CHILTON'S

ELECTRONICS DIAGNOSTIC REPAIR MANUAL

AF837359

Ford Cars and Trucks

Ron Wilson Chilton Sales
P.O. Box 12094
Salem, Oregon 97309
(503) 390-6771

Editor-In-Chief Kerry A. Freeman, S.A.E.
Managing Editor Dean F. Morgantini, S.A.E. □ **Managing Editor** John H. Weise, S.A.E.
Assistant Managing Editor David H. Lee, A.S.E., S.A.E.

Senior Editor Richard J. Rivele, S.A.E. □ **Senior Editor** W. Calvin Settle, Jr., S.A.E.
Project Manager Wayne A. Eiffes, A.S.E., S.A.E.,
Service Editors John M. Baxter, S.A.E., Lawrence C. Braun, S.A.E., A.S.C., Dennis Carroll, Peter M. Conti, Jr., Nick D'Andrea, Ken Grabowski, A.S.E., Martin J. Gunther, Robert McAnally, Michael J. Randazzo, Richard T. Smith, Larry E. Stiles, Anthony Tortorici, A.S.E., S.A.E., Ron Webb
Editorial Consultants Edward K. Shea, S.A.E., Stan Stephenson

Production Manager John J. Cantwell
Art & Production Coordinator Robin S. Miller
Supervisor Mechanical Paste-up Margaret A. Stoner
Mechanical Artists John A. Duncan, Cynthia Fiore

Sales Director Albert M. Kushnerick □ **Assistant** Jacquelyn T. Powers
Regional Sales Managers Joseph Andrews, Jr., David Flaherty, James O. Callahan

OFFICERS
President Lawrence A. Fornasieri
Vice President & General Manager John P. Kushnerick

CHILTON BOOK COMPANY
*ONE OF THE **ABC PUBLISHING COMPANIES,***
*A PART OF **CAPITAL CITIES/ABC, INC.***
Manufactured in USA ©1988 Chilton Book Company • Chilton Way, Radnor, Pa. 19089
ISBN 0–8019–7861–0 Library of Congress Card Catalog 88-48011
1234567890 7654321098

SAFETY NOTICE

Proper service and repair procedures are vital to the safe, reliable operation of all motor vehicles, as well as the personal safety of those performing repairs. This manual outlines procedures for servicing and repairing vehicles using safe, effective methods. The procedures contain many NOTES, CAUTIONS and WARNINGS which should be followed along with standard safety procedures to eliminate the possibilty of personal injury or improper service which could damage the vehicle or compromise its safety.

It is important to note that the repair procedures and techniques, tools and parts for servicng motor vehicles, as well as the skill and experience of the individual performing the work vary widely. It is not possible to anticipate all of the conceivable ways or conditions under which vehicles may be serviced, or to provide cautions as to all of the possible hazards that may result. Standard and accepted safety precautions and equipment should be used when handling toxic or flammable fluids, and safety goggles or other protection should be used during cutting, grinding, chiseling, prying, or any other process that can cause material removal or projectiles.

Some procedures require the use of tools specially designed for a specific purpose. Before substituting another tool or procedure, you must be completely satisfied that neither your personal safety, nor the performance of the vehicle will be endangered

---

Although information in this manual is based on industry sources and is complete as possible at the time of publication, the possibilty exists that some car manufacturers made later changes which could not be included here. While striving for total accuracy, Chilton Book Company cannot assume responsibity for any errors, changes or omissions that may occur in the compilation of this data.

No part of this publication may be reproduced, transmitted or stored in any form or by any means, electronic or mechanical, including photocopy, recording, or by information storage or retrieval system withiout prior written permission from the publisher.

Electronic Diagnosis 1

GENERAL INFORMATION

FORD EEC IV ENGINE CONTROL SYSTEM

General Information

The Ford Electronic Engine Control (EEC) IV system is the fifth generation of engine control systems used on Ford passenger cars and light trucks. EEC IV is similar to previous Ford engine control systems in that the heart of the system is a microcomputer called an Electronic Control Assembly (ECA). The ECA receives data (system inputs) from sensors, switches, relays and other electronic components and issues command signals (system outputs) to various devices in order to control engine operation under a variety of loads and ambient conditions. The ECA is calibrated according to the powertrain, axle ratio and gross vehicle weight (GVW) to optimize fuel economy and driveability while minimizing harmful emissions.

The ECA in the EEC IV system is similar in appearance to earlier control units, except that the calibration module is located within the ECA assembly instead of being attached to the outside as on earlier systems. The harness connectors are edge-card type contacts which provide a more positive connection and allow probing with volt/ohmmeter leads from the rear while connected to simplify diagnosis and testing. The ECA is usually mounted in the passenger compartment, under the front section of the center console, but the location will vary on different models. The ECA could also be located under the right (passenger) seat, or behind the right front (passenger) kick panel, for example.

The EEC IV engine control system is used in conjunction with either a throttle body (CFI) injection, multiport (EFI and SEFI) injection, or feedback carburetor (FBC) fuel delivery system depending on the year, model and powertrain. Although the individual system components vary, the electronic control system operation is basically the same. The major difference is the number and type of output devices being controlled by the ECA.

The EEC IV system electronically controls the fuel injectors or carburetor feedback and mixture control solenoids to maintain a 14.7:1 air/fuel ratio under all driving conditions. This 14.7:1 air/fuel mixture allows the three-way catalytic converter to operate at peak efficiency while getting the most performance and economy from the engine. Ignition spark timing, deceleration fuel cut-off, EGR function (on or off), curb and fast idle speed, evaporative emissions purge, A/C cut-off during wide open throttle, cold engine start and enrichment, electric fuel pump and self-test engine diagnostics are also controlled by the ECA to maintain consistent driveability across a wide range of operating conditions, temperatures and altitudes. The EEC IV system is self-adjusting for operation at high altitude elevations (over 4,000 ft. above sea level) and can actually compensate for engine component wear (a worn timing chain, for example).

The EEC IV engine control system is divided into three major subsystems:

- **Fuel Delivery Subsystem** – which includes the fuel tank and lines, fuel pump and fuel injection or feedback carburetor components. On fuel injected models, this includes the fuel supply manifold, fuel pressure regulator, injectors and fuel filter. On models with feedback carburetors, it includes the carburetor metering circuits, choke assembly and the carburetor body itself.
- **Air Induction Subsystem** – which includes the air cleaner and ducts, intake manifold, carburetor or fuel injection throttle body, throttle air bypass valve, vane airflow meter, turbocharger (if equipped) and related vacuum hoses and air ducts.
- **Electronic Control Subsystem** – which consists of the ECA and its various engine sensors (such as the oxygen sensor and coolant temperature sensor, etc.), along with the wiring harness, relays, fuses, battery and self-diagnostic system. On feedback carburetor models, it would include the throttle position sensor and mixture control solenoid (Carter YFA-1V), duty cycle/feedback solenoid (Motorcraft 2150A), or variable venturi feedback actuator motor (Motorcraft 7200VV). On fuel injected models, it would include the throttle position switch and the wire harness to the fuel injectors, but not the fuel injectors themselves. On turbocharged models, it would include the same multiport fuel injection components, along with such items like overboost pressure switches that are peculiar to turbocharged engines.

EEC IV SYSTEM OPERATION

Crank Mode

The crank mode is entered after initial engine starting, or after engine stall when key is in START. A special operation program is used in the crank mode to aid engine starting. After engine start, one of the run modes is entered and normal engine operation is performed. If the engine stumbles during a run mode, the underspeed mode is entered to help it recover from the stumble and prevent stalling. When cranking the engine, the fuel control is in the open loop mode (no feedback to the ECA) of operation and the ECA sets engine timing at 10–15 degrees BTDC (for the correct timing specification, refer to the Engine Emissions Decal under the hood.

On fuel injected (EFI and SEFI) models, the injectors fire either in a simultaneous, double-fire manner (fires twice every crankshaft revolution), or once per engine revolution in the nor-

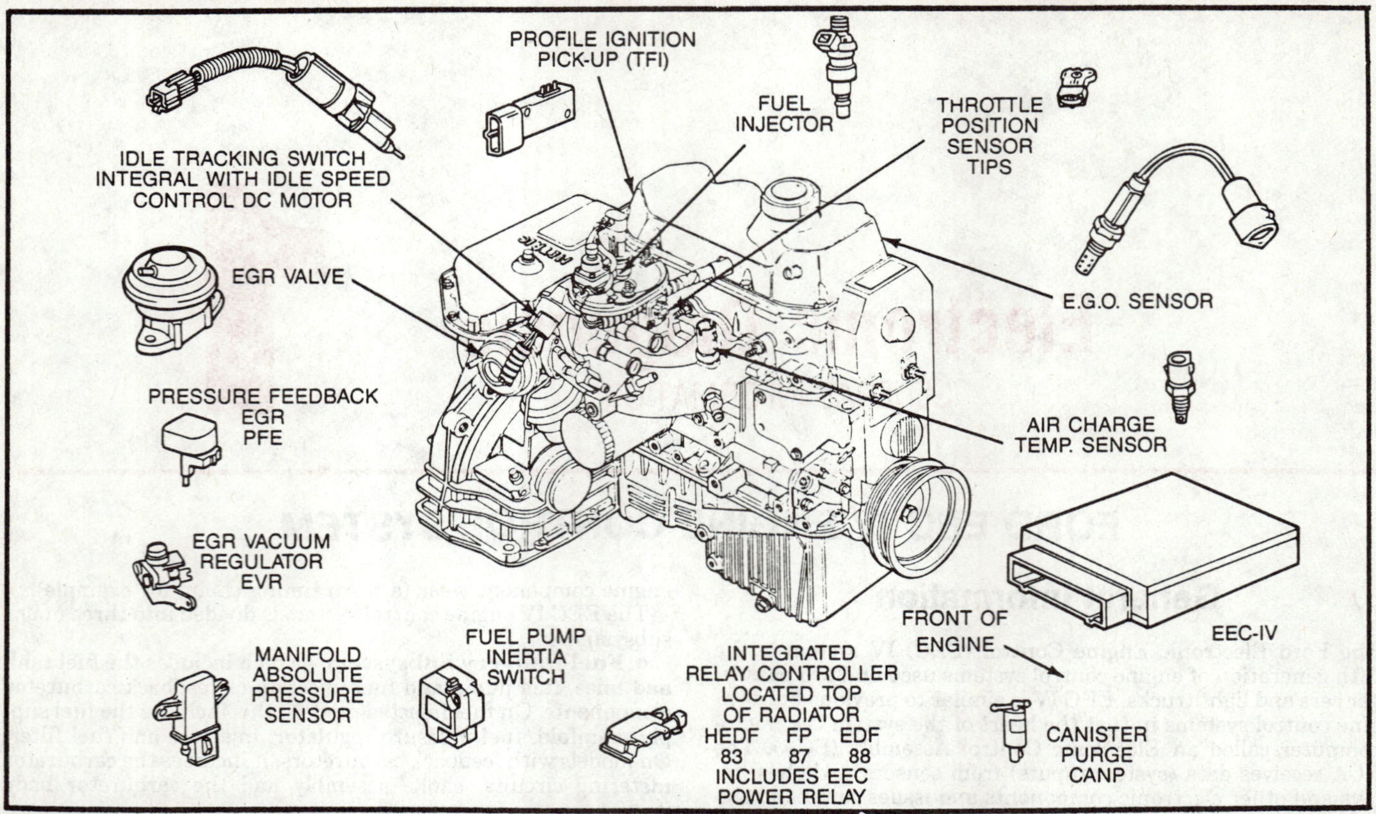

EEC IV fuel injection system components—1.9L CFI engine shown

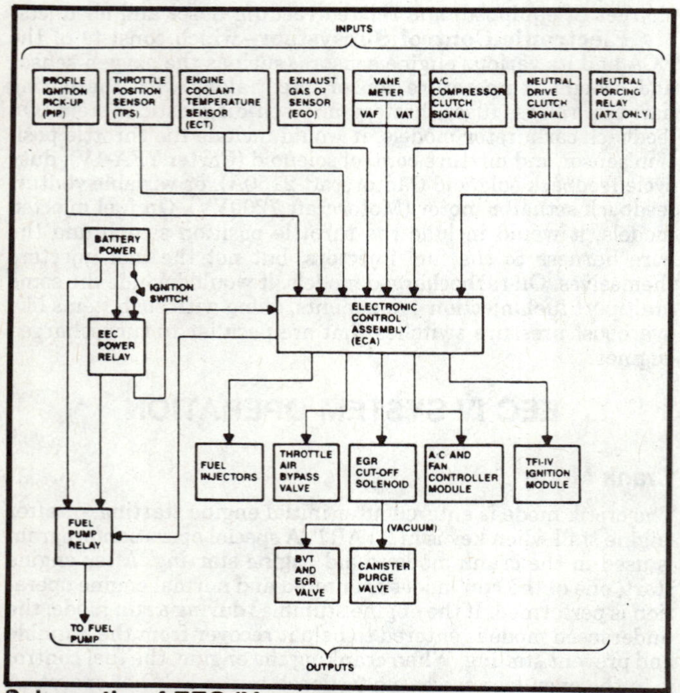

Schematic of EEC IV system operation. The ECA uses the voltage signals from the sensors (inputs) to control various components (outputs) that regulate the operation of the fuel, ignition and emission control systems

mal engine firing order, to provide the base crank air/fuel control. The throttle air bypass valve solenoid is set to open the bypass valve to provide the fast idle/no-touch start.

On models with a feedback carburetor, the air/fuel ratio is controlled through the feedback solenoid. The EGR cut-off solenoid is not energized, so EGR valve if Off.

NOTE: On some engines, the vapor canister purge solenoid is teed with the Thermactor Air Divert (TAD) solenoid, so when thermactor air is upstream, the canister purge if off.

At start-up, the throttle position sensor (TPS) keeps the ECA informed on the position of the throttle plate. When the throttle is kicked after start-up, the ECA will bring the engine down from fast idle by changing the signal to the throttle air bypass valve.

Underspeed Mode

Operation in the underspeed mode (under 500 rpm) is similar to that previously described for the crank mode. The system switches from the underspeed mode to the normal run mode when the required rpm is reached. The underspeed mode is used to provide a good pulse width to the injectors and ignores any signal from the vane meter. During this mode, the vane meter used on fuel injection models flutters and the signal generated would vary with the flutter. Therefore, the vane meter signal is ignored by the ECA in the underspeed mode.

Closed Throttle Mode (Idle or Deceleration)

In the closed throttle mode, the air/fuel ratio is trimmed by either varying the pulse width of the output from the ECA to the injectors, or by varying the duty cycle of the feedback solenoid on carbureted models, to obtain the desired mixture. To calculate what this output signal should be, the ECA evaluates inputs from the ECT sensor, the vane meter, the TPS, the EGO sensor, the PIP sensor and the A/C clutch. These sensors inform the ECA of the various conditions that must be evaluated in order for the ECA to determine the correct air/fuel ratio for the closed-throttle condition present. Therefore, with the input from the EGO sensor, the system is maintained in closed-loop operation at idle. If the EGO sensor fails to switch rich/lean, the ECA programming assumes the EGO sensor has cooled off, and the system goes to open-loop fuel control. Under a deceleration condition, the TPS signal indicates closed throttle and the ECA shut-off fuel for improved fuel economy and emissions. The injectors are turned back on, as required, to prevent engine stalling.

NOTE: The point at which the injectors are turned back on will occur at different rpm's, depending on calibration factors and engine temperature, although the injectors are turned back on if the throttle is opened.

Ignition timing is also determined by ECA using these same inputs. The ECA has a series of tables programmed into the assembly at the factory. These tables provide the ECA with a reference of desired ignition timing for the various operating conditions reflected by the sensor inputs. The throttle air bypass valve position is determined by the ECA as a function of RPM, ECT, A/C On or Off, throttle mode and time since start-up inputs. The signal from the TPS to the ECA indicates that the throttle plate is closed, and the ECA de-energizes the EGR shut-off solenoid to close the EGR valve.

Part Throttle Mode (Cruise)

The air/fuel mixture ratio and ignition timing are calculated in the same manner as previously described for the closed throttle mode. The fuel control system remains in closed-loop during part throttle operation, as long as the EGO sensor is operational. In part throttle operation, the throttle air bypass valve is positioned to provide an electronic dashpot function in the event the throttle is closed. Again, as in the closed throttle mode, the ECA makes this determination based on the inputs from the applicable sensors. The TPS provides the throttle plate position signal to the ECA. With the throttle plate being in a partial open position, the ECA energizes the EGR shut-off solenoid to open the EGR valve.

Wide Open Throttle Mode (WOT)

Control of the air/fuel ratio in WOT mode is the same as in part, or closed throttle situations, except that fuel control switches to open-loop, and the fuel injector pulse width is increased to provide additional fuel enrichment. This pulse width increase is applied as a result of the WOT signal from the TPS to the ECA. This signal from the TPS also causes the ECA to remove the energizing signal from the EGR shut-off solenoid (if present). More spark advance is added in WOT for improved performance.

Cold or Hot Engine Operation

This modified operation changes the normal engine operation output signals, as required, to adjust for uncommon engine operating conditions. These include cold or excessively hot engine.

Limited Operation Strategy (LOS)

In this operation, the ECA provides the necessary output signals to allow the vehicle to "limp home" when an electronic malfunction occurs. The EGR valve is shut-off, the air bypass valve goes to a fixed voltage, timing is locked at the fixed timing (depends on calibration, refer to Engine Emissions Decal on the vehicle), and the injector pulse width is constant.

FORD MICROPROCESSOR CONTROL UNIT (MCU) SYSTEM

General Information

The system is equipped to supply input signals to the microprocessor control unit module (MCU). These inputs originate from the exhaust gas oxygen sensor, cold temperature vacuum switch, idle tracking switch and the rpm tach input from the coil.

The MCU module continuously monitors the input signals and computes the correct operating mode for a given condition. Output signals from the MCU module are applied to control the vacuum solenoid regulator, thermactor bypass control solenoid and the thermactor diverter control solenoid (manual transmission only).

The primary function of the 2.3L MCU module is to control the vacuum regulator solenoid. The MCU module sends ten signals per second to the vacuum regulator solenoid in a timed duty cycle. By varying the "ON" time to the "OFF" time of the cycle, the MCU module is able to maintain or change the air-fuel mixture.

The MCU module does not control ignition timing. The system uses a conventional distributor and coil. The MCU system has three operating modes. These modes are system initialization (start-up), open loop and closed loop.

System Initialization (Start-Up)

The MCU module will initialize when battery power is applied to the computer prior to engine cranking and again immediately with engine starting. During initialization, the duty cycle to the vacuum regulator solenoid is maintained at 50%. After starting, initialization lasts for only a fraction of a second. Then the MCU system goes into the open or closed loop.

Open Loop

The system is in the open loop mode when either the cold temperature vacuum switch or idle tracking switch is activated. In the open loop mode, the MCU module will control the duty cycle with "On" time signals to the vacuum regulator solenoid. These will provide a calibrated air-fuel mixture.

Closed Loop

With the proper signals fro the idle tracking switch and the cold temperature switch, the MCU module changes to the closed loop mode for close range monitoring and control of the air-fuel ratio. The exhaust gas oxygen sensor monitors the exhaust gas to determine if the engine is running rich or lean. This information is used by the MCU module to adjust the carburetor to the air-fuel ratio desired for the operating condition.

Signals from the MCU module, which produce carburetor adjustments, are calibrated. This provides a damping effect to minimize over-correction and abrupt changes.

SYSTEM COMPONENTS

MCU Module

The MCU is the brain of the system. It is a solid-state, programmed micro-computer. It takes the information inputs and uses a program to provide output control signals to control solenoids. The MCU module receives its power from the battery through the ignition switch.

The module is located either in the engine compartment or under the instrument panel to the left of the steering column. It similar to an ignition switch module, but difference in the wiring connectors prevent interchangeability.

With the engine cold or at idle, the air-fuel ratio is at a preset level. When the engine is warm and off idle, the air-fuel mixture is adjusted by a signal from the exhaust gas oxygen sensor. These adjustments are made by an "ON-OFF" control signal from the module. For example, when the sensor signals a rich mixture, control "ON" time is increased, while "ON" time is decreased for a lean mixture.

Cold Temperature Vacuum Switch

The cold temperature switch is a normally closed switch with vacuum controlled through a ported vacuum switch (PVS). The MCU system will enter a closed loop made when the cold temperature switch is activated, provided the vehicle is operating at part throttle.

When the engine is cold, the PVS valve blocks vacuum from the cold temperature vacuum switch. The switch position provides a ground path as a signal to the control unit. When the engine is above 95°F, the PVS valve opens to allow ported vacuum to activate the normally closed switch. With the contacts opened, the ground path signal to the control unit is interrupted.

Idle Tracking switch is a limited switch used to detect the throttle in the idle position. It is normally closed but opens at closed throttle, sending the MCU system into an open loop model. The idle tracking switch is mounted on the rear of the carburetor.

Exhaust Gas Oxygen Sensor

The exhaust gas oxygen sensor is threaded into the exhaust manifold, directly in the path of the exhaust gas stream. The sensor provides information to the computer about the air-fuel ratio, indicated by the oxygen concentration in the exhaust gases.

When it senses a rich mixture, the sensor generates a high voltage signal to the computer. A low voltage is generated when a lean mixture (high oxygen level in the exhaust) is sensed. The voltage signal is used by the MCU module for adjustment of the duty cycle during the closed loop mode of operation.

Vacuum Solenoid Regulator

The vacuum solenoid regulator controls manifold vacuum which has been regulated to 5 in Hg. This solenoid receives ten signals per second from the MCU module. The vacuum applied to the metering rod of the carburetor is controlled by varying the duty cycle ("ON-OFF" ratio of the MCU signal).

At 100% duty cycle, 5 in. Hg. vacuum is applied to the metering rod, producing a full lean condition. At 0% duty cycle, the metering rod diaphragm is vented to the atmosphere, producing a full rich condition.

Tachometer Input

The rpm signal to the computer is taken form the "tach" terminal of the ignition coil. This circuit is identified by the dark green-yellow dot wire. This negative signal is used by the computer to determined the amount of damping to use changing the duty signal to the carburetor. These signals are used only in the closed loop mode of operation.

THERMACTOR AIR SYSTEM

The thermactor air system is used to operate the three-way catalytic converter system. The thermactor air system consists of the air pump, bypass and diverter valves and the air control solenoids.

Thermactor air venting to the atmosphere is controlled by the idle tracking switch to protect the vehicle from overtemperature during extended idling. Venting begins after $2\text{-}2^1{}_2$ minutes of uninterrupted idle. To reduce the excessive amounts of HC and CO during warm-up, thermactor air is directed upstream during this period. The thermactor air valves are located at the right front of the engine.

Thermactor Air Pump

The air pump in an impeller-type centrifugal air filter fan. Heavier than air contaminants are thrown fro the air intake by centrifugal force. This type of air pump does not have a pressure relief valve because this function is controlled by the thermactor bypass valve.

Thermactor Bypass Solenoid Diverter Solenoid

The bypass valve operation and diverter valve operation for non-turbocharged engine with manual transmissions are controlled by the thermactor air control solenoids. The solenoids route vacuum in three possible directions: bypass, upstream or downstream.

In the bypass mode, air is vented to the atmosphere. In the upstream mode, air is injected into the exhaust manifold. In the downstream mode, air is directed into the three-way catalyst between the two catalyst stages.

ON-DEMAND SERVICE CODES

SERVICE CODE	EXPLANATION OF SERVICE CODE	POSSIBLE CAUSES OF CONCERN	POSSIBLE SYMPTOMS
12 KOER	Indicates the system is not capable of raising engine speed above curb idle.	Non-EEC: — Engine running rough/missing. — Throttle linkage binding. — Improper vehicle prep. (e.g., warm-up). EEC: — ISC motor/TKS system/ISC solenoid problems. — Wire harness problems. — ECA problems.	— Rough idle or stalls due to lack of rpm increase with added loads (e.g., Power Steering lock or A/C "on").
13 KOER	Indicates that the engine did not return to a specified lower rpm prior to entering the "goose" test portion of Quick Test.	Non-EEC: — Improper curb idle set. — Throttle/TVS linkage binding. — Improper throttle stop set. — Improper cruise control set. EEC: — ISC motor/ISC solenoid problems. — Idle tracking switch problems. — Wire harness problems. — ECA problems.	— Idle speed concerns (may be accompanied by code 58).
15 KOEO	Indicates an ECA failure.	EEC: — ECA problems.	— Erratic operation or no start.
16 KOER	Indicates that the fuel system has been driven "lean" until the rpm drops, but the EGO sensor continues to indicate "rich."	EEC: — Contaminated EGO sensor. — Wire harness problems. — ECA problems.	— System may correct "lean" inducing stumbles or hesitation.
17 KOER	Indicates that with thermactor air upstream the fuel system has been driven "lean" until the rpm drops, but the EGO sensor continues to indicate "rich" (5.0L CFI only)	EEC: — Contaminated or disconnected EGO sensor. — Wire harness problems. — ECA problems.	— System may correct "rich" causing reduced MPG or black smoke.
21 KOEO	Indicates that the engine coolant temperature is out of range.	Non-EEC: — Engine not up to operating temperature or, in the case of a no-start, test performed in cool ambient conditions. — Engine over operating temperature. • Low coolant. • Stuck thermostat. • Cooling fan problems. EEC: — ECT sensor problems. — Wire harness problems. — ECA problems.	— Reduced MPG. — Rough idle. — Improper idle speed. — Detonation.
21 KOER	Indicates that the engine coolant is not at normal operating temperature.	Non-EEC: — Engine not warmed up. — Thermostat stuck open. — Low coolant level. — Coolant fan not operating. — Thermostat stuck closed. — Radiator blockage (internal or external). — Improper spark timing. — Fuel problems. EEC: — ECT sensor problems. — Wire harness problems. — ECA problems.	— Reduced MPG. — Rough idle. — Improper idle speed. — Detonation.

KOEO: Key On Engine Off.
KOER: Key On Engine Running.

ON-DEMAND SERVICE CODES

SERVICE CODE	EXPLANATION OF SERVICE CODE	POSSIBLE CAUSES OF CONCERN	POSSIBLE SYMPTOMS
22 KOEO	Indicates that the MAP/BP sensor is out of range. The sensor(s) should read atmospheric pressure.	Non-EEC: — Unusually high atmospheric pressure. — Vacuum trapped at MAP sensor. EEC: — MAP/BP sensor problems. — Wire harness problems. — ECA problems.	— No-start. — Stalls. — Detonation. — Reduced MPG. — Loss of power.
22 KOER	Indicates that the MAP/BP sensor is not at normal vacuum levels for Quick Test. The MAP sensor should indicate engine manifold vacuum. The BP sensor should indicate atmospheric pressure.	Non-EEC: — Base engine problems (MAP). • Compression problems. • Improper timing. • Vacuum leaks. • Excess EGR. — Vacuum line connected to BP sensor (BP). EEC: — MAP/BP sensor problems. — Wire harness problems. — ECA problems.	— Stalls. — Detonation. — Reduced MPG. — Loss of power.
23 KOEO	Indicates that the TP sensor is not at the proper closed throttle position. Failure may occur either above or below the proper closed throttle position.	Non-EEC: — Improper base adjustment of curb set, ISC motor, TSP, TKS, or cruise control linkage. EEC: — TKS/ISC motor problems. — Wire harness problems. — TP sensor problems. — ECA problems.	— No-start (EFI/CFI) — Hesitation. — Stalls. — Low/high idle. — Poor performance. — Reduced MPG.
23 KOER	Indicates that the TP sensor is not at the normal throttle position for Quick Test conditions.	Non-EEC: Improper base adjustment of curb set, ISC motor, TSP, TKS, or cruise control linkage. EEC: — TKS/ISC motor problems. — Wire harness problems. — TP sensor problems. — ECA problems.	— Hesitation. — Stalls. — Low/high idle. — Poor performance. — Reduced MPG.
24 KOEO	Indicates that the Air Charge/ Vane Air temperature is out of range.	Non-EEC: — Vehicle testing performed in ambient temperature less than +50 deg. F. — Improper air cleaner duct/door operation. EEC: — ACT sensor problems. — Wire harness problems. — ECA problems.	— Poor idle. — Reduced MPG.
24 KOER	Indicates that the ACT/VAT sensor is not at normal engine operating temperature.	Non-EEC: — Improper operation of the air cleaner duct door. — Cooling system problems. — Base timing problems. EEC: — ACT/VAT sensor problems. — Wire harness problems. — ECA problems.	— Poor idle. — Reduced MPG.
25 KOER	Indicates that knock was not sensed during the "goose" test.	Non-EEC: — Base timing problems. EEC: — Knock sensor problems. — Wire harness problems. — ECA problems.	— Detonation. — Poor performance.

KOEO: Key On Engine Off.
KOER: Key On Engine Running.

ON-DEMAND SERVICE CODES

SERVICE CODE	EXPLANATION OF SERVICE CODE	POSSIBLE CAUSES OF CONCERN	POSSIBLE SYMPTOMS
26 KOEO	Indicates that the VAF meter is not in its closed position.	Non-EEC: — Obstruction in VAF meter. EEC: — VAF Meter problems. — Wire harness problems. — ECA problems.	— No-start. — Stalls. — Runs rough. — Hesitates.
26 KOER	Indicates that the VAF meter is not at the normal position for Quick Test conditions.	Non-EEC: — Unmetered air leaks. — Improper idle speeds. — Engine not at normal operating temperature. EEC: — VAF meter problems. — Wire harness problems. — ECA problems.	— Stalls. — Runs rough. — Hesitates.
31 KOEO/ KOER	Indicates that the EGR valve is not in its normal closed position.	Non-EEC: — Sticking/damaged EGR valve. — Vacuum trapped at EGR valve. EEC: — EGR solenoid problems. — EVP sensor problems. — Wire harness problems. — ECA problems.	— Stalls. — Runs rough. — Dies at idle. — Detonation.
32 KOER	Indicates that the system is not able to open and maintain a specified EGR valve position.	Non-EEC: — Stuck or damaged EGR valve. — Vacuum leaks. EEC: — EGR solenoid problems. — EVP sensor problems. — Wire harness problems. — ECA problems.	— Detonation. — Poor performance.
33 KOER	Indicates that the EGR valve has not returned to its normal closed position after the EGR test.	Non-EEC: — Stuck or damaged EGR valve. EEC: — EGR solenoid problems. — EVP sensor problems.	— Decel stall. — Runs rough.
34 KOER	Indicates that, with engine rpm elevated and stabilized, a specified rpm drop did not occur when EGR was turned "on."	Non-EEC: — Stuck or damaged EGR valve. — BVT/EGR problems. — Vacuum leaks. — Improper exhaust back-pressure EEC: — EGR solenoid problems. — Wire harness problems. — ECA problems.	— Detonation.
35 KOER	Indicates that the engine rpm is too low for the EGR test.	Non-EEC: — Base engine problems. EEC: — ISC problems.	— May run rough. — Lacks power.
41* KOER	Indicates that the EGO sensor output voltage is always less than 0.5 volts ("lean") during the fuel test.	Non-EEC: — Improper fuel delivery. — Carburetor/Throttle-Body injector problem. — Vacuum leaks. — Unmetered air. — Thermactor air is always upstream. EEC: — EGO sensor problems. — Wire harness problems. — ECA problems.	— Runs rough. — Stalls. — Hesitates. — Runs "lean". — May correct "rich" if sensor or harness related. • Spark plugs fouled. • Reduced MPG.

KOEO: Key On Engine Off.
KOER: Key On Engine Running.
*On 3.8L engines this condition refers to the left bank.

ON-DEMAND SERVICE CODES

SERVICE CODE	EXPLANATION OF SERVICE CODE	POSSIBLE CAUSES OF CONCERN	POSSIBLE SYMPTOMS
42* KOER	Indicates that the EGO sensor output voltage is always greater than 0.5 volts ("rich") during the fuel test.	Non-EEC: — Improper fuel delivery. — Carburetor/Throttle-Body injector problems. — Obstructed air intake. — Ignition system problems. — Cannister purge problems. EEC: — EGO sensor problems. — Wire harness problems. — ECA problems.	— Runs rough. — Runs "rich." — Reduced MPG. — Spark plugs fouled. — May correct "lean" if sensor or harness related. ● Hesitations. ● Stalls.
43* KOER	Indicates that the EGO sensor has cooled down and may not have given the proper responses during Quick Test.	Non-EEC: — Vehicle not properly prepared for test (Run at 2000 rpm for 2 minutes prior to test). — Engine below operating temperature.	— Usually no drive complaint.
44* KOER	Indicates that there is a thermactor problem.	Non-EEC: — Air pump problems. — Thermactor valve problems. EEC: — Thermactor solenoid problems. — Wire harness problems. — ECA problems.	— Usually no drive complaint.
45* KOER	Indicates that thermactor air is always upstream during Quick Test.	Non-EEC: — Air pump problems. — Thermactor valve problems. EEC: — Thermactor solenoid problems. — Wire harness problems. — ECA problems.	— Catalyst over temperature.
46* KOER	Indicates that the system is unable to bypass (vent to atmosphere) thermactor air.	Non-EEC: — Air pump problems. — Thermactor valve problems. EEC: — Thermactor solenoid problems. — Wire harness problems. — ECA problems.	— Catalyst over temperature.
47* KOER	Indicates that, even though thermactor air is upstream and fuel control is max. "lean", the EGO sensor indicates "rich".	Non-EEC: — Improper fuel delivery. — Carburetor/Throttle-Body injector problems. — Obstructed air intake. — Ignition system problems. EEC: — EGO sensor problems. — Wire harness problems. — ECA problems.	— Runs "rich". — Runs rough. — Reduced MPG. — Spark plugs fouled. — May correct "lean" if sensor or harness related. ● Hesitations. ● Stalls.

KOEO: Key On Engine Off.
KOER: Key On Engine Running.
*On 3.8L engines this condition refer to the left bank.

ON-DEMAND SERVICE CODES

SERVICE CODE	EXPLANATION OF SERVICE CODE	POSSIBLE CAUSES OF CONCERN	POSSIBLE SYMPTOMS
48 KOER	Indicates that the system does not have proper side-to-side fuel control. (3.8L only.)	Non-EEC: — Catalyst blockage. — Stuck injector. — Plugged injector. EEC: — Wire harness problems. • Injector connections reversed. • EGO connections reversed. • One injector disconnected. • One EGO disconnected. — EGO sensor problems. — ECA problems.	— Stumbles. — Stalls. — Reduce MPG. — Black smoke.
51 KOEO	Indicates that the ECT signal failed at the high end (approximately 5.0 volts). Failure mode indicates −40 deg. F.	EEC: — ECT sensor problems. — Wire harness problems. — ECA problems.	— Hard to start Hot. — Black smoke. — Reduced MPG.
53 KOEO	Indicates that the TP signal has failed at the high end (approximately 5.0 volts). Failure mode indicates WOT.	EEC: — Wire harness problems. • Open signal return circuit. • TP signal shorted to VREF. — TP sensor problems. — ECA problems.	— No start (EFI/CFI). — Poor part throttle performance.
54 KOEO	Indicates that the ACT signal has failed at the high end (approximately 5.0 volts). Failure mode indicates −40 deg. F.	EEC: — Wire harness problems. • Vref open. • ACT signal shorted to signal return. • ACT signal open. — ACT sensor problems. — ECA problems.	— Reduced MPG. — Black smoke.
56 KOEO	Indicates that the VAF signal has failed at the high end (approximately 5.0 volts). Failure mode indicates wide open throttle.	EEC: — Wire harness problems. — VAF sensor problems. — ECA problems.	— Black smoke.
58 KOER	Indicates that the ITS is not in contact with the throttle lever with the ISC motor extended.	Non-EEC: — Improper throttle plate stop adjustment. — Throttle linkage binding. — Cruise control misadjusted. — Improper ISC adjustment. EEC: — ITS problems. — Wire harness problems. — ECA problems.	— Improper idle speeds. — Stalls.
61 KOEO	Indicates that the ECT signal has failed at the low end (approximately 0.0 volts). Failure mode indicates +240 deg. F.	EEC: — ECT sensor problems. — Wire harness problems. — ECA problems.	— Hard start/no start (cold).
63 KOEO	Indicates that the TP signal has failed at the low end (approximately 0.0 volts). Failure mode indicates closed throttle.	EEC: — Wire harness problems. • Vref open. • TP signal shorted to signal return. • TP signal open. — TP sensor problems. — ECA problems.	— Runs rough. — Stumbles. — Stalls.

KOEO: Key On Engine Off.
KOER: Key On Engine Running.

ON-DEMAND SERVICE CODES

SERVICE CODE	EXPLANATION OF SERVICE CODE	POSSIBLE CAUSES OF CONCERN	POSSIBLE SYMPTOMS
64 KOEO	Indicates that the ACT/VAT signal has failed at the low end (approximately 0.0 volts). Failure mode indicates +240 deg.	EEC: — ACT/VAT sensor problems. — Wire harness problems. — ECA problems.	— Runs rough. — Stumbles. — Stalls.
66 KOEO	Indicates that the VAF signal has failed at the low end (approximately 0.0 volts). Failure mode indicates closed throttle.	EEC: — Wire harness problems. — VAF sensor problems. — ECA problems.	— Stumbles — Stalls.
67 KOEO	Indicates that the system is receiving an improper neutral/drive or A/C clutch status input.	Non-EEC: — Vehicle in gear or A/C "on." EEC: — Neutral/drive switch problems. — Wire harness problems. — ECA problems.	— Improper ISC. — Stalls.
68 KOEO	Indicates that the ITS is in contact with the throttle lever with the ISC motor retracted.	Non-EEC: — Improper ISC motor adjustment. EEC: — ITS problems. — Wire harness problems. — ECA problems.	— Improper idle speed control. — Stalls.
72 KOER	Indicates that the MAP sensor has not detected a sufficient manifold vacuum change during the "goose" test.	Non-EEC: — Vacuum leaks. — Base engine problems. EEC: — Map sensor problems. — ECA problems.	— Stumbles. — Hesitations. — Stalls.
73 KOER	Indicates that the system has not detected a sufficient TP change during the "goose" test.	EEC: — TP sensor stuck at WOT. — TP sensor not tracking throttle shaft. — ECA problems.	— Stumbles. — Hesitations.
76 KOER	Indicates that the system has not detected a sufficient VAF change during the "goose" test.	Non-EEC: — Unmetered air entering engine. EEC: — VAF meter sticking. — ECA problems.	— Stumbles. — Hesitations.
77 KOER	Indicates that the operator did not do the "goose" test.	Non-EEC: — Operator did not do a brief WOT. EEC: — ECA problems.	
81 KOEO	Indicates a TAD circuit fault. (For 2.8L only, code indicates a TAB circuit fault.)	EEC: — Open/shorted TAD solenoid. — Wire harness problems. — ECA problems.	— Thermactor service code KOER.
82 KOEO	Indicates a TAB circuit fault. (For 2.8L only, code indicates a TAD circuit fault.)	EEC: — Open/shorted TAB solenoid. — Wire harness problems. — ECA problems.	— Thermactor service code KOER.
83 KOEO	Indicates an EGRC circuit fault.	EEC: — Open/shorted EGRC solenoid. — Wire harness problems. — ECA problems.	— No EGR control. — Detonation.
84 KOEO	Indicates an EGRV circuit fault.	EEC: — Open/shorted EGRV solenoid. — Wire harness problems. — ECA problems.	— Improper EGR control. — Stalls. — Runs rough.
85 KOEO	Indicates a CANP circuit fault.	EEC: — Open/shorted CANP valve. — Wire harness problems. — ECA problems.	— Customer complaints of gasoline odor.
86 KOEO	Indicates a WAC circuit fault.	EEC: — Open/shorted WAC controller. — Wire harness problems. — ECA problems.	— No A/C operation. — No WOT-A/C cutoff.

KOEO: Key On Engine Off.
KOER: Key On Engine Running.

ON-DEMAND SERVICE CODES

SERVICE CODE	EXPLANATION OF SERVICE CODE	POSSIBLE CAUSES OF CONCERN	POSSIBLE SYMPTOMS
87 KOEO	Indicates a fuel pump relay circuit fault	EEC: — Inertia switch problems. — Open/short fuel pump relay. — Wire harness problems. — ECA problems.	— No start. — Pump runs with key in "off" position.
	or a TCP circuit fault (2.8L only).	EEC: — TCP solenoid problems. — Wiring harness problems. — ECA problems.	— Poor cold start/driveaway.
88 KOEO	Indicates a TKS circuit fault (5.0L CFI only)	EEC: — Open/shorted TKS solenoid. — Wire harness problems. — ECA problems.	— High idle speeds. — Stalls (low idle speeds).
	or a VVC circuit fault (2.8L only).	EEC: — VVC relay problems. — Wiring harness problems. — ECA problems.	— Poor cold start/driveaway. — Spark plug loading. — Reduced MPG. — Black smoke.
89 KOEO	Indicates an exhaust heat control circuit fault.	EEC: — Open/shorted EHC solenoid. — Wire harness problems. — ECA problems.	— Stumbles. — Hesitations cold.
91 KOER	Indicates that the right EGO sensor output voltage is always less than 0.5 volts ("lean") during the fuel test (3.8L only).	Non-EEC: — Improper fuel delivery. — Carburetor/Throttle-Body injector problems. — Vacuum leaks. — Unmetered air. — Thermactor air is always upstream. EEC: — EGO sensor problems. — Wire harness problems. — ECA problems.	— Runs rough. — Stalls. — Hesitates. — Runs "lean". — May correct "rich" if sensor or harness related. • Spark plugs fouled. • Reduced MPG.
92 KOER	Indicates that the right EGO sensor output voltage is always greater than 0.5 volts ("rich") during the fuel test (3.8L only).	Non-EEC: — Improper fuel delivery. — Carburetor/Throttle-Body injector problems. — Obstructed air intake. — Ignition system problems. EEC: — EGO sensor problems. — Wire harness problems. — ECA problems.	— Runs rough. — Runs "rich". — Reduced MPG. — Spark plugs fouled. — May correct "lean" if sensor or harness related. • Hesitations. • Stalls.
93 KOER	Indicates that the right EGO sensor has cooled down and may not have given the proper responses during Quick Test (3.8L only).	Non-EEC: — Vehicle not properly prepared for test (Run at 2000 rpm for 2 minutes prior to test). — Engine below operating temperature.	— Usually no drive complaint.
94 KOER	Indicates that there is a thermactor problem on the right bank (3.8L only).	Non-EEC: — Air pump problems. — Thermactor valve problems. EEC: — Thermactor solenoid problems. — Wire harness problems. — ECA problems.	— Usually no drive complaint.
95 KOER	Indicates that the right bank thermactor air is always upstream during Quick Test (3.8L only).	Non-EEC: — Air pump problems. — Thermactor valve problems. EEC: — Thermactor solenoid problems. — Wire harness problems. — ECA problems.	— Catalyst over temperature.

KOEO: Key On Engine Off.
KOER: Key On Engine Running.

ON-DEMAND SERVICE CODES

SERVICE CODE	EXPLANATION OF SERVICE CODE	POSSIBLE CAUSES OF CONCERN	POSSIBLE SYMPTOMS
96 KOER	Indicates that the system is unable to bypass (vent to atmosphere) thermactor air on the right bank (3.8L only).	Non-EEC: — Air pump problems. — Thermactor valve problems. EEC: — Thermactor solenoid problems. — Wire harness problems. — ECA problems.	— Catalyst over temperature.
97 KOER	Indicates that, even though thermactor air is upstream and fuel control is max. "lean," the right EGO sensor indicates "rich" (3.8L only).	Non-EEC: — Improper fuel delivery. — Carburetor/Throttle-Body injector problems. — Obstructed air intake. — Ignition system problems. EEC: — EGO sensor problems. — Wire harness problems. — ECA problems.	— Runs "rich." — Runs rough. — Reduced MPG. — Spark plugs fouled. — May correct "lean" if sensor or harness related. • Hesitations. • Stalls.

KOEO: Key On Engine Off.
KOER: Key On Engine Running.

CONTINUOUS CODES

SERVICE CODE	EXPLANATION OF SERVICE CODE	POSSIBLE CAUSES OF CONCERN	POSSIBLE SYMPTOMS
31 Cont.	Indicates that the EVP sensor signal has been off scale at the high or low end during recent operation.	Non-EEC: — EGR valve sticks at wide open. EEC: — EGR over travel (normal condition on some applications, refer to diagnostics manual). — EVP sensor problems. — Wire harness problems. — ECA problems.	— Detonation. — Hard to start. — No start. — Runs rough. — Stalls. — Dies at idle.
53 Cont.	Indicates that the TP sensor signal has been off scale at the high end during recent operation.	EEC: — Wire harness problems. • Open signal return. • Signal short circuit to Vref. — TP sensor problems. — ECA problems.	— No start (CFI/EFI). — Detonation. — Stumbles. — Poor performance.
54 Cont.	Indicates that the ACT/VAT sensor signal has been off scale at the high end during recent operation.	EEC: — ACT/VAT sensor problems. — Wire harness open circuit. — ECA problems.	— Runs rough. — Black smoke. — Reduced MPG.
56 Cont.	Indicates that the VAF meter signal has been off scale at the high end during recent operation.	EEC: — VAF sensor problems. — Wire harness problems. — ECA problems.	— No start. — Stumbles. — Stalls. — Runs only at WOT.
61 Cont.	Indicates that the ECT sensor signal has been off scale at the low end during recent operation.	EEC: — ECT sensor problems. — Wire harness shorted to ground. — ECA problems.	— Runs rough. — Stalls. — Erratic idle.
63 Cont.	Indicates that the TP sensor signal has been off scale at the low end during recent operation.	EEC: — Wire harness problems. • Vref open. • Signal line open. • Signal shorted to sensor return. — TP sensor problems. — ECA problems.	— Poor performance. — Stumbles. — Detonation.
64 Cont.	Indicates that the ACT/VAT sensor signal has been off scale at the low end during recent operation.	EEC: — ACT/VAT sensor problems. — Wire harness shorted to ground. — ECA problems.	— Runs rough. — Stalls. — Erratic idle.
65 Cont.	Indicates that during recent operation a charging system over-voltage condition (greater than 17.5 v) has occurred.	Non-EEC: — Voltage regulator problems. — Alternator problems. EEC: — Wire harness problems. — ECA problems.	— Lamp burnout. — Electronic-component burnout (e.g., TFI, ECA, Radio, etc.)
66 Cont.	Indicates that the VAF meter signal has been off scale at the low end during recent operation.	EEC: — Wire harness problems. • Vref circuit open. • Signal circuit open. • Signal circuit shorted to signal return. — VAF meter problems. — ECA problems.	— No start. — Stalls. — Lacks power. — Runs only at WOT.

Cont: Continuous Test.

CONTINUOUS CODES

SERVICE CODE	EXPLANATION OF SERVICE CODE	POSSIBLE CAUSES OF CONCERN	POSSIBLE SYMPTOMS
13 Cont.	Indicates that during recent operation, while in the normal operating mode an ISC command to extend the ISC motor shaft occurred without a corresponding TP sensor change.	Non-EEC: — Throttle shaft binding. — Improper ISC adjustment. EEC: — ISC motor problems. — TP sensor problems. — Wire harness problems. — ECA problems.	— Improper idle speed. — Possible stalls.
14 Cont.	Indicates that erratic operation or intermittent loss of PIP information to the ECA has occurred during recent operation.	Non-EEC: — Distributor pickup problems. — TFI module problems. EEC: — Wire harness problems. — ECA problems.	— Engine miss. — Surge. — Rough idle. — Stall.
15 Cont.	Indicates loss of Keep-Alive-Memory during recent operation.	EEC: — Loss of Keep-Alive-Memory battery power. — ECA problems.	— No codes in memory.
18 Cont.	Indicates that during recent operation the ECA received a PIP signal input without receiving a corresponding tach signal input.	Non-EEC: — Ignition coil problems. — TFI module problems. — Vehicle wire harness. EEC: — Wire harness problems. — ECA problems.	— No start. — Stalls. — Idles/runs rough.
21 Cont.	Indicates that during a single drive cycle the engine has reached normal operating temperature and then has cooled down.	Non-EEC: — Thermostat failure. — Coolant loss. EEC: — ECT sensor problems. — Wire harness problems. — ECA problems.	— Reduced MPG. — Heater output loss.
22 Cont.	Indicates that a gross MAP/BP sensor signal error has occurred during recent operation.	Non-EEC: Basic engine problems. • Compression problems. • Improper timing. • Vacuum leaks. EEC: — MAP/BP sensor problems. — Wire harness problems. — ECA problems.	— Detonation. — Poor performance. — Reduced MPG. — Surge. — Stumbles.

Cont: Continuous Test.

Electronic Diagnosis

1984 FORD CARS & TRUCKS
EEC-IV & MCU SYSTEMS

2

When To Use This Diagnostic Procedure

This diagnostic procedure is only used on Electronic Engine Control (EEC-IV) equipped vehicles when the Diagnostic Routines, Section 2 instruct you to perform EEC-IV Diagnostics. It is divided into two test formats. **Quick Test,** a functional EEC-IV system test, and **Pinpoint Tests,** a number of specific component tests.

To test and service the EEC-IV subsystem, perform the Quick Test first. If the vehicle passes all three phases of the Quick Test: Key On, Engine Off, Engine Running, and Continuous Testing without running any Pinpoint Tests, you will know the EEC-IV subsystem is OK and the vehicle's problem exists somewhere else other than in the EEC-IV subsystem. However, if a Step of the Quick Test fails, perform only the Pinpoint Tests specified by the failed Step. Do not begin any Pinpoint Test without following the instructions at the beginning of the Section.

After all tests and services have been completed, repeat the entire Quick Test to make sure the EEC-IV subsystem works properly.

Test Equipment

EQUIPMENT REQUIRED:

- Rotunda Self-Test Automatic Readout (STAR), No. 07-0004 with cable assembly No. 07-0010. Refer to STAR operation.
- Analog volt-ohmmeter (VOM), 0 to 20v DC, (alternate to STAR). Refer to Appendix A.

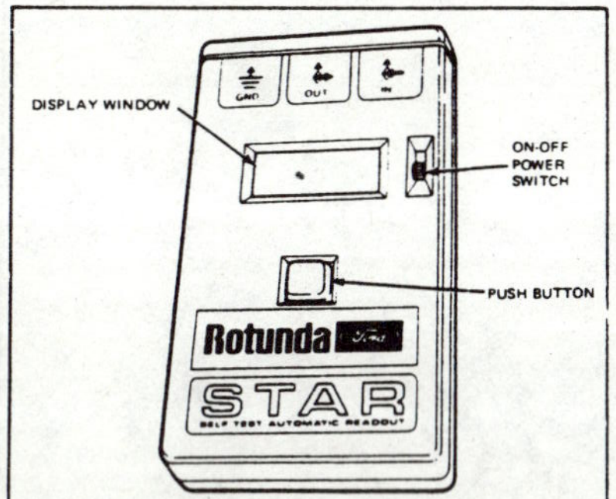

Automatic Readout (STAR) Tester

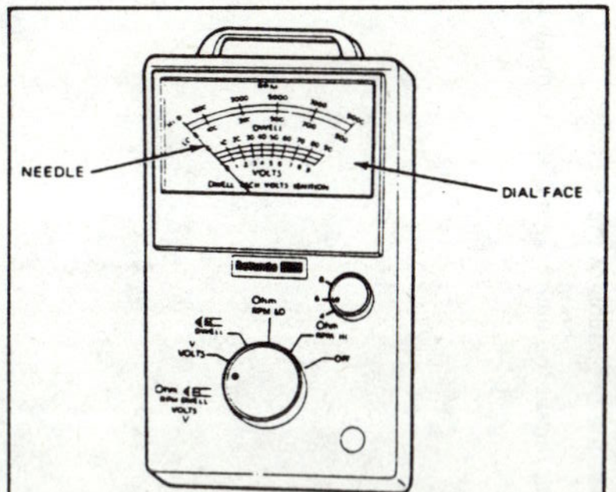

Analog Voltmeter (VOM)

- Jumper wire.
- Vacuum gauge, Rotunda 59-0008 or equivalent. Range 0-30 in. Hg. Resolution 1 in. Hg.

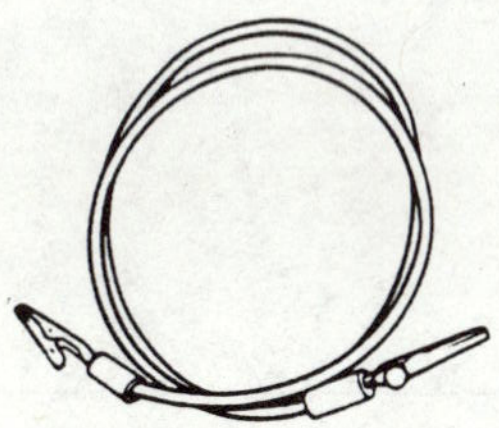

Jumper Wire

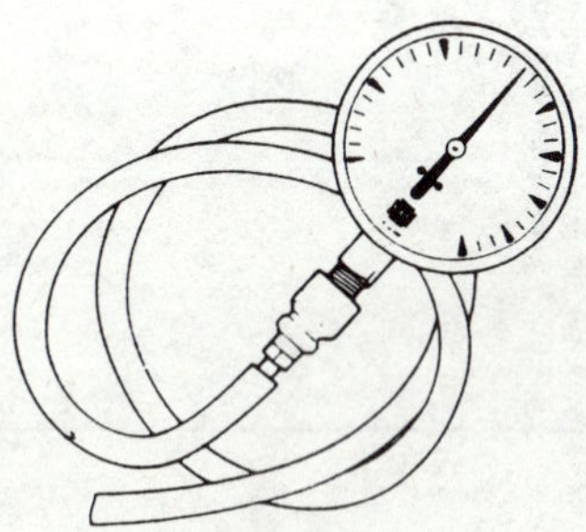

Vacuum Gauge

Test Equipment

- Tachometer, Rotunda No. 59-0010 or equivalent. Range 0-6,000 rpm. Accuracy ± 40 rpm. Resolution 20 rpm.
- Breakout Box, Rotunda T83L-50 EEC-IV or equivalent.

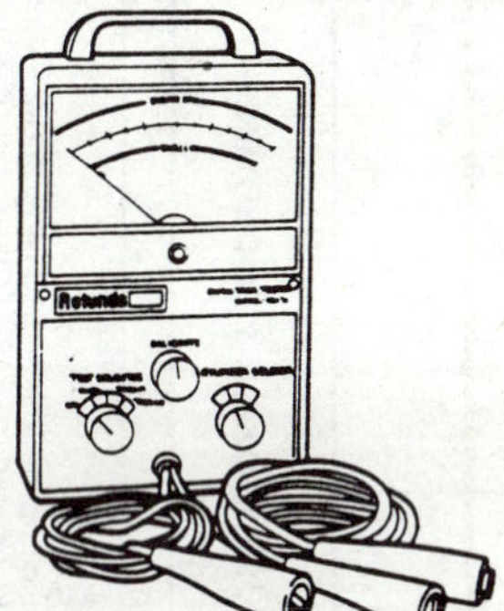

Tachometer

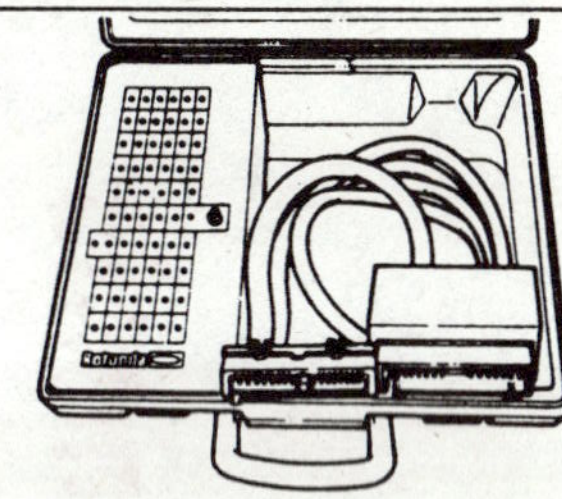

Breakout Box

- Vacuum pump, Rotunda No. 21-0014 or equivalent. Range 0-25 in. Hg.
- Digital volt-ohmmeter (DVOM), Rotunda No. 15-0031 or equivalent. Input impedance 10 Megaohm minimum.

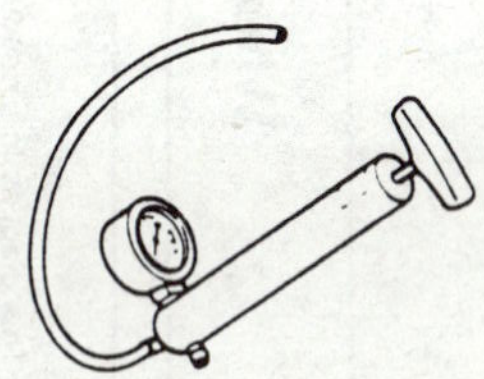

Vacuum Pump

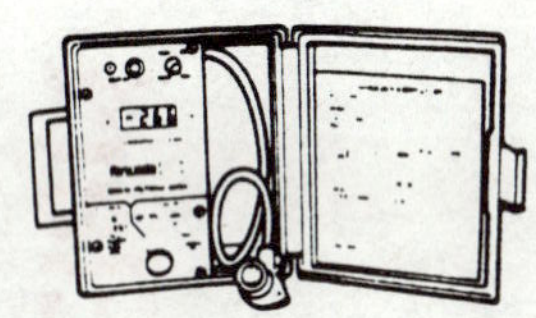

Digital Volt-Ohmmeter (DVOM)

Test Equipment

- Electronic Fuel Injection Pressure Gauge EFI/CFI only, Rotunda No. T80L-9974-A or equivalent.
- Spark tester Tool D81P-6666-A or equivalent.

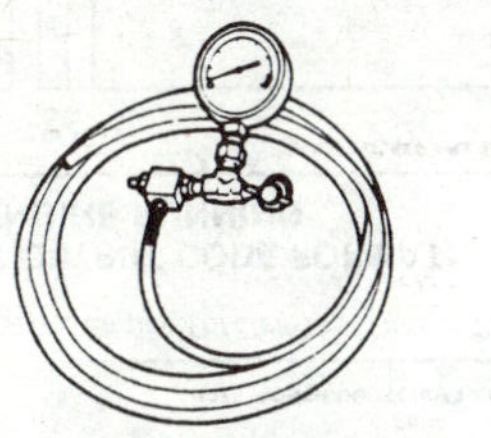
EFI Pressure Gauge

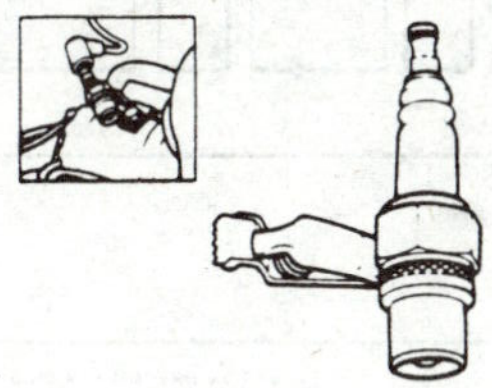
Spark Tester

- Timing light, Rotunda Tool No. 27-0001 or equivalent.
- Non-powered Test lamp.

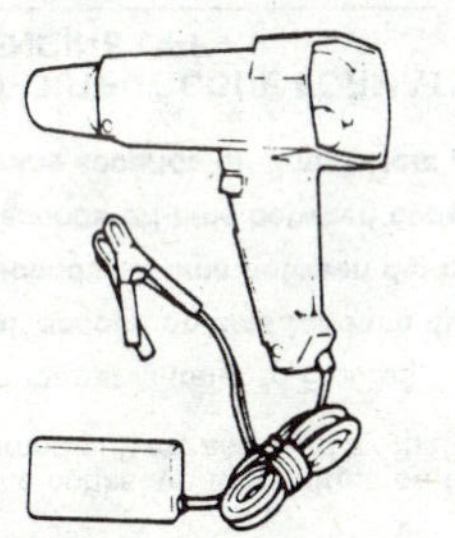
Timing Light

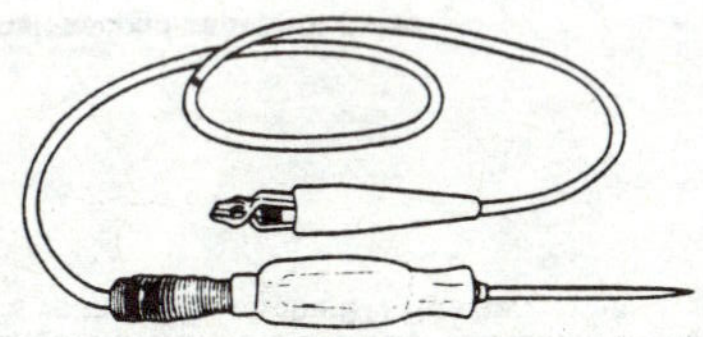
Non-Powered Test-Lamp

Quick Test Description

The Quick Test is a functional test of the EEC-IV system consisting of basic Test Steps (described below). These Steps must be carefully followed, otherwise mis-diagnosis, or the replacement of non-faulty components may result.

Quick Test Steps

1. Visual Check and Vehicle Preparation
   - Checks for obvious faults.
   - Properly prepares the vehicle for testing.
2. Equipment Hook-up
   - Ensures that the proper equipment for gathering test data is ready, prior to testing.
3. Key On, Engine Off Test
   - Is a static check of processor inputs and outputs.
4. Computed Timing Check
   - Verifies the system ability to compute and maintain a fixed spark timing during Self-Test.
5. Engine Running Self-Test
   - Is a dynamic check with the engine in operation.
6. Continuous Self-Test
   - Checks the sensor inputs for opens and shorts while the vehicle is in operation.

The Key On, Engine Off and Engine Running Self-Tests are intended to detect hard faults only, not intermittent faults. Intermittent faults are detected by Continuous Self-Test.

NOTE: A complete Quick Test description is in Appendix A.

Service Code Format

Key On, Engine Off Test

- Fast Codes
- On Demand Code
- Separator Code
- Continuous Code

Engine Running Test

- ID Codes
- Dynamic Response Code
- Fast Code
- On Demand Code

"On Demand" code is information collected from a fault present at the time of testing.

"Continuous" code is information collected and stored from a fault detected in a previous drive-cycle, the fault may not be present at this time.

WARNING: Anyone who departs from the instruction provided in this publication must first establish that he compromises neither his personal safety nor the vehicle integrity by his choice of methods, tools or parts.

Service Code Format

SERVICE CODES

The EEC-IV system communicates service information to the outside world by way of the Self-Test service codes. These service codes are two-digit numbers representing the results of Self-Test.

The service codes are transmitted on the Self-Test output (found in the Self-Test connector) in the form of timed pulses, and read by the technician on a voltmeter or on the STAR tester.

The pulse format is (refer to Figures 1 and 2):

- **One-half second on-time for each digit.**
- **Two seconds off-time between digits.**
- **Four seconds off-time between codes.**
- **Six to nine seconds off-time before and after the half-second separator pulse.**

SELF-TEST OUTPUT CODE FORMAT*
KEY ON, ENGINE OFF

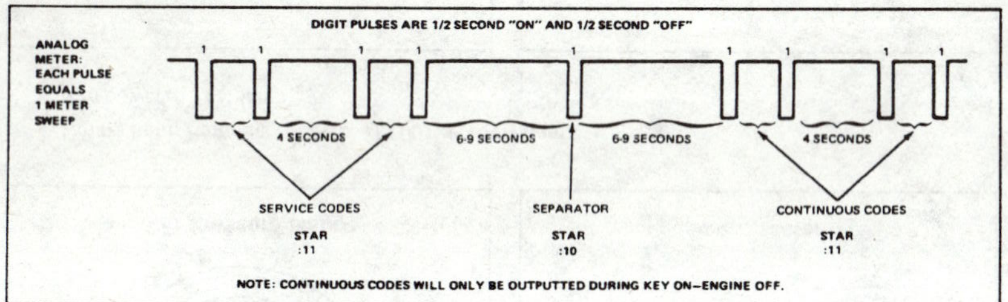

Figure 1 Key On, Engine Off Self-Test Output Code Format

SELF-TEST OUTPUT CODE FORMAT*
KEY ON, ENGINE RUNNING

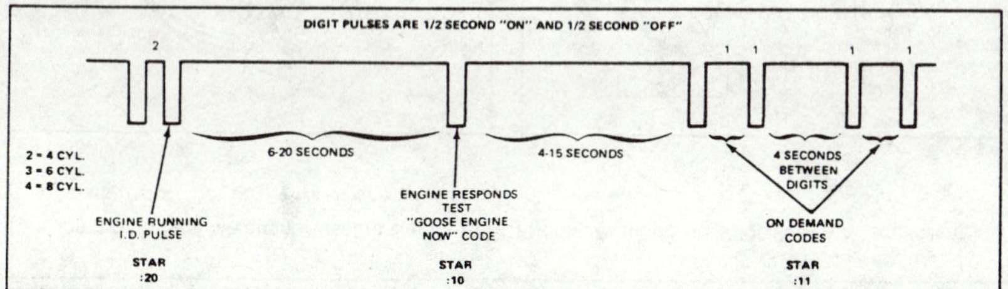

Figure 2 Key On, Engine Running Self-Test Output Code Format

***For a detailed explanation of Self-Test output code format, refer to Appendix B.**

Quick Test

1.6L EFI
1.6L EFI TC
2.3L EFI TC

1.0 VISUAL CHECK AND VEHICLE PREPARATION

Correct test results for the Quick Test are dependent on the proper operation of related non-EEC components systems. It may be necessary to correct faults in these areas before EEC will pass QuickTest. Refer to Diagnostic Routines, Section 2 for service.

Before hooking up any equipment to diagnose the EEC system, make the following checks:

1. Verify the condition of air cleaner and ducting. These components may be removed and reinstalled as necessary for service and/or inspection.
2. Check all engine vacuum hoses for:
   - Leaks or pinched hoses.
   - Proper emission routing per Vehicle Emission Control Information (VECI) Decal.
3. Check the EEC system wiring harness electrical connections for:
   - Proper connections.
   - Loose or detached connectors, wires and terminals.
   - Corrosion.
   - Proper routing of harness.

It may be necessary to disconnect or disassemble the connector assembly to perform some of the inspections. (Note the location of each pin before disassembly).

4. Check processor, sensors and actuators for physical damage.
5. Perform all safety steps required to start and run operational vehicle tests.
6. Apply the emergency brake. Place shift lever in Park; Neutral for manual transmission.
7. **Turn Off all electrical loads** such as the radio, lamps, air conditioner, etc. Be sure doors are closed whenever readings are made.
8. Verify engine coolant is at the specified level.
9. Start engine and idle until the upper radiator hose is hot and pressurized and the throttle is off fast idle. While the engine is operating, check for leaks around the exhaust manifold, EGO sensor, and vacuum hose connections.

   NOTE: If the engine will not start or starts and stalls, continue with this Step 1.0 and on through Steps 2.0 and 3.0 of the Quick Test. If the engine idles rough and/or runs rough, go to Quick Test Step 7.0 , Diagnosis by Symptom.
10. Turn ignition key Off.
11. Service items as required, and proceed to equipment hook-up, Step 2.0 .

Quick Test

1.6L EFI
1.6L EFI TC
2.3L EFI TC

2.0 EQUIPMENT HOOKUP

Using the STAR tester:

- Turn ignition key Off.
- Connect the color-coded adapter cable leads to the STAR tester.
- Connect the adapter cable's two service connectors to the vehicle's appropriate Self-Test connectors.
- Connect the timing light.

Using analog voltmeter:

- Turn ignition key Off.
- Connect a jumper wire from Self-Test Input (STI) to Pin 2, Signal Return on the Self-Test connector (refer to the diagram below).
- Set analog VOM on a DC voltage range to read from 0 to 15 volts DC. Connect VOM from battery (+) to Pin 4 Self-Test Output (STO) in the Self-Test connector.
- Connect the timing light.

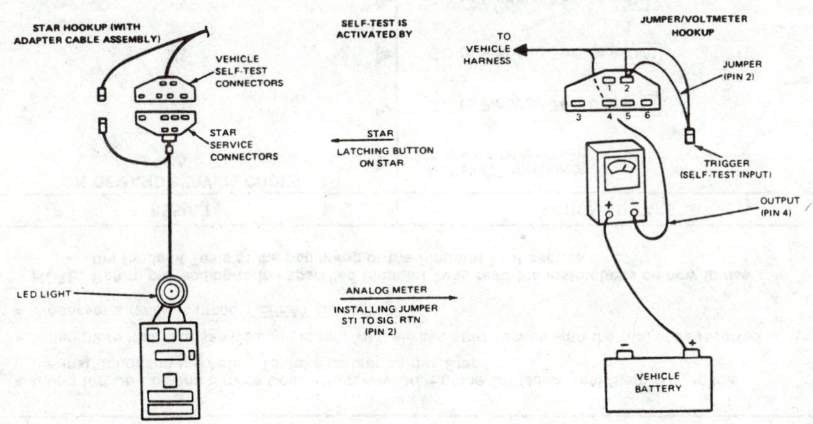

Quick Test

1.6L EFI
1.6L EFI TC
2.3L EFI TC

3.0 KEY ON, ENGINE OFF SELF-TEST

Correct test results for the Quick Test are dependent on the proper operation of related non-EEC-IV components/systems. It may be necessary to correct any faults in these areas before EEC-IV will pass the Quick Test. Refer to the Diagnostic Routines, Section 2 for proper servicing.

- Verify that the vehicle has been properly prepared per Quick Test Steps [1.0] and [2.0] .
- Place transmission in Neutral or Park. Set parking brake. Place A/C or heater control to Off.
- Activate Self-Test.
- Place key in Run to start Self-Test.

NOTE: Do not depress throttle during Self-Test.

- Observe and record all Service Codes indicated. One of the following outputs will occur. (Continuous Service Codes recorded in this Step will be referenced later in Quick Test Step [6.0]).

RESULT				ACTION TO TAKE
CODE FORMAT				
ON DEMAND	SEPARATOR[1]	CONTINUOUS		
11	1(0)	11	▶	Key On, Engine Off **on demand** and **continuous tests** indicate a pass. Go to Quick Test Step [4.0] . If vehicle is a no start, go directly to Pinpoint Test Step [A1] .
ANY CODE(S)	1(0)	11	▶	Key On, Engine Off **on demand** test indicates a fault. Record on demand code(s). Go to Quick Test Step [3.1] .
ANY CODE(S)	1(0)	ANY CODE(S)	▶	Key On, Engine Off **on demand** and **continuous tests** indicate a fault. Record on demand and continuous codes. Go to Quick Test Step [3.1] .
11	1(0)	ANY CODE(S)	▶	**Continuous** test indicates a fault. Record continuous code. Go to Quick Test Step [4.0] .
NO CODES OUTPUTTED			▶	Go to Pinpoint Test Step [Q1] .

[1] **Star tester displays a 10 in place of a 1 for the single digit separator code, therefore the separator code is expressed in the results column as 10.**

Quick Test

1.6L EFI
1.6L EFI TC
2.3L EFI TC

3.1 KEY ON, ENGINE OFF SELF-TEST

- Using the On Demand Service Codes from Key On, Engine Off Quick Test Step 3.0 follow the instructions in the Action To Take column in this Step.
- When more than one service code is received, always start service with the first code received.
- Whenever a repair is made, REPEAT Quick Test.

NOTE: **Before proceeding to the specified Pinpoint Test, read the instructions on how to use the Pinpoint Tests at the beginning of the Pinpoint Test section.**

RESULT		ACTION TO TAKE
ON DEMAND SERVICE CODES		
15	▶	REPLACE processor. REPEAT Quick Test.
21	▶	GO to Pinpoint Test Step [DE1].
22	▶	GO to Pinpoint Test Step [DC1].
23	▶	GO to Pinpoint Test Step [DH1].
24	▶	GO to Pinpoint Test Step [DA1].
26	▶	GO to Pinpoint Test Step [DK1].
51	▶	GO to Pinpoint Test Step [DE10].
53	▶	GO to Pinpoint Test Step [DH20].
54	▶	GO to Pinpoint Test Step [DA10].
56	▶	GO to Pinpoint Test Step [DK10].
61	▶	GO to Pinpoint Test Step [DE20].
63	▶	GO to Pinpoint Test Step [DH30].
64	▶	GO to Pinpoint Test Step [DA20].
66	▶	GO to Pinpoint Test Step [DK20].

Quick Test

1.6L EFI
1.6L EFI TC
2.3L EFI TC

3.1 KEY ON, ENGINE OFF SELF-TEST

RESULT		ACTION TO TAKE
ON DEMAND SERVICE CODES		
67	▶	GO to Pinpoint Test Step [FA1].
NO CODES/ANY CODE NOT LISTED ABOVE	▶	GO to Pinpoint Test Step [Q1].

TEST STEP	RESULT		ACTION TO TAKE
4.0 CHECK TIMING			
• If vehicle is a no start, go directly to Pinpoint Test [A] NO STARTS.	Vehicle stalls while testing.	▶	GO to Pinpoint Test Step [Q30].
• Key Off, wait 10 seconds.			
• Verify Self-Test trigger has been activated.			
• Restart engine and check timing while in Self-Test. You have 2 minutes from the time the last code is outputted to check Self-Test timing.	Timing is not base plus 20° BTDC ±3°	▶	GO to Pinpoint Test Step [P1].
• Correct Self-Test timing equals base plus 20° BTDC ± 3° (see VECI decal for correct base timing).	Timing is base plus 20° BTDC ±3°	▶	GO to Quick Test Step [5.0].
Example: If base timing is 10° BTDC then initial timing during Self-Test is: (10° + 20°) ± 3° = 30° ± 3° therefore correct timing is 27-33° BTDC.			

Quick Test

1.6L EFI
1.6L EFI TC
2.3L EFI TC

5.0 ENGINE RUNNING SELF-TEST

NOTE: If vehicle is a no start, go directly to Pinpoint Test [A1].

- Deactivate Self-Test.
- Start and Run engine at greater than 2000 rpm for 2 minutes. This warms up EGO sensor.
- Turn engine Off, wait 10 seconds.
- Activate Self-Test.
- Start engine. The Engine Running Test will progress as follows:
  —Engine ID code. ①
  —Run test.
  —When the dynamic response ready code occurs, perform a brief WOT. ②
  —Engine running service codes.
- Observe and record all service codes. One of the following outputs will occur.
- If engine stalls in Self-Test, Go to Pinpoint Test Step [Q 30].

RESULT				ACTION TO TAKE
CODE FORMAT				
ENGINE ID	DYNAMIC RESPONSE	ON DEMAND		
2(0)	1(0)	11	▶	Engine Running On Demand Test indicates a pass. If the symptom was of an intermittent nature, Go to Quick Test Step [6.0]. If the drive symptom is currently present, Go to Quick Test Step [7.0]. Otherwise, testing is complete, EEC-IV System is OK.
2(0)	1(0)	ANY CODE(S)	▶	Engine Running On Demand Test indicates a fault. Go to Quick Test Step [5.1].
NO CODES OUTPUTTED			▶	Repeat Self-Test and verify no service codes are present then Go to Pinpoint Test Step [Q1].

① Engine ID code is equal to half of the number of cylinders (a code of 2 equals a 4 cylinder engine) except for STAR tester which adds a zero to all single digit readings (20 equals a 4 cylinder engine).

② STAR displays a 10 in place of a 1 for single digit separator code, therefore, the Dynamic Response code is expressed in the results column as 10.

Quick Test

1.6L EFI
1.6L EFI TC
2.3L EFI TC

5.1 ENGINE RUNNING SELF-TEST

- Using the Service Codes from Engine Running Self-Test Step [5.0] follow the instructions in the Action To Take column in this Step.
- When more than one service code is received, always start service with the first code received.
- Whenever a repair is made, REPEAT Quick Test.

RESULT		ACTION TO TAKE
ENGINE RUNNING SERVICE CODES 12	▶	GO to Pinpoint Test Step [KE1].
13	▶	GO to Pinpoint Test Step [KE8].
21	▶	GO to Pinpoint Test Step [DE1].
22	▶	GO to Pinpoint Test Step [DC1].
23	▶	GO to Pinpoint Test Step [DH1].
24	▶	GO to Pinpoint Test Step [DA1].
25	▶	GO to Pinpoint Test Step [DG1].
26	▶	GO to Pinpoint Test Step [DK1].
34	▶	GO to Pinpoint Test Step [KA1].
41	▶	GO to Pinpoint Test Step [HA8].
42	▶	GO to Pinpoint Test Step [HA5].
73	▶	GO to Pinpoint Test Step [DH40].
76	▶	GO to Pinpoint Test Step [DK30].
77	▶	GO to Pinpoint Test Step [M1].
NO CODES/ANY CODES NOT LISTED ABOVE	▶	GO to Pinpoint Test Step [Q1].

Quick Test

1.6L EFI
1.6L EFI TC
2.3L EFI TC

TEST STEP	RESULT ▶	ACTION TO TAKE
6.0 CONTINUOUS TESTING		
• Do the results of both on demand tests (Key On, Engine Off and Engine Running) result in a pass code 11?	Yes ▶	Go to [6.1].
	No ▶	RETURN to Step [1.0] of Quick Test and make necessary repairs, indicated in Test Steps [3.0] and [5.0] before continuing.
6.1 CLEAR CONTINUOUS CODES		
• Prepare the vehicle for Key On, Engine Off Quick Test per Steps [1.0] [2.0] and [3.0]. • Perform Key On, Engine Off Quick Test. When the service codes begin, exit the Self-Test program (via unlatching STAR or removing the jumper from STI to Signal Return). Exiting Quick Test during code output will clear all codes stored in the continuous memory.	▶	Go to [6.2].
6.2 DETERMINE CODES TO BE TESTED		
• Refer to Continuous Codes recorded in Quick Test Step [3.0]. • Have any of these codes been repaired either in Key On, Engine Off or Engine Running on Demand Tests?	Yes ▶	Disregard codes already serviced. For remaining codes, Go to [6.3].
	No ▶	Go to [6.3].

Quick Test

1.6L EFI
1.6L EFI TC
2.3L EFI TC

6.3 CONTINUOUS MONITOR MODE (WIGGLE TEST)

- Verify proper test equipment hook up per Quick Test Step [2.0].
- **Self-Test de-activated. Do not activate Self-Test unless instructed.**
- Using the service codes determined in Step [6.2], follow the Action To Take instructions. (While in the appropriate Pinpoint Test, one or both of the continuous monitor modes described below may have to be used to find the intermittent fault).
- **Key On, Engine Off**
  With Self-Test de-activated, turn the ignition key to the On position. You are now in the continuous monitor mode (wiggle TEST mode). STO will be activated whenever a fault is detected. A fault is indicated when the STAR LED turns off or VOM deflection is 10.5 V or greater.
- **Engine Running**
  Activate Self-Test and perform engine running Quick Test. After service code output has been completed, do not turn engine off or de-activate Quick Test. Approximately 2 minutes after Code 11 has been outputted the system will enter and remain in continuous monitor mode until Self-Test is de-activated or engine is turned Off. STO will be activated whenever a fault is detected. A fault is indicated when the STAR LED turns off or VOM deflection is 10.5 V or greater.
- Refer to Appendix C for a detailed description of Continuous Monitor.

RESULT		ACTION TO TAKE
CONTINUOUS SERVICE CODES		
11	▶	If drive symptom is currently present, Go to Quick Test Step [7.0], otherwise testing is complete.
14	▶	GO to Pinpoint Test Step [Y1].
18	▶	GO to Pinpoint Test Step [N1].
21	▶	GO to Pinpoint Test Step [DE90].
22	▶	GO to Pinpoint Test Step [DC90].
41	▶	GO to Pinpoint Test Step [HA11].
42	▶	GO to Pinpoint Test Step [HA11].
51	▶	GO to Pinpoint Test Step [DE91].
53	▶	GO to Pinpoint Test Step [DH90].
54	▶	GO to Pinpoint Test Step [DA90].
56	▶	GO to Pinpoint Test Step [DK90].
61	▶	GO to Pinpoint Test Step [DE94].
63	▶	GO to Pinpoint Test Step [DH94].
64	▶	GO to Pinpoint Test Step [DA93].
66	▶	GO to Pinpoint Test Step [DK93].

Quick Test

1.6L EFI
1.6L EFI TC
2.3L EFI TC

TEST STEP	RESULT	ACTION TO TAKE
7.0 DIAGNOSTICS BY SYMPTOM		
• Engine stalls. • Stalls in Self-Test. • Runs rough. • Misses.		• Poor power/ground connections. • Ignition system distributor cap, rotor, wires, coil, plugs. • Base engine valves, cam timing, compression etc. • Fuel delivery, Pinpoint Test Steps HA1 thru HA4 .
• Detonation/spark knock.	2.3L EFI/TC ▶ 1.6L EFI ▶	GO to Pinpoint Test Step DG1 . GO to Pinpoint Test Step KA1 .
• High idle speeds on each restart may be accompanied by detonation for up to 3-5 minutes after a restart.	2.3L EFI/TC ▶ 1.6L EFI ▶	GO to Pinpoint Test Step DG1 . GO to Pinpoint Test Step KA1 .
• Lack of fast idle assist with A/C On.		GO to Pinpoint Test Step FA .
• Gasoline fumes under hood.		GO to Pinpoint Test Step KD1 .
• Shift and light always On or Off.		GO to Pinpoint Test Step KL1 .
• A/C does not cut-off under WOT conditions.		GO to Pinpoint Test Step KM1 .
• Boost cannot exceed 9-11 psi in premium position of octane switch or exceed 9-11 psi in regular position of octane switch 2.3L SVO engine only.		GO to Pinpoint Test Step KN1 .
• Stumble after hot re-start 2.3L and 2.3L SVO engines (HEGO).		GO to Pinpoint Test Step HA21 .

Electrical Schematic

1.6L EFI
1.6L EFI TC

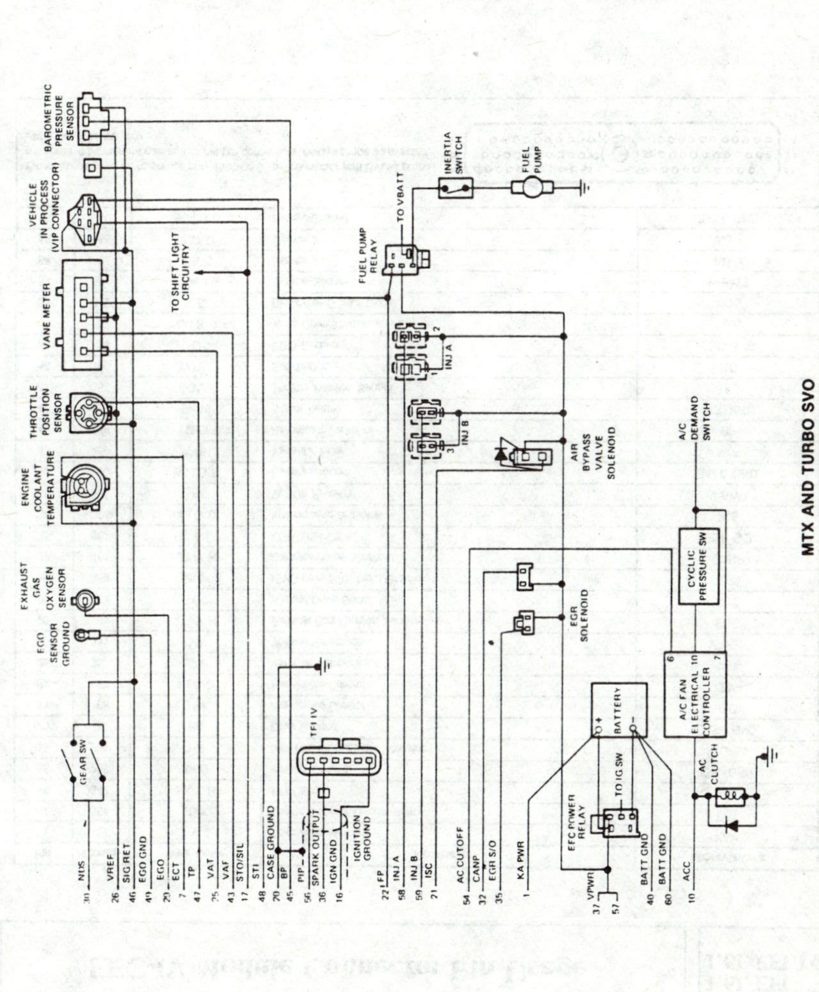

Electrical Schematic

1.6L EFI

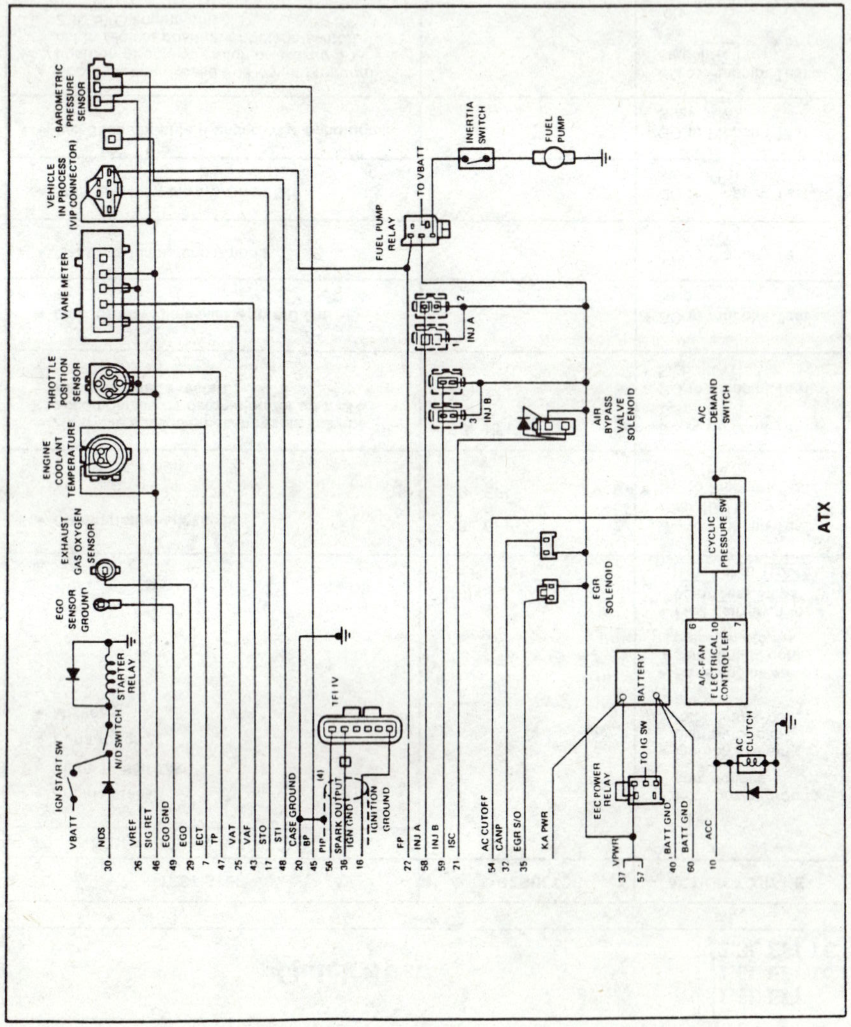

EEC-IV Module Connector Pin Usage

1.6L EFI
1.6L EFI TC

Pin	Circuit	Wire Color	Application	Abbreviations
1	37	Y	Keep Alive Memory	KAM
7	354	LG/Y	Engine Coolant Temperature	ECT
10	347	BK/Y H	A/C Clutch	ACC
16	259	BK/O	Ignition Ground	IGN. GND.
17	201	T/R	Self-Test Output/Shift Indicator Light	STO/SIL
20	57	BK	Case Ground	CASE GND.
21	68	O/BK	Idle Speed Control	ISC
22	97	T/LG D	Fuel Pump Control	FP
25	357	LG/P	Vane Air Temperature	VAT
26	351	O/W	Voltage Reference	V. REF.
29	94	DG/P H	Exhaust Gas Oxygen Sensor	EGO
30	614	GY-O	Neutral Drive Switch	NDS
30	359	BK/W	(SVO Only NDS Input Grounded)	NDS
32	101	GY-Y H	Canister Purge Solenoid	CANP
35	362	Y	EGR Shut Off	EGR S/O
36	324	Y/LG D	Ignition Module Signal	IMS
37	361	R	Vehicle Power	V PWR
40	60	BK-LG D	Battery Ground	BATT. GND.
43	200	W-BK	Vane Air Flow	VAF
45	358	LG/BK	Barometric Pressure	BP
46	359	BK/W	Signal Return	SIG. RET.
47	355	DG/LG	Throttle Position Sensor	TPS
48	100	W/R D	Self-Test In	STI
49	89	O	EGO Ground	EGO GND.
54	73	O-LB H	Wide Open Throttle A/C Cut-Off	WAC
56	349	DB	Distributor Output Signal	DOS
57	361	Red	Vehicle Battery	V BATT.
58	95	T/R D	Injector A	INJ. A
59	96	T/O D	Injector B	INJ. B
60	60	BK-LG D	Battery Ground	BATT. GND.

Pin locations given for reference only. Probing 60 pin connector with DVOM probe will result in permanent damage to the pin connectors. Always probe as directed, using the Breakout Box.

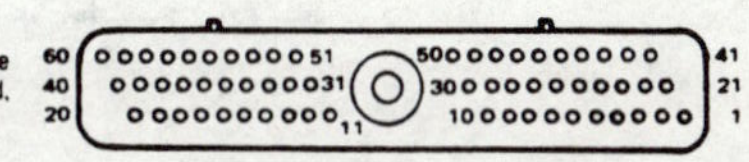

Electrical Schematic

2.3L EFI TC

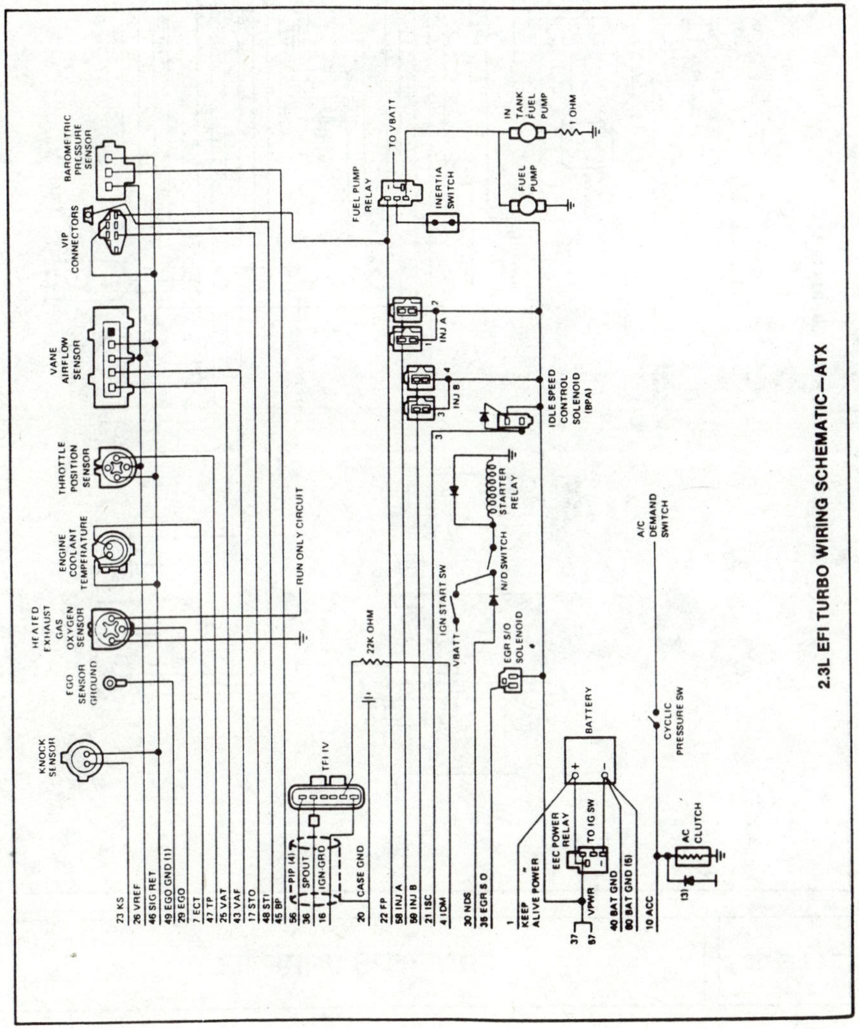

2.3L EFI TURBO WIRING SCHEMATIC—ATX

EEC-IV Module Connector Pin Usage

2.3L EFI TC

Pin	Circuit	Wire Color	Application	Abbreviations
1	37	Y	Keep Alive Memory	KAM
4	11	DG-Y D	Ignition Diagnostic Monitor	IDM
7	354	LG/Y	Engine Coolant Temperature	ECT
10	348	LG/P H	A/C Clutch	ACC
16	259	BK/LG D	Ignition Ground	IGN. GND.
17	201	T/R	Self-Test Output	STO
20	60	BK/LG H	Case Ground	CASE GND.
21	68	O/BK	Idle Speed Control	ISC
22	97	T/LG D	Fuel Pump Control	FP
23	310	Y/R D	Knock Sensor	KS
25	357	LG/P	Vane Air Temperature	VAT
26	351	O/W	Voltage Reference	V. REF.
29	94	DG/P H	Heated Exhaust Gas Oxygen Sensor	HEGO
30	60	BK/LG H	Neutral Drive Switch	NDS
35	362	Y	EGR Shut Off	EGR S/O
36	324	Y/LG D	Ignition Module Signal	IMS
37	361	R	Vehicle Power	V PWR
40	60	BK/LG H	Battery Ground	BATT. GND.
43	200	LG/P	Vane Air Flow	VAF
45	358	LG/BK	Barometric Pressure	BP
46	359	BK/W	Signal Return	SIG. RET.
47	355	DG/LG	Throttle Position Sensor	TPS
48	100	W/R D	Self-Test In	STI
49	89	O	EGO Ground	EGO GND.
56	349	DG	Distributor Output Signal	DOS
57	361	Red	Vehicle Power	V PWR
58	95	T/R D	Injector A	INJ. A
59	96	T/O D	Injector B	INJ. B
60	60	BK/LG H	Battery Ground	BATT. GND.

Pin locations given for reference only. Probing 60 pin connector with DVOM probe will result in permanent damage to the pin connectors. Always probe as directed using the Breakout Box.

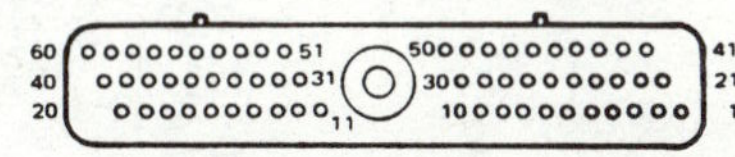

2.3L EFI TC

Electrical Schematic

2.3L EFI TURBO WIRING SCHEMATIC—SVO

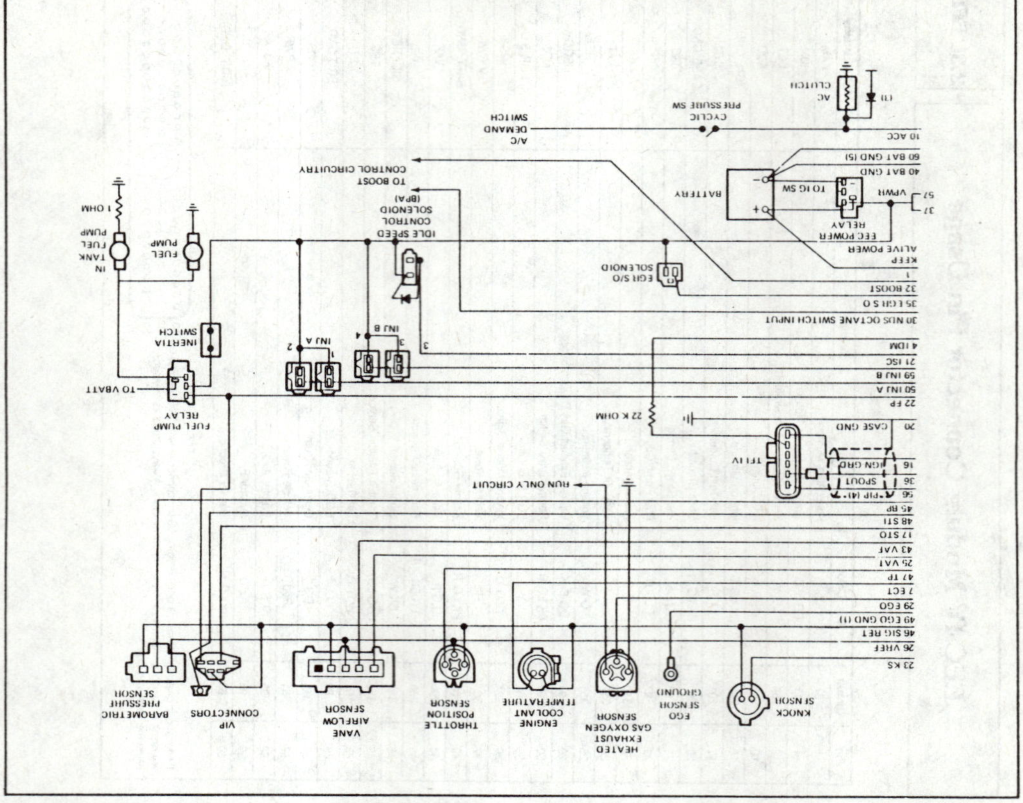

2.3L EFI TC

Electrical Schematic

2.3L EFI TURBO WIRING SCHEMATIC—MTX

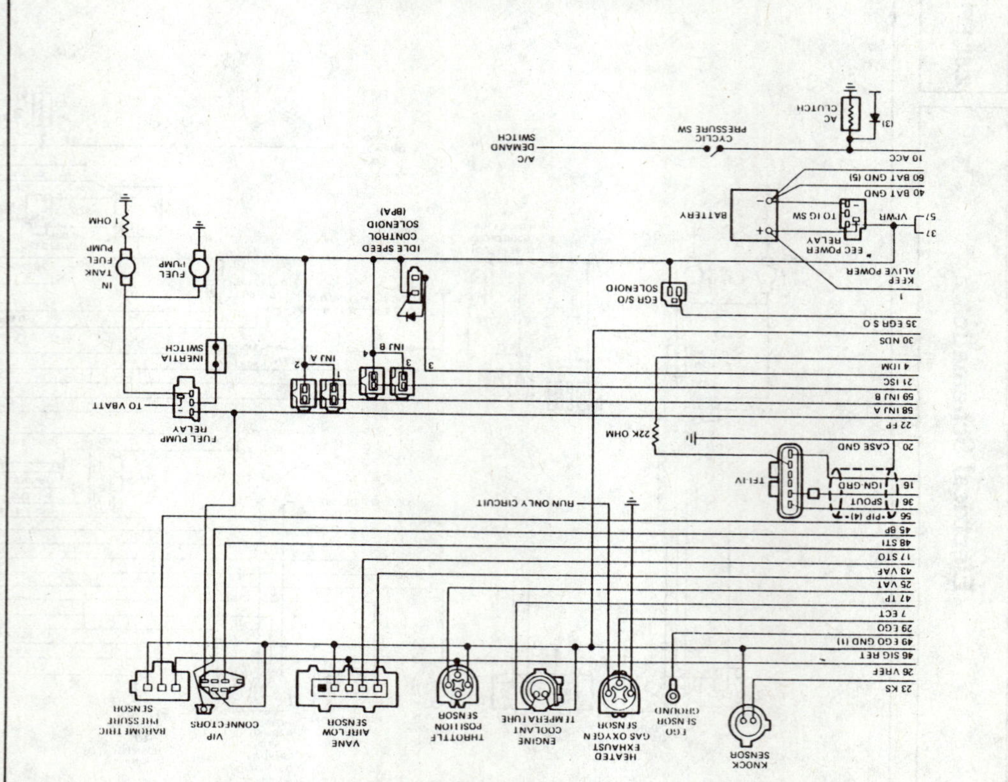

EEC-IV Module Connector Pin Usage

2.3L EFI TC

Pin	Circuit	Wire Color	Application	Abbreviations
1	37	Y	Keep Alive Memory	KAM
4	11	DG-Y D	Ignition Diagnostic Monitor	IDM
7	354	LG Y	Engine Coolant Temperature	ECT
10	347	BK Y H	A C Clutch	ACC
16	259	BK LG D	Ignition Ground	IGN. GND.
17	201	T R	Self-Test Output	STO
20	57	BK	Case Ground	CASE GND.
21	67	O BK	Idle Speed Control	ISC
22	97	T LG D	Fuel Pump Control	FP
23	310	Y R D	Knock Sensor	KS
25	357	LG P	Vane Air Temperature	VAT
26	351	O W	Voltage Reference	V. REF
29	94	DG P H	Heated Exhaust Gas Oxygen Sensor	HEGO
30	376	Y R	Neutral Drive Switch	NDS
32	544	P	Boost SVO Only	BOOST
35	362	Y	EGR Shut Off	EGR S O
36	324	Y LG D	Ignition Module Signal	IMS
37	361	R	Vehicle Power	V PWR
40	57	BK	Battery Ground	BATT. GND.
43	200	LG P	Vane Air Flow	VAF
45	358	LG BK	Barometric Pressure	BP
46	359	BK W	Signal Return	SIG. RET.
47	355	DG LG	Throttle Position Sensor	TPS
48	100	W R D	Self-Test In	STI
49	89	O	EGO Ground	EGO GND.
56	349	DG	Distributor Output Signal	DOS
57	361	Red	Vehicle Power	V PWR
58	95	T R D	Injector A	INJ. A
59	96	T O D	Injector B	INJ. B
60	57	BK	Battery Ground	BATT. GND.

Pin locations given for reference only. Probing 60 pin connector with DVOM probe will result in permanent damage to the pin connectors. Always probe as directed, using the Breakout Box.

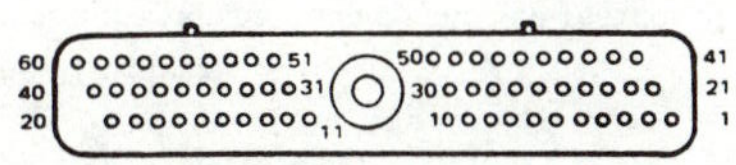

Quick Test

2.3L OHC

1.0 VISUAL CHECK AND VEHICLE PREPARATION

Correct test results for the Quick Test are dependent on the proper operation of related non-EEC components systems. It may be necessary to correct faults in these areas before EEC will pass QuickTest. Refer to Diagnostic Routines, Section 2 for service.

Before hooking up any equipment to diagnose the EEC system, make the following checks:

1. Verify the condition of air cleaner and ducting. These components may be removed and reinstalled as necessary for service and/or inspection.
2. Check all engine vacuum hoses for:
   - Leaks or pinched hoses.
   - Proper emission routing per Vehicle Emission Control Information (VECI) Decal.
3. Check the EEC system wiring harness electrical connections for:
   - Proper connections.
   - Loose or detached connectors, wires and terminals.
   - Corrosion.
   - Proper routing of harness.

It may be necessary to disconnect or disassemble the connector assembly to perform some of the inspections. (Note the location of each pin before disassembly).

4. Check processor, sensors and actuators for physical damage.
5. Perform all safety steps required to start and run operational vehicle tests.
6. Apply the emergency brake. Place shift lever in Park; Neutral for manual transmission.
7. **Turn Off all electrical loads** such as the radio, lamps, air conditioner, etc. Be sure doors are closed whenever readings are made.
8. Verify engine coolant is at the specified level.
9. Start engine and idle until the upper radiator hose is hot and pressurized and the throttle is off fast idle. While the engine is operating, check for leaks around the exhaust manifold, EGO sensor, and vacuum hose connections.

   NOTE: If the engine will not start or starts and stalls, continue with this Step 1.0 and on through Steps 2.0 and 3.0 of the Quick Test. If the engine idles rough and/or runs rough, go to Quick Test Step 7.0 , Diagnosis by Symptom.

10. Turn ignition key Off.
11. Service items as required, and proceed to equipment hook-up, Step 2.0 .

Quick Test — 2.3L OHC

2.0 EQUIPMENT HOOKUP

Using the STAR tester:

- Turn ignition key Off.
- Connect the color-coded adapter cable leads to the STAR tester.
- Connect the adapter cable's two service connectors to the vehicle's appropriate Self-Test connectors.
- Connect the timing light.

Using analog voltmeter:

- Turn ignition key Off.
- Connect a jumper wire from Self-Test Input (STI) to Pin 2, Signal Return on the Self-Test connector (refer to the diagram below).
- Set analog VOM on a DC voltage range to read from 0 to 15 volts DC. Connect VOM from battery (+) to Pin 4 Self-Test Output (STO) in the Self-Test connector.
- Connect the timing light.

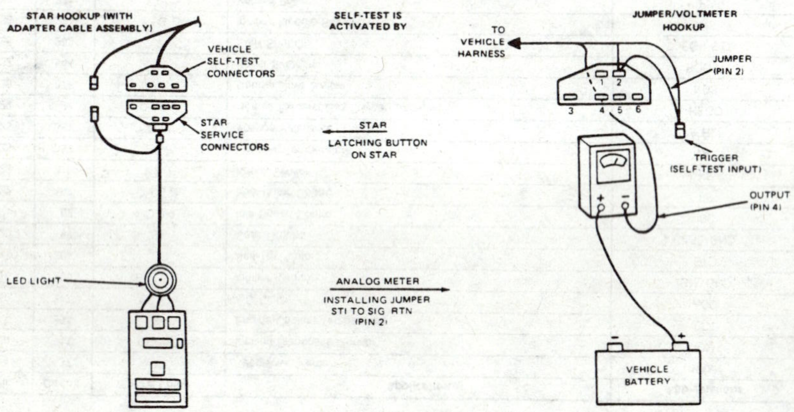

Quick Test — 2.3L OHC

3.0 KEY ON, ENGINE OFF SELF-TEST

Correct test results for the Quick Test are dependent on the proper operation of related non-EEC-IV components/systems. It may be necessary to correct any faults in these areas before EEC-IV will pass the Quick Test. Refer to the Diagnostic Routines, Section 2 for proper servicing.

- Verify that the vehicle has been properly prepared per Quick Test Steps [1.0] and [2.0].
- Place transmission in Neutral or Park. Set parking brake. Place A/C or heater control to Off.
- Activate Self-Test.
- Place key in Run to start Self-Test.

NOTE: Do not depress throttle during Self-Test.

- Observe and record all Service Codes indicated. One of the following outputs will occur. (Continuous Service Codes recorded in this Step will be referenced later in Quick Test Step [6.0]).

RESULT: CODE FORMAT — ON DEMAND	SEPARATOR①	CONTINUOUS		ACTION TO TAKE
11	1(0)	11	▶	Key On, Engine Off **on demand** and **continuous tests** indicate a pass. Go to Quick Test Step [4.0]. If vehicle is a no start, go directly to Pinpoint Test Step [A1].
ANY CODE(S)	1(0)	11	▶	Key On, Engine Off **on demand** test indicates a fault. Record on demand code(s). Go to Quick Test Step [3.1].
ANY CODE(S)	1(0)	ANY CODE(S)	▶	Key On, Engine Off **on demand** and **continuous tests** indicate a fault. Record on demand and continuous codes. Go to Quick Test Step [3.1].
11	1(0)	ANY CODE(S)	▶	**Continuous** test indicates a fault. Record continuous code. Go to Quick Test Step [4.0].
NO CODES OUTPUTTED			▶	Go to Pinpoint Test Step [Q1].

①Star tester displays a 10 in place of a 1 for the single digit separator code, therefore the separator code is expressed in the results column as 10.

Quick Test — 2.3L OHC

3.1 KEY ON, ENGINE OFF SELF-TEST

- Using the On Demand Service Codes from Key On, Engine Off Quick Test Step 3.0 follow the instructions in the Action To Take column in this Step.
- When more than one service code is received, always start service with the first code received.
- Whenever a repair is made, REPEAT Quick Test.

NOTE: **Before proceeding to the specified Pinpoint Test, read the instructions on how to use the Pinpoint Tests at the beginning of the Pinpoint Test section.**

RESULT		ACTION TO TAKE
ON DEMAND SERVICE CODES 15	▶	REPLACE processor. REPEAT Quick Test.
21	▶	GO to Pinpoint Test Step [DE1].
22	▶	GO to Pinpoint Test Step [DF4].
23	▶	GO to Pinpoint Test Step [DH1].
51	▶	GO to Pinpoint Test Step [DE10].
53	▶	GO to Pinpoint Test Step [DH20].
61	▶	GO to Pinpoint Test Step [DE20].
63	▶	GO to Pinpoint Test Step [DH30].
67	▶	GO to Pinpoint Test Step [FA1].
68	▶	GO to Pinpoint Test Step [FB5].
NO CODES OR ANY CODE NOT LISTED ABOVE	▶	GO to Pinpoint Test Step [Q1].

Quick Test — 2.3L OHC

TEST STEP	RESULT		ACTION TO TAKE
4.0 CHECK TIMING			
• If vehicle is a no start, go directly to Pinpoint Test [A] NO STARTS. • Key Off, wait 10 seconds. • Verify Self-Test trigger has been activated.	Vehicle stalls while testing.	▶	GO to Pinpoint Test Step [Q30].
• Restart engine and check timing while in Self-Test. You have 2 minutes from the time the last code is outputted to check Self-Test timing. • Correct Self-Test timing equals base plus 20° BTDC ± 3° (see VECI decal for correct base timing). **Example:** If base timing is 10° BTDC then initial timing during Self-Test is: (10° + 20°) ± 3° = 30° ± 3° therefore correct timing is 27-33° BTDC.	Timing is not base plus 20° BTDC ± 3°	▶	GO to Pinpoint Test Step [P1].
	Timing is base plus 20° BTDC ± 3°	▶	GO to Quick Test Step [5.0].

Quick Test — 2.3L OHC

5.0 ENGINE RUNNING SELF-TEST

NOTE: If vehicle is a no start, go directly to Pinpoint Test [A1].

- Deactivate Self-Test.
- Start and Run engine at greater than 2000 rpm for 2 minutes. This warms up EGO sensor.
- Turn engine Off, wait 10 seconds.
- Activate Self-Test.
- Start engine. The Engine Running Test will progress as follows:
  - —Engine ID code. (1)
  - —Run test.
  - —When the dynamic response ready code occurs, perform a brief WOT. (2)
  - —Engine running service codes.
- Observe and record all service codes. One of the following outputs will occur.
- If engine stalls in Self-Test, Go to Pinpoint Test Step [Q 30].

RESULT: CODE FORMAT — ENGINE ID	DYNAMIC RESPONSE	ON DEMAND		ACTION TO TAKE
2(0)	1(0)	11	▶	Engine Running On Demand Test indicates a pass. If the symptom was of an intermittent nature, Go to Quick Test Step [6.0]. If the drive symptom is currently present, Go to Quick Test Step [7.0]. Otherwise, testing is complete, EEC-IV System is OK.
2(0)	1(0)	ANY CODE(S)	▶	Engine Running On Demand Test indicates a fault. Go to Quick Test Step [5.1].
NO CODES OUTPUTTED			▶	Repeat Self-Test and verify no service codes are present then Go to Pinpoint Test Step [Q1].

(1) Engine ID code is equal to half of the number of cylinders (a code of 2 equals a 4 cylinder engine) except for STAR tester which adds a zero to all single digit readings (20 equals a 4 cylinder engine).

(2) STAR displays a 10 in place of a 1 for single digit separator code, therefore, the Dynamic Response code is expressed in the results column as 10.

Quick Test — 2.3L OHC

5.1 ENGINE RUNNING SELF-TEST

- Using the Service Codes from Engine Running Self-Test Step [5.0] follow the instructions in the Action To Take column in this Step.
- When more than one service code is received, always start service with the first code received.
- Whenever a repair is made, REPEAT Quick Test.

RESULT: ENGINE RUNNING SERVICE CODES		ACTION TO TAKE
12	▶	GO to Pinpoint Test Step [KF1].
13	▶	GO to Pinpoint Test Step [KF20].
21	▶	GO to Pinpoint Test Step [DE1].
22	▶	GO to Pinpoint Test Step [DF1].
23	▶	GO to Pinpoint Test Step [DH3].
41	▶	GO to Pinpoint Test Step [HB1].
42	▶	GO to Pinpoint Test Step [HB20].
43	▶	GO to Pinpoint Test Step [HB40].
44	▶	GO to Pinpoint Test Step [KC1].
45	▶	GO to Pinpoint Test Step [KC1].
46	▶	GO to Pinpoint Test Step [KC1].
58	▶	GO to Pinpoint Test Step [FB1].
72	▶	GO to Pinpoint Test Step [DF10].

Quick Test — 2.3L OHC

5.1 ENGINE RUNNING SELF-TEST (CONTINUED)

RESULT		ACTION TO TAKE
ENGINE RUNNING SERVICE CODES 73	▶	GO to Pinpoint Test Step DH40.
77	▶	GO to Pinpoint Test Step M1.
NO CODES/ANY CODE NOT LISTED ABOVE	▶	GO to Pinpoint Test Step Q1.

TEST STEP	RESULT		ACTION TO TAKE
6.0 CONTINUOUS TESTING			
• Do the results of both on demand tests (Key On, Engine Off and Engine Running) result in a pass code 11?	Yes	▶	Go to 6.1.
	No	▶	RETURN to Step 1.0 of Quick Test and make necessary repairs, indicated in Test Steps 3.0 and 5.0 before continuing.
6.1 CLEAR CONTINUOUS CODES			
• Prepare the vehicle for Key On, Engine Off Quick Test per Steps 1.0 2.0 and 3.0. • Perform Key On, Engine Off Quick Test. When the service codes begin, exit the Self-Test program (via unlatching STAR or removing the jumper from STI to Signal Return). Exiting Quick Test during code output will clear all codes stored in the continuous memory.		▶	Go to 6.2.
6.2 DETERMINE CODES TO BE TESTED			
• Refer to Continuous Codes recorded in Quick Test Step 3.0. • Have any of these codes been repaired either in Key On, Engine Off or Engine Running on Demand Tests?	Yes	▶	Disregard codes already serviced. For remaining codes, Go to 6.3.
	No	▶	Go to 6.3.

Quick Test — 2.3L OHC

6.3 CONTINUOUS MONITOR MODE (WIGGLE TEST)

- Verify proper test equipment hook up per Quick Test Step 2.0.
- **Self-Test de-activated. Do not activate Self-Test unless instructed.**
- Using the service codes determined in Step 6.2, follow the Action To Take instructions.

  (While in the appropriate Pinpoint Test, one or both of the continuous monitor modes described below may have to be used to find the intermittent fault).
- **Key On, Engine Off**
  With Self-Test de-activated, turn the ignition key to the On position. You are now in the continuous monitor mode (wiggle TEST mode). STO will be activated whenever a fault is detected. A fault is indicated when the STAR LED turns off or VOM deflection is 10.5 V or greater.
- **Engine Running**
  Activate Self-Test and perform engine running Quick Test. After service code output has been completed, do not turn engine off or de-activate Quick Test. Approximately 2 minutes after Code 11 has been outputted the system will enter and remain in continuous monitor mode until Self-Test is de-activated or engine is turned Off. STO will be activated whenever a fault is detected. A fault is indicated when the STAR LED turns off or VOM deflection is 10.5 V or greater.
- Refer to Appendix C for a detailed description of Continuous Monitor.

RESULT		ACTION TO TAKE
CONTINUOUS SERVICE CODES 11	▶	If drive symptom is currently present GO to Quick Test Step 7.0, otherwise testing complete.
15	▶	GO to Pinpoint Test Step Q6.
21	▶	GO to Pinpoint Test Step DE90.
22	▶	GO to Pinpoint Test Step DF90.
51	▶	GO to Pinpoint Test Step DE91.
61	▶	GO to Pinpoint Test Step DE94.
63	▶	GO to Pinpoint Test Step DH94.
NO CODES/ CODES NOT LISTED ABOVE	▶	GO to Pinpoint Test Step Q1.

Quick Test — 2.3L OHC

TEST STEP	RESULT	ACTION TO TAKE
7.0 DIAGNOSTICS BY SYMPTOM		
• Engine stalls. • Stalls in Self-Test. • Runs rough. • Misses.		• Poor power/ground connections. • Ignition system distributor cap, rotor, wires, coil, plugs. • Base engine valves, cam timing, compression etc. • EGR Valve.
• Shift indicator light always On or always Off.		GO to Pinpoint Test Step KL1 .
• A/C does not cut-off during WOT operation. • A/C not functioning.		GO to Pinpoint Test Step KM1 .
• High idle speed on hot start. • Low idle speed on cold start. • Engine diesels.		GO to Pinpoint Test Step FC1 .
• Gasoline fumes under hood.		GO to Pinpoint Test Step KD1 .

Electrical Schematic — 2.3L OHC

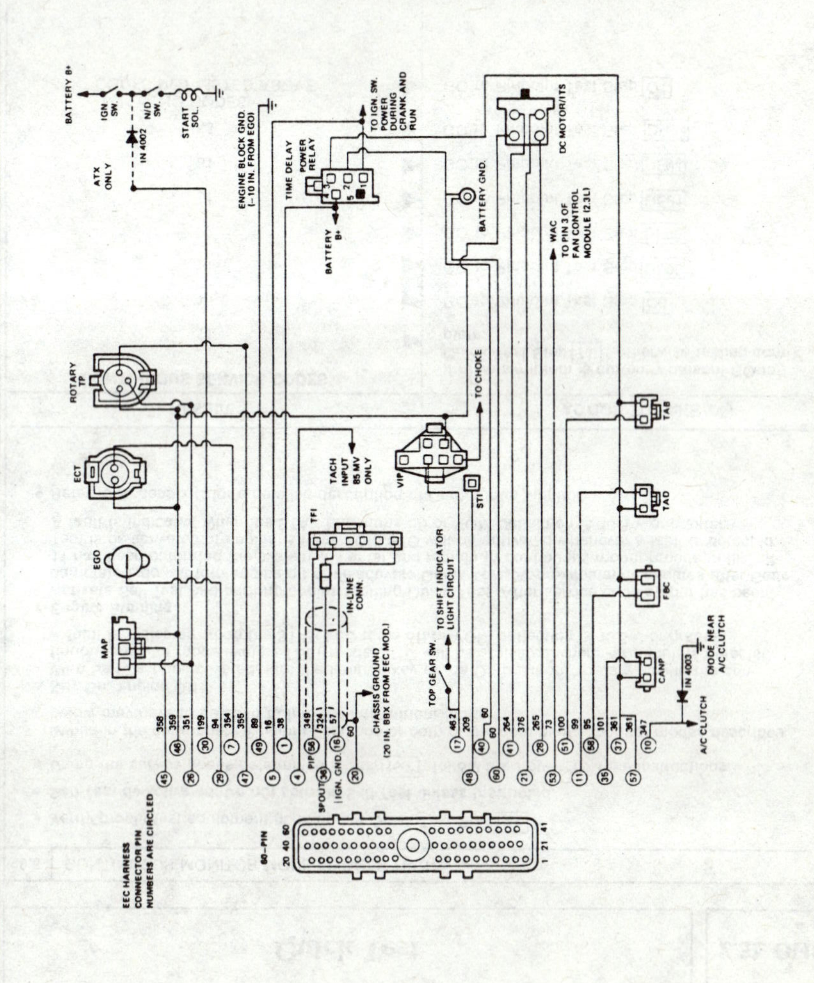

EEC-IV Module Connector Pin Usage

2.3L OHC

Pin	Circuit	Wire Color	Application	Abbreviations
1	38	BK-O	Keep Alive Memory	KAM
5	16	R-LG	Key Power	KEY PWR
7	354	LG-Y	Engine Coolant Temperature	ECT
10	347	BK-Y	A/C Clutch	ACC
11	99	LG-BK D	Thermactor Air Diverter	TAD
16	57	BK	Ignition Ground	IGN GND
17	462	P	Self-Test Output/Shift Ind. Light	STO/SIL
20	60	BK-LG D	Case Ground	CASE GND
21	376	Y-R	Idle Speed Control	ISC
26	351	O-W	Voltage Reference	V. REF.
28	265	LG-W	Idle Tracking Switch	ITS
29	94	DG-P H	Exhaust Gas Oxygen Sensor	EGO
30	199	LB-Y	Neutral Drive Switch	NDS
35	101	GY-Y H	Canister Purge Solenoid	CANP
36	324	Y-LG D	Spark Output	SPOUT
37	361	R	Vehicle Power	V. PWR.
40	60	BK-LG D	Battery Ground	BATT GND
41	264	W-LB	Throttle Position Sensor	TP
45	358	LG-BK	Manifold Absolute Pressure	MAP
46	359	BK-W	Signal Return	SIG RET
47	355	DG-LG	Throttle Position Sensor	TPS
48	209	W-R	Self-Test Input	STI
49	89	O	EGO Ground	EGO GND
51	100	W-R-D	Thermactor Air Bypass Solenoid	TAB
53	73	O-LB H	Vehicle Power	V. PWR.
56	349	DB	Profile Ignition Pick-Up	PIP
57	361	R	Vehicle Power	V. PWR.
58	95	T-R D	Feedback Carburetor Solenoid	FBC
60	60	BK-LG D	Battery Ground	BATT GND

Pin locations given for reference only. Probing 60 pin connector with DVOM probe will result in permanent damage to the pin connectors. Always probe as directed, using the Breakout Box.

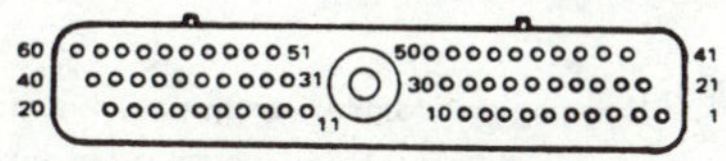

Quick Test

2.3L HSC

1.0 VISUAL CHECK AND VEHICLE PREPARATION

Correct test results for the Quick Test are dependent on the proper operation of related non-EEC components systems. It may be necessary to correct faults in these areas before EEC will pass QuickTest. Refer to Diagnostic Routines, Section 2 for service.

Before hooking up any equipment to diagnose the EEC system, make the following checks:

1. Verify the condition of air cleaner and ducting. These components may be removed and reinstalled as necessary for service and/or inspection.
2. Check all engine vacuum hoses for:
   - Leaks or pinched hoses.
   - Proper emission routing per Vehicle Emission Control Information (VECI) Decal.
3. Check the EEC system wiring harness electrical connections for:
   - Proper connections.
   - Loose or detached connectors, wires and terminals.
   - Corrosion.
   - Proper routing of harness.

It may be necessary to disconnect or disassemble the connector assembly to perform some of the inspections. (Note the location of each pin before disassembly).

4. Check processor, sensors and actuators for physical damage.
5. Perform all safety steps required to start and run operational vehicle tests.
6. Apply the emergency brake. Place shift lever in Park; Neutral for manual transmission.
7. **Turn Off all electrical loads** such as the radio, lamps, air conditioner, etc. Be sure doors are closed whenever readings are made.
8. Verify engine coolant is at the specified level.
9. Start engine and idle until the upper radiator hose is hot and pressurized and the throttle is off fast idle. While the engine is operating, check for leaks around the exhaust manifold, EGO sensor, and vacuum hose connections.

   NOTE: If the engine will not start or starts and stalls, continue with this Step 1.0 and on through Steps 2.0 and 3.0 of the Quick Test. If the engine idles rough and/or runs rough, go to Quick Test Step 7.0 , Diagnosis by Symptom.

10. Turn ignition key Off.
11. Service items as required, and proceed to equipment hook-up, Step 2.0 .

Quick Test — 2.3L HSC

2.0 EQUIPMENT HOOKUP

Using the STAR tester:

- Turn ignition key Off.
- Connect the color-coded adapter cable leads to the STAR tester.
- Connect the adapter cable's two service connectors to the vehicle's appropriate Self-Test connectors.
- Connect the timing light.

Using analog voltmeter:

- Turn ignition key Off.
- Connect a jumper wire from Self-Test Input (STI) to Pin 2, Signal Return on the Self-Test connector (refer to the diagram below).
- Set analog VOM on a DC voltage range to read from 0 to 15 volts DC. Connect VOM from battery (+) to Pin 4 Self-Test Output (STO) in the Self-Test connector.
- Connect the timing light.

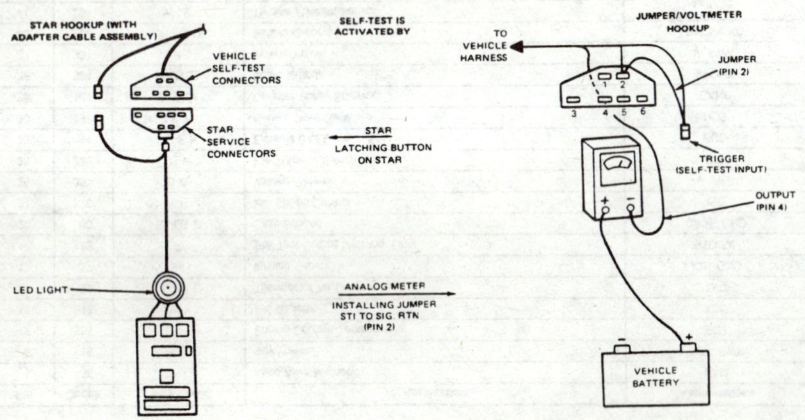

Quick Test — 2.3L HSC

3.0 KEY ON, ENGINE OFF SELF-TEST

Correct test results for the Quick Test are dependent on the proper operation of related non-EEC-IV components/systems. It may be necessary to correct any faults in these areas before EEC-IV will pass the Quick Test. Refer to the Diagnostic Routines, Section 2 for proper servicing.

- Verify that the vehicle has been properly prepared per Quick Test Steps [1.0] and [2.0] .
- Place transmission in Neutral or Park. Set parking brake. Place A/C or heater control to Off.
- Activate Self-Test.
- Place key in Run to start Self-Test.

NOTE: Do not depress throttle during Self-Test.

- Observe and record all Service Codes indicated. One of the following outputs will occur. (Continuous Service Codes recorded in this Step will be referenced later in Quick Test Step [6.0]).

RESULT: CODE FORMAT — ON DEMAND	SEPARATOR①	CONTINUOUS		ACTION TO TAKE
11	1(0)	11	▶	Key On, Engine Off **on demand** and **continuous tests** indicate a pass. Go to Quick Test Step [4.0] . If vehicle is a no start, go directly to Pinpoint Test Step [A1] .
ANY CODE(S)	1(0)	11	▶	Key On, Engine Off **on demand** test indicates a fault. Record on demand code(s). Go to Quick Test Step [3.1] .
ANY CODE(S)	1(0)	ANY CODE(S)	▶	Key On, Engine Off **on demand** and **continuous tests** indicate a fault. Record on demand and continuous codes. Go to Quick Test Step [3.1] .
11	1(0)	ANY CODE(S)	▶	**Continuous** test indicates a fault. Record continuous code. Go to Quick Test Step [4.0] .
NO CODES OUTPUTTED			▶	Go to Pinpoint Test Step [Q1] .

① **Star tester displays a 10 in place of a 1 for the single digit separator code, therefore the separator code is expressed in the results column as 10.**

Quick Test — 2.3L HSC

3.1 KEY ON, ENGINE OFF SELF-TEST

- Using the On Demand Service Codes from Key On, Engine Off Quick Test Step [3.0] follow the instructions in the Action To Take column in this Step.
- When more than one service code is received, always start service with the first code received.
- Whenever a repair is made, REPEAT Quick Test.

NOTE: **Before proceeding to the specified Pinpoint Test, read the instructions on how to use the Pinpoint Tests at the beginning of the Pinpoint Test section.**

RESULT		ACTION TO TAKE
ON DEMAND SERVICE CODES		
15	▶	REPLACE processor. REPEAT Quick Test.
21	▶	GO to Pinpoint Test Step [DE1].
22	▶	GO to Pinpoint Test Step [DF4].
23	▶	GO to Pinpoint Test Step [DH1].
31	▶	GO to Pinpoint Test Step [DD40].
51	▶	GO to Pinpoint Test Step [DE10].
53	▶	GO to Pinpoint Test Step [DH20].
61	▶	GO to Pinpoint Test Step [DE20].
63	▶	GO to Pinpoint Test Step [DH30].
24	▶	GO to Pinpoint Test Step [DB1].
54	▶	GO to Pinpoint Test Step [DB10].
64	▶	GO to Pinpoint Test Step [DB20].
NO CODES OR ANY CODE NOT LISTED ABOVE	▶	GO to Pinpoint Test Step [Q1].

Quick Test — 2.3L HSC

TEST STEP	RESULT		ACTION TO TAKE
4.0 CHECK TIMING			
• If vehicle is a no start, go directly to Pinpoint Test [A] NO STARTS.	Vehicle stalls while testing.	▶	GO to Pinpoint Test Step [Q30].
• Key Off, wait 10 seconds.			
• Verify Self-Test trigger has been activated.			
• Restart engine and check timing while in Self-Test. You have 2 minutes from the time the last code is outputted to check Self-Test timing.	Timing is not base plus 20° BTDC ± 3°	▶	GO to Pinpoint Test Step [P1].
• Correct Self-Test timing equals base plus 20° BTDC ± 3° (see VECI decal for correct base timing). **Example:** If base timing is 10° BTDC then initial timing during Self-Test is: (10° + 20°) ± 3° = 30° ± 3° therefore correct timing is 27-33° BTDC.	Timing is base plus 20° BTDC ± 3°	▶	GO to Quick Test Step [5.0].

Quick Test — 2.3L HSC

5.0 ENGINE RUNNING SELF-TEST

NOTE: If vehicle is a no start, go directly to Pinpoint Test [A1] .

- Deactivate Self-Test.
- Start and Run engine at greater than 2000 rpm for 2 minutes. This warms up EGO sensor.
- Turn engine Off, wait 10 seconds.
- Activate Self-Test.
- Start engine. The Engine Running Test will progress as follows:
  - —Engine ID code.①
  - —Run test.
  - —When the dynamic response ready code occurs, perform a brief WOT. ②
  - —Engine running service codes.
- Observe and record all service codes. One of the following outputs will occur.
- If engine stalls in Self-Test, Go to Pinpoint Test Step [Q 30] .
- Before performing Engine Running Self-Test: Disconnect the vacuum line from canister purge shut-off valve to the canister and plug the canister purge shut-off valve.
- For new (green) engines, less than 1000 miles: If any 40 series service code(s) are received, the following procedure **must** be implemented for the Engine Running Test:
  1. With STI grounded, start the vehicle and immediately increase the rpm to 1750-2500 rpm.
  2. Once desired speed is reached, hold the throttle in that position for the remainder of the test. **Do not** compensate for rpm changes during the test.
  3. When the goose pulse is received, goose the throttle from the elevated rpm then return to idle. **Do not** return to idle before goosing the throttle.

RESULT: CODE FORMAT — ENGINE ID	DYNAMIC RESPONSE	ON DEMAND		ACTION TO TAKE
2(0)	1(0)	11	▶	Engine Running On Demand Test indicates a pass. If the symptom was of an intermittent nature, Go to Quick Test Step [6.0] . If the drive symptom is currently present, Go to Quick Test Step [7.0] . Otherwise, testing is complete, EEC-IV System is OK. Reconnect canister purge.
2(0)	1(0)	ANY CODE(S)	▶	Engine Running On Demand Test indicates a fault. Go to Quick Test Step [5.1] .
NO CODES OUTPUTTED			▶	Repeat Self-Test and verify no service codes are present then Go to Pinpoint Test Step [Q1] .

① Engine ID code is equal to half of the number of cylinders (a code of 2 equals a 4 cylinder engine) except for STAR tester which adds a zero to all single digit readings (20 equals a 4 cylinder engine).

② STAR displays a 10 in place of a 1 for single digit separator code, therefore, the Dynamic Response code is expressed in the results column as 10.

Quick Test — 2.3L HSC

5.1 ENGINE RUNNING SELF-TEST

- Using the Service Codes from Engine Running Self-Test Step [5.0] follow the instructions in the Action To Take column in this Step.
- When more than one service code is received, always start service with the first code received.
- Whenever a repair is made, REPEAT Quick Test.

RESULT: ENGINE RUNNING SERVICE CODES		ACTION TO TAKE
12	▶	GO to Pinpoint Test Step [KG1] .
21	▶	GO to Pinpoint Test Step [DE1] .
22	▶	GO to Pinpoint Test Step [DF1] .
23	▶	GO to Pinpoint Test Step [DH] .
24	▶	GO to Pinpoint Test Step [DB] .
31	▶	GO to Pinpoint Test Step [DD1] .
32	▶	GO to Pinpoint Test Step [DD10] .
33	▶	GO to Pinpoint Test Step [DD10] .
34	▶	GO to Pinpoint Test Step [DD10] .
35	▶	GO to Pinpoint Test Step [DD30] .
41	▶	GO to Pinpoint Test Step [HB1] .
42	▶	GO to Pinpoint Test Step [HB20] .
43	▶	GO to Pinpoint Test Step [HB40] .
44	▶	GO to Pinpoint Test Step [KC1] .
45	▶	GO to Pinpoint Test Step [KC1] .

Quick Test — 2.3L HSC

5.1 ENGINE RUNNING SELF-TEST (CONTINUED)

RESULT		ACTION TO TAKE
ENGINE RUNNING SERVICE CODES 46	▶	GO to Pinpoint Test Step [KC1].
72	▶	GO to Pinpoint Test Step [DF10].
73	▶	GO to Pinpoint Test Step [DH40].
77	▶	GO to Pinpoint Test Step [M1].
NO CODES/ CODES NOT LISTED ABOVE	▶	GO to Pinpoint Test Step [Q1].

TEST STEP	RESULT		ACTION TO TAKE
6.0 CONTINUOUS TESTING			
• Do the results of both on demand tests (Key On, Engine Off and Engine Running) result in a pass code 11?	Yes	▶	Go to [6.1].
	No	▶	RETURN to Step [1.0] of Quick Test and make necessary repairs, indicated in Test Steps [3.0] and [5.0] before continuing.
6.1 CLEAR CONTINUOUS CODES			
• Prepare the vehicle for Key On, Engine Off Quick Test per Steps [1.0] [2.0] and [3.0]. • Perform Key On, Engine Off Quick Test. When the service codes begin, exit the Self-Test program (via unlatching STAR or removing the jumper from STI to Signal Return). Exiting Quick Test during code output will clear all codes stored in the continuous memory.		▶	Go to [6.2].
6.2 DETERMINE CODES TO BE TESTED			
• Refer to Continuous Codes recorded in Quick Test Step [3.0]. • Have any of these codes been repaired either in Key On, Engine Off or Engine Running on Demand Tests?	Yes	▶	Disregard codes already serviced. For remaining codes. Go to [6.3].
	No	▶	Go to [6.3].

Quick Test — 2.3L HSC

6.3 CONTINUOUS MONITOR MODE (WIGGLE TEST)

- Verify proper test equipment hook up per Quick Test Step [2.0].
- **Self-Test de-activated. Do not activate Self-Test unless instructed.**
- Using the service codes determined in Step [6.2], follow the Action To Take instructions. (While in the appropriate Pinpoint Test, one or both of the continuous monitor modes described below may have to be used to find the intermittent fault).
- **Key On, Engine Off**
  With Self-Test de-activated, turn the ignition key to the On position. You are now in the continuous monitor mode (wiggle TEST mode). STO will be activated whenever a fault is detected. A fault is indicated when the STAR LED turns off or VOM deflection is 10.5 V or greater.
- **Engine Running**
  Activate Self-Test and perform engine running Quick Test. After service code output has been completed, do not turn engine off or de-activate Quick Test. Approximately 2 minutes after Code 11 has been outputted the system will enter and remain in continuous monitor mode until Self-Test is de-activated or engine is turned Off. STO will be activated whenever a fault is detected. A fault is indicated when the STAR LED turns off or VOM deflection is 10.5 V or greater.
- Refer to Appendix C for a detailed description of Continuous Monitor.

RESULT		ACTION TO TAKE
CONTINUOUS SERVICE CODES 11	▶	**If drive symptom is currently present GO to Quick Test Step [7.0], otherwise testing complete.**
54	▶	GO to Pinpoint Test Step [DB90].
64	▶	GO to Pinpoint Test Step [DB93].
61	▶	GO to Pinpoint Test Step [DE94].
21	▶	GO to Pinpoint Test Step [DE90].
51	▶	GO to Pinpoint Test Step [DE91].
22	▶	GO to Pinpoint Test Step [DF90].
31	▶	GO to Pinpoint Test Step [DL90].
63	▶	GO to Pinpoint Test Step [DH94].
41	▶	GO to Pinpoint Test Step [HB90].
15	▶	GO to Pinpoint Test Step [Q6].
53	▶	GO to Pinpoint Test Step [DH90].
NO CODES/ CODES NOT LISTED ABOVE	▶	GO to Pinpoint Test Step [Q1].

Quick Test — 2.3L HSC

TEST STEP	RESULT	ACTION TO TAKE
7.0 DIAGNOSTICS BY SYMPTOM • Engine stalls. • Stalls in Self-Test. • Runs rough. • Misses. • Lack of power.		• Poor power/ground connections. • Ignition system distributor cap, rotor, wires, coil, plugs. • Base engine valves, cam timing, compression etc. • EGR Valve, Section 6. • Fuel delivery, Shop Manual, Group 24.
• Shift indicator light always On or always Off.		GO to Pinpoint Test Step KL1 .
• Low idle or runs rough with A/C On.		GO to Pinpoint Test Step KF1 .
• High or low idle speeds on start up. • Engine Diesels.		GO to Pinpoint Test Step FC1 .
• Gasoline fume odors.		GO to Pinpoint Test Step KD1 .

Electrical Schematic — 2.3L HSC

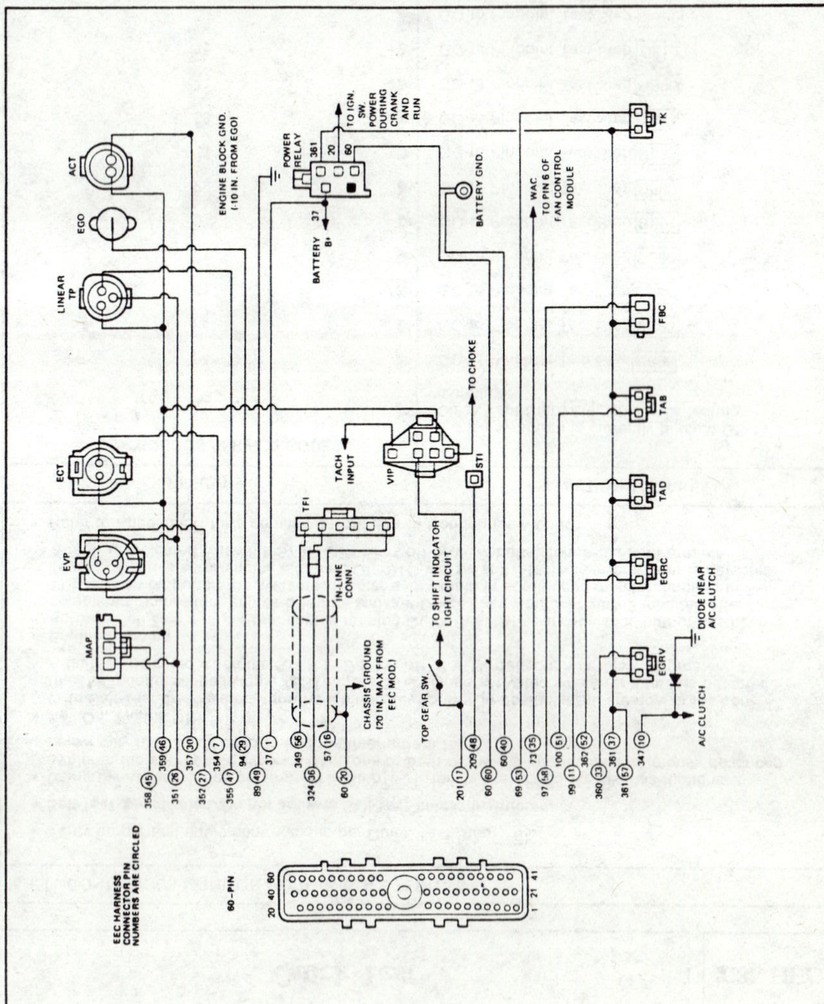

EEC-IV Module Connector Pin Usage — 2.3L HSC

Pin	Circuit	Wire Color	Application	Abbreviations
1	37	Y	Keep Alive Power	KAPWR
7	354	LG-Y	Engine Coolant Temperature Sensor	ECT
10	347	BK-YH	Air Conditioning Clutch	ACC
11	99	LG-B D	Thermactor Air Diverter Solenoid	TAD
16	57	BK-O	Ignition Ground	IGN GRD
17	201	T-R	Self-Test Output	STO
20	60	BK-LG D	Ground	GRD
26	351	O-W	Vehicle Reference Voltage	V REF
27	352	BR-LG	EGR Valve Position Sensor	EVP
29	94	DG-P H	Exhaust Gas Oxygen Sensor	EGO
30	357	LG-P	Air Charge Temperature Sensor	ACT
33	360	DG	EGR Vent Solenoid	EGR V
35	73	O-LB H	Wide Open Throttle A C Cut Off	WAC
36	324	Y-LG D	Ignition Module Signal	SPOUT
37	361	R	Vehicle Power	V PWR
40	60	BK-LG D	Ground	GRD
45	358	LG-BK	Manifold Absolute Pressure Sensor	MAP
46	359	BK-W	Signal Returned	SIG. RTN.
47	355	DG-LG	Throttle Position Sensor	TP
48	209	W-R	Self-Test Input	STI
49	89	O	Exhaust Gas Oxygen Ground	EGO GRD
51	100	W-R D	Thermactor Air Bypass	TAB
52	362	Y	EGR Control Solenoid	EGR C
53	69	R-LG	Throttle Kicker	TK
56	349	DB	Profile Ignition Pick-Up	PIP
57	361	R	Vehicle Power	V PWR
58	97	T-LG D	Feedback Control Solenoid	FBC
60	60	BK-LG D	Ground	GRD

Quick Test — 3.8L CFI

1.0 VISUAL CHECK AND VEHICLE PREPARATION

Correct test results for the Quick Test are dependent on the proper operation of related non-EEC components systems. It may be necessary to correct faults in these areas before EEC will pass QuickTest. Refer to Diagnostic Routines, Section 2 for service.

Before hooking up any equipment to diagnose the EEC system, make the following checks:

1. Verify the condition of air cleaner and ducting. These components may be removed and reinstalled as necessary for service and/or inspection.
2. Check all engine vacuum hoses for:
   - Leaks or pinched hoses.
   - Proper emission routing per Vehicle Emission Control Information (VECI) Decal.
3. Check the EEC system wiring harness electrical connections for:
   - Proper connections.
   - Loose or detached connectors, wires and terminals.
   - Corrosion.
   - Proper routing of harness.

It may be necessary to disconnect or disassemble the connector assembly to perform some of the inspections. (Note the location of each pin before disassembly).

4. Check processor, sensors and actuators for physical damage.
5. Perform all safety steps required to start and run operational vehicle tests.
6. Apply the emergency brake. Place shift lever in Park; Neutral for manual transmission.
7. **Turn Off all electrical loads** such as the radio, lamps, air conditioner, etc. Be sure doors are closed whenever readings are made.
8. Verify engine coolant is at the specified level.
9. Start engine and idle until the upper radiator hose is hot and pressurized and the throttle is off fast idle. While the engine is operating, check for leaks around the exhaust manifold, EGO sensor, and vacuum hose connections.

   NOTE: If the engine will not start or starts and stalls, continue with this Step 1.0 and on through Steps 2.0 and 3.0 of the Quick Test. If the engine idles rough and/or runs rough, go to Quick Test Step 7.0 , Diagnosis by Symptom.

10. Turn ignition key Off.
11. Service items as required, and proceed to equipment hook-up, Step 2.0

Quick Test — 3.8L CFI

2.0 EQUIPMENT HOOKUP

Using the STAR tester:

- Turn ignition key Off.
- Connect the color-coded adapter cable leads to the STAR tester.
- Connect the adapter cable's two service connectors to the vehicle's appropriate Self-Test connectors.
- Connect the timing light.

Using analog voltmeter:

- Turn ignition key Off.
- Connect a jumper wire from Self-Test Input (STI) to Pin 2, Signal Return on the Self-Test connector (refer to the diagram below).
- Set analog VOM on a DC voltage range to read from 0 to 15 volts DC. Connect VOM from battery (+) to Pin 4 Self-Test Output (STO) in the Self-Test connector.
- Connect the timing light.

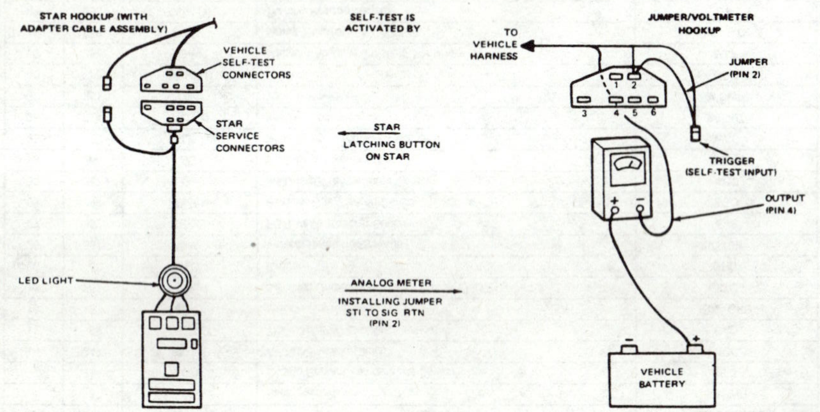

Quick Test — 3.8L CFI

3.0 KEY ON, ENGINE OFF SELF-TEST

Correct test results for the Quick Test are dependent on the proper operation of related non-EEC-IV components/systems. It may be necessary to correct any faults in these areas before EEC-IV will pass the Quick Test. Refer to the Diagnostic Routines, Section 2 for proper servicing.

- Verify that the vehicle has been properly prepared per Quick Test Steps [1.0] and [2.0].
- Place transmission in Neutral or Park. Set parking brake. Place A/C or heater control to Off.
- Activate Self-Test.
- Place key in Run to start Self-Test.

NOTE: Do not depress throttle during Self-Test.

- Observe and record all Service Codes indicated. One of the following outputs will occur. (Continuous Service Codes recorded in this Step will be referenced later in Quick Test Step [6.0]).

RESULT: CODE FORMAT — ON DEMAND	SEPARATOR①	CONTINUOUS		ACTION TO TAKE
11	1(0)	11	▶	Key On, Engine Off **on demand** and **continuous tests** indicate a pass. Go to Quick Test Step [4.0]. If vehicle is a no start, go directly to Pinpoint Test Step [A1].
ANY CODE(S)	1(0)	11	▶	Key On, Engine Off **on demand** test indicates a fault. Record on demand code(s). Go to Quick Test Step [3.1].
ANY CODE(S)	1(0)	ANY CODE(S)	▶	Key On, Engine Off **on demand** and **continuous tests** indicate a fault. Record on demand and continuous codes. Go to Quick Test Step [3.1].
11	1(0)	ANY CODE(S)	▶	**Continuous** test indicates a fault. Record continuous code. Go to Quick Test Step [4.0].
NO CODES OUTPUTTED			▶	Go to Pinpoint Test Step [Q1].

①**Star tester displays a 10 in place of a 1 for the single digit separator code, therefore the separator code is expressed in the results column as 10.**

Quick Test — 3.8L CFI

3.1 KEY ON, ENGINE OFF SELF-TEST

- Using the On Demand Service Codes from Key On, Engine Off Quick Test Step [3.0] follow the instructions in the Action To Take column in this Step.
- When more than one service code is received, always start service with the first code received.
- Whenever a repair is made, REPEAT Quick Test.

NOTE: Before proceeding to the specified Pinpoint Test, read the instructions on how to use the Pinpoint Tests at the beginning of the Pinpoint Test section.

RESULT ON DEMAND SERVICE CODES		ACTION TO TAKE
15	▶	REPLACE processor. REPEAT Quick Test.
21	▶	GO to Pinpoint Test Step [DE1].
22	▶	GO to Pinpoint Test Step [DF4].
23	▶	GO to Pinpoint Test Step [DH1].
24	▶	GO to Pinpoint Test Step [DB1].
31	▶	GO to Pinpoint Test Step [DL40].
51	▶	GO to Pinpoint Test Step [DE10].
53	▶	GO to Pinpoint Test Step [DH20].
54	▶	GO to Pinpoint Test Step [DB10].
61	▶	GO to Pinpoint Test Step [DE20].
63	▶	GO to Pinpoint Test Step [DH30].
64	▶	GO to Pinpoint Test Step [DB20].
65	▶	GO to Pinpoint Test Step [FC1].
68	▶	GO to Pinpoint Test Step [FB1].

Quick Test — 3.8L CFI

3.1 KEY ON, ENGINE OFF SELF-TEST (CONTINUED)

RESULT ON DEMAND SERVICE CODES		ACTION TO TAKE
81	▶	GO to Pinpoint Test Step [UA1].
82	▶	GO to Pinpoint Test Step [UA1].
83	▶	GO to Pinpoint Test Step [UC1].
84	▶	GO to Pinpoint Test Step [UC1].
86	▶	GO to Pinpoint Test Step [UJ1].
87	▶	GO to Pinpoint Test Step [J1].
89	▶	GO to Pinpoint Test Step [UL1].
NO CODES OR ANY CODE NOT LISTED ABOVE	▶	GO to Pinpoint Test Step [Q1].

TEST STEP	RESULT	▶	ACTION TO TAKE
4.0 CHECK TIMING			
• If vehicle is a no start, go directly to Pinpoint Test [A] NO STARTS. • Key Off, wait 10 seconds. • Verify Self-Test trigger has been activated. • Restart engine and check timing while in Self-Test. You have 2 minutes from the time the last code is outputted to check Self-Test timing. • Correct Self-Test timing equals base plus 20° BTDC ± 3° (see VECI decal for correct base timing). **Example:** If base timing is 10° BTDC then initial timing during Self-Test is: (10° + 20°) ± 3° = 30° ± 3° therefore correct timing is 27-33° BTDC.	Vehicle stalls while testing.	▶	GO to Pinpoint Test Step [Q30].
	Timing is not base plus 20° BTDC ± 3°	▶	GO to Pinpoint Test Step [P1].
	Timing is base plus 20° BTDC ± 3°	▶	GO to Quick Test Step [5.0].

Quick Test — 3.8L CFI

5.0 ENGINE RUNNING SELF-TEST

NOTE: If vehicle is a no start, go directly to Pinpoint Test [A1].

- Deactivate Self-Test.
- Start and Run engine at greater than 2000 rpm for 2 minutes. This warms up EGO sensor.
- Turn engine Off, wait 10 seconds.
- Activate Self-Test.
- Start engine. The Engine Running Test will progress as follows:
  - —Engine ID code.(1)
  - —Run test.
  - —When the dynamic response ready code occurs, perform a brief WOT.(2)
  - —Engine running service codes.
- Observe and record all service codes. One of the following outputs will occur.
- If engine stalls in Self-Test, Go to Pinpoint Test Step [Q 30].

RESULT: CODE FORMAT — ENGINE ID	DYNAMIC RESPONSE	ON DEMAND		ACTION TO TAKE
2(0)	1(0)	11	▶	Engine Running On Demand Test indicates a pass. If the symptom was of an intermittent nature, Go to Quick Test Step [6.0]. If the drive symptom is currently present, Go to Quick Test Step [7.0]. Otherwise, testing is complete, EEC-IV System is OK.
2(0)	1(0)	ANY CODE(S)	▶	Engine Running On Demand Test indicates a fault. Go to Quick Test Step [5.1].
NO CODES OUTPUTTED			▶	Repeat Self-Test and verify no service codes are present then Go to Pinpoint Test Step [Q1].

(1) Engine ID code is equal to half of the number of cylinders (a code of 2 equals a 4 cylinder engine) except for STAR tester which adds a zero to all single digit readings (20 equals a 4 cylinder engine).

(2) STAR displays a 10 in place of a 1 for single digit separator code, therefore, the Dynamic Response code is expressed in the results column as 10.

Quick Test — 3.8L CFI

5.1 ENGINE RUNNING SELF-TEST

- Using the Service Codes from Engine Running Self-Test Step [5.0] follow the instructions in the Action To Take column in this Step.
- When more than one service code is received, always start service with the first code received.
- Whenever a repair is made, REPEAT Quick Test.

RESULT: ENGINE RUNNING SERVICE CODES		ACTION TO TAKE
12	▶	GO to Pinpoint Test Step [KF1].
13	▶	GO to Pinpoint Test Step [KF20].
16	▶	GO to Pinpoint Test Step [HC1].
17	▶	GO to Pinpoint Test Step [HC1].
21	▶	GO to Pinpoint Test Step [DE1].
22	▶	GO to Pinpoint Test Step [DF1].
23	▶	GO to Pinpoint Test Step [DH3].
24	▶	GO to Pinpoint Test Step [DB1].
25	▶	GO to Pinpoint Test Step [DG1].
31	▶	GO to Pinpoint Test Step [DL1].
32	▶	GO to Pinpoint Test Step [DL10].
33	▶	GO to Pinpoint Test Step [DL10].
34	▶	GO to Pinpoint Test Step [DL10].
35	▶	GO to Pinpoint Test Step [DL30].

Quick Test — 3.8L CFI

5.1 ENGINE RUNNING SELF-TEST (CONTINUED)

RESULT		ACTION TO TAKE
ENGINE RUNNING SERVICE CODES 41	▶	GO to Pinpoint Test Step [HC1].
42	▶	GO to Pinpoint Test Step [HC1].
43	▶	GO to Pinpoint Test Step [HC1].
44	▶	GO to Pinpoint Test Step [KS1].
45	▶	GO to Pinpoint Test Step [KS1].
46	▶	GO to Pinpoint Test Step [KS1].
48	▶	GO to Pinpoint Test Step [HC1].
58	▶	GO to Pinpoint Test Step [FB1].
72	▶	GO to Pinpoint Test Step [DF10].
73	▶	GO to Pinpoint Test Step [DH40].
77	▶	GO to Pinpoint Test Step [M1].
91	▶	GO to Pinpoint Test Step [HC1].
92	▶	GO to Pinpoint Test Step [HC1].
93	▶	GO to Pinpoint Test Step [HC1].
94	▶	GO to Pinpoint Test Step [HC1].
95	▶	GO to Pinpoint Test Step [HC1].
96	▶	GO to Pinpoint Test Step [HC1].

Quick Test — 3.8L CFI

5.1 ENGINE RUNNING SELF-TEST (CONTINUED)

RESULT		ACTION TO TAKE
ON DEMAND SERVICE CODES 97	▶	GO to Pinpoint Test Step [HC1].
98	▶	GO to Pinpoint Test Step [HC1].
ANY CODE NOT LISTED ABOVE	▶	GO to Pinpoint Test Step [Q1].

TEST STEP	RESULT	▶	ACTION TO TAKE
6.0 CONTINUOUS TESTING			
• Do the results of both on demand tests (Key On, Engine Off and Engine Running) result in a pass code 11?	Yes	▶	Go to [6.1].
	No	▶	RETURN to Step [1.0] of Quick Test and make necessary repairs, indicated in Test Steps [3.0] and [5.0] before continuing.
6.1 CLEAR CONTINUOUS CODES			
• Prepare the vehicle for Key On, Engine Off Quick Test per Steps [1.0] [2.0] and [3.0]. • Perform Key On, Engine Off Quick Test. When the service codes begin, exit the Self-Test program (via unlatching STAR or removing the jumper from STI to Signal Return). Exiting Quick Test during code output will clear all codes stored in the continuous memory.		▶	Go to [6.2].
6.2 DETERMINE CODES TO BE TESTED			
• Refer to Continuous Codes recorded in Quick Test Step [3.0]. • Have any of these codes been repaired either in Key On, Engine Off or Engine Running on Demand Tests?	Yes	▶	Disregard codes already serviced. For remaining codes. Go to [6.3].
	No	▶	Go to [6.3].

Quick Test — 3.8L CFI

6.3 CONTINUOUS MONITOR MODE (WIGGLE TEST)

- Verify proper test equipment hook up per Quick Test Step [2.0].
- **Self-Test de-activated. Do not activate Self-Test unless instructed.**
- Using the service codes determined in Step [6.2], follow the Action To Take instructions. (While in the appropriate Pinpoint Test, one or both of the continuous monitor modes described below may have to be used to find the intermittent fault).
- **Key On, Engine Off**
  With Self-Test de-activated, turn the ignition key to the On position. You are now in the continuous monitor mode (wiggle TEST mode). STO will be activated whenever a fault is detected. A fault is indicated when the STAR LED turns off or VOM deflection is 10.5 V or greater.
- **Engine Running**
  Activate Self-Test and perform engine running Quick Test. After service code output has been completed, do not turn engine off or de-activate Quick Test. Approximately 2 minutes after Code 11 has been outputted the system will enter and remain in continuous monitor mode until Self-Test is de-activated or engine is turned Off. STO will be activated whenever a fault is detected. A fault is indicated when the STAR LED turns off or VOM deflection is 10.5 V or greater.
- Refer to Appendix C for a detailed description of Continuous Monitor.

RESULT		ACTION TO TAKE
CONTINUOUS SERVICE CODES		
11	▶	If drive symptom is currently present GO to Quick Test Step [7.0], otherwise testing complete.
13	▶	GO to Pinpoint Test Step [KF20].
14	▶	GO to Pinpoint Test Step [Y1].
15	▶	GO to Pinpoint Test Step [Q6].
21	▶	GO to Pinpoint Test Step [DE90].
22	▶	GO to Pinpoint Test Step [DC90].
31	▶	GO to Pinpoint Test Step [DL90].
51	▶	GO to Pinpoint Test Step [DE91].
53	▶	GO to Pinpoint Test Step [DH90].
54	▶	GO to Pinpoint Test Step [DB90].
61	▶	GO to Pinpoint Test Step [DE94].
63	▶	GO to Pinpoint Test Step [DH94].
64	▶	GO to Pinpoint Test Step [DB93].
65	▶	GO to Pinpoint Test Step [FC1].
NO CODES/CODES NOT LISTED ABOVE	▶	GO to Pinpoint Test Step [Q1].

Quick Test — 3.8L CFI

TEST STEP	RESULT	ACTION TO TAKE
7.0 DIAGNOSTICS BY SYMPTOM		
• Engine runs rough or miss-fires. • Engine stalls.		• Poor power/ground connections. • Ignition system distributor cap, rotor, wires, coil, plugs. • Base engine valves, cam timing, compression etc. • Fuel delivery. • EGR value function.
• Detonation/spark knock. • Rough idle.		• EGR valve function. • Go to Pinpoint Test Step [KA1].
• Lack of fast idle with A/C On.		• Go to Pinpoint Test Step [KF4].
• Gasoline fumes under hood.		• Go to Pinpoint Test Step [KD1].
• Shift indicator light always On or always Off.		• Go to Pinpoint Test Step [KL1].
• Low idle on code start-up. • High idle on hot start-up. • Engine diesels.		• Go to Pinpoint Test Step [FC1].

Electrical Schematic

3.8L CFI

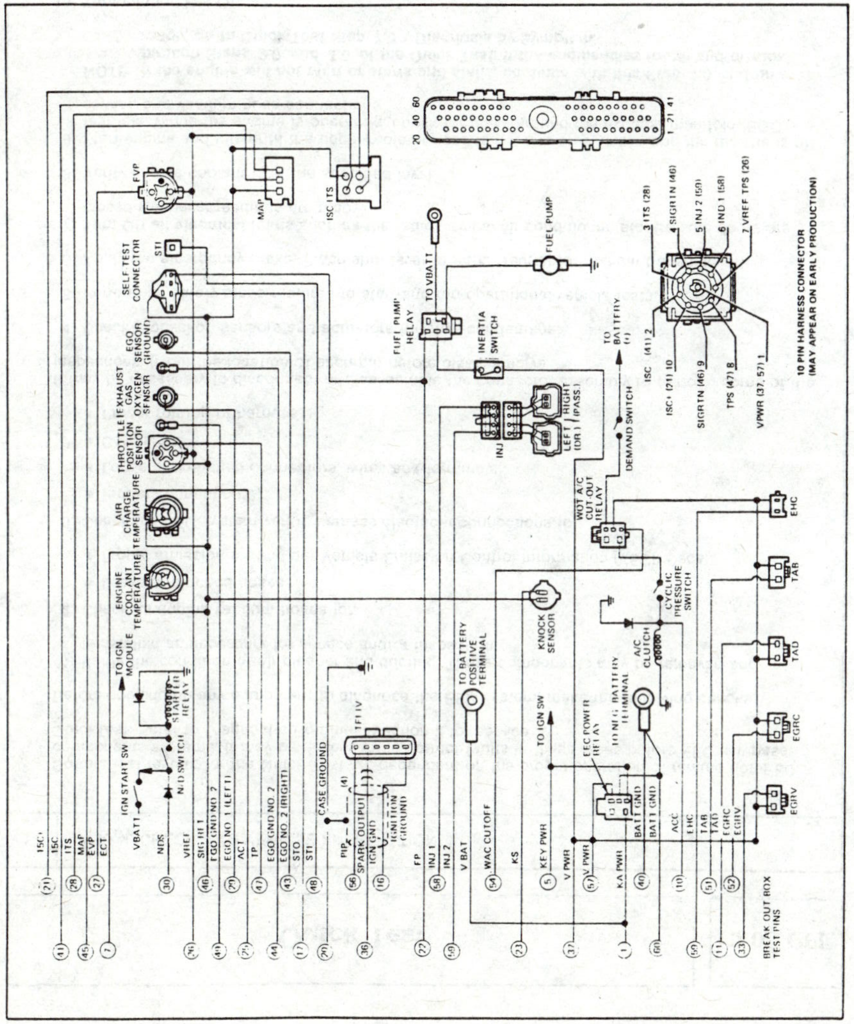

EEC-IV Module Connector Pin Usage

3.8L CFI

Pin	Circuit	Wire Color	Application	Abbreviations
1	38	BK/O	Keep Alive Power	KAPWR
5	16	R/LG	Key Power	KPWR
7	354	LG/Y	Engine Coolant Temperature	ECT
10	347	BK/Y	A/C Clutch	ACC
11	99	LG/BK	Air Management — TAD	AM2
16	259	BK/O	Ignition Ground	IGN. GND.
17	382	Y/B	Self-Test Output	STO
20	57	BK	Case Ground	CSE. GND.
21	376	Y/R	Idle Speed Control (LT MQ, M C)	ISC+
	382	Y/BK	Idle Speed Control (T/X)	ISC+
22	97	T/LG	Fuel Pump	FP
23	310	Y/R	Knock Sensor (M C Only)	KS
25	357	LG/P	Air Charge Temperature	ACT
26	351	O/W	Reference Voltage	V. REF.
27	352	BR/LG	EGR Valve Position	EVP
28	265	LG/W	Idle Tracking Switch	ITS
29	94	DG/P	Exhaust Gas Oxygen Sensor — Left	EGO 1
30	199	LB/Y	Neutral Drive Switch	NDS
33	360	DG	Exhaust Gas Recirculation Vent	EGRV
36	324	Y/LG	Spark Output	SPOUT
37	361	R	Vehicle Power	V. PWR.
40	60	BK/LG	Power Ground	PWR. GND.
41	264	W/LB	Idle Speed Control	ISC –
43	90	DB/LG	Exhaust Gas Oxygen Sensor — Right	EGO 2
44	91	GY/R	Exhaust Gas Oxygen Ground — Right	EGOG 2
45	358	LG/BK	Manifold Absolute Pressure (LTD/MQ AND T/X)	MAP
	356	DB/LG	Manifold Absolute Pressure (M/C)	MAP
46	359	BK/W	Signal Return	SIG RTN
47	355	DG LG	Throttle Position Sensor	TP
48	209	W/R	Self-Test Input	STI
49	89	O	Exhaust Gas Oxygen Ground — Left	EGOG 1
51	100	W/R	Air Management — Tab	AM1
52	362	Y	Exhaust Gas Recirculation Control	EGRC
54	73	O/LB	W.O.T. A/C Cut Off	WAC
55	377	W	Exhaust Heat Control	EHC
56	349	DB	Profile Ignition Pick-Up	PIP
57	361	R	Vehicle Power	V. PWR.
58	95	T/R	Injector 1	INJ 1
59	96	T/O	Injector 2	INJ 2
60	60	BL/LG	Power Ground	PWR. GND.

Pin locations given for reference only. Probing 60 pin connector with DVOM probe will result in permanent damage to the pin connectors. Always probe as directed, using the Breakout Box.

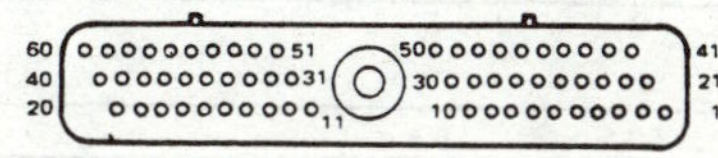

Quick Test — 5.0L CFI

1.0 VISUAL CHECK AND VEHICLE PREPARATION

Correct test results for the Quick Test are dependent on the proper operation of related non-EEC components systems. It may be necessary to correct faults in these areas before EEC will pass QuickTest. Refer to Diagnostic Routines, Section 2 for service.

Before hooking up any equipment to diagnose the EEC system, make the following checks:

1. Verify the condition of air cleaner and ducting. These components may be removed and reinstalled as necessary for service and/or inspection.
2. Check all engine vacuum hoses for:
   - Leaks or pinched hoses.
   - Proper emission routing per Vehicle Emission Control Information (VECI) Decal.
3. Check the EEC system wiring harness electrical connections for:
   - Proper connections.
   - Loose or detached connectors, wires and terminals.
   - Corrosion.
   - Proper routing of harness.

It may be necessary to disconnect or disassemble the connector assembly to perform some of the inspections. (Note the location of each pin before disassembly).

4. Check processor, sensors and actuators for physical damage.
5. Perform all safety steps required to start and run operational vehicle tests.
6. Apply the emergency brake. Place shift lever in Park; Neutral for manual transmission.
7. **Turn Off all electrical loads** such as the radio, lamps, air conditioner, etc. Be sure doors are closed whenever readings are made.
8. Verify engine coolant is at the specified level.
9. Start engine and idle until the upper radiator hose is hot and pressurized and the throttle is off fast idle. While the engine is operating, check for leaks around the exhaust manifold, EGO sensor, and vacuum hose connections.

   NOTE: If the engine will not start or starts and stalls, continue with this Step 1.0 and on through Steps 2.0 and 3.0 of the Quick Test. If the engine idles rough and/or runs rough, go to Quick Test Step 7.0 , Diagnosis by Symptom.

10. Turn ignition key Off.
11. Service items as required, and proceed to equipment hook-up, Step 2.0 .

Quick Test — 5.0L CFI

2.0 EQUIPMENT HOOKUP

Using the STAR tester:

- Turn ignition key Off.
- Connect the color-coded adapter cable leads to the STAR tester.
- Connect the adapter cable's two service connectors to the vehicle's appropriate Self-Test connectors.
- Connect the timing light.

Using analog voltmeter:

- Turn ignition key Off.
- Connect a jumper wire from Self-Test Input (STI) to Pin 2, Signal Return on the Self-Test connector (refer to the diagram below).
- Set analog VOM on a DC voltage range to read from 0 to 15 volts DC. Connect VOM from battery (+) to Pin 4 Self-Test Output (STO) in the Self-Test connector.
- Connect the timing light.

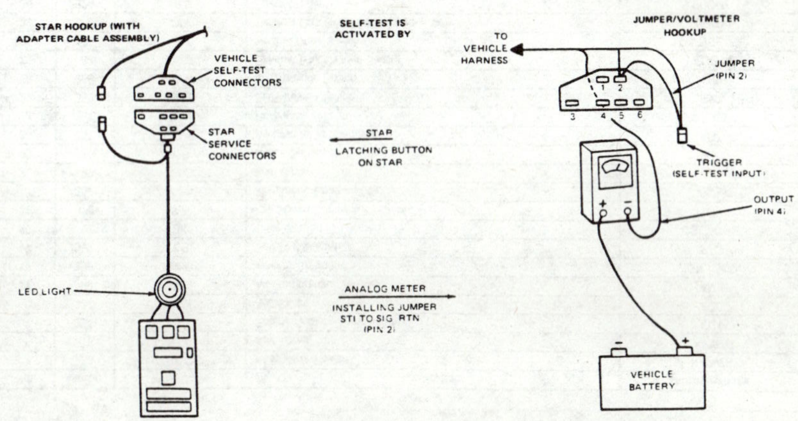

Quick Test — 5.0L CFI

3.0 KEY ON, ENGINE OFF SELF-TEST

Correct test results for the Quick Test are dependent on the proper operation of related non-EEC-IV components/systems. It may be necessary to correct any faults in these areas before EEC-IV will pass the Quick Test. Refer to the Diagnostic Routines, Section 2 for proper servicing.

- Verify that the vehicle has been properly prepared per Quick Test Steps [1.0] and [2.0] .
- Place transmission in Neutral or Park. Set parking brake. Place A/C or heater control to Off.
- Activate Self-Test.
- Place key in Run to start Self-Test.

NOTE: Do not depress throttle during Self-Test.

- Observe and record all Service Codes indicated. One of the following outputs will occur. (Continuous Service Codes recorded in this Step will be referenced later in Quick Test Step [6.0]).

RESULT: CODE FORMAT — ON DEMAND	SEPARATOR①	CONTINUOUS		ACTION TO TAKE
11	1(0)	11	▶	Key On, Engine Off **on demand** and **continuous tests** indicate a pass. Go to Quick Test Step [4.0] . If vehicle is a no start, go directly to Pinpoint Test Step [A1] .
ANY CODE(S)	1(0)	11	▶	Key On, Engine Off **on demand** test indicates a fault. Record on demand code(s). Go to Quick Test Step [3.1] .
ANY CODE(S)	1(0)	ANY CODE(S)	▶	Key On, Engine Off **on demand** and **continuous tests** indicate a fault. Record on demand and continuous codes. Go to Quick Test Step [3.1] .
11	1(0)	ANY CODE(S)	▶	**Continuous** test indicates a fault. Record continuous code. Go to Quick Test Step [4.0] .
NO CODES OUTPUTTED			▶	Go to Pinpoint Test Step [Q1] .

①Star tester displays a 10 in place of a 1 for the single digit separator code, therefore the separator code is expressed in the results column as 10.

Quick Test — 5.0L CFI

3.1 KEY ON, ENGINE OFF SELF-TEST

- Using the On Demand Service Codes from Key On, Engine Off Quick Test Step [3.0] follow the instructions in the Action To Take column in this Step.
- When more than one service code is received, always start service with the first code received.
- Whenever a repair is made, REPEAT Quick Test.

NOTE: Before proceeding to the specified Pinpoint Test, read the instructions on how to use the Pinpoint Tests at the beginning of the Pinpoint Test section.

RESULT: ON DEMAND SERVICE CODES		ACTION TO TAKE
15	▶	REPLACE processor. REPEAT Quick Test.
21	▶	GO to Pinpoint Test Step [DE1] .
22	▶	GO to Pinpoint Test Step [DF4] .
23	▶	GO to Pinpoint Test Step [DH1] .
24	▶	GO to Pinpoint Test Step [DB1] .
31	▶	GO to Pinpoint Test Step [DL40] .
51	▶	GO to Pinpoint Test Step [DE10] .
53	▶	GO to Pinpoint Test Step [DH20] .
54	▶	GO to Pinpoint Test Step [DB10] .
61	▶	GO to Pinpoint Test Step [DE20] .
63	▶	GO to Pinpoint Test Step [DH30] .
64	▶	GO to Pinpoint Test Step [DB20] .
81	▶	GO to Pinpoint Test Step [UA1] .
82	▶	GO to Pinpoint Test Step [UA1] .

Quick Test — 5.0L CFI

3.1 KEY ON, ENGINE OFF SELF-TEST (CONTINUED)

RESULT		ACTION TO TAKE
ON DEMAND SERVICE CODES 83	▶	GO to Pinpoint Test Step [UC1].
84	▶	GO to Pinpoint Test Step [UC1].
85	▶	GO to Pinpoint Test Step [UE1].
86	▶	GO to Pinpoint Test Step [UJ1].
87	▶	GO to Pinpoint Test Step [J1].
88	▶	GO to Pinpoint Test Step [UK1].
NO CODES/CODES NOT LISTED ABOVE	▶	GO to Pinpoint Test Step [Q1].

TEST STEP	RESULT		ACTION TO TAKE
4.0 CHECK TIMING			
• If vehicle is a no start, go directly to Pinpoint Test [A] NO STARTS. • Key Off, wait 10 seconds. • Verify Self-Test trigger has been activated. • Restart engine and check timing while in Self-Test. You have 2 minutes from the time the last code is outputted to check Self-Test timing. • Correct Self-Test timing equals base plus 20° BTDC ± 3° (see VECI decal for correct base timing). **Example:** If base timing is 10° BTDC then initial timing during Self-Test is: (10° + 20°) ± 3° = 30° ± 3° therefore correct timing is 27-33° BTDC.	Vehicle stalls while testing.	▶	GO to Pinpoint Test Step [Q30].
	Timing is not base plus 20° BTDC ± 3°	▶	GO to Pinpoint Test Step [P1].
	Timing is base plus 20° BTDC ± 3°	▶	GO to Quick Test Step [5.0].

Quick Test — 5.0L CFI

5.0 ENGINE RUNNING SELF-TEST

NOTE: If vehicle is a no start, go directly to Pinpoint Test [A1].

- Deactivate Self-Test.
- Start and Run engine at greater than 2000 rpm for 2 minutes. This warms up EGO sensor.
- Turn engine Off, wait 10 seconds.
- Activate Self-Test.
- Start engine. The Engine Running Test will progress as follows:
  —Engine ID code. ①
  —Run test.
  —When the dynamic response ready code occurs, perform a brief WOT. ②
  —Engine running service codes.
- Observe and record all service codes. One of the following outputs will occur.
- If engine stalls in Self-Test, Go to Pinpoint Test Step [Q 30]

RESULT: CODE FORMAT — ENGINE ID	DYNAMIC RESPONSE	ON DEMAND		ACTION TO TAKE
2(0)	1(0)	11	▶	Engine Running On Demand Test indicates a pass. If the symptom was of an intermittent nature, Go to Quick Test Step [6.0]. If the drive symptom is currently present, Go to Quick Test Step [7.0]. Otherwise, testing is complete, EEC-IV System is OK.
2(0)	1(0)	ANY CODE(S)	▶	Engine Running On Demand Test indicates a fault. Go to Quick Test Step [5.1].
NO CODES OUTPUTTED			▶	Repeat Self-Test and verify no service codes are present then Go to Pinpoint Test Step [Q1].

① **Engine ID code is equal to half of the number of cylinders (a code of 2 equals a 4 cylinder engine) except for STAR tester which adds a zero to all single digit readings (20 equals a 4 cylinder engine).**

② **STAR displays a 10 in place of a 1 for single digit separator code, therefore, the Dynamic Response code is expressed in the results column as 10.**

Quick Test — 5.0L CFI

5.1 ENGINE RUNNING SELF-TEST

- Using the Service Codes from Engine Running Self-Test Step [5.0] follow the instructions in the Action To Take column in this Step.
- When more than one service code is received, always start service with the first code received.
- Whenever a repair is made, REPEAT Quick Test.

RESULT		ACTION TO TAKE
ENGINE RUNNING SERVICE CODES 12	▶	GO to Pinpoint Test Step [KT1].
13	▶	GO to Pinpoint Test Step [KT1].
16	▶	GO to Pinpoint Test Step [HC8].
17	▶	GO to Pinpoint Test Step [HC8].
21	▶	GO to Pinpoint Test Step [DE1].
22	▶	GO to Pinpoint Test Step [DF1].
23	▶	GO to Pinpoint Test Step [DH3].
24	▶	GO to Pinpoint Test Step [DB1].
31	▶	GO to Pinpoint Test Step [DL1].
32	▶	GO to Pinpoint Test Step [DL10].
33	▶	GO to Pinpoint Test Step [DL10].
34	▶	GO to Pinpoint Test Step [DL10].
35	▶	GO to Pinpoint Test Step [DL30].

Quick Test — 5.0L CFI

5.1 ENGINE RUNNING SELF-TEST

RESULT		ACTION TO TAKE
ENGINE RUNNING SERVICE CODES 41	▶	GO to Pinpoint Test Step [HC5].
42	▶	GO to Pinpoint Test Step [HC11].
43	▶	GO to Pinpoint Test Step [HC1].
44	▶	GO to Pinpoint Test Step [KS1].
45	▶	GO to Pinpoint Test Step [KS1].
46	▶	GO to Pinpoint Test Step [KS1].
47	▶	GO to Pinpoint Test Step [HC11].
72	▶	GO to Pinpoint Test Step [DF10].
73	▶	GO to Pinpoint Test Step [DH40].
77	▶	GO to Pinpoint Test Step [M1].
NO CODES/CODES NOT LISTED ABOVE	▶	GO to Pinpoint Test Step [Q1].

Quick Test — 5.0L CFI

TEST STEP	RESULT ▶	ACTION TO TAKE
6.0 CONTINUOUS TESTING		
• Do the results of both on demand tests (Key On, Engine Off and Engine Running) result in a pass code 11?	Yes ▶	Go to [6.1].
	No ▶	RETURN to Step [1.0] of Quick Test and make necessary repairs, indicated in Test Steps [3.0] and [5.0] before continuing.
6.1 CLEAR CONTINUOUS CODES		
• Prepare the vehicle for Key On, Engine Off Quick Test per Steps [1.0] [2.0] and [3.0]. • Perform Key On, Engine Off Quick Test. When the service codes begin, exit the Self-Test program (via unlatching STAR or removing the jumper from STI to Signal Return). Exiting Quick Test during code output will clear all codes stored in the continuous memory.	▶	Go to [6.2].
6.2 DETERMINE CODES TO BE TESTED		
• Refer to Continuous Codes recorded in Quick Test Step [3.0]. • Have any of these codes been repaired either in Key On, Engine Off or Engine Running on Demand Tests?	Yes ▶	Disregard codes already serviced. For remaining codes. Go to [6.3].
	No ▶	Go to [6.3].

Quick Test — 5.0L CFI

6.3 CONTINUOUS MONITOR MODE (WIGGLE TEST)

- Verify proper test equipment hook up per Quick Test Step [2.0].
- **Self-Test de-activated. Do not activate Self-Test unless instructed.**
- Using the service codes determined in Step [6.2], follow the Action To Take instructions.

  (While in the appropriate Pinpoint Test, one or both of the continuous monitor modes described below may have to be used to find the intermittent fault).
- **Key On, Engine Off**
  With Self-Test de-activated, turn the ignition key to the On position. You are now in the continuous monitor mode (wiggle TEST mode). STO will be activated whenever a fault is detected. A fault is indicated when the STAR LED turns off or VOM deflection is 10.5 V or greater.
- **Engine Running**
  Activate Self-Test and perform engine running Quick Test. After service code output has been completed, do not turn engine off or de-activate Quick Test. Approximately 2 minutes after Code 11 has been outputted the system will enter and remain in continuous monitor mode until Self-Test is de-activated or engine is turned Off. STO will be activated whenever a fault is detected. A fault is indicated when the STAR LED turns off or VOM deflection is 10.5 V or greater.
- Refer to Appendix C for a detailed description of Continuous Monitor.

RESULT CONTINUOUS SERVICE CODES		ACTION TO TAKE
11	▶	If drive symptom is currently present GO to Quick Test Step [7.0], otherwise testing complete.
14	▶	GO to Pinpoint Test Step [Y1].
15	▶	GO to Pinpoint Test Step [Q6].
18	▶	GO to Pinpoint Test Step [N1].
21	▶	GO to Pinpoint Test Step [DE90].
22	▶	GO to Pinpoint Test Step [DF90].
31	▶	GO to Pinpoint Test Step [DL90].
41	▶	GO to Pinpoint Test Step [HC5].
42	▶	GO to Pinpoint Test Step [HC8].
51	▶	GO to Pinpoint Test Step [DE91].
54	▶	GO to Pinpoint Test Step [DA90].
61	▶	GO to Pinpoint Test Step [DE94].
64	▶	GO to Pinpoint Test Step [DA93].

Quick Test — 5.0L CFI

TEST STEP	RESULT	ACTION TO TAKE
7.0 DIAGNOSTICS BY SYMPTOM		
• Engine runs rough or miss fires. • Engine stalls. • Lack of power.		• Poor power/ground connections. • Ignition system distributor cap, rotor, wires, coil, plugs. • Base engine valves, cam timing, compression etc. • GO to Fuel control HC11. • EGR Valve Function.
• Detonation/spark knock. • Rough idle.		• EGR Valve Function. • GO to Pinpoint Test Step DL1
• Lack of fast idle with A/C On.		• GO to Pinpoint Test Step KT4.
• Gasoline fumes under hood.		• GO to Pinpoint Test Step KD1.

Electrical Schematic — 5.0L CFI

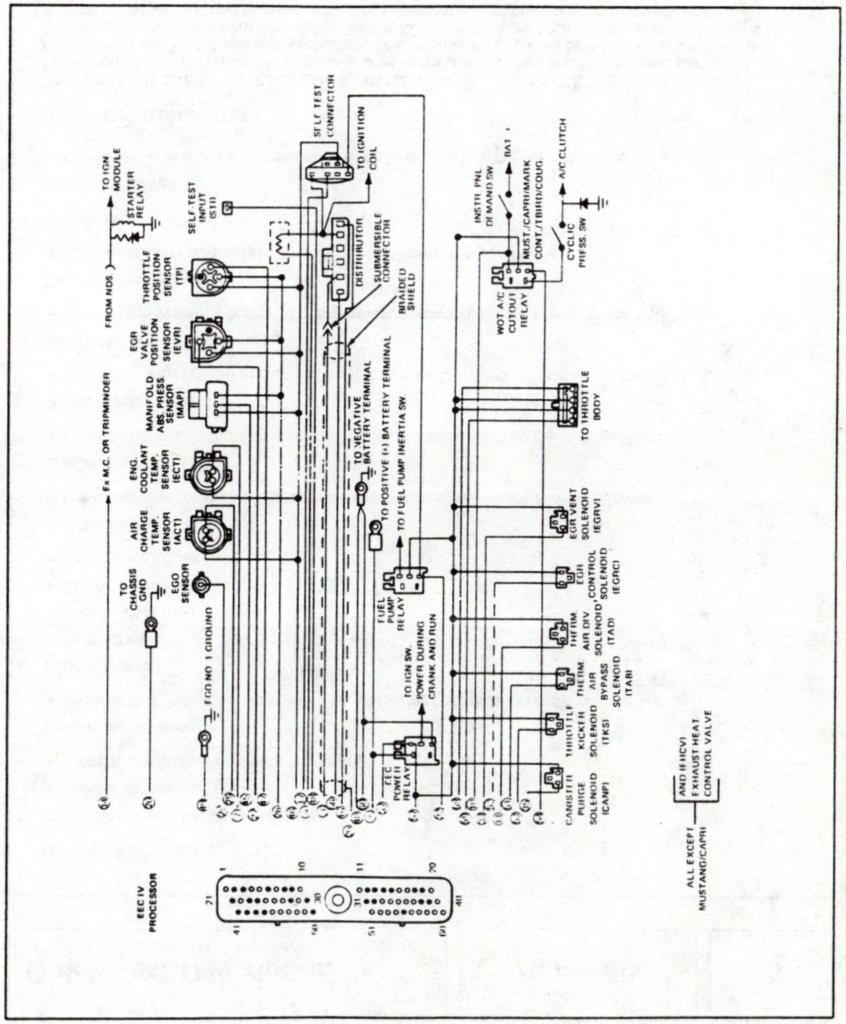

EEC-IV Module Connector Pin Usage — 5.0L CFI

Pin	Circuit	Wire Color	Application	Abbreviations
1	37	Y	Keep Alive Power	KAPWR
4	11	DG-Y D	Ignition Diagnostic Monitor	IDM
7	354	LG-Y	Engine Coolant Temperature Sensor	ECT
10	347	BK-Y H	A C Clutch Compressor Signal	ACC
11	99	LG-BK D	Thermactor Air Diverter Solenoid	TAD
16	57	BK	Ignition Ground	IGN. GND.
17	201	T-R	Self-Test Output	STO
20	60	W-LB H	Case Ground	CSE. GND.
22	97	T-LG D	Fuel Pump Relay	FP
25	357	LG-P	Air Charge Temperature Sensor	ACT
26	351	O-W	Reference Voltage	V. REF
27	352	BR-LG	EGR Valve Position Sensor	EVP
29	94	DG-P H	Exhaust Gas Oxygen Sensor	EGO
33	360	DG	EGR Vent Solenoid	EGRV
34	305	LB-P H	Data Output Link For Tripminder	DOL
35	235	R-BK	Canister Purge Solenoid	CANP
36	324	Y-LG D	Spark Output	SPOUT
37	361	R	Vehicle Power	V. PWR
40	60	BK-LG D	Power Ground	PWR. GND.
45	358	LG-BK	Manifold Absolute Pressure Sensor	MAP
46	359	BK-W	Signal Return (Ground)	SIG. RTN.
47	355	DG-LG	Throttle Position Sensor	TP
48	200	W-BK	Self-Test Input	STI
49	89	O	EGO Ground	EGO GND.
51	100	W-R D	Thermactor Air Bypass Solenoid	TAB
52	362	Y	EGR Control Solenoid	EGRC
53	69	R-LG	Throttle Kicker Solenoid	TK
54	331	R	WOT A C Cutoff	WAC
56	349	DB	Profile Ignition Pick-Up	PIP
57	361	R	Vehicle Power	V. PWR
58	95	T-R D	Fuel Injector	INJ
59	96	T-O D	Fuel Injector	INJ
60	60	BK-LG D	Power Ground	PWR. GND.

Pin locations given for reference only. Probing 60 pin connector with DVOM probe will result in permanent damage to the pin connectors. Always probe as directed using the Breakout Box.

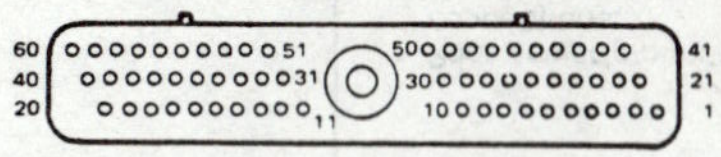

Quick Test Description — Appendix A

QUICK TEST STEPS

Visual Check and Vehicle Preparation

- Checks for obvious faults
- Properly prepares the vehicle for testing.

Equipment Hookup

- Ensures that the proper equipment for gathering test data is ready, prior to testing.

Service Codes

- Fast codes
- Engine identification codes
- Engine service code
- Continuous codes

Key On, Engine Off Self-Test

- Checks processor inputs against calibrated sensor values for Key On, Engine Off.

Continuous Self-Test

- Checks the sensor inputs for opens and shorts while the vehicle is in operation.

Output Cycling Test

- Processor activates all outputs when the throttle is cycled for additional diagnostics.

Computed Timing Check

- Verifies the system's ability to compute and maintain engine timing in the Self-Test.

Engine Running Test

- Checks sensors and actuators under actual operating conditions.

Wiggle Test

Pinpoint Routines

- Corrective action to be entered only when instructed by the functional test.

SELF-TEST DESCRIPTION

The Self-Test is divided into three specialized tests: Key On, Engine Off, Engine Running, and Continuous Testing. The Self-Test is not a conclusive test by itself, but is used as a part of the functional Quick-Test diagnostic procedure. The processor stores the Self-Test program in its permanent memory. When activated, it checks the EEC-IV system by testing its memory integrity and processing capability, and verifies that various sensors and actuators are connected and operating properly.

The Key On, Engine Off and Engine Running tests are functional tests which only detect faults present at the time of the Self-Test. Continuous testing is an ongoing test that stores fault information for retrieval at a later time (during the Self-Test).

Reading Codes—Analog Voltmeter

When a service code is reported on the analog voltmeter for a function test, it will represent itself as a pulsing or sweeping movement of the voltmeter's needle across the dial face of the voltmeter (Figure 3). Therefore, a single-digit number of three will be reported by three needle pulses (sweeps). However, as previously stated, a service code is represented by a two-digit number, such as 2-3. As a result, the Self-Tests service code of 2-3 will appear on the voltmeter as two needle pulses (sweeps), then, after a two-second pause, the needle will pulse (sweep) three times.

The continuous testing codes are separated from the functional codes by a six-second delay, a single half-second sweep, and another six-second delay. They are produced on the voltmeter in the same manner as the functional codes.

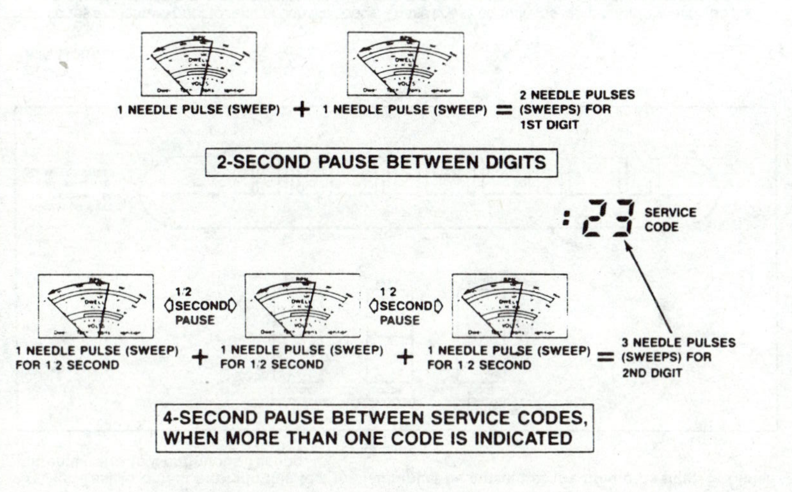

Figure 3 Analog Voltmeter Functional Service Code

Quick Test Description | Appendix | A

Reading Codes—Self-Test Automatic Readout (STAR) Rotunda Model 07-0004 or Equivalent

After hooking up the STAR tester and turning on its power switch, the tester will run a display check and the numerals 88 will begin to flash in the display window (Figure 4). A steady 00 will then appear to signify that the STAR tester is ready to start the Self-Test and receive the test's service codes.

To receive the service codes, press the pushbutton at the front of the STAR tester. The button will latch down, and a colon will appear in the display window in front of the 00 numerals. The colon **must** be displayed to receive the service codes.

If for any reason the technician wishes to clear the display window during the Self-Test, he must turn off the vehicle's engine, press the tester's pushbutton once to unlatch it (colon will disappear), then press the button again to latch down the button (colon will appear again). Every time the STAR tester is turned off, the low battery indicator (LO BAT) should show briefly at the upper left corner of the tester's display window. If the LO BAT indicator shows steadily at any other time during the operation of the STAR tester with any service code, turn its power switch to Off and replace the 9-volt battery in the tester.

The STAR tester will display the last service code received, even after disconnecting it from the vehicle. It will hold the service code on the display until the power is turned off or the pushbutton is unlatched and relatched.

WARNING: Anyone who departs from the instructions provided in this publication must first establish that he compromises neither his personal safety nor the vehicle integrity by his choice of methods, tools, or parts.

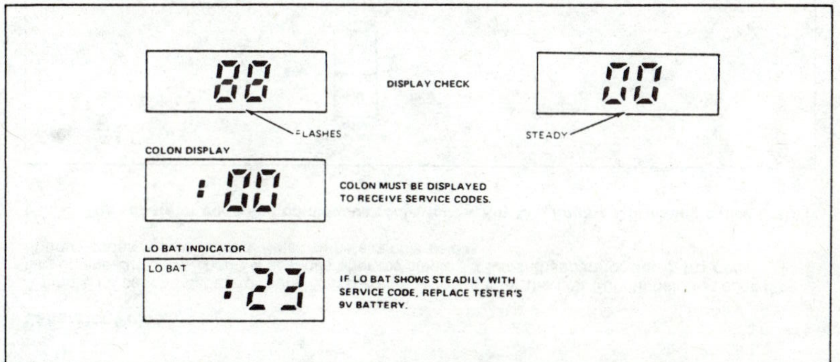

Figure 4 Self-Test Output Code Format

Key On, Engine Off

At this time, a test of the EEC-IV system is conducted with power applied and engine at rest.

For Self-Test to detect errors in this test the fault must be present at the time of testing. For intermittents, refer to Continuous Testing.

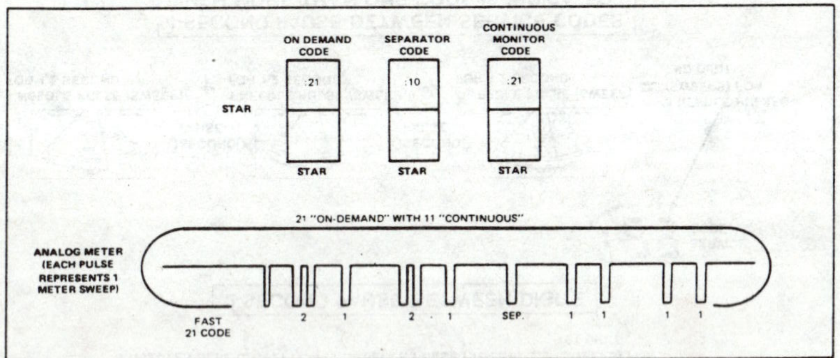

Fast Codes

Fast codes are issued prior to regular service codes. These codes contain the identical information as the regular service codes but are transmitted at 100 times the normal rate. These codes are interpreted by special equipment at the Automotive Assembly Division. They serve no purpose for the field.

For this reason the STAR has been designed to disregard these codes. Some meters in service detect these codes as a short burst of information (slight meter deflection).

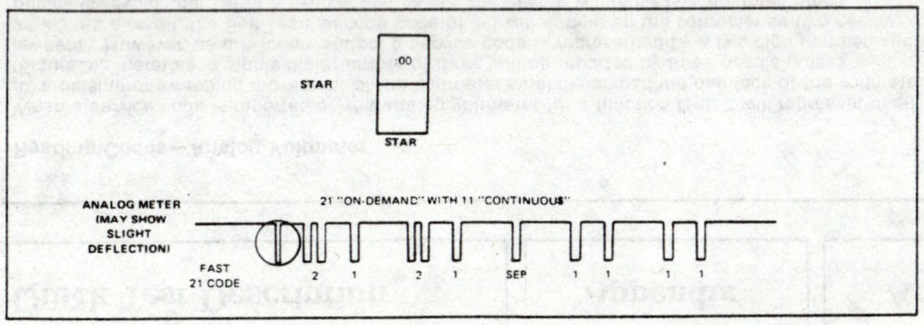

Self-Test Code Output Format | Appendix | B

Separator Pulse

A single 1/2 second separator pulse is issued 6-9 seconds after the last Functional Test code (Key On, Engine Off only). Then 6-9 seconds after the single 1/2 second separator pulse, the continuous codes will be issued. Refer to the example below:

NOTE: The separator code and continuous codes follow Key On, Engine Off testing codes ONLY.

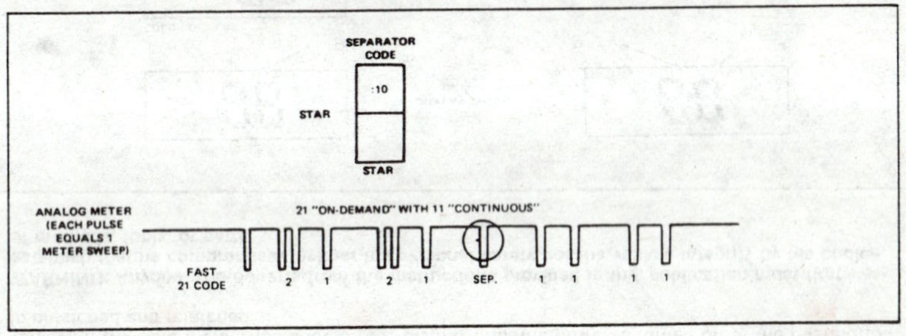

Continuous Codes

Continuous codes are issued as a result of information stored during continuous monitor testing, while the vehicle was in normal operation. These codes are displayed only during the Key On, Engine Off testing and after the separator code.

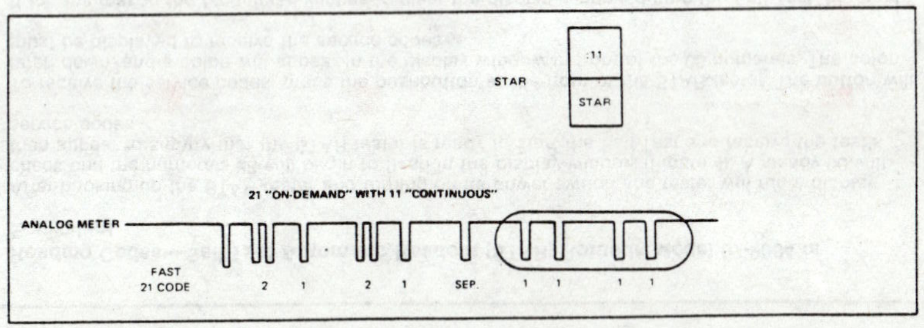

Engine Running Test

At this time, a test of the EEC-IV system is conducted with the engine running. The sensors are checked under actual operating conditions and at normal operating temperatures. The actuators are exercised and checked for corresponding results.

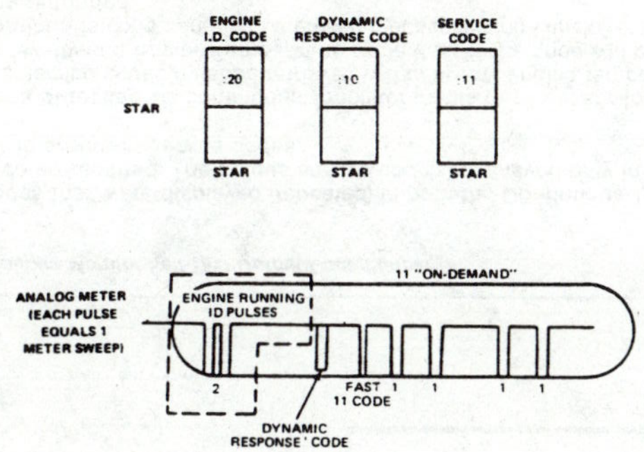

Self-Test Code Output Format | Appendix | B

Engine Identification Codes (ID Codes)

Engine ID codes are issued at the beginning of the engine running test and are one-digit numbers represented by the number of pulses sent out. The engine ID code is equal to 1/2 the number of engine cylinder (i.e, 2 pulses = 4 cylinders). These codes are used to verify the proper processor is installed and that the Self-Test has been entered.

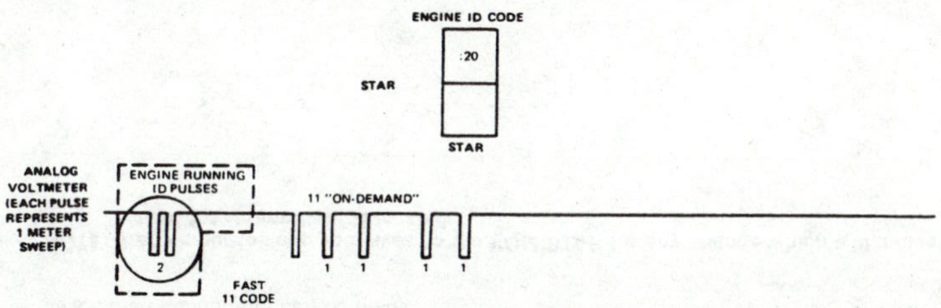

Dynamic Response Check

The dynamic response check verifies the movement of the TP, VAF, and MAP sensors during the brief Wide Open Throttle (WOT) performed during the Engine Running test. The signal for the operator to perform the brief WOT is a single pulse or 10 code on the STAR.

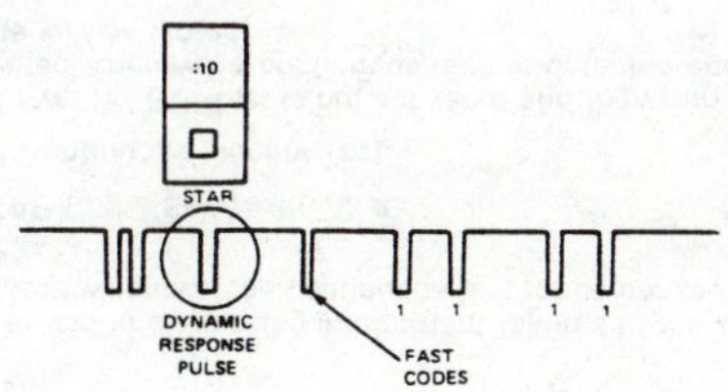

Continuous Monitor Test

This test is intended as an aid in diagnosing intermittent failures in the sensor input circuits. The Self-Test output is energized whenever the continuous test mode senses a fault and de-energized when the system is OK.

- Connect a VOM/STAR to the Self-Test output.
- Key On, Engine Off Continuous Monitor Test.

  To enter this mode, "verify" Self-Test is not activated and then turn the key to run. Self-Test output will be activated whenever a continuous fault is detected and if the duration is long enough, a fault code will be stored.
- Engine Running Continuous Monitor Test.

  This mode of testing is entered approximately 2 minutes after the service codes from the Engine Running portion of Quick Test (Step 5.0) because the engine running test can be entered only once per ignition cycle. An alternate quick method (eliminates waiting for Self-Test to complete its cycle) is to enter the running test, exit, and re-enter (do not shut the engine off). This will put you in the Engine Running Continuous Monitor Test.

  The Continuous Monitor Test will allow the technician to enter this mode of Self-Test and then attempt to re-create the intermittent failure (tap, move, wiggle the harness and/or the suspected sensor).

  If the VOM/STAR indicates a fault (short or open) the corresponding service code will be stored. Now with the knowledge of the affected circuits, a close check of the harness and associated connectors can be made.

NOTE: Remember to keep your eyes on the VOM/STAR for any change which will indicate where the intermittent is located.

Continuous Monitor Test | Appendix | C

The Continuous Monitor Test service codes are separated from the Quick Test Key On, Engine Off codes by a single separator pulse, Figure 5.

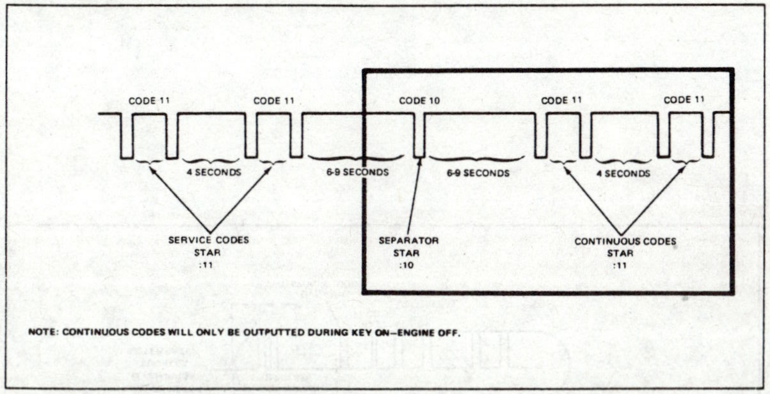

Figure 5 Continuous Monitor Self-Test Output Code Format

The service codes that were displayed (repeated) in both the Continuous Test and On Demand Test are considered repaired. The unique service code(s) displayed only in Continuous Test are considered the intermittent service codes.

During this mode of testing we continuously monitor inputs to processor for opens and shorts. The continuous service codes must be retrieved within twenty engine temperature warm up cycles. On the twenty-first Engine Temperature cycle the service code will be automatically erased. The continuous code can also be erased by deactivating Quick Test, while the service codes are being outputted.

Continuous Monitor Test	Appendix	C

EXAMPLE: How to Use the Continuous Test (Systematically)

If a service code 51 was displayed: Engine Coolant Temperature Sensor:

- Visually inspect the sensor **very closely.**
- Enter Continuous Monitor test.
- Lightly TAP on the sensor.
- Push/pull on the sensor harness connector (do not disconnect, yet).
- Test and Wiggle (shake) harness vigorously working from the sensor connector toward the dash panel and from the dash panel to the processor in short sections.
- If VOM has not given a positive indication of an intermittent, disconnect the sensor from the harness as carefully as possible. Remove terminals from the connector. Visually inspect terminals at both ends for corrosion, bad crimps, improperly seated terminals, etc.
- Reconnect after inspection.
- Disconnect processor from harness as carefully as possible.
- Inspect terminals.
- Only remove terminals associated with the sensor being inspected.
- If the VOM does not give a positive intermittent indication, reconnect the connector and erase the Continuous Test service codes.
- To erase the continuous service codes:
  - —Initiate Key On, Engine Off Self-Test.
  - —Remove jumper from the Self-Test input terminal as soon as the **first** service code is received (even if an "11" is the first code).
  - —Rerun Self-Test with jumper to verify services have been erased.

Continuous Monitor Test	Appendix	C

The diagrams below indicate common three wire sensor failure modes. Use them as a guide for servicing 50 or 60 Series Service Codes.

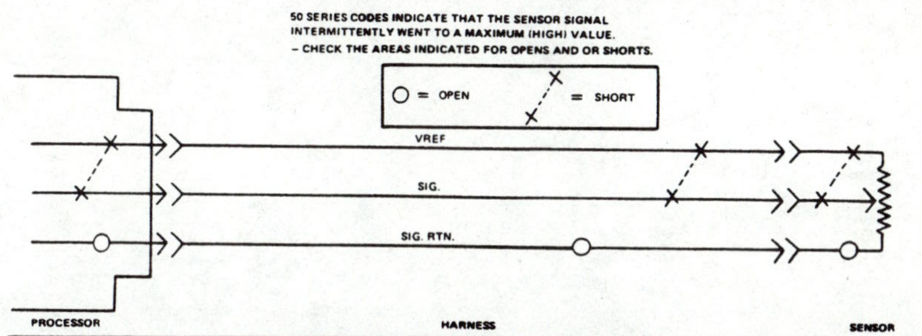

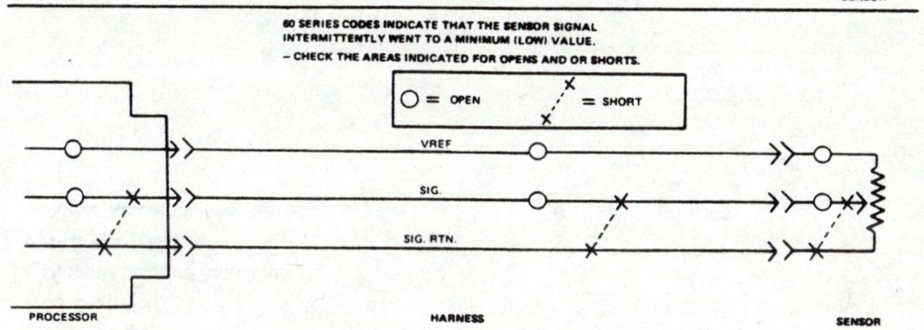

Continuous Monitor Test | Appendix | C

The diagrams below indicate common two wire sensor failure. Use them as a guide for servicing 50 or 60 Series Service Codes.

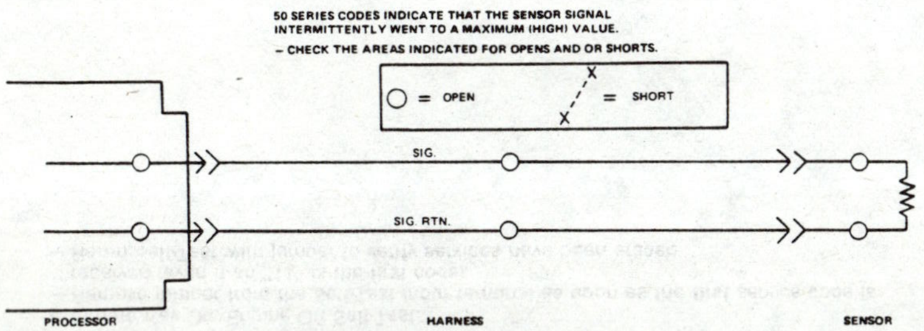

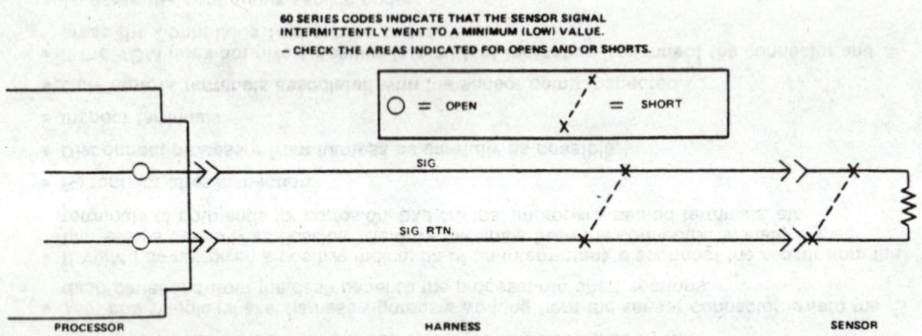

Output State Check | Appendix | D

The output state check aids in servicing output actuators associated with the EEC-IV system. It enables the technician to energize and de-energize most of the system output actuators (turn EGR solenoid and Self-Test Output (STO) On and Off) on command. This mode is entered after all codes have been received from Key On, Engine Off. On Demand and Continuous Testing. At this time, leave Self-Test activated and depress the throttle. Each time the throttle is depressed the output actuators will change state (On to Off or Off to On).

- Enter Self-Test.
- Code Output Ends.
- Do Brief WOT.
- Output Actuators Actuated.
- Next Brief WOT.
- Output Actuators Deactivated.

No Start Problem (All) | Pinpoint Test | A

A

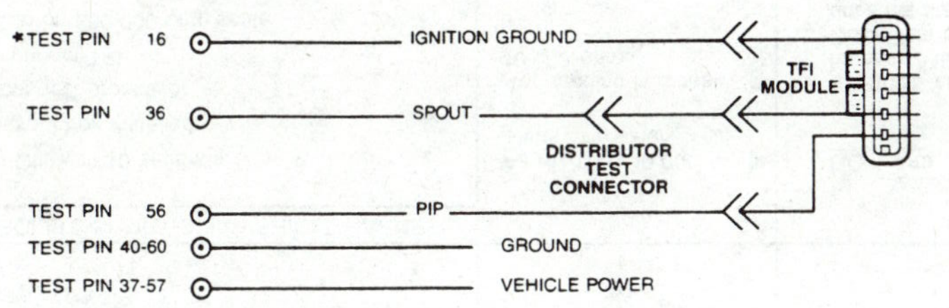

***TEST PIN LOCATED ON BREAKOUT BOX.
ALL CONNECTIONS VIEWED INTO MATING SURFACE.**

NOTE: When Breakout box is installed, ensure that timing switch is in "Computed" position unless otherwise noted.

STOP-WARNING

You may have reached this point because of a fault in a non-EEC System and/or component. To prevent the replacement of good components, be aware that the following non-EEC areas may be at fault:

- Fuel: quantity and quality.
- Ignition: general condition, moisture, cracks, damage, etc.
- Engine: internal, valves, timing belt, camshaft.
- Starter and battery circuit.

This pinpoint test is intended to diagnose only the following:

- Spark (as related to EEC-IV).
- Circuits: pip, spout, ignition ground, vehicle power.

No Start Problem (All) | Pinpoint Test | A

**WARNING: Stop this test at the first sign of a fuel leak and service as required.
CAUTION: No open flame — No smoking during fuel delivery checks.**

TEST STEP	RESULT ▶	ACTION TO TAKE
A1 ATTEMPT TO START ENGINE		
	Engine cranks, but does not start, or stalls out ▶	GO to A2.
	Engine does not crank ▶	REFER to Shop Manual, Group 28.
A2 CHECK FOR SPARK AT PLUGS		
• Disconnect the spark plug wire to any accessible cylinder. • Connect spark tester between spark plug wire and engine ground. • Crank engine and check for spark. • Reconnect the spark plug wire to the spark plug.	Spark ▶	GO to A13.
	No spark ▶	GO to A3.
A3 CHECK FOR SPARK AT COIL		
• Remove high tension coil wire from distributor and install spark tester. • Check for spark while cranking. • Reconnect high tension coil wire to distributor.	Spark ▶	REFER to Section 15, Part 2 for TFI Diagnosis for cap, rotor, wires.
	No spark ▶	GO to A4.
A4 CHECK FOR POWER TO PROCESSOR		
• Key Off, wait 10 seconds. • Disconnect processor 60 pin connector and inspect for damaged pins, corrosion, loose wires. Service as necessary. • Install Breakout box. • Key On, Engine Off. • DVOM on 20 volt scale. • Measure voltage from test Pin 37 to test Pin 40 and test Pin 57 to test Pin 60. • Ensure timing switch is in "Computed" position on Breakout box.	Either voltage reading less than 10.5V ▶	GO to Pinpoint Test B Step B1.
	Both voltage reading 10.5V or greater ▶	GO to A5.

No Start Problem (All) — Pinpoint Test A

TEST STEP	RESULT ▶	ACTION TO TAKE
A5 HARNESS CHECK IGNITION GROUND		
• Breakout box installed. • Key Off, wait 10 seconds. • DVOM on 200 ohm scale. • Measure from test Pin 16 to TFI connector (ign. gnd).	Less than 5 ohms ▶	GO to A6.
	5 ohms or greater ▶	SERVICE harness as necessary. REPEAT Quick Test.
A6 ISOLATION OF PROBLEM TO SPOUT CIRCUIT		
• Breakout box installed. • Processor connected. • Timing switch to "Dist" position on Breakout box. • Attempt to start vehicle. • Does the vehicle start?	Yes ▶	GO to A10.
	No ▶	GO to A7.
A7 SPOUT SIGNAL CHECK		
• Breakout box installed. • Timing switch to "Computed" position on Breakout box. • DVOM on 20 volt scale. • Measure voltage from Pin 36 to chassis ground, during crank.	Less than 3.0V or greater than 6.0V ▶	GO to A8.
	Between 3.0V and 6.0V ▶	EEC OK, REFER to Section 15, for TFI diagnosis.
A8 ISOLATE SHORT(S) IN VEHICLE HARNESS		
• Key Off, wait 10 seconds. • Breakout box installed. • Disconnect processor. • Disconnect TFI. • DVOM on 200,000 ohm scale. • Measure resistance from test Pin 36 to test Pins 16, 20, 26, 40, 60 (short to ground), 37, 57 (short to power) and 56 (short to pip).	All readings 10,000 ohms or greater ▶	GO to A9.
	Any reading less than 10,000 ohms ▶	SERVICE short in harness. REPEAT Quick Test if vehicle does not start. GO to A9.

No Start Problem (All) — Pinpoint Test A

TEST STEP	RESULT ▶	ACTION TO TAKE
A9 ISOLATE SHORT(S) IN PROCESSOR		
• Key Off, wait 10 seconds. • Breakout box installed. • Reconnect processor. • TFI disconnected. • DVOM on 200 ohm scale. • Measure resistance from test Pin 36 to test Pins 37 and 57 (short to power) also test Pins 40 and 60 (short to ground).	All readings 5.0 ohms or greater. ▶	GO to A10.
	Any reading less than 5.0 ohms ▶	REPLACE processor. REPEAT Quick Test.
A10 PIP SIGNAL CHECK		
• Breakout box installed. • DVOM to 20 volt range. • Measure voltage at test Pin 56 to test Pin 16. • Crank engine, record reading.	Between 3.0V and 6.0V ▶	REMOVE Breakout box. REPLACE processor. REPEAT Quick Test.
	Less than 3.0V or greater than 6.0V ▶	GO to A11.
A11 CONTINUITY OF PIP CIRCUIT		
• Breakout box installed. • Key Off, wait 10 seconds. • DVOM on 200 ohm scale. • Disconnect processor. • Connect DVOM from test Pin 56 to TFI connector, PIP circuit. • Measure resistance.	Less than 5 ohms ▶	GO to A12.
	5 ohms or greater ▶	SERVICE open PIP circuit. REPEAT Quick Test.

No Start Problem (All) — Pinpoint Test A

TEST STEP	RESULT ▶	ACTION TO TAKE
A12 CHECK PIP CIRCUIT FOR SHORTS		
• Breakout box installed. • Processor disconnected. • Key Off. • Disconnect TFI connector. • DVOM on 200,000 ohm scale. • Measure resistance from test Pin 56 to test Pins 16, 20, 26, 40, 60 (shorts to ground) and test Pins 37 and 57 (shorts to power) and test Pin 36 (short to spout).	Any resistance less than 10,000 ohms ▶	SERVICE PIP circuit. REPEAT Quick Test.
	All resistance greater than 10,000 ohms ▶	REFER to Section 15 for TFI diagnosis.
A13 SPOUT SIGNAL VERIFICATION		
• Key Off, wait 10 seconds. • Disconnect processors 60 pin connector and inspect for damaged pins, corrosion, loose wires. Service as necessary. • Install Breakout box. • Processor connected. • DVOM on 20 volt scale. • Measure voltage from test Pin 36 to chassis ground, during crank. • Ensure timing switch is in "Computed" position on Breakout box.	Between 3.0V and 6.0V ▶	FBC vehicles. REFER to Shop Manual Group 24. CFI and EFI vehicles GO to A21.
	Less than 3.0V or greater than 6.0V ▶	GO to A10.

No Start Problem (CFI/EFI) — Pinpoint Test A

A

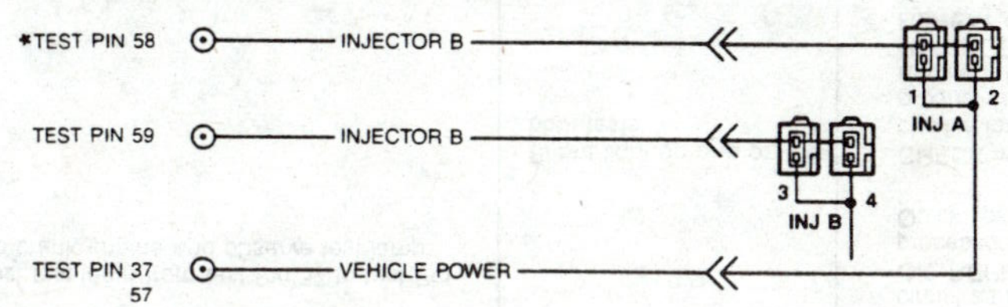

*TEST PIN LOCATED ON BREAKOUT BOX.
ALL CONNECTIONS VIEWED INTO MATING SURFACE.

TEST STEP	RESULT ▶	ACTION TO TAKE
A21 FUEL PUMP CHECK		
• **No smoking nearby.** • Disconnect all injector electrical connections at the injectors. • Connect pressure gauge to Schraeder valve. • Note initial pressure reading. • Observe pressure gauge as you pressurize fuel system. (Turn key to Run for 1 second, then turn key to Off. Wait 10 seconds. Repeat 5 times). • Turn key Off. Wait 10 seconds. • Reconnect all injectors.	PRESSURE GAUGE READING: Increased ▶	GO to A23.
	Did not increase ▶	TURN key Off, and CONTINUE to A22.

WARNING: If fuel starts leaking, turn key Off immediately. No smoking.

No Start Problem (CFI/EFI) — Pinpoint Test A

TEST STEP	RESULT ▶	ACTION TO TAKE
A22 INERTIA SWITCH CHECK		
• Key Off. • Fuel pressure gauge installed. • Locate fuel pump inertia switch. Refer to Owner's Manual for location. • Push the button of inertia switch to reset to On. **NOTE: If switch will not reset to On, replace Inertia Switch and repeat Step A21. If switch button was On, GO to J1.** • Observe pressure gauge as you pressurize fuel system. (Turn key to Run for 1 second, then turn key to Off. Wait 10 seconds.) Repeat 5 times.	PRESSURE GAUGE READING: Increase ▶ No increase ▶	 REPEAT Quick Test GO to J1.
A23 FUEL DELIVERY CHECK		
• Fuel pressure gauge installed. • Pressurized fuel system as per Step A21. • Turn key Off. • Wait for pressure to become steady — approximately 2 to 5 seconds. • Read pressure gauge.	Less than 45 psi and 35 psi or greater ▶	GO to A24.
	Less than 35 psi and/or 45 psi or greater ▶	REFER to Shop Manual Group 24 for CFI/EFI injectors and regulator, and/or Section 24-35 for electric fuel pump Diagnosis.
A24 FUEL PRESSURE LEAK DOWN CHECK		
• Fuel pressure gauge installed. • Key Off. • Wait 2 minutes after pressure gauge reading in Step A23, then note drop in gauge reading.	Less than 4 psi drop in 2 minutes ▶	GO to A25.
	4 psi drop or greater in 2 minutes ▶	Immediately CHECK for hydraulic lock up or fouled spark plugs. REFER to Shop Manual Group 24 for CFI/EFI injector and/or fuel regulator service.

No Start Problem (CFI/EFI) — Pinpoint Test A

TEST STEP	RESULT ▶	ACTION TO TAKE
A25 FUEL DELIVERY TEST		
• Key Off. • Fuel pressure gauge installed. • Pressurize fuel system as per Step A21. • Locate and disconnect the fuel pump relay. • Crank engine for 5 seconds. • Take pressure reading at the end of 5 second crank. **NOTE: Verify fuel quality; air and/or water will also pressurize and look like acceptable fuel delivery.**	PRESSURE GAUGE READING: Pressure is approximately 10-20 psi at the end of 5 second crank cycle. Refer to note below ▶	The EEC system is not the fault of the No Start. Fuel and spark are present. REFER to Section 2 for other No Start Routines. If complaint was runs rough or misses CONTINUE to A26.
	Pressure is greater or less than specified ▶	GO to A26.

NOTE: The colder the engine, the greater the pressure drop (i.e., an engine coolant temperature of 200°F equals approximately a 10 psi drop in 5 seconds. 60°F equals approximately a 20 psi drop in 5 seconds).

TEST STEP	RESULT ▶	ACTION TO TAKE
A26 INJECTOR DRIVE SIGNAL CHECK		
• Key Off, wait 10 seconds. • Breakout box installed. • Use a standard non-powered test lamp. • Connect test lamp from test Pin 37 to test Pin 58 and crank engine observe test lamp. • Connect test lamp from test Pin 37 to test Pin 59 and crank engine and observe test lamp.	Dim glow at light on both tests ▶	GO to A27.
	No light on one or both tests ▶	VERIFY 12V battery power at test Pin 37. SERVICE power circuit as required. If OK, REPLACE processor. REPEAT Quick Test.
	Bright light on one or both tests ▶	CHECK injector circuits for shorts to ground. SERVICE as required. If OK, REPLACE processor. REPEAT Quick Test.

No Start Problem (CFI/EFI) — Pinpoint Test A

TEST STEP	RESULT ▶	ACTION TO TAKE
A27 SERVICING THE SUSPECT INJECTOR CIRCUIT		
• Key Off, wait 10 seconds. • Locate and disconnect fuel pump relay. • DVOM on 20 volt scale. • With DVOM measure the average voltage between test Pin 58 and test Pin 37 during cranking mode. • Also measure average voltage between test Pin 59 and test Pin 37. • After measuring and recording voltages, reconnect fuel pump relay.	1.0V or greater difference between injector circuits ▶	DISCONNECT processor connector and INSPECT for corrosion, damaged pins, etc. RECONNECT processor and RERUN Step A27. If condition still exists, REPLACE processor and REPEAT Quick Test.
	Less than 1.0V difference between injector circuits ▶	GO to A28.
A28 RESISTANCE CHECK OF INJECTOR HARNESS		
• Processor disconnected. • Key Off. • DVOM on 20 ohm scale. • Disconnect injector(s). **EFI:** Disconnect injectors 2, 3 and 4. **CFI:** Disconnect injector No. 2. • Through the vehicle harness, measure the resistance of injector No. 1 (still connected) from test Pin 58 to test Pin 37. Record resistance reading. • Disconnect injector No. 1 and connect injector No. 2. • Through the vehicle harness, measure the resistance of injector No. 2 from: **CFI:** Test Pin 59 to test Pin 37. **EFI:** Test Pin 58 to test Pin 37. Record resistance reading. **EFI ONLY:** • Disconnect injector No. 2 and connect injector No. 3. • Record resistance measured from test Pin 59 to test Pin 37. • Disconnect injector No. 3 and connect injector No. 4. • Record resistance measured from test Pin 59 to test Pin 37. • Injector resistances are 2.0 ohms to 3.5 ohms.	Yes ▶	GO to A29.
	No ▶	CHECK the harness on suspect injector(s). If OK, REPLACE suspect injector(s). RECONNECT all injectors and REPEAT Quick Test.

No Start Problem (CFI/EFI) — Pinpoint Test A

TEST STEP	RESULT ▶	ACTION TO TAKE
A29 ISOLATING A FAULTY INJECTOR		
• Key Off. • Fuel pressure gauge installed. • Electrically disconnect all injectors at intake manifold. • Pressurize the fuel system as per Step A21. • Electrically connect injector for cylinder No. 1. • Crank the engine for 5 seconds. • Observe fuel pressure gauge and record reading immediately after the 5 second crank. • Disconnect injector No. 1 and repressurize the fuel system. • Connect injector No. 2 and crank the engine for 5 seconds. Immediately observe and record pressure. • Disconnect injector No. 2 and repressurize the fuel system. • Connect injector No. 3 and crank the engine for 5 seconds. Immediately observe and record pressure. • Disconnect injector No. 3 and repressurize the fuel system. • Connect injector No. 4 and crank the engine for 5 seconds. Immediately observe and record pressure.	All readings not within 4 psi of each other ▶	REPLACE injector(s) that is/are not within 4 psi of each other. Connect all injectors and REPEAT Quick Test.
	All four readings are within 4 psi of each other ▶	VERIFY all the non-EEC Diagnostic Routines that apply to your drive complaint have been properly performed, then: DISCONNECT processor and INSPECT all Pins for corrosion, damage, etc., and RETEST. If fault is still present, REPLACE processor and REPEAT Quick Test.

Vehicle Battery — Pinpoint Test — B

B

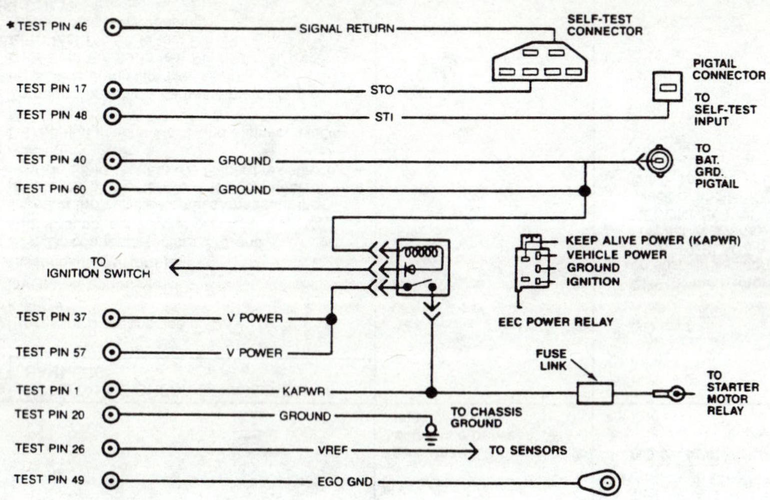

* TEST PINS LOCATED ON BREAKOUT BOX.
ALL HARNESS CONNECTORS VIEWED INTO MATING SURFACE.

STOP-WARNING

You may have reached this point because of a fault in a non-EEC System and/or component. To prevent the replacement of good components, be aware that the following non-EEC areas may be at fault:

- Ignition switch.
- Battery Cables.
- Alternator.
- Voltage Regulator.
- Ground Straps.

This pinpoint test is intended to diagnose only the following:

- Processor.
- Harness circuits: Signal Return, STO, STI, Ground, V Power, KAPWR, VREF, Ignition.
- Battery Voltage.
- Power Relay.

Vehicle Battery — Pinpoint Test — B

TEST STEP	RESULT ▶	ACTION TO TAKE
B1 BATTERY VOLTAGE CHECK		
• Key On, Engine Off. • DVOM on 20 volt scale. • Measure voltage across battery terminals.	10.5V or greater ▶	GO to B2.
	Less than 10.5V ▶	SERVICE discharged battery, REFER to Shop Manual, Group 31.
B2 BATTERY POWER GROUND CHECK		
• Key On, Engine Off. • Processor connected. • DVOM on 20 volt range. • Measure voltage from battery negative post to signal return circuit in the Self-Test connector.	Less than 0.5V ▶	GO to B3.
	0.5V or greater ▶	GO to B4.
B3 12 VOLT BATTERY POWER CHECK AT PROCESSOR		
• Breakout box installed. • Key On, Engine Off. • Processor connected. • DVOM on 20 volt scale. • Measure voltage from battery positive post to test Pin 37 and test Pin 57.	Both readings less than 0.5V ▶	GO to C2.
	One or both readings 0.5V or greater ▶	GO to B7.
B4 GROUND FAULT ISOLATION		
• Breakout box installed. • Key On, Engine Off. • Processor connected. • DVOM on 20 volt scale. • Measure voltage from battery negative post to test Pin 40 and test Pin 60.	Both readings less than 0.5V ▶	GO to B5.
	One or both readings 0.5V or greater ▶	Circuit(s) with greater than 0.5V has high resistance or open. CORRECT faulty ground circuit. REPEAT Quick Test.

Vehicle Battery — Pinpoint Test — B

TEST STEP	RESULT ▶	ACTION TO TAKE
B5 PROCESSOR GROUND FAULT ISOLATION		
• Breakout box installed. • Key Off, wait 10 seconds. • Processor connected. • DVOM on 200 ohm scale. • Measure resistance from test Pin 46 to test Pin 40 and from test Pin 46 to test Pin 60.	Both readings less than 5 ohms ▶	GO to B6.
	One or both readings 5 ohms or greater ▶	DISCONNECT processor connector and INSPECT for corrosion, damaged pins, etc. SERVICE as necessary and RETEST. If fault is still present, REPLACE processor. REPEAT Quick Test.
B6 HARNESS CHECK (SIGNAL RETURN)		
• Breakout box installed. • Key Off, wait 10 seconds. • Processor connected. • DVOM on 200 ohm scale. • Measure resistance from test Pin 46 to self test connector Signal Return circuit.	Less than 5 ohms ▶	System OK. RUN Quick Test.
	5 ohms or greater ▶	CORRECT cause of resistance in the harness Signal Return circuit. REPEAT Quick Test.
B7 12 VOLT BATTERY POWER FAULT ISOLATION		
• Key On, Engine Off. • Processor connected. • DVOM on 20 volt scale. • Measure voltage by connecting the DVOM negative lead to battery negative post. Connect DVOM positive lead to KAPWR circuit at EEC power relay.	10.5V or greater ▶	GO to B8.
	Less than 10.5V ▶	CHECK KAPWR and V Power circuits for shorts to ground and KAPWR circuit from power relay to battery positive post for opens. SERVICE as necessary. REPEAT Quick Test.

Vehicle Battery — Pinpoint Test — B

TEST STEP	RESULT ▶	ACTION TO TAKE
B8 12 VOLT BATTERY POWER FAULT ISOLATION		
• Key On, Engine Off. • Processor connected. • DVOM on 20 volt scale. • Measure voltage by connecting the DVOM negative lead to battery negative post. Connect DVOM positive lead to Ignition circuit at EEC power relay.	10.5V or greater ▶	GO to B9.
	Less than 10.5V ▶	CHECK 10 amp fuse, location 18 in fuse panel. CHECK for open in ignition switch circuits. SERVICE as necessary. REPEAT Quick Test.
B9 12 VOLT BATTERY POWER FAULT ISOLATION		
• Key On, Engine Off. • Processor connected. • DVOM on 20 volt scale. • Measure voltage by connecting the DVOM negative lead to battery negative post. Connect DVOM positive lead to ground circuit at EEC power relay.	Less than 0.5V ▶	GO to B10.
	0.5V or greater ▶	SERVICE open or ground in Ground Circuit. REPEAT Quick Test.
B10 12 VOLT BATTERY POWER FAULT ISOLATION		
• Key On, Engine Off. • Processor connected. • DVOM on 20 volt scale. • Measure voltage by connecting the DVOM negative lead to battery ground. Connect DVOM positive lead to V-Power circuit at EEC power relay.	10.5V or greater ▶	SERVICE short to ground or open in V-Power circuit from EEC power relay to processor connector test Pins 37 and 57. REPEAT Quick Test.
	Less than 10.5V ▶	REPLACE power relay. REPEAT Quick Test.

Reference Voltage — Pinpoint Test C

C

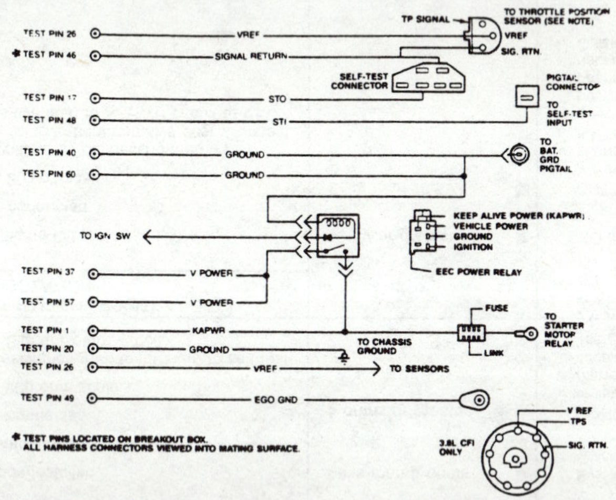

NOTE: For 3.8L CFI a 10-Pin connector was used for the first 60 days of production. Either connector may be encountered. Test pin assignments did not change on Breakout box. If TP sensor must be disconnected for diagnostics, disconnect 10-Pin connector at fuel charging assy.

STOP-WARNING

This pinpoint test is intended to diagnose only the following:

- Processor.
- Sensor harness circuits: Signal Return, STO, STI, Ground, V Power, KAPWR, VREF, Ignition.

Reference Voltage — Pinpoint Test C

TEST STEP	RESULT ▶	ACTION TO TAKE
C1 VEHICLE BATTERY POWER CIRCUIT CHECK		
• Breakout box installed. • Key On, Engine Off. • Processor connected. • DVOM on 20 volt scale. • Connect DVOM positive test lead to test Pin 37 and negative test lead to Signal Rtn. in Self-Test connector.	10.5V or greater ▶	GO to C2.
	Less than 10.5V ▶	GO to B1.
C2 VREF VOLTAGE CHECK		
• Breakout box installed. • Key On, Engine Off. • Processor connected. • DVOM on 20 volt scale. • Measure voltage with DVOM from test Pin 26 to test Pin 46.	6.0V or greater ▶	GO to C3.
	4.0V or less ▶	GO to C4.
	Greater than 4.0V, less than 6.0V ▶	Reference voltage OK, REPEAT Quick Test.
C3 CHECK FOR SHORT TO PROCESSOR POWER VREF		
• Key Off, wait 10 seconds. • Breakout box installed. • Disconnect processor. • Key On, Engine Off. • DVOM on 20 volt scale. • Check for voltage from test Pin 26 to battery ground and from test Pin 46 to battery ground.	Less than 0.5V ▶	REPLACE processor REPEAT Quick Test.
	0.5V or greater ▶	SERVICE short to battery power in EEC harness. REPEAT Quick Test.
C4 CHECK FOR SHORTED THROTTLE POSITION SENSOR		
• Key Off, wait 10 seconds. • Breakout box installed. • Processor connected. • Disconnect Throttle Position (TP) sensor from vehicle harness. • Key On, Engine Off. • DVOM on 20 volt scale. • Measure voltage with DVOM from test Pin 26 to test Pin 46.	Less than 4.0V ▶	Vehicles equipped with EVP sensor, GO to C5. All other vehicles, GO to C6.
	4.0V or greater ▶	REPLACE TP sensor. REPEAT Quick Test.

Reference Voltage — Pinpoint Test C

TEST STEP	RESULT ▶	ACTION TO TAKE
C5 CHECK FOR SHORTED EGR VALVE POSITION (EVP) SENSOR		
• Key Off, Wait 10 seconds. • Breakout box installed. • Processor connected. • Disconnect EGR valve position (EVP) sensor. • Key On, Engine Off. • DVOM on 20 volt scale. • Measure voltage with DVOM from test Pin 26 to test Pin 46.	Less than 4.0V ▶ 4.0V or greater ▶	GO to C6. REPLACE EVP sensor. REPEAT Quick Test.
C6 CHECK FOR SHORTED MAP/BP SENSOR		
• Key Off, wait 10 seconds. • Breakout box installed. • Processor connected. • Disconnect MAP/BP sensor. • Key On, Engine Off. • DVOM on 20 volt scale. • Measure voltage with DVOM from test Pin 26 to test Pin 46.	Less than 4.0V ▶ 4.0V or greater ▶	Vehicles equipped with VAF sensor. GO to C7. All other vehicles. GO to C8. REPLACE MAP/BP sensor. REPEAT Quick Test.
C7 CHECK FOR SHORTED VANE AIR METER (VAF) SENSOR		
• Key Off, wait 10 seconds. • Breakout box installed. • Processor connected. • Disconnect vane air meter (VAF) sensor. • Key On, Engine Off. • DVOM on 20 volt scale. • Measure voltage with DVOM from test Pin 26 to test Pin 46.	Less than 4.0V ▶ 4.0V or greater ▶	GO to C8. REPLACE VAF sensor and REPEAT Quick Test.

Reference Voltage — Pinpoint Test C

TEST STEP	RESULT ▶	ACTION TO TAKE
C8 SHORT TO GROUND IN VREF		
• Breakout box installed. • Processor disconnected. • Key Off, wait 10 seconds. • Disconnect TP and MAP/BP, EVP and VAF, if so equipped. • DVOM on 200 ohm scale. • Measure continuity from test Pin 26 to test Pins 20, 40, 46 and 60.	Less than 5 ohms ▶ 5 ohms or greater ▶	SERVICE short to ground in VREF. REPEAT Quick Test. GO to C9.
C9 CHECK VREF FOR CONTINUITY		
• Breakout box installed. • Processor disconnected. • TP, MAP/BP, EVP and VAF sensors disconnected. • Key Off. • DVOM on 200 ohm scale. • Measure continuity on all above referenced sensors from test Pin 26 to VREF circuit on the sensor harness connectors.	Less than 5 ohms on all readings ▶ 5 ohms or greater on any reading ▶	RECONNECT sensors. REPLACE processor. REPEAT Quick Test. SERVICE open in VREF. REPEAT Quick Test.

Vane Air Temperature Sensor (VAT) — Pinpoint Test — DA

DA

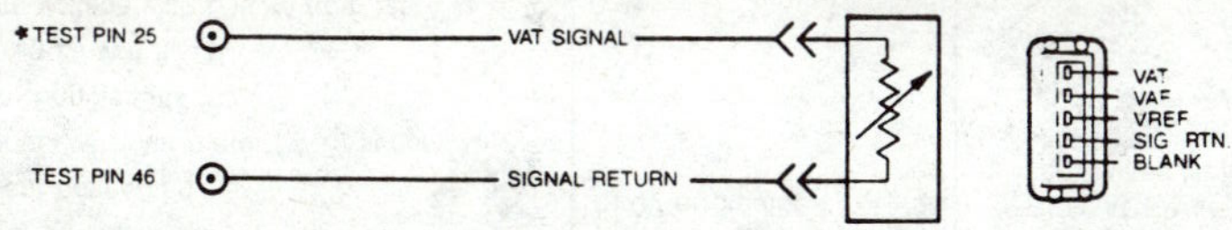

***NOTE: AMBIENT TEMP. MUST BE GREATER THAN 50°F TO PASS THIS TEST.**

TYPICAL RESISTANCE BETWEEN TEST POINTS 25 & 46	5800Ω	2700Ω	300Ω	180Ω	125Ω
AT TEMPERATURE	32°F	65°F	185°F	220°F	240°F

***TEST PIN LOCATED ON BREAKOUT BOX.**
ALL CONNECTIONS VIEWED INTO MATING SURFACE.

STOP-WARNING

You may have reached this point because of a fault in a non-EEC System and/or component. To prevent the replacement of good components, be aware that the following non-EEC areas may be at fault:

- Test performed in unusually low (cold) or high (hot) ambient conditions.
- Ambient temperature must be greater than 50°F for this test.

This pinpoint test is intended to diagnose only the following:

- VAT sensor.
- Circuits: VAT, and Signal Return.
- Vehicle harness.
- Processor assembly.

Vane Air Temperature Sensor (VAT) — Pinpoint Test — DA

TEST STEP	RESULT ▶	ACTION TO TAKE
DA1 SERVICE CODE 24: CHECK VAT SIGNAL TO SIGNAL RETURN VOLTAGE		
• Key Off, wait 10 seconds. • Disconnect harness from VAT sensor. Inspect for damaged pins, corrosion, loose wires, etc. SERVICE as necessary. • DVOM on 20 volt scale. • Key On, Engine Off. • Measure voltage of vane meter harness connector VAT signal to Signal Return.	Reading is between 4.0V and 6.0V ▶ Reading is less than 4.0V ▶ Reading is greater than 6.0V ▶	GO to DA4. GO to DA2. GO to Pinpoint Test Step C1.
DA2 CONTINUITY CHECK OF CIRCUITS VAT AND SIGNAL RETURN		
• Key Off, wait 10 seconds. • Harness disconnected from vane meter. • Disconnect processor 60 pin connector and inspect for damaged pins, loose wires, corrosion, etc. Service as necessary. • Install Breakout box to harness, leave processor disconnected. • DVOM on 200 ohm scale. • Measure resistance of harness connector VAT signal to test Pin 25. • Measure resistance of harness connector signal return to test Pin 46.	Both readings are less than 5 ohms ▶ Either reading is 5 ohms or greater ▶	GO to DA3. CORRECT open circuits, REMOVE Breakout box. RECONNECT harness to processor and vane meter. RERUN Quick Test.
DA3 CHECK VAT CIRCUITS FOR SHORTS		
• Key Off. • Harness disconnected from vane meter and processor: Breakout box installed. • DVOM on 200,000 ohms scale. • Measure resistance from test Pin 25 to test Pins 40, 46 and 60.	Any reading less than 10,000 ohms ▶ All readings are 10,000 ohms or greater ▶	CORRECT circuits with short. REMOVE Breakout box. RECONNECT harness to vane meter and processor. RERUN Quick Test. GO to DA4.

Vane Air Temperature Sensor (VAT) — Pinpoint Test — DA

TEST STEP	RESULT ▶	ACTION TO TAKE
DA4 VAT SENSOR CHECK		
NOTE: Ambient temperature must be greater than 50°F for this test. • Key Off, wait 10 seconds. • Harness disconnected from the vane meter. • DVOM on 200,000 ohm scale. • Measure resistance of VAT sensor, VAT signal Pin to Signal Return Pin. • Is reading from 125 ohms (240°F) to 3700 ohms (50°F)?	Yes ▶	REPLACE processor. RECONNECT harness to vane meter. RERUN Quick Test.
	No ▶	REPLACE vane meter. RERUN Quick Test.
DA10 SERVICE CODE 54: INDUCE OPPOSITE CODE		
• Key Off, wait 10 seconds. • Disconnect harness from vane meter. Inspect for damaged pins, corrosion, loose wires, etc. Service as necessary. • Insert a jumper wire (paper clip) between harness connector VAT Signal and harness connector Signal Return. • Run Key On, Engine Off Quick Test. • Is Code 64 present?	Yes ▶	REPLACE vane meter. REMOVE jumper wire. CONNECT harness to vane meter. RERUN Quick Test.
	No ▶	REMOVE jumper wire. GO to DA11.
DA11 CHECK CONTINUITY OF VAT SIGNAL AND SIGNAL RETURN		
• Key Off, wait 10 seconds. • Harness disconnected from vane meter, jumper wire removed. • Disconnect processor 60 pin connector. Inspect for damaged pins, corrosion, loose wires, etc. Service as necessary. • Install Breakout box leaving processor disconnected. • DVOM on 200 ohm scale. • Measure resistance of harness connector VAT Signal to test Pin 25. • Measure resistance of harness connector Signal Return to test Pin 46.	Both readings are less than 5 ohms ▶	REPLACE processor. REMOVE Breakout box. RECONNECT harness to vane meter and processor. RERUN Quick Test.
	Either reading is 5 ohms or greater ▶	CORRECT open circuit. REMOVE Breakout box. RECONNECT harness to vane meter and processor. RERUN Quick Test.

Vane Air Temperature Sensor (VAT) — Pinpoint Test — DA

TEST STEP	RESULT ▶	ACTION TO TAKE
DA20 SERVICE CODE 64: INDUCE OPPOSITE CODE		
• Key Off, wait 10 seconds. • Disconnect harness from vane meter. Inspect for damaged pins, corrosion, loose wires, etc. Service as necessary. • Run Key On, Engine Off Quick Test. • Is code 54 present?	Yes ▶	REPLACE vane meter. RECONNECT harness to vane meter. RERUN Quick Test.
	No ▶	GO to DA21.
DA21 CHECK VAT SIGNAL FOR SHORTS		
• Key Off, wait 10 seconds. • Harness disconnected from vane meter. • Disconnect processor 60 pin connector. Inspect for damaged pins, corrosion, loose wires, etc. Service as necessary. • Install Breakout box leaving processor disconnected. • DVOM on 200,000 ohm scale. • Measure resistance of Test Pin 25 to Test Pins 40, 46 and 60.	Any reading less than 10,000 ohms ▶	CORRECT circuit shorts. REMOVE Breakout box. RECONNECT processor and vane meter. RERUN Quick Test.
	All readings are 10,000 ohms or greater ▶	REPLACE processor. REMOVE Breakout box. RECONNECT processor. RERUN Quick Test.

Vane Air Temperature — Pinpoint Test — DA

TEST STEP	RESULT ▶	ACTION TO TAKE
DA90 SERVICE CODE 54: CONTINUOUS TEST: CHECK VAT SENSOR		
• Using continuous monitor mode, observe VOM or STAR LED for indication of a fault while performing the following: • Lightly tap on VAT sensor (simulate road shock). • Wiggle VAT connector. • Is a fault indicated?	Yes ▶	DISCONNECT and INSPECT connectors. If connector and terminals are good, REPLACE VAT sensor. REPEAT Quick Test.
	No ▶	GO to DA91.

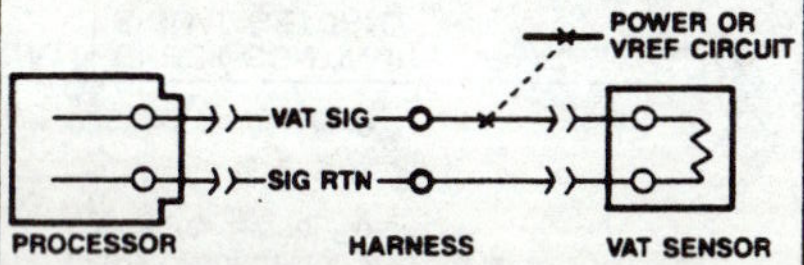

TEST STEP	RESULT ▶	ACTION TO TAKE
DA91 CHECK EEC-IV HARNESS		
• Observe VOM or STAR LED for a fault indication while performing the following: • Referring to the illustration in Step DA90, grasp the harness closest to the sensor connector. Wiggle, shake or bend a small section of the EEC-IV system harness while working your way to the dash panel. Also wiggle, shake or bend the EEC-IV harness from the dash panel to the processor. • Is fault indicated?	Yes ▶	ISOLATE fault and make necessary service. REPEAT Quick Test.
	No ▶	GO to DA92.
DA92 CHECK PROCESSOR AND HARNESS CONNECTORS		
• Key Off, wait 10 seconds. • Disconnect processor 60 pin connector. • Inspect both connectors and connector terminals for obvious damage or faults. • Connectors and terminals are OK.	No ▶	SERVICE as necessary. REPEAT Quick Test.
	Yes ▶	Unable to duplicate fault at this time. Continuous code 54 testing complete.

Vane Air Temperature — Pinpoint Test — DA

TEST STEP	RESULT ▶	ACTION TO TAKE
DA93 SERVICE CODE 64: CONTINUOUS TEST: CHECK VAT SENSOR		
• Using continuous monitor mode, observe VOM or STAR LED for indication of a fault while performing the following: • Lightly tap on VAT sensor (simulate road shock). • Wiggle VAT connector. • Is fault indicated?	Yes ▶	DISCONNECT and INSPECT connectors. If connector and terminals are good, REPLACE VAT sensor. REPEAT Quick Test.
	No ▶	GO to DA94.

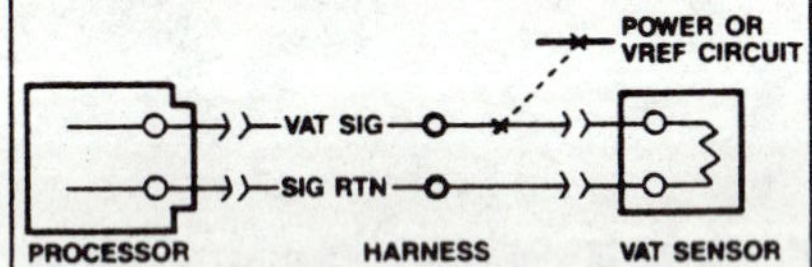

TEST STEP	RESULT ▶	ACTION TO TAKE
DA94 CHECK EEC-IV HARNESS		
• Observe VOM or STAR LED for a fault indication while performing the following: • Referring to the illustration in Step DA93, grasp the harness closest to the sensor connector. Wiggle, shake or bend a small section of the EEC-IV system harness while working your way to the dash panel. Also wiggle, shake or bend the EEC-IV harness from the dash panel to the processor. • Is fault indicated?	Yes ▶	ISOLATE fault and make necessary service. REPEAT Quick Test.
	No ▶	GO to DA95.
DA95 CHECK PROCESSOR AND HARNESS CONNECTORS		
• Key Off, wait 10 seconds. • Disconnect processor 60 pin connector. • Inspect both connectors and connector terminals for obvious damage or faults. • Connectors and terminals are OK.	No ▶	SERVICE as necessary. REPEAT Quick Test.
	Yes ▶	Unable to duplicate fault at this time. Continuous code 64 testing complete.

Air Charge Temperature Sensor (ACT) | Pinpoint Test | DB

DB

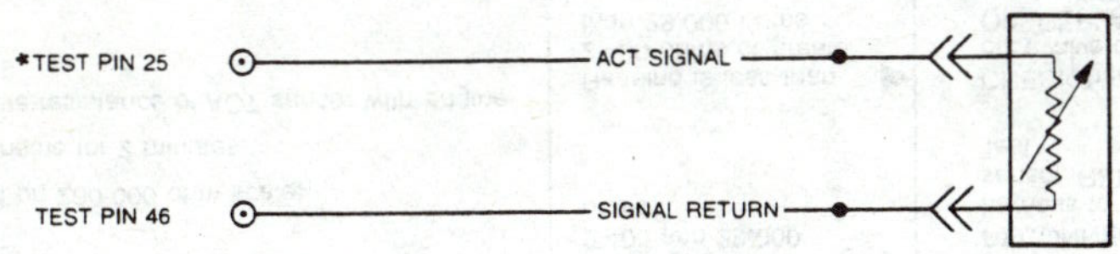

***TEST PIN LOCATED ON BREAKOUT BOX.**
ALL CONNECTIONS VIEWED INTO MATING SURFACE.

NOTE: Ambient temperature must be greater than 50°F for this test.

STOP-WARNING

You may have reached this point because of a fault in a non-EEC System and/or component. To prevent the replacement of good components, be aware that the following non-EEC areas may be at fault:

- Cooling system.
- Air cleaner duct problems.
- Improper engine oil level.

This pinpoint test is intended to diagnose only the following:

- ACT sensor.
- Harness circuits: ACT Signal and Signal Return.
- Processor assembly.

Air Charge Temperature Sensor (ACT) | Pinpoint Test | DB

TEST STEP	RESULT ▶	ACTION TO TAKE
DB1 SERVICE CODE 24: CHECK PROPER INSTALLATION		
• For vehicle with ACT mounted in intake manifold. GO to Step DB2.	Yes ▶	GO to DB2.
• Is ACT mounted properly in air cleaner?	No ▶	INSTALL ACT properly. RERUN Quick Test.
DB2 CHECK ACT SIGNAL TO SIGNAL RETURN VOLTAGE		
• Key Off, wait 10 seconds. • Disconnect harness from ACT sensor. Inspect for damaged pins, corrosion, loose wires, etc. Service as necessary. • DVOM on 20 volt scale. • Key On, Engine Off. • Measure voltage from ACT harness connector ACT Signal to Signal Return.	Reading is between 4.0V and 6.0V ▶	GO to DB5.
	Reading is less than 4.0V ▶	GO to DB3.
	Reading is greater than 6.0V ▶	GO to Pinpoint Test Step C1.
DB3 CONTINUITY CHECK OF ACT SIGNAL AND SIGNAL RETURN		
• Key Off, wait 10 seconds. • Harness disconnected from ACT sensor. • Disconnect processor 60 pin connector. Inspect for damaged pins, corrosion, loose wires, etc. Service as necessary. • Install Breakout box leaving processor disconnected. • DVOM on 200 ohm scale. • Measure resistance of harness connector ACT Signal to test Pin 26. • Measure resistance of harness connector Signal Return to test Pin 46.	Both readings are less than 5 ohms ▶	GO to DB4.
	Either reading is 5 ohms or greater ▶	CORRECT circuit opens. REMOVE Breakout box. RECONNECT processor and ACT sensor. RERUN Quick Test.

Air Charge Temperature Sensor (ACT) — Pinpoint Test — DB

TEST STEP	RESULT ▶	ACTION TO TAKE
DB4 CHECK ACT SIGNAL FOR SHORTS		
• Key Off. Harness disconnected from ACT sensor. Breakout box installed. Processor disconnected. • DVOM on 200,000 ohm scale. • Measure resistance of test Pin 25 to test Pins 40, 46 and 60.	Any reading less than 10,000 ohms ▶	CORRECT circuit shorts. REMOVE Breakout box. RECONNECT processor and ACT sensor. RERUN Quick Test.
	All readings are 10,000 ohms or greater ▶	GO to DB5.
DB5 CHECK ACT SENSOR — ENGINE OFF		
• Key Off, wait 10 seconds. • Harness disconnected from ACT sensor. • DVOM on 200,000 ohm scale. • Measure resistance of ACT sensor. **NOTE: Make sure engine is warmed up prior to this test.**	Reading is between 1,100 and 58,000 ohms (approximately 50°F) ▶	GO to DB6.
	Reading is less than 1,100 ohms or greater than 58,000 ▶	CHECK heat stove duct valve operation. If OK, REPLACE ACT sensor. RECONNECT harness to ACT sensor. RERUN Quick Test.
DB6 CHECK ACT SENSOR — ENGINE RUNNING		
• Key Off. Harness disconnected from ACT sensor. • DVOM on 200,000 ohm scale. • Run engine for 2 minutes. • Measure resistance of ACT sensor with engine running.	Reading is between 2,400 and 29,000 ohms ▶	REPLACE processor. RECONNECT harness to ACT sensor. RERUN Quick Test.
	Reading is less than 2,400 ohms or greater than 29,000 ohms ▶	CHECK heat stove duct valve operation. If OK, REPLACE ACT sensor. RERUN Quick Test.

Air Charge Temperature Sensor (ACT) — Pinpoint Test — DB

TEST STEP	RESULT ▶	ACTION TO TAKE
DB10 SERVICE CODE 54: INDUCE OPPOSITE CODE		
• Key Off, wait 10 seconds. • Disconnect harness from ACT sensor. Inspect for damaged pins, corrosion, loose wires, etc. Service as necessary. • Insert a jumper wire between ACT harness connector ACT Signal to Signal Return. • Run Key On, Engine Off Quick Test. • Is code 64 present?	Yes ▶	REPLACE ACT sensor. REMOVE jumper wire. RECONNECT harness to ACT sensor. RERUN Quick Test.
	No ▶	REMOVE jumper wire. GO to DB11.
DB11 CONTINUITY CHECK ACT SIGNAL AND SIGNAL RETURN		
• Key Off, wait 10 seconds. • Harness disconnected from ACT sensor. • Disconnect processor 60 pin connector. Inspect for damaged pins, corrosion, loose wires, etc. Service as necessary. • Install Breakout box. Leave processor disconnected. • DVOM on 200 ohm scale. • Measure resistance of ACT harness connector ACT Signal to test Pin 25. • Measure resistance of ACT harness connector Signal Return to test Pin 46.	Both readings are less than 5 ohms ▶	REPLACE processor. REMOVE Breakout box. RECONNECT processor and ACT sensor. RERUN Quick Test.
	Either reading is 5 ohms or greater ▶	Correct circuit opens. REMOVE Breakout box. RECONNECT processor and ACT sensor. RERUN Quick Test.

Air Charge Temperature Sensor (ACT) | Pinpoint Test | DB

TEST STEP	RESULT ▶	ACTION TO TAKE
DB20 SERVICE CODE 64: INDUCE OPPOSITE CODE		
• Key Off, wait 10 seconds. • Disconnect harness from ACT sensor. Inspect for damaged pins, corrosion, loose wires, etc. Service as necessary. • Run Key On, Engine Off Quick Test. • Is code 54 present?	Yes ▶	REPLACE ACT sensor. RECONNECT harness to ACT sensor. RERUN Quick Test.
	No ▶	GO to DB21.
DB21 CHECK ACT SIGNAL FOR SHORTS		
• Key Off, wait 10 seconds. • Harness disconnected from ACT sensor. • Disconnect processor 60 pin connector. Inspect for damaged pins, corrosion, loose wires, etc. Service as necessary. • DVOM on 200,000 ohm scale. • Measure resistance of test Pin 25 to test Pins 40, 46 and 60.	Any reading less than 10,000 ohms ▶	CORRECT circuit shorts. REMOVE Breakout box. RECONNECT processor and ACT sensor. RERUN Quick Test.
	All readings are 10,000 ohms or greater ▶	REPLACE processor. REMOVE Breakout box. RECONNECT processor and ACT sensor. RERUN Quick Test.

Air Charge Temperature (ACT) | Pinpoint Test | DB

TEST STEP	RESULT ▶	ACTION TO TAKE
DB90 SERVICE CODE 54: CONTINUOUS TEST: CHECK ACT SENSOR		
• Using continuous monitor mode, observe VOM or STAR LED for indication of a fault while performing the following: • Lightly tap on ACT sensor (simulate road shock). • Wiggle ACT connector. • Is a fault indicated?	Yes ▶	DISCONNECT and INSPECT connectors. If connector and terminals are good, REPLACE ACT sensor. REPEAT Quick Test.
	No ▶	GO to DB91.

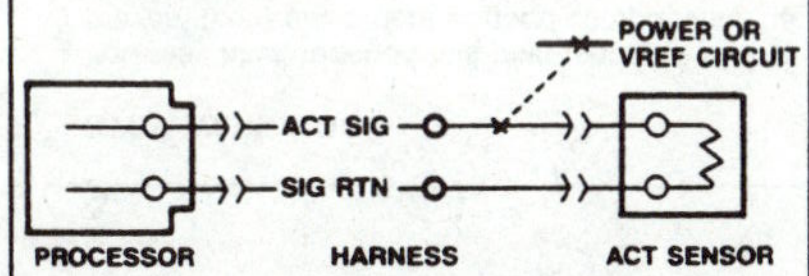

TEST STEP	RESULT ▶	ACTION TO TAKE
DB91 CHECK EEC-IV HARNESS		
• Observe VOM or STAR LED for a fault indication while performing the following: • Referring to the illustration in Step DB90, grasp the harness closest to the sensor connector. Wiggle, shake or bend a small section of the EEC-IV system harness while working your way to the dash panel. Also wiggle, shake or bend the EEC-IV harness from the dash panel to the processor. • Is fault indicated?	Yes ▶	ISOLATE fault and make necessary repairs. REPEAT Quick Test.
	No ▶	GO to DB92.
DB92 CHECK PROCESSOR AND HARNESS CONNECTORS		
• Key Off, wait 10 seconds. • Disconnect processor 60 pin connector. • Inspect both connectors and connector terminals for obvious damage or faults. • Connectors and terminals are OK.	No ▶	SERVICE as necessary. REPEAT Quick Test.
	Yes ▶	Unable to duplicate fault at this time. Continuous code 54 testing complete.

Air Charge Temperature (ACT) — Pinpoint Test — DB

TEST STEP	RESULT ▶	ACTION TO TAKE
DB93 SERVICE CODE 64: CONTINUOUS TEST: CHECK ACT SENSOR		
• Using continuous monitor mode, observe VOM or STAR LED for fault while performing the following: • Lightly tap on ACT sensor (simulate road shock). • Wiggle ACT connector. • Is fault indicated?	Yes ▶	DISCONNECT and INSPECT connectors. If connector and terminals are good, REPLACE ACT sensor. REPEAT Quick Test.
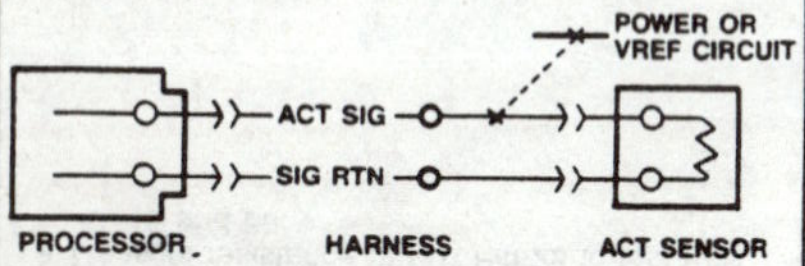	No ▶	GO to DB94.
DB94 CHECK EEC-IV HARNESS		
• Observe VOM or STAR LED for a fault indication while performing the following: • Referring to the illustration in Step DB93, grasp the harness closest to the sensor connector. Wiggle, shake or bend a small section of the EEC-IV system harness while working your way to the dash panel. Also wiggle, shake or bend the EEC-IV harness from the dash panel to the processor.	Yes ▶	ISOLATE fault and make necessary repairs. REPEAT Quick Test.
• Is fault indicated?	No ▶	GO to DB95.
DB95 CHECK PROCESSOR AND HARNESS CONNECTORS		
• Key Off, wait 10 seconds. • Disconnect processor 60 pin connector. • Inspect both connectors and connector terminals for obvious damage or faults.	No ▶	SERVICE as necessary. REPEAT Quick Test.
• Connectors and terminals are OK.	Yes ▶	Unable to duplicate fault at this time. Continuous code 64 testing complete.

Barometric Pressure Sensor (BP) — Pinpoint Test — DC

DC

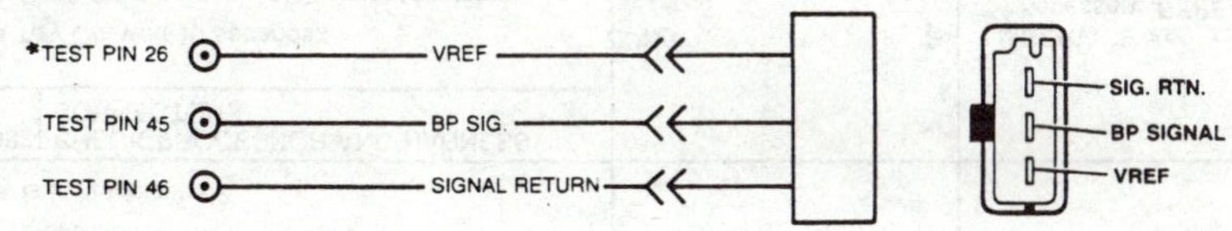

*TEST PIN LOCATED ON BREAKOUT BOX.
ALL HARNESS CONNECTORS VIEWED INTO MATING SURFACE.

STOP-WARNING

You may have reached this point because of a fault in a non-EEC System and/or component. To prevent the replacement of good components, be aware that the following non-EEC areas may be at fault:

- Unusually high or low atmospheric pressure.

This pinpoint test is intended to diagnose only the following:

- An active BP sensor (will not detect on out of range sensor).
- Harness circuits: VREF, BP Signal, and Signal Return.
- Processor assembly.

Barometric Pressure Sensor (BP) — Pinpoint Test — DC

TEST STEP	RESULT ▶	ACTION TO TAKE
DC1 SERVICE CODE 22: VOLTAGE CHECK VREF		
NOTE: The output of the BP sensor is digital in nature and cannot be monitored without special equipment. Therefore, perform a harness check before substituting the BP sensor. • Key Off, wait 10 seconds. • Disconnect harness from BP sensor. Inspect for damaged pins, corrosion, loose wires, etc. Service as necessary. • DVOM on 20 volt scale. • Key On, Engine Off. • Measure voltage from BP sensor harness connector VREF to battery negative terminal.	Reading is between 4.0V and 6.0V ▶	GO to DC2.
	Reading is less than 4.0V ▶	SERVICE open in VREF. If reading is still less than 4.0 volts, REPLACE processor. RECONNECT processor and BP sensor. RERUN Quick Test.
	Reading is greater than 6.0V ▶	GO to Pinpoint Test Step C1.
DC2 SIGNAL RETURN CHECK		
• Key On, Engine Off. • Harness disconnected from BP sensor. • DVOM on 20 volt scale. • Measure voltage from BP sensor harness connector signal return to battery positive post.	Reading is less than 10.0V ▶	SERVICE open in signal return. RECONNECT BP sensor. RERUN Quick Test.
	Reading is 10.0V or greater ▶	GO to Step DC3.
DC3 CHECK CONTINUITY OF BP SIGNAL		
• Key Off, wait 10 seconds. • Harness disconnected from BP sensor. • Disconnect processor 60 pin connector. Inspect for damaged pins, corrosion, loose wires, etc. Service as necessary. • Install Breakout box leaving processor disconnected. • DVOM on 200 ohm scale. • Measure resistance between BP sensor harness connector BP signal and test Pin 45.	Reading is less than 5 ohms ▶	REMOVE Breakout box. GO to DC4.
	Reading is 5 ohms or greater ▶	SERVICE open in BP signal. REMOVE Breakout box. RECONNECT processor and BP sensor. RERUN Quick Test.

Barometric Pressure Sensor (BP) — Pinpoint Test — DC

TEST STEP	RESULT ▶	ACTION TO TAKE
DC4 CHECK BP SIGNAL FOR SHORTS		
• Key Off. • Harness disconnected from BP sensor and processor. • DVOM on 200,000 ohm scale. • Measure resistance between BP sensor harness, BP signal and Signal Return. • Measure resistance between BP sensor harness, BP signal and VREF.	Either reading is less than 10,000 ohms ▶	SERVICE circuit shorts. RECONNECT BP sensor and processor. RERUN Quick Test.
	Both readings are 10,000 ohms or greater ▶	RECONNECT harness to processor and substitute BP sensor. RUN Quick Test. If code 22 is still present, REPLACE processor using original BP sensor. RERUN Quick Test.

Barometric Pressure Sensor (BP) — Pinpoint Test — DC

TEST STEP	RESULT	ACTION TO TAKE
DC90 SERVICE CODE 22: CONTINUOUS TEST: EXERCISE BP SENSOR		
• Using continuous monitor mode, observe VOM or STAR LED for indication of a fault while performing the following: • Connect a vacuum pump to the BP sensor. • Slowly apply 84 kPa (25 in. Hg.) vacuum to the sensor. • Slowly bleed vacuum off the BP sensor. • Lightly tap on BP sensor (simulate road shock). • Wiggle BP connector. • Is fault indicated?	Yes ▶	DISCONNECT and INSPECT connectors. If connector and terminals are good, REPLACE BP sensor. REPEAT Quick Test.
	No ▶	GO to DC91.
DC91 CHECK EEC-IV HARNESS		
• Observe VOM or STAR LED for a fault indication while performing the following: • Referring to the illustration in Step DC90, grasp the harness closest to the sensor connector. Wiggle, shake or bend a small section of the EEC-IV system harness while working your way to the dash panel. Also wiggle, shake or bend the EEC-IV harness from the dash panel to the processor. • Is fault indicated?	Yes ▶	ISOLATE fault and make necessary repairs. REPEAT Quick Test.
	No ▶	GO to DC92.
DC92 CHECK PROCESSOR AND HARNESS CONNECTORS		
• Key Off, wait 10 seconds. • Disconnect processor 60 pin connector. • Inspect both connectors and connector terminals for obvious damage or faults. • Connectors and terminals are OK.	No ▶	SERVICE as necessary. REPEAT Quick Test.
	Yes ▶	Unable to duplicate fault at this time. Continuous code 22 testing complete.

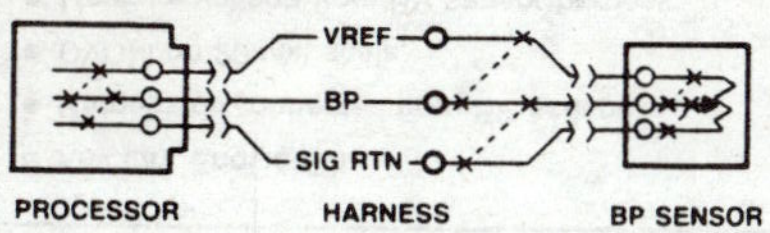

EGR Valve Position Sensor (EVP) — Pinpoint Test — DD

DD

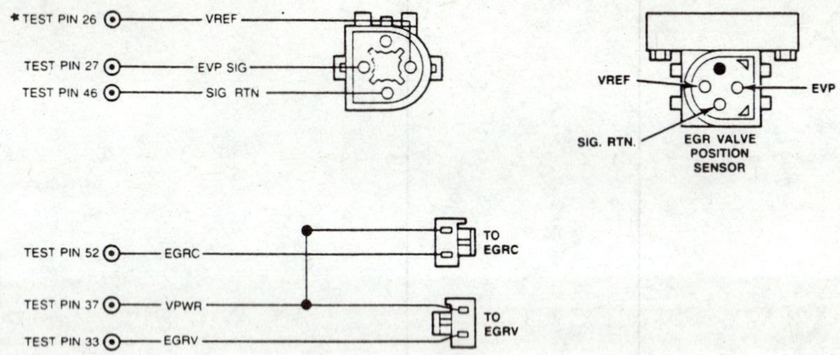

*TEST PINS LOCATED ON BREAKOUT BOX.
ALL HARNESS CONNECTORS VIEWED INTO MATING SURFACE.

STOP-WARNING

You may have reached this point because of a fault in a non-EEC System and/or component. To prevent the replacement of good components, be aware that the following non-EEC area may be at fault:

- Damaged EGR valve.

This Pinpoint Test is intended to diagnose only the following:

- EVP sensor.
- Harness circuits: EVP, Signal Return, VREF, EGRV, EGRC, VPWR.
- EGR solenoids.
- EGR Valve assembly.
- Processor assembly.

EGR Valve Position Sensor (EVP) — Pinpoint Test — DD

TEST STEP	RESULT ▶	ACTION TO TAKE
FAULT CODE 31		
DD1 RUN ENGINE RUNNING QUICK TEST WITH EGR VACUUM SIGNAL LINE DISCONNECTED AT EGR VALVE		
• Key Off, wait 10 seconds. • Disconnect EGR vacuum line at EGR valve and cap EGR vacuum line. • Run Engine Running Quick Test. • Check for code 31.	Code 31 present ▶	GO to DD2.
	No code 31 present, but codes 32, 34 are present ▶	GO to DD10.
DD2 CHECK EVP SENSOR RESISTANCE		
• Key Off, wait 10 seconds. • EGR signal line disconnected and capped. • Disconnect EVP sensor. Inspect for damaged pins, corrosion, and pins pushed out. Service as necessary. • DVOM on 200,000 ohm scale. • Measure resistance of EVP sensor EVP SIG to sensor VREF and sensor Signal Return to sensor VREF.	Both readings between 100 and 5500 ohms ▶	GO to DD3.
	Either reading less than 100 ohms or greater than 5500 ohms ▶	REPLACE EVP sensor. RECONNECT vacuum line and harness. RERUN Quick Test.
DD3 CHECK EVP MOVEMENT		
• Key Off, vacuum signal line disconnected and capped, harness disconnected at EVP sensor. • DVOM on 200,000 ohm scale. • Connect vacuum pump to EGR valve. • Measure resistance of EVP sensor EVP SIG to VREF while gradually increasing vacuum to 33 kPa (10 in. Hg). • Observe resistance as vacuum increases.	Reading gradually decreases from no greater than 5500 ohms to no less than 100 ohms ▶	GO to DD4.
	Reading is less than 100 ohms or greater than 5500 ohms ▶	REPLACE EVP sensor. RECONNECT signal line and harness. RERUN Quick Test.
	Reading does not decrease or unable to hold vacuum ▶	GO to DD16.

EGR Valve Position Sensor (EVP) — Pinpoint Test — DD

TEST STEP	RESULT ▶	ACTION TO TAKE
DD4 MEASURE VREF TO SIGNAL RETURN VOLTAGE		
• Key On, Engine Off. • Vacuum signal line disconnected and capped, harness disconnected from EVP sensor. • DVOM on 20V scale. • Measure voltage from EVP harness connector VREF to Signal Return.	Reading is between 4 and 6V ▶	GO to DD5.
	Reading is less than 4V ▶	GO to DD7.
	Reading is greater than 6V ▶	GO to Pinpoint Test Step C1.
DD5 MEASURE VREF TO SIGNAL INPUT VOLTAGE		
• Key On, Engine Off. • Reconnect EGR vacuum signal line to EGR valve. • Harness disconnected from EVP sensor. • DVOM on 20V scale. • Measure voltage from EVP sensor harness connector, VREF to EVP signal.	Reading is between 4 and 6V ▶	GO to DD6.
	Voltage is less than 4V ▶	GO to DD7.
DD6 CHECK EVP SIGNAL FOR SHORTS TO GROUND		
• Key Off, wait 10 seconds. Harness disconnected from EVP sensor. • Disconnect harness from processor 60 Pin connector and inspect for damaged pins, corrosion, loose wires. Service as necessary. • DVOM on 200,000 ohm scale. • Measure resistance of EVP sensor harness connector EVP Signal to Signal Return and EVP signal to battery negative terminal.	Either reading is less than 10,000 ohms ▶	SERVICE circuit short. RECONNECT processor and EVP sensor. RERUN Quick Test.
	Both readings are 10,000 ohms or greater ▶	REPLACE processor. RECONNECT processor and EVP sensor. RERUN Quick Test.

EGR Valve Position Sensor (EVP) — Pinpoint Test — DD

TEST STEP	RESULT ▶	ACTION TO TAKE
DD7 MEASURE CONTINUITY OF VREF, EVP SIGNAL AND SIGNAL RETURN		
• Key Off, wait 10 seconds. • Harness disconnected from EVP sensor. • Disconnect processor 60 Pin connector and inspect for damaged pins, corrosion, loose wires. Service as necessary. • Install Breakout box leaving processor disconnected. • DVOM on 200 ohm scale. • Measure continuity from test Pin 46 to EVP harness connector Signal Return and, from test Pin 26 to EVP harness connector VREF and from test Pin 27 to EVP harness connector EVP signal.	All readings are less than 5 ohms ▶ Any reading 5 ohms or greater ▶	GO to DD8. SERVICE open circuit(s). REMOVE Breakout box. RECONNECT processor and EVP sensor. RERUN Quick Test.
DD8 CHECK EVP SIGNAL FOR SHORTS TO POWER		
• Key Off, harness disconnected from EVP sensor. Breakout box installed with processor not connected. • DVOM on 200,000 ohm scale. • Measure resistance of test Pin 27 to test Pins 37 and 57.	Either reading less than 10,000 ohms ▶ Both readings 10,000 ohms or greater ▶	SERVICE short circuit. REMOVE Breakout box. RECONNECT processor and EVP sensor. RERUN Quick Test. REPLACE processor. REMOVE Breakout box. RECONNECT processor and EVP sensor. RERUN Quick Test.
FAULT CODES 32, 33 AND 34		
DD10 MEASURE EGR VENT AND EGR CONTROL SOLENOID RESISTANCE		
• Key Off, wait 10 seconds. • Disconnect harness from EGRC and EGRV solenoids. Inspect for damaged pins, corrosion, and pins pushed out. Service as necessary. • DVOM on 200 ohm scale. • Measure resistance of both solenoids.	Both readings between 30 and 70 ohms ▶ Either reading less than 30 ohms or greater than 70 ohms ▶	RECONNECT harness to both solenoids and GO to DD11. REPLACE EGR solenoid assembly. RECONNECT harness and RERUN Quick Test.

EGR Valve Position Sensor (EVP) — Pinpoint Test — DD

TEST STEP	RESULT ▶	ACTION TO TAKE
DD11 OUTPUT STATE CHECK (REFER TO APPENDIX)		
• Key Off, wait 10 seconds. • DVOM on 20V scale. • Connect DVOM negative test lead to STO and positive test lead to battery positive. • Jumper STI to signal return. • Perform Key On, Engine Off Self-Test until the completion of the Continuous Test Codes. • DVOM will indicate zero volts. • Depress and release the throttle. • Did DVOM reading change to a high voltage reading?	NO ▶ YES ▶	DEPRESS throttle to WOT and RELEASE. If STO voltage does not go high, GO to Pinpoint Test Step Q40. Remain in output state check and GO to DD12.
DD12 CHECK EGR SOLENOIDS FOR ELECTRICAL CYCLING		
• Key On, Engine Off. • In output state check. • DVOM on 20V scale. • Connect DVOM to EGRV circuits VPWR and EGRV signal. • While observing DVOM, depress and release the throttle several times to cycle output On and Off. • Repeat for EGRC circuits VPWR and EGRC signal.	Both solenoid outputs cycle On and Off ▶ Either output does not cycle On and Off ▶	REMAIN in output state check. GO to DD13. Exit output state check. GO to DD17.

EGR Valve Position Sensor (EVP) — Pinpoint Test — DD

TEST STEP	RESULT ▶	ACTION TO TAKE
DD13 CHECK EGR SOLENOIDS FOR VACUUM CYCLING		
• Key On. Engine Off. • In output state check. • Disconnect and cap vacuum line from bottom port of EGRC solenoid and connect a vacuum pump. • Connect a vacuum gauge in the common output (top) vacuum line to EGR valve. • Disconnect but **do not** cap vacuum vent line from EGRV solenoid. • While cycling outputs On and Off (by depressing and releasing throttle), observe vacuum gauge at the output. Maintain vacuum at source.	Vacuum output cycles On and Off in less than 2 seconds ▶	RECONNECT all vacuum lines. GO to DD14.
	Vacuum does not cycle On and Off in less than 2 seconds ▶	REPLACE solenoid assembly. RECONNECT all vacuum lines. RERUN Quick Test.
DD14 CHECK VACUUM LINES		
• Key Off, wait 10 seconds. • Vacuum lines reconnected. • Check entire **EEC** vacuum line system per VECI emission schematic decal for kinks, cracks, obstructions or leaks.	Vacuum lines OK ▶	GO to DD15.
	Vacuum lines not OK ▶	SERVICE as necessary. RERUN Quick Test.
DD15 CHECK EVP RESISTANCE WHILE APPLYING VACUUM TO EGR VALVE		
• Key Off. • Disconnect harness from EVP sensor. Inspect for damaged pins, corrosion, and pins pushed out. Service as necessary. • DVOM on 200,000 ohm scale. • Disconnect vacuum line at EGR valve. • Connect vacuum pump to EGR valve. • Measure resistance of EVP sensor EVP Signal Pin to VREF Pin while increasing vacuum to 33 kPa (10 in. Hg). • Observe resistance as vacuum increases.	Reading does not decrease gradually ▶	GO to DD16.
	Reading gradually decreases from no more than 5,500 ohms to no less than 100 ohms ▶	REPLACE processor. RECONNECT EVP sensor and EGR vacuum line. RERUN Quick Test.

EGR Valve Position Sensor (EVP) — Pinpoint Test — DD

TEST STEP	RESULT ▶	ACTION TO TAKE
DD16 MANUALLY EXERCISE EVP SENSOR		
• Key Off, harness disconnected from EVP sensor. • Remove EVP sensor from EGR valve. • Measure resistance from EVP sensor Signal Return Pin to EVP sensor VREF Pin while gradually applying pressure to EVP sensor shaft. • Observe resistance as shaft is slowly pushed in. • Look for sudden changes in resistance readings.	Reading decreases smoothly from no more than 5.500 ohms to no less than 100 ohms ▶	REPLACE EGR valve assembly. RECONNECT EVP sensor and EGR supply vacuum line. RERUN Quick Test.
	Reading does not decrease smoothly or changes abruptly ▶	REPLACE EVP sensor. RECONNECT harness and EGR supply vacuum line. RERUN Quick Test.
DD17 CHECK FOR VOLTAGE ON VEHICLE POWER CIRCUIT		
• Disconnect EGR vent and EGR control solenoids from harness. • Key On, Engine Off. • DVOM on 20V scale. • Measure voltage from battery negative terminal to VPWR circuit on both EGR solenoids.	Either reading is less than 10.5V ▶	SERVICE harness circuit open. RERUN Quick Test.
	Both readings are 10.5V or greater ▶	GO to DD18.
DD18 CHECK CONTINUITY OF EGRV AND EGRC CIRCUITS		
• Key Off, wait 10 seconds. • EGR vent and EGR control solenoid disconnected from harness. • Disconnect processor 60 Pin connector and inspect for damaged pins, corrosion, loose wires. Service as necessary. • Install Breakout box to processor harness connector. Leave processor disconnected. • DVOM on 200 ohm scale. • Measure continuity from test Pin 33 to EGRV solenoid harness connector EGRV signal. • Measure continuity from test Pin 52 to EGRC solenoid harness connector EGRC signal.	Both readings less than 5 ohms ▶	GO to DD19.
	Either reading 5 ohms or greater ▶	SERVICE open circuit. REMOVE Breakout box. RECONNECT harness to processor. RERUN Quick Test.

EGR Valve Position Sensor (EVP) — Pinpoint Test — DD

TEST STEP	RESULT ▶	ACTION TO TAKE
DD19 CHECK EGRV AND EGRC SIGNALS FOR SHORTS TO POWER		
• EGR vent and EGR control solenoids disconnected from harness. • Key Off. Breakout box installed. Processor disconnected. • DVOM on 200,000 ohm scale. • Measure resistance from test Pin 33 to test Pins 40, 46 and 57. • Measure resistance from test Pin 52 to test Pins 40, 46 and 57.	Any reading less than 10,000 ohms ▶	SERVICE circuit short. REMOVE Breakout box. RECONNECT harness to processor. RERUN Quick Test.
	All readings 10,000 ohms or greater ▶	REPLACE processor. REMOVE Breakout box. RECONNECT harness to processor. RERUN Quick Test.
FAULT CODE 35		
DD30 RPM TOO LOW FOR EGR TEST		
• Check for codes 12 and 13.	Code 12 present ▶	Vehicles equipped with DC motor control, GO to KF1. Vehicles equipped with throttle kicker, GO to KG1.
	Code 13 present ▶	Vehicles equipped with DC motor control, GO to KF20. Vehicles equipped with throttle kicker, GO to KG1.
	Codes 12 and 13 not present ▶	GO to DD31.
DD31 RETEST AT 1,500 RPM		
• Key Off, wait 10 seconds. • Install tachometer. • Perform Key On, Engine Running Quick Test while maintaining 1,500 rpm. • Record Engine Running service codes. • Check for code 35.	Code 35 present ▶	REPLACE processor. RERUN Quick Test.
	Code 35 not present ▶	REPEAT Quick Test. SERVICE codes as necessary.

EGR Valve Position Sensor (EVP) — Pinpoint Test — DD

TEST STEP	RESULT ▶	ACTION TO TAKE
DD40 VREF CIRCUIT TO GROUND		
• Key Off, wait 10 seconds. • Disconnect EVP sensor and inspect for damaged pins, corrosion, loose wires. Service as necessary. • DVOM on 20V scale. • Key On, Engine Off. • Measure voltage from VREF (at EVP sensor harness connector) to battery ground.	Voltage readings are between 4V and 6V ▶	GO to DD41.
	Voltage readings are less than 4.0V ▶	GO to DD7.
	Voltage readings are 6.0V or greater ▶	GO to Pinpoint Test step C1.
DD41 SIGNAL RETURN TO VREF		
• Key On. • EVP harness disconnected. • DVOM on 20V scale. • Measure voltage from VREF to signal return at the EVP harness connector.	Voltage is reading between 4V and 6V ▶	GO to DD42.
	Voltage readings are less than 4.0V ▶	SERVICE open in signal return. RERUN Key On, Engine Off Quick Test.
DD42 EVP CIRCUIT VOLTAGE CHECK		
• Key Off, (wait 10 seconds). • Disconnect processor 60 Pin connector and inspect for damaged pins, corrosion, loose wires. Service as necessary. • Install Breakout box and reconnect processor. • Connect EVP sensor. • DVOM on 20V scale. • Key On, Engine Off. • Measure voltage from test Pin 27 to test Pin 46.	Less than .5V ▶	GO to DD43.
	1.25V or greater ▶	GO to DD46.

EGR Valve Position Sensor (EVP) — Pinpoint Test — DD

TEST STEP	RESULT ▶	ACTION TO TAKE
DD43 EVP SIGNAL SHORT TO GROUND		
• Key Off, wait 10 seconds. • Disconnect processor. • Disconnect EVP sensor. • Breakout box installed. • DVOM on 200 ohm scale. • Measure resistance from test Pin 27 to test Pins 20, 40, 46, and 60.	5.0 ohms or greater ▶ Less than 5 ohms ▶	GO to DD44. SERVICE short to ground in EVP signal circuit and REPEAT Quick Test.
DD44 EVP SIGNAL CIRCUIT FOR OPENS		
• Key Off. • Processor disconnected. • EVP sensor disconnected. • Breakout box installed. • DVOM on 200 ohm scale. • Measure resistance from test Pin 27 to EVP signal circuit at EVP connector harness.	5 ohms or greater ▶ Less than 5 ohms ▶	SERVICE open in EVP signal circuit. REPEAT Quick Test. GO to DD45.
DD45 SUBSTITUTE EVP SENSOR		
• Key Off. • Remove and replace present EVP sensor with known good sensor. Reconnect EVP sensor. • Remove Breakout box. • Reconnect processor. • Rerun Key On, Engine Off Quick Test. • Is code 31 present?	Yes ▶ No ▶	REPLACE processor. RECONNECT original EVP sensor. REPEAT Quick Test. Original EVP sensor at fault. INSTALL new EVP sensor. REPEAT Quick Test.

EGR Valve Position Sensor (EVP) — Pinpoint Test — DD

TEST STEP	RESULT ▶	ACTION TO TAKE
DD46 EVP SIGNAL SHORT TO POWER		
• Key Off, wait 10 seconds. • Disconnect processor. • Disconnect EVP sensor. • Breakout box installed. • DVOM on 200 ohm scale. • Measure resistance from test Pin 27 to test Pins 26, 37 and 57.	5 ohms or greater ▶ Less than 5 ohms ▶	GO to DD47. SERVICE short to power. REPEAT Quick Test.
DD47 SUBSTITUTE EVP SENSOR AND EGR VALVE		
• Key Off, wait 10 seconds. • Electrically connect known good EVP sensor and EGR valve assembly. • Remove Breakout box. • Reconnect processor. • Perform Key On, Engine Off Quick Test. • Is code 31 present?	Yes ▶ No ▶	REPLACE processor. Connect original EVP sensor. REPEAT Quick Test. GO to DD48.
DD48 EVP SENSOR CHECK		
• Key Off, wait 10 seconds. • Install original EVP sensor on known good EGR valve. • EVP sensor connected. • Rerun Key On, Engine Off Quick Test. • Is code 31 present?	Yes ▶ No ▶	INSTALL new EVP sensor. REPEAT Quick Test. REFER to EGR systems, Section 6.

EGR Valve Position Sensor (EVP) — Pinpoint Test — DD

TEST STEP	RESULT ▶	ACTION TO TAKE
DD90 SERVICE CODE 31 CONTINUOUS TEST: EXERCISE EVP SENSOR		
• Using continuous monitor mode, observe VOM or STAR LED for indication of a fault while performing the following: • Connect a vacuum pump to the EGR valve. • Very slowly apply 20 kPa (6 in. Hg) vacuum to the EGR valve. • Slowly bleed vacuum off the EGR valve and lightly tap on EVP sensor (simulate road shock). • Wiggle EVP sensor connector. • Is a fault indicated?	Yes ▶	GO to DD91.
	No ▶	GO to DD92.
DD91 MEASURE EVP SIGNAL VOLTAGE WHILE EXERCISING EVP SENSOR		
• Key Off, wait 10 seconds. • Disconnect processor 60 Pin connector and inspect for damaged pins, corrosion, loose wires. Service as necessary. • Install Breakout box and reconnect processor. • VOM or STAR LED still connected to STO as in previous Step. • Connect a DVOM from test Pin 27 to test Pin 46. • DVOM on 20V scale. • Key On, Engine Off. • While observing DVOM, repeat step **DD90**. • Does the fault occur below 4.25V?	Yes ▶	DISCONNECT and INSPECT connector. If connector and terminals are good, REPLACE EVP sensor. REPEAT Quick Test.
	No ▶	EGR valve overshoot may have caused continuous code 31. Sensor service is not required. To verify harness integrity, GO to DD92.

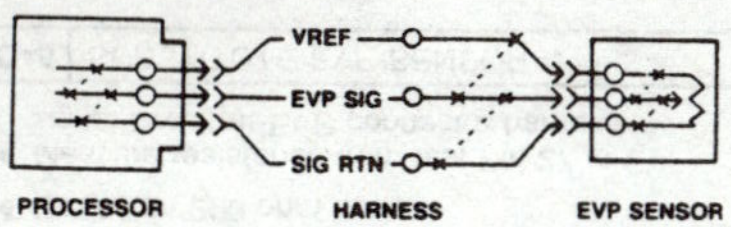

EGR Valve Position Sensor (EVP) — Pinpoint Test — DD

TEST STEP	RESULT ▶	ACTION TO TAKE
DD92 CHECK EEC-IV HARNESS		
• Observe VOM or STAR LED for a fault indication while performing the following: • Referring to the illustration in Step DD90, grasp the harness closest to the sensor connector. Wiggle, shake or bend a small section of the EEC-IV system harness while working your way to the dash panel. Also wiggle, shake or bend the EEC-IV harness from the dash panel to the processor. • Is a fault indicated?	Yes ▶	ISOLATE fault and make necessary repairs. REFER to appropriate figure. REPEAT Quick Test.
	No ▶	GO to DD93.
DD93 CHECK PROCESSOR AND HARNESS CONNECTORS		
• Key Off, wait 10 seconds. • Disconnect processor 60 Pin connector. • Inspect both connectors and connector terminals for obvious damage or faults. • Are connectors and terminals OK?	No ▶	SERVICE as necessary. REPEAT Quick Test.
	Yes ▶	Unable to duplicate fault at this time. Continuous code 31 testing complete.

Engine Coolant Temperature Sensor (ECT) | Pinpoint Test | DE

DE

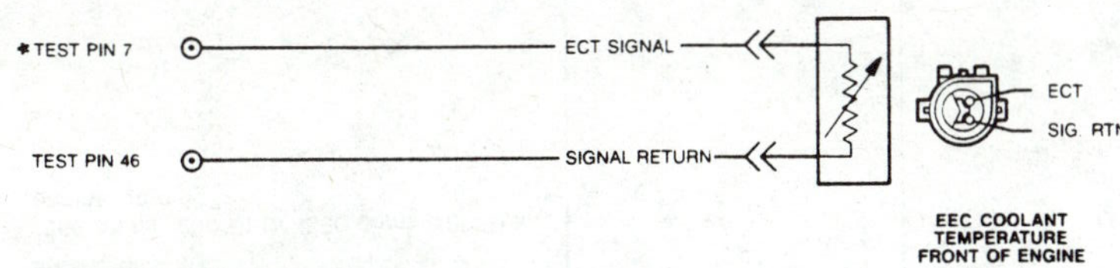

NOTE: TO PASS THIS TEST, ENGINE COOLANT TEMPERATURE MUST BE:
KEY ON, ENGINE OFF (50°F TO 240°F)
ENGINE RUNNING (180°F TO 240°F)

TYPICAL RESISTANCE BETWEEN TEST POINTS 7 & 46	58.750Ω	40.500Ω	3600Ω	1840Ω
AT TEMPERATURE	50°F	65°F	180°F	220°F

*** TEST PIN LOCATED ON BREAKOUT BOX.**
ALL CONNECTIONS VIEWED INTO MATING SURFACE.

STOP-WARNING

You may have reached this point because of a fault in a non-EEC System and/or component. To prevent the replacement of good components, be aware that the following non-EEC areas may be at fault:

- Coolant level.
- Oil level.
- Blocked or obstructed air flow.
- Engine not at normal operating temperature.
- Electro drive cooling fan.

This pinpoint test is intended to diagnose only the following:

- ECT sensor.
- Harness sensor circuits: ECT and Signal Return.
- Processor assembly.

Engine Coolant Temperature Sensor (ECT) | Pinpoint Test | DE

TEST STEP	RESULT ▶	ACTION TO TAKE
DE1 SERVICE CODE 21: CHECK ENGINE OPERATING TEMPERATURE		
• Run engine for 2 minutes at 2.000 rpm. • Check that upper radiator hose is hot and pressurized. • Rerun Quick Test.	Vehicle stalls ▶	Do not service code 21 at this time. REFER to diagnosis by symptoms.
	Code 21 present ▶	GO to DE2.
	Code 21 not present ▶	SERVICE other codes as necessary.
DE2 CHECK ECT AND SIG. RTN. VOLTAGE		
• Key Off, wait 10 seconds. • Disconnect harness from ECT sensor. Inspect for damaged pins, corrosion, loose wires, etc. Service as necessary. • DVOM on 20 volt scale. • Key On, Engine Off. • Measure voltage between ECT sensor harness connector ECT signal to signal return.	Reading is between 4.0V and 6.0V ▶	GO to DE5.
	Reading is less than 4.0V ▶	GO to DE3.
	Reading is greater than 6.0V ▶	GO to Pinpoint Test Step C1.
DE3 CHECK CONTINUITY OF ECT SIG. AND SIG. RTN.		
• Key Off, wait 10 seconds. • Harness disconnected from ECT sensor. • Disconnect processor 60 pin connector — inspect for damaged pins, corrosion, loose wires, etc. Service as necessary. • Install Breakout box to harness. Leave processor disconnected. • DVOM on 200 ohm scale. • Measure resistance of ECT sensor harness connector ECT signal to test Pin 7. • Measure resistance of ECT sensor harness connector Signal Return to test Pin 46.	Both readings are less than 5 ohms ▶	GO to DE4.
	Either reading is 5 ohms or greater ▶	SERVICE circuit opens. REMOVE Breakout box. RECONNECT processor and ECT sensor. RERUN Quick Test.

Engine Coolant Temperature Sensor (ECT) — Pinpoint Test — DE

TEST STEP	RESULT ▶	ACTION TO TAKE
DE4 CHECK ECT SIGNAL FOR SHORTS		
• Key Off. • Harness disconnected from ECT sensor. Breakout box installed. Processor disconnected. • DVOM on 200,000 ohm scale. • Measure resistance from test Pin 7 to test Pins 40 and 60.	Either reading is less than 10,000 ohms ▶	SERVICE circuit shorts. REMOVE Breakout box. RECONNECT processor and ECT sensor. RERUN Quick Test.
	Both readings are 10,000 ohms or greater ▶	GO to DE5.
DE5 ECT SENSOR CHECK		
NOTE: Engine may have cooled down. Always warm engine before taking ECT resistance measurement. • Key Off, wait 10 seconds. • Harness disconnected from ECT sensor. • DVOM on 200,000 ohm scale. • Make a resistance measurement of the ECT sensor. • Is the resistance reading: — 1300 ohms (240°F) to 7700 ohms (140°F) for engine off? — 1550 ohms (230°F) to 4550 ohms (180°F) for engine running?	Yes ▶	REPLACE processor. RECONNECT harness to ECT sensor. RERUN Quick Test.
	No ▶	REPLACE ECT sensor. RECONNECT harness to ECT sensor. RERUN Quick Test.

Engine Coolant Temperature Sensor (ECT) — Pinpoint Test — DE

TEST STEP	RESULT ▶	ACTION TO TAKE
DE10 SERVICE CODE 51: INDUCE OPPOSITE CODE		
• Key Off, wait 10 seconds. • Disconnect harness from ECT sensor. Inspect for damaged pins, corrosion, loose wires, etc. Service as necessary. • Insert a jumper wire between ECT sensor harness connector ECT Signal and Signal Return. • Run Key On, Engine Off Quick Test. • Is code 61 present?	Yes ▶	REPLACE ECT sensor. REMOVE jumper wire. RECONNECT ECT sensor. RERUN Quick Test.
	No ▶	GO to DE11.
DE11 CHECK CONTINUITY OF ECT SIGNAL AND SIGNAL RETURN		
• Key Off, wait 10 seconds. • Harness disconnected from ECT sensor, jumper wire removed. • Disconnect processor 60 pin connector. Inspect for damaged pins, corrosion, loose wires, etc. Service as necessary. • Install Breakout box to harness leaving processor disconnected. • DVOM on 200 ohm scale. • Measure resistance between ECT sensor harness connector ECT signal and test Pin 7. • Measure resistance between ECT sensor harness connector Signal Return and test Pin 46.	Both readings are less than 5 ohms ▶	REPLACE processor. REMOVE Breakout box. RECONNECT processor and ECT sensor. RERUN Quick Test.
	Either reading is 5 ohms or greater ▶	SERVICE open circuit(s). REMOVE Breakout box. RECONNECT processor and ECT sensor. RERUN Quick Test.

Engine Coolant Temperature Sensor (ECT) — Pinpoint Test — DE

TEST STEP	RESULT ▶	ACTION TO TAKE
DE20 SERVICE CODE 61: INDUCE OPPOSITE CODE		
• Key Off, wait 10 seconds. • Disconnect harness from ECT sensor. Inspect for damaged pins, corrosion, loose wires, etc. Service as necessary. • Run Key On, Engine Off Quick Test. • Is code 51 present?	Yes ▶	REPLACE ECT sensor. RECONNECT ECT sensor. RERUN Quick Test.
	No ▶	GO to DE21.
DE21 CHECK ECT SIGNAL FOR SHORT		
• Key Off, wait 10 seconds. • Harness disconnected from ECT sensor. • Disconnect processor 60 pin connector. Inspect for damaged pins, corrosion, loose wires, etc. Service as necessary. • Install Breakout box, leave processor disconnected. • DVOM on 200,000 ohm scale. • Measure resistance of test Pin 7 to test Pins 40 and 60.	Either reading is less than 10,000 ohms ▶	SERVICE circuit shorts. REMOVE Breakout box. RECONNECT processor and ECT sensor. RERUN Quick Test.
	Both readings are 10,000 ohms or greater ▶	REPLACE processor. REMOVE Breakout box. RECONNECT processor and ECT harness. RERUN Quick Test.

Engine Coolant Temperature Sensor (ECT) — Pinpoint Test — DE

TEST STEP	RESULT ▶	ACTION TO TAKE
DE90 SERVICE CODE 21 CONTINUOUS TEST: TEST DRIVE VEHICLE		
• Key Off and wait 10 seconds. • Disconnect all Self-Test equipment and prepare vehicle for test drive. • Drive vehicle. Try to simulate different drive modes or mode in which drive complaint is noticed. Attempt to maintain drive complaint mode for one minute or more, if possible. • Upon completion of drive evaluation, repeat Key On, Engine Off Self-Test. • Is code 21 present in the continuous test results?	Yes ▶	VERIFY thermostat operating properly. If OK, REPLACE ECT sensor. REPEAT Quick Test.
	No ▶	Unable to duplicate fault. Code 21 testing complete.

Engine Coolant Temperature (ECT) — Pinpoint Test — DE

TEST STEP	RESULT ▶	ACTION TO TAKE
DE91 SERVICE CODE 51 CONTINUOUS TEST: CHECK ECT SENSOR		
• Using continuous monitor mode, observe VOM or STAR LED for indication of a fault while performing the following: • Lightly tap on ECT sensor (simulate road shock). • Wiggle ECT connector. • Is fault indicated?	Yes ▶	DISCONNECT and INSPECT connectors. If connector and terminals are good, REPLACE ECT sensor. REPEAT Quick Test.
	No ▶	GO to DE92.

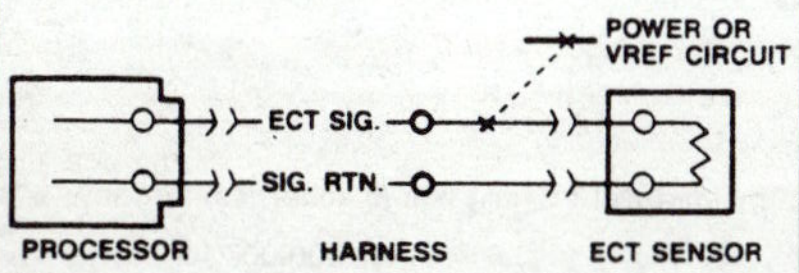

TEST STEP	RESULT ▶	ACTION TO TAKE
DE92 CHECK EEC-IV HARNESS		
• Observe VOM or STAR LED for a fault indication while performing the following: • Referring to the illustration in Step DE91, grasp the harness closest to the sensor connector. Wiggle, shake or bend a small section of the EEC-IV system harness while working your way to the dash panel. Also wiggle, shake or bend the EEC-IV harness from the dash panel to the processor. • Is fault indicated?	Yes ▶	ISOLATE fault and make necessary repairs. REPEAT Quick Test.
	No ▶	GO to DE93.
DE93 CHECK PROCESSOR AND HARNESS CONNECTORS		
• Key Off, wait 10 seconds. • Disconnect processor 60 pin connector. • Inspect both connectors and connector terminals for obvious damage or faults. • Connectors and terminals are OK.	No ▶	SERVICE as necessary. REPEAT Quick Test.
	Yes ▶	Unable to duplicate fault at this time. Continuous code 51 testing complete.

Engine Coolant Temperature (ECT) — Pinpoint Test — DE

TEST STEP	RESULT ▶	ACTION TO TAKE
DE94 SERVICE CODE 61 CONTINUOUS TEST: CHECK ECT SENSOR		
• Using continuous monitor mode, observe VOM or STAR LED for indication of a fault while performing the following: • Lightly tap on ECT sensor (simulate road shock). • Wiggle ECT connector. Voltage greater than 5 volts or STAR LED Off. • Is fault indicated?	Yes ▶	DISCONNECT and INSPECT connectors. If connector and terminals are good, REPLACE ECT sensor. REPEAT Quick Test.
	No ▶	GO to DE95.

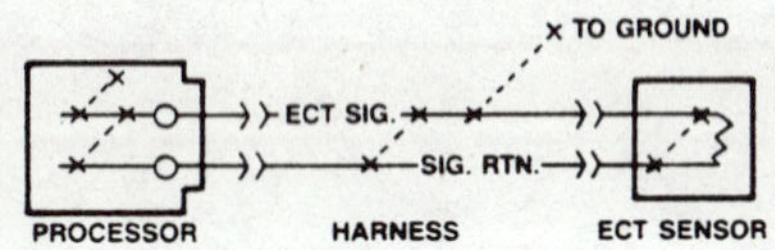

TEST STEP	RESULT ▶	ACTION TO TAKE
DE95 CHECK EEC-IV HARNESS		
• Observe VOM or STAR LED for a fault indication while performing the following: • Referring to the illustration in Step DE94, grasp the harness closest to the sensor connector. Wiggle, shake or bend a small section of the EEC-IV system harness while working your way to the dash panel. Also wiggle, shake or bend the EEC-IV harness from the dash panel to the processor. • Is fault indicated?	Yes ▶	ISOLATE fault and make necessary repairs. REPEAT Quick Test.
	No ▶	GO to DE96.
DE96 CHECK PROCESSOR AND HARNESS CONNECTORS		
• Key Off, wait 10 seconds. • Disconnect processor 60 pin connector. • Inspect both connectors and connector terminals for obvious damage or faults. • Connectors and terminals are OK.	No ▶	SERVICE as necessary. REPEAT Quick Test.
	Yes ▶	Unable to duplicate fault at this time. Continuous code 61 testing complete.

Manifold Absolute Pressure Sensor (MAP) — Pinpoint Test — DF

DF

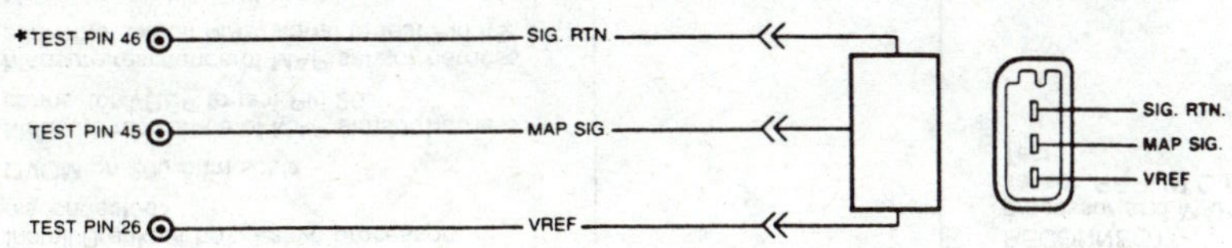

*TEST PINS LOCATED ON BREAKOUT BOX.
ALL HARNESS CONNECTORS VIEWED INTO MATING SURFACE.

STOP-WARNING

You may have reached this point because of a fault in a non-EEC System and/or component. To prevent the replacement of good components, be aware that the following non-EEC areas may be at fault:

- Unusually high/low atmospheric barometer reading.
- Kinked or obstructed vacuum lines.
- Basic engine (valves, vacuum leaks, timing, carburetor, etc.)
- Key On, Engine Off — MAP sensor must **not** see vacuum.
- Engine Running — MAP sensor must see actual manifold vacuum.

This Pinpoint Test is intended to diagnose only the following:

- MAP sensor.
- Harness circuits: VREF, MAP Signal, and Signal Return.
- Processor assembly.

MAP Absolute Pressure Sensor (MAP) — Pinpoint Test — DF

TEST STEP	RESULT ▶	ACTION TO TAKE
FAULT CODE 22		
DF1 CHECK FOR VACUUM LEAKS		
• From engine running code only. • Key Off, wait 10 seconds. • Verify all accessories are off. • Check vacuum line to MAP sensor for leaks, restrictions, kinks, etc. • Check for any engine system vacuum leaks. Refer to VECI decal.	Faults or leaks found ▶	SERVICE as necessary. RERUN Quick Test.
	Faults or leaks not found ▶	GO to DF2.
DF2 CHECK MAP SENSOR		
• Key Off, wait 10 seconds. • Disconnect vacuum line from MAP sensor. • Install vacuum pump to MAP sensor. • Apply 70 kPa (20 in. Hg) vacuum to MAP sensor. • Does MAP sensor hold vacuum?	Yes ▶	REMOVE vacuum pump. RECONNECT vacuum line to MAP sensor. GO to DF3.
	No ▶	REPLACE MAP sensor. CONNECT vacuum line to MAP sensor. RERUN Quick Test.
DF3 CHECK FOR OTHER FAULT CODES		
• Check for engine running, codes 31 or 41.	Code 31 present ▶	For 2.8L truck and 3.8L, 5.0L CFI: GO to Pinpoint Test Step DL1. All others: GO to Pinpoint Test Step DD1.
	Code 41 present ▶	For 1.6L, 2.3L EFI: GO to Pinpoint Test Step HA1. For 5.0L, 3.8L CFI: GO to Pinpoint Test Step HC1. All others: (FBC) GO to Pinpoint Test Step HB1.
	Codes 31 and 41 not present ▶	GO to DF4.

Manifold Absolute Pressure Sensor — Pinpoint Test — DF

TEST STEP	RESULT ▶	ACTION TO TAKE
DF4 CHECK VOLTAGE VREF TO SIGNAL RETURN		
• Key Off, wait 10 seconds. • Disconnect harness from MAP sensor. Inspect for damaged pins, corrosion, loose wires, etc. Service as necessary. • DVOM on 20V scale. • Key On, Engine Off. • Measure voltage from MAP sensor harness connector VREF to Signal Return.	Reading is between 4.0 and 6.0V ▶	GO to DF5.
	Reading is less than 4.0V ▶	GO to DF6.
	Reading is greater than 6.0V ▶	GO to Pinpoint Test Step C1.
DF5 CHECK VOLTAGE MAP SIGNAL TO SIGNAL RETURN		
• Key On, Engine Off. Harness disconnected from MAP sensor. • DVOM on 20V scale. • Measure voltage from MAP sensor harness connector MAP signal to Signal Return.	Reading is less than 4.0V ▶	GO to DF6.
	Reading is 4.0V or greater ▶	GO to DF8.
DF6 CHECK CONTINUITY OF VREF MAP SIGNAL AND SIGNAL RETURN		
• Key Off, wait 10 seconds. • Harness disconnected from MAP sensor. • Disconnect processor 60 Pin connector. Inspect for damaged pins, corrosion, loose wires, etc. Service as necessary. • Install Breakout box. Leave processor disconnected. • DVOM on 200 ohm scale. • Measure resistance of MAP sensor harness connector VREF to test Pin 26. • Measure resistance of MAP sensor harness connector circuit MAP signal to test Pin 45. • Measure resistance of MAP sensor harness connector Signal Return to test Pin 46.	All readings are less than 5 ohms ▶	GO to DF7.
	Any reading is 5 ohms or greater ▶	SERVICE circuit opens. REMOVE Breakout box. RECONNECT processor and MAP sensor. RERUN Quick Test.

Manifold Absolute Pressure Sensor (MAP) — Pinpoint Test — DF

TEST STEP	RESULT ▶	ACTION TO TAKE
DF7 CHECK VREF AND MAP SIGNAL FOR SHORT TO GROUND		
• Key Off, harness disconnected from MAP sensor. Breakout box installed. Processor disconnected. • DVOM on 200 ohm scale. • Measure resistance of test Pin 26 to test Pins 40, 46 and 60. • Measure resistance of test Pin 45 to test Pins 40, 46 and 60.	Any reading less than 100 ohms ▶	SERVICE circuit shorts. REMOVE Breakout box. RECONNECT processor and MAP sensor. RERUN Quick Test.
	All readings are 100 ohms or greater ▶	GO to DF9.
DF8 CHECK VREF AND MAP SIGNAL FOR SHORTS TO POWER		
• Key Off, wait 10 seconds. • Disconnect processor 60 Pin connector and inspect for damaged pins, corrosion, loose wires. Service as necessary. • Install Breakout box, leave processor disconnected. • Harness disconnected from MAP sensor. • DVOM on 200,000 ohm scale. • Measure resistance of test Pin 45 to test Pins 26, 37 and 57.	Any reading less than 10,000 ohms ▶	SERVICE circuit shorts. REMOVE Breakout box. RECONNECT processor and MAP sensor. RERUN Quick Test.
	All readings are 10,000 ohms or greater ▶	GO to DF9.
DF9 SUBSTITUTE MAP SENSOR		
• Key Off. • Remove Breakout box. Reconnect processor. Connect harness and vacuum line to a substitute MAP sensor. • Rerun quick test. • Is code 22 present?	Yes ▶	REPLACE processor. CONNECT harness and vacuum to original MAP sensor. RERUN Quick Test.
	No ▶	INSTALL substitute MAP sensor. RERUN Quick Test.

Manifold Absolute Pressure Sensor (MAP) — Pinpoint Test — DF

TEST STEP	RESULT ▶	ACTION TO TAKE
FAULT CODE 72		
DF10 CHECK THAT VACUUM TO MAP SENSOR DECREASES DURING DYNAMIC RESPONSE		
• Key Off, wait 10 seconds. • Tee a vacuum gauge in the intake manifold vacuum line at the MAP sensor. • Perform Engine Running Quick Test while observing vacuum. • Record engine service codes. • Did vacuum decrease by more than 30 kPa (9 in. Hg) during dynamic response test? • Is code 72 present?	Vacuum decrease is 30 kPa (9 in. Hg) or greater and code 72 is not present ▶	DISCONNECT vacuum equipment and service other codes as necessary.
	Vacuum decrease is 30 kPa (9 in. Hg) or greater and code 72 is present ▶	REPLACE MAP sensor and RERUN Quick Test.
	Vacuum decrease is less than 30 kPa (9 in. Hg) ▶	GO to DF11.
DF11 CHECK VACUUM LINES		
• Check vacuum line for proper routing. Refer to VECI decal. Check MAP sensor vacuum line for kinks or blockage. • Vacuum lines are OK.	Yes ▶	EEC-IV system OK. REFER to shop manual, group 21 for probable subjects affecting engine vacuum.
	No ▶	SERVICE as necessary and REPEAT DF10.

Manifold Absolute Pressure Sensor (MAP) — Pinpoint Test — DF

TEST STEP	RESULT ▶	ACTION TO TAKE
DF90 SERVICE CODE 22: CONTINUOUS TEST: EXERCISE MAP SENSOR		
• Using continuous monitor mode, observe VOM or STAR LED for indication of a fault while performing the following: • Connect a vacuum pump to the MAP sensor. • Slowly apply 84 kPa (25 in. Hg.) vacuum to the sensor. • Slowly bleed vacuum off the MAP sensor. • Lightly tap on MAP sensor (simulate road shock). • Wiggle MAP connector. • Is fault indicated?	Yes ▶	DISCONNECT and INSPECT connectors. If connector and terminals are good, REPLACE sensor. REPEAT Quick Test.
	No ▶	GO to DF91.

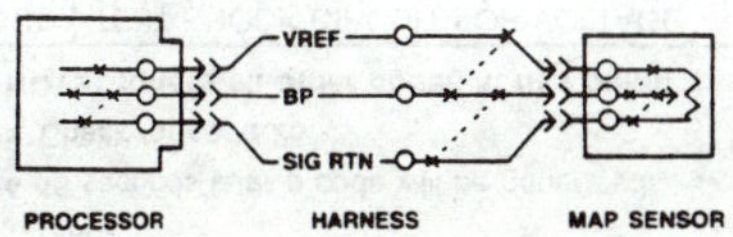

TEST STEP	RESULT ▶	ACTION TO TAKE
DF91 CHECK EEC-IV HARNESS		
• Observe VOM or STAR LED for a fault indication while performing the following: • Referring to the illustration in Step DF90, grasp the harness closest to the sensor connector. Wiggle, shake or bend a small section of the EEC-IV system harness while working your way to the dash panel. Also wiggle, shake or bend the EEC-IV harness from the dash panel to the processor. • Is a fault indicated?	Yes ▶	ISOLATE fault and make necessary repairs. REPEAT Quick Test.
	No ▶	GO to DF92.
DC92 CHECK PROCESSOR AND HARNESS CONNECTORS		
• Key Off, wait 10 seconds. • Disconnect processor 60 Pin connector. • Inspect both connectors and connector terminals for obvious damage or faults. • Are connectors and terminals OK?	No ▶	SERVICE as necessary. REPEAT Quick Test.
	Yes ▶	Unable to duplicate fault at this time. Continuous code 22 testing complete.

Knock Sensor — Pinpoint Test — DG

DG

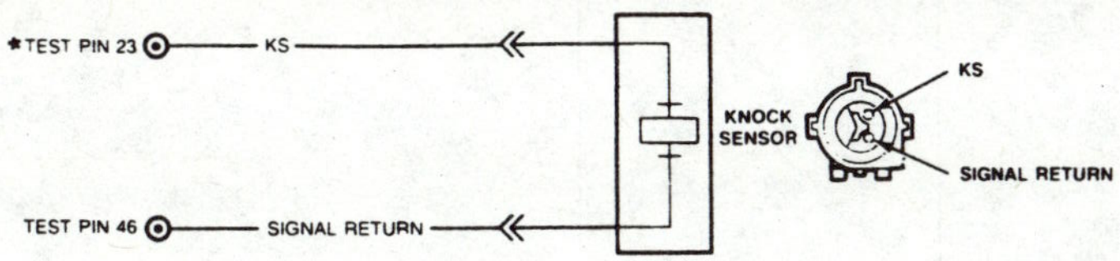

*TEST PINS LOCATED ON BREAKOUT BOX.
ALL HARNESS CONNECTORS VIEWED INTO MATING SURFACE.

STOP-WARNING

You may have reached this point because of a fault in a non-EEC System and/or component. To prevent the replacement of good components, be aware that the following non-EEC areas may be at fault:

- Fuel (quality).
- Basic engine.
- Spark timing.

This Pinpoint Test is intended to diagnose only the following:

- Knock sensor.
- Harness circuits: KS and Signal Return.
- Processor assembly.

Knock Sensor — Pinpoint Test — DG

TEST STEP	RESULT ▶	ACTION TO TAKE
DG1 SERVICE CODE 25: GENERATE KNOCK MANUALLY		
NOTE: With knock conditions sensitive to fuel, altitude and weather, perform Step DG1 before servicing any components. • Vehicle prepared to run (Engine Running Self-Test). • Equipment needed: 4 oz. hammer. • Prepare to rap/tap on exhaust manifold, directly above the knock sensor, when the Dynamic Response Signal is given. **NOTE: There is no need to actually depress throttle at this point.** • Perform Engine Running Self-Test. • Rap moderately on exhaust manifold when meter indicates Dynamic Response Test is ready. • 15 seconds later a code will be generated. • Check for code 25. **NOTE: Ignore all other codes at this point.**	No ▶	Knock system OK. REPEAT Engine Running Self-Test and SERVICE any other codes from that test.
	Yes ▶ **NOTE: Service code 25 may be received whenever the engine is not tapped**	GO to DG2.
DG2 TEST KNOCK CIRCUIT FOR VOLTAGE		
• Key Off. wait 10 seconds. • Disconnect knock sensor connector and inspect. • Set DVOM on 20V scale. • Key On. Engine Off. • Measure voltage between harness connector circuit KS and signal return.	Voltage is between 1 and 4V ▶	GO to DG6.
	Voltage is less than 1V ▶	GO to DG3.
	Voltage is greater than 4V ▶	GO to DG5.
DG3 CHECK CONTINUITY OF KS AND SIGNAL RETURN CIRCUITS		
• Key Off. wait 10 seconds. • Disconnect processor 60 Pin connector and inspect for damaged pins, corrosion, loose wires. Service as necessary. • Connect Breakout box to harness. Leave processor disconnected. • Knock sensor disconnected. • DVOM on 200 ohm scale. • Measure continuity of Signal Return at harness to test Pin 46 and KS at harness to test Pin 23.	Both resistances less than 5 ohms ▶	GO to DG4.
	Either resistance 5 ohms or greater ▶	SERVICE open circuit. REPEAT Quick Test.

Knock Sensor — Pinpoint Test — DG

TEST STEP	RESULT ▶	ACTION TO TAKE
DG4 CHECK KS CIRCUIT FOR SHORT TO GROUND		
• Key Off, wait 10 seconds. • Breakout box installed. • Processor disconnected. • Knock sensor disconnected. • DVOM on 200,000 ohm scale. • Measure resistance of KS circuit to test Pins 40, 46 and 60.	All resistances 10,000 ohms or greater ▶	GO to DG6.
	Any resistance less than 10,000 ohms ▶	SERVICE harness short. REPEAT Quick Test.
DG5 CHECK KS CIRCUIT FOR SHORT TO VOLTAGE		
• Key Off, wait 10 seconds. • Disconnect processor 60 Pin connector and inspect for damaged pins, corrosion, loose wires. Service as necessary. • Connect Breakout box to harness. Leave processor disconnected. • Knock sensor disconnected. • Key On, Engine Off. • DVOM on 20V scale. • Measure voltage from test Pin 23 to test Pin 40.	0.5V or greater ▶	SERVICE harness short to power. REPEAT Quick Test.
	Less than 0.5V ▶	GO to DG6.
DG6 TEST PROCESSOR WITH SUBSTITUTE KNOCK SENSOR		
• Key Off, wait 10 seconds. • Remove Breakout box and reconnect processor. • Equipment Required: — Equivalent knock sensor (same part number). — 4 oz. hammer. • Plug substitute sensor in harness (do not install). • Perform Engine Running Self-Test. • Lightly tap knock sensor when Dynamic Test Ready Signal is given. • 15 seconds later, a code will be generated. • Check for code 25. **NOTE: Ignore all other codes at this time.**	No ▶	INSTALL new knock sensor. REPEAT Quick Test.
	Yes ▶	REPLACE processor and REMOVE substitute sensor. REPEAT Quick Test with original sensor.

Throttle Position Sensor — Pinpoint Test — DH

DH

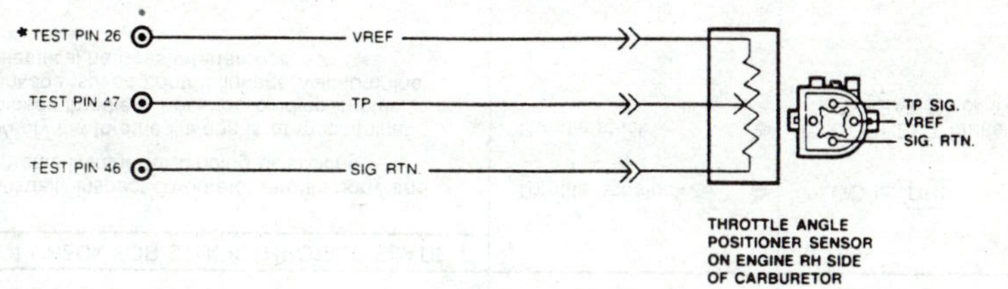

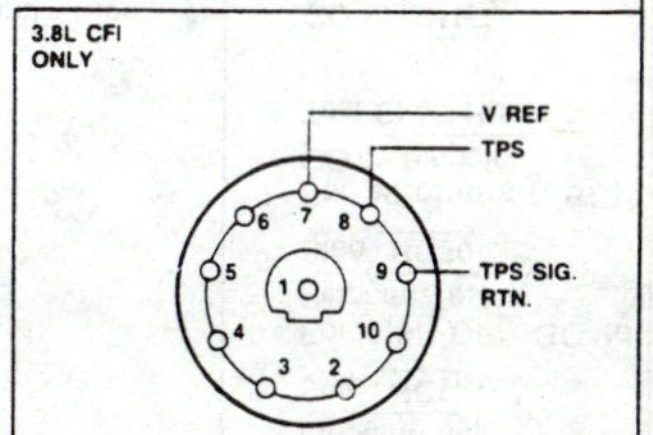

NOTE: FOR 3.8L CFI A 10-PIN CONNECTOR WAS USED FOR THE FIRST 60 DAYS OF PRODUCTION. EITHER CONNECTOR MAY BE ENCOUNTERED. TEST PIN ASSIGNMENTS DID NOT CHANGE ON BREAKOUT BOX. IF TP SENSOR MUST BE DISCONNECTED FOR DIAGNOSTICS, DISCONNECT THE 10 PIN CONNECTOR AT THE FUEL CHARGE ASSY.

* TEST PINS LOCATED ON BREAKOUT BOX.
ALL HARNESS CONNECTORS VIEWED INTO MATING SURFACE.

STOP-WARNING

You may have reached this point because of a fault in a non-EEC System and/or component. To prevent the replacement of good components, be aware that the following non-EEC areas may be at fault:

- Idle speeds/throttle stop adjustment.
- Binding throttle shaft linkage or speed control linkage.
- Choke/high cam system, if equipped.

This Pinpoint Test is intended to diagnose only the following:

- TP sensor.
- Sensor harness circuits: VREF, TP Signal, and Signal Return.
- Processor assembly.

Throttle Position Sensor — Pinpoint Test — DH

TEST STEP	RESULT	ACTION TO TAKE
DH1 SERVICE CODE 23: YOU HAVE BEEN DIRECTED HERE FROM KEY ON, ENGINE OFF TEST. THE FOLLOWING CHECK MUST BE MADE BEFORE SERVICING THIS CODE		
• Check for code 68.	Code 68 present ▶	GO to Pinpoint Test Step FB1.
	Code 68 not present ▶	GO to DH2.
DH2 ATTEMPT TO CLOSE THROTTLE COMPLETELY		
• Lightly press throttle lever toward actuator tip of idle speed motor while performing Key On, Engine Off Self-Test. • Check for service code 23.	Code 23 present ▶	GO to DH5.
	Code 23 not present ▶	TP sensor slightly out of adjustment but do not attempt to adjust. EEC system functioning properly.
DH3 SERVICE CODE 23: YOU HAVE BEEN DIRECTED HERE FROM ENGINE RUNNING TEST. THE FOLLOWING CHECK MUST BE MADE BEFORE SERVICING THIS CODE		
• Check for code 58.	Code 58 present ▶	GO to Pinpoint Test Step FB1.
	Code 58 not present ▶	GO to DH4.

Throttle Position Sensor — Pinpoint Test — DH

TEST STEP	RESULT	ACTION TO TAKE
DH4 REPAIR FOLLOWING ENGINE RUNNING SERVICE CODES BEFORE CONTINUING		
• Check for codes 31 or 41 present in Engine Running Test.	Code 31 present ▶	For 2.8L truck, 3.8L and 5.0L CFI: GO to Pinpoint Test Step DL1. All others: GO to Pinpoint Test Step DD1.
	Code 41 present ▶	For 1.6L, 2.3L EFI: GO to Pinpoint Test Step HA1. For 5.0L CFI: GO to Pinpoint Test Step HC1. For 3.8L CFI: GO to Pinpoint Test Step HC20. For all others (FBC): GO to Pinpoint Test Step HB1.
	Codes 31 and 41 are not present ▶	GO to DH5.
DH5 CHECK FOR STUCK THROTTLE PLATE		
• Visually inspect carburetor/throttle body and throttle linkage for binding or sticking.	Throttle not stuck ▶	GO to DH6.
• Verify the throttle linkage is at mechanical closed throttle. Check for: binding throttle linkage, speed control linkage, vacuum line electrical harness interference, etc.	Throttle stuck ▶	CORRECT faults. REPEAT Quick Test.

Throttle Position Sensor — Pinpoint Test — DH

TEST STEP	RESULT ▶	ACTION TO TAKE
DH6 GENERATE CODE 63		
• Refer to illustration DH. • Key Off, wait 10 seconds. • Disconnect TP sensor harness connector at the throttle body. Inspect for damaged pins, corrosion, loose wires, etc. Service as necessary. • Perform Key On, Engine Off self-test and record codes. **NOTE: Ignore all other codes at this time.**	Code 63 present ▶	REPLACE TP sensor, REFER to Shop Manual Group 24 and REPEAT Quick Test except for Calibration 4-54T 5.0L truck and 2.3L HSC engine. GO to DH20.
	Code 63 not present ▶	REPLACE processor, REPEAT Quick Test, except for Calibration 4-54T (5.0L truck) and 2.3 HSC engine. GO to DH20.
DH20 SERVICE CODE 53. CHECK SIGNAL RETURN FOR OPEN		
• Refer to illustration DH. • Key Off, wait 10 seconds. • Disconnect TP harness connector at throttle body. Inspect for damaged pins, corrosion, loose wires. Service as necessary. • DVOM on 20V scale. • Connect DVOM test leads to battery positive terminal and signal return at harness connector. • Observe voltage reading.	10V or greater ▶	GO to DH21.
	Less than 10V ▶	GO to DH23.
DH21 CHECK TP SIGNAL LINE FOR VOLTAGE		
• Key Off, TP sensor disconnected. • DVOM on 20V scale. • Connect DVOM test leads to TP and signal return at harness connector. • Key On, Engine Off. • Measure voltage on TP circuit.	Voltage less than 4V ▶	GO to DH22.
	Voltage 4V or greater ▶	GO to DH24.

Throttle Position Sensor — Pinpoint Test — DH

TEST STEP	RESULT ▶	ACTION TO TAKE
DH22 GENERATE CODE 63		
• Turn Key Off, wait 10 seconds. • With TP harness disconnected, perform Key On, Engine Off Self-Test. • Check for code 63. **NOTE: Ignore all other codes at this time.**	Code 63 present ▶	REPLACE TP sensor, REFER to Shop Manual Group 24 and REPEAT Quick Test.
	Code 63 not present ▶	REPLACE processor. REPEAT Quick Test.
DH23 CHECK SIG. RET. CIRCUIT FOR OPEN		
• Key Off, wait 10 seconds, TP harness disconnected. • DVOM on 200 ohm scale. • Disconnect processor 60 Pin connector and inspect for damaged pins, corrosion, loose wires. Service as necessary. • Install Breakout box, leave processor disconnected. • Measure continuity of Signal Return to test Pin 46.	DVOM reading less than 5 ohms ▶	REPLACE processor. REPEAT Quick Test.
	DVOM reading 5 ohms or greater ▶	SERVICE open circuit. REPEAT Quick Test.
DH24 CHECK TP SIGNAL FOR SHORT TO POWER		
• Key Off, wait 10 seconds, TP harness disconnected. • DVOM on 200,000 ohm scale. • Disconnect processor 60 Pin connector and inspect for damaged pins, corrosion, loose wires. Service as necessary. • Install Breakout box, leave processor disconnected. • Measure resistance of test Pin 47 to test Pins 26 and 57.	Either resistance is less than 10,000 ohms ▶	SERVICE harness short. REPEAT Quick Test.
	Both resistances are 10,000 ohms or greater ▶	REPLACE processor. REPEAT Quick Test.

Throttle Position Sensor — Pinpoint Test — DH

TEST STEP	RESULT ▶	ACTION TO TAKE
DH30 SERVICE CODE 63: CHECK FOR VREF TO TP SIGNAL INPUT		
• Refer to illustration DH. • Key Off, wait 10 seconds. • Disconnect TP harness connector at throttle body. Inspect for damaged pins, corrosion, and pins pushed out. Service as necessary. • DVOM on 20V scale. • Key On, Engine Off. • Measure voltage from TP harness VREF circuit to TP signal.	Voltage reading between 4V and 6V ▶ Voltage reading less than 4V ▶ Voltage reading 6V or greater ▶	GO to DH31. GO to DH33. GO to Pinpoint Test Step C1.
DH31 CHECK RESISTANCE OF TP CIRCUIT TO GROUND		
• Key Off, wait 10 seconds, TP harness disconnected. • Disconnect processor 60 Pin connector and inspect for damaged pins, corrosion, loose wires. Service as necessary. • DVOM to 200,000 ohm scale. • Connect one test lead to harness TP signal at TP sensor connector and the other test lead to ground.	DVOM reading is less than 10,000 ohms ▶ DVOM reading is 10,000 ohms or greater ▶	SERVICE circuit short and RETEST. GO to DH32.
DH32 GENERATE CODE 53: 5.0L TRUCK CAL. 4-54T AND 2.3 HSC ONLY		
• Key Off, wait 10 seconds. TP harness disconnected. • Jumper VREF to TP signal. • Perform Key On, Engine Off self-test. • Check for Code 53 Code 23 (5.0L truck — 2.3 HSC) **NOTE: Ignore all other codes at this time.**	Code 53 present (Code 23 present) ▶ Code 53 not present (Code 23 not present) ▶	REPLACE TP sensor, REFER to Shop Manual Group 24 and REPEAT Quick Test. REPLACE processor. REPEAT Quick Test.

Throttle Position Sensor — Pinpoint Test — DH

TEST STEP	RESULT ▶	ACTION TO TAKE
DH33 CHECK CONTINUITY OF TP AND VREF CIRCUITS		
• Key Off. wait 10 seconds. TP harness disconnected. • DVOM on 200 ohm scale. • Disconnect processor 60 Pin connector and inspect for damaged pins. corrosion. loose wires. Service as necessary. • Connect Breakout box. Processor connected to Breakout box. • Check continuity of TP Signal to test Pin 47. Check continuity of VREF to test Pin 26.	Either DVOM reading 5 ohms or greater ▶ Both DVOM readings less than 5 ohms ▶	SERVICE faulty circuit. CONNECT throttle position sensor. REMOVE Breakout box and REPEAT Quick Test. REPLACE processor. CONNECT throttle position sensor. REMOVE Breakout box and REPEAT Quick Test.
DH40 SERVICE CODE 73: TP SENSOR MOVES IN ENGINE RESPONSE TEST		
NOTE: Code 73 indicates the TP Sensor did not exceed 25 percent of its rotation in the Engine Response Check. • Key Off. • Install Breakout box. • DVOM on 20V range. • Connect DVOM to test Pins 47 and 46. • Perform Engine Running Quick Test. Step 5.0. • Verify DVOM reading exceeds 3.5V during brief WOT at Engine Response Check.	DVOM exceeds 3.5V during Engine Response Check ▶ DVOM does not exceed 3.5V during Engine Response Check ▶	REPLACE processor. REPEAT Quick Test. VERIFY TP Sensor is properly attached to throttle body. If OK. REPLACE TP Sensor PER Shop Manual Group 24 and REPEAT Quick Test.

Throttle Position Sensor (TPS) — Pinpoint Test — DH

TEST STEP	RESULT ▶	ACTION TO TAKE
DH90 CONTINUOUS TEST SERVICE CODE 53: EXERCISE TP SENSOR		
• Using continuous monitor mode, observe VOM or STAR LED for indication of a fault while performing the following: • Move throttle slowly to WOT position. • Release throttle slowly to closed position and lightly tap on TP sensor (simulate road shock). • Wiggle TP harness connector. • Is a fault indicated?	Yes ▶	GO to DH91.
	No ▶	GO to DH92.

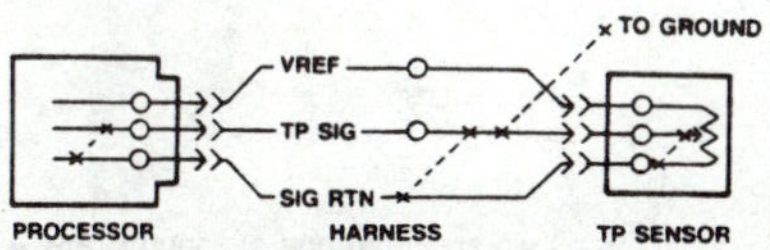

TEST STEP	RESULT ▶	ACTION TO TAKE
DH91 MEASURE THROTTLE POSITION SIGNAL VOLTAGE WHILE EXERCISING TP SENSOR		
• Key Off, wait 10 seconds. • Disconnect processor 60 Pin connector and inspect for damaged pins, corrosion, loose wires. Service as necessary. • Install Breakout box and reconnect processor. • VOM or STAR LED still connected to STO as in previous step. • Connect a DVOM from test Pin 47 to test Pin 46. • DVOM on 20V scale. • Key On, engine Off. • While observing DVOM, repeat step DH90. • Does the fault occur below 4.25V?	Yes ▶	DISCONNECT and INSPECT connectors. If connector and terminals are good, REPLACE TP sensor, REFER to Shop Manual Group 24 and REPEAT Quick Test.
	No ▶	Throttle position sensor overtravel may have caused the continuous code 53. Sensor service is not required. To verify harness integrity, GO to DH92.

Throttle Position Sensor (TPS) — Pinpoint Test — DH

TEST STEP	RESULT ▶	ACTION TO TAKE
DH92 CHECK EEC-IV HARNESS		
• Observe VOM or STAR LED for a fault indication while performing the following: • Referring to the illustration in Step DH90, grasp the harness close to the sensor connector. Wiggle, shake or bend a small section of the EEC-IV system harness while working your way to the dash panel. Also wiggle, shake or bend the EEC-IV harness from the dash panel to the processor. • Is a fault indicated?	Yes ▶	ISOLATE fault and make necessary repairs. REFER to appropriate figure. REPEAT Quick Test.
	No ▶	GO to DH93.
DH93 CHECK PROCESSOR AND HARNESS CONNECTORS		
• Key Off, wait 10 seconds. • Disconnect processor 60 Pin connector. • Inspect both connectors and connector terminals for obvious damage or faults. • Are connectors and terminals OK?	No ▶	SERVICE as necessary. REPEAT Quick Test.
	Yes ▶	Unable to duplicate fault at this time. Continuous code 53 testing complete.
DH94 CONTINUOUS TEST SERVICE CODE 63: EXERCISE TP SENSOR		
• Using continuous monitor mode, observe VOM or STAR LED for indication of a fault while performing the following: • Move throttle slowly to WOT position. • Release throttle slowly to closed condition. • Lightly tap on TP sensor (simulate road shock). • Wiggle TP harness connector. • Is a fault indicated?	Yes ▶	DISCONNECT and INSPECT connectors. If connector and terminals are good, REPLACE TP sensor, REFER to Shop Manual Group 24 and REPEAT Quick Test.
	No ▶	GO to DH95.

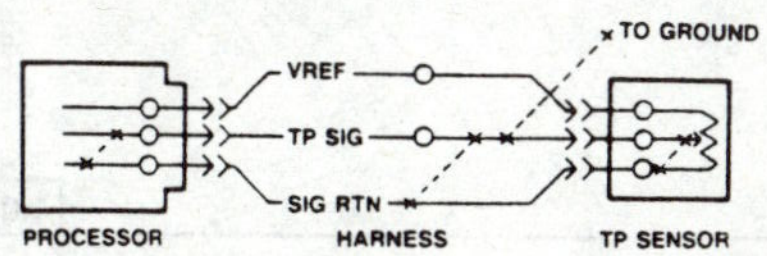

Throttle Position Sensor (TPS) — Pinpoint Test — DH

TEST STEP	RESULT ▶	ACTION TO TAKE
DH95 CHECK EEC-IV HARNESS		
• Observe VOM or STAR LED for a fault indication while performing the following: • Referring to the illustration in Step DH94 grasp the harness close to the sensor connector. Wiggle, shake or bend a small section of the EEC-IV system harness while working your way to the dash panel. Also wiggle, shake or bend the EEC-IV harness from the dash panel to the processor. • Is a fault indicated?	Yes ▶	ISOLATE fault and make necessary repairs. REFER to appropriate figure. REPEAT Quick Test.
	No ▶	Go to DH96.
DH96 CHECK PROCESSOR AND HARNESS CONNECTORS		
• Key Off, wait 10 seconds. • Disconnect processor 60 Pin connector. • Inspect both connectors and connector temrinals for obvious damage or faults. • Are connectors and terminals OK?	No ▶	SERVICE as necessary. REPEAT Quick Test.
	Yes ▶	Unable to duplicate fault at this time. Continuous code 63 testing complete.

Vane Air Flow Sensor (VAF) — Pinpoint Test — DK

DK

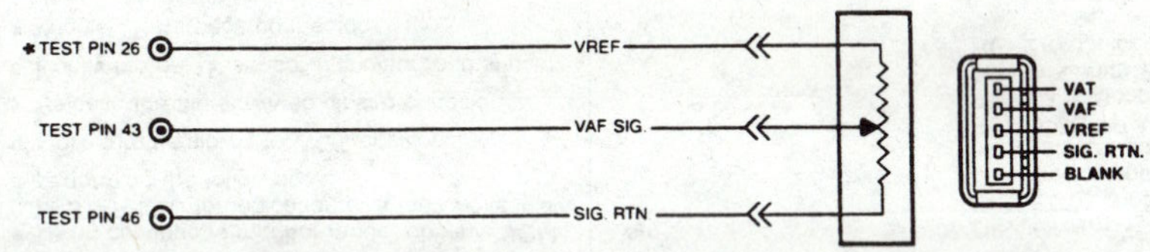

* TEST PIN LOCATED ON BREAKOUT BOX
ALL HARNESS CONNECTORS VIEWED INTO MATING SURFACE.

STOP-WARNING

You may have reached this point because of a fault in a non-EEC System and/or component. To prevent the replacement of good components, be aware that the following non-EEC areas may be at fault:

- Check for unmetered air (air leaks) between VAF meter and throttle body.
- Vacuum leaks.
- Engine sealing (PCV sealing, CANP, valve cover seal dipstick seated).

This Pinpoint Test is intended to diagnose only the following:

- VAF meter.
- Processor.
- Harness circuits: VREF, VAF Sig. and Signal Return.

Vane Air Flow Sensor (VAF) — Pinpoint Test — DK

TEST STEP	RESULT ▶	ACTION TO TAKE
FAULT CODE 26		
DK1 CHECK FOR CONTAMINATION		
NOTE: Code 26 (Key On, Engine Off, or Engine Running) indicates the vane air flow input to the processor is out of closed (Engine Not Running) or at idle limits (Engine Off 0.15-0.50V: Engine at idle 1.50-2.70V). There have been no opens or shorts in the VAF circuit or a code 56 (signal always high) or 66 (signal always low) would have been generated.		
• Key Off, wait 10 seconds. • Remove air cleaner element and check for contamination (oil residue, foreign material, etc.) that may impede VAF sensor vane movement. • Is service code 26 present in Key On, Engine Off portion of Quick Test?	Yes ▶	REINSTALL air cleaner element. REPLACE vane meter. RERUN Quick Test.
	No ▶	GO to DK2.
DK2 VAF SENSOR CHECK		
• Key Off, air cleaner element reinstalled. • Check for unmetered air leaks between vane meter and throttle body. • Disconnect processor 60 Pin connector. Inspect for damaged pins, corrosion, loose wires, etc. Service as necessary. • Install Breakout box. Connect processor to Breakout box harness. • DVOM on 20V scale. • Key On, engine Off. • Place new unsharpened pencil as shown below. • Measure voltage between test Pins 43 and 46. • Does DVOM read between 2.8V and 3.7V?	Yes ▶	Vane meter is capable of outputting an acceptable signal. The VAF code 26 has been caused by incorrect engine speed or an unmetered air leak (vacuum leak). SERVICE as necessary. REMOVE Breakout box. RERUN Quick Test.
	No ▶	REMOVE Breakout box. REPLACE processor. RERUN Quick Test.

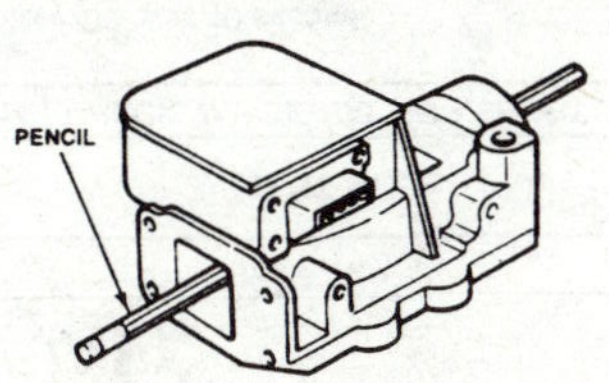

VANE METER

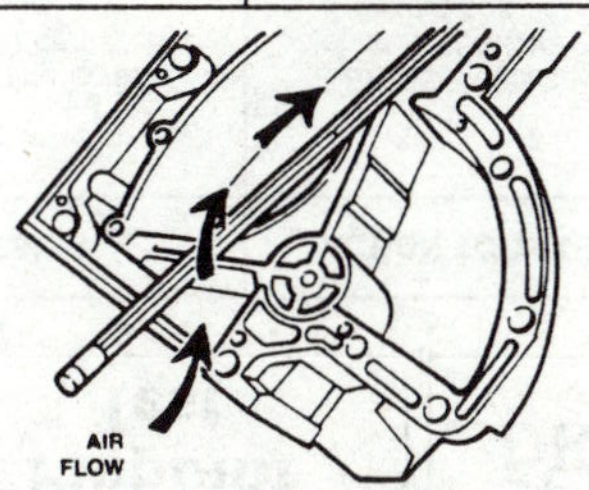

VAF SENSOR AIR VANE

Vane Air Flow Sensor (VAF) — Pinpoint Test — DK

TEST STEP	RESULT ▶	ACTION TO TAKE
FAULT CODE 56		
DK10 CHECK SIGNAL RETURN		
• Key Off, wait 10 seconds. • Disconnect harness from vane meter. Inspect for damaged pins, corrosion, loose wires, etc. Service as necessary. • DVOM on 20V scale. • Measure voltage of vane meter harness connector Signal Return to battery positive terminal.	Reading is less than 10.0V ▶	GO to DK13.
	Reading is 10.0V or greater ▶	GO to DK11.
DK11 CHECK VAF TO SIGNAL RETURN VOLTAGE		
• Key On, Engine Off. Harness disconnected from vane meter. • DVOM on 20V scale. • Measure voltage from vane meter harness connector VAF Signal to Signal Return.	Reading is less than 4.0V ▶	GO to DK12.
	Reading is 4.0V or greater ▶	GO to DK14.
DK12 INDUCE OPPOSITE CODE		
• Key Off, wait 10 seconds. • Harness disconnected from vane meter. • Run Key On, Engine Off Quick Test. • Is code 66 present? **NOTE: Disconnecting vane meter disconnects both VAF and VAT sensors. A code 54 should also be present. Disregard all codes except VAF codes at this time.**	Yes ▶	REPLACE vane meter. RECONNECT harness. RERUN Quick Test.
	No ▶	REPLACE processor. RECONNECT harness to vane meter and processor. RERUN Quick Test.
DK13 CHECK CONTINUITY OF SIGNAL RETURN		
• Key Off. Harness disconnected from vane meter. • Disconnect processor 60 Pin connector. Inspect for damaged pins, corrosion, loose wires, etc. Service as necessary. • Install Breakout box, leave processor disconnected. • DVOM on 200 ohm scale. • Measure resistance of vane meter harness connector Signal Return to test Pin 46.	Reading is less than 5 ohms ▶	REPLACE processor. REMOVE Breakout box. RECONNECT processor and vane meter. RERUN Quick Test.
	Reading is 5 ohms or greater ▶	SERVICE circuit open. REMOVE Breakout box. RECONNECT processor and vane meter. RERUN Quick Test.

Vane Air Flow Sensor (VAF) — Pinpoint Test — DK

TEST STEP	RESULT ▶	ACTION TO TAKE
DK14 CHECK VAF SIGNAL FOR SHORT		
• Key Off, wait 10 seconds. • Harness disconnected from vane meter. • Disconnect processor 60 Pin connector. Inspect for damaged pins, corrosion, loose wires, etc. Service as necessary. • Install Breakout box, leave processor disconnected. • DVOM on 200,000 ohm scale. • Measure resistance from test Pin 43 to test Pins 26 and 57.	Either reading is less than 10,000 ohms ▶	SERVICE circuit shorts. REMOVE Breakout box. RECONNECT processor and vane meter. RERUN Quick Test.
	Both readings are 10,000 ohms or greater ▶	REPLACE processor. REMOVE Breakout box. RECONNECT processor and vane meter. RERUN Quick Test.
FAULT CODE 66		
DK20 CHECK VREF TO SIGNAL RETURN VOLTAGE		
• Key off, wait 10 seconds. • Disconnect harness from vane meter. Inspect for damaged pins, corrosion, loose wires, etc. Service as necessary. • Key On, Engine Off. • DVOM on 20V scale. • Measure voltage from vane meter harness connector VREF to Signal Return.	Reading is between 4.0 and 6.0V ▶	GO to DK21.
	Reading is less than 4.0V ▶	GO to DK24.
	Reading is greater than 6.0V ▶	GO to Pinpoint Test Step C1.
DK21 CHECK VAF SIGNAL FOR SHORT		
• Key Off, wait 10 seconds. • Harness disconnected from vane meter. • DVOM on 200,000 ohm scale. • Disconnect processor 60 Pin connector. Inspect for damaged pins, corrosion, loose wires, etc. Service as necessary. • Measure resistance of vane meter harness connector circuit VAF Signal to Signal Return.	Reading is less than 10,000 ohms ▶	SERVICE circuit shorts. RECONNECT vane meter. RERUN Quick Test.
	Reading is 10,000 ohms or greater ▶	RECONNECT processor. GO to DK22.

Vane Air Flow Sensor (VAF) — Pinpoint Test — DK

TEST STEP	RESULT ▶	ACTION TO TAKE
DK22 INDUCE OPPOSITE CODE		
• Key Off, wait 10 seconds. • Processor reconnected. • Harness disconnected from vane meter. • Install jumper wire in vane meter harness connector VREF to VAF signal. • Perform Key On, Engine Off Quick Test. • Is code 56 present? **NOTE: Disconnecting vane meter disconnects both VAF and VAT sensors. Code 54 should also be present. Disregard all codes except VAF codes at this time.**	Yes ▶	REPLACE vane meter. REMOVE jumper wire. RECONNECT vane meter. RERUN Quick Test.
	No ▶	REMOVE jumper wire and GO to DK23.
DK23 CHECK CONTINUITY OF VAF SIGNAL		
• Key Off, harness disconnected from vane meter. • Disconnect processor 60 Pin connector. Inspect for damaged pins, corrosion, loose wires, etc. Service as necessary. • Install Breakout box, leave processor disconnected. • DVOM on 200 ohm scale. • Measure resistance from vane meter harness connector VAF signal test Pin 43.	Reading is less than 5 ohms ▶	REPLACE processor. REMOVE Breakout box. RECONNECT processor and vane meter. RERUN Quick Test.
	Reading is 5 ohms or greater ▶	SERVICE circuit open. REMOVE Breakout box. RECONNECT processor and vane meter. RERUN Quick Test.
DK24 CHECK CONTINUITY OF VREF		
• Key Off, wait 10 seconds. • Harness disconnected from vane meter. • Disconnect processor 60 Pin connector. Inspect for damaged pins, corrosion, loose wires, etc. Service as necessary. • Install Breakout box, leave processor disconnected. • DVOM on 200 ohm scale. • Measure resistance from vane meter harness connector VREF to test Pin 26.	Reading is less than 5 ohms ▶	REPLACE processor. REMOVE Breakout box. RECONNECT vane meter and processor. RERUN Quick Test.
	Reading is 5 ohms or greater ▶	SERVICE circuit open. REMOVE Breakout box. RECONNECT processor and vane meter. RERUN Quick Test.

Vane Air Flow Sensor (VAF) — Pinpoint Test — DK

TEST STEP	RESULT ▶	ACTION TO TAKE
FAULT CODE 76		
DK30 CHECK FOR VOLTAGE INCREASE IN VAF SIGNAL TO SIGNAL RETURN		
NOTE: A sharp snap of the throttle may not be sufficient to pass this test. Be sure to move throttle to WOT and return.		
• Key Off, wait 10 seconds. • Disconnect processor 60 Pin connector. Inspect for damaged pins, corrosion, loose wires, etc. Service as necessary. • Install Breakout box. Connect processor to Breakout box harness. • DVOM on 20V scale. • Connect DVOM to test Pins 43 and 46. • Perform Engine Running Quick Test while monitoring DVOM. • After dynamic response prompt code 1(0) operator does a brief WOT. DVOM should increase more than 2.0V from reading before WOT. • Observe service codes at end of test.	DVOM increased more than 2.0V and code 76 is still present ▶	REPLACE processor. REMOVE Breakout box. RERUN Quick Test.
	DVOM did not increase more than 2.0V ▶	CHECK air cleaner duct for obstruction. If OK, REPLACE vane meter.
	DVOM increased more than 2.0V and code 76 is not present ▶	Vane meter is OK, SERVICE other codes as necessary.

Vane Air Flow — Pinpoint Test — DK

TEST STEP	RESULT ▶	ACTION TO TAKE
DK90 SERVICE CODE 56 CONTINUOUS TEST: CHECK VAF SENSOR		
• Using continuous monitor mode, observe VOM or STAR LED for indication of a fault while performing the following: • Lightly tap on VAF sensor (simulate road shock). • Wiggle VAF connector. • Is a fault indicated?	Yes ▶	DISCONNECT and INSPECT connectors. If connector and terminals are good, REPLACE VAF sensor. REPEAT Quick Test.
	No ▶	GO to DK91.
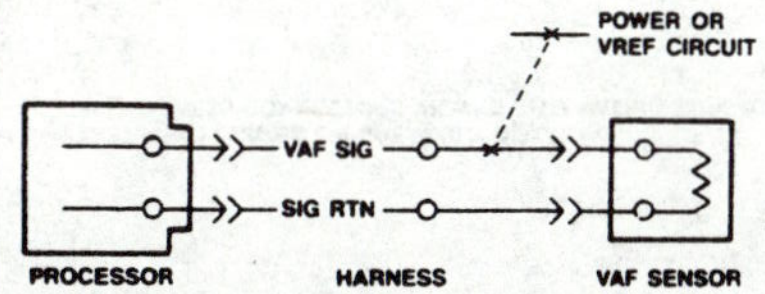		
DK91 CHECK EEC-IV HARNESS		
• Observe VOM or STAR LED for a fault indication while performing the following: • Referring to the illustration in Step DK90, grasp the harness close to the sensor connector. Wiggle, shake or bend a small section of the EEC-IV system harness while working your way to the dash panel. Also wiggle, shake or bend the EEC-IV harness from the dash panel to the processor. • Is a fault indicated?	Yes ▶	ISOLATE fault and make necessary repairs. REFER to appropriate figure. REPEAT Quick Test.
	No ▶	GO to DK92.
DK92 CHECK PROCESSOR AND HARNESS CONNECTORS		
• Key Off, wait 10 seconds. • Disconnect processor 60 Pin connector. • Inspect both connectors and connector terminals for obvious damage or faults. • Are connectors and terminals OK?	No ▶	SERVICE as necessary. REPEAT Quick Test.
	Yes ▶	Unable to duplicate fault at this time. Continuous code 56 testing complete.

Vane Air Flow — Pinpoint Test — DK

TEST STEP	RESULT ▶	ACTION TO TAKE
DK93 SERVICE CODE 66 CONTINUOUS TEST: CHECK VAF SENSOR		
• Using continuous monitor mode, observe VOM or STAR LED for indication of a fault while performing the following: • Lightly tap on VAF sensor (simulate road shock). • Wiggle ACT connector. • Is a fault indicated?	Yes ▶	DISCONNECT and INSPECT connectors. If connector and terminals are good, REPLACE sensor. REPEAT Quick Test.
	No ▶	GO to DK94.
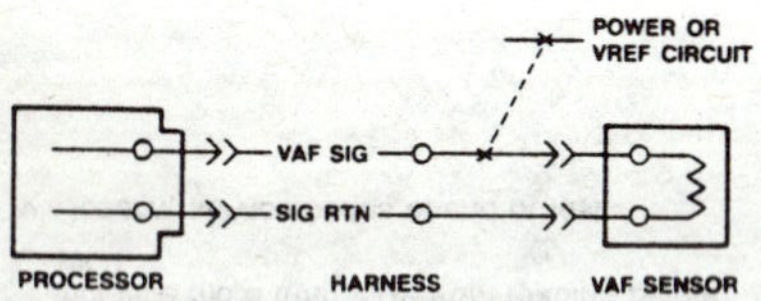		
DK94 CHECK EEC-IV HARNESS		
• Observe VOM or STAR LED for a fault indication while performing the following: • Referring to the illustration in Step DK93 grasp the harness close to the sensor connector. Wiggle, shake or bend a small section of the EEC-IV system harness while working your way to the dash panel. Also wiggle, shake or bend the EEC-IV harness from the dash panel to the processor. • Is a fault indicated?	Yes ▶	ISOLATE fault and make necessary repairs. REFER to appropriate figure. REPEAT Quick Test.
	No ▶	Go to DK95.
DK95 CHECK PROCESSOR AND HARNESS CONNECTORS		
• Key Off, wait 10 seconds. • Disconnect processor 60 Pin connector. • Inspect both connectors and connector terminals for obvious damage or faults. • Are connectors and terminals OK?	No ▶	SERVICE as necessary. REPEAT Quick Test.
	Yes ▶	Unable to duplicate fault at this time. Continuous code 66 testing complete.

EGR Valve Position — Pinpoint Test — DL

DL

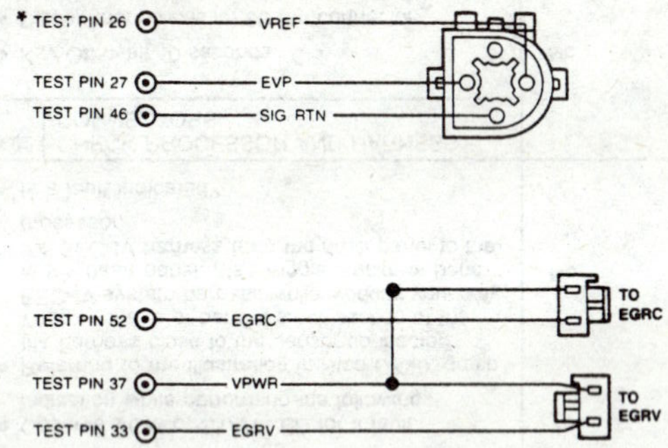

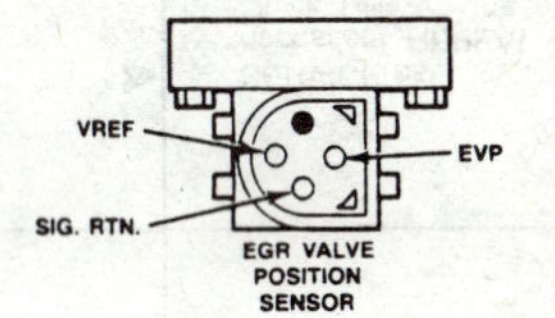

*TEST PINS LOCATED ON BREAKOUT BOX.
ALL HARNESS CONNECTORS VIEWED INTO MATING SURFACE.

STOP-WARNING

You may have reached this point because of a fault in a non-EEC System and/or component. To prevent the replacement of good components, be aware that the following non-EEC area may be at fault:

- Damaged EGR valve.

This Pinpoint Test is intended to diagnose only the following:

- EVP sensor.
- Harness circuits: VREF, EVP, Signal Return, EGRC, VPWR, EGRV.
- EGR solenoids.
- EGR valve assembly.
- Processor assembly.

EGR Valve Position Sensor (EVP) — Pinpoint Test — DL

TEST STEP	RESULT ▶	ACTION TO TAKE
FAULT CODE 31		
DL1 RUN ENGINE RUNNING QUICK TEST WITH EGR VACUUM SIGNAL LINE DISCONNECTED AT EGR VALVE		
• Key Off, wait 10 seconds. • Disconnect EGR vacuum line at EGR valve and cap EGR vacuum line. • Perform Engine Running Quick Test. • Check for code 31.	Code 31 present ▶	GO to DL2.
	No code 31 present, but codes 32 and 34 are present ▶	GO to DL10.
DL2 CHECK EVP SENSOR RESISTANCE		
• Key Off, wait 10 seconds. • EGR signal line disconnected and capped. • Disconnect EVP sensor. Inspect for damaged pins, corrosion, and pins pushed out. Service as necessary. • DVOM on 200,000 ohm scale. • Measure resistance from EVP circuit to VREF circuit and from Signal Return circuit to VREF circuit.	Both readings between 100 ohms and 5,500 ohms ▶	GO to DL3.
	Either reading below 100 ohms or greater than 5,500 ohms ▶	REPLACE EVP sensor. RECONNECT vacuum line and harness. RERUN Quick Test.
DL3 CHECK EVP MOVEMENT		
• Key Off, vacuum signal line disconnected and capped, harness disconnected at EVP sensor. • DVOM on 200,000 ohm scale. • Connect vacuum pump to EGR valve. • Measure resistance from EVP to VREF circuit while gradually increasing vacuum to 33 kPa (10 in. Hg). • Observe resistance as vacuum increases.	Reading gradually decreases from no greater than 5,500 ohms to no less than 100 ohms ▶	GO to DL4.
	Reading is less than 100 ohms or greater than 5,500 ohms ▶	REPLACE EVP sensor. RECONNECT signal line and harness. RERUN Quick Test.
	Reading does not decrease or unable to hold vacuum ▶	GO to DL13.

EGR Valve Position Sensor (EVP) — Pinpoint Test — DL

TEST STEP	RESULT ▶	ACTION TO TAKE
DL4 MEASURE VREF TO SIGNAL RETURN VOLTAGE		
• Key On, Engine Off. Vacuum signal line disconnected and capped, harness disconnected from EVP sensor. • DVOM on 20V scale. • Measure voltage from harness connector VREF to harness connector Signal Return.	Reading is between 4V and 6V ▶	GO to DL5.
	Reading is less than 4V ▶	GO to DL7.
	Reading is greater than 6V ▶	GO to Pinpoint Test Step C1.
DL5 MEASURE VREF TO SIGNAL INPUT VOLTAGE		
• Key On, Engine Off. • Reconnect EGR vacuum signal line to EGR valve. • Harness disconnected from EVP sensor. • DVOM on 20V scale. • Measure voltage from sensor connector VREF to sensor connector EVP.	Reading is between 4V and 6V ▶	GO to DL6.
	Voltage is less than 4V ▶	GO to DL7.
DL6 CHECK EVP SIGNAL FOR SHORTS TO GROUND		
• Key Off, wait 10 seconds. Harness disconnected from EVP sensor. • Disconnect harness from processor 60 Pin connector and inspect for damaged pins, corrosion, loose wires. Service as necessary. • DVOM on 200,000 ohm scale. • Measure resistance from sensor harness connector EVP to harness connector Signal Return and from harness connector EVP to battery negative terminal.	Either reading is less than 10,000 ohms ▶	SERVICE circuit short. RECONNECT processor and EVP sensor. RERUN Quick Test.
	Both readings 10,000 ohms or greater ▶	REPLACE processor. RECONNECT processor and EVP sensor. RERUN Quick Test.

EGR Valve Position Sensor (EVP) — Pinpoint Test — DL

TEST STEP	RESULT ▶	ACTION TO TAKE
DL7 MEASURE CONTINUITY OF VREF, EVP AND SIGNAL RETURN CIRCUITS		
• Key Off, wait 10 seconds. • Harness disconnected from EVP sensor. • Disconnect harness from processor 60 Pin connector and inspect for damaged pins, corrosion, loose wires. Service as necessary. • Install Breakout box leaving processor disconnected. • DVOM on 200 ohm scale. • Measure continuity from test Pin 46 to harness connector Signal Return, test Pin 26 to harness connector VREF and test Pin 27 to harness connector EVP.	All readings are less than 5 ohms ▶	GO to DL8.
	Any reading of 5 ohms or greater ▶	SERVICE open circuit(s). REMOVE Breakout box. RECONNECT processor and EVP sensor. RERUN Quick Test.
DL8 CHECK EVP SIGNAL FOR SHORTS TO POWER		
• Key Off, harness disconnected from EVP sensor. Breakout box installed with processor not connected. • DVOM on 200,000 ohm scale. • Measure resistance from test Pin 27 to test Pins 37 and 57.	Either reading less than 10,000 ohms ▶	SERVICE short circuit. REMOVE Breakout box. RECONNECT processor and EVP sensor. RERUN Quick Test.
	Both readings 10,000 ohms or greater ▶	REPLACE processor. REMOVE Breakout box. RECONNECT processor and EVP sensor. RERUN Quick Test.

EGR Valve Position Sensor (EVP) — Pinpoint Test — DL

TEST STEP	RESULT ▶	ACTION TO TAKE
FAULT CODES 32, 33 AND 34		
DL10 OUTPUT STATE CHECK (REFER TO APPENDIX)		
• Key Off, wait 10 seconds. • DVOM on 20V scale. • Connect DVOM negative test lead to STO and positive test lead to battery positive. • Jumper STI to signal return. • Perform Key On, Engine Off Self-Test until the completion of the Continuous Test Codes. • DVOM will indicate zero volts. • Depress and release the throttle. • Did DVOM reading change to a high voltage reading?	NO ▶	DEPRESS throttle to WOT and RELEASE. If STO voltage does not go high, GO to Pinpoint Test Step Q40.
	YES ▶	REMAIN in output state check. GO to DL11.
DL11 CHECK EGR VENT AND EGR CONTROL SOLENOIDS FOR VACUUM CYCLING		
• Key On, Engine Off. • In output state check. • Disconnect and cap vacuum line from bottom port of EGRC solenoid and connect vacuum pump. • Connect a vacuum gauge in the common output (top) vacuum line to EGR valve. • Disconnect, but **do not** cap the vacuum vent line from the EGRV solenoid. • While cycling outputs On and Off (by depressing and releasing throttle) observe vacuum gauge at the output. Maintain vacuum at source.	Vacuum output cycles On and Off in less than 2 seconds ▶	RECONNECT all vacuum lines. GO to DL12.
	Vacuum does not cycle On and Off in less than 2 seconds ▶	REPLACE solenoid assembly. RECONNECT all vacuum lines. RERUN Quick Test.
DL12 CHECK VACUUM LINES		
• Key Off, wait 10 seconds. • Vacuum lines reconnected. • Check entire EEC vacuum line system per VECI schematic decal for leaks, kinks, cracks or obstructions.	Vacuum lines OK ▶	EEC-IV system OK. REFER to Section 29 for EGR system diagnosis.
	Vacuum lines are not OK ▶	SERVICE as necessary. RERUN Quick Test.

EGR Valve Position Sensor (EVP) — Pinpoint Test — DL

TEST STEP	RESULT ▶	ACTION TO TAKE
DL13 MANUALLY EXERCISE EVP SENSOR		
• Key Off, harness disconnected. • Remove EVP sensor from EGR valve. • Measure resistance from EVP sensor Signal Return Pin to EVP sensor VREF Pin while gradually applying pressure to EVP sensor shaft. • Observe resistance as shaft is slowly pushed in. • Look for sudden changes in resistance readings.	Reading decreases smoothly from no more than 5,500 ohms to no less than 100 ohms ▶	REPLACE EGR valve assembly. RECONNECT EVP sensor and EGR supply vacuum line. RERUN Quick Test.
	Reading does not decrease smoothly or changes abruptly ▶	REPLACE EVP sensor. RECONNECT harness and EGR supply vacuum line. RERUN Quick Test.
FAULT CODE 35		
DL30 RPM TOO LOW FOR EGR TEST		
• Check for codes 12 and 13.	Code 12 present	Vehicles equipped with DC motor control GO to KF1. Vehicles equipped with throttle kicker GO to KT1.
	Code 13 present ▶	Vehicles equipped with DC motor control GO to KF20. Vehicles equipped with throttle kicker GO to KT1.
	Codes 12 and 13 not present ▶	GO to DL31.
DL31 RETEST AT 1500 RPM		
• Key Off, wait 10 seconds. • Install tachometer. • Perform Key On, Engine Running Quick Test while maintaining 1500 rpm. • Record engine running service codes. • Check for code 35.	Code 35 present ▶	REPLACE processor. RERUN Quick Test.
	Code 35 not present ▶	REPEAT Quick Test. SERVICE codes as necessary.

EGR Valve Position Sensor (EVP) — Pinpoint Test — DL

TEST STEP	RESULT ▶	ACTION TO TAKE
DL40 VREF CIRCUIT TO GROUND		
• Key Off, wait 10 seconds. • Disconnect the EVP sensor and inspect for damaged pins, corrosion, loose wires. Service as necessary. • DVOM on 20V scale. • Key On, Engine Off. • Measure voltage from VREF (at EVP sensor harness connector) to battery ground.	Voltage readings between 4V and 6V ▶	GO to DL41.
	Voltage readings are less than 4.0V ▶	GO to DL7.
	Voltage readings are greater than 6.0V ▶	GO to Pinpoint Test Step C1.
DL41 SIGNAL RETURN TO VREF		
• Key On. • EVP harness disconnected. • DVOM on 20V scale. • Measure voltage from VREF to signal return at the EVP harness connector.	Voltage readings are between 4V and 6V ▶	GO to DL42.
	Voltage readings are less than 4.0V ▶	SERVICE open in signal return. RERUN Key On, Engine Off Self-Test.
DL42 EVP CIRCUIT VOLTAGE CHECK		
• Key Off, wait 10 seconds. • Disconnect processor 60 Pin connector and inspect for damaged pins, corrosion, loose wires. Service as necessary. • Install the Breakout box, reconnect processor. • Connect EVP sensor. • DVOM on 20V scale. • Key On, Engine Off. • Measure voltage from test Pin 46 to test Pin 27.	Less than 0.5V ▶	GO to DL43.
	1.25V or greater ▶	GO to DL46.
DL43 EVP SIGNAL SHORT TO GROUND		
• Key Off, wait 10 seconds. • Disconnect processor. • Disconnect EVP sensor. • Breakout box installed. • DVOM on 200 ohm scale. • Measure resistance from test Pin 27 to test Pins 20, 40, 46 and 60.	5.0 ohms or greater ▶	GO to DL44.
	Less than 5.0 ohms ▶	SERVICE short to ground in EVP signal circuit. REPEAT Quick Test.

EGR Valve Position Sensor (EVP) — Pinpoint Test — DL

TEST STEP	RESULT	ACTION TO TAKE
DL44 EVP SIGNAL CIRCUIT FOR OPENS		
• Key Off. • Processor disconnected. • EVP sensor disconnected. • Breakout box installed. • DVOM on 200 ohm scale. • Measure resistance from test Pin 27 to EVP signal circuit at EVP connector harness.	5 ohms or greater ▶ Less than 5 ohms ▶	SERVICE open in EVP signal circuit. REPEAT Quick Test. GO to DL45.
DL45 SUBSTITUTE EVP SENSOR		
• Key Off. • Remove and replace present EVP sensor with known good sensor. Reconnect EVP sensor. • Remove Breakout box. • Reconnect processor. • Rerun Key On, Engine Off Quick Test. • Is code 31 present?	Yes ▶ No ▶	REPLACE processor and REINSTALL original EVP sensor. Testing completed.
DL46 EVP SIGNAL SHORT TO POWER		
• Key Off, wait 10 seconds. • Disconnect processor. • Disconnect EVP sensor. • Breakout box installed. • DVOM on 200 ohm scale. • Measure resistance from test Pin 27 to test Pins 26, 37 and 57.	5 ohms or greater ▶ Less than 5 ohms ▶	GO to DL47. SERVICE short to power. RERUN Key On, Engine Off Quick Test.
DL47 SUBSTITUTE EVP SENSOR AND EGR VALVE		
• Key Off, wait 10 seconds. • Substitute a known good sensor and EGR valve assembly. • Sensor connected. • Remove Breakout box. • Reconnect processor. • Rerun Key On, Engine Off Quick Test. • Is code 31 present?	Yes ▶ No ▶	REPLACE processor. REINSTALL original EVP sensor and EGR valve assembly. GO to DL48.

EGR Valve Position Sensor (EVP) — Pinpoint Test — DL

TEST STEP	RESULT	ACTION TO TAKE
DL48 EVP SENSOR CHECK		
• Key Off, wait 10 seconds. • Install original EVP sensor. • EVP sensor connected. • Rerun Key On, Engine Off Quick Test. • Is code 31 present?	Yes ▶ No ▶	REPLACE EVP sensor. Testing completed. Testing completed.

EGR Valve Position Sensor (EVP) — Pinpoint Test DL

TEST STEP	RESULT ▶	ACTION TO TAKE
DL90 SERVICE CODE 31 CONTINUOUS TEST: EXERCISE EVP SENSOR		
• Using continuous monitor mode, observe VOM or STAR LED for indication of a fault while performing the following: • Connect a vacuum pump to the EGR valve. • Very slowly apply 20 kPa (6 in. Hg) vacuum to the EGR valve. • Slowly bleed vacuum off the EGR valve and lightly tap on EVP sensor (simulate road shock). • Wiggle EVP sensor connector. • Is a fault indicated?	Yes ▶	GO to DL91.
	No ▶	GO to DL92.
DL91 MEASURE EVP SIGNAL VOLTAGE WHILE EXERCISING EVP SENSOR		
• Key Off, wait 10 seconds. • Disconnect processor 60 Pin connector and inspect for damaged pins, corrosion, loose wires. Service as necessary. • Install Breakout box and reconnect processor. • VOM or STAR LED still connected to STO as in previous step. • Connect a DVOM from test Pin 27 to test Pin 46. • DVOM on 20V scale. • Key On, Engine Off. • While observing DVOM, repeat step DL90. • Does the fault occur below 4.25V?	Yes ▶	DISCONNECT and INSPECT connector. If connector and terminals are good, REPLACE EVP sensor. REPEAT Quick Test.
	No ▶	EGR valve overshoot may have caused continuous code 31. Sensor service is not required. To verify harness integrity, GO to DL92.

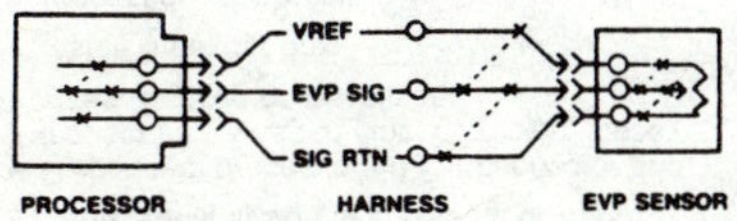

EGR Valve Position Sensor (EVP) — Pinpoint Test DL

TEST STEP	RESULT ▶	ACTION TO TAKE
DL92 CHECK EEC-IV HARNESS		
• Observe VOM or STAR LED for a fault indication while performing the following: • Referring to the illustration in Step DL90, grasp the harness close to the sensor connector. Wiggle, shake or bend a small section of the EEC-IV system harness while working your way to the dash panel. Also wiggle, shake or bend the EEC-IV harness from the dash panel to the processor. • Is a fault indicated?	Yes ▶	ISOLATE fault and make necessary repairs. REFER to appropriate figure. REPEAT Quick Test.
	No ▶	GO to DL93.
DL93 CHECK PROCESSOR AND HARNESS CONNECTORS		
• Key Off, wait 10 seconds. • Disconnect processor 60 Pin connector. • Inspect both connectors and connector terminals for obvious damage or faults. • Are connectors and terminals OK?	No ▶	SERVICE as necessary. REPEAT Quick Test.
	Yes ▶	Unable to duplicate fault at this time. Continuous Code 31 testing complete.

Neutral Drive Switch A/C Input — Pinpoint Test — FA

TEST STEP	RESULT ▶	ACTION TO TAKE
FA1 CODE 67 SYSTEM IDENTIFICATION		
1.6L MTX		GO to FA10.
2.3L Turbo, M/T and 1.6L Turbo, SVO		GO to FA15.
2.3L SVO Turbo Octane Switch		GO to KN15.
4.9L M/T, Truck		GO to FA25.
All other systems		GO to FA2.
FA2 NEUTRAL DRIVE INPUT CHECK		
• Key Off, wait 10 seconds. • Verify heater control is in Off position, if so equipped. • Verify transmission is in Neutral or Park. • Disconnect processor 60 Pin connector and inspect for damaged pins, corrosion, loose wires. Service as necessary. • Install Breakout box. • Processor connected. • Key On, Engine Off. • DVOM on 20V scale. • Measure voltage from test Pin 30 (Neutral Drive circuit) to chassis ground.	Less than .5V ▶ .5V or greater ▶	GO to FA4. GO to FA3.

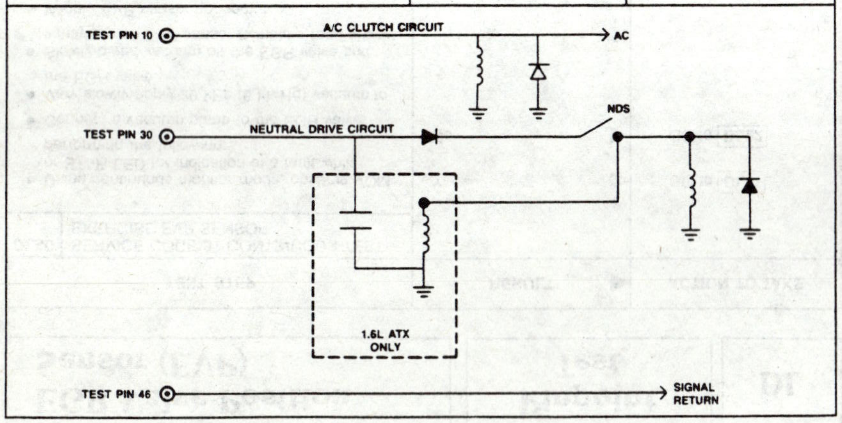

Neutral Drive Switch A/C Input — Pinpoint Test — FA

TEST STEP	RESULT ▶	ACTION TO TAKE
FA3 NEUTRAL DRIVE SWITCH CHECK		
• Key Off, wait 10 seconds. • Breakout box installed. • DVOM on 200 ohm scale. • Locate the Neutral Drive switch. • Disconnect vehicle harness from the Neutral Drive switch and measure resistance across the two contacts.	Less than 5 ohms ▶ 5 ohms or greater ▶	SERVICE open in vehicle harness Neutral Drive circuit. REPEAT Quick Test. REPLACE Neutral Drive switch. REPEAT Quick Test.
FA4 A/C INPUT CHECK		
• Breakout box installed. • Key On, Engine Off. • DVOM on 20V scale. • A/C control Off. Measure voltage at test Pin 10 A/C clutch circuit to chassis ground.	1.0V or greater ▶ Less than 1.0V ▶	SERVICE short to power in A/C clutch circuit. REPEAT Quick Test. REPLACE processor. REPEAT Quick Test.
FA10 NEUTRAL/CLUTCH INPUT CHECK 1.6L, MTX		
• Key Off, wait 10 seconds. • Verify A/C is Off, if so equipped. • Verify transaxle is in Neutral and clutch is released. • Disconnect processor 60 Pin connector and inspect for damaged pins, corrosion, loose wires. Service as necessary. • Install Breakout box. • Connect processor. • DVOM on 200 ohm scale. • Measure resistance from test Pin 30 (A/C clutch circuit) to test Pin 46 (Signal Return circuit).	Less than 5 ohms ▶ 5 ohms or greater ▶	GO to FA4. GO to FA11.

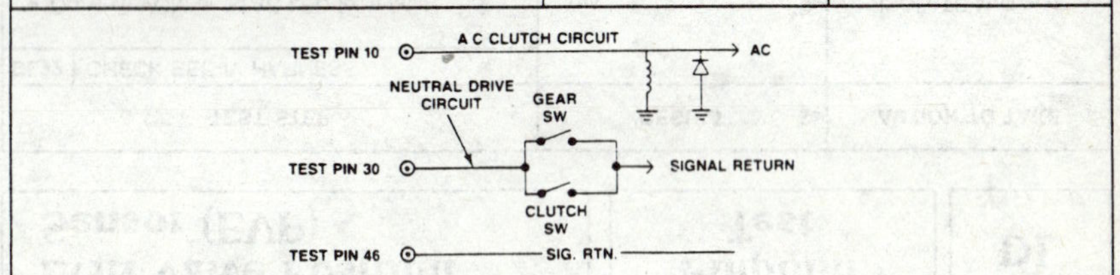

Neutral Drive Switch A/C Input — Pinpoint Test — FA

TEST STEP	RESULT ▶	ACTION TO TAKE
FA11 NEUTRAL/CLUTCH SWITCH CHECK — 1.6L, MTX		
• Key Off. • DVOM on 200 ohm scale. • Breakout box installed. • Locate neutral switch (on transaxle) and clutch switch (under dash). • Disconnect vehicle harness at both switches and measure resistance across each switch.	Less than 5 ohms at both switches ▶	SERVICE open(s) in vehicle, harness Neutral Drive and Signal Return circuits. REPEAT Quick Test.
	5 ohms or greater at either or both switches ▶	REPLACE open switch(es). RECONNECT harness and REPEAT Quick Test.
FA15 NEUTRAL INPUT CHECK — 2.3L TC M/T AND 1.6L TC SVO		
• Key Off, wait 10 seconds. • Verify A/C is Off, if so equipped. • Disconnect processor 60 Pin connector and inspect for damaged pins, corrosion, loose wires. Service as necessary. • Install Breakout box. • Connect processor. • DVOM on 200 ohm scale. • Measure resistance from test Pin 30 (Neutral input circuit) to test Pin 46 (Signal Return circuit).	Less than 5 ohms ▶	REPLACE processor and REPEAT Quick Test.
	5 ohms or greater ▶	SERVICE open in Neutral input or Signal Return circuit. REPEAT Quick Test.

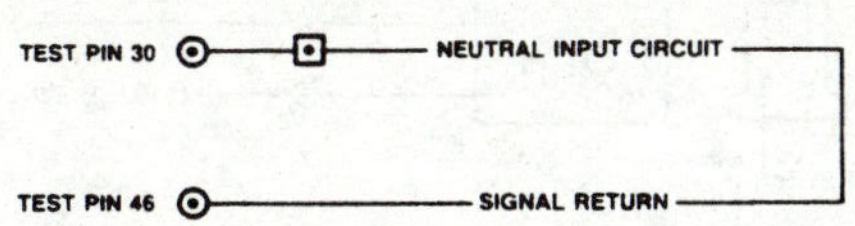

Neutral Drive Switch A/C Input — Pinpoint Test — FA

TEST STEP	RESULT ▶	ACTION TO TAKE
FA25 NEUTRAL INPUT CHECK — 4.9L M/T. TRUCK		
• Key Off, wait 10 seconds. • Verify A/C is Off, if so equipped. • Disconnect processor 60 Pin connector and inspect for damaged pins, corrosion, loose wires. Service as necessary. • Install Breakout box. • Connect processor. • DVOM on 200 ohm scale. • Measure resistance from test Pin 46 (Neutral input circuit) to VIP connector.	Less than 5 ohms ▶	REPLACE processor. REPEAT Quick Test.
	5 ohms or greater ▶	SERVICE open in Neutral input circuit to VIP tester. REPEAT Quick Test.

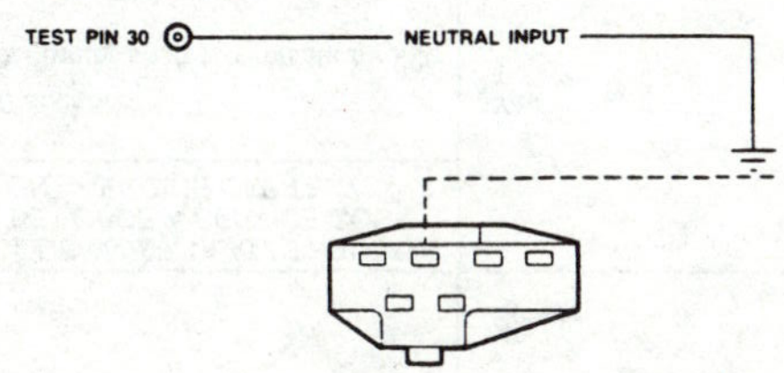

Idle Tracking Switch — Pinpoint Test — FB

FB

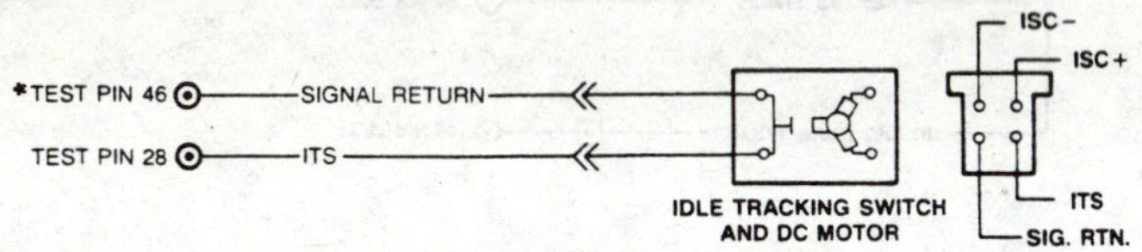

NOTE: FOR 3.8L CFI A 10-PIN CONNECTOR WAS USED FOR THE FIRST 60 DAYS OF PRODUCTION. EITHER CONNECTOR MAY BE ENCOUNTERED. TEST PIN. ASSIGNMENTS DID NOT CHANGE ON BREAKOUT BOX. IF ITS MUST BE DISCONNECTED FOR DIAGNOSTICS, DISCONNECT THE 10-PIN CONNECTOR AT THE FUEL CHARGE ASSY.

3.8L CFI ONLY

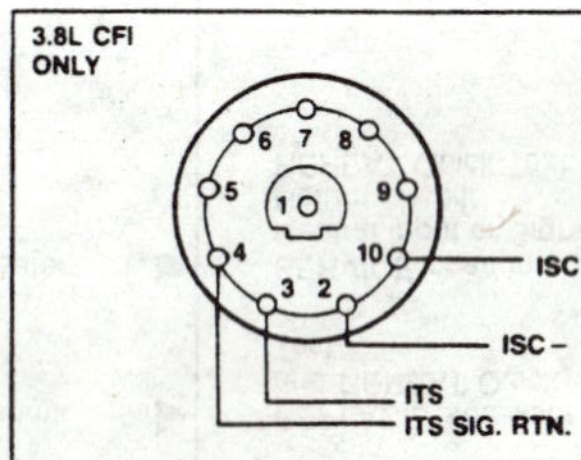

*** TEST PIN LOCATED ON BREAKOUT BOX**
ALL HARNESS CONNECTORS VIEWED INTO MATING SURFACE.

STOP-WARNING

You may have reached this point because of a fault in a non-EEC System and/or component. To prevent the replacement of good components, be aware that the following non-EEC areas may be at fault:

- Anti-diesel speed too high.
- Throttle sticking or on high cam.

This Pinpoint Test is intended to diagnose only the following:

- DC motor.
- Idle tracking switch.
- Harness circuits: ITS and Signal Return.
- Processor assembly.

Idle Tracking Switch — Pinpoint Test — FB

TEST STEP	RESULT ▶	ACTION TO TAKE
FB1 SERVICE CODES 58 AND 68: ITS FAULT ISOLATION		
• Is code 58 or 68 present?	Service code 58 ▶	GO to FB2.
	Service code 68 ▶	GO to FB5.
FB2 SERVICE CODE 58: SIMULATE THROTTLE CONTACT TO FORCE A SERVICE CODE 68 IN KEY ON — ENGINE OFF TEST		
• Key Off, wait 10 seconds. • Press on ISC DC motor shaft to simulate throttle contact. • With a force pushing on DC motor shaft, perform Key On, Engine Off Self-Test. **NOTE: Ignore all other codes at this point.** • Is code 68 present?	Yes ▶ No ▶	GO to Pinpoint Test Step KF4. GO to FB3.
FB3 IDLE TRACKING SWITCH ALWAYS "NOT TRACKING" (CLOSED)		
• Key Off, wait 10 seconds. • Disconnect harness from ISC. • Perform Key On, Engine Off self-test. **NOTE: Ignore all other codes at this point.** • Is code 68 present?	Yes ▶ No ▶	REPLACE DC motor. REPEAT Quick Test. GO to FB4.

Idle Tracking Switch — Pinpoint Test — FB

TEST STEP	RESULT ▶	ACTION TO TAKE
FB4 CHECK ITS CIRCUIT FOR SHORT TO GROUND		
• Key Off, wait 10 seconds. • Disconnect processor 60 Pin connector and inspect for damaged pins, corrosion, loose wires. Service as necessary. • Connect Breakout box. Leave processor disconnected. • Harness disconnected at ISC. • DVOM on 200,000 ohm scale. • Measure resistance of test Pin 28 to test Pins 40, 46 and 60.	Any resistance less than 10,000 ohms ▶ All resistances 10,000 ohms or more ▶	SERVICE short to ground. REPEAT Quick Test. REPLACE processor. REPEAT Quick Test.
FB5 SERVICE CODE 68: SIMULATE NO THROTTLE CONTACT TO FORCE A CODE 11 IN KEY ON — ENGINE OFF TEST		
• Key Off, wait 10 seconds. • Move throttle away from ISC DC motor shaft. • Perform Key On, Engine Off Self-Test. • Is code 68 present? **NOTE: Ignore all other codes at this point.**	Yes ▶ No ▶	GO to FB6. GO to Pinpoint Test Step KF4.

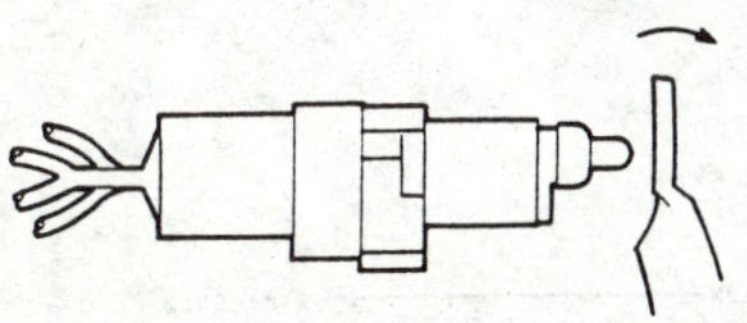

Idle Tracking Switch — Pinpoint Test — FB

TEST STEP	RESULT ▶	ACTION TO TAKE
FB6 SIMULATE A CLOSED CONTACT CONDITION IN KEY ON — ENGINE OFF TEST		
• Key Off, wait 10 seconds. • Disconnect harness from ISC. • Connect jumper between ISC harness connector Pins 3 and 4. • Install jumper as shown. • Perform Key On, Engine Off Self-Test. • Is code 68 present?	Yes ▶ No ▶	GO to FB7. REPLACE ISC DC motor. REPEAT Quick Test.

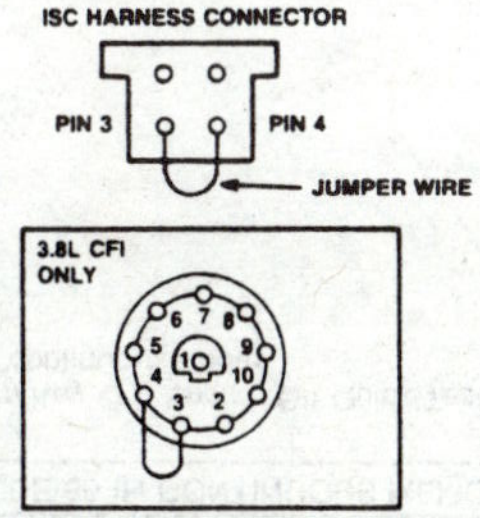

(REFER TO NOTE ON ILLUSTRATION FB)

CAUTION: Do not short any other pins.

TEST STEP	RESULT ▶	ACTION TO TAKE
FB7 CHECK ITS AND SIGNAL RETURN CIRCUITS FOR CONTINUITY		
• Key Off, wait 10 seconds. • Disconnect processor 60 Pin connector and inspect for damaged pins, corrosion, loose wires. Service as necessary. • Connect Breakout box to harness. Leave processor disconnected. • DVOM on 200 ohm scale. • Measure continuity of test Pin 46 to Signal Return circuit and test Pin 28 to ITS circuit.	Either resistance 5 ohms or greater ▶ Both resistances less than 5 ohms ▶	SERVICE faulty circuit. REPEAT Quick Test. REPLACE processor. REPEAT Quick Test.

Key Power Check | Pinpoint Test | FC

FC

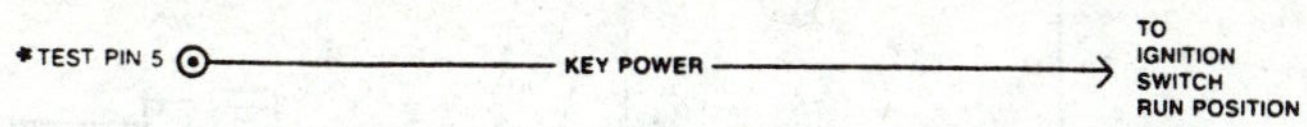

*TEST PINS LOCATED ON BREAKOUT BOX.
ALL HARNESS CONNECTORS VIEWED INTO MATING SURFACE.

STOP-WARNING

You may have reached this point because of a fault in a non-EEC System and/or component. To prevent the replacement of good components, be aware that the following non-EEC areas may be at fault:

- Charging system overvoltage.
- Battery charger connected with engine running.
- Jump starting.

This Pinpoint Test is intended to diagnose only the following:

- Harness circuit: key power.
- Processor.

Key Power Check | Pinpoint Test | FC

TEST STEP	RESULT ▶	ACTION TO TAKE
FC1 BATTERY VOLTAGE GREATER THAN 17.5V		
• Key Off, wait 10 seconds. • Install Breakout box with processor connected. • DVOM at 20V range and connected to test Pins 5 and 60. • Perform Engine Running Quick Test. Observe DVOM during test, record service codes.	DVOM exceeds 17.5V during Quick Test ▶	CORRECT charging system for over-voltage condition.
	DVOM remains below 17.5V and code 65 is present ▶	REPLACE processor and REPEAT Quick Test.
	DVOM remains below 17.5V and code 65 is not present ▶	GO to FC2.
FC2 CODE 65 IN CONTINUOUS MEMORY		
• Perform Key On, Engine Off Quick Test and record continuous codes.	Code 65 not present ▶	TESTING complete. CHARGING system OK at this time.
	Code 65 present ▶	CHECK charging system per Shop Manual Group 31 for cause of intermittent overcharging (greater than 17.5V).

Fuel Control | Pinpoint Test | HA

HA

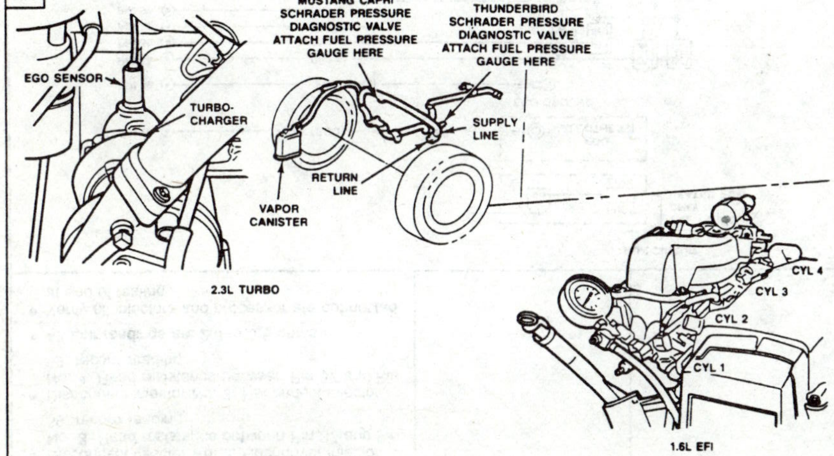

STOP-WARNING

You may have reached this point because of a fault in a non-EEC System and/or component. To prevent the replacement of good components, be aware that the following non-EEC areas may be at fault:

- Ignition Coil
- Distributor Cap
- Distributor Rotor
- Fouled Spark Plugs
- Spark Plug Wires
- CANP Problems
- PCV Valves (see note below)
- EGR Valve and Gasket
- Air Filter
- Fuel Contamination, Engine Oil
- Poor Power Ground
- Fuel Pressure
- Manifold Leaks, Intake/Exhaust
- Engine Not at Normal Operating Temperatures

This Pinpoint Test is intended to diagnose only the following:

- EGO Sensor
- Harness Circuits EGO Grd., EGO, INJA, INJB, V PWR
- EGO Sensor Connection
  — Code 42 start at HA5
  — Code 41 start at HA8
- Vacuum Systems
- Fuel Injectors
- Processor Assembly

NOTE: Fuel contaminated engine oil may affect 4-1, 4-2 Service Codes. If this is suspected, remove the PCV from the valve cover and repeat the Quick Test. If the problem is corrected, then change the engine oil and filter.

Fuel Control | Pinpoint Test | HA

TEST STEP	RESULT ▶	ACTION TO TAKE
HA1 FUEL PRESSURE CHECK		
• Key Off, wait 10 seconds. • Install fuel pressure gauge. • Start and run engine. Fuel pressure must be 172-310 kPa (25-45 psi). **For No Starts:** • If engine will not run, cycle the key from Off to On several times. • Fuel pressure must remain at 276 ± 34 kPa (40 ± 5 psi) for 60 seconds after final key Off.	Fuel pressure is within specification ▶	GO to HA2.
	Fuel pressure is not within specification ▶	REFER to the Shop Manual Group 24 for electric fuel pump and fuel pressure regulator check.

Fuel Control — Pinpoint Test — HA

TEST STEP	RESULT ▶	ACTION TO TAKE
HA2 HARNESS INJECTOR RESISTANCE CHECK		
• Key Off, wait 10 seconds. • Disconnect processor 60 Pin connector and inspect for damaged pins, corrosion, loose wires. Service as necessary. • Connect Breakout box to harness. Leave processor disconnected. • Key Off. DVOM on 200 ohm scale. Disconnect injectors for cylinders 2, 3 and 4 electrically. • Through the vehicle harness, measure the resistance of injector No. 1 between Pin 37 and Pin 58, record reading. • Disconnect injector No. 1. Reconnect injector No. 2. Read resistance between Pin 37 and Pin 58, record reading. • Disconnect injector No. 2. Reconnect injector No. 3. Read resistance between Pin 37 and Pin 59, record reading. • Disconnect injector No. 3. Reconnect injector No. 4. Read resistance between Pin 37 and Pin 59, record reading. • All four readings are 2.0 to 3.5 ohms. • Verify all injectors and processor are connected at end of testing.	Yes ▶ No ▶	GO to HA3 SERVICE the harness/connectors on the suspect injector for opens or shorts. If OK, REPLACE injector. REPEAT Quick Test.

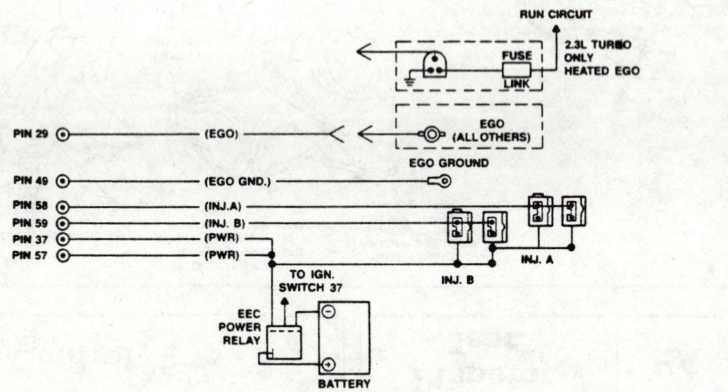

Fuel Control — Pinpoint Test — HA

TEST STEP	RESULT ▶	ACTION TO TAKE
HA3 INJECTOR DRIVE SIGNAL CHECK		
Requires standard non-powered 12 volt test lamp. • Key Off. • Breakout box installed. • Connect processor to Breakout box. • Refer to illustration HA2. • Connect test lamp between Breakout box Pin 37 and Pin 58. • Crank or start engine. • Repeat above test between Breakout box Pin 37 and Pin 59.	Dim glow at light on both tests ▶ No light on one or both tests ▶ Bright light on one or both tests ▶	GO to HA4. VERIFY 12-volt battery power at Pins 37 and 57. If OK, REPLACE processor. REPEAT Quick Test. CHECK circuits 95 and 96 for shorts to ground. If OK, REPLACE processor. REPEAT Quick Test.
HA4 INJECTOR BALANCE TEST		
• Connect tachometer to engine. Run engine at 2000 rpm. It may be necessary to disconnect ISC and use throttle body stop screw to set engine speed. • Disconnect and reconnect the injectors one at a time; note rpm drop for each injector. • Does each injector produce at least a 150 rpm drop? • Reset curb idle. Refer to VECI decal. Verify ISC connection at end of testing.	Yes ▶ No ▶	Fuel delivery OK. Problem is with area common to all cylinders, i.e.: Air/vacuum leak, fuel contamination, EGR. REPLACE faulty injector. REPEAT Quick Test.

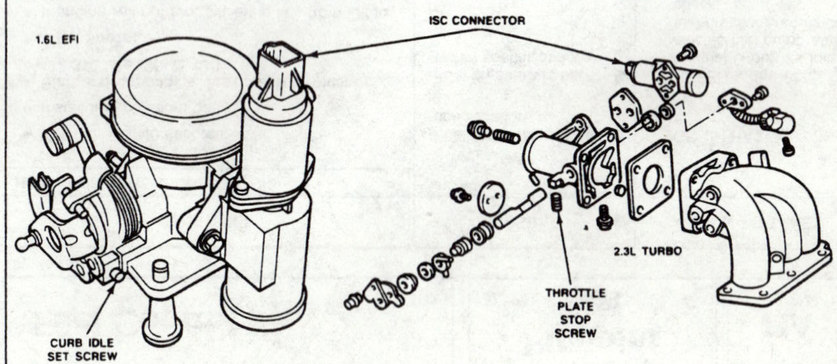

Fuel Control — Pinpoint Test — HA

TEST STEP	RESULT ▶	ACTION TO TAKE
HA5 FUEL CONTROL — ALWAYS RICH: CODE 42		
• Key Off, wait 10 seconds. • **NOTE: Non-EEC areas could cause a code 42. Check for: — Fuel contaminated engine oil — Ignition caused misfire — CANP problems** • Refer to illustration HA. • Disconnect vehicle harness at the EGO sensor. Using a jumper wire, ground vehicle harness EGO circuit at the EGO sensor to the engine block. • Repeat Engine Running Self-Test. • Service code 41 results.	Yes ▶ No ▶	GO to HA7. GO to HA6.
HA6 HARNESS CHECK		
• Key Off, wait 10 seconds. • Refer to illustration HA. • Install Breakout box. • Check continuity of EGO ground circuit between Breakout box Pin 49 and EGO ground at engine block. • Check continuity of EGO circuit between Breakout box Pin 29 and EGO harness connector. • Both circuits less than 5 ohms.	Yes ▶ No ▶	DISCONNECT processor connector. INSPECT for damage or corrosion. If OK, REPLACE processor. REPEAT Quick Test. CORRECT harness circuit with resistance greater than 5 ohms. REPEAT Quick Test.

Fuel Control — Pinpoint Test — HA

TEST STEP	RESULT ▶	ACTION TO TAKE
HA7 EGO CHECK		
• Refer to illustration HA. • DVOM on 20V scale. • With EGO sensor disconnected from the harness, connect a DVOM from EGO sensor to engine ground. • Run the engine at 2000 rpm for 2 minutes. While observing DVOM, disconnect the manifold vacuum hose indicated below. • Meter indicates less than 0.4V.	Yes ▶ No ▶	EGO sensor OK. GO to HA1. REPLACE EGO sensor. REPEAT Quick Test.

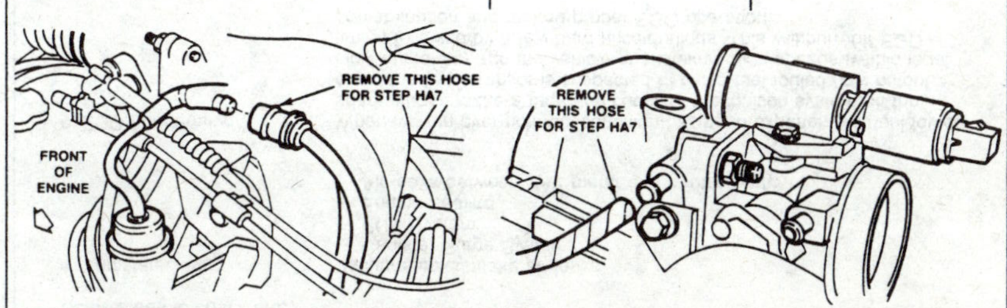

TEST STEP	RESULT ▶	ACTION TO TAKE
HA8 FUEL CONTROL — ALWAYS LEAN: CODE 41		
NOTE: Vacuum/air leaks in non-EEC areas could cause a code 41. Check for: — Leaking vacuum actuator (eg: A/C control motor) — Engine sealing — EGR system — PCV system — Unmetered air leak between air meter and throttle body — Lead contaminated EGO sensor • Refer to illustration HA. • Key Off. DVOM on 20V scale. Disconnect EGO sensor from vehicle harness. Connect DVOM to EGO sensor and engine ground. Remove air cleaner to gain access to air meter inlet. Using a standard wood lead pencil, prop the air meter door part-way open. • Start the engine and run at approximately 2000 rpm for 2 minutes. • Does the DVOM read greater than 0.5V at the end of 2 minutes?	Yes ▶ No ▶	GO to HA9. REPLACE EGO sensor. REPEAT Quick Test.

Fuel Control — Pinpoint Test — HA

TEST STEP	RESULT ▶	ACTION TO TAKE
HA9 HARNESS CHECK EGO CIRCUITS		
• Key Off, wait 10 seconds. • Install Breakout box. Processor disconnected. • Check continuity of EGO circuit from Breakout box Pin 49 to engine block ground. Also EGO circuit from EGO connector to Breakout box Pin 29. • Both circuits less than 5 ohms.	Yes ▶	GO to HA10.
	No ▶	SERVICE/CORRECT as necessary. REPEAT Quick Test.
HA10 CHECK EGO CIRCUIT FOR SHORT TO GROUND		
• Key Off. • Breakout box installed. Processor disconnected. • DVOM to 200,000 ohms range. • Disconnect EGO sensor at harness. • Measure the resistance between Breakout box Pin 29 and Pin 40.	Reading is 10,000 ohms or greater ▶	EGO input circuit OK. DISCONNECT processor connector. INSPECT for corrosion or damaged pins. If OK, REPLACE processor. REPEAT Quick Test.
	Reading is less than 10,000 ohms ▶	CORRECT cause of resistance to ground. REPEAT Quick Test.

Fuel Control — Pinpoint Test — HA

HA11 CONTINUOUS TESTING: CODE 41/42

41 — EGO indicated the fuel system was lean for more than 15 seconds when the system should have been in closed loop fuel control.

42 — EGO indicated the fuel system was rich for more than 15 seconds when the system should have been in closed loop fuel control.

*CLOSED LOOP — Fuel control under the influence of the EGO sensor.

*OPEN LOOP — Fuel control NOT under the influence of the EGO sensor.

Before attempting to correct a fuel control code, 41/42, diagnose all other drive complaints first, eg., rough idle, misses, etc.

NOTE: The fuel control code may help in this diagnosis.

Using the fuel control service code, isolate the cause of the fuel control problem.

Some areas to check are:

- Unmetered Air: Vacuum leaks/intake air leaks.
  - — Canister Purge System.
  - — PCV System.
  - — Engine sealing.
  - — Air leaks between VAF meter and throttle body.
- EGO Fuel Fouled: Whenever an over-rich fuel condition has been experienced (fuel fouled spark plugs). make a thorough check of the ignition system. In the event the EGO sensor is suspected of being fuel fouled (low output, slow response), run the vehicle at sustained high speeds (within legal limits) followed by a few hard accelerations. This will burn off EGO contamination and restore proper EGO operation.
- Fuel Pressure: Perform Pinpoint Test Step HA1.
- Ignition System: Always in default spark (10 degrees). Refer to Quick Test Step 4.0.
- Improper Fueling: Lead fouled EGO sensor.
- TP Sensor: Not moving (mechanical damage). Connect DVOM to test Pin 47 and to test Pin 46. Key to Run. Observe DVOM while moving the throttle. Reading must increase with increase in throttle opening. If not correct, REPLACE as necessary.
- If at this point, the drive concern is still present, perform Steps HA2 through HA4 only.

Fuel Control — Pinpoint Test HA

TEST STEP	RESULT ▶	ACTION TO TAKE
HA21 CHECK HEATER ELEMENT RESISTANCE ON HEATED EGO		
• Key Off, Engine Off. • DVOM on 200 ohm scale. • Measure resistance between RUN circuit and ground on heated EGO connector at room temperature.	Reading between 2.5 and 5.0 ohms ▶	GO to HA22.
	Reading less than 2.5 or greater than 5.0 ohms ▶	REPLACE HEGO sensor.
HA22 CHECK FOR POWER AT HEGO HARNESS CONNECTOR		
• Key On, Engine Off. • DVOM on 20V scale. • Connect positive lead to run circuit and negative lead to ground on HEGO harness connector.	Reading 10.5V or greater ▶	HEGO sensor system OK, RETURN to Section 2.
	Reading less than 10.5V ▶	GO to HA23.
HA23 CHECK CONTINUITY OF GROUND TO HEGO CONNECTOR		
• Key Off, wait 10 seconds. • DVOM on 200 ohm scale. • Measure continuity of ground circuit from HEGO connector to battery ground.	Reading less than 5.0 ohms ▶	SERVICE open in run circuit.
	Reading 5.0 ohms or greater ▶	SERVICE open in ground circuit.

Fuel Control (FBC) — Pinpoint Test HB

HB

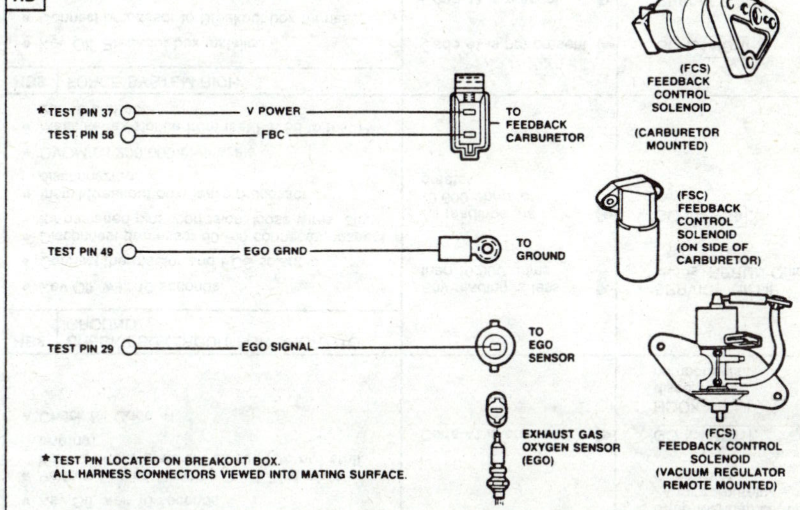

* TEST PIN LOCATED ON BREAKOUT BOX.
ALL HARNESS CONNECTORS VIEWED INTO MATING SURFACE.

STOP-WARNING

You may have reached this point because of a fault in a non-EEC System and/or component. To prevent the replacement of good components, be aware that the following non-EEC areas may be at fault:

- Fouled Spark Plugs
- Distributor Cap and Rotor
- Ignition Coil
- Spark Plug Wires
- CANP Problems
- PCV Valve
- Oil Contamination
- EGR Valve and Gasket
- Air Filter
- Fuel Filter
- Fuel Pump and Pressure
- Manifold Leak, Intake Exhaust
- Engine Not at Normal Operating Temperature
- Carburetor
- Choke, Sticking Electrical

This Pinpoint Test is intended to diagnose only the following:

- EGO Sensor Connection
- EGO Sensor
- Harness Circuits: V Pwer, FBC, EGO Control, EGO Signal
- Impact of Thermactor Air
- FBC Solenoid
- Vacuum System (EEC Only)
- Choke Circuit
- Processor Assembly

Fuel Control (FBC) — Pinpoint Test HB

TEST STEP	RESULT ▶	ACTION TO TAKE
FAULT CODE 41		
HB1 VERIFY VEHICLE PREPARATION		
• Did vehicle run for more than 2 minutes before performing Quick Test?	Yes ▶	GO to HB2.
• For 2.3L HSC verify canister purge shut-off valve was plugged per Quick Test Step 5.1.	No ▶	PREPARE vehicle, RERUN Quick Test.
HB2 CHECK FOR OBVIOUS FAULTS		
• Check EGO sensor connection. • Check EGO ground for good connection. • Check for intake or exhaust manifold leaks. • Are all OK?	Yes ▶	For 2.3L HSC engine, GO to HB3. All others GO to HB4.
	No ▶	SERVICE fault. RERUN Quick Test.
HB3 CHECK VACUUM		
• Key Off. • Check all FCS vacuum hoses to carburetor for leaks, blockages, and proper connection. • Refer to VECI decal. • Are all vacuum hoses OK?	Yes ▶	GO to HB4.
	No ▶	SERVICE or REPLACE faulty hoses. RERUN Quick Test.
HB4 CHECK EGO SIGNAL FOR SHORT		
• Key Off, wait 10 seconds. • Disconnect harness from EGO sensor and processor. • DVOM on 200,000 ohm scale. • Measure resistance of EGO sensor harness connector EGO signal to vehicle battery negative terminal.	Reading is less than 10,000 ohms ▶	SERVICE circuit short. RERUN Quick Test.
	Reading is 10,000 ohms or greater ▶	CONNECT EGO processor and sensor. GO to HB5.
HB5 TRY TO FORCE RICH BY REMOVING AIR SUPPLY		
• Key Off. • Reconnect EGO sensor and processor to harness. • Disconnect thermactor air supply hose at the air pump and cap the hose. • Run engine running Quick Test. **NOTE: Because thermactor air supply is disconnected a Code 11 does not indicate a pass. Treat Code 11 as a Code 44.**	Code 41 is present ▶	GO to HB6. Leave thermactor disconnected.
	Code 44 is present ▶	RETURN to Quick Test Step 5.1.
	Ignore all other service codes at this time	

Fuel Control (FBC) — Pinpoint Test HB

TEST STEP	RESULT ▶	ACTION TO TAKE
HB6 CHECK FCS SOLENOID		
• Key Off, wait 10 seconds. • Thermactor disconnected. • Disconnect harness from FCS solenoid. • Perform Engine Running Quick Test. • Check for Code 41. **NOTE: Ignore all other codes at this time.**	Code 41 is not present ▶	GO to HB8.
	Code 41 is present ▶	GO to HB7.
HB7 FORCE SYSTEM RICH		
• Key Off, wait 10 seconds. • FCS and thermactor disconnected. • Run for 2 minutes at part throttle. • Key Off, wait 10 seconds. • Perform Engine Running Quick Test while holding choke plate 3/4 closed **(Do not stall engine).** • Check for Code 41.	Code 41 is not present ▶	REPLACE FCS. RERUN Quick Test if Code 41 is still present. REFER to Shop Manual, Group 24 for carburetor diagnostics.
	Code 41 is present ▶	GO to HB11. HOOK-UP all disconnected components.
HB8 CHECK FBC CIRCUIT FOR SHORT TO GROUND		
• Key Off, wait 10 seconds. • Connect thermactor and FCS solenoid. • Disconnect processor 60 Pin connector, inspect for damaged pins, corrosion, loose wires, etc. • Install Breakout box, leave processor disconnected. • DVOM on 200,000 ohm scale. • Measure resistance from test Pin 58 to test Pins 20, 40 and 46.	Any reading is less than 10,000 ohms ▶	SERVICE circuit shorts. RERUN Quick Test.
	All readings are 10,000 ohms or greater ▶	GO to HB9.
HB9 FORCE SYSTEM RICH		
• Key Off, Breakout box installed. • Connect processor to Breakout box harness. • Run vehicle for 2 minutes at part throttle. • Key Off, wait 10 seconds. • Perform Engine Running Quick Test while holding choke plate 3 4 closed **(Do not stall engine).** • Check for Code 41.	Code 41 is not present ▶	GO to HB10.
	Code 41 is present ▶	REPLACE FCS solenoid. REMOVE Breakout box. RECONNECT FCS and processor. RERUN Quick Test.

Fuel Control (FBC) — Pinpoint Test — HB

TEST STEP	RESULT ▶	ACTION TO TAKE
HB10 CHECK FCS SOLENOID		
• Key Off, wait 10 seconds. • Breakout box installed with processor connected. • Disconnect harness from FCS solenoid. • DVOM on 2,000 ohm scale. • Measure resistance between FCS FBC circuit terminal and battery negative terminal. • Measure resistance between FCS V power circuit terminal and battery negative terminal.	Either reading is less than 1,000 ohms ▶	REPLACE FCS solenoid. REMOVE Breakout box. RECONNECT processor and FCS. RERUN Quick Test.
	Both readings are 1,000 ohms or greater ▶	REMOVE Breakout box. REPLACE processor. RECONNECT FCS and processor. RERUN Quick Test.
HB11 CHECK EGO CIRCUIT FOR SHORT		
• Key Off, wait 10 seconds. • Thermactor and FCS reconnected. • Disconnect harness from EGO sensor. • DVOM on 200,000 ohm scale. • Measure resistance of EGO harness connector EGO circuit to battery negative terminal.	Reading is less than 10,000 ohms ▶	REPLACE processor. RECONNECT EGO sensor and processor. RERUN Quick Test.
	Reading is 10,000 ohms or greater ▶	GO to HB12.
HB12 FORCE EGO RICH		
• Key Off, EGO sensor disconnected. • Run engine for 2 minutes. • DVOM on 20V scale. • With engine running hold choke plate 3/4 closed **(Do not stall engine).** • Measure voltage from EGO sensor pig tail terminal to battery negative terminal.	Reading is less than .45V ▶	REPLACE EGO sensor. RERUN Quick Test.
	Reading is .45V or greater ▶	GO to HB13.
HB13 CHECK EGO CIRCUIT FOR SHORT TO VREF		
• Key On, Engine Off, EGO disconnected. • DVOM on 20V scale. • Measure voltage from harness EGO circuit to battery negative terminal.	Reading is less than 2.0V ▶	GO to HB14.
	Reading is 2.0V or greater ▶	SERVICE circuit short. RECONNECT EGO sensor. RERUN Quick Test.

Fuel Control (FBC) — Pinpoint Test — HB

TEST STEP	RESULT ▶	ACTION TO TAKE
HB14 CHECK CONTINUITY EGO AND EGO GROUND CIRCUITS		
• Key Off, wait 10 seconds. • EGO sensor disconnected. • Disconnect processor 60 Pin connector, inspect for damaged pins, corrosion, loose wires, etc. Service as necessary. • Install Breakout box, leave processor disconnected. • DVOM on 200 ohm scale. • Measure resistance from test Pin 29 to EGO sensor harness EGO circuit. • Measure resistance from test Pin 49 to test Pins 40 and 60.	All readings are less than 5 ohms ▶	REPLACE processor. REMOVE Breakout box. RECONNECT EGO and processor. RERUN Quick Test.
	Any reading is 5 ohms or greater ▶	SERVICE circuit open. REMOVE Breakout box. RECONNECT processor and EGO. RERUN Quick Test.
FAULT CODES 42/47		
HB20 CHECK CHOKE		
• Does choke function properly without sticking or binding? **NOTE: For 2.3L HSC engine verify canister purge shut-off valve was plugged per Quick Test Step 5.1.**	No ▶	SERVICE choke as required. REFER to section 4. RERUN Quick Test.
	Yes ▶	GO to HB21.
HB21 GENERATE KNOWN FAULT		
• Run engine for 2 minutes at part throttle. • Key Off, wait 10 seconds. • Disconnect harness from EGO sensor. Jumper harness EGO circuit to ground. • Rerun Engine Running Quick Test. • Check for Code 41, ignore all other codes at this time.	Code 41 is present ▶	For 2.3L HSC engine, GO to HB22. All others, GO to HB23.
	Code 41 is not present ▶	GO to HB30.
HB22 CHECK VACUUM HOSES		
• Check all vacuum hoses from the FCS solenoid to the carburetor for leaks, blockages, and proper connection. • Refer to VECI decal. • Are vacuum hoses OK?	Yes ▶	RECONNECT all hoses to carburetor. GO to HB23.
	No ▶	REMOVE jumper wire and SERVICE as necessary. RERUN Quick Test.

Fuel Control (FBC) — Pinpoint Test HB

TEST STEP	RESULT ▶	ACTION TO TAKE
HB23 FORCE SYSTEM LEAN		
• Key Off, wait 10 seconds. • Reconnect EGO sensor. • Disconnect a 7/32 vacuum line at the intake manifold vacuum tree. • Perform Engine Running Quick Test. If engine stalls RERUN test while manually maintaining approximately 1500 rpm. • Is code 42/47 still present? Disregard other codes at this time.	Yes ▶	REPLACE EGO sensor. RECONNECT vacuum hose. RERUN Quick Test.
	No ▶	GO to HB24.
HB24 CHECK FCS SOLENOID		
• Key Off, wait 10 seconds. • Vacuum hose reconnected. • Disconnect FCS solenoid. • DVOM on 200 ohm scale. • Measure resistance between terminals on the FCS solenoid. • Is resistance within the following ranges: 15 to 30 ohms: 2150 carburetor 30 to 60 ohms: All other	Yes ▶	RECONNECT FCS. GO to HB25.
	No ▶	REPLACE FCS. RERUN Quick Test.
HB25 REFER TO APPENDIX, ENTER OUTPUT STATE CHECK		
• Key Off, wait 10 seconds. • DVOM on 20V scale. • Connect DVOM negative test lead to STO and positive test lead to battery positive. • Jumper STI to signal return. • Perform Key On, Engine Off Self-Test until the completion of the Continuous Test Codes. • DVOM will indicate 0 volts. • Depress and release the throttle. • Did DVOM reading change to a high voltage reading?	Yes ▶	REMAIN in output state check. GO to HB26.
	No ▶	DEPRESS throttle to WOT and RELEASE. If STO voltage does not go high GO to Pinpoint Test Step Q40.

Fuel Control (FBC) — Pinpoint Test HB

TEST STEP	RESULT ▶	ACTION TO TAKE
HB26 CHECK FCS OPERATION		
• Key On, in output state check. • DVOM on 20V scale. • Connect DVOM to FBC circuit on the FCS connector and to battery negative post. • While observing DVOM, depress and release the throttle several times to cycle output On and Off. • Does FCS cycle On and Off?	Yes ▶	2.3L HSC engine: GO to HB27.
	▶	All others: EXIT output state check. GO to Pinpoint Test Step KD1.
	No ▶	EXIT output state check. GO to HB28.
HB27 CHECK SOLENOID VACUUM CYCLING		
• Key On, Engine Off. • Disconnect input vacuum source from FCS and plug. • Connect a vacuum pump to FCS source vacuum port. • Disconnect the idle vacuum line from FCS. • Connect vacuum gauge to FCS idle vacuum port. • Apply 20 in. Hg while cycling outputs On and Off (by depressing and releasing throttle). • Observe the vacuum gauge.	Vacuum output cycles On and Off in less than 2 seconds ▶	RECONNECT all vacuum lines. REFER to Shop Manual, Group 24.
	Vacuum output does not cycle On or Off in less than 2 seconds ▶	REPLACE FCS. RECONNECT all vacuum lines. RERUN Quick Test.
HB28 CHECK V PWR CIRCUIT FOR POWER		
• Key Off, wait 10 seconds. • Deactivate Quick Test. • Disconnect harness from FCS. • DVOM on 20V scale. • Key On, Engine Off. • Measure voltage of FCS harness connector V PWR circuit to battery negative terminal.	Reading is less than 10V ▶	RECONNECT FCS solenoid. SERVICE circuit open. RERUN Quick Test.
	Reading is 10V or greater ▶	GO to HB29.

Fuel Control (FBC) — Pinpoint Test — HB

TEST STEP	RESULT ▶	ACTION TO TAKE
HB29 CHECK CONTINUITY OF FBC CIRCUIT		
• Key Off, wait 10 seconds. • FCS disconnected. • Disconnect processor 60 Pin connector, inspect connectors for damaged pins, corrosion, loose wires, etc. Service as necessary. • Install Breakout box, leave processor disconnected. • DVOM on 200 ohm scale. • Measure resistance from FCS harness connector FBC circuit to test Pin 58.	Reading is less than 5 ohms ▶	REMOVE Breakout box. RECONNECT FCS solenoid. REPLACE processor and RERUN Quick Test.
	Reading is 5 ohms or greater ▶	SERVICE circuit open. REMOVE Breakout box. RECONNECT processor and FCS solenoid. RERUN Quick Test.
HB30 CHECK EGO CIRCUIT FOR SHORT TO POWER		
• Key Off, wait 10 seconds. • Reconnect EGO sensor. • Disconnect processor 60 pin connector. Inspect for damaged pins, corrosion, loose wires, etc. Service as necessary. • Install Breakout box, leave processor disconnected. • DVOM on 20V scale. • Key On, Engine Off. • Measure voltage from test Pin 29 to battery negative terminal.	Reading is less than .4V ▶	REMOVE Breakout box. REPLACE processor. RERUN Quick Test. If Code 42 is still present GO to Pinpoint Test Step KD1. Except 2.3L HSC engine: GO to Shop Manual, Group 24.
	Reading is .4V or greater ▶	SERVICE circuit short. REMOVE Breakout box. RECONNECT processor. RERUN Quick Test.

Fuel Control (FBC) — Pinpoint Test — HB

TEST STEP	RESULT ▶	ACTION TO TAKE
FAULT CODE 43		
HB40 WARM ENGINE		
• Run vehicle at 2000 rpm for 2 minutes. • Key Off, wait 10 seconds. • Perform Engine Running Quick Test.	Code 43 is not present ▶	SERVICE other codes as necessary.
	Code 43 is present ▶	GO to HB41.
HB41 CHECK EXHAUST LEAKS		
• Are there any exhaust leaks at manifold or inlet pipe?	Yes ▶	SERVICE as necessary. RERUN Quick Test.
	No ▶	GO to HB42.
HB42 CHECK IDLE QUALITY		
• Start vehicle, let engine idle. • Does engine idle quality deteriorate and remain poor during and after the Quick Test?	Yes ▶	REFER to Carburetors, Section 4.
	No ▶	REPLACE EGO sensor. RERUN Quick Test.

Fuel Control — CFI | Pinpoint Test | HC

STOP-WARNING

You may have reached this point because of a fault in a non-EEC System and/or component. To prevent the replacement of good components, be aware that the following non-EEC areas may be at fault:

- Ignition Coil
- Distributor Cap
- Distributor Rotor
- Fouled Spark Plugs
- Spark Plug Wires
- CANP Problems
- PCV Valves (see note below)
- EGR Valve and Gasket
- Air Filter
- Fuel Contamination, Engine Oil
- Poor Power Ground
- Fuel Pressure
- Manifold Leaks, Intake/Exhaust
- Engine Not at Normal Operating Temperatures

This Pinpoint Test is intended to diagnose only the following:

- EGO Sensor
- EGO Signal and Ground Circuits
- EGO Sensor Connection
- Vacuum Systems
- Fuel Injectors
- Processor Assembly
- Thermactor Impact

NOTE: Fuel-contaminated engine oil may affect 4-1, 4-2 and 4-3 Service Codes, so if it is suspected, remove the PCV from the valve cover, and rerun the Quick Test. If the problem is corrected, then change the engine oil and filter.

TEST STEP	ACTION TO TAKE
HC1 SYSTEM IDENTIFICATION	
Is engine 5.0L CFI? ▶	GO to HC2.
Is engine 3.8L CFI? ▶	GO to HC20.

Fuel Control — 5.0L CFI | Pinpoint Test | HC

HC2

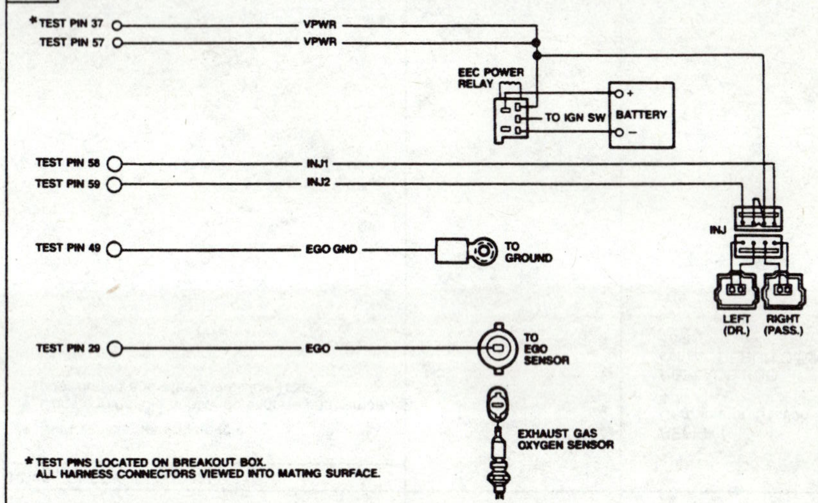

* TEST PINS LOCATED ON BREAKOUT BOX.
ALL HARNESS CONNECTORS VIEWED INTO MATING SURFACE.

TEST STEP	RESULT ▶	ACTION TO TAKE
HC2 WARM ENGINE		
• Run vehicle at 2000 rpm for 2 minutes. • Key Off, wait 10 seconds. • Perform Engine Running Quick Test.	Code 43 is not present ▶	SERVICE other codes as necessary.
	Code 43 is present ▶	GO to HC3.
HC3 CHECK EXHAUST LEAKS		
• Are there any exhaust leaks at manifold or inlet pipe?	Yes ▶	SERVICE as necessary. RERUN Quick Test.
	No ▶	GO to HC4.

Fuel Control — 5.0L CFI | Pinpoint Test | HC

TEST STEP	RESULT ▶	ACTION TO TAKE
FAULT CODE 43		
HC4 CHECK IDLE QUALITY		
• Start vehicle, let engine idle. • Does engine idle quality deteriorate and remain poor during and after the Quick Test?	Yes ▶ No ▶	REFER to EFI/CFI, Section 4. REPLACE EGO sensor. RERUN Quick Test.
HC5 FUEL CONTROL — ALWAYS LEAN, CODE 41		
• Key Off. DVOM to 20V range. Disconnect EGO sensor from vehicle harness. Connect DVOM to EGO sensor and engine ground. • Start the engine and run at approximately 2000 rpm for 2 minutes. • Does the DVOM read greater than 0.5V at the end of 2 minutes?	Yes ▶ No ▶	GO to HC6. REPLACE EGO sensor. REPEAT Quick Test.
HC6 EGO AND EGO GROUND CONTINUITY CHECK		
• Key Off, wait 10 seconds, EGO disconnected. • DVOM on 2000 ohm scale. • Disconnect processor 60 Pin connector and inspect for damaged pins, corrosion, loose wires. Service as necessary. • Connect Breakout box. Leave processor disconnected. • Measure continuity from test Pin 29 to EGO sensor connector. • Measure continuity from test Pin 49 to test Pins 40 and 60.	Both resistance readings less than 5 ohms ▶ Either resistance reading 5 ohms or greater ▶	GO to HC7. SERVICE harness open in ground circuits. REPEAT Quick Test.
HC7 CHECK EGO CIRCUIT FOR SHORT TO GROUND		
• Key Off. • Breakout box installed. Processor disconnected. • DVOM to 200,000 ohms range. • Disconnect EGO sensor at harness. • Measure the resistance between Breakout box Pin 29 and Pin 40.	Reading is 10,000 ohms or greater ▶ Reading is less than 10,000 ohms ▶	EGO input circuit OK. GO to HC11. CORRECT cause of resistance to ground. REPEAT Quick Test.

Fuel Control — 5.0L CFI | Pinpoint Test | HC

TEST STEP	RESULT ▶	ACTION TO TAKE
HC8 TRY TO FORCE A CODE 41		
• Key Off. • Disconnect vehicle harness at the EGO sensor. Using a jumper wire, ground EGO circuit in the harness connector to the engine block. • Rerun Engine Running Test. • Does Service Code 41 result? • Remove jumper wire when completed.	Yes ▶ No ▶	GO to HC10. GO to HC9.
HC9 HARNESS CONTINUITY CHECK		
• Key Off, wait 10 seconds, EGO sensor disconnected. • DVOM on 2000 ohm scale. • Disconnect processor 60 Pin connector and inspect for damaged pins, corrosion, loose wires. Service as necessary. • Connect Breakout box. Leave processor disconnected. • Measure continuity from test Pin 29 to EGO sensor connector. • Measure continuity from test Pin 49 to EGO ground terminal on the engine.	Both resistance readings less than 5 ohms ▶ Either resistance reading 5 ohms or greater ▶	DISCONNECT module harness connector. INSPECT for damage or corrosion. If OK, REPLACE processor. SERVICE harness open circuit. REPEAT Quick Test.
HC10 FORCED LEAN EGO TEST		
• Key Off. • DVOM on 20V range. • Disconnect the PCV hose at the intake manifold port. • With EGO sensor disconnected from the harness, connect a DVOM from EGO sensor to engine ground. • Run the engine at 2000 rpm for 2 minutes while observing DVOM.	Less than 0.4V ▶ 0.4V or greater ▶	EGO sensor OK. GO to HC11. REPLACE EGO sensor and RETEST.
HC11 FUEL PRESSURE CHECK		
• Install fuel pressure gauge. • Start and Run engine. Observe gauge pressure reading.	Fuel pressure is within 24-45 psi ▶ Fuel pressure is less than 24 psi or greater than 45 psi ▶	GO to HC12. REFER to the Shop manual for electric fuel pump and fuel regulator check.

Fuel Control — 5.0L CFI | Pinpoint Test | HC

TEST STEP	RESULT	ACTION TO TAKE
HC12 CHECK RESISTANCE OF INJECTORS HARNESS AND PROCESSOR CONNECTION		
• Key Off. • DVOM on 200 ohm range. • Disconnect processor 60 Pin connector and inspect for damaged pins, corrosion, loose wires. Service as necessary. • Using DVOM test leads measure resistance from test Pins 57 to 58 and test Pins 57 to 59.	Resistance is less than 1.5 ohms or greater than 3.5 ohms ▶	SERVICE the harness/connectors on the suspect injector for opens or shorts. If OK, REPLACE injector.
	Resistance is 1.5-3.5 ohms ▶	GO to HC13.
HC13 INJECTOR DRIVE SIGNAL CHECK		
• Requires standard non-powered 12V test lamp. • Key Off, Breakout box installed, reconnect processor. • Connect test lamp from test Pin 57 to test Pin 59. • Crank or start engine. • Repeat above test from test Pin 57 to test Pin 58. • How brightly does the test lamp glow?	Dim glow at light on both tests ▶	Normal. GO to HC14.
	No light on one or both tests ▶	VERIFY 12V battery power at test Pin 57. If OK, REPLACE processor.
	Bright light on one or both tests ▶	CHECK both injector circuits for shorts to ground. If OK, REPLACE processor.
HC14 INJECTOR BALANCE TEST		
• Connect tachometer to engine. Run engine at 2000 rpm. May be necessary to disconnect TK solenoid and use throttle body stop screw to set engine speed. • Disconnect and reconnect the injectors one at a time; note rpm drop for each injector. • Does each injector produce at least a 100 rpm drop? • Reset curb idle and verify TK connection at end of testing.	Yes ▶	Fuel delivery OK. Problem is with area common to all cylinders, i.e.: Air vac leak, fuel contamination, EGR.
	No ▶	REPLACE suspect injector.

Fuel Control — 3.8L CFI | Pinpoint Test | HC

HC20A GENERAL INSTRUCTIONS

STOP-WARNING

You may have reached this point because of a fault in a non-EEC System and/or component. To prevent the replacement of good components. be aware that the following non-EEC areas may be at fault:

- Ignition Coil
- Distributor Cap
- Distributor Rotor
- Fouled Spark Plugs
- Spark Plug Wires
- CANP Problems
- PCV Valves (see note below)
- EGR Valve and Gasket
- Air Filter
- Fuel Contamination, Engine Oil
- Poor Power Ground
- Fuel Pressure
- Manifold Leaks, Intake/Exhaust
- Engine Not at Normal Operating Temperatures

This Pinpoint Test is intended to diagnose only the following:

- EGO Sensors
- EGO Signals and Ground Circuits
- EGO Sensor Connections
- Fuel Injectors
- Processor Assembly

NOTE: Fuel-contaminated engine oil may affect 4-1, 4-2 and Service Codes, so if it is suspected, remove the PCV from the valve cover, and rerun the Quick Test. If the problem is corrected, then change the engine oil and filter.

Throttle Body — 3.8L CFI | Pinpoint Test | HC

HC20B REFERENCE ILLUSTRATION, 10-PIN CONNECTOR THROTTLE BODY

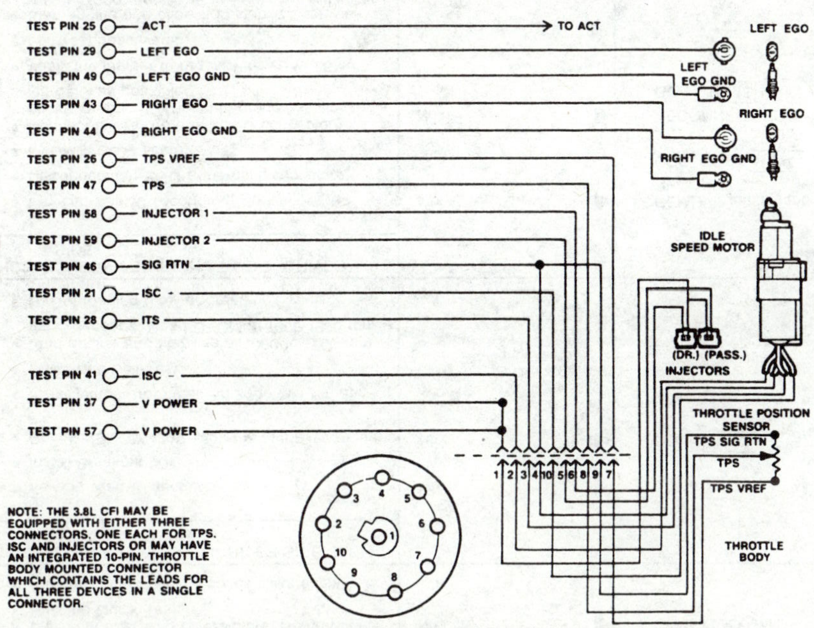

NOTE: THE 3.8L CFI MAY BE EQUIPPED WITH EITHER THREE CONNECTORS, ONE EACH FOR TPS, ISC AND INJECTORS OR MAY HAVE AN INTEGRATED 10-PIN, THROTTLE BODY MOUNTED CONNECTOR WHICH CONTAINS THE LEADS FOR ALL THREE DEVICES IN A SINGLE CONNECTOR.

10-PIN CONNECTOR LOOKING INTO MATING SURFACE.

NOTE: When servicing the 10-pin connector, care must be taken that the red retaining ring is not lost or damaged. Care must also be taken that pins are placed into the proper holes on reassembly.

Fuel Control — 3.8L CFI | Pinpoint Test | HC

HC20C REFERENCE DRAWING — SEPARATE CONNECTORS FOR TPS-ISC AND INJECTORS

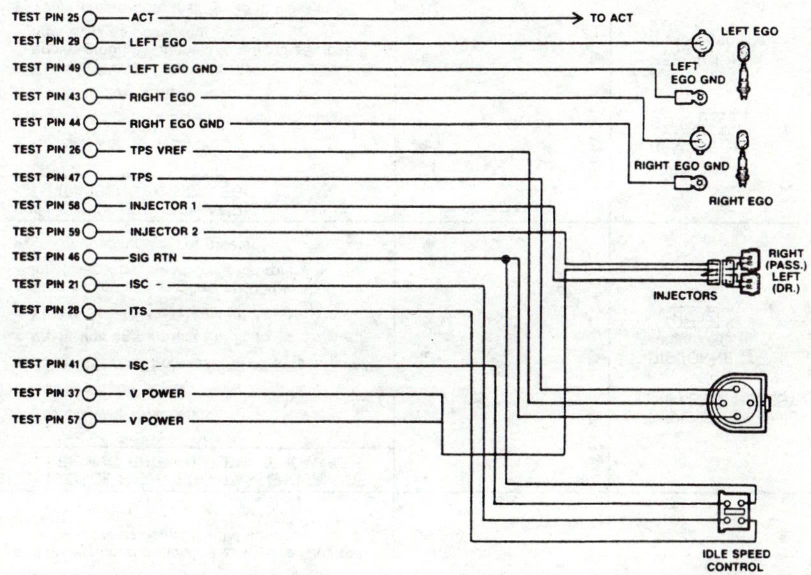

NOTE: The 3.8L CFI may be equipped with either three connectors, one each for TPS, ISC and injectors or may have an integrated 10-pin, throttle body mounted connector which contains the leads for all three devices in a single connector.

Fuel Control — 3.8L CFI | Pinpoint Test | HC

TEST STEP	RESULT ▶	ACTION TO TAKE
HC21 VERIFY ENGINE IS AT TEMPERATURE		
• Run engine at 2000 rpm for two minutes (Upper hose hot and pressurized?). • Engine at operating temperature. Rerun Engine Running Quick Test. • Are fuel control service codes still present?	Yes ▶ No ▶	GO to HC22. GO to Quick Test Step 5.
HC22 FORCE LEFT EGO (DRIVER SIDE) RICH AT BREAKOUT BOX		
• Key Off, wait 10 seconds. • Install Breakout box. • Connect DVOM (20V scale) to test Pins 29 and 49. • Install jumper from test Pin 25 to test Pin 26. • Deactivate Self-Test. • Start engine and stabilize at 1000 to 1200 rpm for 15 seconds. Read DVOM at 1000-1200 rpm. • DVOM reading 0.5V or greater?	Yes ▶ No ▶	GO to HC24. GO to HC23.
HC23 CHECK FOR RICH AT LEFT (DRIVER SIDE) EGO		
• Key Off, wait 10 seconds. • Disconnect left EGO at harness connector. • DVOM on 20V scale. • Connect DVOM from left EGO connector to engine ground. (Connect to ground as close to left EGO as possible.) • Verify the jumper from test Pins 25 to 26. • Self-Test deactivated. • Start engine and allow it to stabilize between 1000 to 1200 rpm for 15 seconds. Read DVOM at 1000-1200 rpm. • DVOM reading 0.5V or greater?	Yes ▶ No ▶	REMOVE jumper and RECONNECT EGO. GO to HC34. RECONNECT EGO. GO to HC25.

Fuel Control — 3.8L CFI | Pinpoint Test | HC

TEST STEP	RESULT ▶	ACTION TO TAKE
HC24 FORCE RIGHT EGO (PASSENGER SIDE) RICH AT BREAKOUT BOX (FROM HC22)		
• Key Off, wait 10 seconds. • DVOM on 20V scale. • Connect DVOM from test Pin 43 to test Pin 44. • Verify jumper from test Pin 25 to test Pin 26. • Ensure Self-Test is deactivated. • Start engine and stabilize at 1000 to 1200 rpm for 15 seconds. Read DVOM at rpm. • DVOM reading 0.5V or greater?	Yes ▶ No ▶	REMOVE jumper and Go to HC27. GO to HC26.
HC25 FORCE RIGHT EGO (PASSENGER SIDE) RICH AT BREAKOUT BOX (FROM HC23)		
• Key Off, wait 10 seconds. • DVOM on 20V scale. • Connect DVOM from test Pin 43 to test Pin 44. • Verify jumper from test Pin 25 to Pin 26. • Ensure Self-Test deactivated. • Start engine and stabilize at 1000 to 1200 rpm for 15 seconds. Read DVOM at rpm. • DVOM reading 0.5V or greater?	Yes ▶ No ▶	REPLACE left (driver side) EGO. RETEST. DISCONNECT jumper. GO to HC32.
HC26 CHECK FOR RICH AT RIGHT (PASSENGER SIDE) EGO		
• Key Off, wait 10 seconds. • Disconnect right EGO at the harness connector. • DVOM to 20V scale. • Connect DVOM from right EGO connector to engine ground. (Connect to ground as close to right EGO as possible.) • Verify jumper from test Pin 25 to test Pin 26. • Verify Self-Test deactivated. • Start engine and allow it to stabilize between 1000 to 1200 rpm for 15 seconds. Read DVOM at rpm. • DVOM reading 0.5V or greater?	Yes ▶ No ▶	REMOVE jumper. GO to HC35. REPLACE right (passenger side) EGO.

Fuel Control — 3.8L CFI | Pinpoint Test | HC

TEST STEP	RESULT ▶	ACTION TO TAKE
HC27 CHECK LEFT EGO (DRIVER SIDE) LEAN AT BREAKOUT BOX		
• Key Off, wait 10 seconds. • Connect DVOM (20V scale) to test Pins 29 to 49. • Ensure Self-Test deactivated. • Start engine and run at 2000 rpm for 30 seconds to ensure engine is at operating temperature. • Shut engine Off and wait 10 seconds. • Restart engine and monitor EGO voltage on DVOM. • Does EGO voltage stabilize at 0.5V or less within 15 seconds?	Yes ▶ No ▶	GO to **HC29**. GO to **HC28**.
HC28 CHECK FOR LEAN AT LEFT (DRIVER SIDE) EGO		
• Key Off, wait 10 seconds. • Disconnect left EGO at harness connector. • DVOM on 20V scale. • Connect DVOM from left EGO connector to engine ground (connect to ground as close to left EGO as possible). • Verify Self-Test deactivated. • Start engine and run at 2000 rpm for 30 seconds. • Shut engine Off, wait 10 seconds. • Restart engine, monitor EGO voltage. • Does EGO voltage stabilize at 0.5V or less within 15 seconds?	Yes ▶ No ▶	RECONNECT EGO. GO to **HC34**. GO to **HC30**.

Fuel Control — 3.8L CFI | Pinpoint Test | HC

TEST STEP	RESULT ▶	ACTION TO TAKE
HC29 FORCE RIGHT EGO (PASSENGER SIDE) LEAN AT BREAKOUT BOX (FROM HC27)		
• Key Off, wait 10 seconds. • Connect DVOM (20V scale) to test Pins 43 and 44. • Ensure Self-Test deactivated. • Start engine and run at 2000 rpm for 30 seconds to ensure engine is at operating temperature. • Shut engine Off and wait 10 seconds. • Restart engine and monitor EGO voltage on DVOM. • Does EGO voltage stabilize at 0.5V or less within 30 seconds? **NOTE: DVOM reading must be taken within 30 seconds of re-start.**	Yes ▶ No ▶	REPLACE processor, RETEST. REPLACE (passenger side) right EGO and RETEST.
HC30 FORCE RIGHT EGO (PASSENGER SIDE) LEAN AT BREAKOUT BOX (FROM HC28)		
• Key Off, wait 10 seconds. • Connect DVOM (20V scale) to test Pins 43 and 44. • Verify Self-Test deactivated. • Start engine. Run at 2000 rpm for 30 seconds to ensure engine is at operating temperature. • Shut engine Off, wait 10 seconds. • Restart engine, monitor EGO voltage on DVOM. • Does EGO voltage stabilize at 0.5V or less within 30 seconds? **NOTE: DVOM reading must be taken within 30 seconds of re-start.**	Yes ▶ No ▶	REPLACE left (driver side) EGO and RETEST. GO to **HC31**.
HC31 CHECK FOR LEAN AT RIGHT (PASSENGER SIDE) EGO		
• Key Off, wait 10 seconds. • Disconnect right EGO at harness connector. • DVOM on 20V scale. • Connect DVOM from right EGO connector to engine ground. (Connect to ground as close to right EGO as possible.) • Ensure Self-Test deactivated. • Start engine and run at 2000 rpm for 30 seconds. • Shut engine Off, wait 10 seconds. • Restart engine, monitor EGO voltage on DVOM. • Does EGO voltage stabilize at 0.5V or less within 30 seconds? **NOTE: DVOM reading must be taken within 30 seconds of re-start.**	Yes ▶ No ▶	RECONNECT EGO. GO to **HC35**. TURN engine Off. LEAVE EGO and DVOM connected. GO to **HC33**.

Fuel Control — 3.8L CFI | Pinpoint Test | HC

TEST STEP	RESULT ▶	ACTION TO TAKE
HC32 VERIFY SYSTEM IS RICH		
• Key Off, wait 10 seconds. • Verify both left and right EGOs are connected. • Install jumper between test Pins 25 and 26. • Verify Self-Test deactivated. • Start engine and stabilize at 1500 rpm for 30 seconds. • Black smoke present in exhaust?	Yes ▶ No ▶	REPLACE both EGOs and RETEST. GO to HC40.
HC33 VERIFY SYSTEM IS LEAN		
• Verify right EGO disconnected and DVOM connected to EGO sensor connector and ground. • Remove air cleaner cover. • Start engine, wait 10 seconds. • Run engine at 2000 rpm for 60 seconds. • Turn engine Off (key Off). • When engine stops, disconnect both injectors at the injector harness. • Push accelerator to the floor and hold it there. • Crank engine for 10 seconds and monitor EGO voltage. • EGO voltage less than 0.5V?	Yes ▶ No ▶	RECONNECT EGO. CHECK fuel delivery. GO to HC40. REPLACE both EGO sensors and RETEST.
HC34 CHECK LEFT EGO HARNESS FOR SHORTS OR OPEN		
• Engine Off, wait 10 seconds. • DVOM to 200 ohm scale. • Connect DVOM to left EGO connector and test Pin 29. • Resistance less than 5 ohms?	Yes ▶ No ▶	REPLACE processor and RETEST. SERVICE harness and RETEST.
HC35 CHECK RIGHT EGO HARNESS FOR SHORT OR OPEN		
• Engine Off, wait 10 seconds. • DVOM to 200 ohm scale. • Connect DVOM to right EGO connector and test Pin 43. • Resistance less than 5 ohms?	Yes ▶ No ▶	REPLACE processor and RETEST. SERVICE harness and RETEST.

Fuel Control — 3.8L CFI | Pinpoint Test | HC

TEST STEP	RESULT ▶	ACTION TO TAKE
HC40 FUEL PRESSURE CHECK		
• Install fuel pressure gauge. • Start and run engine. Fuel pressure must be 241-310 kPa (35-45 psi). • Fuel pressure within specification? **For no starts:** • If engine will not run cycle the key from Off to On several times. • Fuel pressure must remain at 276 ± 34 kPa (40 psi ± 5 psi) for 60 seconds after final key Off.	Yes ▶ No ▶	GO to HC41. REFER to Shop manual Group 24 for electric fuel pump.
HC41 CHECK FOR INJECTOR LEAK DOWN		
• Observe pressure drop 2 minutes after performing Step HC40. • Does pressure drop 4 psi or less?	Yes ▶ No ▶	GO to HC42. REFER to Shop manual Group 24 for service.
HC42 CHECK LEFT INJECTOR PRESSURE		
• Disconnect the left (driver side) injector connector. • Ensure pressurized fuel system. • Disable fuel pump by disconnecting inertia switch or fuel pump relay. • Observe pressure gauge reading. • Crank engine for two seconds. • Turn key to Off. • Wait 5 seconds, then observe pressure drop. • Is drop in pressure between 14 and 110 kPa (2 to 16 psi)?	Yes ▶ No ▶	GO to HC43. GO to HC46.
HC43 CHECK RIGHT INJECTOR PRESSURE		
• Reactivate fuel pump. • Connect left injector and disconnect right (passenger side) injector. • Repeat HC42 for right injector. • Is drop in pressure between 14 and 110 kPa (2 to 16 psi)?	Yes ▶ No ▶	GO to HC44. GO to HC46.

Fuel Control — 3.8L CFI — Pinpoint Test — HC

TEST STEP	RESULT ▶	ACTION TO TAKE
HC44 CHECK FUEL PRESSURE REGULATOR		
• Reconnect injector. • Reconnect inertia switch or fuel pump relay. • Start engine, let idle. • Observe pressure gauge. • Is pressure 241 to 310 kPa (35 to 45 psi)?	Yes ▶	GO to HC45.
	No ▶	CHECK to ensure that fuel lines; supply and return, are free from kinks. If OK, GO to Shop manual Group 24 to check pressure regulator and throttle body for proper operation.
HC45 INJECTOR BALANCE CHECK		
• Connect tachometer to engine. • Disconnect and plug hose to EGR valve. • Run engine at 2000 rpm. (It may be necessary to disconnect idle speed control and use throttle body stop screw to set engine speed.) • Disconnect left injector by unplugging the injector connector. • Note rpm when engine rpm stabilizes. • Reconnect left injector and stabilize rpm at 2000 rpm. • Disconnect right injector as in above Step. • Note rpm when engine rpm stabilizes. • Reconnect right injector and allow engine to stabilize at idle. • Note the difference between the two rpm readings. • Does the rpm drop at least 150 rpm for each injector?	Yes ▶	RECONNECT EGR hose. Fuel system OK. REFER to diagnostic routines for other fuel related concerns (fouled plugs, fuel-contaminated engine oil).
	No ▶	RECONNECT EGR hose. REPLACE both injectors.
HC46 ISOLATE SUSPECT INJECTOR		
• Key Off, wait for 10 seconds. • Disconnect both injectors from the injector harness. • Reconnect the injector connectors as follows: A. Connect left connector to right (passenger side) injector. B. Connect right connector to left (driver side) injector. • Rerun Self-Test. • Suspect injector still fails.	Yes ▶	RESTORE connectors to proper injectors. REPLACE injector. RETEST.
	No ▶	GO to HC47.

Fuel Control — 3.8L CFI — Pinpoint Test — HC

TEST STEP	RESULT ▶	ACTION TO TAKE
HC47 CHECK FOR SHORT TO BATTERY AT INJECTOR CONNECTORS		
• Key Off, wait 10 seconds. • Disconnect injector connectors. • DVOM to 20V scale. • Connect DVOM between injector connector and ground. • DVOM reads less than 1.0V? **NOTE: This step must be run on both left and right injectors.**		RECONNECT injector connectors.
	Yes ▶	GO to HC48.
	No ▶	SERVICE short and RETEST.
HC48 CHECK LEFT AND RIGHT INJECTOR FOR SHORT TO GROUND		
• Key Off, wait 10 seconds. • Verify injectors are connected. • Verify Breakout box installed. • DVOM to 200 ohm range. • Connect DVOM between test Pin 60 and: — For left injector test Pin 58 — For right injector test Pin 59 • DVOM reads less than 5 ohms?	Yes ▶	GO to HC49.
	No ▶	GO to HC51.
HC49 ISOLATE SHORT TO GROUND		
• Disconnect the harness connector at the throttle body. • DVOM on 200 ohm scale. • Disconnect injector harness connectors from injectors. • Connect DVOM from injector 1 of the throttle body connector to ground. Note reading. • Connect DVOM from injector 2 of the throttle body connector to ground. • Is resistance on either or both injectors less than 5 ohms?	Yes ▶	CORRECT appropriate short or REPLACE appropriate injector.
	No ▶	GO to HC50.

Fuel Control — 3.8L CFI | Pinpoint Test | HC

TEST STEP	RESULT ▶	ACTION TO TAKE
HC50 SHORT IN HARNESS OR PROCESSOR		
• Verify harness connection to the throttle body is disconnected. • Disconnect processor from Breakout box. • DVOM on 200 ohm scale. • Connect DVOM from injector 1 of the harness connector to test Pin 58. Note reading. • Connect DVOM from injector 2 of the harness connector to test Pin 59. Note reading. • Either or both readings less than 5 ohms?	Yes ▶ No ▶	SERVICE appropriate short. RECHECK. REPLACE processor.
HC51 CHECK FOR OPEN INJECTORS		
• Verify injectors are disconnected. • DVOM to 200 ohm scale. • Connect DVOM across the terminals of each injector. Note readings. • Is resistance less than 1.5 ohms or greater than 2.5 ohms?	Yes ▶ No ▶	REPLACE appropriate injector and RETEST. GO to HC52.
HC52 CHECK HARNESS CONTINUITY		
• Verify that the throttle body harness connector is disconnected. • DVOM on 200 ohm scale. • Connect DVOM from injector 1 to test Pin 58. Note reading. • Connect DVOM from injector 2 to test Pin 59. Note reading. • Resistance less than 2 ohms?	Yes ▶ No ▶	RECONNECT harness to throttle body assembly. RECONNECT processor to Breakout box. GO to HC53. SERVICE appropriate harness wire and RETEST.

Fuel Control — 3.8L CFI | Pinpoint Test | HC

TEST STEP	RESULT ▶	ACTION TO TAKE
HC53 CHECK FOR SHORT BETWEEN INJECTOR INPUTS AT PROCESSOR		
• Verify processor connected to Breakout box. • Verify throttle body harness connector is connected. • DVOM to 2000 ohm scale. • Verify injectors are not connected to harness. • Connect DVOM from test Pin 58 to test Pin 59. • Resistance less than 10 kilohms?	Yes ▶ No ▶	GO to HC54. REPLACE processor. RETEST.
HC54 CHECK FOR SHORT TO INJECTOR CIRCUITS AT THROTTLE BODY		
• Verify injectors are not connected to harness. • Disconnect processor from Breakout box. • Disconnect harness connector from throttle body. • DVOM on 2000 ohm scale. • Connect DVOM from injector 1 to injector 2 at the throttle body connector. • Resistance less than 10 kilohms?	Yes ▶ No ▶	CORRECT short and RETEST. REPLACE processor.

Fuel Pump Circuit — (Inertia Switch) | Pinpoint Test | J

J

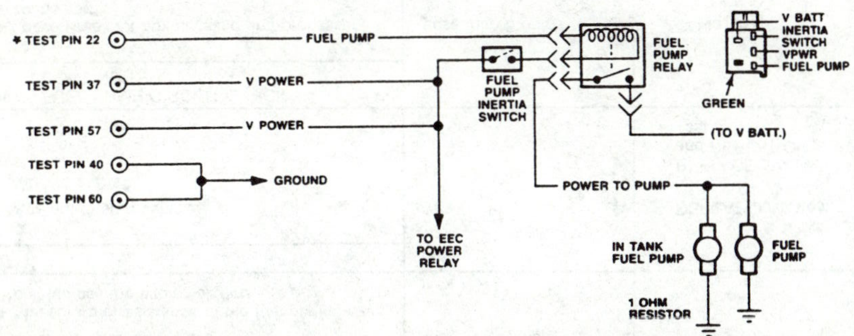

* TEST PINS LOCATED ON BREAKOUT BOX.
ALL HARNESS CONNECTORS VIEWED INTO MATING SURFACE.

STOP-WARNING

You may have reached this point because of a fault in a non-EEC System and/or component. To prevent the replacement of good components, be aware that the following non-EEC areas may be at fault:

- Fuel Lines
- Fuel Filters
- Throttle Body
- Contaminated Fuel
- Fuel Pump

This Pinpoint Test is intended to diagnose only the following:

- Fuel Pump Relay
- Harness Circuits: V Batt., V Power, F.P., Ground and Power to Pump(s)
- Processor Assembly

Fuel Pump Circuit — (Inertia Switch) | Pinpoint Test | J

TEST STEP	RESULT ▶	ACTION TO TAKE
J1 NO FUEL PUMP PRESSURE		
• Fuel pressure gauge installed. • Check if fuel pump runs. • Cycle key from Off to Run, repeat several times. (Do not enter start mode.) Fuel pump should run briefly each time the key enters Run.	Yes ▶	GO to Shop Manual Group 24 — Electric Fuel Pump.
	No ▶	GO to [J2].
J2 CHECK FOR V POWER TO PROCESSOR		
• Key Off, wait 10 seconds. • Disconnect processor 60 Pin connector and inspect for damaged pins, corrosion, loose wires. Service as necessary. • Install Breakout box and reconnect processor. • Key On, engine Off. • DVOM on 20V scale. • Measure voltage from test Pin 37 to test Pin 40 and test Pin 57 to test Pin 60.	Either voltage reading is less than 10.5V ▶	GO to [B1].
	Both voltage readings are 10.5V or greater ▶	GO to [J3].
J3 POWER TO PUMP(S) CHECK		
• Key On, engine Off. • Leave Breakout box installed and processor connected. • Locate fuel pump relay. • DVOM on 20V scale. • With DVOM measure voltage from chassis ground to power to pump(s) circuit at fuel pump relay during crank mode.	8.0V or greater during crank ▶	GO to Shop Manual Group 24 — Electric Fuel Pump.
	Less than 8.0V during crank ▶	GO to [J4].
J4 FUEL PUMP CIRCUIT CHECK TO V BATT.		
• Key On, engine Off. • Leave Breakout box installed and processor connected. • Locate fuel pump relay. • DVOM on 20V scale. • With DVOM measure voltage from chassis ground to V Batt. at fuel pump relay.	10.5V or greater ▶	GO to [J5].
	Less than 10.5V ▶	SERVICE open in V Batt. between fuel pump relay and vehicle battery positive post, then GO to [J13].

Fuel Pump Circuit — (Inertia Switch) | Pinpoint Test | J

TEST STEP	RESULT ▶	ACTION TO TAKE
J5 FUEL PUMP CIRCUIT CHECK		
• Key On, engine Off. • Leave Breakout box installed and processor connected. • Locate fuel pump relay. • DVOM on 20V scale. • With DVOM measure voltage from chassis ground to fuel pump circuit at the fuel pump relay.	10.5V or greater ▶ Less than 10.5V ▶	GO to J6. GO to J9.
J6 RESISTANCE CHECK OF FUEL PUMP RELAY		
• Key Off, wait 10 seconds. • Leave Breakout box installed and processor connected. • Locate and disconnect fuel pump relay. • DVOM on 200 ohm scale. • Measure the resistance of the fuel pump relay coil (across the two small pins).	Reading is 50-100 ohms ▶ Reading is less than 50 ohms or greater than 100 ohms ▶	GO to J8. REPLACE fuel pump relay and GO to J7.
J7		
• Attempt to start vehicle. • Vehicle starts?	Yes ▶ No ▶	REPEAT Quick Test. REPLACE processor and REPEAT Quick Test.
J8 FUEL PUMP CIRCUIT CHECK CONTINUITY		
• Leave Breakout box installed and processor connected. • Key Off, wait 10 seconds. • DVOM on 200 ohm scale. • Measure resistance on fuel pump circuit from pump relay to test Pin 22.	Less than 5 ohms ▶ 5 ohms or greater ▶	GO to J11. SERVICE open in fuel pump circuit and REPEAT Quick Test.

Fuel Pump Circuit — (Inertia Switch) | Pinpoint Test | J

TEST STEP	RESULT ▶	ACTION TO TAKE
J9 FUEL PUMP CIRCUIT CHECK (V POWER CIRCUIT)		
• Leave Breakout box installed and processor connected. • Key Off. • Locate the inertia switch and fuel pump relay. • DVOM on 200 ohm scale. • Measure resistance of V Power circuit between inertia switch and fuel pump relay.	5 ohms or greater ▶ Less than 5 ohms ▶	SERVICE open in V Power circuit. REPEAT Quick Test. GO to J10.
J10 INERTIA SWITCH CHECK		
• Leave Breakout box installed and processor connected. • Key On, engine Off. • DVOM on 20V scale. • Measure voltage at inertia switch input V Power circuit to ground.	10.5V or greater ▶ Less than 10.5V ▶	REPLACE inertia switch and REPEAT Quick Test. SERVICE open from EEC power relay to inertia switch. REPEAT Quick Test.
J11 CHECK FOR SHORT TO POWER		
• Key Off, wait 10 seconds. • Leave Breakout box installed and disconnect processor. • Fuel pump relay disconnected. • DVOM on 200,000 ohm scale. • Measure resistance from test Pin 22 to test Pins 37 and 57.	10,000 ohms or greater ▶ Less than 10,000 ohms ▶	GO to J12. SERVICE short to power in the fuel pump circuit. GO to J7.
J12 CHECK FOR SHORT TO GROUND		
• Key Off. • Leave Breakout box installed and processor disconnected. • Fuel pump relay disconnected. • DVOM on 200,000 ohm scale. • Measure resistance from test Pin 22 to test Pins 40 and 60.	10,000 ohms or greater ▶ Less than 10,000 ohms ▶	REPLACE processor and REPEAT Quick Test. SERVICE short to ground in the fuel pump circuit. REPEAT Quick Test.

Fuel Pump Circuit — (Inertia Switch) | Pinpoint Test | J

TEST STEP	RESULT ▶	ACTION TO TAKE
J13 CHECK POWER TO PUMP(S) CIRCUIT — SHORT TO GROUND		
• Key Off, wait 10 seconds. • Fuel pump relay disconnected. • Fuel pump(s) disconnected. • DVOM on 200,000 ohm scale. • Measure resistance from power to pump(s) circuit to chassis ground.	Less than 10,000 ohms ▶	SERVICE short to ground in power to pump(s) circuit. RECONNECT all components. REPEAT Quick Test.
	10,000 ohms or greater ▶	RECONNECT all components. REPEAT Quick Test.

EGR On/Off Control | Pinpoint Test | KA

KA

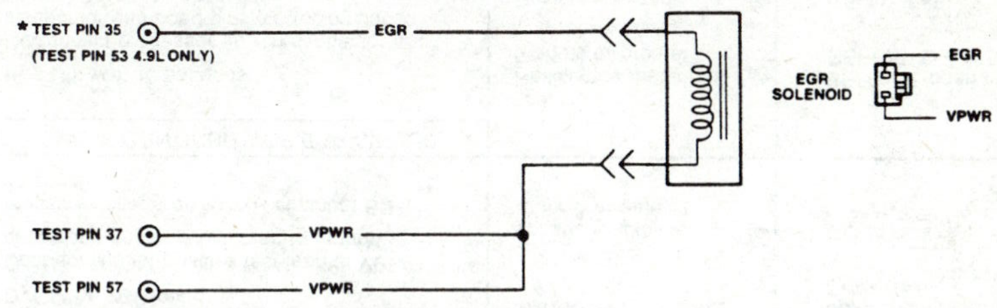

*TEST PINS LOCATED ON BREAKOUT BOX.
ALL HARNESS CONNECTORS VIEWED INTO MATING SURFACE.

STOP-WARNING

You may have reached this point because of a fault in a non-EEC System and/or component. To prevent the replacement of good components, be aware that the following non-EEC areas may be at fault:

- Air or Vacuum Leaks
- EGR Flow Restrictions
- EGR Value

NOTE: Code 34 may be the result of high volume exhaust vent system (reduces backpressure). If this is suspected, perform the test in a well-ventilated area without exhaust vent connected.

This Pinpoint Test is intended to diagnose only the following:

- Circuits: EGR and VPWR
- EGR Solenoid
- Presence of Manifold Vacuum
- Processor Assembly

EGR On/Off Control — Pinpoint Test — KA

TEST STEP	RESULT ▶	ACTION TO TAKE
KA1 MEASURE EGR SOLENOID RESISTANCE		
• Key Off, wait 10 seconds. • DVOM on 200 ohm scale. • Disconnect EGR solenoid connector and measure solenoid resistance.	Resistance is between 65 and 110 ohms ▶	CONNECT EGR solenoid. GO to KA2.
	Resistance is less than 65 ohms or greater than 110 ohms ▶	REPLACE EGR solenoid. REPEAT Quick Test.
KA2 ENTER OUTPUT STATE CHECK (REFER TO APPENDIX)		
• Key Off, wait 10 seconds. • DVOM on 20V scale. • Connect DVOM negative test lead to STO and positive test lead to battery positive. • Jumper STI to signal return. • Perform Key On, Engine Off Self-Test until the completion of the Continuous Test Codes. • DVOM will indicate zero volts. • Depress and release the throttle. • Did DVOM reading change to a high voltage reading?	Yes ▶	REMAIN in Output State Check. GO to KA3.
	No ▶	DEPRESS throttle to WOT and release. If STO voltage does not go high, GO to Pinpoint Test Step Q40. Leave equipment hooked up.
KA3 CHECK EGR ON/OFF CONTROL SOLENOID ELECTRICAL OPERATION		
• DVOM on 20V scale. • Connect DVOM positive test lead to VPWR circuit on EGR solenoid and negative test lead to EGR output circuit. • While observing DVOM, depress and release the throttle several times to cycle output On and Off. • EGR output cycles On and Off.	Yes ▶	GO to KA4.
	No ▶	GO to KA6.

EGR On/Off Control — Pinpoint Test — KA

TEST STEP	RESULT ▶	ACTION TO TAKE
KA4 CHECK SOLENOID FOR VACUUM CYCLING		
• Install vacuum pump to the solenoid vacuum supply port and install a vacuum gauge to the output port. Apply 6 in. Hg minimum. • While cycling outputs On and Off (by depressing and releasing throttle) observe the vacuum gauge at the output. **NOTE: Maintain vacuum at source.**	Vacuum output cycles On and Off ▶	GO to KA5.
	Vacuum output does not cycle On and Off ▶	REPLACE solenoid. REPEAT Quick Test.
KA5 CHECK MANIFOLD VACUUM LINES FOR BLOCKAGE OR LEAKS		
• With vacuum lines disconnected at solenoid, check for vacuum. • Start engine.	Vacuum present ▶	EEC-IV system OK. REFER to Section 6.
	No vacuum present ▶	SERVICE vacuum source blockage or leak. REPEAT Quick Test.
KA6 CHECK VOLTAGE OF VPWR CIRCUIT		
• Key On, Engine Off. • DVOM on 20V scale. • Connect DVOM positive test lead to VPWR circuit and negative test lead to ground. • Measure voltage on VPWR circuit at EGR solenoid.	Voltage reading is less than 10.5V ▶	SERVICE harness open circuit. REPEAT Quick Test.
	Voltage reading is 10.5V or greater ▶	GO to KA7.
KA7 CHECK CONTINUITY OF EGR CIRCUIT		
• Key Off, wait 10 seconds. • Disconnect processor 60 Pin connector and inspect for damaged pins, corrosion, loose wires. Service as necessary. • Connect Breakout box to harness. Leave processor disconnected. • DVOM on 200 ohm scale. • Measure continuity from test Pin 35* to EGR circuit at harness connector. *Test Pin 53 for 4.9L engine, truck only.	Resistance reading is 5 ohms or greater ▶	SERVICE open circuit. REPEAT Quick Test.
	Resistance reading is less than 5 ohms ▶	GO to KA8.

EGR On/Off Control — Pinpoint Test — KA

TEST STEP	RESULT ▶	ACTION TO TAKE
KA8 CHECK FOR SHORT TO POWER		
• Key Off, wait 10 seconds. • DVOM on 200,000 ohm scale. • Leave Breakout box installed and processor disconnected. • With EGR solenoid disconnected, measure resistance from test Pin 35* to test Pins 37 and 57. *Test Pin 53 for 4.9L engine, truck only.	Resistance reading is 10,000 ohms or greater ▶	GO to KA9.
	Resistance reading is less than 10,000 ohms ▶	SERVICE short to power. REPEAT Quick Test.
KA9 CHECK FOR SHORT TO GROUND		
• Key Off, wait 10 seconds. • Leave Breakout box installed and processor disconnected. • DVOM on 200,000 ohm scale. • Measure resistance of test Pin 35* to test Pins 40, 46 and 60. *Test Pin 53 for 4.9L engine, truck only.	Resistance reading is less than 10,000 ohms ▶	SERVICE short to ground. REPEAT Quick Test.
	Resistance reading is 10,000 ohms or greater ▶	REPLACE processor. REPEAT Quick Test.

Air Management System — Pinpoint Test — KC

KC

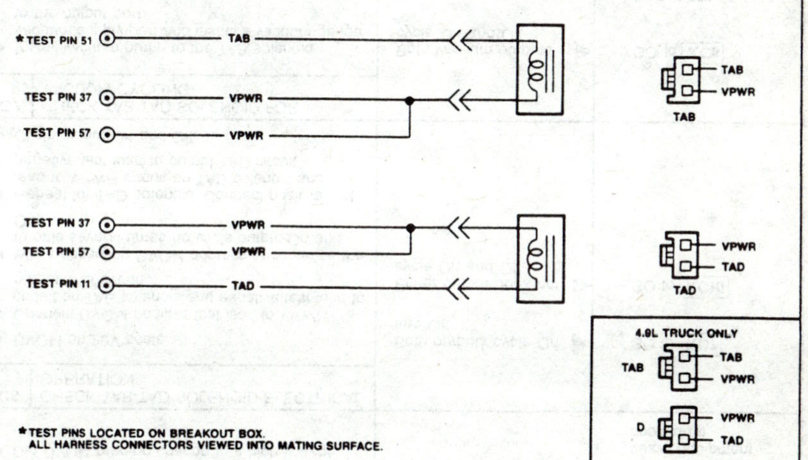

*TEST PINS LOCATED ON BREAKOUT BOX.
ALL HARNESS CONNECTORS VIEWED INTO MATING SURFACE.

STOP-WARNING

You may have reached this point because of a fault in a non-EEC System and/or component. To prevent the replacement of good components, be aware that the following non-EEC areas may be at fault:

- Thermactor System
  - — Belt
  - — Pump
  - — Valve

This Pinpoint Test is intended to diagnose only the following:

- TAB and TAD Solenoid Valve Assemblies
- Harness Circuits: TAB, TAD and VPWR
- Vacuum Supply
- Processor Assembly

Air Management System — Pinpoint Test — KC

TEST STEP	RESULT ▶	ACTION TO TAKE
KC1 SERVICE CODES 44, 45 AND 46: VERIFY VACUUM LINE ROUTING		
• Verify proper vacuum line routing to the TAB/TAD solenoids and to the bypass diverter valve. Refer to VECI decal. • Check for kinked or blocked vacuum lines. • Check for kinked or blocked air hoses. • Check for disconnected vacuum lines. • Are visual checks satisfactory?	No ▶	SERVICE routing or faults. REPEAT Quick Test.
	Yes ▶	Service code 44, GO to KC4. Service code 45, GO to KC2. Service code 46, GO to KC3.
KC2 ATTEMPT TO ELIMINATE SERVICE CODE 45 (TAD ONLY)		
• Disconnect vacuum line on diverter valve and cap vacuum line. • Key Off, wait 10 seconds. • Repeat Engine Running Self-Test and record service codes. • Is code 45 present?	Yes ▶	EEC-IV system OK. REFER to Section 3 for diverter valve or check valve diagnostics.
	No ▶	GO to KC4.
KC3 ATTEMPT TO ELIMINATE SERVICE CODE 46 (TAB ONLY)		
• Disconnect vacuum line on bypass valve and cap vacuum line. • Key Off, wait 10 seconds. • Repeat Engine Running Self-Test and record codes. • Is code 46 present?	Yes ▶	EEC-IV system OK. REFER to Section 3 for bypass valve diagnostics.
	No ▶	GO to KC4.
KC4 MEASURE TAB TAD SOLENOID RESISTANCE		
• Key Off, wait 10 seconds. • DVOM on 200 ohm scale. • Disconnect TAB solenoid connector and measure solenoid resistance. • Disconnect TAD solenoid connector and measure solenoid resistance.	Both resistances are between 50 and 100 ohms ▶	CONNECT TAB/TAD solenoids. GO to KC5.
	Either resistance is less than 50 ohms or greater than 100 ohms ▶	REPLACE TAB TAD solenoid assembly. REPEAT Quick Test.

Air Management System — Pinpoint Test — KC

TEST STEP	RESULT ▶	ACTION TO TAKE
KC5 ENTER OUTPUT STATE CHECK (REFER TO APPENDIX)		
• Key Off. wait 10 seconds. • DVOM on 20V scale. • Connect DVOM negative test lead to STO and positive test lead to battery positive. • Jumper STI to signal return. • Perform Key On, Engine Off Self-Test until the completion of the Continuous Test Codes. • DVOM will indicate zero volts. • Depress and release the throttle. • Did DVOM reading change to a high voltage reading?	Yes ▶	REMAIN in Output State Check. GO to KC6.
	No ▶	DEPRESS throttle to WOT and RELEASE. If STO voltage does not go high, GO to Pinpoint Test Step Q40. Leave equipment hooked up.
KC6 CHECK TAB/TAD SOLENOID ELECTRICAL OPERATION		
• DVOM on 20V scale. • Connect DVOM positive test lead to VPWR circuit on TAB solenoid and negative test lead to output TAB circuit. • While observing DVOM depress and release the throttle several times (to cycle output On and Off). • Repeat for TAD solenoid. Connect positive test lead to VPWR circuit on TAD solenoid and negative test lead to output TAD circuit. • Cycle output On and Off.	Both outputs cycle On and Off ▶	GO to KC7.
	Either output does not cycle On and Off ▶	GO to KC10.
KC7 CHECK TAB TAD SOLENOID FOR VACUUM CYCLING		
• Install vacuum pump to the TAB solenoid vacuum supply port and install a vacuum gauge to the output port. • While cycling outputs On and Off (by depressing and releasing throttle), observe the vacuum gauge at the output. **NOTE: Maintain vacuum at source.** • Repeat for TAD solenoid. Connect vacuum pump to the TAD solenoid vacuum supply port and connect a vacuum gauge to the output port. • Cycle output On and Off.	Both vacuum outputs cycle On and Off ▶	GO to KC8.
	Either vacuum output does not cycle On and Off ▶	REPLACE solenoid assembly. REPEAT Quick Test.

Air Management System — Pinpoint Test — KC

TEST STEP	RESULT ▶	ACTION TO TAKE
KC8 CHECK TAB TAD VACUUM		
• Disconnect vacuum line from TAB solenoid to thermactor valve. • Inspect vacuum line for blockage, cracks, kinks (e.g. connect vacuum pump and gauge and verify vacuum line holds vacuum, or blow into vacuum line to verify no restrictions). • Repeat this Step for the TAD vacuum line. • Thermactor vacuum lines OK?	Yes ▶	GO to KC9.
	No ▶	SERVICE as required. REPEAT Quick Test.
KC9 CHECK MANIFOLD VACUUM LINES FOR BLOCKAGE OR LEAKS		
• With vacuum lines disconnected at TAD/TAB solenoids, check for vacuum. • Start engine.	Vacuum present ▶	EEC-IV system OK. REFER to Section 3 for thermactor valve and air pump diagnostics.
	No vacuum present ▶	SERVICE vacuum source blockage or leak. REPEAT Quick Test.
KC10 CHECK VOLTAGE OF VPWR CIRCUIT		
• Key On, engine Off. • DVOM on 20V scale. • Positive DVOM test lead to VPWR circuit and negative test lead to ground. • Measure voltage on TAB solenoid VPWR circuit. • Repeat for TAD solenoid.	Both voltage readings are greater than 10.5V ▶	GO to KC11.
	Either voltage reading is less than 10.5V ▶	SERVICE harness circuit open. REPEAT Quick Test.
KC11 CHECK CONTINUITY OF TAB AND TAD CIRCUITS		
• Key Off, wait 10 seconds. • Disconnect processor 60 Pin connector and inspect for damaged pins, corrosion, loose wires. Service as necessary. • Connect Breakout box to harness. Leave processor disconnected. • DVOM on 200 ohm scale. • Measure continuity of test Pin 51 to TAB circuit at harness connector. • Measure continuity of test Pin 11 to TAD circuit at harness connector.	Both resistance readings are less than 5 ohms ▶	GO to KC12.
	Either resistance reading is 5 ohms or greater ▶	SERVICE harness open circuit. REPEAT Quick Test.

Air Management System — Pinpoint Test — KC

TEST STEP	RESULT ▶	ACTION TO TAKE
KC12 CHECK FOR SHORT TO POWER		
• Key Off, wait 10 seconds. • DVOM on 200,000 ohm. • Leave Breakout box installed and processor disconnected. • With TAB/TAD solenoids disconnected, measure resistance from test Pins 51 and/or 11 to test Pins 37 and 57.	Resistance reading is 10,000 ohms or greater ▶	GO to KC13.
	Resistance reading is less than 10,000 ohms ▶	SERVICE short to power. REPEAT Quick Test.
KC13 CHECK FOR SHORT TO GROUND		
• Key Off, wait 10 seconds. • DVOM on 200,000 ohm scale. • Leave Breakout box installed and processor disconnected. • With TAB/TAD solenoids disconnected, measure resistance from test Pins 51 and/or 11 to test Pins 40, 46 and 60.	Resistance reading is less than 10,000 ohms ▶	SERVICE short to ground. REPEAT Quick Test.
	Resistance reading is 10,000 ohms or greater ▶	REPLACE processor. REPEAT Quick Test.

Canister Purge — Pinpoint Test — KD

KD

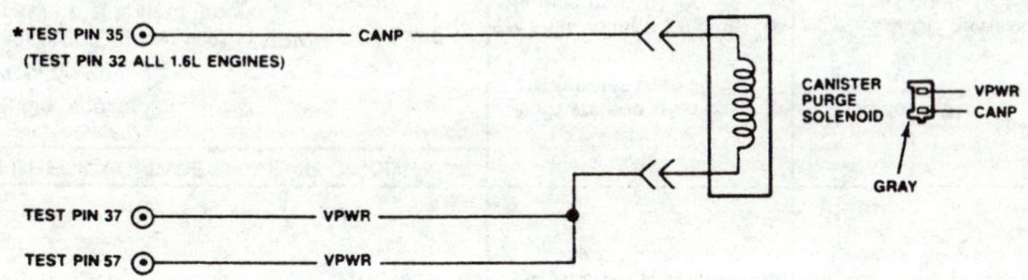

*TEST PINS LOCATED ON BREAKOUT BOX.
ALL HARNESS CONNECTORS VIEWED INTO MATING SURFACE.

STOP-WARNING

This Pinpoint Test is intended to diagnose only the following:

- Harness Circuits: CANP and VPWR
- Processor Assembly

Canister Purge — Pinpoint Test — KD

TEST STEP	RESULT ▶	ACTION TO TAKE
KD1 MEASURE CANP SOLENOID RESISTANCE		
• Key Off, wait 10 seconds. • DVOM on 200 ohm scale. • Disconnect CANP solenoid connector and measure solenoid resistance.	Resistance is between 40 and 90 ohms ▶	CONNECT CANP solenoid. GO to KD2.
	Resistance is less than 40 ohms or greater than 90 ohms ▶	REPLACE CANP solenoid. REPEAT Quick Test.
KD2 ENTER OUTPUT STATE CHECK (REFER TO APPENDIX)		
• Key Off, wait 10 seconds. • DVOM on 20V scale. • Connect DVOM negative test lead to STO and positive test lead to battery positive. • Jumper STI to signal return. • Perform Key On, Engine Off Self-Test until the completion of the Continuous Test Codes. • DVOM will indicate zero volts. • Depress and release the throttle. • Did DVOM reading change to a high voltage reading?	Yes ▶	REMAIN in Output State Check. GO to KD3.
	No ▶	DEPRESS throttle to WOT and release. If STO voltage does not go high, GO to Pinpoint Test Step Q40. Leave equipment hooked up.
KD3 CHECK CANISTER PURGE (CANP) SOLENOID ELECTRICAL OPERATION		
• DVOM on 20V scale. • Connect DVOM positive test lead to VPWR on CANP solenoid and negative test lead to CANP output circuit. • While observing DVOM depress and release the throttle several times (to cycle output On and Off). • CANP output cycles On and Off?	Yes ▶	GO to KD4.
	No ▶	REMOVE jumper. GO to KD5.

Canister Purge — Pinpoint Test — KD

TEST STEP	RESULT	ACTION TO TAKE
KD4 CHECK CANISTER PURGE SOLENOID FOR HOLDING VACUUM		
• Key Off, wait 10 seconds. • Remove jumper from STI to Signal Return. • Disconnect vacuum hose at canister purge solenoid at PCV side. • Apply 16 in. Hg to CANP solenoid. • Does CANP solenoid hold vacuum for 20 seconds?	Yes ▶	EEC-IV system OK. Refer to Shop Manual Group 24.
	No ▶	REPLACE CANP solenoid. REPEAT Quick Test. If code 42 is still present, GO to Section 4.
KD5 CHECK VOLTAGE OF VPWR CIRCUIT		
• Key On, engine Off. • DVOM on 20V scale. • Connect DVOM positive test lead to VPWR and negative test lead to ground. • Measure voltage on CANP solenoid.	Voltage reading is less than 10.5V ▶	SERVICE harness open circuit. REPEAT Quick Test.
	Voltage reading is 10.5V or greater ▶	GO to KD6.
KD6 CHECK CONTINUITY OF CANP CIRCUIT		
• Key Off, wait 10 seconds. • Disconnect processor and inspect both 60 Pin connectors and inspect for damaged pins, corrosion, loose wires. Service as necessary. • Connect Breakout box to harness. Leave processor disconnected. • DVOM on 200 ohm scale. • Measure continuity from test Pin 35* to CANP at harness connector. *Test Pin 32 for all 1.6L engines.	Resistance reading is 5 ohms or greater ▶	SERVICE open circuit. REPEAT Quick Test.
	Resistance reading is less than 5 ohms ▶	GO to KD7.

Canister Purge — Pinpoint Test — KD

TEST STEP	RESULT	ACTION TO TAKE
KD7 CHECK FOR SHORT TO POWER		
• Key Off, wait 10 seconds. • DVOM on 200,000 ohm scale. • Leave Breakout box to harness. Leave processor disconnected. • With CANP solenoid disconnected, measure resistance from test Pin 35* to test Pins 37 and 57. *Test Pin 32 for all 1.6L engine applications.	Resistance reading is 10,000 ohms or greater ▶	GO to KD8.
	Resistance reading is less than 10,000 ohms ▶	SERVICE short to power. REPEAT Quick Test.
KD8 CHECK FOR SHORT TO GROUND		
• Key Off, wait 10 seconds. • Leave Breakout box installed and processor disconnected. • DVOM on 200,000 ohm scale. • Measure resistance of test Pin 35* to test Pins 40, 46 and 60. *Test Pin 32 for all 1.6L engine applications.	Resistance reading is less than 10,000 ohms ▶	SERVICE short to ground. REPEAT Quick Test.
	Resistance reading is 10,000 ohms or greater ▶	REMOVE Breakout box. REPLACE processor. RERUN Quick Test.

Idle Speed Control (By-Pass Air) | Pinpoint Test | KE

KE

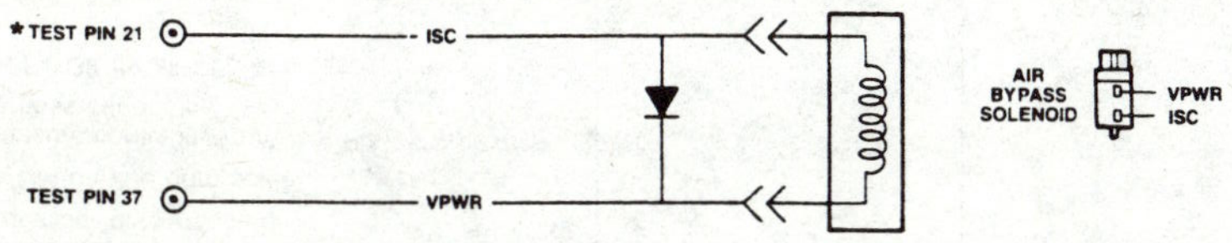

*TEST PINS LOCATED ON BREAKOUT BOX.
ALL HARNESS CONNECTORS VIEWED INTO MATING SURFACE.

CAUTION: If the engine exhibits rough running and/or idle, correct these conditions before diagnosing Idle Speed Control (ISC). Rough running/misses may be caused by:

- **Ignition System**
  Refer to Section 15
- **Fuel System**
  Refer to Pinpoint Test Steps HA1 through HA4
- **EGR System, Section 6**

STOP-WARNING

You may have reached this point because of a fault in a non-EEC System and/or component. To prevent the replacement of good components, be aware that the following non-EEC areas may be at fault:

- Engine not up to operating temperature
- Engine over operating temperature
- Improper Idle Speed Throttle Stop Adjustment
- A/C input (electrical problem)
- Throttle Speed Control Linkage

This Pinpoint Test is intended to diagnose only the following:

- RPM in Self-Test only
- ISC Actuator
- Harness Circuits ISC and VPWR
- Processor Assembly

Idle Speed Control (By-Pass Air) | Pinpoint Test | KE

TEST STEP	RESULT ▶	ACTION TO TAKE
KE1 MEASURE ISC SOLENOID RESISTANCE		
• Key Off. • DVOM on 200 ohm scale. • Disconnect ISC solenoid connector and measure solenoid resistance.	Resistance is between 7 and 13 ohms ▶	GO to KE2.
	Resistance is less than 7 ohms or greater than 13 ohms ▶	REPLACE ISC solenoid. REPEAT Quick Test.
KE2 ISC SHORT TO CASE (GROUND) CHECK		
• Key Off. • DVOM on 200,000 ohm scale. • Leave ISC harness disconnected. • Check for short from either ISC pin to ISC housing.	10,000 ohms or greater ▶	GO to KE3.
	Less than 10,000 ohms ▶	REPLACE ISC solenoid. REPEAT Quick Test.
KE3 CHECK VOLTAGE OF VPWR CIRCUIT		
• Key On. Engine Off. • DVOM on 20V scale. • Leave ISC harness disconnected. • Connect DVOM positive test lead to VPWR circuit and negative test lead to ground. • Measure voltage on ISC solenoid VPWR circuit.	Voltage reading is less than 10.5V ▶	SERVICE harness open circuit. REPEAT Quick Test.
	Voltage reading is 10.5V or greater ▶	GO to KE4.
KE4 CHECK CONTINUITY OF ISC CIRCUIT		
• Key Off. • Leave ISC harness disconnected. • Disconnect processor and inspect both 60 Pin connectors for damaged pins, corrosion, loose wires. Service as necessary. • Connect Breakout box to harness. Leave processor disconnected. • DVOM on 200 ohm scale. • Measure continuity of test Pin 21 to ISC circuit at harness connector.	Resistance reading is 5 ohms or greater ▶	SERVICE open circuit. REPEAT Quick Test.
	Resistance reading is less than 5 ohms ▶	GO to KE5.

Idle Speed Control (By-Pass Air) — Pinpoint Test — KE

TEST STEP	RESULT ▶	ACTION TO TAKE
KE5 CHECK FOR SHORT TO POWER		
• Key Off, wait 10 seconds. • DVOM on 200,000 ohm scale. • Leave Breakout box to harness. Leave processor disconnected. • Leave ISC solenoid disconnected, measure resistance from test Pin 21 to test Pin 37.	Resistance reading is 10,000 ohms or greater ▶ Resistance reading is less than 10,000 ohms ▶	GO to KE6. SERVICE short to power. REPEAT Quick Test.
KE6 CHECK FOR SHORT TO GROUND		
• Key Off, wait 10 seconds. • Leave Breakout box installed and processor disconnected. • DVOM on 200,000 ohm scale. • Measure resistance of test Pin 21 to test Pins 40, 46 and 60.	Resistance reading is less than 10,000 ohms ▶ Resistance reading is 10,000 ohms or greater ▶	SERVICE short to ground. REPEAT Quick Test. GO to KE7.
KE7 CHECK FOR ISC SIGNAL AT THE PROCESSOR		
• Connect processor. • Connect ISC actuator. • Vehicle prepared for Quick Test. • DVOM on 20V scale. • Connect DVOM between test Pin 21 and test Pin 40. • Start engine and observe the DVOM during Self-Test.	Meter reading varies during Quick Test ▶ Meter reading does not vary during Quick Test ▶	REPLACE ISC actuator. REPEAT Quick Test. DISCONNECT processor and INSPECT for bent, damaged, corroded pins, etc. CORRECT as necessary. If OK, REPLACE processor. REPEAT Quick Test.
KE8 CODE 13: RPM DOES NOT DROP TO LESS THAN 1500 AT END OF TEST		
• Disconnect ISC harness connector. • Connect engine tachometer. • Repeat Engine Running Quick Test. • At end of test, record service codes for future use. • Does rpm remain below 1500 rpm during test?	Yes ▶ No ▶	REPLACE processor. REPEAT Quick Test. CHECK engine vacuum hoses. REFER to VECI decal. CHECK that throttle plates are fully closed. CHECK throttle linkage and or speed control linkage for binding. If OK, REPLACE ISC actuator. REPEAT Quick Test.

Idle Speed Control (DC Motor Control) — Pinpoint Test — KF

KF

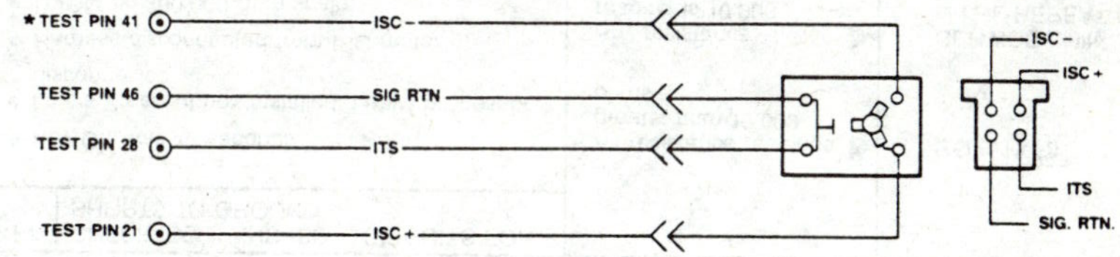

NOTE: FOR 3.8L CFI A 10-PIN CONNECTOR WAS USED FOR THE FIRST 60 DAYS OF PRODUCTION. EITHER CONNECTOR MAY BE ENCOUNTERED. TEST PIN ASSIGNMENTS DID NOT CHANGE ON BREAKOUT BOX.

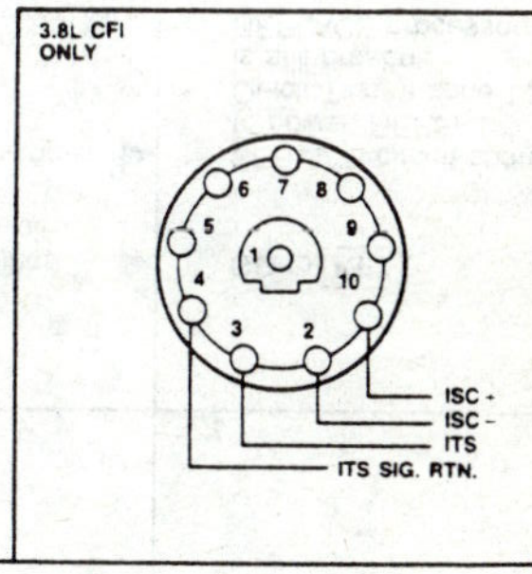

★ TEST PINS LOCATED ON BREAKOUT BOX.
ALL HARNESS CONNECTORS VIEWED INTO MATING SURFACE.

STOP-WARNING

You may have reached this point because of a fault in a non-EEC System and/or component. To prevent the replacement of good components, be aware that the following non-EEC areas may be at fault:

- Anti-Diesel Speed too High
- Basic Engine
- Vacuum Leaks
- Throttle Sticking or on High Cam

This Pinpoint Test is intended to diagnose only the following:

- DC Motor
- Idle Tracking Switch
- Harness circuits: ISC+, ISC−, Signal Return and ITS
- Processor Assembly

RPM Test (DC Motor Control) — Pinpoint Test — KF

TEST STEP	RESULT ▶	ACTION TO TAKE
KF1 DC MOTOR CONTROL FAULT ISOLATION: SERVICE CODE 12		
• Is code 58 present?	Yes ▶	GO to Pinpoint Test Step FB1.
	No ▶	GO to KF2.
KF2 CHECK IF CODE 31 IS PRESENT		
• Is code 31 present?	Yes ▶	For 2.8L Truck and 3.8L, 5.0L CFI engines, GO to Pinpoint Test Step DL1. All others, GO to Pinpoint Test Step DD1.
	No ▶	GO to KF3.
KF3 CHECK IF CODE 41 IS PRESENT		
• Is code 41 present?	Yes ▶	For 1.6L, 2.3L EFI engines, GO to Pinpoint Test Step HA1. For 5.0L CFI engine, GO to Pinpoint Test Step HC1. For 3.8L CFI engine, GO to Pinpoint Test Step HC20. All others, (FBC) GO to Pinpoint Test Step HB1.
	No ▶	GO to KF4.
KF4 CHECK FOR THROTTLE STICKING		
• Check throttle plates and linkages for binding. • Check speed control for binding (if so equipped).	Yes ▶	SERVICE as necessary. REPEAT Quick Test.
	No ▶	GO to KF5.

RPM Test (DC Motor Control) — Pinpoint Test — KF

TEST STEP	RESULT ▶	ACTION TO TAKE
KF5 CHECK TP AND ISC CIRCUITS FOR CONTINUITY		
• Key Off, wait 10 seconds. • Disconnect processor 60 Pin connector and inspect for damaged pins, corrosion, loose wires. Service as necessary. • Connect Breakout box. Leave processor disconnected. • Disconnect harness connector for ISC motor. • DVOM on 200 ohm scale. • Check continuity of test Pin 41 to ISC− circuit and test Pin 21 to ISC+ circuit at harness connector.	Both resistance readings are less than 5 ohms ▶	GO to KF6.
	Either resistance reading is 5 ohms or greater ▶	SERVICE faulty circuits. REPEAT Quick Test.
KF6 CHECK ISC+ AND ISC− CIRCUITS FOR SHORTS TO POWER		
• Key On, engine Off. • Leave Breakout box installed. Leave processor disconnected. • Harness disconnected from ISC motor. • DVOM on 20V scale. • Measure voltage of test Pin 41 to test Pins 40 and 60. Measure voltage of test Pin 21 to test Pins 40 and 60.	All voltage readings are less than 1 volt ▶	GO to KF7.
	Any voltage reading is 1 volt or more ▶	SERVICE circuit short to power. REPEAT Quick Test. If code 12 is still present, REPLACE processor.
KF7 CHECK ISC+ AND ISC− CIRCUITS FOR SHORTS TO GROUND		
• Key Off, wait 10 seconds. • Leave Breakout box installed. Leave processor disconnected. • Harness disconnected from ISC motor. • DVOM on 200,000 ohm scale. • Measure resistance of test Pin 41 to test Pins 40, 46 and 60. Measure resistance of test Pin 21 to test Pins 40, 46 and 60.	All resistance readings greater than 10,000 ohms ▶	GO to KF8.
	Any resistance reading is 10,000 ohms or less ▶	SERVICE faulty circuit. REPEAT Quick Test. If code 12 is still present, REPLACE processor.

RPM Test (DC Motor Control) | Pinpoint Test | KF

TEST STEP	RESULT ▶	ACTION TO TAKE
KF8 PREPARE IDLE SPEED CONTROL DC MOTOR FOR CHECK OF OPERATION		
• Key Off, wait 10 seconds. • Leave Breakout box installed. Leave processor disconnected. • Harness disconnected from ISC motor. • DVOM on 20V scale. • Connect DVOM positive test lead to test Pin 1 and negative test lead to ground.	Voltage reading is 10.5V or greater ▶	GO to KF9.
	Voltage reading is less than 10.5V ▶	SERVICE open in circuit 37. REPEAT Quick Test. If code 12 is still present, GO to KF9.
KF9 PREPARE ISC DC MOTOR FOR OPERATION CHECK		
• Key Off, wait 10 seconds. • Leave Breakout box installed. Leave processor disconnected. • Connect ISC DC motor to harness connector. • Jumper test Pin 21 to test Pin 1 and test Pin 41 to ground (to extend DC motor). • Does DC motor shaft extend 5 cm (2 inches) or more?	Yes ▶	GO to KF10.
	No ▶	REPLACE DC motor. REPEAT Quick Test.
KF10 RETRACT ISC DC MOTOR		
• Key Off, wait 10 seconds. • Leave Breakout box installed. Leave processor disconnected. • Jumper test Pin 41 to test Pin 1 and test Pin 21 to ground (to retract DC motor). • Does ISC DC motor retract?	Yes ▶	GO to KF11.
	No ▶	REPLACE DC motor. REPEAT Quick Test.
KF11 EXTEND ISC DC MOTOR		
• Key Off, wait 10 seconds. • Leave Breakout box installed. Leave processor disconnected. • Jumper test Pin 21 to test Pin 1 and test Pin 41 to ground (to extend DC motor). • Does ISC DC motor extend?	No ▶	REPLACE DC motor. REPEAT Quick Test.
	Yes ▶	REPLACE processor. REPEAT Quick Test.

RPM Test (DC Motor Control) | Pinpoint Test | KF

TEST STEP	RESULT ▶	ACTION TO TAKE
KF20 RPM NOT RETURNING TO NORMAL SPEED: CODE 13		
• Prepare system for normal engine operation. • De-activate Self-Test. • Air conditioner Off. • Run engine. Alternate between 30 seconds at idle and 5 seconds at part throttle modes for 3 minutes. • Is idle speed erratic?	No ▶	GO to KF21
	Yes ▶	CHECK for: • Vacuum leaks • Code 22 • Code 31 • Code 41 • Code 58 SERVICE above codes as necessary before continuing.
KF21 PERFORM ENGINE RUNNING SELF-TEST		
• Connect Self-Test equipment. • Perform Engine Running Self-Test. • Is code 13 present?	No ▶	SERVICE other codes as necessary.
	Yes ▶	GO to KF22.
KF22 CHECK FOR THROTTLE STICKING OR CHOKE POSITION		
• Inspect throttle mechanisms (plate linkage) for sticking or binding. • Inspect choke for sticking or binding. • Inspect for throttle stuck on high cam. • Are above inspections satisfactory?	No ▶	SERVICE fault. REPEAT Quick Test.
	Yes ▶	GO to KF23.
KF23 PERFORM KEY ON — ENGINE OFF TEST. OBSERVE ON DEMAND CODES		
• Perform Key On, Engine Off Test. Record On Demand codes. • Key On, Engine Off. • Is code 68 present?	Yes ▶	GO to Pinpoint Test Step FB5.
	No ▶	REFER to Section 4 for anti-diesel speed adjustment.

Idle Speed Control (Throttle Kicker) — Pinpoint Test — KG

KG

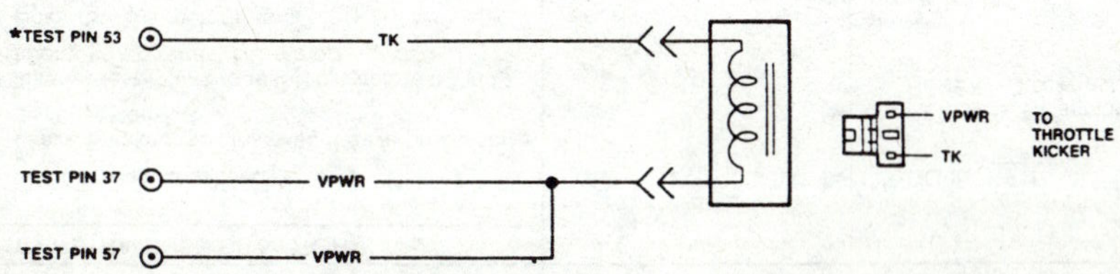

*TEST PINS LOCATED ON BREAKOUT BOX.
ALL HARNESS CONNECTORS VIEWED INTO MATING SURFACE.

STOP-WARNING

You may have reached this point because of a fault in a non-EEC System and/or component. To prevent the replacement of good components, be aware that the following non-EEC areas may be at fault:

- Engine Running Below Operating Temperature
- Engine Running Above Operating Temperature
- Improper Idle Speeds
- A/C Input (Electrical Problems)

This Pinpoint Test is intended to diagnose only the following:

- TK Solenoid
- TK Actuator
- Harness circuits: TK and VPWR
- Processor Assembly
- Vacuum Source
- Actuator Check

Idle Speed Control (Throttle Kicker) — Pinpoint Test — KG

TEST STEP	RESULT ▶	ACTION TO TAKE
KG1 VERIFY LINE ROUTINGS AND VISUAL CHECKS		
• Verify proper vacuum line routing to the TK solenoid. Refer to VECI decal. • Check for kinked or blocked vacuum lines. • Check for kinked or blocked air hoses. • Check for disconnected vacuum lines. • Are visual checks satisfactory?	No ▶	SERVICE routing or faults. REPEAT Quick Test.
	Yes ▶	Service codes 12 or 13. GO to KG2.
KG2 VACUUM CHECK		
• Apply 20 in Hg vacuum to throttle kicker. • Verify vacuum holds.	Yes ▶	GO to KG3.
	No ▶	REPLACE solenoid. REPEAT Quick Test.
KG3 ATTEMPT TO ELIMINATE SERVICE CODE 12 OR 13		
• While vacuum is applied to throttle kicker, rerun Engine Running Self-Test. • Is service code 12 or 13 present?	Yes ▶	EEC-IV system OK. REFER to Section 4 for idle adjustment.
	No ▶	GO to KG4.
KG4 MEASURE TK SOLENOID RESISTANCE		
• Key Off, wait 10 seconds. • DVOM on 200 ohm scale. • Disconnect TK solenoid connector and measure solenoid resistance.	Resistance is between 50-100 ohms ▶	CONNECT TK solenoid. GO to KG5.
	Resistance is less than 50 ohms or greater than 100 ohms ▶	REPLACE TK solenoid. REPEAT Quick Test.

Idle Speed Control (Throttle Kicker) — Pinpoint Test — KG

TEST STEP	RESULT ▶	ACTION TO TAKE
KG5 ENTER OUTPUT STATE CHECK (REFER TO APPENDIX)		
• Key Off, wait 10 seconds. • DVOM on 20V scale. • Connect DVOM negative test lead to STO and positive test lead to battery positive. • Jumper STI to signal return. • Perform Key On, Engine Off Self-Test until the completion of the Continuous Test Codes. • DVOM will indicate zero volts. • Depress and release the throttle. • Did DVOM reading change to a high voltage reading?	Yes ▶	REMAIN in Output State Check. GO to KG6.
	No ▶	DEPRESS throttle to WOT and release. If STO voltage does not go high, GO to Pinpoint Test Step Q40. Leave equipment hooked up.
KG6 CHECK TK SOLENOID ELECTRICAL OPERATION		
• DVOM on 20V scale. • Connect DVOM positive test lead to circuit VPWR on solenoid and negative test lead to output circuit TK on solenoid. • While observing DVOM, depress and release the throttle several times to cycle output On and Off. • TK output cycles On and Off?	Yes ▶	GO to KG7.
	No ▶	GO to KG9.
KG7 CHECK TK SOLENOID FOR VACUUM CYCLING		
• Install vacuum pump to the TK solenoid vacuum supply port and install a vacuum gauge to the output port. Apply 6 in Hg minimum. • While cycling outputs On and Off (by depressing and releasing throttle) observe the vacuum gauge at the output. **NOTE: Maintain vacuum at source.**	Vacuum output cycles On and Off ▶	GO to KG8.
	Vacuum output does not cycle On and Off ▶	REPLACE solenoid. REPEAT Quick Test.

Idle Speed Control (Throttle Kicker) — Pinpoint Test — KG

TEST STEP	RESULT ▶	ACTION TO TAKE
KG8 CHECK MANIFOLD VACUUM LINES FOR BLOCKAGE OR LEAKS		
• With vacuum lines disconnected at TK solenoid, check for vacuum. • Start engine.	Vacuum present ▶	EEC-IV system OK. REFER to Section 4 for idle adjustment.
	No vacuum present ▶	SERVICE vacuum source blockage or leak. REPEAT Quick Test.
KG9 CHECK VOLTAGE OF VPWR CIRCUIT		
• Key On, Engine Off. • DVOM on 20V scale. • Connect DVOM positive test lead to VPWR circuit and negative test lead to ground. • Measure voltage on TK solenoid VPWR circuit.	Voltage reading is 10.5V or less ▶	SERVICE harness open circuit. REPEAT Quick Test.
	Voltage reading is 10.5V or greater ▶	GO to KG10.
KG10 CHECK CONTINUITY OF TK CIRCUIT		
• Key Off, wait 10 seconds. • Disconnect processor and inspect both 60 Pin connectors. • Connect Breakout box to harness, leave processor disconnected. • DVOM on 200 ohm scale. • Measure continuity of test Pin 53 to TK circuit at harness connector.	Resistance reading is 5 ohms or greater ▶	SERVICE harness open circuit. REPEAT Quick Test.
	Resistance reading is less than 5 ohms ▶	GO to KG11.
KG11 CHECK FOR SHORT TO POWER		
• Key Off, wait 10 seconds. • DVOM on 200,000 ohm scale. • Leave Breakout box installed and processor disconnected. • With TK solenoid disconnected, measure resistance from test Pin 53 to test Pins 37 and 57.	Resistance reading is 10,000 ohms or greater ▶	GO to KG12.
	Resistance reading is less than 10,000 ohms ▶	SERVICE short to power. REPEAT Quick Test.

Idle Speed Control (Throttle Kicker) — Pinpoint Test — KG

TEST STEP	RESULT ▶	ACTION TO TAKE
KG12 CHECK FOR SHORT TO GROUND		
• Key Off, wait 10 seconds. • Leave Breakout box installed and processor disconnected. • DVOM on 200,000 ohm scale. • Measure resistance of test Pin 53 to test Pins 40, 46 and 60.	Resistance reading is less than 10,000 ohms ▶	SERVICE short to ground. REPEAT Quick Test.
	Resistance reading is 10,000 ohms or greater ▶	REPLACE processor. REPEAT Quick Test.

Variable Voltage Choke (VVC) — Pinpoint Test — KJ

KJ

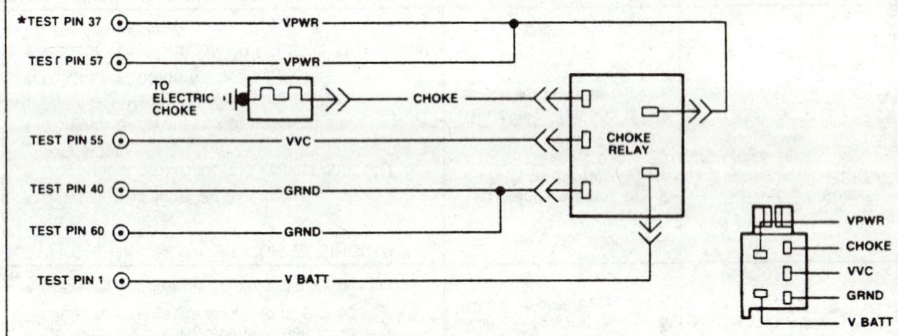

*TEST PINS LOCATED ON BREAKOUT BOX.
ALL HARNESS CONNECTORS VIEWED INTO MATING SURFACE.

STOP-WARNING

This Pinpoint Test is intended to diagnose only the following:

- Harness circuits: VPWR, Choke, VVC, Grnd., and V Batt.
- Choke Relay

Variable Voltage Choke (VVC) — Pinpoint Test — KJ

TEST STEP	RESULT ▶	ACTION TO TAKE
KJ1 MEASURE CHOKE VOLTAGE		
• DVOM on 20V scale. • Perform Engine Running Self-Test. • Measure choke voltage.	Choke voltage is 10.5V or greater ▶	CHECK choke plates for binding and sticking. EEC-IV system OK. REFER to Section 4 for choke diagnostics.
	Choke voltage is less than 10.5V ▶	GO to KJ2.
KJ2 CHECK FOR VOLTAGE ON V BATT CIRCUIT		
• Key On, Engine Off. • DVOM on 20V scale. • Connect DVOM positive test lead to V Batt. at relay harness connector and negative test lead to ground at relay harness connector.	Voltage is less than 10.5V ▶	SERVICE open in choke power circuit. REPEAT Quick Test.
	Voltage is 10.5V or greater ▶	GO to KJ3.
KJ3 CHECK FOR SHORT IN GROUND CIRCUIT		
• Key On, Engine Off. • DVOM on 20V scale. • Measure voltage with DVOM from test Pins 40 and 60 to negative battery post.	Less than 1.0V on both circuits ▶	Ground OK. GO to KJ4.
	1.0V or greater on one or both circuits ▶	CORRECT faulty ground circuit (high resistance indicated by voltage reading). REPEAT Quick Test.

Variable Voltage Choke (VVC) — Pinpoint Test — KJ

TEST STEP	RESULT ▶	ACTION TO TAKE
KJ4 ENTER OUTPUT STATE CHECK (REFER TO APPENDIX)		
• Key Off, wait 10 seconds. • DVOM on 20V scale. • Connect DVOM negative test lead to STO and positive test lead to battery positive. • Jumper STI to signal return. • Perform Key On, Engine Off Self-Test until the completion of the Continuous Test Codes. • DVOM will indicate zero volts. • Depress and release the throttle. • Did DVOM reading change to a high voltage reading?	Yes ▶	REMAIN in Output State Check. GO to KJ5.
	No ▶	DEPRESS throttle to WOT and RELEASE. If STO voltage does not go high, GO to Pinpoint Test Step Q40.
KJ5 CHECK VVC OUTPUT ELECTRICAL OPERATION		
• Key On, Engine Off. • DVOM on 20V scale. • Connect DVOM positive test lead to V Batt. on VVC relay and negative test lead to VVC output at choke. • While observing DVOM depress and release the throttle several times to cycle output On and Off. • VVC output cycles On and Off?	Yes ▶	EEC-IV system OK. REFER to Section 4 for choke diagnostics.
	No ▶	REPLACE choke relay. REPEAT Quick Test.

Temperature Compensated Pump — Pinpoint Test — KK

KK

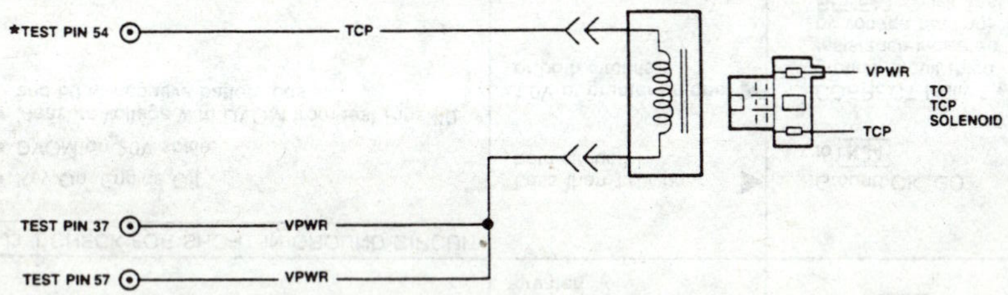

*TEST PINS LOCATED ON BREAKOUT BOX.
ALL HARNESS CONNECTORS VIEWED INTO MATING SURFACE.

STOP-WARNING

This Pinpoint Test is intended to diagnose only the following:

- Harness circuits: TCP and VPWR
- TCP Solenoid
- Processor Assembly

Temperature Compensated Pump — Pinpoint Test — KK

TEST STEP	RESULT ▶	ACTION TO TAKE
KK1 MEASURE TCP SOLENOID RESISTANCE		
• Key Off, wait 10 seconds. • DVOM on 200 ohm scale. • Disconnect TCP solenoid connector and measure solenoid resistance.	Resistance is between 50-100 ohms ▶	CONNECT TCP solenoid. GO to KK2.
	Resistance is less than 50 ohms or greater than 100 ohms ▶	REPLACE TCP solenoid. REPEAT Quick Test.
KK2 ENTER OUTPUT STATE CHECK (REFER TO APPENDIX)		
• Key Off, wait 10 seconds. • DVOM on 20V scale. • Connect DVOM negative test lead to STO and positive test lead to battery positive. • Jumper STI to signal return. • Perform Key On, Engine Off Self-Test until the completion of the Continuous Test Codes. • DVOM will indicate zero volts. • Depress and release the throttle. • Did DVOM reading change to a high voltage reading?	Yes ▶	REMAIN in Output State Check. GO to KK3.
	No ▶	DEPRESS throttle to WOT and release. If STO voltage does not go high, GO to Pinpoint Test Step Q40. Leave equipment hooked up.
KK3 CHECK TCP SOLENOID ELECTRICAL OPERATION		
• DVOM on 20V scale. • Connect DVOM positive test lead to circuit VPWR on TCP solenoid and negative test lead to output circuit TCP. • While observing DVOM depress and release the throttle several times to cycle output On and Off. • Does TCP output cycle On and Off?	Yes ▶	GO to KK4.
	No ▶	GO to KK6.

Temperature Compensated Pump — Pinpoint Test KK

TEST STEP	RESULT ▶	ACTION TO TAKE
KK4 CHECK TCP SOLENOID FOR VACUUM CYCLING		
• Install vacuum pump to the TCP solenoid vacuum supply port and install a vacuum gauge to the output port. Apply 6 in Hg minimum. • While cycling outputs On and Off (by depressing and releasing throttle) observe the vacuum gauge at the output. **NOTE: Maintain vacuum at source.**	Vacuum output cycles On and Off ▶ Vacuum output does not cycle On and Off ▶	GO to KK5. REPLACE solenoid. REPEAT Quick Test.
KK5 CHECK MANIFOLD VACUUM LINES FOR BLOCKAGE OR LEAKS		
• With vacuum lines disconnected at TCP solenoid, check for vacuum. • Start engine.	Vacuum present ▶ No vacuum present ▶	EEC-IV system OK. REFER to Section 11 for TCP diagnostics. SERVICE vacuum source blockage or leak. REPEAT Quick Test.
KK6 CHECK VOLTAGE OF VPWR CIRCUIT		
• Key On, Engine Off. • DVOM on 20V scale. • Connect DVOM positive test lead to VPWR circuit and negative test lead to ground. • Measure voltage on TCP solenoid VPWR circuit.	Voltage readings is 10.5V or less ▶ Voltage reading is 10.5V or greater ▶	SERVICE harness open circuit. REPEAT Quick Test. GO to KK7.
KK7 CHECK CONTINUITY OF TCP CIRCUIT		
• Key Off, wait 10 seconds. • Disconnect processor and inspect both 60 Pin connectors. • Connect Breakout box to harness, leave processor disconnected. • DVOM on 200 ohm scale. • Measure continuity of test Pin 54 to TCP circuit at harness connector.	Resistance reading is 5 ohms or greater ▶ Resistance reading is less than 5 ohms ▶	SERVICE harness open circuit. REPEAT Quick Test. GO to KK8.

Temperature Compensated Pump — Pinpoint Test KK

TEST STEP	RESULT ▶	ACTION TO TAKE
KK8 CHECK FOR SHORT TO POWER		
• Key Off, wait 10 seconds. • DVOM on 200,000 ohm scale. • Leave Breakout box installed and processor disconnected. • With TCP solenoid disconnected, measure resistance from test Pin 54 to test Pins 37 and 57.	Resistance reading is 10,000 ohms or greater ▶ Resistance reading is less than 10,000 ohms ▶	GO to KK9. SERVICE short to power. REPEAT Quick Test.
KK9 CHECK FOR SHORT TO GROUND		
• Key Off, wait 10 seconds. • Leave Breakout box installed and processor disconnected. • DVOM on 200,000 ohm scale. • Measure resistance of test Pin 54 to test Pin 40, 46 and 60.	Resistance reading is less than 10,000 ohms ▶ Resistance reading is 10,000 ohms or greater ▶	SERVICE short to ground. REPEAT Quick Test. REPLACE processor. REPEAT Quick Test.

Shift Indicator Light (SIL) — Pinpoint Test — KL

KL

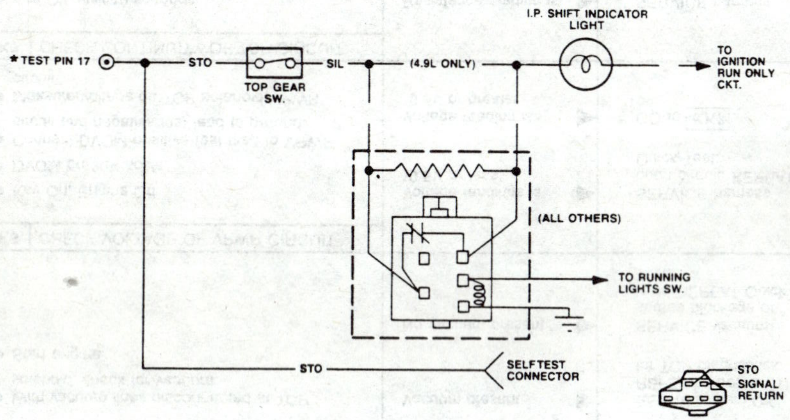

*TEST PINS LOCATED ON BREAKOUT BOX.
ALL HARNESS CONNECTORS VIEWED INTO MATING SURFACE.

STOP-WARNING

This Pinpoint Test is intended to diagnose only the following:

- Top Gear Switch
- Shift Indicator Light Bulb and Fuse
- Harness Circuits: SIL and STO

Shift Indicator Light (SIL) — Pinpoint Test — KL

TEST STEP	RESULT ▶	ACTION TO TAKE
KL1 DETERMINE TYPE OF FAULT		
NOTE: **Before proceeding with this Pinpoint Test, Quick Test Steps 1 through 5 must be completed and all fault codes serviced. Code 11 must be present for both Key On, Engine Off and Engine Running Quick Tests.**		
• Verify SIL fault by drive evaluating vehicle.	SIL is always Off ▶	GO to KL2.
	SIL is not always Off ▶	GO to KL4.
KL2 CHECK CONTINUITY OF STO CIRCUIT TO TOP GEAR SWITCH		
• Key Off, wait 10 seconds. • Disconnect harness from top gear switch. Inspect for damaged pins, corrosion, loose wires, etc. Service as necessary. • DVOM on 200 ohm scale. • Measure resistance of top gear switch harness connector STO circuit to Self-Test connector.	Reading is less than 5 ohms ▶	GO to KL3.
	Reading is 5 ohms or greater ▶	SERVICE circuit open. RECONNECT top gear switch. RE-EVALUATE SIL function.
KL3 CHECK TOP GEAR SWITCH, CONTINUITY TRANSMISSION IN FIRST GEAR		
• Key Off, harness disconnected from top gear switch. • **Transmission in first gear.** • DVOM on 200 ohm scale. • Measure resistance of top gear switch.	Reading is less than 5 ohms ▶	CHECK fuse and bulb, if OK SERVICE circuit open in SIL circuit and Ign. RECONNECT top gear switch. RE-EVALUATE SIL.
	Reading is 5 ohms or greater ▶	REPLACE top gear switch. RE-EVALUATE SIL.
KL4 CHECK TOP GEAR SWITCH, TRANSMISSION IN TOP GEAR		
• Key Off, wait 10 seconds. • Disconnect harness from top gear switch. Inspect for damaged pins, corrosion, loose wires, etc. Service as necessary. • DVOM on 200,000 ohm scale. • **Transmission in top gear.** • Measure resistance of top gear switch.	Reading is less than 10,000 ohms ▶	REPLACE top gear switch. RE-EVALUATE SIL.
	Reading is 10,000 ohms or greater ▶	SERVICE short in SIL circuit. RECONNECT top gear switch. RE-EVALUATE SIL.

WOT A/C Cutoff (WAC) | Pinpoint Test | KM

KM | 1.6L AND 2.3L HSC ENGINES

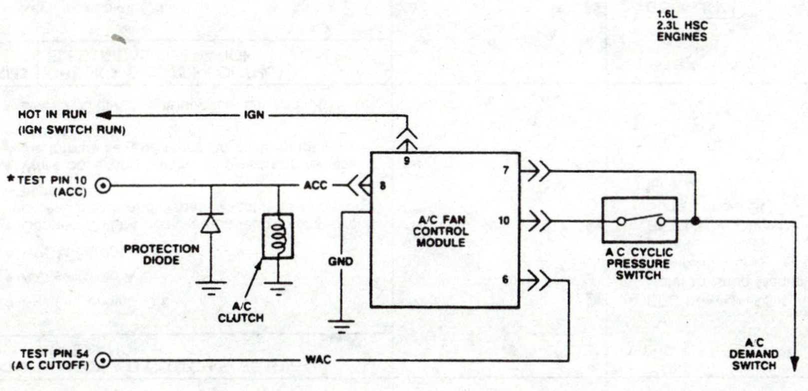

*TEST PINS LOCATED ON BREAKOUT BOX.
ALL HARNESS CONNECTORS VIEWED INTO MATING SURFACE.

STOP-WARNING

This Pinpoint Test is intended to diagnose only the following:

- Harness Circuits: Ign.. Acc.. Gnd.. WAC. and A/C Demand Switch
- A/C Fan Electrical Controller
- Processor Assembly

WOT A/C Cutoff (WAC) | Pinpoint Test | KM

KM | 2.3L OHC ENGINE

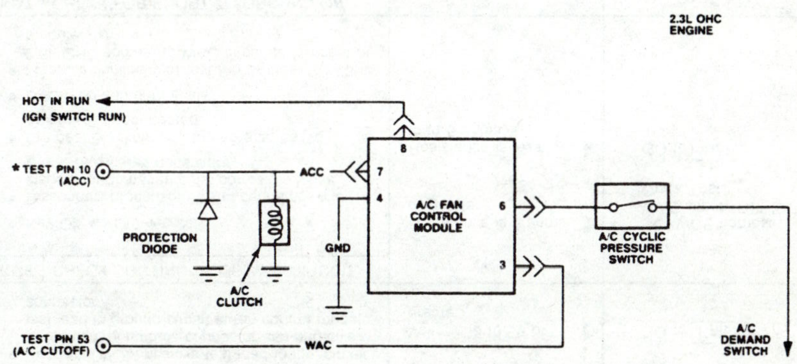

*TEST PINS LOCATED ON BREAKOUT BOX.
ALL HARNESS CONNECTORS VIEWED INTO MATING SURFACE.

STOP-WARNING

This Pinpoint Test is intended to diagnose only the following:

- Harness Circuits: Ign.. Acc.. Gnd.. WAC. and A C Demand Switch
- A C Fan Electrical Controller
- Processor Assembly

WOT A/C Cutoff (WAC) — Pinpoint Test — KM

TEST STEP	RESULT ▶	ACTION TO TAKE
KM1 ENTER OUTPUT STATE CHECK (REFER TO APPENDIX)		
• Key Off, wait 10 seconds. • DVOM on 20V scale. • Connect DVOM negative test lead to STO and positive test lead to battery positive. • Jumper STI to signal return. • Perform Key On, Engine Off Self-Test until the completion of the Continuous Test Codes. • DVOM will indicate zero volts. • Depress and release the throttle. • Did DVOM reading change to a high voltage reading?	Yes ▶ No ▶	REMAIN in Output State Check. GO to KM2. DEPRESS throttle to WOT and RELEASE. If STO voltage does not go high, GO to Pinpoint Test Step Q40. Leave equipment hooked up.
KM2 CHECK A/C CLUTCH ELECTRICAL OPERATION		
• Key On, engine Off. • A/C switch to A/C. • DVOM on 20V scale. • Connect DVOM positive test lead to acc. circuit on A/C clutch and negative test lead to battery negative. • While observing DVOM, depress and release the throttle several times (to cycle output On and Off). • Does A/C clutch output cycle On and Off?	Yes ▶ No ▶	EEC-IV system OK. Refer to Shop Manual Group 36. REMAIN in Output State Check. GO to KM3.
KM3 CHECK A/C CUTOFF OUTPUT ELECTRICAL OPERATION		
• Key On, engine Off. • DVOM on 20V scale. • Connect DVOM positive test lead to ign. circuit on fan control module and negative lead to WAC circuit on fan control module. • While observing DVOM, depress and release the throttle several times to cycle the output On and Off. • Does A/C cutoff cycle On and Off?	Yes ▶ No ▶	GO to KM4. GO to KM5.

WOT A/C Cutoff (WAC) — Pinpoint Test — KM

TEST STEP	RESULT ▶	ACTION TO TAKE
KM4 CHECK CONTINUITY OF GROUND CIRCUIT TO FAN CONTROL MODULE		
• Key Off, wait 10 seconds. • DVOM on 200 ohm scale. • Measure continuity of ground circuit to fan control module connector.	Resistance reading is 5 ohms or greater ▶ Resistance reading is less than 5 ohms ▶	SERVICE harness circuit. REPEAT Quick Test. REPLACE fan control module. REPEAT Quick Test.
KM5 CHECK FOR IGN. CIRCUIT VOLTAGE		
• Key On, engine Off. • DVOM on 20V scale. • Connect DVOM positive test lead to ign. circuit at fan control module connector and negative test lead to ground circuit at fan control module connector.	Voltage is less than 10.5V ▶ Voltage is 10.5V or greater ▶	SERVICE open in ign. circuit. REPEAT Quick Test. GO to KM6.
KM6 CHECK CONTINUITY OF WAC CIRCUIT		
• Key Off, wait 10 seconds. • Disconnect processor 60 Pin connector and inspect for damaged pins, corrosion, loose wires. Service as necessary. • Connect Breakout box to harness. Leave processor disconnected. • DVOM on 200 ohm scale. • Measure continuity of test Pin 54 (test Pin 53 for 2.3L OHC engine) to WAC circuit at fan control module.	Resistance is 5 ohms or greater ▶ Resistance is less than 5 ohms ▶	SERVICE harness WAC circuit. REPEAT Quick Test. GO to KM7.
KM7 CHECK FOR SHORT TO GROUND ON WAC CIRCUIT		
• Key Off, wait 10 seconds. • Leave Breakout box installed and processor disconnected. • DVOM on 200,000 ohm scale. • Measure resistance of test Pin 54 (test Pin 53 for 2.3L OHC engine) to test Pins 40, 46 and 60.	Resistance reading is less than 10,000 ohms ▶ Resistance reading is 10,000 ohms or greater ▶	GO to KM8. REPLACE processor. REPEAT Quick Test.

WOT A/C Cutoff (WAC) — Pinpoint Test — KM

TEST STEP		RESULT ▶	ACTION TO TAKE
KM8	CHECK FOR SHORT TO GROUND WITH FAN CONTROL MODULE DISCONNECTED		
	• Key Off, wait 10 seconds. • Leave Breakout box installed and processor disconnected. • DVOM on 200,000 ohm scale. • Disconnect fan control module. • Measure resistance of test Pin 54 (test Pin 53 for 2.3L OHC engine) to test Pins 40, 46 and 60.	All resistance readings 10,000 ohms or greater ▶	REPLACE fan control module. REPEAT Quick Test.
		Any resistance reading is less than 10,000 ohms ▶	SERVICE harness short. REPEAT Quick Test.

Turbo Boost — Pinpoint Test — KN

KN

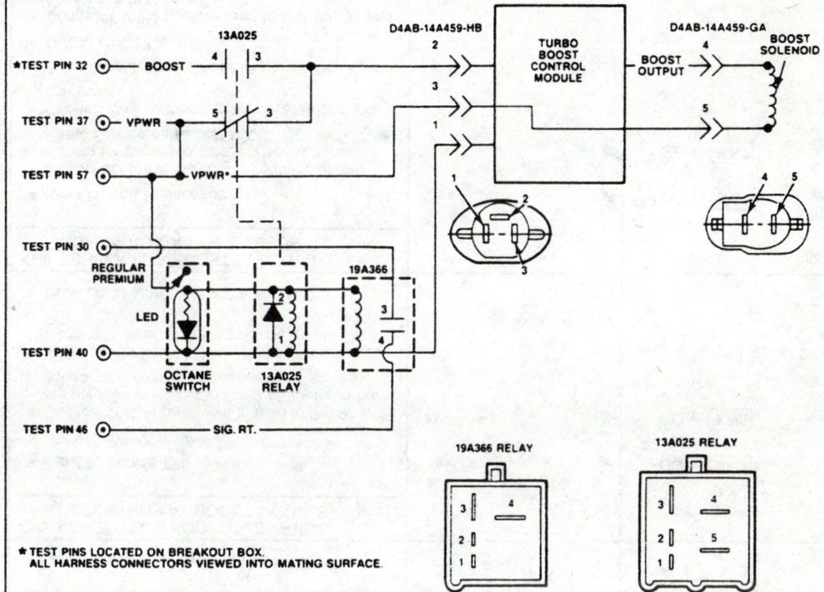

★ TEST PINS LOCATED ON BREAKOUT BOX.
ALL HARNESS CONNECTORS VIEWED INTO MATING SURFACE.

STOP-WARNING

You may have reached this point because of a fault in a non-EEC System and or component. To prevent the replacement of good components, be aware that the following non-EEC areas may be at fault:

- Contaminated Fuel (premium fuel required).
- Octane Switch must be in the premium position.

This Pinpoint Test is intended to diagnose only the following:

- Harness circuits: VPWR, Boost, Grd., Signal Return, Octane Switch, Input (NOS), and Boost Output from the Turbo Boost Control Module.

NOTE: The boost system is protected electrically by a fuse. Refer to the Owner's Manual for checkout and replacement before proceeding with the Pinpoint Tests.

Turbo Boost — Pinpoint Test — KN

TEST STEP	RESULT ▶	ACTION TO TAKE
KN1 TEST FOR CODE 67		
• Place octane switch in regular position. • Perform Key On, Engine Off Self-Test. • Is code 67 present?	Yes ▶	GO to KN2.
	No ▶	GO to KN15.
KN2 MEASURE BOOST SOLENOID RESISTANCE		
• Key Off, wait 10 seconds. • DVOM on 200 ohm scale. • Disconnect boost solenoid connector and measure the solenoid resistance.	Resistance is between 1.0-2.0 ohms ▶	GO to KN3.
	Resistance is less than 1.0 or greater than 2.0 ohms ▶	REPLACE solenoid. RETEST.
KN3 ENTER OUTPUT STATE CHECK (REFER TO APPENDIX)		
• Place octane switch in premium position. • Key Off, wait 10 seconds. • DVOM on 20V scale. • Connect DVOM negative test lead to STO and positive test lead to battery positive. • Jumper STI to signal return. • Perform Key On, Engine Off Self-Test until the completion of the Continuous Test Codes. • DVOM will indicate zero volts. • Depress and release the throttle. • Did DVOM reading change to a high voltage reading?	Yes ▶	REMAIN in Output State Check. GO to KN4.
	No ▶	DEPRESS throttle to WOT and release. If STO voltage does not go high, GO to Pinpoint Test Step Q40. Leave equipment hooked up.

Turbo Boost — Pinpoint Test — KN

TEST STEP	RESULT ▶	ACTION TO TAKE
KN4 CHECK BOOST OUTPUT ELECTRICAL OPERATION		
• Key On, Engine Off. • DVOM on 20V scale. • Connect DVOM positive test lead to circuit VPWR on boost connector and negative test lead to boost output on boost connector. • While observing DVOM, depress and release the throttle several times to cycle output On and Off. • Does boost output cycle On and Off?	Yes ▶	GO to KN21.
	No ▶	GO to KN5.
KN5 CHECK FOR BOOST ELECTRICAL OPERATION AT BOOST MODULE		
• Key On, Engine Off. • DVOM on 20V scale. • Connect DVOM positive test lead to Pin 2, and the negative test lead to Pin 3 of the turbo boost control module harness connector. • While observing DVOM, depress and release the throttle several times to cycle output On and Off. • Does boost output cycle On and Off?	Yes ▶	GO to KN8.
	No ▶	GO to KN6.
KN6 CHECK FOR BOOST ELECTRICAL OPERATION AT PROCESSOR		
• Key Off, wait 10 seconds. • Disconnect processor 60 Pin connector and inspect for damaged pins, corrosion, loose wires. Service as necessary. • Connect Breakout box to harness. Reconnect processor. • Enter Output State Check (Refer to Step KN3). • DVOM on 20V scale. • Connect DVOM positive lead to test Pin 37 and negative lead to test Pin 32. • While observing DVOM, depress and release the throttle several times to cycle output On and Off. • Does boost output cycle On and Off?	Yes ▶	GO to KN11.
	No ▶	GO to KN7.

Turbo Boost — Pinpoint Test — KN

TEST STEP	RESULT ▶	ACTION TO TAKE
KN7 CHECK FOR SHORT TO POWER AND GROUND		
• Key Off, wait 10 seconds. • Remove jumper from STI and Sig. Rtn. • Breakout box connected. • DVOM to 200,000 ohm scale. • Measure resistance from test Pin 32 to test Pins 37, 46 and 60.	Any reading less than 10,000 ohms ▶	SERVICE short. REPEAT Quick Test.
	All readings 10,000 ohms or greater ▶	REPLACE processor. REPEAT Quick Test.
KN8 CHECK FOR POWER AND GROUND AT TURBO BOOST CONTROL MODULE		
• Key On, Engine Off. • DVOM on 20V scale. • Connect positive test lead to Pins 3 and 5 and negative lead to Pin 1 of turbo boost control module harness.	Voltage reading is 10.5V or greater ▶	GO to KN9.
	Voltage reading is less than 10.5V ▶	SERVICE open in harness. REPEAT Quick Test.
KN9 CHECK CONTINUITY OF BOOST OUTPUT AND VPWR FROM TURBO BOOST MODULE TO BOOST SOLENOID		
• Key Off, wait 10 seconds. • DVOM on 200 ohm scale. • Measure resistance from boost solenoid connector, boost output to Pin 4 and VPWR at the boost solenoid connector to Pin 5 at the turbo boost module connector.	All readings less than 5.0 ohms ▶	GO to KN10.
	Any reading greater than 5.0 ohms ▶	SERVICE open in harness. REPEAT Quick Test.
KN10 CHECK FOR SHORT TO POWER IN BOOST OUTPUT HARNESS		
• Key Off, wait 10 seconds. • DVOM on 200,000 ohm scale. • Disconnect turbo boost control module and boost solenoid. • Measure resistance between Pins 4 and 5 of turbo boost control module harness connector.	Reading is 10,000 ohms or greater ▶	REPLACE turbo boost control module. REPEAT Quick Test.
	Reading is less than 10,000 ohms ▶	SERVICE short in harness. REPEAT Quick Test.

Turbo Boost — Pinpoint Test — KN

TEST STEP	RESULT ▶	ACTION TO TAKE
KN11 CHECK CONTINUITY OF BOOST FROM RELAY TO BREAKOUT BOX		
• Key Off, wait 10 seconds. • DVOM on 200 ohm scale. • Measure resistance from Pin 4 of boost relay to test Pin 32.	Reading is less than 5.0 ohms ▶	GO to KN12.
	Reading is greater than 5.0 ohms ▶	SERVICE open in harness. REPEAT Quick Test.
KN12 CHECK CONTINUITY OF BOOST FROM BOOST RELAY TO TURBO BOOST CONTROL MODULE		
• Key Off, wait 10 seconds. • DVOM on 200 ohm scale. • Measure resistance from Pin 3 of boost relay to Pin 2 of turbo boost control module.	Reading is less than 5.0 ohms ▶	GO to KN13.
	Reading is 5.0 ohms or greater ▶	SERVICE open in harness. REPEAT Quick Test.
KN13 CHECK FOR SHORT TO POWER AND GROUND		
• Key Off, wait 10 seconds. • Breakout box installed. • DVOM to 200,000 ohm scale. • Disconnect turbo boost control module and boost relay. • Measure resistance from Pin 3 of boost relay to test Pins 37, 40 and 46.	Any reading is less than 10,000 ohms ▶	SERVICE short in harness. REPEAT Quick Test.
	All readings are 10,000 ohms or greater ▶	GO to KN14.
KN14 CHECK FOR POWER AND GROUND AT THE BOOST RELAY		
• Key On, engine Off. • Octane switch in premium position. • DVOM to 20V scale. • Connect positive lead to Pin 2 and negative lead to Pin 1 of the turbo boost relay.	Voltage reading is 10.5V or greater ▶	REPLACE boost relay. REPEAT Quick Test.
	Voltage reading is less than 10.5V ▶	SERVICE open in harness. REPEAT Quick Test.

Turbo Boost — Pinpoint Test — KN

TEST STEP	RESULT	ACTION TO TAKE
KN15 CHECK OCTANE SWITCH INPUT FOR INPUT CHANGE		
• Key Off, wait 10 seconds. • Disconnect processor 60 Pin connector and inspect for damaged pins, corrosion, loose wires. Service as necessary. • Install Breakout box. Reconnect processor. • DVOM to 20V scale. • Connect positive test lead to test Pin 30 and negative test lead to test Pin 46. • Key On, engine Off. • Cycle octane switch several times while observing DVOM. • Does voltage change from 0V to 5V?	Yes ▶ No ▶	REPLACE processor. REPEAT Quick Test. GO to KN16.
KN16 CHECK CONTINUITY BETWEEN PROCESSOR AND OCTANE SWITCH RELAY OCTANE SWITCH		
• Key Off, wait 10 seconds. • Breakout box installed. • DVOM on 200 ohm scale. • Measure resistance between test Pin 30 and Pin 3 on the octane switch relay.	Reading is less than 5.0 ohms ▶ Reading is 5.0 ohms or greater ▶	GO to KN17. SERVICE open in harness. REPEAT Quick Test.
KN17 CHECK CONTINUITY BETWEEN PROCESSOR AND OCTANE SWITCH RELAY SIGNAL RETURN		
• Key Off, wait 10 seconds. • Breakout box installed. • DVOM on 200 ohm scale. • Measure resistance between test Pin 46 and Pin 4 on the octane switch relay.	Reading is less than 5.0 ohms ▶ Reading is 5.0 ohms or greater ▶	GO to KN18. SERVICE open in harness. REPEAT Quick Test.
KN18 CHECK FOR POWER AND GROUND AT THE OCTANE SWITCH RELAY		
• Key On, engine Off. • DVOM on 20V scale. • Octane switch in premium position. • Measure voltage between Pin 1 and Pin 2 on the octane switch relay harness connector.	Voltage reading is 10.5V or greater ▶ Voltage reading is less than 10.5V ▶	REPLACE octane switch relay. REPEAT Quick Test. GO to KN19.

Turbo Boost — Pinpoint Test — KN

TEST STEP	RESULT	ACTION TO TAKE
KN19 CHECK CONTINUITY FROM OCTANE SWITCH RELAY TO BATTERY GROUND		
• Key On, Engine Off. • DVOM on 200 ohm scale. • Measure resistance between Pin 1 on octane switch relay and battery ground.	Reading is less than 5.0 ohms ▶ Reading is 5.0 ohms or greater ▶	GO to KN20. SERVICE open in harness. REPEAT Quick Test.
KN20 CHECK FOR POWER AT OCTANE SWITCH		
• Key On, Engine Off. • DVOM on 20V scale. • Connect positive lead to input of octane switch and negative lead to battery ground. • Measure voltage.	Voltage reading is 10.5V or greater ▶ Voltage reading is less than 10.5V ▶	REPLACE octane switch. REPEAT Quick Test. SERVICE open circuit. REPEAT Quick Test.
KN21 CHECK BOOST SOLENOID FUNCTION		
• Remain in output state check. • Place octane switch in the premium position. • Disconnect black solenoid vacuum hose at the turbocharger end. **(Not the black and yellow stripe hose.)** • Attach a vacuum pump to the hose. • Depress the throttle once to cycle the solenoid closed. • Apply vacuum to the solenoid. • Depress the throttle once again to cycle the solenoid open and release trapped vacuum. • Did the solenoid hold and then release vacuum?	Yes ▶ No ▶	EEC-IV system OK. REFER to Shop Manual for boost diagnostics. REPLACE solenoid. RETEST.

Exhaust Heat Control — Pinpoint Test — KP

KP

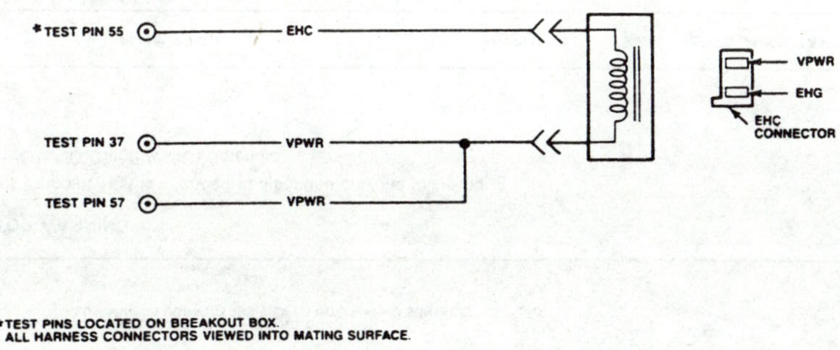

STOP-WARNING

This Pinpoint Test is intended to diagnose only the following:

- Exhaust Heat Solenoid
- Vacuum Routing

Exhaust Heat Control — Pinpoint Test — KP

TEST STEP	RESULT ▶	ACTION TO TAKE
KP1 ENTER OUTPUT STATE CHECK (REFER TO APPENDIX)		
• Key Off, wait 10 seconds. • DVOM on 20V scale. • Connect DVOM negative test lead to STO and positive test lead to battery positive. • Jumper STI to signal return. • Perform Key On, Engine Off Self-Test until the completion of the Continuous Test Codes. • DVOM will indicate zero volts. • Depress and release the throttle. • Did DVOM reading change to a high voltage reading?	Yes ▶ No ▶	REMAIN in Output State Check. GO to KP3. DEPRESS throttle to WOT and release. If STO voltage does not go high, GO to Pinpoint Test Step Q40. Leave equipment hooked up.
KP2 CHECK EXHAUST HEAT SOLENOID FOR VACUUM CYCLING		
• Install vacuum pump to the exhaust heat solenoid vacuum supply port and install a vacuum gauge to the output port. Apply 6 in Hg minimum. • While cycling outputs On and Off (by depressing and releasing throttle) observe the vacuum gauge at the output. **NOTE: Maintain vacuum at source.**	Vacuum output cycles On and Off ▶ Vacuum output does not cycle On and Off ▶	GO to KP3. REPLACE solenoid. REPEAT Quick Test.
KP3 CHECK MANIFOLD VACUUM LINES FOR BLOCKAGE OR LEAKS		
• With vacuum lines disconnected at exhaust heat solenoid, check for vacuum. • Start engine.	Vacuum present ▶ No vacuum present ▶	EEC-IV system OK. REFER to Section 5 for exhaust heat diagnostics. SERVICE vacuum source blockage or leak. REPEAT Quick Test.

WOT A/C Cutoff (WAC) — Pinpoint Test — KR

KR

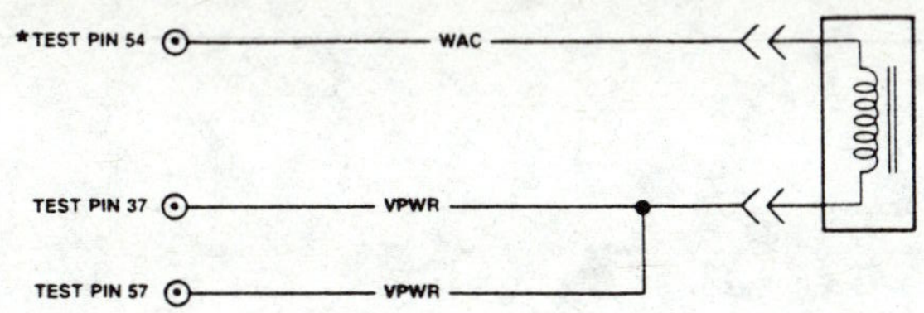

*TEST PINS LOCATED ON BREAKOUT BOX.
ALL HARNESS CONNECTORS VIEWED INTO MATING SURFACE.

STOP-WARNING

This Pinpoint Test is intended to diagnose only the following:

- WOT A/C Cutoff (WAC) Solenoid

TEST STEP	RESULT	ACTION TO TAKE
KR1 CYCLE WAC RELAY		
• Key Off, wait 10 seconds. • Disconnect processor 60 Pin connector and inspect for damaged pins, corrosion, loose wires. Service as necessary. • Connect Breakout box to harness. Leave processor disconnected. • Key On, Engine Off. • DVOM on 20V scale. • Connect DVOM negative test lead to ground and positive test lead to A/C clutch. • Turn A/C demand switch to On. • Jumper test Pin 54 to test Pin 60. • DVOM will indicate zero volts. • Depress and release throttle. • Did DVOM change to a high voltage reading?	Yes ▶	EEC-IV system is OK. REFER to Shop Manual Group 36 for A/C system diagnostics.
	No ▶	REPLACE WAC relay. REPEAT Quick Test.

Air Management System — Pinpoint Test — KS

KS

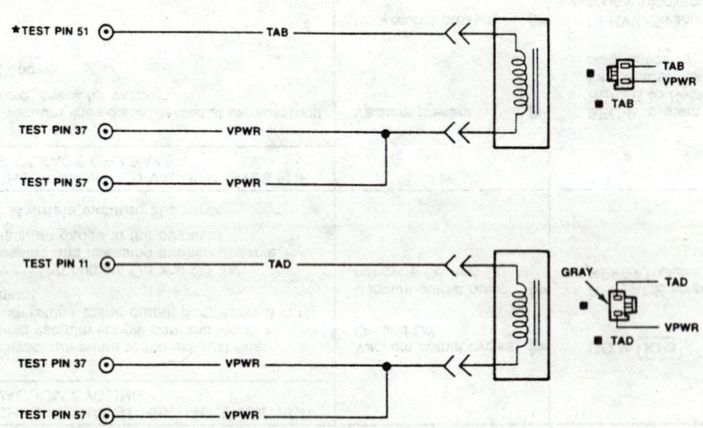

*TEST PINS LOCATED ON BREAKOUT BOX.
ALL HARNESS CONNECTORS VIEWED INTO MATING SURFACE.

STOP-WARNING

You may have reached this point because of a fault in a non-EEC System and/or component. To prevent the replacement of good components, be aware that the following non-EEC areas may be at fault:

- Thermactor System
  - — Belt
  - — Pump
  - — Valve

This Pinpoint Test is intended to diagnose only the following:

- TAB and TAD Solenoid Valve Assemblies
- Vacuum Supply

Air Management System — Pinpoint Test — KS

TEST STEP	RESULT ▶	ACTION TO TAKE
KS1 SERVICE CODES: 44, 45 AND 46: VERIFY VACUUM LINE ROUTING		
• Verify proper vacuum line routing to the TAB TAD solenoids and to the bypass diverter valve. Refer to VECI decal. • Check for kinked or blocked vacuum lines. • Check for kinked or blocked air hoses. • Check for disconnected vacuum lines. • Are visual checks satifactory?	No ▶	SERVICE routing or faults. REPEAT Quick Test.
	Yes ▶	Service code 44: GO to KS4. Service code 45: GO to KS2. Service code 46: GO to KS3.
KS2 ATTEMPT TO ELIMINATE SERVICE CODE 45 (TAD ONLY)		
• Disconnect vacuum line on diverter valve and cap vacuum line. • Key Off, wait 10 seconds. • Repeat Engine Running Self-Test and record service codes. • Is code 45 present?	Yes ▶	EEC-IV system OK. REFER to Section 2 for diverter valve or check valve diagnostics.
	No ▶	GO to KS4.
KS3 ATTEMPT TO ELIMINATE SERVICE CODE 46 (TAB ONLY)		
• Disconnect vacuum line on bypass valve and cap vacuum line. • Key Off, wait 10 seconds. • Repeat Engine Running Self-Test and record codes. • Is code 46 present?	Yes ▶	EEC-IV system OK. REFER to Section 3 for bypass valve diagnostics.
	No ▶	GO to KS4

Air Management System — Pinpoint Test — KS

TEST STEP	RESULT ▶	ACTION TO TAKE
KS4 ENTER OUTPUT STATE CHECK (REFER TO APPENDIX)		
• Key Off, wait 10 seconds. • DVOM on 20V scale. • Connect DVOM negative test lead to STO and positive test lead to battery positive. • Jumper STI to signal return. • Perform Key On, Engine Off Self-Test until the completion of the Continuous Test Codes. • DVOM will indicate zero volts. • Depress and release the throttle. • Did DVOM reading change to a high voltage reading?	Yes ▶	REMAIN in Output State Check. GO to KS5.
	No ▶	DEPRESS throttle to WOT and RELEASE. If STO voltage does not go high, GO to Pinpoint Test Step Q40. Leave equipment hooked up.
KS5 CHECK TAB TAD SOLENOIDS FOR VACUUM CYCLING		
• Install vacuum pump to the TAB solenoid vacuum supply port and install a vacuum gauge to the output port. • While cycling outputs On and Off (by depressing and releasing throttle) observe the vacuum gauge at the output. **NOTE: Maintain vacuum at source.** • Repeat for TAD solenoid. Connect vacuum pump to the TAD solenoid vacuum supply port and connect a vacuum gauge to the output port. • Cycle output On and Off.	Both vacuum outputs cycle On and Off ▶	GO to KS6.
	Either vacuum output does not cycle On and Off ▶	REPLACE solenoid assembly. REPEAT Quick Test.
KS6 CHECK TAB TAD VACUUMS		
• Disconnect vacuum line from TAB solenoid to thermactor valve. • Inspect vacuum line for blockage, cracks, kinks (e.g. connect vacuum pump and gauge and verify vacuum line holds vacuum, or blow into vacuum line to verify no restrictions). • Repeat this Step for the TAD vacuum line. • Are thermactor vacuum lines OK?	Yes ▶	GO to KS7.
	No ▶	SERVICE as required. REPEAT Quick Test.

Air Management System | Pinpoint Test | KS

TEST STEP	RESULT ▶	ACTION TO TAKE
KS7 CHECK MANIFOLD VACUUM LINES FOR BLOCKAGE OR LEAKS		
• With vacuum lines disconnected at TAD/TAB solenoids, check for vacuum. • Start engine.	Vacuum present ▶	EEC-IV system OK. REFER to Section 3 for thermactor valve and air pump diagnostics.
	No vacuum present ▶	SERVICE vacuum source blockage or leak. REPEAT Quick Test.

Idle Speed Control (Throttle Kicker) | Pinpoint Test | KT

KT

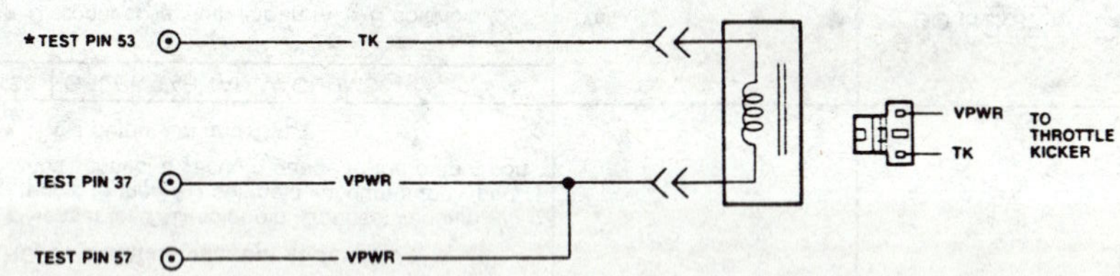

*TEST PINS LOCATED ON BREAKOUT BOX.
ALL HARNESS CONNECTORS VIEWED INTO MATING SURFACE.

STOP-WARNING

You may have reached this point because of a fault in a non-EEC System and/or component. To prevent the replacement of good components, be aware that the following non-EEC areas may be at fault:

- Engine Not At Operating Temperature.
- Engine Above Normal Operating Temperature.
- Improper Idle Speeds.
- A/C Input (Electrical Problems).

This Pinpoint Test is intended to diagnose only the following:

- TK Solenoid.
- TK Actuator.
- Harness circuits: TK and VPWR.
- Vacuum Source.
- Actuator Check.

Idle Speed Control (Throttle Kicker) — Pinpoint Test KT

TEST STEP	RESULT ▶	ACTION TO TAKE
KT1 VERIFY LINE ROUTINGS AND VISUAL CHECKS		
• Verify proper vacuum line routing to the TK solenoid. Refer to VECI decal. • Check for kinked or blocked vacuum lines. • Check for kinked or blocked air hoses. • Check for disconnected vacuum lines. • Are visual checks satisfactory?	No ▶	SERVICE routing or faults. REPEAT Quick Test.
	Yes ▶	Service codes 12 or 13: GO to KT2.
KT2 VACUUM CHECK		
• Apply 20 in-Hg vacuum to throttle kicker. • Verify vacuum holds.	Yes ▶	GO to KT3.
	No ▶	REPLACE solenoid. REPEAT Quick Test.
KT3 ATTEMPT TO ELIMINATE SERVICE CODE 12 OR 13		
• While vacuum is applied to throttle kicker, rerun Engine Running Self-Test. • Is service code 12 or 13 present?	Yes ▶	GO to Section 4 for idle adjustment, EEC OK.
	No ▶	GO to KT4.
KT4 ENTER OUTPUT STATE CHECK (REFER TO APPENDIX)		
• Key Off, wait 10 seconds. • DVOM on 20V scale. • Connect DVOM negative test lead to STO and positive test lead to battery positive. • Jumper STI to signal return. • Perform Key On, Engine Off Self-Test until the completion of the Continuous Test Codes. • DVOM will indicate zero volts. • Depress and release the throttle. • Did DVOM reading change to a high voltage reading?	Yes ▶	REMAIN in Output State Check. GO to KT5.
	No ▶	DEPRESS throttle to WOT and release. If STO voltage does not go high, GO to Pinpoint Test Step Q40. Leave equipment hooked up.

Idle Speed Control (Throttle Kicker) — Pinpoint Test KT

TEST STEP	RESULT ▶	ACTION TO TAKE
KT5 CHECK TK SOLENOID FOR VACUUM CYCLING		
• Install vacuum pump to the TK solenoid vacuum supply port and install a vacuum gauge to the output port. Apply 6 in-Hg minimum. • While cycling outputs On and Off (by depressing and releasing throttle) observe the vacuum gauge at the output. **NOTE: Maintain vacuum at source.**	Vacuum output cycles On and Off ▶	GO to KT6.
	Vacuum output does not cycle On and Off ▶	REPLACE solenoid. REPEAT Quick Test.
KT6 CHECK MANIFOLD VACUUM LINES FOR BLOCKAGE OR LEAKS		
• With vacuum lines disconnected at TK solenoid, check for vacuum. • Start engine.	Vacuum present ▶	EEC-IV system OK. REFER to Section 4 for Idle Adjustment.
	No vacuum present ▶	SERVICE vacuum source blockage or leak. REPEAT Quick Test.

Dynamic Response Test — Pinpoint Test — M

M

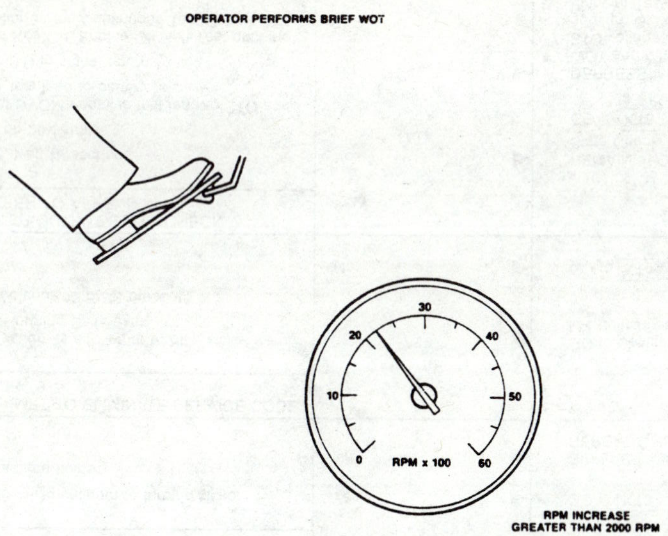

STOP-WARNING

You may have reached this point because of a fault in a non-EEC System and/or component. To prevent the replacement of good components, be aware that the following non-EEC areas may be at fault:

- Operator did not perform a brief WOT after dynamic response code.
- Mechanical engine problems: engine did not achieve greater than 2000 rpm.

This Pinpoint Test is intended to diagnose only the following:

- Throttle movement (greater than 3/4 throttle).
- Vane Air Flow (greater than 50% open).
- RPM increase (greater than 2000 rpm).

Dynamic Response Test — Pinpoint Test — M

TEST STEP	RESULT ▶	ACTION TO TAKE
M1 CODE 77: SYSTEM FAILED TO RECOGNIZE BRIEF WOT		
NOTE: A brief snap of the throttle may not be sufficient to pass this test. Be sure to go to WOT and return.		
• Repeat Engine Running Test of Quick Test. Be sure operator is familiar with the engine running format which proceeds as follows: — With Self-Test activated restart the engine. — ID Code 2 (0) start of test. — Dynamic response Code 1 (0) perform brief WOT. — Testing over. — Service code output begins.	Code 77 still present ▶	REPLACE processor. REPEAT Quick Test.
	No Code 77 ▶	Dynamic Response Test passed. SERVICE any other service code(s) received as necessary.

Ignition Diagnostic Monitor (IDM) — Pinpoint Test — N

N

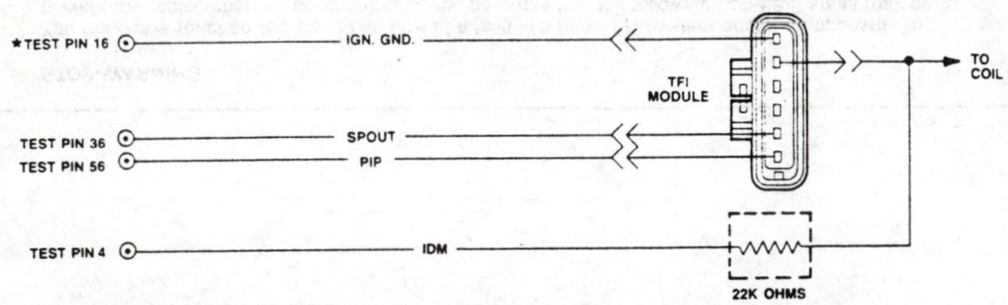

*TEST PINS LOCATED ON BREAKOUT BOX.
ALL HARNESS CONNECTORS VIEWED INTO MATING SURFACE.

STOP-WARNING

You may have reached this point because of a fault in a non-EEC System and or component. To prevent the replacement of good components, be aware that the following non-EEC areas may be at fault:

- Ignition module.
- Ignition coil.
- Spark plugs and high tension cables.
- Distributor and PIP sensor.

This Pinpoint Test is intended to diagnose only the following:

- Harness circuits: ignition ground, spout, PIP, IDM.

Ignition Diagnostic Monitor (IDM) — Pinpoint Test — N

TEST STEP	RESULT ▶	ACTION TO TAKE
N1 CHECK TFI MODULE		
• Key Off, wait 10 seconds. • Deactivate Self-Test. • Connect VOM or STAR per Quick Test Step 2.0. • Enter Engine Running Continuous Monitor Test (as instructed in Quick Test Step 6.3).	Yes ▶	DISCONNECT and INSPECT connectors. If connector and terminals are good, GO to Section 15 TFI Ignition Diagnostics.
• Observe VOM or STAR LED for indication of a fault while performing the following: • Lightly tap on TFI module (simulate road shock). • Wiggle TFI connector. • Is a fault indicated?	No ▶	GO to N2.
N2 CHECK EEC-IV HARNESS		
• While still in continuous monitor test from step N1 observe VOM or STAR LED for a fault indication while performing the following:	Yes ▶	ISOLATE fault and make necessary repairs. REPEAT Quick Test.
• While looking for faults listed in the table below, grasp the harness close to the TFI connector. Wiggle, shake or bend a small section of the EEC-IV system harness while working your way to the dash panel. Also wiggle, shake or bend the EEC-IV harness from the dash panel to the processor. Do this test on the circuits listed one at a time if needed to locate a faulty circuit. FAULT — BREAKOUT BOX NO. PIP shorted to ground — Test Pin 56 Spout shorted to ground — Test Pin 36 Ign. ground open — Test Pin 16 IDM open or shorted to ground, power — Test Pin 4 • Is a fault indicated?	No ▶	GO to N3.
N3 CHECK PROCESSOR AND HARNESS CONNECTORS		
• Key Off, wait 10 seconds. • Disconnect processor 60 Pin connector and inspect for damaged pins, corrosion, loose wires.	No ▶	SERVICE as necessary. REPEAT Quick Test.
• Are connectors and terminals OK? • Reconnect processor when this step is completed.	Yes ▶	Unable to DUPLICATE an IDM fault in the EEC-IV System, for further diagnosis GO to Section 15 TFI Ignition System Diagnostics.

Spark Timing Check — Pinpoint Test — P

P

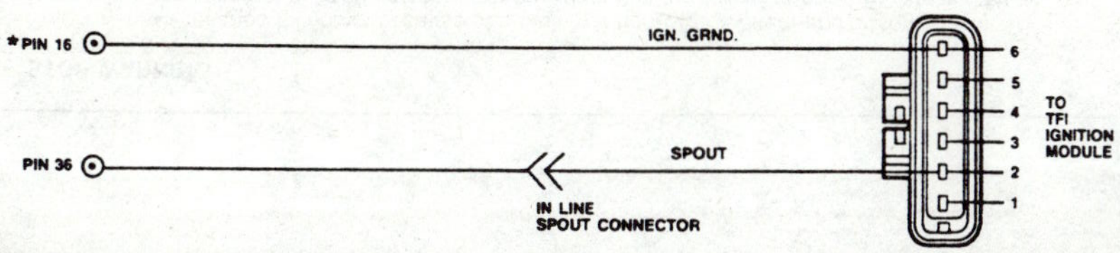

*TEST PINS LOCATED ON BREAKOUT BOX.
ALL HARNESS CONNECTORS VIEWED INTO MATING SURFACE.

STOP-WARNING

You may have reached this point because of a fault in a non-EEC System and/or component. To prevent the replacement of good components, be aware that the following non-EEC areas may be at fault:

- Base Engine.
- PIP Sensor.
- TFI Module.

This Pinpoint Test is intended to diagnose only the following:

- Harness Spout Circuit.
- Base Timing.
- Processor Assembly.

Spark Timing Check — Pinpoint Test — P

TEST STEP	RESULT ▶	ACTION TO TAKE
P1 CHECK SPARK TIMING		
• Verify that Self-Test is activated during Engine Running Test.	Yes ▶	GO to Quick Test Step 5.0.
• Check timing while performing Engine Running Test. Record value. • Timing base plus 20 degrees ± 3 degrees? **NOTE: Self-Test locks the timing at 20 degrees plus base throughout this test and for two minutes after the last service code is outputted. Timing check must be made during this time period. Self-Test timing is base + 20 degrees (±3 degrees) BTDC. (See VECI decal for base value.)**	No ▶	GO to P2.
P2 CHECK SPARK OUTPUT (SPOUT) CIRCUIT TO THE TFI MODULE		
• Locate spout connector and open this connection.	Yes ▶	GO to P3.
• Start engine. • Check base timing. • Timing base ±3 degrees BTDC. **NOTE: Base timing must be checked to decal value.**	No ▶	RECONNECT spout connector. ADJUST timing. REFER to Section 15 for engine timing instructions. After timing is reset, PERFORM Quick Test Step 4.
P3 CHECK FOR POWER TO PROCESSOR		
• Key Off, wait 10 seconds. • Disconnect processor 60 Pin connector and inspect for damaged pins, corrosion, loose wires. Service as necessary. • Install Breakout Box. • Key On, Engine Off.	Either voltage reading less than 10.5V ▶	GO to Pinpoint Test Step B1.
• DVOM on 20V scale. • Measure voltage from test Pin 37 to test Pin 40 and test Pin 57 to test Pin 60.	Both voltage readings 10.5V or greater ▶	GO to P4.

Spark Timing Check — Pinpoint Test P

TEST STEP	RESULT ▶	ACTION TO TAKE
P4 CHECK HARNESS FOR CONTINUITY		
• Key Off, wait 10 seconds. • DVOM on 200 ohm scale. • Measure spout circuit from test Pin 36 to spout pin (pin 2) on the TFI harness connector.	DVOM reading 5 ohms or less ▶	GO to P5.
	DVOM reading greater than 5 ohms ▶	REPAIR open circuit. CONNECT spout test connector. CHECK timing per P1.
P5 HARNESS CHECK — IGNITION GROUND		
• Breakout box installed. • Key Off. • DVOM on 200 ohm scale. • Measure from test Pin 16 to TFI connector Pin 6 (ign. grnd.).	Less than 5 ohms ▶	GO to Pinpoint Test Step A7.
	5 ohms or greater ▶	SERVICE harness as necessary. REPEAT Quick Test.

No Codes/Invalid Codes — Pinpoint Test Q

Q

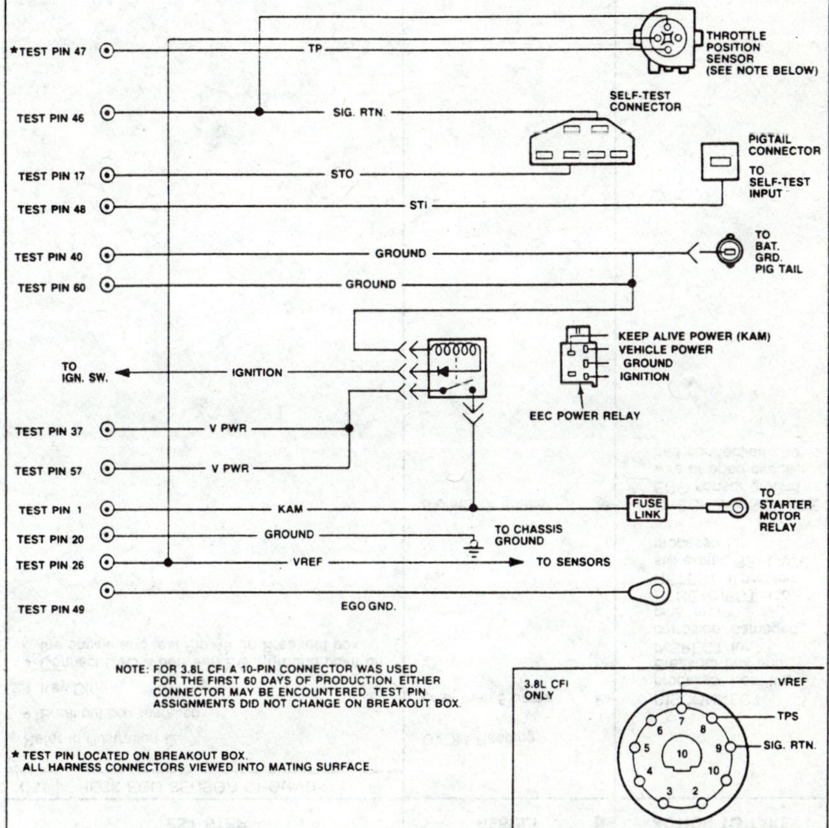

NOTE: FOR 3.8L CFI A 10-PIN CONNECTOR WAS USED FOR THE FIRST 60 DAYS OF PRODUCTION. EITHER CONNECTOR MAY BE ENCOUNTERED. TEST PIN ASSIGNMENTS DID NOT CHANGE ON BREAKOUT BOX.

* TEST PIN LOCATED ON BREAKOUT BOX.
ALL HARNESS CONNECTORS VIEWED INTO MATING SURFACE.

STOP-WARNING

This Pinpoint Test is intended to diagnose only the following:

- Processor.
- Harness Circuits: Signal Return, STO, STI, Ground, V PWR, KAM, VREF, Ignition.

No Codes/Invalid Codes — Pinpoint Test Q

TEST STEP	RESULT ▶	ACTION TO TAKE
Q1 NO CODES, INVALID CODES		
Refer to illustration Q. • Key Off, wait 10 seconds. • DVOM on 20V scale. • Disconnect TP sensor. • Key On, Engine Off. • Measure voltage at TP harness connector — VREF to signal return.	6.0V and greater ▶	GO to Pinpoint Test Step C2.
	Less than 4.0V ▶	GO to Pinpoint Test Step C1.
	4.0V to 6.0V ▶	RECONNECT TP sensor. GO to Q2.
Q2 BATTERY POWER GROUND CHECK (NEGATIVE SUPPLY)		
• Refer to illustration Q. • Key On, Engine Off. • DVOM to 20V scale. Take a voltage reading from battery negative (−) post to Signal Return, in the Self-Test connector.	Less than 0.5V ▶	GO to Q3.
	0.5V or greater ▶	GO to Pinpoint Test Step B4.
Q3 SELF-TEST INPUT CHECK		
Refer to illustration Q. • Key Off, wait 10 seconds. • Disconnect processor 60 pin connector and inspect for damaged pins, corrosion, loose wires. Repair as necessary. • Set DVOM to 200 ohm scale. • Check resistance from Self-Test input. At the Self-Test pigtail connector to Breakout box test Pin 48.	Less than 5 ohms ▶	GO to Q4.
	5 ohms or greater ▶	CORRECT open in circuit.
Q4 SELF-TEST OUTPUT CIRCUIT CHECK		
Refer to illustration Q. • Breakout box installed. • DVOM to 200 ohm scale. • Check resistance from Self-Test output at Self-Test connector to test Pin 17 of Breakout box.	5 ohms or greater ▶	CORRECT open in circuit.
	Less than 5 ohms ▶	GO to Q5.

No Codes/Invalid Codes — Pinpoint Test Q

TEST STEP	RESULT ▶	ACTION TO TAKE
Q5 CHECK EGO SENSOR GROUND		
Refer to illustration Q. • Breakout box installed. • Key Off. • Connect DVOM between EGO ground point on the engine and test Pin 49 on Breakout box.	DVOM Reading Less than 5 ohms ▶	DISCONNECT processor from the Breakout box and INSPECT for corrosion, damaged pins, etc. RECONNECT and RETEST. If problem still exists, REPLACE processor.
	5 ohms or greater ▶	CHECK and SERVICE EGO sensor ground wire or open circuit bad connection.

Invalid Codes — Continuous Test | Pinpoint Test | Q

TEST STEP	RESULT ▶	ACTION TO TAKE
Q6 CONDITIONS FOR CONTINUOUS CODE 15		
• Power interruption to Keep Alive Memory (KAM) Pin 1 may result in a service code being outputted.* • Clear continuous codes (use procedure described in Quick Test Step 6.) • Repeat Quick Test Step 3.0 through Continuous Codes. • Code 15 present on retest? *NOTE: **Anytime power is interrupted to the processor, a code 15 may be outputted the first time Self-Test is run after restoration of power. Repeat Self-Test to ensure correct diagnosis.**	Yes ▶	GO to Q7.
	No ▶	Test complete.
Q7 INSPECT ENGINE COMPARTMENT WIRING FOR PROPER ROUTING		
• Are any EEC components or EEC wiring close to ignition components or wires (High Electrical Energy Sources)? If EEC wiring close, reroute and rerun Self-Test. • Is code 15 still present?	Yes ▶	GO to Q8.
	No ▶	Test complete.
Q8 CHECK POWER CIRCUIT TO KEEP ALIVE MEMORY		
• Check power circuit to Keep Alive Memory for voltage. • Key Off, wait 10 seconds. • Disconnect processor and inspect both 60 Pin connectors. • Connect Breakout box to harness, leave processor disconnected. • DVOM on 20V scale. • Connect positive test lead to test Pin 1 and negative test lead to test Pin 46. • Key On. • Observe voltage reading.	DVOM reading less than 10 volts ▶	SERVICE open to KAM circuit. REPEAT Quick Test.
	DVOM reading 10V or greater ▶	GO to Q9.
Q9 CHECK INVALID CODE		
• Key On, engine Off. • Activate Self-Test. • Is code 15 present?	Yes ▶	REPLACE processor. REPEAT Quick Test.
	No ▶	DISCONNECT Breakout box. REPEAT Self-Test.*

Engine Stalls | Pinpoint Test | Q

TEST STEP	RESULT ▶	ACTION TO TAKE
Q30 DIAGNOSTIC ROUTINES COMPLETE		
• Have diagnostic routines in Section 2 been completed?	Yes ▶	GO to Q31.
	No ▶	GO to Diagnostic Routines, Section 2.
Q31 ATTEMPT TO BRING ENGINE TO OPERATING TEMPERATURE		
• Key Off, wait 10 seconds. • Install tachometer. • Try to maintain engine at 2,000 rpm for 2 minutes. • Does engine stall?	Yes ▶	GO to Pinpoint Test Step A13.
	No ▶	GO to Q32.
Q32 MAINTAIN ENGINE OPERATION DURING ENGINE RUNNING SELF-TEST		
• Engine at operating temperature. • Tachometer installed. • Key Off, wait 10 seconds. • Activate Quick Test. • Perform Engine Running Self-Test while maintaining 2,000 rpm.	Code 11 ▶	GO to Quick Test Step 5.0.
	Any service code ▶	GO to Quick Test Step 5.1.
	No codes ▶	GO to Pinpoint Test Step Q1.

Output State Check Not Functioning — Pinpoint Test Q

TEST STEP	RESULT ▶	ACTION TO TAKE
Q40 CHECK FOR CODES 23, 53, 63 OR 68		
• Key Off, wait 10 seconds. • Perform Key On, Engine Off Self-Test. • Leave Key On to enter Output State Check. • Are codes 23, 53, 63 or 68 present in the On Demand Test results?	Yes ▶	GO to Quick Test Step 3.1 and SERVICE appropriate code as instructed.
	No ▶	Leave Key On, GO to Q41.
Q41 ATTEMPT TO ENTER OUTPUT STATE CHECK (REFER TO APPENDIX)		
• Key On, engine Off in Output State Check. • DVOM on 20V scale. • Connect DVOM leads to STO circuit of Self-Test connector and battery positive terminal. • Depress throttle to wide open and release. • Verify throttle is in closed position. • Is DVOM reading greater than 10V?	Yes ▶	CHECK throttle and throttle linkages for sticking and binding. SERVICE as necessary. REPEAT Quick Test.
	No ▶	REPLACE TP sensor. REPEAT Quick Test.

Output Circuit Check — (Thermactor) — Pinpoint Test UA

UA

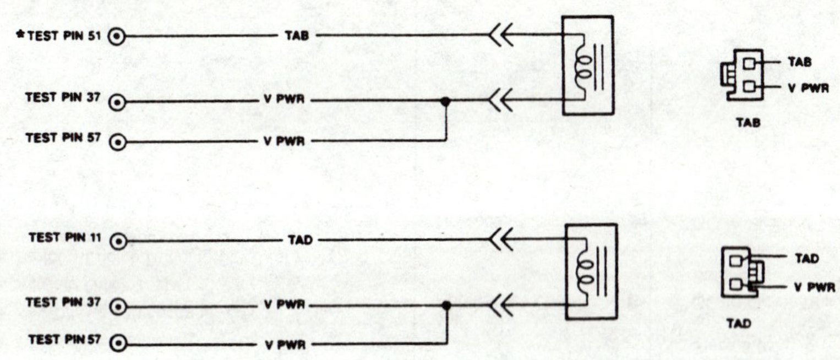

*TEST PINS LOCATED ON BREAKOUT BOX.
ALL HARNESS CONNECTORS VIEWED INTO MATING SURFACE.

STOP-WARNING

You may have reached this point because of a fault in a non-EEC System and/or component. To prevent the replacement of good components, be aware that the following non-EEC areas may be at fault:

- Thermactor System.
  - Belt.
  - Pump.
  - Valve.

This Pinpoint Test is intended to diagnose only the following:

- TAB and TAD Solenoid Valve Assemblies.
- Harness Circuits: TAB, TAD and V PWR.
- Processor Assembly.

Output Circuit Check — (Thermactor) | Pinpoint Test | UA

TEST STEP	RESULT ▶	ACTION TO TAKE
UA1 MEASURE TAB/TAD SOLENOID RESISTANCE		
• Key Off, wait 10 seconds. • DVOM on 200 ohm scale. • Disconnect TAB solenoid connector and measure solenoid resistance. • Disconnect TAD solenoid connector and measure solenoid resistance.	Both resistances are between 50 and 100 ohms ▶	CONNECT TAB/TAD solenoids. GO to UA2.
	Either resistance is less than 50 ohms or greater than 100 ohms ▶	REPLACE TAB/TAD solenoid assembly. REPEAT Quick Test.
UA2 CHECK VOLTAGE OF V PWR CIRCUIT		
• Key On, Engine Off. • DVOM on 20V scale. • Positive DVOM test lead to V PWR circuit and negative test lead to ground. • Measure voltage on TAB solenoid V PWR circuit. • Repeat for TAD solenoid.	Both voltage readings 10.5V or greater ▶	GO to UA3.
	Either voltage reading is less than 10.5V ▶	SERVICE harness circuit open. REPEAT Quick Test.
UA3 CHECK CONTINUITY OF TAB AND TAD CIRCUITS		
• Key Off, wait 10 seconds. • Disconnect processor 60 Pin connector and inspect for damaged pins, corrosion, loose wires. Service as necessary. • Connect Breakout box to harness. Leave processor disconnected. • DVOM on 200 ohm scale. • Measure continuity of test Pin 51 to TAB circuit at harness connector. • Measure continuity of test Pin 11 to TAD circuit at harness connector.	Both resistance readings are less than 5 ohms ▶	GO to UA4.
	Either resistance reading is 5 ohms or greater ▶	SERVICE harness open circuit. REPEAT Quick Test.
UA4 CHECK FOR SHORT TO POWER		
• Key Off, wait 10 seconds. • DVOM on 200,000 ohm scale. • Leave Breakout box installed and processor disconnected. • With TAB/TAD solenoid disconnected, measure resistance from test Pins 51 and/or 11 to test Pins 37 and 57.	Resistance reading is 10,000 ohms or greater ▶	GO to UA5.
	Resistance reading is less than 10,000 ohms ▶	SERVICE short to power. REPEAT Quick Test.

Output Circuit Check — (Thermactor) | Pinpoint Test | UA

TEST STEP	RESULT ▶	ACTION TO TAKE
UA5 CHECK FOR SHORT TO GROUND		
• Key Off, wait 10 seconds. • DVOM on 200,000 ohm scale. • Leave Breakout box installed and processor disconnected. • With TAB/TAD solenoids disconnected, measure resistance from test Pins 51 and/or 11 to test Pins 40, 46 and 60.	Resistance reading is less than 10,000 ohms ▶	SERVICE short to ground. REPEAT Quick Test.
	Resistance reading is 10,000 ohms or greater ▶	REPLACE processor. REPEAT Quick Test.

Output Circuit Check — (EGRV/EGRC) | Pinpoint Test | UC

UC

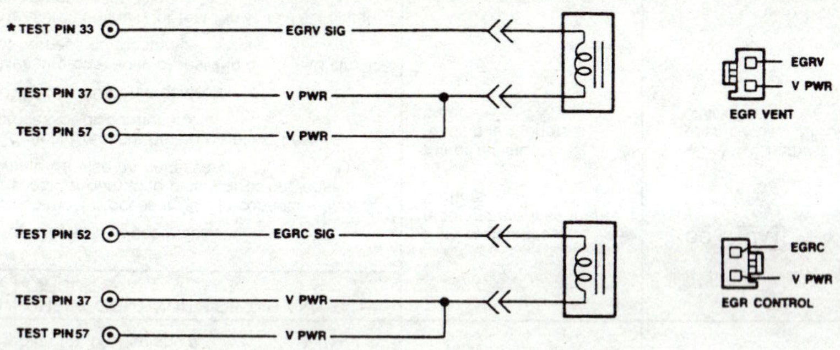

* TEST PINS LOCATED ON BREAKOUT BOX.
ALL HARNESS CONNECTORS VIEWED INTO MATING SURFACE.

STOP-WARNING

You may have reached this point because of a fault in a non-EEC System and/or component. To prevent the replacement of good components, be aware that the following non-EEC areas may be at fault:

- EGR Valve.
  — Vacuum Hose.

This Pinpoint Test is intended to diagnose only the following:

- EGRV and EGRC Solenoid Valve Assemblies.
- Harness Circuits: EGRV, EGRC and V PWR.
- Processor Assembly.

Output Circuit Check — (EGRV/EGRC) | Pinpoint Test | UC

TEST STEP	RESULT ▶	ACTION TO TAKE
UC1 SERVICE CODES 83 AND 84: MEASURE EGRV EGRC SOLENOID RESISTANCE		
• Key Off, wait 10 seconds. • DVOM on 200 ohm scale. • Disconnect EGRV solenoid connector and measure solenoid resistance. • Disconnect EGRC solenoid connector and measure solenoid resistance.	Both resistances are between 30 and 70 ohms ▶	CONNECT EGRC/EGRV solenoids. GO to UC2.
	Either resistance is less than 30 ohms or greater than 70 ohms ▶	REPLACE EGRC/EGRV solenoid assembly. REPEAT Quick Test.
UC2 CHECK VOLTAGE OF V PWR		
• Key On, Engine Off. • DVOM on 20V scale. • Positive DVOM test lead to V PWR circuit and negative test lead to ground. • Measure voltage on EGRV solenoid V PWR circuit. • Repeat for EGRC solenoid.	Both voltage readings are greater than 10.5V ▶	GO to UC3.
	Either voltage reading is less than 10.5V ▶	SERVICE harness circuit open. REPEAT Quick Test.
UC3 CHECK CONTINUITY OF EGRV AND EGRC CIRCUITS		
• Key Off, wait 10 seconds. • Disconnect processor 60 Pin connector and inspect for damaged pins, corrosion, loose wires. Service as necessary. • Connect Breakout box to harness. Leave processor disconnected. • DVOM on 200 ohm scale. • Measure continuity of test Pin 33 to harness EGRV circuit at EGRV solenoid. • Measure continuity of test Pin 52 to harness EGRC circuit at EGRC solenoid.	Both resistance readings are less than 5 ohms ▶	GO to UC4.
	Either resistance reading is 5 ohms or greater ▶	SERVICE harness open circuit. REPEAT Quick Test.
UC4 CHECK FOR SHORT TO POWER		
• Key Off, wait 10 seconds. • DVOM on 200,000 ohm scale. • Leave Breakout box installed and processor disconnected. • With EGRV EGRC solenoids disconnected, measure resistance from test Pins 33 and or 52 to test Pins 37 and 57.	Resistance reading is less than 10,000 ohms ▶	SERVICE short to power and REPEAT Quick Test.
	Resistance reading is 10,000 ohms or greater ▶	GO to UC5.

Output Circuit Check — (EGRV/EGRC) | Pinpoint Test | UC

TEST STEP	RESULT ▶	ACTION TO TAKE
UC5 CHECK FOR SHORT TO GROUND		
• Key Off, wait 10 seconds. • DVOM on 200,000 ohm scale. • Leave Breakout box installed and processor disconnected. • With EGRV/EGRC solenoids disconnected, measure resistance from test Pins 33 and/or 52 to test Pins 40, 46 and 60.	Resistance reading is less than 10,000 ohms ▶	SERVICE short to ground. REPEAT Quick Test.
	Resistance reading is 10,000 ohms or greater ▶	REPLACE processor. REPEAT Quick Test.

Output Circuit Check Canister Purge | Pinpoint Test | UE

UE

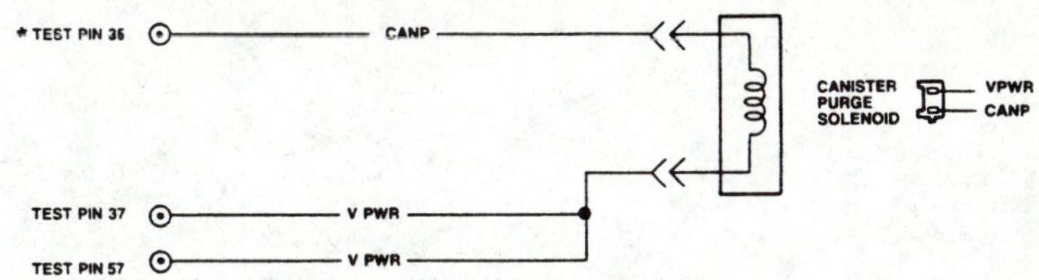

* TEST PINS LOCATED ON BREAKOUT BOX.
ALL HARNESS CONNECTORS VIEWED INTO MATING SURFACE.

STOP-WARNING

This Pinpoint Test is intended to diagnose only the following:

- Harness Circuits: CANP and V PWR.
- Processor Assembly.

TEST STEP	RESULT ▶	ACTION TO TAKE
UE1 MEASURE CANP SOLENOID RESISTANCE		
• Key Off, wait 10 seconds. • DVOM on 200 ohm scale. • Disconnect CANP solenoid connector and measure solenoid resistance.	Resistance is between 40 and 90 ohms ▶	CONNECT CANP solenoid. GO to UE2.
	Resistance is less than 40 ohms or greater than 90 ohms ▶	REPLACE CANP solenoid. REPEAT Quick Test.

Output Circuit Check — Canister Purge | Pinpoint Test | UE

TEST STEP	RESULT ▶	ACTION TO TAKE
UE2 CHECK VOLTAGE OF V PWR CIRCUIT		
• Key On. Engine Off. • DVOM on 20V scale. • Connect DVOM positive test lead to V PWR circuit and negative test lead to ground. • Measure voltage on CANP solenoid V PWR circuit.	Voltage reading is less than 10.5V ▶	SERVICE harness open circuit. REPEAT Quick Test.
	Voltage reading is 10.5V or greater ▶	GO to UE3.
UE3 CHECK CONTINUITY OF CANP CIRCUIT		
• Key Off, wait 10 seconds. • Disconnect processor 60 Pin connector and inspect for damaged pins, corrosion, loose wires. Service as necessary. • Connect Breakout box to harness. Leave processor disconnected. • DVOM on 200 ohm scale. • Measure continuity of test Pin 35 to CANP circuit at harness connector.	Resistance reading is 5 ohms or greater ▶	SERVICE open circuit. REPEAT Quick Test.
	Resistance reading is less than 5 ohms ▶	GO to UE4.
UE4 CHECK FOR SHORT TO POWER		
• Key Off, wait 10 seconds. • DVOM on 200,000 ohm scale. • Leave Breakout box installed and processor disconnected. • With CANP solenoid disconnected, measure resistance from test Pin 35 to test Pins 37 and 57.	Resistance reading is 10,000 ohms or greater ▶	GO to UE5.
	Resistance reading is less than 10,000 ohms ▶	SERVICE short to power. REPEAT Quick Test.
UE5 CHECK FOR SHORT TO GROUND		
• Key Off, wait 10 seconds. • Leave Breakout box installed and processor disconnected. • DVOM on 200,000 ohm scale. • Measure resistance of test Pin 35 to test Pins 40, 46 and 60.	Resistance reading is less than 10,000 ohms ▶	SERVICE short to ground. REPEAT Quick Test.
	Resistance reading is 10,000 ohms or greater ▶	REPLACE processor. REPEAT Quick Test.

Output Circuit Check — Temperature Compensated Fuel Pump | Pinpoint Test | UG

UG

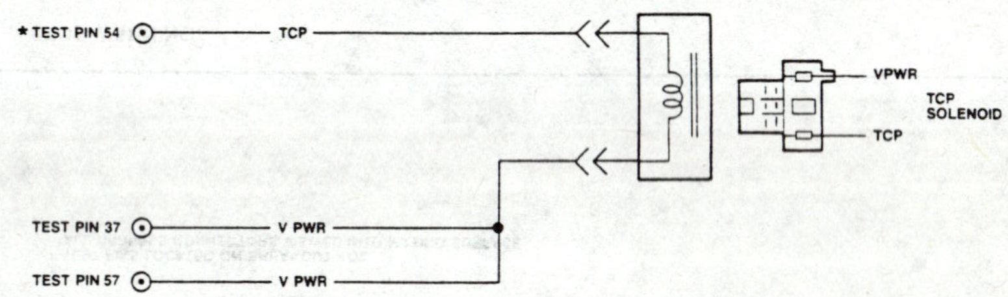

* TEST PINS LOCATED ON BREAKOUT BOX.
ALL HARNESS CONNECTORS VIEWED INTO MATING SURFACE.

STOP-WARNING

This Pinpoint Test is intended to diagnose only the following:

- Harness Circuits: V PWR and TCP.
- TCP Solenoid.
- Processor Assembly.

Output Circuit Check — Temperature Compensated Fuel Pump | Pinpoint Test | UG

TEST STEP	RESULT	▶	ACTION TO TAKE
UG1 MEASURE TCP SOLENOID RESISTANCE			
• Key Off, wait 10 seconds. • DVOM on 200 ohm scale. • Disconnect TCP solenoid connector and measure solenoid resistance.	Resistance is between 50 and 100 ohms	▶	CONNECT TCP solenoid. GO to UG2.
	Resistance is less than 50 ohms or greater than 100 ohms	▶	REPLACE TCP solenoid. REPEAT Quick Test.
UG2 CHECK VOLTAGE OF V PWR CIRCUIT			
• Key On, Engine Off. • DVOM on 20V scale. • Connect DVOM positive test lead to V PWR circuit and negative test lead to ground. • Measure voltage on TCP solenoid V PWR circuit.	Voltage reading is less than 10.5V	▶	SERVICE harness open circuit. REPEAT Quick Test.
	Voltage reading is 10.5V or greater	▶	GO to UG3.
UG3 CHECK CONTINUITY OF TCP CIRCUIT			
• Key Off, wait 10 seconds. • Disconnect processor 60 Pin connector and inspect for damaged pins, corrosion, loose wires. Service as necessary. • Connect Breakout box to harness. Leave processor disconnected. • DVOM on 200 ohm scale. • Measure continuity of test Pin 54 to TCP circuit at harness connector.	Resistance readings are 5 ohms or greater	▶	SERVICE harness open circuit. REPEAT Quick Test.
	Resistance reading is less than 5 ohms	▶	GO to UG4.
UG4 CHECK FOR SHORT TO POWER			
• Key Off, wait 10 seconds. • DVOM on 200,000 ohm scale. • Leave Breakout box installed and processor disconnected. • With TCP solenoid disconnected, measure resistance from test Pin 54 to test Pins 37 and 57.	Resistance reading is 10,000 ohms or greater	▶	GO to UG5.
	Resistance reading less than 10,000 ohms	▶	SERVICE short to power. REPEAT Quick Test.

Output Circuit Check — Temperature Compensated Fuel Pump | Pinpoint Test | UG

TEST STEP	RESULT	▶	ACTION TO TAKE
UG5 CHECK FOR SHORT TO GROUND			
• Key Off, wait 10 seconds. • Leave Breakout box installed and processor disconnected. • DVOM on 200,000 ohm scale. • Measure resistance of test Pin 54 to test Pins 40, 46 and 60.	Resistance readings are less than 10,000 ohms	▶	SERVICE short to ground. REPEAT Quick Test.
	Resistance readings are 10,000 ohms or greater	▶	REPLACE processor. REPEAT Quick Test.

Output Circuit Check — Variable Voltage Choke (VVC) | Pinpoint Test | UH

UH

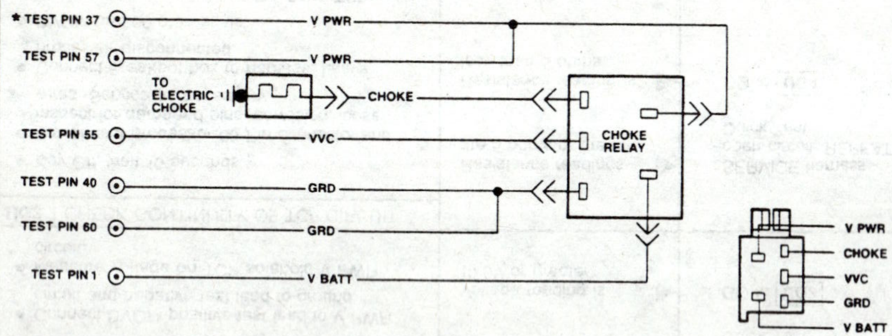

*TEST PINS LOCATED ON BREAKOUT BOX.
ALL HARNESS CONNECTORS VIEWED INTO MATING SURFACE.

STOP-WARNING

This Pinpoint Test is intended to diagnose only the following:

- Harness Circuits: VVC, V PWR and Choke.
- Choke Relay.
- Processor Assembly.

Output Circuit Check — Variable Voltage Choke (VVC) | Pinpoint Test | UH

TEST STEP	RESULT	▶	ACTION TO TAKE
UH1 MEASURE CHOKE VOLTAGE			
• DVOM on 20V scale. • Perform Engine Running Self-Test. • Measure choke voltage.	Choke voltage is 10.5V or greater	▶	CHECK choke plates for binding and sticking. EEC-IV system OK. REFER to Section 4, for choke diagnostics.
	Choke voltage is less than 10.5V	▶	GO to UH2.
UH2 ENTER OUTPUT STATE CHECK (REFER TO APPENDIX)			
• Key Off, wait 10 seconds. • DVOM on 20V scale. • Connect DVOM negative test lead to STO and positive test lead to battery positive. • Jumper STI to signal return. • Perform Key On, Engine Off Self-Test until the completion of the Continuous Test Codes. • DVOM will indicate zero volts. • Depress and release the throttle. • Did DVOM reading change to a high voltage reading?	Yes	▶	REMAIN in Output State Check. GO to UH3.
	No	▶	DEPRESS throttle to WOT and RELEASE. If STO voltage does not go high, GO to Pinpoint Test Step Q40.
UH3 CHECK VVC OUTPUT ELECTRICAL OPERATION			
• Key On, Engine Off. • DVOM on 20V scale. • Connect DVOM positive test lead to V PWR on VVC relay and negative test lead to VVC output circuit VVC. • While observing DVOM, depress and release the throttle several times to cycle output On and Off. • Does VVC output cycle On and Off?	Yes	▶	GO to UH4.
	No	▶	REMAIN in Output State Check. GO to UH5.

Output Circuit Check — Variable Voltage Choke (VVC) | Pinpoint Test | UH

TEST STEP	RESULT ▶	ACTION TO TAKE
UH4 CHECK CONTINUITY OF GROUND CIRCUIT AND CHOKE CIRCUIT		
• Key Off, wait 10 seconds. • DVOM on 200 ohm scale. • Measure continuity of ground circuit from the relay harness connector. • Measure continuity of the choke circuit from the relay harness connector to the choke.	Either resistance reading is 5 ohms or greater ▶	SERVICE harness circuit. REPEAT Quick Test.
	Both resistance readings are less than 5 ohms ▶	REPLACE choke relay. REPEAT Quick Test.
UH5 CHECK FOR VOLTAGE ON V PWR CIRCUIT		
• Key On, Engine Off. • DVOM on 20V scale. • Connect DVOM positive test lead to V PWR at relay harness connector and negative test lead to the ground circuit at the relay harness connector.	Voltage is less than 10.5V ▶	SERVICE open in choke power circuit. REPEAT Quick Test.
	Voltage is 10.5V or greater ▶	GO to UH6.
UH6 CHECK CONTINUITY OF VVC CIRCUIT		
• Key Off, wait 10 seconds. • Disconnect processor 60 Pin connector and inspect for damaged pins, corrosion, loose wires. Service as necessary. • Connect Breakout box to harness. Leave processor disconnected. • DVOM on 200 ohm scale. • Measure continuity of test Pin 55 to VVC circuit at the relay harness connector.	Resistance is 5 ohms or greater ▶	SERVICE harness circuit. REPEAT Quick Test.
	Resistance is less than 5 ohms ▶	GO to UH7
UH7 CHECK FOR SHORT TO POWER		
• Key Off, wait 10 seconds. • DVOM on 200,000 ohm scale. • Leave Breakout box installed and processor disconnected. • With VVC relay disconnected, measure resistance from test Pin 55 to test Pins 1, 37 and 57.	Resistance reading is 10,000 ohms or greater ▶	GO to UH8
	Resistance reading is less than 10,000 ohms ▶	SERVICE short to power. REPEAT Quick Test.

Output Circuit Check — Variable Voltage Choke (VVC) | Pinpoint Test | UH

TEST STEP	RESULT ▶	ACTION TO TAKE
UH8 CHECK FOR SHORT TO GROUND ON VVC CIRCUIT		
• Key Off, wait 10 seconds. • DVOM on 200,000 ohm scale. • VVC relay connected. • Leave Breakout box installed and processor disconnected. • Measure resistance of test Pin 55 to test Pins 40, 46 and 60.	Any resistance readings 10,000 ohms or greater ▶	GO to UH9
	All resistance readings less than 10,000 ohms ▶	REPLACE processor. REPEAT Quick Test.
UH9 CHECK FOR SHORT TO GROUND WITH VVC RELAY DISCONNECTED		
• Key Off, wait 10 seconds. • Leave Breakout box installed and processor disconnected. • DVOM on 200,000 ohm scale. • Disconnect VVC relay. • Measure resistance from test Pin 55 to test Pins 40, 46 and 60.	All resistance readings are 10,000 ohms or greater ▶	REPLACE choke relay. REPEAT Quick Test.
	Any resistance reading is less than 10,000 ohms ▶	SERVICE harness short. REPEAT Quick Test.

Output Circuit Check — WOT A/C Cut-Off | Pinpoint Test | UJ

UJ

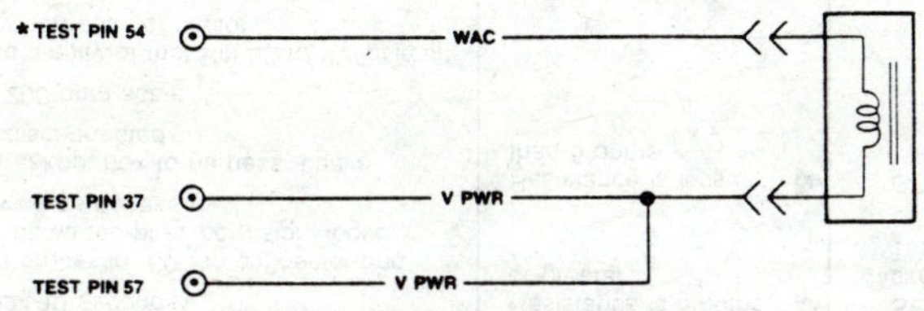

* TEST PINS LOCATED ON BREAKOUT BOX.
ALL HARNESS CONNECTORS VIEWED INTO MATING SURFACE.

STOP-WARNING

This Pinpoint Test is intended to diagnose only the following:

- Harness Circuits: WAC and V PWR.
- WOT A/C Cut-Off Solenoid.
- Processor Assembly.

TEST STEP	RESULT ▶	ACTION TO TAKE
UJ1 MEASURE WAC SOLENOID RESISTANCE		
• Key Off, wait 10 seconds. • DVOM on 200 ohm scale. • Disconnect WAC solenoid connector and measure solenoid resistance.	Resistance is between 50 and 100 ohms ▶	CONNECT WAC solenoid. GO to UJ2.
	Resistance is less than 50 ohms or greater than 100 ohms ▶	REPLACE WAC solenoid. REPEAT Quick Test.

Output Circuit Check — WOT A/C Cut-Off | Pinpoint Test | UJ

TEST STEP	RESULT ▶	ACTION TO TAKE
UJ2 CHECK VOLTAGE OF V PWR CIRCUIT		
• Key On, Engine Off. • DVOM on 20V scale. • Connect DVOM positive test lead to V PWR circuit and negative test lead to ground. • Measure voltage on WAC solenoid V PWR circuit.	Voltage reading is less than 10.5V ▶	SERVICE harness open circuit. REPEAT Quick Test.
	Voltage reading is 10.5V or greater ▶	GO to UJ3.
UJ3 CHECK CONTINUITY OF WAC CIRCUIT		
• Key Off, wait 10 seconds. • Disconnect processor 60 Pin connector and inspect for damaged pins, corrosion, loose wires. Service as necessary. • Connect Breakout box to harness. Leave processor disconnected. • DVOM on 200 ohm scale. • Measure continuity of test Pin 54 to WAC circuit at harness connector.	Resistance readings are 5 ohms or greater ▶	SERVICE harness open circuit. REPEAT Quick Test.
	Resistance reading is less than 5 ohms ▶	GO to UJ4.
UJ4 CHECK FOR SHORT TO POWER		
• Key Off, wait 10 seconds. • DVOM on 200,000 ohm scale. • Leave Breakout box installed and processor disconnected. • With WAC solenoid disconnected, measure resistance from test Pin 54 to test Pins 37 and 57.	Resistance reading is 10,000 ohms or greater ▶	GO to UJ5.
	Resistance reading is less than 10,000 ohms ▶	SERVICE short to power. REPEAT Quick Test.
UJ5 CHECK FOR SHORT TO GROUND		
• Key Off, wait 10 seconds. • Leave Breakout box installed and processor disconnected. • DVOM on 200,000 ohm scale. • Measure resistance of test Pin 54 to test Pins 40, 46 and 60.	Resistance readings are less than 10,000 ohms ▶	SERVICE short to ground. REPEAT Quick Test.
	Resistance readings are 10,000 ohms or greater ▶	REPLACE processor. REPEAT Quick Test.

Output Circuit Check — Idle Speed Control (Throttle Kicker) | Pinpoint Test | UK

UK

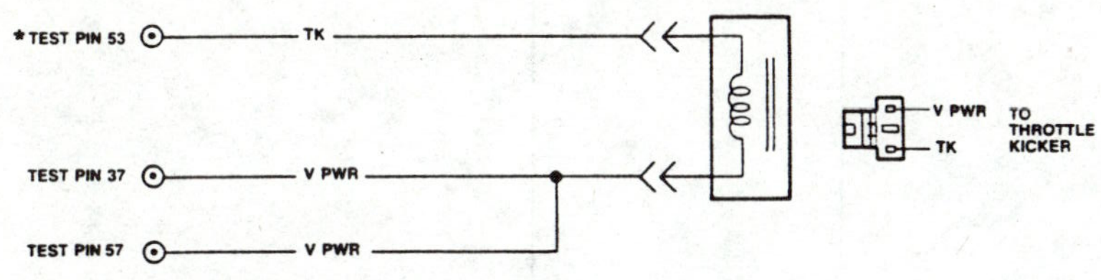

*TEST PINS LOCATED ON BREAKOUT BOX.
ALL HARNESS CONNECTORS VIEWED INTO MATING SURFACE.

STOP-WARNING

You may have reached this point because of a fault in a non-EEC System and/or component. To prevent the replacement of good components, be aware that the following non-EEC areas may be at fault:

- Engine not at operating temperature.
- Engine above operating temperature.
- Improper Idle Speeds.
- A/C Input (Electrical Problems).

This Pinpoint Test is intended to diagnose only the following:

- TK Solenoid.
- TK Actuator.
- Harness Circuits: TK and V PWR.
- Processor Assembly.

Output Circuit Check — Idle Speed Control (Throttle Kicker) | Pinpoint Test | UK

TEST STEP	RESULT ▶	ACTION TO TAKE
UK1 MEASURE TK SOLENOID RESISTANCE		
• Key Off, wait 10 seconds. • DVOM on 200 ohm scale. • Disconnect TK solenoid connector and measure solenoid resistance.	Resistance is between 50 and 100 ohms ▶	CONNECT TK solenoid. GO to **UK2**.
	Resistance is less than 50 ohms or greater than 100 ohms ▶	REPLACE TK solenoid. REPEAT Quick Test.
UK2 CHECK VOLTAGE OF V PWR CIRCUIT		
• Key On, Engine Off. • DVOM on 20V scale. • Connect DVOM positive test lead to V PWR circuit and negative test lead to ground. • Measure voltage on TK solenoid V PWR circuit.	Voltage reading is 10.5V or less ▶	SERVICE harness open circuit. REPEAT Quick Test.
	Voltage reading is 10.5V or greater ▶	GO to **UK3**.
UK3 CHECK CONTINUITY OF TK CIRCUIT		
• Key Off, wait 10 seconds. • Disconnect processor 60 Pin connector and inspect for damaged pins, corrosion, loose wires. Service as necessary. • Connect Breakout box to harness. Leave processor disconnected. • DVOM on 200 ohm scale. • Measure continuity of test Pin 53 to TK circuit at harness connector.	Resistance readings are 5 ohms or greater ▶	SERVICE harness open circuit. REPEAT Quick Test.
	Resistance reading is less than 5 ohms ▶	GO to **UK4**.
UK4 CHECK FOR SHORT TO POWER		
• Key Off, wait 10 seconds. • DVOM on 200,000 ohm scale. • Leave Breakout box installed. Leave processor disconnected. • With TK solenoid disconnected, measure resistance from test Pin 53 to test Pins 37 and 57.	Resistance reading is 10,000 ohms or greater ▶	GO to **UK5**.
	Resistance reading is less than 10,000 ohms ▶	SERVICE short to power. REPEAT Quick Test.

Output Circuit Check — Idle Speed Control (Throttle Kicker) | Pinpoint Test | UK

TEST STEP	RESULT ▶	ACTION TO TAKE
UK5 CHECK FOR SHORT TO GROUND		
• Key Off, wait 10 seconds. • Leave Breakout box installed and processor disconnected. • DVOM on 200,000 ohm scale. • Measure resistance of test Pin 53 to test Pins 40, 46 and 60.	Resistance readings are less than 10,000 ohms ▶	SERVICE short to ground. REPEAT Quick Test.
	Resistance readings are 10,000 ohms or greater ▶	REPLACE processor. REPEAT Quick Test.

Output Circuit Check — Exhaust Heat Control | Pinpoint Test | UL

UL

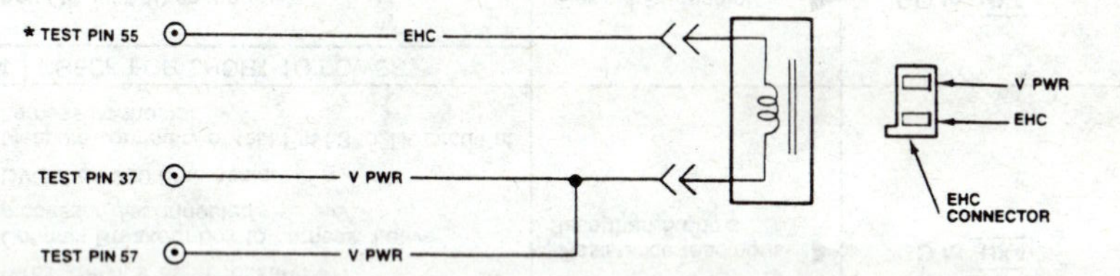

*TEST PINS LOCATED ON BREAKOUT BOX.
ALL HARNESS CONNECTORS VIEWED INTO MATING SURFACE.

STOP-WARNING

This Pinpoint Test is intended to diagnose only the following:

- Harness Circuits: EHC and V PWR.
- Exhaust Heat Solenoid.
- Processor Assembly.

TEST STEP	RESULT ▶	ACTION TO TAKE
UL1 MEASURE EHC SOLENOID RESISTANCE		
• Key Off, wait 10 seconds. • DVOM on 200 ohm scale. • Disconnect exhaust heat solenoid connector and measure solenoid resistance.	Resistance is between 50 and 100 ohms ▶	CONNECT exhaust heat solenoid. GO to UL2.
	Resistance is less than 50 ohms or greater than 100 ohms ▶	REPLACE exhaust heat solenoid. REPEAT Quick Test.

Output Circuit Check — Exhaust Heat Control — Pinpoint Test UL

TEST STEP	RESULT ▶	ACTION TO TAKE
UL2 CHECK VOLTAGE OF V PWR CIRCUIT		
• Key On. Engine Off. • DVOM on 20V scale. • Connect DVOM positive test lead to V PWR circuit and negative test lead to ground. • Measure voltage on EHC solenoid V PWR circuit.	Voltage reading is less than 10.5V ▶	SERVICE harness open circuit. REPEAT Quick Test.
	Voltage reading is 10.5V or greater ▶	GO to UL3.
UL3 CHECK CONTINUITY OF EHC CIRCUIT		
• Key Off, wait 10 seconds. • Disconnect processor 60 Pin connector and inspect for damaged pins, corrosion, loose wires. Service as necessary. • Connect Breakout box to harness. Leave processor disconnected. • DVOM on 200 ohm scale. • Measure continuity of test Pin 55 to EHC circuit at harness connector.	Resistance readings are 5 ohms or greater ▶	SERVICE harness open circuit. REPEAT Quick Test.
	Resistance reading is less than 5 ohms ▶	GO to UL4.
UL4 CHECK FOR SHORT TO POWER		
• Key Off, wait 10 seconds. • DVOM on 200,000 ohm scale. • Leave Breakout box installed and processor disconnected. • With exhaust heat solenoid disconnected, measure resistance from test Pin 55 to test Pins 37 and 57.	Resistance reading is 10,000 ohms or greater ▶	GO to UL5.
	Resistance reading is less than 10,000 ohms ▶	SERVICE short to power. REPEAT Quick Test.
UL5 CHECK FOR SHORT TO GROUND		
• Key Off, wait 10 seconds. • Leave Breakout box installed and processor disconnected. • DVOM on 200,000 ohm scale. • Measure resistance of test Pin 55 to test Pins 40, 46 and 60.	Resistance readings are less than 10,000 ohms ▶	SERVICE short to ground. REPEAT Quick Test.
	Resistance readings are 10,000 ohms or greater ▶	REPLACE processor. REPEAT Quick Test.

Erratic Ignition — Pinpoint Test Y

TEST STEP	RESULT ▶	ACTION TO TAKE
Y1 CODE 14: ERRATIC IGNITION		
NOTE: Code 14 indicates two successive erratic profile ignition pick-up (PIP) pulses occurred, resulting in a possible engine miss or stall. • Check EEC-IV and ignition systems harnesses for: — Loose wires/connectors. — Arcing secondary ignition components (coil, cap, rotor, wires, plugs, etc). — On-board transmitter (2-way radio).* • Are any of the above present? *Verify all 2-way radio installations. Carefully follow manufacturer's installation instructions regarding the routing of antenna and power leads.	Yes ▶	SERVICE as necessary. REPEAT Quick Test.
	No ▶	GO to Y2.
Y2 CHECK DISTRIBUTOR		
• Key Off, wait 10 seconds. • Deactivate Self-Test. • Connect VOM or STAR per Quick Test Step 2.0. • Enter Engine Running Continuous Monitor Test (as instructed in Quick Test Step 6.3). • Observe VOM or STAR LED for indication of a fault while performing the following: • Lightly tap on TFI module and distributor. • Wiggle TFI connector. • Is a fault indicated?	Yes ▶	DISCONNECT and INSPECT connectors. If connector and terminals are good, GO to Section 15 Ignition Diagnostics.
	No ▶	GO to Y3.
Y3 CHECK EEC-IV HARNESS		
• While still in Continuous Monitor Test from Step Y2 observe VOM or STAR LED for a fault indication while performing the following: • While looking for faults grasp the harness close to the TFI connector. Wiggle, shake or bend a small section of the ignition and EEC-IV systems harness while working your way to the other components and dash panel. Also wiggle, shake or bend the EEC-IV harness from the dash panel to the processor. Isolate the PIP circuit if needed for this test. • Is a fault indicated?	Yes ▶	ISOLATE fault and make necessary repairs. REPEAT Quick Test.
	No ▶	GO to Y4.

Erratic Ignition	Pinpoint Test	Y

TEST STEP	RESULT ▶	ACTION TO TAKE
Y4 CHECK PROCESSOR AND HARNESS CONNECTORS		
• Key Off, wait 10 seconds. • Disconnect processor 60 Pin connector and inspect for damaged pins, corrosion, loose wires. • Are connectors and terminals OK? • Reconnect processor when this Step is completed.	No ▶	SERVICE as necessary. REPEAT Quick Test.
	Yes ▶	Unable to duplicate an erratic ignition fault in the EEC-IV System, for further diagnosis GO to Section 15 Ignition System Diagnostics.

MCU System Description

V-EIGHT CYLINDER MCU SYSTEM

The MCU is basically the same on all V8 cylinder engines. The heart of the system is the fuel control system. This is necessary to keep the air-fuel (A/F) ratio at a proper chemical balance (14.7) to obtain maximum catalyst efficiency. The fuel control "loop" consists of an Exhaust Gas Oxygen (EGO) sensor, Microprocessor Control Unit (MCU), and carburetor with Feedback Carburetor Actuator (FBCA).

The EGO sensor senses whether the exhaust gas is rich or lean of a proper chemical balance. This signal is sent to the MCU module, which sends a varying signal to the Feedback Carburetor Actuator (FBCA) to move the air-fuel ratio back to the proper chemical balance. This operation is called "closed loop" operation.

The other mode of operation is called "open loop". In this mode, the MCU module sends out a fixed signal to the Feedback Carburetor Actuator (FBCA). During this time, the input from the EGO sensor is ignored, thus opening the loop.

The determining factor when the system goes into open or closed loop is based upon information from the switch inputs, which senses parameters such as coolant temperature, manifold vacuum, and rpm. Generally, the vehicle will be in closed loop when the vehicle is at operating temperature and at a steady part throttle cruise.

Other functions controlled by the MCU are Canister Purge (CANP), Thermactor Air Diverter (TAD), Thermactor Air Bypass (TAB), Throttle Kicker (TK), and Spark Retard System, or spark retard through the Knock sensor.

How To Use The Diagnostic Procedure

ANALOG VOLTMETER

When a Service Code is reported on the Analog Voltmeter, it will represent itself as a pulsing or sweeping movement of the voltmeter's needle across the dial face of the voltmeter. Therefore, a single-digit number of three will be reported by three needle pulses (sweeps). However, as previously stated, a Service Code is represented by a two-digit number, such as 3-2. So as a result, the Self-Test's Service Code of 3-2 will appear on the voltmeter as three needle pulses (sweeps), then after a two second pause, the needle will pulse (sweep) twice more.

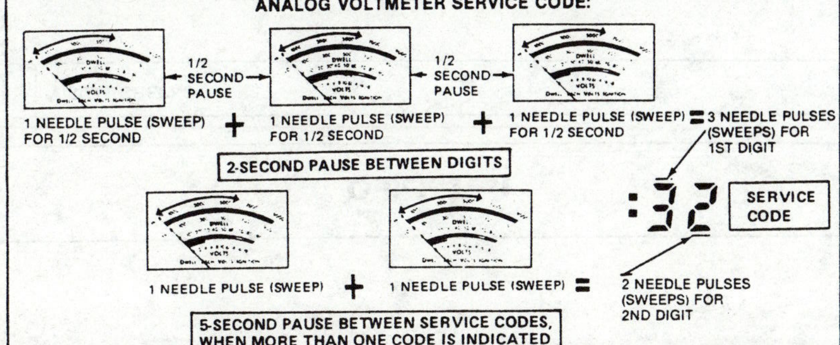

How To Use The Diagnostic Procedure

SELF-TEST AUTOMATIC READOUT (STAR)
Rotunda Model No. 07-0004 or Equivalent

After hooking up the STAR tester and turning on its power switch, the tester will run a display check: the numerals 88 will begin to flash in the display window, and then a steady 00 will appear to signify that the STAR tester is ready to start the Self-Test and receive the test's Service Codes.

To receive the Service Codes, press the pushbutton at the front of the STAR tester. The button will latch down, and a colon will appear in the display window in front of the :00 numerals. The colon MUST be displayed to receive the Service Codes.

If for any reason the technician wishes to clear the display window during the Self-Test, he must turn off the vehicle's engine, press the tester's pushbutton once to unlatch it (colon will disappear), then press the button again to latch down the button (colon will appear again). Every time the STAR tester is turned off, the low battery indicator (LO BAT) should show briefly at the upper left-hand corner of the tester's display window. If the LO BAT indicator shows steadily at any other time during the operation of the STAR tester with any Service Code, turn its power switch to OFF, and replace the 9-volt battery in the tester.

The STAR tester will display the last Service Code received, even after disconnecting it from the vehicle. It will hold the Service Code on the display until the power is turned off or the pushbutton is unlatched and relatched.

Anyone who departs from the instructions provided in this publication must first establish that he compromises neither his personal safety nor the vehicle integrity by his choice of methods, tools, or parts.

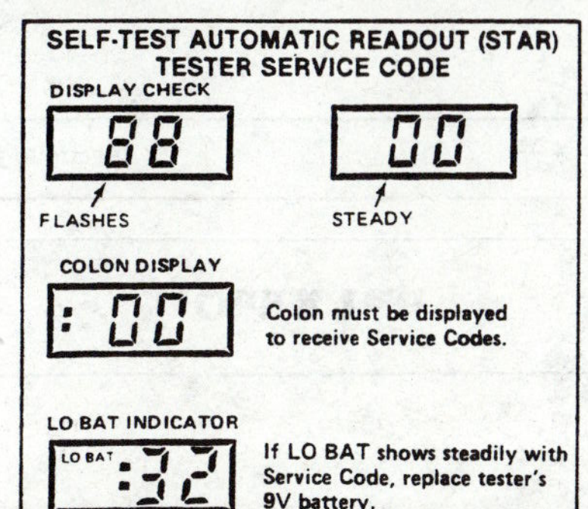

Quick Test

1	VISUAL CHECKS

A. CHECK VACUUM LINES FOR CRACKED OR BROKEN HOSES AND PROPER CONNECTIONS.

B. CHECK THE ELECTRICAL HARNESS FOR:

- Broken, detached, or frayed wires/insulation.
- Terminals not completely seated in the connectors.
- Corrosion on terminals.

Quick Test

1	VISUAL CHECKS (Continued)

C. Verify that coolant level is to specification.

2	PRE-TEST CONDITIONS

A. Transmission in Park or Neutral.

Quick Test

2	PRE-TEST CONDITIONS (Continued)

B. Set parking brake. Block the wheels.

C. Turn Accessories Off.

Quick Test

2	PRE-TEST CONDITIONS (Continued)

D. Make sure that the vacuum hoses are connected to the air cleaner.

E. When using an analog voltmeter, make sure there is power to the choke (engine must be running).

10 to 12 Volts D.C. Scale

Quick Test

2 PRE-TEST CONDITIONS (Continued)

F. Start and idle the engine until the radiator upper hose is hot and pressurized.

If vehicle will not start, refer to No Start, Pinpoint Test 1.

3 SET-UP PROCEDURE

A. Turn the ignition key Off.

B. Locate the Self-Test connector near MCU module (left or right fender apron as viewed from driver's seat).

Quick Test

3 SET-UP PROCEDURE (Continued)

- If you are using an analog VOM during the Quick Test, connect a jumper wire or paper clip from Self-Test Trigger (Circuit 201) to ground (Circuit 60) on Self-Test connector.
- Then connect an analog voltmeter from battery (+) to Self-Test Output on the Self-Test connector.
- Set voltmeter on DC volt range to read from 0 to 15 VDC.

- However, if you are using a STAR tester for the Quick Test:
- Connect black lead to the battery (—) post.
- Connect red lead to the Self-Test Output terminal on the Self-Test connector.
- Connect white lead to the Self-Test Trigger.
- Refer to back of STAR tester for additional instructions.

Quick Test

3	SET-UP PROCEDURE (Continued)

C. Remove PCV from breather cap on valve cover.

D. Tee a vacuum gauge to Canister Purge Solenoid Valve hose on the carbon canister side of the hose.

Quick Test

3	SET-UP PROCEDURE (Continued)

E. On 5.8L engine calibrations with a vacuum delay valve, there is a Tee and restrictor in the Thermactor Bypass Vacuum Control line. The restrictor must be uncapped during the Quick Test. Recap it after the Quick Test is completed.

Quick Test — Test 4

TEST STEP	RESULT ▶	ACTION TO TAKE
4.0 KEY ON, ENGINE OFF QUICK TEST		
• Key Off.	Service Code: 1-1 ▶	CONTINUE to Quick Test 5.
• Turn key to On with engine Off.	Service Codes: ▶	See note below.
• Observe voltmeter.	5-1 ▶	Pinpoint Test 4, open switch.
• Start service with first code received.	6-5 ▶	Pinpoint Test 5, closed switch.
	Steady high or low voltage ▶	Pinpoint Test 14, No Self-Test Output.
	Erroneous Codes ▶	Pinpoint Test 14, No Self-Test Output.

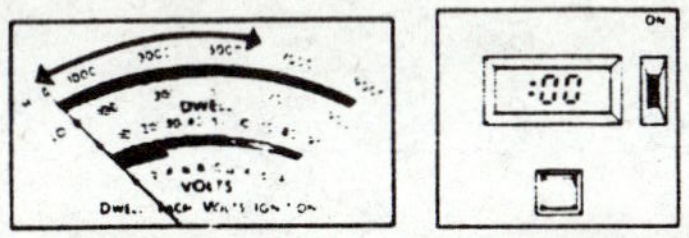

VOLTMETER

STAR TESTER

Display :00

NOTE: ONE QUICK INITIALIZATION PULSE MAY OCCUR ON THE VOLTMETER IMMEDIATELY AFTER TURNING THE IGNITION KEY ON. THE OUTPUT SERVICE CODE WILL OCCUR IN APPROXIMATELY FIVE SECONDS.

*Before proceeding with appropriate Pinpoint Test, read note on Pinpoint Test usage on page 27-21.

Quick Test — Test 5

TEST STEP	RESULT ▶	ACTION TO TAKE
5.0 ENGINE RUNNING QUICK TEST	Service Code: 1-1* for vehicles below 4000 ft. elevation. OR 6-2* for vehicles ABOVE 4000 ft. elevation. ▶	For vehicles with drive complaint of detonation (WOT only) and/or poor performance and equipped with Universal Ignition Modules, CONTINUE to Test 6. MCU system OK.
• Verify that engine is at normal operating temperature.		
• Run engine at 2000 rpm for 2 minutes, then turn key Off. Immediately restart engine and idle.		
• Observe voltmeter and vacuum gauge for initialization pulses after restarting the engine. The throttle kicker will also extend at this time (rpm increases) and remain on throughout test.	For the following results, go to ▶	Pinpoint Test Number:
• **Vehicles with Knock Sensor:** When 4 initial pulses occur, immediately simulate spark knock by placing a 3/8 inch extension on manifold near base of knock sensor. Tap extension end lightly with light (4 oz.) hammer for approximately 15 seconds.	No start ▶	Test 1, No Start
	No initialization pulses on voltmeter or erroneous codes ▶	Test 14, No Self-Test Output
	No pulses on vacuum gauge ▶	Test 14, No Self-Test Output
OR	No Throttle Kicker extension or rpm increase ▶	Test 8, RPM Out of Specification
• After the 4 initial pulses appear, observe and record voltmeter pulses (service codes). Within 90 seconds, the service codes are complete and the throttle kicker will retract (rpm decreases).	Service Codes: 1-2 ▶	Test 8, RPM Out of Specification
	2-5 ▶	Test 13, Knock Sensor
• Always start service with first code received.	4-1 ▶	Test 6, Fuel Lean
	4-2 ▶	Test 7, Fuel Rich
	4-4 ▶	Test 10, Thermactor
	4-5 ▶	Test 11, Always Upstream
	4-6 ▶	Test 12, Alway Not Bypassing
	5-3 or 5-4 ▶	Test 2, Open Switch
	5-1 or 5-5 ▶	Test 4, Open Switch
	6-2* below 4000 ft. 1-1* ABOVE 4000 ft. ▶	Test 15.
	6-1 or 6-5 ▶	Test 5, Closed Switch

*Before proceeding with appropriate Pinpoint Test, read important note on Pinpoint Test usage on page 27-21.

Quick Test — Test 6

TEST STEP	RESULT ▶	ACTION TO TAKE
6.0 SPARK KNOCK, ONLY AT WIDE OPEN THROTTLE, AND/OR POOR PERFORMANCE/MILEAGE. PERFORM STEP 6.3 ON VEHICLES WITH KNOCK SENSOR ONLY		
• Disconnect knock sensor. • Verify initial timing per Emission decal, correct timing if necessary. • With vehicle at normal operating temperature, disconnect the two wire ignition module connector (contains yellow circuit 324 and black/white circuit 60 wires) from vehicle harness, and jumper the wires together at the ignition module. • Read and note ignition timing. • Remove jumper wire, and reread ignition timing.	Retards 16-20 degrees ▶	RECONNECT ignition module, and go to Step [6.1]
	Does not retard 16-20 degrees ▶	REPLACE ignition module, and repeat all of Step [6.0]

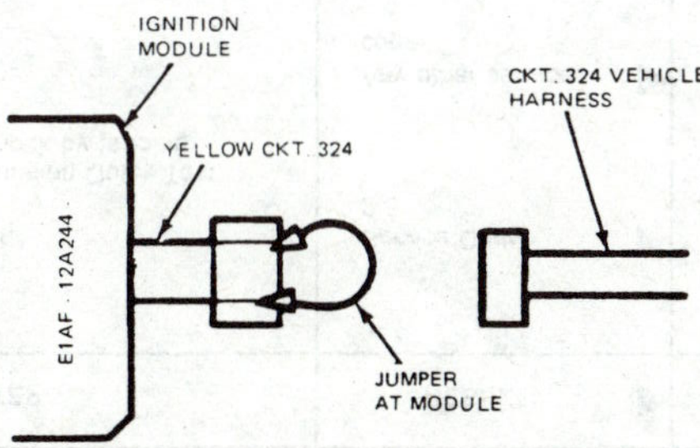

Quick Test — Test 6

TEST STEP	RESULT ▶	ACTION TO TAKE
6.1 • With vehicle at normal operating temperature, set high cam on lowest step that will maintain an engine speed greater than 2400 rpm. • Read and note ignition timing. • Remove and plug the vacuum hose from zone vacuum switch. Reread timing.	Retards more than 5 degrees ▶	If a system has a knock sensor, CONTINUE to Step [6.3], otherwise MCU System, OK; testing completed.
	Retards less than 5 degrees ▶	CONTINUE to Step [6.2]

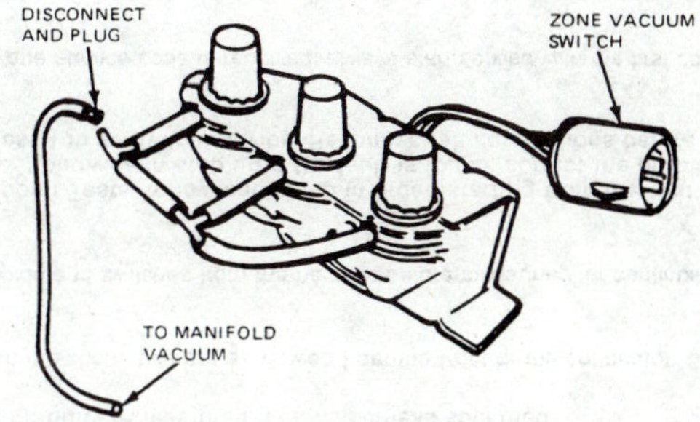

TEST STEP	RESULT ▶	ACTION TO TAKE
6.2 While still at 2400 rpm, disconnect ignition module (connector with yellow and black/white wires).	Timing retards ▶	CHECK circuit 324 for short to ground. If OK, REPLACE MCU and retest.
	Timing does not retard ▶	CHECK circuit 324 for open. If OK, REPLACE MCU and retest

Quick Test — Test 6

TEST STEP	RESULT ▶	ACTION TO TAKE
6.3 • Reconnect knock sensor. • Rerun Step 5, Engine Running Quick Test, but DO NOT simulate knock by tapping sensor.	Service Code: 2-5 ▶	MCU. OK; testing completed
	Any other service code ▶	REPLACE knock sensor and retest

Pinpoint Test

INSTRUCTIONS FOR USING THE PINPOINT TESTS

Important note on Pinpoint Test usage:

- Do not measure voltage or resistance at the MCU module or connect any test lamps to it, unless otherwise specified.
- The symbol "R" indicates resistance: that is. R = 100 ohms means the resistance is 100 ohms.
- Isolate both ends of a circuit and turn key Off whenever checking for shorts (R less than 1000 ohms) or continuity (R less than 5 ohms), unless specified.
- Connectors shown in schematics are as viewed from the front of the connector. See diagram below.
- Disconnect solenoids and switches from the harness before measuring for continuity or resistance.
- In using the Pinpoint Tests, follow each step in order, starting from the first step in the appropriate routine. Follow each step until the fault is found; correct the problem, then proceed to Quick Test 4 to verify the service. Reconnect all connections before conducting the Quick Test.
- When more than one service code is received. always start service with the first code received.

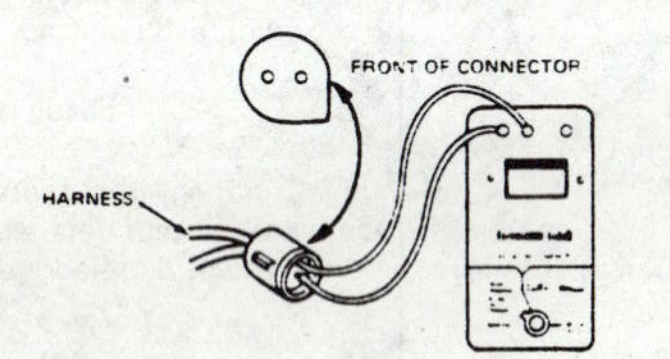

No Start — Pinpoint Test 1

TEST STEP	RESULT	▶	ACTION TO TAKE
1.0 NO START — DETECTS MCU RELATED FAULTS ONLY			
1.1 • Key Off. • Disconnect the vehicle harness from the ignition module and the horseshoe connector from the coil. • Connect an ohmmeter between circuit 11 and engine block ground. • Measure resistance.	R is more than 1000 ohms.	▶	MCU System OK, refer to appropriate Routine, Section 2.
	R is 1000 ohms or less.	▶	CONTINUE to Step 1.2

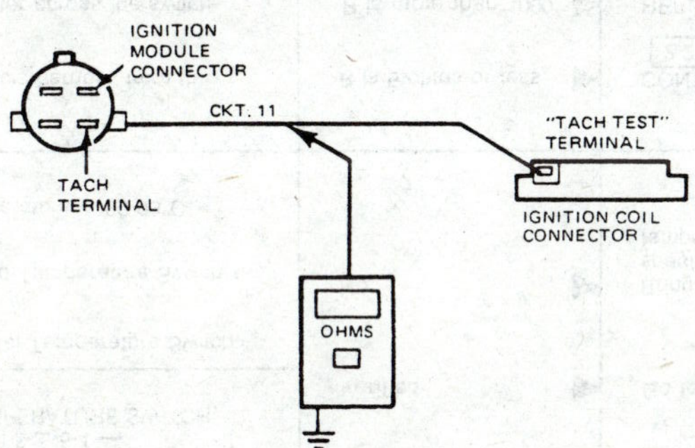

No Start — Pinpoint Test 1

TEST STEP	RESULT	▶	ACTION TO TAKE
1.2 • Disconnect the vehicle harness from the MCU. • Check circuit 11 for short to ground.	R is more than 1000 ohms.	▶	REPLACE MCU, and Retest.
	R is 1000 ohms or less.	▶	SERVICE or REPLACE circuit 11, and Retest.

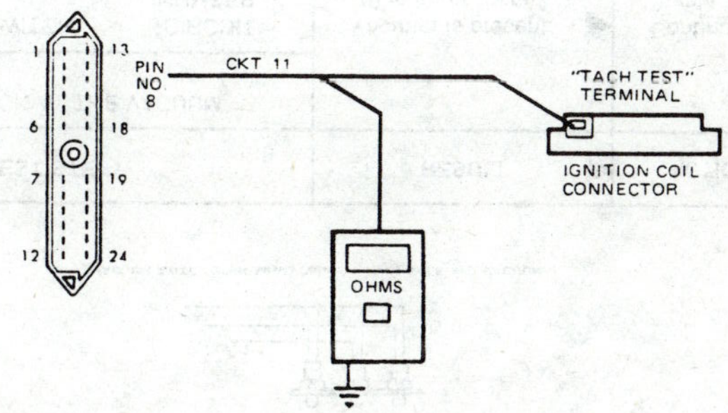

Temperature Switch — Pinpoint Test 2

TEST STEP	RESULT	ACTION TO TAKE
2.0 SERVICE CODE(S) 5-3, 5-4 — ELECTRICAL TEMPERATURE SWITCH, OPEN • CODE 5-3 is for Dual Temperature Switch, circuit 308. • CODE 5-4 is for Mid-Temperature Switch, circuit 309. • Verify coolant temperature is 60-93°C (140-200°F).	Verified ▶	Go to next Step 2.1
	No ▶	Bring engine coolant to specified temperature.
2.1 • Disconnect the vehicle harness from the switch. • Connect an ohmmeter across the switch terminals.	R is 5 ohms or less. ▶	CONTINUE to Step 2.2
	R is more than 1000 ohms. ▶	REPLACE switch
2.2 • Connect one lead of DVOM to Circuit 308 (Dual) or Circuit 309 (MID) and the other lead to engine ground.	R is 1,000 ohms or less. ▶	Service or replace circuit.
	R is more than 1000 ohms. ▶	Continue to next step 2.3
2.3 • Disconnect vehicle harness from MCU. • Check circuits 60 and 308 (Dual), 309 (MID) for continuity.	R is 5 ohms or less. ▶	Replace MCU.
	R is more than 1000 ohms. ▶	Service or repair circuit. Retest

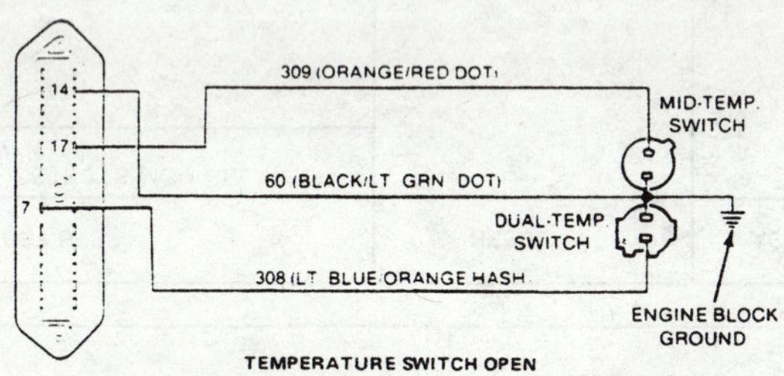

TEMPERATURE SWITCH OPEN
(Service Codes, 5-3 DUAL-TEMPERATURE, and 5-4, MID-TEMPERATURE)

Zoned Vacuum Switch — Pinpoint Test 4

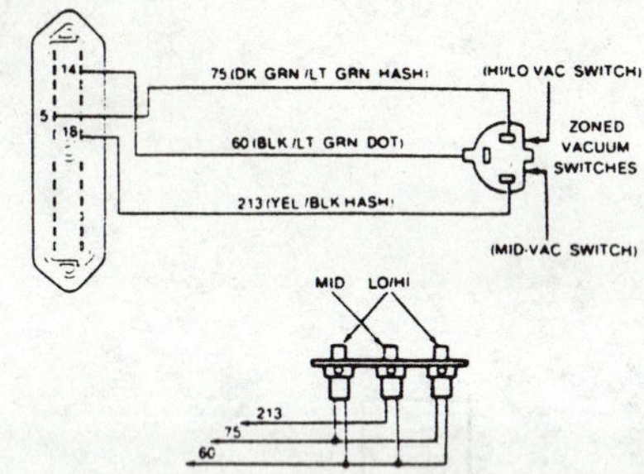

VACUUM SWITCH OPEN (Service Codes 5-1, Hi-Lo, or 5-5, MID VACUUM)

TEST STEP	RESULT	ACTION TO TAKE
4.0 VACUUM CHECK AT THE VACUUM SWITCHES. CODE: 5-1, SWITCH: Hi/Low, CIRCUIT NUMBER: 75 CODE: 5-5, SWITCH: Mid VAC, CIRCUIT NUMBER: 213 • Verify manifold vacuum is present at switches during idle condition.	Vacuum is present. (10 in hg or more) ▶	Continue to next Step 4.1
	No vacuum or less than 10. ▶	Check vacuum line for pinched, plugged, cracked conditions, or basic mechanical problem.
4.1 VACUUM SWITCH BLEED-DOWN TEST • Key Off. • Apply vacuum to the switch assembly.	Holds vacuum. ▶	For a service code 5-1 go to 4.2. For a service code 5-5 go to 4.4
	Does not hold vacuum. ▶	Check vacuum hoses for leaks. If no leaks are found replace the switch assembly and retest.

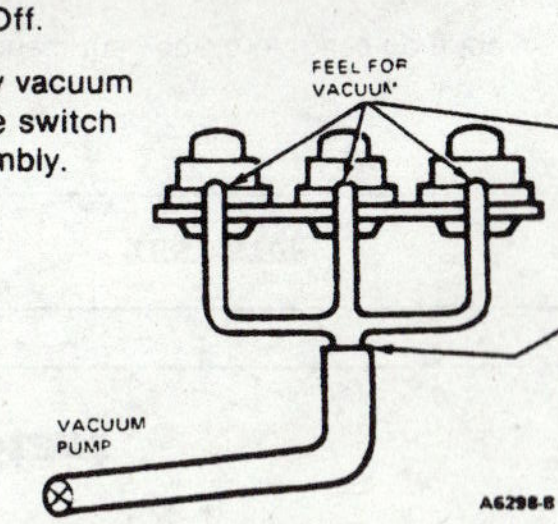

Zoned Vacuum Switch | Pinpoint | Test 4

TEST STEP	RESULT ▶	ACTION TO TAKE
4.2 HI/LOW VACUUM SWITCH		
• Key Off. • Disconnect the vehicle harness from the vacuum switch connector. For Hi/Low, do not apply vacuum to switch. • Check resistance of Hi/Low vacuum switch circuit 75 to circuit 60.	R is 5 ohms or less. ▶	CONTINUE to Step 4.3
	R is more than 5 ohms. ▶	REPLACE switch assembly. Retest.
4.3		
• Disconnect the 24 pin connector from MCU module.	R is 5 ohms or less. ▶	REPLACE MCU and retest.
• Check circuits 60 and 75 of the vehicle harness for continuity.	R is more than 5 ohms. ▶	SERVICE or REPLACE circuit and retest.

Zoned Vacuum Switch | Pinpoint | Test 4

TEST STEP	RESULT ▶	ACTION TO TAKE
4.4 MID VACUUM SWITCH		
• Key Off. • Disconnect the vehicle harness from the vacuum switch connector. • For MIDVAC, apply minimum of (15 inches Hg.) of vacuum to switch.	R is 5 ohms or less. ▶	RECONNECT vacuum hose. CONTINUE to Step 4.5
• Check resistance of MIDVAC vacuum switch, circuit 213 to circuit 60.	R is more than 5 ohms. ▶	REPLACE switch assembly. Retest.
4.5		
• Disconnect the 24 pin connector from MCU module.	R is 5 ohms or less. ▶	REPLACE MCU. Retest.
• Check circuits 60 and 213 of the vehicle harness for continuity.	R is more than 5 ohms. ▶	SERVICE or REPLACE circuit. Retest.

Zoned Vacuum Switch — Pinpoint Test 5

TEST STEP	RESULT	▶	ACTION TO TAKE
5.0 SERVICE CODE 6-5, VACUUM SWITCH, CLOSED			
• Verify engine is at normal operating temperature. If not, refer to Section 2, Routines 216 and 217. Verify that the vacuum hoses are connected to the switch; if not, connect and continue.	Verified	▶	GO to Step 5.1
	No	▶	Bring engine to normal operating condition.
5.1			
• Key Off.			
• Disconnect vehicle harness from vacuum switch connector.	R is more than 1000 ohms.	▶	RECONNECT vacuum hose. CONTINUE to Step 5.2
• Connect ohmmeter across the MIDVAC switch terminals at the switch connector. Must measure from circuit 213 to circuit 60.	R is 1000 ohms or less.	▶	REPLACE switch assembly. Retest.
5.2			
• Disconnect the vehicle harness from the MCU.	R is more than 1000 ohms.	▶	REPLACE MCU. Retest.
• Connect one lead of ohmmeter to the vehicle harness circuit 213 and the other lead to engine block ground.	R is 1000 ohms or less.	▶	SERVICE or REPLACE circuit. Retest.

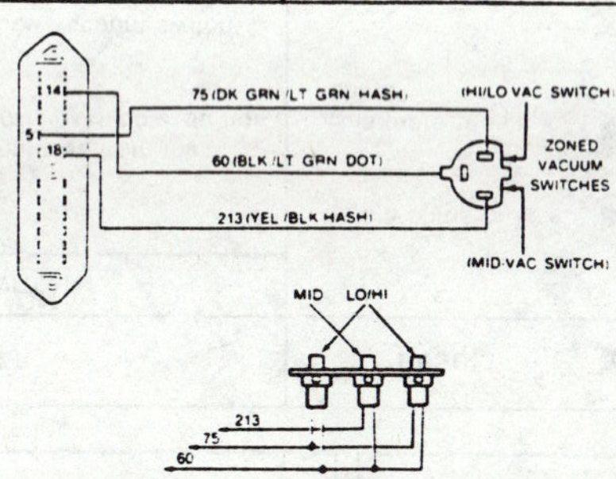

VACUUM SWITCH OPEN (Service Codes 5-1, Hi-Lo, or 5-5, MID VACUUM)

Fuel Always Lean — Pinpoint Test 6

6.0 SERVICE CODE 4-1

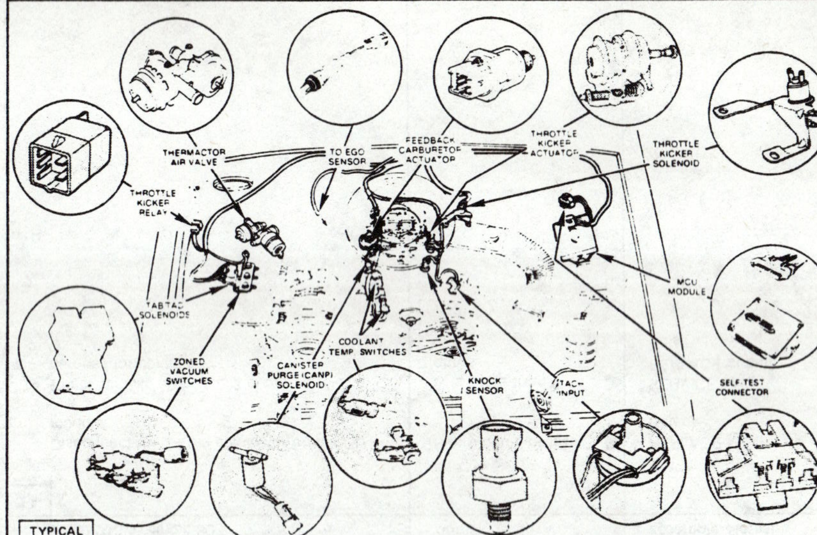

TYPICAL

Components controlled by the MCU System in this Pinpoint Test are the Feedback Carburetor Actuator, the Ego Sensor, Thermactor Air Valve, the TAB/TAD Solenoids, and the MCU Module. The components not directly controlled by the MCU System during this Pinpoint Test but could possibly cause a Service Code fault are as follows:

- Malfunctioning Carburetor
- Low Fuel Pressure
  - —Fuel Line Kink
  - —Fuel Pump Malfunction
- Vacuum Leaks
  - —EGR Gasket
  - —Head Gasket
  - —Cracked Hoses, etc.
- Canister Purge Solenoid
- Bowl Vent
- PCV

NOTE: Virtually anything that adds air or restricts fuel should be investigated.

Fuel Always Lean — Pinpoint Test 6

TEST STEP	RESULT	ACTION TO TAKE
6.0 SERVICE CODE 4-1—FUEL ALWAYS LEAN (CONTINUED)		
• Run engine at 2000 rpm for 60 seconds, to ensure that the EGO sensor is at an operating temperature.	Service code 4-1 still present. ▶	Continue to Pinpoint Test 6.1
• Rerun Engine Running Self-Test.	No service code 4-1 present. ▶	SERVICE the next service recorded from this self-test.

Fuel Always Lean — Pinpoint Test 6

TEST STEP	RESULT	ACTION TO TAKE
6.1 EGO SENSOR CHECK		
CAUTION: To prevent damage to the EGO sensor, never connect an ohmmeter to the EGO sensor; use a voltmeter with an input impedence of 10 megohms or greater.		
• Key Off. • Deactivate this vehicle self test. • Disconnect the vehicle harness from the EGO sensor. • Connect a jumper wire between the vehicle harness connector and the EGO sensor connector. • Connect a voltmeter between engine block ground and the metal clip of the jumper wire. Verify a good connection exists with the jumper wire between the two connectors. • Engine On, rpm between 2000-2300. • Disconnect zoned vacuum switches from the engine vacuum source, and connect a hand vacuum pump to the switches. • Apply 34-41 kPa (10-12 In.-Hg.) to the zoned vacuum switches. • Wait 2 minutes, then observe the voltmeter.	Voltage is steady, then alternating above and below 0.50 volts ▶	GO to Step 6.14
	Voltage remained under 0.50 volts ▶	CONTINUE to Step 6.2
	Voltage remained above 0.50 volts ▶	GO to Step 6.6

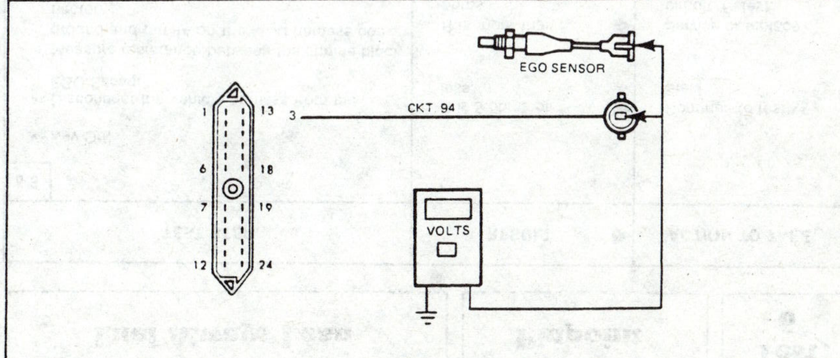

Fuel Always Lean — Pinpoint Test 6

TEST STEP	RESULT	ACTION TO TAKE
6.2 • Key Off. • Reconnect EGO sensor to to the vehicle harness. • Disconnect the thermactor air hose at the air pump. • Disconnect the vehicle harness connector from the MCU. • Connect a voltmeter between circuit 60 and circuit 94 of the harness connector. • Start engine. Run engine at idle. • Place carburetor on high idle cam. • Force system rich by depressing the CVR rod. and observe voltmeter.	Voltage is 0.5 volts or more ▶	GO to Step 6.6
	Voltage is less than 0.5 volts. ▶	CONTINUE to Step 6.3

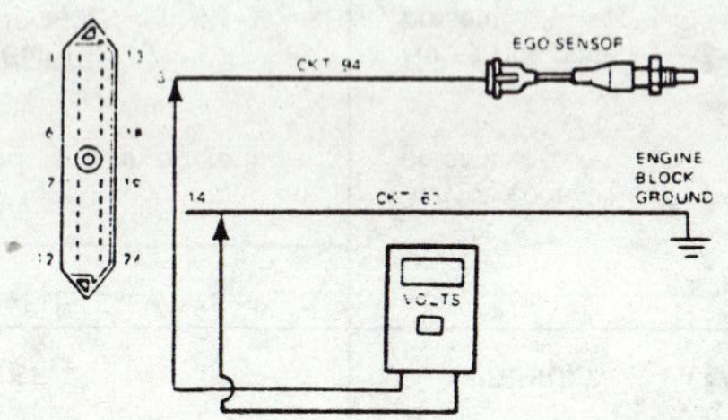

Fuel Always Lean — Pinpoint Test 6

TEST STEP	RESULT	ACTION TO TAKE
6.3 • Key Off. • Disconnect the vehicle harness from the EGO Sensor. • Measure resistance between the engine block ground and pin 14 on the MCU harness connector.	R is 5 ohms or less. ▶	Continue to test Step 6.4
	R is more than 5 ohms. ▶	Service or replace circuit. Retest.
6.4 • Check continuity of circuit 94.	R is 5 ohms or less. ▶	Continue to test Step 6.5
	R is more than 5 ohms. ▶	Service or replace circuit 94. Retest.
6.5 • Connect an ohmmeter between the engine block ground and circuit 94. • Measure the resistance.	R is 1,000 ohms or less. ▶	SERVICE or replace circuit 94. Retest.
	R is more than 1,000 ohms. ▶	REPLACE EGO sensor. Retest.

Fuel Always Lean — Pinpoint Test 6

TEST STEP	RESULT	▶ ACTION TO TAKE
6.6 FEEDBACK CARBURETOR ACTUATOR MOTOR (FBCA)		
• Key Off. • Apply 34-41 kPa (10-12 In.-Hg.) to the zoned vacuum switches. • Reconnect the vehicle harness connector to the MCU and EGO sensor. • Disconnect the vehicle harness from the FBCA motor. • Connect a known good FBCA motor, shaft fully depressed, to the vehicle harness. Do not remove existing FBCA motor from carburetor. • Start engine and set on low speed step of fast idle cam. • While observing the FBCA motor shaft, remove the PCV hose from the carburetor.* • Within two minutes, the known good FBCA Motor shaft should fully extend.	FBCA Motor shaft fully extends.	RECONNECT the PCV hose to the carburetor and the thermactor hose to the pump. GO to Step 6.10
	FBCA Motor shaft does not extend	RECONNECT the PCV hose to the carburetor. CONTINUE to Step 6.7

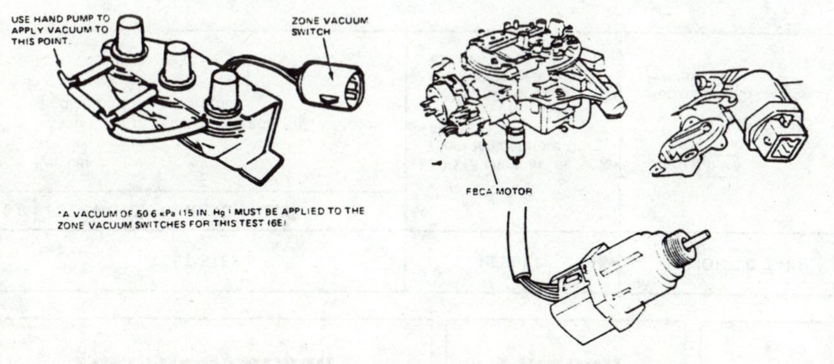

*Because of the vacuum leak caused by this Test Step, the vehicle may run rough.

Fuel Always Lean — Pinpoint Test 6

TEST STEP	RESULT	▶ ACTION TO TAKE
6.7 FBCA MOTOR HARNESS CHECK		
• Key Off. • Disconnect the vehicle harness from the MCU and the FBCA motor. • With an ohmmeter, check the continuity of circuit 687, 95, 96, 97 and 98 of the vehicle harness.	R is 5 ohms or less for each circuit.	CONTINUE to Step 6.8
	R is more than 5 ohms in any one of the circuits.	SERVICE or REPLACE the circuit(s). Retest.

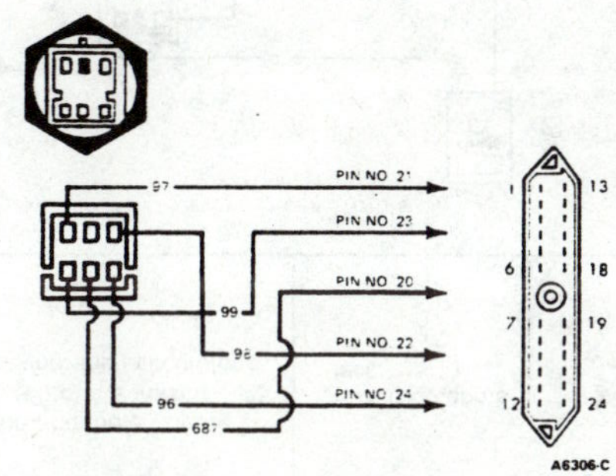

Fuel Always Lean | Pinpoint | Test 6

TEST STEP	RESULT ▶	ACTION TO TAKE
6.8 FBCA MOTOR WINDING CHECK		
• Key Off. • Measure the resistance of each FBCA Motor winding on the carburetor.	50-175 ohms at room temperature or above. ▶	Go to 6.9
	Not 50-175 ohms at room temperature or above. ▶	Replace FBCA Motor. Retest.

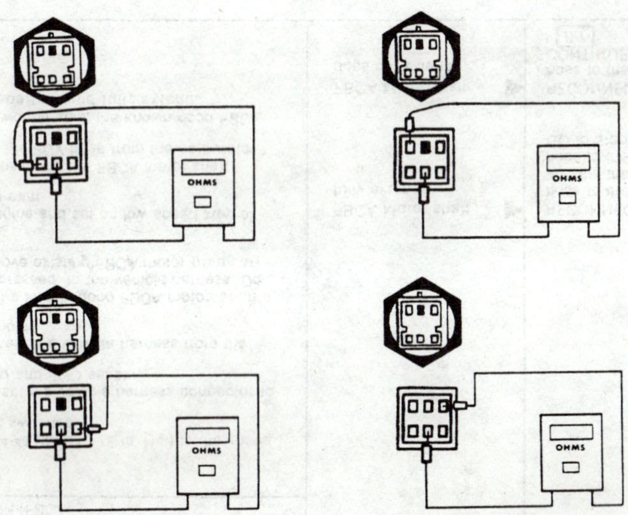

Fuel Always Lean | Pinpoint | Test 6

TEST STEP	RESULT ▶	ACTION TO TAKE
6.9		
• Connect one lead of ohmmeter to engine block ground. • With the other lead of the ohmmeter, probe each circuit (95, 96, 97, 98) of the vehicle harness connector, checking each circuit for a short to ground.	R is more than 1000 ohms. ▶	Replace MCU. Retest.
	R is 1000 ohms or less. ▶	SERVICE or REPLACE wire(s). Retest.

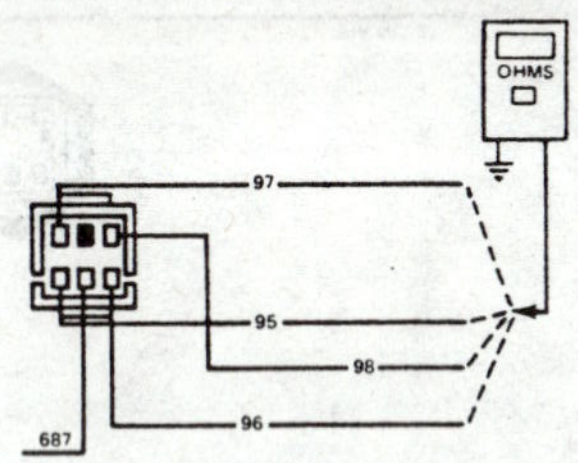

TEST STEP	RESULT ▶	ACTION TO TAKE
6.10 FBCA MOTOR		
• Key Off. • Connect an ohmmeter lead to pin 687 of the original FBCA motor in the carburetor and the other lead to the FBCA motor case.	R is more than 190,000 ohms. ▶	CONTINUE to Step 6.11
	R is 190,000 ohms or less. ▶	REPLACE FBCA motor. Retest.

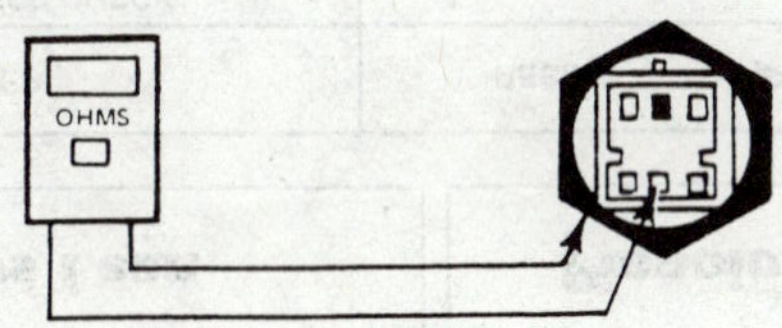

Fuel Always Lean — Pinpoint — Test 6

TEST STEP	RESULT	►	ACTION TO TAKE
6.11 • Disconnect FBCA motor from the carburetor. • Check the pintle for binding or damage. • Check the FBCA motor for physical damage, and check for movement by turning the key On.	FBCA motor, and pintle, OK.	►	RECONNECT the FBCA motor to the carburetor and vehicle harness. GO to [6.12]
	FBCA pintle, binding.	►	REPLACE FBCA pintle and retest.
	FBCA motor, physically damaged or no movement.	►	Replace FBCA motor and retest.
6.12 THERMACTOR AIR DIVERTER SOLENOID (TAD) • Key Off. • Disconnect the vehicle harness from the TAD solenoid. • Connect an ohmmeter across the TAD solenoid terminals. • Observe ohmmeter.	R is 50-100 ohms	►	CONTINUE to Step [6.13]
	R less than 50 ohms or greater than 100 ohms.	►	REPLACE TAD solenoid. Retest.

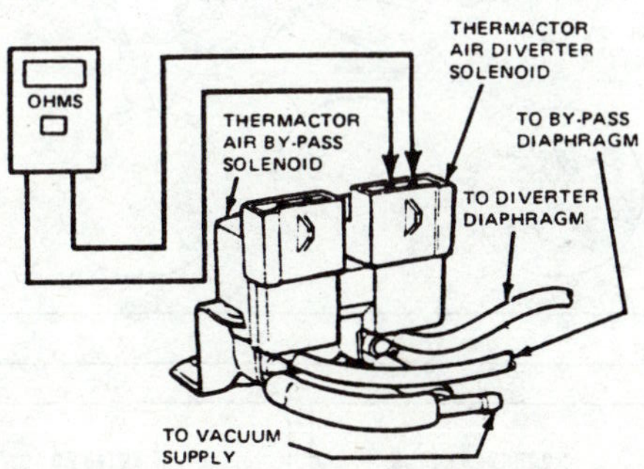

Fuel Always Lean — Pinpoint — Test 6

TEST STEP	RESULT	►	ACTION TO TAKE
6.13 • Disconnect the vehicle harness connector from the MCU. • Connect one lead of ohmmeter to circuit 200 of the vehicle harness and the other lead to engine block ground.	R is more than 1000 ohms.	►	RECONNECT the vehicle harness to the MCU. CONTINUE to Step [6.14]
	R is 1000 ohms or less.	►	SERVICE or REPLACE circuit 200. Retest.

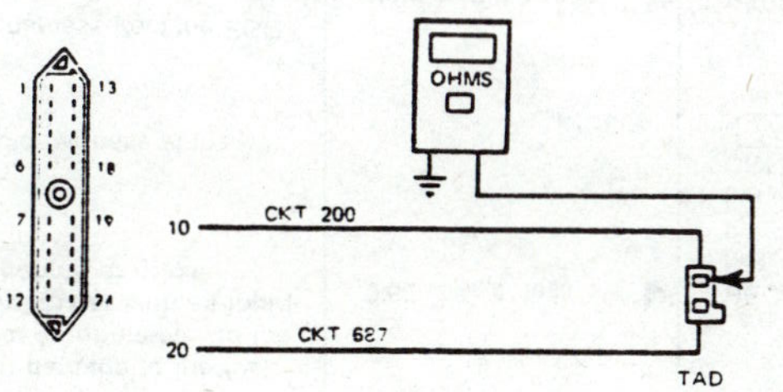

A6311-A

TEST STEP	RESULT	►	ACTION TO TAKE
6.14 • Check Thermactor Air Diverter valve, TAD solenoids, and connections. Refer to Section 10. • If Service Code 4-1 and 4-5 were both displayed during the Quick Test, perform Pinpoint Test 11.	No problem exists	►	CHECK for carburetor leaks[1], carburetor feedback[2], bowl vent hose leaks[2], and the in-line bowl vent solenoid for proper operation[2]. [1]Per Group 24. [2]Per Section 3.
	Problem exists	►	CORRECT problems. Retest.

Fuel Always Rich — Pinpoint Test 7

7.0 SERVICE CODE 4-2

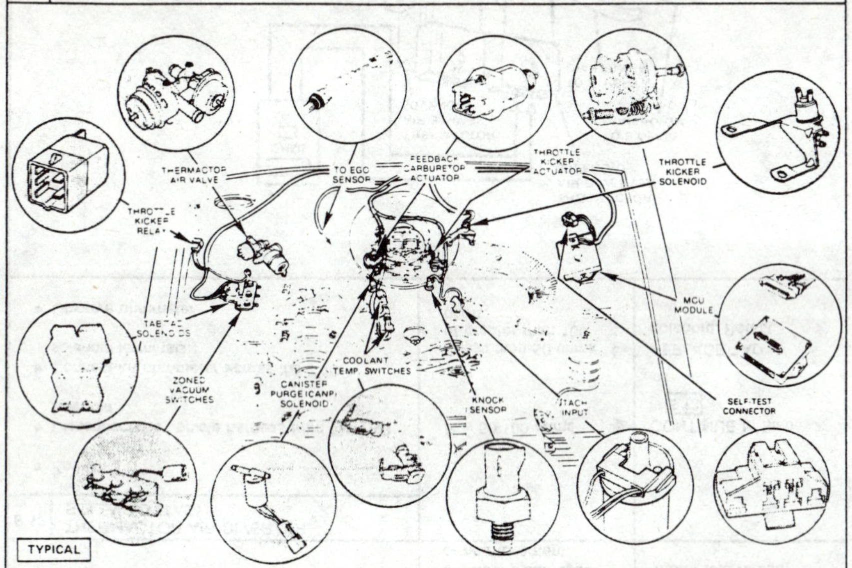

Components controlled by the MCU System in this Pinpoint Test are the EGO Sensor, the Feedback Carburetor Actuator, and the MCU Module. The components not directly controlled by the MCU System during this Pinpoint Test but could possibly cause a Service Code fault are as follows:

- Malfunctioning Carburetor
  - —High Float Level
  - —No Power To Choke Cap
- Misfiring Spark Plugs
  - —Coil
  - —Distributor
- Restricted Air Cleaner
- Restricted PVS
- Air Pump Not Working
  - —Malfunctioning Pulley
  - —Broken Belt(s)
- Throttle Kicker Always On
- PCV

NOTE: Virtually anything that adds fuel or restricts air should be investigated.

Fuel Always Rich — Pinpoint Test 7

TEST STEP	RESULT ▶	ACTION TO TAKE
7.0 SERVICE CODE 4-2—FUEL ALWAYS RICH (CONTINUED)		
• Disconnect the jumper wire from Self-Test Trigger input circuits 201 and 60. • Vehicle must be at normal operating temperature. • Start engine, run at idle. • Verify that there is power to the choke cap, cold enrichment circuit is operating properly, and choke is Off.	System, OK. ▶	Continue to Step 7.1
	System, not OK. ▶	Service as required. Retest.
7.1 **CAUTION: To prevent damage to the EGO sensor, never connect an ohmmeter to the EGO sensor; use a voltmeter with an input impedence of 10 megohms or greater.** • Key Off. • Disconnect the thermactor air hose at the pump. • Deactivate self test. • Disconnect the vehicle harness from the EGO sensor. • Connect a voltmeter between the EGO sensor and engine block ground. • Disconnect the PCV hose from the carburetor, creating a vacuum leak. • Run engine at 2400 rpm for 60 seconds, then observe voltmeter.	0.50 volts or less. ▶	RECONNECT PCV hose. RECONNECT the EGO sensor to the vehicle harness. CONTINUE to Step 7.2
	Voltage more than 0.50 volts. ▶	REPLACE EGO sensor. Retest.

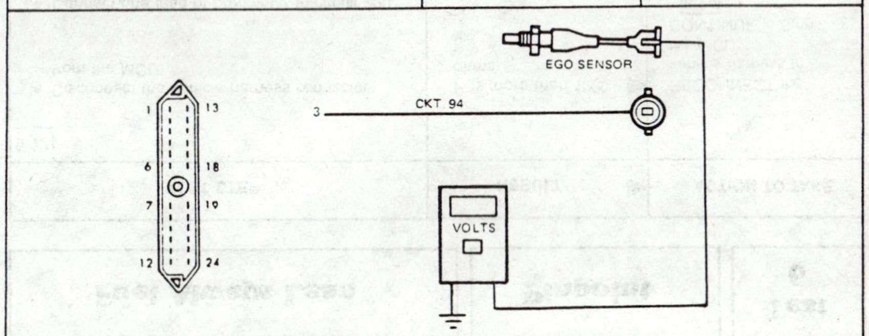

Fuel Always Rich — Pinpoint — Test 7

TEST STEP	RESULT ▶	ACTION TO TAKE
7.2 • Key Off. • Disconnect the vehicle harness from the FBCA Motor, connect a known good FBCA motor (shaft fully depressed) to harness, but do not remove the original FBCA motor from the carburetor. • Key to Run.	FBCA motor shaft extends fully, then may retract slightly. ▶	CONTINUE to Step 7.3
	Shaft does not extend fully. ▶	CONTINUE to Step 7.4

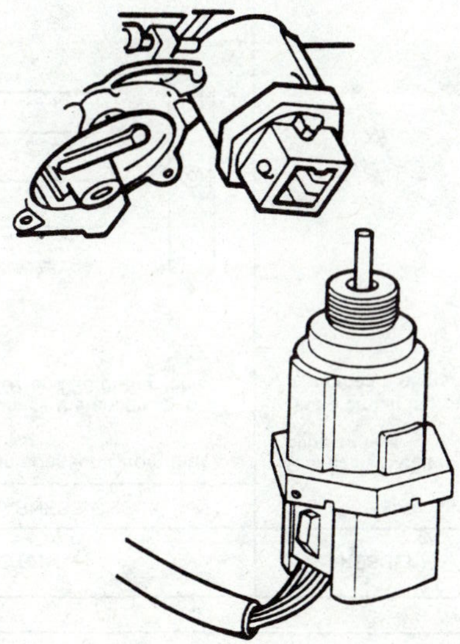

Fuel Always Rich — Pinpoint — Test 7

TEST STEP	RESULT ▶	ACTION TO TAKE
7.3 • Apply a vacuum of 33-40 kPa (10-12 inches Hg.) to the zone vacuum switches for this test. • Start vehicle, place on high cam. • Allow 120 seconds, then force system rich by depressing CVR rod. • Observe remote FBCA motor shaft.	Shaft retracts until flush with motor ▶	Key Off. RECONNECT Thermactor air hose to the pump. GO to Step 7.6
	Shaft does not retract ▶	Key Off. DISCONNECT the known good FBCA motor from the harness. CONTINUE to Step 7.4

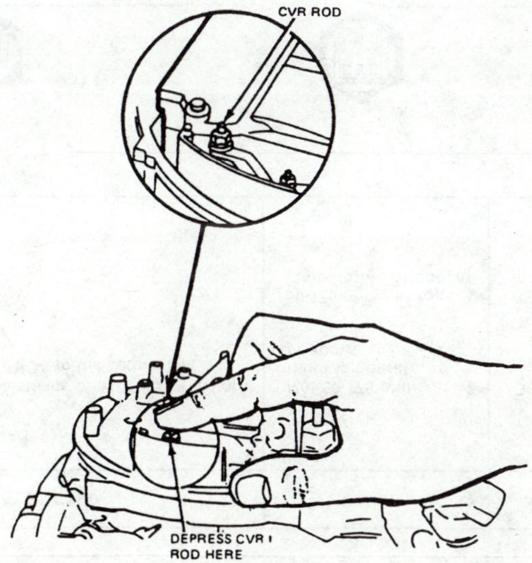

Fuel Always Rich — Pinpoint — Test 7

TEST STEP	RESULT ▶	ACTION TO TAKE
7.4 FBCA MOTOR HARNESS CHECK		
• Disconnect the 24 pin connector from the MCU module. • With an ohmmeter, check the continuity of circuits 687, 95, 96, 97 and 98 of the vehicle harness.	R of each circuit is 5 ohms or less. ▶	CONTINUE to Step 7.5
	R of a circuit is more than 5 ohms. ▶	SERVICE or REPLACE circuit. Retest.

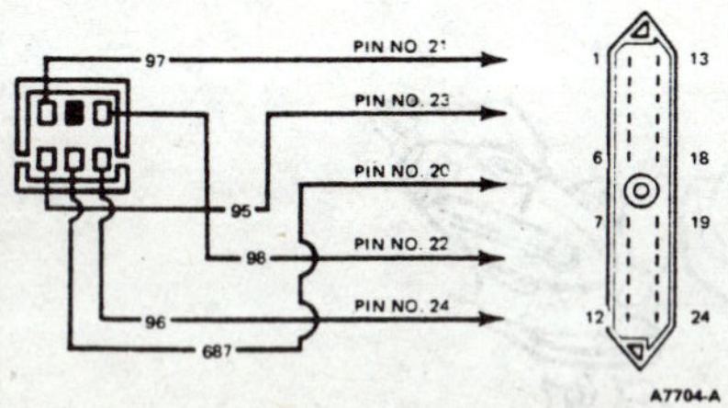

TEST STEP	RESULT ▶	ACTION TO TAKE
7.5 CHECK FBCA MOTOR HARNESS FOR A SHORT TO GROUND		
• Connect one lead of ohmmeter to engine block ground, and with the other lead, probe circuits 95, 96, 97 and 98 of the vehicle harness connector, checking each circuit for a short to ground.	R of each circuit is more than 1000 ohms. ▶	GO to 7.6
	R of each circuit is 1,000 ohms or less. ▶	SERVICE or REPLACE circuit. Retest.

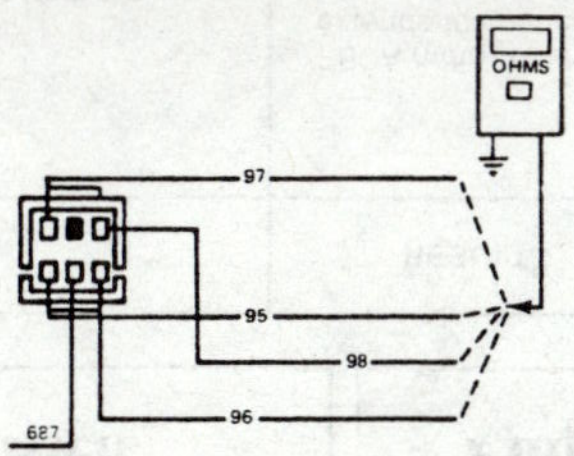

Fuel Always Rich — Pinpoint — Test 7

TEST STEP	RESULT ▶	ACTION TO TAKE
7.6		
• Measure the resistance of each FBCA motor winding to the FBCA in the carburetor.	Not 50-175 ohms at room temperature or above. ▶	GO to 7.7
	50-175 ohms at room temperature or above. ▶	REPLACE FBCA motor and retest.

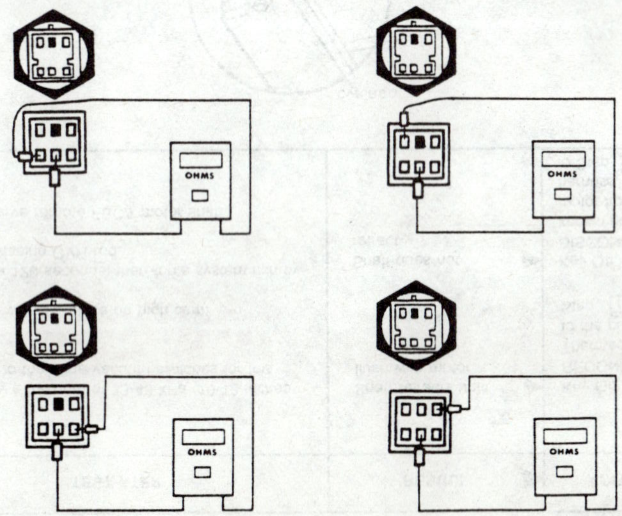

Fuel Always Rich — Pinpoint — Test 7

TEST STEP	RESULT ▶	ACTION TO TAKE
7.7 FBCA MOTOR		
• Key Off. • Connect an ohmmeter lead to pin 687 at the original FBCA motor in the carburetor and the other lead to the FBCA motor outside case.	R is more than 190,000 ohms. ▶	CONTINUE to Step 7.8
	R is 190,000 ohms or less ▶	REPLACE FBCA. Retest.

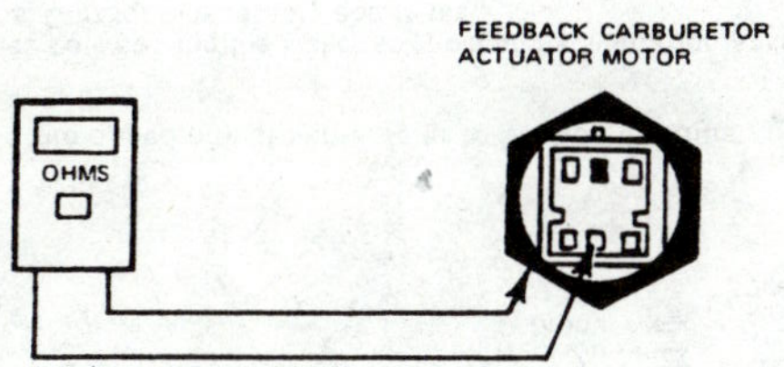

Fuel Always Rich — Pinpoint — Test 7

TEST STEP	RESULT ▶	ACTION TO TAKE
7.8		
• Disconnect the FBCA motor from the carburetor. • Check the pintle for binding or damage. • Check the FBCA motor for physical damage and check for movement by turning the key On.	FBCA motor and pintle, OK. ▶	Check Evaporative Emission Control system per Section 7. Check all vacuum lines for leaks, wrong or bad connections per Emissions Decal. Check bowl vent per Section 3. Check crankcase for fuel contamination by rerunning Quick Tests with PCV valve removed from the valve cover. If, the above items are OK, replace MCU and retest.
	FBCA pintle, binding. ▶	REPLACE the FBCA pintle. Retest.
	FBCA motor physically damaged or no movement. ▶	REPLACE FBCA motor and retest.

RPM Check | Pinpoint | Test 8

8.0 SERVICE CODE 1-2

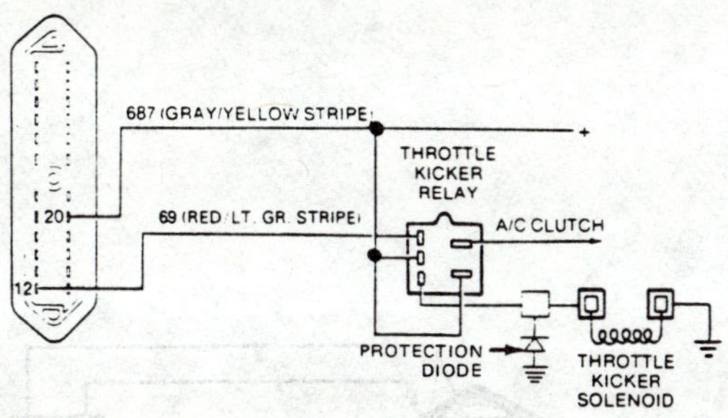

- Check carburetor idle speed adjustments, refer to Section 4. Adjust, if necessary, and re-test.
- Check vacuum lines between throttle kicker solenoid and actuator for leaks, blockage, or improper connections. Correct if necessary and re-test.

TEST STEP	RESULT	► ACTION TO TAKE
8.1		
• Disconnect self test input. • "T"-in a vacuum gauge to the throttle kicker. • Start engine, run at idle. • Turn A/C On. Vacuum should be present. • Turn A/C Off, no vacuum present.	Vacuum switches.	► Key Off. GO to Step 8.2
	No vacuum, or no vacuum switching.	► CHECK vacuum line for leaks, blockage, or wrong connections. If OK continue to 8.3

RPM Check | Pinpoint | Test 8

TEST STEP	RESULT	► ACTION TO TAKE
8.2		
• Key Off. • Check throttle kicker actuator for proper function. Apply at least 33 kPa (10 inches Hg.) vacuum to actuator vacuum line. Observe actuator arm.	Arm extends and holds	► GO to Step 8.3
	Arm does not extend and hold	► CHECK actuator. SERVICE or REPLACE and re-test
8.3		
• Check throttle kicker solenoid resistance (harness disconnected).	R is between 45 and 90 ohms	► GO to Step 8.4
	R less than 45 ohms or greater than 90 ohms.	► REPLACE solenoid. Re-test
8.4		
• Check circuits between throttle kicker relay connector and throttle kicker solenoid, and between the solenoid and ground, for continuity.	R is less than 5 ohms	► GO to Step 8.5
	R is more than 5 ohms.	► Service or replace open circuit in harness. Retest.

RPM Check — Pinpoint Test 8

TEST STEP	RESULT ▶	ACTION TO TAKE
8.5		
• MCU and relay disconnected. • Check circuits 69 and 687 for continuity between MCU and throttle kicker relay connector.	R is 5 ohms or less. ▶	GO to Step 8.6
	R is more than 5 ohms. ▶	SERVICE open circuit. Re-test
8.6		
• Check circuit 69 for isolation from ground.	R is more than 1,000 ohms. ▶	GO to Step 8.7
	R is 1,000 ohms or less. ▶	SERVICE grounded circuit 69. Re-test
8.7		
• Reconnect throttle kicker relay and throttle kicker solenoid. • Key On, Engine Off, A/C Off. • Check throttle kicker relay function when energized. Connect voltmeter across solenoid terminals. Ground circuit 69 at MCU connector, Pin 12, and observe voltage.	Voltage is more than 10 volts. ▶	Verify that circuit 69 is not shorted to 687. REPAIR harness if necessary then replace MCU.
	Voltage is less than 10 volts. ▶	Replace the relay. Retest.

Thermactor Check — Pinpoint Test 10

10.0 SERVICE CODE 4-4

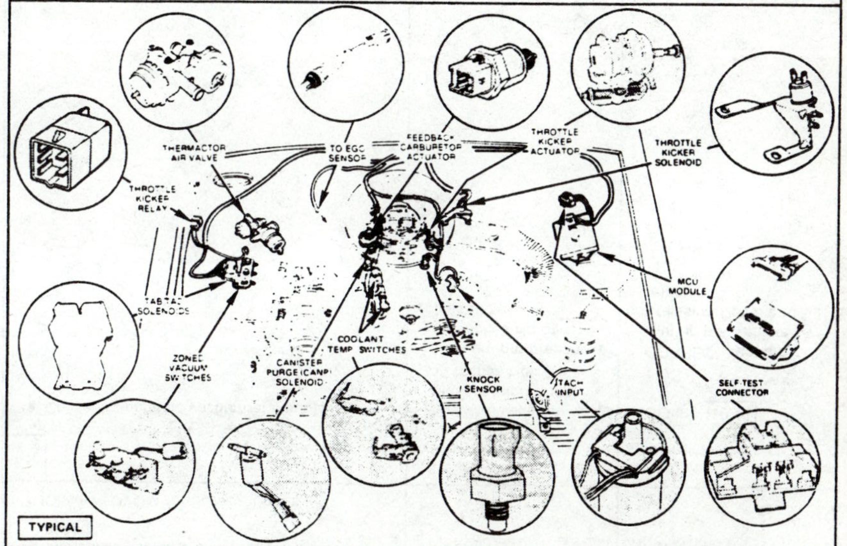

Components controlled by the MCU System in this Pinpoint Test are the TAB/TAD Solenoids, the Thermactor Air Valve, the EGO Sensor, and the MCU Module. The components not directly controlled by the MCU System during this Pinpoint Test but could possibly cause a Service Code fault are as follows:

- Retard Delay Valve (RDV)
- Temperature Vacuum Switch (TVS)
- Malfunctioning Carburetor
  - —High Float Level
  - —No Power To Choke Plate
- Misfiring Spark Plugs
  - —Coil
  - —Distributor
- Restricted Air Cleaner
- Restricted PVS
- Air Pump Not Working
  - —Malfunctioning Pulley
  - —Broken Belt(s)
- Throttle Kicker Always On
- PCV

NOTE: Virtually anything that adds fuel, restricts air, or inhibits Thermactor System from switching air upstream to the exhaust manifold should be investigated.

Thermactor Check — Pinpoint Test 10

10.0 SERVICE CODE 4-4 (CONTINUED)

TYPICAL THERMACTOR SYSTEM

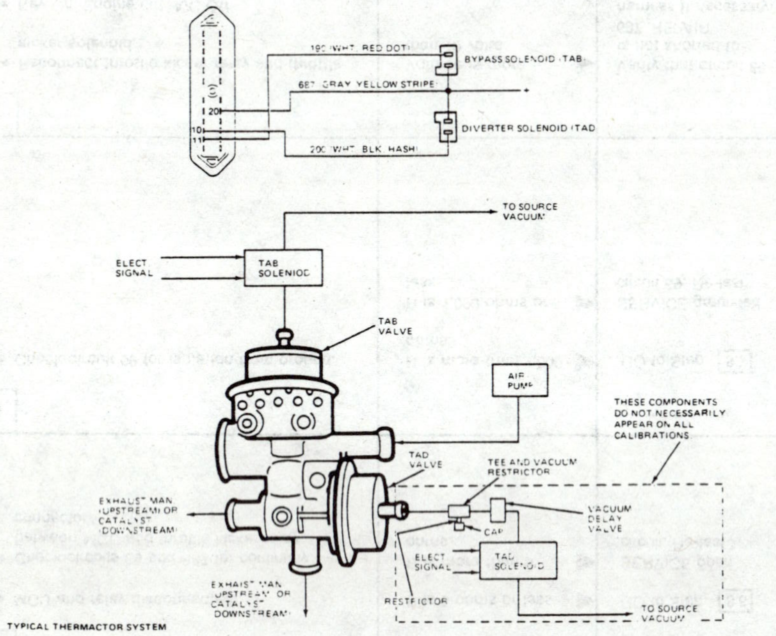

TYPICAL THERMACTOR SYSTEM

Thermactor Check — Pinpoint Test 10

TEST STEP	RESULT	ACTION TO TAKE
10.01 THERMACTOR CHECK		
This Step is for police vehicles only; all other vehicles, go to Pinpoint Test 10, Step 10.10 • Bring engine to operating temperature. • Verify upper radiator hose is hot and pressurized. • Turn engine off. • Restart engine at curb idle. • Immediately initiate Self-Test within 10 seconds. • Observe DVOM for Service Codes.	Any other Service Codes ▶	Go to the appropriate Pinpoint Test.
	Service Code 4-4 ▶	GO to 10.10
	Service Code 1-1 ▶	Continue with Quick Test.
10.10		
• Check for manifold vacuum at input to TAB-TAD solenoids.	Manifold vacuum present (more than 10 inches Hg). ▶	GO to 10.11
	Vacuum not present (10 inches Hg or less). ▶	CORRECT vacuum routing per Vehicle Emission Decal and retest.

Thermactor Check — Pinpoint Test 10

TEST STEP	RESULT	ACTION TO TAKE
10.11 • Remove the vacuum hose from the TAB valve, and connect a vacuum gauge to hose. Start engine, increase rpm to above 2,000 rpm for 2 minutes. Shut off, then restart and initiate Self-Test. Observe gauge during Self-Test.	Vacuum pulses above and below 17 kPa (5 inches Hg.) ▶	GO to 10.12
	Vacuum always above 17 kPa (5 inches Hg.) ▶	GO to 10.20
	Vacuum always below 17 kPa (5 inches Hg.) ▶	GO to 10.30
10.12 • Reconnect hose to TAB valve. Remove vacuum hose from the TAD valve and connect a vacuum gauge to hose. Start/re-start engine, increase the rpm to above 2,000 rpm for 2 minutes. Shut off, then restart and initiate Self-Test. Observe gauge during test.	Vacuum pulses above and below 17 kPa (5 inches Hg.) ▶	GO to Thermactor Air System Routine, Section 10.
	Vacuum always above 17 kPa (5 inches Hg.) ▶	GO to 10.40
	Vacuum always below 17 kPa (5 inches Hg.) ▶	GO to 10.50

Thermactor Check — Pinpoint Test 10

TEST STEP	RESULT	ACTION TO TAKE
10.20 • Key Off. • Check circuit 190 for a short to ground. Measure resistance.	R is more than 1,000 ohms. ▶	GO to 10.22
	R is 1,000 ohms or less. ▶	SERVICE short. Retest.
10.21 • Measure the TAB solenoid resistance. Should be between 50-110 ohms.	R within range ▶	GO to 10.32
	R not within range ▶	REPLACE TAB solenoid. Retest. If Service Code 4-4 is still present, replace MCU.
10.22 • Check at the TAB solenoid output to make sure vacuum is not present when the solenoid is de-energized (vehicle harness disconnected, engine at idle). • Disconnect the TAB solenoid electrically. • Start engine, run at idle.	Vacuum is present ▶	REPLACE TAB solenoid. Retest.
	Vacuum is not present ▶	REPLACE MCU module. Retest.

Thermactor Check — Pinpoint Test 10

TEST STEP	RESULT	ACTION TO TAKE
10.30 • Key Off. • Check the vacuum hose between the TAB valve and TAB solenoid for leakage or blockage.	Leakage or blockage found	SERVICE as required. Retest.
	No problem found	GO to 10.21
10.32 • Check the TAB solenoid output to make sure vacuum is not present when the solenoid is energized. • Jumper the TAB solenoid circuit 190 to ground. • Start engine, run at idle.	Vacuum present.	GO to 10.33
	Vacuum is not present.	REPLACE TAB solenoid. Retest if Service Code 4-4 present, continue to 10.33.
10.33 • Key Off. • Check circuits 687 and 190 for continuity between the TAB solenoid and MCU module. Measure resistance.	R is more than 5 ohms.	SERVICE circuit. Retest.
	R is 5 ohms or less.	REPLACE MCU module. Retest.

Thermactor Check — Pinpoint Test 10

TEST STEP	RESULT	ACTION TO TAKE
10.40 • Key Off. • Check circuit 200 for a short to ground. Measure resistance.	R is more than 1,000 ohms.	GO to 10.41
	R is 1,000 ohms or less.	SERVICE short
10.41 • Check at the TAD solenoid output to make sure vacuum is not present when the solenoid is de-energized (vehicle harness disconnected, engine at idle). • Disconnect the TAD solenoid electrically. • Start engine, run at idle.	Vacuum is present	REPLACE TAD solenoid. Retest.
	Vacuum is not present	REPLACE MCU. Retest.
10.50 • Check the vacuum hoses between the TAD valve and TAD solenoid for leakage or blockage.	Leakage or blockage found	SERVICE as required. Retest.
	OK	GO to 10.52

Thermactor Check — Pinpoint Test 10

TEST STEP	RESULT ▶	ACTION TO TAKE
10.52 • Key Off. • Measure the TAD solenoid resistance. Should be between 50-110 ohms.	R in range ▶	GO to 10.53
	R not in range ▶	REPLACE solenoid. Retest if Service Code 4-4 is still present go to 10.54.
10.53 • Check at the TAD solenoid output to make sure vacuum is present when the solenoid is energized by 12V. • Jumper the TAD solenoid circuit 200 to ground. • Start engine, run at idle.	Vacuum present ▶	GO to 10.54
	Vacuum not present ▶	REPLACE solenoid. Retest.
10.54 • Key Off. • Check circuits 687 and 200 for continuity between TAD solenoid and MCU module. Measure resistance.	R is more than 5 ohms. ▶	SERVICE circuit. Retest
	R is 5 ohms or less. ▶	REPLACE MCU module. Retest.

Thermactor Air Switching — Pinpoint Test 11

11.0 SERVICE CODE 4-5 THERMACTOR AIR ALWAYS UPSTREAM

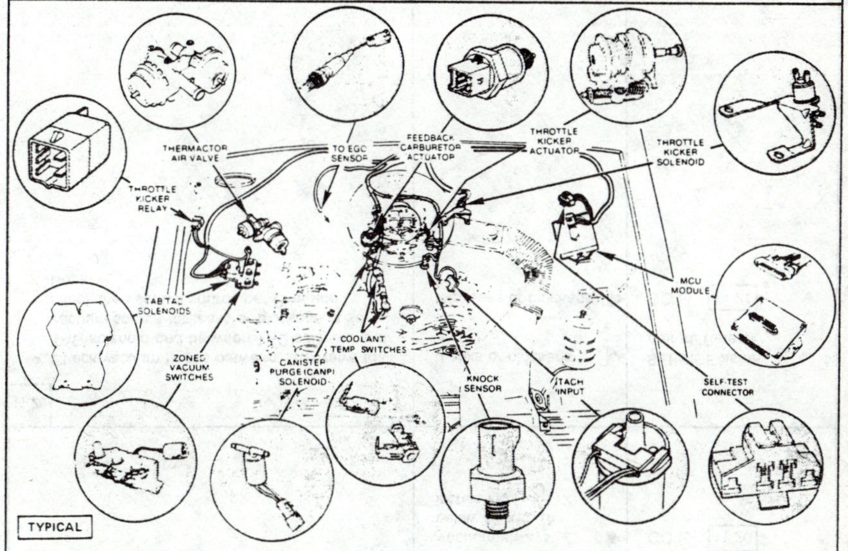

Components controlled by the MCU System in this Pinpoint Test are the TAB/TAD Solenoids, the Thermactor Air Valve, and the MCU Module. The component not directly controlled by the MCU System during this Pinpoint Test but could possibly cause a Service Code fault are as follows:

- Broken, Loose Belts
- Crimped, Pinched or Missing Hoses, Plugged Restrictor
- Malfunctioning Vacuum Delay Valve

NOTE: Virtually anything that inhibits the Thermactor System from switching air downstream to the catalyst should be investigated.

Thermactor Air Switching — Pinpoint — Test 11

11.0 SERVICE CODE 4-5 (Continued)

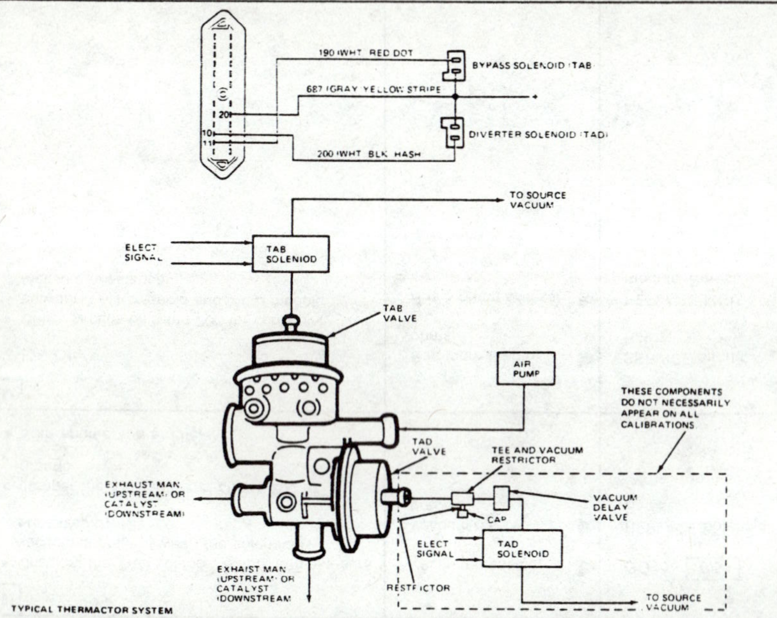

TYPICAL THERMACTOR SYSTEM

TEST STEP	RESULT ▶	ACTION TO TAKE
11.081		
• Verify the restrictor Teed into the TAD vacuum line was uncapped and restrictor was free from contamination.	Restrictor OK ▶	GO to 11.091
	Problem with restrictor ▶	SERVICE as required and retest
11.091		
• Cap vacuum restrictor, and remove vacuum delay valve from TAD vacuum control line. Reconnect system without delay valve and retest.	No Code 4-5 ▶	System OK. SERVICE vacuum delay valve per Section 3.
	Code 4-5 ▶	GO to 11.10

Thermactor Air Switching — Pinpoint — Test 11

TEST STEP	RESULT ▶	ACTION TO TAKE
11.10		
• Remove vacuum hose from the TAD valve, and connect a vacuum gauge to hose. Start/re-start engine, increase rpm to above 2,000 for 2 minutes. Shut off then restart and initiate self test. Observe vacuum gauge during test.	Vacuum pulses above and below 17 kPa (5 inches Hg.) ▶	MCU system, OK; so CHECK thermactor air, Section 10.
	Vacuum always above 17 kPa (5 inches Hg.) ▶	GO to 11.30
	Vacuum always below 17 kPa (5 inches Hg.) ▶	GO to 11.20
11.20		
• Check vacuum hoses between TAD valve to TAD solenoid and between TAD solenoid and vacuum source for leaks or blockage. Also check the vacuum source per Emission Decal.	Leaks or blockage found ▶	SERVICE as required. Retest.
	No leaks or blockage ▶	GO to 11.21

Thermactor Air Switching — Pinpoint Test 11

TEST STEP	RESULT	ACTION TO TAKE
11.21 • Key Off. • Measure the TAD solenoid resistance. • Should be between 50-110 ohms.	R within range ▶	GO to 11.22
	R not within range ▶	REPLACE solenoid. Retest.
11.22 • Check at the TAD solenoid output that vacuum is present when the solenoid is energized by 12V. • Jumper the TAD solenoid circuit 200 to ground. • Start engine, run at idle.	Vacuum present ▶	GO to 11.23
	Vacuum not present ▶	REPLACE solenoid. Retest.
11.23 • Key Off. • Check continuity at circuits 200 and 687 between the MCU module and the TAD solenoid for continuity. Measure resistance.	R is 5 ohms or less. ▶	REPLACE MCU module. Retest.
	R is more than 5 ohms. ▶	SERVICE circuit. Retest.
11.30 • Check at the TAD solenoid output that vacuum is not present when the solenoid is de-energized. • Disconnect the TAD solenoid electrically. • Start engine. run at idle.	Vacuum present ▶	REPLACE solenoid. Retest.
	Vacuum not present ▶	GO to 11.31
11.31 • Key Off. • Check circuit 200 for a short to ground. Should not be shorted. Measure resistance.	R is 1,000 ohms or less. ▶	SERVICE short. Retest.
	R is more than 1,000 ohms. ▶	REPLACE MCU module. Retest.

Thermactor Air Bypass — Pinpoint Test 12

12.0 SERVICE CODE 4-6 THERMACTOR AIR ALWAYS NOT BYPASSING

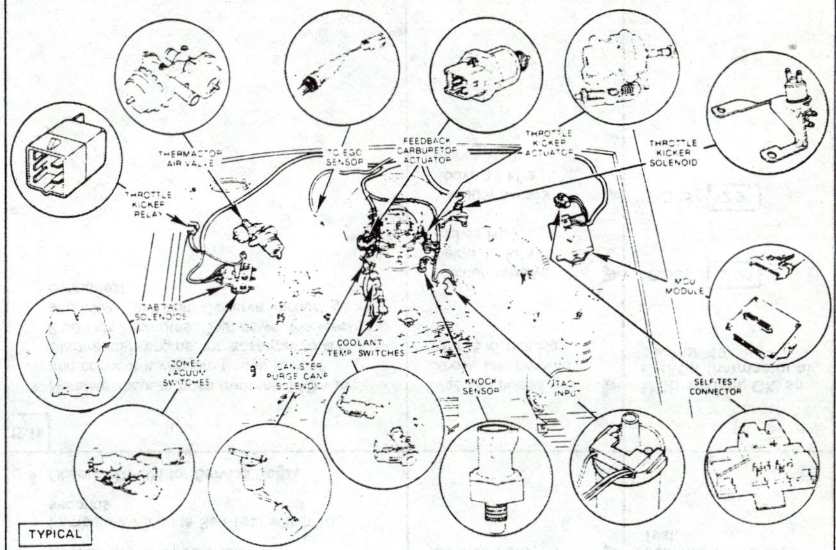

TYPICAL

Components controlled by the MCU System in this Pinpoint Test are the TAB/TAD Solenoids, the Thermactor Air Valve, and the MCU Module. The component not directly controlled by the MCU System during this Pinpoint Test but could possibly cause a Service Code fault are as follows:

- Broken, Loose Belts
- Crimped, Pinched or Missing Hoses, Plugged Restrictor
- Malfunctiong Vacuum Delay Valve

NOTE: Virtually anything that inhibits the Thermactor System from switching air downstream to the catalyst should be investigated.

Thermactor Air Bypass — Pinpoint — Test 12

12.0 SERVICE CODE 4-6 (CONTINUED)

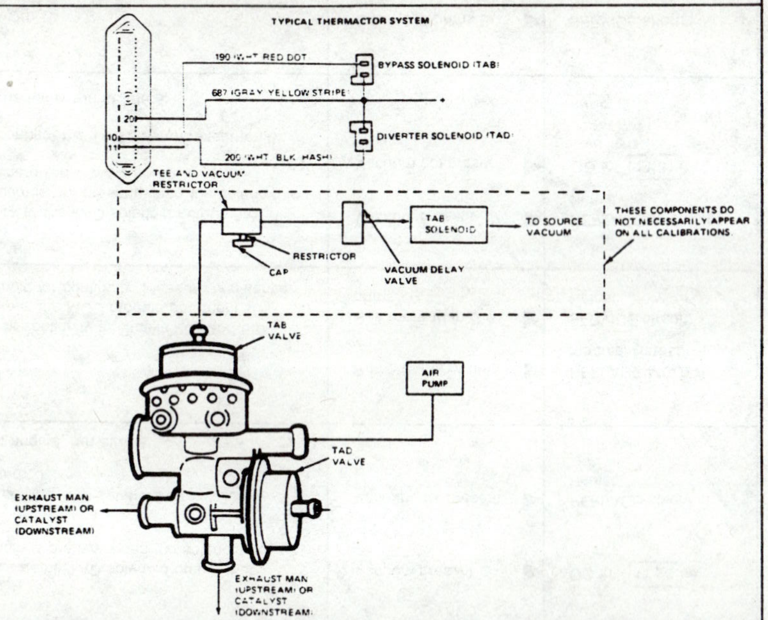

TEST STEP	RESULT	▶	ACTION TO TAKE
12.01			
• Verify the restrictor Teed into the TAB vacuum line was uncapped and restrictor was free from contamination.	Restrictor, OK	▶	GO to [12.02]
	Problem with restrictor	▶	SERVICE as required and retest.
12.02			
• Cap vacuum restrictor, and remove vacuum delay valve from TAB vacuum control line. Reconnect system without delay valve and retest.	No Code 4-6	▶	System, OK. SERVICE vacuum delay valve per Section 3.
	Code 4-6	▶	GO to [12.10]

Thermactor Air Bypass — Pinpoint — Test 12

TEST STEP	RESULT	▶	ACTION TO TAKE
12.03 THERMACTOR CHECK			
This Step is for police vehicles only; all other vehicles, go to Pinpoint Test 12, Step [12.10]	Any other Service Codes.	▶	Go to the appropriate Pinpoint Test.
• Bring engine to operating temperature. • Verify upper radiator hose is hot and pressurized.	Service Code 4-6.	▶	GO to [12.10]
• Turn engine Off. • Restart engine at curb idle.	Service Code 1-1.	▶	Continue with Quick Test.
• Immediately initiate Self-Test within 10 seconds. • Observe DVOM for Service Codes.			
12.10			
• Remove vacuum hose from the TAB valve, and connect a vacuum gauge to hose. Start/re-start engine, increase rpm to above 2,000 for 2 minutes. Shut down, then restart and initiate self test. Observe vacuum gauge during test.	Vacuum pulses above and below 17 kPa (5 inches Hg.)	▶	MCU system, OK; so CHECK thermactor air, Section 10.
	Vacuum always below 17 kPa (5 inches Hg.)	▶	GO to [12.20]
	Vacuum always above 17 kPa (5 inches Hg.)	▶	GO to [12.30]

Thermactor Air Bypass — Pinpoint Test 12

TEST STEP	RESULT	ACTION TO TAKE
12.20 • Check vacuum hoses between TAB valve to TAB solenoid and between TAB solenoid and vacuum source for leaks or blockage.	Leaks or blockage found ▶	SERVICE as required. Retest.
	OK ▶	GO to 12.21
12.21 • Check at the TAB solenoid output to make sure vacuum is present when the solenoid is energized. • Jumper the TAB solenoid circuit 190 to ground. • Start engine, run at idle.	Vacuum ▶	GO to 12.22
	No vacuum ▶	REPLACE solenoid. Retest.
12.22 • Check circuit 190 for continuity. Measure resistance.	R is more than 10 ohms. ▶	SERVICE harness. Retest.
	R is 10 ohms or less. ▶	REPLACE MCU module. Retest.
12.30 • Check at the TAB solenoid output to make sure vacuum is not present when the solenoid is de-energized. • Disconnect the TAB solenoid electrically. • Start engine, run at idle.	Vacuum not present ▶	GO to 12.31
	Vacuum present ▶	REPLACE solenoid. Retest.
12.31 • Check circuit 190 for a short to ground. Measure resistance.	R is 1,000 ohms or less. ▶	SERVICE short. Retest.
	R is more than 1,000 ohms. ▶	REPLACE MCU module. Retest.

Knock Detection — Pinpoint Test 13

TEST STEP	RESULT	ACTION TO TAKE
13.0 SERVICE CODE 2-5—KNOCK DETECTION SYSTEM INOPERATIVE		
• Verify that the intake manifold was tapped with a steel object within two inches of the knock sensor.	No ▶	RETEST per Step 5.0 Engine Running Quick Test.
• Disconnect the jumper wire from the Self-Test Trigger input circuit 201.	Yes ▶	CONTINUE to next Step 13.1
13.1 • Key Off. • Disconnect the vehicle harness from the knock sensor. • Using a torque wrench with a 1⅛ inch deep socket, verify sensor is tightened at 16-24 N•m (12-18 lb-ft). Do not tighten more than 24 N•m (18 lb-ft).	Torque at 16-24 N•m (12-18 lb-ft) ▶	CONTINUE to 13.2
	Torque not at 16-24 N•m (12-18 lb-ft) ▶	TIGHTEN to 16-24 N•m (12-18 lb-ft)

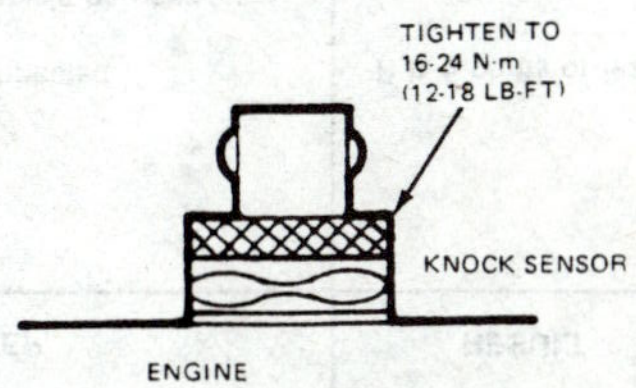

Knock Detection — Pinpoint Test 13

TEST STEP	RESULT	ACTION TO TAKE
13.2 • Disconnect the vehicle harness from the MCU and knock sensor. • Connect one lead of ohmmeter to circuit 310 of the vehicle harness and the other lead to engine block.	R is more than 1000 ohms. ▶	CONTINUE to Step 13.3
	R is 1,000 ohms or less. ▶	SERVICE or REPLACE circuit 310 for short to ground. Retest.

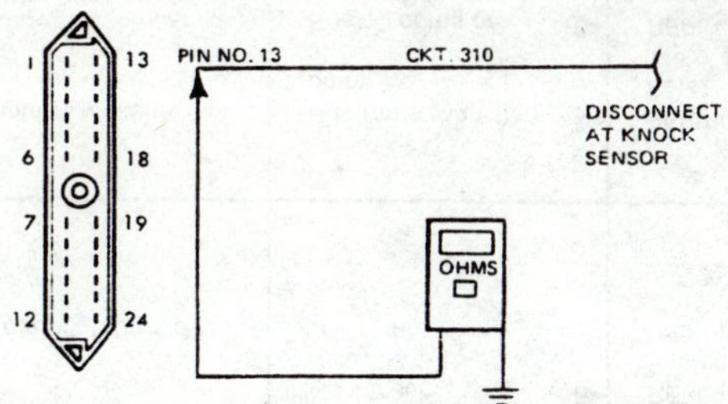

Knock Detection — Pinpoint Test 13

TEST STEP	RESULT	ACTION TO TAKE
13.3 • Key Off. • Leave knock sensor disconnected. • Check circuit 310 of the vehicle harness for continuity.	R is 5 ohms or less. ▶	CONTINUE to Step 13.4
	R is more than 5 ohms. ▶	SERVICE or REPLACE circuit 310. Retest.

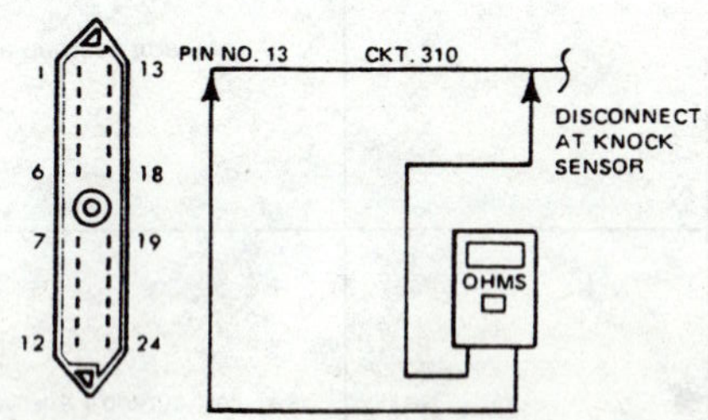

Knock Detection — Pinpoint Test 13

TEST STEP	RESULT ▶	ACTION TO TAKE
13.4 • Check Circuit 60 of the vehicle harness for continuity between the knock sensor and MCU.	R is 5 ohms or less. ▶	RECONNECT the vehicle harness connector to the MCU. GO to Step 13.5
	R is more than 5 ohms. ▶	SERVICE or REPLACE circuit 60. Retest.

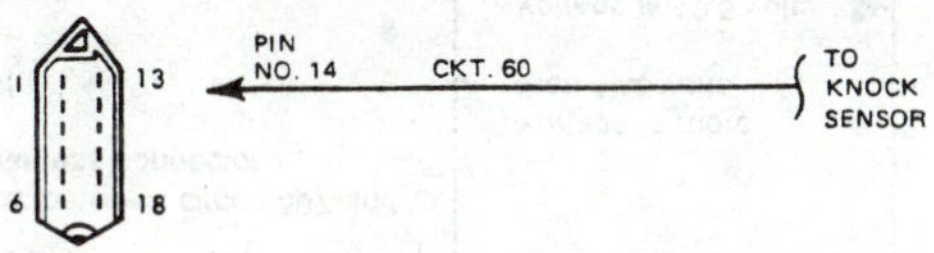

Knock Detection — Pinpoint Test 13

TEST STEP	RESULT ▶	ACTION TO TAKE
13.5 • Reconnect the jumper to circuits 201 and 60 of the VIP connector. • Connect a circuit test lamp to the positive terminal of the battery. • Disconnect the knock sensor from the vehicle harness. • Go to Step 5.0, Engine Running Quick Test, and follow instructions. • When the four initialization pulses occur, continuously tap circuit 310 of the vehicle harness connector, which mates to the knock sensor connector, with the probe end of the circuit tester lamp for 5 seconds. • Observe voltmeter pulses.	Service Code 2-5 ▶	REPLACE MCU. Retest.
	Any other Service Code ▶	REPLACE Knock Sensor. Retest.

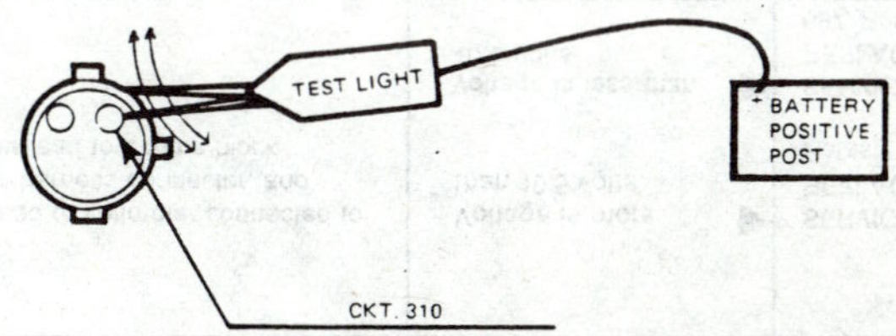

No Self-Test Output — Pinpoint Test 14

TEST STEP	RESULT ▶	ACTION TO TAKE
14.0 NO SELF-TEST OUTPUT • Disconnect the jumper wire from the Self-Test Trigger input circuits 201 and 60. • Disconnect the harness connector from the MCU. • Connect a voltmeter between circuit 687 and 60 of the vehicle harness connector. • Key On, engine Off.	Voltage is more than 10.5 volts.	GO to Step 14.2
	Voltage is 10.5 volts or less. ▶	CONTINUE to Step 14.1

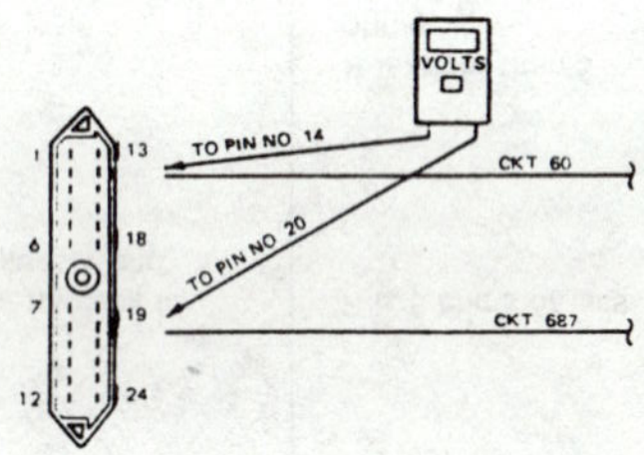

No Self-Test Output — Pinpoint Test 14

TEST STEP	RESULT ▶	ACTION TO TAKE
14.1 • Leave positive lead of voltmeter connected to circuit 687 of the harness connector, and connect the other lead to engine block ground.	Voltage is more than 10.5 volts. ▶	SERVICE or REPLACE circuit 60. Retest.
	Voltage is less than 10.5 volts. ▶	SERVICE or REPLACE circuit 687. Retest.

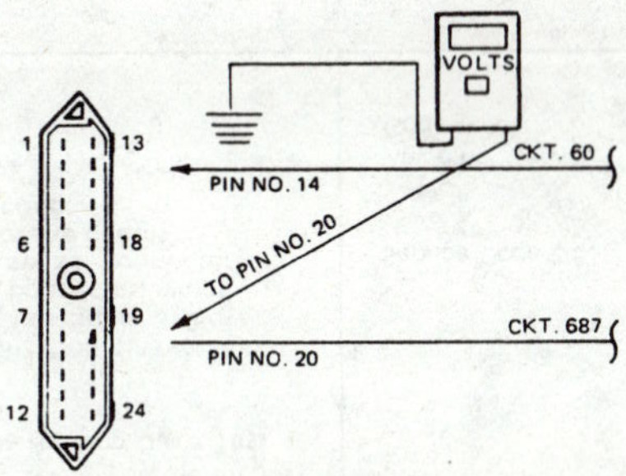

TEST STEP	RESULT ▶	ACTION TO TAKE
14.2 • Key Off. • Check circuit 60 Pin No. 14 for continuity between the Self-Test connector end and engine block ground.	R is 1,000 ohms or less. ▶	CONTINUE to Step 14.3
	R is more than 1,000 ohms. ▶	SERVICE or REPLACE circuit 60. Retest.

No Self-Test Output | Pinpoint | Test 14

TEST STEP	RESULT ▶	ACTION TO TAKE
14.3		
• Check circuit 201 for continuity between the Self Test connector and the vehicle harness connector that mates with the MCU, Pin No. 19.	R is 5 ohms or less. ▶	CONTINUE to Step 14.4
	R is more than 5 ohms. ▶	SERVICE or REPLACE circuit 201. Retest.
14.4 TACH LEAD		
• Connect ohmmeter between Tach terminal of the coil horseshoe connector and circuit 11 of the vehicle harness connector that mates with the MCU connector, Pin No. 8. • Observe ohmmeter.	R is 5 ohms or less. ▶	CONTINUE to Step 14.5
	R is more than 5 ohms. ▶	SERVICE or REPLACE circuit 11. Retest.

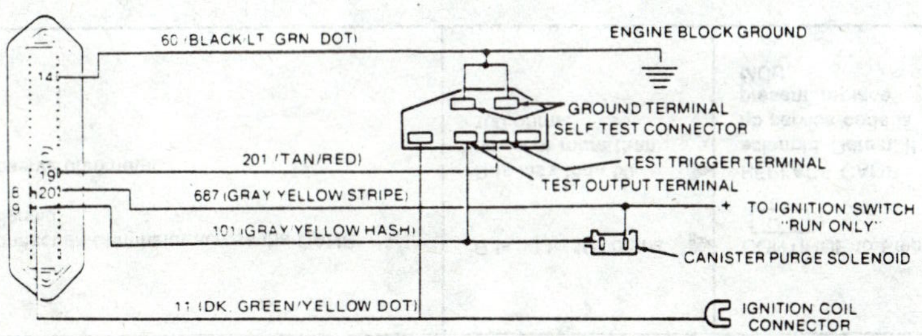

No Self-Test Output | Pinpoint | Test 14

TEST STEP	RESULT ▶	ACTION TO TAKE
14.5		
• Disconnect the vehicle harness from the CANP solenoid and the MCU. • Check circuits 687 and 101 for continuity. • Check circuit 101 to the Self Test connector for continuity.	R of each circuit is 5 ohms or less. ▶	CONTINUE to Step 14.6
	R of a circuit is more than 5 ohms. ▶	SERVICE or REPLACE circuit. Retest.

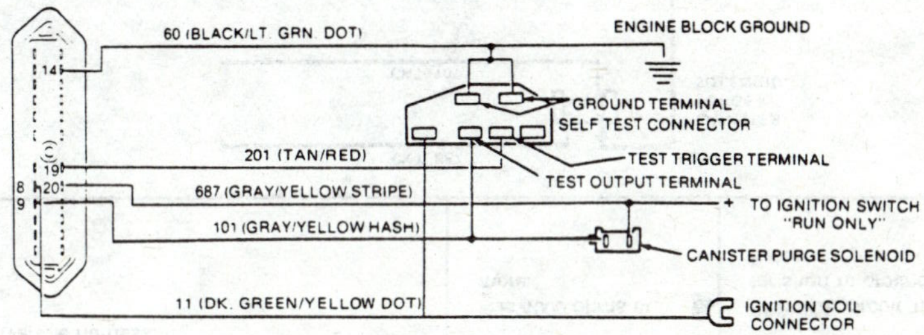

No Self-Test Output — Pinpoint Test 14

TEST STEP	RESULT ▶	ACTION TO TAKE
14.6 • Connect an ohmmeter across the CANP solenoid. • Observe ohmmeter.	R is 50 to 100 ohms. ▶	CONTINUE to Step 14.7
	R is less than 50 ohms or more than 100 ohms. ▶	REPLACE CANP solenoid. Retest, if no service code is present, replace MCU.

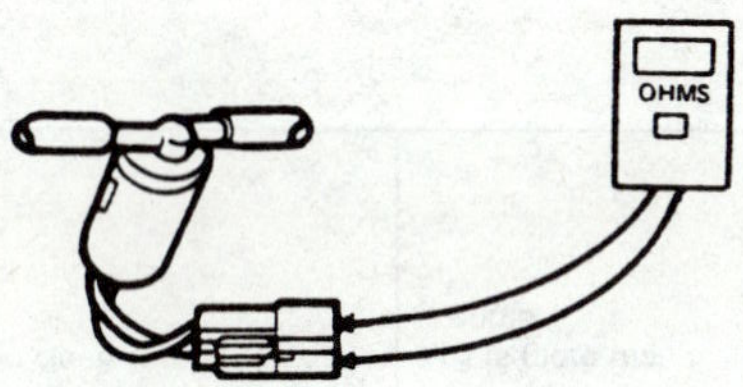

No Self-Test Output — Pinpoint Test 14

TEST STEP	RESULT ▶	ACTION TO TAKE
14.7 • Connect one lead of ohmmeter to engine block ground and the other lead to circuit 101 of vehicle harness.	R is less than 1000 ohms. ▶	CONTINUE to Step 14.8
	R is 1000 ohms or more. ▶	SERVICE circuit 101 for short to ground. Retest.

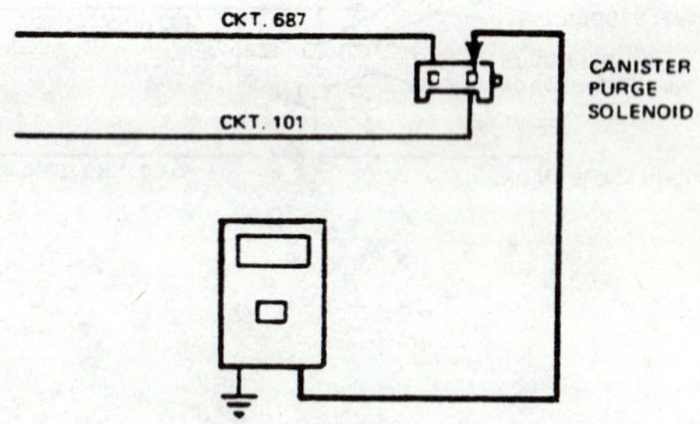

TEST STEP	RESULT ▶	ACTION TO TAKE
14.8 • Check to make sure the Canister Purge solenoid passes vacuum when it is energized (12V) and blocks vacuum when it is de-energized. Check housing and hose for leaks. • Make sure the CANP hose is not cracked, pinched, or plugged.	Solenoid blocks and passes vacuum as indicated ▶	REPLACE MCU. Retest.
	Solenoid does not pass or block vacuum as indicated. Solenoid housing leaks ▶	REPLACE solenoid. Retest.

Barometric Pressure Switch	Pinpoint	Test 15

TEST STEP	RESULT ▶	ACTION TO TAKE
15.0 SERVICE CODE 6-2		
• Make sure the Barometric Pressure switch connector is properly mated with the wiring harness connector. If it is not, connect and retest.	R is 10 ohms or less above 4,000 feet. ▶	Continue to 15.1
• Determine local altitude. If below 3,500 feet, the Barometric Pressure switch is normally open; if above 4,000 feet, the switch is normally closed; and if between 3,500 and 4,000 feet, do not test altitude switch the switch may be in either condition.	R is more than 10 ohms, above 4,000 feet. ▶	Replace switch and retest.
• Disconnect the Barometric Pressure Switch from the vehicle harness.	R is more than 1,000 ohms, below 3,500 feet. ▶	Continue to 15.1
• Connect an ohmmeter across the switch terminals.	R is 1000 ohms or less, below 3,500 feet. ▶	Replace switch and retest.
15.1		
• Disconnect the vehicle harness from the MCU Module.	R is 10 ohms or less. ▶	Replace MCU module and retest.
• Check circuit 60, Pin No. 14 and Pin No. 6 of the MCU module connector, for continuity.	R is more than 10 ohms. ▶	Service or replace circuit. Retest.

Schematics—Electrical

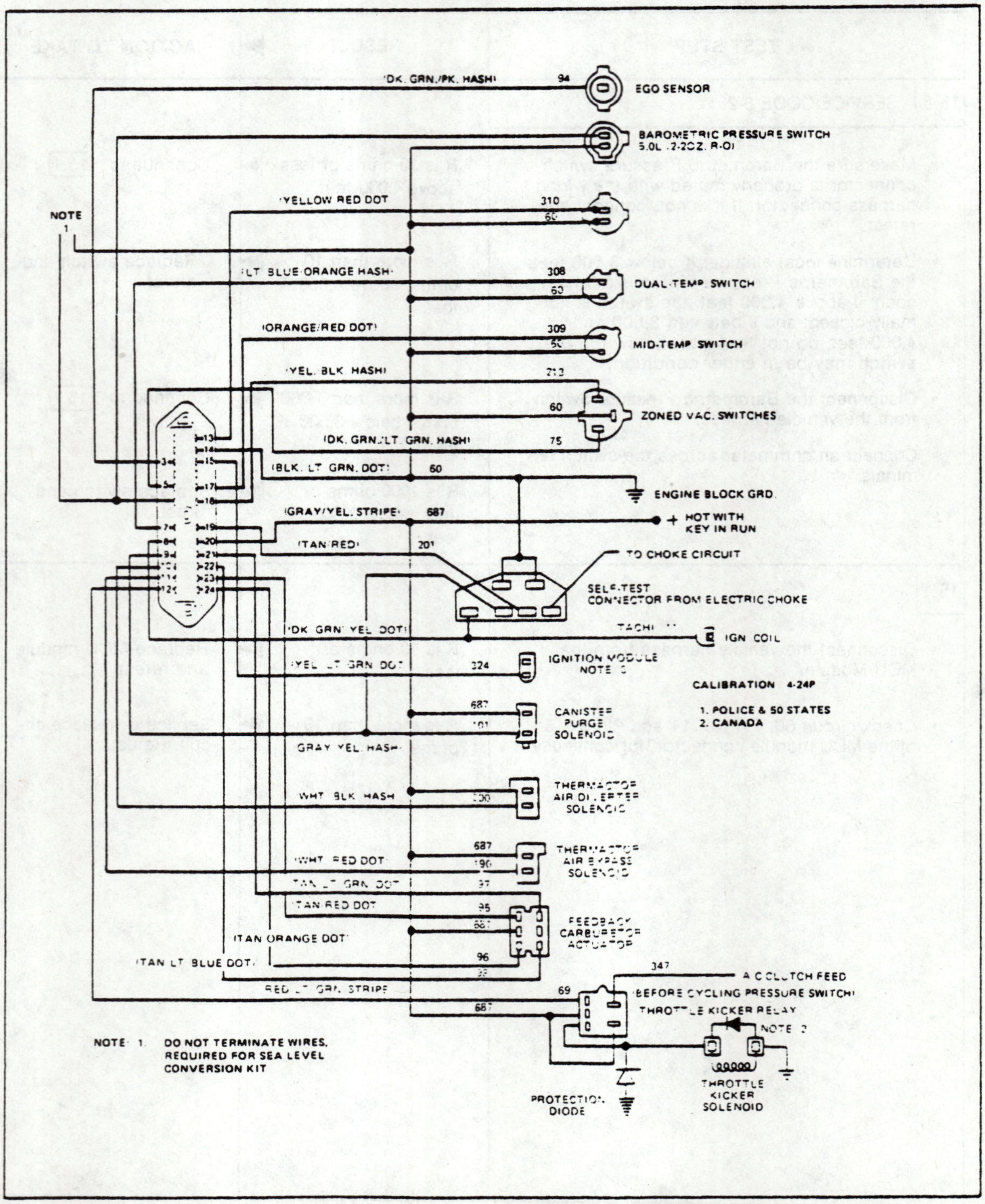

When To Use This Diagnostic Procedure

This diagnostic procedure is only used on Electronic Engine Control (EEC-IV) equipped vehicles when the Diagnostic Routines. Section 2 instruct you to perform EEC-IV Diagnostics. It is divided into two test formats. **Quick Test,** a functional EEC-IV system test, and **Pinpoint Tests,** a number of specific component tests.

To test and service the EEC-IV subsystem, perform the Quick Test first. If the vehicle passes all three phases of the Quick Test: Key On, Engine Off, Engine Running, and Continuous Testing without running any Pinpoint Tests, you will know the EEC-IV subsystem is OK and the vehicle's problem exists somewhere else other than in the EEC-IV subsystem. However, if a Step of the Quick Test fails, perform only the Pinpoint Tests specified by the failed Step. Do not begin any Pinpoint Test without following the instructions at the beginning of the Section.

After all tests and services have been completed, repeat the entire Quick Test to make sure the EEC-IV subsystem works properly.

Test Equipment

EQUIPMENT REQUIRED:

- Rotunda Self-Test Automatic Readout (STAR), No. 07-0004 with cable assembly No. 07-0010. Refer to STAR operation.
- Analog volt-ohmmeter (VOM), 0 to 20v DC, (alternate to STAR). Refer to Appendix A.

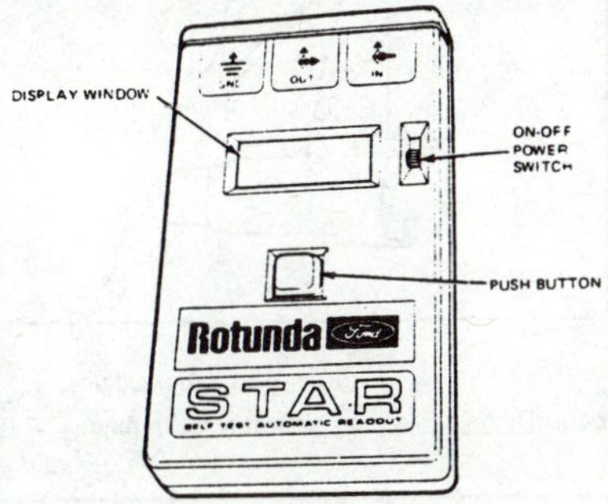

Automatic Readout (STAR) Tester

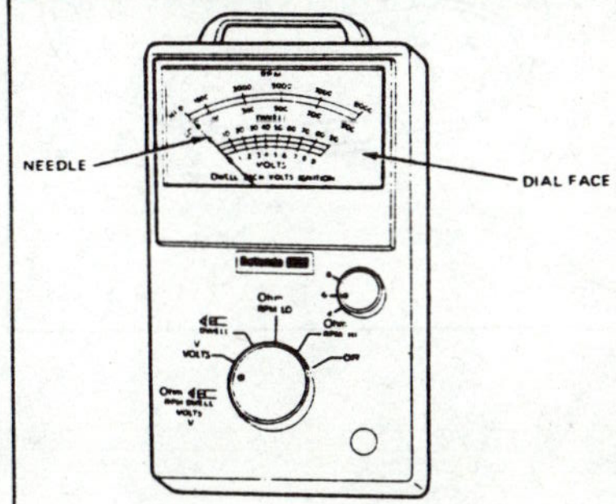

Analog Voltmeter (VOM)

- Jumper wire.
- Vacuum gauge, Rotunda 59-0008 or equivalent. Range 0-30 in. Hg. Resolution 1 in. Hg.

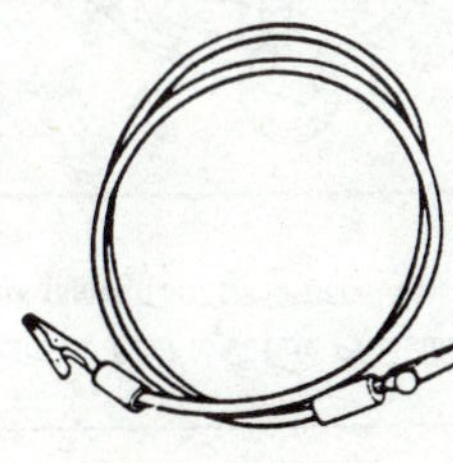

Jumper Wire

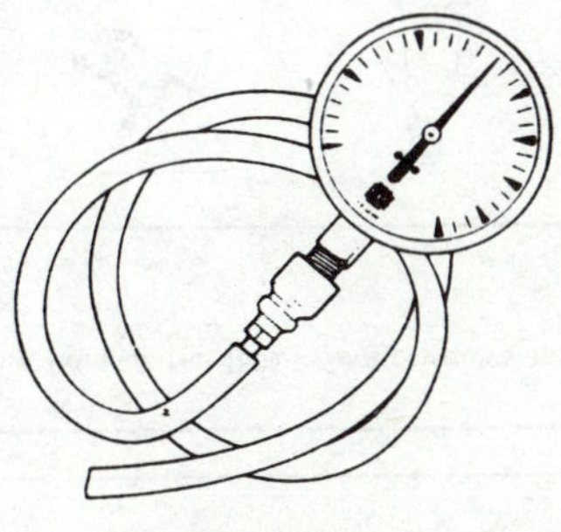

Vacuum Gauge

Test Equipment

- Tachometer, Rotunda No. 59-0010 or equivalent. Range 0-6,000 rpm. Accuracy ± 40 rpm. Resolution 20 rpm.
- Breakout Box, Rotunda T83L-50 EEC-IV or equivalent.

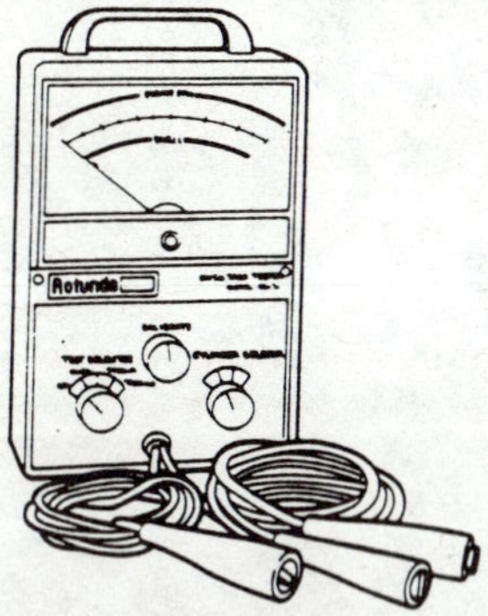

Tachometer

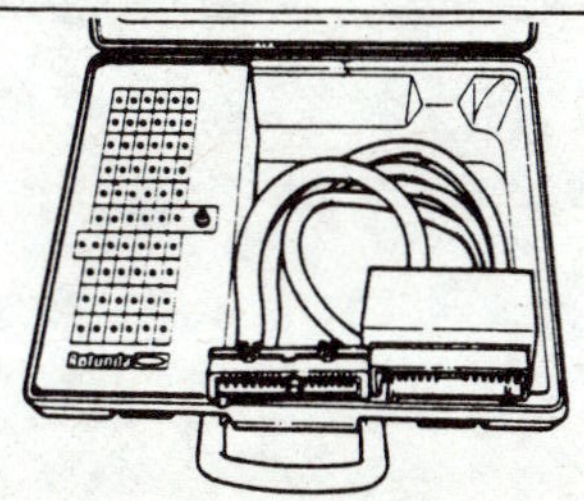

Breakout Box

- Vacuum pump, Rotunda No. 21-0014 or equivalent. Range 0-25 in. Hg.
- Digital volt-ohmmeter (DVOM), Rotunda No. 15-0031 or equivalent. Input impedance 10 Megaohm minimum.

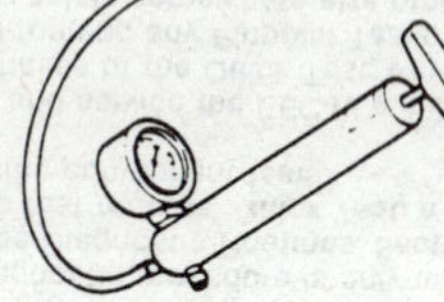

Vacuum Pump

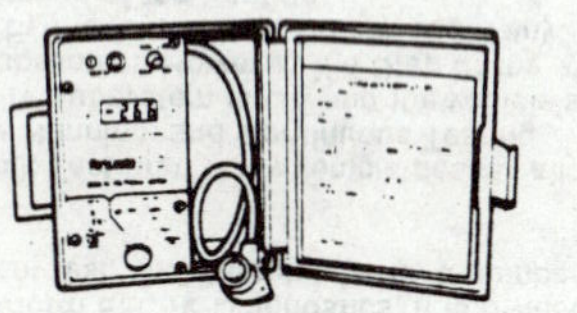

Digital Volt-Ohmmeter (DVOM)

Test Equipment

- Electronic Fuel Injection Pressure Gauge EFI/CFI only, Rotunda No. T80L-9974-A or equivalent.
- Spark tester Tool D81P-6666-A or equivalent.

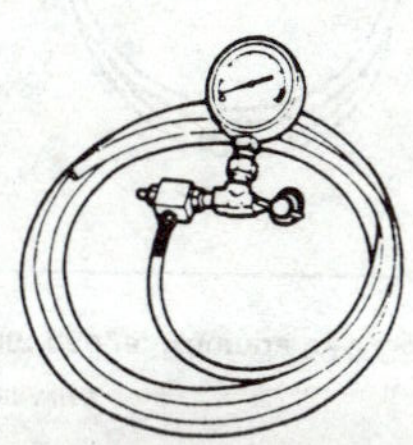

EFI Pressure Gauge

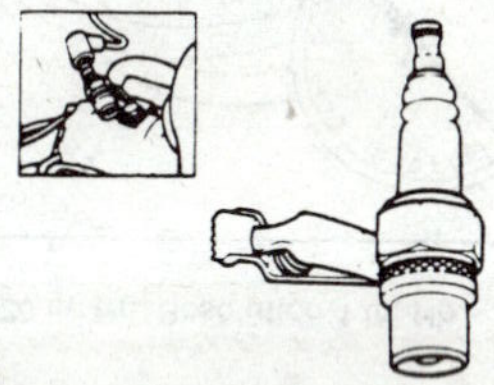

Spark Tester

- Timing light, Rotunda Tool No. 27-0001 or equivalent.
- Non-powered Test lamp.

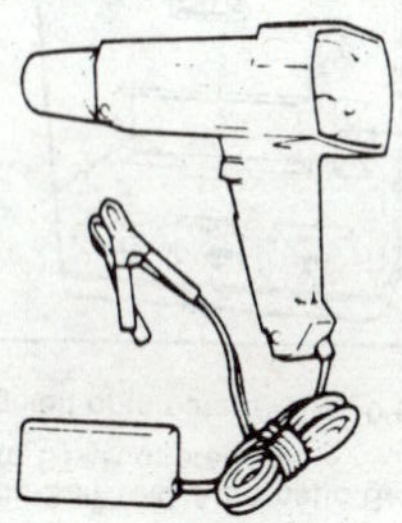

Timing Light

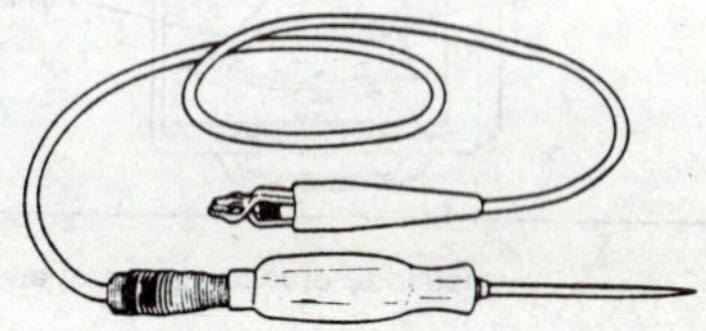

Non-Powered Test-Lamp

Quick Test Description

The Quick Test is a functional test of the EEC-IV system consisting of basic Test Steps (described below). These Steps must be carefully followed, otherwise mis-diagnosis, or the replacement of non-faulty components may result.

Quick Test Steps

1. Visual Check and Vehicle Preparation
   - Checks for obvious faults.
   - Properly prepares the vehicle for testing.
2. Equipment Hook-up
   - Ensures that the proper equipment for gathering test data is ready, prior to testing.
3. Key On, Engine Off Test
   - Is a static check of processor inputs and outputs.
4. Computed Timing Check
   - Verifies the system ability to compute and maintain a fixed spark timing during Self-Test.
5. Engine Running Self-Test
   - Is a dynamic check with the engine in operation.
6. Continuous Self-Test
   - Checks the sensor inputs for opens and shorts while the vehicle is in operation.

The Key On, Engine Off and Engine Running Self-Tests are intended to detect hard faults only, not intermittent faults. Intermittent faults are detected by Continuous Self-Test.

NOTE: A complete Quick Test description is in Appendix A.

Service Code Format

Key On, Engine Off Test

- Fast Codes
- On Demand Code
- Separator Code
- Continuous Code

Engine Running Test

- ID Codes
- Dynamic Response Code
- Fast Code
- On Demand Code

"On Demand" code is information collected from a fault present at the time of testing.

"Continuous" code is information collected and stored from a fault detected in a previous drive-cycle, the fault may not be present at this time.

WARNING: Anyone who departs from the instruction provided in this publication must first establish that he compromises neither his personal safety nor the vehicle integrity by his choice of methods, tools or parts.

Service Code Format

SERVICE CODES

The EEC-IV system communicates service information to the outside world by way of the Self-Test service codes. These service codes are two-digit numbers representing the results of Self-Test.

The service codes are transmitted on the Self-Test output (found in the Self-Test connector) in the form of timed pulses, and read by the technician on a voltmeter or on the STAR tester.

The pulse format is (refer to Figures 1 and 2):

- **One-half second on-time for each digit.**
- **Two seconds off-time between digits.**
- **Four seconds off-time between codes.**
- **Six to nine seconds off-time before and after the half-second separator pulse.**

SELF-TEST OUTPUT CODE FORMAT*
KEY ON, ENGINE OFF

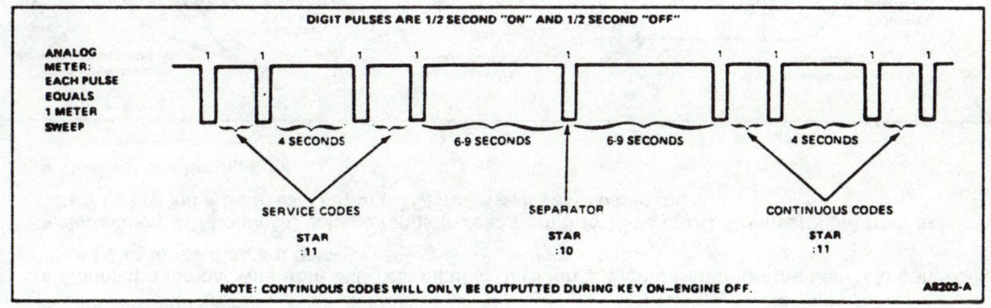

Figure 1 Key On, Engine Off Self-Test Output Code Format

SELF-TEST OUTPUT CODE FORMAT*
KEY ON, ENGINE RUNNING

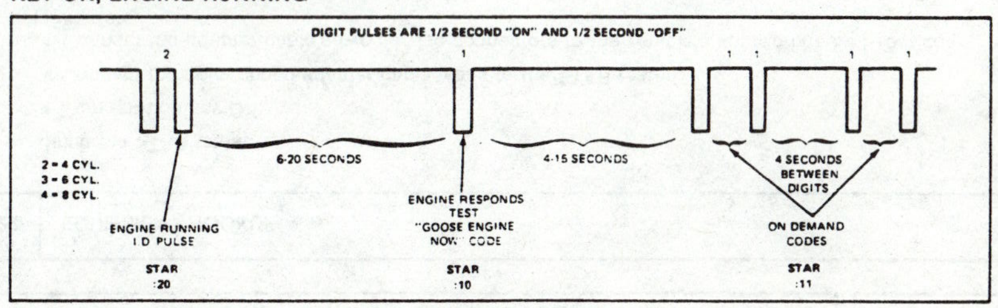

Figure 2 Key On, Engine Running Self-Test Output Code Format

***For a detailed explanation of Self-Test output code format, refer to Appendix B.**

Quick Test — 2.8L FBC

1.0 VISUAL CHECK AND VEHICLE PREPARATION

Correct test results for the Quick Test are dependent on the proper operation of related non-EEC components systems. It may be necessary to correct faults in these areas before EEC will pass QuickTest. Refer to Diagnostic Routines, Section 2 for service.

Before hooking up any equipment to diagnose the EEC system, make the following checks:

1. Verify the condition of air cleaner and ducting. These components may be removed and reinstalled as necessary for service and/or inspection.
2. Check all engine vacuum hoses for:
   - Leaks or pinched hoses.
   - Proper emission routing per Vehicle Emission Control Information (VECI) Decal.
3. Check the EEC system wiring harness electrical connections for:
   - Proper connections.
   - Loose or detached connectors, wires and terminals.
   - Corrosion.
   - Proper routing of harness.

It may be necessary to disconnect or disassemble the connector assembly to perform some of the inspections. (Note the location of each pin before disassembly).

4. Check processor, sensors and actuators for physical damage.
5. Perform all safety steps required to start and run operational vehicle tests.
6. Apply the emergency brake. Place shift lever in Park; Neutral for manual transmission.
7. **Turn Off all electrical loads** such as the radio, lamps, air conditioner, etc. Be sure doors are closed whenever readings are made.
8. Verify engine coolant is at the specified level.
9. Start engine and idle until the upper radiator hose is hot and pressurized and the throttle is off fast idle. While the engine is operating, check for leaks around the exhaust manifold, EGO sensor, and vacuum hose connections.

   NOTE: If the engine will not start or starts and stalls, continue with this Step 1.0 and on through Steps 2.0 and 3.0 of the Quick Test. If the engine idles rough and/or runs rough, go to Quick Test Step 7.0 , Diagnosis by Symptom.
10. Turn ignition key Off.
11. Service items as required, and proceed to equipment hook-up. Step 2.0

Quick Test — 2.8L FBC

2.0 EQUIPMENT HOOKUP

Using the STAR tester:

- Turn ignition key Off.
- Connect the color-coded adapter cable leads to the STAR tester.
- Connect the adapter cable's two service connectors to the vehicle's appropriate Self-Test connectors.
- Connect the timing light.

Using analog voltmeter:

- Turn ignition key Off.
- Connect a jumper wire from Self-Test Input (STI) to Pin 2, Signal Return on the Self-Test connector (refer to the diagram below).
- Set analog VOM on a DC voltage range to read from 0 to 15 volts DC. Connect VOM from battery (+) to Pin 4 Self-Test Output (STO) in the Self-Test connector.
- Connect the timing light.

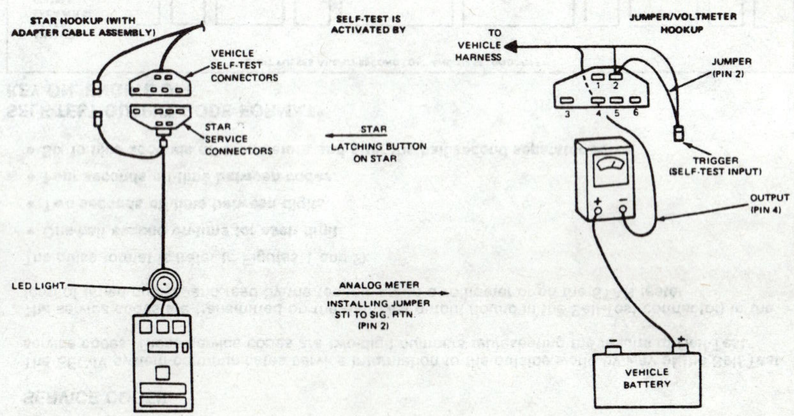

Quick Test — 2.8L FBC

3.0 KEY ON, ENGINE OFF SELF-TEST

Correct test results for the Quick Test are dependent on the proper operation of related non-EEC-IV components/systems. It may be necessary to correct any faults in these areas before EEC-IV will pass the Quick Test. Refer to the Diagnostic Routines, Section 2 for proper servicing.

- Verify that the vehicle has been properly prepared per Quick Test Steps [1.0] and [2.0].
- Place transmission in Neutral or Park. Set parking brake. Place A/C or heater control to Off.
- Activate Self-Test.
- Place key in Run to start Self-Test.

NOTE: Do not depress throttle during Self-Test.

- Observe and record all Service Codes indicated. One of the following outputs will occur. (Continuous Service Codes recorded in this Step will be referenced later in Quick Test Step [6.0]).

RESULT: CODE FORMAT — ON DEMAND	SEPARATOR①	CONTINUOUS		ACTION TO TAKE
11	1(0)	11	▶	Key On, Engine Off **on demand** and **continuous tests** indicate a pass. Go to Quick Test Step [4.0]. If vehicle is a no start, go directly to Pinpoint Test Step [A1].
ANY CODE(S)	1(0)	11	▶	Key On, Engine Off **on demand** test indicates a fault. Record on demand code(s). Go to Quick Test Step [3.1].
ANY CODE(S)	1(0)	ANY CODE(S)	▶	Key On, Engine Off **on demand** and **continuous tests** indicate a fault. Record on demand and continuous codes. Go to Quick Test Step [3.1].
11	1(0)	ANY CODE(S)	▶	**Continuous** test indicates a fault. Record continuous code. Go to Quick Test Step [4.0].
NO CODES OUTPUTTED			▶	Go to Pinpoint Test Step [Q1].

①Star tester displays a 10 in place of a 1 for the single digit separator code, therefore the separator code is expressed in the results column as 10.

Quick Test — 2.8L FBC

3.1 KEY ON, ENGINE OFF SELF-TEST

- Using the On Demand Service Codes from Key On, Engine Off Quick Test Step [3.0] follow the instructions in the Action To Take column in this Step.
- When more than one service code is received, always start service with the first code received.
- Whenever a repair is made, REPEAT Quick Test.

NOTE: **Before proceeding to the specified Pinpoint Test, read the instructions on how to use the Pinpoint Tests at the beginning of the Pinpoint Test section.**

RESULT: ON DEMAND SERVICE CODES		ACTION TO TAKE
15	▶	REPLACE processor. REPEAT Quick Test.
21	▶	GO to Pinpoint Test Step [DE1].
22	▶	GO to Pinpoint Test Step [DF4].
23	▶	GO to Pinpoint Test Step [DH1].
24	▶	GO to Pinpoint Test Step [DB1].
31	▶	GO to Pinpoint Test Step [DL40].
51	▶	GO to Pinpoint Test Step [DE10].
53	▶	GO to Pinpoint Test Step [DH20].
54	▶	GO to Pinpoint Test Step [DB10].
61	▶	GO to Pinpoint Test Step [DE20].
63	▶	GO to Pinpoint Test Step [DH30].
64	▶	GO to Pinpoint Test Step [DB20].
67	▶	GO to Pinpoint Test Step [FA2].
68	▶	GO to Pinpoint Test Step [FB1].

Quick Test — 2.8L FBC

3.1 KEY ON, ENGINE OFF SELF-TEST (CONTINUED)

RESULT	ACTION TO TAKE
ON DEMAND SERVICE CODES 81	GO to Pinpoint Test Step [UA1].
82	
83	GO to Pinpoint Test Step [UC1].
84	GO to Pinpoint Test Step [UC1].
85	GO to Pinpoint Test Step [UE1].
87	GO to Pinpoint Test Step [UG1].
88	GO to Pinpoint Test Step [UH1].
NO CODES OR ANY CODE NOT LISTED ABOVE	GO to Pinpoint Test Step [Q1].

4.0 CHECK TIMING

TEST STEP	RESULT ▶	ACTION TO TAKE
• If vehicle is a no start, go directly to Pinpoint Test [A] NO STARTS. • Key Off, wait 10 seconds. • Verify Self-Test trigger has been activated.	Vehicle stalls while testing. ▶	GO to Pinpoint Test Step [Q30].
• Restart engine and check timing while in Self-Test. You have 2 minutes from the time the last code is outputted to check Self-Test timing.	Timing is not base plus 20° BTDC ±3° ▶	GO to Pinpoint Test Step [P1].
• Correct Self-Test timing equals base plus 20° BTDC ± 3° (see VECI decal for correct base timing).	Timing is base plus 20° BTDC ±3° ▶	GO to Quick Test Step [5.0].

Example:

If base timing is 10° BTDC then initial timing during Self-Test is:

(10° + 20°) ± 3° = 30° ± 3° therefore correct timing is 27-33° BTDC.

Quick Test — 2.8L FBC

5.0 ENGINE RUNNING SELF-TEST

NOTE: If vehicle is a no start, go directly to Pinpoint Test [A1].

- Deactivate Self-Test.
- Start and Run engine at greater than 2000 rpm for 2 minutes. This warms up EGO sensor.
- Turn engine Off, wait 10 seconds.
- Activate Self-Test.
- Start engine. The Engine Running Test will progress as follows:
  - —Engine ID code. ①
  - —Run test.
  - —When the dynamic response ready code occurs, perform a brief WOT. ②
  - —Engine running service codes.
- Observe and record all service codes. One of the following outputs will occur.
- If engine stalls in Self-Test, Go to Pinpoint Test Step [Q 30].

RESULT: CODE FORMAT — ENGINE ID	DYNAMIC RESPONSE	ON DEMAND		ACTION TO TAKE
2(0)	1(0)	11	▶	Engine Running On Demand Test indicates a pass. If the symptom was of an intermittent nature, Go to Quick Test Step [6.0]. If the drive symptom is currently present, Go to Quick Test Step [7.0]. Otherwise, testing is complete, EEC-IV System is OK.
2(0)	1(0)	ANY CODE(S)	▶	Engine Running On Demand Test indicates a fault. Go to Quick Test Step [5.1].
NO CODES OUTPUTTED			▶	Repeat Self-Test and verify no service codes are present then Go to Pinpoint Test Step [Q1].

① **Engine ID code is equal to half of the number of cylinders (a code of 2 equals a 4 cylinder engine) except for STAR tester which adds a zero to all single digit readings (20 equals a 4 cylinder engine).**

② **STAR displays a 10 in place of a 1 for single digit separator code, therefore, the Dynamic Response code is expressed in the results column as 10.**

Quick Test — 2.8L FBC

5.1 ENGINE RUNNING SELF-TEST

- Using the Service Codes from Engine Running Self-Test Step [5.0] follow the instructions in the Action To Take column in this Step.
- When more than one service code is received, always start service with the first code received.
- Whenever a repair is made, REPEAT Quick Test.

RESULT		ACTION TO TAKE
ENGINE RUNNING SERVICE CODES		
12	▶	GO to Pinpoint Test Step [KF1].
13	▶	GO to Pinpoint Test Step [KF20].
21	▶	GO to Pinpoint Test Step [DE1].
22	▶	GO to Pinpoint Test Step [DF1].
23	▶	GO to Pinpoint Test Step [DH1].
24	▶	GO to Pinpoint Test Step [DB1].
25	▶	GO to Pinpoint Test Step [DG1].
31	▶	GO to Pinpoint Test Step [DL1].
32	▶	GO to Pinpoint Test Step [DL10].
33	▶	GO to Pinpoint Test Step [DL10].
34	▶	GO to Pinpoint Test Step [DL10].
35	▶	GO to Pinpoint Test Step [DL30].
41	▶	GO to Pinpoint Test Step [HB1].
42	▶	GO to Pinpoint Test Step [HB20].

Quick Test — 2.8L FBC

5.1 ENGINE RUNNING SELF-TEST (CONTINUED)

RESULT		ACTION TO TAKE
ENGINE RUNNING SERVICE CODES		
43	▶	GO to Pinpoint Test Step [HB1].
44	▶	GO to Pinpoint Test Step [KS1].
45	▶	GO to Pinpoint Test Step [KS1].
46	▶	GO to Pinpoint Test Step [KS1].
47	▶	GO to Pinpoint Test Step [HB20].
58	▶	GO to Pinpoint Test Step [FB1].
72	▶	GO to Pinpoint Test Step [DF10].
73	▶	GO to Pinpoint Test Step [DH40].
77	▶	GO to Pinpoint Test Step [M1].
INVALID CODES	▶	REPEAT Engine Running Self-Test and verify codes. GO to Pinpoint Test Step [Q1].

Quick Test — 2.8L FBC

TEST STEP	RESULT ▶	ACTION TO TAKE
6.0 CONTINUOUS TESTING		
• Do the results of both on demand tests (Key On, Engine Off and Engine Running) result in a pass code 11?	Yes ▶	Go to [6.1].
	No ▶	RETURN to Step [1.0] of Quick Test and make necessary repairs, indicated in Test Steps [3.0] and [5.0] before continuing.
6.1 CLEAR CONTINUOUS CODES		
• Prepare the vehicle for Key On, Engine Off Quick Test per Steps [1.0] [2.0] and [3.0]. • Perform Key On, Engine Off Quick Test. When the service codes begin, exit the Self-Test program (via unlatching STAR or removing the jumper from STI to Signal Return). Exiting Quick Test during code output will clear all codes stored in the continuous memory.	▶	Go to [6.2].
6.2 DETERMINE CODES TO BE TESTED		
• Refer to Continuous Codes recorded in Quick Test Step [3.0]. • Have any of these codes been repaired either in Key On, Engine Off or Engine Running on Demand Tests?	Yes ▶	Disregard codes already serviced. For remaining codes, Go to [6.3].
	No ▶	Go to [6.3].

Quick Test — 2.8L FBC

6.3 CONTINUOUS MONITOR MODE (WIGGLE TEST)

- Verify proper test equipment hook up per Quick Test Step [2.0].
- **Self-Test de-activated. Do not activate Self-Test unless instructed.**
- Using the service codes determined in Step [6.2], follow the Action To Take instructions. (While in the appropriate Pinpoint Test, one or both of the continuous monitor modes described below may have to be used to find the intermittent fault).
- **Key On, Engine Off**
  With Self-Test de-activated, turn the ignition key to the On position. You are now in the continuous monitor mode (wiggle TEST mode). STO will be activated whenever a fault is detected. A fault is indicated when the STAR LED turns off or VOM deflection is 10.5 V or greater.
- **Engine Running**
  Activate Self-Test and perform engine running Quick Test. After service code output has been completed, do not turn engine off or de-activate Quick Test. Approximately 2 minutes after Code 11 has been outputted the system will enter and remain in continuous monitor mode until Self-Test is de-activated or engine is turned Off. STO will be activated whenever a fault is detected. A fault is indicated when the STAR LED turns off or VOM deflection is 10.5 V or greater.
- Refer to Appendix C for a detailed description of Continuous Monitor.

RESULT	ACTION TO TAKE
CONTINUOUS SERVICE CODES	
11 ▶	If drive symptom is currently present GO to Quick Test Step [7.0], otherwise testing complete.
15 ▶	GO to Pinpoint Test Step [Q6].
21 ▶	GO to Pinpoint Test Step [DE90].
22 ▶	GO to Pinpoint Test Step [DF90].
31 ▶	GO to Pinpoint Test Step [DD90].
51 ▶	GO to Pinpoint Test Step [DE91].
53 ▶	GO to Pinpoint Test Step [DH90].
61 ▶	GO to Pinpoint Test Step [DE94].
63 ▶	GO to Pinpoint Test Step [DH94].
54 ▶	GO to Pinpoint Test Step [DB90].
64 ▶	GO to Pinpoint Test Step [DB93].
NO CODES/CODES NOT LISTED ABOVE ▶	GO to Pinpoint Test Step [Q1].

Quick Test — 2.8L FBC

TEST STEP		RESULT	ACTION TO TAKE
7.0	DIAGNOSTICS BY SYMPTOM		
• Engine stalls. • Stalls in Self-Test. • Runs rough. • Misses. • Lack of power.			• Poor power/ground connections. • Ignition system distributor cap, rotor, wires, coil, plugs. • Base engine valves, cam timing, compression etc. • EGR Valve, Section 6. • Fuel Delivery, Section II or Shop Manual, Group 24.
• Shift indicator light always On or Always Off.			• GO to Pinpoint Test Step KL1 .
• A/C does not cut-off during WOT operation. • A/C not functioning.			• GO to Pinpoint Test Step KG1 .
• High or low idle speeds on start up. • Engine diesels.			• GO to Pinpoint Test Step FC1 .
• Gasoline fume odors.			• GO to Section 7 or Shop Manual, Group 24.
• Lack of fast idle with A/C On.			• GO to KG1 .

Electrical Schematic — 2.8L FBC

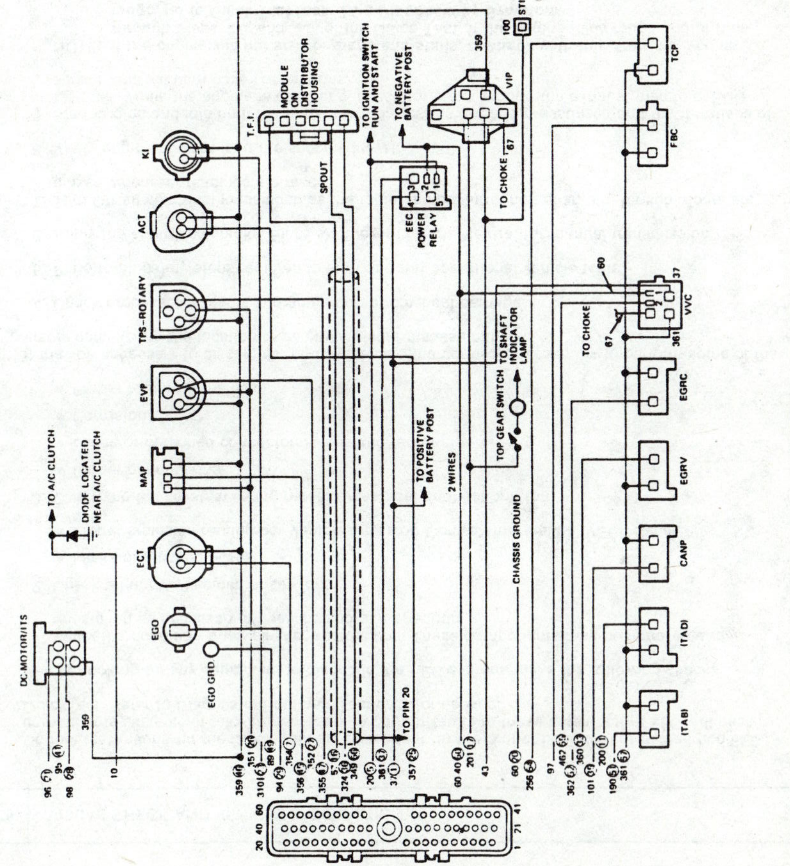

EEC-IV Module Connector Pin Usage

2.8L FBC

Pin	Circuit	Wire Color	Application	Abbreviations
1	37	Y	Keep Alive Memory	KAM
5	20	W-LB H	Ignition Run Start	—
7	354	LG-Y	Engine Coolant Sensor	ECT
10		-	A C Clutch	ACC
11	200	W-BK	Thermactor Air Diverter	TAD
16	57	BK	Ignition Ground	IGN. GRD.
17	201	T-R	Self-Test Output	STO
20	60	BK-LG D	Chassis Ground	GRD.
21	96	T-O D	Idle Speed Control	ISC
23	310	Y-R D	Knock Sensor	KI
25	357	LG-P	Air Charge Temperature Sensor	ACT
26	351	O-W	Reference Voltage	V. REF.
27	352	BR-LG	EGR Valve Position Sensor	EVP
28	98	T-LB D	Idle Tracking Switch	ITS
29	94	DG-P H	Exhaust Gas Oxygen Sensor	EGO
33	360	DG	EGR Vent Solenoid	EGRV
35	101	GY-Y H	Canister Purge	CANP
36	324	Y-LG D	Ignition Module Signal	SPOUT
37	361	R	Vehicle Power	V PWR
41	95	T-R D	Idle Speed Control	ISC
45	356	DB-LG	Manifold Absolute Pressure Sensor	MAP
46	359	BK-W	Signal Return	SIG. RTN.
47	355	DG-LG	Throttle Position Sensor	TPS
49	89	O	Exhaust Gas Oxygen Sensor Ground	EGO GRD
51	190	W-R D	Thermactor Air Bypass	TAB
52	362	Y	EGR Control Solenoid	EGRC
54	256	O-W	Temperature Compensation Pump	TCP
55	462	P	Variable Voltage Choke	VVC
56	349	DB	Profile Ignition Pick-Up	PIP
57	361	R	Vehicle Power	V POWER
60	60	BK-LG D	Vehicle Ground	GRD

Quick Test

4.9L FBC

1.0 VISUAL CHECK AND VEHICLE PREPARATION

Correct test results for the Quick Test are dependent on the proper operation of related non-EEC components systems. It may be necessary to correct faults in these areas before EEC will pass QuickTest. Refer to Diagnostic Routines, Section 2 for service.

Before hooking up any equipment to diagnose the EEC system, make the following checks:

1. Verify the condition of air cleaner and ducting. These components may be removed and reinstalled as necessary for service and/or inspection.
2. Check all engine vacuum hoses for:
   - Leaks or pinched hoses.
   - Proper emission routing per Vehicle Emission Control Information (VECI) Decal.
3. Check the EEC system wiring harness electrical connections for:
   - Proper connections.
   - Loose or detached connectors, wires and terminals.
   - Corrosion.
   - Proper routing of harness.

It may be necessary to disconnect or disassemble the connector assembly to perform some of the inspections. (Note the location of each pin before disassembly).

4. Check processor, sensors and actuators for physical damage.
5. Perform all safety steps required to start and run operational vehicle tests.
6. Apply the emergency brake. Place shift lever in Park; Neutral for manual transmission.
7. **Turn Off all electrical loads** such as the radio, lamps, air conditioner, etc. Be sure doors are closed whenever readings are made.
8. Verify engine coolant is at the specified level.
9. Start engine and idle until the upper radiator hose is hot and pressurized and the throttle is off fast idle. While the engine is operating, check for leaks around the exhaust manifold, EGO sensor, and vacuum hose connections.

   NOTE: If the engine will not start or starts and stalls, continue with this Step 1.0 and on through Steps 2.0 and 3.0 of the Quick Test. If the engine idles rough and/or runs rough, go to Quick Test Step 7.0 , Diagnosis by Symptom.

10. Turn ignition key Off.
11. Service items as required, and proceed to equipment hook-up, Step 2.0 .

Quick Test — 4.9L FBC

2.0 EQUIPMENT HOOKUP

Using the STAR tester:

- Turn ignition key Off.
- Connect the color-coded adapter cable leads to the STAR tester.
- Connect the adapter cable's two service connectors to the vehicle's appropriate Self-Test connectors.
- Connect the timing light.

Using analog voltmeter:

- Turn ignition key Off.
- Connect a jumper wire from Self-Test Input (STI) to Pin 2, Signal Return on the Self-Test connector (refer to the diagram below).
- Set analog VOM on a DC voltage range to read from 0 to 15 volts DC. Connect VOM from battery (+) to Pin 4 Self-Test Output (STO) in the Self-Test connector.
- Connect the timing light.

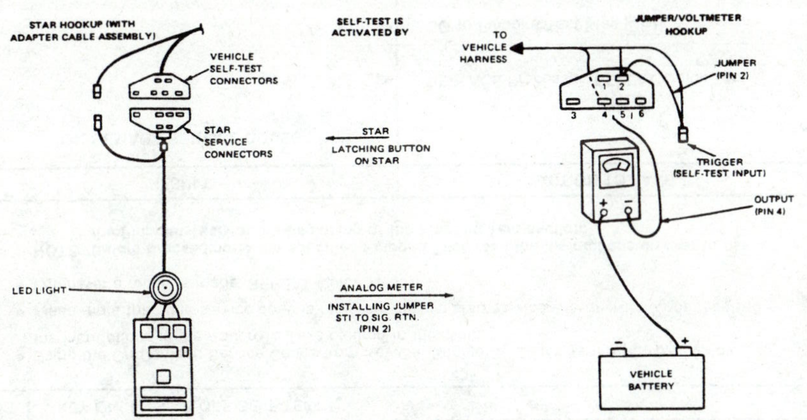

Quick Test — 4.9L FBC

3.0 KEY ON, ENGINE OFF SELF-TEST

Correct test results for the Quick Test are dependent on the proper operation of related non-EEC-IV components/systems. It may be necessary to correct any faults in these areas before EEC-IV will pass the Quick Test. Refer to the Diagnostic Routines, Section 2 for proper servicing.

- Verify that the vehicle has been properly prepared per Quick Test Steps [1.0] and [2.0].
- Place transmission in Neutral or Park. Set parking brake. Place A/C or heater control to Off.
- Activate Self-Test.
- Place key in Run to start Self-Test.

  NOTE: Do not depress throttle during Self-Test.
- Observe and record all Service Codes indicated. One of the following outputs will occur. (Continuous Service Codes recorded in this Step will be referenced later in Quick Test Step [6.0]).

RESULT — CODE FORMAT: ON DEMAND	SEPARATOR[1]	CONTINUOUS		ACTION TO TAKE
11	1(0)	11	▶	Key On, Engine Off **on demand** and **continuous tests** indicate a pass. Go to Quick Test Step [4.0]. If vehicle is a no start, go directly to Pinpoint Test Step [A1].
ANY CODE(S)	1(0)	11	▶	Key On, Engine Off **on demand** test indicates a fault. Record on demand code(s). Go to Quick Test Step [3.1].
ANY CODE(S)	1(0)	ANY CODE(S)	▶	Key On, Engine Off **on demand** and **continuous tests** indicate a fault. Record on demand and continuous codes. Go to Quick Test Step [3.1].
11	1(0)	ANY CODE(S)	▶	**Continuous** test indicates a fault. Record continuous code. Go to Quick Test Step [4.0].
NO CODES OUTPUTTED			▶	Go to Pinpoint Test Step [Q1].

[1] **Star tester displays a 10 in place of a 1 for the single digit separator code, therefore the separator code is expressed in the results column as 10.**

Quick Test — 4.9L FBC

3.1 KEY ON, ENGINE OFF SELF-TEST

- Using the On Demand Service Codes from Key On, Engine Off Quick Test Step 3.0 follow the instructions in the Action To Take column in this Step.
- When more than one service code is received, always start service with the first code received.
- Whenever a repair is made, REPEAT Quick Test.

NOTE: Before proceeding to the specified Pinpoint Test, read the instructions on how to use the Pinpoint Tests at the beginning of the Pinpoint Test section.

RESULT		ACTION TO TAKE
ON DEMAND SERVICE CODES		
15	▶	REPLACE Processor. REPEAT Quick Test.
21	▶	GO to Pinpoint Test Step [DE1].
22	▶	GO to Pinpoint Test Step [DF4].
23	▶	GO to Pinpoint Test Step [DH1].
51	▶	GO to Pinpoint Test Step [DE10].
53	▶	GO to Pinpoint Test Step [DH20].
61	▶	GO to Pinpoint Test Step [DE20].
63	▶	GO to Pinpoint Test Step [DH30].
67	▶	GO to Pinpoint Test Step [FA1].
68	▶	GO to Pinpoint Test Step [FB5].
NO CODES OR ANY CODE NOT LISTED ABOVE	▶	GO to Pinpoint Test Step [Q1].

Quick Test — 4.9L FBC

TEST STEP	RESULT		ACTION TO TAKE
4.0 CHECK TIMING			
• If vehicle is a no start, go directly to Pinpoint Test [A] NO STARTS.	Vehicle stalls while testing.	▶	GO to Pinpoint Test Step [Q30].
• Key Off, wait 10 seconds.			
• Verify Self-Test trigger has been activated.			
• Restart engine and check timing while in Self-Test. You have 2 minutes from the time the last code is outputted to check Self-Test timing.	Timing is not base plus 20° BTDC ±3°	▶	GO to Pinpoint Test Step [P1].
• Correct Self-Test timing equals base plus 20° BTDC ± 3° (see VECI decal for correct base timing).	Timing is base plus 20° BTDC ±3°	▶	GO to Quick Test Step [5.0].
Example: If base timing is 10° BTDC then initial timing during Self-Test is: (10° + 20°) ± 3° = 30° ± 3° therefore correct timing is 27-33° BTDC.			

Quick Test — 4.9L FBC

5.0 ENGINE RUNNING SELF-TEST

NOTE: If vehicle is a no start, go directly to Pinpoint Test [A1].

- Deactivate Self-Test.
- Start and Run engine at greater than 2000 rpm for 2 minutes. This warms up EGO sensor.
- Turn engine Off, wait 10 seconds.
- Activate Self-Test.
- Start engine. The Engine Running Test will progress as follows:
  - —Engine ID code.①
  - —Run test.
  - —When the dynamic response ready code occurs, perform a brief WOT.②
  - —Engine running service codes.
- Observe and record all service codes. One of the following outputs will occur.
- If engine stalls in Self-Test, Go to Pinpoint Test Step [Q 30].

RESULT: CODE FORMAT — ENGINE ID	DYNAMIC RESPONSE	ON DEMAND		ACTION TO TAKE
2(0)	1(0)	11	▶	Engine Running On Demand Test indicates a pass. If the symptom was of an intermittent nature, Go to Quick Test Step [6.0]. If the drive symptom is currently present, Go to Quick Test Step [7.0]. Otherwise, testing is complete. EEC-IV System is OK.
2(0)	1(0)	ANY CODE(S)	▶	Engine Running On Demand Test indicates a fault. Go to Quick Test Step [5.1].
NO CODES OUTPUTTED			▶	Repeat Self-Test and verify no service codes are present then Go to Pinpoint Test Step [Q1].

① **Engine ID code is equal to half of the number of cylinders (a code of 2 equals a 4 cylinder engine) except for STAR tester which adds a zero to all single digit readings (20 equals a 4 cylinder engine).**

② **STAR displays a 10 in place of a 1 for single digit separator code, therefore, the Dynamic Response code is expressed in the results column as 10.**

Quick Test — 4.9L FBC

5.1 ENGINE RUNNING SELF-TEST

- Using the Service Codes from Engine Running Self-Test Step [5.0] follow the instructions in the Action To Take column in this Step.
- When more than one service code is received, always start service with the first code received.
- Whenever a repair is made, REPEAT Quick Test.

RESULT: ENGINE RUNNING SERVICE CODES		ACTION TO TAKE
12	▶	GO to Pinpoint Test Step [KF1].
13	▶	GO to Pinpoint Test Step [KF20].
21	▶	GO to Pinpoint Test Step [DE1].
22	▶	GO to Pinpoint Test Step [DF1].
23	▶	GO to Pinpoint Test Step [DH3].
41	▶	GO to Pinpoint Test Step [HB1].
42	▶	GO to Pinpoint Test Step [HB20].
43	▶	GO to Pinpoint Test Step [HB40].
44	▶	GO to Pinpoint Test Step [KC1].
45	▶	GO to Pinpoint Test Step [KC1].
46	▶	GO to Pinpoint Test Step [KC1].
58	▶	GO to Pinpoint Test Step [FB1].
72	▶	GO to Pinpoint Test Step [DF10].

Quick Test — 4.9L FBC

5.1 ENGINE RUNNING SELF-TEST (CONTINUED)

RESULT		ACTION TO TAKE
ENGINE RUNNING SERVICE CODES		
73	▶	GO to Pinpoint Test Step [DH40].
77	▶	GO to Pinpoint Test Step [M1].
NO CODES/ANY CODE NOT LISTED ABOVE	▶	GO to Pinpoint Test Step [Q1].

TEST STEP	RESULT		ACTION TO TAKE
6.0 CONTINUOUS TESTING			
• Do the results of both on demand tests (Key On, Engine Off and Engine Running) result in a pass code 11?	Yes	▶	Go to [6.1].
	No	▶	RETURN to Step [1.0] of Quick Test and make necessary repairs, indicated in Test Steps [3.0] and [5.0] before continuing.
6.1 CLEAR CONTINUOUS CODES			
• Prepare the vehicle for Key On, Engine Off Quick Test per Steps [1.0] [2.0] and [3.0]. • Perform Key On, Engine Off Quick Test. When the service codes begin, exit the Self-Test program (via unlatching STAR or removing the jumper from STI to Signal Return). Exiting Quick Test during code output will clear all codes stored in the continuous memory.		▶	Go to [6.2].
6.2 DETERMINE CODES TO BE TESTED			
• Refer to Continuous Codes recorded in Quick Test Step [3.0]. • Have any of these codes been repaired either in Key On, Engine Off or Engine Running on Demand Tests?	Yes	▶	Disregard codes already serviced. For remaining codes, Go to [6.3].
	No	▶	Go to [6.3].

Quick Test — 4.9L FBC

6.3 CONTINUOUS MONITOR MODE (WIGGLE TEST)

- Verify proper test equipment hook up per Quick Test Step [2.0].
- **Self-Test de-activated. Do not activate Self-Test unless instructed.**
- Using the service codes determined in Step [6.2], follow the Action To Take instructions. (While in the appropriate Pinpoint Test, one or both of the continuous monitor modes described below may have to be used to find the intermittent fault).
- **Key On, Engine Off**
  With Self-Test de-activated, turn the ignition key to the On position. You are now in the continuous monitor mode (wiggle TEST mode). STO will be activated whenever a fault is detected. A fault is indicated when the STAR LED turns off or VOM deflection is 10.5 V or greater.
- **Engine Running**
  Activate Self-Test and perform engine running Quick Test. After service code output has been completed, do not turn engine off or de-activate Quick Test. Approximately 2 minutes after Code 11 has been outputted the system will enter and remain in continuous monitor mode until Self-Test is de-activated or engine is turned Off. STO will be activated whenever a fault is detected. A fault is indicated when the STAR LED turns off or VOM deflection is 10.5 V or greater.

  Refer to Appendix C for a detailed description of Continuous Monitor.

RESULT		ACTION TO TAKE
CONTINUOUS SERVICE CODES		
11	▶	If drive symptom is currently present GO to Quick Test Step [7.0], otherwise testing complete.
15	▶	GO to Pinpoint Test Step [Q6].
21	▶	GO to Pinpoint Test Step [DE90].
22	▶	GO to Pinpoint Test Step [DF90].
41	▶	GO to Pinpoint Test Step [HB1].
42	▶	GO to Pinpoint Test Step [HB20].
51	▶	GO to Pinpoint Test Step [DE91].
53	▶	GO to Pinpoint Test Step [DH90].
61	▶	GO to Pinpoint Test Step [DE94].
63	▶	GO to Pinpoint Test Step [DH94].
NO CODES/CODES NOT LISTED	▶	GO to Pinpoint Test Step [Q1].

Quick Test — 4.9L FBC

TEST STEP	RESULT	ACTION TO TAKE
7.0 DIAGNOSTICS BY SYMPTOM		
• Engine runs rough or miss-fires. • Engine stalls.		• Poor power/ground connections. • Ignition system distributor cap, rotor, wires, coil, plugs. • Base engine valves, cam timing, compression etc. • Fuel delivery. • EGR value function.
• Detonation/spark knock. • Rough idle.		• EGR valve function. • Go to Pinpoint Test Step KA1 .
• Lack of fast idle with A/C On.		• Go to Pinpoint Test Step KF4 .
• Gasoline fumes under hood.		• Go to Pinpoint Test Step KD1 .
• Shift indicator light always On or always Off.		• Go to Pinpoint Test Step KL1 .
• Low idle on code start-up. • High idle on hot start-up. • Engine diesels.		• Go to Pinpoint Test Step FC1 .

Electrical Schematic — 4.9L FBC

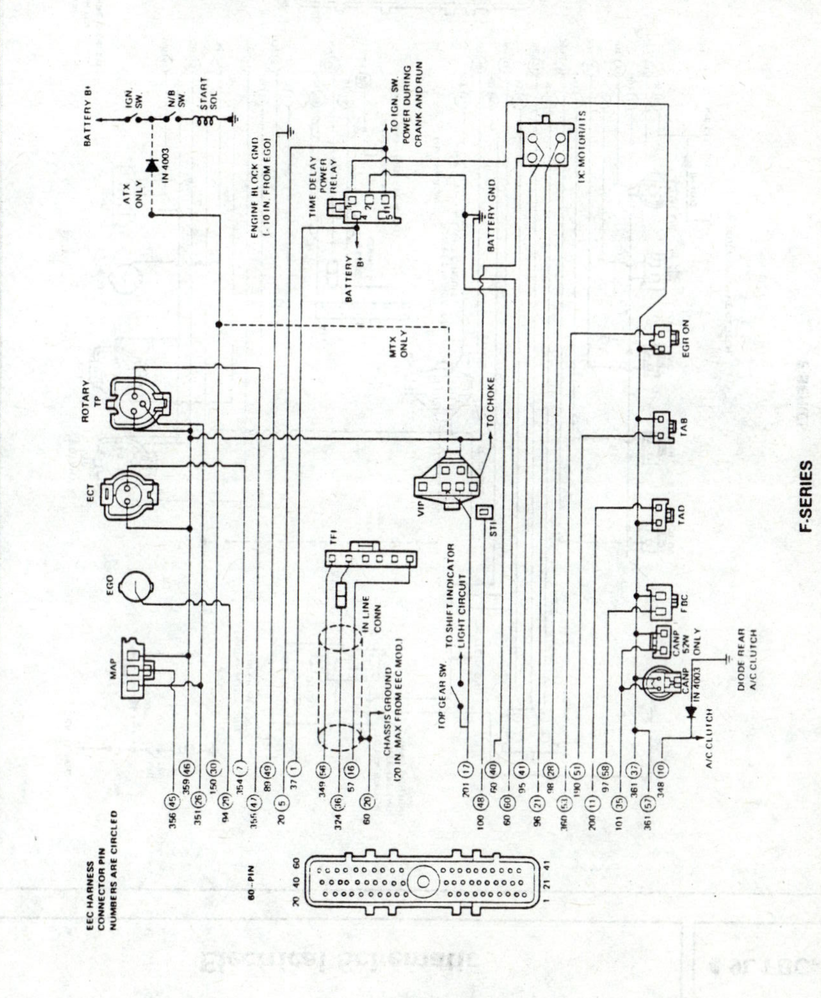

EEC-IV Module Connector Pin Usage

4.9L FBC

F-SERIES

Pin	Circuit	Wire Color	Application	Abbreviations
1	37	Y	Keep Alive Power	KA PWR
5	20	W/LB H	Key Power	K PWR
7	354	LG/Y	Engine Coolant Temperature	ECT
10	348	LG/PH	A/C Clutch	ACC
11	200	W/BK	Thermactor Air Diverter	TAD
16	57	BK	Case Ground	CASE GND.
17	201	T/R	Self-Test Output/Shift Ind. Light	STO/SIL
20	60	BK/LG D	Ignition Ground	IGN. GND.
21	96	T/O D	Idle Speed Control	ISC
26	351	O/W	Voltage Reference	V. REF.
28	98	T/LB D	Idle Tracking Switch	ITS
29	94	DG/P H	Exhaust Gas Oxygen Sensor	EGO
30	150	DG/W	Neutral Drive Switch	N/D
35	101	GY/Y H	Canister Purge	CANP
36	324	Y/LG D	Spark Output	SPOUT
37	361	R	Vehicle Power	V PWR
40	60	BK/LG D	Battery Ground	BATT. GND.
41	95	T/R D	Idle Speed Control	ISC
45	356	DB/LG	Manifold Absolute Pressure	MAP
46	359	BK/W	Signal Return	SIG. RTN.
47	355	DG/LB	Throttle Position Sensor	TPS
48	100	W/R D	Self-Test Input	STI
49	89	O	Exhaust Gas Oxygen Ground	EGO GND.
51	190	W/R D	Thermactor Air Bypass	TAB
53	360	DG	Exhaust Gas Recirculate Shut Off	EGR S/O
56	349	DB	Profile Ignition Pick-Up	PIP
57	361	R	Vehicle Power	V PWR
58	97	T/LG D	Feedback Control	FBC
60	60	BK/LG D	Battery Ground	BATT. GND.

Pin locations given for reference only. Probing 60 pin connector with DVOM probe will result in permanent damage to the pin connectors. Always probe as directed, using the Breakout Box.

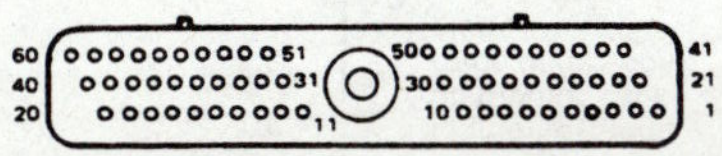

Electrical Schematic

4.9L FBC

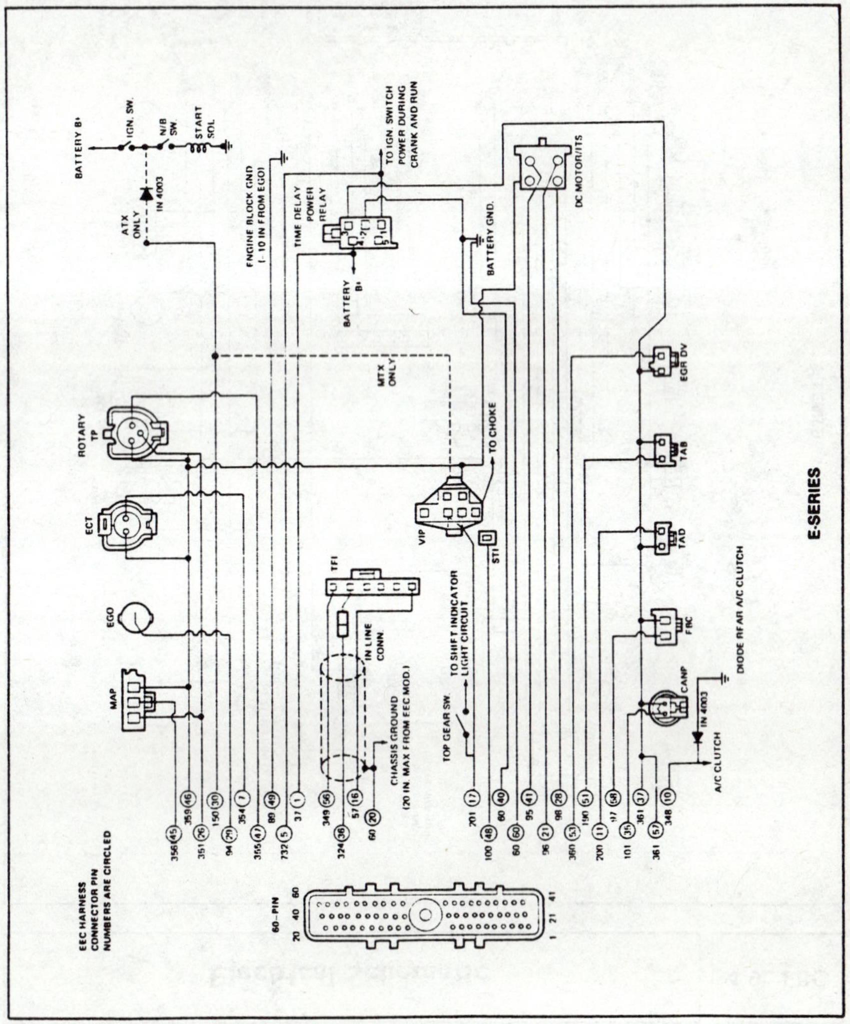

EEC-IV Module Connector Pin Usage — 4.9L FBC

E-SERIES

Pin	Circuit	Wire Color	Application	Abbreviations
1	37	BK/Y	Keep Alive Power	KA PWR
5	732	W	Key Power	K PWR
7	354	GR/Y	Engine Coolant Temperature	ECT
10	348	GR/W	A/C Clutch	ACC
11	200	BL	Thermactor Air Diverter	TAD
16	57	BK	Case Ground	CASE GND.
17	201	BL/Y	Self-Test Output	STO
20	60	BK/W	Ignition Ground	IGN. GND.
21	96	O/BK	Idle Speed Control	ISC
26	351	R	Voltage Reference	V. REF.
28	98	BK	Idle Tracking Switch	ITS
29	94	GR/P	Exhaust Gas Oxygen Sensor	EGO
30	150	R	Neutral Drive Switch	N/D
35	101	GY	Canister Purge	CANP
36	324	Y/GR	Spark Output	SPOUT
37	361	R	Vehicle Power	V PWR
40	60	BK/W	Battery Ground	BATT. GND.
41	95	W	Idle Speed Control	ISC
45	356	BL/GR	Manifold Absolute Pressure	MAP
46	359	BK/W	Signal Return	SIG. RTN.
47	355	GR	Throttle Position Sensor	TPS
48	100	BK/R	Self-Test Input	STI
49	89	O	Exhaust Gas Oxygen Ground	EGO GND.
51	190	Y	Thermactor Air Bypass	TAB
53	360	Y/BL	Exhaust Gas Recirculate Shut Off	EGR S/O
56	349	BR	Profile Ignition Pick-Up	PIP
57	361	R	Vehicle Power	V PWR
58	97	O	Feed Back Control	FBC
60	60	BK/W	Battery Ground	BATT. GND.

Pin locations given for reference only. Probing 60 pin connector with DVOM probe will result in permanent damage to the pin connectors. Always probe as directed using the Breakout Box.

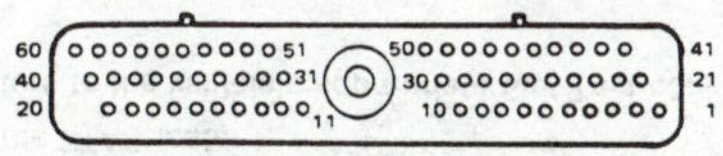

Quick Test — 5.0L 5.8L FBC

1.0 VISUAL CHECK AND VEHICLE PREPARATION

Correct test results for the Quick Test are dependent on the proper operation of related non-EEC components systems. It may be necessary to correct faults in these areas before EEC will pass QuickTest. Refer to Diagnostic Routines, Section 2 for service.

Before hooking up any equipment to diagnose the EEC system, make the following checks:

1. Verify the condition of air cleaner and ducting. These components may be removed and reinstalled as necessary for service and/or inspection.
2. Check all engine vacuum hoses for:
   - Leaks or pinched hoses.
   - Proper emission routing per Vehicle Emission Control Information (VECI) Decal.
3. Check the EEC system wiring harness electrical connections for:
   - Proper connections.
   - Loose or detached connectors, wires and terminals.
   - Corrosion.
   - Proper routing of harness.

It may be necessary to disconnect or disassemble the connector assembly to perform some of the inspections. (Note the location of each pin before disassembly).

4. Check processor, sensors and actuators for physical damage.
5. Perform all safety steps required to start and run operational vehicle tests.
6. Apply the emergency brake. Place shift lever in Park; Neutral for manual transmission.
7. **Turn Off all electrical loads** such as the radio, lamps, air conditioner, etc. Be sure doors are closed whenever readings are made.
8. Verify engine coolant is at the specified level.
9. Start engine and idle until the upper radiator hose is hot and pressurized and the throttle is off fast idle. While the engine is operating, check for leaks around the exhaust manifold, EGO sensor, and vacuum hose connections.

   NOTE: If the engine will not start or starts and stalls, continue with this Step 1.0 and on through Steps 2.0 and 3.0 of the Quick Test. If the engine idles rough and/or runs rough, go to Quick Test Step 7.0 , Diagnosis by Symptom.

10. Turn ignition key Off.
11. Service items as required, and proceed to equipment hook-up. Step 2.0

Quick Test

5.0L
5.8L FBC

2.0 EQUIPMENT HOOKUP

Using the STAR tester:

- Turn ignition key Off.
- Connect the color-coded adapter cable leads to the STAR tester.
- Connect the adapter cable's two service connectors to the vehicle's appropriate Self-Test connectors.
- Connect the timing light.

Using analog voltmeter:

- Turn ignition key Off.
- Connect a jumper wire from Self-Test Input (STI) to Pin 2, Signal Return on the Self-Test connector (refer to the diagram below).
- Set analog VOM on a DC voltage range to read from 0 to 15 volts DC. Connect VOM from battery (+) to Pin 4 Self-Test Output (STO) in the Self-Test connector.
- Connect the timing light.

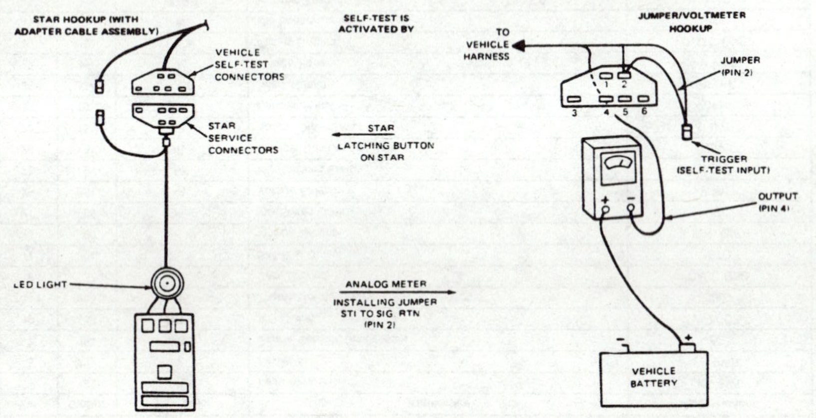

Quick Test

5.0L
5.8L FBC

3.0 KEY ON, ENGINE OFF SELF-TEST

Correct test results for the Quick Test are dependent on the proper operation of related non-EEC-IV components/systems. It may be necessary to correct any faults in these areas before EEC-IV will pass the Quick Test. Refer to the Diagnostic Routines, Section 2 for proper servicing.

- Verify that the vehicle has been properly prepared per Quick Test Steps [1.0] and [2.0].
- Place transmission in Neutral or Park. Set parking brake. Place A/C or heater control to Off.
- Activate Self-Test.
- Place key in Run to start Self-Test.

NOTE: Do not depress throttle during Self-Test.

- Observe and record all Service Codes indicated. One of the following outputs will occur. (Continuous Service Codes recorded in this Step will be referenced later in Quick Test Step [6.0]).

RESULT: CODE FORMAT — ON DEMAND	SEPARATOR①	CONTINUOUS		ACTION TO TAKE
11	1(0)	11	▶	Key On, Engine Off **on demand** and **continuous tests** indicate a pass. Go to Quick Test Step [4.0]. If vehicle is a no start, go directly to Pinpoint Test Step [A1].
ANY CODE(S)	1(0)	11	▶	Key On, Engine Off **on demand** test indicates a fault. Record on demand code(s). Go to Quick Test Step [3.1].
ANY CODE(S)	1(0)	ANY CODE(S)	▶	Key On, Engine Off **on demand** and **continuous tests** indicate a fault. Record on demand and continuous codes. Go to Quick Test Step [3.1].
11	1(0)	ANY CODE(S)	▶	**Continuous** test indicates a fault. Record continuous code. Go to Quick Test Step [4.0].
NO CODES OUTPUTTED			▶	Go to Pinpoint Test Step [Q1].

①Star tester displays a 10 in place of a 1 for the single digit separator code, therefore the separator code is expressed in the results column as 10.

Quick Test

5.0L 5.8L FBC

3.1 KEY ON, ENGINE OFF SELF-TEST

- Using the On Demand Service Codes from Key On, Engine Off Quick Test Step 3.0 follow the instructions in the Action To Take column in this Step.
- When more than one service code is received, always start service with the first code received.
- Whenever a repair is made, REPEAT Quick Test.

NOTE: **Before proceeding to the specified Pinpoint Test, read the instructions on how to use the Pinpoint Tests at the beginning of the Pinpoint Test section.**

RESULT	ACTION TO TAKE
ON DEMAND SERVICE CODES 15 ▶	REPLACE processor. REPEAT Quick Test.
21 ▶	GO to Pinpoint Test Step [DE1].
22 ▶	GO to Pinpoint Test Step [DF4].
23 ▶	GO to Pinpoint Test Step [DH2].
31 ▶	GO to Pinpoint Test Step [DD40].
51 ▶	GO to Pinpoint Test Step [DE10].
53 ▶	GO to Pinpoint Test Step [DH20].
61 ▶	GO to Pinpoint Test Step [DE20].
63 ▶	GO to Pinpoint Test Step [DH30].
NO CODES OR ANY CODE NOT LISTED ABOVE ▶	GO to Pinpoint Test Step [Q1].

Quick Test

5.0L 5.8L FBC

TEST STEP	RESULT ▶	ACTION TO TAKE
4.0 CHECK TIMING		
• If vehicle is a no start, go directly to Pinpoint Test [A] NO STARTS.	Vehicle stalls while testing. ▶	GO to Pinpoint Test Step [Q30].
• Key Off, wait 10 seconds.		
• Verify Self-Test trigger has been activated.		
• Restart engine and check timing while in Self-Test. You have 2 minutes from the time the last code is outputted to check Self-Test timing.	Timing is not base plus 20° BTDC ±3° ▶	GO to Pinpoint Test Step [P1].
• Correct Self-Test timing equals base plus 20° BTDC ± 3° (see VECI decal for correct base timing).	Timing is base plus 20° BTDC ±3° ▶	GO to Quick Test Step [5.0].
Example: If base timing is 10° BTDC then initial timing during Self-Test is: (10° + 20°) ± 3° = 30° ± 3° therefore correct timing is 27-33° BTDC.		

Quick Test

5.0L
5.8L FBC

5.0 ENGINE RUNNING SELF-TEST

NOTE: If vehicle is a no start, go directly to Pinpoint Test [A1].

- Deactivate Self-Test.
- Start and Run engine at greater than 2000 rpm for 2 minutes. This warms up EGO sensor.
- Turn engine Off, wait 10 seconds.
- Activate Self-Test.
- Start engine. The Engine Running Test will progress as follows:
  - —Engine ID code. (1)
  - —Run test.
  - —When the dynamic response ready code occurs, perform a brief WOT. (2)
  - —Engine running service codes.
- Observe and record all service codes. One of the following outputs will occur.
- If engine stalls in Self-Test, Go to Pinpoint Test Step [Q 30].

RESULT: CODE FORMAT — ENGINE ID	DYNAMIC RESPONSE	ON DEMAND		ACTION TO TAKE
2(0)	1(0)	11	▶	Engine Running On Demand Test indicates a pass. If the symptom was of an intermittent nature, Go to Quick Test Step [6.0]. If the drive symptom is currently present, Go to Quick Test Step [7.0]. Otherwise, testing is complete. EEC-IV System is OK.
2(0)	1(0)	ANY CODE(S)	▶	Engine Running On Demand Test indicates a fault. Go to Quick Test Step [5.1].
NO CODES OUTPUTTED			▶	Repeat Self-Test and verify no service codes are present then Go to Pinpoint Test Step [Q1].

(1) Engine ID code is equal to half of the number of cylinders (a code of 2 equals a 4 cylinder engine) except for STAR tester which adds a zero to all single digit readings (20 equals a 4 cylinder engine).

(2) STAR displays a 10 in place of a 1 for single digit separator code, therefore, the Dynamic Response code is expressed in the results column as 10.

Quick Test

5.0L
5.8L FBC

5.1 ENGINE RUNNING SELF-TEST

- Using the Service Codes from Engine Running Self-Test Step [5.0] follow the instructions in the Action To Take column in this Step.
- When more than one service code is received, always start service with the first code received.
- Whenever a repair is made, REPEAT Quick Test.

RESULT: ENGINE RUNNING SERVICE CODES		ACTION TO TAKE
12	▶	GO to Pinpoint Test Step [KG1].
16	▶	GO to Pinpoint Test Step [HB21].
21	▶	GO to Pinpoint Test Step [DE1].
22	▶	GO to Pinpoint Test Step [DF1].
23	▶	GO to Pinpoint Test Step [DH4].
31	▶	GO to Pinpoint Test Step [DD1].
32	▶	GO to Pinpoint Test Step [DD10].
33	▶	GO to Pinpoint Test Step [DD10].
34	▶	GO to Pinpoint Test Step [DD10].
35	▶	GO to Pinpoint Test Step [DD30].
41	▶	GO to Pinpoint Test Step [HB1].
42	▶	GO to Pinpoint Test Step [HB20].
43	▶	GO to Pinpoint Test Step [HB40].
44	▶	GO to Pinpoint Test Step [KC1].
45	▶	GO to Pinpoint Test Step [KC1].

Quick Test

5.0L 5.8L FBC

5.1 ENGINE RUNNING SELF-TEST (CONTINUED)

RESULT	ACTION TO TAKE
ENGINE RUNNING SERVICE CODES	
46 ▶	GO to Pinpoint Test Step [KC1].
72 ▶	GO to Pinpoint Test Step [DF10].
73 ▶	GO to Pinpoint Test Step [DH40].
77 ▶	GO to Pinpoint Test Step [M1].
NO CODES/CODES NOT LISTED ABOVE ▶	GO to Pinpoint Test Step [Q1].

TEST STEP	RESULT ▶	ACTION TO TAKE
6.0 CONTINUOUS TESTING		
• Do the results of both on demand tests (Key On, Engine Off and Engine Running) result in a pass code 11?	Yes ▶	Go to [6.1].
	No ▶	RETURN to Step [1.0] of Quick Test and make necessary repairs, indicated in Test Steps [3.0] and [5.0] before continuing.
6.1 CLEAR CONTINUOUS CODES		
• Prepare the vehicle for Key On, Engine Off Quick Test per Steps [1.0] [2.0] and [3.0]. • Perform Key On, Engine Off Quick Test. When the service codes begin, exit the Self-Test program (via unlatching STAR or removing the jumper from STI to Signal Return). Exiting Quick Test during code output will clear all codes stored in the continuous memory.	▶	Go to [6.2].
6.2 DETERMINE CODES TO BE TESTED		
• Refer to Continuous Codes recorded in Quick Test Step [3.0]. • Have any of these codes been repaired either in Key On, Engine Off or Engine Running on Demand Tests?	Yes ▶	Disregard codes already serviced. For remaining codes, Go to [6.3].
	No ▶	Go to [6.3].

Quick Test

5.0L 5.8L FBC

6.3 CONTINUOUS MONITOR MODE (WIGGLE TEST)

- Verify proper test equipment hook up per Quick Test Step [2.0].
- **Self-Test de-activated. Do not activate Self-Test unless instructed.**
- Using the service codes determined in Step [6.2], follow the Action To Take instructions. (While in the appropriate Pinpoint Test, one or both of the continuous monitor modes described below may have to be used to find the intermittent fault).
- **Key On, Engine Off**
  With Self-Test de-activated, turn the ignition key to the On position. You are now in the continuous monitor mode (wiggle TEST mode). STO will be activated whenever a fault is detected. A fault is indicated when the STAR LED turns off or VOM deflection is 10.5 V or greater.
- **Engine Running**
  Activate Self-Test and perform engine running Quick Test. After service code output has been completed, do not turn engine off or de-activate Quick Test. Approximately 2 minutes after Code 11 has been outputted the system will enter and remain in continuous monitor mode until Self-Test is de-activated or engine is turned Off. STO will be activated whenever a fault is detected. A fault is indicated when the STAR LED turns off or VOM deflection is 10.5 V or greater.
- Refer to Appendix C for a detailed description of Continuous Monitor.

RESULT	ACTION TO TAKE
CONTINUOUS SERVICE CODES	
11 ▶	If drive symptom is currently present GO to Quick Test Step [7.0], otherwise testing complete.
15 ▶	GO to Pinpoint Test Step [Q6].
21 ▶	GO to Pinpoint Test Step [DE90].
22 ▶	GO to Pinpoint Test Step [DF90].
31 ▶	GO to Pinpoint Test Step [DD90].
41 ▶	GO to Pinpoint Test Step [HB1].
42 ▶	GO to Pinpoint Test Step [HB20].
51 ▶	GO to Pinpoint Test Step [DE91].
61 ▶	GO to Pinpoint Test Step [DE94].
63 ▶	GO to Pinpoint Test Step [DH94].
NO CODES/CODES NOT LISTED ABOVE ▶	GO to Pinpoint Test Step [Q1].

Quick Test

5.0L 5.8L FBC

TEST STEP	RESULT	ACTION TO TAKE
7.0 DIAGNOSTICS BY SYMPTOM		
• Runs rough. • Misses. • Lack of power.		• EGR system, Section 6. • Ignition system distributor cap, rotor, wires, coil, plugs: Section 15. • Base engine valves, cam timing, compression etc. Refer to Shop Manual, Group 21. • Fuel delivery, refer to Section 11 or Shop Manual, Group 24.
• Lack of fast idle with A/C On. • Fast idle speed.		GO to Pinpoint Test Step KG1 .
• Gas fume odors present under hood.		GO to Pinpoint Test Step KD1 and/or Section 7.

Electrical Schematic

5.0L 5.8L FBC

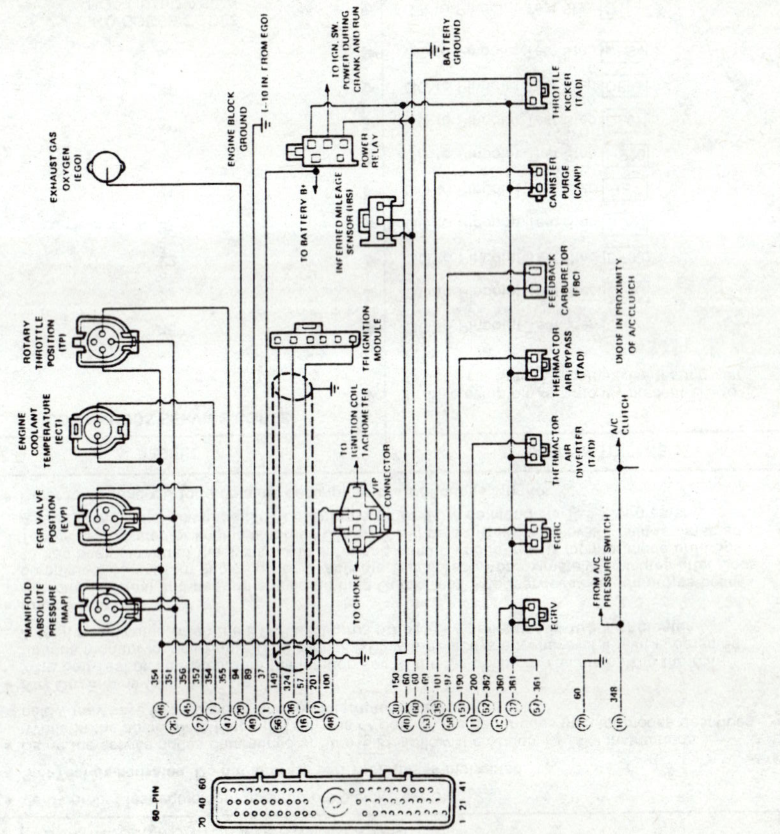

F-SERIES

EEC-IV Module Connector Pin Usage

5.0L 5.8L FBC

F-SERIES

Pin	Circuit	Wire Color	Application	Abbreviations
1	37	Y	Keep Alive Memory	KAM
7	354	LG-Y	Engine Coolant Temperature	ECT
10	348	LG-P H	A C Clutch	ACC
11	200	W-BK	Thermactor Air Diverter	TAD
16	57	BK	Ignition Ground	IGN GND
17	201	T-R	Self-Test Output	STO
20	60	BK-LG D	Case Ground	CASE GND
26	351	O-W	Voltage Reference	V REF
27	352	BR-LG	EGR Valve Position	EVP
29	94	DG-P H	Exhaust Gas Oxygen Sensor	EGO
30	150	DG-W	Inferred Mileage Sensor	IMS
33	360	DG	EGR Vent	EGRV
35	101	GY-Y H	Canister Purge	CANP
36	324	Y LG D	Spark Output	SPOUT
37	361	R	Vehicle Power	V PWR
40	60	BK-LG D	Battery Ground	BATT GND
45	356	DB-LG	Manifold Absolute Pressure	MAP
46	359	BK-W	Signal Return	SIG RET
47	355	DG-L	Throttle Position Sensor	TPS
48	100	W-R D	Self-Test In	STI
49	89	O	EGO Ground (Engine Block)	EGO GND
51	190	W-R D	Thermactor Air Bypass	TAB
52	362	Y	EGR Control	EGRC
53	69	R-LG	Throttle Kicker	TKS
56	349	DB	Profile Ignition Pick-Up	PIP
57	361	R	Vehicle Power	V PWR
58	97	T-LG D	Feedback Carburetor	FBC
60	60	BK-LG D	Battery Ground	BATT GND

Electrical Schematic

5.0L 5.8L FBC

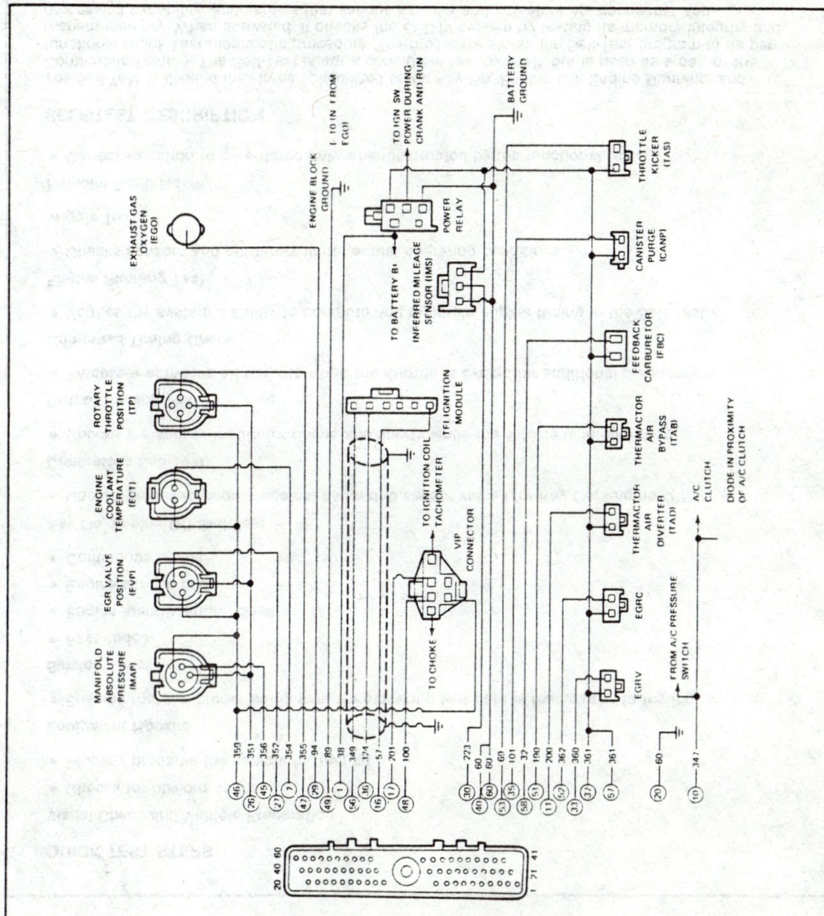

ECONOLINE

EEC-IV Module Connector Pin Usage

5.0L 5.8L FBC

ECONOLINE

Pin	Circuit	Wire Color	Application	Abbreviations
1	38	BK-R	Keep Alive Memory	KAM
7	354	GR-Y	Engine Coolant Temperature	ECT
10	347	BK	A C Clutch	ACC
11	200	BL	Thermactor Air Diverter	TAD
16	57	BK	Ignition Ground	IGN GND
17	201	BL-Y	Self-Test Output	STO
20	60	BK-W	Case Ground	CASE GND
26	351	R	Voltage Reference	V. REF
27	352	DR-GR	EGR Valve Position	EVP
29	94	GR-P	Exhaust Gas Oxygen Sensor	EGO
30	223	O-GR	Inferred Mileage Sensor	IMS
33	360	Y-BL	EGR Vent	EGRV
35	101	GY	Canister Purge	CANP
36	324	Y-GR	Spark Output	SPOUT
37	361	R	Vehicle Power	V PWR
40	60	BK-W	Battery Ground	BATT GND
45	356	BL-GR	Manifold Absolute Pressure	MAP
46	359	BK-W	Signal Return	SIG RET.
47	355	GR	Throttle Position Sensor	TPS
48	100	BK-R	Self-Test In	STI
49	89	O	EGO Ground (Engine Block)	EGO GND
51	190	Y	Thermactor Air Bypass	TAB
52	362	Y	EGR Control	EGRC
53	69	R-GR	Throttle Kicker	TKS
56	349	BR	Profile Ignition Pick-Up	PIP
57	361	R	Vehicle Power	V PWR
58	32	R-BL	Feedback Carburetor	FBC
60	60	BK-W	Battery Ground	BATT GND

Quick Test Description

Appendix **A**

QUICK TEST STEPS

Visual Check and Vehicle Preparation

- Checks for obvious faults
- Properly prepares the vehicle for testing.

Equipment Hookup

- Ensures that the proper equipment for gathering test data is ready, prior to testing.

Service Codes

- Fast codes
- Engine identification codes
- Engine service code
- Continuous codes

Key On, Engine Off Self-Test

- Checks processor inputs against calibrated sensor values for Key On, Engine Off.

Continuous Self-Test

- Checks the sensor inputs for opens and shorts while the vehicle is in operation.

Output Cycling Test

- Processor activates all outputs when the throttle is cycled for additional diagnostics.

Computed Timing Check

- Verifies the system's ability to compute and maintain engine timing in the Self-Test.

Engine Running Test

- Checks sensors and actuators under actual operating conditions.

Wiggle Test

Pinpoint Routines

- Corrective action to be entered only when instructed by the functional test.

SELF-TEST DESCRIPTION

The Self-Test is divided into three specialized tests: Key On, Engine Off. Engine Running, and Continuous Testing. The Self-Test is not a conclusive test by itself, but is used as a part of the functional Quick-Test diagnostic procedure. The processor stores the Self-Test program in its permanent memory. When activated, it checks the EEC-IV system by testing its memory integrity and processing capability, and verifies that various sensors and actuators are connected and operating properly.

The Key On, Engine Off and Engine Running tests are functional tests which only detect faults present at the time of the Self-Test. Continuous testing is an ongoing test that stores fault information for retrieval at a later time (during the Self-Test).

Reading Codes—Analog Voltmeter

When a service code is reported on the analog voltmeter for a function test, it will represent itself as a pulsing or sweeping movement of the voltmeter's needle across the dial face of the voltmeter (Figure 3). Therefore, a single-digit number of three will be reported by three needle pulses (sweeps). However, as previously stated, a service code is represented by a two-digit number, such as 2-3. As a result, the Self-Tests service code of 2-3 will appear on the voltmeter as two needle pulses (sweeps), then, after a two-second pause, the needle will pulse (sweep) three times.

The continuous testing codes are separated from the functional codes by a six-second delay, a single half-second sweep, and another six-second delay. They are produced on the voltmeter in the same manner as the functional codes.

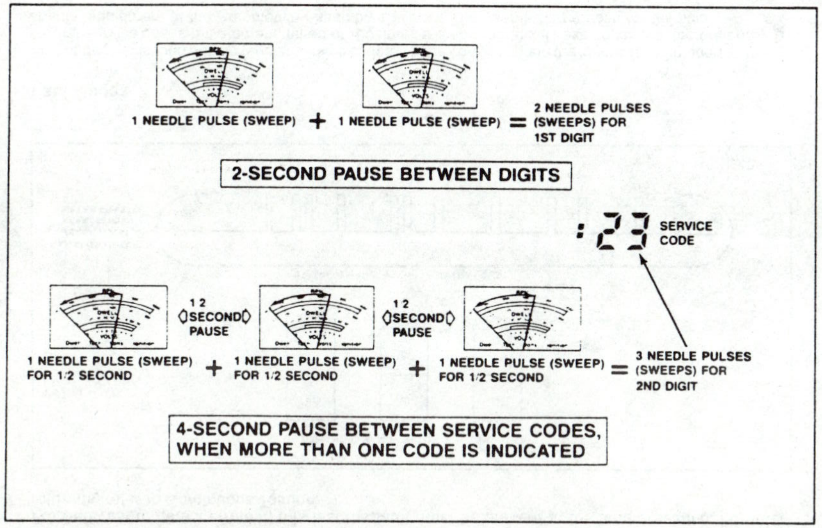

Figure 3 Analog Voltmeter Functional Service Code

Quick Test Description | Appendix | A

Reading Codes—Self-Test Automatic Readout (STAR) Rotunda Model 07-0004 or Equivalent

After hooking up the STAR tester and turning on its power switch, the tester will run a display check and the numerals 88 will begin to flash in the display window (Figure 4). A steady 00 will then appear to signify that the STAR tester is ready to start the Self-Test and receive the test's service codes.

To receive the service codes, press the pushbutton at the front of the STAR tester. The button will latch down, and a colon will appear in the display window in front of the 00 numerals. The colon **must** be displayed to receive the service codes.

If for any reason the technician wishes to clear the display window during the Self-Test, he must turn off the vehicle's engine, press the tester's pushbutton once to unlatch it (colon will disappear), then press the button again to latch down the button (colon will appear again). Every time the STAR tester is turned off, the low battery indicator (LO BAT) should show briefly at the upper left corner of the tester's display window. If the LO BAT indicator shows steadily at any other time during the operation of the STAR tester with any service code, turn its power switch to Off and replace the 9-volt battery in the tester.

The STAR tester will display the last service code received, even after disconnecting it from the vehicle. It will hold the service code on the display until the power is turned off or the pushbutton is unlatched and relatched.

WARNING: Anyone who departs from the instructions provided in this publication must first establish that he compromises neither his personal safety nor the vehicle integrity by his choice of methods, tools, or parts.

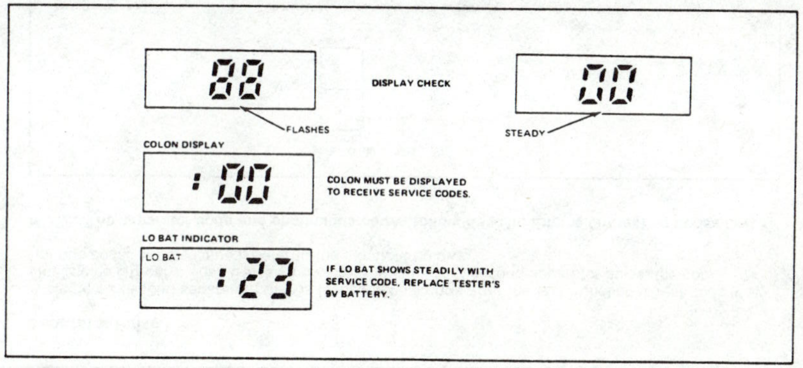

Figure 4 Self-Test Output Code Format

Key On, Engine Off

At this time, a test of the EEC-IV system is conducted with power applied and engine at rest.

For Self-Test to detect errors in this test the fault must be present at the time of testing. For intermittents, refer to Continuous Testing.

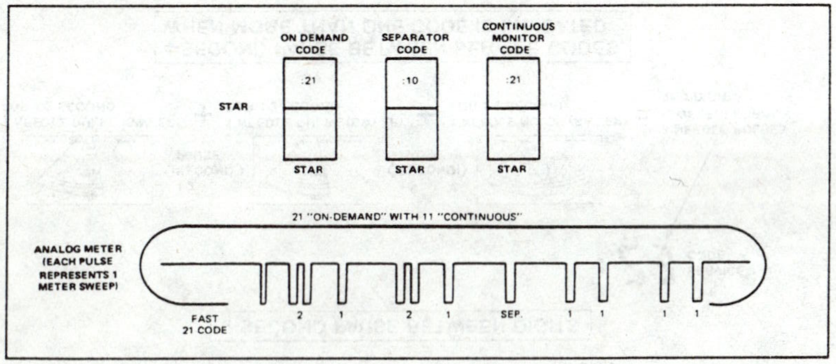

Fast Codes

Fast codes are issued prior to regular service codes. These codes contain the identical information as the regular service codes but are transmitted at 100 times the normal rate. These codes are interpreted by special equipment at the Automotive Assembly Division. They serve no purpose for the field.

For this reason the STAR has been designed to disregard these codes. Some meters in service detect these codes as a short burst of information (slight meter deflection).

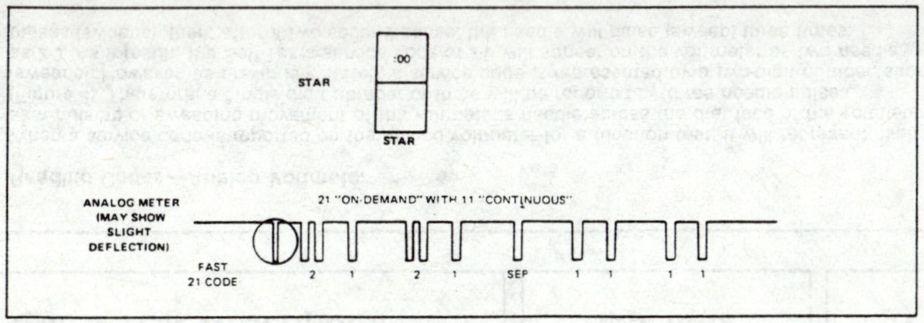

Self-Test Code Output Format | Appendix | B

Separator Pulse

A single 1/2 second separator pulse is issued 6-9 seconds after the last Functional Test code (Key On, Engine Off only). Then 6-9 seconds after the single 1/2 second separator pulse, the continuous codes will be issued. Refer to the example below:

NOTE: The separator code and continuous codes follow Key On, Engine Off testing codes ONLY.

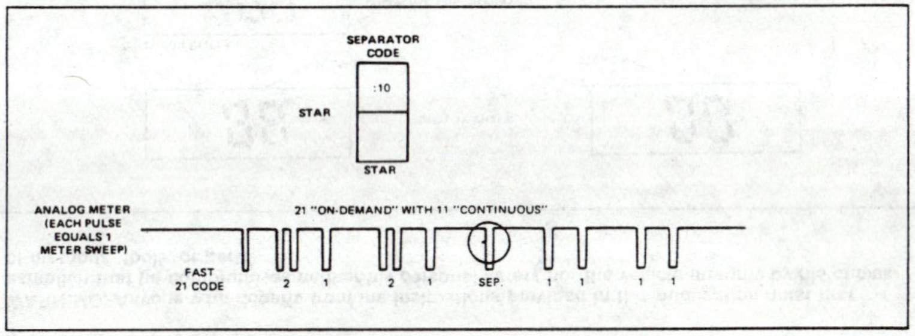

Continuous Codes

Continuous codes are issued as a result of information stored during continuous monitor testing, while the vehicle was in normal operation. These codes are displayed only during the Key On, Engine Off testing and after the separator code.

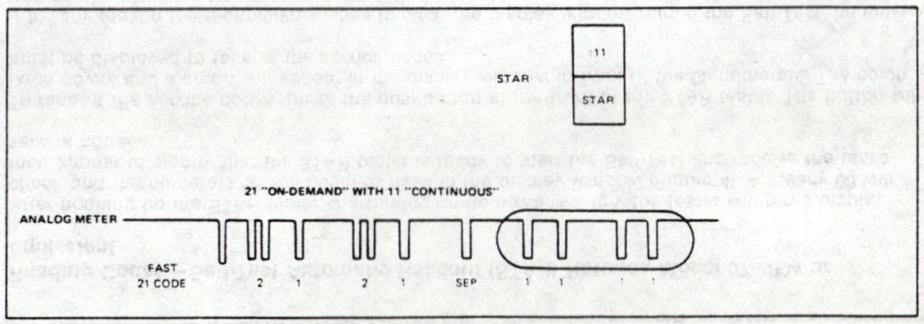

Self-Test Code Output Format	Appendix	B

Engine Running Test

At this time, a test of the EEC-IV system is conducted with the engine running. The sensors are checked under actual operating conditions and at normal operating temperatures. The actuators are exercised and checked for corresponding results.

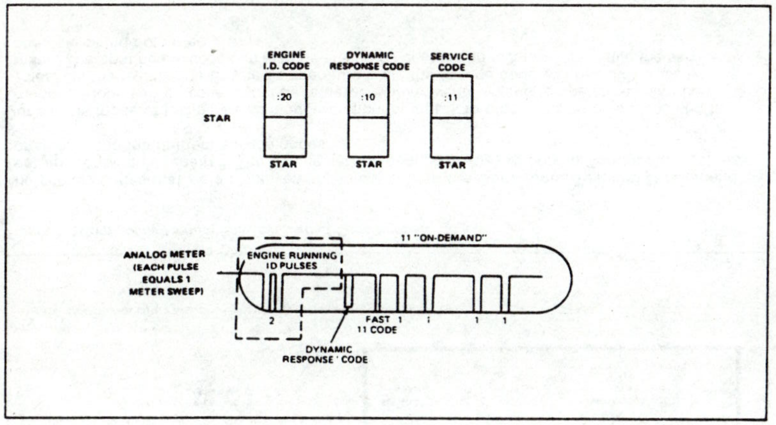

Self-Test Code Output Format	Appendix	B

Engine Identification Codes (ID Codes)

Engine ID codes are issued at the beginning of the engine running test and are one-digit numbers represented by the number of pulses sent out. The engine ID code is equal to 1/2 the number of engine cylinder (i.e, 2 pulses = 4 cylinders). These codes are used to verify the proper processor is installed and that the Self-Test has been entered.

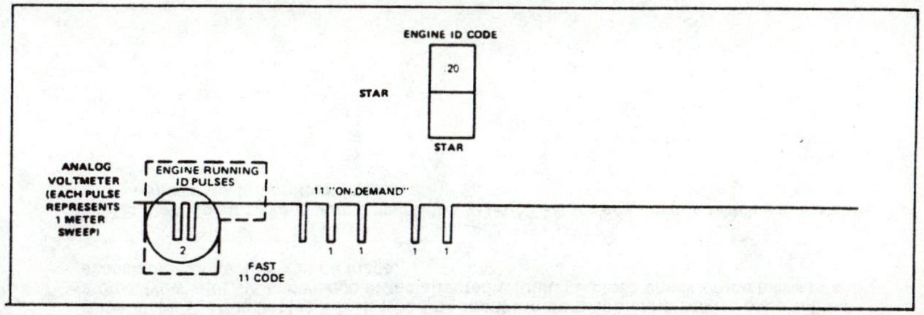

Dynamic Response Check

The dynamic response check verifies the movement of the TP, VAF, and MAP sensors during the brief Wide Open Throttle (WOT) performed during the Engine Running test. The signal for the operator to perform the brief WOT is a single pulse or 10 code on the STAR.

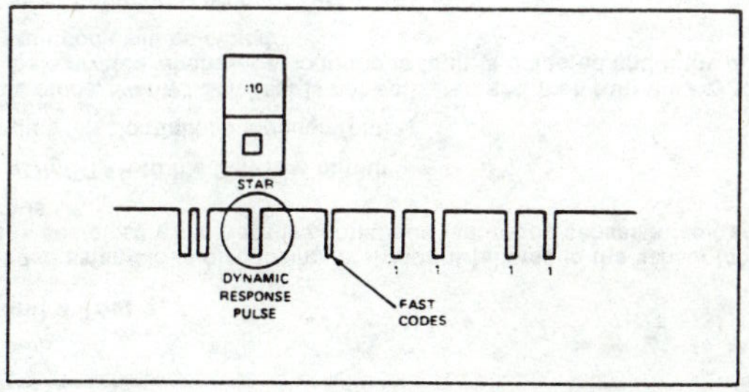

Continuous Monitor Test

This test is intended as an aid in diagnosing intermittent failures in the sensor input circuits. The Self-Test output is energized whenever the continuous test mode senses a fault and de-energized when the system is OK.

- Connect a VOM/STAR to the Self-Test output.
- Key On, Engine Off Continuous Monitor Test.

  To enter this mode, "verify" Self-Test is not activated and then turn the key to run. Self-Test output will be activated whenever a continuous fault is detected and if the duration is long enough, a fault code will be stored.

- Engine Running Continuous Monitor Test.

  This mode of testing is entered approximately 2 minutes after the service codes from the Engine Running portion of Quick Test (Step 5.0) because the engine running test can be entered only once per ignition cycle. An alternate quick method (eliminates waiting for Self-Test to complete its cycle) is to enter the running test, exit, and re-enter (do not shut the engine off). This will put you in the Engine Running Continuous Monitor Test.

  The Continuous Monitor Test will allow the technician to enter this mode of Self-Test and then attempt to re-create the intermittent failure (tap, move, wiggle the harness and/or the suspected sensor).

  If the VOM/STAR indicates a fault (short or open) the corresponding service code will be stored. Now with the knowledge of the affected circuits, a close check of the harness and associated connectors can be made.

NOTE: Remember to keep your eyes on the VOM/STAR for any change which will indicate where the intermittent is located.

Continuous Monitor Test | Appendix | C

The Continuous Monitor Test service codes are separated from the Quick Test Key On, Engine Off codes by a single separator pulse, Figure 5.

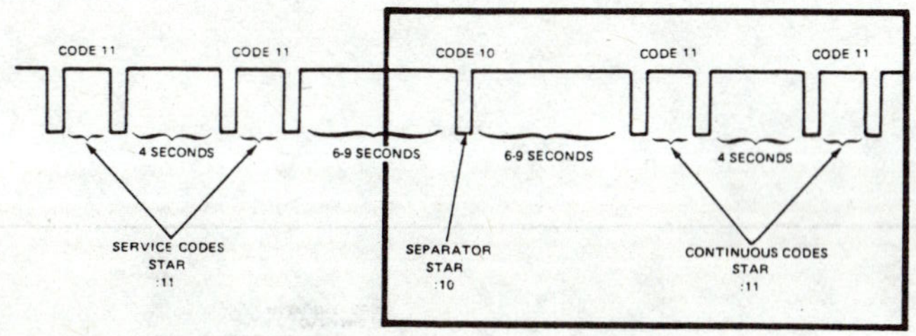

NOTE: CONTINUOUS CODES WILL ONLY BE OUTPUTTED DURING KEY ON—ENGINE OFF.

Figure 5 Continuous Monitor Self-Test Output Code Format

The service codes that were displayed (repeated) in both the Continuous Test and On Demand Test are considered repaired. The unique service code(s) displayed only in Continuous Test are considered the intermittent service codes.

During this mode of testing we continuously monitor inputs to processor for opens and shorts. The continuous service codes must be retrieved within twenty engine temperature warm up cycles. On the twenty-first Engine Temperature cycle the service code will be automatically erased. The continuous code can also be erased by deactivating Quick Test, while the service codes are being outputted.

Continuous Monitor Test | Appendix | C

EXAMPLE: How to Use the Continuous Test (Systematically)

If a service code 51 was displayed: Engine Coolant Temperature Sensor:

- Visually inspect the sensor **very closely.**
- Enter Continuous Monitor test.
- Lightly TAP on the sensor.
- Push/pull on the sensor harness connector (do not disconnect, yet).
- Test and Wiggle (shake) harness vigorously working from the sensor connector toward the dash panel and from the dash panel to the processor in short sections.
- If VOM has not given a positive indication of an intermittent, disconnect the sensor from the harness as carefully as possible. Remove terminals from the connector. Visually inspect terminals at both ends for corrosion, bad crimps, improperly seated terminals, etc.
- Reconnect after inspection.
- Disconnect processor from harness as carefully as possible.
- Inspect terminals.
- Only remove terminals associated with the sensor being inspected.
- If the VOM does not give a positive intermittent indication, reconnect the connector and erase the Continuous Test service codes.
- To erase the continuous service codes:
  — Initiate Key On, Engine Off Self-Test.
  — Remove jumper from the Self-Test input terminal as soon as the **first** service code is received (even if an "11" is the first code).
  — Rerun Self-Test with jumper to verify services have been erased.

Continuous Monitor Test | Appendix | C

The diagrams below indicate common three wire sensor failure modes. Use them as a guide for servicing 50 or 60 Series Service Codes.

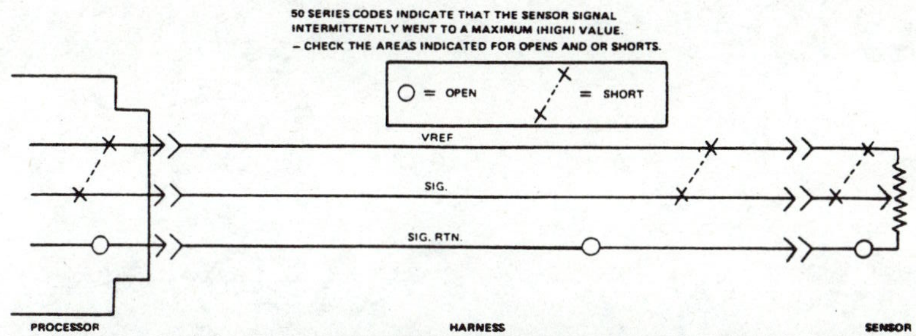

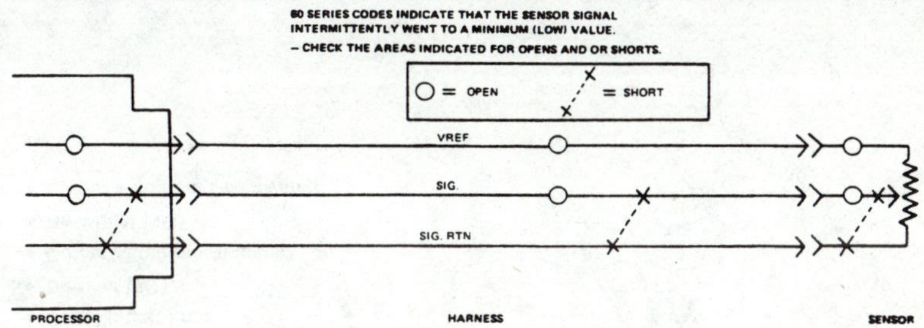

Continuous Monitor Test | Appendix | C

The diagrams below indicate common two wire sensor failure. Use them as a guide for servicing 50 or 60 Series Service Codes.

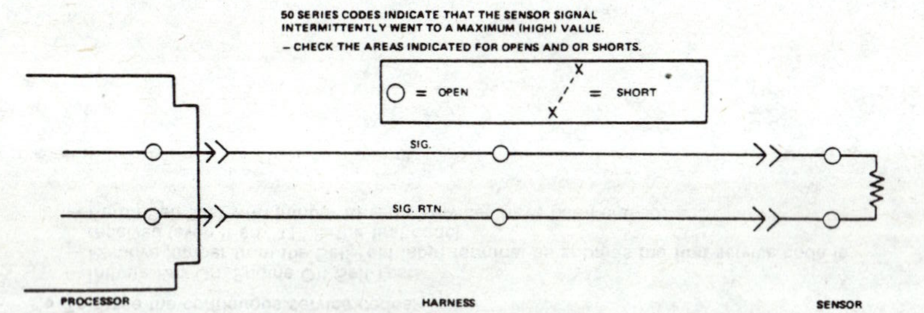

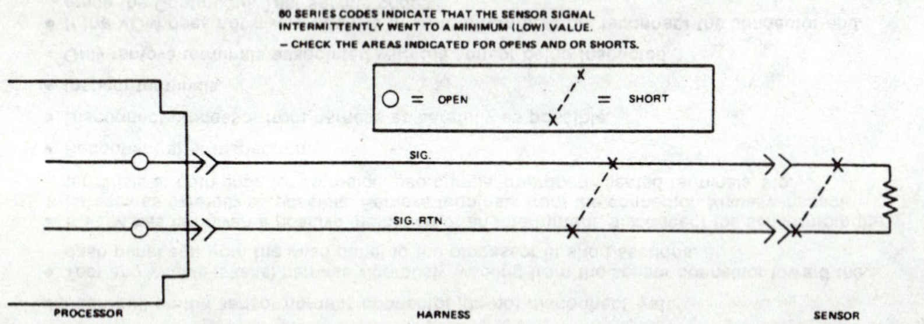

Output State Check | Appendix | D

The output state check aids in servicing output actuators associated with the EEC-IV system. It enables the technician to energize and de-energize most of the system output actuators (turn EGR solenoid and Self-Test Output (STO) On and Off) on command. This mode is entered after all codes have been received from Key On, Engine Off. On Demand and Continuous Testing. At this time, leave Self-Test activated and depress the throttle. Each time the throttle is depressed the output actuators will change state (On to Off or Off to On).

- Enter Self-Test.
- Code Output Ends.
- Do Brief WOT.
- Output Actuators Actuated.
- Next Brief WOT.
- Output Actuators Deactivated.

No Start Problem (All) | Pinpoint Test | A

A

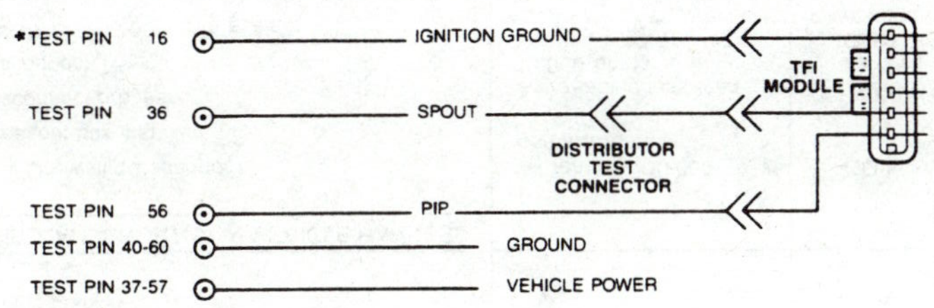

*TEST PIN LOCATED ON BREAKOUT BOX.
ALL CONNECTIONS VIEWED INTO MATING SURFACE.

NOTE: When Breakout box is installed, ensure that timing switch is in "Computed" position unless otherwise noted.

STOP-WARNING

You may have reached this point because of a fault in a non-EEC System and/or component. To prevent the replacement of good components, be aware that the following non-EEC areas may be at fault:

- Fuel: quantity and quality.
- Ignition: general condition, moisture, cracks, damage, etc.
- Engine: internal, valves, timing belt, camshaft.
- Starter and battery circuit.

This pinpoint test is intended to diagnose only the following:

- Spark (as related to EEC-IV).
- Circuits: pip, spout, ignition ground, vehicle power.

No Start Problem (All) | Pinpoint Test | A

WARNING: Stop this test at the first sign of a fuel leak and service as required.
CAUTION: No open flame — No smoking during fuel delivery checks.

TEST STEP	RESULT ▶	ACTION TO TAKE
A1 ATTEMPT TO START ENGINE	Engine cranks, but does not start, or stalls out ▶	GO to A2.
	Engine does not crank ▶	REFER to Shop Manual, Group 28.
A2 CHECK FOR SPARK AT PLUGS • Disconnect the spark plug wire to any accessible cylinder. • Connect spark tester between spark plug wire and engine ground. • Crank engine and check for spark. • Reconnect the spark plug wire to the spark plug.	Spark ▶	GO to A13.
	No spark ▶	GO to A3.
A3 CHECK FOR SPARK AT COIL • Remove high tension coil wire from distributor and install spark tester. • Check for spark while cranking. • Reconnect high tension coil wire to distributor.	Spark ▶	REFER to Section 15, Part 2 for TFI Diagnosis for cap, rotor, wires.
	No spark ▶	GO to A4.
A4 CHECK FOR POWER TO PROCESSOR • Key Off, wait 10 seconds. • Disconnect processor 60 pin connector and inspect for damaged pins, corrosion, loose wires. Service as necessary. • Install Breakout box. • Key On, Engine Off. • DVOM on 20 volt scale. • Measure voltage from test Pin 37 to test Pin 40 and test Pin 57 to test Pin 60. • Ensure timing switch is in "Computed" position on Breakout box.	Either voltage reading less than 10.5V ▶	GO to Pinpoint Test B Step B1.
	Both voltage reading 10.5V or greater ▶	GO to A5.

No Start Problem (All) — Pinpoint Test A

TEST STEP	RESULT ▶	ACTION TO TAKE
A5 HARNESS CHECK IGNITION GROUND		
• Breakout box installed. • Key Off, wait 10 seconds. • DVOM on 200 ohm scale. • Measure from test Pin 16 to TFI connector (ign. gnd).	Less than 5 ohms ▶	GO to A6.
	5 ohms or greater ▶	SERVICE harness as necessary. REPEAT Quick Test.
A6 ISOLATION OF PROBLEM TO SPOUT CIRCUIT		
• Breakout box installed. • Processor connected. • Timing switch to "Dist" position on Breakout box. • Attempt to start vehicle. • Does the vehicle start?	Yes ▶	GO to A10.
	No ▶	GO to A7.
A7 SPOUT SIGNAL CHECK		
• Breakout box installed. • Timing switch to "Computed" position on Breakout box. • DVOM on 20 volt scale. • Measure voltage from Pin 36 to chassis ground, during crank.	Less than 3.0V or greater than 6.0V ▶	GO to A8.
	Between 3.0V and 6.0V ▶	EEC OK, REFER to Section 15, for TFI diagnosis.
A8 ISOLATE SHORT(S) IN VEHICLE HARNESS		
• Key Off, wait 10 seconds. • Breakout box installed. • Disconnect processor. • Disconnect TFI. • DVOM on 200,000 ohm scale. • Measure resistance from test Pin 36 to test Pins 16, 20, 26, 40, 60 (short to ground), 37, 57 (short to power) and 56 (short to pip).	All readings 10,000 ohms or greater ▶	GO to A9.
	Any reading less than 10,000 ohms ▶	SERVICE short in harness. REPEAT Quick Test if vehicle does not start. GO to A9.

No Start Problem (All) — Pinpoint Test A

TEST STEP	RESULT ▶	ACTION TO TAKE
A9 ISOLATE SHORT(S) IN PROCESSOR		
• Key Off, wait 10 seconds. • Breakout box installed. • Reconnect processor. • TFI disconnected. • DVOM on 200 ohm scale. • Measure resistance from test Pin 36 to test Pins 37 and 57 (short to power) also test Pins 40 and 60 (short to ground).	All readings 5.0 ohms or greater. ▶	GO to A10.
	Any reading less than 5.0 ohms ▶	REPLACE processor. REPEAT Quick Test.
A10 PIP SIGNAL CHECK		
• Breakout box installed. • DVOM to 20 volt range. • Measure voltage at test Pin 56 to test Pin 16. • Crank engine, record reading.	Between 3.0V and 6.0V ▶	REMOVE Breakout box. REPLACE processor. REPEAT Quick Test.
	Less than 3.0V or greater than 6.0V ▶	GO to A11.
A11 CONTINUITY OF PIP CIRCUIT		
• Breakout box installed. • Key Off, wait 10 seconds. • DVOM on 200 ohm scale. • Disconnect processor. • Connect DVOM from test Pin 56 to TFI connector, PIP circuit. • Measure resistance.	Less than 5 ohms ▶	GO to A12.
	5 ohms or greater ▶	SERVICE open PIP circuit. REPEAT Quick Test.

No Start Problem (All) — Pinpoint Test A

TEST STEP	RESULT ▶	ACTION TO TAKE
A12 CHECK PIP CIRCUIT FOR SHORTS		
• Breakout box installed. • Processor disconnected. • Key Off. • Disconnect TFI connector. • DVOM on 200,000 ohm scale. • Measure resistance from test Pin 56 to test Pins 16, 20, 26, 40, 60 (shorts to ground) and test Pins 37 and 57 (shorts to power) and test Pin 36 (short to spout).	Any resistance less than 10,000 ohms ▶	SERVICE PIP circuit. REPEAT Quick Test.
	All resistance greater than 10,000 ohms ▶	REFER to Section 15 for TFI diagnosis.
A13 SPOUT SIGNAL VERIFICATION		
• Key Off, wait 10 seconds. • Disconnect processors 60 pin connector and inspect for damaged pins, corrosion, loose wires. Service as necessary. • Install Breakout box. • Processor connected. • DVOM on 20 volt scale. • Measure voltage from test Pin 36 to chassis ground, during crank. • Ensure timing switch is in "Computed" position on Breakout box.	Between 3.0V and 6.0V ▶	FBC vehicles. REFER to Shop Manual Group 24. CFI and EFI vehicles GO to **A21**.
	Less than 3.0V or greater than 6.0V ▶	GO to **A10**.

No Start Problem (CFI/EFI) — Pinpoint Test A

A

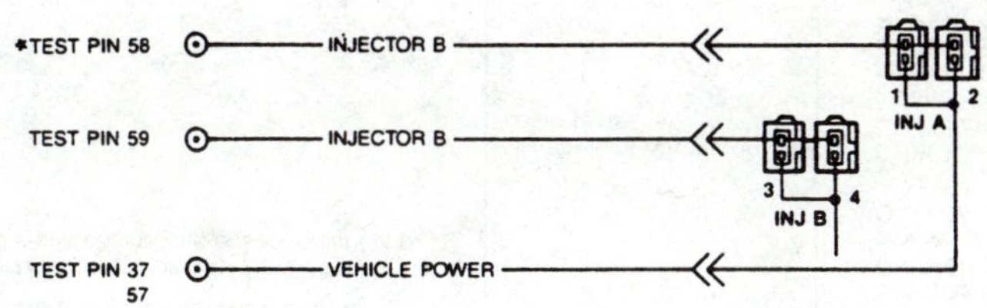

*TEST PIN LOCATED ON BREAKOUT BOX.
ALL CONNECTIONS VIEWED INTO MATING SURFACE.

TEST STEP	RESULT ▶	ACTION TO TAKE
A21 FUEL PUMP CHECK		
• **No smoking nearby.** • Disconnect all injector electrical connections at the injectors. • Connect pressure gauge to Schraeder valve. • Note initial pressure reading. • Observe pressure gauge as you pressurize fuel system. (Turn key to Run for 1 second, then turn key to Off. Wait 10 seconds. Repeat 5 times). • Turn key Off. Wait 10 seconds. • Reconnect all injectors.	PRESSURE GAUGE READING: Increased ▶	GO to **A23**.
	Did not increase ▶	TURN key Off, and CONTINUE to **A22**.

WARNING: If fuel starts leaking, turn key Off immediately. No smoking.

No Start Problem (CFI/EFI) — Pinpoint Test A

TEST STEP	RESULT ▶	ACTION TO TAKE
A22 INERTIA SWITCH CHECK		
• Key Off. • Fuel pressure gauge installed. • Locate fuel pump inertia switch. Refer to Owner's Manual for location. • Push the button of inertia switch to reset to On. **NOTE: If switch will not reset to On, replace Inertia Switch and repeat Step A21. If switch button was On, GO to J1.** • Observe pressure gauge as you pressurize fuel system. (Turn key to Run for 1 second, then turn key to Off. Wait 10 seconds.) Repeat 5 times.	PRESSURE GAUGE READING: Increase ▶	REPEAT Quick Test
	No increase ▶	GO to J1.
A23 FUEL DELIVERY CHECK		
• Fuel pressure gauge installed. • Pressurized fuel system as per Step A21. • Turn key Off. • Wait for pressure to become steady — approximately 2 to 5 seconds. • Read pressure gauge.	Less than 45 psi and 35 psi or greater ▶	GO to A24.
	Less than 35 psi and/or 45 psi or greater ▶	REFER to Shop Manual Group 24 for CFI/EFI injectors and regulator, and/or Section 24-35 for electric fuel pump Diagnosis.
A24 FUEL PRESSURE LEAK DOWN CHECK		
• Fuel pressure gauge installed. • Key Off. • Wait 2 minutes after pressure gauge reading in Step A23, then note drop in gauge reading.	Less than 4 psi drop in 2 minutes ▶	GO to A25.
	4 psi drop or greater in 2 minutes ▶	Immediately CHECK for hydraulic lock up or fouled spark plugs. REFER to Shop Manual Group 24 for CFI/EFI injector and/or fuel regulator service.

No Start Problem (CFI/EFI) — Pinpoint Test A

TEST STEP	RESULT ▶	ACTION TO TAKE
A25 FUEL DELIVERY TEST		
• Key Off. • Fuel pressure gauge installed. • Pressurize fuel system as per Step A21. • Locate and disconnect the fuel pump relay. • Crank engine for 5 seconds. • Take pressure reading at the end of 5 second crank. **NOTE: Verify fuel quality; air and/or water will also pressurize and look like acceptable fuel delivery.**	PRESSURE GAUGE READING: Pressure is approximately 10-20 psi at the end of 5 second crank cycle. Refer to note below ▶	The EEC system is not the fault of the No Start. Fuel and spark are present. REFER to Section 2 for other No Start Routines. If complaint was runs rough or misses CONTINUE to A26.
	Pressure is greater or less than specified ▶	GO to A26.

NOTE: The colder the engine, the greater the pressure drop (i.e., an engine coolant temperature of 200°F equals approximately a 10 psi drop in 5 seconds. 60°F equals approximately a 20 psi drop in 5 seconds).

TEST STEP	RESULT ▶	ACTION TO TAKE
A26 INJECTOR DRIVE SIGNAL CHECK		
• Key Off, wait 10 seconds. • Breakout box installed. • Use a standard non-powered test lamp. • Connect test lamp from test Pin 37 to test Pin 58 and crank engine observe test lamp. • Connect test lamp from test Pin 37 to test Pin 59 and crank engine and observe test lamp.	Dim glow at light on both tests ▶	GO to A27.
	No light on one or both tests ▶	VERIFY 12V battery power at test Pin 37. SERVICE power circuit as required. If OK, REPLACE processor. REPEAT Quick Test.
	Bright light on one or both tests ▶	CHECK injector circuits for shorts to ground. SERVICE as required. If OK, REPLACE processor. REPEAT Quick Test.

No Start Problem (CFI/EFI) — Pinpoint Test A

TEST STEP	RESULT ▶	ACTION TO TAKE
A27 SERVICING THE SUSPECT INJECTOR CIRCUIT		
• Key Off, wait 10 seconds. • Locate and disconnect fuel pump relay. • DVOM on 20 volt scale. • With DVOM measure the average voltage between test Pin 58 and test Pin 37 during cranking mode. • Also measure average voltage between test Pin 59 and test Pin 37. • After measuring and recording voltages, reconnect fuel pump relay.	1.0V or greater difference between injector circuits ▶	DISCONNECT processor connector and INSPECT for corrosion, damaged pins, etc. RECONNECT processor and RERUN Step A27. If condition still exists, REPLACE processor and REPEAT Quick Test.
	Less than 1.0V difference between injector circuits ▶	GO to A28.
A28 RESISTANCE CHECK OF INJECTOR HARNESS		
• Processor disconnected. • Key Off. • DVOM on 20 ohm scale. • Disconnect injector(s). **EFI:** Disconnect injectors 2, 3 and 4. **CFI:** Disconnect injector No. 2. • Through the vehicle harness, measure the resistance of injector No. 1 (still connected) from test Pin 58 to test Pin 37. Record resistance reading. • Disconnect injector No. 1 and connect injector No. 2. • Through the vehicle harness, measure the resistance of injector No. 2 from: **CFI:** Test Pin 59 to test Pin 37. **EFI:** Test Pin 58 to test Pin 37. Record resistance reading. **EFI ONLY:** • Disconnect injector No. 2 and connect injector No. 3. • Record resistance measured from test Pin 59 to test Pin 37. • Disconnect injector No. 3 and connect injector No. 4. • Record resistance measured from test Pin 59 to test Pin 37. • Injector resistances are 2.0 ohms to 3.5 ohms.	Yes ▶	GO to A29.
	No ▶	CHECK the harness on suspect injector(s). If OK, REPLACE suspect injector(s). RECONNECT all injectors and REPEAT Quick Test.

No Start Problem (CFI/EFI) — Pinpoint Test A

TEST STEP	RESULT ▶	ACTION TO TAKE
A29 ISOLATING A FAULTY INJECTOR		
• Key Off. • Fuel pressure gauge installed. • Electrically disconnect all injectors at intake manifold. • Pressurize the fuel system as per Step A21. • Electrically connect injector for cylinder No. 1. • Crank the engine for 5 seconds. • Observe fuel pressure gauge and record reading immediately after the 5 second crank. • Disconnect injector No. 1 and repressurize the fuel system. • Connect injector No. 2 and crank the engine for 5 seconds. Immediately observe and record pressure. • Disconnect injector No. 2 and repressurize the fuel system. • Connect injector No. 3 and crank the engine for 5 seconds. Immediately observe and record pressure. • Disconnect injector No. 3 and repressurize the fuel system. • Connect injector No. 4 and crank the engine for 5 seconds. Immediately observe and record pressure.	All readings not within 4 psi of each other ▶	REPLACE injector(s) that is/are not within 4 psi of each other. Connect all injectors and REPEAT Quick Test.
	All four readings are within 4 psi of each other ▶	VERIFY all the non-EEC Diagnostic Routines that apply to your drive complaint have been properly performed, then: DISCONNECT processor and INSPECT all Pins for corrosion, damage, etc., and RETEST. If fault is still present, REPLACE processor and REPEAT Quick Test.

Vehicle Battery — Pinpoint Test — B

B

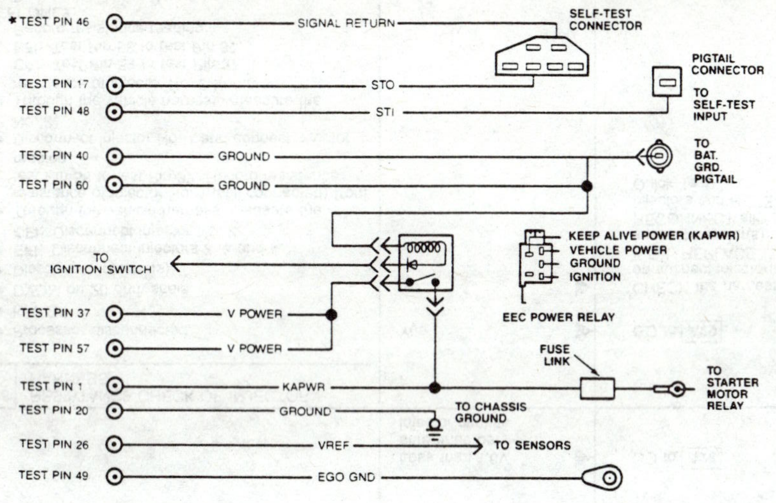

*TEST PINS LOCATED ON BREAKOUT BOX.
ALL HARNESS CONNECTORS VIEWED INTO MATING SURFACE.

STOP-WARNING

You may have reached this point because of a fault in a non-EEC System and or component. To prevent the replacement of good components, be aware that the following non-EEC areas may be at fault:

- Ignition switch.
- Battery Cables.
- Alternator.
- Voltage Regulator.
- Ground Straps.

This pinpoint test is intended to diagnose only the following:

- Processor.
- Harness circuits: Signal Return, STO, STI, Ground, V Power, KAPWR, VREF, Ignition.
- Battery Voltage.
- Power Relay.

Vehicle Battery — Pinpoint Test — B

TEST STEP	RESULT	►	ACTION TO TAKE
B1 BATTERY VOLTAGE CHECK			
• Key On, Engine Off. • DVOM on 20 volt scale. • Measure voltage across battery terminals.	10.5V or greater	►	GO to B2.
	Less than 10.5V	►	SERVICE discharged battery, REFER to Shop Manual, Group 31.
B2 BATTERY POWER GROUND CHECK			
• Key On, Engine Off. • Processor connected. • DVOM on 20 volt range. • Measure voltage from battery negative post to signal return circuit in the Self-Test connector.	Less than 0.5V	►	GO to B3.
	0.5V or greater	►	GO to B4.
B3 12 VOLT BATTERY POWER CHECK AT PROCESSOR			
• Breakout box installed. • Key On, Engine Off. • Processor connected. • DVOM on 20 volt scale. • Measure voltage from battery positive post to test Pin 37 and test Pin 57.	Both readings less than 0.5V	►	GO to C2.
	One or both readings 0.5V or greater	►	GO to B7.
B4 GROUND FAULT ISOLATION			
• Breakout box installed. • Key On, Engine Off. • Processor connected. • DVOM on 20 volt scale. • Measure voltage from battery negative post to test Pin 40 and test Pin 60.	Both readings less than 0.5V	►	GO to B5.
	One or both readings 0.5V or greater	►	Circuit(s) with greater than 0.5V has high resistance or open. CORRECT faulty ground circuit. REPEAT Quick Test.

Vehicle Battery — Pinpoint Test B

TEST STEP	RESULT ▶	ACTION TO TAKE
B5 PROCESSOR GROUND FAULT ISOLATION		
• Breakout box installed. • Key Off, wait 10 seconds. • Processor connected. • DVOM on 200 ohm scale. • Measure resistance from test Pin 46 to test Pin 40 and from test Pin 46 to test Pin 60.	Both readings less than 5 ohms ▶	GO to B6.
	One or both readings 5 ohms or greater ▶	DISCONNECT processor connector and INSPECT for corrosion, damaged pins. etc. SERVICE as necessary and RETEST. If fault is still present, REPLACE processor. REPEAT Quick Test.
B6 HARNESS CHECK (SIGNAL RETURN)		
• Breakout box installed. • Key Off, wait 10 seconds. • Processor connected. • DVOM on 200 ohm scale. • Measure resistance from test Pin 46 to self test connector Signal Return circuit.	Less than 5 ohms ▶	System OK. RUN Quick Test.
	5 ohms or greater ▶	CORRECT cause of resistance in the harness Signal Return circuit. REPEAT Quick Test.
B7 12 VOLT BATTERY POWER FAULT ISOLATION		
• Key On, Engine Off. • Processor connected. • DVOM on 20 volt scale. • Measure voltage by connecting the DVOM negative lead to battery negative post. Connect DVOM positive lead to KAPWR circuit at EEC power relay.	10.5V or greater ▶	GO to B8.
	Less than 10.5V ▶	CHECK KAPWR and V Power circuits for shorts to ground and KAPWR circuit from power relay to battery positive post for opens. SERVICE as necessary. REPEAT Quick Test.

Vehicle Battery — Pinpoint Test B

TEST STEP	RESULT ▶	ACTION TO TAKE
B8 12 VOLT BATTERY POWER FAULT ISOLATION		
• Key On, Engine Off. • Processor connected. • DVOM on 20 volt scale. • Measure voltage by connecting the DVOM negative lead to battery negative post. Connect DVOM positive lead to Ignition circuit at EEC power relay.	10.5V or greater ▶	GO to B9.
	Less than 10.5V ▶	CHECK 10 amp fuse, location 18 in fuse panel. CHECK for open in ignition switch circuits. SERVICE as necessary. REPEAT Quick Test.
B9 12 VOLT BATTERY POWER FAULT ISOLATION		
• Key On, Engine Off. • Processor connected. • DVOM on 20 volt scale. • Measure voltage by connecting the DVOM negative lead to battery negative post. Connect DVOM positive lead to ground circuit at EEC power relay.	Less than 0.5V ▶	GO to B10.
	0.5V or greater ▶	SERVICE open or ground in Ground Circuit. REPEAT Quick Test.
B10 12 VOLT BATTERY POWER FAULT ISOLATION		
• Key On, Engine Off. • Processor connected. • DVOM on 20 volt scale. • Measure voltage by connecting the DVOM negative lead to battery ground. Connect DVOM positive lead to V-Power circuit at EEC power relay.	10.5V or greater ▶	SERVICE short to ground or open in V-Power circuit from EEC power relay to processor connector test Pins 37 and 57. REPEAT Quick Test.
	Less than 10.5V ▶	REPLACE power relay. REPEAT Quick Test.

Reference Voltage — Pinpoint Test C

C

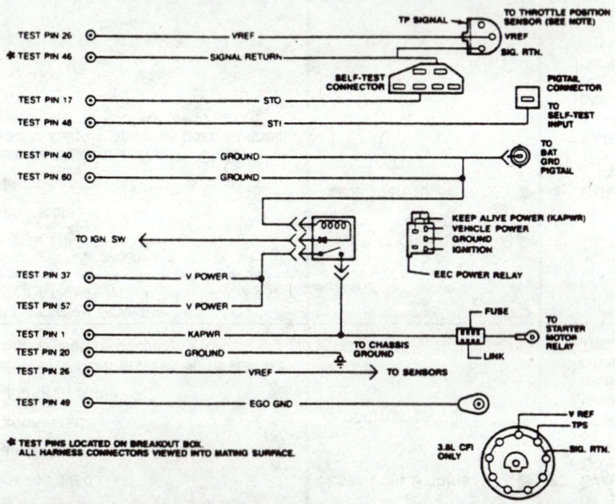

NOTE: For 3.8L CFI a 10-Pin connector was used for the first 60 days of production. Either connector may be encountered. Test pin assignments did not change on Breakout box. If TP sensor must be disconnected for diagnostics, disconnect 10-Pin connector at fuel charging assy.

STOP-WARNING

This pinpoint test is intended to diagnose only the following:

- Processor.
- Sensor harness circuits: Signal Return, STO, STI, Ground, V Power, KAPWR, VREF, Ignition.

Reference Voltage — Pinpoint Test C

TEST STEP	RESULT	▶	ACTION TO TAKE
C1 VEHICLE BATTERY POWER CIRCUIT CHECK			
• Breakout box installed. • Key On, Engine Off. • Processor connected. • DVOM on 20 volt scale. • Connect DVOM positive test lead to test Pin 37 and negative test lead to Signal Rtn. in Self-Test connector.	10.5V or greater	▶	GO to C2.
	Less than 10.5V	▶	GO to B1.
C2 VREF VOLTAGE CHECK			
• Breakout box installed. • Key On, Engine Off. • Processor connected. • DVOM on 20 volt scale. • Measure voltage with DVOM from test Pin 26 to test Pin 46.	6.0V or greater	▶	GO to C3.
	4.0V or less	▶	GO to C4.
	Greater than 4.0V, less than 6.0V	▶	Reference voltage OK, REPEAT Quick Test.
C3 CHECK FOR SHORT TO PROCESSOR POWER VREF			
• Key Off, wait 10 seconds. • Breakout box installed. • Disconnect processor. • Key On, Engine Off. • DVOM on 20 volt scale. • Check for voltage from test Pin 26 to battery ground and from test Pin 46 to battery ground.	Less than 0.5V	▶	REPLACE processor REPEAT Quick Test.
	0.5V or greater	▶	SERVICE short to battery power in EEC harness. REPEAT Quick Test.
C4 CHECK FOR SHORTED THROTTLE POSITION SENSOR			
• Key Off, wait 10 seconds. • Breakout box installed. • Processor connected. • Disconnect Throttle Position (TP) sensor from vehicle harness. • Key On, Engine Off. • DVOM on 20 volt scale. • Measure voltage with DVOM from test Pin 26 to test Pin 46.	Less than 4.0V	▶	Vehicles equipped with EVP sensor, GO to C5. All other vehicles, GO to C6.
	4.0V or greater	▶	REPLACE TP sensor. REPEAT Quick Test.

Reference Voltage — Pinpoint Test C

TEST STEP	RESULT ▶	ACTION TO TAKE
C5 CHECK FOR SHORTED EGR VALVE POSITION (EVP) SENSOR		
• Key Off, Wait 10 seconds. • Breakout box installed. • Processor connected. • Disconnect EGR valve position (EVP) sensor. • Key On, Engine Off. • DVOM on 20 volt scale. • Measure voltage with DVOM from test Pin 26 to test Pin 46.	Less than 4.0V ▶ 4.0V or greater ▶	GO to C6. REPLACE EVP sensor. REPEAT Quick Test.
C6 CHECK FOR SHORTED MAP/BP SENSOR		
• Key Off, wait 10 seconds. • Breakout box installed. • Processor connected. • Disconnect MAP/BP sensor. • Key On, Engine Off. • DVOM on 20 volt scale. • Measure voltage with DVOM from test Pin 26 to test Pin 46.	Less than 4.0V ▶ 4.0V or greater ▶	Vehicles equipped with VAF sensor, GO to C7. All other vehicles, GO to C8. REPLACE MAP/BP sensor. REPEAT Quick Test.
C7 CHECK FOR SHORTED VANE AIR METER (VAF) SENSOR		
• Key Off, wait 10 seconds. • Breakout box installed. • Processor connected. • Disconnect vane air meter (VAF) sensor. • Key On, Engine Off. • DVOM on 20 volt scale. • Measure voltage with DVOM from test Pin 26 to test Pin 46.	Less than 4.0V ▶ 4.0V or greater ▶	GO to C8. REPLACE VAF sensor and REPEAT Quick Test.

Reference Voltage — Pinpoint Test C

TEST STEP	RESULT ▶	ACTION TO TAKE
C8 SHORT TO GROUND IN VREF		
• Breakout box installed. • Processor disconnected. • Key Off, wait 10 seconds. • Disconnect TP and MAP/BP, EVP and VAF, if so equipped. • DVOM on 200 ohm scale. • Measure continuity from test Pin 26 to test Pins 20, 40, 46 and 60.	Less than 5 ohms ▶ 5 ohms or greater ▶	SERVICE short to ground in VREF. REPEAT Quick Test. GO to C9.
C9 CHECK VREF FOR CONTINUITY		
• Breakout box installed. • Processor disconnected. • TP, MAP/BP, EVP and VAF sensors disconnected. • Key Off. • DVOM on 200 ohm scale. • Measure continuity on all above referenced sensors from test Pin 26 to VREF circuit on the sensor harness connectors.	Less than 5 ohms on all readings ▶ 5 ohms or greater on any reading ▶	RECONNECT sensors. REPLACE processor. REPEAT Quick Test. SERVICE open in VREF. REPEAT Quick Test.

Vane Air Temperature Sensor (VAT) | Pinpoint Test | DA

DA

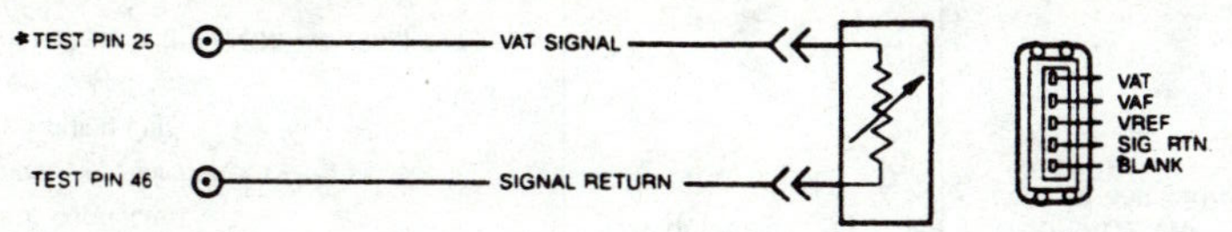

*NOTE: AMBIENT TEMP. MUST BE GREATER THAN 50°F TO PASS THIS TEST.

TYPICAL RESISTANCE BETWEEN TEST POINTS 25 & 46	5800Ω	2700Ω	300Ω	180Ω	125Ω
AT TEMPERATURE	32°F	65°F	185°F	220°F	240°F

*TEST PIN LOCATED ON BREAKOUT BOX.
ALL CONNECTIONS VIEWED INTO MATING SURFACE.

STOP-WARNING

You may have reached this point because of a fault in a non-EEC System and/or component. To prevent the replacement of good components, be aware that the following non-EEC areas may be at fault:

- Test performed in unusually low (cold) or high (hot) ambient conditions.
- Ambient temperature must be greater than 50°F for this test.

This pinpoint test is intended to diagnose only the following:

- VAT sensor.
- Circuits: VAT, and Signal Return.
- Vehicle harness.
- Processor assembly.

Vane Air Temperature Sensor (VAT) | Pinpoint Test | DA

TEST STEP	RESULT ▶	ACTION TO TAKE
DA1 SERVICE CODE 24: CHECK VAT SIGNAL TO SIGNAL RETURN VOLTAGE		
• Key Off, wait 10 seconds. • Disconnect harness from VAT sensor. Inspect for damaged pins, corrosion, loose wires, etc. SERVICE as necessary. • DVOM on 20 volt scale. • Key On, Engine Off. • Measure voltage of vane meter harness connector VAT signal to Signal Return.	Reading is between 4.0V and 6.0V ▶ Reading is less than 4.0V ▶ Reading is greater than 6.0V ▶	GO to DA4. GO to DA2. GO to Pinpoint Test Step C1.
DA2 CONTINUITY CHECK OF CIRCUITS VAT AND SIGNAL RETURN		
• Key Off, wait 10 seconds. • Harness disconnected from vane meter. • Disconnect processor 60 pin connector and inspect for damaged pins, loose wires, corrosion, etc. Service as necessary. • Install Breakout box to harness, leave processor disconnected. • DVOM on 200 ohm scale. • Measure resistance of harness connector VAT signal to test Pin 25. • Measure resistance of harness connector signal return to test Pin 46.	Both readings are less than 5 ohms ▶ Either reading is 5 ohms or greater ▶	GO to DA3. CORRECT open circuits, REMOVE Breakout box. RECONNECT harness to processor and vane meter. RERUN Quick Test.
DA3 CHECK VAT CIRCUITS FOR SHORTS		
• Key Off. • Harness disconnected from vane meter and processor; Breakout box installed. • DVOM on 200,000 ohms scale. • Measure resistance from test Pin 25 to test Pins 40, 46 and 60.	Any reading less than 10,000 ohms ▶ All readings are 10,000 ohms or greater ▶	CORRECT circuits with short. REMOVE Breakout box. RECONNECT harness to vane meter and processor. RERUN Quick Test. GO to DA4.

Vane Air Temperature Sensor (VAT) — Pinpoint Test — DA

TEST STEP	RESULT ▶	ACTION TO TAKE
DA4 VAT SENSOR CHECK		
NOTE: Ambient temperature must be greater than 50°F for this test. • Key Off, wait 10 seconds. • Harness disconnected from the vane meter. • DVOM on 200,000 ohm scale. • Measure resistance of VAT sensor, VAT signal Pin to Signal Return Pin. • Is reading from 125 ohms (240°F) to 3700 ohms (50°F)?	Yes ▶	REPLACE processor. RECONNECT harness to vane meter. RERUN Quick Test.
	No ▶	REPLACE vane meter. RERUN Quick Test.
DA10 SERVICE CODE 54: INDUCE OPPOSITE CODE		
• Key Off, wait 10 seconds. • Disconnect harness from vane meter. Inspect for damaged pins, corrosion, loose wires, etc. Service as necessary. • Insert a jumper wire (paper clip) between harness connector VAT Signal and harness connector Signal Return. • Run Key On, Engine Off Quick Test. • Is Code 64 present?	Yes ▶	REPLACE vane meter. REMOVE jumper wire. CONNECT harness to vane meter. RERUN Quick Test.
	No ▶	REMOVE jumper wire. GO to DA11.
DA11 CHECK CONTINUITY OF VAT SIGNAL AND SIGNAL RETURN		
• Key Off, wait 10 seconds. • Harness disconnected from vane meter, jumper wire removed. • Disconnect processor 60 pin connector. Inspect for damaged pins, corrosion, loose wires, etc. Service as necessary. • Install Breakout box leaving processor disconnected. • DVOM on 200 ohm scale. • Measure resistance of harness connector VAT Signal to test Pin 25. • Measure resistance of harness connector Signal Return to test Pin 46.	Both readings are less than 5 ohms ▶	REPLACE processor. REMOVE Breakout box. RECONNECT harness to vane meter and processor. RERUN Quick Test.
	Either reading is 5 ohms or greater ▶	CORRECT open circuit. REMOVE Breakout box. RECONNECT harness to vane meter and processor. RERUN Quick Test.

Vane Air Temperature Sensor (VAT) — Pinpoint Test — DA

TEST STEP	RESULT ▶	ACTION TO TAKE
DA20 SERVICE CODE 64: INDUCE OPPOSITE CODE		
• Key Off, wait 10 seconds. • Disconnect harness from vane meter. Inspect for damaged pins, corrosion, loose wires, etc. Service as necessary. • Run Key On, Engine Off Quick Test. • Is code 54 present?	Yes ▶	REPLACE vane meter. RECONNECT harness to vane meter. RERUN Quick Test.
	No ▶	GO to DA21.
DA21 CHECK VAT SIGNAL FOR SHORTS		
• Key Off, wait 10 seconds. • Harness disconnected from vane meter. • Disconnect processor 60 pin connector. Inspect for damaged pins, corrosion, loose wires, etc. Service as necessary. • Install Breakout box leaving processor disconnected. • DVOM on 200,000 ohm scale. • Measure resistance of Test Pin 25 to Test Pins 40, 46 and 60.	Any reading less than 10,000 ohms ▶	CORRECT circuit shorts. REMOVE Breakout box. RECONNECT processor and vane meter. RERUN Quick Test.
	All readings are 10,000 ohms or greater ▶	REPLACE processor. REMOVE Breakout box. RECONNECT processor. RERUN Quick Test.

Vane Air Temperature — Pinpoint Test DA

TEST STEP	RESULT ▶	ACTION TO TAKE
DA90 SERVICE CODE 54: CONTINUOUS TEST: CHECK VAT SENSOR		
• Using continuous monitor mode, observe VOM or STAR LED for indication of a fault while performing the following: • Lightly tap on VAT sensor (simulate road shock). • Wiggle VAT connector. • Is a fault indicated?	Yes ▶	DISCONNECT and INSPECT connectors. If connector and terminals are good, REPLACE VAT sensor. REPEAT Quick Test.
	No ▶	GO to DA91.

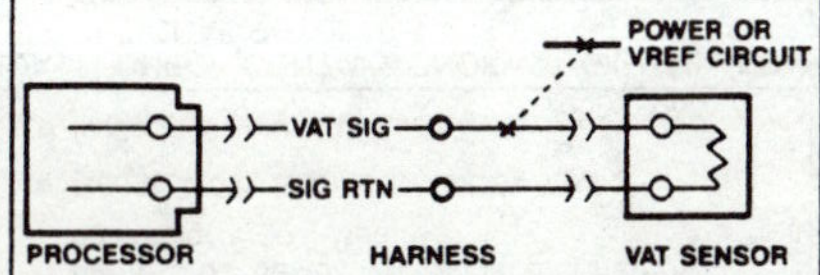

TEST STEP	RESULT ▶	ACTION TO TAKE
DA91 CHECK EEC-IV HARNESS		
• Observe VOM or STAR LED for a fault indication while performing the following: • Referring to the illustration in Step DA90, grasp the harness closest to the sensor connector. Wiggle, shake or bend a small section of the EEC-IV system harness while working your way to the dash panel. Also wiggle, shake or bend the EEC-IV harness from the dash panel to the processor. • Is fault indicated?	Yes ▶	ISOLATE fault and make necessary service. REPEAT Quick Test.
	No ▶	GO to DA92.
DA92 CHECK PROCESSOR AND HARNESS CONNECTORS		
• Key Off, wait 10 seconds. • Disconnect processor 60 pin connector. • Inspect both connectors and connector terminals for obvious damage or faults. • Connectors and terminals are OK.	No ▶	SERVICE as necessary. REPEAT Quick Test.
	Yes ▶	Unable to duplicate fault at this time. Continuous code 54 testing complete.

Vane Air Temperature — Pinpoint Test DA

TEST STEP	RESULT ▶	ACTION TO TAKE
DA93 SERVICE CODE 64: CONTINUOUS TEST: CHECK VAT SENSOR		
• Using continuous monitor mode, observe VOM or STAR LED for indication of a fault while performing the following: • Lightly tap on VAT sensor (simulate road shock). • Wiggle VAT connector. • Is fault indicated?	Yes ▶	DISCONNECT and INSPECT connectors. If connector and terminals are good, REPLACE VAT sensor. REPEAT Quick Test.
	No ▶	GO to DA94.

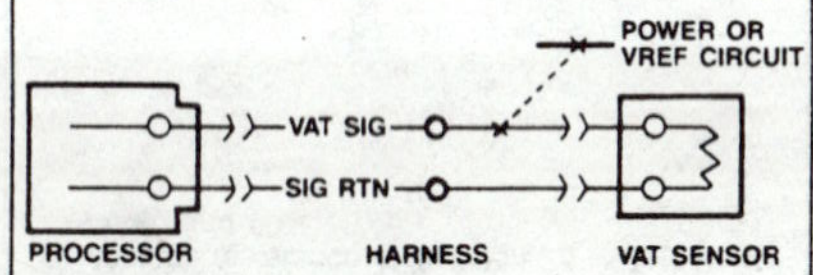

TEST STEP	RESULT ▶	ACTION TO TAKE
DA94 CHECK EEC-IV HARNESS		
• Observe VOM or STAR LED for a fault indication while performing the following: • Referring to the illustration in Step DA93, grasp the harness closest to the sensor connector. Wiggle, shake or bend a small section of the EEC-IV system harness while working your way to the dash panel. Also wiggle, shake or bend the EEC-IV harness from the dash panel to the processor. • Is fault indicated?	Yes ▶	ISOLATE fault and make necessary service. REPEAT Quick Test.
	No ▶	GO to DA95.
DA95 CHECK PROCESSOR AND HARNESS CONNECTORS		
• Key Off, wait 10 seconds. • Disconnect processor 60 pin connector. • Inspect both connectors and connector terminals for obvious damage or faults. • Connectors and terminals are OK.	No ▶	SERVICE as necessary. REPEAT Quick Test.
	Yes ▶	Unable to duplicate fault at this time. Continuous code 64 testing complete.

Air Charge Temperature Sensor (ACT) — Pinpoint Test — DB

DB

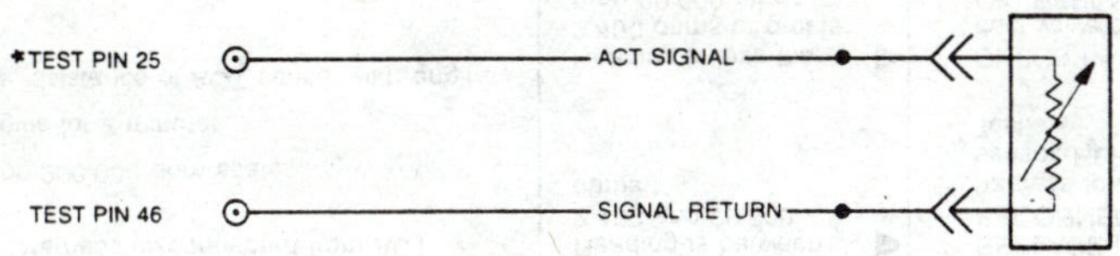

***TEST PIN LOCATED ON BREAKOUT BOX.**
ALL CONNECTIONS VIEWED INTO MATING SURFACE.

NOTE: Ambient temperature must be greater than 50°F for this test.

STOP-WARNING

You may have reached this point because of a fault in a non-EEC System and/or component. To prevent the replacement of good components, be aware that the following non-EEC areas may be at fault:

- Cooling system.
- Air cleaner duct problems.
- Improper engine oil level.

This pinpoint test is intended to diagnose only the following:

- ACT sensor.
- Harness circuits: ACT Signal and Signal Return.
- Processor assembly.

Air Charge Temperature Sensor (ACT) — Pinpoint Test — DB

TEST STEP	RESULT ▶	ACTION TO TAKE
DB1 SERVICE CODE 24: CHECK PROPER INSTALLATION		
• For vehicle with ACT mounted in intake manifold. GO to Step DB2. • Is ACT mounted properly in air cleaner?	Yes ▶	GO to DB2.
	No ▶	INSTALL ACT properly. RERUN Quick Test.
DB2 CHECK ACT SIGNAL TO SIGNAL RETURN VOLTAGE		
• Key Off, wait 10 seconds. • Disconnect harness from ACT sensor. Inspect for damaged pins, corrosion, loose wires, etc. Service as necessary. • DVOM on 20 volt scale. • Key On, Engine Off. • Measure voltage from ACT harness connector ACT Signal to Signal Return.	Reading is between 4.0V and 6.0V ▶	GO to DB5.
	Reading is less than 4.0V ▶	GO to DB3.
	Reading is greater than 6.0V ▶	GO to Pinpoint Test Step C1.
DB3 CONTINUITY CHECK OF ACT SIGNAL AND SIGNAL RETURN		
• Key Off, wait 10 seconds. • Harness disconnected from ACT sensor. • Disconnect processor 60 pin connector. Inspect for damaged pins, corrosion, loose wires, etc. Service as necessary. • Install Breakout box leaving processor disconnected. • DVOM on 200 ohm scale. • Measure resistance of harness connector ACT Signal to test Pin 26. • Measure resistance of harness connector Signal Return to test Pin 46.	Both readings are less than 5 ohms ▶	GO to DB4.
	Either reading is 5 ohms or greater ▶	CORRECT circuit opens. REMOVE Breakout box. RECONNECT processor and ACT sensor. RERUN Quick Test.

Air Charge Temperature Sensor (ACT) — Pinpoint Test DB

TEST STEP	RESULT ▶	ACTION TO TAKE
DB4 CHECK ACT SIGNAL FOR SHORTS		
• Key Off. Harness disconnected from ACT sensor. Breakout box installed. Processor disconnected. • DVOM on 200,000 ohm scale. • Measure resistance of test Pin 25 to test Pins 40, 46 and 60.	Any reading less than 10,000 ohms ▶	CORRECT circuit shorts. REMOVE Breakout box. RECONNECT processor and ACT sensor. RERUN Quick Test.
	All readings are 10,000 ohms or greater ▶	GO to DB5.
DB5 CHECK ACT SENSOR — ENGINE OFF		
• Key Off, wait 10 seconds. • Harness disconnected from ACT sensor. • DVOM on 200,000 ohm scale. • Measure resistance of ACT sensor. **NOTE: Make sure engine is warmed up prior to this test.**	Reading is between 1,100 and 58,000 ohms (approximately 50°F) ▶	GO to DB6.
	Reading is less than 1,100 ohms or greater than 58,000 ▶	CHECK heat stove duct valve operation. If OK, REPLACE ACT sensor. RECONNECT harness to ACT sensor. RERUN Quick Test.
DB6 CHECK ACT SENSOR — ENGINE RUNNING		
• Key Off. Harness disconnected from ACT sensor. • DVOM on 200,000 ohm scale. • Run engine for 2 minutes. • Measure resistance of ACT sensor with engine running.	Reading is between 2,400 and 29,000 ohms ▶	REPLACE processor. RECONNECT harness to ACT sensor. RERUN Quick Test.
	Reading is less than 2,400 ohms or greater than 29,000 ohms ▶	CHECK heat stove duct valve operation. If OK, REPLACE ACT sensor. RERUN Quick Test.

Air Charge Temperature Sensor (ACT) — Pinpoint Test DB

TEST STEP	RESULT ▶	ACTION TO TAKE
DB10 SERVICE CODE 54: INDUCE OPPOSITE CODE		
• Key Off, wait 10 seconds. • Disconnect harness from ACT sensor. Inspect for damaged pins, corrosion, loose wires, etc. Service as necessary. • Insert a jumper wire between ACT harness connector ACT Signal to Signal Return. • Run Key On, Engine Off Quick Test. • Is code 64 present?	Yes ▶	REPLACE ACT sensor. REMOVE jumper wire. RECONNECT harness to ACT sensor. RERUN Quick Test.
	No ▶	REMOVE jumper wire. GO to DB11.
DB11 CONTINUITY CHECK ACT SIGNAL AND SIGNAL RETURN		
• Key Off, wait 10 seconds. • Harness disconnected from ACT sensor. • Disconnect processor 60 pin connector. Inspect for damaged pins, corrosion, loose wires, etc. Service as necessary. • Install Breakout box. Leave processor disconnected. • DVOM on 200 ohm scale. • Measure resistance of ACT harness connector ACT Signal to test Pin 25. • Measure resistance of ACT harness connector Signal Return to test Pin 46.	Both readings are less than 5 ohms ▶	REPLACE processor. REMOVE Breakout box. RECONNECT processor and ACT sensor. RERUN Quick Test.
	Either reading is 5 ohms or greater ▶	Correct circuit opens. REMOVE Breakout box. RECONNECT processor and ACT sensor. RERUN Quick Test.

Air Charge Temperature Sensor (ACT) — Pinpoint Test — DB

TEST STEP	RESULT ▶	ACTION TO TAKE
DB20 SERVICE CODE 64: INDUCE OPPOSITE CODE		
• Key Off, wait 10 seconds. • Disconnect harness from ACT sensor. Inspect for damaged pins, corrosion, loose wires, etc. Service as necessary. • Run Key On, Engine Off Quick Test. • Is code 54 present?	Yes ▶	REPLACE ACT sensor. RECONNECT harness to ACT sensor. RERUN Quick Test.
	No ▶	GO to DB21.
DB21 CHECK ACT SIGNAL FOR SHORTS		
• Key Off, wait 10 seconds. • Harness disconnected from ACT sensor. • Disconnect processor 60 pin connector. Inspect for damaged pins, corrosion, loose wires, etc. Service as necessary. • DVOM on 200,000 ohm scale. • Measure resistance of test Pin 25 to test Pins 40, 46 and 60.	Any reading less than 10,000 ohms ▶	CORRECT circuit shorts. REMOVE Breakout box. RECONNECT processor and ACT sensor. RERUN Quick Test.
	All readings are 10,000 ohms or greater ▶	REPLACE processor. REMOVE Breakout box. RECONNECT processor and ACT sensor. RERUN Quick Test.

Air Charge Temperature (ACT) — Pinpoint Test — DB

TEST STEP	RESULT ▶	ACTION TO TAKE
DB90 SERVICE CODE 54: CONTINUOUS TEST: CHECK ACT SENSOR		
• Using continuous monitor mode, observe VOM or STAR LED for indication of a fault while performing the following: • Lightly tap on ACT sensor (simulate road shock). • Wiggle ACT connector. • Is a fault indicated?	Yes ▶	DISCONNECT and INSPECT connectors. If connector and terminals are good, REPLACE ACT sensor. REPEAT Quick Test.
	No ▶	GO to DB91.

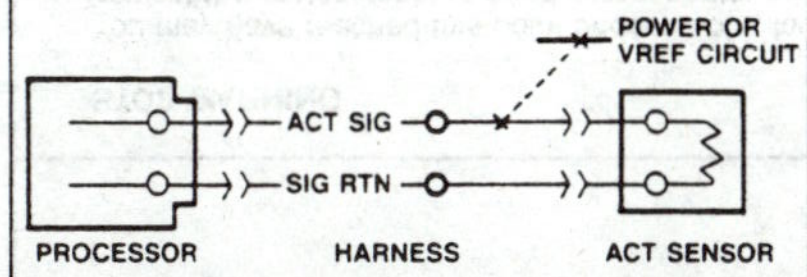

TEST STEP	RESULT ▶	ACTION TO TAKE
DB91 CHECK EEC-IV HARNESS		
• Observe VOM or STAR LED for a fault indication while performing the following: • Referring to the illustration in Step DB90, grasp the harness closest to the sensor connector. Wiggle, shake or bend a small section of the EEC-IV system harness while working your way to the dash panel. Also wiggle, shake or bend the EEC-IV harness from the dash panel to the processor. • Is fault indicated?	Yes ▶	ISOLATE fault and make necessary repairs. REPEAT Quick Test.
	No ▶	GO to DB92.
DB92 CHECK PROCESSOR AND HARNESS CONNECTORS		
• Key Off, wait 10 seconds. • Disconnect processor 60 pin connector. • Inspect both connectors and connector terminals for obvious damage or faults. • Connectors and terminals are OK.	No ▶	SERVICE as necessary. REPEAT Quick Test.
	Yes ▶	Unable to duplicate fault at this time. Continuous code 54 testing complete.

Air Charge Temperature (ACT) Pinpoint Test DB

TEST STEP	RESULT ▶	ACTION TO TAKE
DB93 SERVICE CODE 64: CONTINUOUS TEST: CHECK ACT SENSOR		
• Using continuous monitor mode, observe VOM or STAR LED for fault while performing the following: • Lightly tap on ACT sensor (simulate road shock). • Wiggle ACT connector. • Is fault indicated?	Yes ▶	DISCONNECT and INSPECT connectors. If connector and terminals are good, REPLACE ACT sensor. REPEAT Quick Test.
	No ▶	GO to DB94.
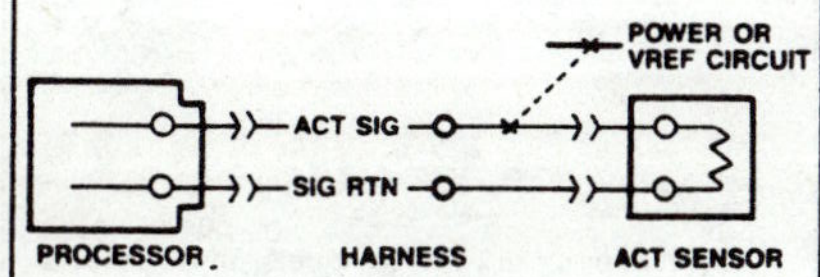		
DB94 CHECK EEC-IV HARNESS		
• Observe VOM or STAR LED for a fault indication while performing the following: • Referring to the illustration in Step DB93, grasp the harness closest to the sensor connector. Wiggle, shake or bend a small section of the EEC-IV system harness while working your way to the dash panel. Also wiggle, shake or bend the EEC-IV harness from the dash panel to the processor. • Is fault indicated?	Yes ▶	ISOLATE fault and make necessary repairs. REPEAT Quick Test.
	No ▶	GO to DB95.
DB95 CHECK PROCESSOR AND HARNESS CONNECTORS		
• Key Off, wait 10 seconds. • Disconnect processor 60 pin connector. • Inspect both connectors and connector terminals for obvious damage or faults. • Connectors and terminals are OK.	No ▶	SERVICE as necessary. REPEAT Quick Test.
	Yes ▶	Unable to duplicate fault at this time. Continuous code 64 testing complete.

Barometric Pressure Sensor (BP) Pinpoint Test DC

DC

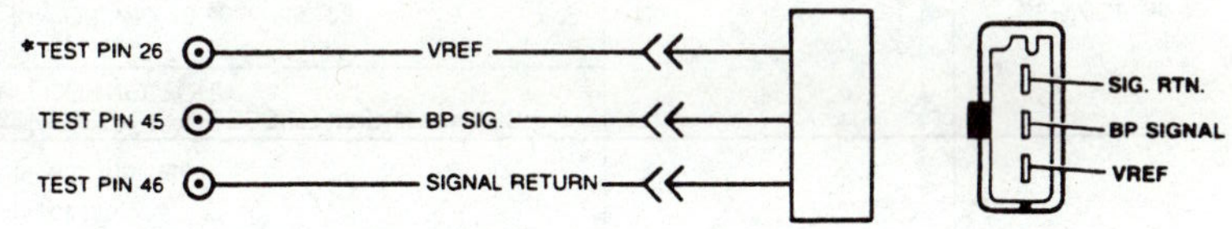

*TEST PIN LOCATED ON BREAKOUT BOX.
ALL HARNESS CONNECTORS VIEWED INTO MATING SURFACE.

STOP-WARNING

You may have reached this point because of a fault in a non-EEC System and/or component. To prevent the replacement of good components, be aware that the following non-EEC areas may be at fault:

- Unusually high or low atmospheric pressure.

This pinpoint test is intended to diagnose only the following:

- An active BP sensor (will not detect on out of range sensor).
- Harness circuits: VREF, BP Signal, and Signal Return.
- Processor assembly.

Barometric Pressure Sensor (BP) — Pinpoint Test — DC

TEST STEP	RESULT ▶	ACTION TO TAKE
DC1 SERVICE CODE 22: VOLTAGE CHECK VREF		
NOTE: The output of the BP sensor is digital in nature and cannot be monitered without special equipment. Therefore, perform a harness check before substituting the BP sensor. • Key Off, wait 10 seconds • Disconnect harness from BP sensor. Inspect for damaged pins, corrosion, loose wires, etc. Service as necessary. • DVOM on 20 volt scale. • Key On, Engine Off. • Measure voltage from BP sensor harness connector VREF to battery negative terminal.	Reading is between 4.0V and 6.0V ▶	GO to DC2.
	Reading is less than 4.0V ▶	SERVICE open in VREF. If reading is still less than 4.0 volts, REPLACE processor. RECONNECT processor and BP sensor. RERUN Quick Test.
	Reading is greater than 6.0V ▶	GO to Pinpoint Test Step C1.
DC2 SIGNAL RETURN CHECK		
• Key On, Engine Off. • Harness disconnected from BP sensor. • DVOM on 20 volt scale. • Measure voltage from BP sensor harness connector signal return to battery positive post.	Reading is less than 10.0V ▶	SERVICE open in signal return. RECONNECT BP sensor. RERUN Quick Test.
	Reading is 10.0V or greater ▶	GO to Step DC3.
DC3 CHECK CONTINUITY OF BP SIGNAL		
• Key Off, wait 10 seconds. • Harness disconnected from BP sensor. • Disconnect processor 60 pin connector. Inspect for damaged pins, corrosion, loose wires, etc. Service as necessary. • Install Breakout box leaving processor disconnected. • DVOM on 200 ohm scale. • Measure resistance between BP sensor harness connector BP signal and test Pin 45.	Reading is less than 5 ohms ▶	REMOVE Breakout box. GO to DC4.
	Reading is 5 ohms or greater ▶	SERVICE open in BP signal. REMOVE Breakout box. RECONNECT processor and BP sensor. RERUN Quick Test.

Barometric Pressure Sensor (BP) — Pinpoint Test — DC

TEST STEP	RESULT ▶	ACTION TO TAKE
DC4 CHECK BP SIGNAL FOR SHORTS		
• Key Off. • Harness disconnected from BP sensor and processor. • DVOM on 200,000 ohm scale. • Measure resistance between BP sensor harness, BP signal and Signal Return. • Measure resistance between BP sensor harness, BP signal and VREF.	Either reading is less than 10,000 ohms ▶	SERVICE circuit shorts. RECONNECT BP sensor and processor. RERUN Quick Test.
	Both readings are 10,000 ohms or greater ▶	RECONNECT harness to processor and substitute BP sensor. RUN Quick Test. If code 22 is still present, REPLACE processor using original BP sensor. RERUN Quick Test.

Barometric Pressure Sensor (BP) — Pinpoint Test — DC

TEST STEP	RESULT	▶ ACTION TO TAKE
DC90 SERVICE CODE 22: CONTINUOUS TEST: EXERCISE BP SENSOR		
• Using continuous monitor mode, observe VOM or STAR LED for indication of a fault while performing the following: • Connect a vacuum pump to the BP sensor. • Slowly apply 84 kPa (25 in. Hg.) vacuum to the sensor. • Slowly bleed vacuum off the BP sensor. • Lightly tap on BP sensor (simulate road shock). • Wiggle BP connector. • Is fault indicated?	Yes	▶ DISCONNECT and INSPECT connectors. If connector and terminals are good, REPLACE BP sensor. REPEAT Quick Test.
	No	▶ GO to DC91.
DC91 CHECK EEC-IV HARNESS		
• Observe VOM or STAR LED for a fault indication while performing the following: • Referring to the illustration in Step DC90, grasp the harness closest to the sensor connector. Wiggle, shake or bend a small section of the EEC-IV system harness while working your way to the dash panel. Also wiggle, shake or bend the EEC-IV harness from the dash panel to the processor. • Is fault indicated?	Yes	▶ ISOLATE fault and make necessary repairs. REPEAT Quick Test.
	No	▶ GO to DC92.
DC92 CHECK PROCESSOR AND HARNESS CONNECTORS		
• Key Off, wait 10 seconds. • Disconnect processor 60 pin connector. • Inspect both connectors and connector terminals for obvious damage or faults. • Connectors and terminals are OK.	No	▶ SERVICE as necessary. REPEAT Quick Test.
	Yes	▶ Unable to duplicate fault at this time. Continuous code 22 testing complete.

Illustration for Step DC90:

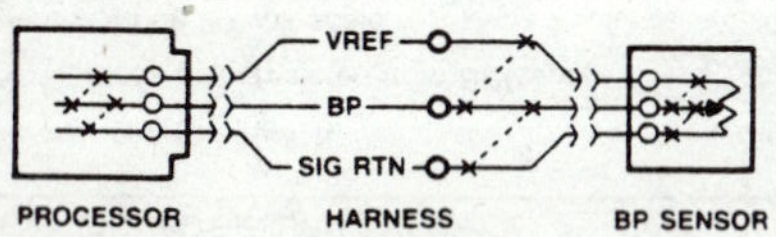

EGR Valve Position Sensor (EVP) — Pinpoint Test — DD

DD

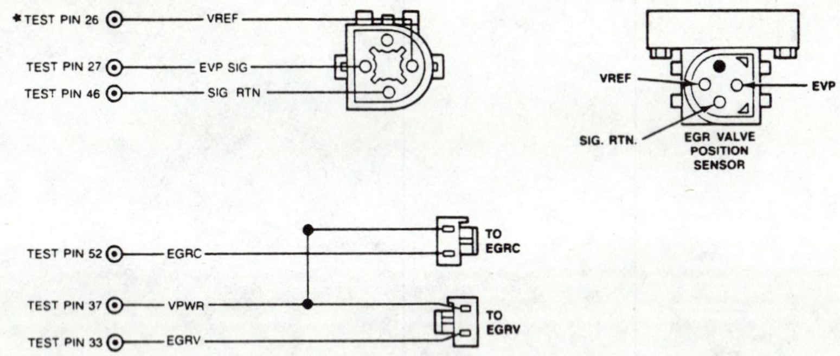

*TEST PINS LOCATED ON BREAKOUT BOX.
ALL HARNESS CONNECTORS VIEWED INTO MATING SURFACE.

STOP-WARNING

You may have reached this point because of a fault in a non-EEC System and/or component. To prevent the replacement of good components, be aware that the following non-EEC area may be at fault:

- Damaged EGR valve.

This Pinpoint Test is intended to diagnose only the following:

- EVP sensor.
- Harness circuits: EVP, Signal Return, VREF, EGRV, EGRC, VPWR.
- EGR solenoids.
- EGR Valve assembly.
- Processor assembly.

EGR Valve Position Sensor (EVP) — Pinpoint Test DD

TEST STEP	RESULT ▶	ACTION TO TAKE
FAULT CODE 31		
DD1 RUN ENGINE RUNNING QUICK TEST WITH EGR VACUUM SIGNAL LINE DISCONNECTED AT EGR VALVE		
• Key Off, wait 10 seconds. • Disconnect EGR vacuum line at EGR valve and cap EGR vacuum line. • Run Engine Running Quick Test. • Check for code 31.	Code 31 present ▶	GO to DD2.
	No code 31 present, but codes 32, 34 are present ▶	GO to DD10.
DD2 CHECK EVP SENSOR RESISTANCE		
• Key Off, wait 10 seconds. • EGR signal line disconnected and capped. • Disconnect EVP sensor. Inspect for damaged pins, corrosion, and pins pushed out. Service as necessary. • DVOM on 200,000 ohm scale. • Measure resistance of EVP sensor EVP SIG to sensor VREF and sensor Signal Return to sensor VREF.	Both readings between 100 and 5500 ohms ▶	GO to DD3.
	Either reading less than 100 ohms or greater than 5500 ohms ▶	REPLACE EVP sensor. RECONNECT vacuum line and harness. RERUN Quick Test.
DD3 CHECK EVP MOVEMENT		
• Key Off, vacuum signal line disconnected and capped, harness disconnected at EVP sensor. • DVOM on 200,000 ohm scale. • Connect vacuum pump to EGR valve. • Measure resistance of EVP sensor EVP SIG to VREF while gradually increasing vacuum to 33 kPa (10 in. Hg). • Observe resistance as vacuum increases.	Reading gradually decreases from no greater than 5500 ohms to no less than 100 ohms ▶	GO to DD4.
	Reading is less than 100 ohms or greater than 5500 ohms ▶	REPLACE EVP sensor. RECONNECT signal line and harness. RERUN Quick Test.
	Reading does not decrease or unable to hold vacuum ▶	GO to DD16.

EGR Valve Position Sensor (EVP) — Pinpoint Test DD

TEST STEP	RESULT ▶	ACTION TO TAKE
DD4 MEASURE VREF TO SIGNAL RETURN VOLTAGE		
• Key On, Engine Off. • Vacuum signal line disconnected and capped, harness disconnected from EVP sensor. • DVOM on 20V scale. • Measure voltage from EVP harness connector VREF to Signal Return.	Reading is between 4 and 6V ▶	GO to DD5.
	Reading is less than 4V ▶	GO to DD7.
	Reading is greater than 6V ▶	GO to Pinpoint Test Step C1.
DD5 MEASURE VREF TO SIGNAL INPUT VOLTAGE		
• Key On, Engine Off. • Reconnect EGR vacuum signal line to EGR valve. • Harness disconnected from EVP sensor. • DVOM on 20V scale. • Measure voltage from EVP sensor harness connector, VREF to EVP signal.	Reading is between 4 and 6V ▶	GO to DD6.
	Voltage is less than 4V ▶	GO to DD7.
DD6 CHECK EVP SIGNAL FOR SHORTS TO GROUND		
• Key Off, wait 10 seconds. Harness disconnected from EVP sensor. • Disconnect harness from processor 60 Pin connector and inspect for damaged pins, corrosion, loose wires. Service as necessary. • DVOM on 200,000 ohm scale. • Measure resistance of EVP sensor harness connector EVP Signal to Signal Return and EVP signal to battery negative terminal.	Either reading is less than 10,000 ohms ▶	SERVICE circuit short. RECONNECT processor and EVP sensor. RERUN Quick Test.
	Both readings are 10,000 ohms or greater ▶	REPLACE processor. RECONNECT processor and EVP sensor. RERUN Quick Test.

EGR Valve Position Sensor (EVP) — Pinpoint Test — DD

TEST STEP	RESULT ▶	ACTION TO TAKE
DD7 MEASURE CONTINUITY OF VREF, EVP SIGNAL AND SIGNAL RETURN		
• Key Off, wait 10 seconds. • Harness disconnected from EVP sensor. • Disconnect processor 60 Pin connector and inspect for damaged pins, corrosion, loose wires. Service as necessary. • Install Breakout box leaving processor disconnected. • DVOM on 200 ohm scale. • Measure continuity from test Pin 46 to EVP harness connector Signal Return and, from test Pin 26 to EVP harness connector VREF and from test Pin 27 to EVP harness connector EVP signal.	All readings are less than 5 ohms ▶	GO to DD8.
	Any reading 5 ohms or greater ▶	SERVICE open circuit(s). REMOVE Breakout box. RECONNECT processor and EVP sensor. RERUN Quick Test.
DD8 CHECK EVP SIGNAL FOR SHORTS TO POWER		
• Key Off, harness disconnected from EVP sensor. Breakout box installed with processor not connected. • DVOM on 200,000 ohm scale. • Measure resistance of test Pin 27 to test Pins 37 and 57.	Either reading less than 10,000 ohms ▶	SERVICE short circuit. REMOVE Breakout box. RECONNECT processor and EVP sensor. RERUN Quick Test.
	Both readings 10,000 ohms or greater ▶	REPLACE processor. REMOVE Breakout box. RECONNECT processor and EVP sensor. RERUN Quick Test.
FAULT CODES 32, 33 AND 34		
DD10 MEASURE EGR VENT AND EGR CONTROL SOLENOID RESISTANCE		
• Key Off, wait 10 seconds. • Disconnect harness from EGRC and EGRV solenoids. Inspect for damaged pins, corrosion, and pins pushed out. Service as necessary. • DVOM on 200 ohm scale. • Measure resistance of both solenoids.	Both readings between 30 and 70 ohms ▶	RECONNECT harness to both solenoids and GO to DD11.
	Either reading less than 30 ohms or greater than 70 ohms ▶	REPLACE EGR solenoid assembly. RECONNECT harness and RERUN Quick Test.

EGR Valve Position Sensor (EVP) — Pinpoint Test — DD

TEST STEP	RESULT ▶	ACTION TO TAKE
DD11 OUTPUT STATE CHECK (REFER TO APPENDIX)		
• Key Off, wait 10 seconds. • DVOM on 20V scale. • Connect DVOM negative test lead to STO and positive test lead to battery positive. • Jumper STI to signal return. • Perform Key On, Engine Off Self-Test until the completion of the Continuous Test Codes. • DVOM will indicate zero volts. • Depress and release the throttle. • Did DVOM reading change to a high voltage reading?	NO ▶	DEPRESS throttle to WOT and RELEASE. If STO voltage does not go high, GO to Pinpoint Test Step Q40.
	YES ▶	Remain in output state check and GO to DD12.
DD12 CHECK EGR SOLENOIDS FOR ELECTRICAL CYCLING		
• Key On, Engine Off. • In output state check. • DVOM on 20V scale. • Connect DVOM to EGRV circuits VPWR and EGRV signal. • While observing DVOM, depress and release the throttle several times to cycle output On and Off. • Repeat for EGRC circuits VPWR and EGRC signal.	Both solenoid outputs cycle On and Off ▶	REMAIN in output state check. GO to DD13.
	Either output does not cycle On and Off ▶	Exit output state check. GO to DD17.

EGR Valve Position Sensor (EVP) — Pinpoint Test DD

TEST STEP	RESULT ▶	ACTION TO TAKE
DD13 CHECK EGR SOLENOIDS FOR VACUUM CYCLING		
• Key On, Engine Off. • In output state check. • Disconnect and cap vacuum line from bottom port of EGRC solenoid and connect a vacuum pump. • Connect a vacuum gauge in the common output (top) vacuum line to EGR valve. • Disconnect but **do not** cap vacuum vent line from EGRV solenoid. • While cycling outputs On and Off (by depressing and releasing throttle), observe vacuum gauge at the output. Maintain vacuum at source.	Vacuum output cycles On and Off in less than 2 seconds ▶	RECONNECT all vacuum lines. GO to DD14.
	Vacuum does not cycle On and Off in less than 2 seconds ▶	REPLACE solenoid assembly. RECONNECT all vacuum lines. RERUN Quick Test.
DD14 CHECK VACUUM LINES		
• Key Off, wait 10 seconds. • Vacuum lines reconnected. • Check entire **EEC** vacuum line system per VECI emission schematic decal for kinks, cracks, obstructions or leaks.	Vacuum lines OK ▶	GO to DD15.
	Vacuum lines not OK ▶	SERVICE as necessary. RERUN Quick Test.
DD15 CHECK EVP RESISTANCE WHILE APPLYING VACUUM TO EGR VALVE		
• Key Off. • Disconnect harness from EVP sensor. Inspect for damaged pins, corrosion, and pins pushed out. Service as necessary. • DVOM on 200,000 ohm scale. • Disconnect vacuum line at EGR valve. • Connect vacuum pump to EGR valve. • Measure resistance of EVP sensor EVP Signal Pin to VREF Pin while increasing vacuum to 33 kPa (10 in. Hg). • Observe resistance as vacuum increases.	Reading does not decrease gradually ▶	GO to DD16.
	Reading gradually decreases from no more than 5,500 ohms to no less than 100 ohms ▶	REPLACE processor. RECONNECT EVP sensor and EGR vacuum line. RERUN Quick Test.

EGR Valve Position Sensor (EVP) — Pinpoint Test DD

TEST STEP	RESULT ▶	ACTION TO TAKE
DD16 MANUALLY EXERCISE EVP SENSOR		
• Key Off, harness disconnected from EVP sensor. • Remove EVP sensor from EGR valve. • Measure resistance from EVP sensor Signal Return Pin to EVP sensor VREF Pin while gradually applying pressure to EVP sensor shaft. • Observe resistance as shaft is slowly pushed in. • Look for sudden changes in resistance readings.	Reading decreases smoothly from no more than 5,500 ohms to no less than 100 ohms ▶	REPLACE EGR valve assembly. RECONNECT EVP sensor and EGR supply vacuum line. RERUN Quick Test.
	Reading does not decrease smoothly or changes abruptly ▶	REPLACE EVP sensor. RECONNECT harness and EGR supply vacuum line. RERUN Quick Test.
DD17 CHECK FOR VOLTAGE ON VEHICLE POWER CIRCUIT		
• Disconnect EGR vent and EGR control solenoids from harness. • Key On, Engine Off. • DVOM on 20V scale. • Measure voltage from battery negative terminal to VPWR circuit on both EGR solenoids.	Either reading is less than 10.5V ▶	SERVICE harness circuit open. RERUN Quick Test.
	Both readings are 10.5V or greater ▶	GO to DD18.
DD18 CHECK CONTINUITY OF EGRV AND EGRC CIRCUITS		
• Key Off, wait 10 seconds. • EGR vent and EGR control solenoid disconnected from harness. • Disconnect processor 60 Pin connector and inspect for damaged pins, corrosion, loose wires. Service as necessary. • Install Breakout box to processor harness connector. Leave processor disconnected. • DVOM on 200 ohm scale. • Measure continuity from test Pin 33 to EGRV solenoid harness connector EGRV signal. • Measure continuity from test Pin 52 to EGRC solenoid harness connector EGRC signal.	Both readings less than 5 ohms ▶	GO to DD19.
	Either reading 5 ohms or greater ▶	SERVICE open circuit. REMOVE Breakout box. RECONNECT harness to processor. RERUN Quick Test.

EGR Valve Position Sensor (EVP) — Pinpoint Test — DD

TEST STEP	RESULT	ACTION TO TAKE
DD19 CHECK EGRV AND EGRC SIGNALS FOR SHORTS TO POWER		
• EGR vent and EGR control solenoids disconnected from harness. • Key Off. Breakout box installed. Processor disconnected. • DVOM on 200,000 ohm scale. • Measure resistance from test Pin 33 to test Pins 40, 46 and 57. • Measure resistance from test Pin 52 to test Pins 40, 46 and 57.	Any reading less than 10,000 ohms ▶	SERVICE circuit short. REMOVE Breakout box. RECONNECT harness to processor. RERUN Quick Test.
	All readings 10,000 ohms or greater ▶	REPLACE processor. REMOVE Breakout box. RECONNECT harness to processor. RERUN Quick Test.
FAULT CODE 35		
DD30 RPM TOO LOW FOR EGR TEST		
• Check for codes 12 and 13.	Code 12 present ▶	Vehicles equipped with DC motor control, GO to KF1. Vehicles equipped with throttle kicker, GO to KG1.
	Code 13 present ▶	Vehicles equipped with DC motor control, GO to KF20. Vehicles equipped with throttle kicker, GO to KG1.
	Codes 12 and 13 not present ▶	GO to DD31.
DD31 RETEST AT 1,500 RPM		
• Key Off, wait 10 seconds. • Install tachometer. • Perform Key On, Engine Running Quick Test while maintaining 1,500 rpm. • Record Engine Running service codes. • Check for code 35.	Code 35 present ▶	REPLACE processor. RERUN Quick Test.
	Code 35 not present ▶	REPEAT Quick Test. SERVICE codes as necessary.

EGR Valve Position Sensor (EVP) — Pinpoint Test — DD

TEST STEP	RESULT	ACTION TO TAKE
DD40 VREF CIRCUIT TO GROUND		
• Key Off, wait 10 seconds. • Disconnect EVP sensor and inspect for damaged pins, corrosion, loose wires. Service as necessary. • DVOM on 20V scale. • Key On, Engine Off. • Measure voltage from VREF (at EVP sensor harness connector) to battery ground.	Voltage readings are between 4V and 6V ▶	GO to DD41.
	Voltage readings are less than 4.0V ▶	GO to DD7.
	Voltage readings are 6.0V or greater ▶	GO to Pinpoint Test step C1.
DD41 SIGNAL RETURN TO VREF		
• Key On. • EVP harness disconnected. • DVOM on 20V scale. • Measure voltage from VREF to signal return at the EVP harness connector.	Voltage is reading between 4V and 6V ▶	GO to DD42.
	Voltage readings are less than 4.0V ▶	SERVICE open in signal return. RERUN Key On, Engine Off Quick Test.
DD42 EVP CIRCUIT VOLTAGE CHECK		
• Key Off, (wait 10 seconds). • Disconnect processor 60 Pin connector and inspect for damaged pins, corrosion, loose wires. Service as necessary. • Install Breakout box and reconnect processor. • Connect EVP sensor. • DVOM on 20V scale. • Key On, Engine Off. • Measure voltage from test Pin 27 to test Pin 46.	Less than .5V ▶	GO to DD43.
	1.25V or greater ▶	GO to DD46.

EGR Valve Position Sensor (EVP) — Pinpoint Test — DD

TEST STEP	RESULT ▶	ACTION TO TAKE
DD43 EVP SIGNAL SHORT TO GROUND		
• Key Off, wait 10 seconds. • Disconnect processor. • Disconnect EVP sensor. • Breakout box installed. • DVOM on 200 ohm scale. • Measure resistance from test Pin 27 to test Pins 20, 40, 46, and 60.	5.0 ohms or greater ▶	GO to DD44.
	Less than 5 ohms ▶	SERVICE short to ground in EVP signal circuit and REPEAT Quick Test.
DD44 EVP SIGNAL CIRCUIT FOR OPENS		
• Key Off. • Processor disconnected. • EVP sensor disconnected. • Breakout box installed. • DVOM on 200 ohm scale. • Measure resistance from test Pin 27 to EVP signal circuit at EVP connector harness.	5 ohms or greater ▶	SERVICE open in EVP signal circuit. REPEAT Quick Test.
	Less than 5 ohms ▶	GO to DD45.
DD45 SUBSTITUTE EVP SENSOR		
• Key Off. • Remove and replace present EVP sensor with known good sensor. Reconnect EVP sensor. • Remove Breakout box. • Reconnect processor. • Rerun Key On, Engine Off Quick Test. • Is code 31 present?	Yes ▶	REPLACE processor. RECONNECT original EVP sensor. REPEAT Quick Test.
	No ▶	Original EVP sensor at fault. INSTALL new EVP sensor. REPEAT Quick Test.

EGR Valve Position Sensor (EVP) — Pinpoint Test — DD

TEST STEP	RESULT ▶	ACTION TO TAKE
DD46 EVP SIGNAL SHORT TO POWER		
• Key Off, wait 10 seconds. • Disconnect processor. • Disconnect EVP sensor. • Breakout box installed. • DVOM on 200 ohm scale. • Measure resistance from test Pin 27 to test Pins 26, 37 and 57.	5 ohms or greater ▶	GO to DD47.
	Less than 5 ohms ▶	SERVICE short to power. REPEAT Quick Test.
DD47 SUBSTITUTE EVP SENSOR AND EGR VALVE		
• Key Off, wait 10 seconds. • Electrically connect known good EVP sensor and EGR valve assembly. • Remove Breakout box. • Reconnect processor. • Perform Key On, Engine Off Quick Test. • Is code 31 present?	Yes ▶	REPLACE processor. Connect original EVP sensor. REPEAT Quick Test.
	No ▶	GO to DD48.
DD48 EVP SENSOR CHECK		
• Key Off, wait 10 seconds. • Install original EVP sensor on known good EGR valve. • EVP sensor connected. • Rerun Key On, Engine Off Quick Test. • Is code 31 present?	Yes ▶	INSTALL new EVP sensor. REPEAT Quick Test.
	No ▶	REFER to EGR systems, Section 6.

EGR Valve Position Sensor (EVP) | Pinpoint Test | DD

TEST STEP	RESULT ▶	ACTION TO TAKE
DD90 SERVICE CODE 31 CONTINUOUS TEST: EXERCISE EVP SENSOR		
• Using continuous monitor mode, observe VOM or STAR LED for indication of a fault while performing the following: • Connect a vacuum pump to the EGR valve. • Very slowly apply 20 kPa (6 in. Hg) vacuum to the EGR valve. • Slowly bleed vacuum off the EGR valve and lightly tap on EVP sensor (simulate road shock). • Wiggle EVP sensor connector. • Is a fault indicated?	Yes ▶ No ▶	GO to DD91. GO to DD92.

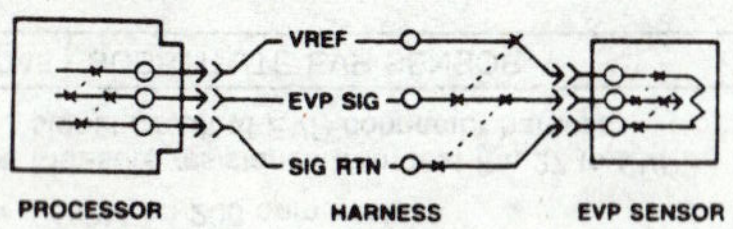

TEST STEP	RESULT ▶	ACTION TO TAKE
DD91 MEASURE EVP SIGNAL VOLTAGE WHILE EXERCISING EVP SENSOR		
• Key Off, wait 10 seconds. • Disconnect processor 60 Pin connector and inspect for damaged pins, corrosion, loose wires. Service as necessary. • Install Breakout box and reconnect processor. • VOM or STAR LED still connected to STO as in previous Step. • Connect a DVOM from test Pin 27 to test Pin 46. • DVOM on 20V scale. • Key On, Engine Off. • While observing DVOM, repeat step **DD90**. • Does the fault occur below 4.25V?	Yes ▶ No ▶	DISCONNECT and INSPECT connector. If connector and terminals are good, REPLACE EVP sensor. REPEAT Quick Test. EGR valve overshoot may have caused continuous code 31. Sensor service is not required. To verify harness integrity, GO to DD92.

EGR Valve Position Sensor (EVP) | Pinpoint Test | DD

TEST STEP	RESULT ▶	ACTION TO TAKE
DD92 CHECK EEC-IV HARNESS		
• Observe VOM or STAR LED for a fault indication while performing the following: • Referring to the illustration in Step DD90, grasp the harness closest to the sensor connector. Wiggle, shake or bend a small section of the EEC-IV system harness while working your way to the dash panel. Also wiggle, shake or bend the EEC-IV harness from the dash panel to the processor. • Is a fault indicated?	Yes ▶ No ▶	ISOLATE fault and make necessary repairs. REFER to appropriate figure. REPEAT Quick Test. GO to DD93.
DD93 CHECK PROCESSOR AND HARNESS CONNECTORS		
• Key Off, wait 10 seconds. • Disconnect processor 60 Pin connector. • Inspect both connectors and connector terminals for obvious damage or faults. • Are connectors and terminals OK?	No ▶ Yes ▶	SERVICE as necessary. REPEAT Quick Test. Unable to duplicate fault at this time. Continuous code 31 testing complete.

Engine Coolant Temperature Sensor (ECT) | Pinpoint Test | DE

DE

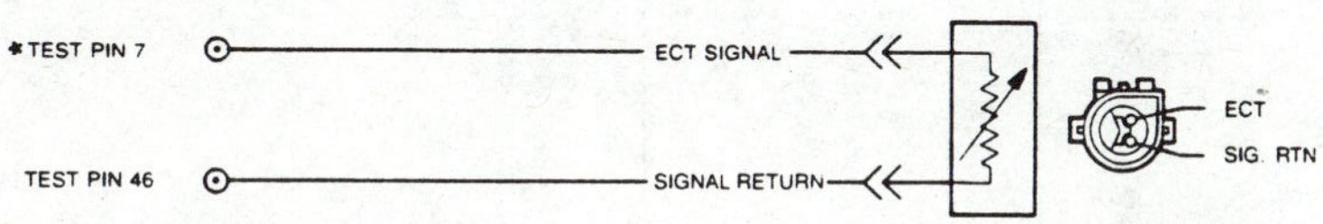

EEC COOLANT TEMPERATURE FRONT OF ENGINE

NOTE: TO PASS THIS TEST, ENGINE COOLANT TEMPERATURE MUST BE:
KEY ON, ENGINE OFF (50°F TO 240°F)
ENGINE RUNNING (180°F TO 240°F)

TYPICAL RESISTANCE BETWEEN TEST POINTS 7 & 46	58,750Ω	40,500Ω	3600Ω	1840Ω
AT TEMPERATURE	50°F	65°F	180°F	220°F

* TEST PIN LOCATED ON BREAKOUT BOX.
ALL CONNECTIONS VIEWED INTO MATING SURFACE.

STOP-WARNING

You may have reached this point because of a fault in a non-EEC System and/or component. To prevent the replacement of good components, be aware that the following non-EEC areas may be at fault:

- Coolant level.
- Oil level.
- Blocked or obstructed air flow.
- Engine not at normal operating temperature.
- Electro drive cooling fan.

This pinpoint test is intended to diagnose only the following:

- ECT sensor.
- Harness sensor circuits: ECT and Signal Return.
- Processor assembly.

Engine Coolant Temperature Sensor (ECT) | Pinpoint Test | DE

TEST STEP	RESULT ▶	ACTION TO TAKE
DE1 SERVICE CODE 21: CHECK ENGINE OPERATING TEMPERATURE		
• Run engine for 2 minutes at 2,000 rpm. • Check that upper radiator hose is hot and pressurized. • Rerun Quick Test.	Vehicle stalls ▶	Do not service code 21 at this time. REFER to diagnosis by symptoms.
	Code 21 present ▶	GO to DE2 .
	Code 21 not present ▶	SERVICE other codes as necessary.
DE2 CHECK ECT AND SIG. RTN. VOLTAGE		
• Key Off, wait 10 seconds. • Disconnect harness from ECT sensor. Inspect for damaged pins, corrosion, loose wires, etc. Service as necessary. • DVOM on 20 volt scale. • Key On, Engine Off. • Measure voltage between ECT sensor harness connector ECT signal to signal return.	Reading is between 4.0V and 6.0V ▶	GO to DE5 .
	Reading is less than 4.0V ▶	GO to DE3 .
	Reading is greater than 6.0V ▶	GO to Pinpoint Test Step C1 .
DE3 CHECK CONTINUITY OF ECT SIG. AND SIG. RTN.		
• Key Off, wait 10 seconds. • Harness disconnected from ECT sensor. • Disconnect processor 60 pin connector — inspect for damaged pins, corrosion, loose wires, etc. Service as necessary. • Install Breakout box to harness. Leave processor disconnected. • DVOM on 200 ohm scale. • Measure resistance of ECT sensor harness connector ECT signal to test Pin 7. • Measure resistance of ECT sensor harness connector Signal Return to test Pin 46.	Both readings are less than 5 ohms ▶	GO to DE4 .
	Either reading is 5 ohms or greater ▶	SERVICE circuit opens. REMOVE Breakout box. RECONNECT processor and ECT sensor. RERUN Quick Test.

Engine Coolant Temperature Sensor (ECT) — Pinpoint Test — DE

TEST STEP	RESULT	ACTION TO TAKE
DE4 CHECK ECT SIGNAL FOR SHORTS • Key Off. • Harness disconnected from ECT sensor. Breakout box installed. Processor disconnected. • DVOM on 200,000 ohm scale. • Measure resistance from test Pin 7 to test Pins 40 and 60.	Either reading is less than 10,000 ohms ▶	SERVICE circuit shorts. REMOVE Breakout box. RECONNECT processor and ECT sensor. RERUN Quick Test.
	Both readings are 10,000 ohms or greater ▶	GO to DE5.
DE5 ECT SENSOR CHECK **NOTE: Engine may have cooled down. Always warm engine before taking ECT resistance measurement.** • Key Off, wait 10 seconds. • Harness disconnected from ECT sensor. • DVOM on 200,000 ohm scale. • Make a resistance measurement of the ECT sensor. • Is the resistance reading: — 1300 ohms (240°F) to 7700 ohms (140°F) for engine off? — 1550 ohms (230°F) to 4550 ohms (180°F) for engine running?	Yes ▶	REPLACE processor. RECONNECT harness to ECT sensor. RERUN Quick Test.
	No ▶	REPLACE ECT sensor. RECONNECT harness to ECT sensor. RERUN Quick Test.

Engine Coolant Temperature Sensor (ECT) — Pinpoint Test — DE

TEST STEP	RESULT	ACTION TO TAKE
DE10 SERVICE CODE 51: INDUCE OPPOSITE CODE • Key Off, wait 10 seconds. • Disconnect harness from ECT sensor. Inspect for damaged pins, corrosion, loose wires, etc. Service as necessary. • Insert a jumper wire between ECT sensor harness connector ECT Signal and Signal Return. • Run Key On, Engine Off Quick Test. • Is code 61 present?	Yes ▶	REPLACE ECT sensor. REMOVE jumper wire. RECONNECT ECT sensor. RERUN Quick Test.
	No ▶	GO to DE11.
DE11 CHECK CONTINUITY OF ECT SIGNAL AND SIGNAL RETURN • Key Off, wait 10 seconds. • Harness disconnected from ECT sensor, jumper wire removed. • Disconnect processor 60 pin connector. Inspect for damaged pins, corrosion, loose wires, etc. Service as necessary. • Install Breakout box to harness leaving processor disconnected. • DVOM on 200 ohm scale. • Measure resistance between ECT sensor harness connector ECT signal and test Pin 7. • Measure resistance between ECT sensor harness connector Signal Return and test Pin 46.	Both readings are less than 5 ohms ▶	REPLACE processor. REMOVE Breakout box. RECONNECT processor and ECT sensor. RERUN Quick Test.
	Either reading is 5 ohms or greater ▶	SERVICE open circuit(s). REMOVE Breakout box. RECONNECT processor and ECT sensor. RERUN Quick Test.

Engine Coolant Temperature Sensor (ECT) — Pinpoint Test — DE

TEST STEP	RESULT ▶	ACTION TO TAKE
DE20 SERVICE CODE 61: INDUCE OPPOSITE CODE		
• Key Off, wait 10 seconds. • Disconnect harness from ECT sensor. Inspect for damaged pins, corrosion, loose wires, etc. Service as necessary. • Run Key On, Engine Off Quick Test. • Is code 51 present?	Yes ▶	REPLACE ECT sensor. RECONNECT ECT sensor. RERUN Quick Test.
	No ▶	GO to DE21.
DE21 CHECK ECT SIGNAL FOR SHORT		
• Key Off, wait 10 seconds. • Harness disconnected from ECT sensor. • Disconnect processor 60 pin connector. Inspect for damaged pins, corrosion, loose wires, etc. Service as necessary. • Install Breakout box, leave processor disconnected. • DVOM on 200,000 ohm scale. • Measure resistance of test Pin 7 to test Pins 40 and 60.	Either reading is less than 10,000 ohms ▶	SERVICE circuit shorts. REMOVE Breakout box. RECONNECT processor and ECT sensor. RERUN Quick Test.
	Both readings are 10,000 ohms or greater ▶	REPLACE processor. REMOVE Breakout box. RECONNECT processor and ECT harness. RERUN Quick Test.

Engine Coolant Temperature Sensor (ECT) — Pinpoint Test — DE

TEST STEP	RESULT ▶	ACTION TO TAKE
DE90 SERVICE CODE 21 CONTINUOUS TEST: TEST DRIVE VEHICLE		
• Key Off and wait 10 seconds. • Disconnect all Self-Test equipment and prepare vehicle for test drive. • Drive vehicle. Try to simulate different drive modes or mode in which drive complaint is noticed. Attempt to maintain drive complaint mode for one minute or more, if possible. • Upon completion of drive evaluation, repeat Key On, Engine Off Self-Test. • Is code 21 present in the continuous test results?	Yes ▶	VERIFY thermostat operating properly. If OK, REPLACE ECT sensor. REPEAT Quick Test.
	No ▶	Unable to duplicate fault. Code 21 testing complete.

Engine Coolant Temperature (ECT) | Pinpoint Test | DE

TEST STEP	RESULT ▶	ACTION TO TAKE
DE91 SERVICE CODE 51 CONTINUOUS TEST: CHECK ECT SENSOR		
• Using continuous monitor mode, observe VOM or STAR LED for indication of a fault while performing the following: • Lightly tap on ECT sensor (simulate road shock). • Wiggle ECT connector. • Is fault indicated?	Yes ▶	DISCONNECT and INSPECT connectors. If connector and terminals are good, REPLACE ECT sensor. REPEAT Quick Test.
	No ▶	GO to DE92.
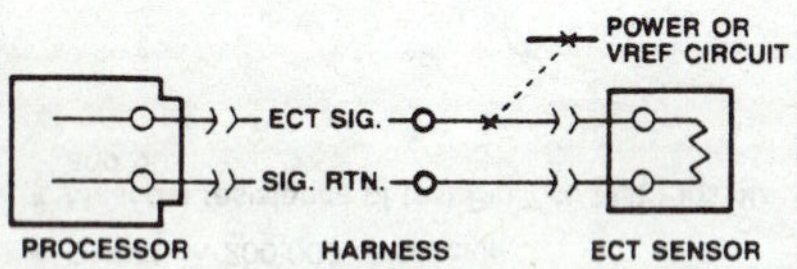		
DE92 CHECK EEC-IV HARNESS		
• Observe VOM or STAR LED for a fault indication while performing the following: • Referring to the illustration in Step DE91, grasp the harness closest to the sensor connector. Wiggle, shake or bend a small section of the EEC-IV system harness while working your way to the dash panel. Also wiggle, shake or bend the EEC-IV harness from the dash panel to the processor. • Is fault indicated?	Yes ▶	ISOLATE fault and make necessary repairs. REPEAT Quick Test.
	No ▶	GO to DE93.
DE93 CHECK PROCESSOR AND HARNESS CONNECTORS		
• Key Off, wait 10 seconds. • Disconnect processor 60 pin connector. • Inspect both connectors and connector terminals for obvious damage or faults. • Connectors and terminals are OK.	No ▶	SERVICE as necessary. REPEAT Quick Test.
	Yes ▶	Unable to duplicate fault at this time. Continuous code 51 testing complete.

Engine Coolant Temperature (ECT) | Pinpoint Test | DE

TEST STEP	RESULT ▶	ACTION TO TAKE
DE94 SERVICE CODE 61 CONTINUOUS TEST: CHECK ECT SENSOR		
• Using continuous monitor mode, observe VOM or STAR LED for indication of a fault while performing the following: • Lightly tap on ECT sensor (simulate road shock). • Wiggle ECT connector. Voltage greater than 5 volts or STAR LED Off. • Is fault indicated?	Yes ▶	DISCONNECT and INSPECT connectors. If connector and terminals are good, REPLACE ECT sensor. REPEAT Quick Test.
	No ▶	GO to DE95.
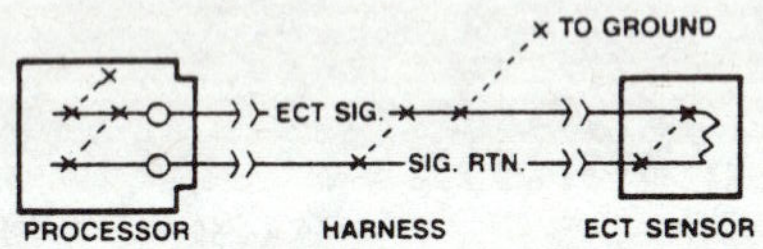		
DE95 CHECK EEC-IV HARNESS		
• Observe VOM or STAR LED for a fault indication while performing the following: • Referring to the illustration in Step DE94, grasp the harness closest to the sensor connector. Wiggle, shake or bend a small section of the EEC-IV system harness while working your way to the dash panel. Also wiggle, shake or bend the EEC-IV harness from the dash panel to the processor. • Is fault indicated?	Yes ▶	ISOLATE fault and make necessary repairs. REPEAT Quick Test.
	No ▶	GO to DE96.
DE96 CHECK PROCESSOR AND HARNESS CONNECTORS		
• Key Off, wait 10 seconds. • Disconnect processor 60 pin connector. • Inspect both connectors and connector terminals for obvious damage or faults. • Connectors and terminals are OK.	No ▶	SERVICE as necessary. REPEAT Quick Test.
	Yes ▶	Unable to duplicate fault at this time. Continuous code 61 testing complete.

Manifold Absolute Pressure Sensor (MAP) | Pinpoint Test | DF

DF

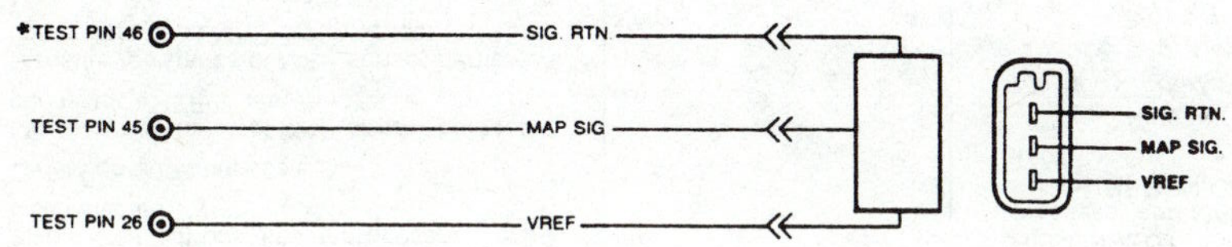

*TEST PINS LOCATED ON BREAKOUT BOX.
ALL HARNESS CONNECTORS VIEWED INTO MATING SURFACE.

STOP-WARNING

You may have reached this point because of a fault in a non-EEC System and/or component. To prevent the replacement of good components, be aware that the following non-EEC areas may be at fault:

- Unusually high/low atmospheric barometer reading.
- Kinked or obstructed vacuum lines.
- Basic engine (valves, vacuum leaks, timing, carburetor, etc.)
- Key On, Engine Off — MAP sensor must **not** see vacuum.
- Engine Running — MAP sensor must see actual manifold vacuum.

This Pinpoint Test is intended to diagnose only the following:

- MAP sensor.
- Harness circuits: VREF, MAP Signal, and Signal Return.
- Processor assembly.

MAP Absolute Pressure Sensor (MAP) | Pinpoint Test | DF

TEST STEP	RESULT ▶	ACTION TO TAKE
FAULT CODE 22		
DF1 CHECK FOR VACUUM LEAKS		
• From engine running code only. • Key Off, wait 10 seconds. • Verify all accessories are off. • Check vacuum line to MAP sensor for leaks, restrictions, kinks, etc. • Check for any engine system vacuum leaks. Refer to VECI decal.	Faults or leaks found ▶	SERVICE as necessary. RERUN Quick Test.
	Faults or leaks not found ▶	GO to DF2.
DF2 CHECK MAP SENSOR		
• Key Off, wait 10 seconds. • Disconnect vacuum line from MAP sensor. • Install vacuum pump to MAP sensor. • Apply 70 kPa (20 in. Hg) vacuum to MAP sensor. • Does MAP sensor hold vacuum?	Yes ▶	REMOVE vacuum pump. RECONNECT vacuum line to MAP sensor. GO to DF3.
	No ▶	REPLACE MAP sensor. CONNECT vacuum line to MAP sensor. RERUN Quick Test.
DF3 CHECK FOR OTHER FAULT CODES		
• Check for engine running, codes 31 or 41.	Code 31 present ▶	For 2.8L truck and 3.8L, 5.0L CFI: GO to Pinpoint Test Step DL1. All others: GO to Pinpoint Test Step DD1.
	Code 41 present ▶	For 1.6L, 2.3L EFI: GO to Pinpoint Test Step HA1. For 5.0L, 3.8L CFI: GO to Pinpoint Test Step HC1. All others: (FBC) GO to Pinpoint Test Step HB1.
	Codes 31 and 41 not present ▶	GO to DF4.

Manifold Absolute Pressure Sensor — Pinpoint Test — DF

TEST STEP	RESULT	►	ACTION TO TAKE
DF4 CHECK VOLTAGE VREF TO SIGNAL RETURN			
• Key Off, wait 10 seconds. • Disconnect harness from MAP sensor. Inspect for damaged pins, corrosion, loose wires, etc. Service as necessary. • DVOM on 20V scale. • Key On, Engine Off. • Measure voltage from MAP sensor harness connector VREF to Signal Return.	Reading is between 4.0 and 6.0V	►	GO to DF5.
	Reading is less than 4.0V	►	GO to DF6.
	Reading is greater than 6.0V	►	GO to Pinpoint Test Step C1.
DF5 CHECK VOLTAGE MAP SIGNAL TO SIGNAL RETURN			
• Key On, Engine Off. Harness disconnected from MAP sensor. • DVOM on 20V scale. • Measure voltage from MAP sensor harness connector MAP signal to Signal Return.	Reading is less than 4.0V	►	GO to DF6.
	Reading is 4.0V or greater	►	GO to DF8.
DF6 CHECK CONTINUITY OF VREF MAP SIGNAL, AND SIGNAL RETURN			
• Key Off, wait 10 seconds. • Harness disconnected from MAP sensor. • Disconnect processor 60 Pin connector. Inspect for damaged pins, corrosion, loose wires, etc. Service as necessary. • Install Breakout box. Leave processor disconnected. • DVOM on 200 ohm scale. • Measure resistance of MAP sensor harness connector VREF to test Pin 26. • Measure resistance of MAP sensor harness connector circuit MAP signal to test Pin 45. • Measure resistance of MAP sensor harness connector Signal Return to test Pin 46.	All readings are less than 5 ohms	►	GO to DF7.
	Any reading is 5 ohms or greater	►	SERVICE circuit opens. REMOVE Breakout box. RECONNECT processor and MAP sensor. RERUN Quick Test.

Manifold Absolute Pressure Sensor (MAP) — Pinpoint Test — DF

TEST STEP	RESULT	►	ACTION TO TAKE
DF7 CHECK VREF AND MAP SIGNAL FOR SHORT TO GROUND			
• Key Off, harness disconnected from MAP sensor. Breakout box installed. Processor disconnected. • DVOM on 200 ohm scale. • Measure resistance of test Pin 26 to test Pins 40, 46 and 60. • Measure resistance of test Pin 45 to test Pins 40, 46 and 60.	Any reading less than 100 ohms	►	SERVICE circuit shorts. REMOVE Breakout box. RECONNECT processor and MAP sensor. RERUN Quick Test.
	All readings are 100 ohms or greater	►	GO to DF9.
DF8 CHECK VREF AND MAP SIGNAL FOR SHORTS TO POWER			
• Key Off, wait 10 seconds. • Disconnect processor 60 Pin connector and inspect for damaged pins, corrosion, loose wires. Service as necessary. • Install Breakout box, leave processor disconnected. • Harness disconnected from MAP sensor. • DVOM on 200,000 ohm scale. • Measure resistance of test Pin 45 to test Pins 26, 37 and 57.	Any reading less than 10,000 ohms	►	SERVICE circuit shorts. REMOVE Breakout box. RECONNECT processor and MAP sensor. RERUN Quick Test.
	All readings are 10,000 ohms or greater	►	GO to DF9.
DF9 SUBSTITUTE MAP SENSOR			
• Key Off. • Remove Breakout box. Reconnect processor. Connect harness and vacuum line to a substitute MAP sensor. • Rerun quick test. • Is code 22 present?	Yes	►	REPLACE processor. CONNECT harness and vacuum to original MAP sensor. RERUN Quick Test.
	No	►	INSTALL substitute MAP sensor. RERUN Quick Test.

Manifold Absolute Pressure Sensor (MAP) — Pinpoint Test DF

TEST STEP	RESULT ▶	ACTION TO TAKE
FAULT CODE 72		
DF10 CHECK THAT VACUUM TO MAP SENSOR DECREASES DURING DYNAMIC RESPONSE		
• Key Off, wait 10 seconds. • Tee a vacuum gauge in the intake manifold vacuum line at the MAP sensor. • Perform Engine Running Quick Test while observing vacuum. • Record engine service codes. • Did vacuum decrease by more than 30 kPa (9 in. Hg) during dynamic response test? • Is code 72 present?	Vacuum decrease is 30 kPa (9 in. Hg) or greater and code 72 is not present ▶	DISCONNECT vacuum equipment and service other codes as necessary.
	Vacuum decrease is 30 kPa (9 in. Hg) or greater and code 72 is present ▶	REPLACE MAP sensor and RERUN Quick Test.
	Vacuum decrease is less than 30 kPa (9 in. Hg) ▶	GO to DF11.
DF11 CHECK VACUUM LINES		
• Check vacuum line for proper routing. Refer to VECI decal. Check MAP sensor vacuum line for kinks or blockage. • Vacuum lines are OK.	Yes ▶	EEC-IV system OK. REFER to shop manual, group 21 for probable subjects affecting engine vacuum.
	No ▶	SERVICE as necessary and REPEAT DF10.

Manifold Absolute Pressure Sensor (MAP) — Pinpoint Test DF

TEST STEP	RESULT ▶	ACTION TO TAKE
DF90 SERVICE CODE 22: CONTINUOUS TEST: EXERCISE MAP SENSOR		
• Using continuous monitor mode, observe VOM or STAR LED for indication of a fault while performing the following: • Connect a vacuum pump to the MAP sensor. • Slowly apply 84 kPa (25 in. Hg.) vacuum to the sensor. • Slowly bleed vacuum off the MAP sensor. • Lightly tap on MAP sensor (simulate road shock). • Wiggle MAP connector. • Is fault indicated?	Yes ▶	DISCONNECT and INSPECT connectors. If connector and terminals are good, REPLACE sensor. REPEAT Quick Test.
	No ▶	GO to DF91.
DF91 CHECK EEC-IV HARNESS		
• Observe VOM or STAR LED for a fault indication while performing the following: • Referring to the illustration in Step DF90, grasp the harness closest to the sensor connector. Wiggle, shake or bend a small section of the EEC-IV system harness while working your way to the dash panel. Also wiggle, shake or bend the EEC-IV harness from the dash panel to the processor. • Is a fault indicated?	Yes ▶	ISOLATE fault and make necessary repairs. REPEAT Quick Test.
	No ▶	GO to DF92.
DC92 CHECK PROCESSOR AND HARNESS CONNECTORS		
• Key Off, wait 10 seconds. • Disconnect processor 60 Pin connector. • Inspect both connectors and connector terminals for obvious damage or faults. • Are connectors and terminals OK?	No ▶	SERVICE as necessary. REPEAT Quick Test.
	Yes ▶	Unable to duplicate fault at this time. Continuous code 22 testing complete.

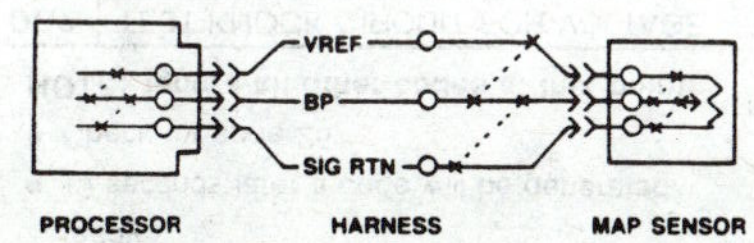

Knock Sensor — Pinpoint Test — DG

DG

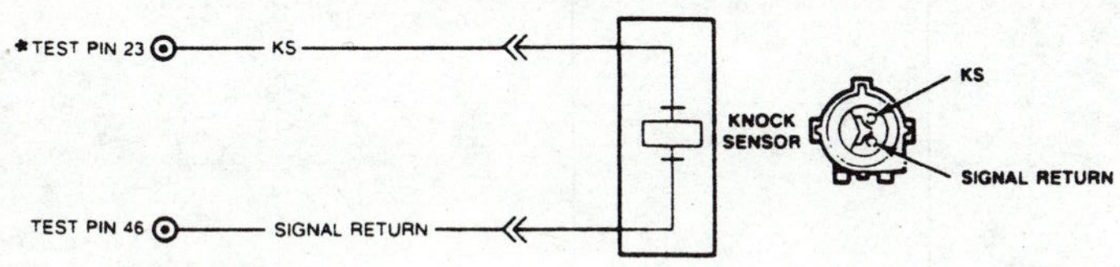

*TEST PINS LOCATED ON BREAKOUT BOX.
ALL HARNESS CONNECTORS VIEWED INTO MATING SURFACE.

STOP-WARNING

You may have reached this point because of a fault in a non-EEC System and/or component. To prevent the replacement of good components, be aware that the following non-EEC areas may be at fault:

- Fuel (quality).
- Basic engine.
- Spark timing.

This Pinpoint Test is intended to diagnose only the following:

- Knock sensor.
- Harness circuits: KS and Signal Return.
- Processor assembly.

Knock Sensor — Pinpoint Test — DG

TEST STEP	RESULT ▶	ACTION TO TAKE
DG1 SERVICE CODE 25: GENERATE KNOCK MANUALLY		
NOTE: With knock conditions sensitive to fuel, altitude and weather, perform Step DG1 before servicing any components. • Vehicle prepared to run (Engine Running Self-Test). • Equipment needed: 4 oz. hammer. • Prepare to rap tap on exhaust manifold, directly above the knock sensor, when the Dynamic Response Signal is given. **NOTE: There is no need to actually depress throttle at this point.** • Perform Engine Running Self-Test. • Rap moderately on exhaust manifold when meter indicates Dynamic Response Test is ready. • 15 seconds later a code will be generated. • Check for code 25. **NOTE: Ignore all other codes at this point.**	No ▶ Yes ▶ **NOTE: Service code 25 may be received whenever the engine is not tapped**	Knock system OK. REPEAT Engine Running Self-Test and SERVICE any other codes from that test. GO to DG2.
DG2 TEST KNOCK CIRCUIT FOR VOLTAGE		
• Key Off, wait 10 seconds. • Disconnect knock sensor connector and inspect. • Set DVOM on 20V scale. • Key On. Engine Off. • Measure voltage between harness connector circuit KS and signal return.	Voltage is between 1 and 4V ▶ Voltage is less than 1V ▶ Voltage is greater than 4V ▶	GO to DG6. GO to DG3. GO to DG5.
DG3 CHECK CONTINUITY OF KS AND SIGNAL RETURN CIRCUITS		
• Key Off, wait 10 seconds. • Disconnect processor 60 Pin connector and inspect for damaged pins, corrosion, loose wires. Service as necessary. • Connect Breakout box to harness. Leave processor disconnected. • Knock sensor disconnected. • DVOM on 200 ohm scale. • Measure continuity of Signal Return at harness to test Pin 46 and KS at harness to test Pin 23.	Both resistances less than 5 ohms ▶ Either resistance 5 ohms or greater ▶	GO to DG4. SERVICE open circuit. REPEAT Quick Test.

Knock Sensor — Pinpoint Test — DG

TEST STEP	RESULT ▶	ACTION TO TAKE
DG4 CHECK KS CIRCUIT FOR SHORT TO GROUND		
• Key Off, wait 10 seconds. • Breakout box installed. • Processor disconnected. • Knock sensor disconnected. • DVOM on 200,000 ohm scale. • Measure resistance of KS circuit to test Pins 40, 46 and 60.	All resistances 10,000 ohms or greater ▶ Any resistance less than 10,000 ohms ▶	GO to DG6. SERVICE harness short. REPEAT Quick Test.
DG5 CHECK KS CIRCUIT FOR SHORT TO VOLTAGE		
• Key Off, wait 10 seconds. • Disconnect processor 60 Pin connector and inspect for damaged pins, corrosion, loose wires. Service as necessary. • Connect Breakout box to harness. Leave processor disconnected. • Knock sensor disconnected. • Key On, Engine Off. • DVOM on 20V scale. • Measure voltage from test Pin 23 to test Pin 40.	0.5V or greater ▶ Less than 0.5V ▶	SERVICE harness short to power. REPEAT Quick Test. GO to DG6.
DG6 TEST PROCESSOR WITH SUBSTITUTE KNOCK SENSOR		
• Key Off, wait 10 seconds. • Remove Breakout box and reconnect processor. • Equipment Required: — Equivalent knock sensor (same part number). — 4 oz. hammer. • Plug substitute sensor in harness (do not install). • Perform Engine Running Self-Test. • Lightly tap knock sensor when Dynamic Test Ready Signal is given. • 15 seconds later, a code will be generated. • Check for code 25. **NOTE: Ignore all other codes at this time.**	No ▶ Yes ▶	INSTALL new knock sensor. REPEAT Quick Test. REPLACE processor and REMOVE substitute sensor. REPEAT Quick Test with original sensor.

Throttle Position Sensor — Pinpoint Test — DH

DH

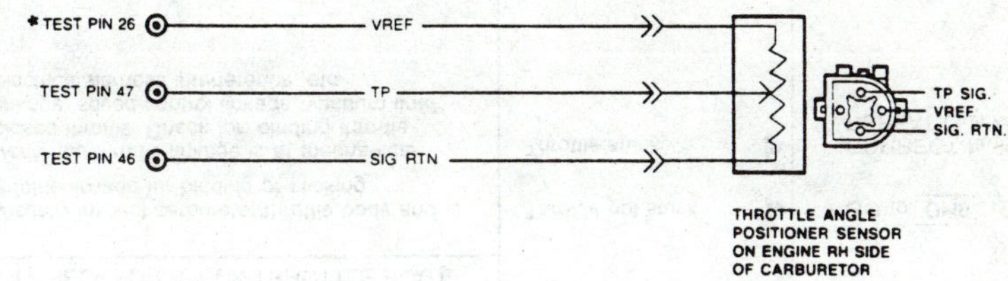

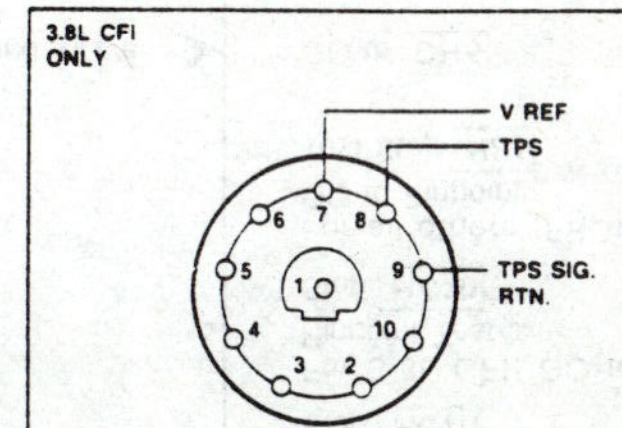

NOTE: FOR 3.8L CFI A 10-PIN CONNECTOR WAS USED FOR THE FIRST 60 DAYS OF PRODUCTION. EITHER CONNECTOR MAY BE ENCOUNTERED. TEST PIN ASSIGNMENTS DID NOT CHANGE ON BREAKOUT BOX. IF TP SENSOR MUST BE DISCONNECTED FOR DIAGNOSTICS, DISCONNECT THE 10 PIN CONNECTOR AT THE FUEL CHARGE ASSY.

*TEST PINS LOCATED ON BREAKOUT BOX. ALL HARNESS CONNECTORS VIEWED INTO MATING SURFACE.

STOP-WARNING

You may have reached this point because of a fault in a non-EEC System and/or component. To prevent the replacement of good components, be aware that the following non-EEC areas may be at fault:

- Idle speeds/throttle stop adjustment.
- Binding throttle shaft/linkage or speed control linkage.
- Choke/high cam system, if equipped.

This Pinpoint Test is intended to diagnose only the following:

- TP sensor.
- Sensor harness circuits: VREF, TP Signal, and Signal Return.
- Processor assembly.

Throttle Position Sensor — Pinpoint Test — DH

TEST STEP	RESULT ▶	ACTION TO TAKE
DH1 SERVICE CODE 23: YOU HAVE BEEN DIRECTED HERE FROM KEY ON, ENGINE OFF TEST. THE FOLLOWING CHECK MUST BE MADE BEFORE SERVICING THIS CODE		
• Check for code 68.	Code 68 present ▶	GO to Pinpoint Test Step FB1.
	Code 68 not present ▶	GO to DH2.
DH2 ATTEMPT TO CLOSE THROTTLE COMPLETELY		
• Lightly press throttle lever toward actuator tip of idle speed motor while performing Key On, Engine Off Self-Test. • Check for service code 23.	Code 23 present ▶	GO to DH5.
	Code 23 not present ▶	TP sensor slightly out of adjustment but do not attempt to adjust. EEC system functioning properly.
DH3 SERVICE CODE 23: YOU HAVE BEEN DIRECTED HERE FROM ENGINE RUNNING TEST. THE FOLLOWING CHECK MUST BE MADE BEFORE SERVICING THIS CODE		
• Check for code 58.	Code 58 present ▶	GO to Pinpoint Test Step FB1.
	Code 58 not present ▶	GO to DH4.

Throttle Position Sensor — Pinpoint Test — DH

TEST STEP	RESULT ▶	ACTION TO TAKE
DH4 REPAIR FOLLOWING ENGINE RUNNING SERVICE CODES BEFORE CONTINUING		
• Check for codes 31 or 41 present in Engine Running Test.	Code 31 present ▶	For 2.8L truck, 3.8L and 5.0L CFI: GO to Pinpoint Test Step DL1. All others: GO to Pinpoint Test Step DD1.
	Code 41 present ▶	For 1.6L, 2.3L EFI: GO to Pinpoint Test Step HA1. For 5.0L CFI: GO to Pinpoint Test Step HC1. For 3.8L CFI: GO to Pinpoint Test Step HC20. For all others (FBC): GO to Pinpoint Test Step HB1.
	Codes 31 and 41 are not present ▶	GO to DH5.
DH5 CHECK FOR STUCK THROTTLE PLATE		
• Visually inspect carburetor/throttle body and throttle linkage for binding or sticking. • Verify the throttle linkage is at mechanical/closed throttle. Check for: binding throttle linkage, speed control linkage, vacuum line/electrical harness interference, etc.	Throttle not stuck ▶	GO to DH6.
	Throttle stuck ▶	CORRECT faults. REPEAT Quick Test.

Throttle Position Sensor — Pinpoint Test — DH

TEST STEP	RESULT ▶	ACTION TO TAKE
DH6 GENERATE CODE 63		
• Refer to illustration DH. • Key Off. wait 10 seconds. • Disconnect TP sensor harness connector at the throttle body. Inspect for damaged pins, corrosion, loose wires, etc. Service as necessary. • Perform Key On, Engine Off self-test and record codes. **NOTE: Ignore all other codes at this time.**	Code 63 present ▶	REPLACE TP sensor. REFER to Shop Manual Group 24 and REPEAT Quick Test except for Calibration 4-54T 5.0L truck and 2.3L HSC engine. GO to DH20.
	Code 63 not present ▶	REPLACE processor. REPEAT Quick Test. except for Calibration 4-54T (5.0L truck) and 2.3 HSC engine. GO to DH20.
DH20 SERVICE CODE 53. CHECK SIGNAL RETURN FOR OPEN		
• Refer to illustration DH. • Key Off, wait 10 seconds. • Disconnect TP harness connector at throttle body. Inspect for damaged pins, corrosion, loose wires. Service as necessary. • DVOM on 20V scale. • Connect DVOM test leads to battery positive terminal and signal return at harness connector. • Observe voltage reading.	10V or greater ▶	GO to DH21.
	Less than 10V ▶	GO to DH23.
DH21 CHECK TP SIGNAL LINE FOR VOLTAGE		
• Key Off. TP sensor disconnected. • DVOM on 20V scale. • Connect DVOM test leads to TP and signal return at harness connector. • Key On, Engine Off. • Measure voltage on TP circuit.	Voltage less than 4V ▶	GO to DH22.
	Voltage 4V or greater ▶	GO to DH24.

Throttle Position Sensor — Pinpoint Test — DH

TEST STEP	RESULT ▶	ACTION TO TAKE
DH22 GENERATE CODE 63		
• Turn Key Off, wait 10 seconds. • With TP harness disconnected, perform Key On, Engine Off Self-Test. • Check for code 63. **NOTE: Ignore all other codes at this time.**	Code 63 present ▶	REPLACE TP sensor, REFER to Shop Manual Group 24 and REPEAT Quick Test.
	Code 63 not present ▶	REPLACE processor. REPEAT Quick Test.
DH23 CHECK SIG. RET. CIRCUIT FOR OPEN		
• Key Off, wait 10 seconds, TP harness disconnected. • DVOM on 200 ohm scale. • Disconnect processor 60 Pin connector and inspect for damaged pins, corrosion, loose wires. Service as necessary. • Install Breakout box, leave processor disconnected. • Measure continuity of Signal Return to test Pin 46.	DVOM reading less than 5 ohms ▶	REPLACE processor. REPEAT Quick Test.
	DVOM reading 5 ohms or greater ▶	SERVICE open circuit. REPEAT Quick Test.
DH24 CHECK TP SIGNAL FOR SHORT TO POWER		
• Key Off, wait 10 seconds, TP harness disconnected. • DVOM on 200,000 ohm scale. • Disconnect processor 60 Pin connector and inspect for damaged pins, corrosion, loose wires. Service as necessary. • Install Breakout box, leave processor disconnected. • Measure resistance of test Pin 47 to test Pins 26 and 57.	Either resistance is less than 10,000 ohms ▶	SERVICE harness short. REPEAT Quick Test.
	Both resistances are 10,000 ohms or greater ▶	REPLACE processor. REPEAT Quick Test.

Throttle Position Sensor — Pinpoint Test — DH

TEST STEP	RESULT ▶	ACTION TO TAKE
DH30 SERVICE CODE 63: CHECK FOR VREF TO TP SIGNAL INPUT		
• Refer to illustration DH. • Key Off, wait 10 seconds. • Disconnect TP harness connector at throttle body. Inspect for damaged pins, corrosion, and pins pushed out. Service as necessary. • DVOM on 20V scale. • Key On, Engine Off. • Measure voltage from TP harness VREF circuit to TP signal.	Voltage reading between 4V and 6V ▶	GO to DH31.
	Voltage reading less than 4V ▶	GO to DH33.
	Voltage reading 6V or greater ▶	GO to Pinpoint Test Step C1.
DH31 CHECK RESISTANCE OF TP CIRCUIT TO GROUND		
• Key Off, wait 10 seconds, TP harness disconnected. • Disconnect processor 60 Pin connector and inspect for damaged pins, corrosion, loose wires. Service as necessary. • DVOM to 200,000 ohm scale. • Connect one test lead to harness TP signal at TP sensor connector and the other test lead to ground.	DVOM reading is less than 10,000 ohms ▶	SERVICE circuit short and RETEST.
	DVOM reading is 10,000 ohms or greater ▶	GO to DH32.
DH32 GENERATE CODE 53: 5.0L TRUCK CAL. 4-54T AND 2.3 HSC ONLY		
• Key Off, wait 10 seconds. TP harness disconnected. • Jumper VREF to TP signal. • Perform Key On, Engine Off self-test. • Check for Code 53 Code 23 (5.0L truck — 2.3 HSC) **NOTE: Ignore all other codes at this time.**	Code 53 present (Code 23 present) ▶	REPLACE TP sensor, REFER to Shop Manual Group 24 and REPEAT Quick Test.
	Code 53 not present (Code 23 not present) ▶	REPLACE processor. REPEAT Quick Test.

Throttle Position Sensor — Pinpoint Test — DH

TEST STEP	RESULT ▶	ACTION TO TAKE
DH33 CHECK CONTINUITY OF TP AND VREF CIRCUITS		
• Key Off, wait 10 seconds. TP harness disconnected. • DVOM on 200 ohm scale. • Disconnect processor 60 Pin connector and inspect for damaged pins, corrosion, loose wires. Service as necessary. • Connect Breakout box. Processor connected to Breakout box. • Check continuity of TP Signal to test Pin 47. Check continuity of VREF to test Pin 26.	Either DVOM reading 5 ohms or greater ▶	SERVICE faulty circuit. CONNECT throttle position sensor. REMOVE Breakout box and REPEAT Quick Test.
	Both DVOM readings less than 5 ohms ▶	REPLACE processor. CONNECT throttle position sensor. REMOVE Breakout box and REPEAT Quick Test.
DH40 SERVICE CODE 73: TP SENSOR MOVES IN ENGINE RESPONSE TEST		
NOTE: Code 73 indicates the TP Sensor did not exceed 25 percent of its rotation in the Engine Response Check. • Key Off. • Install Breakout box. • DVOM on 20V range. • Connect DVOM to test Pins 47 and 46. • Perform Engine Running Quick Test. Step 5.0. • Verify DVOM reading exceeds 3.5V during brief WOT at Engine Response Check.	DVOM exceeds 3.5V during Engine Response Check ▶	REPLACE processor. REPEAT Quick Test.
	DVOM does not exceed 3.5V during Engine Response Check ▶	VERIFY TP Sensor is properly attached to throttle body. If OK, REPLACE TP Sensor PER Shop Manual Group 24 and REPEAT Quick Test.

Throttle Position Sensor (TPS) — Pinpoint Test — DH

TEST STEP	RESULT	▶ ACTION TO TAKE
DH90 CONTINUOUS TEST SERVICE CODE 53: EXERCISE TP SENSOR		
• Using continuous monitor mode, observe VOM or STAR LED for indication of a fault while performing the following: • Move throttle slowly to WOT position. • Release throttle slowly to closed position and lightly tap on TP sensor (simulate road shock). • Wiggle TP harness connector. • Is a fault indicated?	Yes ▶ No ▶	GO to DH91. GO to DH92.

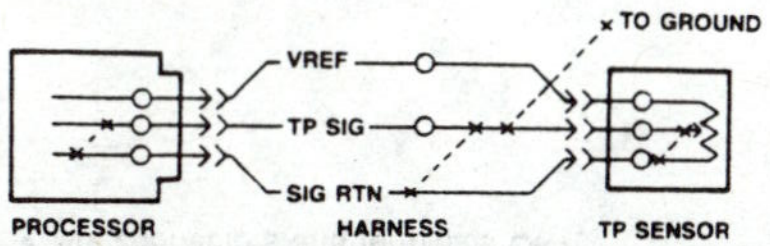

TEST STEP	RESULT	▶ ACTION TO TAKE
DH91 MEASURE THROTTLE POSITION SIGNAL VOLTAGE WHILE EXERCISING TP SENSOR		
• Key Off, wait 10 seconds. • Disconnect processor 60 Pin connector and inspect for damaged pins, corrosion, loose wires. Service as necessary. • Install Breakout box and reconnect processor. • VOM or STAR LED still connected to STO as in previous step. • Connect a DVOM from test Pin 47 to test Pin 46. • DVOM on 20V scale. • Key On, engine Off. • While observing DVOM, repeat step DH90. • Does the fault occur below 4.25V?	Yes ▶	DISCONNECT and INSPECT connectors. If connector and terminals are good, REPLACE TP sensor, REFER to Shop Manual Group 24 and REPEAT Quick Test.
	No ▶	Throttle position sensor overtravel may have caused the continuous code 53. Sensor service is not required. To verify harness integrity, GO to DH92.

Throttle Position Sensor (TPS) — Pinpoint Test — DH

TEST STEP	RESULT	▶ ACTION TO TAKE
DH92 CHECK EEC-IV HARNESS		
• Observe VOM or STAR LED for a fault indication while performing the following: • Referring to the illustration in Step DH90, grasp the harness close to the sensor connector. Wiggle, shake or bend a small section of the EEC-IV system harness while working your way to the dash panel. Also wiggle, shake or bend the EEC-IV harness from the dash panel to the processor. • Is a fault indicated?	Yes ▶	ISOLATE fault and make necessary repairs. REFER to appropriate figure. REPEAT Quick Test.
	No ▶	GO to DH93.
DH93 CHECK PROCESSOR AND HARNESS CONNECTORS		
• Key Off, wait 10 seconds. • Disconnect processor 60 Pin connector. • Inspect both connectors and connector terminals for obvious damage or faults. • Are connectors and terminals OK?	No ▶	SERVICE as necessary. REPEAT Quick Test.
	Yes ▶	Unable to duplicate fault at this time. Continuous code 53 testing complete.
DH94 CONTINUOUS TEST SERVICE CODE 63: EXERCISE TP SENSOR		
• Using continuous monitor mode, observe VOM or STAR LED for indication of a fault while performing the following: • Move throttle slowly to WOT position. • Release throttle slowly to closed condition. • Lightly tap on TP sensor (simulate road shock). • Wiggle TP harness connector. • Is a fault indicated?	Yes ▶	DISCONNECT and INSPECT connectors. If connector and terminals are good, REPLACE TP sensor, REFER to Shop Manual Group 24 and REPEAT Quick Test.
	No ▶	GO to DH95.

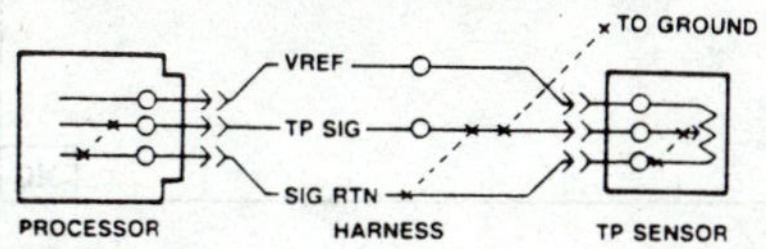

Throttle Position Sensor (TPS) — Pinpoint Test — DH

TEST STEP	RESULT	▶	ACTION TO TAKE
DH95 CHECK EEC-IV HARNESS			
• Observe VOM or STAR LED for a fault indication while performing the following: • Referring to the illustration in Step DH94 grasp the harness close to the sensor connector. Wiggle, shake or bend a small section of the EEC-IV system harness while working your way to the dash panel. Also wiggle, shake or bend the EEC-IV harness from the dash panel to the processor. • Is a fault indicated?	Yes	▶	ISOLATE fault and make necessary repairs. REFER to appropriate figure. REPEAT Quick Test.
	No	▶	Go to DH96.
DH96 CHECK PROCESSOR AND HARNESS CONNECTORS			
• Key Off, wait 10 seconds. • Disconnect processor 60 Pin connector. • Inspect both connectors and connector temrinals for obvious damage or faults. • Are connectors and terminals OK?	No	▶	SERVICE as necessary. REPEAT Quick Test.
	Yes	▶	Unable to duplicate fault at this time. Continuous code 63 testing complete.

Vane Air Flow Sensor (VAF) — Pinpoint Test — DK

DK

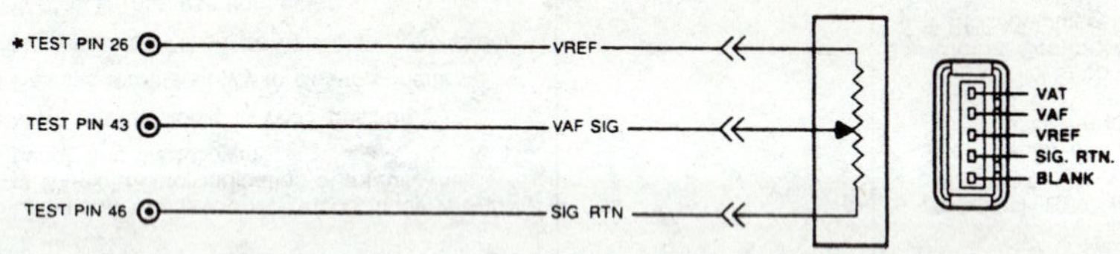

* TEST PIN LOCATED ON BREAKOUT BOX
ALL HARNESS CONNECTORS VIEWED INTO MATING SURFACE.

STOP-WARNING

You may have reached this point because of a fault in a non-EEC System and/or component. To prevent the replacement of good components, be aware that the following non-EEC areas may be at fault:

- Check for unmetered air (air leaks) between VAF meter and throttle body.
- Vacuum leaks.
- Engine sealing (PCV sealing, CANP, valve cover seal dipstick seated).

This Pinpoint Test is intended to diagnose only the following:

- VAF meter.
- Processor.
- Harness circuits: VREF, VAF Sig. and Signal Return.

Vane Air Flow Sensor (VAF) — Pinpoint Test — DK

TEST STEP	RESULT ▶	ACTION TO TAKE
FAULT CODE 26		
DK1 CHECK FOR CONTAMINATION		
NOTE: Code 26 (Key On, Engine Off, or Engine Running) indicates the vane air flow input to the processor is out of closed (Engine Not Running) or at idle limits (Engine Off 0.15-0.50V: Engine at idle 1.50-2.70V). There have been no opens or shorts in the VAF circuit or a code 56 (signal always high) or 66 (signal always low) would have been generated.		
• Key Off, wait 10 seconds. • Remove air cleaner element and check for contamination (oil residue, foreign material, etc.) that may impede VAF sensor vane movement. • Is service code 26 present in Key On, Engine Off portion of Quick Test?	Yes ▶	REINSTALL air cleaner element. REPLACE vane meter. RERUN Quick Test.
	No ▶	GO to DK2.
DK2 VAF SENSOR CHECK		
• Key Off, air cleaner element reinstalled. • Check for unmetered air leaks between vane meter and throttle body. • Disconnect processor 60 Pin connector. Inspect for damaged pins, corrosion, loose wires, etc. Service as necessary. • Install Breakout box. Connect processor to Breakout box harness. • DVOM on 20V scale. • Key On, engine Off. • Place new unsharpened pencil as shown below. • Measure voltage between test Pins 43 and 46. • Does DVOM read between 2.8V and 3.7V?	Yes ▶	Vane meter is capable of outputting an acceptable signal. The VAF code 26 has been caused by incorrect engine speed or an unmetered air leak (vacuum leak). SERVICE as necessary. REMOVE Breakout box. RERUN Quick Test.
	No ▶	REMOVE Breakout box. REPLACE processor. RERUN Quick Test.

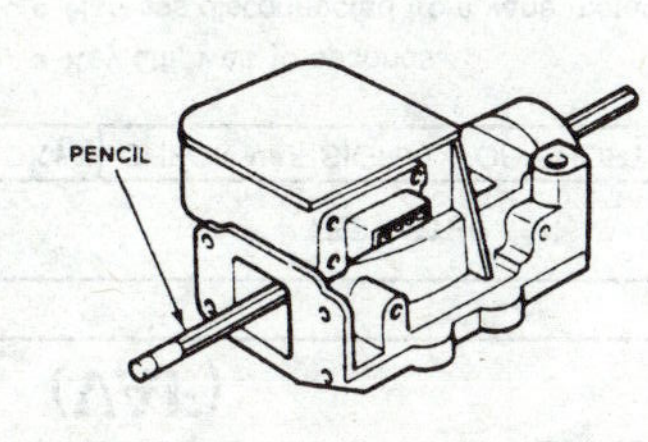

VANE METER

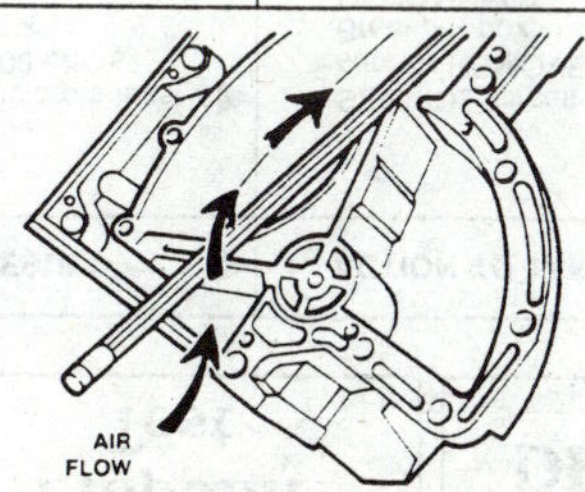

VAF SENSOR AIR VANE

Vane Air Flow Sensor (VAF) — Pinpoint Test — DK

TEST STEP	RESULT ▶	ACTION TO TAKE
FAULT CODE 56		
DK10 CHECK SIGNAL RETURN		
• Key Off, wait 10 seconds. • Disconnect harness from vane meter. Inspect for damaged pins, corrosion, loose wires, etc. Service as necessary. • DVOM on 20V scale. • Measure voltage of vane meter harness connector Signal Return to battery positive terminal.	Reading is less than 10.0V ▶	GO to DK13.
	Reading is 10.0V or greater ▶	GO to DK11.
DK11 CHECK VAF TO SIGNAL RETURN VOLTAGE		
• Key On, Engine Off. Harness disconnected from vane meter. • DVOM on 20V scale. • Measure voltage from vane meter harness connector VAF Signal to Signal Return.	Reading is less than 4.0V ▶	GO to DK12.
	Reading is 4.0V or greater ▶	GO to DK14.
DK12 INDUCE OPPOSITE CODE		
• Key Off, wait 10 seconds. • Harness disconnected from vane meter. • Run Key On, Engine Off Quick Test. • Is code 66 present?	Yes ▶	REPLACE vane meter. RECONNECT harness. RERUN Quick Test.
NOTE: Disconnecting vane meter disconnects both VAF and VAT sensors. A code 54 should also be present. Disregard all codes except VAF codes at this time.	No ▶	REPLACE processor. RECONNECT harness to vane meter and processor. RERUN Quick Test.
DK13 CHECK CONTINUITY OF SIGNAL RETURN		
• Key Off. Harness disconnected from vane meter. • Disconnect processor 60 Pin connector. Inspect for damaged pins, corrosion, loose wires, etc. Service as necessary. • Install Breakout box, leave processor disconnected. • DVOM on 200 ohm scale. • Measure resistance of vane meter harness connector Signal Return to test Pin 46.	Reading is less than 5 ohms ▶	REPLACE processor. REMOVE Breakout box. RECONNECT processor and vane meter. RERUN Quick Test.
	Reading is 5 ohms or greater ▶	SERVICE circuit open. REMOVE Breakout box. RECONNECT processor and vane meter. RERUN Quick Test.

Vane Air Flow Sensor (VAF) — Pinpoint Test — DK

TEST STEP	RESULT ▶	ACTION TO TAKE
DK14 CHECK VAF SIGNAL FOR SHORT		
• Key Off, wait 10 seconds. • Harness disconnected from vane meter. • Disconnect processor 60 Pin connector. Inspect for damaged pins, corrosion, loose wires, etc. Service as necessary. • Install Breakout box, leave processor disconnected. • DVOM on 200,000 ohm scale. • Measure resistance from test Pin 43 to test Pins 26 and 57.	Either reading is less than 10,000 ohms ▶	SERVICE circuit shorts. REMOVE Breakout box. RECONNECT processor and vane meter. RERUN Quick Test.
	Both readings are 10,000 ohms or greater ▶	REPLACE processor. REMOVE Breakout box. RECONNECT processor and vane meter. RERUN Quick Test.
FAULT CODE 66		
DK20 CHECK VREF TO SIGNAL RETURN VOLTAGE		
• Key off, wait 10 seconds. • Disconnect harness from vane meter. Inspect for damaged pins, corrosion, loose wires, etc. Service as necessary. • Key On, Engine Off. • DVOM on 20V scale. • Measure voltage from vane meter harness connector VREF to Signal Return.	Reading is between 4.0 and 6.0V ▶	GO to **DK21**.
	Reading is less than 4.0V ▶	GO to **DK24**.
	Reading is greater than 6.0V ▶	GO to Pinpoint Test Step **C1**.
DK21 CHECK VAF SIGNAL FOR SHORT		
• Key Off, wait 10 seconds. • Harness disconnected from vane meter. • DVOM on 200,000 ohm scale. • Disconnect processor 60 Pin connector. Inspect for damaged pins, corrosion, loose wires, etc. Service as necessary. • Measure resistance of vane meter harness connector circuit VAF Signal to Signal Return.	Reading is less than 10,000 ohms ▶	SERVICE circuit shorts. RECONNECT vane meter. RERUN Quick Test.
	Reading is 10,000 ohms or greater ▶	RECONNECT processor. GO to **DK22**.

Vane Air Flow Sensor (VAF) — Pinpoint Test — DK

TEST STEP	RESULT ▶	ACTION TO TAKE
DK22 INDUCE OPPOSITE CODE		
• Key Off, wait 10 seconds. • Processor reconnected. • Harness disconnected from vane meter. • Install jumper wire in vane meter harness connector VREF to VAF signal. • Perform Key On, Engine Off Quick Test. • Is code 56 present? **NOTE: Disconnecting vane meter disconnects both VAF and VAT sensors. Code 54 should also be present. Disregard all codes except VAF codes at this time.**	Yes ▶	REPLACE vane meter. REMOVE jumper wire. RECONNECT vane meter. RERUN Quick Test.
	No ▶	REMOVE jumper wire and GO to **DK23**.
DK23 CHECK CONTINUITY OF VAF SIGNAL		
• Key Off, harness disconnected from vane meter. • Disconnect processor 60 Pin connector. Inspect for damaged pins, corrosion, loose wires, etc. Service as necessary. • Install Breakout box, leave processor disconnected. • DVOM on 200 ohm scale. • Measure resistance from vane meter harness connector VAF signal test Pin 43.	Reading is less than 5 ohms ▶	REPLACE processor. REMOVE Breakout box. RECONNECT processor and vane meter. RERUN Quick Test.
	Reading is 5 ohms or greater ▶	SERVICE circuit open. REMOVE Breakout box. RECONNECT processor and vane meter. RERUN Quick Test.
DK24 CHECK CONTINUITY OF VREF		
• Key Off, wait 10 seconds. • Harness disconnected from vane meter. • Disconnect processor 60 Pin connector. Inspect for damaged pins, corrosion, loose wires, etc. Service as necessary. • Install Breakout box, leave processor disconnected. • DVOM on 200 ohm scale. • Measure resistance from vane meter harness connector VREF to test Pin 26.	Reading is less than 5 ohms ▶	REPLACE processor. REMOVE Breakout box. RECONNECT vane meter and processor. RERUN Quick Test.
	Reading is 5 ohms or greater ▶	SERVICE circuit open. REMOVE Breakout box. RECONNECT processor and vane meter. RERUN Quick Test.

Vane Air Flow Sensor (VAF) — Pinpoint Test — DK

TEST STEP	RESULT ▶	ACTION TO TAKE
FAULT CODE 76		
DK30 CHECK FOR VOLTAGE INCREASE IN VAF SIGNAL TO SIGNAL RETURN		
NOTE: A sharp snap of the throttle may not be sufficient to pass this test. Be sure to move throttle to WOT and return.		
• Key Off, wait 10 seconds. • Disconnect processor 60 Pin connector. Inspect for damaged pins, corrosion, loose wires, etc. Service as necessary. • Install Breakout box. Connect processor to Breakout box harness. • DVOM on 20V scale. • Connect DVOM to test Pins 43 and 46. • Perform Engine Running Quick Test while monitoring DVOM. • After dynamic response prompt code 1(0) operator does a brief WOT. DVOM should increase more than 2.0V from reading before WOT. • Observe service codes at end of test.	DVOM increased more than 2.0V and code 76 is still present ▶	REPLACE processor. REMOVE Breakout box. RERUN Quick Test.
	DVOM did not increase more than 2.0V ▶	CHECK air cleaner duct for obstruction. If OK, REPLACE vane meter.
	DVOM increased more than 2.0V and code 76 is not present ▶	Vane meter is OK, SERVICE other codes as necessary.

Vane Air Flow — Pinpoint Test — DK

TEST STEP	RESULT ▶	ACTION TO TAKE
DK90 SERVICE CODE 56 CONTINUOUS TEST: CHECK VAF SENSOR		
• Using continuous monitor mode, observe VOM or STAR LED for indication of a fault while performing the following: • Lightly tap on VAF sensor (simulate road shock). • Wiggle VAF connector. • Is a fault indicated?	Yes ▶	DISCONNECT and INSPECT connectors. If connector and terminals are good, REPLACE VAF sensor. REPEAT Quick Test.
	No ▶	GO to DK91.
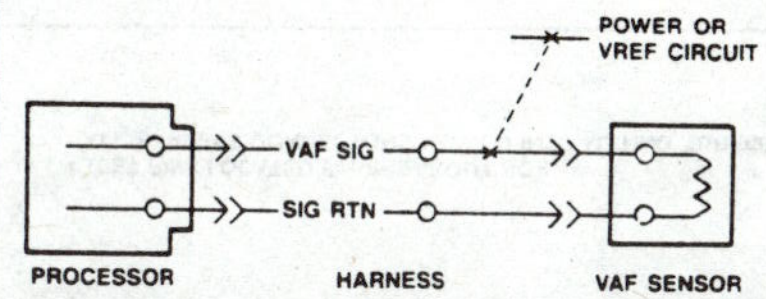		
DK91 CHECK EEC-IV HARNESS		
• Observe VOM or STAR LED for a fault indication while performing the following: • Referring to the illustration in Step DK90, grasp the harness close to the sensor connector. Wiggle, shake or bend a small section of the EEC-IV system harness while working your way to the dash panel. Also wiggle, shake or bend the EEC-IV harness from the dash panel to the processor. • Is a fault indicated?	Yes ▶	ISOLATE fault and make necessary repairs. REFER to appropriate figure. REPEAT Quick Test.
	No ▶	GO to DK92.
DK92 CHECK PROCESSOR AND HARNESS CONNECTORS		
• Key Off, wait 10 seconds. • Disconnect processor 60 Pin connector. • Inspect both connectors and connector terminals for obvious damage or faults. • Are connectors and terminals OK?	No ▶	SERVICE as necessary. REPEAT Quick Test.
	Yes ▶	Unable to duplicate fault at this time. Continuous code 56 testing complete.

Vane Air Flow — Pinpoint Test — DK

TEST STEP	RESULT ▶	ACTION TO TAKE
DK93 SERVICE CODE 66 CONTINUOUS TEST: CHECK VAF SENSOR		
• Using continuous monitor mode, observe VOM or STAR LED for indication of a fault while performing the following: • Lightly tap on VAF sensor (simulate road shock). • Wiggle ACT connector. • Is a fault indicated?	Yes ▶	DISCONNECT and INSPECT connectors. If connector and terminals are good, REPLACE sensor. REPEAT Quick Test.
	No ▶	GO to DK94.
DK94 CHECK EEC-IV HARNESS		
• Observe VOM or STAR LED for a fault indication while performing the following: • Referring to the illustration in Step DK93 grasp the harness close to the sensor connector. Wiggle, shake or bend a small section of the EEC-IV system harness while working your way to the dash panel. Also wiggle, shake or bend the EEC-IV harness from the dash panel to the processor. • Is a fault indicated?	Yes ▶	ISOLATE fault and make necessary repairs. REFER to appropriate figure. REPEAT Quick Test.
	No ▶	Go to DK95.
DK95 CHECK PROCESSOR AND HARNESS CONNECTORS		
• Key Off, wait 10 seconds. • Disconnect processor 60 Pin connector. • Inspect both connectors and connector terminals for obvious damage or faults. • Are connectors and terminals OK?	No ▶	SERVICE as necessary. REPEAT Quick Test.
	Yes ▶	Unable to duplicate fault at this time. Continuous code 66 testing complete.

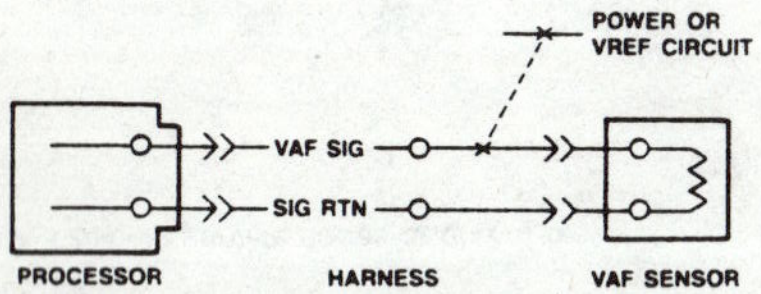

EGR Valve Position — Pinpoint Test — DL

DL

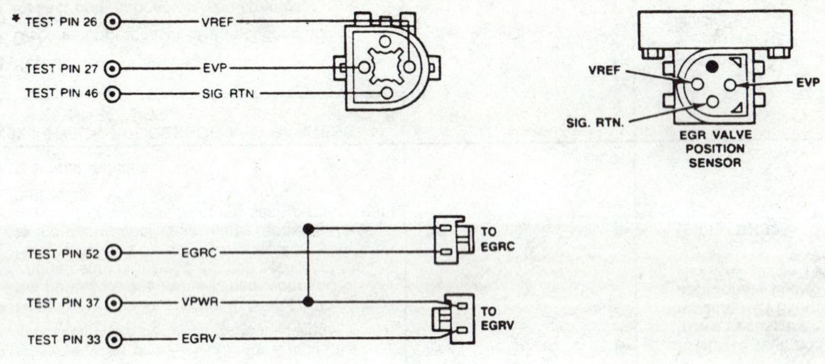

*TEST PINS LOCATED ON BREAKOUT BOX.
ALL HARNESS CONNECTORS VIEWED INTO MATING SURFACE.

STOP-WARNING

You may have reached this point because of a fault in a non-EEC System and/or component. To prevent the replacement of good components, be aware that the following non-EEC area may be at fault:

- Damaged EGR valve.

This Pinpoint Test is intended to diagnose only the following:

- EVP sensor.
- Harness circuits: VREF, EVP, Signal Return, EGRC, VPWR, EGRV.
- EGR solenoids.
- EGR valve assembly.
- Processor assembly.

EGR Valve Position Sensor (EVP) — Pinpoint Test — DL

TEST STEP	RESULT ▶	ACTION TO TAKE
FAULT CODE 31		
DL1 RUN ENGINE RUNNING QUICK TEST WITH EGR VACUUM SIGNAL LINE DISCONNECTED AT EGR VALVE		
• Key Off, wait 10 seconds. • Disconnect EGR vacuum line at EGR valve and cap EGR vacuum line. • Perform Engine Running Quick Test. • Check for code 31.	Code 31 present ▶	GO to DL2.
	No code 31 present, but codes 32 and 34 are present ▶	GO to DL10.
DL2 CHECK EVP SENSOR RESISTANCE		
• Key Off, wait 10 seconds. • EGR signal line disconnected and capped. • Disconnect EVP sensor. Inspect for damaged pins, corrosion, and pins pushed out. Service as necessary. • DVOM on 200,000 ohm scale. • Measure resistance from EVP circuit to VREF circuit and from Signal Return circuit to VREF circuit.	Both readings between 100 ohms and 5,500 ohms ▶	GO to DL3.
	Either reading below 100 ohms or greater than 5,500 ohms ▶	REPLACE EVP sensor. RECONNECT vacuum line and harness. RERUN Quick Test.
DL3 CHECK EVP MOVEMENT		
• Key Off, vacuum signal line disconnected and capped, harness disconnected at EVP sensor. • DVOM on 200,000 ohm scale. • Connect vacuum pump to EGR valve. • Measure resistance from EVP to VREF circuit while gradually increasing vacuum to 33 kPa (10 in. Hg). • Observe resistance as vacuum increases.	Reading gradually decreases from no greater than 5,500 ohms to no less than 100 ohms ▶	GO to DL4.
	Reading is less than 100 ohms or greater than 5,500 ohms ▶	REPLACE EVP sensor. RECONNECT signal line and harness. RERUN Quick Test.
	Reading does not decrease or unable to hold vacuum ▶	GO to DL13.

EGR Valve Position Sensor (EVP) — Pinpoint Test — DL

TEST STEP	RESULT ▶	ACTION TO TAKE
DL4 MEASURE VREF TO SIGNAL RETURN VOLTAGE		
• Key On, Engine Off. Vacuum signal line disconnected and capped, harness disconnected from EVP sensor. • DVOM on 20V scale. • Measure voltage from harness connector VREF to harness connector Signal Return.	Reading is between 4V and 6V ▶	GO to DL5.
	Reading is less than 4V ▶	GO to DL7.
	Reading is greater than 6V ▶	GO to Pinpoint Test Step C1.
DL5 MEASURE VREF TO SIGNAL INPUT VOLTAGE		
• Key On, Engine Off. • Reconnect EGR vacuum signal line to EGR valve. • Harness disconnected from EVP sensor. • DVOM on 20V scale. • Measure voltage from sensor connector VREF to sensor connector EVP.	Reading is between 4V and 6V ▶	GO to DL6.
	Voltage is less than 4V ▶	GO to DL7.
DL6 CHECK EVP SIGNAL FOR SHORTS TO GROUND		
• Key Off, wait 10 seconds. Harness disconnected from EVP sensor. • Disconnect harness from processor 60 Pin connector and inspect for damaged pins, corrosion, loose wires. Service as necessary. • DVOM on 200,000 ohm scale. • Measure resistance from sensor harness connector EVP to harness connector Signal Return and from harness connector EVP to battery negative terminal.	Either reading is less than 10,000 ohms ▶	SERVICE circuit short. RECONNECT processor and EVP sensor. RERUN Quick Test.
	Both readings 10,000 ohms or greater ▶	REPLACE processor. RECONNECT processor and EVP sensor. RERUN Quick Test.

EGR Valve Position Sensor (EVP) — Pinpoint Test — DL

TEST STEP	RESULT ▶	ACTION TO TAKE
DL7 MEASURE CONTINUITY OF VREF, EVP AND SIGNAL RETURN CIRCUITS		
• Key Off, wait 10 seconds. • Harness disconnected from EVP sensor. • Disconnect harness from processor 60 Pin connector and inspect for damaged pins, corrosion, loose wires. Service as necessary. • Install Breakout box leaving processor disconnected. • DVOM on 200 ohm scale. • Measure continuity from test Pin 46 to harness connector Signal Return, test Pin 26 to harness connector VREF and test Pin 27 to harness connector EVP.	All readings are less than 5 ohms ▶	GO to DL8.
	Any reading of 5 ohms or greater ▶	SERVICE open circuit(s). REMOVE Breakout box. RECONNECT processor and EVP sensor. RERUN Quick Test.
DL8 CHECK EVP SIGNAL FOR SHORTS TO POWER		
• Key Off, harness disconnected from EVP sensor. Breakout box installed with processor not connected. • DVOM on 200,000 ohm scale. • Measure resistance from test Pin 27 to test Pins 37 and 57.	Either reading less than 10,000 ohms ▶	SERVICE short circuit. REMOVE Breakout box. RECONNECT processor and EVP sensor. RERUN Quick Test.
	Both readings 10,000 ohms or greater ▶	REPLACE processor. REMOVE Breakout box. RECONNECT processor and EVP sensor. RERUN Quick Test.

EGR Valve Position Sensor (EVP) — Pinpoint Test — DL

TEST STEP	RESULT ▶	ACTION TO TAKE
FAULT CODES 32, 33 AND 34		
DL10 OUTPUT STATE CHECK (REFER TO APPENDIX)		
• Key Off, wait 10 seconds. • DVOM on 20V scale. • Connect DVOM negative test lead to STO and positive test lead to battery positive. • Jumper STI to signal return. • Perform Key On, Engine Off Self-Test until the completion of the Continuous Test Codes. • DVOM will indicate zero volts. • Depress and release the throttle. • Did DVOM reading change to a high voltage reading?	NO ▶	DEPRESS throttle to WOT and RELEASE. If STO voltage does not go high, GO to Pinpoint Test Step Q40.
	YES ▶	REMAIN in output state check. GO to DL11.
DL11 CHECK EGR VENT AND EGR CONTROL SOLENOIDS FOR VACUUM CYCLING		
• Key On, Engine Off. • In output state check. • Disconnect and cap vacuum line from bottom port of EGRC solenoid and connect vacuum pump. • Connect a vacuum gauge in the common output (top) vacuum line to EGR valve. • Disconnect, but **do not** cap the vacuum vent line from the EGRV solenoid. • While cycling outputs On and Off (by depressing and releasing throttle) observe vacuum gauge at the output. Maintain vacuum at source.	Vacuum output cycles On and Off in less than 2 seconds ▶	RECONNECT all vacuum lines. GO to DL12.
	Vacuum does not cycle On and Off in less than 2 seconds ▶	REPLACE solenoid assembly. RECONNECT all vacuum lines. RERUN Quick Test.
DL12 CHECK VACUUM LINES		
• Key Off, wait 10 seconds. • Vacuum lines reconnected. • Check entire EEC vacuum line system per VECI schematic decal for leaks, kinks, cracks or obstructions.	Vacuum lines OK ▶	EEC-IV system OK. REFER to Section 29 for EGR system diagnosis.
	Vacuum lines are not OK ▶	SERVICE as necessary. RERUN Quick Test.

EGR Valve Position Sensor (EVP) — Pinpoint Test — DL

TEST STEP	RESULT ▶	ACTION TO TAKE
DL13 MANUALLY EXERCISE EVP SENSOR		
• Key Off, harness disconnected. • Remove EVP sensor from EGR valve. • Measure resistance from EVP sensor Signal Return Pin to EVP sensor VREF Pin while gradually applying pressure to EVP sensor shaft. • Observe resistance as shaft is slowly pushed in. • Look for sudden changes in resistance readings.	Reading decreases smoothly from no more than 5,500 ohms to no less than 100 ohms ▶	REPLACE EGR valve assembly. RECONNECT EVP sensor and EGR supply vacuum line. RERUN Quick Test.
	Reading does not decrease smoothly or changes abruptly ▶	REPLACE EVP sensor. RECONNECT harness and EGR supply vacuum line. RERUN Quick Test.
FAULT CODE 35		
DL30 RPM TOO LOW FOR EGR TEST		
• Check for codes 12 and 13.	Code 12 present	Vehicles equipped with DC motor control GO to KF1. Vehicles equipped with throttle kicker GO to KT1.
	Code 13 present ▶	Vehicles equipped with DC motor control GO to KF20. Vehicles equipped with throttle kicker GO to KT1.
	Codes 12 and 13 not present ▶	GO to DL31.
DL31 RETEST AT 1500 RPM		
• Key Off, wait 10 seconds. • Install tachometer. • Perform Key On, Engine Running Quick Test while maintaining 1500 rpm. • Record engine running service codes. • Check for code 35.	Code 35 present ▶	REPLACE processor. RERUN Quick Test.
	Code 35 not present ▶	REPEAT Quick Test. SERVICE codes as necessary.

EGR Valve Position Sensor (EVP) — Pinpoint Test — DL

TEST STEP	RESULT ▶	ACTION TO TAKE
DL40 VREF CIRCUIT TO GROUND		
• Key Off, wait 10 seconds. • Disconnect the EVP sensor and inspect for damaged pins, corrosion, loose wires. Service as necessary. • DVOM on 20V scale. • Key On, Engine Off. • Measure voltage from VREF (at EVP sensor harness connector) to battery ground.	Voltage readings between 4V and 6V ▶	GO to DL41.
	Voltage readings are less than 4.0V ▶	GO to DL7.
	Voltage readings are greater than 6.0V ▶	GO to Pinpoint Test Step C1.
DL41 SIGNAL RETURN TO VREF		
• Key On. • EVP harness disconnected. • DVOM on 20V scale. • Measure voltage from VREF to signal return at the EVP harness connector.	Voltage readings are between 4V and 6V ▶	GO to DL42.
	Voltage readings are less than 4.0V ▶	SERVICE open in signal return. RERUN Key On, Engine Off Self-Test.
DL42 EVP CIRCUIT VOLTAGE CHECK		
• Key Off, wait 10 seconds. • Disconnect processor 60 Pin connector and inspect for damaged pins, corrosion, loose wires. Service as necessary. • Install the Breakout box, reconnect processor. • Connect EVP sensor. • DVOM on 20V scale. • Key On, Engine Off. • Measure voltage from test Pin 46 to test Pin 27.	Less than 0.5V ▶	GO to DL43.
	1.25V or greater ▶	GO to DL46.
DL43 EVP SIGNAL SHORT TO GROUND		
• Key Off, wait 10 seconds. • Disconnect processor. • Disconnect EVP sensor. • Breakout box installed. • DVOM on 200 ohm scale. • Measure resistance from test Pin 27 to test Pins 20, 40, 46 and 60.	5.0 ohms or greater ▶	GO to DL44.
	Less than 5.0 ohms ▶	SERVICE short to ground in EVP signal circuit. REPEAT Quick Test.

EGR Valve Position Sensor (EVP) — Pinpoint Test — DL

TEST STEP	RESULT ▶	ACTION TO TAKE
DL44 EVP SIGNAL CIRCUIT FOR OPENS		
• Key Off. • Processor disconnected. • EVP sensor disconnected. • Breakout box installed. • DVOM on 200 ohm scale. • Measure resistance from test Pin 27 to EVP signal circuit at EVP connector harness.	5 ohms or greater ▶ Less than 5 ohms ▶	SERVICE open in EVP signal circuit. REPEAT Quick Test. GO to DL45.
DL45 SUBSTITUTE EVP SENSOR		
• Key Off. • Remove and replace present EVP sensor with known good sensor. Reconnect EVP sensor. • Remove Breakout box. • Reconnect processor. • Rerun Key On, Engine Off Quick Test. • Is code 31 present?	Yes ▶ No ▶	REPLACE processor and REINSTALL original EVP sensor. Testing completed.
DL46 EVP SIGNAL SHORT TO POWER		
• Key Off, wait 10 seconds. • Disconnect processor. • Disconnect EVP sensor. • Breakout box installed. • DVOM on 200 ohm scale. • Measure resistance from test Pin 27 to test Pins 26, 37 and 57.	5 ohms or greater ▶ Less than 5 ohms ▶	GO to DL47. SERVICE short to power. RERUN Key On, Engine Off Quick Test.
DL47 SUBSTITUTE EVP SENSOR AND EGR VALVE		
• Key Off, wait 10 seconds. • Substitute a known good sensor and EGR valve assembly. • Sensor connected. • Remove Breakout box. • Reconnect processor. • Rerun Key On, Engine Off Quick Test. • Is code 31 present?	Yes ▶ No ▶	REPLACE processor. REINSTALL original EVP sensor and EGR valve assembly. GO to DL48.

EGR Valve Position Sensor (EVP) — Pinpoint Test — DL

TEST STEP	RESULT ▶	ACTION TO TAKE
DL48 EVP SENSOR CHECK		
• Key Off, wait 10 seconds. • Install original EVP sensor. • EVP sensor connected. • Rerun Key On, Engine Off Quick Test. • Is code 31 present?	Yes ▶ No ▶	REPLACE EVP sensor. Testing completed. Testing completed.

EGR Valve Position Sensor (EVP) — Pinpoint Test — DL

TEST STEP	RESULT ▶	ACTION TO TAKE
DL90 SERVICE CODE 31 CONTINUOUS TEST: EXERCISE EVP SENSOR		
• Using continuous monitor mode, observe VOM or STAR LED for indication of a fault while performing the following: • Connect a vacuum pump to the EGR valve. • Very slowly apply 20 kPa (6 in. Hg) vacuum to the EGR valve. • Slowly bleed vacuum off the EGR valve and lightly tap on EVP sensor (simulate road shock). • Wiggle EVP sensor connector. • Is a fault indicated?	Yes ▶ No ▶	GO to DL91. GO to DL92.
DL91 MEASURE EVP SIGNAL VOLTAGE WHILE EXERCISING EVP SENSOR		
• Key Off, wait 10 seconds. • Disconnect processor 60 Pin connector and inspect for damaged pins, corrosion, loose wires. Service as necessary. • Install Breakout box and reconnect processor. • VOM or STAR LED still connected to STO as in previous step. • Connect a DVOM from test Pin 27 to test Pin 46. • DVOM on 20V scale. • Key On, Engine Off. • While observing DVOM, repeat step DL90. • Does the fault occur below 4.25V?	Yes ▶ No ▶	DISCONNECT and INSPECT connector. If connector and terminals are good, REPLACE EVP sensor. REPEAT Quick Test. EGR valve overshoot may have caused continuous code 31. Sensor service is not required. To verify harness integrity, GO to DL92.

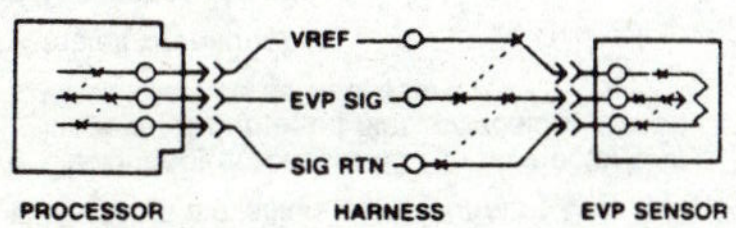

EGR Valve Position Sensor (EVP) — Pinpoint Test — DL

TEST STEP	RESULT ▶	ACTION TO TAKE
DL92 CHECK EEC-IV HARNESS		
• Observe VOM or STAR LED for a fault indication while performing the following: • Referring to the illustration in Step DL90, grasp the harness close to the sensor connector. Wiggle, shake or bend a small section of the EEC-IV system harness while working your way to the dash panel. Also wiggle, shake or bend the EEC-IV harness from the dash panel to the processor. • Is a fault indicated?	Yes ▶ No ▶	ISOLATE fault and make necessary repairs. REFER to appropriate figure. REPEAT Quick Test. GO to DL93.
DL93 CHECK PROCESSOR AND HARNESS CONNECTORS		
• Key Off, wait 10 seconds. • Disconnect processor 60 Pin connector. • Inspect both connectors and connector terminals for obvious damage or faults. • Are connectors and terminals OK?	No ▶ Yes ▶	SERVICE as necessary. REPEAT Quick Test. Unable to duplicate fault at this time. Continuous Code 31 testing complete.

Neutral Drive Switch A/C Input — Pinpoint Test — FA

TEST STEP	RESULT	▶	ACTION TO TAKE
FA1 CODE 67 SYSTEM IDENTIFICATION			
1.6L MTX			GO to FA10.
2.3L Turbo, M/T and 1.6L Turbo, SVO			GO to FA15.
2.3L SVO Turbo Octane Switch			GO to KN15.
4.9L M/T, Truck			GO to FA25.
All other systems			GO to FA2.
FA2 NEUTRAL DRIVE INPUT CHECK			
• Key Off, wait 10 seconds. • Verify heater control is in Off position, if so equipped. • Verify transmission is in Neutral or Park. • Disconnect processor 60 Pin connector and inspect for damaged pins, corrosion, loose wires. Service as necessary. • Install Breakout box. • Processor connected. • Key On, Engine Off. • DVOM on 20V scale. • Measure voltage from test Pin 30 (Neutral Drive circuit) to chassis ground.	Less than .5V	▶	GO to FA4.
	.5V or greater	▶	GO to FA3.

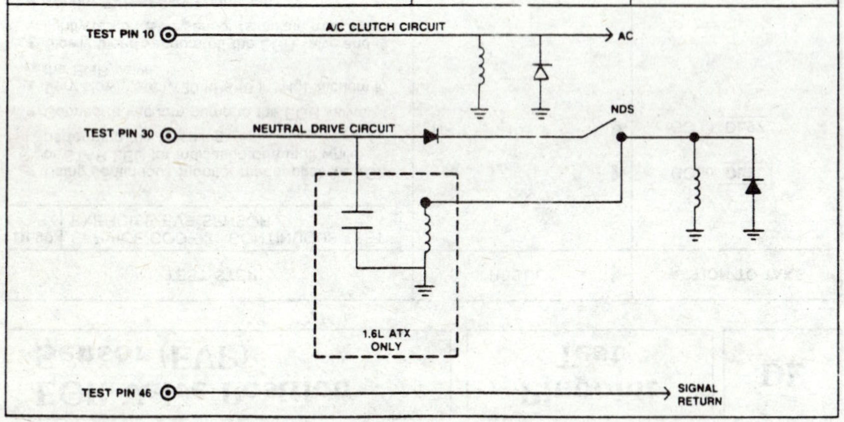

Neutral Drive Switch A/C Input — Pinpoint Test — FA

TEST STEP	RESULT	▶	ACTION TO TAKE
FA3 NEUTRAL DRIVE SWITCH CHECK			
• Key Off, wait 10 seconds. • Breakout box installed. • DVOM on 200 ohm scale. • Locate the Neutral Drive switch. • Disconnect vehicle harness from the Neutral Drive switch and measure resistance across the two contacts.	Less than 5 ohms	▶	SERVICE open in vehicle harness Neutral Drive circuit. REPEAT Quick Test.
	5 ohms or greater	▶	REPLACE Neutral Drive switch. REPEAT Quick Test.
FA4 A/C INPUT CHECK			
• Breakout box installed. • Key On, Engine Off. • DVOM on 20V scale. • A/C control Off. Measure voltage at test Pin 10 A/C clutch circuit to chassis ground.	1.0V or greater	▶	SERVICE short to power in A/C clutch circuit. REPEAT Quick Test.
	Less than 1.0V	▶	REPLACE processor. REPEAT Quick Test.
FA10 NEUTRAL/CLUTCH INPUT CHECK 1.6L MTX			
• Key Off, wait 10 seconds. • Verify A/C is Off, if so equipped. • Verify transaxle is in Neutral and clutch is released. • Disconnect processor 60 Pin connector and inspect for damaged pins, corrosion, loose wires. Service as necessary. • Install Breakout box. • Connect processor. • DVOM on 200 ohm scale. • Measure resistance from test Pin 30 (A C clutch circuit) to test Pin 46 (Signal Return circuit).	Less than 5 ohms	▶	GO to FA4.
	5 ohms or greater	▶	GO to FA11.

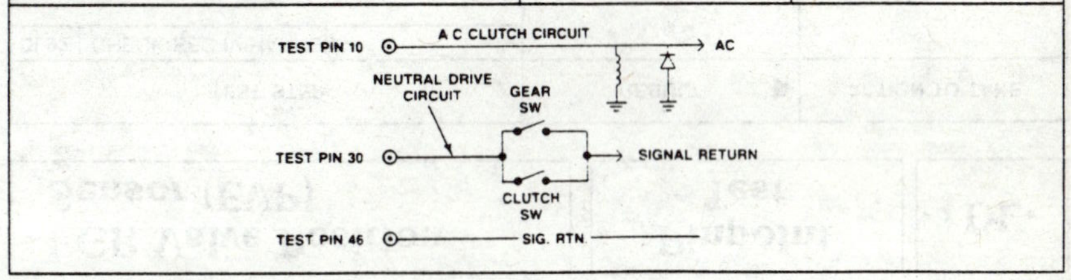

Neutral Drive Switch A/C Input — Pinpoint Test — FA

TEST STEP	RESULT ▶	ACTION TO TAKE
FA11 NEUTRAL/CLUTCH SWITCH CHECK — 1.6L. MTX		
• Key Off. • DVOM on 200 ohm scale. • Breakout box installed. • Locate neutral switch (on transaxle) and clutch switch (under dash). • Disconnect vehicle harness at both switches and measure resistance across each switch.	Less than 5 ohms at both switches ▶	SERVICE open(s) in vehicle. harness Neutral Drive and Signal Return circuits. REPEAT Quick Test.
	5 ohms or greater at either or both switches ▶	REPLACE open switch(es). RECONNECT harness and REPEAT Quick Test.
FA15 NEUTRAL INPUT CHECK — 2.3L TC M/T AND 1.6L TC SVO		
• Key Off, wait 10 seconds. • Verify A/C is Off, if so equipped. • Disconnect processor 60 Pin connector and inspect for damaged pins, corrosion, loose wires. Service as necessary. • Install Breakout box. • Connect processor. • DVOM on 200 ohm scale. • Measure resistance from test Pin 30 (Neutral input circuit) to test Pin 46 (Signal Return circuit).	Less than 5 ohms ▶	REPLACE processor and REPEAT Quick Test.
	5 ohms or greater ▶	SERVICE open in Neutral input or Signal Return circuit. REPEAT Quick Test.

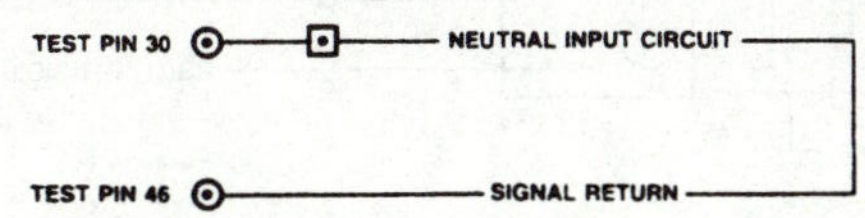

Neutral Drive Switch A/C Input — Pinpoint Test — FA

TEST STEP	RESULT ▶	ACTION TO TAKE
FA25 NEUTRAL INPUT CHECK — 4.9L M T, TRUCK		
• Key Off, wait 10 seconds. • Verify A/C is Off, if so equipped. • Disconnect processor 60 Pin connector and inspect for damaged pins, corrosion, loose wires. Service as necessary. • Install Breakout box. • Connect processor. • DVOM on 200 ohm scale. • Measure resistance from test Pin 46 (Neutral input circuit) to VIP connector.	Less than 5 ohms ▶	REPLACE processor. REPEAT Quick Test.
	5 ohms or greater ▶	SERVICE open in Neutral input circuit to VIP tester. REPEAT Quick Test.

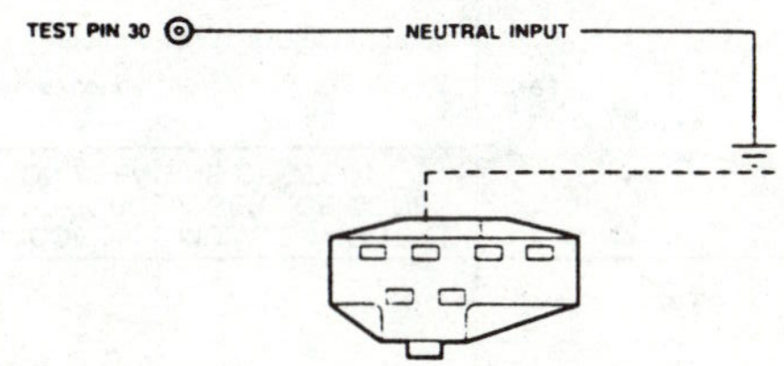

Idle Tracking Switch — Pinpoint Test — FB

FB

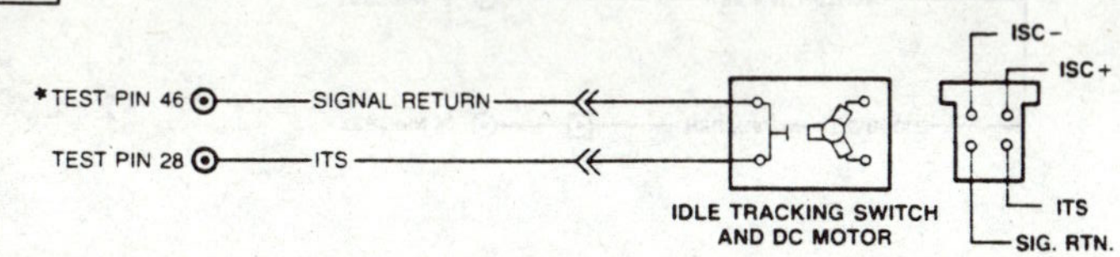

NOTE: FOR 3.8L CFI A 10-PIN CONNECTOR WAS USED FOR THE FIRST 60 DAYS OF PRODUCTION. EITHER CONNECTOR MAY BE ENCOUNTERED. TEST PIN ASSIGNMENTS DID NOT CHANGE ON BREAKOUT BOX. IF ITS MUST BE DISCONNECTED FOR DIAGNOSTICS. DISCONNECT THE 10-PIN CONNECTOR AT THE FUEL CHARGE ASSY.

3.8L CFI ONLY

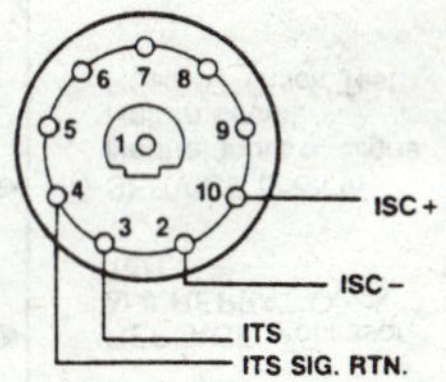

* TEST PIN LOCATED ON BREAKOUT BOX
ALL HARNESS CONNECTORS VIEWED INTO MATING SURFACE.

STOP-WARNING

You may have reached this point because of a fault in a non-EEC System and/or component. To prevent the replacement of good components. be aware that the following non-EEC areas may be at fault:

- Anti-diesel speed too high.
- Throttle sticking or on high cam.

This Pinpoint Test is intended to diagnose only the following:

- DC motor.
- Idle tracking switch.
- Harness circuits: ITS and Signal Return.
- Processor assembly.

Idle Tracking Switch — Pinpoint Test — FB

TEST STEP	RESULT ▶	ACTION TO TAKE
FB1 SERVICE CODES 58 AND 68: ITS FAULT ISOLATION		
• Is code 58 or 68 present?	Service code 58 ▶	GO to FB2.
	Service code 68 ▶	GO to FB5.
FB2 SERVICE CODE 58: SIMULATE THROTTLE CONTACT TO FORCE A SERVICE CODE 68 IN KEY ON — ENGINE OFF TEST		
• Key Off. wait 10 seconds. • Press on ISC DC motor shaft to simulate throttle contact. • With a force pushing on DC motor shaft, perform Key On. Engine Off Self-Test. **NOTE: Ignore all other codes at this point.** • Is code 68 present?	Yes ▶	GO to Pinpoint Test Step KF4.
	No ▶	GO to FB3.
FB3 IDLE TRACKING SWITCH ALWAYS "NOT TRACKING" (CLOSED)		
• Key Off, wait 10 seconds. • Disconnect harness from ISC. • Perform Key On. Engine Off self-test. **NOTE: Ignore all other codes at this point.** • Is code 68 present?	Yes ▶	REPLACE DC motor. REPEAT Quick Test.
	No ▶	GO to FB4.

Idle Tracking Switch — Pinpoint Test — FB

TEST STEP	RESULT ▶	ACTION TO TAKE
FB4 CHECK ITS CIRCUIT FOR SHORT TO GROUND		
• Key Off, wait 10 seconds. • Disconnect processor 60 Pin connector and inspect for damaged pins, corrosion, loose wires. Service as necessary. • Connect Breakout box. Leave processor disconnected. • Harness disconnected at ISC. • DVOM on 200,000 ohm scale. • Measure resistance of test Pin 28 to test Pins 40, 46 and 60.	Any resistance less than 10,000 ohms ▶ All resistances 10,000 ohms or more ▶	SERVICE short to ground. REPEAT Quick Test. REPLACE processor. REPEAT Quick Test.
FB5 SERVICE CODE 68: SIMULATE NO THROTTLE CONTACT TO FORCE A CODE 11 IN KEY ON — ENGINE OFF TEST		
• Key Off, wait 10 seconds. • Move throttle away from ISC DC motor shaft. • Perform Key On, Engine Off Self-Test. • Is code 68 present? **NOTE: Ignore all other codes at this point.**	Yes ▶ No ▶	GO to FB6. GO to Pinpoint Test Step KF4.

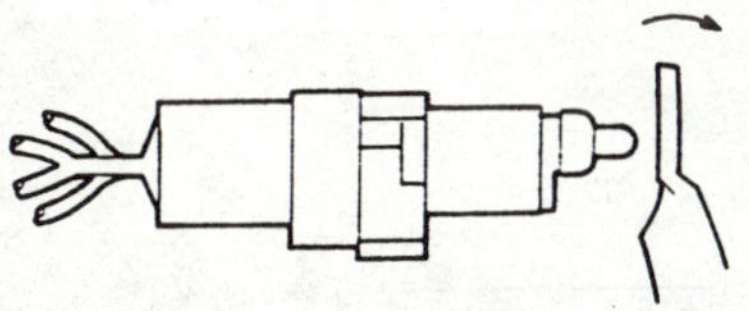

Idle Tracking Switch — Pinpoint Test — FB

TEST STEP	RESULT ▶	ACTION TO TAKE
FB6 SIMULATE A CLOSED CONTACT CONDITION IN KEY ON — ENGINE OFF TEST		
• Key Off, wait 10 seconds. • Disconnect harness from ISC. • Connect jumper between ISC harness connector Pins 3 and 4. • Install jumper as shown. • Perform Key On, Engine Off Self-Test. • Is code 68 present?	Yes ▶ No ▶	GO to FB7. REPLACE ISC DC motor. REPEAT Quick Test.

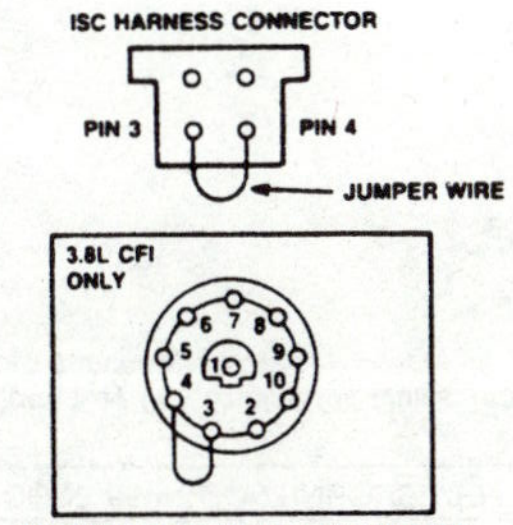

(REFER TO NOTE ON ILLUSTRATION FB)

CAUTION: Do not short any other pins.

TEST STEP	RESULT ▶	ACTION TO TAKE
FB7 CHECK ITS AND SIGNAL RETURN CIRCUITS FOR CONTINUITY		
• Key Off, wait 10 seconds. • Disconnect processor 60 Pin connector and inspect for damaged pins, corrosion, loose wires. Service as necessary. • Connect Breakout box to harness. Leave processor disconnected. • DVOM on 200 ohm scale. • Measure continuity of test Pin 46 to Signal Return circuit and test Pin 28 to ITS circuit.	Either resistance 5 ohms or greater ▶ Both resistances less than 5 ohms ▶	SERVICE faulty circuit. REPEAT Quick Test. REPLACE processor. REPEAT Quick Test.

Key Power Check — Pinpoint Test — FC

FC

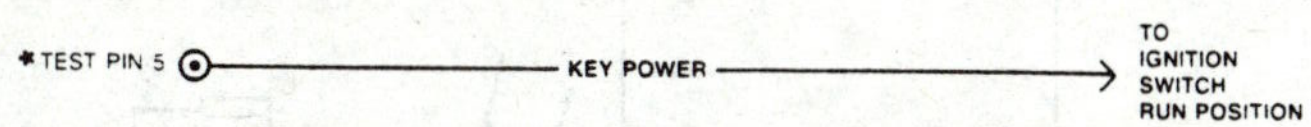

*TEST PINS LOCATED ON BREAKOUT BOX.
ALL HARNESS CONNECTORS VIEWED INTO MATING SURFACE.

STOP-WARNING

You may have reached this point because of a fault in a non-EEC System and/or component. To prevent the replacement of good components, be aware that the following non-EEC areas may be at fault:

- Charging system overvoltage.
- Battery charger connected with engine running.
- Jump starting.

This Pinpoint Test is intended to diagnose only the following:

- Harness circuit: key power.
- Processor.

Key Power Check — Pinpoint Test — FC

TEST STEP	RESULT ▶	ACTION TO TAKE
FC1 BATTERY VOLTAGE GREATER THAN 17.5V		
• Key Off, wait 10 seconds. • Install Breakout box with processor connected. • DVOM at 20V range and connected to test Pins 5 and 60. • Perform Engine Running Quick Test. Observe DVOM during test, record service codes.	DVOM exceeds 17.5V during Quick Test ▶	CORRECT charging system for over-voltage condition.
	DVOM remains below 17.5V and code 65 is present ▶	REPLACE processor and REPEAT Quick Test.
	DVOM remains below 17.5V and code 65 is not present ▶	GO to FC2.
FC2 CODE 65 IN CONTINUOUS MEMORY		
• Perform Key On, Engine Off Quick Test and record continuous codes.	Code 65 not present ▶	TESTING complete. CHARGING system OK at this time.
	Code 65 present ▶	CHECK charging system per Shop Manual Group 31 for cause of intermittent overcharging (greater than 17.5V).

Fuel Control | Pinpoint Test | HA

HA

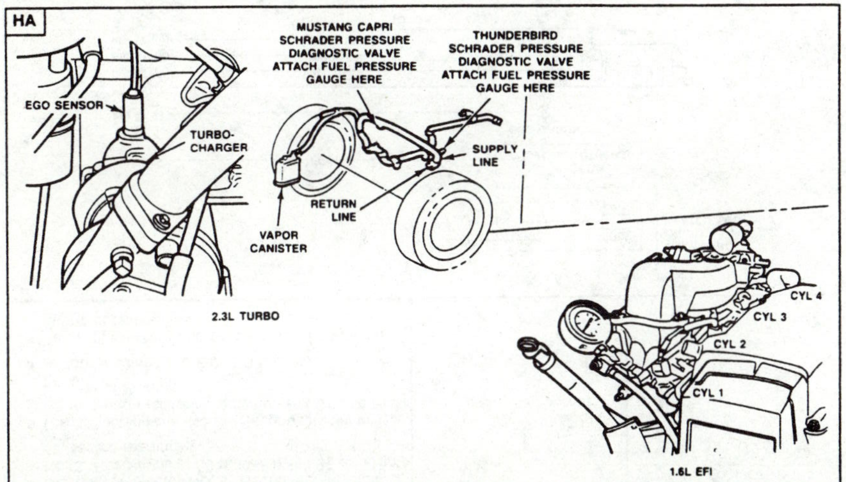

STOP-WARNING

You may have reached this point because of a fault in a non-EEC System and/or component. To prevent the replacement of good components, be aware that the following non-EEC areas may be at fault:

- Ignition Coil
- Distributor Cap
- Distributor Rotor
- Fouled Spark Plugs
- Spark Plug Wires
- CANP Problems
- PCV Valves (see note below)
- EGR Valve and Gasket
- Air Filter
- Fuel Contamination, Engine Oil
- Poor Power Ground
- Fuel Pressure
- Manifold Leaks, Intake/Exhaust
- Engine Not at Normal Operating Temperatures

This Pinpoint Test is intended to diagnose only the following:

- EGO Sensor
- Harness Circuits EGO Grd., EGO, INJA, INJB, V PWR
- EGO Sensor Connection
  - — Code 42 start at HA5
  - — Code 41 start at HA8
- Vacuum Systems
- Fuel Injectors
- Processor Assembly

NOTE: Fuel contaminated engine oil may affect 4-1, 4-2 Service Codes. If this is suspected, remove the PCV from the valve cover and repeat the Quick Test. If the problem is corrected, then change the engine oil and filter.

Fuel Control | Pinpoint Test | HA

TEST STEP	RESULT ▶	ACTION TO TAKE
HA1 FUEL PRESSURE CHECK		
• Key Off, wait 10 seconds. • Install fuel pressure gauge. • Start and run engine. Fuel pressure must be 172-310 kPa (25-45 psi). **For No Starts:** • If engine will not run, cycle the key from Off to On several times. • Fuel pressure must remain at 276 ± 34 kPa (40 ± 5 psi) for 60 seconds after final key Off.	Fuel pressure is within specification ▶	GO to HA2 .
	Fuel pressure is not within specification ▶	REFER to the Shop Manual Group 24 for electric fuel pump and fuel pressure regulator check.

Fuel Control — Pinpoint Test — HA

TEST STEP	RESULT ▶	ACTION TO TAKE
HA2 HARNESS INJECTOR RESISTANCE CHECK		
• Key Off, wait 10 seconds. • Disconnect processor 60 Pin connector and inspect for damaged pins, corrosion, loose wires. Service as necessary. • Connect Breakout box to harness. Leave processor disconnected. • Key Off. DVOM on 200 ohm scale. Disconnect injectors for cylinders 2, 3 and 4 electrically. • Through the vehicle harness, measure the resistance of injector No. 1 between Pin 37 and Pin 58, record reading. • Disconnect injector No. 1. Reconnect injector No. 2. Read resistance between Pin 37 and Pin 58, record reading. • Disconnect injector No. 2. Reconnect injector No. 3. Read resistance between Pin 37 and Pin 59, record reading. • Disconnect injector No. 3. Reconnect injector No. 4. Read resistance between Pin 37 and Pin 59, record reading. • All four readings are 2.0 to 3.5 ohms. • Verify all injectors and processor are connected at end of testing.	Yes ▶ No ▶	GO to HA3 SERVICE the harness connectors on the suspect injector for opens or shorts. If OK, REPLACE injector. REPEAT Quick Test.

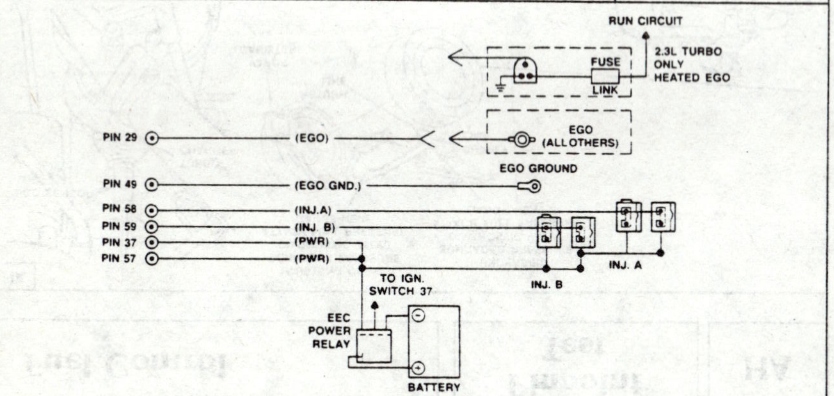

Fuel Control — Pinpoint Test — HA

TEST STEP	RESULT ▶	ACTION TO TAKE
HA3 INJECTOR DRIVE SIGNAL CHECK		
Requires standard non-powered 12 volt test lamp. • Key Off. • Breakout box installed. • Connect processor to Breakout box. • Refer to illustration HA2. • Connect test lamp between Breakout box Pin 37 and Pin 58. • Crank or start engine. • Repeat above test between Breakout box Pin 37 and Pin 59.	Dim glow at light on both tests ▶ No light on one or both tests ▶ Bright light on one or both tests ▶	GO to HA4 VERIFY 12-volt battery power at Pins 37 and 57. If OK, REPLACE processor. REPEAT Quick Test. CHECK circuits 95 and 96 for shorts to ground. If OK, REPLACE processor. REPEAT Quick Test.
HA4 INJECTOR BALANCE TEST		
• Connect tachometer to engine. Run engine at 2000 rpm. It may be necessary to disconnect ISC and use throttle body stop screw to set engine speed. • Disconnect and reconnect the injectors one at a time; note rpm drop for each injector. • Does each injector produce at least a 150 rpm drop? • Reset curb idle. Refer to VECI decal. Verify ISC connection at end of testing.	Yes ▶ No ▶	Fuel delivery OK. Problem is with area common to all cylinders, i.e.: Air/vacuum leak, fuel contamination, EGR. REPLACE faulty injector. REPEAT Quick Test.

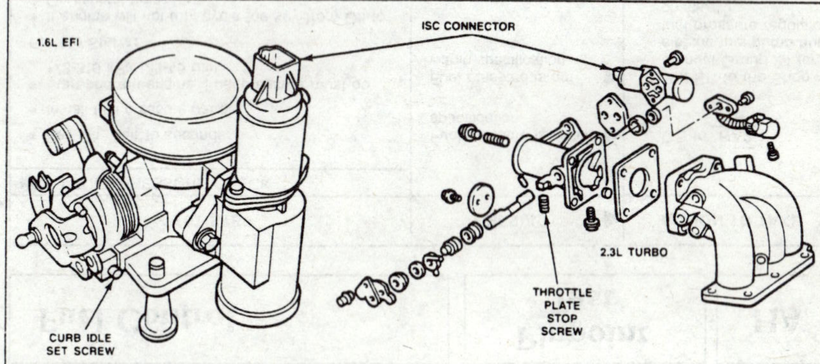

Fuel Control — Pinpoint Test — HA

TEST STEP	RESULT ▶	ACTION TO TAKE
HA5 FUEL CONTROL — ALWAYS RICH: CODE 42		
• Key Off, wait 10 seconds. • **NOTE: Non-EEC areas could cause a code 42. Check for:** **— Fuel contaminated engine oil** **— Ignition caused misfire** **— CANP problems** • Refer to illustration HA. • Disconnect vehicle harness at the EGO sensor. Using a jumper wire, ground vehicle harness EGO circuit at the EGO sensor to the engine block. • Repeat Engine Running Self-Test. • Service code 41 results.	Yes ▶ No ▶	GO to HA7. GO to HA6.
HA6 HARNESS CHECK		
• Key Off, wait 10 seconds. • Refer to illustration HA. • Install Breakout box. • Check continuity of EGO ground circuit between Breakout box Pin 49 and EGO ground at engine block. • Check continuity of EGO circuit between Breakout box Pin 29 and EGO harness connector. • Both circuits less than 5 ohms.	Yes ▶ No ▶	DISCONNECT processor connector. INSPECT for damage or corrosion. If OK, REPLACE processor. REPEAT Quick Test. CORRECT harness circuit with resistance greater than 5 ohms. REPEAT Quick Test.

Fuel Control — Pinpoint Test — HA

TEST STEP	RESULT ▶	ACTION TO TAKE
HA7 EGO CHECK		
• Refer to illustration HA. • DVOM on 20V scale. • With EGO sensor disconnected from the harness, connect a DVOM from EGO sensor to engine ground. • Run the engine at 2000 rpm for 2 minutes. While observing DVOM, disconnect the manifold vacuum hose indicated below. • Meter indicates less than 0.4V.	Yes ▶ No ▶	EGO sensor OK. GO to HA1. REPLACE EGO sensor. REPEAT Quick Test.

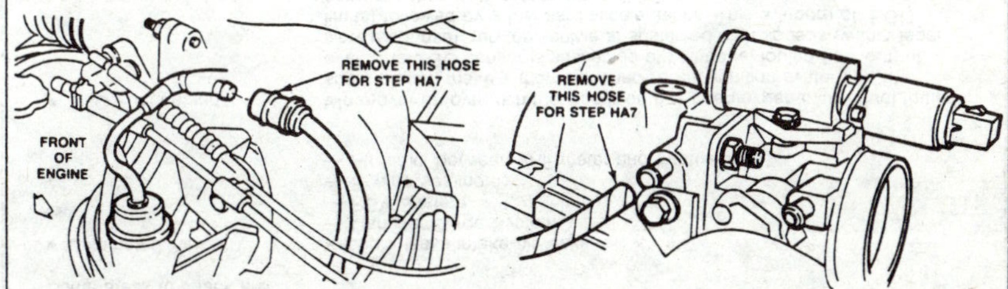

TEST STEP	RESULT ▶	ACTION TO TAKE
HA8 FUEL CONTROL — ALWAYS LEAN: CODE 41		
NOTE: Vacuum/air leaks in non-EEC areas could cause a code 41. Check for: **— Leaking vacuum actuator (eg: A/C control motor)** **— Engine sealing** **— EGR system** **— PCV system** **— Unmetered air leak between air meter and throttle body** **— Lead contaminated EGO sensor** • Refer to illustration HA. • Key Off. DVOM on 20V scale. Disconnect EGO sensor from vehicle harness. Connect DVOM to EGO sensor and engine ground. Remove air cleaner to gain access to air meter inlet. Using a standard wood lead pencil, prop the air meter door part-way open. • Start the engine and run at approximately 2000 rpm for 2 minutes. • Does the DVOM read greater than 0.5V at the end of 2 minutes?	Yes ▶ No ▶	GO to HA9. REPLACE EGO sensor. REPEAT Quick Test.

Fuel Control — Pinpoint Test — HA

TEST STEP	RESULT ▶	ACTION TO TAKE
HA9 HARNESS CHECK EGO CIRCUITS		
• Key Off, wait 10 seconds. • Install Breakout box. Processor disconnected. • Check continuity of EGO circuit from Breakout box Pin 49 to engine block ground. Also EGO circuit from EGO connector to Breakout box Pin 29. • Both circuits less than 5 ohms.	Yes ▶	GO to HA10.
	No ▶	SERVICE/CORRECT as necessary. REPEAT Quick Test.
HA10 CHECK EGO CIRCUIT FOR SHORT TO GROUND		
• Key Off. • Breakout box installed. Processor disconnected. • DVOM to 200,000 ohms range. • Disconnect EGO sensor at harness. • Measure the resistance between Breakout box Pin 29 and Pin 40.	Reading is 10,000 ohms or greater ▶	EGO input circuit OK. DISCONNECT processor connector. INSPECT for corrosion or damaged pins. If OK, REPLACE processor. REPEAT Quick Test.
	Reading is less than 10,000 ohms ▶	CORRECT cause of resistance to ground. REPEAT Quick Test.

Fuel Control — Pinpoint Test — HA

HA11 CONTINUOUS TESTING: CODE 41/42

41 — EGO indicated the fuel system was lean for more than 15 seconds when the system should have been in closed loop fuel control.

42 — EGO indicated the fuel system was rich for more than 15 seconds when the system should have been in closed loop fuel control.

*CLOSED LOOP — Fuel control under the influence of the EGO sensor.

*OPEN LOOP — Fuel control NOT under the influence of the EGO sensor.

Before attempting to correct a fuel control code, 41/42, diagnose all other drive complaints first, eg., rough idle, misses, etc.

NOTE: The fuel control code may help in this diagnosis.

Using the fuel control service code, isolate the cause of the fuel control problem.

Some areas to check are:

- Unmetered Air: Vacuum leaks/intake air leaks.
  - — Canister Purge System.
  - — PCV System.
  - — Engine sealing.
  - — Air leaks between VAF meter and throttle body.
- EGO Fuel Fouled: Whenever an over-rich fuel condition has been experienced (fuel fouled spark plugs), make a thorough check of the ignition system. In the event the EGO sensor is suspected of being fuel fouled (low output, slow response), run the vehicle at sustained high speeds (within legal limits) followed by a few hard accelerations. This will burn off EGO contamination and restore proper EGO operation.
- Fuel Pressure: Perform Pinpoint Test Step HA1.
- Ignition System: Always in default spark (10 degrees). Refer to Quick Test Step 4.0.
- Improper Fueling: Lead fouled EGO sensor.
- TP Sensor: Not moving (mechanical damage). Connect DVOM to test Pin 47 and to test Pin 46. Key to Run. Observe DVOM while moving the throttle. Reading must increase with increase in throttle opening. If not correct, REPLACE as necessary.
- If at this point, the drive concern is still present, perform Steps HA2 through HA4 only.

Fuel Control — Pinpoint Test — HA

TEST STEP	RESULT ▶	ACTION TO TAKE
HA21 CHECK HEATER ELEMENT RESISTANCE ON HEATED EGO		
• Key Off. Engine Off. • DVOM on 200 ohm scale. • Measure resistance between RUN circuit and ground on heated EGO connector at room temperature.	Reading between 2.5 and 5.0 ohms ▶	GO to HA22.
	Reading less than 2.5 or greater than 5.0 ohms ▶	REPLACE HEGO sensor.
HA22 CHECK FOR POWER AT HEGO HARNESS CONNECTOR		
• Key On, Engine Off. • DVOM on 20V scale. • Connect positive lead to run circuit and negative lead to ground on HEGO harness connector.	Reading 10.5V or greater ▶	HEGO sensor system OK, RETURN to Section 2.
	Reading less than 10.5V ▶	GO to HA23.
HA23 CHECK CONTINUITY OF GROUND TO HEGO CONNECTOR		
• Key Off, wait 10 seconds. • DVOM on 200 ohm scale. • Measure continuity of ground circuit from HEGO connector to battery ground.	Reading less than 5.0 ohms ▶	SERVICE open in run circuit.
	Reading 5.0 ohms or greater ▶	SERVICE open in ground circuit.

Fuel Control (FBC) — Pinpoint Test — HB

HB

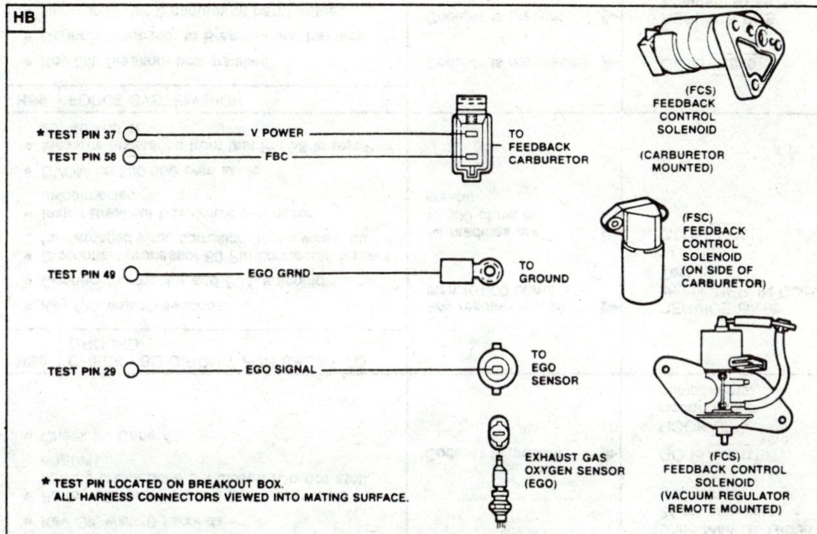

* TEST PIN LOCATED ON BREAKOUT BOX.
ALL HARNESS CONNECTORS VIEWED INTO MATING SURFACE.

STOP-WARNING

You may have reached this point because of a fault in a non-EEC System and or component. To prevent the replacement of good components, be aware that the following non-EEC areas may be at fault:

- Fouled Spark Plugs
- Distributor Cap and Rotor
- Ignition Coil
- Spark Plug Wires
- CANP Problems
- PCV Valve
- Oil Contamination
- EGR Valve and Gasket
- Air Filter
- Fuel Filter
- Fuel Pump and Pressure
- Manifold Leak, Intake Exhaust
- Engine Not at Normal Operating Temperature
- Carburetor
- Choke, Sticking Electrical

This Pinpoint Test is intended to diagnose only the following:

- EGO Sensor Connection
- EGO Sensor
- Harness Circuits: V Pwer, FBC, EGO Control, EGO Signal
- Impact of Thermactor Air
- FBC Solenoid
- Vacuum System (EEC Only)
- Choke Circuit
- Processor Assembly

Fuel Control (FBC) — Pinpoint Test — HB

TEST STEP	RESULT ▶	ACTION TO TAKE
FAULT CODE 41		
HB1 VERIFY VEHICLE PREPARATION		
• Did vehicle run for more than 2 minutes before performing Quick Test?	Yes ▶	GO to HB2.
• For 2.3L HSC verify canister purge shut-off valve was plugged per Quick Test Step 5.1.	No ▶	PREPARE vehicle. RERUN Quick Test.
HB2 CHECK FOR OBVIOUS FAULTS		
• Check EGO sensor connection. • Check EGO ground for good connection. • Check for intake or exhaust manifold leaks. • Are all OK?	▶	For 2.3L HSC engine, GO to HB3.
	Yes ▶	All others GO to HB4.
	No ▶	SERVICE fault. RERUN Quick Test.
HB3 CHECK VACUUM		
• Key Off. • Check all FCS vacuum hoses to carburetor for leaks, blockages, and proper connection. • Refer to VECI decal. • Are all vacuum hoses OK?	Yes ▶	GO to HB4.
	No ▶	SERVICE or REPLACE faulty hoses. RERUN Quick Test.
HB4 CHECK EGO SIGNAL FOR SHORT		
• Key Off, wait 10 seconds. • Disconnect harness from EGO sensor and processor. • DVOM on 200,000 ohm scale. • Measure resistance of EGO sensor harness connector EGO signal to vehicle battery negative terminal.	Reading is less than 10,000 ohms ▶	SERVICE circuit short. RERUN Quick Test.
	Reading is 10,000 ohms or greater ▶	CONNECT EGO processor and sensor. GO to HB5.
HB5 TRY TO FORCE RICH BY REMOVING AIR SUPPLY		
• Key Off. • Reconnect EGO sensor and processor to harness. • Disconnect thermactor air supply hose at the air pump and cap the hose. • Run engine running Quick Test. **NOTE: Because thermactor air supply is disconnected a Code 11 does not indicate a pass. Treat Code 11 as a Code 44.**	Code 41 is present ▶	GO to HB6. Leave thermactor disconnected.
	Code 44 is present ▶	RETURN to Quick Test Step 5.1.
	Ignore all other service codes at this time	

Fuel Control (FBC) — Pinpoint Test — HB

TEST STEP	RESULT ▶	ACTION TO TAKE
HB6 CHECK FCS SOLENOID		
• Key Off, wait 10 seconds. • Thermactor disconnected. • Disconnect harness from FCS solenoid. • Perform Engine Running Quick Test. • Check for Code 41. **NOTE: Ignore all other codes at this time.**	Code 41 is not present ▶	GO to HB8.
	Code 41 is present ▶	GO to HB7.
HB7 FORCE SYSTEM RICH		
• Key Off, wait 10 seconds. • FCS and thermactor disconnected. • Run for 2 minutes at part throttle. • Key Off, wait 10 seconds. • Perform Engine Running Quick Test while holding choke plate 3/4 closed **(Do not stall engine).** • Check for Code 41.	Code 41 is not present ▶	REPLACE FCS. RERUN Quick Test if Code 41 is still present. REFER to Shop Manual, Group 24 for carburetor diagnostics.
	Code 41 is present ▶	GO to HB11. HOOK-UP all disconnected components.
HB8 CHECK FBC CIRCUIT FOR SHORT TO GROUND		
• Key Off, wait 10 seconds. • Connect thermactor and FCS solenoid. • Disconnect processor 60 Pin connector, inspect for damaged pins, corrosion, loose wires, etc. • Install Breakout box, leave processor disconnected. • DVOM on 200,000 ohm scale. • Measure resistance from test Pin 58 to test Pins 20, 40 and 46.	Any reading is less than 10,000 ohms ▶	SERVICE circuit shorts. RERUN Quick Test.
	All readings are 10,000 ohms or greater ▶	GO to HB9.
HB9 FORCE SYSTEM RICH		
• Key Off, Breakout box installed. • Connect processor to Breakout box harness. • Run vehicle for 2 minutes at part throttle. • Key Off, wait 10 seconds. • Perform Engine Running Quick Test while holding choke plate 3/4 closed **(Do not stall engine).** • Check for Code 41.	Code 41 is not present ▶	GO to HB10.
	Code, 41 is present ▶	REPLACE FCS solenoid. REMOVE Breakout box. RECONNECT FCS and processor. RERUN Quick Test.

Fuel Control (FBC) — Pinpoint Test HB

TEST STEP	RESULT ▶	ACTION TO TAKE
HB10 CHECK FCS SOLENOID		
• Key Off, wait 10 seconds. • Breakout box installed with processor connected. • Disconnect harness from FCS solenoid. • DVOM on 2,000 ohm scale. • Measure resistance between FCS FBC circuit terminal and battery negative terminal. • Measure resistance between FCS V power circuit terminal and battery negative terminal.	Either reading is less than 1,000 ohms ▶	REPLACE FCS solenoid. REMOVE Breakout box. RECONNECT processor and FCS. RERUN Quick Test.
	Both readings are 1,000 ohms or greater ▶	REMOVE Breakout box. REPLACE processor. RECONNECT FCS and processor. RERUN Quick Test.
HB11 CHECK EGO CIRCUIT FOR SHORT		
• Key Off, wait 10 seconds. • Thermactor and FCS reconnected. • Disconnect harness from EGO sensor. • DVOM on 200,000 ohm scale. • Measure resistance of EGO harness connector EGO circuit to battery negative terminal.	Reading is less than 10,000 ohms ▶	REPLACE processor. RECONNECT EGO sensor and processor. RERUN Quick Test.
	Reading is 10,000 ohms or greater ▶	GO to HB12.
HB12 FORCE EGO RICH		
• Key Off, EGO sensor disconnected. • Run engine for 2 minutes. • DVOM on 20V scale. • With engine running hold choke plate 3 4 closed **(Do not stall engine).** • Measure voltage from EGO sensor pig tail terminal to battery negative terminal.	Reading is less than .45V ▶	REPLACE EGO sensor. RERUN Quick Test.
	Reading is .45V or greater ▶	GO to HB13.
HB13 CHECK EGO CIRCUIT FOR SHORT TO VREF		
• Key On, Engine Off, EGO disconnected. • DVOM on 20V scale. • Measure voltage from harness EGO circuit to battery negative terminal.	Reading is less than 2.0V ▶	GO to HB14.
	Reading is 2.0V or greater ▶	SERVICE circuit short. RECONNECT EGO sensor. RERUN Quick Test.

Fuel Control (FBC) — Pinpoint Test HB

TEST STEP	RESULT ▶	ACTION TO TAKE
HB14 CHECK CONTINUITY EGO AND EGO GROUND CIRCUITS		
• Key Off, wait 10 seconds. • EGO sensor disconnected. • Disconnect processor 60 Pin connector, inspect for damaged pins, corrosion, loose wires, etc. Service as necessary. • Install Breakout box, leave processor disconnected. • DVOM on 200 ohm scale. • Measure resistance from test Pin 29 to EGO sensor harness EGO circuit. • Measure resistance from test Pin 49 to test Pins 40 and 60.	All readings are less than 5 ohms ▶	REPLACE processor. REMOVE Breakout box. RECONNECT EGO and processor. RERUN Quick Test.
	Any reading is 5 ohms or greater ▶	SERVICE circuit open. REMOVE Breakout box. RECONNECT processor and EGO. RERUN Quick Test.
FAULT CODES 42/47		
HB20 CHECK CHOKE		
• Does choke function properly without sticking or binding? **NOTE: For 2.3L HSC engine verify canister purge shut-off valve was plugged per Quick Test Step 5.1.**	No ▶	SERVICE choke as required. REFER to section 4. RERUN Quick Test.
	Yes ▶	GO to HB21.
HB21 GENERATE KNOWN FAULT		
• Run engine for 2 minutes at part throttle. • Key Off, wait 10 seconds. • Disconnect harness from EGO sensor. Jumper harness EGO circuit to ground. • Rerun Engine Running Quick Test. • Check for Code 41, ignore all other codes at this time.	Code 41 is present ▶	For 2.3L HSC engine, GO to HB22. All others, GO to HB23.
	Code 41 is not present ▶	GO to HB30.
HB22 CHECK VACUUM HOSES		
• Check all vacuum hoses from the FCS solenoid to the carburetor for leaks, blockages, and proper connection. • Refer to VECI decal. • Are vacuum hoses OK?	Yes ▶	RECONNECT all hoses to carburetor. GO to HB23.
	No ▶	REMOVE jumper wire and SERVICE as necessary. RERUN Quick Test.

Fuel Control (FBC) — Pinpoint Test — HB

TEST STEP	RESULT ▶	ACTION TO TAKE
HB23 FORCE SYSTEM LEAN		
• Key Off, wait 10 seconds. • Reconnect EGO sensor. • Disconnect a 7/32 vacuum line at the intake manifold vacuum tree. • Perform Engine Running Quick Test. If engine stalls RERUN test while manually maintaining approximately 1500 rpm. • Is code 42/47 still present? Disregard other codes at this time.	Yes ▶	REPLACE EGO sensor. RECONNECT vacuum hose. RERUN Quick Test.
	No ▶	GO to HB24.
HB24 CHECK FCS SOLENOID		
• Key Off, wait 10 seconds. • Vacuum hose reconnected. • Disconnect FCS solenoid. • DVOM on 200 ohm scale. • Measure resistance between terminals on the FCS solenoid. • Is resistance within the following ranges: 15 to 30 ohms: 2150 carburetor 30 to 60 ohms: All other	Yes ▶	RECONNECT FCS. GO to HB25.
	No ▶	REPLACE FCS. RERUN Quick Test.
HB25 REFER TO APPENDIX, ENTER OUTPUT STATE CHECK		
• Key Off, wait 10 seconds. • DVOM on 20V scale. • Connect DVOM negative test lead to STO and positive test lead to battery positive. • Jumper STI to signal return. • Perform Key On, Engine Off Self-Test until the completion of the Continuous Test Codes. • DVOM will indicate 0 volts. • Depress and release the throttle. • Did DVOM reading change to a high voltage reading?	Yes ▶	REMAIN in output state check. GO to HB26.
	No ▶	DEPRESS throttle to WOT and RELEASE. If STO voltage does not go high GO to Pinpoint Test Step Q40.

Fuel Control (FBC) — Pinpoint Test — HB

TEST STEP	RESULT ▶	ACTION TO TAKE
HB26 CHECK FCS OPERATION		
• Key On. in output state check. • DVOM on 20V scale. • Connect DVOM to FBC circuit on the FCS connector and to battery negative post. • While observing DVOM, depress and release the throttle several times to cycle output On and Off. • Does FCS cycle On and Off?	Yes ▶	2.3L HSC engine: GO to HB27.
	▶	All others: EXIT output state check. GO to Pinpoint Test Step KD1.
	No ▶	EXIT output state check. GO to HB28.
HB27 CHECK SOLENOID VACUUM CYCLING		
• Key On, Engine Off. • Disconnect input vacuum source from FCS and plug. • Connect a vacuum pump to FCS source vacuum port. • Disconnect the idle vacuum line from FCS. • Connect vacuum gauge to FCS idle vacuum port. • Apply 20 in. Hg while cycling outputs On and Off (by depressing and releasing throttle). • Observe the vacuum gauge.	Vacuum output cycles On and Off in less than 2 seconds ▶	RECONNECT all vacuum lines. REFER to Shop Manual, Group 24.
	Vacuum output does not cycle On or Off in less than 2 seconds ▶	REPLACE FCS. RECONNECT all vacuum lines. RERUN Quick Test.
HB28 CHECK V PWR CIRCUIT FOR POWER		
• Key Off, wait 10 seconds. • Deactivate Quick Test. • Disconnect harness from FCS. • DVOM on 20V scale. • Key On, Engine Off. • Measure voltage of FCS harness connector V PWR circuit to battery negative terminal.	Reading is less than 10V ▶	RECONNECT FCS solenoid. SERVICE circuit open. RERUN Quick Test.
	Reading is 10V or greater ▶	GO to HB29.

Fuel Control (FBC) — Pinpoint Test — HB

TEST STEP	RESULT ▶	ACTION TO TAKE
HB29 CHECK CONTINUITY OF FBC CIRCUIT		
• Key Off, wait 10 seconds. • FCS disconnected. • Disconnect processor 60 Pin connector, inspect connectors for damaged pins, corrosion, loose wires, etc. Service as necessary. • Install Breakout box, leave processor disconnected. • DVOM on 200 ohm scale. • Measure resistance from FCS harness connector FBC circuit to test Pin 58.	Reading is less than 5 ohms ▶	REMOVE Breakout box. RECONNECT FCS solenoid. REPLACE processor and RERUN Quick Test.
	Reading is 5 ohms or greater ▶	SERVICE circuit open. REMOVE Breakout box. RECONNECT processor and FCS solenoid. RERUN Quick Test.
HB30 CHECK EGO CIRCUIT FOR SHORT TO POWER		
• Key Off, wait 10 seconds. • Reconnect EGO sensor. • Disconnect processor 60 pin connector. Inspect for damaged pins, corrosion, loose wires, etc. Service as necessary. • Install Breakout box, leave processor disconnected. • DVOM on 20V scale. • Key On, Engine Off. • Measure voltage from test Pin 29 to battery negative terminal.	Reading is less than .4V ▶	REMOVE Breakout box. REPLACE processor. RERUN Quick Test. If Code 42 is still present GO to Pinpoint Test Step KD1. Except 2.3L HSC engine: GO to Shop Manual, Group 24.
	Reading is .4V or greater ▶	SERVICE circuit short. REMOVE Breakout box. RECONNECT processor. RERUN Quick Test.

Fuel Control (FBC) — Pinpoint Test — HB

TEST STEP	RESULT ▶	ACTION TO TAKE
FAULT CODE 43		
HB40 WARM ENGINE		
• Run vehicle at 2000 rpm for 2 minutes. • Key Off, wait 10 seconds. • Perform Engine Running Quick Test.	Code 43 is not present ▶	SERVICE other codes as necessary.
	Code 43 is present ▶	GO to HB41.
HB41 CHECK EXHAUST LEAKS		
• Are there any exhaust leaks at manifold or inlet pipe?	Yes ▶	SERVICE as necessary. RERUN Quick Test.
	No ▶	GO to HB42.
HB42 CHECK IDLE QUALITY		
• Start vehicle, let engine idle. • Does engine idle quality deteriorate and remain poor during and after the Quick Test?	Yes ▶	REFER to Carburetors, Section 4.
	No ▶	REPLACE EGO sensor. RERUN Quick Test.

Fuel Control — CFI | Pinpoint Test | HC

STOP-WARNING

You may have reached this point because of a fault in a non-EEC System and/or component. To prevent the replacement of good components, be aware that the following non-EEC areas may be at fault:

- Ignition Coil
- Distributor Cap
- Distributor Rotor
- Fouled Spark Plugs
- Spark Plug Wires
- CANP Problems
- PCV Valves (see note below)
- EGR Valve and Gasket
- Air Filter
- Fuel Contamination, Engine Oil
- Poor Power Ground
- Fuel Pressure
- Manifold Leaks, Intake/Exhaust
- Engine Not at Normal Operating Temperatures

This Pinpoint Test is intended to diagnose only the following:

- EGO Sensor
- EGO Signal and Ground Circuits
- EGO Sensor Connection
- Vacuum Systems
- Fuel Injectors
- Processor Assembly
- Thermactor Impact

NOTE: Fuel-contaminated engine oil may affect 4-1, 4-2 and 4-3 Service Codes, so if it is suspected, remove the PCV from the valve cover, and rerun the Quick Test. If the problem is corrected, then change the engine oil and filter.

TEST STEP		ACTION TO TAKE
HC1 SYSTEM IDENTIFICATION		
Is engine 5.0L CFI?	▶	GO to HC2.
Is engine 3.8L CFI?	▶	GO to HC20.

Fuel Control — 5.0L CFI | Pinpoint Test | HC

HC2

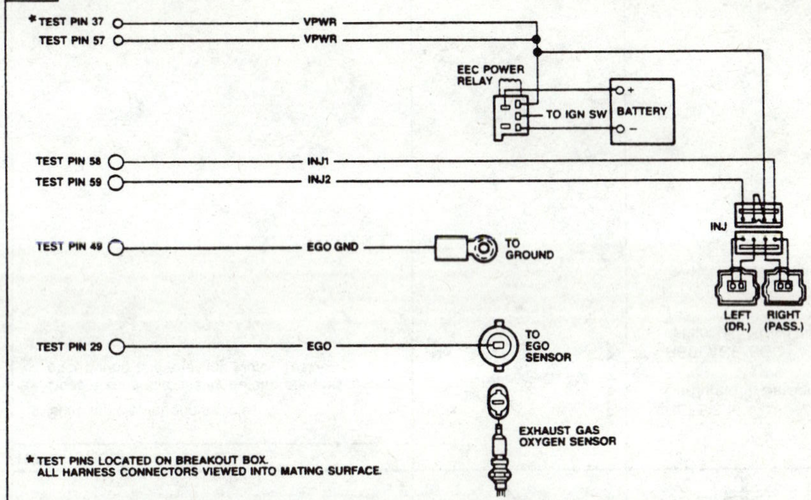

TEST STEP	RESULT		ACTION TO TAKE
HC2 WARM ENGINE			
• Run vehicle at 2000 rpm for 2 minutes. • Key Off, wait 10 seconds. • Perform Engine Running Quick Test.	Code 43 is not present	▶	SERVICE other codes as necessary.
	Code 43 is present	▶	GO to HC3.
HC3 CHECK EXHAUST LEAKS			
• Are there any exhaust leaks at manifold or inlet pipe?	Yes	▶	SERVICE as necessary. RERUN Quick Test.
	No	▶	GO to HC4.

Fuel Control — 5.0L CFI | Pinpoint Test | HC

TEST STEP	RESULT ▶	ACTION TO TAKE
FAULT CODE 43		
HC4 CHECK IDLE QUALITY		
• Start vehicle. let engine idle. • Does engine idle quality deteriorate and remain poor during and after the Quick Test?	Yes ▶	REFER to EFI/CFI, Section 4.
	No ▶	REPLACE EGO sensor. RERUN Quick Test.
HC5 FUEL CONTROL — ALWAYS LEAN. CODE 41		
• Key Off. DVOM to 20V range. Disconnect EGO sensor from vehicle harness. Connect DVOM to EGO sensor and engine ground. • Start the engine and run at approximately 2000 rpm for 2 minutes. • Does the DVOM read greater than 0.5V at the end of 2 minutes?	Yes ▶	GO to HC6.
	No ▶	REPLACE EGO sensor. REPEAT Quick Test.
HC6 EGO AND EGO GROUND CONTINUITY CHECK		
• Key Off, wait 10 seconds, EGO disconnected. • DVOM on 2000 ohm scale. • Disconnect processor 60 Pin connector and inspect for damaged pins, corrosion, loose wires. Service as necessary. • Connect Breakout box. Leave processor disconnected. • Measure continuity from test Pin 29 to EGO sensor connector. • Measure continuity from test Pin 49 to test Pins 40 and 60.	Both resistance readings less than 5 ohms ▶	GO to HC7.
	Either resistance reading 5 ohms or greater ▶	SERVICE harness open in ground circuits. REPEAT Quick Test.
HC7 CHECK EGO CIRCUIT FOR SHORT TO GROUND		
• Key Off. • Breakout box installed. Processor disconnected. • DVOM to 200,000 ohms range. • Disconnect EGO sensor at harness. • Measure the resistance between Breakout box Pin 29 and Pin 40.	Reading is 10,000 ohms or greater ▶	EGO input circuit OK. GO to HC11.
	Reading is less than 10,000 ohms ▶	CORRECT cause of resistance to ground. REPEAT Quick Test.

Fuel Control — 5.0L CFI | Pinpoint Test | HC

TEST STEP	RESULT ▶	ACTION TO TAKE
HC8 TRY TO FORCE A CODE 41		
• Key Off. • Disconnect vehicle harness at the EGO sensor. Using a jumper wire, ground EGO circuit in the harness connector to the engine block. • Rerun Engine Running Test. • Does Service Code 41 result? • Remove jumper wire when completed.	Yes ▶	GO to HC10.
	No ▶	GO to HC9.
HC9 HARNESS CONTINUITY CHECK		
• Key Off. wait 10 seconds, EGO sensor disconnected. • DVOM on 2000 ohm scale. • Disconnect processor 60 Pin connector and inspect for damaged pins, corrosion, loose wires. Service as necessary. • Connect Breakout box. Leave processor disconnected. • Measure continuity from test Pin 29 to EGO sensor connector. • Measure continuity from test Pin 49 to EGO ground terminal on the engine.	Both resistance readings less than 5 ohms ▶	DISCONNECT module harness connector. INSPECT for damage or corrosion. If OK, REPLACE processor.
	Either resistance reading 5 ohms or greater ▶	SERVICE harness open circuit. REPEAT Quick Test.
HC10 FORCED LEAN EGO TEST		
• Key Off. • DVOM on 20V range. • Disconnect the PCV hose at the intake manifold port. • With EGO sensor disconnected from the harness, connect a DVOM from EGO sensor to engine ground. • Run the engine at 2000 rpm for 2 minutes while observing DVOM.	Less than 0.4V ▶	EGO sensor OK. GO to HC11.
	0.4V or greater ▶	REPLACE EGO sensor and RETEST.
HC11 FUEL PRESSURE CHECK		
• Install fuel pressure gauge. • Start and Run engine. Observe gauge pressure reading.	Fuel pressure is within 24-45 psi ▶	GO to HC12.
	Fuel pressure is less than 24 psi or greater than 45 psi ▶	REFER to the Shop manual for electric fuel pump and fuel regulator check.

Fuel Control — 5.0L CFI | Pinpoint Test | HC

TEST STEP	RESULT	ACTION TO TAKE
HC12 CHECK RESISTANCE OF INJECTORS HARNESS AND PROCESSOR CONNECTION		
• Key Off. • DVOM on 200 ohm range. • Disconnect processor 60 Pin connector and inspect for damaged pins, corrosion, loose wires. Service as necessary.	Resistance is less than 1.5 ohms or greater than 3.5 ohms ▶	SERVICE the harness/connectors on the suspect injector for opens or shorts. If OK, REPLACE injector.
• Using DVOM test leads measure resistance from test Pins 57 to 58 and test Pins 57 to 59.	Resistance is 1.5-3.5 ohms ▶	GO to HC13.
HC13 INJECTOR DRIVE SIGNAL CHECK		
• Requires standard non-powered 12V test lamp. • Key Off, Breakout box installed, reconnect processor.	Dim glow at light on both tests ▶	Normal. GO to HC14.
• Connect test lamp from test Pin 57 to test Pin 59. • Crank or start engine. • Repeat above test from test Pin 57 to test Pin 58.	No light on one or both tests ▶	VERIFY 12V battery power at test Pin 57. If OK, REPLACE processor.
• How brightly does the test lamp glow?	Bright light on one or both tests ▶	CHECK both injector circuits for shorts to ground. If OK, REPLACE processor.
HC14 INJECTOR BALANCE TEST		
• Connect tachometer to engine. Run engine at 2000 rpm. May be necessary to disconnect TK solenoid and use throttle body stop screw to set engine speed. • Disconnect and reconnect the injectors one at a time; note rpm drop for each injector.	Yes ▶	Fuel delivery OK. Problem is with area common to all cylinders, i.e.: Air/vac leak, fuel contamination, EGR.
• Does each injector produce at least a 100 rpm drop? • Reset curb idle and verify TK connection at end of testing.	No ▶	REPLACE suspect injector.

Fuel Control — 3.8L CFI | Pinpoint Test | HC

HC20A GENERAL INSTRUCTIONS

STOP-WARNING

You may have reached this point because of a fault in a non-EEC System and/or component. To prevent the replacement of good components, be aware that the following non-EEC areas may be at fault:

- Ignition Coil
- Distributor Cap
- Distributor Rotor
- Fouled Spark Plugs
- Spark Plug Wires
- CANP Problems
- PCV Valves (see note below)
- EGR Valve and Gasket
- Air Filter
- Fuel Contamination, Engine Oil
- Poor Power Ground
- Fuel Pressure
- Manifold Leaks, Intake/Exhaust
- Engine Not at Normal Operating Temperatures

This Pinpoint Test is intended to diagnose only the following:

- EGO Sensors
- EGO Signals and Ground Circuits
- EGO Sensor Connections
- Fuel Injectors
- Processor Assembly

NOTE: Fuel-contaminated engine oil may affect 4-1, 4-2 and Service Codes, so if it is suspected, remove the PCV from the valve cover, and rerun the Quick Test. If the problem is corrected, then change the engine oil and filter.

Throttle Body — 3.8L CFI | Pinpoint Test | HC

HC20B REFERENCE ILLUSTRATION, 10-PIN CONNECTOR THROTTLE BODY

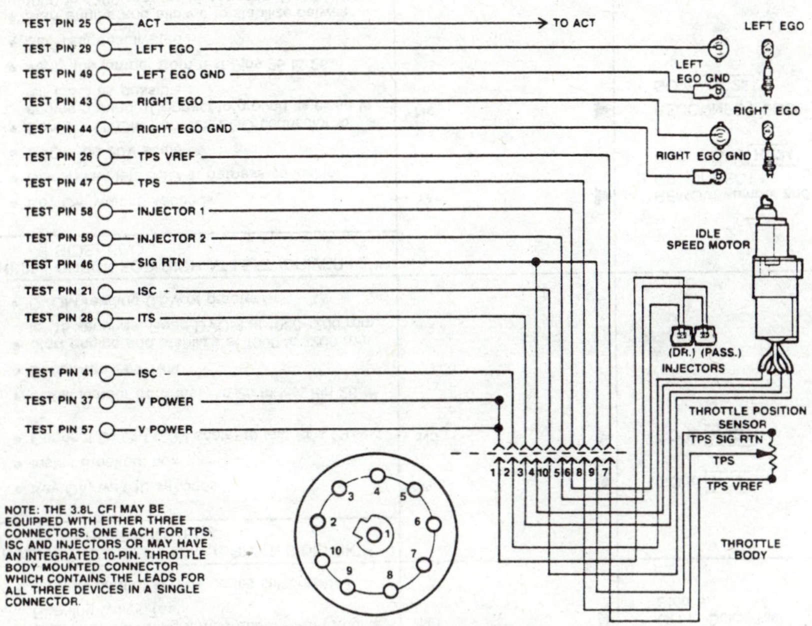

10-PIN CONNECTOR LOOKING INTO MATING SURFACE.

NOTE: When servicing the 10-pin connector, care must be taken that the red retaining ring is not lost or damaged. Care must also be taken that pins are placed into the proper holes on reassembly.

Fuel Control — 3.8L CFI | Pinpoint Test | HC

HC20C REFERENCE DRAWING — SEPARATE CONNECTORS FOR TPS-ISC AND INJECTORS

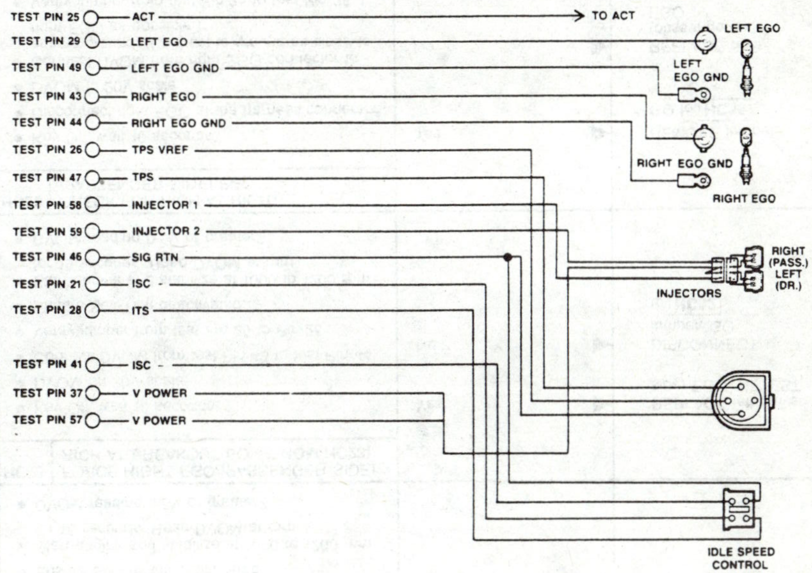

NOTE: The 3.8L CFI may be equipped with either three connectors, one each for TPS, ISC and injectors or may have an integrated 10-pin, throttle body mounted connector which contains the leads for all three devices in a single connector.

Fuel Control — 3.8L CFI | Pinpoint Test | HC

TEST STEP	RESULT ▶	ACTION TO TAKE
HC21 VERIFY ENGINE IS AT TEMPERATURE		
• Run engine at 2000 rpm for two minutes (Upper hose hot and pressurized?). • Engine at operating temperature. Rerun Engine Running Quick Test. • Are fuel control service codes still present?	Yes ▶ No ▶	GO to HC22. GO to Quick Test Step 5.
HC22 FORCE LEFT EGO (DRIVER SIDE) RICH AT BREAKOUT BOX		
• Key Off, wait 10 seconds. • Install Breakout box. • Connect DVOM (20V scale) to test Pins 29 and 49. • Install jumper from test Pin 25 to test Pin 26. • Deactivate Self-Test. • Start engine and stabilize at 1000 to 1200 rpm for 15 seconds. Read DVOM at 1000-1200 rpm. • DVOM reading 0.5V or greater?	Yes ▶ No ▶	GO to HC24. GO to HC23.
HC23 CHECK FOR RICH AT LEFT (DRIVER SIDE) EGO		
• Key Off, wait 10 seconds. • Disconnect left EGO at harness connector. • DVOM on 20V scale. • Connect DVOM from left EGO connector to engine ground. (Connect to ground as close to left EGO as possible.) • Verify the jumper from test Pins 25 to 26. • Self-Test deactivated. • Start engine and allow it to stabilize between 1000 to 1200 rpm for 15 seconds. Read DVOM at 1000-1200 rpm. • DVOM reading 0.5V or greater?	Yes ▶ No ▶	REMOVE jumper and RECONNECT EGO. GO to HC34. RECONNECT EGO. GO to HC25.

Fuel Control — 3.8L CFI | Pinpoint Test | HC

TEST STEP	RESULT ▶	ACTION TO TAKE
HC24 FORCE RIGHT EGO (PASSENGER SIDE) RICH AT BREAKOUT BOX (FROM HC22)		
• Key Off, wait 10 seconds. • DVOM on 20V scale. • Connect DVOM from test Pin 43 to test Pin 44. • Verify jumper from test Pin 25 to test Pin 26. • Ensure Self-Test is deactivated. • Start engine and stabilize at 1000 to 1200 rpm for 15 seconds. Read DVOM at rpm. • DVOM reading 0.5V or greater?	Yes ▶ No ▶	REMOVE jumper and Go to HC27. GO to HC26.
HC25 FORCE RIGHT EGO (PASSENGER SIDE) RICH AT BREAKOUT BOX (FROM HC23)		
• Key Off, wait 10 seconds. • DVOM on 20V scale. • Connect DVOM from test Pin 43 to test Pin 44. • Verify jumper from test Pin 25 to Pin 26. • Ensure Self-Test deactivated. • Start engine and stabilize at 1000 to 1200 rpm for 15 seconds. Read DVOM at rpm. • DVOM reading 0.5V or greater?	Yes ▶ No ▶	REPLACE left (driver side) EGO. RETEST. DISCONNECT jumper. GO to HC32.
HC26 CHECK FOR RICH AT RIGHT (PASSENGER SIDE) EGO		
• Key Off, wait 10 seconds. • Disconnect right EGO at the harness connector. • DVOM to 20V scale. • Connect DVOM from right EGO connector to engine ground. (Connect to ground as close to right EGO as possible.) • Verify jumper from test Pin 25 to test Pin 26. • Verify Self-Test deactivated. • Start engine and allow it to stabilize between 1000 to 1200 rpm for 15 seconds. Read DVOM at rpm. • DVOM reading 0.5V or greater?	Yes ▶ No ▶	REMOVE jumper. GO to HC35. REPLACE right (passenger side) EGO.

Fuel Control — 3.8L CFI | Pinpoint Test | HC

TEST STEP	RESULT ▶	ACTION TO TAKE
HC27 CHECK LEFT EGO (DRIVER SIDE) LEAN AT BREAKOUT BOX		
• Key Off, wait 10 seconds. • Connect DVOM (20V scale) to test Pins 29 to 49. • Ensure Self-Test deactivated. • Start engine and run at 2000 rpm for 30 seconds to ensure engine is at operating temperature. • Shut engine Off and wait 10 seconds. • Restart engine and monitor EGO voltage on DVOM. • Does EGO voltage stabilize at 0.5V or less within 15 seconds?	Yes ▶ No ▶	GO to HC29. GO to HC28.
HC28 CHECK FOR LEAN AT LEFT (DRIVER SIDE) EGO		
• Key Off, wait 10 seconds. • Disconnect left EGO at harness connector. • DVOM on 20V scale. • Connect DVOM from left EGO connector to engine ground (connect to ground as close to left EGO as possible). • Verify Self-Test deactivated. • Start engine and run at 2000 rpm for 30 seconds. • Shut engine Off, wait 10 seconds. • Restart engine, monitor EGO voltage. • Does EGO voltage stabilize at 0.5V or less within 15 seconds?	Yes ▶ No ▶	RECONNECT EGO. GO to HC34. GO to HC30.

Fuel Control — 3.8L CFI | Pinpoint Test | HC

TEST STEP	RESULT ▶	ACTION TO TAKE
HC29 FORCE RIGHT EGO (PASSENGER SIDE) LEAN AT BREAKOUT BOX (FROM HC27)		
• Key Off, wait 10 seconds. • Connect DVOM (20V scale) to test Pins 43 and 44. • Ensure Self-Test deactivated. • Start engine and run at 2000 rpm for 30 seconds to ensure engine is at operating temperature. • Shut engine Off and wait 10 seconds. • Restart engine and monitor EGO voltage on DVOM. • Does EGO voltage stabilize at 0.5V or less within 30 seconds? **NOTE: DVOM reading must be taken within 30 seconds of re-start.**	Yes ▶ No ▶	REPLACE processor, RETEST. REPLACE (passenger side) right EGO and RETEST.
HC30 FORCE RIGHT EGO (PASSENGER SIDE) LEAN AT BREAKOUT BOX (FROM HC28)		
• Key Off, wait 10 seconds. • Connect DVOM (20V scale) to test Pins 43 and 44. • Verify Self-Test deactivated. • Start engine. Run at 2000 rpm for 30 seconds to ensure engine is at operating temperature. • Shut engine Off, wait 10 seconds. • Restart engine, monitor EGO voltage on DVOM. • Does EGO voltage stabilize at 0.5V or less within 30 seconds? **NOTE: DVOM reading must be taken within 30 seconds of re-start.**	Yes ▶ No ▶	REPLACE left (driver side) EGO and RETEST. GO to HC31.
HC31 CHECK FOR LEAN AT RIGHT (PASSENGER SIDE) EGO		
• Key Off, wait 10 seconds. • Disconnect right EGO at harness connector. • DVOM on 20V scale. • Connect DVOM from right EGO connector to engine ground. (Connect to ground as close to right EGO as possible.) • Ensure Self-Test deactivated. • Start engine and run at 2000 rpm for 30 seconds. • Shut engine Off, wait 10 seconds. • Restart engine, monitor EGO voltage on DVOM. • Does EGO voltage stabilize at 0.5V or less within 30 seconds? **NOTE: DVOM reading must be taken within 30 seconds of re-start.**	Yes ▶ No ▶	RECONNECT EGO. GO to HC35. TURN engine Off. LEAVE EGO and DVOM connected. GO to HC33.

Fuel Control — 3.8L CFI | Pinpoint Test | HC

TEST STEP	RESULT ▶	ACTION TO TAKE
HC32 VERIFY SYSTEM IS RICH		
• Key Off, wait 10 seconds. • Verify both left and right EGOs are connected. • Install jumper between test Pins 25 and 26. • Verify Self-Test deactivated. • Start engine and stabilize at 1500 rpm for 30 seconds. • Black smoke present in exhaust?	Yes ▶	REPLACE both EGOs and RETEST.
	No ▶	GO to HC40.
HC33 VERIFY SYSTEM IS LEAN		
• Verify right EGO disconnected and DVOM connected to EGO sensor connector and ground. • Remove air cleaner cover. • Start engine, wait 10 seconds. • Run engine at 2000 rpm for 60 seconds. • Turn engine Off (key Off). • When engine stops, disconnect both injectors at the injector harness. • Push accelerator to the floor and hold it there. • Crank engine for 10 seconds and monitor EGO voltage. • EGO voltage less than 0.5V?	Yes ▶	RECONNECT EGO. CHECK fuel delivery. GO to HC40.
	No ▶	REPLACE both EGO sensors and RETEST.
HC34 CHECK LEFT EGO HARNESS FOR SHORTS OR OPEN		
• Engine Off, wait 10 seconds. • DVOM to 200 ohm scale. • Connect DVOM to left EGO connector and test Pin 29. • Resistance less than 5 ohms?	Yes ▶	REPLACE processor and RETEST.
	No ▶	SERVICE harness and RETEST.
HC35 CHECK RIGHT EGO HARNESS FOR SHORT OR OPEN		
• Engine Off, wait 10 seconds. • DVOM to 200 ohm scale. • Connect DVOM to right EGO connector and test Pin 43. • Resistance less than 5 ohms?	Yes ▶	REPLACE processor and RETEST.
	No ▶	SERVICE harness and RETEST.

Fuel Control — 3.8L CFI | Pinpoint Test | HC

TEST STEP	RESULT ▶	ACTION TO TAKE
HC40 FUEL PRESSURE CHECK		
• Install fuel pressure gauge. • Start and run engine. Fuel pressure must be 241-310 kPa (35-45 psi). • Fuel pressure within specification? **For no starts:** • If engine will not run cycle the key from Off to On several times. • Fuel pressure must remain at 276 ± 34 kPa (40 psi ± 5 psi) for 60 seconds after final key Off.	Yes ▶	GO to HC41.
	No ▶	REFER to Shop manual Group 24 for electric fuel pump.
HC41 CHECK FOR INJECTOR LEAK DOWN		
• Observe pressure drop 2 minutes after performing Step HC40. • Does pressure drop 4 psi or less?	Yes ▶	GO to HC42.
	No ▶	REFER to Shop manual Group 24 for service.
HC42 CHECK LEFT INJECTOR PRESSURE		
• Disconnect the left (driver side) injector connector. • Ensure pressurized fuel system. • Disable fuel pump by disconnecting inertia switch or fuel pump relay. • Observe pressure gauge reading. • Crank engine for two seconds. • Turn key to Off. • Wait 5 seconds, then observe pressure drop. • Is drop in pressure between 14 and 110 kPa (2 to 16 psi)?	Yes ▶	GO to HC43.
	No ▶	GO to HC46.
HC43 CHECK RIGHT INJECTOR PRESSURE		
• Reactivate fuel pump. • Connect left injector and disconnect right (passenger side) injector. • Repeat HC42 for right injector. • Is drop in pressure between 14 and 110 kPa (2 to 16 psi)?	Yes ▶	GO to HC44.
	No ▶	GO to HC46.

Fuel Control — 3.8L CFI | Pinpoint Test | HC

TEST STEP	RESULT ▶	ACTION TO TAKE
HC44 CHECK FUEL PRESSURE REGULATOR		
• Reconnect injector. • Reconnect inertia switch or fuel pump relay. • Start engine, let idle. • Observe pressure gauge. • Is pressure 241 to 310 kPa (35 to 45 psi)?	Yes ▶	GO to HC45.
	No ▶	CHECK to ensure that fuel lines; supply and return, are free from kinks. If OK, GO to Shop manual Group 24 to check pressure regulator and tnrottle body for proper operation.
HC45 INJECTOR BALANCE CHECK		
• Connect tachometer to engine. • Disconnect and plug hose to EGR valve. • Run engine at 2000 rpm. (It may be necessary to disconnect idle speed control and use throttle body stop screw to set engine speed.) • Disconnect left injector by unplugging the injector connector. • Note rpm when engine rpm stabilizes. • Reconnect left injector and stabilize rpm at 2000 rpm. • Disconnect right injector as in above Step. • Note rpm when engine rpm stabilizes. • Reconnect right injector and allow engine to stabilize at idle. • Note the difference between the two rpm readings. • Does the rpm drop at least 150 rpm for each injector?	Yes ▶	RECONNECT EGR hose. Fuel system OK. REFER to diagnostic routines for other fuel related concerns (fouled plugs, fuel-contaminated engine oil).
	No ▶	RECONNECT EGR hose. REPLACE both injectors.
HC46 ISOLATE SUSPECT INJECTOR		
• Key Off, wait for 10 seconds. • Disconnect both injectors from the injector harness. • Reconnect the injector connectors as follows: A. Connect left connector to right (passenger side) injector. B. Connect right connector to left (driver side) injector. • Rerun Self-Test. • Suspect injector still fails.	Yes ▶	RESTORE connectors to proper injectors. REPLACE injector. RETEST.
	No ▶	GO to HC47.

Fuel Control — 3.8L CFI | Pinpoint Test | HC

TEST STEP	RESULT ▶	ACTION TO TAKE
HC47 CHECK FOR SHORT TO BATTERY AT INJECTOR CONNECTORS		
• Key Off, wait 10 seconds. • Disconnect injector connectors. • DVOM to 20V scale. • Connect DVOM between injector connector and ground. • DVOM reads less than 1.0V? NOTE: **This step must be run on both left and right injectors.**		RECONNECT injector connectors.
	Yes ▶	GO to HC48.
	No ▶	SERVICE short and RETEST.
HC48 CHECK LEFT AND RIGHT INJECTOR FOR SHORT TO GROUND		
• Key Off, wait 10 seconds. • Verify injectors are connected. • Verify Breakout box installed. • DVOM to 200 ohm range. • Connect DVOM between test Pin 60 and: — For left injector test Pin 58 — For right injector test Pin 59 • DVOM reads less than 5 ohms?	Yes ▶	GO to HC49.
	No ▶	GO to HC51.
HC49 ISOLATE SHORT TO GROUND		
• Disconnect the harness connector at the throttle body. • DVOM on 200 ohm scale. • Disconnect injector harness connectors from injectors. • Connect DVOM from injector 1 of the throttle body connector to ground. Note reading. • Connect DVOM from injector 2 of the throttle body connector to ground. • Is resistance on either or both injectors less than 5 ohms?	Yes ▶	CORRECT appropriate short or REPLACE appropriate injector.
	No ▶	GO to HC50.

Fuel Control — 3.8L CFI | Pinpoint Test | HC

TEST STEP	RESULT ▶	ACTION TO TAKE
HC50 SHORT IN HARNESS OR PROCESSOR		
• Verify harness connection to the throttle body is disconnected. • Disconnect processor from Breakout box. • DVOM on 200 ohm scale. • Connect DVOM from injector 1 of the harness connector to test Pin 58. Note reading. • Connect DVOM from injector 2 of the harness connector to test Pin 59. Note reading. • Either or both readings less than 5 ohms?	Yes ▶	SERVICE appropriate short. RECHECK.
	No ▶	REPLACE processor.
HC51 CHECK FOR OPEN INJECTORS		
• Verify injectors are disconnected. • DVOM to 200 ohm scale. • Connect DVOM across the terminals of each injector. Note readings. • Is resistance less than 1.5 ohms or greater than 2.5 ohms?	Yes ▶	REPLACE appropriate injector and RETEST.
	No ▶	GO to HC52.
HC52 CHECK HARNESS CONTINUITY		
• Verify that the throttle body harness connector is disconnected. • DVOM on 200 ohm scale. • Connect DVOM from injector 1 to test Pin 58. Note reading. • Connect DVOM from injector 2 to test Pin 59. Note reading. • Resistance less than 2 ohms?	Yes ▶	RECONNECT harness to throttle body assembly. RECONNECT processor to Breakout box. GO to HC53.
	No ▶	SERVICE appropriate harness wire and RETEST.

Fuel Control — 3.8L CFI | Pinpoint Test | HC

TEST STEP	RESULT ▶	ACTION TO TAKE
HC53 CHECK FOR SHORT BETWEEN INJECTOR INPUTS AT PROCESSOR		
• Verify processor connected to Breakout box. • Verify throttle body harness connector is connected. • DVOM to 2000 ohm scale. • Verify injectors are not connected to harness. • Connect DVOM from test Pin 58 to test Pin 59. • Resistance less than 10 kilohms?	Yes ▶	GO to HC54.
	No ▶	REPLACE processor. RETEST.
HC54 CHECK FOR SHORT TO INJECTOR CIRCUITS AT THROTTLE BODY		
• Verify injectors are not connected to harness. • Disconnect processor from Breakout box. • Disconnect harness connector from throttle body. • DVOM on 2000 ohm scale. • Connect DVOM from injector 1 to injector 2 at the throttle body connector. • Resistance less than 10 kilohms?	Yes ▶	CORRECT short and RETEST.
	No ▶	REPLACE processor.

Fuel Pump Circuit — (Inertia Switch) | Pinpoint Test | J

J

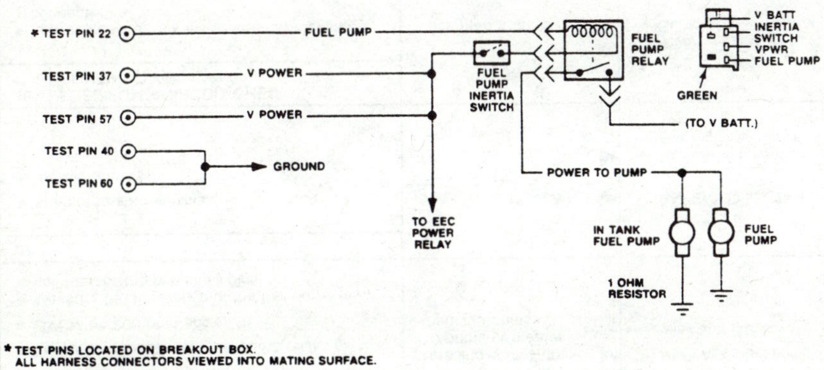

* TEST PINS LOCATED ON BREAKOUT BOX.
ALL HARNESS CONNECTORS VIEWED INTO MATING SURFACE.

STOP-WARNING

You may have reached this point because of a fault in a non-EEC System and/or component. To prevent the replacement of good components, be aware that the following non-EEC areas may be at fault:

- Fuel Lines
- Fuel Filters
- Throttle Body
- Contaminated Fuel
- Fuel Pump

This Pinpoint Test is intended to diagnose only the following:

- Fuel Pump Relay
- Harness Circuits: V Batt., V Power, F.P., Ground and Power to Pump(s)
- Processor Assembly

Fuel Pump Circuit — (Inertia Switch) | Pinpoint Test | J

TEST STEP	RESULT ▶	ACTION TO TAKE
J1 NO FUEL PUMP PRESSURE		
• Fuel pressure gauge installed. • Check if fuel pump runs. • Cycle key from Off to Run, repeat several times. (Do not enter start mode.) Fuel pump should run briefly each time the key enters Run.	Yes ▶	GO to Shop Manual Group 24 — Electric Fuel Pump.
	No ▶	GO to [J2].
J2 CHECK FOR V POWER TO PROCESSOR		
• Key Off, wait 10 seconds. • Disconnect processor 60 Pin connector and inspect for damaged pins, corrosion, loose wires. Service as necessary. • Install Breakout box and reconnect processor. • Key On, engine Off. • DVOM on 20V scale. • Measure voltage from test Pin 37 to test Pin 40 and test Pin 57 to test Pin 60.	Either voltage reading is less than 10.5V ▶	GO to [B1].
	Both voltage readings are 10.5V or greater ▶	GO to [J3].
J3 POWER TO PUMP(S) CHECK		
• Key On, engine Off. • Leave Breakout box installed and processor connected. • Locate fuel pump relay. • DVOM on 20V scale. • With DVOM measure voltage from chassis ground to power to pump(s) circuit at fuel pump relay during crank mode.	8.0V or greater during crank ▶	GO to Shop Manual Group 24 — Electric Fuel Pump.
	Less than 8.0V during crank ▶	GO to [J4].
J4 FUEL PUMP CIRCUIT CHECK TO V BATT.		
• Key On, engine Off. • Leave Breakout box installed and processor connected. • Locate fuel pump relay. • DVOM on 20V scale. • With DVOM measure voltage from chassis ground to V Batt. at fuel pump relay.	10.5V or greater ▶	GO to [J5].
	Less than 10.5V ▶	SERVICE open in V Batt. between fuel pump relay and vehicle battery positive post, then GO to [J13].

Fuel Pump Circuit — (Inertia Switch) | Pinpoint Test | J

TEST STEP	RESULT ▶	ACTION TO TAKE
J5 FUEL PUMP CIRCUIT CHECK		
• Key On, engine Off. • Leave Breakout box installed and processor connected. • Locate fuel pump relay. • DVOM on 20V scale. • With DVOM measure voltage from chassis ground to fuel pump circuit at the fuel pump relay.	10.5V or greater ▶ Less than 10.5V ▶	GO to J6. GO to J9.
J6 RESISTANCE CHECK OF FUEL PUMP RELAY		
• Key Off, wait 10 seconds. • Leave Breakout box installed and processor connected. • Locate and disconnect fuel pump relay. • DVOM on 200 ohm scale. • Measure the resistance of the fuel pump relay coil (across the two small pins).	Reading is 50-100 ohms ▶ Reading is less than 50 ohms or greater than 100 ohms ▶	GO to J8. REPLACE fuel pump relay and GO to J7.
J7		
• Attempt to start vehicle. • Vehicle starts?	Yes ▶ No ▶	REPEAT Quick Test. REPLACE processor and REPEAT Quick Test.
J8 FUEL PUMP CIRCUIT CHECK CONTINUITY		
• Leave Breakout box installed and processor connected. • Key Off, wait 10 seconds. • DVOM on 200 ohm scale. • Measure resistance on fuel pump circuit from pump relay to test Pin 22.	Less than 5 ohms ▶ 5 ohms or greater ▶	GO to J11. SERVICE open in fuel pump circuit and REPEAT Quick Test.

Fuel Pump Circuit — (Inertia Switch) | Pinpoint Test | J

TEST STEP	RESULT ▶	ACTION TO TAKE
J9 FUEL PUMP CIRCUIT CHECK (V POWER CIRCUIT)		
• Leave Breakout box installed and processor connected. • Key Off. • Locate the inertia switch and fuel pump relay. • DVOM on 200 ohm scale. • Measure resistance of V Power circuit between inertia switch and fuel pump relay.	5 ohms or greater ▶ Less than 5 ohms ▶	SERVICE open in V Power circuit. REPEAT Quick Test. GO to J10.
J10 INERTIA SWITCH CHECK		
• Leave Breakout box installed and processor connected. • Key On, engine Off. • DVOM on 20V scale. • Measure voltage at inertia switch input V Power circuit to ground.	10.5V or greater ▶ Less than 10.5V ▶	REPLACE inertia switch and REPEAT Quick Test. SERVICE open from EEC power relay to inertia switch. REPEAT Quick Test.
J11 CHECK FOR SHORT TO POWER		
• Key Off, wait 10 seconds. • Leave Breakout box installed and disconnect processor. • Fuel pump relay disconnected. • DVOM on 200,000 ohm scale. • Measure resistance from test Pin 22 to test Pins 37 and 57.	10,000 ohms or greater ▶ Less than 10,000 ohms ▶	GO to J12. SERVICE short to power in the fuel pump circuit. GO to J7.
J12 CHECK FOR SHORT TO GROUND		
• Key Off. • Leave Breakout box installed and processor disconnected. • Fuel pump relay disconnected. • DVOM on 200,000 ohm scale. • Measure resistance from test Pin 22 to test Pins 40 and 60.	10,000 ohms or greater ▶ Less than 10,000 ohms ▶	REPLACE processor and REPEAT Quick Test. SERVICE short to ground in the fuel pump circuit. REPEAT Quick Test.

Fuel Pump Circuit — (Inertia Switch) | Pinpoint Test | J

TEST STEP	RESULT ▶	ACTION TO TAKE
J13 CHECK POWER TO PUMP(S) CIRCUIT — SHORT TO GROUND		
• Key Off, wait 10 seconds. • Fuel pump relay disconnected. • Fuel pump(s) disconnected. • DVOM on 200,000 ohm scale. • Measure resistance from power to pump(s) circuit to chassis ground.	Less than 10,000 ohms ▶	SERVICE short to ground in power to pump(s) circuit. RECONNECT all components. REPEAT Quick Test.
	10,000 ohms or greater ▶	RECONNECT all components. REPEAT Quick Test.

EGR On/Off Control | Pinpoint Test | KA

KA

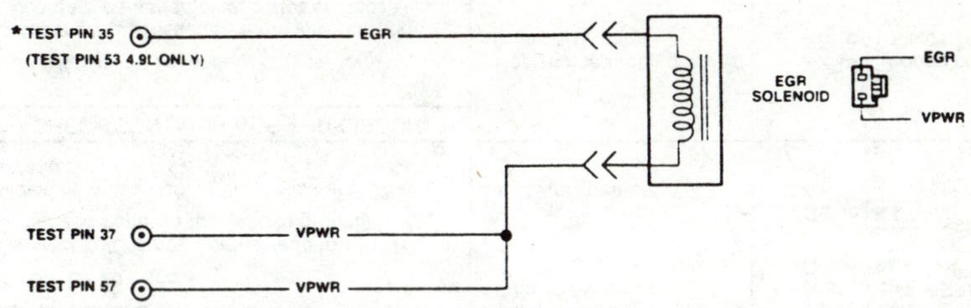

* TEST PINS LOCATED ON BREAKOUT BOX.
ALL HARNESS CONNECTORS VIEWED INTO MATING SURFACE.

STOP-WARNING

You may have reached this point because of a fault in a non-EEC System and or component. To prevent the replacement of good components, be aware that the following non-EEC areas may be at fault:

- Air or Vacuum Leaks
- EGR Flow Restrictions
- EGR Value

NOTE: Code 34 may be the result of high volume exhaust vent system (reduces backpressure). If this is suspected, perform the test in a well-ventilated area without exhaust vent connected.

This Pinpoint Test is intended to diagnose only the following:

- Circuits: EGR and VPWR
- EGR Solenoid
- Presence of Manifold Vacuum
- Processor Assembly

EGR On/Off Control — Pinpoint Test — KA

TEST STEP	RESULT ▶	ACTION TO TAKE
KA1 MEASURE EGR SOLENOID RESISTANCE		
• Key Off, wait 10 seconds. • DVOM on 200 ohm scale. • Disconnect EGR solenoid connector and measure solenoid resistance.	Resistance is between 65 and 110 ohms ▶	CONNECT EGR solenoid. GO to KA2.
	Resistance is less than 65 ohms or greater than 110 ohms ▶	REPLACE EGR solenoid. REPEAT Quick Test.
KA2 ENTER OUTPUT STATE CHECK (REFER TO APPENDIX)		
• Key Off, wait 10 seconds. • DVOM on 20V scale. • Connect DVOM negative test lead to STO and positive test lead to battery positive. • Jumper STI to signal return. • Perform Key On, Engine Off Self-Test until the completion of the Continuous Test Codes. • DVOM will indicate zero volts. • Depress and release the throttle. • Did DVOM reading change to a high voltage reading?	Yes ▶	REMAIN in Output State Check. GO to KA3.
	No ▶	DEPRESS throttle to WOT and release. If STO voltage does not go high, GO to Pinpoint Test Step Q40. Leave equipment hooked up.
KA3 CHECK EGR ON/OFF CONTROL SOLENOID ELECTRICAL OPERATION		
• DVOM on 20V scale. • Connect DVOM positive test lead to VPWR circuit on EGR solenoid and negative test lead to EGR output circuit. • While observing DVOM, depress and release the throttle several times to cycle output On and Off. • EGR output cycles On and Off.	Yes ▶	GO to KA4.
	No ▶	GO to KA6.

EGR On/Off Control — Pinpoint Test — KA

TEST STEP	RESULT ▶	ACTION TO TAKE
KA4 CHECK SOLENOID FOR VACUUM CYCLING		
• Install vacuum pump to the solenoid vacuum supply port and install a vacuum gauge to the output port. Apply 6 in. Hg minimum. • While cycling outputs On and Off (by depressing and releasing throttle) observe the vacuum gauge at the output. **NOTE: Maintain vacuum at source.**	Vacuum output cycles On and Off ▶	GO to KA5.
	Vacuum output does not cycle On and Off ▶	REPLACE solenoid. REPEAT Quick Test.
KA5 CHECK MANIFOLD VACUUM LINES FOR BLOCKAGE OR LEAKS		
• With vacuum lines disconnected at solenoid, check for vacuum. • Start engine.	Vacuum present ▶	EEC-IV system OK. REFER to Section 6.
	No vacuum present ▶	SERVICE vacuum source blockage or leak. REPEAT Quick Test.
KA6 CHECK VOLTAGE OF VPWR CIRCUIT		
• Key On, Engine Off. • DVOM on 20V scale. • Connect DVOM positive test lead to VPWR circuit and negative test lead to ground. • Measure voltage on VPWR circuit at EGR solenoid.	Voltage reading is less than 10.5V ▶	SERVICE harness open circuit. REPEAT Quick Test.
	Voltage reading is 10.5V or greater ▶	GO to KA7.
KA7 CHECK CONTINUITY OF EGR CIRCUIT		
• Key Off, wait 10 seconds. • Disconnect processor 60 Pin connector and inspect for damaged pins, corrosion, loose wires. Service as necessary. • Connect Breakout box to harness. Leave processor disconnected. • DVOM on 200 ohm scale. • Measure continuity from test Pin 35* to EGR circuit at harness connector. *Test Pin 53 for 4.9L engine, truck only.	Resistance reading is 5 ohms or greater ▶	SERVICE open circuit. REPEAT Quick Test.
	Resistance reading is less than 5 ohms ▶	GO to KA8.

EGR On/Off Control — Pinpoint Test — KA

TEST STEP	RESULT ▶	ACTION TO TAKE
KA8 CHECK FOR SHORT TO POWER		
• Key Off, wait 10 seconds. • DVOM on 200,000 ohm scale. • Leave Breakout box installed and processor disconnected. • With EGR solenoid disconnected, measure resistance from test Pin 35* to test Pins 37 and 57. *Test Pin 53 for 4.9L engine, truck only.	Resistance reading is 10,000 ohms or greater ▶	GO to KA9.
	Resistance reading is less than 10,000 ohms ▶	SERVICE short to power. REPEAT Quick Test.
KA9 CHECK FOR SHORT TO GROUND		
• Key Off, wait 10 seconds. • Leave Breakout box installed and processor disconnected. • DVOM on 200,000 ohm scale. • Measure resistance of test Pin 35* to test Pins 40, 46 and 60. *Test Pin 53 for 4.9L engine, truck only.	Resistance reading is less than 10,000 ohms ▶	SERVICE short to ground. REPEAT Quick Test.
	Resistance reading is 10,000 ohms or greater ▶	REPLACE processor. REPEAT Quick Test.

Air Management System — Pinpoint Test — KC

KC

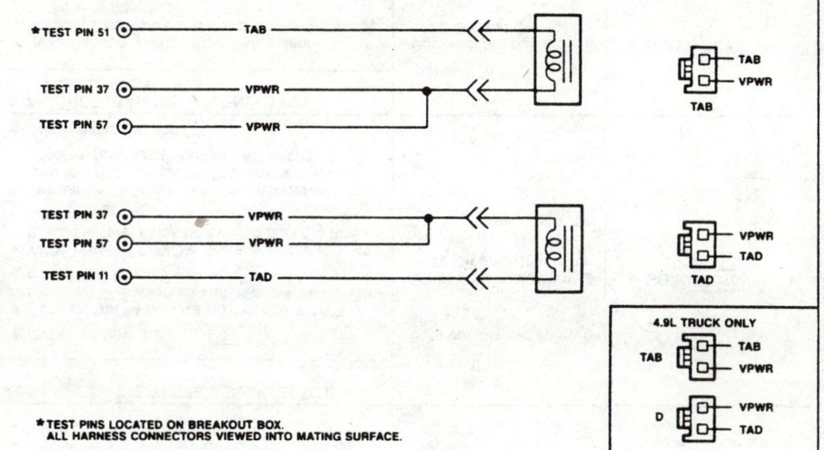

STOP-WARNING

You may have reached this point because of a fault in a non-EEC System and/or component. To prevent the replacement of good components, be aware that the following non-EEC areas may be at fault:

- Thermactor System
  - — Belt
  - — Pump
  - — Valve

This Pinpoint Test is intended to diagnose only the following:

- TAB and TAD Solenoid Valve Assemblies
- Harness Circuits: TAB, TAD and VPWR
- Vacuum Supply
- Processor Assembly

Air Management System — Pinpoint Test — KC

TEST STEP	RESULT ▶	ACTION TO TAKE
KC1 SERVICE CODES 44, 45 AND 46: VERIFY VACUUM LINE ROUTING		
• Verify proper vacuum line routing to the TAB/TAD solenoids and to the bypass diverter valve. Refer to VECI decal. • Check for kinked or blocked vacuum lines. • Check for kinked or blocked air hoses. • Check for disconnected vacuum lines. • Are visual checks satisfactory?	No ▶	SERVICE routing or faults. REPEAT Quick Test.
	Yes ▶	Service code 44, GO to **KC4**. Service code 45, GO to **KC2**. Service code 46, GO to **KC3**.
KC2 ATTEMPT TO ELIMINATE SERVICE CODE 45 (TAD ONLY)		
• Disconnect vacuum line on diverter valve and cap vacuum line. • Key Off, wait 10 seconds. • Repeat Engine Running Self-Test and record service codes. • Is code 45 present?	Yes ▶	EEC-IV system OK. REFER to Section 3 for diverter valve or check valve diagnostics.
	No ▶	GO to **KC4**.
KC3 ATTEMPT TO ELIMINATE SERVICE CODE 46 (TAB ONLY)		
• Disconnect vacuum line on bypass valve and cap vacuum line. • Key Off, wait 10 seconds. • Repeat Engine Running Self-Test and record codes. • Is code 46 present?	Yes ▶	EEC-IV system OK. REFER to Section 3 for bypass valve diagnostics.
	No ▶	GO to **KC4**.
KC4 MEASURE TAB/TAD SOLENOID RESISTANCE		
• Key Off, wait 10 seconds. • DVOM on 200 ohm scale. • Disconnect TAB solenoid connector and measure solenoid resistance. • Disconnect TAD solenoid connector and measure solenoid resistance.	Both resistances are between 50 and 100 ohms ▶	CONNECT TAB/TAD solenoids. GO to **KC5**.
	Either resistance is less than 50 ohms or greater than 100 ohms ▶	REPLACE TAB/TAD solenoid assembly. REPEAT Quick Test.

Air Management System — Pinpoint Test — KC

TEST STEP	RESULT ▶	ACTION TO TAKE
KC5 ENTER OUTPUT STATE CHECK (REFER TO APPENDIX)		
• Key Off, wait 10 seconds. • DVOM on 20V scale. • Connect DVOM negative test lead to STO and positive test lead to battery positive. • Jumper STI to signal return. • Perform Key On, Engine Off Self-Test until the completion of the Continuous Test Codes. • DVOM will indicate zero volts. • Depress and release the throttle. • Did DVOM reading change to a high voltage reading?	Yes ▶	REMAIN in Output State Check. GO to **KC6**.
	No ▶	DEPRESS throttle to WOT and RELEASE. If STO voltage does not go high, GO to Pinpoint Test Step **Q40**. Leave equipment hooked up.
KC6 CHECK TAB/TAD SOLENOID ELECTRICAL OPERATION		
• DVOM on 20V scale. • Connect DVOM positive test lead to VPWR circuit on TAB solenoid and negative test lead to output TAB circuit. • While observing DVOM depress and release the throttle several times (to cycle output On and Off). • Repeat for TAD solenoid. Connect positive test lead to VPWR circuit on TAD solenoid and negative test lead to output TAD circuit. • Cycle output On and Off.	Both outputs cycle On and Off ▶	GO to **KC7**.
	Either output does not cycle On and Off ▶	GO to **KC10**.
KC7 CHECK TAB/TAD SOLENOID FOR VACUUM CYCLING		
• Install vacuum pump to the TAB solenoid vacuum supply port and install a vacuum gauge to the output port. • While cycling outputs On and Off (by depressing and releasing throttle), observe the vacuum gauge at the output. **NOTE: Maintain vacuum at source.** • Repeat for TAD solenoid. Connect vacuum pump to the TAD solenoid vacuum supply port and connect a vacuum gauge to the output port. • Cycle output On and Off.	Both vacuum outputs cycle On and Off ▶	GO to **KC8**.
	Either vacuum output does not cycle On and Off ▶	REPLACE solenoid assembly. REPEAT Quick Test.

Air Management System — Pinpoint Test — KC

TEST STEP	RESULT ▶	ACTION TO TAKE
KC8 CHECK TAB TAD VACUUM		
• Disconnect vacuum line from TAB solenoid to thermactor valve. • Inspect vacuum line for blockage, cracks, kinks (e.g. connect vacuum pump and gauge and verify vacuum line holds vacuum, or blow into vacuum line to verify no restrictions). • Repeat this Step for the TAD vacuum line. • Thermactor vacuum lines OK?	Yes ▶	GO to KC9.
	No ▶	SERVICE as required. REPEAT Quick Test.
KC9 CHECK MANIFOLD VACUUM LINES FOR BLOCKAGE OR LEAKS		
• With vacuum lines disconnected at TAD/TAB solenoids, check for vacuum. • Start engine.	Vacuum present ▶	EEC-IV system OK. REFER to Section 3 for thermactor valve and air pump diagnostics.
	No vacuum present ▶	SERVICE vacuum source blockage or leak. REPEAT Quick Test.
KC10 CHECK VOLTAGE OF VPWR CIRCUIT		
• Key On, engine Off. • DVOM on 20V scale. • Positive DVOM test lead to VPWR circuit and negative test lead to ground. • Measure voltage on TAB solenoid VPWR circuit. • Repeat for TAD solenoid.	Both voltage readings are greater than 10.5V ▶	GO to KC11.
	Either voltage reading is less than 10.5V ▶	SERVICE harness circuit open. REPEAT Quick Test.
KC11 CHECK CONTINUITY OF TAB AND TAD CIRCUITS		
• Key Off, wait 10 seconds. • Disconnect processor 60 Pin connector and inspect for damaged pins, corrosion, loose wires. Service as necessary. • Connect Breakout box to harness. Leave processor disconnected. • DVOM on 200 ohm scale. • Measure continuity of test Pin 51 to TAB circuit at harness connector. • Measure continuity of test Pin 11 to TAD circuit at harness connector.	Both resistance readings are less than 5 ohms ▶	GO to KC12.
	Either resistance reading is 5 ohms or greater ▶	SERVICE harness open circuit. REPEAT Quick Test.

Air Management System — Pinpoint Test — KC

TEST STEP	RESULT ▶	ACTION TO TAKE
KC12 CHECK FOR SHORT TO POWER		
• Key Off, wait 10 seconds. • DVOM on 200,000 ohm. • Leave Breakout box installed and processor disconnected. • With TAB/TAD solenoids disconnected, measure resistance from test Pins 51 and/or 11 to test Pins 37 and 57.	Resistance reading is 10,000 ohms or greater ▶	GO to KC13.
	Resistance reading is less than 10,000 ohms ▶	SERVICE short to power. REPEAT Quick Test.
KC13 CHECK FOR SHORT TO GROUND		
• Key Off, wait 10 seconds. • DVOM on 200,000 ohm scale. • Leave Breakout box installed and processor disconnected. • With TAB/TAD solenoids disconnected, measure resistance from test Pins 51 and/or 11 to test Pins 40, 46 and 60.	Resistance reading is less than 10,000 ohms ▶	SERVICE short to ground. REPEAT Quick Test.
	Resistance reading is 10,000 ohms or greater ▶	REPLACE processor. REPEAT Quick Test.

Canister Purge — Pinpoint Test — KD

KD

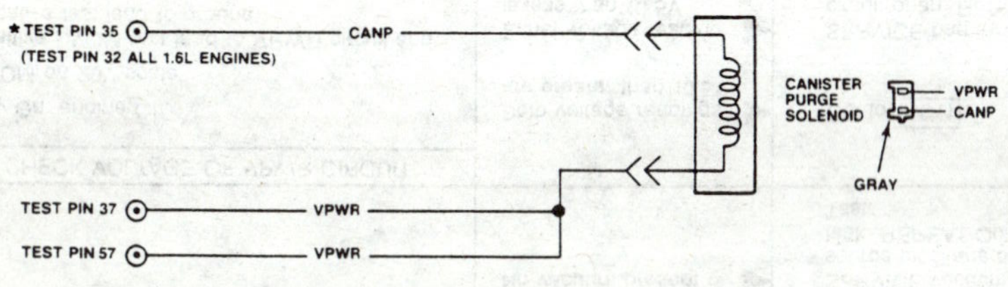

*TEST PINS LOCATED ON BREAKOUT BOX.
ALL HARNESS CONNECTORS VIEWED INTO MATING SURFACE.

STOP-WARNING

This Pinpoint Test is intended to diagnose only the following:

- Harness Circuits: CANP and VPWR
- Processor Assembly

Canister Purge — Pinpoint Test — KD

TEST STEP	RESULT ▶	ACTION TO TAKE
KD1 MEASURE CANP SOLENOID RESISTANCE		
• Key Off, wait 10 seconds. • DVOM on 200 ohm scale. • Disconnect CANP solenoid connector and measure solenoid resistance.	Resistance is between 40 and 90 ohms ▶	CONNECT CANP solenoid. GO to KD2.
	Resistance is less than 40 ohms or greater than 90 ohms ▶	REPLACE CANP solenoid. REPEAT Quick Test.
KD2 ENTER OUTPUT STATE CHECK (REFER TO APPENDIX)		
• Key Off, wait 10 seconds. • DVOM on 20V scale. • Connect DVOM negative test lead to STO and positive test lead to battery positive. • Jumper STI to signal return. • Perform Key On, Engine Off Self-Test until the completion of the Continuous Test Codes. • DVOM will indicate zero volts. • Depress and release the throttle. • Did DVOM reading change to a high voltage reading?	Yes ▶	REMAIN in Output State Check. GO to KD3.
	No ▶	DEPRESS throttle to WOT and release. If STO voltage does not go high, GO to Pinpoint Test Step Q40. Leave equipment hooked up.
KD3 CHECK CANISTER PURGE (CANP) SOLENOID ELECTRICAL OPERATION		
• DVOM on 20V scale. • Connect DVOM positive test lead to VPWR on CANP solenoid and negative test lead to CANP output circuit. • While observing DVOM depress and release the throttle several times (to cycle output On and Off). • CANP output cycles On and Off?	Yes ▶	GO to KD4.
	No ▶	REMOVE jumper. GO to KD5.

Canister Purge — Pinpoint Test — KD

TEST STEP	RESULT ▶	ACTION TO TAKE
KD4 CHECK CANISTER PURGE SOLENOID FOR HOLDING VACUUM		
• Key Off, wait 10 seconds. • Remove jumper from STI to Signal Return. • Disconnect vacuum hose at canister purge solenoid at PCV side. • Apply 16 in. Hg to CANP solenoid. • Does CANP solenoid hold vacuum for 20 seconds?	Yes ▶	EEC-IV system OK. Refer to Shop Manual Group 24.
	No ▶	REPLACE CANP solenoid. REPEAT Quick Test. If code 42 is still present, GO to Section 4.
KD5 CHECK VOLTAGE OF VPWR CIRCUIT		
• Key On, engine Off. • DVOM on 20V scale. • Connect DVOM positive test lead to VPWR and negative test lead to ground. • Measure voltage on CANP solenoid.	Voltage reading is less than 10.5V ▶	SERVICE harness open circuit. REPEAT Quick Test.
	Voltage reading is 10.5V or greater ▶	GO to KD6.
KD6 CHECK CONTINUITY OF CANP CIRCUIT		
• Key Off, wait 10 seconds. • Disconnect processor and inspect both 60 Pin connectors and inspect for damaged pins, corrosion, loose wires. Service as necessary. • Connect Breakout box to harness. Leave processor disconnected. • DVOM on 200 ohm scale. • Measure continuity from test Pin 35* to CANP at harness connector. *Test Pin 32 for all 1.6L engines.	Resistance reading is 5 ohms or greater ▶	SERVICE open circuit. REPEAT Quick Test.
	Resistance reading is less than 5 ohms ▶	GO to KD7.

Canister Purge — Pinpoint Test — KD

TEST STEP	RESULT ▶	ACTION TO TAKE
KD7 CHECK FOR SHORT TO POWER		
• Key Off, wait 10 seconds. • DVOM on 200,000 ohm scale. • Leave Breakout box to harness. Leave processor disconnected. • With CANP solenoid disconnected, measure resistance from test Pin 35* to test Pins 37 and 57. *Test Pin 32 for all 1.6L engine applications.	Resistance reading is 10,000 ohms or greater ▶	GO to KD8.
	Resistance reading is less than 10,000 ohms ▶	SERVICE short to power. REPEAT Quick Test.
KD8 CHECK FOR SHORT TO GROUND		
• Key Off, wait 10 seconds. • Leave Breakout box installed and processor disconnected. • DVOM on 200,000 ohm scale. • Measure resistance of test Pin 35* to test Pins 40, 46 and 60. *Test Pin 32 for all 1.6L engine applications.	Resistance reading is less than 10,000 ohms ▶	SERVICE short to ground. REPEAT Quick Test.
	Resistance reading is 10,000 ohms or greater ▶	REMOVE Breakout box. REPLACE processor. RERUN Quick Test.

Idle Speed Control (By-Pass Air) — Pinpoint Test — KE

KE

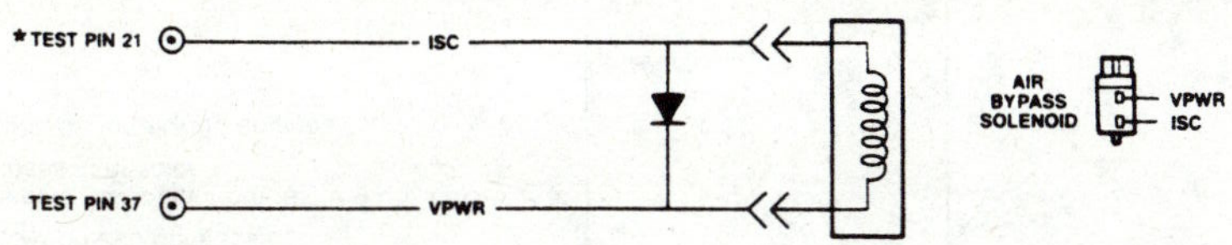

*TEST PINS LOCATED ON BREAKOUT BOX.
ALL HARNESS CONNECTORS VIEWED INTO MATING SURFACE.

CAUTION: If the engine exhibits rough running and/or idle, correct these conditions before diagnosing Idle Speed Control (ISC). Rough running/misses may be caused by:

- **Ignition System**
  Refer to Section 15
- **Fuel System**
  Refer to Pinpoint Test Steps HA1 through HA4
- **EGR System, Section 6**

STOP-WARNING

You may have reached this point because of a fault in a non-EEC System and/or component. To prevent the replacement of good components, be aware that the following non-EEC areas may be at fault:

- Engine not up to operating temperature
- Engine over operating temperature
- Improper Idle Speed Throttle Stop Adjustment
- A/C input (electrical problem)
- Throttle Speed Control Linkage

This Pinpoint Test is intended to diagnose only the following:

- RPM in Self-Test only
- ISC Actuator
- Harness Circuits ISC and VPWR
- Processor Assembly

Idle Speed Control (By-Pass Air) — Pinpoint Test — KE

TEST STEP	RESULT ▶	ACTION TO TAKE
KE1 MEASURE ISC SOLENOID RESISTANCE		
• Key Off. • DVOM on 200 ohm scale. • Disconnect ISC solenoid connector and measure solenoid resistance.	Resistance is between 7 and 13 ohms ▶	GO to KE2.
	Resistance is less than 7 ohms or greater than 13 ohms ▶	REPLACE ISC solenoid. REPEAT Quick Test.
KE2 ISC SHORT TO CASE (GROUND) CHECK		
• Key Off. • DVOM on 200,000 ohm scale. • Leave ISC harness disconnected. • Check for short from either ISC pin to ISC housing.	10,000 ohms or greater ▶	GO to KE3.
	Less than 10,000 ohms ▶	REPLACE ISC solenoid. REPEAT Quick Test.
KE3 CHECK VOLTAGE OF VPWR CIRCUIT		
• Key On, Engine Off. • DVOM on 20V scale. • Leave ISC harness disconnected. • Connect DVOM positive test lead to VPWR circuit and negative test lead to ground. • Measure voltage on ISC solenoid VPWR circuit.	Voltage reading is less than 10.5V ▶	SERVICE harness open circuit. REPEAT Quick Test.
	Voltage reading is 10.5V or greater ▶	GO to KE4.
KE4 CHECK CONTINUITY OF ISC CIRCUIT		
• Key Off. • Leave ISC harness disconnected. • Disconnect processor and inspect both 60 Pin connectors for damaged pins, corrosion, loose wires. Service as necessary. • Connect Breakout box to harness. Leave processor disconnected. • DVOM on 200 ohm scale. • Measure continuity of test Pin 21 to ISC circuit at harness connector.	Resistance reading is 5 ohms or greater ▶	SERVICE open circuit. REPEAT Quick Test.
	Resistance reading is less than 5 ohms ▶	GO to KE5.

Idle Speed Control (By-Pass Air) — Pinpoint Test — KE

TEST STEP	RESULT ▶	ACTION TO TAKE
KE5 CHECK FOR SHORT TO POWER		
• Key Off, wait 10 seconds. • DVOM on 200,000 ohm scale. • Leave Breakout box to harness. Leave processor disconnected. • Leave ISC solenoid disconnected, measure resistance from test Pin 21 to test Pin 37.	Resistance reading is 10,000 ohms or greater ▶	GO to KE6.
	Resistance reading is less than 10,000 ohms ▶	SERVICE short to power. REPEAT Quick Test.
KE6 CHECK FOR SHORT TO GROUND		
• Key Off, wait 10 seconds. • Leave Breakout box installed and processor disconnected. • DVOM on 200,000 ohm scale. • Measure resistance of test Pin 21 to test Pins 40, 46 and 60.	Resistance reading is less than 10,000 ohms ▶	SERVICE short to ground. REPEAT Quick Test.
	Resistance reading is 10,000 ohms or greater ▶	GO to KE7.
KE7 CHECK FOR ISC SIGNAL AT THE PROCESSOR		
• Connect processor. • Connect ISC actuator. • Vehicle prepared for Quick Test. • DVOM on 20V scale. • Connect DVOM between test Pin 21 and test Pin 40. • Start engine and observe the DVOM during Self-Test.	Meter reading varies during Quick Test ▶	REPLACE ISC actuator. REPEAT Quick Test.
	Meter reading does not vary during Quick Test ▶	DISCONNECT processor and INSPECT for bent, damaged, corroded pins, etc. CORRECT as necessary. If OK, REPLACE processor. REPEAT Quick Test.
KE8 CODE 13: RPM DOES NOT DROP TO LESS THAN 1500 AT END OF TEST		
• Disconnect ISC harness connector. • Connect engine tachometer. • Repeat Engine Running Quick Test. • At end of test, record service codes for future use. • Does rpm remain below 1500 rpm during test?	Yes ▶	REPLACE processor. REPEAT Quick Test.
	No ▶	CHECK engine vacuum hoses. REFER to VECI decal. CHECK that throttle plates are fully closed. CHECK throttle linkage and/or speed control linkage for binding. If OK, REPLACE ISC actuator. REPEAT Quick Test.

Idle Speed Control (DC Motor Control) — Pinpoint Test — KF

KF

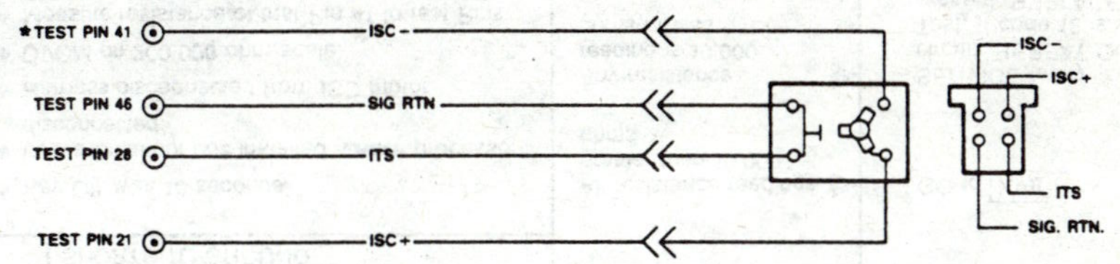

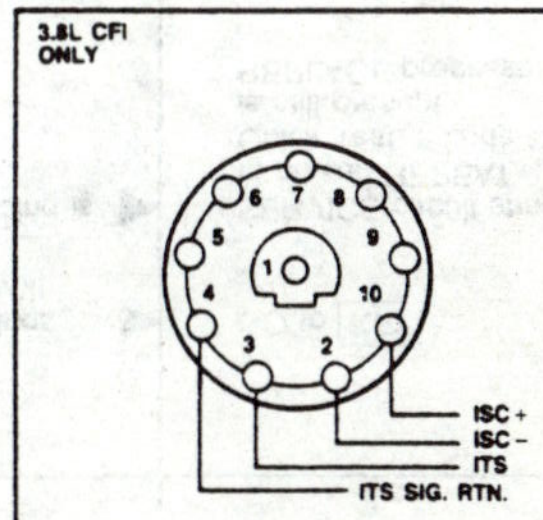

NOTE: FOR 3.8L CFI A 10-PIN CONNECTOR WAS USED FOR THE FIRST 60 DAYS OF PRODUCTION. EITHER CONNECTOR MAY BE ENCOUNTERED. TEST PIN ASSIGNMENTS DID NOT CHANGE ON BREAKOUT BOX.

* TEST PINS LOCATED ON BREAKOUT BOX.
ALL HARNESS CONNECTORS VIEWED INTO MATING SURFACE.

STOP-WARNING

You may have reached this point because of a fault in a non-EEC System and/or component. To prevent the replacement of good components, be aware that the following non-EEC areas may be at fault:

- Anti-Diesel Speed too High
- Basic Engine
- Vacuum Leaks
- Throttle Sticking or on High Cam

This Pinpoint Test is intended to diagnose only the following:

- DC Motor
- Idle Tracking Switch
- Harness circuits: ISC +, ISC −, Signal Return and ITS
- Processor Assembly

RPM Test (DC Motor Control) | Pinpoint Test | KF

TEST STEP	RESULT	ACTION TO TAKE
KF1 DC MOTOR CONTROL FAULT ISOLATION: SERVICE CODE 12		
• Is code 58 present?	Yes ▶	GO to Pinpoint Test Step FB1.
	No ▶	GO to KF2.
KF2 CHECK IF CODE 31 IS PRESENT		
• Is code 31 present?	Yes ▶	For 2.8L Truck and 3.8L, 5.0L CFI engines, GO to Pinpoint Test Step DL1. All others, GO to Pinpoint Test Step DD1.
	No ▶	GO to KF3.
KF3 CHECK IF CODE 41 IS PRESENT		
• Is code 41 present?	Yes ▶	For 1.6L, 2.3L EFI engines, GO to Pinpoint Test Step HA1. For 5.0L CFI engine, GO to Pinpoint Test Step HC1. For 3.8L CFI engine, GO to Pinpoint Test Step HC20. All others, (FBC) GO to Pinpoint Test Step HB1.
	No ▶	GO to KF4.
KF4 CHECK FOR THROTTLE STICKING		
• Check throttle plates and linkages for binding. • Check speed control for binding (if so equipped).	Yes ▶	SERVICE as necessary. REPEAT Quick Test.
	No ▶	GO to KF5.

RPM Test (DC Motor Control) | Pinpoint Test | KF

TEST STEP	RESULT	ACTION TO TAKE
KF5 CHECK TP AND ISC CIRCUITS FOR CONTINUITY		
• Key Off, wait 10 seconds. • Disconnect processor 60 Pin connector and inspect for damaged pins, corrosion, loose wires. Service as necessary. • Connect Breakout box. Leave processor disconnected. • Disconnect harness connector for ISC motor. • DVOM on 200 ohm scale. • Check continuity of test Pin 41 to ISC− circuit and test Pin 21 to ISC+ circuit at harness connector.	Both resistance readings are less than 5 ohms ▶ Either resistance reading is 5 ohms or greater ▶	GO to KF6. SERVICE faulty circuits. REPEAT Quick Test.
KF6 CHECK ISC+ AND ISC− CIRCUITS FOR SHORTS TO POWER		
• Key On, engine Off. • Leave Breakout box installed. Leave processor disconnected. • Harness disconnected from ISC motor. • DVOM on 20V scale. • Measure voltage of test Pin 41 to test Pins 40 and 60. Measure voltage of test Pin 21 to test Pins 40 and 60.	All voltage readings are less than 1 volt ▶ Any voltage reading is 1 volt or more ▶	GO to KF7. SERVICE circuit short to power. REPEAT Quick Test. If code 12 is still present, REPLACE processor.
KF7 CHECK ISC+ AND ISC− CIRCUITS FOR SHORTS TO GROUND		
• Key Off, wait 10 seconds. • Leave Breakout box installed. Leave processor disconnected. • Harness disconnected from ISC motor. • DVOM on 200,000 ohm scale. • Measure resistance of test Pin 41 to test Pins 40, 46 and 60. Measure resistance of test Pin 21 to test Pins 40, 46 and 60.	All resistance readings greater than 10,000 ohms ▶ Any resistance reading is 10,000 ohms or less ▶	GO to KF8. SERVICE faulty circuit. REPEAT Quick Test. If code 12 is still present, REPLACE processor.

RPM Test (DC Motor Control) — Pinpoint Test — KF

TEST STEP	RESULT ▶	ACTION TO TAKE
KF8 PREPARE IDLE SPEED CONTROL DC MOTOR FOR CHECK OF OPERATION		
• Key Off, wait 10 seconds. • Leave Breakout box installed. Leave processor disconnected. • Harness disconnected from ISC motor. • DVOM on 20V scale. • Connect DVOM positive test lead to test Pin 1 and negative test lead to ground.	Voltage reading is 10.5V or greater ▶	GO to KF9.
	Voltage reading is less than 10.5V ▶	SERVICE open in circuit 37. REPEAT Quick Test. If code 12 is still present, GO to KF9.
KF9 PREPARE ISC DC MOTOR FOR OPERATION CHECK		
• Key Off, wait 10 seconds. • Leave Breakout box installed. Leave processor disconnected. • Connect ISC DC motor to harness connector. • Jumper test Pin 21 to test Pin 1 and test Pin 41 to ground (to extend DC motor). • Does DC motor shaft extend 5 cm (2 inches) or more?	Yes ▶	GO to KF10.
	No ▶	REPLACE DC motor. REPEAT Quick Test.
KF10 RETRACT ISC DC MOTOR		
• Key Off, wait 10 seconds. • Leave Breakout box installed. Leave processor disconnected. • Jumper test Pin 41 to test Pin 1 and test Pin 21 to ground (to retract DC motor). • Does ISC DC motor retract?	Yes ▶	GO to KF11.
	No ▶	REPLACE DC motor. REPEAT Quick Test.
KF11 EXTEND ISC DC MOTOR		
• Key Off, wait 10 seconds. • Leave Breakout box installed. Leave processor disconnected. • Jumper test Pin 21 to test Pin 1 and test Pin 41 to ground (to extend DC motor). • Does ISC DC motor extend?	No ▶	REPLACE DC motor. REPEAT Quick Test.
	Yes ▶	REPLACE processor. REPEAT Quick Test.

RPM Test (DC Motor Control) — Pinpoint Test — KF

TEST STEP	RESULT ▶	ACTION TO TAKE
KF20 RPM NOT RETURNING TO NORMAL SPEED: CODE 13		
• Prepare system for normal engine operation. • De-activate Self-Test. • Air conditioner Off. • Run engine. Alternate between 30 seconds at idle and 5 seconds at part throttle modes for 3 minutes. • Is idle speed erratic?	No ▶	GO to KF21.
	Yes ▶	CHECK for: • Vacuum leaks • Code 22 • Code 31 • Code 41 • Code 58 SERVICE above codes as necessary before continuing.
KF21 PERFORM ENGINE RUNNING SELF-TEST		
• Connect Self-Test equipment. • Perform Engine Running Self-Test. • Is code 13 present?	No ▶	SERVICE other codes as necessary.
	Yes ▶	GO to KF22.
KF22 CHECK FOR THROTTLE STICKING OR CHOKE POSITION		
• Inspect throttle mechanisms (plate linkage) for sticking or binding. • Inspect choke for sticking or binding. • Inspect for throttle stuck on high cam. • Are above inspections satisfactory?	No ▶	SERVICE fault. REPEAT Quick Test.
	Yes ▶	GO to KF23.
KF23 PERFORM KEY ON — ENGINE OFF TEST. OBSERVE ON DEMAND CODES		
• Perform Key On, Engine Off Test. Record On Demand codes. • Key On, Engine Off. • Is code 68 present?	Yes ▶	GO to Pinpoint Test Step FB5.
	No ▶	REFER to Section 4 for anti-diesel speed adjustment.

Idle Speed Control (Throttle Kicker) — Pinpoint Test — KG

KG

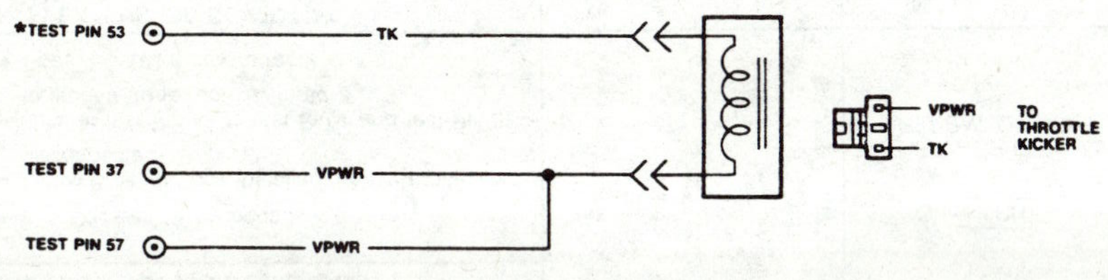

*TEST PINS LOCATED ON BREAKOUT BOX.
ALL HARNESS CONNECTORS VIEWED INTO MATING SURFACE.

STOP-WARNING

You may have reached this point because of a fault in a non-EEC System and/or component. To prevent the replacement of good components, be aware that the following non-EEC areas may be at fault:

- Engine Running Below Operating Temperature
- Engine Running Above Operating Temperature
- Improper Idle Speeds
- A/C Input (Electrical Problems)

This Pinpoint Test is intended to diagnose only the following:

- TK Solenoid
- TK Actuator
- Harness circuits: TK and VPWR
- Processor Assembly
- Vacuum Source
- Actuator Check

Idle Speed Control (Throttle Kicker) — Pinpoint Test — KG

TEST STEP	RESULT ▶	ACTION TO TAKE
KG1 VERIFY LINE ROUTINGS AND VISUAL CHECKS		
• Verify proper vacuum line routing to the TK solenoid. Refer to VECI decal. • Check for kinked or blocked vacuum lines. • Check for kinked or blocked air hoses. • Check for disconnected vacuum lines. • Are visual checks satisfactory?	No ▶	SERVICE routing or faults. REPEAT Quick Test.
	Yes ▶	Service codes 12 or 13. GO to KG2.
KG2 VACUUM CHECK		
• Apply 20 in Hg vacuum to throttle kicker. • Verify vacuum holds.	Yes ▶	GO to KG3.
	No ▶	REPLACE solenoid. REPEAT Quick Test.
KG3 ATTEMPT TO ELIMINATE SERVICE CODE 12 OR 13		
• While vacuum is applied to throttle kicker, rerun Engine Running Self-Test. • Is service code 12 or 13 present?	Yes ▶	EEC-IV system OK. REFER to Section 4 for idle adjustment.
	No ▶	GO to KG4.
KG4 MEASURE TK SOLENOID RESISTANCE		
• Key Off, wait 10 seconds. • DVOM on 200 ohm scale. • Disconnect TK solenoid connector and measure solenoid resistance.	Resistance is between 50-100 ohms ▶	CONNECT TK solenoid. GO to KG5.
	Resistance is less than 50 ohms or greater than 100 ohms ▶	REPLACE TK solenoid. REPEAT Quick Test.

Idle Speed Control (Throttle Kicker) — Pinpoint Test — KG

TEST STEP	RESULT ▶	ACTION TO TAKE
KG5 ENTER OUTPUT STATE CHECK (REFER TO APPENDIX)		
• Key Off, wait 10 seconds. • DVOM on 20V scale. • Connect DVOM negative test lead to STO and positive test lead to battery positive. • Jumper STI to signal return. • Perform Key On. Engine Off Self-Test until the completion of the Continuous Test Codes. • DVOM will indicate zero volts. • Depress and release the throttle. • Did DVOM reading change to a high voltage reading?	Yes ▶	REMAIN in Output State Check. GO to KG6.
	No ▶	DEPRESS throttle to WOT and release. If STO voltage does not go high, GO to Pinpoint Test Step Q40. Leave equipment hooked up.
KG6 CHECK TK SOLENOID ELECTRICAL OPERATION		
• DVOM on 20V scale. • Connect DVOM positive test lead to circuit VPWR on solenoid and negative test lead to output circuit TK on solenoid. • While observing DVOM, depress and release the throttle several times to cycle output On and Off. • TK output cycles On and Off?	Yes ▶	GO to KG7.
	No ▶	GO to KG9.
KG7 CHECK TK SOLENOID FOR VACUUM CYCLING		
• Install vacuum pump to the TK solenoid vacuum supply port and install a vacuum gauge to the output port. Apply 6 in Hg minimum. • While cycling outputs On and Off (by depressing and releasing throttle) observe the vacuum gauge at the output. **NOTE: Maintain vacuum at source.**	Vacuum output cycles On and Off ▶	GO to KG8.
	Vacuum output does not cycle On and Off ▶	REPLACE solenoid. REPEAT Quick Test.

Idle Speed Control (Throttle Kicker) — Pinpoint Test — KG

TEST STEP	RESULT ▶	ACTION TO TAKE
KG8 CHECK MANIFOLD VACUUM LINES FOR BLOCKAGE OR LEAKS		
• With vacuum lines disconnected at TK solenoid, check for vacuum. • Start engine.	Vacuum present ▶	EEC-IV system OK. REFER to Section 4 for idle adjustment.
	No vacuum present ▶	SERVICE vacuum source blockage or leak. REPEAT Quick Test.
KG9 CHECK VOLTAGE OF VPWR CIRCUIT		
• Key On, Engine Off. • DVOM on 20V scale. • Connect DVOM positive test lead to VPWR circuit and negative test lead to ground. • Measure voltage on TK solenoid VPWR circuit.	Voltage reading is 10.5V or less ▶	SERVICE harness open circuit. REPEAT Quick Test.
	Voltage reading is 10.5V or greater ▶	GO to KG10.
KG10 CHECK CONTINUITY OF TK CIRCUIT		
• Key Off, wait 10 seconds. • Disconnect processor and inspect both 60 Pin connectors. • Connect Breakout box to harness, leave processor disconnected. • DVOM on 200 ohm scale. • Measure continuity of test Pin 53 to TK circuit at harness connector.	Resistance reading is 5 ohms or greater ▶	SERVICE harness open circuit. REPEAT Quick Test.
	Resistance reading is less than 5 ohms ▶	GO to KG11.
KG11 CHECK FOR SHORT TO POWER		
• Key Off, wait 10 seconds. • DVOM on 200,000 ohm scale. • Leave Breakout box installed and processor disconnected. • With TK solenoid disconnected, measure resistance from test Pin 53 to test Pins 37 and 57.	Resistance reading is 10,000 ohms or greater ▶	GO to KG12.
	Resistance reading is less than 10,000 ohms ▶	SERVICE short to power. REPEAT Quick Test.

Idle Speed Control (Throttle Kicker) — Pinpoint Test — KG

TEST STEP	RESULT ▶	ACTION TO TAKE
KG12 CHECK FOR SHORT TO GROUND		
• Key Off, wait 10 seconds. • Leave Breakout box installed and processor disconnected. • DVOM on 200,000 ohm scale. • Measure resistance of test Pin 53 to test Pins 40, 46 and 60.	Resistance reading is less than 10,000 ohms ▶	SERVICE short to ground. REPEAT Quick Test.
	Resistance reading is 10,000 ohms or greater ▶	REPLACE processor. REPEAT Quick Test.

Variable Voltage Choke (VVC) — Pinpoint Test — KJ

KJ

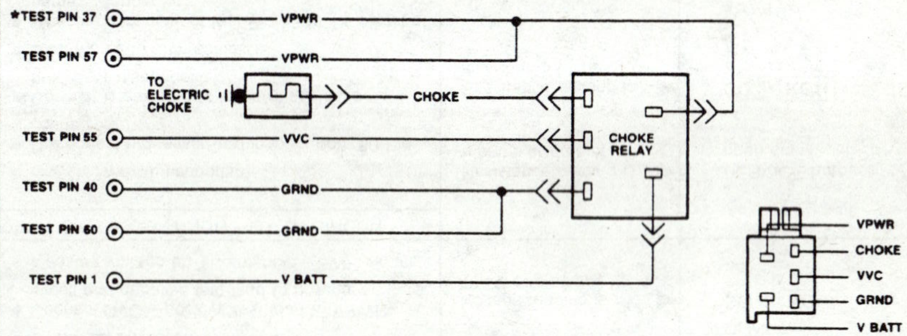

*TEST PINS LOCATED ON BREAKOUT BOX.
ALL HARNESS CONNECTORS VIEWED INTO MATING SURFACE.

STOP-WARNING

This Pinpoint Test is intended to diagnose only the following:

- Harness circuits: VPWR, Choke, VVC, Grnd., and V Batt.
- Choke Relay

Variable Voltage Choke (VVC) — Pinpoint Test KJ

TEST STEP	RESULT	ACTION TO TAKE
KJ1 MEASURE CHOKE VOLTAGE • DVOM on 20V scale. • Perform Engine Running Self-Test. • Measure choke voltage.	Choke voltage is 10.5V or greater ▶	CHECK choke plates for binding and sticking. EEC-IV system OK. REFER to Section 4 for choke diagnostics.
	Choke voltage is less than 10.5V ▶	GO to KJ2.
KJ2 CHECK FOR VOLTAGE ON V BATT CIRCUIT • Key On, Engine Off. • DVOM on 20V scale. • Connect DVOM positive test lead to V Batt. at relay harness connector and negative test lead to ground at relay harness connector.	Voltage is less than 10.5V ▶	SERVICE open in choke power circuit. REPEAT Quick Test.
	Voltage is 10.5V or greater ▶	GO to KJ3.
KJ3 CHECK FOR SHORT IN GROUND CIRCUIT • Key On, Engine Off. • DVOM on 20V scale. • Measure voltage with DVOM from test Pins 40 and 60 to negative battery post.	Less than 1.0V on both circuits ▶	Ground OK. GO to KJ4.
	1.0V or greater on one or both circuits ▶	CORRECT faulty ground circuit (high resistance indicated by voltage reading). REPEAT Quick Test.

Variable Voltage Choke (VVC) — Pinpoint Test KJ

TEST STEP	RESULT	ACTION TO TAKE
KJ4 ENTER OUTPUT STATE CHECK (REFER TO APPENDIX) • Key Off, wait 10 seconds. • DVOM on 20V scale. • Connect DVOM negative test lead to STO and positive test lead to battery positive. • Jumper STI to signal return. • Perform Key On, Engine Off Self-Test until the completion of the Continuous Test Codes. • DVOM will indicate zero volts. • Depress and release the throttle. • Did DVOM reading change to a high voltage reading?	Yes ▶	REMAIN in Output State Check. GO to KJ5.
	No ▶	DEPRESS throttle to WOT and RELEASE. If STO voltage does not go high, GO to Pinpoint Test Step Q40.
KJ5 CHECK VVC OUTPUT ELECTRICAL OPERATION • Key On, Engine Off. • DVOM on 20V scale. • Connect DVOM positive test lead to V Batt. on VVC relay and negative test lead to VVC output at choke. • While observing DVOM depress and release the throttle several times to cycle output On and Off. • VVC output cycles On and Off?	Yes ▶	EEC-IV system OK. REFER to Section 4 for choke diagnostics.
	No ▶	REPLACE choke relay. REPEAT Quick Test.

Temperature Compensated Pump — Pinpoint Test — KK

KK

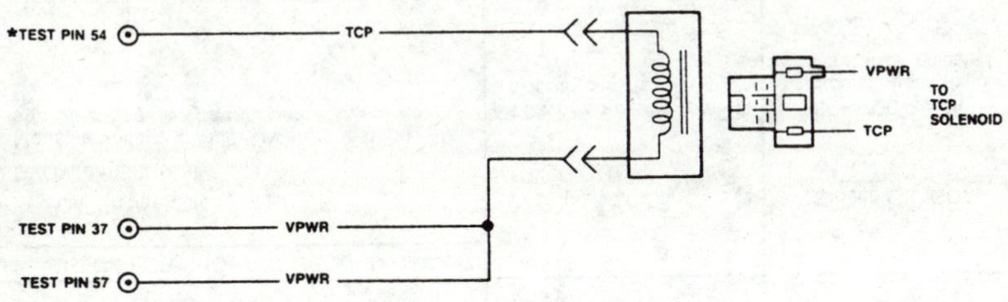

*TEST PINS LOCATED ON BREAKOUT BOX.
ALL HARNESS CONNECTORS VIEWED INTO MATING SURFACE.

STOP-WARNING

This Pinpoint Test is intended to diagnose only the following:

- Harness circuits: TCP and VPWR
- TCP Solenoid
- Processor Assembly

Temperature Compensated Pump — Pinpoint Test — KK

TEST STEP	RESULT ▶	ACTION TO TAKE
KK1 MEASURE TCP SOLENOID RESISTANCE		
• Key Off, wait 10 seconds. • DVOM on 200 ohm scale. • Disconnect TCP solenoid connector and measure solenoid resistance.	Resistance is between 50-100 ohms ▶	CONNECT TCP solenoid. GO to KK2.
	Resistance is less than 50 ohms or greater than 100 ohms ▶	REPLACE TCP solenoid. REPEAT Quick Test.
KK2 ENTER OUTPUT STATE CHECK (REFER TO APPENDIX)		
• Key Off, wait 10 seconds. • DVOM on 20V scale. • Connect DVOM negative test lead to STO and positive test lead to battery positive. • Jumper STI to signal return. • Perform Key On, Engine Off Self-Test until the completion of the Continuous Test Codes. • DVOM will indicate zero volts. • Depress and release the throttle. • Did DVOM reading change to a high voltage reading?	Yes ▶	REMAIN in Output State Check. GO to KK3.
	No ▶	DEPRESS throttle to WOT and release. If STO voltage does not go high, GO to Pinpoint Test Step Q40. Leave equipment hooked up.
KK3 CHECK TCP SOLENOID ELECTRICAL OPERATION		
• DVOM on 20V scale. • Connect DVOM positive test lead to circuit VPWR on TCP solenoid and negative test lead to output circuit TCP. • While observing DVOM depress and release the throttle several times to cycle output On and Off. • Does TCP output cycle On and Off?	Yes ▶	GO to KK4.
	No ▶	GO to KK6.

Temperature Compensated Pump — Pinpoint Test — KK

TEST STEP	RESULT ▶	ACTION TO TAKE
KK4 CHECK TCP SOLENOID FOR VACUUM CYCLING		
• Install vacuum pump to the TCP solenoid vacuum supply port and install a vacuum gauge to the output port. Apply 6 in Hg minimum. • While cycling outputs On and Off (by depressing and releasing throttle) observe the vacuum gauge at the output. **NOTE: Maintain vacuum at source.**	Vacuum output cycles On and Off ▶	GO to KK5.
	Vacuum output does not cycle On and Off ▶	REPLACE solenoid. REPEAT Quick Test.
KK5 CHECK MANIFOLD VACUUM LINES FOR BLOCKAGE OR LEAKS		
• With vacuum lines disconnected at TCP solenoid, check for vacuum. • Start engine.	Vacuum present ▶	EEC-IV system OK. REFER to Section 11 for TCP diagnostics.
	No vacuum present ▶	SERVICE vacuum source blockage or leak. REPEAT Quick Test.
KK6 CHECK VOLTAGE OF VPWR CIRCUIT		
• Key On, Engine Off. • DVOM on 20V scale. • Connect DVOM positive test lead to VPWR circuit and negative test lead to ground. • Measure voltage on TCP solenoid VPWR circuit.	Voltage readings is 10.5V or less ▶	SERVICE harness open circuit. REPEAT Quick Test.
	Voltage reading is 10.5V or greater ▶	GO to KK7.
KK7 CHECK CONTINUITY OF TCP CIRCUIT		
• Key Off, wait 10 seconds. • Disconnect processor and inspect both 60 Pin connectors. • Connect Breakout box to harness, leave processor disconnected. • DVOM on 200 ohm scale. • Measure continuity of test Pin 54 to TCP circuit at harness connector.	Resistance reading is 5 ohms or greater ▶	SERVICE harness open circuit. REPEAT Quick Test.
	Resistance reading is less than 5 ohms ▶	GO to KK8.

Temperature Compensated Pump — Pinpoint Test — KK

TEST STEP	RESULT ▶	ACTION TO TAKE
KK8 CHECK FOR SHORT TO POWER		
• Key Off, wait 10 seconds. • DVOM on 200,000 ohm scale. • Leave Breakout box installed and processor disconnected. • With TCP solenoid disconnected, measure resistance from test Pin 54 to test Pins 37 and 57.	Resistance reading is 10,000 ohms or greater ▶	GO to KK9.
	Resistance reading is less than 10,000 ohms ▶	SERVICE short to power. REPEAT Quick Test.
KK9 CHECK FOR SHORT TO GROUND		
• Key Off, wait 10 seconds. • Leave Breakout box installed and processor disconnected. • DVOM on 200,000 ohm scale. • Measure resistance of test Pin 54 to test Pin 40, 46 and 60.	Resistance reading is less than 10,000 ohms ▶	SERVICE short to ground. REPEAT Quick Test.
	Resistance reading is 10,000 ohms or greater ▶	REPLACE processor. REPEAT Quick Test.

Shift Indicator Light (SIL) | Pinpoint Test | KL

KL

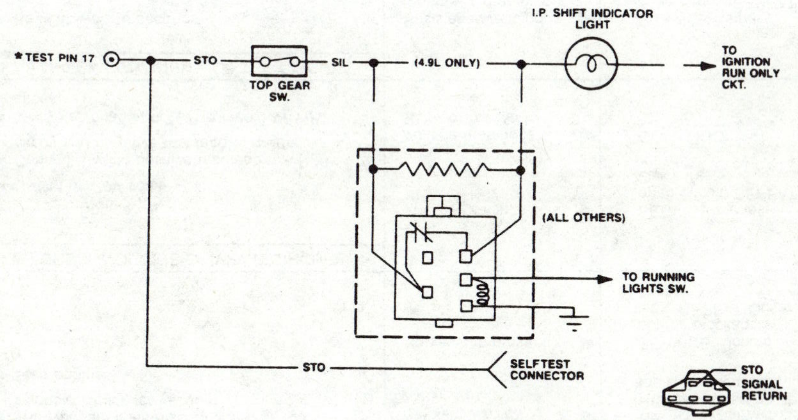

*TEST PINS LOCATED ON BREAKOUT BOX.
ALL HARNESS CONNECTORS VIEWED INTO MATING SURFACE.

STOP-WARNING

This Pinpoint Test is intended to diagnose only the following:

- Top Gear Switch
- Shift Indicator Light Bulb and Fuse
- Harness Circuits: SIL and STO

Shift Indicator Light (SIL) | Pinpoint Test | KL

TEST STEP	RESULT ▶	ACTION TO TAKE
KL1 DETERMINE TYPE OF FAULT		
NOTE: Before proceeding with this Pinpoint Test, Quick Test Steps 1 through 5 must be completed and all fault codes serviced. Code 11 must be present for both Key On, Engine Off and Engine Running Quick Tests.		
• Verify SIL fault by drive evaluating vehicle.	SIL is always Off ▶	GO to KL2.
	SIL is not always Off ▶	GO to KL4.
KL2 CHECK CONTINUITY OF STO CIRCUIT TO TOP GEAR SWITCH		
• Key Off, wait 10 seconds. • Disconnect harness from top gear switch. Inspect for damaged pins, corrosion, loose wires, etc. Service as necessary. • DVOM on 200 ohm scale. • Measure resistance of top gear switch harness connector STO circuit to Self-Test connector.	Reading is less than 5 ohms ▶	GO to KL3.
	Reading is 5 ohms or greater ▶	SERVICE circuit open. RECONNECT top gear switch. RE-EVALUATE SIL function.
KL3 CHECK TOP GEAR SWITCH, CONTINUITY TRANSMISSION IN FIRST GEAR		
• Key Off, harness disconnected from top gear switch. • **Transmission in first gear.** • DVOM on 200 ohm scale. • Measure resistance of top gear switch.	Reading is less than 5 ohms ▶	CHECK fuse and bulb, if OK SERVICE circuit open in SIL circuit and Ign. RECONNECT top gear switch. RE-EVALUATE SIL.
	Reading is 5 ohms or greater ▶	REPLACE top gear switch. RE-EVALUATE SIL.
KL4 CHECK TOP GEAR SWITCH, TRANSMISSION IN TOP GEAR		
• Key Off, wait 10 seconds. • Disconnect harness from top gear switch. Inspect for damaged pins, corrosion, loose wires, etc. Service as necessary. • DVOM on 200,000 ohm scale. • **Transmission in top gear.** • Measure resistance of top gear switch.	Reading is less than 10,000 ohms ▶	REPLACE top gear switch. RE-EVALUATE SIL.
	Reading is 10,000 ohms or greater ▶	SERVICE short in SIL circuit. RECONNECT top gear switch. RE-EVALUATE SIL.

WOT A/C Cutoff (WAC) | Pinpoint Test | KM

KM | 1.6L AND 2.3L HSC ENGINES

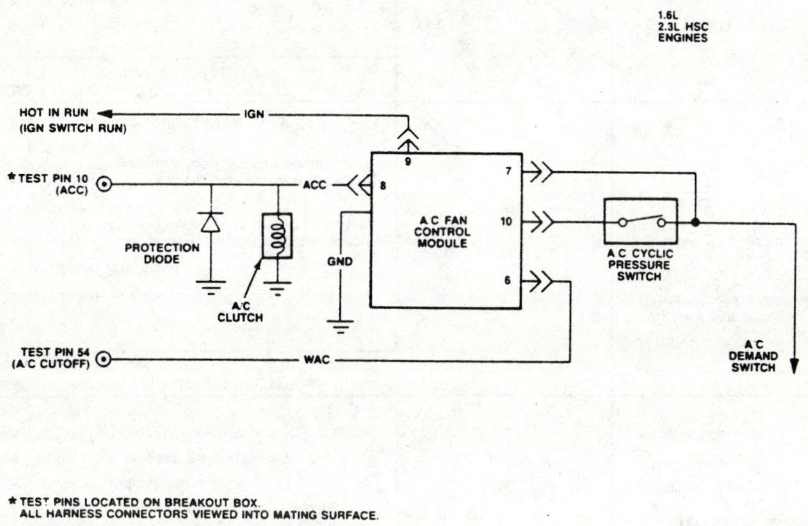

STOP-WARNING

This Pinpoint Test is intended to diagnose only the following:

- Harness Circuits: Ign., Acc., Gnd., WAC, and A/C Demand Switch
- A/C Fan Electrical Controller
- Processor Assembly

WOT A/C Cutoff (WAC) | Pinpoint Test | KM

KM | 2.3L OHC ENGINE

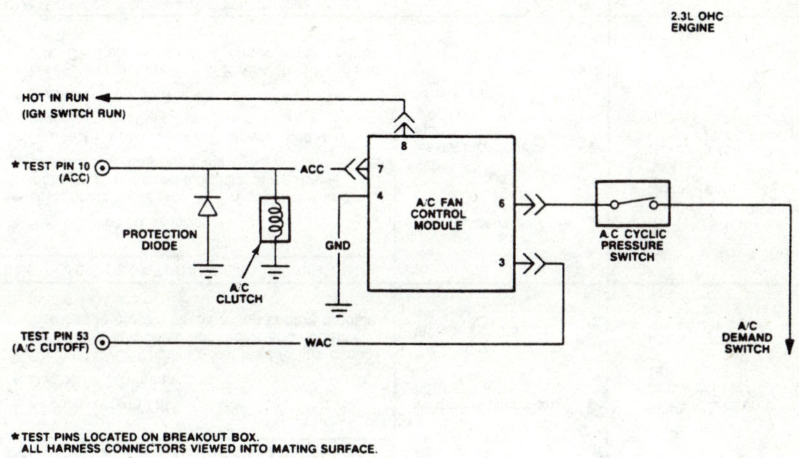

STOP-WARNING

This Pinpoint Test is intended to diagnose only the following:

- Harness Circuits: Ign., Acc., Gnd., WAC, and A/C Demand Switch
- A/C Fan Electrical Controller
- Processor Assembly

WOT A/C Cutoff (WAC) — Pinpoint Test — KM

TEST STEP	RESULT ▶	ACTION TO TAKE
KM1 ENTER OUTPUT STATE CHECK (REFER TO APPENDIX)		
• Key Off, wait 10 seconds. • DVOM on 20V scale. • Connect DVOM negative test lead to STO and positive test lead to battery positive. • Jumper STI to signal return. • Perform Key On, Engine Off Self-Test until the completion of the Continuous Test Codes. • DVOM will indicate zero volts. • Depress and release the throttle. • Did DVOM reading change to a high voltage reading?	Yes ▶	REMAIN in Output State Check. GO to KM2.
	No ▶	DEPRESS throttle to WOT and RELEASE. If STO voltage does not go high, GO to Pinpoint Test Step Q40. Leave equipment hooked up.
KM2 CHECK A/C CLUTCH ELECTRICAL OPERATION		
• Key On, engine Off. • A/C switch to A/C. • DVOM on 20V scale. • Connect DVOM positive test lead to acc. circuit on A/C clutch and negative test lead to battery negative. • While observing DVOM, depress and release the throttle several times (to cycle output On and Off). • Does A/C clutch output cycle On and Off?	Yes ▶	EEC-IV system OK. Refer to Shop Manual Group 36.
	No ▶	REMAIN in Output State Check. GO to KM3.
KM3 CHECK A/C CUTOFF OUTPUT ELECTRICAL OPERATION		
• Key On, engine Off. • DVOM on 20V scale. • Connect DVOM positive test lead to ign. circuit on fan control module and negative lead to WAC circuit on fan control module. • While observing DVOM, depress and release the throttle several times to cycle the output On and Off. • Does A/C cutoff cycle On and Off?	Yes ▶	GO to KM4.
	No ▶	GO to KM5.

WOT A/C Cutoff (WAC) — Pinpoint Test — KM

TEST STEP	RESULT ▶	ACTION TO TAKE
KM4 CHECK CONTINUITY OF GROUND CIRCUIT TO FAN CONTROL MODULE		
• Key Off, wait 10 seconds. • DVOM on 200 ohm scale. • Measure continuity of ground circuit to fan control module connector.	Resistance reading is 5 ohms or greater ▶	SERVICE harness circuit. REPEAT Quick Test.
	Resistance reading is less than 5 ohms ▶	REPLACE fan control module. REPEAT Quick Test.
KM5 CHECK FOR IGN. CIRCUIT VOLTAGE		
• Key On, engine Off. • DVOM on 20V scale. • Connect DVOM positive test lead to ign. circuit at fan control module connector and negative test lead to ground circuit at fan control module connector.	Voltage is less than 10.5V ▶	SERVICE open in ign. circuit. REPEAT Quick Test.
	Voltage is 10.5V or greater ▶	GO to KM6.
KM6 CHECK CONTINUITY OF WAC CIRCUIT		
• Key Off, wait 10 seconds. • Disconnect processor 60 Pin connector and inspect for damaged pins, corrosion, loose wires. Service as necessary. • Connect Breakout box to harness. Leave processor disconnected. • DVOM on 200 ohm scale. • Measure continuity of test Pin 54 (test Pin 53 for 2.3L OHC engine) to WAC circuit at fan control module.	Resistance is 5 ohms or greater ▶	SERVICE harness WAC circuit. REPEAT Quick Test.
	Resistance is less than 5 ohms ▶	GO to KM7.
KM7 CHECK FOR SHORT TO GROUND ON WAC CIRCUIT		
• Key Off, wait 10 seconds. • Leave Breakout box installed and processor disconnected. • DVOM on 200,000 ohm scale. • Measure resistance of test Pin 54 (test Pin 53 for 2.3L OHC engine) to test Pins 40, 46 and 60.	Resistance reading is less than 10,000 ohms ▶	GO to KM8.
	Resistance reading is 10,000 ohms or greater ▶	REPLACE processor. REPEAT Quick Test.

WOT A/C Cutoff (WAC) — Pinpoint Test — KM

TEST STEP	RESULT ▶	ACTION TO TAKE
KM8 CHECK FOR SHORT TO GROUND WITH FAN CONTROL MODULE DISCONNECTED		
• Key Off. wait 10 seconds. • Leave Breakout box installed and processor disconnected. • DVOM on 200.000 ohm scale. • Disconnect fan control module. • Measure resistance of test Pin 54 (test Pin 53 for 2.3L OHC engine) to test Pins 40. 46 and 60.	All resistance readings 10.000 ohms or greater ▶	REPLACE fan control module. REPEAT Quick Test.
	Any resistance reading is less than 10.000 ohms ▶	SERVICE harness short. REPEAT Quick Test.

Turbo Boost — Pinpoint Test — KN

KN

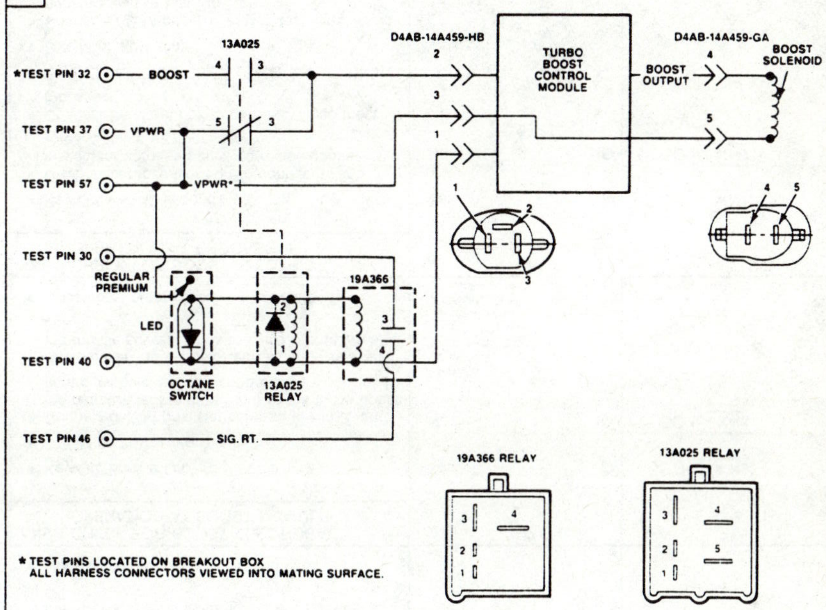

* TEST PINS LOCATED ON BREAKOUT BOX
ALL HARNESS CONNECTORS VIEWED INTO MATING SURFACE.

STOP-WARNING

You may have reached this point because of a fault in a non-EEC System and or component. To prevent the replacement of good components. be aware that the following non-EEC areas may be at fault:

- Contaminated Fuel (premium fuel required).
- Octane Switch must be in the premium position.

This Pinpoint Test is intended to diagnose only the following:

- Harness circuits: VPWR. Boost. Grd.. Signal Return. Octane Switch. Input (NOS). and Boost Output from the Turbo Boost Control Module.

NOTE: The boost system is protected electrically by a fuse. Refer to the Owner's Manual for checkout and replacement before proceeding with the Pinpoint Tests.

Turbo Boost — Pinpoint Test — KN

TEST STEP	RESULT ▶	ACTION TO TAKE
KN1 TEST FOR CODE 67		
• Place octane switch in regular position. • Perform Key On, Engine Off Self-Test. • Is code 67 present?	Yes ▶	GO to KN2.
	No ▶	GO to KN15.
KN2 MEASURE BOOST SOLENOID RESISTANCE		
• Key Off, wait 10 seconds. • DVOM on 200 ohm scale. • Disconnect boost solenoid connector and measure the solenoid resistance.	Resistance is between 1.0-2.0 ohms ▶	GO to KN3.
	Resistance is less than 1.0 or greater than 2.0 ohms ▶	REPLACE solenoid. RETEST.
KN3 ENTER OUTPUT STATE CHECK (REFER TO APPENDIX)		
• Place octane switch in premium position. • Key Off, wait 10 seconds. • DVOM on 20V scale. • Connect DVOM negative test lead to STO and positive test lead to battery positive. • Jumper STI to signal return. • Perform Key On, Engine Off Self-Test until the completion of the Continuous Test Codes. • DVOM will indicate zero volts. • Depress and release the throttle. • Did DVOM reading change to a high voltage reading?	Yes ▶	REMAIN in Output State Check. GO to KN4.
	No ▶	DEPRESS throttle to WOT and release. If STO voltage does not go high, GO to Pinpoint Test Step Q40. Leave equipment hooked up.

Turbo Boost — Pinpoint Test — KN

TEST STEP	RESULT ▶	ACTION TO TAKE
KN4 CHECK BOOST OUTPUT ELECTRICAL OPERATION		
• Key On, Engine Off. • DVOM on 20V scale. • Connect DVOM positive test lead to circuit VPWR on boost connector and negative test lead to boost output on boost connector. • While observing DVOM, depress and release the throttle several times to cycle output On and Off. • Does boost output cycle On and Off?	Yes ▶	GO to KN21.
	No ▶	GO to KN5.
KN5 CHECK FOR BOOST ELECTRICAL OPERATION AT BOOST MODULE		
• Key On, Engine Off. • DVOM on 20V scale. • Connect DVOM positive test lead to Pin 2, and the negative test lead to Pin 3 of the turbo boost control module harness connector. • While observing DVOM, depress and release the throttle several times to cycle output On and Off. • Does boost output cycle On and Off?	Yes ▶	GO to KN8.
	No ▶	GO to KN6.
KN6 CHECK FOR BOOST ELECTRICAL OPERATION AT PROCESSOR		
• Key Off, wait 10 seconds. • Disconnect processor 60 Pin connector and inspect for damaged pins, corrosion, loose wires. Service as necessary. • Connect Breakout box to harness. Reconnect processor. • Enter Output State Check (Refer to Step KN3). • DVOM on 20V scale. • Connect DVOM positive lead to test Pin 37 and negative lead to test Pin 32. • While observing DVOM, depress and release the throttle several times to cycle output On and Off. • Does boost output cycle On and Off?	Yes ▶	GO to KN11.
	No ▶	GO to KN7.

Turbo Boost — Pinpoint Test — KN

TEST STEP	RESULT ▶	ACTION TO TAKE
KN7 CHECK FOR SHORT TO POWER AND GROUND		
• Key Off, wait 10 seconds. • Remove jumper from STI and Sig. Rtn. • Breakout box connected. • DVOM to 200,000 ohm scale. • Measure resistance from test Pin 32 to test Pins 37, 46 and 60.	Any reading less than 10,000 ohms ▶	SERVICE short. REPEAT Quick Test.
	All readings 10,000 ohms or greater ▶	REPLACE processor. REPEAT Quick Test.
KN8 CHECK FOR POWER AND GROUND AT TURBO BOOST CONTROL MODULE		
• Key On, Engine Off. • DVOM on 20V scale. • Connect positive test lead to Pins 3 and 5 and negative lead to Pin 1 of turbo boost control module harness.	Voltage reading is 10.5V or greater ▶	GO to KN9.
	Voltage reading is less than 10.5V ▶	SERVICE open in harness. REPEAT Quick Test.
KN9 CHECK CONTINUITY OF BOOST OUTPUT AND VPWR FROM TURBO BOOST MODULE TO BOOST SOLENOID		
• Key Off, wait 10 seconds. • DVOM on 200 ohm scale. • Measure resistance from boost solenoid connector, boost output to Pin 4 and VPWR at the boost solenoid connector to Pin 5 at the turbo boost module connector.	All readings less than 5.0 ohms ▶	GO to KN10.
	Any reading greater than 5.0 ohms ▶	SERVICE open in harness. REPEAT Quick Test.
KN10 CHECK FOR SHORT TO POWER IN BOOST OUTPUT HARNESS		
• Key Off, wait 10 seconds. • DVOM on 200,000 ohm scale. • Disconnect turbo boost control module and boost solenoid. • Measure resistance between Pins 4 and 5 of turbo boost control module harness connector.	Reading is 10,000 ohms or greater ▶	REPLACE turbo boost control module. REPEAT Quick Test.
	Reading is less than 10,000 ohms ▶	SERVICE short in harness. REPEAT Quick Test.

Turbo Boost — Pinpoint Test — KN

TEST STEP	RESULT ▶	ACTION TO TAKE
KN11 CHECK CONTINUITY OF BOOST FROM RELAY TO BREAKOUT BOX		
• Key Off, wait 10 seconds. • DVOM on 200 ohm scale. • Measure resistance from Pin 4 of boost relay to test Pin 32.	Reading is less than 5.0 ohms ▶	GO to KN12.
	Reading is greater than 5.0 ohms ▶	SERVICE open in harness. REPEAT Quick Test.
KN12 CHECK CONTINUITY OF BOOST FROM BOOST RELAY TO TURBO BOOST CONTROL MODULE		
• Key Off, wait 10 seconds. • DVOM on 200 ohm scale. • Measure resistance from Pin 3 of boost relay to Pin 2 of turbo boost control module.	Reading is less than 5.0 ohms ▶	GO to KN13.
	Reading is 5.0 ohms or greater ▶	SERVICE open in harness. REPEAT Quick Test.
KN13 CHECK FOR SHORT TO POWER AND GROUND		
• Key Off, wait 10 seconds. • Breakout box installed. • DVOM to 200,000 ohm scale. • Disconnect turbo boost control module and boost relay. • Measure resistance from Pin 3 of boost relay to test Pins 37, 40 and 46.	Any reading is less than 10,000 ohms ▶	SERVICE short in harness. REPEAT Quick Test.
	All readings are 10,000 ohms or greater ▶	GO to KN14.
KN14 CHECK FOR POWER AND GROUND AT THE BOOST RELAY		
• Key On, engine Off. • Octane switch in premium position. • DVOM to 20V scale. • Connect positive lead to Pin 2 and negative lead to Pin 1 of the turbo boost relay.	Voltage reading is 10.5V or greater ▶	REPLACE boost relay. REPEAT Quick Test.
	Voltage reading is less than 10.5V ▶	SERVICE open in harness. REPEAT Quick Test.

Turbo Boost — Pinpoint Test — KN

TEST STEP	RESULT ▶	ACTION TO TAKE
KN15 CHECK OCTANE SWITCH INPUT FOR INPUT CHANGE		
• Key Off, wait 10 seconds. • Disconnect processor 60 Pin connector and inspect for damaged pins, corrosion, loose wires. Service as necessary. • Install Breakout box. Reconnect processor. • DVOM to 20V scale. • Connect positive test lead to test Pin 30 and negative test lead to test Pin 46. • Key On, engine Off. • Cycle octane switch several times while observing DVOM. • Does voltage change from 0V to 5V?	Yes ▶ No ▶	REPLACE processor. REPEAT Quick Test. GO to KN16.
KN16 CHECK CONTINUITY BETWEEN PROCESSOR AND OCTANE SWITCH RELAY OCTANE SWITCH		
• Key Off, wait 10 seconds. • Breakout box installed. • DVOM on 200 ohm scale. • Measure resistance between test Pin 30 and Pin 3 on the octane switch relay.	Reading is less than 5.0 ohms ▶ Reading is 5.0 ohms or greater ▶	GO to KN17. SERVICE open in harness. REPEAT Quick Test.
KN17 CHECK CONTINUITY BETWEEN PROCESSOR AND OCTANE SWITCH RELAY SIGNAL RETURN		
• Key Off, wait 10 seconds. • Breakout box installed. • DVOM on 200 ohm scale. • Measure resistance between test Pin 46 and Pin 4 on the octane switch relay.	Reading is less than 5.0 ohms ▶ Reading is 5.0 ohms or greater ▶	GO to KN18. SERVICE open in harness. REPEAT Quick Test.
KN18 CHECK FOR POWER AND GROUND AT THE OCTANE SWITCH RELAY		
• Key On, engine Off. • DVOM on 20V scale. • Octane switch in premium position. • Measure voltage between Pin 1 and Pin 2 on the octane switch relay harness connector.	Voltage reading is 10.5V or greater ▶ Voltage reading is less than 10.5V ▶	REPLACE octane switch relay. REPEAT Quick Test. GO to KN19.

Turbo Boost — Pinpoint Test — KN

TEST STEP	RESULT ▶	ACTION TO TAKE
KN19 CHECK CONTINUITY FROM OCTANE SWITCH RELAY TO BATTERY GROUND		
• Key On, Engine Off. • DVOM on 200 ohm scale. • Measure resistance between Pin 1 on octane switch relay and battery ground.	Reading is less than 5.0 ohms ▶ Reading is 5.0 ohms or greater ▶	GO to KN20. SERVICE open in harness. REPEAT Quick Test.
KN20 CHECK FOR POWER AT OCTANE SWITCH		
• Key On, Engine Off. • DVOM on 20V scale. • Connect positive lead to input of octane switch and negative lead to battery ground. • Measure voltage.	Voltage reading is 10.5V or greater ▶ Voltage reading is less than 10.5V ▶	REPLACE octane switch. REPEAT Quick Test. SERVICE open circuit. REPEAT Quick Test.
KN21 CHECK BOOST SOLENOID FUNCTION		
• Remain in output state check. • Place octane switch in the premium position. • Disconnect black solenoid vacuum hose at the turbocharger end. **(Not the black and yellow stripe hose.)** • Attach a vacuum pump to the hose. • Depress the throttle once to cycle the solenoid closed. • Apply vacuum to the solenoid. • Depress the throttle once again to cycle the solenoid open and release trapped vacuum. • Did the solenoid hold and then release vacuum?	Yes ▶ No ▶	EEC-IV system OK. REFER to Shop Manual for boost diagnostics. REPLACE solenoid. RETEST.

Exhaust Heat Control — Pinpoint Test — KP

KP

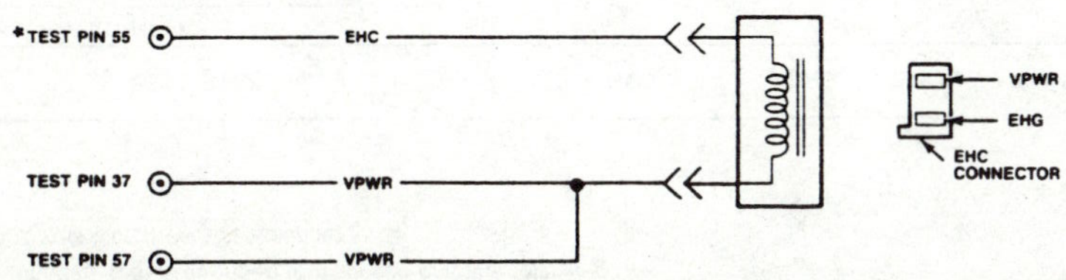

*TEST PINS LOCATED ON BREAKOUT BOX.
ALL HARNESS CONNECTORS VIEWED INTO MATING SURFACE.

STOP-WARNING

This Pinpoint Test is intended to diagnose only the following:

- Exhaust Heat Solenoid
- Vacuum Routing

Exhaust Heat Control — Pinpoint Test — KP

TEST STEP	RESULT ▶	ACTION TO TAKE
KP1 ENTER OUTPUT STATE CHECK (REFER TO APPENDIX)		
• Key Off. wait 10 seconds. • DVOM on 20V scale. • Connect DVOM negative test lead to STO and positive test lead to battery positive. • Jumper STI to signal return. • Perform Key On. Engine Off Self-Test until the completion of the Continuous Test Codes. • DVOM will indicate zero volts. • Depress and release the throttle. • Did DVOM reading change to a high voltage reading?	Yes ▶	REMAIN in Output State Check. GO to KP3.
	No ▶	DEPRESS throttle to WOT and release. If STO voltage does not go high. GO to Pinpoint Test Step Q40. Leave equipment hooked up.
KP2 CHECK EXHAUST HEAT SOLENOID FOR VACUUM CYCLING		
• Install vacuum pump to the exhaust heat solenoid vacuum supply port and install a vacuum gauge to the output port. Apply 6 in Hg minimum. • While cycling outputs On and Off (by depressing and releasing throttle) observe the vacuum gauge at the output. **NOTE: Maintain vacuum at source.**	Vacuum output cycles On and Off ▶	GO to KP3.
	Vacuum output does not cycle On and Off ▶	REPLACE solenoid. REPEAT Quick Test.
KP3 CHECK MANIFOLD VACUUM LINES FOR BLOCKAGE OR LEAKS		
• With vacuum lines disconnected at exhaust heat solenoid, check for vacuum. • Start engine.	Vacuum present ▶	EEC-IV system OK. REFER to Section 5 for exhaust heat diagnostics.
	No vacuum present ▶	SERVICE vacuum source blockage or leak. REPEAT Quick Test.

WOT A/C Cutoff (WAC) — Pinpoint Test — KR

KR

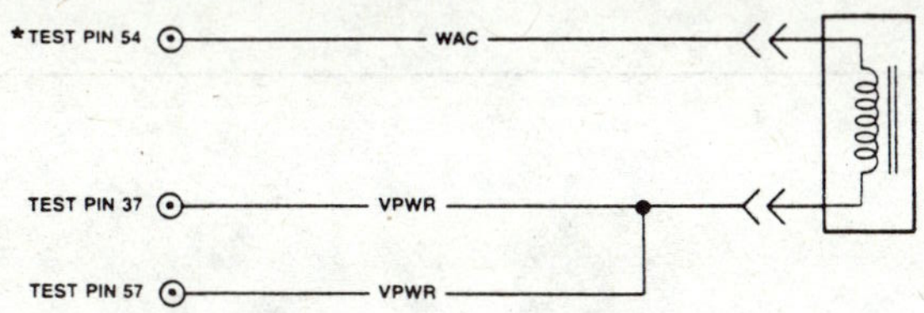

*TEST PINS LOCATED ON BREAKOUT BOX.
ALL HARNESS CONNECTORS VIEWED INTO MATING SURFACE.

STOP-WARNING

This Pinpoint Test is intended to diagnose only the following:

- WOT A/C Cutoff (WAC) Solenoid

TEST STEP	RESULT ▶	ACTION TO TAKE
KR1 CYCLE WAC RELAY		
• Key Off, wait 10 seconds. • Disconnect processor 60 Pin connector and inspect for damaged pins, corrosion, loose wires. Service as necessary. • Connect Breakout box to harness. Leave processor disconnected. • Key On, Engine Off. • DVOM on 20V scale. • Connect DVOM negative test lead to ground and positive test lead to A C clutch. • Turn A C demand switch to On. • Jumper test Pin 54 to test Pin 60. • DVOM will indicate zero volts. • Depress and release throttle. • Did DVOM change to a high voltage reading?	Yes ▶ No ▶	EEC-IV system is OK. REFER to Shop Manual Group 36 for A C system diagnostics. REPLACE WAC relay. REPEAT Quick Test.

Air Management System — Pinpoint Test — KS

KS

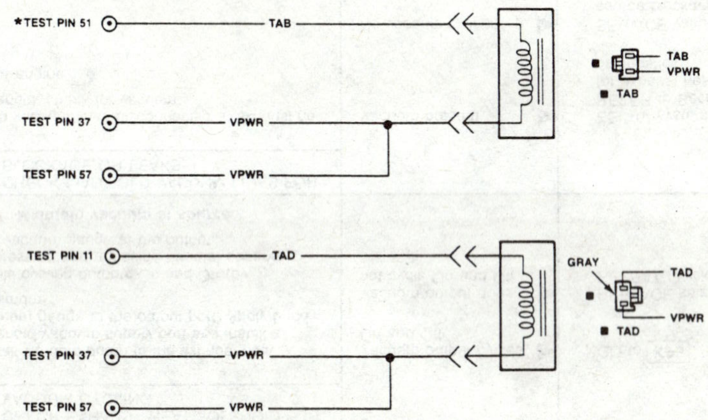

*TEST PINS LOCATED ON BREAKOUT BOX.
ALL HARNESS CONNECTORS VIEWED INTO MATING SURFACE.

STOP-WARNING

You may have reached this point because of a fault in a non-EEC System and/or component. To prevent the replacement of good components, be aware that the following non-EEC areas may be at fault:

- Thermactor System
  - — Belt
  - — Pump
  - — Valve

This Pinpoint Test is intended to diagnose only the following:

- TAB and TAD Solenoid Valve Assemblies
- Vacuum Supply

Air Management System — Pinpoint Test — KS

TEST STEP	RESULT ▶	ACTION TO TAKE
KS1 SERVICE CODES: 44, 45 AND 46: VERIFY VACUUM LINE ROUTING		
• Verify proper vacuum line routing to the TAB TAD solenoids and to the bypass diverter valve. Refer to VECI decal. • Check for kinked or blocked vacuum lines. • Check for kinked or blocked air hoses. • Check for disconnected vacuum lines. • Are visual checks satifactory?	No ▶	SERVICE routing or faults. REPEAT Quick Test.
	Yes ▶	Service code 44: GO to KS4. Service code 45: GO to KS2. Service code 46: GO to KS3.
KS2 ATTEMPT TO ELIMINATE SERVICE CODE 45 (TAD ONLY)		
• Disconnect vacuum line on diverter valve and cap vacuum line. • Key Off, wait 10 seconds. • Repeat Engine Running Self-Test and record service codes. • Is code 45 present?	Yes ▶	EEC-IV system OK. REFER to Section 2 for diverter valve or check valve diagnostics.
	No ▶	GO to KS4.
KS3 ATTEMPT TO ELIMINATE SERVICE CODE 46 (TAB ONLY)		
• Disconnect vacuum line on bypass valve and cap vacuum line. • Key Off, wait 10 seconds. • Repeat Engine Running Self-Test and record codes. • Is code 46 present?	Yes ▶	EEC-IV system OK. REFER to Section 3 for bypass valve diagnostics.
	No ▶	GO to KS4.

Air Management System — Pinpoint Test — KS

TEST STEP	RESULT ▶	ACTION TO TAKE
KS4 ENTER OUTPUT STATE CHECK (REFER TO APPENDIX)		
• Key Off, wait 10 seconds. • DVOM on 20V scale. • Connect DVOM negative test lead to STO and positive test lead to battery positive. • Jumper STI to signal return. • Perform Key On, Engine Off Self-Test until the completion of the Continuous Test Codes. • DVOM will indicate zero volts. • Depress and release the throttle. • Did DVOM reading change to a high voltage reading?	Yes ▶	REMAIN in Output State Check. GO to KS5.
	No ▶	DEPRESS throttle to WOT and RELEASE. If STO voltage does not go high, GO to Pinpoint Test Step Q40. Leave equipment hooked up.
KS5 CHECK TAB TAD SOLENOIDS FOR VACUUM CYCLING		
• Install vacuum pump to the TAB solenoid vacuum supply port and install a vacuum gauge to the output port. • While cycling outputs On and Off (by depressing and releasing throttle) observe the vacuum gauge at the output. **NOTE: Maintain vacuum at source.** • Repeat for TAD solenoid. Connect vacuum pump to the TAD solenoid vacuum supply port and connect a vacuum gauge to the output port. • Cycle output On and Off.	Both vacuum outputs cycle On and Off ▶	GO to KS6.
	Either vacuum output does not cycle On and Off ▶	REPLACE solenoid assembly. REPEAT Quick Test.
KS6 CHECK TAB TAD VACUUMS		
• Disconnect vacuum line from TAB solenoid to thermactor valve. • Inspect vacuum line for blockage, cracks, kinks (e.g. connect vacuum pump and gauge and verify vacuum line holds vacuum, or blow into vacuum line to verify no restrictions). • Repeat this Step for the TAD vacuum line. • Are thermactor vacuum lines OK?	Yes ▶	GO to KS7.
	No ▶	SERVICE as required. REPEAT Quick Test.

Air Management System — Pinpoint Test — KS

TEST STEP	RESULT	▶	ACTION TO TAKE
KS7 CHECK MANIFOLD VACUUM LINES FOR BLOCKAGE OR LEAKS			
• With vacuum lines disconnected at TAD/TAB solenoids. check for vacuum. • Start engine.	Vacuum present	▶	EEC-IV system OK. REFER to Section 3 for thermactor valve and air pump diagnostics.
	No vacuum present	▶	SERVICE vacuum source blockage or leak. REPEAT Quick Test.

Idle Speed Control (Throttle Kicker) — Pinpoint Test — KT

KT

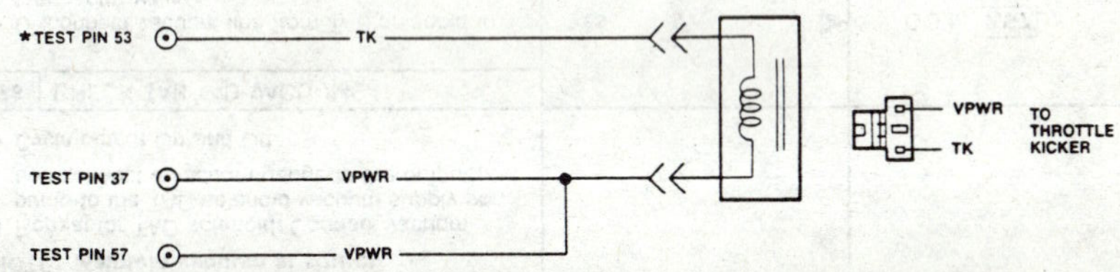

*TEST PINS LOCATED ON BREAKOUT BOX.
ALL HARNESS CONNECTORS VIEWED INTO MATING SURFACE.

STOP-WARNING

You may have reached this point because of a fault in a non-EEC System and/or component. To prevent the replacement of good components, be aware that the following non-EEC areas may be at fault:

- Engine Not At Operating Temperature.
- Engine Above Normal Operating Temperature.
- Improper Idle Speeds.
- A/C Input (Electrical Problems).

This Pinpoint Test is intended to diagnose only the following:

- TK Solenoid.
- TK Actuator.
- Harness circuits: TK and VPWR.
- Vacuum Source.
- Actuator Check.

Idle Speed Control (Throttle Kicker) — Pinpoint Test — KT

TEST STEP	RESULT	ACTION TO TAKE
KT1 VERIFY LINE ROUTINGS AND VISUAL CHECKS		
• Verify proper vacuum line routing to the TK solenoid. Refer to VECI decal. • Check for kinked or blocked vacuum lines. • Check for kinked or blocked air hoses. • Check for disconnected vacuum lines. • Are visual checks satisfactory?	No ▶ Yes ▶	SERVICE routing or faults. REPEAT Quick Test. Service codes 12 or 13: GO to KT2.
KT2 VACUUM CHECK		
• Apply 20 in-Hg vacuum to throttle kicker. • Verify vacuum holds.	Yes ▶ No ▶	GO to KT3. REPLACE solenoid. REPEAT Quick Test.
KT3 ATTEMPT TO ELIMINATE SERVICE CODE 12 OR 13		
• While vacuum is applied to throttle kicker, rerun Engine Running Self-Test. • Is service code 12 or 13 present?	Yes ▶ No ▶	GO to Section 4 for idle adjustment, EEC OK. GO to KT4.
KT4 ENTER OUTPUT STATE CHECK (REFER TO APPENDIX)		
• Key Off, wait 10 seconds. • DVOM on 20V scale. • Connect DVOM negative test lead to STO and positive test lead to battery positive. • Jumper STI to signal return. • Perform Key On, Engine Off Self-Test until the completion of the Continuous Test Codes. • DVOM will indicate zero volts. • Depress and release the throttle. • Did DVOM reading change to a high voltage reading?	Yes ▶ No ▶	REMAIN in Output State Check. GO to KT5. DEPRESS throttle to WOT and release. If STO voltage does not go high, GO to Pinpoint Test Step Q40. Leave equipment hooked up.

Idle Speed Control (Throttle Kicker) — Pinpoint Test — KT

TEST STEP	RESULT	ACTION TO TAKE
KT5 CHECK TK SOLENOID FOR VACUUM CYCLING		
• Install vacuum pump to the TK solenoid vacuum supply port and install a vacuum gauge to the output port. Apply 6 in-Hg minimum. • While cycling outputs On and Off (by depressing and releasing throttle) observe the vacuum gauge at the output. **NOTE: Maintain vacuum at source.**	Vacuum output cycles On and Off ▶ Vacuum output does not cycle On and Off ▶	GO to KT6. REPLACE solenoid. REPEAT Quick Test.
KT6 CHECK MANIFOLD VACUUM LINES FOR BLOCKAGE OR LEAKS		
• With vacuum lines disconnected at TK solenoid, check for vacuum. • Start engine.	Vacuum present ▶ No vacuum present ▶	EEC-IV system OK. REFER to Section 4 for Idle Adjustment. SERVICE vacuum source blockage or leak. REPEAT Quick Test.

Dynamic Response Test | Pinpoint Test | M

M

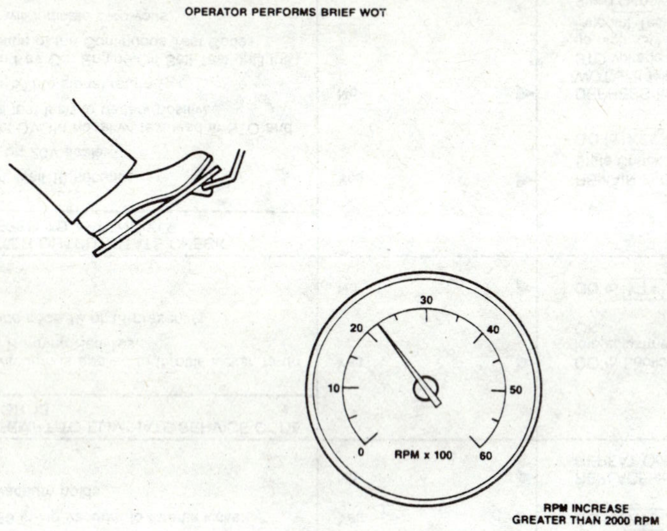

STOP-WARNING

You may have reached this point because of a fault in a non-EEC System and/or component. To prevent the replacement of good components, be aware that the following non-EEC areas may be at fault:

- Operator did not perform a brief WOT after dynamic response code.
- Mechanical engine problems: engine did not achieve greater than 2000 rpm.

This Pinpoint Test is intended to diagnose only the following:

- Throttle movement (greater than 3/4 throttle).
- Vane Air Flow (greater than 50% open).
- RPM increase (greater than 2000 rpm).

Dynamic Response Test | Pinpoint Test | M

TEST STEP	RESULT ▶	ACTION TO TAKE
M1 CODE 77: SYSTEM FAILED TO RECOGNIZE BRIEF WOT		
NOTE: A brief snap of the throttle may not be sufficient to pass this test. Be sure to go to WOT and return.		
• Repeat Engine Running Test of Quick Test. Be sure operator is familiar with the engine running format which proceeds as follows: — With Self-Test activated restart the engine. — ID Code 2 (0) start of test. — Dynamic response Code 1 (0) perform brief WOT. — Testing over. — Service code output begins.	Code 77 still present ▶	REPLACE processor. REPEAT Quick Test.
	No Code 77 ▶	Dynamic Response Test passed. SERVICE any other service code(s) received as necessary.

Ignition Diagnostic Monitor (IDM) | Pinpoint Test | N

N

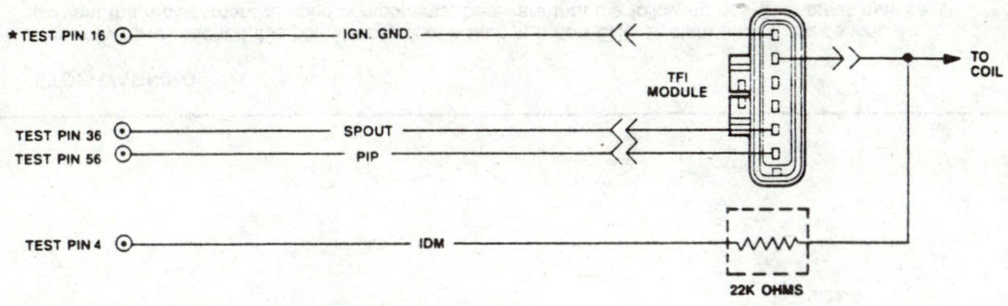

*TEST PINS LOCATED ON BREAKOUT BOX.
ALL HARNESS CONNECTORS VIEWED INTO MATING SURFACE.

STOP-WARNING

You may have reached this point because of a fault in a non-EEC System and/or component. To prevent the replacement of good components, be aware that the following non-EEC areas may be at fault:

- Ignition module.
- Ignition coil.
- Spark plugs and high tension cables.
- Distributor and PIP sensor.

This Pinpoint Test is intended to diagnose only the following:

- Harness circuits: ignition ground, spout, PIP, IDM.

Ignition Diagnostic Monitor (IDM) | Pinpoint Test | N

TEST STEP	RESULT ▶	ACTION TO TAKE
N1 CHECK TFI MODULE		
• Key Off, wait 10 seconds. • Deactivate Self-Test. • Connect VOM or STAR per Quick Test Step 2.0. • Enter Engine Running Continuous Monitor Test (as instructed in Quick Test Step 6.3). • Observe VOM or STAR LED for indication of a fault while performing the following: • Lightly tap on TFI module (simulate road shock). • Wiggle TFI connector. • Is a fault indicated?	Yes ▶ No ▶	DISCONNECT and INSPECT connectors. If connector and terminals are good. GO to Section 15 TFI Ignition Diagnostics. GO to N2.
N2 CHECK EEC-IV HARNESS		
• While still in continuous monitor test from step N1 observe VOM or STAR LED for a fault indication while performing the following: • While looking for faults listed in the table below, grasp the harness close to the TFI connector. Wiggle, shake or bend a small section of the EEC-IV system harness while working your way to the dash panel. Also wiggle, shake or bend the EEC-IV harness from the dash panel to the processor. Do this test on the circuits listed one at a time if needed to locate a faulty circuit. **FAULT** — **BREAKOUT BOX NO.** PIP shorted to ground — Test Pin 56 Spout shorted to ground — Test Pin 36 Ign. ground open — Test Pin 16 IDM open or shorted to ground, power — Test Pin 4 • Is a fault indicated?	Yes ▶ No ▶	ISOLATE fault and make necessary repairs. REPEAT Quick Test. GO to N3.
N3 CHECK PROCESSOR AND HARNESS CONNECTORS		
• Key Off, wait 10 seconds. • Disconnect processor 60 Pin connector and inspect for damaged pins, corrosion, loose wires. • Are connectors and terminals OK? • Reconnect processor when this step is completed.	No ▶ Yes ▶	SERVICE as necessary. REPEAT Quick Test. Unable to DUPLICATE an IDM fault in the EEC-IV System, for further diagnosis GO to Section 15 TFI Ignition System Diagnostics.

Spark Timing Check | Pinpoint Test | P

P

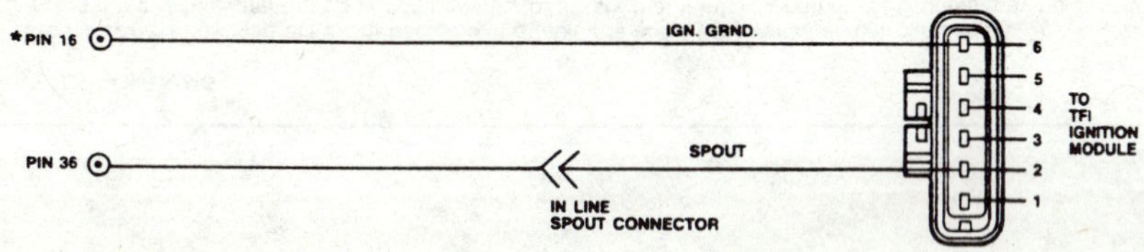

*TEST PINS LOCATED ON BREAKOUT BOX.
ALL HARNESS CONNECTORS VIEWED INTO MATING SURFACE.

STOP-WARNING

You may have reached this point because of a fault in a non-EEC System and/or component. To prevent the replacement of good components, be aware that the following non-EEC areas may be at fault:

- Base Engine.
- PIP Sensor.
- TFI Module.

This Pinpoint Test is intended to diagnose only the following:

- Harness Spout Circuit.
- Base Timing.
- Processor Assembly.

Spark Timing Check | Pinpoint Test | P

TEST STEP	RESULT ▶	ACTION TO TAKE
P1 CHECK SPARK TIMING		
• Verify that Self-Test is activated during Engine Running Test. • Check timing while performing Engine Running Test. Record value. • Timing base plus 20 degrees ± 3 degrees? NOTE: **Self-Test locks the timing at 20 degrees plus base throughout this test and for two minutes after the last service code is outputted. Timing check must be made during this time period. Self-Test timing is base + 20 degrees (±3 degrees) BTDC. (See VECI decal for base value.)**	Yes ▶ No ▶	GO to Quick Test Step 5.0. GO to P2.
P2 CHECK SPARK OUTPUT (SPOUT) CIRCUIT TO THE TFI MODULE		
• Locate spout connector and open this connection. • Start engine. • Check base timing. • Timing base ±3 degrees BTDC. NOTE: **Base timing must be checked to decal value.**	Yes ▶ No ▶	GO to P3. RECONNECT spout connector. ADJUST timing. REFER to Section 15 for engine timing instructions. After timing is reset, PERFORM Quick Test Step 4.
P3 CHECK FOR POWER TO PROCESSOR		
• Key Off, wait 10 seconds. • Disconnect processor 60 Pin connector and inspect for damaged pins, corrosion, loose wires. Service as necessary. • Install Breakout Box. • Key On, Engine Off. • DVOM on 20V scale. • Measure voltage from test Pin 37 to test Pin 40 and test Pin 57 to test Pin 60.	Either voltage reading less than 10.5V ▶ Both voltage readings 10.5V or greater ▶	GO to Pinpoint Test Step B1. GO to P4.

Spark Timing Check — Pinpoint Test P

TEST STEP	RESULT ▶	ACTION TO TAKE
P4 CHECK HARNESS FOR CONTINUITY		
• Key Off, wait 10 seconds. • DVOM on 200 ohm scale. • Measure spout circuit from test Pin 36 to spout pin (pin 2) on the TFI harness connector.	DVOM reading 5 ohms or less ▶	GO to P5.
	DVOM reading greater than 5 ohms ▶	REPAIR open circuit. CONNECT spout test connector. CHECK timing per P1.
P5 HARNESS CHECK — IGNITION GROUND		
• Breakout box installed. • Key Off. • DVOM on 200 ohm scale. • Measure from test Pin 16 to TFI connector Pin 6 (ign. grnd.).	Less than 5 ohms ▶	GO to Pinpoint Test Step A7.
	5 ohms or greater ▶	SERVICE harness as necessary. REPEAT Quick Test.

No Codes/Invalid Codes — Pinpoint Test Q

Q

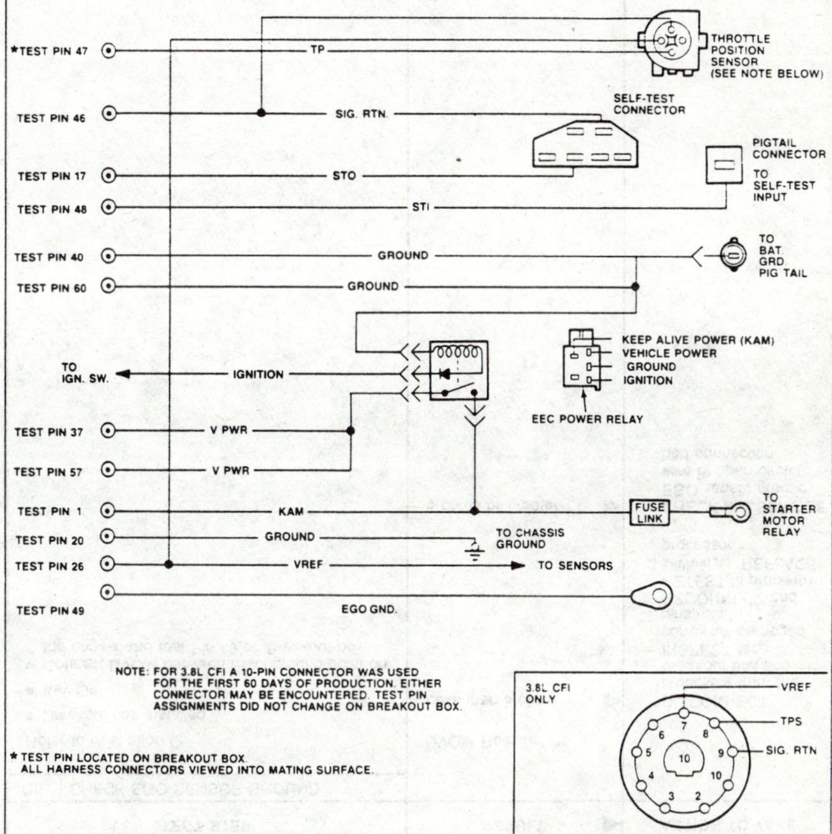

STOP-WARNING

This Pinpoint Test is intended to diagnose only the following:

- Processor.
- Harness Circuits: Signal Return, STO, STI, Ground, V PWR, KAM, VREF, Ignition.

No Codes/Invalid Codes — Pinpoint Test — Q

TEST STEP	RESULT ▶	ACTION TO TAKE
Q1 — NO CODES. INVALID CODES		
Refer to illustration Q. • Key Off. wait 10 seconds. • DVOM on 20V scale. • Disconnect TP sensor. • Key On. Engine Off. • Measure voltage at TP harness connector — VREF to signal return.	6.0V and greater ▶	GO to Pinpoint Test Step C2.
	Less than 4.0V ▶	GO to Pinpoint Test Step C1.
	4.0V to 6.0V ▶	RECONNECT TP sensor. GO to Q2.
Q2 — BATTERY POWER GROUND CHECK (NEGATIVE SUPPLY)		
• Refer to illustration Q. • Key On. Engine Off. • DVOM to 20V scale. Take a voltage reading from battery negative (−) post to Signal Return, in the Self-Test connector.	Less than 0.5V ▶	GO to Q3.
	0.5V or greater ▶	GO to Pinpoint Test Step B4.
Q3 — SELF-TEST INPUT CHECK		
Refer to illustration Q. • Key Off. wait 10 seconds. • Disconnect processor 60 pin connector and inspect for damaged pins. corrosion. loose wires. Repair as necessary. • Set DVOM to 200 ohm scale. • Check resistance from Self-Test input. At the Self-Test pigtail connector to Breakout box test Pin 48.	Less than 5 ohms ▶	GO to Q4.
	5 ohms or greater ▶	CORRECT open in circuit.
Q4 — SELF-TEST OUTPUT CIRCUIT CHECK		
Refer to illustration Q. • Breakout box installed. • DVOM to 200 ohm scale. • Check resistance from Self-Test output at Self-Test connector to test Pin 17 of Breakout box.	5 ohms or greater ▶	CORRECT open in circuit.
	Less than 5 ohms ▶	GO to Q5.

No Codes/Invalid Codes — Pinpoint Test — Q

TEST STEP	RESULT ▶	ACTION TO TAKE
Q5 — CHECK EGO SENSOR GROUND		
Refer to illustration Q. • Breakout box installed. • Key Off. • Connect DVOM between EGO ground point on the engine and test Pin 49 on Breakout box.	DVOM Reading Less than 5 ohms ▶	DISCONNECT processor from the Breakout box and INSPECT for corrosion, damaged pins, etc. RECONNECT and RETEST. If problem still exists, REPLACE processor.
	5 ohms or greater ▶	CHECK and SERVICE EGO sensor ground wire or open circuit bad connection.

Invalid Codes — Continuous Test — Pinpoint Test Q

TEST STEP	RESULT ▶	ACTION TO TAKE
Q6 CONDITIONS FOR CONTINUOUS CODE 15		
• Power interruption to Keep Alive Memory (KAM) Pin 1 may result in a service code being outputted.* • Clear continuous codes (use procedure described in Quick Test Step 6.) • Repeat Quick Test Step 3.0 through Continuous Codes. • Code 15 present on retest? *NOTE: **Anytime power is interrupted to the processor, a code 15 may be outputted the first time Self-Test is run after restoration of power. Repeat Self-Test to ensure correct diagnosis.**	Yes ▶ No ▶	GO to Q7. Test complete.
Q7 INSPECT ENGINE COMPARTMENT WIRING FOR PROPER ROUTING		
• Are any EEC components or EEC wiring close to ignition components or wires (High Electrical Energy Sources)? If EEC wiring close, reroute and rerun Self-Test. • Is code 15 still present?	Yes ▶ No ▶	GO to Q8. Test complete.
Q8 CHECK POWER CIRCUIT TO KEEP ALIVE MEMORY		
• Check power circuit to Keep Alive Memory for voltage. • Key Off, wait 10 seconds. • Disconnect processor and inspect both 60 Pin connectors. • Connect Breakout box to harness, leave processor disconnected. • DVOM on 20V scale. • Connect positive test lead to test Pin 1 and negative test lead to test Pin 46. • Key On. • Observe voltage reading.	DVOM reading less than 10 volts ▶ DVOM reading 10V or greater ▶	SERVICE open to KAM circuit. REPEAT Quick Test. GO to Q9.
Q9 CHECK INVALID CODE		
• Key On, engine Off. • Activate Self-Test. • Is code 15 present?	Yes ▶ No ▶	REPLACE processor. REPEAT Quick Test. DISCONNECT Breakout box. REPEAT Self-Test.*

Engine Stalls — Pinpoint Test Q

TEST STEP	RESULT ▶	ACTION TO TAKE
Q30 DIAGNOSTIC ROUTINES COMPLETE		
• Have diagnostic routines in Section 2 been completed?	Yes ▶ No ▶	GO to Q31. GO to Diagnostic Routines, Section 2.
Q31 ATTEMPT TO BRING ENGINE TO OPERATING TEMPERATURE		
• Key Off, wait 10 seconds. • Install tachometer. • Try to maintain engine at 2,000 rpm for 2 minutes. • Does engine stall?	Yes ▶ No ▶	GO to Pinpoint Test Step A13. GO to Q32.
Q32 MAINTAIN ENGINE OPERATION DURING ENGINE RUNNING SELF-TEST		
• Engine at operating temperature. • Tachometer installed. • Key Off, wait 10 seconds. • Activate Quick Test. • Perform Engine Running Self-Test while maintaining 2,000 rpm.	Code 11 ▶ Any service code ▶ No codes ▶	GO to Quick Test Step 5.0. GO to Quick Test Step 5.1. GO to Pinpoint Test Step Q1.

Output State Check Not Functioning — Pinpoint Test Q

TEST STEP	RESULT ▶	ACTION TO TAKE
Q40 CHECK FOR CODES 23, 53, 63 OR 68		
• Key Off, wait 10 seconds. • Perform Key On, Engine Off Self-Test. • Leave Key On to enter Output State Check. • Are codes 23, 53, 63 or 68 present in the On Demand Test results?	Yes ▶	GO to Quick Test Step 3.1 . and SERVICE appropriate code as instructed.
	No ▶	Leave Key On, GO to Q41 .
Q41 ATTEMPT TO ENTER OUTPUT STATE CHECK (REFER TO APPENDIX)		
• Key On, engine Off in Output State Check. • DVOM on 20V scale. • Connect DVOM leads to STO circuit of Self-Test connector and battery positive terminal. • Depress throttle to wide open and release. • Verify throttle is in closed position. • Is DVOM reading greater than 10V?	Yes ▶	CHECK throttle and throttle linkages for sticking and binding. SERVICE as necessary. REPEAT Quick Test.
	No ▶	REPLACE TP sensor. REPEAT Quick Test.

Output Circuit Check — (Thermactor) — Pinpoint Test UA

UA

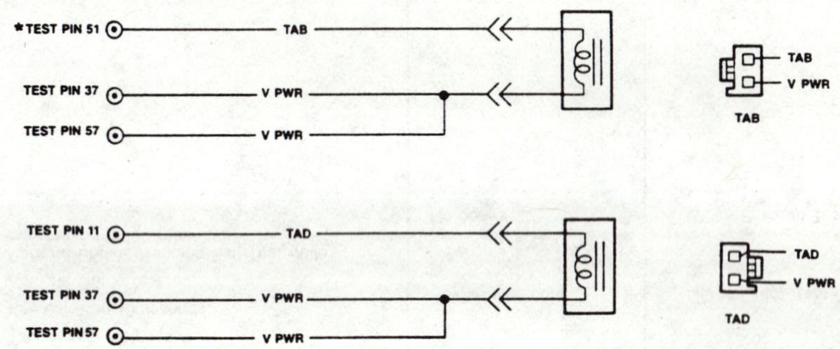

*TEST PINS LOCATED ON BREAKOUT BOX.
ALL HARNESS CONNECTORS VIEWED INTO MATING SURFACE.

STOP-WARNING

You may have reached this point because of a fault in a non-EEC System and/or component. To prevent the replacement of good components, be aware that the following non-EEC areas may be at fault:

- Thermactor System.
  - — Belt.
  - — Pump.
  - — Valve.

This Pinpoint Test is intended to diagnose only the following:

- TAB and TAD Solenoid Valve Assemblies.
- Harness Circuits: TAB, TAD and V PWR.
- Processor Assembly.

Output Circuit Check — (Thermactor) | Pinpoint Test | UA

TEST STEP	RESULT ▶	ACTION TO TAKE
UA1 MEASURE TAB/TAD SOLENOID RESISTANCE		
• Key Off, wait 10 seconds. • DVOM on 200 ohm scale. • Disconnect TAB solenoid connector and measure solenoid resistance. • Disconnect TAD solenoid connector and measure solenoid resistance.	Both resistances are between 50 and 100 ohms ▶	CONNECT TAB/TAD solenoids. GO to UA2.
	Either resistance is less than 50 ohms or greater than 100 ohms ▶	REPLACE TAB/TAD solenoid assembly. REPEAT Quick Test.
UA2 CHECK VOLTAGE OF V PWR CIRCUIT		
• Key On, Engine Off. • DVOM on 20V scale. • Positive DVOM test lead to V PWR circuit and negative test lead to ground. • Measure voltage on TAB solenoid V PWR circuit. • Repeat for TAD solenoid.	Both voltage readings 10.5V or greater ▶	GO to UA3.
	Either voltage reading is less than 10.5V ▶	SERVICE harness circuit open. REPEAT Quick Test.
UA3 CHECK CONTINUITY OF TAB AND TAD CIRCUITS		
• Key Off, wait 10 seconds. • Disconnect processor 60 Pin connector and inspect for damaged pins, corrosion, loose wires. Service as necessary. • Connect Breakout box to harness. Leave processor disconnected. • DVOM on 200 ohm scale. • Measure continuity of test Pin 51 to TAB circuit at harness connector. • Measure continuity of test Pin 11 to TAD circuit at harness connector.	Both resistance readings are less than 5 ohms ▶	GO to UA4.
	Either resistance reading is 5 ohms or greater ▶	SERVICE harness open circuit. REPEAT Quick Test.
UA4 CHECK FOR SHORT TO POWER		
• Key Off, wait 10 seconds. • DVOM on 200,000 ohm scale. • Leave Breakout box installed and processor disconnected. • With TAB/TAD solenoid disconnected, measure resistance from test Pins 51 and/or 11 to test Pins 37 and 57.	Resistance reading is 10,000 ohms or greater ▶	GO to UA5.
	Resistance reading is less than 10,000 ohms ▶	SERVICE short to power. REPEAT Quick Test.

Output Circuit Check — (Thermactor) | Pinpoint Test | UA

TEST STEP	RESULT ▶	ACTION TO TAKE
UA5 CHECK FOR SHORT TO GROUND		
• Key Off, wait 10 seconds. • DVOM on 200,000 ohm scale. • Leave Breakout box installed and processor disconnected. • With TAB/TAD solenoids disconnected, measure resistance from test Pins 51 and/or 11 to test Pins 40, 46 and 60.	Resistance reading is less than 10,000 ohms ▶	SERVICE short to ground. REPEAT Quick Test.
	Resistance reading is 10,000 ohms or greater ▶	REPLACE processor. REPEAT Quick Test.

Output Circuit Check — (EGRV/EGRC) | Pinpoint Test | UC

UC

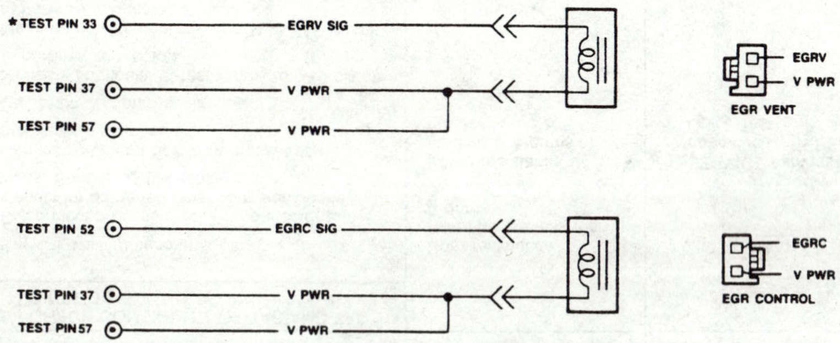

*TEST PINS LOCATED ON BREAKOUT BOX.
ALL HARNESS CONNECTORS VIEWED INTO MATING SURFACE.

STOP-WARNING

You may have reached this point because of a fault in a non-EEC System and/or component. To prevent the replacement of good components, be aware that the following non-EEC areas may be at fault:

- EGR Valve.
  — Vacuum Hose.

This Pinpoint Test is intended to diagnose only the following:

- EGRV and EGRC Solenoid Valve Assemblies.
- Harness Circuits: EGRV, EGRC and V PWR.
- Processor Assembly.

Output Circuit Check — (EGRV/EGRC) | Pinpoint Test | UC

TEST STEP	RESULT ▶	ACTION TO TAKE
UC1 SERVICE CODES 83 AND 84: MEASURE EGRV/EGRC SOLENOID RESISTANCE		
• Key Off, wait 10 seconds. • DVOM on 200 ohm scale. • Disconnect EGRV solenoid connector and measure solenoid resistance. • Disconnect EGRC solenoid connector and measure solenoid resistance.	Both resistances are between 30 and 70 ohms ▶	CONNECT EGRC/EGRV solenoids. GO to UC2.
	Either resistance is less than 30 ohms or greater than 70 ohms ▶	REPLACE EGRC/EGRV solenoid assembly. REPEAT Quick Test.
UC2 CHECK VOLTAGE OF V PWR		
• Key On, Engine Off. • DVOM on 20V scale. • Positive DVOM test lead to V PWR circuit and negative test lead to ground. • Measure voltage on EGRV solenoid V PWR circuit. • Repeat for EGRC solenoid.	Both voltage readings are greater than 10.5V ▶	GO to UC3.
	Either voltage reading is less than 10.5V ▶	SERVICE harness circuit open. REPEAT Quick Test.
UC3 CHECK CONTINUITY OF EGRV AND EGRC CIRCUITS		
• Key Off, wait 10 seconds. • Disconnect processor 60 Pin connector and inspect for damaged pins, corrosion, loose wires. Service as necessary. • Connect Breakout box to harness. Leave processor disconnected. • DVOM on 200 ohm scale. • Measure continuity of test Pin 33 to harness EGRV circuit at EGRV solenoid. • Measure continuity of test Pin 52 to harness EGRC circuit at EGRC solenoid.	Both resistance readings are less than 5 ohms ▶	GO to UC4.
	Either resistance reading is 5 ohms or greater ▶	SERVICE harness open circuit. REPEAT Quick Test.
UC4 CHECK FOR SHORT TO POWER		
• Key Off, wait 10 seconds. • DVOM on 200,000 ohm scale. • Leave Breakout box installed and processor disconnected. • With EGRV/EGRC solenoids disconnected, measure resistance from test Pins 33 and/or 52 to test Pins 37 and 57.	Resistance reading is less than 10,000 ohms ▶	SERVICE short to power and REPEAT Quick Test.
	Resistance reading is 10,000 ohms or greater ▶	GO to UC5.

Output Circuit Check — (EGRV/EGRC) | Pinpoint Test | UC

TEST STEP	RESULT ▶	ACTION TO TAKE
UC5 CHECK FOR SHORT TO GROUND		
• Key Off, wait 10 seconds. • DVOM on 200,000 ohm scale. • Leave Breakout box installed and processor disconnected. • With EGRV/EGRC solenoids disconnected, measure resistance from test Pins 33 and/or 52 to test Pins 40, 46 and 60.	Resistance reading is less than 10,000 ohms ▶	SERVICE short to ground. REPEAT Quick Test.
	Resistance reading is 10,000 ohms or greater ▶	REPLACE processor. REPEAT Quick Test.

Output Circuit Check Canister Purge | Pinpoint Test | UE

UE

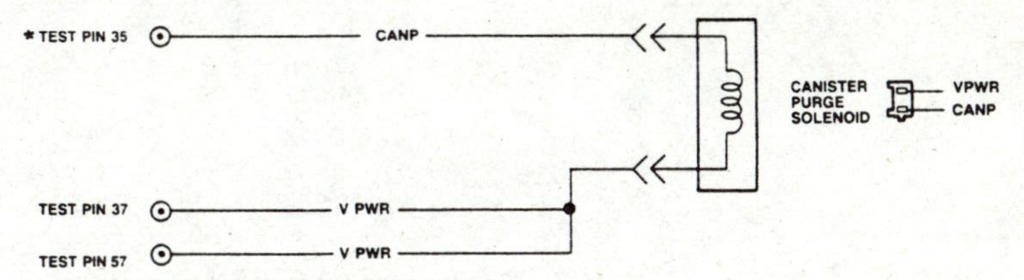

*TEST PINS LOCATED ON BREAKOUT BOX.
ALL HARNESS CONNECTORS VIEWED INTO MATING SURFACE.

STOP-WARNING

This Pinpoint Test is intended to diagnose only the following:

- Harness Circuits: CANP and V PWR.
- Processor Assembly.

TEST STEP	RESULT ▶	ACTION TO TAKE
UE1 MEASURE CANP SOLENOID RESISTANCE		
• Key Off, wait 10 seconds. • DVOM on 200 ohm scale. • Disconnect CANP solenoid connector and measure solenoid resistance.	Resistance is between 40 and 90 ohms ▶	CONNECT CANP solenoid. GO to UE2.
	Resistance is less than 40 ohms or greater than 90 ohms ▶	REPLACE CANP solenoid. REPEAT Quick Test.

Output Circuit Check — Canister Purge | Pinpoint Test | UE

TEST STEP	RESULT ▶	ACTION TO TAKE
UE2 CHECK VOLTAGE OF V PWR CIRCUIT		
• Key On, Engine Off. • DVOM on 20V scale. • Connect DVOM positive test lead to V PWR circuit and negative test lead to ground. • Measure voltage on CANP solenoid V PWR circuit.	Voltage reading is less than 10.5V ▶	SERVICE harness open circuit. REPEAT Quick Test.
	Voltage reading is 10.5V or greater ▶	GO to UE3.
UE3 CHECK CONTINUITY OF CANP CIRCUIT		
• Key Off, wait 10 seconds. • Disconnect processor 60 Pin connector and inspect for damaged pins, corrosion, loose wires. Service as necessary. • Connect Breakout box to harness. Leave processor disconnected. • DVOM on 200 ohm scale. • Measure continuity of test Pin 35 to CANP circuit at harness connector.	Resistance reading is 5 ohms or greater ▶	SERVICE open circuit. REPEAT Quick Test.
	Resistance reading is less than 5 ohms ▶	GO to UE4.
UE4 CHECK FOR SHORT TO POWER		
• Key Off, wait 10 seconds. • DVOM on 200,000 ohm scale. • Leave Breakout box installed and processor disconnected. • With CANP solenoid disconnected, measure resistance from test Pin 35 to test Pins 37 and 57.	Resistance reading is 10,000 ohms or greater ▶	GO to UE5.
	Resistance reading is less than 10,000 ohms ▶	SERVICE short to power. REPEAT Quick Test.
UE5 CHECK FOR SHORT TO GROUND		
• Key Off, wait 10 seconds. • Leave Breakout box installed and processor disconnected. • DVOM on 200,000 ohm scale. • Measure resistance of test Pin 35 to test Pins 40, 46 and 60.	Resistance reading is less than 10,000 ohms ▶	SERVICE short to ground. REPEAT Quick Test.
	Resistance reading is 10,000 ohms or greater ▶	REPLACE processor. REPEAT Quick Test.

Output Circuit Check — Temperature Compensated Fuel Pump | Pinpoint Test | UG

UG

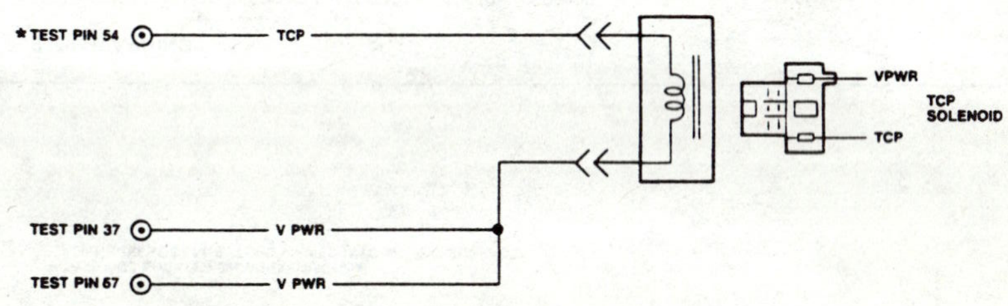

* TEST PINS LOCATED ON BREAKOUT BOX.
ALL HARNESS CONNECTORS VIEWED INTO MATING SURFACE.

STOP-WARNING

This Pinpoint Test is intended to diagnose only the following:

- Harness Circuits: V PWR and TCP.
- TCP Solenoid.
- Processor Assembly.

Output Circuit Check — Temperature Compensated Fuel Pump | Pinpoint Test | UG

TEST STEP	RESULT ▶	ACTION TO TAKE
UG1 MEASURE TCP SOLENOID RESISTANCE		
• Key Off, wait 10 seconds. • DVOM on 200 ohm scale. • Disconnect TCP solenoid connector and measure solenoid resistance.	Resistance is between 50 and 100 ohms ▶	CONNECT TCP solenoid. GO to UG2.
	Resistance is less than 50 ohms or greater than 100 ohms ▶	REPLACE TCP solenoid. REPEAT Quick Test.
UG2 CHECK VOLTAGE OF V PWR CIRCUIT		
• Key On, Engine Off. • DVOM on 20V scale. • Connect DVOM positive test lead to V PWR circuit and negative test lead to ground. • Measure voltage on TCP solenoid V PWR circuit.	Voltage reading is less than 10.5V ▶	SERVICE harness open circuit. REPEAT Quick Test.
	Voltage reading is 10.5V or greater ▶	GO to UG3.
UG3 CHECK CONTINUITY OF TCP CIRCUIT		
• Key Off, wait 10 seconds. • Disconnect processor 60 Pin connector and inspect for damaged pins, corrosion, loose wires. Service as necessary. • Connect Breakout box to harness. Leave processor disconnected. • DVOM on 200 ohm scale. • Measure continuity of test Pin 54 to TCP circuit at harness connector.	Resistance readings are 5 ohms or greater ▶	SERVICE harness open circuit. REPEAT Quick Test.
	Resistance reading is less than 5 ohms ▶	GO to UG4.
UG4 CHECK FOR SHORT TO POWER		
• Key Off, wait 10 seconds. • DVOM on 200,000 ohm scale. • Leave Breakout box installed and processor disconnected. • With TCP solenoid disconnected, measure resistance from test Pin 54 to test Pins 37 and 57.	Resistance reading is 10,000 ohms or greater ▶	GO to UG5.
	Resistance reading less than 10,000 ohms ▶	SERVICE short to power. REPEAT Quick Test.

Output Circuit Check — Temperature Compensated Fuel Pump | Pinpoint Test | UG

TEST STEP	RESULT ▶	ACTION TO TAKE
UG5 CHECK FOR SHORT TO GROUND		
• Key Off, wait 10 seconds. • Leave Breakout box installed and processor disconnected. • DVOM on 200,000 ohm scale. • Measure resistance of test Pin 54 to test Pins 40, 46 and 60.	Resistance readings are less than 10,000 ohms ▶	SERVICE short to ground. REPEAT Quick Test.
	Resistance readings are 10,000 ohms or greater ▶	REPLACE processor. REPEAT Quick Test.

Output Circuit Check — Variable Voltage Choke (VVC)

Pinpoint Test **UH**

UH

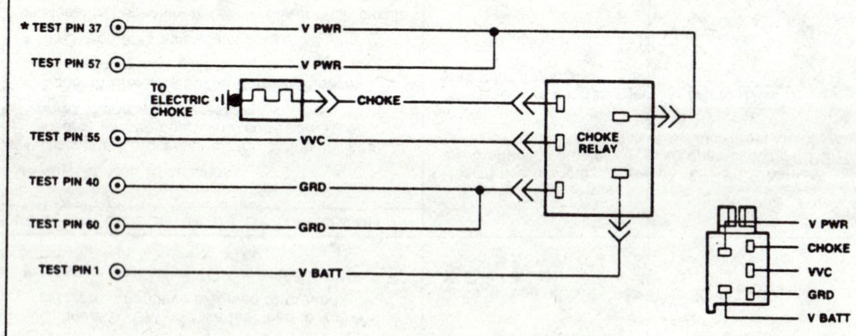

*TEST PINS LOCATED ON BREAKOUT BOX.
ALL HARNESS CONNECTORS VIEWED INTO MATING SURFACE.

STOP-WARNING

This Pinpoint Test is intended to diagnose only the following:

- Harness Circuits: VVC, V PWR and Choke.
- Choke Relay.
- Processor Assembly.

Output Circuit Check — Variable Voltage Choke (VVC)

Pinpoint Test **UH**

TEST STEP	RESULT	ACTION TO TAKE
UH1 MEASURE CHOKE VOLTAGE		
• DVOM on 20V scale. • Perform Engine Running Self-Test. • Measure choke voltage.	Choke voltage is 10.5V or greater ▶	CHECK choke plates for binding and sticking. EEC-IV system OK. REFER to Section 4, for choke diagnostics.
	Choke voltage is less than 10.5V ▶	GO to UH2.
UH2 ENTER OUTPUT STATE CHECK (REFER TO APPENDIX)		
• Key Off, wait 10 seconds. • DVOM on 20V scale. • Connect DVOM negative test lead to STO and positive test lead to battery positive. • Jumper STI to signal return. • Perform Key On, Engine Off Self-Test until the completion of the Continuous Test Codes. • DVOM will indicate zero volts. • Depress and release the throttle. • Did DVOM reading change to a high voltage reading?	Yes ▶	REMAIN in Output State Check. GO to UH3.
	No ▶	DEPRESS throttle to WOT and RELEASE. If STO voltage does not go high, GO to Pinpoint Test Step Q40.
UH3 CHECK VVC OUTPUT ELECTRICAL OPERATION		
• Key On, Engine Off. • DVOM on 20V scale. • Connect DVOM positive test lead to V PWR on VVC relay and negative test lead to VVC output circuit VVC. • While observing DVOM, depress and release the throttle several times to cycle output On and Off. • Does VVC output cycle On and Off?	Yes ▶	GO to UH4.
	No ▶	REMAIN in Output State Check. GO to UH5.

Output Circuit Check — Variable Voltage Choke (VVC) | Pinpoint Test | UH

TEST STEP	RESULT ▶	ACTION TO TAKE
UH4 CHECK CONTINUITY OF GROUND CIRCUIT AND CHOKE CIRCUIT		
• Key Off, wait 10 seconds. • DVOM on 200 ohm scale. • Measure continuity of ground circuit from the relay harness connector. • Measure continuity of the choke circuit from the relay harness connector to the choke.	Either resistance reading is 5 ohms or greater ▶ Both resistance readings are less than 5 ohms ▶	SERVICE harness circuit. REPEAT Quick Test. REPLACE choke relay. REPEAT Quick Test.
UH5 CHECK FOR VOLTAGE ON V PWR CIRCUIT		
• Key On, Engine Off. • DVOM on 20V scale. • Connect DVOM positive test lead to V PWR at relay harness connector and negative test lead to the ground circuit at the relay harness connector.	Voltage is less than 10.5V ▶ Voltage is 10.5V or greater ▶	SERVICE open in choke power circuit. REPEAT Quick Test. GO to UH6.
UH6 CHECK CONTINUITY OF VVC CIRCUIT		
• Key Off, wait 10 seconds. • Disconnect processor 60 Pin connector and inspect for damaged pins, corrosion, loose wires. Service as necessary. • Connect Breakout box to harness. Leave processor disconnected. • DVOM on 200 ohm scale. • Measure continuity of test Pin 55 to VVC circuit at the relay harness connector.	Resistance is 5 ohms or greater ▶ Resistance is less than 5 ohms ▶	SERVICE harness circuit. REPEAT Quick Test. GO to UH7.
UH7 CHECK FOR SHORT TO POWER		
• Key Off, wait 10 seconds. • DVOM on 200,000 ohm scale. • Leave Breakout box installed and processor disconnected. • With VVC relay disconnected, measure resistance from test Pin 55 to test Pins 1, 37 and 57.	Resistance reading is 10,000 ohms or greater ▶ Resistance reading is less than 10,000 ohms ▶	GO to UH8. SERVICE short to power. REPEAT Quick Test.

Output Circuit Check — Variable Voltage Choke (VVC) | Pinpoint Test | UH

TEST STEP	RESULT ▶	ACTION TO TAKE
UH8 CHECK FOR SHORT TO GROUND ON VVC CIRCUIT		
• Key Off, wait 10 seconds. • DVOM on 200,000 ohm scale. • VVC relay connected. • Leave Breakout box installed and processor disconnected. • Measure resistance of test Pin 55 to test Pins 40, 46 and 60.	Any resistance readings 10,000 ohms or greater ▶ All resistance readings less than 10,000 ohms ▶	GO to UH9. REPLACE processor. REPEAT Quick Test.
UH9 CHECK FOR SHORT TO GROUND WITH VVC RELAY DISCONNECTED		
• Key Off, wait 10 seconds. • Leave Breakout box installed and processor disconnected. • DVOM on 200,000 ohm scale. • Disconnect VVC relay. • Measure resistance from test Pin 55 to test Pins 40, 46 and 60.	All resistance readings are 10,000 ohms or greater ▶ Any resistance reading is less than 10,000 ohms ▶	REPLACE choke relay. REPEAT Quick Test. SERVICE harness short. REPEAT Quick Test.

Output Circuit Check — WOT A/C Cut-Off | Pinpoint Test | UJ

UJ

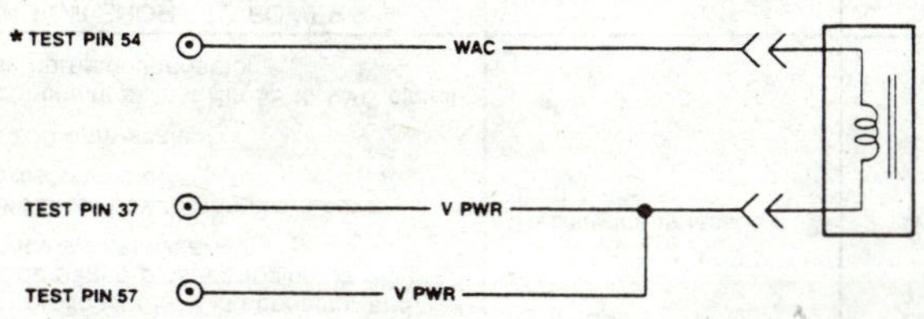

* TEST PINS LOCATED ON BREAKOUT BOX.
ALL HARNESS CONNECTORS VIEWED INTO MATING SURFACE.

STOP-WARNING

This Pinpoint Test is intended to diagnose only the following:

- Harness Circuits: WAC and V PWR.
- WOT A/C Cut-Off Solenoid.
- Processor Assembly.

TEST STEP	RESULT ▶	ACTION TO TAKE
UJ1 MEASURE WAC SOLENOID RESISTANCE		
• Key Off, wait 10 seconds. • DVOM on 200 ohm scale. • Disconnect WAC solenoid connector and measure solenoid resistance.	Resistance is between 50 and 100 ohms ▶	CONNECT WAC solenoid. GO to UJ2.
	Resistance is less than 50 ohms or greater than 100 ohms ▶	REPLACE WAC solenoid. REPEAT Quick Test.

Output Circuit Check — WOT A/C Cut-Off | Pinpoint Test | UJ

TEST STEP	RESULT ▶	ACTION TO TAKE
UJ2 CHECK VOLTAGE OF V PWR CIRCUIT		
• Key On, Engine Off. • DVOM on 20V scale. • Connect DVOM positive test lead to V PWR circuit and negative test lead to ground. • Measure voltage on WAC solenoid V PWR circuit.	Voltage reading is less than 10.5V ▶	SERVICE harness open circuit. REPEAT Quick Test.
	Voltage reading is 10.5V or greater ▶	GO to UJ3.
UJ3 CHECK CONTINUITY OF WAC CIRCUIT		
• Key Off, wait 10 seconds. • Disconnect processor 60 Pin connector and inspect for damaged pins, corrosion, loose wires. Service as necessary. • Connect Breakout box to harness. Leave processor disconnected. • DVOM on 200 ohm scale. • Measure continuity of test Pin 54 to WAC circuit at harness connector.	Resistance readings are 5 ohms or greater ▶	SERVICE harness open circuit. REPEAT Quick Test.
	Resistance reading is less than 5 ohms ▶	GO to UJ4.
UJ4 CHECK FOR SHORT TO POWER		
• Key Off, wait 10 seconds. • DVOM on 200,000 ohm scale. • Leave Breakout box installed and processor disconnected. • With WAC solenoid disconnected, measure resistance from test Pin 54 to test Pins 37 and 57.	Resistance reading is 10,000 ohms or greater ▶	GO to UJ5.
	Resistance reading is less than 10,000 ohms ▶	SERVICE short to power. REPEAT Quick Test.
UJ5 CHECK FOR SHORT TO GROUND		
• Key Off, wait 10 seconds. • Leave Breakout box installed and processor disconnected. • DVOM on 200,000 ohm scale. • Measure resistance of test Pin 54 to test Pins 40, 46 and 60.	Resistance readings are less than 10,000 ohms ▶	SERVICE short to ground. REPEAT Quick Test.
	Resistance readings are 10,000 ohms or greater ▶	REPLACE processor. REPEAT Quick Test.

Output Circuit Check — Idle Speed Control (Throttle Kicker) | Pinpoint Test | UK

UK

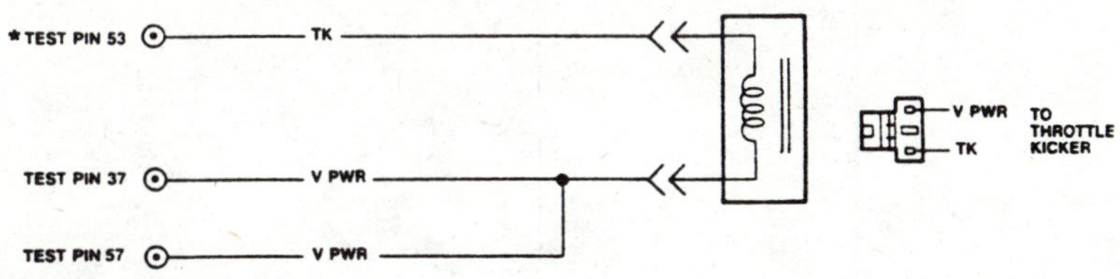

*TEST PINS LOCATED ON BREAKOUT BOX.
ALL HARNESS CONNECTORS VIEWED INTO MATING SURFACE.

STOP-WARNING

You may have reached this point because of a fault in a non-EEC System and or component. To prevent the replacement of good components, be aware that the following non-EEC areas may be at fault:

- Engine not at operating temperature.
- Engine above operating temperature.
- Improper Idle Speeds.
- A/C Input (Electrical Problems).

This Pinpoint Test is intended to diagnose only the following:

- TK Solenoid.
- TK Actuator.
- Harness Circuits: TK and V PWR.
- Processor Assembly.

Output Circuit Check — Idle Speed Control (Throttle Kicker) | Pinpoint Test | UK

TEST STEP	RESULT ▶	ACTION TO TAKE
UK1 MEASURE TK SOLENOID RESISTANCE		
• Key Off, wait 10 seconds. • DVOM on 200 ohm scale. • Disconnect TK solenoid connector and measure solenoid resistance.	Resistance is between 50 and 100 ohms ▶	CONNECT TK solenoid. GO to UK2.
	Resistance is less than 50 ohms or greater than 100 ohms ▶	REPLACE TK solenoid. REPEAT Quick Test.
UK2 CHECK VOLTAGE OF V PWR CIRCUIT		
• Key On, Engine Off. • DVOM on 20V scale. • Connect DVOM positive test lead to V PWR circuit and negative test lead to ground. • Measure voltage on TK solenoid V PWR circuit.	Voltage reading is 10.5V or less ▶	SERVICE harness open circuit. REPEAT Quick Test.
	Voltage reading is 10.5V or greater ▶	GO to UK3.
UK3 CHECK CONTINUITY OF TK CIRCUIT		
• Key Off, wait 10 seconds. • Disconnect processor 60 Pin connector and inspect for damaged pins, corrosion, loose wires. Service as necessary. • Connect Breakout box to harness. Leave processor disconnected. • DVOM on 200 ohm scale. • Measure continuity of test Pin 53 to TK circuit at harness connector.	Resistance readings are 5 ohms or greater ▶	SERVICE harness open circuit. REPEAT Quick Test.
	Resistance reading is less than 5 ohms ▶	GO to UK4.
UK4 CHECK FOR SHORT TO POWER		
• Key Off, wait 10 seconds. • DVOM on 200,000 ohm scale. • Leave Breakout bcx installed. Leave processor disconnected. • With TK solenoid disconnected, measure resistance from test Pin 53 to test Pins 37 and 57.	Resistance reading is 10,000 ohms or greater ▶	GO to UK5.
	Resistance reading is less than 10,000 ohms ▶	SERVICE short to power. REPEAT Quick Test.

Output Circuit Check — Idle Speed Control (Throttle Kicker) | Pinpoint Test | UK

TEST STEP	RESULT ▶	ACTION TO TAKE
UK5 CHECK FOR SHORT TO GROUND		
• Key Off, wait 10 seconds. • Leave Breakout box installed and processor disconnected. • DVOM on 200,000 ohm scale. • Measure resistance of test Pin 53 to test Pins 40, 46 and 60.	Resistance readings are less than 10,000 ohms ▶	SERVICE short to ground. REPEAT Quick Test.
	Resistance readings are 10,000 ohms or greater ▶	REPLACE processor. REPEAT Quick Test.

Output Circuit Check — Exhaust Heat Control | Pinpoint Test | UL

UL

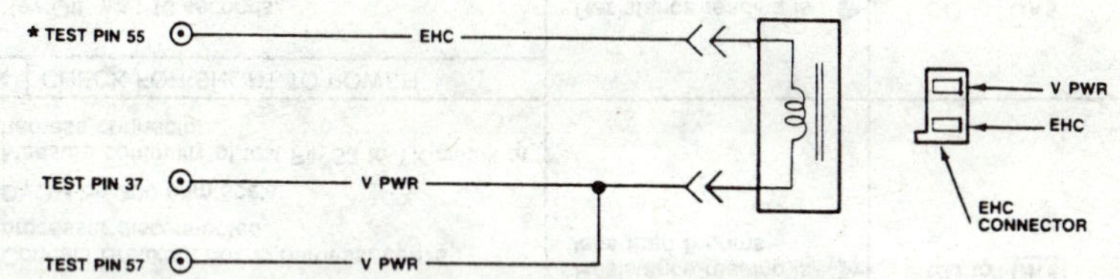

* TEST PINS LOCATED ON BREAKOUT BOX.
ALL HARNESS CONNECTORS VIEWED INTO MATING SURFACE.

STOP-WARNING

This Pinpoint Test is intended to diagnose only the following:

- Harness Circuits: EHC and V PWR.
- Exhaust Heat Solenoid.
- Processor Assembly.

TEST STEP	RESULT ▶	ACTION TO TAKE
UL1 MEASURE EHC SOLENOID RESISTANCE		
• Key Off, wait 10 seconds. • DVOM on 200 ohm scale. • Disconnect exhaust heat solenoid connector and measure solenoid resistance.	Resistance is between 50 and 100 ohms ▶	CONNECT exhaust heat solenoid. GO to UL2.
	Resistance is less than 50 ohms or greater than 100 ohms ▶	REPLACE exhaust heat solenoid. REPEAT Quick Test.

Output Circuit Check — Exhaust Heat Control | Pinpoint Test | UL

TEST STEP	RESULT ▶	ACTION TO TAKE
UL2 CHECK VOLTAGE OF V PWR CIRCUIT		
• Key On, Engine Off. • DVOM on 20V scale. • Connect DVOM positive test lead to V PWR circuit and negative test lead to ground. • Measure voltage on EHC solenoid V PWR circuit.	Voltage reading is less than 10.5V ▶	SERVICE harness open circuit. REPEAT Quick Test.
	Voltage reading is 10.5V or greater ▶	GO to UL3.
UL3 CHECK CONTINUITY OF EHC CIRCUIT		
• Key Off, wait 10 seconds. • Disconnect processor 60 Pin connector and inspect for damaged pins, corrosion, loose wires. Service as necessary. • Connect Breakout box to harness. Leave processor disconnected. • DVOM on 200 ohm scale. • Measure continuity of test Pin 55 to EHC circuit at harness connector.	Resistance readings are 5 ohms or greater ▶	SERVICE harness open circuit. REPEAT Quick Test.
	Resistance reading is less than 5 ohms ▶	GO to UL4.
UL4 CHECK FOR SHORT TO POWER		
• Key Off, wait 10 seconds. • DVOM on 200,000 ohm scale. • Leave Breakout box installed and processor disconnected. • With exhaust heat solenoid disconnected, measure resistance from test Pin 55 to test Pins 37 and 57.	Resistance reading is 10,000 ohms or greater ▶	GO to UL5.
	Resistance reading is less than 10,000 ohms ▶	SERVICE short to power. REPEAT Quick Test.
UL5 CHECK FOR SHORT TO GROUND		
• Key Off, wait 10 seconds. • Leave Breakout box installed and processor disconnected. • DVOM on 200,000 ohm scale. • Measure resistance of test Pin 55 to test Pins 40, 46 and 60.	Resistance readings are less than 10,000 ohms ▶	SERVICE short to ground. REPEAT Quick Test.
	Resistance readings are 10,000 ohms or greater ▶	REPLACE processor. REPEAT Quick Test.

Erratic Ignition | Pinpoint Test | Y

TEST STEP	RESULT ▶	ACTION TO TAKE
Y1 CODE 14: ERRATIC IGNITION		
NOTE: Code 14 indicates two successive erratic profile ignition pick-up (PIP) pulses occurred, resulting in a possible engine miss or stall. • Check EEC-IV and ignition systems harnesses for: — Loose wires/connectors. — Arcing secondary ignition components (coil, cap, rotor, wires, plugs, etc). — On-board transmitter (2-way radio).* • Are any of the above present? *Verify all 2-way radio installations. Carefully follow manufacturer's installation instructions regarding the routing of antenna and power leads.	Yes ▶	SERVICE as necessary. REPEAT Quick Test.
	No ▶	GO to Y2.
Y2 CHECK DISTRIBUTOR		
• Key Off, wait 10 seconds. • Deactivate Self-Test. • Connect VOM or STAR per Quick Test Step 2.0. • Enter Engine Running Continuous Monitor Test (as instructed in Quick Test Step 6.3). • Observe VOM or STAR LED for indication of a fault while performing the following: • Lightly tap on TFI module and distributor. • Wiggle TFI connector. • Is a fault indicated?	Yes ▶	DISCONNECT and INSPECT connectors. If connector and terminals are good, GO to Section 15 Ignition Diagnostics.
	No ▶	GO to Y3.
Y3 CHECK EEC-IV HARNESS		
• While still in Continuous Monitor Test from Step Y2 observe VOM or STAR LED for a fault indication while performing the following: • While looking for faults grasp the harness close to the TFI connector. Wiggle, shake or bend a small section of the ignition and EEC-IV systems harness while working your way to the other components and dash panel. Also wiggle, shake or bend the EEC-IV harness from the dash panel to the processor. Isolate the PIP circuit if needed for this test. • Is a fault indicated?	Yes ▶	ISOLATE fault and make necessary repairs. REPEAT Quick Test.
	No ▶	GO to Y4.

Erratic Ignition

Pinpoint Test Y

TEST STEP	RESULT ▶	ACTION TO TAKE
Y4 CHECK PROCESSOR AND HARNESS CONNECTORS		
• Key Off, wait 10 seconds. • Disconnect processor 60 Pin connector and inspect for damaged pins, corrosion, loose wires. • Are connectors and terminals OK? • Reconnect processor when this Step is completed.	No ▶	SERVICE as necessary. REPEAT Quick Test.
	Yes ▶	Unable to duplicate an erratic ignition fault in the EEC-IV System, for further diagnosis GO to Section 15 Ignition System Diagnostics.

MCU System Description

2.0/2.3 LIGHT TRUCK IN-LINE FOUR CYLINDER MCU SYSTEM

The MCU is basically the same on all in-line 4-cylinder engines. The heart of the system is the fuel control system. This is necessary to keep the air-fuel (A/F) ratio at proper chemical balance (14.7) to obtain maximum catalyst efficiency. The fuel control "Loop" consists of an Exhaust Gas Oxygen (EGO) sensor, Microprocessor Control Unit (MCU), and Fuel Control Solenoid.

The EGO sensor senses whether the exhaust gas is rich or lean of proper chemical balance. This signal is sent to the MCU module, which sends a varying signal to the fuel control solenoid to move the air-fuel ratio back to the proper chemical balance. This operation is called "closed loop" operation.

The other mode of operation is called "open loop." In this mode, the MCU module sends out a fixed signal to the fuel control solenoid. During this time the input from the EGO sensor is ignored, thus opening the loop.

The determining factor when the system goes into open or closed loop is based upon information from the switch inputs, which sense coolant temperature, manifold vacuum, and throttle position. Generally, the vehicle will be in closed loop when the vehicle is at operating temperature and at a steady part throttle cruise.

Other functions controlled by the MCU module, by means of vacuum solenoids, are the Thermactor Air Bypass (TAB) valve and the Thermactor Air Diverter (TAD) valve. Also controlled, but not on all calibrations, are Canister Purge and the Spark Retard Solenoid.

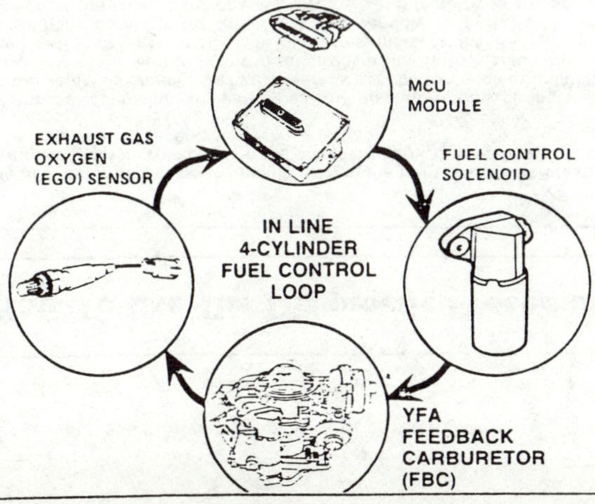

How To Use The Diagnostic Procedure

DESCRIPTION

The Microprocessor Control Unit has the capability to diagnose a malfunction within its own system. It advises the technician where to look for MCU problems through the use of a Self-Test computer program built into the MCU module. When triggered, the Self-Test simulates a variety of engine operating conditions, so that operating conditions can be detected.

A received response to the initiated Self-Test is reported to the technician as a pulsing series of voltage outputs that will appear on one of two testing devices: an Analog Voltmeter or Self-Test Automatic Readout (STAR) tester.

The series of voltage pulses are called a Service Code, representing a two-digit number. It's the two-digit Service Code that advises the technician of the nature of the MCU problem. Hooking up the Analog Voltmeter or STAR tester to the MCU System will be described and shown later.

After performing the "Self-Test", the technician must follow all diagnostic procedures in the Engine/Emissions Diagnosis Manual prior to replacing any components. This will help assure service reliability.

How To Use The Diagnostic Procedure

ANALOG VOLTMETER

When a Service Code is reported on the Analog Voltmeter, it will represent itself as a pulsing or sweeping movement of the voltmeter's needle across the dial face of the voltmeter. Therefore, a single-digit number of three will be reported by three needle pulses (sweeps). However, as previously stated, a Service Code is represented by a two-digit number, such as 3-2. So as a result, the Self-Test's Service Code of 3-2 will appear on the voltmeter as three needle pulses (sweeps), then after a two second pause, the needle will pulse (sweep) twice more.

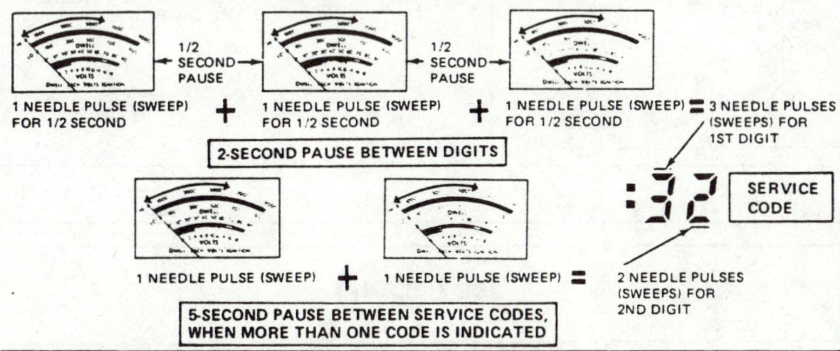

SELF-TEST AUTOMATIC READOUT (STAR)
Rotunda Model No. 07-0004 or Equivalent

After hooking up the STAR tester and turning on its power switch, the tester will run a display check: the numerals 88 will begin to flash in the display window, and then a steady 00 will appear to signify that the STAR tester is ready to start the Self-Test and receive the test's Service Codes.

How To Use The Diagnostic Procedure

To receive the Service Codes, press the pushbutton at the front of the STAR tester. The button will latch down, and a colon will appear in the display window in front of the :00 numerals. The colon MUST be displayed to receive the Service Codes.

If for any reason the technician wishes to clear the display window during the Self-Test, he must turn off the vehicle's engine, press the tester's pushbutton once to unlatch it (colon will disappear), then press the button again to latch down the button (colon will appear again). Every time the STAR tester is turned off, the low battery indicator (LO BAT) should show briefly at the upper left-hand corner of the tester's display window. If the LO BAT indicator shows steadily at any other time during the operation of the STAR tester with any Service Code, turn its power switch to OFF, and replace the 9-volt battery in the tester.

The STAR tester will display the last Service Code received, even after disconnecting it from the vehicle. It will hold the Service Code on the display until the power is turned off or the pushbutton is unlatched and relatched.

Anyone who departs from the instructions provided in this publication must first establish that he compromises neither his personal safety nor the vehicle integrity by his choice of methods, tools, or parts.

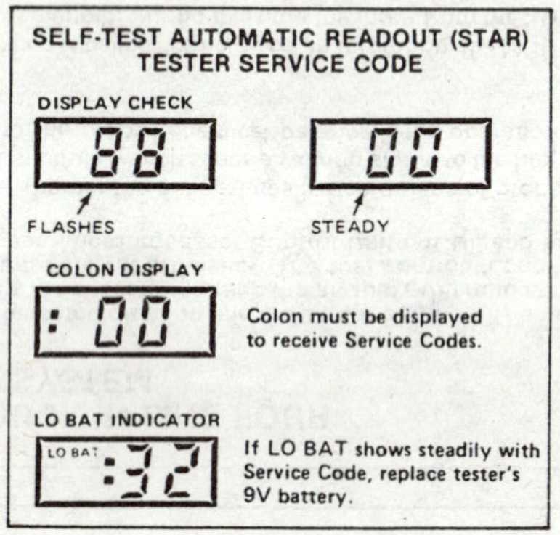

Quick Test

1	VISUAL CHECKS

A. CAREFULLY REMOVE THE ENGINE CONTROL HOUSING TOP.

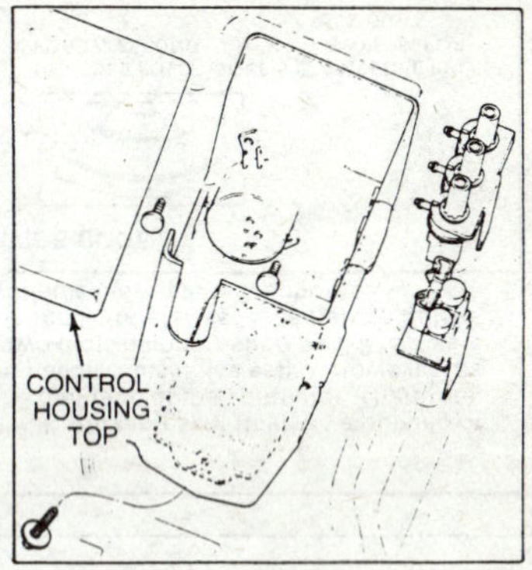

B. CHECK VACUUM LINES FOR CRACKED OR BROKEN HOSES AND PROPER CONNECTIONS PER VACUUM SCHEMATIC.

C. CHECK THE ELECTRICAL HARNESS FOR:

- Broken, detached, or frayed wires/insulation.
- Corrosion on terminals.
- Before reading continuity or shorts, verify terminals are completely seated in the connector and wires are properly crimped to the terminals.

Quick Test

1 VISUAL CHECKS (Continued)

D. Verify that coolant level is to specification.

2 PRE-TEST CONDITIONS

A. Transmission in Park or Neutral.

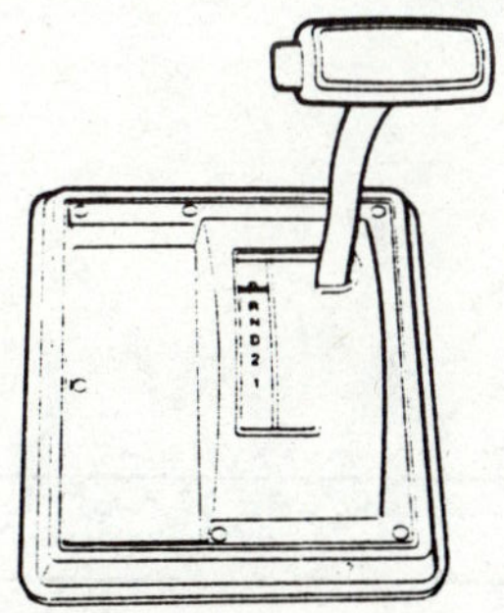

Quick Test

2 PRE-TEST CONDITIONS (Continued)

B. Set parking brake. Block the wheels.

C. Turn accessories off.

Quick Test

2 PRE-TEST CONDITIONS (Continued)

D. Make sure that the vacuum hoses are connected to the air cleaner.

E. Using an analog voltmeter, make sure there is power to the choke (engine must be running).

Quick Test

2 PRE-SET CONDITIONS (Continued)

F. Start and idle the engine until the radiator upper hose is hot and pressurized.

If vehicle will not start, see No Start, Pinpoint Test 1.0.

3 SET-UP PROCEDURE

A. Turn the ignition key Off.

B. Locate the Self-Test connector near MCU module (left fender apron as viewed from driver's seat).

Quick Test

3	SET-UP PROCEDURE (Continued)

- If you are using an analog VOM during the Quick Test, connect a jumper wire or paper clip from Self-Test Trigger (circuit 201) to ground (circuit 60) on Self-Test connector.
- Then connect an analog voltmeter from battery (+) to Self-Test Output on the Self-Test connector.
- Set voltmeter on DC volt range to read from o to 15 VDC.

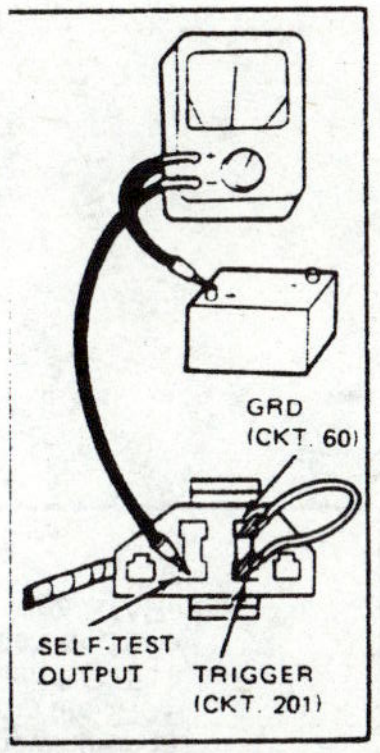

- However, if you are using a STAR tester for the Quick Test:
- Connect black lead to the battery (—) post.
- Connect red lead to the Self-Test Ouput terminal on the Self-Test connector.
- Connect white lead to the Self-Test Trigger.
- Refer to back of STAR tester for additional instructions.

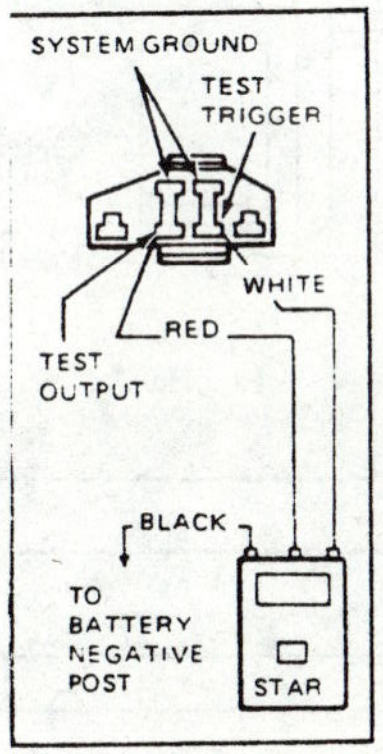

Quick Test

3	SET-UP PROCEDURE (Continued)

C. On engines with a vacuum purge valve. disable the canister purge system during this test.

- Remove the hose from the canister control valve that goes to the carbon canister (this simulates a clean carbon canister).
- Do not plug the hose.
- Connect the hose after testing and servicing is completed.

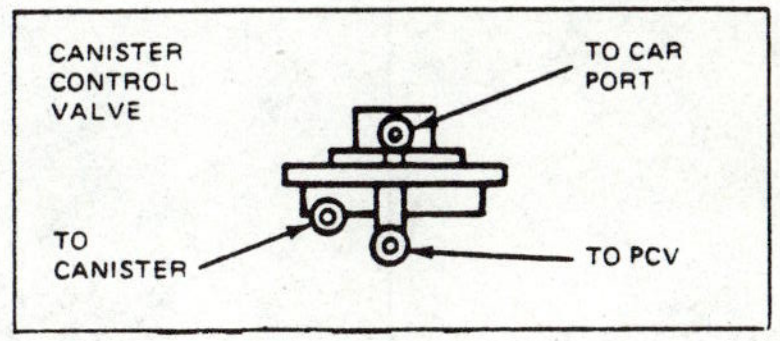

D. Make sure the carburetor throttle linkage is off high cam, and the choke is open.

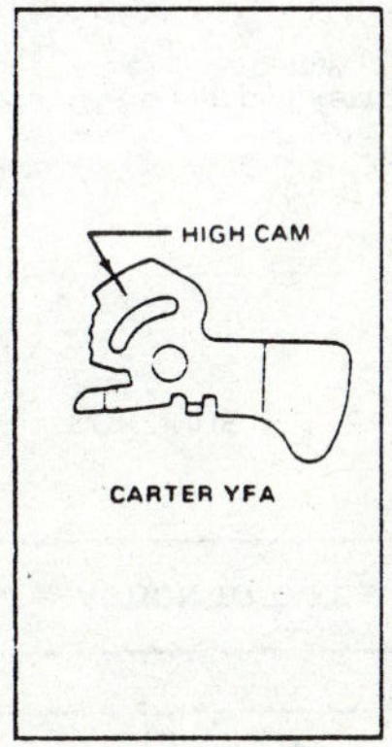

Quick Test

3 SET-UP PROCEDURE (Continued)

E.

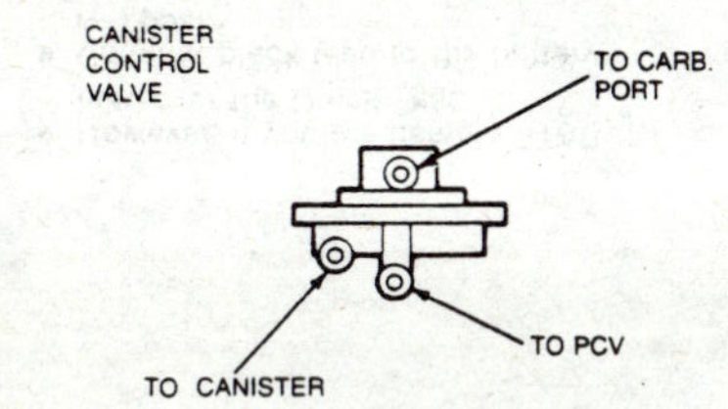

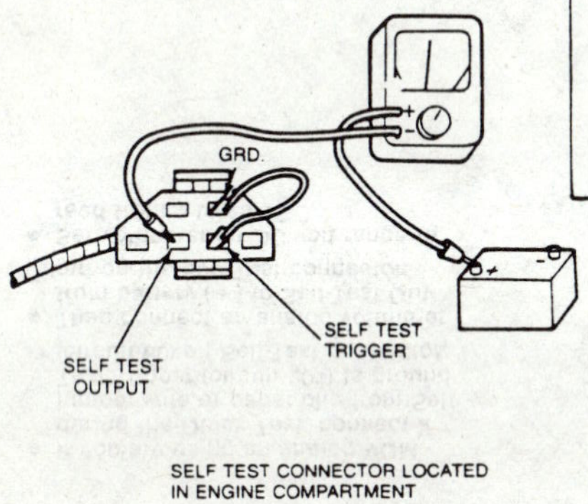

- Key OFF.
- Connect STAR to self-test connector as shown:

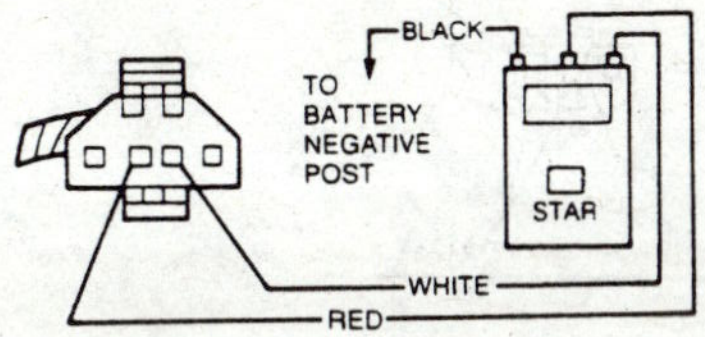

- Turn STAR unit ON. Display should show [00]
- Press STAR pushbutton to latch it. A colon (:) MUST be showing on the display. For more complete STAR instructions, see Sec. II. NOTE: If pushbutton will not latch or display doesn't show [:00] see STAR decal for service.

Quick Test

Test 4

TEST STEP	RESULT	▶	ACTION TO TAKE
4.0 KEY ON, ENGINE OFF TEST			
• Start engine. • Run engine until upper radiator hose is hot and pressurized. • Turn the ignition key to Off, then turn to On with engine off.	Service Code 1-1, OK.	▶	CONTINUE to Quick Test 5.
	Failure Codes:		
	5-1 Electric Temperature Switch	▶	GO to Pinpoint Test 2. See note below.
	5-2 } Idle Decel 5-3 } WOT Crowd 5-5 } Road Load	▶	GO to Pinpoint Test 4. See note below.
	Steady high or low reading or STAR CODE showing :00 steadily	▶	GO to Pinpoint Test 14. See note below.
	Any code not listed	▶	RERUN Quick Test 4.

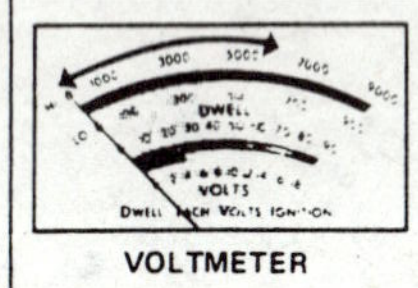

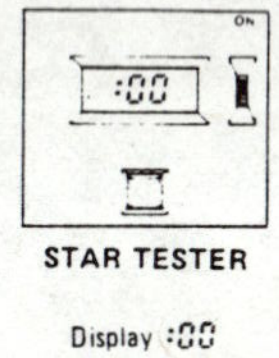

NOTE: One quick initialization pulse may occur on the voltmeter immediately after turning the ignition key On. The output service code will occur in approximately five seconds.

- Observe and record Service Codes shown on the voltmeter or STAR tester.
- Start Service with the first Code received and proceed with each successive code in the order received.

*Before proceeding with appropriate Pinpoint Test, read note on Pinpoint Test usage, located at the beginning of Pinpoint Tests.

Quick Test — Test 5

TEST STEP	RESULT ▶	ACTION TO TAKE
5.0 MCU SYSTEM RUNNING TEST		
VOM • Start or re-start engine. • Within 20 seconds of re-start, increase rpm (2,500 rpm) until VOM pulses 2, then maintain that rpm until the service code is received, (ignore any pulses longer than 1 second).	Service Code 1-1, OK. ▶	CONTINUE Quick Test
• Observe VOM. Return to idle as soon as the output begins. Test will run between 10-40 seconds. **NOTE: If VOM does not pulse or pulses more than 2 times in the first 7 seconds, GO to Pinpoint Test 15.**	Failure Codes: 6-2 Idle Decel 6-3 WOT Crowd 6-5 Road Load ▶	GO to Pinpoint Test 4. See note below.
or STAR • Start or re-start engine. • Within 20 seconds of re-start, increase rpm (2,500 rpm) until STAR displays Code 20, hold that rpm.	4-1 Lean Fuel ▶	GO to Pinpoint Test 8. See note below.
	4-2 Rich Fuel ▶	GO to Pinpoint Test 9. See note below.
• Observe STAR display. Return to idle as soon as code changes. Test will run 10-40 seconds. **NOTE: If code does not change or code other than 10 or 20 appears in first seven seconds, GO to Pinpoint Test 15.**	4.4 Thermactor System ▶	GO to Pinpoint Test 10. See note below.
	4-5 Thermactor Air Diverter ▶	GO to Pinpoint Test 11. See note below.
	4-6 Thermactor Air Bypass ▶	GO to Pinpoint Test 12. See note below.
	3-3 Running Test Not Initiated ▶	GO to Pinpoint Test 13. See note below.
	Any code not listed ▶	REPEAT Quick Test 5.
	Steady high or low ▶	GO to Pinpoint Test 14. See note below.

*Before proceeding with appropriate Pinpoint Test, read note on Pinpoint Test usage, located at the beginning of Pinpoint Tests.

Quick Test — Test 6

TEST STEP	RESULT ▶	ACTION TO TAKE
6.0 MISCELLANEOUS TEST		
6.1 CANISTER PURGE CHECK		
• Tee in vacuum gauge between the Canister Purge Solenoid and Canister. • Engine OFF. Re-start engine.	Vacuum gauge pulses three times up to 0-3 kPa (0-1 In-Hg)	
• Increase RPM to 2,500 RPM initiate Self-Test, and observe vacuum gauge during the initialization pulses.	Vacuum always high or low	GO to Pinpoint Test [16]

*Before proceeding with appropriate Pinpoint Test, read note on Pinpoint Test usage, located at the beginning of Pinpoint Tests.

Pinpoint Tests

INSTRUCTIONS FOR USING THE PINPOINT TESTS

- Do not measure voltage or resistance at the MCU module or connect any test lights to it, unless otherwise specified.
- The symbol "R" indicates resistance; that is, R = 100 ohms means the resistance is 100 ohms.
- Isolate both ends of a circuit, and turn key Off whenever checking for shorts (R less than 1,000 ohms) or continuity (R less than 5 ohms), unless specified.
- Connectors shown in schematics are as viewed from the **front** of the connector. Refer to diagram below.
- Disconnect solenoids and switches from the harness before measuring for continuity, resistance, or energizing by way of 12V source.
- In using the Pinpoint Tests, follow each step in order, starting from the first step in the appropriate Test. Follow each step until the fault is found; correct the problem, then proceed to Quick Test to verify the service. Reconnect all connections before conducting the Quick Test.
- Before measuring continuity or shorts while performing any of the test steps, verify that terminals are completely seated in the connector and wires are properly crimped to terminals.
- Always verify that vacuum hoses are routed properly per the vacuum schematic.

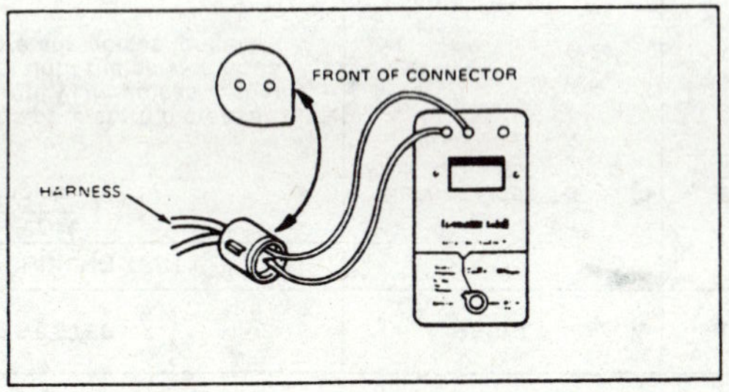

No Start | Pinpoint | Test 1

TEST STEP	RESULT ▶	ACTION TO TAKE
1.0 NO START		
• No start detects MCU faults only.		
1.1		
• Check Tach lead (circuit 11) for a short to ground. For this check, leave harness connected to MCU, and disconnect the coil horseshoe connector and ignition module from the harness. • Measure resistance.	R 1,000 ohms or less. ▶	GO to 1.2
	R more than 1,000 ohms. ▶	MCU system is OK. Refer to Routines, Section 2.
1.2		
• To isolate fault, disconnect MCU module from harness, and again check for short to ground in circuit 11. • Measure resistance.	R 1,000 ohms or less. ▶	SERVICE or replace circuit 11. Restest.
	R more than 1,000 ohms. ▶	REPLACE MCU module. Retest.

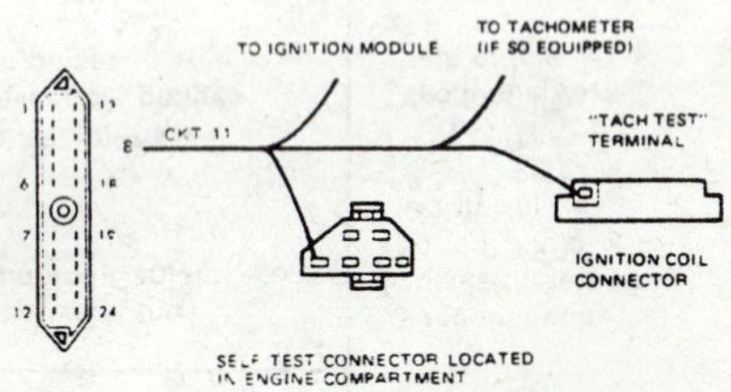

Low Temperature Switch — Pinpoint Test 2

TEST STEP	RESULT	▶ ACTION TO TAKE
2.0 SWITCH NO. 1 LOW TEMPERATURE SWITCH (SERVICE CODE 5-1)		
2.1 • Verify water temperature was above 95°F during Self-Test.	Yes	▶ GO to 2.2
	No	▶ Restart vehicle, allow to run at idle for two minutes.
2.2 • Disconnect the vehicle harness from the switch.	R 5 ohms or less.	▶ GO to 2.3
• Check contacts of low temperature switch; should be closed above 95°F. Measure resistance.	R more than 5 ohms.	▶ REPLACE low temperature switch. Retest.
2.3 • Disconnect the vehicle harness from the MCU.	R 5 ohms or less.	▶ GO to 2.4
• Check circuits 75 and 60 for continuity. Measure resistance.	R more than 5 ohms.	▶ SERVICE or replace circuit. Retest.
2.4 • Connect one lead of VOM to circuit 75 and the other lead to engine ground.	R is more than 1,000 ohms.	▶ REPLACE MCU. Retest.
• Measure resistance.	R is 1,000 ohms or less.	▶ REPAIR or REPLACE circuit. Retest.

Vacuum Switches — Pinpoint Test 4

4.0 VACUUM SWITCH

CODE	SWITCH NUMBER IS	NAME/FUNCTION	CIRCUIT NUMBER IS +
52 or 62	2	Idle Decel	189
53 or 63	3	WOT Crowd	308
55 or 65	5	Road Load	356

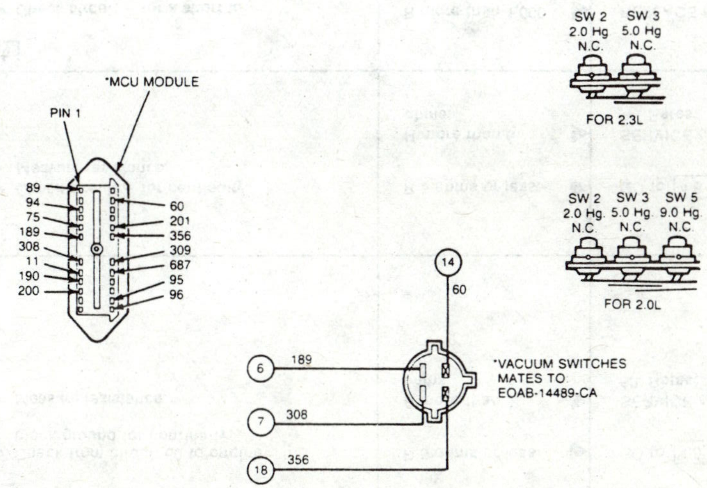

Locate vacuum switch by referring to proper vacuum schematic and comparing to vacuum routings on vehicle.

Use chart to identify proper circuit number for

Vacuum Switches — Pinpoint Test 4

TEST STEP	RESULT ►	ACTION TO TAKE
4.1 • Verify vacuum is properly routed to the appropriate switch. • Verify that there is proper amount of vacuum at switch. • For switches 3 & 5 check with engine at idle. • For switch 2 check with engine at 2500 RPM.	For switches 2,3 and 5 with 32 kPa (10 In. Hg.) or more of vacuum. ►	GO to 4.2
	Proper vacuum not present. ►	CHECK vacuum lines for blockage or leakage. Also check TVS, PVS, ETC if appropriate.
4.2 • Check vacuum switch contacts. Contacts should have continuity without vacuum. Measure resistance of switch contacts.	R 5 ohms or less. ►	GO to 4.3
	R more than 5 ohms. ►	REPLACE vacuum switch. Retest.
4.3 • Check vacuum switch contacts with vacuum applied to switch. Measure resistance of switch contacts. (Contacts should be open when vacuum is 32 kPa (10 In. Hg.) or more.	R more than 5 ohms. ►	GO to 4.4
	R 5 ohms or less. ►	REPLACE vacuum switch. Retest.

Vacuum Switches — Pinpoint Test 4

TEST STEP	RESULT ►	ACTION TO TAKE
4.4 • Check from circuit 60 to engine block ground for continuity. • Measure resistance.	R 5 ohms or less. ►	GO to 4.5
	R more than 5 ohms. ►	SERVICE circuit 60. Retest.
4.5 • Check circuit + for continuity. Measure resistance.	R 5 ohms or less. ►	GO to 4.6
	R more than 5 ohms. ►	SERVICE circuit +. Retest.
4.6 • Check circuit + for a short to ground. Measure resistance.	R more than 1,000 ohms. ►	REPLACE MCU module. Retest.
	R 1,000 ohms or less. ►	SERVICE short in circuit +. Retest.

Fuel Always Lean | Pinpoint | Test 8

8.0 SERVICE CODE 4-1

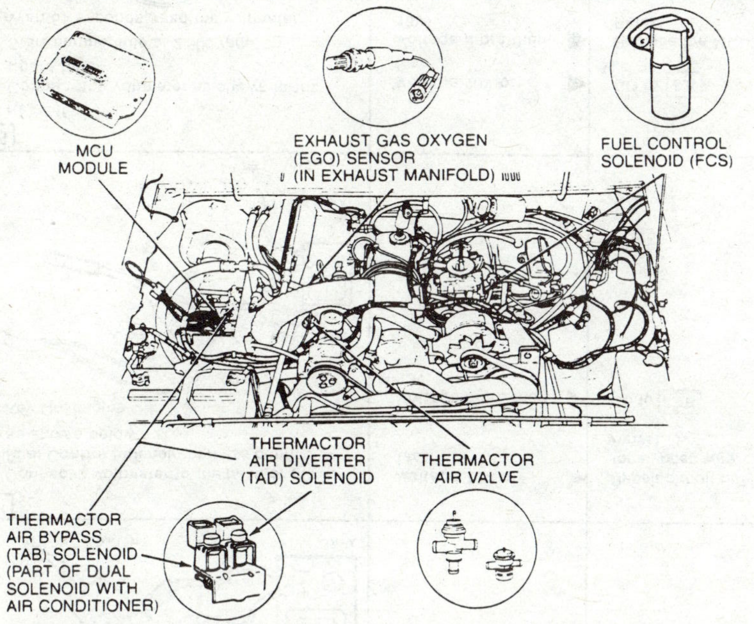

Components controlled by the MCU System in this Pinpoint Test are the EGO Sensor, the Fuel Control Solenoid, the MCU Module, and the Thermactor Air Valve. The components that are not directly controlled by the MCU System during this Pinpoint Test but could possibly cause a Service Code fault are as follows:

- Malfunctioning Carburetor
- Low Fuel Pressure
  - — Fuel Line Kink
  - — Fuel Pump Malfunction
- Vacuum Leaks
  - — EGR Valve and Gasket
  - — Head Gasket
  - — Cracked Hoses, etc.
- Bowl Vent
- Canister Purge Solenoid

NOTE: Virtually anything that adds air or restricts fuel should be investigated.

Fuel Always Lean | Pinpoint | Test 8

8.0 SERVICE CODE 4-1

- Disconnect Self-Test trigger jumper.

OR

- Press STAR pushbutton to unlatch, colon disappears from display.

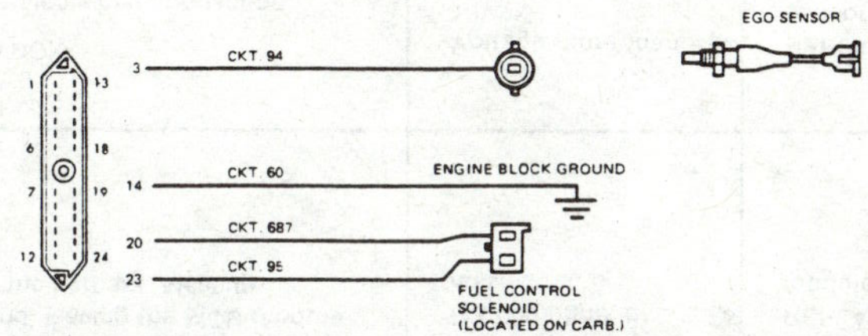

TEST STEP	RESULT	▶	ACTION TO TAKE
8.1 • Key Off. • Disconnect MCU module, FCS, and EGO sensor from vehicle harness. • Check circuits 94 and 95 for a short to ground. Measure resistance with all accessories Off.	R more than 1,000 ohms.	▶	GO to 8.2
	R 1,000 ohms or less.	▶	SERVICE short. Retest.
8.2 • Check circuits 94 and 95 for continuity. Measure resistance. • Check circuit 60 for continuity. Measure resistance.	R 5 ohms or less.	▶	GO to 8.3
	R more than 5 ohms.	▶	SERVICE circuit #94, #95 or #60 as required. Retest.

Fuel Always Lean — Pinpoint Test 8

TEST STEP	RESULT ▶	ACTION TO TAKE
8.3 • Reconnect MCU module and EGO sensor. • Check the resistance of the FCS. Should be 28-66 ohms.	R within range. ▶	GO to 8.4
	R not within range. ▶	REPLACE FCS. Retest.

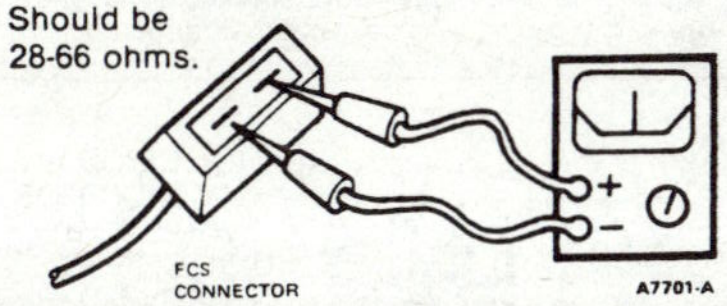

TEST STEP	RESULT ▶	ACTION TO TAKE
8.4 • Connect a voltmeter to the back of the Fuel Control Solenoid harness connector as shown below. • Key On, engine Off.	Voltage 10V or less. ▶	Repair circuit 687 for an open wire. Retest.
	Voltage more than 10V. ▶	Go to 8.4B

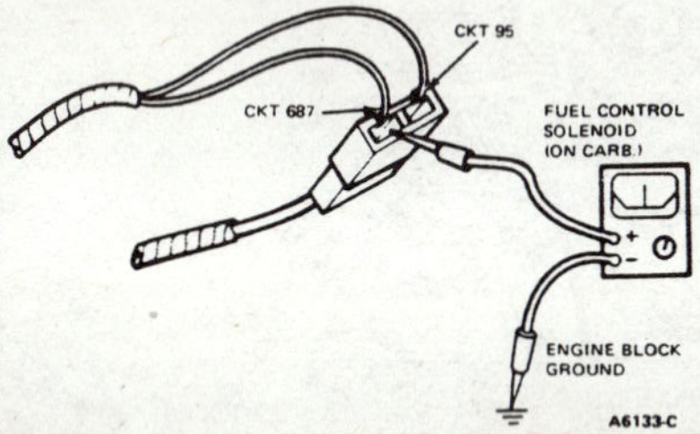

TEST STEP	RESULT ▶	ACTION TO TAKE
8.4B • Key Off. • Connect the voltmeter as shown in the figure below. • Start engine, idle at 2,500 rpm. • Wait 55 seconds, read the voltmeter.	Voltage 10V or less. ▶	GO to 8.5
	Voltage more than 10V. ▶	Replace the MCU. Retest.

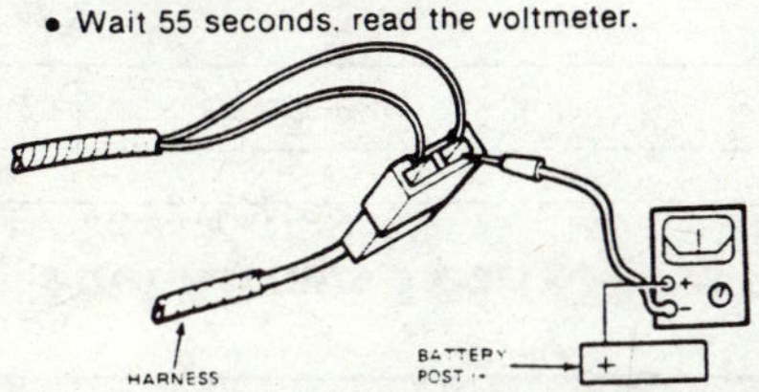

Fuel Always Lean — Pinpoint Test 8

TEST STEP	RESULT ▶	ACTION TO TAKE
8.5 • EGO and Module Check: Disconnect the Thermactor air supply hose from the air pump, and plug the hose. Set the RPM to 2,500, and hold choke 3/4 closed to force the system rich. With the VOM still connected to the Fuel Control Solenoid, observe the voltage after 55 seconds.	Voltage 10V or less. ▶	GO to 8.6
	Voltage more than 10V. ▶	MCU system is OK. Return to diagnostic routines for drive complaint observed.
8.6 • Module short check. Turn key Off, and disconnect the harness from the EGO sensor. Check circuit 94 for a short to ground, leaving the MCU module connected to the harness. Measure resistance.	R more than 1,000 ohms. ▶	GO to 8.7
	R 1,000 ohms or less. ▶	REPLACE MCU module. Retest.
8.7 FAULT ISOLATION: CAUTION: In this step when connecting the jumper to B+, make sure the other end does not come into contact with ground, or violent short will occur.	Voltage more than 10V. ▶	REPLACE EGO sensor. Retest.
• Disconnect harness from EGO sensor. With engine at idle (important), connect a jumper wire between the harness connector that mates with the EGO sensor and battery positive (+) terminal. Set RPM at 2,500 to 2,800, and measure the voltage after 55 seconds at the Fuel Control Solenoid.	Voltage 10V or less. ▶	REPLACE MCU module. Retest.

Fuel Always Rich	Pinpoint	Test 9

9.0 SERVICE CODE 4-2

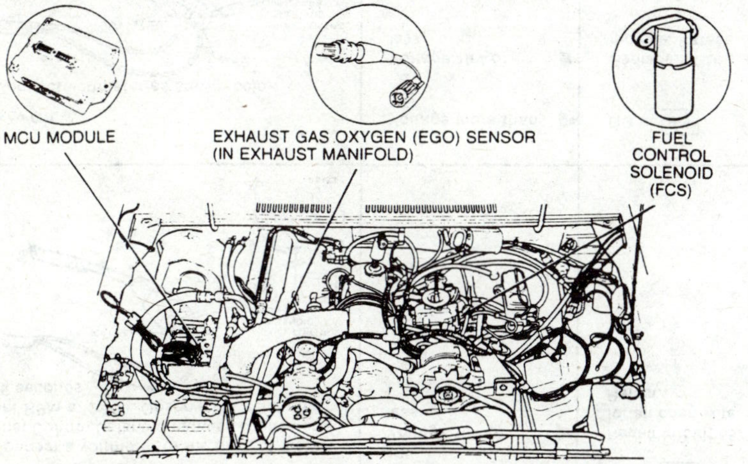

Components controlled by the MCU System in this Pinpoint Test are the EGO Sensor, the MCU Module, and the Fuel Control Solenoid. The components that are not directly controlled by the MCU System during this Pinpoint Test but could possibly cause a Service Code fault are as follows:

- Malfunctioning Carburetor
  - — High Float Level
  - — No Power to Choke Plate
- Spark Plugs Misfiring
  - — Coil
  - — Distributor
- Restricted Air Cleaner
- Restricted PVS
- Air Pump Not Working
  - — Malfunctioning Pulley
  - — Broken Belt(s)
- Throttle Kicker Always On

NOTE: Virtually anything that adds fuel or restricts air should be investigated.

Fuel Always Rich	Pinpoint	Test 9

9.0 SERVICE CODE 4-2

- Disconnect Self-Test trigger jumper.

OR

- Press STAR pushbutton to unlatch, colon disappears from display.

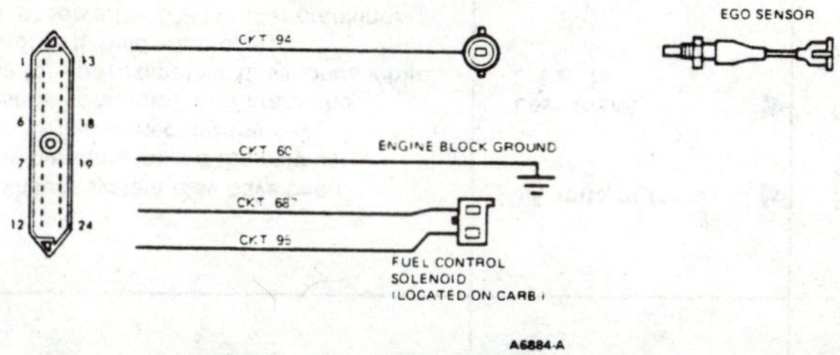

TEST STEP	RESULT ▶	ACTION TO TAKE
9.1		
• Check choke plate for sticking or binding.	Sticking or binding ▶	SERVICE as required. Retest.
	No problem ▶	GO to 9.2
9.2		
• Check circuits 687 and 95 for continuity. Measure resistance.	R 5 ohms or less. ▶	GO to 9.3
	R more than 5 ohms. ▶	SERVICE circuit. Retest.

Fuel Always Rich | Pinpoint | Test 9

9.3

- Check the resistance of the Control Solenoid. Should be 28-66 ohms.

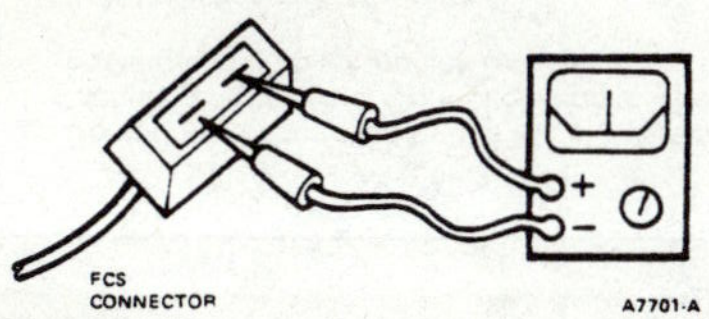

RESULT	▶	ACTION TO TAKE
"R" within range.	▶	GO to 9.4
"R" not within range	▶	REPLACE FCS. Retest.

9.4

- Key On, engine Off.
- Connect a voltmeter to the back of the Fuel Control Solenoid harness connector. Set RPM at 2,500. Observe the VOM after 55 seconds.

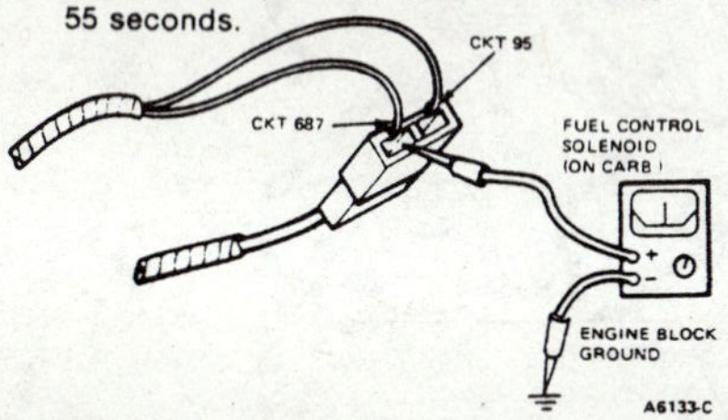

RESULT	▶	ACTION TO TAKE
Voltage more than 10V.	▶	GO to 9.4B
Voltage 10V or less.	▶	Repair circuit 687 for an open wire. Retest.

9.4B

- Key Off.
- Connect voltmeter as shown below.

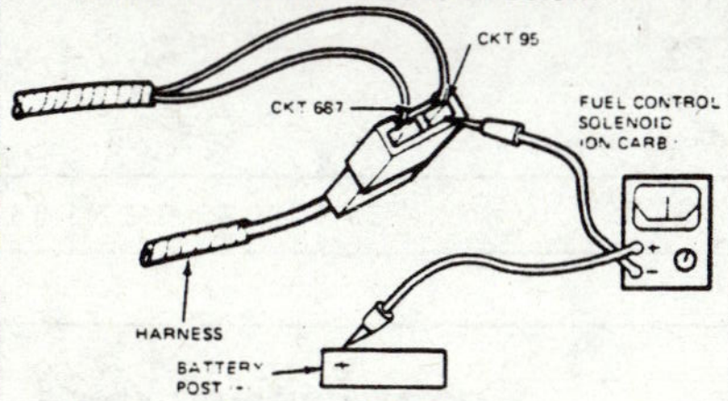

- Start engine, run at 2,500 rpm.
- Wait 55 seconds, read the voltmeter.

RESULT	▶	ACTION TO TAKE
Voltage more than 10V.	▶	GO to 9.5
Voltage 10V or less.	▶	Replace MCU module. Retest.

Fuel Always Rich | Pinpoint | Test 9

TEST STEP	RESULT	▶	ACTION TO TAKE
9.5 • EGO sensor check: Disconnect the EGO sensor from the harness. Connect a digital voltmeter (input impedence 10 megohm or greater, such as Rotunda T79L-50-DVOM-1 or equivalent, between the EGO sensor and ground. Place DVOM in the volts position. Start engine, and run it at 2,000 RPM for 1 minute to warm up the EGO sensor. Turn engine Off, and immediately observe the DVOM.	0.4 volts or more	▶	GO to 9.6
	Less than 0.4 volts	▶	MCU System OK. Return to Diagnostic Routines for drive complaint observed.
9.6 • The exhaust system may have been rich. Immediately purge the exhaust system by disconnecting the coil horseshoe connector, and crank the engine for approximately 10 seconds while the throttle is held wide open. Again, observe the DVOM after cranking.	0.4 volts or more	▶	REPLACE the EGO sensor. Retest.
	Less than 0.4 volts	▶	CHECK for rich carburetor, Section 4.

Thermactor Check	Pinpoint	Test 10

10.0 SERVICE CODE 4-4

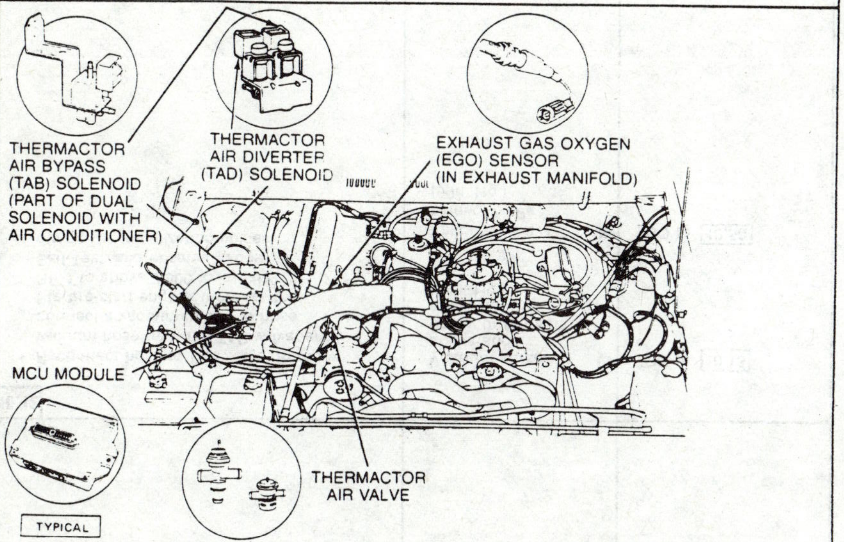

Components controlled by the MCU System in this Pinpoint Test are the Thermactor Air Bypass and Diverter, Thermactor Air Valve, EGO Sensor, Carburetor, and the MCU Module. The components not directly controlled by the MCU System during the Pinpoint Test but could possibly cause a Service Code fault are as follows:

- Malfunctioning Carburetor
  - — High Float Level
  - — No Power to Choke Plate
- Misfiring Spark Plugs
  - — Coil
  - — Distributor
- Restricted Air Cleaner
- Restricted PVS
- Air Pump Not Working
  - — Malfunctioning Pulley
  - — Broken Belt(s)

NOTE: Virtually anything that adds fuel, restricts air, or inhibits the Thermactor system from switching air upstream to the exhaust manifold should be investigated.

Thermactor Check	Pinpoint	Test 10

10.0 SERVICE CODE 4-4 (Continued)

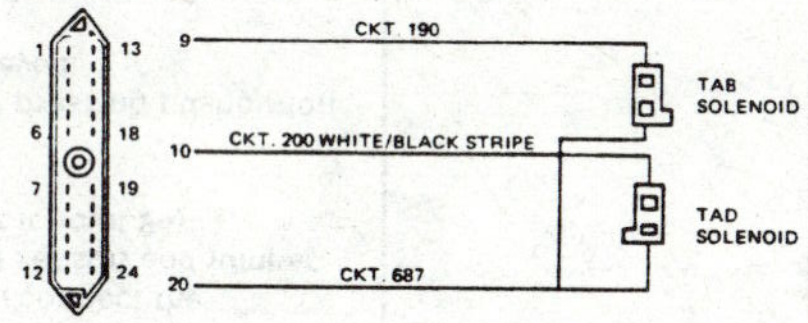

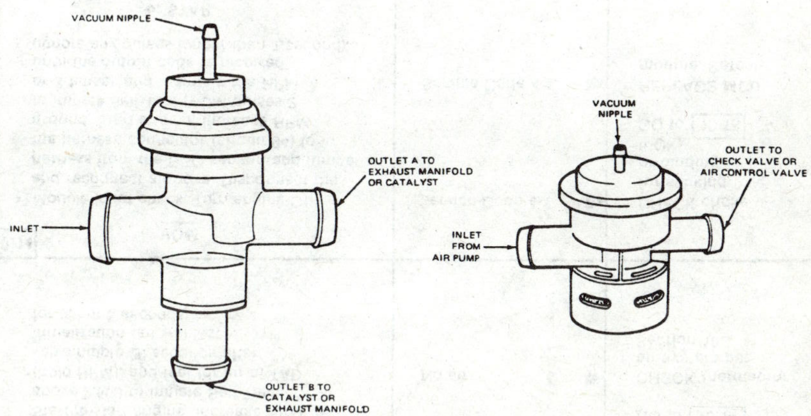

Typical Thermactor System

Thermactor Check — Pinpoint Test 10

TEST STEP	RESULT	▶	ACTION TO TAKE
10.11 • Remove the vacuum hose from the TAB valve, and connect a vacuum gauge to hose. Start engine, increase RPM to above 2,500 to initiate the Self-Test, and observe the vacuum gauge during the initialization pulses.	Vacuum pulses above and below or remains above 17 kPa (5 In. Hg.)	▶	GO to 10.12
	Vacuum always below 17 kPa (5 In. Hg.)	▶	GO to 10.30
10.12 • Reconnect hose to TAB valve. Remove vacuum hose from the TAD valve, and connect a vacuum gauge to hose. Start/re-start engine, increase the RPM to above 2,500 to initiate Self-Test, and observe gauge during the initialization pulses.	Vacuum pulses above and below or remains above 17 kPa (5 In. Hg.)	▶	GO to 10.13
	Vacuum always below 17 kPa (5 In. Hg.)	▶	GO to 10.50

Thermactor Check — Pinpoint Test 10

TEST STEP	RESULT	▶	ACTION TO TAKE
10.13 • Reconnect hose to TAD valve. Remove upstream air hose at TAD valve. Start/re-start engine. increase RPM to above 2,500 to initiate Self-Test (hold RPM), and feel for air at TAD valve nipple 20 seconds after initialization (air will last for about 6 seconds).	Air	▶	Thermactor air system is OK GO to 10.14
	No air	▶	CHECK thermactor air system per Section 10.
10.14 <u>VOM</u> • Module input check. Turn engine Off, and reconnect air hose. Disconnect the harness from the EGO sensor, and jumper the harness connector (circuit 94) to ground. Start engine, increase RPM to initiate Self-Test (VOM pulses 2 or 3 times), and maintain the RPM until the output code is received (ignore any pulses longer than 1 second). or <u>STAR</u> • Module input check. Turn engine Off, and reconnect air hose. Disconnect the harness from the EGO sensor, and jumper the harness connector (circuit 94) to ground. • Clear STAR display by pressing pushbutton twice (colon still displayed). • Start engine until STAR displays Code 20 or 30, then hold that RPM. Observe the STAR display, return to idle as soon as Code changes. Test will run 10-40 seconds.	Service Code 4-1.	▶	CHECK choke for sticking or binding. If OK, GO to 10.15
	Service Code 4-4.	▶	REPLACE MCU module. Retest.

Thermactor Check — Pinpoint Test 10

TEST STEP	RESULT ►	ACTION TO TAKE
10.15 • EGO sensor check. Disconnect the EGO sensor from the harness. Connect a digital voltmeter (input impedence 10 megohm or greater, Rotunda T79L-50-DVOM-1 or equivalent, between the EGO sensor and ground. Place DVOM in the volts position. Start engine, and run it at 2,000 RPM for 1 minute to warm up the EGO sensor. Turn engine Off, and immediately observe the DVOM.	0.4 volts or more ►	GO to 10.16
	Less than 0.4 volts ►	CHECK for rich carburetor Section 4.
10.16 • The exhaust system may have been rich. Immediately purge the exhaust system by disconnecting the coil horseshoe connector, and crank the engine for approximately 10 seconds while the throttle is held wide open. Again, observe the DVOM after cranking.	0.4 volts or more ►	REPLACE the EGO sensor. Retest.
	Less than 0.4 volts ►	MCU system OK. Return to Diagnostic Routines for drive complaints observed.
10.30 • Check the vacuum hose between the TAB valve and TAB solenoid and between the vacuum source and TAB solenoid for leaks, blockage or misconnections. Also check vacuum source.	Leakage or blockage found ►	SERVICE as required. Retest.
	No problem found ►	GO to 10.31
10.31 • Check Retard Delay Valve (RDV) for proper operation and installation and Temperature Vacuum Switch (TVS) for proper operation, if applicable.	RDV or TVS not OK ►	SERVICE as required. Retest.
	RDV and TVS, OK. ►	GO to 10.32

Thermactor Check — Pinpoint Test 10

TEST STEP	RESULT ►	ACTION TO TAKE
10.32 • Check at the TAB solenoid output to make sure vacuum is present when the solenoid is energized (12V).	Vacuum present ►	GO to 10.33
	Vacuum not present ►	REPLACE TAB solenoid. Retest.
10.33 • Check circuits 687 and 190 for continuity between the TAB solenoid and MCU module. Measure resistance.	R more than 5 ohms. ►	SERVICE circuit as required. Retest.
	R 5 ohms or less. ►	REPLACE MCU module. Retest.
10.50 • Check the vacuum hoses between the TAD valve and TAD solenoid and between the vacuum source and the TAD solenoid for leakage or blocks. Also check vacuum source.	Leakage or blockage found ►	SERVICE as required. Retest.
	OK ►	GO to 10.51
10.51 • Check RDV for proper operation and installation and TVS for proper operation, if applicable.	RDV or TVS not OK ►	SERVICE as required. Retest.
	RDV and TVS, OK. ►	GO to 10.52

Thermactor Check — Pinpoint Test 10

TEST STEP	RESULT	ACTION TO TAKE
10.52		
• Measure the TAD solenoid resistance. Should be between 50-110 ohms.	"R" in range ▶	GO to 10.53
	"R" not in range ▶	REPLACE solenoid. Retest.
10.53		
• Check at the TAD solenoid output to make sure vacuum is present when the solenoid is energized (12V).	Vacuum present ▶	GO to 10.54
	Vacuum not present ▶	REPLACE solenoid. Retest.
10.54		
• Check circuits 687 and 200 for continuity between TAD solenoid and MCU module. Measure resistance.	R more than 5 ohms. ▶	SERVICE circuit as required. Retest.
	R 5 ohms or less. ▶	REPLACE MCU module. Retest.

Thermactor Air Switching — Pinpoint Test 11

11.0 SERVICE CODE 4-5 THERMACTOR AIR ALWAYS UPSTREAM

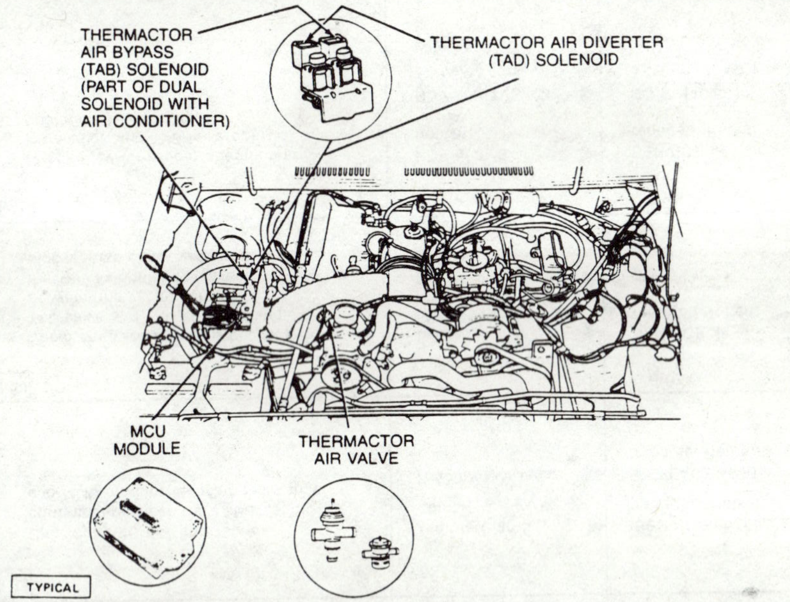

Components controlled by the MCU System in this Pinpoint Test are the Thermactor Air Diverter, Thermactor Air Valve, and the MCU Module. The components that are not directly controlled by the MCU System during this Pinpoint Test but could possibly cause a Service Code fault are as follows:

- Belts Broken or Loose
- Hoses Pinched, Crimped, Missing, Plugged Restrictor
- Malfunctioning Vacuum Delay Valve

NOTE: Virtually anything that inhibits the Thermactor system from switching air downstream to the catalyst should be investigated.

Thermactor Air Switching	Pinpoint	Test 11

11.0 SERVICE CODE 4-5 (Continued)

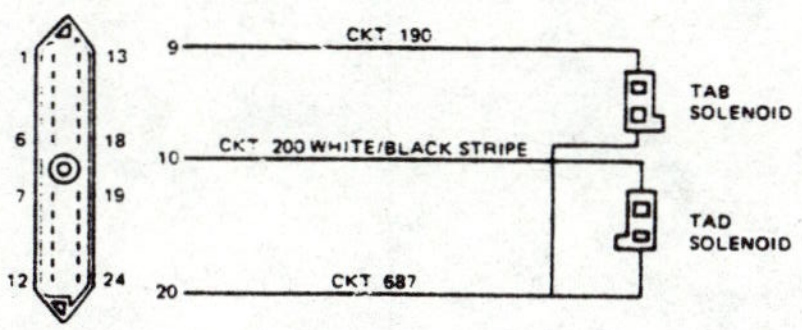

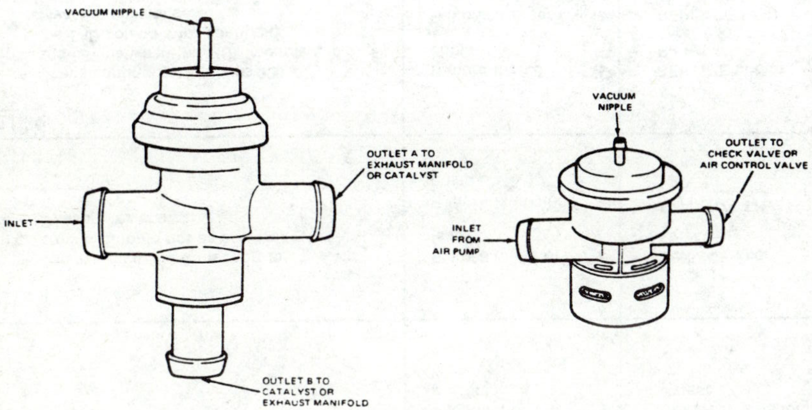

Typical Thermactor System

Thermactor Air Switching	Pinpoint	Test 11

TEST STEP	RESULT ▶	ACTION TO TAKE
11.10		
• Remove vacuum hose from the TAD valve, and connect a vacuum gauge to hose. Start/re-start engine, increase RPM to above 2,500 to initiate Self-Test, and observe gauge during the initialization pulses.	Vacuum pulses above and below 17 kPa (5 In. Hg.) ▶	MCU system OK, CHECK thermactor air in Section 10.
	Vacuum always above 17 kPa (5 In. Hg.) or vacuum always below 17 kPa (5 In. Hg.) ▶	GO to 11.20
11.20		
• Check vacuum hoses between TAD valve to TAD solenoid and between TAD solenoid and vacuum source for leaks or blockage. Also check the vacuum source.	Leaks or blockage found ▶	SERVICE as required. Retest.
	No leaks or blockage ▶	GO to 11.21
11.21		
• Measure the TAD solenoid resistance. • Should be between 50-110 ohms.	"R" within range ▶	GO to 11.22
	"R" not within range ▶	REPLACE solenoid. Retest.

Thermactor Air Switching — Pinpoint Test 11

TEST STEP	RESULT	ACTION TO TAKE
11.22		
• Check at the TAD solenoid output that vacuum is present when the solenoid is energized (12V).	Vacuum present ▶	GO to 11.23
	Vacuum not present ▶	REPLACE solenoid. Retest.
11.23		
• Check circuit 200 for a short to ground. Should not be shorted. Measure resistance.	R 1,000 ohms or less. ▶	SERVICE short. Retest.
	R more than 1,000 ohms. ▶	GO to 11.24
11.24		
• Check continuity at circuits 200 and 687 between the MCU module and the TAD solenoid for continuity. Measure resistance.	R more than 5 ohms. ▶	SERVICE circuit. Retest.
	R 5 ohms or less. ▶	REPLACE MCU module. Retest.

Thermactor Air Bypass — Pinpoint Test 12

12.0 SERVICE CODE 4-6 THERMACTOR AIR ALWAYS NOT BYPASSING

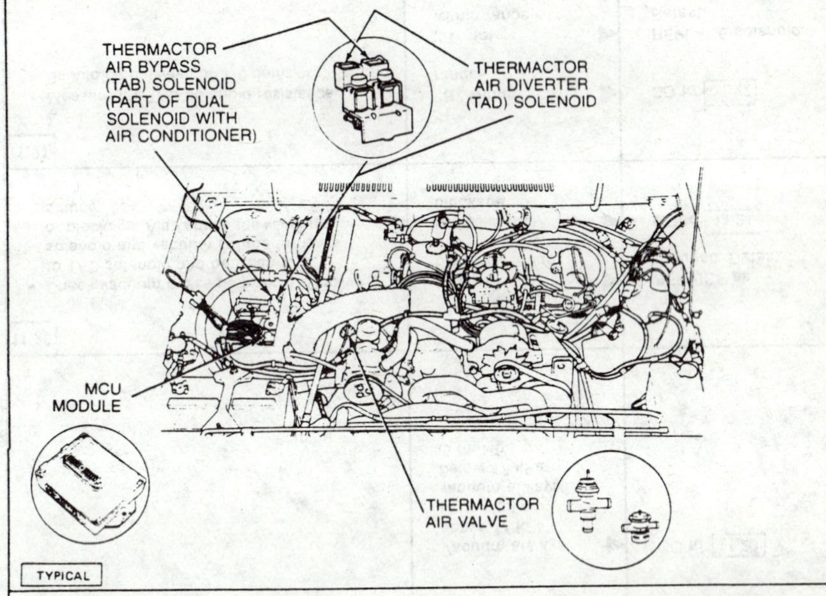

Components controlled by the MCU System in this Pinpoint Test are the Thermactor Air Bypass Solenoid, Thermactor Air Valve, and the MCU Module. The components that are not directly controlled by the MCU System during this Pinpoint Test but could possibly cause a Service Code fault are as follows:

- Belts Broken or Loose
- Hoses Pinched, Crimped, Missing, Plugged Restrictor
- Malfunctioning Vacuum Delay Valve

NOTE: Virtually anything that inhibits the Thermactor system from switching air downstream to the catalyst should be investigated.

Thermactor Air Bypass | Pinpoint | Test 12

12.0 SERVICE CODE 4-6 (Continued)

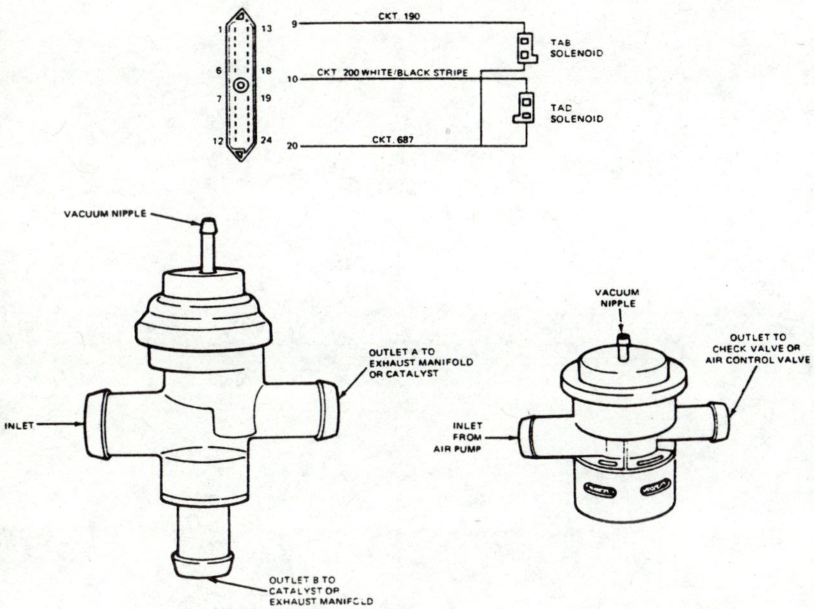

Thermactor Air Bypass | Pinpoint | Test 12

TEST STEP	RESULT ▶	ACTION TO TAKE
12.10 • Remove vacuum hose from the TAB valve, and connect a vacuum gauge to hose. Start/re-start engine, increase RPM to above 2,500 to initiate Self-Test, and observe the gauge during the initialization pulses.	Vacuum pulses above and below 17 kPa (5 In. Hg.) ▶	MCU system OK, CHECK thermactor air, Section 10. GO to 12.21
	Vacuum always below 17 kPa (5 In. Hg.) ▶	GO to 12.21
	Vacuum always above 17 kPa (5 In. Hg.) ▶	GO to 12.30
12.21 • Check at the TAB solenoid output to make sure vacuum is present when the solenoid is energized. • Jumper the TAB solenoid circuit 190 to ground. • Start engine, run at idle.	Vacuum ▶	GO to 12.22
	No Vacuum ▶	REPLACE solenoid. Retest.
12.22 • Check circuit 190 for continuity. Measure resistance.	R is more than 10 ohms. ▶	SERVICE harness. Retest.
	R is 10 ohms or less. ▶	REPLACE MCU module. Retest.
12.30 • Check at the TAB solenoid output to make sure vacuum is not present when the solenoid is de-energized. • Disconnect the TAB solenoid electrical. • Start engine, run at idle.	Vacuum not present. ▶	GO to 12.31
	Vacuum present. ▶	REPLACE solenoid. Retest.
12.31 • Check circuit 190 for a short to ground. Measure resistance.	R is 1,000 ohms or less. ▶	SERVICE short. Retest.
	R is more than 1,000 ohms. ▶	REPLACE MCU module. Retest.

Engine Running Test Not Initiated — Pinpoint Test 13

13.0 SERVICE CODE 3-3

It is necessary to increase the rpm to greater than 2,500 within 20 seconds after start in order to initiate Self-Test. Turn key Off, and repeat Quick Test 5.

Self-Test Not Functional — Pinpoint Test 14

14.0

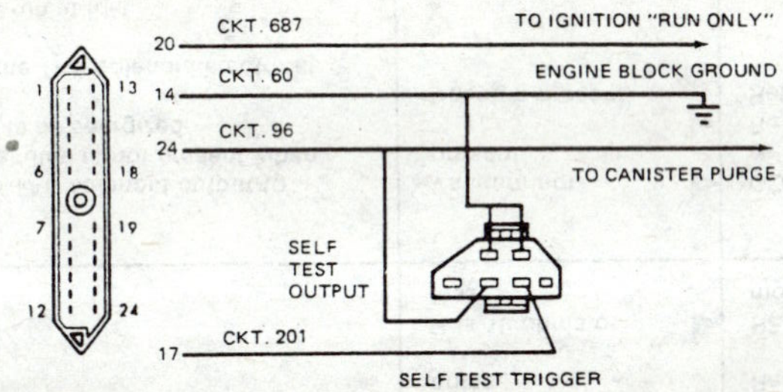

TEST STEP	RESULT	▶ ACTION TO TAKE
14.1 **VOM** • Make sure that the Self-Test trigger jumper and VOM were connected to the correct pins on the Self-Test connector and that good connection existed.	Yes	▶ CONTINUE to next Step 14.2.
or STAR • Make sure STAR unit was properly connected and pushbutton latched (colon displayed) when Self-Test was attempted.	No	▶ Correct the problem and retest.
14.2 • Check circuits 96, 201, and 60 between MCU and Self-Test connector for continuity, and verify circuit 96 is not shorted to ground.	Circuits, OK.	▶ GO to 14.3
	Fault found in circuit.	▶ SERVICE circuit. If circuit 101 is shorted to ground, go to 16.2. Retest.
14.3 • Check power circuit 687 (hot when ignition in Run) and ground circuit 60 to MCU module for continuity. If fuse link is blown, look for a short in the power circuit (687).	687 circuit and circuit 60, OK.	▶ REPLACE MCU module. Retest.
	Faults found in circuits 687 and 60.	▶ SERVICE circuit 687 and circuit 60. Retest.

Tachometer Lead Check — Pinpoint Test 15

15.0

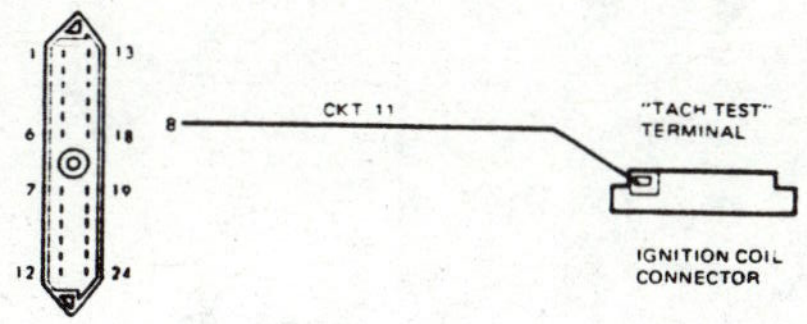

TEST STEP	RESULT ▶	ACTION TO TAKE
15.1 • Check Tach Lead (circuit 11) for continuity between the ignition coil and MCU module.	Circuit open ▶	SERVICE circuit. Retest.
	Circuit OK ▶	REPLACE MCU module. Retest.

Canister Purge Solenoid — Pinpoint Test 16

16.0

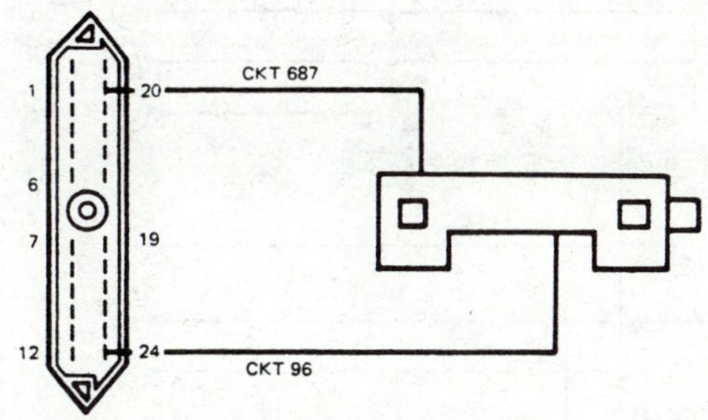

TEST STEP	RESULT ▶	ACTION TO TAKE
16.1 • Check vacuum hoses for leaks or blockage and check vacuum source.	Leaks or blockage ▶	SERVICE as required. Retest.
	No leaks or blockage ▶	GO to 16.2
16.2 • Check to make sure the canister purge solenoid passes vacuum when it is energized (12V) and blocks vacuum when it is de-energized.	Solenoid blocks and passes vacuum as indicated. ▶	GO to 16.3
	Solenoid does not pass or block vacuum as indicated ▶	REPLACE solenoid. Retest.

Canister Purge Solenoid — Pinpoint Test 16

TEST STEP	RESULT ▶	ACTION TO TAKE
16.3 • Check circuits 687 and 96 between the solenoid and the MCU module for continuity. Measure resistance.	R more than 5 ohms. ▶	SERVICE open. Retest.
	R 5 ohms or less. ▶	GO to 16.4
16.4 • Check circuit 96 for short to ground. Measure resistance.	R 1,000 ohms or less. ▶	SERVICE short. Retest.
	R more than 1,000 ohms. ▶	REPLACE MCU module. Retest.

Schematics—Electrical

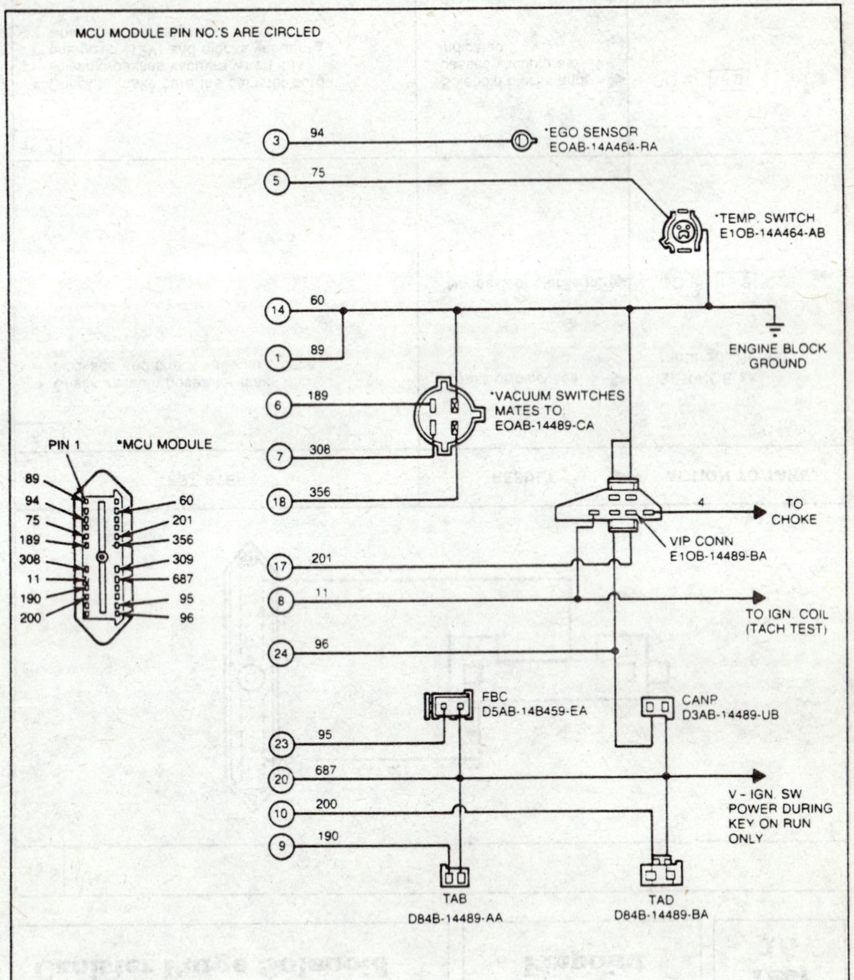

*GREASE REQUIRED

Electronic Diagnosis

3

1985 FORD CARS & TRUCKS
EEC-IV & MCU SYSTEMS

When To Use This Diagnostic Procedure

This diagnostic procedure is only used on Electronic Engine Control (EEC-IV) equipped vehicles when the Diagnostic Routines, Section 2 instruct you to perform EEC-IV Diagnostics. It is divided into two test formats. **Quick Test**, a functional EEC-IV system test, and **Pinpoint Tests**, a number of specific component tests.

To test and service the EEC-IV subsystem, perform the Quick Test first. **If the vehicle passes all three phases of the Quick Test: Key On, Engine Off, Engine Running, and Continuous Testing without running any Pinpoint Tests, you will know the EEC-IV subsystem is OK and the vehicle's problem exists somewhere else other than in the EEC-IV subsystem.** However, if a Step of the Quick Test fails, perform only the Pinpoint Tests specified by the failed Step. Do not begin any Pinpoint Test without following the instructions at the beginning of the Section.

After all tests and services have been completed, repeat the entire Quick Test to make sure the EEC-IV subsystem works properly.

Quick Test Description

The Quick Test is a functional test of the EEC-IV system consisting of basic Test Steps (described below). These Steps must be carefully followed in sequence, otherwise misdiagnosis, or the replacement of non-faulty components may result.

Quick Test Steps

1. Visual Check and Vehicle Preparation
   - Checks for obvious faults.
   - Properly prepares the vehicle for testing.
2. Equipment Hook-up
   - Ensures that the proper equipment for gathering test data is ready, prior to testing.
3. Key On, Engine Off Self-Test
   - Is a static check of processor inputs and outputs.
4. Computed Timing Check
   - Verifies the system ability to compute and maintain a fixed spark timing during Self-Test.
5. Engine Running Self-Test
   - Is a dynamic check with the engine in operation.
6. Continuous Self-Test
   - Checks the sensor inputs for opens and shorts while the vehicle is in operation.

The Key On, Engine Off and Engine Running Self-Tests are intended to detect faults present at the time of testing only, not intermittent faults. Intermittent faults are detected by Continuous Self-Test.

A continuous memory code is information collected and stored from an intermittant fault detected in a previous drive-cycle, the fault may not be present at this time.

WARNING: Anyone who departs from the instruction provided in this publication must first establish that he compromises neither his personal safety nor the vehicle integrity by his choice of methods, tools or parts.

QUICK TEST: Visual Check And Vehicle Preparation — 1.0

Correct test results for the Quick Test are dependent on the proper operation of related non-EEC components systems. It may be necessary to correct faults in these areas before EEC will pass Quick Test. Refer to Diagnostic Routines, Section 2 for service.

Before hooking up any equipment to diagnose the EEC system, make the following checks:

1. Verify the condition of air cleaner and ducting. These components may be removed and reinstalled as necessary for service and/or inspection.
2. Check all engine vacuum hoses for:
   - Leaks or pinched hoses.
   - Proper emission routing per Vehicle Emission Control Information (VECI) Decal.
3. Check the EEC system wiring harness electrical connections for:
   - Proper connections.
   - Loose or detached connectors, wires and terminals.
   - Corrosion.
   - Proper routing of harness.

It may be necessary to disconnect or disassemble the connector assembly to perform some of the inspections. (Note the location of each pin before disassembly.)

4. Check processor, sensors and actuators for physical damage.
5. Perform all safety steps required to start and run operational vehicle tests.
6. Apply the emergency brake. Place shift lever in Park; Neutral for manual transmission.
7. **Turn Off all electrical loads** such as the radio, lamps, air conditioner, etc. Be sure doors are closed whenever readings are made.
8. Verify engine coolant is at the specified level.
9. Start engine and idle until the upper radiator hose is hot and pressurized and the throttle is off fast idle. While the engine is operating, check for leaks around the exhaust manifold, EGO sensor, and vacuum hose connections.

   NOTE: **If engine will not start, starts but stalls, idles rough or runs rough continue through Quick Test Step 3.0. If any of the above conditions are still present after a pass code 11 in the Key On, Engine Off self-test, go to Pinpoint Test Step A2 for no start or diagnostics by symptom in the proper Engine Supplement Section for stalls or roughness.**

10. Turn ignition Key Off.
11. Service items as required, and proceed to equipment hook-up, Step 2.0

QUICK TEST: Equipment Hook-Up — 2.0

Using the STAR tester:

- Turn the ignition key Off.
- Connect the color-coded adapter cable leads to the STAR tester.
- Connect the adapter cable's two service connectors to the vehicle's appropriate Self-Test connectors.
- Connect the timing light.

After equipment hook-up, GO to Quick Test Step 3.0

Using analog voltmeter:

- Turn ignition key Off.
- Connect a jumper wire from Self-Test Input (STI) to Pin 2, Signal Return on the Self-Test connector (refer to the diagram below).
- Set analog VOM on a DC voltage range to read from 0 to 15 volts DC. Connect VOM from battery (+) to Pin 4 Self-Test Output (STO) in the Self-Test connector.
- Connect the timing light.

After equipment hook-up, GO to Quick Test Step 3.0

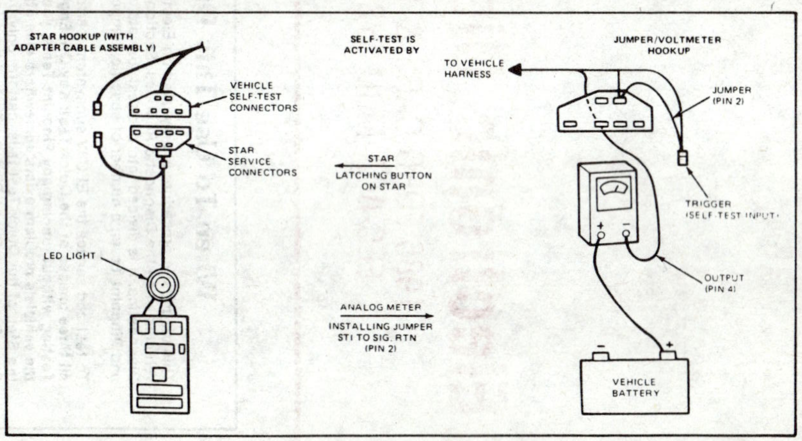

QUICK TEST: Key On, Engine Off Self-Test — 3.0

A CODE OUTPUT

Correct test results for the QUICK TEST are dependent on the proper operation of related non-EEC-IV components systems. It may be necessary to correct any faults in these areas before EEC-IV will pass the QUICK TEST. Refer to the diagnostic routines, section 2 for proper servicing.

- Verify that the vehicle has been properly prepared per QUICK TEST STEPS 1.0 and 2.0.
- Place transmission in neutral or park. Set parking brake. Place A C or heater control to off. Place octane switch in premium position on 2.3L SVO TURBO.
- Activate SELF-TEST.
- Place key in run position to start SELF-TEST.

NOTE: Do not depress throttle during SELF-TEST.

- Observe and record all service codes indicated. One of the following outputs will occur.

NOTE: Continuous memory codes recorded in this step will be used for diagnosis later in QUICK TEST Step 6.0!

RESULTS: Code Format — Key On Engine Off	Separator	Continuous Memory		ACTION TO TAKE
11	1(0)	11	▶	KEY ON ENGINE OFF and CONTINUOUS MEMORY indicate a pass. Go To QUICK TEST STEP 4.0. If vehicle is a no start go directly to PINPOINT TEST STEP A1.
ANY CODE(S)	1(0)	11	▶	KEY ON ENGINE OFF TEST indicates a fault. Record codes and Go To QUICK TEST STEP 3.0B.
ANY CODE(S)	1(0)	ANY CODE(S)	▶	KEY ON ENGINE OFF and CONTINUOUS MEMORY indicate a fault. Record ALL codes. DO NOT REPAIR CONTINUOUS MEMORY CODES AT THIS TIME! KEY ON ENGINE OFF and ENGINE RUNNING codes must be repaired first. Go To QUICK TEST STEP 3.0B.
11	1(0)	ANY CODE(S) EXCEPT 15	▶	CONTINUOUS MEMORY indicates a fault. DO NOT REPAIR THESE CODES AT THIS TIME! KEY ON ENGINE OFF and ENGINE RUNNING codes must be repaired first. Go To QUICK TEST STEP 4.0
11	1(0)	15	▶	Go To Pinpoint Test Step Q10.
NO CODES OUTPUTTED			▶	Repeat SELF-TEST and verify that no service codes are present. Go To Pinpoint Test Step Q1.

*Star tester displays a 10 in place of a 1 for the single digit separator code, therefore the separator code is expressed in the results column as 1(0).

QUICK TEST: Key On, Engine Off Self-Test — 3.0

B RESULTS AND ACTION TO TAKE

- Using the KEY ON ENGINE OFF service codes from QUICK TEST step 3.0A, follow the instructions in the ACTION TO TAKE column in this step.
- When more than one service code is received always start with the first code received.
- Whenever a repair is made REPEAT QUICK TEST!

NOTE: Before proceeding to the specified Pinpoint Test, read the instructions on how to use the Pinpoint Tests at the beginning of the Pinpoint Test section.

RESULTS ▶	ACTION TO TAKE								
						TRUCK			
Key On, Engine Off Service Codes	1.6L EFI 1.6L & 2.3L EFI TC	2.3L OHC FBC	2.3L HSC CFI	3.8L CFI	5.0L CFI	2.8L FBC	4.9L FBC	5.0L 5.8L FBC	2.3L 5.0L EFI
15 GO to Pinpoint Step ▶	Replace Processor — Repeat Quick Test								
21 GO to Pinpoint Step ▶	DE1	DE1	DE1	DE1	DE1	DE1	DE1	DE1	DE1
22 GO to Pinpoint Step ▶	DF1	DF1	DF1	DF1	DF1	DF1	DF1	DF1	DF1
23 GO to Pinpoint Step ▶	DH1	DH1	DH1	DH1	DH1	DH1	DH1	DH1	DH1
24 GO to Pinpoint Step ▶	DA1	*	DB1	DB1	DB1	DB1	*	*	DB1
26 GO to Pinpoint Step ▶	DK1	*	*	*	*	*	*	*	*
31 GO to Pinpoint Step ▶	*	*	DM2	DD2	DD2	DD2	*	DD2	DD2
51 GO to Pinpoint Step ▶	DE10	DE10	DE10	DE10	DE10	DE10	DE10	DE10	DE10
52 GO to Pinpoint Step ▶	*	*	FF1	*	*	*	*	*	FF1
53 GO to Pinpoint Step ▶	DH3	DH3	DH3	DH3	DH3	DH3	DH3	DH3	DH3
54 GO to Pinpoint Step ▶	DA10	*	DB10	DB10	DB10	DB10	*	*	DB10
55 GO to Pinpoint Step ▶	*	FC3	*	*	*	*	FC3	*	*
56 GO to Pinpoint Step ▶	DK10	*	*	*	*	*	*	*	*
61 GO to Pinpoint Step ▶	DE20	DE20	DE20	DE20	DE20	DE20	DE20	DE20	DE20
63 GO to Pinpoint Step ▶	DH10	DH10	DH10	DH10	DH10	DH10	DH10	DH10	DH10
64 GO to Pinpoint Step ▶	DA20	*	DB20	DB20	DB20	DB20	*	*	DB20
66 GO to Pinpoint Step ▶	DK20	*	*	*	*	*	*	*	*
67 GO to Pinpoint Step ▶	FA1	FA1	FA1	FA1	*	FA1	FA1	*	FA1
68 GO to Pinpoint Step ▶	*	KF1	KF1	KF1	*	KF1	KF1	*	*
81 GO to Pinpoint Step ▶	*	*	*	KC8	KC8	KC8	*	*	KC8
82 GO to Pinpoint Step ▶	*	*	*	KC8	KC8	KC8	*	*	KC8
83 GO to Pinpoint Step ▶	*	*	*	DD17	DD17	DD17	*	*	DD17
84 GO to Pinpoint Step ▶	*	*	DM17	DD17	DD17	DD17	*	*	DD17
85 GO to Pinpoint Step ▶	*	*	KD4	*	KD4	KD4	*	*	*
86 GO to Pinpoint Step ▶	*	*	*	KM10	KM10	*	*	*	*
87 GO to Pinpoint Step ▶	*	*	J7	J7	J7	KK5	*	*	J7
88 GO to Pinpoint Step ▶	*	*	*	*	KG8	KJ6	*	*	*
89 GO to Pinpoint Step ▶	*	*	*	KP5	KP5	KR1	*	*	KR1
No Codes Code Not Listed	GO to Pinpoint Test Step Q1								

QUICK TEST: TIMING CHECK — 4.0

TEST STEP	RESULT	ACTION TO TAKE
• If engine is a no start go directly to Pinpoint Test A NO STARTS. • If engine starts but stalls go directly to DIAGNOSTIC BY SYMPTOM in the proper ENGINE SUPPLEMENT SECTION.	Engine stalls during timing check ▶	Go To DIAGNOSTIC BY SYMPTOM in the proper ENGINE SUPPLEMENT SECTION.
• Key off, wait 10 seconds. • Verify SELF-TEST trigger has been activated.	Timing is not base + 20° BTDC ▶	Go To Pinpoint Test Step [P1].
• Restart engine and check timing while in SELF-TEST. You have 2 minutes from the time the last code is outputted to check SELF-TEST timing. • Correct SELF-TEST timing equals base plus 20° BTDC ± 3° (see VECI decal for correct base timing). EXAMPLE: If base timing is 10° BTDC then SELF-TEST timing is: 10° + 20° = 30° ± 3° therefore SELF-TEST timing is 27°-33° BTDC.	Timing is base plus 20° BTDC ▶	Go To QUICK TEST Step [5.0].

QUICK TEST: Engine Running Self-Test — 5.0

A CODE OUTPUT

NOTE: If engine is a no start go directly to Pinpoint Test Step A1. If engine stalls go directly to Diagnostic By Symptom in the proper engine supplement section.

- Deactivate SELF-TEST.
- Start and run engine at greater than 2000 RPM for 2 minutes. This warms up the EGO sensor.
- Turn engine off, wait 10 seconds.
- Activate SELF-TEST.
- Start engine. The ENGINE RUNNING SELF-TEST will progress as follows:
  - — Engine ID code. ①
    Depress and release the brake pedal, only once, after the ID code has been received. This tests the Brake on/off switch on 2.8L and 2.3L EFI trucks.
  - — Run test.
  - — When the dynamic response code occurs, perform a brief WOT. ②
  - — ENGINE RUNNING service codes.
- Observe and record all service codes. One of the following outputs will occur.

NOTE: If engine stalls during Self-Test, go directly to Diagnostic By Symptom in the proper engine supplement section.

RESULTS			ACTION TO TAKE
Code Format			
Engine ID	Dynamic Response	Engine Running	
2(0), 3(0) or 4(0)	1(0)	11 ▶	ENGINE RUNNING SELF-TEST indicates a pass. If the symptom was of an intermittent nature, Go To QUICK TEST STEP 6.0. If the drive symptom is currently present, Go To DIAGNOSTIC BY SYMPTOM in the proper ENGINE SUPPLEMENT SECTION. Otherwise testing is complete, EEC-IV system is OK.
2(0), 3(0) or 4(0)	1(0)	ANY CODES ▶	ENGINE RUNNING SELF-TEST indicates a fault. Go To QUICK TEST STEP 5.0B.
NO CODES OUTPUTTED ▶			Repeat SELF-TEST and verify that no service codes are present, then Go To Pinpoint Test Step Q1.

① Engine ID code is equal to half the number of cylinders (a code of 2 equals a 4 cylinder engine. STAR tester adds a zero to all single digit readings, so a 20 code equals a 4 cylinder engine).
② Star displays a 10 in place of a 1 for the single digit separator code, therefore, the dynamic response code is expressed in the results column as 1(0).

QUICK TEST: Engine Running Self-Test — 5.0

B — RESULTS AND ACTION TO TAKE

- Using the ENGINE RUNNING service codes from QUICK TEST STEP 5.0A, follow the instructions in the ACTION TO TAKE column in this step.
- When more than one service code is received, always start service with the first code received.
- Whenever a repair is made, REPEAT QUICK TEST!

RESULTS ▶	ACTION TO TAKE								
						TRUCK			
Engine Running Service Codes	1.6L EFI 1.6L & 2.3L EFI TC	2.3L OHC FBC	2.3L HSC CFI	3.8L CFI	5.0L CFI	2.8L FBC	4.9L FBC	5.0L 5.8L FBC	2.3L 5.0L EFI
12 GO to Pinpoint Step ▶	KE1	KF16	KF16	KF16	KG1	KF16	KF16	KG1	KE1
13 GO to Pinpoint Step ▶	KE8	KF19	KF19	KF19	*	KF19	KF19	*	KE8
16 GO to Pinpoint Step ▶	*	HB21	*	HD16	HC10	HB21	HB21	HB21	*
17 GO to Pinpoint Step ▶	*	*	*	HD8	HC10	*	*	*	*
21 GO to Pinpoint Step ▶	DE1	DE1	DE1	DE1	DE1	DE1	DE1	DE1	DE1
22 GO to Pinpoint Step ▶	DF1	DF7	DF7	DF7	DF7	DF7	DF7	DF7	DF7
23 GO to Pinpoint Step ▶	DH1	DH1	DH1	DH1	DH1	DH1	DH1	DH1	DH1
24 GO to Pinpoint Step ▶	DA1	*	DB1	DB1	DB1	DB1	*	*	DB1
25 GO to Pinpoint Step ▶	DG1	*	*	DG1	*	DG1	*	*	DG1
26 GO to Pinpoint Step ▶	DK1	*	*	*	*	*	*	*	*
31 GO to Pinpoint Step ▶	*	*	DM1	DD1	DD1	DD1	*	DD1	DD1
32 GO to Pinpoint Step ▶	*	*	DM11	DD11	DD11	DD11	*	DD11	DD11
33 GO to Pinpoint Step ▶	*	*	DM11	DD11	DD11	DD11	*	DD11	DD11
34 GO to Pinpoint Step ▶	KA1	*	DM11	DD11	DD11	DD11	KA1	DD11	DD11
35 GO to Pinpoint Step ▶	*	*	DM30	DD30	DD30	DD30	*	DD30	DD30
41 GO to Pinpoint Step ▶	HA10	HB1	HF5	HD1	HC6	HB1	HB1	HB1	HE10
42 GO to Pinpoint Step ▶	HA7	HB20	HF9	HD16	HC10	HB20	HB20	HB20	HE7
43 GO to Pinpoint Step ▶	*	HB40	*	HD23	HC17	HB40	HB40	HB40	*
44 GO to Pinpoint Step ▶	*	KC1	*	KC1	KC1	KC1	KC1	KC1	KC1
45 GO to Pinpoint Step ▶	*	KC1	*	KC1	KC1	KC1	KC1	KC1	KC1
46 GO to Pinpoint Step ▶	*	KC1	*	KC1	KC1	KC1	KC1	KC1	KC1
47 GO to Pinpoint Step ▶	*	HB20	*	*	HC10	HB20	HB20	HB20	*
48 GO to Pinpoint Step ▶	*	*	*	HD6	*	*	*	*	*
55 GO to Pinpoint Step ▶	*	*	*	FC3	FC3	FC3	*	*	*
58 GO to Pinpoint Step ▶	*	KF12	KF12	KF12	*	KF12	KF12	*	*
65 GO to Pinpoint Step ▶	*	*	*	FC1	FC1	FC1	*	*	*
72 GO to Pinpoint Step ▶	*	DF10	DF10	DF10	DF10	DF10	DF10	DF10	DF10
73 GO to Pinpoint Step ▶	DH20	DH20	DH20	DH20	DH20	DH20	DH20	DH20	DH20
74 GO to Pinpoint Step ▶	*	*	*	*	*	FD1	*	*	FD1
75 GO to Pinpoint Step ▶	*	*	*	*	*	FD4	*	*	FD4
76 GO to Pinpoint Step ▶	DK30	*	*	*	*	*	*	*	*
77 GO to Pinpoint Step ▶	M1	M1	M1	M1	M1	M1	M1	M1	M1
91 GO to Pinpoint Step ▶	*	*	*	HD1	*	*	*	*	*
92 GO to Pinpoint Step ▶	*	*	*	HD19	*	*	*	*	*
93 GO to Pinpoint Step ▶	*	*	*	HD23	*	*	*	*	*
94 GO to Pinpoint Step ▶	*	*	*	KC1	*	*	*	*	*
95 GO to Pinpoint Step ▶	*	*	*	KC1	*	*	*	*	*
96 GO to Pinpoint Step ▶	*	*	*	KC1	*	*	*	*	*
97 GO to Pinpoint Step ▶	*	*	*	HD19	*	*	*	*	*
98 GO to Pinpoint Step ▶	*	*	*	HD12	*	*	*	*	*
No Codes/Code Not Listed	GO to Pinpoint Test Step Q1								

QUICK TEST: Continuous Self-Test — 6.0

TEST STEP	RESULT ▶	ACTION TO TAKE
A — CONTINUOUS MEMORY CODES		
• To ensure proper diagnosis of CONTINUOUS MEMORY codes, QUICK TEST STEPS 1.0 through 5.0 must be successfully completed. • Do the results of both KEY ON ENGINE OFF and ENGINE RUNNING SELF-TEST result in a pass code 11?	Yes ▶	Go To 6.0B.
	No ▶	Return to Step 1.0 of QUICK TEST and make the necessary repairs indicated in QUICK TEST STEPS 3.0 and 5.0 before continuing.
B — CLEAR CONTINUOUS MEMORY CODES		
• Prepare the vehicle for KEY ON ENGINE OFF QUICK TEST per steps 1.0, 2.0, and 3.0. • Perform KEY ON ENGINE OFF SELF-TEST. When the service codes begin, exit the self-test program (via unlatching STAR or removing the jumper wire from STI to Signal Return). Exiting QUICK TEST during code output will clear all codes stored in the CONTINUOUS MEMORY. • Repeat KEY ON ENGINE OFF SELF-TEST. • Does KEY ON ENGINE OFF SELF-TEST output 11-10-11?	Yes ▶	Go To 6.0C.
	No ▶	Check STI circuit for short to ground. Repair as necessary. Repeat Step 6.0B.
C — DETERMINE CODES TO BE TESTED		
• Refer to CONTINUOUS MEMORY CODES recorded in QUICK TEST STEP 3.0A. • Have any of the codes been repaired either in KEY ON ENGINE OFF or ENGINE RUNNING SELF-TESTS?	Yes ▶	Disregard codes already serviced. For remaining codes, Go To 6.0D.
	No ▶	Go To 6.0D.

QUICK TEST: Continuous Self-Test — 6.0

D TEST RESULTS AND ACTION TO TAKE

- YOU SHOULD NOT BE HERE UNLESS ALL PREVIOUS QUICK TEST STEPS HAVE BEEN SUCCESSFULLY COMPLETED!
- Verify proper test equipment hook up per QUICK TEST STEP 2.0.
- SELF-TEST deactivated. Do not activate SELF-TEST unless instructed.
- Using the service codes determined in STEP 6.0C, follow the action to take instructions. (While in the appropriate PINPOINT TEST, one or both of the continuous monitor modes described below may have to be used to find the intermittent fault.)

CONTINUOUS MONITOR (WIGGLE) MODES

- KEY ON ENGINE OFF
  With SELF-TEST deactivated, turn the ignition key to the ON position. You are now in the continuous monitor mode (wiggle test). STO will be activated whenever a fault is detected. A fault is indicated when the STAR LED turns off or the VOM deflection is 10.5V or greater.
- ENGINE RUNNING
  Activate SELF-TEST and perform ENGINE RUNNING QUICK TEST. After service code output has been completed do not turn engine off or deactivate SELF-TEST. Approximately 2 minutes after code 11 has been outputted the system will enter and remain in continuous monitor mode until SELF-TEST is deactivated or the engine is turned off. STO will be activated whenever a fault is detected. A fault is indicated the same way as KEY ON ENGINE OFF.
- Refer to the appendix for a detailed description of the continuous monitor (wiggle test).

RESULTS ▶	ACTION TO TAKE								
						TRUCK			
Continuous Memory Service Codes	1.6L EFI 1.6L & 2.3L EFI TC	2.3L OHC FBC	2.3L HSC CFI	3.8L CFI	5.0L CFI	2.8L FBC	4.9L FBC	5.0L 5.8L FBC	2.3L 5.0L EFI
11 GO to Pinpoint Step ▶	If drive symptom is currently present go to Diagnostic by Symptom in the proper ENGINE SUPPLEMENT section. Otherwise testing is complete.								
13 GO to Pinpoint Step ▶	*	*	KF19	KF19	*	*	*	*	*
14 GO to Pinpoint Step ▶	Y1	Y1	Y1	Y1	Y1	Y1	Y1	Y1	Y1
18 GO to Pinpoint Step ▶	N1	N1	N1	N1	N1	N1	N1	N1	N1
21 GO to Pinpoint Step ▶	DE90	DE90	DE90	DE90	DE90	DE90	DE90	DE90	DE90
22 GO to Pinpoint Step ▶	DF90	DF90	DF90	DF90	DF90	DF90	DF90	DF90	DF90
31 GO to Pinpoint Step ▶	*	*	DM90	DD90	DD90	DD90	*	DD90	DD90
41 GO to Pinpoint Step ▶	HA20	HB1	*	*	HC13	HB1	HB1	HB1	*
42 GO to Pinpoint Step ▶	HA20	HB20	*	*	HC13	HB20	HB20	HB20	HE20
51 GO to Pinpoint Step ▶	DE91	DE91	DE91	DE91	DE91	DE91	DE91	DE91	DE91
53 GO to Pinpoint Step ▶	DH90	DH90	DH90	DH90	DH90	DH90	DH90	DH90	*
54 GO to Pinpoint Step ▶	DA90	*	DB90	DB90	DB90	DB90	*	*	DB90
55 GO to Pinpoint Step ▶	*	*	*	FC3	FC3	*	*	*	*
56 GO to Pinpoint Step ▶	DK90	*	*	*	*	*	*	*	*
61 GO to Pinpoint Step ▶	DE94	DE94	DE94	DE94	DE94	DE94	DE94	DE94	DE94
63 GO to Pinpoint Step ▶	DH94	DH94	DH94	DH94	DH94	DH94	DH94	DH94	DH94
64 GO to Pinpoint Step ▶	DA93	*	DB93	DB93	DB93	DB93	*	*	DB93
65 GO to Pinpoint Step ▶	*	*	*	FC1	FC1	*	*	*	*
66 GO to Pinpoint Step ▶	DK93	*	*	*	*	*	*	*	*
No Codes/Code Not Listed	GO to Pinpoint Test Step Q1								

APPENDIX: Self-Test Description

SELF-TEST DESCRIPTION

The Self-Test is divided into three specialized tests: Key On, Engine Off, Engine Running, and Continuous memory. The Self-Test is not a conclusive test by itself, but is used as a part of the functional Quick-Test diagnostic procedure. The processor stores the Self-Test program in its permanent memory. When activated, it checks the EEC-IV system by testing its memory integrity and processing capability, and verifies that various sensors and actuators are connected and operating properly.

The Key On, Engine Off and Engine Running tests are functional tests which only detect faults present at the time of the Self-Test. Continuous testing is an ongoing test that stores fault information for retrieval at a later time (during the Self-Test).

Key On, Engine Off

At this time, a test of the EEC-IV system is conducted with power applied and engine at rest.

For Self-Test to detect errors in this test the fault must be present at the time of testing. For intermittents, refer to Continuous Memory Codes.

Separator Pulse

A single 1/2 second separator pulse is issued 6-9 seconds after the last Functional Test code (Key On, Engine Off only). Then 6-9 seconds after the single 1/2 second separator pulse, the continuous codes will be issued. Refer to the example below:

NOTE: The separator code and continuous memory codes follow Key On, Engine Off testing codes ONLY.

Continuous Memory Codes

Continuous memory codes are issued as a result of information stored during continuous monitor testing, while the vehicle was in normal operation. These codes are displayed only during the Key On, Engine Off testing and after the separator code. These codes should be used for diagnosis only when Key On, Engine Off and Engine Running self-tests result in code 11 and all Quick Test steps 1.0 through 5.0 have been successfully completed.

Engine Running Test

At this time, a test of the EEC-IV system is conducted with the engine running. The sensors are checked under actual operating conditions and at normal operating temperatures. The actuators are exercised and checked for corresponding results.

Engine Identification Codes (ID Codes)

Engine ID codes are issued at the beginning of the engine running test and are one-digit numbers represented by the number of pulses sent out. The engine ID code is equal to 1/2 the number of engine cylinder (i.e. 2 pulses = 4 cylinders). These codes are used to verify the proper processor is installed and that the Self-Test has been entered.

Dynamic Response Check

The dynamic response check verifies the movement of the TP, VAF, and MAP sensors during the brief Wide Open Throttle (WOT) performed during the Engine Running Test. The signal for the operator to perform the brief WOT is a single pulse or 10 code on the STAR. Also, during the Engine Running Test, the brake pedal must be depressed and released after the ID code to assure operation of the brake on/off switch for 2.8L and 2.3L EFI Light trucks.

APPENDIX: Code Output Format

SERVICE CODES

The EEC-IV system communicates service information to the outside world by way of the Self-Test service codes. These service codes are two-digit numbers representing the results of Self-Test.

The service codes are transmitted on the Self-Test output (found in the Self-Test connector) in the form of timed pulses, and read by the technician on a voltmeter or on the STAR tester.

SELF-TEST OUTPUT CODE FORMAT* KEY ON, ENGINE OFF

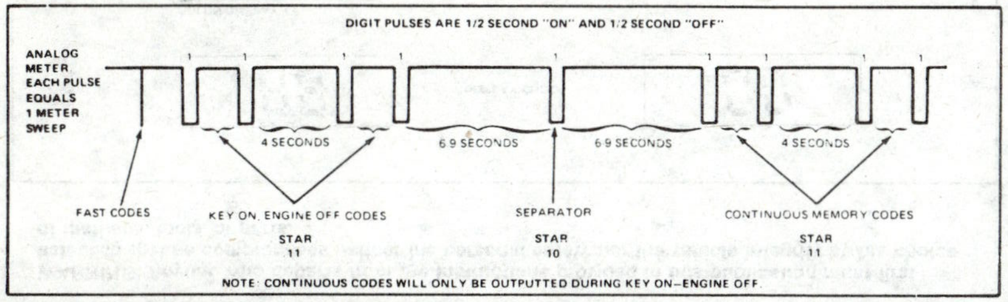

Figure 1 Key On, Engine Off Self-Test Output Code Format

SELF-TEST OUTPUT CODE FORMAT* ENGINE RUNNING

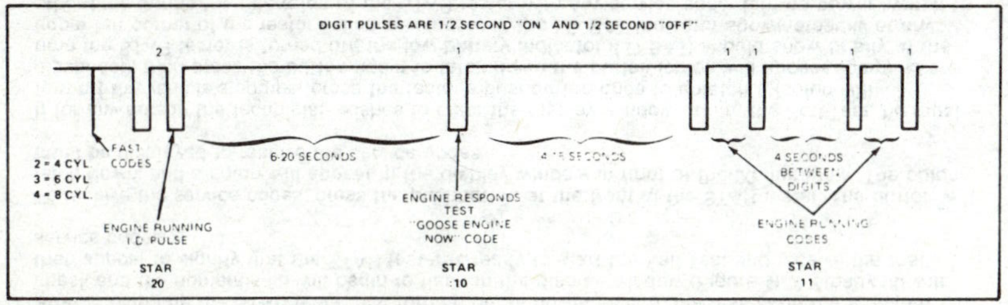

Figure 2 Engine Running Self-Test Output Code Format

Fast Codes

Fast codes are issued prior to regular service codes. These codes contain the identical information as the regular service codes but are transmitted at 100 times the normal rate. These codes are interpreted by special equipment at the Body and Assembly Division. They serve no purpose for the field.

For this reason the STAR has been designed to disregard these codes. Some meters in service detect these codes as a short burst of information (slight meter deflection).

APPENDIX: Self-Test With Analog Voltmeter

Reading Codes—Analog Voltmeter

When a service code is reported on the analog voltmeter for a function test, it will represent itself as a pulsing or sweeping movement of the voltmeter's needle across the dial face of the voltmeter (Figure 3). Therefore, a single-digit number of three will be reported by three needle pulses (sweeps). However, as previously stated, a service code is represented by a two-digit number, such as 2-3. As a result, the Self-Tests service code of 2-3 will appear on the voltmeter as two needle pulses (sweeps), then, after a two-second pause, the needle will pulse (sweep) three times.

The continuous memory codes are separated from the functional codes by a six-second delay, a single half-second sweep, and another six-second delay. They are produced on the voltmeter in the same manner as the functional codes.

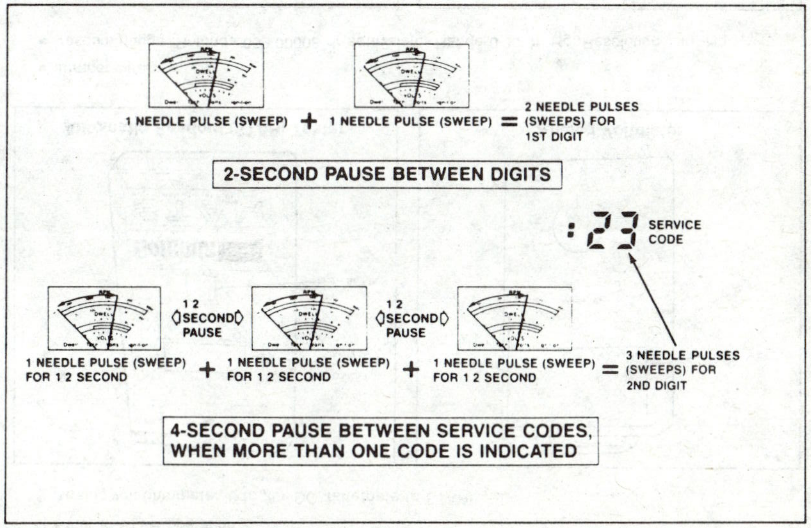

Figure 3 Analog Voltmeter Functional Service Code

APPENDIX: Self-Test With STAR Tester

Reading Codes—Self-Test Automatic Readout (STAR) Rotunda Model 007-00004 or Equivalent

After hooking up the STAR tester and turning on its power switch, the tester will run a display check and the numerals 88 will begin to flash in the display window (Figure 4). A steady 00 will then appear to signify that the STAR tester is ready to start the Self-Test and receive the test's service codes.

To receive the service codes, press the pushbutton at the front of the STAR tester. The button will latch down, and a colon will appear in the display window in front of the 00 numerals. The colon **must** be displayed to receive the service codes.

If for any reason the technician wishes to clear the display window during the Self-Test, he must turn off the vehicle's engine, press the tester's pushbutton once to unlatch it (colon will disappear), then press the button again to latch down the button (colon will appear again). Every time the STAR tester is turned off, the low battery indicator (LO BAT) should show briefly at the upper left corner of the tester's display window. If the LO BAT indicator shows steadily at any other time during the operation of the STAR tester with any service code, turn its power switch to Off and replace the 9-volt battery in the tester.

The STAR tester will display the last service code received, even after disconnecting it from the vehicle. It will hold the service code on the display until the power is turned off or the pushbutton is unlatched and relatched.

WARNING: Anyone who departs from the instructions provided in this publication must first establish that he compromises neither his personal safety nor the vehicle integrity by his choice of methods, tools, or parts.

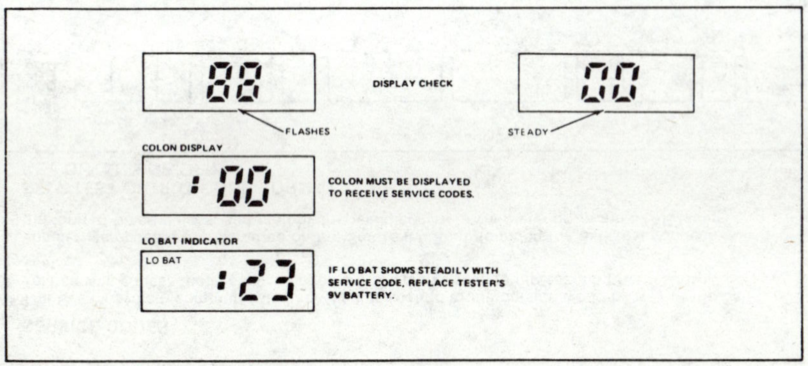

Figure 4 Self-Test Output Code Format

APPENDIX: Test Equipment

EQUIPMENT REQUIRED:

- Rotunda Self-Test Automatic Readout (STAR). No. 007-00004 with cable assembly No. 007-00010. Refer to STAR operation.
- Analog volt-ohmmeter, 0 to 20v DC, (alternate to STAR).

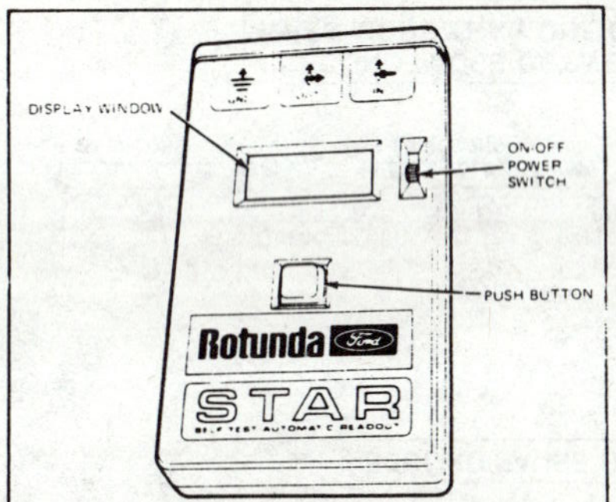

Automatic Readout (STAR) Tester

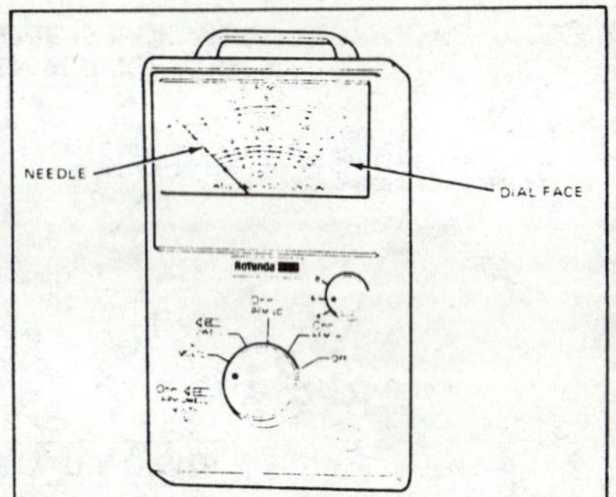

Analog Voltmeter

- Jumper wire.
- Vacuum gauge, Rotunda 059-00008 or equivalent. Range 0-30 in. Hg. Resolution 1 in. Hg.

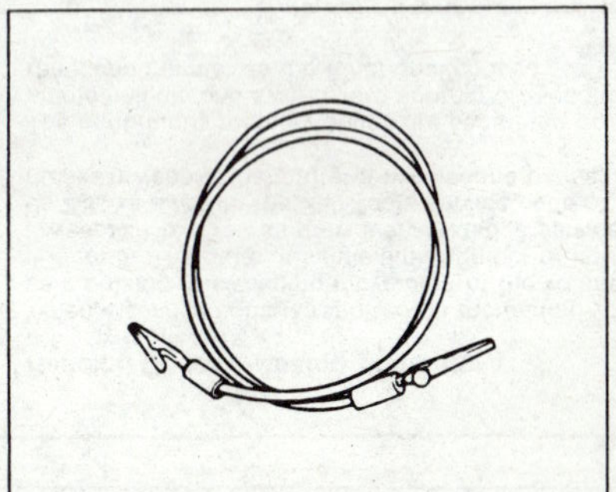

Jumper Wire

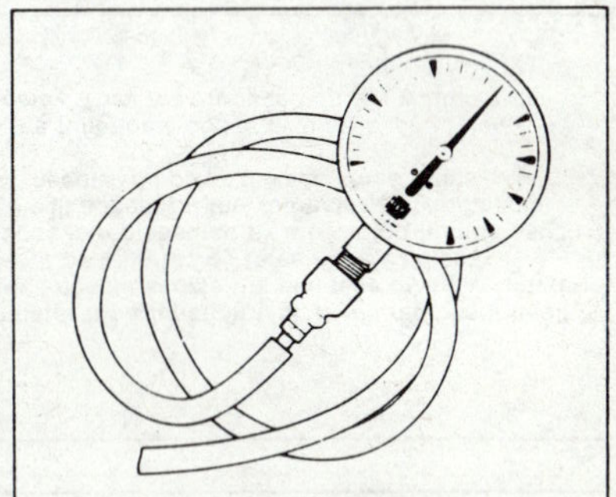

Vacuum Gauge

APPENDIX: Test Equipment (Continued)

- Tachometer, Rotunda No. 059-00010 or equivalent. Range 0-6,000 rpm. Accuracy ± 40 rpm. Resolution 20 rpm.
- Breakout Box, Rotunda 014-00322, Special Service Tool T83L-50-EEC-IV or equivalent.

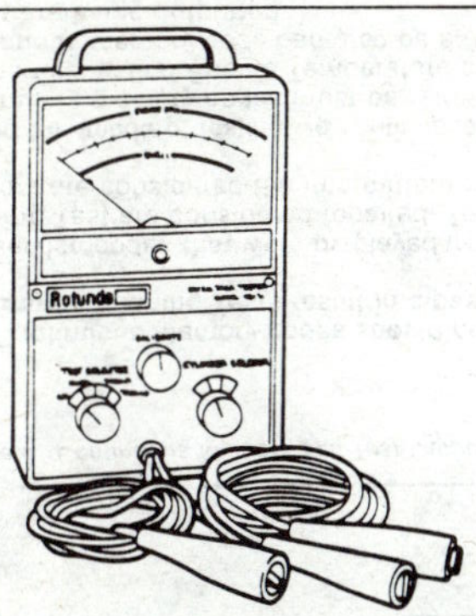

Tachometer

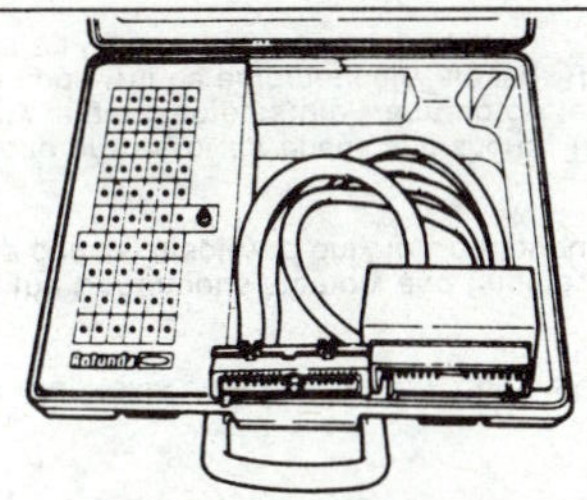

Breakout Box

- Vacuum pump, Rotunda No. 021-00014 or equivalent. Range 0-30 in. Hg.
- Digital volt-ohmmeter, Rotunda No. 014-00407 or equivalent. Input impedance 10 Megaohm minimum.

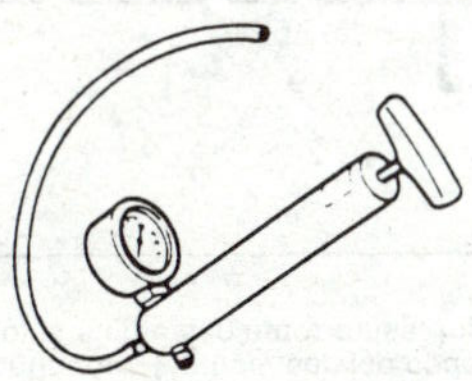

Vacuum Pump

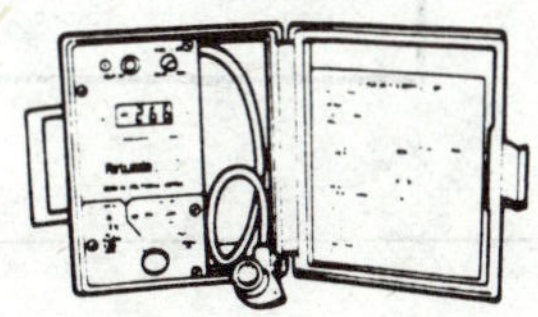

Digital Volt-Ohmmeter

APPENDIX: Test Equipment (Continued)

- Electronic Fuel Injection Pressure Gauge EFI/CFI only. Tool T80L-9974-A or equivalent.
- Spark tester (optional modified spark plug with side electrode removed). Tool D81P-6666-A or equivalent.

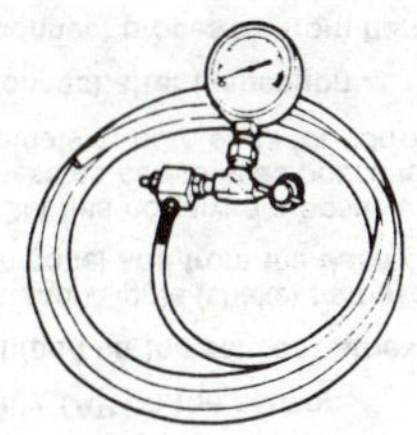

EFI Pressure Gauge

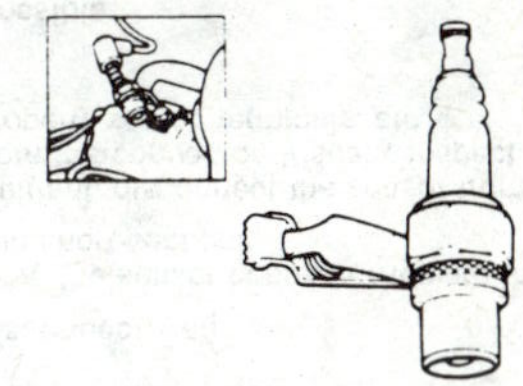

Spark Tester

- Timing light, Rotunda model 059-00006 or equivalent.
- Non-powered Test lamp.

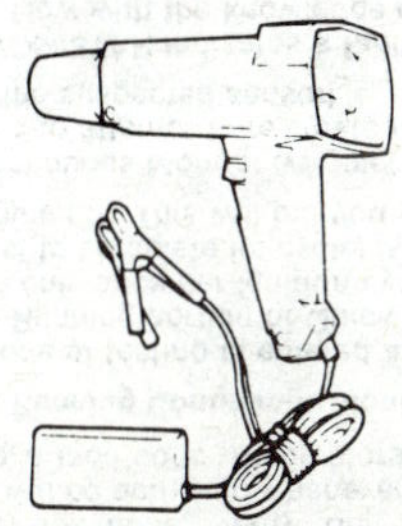

Timing Light

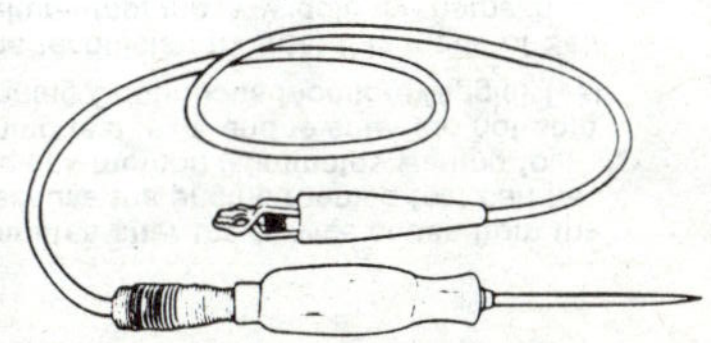

Non-Powered Test-Lamp

APPENDIX: Continuous Self-Test

The Continuous Memory service codes are separated from the Quick Test Key On, Engine Off codes by a single separator pulse, Figure 5.

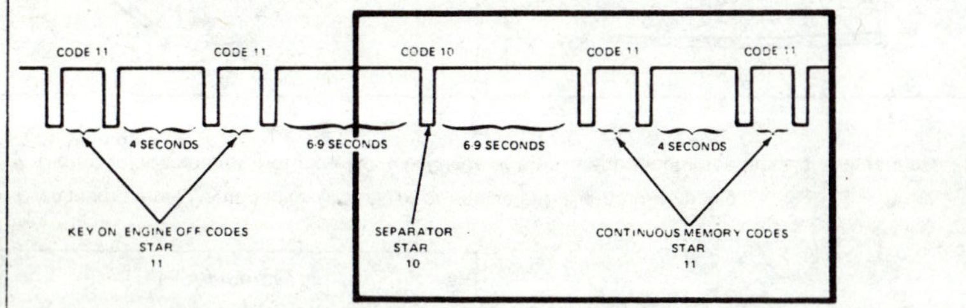

Figure 5 Continuous Memory Self-Test Output Code Format

The Continuous Memory codes should never be used for Diagnosis until the Key On, Engine Off and Engine Running Tests result in a pass code 11.

The service codes that were displayed (repeated) in both the Continuous Memory and Engine Running Test are considered repaired. The unique service code(s) displayed only in Continuous Memory are considered the intermittent service codes.

During this mode of testing we continuously monitor inputs to processor for opens and shorts. The continuous memory codes must be retrieved within twenty engine temperature warm up cycles. On the twenty-first Engine Temperature cycle the service code will be automatically erased. The continuous memory code can also be erased by deactivating Quick Test, while the service codes are being outputted.

Continuous Monitor Test

This test is intended as an aid in diagnosing intermittent failures in the sensor input circuits. The Self-Test output is energized whenever the continuous memory mode senses a fault and de-energized when the system is OK.

- Connect a VOM/STAR to the Self-Test output.

APPENDIX: Continuous Self-Test (Continued)

- **Key On, Engine Off** Continuous Monitor Test, (Wiggle).

  To enter this mode, "verify" Self-Test is not activated and then turn the key to run. Self-Test output will be activated whenever a continuous fault is detected and if the duration is long enough, a fault code will be stored.

- **Engine Running Continuous** Monitor Test, (Wiggle).

  This mode of testing is entered approximately 2 minutes after the service codes from the Engine Running portion of Quick Test (Step 5.0) because the engine running test can be entered only once per ignition cycle. An alternate quick method (eliminates waiting for Self-Test to complete its cycle) is to enter the running test, exit, and re-enter (do not shut the engine off). This will put you in the Engine Running Continuous Monitor (Wiggle) Test.

  The Continuous Monitor (Wiggle) Test will allow the technician to enter this mode of Self-Test and to attempt to re-create the intermittent failure (tap, move, wiggle the harness and/or the suspected sensor).

  If the VOM/STAR indicates a fault (short or open) the corresponding service code will be stored. Now with the knowledge of the affected circuits, a close check of the harness and associated connectors can be made.

NOTE: Remember to keep your eyes on the VOM/STAR for any change which will indicate where the intermittent is located.

EXAMPLE: How to Use the Continuous Monitor Test (Systematically)

If a service code 51 was displayed: Engine Coolant Temperature Sensor:

- Visually inspect the sensor **very closely.**
- Enter Continuous Monitor test.
- Lightly TAP on the sensor.
- Push/pull on the sensor harness connector (do not disconnect, yet).
- Test and Wiggle (shake) harness vigorously working from the sensor connector toward the dash panel and from the dash panel to the processor in short sections.
- If VOM has not given a positive indication of an intermittent, disconnect the sensor from the harness as carefully as possible. Remove terminals from the connector. Visually inspect terminals at both ends for corrosion, bad crimps, improperly seated terminals, etc.
- Reconnect after inspection.
- Disconnect processor from harness as carefully as possible.
- Inspect terminals.
- Only remove terminals associated with the sensor being inspected.
- If the VOM does not give a positive intermittent indication, reconnect the connector and erase the Continuous Test service codes.
- To erase the continuous service codes:
  - —Initiate Key On, Engine Off Self-Test.
  - —Remove jumper from the Self-Test input terminal as soon as the **first** service code is received (even if an "11" is the first code).
  - —Rerun Self-Test with jumper to verify services have been erased.

Diagnostic By Symptom

1.6L EFI
1.6L EFI TC
2.3L EFI TC

SYMPTOM	RESULT	ACTION TO TAKE
7.0 DIAGNOSTICS BY SYMPTOM • Engine stalls. • Stalls in Self-Test. • Runs rough. • Misses. • Always Rich/Lean		• Fuel delivery. Pinpoint Test Steps [HA1] thru [HA5]. • BP test. GO to Pinpoint Test Step [DF20]. • Poor power/ground connections. • Ignition system distributor cap, rotor, wires, coil, plugs. • Base engine valves, cam timing, compression etc.
• Detonation/spark knock.	2.3L EFI/TC ▶ 1.6L EFI ▶	GO to Pinpoint Test Step [DG1]. GO to Pinpoint Test Step [KA1].
• High idle speeds on each restart may be accompanied by detonation for up to 3-5 minutes after a restart.	2.3L EFI/TC ▶ 1.6L EFI ▶	GO to Pinpoint Test Step [DG1]. GO to Pinpoint Test Step [KA1].
• Lack of fast idle assist with A/C On.		GO to Pinpoint Test Step [FA1].
• Gasoline fumes under hood.		GO to Pinpoint Test Step [KD1].
• Shift and light always On or Off.		GO to Pinpoint Test Step [KL1].
• A/C does not cut-off under WOT conditions.		GO to Pinpoint Test Step [KM1].
• Boost control not functioning properly.		GO to Pinpoint Test Step [KN1].

APPENDIX: Output State Check

The output state check aids in servicing output actuators associated with the EEC-IV system. It enables the technician to energize and de-energize most of the system output actuators (turn EGR solenoid and Self-Test Output (STO) On and Off) on command. This mode is entered after all codes have been received from Key On, Engine Off and Continuous Testing. At this time, leave Self-Test activated and depress the throttle. Each time the throttle is depressed the output actuators will change state (On to Off or Off to On).

- Enter Self-Test.
- Code Output Ends.
- Do Brief WOT.
- Output Actuators Actuated.
- Next Brief WOT.
- Output Actuators Deactivated.

EEC-IV Module Connector Pin Usage

1.6L EFI
1.6L EFI TC

Pin	Circuit	Wire Color	Application	Abbreviations
1	37A	Y	Keep Alive Memory	KA PWR
7	354	LG Y	Engine Coolant Temperature	ECT
10	347	BK Y H	A C Clutch	ACC
16	259	BK O	Ignition Ground	IGN GND
17	201A	T R	Self-Test Output Shift Indicator Light	STO SIL
20	57	BK	Case Ground	CSE GND
21	68	O BK	Idle Speed Control	ISC
22	97	T LG D	Fuel Pump	FP
25	357	LG P	Vane Air Temperature	VAT
26	351	O W	Reference Voltage	V REF
29	94	DG P H	Exhaust Gas Oxygen Sensor	EGO
30	614	GY-O	Neutral Drive Switch (Turbo Grounded)	NDS
32	101	GY-Y H	Canister Purge Solenoid	CANP
35	362	Y	Exhaust Gas Recirculation Shut Off	EGR S O
36	324A	Y LG D	Spark Output	SPOUT
37	361	R	Vehicle Power	V PWR
40	60A	BK-LG D	Power Ground	PWR GND
43	200	W-BK	Vane Air Flow	VAF
45	358	LG BK	Barometric Pressure	BP
46	359	BK/W	Signal Return	SIG RET
47	355	DG/LG	Throttle Position Sensor	TPS
48	100	W R D	Self-Test Input	STI
49	89	O	Exhaust Gas Oxygen Ground	EGOG
54	73	O-LB H	Wide Open Throttle A C Cut-Off	WAC
56	349	DB	Profile Ignition Pickup	PIP
57	361	R	Vehicle Power	V PWR
58	95	T/R D	Injector 1 (Bank 1)	INJ 1
59	96	T/O D	Injector 2 (Bank 2)	INJ 2
60	60	BK-LG D	Power Ground	PWR GND

Pin locations given for reference only. Probing 60 pin connector with DVOM probe will result in permanent damage to the pin connectors. Always probe as directed using the Breakout Box.

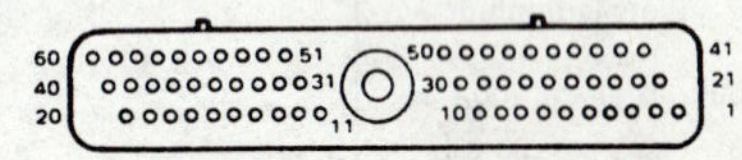

Electrical Schematic

1.6L EFI
1.6L EFI TC

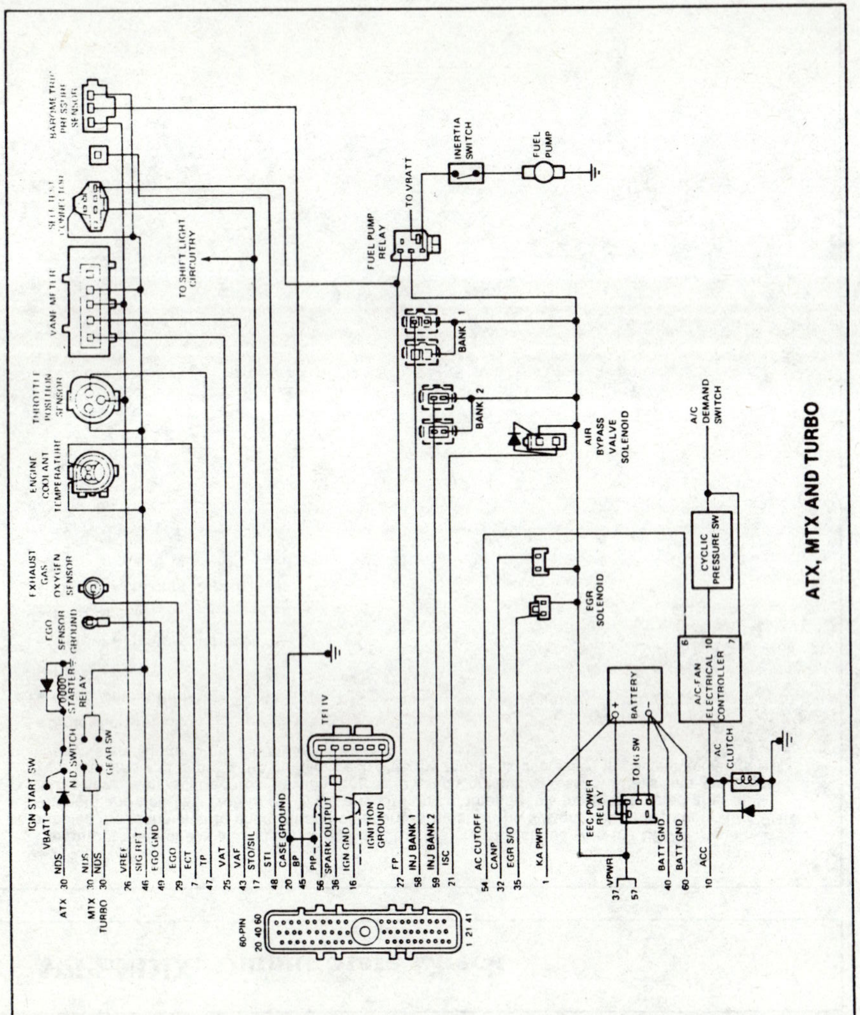

Electrical Schematic

2.3L EFI TC

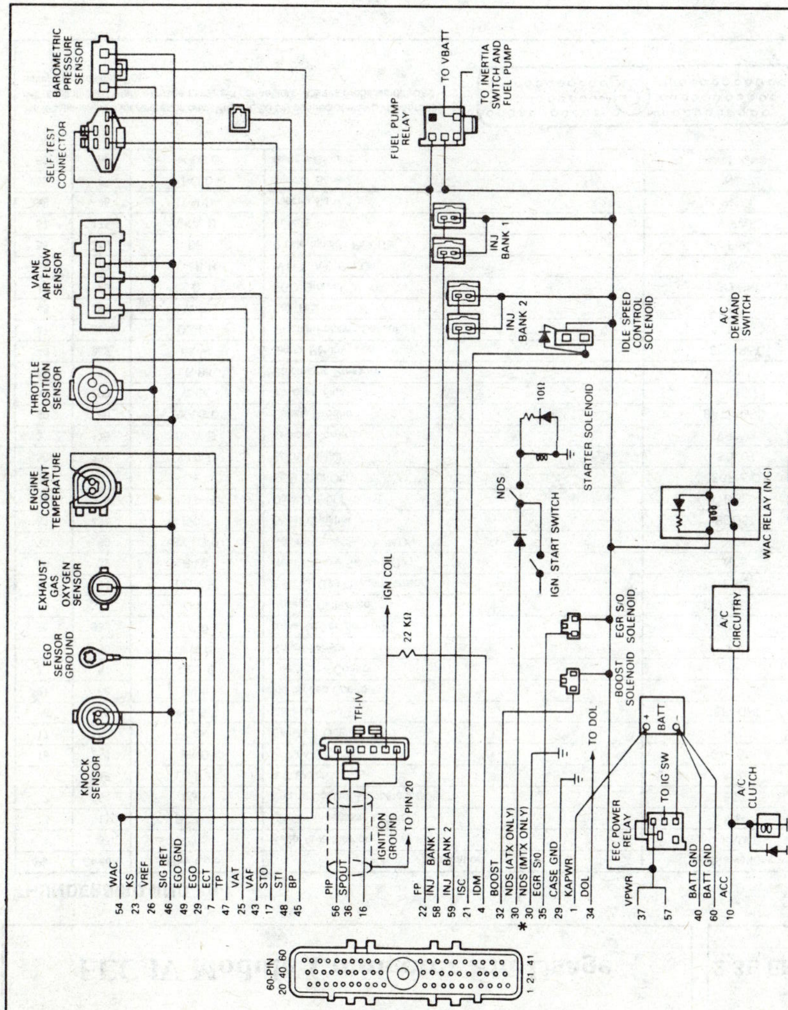

*** NOTE: On Merkur XR4Ti only, calibration with manual transmission requires that Pin 30 to be an "open" circuit.**

EEC-IV Module Connector Pin Usage

2.3L EFI TC

MUSTANG AND SVO MUSTANG

Pin	Circuit	Wire Color	Application	Abbreviations
1	37	Y	Keep Alive Memory	KAM
4	11	DG-Y D	Ignition Diagnostic Monitor	IDM
7	354	LG Y	Engine Coolant Temperature	ECT
10	347	BK-Y H	A C Clutch	ACC
16	259	BK-0	Ignition Ground	IGN GND
17	382	Y-BK	Self-Test Output	STO
20	57	BK	Case Ground	CSE GND
21	67	GY-W H	Idle Speed Control	ISC
22	97	T-LG D	Fuel Pump Control	FP
23	310	Y-R D	Knock Sensor	KS
25	357	LG-P	Vane Air Temperature	VAT
26	351	O-W	Voltage Reference	V REF
29	94	DG-P H	Exhaust Gas Oxygen Sensor	EGO
30	376	Y-R	Neutral Drive Switch	NDS
30	376	Y-R	Octane Switch (SVO)	OCT SW
32	464	BK-P	Boost Control	BOOST
34	305	LB-P H	Data Output Link	DOL
35	362	Y	EGR Shut-Off	EGR S O
36	324	Y-LG D	Spark Output	SPOUT
37	361	R	Vehicle Power	V PWR
40	57	B	Battery Ground	BATT GND
43	200	W-BK	Vane Air Flow	VAF
45	358	LG-BK	Barometric Pressure	BP
46	359	BK-W	Signal Return	SIG RET
47	355	DG-LG	Throttle Position Sensor	T P
48	209	W-R	Self-Test Input	STI
49	89	O	EGO Ground	EGO GND
54	73	O-LB H	W.O.T. A C Cut-Off	WAC
56	349	DB	Profile Ignition Pick-Up	PIP
57	361	R	Vehicle Power	V PWR
58	95	T-R D	Injector Bank 1	INJ 1
59	96	T-O D	Injector Bank 2	INJ 2
60	57	BK	Battery Ground	BATT GND

Pin locations given for reference only. Probing 60 pin connector with DVOM probe will result in permanent damage to the pin connectors. Always probe as directed, using the Breakout Box.

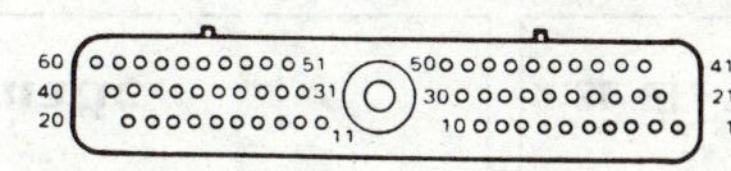

EEC-IV Module Connector Pin Usage — 2.3L EFI TC

THUNDERBIRD AND XR-7

Pin	Circuit	Wire Color	Application	Abbreviations
1	37	Y	Keep Alive Memory	KAM
4	11	DG-Y D	Ignition Diagnostic Monitor	IDM
7	354	LG Y	Engine Coolant Temperature	ECT
10	348	LG-P H	A C Clutch	ACC
16	259	BK-O	Ignition Ground	IGN GND
17	382	Y-BK	Self-Test Output	STO
20	60	BK-LG D	Case Ground	CSE GND
21	68	O-BK	Idle Speed Control	ISC
22	97	T-LG D	Fuel Pump Control	FP
23	310	Y-R D	Knock Sensor	KS
25	357	LG-P	Vane Air Temperature	VAT
26	351	O-W	Voltage Reference	V REF
29	94	DG-P H	Exhaust Gas Oxygen Sensor	EGO
30	32	R-LB	Neutral Drive Switch (ATX)	NDS
30	60	BK-LG D	Neutral Drive Switch (NTX)	NDS
32	464	BK-P	Boost Constrol	BOOST
34	305	LB-P H	Data Output Link	DOL
35	362	Y	EGR Shut-Off	EGR S O
36	324	Y-LG D	Spark Output	SPOUT
37	175	BK-Y D	Vehicle Power	V PWR
40	60	BK-LG D	Battery Ground	BATT GND
43	200	W-BK	Vane Air Flow	VAF
45	358	LG-BK	Barometric Pressure	BP
46	359	BK-W	Signal Return	SIG RET
47	355	DG-LG	Throttle Position Sensor	T P
48	209	W-R	Self-Test Input	STI
49	89	O	EGO Ground	EGO GND
54	73	O-LB H	W.O.T A C Cut-Off	WAC
56	349	DB	Profile Ignition Pick-Up	PIP
57	175	BK-Y D	Vehicle Power	V PWR
58	95	T-R D	Injector Bank 1	INJ 1
59	96	T-O D	Injector Bank 2	INJ 2
60	60	BK-LG D	Battery Ground	BATT GND

Pin locations given for reference only. Probing 60 pin connector with DVOM probe will result in permanent damage to the pin connectors. Always probe as directed, using the Breakout Box

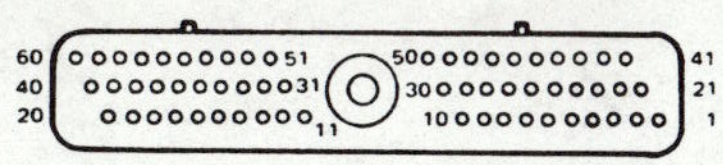

Electrical Schematic — 2.3L EFI TC

MERKUR XR4Ti

Pin	Circuit	Wire Color	Application	Abbreviations
1	37	Y	Keep Alive Memory	KAM
4	11	DG-Y D	Ignition Diagnostic Monitor	IDM
7	354	LG Y	Engine Coolant Temperature	ECT
10	347	BK-Y H	A C Clutch	ACC
16	259	R-O H	Ignition Ground	IGN GND
17	201	T-R	Self-Test Output	STO
20	57	BK	Case Ground	CSE GND
21	67	GY-W H	Idle Speed Control	ISC
22	97	T-LG D	Fuel Pump Control	FP
23	310	Y-R D	Knock Sensor	KS
25	357	LG-P	Vane Air Temperature	VAT
26	351	O-W	Voltage Reference	V. REF
29	94	DG-P H	Exhaust Gas Oxygen Sensor	EGO
30	376	Y-LG D	Neutral Drive Switch	NDS
32	544	P	Boost Control	BOOST
34	139	LG-P	Data Output Link	DOL
35	362	Y	EGR Shut-Off	EGR S O
36	324	Y-LG D	Spark Output	SPOUT
37	361	R	Vehicle Power	V PWR
40	57	BK	Battery Control	BATT. GND
43	200	W-BK	Vane Air Flow	VAF
45	358	LG-BK	Barometric Pressure	BP
46	359	BK-W	Signal Return	SIG RET
47	355	DG-LG	Throttle Position Sensor	T.P.
48	100	W-R D	Self-Test Input	STI
49	89	O	EGO Ground	EGO GND
54	208	GY-BK D	W.O.T. A C Cut-Off	WAC
56	349	DB	Profile Ignition Pick-Up	PIP
57	361	R	Vehicle Power	V PWR
58	95	T-R D	Injector Bank 1	INJ 1
59	96	T-O D	Injector Bank 2	INJ 2
60	57	BK	Battery Ground	BATT. GND

Pin locations given for reference only. Probing 60 pin connector with DVOM probe will result in permanent damage to the pin connectors. Always probe as directed, using the Breakout Box

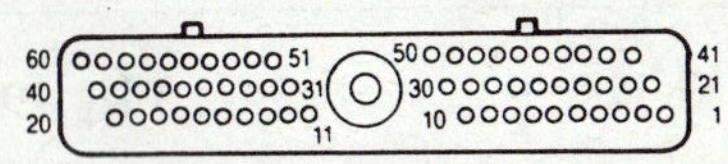

Diagnostic By Symptom

2.3L HSC-CFI

SYMPTOM	RESULT	ACTION TO TAKE
• Engine runs rough or miss fires. • Engine stalls. • Lack of power. • Rough idle.		• GO to Fuel Control [HF1] thru [HF4]. • GO to Pinpoint Test Step [DF20]. • EGR Valve Function. • Ignition system distributor cap, rotor, wires, coil, plugs. • Base engine valves, cam timing, compression etc. • Poor power/ground connections.
• Lack of fast idle with A/C On.		• GO to Pinpoint Test Step [KF9].
• Gasoline fumes under hood.		• GO to Pinpoint Test Step [KD1].
• Stalls during parking maneuvers.		• GO to Pinpoint Test Step [FF3].
• High idle in Drive, automatic transmission only.		• Place transmission in Drive with engine Off. Test neutral drive circuit (test Pin 30) for short to ground. SERVICE as necessary. **NOTE: Circuit should NOT have continuity to ground in Drive.**

Electrical Schematic

2.3L HSC-CFI

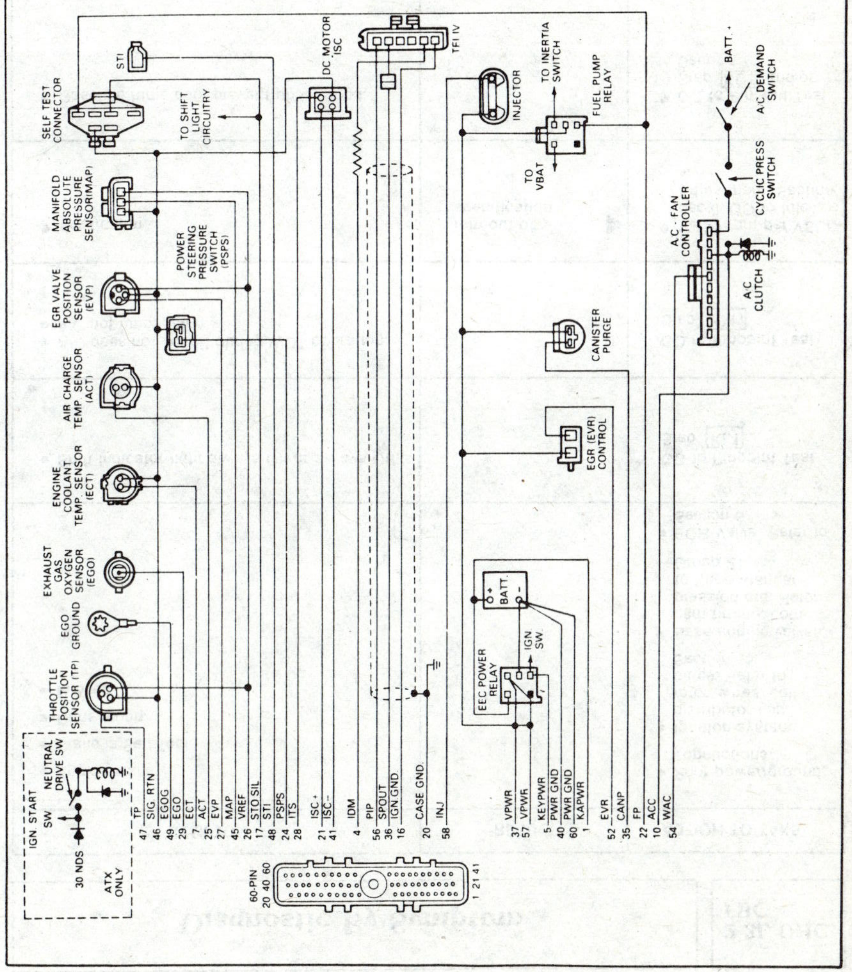

EEC-IV Module Connector Pin Usage

2.3L HSC-CFI

Pin	Circuit	Wire Color	Application	Abbreviations
1	37	Y	Keep Alive Power	KAPWR
4	11	DG-Y D	Ignition Diagnostic Monitor	IDM
5	20	W-LB H	Key Power	KPWR
7	354	LG-Y	Engine Coolant Temperature Sensor	ECT
10	347	BK-Y H	A C Clutch Compressor Signal	ACC
16	259	BK-O	Ignition Ground	IGN GND
17	201	T-R	Self-Test Output Shift Indicator Lite	STO SIL
20	57	BK	Case Ground	CSE GND
21	376	Y R	Idle Speed Control	ISC -
22	95	T-RD	Fuel Pump Relay	FP
24	330	Y-LG S	Power Steering Pressure Switch	PSPS
25	357	LG-P	Air Charge Temperature Sensor	ACT
26	351	O-W	Reference Voltage	V REF
27	352	BR-LG	EGR Valve Position Sensor	EVP
28	265	LG W	Idle Tracking Switch	ITS
29	94	DG P H	Exhaust Gas Oxygen Sensor	EGO
30	614	GY-O	Neutral Drive Switch (A T Only)	NDS
35	101	GR-Y H	Canister Purge Solenoid	CANP
36	324	Y-LG D	Spark Output	SPOUT
37	361	R	Vehicle Power	V PWR
40	60	BK-LG D	Power Ground	PWR GND
41	264	W LB	Idle Speed Control	ISC -
45	358	LG-BK	Manifold Absolute Pressure Sensor	MAP
46	359	BK-W	Signal Return (Ground)	SIG RTN
47	355	DG-LG	Throttle Position Sensor	TP
48	209	W-R	Self-Test Input	STI
49	89	O	EGO Ground	EGO GND
52	362	Y	EGR Valve Regulator Solenoid	EVR
54	73	O-LB H	WOT A C Cutoff	WAC
56	349	DB	Profile Ignition Pick-Up	PIP
57	361	R	Vehicle Power	V PWR
58	97	T-LG D	Fuel Injector	INJ
60	60	BK-LG D	Power Ground	PWR GND

Pin locations given for reference only. Probing 60 pin connector with DVOM probe will result in permanent damage to the pin connectors. Always probe as directed, using the Breakout Box.

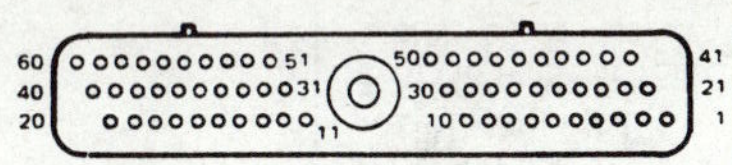

Diagnostic By Symptom

2.3L OHC FBC

SYMPTOM	RESULT	ACTION TO TAKE
• Engine stalls. • Stalls in Self-Test. • Runs rough. • Misses.		• Poor power/ground connections. • Ignition system distributor cap, rotor, wires, coil, plugs. Refer to Section 15. • Base engine valves, cam timing, compression etc. Refer to Shop Manual Group 21. • EGR Valve. Refer to Section 6.
• Shift indicator light always On or always Off.		GO to Pinpoint Test Step [KL1].
• A/C does not cut-off during WOT operation. • A/C not functioning.		GO to Pinpoint Test Step [KM1].
• Erratic rpm.	rpm out of specification ▶	• Verify rpm per VECI decal. GO to Idle Adjustment, Section 4.
• Gasoline fume odor present under hood.		• GO to Pinpont Test Step [KD1] and/or Section 7.

Electrical Schematic

2.3L OHC FBC

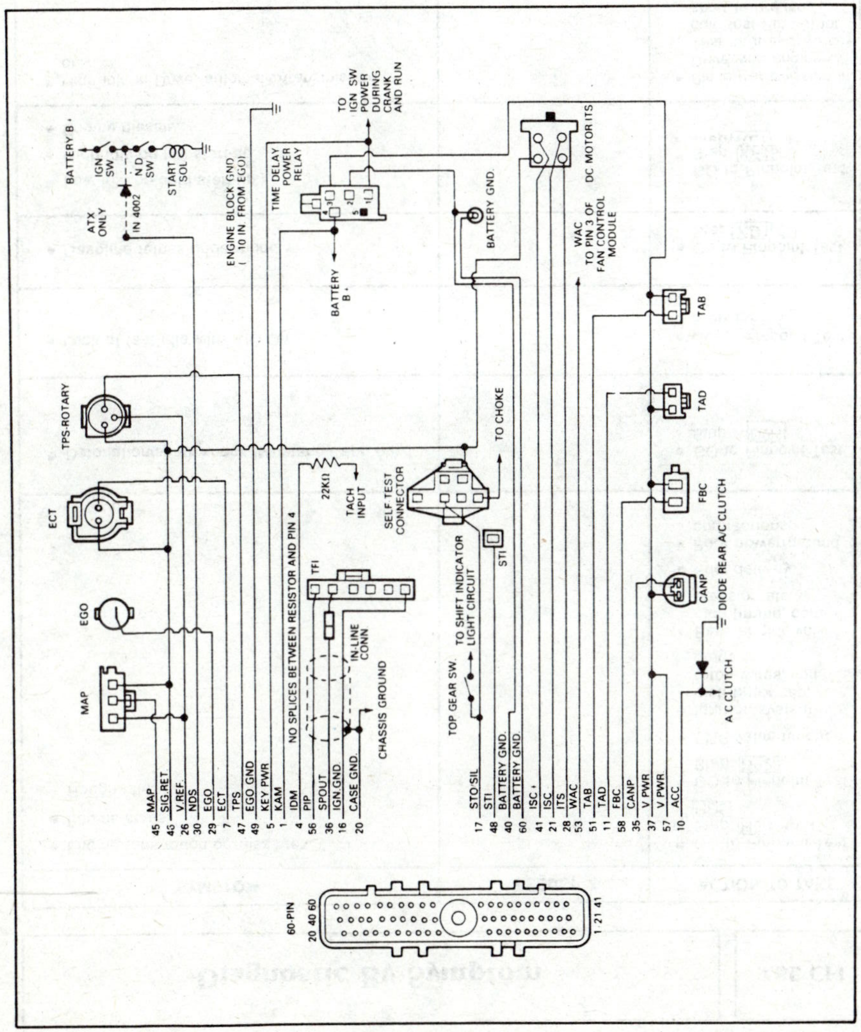

EEC-IV Module Connector Pin Usage

2.3L OHC FBC

Pin	Circuit	Wire Color	Application	Abbreviations
1	38	BK-O	Keep Alive Memory	KAM
4	11		Ignition Diagnostic Monitor	IDM
5	16	R-LG	Key Power	KEY PWR
7	354	LG-Y	Engine Coolant Temperature	ECT
10	347	BK-Y	A C Clutch	ACC
11	99	LG-BK D	Thermactor Air Diverter	TAD
16	57	BK	Ignition Ground	IGN GND
17	462	P	Self-Test Output Shift Ind. Light	STO SIL
20	60	BK-LG D	Case Ground	CASE GND
21	376	Y-R	Idle Speed Control	ISC
26	351	O-W	Voltage Reference	V REF
28	265	LG-W	Idle Tracking Switch	ITS
29	94	DG-P H	Exhaust Gas Oxygen Sensor	EGO
30	199	LB-Y	Neutral Drive Switch	NDS
35	101	GY-Y H	Canister Purge Solenoid	CANP
36	324	Y-LG D	Spark Output	SPOUT
37	361	R	Vehicle Power	V PWR
40	60	BK-LG D	Battery Ground	BATT GND
41	264	W-LB	Idle Speed Control	ISC -
45	358	LG-BK	Manifold Absolute Pressure	MAP
46	359	BK-W	Signal Return	SIG RET
47	355	DG-LG	Throttle Position Sensor	TPS
48	209	W-R	Self-Test Input	STI
49	89	O	EGO Ground	EGO GND
51	100	W-R-D	Thermactor Air Bypass Solenoid	TAB
53	73	O-LB H	Wide Open Throttle A C Cut-Off	WAC
56	349	DB	Profile Ignition Pick-Up	PIP
57	361	R	Vehicle Power	V PWR
58	95	T-R D	Feedback Carburetor Solenoid	FBC
60	60	BK-LG D	Battery Ground	BATT GND

Pin locations given for reference only. Probing 60 pin connector with DVOM probe will result in permanent damage to the pin connectors. Always probe as directed using the Breakout Box

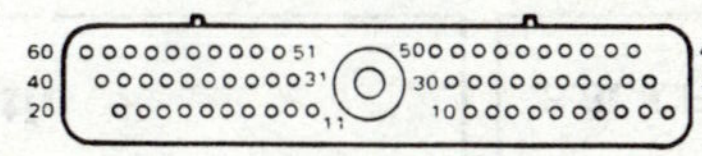

Diagnostic By Symptom

3.8L CFI

SYMPTOM	RESULT	ACTION TO TAKE
• Engine runs rough or miss-fires. • Engine stalls. • Rough idle.		• GO to Pinpoint Test Step HD1 thru HD5. • GO to Pinpoint Test Step DF20. • EGR value function. • Ignition system distributor cap, rotor, wires, coil, plugs. • Base engine valves, cam timing, compression etc. • Fuel delivery. • Poor power/ground connections.
• Detonation/spark knock (Mustang/Capri only).		• GO to Pinpoint Test Step DG1.
• Lack of fast idle with A/C On.		• GO to Pinpoint Test Step KF9.
• Gasoline fumes under hood.		• Go to Pinpoint Test Step KD1.
• Low idle on cold start-up. • High idle on hot start-up. • Engine diesels.		• GO to Pinpoint Test Step KF16.
• High idle in Drive, automatic transmission only.		• Place transmission in Drive with engine Off. Test neutral drive circuit (test Pin 30) for short to ground. SERVICE as necessary. **NOTE: Circuit should NOT have continuity to ground in Drive.**

Electrical Schematic

3.8L CFI

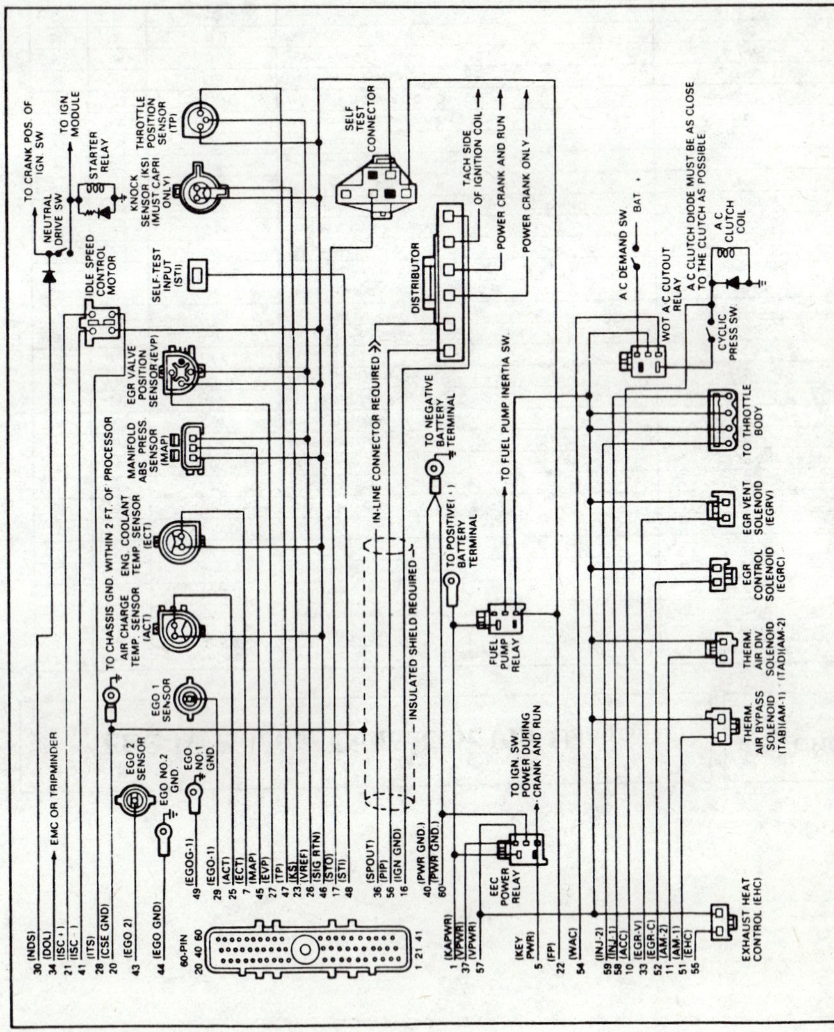

EEC-IV Module Connector Pin Usage

3.8L CFI

Pin	Circuit	Wire Color	Application	Abbreviations
1	38	BK O	Keep Alive Power	KAPWR
5	16	R LG	Key Power	KPWR
7	354	LG Y	Engine Coolant Temperature	ECT
10	347	BK Y	A C Clutch	ACC
11	99	LG BK	Air Management — TAD	AM2
16	259	BK O	Ignition Ground	IGN GND
17	382	Y B	Self-Test Output	STO
20	57	BK	Case Ground	CSE GND
21	376	Y R	Idle Speed Control (LT MQ. M C)	ISC+
	382	Y BK	Idle Speed Control (T X)	ISC+
22	97	T LG	Fuel Pump	FP
23	310	Y R	Knock Sensor (M C Only)	KS
25	357	LG P	Air Charge Temperature	ACT
26	351	O W	Reference Voltage	V REF.
27	352	BR LG	EGR Valve Position	EVP
28	265	LG W	Idle Tracking Switch	ITS
29	94	DG P	Exhaust Gas Oxygen Sensor — Left	EGO 1(L)
30	199	LB Y	Neutral Drive Switch	NDS
33	360	DG	Exhaust Gas Recirculation Vent	EGRV
34	305	LB-PH	Data Output Line	DOL
36	324	Y LG	Spark Output	SPOUT
37	361	R	Vehicle Power	V PWR.
40	60	BK LG	Battery Ground	BATT. GND.
41	264	W LB	Idle Speed Control	ISC −
43	90	DB LG	Exhaust Gas Oxygen Sensor — Right	EGO 2(R)
44	91	GY R	Exhaust Gas Oxygen Ground — Right	EGOG 2(R)
45	358	LG BK	Manifold Absolute Pressure (LTD MQ AND T X)	MAP
	356	DB LG	Manifold Absolute Pressure (M C)	MAP
46	359	BK W	Signal Return	SIG RTN
47	355	DG LG	Throttle Position Sensor	TP
48	209	W R	Self-Test Input	STI
49	89	O	Exhaust Gas Oxygen Ground — Left	EGOG 1(L)
51	100	W R	Air Management — Tab	AM1
52	362	Y	Exhaust Gas Recirculation Control	EGRC
54	73	O LB	W O T A C Cut Off	WAC
55	377	W	Exhaust Heat Control	EHC
56	349	DB	Profile Ignition Pick-Up	PIP
57	361	R	Vehicle Power	V PWR
58	95	T R	Injector 1	INJ 1
59	96	T O	Injector 2	INJ 2
60	60	BL LG	Battery Ground	BATT GND

Pin locations given for reference only. Probing 60 pin connector with DVOM probe will result in permanent damage to the pin connectors. Always probe as directed using the Breakout Box.

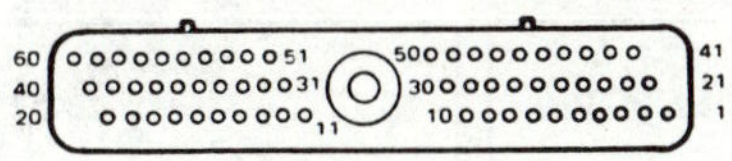

Diagnostic By Symptom

5.0L CFI

SYMPTOM	RESULT	ACTION TO TAKE
• Engine runs rough or miss fires. • Engine stalls. • Lack of power. • Rough idle.		• GO to Fuel control HC1 thru HC5. • GO to Pinpoint Test Step DF20. • EGR Valve Function. • Poor power/ground connections. • Ignition system distributor cap, rotor, wires, coil, plugs. • Base engine valves, cam timing, compression etc.
• Lack of fast idle with A/C On.		• GO to Pinpoint Test Step KG1.
• Gasoline fumes under hood.		• GO to Pinpoint Test Step KD1.
• Stumble after hot restart (5.0L HO only)		• GO to Pinpoint Test Step HC14.

Electrical Schematic — 5.0L CFI

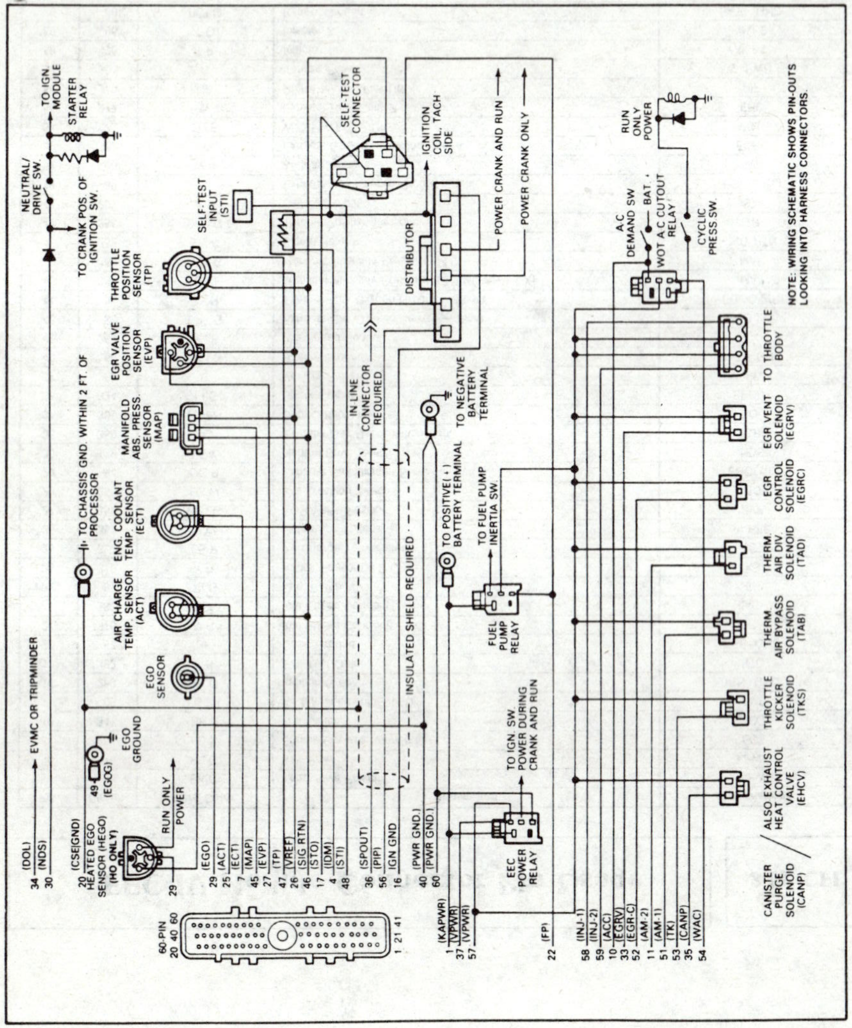

EEC-IV Module Connector Pin Usage — 5.0L CFI

Pin	Circuit	Wire Color	Application	Abbreviations
1	37	Y	Keep Alive Power	KAPWR
4	11	DG-Y D	Ignition Diagnostic Monitor	IDM
7	354	LG-Y	Engine Coolant Temperature Sensor	ECT
10	347	BK-Y H	A/C Clutch Compressor Signal	ACC
11	99	LG-BK D	Thermactor Air Diverter Solenoid	TAD
16	57	BK	Ignition Ground	IGN GND
17	201	T-R	Self-Test Output	STO
20	60	BK-LG D	Case Ground	CSE GND
22	97	T-LG D	Fuel Pump Relay	FP
25	357	LG-P	Air Charge Temperature Sensor	ACT
26	351	O-W	Reference Voltage	V REF
27	352	BR-LG	EGR Valve Position Sensor	EVP
29	94	DG-P H	Exhaust Gas Oxygen Sensor	EGO
30	199	W-PD	Neutral Drive Switch	NDS
33	360	DG	EGR Vent Solenoid	EGRV
34	305	LB-P H	Data Output Link For Tripminder	DOL
35	235	R-BK	Canister Purge Solenoid	CANP
36	324	Y-LG D	Spark Output	SPOUT
37	361	R	Vehicle Power	V PWR
40	60	BK-LG D	Power Ground	PWR GND
45	358	LG-BK	Manifold Absolute Pressure Sensor	MAP
46	359	BK-W	Signal Return (Ground)	SIG RTN
47	355	DG-LG	Throttle Position Sensor	TP
48	200	W-BK	Self-Test Input	STI
49	89	O	EGO Ground	EGO GND
51	100	W-R D	Thermactor Air Bypass Solenoid	TAB
52	362	Y	EGR Control Solenoid	EGRC
53	69	R-LG	Throttle Kicker Solenoid	TK
54	73	O-LB-H	WOT A/C Cutoff	WAC
56	349	DB	Profile Ignition Pick-Up	PIP
57	361	R	Vehicle Power	V PWR
58	95	T-R D	Injector #1	INJ #1
59	96	T-O D	Injector #2	INJ #2
60	60	BK-LG D	Power Ground	PWR GND

Pin locations given for reference only. Probing 60 pin connector with DVOM probe will result in permanent damage to the pin connectors. Always probe as directed, using the Breakout Box.

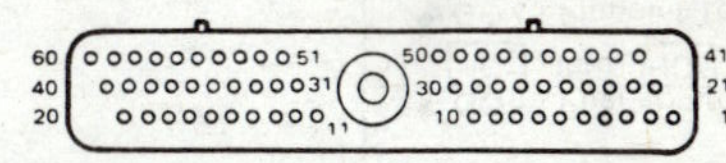

Diagnostic By Symptom

2.8L FBC

SYMPTOM	RESULT	ACTION TO TAKE
• Engine stalls. • Stalls in Self-Test. • Runs rough. • Misses. • Lack of power.	Carburetor icing ▶ Minor lean or rich fuel drive ▶ Choke always On/Off, slow in response ▶	• GO to KH1. • GO to KK1. • Poor power/ground connections. • Ignition system distributor cap, rotor, wires, coil, plugs. • Base engine valves, cam timing, compression etc. • EGR Valve, Section 6. • Fuel Delivery, Section II or Shop Manual, Group 24. • GO to KJ1.
• Shift indicator light always On or Always Off.		• GO to Pinpoint Test Step KL1.
• Erratic rpm.	rpm out of specification ▶	• Verify rpm per VECI decal. GO to Idle Adjustment. Section 4.
• Gasoline fume odors present under hood.		• GO to Section 7 or Shop Manual, Group 24.
• Lack of fast idle with A/C On.		• GO to Pinpoint Test Step FA1 and/or KF1.

Electrical Schematic

2.8L FBC

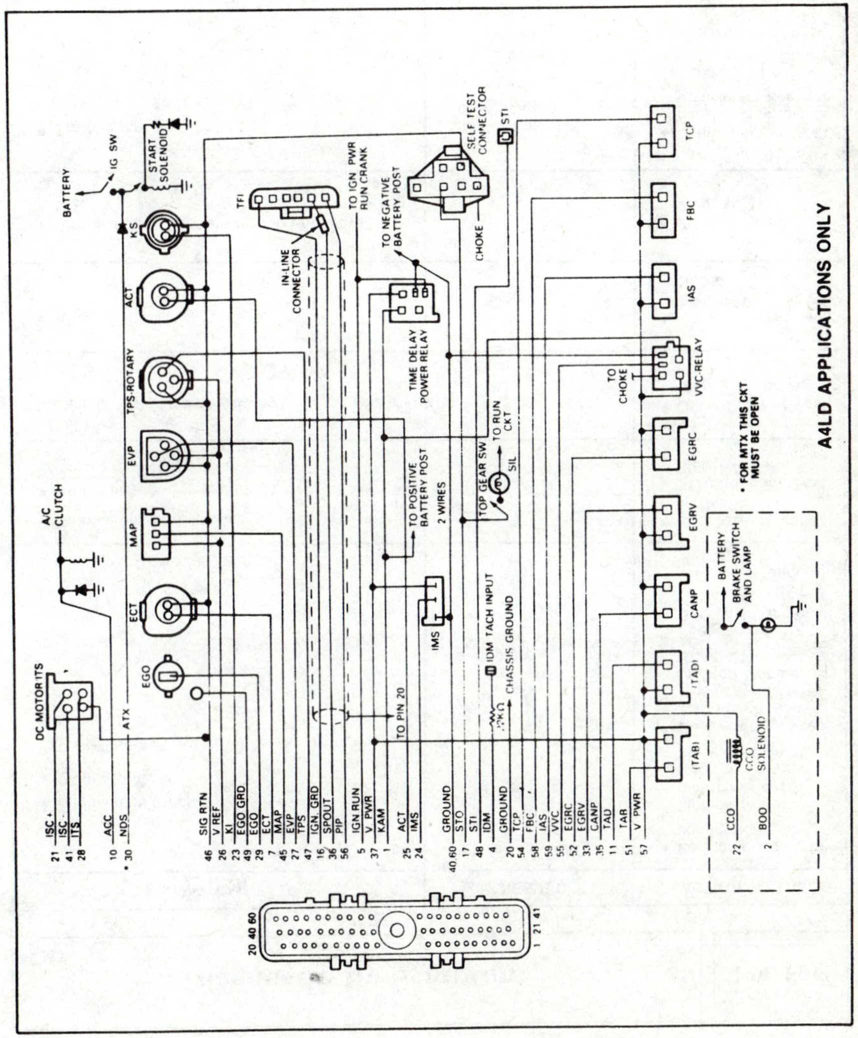

EEC-IV Module Connector Pin Usage — 2.8L FBC

Pin	Circuit	Wire Color	Application	Abbreviations
1	37	Y	Keep Alive Memory	KAM
2	511	DG	Brake On Off	BOO
4	11	DG-Y D	Ignition Diagnostic Monitor	IDM
5	20	W-LB H	Ignition Run Start	—
7	354	LG-Y	Engine Coolant Sensor	ECT
10	347	BK-YH	A C Clutch	ACC
11	200	W-BK	Thermactor Air Diverter	TAD
16	57	BK	Ignition Ground	IGN GRD
17	201	T-R	Self-Test Output	STO
20	60	BK-LG D	Chassis Ground	GRD
21	96	T-O D	Idle Speed Control	ISC
22	237	O-Y	Converter Clutch Override	CCO
23	310	Y-R D	Knock Sensor	KI
24	223	T-LG H	Inferred Milage Sensor	IMS
25	357	LG-P	Air Charge Temperature Sensor	ACT
26	351	O-W	Reference Voltage	V REF
27	352	BR-LG	EGR Valve Position Sensor	EVP
28	98	T-LB D	Idle Tracking Switch	ITS
29	94	DG-P H	Exhaust Gas Oxygen Sensor	EGO
30	150	DG-W	Neutral Drive Switch	NDS
33	360	DG	EGR Vent Solenoid	EGRV
35	101	GY-Y H	Canister Purge	CANP
36	324	Y-LG D	Ignition Module Signal	SPOUT
37	361	R	Vehicle Power	V PWR
40	60	BK-LG D	Vehicle Ground	GND
41	95	T-R D	Idle Speed Control	ISC
45	356	DB-LG	Manifold Absolute Pressure Sensor	MAP
46	359	BK-W	Signal Return	SIG RTN
47	355	DG-LG	Throttle Position Sensor	TPS
48	100	W-R D	Self-Test Input	STI
49	89	O	Exhaust Gas Oxygen Sensor Ground	EGO GRD
51	190	W-R D	Thermactor Air Bypass	TAB
52	362	Y	EGR Control Solenoid	EGRC
54	256	O-W	Temperature Compensation Pump	TCP
55	462	P	Variable Voltage Choke	VVC
56	349	DB	Profile Ignition Pick-Up	PIP
57	361	R	Vehicle Power	V POWER
58	97	T-LG D	Feedback Carburetor	FBC
59	75	DG-LG H	Inlet Air Solenoid	IAS
60	60	BK-LG D	Vehicle Ground	GRD

Pin locations given for reference only. Probing 60 pin connector with DVOM probe will result in permanent damage to the pin connectors. Always probe as directed, using the Breakout Box

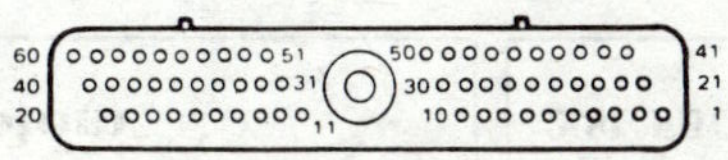

Diagnostic By Symptom — 4.9L FBC

SYMPTOM	RESULT	ACTION TO TAKE
• Engine runs rough or miss-fires. • Engine stalls. • Loss of power. • Misses.		• Poor power/ground connections. • Ignition system distributor cap, rotor, wires, coil, plugs, Section 15. • Base engine valves, cam timing, compression etc., refer to Shop Manual Group 21. • Fuel delivery, Section 11 and/or Group 24. • EGR valve function, Section 6.
• Detonation/spark knock. • Rough idle.		• EGR valve function, Section 6. • GO to Pinpoint Test Step [KA1].
• Lack of fast idle with A/C On.		• GO to Pinpoint Test Step [FA1] and/or [KF1].
• Gasoline fumes under hood.		• GO to Pinpoint Test Step [KD1] and/or Section 7.
• Shift indicator light always On or always Off.		• GO to Pinpoint Test Step [KL1].
• Erractic rpm.	rpm out of specification ▶	• Verify rpm per VECI decal. GO to Idle Adjustment Section 4.

Electrical Schematic — 4.9L FBC

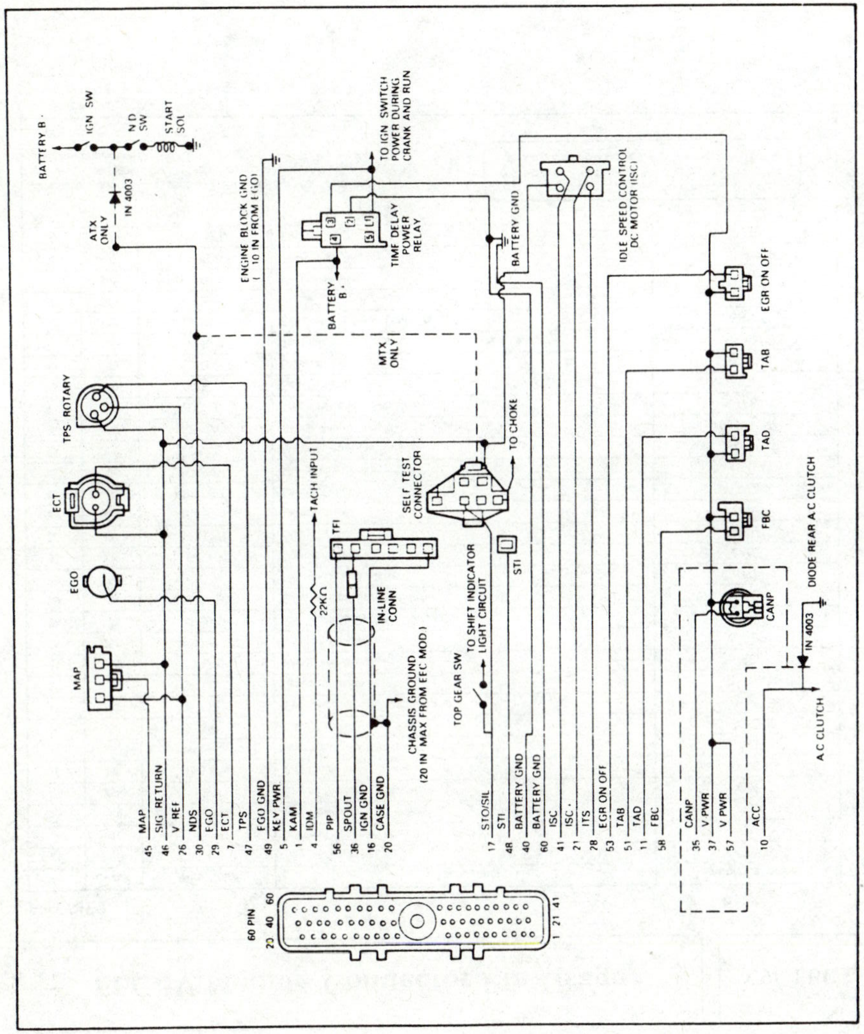

EEC-IV Module Connector Pin Usage — 4.9L FBC

ECONOLINE

Pin	Circuit	Wire Color	Application	Abbreviations
1	37	BK-Y	Keep Alive Power	KA PWR
4	11	BK-Y	Ignition Diagnostic Module	IDM
5	732	W	Key Power	K PWR
7	354	GR Y	Engine Coolant Temperature	ECT
10	348	GR W	A C Clutch	ACC
11	200	BL	Thermactor Air Diverter	TAD
16	57	BK	Case Ground	CASE GND
17	201	BL-Y	Self-Test Output Shift Ind Light	STO SIL
20	60	BK W	Ignition Ground	IGN GND
21	96	O-BK	Idle Speed Control	ISC
26	351	R	Voltage Reference	V. REF
28	98	BK	Idle Tracking Switch	ITS
29	94	GR-P	Exhaust Gas Oxygen Sensor	EGO
30	150	R	Neutral Drive Switch	N D
35	101	GY	Canister Purge	CANP
36	324	Y-GR	Spark Output	SPOUT
37	361	R	Vehicle Power	V PWR
40	60	BK-W	Battery Ground	BATT. GND
41	95	W	Idle Speed Control	ISC
45	356	BL-GR	Manifold Absolute Pressure	MAP
46	359	BK W	Signal Return	SIG. RTN.
47	355	GR	Throttle Position Sensor	TPS
48	100	BK-R	Self-Test Input	STI
49	89	O	Exhaust Gas Oxygen Ground	EGO GND
51	190	Y	Thermactor Air Bypass	TAB
53	360	Y-BL	Exhaust Gas Recirculate Shut Off	EGR S O
56	349	BR	Profile Ignition Pick-Up	PIP
57	361	R	Vehicle Power	V PWR
58	97	O	Feedback Control	FBC
60	60	BK-W	Battery Ground	BATT GND

Pin locations given for reference only. Probing 60 pin connector with DVOM probe will result in permanent damage to the pin connectors. Always probe as directed, using the Breakout Box.

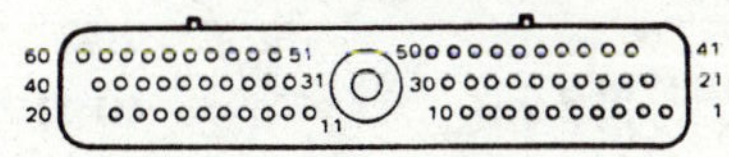

EEC-IV Module Connector Pin Usage — 4.9L FBC

F-SERIES

Pin	Circuit	Wire Color	Application	Abbreviations
1	37	Y	Keep Alive Power	KA PWR
4	11	DG-Y D	Ignition Diagnostic Module	IDM
5	20	W LB H	Key Power	K PWR
7	354	LG Y	Engine Coolant Temperature	ECT
10	348	LG PH	A C Clutch	ACC
11	200	W BK	Thermactor Air Diverter	TAD
16	57	BK	Case Ground	CASE GND
17	201	T R	Self-Test Output Shift Ind Light	STO SIL
20	60	BK LG D	Ignition Ground	IGN GND
21	96	T O D	Idle Speed Control	ISC
26	351	O W	Voltage Reference	V REF
28	98	T LB D	Idle Tracking Switch	ITS
29	94	DG PH	Exhaust Gas Oxygen Sensor	EGO
30	150	DG W	Neutral Drive Switch	N D
35	101	GY Y H	Canister Purge	CANP
36	324	Y LG D	Spark Output	SPOUT
37	361	R	Vehicle Power	V PWR
40	60	BK LG D	Battery Ground	BATT GND
41	95	T R D	Idle Speed Control	ISC
45	356	DB LG	Manifold Absolute Pressure	MAP
46	359	BK W	Signal Return	SIG RTN
47	355	DG LB	Throttle Position Sensor	TPS
48	100	W R D	Self-Test Input	STI
49	89	O	Exhaust Gas Oxygen Ground	EGO GND
51	190	W R D	Thermactor Air Bypass	TAB
53	360	DG	Exhaust Gas Recirculate Shut Off	EGR S O
56	349	DB	Profile Ignition Pick-Up	PIP
57	361	R	Vehicle Power	V PWR
58	97	T LG D	Feedback Control	FBC
60	60	BK LG D	Battery Ground	BATT GND

Pin locations given for reference only. Probing 60 pin connector with DVOM probe will result in permanent damage to the pin connectors. Always probe as directed using the Breakout Box.

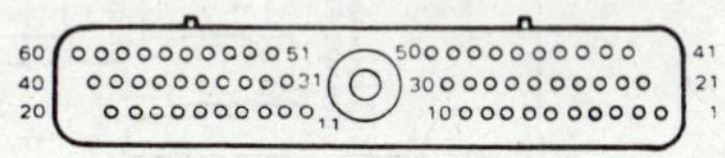

Diagnostic By Symptom — 5.0L 5.8L FBC

SYMPTOM	RESULT	ACTION TO TAKE
• Runs rough. • Misses. • Lack of power. • Engine stalls.		• Poor ground or power connections. • EGR system, Section 6. • Ignition system distributor cap, rotor, wires, coil, plugs: Section 15. • Base engine valves, cam timing, compression etc. Refer to Shop Manual, Group 21. • Fuel delivery, refer to Section 11 or Shop Manual, Group 24.
• Lack of fast idle with A/C On.		GO to Pinpoint Test Step KC1
• Gas fume odors present under hood.		GO to Pinpoint Test Step KD1 and/or Section 7.

Electrical Schematic

**5.0L
5.8L FBC**

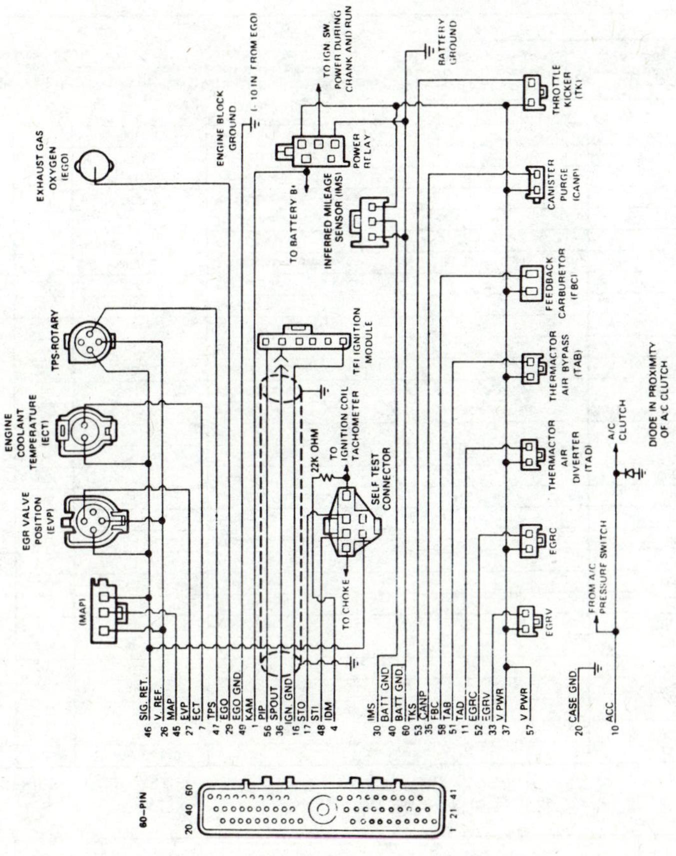

F-SERIES OR ECONOLINE

EEC-IV Module Connector Pin Usage

**5.0L
5.8L FBC**

ECONOLINE

Pin	Circuit	Wire Color	Application	Abbreviations
1	38	BK-R	Keep Alive Memory	KAM
4	11	B-Y	Ignition Diagnostic Module	IDM
7	354	GR-Y	Engine Coolant Temperature	ECT
10	347	BK	A C Clutch	ACC
11	200	BL	Thermactor Air Diverter	TAD
16	57	BK	Ignition Ground	IGN GND
17	201	BL-Y	Self-Test Output	STO
20	60	BK-W	Case Ground	CASE GND
26	351	R	Voltage Reference	V REF
27	352	DR-GR	EGR Valve Position	EVP
29	94	GR-P	Exhaust Gas Oxygen Sensor	EGO
30	223	O-GR	Inferred Mileage Sensor	IMS
33	360	Y-BL	EGR Vent	EGRV
35	101	GY	Canister Purge	CANP
36	324	Y-GR	Spark Out	SPOUT
37	361	R	Vehicle Power	V PWR
40	60	BK-W	Battery Ground	BATT GND
45	356	BL-GR	Manifold Absolute Pressure	MAP
46	359	BK-W	Signal Return	SIG RET
47	355	GR	Throttle Position Sensor	TPS
48	100	BK-R	Self-Test In	STI
49	89	O	EGO Ground (Engine Block)	EGO GND
51	190	Y	Thermactor Air Bypass	TAB
52	362	Y	EGR Control	EGRC
53	69	R-GR	Throttle Kicker	TKS
56	349	BR	Profile Ignition Pick-Up	PIP
57	361	R	Vehicle Power	V PWR
58	32	R-BL	Feedback Carburetor	FBC
60	60	BK-W	Battery Ground	BATT GND

Pin locations given for reference only. Probing 60 pin connector with DVOM probe will result in permanent damage to the pin connectors. Always probe as directed using the Breakout Box.

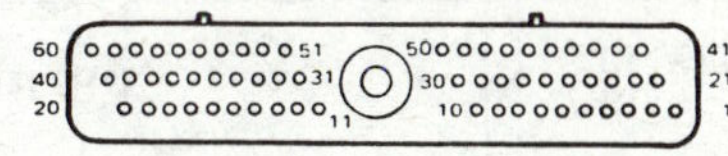

EEC-IV Module Connector Pin Usage

5.0L 5.8L FBC

F-SERIES

Pin	Circuit	Wire Color	Application	Abbreviations
1	37	Y	Keep Alive Memory	KAM
4	11	B-Y	Ignition Diagnostic Module	IDM
7	354	LG-Y	Engine Coolant Temperature	ECT
10	348	LG-P H	A C Clutch	ACC
11	200	W-BK	Thermactor Air Diverter	TAD
16	57	BK	Ignition Ground	IGN GND
17	201	T-R	Self-Test Out	STO
20	60	BK-LG D	Case Ground	CASE GND
26	351	O-W	Voltage Reference	V REF
27	352	BR-LG	EGR Valve Position	EVP
29	94	DG-P H	Exhaust Gas Oxygen Sensor	EGO
30	150	DG-W	Inferred Mileage Sensor	IMS
33	360	DG	EGR Vent	EGRV
35	101	GY-Y H	Canister Purge	CANP
36	324	Y LG D	Spark Output	SPOUT
37	361	R	Vehicle Power	V PWR
40	60	BK-LG D	Battery Ground	BATT. GND
45	356	DB-LG	Manifold Absolute Pressure	MAP
46	359	BK-W	Signal Return	SIG RET
47	355	DG-L	Throttle Position Sensor	TPS
48	100	W-R D	Self-Test In	STI
49	89	O	EGO Ground (Engine Block)	EGO GND
51	190	W-R D	Thermactor Air Bypass	TAB
52	362	Y	EGR Control	EGRC
53	69	R-LG	Throttle Kicker	TKS
56	349	DB	Profile Ignition Pick-Up	PIP
57	361	R	Vehicle Power	V PWR
58	97	T-LG D	Feedback Carburetor	FBC
60	60	BK-LG D	Battery Ground	BATT GND

Pin locations given for reference only. Probing 60 pin connector with DVOM probe will result in permanent damage to the pin connectors. Always probe as directed, using the Breakout Box.

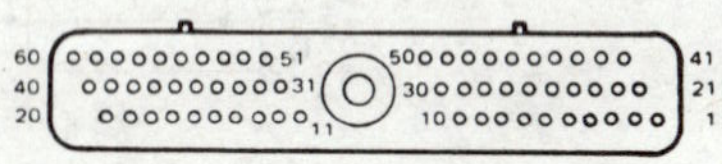

Diagnostic By Symptom

2.3L 5.0L EFI

SYMPTOM	RESULT	ACTION TO TAKE
• Engine stalls. • Stalls in Self-Test. • Runs rough. • Misses. • Always rich/lean.		• Fuel delivery. Pinpoint Test Steps HE1 thru HE5. • Map test GO to Pinpoint Test Step DF20. • Poor power/ground connections. • Ignition system distributor cap, rotor, wires, coil, plugs • Base engine valves, cam timing, compression etc.
• Detonation/spark knock.		GO to Pinpoint Test Step DG1.
• Lack of fast idle assist with A/C On.		GO to Pinpoint Test Step FA1.
• Shift light always On or Off.		GO to Pinpoint Test Step KL1.
• A/C does not cut-off under WOT conditions.		GO to Pinpoint Test Step KM1.
• Surges with A/C on at idle.		GO to Pinpoint Test Step KM20.
• Stumble after hot re-start (HEGO).		GO to Pinpoint Test Step HE21.
• Stalls during parking maneuvers.		GO to Pinpoint Test Step FF3.

Electrical Schematic

2.3L OHC EFI

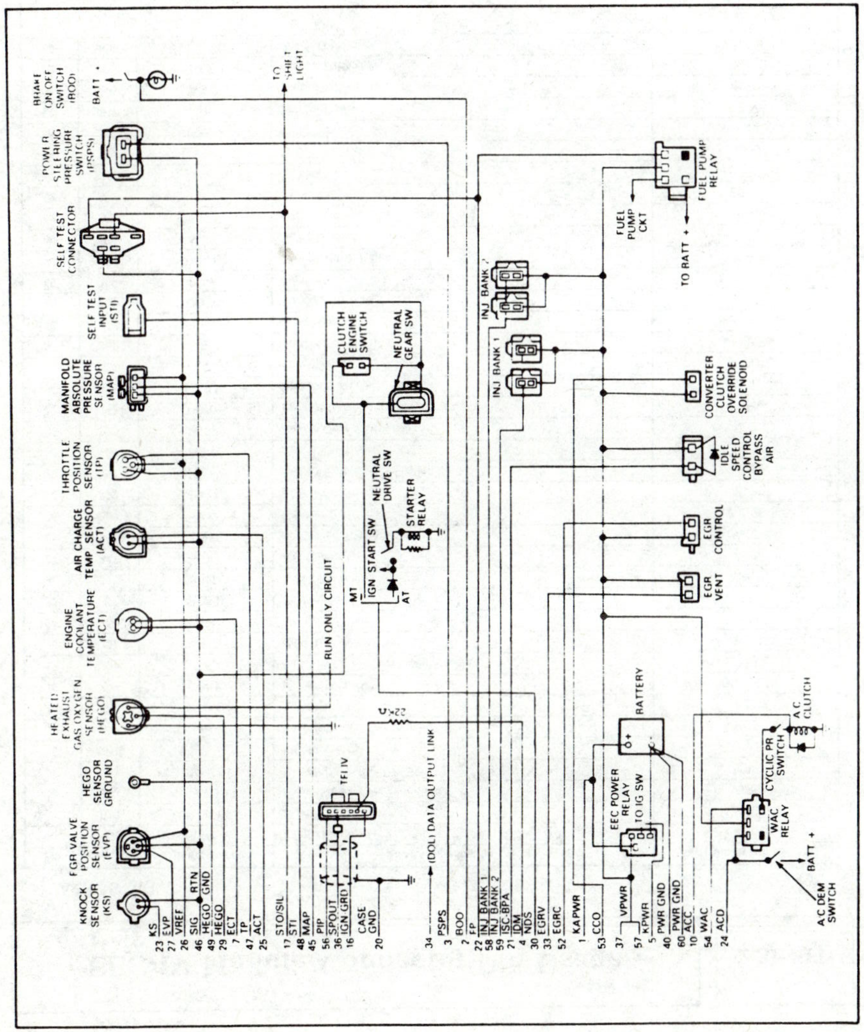

EEC-IV Module Connector Pin Usage

2.3L EFI

RANGER

Pin	Circuit	Wire Color	Application	Abbreviations
1	37A	Y	Keep Alive Power	KAPWR
2	511	LG	Brake On Off	BOO
3	330	Y-LG	Power Steering Pressure Switch	PSPS
4	11	DG-Y D	Ignition Diagnostic Monitor	IDM
5	361C	R	Key Power	KPWR
7	354	LG-Y	Engine Coolant Temperature	ECT
10	347	BK-Y H	A C Clutch	ACC
16	57A	BK	Ignition Ground	IGN GND
17	201	T-R	Self-Test Output Shift Indicator Light	STO SIL
20	57	BK	Case Ground	CSE GND
21	67	GY-W H	Idle Speed Control	ISC
22	97	T-LG D	Fuel Pump	FP
23	99	LG-BK D	Knock Sensor	KS
24	347	BK-Y H	A C Demand	ACD
25	310	Y-R D	Air Charge Temperature	ACT
26	351	O-W	Reference Voltage	V REF
27	352	BR-LG	EGR Valve Position	EVP
29	94	DG-P H	Heated Exhaust Gas Oxygen Sensor	HEGO
30	150	DG-W	Neutral Drive Switch (AT)	NDS
30	200	W-BK	Neutral Drive Switch (MT)	NDS
33	360	DG	Exhaust Gas Recirculation Vent	EGRV
34	305	LB-P H	Data Output Link	DOL
36	324	Y-LG D	Spark Output	SPOUT
37	361	R	Vehicle Power	V PWR
40	60A	BK-LG D	Power Ground	PWR GND
45	356	DB-LG	Manifold Absolute Pressure	MAP
46	359	BK-W	Signal Return	SIG RTN
47	355	DG-LG	Throttle Position Sensor	T P
48	100	W-R D	Self-Test Input	STI
49	89	O	Heated Exhaust Gas Oxygen Sensor Ground	HEGOG
52	362	Y	Exhaust Gas Recirculation Control	EGRC
53	237B	O-Y	Clutch Converter Override	C C O
54	331	R	W O T A C Cut Off	WAC
56	349	DB	Profile Ignition Pick-Up	PIP
57	361B	R	Vehicle Power	V PWR
58	96	T-O D	Injector 1 (Bank 1)	INJ 1
59	95	T-R D	Injector 2 (Bank 2)	INJ 2
60	60B	BK-LG D	Power Ground	PWR GRD

Pin locations given for reference only. Probing 60 pin connector with DVOM probe will result in permanent damage to the pin connectors. Always probe as directed, using the Breakout Box.

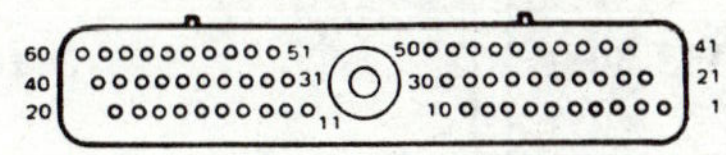

EEC-IV Module Connector Pin Usage

2.3L EFI

AEROSTAR

Pin	Circuit	Wire Color	Application	Abbreviations
1	37A	Y	Keep Alive Power	KAPWR
2	810	R-LG H	Brake On-Off	BOO
3	330	Y-LG	Power Steering Pressure Switch	PSPS
4	11	DG-Y D	Ignition Diagnostic Monitor	IDM
5	361C	R	Key Power	KPWR
7	354	LG-Y	Engine Coolant Temperature	ECT
10	347	BK-Y H	A C Clutch	ACC
16	57A	BK	Ignition Ground	IGN. GND
17	201	T-R	Self-Test Output Shift Indicator Light	STO SIL
20	57	BK	Case Ground	CSE. GND
21	67	GY-W H	Idle Speed Control	ISC
22	97	T-LG D	Fuel Pump	FP
23	310	Y-R D	Knock Sensor	KS
24	347	BK-Y H	A C Demand	ACD
25	357	LG-P	Air Charge Temperature	ACT
26	351	O-W	Reference Voltage	V. REF.
27	352	BR-LG	EGR Valve Position	EVP
29	94	DG-P H	Heated Exhaust Gas Oxygen Sensor	HEGO
30	150	DG-W	Neutral Drive Switch (AT)	NDS
30	200	W-BK	Neutral Drive Switch (MT)	NDS
33	360	DG	Exhaust Gas Recirculation Vent	EGRV
34	305	LB-P H	Data Output Link	DOL
36	324	Y-LG D	Spark Output	SPOUT
37	361	R	Vehicle Power	V. PWR.
40	60A	BK-LG D	Power Ground	PWR. GND.
45	356	DB-LG	Manifold Absolute Pressure	MAP
46	359	BK-W	Signal Return	SIG. RTN
47	355	DG-LG	Throttle Position Sensor	T.P.
48	100	W-R D	Self-Test Input	STI
49	89	O	Heated Exhaust Gas Oxygen Ground	HEGOG
52	362	Y	Exhaust Gas Recirculation Control	EGRC
53	237B	O-Y		
54	331	R	W.O.T. A C Cut Off	WAC
56	349	DB	Profile Ignition Pick-Up	PIP
57	361B	R	Vehicle Power	V. PWR.
58	96	T-O D	Injector 1 (Bank 1)	INJ. 1
59	95	T-R D	Injector 2 (Bank 2)	INJ. 2
60	60B	BK-LG D	Power Ground	PWR. GRD.

Pin locations given for reference only. Probing 60 pin connector with DVOM probe will result in permanent damage to the pin connectors. Always probe as directed, using the Breakout Box.

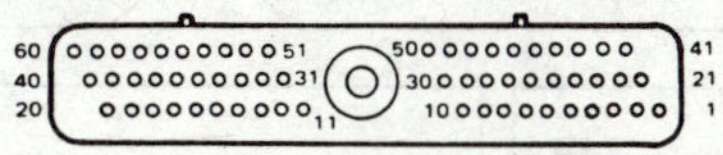

Electrical Schematic

5.0L EFI

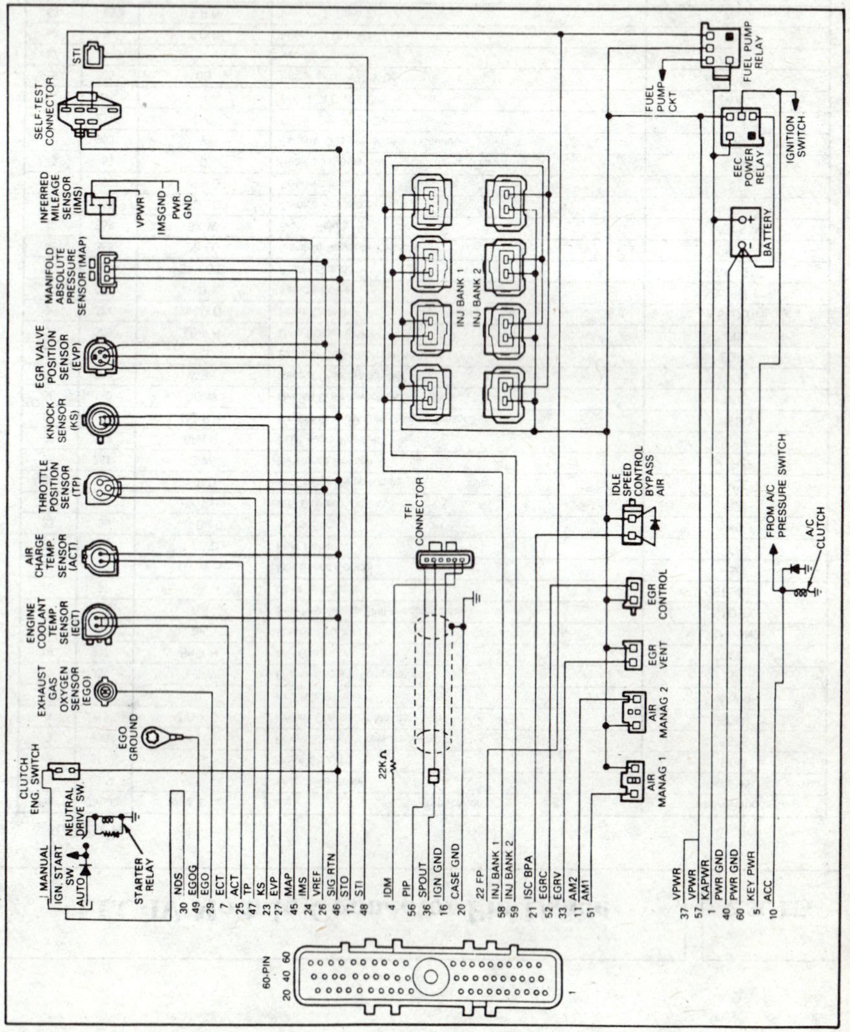

EEC-IV Module Connector Pin Usage

5.0L EFI

Pin	Circuit	Wire Color	Application	Abbreviations
1	37D	Y	Keep Alive Power	KAPWR
4	11D	DG-YD	Ignition Diagnostic Monitor	IDM
5	361R	R	Key Power	KPWR
7	354	LG-Y	Engine Coolant Temperature	ECT
10	348	LG-P/H	A/C Clutch	ACC
11	200	W-BK	Air Management 2	AM-2
16	57	BK	Ignition Ground	IGN GND
17	201	T-R	Self-Test Output	STO
20	57A	BK	Case Ground	CSE GND
21	67	GY-W/H	Idle Speed Control	ISC
22	97B	T-LG/D	Fuel Pump	FP
23	99	LG-BK/D	Knock Sensor	K.S.
24	223	T-LG/H	Inferred Mileage Sensor	IMS
25	310	Y-R/D	Air Charge Temperature	ACT
26	351	O-W	Reference Voltage	V. REF.
27	352	BR-LG	EGR Valve Position	EVP
29	94	DG-P/H	Exhaust Gas Oxygen Sensor	EGO
30	150	DG-W	Neutral Drive Switch	NDS
33	360	DG	Exhaust Gas Recirculation Vent	EGRV
36	324A	Y-LG/D	Spark Output	SPOUT
37	361	R	Vehicle Power	V. PWR
40	60A	BK-LG/D	Power Ground	PWR. GND.
45	356	DB-LG	Manifold Absolute Pressure	MAP
46	359	BW-W	Signal Return	SIG. RTN.
47	355	DG-LG	Throttle Position Sensor	T.P.
48	100	W-R/D	Self-Test Input	S.T.I.
49	89	O	Exhaust Gas Oxygen Ground	EGOG
51	190	W-R/D	Air Management 1	AM-1
52	362	Y	Exhaust Gas Recirculation Control	EGRC
56	349	DB	Profile Ignition Pick-Up	PIP
57	361A	R	Vehicle Power	V. PWR.
58	96	T-O/D	Injector 1 (Bank 1)	INJ. 1
59	95	T-R/D	Injector 2 (Bank 2)	INJ. 2
60	60B	BK-LG/D	Power Ground	PWR. GND.

Pin locations given for reference only. Probing 60 pin connector with DVOM probe will result in permanent damage to the pin connectors. Always probe as directed, using the Breakout Box.

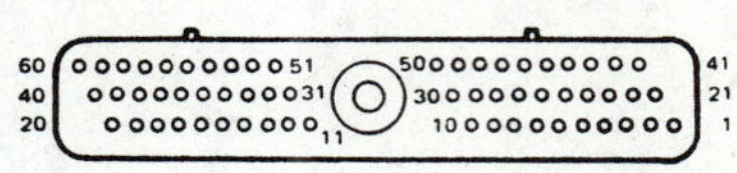

How To Use The Diagnostic Procedure

DESCRIPTION

The Microprocessor Control Unit has the capability to diagnose a malfunction within its own system. It advises the technician where to look for MCU problems through the use of a Self-Test computer program built into the MCU module. When triggered, the Self-Test simulates a variety of engine operating conditions, so that operating conditions can be detected.

A received response to the initiated Self-Test is reported to the technician as a pulsing series of voltage outputs that will appear on one of two testing devices: an Analog Voltmeter or the optional Self-Test Automatic Readout (STAR) tester, Rotunda model 007-00004.

The series of voltage pulses are called a Service Code, representing a two-digit number. It's the two-digit Service Code that advises the technician of the nature of the MCU problem. Hooking up the Analog Voltmeter or STAR tester to the MCU System will be described and shown later.

After performing the "Self-Test," the technician must follow all diagnostic procedures in the Engine/Emissions Diagnosis manual prior to replacing any components. This will help assure service reliability.

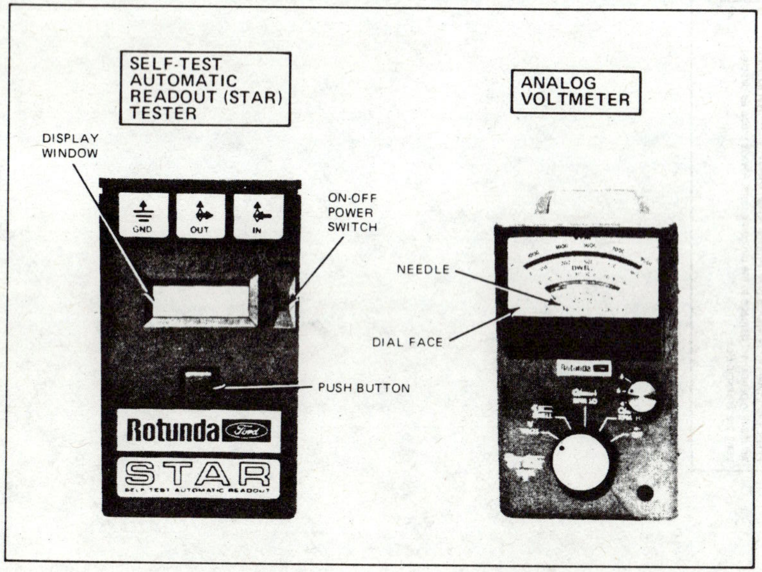

How To Use The Diagnostic Procedure

ANALOG VOLTMETER

When a Service Code is reported on the Analog Voltmeter, it will represent itself as a pulsing or sweeping movement of the voltmeter's needle across the dial face of the voltmeter. Therefore, a single-digit number of three will be reported by three needle pulses (sweeps). However, as previously stated, a Service Code is represented by a two-digit number, such as 3-2. So as a result, the Self-Test's Service Code of 3-2 will appear on the voltmeter as three needle pulses (sweeps), then after a two second pause, the needle will pulse (sweep) twice more.

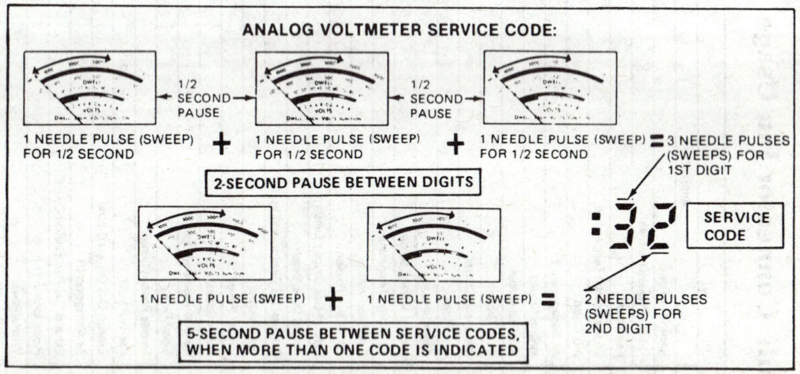

SELF-TEST AUTOMATIC READOUT (STAR)
Rotunda Model No. 007-00004 or Equivalent

After hooking up the STAR tester and turning on its power switch, the tester will run a display check: the numerals 88 will begin to flash in the display window, and then a steady 00 will appear to signify that the STAR tester is ready to start the Self-Test and receive the test's Service Codes.

How To Use The Diagnostic Procedure

To receive the Service Codes, press the pushbutton at the front of the STAR tester. The button will latch down, and a colon will appear in the display window in front of the :00 numerals. The colon MUST be displayed to receive the Service Codes.

If for any reason the technician wishes to clear the display window during the Self-Test, he must turn off the vehicle's engine, press the tester's pushbutton once to unlatch it (colon will disappear), then press the button again to latch down the button (colon will appear again). Every time the STAR tester is turned off, the low battery indicator (LO BAT) should show briefly at the upper left-hand corner of the tester's display window. If the LO BAT indicator shows steadily at any other time during the operation of the STAR tester with any Service Code, turn its power switch to OFF, and replace the 9-volt battery in the tester.

The STAR tester will display the last Service Code received, even after disconnecting it from the vehicle. It will hold the Service Code on the display until the power is turned off or the pushbutton is unlatched and relatched.

Anyone who departs from the instructions provided in this publication must first establish that he compromises neither his personal safety nor the vehicle integrity by his choice of methods, tools, or parts.

SELF-TEST AUTOMATIC READOUT (STAR) TESTER SERVICE CODE

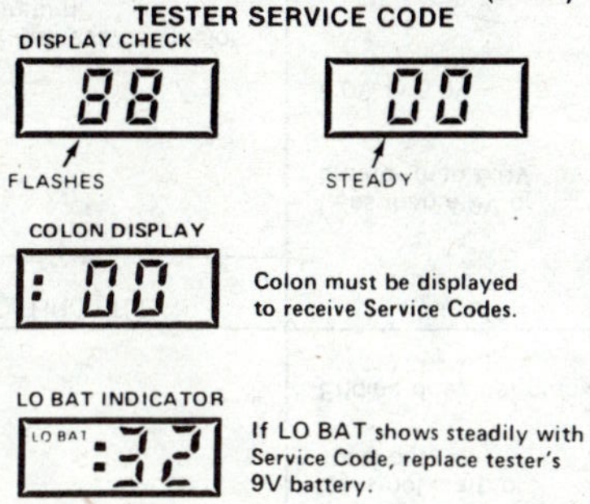

EEC IV No Start

Pinpoint Test A

A

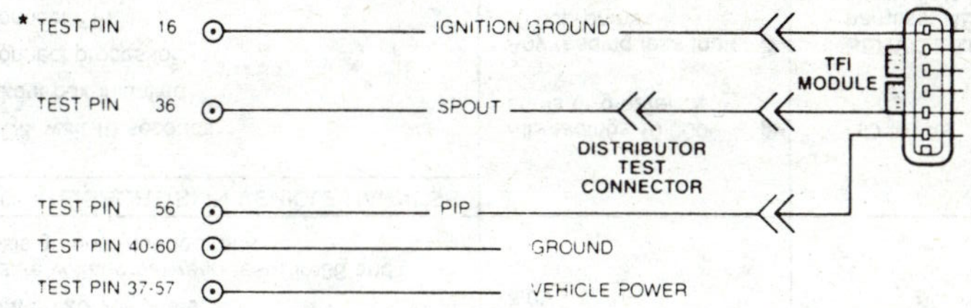

* TEST PIN LOCATED ON BREAKOUT BOX.
ALL CONNECTIONS VIEWED INTO MATING SURFACE.

NOTE: When Breakout box is installed, ensure that timing switch is in "Computed" position unless otherwise noted.

STOP-WARNING

You should enter this Pinpont Test only when Quick Test Steps 1.0 through 3.0 have been successfully completed and the engine is still a no start.

This Pinpoint Test will not diagnose ignition system problems.

To prevent the replacement of good components, be aware that the following non-EEC areas may be at fault:

- Fuel: quantity and quality.
- Ignition: general condition, moisture, cracks, damage, etc.
- Engine: internal, valves, timing belt, camshaft.
- Starter and battery circuit.

This pinpoint test is intended to diagnose only the following:

- Spark (as related to EEC-IV).
- Circuits: pip, spout, ignition ground, vehicle power.

EEC IV No Start — Pinpoint Test A

WARNING: Stop this test at the first sign of a fuel leak and service as required.
CAUTION: No open flame — No smoking during fuel delivery checks.

TEST STEP	RESULT ▶	ACTION TO TAKE
A1 ATTEMPT TO START ENGINE		
	Engine cranks, but does not start, or stalls out ▶	GO to A2.
	Engine does not crank ▶	REFER to Shop Manual
A2 CHECK FOR V REF AT THROTTLE POSITION SENSOR		
• Key Off, wait 10 seconds. • DVOM on 20V scale. • Disconnect TP sensor. • Key On, Engine Off. • Measure voltage at the TP harness connector between VREF and signal return.	Less than 4.0V or greater than 6.0V ▶	GO to Pinpoint Test Step C1.
	4.0V to 6.0V ▶	RECONNECT TP sensor. GO to A3.
A3 CHECK FOR SPARK AT PLUGS		
• Disconnect the spark plug wire to any accessible cylinder. • Connect spark tester between spark plug wire and engine ground. • Crank engine and check for spark. • Reconnect the spark plug wire to the spark plug.	Spark ▶	GO to A13.
	No spark ▶	GO to A4.
A4 CHECK FOR SPARK AT COIL		
• Remove high tension coil wire from distributor and install spark tester. • Check for spark while cranking. • Reconnect high tension coil wire to distributor.	Spark ▶	REFER to Diagnosis for cap, rotor, wires.
	No spark ▶	GO to A5.

EEC IV No Start — Pinpoint Test A

TEST STEP	RESULT ▶	ACTION TO TAKE
A5 HARNESS CHECK (IGNITION GROUND)		
• Key Off, wait 10 seconds. • Install Breakout box. Leave processor disconnected. • DVOM on 200 ohm scale. • Disconnect TFI. • Measure resistance between test Pin 16 and TFI harness connector ign. gnd.	Less than 5 ohms ▶	GO to A6.
	5 ohms or greater ▶	SERVICE harness as necessary. REPEAT Quick Test.
A6 ISOLATION OF PROBLEM TO SPOUT CIRCUIT		
• Breakout box installed. • Connect TFI. • Connect processor. • Timing switch to "Dist" position on Breakout box. • Attempt to start vehicle. • Does the vehicle start?	Yes ▶	GO to A10.
	No ▶	GO to A7.
A7 SPOUT SIGNAL CHECK		
• Breakout box installed. • Timing switch to "Computed" position on Breakout box. • DVOM on 20 volt scale. • Measure voltage between test Pin 36 and chassis ground, during crank.	Less than 3.0V or greater than 6.0V ▶	GO to A8.
	Between 3.0V and 6.0V ▶	EEC OK
A8 ISOLATE SHORT(S) IN VEHICLE HARNESS		
• Key Off, wait 10 seconds. • Breakout box installed. • Disconnect processor. • Disconnect TFI. • DVOM on 200,000 ohm scale. • Measure resistance between test Pin 36 and test Pins 16, 20, 26, 40, 60 (short to ground), 37, 57 (short to power) and 56 (short to pip).	All readings 10,000 ohms or greater ▶	GO to A9.
	Any reading less than 10,000 ohms ▶	SERVICE short in harness. REPEAT Quick Test if vehicle does not start. GO to A9.

EEC IV No Start — Pinpoint Test — A

TEST STEP	RESULT ▶	ACTION TO TAKE
A9 ISOLATE SHORT(S) IN PROCESSOR		
• Key Off, wait 10 seconds. • Breakout box installed • Reconnect processor. • TFI disconnected. • DVOM on 200 ohm scale. • Measure resistance between test Pin 36 and test Pins 37 and 57 (short to power) also, test Pins 40 and 60 (short to ground).	All readings 5.0 ohms or greater ▶ Any reading less than 5.0 ohms ▶	Connect TFI. GO to **A10**. REPLACE processor. REPEAT Quick Test
A10 PIP SIGNAL CHECK		
• Breakout box installed. • DVOM to 20 volt range. • Measure voltage between test Pin 56 and test Pin 16. • Crank engine, record reading.	Between 3.0V and 6.0V ▶ Less than 3.0V or greater than 6.0V ▶	REMOVE Breakout box. REPLACE processor. REPEAT Quick Test. GO to **A11**.
A11 CONTINUITY OF PIP CIRCUIT CHECK		
• Breakout box installed. • Key Off, wait 10 seconds. • DVOM on 200 ohm scale. • Disconnect TFI. • Disconnect processor. • Measure resistance between test Pin 56 and TFI connector, PIP circuit.	Less than 5 ohms ▶ 5 ohms or greater ▶	GO to **A12**. SERVICE open PIP circuit. REPEAT Quick Test

EEC IV No Start — Pinpoint Test — A

TEST STEP	RESULT ▶	ACTION TO TAKE
A12 CHECK PIP CIRCUIT FOR SHORTS		
• Breakout box installed. • Processor disconnected. • Key Off. • Disconnect TFI connector. • DVOM on 200,000 ohm scale. • Measure resistance between test Pin 56 and test Pins 16, 20, 26, 40, 60 (shorts to ground) and test Pins 37 and 57 (shorts to power) and test Pin 36 (short to spout).	Any resistance less than 10,000 ohms ▶ All resistance greater than 10,000 ohms ▶	SERVICE PIP circuit. REPEAT Quick Test. REFER to Section 15 for TFI diagnosis.
A13 SPOUT SIGNAL VERIFICATION		
• Key Off, wait 10 seconds. • Disconnect processors 60 pin connector and inspect for damaged pins, corrosion, loose wires. Service as necessary. • Install Breakout box. • Processor connected. • DVOM on 20 volt scale. • Measure voltage between test Pin 36 and chassis ground, during crank. • Ensure timing switch is in "Computed" position on Breakout box.	Between 3.0V and 6.0V ▶ Less than 3.0V or greater than 6.0V ▶	FBC vehicles. REFER to Shop Manual CFI and EFI vehicles GO to **A21**. GO to **A10**.

EEC IV No Start — Pinpoint Test A

TEST STEP	RESULT ▶	ACTION TO TAKE
A21 FUEL PUMP CHECK		
• **No smoking nearby.** • Disconnect all injector electrical connections at the injectors. • Connect pressure gauge to Schraeder valve, if so equipped. • Note initial pressure reading. • Observe pressure gauge as you pressurize fuel system. (Turn key to Run for 1 second, then turn key to Off. Wait 10 seconds. Repeat 5 times.) • Turn key Off. Wait 10 seconds. • Reconnect all injectors.	PRESSURE GAUGE READING: Increased ▶	1.6L EFI and 1.6L EFI. TC SVO, 2.3L EFI-TC and SVO. GO to HA1. 2.3L HSC-CFI, GO to HF1. 3.8L CFI, GO to HD-1. 5.0L CFI, GO to HC1. 2.3L and 5.0L EFI Truck, GO to HE1.
	Did not increase ▶	TURN key Off, and CONTINUE to A22.
WARNING: If fuel starts leaking, turn key OFF immediately. No smoking.		
A22 INERTIA SWITCH CHECK		
• Key Off. • Fuel pressure gauge installed. • Locate fuel pump inertia switch. Refer to Owner's Manual for location. • Push the button of inertia switch to reset to On. **NOTE: If switch will not reset to On, replace Inertia Switch and repeat Step A21.** **If switch button was On, GO to J1.** • Observe pressure gauge as you pressurize fuel system. (Turn key to Run for 1 second, then turn key to Off. Wait 10 seconds.) Repeat 5 times	PRESSURE GAUGE READING: Increase ▶ No increase ▶	 REPEAT Quick Test GO to J1

Vehicle Battery — Pinpoint Test B

B

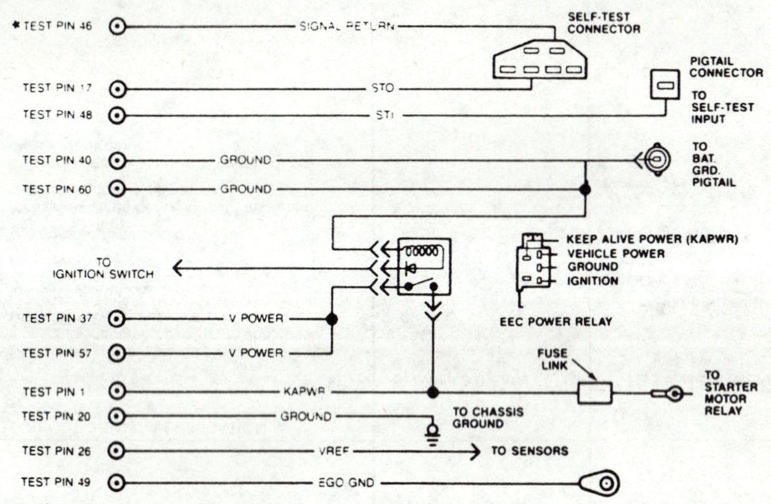

*TEST PINS LOCATED ON BREAKOUT BOX.
ALL HARNESS CONNECTORS VIEWED INTO MATING SURFACE.

STOP-WARNING

You should enter this Pinpoint Test only when directed here from Pinpoint Tests C, J or P.

To prevent the replacement of good components, be aware that the following non-EEC areas may be at fault:

- Ignition switch.
- Battery Cables.
- Alternator.
- Voltage Regulator.
- Ground Straps.

This pinpoint test is intended to diagnose only the following:

- Processor.
- Harness circuits: Signal Return, STO, STI, Ground, V Power, KAPWR, VREF, Ignition.
- Battery Voltage.
- Power Relay.

Vehicle Battery — Pinpoint Test B

TEST STEP	RESULT ▶	ACTION TO TAKE
B1 BATTERY VOLTAGE CHECK		
• Key On, Engine Off. • DVOM on 20 volt scale. • Measure voltage across battery terminals.	10.5V or greater ▶	GO to B2.
	Less than 10.5V ▶	SERVICE discharged battery
B2 BATTERY POWER GROUND CHECK		
• Key On, Engine Off. • Processor connected. • DVOM on 20 volt range. • Measure voltage between battery negative post and Signal Return circuit in the Self-Test connector.	Less than 0.5V ▶	GO to B6.
	0.5V or greater ▶	GO to B3.
B3 GROUND FAULT ISOLATION		
• Breakout box installed. • Key On, Engine Off. • Processor connected. • DVOM on 20 volt scale. • Measure voltage between battery negative post and test Pin 40 and test Pin 60.	Both readings less than 0.5V ▶	GO to B4.
	One or both readings 0.5V or greater ▶	Circuit(s) with greater than 0.5V has high resistance or open. CORRECT faulty ground circuit. REPEAT Quick Test.
B4 PROCESSOR GROUND FAULT ISOLATION		
• Breakout box installed. • Key Off, wait 10 seconds. • Processor connected. • DVOM on 200 ohm scale. • Measure resistance between test Pin 46 and test Pin 40 and between test Pin 46 and test Pin 60.	Both readings less than 5 ohms ▶	GO to B5.
	One or both readings 5 ohms or greater ▶	DISCONNECT processor connector and INSPECT for corrosion, damaged pins, etc. SERVICE as necessary and RETEST. If fault is still present, REPLACE processor. REPEAT Quick Test.

Vehicle Battery — Pinpoint Test B

TEST STEP	RESULT ▶	ACTION TO TAKE
B5 HARNESS CHECK (SIGNAL RETURN)		
• Breakout box installed. • Key Off, wait 10 seconds. • Processor connected. • DVOM on 200 ohm scale. • Measure resistance between test Pin 46 and Signal Return circuit at the Self-Test connector.	Less than 5 ohms ▶	System OK. RUN Quick Test.
	5 ohms or greater ▶	CORRECT cause of resistance in the harness Signal Return circuit. REPEAT Quick Test.
B6 12 VOLT BATTERY POWER FAULT ISOLATION		
• Key On, Engine Off. • Processor connected. • DVOM on 20 volt scale. • Measure voltage between the battery negative post and KAPWR circuit at EEC power relay.	10.5V or greater ▶	GO to B7.
	Less than 10.5V ▶	CHECK KAPWR and V Power circuits for shorts to ground and KAPWR circuit from power relay to battery positive post for opens. SERVICE as necessary. REPEAT Quick Test.
B7 12 VOLT BATTERY POWER FAULT ISOLATION		
• Key On, Engine Off. • Processor connected. • DVOM on 20 volt scale. • Measure voltage between the battery negative post and Ignition circuit at EEC power relay.	10.5V or greater ▶	GO to B8.
	Less than 10.5V ▶	CHECK for open in ignition switch circuits. SERVICE as necessary. REPEAT Quick Test.
B8 12 VOLT BATTERY POWER FAULT ISOLATION		
• Key On, Engine Off. • Processor connected. • DVOM on 20 volt scale. • Measure voltage between the battery negative post and ground circuit at EEC power relay.	Less than 0.5V ▶	GO to B9.
	0.5V or greater ▶	SERVICE open or ground in Ground Circuit. REPEAT Quick Test.

Vehicle Battery — Pinpoint Test B

TEST STEP	RESULT ▶	ACTION TO TAKE
B9 12 VOLT BATTERY POWER FAULT ISOLATION		
• Key On, Engine Off. • Processor connected. • DVOM on 20 volt scale. • Measure voltage between the battery negative post and V-Power circuit at EEC power relay.	10.5V or greater ▶	SERVICE short to ground or open in V-Power circuit from EEC power relay to processor connector test Pins 37 and 57. REPEAT Quick Test.
	Less than 10.5V ▶	REPLACE power relay. REPEAT Quick Test.

Reference Voltage — Pinpoint Test C

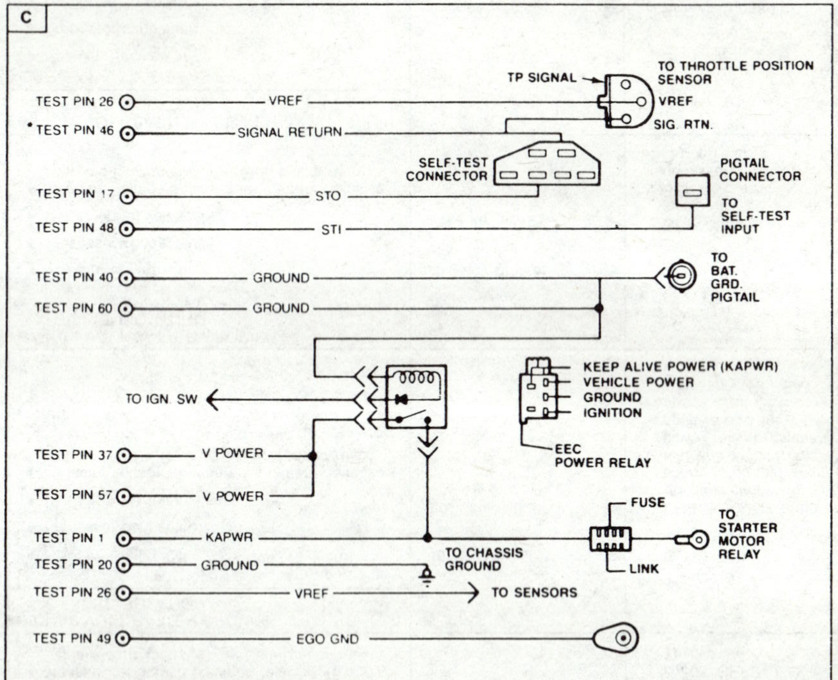

*TEST PINS LOCATED ON BREAKOUT BOX.
ALL HARNESS CONNECTORS VIEWED INTO MATING SURFACE.

STOP-WARNING

You should enter this Pinpoint Test only when a check for VREF has failed in the sensor Pinpoint Tests (D-Series) or Pinpoint Tests A or Q.

This pinpoint test is intended to diagnose only the following:

- Processor.
- Sensor harness circuits: Signal Return, STO, STI, Ground, V Power, KAPWR, VREF, Ignition.

Reference Voltage — Pinpoint Test C

TEST STEP	RESULT ▶	ACTION TO TAKE
C1 VEHICLE BATTERY POWER CIRCUIT CHECK		
• Breakout box installed. • Key On, Engine Off. • Processor connected. • DVOM on 20 volt scale. • Measure voltage between test Pin 37 and Signal Return in Self-Test connector.	10.5V or greater ▶ Less than 10.5V ▶	GO to C2. GO to B1.
C2 VREF VOLTAGE CHECK		
• Breakout box installed. • Key On, Engine Off. • Processor connected. • DVOM on 20 volt scale. • Measure voltage between test Pin 26 and test Pin 46.	6.0V or greater ▶ 4.0V or less ▶ Greater than 4.0V, less than 6.0V ▶	GO to C4. GO to C5. GO to C3.
C3 CHECK VREF AND SIGNAL RETURN FOR CONTINUITY		
• Breakout box installed. • Processor disconnected. • Key Off. • DVOM on 200 ohm scale. • Measure resistance from test Pins 26 and 46 to the faulty VREF sensor harness connector.	Less than 5 ohms on all readings ▶ 5 ohms or greater on any reading ▶	RECONNECT sensors. Reference voltage OK. REPEAT Quick Test. SERVICE open in VREF or Signal Return. REPEAT Quick Test.
C4 CHECK FOR SHORT FROM VREF TO VPWR		
• Key Off, wait 10 seconds. • Breakout box installed. • Disconnect processor. • Key On, Engine Off. • DVOM on 20 volt scale. • Measure voltage between test Pin 26 and battery ground.	Less than 0.5V ▶ 0.5V or greater ▶	REPLACE processor. REPEAT Quick Test. SERVICE short to battery power in EEC harness. REPEAT Quick Test. If condition persists, REPLACE processor.

Reference Voltage — Pinpoint Test C

TEST STEP	RESULT ▶	ACTION TO TAKE
C5 CHECK FOR SHORTED THROTTLE POSITION SENSOR		
• Key Off, wait 10 seconds. • Breakout box installed. • Processor connected. • Disconnect Throttle Position (TP) sensor from vehicle harness. • Key On, Engine Off. • DVOM on 20 volt scale. • Measure voltage between test Pin 26 and test Pin 46.	Less than 4.0V ▶ 4.0V or greater ▶	Vehicles equipped with EVP sensor, GO to C6. All other vehicles, GO to C7. REPLACE TP sensor. REPEAT Quick Test.
C6 CHECK FOR SHORTED EGR VALVE POSITION (EVP) SENSOR		
• Key Off, wait 10 seconds. • Breakout box installed. • Processor connected. • Disconnect EGR valve position (EVP) sensor. • Key On, Engine Off. • DVOM on 20 volt scale. • Measure voltage between test Pin 26 and test Pin 46.	Less than 4.0V ▶ 4.0V or greater ▶	GO to C7. REPLACE EVP sensor. REPEAT Quick Test.
C7 CHECK FOR SHORTED MAP BP SENSOR		
• Key Off, wait 10 seconds. • Breakout box installed. • Processor connected. • Disconnect MAP BP sensor. • Key On, Engine Off. • DVOM on 20 volt scale. • Measure voltage between test Pin 26 and test Pin 46.	Less than 4.0V ▶ 4.0V or greater ▶	Vehicles equipped with VAF sensor, GO to C8. All other vehicles, GO to C9. REPLACE MAP BP sensor. REPEAT Quick Test.

Reference Voltage — Pinpoint Test C

TEST STEP	RESULT ▶	ACTION TO TAKE
C8 CHECK FOR SHORTED VANE AIR METER (VAF) SENSOR		
• Key Off, wait 10 seconds. • Breakout box installed. • Processor connected. • Disconnect vane air meter (VAF) sensor. • Key On, Engine Off. • DVOM on 20 volt scale. • Measure voltage between test Pin 26 and test Pin 46.	Less than 4.0V ▶ 4.0V or greater ▶	GO to C9. REPLACE VAF sensor and REPEAT Quick Test.
C9 SHORT TO GROUND IN VREF		
• Breakout box installed. • Processor disconnected. • Key Off, wait 10 seconds. • Disconnect TP and MAP BP, EVP and VAF, if so equipped. • DVOM on 200 ohm scale. • Measure resistance between test Pin 26 and test Pins 20, 40, 46 and 60.	Less than 5 ohms ▶ 5 ohms or greater ▶	SERVICE short to ground. CONNECT all sensors. REPEAT Quick Test. If original condition still exists, REPLACE processor. RECONNECT sensors. REPLACE processor. REPEAT Quick Test.

Vane Air Temperature Sensor (VAT) — Pinpoint Test DA

DA

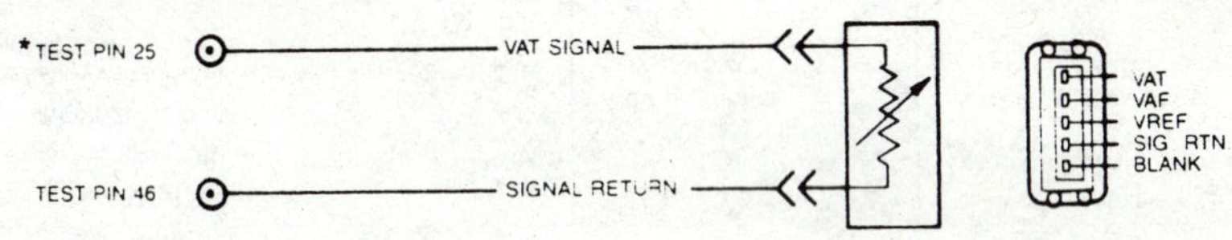

NOTE: AMBIENT TEMP. MUST BE GREATER THAN 50°F TO PASS THIS TEST.

TYPICAL RESISTANCE BETWEEN TEST PINS 25 & 46	5800Ω	2700Ω	300Ω	80Ω	125Ω
AT TEMPERATURE	32°F	65°F	185°F	220°F	240°F

*TEST PIN LOCATED ON BREAKOUT BOX.
ALL CONNECTIONS VIEWED INTO MATING SURFACE.

STOP-WARNING

You should enter this Pinpoint Test only when a service code 24, 54 or 64 is received in Quick Test Step 3.0, 5.0 or 6.0.

To prevent the replacement of good components, be aware that the following non-EEC areas may be at fault:

- Test performed in unusually low (cold) or high (hot) ambient conditions.
- Ambient temperature must be greater than 50°F for this test.

This pinpoint test is intended to diagnose only the following:

- VAT sensor.
- Circuits: VAT, and Signal Return.
- Vehicle harness.
- Processor assembly.

Vane Air Temperature Sensor (VAT) — Pinpoint Test — DA

TEST STEP	RESULT ▶	ACTION TO TAKE
DA1 SERVICE CODE 24: CHECK AMBIENT TEMPERATURE		
• Ambient temperature must be greater than 50°F for this test.	Yes ▶	GO to DA2.
	No ▶	RERUN Quick Test.
DA2 CHECK FOR V REF AT THROTTLE POSITION SENSOR		
• Refer to illustration Q. • Key Off, wait 10 seconds. • DVOM on 20V scale. • Disconnect TP sensor. • Key On, Engine Off. • Measure voltage at the TP harness connector between VREF and signal return.	Less than 4.0V or greater than 6.0V ▶	GO to Pinpoint Test Step C1.
	4.0V to 6.0V ▶	RECONNECT TP sensor. GO to DA3.
DA3 VAT SENSOR CHECK		
NOTE: Ambient temperature must be greater than 50°F for this test. • Key Off, wait 10 seconds. • Harness disconnected from the vane meter. • DVOM on 200,000 ohm scale. • Measure resistance at the VAT sensor between VAT signal and Signal Return. • Is reading from 125 ohms (240°F) to 3700 ohms (50°F)?	Yes ▶	REPLACE processor. RECONNECT harness to vane meter. RERUN Quick Test.
	No ▶	REPLACE vane meter. RERUN Quick Test.

Vane Air Temperature Sensor (VAT) — Pinpoint Test — DA

TEST STEP	RESULT ▶	ACTION TO TAKE
DA10 SERVICE CODE 54: INDUCE OPPOSITE CODE		
• Key Off, wait 10 seconds. • Disconnect harness from vane meter. Inspect for damaged pins, corrosion, loose wires, etc. Service as necessary. • Insert a jumper wire (paper clip) at the vane meter harness connector between VAT Signal and harness connector Signal Return. • Run Key On, Engine Off Quick Test. • Is Code 64 present?	Yes ▶	REPLACE vane meter. REMOVE jumper wire. CONNECT harness to vane meter. RERUN Quick Test.
	No ▶	REMOVE jumper wire. GO to DA11.
DA11 CHECK CONTINUITY OF VAT SIGNAL AND SIGNAL RETURN		
• Key Off, wait 10 seconds. • Harness disconnected from vane meter, jumper wire removed. • Disconnect processor 60 pin connector. Inspect for damaged pins, corrosion, loose wires, etc. Service as necessary. • Install Breakout box leaving processor disconnected. • DVOM on 200 ohm scale. • Measure resistance between VAT signal, at the vane meter harness connector, and test Pin 25. • Measure resistance between signal return, at the vane meter harness connector, and test Pin 46.	Both readings are less than 5 ohms ▶	REPLACE processor. REMOVE Breakout box. RECONNECT harness to vane meter and processor. RERUN Quick Test.
	Either reading is 5 ohms or greater ▶	CORRECT open circuit. REMOVE Breakout box. RECONNECT harness to vane meter and processor. RERUN Quick Test.

Vane Air Temperature Sensor (VAT) — Pinpoint Test — DA

TEST STEP	RESULT ▶	ACTION TO TAKE
DA20 SERVICE CODE 64: INDUCE OPPOSITE CODE		
• Key off, wait 10 seconds. • Disconnect harness from vane meter. Inspect for damaged pins, corrosion, loose wires, etc. Service as necessary. • Run Key On, Engine Off Quick Test. • Is code 54 present?	Yes ▶ No ▶	REPLACE vane meter. RECONNECT harness to vane meter. RERUN Quick Test. GO to DA21.
DA21 CHECK FOR V REF AT THROTTLE POSITION SENSOR		
• Refer to illustration Q. • Key Off, wait 10 seconds. • DVOM on 20V scale. • Disconnect TP sensor. • Key On, Engine Off. • Measure voltage at the TP harness connector between VREF and signal return.	Less than 4.0V or greater than 6.0V ▶ 4.0V to 6.0V ▶	GO to Pinpoint Test Step C1. RECONNECT TP sensor, GO to DA22.
DA22 CHECK VAT SIGNAL FOR SHORTS		
• Key Off, wait 10 seconds. • Harness disconnected from vane meter. • Disconnect processor 60 pin connector. Inspect for damaged pins, corrosion, loose wires, etc. Service as necessary. • Install Breakout box leaving processor disconnected. • DVOM on 200,000 ohm scale. • Measure resistance between Test Pin 25 and Test Pins 40, 46 and 60.	Any reading less than 10,000 ohms ▶ All readings are 10,000 ohms or greater ▶	CORRECT circuit shorts. REMOVE Breakout box. RECONNECT processor and vane meter. RERUN Quick Test. REPLACE processor. REMOVE Breakout box. RECONNECT processor. RERUN Quick Test.

Vane Air Temperature — Pinpoint Test — DA

TEST STEP	RESULT ▶	ACTION TO TAKE
DA90 SERVICE CODE 54: CONTINUOUS TEST: CHECK VAT SENSOR		
• Using continuous monitor mode, observe VOM or STAR LED for indication of a fault while performing the following: • Lightly tap on VAT sensor (simulate road shock). • Wiggle VAT connector. • Is a fault indicated?	Yes ▶ No ▶	DISCONNECT and INSPECT connectors. If connector and terminals are good, REPLACE VAT sensor. REPEAT Quick Test. GO to DA91.

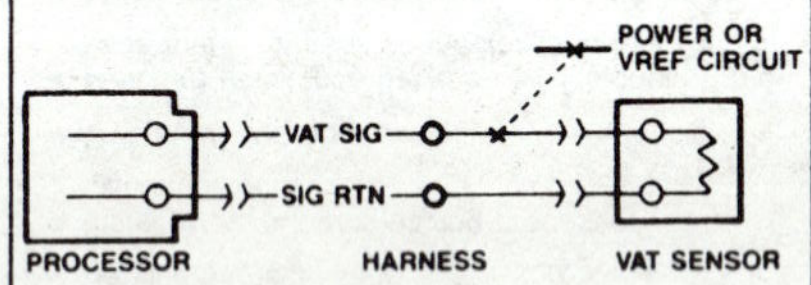

TEST STEP	RESULT ▶	ACTION TO TAKE
DA91 CHECK EEC-IV HARNESS		
• Observe VOM or STAR LED for a fault indication while performing the following: • Referring to the illustration in Step DA90, grasp the harness closest to the sensor connector. Wiggle, shake or bend a small section of the EEC-IV system harness while working your way to the dash panel. Also wiggle, shake or bend the EEC-IV harness from the dash panel to the processor. • Is fault indicated?	Yes ▶ No ▶	ISOLATE fault and make necessary service. REPEAT Quick Test. GO to DA92.
DA92 CHECK PROCESSOR AND HARNESS CONNECTORS		
• Key Off, wait 10 seconds. • Disconnect processor 60 pin connector. • Inspect both connectors and connector terminals for obvious damage or faults. • Connectors and terminals are OK.	No ▶ Yes ▶	SERVICE as necessary. REPEAT Quick Test. Unable to duplicate fault at this time. Continuous code 54 testing complete.

Vane Air Temperature — Pinpoint Test DA

TEST STEP	RESULT ►	ACTION TO TAKE
DA93 SERVICE CODE 64: CONTINUOUS TEST: CHECK VAT SENSOR		
• Using continuous monitor mode, observe VOM or STAR LED for indication of a fault while performing the following: • Lightly tap on VAT sensor (simulate road shock) • Wiggle VAT connector. • Is fault indicated?	Yes ►	DISCONNECT and INSPECT connectors. If connector and terminals are good, REPLACE VAT sensor. REPEAT Quick Test.
	No ►	GO to **DA94**

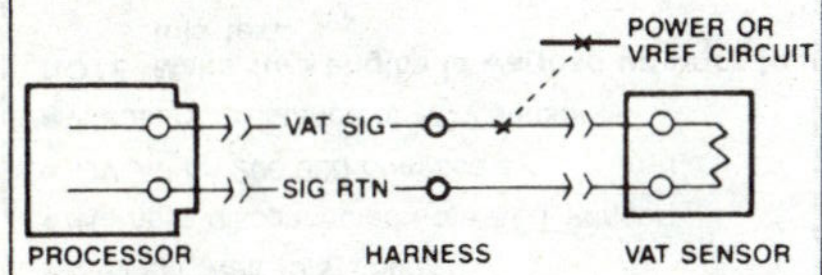

TEST STEP	RESULT ►	ACTION TO TAKE
DA94 CHECK EEC-IV HARNESS		
• Observe VOM or STAR LED for a fault indication while performing the following. • Referring to the illustration in Step DA93, grasp the harness closest to the sensor connector. Wiggle, shake or bend a small section of the EEC-IV system harness while working your way to the dash panel. Also wiggle, shake or bend the EEC-IV harness from the dash panel to the processor. • Is fault indicated?	Yes ►	ISOLATE fault and make necessary service. REPEAT Quick Test.
	No ►	GO to **DA95**
DA95 CHECK PROCESSOR AND HARNESS CONNECTORS		
• Key Off, wait 10 seconds. • Disconnect processor 60 pin connector. • Inspect both connectors and connector terminals for obvious damage or faults. • Connectors and terminals are OK.	No ►	SERVICE as necessary. REPEAT Quick Test.
	Yes ►	Unable to duplicate fault at this time. Continuous code 64 testing complete.

Air Charge Temperature Sensor (ACT) — Pinpoint Test DB

DB

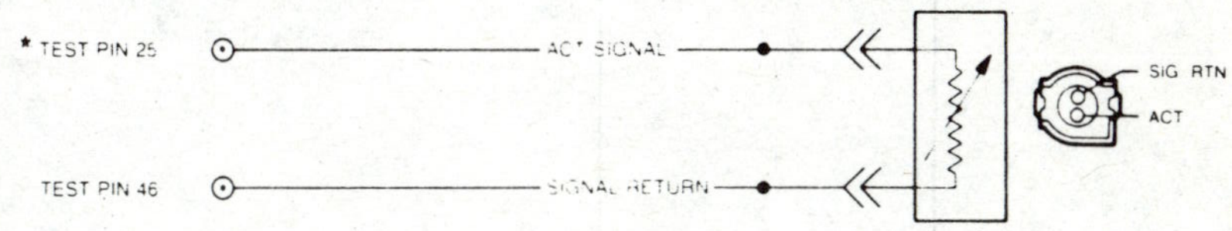

TYPICAL RESISTANCE BETWEEN TEST PINS 25 AND 46	58,750Ω	40,500Ω	3600Ω	1840Ω
AT TEMPERATURE	50 F	65 F	180 F	220 F

* TEST PIN LOCATED ON BREAKOUT BOX.
ALL CONNECTIONS VIEWED INTO MATING SURFACE.

NOTE: Ambient temperature must be greater than 50°F for this test.

STOP-WARNING

You should enter this Pinpont Test only when a service code 24, 54 or 64 is received in Quick Test Step 3.0, 5.0 or 6.0.

To prevent the replacement of good components, be aware that the following non-EEC areas may be at fault:

- Cooling system.
- Air cleaner duct problems
- Improper engine oil level.

This pinpoint test is intended to diagnose only the following:

- ACT sensor.
- Harness circuits: ACT Signal and Signal Return.
- Processor assembly.

Air Charge Temperature Sensor (ACT) — Pinpoint Test — DB

TEST STEP	RESULT ▶	ACTION TO TAKE
DB1 SERVICE CODE 24: CHECK PROPER INSTALLATION		
• For vehicle with ACT mounted in intake manifold. GO to step DB2.	Yes ▶	GO to DB2.
• Is ACT mounted properly in air cleaner?	No ▶	INSTALL ACT properly. RERUN Quick Test.
DB2 CHECK FOR V REF AT THROTTLE POSITION SENSOR		
• Key Off, wait 10 seconds • DVOM on 20V scale. • Disconnect TP sensor. • Key On, Engine Off. • Measure voltage at the TP harness connector between VREF and signal return.	Less than 4.0V or greater than 6.0V ▶	GO to Pinpoint Test Step C1.
	4.0V to 6.0V ▶	RECONNECT TP sensor. GO to DB3.
DB3 CHECK ACT SENSOR — ENGINE OFF		
• Key Off, wait 10 seconds • Harness disconnected from ACT sensor. • DVOM on 200,000 ohm scale • Measure resistance of ACT sensor. **NOTE: Make sure engine is warmed up prior to this test.**	Reading is between 1,100 and 58,000 ohms (approximately 50°F) ▶	GO to DB4.
	Reading is less than 1,100 ohms or greater than 58,000 ▶	CHECK heat stove duct valve operation. If OK, REPLACE ACT sensor. RECONNECT harness to ACT sensor. RERUN Quick Test.
DB4 CHECK ACT SENSOR — ENGINE RUNNING		
• Key Off Harness disconnected from ACT sensor. • DVOM on 200,000 ohm scale. • Run engine for 2 minutes. • Measure resistance of ACT sensor with engine running.	Reading is between 2,400 and 29,000 ohms ▶	REPLACE processor. RECONNECT harness to ACT sensor. RERUN Quick Test.
	Reading is less than 2,400 ohms or greater than 29,000 ohms ▶	CHECK heat stove duct valve operation. If OK, REPLACE ACT sensor. RERUN Quick Test.

Air Charge Temperature Sensor (ACT) — Pinpoint Test — DB

TEST STEP	RESULT ▶	ACTION TO TAKE
DB10 SERVICE CODE 54: INDUCE OPPOSITE CODE		
• Key Off, wait 10 seconds. • Disconnect harness from ACT sensor. Inspect for damaged pins, corrosion, loose wires, etc. Service as necessary. • Insert a jumper wire at the ACT harness connector between ACT Signal to Signal Return. • Run Key On, Engine Off Quick Test. • Is code 64 present?	Yes ▶	REPLACE ACT sensor. REMOVE jumper wire. RECONNECT harness to ACT sensor. RERUN Quick Test.
	No ▶	REMOVE jumper wire. GO to DB11.
DB11 CONTINUITY CHECK ACT SIGNAL AND SIGNAL RETURN		
• Key Off, wait 10 seconds. • Harness disconnected from ACT sensor. • Disconnect processor 60 pin connector. Inspect for damaged pins, corrosion, loose wires, etc. Service as necessary. • Install Breakout box. Leave processor disconnected. • DVOM on 200 ohm scale. • Measure resistance between ACT signal, at the ACT harness connector, and test Pin 25. • Measure resistance between Signal Return, at the ACT harness connector, and test Pin 46.	Both readings are less than 5 ohms ▶	REPLACE processor. REMOVE Breakout box. RECONNECT processor and ACT sensor. RERUN Quick Test.
	Either reading is 5 ohms or greater ▶	CORRECT circuit opens. REMOVE Breakout box. RECONNECT processor and ACT sensor. RERUN Quick Test.

Air Charge Temperature Sensor (ACT) — Pinpoint Test — DB

TEST STEP	RESULT ▶	ACTION TO TAKE
DB20 SERVICE CODE 64: INDUCE OPPOSITE CODE		
• Key Off, wait 10 seconds. • Disconnect harness from ACT sensor. Inspect for damaged pins, corrosion, loose wires, etc. Service as necessary. • Run Key On, Engine Off Quick Test. • Is code 54 present?	Yes ▶ No ▶	REPLACE ACT sensor. RECONNECT harness to ACT sensor. RERUN Quick Test. GO to DB21.
DB21 CHECK FOR V REF AT THROTTLE POSITION SENSOR		
• Key Off, wait 10 seconds. • DVOM on 20V scale. • Disconnect TP sensor. • Key On, Engine Off. • Measure voltage at the TP harness connector between VREF and Signal Return.	Less than 4.0V or greater than 6.0V ▶ 4.0V to 6.0V ▶	GO to Pinpoint Test Step C1. RECONNECT TP sensor. GO to DB22.
DB22 CHECK ACT SIGNAL FOR SHORTS TO GROUND		
• Key Off, wait 10 seconds. • Harness disconnected from ACT sensor. • Disconnect processor 60 pin connector. Inspect for damaged pins, corrosion, loose wires, etc. Service as necessary. • DVOM on 200,000 ohm scale. • Measure resistance between test Pin 25 and test Pins 40, 46 and 60.	Any reading less than 10,000 ohms ▶ All readings are 10,000 ohms or greater ▶	CORRECT circuit shorts. REMOVE Breakout box. RECONNECT processor and ACT sensor. RERUN Quick Test. REPLACE processor. REMOVE Breakout box. RECONNECT processor and ACT sensor. RERUN Quick Test.

Air Charge Temperature (ACT) — Pinpoint Test — DB

TEST STEP	RESULT ▶	ACTION TO TAKE
DB90 SERVICE CODE 54: CONTINUOUS TEST: CHECK ACT SENSOR		
• Using continuous monitor mode, observe VOM or STAR LED for indication of a fault while performing the following: • Lightly tap on ACT sensor (simulate road shock). • Wiggle ACT connector. • Is a fault indicated?	Yes ▶ No ▶	DISCONNECT and INSPECT connectors. If connector and terminals are good, REPLACE ACT sensor. REPEAT Quick Test. GO to DB91.

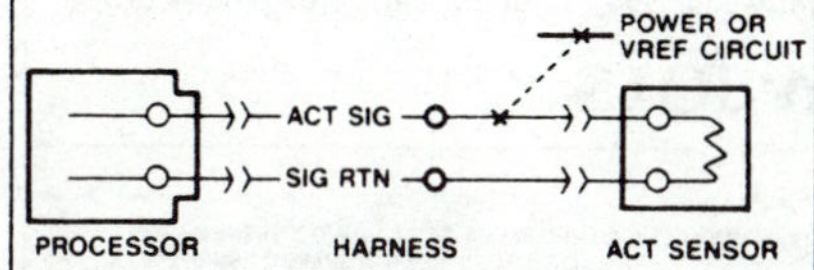

TEST STEP	RESULT ▶	ACTION TO TAKE
DB91 CHECK EEC-IV HARNESS		
• Observe VOM or STAR LED for a fault indication while performing the following: • Referring to the illustration in Step DB90, grasp the harness closest to the sensor connector. Wiggle, shake or bend a small section of the EEC-IV system harness while working your way to the dash panel. Also wiggle, shake or bend the EEC-IV harness from the dash panel to the processor. • Is fault indicated?	Yes ▶ No ▶	ISOLATE fault and make necessary repairs. REPEAT Quick Test. GO to DB92.
DB92 CHECK PROCESSOR AND HARNESS CONNECTORS		
• Key Off, wait 10 seconds. • Disconnect processor 60 pin connector. • Inspect both connectors and connector terminals for obvious damage or faults. • Connectors and terminals are OK.	No ▶ Yes ▶	SERVICE as necessary. REPEAT Quick Test. Unable to duplicate fault at this time. Continuous code 54 testing complete.

Air Charge Temperature (ACT) — Pinpoint Test — DB

TEST STEP	RESULT	▶	ACTION TO TAKE
DB93 SERVICE CODE 64: CONTINUOUS TEST: CHECK ACT SENSOR			
• Using continuous monitor mode, observe VOM or STAR LED for fault while performing the following: • Lightly tap on ACT sensor (simulate road shock). • Wiggle ACT connector. • Is fault indicated?	Yes	▶	DISCONNECT and INSPECT connectors. If connector and terminals are good, REPLACE ACT sensor. REPEAT Quick Test.
	No	▶	GO to DB94.
DB94 CHECK EEC-IV HARNESS			
• Observe VOM or STAR LED for a fault indication while performing the following: • Referring to the illustration in Step DB93, grasp the harness closest to the sensor connector. Wiggle, shake or bend a small section of the EEC-IV system harness while working your way to the dash panel. Also wiggle, shake or bend the EEC-IV harness from the dash panel to the processor. • Is fault indicated?	Yes	▶	ISOLATE fault and make necessary repairs. REPEAT Quick Test.
	No	▶	GO to DB95.
DB95 CHECK PROCESSOR AND HARNESS CONNECTORS			
• Key Off, wait 10 seconds. • Disconnect processor 60 pin connector. • Inspect both connectors and connector terminals for obvious damage or faults. • Connectors and terminals are OK.	No	▶	SERVICE as necessary. REPEAT Quick Test.
	Yes	▶	Unable to duplicate fault at this time. Continuous code 64 testing complete.

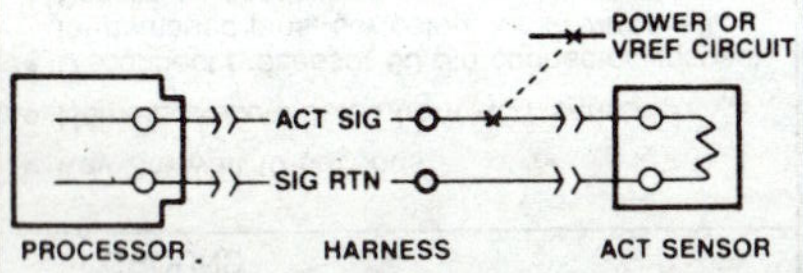

EGR Valve Position Sensor (EVP) Control/Vent (EGRC/EGRV) — Pinpoint Test — DD

DD

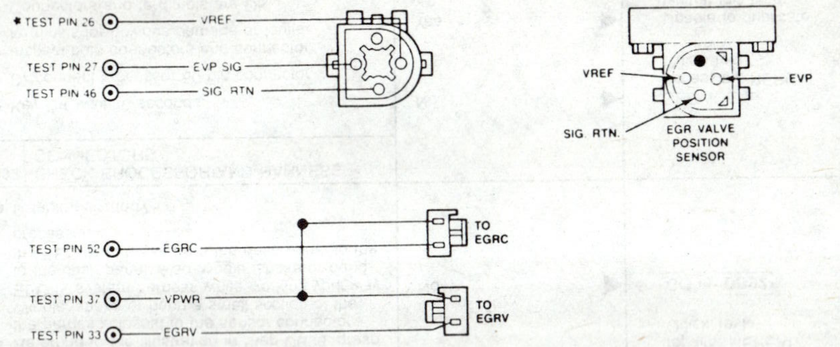

* TEST PINS LOCATED ON BREAKOUT BOX.
ALL HARNESS CONNECTORS VIEWED INTO MATING SURFACE.

STOP-WARNING

You should enter this Pinpoint Test only when a Service Code 31, 32, 33, 34, 35, 83 or 84 is received in Quick Test Step 3.0, 5.0 or 6.0.

To prevent the replacement of good components, be aware that the following non-EEC area may be at fault:

- Damaged EGR valve.

This Pinpoint Test is intended to diagnose only the following:

- EVP sensor.
- Harness circuits: EVP, Signal Return, VREF, EGRV, EGRC, VPWR.
- EGR solenoids.
- EGR Valve assembly.
- Processor assembly.

EGR Valve Position Sensor (EVP) Control/Vent (EGRC/EGRV) — Pinpoint Test DD

TEST STEP	RESULT ▶	ACTION TO TAKE
FAULT CODE 31		
DD1 RUN ENGINE RUNNING QUICK TEST WITH EGR VACUUM SIGNAL LINE DISCONNECTED AT EGR VALVE		
• Key Off, wait 10 seconds. • Disconnect EGR vacuum line at EGR valve and cap EGR vacuum line. • Run Engine Running Quick Test. • Check for code 31.	Code 31 present ▶ No code 31 present, but codes 32, 34 are present ▶	GO to DD2. GO to DD11.
DD2 CHECK EVP RESISTANCE WHILE APPLYING VACUUM TO EGR VALVE		
• Key Off, vacuum signal line disconnected and capped, harness disconnected at EVP sensor. • DVOM on 200,000 ohm scale. • Connect vacuum pump to EGR valve. • Measure resistance at the EVP sensor between EVP SIG and VREF while gradually increasing vacuum to 33 kPa (10 in. Hg). • Observe resistance as vacuum increases.	Reading gradually decreases from no greater than 5500 ohms to no less than 100 ohms ▶ Reading is less than 100 ohms or greater than 5500 ohms ▶ Reading does not decrease or unable to hold vacuum ▶	GO to DD3. REPLACE EVP sensor. RECONNECT signal line and harness. RERUN Quick Test. GO to DD16.
DD3 MEASURE VREF TO SIGNAL RETURN VOLTAGE		
• Key On, Engine Off. • Vacuum signal line disconnected and capped, harness disconnected from EVP sensor. • DVOM on 20V scale. • Measure voltage at the EVP harness connector between VREF and Signal Return.	Reading is between 4 and 6V ▶ Reading is less than 4V or greater than 6V ▶	GO to DD4. GO to Pinpoint Test Step C1.

EGR Valve Position Sensor (EVP) Control/Vent (EGRC/EGRV) — Pinpoint Test DD

TEST STEP	RESULT ▶	ACTION TO TAKE
DD4 MEASURE CONTINUITY OF EVP SIGNAL CIRCUIT		
• Key Off, wait 10 seconds. • Harness disconnected from EVP sensor. • Disconnect processor 60 Pin connector and inspect for damaged pins, corrosion, loose wires. Service as necessary. • Install Breakout box leaving processor disconnected. • DVOM on 200 ohm scale. • Measure resistance between test Pin 27 and EVP signal at the EVP harness connector.	Less than 5 ohms ▶ 5 ohms or greater ▶	GO to DD5. SERVICE open circuit. REMOVE Breakout box. RECONNECT processor and EVP sensor. RERUN Quick Test.
DD5 CHECK EVP SIGNAL FOR SHORTS TO VREF AND SIGNAL RETURN		
• Key Off, harness disconnected from EVP sensor. Breakout box installed with processor not connected. • DVOM on 200,000 ohm scale. • Measure resistance between test Pin 27 and test Pins 26 and 46.	Either reading less than 10,000 ohms ▶ Both readings 10,000 ohms or greater ▶	SERVICE short circuit. REMOVE Breakout box. RECONNECT processor and EVP sensor. RERUN Quick Test. GO to DD6.
DD6 SUBSTITUTE EVP SENSOR AND EGR VALVE		
• Key Off, wait 10 seconds. • Electrically connect known good EVP sensor and EGR valve assembly. • Remove Breakout box. • Reconnect processor. • Perform Key On, Engine Off Quick Test. • Is code 31 present?	Yes ▶ No ▶	REPLACE processor. Connect original EVP sensor and EGR valve assembly. REPEAT Quick Test. GO to DD7.

EGR Valve Position Sensor (EVP) Control/Vent (EGRC/EGRV) — Pinpoint Test DD

TEST STEP	RESULT	ACTION TO TAKE
DD7 EVP SENSOR CHECK		
• Key Off, wait 10 seconds. • Install original EVP sensor on known good EGR valve. • EVP sensor connected. • Rerun Key On, Engine Off Quick Test. • Is code 31 present?	Yes ▶	INSTALL new EVP sensor. REPEAT Quick Test.
	No ▶	REFER to EGR systems
FAULT CODES 32, 33 AND 34		
DD11 OUTPUT STATE CHECK (REFER TO APPENDIX)		
NOTE: Do not use STAR tester for this test step. Use VOM/DVOM. • Key Off, wait 10 seconds. • DVOM on 20V scale. • Connect DVOM negative test lead to STO and positive test lead to battery positive. • Jumper STI to signal return. • Perform Key On, Engine Off Self-Test until the completion of the Continuous Test Codes. • DVOM will indicate zero volts. • Depress and release the throttle • Did DVOM reading change to a high voltage reading?	No ▶	DEPRESS throttle to WOT and RELEASE. If STO voltage does not go high, GO to Pinpoint Test Step Q40.
	Yes ▶	Remain in output state check and GO to DD12.
DD12 CHECK EGR SOLENOIDS FOR ELECTRICAL CYCLING		
• Key On, Engine Off. • In output state check. • DVOM on 20V scale. • Connect DVOM to EGRV circuits VPWR and EGRV signal. • While observing DVOM, depress and release the throttle several times to cycle output On and Off. • Repeat for EGRC circuits VPWR and EGRC signal.	Both solenoid outputs cycle On and Off ▶	REMAIN in output state check. GO to DD13.
	Either output does not cycle On and Off ▶	Exit output state check. GO to DD18.

EGR Valve Position Sensor (EVP) Control/Vent (EGRC/EGRV) — Pinpoint Test DD

TEST STEP	RESULT	ACTION TO TAKE
DD13 CHECK EGR SOLENOIDS FOR VACUUM CYCLING		
• Key On, Engine Off. • In output state check. • Disconnect and cap vacuum line from bottom port of EGRC solenoid and connect a vacuum pump. • Connect a vacuum gauge in the common output (top) vacuum line to EGR valve. • Disconnect but **do not** cap vacuum vent line from EGRV solenoid. • While cycling outputs On and Off (by depressing and releasing throttle), observe vacuum gauge at the output. Maintain vacuum at source.	Vacuum output cycles On and Off in less than 2 seconds ▶	RECONNECT all vacuum lines. GO to DD15.
	Vacuum does not cycle On and Off in less than 2 seconds ▶	GO to DD14.
DD14 CHECK EGR SOLENOID FRESH AIR SUPPLY		
• Key On, Engine Off. • In output state check. • Disconnect and cap vacuum line from bottom port of EGRC solenoid and connect a vacuum pump. • Connect a vacuum gauge in the common output (top) vacuum line to EGR valve. • Disconnect but **do not** cap vacuum vent line from EGRV solenoid. • While cycling outputs On and Off (by depressing and releasing throttle), observe vacuum gauge at the output. Maintain vacuum at source.	Vacuum output cycles On and Off in less than 2 seconds ▶	REPLACE filter if so equipped and/or CHECK vent line for obstructions. SERVICE as necessary. RECONNECT all vacuum lines. RERUN Quick Test.
	Vacuum does not cycle On and Off in less than 2 seconds ▶	REPLACE solenoid assembly. RECONNECT all vacuum lines. RERUN Quick Test.
DD15 CHECK VACUUM LINES		
• Key Off, wait 10 seconds. • Vacuum lines reconnected. • Check entire **EEC** vacuum line system per VECI emission schematic decal for kinks, cracks, obstructions or leaks.	Vacuum lines OK ▶	GO to DD16.
	Vacuum lines not OK ▶	SERVICE as necessary. RERUN Quick Test.

EGR Valve Position Sensor (EVP) Control/Vent (EGRC/EGRV) — Pinpoint Test — DD

TEST STEP	RESULT ▶	ACTION TO TAKE
DD16 CHECK EVP RESISTANCE WHILE APPLYING VACUUM TO EGR VALVE		
• Key Off. • Disconnect harness from EVP sensor. Inspect for damaged pins, corrosion, and pins pushed out. Service as necessary. • DVOM on 200,000 ohm scale. • Disconnect vacuum line at EGR valve. • Connect vacuum pump to EGR valve. • Measure resistance of the EVP sensor between EVP Signal Pin and VREF Pin while increasing vacuum to 33 kPa (10 in. Hg). • Observe resistance as vacuum increases.	Reading does not decrease gradually ▶ Reading gradually decreases from no more than 5,500 ohms to no less than 100 ohms ▶	GO to DD17. REPLACE processor. RECONNECT EVP sensor and EGR vacuum line. RERUN Quick Test.
DD17 MANUALLY EXERCISE EVP SENSOR		
• Key Off, harness disconnected from EVP sensor. • Remove EVP sensor from EGR valve. • Measure resistance of the EVP sensor between EVP Signal Pin and VREF Pin while gradually applying pressure to EVP sensor shaft. • Observe resistance as shaft is slowly pushed in and slowly released. • Look for sudden changes in resistance readings.	Both readings decrease and increase smoothly between 5,500 ohms and 100 ohms ▶ Either reading decreases or increases abruptly between 5,500 ohms and 100 ohms ▶	REPLACE EGR valve assembly. RECONNECT EVP sensor and EGR supply vacuum line. RERUN Quick Test. REPLACE EVP sensor. RECONNECT harness and EGR supply vacuum line. RERUN Quick Test.
DD18 MEASURE EGRV EGRC SOLENOID RESISTANCE		
• Key Off, wait 10 seconds. • DVOM on 200 ohm scale. • Disconnect EGRV solenoid connector and measure solenoid resistance. • Disconnect EGRC solenoid connector and measure solenoid resistance.	Both resistances are between 30 and 70 ohms ▶ Either resistance is less than 30 ohms or greater than 70 ohms ▶	CONNECT EGRC EGRV solenoids. GO to DD19. REPLACE EGRC EGRV solenoid assembly. REPEAT Quick Test.

EGR Valve Position Sensor (EVP) Control/Vent (EGRC/EGRV) — Pinpoint Test — DD

TEST STEP	RESULT ▶	ACTION TO TAKE
DD19 CHECK FOR VOLTAGE ON VEHICLE POWER CIRCUIT		
• Disconnect EGR vent and EGR control solenoids from harness. • Key On, Engine Off. • DVOM on 20V scale. • Measure voltage between battery negative terminal and VPWR circuit on both EGR solenoids.	Either reading is less than 10.5V ▶ Both readings are 10.5V or greater ▶	SERVICE harness circuit open. RERUN Quick Test. GO to DD20.
DD20 CHECK CONTINUITY OF EGRV AND EGRC CIRCUITS		
• Key Off, wait 10 seconds. • EGR vent and EGR control solenoid disconnected from harness. • Disconnect processor 60 Pin connector and inspect for damaged pins, corrosion, loose wires. Service as necessary. • Install Breakout box to processor harness connector. Leave processor disconnected. • DVOM on 200 ohm scale. • Measure resistance between test Pin 33 and EGRV signal at the EGRV solenoid harness connector. • Measure resistance between test Pin 52 and EGRC signal at the EGRC solenoid harness connector.	Both readings less than 5 ohms ▶ Either reading 5 ohms or greater ▶	GO to DD21. SERVICE open circuit. REMOVE Breakout box. RECONNECT harness to processor. RERUN Quick Test.
DD21 CHECK FOR SHORT TO GROUND		
• Key Off, wait 10 seconds. • DVOM on 200,000 ohm scale. • Leave Breakout box installed and processor disconnected. • EGRV EGRC solenoids disconnected. • Measure resistance between test Pins 33 and or 52 and test Pins 40, 46 and 60.	Resistance reading is less than 10,000 ohms ▶ Resistance reading is 10,000 ohms or greater ▶	SERVICE short to ground. REPEAT Quick Test. GO to DD22.

EGR Valve Position Sensor (EVP) Control/Vent (EGRC/EGRV) — Pinpoint Test DD

TEST STEP	RESULT ▶	ACTION TO TAKE
DD22 CHECK EGRV AND EGRC SIGNALS FOR SHORTS TO POWER		
• EGR vent and EGR control solenoids disconnected from harness. • Key Off. Breakout box installed. Processor disconnected. • DVOM on 200,000 ohm scale. • Measure resistance between test Pin 33 and test Pins 37 and 57.	Any reading less than 10,000 ohms ▶	SERVICE circuit short. REMOVE Breakout box. RECONNECT harness to processor. RERUN Quick Test. If code is repeated, REPLACE processor.
• Measure resistance between test Pin 52 and test Pins 37 and 57.	All readings 10,000 ohms or greater ▶	REPLACE processor. REMOVE Breakout box. RECONNECT harness to processor. RERUN Quick Test.
FAULT CODE 35		
DD30 RPM TOO LOW FOR EGR TEST		
• Check for code 12.	Code 12 present ▶	Vehicles equipped with air bypass (EFI), GO to KE1. Vehicles equipped with DC motor control. GO to KF1. Vehicles equipped with throttle kicker, GO to KG1.
	Code 12 not present ▶	GO to DD31.
DD31 RETEST AT 1,500 RPM		
• Key Off, wait 10 seconds. • Install tachometer. • Perform Key On, Engine Running Quick Test while maintaining 1,500 rpm. • Record Engine Running service codes. • Check for code 35.	Code 35 present ▶	REPLACE processor. RERUN Quick Test.
	Code 35 not present ▶	REPEAT Quick Test. SERVICE codes as necessary.

EGR Valve Position Sensor (EVP) Control/Vent (EGRC/EGRV) — Pinpoint Test DD

TEST STEP	RESULT ▶	ACTION TO TAKE
DD90 SERVICE CODE 31 CONTINUOUS TEST: EXERCISE EVP SENSOR		
• Using continuous monitor mode, observe VOM or STAR LED for indication of a fault while performing the following: • Connect a vacuum pump to the EGR valve. • Very slowly apply 20 kPa (6 in. Hg) vacuum to the EGR valve. • Slowly bleed vacuum off the EGR valve and lightly tap on EVP sensor (simulate road shock). • Wiggle EVP sensor connector. • Is a fault indicated? 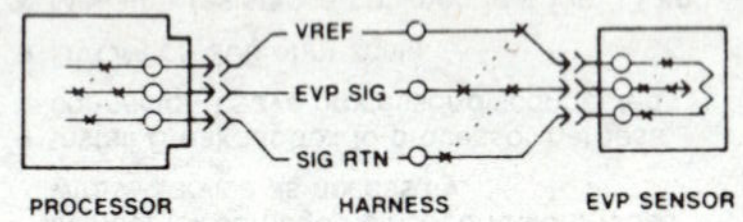 	Yes ▶ No ▶	GO to DD91. GO to DD92.
DD91 MEASURE EVP SIGNAL VOLTAGE WHILE EXERCISING EVP SENSOR		
• Key Off, wait 10 seconds. • Disconnect processor 60 Pin connector and inspect for damaged pins, corrosion, loose wires. Service as necessary. • Install Breakout box and reconnect processor. • VOM or STAR LED still connected to STO as in previous Step. • Connect a DVOM from test Pin 27 to test Pin 46. • DVOM on 20V scale. • Key On, Engine Off. • While observing DVOM, repeat step **DD90**. • Does the fault occur below 4.25V?	Yes ▶	DISCONNECT and INSPECT connector. If connector and terminals are good, REPLACE EVP sensor. REPEAT Quick Test.
	No ▶	EGR valve overshoot may have caused continuous code 31. Sensor service is not required. To verify harness integrity, GO to DD92.

EGR Valve Position Sensor (EVP) Control/Vent (EGRC/EGRV) — Pinpoint Test DD

TEST STEP	RESULT ▶	ACTION TO TAKE
DD92 CHECK EEC-IV HARNESS		
• Observe VOM or STAR LED for a fault indication while performing the following: • Referring to the illustration in Step DD90, grasp the harness closest to the sensor connector. Wiggle, shake or bend a small section of the EEC-IV system harness while working your way to the dash panel. Also wiggle, shake or bend the EEC-IV harness from the dash panel to the processor. • Is a fault indicated?	Yes ▶	ISOLATE fault and make necessary repairs. REFER to appropriate figure. REPEAT Quick Test.
	No ▶	GO to DD93.
DD93 CHECK PROCESSOR AND HARNESS CONNECTORS		
• Key Off, wait 10 seconds. • Disconnect processor 60 Pin connector. • Inspect both connectors and connector terminals for obvious damage or faults. • Are connectors and terminals OK?	No ▶	SERVICE as necessary. REPEAT Quick Test.
	Yes ▶	Unable to duplicate fault at this time. Continuous code 31 testing complete.

Engine Coolant Temperature Sensor (ECT) — Pinpoint Test DE

DE

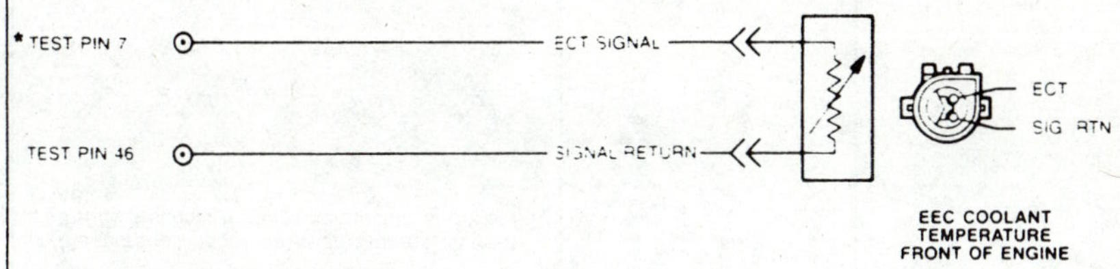

NOTE: TO PASS THIS TEST, ENGINE COOLANT TEMPERATURE MUST BE:
KEY ON, ENGINE OFF (50°F TO 240°F)
ENGINE RUNNING (180°F TO 240°F)

TYPICAL RESISTANCE BETWEEN TEST PINS 7 & 46	58,750Ω	40,500Ω	3600Ω	1840Ω
AT TEMPERATURE	50°F	65°F	180°F	220°F

***TEST PIN LOCATED ON BREAKOUT BOX.**
ALL CONNECTIONS VIEWED INTO MATING SURFACE

STOP-WARNING

You should enter this Pinpont Test only when a service code 21, 51 or 61 is received in Quick Test Step 3.0, 5.0 or 6.0.

To prevent the replacement of good components, be aware that the following non-EEC areas may be at fault:

- Coolant level.
- Oil level.
- Blocked or obstructed air flow.
- Engine not at normal operating temperature
- Electro drive cooling fan.

This pinpoint test is intended to diagnose only the following:

- ECT sensor.
- Harness sensor circuits: ECT and Signal Return
- Processor assembly.

Engine Coolant Temperature Sensor (ECT) — Pinpoint Test — DE

TEST STEP	RESULT ▶	ACTION TO TAKE
DE1 SERVICE CODE 21: CHECK ENGINE OPERATING TEMPERATURE		
• Run engine for 2 minutes at 2,000 rpm. • Check that upper radiator hose is hot and pressurized. • Rerun Quick Test.	Vehicle stalls ▶	Do not service code 21 at this time. REFER to diagnosis by symptoms.
	Code 21 present ▶	GO to DE2.
	Code 21 not present ▶	SERVICE other codes as necessary.
DE2 CHECK FOR V REF AT THROTTLE POSITION SENSOR		
• Refer to illustration Q. • Key Off, wait 10 seconds. • DVOM on 20V scale. • Disconnect TP sensor. • Key On, Engine Off. • Measure voltage at the TP harness connector between VREF and signal return.	Less than 4.0V or greater than 6.0V ▶	GO to Pinpoint Test Step C1.
	4.0V to 6.0V ▶	RECONNECT TP sensor. GO to DE3.
DE3 ECT SENSOR CHECK		
NOTE: Engine may have cooled down. Always warm engine before taking ECT resistance measurement. • Key Off, wait 10 seconds. • Harness disconnected from ECT sensor. • DVOM on 200,000 ohm scale. • Measure resistance of the ECT sensor. • Is the resistance reading: — 1300 ohms (240°F) to 7700 ohms (140°F) for engine off? — 1550 ohms (230°F) to 4550 ohms (180°F) for engine running?	Yes ▶	REPLACE processor. RECONNECT harness to ECT sensor. RERUN Quick Test.
	No ▶	REPLACE ECT sensor. RECONNECT harness to ECT sensor. RERUN Quick Test.

Engine Coolant Temperature Sensor (ECT) — Pinpoint Test — DE

TEST STEP	RESULT ▶	ACTION TO TAKE
DE10 SERVICE CODE 51: INDUCE OPPOSITE CODE		
• Key Off, wait 10 seconds. • Disconnect harness from ECT sensor. Inspect for damaged pins, corrosion, loose wires, etc. Service as necessary. • Insert a jumper wire at the ECT sensor harness connector between ECT Signal and Signal Return. • Run Key On, Engine Off Quick Test. • Is code 61 present?	Yes ▶	REPLACE ECT sensor. REMOVE jumper wire. RECONNECT ECT sensor. RERUN Quick Test.
	No ▶	GO to DE11.
DE11 CHECK CONTINUITY OF ECT SIGNAL AND SIGNAL RETURN		
• Key Off, wait 10 seconds. • Harness disconnected from ECT sensor, jumper wire removed. • Disconnect processor 60 pin connector. Inspect for damaged pins, corrosion, loose wires, etc. Service as necessary. • Install Breakout box to harness leaving processor disconnected. • DVOM on 200 ohm scale. • Measure resistance between ECT signal at the ECT harness connector and test Pin 7. • Measure resistance between Signal Return at the ECT sensor harness connector, and test Pin 46.	Both readings are less than 5 ohms ▶	REPLACE processor. REMOVE Breakout box. RECONNECT processor and ECT sensor. RERUN Quick Test.
	Either reading is 5 ohms or greater ▶	SERVICE open circuit(s). REMOVE Breakout box. RECONNECT processor and ECT sensor. RERUN Quick Test.

Engine Coolant Temperature Sensor (ECT) — Pinpoint Test — DE

TEST STEP	RESULT ▶	ACTION TO TAKE
DE20 SERVICE CODE 61: INDUCE OPPOSITE CODE		
• Key Off, wait 10 seconds. • Disconnect harness from ECT sensor. Inspect for damaged pins, corrosion, loose wires, etc. Service as necessary • Run Key On, Engine Off Quick Test.	Yes ▶	REPLACE ECT sensor. RECONNECT ECT sensor. RERUN Quick Test.
• Is code 51 present?	No ▶	GO to DE21.
DE21 CHECK FOR V REF AT THROTTLE POSITION SENSOR		
• Refer to illustration Q. • Key Off, wait 10 seconds. • DVOM on 20V scale.	Less than 4.0V or greater than 6.0V ▶	GO to Pinpoint Test Step C1.
• Disconnect TP sensor. • Key On, Engine Off. • Measure voltage at the TP harness connector between VREF and signal return.	4.0V to 6.0V ▶	RECONNECT TP sensor. GO to DE22.
DE22 CHECK ECT SIGNAL FOR SHORT		
• Key Off, wait 10 seconds. • Harness disconnected from ECT sensor. • Disconnect processor 60 pin connector. Inspect for damaged pins, corrosion, loose wires, etc Service as necessary.	Either reading is less than 10,000 ohms ▶	SERVICE circuit shorts. REMOVE Breakout box. RECONNECT processor and ECT sensor. RERUN Quick Test.
• Install Breakout box, leave processor disconnected • DVOM on 200,000 ohm scale • Measure resistance between test Pin 7 and test Pins 40, 46 and 60.	Both readings are 10,000 ohms or greater ▶	REPLACE processor. REMOVE Breakout box. RECONNECT processor and ECT harness. RERUN Quick Test.

Engine Coolant Temperature Sensor (ECT) — Pinpoint Test — DE

TEST STEP	RESULT ▶	ACTION TO TAKE
DE90 SERVICE CODE 21 CONTINUOUS TEST: TEST DRIVE VEHICLE		
• Key Off and wait 10 seconds. • Disconnect all Self-Test equipment and prepare vehicle for test drive. • Drive vehicle. Try to simulate different drive modes or mode in which drive complaint is noticed. Attempt to maintain drive complaint mode for one minute or more, if possible. • Upon completion of drive evaluation, repeat Key On, Engine Off Self-Test. • Is code 21 present in the continuous test results?	Yes ▶	VERIFY thermostat operating properly. If OK, REPLACE ECT sensor. REPEAT Quick Test.
	No ▶	Unable to duplicate fault. Code 21 testing complete.

Engine Coolant Temperature (ECT) — Pinpoint Test — DE

TEST STEP	RESULT ▶	ACTION TO TAKE
DE91 SERVICE CODE 51 CONTINUOUS TEST: CHECK ECT SENSOR		
• Using continuous monitor mode, observe VOM or STAR LED for indication of a fault while performing the following: • Lightly tap on ECT sensor (simulate road shock). • Wiggle ECT connector. • Is fault indicated?	Yes ▶	DISCONNECT and INSPECT connectors. If connector and terminals are good, REPLACE ECT sensor. REPEAT Quick Test.
	No ▶	GO to DE92.

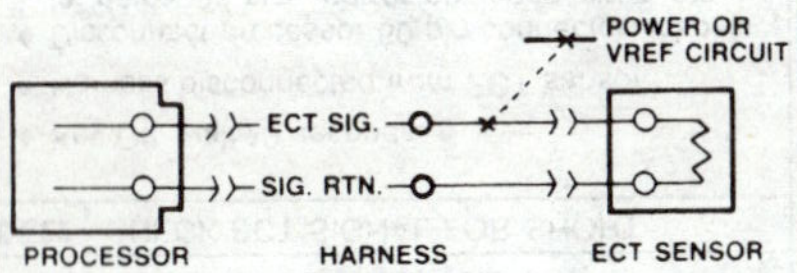

TEST STEP	RESULT ▶	ACTION TO TAKE
DE92 CHECK EEC-IV HARNESS		
• Observe VOM or STAR LED for a fault indication while performing the following: • Referring to the illustration in Step DE91, grasp the harness closest to the sensor connector. Wiggle, shake or bend a small section of the EEC-IV system harness while working your way to the dash panel. Also wiggle, shake or bend the EEC-IV harness from the dash panel to the processor. • Is fault indicated?	Yes ▶	ISOLATE fault and make necessary repairs. REPEAT Quick Test.
	No ▶	GO to DE93.
DE93 CHECK PROCESSOR AND HARNESS CONNECTORS		
• Key Off, wait 10 seconds. • Disconnect processor 60 pin connector. • Inspect both connectors and connector terminals for obvious damage or faults.	No ▶	SERVICE as necessary. REPEAT Quick Test.
• Connectors and terminals are OK.	Yes ▶	Unable to duplicate fault at this time. Continuous code 51 testing complete.

Engine Coolant Temperature (ECT) — Pinpoint Test — DE

TEST STEP	RESULT ▶	ACTION TO TAKE
DE94 SERVICE CODE 61 CONTINUOUS TEST: CHECK ECT SENSOR		
• Using continuous monitor mode, observe VOM or STAR LED for indication of a fault while performing the following: • Lightly tap on ECT sensor (simulate road shock). • Wiggle ECT connector. Voltage greater than 5 volts or STAR LED Off. • Is fault indicated?	Yes ▶	DISCONNECT and INSPECT connectors. If connector and terminals are good, REPLACE ECT sensor. REPEAT Quick Test.
	No ▶	GO to DE95.

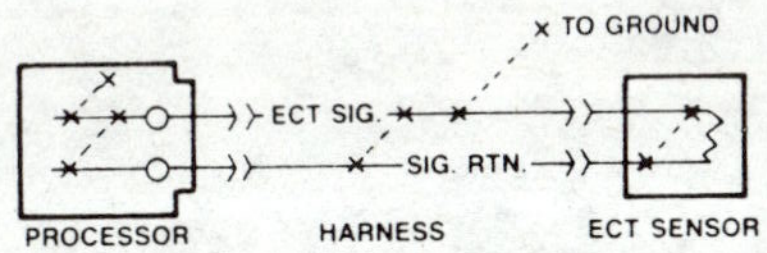

TEST STEP	RESULT ▶	ACTION TO TAKE
DE95 CHECK EEC-IV HARNESS		
• Observe VOM or STAR LED for a fault indication while performing the following: • Referring to the illustration in Step DE94, grasp the harness closest to the sensor connector. Wiggle, shake or bend a small section of the EEC-IV system harness while working your way to the dash panel. Also wiggle, shake or bend the EEC-IV harness from the dash panel to the processor. • Is fault indicated?	Yes ▶	ISOLATE fault and make necessary repairs. REPEAT Quick Test.
	No ▶	GO to DE96.
DE96 CHECK PROCESSOR AND HARNESS CONNECTORS		
• Key Off, wait 10 seconds. • Disconnect processor 60 pin connector. • Inspect both connectors and connector terminals for obvious damage or faults.	No ▶	SERVICE as necessary. REPEAT Quick Test.
• Connectors and terminals are OK.	Yes ▶	Unable to duplicate fault at this time. Continuous code 61 testing complete.

Manifold Absolute Pressure (MAP)/ Barometric Pressure (BP) Sensor — Pinpoint Test — DF

DF

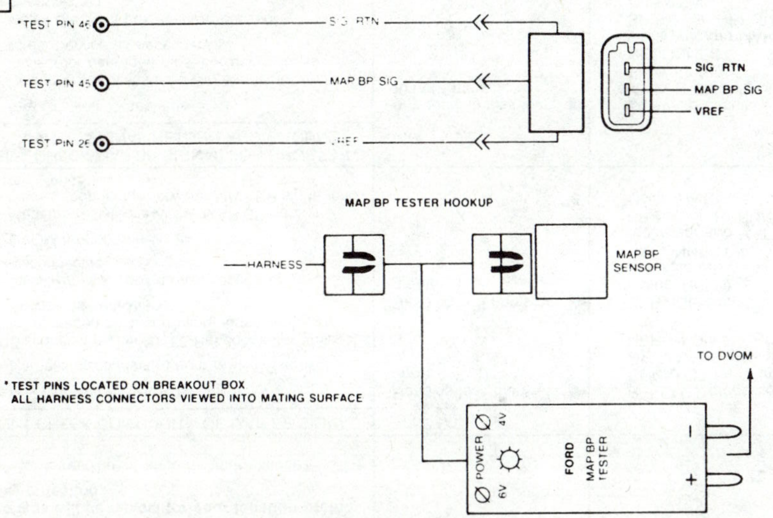

STOP-WARNING

You should enter this Pinpoint Test only when a Service Code 22 or 72 is received in Quick Test Step 3.0, 5.0 or 6.0 or when directed here from Diagnostics by Symptom in the Engine Supplement Section.

To prevent the replacement of good components, be aware that the following non-EEC areas may be at fault:

- Unusually high low atmospheric barometer reading (MAP BP).
- Kinked or obstructed vacuum lines (MAP).
- Basic engine (valves, vacuum leaks, timing, carburetor, EGR valve (MAP), etc.).
- Key On, Engine Off — MAP sensor must **not** see vacuum.
- Engine Running — MAP sensor must see actual manifold vacuum.

This Pinpoint Test is intended to diagnose only the following:

- MAP BP sensor.
- Harness circuits: VREF, MAP BP Signal, and Signal Return.
- Processor assembly.

Manifold Absolute Pressure (MAP)/ Barometric Pressure (BP) Sensor — Pinpoint Test — DF

TEST STEP	RESULT ▶	ACTION TO TAKE
FAULT CODE 22, ENGINE OFF		
DF1 CONNECTING MAP BP TESTER		
• Key Off. • Disconnect the MAP BP sensor from the wire harness. • Connect the MAP BP tester between the wire harness and the MAP BP sensor. • Insert tester banana plugs into DVOM. • Set DVOM to 20V scale. Refer to illustration DF	Tester properly hooked up ▶	GO to **DF2**.
DF2 POWER TO MAP BP SENSOR TEST		
• MAP BP tester connected. • Key On.	(ONLY) Green light, VREF is OK ▶	GO to **DF4**.
	"Less than 4V" light (RED) or no lights, VREF is too low OR "Greater than 6V" light (Red), VREF is too high ▶	GO to **DF3**.
DF3 VREF ISOLATION		
• MAP BP tester connected. • Key On. • MAP BP disconnected. • Repeat Test Step **DF2**.	(ONLY) Green light, VREF is OK ▶	REPLACE MAP BP sensor. REPEAT Quick Test.
	"Less than 4V" light (RED) or no lights, VREF is too low OR "Greater than 6V" light (Red), VREF is too high ▶	REMOVE MAP BP tester. GO to Pinpoint Test Step **C1**.

Manifold Absolute Pressure (MAP)/ Barometric Pressure (BP) Sensor — Pinpoint Test DF

TEST STEP	RESULT ▶	ACTION TO TAKE
DF4 MAP BP TESTER OUTPUT READING		
• MAP tester connected, refer to Note. • Key On. • Approximate Altitude (Ft.) / Voltage Output (+ − .03V): 0 — 1.59 1000 — 1.56 2000 — 1.53 3000 — 1.50 4000 — 1.47 5000 — 1.44 6000 — 1.41 7000 — 1.39 **NOTE: Measure several known good MAP sensors on available vehicles. The measured voltage will be typical for your location on the day of testing.** Is reading in range for your altitude?	Yes ▶ No (Sensor output is out-of-range) ▶	GO to DF5. GO to DF6.
DF5 CHECK CONTINUITY OF MAP BP SIGNAL		
• Key Off, wait 10 seconds. • Harness disconnected from MAP BP sensor. • Disconnect processor 60 Pin connector. Inspect for damaged pins, corrosion, loose wires, etc. Service as necessary. • Install Breakout box. Leave processor disconnected. • DVOM on 200 ohm scale. • Measure resistance at the MAP BP sensor harness connector between MAP BP signal and test Pin 45.	Reading less than 5 ohms ▶ Reading is 5 ohms or greater ▶	REPLACE processor. CONNECT harness and MAP/BP sensor. RERUN Quick Test. SERVICE circuit opens. REMOVE Breakout box. RECONNECT processor and MAP BP sensor. RERUN Quick Test.
DF6 CHECK MAP BP SIGNAL FOR SHORTS TO VREF, SIGNAL RETURN AND GROUND		
• Key Off, wait 10 seconds. • Disconnect processor 60 Pin connector and inspect for damaged pins, corrosion, loose wires. Service as necessary. • Install Breakout box, leave processor disconnected. • Harness disconnected from MAP BP sensor. • DVOM on 200,000 ohm scale. • Measure resistance between test Pin 45 and test Pins 26, 46, 40 and 60.	Any reading less than 10,000 ohms ▶ All readings are 10,000 ohms or greater ▶	SERVICE circuit shorts. REMOVE Breakout box. RECONNECT processor and MAP/BP Sensor. RERUN Quick Test. REPLACE MAP/BP sensor. REMOVE Breakout box. RECONNECT electrical connections. RERUN Quick Test.

Manifold Absolute Pressure (MAP)/ Barometric Pressure (BP) Sensor — Pinpoint Test DF

TEST STEP	RESULT ▶	ACTION TO TAKE
CODE 22 ENGINE RUNNING		
DF7 CHECK MAP SENSOR		
• Key Off, wait 10 seconds. • Disconnect vacuum line from MAP sensor. • Install vacuum pump to MAP sensor. • Apply 18 in. Hg vacuum to MAP sensor. • Does MAP sensor hold vacuum?	Yes ▶ No ▶	RELEASE vacuum. GO to DF8. REPLACE MAP sensor. CONNECT vacuum line to MAP sensor. RERUN Quick Test.
DF8 ATTEMPT TO ELIMINATE CODE 22 (ENGINE RUNNING)		
• Key Off, wait 10 seconds. • Plug MAP vacuum supply hose. • Start engine and maintain 1500 ± 100 engine rpm. • Slowly apply 15 inches Hg vacuum to MAP sensor. • While maintaining rpm, perform Engine Running Quick Test. • Is code 22 still present? **NOTE: Disregard any other codes at this time.**	Yes ▶ No ▶	REPLACE MAP sensor. CONNECT vacuum line to MAP sensor. RERUN Quick Test. INSPECT vacuum supply hose to MAP sensor. SERVICE as necessary. If OK, SERVICE other engine running codes. If none, GO to Diagnostic Routines, Section 2 for a low vacuum problem.
FAULT CODE 72		
DF10 CHECK THAT VACUUM TO MAP SENSOR DECREASES DURING DYNAMIC RESPONSE		
• Key Off, wait 10 seconds. • Tee a vacuum gauge in the intake manifold vacuum line at the MAP sensor. • Perform Engine Running Quick Test while observing vacuum. • Record engine service codes. • Did vacuum decrease by more than 30 kPa (10 in. Hg) during dynamic response test? • Is code 72 present?	Vacuum decrease is 30 kPa (10 in. Hg) or greater and code 72 is not present ▶ Vacuum decrease is 30 kPa (10 in. Hg) or greater and code 72 is present ▶ Vacuum decrease is less than 30 kPa (10 in. Hg) ▶	DISCONNECT vacuum equipment and SERVICE other codes as necessary. REPLACE MAP sensor and RERUN Quick Test. GO to DF11.

Manifold Absolute Pressure (MAP)/ Barometric Pressure (BP) Sensor — Pinpoint Test — DF

TEST STEP	RESULT ▶	ACTION TO TAKE
DF11 CHECK VACUUM LINES		
• Check vacuum line for proper routing. Refer to VECI decal. Check MAP sensor vacuum line for kinks or blockage. • Vacuum lines are OK.	Yes ▶	EEC-IV system OK.
	No ▶	SERVICE as necessary and REPEAT DF10.
DF20 CONNECTING MAP BP TESTER		
• Key Off. • Disconnect the MAP BP sensor from the wire harness. • Connect the MAP BP Tester between the wire harness and the MAP BP sensor. • Plug tester banana plugs into DVOM. • Set DVOM to 20V scale. • Refer to illustration DF.	Tester properly hooked up ▶	GO to DF21.
DF21 MAP BP TESTER OUTPUT READING		
• MAP Tester connected, refer to Note. • Key On. • Approximate Altitude (Ft.) / Voltage Output (+ – .03V): 0 — 1.59 1000 — 1.56 2000 — 1.53 3000 — 1.50 4000 — 1.47 5000 — 1.44 6000 — 1.41 7000 — 1.39 **NOTE: Measure several known good MAP sensors on available vehicles. The measured voltage will be typical for your location on the day of testing.** Is reading in range for your altitude?	Yes ▶	GO to Diagnostic by Symptom.
	No (Sensor output is out-of-range) ▶	REPLACE MAP BP sensor.

Manifold Absolute Pressure (MAP)/ Barometric Pressure (BP) Sensor — Pinpoint Test — DF

TEST STEP	RESULT ▶	ACTION TO TAKE
DF90 SERVICE CODE 22: CONTINUOUS TEST: EXERCISE MAP SENSOR		
• Using continuous monitor mode, observe VOM or STAR LED for indication of a fault while performing the following: • Connect a vacuum pump to the MAP sensor. • Slowly apply 84 kPa (25 in. Hg.) vacuum to the sensor. • Slowly bleed vacuum off the MAP sensor. • Lightly tap on MAP sensor (simulate road shock). • Wiggle MAP connector. • Is fault indicated? 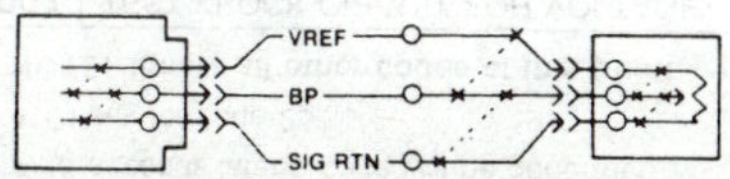	Yes ▶	DISCONNECT and INSPECT connectors. If connector and terminals are good, REPLACE sensor. REPEAT Quick Test.
	No ▶	GO to DF91.
DF91 CHECK EEC-IV HARNESS		
• Observe VOM or STAR LED for a fault indication while performing the following: • Referring to the illustration in Step DF90, grasp the harness closest to the sensor connector. Wiggle, shake or bend a small section of the EEC-IV system harness while working your way to the dash panel. Also wiggle, shake or bend the EEC-IV harness from the dash panel to the processor. • Is a fault indicated?	Yes ▶	ISOLATE fault and make necessary repairs. REPEAT Quick Test.
	No ▶	GO to DF92.
DF92 CHECK PROCESSOR AND HARNESS CONNECTORS		
• Key Off, wait 10 seconds. • Disconnect processor 60 Pin connector. • Inspect both connectors and connector terminals for obvious damage or faults. • Are connectors and terminals OK?	No ▶	SERVICE as necessary. REPEAT Quick Test.
	Yes ▶	Unable to duplicate fault at this time. Continuous code 22 testing complete.

Knock Sensor — Pinpoint Test — DG

DG

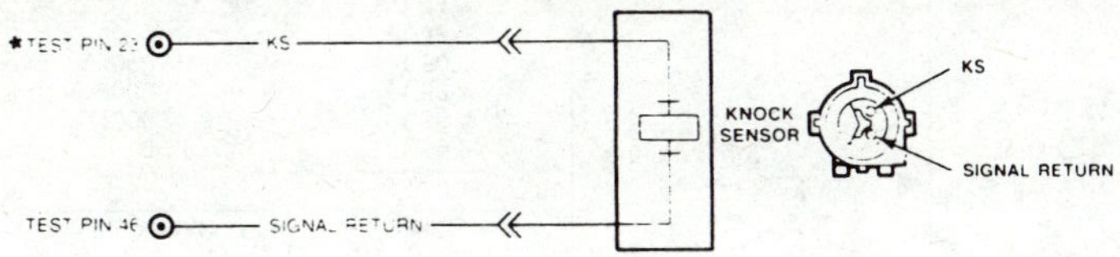

★TEST PINS LOCATED ON BREAKOUT BOX.
ALL HARNESS CONNECTORS VIEWED INTO MATING SURFACE.

STOP-WARNING

You should enter this Pinpoint Test only when a service code 25 is received in Quick Test Step 5.0 or you are directed here from Diagnostics by Symptom in the Engine Supplement Section.

To prevent the replacement of good components, be aware that the following non-EEC areas may be at fault:

- Fuel (quality).
- Basic engine.
- Spark timing.

This Pinpoint Test is intended to diagnose only the following:

- Knock sensor.
- Harness circuits: KS and Signal Return.
- Processor assembly.

Knock Sensor — Pinpoint Test — DG

TEST STEP	RESULT ▶	ACTION TO TAKE
DG1 SERVICE CODE 25: GENERATE KNOCK MANUALLY		
NOTE: With knock conditions sensitive to fuel, altitude and weather, perform Step DG1 before servicing any components. • Vehicle prepared to run (Engine Running Self-Test). • Equipment needed: 4 oz. hammer. • Prepare to rap tap on exhaust manifold, directly above the knock sensor, when the Dynamic Response Signal is given. **NOTE: There is no need to actually depress throttle at this point.** • Perform Engine Running Self-Test. • Rap moderately on exhaust manifold when meter indicates Dynamic Response Test is ready. • 15 seconds later a code will be generated. • Check for code 25. **NOTE: Ignore all other codes at this point.**	No ▶ Yes ▶ **NOTE: Service code 25 may be received whenever the engine is not tapped**	Knock system OK. REPEAT Engine Running Self-Test and SERVICE any other codes from that test. GO to DG2.
DG2 TEST KNOCK CIRCUIT FOR VOLTAGE		
• Key Off, wait 10 seconds. • Disconnect knock sensor connector and inspect. • Set DVOM on 20V scale. • Key On, Engine Off. • Measure voltage at the harness connector between KS and signal return.	Voltage is between 1 and 4V ▶ Voltage is less than 1V ▶ Voltage is greater than 4V ▶	GO to DG6. GO to DG3. GO to DG5.
DG3 CHECK CONTINUITY OF KS AND SIGNAL RETURN CIRCUITS		
• Key Off, wait 10 seconds. • Disconnect processor 60 Pin connector and inspect for damaged pins, corrosion, loose wires. Service as necessary. • Connect Breakout box to harness. Leave processor disconnected. • Knock sensor disconnected. • DVOM on 200 ohm scale. • Measure resistance between Signal Return at the harness and test Pin 46 and KS at the harness to test Pin 23.	Both resistances less than 5 ohms ▶ Either resistance 5 ohms or greater ▶	GO to DG4. SERVICE open circuit. REPEAT Quick Test.

Knock Sensor — Pinpoint Test DG

TEST STEP	RESULT ▶	ACTION TO TAKE
DG4 CHECK KS CIRCUIT FOR SHORT TO GROUND		
• Key Off, wait 10 seconds. • Breakout box installed. • Processor disconnected. • Knock sensor disconnected. • DVOM on 200,000 ohm scale. • Measure resistance between KS at the harness and test Pins 40, 46 and 60.	All resistances 10,000 ohms or greater ▶ Any resistance less than 10,000 ohms ▶	GO to **DG6**. SERVICE harness short. REPEAT Quick Test.
DG5 CHECK KS CIRCUIT FOR SHORT TO VOLTAGE		
• Key Off, wait 10 seconds. • Disconnect processor 60 Pin connector and inspect for damaged pins, corrosion, loose wires. Service as necessary. • Connect Breakout box to harness. Leave processor disconnected. • Knock sensor disconnected. • Key On, Engine Off. • DVOM on 20V scale. • Measure voltage between test Pin 23 and test Pin 40.	0.5V or greater ▶ Less than 0.5V ▶	SERVICE harness short to power. REPEAT Quick Test. GO to **DG6**.
DG6 TEST PROCESSOR WITH SUBSTITUTE KNOCK SENSOR		
• Key Off, wait 10 seconds. • Remove Breakout box and reconnect processor. • Equipment Required: — Equivalent knock sensor (same part number). — 4 oz. hammer. • Plug substitute sensor in harness (do not install). • Perform Engine Running Self-Test. • Lightly tap knock sensor when Dynamic Test Ready Signal is given. • 15 seconds later, a code will be generated. • Check for code 25. **NOTE: Ignore all other codes at this time.**	No ▶ Yes ▶	INSTALL new knock sensor. REPEAT Quick Test. REPLACE processor and REMOVE substitute sensor. REPEAT Quick Test with original sensor.

Throttle Position Sensor (TPS) — Pinpoint Test DH

DH

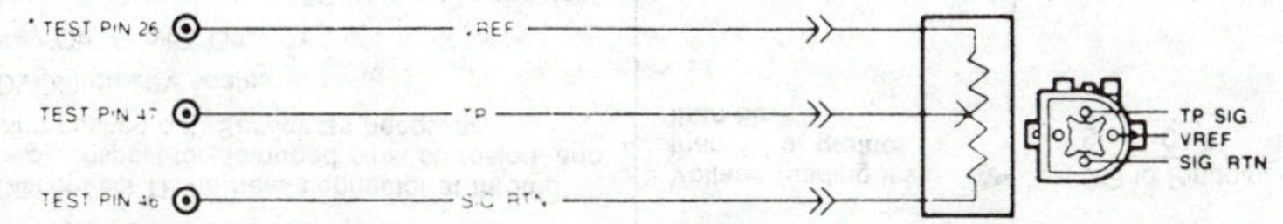

*TEST PINS LOCATED ON BREAKOUT BOX
ALL HARNESS CONNECTORS VIEWED INTO MATING SURFACE

STOP-WARNING

You should enter this Pinpoint Test only when a Service Code 23, 53, 63 or 73 is received in Quick Test Step 3.0, 5.0 or 6.0.

To prevent the replacement of good components, be aware that the following non-EEC areas may be at fault:

- Idle speeds throttle stop adjustment.
- Binding throttle shaft linkage or speed control linkage
- Choke high cam system, if equipped.

This Pinpoint Test is intended to diagnose only the following:

- TP sensor.
- Sensor harness circuits: VREF, TP Signal, and Signal Return.
- Processor assembly.

Throttle Position Sensor (TPS) — Pinpoint Test — DH

TEST STEP	RESULT ▶	ACTION TO TAKE
DH1 SERVICE CODE 23: THE FOLLOWING CHECK MUST BE MADE BEFORE SERVICING THIS CODE		
• Check for code 68: Key On, Engine Off or codes 58, 31 or 41 Engine Running.	Code(s) present ▶	SERVICE as necessary.
	No codes present ▶	GO to DH2.
DH2 CHECK FOR STUCK THROTTLE PLATE		
• Visually inspect carburetor/throttle body and throttle linkage for binding or sticking. • Verify the throttle linkage is at mechanical closed throttle. Check for: binding throttle linkage, speed control linkage, vacuum line electrical harness interference, etc.	Throttle not stuck ▶	GO to DH3.
	Throttle stuck ▶	CORRECT faults. REPEAT Quick Test.
DH3 GENERATE CODE 63		
• Refer to illustration DH. • Key Off, wait 10 seconds. • Disconnect TP sensor harness connector at the throttle body. Inspect for damaged pins, corrosion, loose wires, etc. Service as necessary. • Perform Key On, Engine Off self-test and record codes. **NOTE: Ignore all other codes at this time.**	Code 63 present ▶	GO to DH4.
	Code 63 not present ▶	GO to DH5.
DH4 CHECK VOLTAGE VREF TO SIGNAL RETURN		
• Refer to illustration DH. • Key Off, wait 10 seconds. • Disconnect TP harness connector at throttle body. Inspect for damaged pins, corrosion, and pins pushed out. Service as necessary. • DVOM on 20V scale. • Key On, Engine Off. • Measure voltage at the TP harness connector between VREF and Signal Return.	Voltage reading between 4V and 6V ▶	REPLACE TP sensor. REFER to Section before replacement for adjustment procedures for EFI applications. REPEAT Quick Test.
	Voltage reading less than 4V or greater than 6V ▶	GO to Pinpoint Test Step C1.

Throttle Position Sensor (TPS) — Pinpoint Test — DH

TEST STEP	RESULT ▶	ACTION TO TAKE
DH5 CHECK TP SIGNAL FOR SHORT TO POWER		
• Key Off, wait 10 seconds, TP harness disconnected. • DVOM on 200,000 ohm scale. • Disconnect processor 60 Pin connector and inspect for damaged pins, corrosion, loose wires. Service as necessary. • Install Breakout box, leave processor disconnected. • Measure resistance between test Pin 47 and test Pins 26 and 57.	Either resistance is less than 10,000 ohms ▶	SERVICE harness short. REPEAT Quick Test.
	Both resistances are 10,000 ohms or greater ▶	REPLACE processor. REPEAT Quick Test.
DH10 SERVICE CODE 63: GENERATE CODE 53		
• Key Off, wait 10 seconds, TP harness disconnected. • Jumper VREF to TP signal at TP harness connector. • Perform Key On, Engine Off self-test. **NOTE: If no codes are generated, immediately remove jumper and go directly to DH13.** • Check for Code 53 (Code 23) **NOTE: Ignore all other codes at this time.**	Code 53 present (Code 23 present) ▶	REPLACE TP sensor. REFER to Section before replacement for adjustment procedures for EFI applications and REPEAT Quick Test.
	Code 53 not present (Code 23 not present) ▶	GO to DH11.
DH11 SERVICE CODE 63: CHECK VOLTAGE VREF TO SIGNAL RETURN		
• Refer to illustration DH. • Key Off, wait 10 seconds. • Disconnect TP harness connector at throttle body. Inspect for damaged pins, corrosion, and pins pushed out. Service as necessary. • DVOM on 20V scale. • Key On, Engine Off. • Measure voltage at the TP harness connector between VREF and Signal Return.	Voltage reading between 4V and 6V ▶	GO to DH12.
	Voltage reading less than 4V or greater than 6V ▶	GO to Pinpoint Test Step C1.

Throttle Position Sensor (TPS) — Pinpoint Test — DH

TEST STEP	RESULT	▶ ACTION TO TAKE
DH12 CHECK CONTINUITY OF TP CIRCUIT		
• Key Off, wait 10 seconds, TP harness disconnected. • DVOM on 200 ohm scale. • Disconnect processor 60 Pin connector and inspect for damaged pins, corrosion, loose wires. Service as necessary. • Connect Breakout box. Processor connected to Breakout box. • Measure resistance between TP Signal at the harness connector and test Pin 47.	DVOM reading 5 ohms or greater ▶	SERVICE faulty circuit. CONNECT throttle position sensor. REMOVE Breakout box and REPEAT Quick Test.
	DVOM reading less than 5 ohms ▶	GO to DH13.
DH13 CHECK RESISTANCE OF TP CIRCUIT TO GROUND SIGNAL RETURN		
• Key Off, wait 10 seconds, TP harness disconnected. • Disconnect processor 60 Pin connector and inspect for damaged pins, corrosion, loose wires. Service as necessary. • DVOM on 200,000 ohm scale. • Measure resistance between TP signal at TP harness connector to ground and test Pin 46.	DVOM reading is less than 10,000 ohms ▶	SERVICE circuit short and RETEST.
	DVOM reading is 10,000 ohms or greater ▶	REPLACE processor. CONNECT throttle position sensor. REMOVE Breakout box and REPEAT Quick Test.
DH20 SERVICE CODE 73: TP SENSOR MOVES IN ENGINE RESPONSE TEST		
NOTE: Code 73 indicates the TP Sensor did not exceed 25 percent of its rotation in the Engine Response Check. • Key Off. • Install Breakout box. • DVOM on 20V range. • Connect DVOM to test Pins 47 and 46. • Perform Engine Running Quick Test, Step 5.0. • Verify DVOM reading exceeds 3.5V during brief WOT at Engine Response Check.	DVOM exceeds 3.5V during Engine Response Check ▶	REPLACE processor. REPEAT Quick Test.
	DVOM does not exceed 3.5V during Engine Response Check ▶	VERIFY TP Sensor is properly attached to throttle body. If OK, REPLACE TP Sensor. Refer to Section before replacement for adjustment procedures for EFI applications. REPEAT Quick Test.

Throttle Position Sensor (TPS) — Pinpoint Test — DH

TEST STEP	RESULT	▶ ACTION TO TAKE
DH90 CONTINUOUS TEST SERVICE CODE 53: EXERCISE TP SENSOR		
• Using continuous monitor mode, observe VOM or STAR LED for indication of a fault while performing the following: • Move throttle slowly to WOT position. • Release throttle slowly to closed position and lightly tap on TP sensor (simulate road shock). • Wiggle TP harness connector. • Is a fault indicated?	Yes ▶	GO to DH91.
	No ▶	GO to DH92.

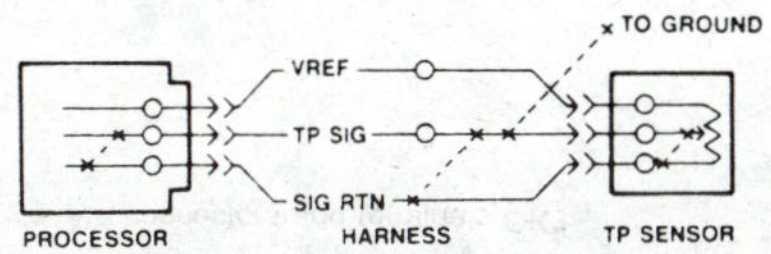

TEST STEP	RESULT	▶ ACTION TO TAKE
DH91 MEASURE THROTTLE POSITION SIGNAL VOLTAGE WHILE EXERCISING TP SENSOR		
• Key Off, wait 10 seconds. • Disconnect processor 60 Pin connector and inspect for damaged pins, corrosion, loose wires. Service as necessary. • Install Breakout box and reconnect processor. • VOM or STAR LED still connected to STO as in previous step. • Connect a DVOM from test Pin 47 to test Pin 46. • DVOM on 20V scale. • Key On, engine Off. • While observing DVOM, repeat step DH90. • Does the fault occur below 4.25V?	Yes ▶	DISCONNECT and INSPECT connectors. If connector and terminals are good, REPLACE TP sensor, REPEAT Quick Test.
	No ▶	Throttle position sensor overtravel may have caused the continuous code 53. Sensor service is not required. To verify harness integrity, GO to DH92.

Throttle Position Sensor (TPS) — Pinpoint Test — DH

TEST STEP	RESULT ▶	ACTION TO TAKE
DH92 CHECK EEC-IV HARNESS		
• Observe VOM or STAR LED for a fault indication while performing the following: • Referring to the illustration in Step DH90, grasp the harness close to the sensor connector. Wiggle, shake or bend a small section of the EEC-IV system harness while working your way to the dash panel. Also wiggle, shake or bend the EEC-IV harness from the dash panel to the processor. • Is a fault indicated?	Yes ▶	ISOLATE fault and make necessary repairs. REFER to appropriate figure. REPEAT Quick Test.
	No ▶	GO to **DH93**.
DH93 CHECK PROCESSOR AND HARNESS CONNECTORS		
• Key Off, wait 10 seconds. • Disconnect processor 60 Pin connector. • Inspect both connectors and connector terminals for obvious damage or faults. • Are connectors and terminals OK?	No ▶	SERVICE as necessary. REPEAT Quick Test.
	Yes ▶	Unable to duplicate fault at this time. Continuous code 53 testing complete.
DH94 CONTINUOUS TEST SERVICE CODE 63: EXERCISE TP SENSOR		
• Using continuous monitor mode, observe VOM or STAR LED for indication of a fault while performing the following: • Move throttle slowly to WOT position. • Release throttle slowly to closed condition. • Lightly tap on TP sensor (simulate road shock). • Wiggle TP harness connector. • Is a fault indicated?	Yes ▶	DISCONNECT and INSPECT connectors. If connector and terminals are good, REPLACE TP sensor, REPEAT Quick Test.
	No ▶	GO to **DH95**.

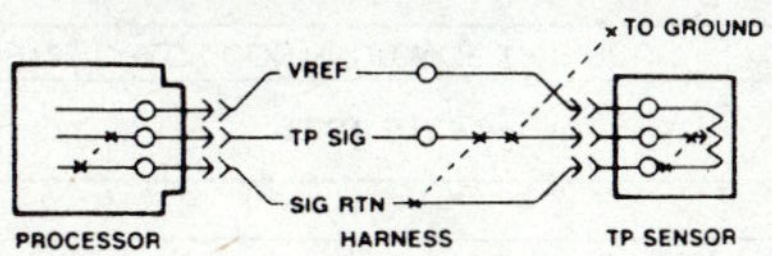

Throttle Position Sensor (TPS) — Pinpoint Test — DH

TEST STEP	RESULT ▶	ACTION TO TAKE
DH95 CHECK EEC-IV HARNESS		
• Observe VOM or STAR LED for a fault indication while performing the following: • Referring to the illustration in Step DH94 grasp the harness close to the sensor connector. Wiggle, shake or bend a small section of the EEC-IV system harness while working your way to the dash panel. Also wiggle, shake or bend the EEC-IV harness from the dash panel to the processor. • Is a fault indicated?	Yes ▶	ISOLATE fault and make necessary repairs. REFER to appropriate figure. REPEAT Quick Test.
	No ▶	Go to **DH96**.
DH96 CHECK PROCESSOR AND HARNESS CONNECTORS		
• Key Off, wait 10 seconds. • Disconnect processor 60 Pin connector. • Inspect both connectors and connector terminals for obvious damage or faults. • Are connectors and terminals OK?	No ▶	SERVICE as necessary. REPEAT Quick Test.
	Yes ▶	Unable to duplicate fault at this time. Continuous code 63 testing complete.

Vane Air Flow Sensor (VAF) — Pinpoint Test — DK

DK

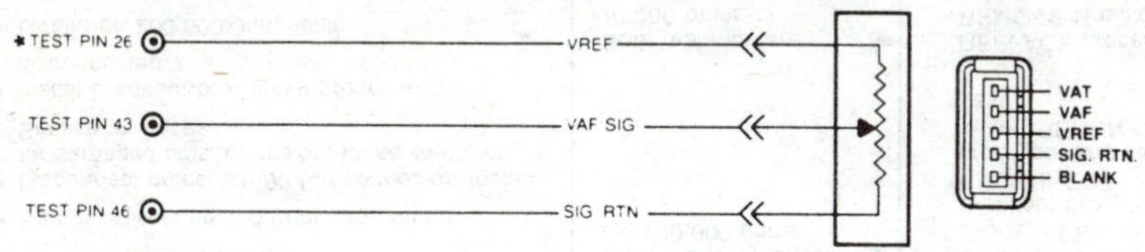

* TEST PIN LOCATED ON BREAKOUT BOX
ALL HARNESS CONNECTORS VIEWED INTO MATING SURFACE.

STOP-WARNING

You should enter this Pinpoint Test only when a service code 26, 56, 66 or 76 is received in Quick Test Step 3.0, 5.0 or 6.0.

To prevent the replacement of good components, be aware that the following non-EEC areas may be at fault:

- Check for unmetered air (air leaks) between VAF meter and throttle body.
- Vacuum leaks.
- Engine sealing (PCV sealing, CANP, valve cover seal dipstick seated).

This Pinpoint Test is intended to diagnose only the following:

- VAF meter.
- Processor.
- Harness circuits: VREF, VAF Sig and Signal Return.

Vane Air Flow Sensor (VAF) — Pinpoint Test — DK

TEST STEP	RESULT ▶	ACTION TO TAKE
FAULT CODE 26		
DK1 CHECK FOR CONTAMINATION		
NOTE: Code 26 (Key On, Engine Off, or Engine Running) indicates the vane air flow input to the processor is out of closed (Engine Not Running) or at idle limits (Engine Off 0.15-0.50V: Engine at idle 1.50-2.70V). There have been no opens or shorts in the VAF circuit or a code 56 (signal always high) or 66 (signal always low) would have been generated.		
• Key Off, wait 10 seconds. • Remove air cleaner element and check for contamination (oil residue, foreign material, etc.) that may impede VAF sensor vane movement. • Is service code 26 present in Key On, Engine Off portion of Quick Test?	Yes ▶	REINSTALL air cleaner element. REPLACE vane meter. RERUN Quick Test.
	No ▶	GO to DK2.
DK2 VAF SENSOR CHECK		
• Key Off, air cleaner element reinstalled. • Check for unmetered air leaks between vane meter and throttle body. • Disconnect processor 60 Pin connector. Inspect for damaged pins, corrosion, loose wires, etc. Service as necessary. • Install Breakout box. Connect processor to Breakout box harness. • DVOM on 20V scale. • Key On, engine Off. • Place new unsharpened pencil as shown below. • Measure voltage between test Pins 43 and 46. • Does DVOM read between 2.8V and 3.7V?	Yes ▶	Vane meter is capable of outputting an acceptable signal. The VAF code 26 has been caused by incorrect engine speed or an unmetered air leak (vacuum leak). SERVICE as necessary. REMOVE Breakout box. RERUN Quick Test.
	No ▶	REMOVE Breakout box. REPLACE vane meter. RERUN Quick Test.

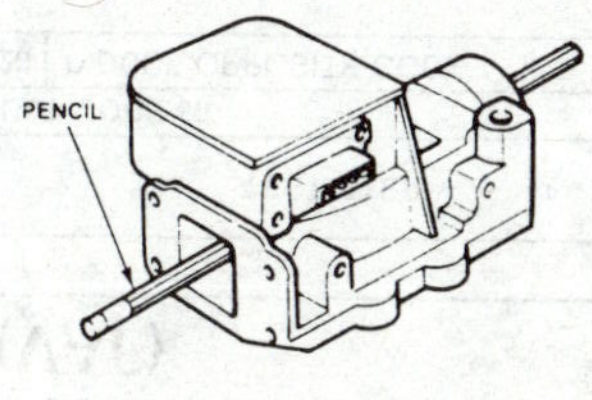

VANE METER

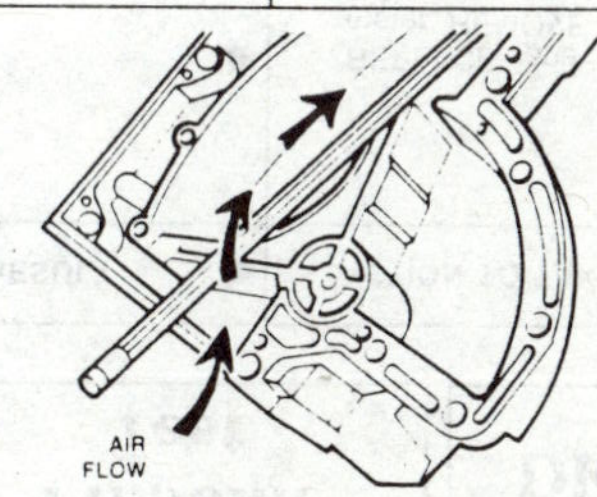

VAF SENSOR AIR VANE

Vane Air Flow Sensor (VAF) — Pinpoint Test — DK

TEST STEP	RESULT ▶	ACTION TO TAKE
FAULT CODE 56		
DK10 INDUCE OPPOSITE CODE		
• Key Off, wait 10 seconds. • Harness disconnected from vane meter. • Run Key On, Engine Off Quick Test. • Is code 66 present? **NOTE: Disconnecting vane meter disconnects both VAF and VAT sensors. A code 54 should also be present. Disregard all codes except VAF codes at this time.**	Yes ▶ No ▶	GO to DK11. GO to DK12.
DK11 CHECK VAF TO SIGNAL RETURN VOLTAGE		
• Key Off, wait 10 seconds. • Disconnect harness from vane meter. Inspect for damaged pins, corrosion, loose wires, etc. Service as necessary. • Key On, engine Off. • DVOM on 20V scale. • Measure voltage at the vane meter harness connector between VREF and Signal Return.	Reading is between 4.0V and 6.0V ▶ Reading is less than 4.0V or greater than 6.0V ▶	REPLACE vane meter. RECONNECT harness. RERUN Quick Test. GO to Pinpoint Test Step C1.
DK12 CHECK VAF SIGNAL FOR SHORT		
• Key Off, wait 10 seconds. • Harness disconnected from vane meter. • Disconnect processor 60 Pin connector. Inspect for damaged pins, corrosion, loose wires, etc. Service as necessary. • Install Breakout box, leave processor disconnected. • DVOM on 200,000 ohm scale. • Measure resistance between test Pin 43 and test Pins 26 and 57.	Either reading is less than 10,000 ohms ▶ Both readings are 10,000 ohms or greater ▶	SERVICE circuit shorts. REMOVE Breakout box. RECONNECT processor and vane meter. RERUN Quick Test. REPLACE processor. REMOVE Breakout box. RECONNECT processor and vane meter. RERUN Quick Test.

Vane Air Flow Sensor (VAF) — Pinpoint Test — DK

TEST STEP	RESULT ▶	ACTION TO TAKE
FAULT CODE 66		
DK20 INDUCE OPPOSITE CODE		
• Key Off, wait 10 seconds. • Disconnect harness from vane meter. • Install jumper wire in vane meter harness connector between VREF and VAF signal. • Perform Key On, Engine Off Quick Test. **NOTE: If no codes are generated, immediately remove jumper and go directly to DK23.** • Is code 56 present? **NOTE: Disconnecting vane meter disconnects both VAF and VAT sensors. Code 54 should also be present. Disregard all codes except VAF codes at this time.**	Yes ▶ No ▶	REPLACE vane meter. REMOVE jumper wire. RECONNECT vane meter. RERUN Quick Test. REMOVE jumper wire and GO to DK21.
DK21 CHECK VREF TO SIGNAL RETURN VOLTAGE		
• Key off, wait 10 seconds. • Harness disconnected from vane meter. • Key On, Engine Off. • DVOM on 20V scale. • Measure voltage at the vane meter harness connector between VREF and Signal Return.	Reading is between 4.0 and 6.0V ▶ Reading is less than 4.0V or greater than 6.0V ▶	GO to DK22. GO to Pinpoint Test Step C1.
DK22 CHECK CONTINUITY OF VAF SIGNAL		
• Key Off, harness disconnected from vane meter. • Disconnect processor 60 Pin connector. Inspect for damaged pins, corrosion, loose wires, etc. Service as necessary. • Install Breakout box; leave processor disconnected. • DVOM on 200 ohm scale. • Measure resistance between VAF signal, at the vane meter harness connector, and test Pin 43.	Reading is less than 5 ohms ▶ Reading is 5 ohms or greater ▶	GO to DK23. SERVICE circuit open. REMOVE Breakout box. RECONNECT processor and vane meter. RERUN Quick Test.

Vane Air Flow Sensor (VAF) — Pinpoint Test — DK

TEST STEP	RESULT ▶	ACTION TO TAKE
DK23 CHECK VAF SIGNAL FOR SHORT		
• Key Off, wait 10 seconds. • Processor disconnected. • Harness disconnected from vane meter. • DVOM on 200,000 ohm scale. • Measure resistance at the vane meter harness connector, from VAF Signal to Signal Return and negative battery terminal.	Reading is less than 10,000 ohms ▶	SERVICE circuit shorts. RECONNECT vane meter. RERUN Quick Test.
	Reading is 10,000 ohms or greater ▶	RECONNECT processor. REMOVE Breakout box. RECONNECT processor and vane meter. RERUN Quick Test.
FAULT CODE 76		
DK30 CHECK FOR VOLTAGE INCREASE IN VAF SIGNAL TO SIGNAL RETURN		
NOTE: A sharp snap of the throttle may not be sufficient to pass this test. Be sure to move throttle to WOT and return.		
• Key Off, wait 10 seconds. • Disconnect processor 60 Pin connector. Inspect for damaged pins, corrosion, loose wires, etc. Service as necessary. • Install Breakout box. Connect processor to Breakout box harness. • DVOM on 20V scale. • Connect DVOM to test Pins 43 and 46. • Perform Engine Running Quick Test while monitoring DVOM. • After dynamic response prompt code 1(0) operator does a brief WOT, DVOM should increase more than 2.0V from reading before WOT. • Observe service codes at end of test.	DVOM increased more than 2.0V and code 76 is still present ▶	REPLACE processor. REMOVE Breakout box. RERUN Quick Test.
	DVOM did not increase more than 2.0V ▶	CHECK air cleaner duct for obstruction. If OK, REPLACE vane meter.
	DVOM increased more than 2.0V and code 76 is not present ▶	Vane meter is OK. SERVICE other codes as necessary.

Vane Air Flow — Pinpoint Test — DK

TEST STEP	RESULT ▶	ACTION TO TAKE
DK90 SERVICE CODE 56 CONTINUOUS TEST: CHECK VAF SENSOR		
• Using continuous monitor mode, observe VOM or STAR LED for indication of a fault while performing the following: • Lightly tap on VAF sensor (simulate road shock). • Wiggle VAF connector. • Is a fault indicated?	Yes ▶	DISCONNECT and INSPECT connectors. If connector and terminals are good, REPLACE VAF sensor. REPEAT Quick Test.
	No ▶	GO to **DK91**.
DK91 CHECK EEC-IV HARNESS		
• Observe VOM or STAR LED for a fault indication while performing the following: • Referring to the illustration in Step DK90, grasp the harness close to the sensor connector. Wiggle, shake or bend a small section of the EEC-IV system harness while working your way to the dash panel. Also wiggle, shake or bend the EEC-IV harness from the dash panel to the processor. • Is a fault indicated?	Yes ▶	ISOLATE fault and make necessary repairs. REFER to appropriate figure. REPEAT Quick Test.
	No ▶	GO to **DK92**.
DK92 CHECK PROCESSOR AND HARNESS CONNECTORS		
• Key Off, wait 10 seconds. • Disconnect processor 60 Pin connector. • Inspect both connectors and connector terminals for obvious damage or faults. • Are connectors and terminals OK?	No ▶	SERVICE as necessary. REPEAT Quick Test.
	Yes ▶	Unable to duplicate fault at this time. Continuous code 56 testing complete.

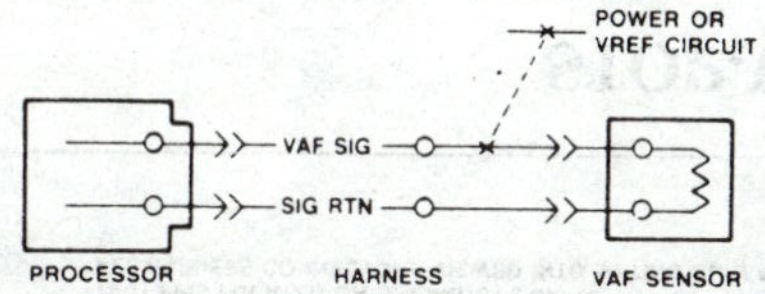

Vane Air Flow — Pinpoint Test — DK

TEST STEP	RESULT ▶	ACTION TO TAKE
DK93 SERVICE CODE 66 CONTINUOUS TEST: CHECK VAF SENSOR		
• Using continuous monitor mode, observe VOM or STAR LED for indication of a fault while performing the following: • Lightly tap on VAF sensor (simulate road shock). • Wiggle ACT connector. • Is a fault indicated?	Yes ▶	DISCONNECT and INSPECT connectors. If connector and terminals are good, REPLACE sensor. REPEAT Quick Test.
	No ▶	GO to DK94.

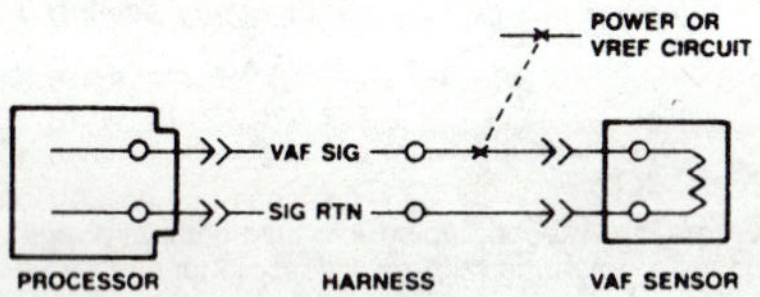

TEST STEP	RESULT ▶	ACTION TO TAKE
DK94 CHECK EEC-IV HARNESS		
• Observe VOM or STAR LED for a fault indication while performing the following: • Referring to the illustration in Step DK93 grasp the harness close to the sensor connector. Wiggle, shake or bend a small section of the EEC-IV system harness while working your way to the dash panel. Also wiggle, shake or bend the EEC-IV harness from the dash panel to the processor. • Is a fault indicated?	Yes ▶	ISOLATE fault and make necessary repairs. REFER to appropriate figure. REPEAT Quick Test.
	No ▶	Go to DK95.
DK95 CHECK PROCESSOR AND HARNESS CONNECTORS		
• Key Off, wait 10 seconds. • Disconnect processor 60 Pin connector. • Inspect both connectors and connector terminals for obvious damage or faults. • Are connectors and terminals OK?	No ▶	SERVICE as necessary. REPEAT Quick Test.
	Yes ▶	Unable to duplicate fault at this time. Continuous code 66 testing complete.

EGR Valve Position Sensor (EVP) EGR Valve Regulator (EVR) — Pinpoint Test — DM

DM

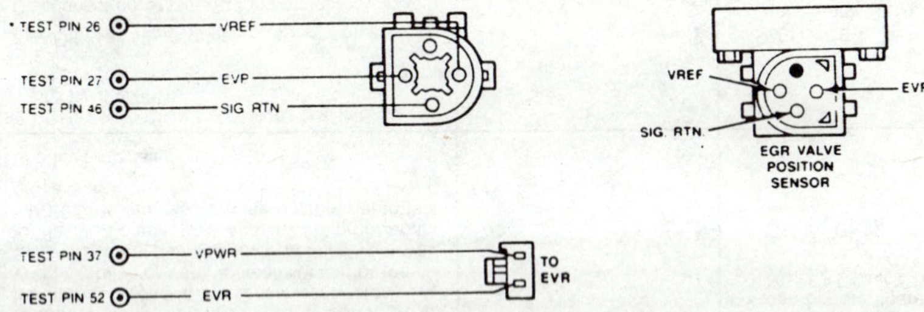

•TEST PINS LOCATED ON BREAKOUT BOX.
ALL HARNESS CONNECTORS VIEWED INTO MATING SURFACE.

STOP-WARNING

You should enter this Pinpoint Test only when a Service Code 31, 32, 33, 34, 35, 83 or 84 is received in Quick Test Step 3.0, 5.0 or 6.0.

To prevent the replacement of good components, be aware that the following non-EEC area may be at fault:

- Damaged EGR valve.

This Pinpoint Test is intended to diagnose only the following:

- EVP sensor.
- Harness circuits: VREF, EVP, Signal Return, EVR, VPWR.
- EVR (EGR valve regulator).
- EGR valve assembly.
- Processor assembly.

EGR Valve Position Sensor (EVP) EGR Valve Regulator (EVR) — Pinpoint Test — DM

TEST STEP	RESULT ▶	ACTION TO TAKE
FAULT CODE 31		
DM1 RUN ENGINE RUNNING QUICK TEST WITH EGR VACUUM SIGNAL LINE DISCONNECTED AT EGR VALVE		
• Key Off, wait 10 seconds. • Disconnect EGR vacuum line at EGR valve and cap EGR vacuum line. • Perform Engine Running Quick Test. • Check for code 31.	Code 31 present ▶	GO to DM2.
	No code 31 present, but codes 32 and 34 are present ▶	GO to DM11.
DM2 CHECK EVP RESISTANCE WHILE APPLYING VACUUM TO EGR VALVE		
• Key Off, wait 10 seconds. • Vacuum signal line disconnected and capped. • Disconnect harness from EVP sensor. • DVOM on 200,000 ohm scale. • Connect vacuum pump to EGR valve. • Measure resistance between EVP and VREF at EVP sensor while gradually increasing vacuum to 33 kPa (10 in. Hg). • Observe resistance as vacuum increases.	Reading gradually decreases from no greater than 5,500 ohms to no less than 100 ohms ▶	GO to DM3.
	Reading is less than 100 ohms or greater than 5,500 ohms ▶	REPLACE EVP sensor. RECONNECT signal line and harness. RERUN Quick Test.
	Reading does not decrease or unable to hold vacuum ▶	GO to DM16.
DM3 MEASURE VREF TO SIGNAL RETURN VOLTAGE		
• Key On, Engine Off. Vacuum signal line disconnected and capped, harness disconnected from EVP sensor. • DVOM on 20V scale. • Measure voltage at the EVP harness connector between VREF and Signal Return.	Reading is between 4V and 6V ▶	GO to DM4.
	Reading is less than 4V or greater than 6V ▶	GO to Pinpoint Test Step C1.

EGR Valve Position Sensor (EVP) EGR Valve Regulator (EVR) — Pinpoint Test — DM

TEST STEP	RESULT ▶	ACTION TO TAKE
DM4 MEASURE CONTINUITY OF EVP SIGNAL		
• Key Off, wait 10 seconds. • Harness disconnected from EVP sensor. • Disconnect harness from processor 60 Pin connector and inspect for damaged pins, corrosion, loose wires. Service as necessary. • Install Breakout box leaving processor disconnected. • DVOM on 200 ohm scale. • Measure resistance between test Pin 27 and EVP signal at the EVP harness connector.	Less than 5 ohms ▶	GO to DM5.
	5 ohms or greater ▶	SERVICE open circuit. REMOVE Breakout box. RECONNECT processor and EVP sensor. RERUN Quick Test.
DM5 CHECK EVP SIGNAL FOR SHORTS TO VREF AND SIGNAL RETURN		
• Key Off, harness disconnected from EVP sensor. Breakout box installed with processor not connected. • DVOM on 200,000 ohm scale. • Measure resistance between test Pin 27 and test Pins 26 and 46.	Any reading less than 10,000 ohms ▶	SERVICE short circuit. REMOVE Breakout box. RECONNECT processor and EVP sensor. RERUN Quick Test.
	All readings 10,000 ohms or greater ▶	GO to DM6.
DM6 SUBSTITUTE EVP SENSOR AND EGR VALVE		
• Key Off, wait 10 seconds. • Electrically connect a known good EVP sensor and EGR valve assembly. • Remove Breakout box. • Reconnect processor. • Perform Key On, Engine Off Quick Test. • Is code 31 present?	Yes ▶	REPLACE processor. CONNECT original EVP sensor and EGR valve assembly. REPEAT Quick Test.
	No ▶	GO to DM7.

EGR Valve Position Sensor (EVP) EGR Valve Regulator (EVR) — Pinpoint Test — DM

TEST STEP	RESULT ▶	ACTION TO TAKE
DM7 EVP SENSOR CHECK		
• Key Off, wait 10 seconds. • Install original EVP sensor on known good EGR valve. • EVP sensor connected. • Rerun Key On, Engine Off Quick Test. • Is code 31 present?	Yes ▶	REPLACE processor. CONNECT original EVP sensor and EGR valve assembly. REPEAT Quick Test.
	No ▶	REFER to EGR systems
FAULT CODES 32, 33 AND 34		
DM11 OUTPUT STATE CHECK (REFER TO APPENDIX)		
NOTE: Do not use STAR tester for this test Step, use VOM/DVOM. • Key Off, wait 10 seconds • DVOM on 20V scale. • Connect DVOM negative test lead to STO and positive test lead to battery positive. • Jumper STI to signal return. • Perform Key On, Engine Off Self-Test until the completion of the Continuous Test Codes. • DVOM will indicate zero volts. • Depress and release the throttle • Did DVOM reading change to a high voltage reading?	No ▶	DEPRESS throttle to WOT and RELEASE. If STO voltage does not go high, GO to Pinpoint Test Step Q40.
	Yes ▶	REMAIN in output state check. GO to DM12.
DM12 CHECK EVR SOLENOID FOR ELECTRICAL CYCLING		
• Key On, Engine Off • In output state check • DVOM on 20V scale • Connect DVOM to EVR circuits VPWR and EVR signal. • While observing DVOM, depress and release the throttle several times to cycle output On and Off.	Solenoid output cycles On and Off ▶	REMAIN in output state check. GO to DM13.
	Ouptut does not cycle On and Off ▶	REMOVE jumper. GO to DM17.

EGR Valve Position Sensor (EVP) EGR Valve Regulator (EVR) — Pinpoint Test — DM

TEST STEP	RESULT ▶	ACTION TO TAKE
DM13 CHECK EVR SOLENOID FOR VACUUM CYCLING		
• Key On, Engine Off. • In output state check. • Disconnect and cap vacuum line from bottom port of EVR solenoid and connect vacuum pump. • Connect a vacuum gauge in the common output (top) vacuum line to EGR valve. • While cycling outputs On and Off (by depressing and releasing throttle) observe vacuum gauge at the output. Maintain vacuum at source.	Vacuum output cycles On and Off in less than 2 seconds ▶	RECONNECT all vacuum lines. GO to DM14.
	Vacuum does not cycle On and Off in less than 2 seconds ▶	CHECK EVR filter for obstructions. REPLACE as necessary. If OK, REPLACE solenoid assembly. RECONNECT all vacuum lines. RERUN Quick Test.
DM14 CHECK VACUUM LINES		
• Key Off, wait 10 seconds. • Vacuum lines reconnected. • Check entire EEC vacuum line system per VECI schematic decal for leaks, kinks, cracks or obstructions.	Vacuum lines OK ▶	GO to DM15.
	Vacuum lines are not OK ▶	SERVICE as necessary. RERUN Quick Test.
DM15 CHECK EVP RESISTANCE WHILE APPLYING VACUUM TO EGR VALVE		
• Key Off. • Disconnect harness from EVP sensor. Inspect for damaged pins, corrosion, and pins pushed out. Service as ncessary. • DVOM on 200,000 ohm scale. • Disconnect vacuum line at EGR valve. • Connect vacuum pump to EGR valve. • Measure resistance at the EVP sensor between EVP Signal Pin and VREF Pin while increasing vacuum to 33 kPa (10 in. Hg). • Observe resistance as vacuum increases.	Reading does not decrease gradually ▶	GO to DM16.
	Reading gradually decreases from no more than 5,500 ohms to no less than 100 ohms ▶	REPLACE processor. RECONNECT EVP sensor and EGR vacuum line. RERUN Quick Test.

EGR Valve Position Sensor (EVP) EGR Valve Regulator (EVR) — Pinpoint Test — DM

TEST STEP	RESULT ▶	ACTION TO TAKE
DM16 MANUALLY EXERCISE EVP SENSOR		
• Key Off, harness disconnected • Remove EVP sensor from EGR valve. • Measure resistance at the EVP sensor between Signal Return Pin and VREF Pin while gradually applying pressure to EVP sensor shaft • Observe resistance as shaft is slowly pushed in and slowly released. • Look for sudden changes in resistance readings.	Both readings decrease and increase smoothly between 5,500 ohms and 100 ohms ▶	REPLACE EGR valve assembly. RECONNECT EVP sensor and EGR supply vacuum line. RERUN Quick Test.
	Either reading decreases or increases abruptly between 5,500 and 100 ohms ▶	REPLACE EVP sensor. RECONNECT harness and EGR supply vacuum line. RERUN Quick Test.
DM17 MEASURE EVR SOLENOID RESISTANCE		
• Key Off, wait 10 seconds. • DVOM on 200 ohm scale. • Disconnect EVR solenoid connector and measure solenoid resistance	Resistance is between 30 and 70 ohms ▶	CONNECT EVR solenoids. GO to **DM18**.
	Resistance is less than 30 ohms or greater than 70 ohms ▶	REPLACE EVR solenoid assembly. REPEAT Quick Test.
DM18 CHECK FOR VOLTAGE ON VEHICLE POWER CIRCUIT		
• Disconnect EVR solenoid from harness • Key On, Engine Off. • DVOM on 20V scale • Measure voltage between battery negative terminal and VPWR circuit on EVR solenoid.	Reading is less than 10.5V ▶	SERVICE harness circuit open. RERUN Quick Test.
	Readings are 10.5V or greater ▶	GO to **DM19**.
DM19 CHECK CONTINUITY OF EVR CIRCUITS		
• Key Off, wait 10 seconds. • EVR solenoid disconnected from harness. • Disconnect processor 60 Pin connector and inspect for damaged pins, corrosion, loose wires. Service as necessary • Install Breakout box to processor harness connector. Leave processor disconnected. • DVOM on 200 ohm scale. • Measure resistance between test Pin 52 and EVR signal at the EVR solenoid harness connector.	Reading less than 5 ohms ▶	GO to **DM20**.
	Reading 5 ohms or greater ▶	SERVICE open circuit. REMOVE Breakout box. RECONNECT harness to processor. RERUN Quick Test.

EGR Valve Position Sensor (EVP) EGR Valve Regulator (EVR) — Pinpoint Test — DM

TEST STEP	RESULT ▶	ACTION TO TAKE
DM20 CHECK FOR SHORT TO GROUND		
• Key Off, wait 10 seconds. • DVOM on 200,000 ohm scale. • Leave Breakout box installed and processor disconnected. • EVR solenoid disconnected. • Measure resistance between test Pin 52 and test Pins 40, 46 and 60.	Resistance reading is less than 10,000 ohms ▶	SERVICE short to ground. REPEAT Quick Test.
	Resistance reading is 10,000 ohms or greater ▶	GO to **DM21**.
DM21 CHECK EVR SIGNAL FOR SHORTS TO POWER		
• EVR solenoid disconnected from harness. • Key Off. Breakout box installed. Processor disconnected. • DVOM on 200,000 ohm scale. • Measure resistance between test Pin 52 and test Pins 37 and 57.	Any reading less than 10,000 ohms ▶	SERVICE circuit short. REMOVE Breakout box. RECONNECT harness to processor. RERUN Quick Test. If code is repeated, REPLACE processor.
	All readings 10,000 ohms or greater ▶	REPLACE processor. REMOVE Breakout box. RECONNECT harness to processor. RERUN Quick Test.
FAULT CODE 35 **DM30** RPM TOO LOW FOR EGR TEST		
• Check for code 12.	Code 12 present ▶	Vehicles equipped with air bypass (EFI) GO to **KE1**. Vehicles equipped with DC motor control GO to **KF1**. Vehicles equipped with throttle kicker GO to **KG1**.
	Code 12 not present ▶	GO to **DM31**.

EGR Valve Position Sensor (EVP) EGR Valve Regulator (EVR) — Pinpoint Test DM

TEST STEP	RESULT ▶	ACTION TO TAKE
DM31 RETEST AT 1500 RPM		
• Key Off, wait 10 seconds. • Install tachometer. • Perform Key On, Engine Running Quick Test while maintaining 1500 rpm. • Record engine running service codes. • Check for code 35.	Code 35 present ▶ Code 35 not present ▶	REPLACE processor. RERUN Quick Test. REPEAT Quick Test. SERVICE codes as necessary.
DM90 SERVICE CODE 31 CONTINUOUS TEST: EXERCISE EVP SENSOR		
• Using continuous monitor mode, observe VOM or STAR LED for indication of a fault while performing the following: • Connect a vacuum pump to the EGR valve. • Very slowly apply 20 kPa (6 in. Hg) vacuum to the EGR valve. • Slowly bleed vacuum off the EGR valve and lightly tap on EVP sensor (simulate road shock). • Wiggle EVP sensor connector. • Is a fault indicated?	Yes ▶ No ▶	GO to DM91. GO to DM92.

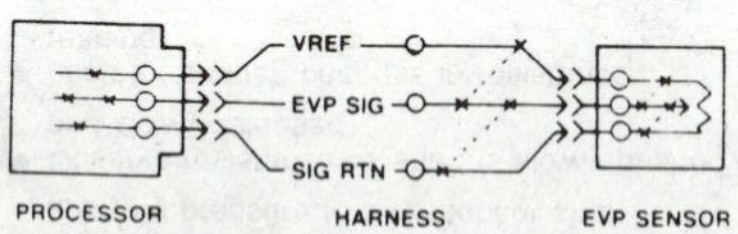

EGR Valve Position Sensor (EVP) EGR Valve Regulator (EVR) — Pinpoint Test DM

TEST STEP	RESULT ▶	ACTION TO TAKE
DM91 MEASURE EVP SIGNAL VOLTAGE WHILE EXERCISING EVP SENSOR		
• Key Off, wait 10 seconds. • Disconnect processor 60 Pin connector and inspect for damaged pins, corrosion, loose wires. Service as necessary. • Install Breakout box and reconnect processor. • VOM or STAR LED still connected to STO as in previous step. • Connect a DVOM from test Pin 27 to test Pin 46. • DVOM on 20V scale. • Key On, Engine Off. • While observing DVOM, repeat step DM90. • Does the fault occur below 4.25V?	Yes ▶ No ▶	DISCONNECT and INSPECT connector. If connector and terminals are good, REPLACE EVP sensor. REPEAT Quick Test. EGR valve overshoot may have caused continuous code 31. Sensor service is not required. To verify harness integrity, GO to DM92.
DM92 CHECK EEC-IV HARNESS		
• Observe VOM or STAR LED for a fault indication while performing the following: • Referring to the illustration in step DM90, grasp the harness close to the sensor connector. Wiggle, shake or bend a small section of the EEC-IV system harness while working your way to the dash panel. Also wiggle, shake or bend the EEC-IV harness from the dash panel to the processor. • Is a fault indicated?	Yes ▶ No ▶	ISOLATE fault and make necessary repairs. REFER to appropriate figure. REPEAT Quick Test. GO to DM93.
DM93 CHECK PROCESSOR AND HARNESS CONNECTORS		
• Key Off, wait 10 seconds. • Disconnect processor 60 Pin connector. • Inspect both connectors and connector terminals for obvious damage or faults. • Are connectors and terminals OK?	No ▶ Yes ▶	SERVICE as necessary. REPEAT Quick Test. Unable to duplicate fault at this time. Continuous Code 31 testing complete.

Neutral Drive Switch A/C Input — Pinpoint Test — FA

TEST STEP	RESULT ▶	ACTION TO TAKE
FA1 CODE 67 SYSTEM IDENTIFICATION		
1.6L MTX, 2.3L EFI MTX Truck, 2.3L CFI MTX		GO to FA10
2.3L Turbo, MTX and 1.6L Turbo		GO to FA15
2.3L SVO Turbo Octane Switch		GO to FA30
4.9L M/T, Truck		GO to FA25
All other systems		GO to FA2

FA2 NEUTRAL DRIVE INPUT CHECK

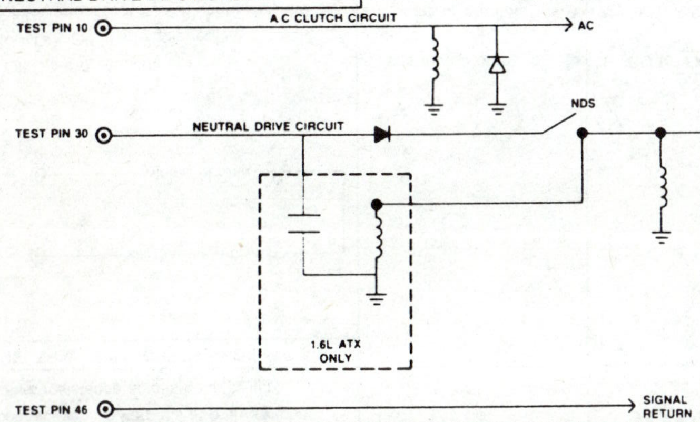

TEST STEP	RESULT ▶	ACTION TO TAKE
• Key Off, wait 10 seconds. • Verify heater control is in Off position, if so equipped. • Verify transmission is in Neutral or Park. • Disconnect processor 60 Pin connector and inspect for damaged pins, corrosion, loose wires. Service as necessary. • Install Breakout box. • Processor connected. • Key On, Engine Off. • DVOM on 20V scale • Measure voltage between test Pin 30 (Neutral Drive circuit) and chassis ground.	Less than 1.0V ▶ 1.0V or greater ▶	GO to FA4. GO to FA3.

Neutral Drive Switch A/C Input — Pinpoint Test — FA

TEST STEP	RESULT ▶	ACTION TO TAKE
FA3 NEUTRAL DRIVE SWITCH CHECK		
• Key Off, wait 10 seconds. • Breakout box installed. • DVOM on 200 ohm scale. • Locate the Neutral Drive switch. • Disconnect vehicle harness from the Neutral Drive switch and measure resistance across the two contacts.	Less than 5 ohms ▶	SERVICE open in vehicle harness Neutral Drive circuit REPEAT Quick Test.
	5 ohms or greater ▶	REPLACE Neutral Drive switch. REPEAT Quick Test.
FA4 A/C INPUT CHECK		
• Breakout box installed. • Key On, Engine Off. • DVOM on 20V scale. • A/C control Off. Measure voltage between test Pin 10 (A/C clutch circuit) and chassis ground. **NOTE: On 2.3L Turbo, check for code 18 in continuous memory.**	1.0V or greater ▶	SERVICE short to power in A/C clutch circuit. REPEAT Quick Test.
	Less than 1.0V ▶	REPLACE processor REPEAT Quick Test.
FA10 NEUTRAL/CLUTCH INPUT CHECK 1.6L, 1.6L TC, 2.3L EFI TRUCK, 2.3L CFI MTX		

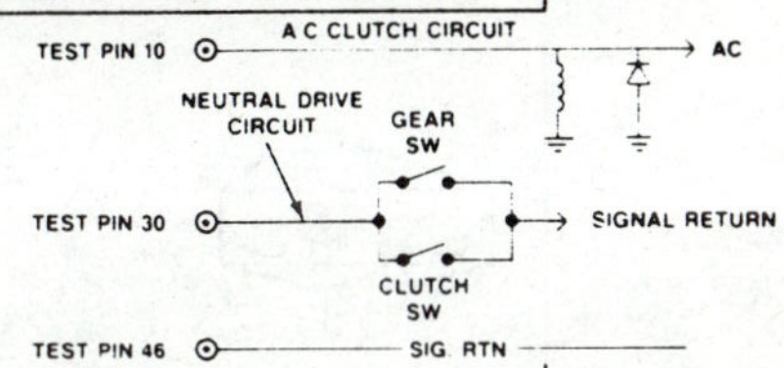

TEST STEP	RESULT ▶	ACTION TO TAKE
• Key Off, wait 10 seconds. • Verify A/C is Off, if so equipped. • Verify transaxle is in Neutral and clutch is released. • Disconnect processor 60 Pin connector and inspect for damaged pins, corrosion, loose wires. Service as necessary. • Install Breakout box. • Connect processor. • DVOM on 200 ohm scale. • Measure resistance between test Pin 30 (Neutral Drive circuit) and test Pin 46 (Signal Return circuit).	Less than 5 ohms ▶ 5 ohms or greater ▶	GO to FA4. GO to FA11.

Neutral Drive Switch A/C Input — Pinpoint Test — FA

TEST STEP	RESULT ▶	ACTION TO TAKE
FA11 NEUTRAL/CLUTCH SWITCH CHECK — 1.6L, 1.6L TC, 2.3L EFI TRUCK, 2.3L CFI MTX		
• Key Off. • DVOM on 200 ohm scale. • Breakout box installed. • Locate neutral switch (on transaxle) and clutch switch (under dash). • Disconnect vehicle harness at both switches. • Measure resistance across each switch.	Less than 5 ohms at both switches ▶ 5 ohms or greater at either or both switches ▶	GO to FA4. REPLACE open switch(es). RECONNECT harness and REPEAT Quick Test.
FA15 NEUTRAL INPUT CHECK — 2.3L TC MTX		
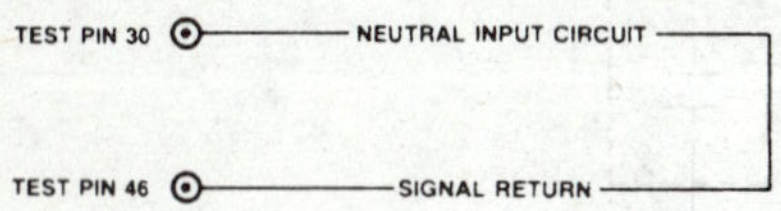		
• Key Off, wait 10 seconds. • Verify A/C is Off, if so equipped. • Disconnect processor 60 Pin connector and inspect for damaged pins, corrosion, loose wires. Service as necessary. • Install Breakout box. • Leave processor disconnected. • DVOM on 200 ohm scale. • Measure resistance between test Pin 30 (Neutral input circuit) and test Pin 46 (Signal Return circuit).	Less than 5 ohms ▶ 5 ohms or greater ▶	GO to FA4. SERVICE Open in Neutral input or Signal Return circuit. REPEAT Quick Test.

Neutral Drive Switch A/C Input — Pinpoint Test — FA

TEST STEP	RESULT ▶	ACTION TO TAKE
FA25 NEUTRAL INPUT CHECK — 4.9L M.T. TRUCK		
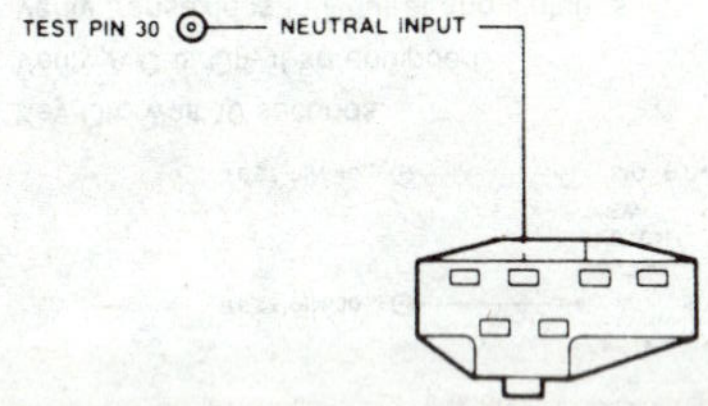		
• Key Off, wait 10 seconds. • Verify A/C is Off, if so equipped. • Disconnect processor 60 Pin connector and inspect for damaged pins, corrosion, loose wires. Service as necessary. • Install Breakout box. • Connect processor. • DVOM on 200 ohm scale. • Measure resistance between test Pin 30 (Neutral input circuit) and VIP connector.	Less than 5 ohms ▶ 5 ohms or greater ▶	GO to FA4. SERVICE open in Neutral input circuit to VIP tester. REPEAT Quick Test.
FA30 CHECK OCTANE SWITCH INPUT FOR INPUT CHANGE		
• Key Off, wait 10 seconds. • Disconnect processor 60 Pin connector and inspect for damaged pins, corrosion, loose wires. Service as necessary. • Install Breakout box. Reconnect processor. • DVOM to 20V scale. • Connect positive test lead to test Pin 30 and negative test lead to test Pin 46. • Key On, engine Off. • Cycle octane switch several times while observing DVOM. • Does voltage change from 0V to 5V?	Yes ▶ No ▶	REPLACE processor. REPEAT Quick Test EEC-IV system OK.

Key Power Check | Pinpoint Test | FC

FC

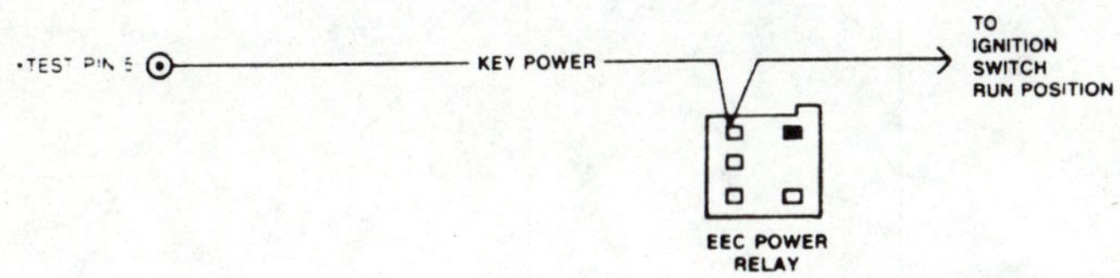

•TEST PINS LOCATED ON BREAKOUT BOX
ALL HARNESS CONNECTORS VIEWED INTO MATING SURFACE

STOP-WARNING

You should enter this Pinpoint Test only when a Service Code 55 or 65 is received in Quick Test Step 3.0, 5.0 or 6.0.

To prevent the replacement of good components, be aware that the following non-EEC areas may be at fault:

- Charging system overvoltage
- Battery charger connected with engine running.
- Jump starting.

This Pinpoint Test is intended to diagnose only the following:

- Harness circuit: key power.
- Processor.

Key Power Check | Pinpoint Test | FC

TEST STEP	RESULT ▶	ACTION TO TAKE
FC1 BATTERY VOLTAGE GREATER THAN 17.5V		
• Key Off, wait 10 seconds. • Install Breakout box with processor connected. • DVOM at 20V range and connected to test Pins 5 and 60. • Perform Engine Running Quick Test. Observe DVOM during test, record service codes.	DVOM exceeds 17.5V during Quick Test ▶	CORRECT charging system for over-voltage condition.
	DVOM remains below 17.5V and code 65 is present ▶	REPLACE processor and REPEAT Quick Test.
	DVOM remains below 17.5V and code 65 is not present ▶	GO to FC2.
FC2 CODE 65 IN CONTINUOUS MEMORY		
• Perform Key On, Engine Off Quick Test and record continuous codes.	Code 65 not present ▶	TESTING complete. CHARGING system OK at this time.
	Code 65 present ▶	CHECK charging system per Shop Manual
FC3 BATTERY VOLTAGE LESS THAN 7.5V		
• Key Off, wait 10 seconds. • Disconnect processor 60 Pin connector and inspect for damaged pins, corrosion, loose wires. Service as necessary. • Install Breakout box to processor harness connector. • Processor connected. • DVOM on 20V scale. • Measure voltage between test Pin 5 and test Pin 60. • Perform Engine Running Quick Test.	Voltage below 7.5V during Quick Test ▶	GO to FC4.
	Voltage remains above 7.5V and code 55 is present ▶	REPLACE processor. REPEAT Quick Test.
	Voltage remains above 7.5V and code 55 is not present ▶	GO to FC5.

Key Power Check — Pinpoint Test FC

TEST STEP	RESULT ▶	ACTION TO TAKE
FC4 KEY POWER CIRCUIT CHECK		
• Key Off. • Breakout box installed, processor connected. • DVOM at 200 ohm scale. • Measure resistance between test Pin 5 and key power terminal of EEC power relay.	5 ohms or less ▶	CORRECT charging system for under-voltage condition.
	Greater than 5 ohms ▶	SERVICE open in key power circuit. REPEAT Quick Test.
FC5 CODE 55 IN CONTINUOUS MEMORY		
• Perform Key On, Engine Off Quick Test and record continuous codes.	Code 55 not present ▶	TESTING complete. CHARGING system OK at this time.
	Code 55 present ▶	CHECK charging system per Shop Manual

Brake On/Off (BOO) — Pinpoint Test FD

FD

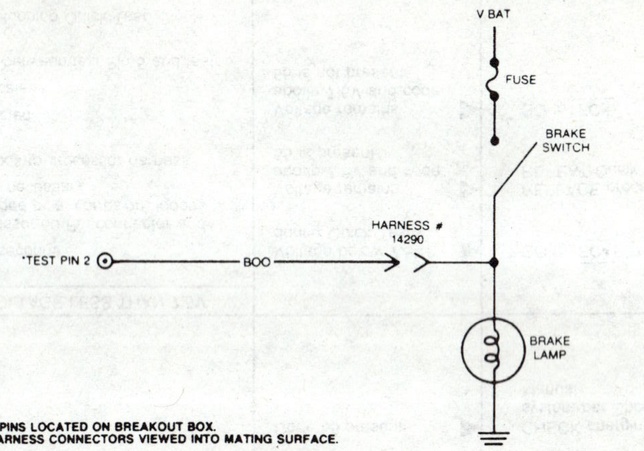

*TEST PINS LOCATED ON BREAKOUT BOX.
ALL HARNESS CONNECTORS VIEWED INTO MATING SURFACE.

STOP-WARNING

You should enter this Pinpoint Test only when a Service Code 74 or 75 is received in Quick Test Step 5.0.

To prevent the replacement of good components, be aware that the following non-EEC areas may be at fault:

- Brake lamp, brake switch, and fuse.

This pinpoint test is intended to diagnose only the following:

- BOO circuit.
- Processor assembly.

Brake On/Off (BOO) — Pinpoint Test — FD

TEST STEP	RESULT ▶	ACTION TO TAKE
FD1 SERVICE CODE 74		
• Did you press brake during the Engine Running Quick Test?	Yes ▶	GO to FD2.
	No ▶	RERUN Engine Running Quick Test. PRESS brake once during test.
FD2 BOO CIRCUIT CYCLING		
• Key Off, wait 10 seconds. • Disconnect processor 60 Pin connector. Inspect for damaged pins, corrosion, loose wires, etc. Service as necessary. • Install Breakout box, leave processor disconnected. • DVOM on 20V scale. • Measure voltage between test Pin 2 and test Pin 40 while depressing and releasing brake. • Does the voltage cycle?	Yes ▶	REPLACE processor. RETEST.
	No ▶	GO to FD3.
FD3 BOO CIRCUIT SHORT TO GROUND		
• Key Off. • Breakout box installed. • Processor disconnected. • DVOM on 200 Ohm scale. • Disconnect BOO circuit from 14290 harness (12 pin connector). • Measure resistance between test Pin 2 and ground. • Is resistance reading greater than 5 ohms?	Yes ▶	SERVICE BOO circuit short to ground.
	No ▶	GO to Shop manual.

Brake On/Off (BOO) — Pinpoint Test — FD

TEST STEP	RESULT ▶	ACTION TO TAKE
FD4 BOO CIRCUIT CYCLING CODE 75		
• Key Off, wait 10 seconds. • Disconnect processor 60 Pin connector. Inspect for damaged pins, corrosion, loose wires, etc. Service as necessary. • Install Breakout box, leave processor disconnected. • DVOM on 20V scale. • Measure voltage between test Pin 2 and test Pin 40 while depressing and releasing brake. • Does the voltage cycle?	Yes ▶	REPLACE processor. RETEST.
	No ▶	GO to FD5.
FD5 BOO CIRCUIT SHORT TO POWER		
• Key Off. • Breakout box installed. • Processor disconnected. • DVOM on 200 Ohm scale. • Disconnect BOO circuit from 14290 harness (12 pin connector). • Measure resistance between test Pin 2 and ground. • Is resistance reading greater than 5 ohms?	Yes ▶	SERVICE BOO circuit short to power.
	No ▶	SERVICE open in BOO circuit.

Power Steering Pressure Switch (PSPS) — Pinpoint Test — FF

FF

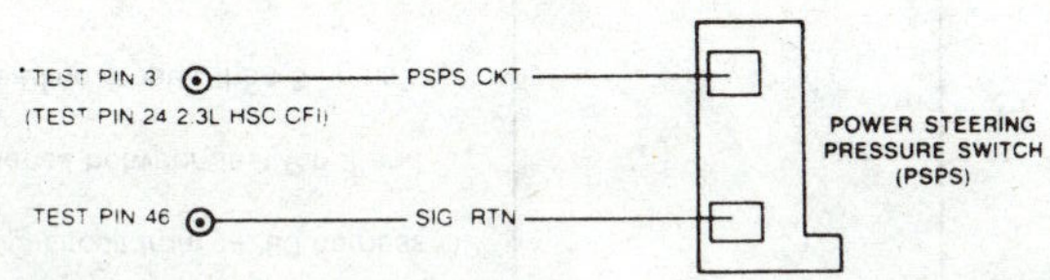

*TEST PINS LOCATED ON BREAKOUT BOX.
ALL HARNESS CONNECTORS VIEWED INTO MATING SURFACE.

STOP-WARNING

You should enter this Pinpoint Test only when a Service Code 52 is received in Quick Test Step 3.0 or if you are directed here from Diagnostics by Symptom in the Engine Supplement Section.

To prevent the replacement of good components, be aware that the following non-EEC areas may be at fault:

- Idle speeds throttle stop adjustment.
- Binding throttle shaft linkage or speed control linkage.

This Pinpoint Test is intended to diagnose only the following:

- Power steering pressure switch.
- Switch harness circuits: PSPS Signal, and Signal Return.
- Processor assembly

Power Steering Pressure Switch (PSPS) — Pinpoint Test — FF

TEST STEP	RESULT ▶	ACTION TO TAKE
FF1 ATTEMPT TO ELIMINATE CODE 52		
• Key Off, wait 10 seconds. • Disconnect PSPS. • Jumper PSPS circuit to Signal Return at harness. • Rerun Key On, Engine Off Quick Test. • Is code 52 still present?	Yes ▶	GO to **FF2**.
	No ▶	REPLACE PSPS. REPEAT Quick Test.
FF2 PSPS HARNESS CHECK		
• Key Off, wait 10 seconds. • Disconnect harness from processor 60 Pin connector and inspect for damaged pins, corrosion, loose wires. Service as necessary. • PSPS disconnected. • Processor disconnected. • Breakout box installed. • DVOM on 200 ohm scale. • Measure resistance between test Pin 46 and Signal Return at the PSPS connector and between test Pin 3 (2.3L HSC CFI test Pin 24) and PSPS circuit at the PSPS harness connector. • Are both readings less than 5 ohms?	Yes ▶	REPLACE processor. REPEAT Quick Test.
	No ▶	SERVICE open in circuit. REPEAT Quick Test.
FF3 SWITCH INTEGRITY		
• Install tachometer. • Start engine, allow to idle in Neutral/Park. • Disconnect PSPS at switch. • Does rpm increase?	Yes ▶	REPLACE PSPS.
	No ▶	GO to **FF4**.

Power Steering Pressure Switch (PSPS) — Pinpoint Test — FF

TEST STEP	RESULT	▶	ACTION TO TAKE
FF4 PSPS HARNESS CHECK			
• Key Off, wait 10 seconds. • Disconnect harness from processor 60 Pin connector and inspect for damaged pins, corrosion, loose wires. Service as necessary. • PSPS disconnected. • Processor disconnected. • Breakout box installed. • DVOM on 200 ohm scale. • Measure resistance between test Pin 3 (2.3L HSC CFI test Pin 24) and test Pin 46. • Is reading less than 5 ohms?	Yes	▶	SERVICE short in harness.
	No	▶	REPLACE processor.

Fuel Control — EFI — Pinpoint Test — HA

HA

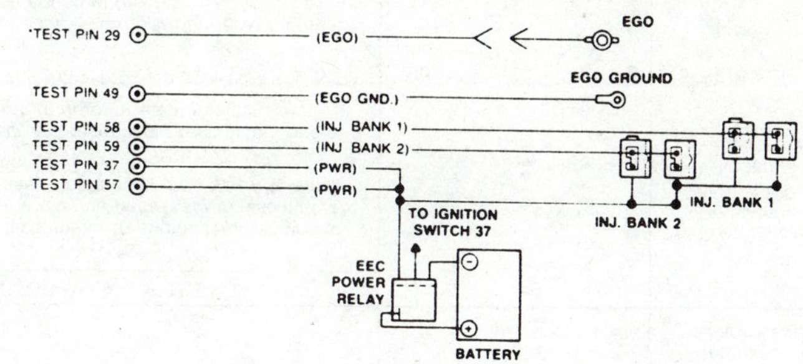

*TEST PINS LOCATED ON BREAKOUT BOX.
ALL HARNESS CONNECTORS VIEWED INTO MATING SURFACE.

STOP-WARNING

You should enter this Pinpoint Test Only when a Service Code 41 or 42 is received in Quick Test Step 5.0 or 6.0 or when directed here from Pinpoint Test A or Diagnostics by Symptom in the Engine Supplement Section.

To prevent the replacement of good components, be aware that the following non-EEC areas may be at fault:

- Ignition Coil
- Distributor Cap
- Distributor Rotor
- Fouled Spark Plugs
- Spark Plug Wires
- CANP Problems
- PCV Valves (see note below)
- EGR Valve and Gasket
- Air Filter
- Fuel Contamination, Engine Oil
- Poor Power Ground
- Fuel Pressure
- Manifold Leaks, Intake/Exhaust
- Engine Not at Normal Operating Temperatures

This Pinpoint Test is intended to diagnose only the following:

- EGO Sensor
- Harness Circuits EGO Grd., EGO, INJ BANK 1, INJ BANK 2, V PWR
- EGO Sensor Connection
  - — Code 42 start at HA7
  - — Code 41 start at HA10
- Vacuum Systems
- Fuel Injectors
- Processor Assembly

NOTE: Fuel contaminated engine oil may affect 41, 42 Service Codes. If this is suspected, remove the PCV from the valve cover and repeat the Quick Test. If the problem is corrected, then change the engine oil and filter.

Fuel Control — EFI | Pinpoint Test | HA

TEST STEP	RESULT ▶	ACTION TO TAKE
HA1 FUEL PRESSURE CHECK		
• Key Off, wait 10 seconds. • Install fuel pressure gauge. • Start and run engine. Fuel pressure must be 172-310 kPa (25-45 psi). • Fuel pressure must remain at 276 ± 34 kPa (30 ± 5 psi) for 60 seconds after final key Off. **For No Starts:** • If engine will not run, cycle the key from Off to On several times.	Fuel pressure is within specification ▶	GO to HA2.
	Fuel pressure is not within specification ▶	REFER to the Shop Manual for electric fuel pump and fuel pressure regulator check.
HA2 FUEL DELIVERY TEST		
• Key Off. • Fuel pressure gauge installed. • Pressurize fuel system as per Step HA1. • Locate and disconnect the fuel pump relay. • Crank engine for 5 seconds. • Take pressure reading at the end of 5 second crank. **NOTE: Verify fuel quality; air and/or water will also pressurize and look like acceptable fuel delivery.**	PRESSURE GAUGE READING: Pressure is approximately 10-20 psi at the end of 5 second crank cycle. Refer to note below. ▶	The EEC system is not the fault of the No Start. Fuel and spark are present. REFER to Section for other No Start Routines. If complaint was runs rough/misses or fuel code GO to HA3.
	Pressure is greater or less than specified ▶	GO to HA3.

NOTE: The colder the engine, the greater the pressure drop (i.e., an engine coolant temperature of 200°F equals approximately a 10 psi drop in 5 seconds. 60°F equals approximately a 20 psi drop in 5 seconds).

TEST STEP	RESULT ▶	ACTION TO TAKE
HA3 HARNESS INJECTOR RESISTANCE CHECK		
• Key Off, wait 10 seconds. • Disconnect processor 60 Pin connector and inspect for damaged pins, corrosion, loose wires. Service as necessary. • Connect Breakout box to harness. Leave processor disconnected. • DVOM on 200 ohm scale. • Through the vehicle harness, measure the resistance of injector Bank 1 between test Pin 37 and test Pin 58, record reading. • Through the vehicle harness, measure the resistance of injector Bank 2 between test Pin 37 and test Pin 59, record reading. • Are both readings between 1.2 ohms and 1.8 ohms?	Yes ▶	GO to HA5.
	No ▶	GO to HA4.

Fuel Control — EFI | Pinpoint Test | HA

TEST STEP	RESULT ▶	ACTION TO TAKE
HA4 ISOLATE FAULTY INJECTOR CIRCUIT		
• Key Off. • Breakout box installed. • Leave processor disconnected. • DVOM on 200 ohm scale. • Disconnect all injectors on suspect Bank. Measure resistance of each injector by connecting one injector at a time, and reading the resistance between test Pins 37 and 58 for Bank 1 or test Pins 37 and 59 for Bank 2. • Are all readings between 2.0 and 2.7 ohms?	Yes ▶	GO to HA5.
	No ▶	SERVICE the harness connectors on the suspect injector for opens or shorts. If OK, REPLACE injector. REPEAT Quick Test.
HA5 INJECTOR DRIVE SIGNAL CHECK		
Requires standard non-powered 12 volt test lamp. • Key Off. • Breakout box installed. • Connect processor to Breakout box. • Connect test lamp between test Pin 37 and test Pin 58. • Crank or start engine. • Repeat above test between test Pin 37 and test Pin 59.	Dim glow at light on both tests ▶	GO to HA6.
	No light on one or both tests ▶	VERIFY 12-volt battery power at Pins 37 and 57. If OK, REPLACE processor. REPEAT Quick Test.
	Bright light on one or both tests ▶	CHECK circuits Bank 1 and Bank 2 for shorts to ground. If OK, REPLACE processor. REPEAT Quick Test.
HA6 INJECTOR BALANCE TEST		
• Connect tachometer to engine. Run engine at 2000 rpm. It may be necessary to disconnect ISC and use throttle body stop screw to set engine speed. • Disconnect and reconnect the injectors one at a time; note rpm drop for each injector. • Does each injector produce at least a 150 rpm drop? • Reset curb idle. Refer to VECI decal. Verify ISC connection at end of testing.	Yes ▶	Fuel delivery OK. Problem is with area common to all cylinders, i.e.: Air vacuum leak, fuel contamination, EGR.
	No ▶	REPLACE faulty injector. REPEAT Quick Test.

Fuel Control — EFI | Pinpoint Test | HA

TEST STEP	RESULT ▶	ACTION TO TAKE
HA7 FUEL CONTROL — ALWAYS RICH: CODE 42		
• Key Off, wait 10 seconds. • **NOTE: Non-EEC areas could cause a code 42. Check for: — Fuel contaminated engine oil — Ignition caused misfire — CANP problems** • Disconnect vehicle harness at the EGO sensor. Using a jumper wire, ground vehicle harness EGO circuit at the EGO sensor to the engine block. • Repeat Engine Running Self-Test. • Service code 41 results.	Yes ▶ No ▶	GO to HA9. GO to HA8.
HA8 HARNESS CHECK		
• Key Off, wait 10 seconds. • Install Breakout box. • Measure resistance of EGO ground circuit between test Pin 49 and EGO ground at engine block. • Measure resistance of EGO circuit between test Pin 29 and EGO harness connector. • Both circuits less than 5 ohms.	Yes ▶ No ▶	DISCONNECT processor connector. INSPECT for damage or corrosion. If OK, REPLACE processor. REPEAT Quick Test. CORRECT harness circuit with resistance greater than 5 ohms. REPEAT Quick Test.

Fuel Control — EFI | Pinpoint Test | HA

TEST STEP	RESULT ▶	ACTION TO TAKE
HA9 EGO CHECK		
• DVOM on 20V scale. • With EGO sensor disconnected from the harness, connect a DVOM from EGO sensor to engine ground. • Run the engine at 2000 rpm for 1 minute. While observing DVOM, disconnect the manifold vacuum hose indicated below. • Meter indicates less than 0.4V.	Yes ▶ No ▶	EGO sensor OK. GO to HA1. REPLACE EGO sensor. REPEAT Quick Test.

1.6L EFI

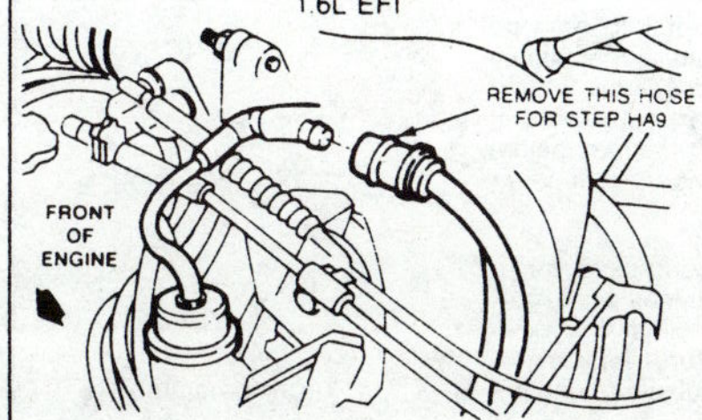

2.3L TURBO

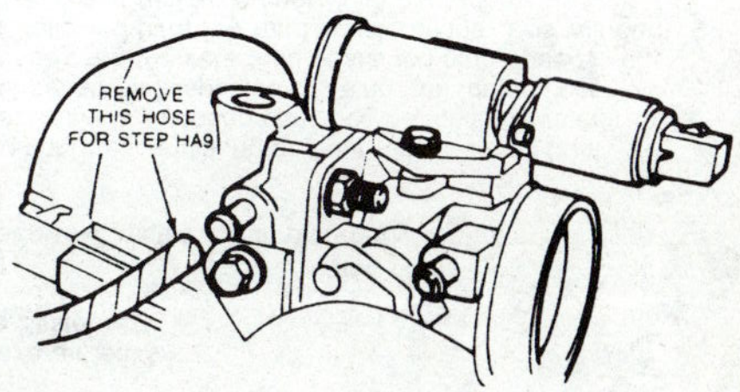

TEST STEP	RESULT ▶	ACTION TO TAKE
HA10 FUEL CONTROL — ALWAYS LEAN: CODE 41		
NOTE: Vacuum air leaks in non-EEC areas could cause a code 41. Check for: — Leaking vacuum actuator (eg: A C control motor) — Engine sealing — EGR system — PCV system — Unmetered air leak between air meter and throttle body — Lead contaminated EGO sensor • Key Off. DVOM on 20V scale. Disconnect EGO sensor from vehicle harness. Connect DVOM to EGO sensor and engine ground. Remove air cleaner to gain access to air meter inlet. Using a standard wood lead pencil, prop the air meter door part-way open. • Start the engine and run at approximately 2000 rpm for 2 minutes. • Does the DVOM read greater than 0.5V at the end of 1 minute?	Yes ▶ No ▶	GO to HA11. REPLACE EGO sensor. REPEAT Quick Test.

Fuel Control — EFI | Pinpoint Test | HA

TEST STEP	RESULT ▶	ACTION TO TAKE
HA11 HARNESS CHECK EGO CIRCUITS		
• Key Off. • Install Breakout box. Processor disconnected. • Measure resistance of EGO circuit between test Pin 49 and engine block ground. Also EGO circuit between EGO connector and test Pin 29. • Both circuits less than 5 ohms.	Yes ▶	GO to HA12.
	No ▶	SERVICE/CORRECT as necessary. REPEAT Quick Test.
HA12 CHECK EGO CIRCUIT FOR SHORT TO GROUND		
• Key Off. • Breakout box installed. Processor disconnected. • DVOM to 200,000 ohms range. • Measure the resistance between test Pin 29 and test Pin 40.	Reading is 10,000 ohms or greater ▶	GO to HA13.
	Reading is less than 10,000 ohms ▶	CORRECT cause of resistance to ground. REPEAT Quick Test.
HA13 ATTEMPT TO ELIMINATE CODE 41		
• Key Off. • Reconnect EGO sensor. Remove air cleaner to gain access to air meter inlet. Using a standard wood lead pencil, prop the air meter door part-way open. • Start the engine and run at approximately 2000 rpm for 2 minutes. • Is code 41 present?	Yes ▶	INSPECT for corrosion or damaged pins. If OK, REPLACE processor. REPEAT Quick Test.
	No ▶	EGO input circuit OK. GO to HA1.

Fuel Control — EFI | Pinpoint Test | HA

HA20 CONTINUOUS TESTING: CODE 41 42

41 — EGO indicated the fuel system was lean for more than 15 seconds when the system should have been in closed loop fuel control.

42 — EGO indicated the fuel system was rich for more than 15 seconds when the system should have been in closed loop fuel control.

*CLOSED LOOP — Fuel control under the influence of the EGO sensor.

*OPEN LOOP — Fuel control NOT under the influence of the EGO sensor.

Before attempting to correct a fuel control code, 41/42, diagnose all other drive complaints first, eg., rough idle, misses, etc.

NOTE: The fuel control code may help in this diagnosis.

Using the fuel control service code, isolate the cause of the fuel control problem.

Some areas to check are:

- Unmetered Air: Vacuum leaks/intake air leaks.
  - — Canister Purge System.
  - — PCV System.
  - — Engine sealing.
  - — Air leaks between VAF meter and throttle body.
- EGO Fuel Fouled: Whenever an over-rich fuel condition has been experienced (fuel fouled spark plugs), make a thorough check of the ignition system. In the event the EGO sensor is suspected of being fuel fouled (low output, slow response), run the vehicle at sustained high speeds (within legal limits) followed by a few hard accelerations. This will burn off EGO contamination and restore proper EGO operation.
- Fuel Pressure: Perform Pinpoint Test Step HA1.
- Ignition System: Always in default spark (10 degrees). Refer to Quick Test Step 4.0.
- Improper Fueling: Lead fouled EGO sensor.
- TP Sensor: Not moving (mechanical damage). Connect DVOM to test Pin 47 and to test Pin 46. Key to Run. Observe DVOM while moving the throttle. Reading must increase with increase in throttle opening. If not correct, REPLACE as necessary.
- If at this point, the drive concern is still present, perform Steps HA2 through HA5 only.

Fuel Control (FBC) — Pinpoint Test — HB

HB

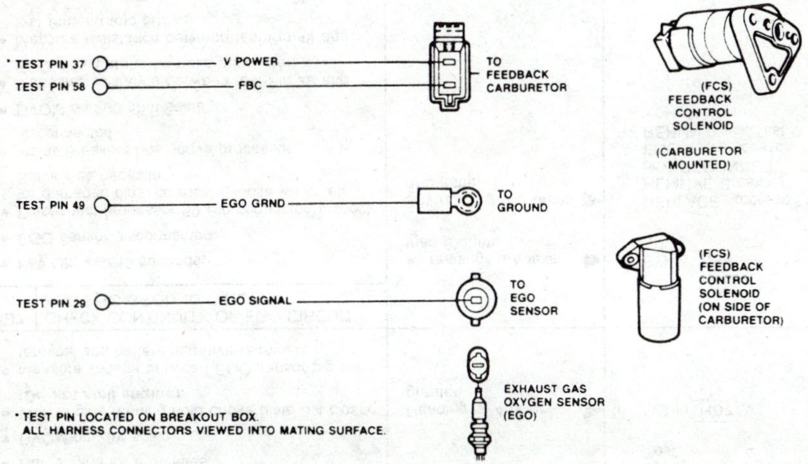

* TEST PIN LOCATED ON BREAKOUT BOX.
ALL HARNESS CONNECTORS VIEWED INTO MATING SURFACE.

STOP-WARNING

You should enter this Pinpoint Test only when a Service Code 16, 41, 42, 43 or 47 in Quick Test Step 5.0 or 6.0 is received or when directed here from Diagnostics by Symptom in the Engine Supplement Section.

To prevent the replacement of good components, be aware that the following non-EEC areas may be at fault:

- Fouled Spark Plugs
- Distributor Cap and Rotor
- Ignition Coil
- Spark Plug Wires
- CANP Problems
- PCV Valve
- Oil Contamination
- EGR Valve and Gasket
- Air Filter
- Fuel Filter
- Fuel Pump and Pressure
- Manifold Leak, Intake Exhaust
- Engine Not at Normal Operating Temperature
- Carburetor
- Choke, Sticking Electrical

This Pinpoint Test is intended to diagnose only the following:

- EGO Sensor Connection
- EGO Sensor
- Harness Circuits: V Pwer, FBC, EGO Control, EGO Signal
- Impact of Thermactor Air
- FBC Solenoid
- Vacuum System (EEC Only)
- Choke Circuit
- Processor Assembly

Fuel Control (FBC) — Pinpoint Test — HB

TEST STEP	RESULT ▶	ACTION TO TAKE
FAULT CODE 41		
HB1 VERIFY VEHICLE PREPARATION		
• Did vehicle run for more than 2 minutes before performing Quick Test?	Yes ▶	GO to HB2.
	No ▶	PREPARE vehicle, RERUN Quick Test.
HB2 TRY TO ELIMINATE CODE 41 BY REMOVING THERMACTOR AIR SUPPLY		
• Key Off. • Reconnect EGO sensor and processor to harness. • Disconnect thermactor air supply hose at the air pump and cap the hose. • Run engine running Quick Test.	Code 41 is present ▶	GO to HB3. Leave thermactor disconnected.
	Code 44 is present ▶	RETURN to Quick Test Step 5.0B.
	Code 11 or 42 is present ▶	Handle as Code 44. RETURN to Quick Test Step 5.0B.
HB3 CHECK FCS SOLENOID		
• Key Off, wait 10 seconds. • Thermactor disconnected. • Disconnect harness from FCS solenoid. • Perform Engine Running Quick Test. • Check for Code 41. **NOTE: Ignore all other codes at this time.**	Code 41 is not present ▶	GO to HB5.
	Code 41 is present ▶	GO to HB4.
HB4 FORCE SYSTEM RICH		
• Key Off, wait 10 seconds. • FCS and thermactor disconnected. • Run for 2 minutes at part throttle. • Key Off, wait 10 seconds. • Perform Engine Running Quick Test while holding choke plate 3/4 closed **(Do not stall engine).** • Check for Code 41.	Code 41 is not present ▶	REPLACE FCS. RERUN Quick Test if Code 41 is still present. REFER to Shop Manual, for carburetor diagnostics.
	Code 41 is present ▶	GO to HB6. HOOK-UP all disconnected components.

Fuel Control (FBC) — Pinpoint Test — HB

TEST STEP	RESULT ▶	ACTION TO TAKE
HB5 CHECK FBC CIRCUIT FOR SHORT TO GROUND		
• Key Off, wait 10 seconds. • Connect thermactor. • Leave FCS solenoid disconnected. • Disconnect processor 60 Pin connector, inspect for damaged pins, corrosion, loose wires, etc. • Install Breakout box, leave processor disconnected. • DVOM on 200,000 ohm scale. • Measure resistance between test Pin 58 and test Pins 20, 40 and 46.	Any reading is less than 10,000 ohms ▶	SERVICE circuit shorts. RERUN Quick Test.
	All readings are 10,000 ohms or greater ▶	REPLACE processor, RERUN Quick Test. HOOK-UP all disconnected components.
HB6 FORCE EGO RICH		
• Key Off, EGO sensor disconnected. • Run engine for 2 minutes. • DVOM on 20V scale. • With engine running hold choke plate 3/4 closed **(Do not stall engine).** • Measure voltage between EGO sensor pig tail terminal and battery negative terminal.	Reading is less than .45V ▶	REPLACE EGO sensor. RERUN Quick Test.
	Reading is .45V or greater ▶	GO to HB7.
HB7 CHECK CONTINUITY OF EGO CIRCUIT AND EGO GROUND CIRCUIT		
• Key Off, wait 10 seconds. • EGO sensor disconnected. • Disconnect processor 60 Pin connector, inspect for damaged pins, corrosion, loose wires, etc. Service as necessary. • Install Breakout box, leave processor disconnected. • DVOM on 200 ohm scale. • Measure resistance between test Pin 29 and EGO circuit at the harness connector. • Measure resistance between test Pin 49 and test Pins 40 and 60.	All readings are less than 5 ohms ▶	GO to HB8.
	Any reading is 5 ohms or greater ▶	REPLACE processor. REMOVE Breakout box. RECONNECT EGO and processor. RERUN Quick Test.

Fuel Control (FBC) — Pinpoint Test — HB

TEST STEP	RESULT ▶	ACTION TO TAKE
HB8 CHECK EGO CIRCUIT FOR SHORT		
• Key Off, wait 10 seconds. • Thermactor and FCS reconnected. • Disconnect harness from EGO sensor. • DVOM on 200,000 ohm scale. • Measure resistance between EGO circuit at EGO harness connector and battery negative terminal.	Reading is less than 10,000 ohms ▶	SERVICE circuit short. REMOVE Breakout box. RECONNECT processor and EGO sensor. RERUN Quick Test.
	Reading is 10,000 ohms or greater ▶	REPLACE processor. RECONNECT EGO sensor and processor. RERUN Quick Test.
FAULT CODES 42/47		
HB20 CHECK CHOKE		
• Does choke function properly without sticking or binding?	No ▶	SERVICE choke as required. RERUN Quick Test.
	Yes ▶	GO to HB21.
HB21 GENERATE KNOWN FAULT		
• Run engine for 2 minutes at part throttle. • Key Off, wait 10 seconds. • Disconnect harness from EGO sensor. Jumper harness EGO circuit to ground. • Rerun Engine Running Quick Test. • Check for Code 41. Ignore all other codes at this time.	Code 41 is present ▶	GO to HB22.
	Code 41 is not present ▶	GO to HB30.
HB22 EGO CHECK		
• Key Off, wait 10 seconds. • Disconnect 7/32 vacuum line. • DVOM on 20V scale. • With EGO sensor disconnected from the harness, connect a DVOM from EGO sensor to engine ground. • Run the engine at 2000 rpm for 2 minutes while observing DVOM. • Does meter indicate less than 0.4V?	Yes ▶	EGO sensor OK. GO to HB23.
	No ▶	REPLACE EGO sensor. RECONNECT vacuum hose. REPEAT Quick Test.

Fuel Control (FBC) — Pinpoint Test — HB

TEST STEP	RESULT ▶	ACTION TO TAKE
HB23 CHECK FCS SOLENOID		
• Key Off, wait 10 seconds. • Vacuum hose reconnected. • Disconnect FCS solenoid. • DVOM on 200 ohm scale. • Measure resistance between terminals on the FCS solenoid. • Is resistance within the following ranges: 15 to 30 ohms: 2150 carburetor 30 to 60 ohms: YFA carburetor	Yes ▶	RECONNECT FCS. GO to HB24.
	No ▶	REPLACE FCS. RERUN Quick Test.
HB24 CHECK V PWR CIRCUIT FOR POWER		
• Key Off, wait 10 seconds. • Deactivate Quick Test. • Disconnect harness from FCS. • DVOM on 20V scale. • Key On. Engine Off. • Measure voltage between VPWR circuit at FCS harness connector and battery negative terminal.	Reading is less than 10V ▶	RECONNECT FCS solenoid. SERVICE circuit open. RERUN Quick Test.
	Reading is 10V or greater ▶	GO to HB25.
HB25 CHECK FOR SHORT TO POWER		
• Key Off, wait 10 seconds. • Disconnect processor 60 Pin connector, inspect connectors for damaged pins, corrosion, loose wires, etc. Service as necessary. • Install Breakout box, leave processor disconnected. • DVOM on 200,000 ohm scale. • Leave FCS solenoid disconnected, measure resistance between test Pin 58 and test Pin 37.	Resistance reading is 10,000 ohms or greater ▶	GO to HB26.
	Resistance reading is less than 10,000 ohms ▶	SERVICE short to power. REPEAT Quick Test.
HB26 CHECK CONTINUITY OF FBC CIRCUIT		
• Key Off, wait 10 seconds. • Breakout box installed. • FCS disconnected. • DVOM on 200 ohm scale. • Measure resistance at the FCS harness connector between FBC circuit and test Pin 58.	Reading is less than 5 ohms ▶	REMOVE Breakout box. RECONNECT FCS solenoid. REPLACE processor and RERUN Quick Test.
	Reading is 5 ohms or greater ▶	SERVICE circuit open. REMOVE Breakout box. RECONNECT processor and FCS solenoid. RERUN Quick Test.

Fuel Control (FBC) — Pinpoint Test — HB

TEST STEP	RESULT ▶	ACTION TO TAKE
HB30 CHECK CANISTER PURGE		
• Key Off, wait 10 seconds. • Disconnect vacuum hose between canister purge solenoid and canister, plug both ends. • Rerun Engine Running Self-Test. • Is service code 42 still present?	Yes ▶	GO to HB31.
	No ▶	GO to KD1.
HB31 HARNESS CHECK		
• Key Off, wait 10 seconds. • Install Breakout box. • Measure resistance between test Pin 49 and EGO ground at engine block. • Measure resistance between test Pin 29 and EGO harness connector. • Both circuits less than 5 ohms.	Yes ▶	DISCONNECT processor connector. INSPECT for damage or corrosion. If OK, REPLACE processor. REPEAT Quick Test.
	No ▶	CORRECT harness circuit with resistance greater than 5 ohms. REPEAT Quick Test.
FAULT CODE 43		
HB40 WARM ENGINE		
• Run vehicle at 2000 rpm for 2 minutes. • Key Off, wait 10 seconds. • Perform Engine Running Quick Test.	Code 43 is not present ▶	SERVICE other codes as necessary.
	Code 43 is present ▶	GO to HB41.
HB41 CHECK EXHAUST LEAKS		
• Are there any exhaust leaks at manifold or inlet pipe?	Yes ▶	SERVICE as necessary. RERUN Quick Test.
	No ▶	GO to HB42.
HB42 CHECK IDLE QUALITY		
• Start vehicle, let engine idle. • Does engine idle quality deteriorate and remain poor during and after the Quick Test?	Yes ▶	REFER to Carburetors.
	No ▶	REPLACE EGO sensor. RERUN Quick Test.

Fuel Control — 5.0L CFI | Pinpoint Test | HC

HC

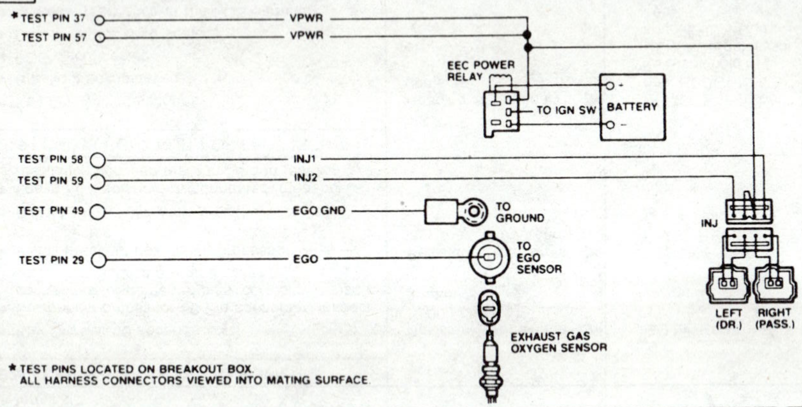

* TEST PINS LOCATED ON BREAKOUT BOX.
ALL HARNESS CONNECTORS VIEWED INTO MATING SURFACE.

STOP-WARNING

You should enter this Pinpoint Test only when a Service Code 41, 42, 43, 16, 17 or 47 in Quick Test Step 5.0 or 6.0 is received or when directed here from Pinpoint Test A or Diagnostics by Symptom in the Engine Supplement Section.

To prevent the replacement of good components, be aware that the following non-EEC areas may be at fault:

- Ignition Coil
- Distributor Cap
- Distributor Rotor
- Fouled Spark Plugs
- Spark Plug Wires
- CANP Problems
- PCV Valves (see note below)
- EGR Valve and Gasket
- Air Filter
- Fuel Contamination, Engine Oil
- Poor Power Ground
- Fuel Pressure
- Manifold Leaks, Intake Exhaust
- Engine Not at Normal Operating Temperatures

This Pinpoint Test is intended to diagnose only the following:

- EGO Sensor
- EGO Signal and Ground Circuits
- EGO Sensor Connection
- Vacuum Systems
- Fuel Injectors
- Processor Assembly
- Thermactor Impact

NOTE: **Fuel-contaminated engine oil may affect 41, 42 and 43 Service Codes, so if it is suspected, remove the PCV from the valve cover, and rerun the Quick Test. If the problem is corrected, then change the engine oil and filter.**

Fuel Control — 5.0L CFI | Pinpoint Test | HC

TEST STEP	RESULT ▶	ACTION TO TAKE
HC1 FUEL PRESSURE CHECK		
• Install fuel pressure gauge. • Start and run engine. Fuel pressure must be 241-310 kPa (35-45 psi). • Fuel pressure must remain at 276 ± 34 kPa (40 psi ± 5 psi) for 60 seconds after final key Off. • Is fuel pressure within specification? **For No Starts:** • If engine will not run cycle the key from Off to On several times.	Yes ▶ No ▶	GO to HC2. REFER to Shop Manual for electric fuel pump.
HC2 FUEL DELIVERY TEST		
• Key Off. • Fuel pressure gauge installed. • Pressurize fuel system as per Step HC1. • Locate and disconnect the fuel pump relay. • Crank engine for 5 seconds. • Take pressure reading at the end of 5 second crank. NOTE: **Verify fuel quality; air and/or water will also pressurize and look like acceptable fuel delivery.**	PRESSURE GAUGE READING: Pressure is approximately 10-20 psi at the end of 5 second crank cycle. Refer to note below ▶ Pressure is greater or less than specified ▶	The EEC system is not the fault of the No Start. Fuel and spark are present. REFER to Section for other No Start Routines. If complaint was runs rough, misses or fuel code, GO to HC3. GO to HC3.
NOTE: **The colder the engine, the greater the pressure drop (i.e., an engine coolant temperature of 200°F equals approximately a 10 psi drop in 5 seconds. 60°F equals approximately a 20 psi drop in 5 seconds).**		
HC3 HARNESS INJECTOR RESISTANCE CHECK		
• Key Off. • DVOM on 200 ohm range. • Disconnect processor 60 Pin connector and inspect for damaged pins, corrosion, loose wires. Service as necessary. • Install Breakout box, processor disconnected. • Measure resistance between test Pins 57 and 58 and between test Pins 57 and 59.	Resistance is less than 1.5 ohms or greater than 3.5 ohms ▶ Resistance is 1.5-3.5 ohms ▶	SERVICE the harness/connectors on the suspect injector for opens or shorts. If OK, REPLACE injector. GO to HC4.

Fuel Control — 5.0L CFI — Pinpoint Test — HC

TEST STEP	RESULT ▶	ACTION TO TAKE
HC4 INJECTOR DRIVE SIGNAL CHECK		
Requires standard non-powered 12 volt test lamp. • Key Off. • Breakout box installed. • Connect processor to Breakout box. • Refer to illustration HC. • Connect test lamp between test Pin 37 and test Pin 58. • Crank or start engine. • Repeat above test between test Pin 37 and test Pin 59.	Dim glow a light on both tests ▶ No light on one or both tests ▶ Bright light on one or both tests ▶	GO to HC5. VERIFY 12-volt battery power at Pins 37 and 57. If OK, REPLACE processor. REPEAT Quick Test. CHECK injector circuits 1 and 2 for shorts to ground. If OK, REPLACE processor. REPEAT Quick Test.
HC5 INJECTOR BALANCE TEST		
• Connect tachometer to engine. Run engine at 2000 rpm. It may be necessary to disconnect ISC and use throttle body stop screw to set engine speed. • Disconnect and reconnect the injectors one at a time; note rpm drop for each injector. • Does each injector produce at least a 150 rpm drop? • Reset curb idle. Refer to VECI decal. Verify ISC connection at end of testing.	Yes ▶ No ▶	Fuel delivery OK. Problem is with area common to all cylinders, i.e.: Air vacuum leak, fuel contamination, EGR. REPLACE faulty injector. REPEAT Quick Test.
HC6 FUEL CONTROL — ALWAYS LEAN:		
NOTE: Vacuum air leaks in non-EEC areas could cause a lean code. Check for: **— Leaking vacuum actuator (eg: A/C control motor)** **— Engine sealing** **— EGR system** **— PCV system** **— Lead contaminated EGO sensor** • Refer to illustration HC. • Key Off. DVOM on 20V scale. Disconnect EGO sensor from vehicle harness. Connect DVOM to EGO sensor and engine ground. Disconnect ACT sensor. • Start the engine and run at approximately 2000 rpm. Does the DVOM read greater than 0.5V within 1 minute?	Yes ▶ No ▶	GO to HC7. REPLACE EGO sensor. REPEAT Quick Test.

Fuel Control — 5.0L CFI — Pinpoint Test — HC

TEST STEP	RESULT ▶	ACTION TO TAKE
HC7 HARNESS CHECK EGO CIRCUITS		
• Key Off, wait 10 seconds. • Disconnect processor 60 pin connector and inspect for damaged pins, corrosion, loose wires. Service as necessary. • Install Breakout box. Processor disconnected. • Measure resistance between test Pin 49 and engine block ground and between EGO connector and test Pin 29. • Are both circuits less than 5 ohms?	Yes ▶ No ▶	GO to HC8. SERVICE CORRECT as necessary. REPEAT Quick Test.
HC8 CHECK EGO CIRCUIT FOR SHORT TO GROUND		
• Key Off. • Breakout box installed. Processor disconnected. • DVOM to 200,000 ohms range. • Disconnect EGO sensor at harness. • Measure resistance between test Pin 29 and test Pin 40.	Reading is 10,000 ohms or greater ▶ Reading is less than 10,000 ohms ▶	GO to HC9. CORRECT cause of resistance to ground. REPEAT Quick Test.
HC9 ATTEMPT TO ELIMINATE LEAN CODE		
• Key Off. • Reconnect EGO sensor. • Make certain that the ACT sensor is still disconnected. • Start the engine and run at approximately 2000 rpm for 1 minute, then, allow engine to return to idle. • Enter self-test. • Is code 41 still present? **NOTE: Disregard any other output codes at this time.**	Yes ▶ No ▶	INSPECT for corrosion or damaged pins. If OK, REPLACE processor. REPEAT Quick Test. EGO input circuit OK. RECONNECT ACT sensor. GO to HC1.

Fuel Control — 5.0L CFI — Pinpoint Test — HC

TEST STEP	RESULT ▶	ACTION TO TAKE
HC10 FUEL CONTROL — ALWAYS RICH		
• Key Off, wait 10 seconds. • **NOTE: Non-EEC areas could cause a code 42. Check for: — Fuel contaminated engine oil — Ignition caused misfire — CANP problems** • Refer to illustration HC. • Disconnect vehicle harness at the EGO sensor. Using a jumper wire, ground vehicle harness EGO circuit at the EGO sensor to the engine block. • Repeat Engine Running Self-Test. • Service code 41 results.	Yes ▶ No ▶	GO to HC12. GO to HC11.
HC11 HARNESS CHECK		
• Key Off, wait 10 seconds. • Refer to illustration HC. • Install Breakout box. • Measure resistance between test Pin 49 and EGO ground at engine block. • Measure resistance between test Pin 29 and EGO harness connector. • Are both circuits less than 5 ohms?	Yes ▶ No ▶	DISCONNECT processor connector. INSPECT for damage or corrosion. If OK, REPLACE processor. REPEAT Quick Test. CORRECT harness circuit with resistance greater than 5 ohms. REPEAT Quick Test.
HC12 EGO CHECK		
• Refer to illustration HC. • DVOM on 20V scale. • With EGO sensor disconnected from the harness, connect a DVOM from EGO sensor to engine ground. • Disconnect PCV hose. • Start engine and run at approximately 2000 rpm. Does DVOM read less than 0.4V within 1 minute?	Yes ▶ No ▶	EGO sensor OK. GO to HC1. REPLACE EGO sensor. REPEAT Quick Test.

Fuel Control — 5.0L CFI — Pinpoint Test — HC

HC13 CONTINUOUS TESTING: CODE 41/42

41 — EGO indicated the fuel system was lean for more than 15 seconds when the system should have been in closed loop fuel control.

42 — EGO indicated the fuel system was rich for more than 15 seconds when the system should have been in closed loop fuel control.

*CLOSED LOOP — Fuel control under the influence of the EGO sensor.

*OPEN LOOP — Fuel control NOT under the influence of the EGO sensor.

Before attempting to correct a fuel control code, 41/42, diagnose all other drive complaints first, eg., rough idle, misses, etc.

NOTE: The fuel control code may help in this diagnosis.

Using the fuel control service code, isolate the cause of the fuel control problem.

Some areas to check are:

- Unmetered Air: Vacuum leaks/intake air leaks.
  - Canister Purge System.
  - PCV System.
  - Engine sealing.
  - Air leaks between VAF meter and throttle body.
- EGO Fuel Fouled: Whenever an over-rich fuel condition has been experienced (fuel fouled spark plugs), make a thorough check of the ignition system. In the event the EGO sensor is suspected of being fuel fouled (low output, slow response), run the vehicle at sustained high speeds (within legal limits) followed by a few hard accelerations. This will burn off EGO contamination and restore proper EGO operation.
- Fuel Pressure: Perform Pinpoint Test Step HC1.
- Ignition System: Always in default spark (10 degrees). Refer to Quick Test Step 4.0.
- Improper Fueling: Lead fouled EGO sensor.
- TP Sensor: Not moving (mechanical damage). Connect DVOM to test Pin 47 and to test Pin 46. Key to Run. Observe DVOM while moving the throttle. Reading must increase with increase in throttle opening. If not correct, REPLACE as necessary.
- If at this point, the drive concern is still present, perform Steps **HC2** through **HC5** only.

Fuel Control — 5.0L CFI | Pinpoint Test | HC

TEST STEP	RESULT ▶	ACTION TO TAKE
HC14 CHECK HEATER ELEMENT RESISTANCE ON HEATED EGO		
• Key Off, Engine Off. • DVOM on 200 ohm scale. • Measure resistance between Run circuit and ground on heated EGO connector at room temperature.	Reading between 2.5 and 5.0 ohms ▶	GO to HC15.
	Reading less than 2.5 or greater than 5.0 ohms ▶	REPLACE HEGO sensor
HC15 CHECK FOR POWER AT HEGO HARNESS CONNECTOR		
• Key On, Engine Off. • DVOM on 20V scale. • Connect positive lead to run circuit and negative lead to ground on HEGO harness connector.	Reading 10.5V or greater ▶	HEGO sensor system OK, RETURN to Section 2.
	Reading less than 10.5V ▶	GO to HC16.
HC16 CHECK CONTINUITY OF GROUND TO HEGO CONNECTOR		
• Key Off, wait 10 seconds. • DVOM on 200 ohm scale. • Measure resistance of ground circuit between HEGO connector and battery ground.	Reading less than 5.0 ohms ▶	SERVICE open in run circuit.
	Reading 5.0 ohms or greater ▶	SERVICE open in ground circuit.
HC17 WARM ENGINE, CODE 43		
• Run vehicle at 2000 rpm for 2 minutes. • Key Off, wait 10 seconds. • Perform Engine Running Quick Test.	Code 43 is not present ▶	SERVICE other codes as necessary.
	Code 43 is present ▶	GO to HC18.
HC18 CHECK EXHAUST LEAKS, CODE 43		
• Are there any exhaust leaks at manifold or inlet pipe?	Yes ▶	SERVICE as necessary. RERUN Quick Test.
	No ▶	GO to HC19.
HC19 CHECK IDLE QUALITY, CODE 43		
• Start vehicle, let engine idle. • Does engine idle quality deteriorate and remain poor during and after the Quick Test?	Yes ▶	REFER to EFI/CFI, Section 4.
	No ▶	REPLACE EGO sensor. RERUN Quick Test.

Fuel Control — 3.8L CFI | Pinpoint Test | HD

HD REFERENCE DRAWING

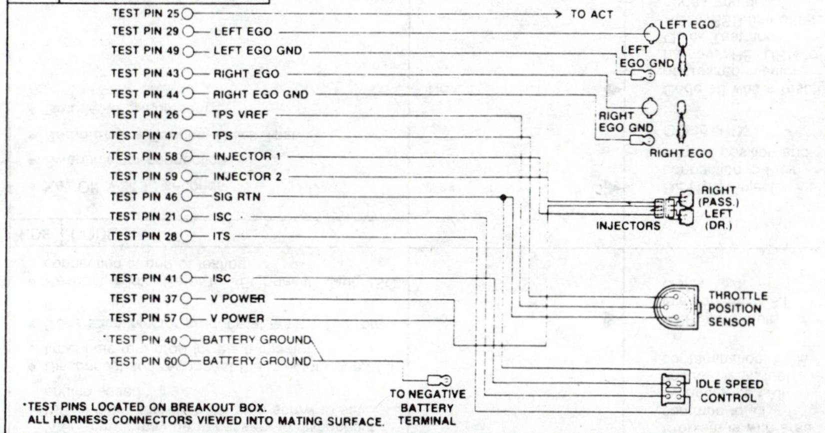

*TEST PINS LOCATED ON BREAKOUT BOX.
ALL HARNESS CONNECTORS VIEWED INTO MATING SURFACE.

STOP-WARNING

You should enter this Pinpoint Test only when a Service Code 16, 17, 41, 91, 42, 92, 43, 93, 97, 98 or 48 is received in Quick Test Step 5.0 or 6.0 or when directed here from Pinpoint Test A or by Diagnostics by Symptom in the Engine Supplement Section.

To prevent the replacement of good components, be aware that the following non-EEC areas may be at fault:

- Ignition Coil
- Distributor Cap
- Distributor Rotor
- Fouled Spark Plugs
- Spark Plug Wires
- CANP Problems
- PCV Valves (see note below)
- EGR Valve and Gasket
- Air Filter
- Fuel Contamination, Engine Oil
- Poor Power Ground
- Fuel Pressure
- Manifold Leaks, Intake Exhaust
- Engine Not at Normal Operating Temperatures

This Pinpoint Test is intended to diagnose only the following:

- EGO Sensors
- EGO Signals and Ground Circuits
- EGO Sensor Connections
- Fuel Injectors
- Processor Assembly

NOTE: Fuel-contaminated engine oil may affect 41, 91, 42, 92, 43 and 93 Service Codes, so if it is suspected, remove the PCV from the valve cover, and rerun the Quick Test. If the problem is corrected, then change the engine oil and filter.

Fuel Control — 3.8L CFI | Pinpoint Test | HD

<table>
<tr><th>TEST STEP</th><th>RESULT ▶</th><th>ACTION TO TAKE</th></tr>
<tr><td colspan="3">HD1 FUEL PRESSURE CHECK</td></tr>
<tr><td rowspan="2">• Install fuel pressure gauge.
• Start and run engine. Fuel pressure must be 241-310 kPa (35-45 psi).
• Fuel pressure must remain at 276 ± 34 kPa (40 psi ± 5 psi) for 60 seconds after final key Off.
• Is fuel pressure within specification?
For No Starts:
• If engine will not run cycle the key from Off to On several times.</td><td>Yes ▶</td><td>GO to HD2.</td></tr>
<tr><td>No ▶</td><td>REFER to Shop Manual for electric fuel pump.</td></tr>
<tr><td colspan="3">HD2 FUEL DELIVERY TEST</td></tr>
<tr><td rowspan="2">• Key Off.
• Fuel pressure gauge installed.
• Pressurize fuel system as per Step HD1.
• Locate and disconnect the fuel pump relay.
• Crank engine for 5 seconds.
• Take pressure reading at the end of 5 second crank.
NOTE: Verify fuel quality; air and/or water will also pressurize and look like acceptable fuel delivery.</td><td>PRESSURE GAUGE READING:
Pressure is approximately 10-20 psi at the end of 5 second crank cycle. Refer to note below ▶</td><td>The EEC system is not the fault of the No Start. Fuel and spark are present. REFER to Section for other No Start Routines. If complaint was runs rough, misses or fuel code, GO to HD3.</td></tr>
<tr><td>Pressure is greater or less than specified ▶</td><td>GO to HD3.</td></tr>
<tr><td colspan="3">NOTE: The colder the engine, the greater the pressure drop (i.e., an engine coolant temperature of 200°F equals approximately a 10 psi drop in 5 seconds. 60°F equals approximately a 20 psi drop in 5 seconds).</td></tr>
<tr><td colspan="3">HD3 HARNESS INJECTOR RESISTANCE CHECK</td></tr>
<tr><td rowspan="2">• Key Off.
• DVOM on 200 ohm range.
• Disconnect processor 60 Pin connector and inspect for damaged pins, corrosion, loose wires. Service as necessary.
• Install Breakout box, processor disconnected.
• Measure resistance between test Pins 57 and 58 and between test Pins 57 and 59.</td><td>Resistance is less than 1.5 ohms or greater than 3.5 ohms ▶</td><td>SERVICE the harness connectors on the suspect injector for opens or shorts. If OK, REPLACE injector.</td></tr>
<tr><td>Resistance is 1.5-3.5 ohms ▶</td><td>GO to HD4.</td></tr>
</table>

Fuel Control — 3.8L CFI | Pinpoint Test | HD

<table>
<tr><th>TEST STEP</th><th>RESULT ▶</th><th>ACTION TO TAKE</th></tr>
<tr><td colspan="3">HD4 INJECTOR DRIVE SIGNAL CHECK</td></tr>
<tr><td rowspan="3">Requires standard non-powered 12 volt test lamp.
• Key Off.
• Breakout box installed.
• Connect processor to Breakout box.
• Refer to illustration HD.
• Connect test lamp between test Pin 37 and Pin 58.
• Crank or start engine.
• Repeat above test between test Pin 37 and Pin 59.</td><td>Dim glow at light on both tests ▶</td><td>GO to HD5.</td></tr>
<tr><td>No light on one or both tests ▶</td><td>VERIFY 12-volt battery power at Pins 37 and 57. If OK, REPLACE processor. REPEAT Quick Test.</td></tr>
<tr><td>Bright light on one or both tests ▶</td><td>CHECK injector circuits 1 and 2 for shorts to ground. If OK, REPLACE processor. REPEAT Quick Test.</td></tr>
<tr><td colspan="3">HD5 INJECTOR BALANCE TEST</td></tr>
<tr><td rowspan="2">• Connect tachometer to engine. Run engine at 2000 rpm. It may be necessary to disconnect ISC and use throttle body stop screw to set engine speed.
• Disconnect and reconnect the injectors one at a time: note rpm drop for each injector.
• Does each injector produce at least a 150 rpm drop?
• Reset curb idle. Refer to VECI decal. Verify ISC connection at end of testing.</td><td>Yes ▶</td><td>Fuel delivery OK. Problem is with area common to all cylinders, i.e.: Air/vacuum leak, fuel contamination, EGR.</td></tr>
<tr><td>No ▶</td><td>REPLACE faulty injector. REPEAT Quick Test.</td></tr>
<tr><td colspan="3">HD6 CODE: 48</td></tr>
<tr><td rowspan="2">• Key Off, wait 10 seconds.
• Swap injector connection.
• Rerun Engine Running Quick Test.
• Is code 48 still present?</td><td>Yes ▶</td><td>RETURN injector connection to their original position and GO to HD7.</td></tr>
<tr><td>No ▶</td><td>Code 48 was a result of crossed injector harness. RETURN to Quick Test and ADDRESS any other codes and or conditions.</td></tr>
</table>

Fuel Control — 3.8L CFI — Pinpoint Test — HD

TEST STEP	RESULT	▶	ACTION TO TAKE
HD7 EGO VERIFICATION			
• Key Off, wait 10 seconds. • Disconnect left EGO at EGO sensor. • Rerun Engine Running Quick Test. • Is code 48 still present?	Yes	▶	Code 48 is a result of an open in left EGO circuit. GO to HD8.
	No	▶	Original code 48 is a result of an open (lean) in the right EGO circuit. RECONNECT left EGO and GO to HD12.
HD8 FUEL CONTROL — ALWAYS LEAN: LEFT EGO			
NOTE: Vacuum/air leaks in non-EEC areas could cause a code. Check for: **— Leaking vacuum actuator (eg: A/C control motor)** **— Engine sealing** **— EGR system** **— PCV system** **— Lead contaminated EGO sensor** • Refer to illustration HD. • Key Off. DVOM on 20V scale. Disconnect left EGO sensor from vehicle harness. Connect DVOM to EGO sensor and engine ground. Disconnect ACT sensor. • Start the engine and run at approximately 2000 rpm. Does the DVOM read greater than 0.5V within 1 minute?	Yes	▶	GO to HD9.
	No	▶	REPLACE EGO sensor. REPEAT Quick Test.
HD9 HARNESS CHECK EGO CIRCUITS (LEFT)			
• Key Off, wait 10 seconds. • Disconnect processor 60 pin connector and inspect for damaged pins, corrosion, loose wires. Service as necessary. • Install Breakout box. Processor disconnected. • Measure resistance between test Pin 49 and engine block ground and between left EGO connector and test Pin 29. • Are both circuits less than 5 ohms?	Yes	▶	GO to HD10.
	No	▶	SERVICE CORRECT as necessary. REPEAT Quick Test.

Fuel Control — 3.8L CFI — Pinpoint Test — HD

TEST STEP	RESULT	▶	ACTION TO TAKE
HD10 CHECK EGO CIRCUIT FOR SHORT TO GROUND (LEFT)			
• Key Off. • Breakout box installed. Processor disconnected. • DVOM to 200,000 ohms range. • Measure resistance between test Pin 29 (left EGO) and test Pin 40.	Reading is 10,000 ohms or greater	▶	GO to HD11.
	Reading is less than 10,000 ohms	▶	CORRECT cause of resistance to ground. REPEAT Quick Test.
HD11 ATTEMPT TO ELIMINATE LEFT EGO LEAN CODES			
• Key Off. • Reconnect EGO sensor. • Make certain that the ACT sensor is still disconnected. • Start the engine and run at approximately 2000 rpm for 1 minute. Allow engine to return to idle. • Enter Engine Running Quick Test. • Is code 41 still present? **NOTE: Disregard any other output codes at this time.**	Yes	▶	INSPECT for corrosion or damaged pins. If OK, REPLACE processor. REPEAT Quick Test.
	No	▶	EGO input circuit OK. RECONNECT ACT sensor. GO to HD1.
HD12 FUEL CONTROL — ALWAYS LEAN: RIGHT EGO			
NOTE: Vacuum/air leaks in non-EEC areas could cause a code. Check for: **— Leaking vacuum actuator (eg: A/C control motor)** **— Engine sealing** **— EGR system** **— PCV system** • Refer to illustration HD. • Key Off. DVOM on 20V scale. Disconnect right EGO sensor from vehicle harness. Connect DVOM to EGO sensor and engine ground. Disconnect ACT sensor. • Start the engine and run at approximately 2000 rpm. Does the DVOM read greater than 0.5V within 1 minute?	Yes	▶	GO to HD13.
	No	▶	REPLACE EGO sensor. REPEAT Quick Test.

Fuel Control — 3.8L CFI | Pinpoint Test | HD

TEST STEP	RESULT ▶	ACTION TO TAKE
HD13 HARNESS CHECK EGO CIRCUITS (RIGHT)		
• Key Off, wait 10 seconds. • Disconnect processor 60 Pin connector and inspect for damaged pins, corrosion, loose wires. Service as necessary. • Install Breakout box. Processor disconnected. • Measure resistance between test Pin 44 and engine block ground and between right EGO connector and test Pin 43. • Are both circuits less than 5 ohms?	Yes ▶ No ▶	GO to HD14. SERVICE CORRECT as necessary. REPEAT Quick Test.
HD14 CHECK EGO CIRCUIT FOR SHORT TO GROUND (RIGHT)		
• Key Off. • Breakout box installed. Processor disconnected. • DVOM to 200,000 ohms range. • Disconnect right EGO sensor at harness. • Measure resistance between test Pin 43 and test Pin 40.	Reading is 10,000 ohms or greater ▶ Reading is less than 10,000 ohms ▶	GO to HD15. CORRECT cause of resistance to ground. REPEAT Quick Test.
HD15 ATTEMPT TO ELIMINATE RIGHT LEAN CODE		
• Key Off. • Reconnect EGO sensor. • Make certain that the ACT sensor is still disconnected. • Start the engine and run at approximately 2000 rpm for 1 minute. Allow engine to return to idle. • Enter Engine Running Self-Test. • Is code 91 still present? **NOTE: Disregard any other code output at this time.**	Yes ▶ No ▶	INSPECT for corrosion or damaged pins. If OK, REPLACE processor. REPEAT Quick Test. EGO input circuit OK. RECONNECT ACT sensor. GO to HD1.

Fuel Control — 3.8L CFI | Pinpoint Test | HD

TEST STEP	RESULT ▶	ACTION TO TAKE
HD16 FUEL CONTROL — ALWAYS RICH: LEFT		
• Key Off, wait 10 seconds. • **NOTE: Non-EEC areas could cause a rich code. Check for: — Fuel contaminated engine oil — Ignition caused misfire — CANP problems** • Refer to illustration HD. • Disconnect vehicle harness at the left EGO sensor. Using a jumper wire, ground vehicle harness EGO circuit at the EGO sensor to the engine block. • Repeat Engine Running Self-Test. • Service code 41/48 results.	Yes ▶ No ▶	GO to HD18. GO to HD17.
HD17 HARNESS CHECK (LEFT)		
• Key Off, wait 10 seconds. • Refer to illustration HD. • Disconnect processor 60 Pin connector and inspect for damaged pins, corrosion, loose wires. Service as necessary. • Install Breakout box. • Measure resistance between test Pin 49 and left EGO ground at engine block. • Measure resistance between test Pin 29 and left EGO harness connector. • Are both circuits less than 5 ohms?	Yes ▶ No ▶	DISCONNECT processor connector. INSPECT for damage or corrosion. If OK, REPLACE processor. REPEAT Quick Test. CORRECT harness circuit with resistance greater than 5 ohms. REPEAT Quick Test.
HD18 EGO CHECK (LEFT)		
• Refer to illustration HD. • DVOM on 20V scale. • With EGO sensor disconnected from the harness, connect a DVOM from EGO sensor to engine ground. • Disconnect PCV hose. • Start engine and run at approximately 2000 rpm. Does the DVOM read less than 0.4V within 2 minutes?	Yes ▶ No ▶	EGO sensor OK. GO to HD1. REPLACE EGO sensor. REPEAT Quick Test.

Fuel Control — 3.8L CFI | Pinpoint Test | HD

TEST STEP	RESULT ▶	ACTION TO TAKE
HD19 FUEL CONTROL — ALWAYS RICH: RIGHT		
• Key Off, wait 10 seconds. • **NOTE: Non-EEC areas could cause a rich code. Check for:** **— Fuel contaminated engine oil** **— Ignition caused misfire** **— CANP problems** • Refer to illustration HD. • Disconnect vehicle harness at the right EGO sensor. Using a jumper wire, ground vehicle harness EGO circuit at the EGO sensor to the engine block. • Repeat Engine Running Self-Test. • Service code 91 results.	Yes ▶ No ▶	GO to HD21. GO to HD20.
HD20 HARNESS CHECK (RIGHT)		
• Key Off, wait 10 seconds. • Refer to illustration HD. • Install Breakout box. • Measure resistance between test Pin 44 and right EGO ground at engine block. • Measure resistance between test Pin 43 and right EGO harness connector. • Are both circuits less than 5 ohms?	Yes ▶ No ▶	DISCONNECT processor connector. INSPECT for damage or corrosion. If OK, REPLACE processor. REPEAT Quick Test. CORRECT harness circuit with resistance greater than 5 ohms. REPEAT Quick Test.
HD21 EGO CHECK		
• Refer to illustration HD. • DVOM on 20V scale. • With EGO sensor disconnected from the harness, connect a DVOM from EGO sensor to engine ground. • Disconnect PCV hose. • Start engine and run at approximately 2000 rpm. Does the DVOM read less than 0.4V within 1 minute?	Yes ▶ No ▶	EGO sensor OK. GO to HD1. REPLACE EGO sensor. REPEAT Quick Test.

Fuel Control — 3.8L CFI | Pinpoint Test | HD

HD22 CONTINUOUS TESTING: CODE 41 42

41 — EGO indicated the fuel system was lean for more than 15 seconds when the system should have been in closed loop fuel control.

42 — EGO indicated the fuel system was rich for more than 15 seconds when the system should have been in closed loop fuel control.

*CLOSED LOOP — Fuel control under the influence of the EGO sensor.

*OPEN LOOP — Fuel control NOT under the influence of the EGO sensor.

Before attempting to correct a fuel control code, 41 42, diagnose all other drive complaints first, eg., rough idle, misses, etc.

NOTE: The fuel control code may help in this diagnosis.

Using the fuel control service code, isolate the cause of the fuel control problem.

Some areas to check are:

- Unmetered Air: Vacuum leaks intake air leaks.
  - — Canister Purge System.
  - — PCV System.
  - — Engine sealing.
  - — Air leaks between VAF meter and throttle body.
- EGO Fuel Fouled: Whenever an over-rich fuel condition has been experienced (fuel fouled spark plugs), make a thorough check of the ignition system. In the event the EGO sensor is suspected of being fuel fouled (low output, slow response), run the vehicle at sustained high speeds (within legal limits) followed by a few hard accelerations. This will burn off EGO contamination and restore proper EGO operation.
- Fuel Pressure: Perform Pinpoint Test Step HD1.
- Ignition System: Always in default spark (10 degrees). Refer to Quick Test Step 4.0.
- Improper Fueling: Lead fouled EGO sensor.
- TP Sensor: Not moving (mechanical damage). Connect DVOM to test Pin 47 and to test Pin 46. Key to Run. Observe DVOM while moving the throttle. Reading must increase with increase in throttle opening. If not correct, REPLACE as necessary.
- If at this point, the drive concern is still present, perform Steps HD2 through HD5 only.

Fuel Control — 3.8L CFI — Pinpoint Test HD

TEST STEP	RESULT ▶	ACTION TO TAKE
HD23 WARM ENGINE, CODE 43/93		
• Run vehicle at 2000 rpm for 2 minutes. • Key Off, wait 10 seconds. • Perform Engine Running Quick Test.	Code 43 is not present ▶	SERVICE other codes as necessary.
	Code 43 is present ▶	GO to HD24.
HD24 CHECK EXHAUST LEAKS, CODE 43/93		
• Are there any exhaust leaks at manifold or inlet pipe?	Yes ▶	SERVICE as necessary. RERUN Quick Test.
	No ▶	GO to HD25.
HD25 CHECK IDLE QUALITY, CODE 43/93		
• Start vehicle, let engine idle. • Does engine idle quality deteriorate and remain poor during and after the Quick Test?	Yes ▶	REFER to EFI/CFI,
	No ▶	REPLACE EGO sensor. RERUN Quick Test.

Fuel Control — EFI Truck — Pinpoint Test HE

HE

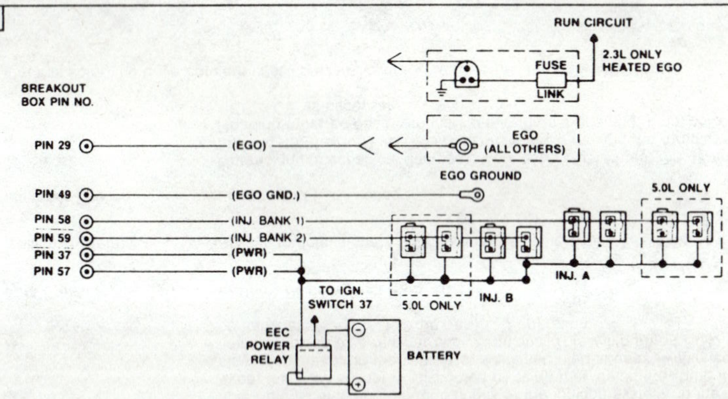

STOP-WARNING

You should enter this Pinpoint Test only when a Service Code 41 or 42 is received in Quick Test Step 5.0 or 6.0 or when directed here from Pinpoint Test A or Diagnostics by Symptom in the Engine Supplement Section.

To prevent the replacement of good components, be aware that the following non-EEC areas may be at fault:

- Ignition Coil
- Distributor Cap
- Distributor Rotor
- Fouled Spark Plugs
- Spark Plug Wires
- CANP Problems
- PCV Valves (see note below)
- EGR Valve and Gasket
- Air Filter
- Fuel Contamination, Engine Oil
- Poor Power Ground
- Fuel Pressure
- Manifold Leaks, Intake/Exhaust
- Engine Not at Normal Operating Temperatures

This Pinpoint Test is intended to diagnose only the following:

- EGO Sensor
- Harness Circuits EGO Grd., EGO, Inj. Bank 1, Inj. Bank 2, V PWR
- EGO Sensor Connection
  - — Code 42 start at **HE7**
  - — Code 41 start at **HE10**
- Vacuum Systems
- Fuel Injectors
- Processor Assembly

NOTE: Fuel contaminated engine oil may affect 41 and 42 Service Codes. If this is suspected, remove the PCV from the valve cover and repeat the Quick Test. If the problem is corrected, then change the engine oil and filter.

Fuel Control — EFI Truck | Pinpoint Test | HE

TEST STEP	RESULT ▶	ACTION TO TAKE
HE1 FUEL PRESSURE CHECK		
• Key Off, wait 10 seconds. • Install fuel pressure gauge. • Start and run engine. Fuel pressure must be 172-310 kPa (25-45 psi). • Fuel pressure must remain at 276 ± 34 kPa (30 ± 5 psi) for 60 seconds after final key Off. **For No Starts:** • If engine will not run, cycle the key from Off to On several times.	Fuel pressure is within specification ▶ Fuel pressure is not within specification ▶	GO to HE2. REFER to the Shop Manual for electric fuel pump and fuel pressure regulator check.
HE2 FUEL DELIVERY TEST		
• Key Off. • Fuel pressure gauge installed. • Pressurize fuel system as per Step HE1. • Locate and disconnect the fuel pump relay. • Crank engine for 5 seconds. • Take pressure reading at the end of 5 second crank. **NOTE: Verify fuel quality; air and/or water will also pressurize and look like acceptable fuel delivery.**	PRESSURE GAUGE READING: Pressure is approximately 10-20 psi at the end of 5 second crank cycle. Refer to note below ▶ Pressure is greater or less than specified ▶	 The EEC system is not the fault of the No Start. Fuel and spark are present. REFER to Section for other No Start Routines. If complaint was runs rough, misses or fuel code, GO to HE3. GO to HE3.

NOTE: The colder the engine, the greater the pressure drop (i.e., an engine coolant temperature of 200°F equals approximately a 10 psi drop in 5 seconds. 60°F equals approximately a 20 psi drop in 5 seconds).

Fuel Control — EFI Truck | Pinpoint Test | HE

TEST STEP	RESULT ▶	ACTION TO TAKE
HE3 HARNESS INJECTOR RESISTANCE CHECK		
• Key Off, wait 10 seconds. • Disconnect processor 60 pin connector and inspect for damaged pins, corrosion, loose wires. Service as necessary. • Connect Breakout box to harness. Leave processor disconnected. • DVOM on 200 ohm scale. • Through the vehicle harness, measure the resistance of injector Bank 1 between test Pin 37 and test Pin 58, record reading. • Through the vehicle harness, measure the resistance of injector Bank 2 between test Pin 37 and test Pin 59, record reading. • Are both readings between 3.5 and 6.0 ohms for 5.0L EFI, and 7.5 and 10.0 ohms for 2.3L EFI?	Yes ▶ No ▶	GO to HE5. GO to HE4.
HE4 ISOLATE FAULTY INJECTOR CIRCUIT		
• Key Off. • Breakout box installed. • Leave processor disconnected. • DVOM on 200 ohm scale. • Disconnect all injectors on suspect Bank. Measure the resistance of each injector by connecting one injector at a time, and reading resistance between test Pin 37 and 58 for Bank 1 or test Pin 37 and 59 for Bank 2. • Are all readings between 15.0 and 19.0 ohms?	Yes ▶ No ▶	GO to HE5. SERVICE the harness/connectors on the suspect injector for opens or shorts. If OK, REPLACE injector. REPEAT Quick Test.
HE5 INJECTOR DRIVE SIGNAL CHECK		
Requires standard non-powered 12-volt test lamp. • Key Off. • Breakout box installed. • Connect processor to Breakout box. • Connect test lamp between test Pin 37 and test Pin 58. • Crank or start engine. • Repeat above test between test Pin 37 and test Pin 59.	Dim glow at light on both tests ▶ No light on one or both tests ▶ Bright light on one or both tests ▶	GO to HE6. VERIFY 12-volt battery power at Pins 37 and 57. If OK, REPLACE processor. REPEAT Quick Test. CHECK circuits Bank 1 and Bank 2 for shorts to ground. If OK, REPLACE processor. REPEAT Quick Test.

Fuel Control — EFI Truck | Pinpoint Test | HE

TEST STEP	RESULT ▶	ACTION TO TAKE
HE6 INJECTOR BALANCE TEST		
• Connect tachometer to engine. Run engine at 2000 rpm. Disconnect ISC and use throttle body stop screw to set engine speed. • Disconnect and reconnect the injectors one at a time; note rpm drop for each injector. • Does each injector produce at least a 150 rpm drop for a 2.3L EFI and at least a 50 rpm drop for a 5.0L EFI? • Reset curb idle. Refer to VECI decal. Verify ISC connection at end of testing.	Yes ▶	Fuel delivery OK. Problem is with area common to all cylinders, i.e.: Air/vacuum leak, fuel contamination, EGR.
	No ▶	REPLACE faulty injector. REPEAT Quick Test.
HE7 FUEL CONTROL — ALWAYS RICH: CODE 42		
• Key Off, wait 10 seconds. • **NOTE: Non-EEC areas could cause a code 42. Check for: — Fuel contaminated engine oil — Ignition caused misfire — CANP problems** • Disconnect vehicle harness at the EGO sensor. Using a jumper wire, ground vehicle harness EGO circuit at the EGO sensor to the engine block. • Repeat Engine Running Self-Test. • Service code 41 results.	Yes ▶	GO to HE9.
	No ▶	GO to HE8.
HE8 EGO HARNESS CHECK		
• Key Off, wait 10 seconds. • Install Breakout box. • Measure resistance between test Pin 49 and EGO ground at engine block. • Measure resistance between test Pin 29 and EGO harness connector. • Are both circuits less than 5 ohms?	Yes ▶	DISCONNECT processor connector. INSPECT for damage or corrosion. If OK, REPLACE processor. REPEAT Quick Test.
	No ▶	CORRECT harness circuit with resistance greater than 5 ohms. REPEAT Quick Test.

Fuel Control — EFI Truck | Pinpoint Test | HE

TEST STEP	RESULT ▶	ACTION TO TAKE
HE9 EGO CHECK		
• DVOM on 20V scale. • With EGO sensor disconnected from the harness, connect a DVOM from EGO sensor to engine ground. • Disconnect PCV hose. • Start engine and run at approximately 2000 rpm. Does the DVOM read less than 0.4V within 1 minute?	Yes ▶	EGO sensor OK. GO to HE1.
	No ▶	REPLACE EGO sensor. REPEAT Quick Test.
HE10 FUEL CONTROL — ALWAYS LEAN: CODE 41		
NOTE: Vacuum/air leaks in non-EEC areas could cause a code 41. Check for: — Leaking vacuum actuator (eg: A/C control motor) — Engine sealing — EGR system — PCV system — Unmetered air leak between air meter and throttle body — Lead contaminated EGO sensor • Key Off. DVOM on 20V scale. Disconnect EGO sensor from vehicle harness. Connect DVOM to EGO sensor and engine ground. • Disconnect ACT sensor. • Start the engine and run at approximately 2000 rpm. Does the DVOM read greater than 0.5V within 1 minute?	Yes ▶	GO to HE11.
	No ▶	REPLACE EGO sensor. REPEAT Quick Test.
HE11 HARNESS CHECK EGO CIRCUITS		
• Key Off. • Install Breakout box. Processor disconnected. • Measure resistance between test Pin 49 and engine block ground and between EGO connector and test Pin 29. • Are both circuits less than 5 ohms?	Yes ▶	GO to HE12.
	No ▶	SERVICE CORRECT as necessary. REPEAT Quick Test.

Fuel Control — EFI Truck | Pinpoint Test | HE

TEST STEP	RESULT ▶	ACTION TO TAKE
HE12 CHECK EGO CIRCUIT FOR SHORT TO GROUND		
• Key Off. • Breakout box installed. Processor disconnected. • DVOM to 200,000 ohms range. • Measure resistance between test Pin 29 and test Pin 40.	Reading is 10,000 ohms or greater ▶	GO to HE13.
	Reading is less than 10,000 ohms ▶	CORRECT cause of resistance to ground. REPEAT Quick Test.
HE13 ATTEMPT TO ELIMINATE CODE 41		
• Key Off. • Reconnect EGO sensor. • Make certain that ACT sensor is still disconnected. • Start the engine and run at approximately 2000 rpm for 1 minute. Allow engine to return to idle. • Perform Engine Running Self-Test. • Is code 41 still present? **NOTE: Disregard any other code output at this time.**	Yes ▶	INSPECT for corrosion or damaged pins. If OK, REPLACE processor. REPEAT Quick Test.
	No ▶	RECONNECT ACT sensor. EGO input circuit OK. GO to HE1.

Fuel Control — EFI Truck | Pinpoint Test | HE

HE20 CONTINUOUS TESTING: CODE 41 42

41 — EGO indicated the fuel system was lean for more than 15 seconds when the system should have been in closed loop fuel control.

42 — EGO indicated the fuel system was rich for more than 15 seconds when the system should have been in closed loop fuel control.

*CLOSED LOOP — Fuel control under the influence of the EGO sensor.

*OPEN LOOP — Fuel control NOT under the influence of the EGO sensor.

Before attempting to correct a fuel control code. 41 42. diagnose all other drive complaints first. eg., rough idle, misses, etc.

NOTE: The fuel control code may help in this diagnosis.

Using the fuel control service code, isolate the cause of the fuel control problem.

Some areas to check are:

- Unmetered Air: Vacuum leaks intake air leaks.
  - — Canister Purge System.
  - — PCV System.
  - — Engine sealing.
  - — Air leaks between VAF meter and throttle body.
- EGO Fuel Fouled: Whenever an over-rich fuel condition has been experienced (fuel fouled spark plugs), make a thorough check of the ignition system. In the event the EGO sensor is suspected of being fuel fouled (low output, slow response), run the vehicle at sustained high speeds (within legal limits) followed by a few hard accelerations. This will burn off EGO contamination and restore proper EGO operation.
- Fuel Pressure: Perform Pinpoint Test Step HE1.
- Ignition System: Always in default spark (10 degrees). Refer to Quick Test Step 4.0.
- Improper Fueling: Lead fouled EGO sensor.
- TP Sensor: Not moving (mechanical damage). Connect DVOM to test Pin 47 and to test Pin 46. Key to Run. Observe DVOM while moving the throttle. Reading must increase with increase in throttle opening. If not correct, REPLACE as necessary.
- If at this point, the drive concern is still present, perform Steps HE2 through HE5 only.

Fuel Control — EFI Truck | Pinpoint Test | HE

TEST STEP	RESULT ▶	ACTION TO TAKE
HE21 CHECK HEATER ELEMENT RESISTANCE ON HEATED EGO		
• Key Off, Engine Off. • DVOM on 200 ohm scale. • Measure resistance between Run circuit and ground on heated EGO connector at room temperature.	Reading between 2.5 and 5.0 ohms ▶	GO to HE22.
	Reading less than 2.5 or greater than 5.0 ohms ▶	REPLACE HEGO sensor.
HE22 CHECK FOR POWER AT HEGO HARNESS CONNECTOR		
• Key On, Engine Off. • DVOM on 20V scale. • Connect positive lead to run circuit and negative lead to ground on HEGO harness connector.	Reading 10.5V or greater ▶	HEGO sensor system OK, RETURN to Section 2.
	Reading less than 10.5V ▶	GO to HE23.
HE23 CHECK CONTINUITY OF GROUND TO HEGO CONNECTOR		
• Key Off, wait 10 seconds. • DVOM on 200 ohm scale. • Measure resistance of ground circuit from HEGO connector to battery ground.	Reading less than 5.0 ohms ▶	SERVICE open in run circuit.
	Reading 5.0 ohms or greater ▶	SERVICE open in ground circuit.

Fuel Control — 2.3L (HSC) CFI | Pinpoint Test | HF

HF

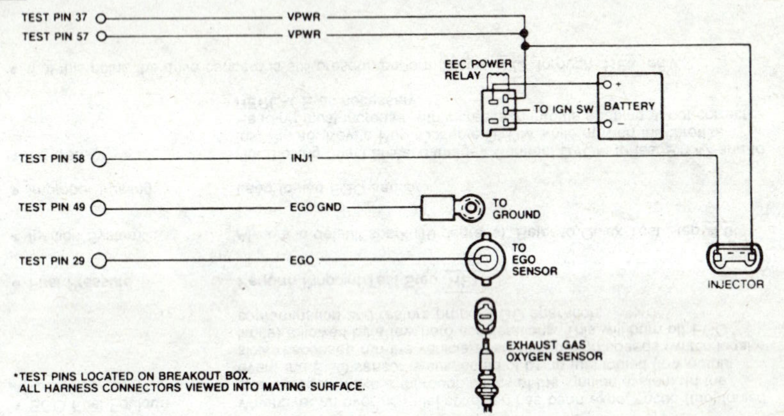

*TEST PINS LOCATED ON BREAKOUT BOX.
ALL HARNESS CONNECTORS VIEWED INTO MATING SURFACE.

STOP-WARNING

You should enter this Pinpoint Test only when a Service Code 41 or 42 is received in Quick Test Step 5.0 or 6.0 or when directed here from Pinpoint Test A or Diagnostics by Symptom in the Engine Supplement Section.

To prevent the replacement of good components, be aware that the following non-EEC areas may be at fault:

- Ignition Coil
- Distributor Cap
- Distributor Rotor
- Fouled Spark Plugs
- Spark Plug Wires
- CANP Problems
- PCV Valves (see note below)
- EGR Valve and Gasket
- Air Filter
- Fuel Contamination, Engine Oil
- Poor Power Ground
- Fuel Pressure
- Manifold Leaks, Intake Exhaust
- Engine Not at Normal Operating Temperatures

This Pinpoint Test is intended to diagnose only the following:

- EGO Sensor
- EGO Signal and Ground Circuit
- EGO Sensor Connection
- Fuel Injector
- Processor Assembly

NOTE: Fuel contaminated engine oil may affect 41 and 42 Service Codes, so if it is suspected, remove the PCV from the valve cover, and rerun the Quick Test. If the problem is corrected, then change the engine oil and filter.

Fuel Control — 2.3L (HSC) CFI — Pinpoint Test — HF

TEST STEP	RESULT ▶	ACTION TO TAKE
HF1 FUEL PRESSURE CHECK		
• Install fuel pressure gauge. • Start and run engine. Fuel pressure must be between 13 and 16 psi. • Fuel pressure must remain between 13.0 and 16.0 psi for 60 seconds after final key Off. • Is fuel pressure within specification? **For No Starts:** • If engine will not run cycle the key from Off to On several times.	Yes ▶ No ▶	GO to HF2. REFER to Shop Manual for electric fuel pump.
HF2 FUEL DELIVERY TEST		
• Key Off. • Fuel pressure gauge installed. • Pressurize fuel system as per Step HF1. • Locate and disconnect the fuel pump relay. • Crank engine for 5 seconds. • Take pressure reading at the end of 5 second crank. **NOTE: Verify fuel quality; air and/or water will also pressurize and look like acceptable fuel delivery.**	PRESSURE GAUGE READING: Pressure drops approximately 5-8 psi at the end of 5 second crank cycle. Refer to note below ▶ Pressure is greater or less than specified ▶	 The EEC system is not the fault of the No Start. Fuel and spark are present. REFER to Section for other No Start Routines. If complaint was runs rough or misses, CONTINUE to HF3. GO to HF3.
NOTE: The colder the engine, the greater the pressure drop (i.e., an engine coolant temperature of 200°F equals approximately a 5 psi drop in 5 seconds. 60°F equals approximately a 8 psi drop in 5 seconds).		
HF3 HARNESS INJECTOR RESISTANCE CHECK		
• Key Off. • DVOM on 200 ohm range. • Disconnect processor 60 Pin connector and inspect for damaged pins, corrosion, loose wires. Service as necessary. • Install Breakout box, processor disconnected. • Measure resistance between test Pins 57 and 58.	Resistance is less than 1.0 ohms or greater than 2.0 ohms ▶ Resistance is 1.0 or 2.0 ohms ▶	SERVICE the harness/connectors on the suspect injector for opens or shorts. If OK, REPLACE injector. GO to HF4.

Fuel Control — 2.3L (HSC) CFI — Pinpoint Test — HF

TEST STEP	RESULT ▶	ACTION TO TAKE
HF4 INJECTOR DRIVE SIGNAL CHECK		
Requires standard non-powered 12-volt test lamp. • Key Off. • Breakout box installed. • Connect processor to Breakout box. • Refer to illustration HF. • Connect test lamp between test Pin 37 and test Pin 58. • Crank or start engine.	Dim glow at light ▶ No light ▶ Bright light ▶	Fuel delivery OK. Problem is with area common to all cylinders, i.e., air/vacuum leak, fuel contamination, EGR. VERIFY 12-volt battery power at Pins 37 and 57. If OK, REPLACE processor. REPEAT Quick Test. CHECK injector circuit for shorts to ground. If OK, REPLACE processor. REPEAT Quick Test.
HF5 FUEL CONTROL — ALWAYS LEAN: CODE 41		
NOTE: Vacuum/air leaks in non-EEC areas could cause a code 41. Check for: **— Leaking vacuum actuator (eg: A/C control motor)** **— Engine sealing** **— EGR system** **— PCV system** **— Lead contaminated EGO sensor** • Refer to illustration HF. • Key Off. DVOM on 20V scale. Disconnect EGO sensor from vehicle harness. Connect DVOM to EGO sensor and engine ground. Disconnect ACT sensor. • Start the engine and run at approximately 2000 rpm. Does the DVOM read greater than 0.5V within 1 minute?	Yes ▶ No ▶	GO to HF6. REPLACE EGO sensor. REPEAT Quick Test.
HF6 HARNESS CHECK EGO CIRCUITS		
• Key Off, wait 10 seconds. • Disconnect processor 60 Pin connector and inspect for damaged pins, corrosion, loose wires. Service as necessary. • Install Breakout box. Processor disconnected. • Measure resistance between test Pin 49 and engine block ground and between EGO connector and test Pin 29. • Are both circuits less than 5 ohms?	Yes ▶ No ▶	GO to HF7. SERVICE CORRECT as necessary. REPEAT Quick Test.

Fuel Control — 2.3L (HSC) CFI — Pinpoint Test — HF

TEST STEP	RESULT ▶	ACTION TO TAKE
HF7 CHECK EGO CIRCUIT FOR SHORT TO GROUND		
• Key Off. • Breakout box installed. Processor disconnected. • DVOM to 200,000 ohms range. • Disconnect EGO sensor at harness. • Measure resistance between test Pin 29 and test Pin 40.	Reading is 10,000 ohms or greater ▶	GO to HF8.
	Reading is less than 10,000 ohms ▶	CORRECT cause of resistance to ground. REPEAT Quick Test.
HF8 ATTEMPT TO ELIMINATE CODE 41		
• Key Off. • Reconnect EGO sensor. • Make certain that the ACT sensor is still disconnected. • Start the engine and run at approximately 2000 rpm for 1 minute, allow engine to return to idle. • Enter Self-Test. • Is code 41 still present? **NOTE: Disregard any other code output at this time.**	Yes ▶	INSPECT for corrosion or damaged pins. If OK, REPLACE processor. REPEAT Quick Test.
	No ▶	EGO input circuit OK. RECONNECT ACT sensor. GO to HF1.
HF9 FUEL CONTROL — ALWAYS RICH: CODE 42		
• Key Off, wait 10 seconds. • **NOTE: Non-EEC areas could cause a code 42.** **Check for:** **— Fuel contaminated engine oil** **— Ignition caused misfire** **— CANP problems** • Refer to illustration HF. • Disconnect vehicle harness at the EGO sensor. Using a jumper wire, ground vehicle harness EGO circuit at the EGO sensor to the engine block. • Repeat Engine Running Self-Test. • Service code 41 results.	Yes ▶	GO to HF11.
	No ▶	GO to HF10.

Fuel Control — 2.3L (HSC) CFI — Pinpoint Test — HF

TEST STEP	RESULT ▶	ACTION TO TAKE
HF10 HARNESS CHECK		
• Key Off, wait 10 seconds. • Refer to illustration HF. • Disconnect processor 60 Pin connector and inspect for damaged pins, corrosion, loose wires. Service as necessary. • Install Breakout box. • Measure resistance between test Pin 49 and EGO ground at engine block. • Measure resistance between test Pin 29 and EGO harness connector. • Are both circuits less than 5 ohms?	Yes ▶	DISCONNECT processor connector. INSPECT for damage or corrosion. If OK, REPLACE processor. REPEAT Quick Test.
	No ▶	CORRECT harness circuit with resistance greater than 5 ohms. REPEAT Quick Test.
HF11 EGO CHECK		
• Refer to illustration HF. • DVOM on 20V scale. • With EGO sensor disconnected from the harness, connect a DVOM from EGO sensor to engine ground. • Disconnect PCV hose. • Start engine and run at approximately 2000 rpm. Does the DVOM read less than 0.4V within 1 minute?	Yes ▶	EGO sensor OK. GO to HF1.
	No ▶	REPLACE EGO sensor. REPEAT Quick Test.

Fuel Control — 2.3L (HSC) CFI | Pinpoint Test | HF

HF12 CONTINUOUS TESTING: CODE 41/42

41 — EGO indicated the fuel system was lean for more than 15 seconds when the system should have been in closed loop fuel control.

42 — EGO indicated the fuel system was rich for more than 15 seconds when the system should have been in closed loop fuel control.

*CLOSED LOOP — Fuel control under the influence of the EGO sensor.

*OPEN LOOP — Fuel control NOT under the influence of the EGO sensor.

Before attempting to correct a fuel control code, 41/42, diagnose all other drive complaints first. eg., rough idle, misses, etc.

NOTE: The fuel control code may help in this diagnosis.

Using the fuel control service code, isolate the cause of the fuel control problem.

Some areas to check are:

- Unmetered Air: Vacuum leaks/intake air leaks.
  - — Canister Purge System.
  - — PCV System.
  - — Engine sealing.
  - — Air leaks between VAF meter and throttle body.
- EGO Fuel Fouled: Whenever an over-rich fuel condition has been experienced (fuel fouled spark plugs), make a thorough check of the ignition system. In the event the EGO sensor is suspected of being fuel fouled (low output, slow response), run the vehicle at sustained high speeds (within legal limits) followed by a few hard accelerations. This will burn off EGO contamination and restore proper EGO operation.
- Fuel Pressure: Perform Pinpoint Test Step HF1.
- Ignition System: Always in default spark (10 degrees). Refer to Quick Test Step 4.0.
- Improper Fueling: Lead fouled EGO sensor.
- TP Sensor: Not moving (mechanical damage). Connect DVOM to test Pin 47 and to test Pin 46. Key to Run. Observe DVOM while moving the throttle. Reading must increase with increase in throttle opening. If not correct, REPLACE as necessary.
- If at this point, the drive concern is still present, perform Steps HF2 through HF3 only.

Fuel Control — 2.3L (HSC) CFI | Pinpoint Test | HF

TEST STEP	RESULT ▶	ACTION TO TAKE
HF13 WARM ENGINE, CODE 43		
• Run vehicle at 2000 rpm for 2 minutes. • Key Off, wait 10 seconds. • Perform Engine Running Quick Test.	Code 43 is not present ▶	SERVICE other codes as necessary.
	Code 43 is present ▶	GO to HF14.
HF14 CHECK EXHAUST LEAKS, CODE 43		
• Are there any exhaust leaks at manifold or inlet pipe?	Yes ▶	SERVICE as necessary. RERUN Quick Test.
	No ▶	GO to HF15.
HF15 CHECK IDLE QUALITY, CODE 43		
• Start vehicle, let engine idle. • Does engine idle quality deteriorate and remain poor during and after the Quick Test?	Yes ▶	REFER to EFI/CFI.
	No ▶	REPLACE EGO sensor. RERUN Quick Test.

Fuel Pump Circuit — (Inertia Switch) | Pinpoint Test | J

J

TYPE I — VEHICLE APPLICATION: LINCOLN TOWN CAR, FORD CROWN VICTORIA/MERCURY GRAND MARQUIS, TEMPO/TOPAZ, ESCORT/LYNX AND EXP

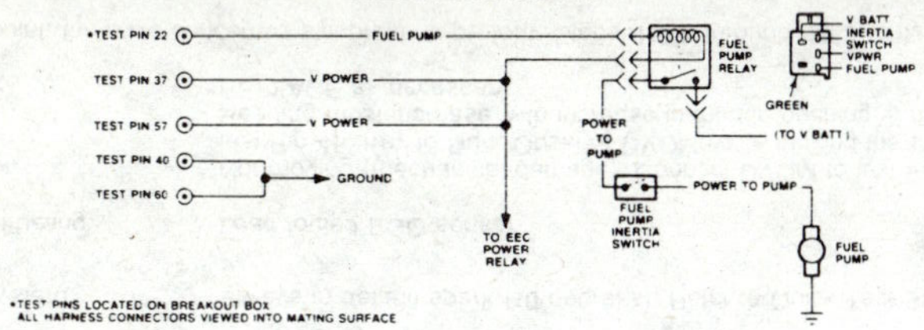

TYPE II — VEHICLE APPLICATION: MARK VII/CONTINENTAL, THUNDERBIRD COUGAR, LTD/MARQUIS, MUSTANG/CAPRI, RANGER, E-SERIES AND F-SERIES

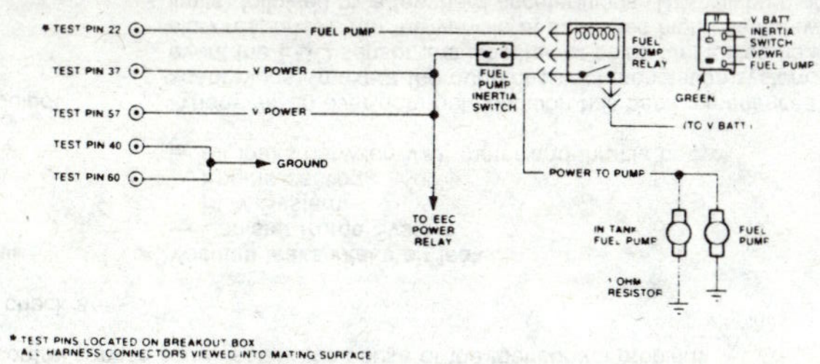

STOP-WARNING

You should enter this Pinpoint Test only when a service code 87 is received in Quick Test Step 3.0 or you are directed here from Pinpoint Test Step A22.

To prevent the replacement of good components, be aware that the following non-EEC areas may be at fault:

- Fuel Lines
- Fuel Filters
- Throttle Body
- Contaminated Fuel
- Fuel Pump

This Pinpoint Test is intended to diagnose only the following:

- Fuel Pump Relay
- Harness Circuits: V Batt., V Power, F.P., Ground and Power to Pump(s)
- Processor Assembly

Fuel Pump Circuit — (Inertia Switch) | Pinpoint Test | J

TEST STEP	RESULT ▶	ACTION TO TAKE
J1 NO FUEL PUMP PRESSURE		
• Fuel pressure gauge installed. • Check if fuel pump runs. • Cycle key from Off to Run, repeat several times. (Do not enter start mode.) Fuel pump should run briefly each time the key enters Run.	Yes ▶ No ▶	GO to Shop Manual Electric Fuel Pump. GO to J2.
J2 CHECK FOR V POWER TO PROCESSOR		
• Key Off, wait 10 seconds. • Disconnect processor 60 Pin connector and inspect for damaged pins, corrosion, loose wires. Service as necessary. • Install Breakout box and reconnect processor. • Key On, engine Off. • DVOM on 20V scale. • Measure voltage between test Pin 37 and test Pin 40 and between test Pin 57 and test Pin 60.	Either voltage reading is less than 10.5V ▶ Both voltage readings are 10.5V or greater ▶	GO to B1. GO to J3.
J3 RESISTANCE CHECK OF FUEL PUMP INERTIA SWITCH		
• Key Off, wait 10 seconds. • Leave Breakout box installed and processor connected. • Locate and disconnect fuel pump inertia switch. • DVOM on 200 ohm scale. • Measure the resistance of the fuel pump inertia switch.	Less than 5.0 ohms ▶ 5.0 ohms or greater ▶	GO to J4. REPLACE fuel pump inertia switch. REPEAT Quick Test.
J4 POWER TO PUMP(S) CHECK		
• Key On, engine Off. • Leave Breakout box installed and processor connected. • Locate fuel pump relay. • DVOM on 20V scale. • Measure voltage between chassis ground and power-to-pump(s) circuit at fuel pump inertia switch during crank mode.	8.0V or greater during crank ▶ Less than 8.0V during crank ▶	GO to J5. GO to J6.

Fuel Pump Circuit — (Inertia Switch) — Pinpoint Test J

TEST STEP	RESULT ▶	ACTION TO TAKE
J5 POWER AT FUEL PUMP(S) CHECK • Key On, engine Off. • Leave Breakout box installed and processor connected. • Locate fuel pump(s). • DVOM on 20V scale. • Measure voltage between chassis ground and power-to-pump(s) circuit at fuel pump during crank mode.	8.0V or greater during crank ▶	GO to Shop Manual Electric Fuel Pump.
	Less than 8.0V during crank ▶	SERVICE open in power to the pump(s) circuit. REPEAT Quick Test.
J6 FUEL PUMP CIRCUIT CHECK TO V BATT. • Key On, engine Off. • Leave Breakout box installed and processor connected. • Locate fuel pump relay. • DVOM on 20V scale. • Measure voltage between chassis ground and V Batt. at the fuel pump relay.	10.5V or greater ▶	GO to J7.
	Less than 10.5V ▶	SERVICE open in V Batt. between fuel pump relay and vehicle battery positive post. REPEAT Quick Test.
J7 V POWER TO FUEL PUMP RELAY CHECK • Key On, engine Off. • Leave Breakout box installed and processor connected. • Locate fuel pump relay. • DVOM on 20V scale. • Measure voltage between chassis ground and V Power circuit at the fuel pump relay.	10.5V or greater ▶	GO to J8.
	Less than 10.5V ▶	SERVICE open in V Power circuit between the processor and the fuel pump relay. REPEAT Quick Test.
J8 FUEL PUMP CIRCUIT CHECK CONTINUITY • Leave Breakout box installed and processor connected. • Key Off, wait 10 seconds. • DVOM on 200 ohm scale. • Measure resistance between fuel pump circuit at the pump relay and test Pin 22.	Less than 5 ohms ▶	GO to J9.
	5 ohms or greater ▶	SERVICE open in fuel pump circuit and REPEAT Quick Test.

Fuel Pump Circuit — (Inertia Switch) — Pinpoint Test J

TEST STEP	RESULT ▶	ACTION TO TAKE
J9 CHECK FOR SHORT TO GROUND • Key Off. • Leave Breakout box installed and processor disconnected. • Fuel pump relay disconnected. • DVOM on 200,000 ohm scale. • Measure resistance between test Pin 22 and test Pins 40 and 60.	10,000 ohms or greater ▶	GO to J10.
	Less than 10,000 ohms ▶	SERVICE short in the fuel pump circuit. REPEAT Quick Test.
J10 CHECK FOR SHORT TO POWER • Key Off, wait 10 seconds. • Leave Breakout box installed and disconnect processor. • Fuel pump relay disconnected. • DVOM on 200,000 ohm scale. • Measure resistance between test Pin 22 and test Pins 37 and 57.	10,000 ohms or greater ▶	RECONNECT fuel pump relay. GO to J11.
	Less than 10,000 ohms ▶	SERVICE short to power in the fuel pump circuit. RECONNECT processor, ATTEMPT to start behicle. If vehicle fails to start, REPLACE processor. REPEAT Quick Test.
J11 FINAL SYSTEM CHECK • Leave Breakout box installed and processor disconnected. • Connect jumper wire to test Pin 22 and to test Pin 40 or 60. • DVOM on 20V scale. • Key On, engine Off. • Measure voltage between chassis ground and power-to-pump(s) circuit at fuel pump relay.	10.5V or greater ▶	REPLACE processor. REPEAT Quick Test.
	Less than 10.5V ▶	REPLACE fuel pump relay. RECONNECT processor and REPEAT Quick Test.

EGR On/Off Control — Pinpoint Test — KA

KA

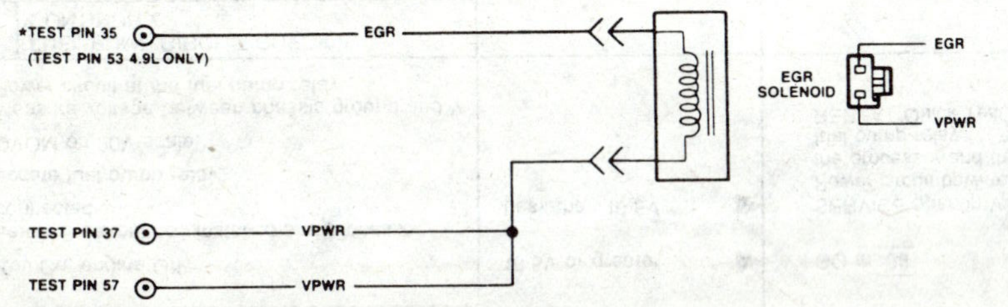

*TEST PINS LOCATED ON BREAKOUT BOX.
ALL HARNESS CONNECTORS VIEWED INTO MATING SURFACE.

STOP-WARNING

You should enter this Pinpoint Test only when a service code 34 is received in Quick Test Step 5.0.

To prevent the replacement of good components, be aware that the following non-EEC areas may be at fault:

- Air or Vacuum Leaks
- EGR Flow Restrictions
- EGR Value

NOTE: Code 34 may be the result of high volume exhaust vent system (reduces backpressure). If this is suspected, perform the test in a well-ventilated area without exhaust vent connected.

This Pinpoint Test is intended to diagnose only the following:

- Circuits: EGR and VPWR
- EGR Solenoid
- Presence of Manifold Vacuum
- Processor Assembly

EGR On/Off Control — Pinpoint Test — KA

TEST STEP	RESULT ▶	ACTION TO TAKE
KA1 ENTER OUTPUT STATE CHECK (REFER TO APPENDIX)		
NOTE: Do not use STAR tester for this step, use a VOM/DVOM. • Key Off, wait 10 seconds. • DVOM on 20V scale. • Connect DVOM negative test lead to STO and positive test lead to battery positive. • Jumper STI to signal return. • Perform Key On, Engine Off Self-Test until the completion of the Continuous Test Codes. • DVOM will indicate zero volts. • Depress and release the throttle. • Did DVOM reading change to a high voltage reading?	Yes ▶	REMAIN in Output State Check. GO to KA2.
	No ▶	DEPRESS throttle to WOT and release. If STO voltage does not go high, GO to Pinpoint Test Step Q40. Leave equipment hooked up.
KA2 CHECK EGR ON/OFF CONTROL SOLENOID ELECTRICAL OPERATION		
• DVOM on 20V scale. • Connect DVOM positive test lead to VPWR circuit on EGR solenoid and negative test lead to EGR output circuit. • While observing DVOM, depress and release the throttle several times to cycle output On and Off. • EGR output cycles On and Off.	Yes ▶	GO to KA3.
	No ▶	REMOVE jumper. GO to KA5.
KA3 CHECK SOLENOID FOR VACUUM CYCLING		
• Install vacuum pump to the solenoid vacuum supply port and install a vacuum gauge to the output port. Apply 6 in. Hg minimum. • While cycling outputs On and Off (by depressing and releasing throttle) observe the vacuum gauge at the output. **NOTE: Maintain vacuum at source.**	Vacuum output cycles On and Off ▶	GO to KA4.
	Vacuum output does not cycle On and Off ▶	REPLACE solenoid. REPEAT Quick Test.

EGR On/Off Control — Pinpoint Test — KA

TEST STEP	RESULT ▶	ACTION TO TAKE
KA4 CHECK MANIFOLD VACUUM LINES FOR BLOCKAGE OR LEAKS		
• Vacuum lines disconnected at solenoid. • Start engine. • Check for vacuum.	Vacuum present ▶	EEC-IV system OK.
	No vacuum present ▶	SERVICE vacuum source blockage or leak. REPEAT Quick Test.
KA5 MEASURE EGR SOLENOID RESISTANCE		
• Key Off, wait 10 seconds. • DVOM on 200 ohm scale. • Disconnect EGR solenoid connector and measure solenoid resistance.	Resistance is between 65 and 110 ohms ▶	CONNECT EGR solenoid. GO to KA6.
	Resistance is less than 65 ohms or greater than 110 ohms ▶	REPLACE EGR solenoid. REPEAT Quick Test.
KA6 CHECK VOLTAGE OF VPWR CIRCUIT		
• Key On, Engine Off. • DVOM on 20V scale. • Measure voltage at the EGR solenoid harness connector between VPWR circuit and ground.	Voltage reading is less than 10.5V ▶	SERVICE harness open circuit. REPEAT Quick Test.
	Voltage reading is 10.5V or greater ▶	GO to KA7.
KA7 CHECK CONTINUITY OF EGR CIRCUIT		
• Key Off, wait 10 seconds. • Disconnect processor 60 Pin connector and inspect for damaged pins, corrosion, loose wires. Service as necessary. • Connect Breakout box to harness. Leave processor disconnected. • DVOM on 200 ohm scale. • Measure resistance between test Pin 35* and EGR circuit at harness connector. *Test Pin 53 for 4.9L engine, truck only.	Resistance reading is 5 ohms or greater ▶	SERVICE open circuit. REPEAT Quick Test.
	Resistance reading is less than 5 ohms ▶	GO to KA8.

EGR On/Off Control — Pinpoint Test — KA

TEST STEP	RESULT ▶	ACTION TO TAKE
KA8 CHECK FOR SHORT TO GROUND		
• Key Off, wait 10 seconds. • Leave Breakout box installed and processor disconnected. • Disconnect EGR solenoid. • DVOM on 200,000 ohm scale. • Measure resistance between test Pin 35* and test Pins 40, 46 and 60. *Test Pin 53 for 4.9L engine, truck only.	Resistance reading is less than 10,000 ohms ▶	SERVICE short to ground. REPEAT Quick Test.
	Resistance reading is 10,000 ohms or greater ▶	GO to KA9.
KA9 CHECK FOR SHORT TO POWER		
• Key Off, wait 10 seconds. • DVOM on 200,000 ohm scale. • Leave Breakout box installed and processor disconnected. • EGR solenoid disconnected. • Measure resistance between test Pin 35* and test Pins 37 and 57. *Test Pin 53 for 4.9L engine, truck only.	Resistance reading is 10,000 ohms or greater ▶	REPLACE Processor. REPEAT Quick Test.
	Resistance reading is less than 10,000 ohms ▶	SERVICE short to power. REPEAT Quick Test. If code is repeated, REPLACE processor.

Air Management System — Pinpoint Test — KC

KC

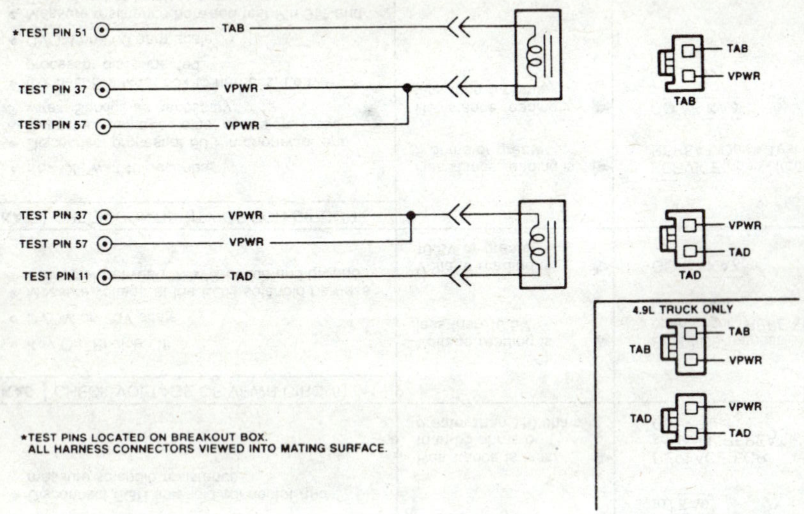

*TEST PINS LOCATED ON BREAKOUT BOX.
ALL HARNESS CONNECTORS VIEWED INTO MATING SURFACE.

STOP-WARNING

You should enter this Pinpoint Test only when a service code 44, 45, 46, 94, 95, 96, 81 or 82 is received in Quick Test Step 3.0 or 5.0.

To prevent the replacement of good components, be aware that the following non-EEC areas may be at fault:

- Thermactor System
  - — Belt
  - — Pump
  - — Valve

This Pinpoint Test is intended to diagnose only the following:

- TAB and TAD Solenoid Valve Assemblies
- Harness Circuits: TAB, TAD and VPWR
- Vacuum Supply
- Processor Assembly

Air Management System — Pinpoint Test — KC

TEST STEP	RESULT ▶	ACTION TO TAKE
KC1 SERVICE CODES 44 (94), 45 (95) AND 46 (96): VERIFY VACUUM LINE ROUTING		
• Verify proper vacuum line routing to the TAB TAD solenoids and to the bypass diverter valve. Refer to VECI decal. • Check for kinked or blocked vacuum lines. • Check for kinked or blocked air hoses. • Check for disconnected vacuum lines. • Are visual checks satisfactory?	No ▶	SERVICE routing or faults. REPEAT Quick Test.
	Yes ▶	Service code 44 (94), GO to **KC4**. Service code 45 (95), GO to **KC2**. Service code 46 (96), GO to **KC3**.
KC2 ATTEMPT TO ELIMINATE SERVICE CODE 45 (95) (TAD ONLY)		
• Disconnect vacuum line on diverter valve and cap vacuum line. • Key Off, wait 10 seconds. • Repeat Engine Running Self-Test and record service codes. • Is code 45 (95) present?	Yes ▶	EEC-IV system OK. REFER to Section for diverter valve or check valve diagnostics.
	No ▶	GO to **KC4**.
KC3 ATTEMPT TO ELIMINATE SERVICE CODE 46 (96) (TAB ONLY)		
• Disconnect vacuum line on bypass valve and cap vacuum line. • Key Off, wait 10 seconds. • Repeat Engine Running Self-Test and record codes. • Is code 46 (96) present?	Yes ▶	EEC-IV system OK. REFER to Section for bypass valve diagnostics.
	No ▶	GO to **KC4**.

Air Management System — Pinpoint Test — KC

TEST STEP	RESULT ▶	ACTION TO TAKE
KC4 ENTER OUTPUT STATE CHECK (REFER TO APPENDIX)		
NOTE: Do not use STAR tester for this Step, use a VOM/DVOM. • Key Off, wait 10 seconds. • DVOM on 20V scale. • Connect DVOM negative test lead to STO and positive test lead to battery positive. • Jumper STI to signal return. • Perform Key On, Engine Off Self-Test until the completion of the Continuous Test Codes. • DVOM will indicate zero volts. • Depress and release the throttle. • Did DVOM reading change to a high voltage reading?	Yes ▶	REMAIN in Output State Check. GO to KC5.
	No ▶	DEPRESS throttle to WOT and RELEASE. If STO voltage does not go high, GO to Pinpoint Test Step Q40. Leave equipment hooked up.
KC5 CHECK TAB/TAD SOLENOID ELECTRICAL OPERATION		
• DVOM on 20V scale. • Connect DVOM positive test lead to VPWR circuit on TAB solenoid and negative test lead to output TAB circuit. • While observing DVOM depress and release the throttle several times (to cycle output On and Off). • Repeat for TAD solenoid. Connect positive test lead to VPWR circuit on TAD solenoid and negative test lead to output TAD circuit. • Cycle output On and Off.	Both outputs cycle On and Off ▶	GO to KC6.
	Either output does not cycle On and Off ▶	REMOVE jumper. GO to KC8.
KC6 CHECK TAB/TAD SOLENOID FOR VACUUM CYCLING		
• Install vacuum pump to the TAB solenoid vacuum supply port and install a vacuum gauge to the output port. • While cycling outputs On and Off (by depressing and releasing throttle), observe the vacuum gauge at the output. **NOTE: Maintain vacuum at source.** • Repeat for TAD solenoid. Connect vacuum pump to the TAD solenoid vacuum supply port and connect a vacuum gauge to the output port. • Cycle output On and Off.	Both vacuum outputs cycle On and Off ▶	GO to KC7.
	Either vacuum output does not cycle On and Off ▶	REPLACE solenoid assembly. REPEAT Quick Test.

Air Management System — Pinpoint Test — KC

TEST STEP	RESULT ▶	ACTION TO TAKE
KC7 CHECK MANIFOLD VACUUM LINES FOR BLOCKAGE OR LEAKS		
• Vacuum lines disconnected at TAD/TAB solenoids. • Start engine. • Check for vacuum.	Vacuum present ▶	EEC-IV system OK. REFER to Section for thermactor valve and air pump diagnostics.
	No vacuum present ▶	SERVICE vacuum source blockage or leak. REPEAT Quick Test.
KC8 MEASURE TAB/TAD SOLENOID RESISTANCE		
• Key Off, wait 10 seconds. • DVOM on 200 ohm scale. • Disconnect TAB solenoid connector and measure solenoid resistance. • Disconnect TAD solenoid connector and measure solenoid resistance.	Both resistances are between 50 and 100 ohms ▶	CONNECT TAB/TAD solenoids. GO to KC9.
	Either resistance is less than 50 ohms or greater than 100 ohms ▶	REPLACE TAB/TAD solenoid assembly. REPEAT Quick Test.
KC9 CHECK VOLTAGE OF VPWR CIRCUIT		
• Key On, engine Off. • DVOM on 20V scale. • Measure voltage between TAB solenoid VPWR circuit and ground. • Repeat for TAD solenoid.	Both voltage readings are greater than 10.5V ▶	GO to KC10.
	Either voltage reading is less than 10.5V ▶	SERVICE harness circuit open. REPEAT Quick Test.
KC10 CHECK CONTINUITY OF TAB AND TAD CIRCUITS		
• Key Off, wait 10 seconds. • Disconnect processor 60 Pin connector and inspect for damaged pins, corrosion, loose wires. Service as necessary. • Connect Breakout box to harness. Leave processor disconnected. • DVOM on 200 ohm scale. • Measure resistance between test Pin 51 and TAB circuit at harness connector. • Measure resistance between test Pin 11 and TAD circuit at harness connector.	Both resistance readings are less than 5 ohms ▶	GO to KC11.
	Either resistance reading is 5 ohms or greater ▶	SERVICE harness open circuit. REPEAT Quick Test.

Air Management System — Pinpoint Test — KC

TEST STEP	RESULT ▶	ACTION TO TAKE
KC11 CHECK FOR SHORT TO GROUND • Key Off, wait 10 seconds. • DVOM on 200,000 ohm. • Leave Breakout box installed and processor disconnected. • Disconnect TAB/TAD solenoids. • Measure resistance between test Pins 51 and or 11 and test Pins 37 and 57.	Resistance reading is 10,000 ohms or greater ▶	GO to KC12.
	Resistance reading is less than 10,000 ohms ▶	SERVICE short to power. REPEAT Quick Test.
KC12 CHECK FOR SHORT TO GROUND • Key Off, wait 10 seconds. • DVOM on 200,000 ohm scale. • Leave Breakout box installed and processor disconnected. • TAB/TAD solenoids disconnected. • Measure resistance between test Pins 51 and or 11 and test Pins 40, 46 and 60.	Resistance reading is less than 10,000 ohms ▶	SERVICE circuit short. REMOVE Breakout box. RECONNECT harness to processor. RERUN Quick Test. If code is repeated, REPLACE processor.
	Resistance reading is 10,000 ohms or greater ▶	REPLACE processor. REMOVE Breakout box. RECONNECT harness to processor. RERUN Quick Test.

Canister Purge (CANP) — Pinpoint Test — KD

KD

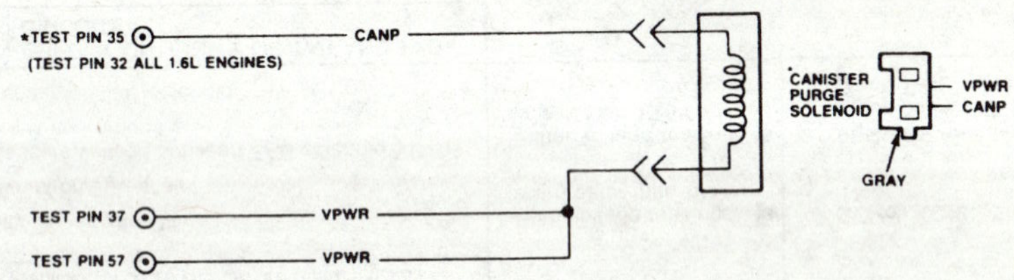

*TEST PINS LOCATED ON BREAKOUT BOX.
ALL HARNESS CONNECTORS VIEWED INTO MATING SURFACE.

STOP-WARNING

You should enter this Pinpoint Test only when a service code 85 is received in Quick Test Step 3.0 or when you are directed here from a fuel control Pinpoint Test and Diagnostics by Symptom in the Engine Supplement Section.

This Pinpoint Test is intended to diagnose only the following:

- Harness Circuits: CANP and VPWR
- Processor Assembly

Canister Purge (CANP) Pinpoint Test KD

TEST STEP	RESULT ▶	ACTION TO TAKE
KD1 ENTER OUTPUT STATE CHECK (REFER TO APPENDIX)		
NOTE: Do not use STAR tester for this step, use VOM DVOM. • Key Off, wait 10 seconds. • DVOM on 20V scale. • Connect DVOM negative test lead to STO and positive test lead to battery positive. • Jumper STI to signal return. • Perform Key On, Engine Off Self-Test until the completion of the Continuous Test Codes. • DVOM will indicate zero volts. • Depress and release the throttle. • Did DVOM reading change to a high voltage reading?	Yes ▶	REMAIN in Output State Check. GO to KD2.
	No ▶	DEPRESS throttle to WOT and release. If STO voltage does not go high, GO to Pinpoint Test Step Q40. Leave equipment hooked up.
KD2 CHECK CANISTER PURGE (CANP) SOLENOID ELECTRICAL OPERATION		
• DVOM on 20V scale. • Connect DVOM positive test lead to VPWR on CANP solenoid and negative test lead to CANP output circuit. • While observing DVOM depress and release the throttle several times (to cycle output On and Off). • CANP output cycles On and Off?	Yes ▶	GO to KD3.
	No ▶	REMOVE jumper. GO to KD4.
KD3 CHECK CANISTER PURGE SOLENOID FOR HOLDING VACUUM		
• Key Off, wait 10 seconds. • Remove jumper from STI to Signal Return. • Disconnect vacuum hose at canister purge solenoid at PCV side. • Apply 16 in. Hg to CANP solenoid. • Does CANP solenoid hold vacuum for 20 seconds?	Yes ▶	EEC-IV system OK.
	No ▶	REPLACE CANP solenoid. REPEAT Quick Test. If code 42 is still present, GO to Section

Canister Purge (CANP) Pinpoint Test KD

TEST STEP	RESULT ▶	ACTION TO TAKE
KD4 MEASURE CANP SOLENOID RESISTANCE		
• Key Off, wait 10 seconds. • DVOM on 200 ohm scale. • Disconnect CANP solenoid connector and measure solenoid resistance.	Resistance is between 40 and 90 ohms ▶	CONNECT CANP solenoid. GO to KD5.
	Resistance is less than 40 ohms or greater than 90 ohms ▶	REPLACE CANP solenoid. REPEAT Quick Test.
KD5 CHECK VOLTAGE OF VPWR CIRCUIT		
• Key On, engine Off. • DVOM on 20V scale. • Measure voltage at the CANP solenoid harness connector between VPWR and ground.	Voltage reading is less than 10.5V ▶	SERVICE harness open circuit. REPEAT Quick Test.
	Voltage reading is 10.5V or greater ▶	GO to KD6.
KD6 CHECK CONTINUITY OF CANP CIRCUIT		
• Key Off, wait 10 seconds. • Disconnect processor and inspect both 60 Pin connectors and inspect for damaged pins, corrosion, loose wires. Service as necessary. • Connect Breakout box to harness. Leave processor disconnected. • DVOM on 200 ohm scale. • Measure resistance between test Pin 35* and CANP at harness connector. *Test Pin 32 for all 1.6L engines.	Resistance reading is 5 ohms or greater ▶	SERVICE open circuit. REPEAT Quick Test.
	Resistance reading is less than 5 ohms ▶	GO to KD7.

Canister Purge (CANP) — Pinpoint Test — KD

TEST STEP	RESULT ▶	ACTION TO TAKE
KD7 CHECK FOR SHORT TO GROUND		
• Key Off, wait 10 seconds. • Leave Breakout box to harness. Leave processor disconnected. • Disconnect CANP solenoid. • DVOM on 200,000 ohm scale. • Measure resistance between test Pin 35* and test Pins 40, 46 and 60. *Test Pin 32 for all 1.6L engine applications.	Resistance reading is 10,000 ohms or greater ▶	GO to KD8
	Resistance reading is less than 10,000 ohms ▶	SERVICE short to power. REPEAT Quick Test.
KD8 CHECK FOR SHORT TO GROUND		
• Key Off, wait 10 seconds. • DVOM on 200,000 ohm scale. • Leave Breakout box installed and processor disconnected. • CANP solenoid disconnected. • Measure resistance between test Pin 35* and test Pins 37 and 57. *Test Pin 32 for all 1.6L engine applications.	Resistance reading is less than 10,000 ohms ▶	SERVICE short to ground. REPEAT Quick Test. If code is repeated, REPLACE processor. RERUN Quick Test.
	Resistance reading is 10,000 ohms or greater ▶	REMOVE Breakout box. REPLACE processor. RERUN Quick Test.

Idle Speed Control (By-Pass Air) — Pinpoint Test — KE

KE

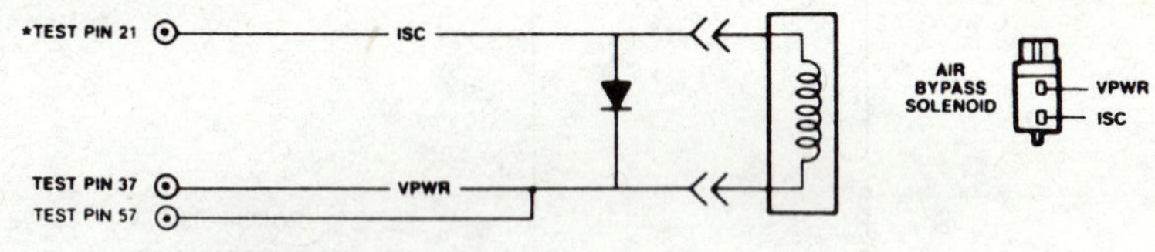

*TEST PINS LOCATED ON BREAKOUT BOX.
ALL HARNESS CONNECTORS VIEWED INTO MATING SURFACE.

STOP-WARNING

You should enter this Pinpoint Test only when a service code 12 or 13 is received in Quick Test Step 5.0.

CAUTION: If the engine exhibits rough running and/or idle, correct these conditions before diagnosing Idle Speed Control (ISC). Rough running/misses may be caused by:

- **Ignition System**
  Refer to Section 15
- **Fuel System**
  Refer to Pinpoint Test Steps HA1 through HA4
- **EGR System, Section 6**

To prevent the replacement of good components, be aware that the following non-EEC areas may be at fault:

- Engine not up to operating temperature
- Engine over operating temperature
- Improper Idle Speed Throttle Stop Adjustment
- A/C input (electrical problem)
- Throttle Speed Control Linkage

This Pinpoint Test is intended to diagnose only the following:

- RPM in Self-Test only
- ISC Actuator
- Harness Circuits ISC and VPWR
- Processor Assembly

Idle Speed Control (By-Pass Air) — Pinpoint Test — KE

TEST STEP	RESULT	ACTION TO TAKE
KE1 MEASURE ISC SOLENOID RESISTANCE		
• Key Off. • DVOM on 200 ohm scale. • Disconnect ISC solenoid connector and measure solenoid resistance.	Resistance is between 7 and 13 ohms ▶	GO to **KE2**.
	Resistance is less than 7 ohms or greater than 13 ohms ▶	REPLACE ISC solenoid. REPEAT Quick Test.
KE2 ISC SHORT TO CASE (GROUND) CHECK		
• Key Off. • DVOM on 200,000 ohm scale. • ISC solenoid disconnected. • Measure resistance from either ISC pin to ISC housing.	10,000 ohms or greater ▶	GO to **KE3**.
	Less than 10,000 ohms ▶	REPLACE ISC solenoid. REPEAT Quick Test.
KE3 CHECK VOLTAGE OF VPWR CIRCUIT		
• Key On, Engine Off. • DVOM on 20V scale. • ISC solenoid disconnected. • Measure voltage at the ISC solenoid harness connector between VPWR and ground.	Voltage reading is less than 10.5V ▶	SERVICE harness open circuit. REPEAT Quick Test.
	Voltage reading is 10.5V or greater ▶	GO to **KE4**.
KE4 CHECK CONTINUITY OF ISC CIRCUIT		
• Key Off. • ISC solenoid disconnected. • Disconnect processor and inspect both 60 Pin connectors for damaged pins, corrosion, loose wires. Service as necessary. • Connect Breakout box to harness. Leave processor disconnected. • DVOM on 200 ohm scale. • Measure resistance between test Pin 21 and ISC circuit at harness connector.	Resistance reading is 5 ohms or greater ▶	SERVICE open circuit. REPEAT Quick Test.
	Resistance reading is less than 5 ohms ▶	GO to **KE5**.

Idle Speed Control (By-Pass Air) — Pinpoint Test — KE

TEST STEP	RESULT	ACTION TO TAKE
KE5 CHECK FOR SHORT TO GROUND		
• Key Off, wait 10 seconds. • Leave Breakout box installed and processor disconnected. • ISC solenoid disconnected. • DVOM on 200,000 ohm scale. • Measure resistance between test Pin 21 and test Pins 40, 46 and 60.	Resistance reading is less than 10,000 ohms ▶	SERVICE short to ground. REPEAT Quick Test.
	Resistance reading is 10,000 ohms or greater ▶	GO to **KE6**.
KE6 CHECK FOR SHORT TO POWER		
• Key Off, wait 10 seconds. • DVOM on 200,000 ohm scale. • ISC solenoid disconnected. • Leave Breakout box connected to harness. Leave processor disconnected. • Measure resistance between test Pin 21 and test Pin 37.	Resistance reading is 10,000 ohms or greater ▶	GO to **KE7**.
	Resistance reading is less than 10,000 ohms ▶	SERVICE short to power. REPEAT Quick Test.
KE7 CHECK FOR ISC SIGNAL AT THE PROCESSOR		
• Connect processor. • Connect ISC actuator. • Vehicle prepared for Quick Test. • DVOM on 20V scale. • Connect DVOM between test Pin 21 and test Pin 40. • Start engine and observe the DVOM during Self-Test.	Meter reading varies during Quick Test ▶	REPLACE ISC actuator. REPEAT Quick Test.
	Meter reading does not vary during Quick Test ▶	DISCONNECT processor and INSPECT for bent, damaged, corroded pins, etc. CORRECT as necessary. If OK, REPLACE processor. REPEAT Quick Test.
KE8 **CODE 13:** RPM DOES NOT DROP TO LESS THAN 1500 AT END OF TEST		
• Disconnect ISC harness connector. • Connect engine tachometer. • Repeat Engine Running Quick Test. • At end of test, record service codes for future use. • Does rpm remain below 1500 rpm during test?	Yes ▶	REPLACE processor. REPEAT Quick Test.
	No ▶	CHECK engine vacuum hoses. REFER to VECI decal. VERIFY curb idle. CHECK that throttle plates are fully closed. CHECK throttle linkage and/or speed control linkage for binding. If OK, REPLACE ISC actuator. REPEAT Quick Test.

Idle Speed Control (DC Motor/Idle Tracking Switch Assembly) — Pinpoint Test — KF

KF

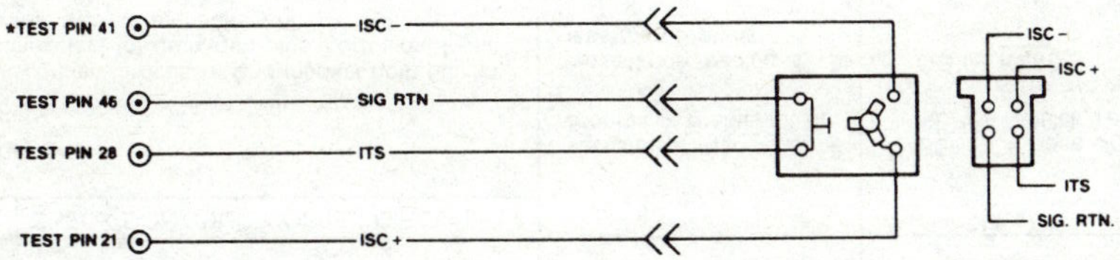

*TEST PINS LOCATED ON BREAKOUT BOX.
ALL HARNESS CONNECTORS VIEWED INTO MATING SURFACE.

STOP-WARNING

You should enter this Pinpoint Test only when a Service Code 12, 13, 58 or 68 is received in Quick Test Step 3.0, 5.0 or 6.0 or when directed here from Diagnostics by Symptom in the Engine Supplement Section.

To prevent the replacement of good components, be aware that the following non-EEC areas may be at fault:

- Anti-Diesel Speed too High
- Basic Engine
- Vacuum Leaks
- Throttle Sticking or on High Cam

This Pinpoint Test is intended to diagnose only the following:

- DC Motor
- Idle Tracking Switch
- Harness circuits: ISC +, ISC –, Signal Return and ITS
- Processor Assembly

Idle Speed Control (DC Motor/Idle Tracking Switch Assembly) — Pinpoint Test — KF

TEST STEP	RESULT ▶	ACTION TO TAKE
KF1 SERVICE CODE 68: SIMULATE NO THROTTLE CONTACT TO FORCE A CODE 11 IN KEY ON — ENGINE OFF TEST		
• Key Off, wait 10 seconds. • Move throttle away from ISC DC motor shaft. • Perform Key On, Engine Off Self-Test. • Is code 68 present? **NOTE: Ignore all other codes at this point.**	Yes ▶	GO to KF3.
	No ▶	GO to Pinpoint Test Step KF2.
KF2 RETRACT ISC DC MOTOR		
• Key Off, wait 10 seconds. • Leave Breakout box installed. Leave processor disconnected. • Jumper test Pin 41 to test Pin 1 and test Pin 21 to ground (to retract DC motor). • Does ISC DC motor retract?	Yes ▶	GO to ISC Adjustments
	No ▶	GO to KF5.
KF3 SIMULATE A CLOSED CONTACT CONDITION IN KEY ON — ENGINE OFF TEST		
• Key Off, wait 10 seconds. • Disconnect harness from ISC. • Connect jumper between ISC harness connector Pins 3 and 4. • Install jumper as shown. • Perform Key On, Engine Off Self-Test. • Is code 68 present?	Yes ▶	GO to KF4.
	No ▶	REPLACE ISC DC motor. REPEAT Quick Test.

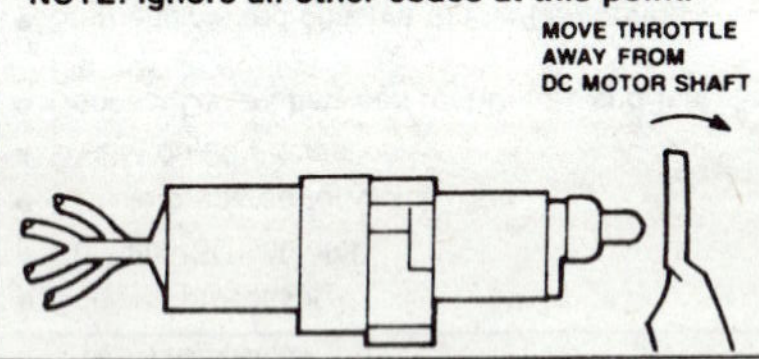

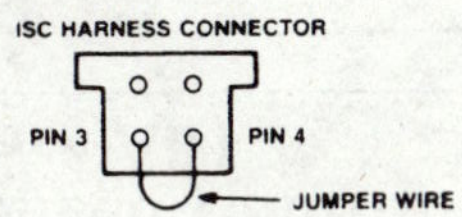

CAUTION: Do not short any other pins.

Idle Speed Control (DC Motor/Idle Tracking Switch Assembly) — Pinpoint Test KF

TEST STEP	RESULT	ACTION TO TAKE
KF4 CHECK ITS AND SIGNAL RETURN CIRCUITS FOR CONTINUITY		
• Key Off, wait 10 seconds. • Disconnect processor 60 Pin connector and inspect for damaged pins, corrosion, loose wires. Service as necessary. • Connect Breakout box to harness. Leave processor disconnected. • DVOM on 200 ohm scale. • Measure resistance between test Pin 46 and Signal Return circuit at ISC harness connector and between test Pin 28 and ITS circuit at ISC harness connector.	Either resistance 5 ohms or greater ▶	SERVICE faulty circuit. REPEAT Quick Test.
	Both resistances less than 5 ohms ▶	REPLACE processor. REPEAT Quick Test.
KF5 CHECK TP AND ISC CIRCUITS FOR CONTINUITY		
• Key Off, wait 10 seconds. • Disconnect processor 60 Pin connector and inspect for damaged pins, corrosion, loose wires. Service as necessary. • Connect Breakout box. Leave processor disconnected. • Disconnect harness connector for ISC motor. • DVOM on 200 ohm scale. • Measure resistance between test Pin 41 and ISC – circuit and between test Pin 21 and ISC + circuit at harness connector.	Both resistance readings are less than 5 ohms ▶	GO to **KF6**.
	Either resistance reading is 5 ohms or greater ▶	SERVICE faulty circuits. REPEAT Quick Test.
KF6 CHECK ISC + AND ISC – CIRCUITS FOR SHORTS TO POWER		
• Key On, Engine Off. • Leave Breakout box installed. Leave processor disconnected. • Harness disconnected from ISC motor. • DVOM on 20V scale. • Measure voltage between test Pin 41 and test Pins 40 and 60. Measure voltage between test Pin 21 and test Pins 40 and 60.	All voltage readings are less than 1 volt ▶	GO to **KF7**.
	Any voltage reading is 1 volt or more ▶	SERVICE circuit short to power. REPEAT Quick Test. If code 12 is still present, REPLACE processor.

Idle Speed Control (DC Motor/Idle Tracking Switch Assembly) — Pinpoint Test KF

TEST STEP	RESULT	ACTION TO TAKE
KF7 CHECK ISC – AND ISC – CIRCUITS FOR SHORTS TO GROUND		
• Key Off, wait 10 seconds. • Leave Breakout box installed. Leave processor disconnected. • Harness disconnected from ISC motor. • DVOM on 200,000 ohm scale. • Measure resistance between test Pin 41 and test Pins 40, 46 and 60. Measure resistance between test Pin 21 and test Pins 40, 46 and 60.	All resistance readings greater than 10,000 ohms ▶	GO to **KF8**.
	Any resistance reading is 10,000 ohms or less ▶	SERVICE faulty circuit. REPEAT Quick Test. If code 12 is still present, REPLACE processor.
KF8 PREPARE IDLE SPEED CONTROL DC MOTOR FOR OPERATION		
• Key Off, wait 10 seconds. • Leave Breakout box installed. Leave processor disconnected. • Harness disconnected from ISC motor. • DVOM on 20V scale. • Connect DVOM positive test lead to test Pin 1 and negative test lead to ground.	Voltage reading is 10.5V or greater ▶	GO to **KF9**.
	Voltage reading is less than 10.5V ▶	SERVICE open in KAPWR circuit. REPEAT Quick Test. If code 12 is still present, GO to **KF9**.
KF9 ISC DC MOTOR OPERATION CHECK		
• Key Off, wait 10 seconds. • Leave Breakout box installed. Leave processor disconnected. • Connect ISC DC motor to harness connector. • Jumper test Pin 21 to test Pin 1 and test Pin 41 to ground (to extend DC motor). • Does DC motor shaft extend 5 cm (2 inches) or more? 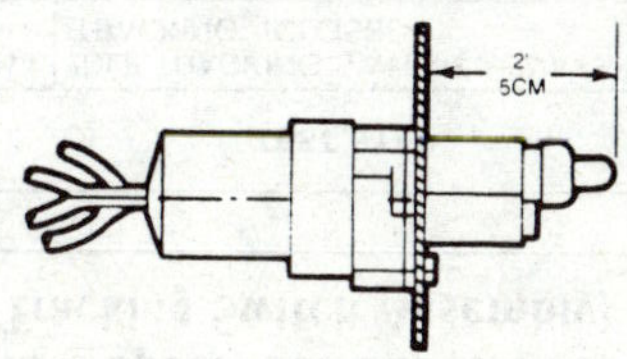	Yes ▶	GO to **KF10**.
	No ▶	REPLACE DC motor. REPEAT Quick Test.

Idle Speed Control (DC Motor/Idle Tracking Switch Assembly) — Pinpoint Test — KF

TEST STEP	RESULT ▶	ACTION TO TAKE
KF10 RETRACT ISC DC MOTOR		
• Key Off, wait 10 seconds. • Leave Breakout box installed. Leave processor disconnected. • Jumper test Pin 41 to test Pin 1 and test Pin 21 to ground (to retract DC motor). • Does ISC DC motor retract?	Yes ▶ No ▶	GO to KF11. REPLACE DC motor. REPEAT Quick Test.
KF11 EXTEND ISC DC MOTOR		
• Key Off, wait 10 seconds. • Leave Breakout box installed. Leave processor disconnected. • Jumper test Pin 21 to test Pin 1 and test Pin 41 to ground (to extend DC motor). • Does ISC DC motor extend?	No ▶ Yes ▶	REPLACE DC motor. REPEAT Quick Test. REPLACE processor. REPEAT Quick Test.
KF12 SERVICE CODE 58: SIMULATE THROTTLE CONTACT TO FORCE A SERVICE CODE 68 IN KEY ON — ENGINE OFF TEST		
• Key Off, wait 10 seconds. • Press on ISC DC motor shaft to simulate throttle contact. • With a force pushing on DC motor shaft, perform Key On, Engine Off Self-Test. **NOTE: Ignore all other codes at this point.** • Is code 68 present?	Yes ▶ No ▶	GO to Pinpoint Test Step KF5. GO to KF13.
KF13 RETRACT ISC DC MOTOR		
• Key Off, wait 10 seconds. • Leave Breakout box installed. Leave processor disconnected. • Jumper test Pin 41 to test Pin 1 and test Pin 21 to ground (to retract DC motor). • Does ISC DC motor retract?	Yes ▶ No ▶	GO to ISC Adjustments. GO to KF14.

Idle Speed Control (DC Motor/Idle Tracking Switch Assembly) — Pinpoint Test — KF

TEST STEP	RESULT ▶	ACTION TO TAKE
KF14 IDLE TRACKING SWITCH ALWAYS "NOT TRACKING" (CLOSED)		
• Key Off, wait 10 seconds. • Disconnect harness from ISC. • Perform Key On, Engine Off self-test. **NOTE: Ignore all other codes at this point.** • Is code 68 present?	Yes ▶ No ▶	REPLACE DC motor. REPEAT Quick Test. GO to KF15.
KF15 CHECK ITS CIRCUIT FOR SHORT TO GROUND		
• Key Off, wait 10 seconds. • Disconnect processor 60 Pin connector and inspect for damaged pins, corrosion, loose wires. Service as necessary. • Connect Breakout box. Leave processor disconnected. • Harness disconnected at ISC. • DVOM on 200,000 ohm scale. • Measure resistance between test Pin 28 and test Pins 40, 46 and 60.	Any resistance less than 10,000 ohms ▶ All resistances 10,000 ohms or more ▶	SERVICE short to ground. REPEAT Quick Test. REPLACE processor. REPEAT Quick Test.
KF16 DC MOTOR CONTROL FAULT ISOLATION: SERVICE CODE 12		
• Are service codes 58, 68, 31 or 41 present?	Yes ▶ No ▶	GO to Quick Test Step 5.0 for appropriate Pinpoint Test. GO to KF17.
KF17 CHECK FOR THROTTLE STICKING		
• Check throttle plates and linkages for binding. • Check speed control for binding (if so equipped).	Yes ▶ No ▶	SERVICE as necessary. REPEAT Quick Test. GO to KF18.

Idle Speed Control (DC Motor/Idle Tracking Switch Assembly) — Pinpoint Test — KF

TEST STEP	RESULT ▶	ACTION TO TAKE
KF18 EXTEND ISC DC MOTOR		
• Key Off, wait 10 seconds. • Leave Breakout box installed. Leave processor disconnected. • Jumper test Pin 21 to test Pin 1 and test Pin 41 to ground (to extend DC motor). • Does ISC DC motor extend?	No ▶ Yes ▶	REPLACE DC motor. REPEAT Quick Test. REPLACE processor. REPEAT Quick Test.
KF19 RPM NOT RETURNING TO NORMAL SPEED: CODE 13		
• Prepare system for normal engine operation. • De-activate Self-Test. • Air conditioner Off. • Run engine. Alternate between 30 seconds at idle and 5 seconds at part throttle modes for 3 minutes. • Is idle speed erratic?	No ▶ Yes ▶	GO to KF20. CHECK for: • Vacuum leaks • Code 22 • Code 31 • Code 41 • Code 58 SERVICE above codes as necessary before continuing.
KF20 CHECK FOR THROTTLE STICKING OR CHOKE POSITION		
• Inspect throttle mechanisms (plate linkage) for sticking or binding. • Inspect choke for sticking or binding. • Inspect for throttle stuck on high cam. • Are above inspections satisfactory?	No ▶ Yes ▶	SERVICE fault. REPEAT Quick Test. GO to KF21.
KF21 PERFORM KEY ON — ENGINE OFF TEST. OBSERVE ON DEMAND CODES		
• Perform Key On, Engine Off Test. Record On Demand codes. • Key On, Engine Off. • Is code 68 present?	Yes ▶ No ▶	GO to Pinpoint Test Step KF1. REFER to Section for anti-diesel speed adjustment.

Idle Speed Control (Throttle Kicker) — Pinpoint Test — KG

KG

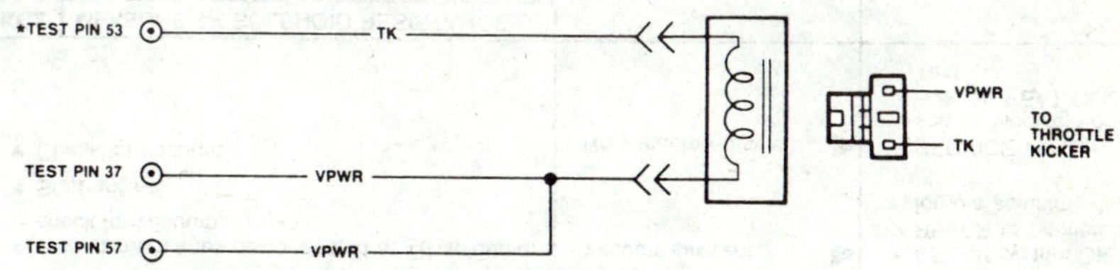

*TEST PINS LOCATED ON BREAKOUT BOX.
ALL HARNESS CONNECTORS VIEWED INTO MATING SURFACE.

STOP-WARNING

You should enter this Pinpoint Test only when a service code 12 or 88 is received in Quick Test Step 3.0 or 5.0.

To prevent the replacement of good components, be aware that the following non-EEC areas may be at fault:

- Engine Running Below Operating Temperature
- Engine Running Above Operating Temperature
- Improper Idle Speeds
- A/C Input (Electrical Problems)

This Pinpoint Test is intended to diagnose only the following:

- TK Solenoid
- TK Actuator
- Harness circuits: TK and VPWR
- Processor Assembly
- Vacuum Source
- Actuator Check

Idle Speed Control (Throttle Kicker) — Pinpoint Test — KG

TEST STEP	RESULT ▶	ACTION TO TAKE
KG1 VERIFY LINE ROUTINGS AND VISUAL CHECKS		
• Verify proper vacuum line routing to the TK solenoid. Refer to VECI decal. • Check for kinked or blocked vacuum lines. • Check for kinked or blocked air hoses. • Check for disconnected vacuum lines. • Are visual checks satisfactory?	No ▶	SERVICE routing or faults. REPEAT Quick Test.
	Yes ▶	Service code 12. GO to KG2.
KG2 VACUUM CHECK		
• Apply 20 in Hg vacuum to throttle kicker. • Verify vacuum holds.	Yes ▶	GO to KG3.
	No ▶	REPLACE solenoid. REPEAT Quick Test.
KG3 ATTEMPT TO ELIMINATE SERVICE CODE 12		
• While vacuum is applied to throttle kicker, rerun Engine Running Self-Test. • Is service code 12 present?	Yes ▶	EEC-IV system OK. REFER to Section 4 for idle adjustment.
	No ▶	GO to KG4.
KG4 ENTER OUTPUT STATE CHECK (REFER TO APPENDIX)		
NOTE: Do not use STAR tester for this Step, use VOM/DVOM. • Key Off, wait 10 seconds. • DVOM on 20V scale. • Connect DVOM negative test lead to STO and positive test lead to battery positive. • Jumper STI to signal return. • Perform Key On, Engine Off Self-Test until the completion of the Continuous Test Codes. • DVOM will indicate zero volts. • Depress and release the throttle. • Did DVOM reading change to a high voltage reading?	Yes ▶	REMAIN in Output State Check. GO to KG5.
	No ▶	DEPRESS throttle to WOT and release. If STO voltage does not go high, GO to Pinpoint Test Step Q40. Leave equipment hooked up.

Idle Speed Control (Throttle Kicker) — Pinpoint Test — KG

TEST STEP	RESULT ▶	ACTION TO TAKE
KG5 CHECK TK SOLENOID ELECTRICAL OPERATION		
• DVOM on 20V scale. • Connect DVOM positive test lead to circuit VPWR on solenoid and negative test lead to output circuit TK on solenoid. • While observing DVOM, depress and release the throttle several times to cycle output On and Off. • TK output cycles On and Off?	Yes ▶	GO to KG6.
	No ▶	REMOVE jumper. GO to KG8.
KG6 CHECK TK SOLENOID FOR VACUUM CYCLING		
• Install vacuum pump to the TK solenoid vacuum supply port and install a vacuum gauge to the output port. Apply 6 in Hg minimum.	Vacuum output cycles On and Off ▶	GO to KG7.
• While cycling outputs On and Off (by depressing and releasing throttle) observe the vacuum gauge at the output. **NOTE: Maintain vacuum at source.**	Vacuum output does not cycle On and Off ▶	REPLACE solenoid. REPEAT Quick Test.
KG7 CHECK MANIFOLD VACUUM LINES FOR BLOCKAGE OR LEAKS		
• With vacuum lines disconnected at TK solenoid, check for vacuum. • Start engine. • Check for vacuum.	Vacuum present ▶	EEC-IV system OK. REFER to Section 4 for idle adjustment.
	No vacuum present ▶	SERVICE vacuum source blockage or leak. REPEAT Quick Test.
KG8 MEASURE TK SOLENOID RESISTANCE		
• Key Off, wait 10 seconds. • DVOM on 200 ohm scale. • Disconnect TK solenoid connector and measure solenoid resistance.	Resistance is between 50-100 ohms ▶	CONNECT TK solenoid. GO to KG9.
	Resistance is less than 50 ohms or greater than 100 ohms ▶	REPLACE TK solenoid. REPEAT Quick Test.

Idle Speed Control (Throttle Kicker) — Pinpoint Test KG

TEST STEP	RESULT ▶	ACTION TO TAKE
KG9 CHECK VOLTAGE OF VPWR CIRCUIT		
• Key On, Engine Off. • DVOM on 20V scale. • Measure voltage at the TK solenoid harness connector, between VPWR and ground.	Voltage reading is 10.5V or less ▶	SERVICE harness open circuit. REPEAT Quick Test.
	Voltage reading is 10.5V or greater ▶	GO to KG10.
KG10 CHECK CONTINUITY OF TK CIRCUIT		
• Key Off, wait 10 seconds. • Disconnect processor and inspect both 60 Pin connectors. • Connect Breakout box to harness, leave processor disconnected. • DVOM on 200 ohm scale. • Measure resistance between test Pin 53 and TK circuit at harness connector.	Resistance reading is 5 ohms or greater ▶	SERVICE harness open circuit. REPEAT Quick Test.
	Resistance reading is less than 5 ohms ▶	GO to KG11.
KG11 CHECK FOR SHORT TO GROUND		
• Key Off, wait 10 seconds. • Leave Breakout box installed and processor disconnected. • Disconnect TK solenoid. • DVOM on 200,000 ohm scale. • Measure resistance between test Pin 53 and test Pins 40, 46 and 60.	Resistance reading is less than 10,000 ohms ▶	SERVICE short to ground. REPEAT Quick Test.
	Resistance reading is 10,000 ohms or greater ▶	GO to KG12.
KG12 CHECK FOR SHORT TO POWER		
• Key Off, wait 10 seconds. • DVOM on 200,000 ohm scale. • Leave Breakout box installed and processor disconnected. • TK solenoid disconnected. • Measure resistance between test Pin 53 and test Pins 37 and 57.	Resistance reading is 10,000 ohms or greater ▶	REPLACE processor. REMOVE Breakout box. RECONNECT harness to processor. RERUN Quick Test.
	Resistance reading is less than 10,000 ohms ▶	SERVICE short to power. REPEAT Quick Test. If code is repeated, REPLACE processor.

Inlet Air Solenoid (IAS) — Pinpoint Test KH

KH

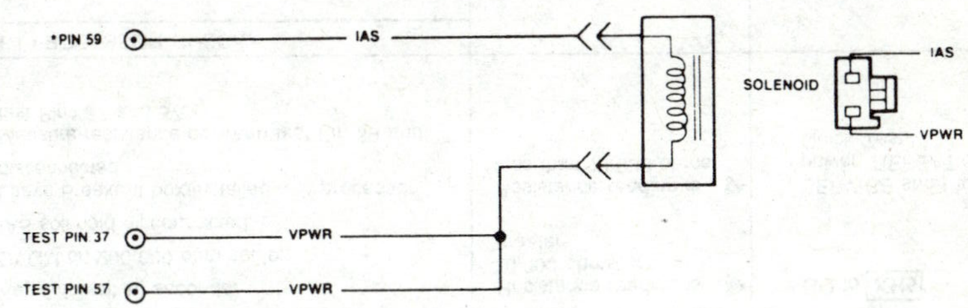

*TEST PINS LOCATED ON BREAKOUT BOX.
ALL HARNESS CONNECTORS VIEWED INTO MATING SURFACE.

STOP-WARNING

You should enter this Pinpoint Test only when directed here from Diagnostics by Symptom in the Engine Supplement Section.

To prevent the replacement of good components, be aware that the following non-EEC areas may be at fault:

- Air or Vacuum Leaks
- Heat Riser and Stovepipe

This Pinpoint Test is intended to diagnose only the following:

- Circuits IAS and VPWR
- IAS Solenoid
- Presence of Manifold Vacuum
- Processor Assembly

Inlet Air Solenoid (IAS) — Pinpoint Test — KH

TEST STEP	RESULT ▶	ACTION TO TAKE
KH1 SIMULATE SUB-ZERO AMBIENT TEMPERATURE		
• Locate, then disconnect ACT sensor. • Start vehicle. • Does inlet air door close on air cleaner inlet air duct?	Yes ▶	IAS is functioning. GO to Diagnostic Routines.
	No ▶	GO to KH2.
KH2 VACUUM CHECK		
• Key Off. • Disconnect both vacuum hoses from IAS. • Connect both vacuum hoses. • Key On, start vehicle. • Does inlet air door close?	Yes ▶	RECONNECT source vacuum hose. GO to KH4.
	No ▶	GO to KH3.
KH3 DUCT AND VALVE CHECK		
• Key Off. • Connect vacuum pump to inlet air valve vacuum hose. • Apply vacuum. • Does inlet air door close?	Yes ▶	VERIFY proper source vacuum.
	No ▶	GO to inlet air duct and valve assembly.
KH4 MEASURE IAS SOLENOID RESISTANCE		
• Key Off, wait 10 seconds. • DVOM on 200 ohm scale. • Disconnect IAS solenoid connector. • Measure solenoid resistance.	Resistance is between 65 and 110 ohms ▶	CONNECT IAS solenoid. GO to KH5.
	Resistance is less than 65 ohms or greater than 110 ohms ▶	REPLACE IAS solenoid. REPEAT Quick Test.
KH5 CHECK VOLTAGE OF VPWR CIRCUIT		
• Key On, Engine Off. • DVOM on 20V scale. • Measure voltage between VPWR circuit at IAS solenoid and to ground.	Voltage reading is less than 10.5V ▶	SERVICE harness open circuit. REPEAT Quick Test.
	Voltage reading is 10.5V or greater ▶	GO to KH6.

Inlet Air Solenoid (IAS) — Pinpoint Test — KH

TEST STEP	RESULT ▶	ACTION TO TAKE
KH6 CHECK CONTINUITY OF IAS CIRCUIT		
• Key Off, wait 10 seconds. • Disconnect processor 60 Pin connector and inspect for damaged pins, corrosion, loose wires. Service as necessary. • Connect Breakout box to harness. Leave processor disconnected. • DVOM on 200 ohm scale. • Measure resistance between test Pin 59 and IAS circuit at harness connector.	Resistance reading is 5 ohms or greater ▶	SERVICE open circuit. REPEAT Quick Test.
	Resistance reading is less than 5 ohms ▶	GO to KH7.
KH7 CHECK FOR SHORT TO GROUND		
• Key Off, wait 10 seconds. • Leave Breakout box installed and processor disconnected. • Disconnect IAS solenoid. • DVOM on 200,000 ohm scale. • Measure resistance between test Pin 59 and test Pins 40, 46 and 60.	Resistance reading is less than 10,000 ohms ▶	SERVICE short to ground. REPEAT Quick Test.
	Resistance reading is 10,000 ohms or greater ▶	GO to KH8.
KH8 CHECK FOR SHORT TO POWER		
• Key Off, wait 10 seconds. • DVOM on 200,000 ohm scale. • IAS solenoid disconnected. • Leave Breakout box installed and processor disconnected. • Measure resistance between test Pin 59 and test Pins 37 and 57.	Resistance reading is 10,000 ohms or greater ▶	GO to KH9.
	Resistance reading is less than 10,000 ohms ▶	SERVICE short to power. REPEAT Quick Test.
KH9 PROCESSOR CHECK		
• Key Off. • Leave Breakout box installed and processor disconnected. • Connect a jumper between test Pin 59 and test Pin 40. • Key On, start vehicle. • Does inlet air door close?	Yes ▶	REPLACE processor.
	No ▶	REPLACE IAS.

Variable Voltage Choke (VVC) — Pinpoint Test — KJ

KJ

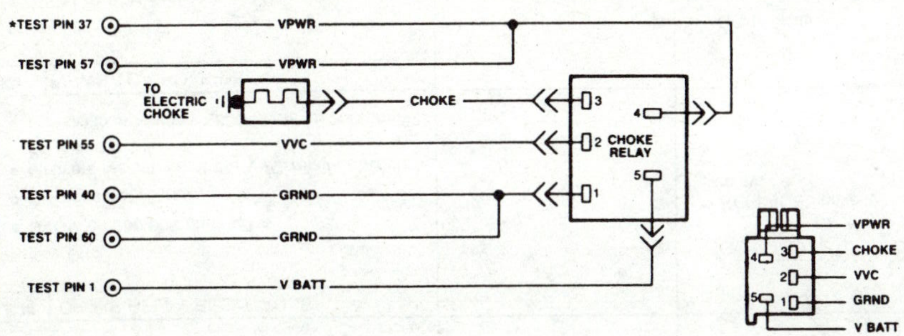

*TEST PINS LOCATED ON BREAKOUT BOX.
ALL HARNESS CONNECTORS VIEWED INTO MATING SURFACE.

STOP-WARNING

You should enter this Pinpoint Test only when a service code 88 is received in Quick Test Step 3.0 or when directed here from Diagnostics by Symptom in the Engine Supplement Section.

This Pinpoint Test is intended to diagnose only the following:

- Harness circuits: VPWR, Choke, VVC, Grnd., and V Batt.
- Choke Relay

Variable Voltage Choke (VVC) — Pinpoint Test — KJ

TEST STEP	RESULT ▶	ACTION TO TAKE
KJ1 ENTER OUTPUT STATE CHECK (REFER TO APPENDIX)		
NOTE: Do not use STAR tester for this step, use VOM/DVOM. • Key Off, wait 10 seconds. • DVOM on 20V scale. • Connect DVOM negative test lead to STO and positive test lead to battery positive. • Jumper STI to signal return. • Perform Key On, Engine Off Self-Test until the completion of the Continuous Test Codes. • DVOM will indicate zero volts. • Depress and release the throttle. • Did DVOM reading change to a high voltage reading?	Yes ▶ No ▶	REMAIN in Output State Check. GO to KJ2. DEPRESS throttle to WOT and RELEASE. If STO voltage does not go high, GO to Pinpoint Test Step Q40.
KJ2 CHECK VVC OUTPUT ELECTRICAL OPERATION		
• Key On, Engine Off. • DVOM on 20V scale. • Connect DVOM positive test lead to choke circuit on choke cap and negative test lead to ground. • While observing DVOM depress and release the throttle several times to cycle output On and Off. • Choke output cycles On and Off?	Yes ▶ No ▶	GO to Section 4. REMOVE jumper. GO to KJ3.
KJ3 CHECK CONTINUITY OF GROUND CIRCUIT		
• Key Off, wait 10 seconds. • DVOM on 200 ohm scale. • Measure resistance between battery negative terminal and Pin 1 at the relay harness connector.	Resistance reading is 5 ohms or greater ▶ Resistance reading is less than 5 ohms ▶	SERVICE harness circuit. REPEAT Quick Test. GO to KJ4.

Variable Voltage Choke (VVC) — Pinpoint Test — KJ

TEST STEP	RESULT ▶	ACTION TO TAKE
KJ4 CONTINUITY OF CHOKE RELAY		
• Key Off, wait 10 seconds. • Disconnect VVC relay. • DVOM on 200 ohm scale. • Measure resistance between Pin 3 of the VVC relay connector and the choke connector. • Is the reading greater than 5 ohms?	Yes ▶	SERVICE open in choke circuit.
	No ▶	GO to KJ5.
KJ5 CHECK FOR VOLTAGE ON V BATT AND VPWR CIRCUITS		
• Key On, Engine Off. • DVOM on 20V scale. • VVC relay disconnected. • Measure voltage between Pin 1 and Pin 5 and Pin 1 and Pin 4 of VVC relay harness connector.	Voltage is less than 10.5V ▶	SERVICE open in choke power circuit. REPEAT Quick Test.
	Voltage is 10.5V or greater ▶	REPLACE choke relay.
KJ6 CHOKE RELAY RESISTANCE		
• Key Off. • DVOM on 200,000 ohm scale. • Disconnect VVC relay. • Measure resistance at the VVC relay between Pin 2 and Pin 4. • Is resistance reading 1,800 ohms ± 200 ohms?	Yes ▶	GO to KJ7.
	No ▶	REPLACE choke relay.
KJ7 VPWR TO CHOKE RELAY		
• Key Off. • DVOM on 20V scale. • VVC relay disconnected. • Key On, engine Off. • Measure voltage between Pin 4 at the VVC relay connector and engine block ground. • Is voltage greater than 10V?	Yes ▶	GO to KJ8.
	No ▶	SERVICE VPWR circuit and RETEST.

Variable Voltage Choke (VVC) — Pinpoint Test — KJ

TEST STEP	RESULT ▶	ACTION TO TAKE
KJ8 CHECK CONTINUITY OF VVC CIRCUIT		
• Key Off, wait 10 seconds. • Disconnect processor 60 Pin connector and inspect for damaged pins, corrosion, loose wires. Service as necessary. • Connect Breakout box to harness. Leave processor disconnected. • VVC relay disconnected. • DVOM on 200 ohm scale. • Measure resistance between test Pin 55 and VVC circuit at the relay harness connector.	Resistance is 5 ohms or greater ▶	SERVICE harness circuit. REPEAT Quick Test.
	Resistance is less than 5 ohms ▶	GO to KJ9.
KJ9 CHECK FOR SHORT TO GROUND ON VVC CIRCUIT		
• Key Off, wait 10 seconds. • DVOM on 200,000 ohm scale. • VVC relay disconnected. • Leave Breakout box installed and processor disconnected. • Measure resistance between Pin 55 to test Pins 40, 46 and 60.	All resistance readings 10,000 ohms or greater ▶	GO to KJ10.
	Any resistance readings less than 10,000 ohms ▶	SERVICE harness short(s). REPEAT Quick Test.
KJ10 CHECK FOR SHORT TO POWER		
• Key Off, wait 10 seconds. • VVC relay disconnected. • DVOM on 200,000 ohm scale. • Leave Breakout box installed and processor disconnected. • Measure resistance from test Pin 55 to test Pins 1, 37 and 57.	Resistance reading is 10,000 ohms or greater ▶	REPLACE processor. REPEAT Quick Test.
	Resistance reading is less than 10,000 ohms ▶	SERVICE short to power. REPEAT Quick Test. If code is repeated, REPLACE processor.

Temperature Compensated Pump (TCP) — Pinpoint Test — KK

KK

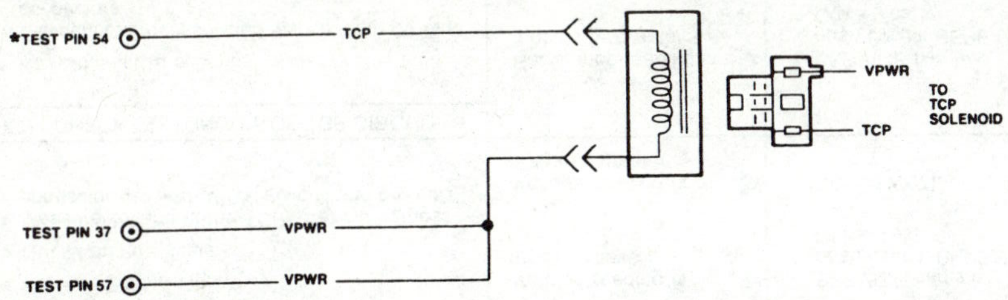

* TEST PINS LOCATED ON BREAKOUT BOX.
ALL HARNESS CONNECTORS VIEWED INTO MATING SURFACE.

STOP-WARNING

You should enter this Pinpoint Test only when a service code 87 is received in Quick Test Step 3.0 or when directed here from Diagnostics by Symptom in the Engine Supplement Section.

This Pinpoint Test is intended to diagnose only the following:

- Harness circuits: TCP and VPWR
- TCP Solenoid
- Processor Assembly

Temperature Compensated Pump (TCP) — Pinpoint Test — KK

TEST STEP	RESULT ▶	ACTION TO TAKE
KK1 ENTER OUTPUT STATE CHECK (REFER TO APPENDIX)		
NOTE: Do not use STAR tester for this Step, use VOM/DVOM. • Key Off, wait 10 seconds. • DVOM on 20V scale. • Connect DVOM negative test lead to STO and positive test lead to battery positive. • Jumper STI to signal return. • Perform Key On, Engine Off Self-Test until the completion of the Continuous Test Codes. • DVOM will indicate zero volts. • Depress and release the throttle. • Did DVOM reading change to a high voltage reading?	Yes ▶	REMAIN in Output State Check. GO to KK2.
	No ▶	DEPRESS throttle to WOT and release. If STO voltage does not go high, GO to Pinpoint Test Step Q40. Leave equipment hooked up.
KK2 CHECK TCP SOLENOID ELECTRICAL OPERATION		
• DVOM on 20V scale. • Connect DVOM positive test lead to circuit VPWR on TCP solenoid and negative test lead to output circuit TCP. • While observing DVOM depress and release the throttle several times to cycle output On and Off. • Does TCP output cycle On and Off?	Yes ▶	GO to KK3.
	No ▶	REMOVE jumper to STI. GO to KK5.
KK3 CHECK TCP SOLENOID FOR VACUUM CYCLING		
• Install vacuum pump to the TCP solenoid vacuum supply port and install a vacuum gauge to the output port. Apply 6 in Hg minimum. • While cycling outputs On and Off (by depressing and releasing throttle) observe the vacuum gauge at the output. **NOTE: Maintain vacuum at source.**	Vacuum output cycles On and Off ▶	GO to KK4.
	Vacuum output does not cycle On and Off ▶	REPLACE solenoid. REPEAT Quick Test.

Temperature Compensated Pump (TCP) — Pinpoint Test KK

TEST STEP	RESULT ▶	ACTION TO TAKE
KK4 CHECK MANIFOLD VACUUM LINES FOR BLOCKAGE OR LEAKS		
• With vacuum lines disconnected at TCP solenoid, check for vacuum. • Start engine. • Check for vacuum.	Vacuum present ▶	EEC-IV system OK. REFER to Section for TCP diagnostics.
	No vacuum present ▶	SERVICE vacuum source blockage or leak. REPEAT Quick Test.
KK5 MEASURE TCP SOLENOID RESISTANCE		
• Key Off, wait 10 seconds. • DVOM on 200 ohm scale. • Disconnect TCP solenoid connector and measure solenoid resistance.	Resistance is between 50-100 ohms ▶	CONNECT TCP solenoid. GO to KK6.
	Resistance is less than 50 ohms or greater than 100 ohms ▶	REPLACE TCP solenoid. REPEAT Quick Test.
KK6 CHECK VOLTAGE OF VPWR CIRCUIT		
• Key On, Engine Off. • DVOM on 20V scale. • Measure voltage at the TCP solenoid harness connector between VPWR circuit and ground.	Voltage reading is 10.5V or less ▶	SERVICE harness open circuit. REPEAT Quick Test.
	Voltage reading is 10.5V or greater ▶	GO to KK7.
KK7 CHECK CONTINUITY OF TCP CIRCUIT		
• Key Off, wait 10 seconds. • Disconnect processor and inspect both 60 Pin connectors. • Connect Breakout box to harness, leave processor disconnected. • DVOM on 200 ohm' scale. • Measure resistance between test Pin 54 and TCP circuit at harness connector.	Resistance reading is 5 ohms or greater ▶	SERVICE harness open circuit. REPEAT Quick Test.
	Resistance reading is less than 5 ohms ▶	GO to KK8.

Temperature Compensated Pump (TCP) — Pinpoint Test KK

TEST STEP	RESULT ▶	ACTION TO TAKE
KK8 CHECK FOR SHORT TO GROUND		
• Key Off, wait 10 seconds. • Leave Breakout box installed and processor disconnected. • Disconnect TCP solenoid. • DVOM on 200,000 ohm scale. • Measure resistance between test Pin 54 and test Pin 40, 46 and 60.	Resistance reading is less than 10,000 ohms ▶	SERVICE short to ground. REPEAT Quick Test.
	Resistance reading is 10,000 ohms or greater ▶	GO to KK9.
KK9 CHECK FOR SHORT TO POWER		
• Key Off, wait 10 seconds. • DVOM on 200,000 ohm scale. • Leave Breakout box installed and processor disconnected. • TCP solenoid disconnected. • Measure resistance between test Pin 54 and test Pins 37 and 57.	Resistance reading is 10,000 ohms or greater ▶	REPLACE processor. RERUN Quick Test.
	Resistance reading is less than 10,000 ohms ▶	SERVICE short to power. REPEAT Quick Test. If code is repeated, REPLACE processor.

Shift Indicator Light (SIL) — Pinpoint Test — KL

KL

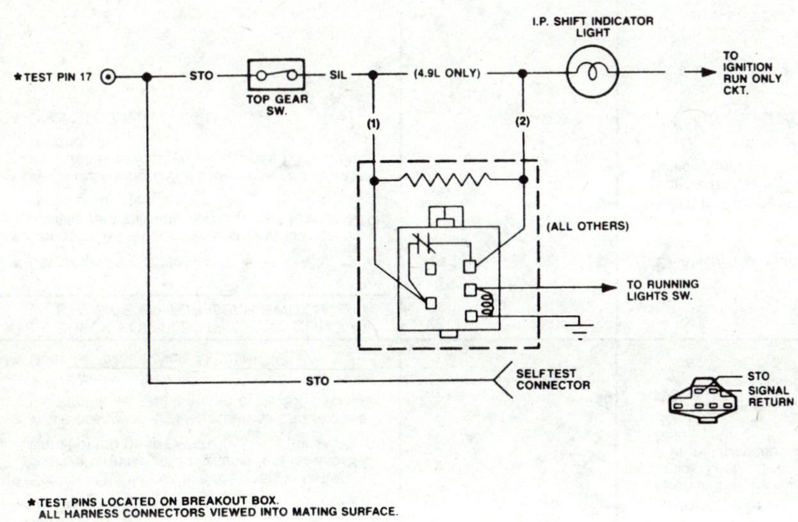

★ TEST PINS LOCATED ON BREAKOUT BOX.
ALL HARNESS CONNECTORS VIEWED INTO MATING SURFACE.

STOP-WARNING

You should enter this Pinpoint Test only when directed here from Diagnostics by Symptom in the Engine Supplement Section or from Pinpoint Test Step Q6.

This Pinpoint Test is intended to diagnose only the following:

- Top Gear Switch
- Shift Indicator Light Bulb and Fuse
- Harness Circuits: SIL and STO

Shift Indicator Light (SIL) — Pinpoint Test — KL

TEST STEP	RESULT ▶	ACTION TO TAKE
KL1 ENTER OUTPUT STATE CHECK IN NEUTRAL (REFER TO APPENDIX)		
NOTE: Do not use STAR tester for this Step, use VOM/DVOM. • Key Off, wait 10 seconds. • Transmission in neutral. • DVOM on 20V scale. • Connect DVOM negative test lead to STO and positive test lead to battery positive. • Jumper STI to signal return. • Perform Key On, Engine Off Self-Test until the completion of the Continuous Test Codes. • DVOM will indicate zero volts. • Depress and release the throttle. • Did DVOM reading change to a high voltage reading?	Yes ▶ No ▶	REMAIN in output state check, Go to KL2. DEPRESS throttle to WOT and release. If STO voltage does not go high, GO to Pinpoint Test Step KL7. Leave equipment hooked up.
KL2 CHECK FOR ELECTRICAL CYCLING AT DIMMER RELAY INPUT		
• If 4.9L truck go to step KL5. • DVOM on 20V scale. • Connect DVOM positive test lead to battery positive terminal and negative test lead to Pin 1 on dimmer relay. • While observing DVOM, depress and release the throttle several times to cycle Pin 1 On and Off. • Does Pin 1 cycle On and Off?	Yes ▶ No ▶	GO to KL3. GO to KL5.
KL3 CHECK FOR ELECTRICAL CYCLING AT DIMMER RELAY OUTPUT		
• DVOM on 20V scale. • Connect DVOM positive test lead to battery positive terminal and negative test lead to Pin 2 on dimmer relay. • While observing DVOM, depress and release the throttle several times to cycle Pin 2 On and Off. • Does Pin 2 cycle On and Off?	Yes ▶ No ▶	REMOVE jumper, GO to KL4. REMOVE jumper. REPLACE dimmer relay.

Shift Indicator Light (SIL) — Pinpoint Test — KL

TEST STEP	RESULT ▶	ACTION TO TAKE
KL4 CHECK CONTINUITY BETWEEN DIMMER RELAY OUTPUT AND SIL BULB		
• Key Off, wait 10 seconds. • DVOM on 200 ohm scale. • Measure resistance between Pin 2 of dimmer relay and SIL bulb.	Resistance reading is 5 ohms or greater ▶	SERVICE harness open circuit. REPEAT Quick Test.
	Resistance reading is less than 5 ohms ▶	REPLACE SIL bulb.
KL5 CHECK FOR ELECTRICAL CYCLING AT SIL SIDE OF TOP GEAR SWITCH		
• DVOM on 20V scale. • Connect DVOM positive test lead to battery positive terminal and negative test lead to SIL circuit of top gear switch. • While observing DVOM, depress and release the throttle several times to cycle SIL circuit On and Off. • Does SIL circuit cycle On and Off?	Yes ▶	SERVICE harness open circuit. REPEAT Quick Test. 4.9L only, change SIL bulb.
	No ▶	GO to **KL6**.
KL6 CHECK FOR ELECTRICAL CYCLING AT STO SIDE OF TOP GEAR SWITCH		
• DVOM on 20V scale. • Connect DVOM positive test lead to battery positive terminal and negative test lead to STO circuit of top gear switch. • While observing DVOM, depress and release the throttle several times to cycle STO circuit On and Off. • Does STO circuit cycle On and Off?	Yes ▶	CHANGE top gear switch.
	No ▶	SERVICE harness open circuit. REPEAT Quick Test.

Shift Indicator Light (SIL) — Pinpoint Test — KL

TEST STEP	RESULT ▶	ACTION TO TAKE
KL7 ENTER OUTPUT STATE CHECK IN TOP GEAR (REFER TO APPENDIX)		
NOTE: Do not use STAR tester for this Step, use VOM/DVOM. • Key Off, wait 10 seconds. • Transmission in top gear. • DVOM on 20V scale. • Connect DVOM negative test lead to STO and positive test lead to battery positive. • Jumper STI to signal return. • Perform Key On, Engine Off Self-Test until the completion of the Continuous Test Codes. • DVOM will indicate zero volts. • Depress and release the throttle. • Did DVOM reading change to a high voltage reading?	Yes ▶	CHECK SIL bulb and fuse 15. If OK, SERVICE short to ground in SIL circuit.
	No ▶	GO to Pinpoint Test Step **Q40**.

WOT A/C Cutoff (WAC) A/C Demand — Pinpoint Test — KM

KM 2.3L EFI, 3.8L AND 5.0L CFI ENGINES

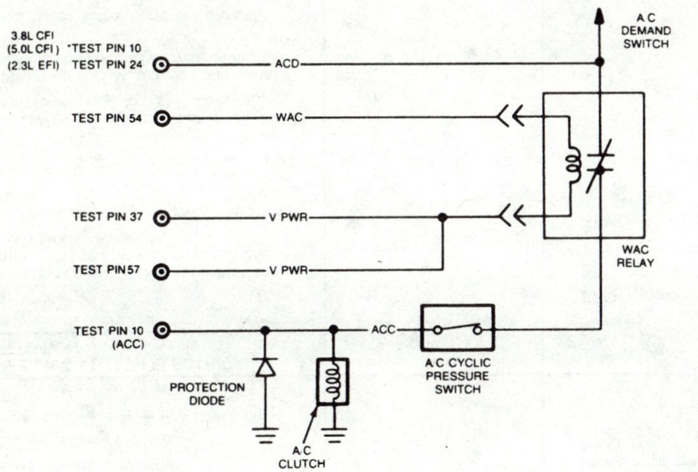

STOP-WARNING

You should enter this Pinpoint Test only when a Service Code 36 or 86 is received in Quick Test Step 3.0 or 5.0 or when directed here from Diagnostics by Symptom in the Engine Supplement Section.

This Pinpoint Test is intended to diagnose only the following:

- Harness Circuits: WAC and V PWR
- WOT A/C Cut-Off Relay
- Processor Assembly
- A/C Demand Switch

WOT A/C Cutoff (WAC) A/C Demand — Pinpoint Test — KM

KM 2.3L OHC, 2.3L EFI TC, 2.3L CFI ENGINES

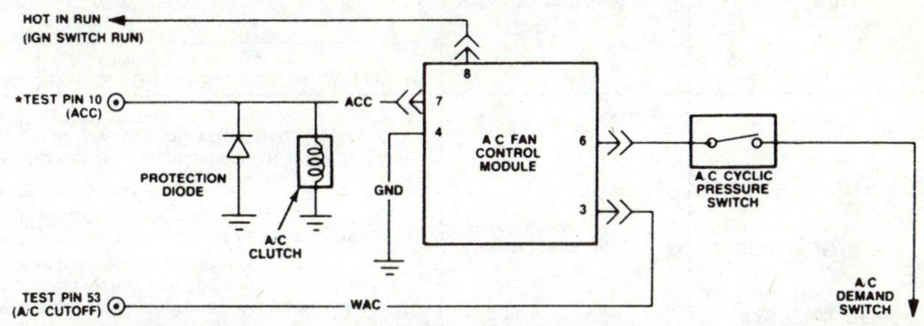

*TEST PINS LOCATED ON BREAKOUT BOX.
ALL HARNESS CONNECTORS VIEWED INTO MATING SURFACE.

KM 1.6L EFI, 1.6L EFI TC ENGINES

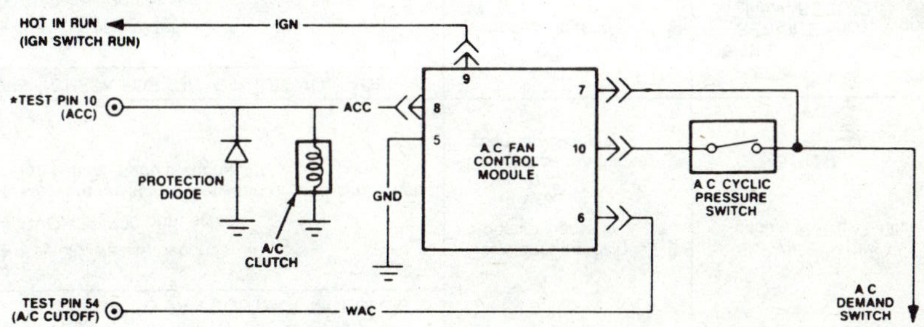

*TEST PINS LOCATED ON BREAKOUT BOX.
ALL HARNESS CONNECTORS VIEWED INTO MATING SURFACE.

WOT A/C Cutoff (WAC) A/C Demand — Pinpoint Test — KM

TEST STEP	RESULT ▶	ACTION TO TAKE
KM1 ENTER OUTPUT STATE CHECK (REFER TO APPENDIX)		
NOTE: Do not use STAR tester for this Step, use VOM/DVOM. • Key Off, wait 10 seconds. • DVOM on 20V scale. • Connect DVOM negative test lead to STO and positive test lead to battery positive. • Jumper STI to signal return. • Perform Key On, Engine Off Self-Test until the completion of the Continuous Test Codes. • DVOM will indicate zero volts. • Depress and release the throttle. • Did DVOM reading change to a high voltage reading?	Yes ▶	REMAIN in Output State Check. GO to KM2.
	No ▶	DEPRESS throttle to WOT and RELEASE. If STO voltage does not go high, GO to Pinpoint Test Step Q40. Leave equipment hooked up.
KM2 CHECK A/C CLUTCH ELECTRICAL OPERATION		
• Key On, engine Off. • Disconnect processor 60 Pin connector and inspect for damaged pins, corrosion, loose wires. Service as necessary. • Connect Breakout box to harness. Reconnect processor. • A/C switch to A/C. • DVOM on 20V scale. • Connect DVOM positive test lead to test Pin 37 and negative test lead to test Pin 54. • While observing DVOM, depress and release the throttle several times (to cycle output On and Off). • Does A/C clutch output cycle On and Off?	Yes ▶	EEC-IV system OK.
	No ▶	REMOVE jumper. GO to KM3.
KM3 FAN CONTROL MODULE OR WAC RELAY FAULT ISOLATION		
• A/C fan/control module applications 1.6L EFI, 1.6L EFI Turbo, 2.3L CFI.	A/C fan control ▶	GO to KM4.
• All other systems with WAC relay.	WAC relay ▶	GO to KM10.

WOT A/C Cutoff (WAC) A/C Demand — Pinpoint Test — KM

TEST STEP	RESULT ▶	ACTION TO TAKE
KM4 CHECK CONTINUITY OF GROUND CIRCUIT TO FAN CONTROL MODULE		
• Key Off, wait 10 seconds. • DVOM on 200 ohm scale. • Measure resistance between ground circuit and fan control module connector.	Resistance reading is 5 ohms or greater ▶	SERVICE harness circuit. REPEAT Quick Test.
	Resistance reading is less than 5 ohms ▶	GO to KM5.
KM5 CHECK FOR IGN. CIRCUIT VOLTAGE		
• Key On, engine Off. • DVOM on 20V scale. • Measure voltage between ignition circuit and ground circuit on the fan control module connector.	Voltage is less than 10.5V ▶	SERVICE open in ign. circuit. REPEAT Quick Test.
	Voltage is 10.5V or greater ▶	GO to KM6.
KM6 CHECK CONTINUITY OF WAC CIRCUIT		
• Key Off, wait 10 seconds. • Disconnect processor 60 Pin connector and inspect for damaged pins, corrosion, loose wires. Service as necessary. • Connect Breakout box to harness. Leave processor disconnected. • DVOM on 200 ohm scale. • Measure resistance between test Pin 54 (test Pin 53 for 2.3L OHC engine) and WAC circuit at fan control module.	Resistance is 5 ohms or greater ▶	SERVICE harness WAC circuit. REPEAT Quick Test.
	Resistance is less than 5 ohms ▶	GO to KM7.
KM7 CHECK FOR SHORT TO GROUND ON WAC CIRCUIT		
• Key Off, wait 10 seconds. • Leave Breakout box installed and processor disconnected. • DVOM on 200,000 ohm scale. • Measure resistance between test Pin 54 (test Pin 53 for 2.3L OHC engine) and test Pins 40, 46 and 60.	Resistance reading is less than 10,000 ohms ▶	GO to KM8.
	Resistance reading is 10,000 ohms or greater ▶	REPLACE processor. REPEAT Quick Test.

WOT A/C Cutoff (WAC) A/C Demand — Pinpoint Test — KM

TEST STEP	RESULT ▶	ACTION TO TAKE
KM8 CHECK FOR SHORT TO GROUND WITH FAN CONTROL MODULE DISCONNECTED		
• Key Off, wait 10 seconds. • Leave Breakout box installed and processor disconnected. • DVOM on 200,000 ohm scale. • Disconnect fan control module. • Measure resistance between test Pin 54 (test Pin 53 for 2.3L OHC engine) and test Pins 40, 46 and 60.	All resistance readings 10,000 ohms or greater ▶	REPLACE fan control module. REPEAT Quick Test.
	Any resistance reading is less than 10,000 ohms ▶	SERVICE harness short. REPEAT Quick Test.
KM10 MEASURE WAC RELAY RESISTANCE		
• Key Off, wait 10 seconds. • DVOM on 200 ohm scale. • Disconnect WAC relay connector and measure relay resistance.	Resistance is between 50 and 70 ohms ▶	CONNECT WAC solenoid. GO to **KM11**.
	Resistance is less than 50 ohms or greater than 70 ohms ▶	REPLACE WAC relay. REPEAT Quick Test.
KM11 CHECK VOLTAGE OF V PWR CIRCUIT		
• Key On, Engine Off. • DVOM on 20V scale. • Connect DVOM positive test lead to V PWR circuit and negative test lead to ground. • Measure voltage on WAC relay V PWR circuit.	Voltage reading is less than 10.5V ▶	SERVICE harness open circuit. REPEAT Quick Test.
	Voltage reading is 10.5V or greater ▶	GO to **KM12**.
KM12 CHECK CONTINUITY OF WAC CIRCUIT		
• Key Off, wait 10 seconds. • Disconnect processor 60 Pin connector and inspect for damaged pins, corrosion, loose wires. Service as necessary. • Connect Breakout box to harness. Leave processor disconnected. • DVOM on 200 ohm scale. • Measure resistance between test Pin 54 and WAC circuit at harness connector.	Resistance readings are 5 ohms or greater ▶	SERVICE harness open circuit. REPEAT Quick Test.
	Resistance reading is less than 5 ohms ▶	GO to **KM13**.

WOT A/C Cutoff (WAC) A/C Demand — Pinpoint Test — KM

TEST STEP	RESULT ▶	ACTION TO TAKE
KM13 CHECK FOR SHORT TO GROUND		
• Key Off, wait 10 seconds. • Leave Breakout box installed and processor disconnected. • DVOM on 200,000 ohm scale. • Measure resistance between test Pin 54 and test Pins 40, 46 and 60.	Resistance readings are less than 10,000 ohms ▶	SERVICE short to ground. REPEAT Quick Test.
	Resistance readings are 10,000 ohms or greater ▶	GO to **KM14**.
KM14 CHECK FOR SHORT TO POWER		
• Key Off, wait 10 seconds. • DVOM on 200,000 ohm scale. • Leave Breakout box installed and processor disconnected. • WAC solenoid disconnected. • Measure resistance between test Pin 54 and test Pins 37 and 57.	Resistance reading is 10,000 ohms or greater ▶	REPLACE processor. REPEAT Quick Test.
	Resistance reading is less than 10,000 ohms ▶	SERVICE short to power. REPEAT Quick Test.
KM20 CYCLE A/C DEMAND SWITCH		
• Key Off, wait 10 seconds. • Disconnect processor 60 Pin connector and inspect for damaged pins, corrosion, loose wires. Service as necessary. • Connect Breakout box to harness. Leave processor disconnected. • DVOM on 20V scale. • Measure voltage between test Pin 10 (and Pin 24 for 2.3L EFI truck) and test Pin 40. • Does output cycle 4.0-10.5V when A/C switch is cycled?	Yes ▶	REPLACE processor. REPEAT Quick Test.
	No ▶	GO to **KM21**.
KM21 CHECK CONTINUITY OF ACC/ACD CIRCUIT		
• Key Off, wait 10 seconds. • DVOM on 200 ohm scale. • Measure resistance between ACC/ACD test Pin 10 (ACD Pin 24, 2.3L EFI truck) and A/C clutch and A/C demand switch.	Resistance reading is 5 ohms or greater ▶	SERVICE harness circuit. REPEAT Quick Test.
	Resistance reading is less than 5 ohms ▶	EEC-IV system OK.

Turbo Boost — Pinpoint Test — KN

KN

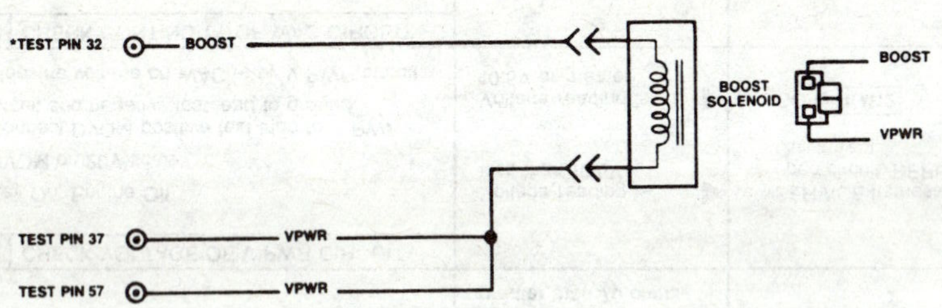

*TEST PINS LOCATED ON BREAKOUT BOX.
ALL HARNESS CONNECTORS VIEWED INTO MATING SURFACE.

STOP-WARNING

You should enter this Pinpoint Test only when directed here from Diagnostics by Symptom in the Engine Supplement Section.

This Pinpoint Test is intended to diagnose only the following:

- Harness circuits: VPWR, Boost

Turbo Boost — Pinpoint Test — KN

TEST STEP	RESULT	ACTION TO TAKE
KN1 ENTER OUPUT STATE CHECK (REFER TO APPENDIX)		
NOTE: Do not use STAR tester for this Step, use VOM/DVOM. • Key Off, wait 10 seconds. • DVOM on 20V scale. • Connect DVOM negative test lead to STO and positive test lead to battery positive. • Jumper STI to signal return. • Perform Key On, Engine Off Self-Test until the completion of the Continuous Test Codes. • DVOM will indicate zero volts. • Depress and release the throttle. • Did DVOM reading change to a high voltage reading?	Yes ▶	REMAIN in Output State Check. GO to KN2.
	No ▶	DEPRESS throttle to WOT and release. If STO voltage does not go high, GO to Pinpoint Test Step Q40. Leave equipment hooked up.
KN2 CHECK BOOST OUTPUT ELECTRICAL OPERATION		
• Key On, Engine Off. • DVOM on 20V scale. • Connect DVOM positive test lead to circuit VPWR on boost connector and negative test lead to boost output on boost connector. • While observing DVOM, depress and release the throttle several times to cycle output On and Off. • Does boost output cycle On and Off?	Yes ▶	GO to KN3.
	No ▶	REMOVE jumper. GO to KN4.

Turbo Boost — Pinpoint Test — KN

TEST STEP	RESULT ▶	ACTION TO TAKE
KN3 CHECK BOOST SOLENOID FUNCTION		
• Remain in output state check. • Disconnect black solenoid vacuum hose at the turbocharger end. **(Not the black and yellow stripe hose.)** • Attach a vacuum pump to the hose. • Depress the throttle once to cycle the solenoid closed. • Apply vacuum to the solenoid. • Depress the throttle once again to cycle the solenoid open and release trapped vacuum. • Did the solenoid hold and then release vacuum?	Yes ▶ No ▶	EEC-IV system OK. REFER to Shop Manual for boost diagnostics. REPLACE solenoid. RETEST.
KN4 MEASURE BOOST SOLENOID RESISTANCE		
• Key Off, wait 10 seconds. • DVOM on 200 ohm scale. • Disconnect boost solenoid connector and measure the solenoid resistance.	Resistance is between 65 and 110 ohms ▶ Resistance is less than 65 ohms or greater than 110 ohms ▶	CONNECT BOOST solenoid. GO to KN5. REPLACE BOOST solenoid. REPEAT Quick Test.
KN5 CHECK VOLTAGE OF VPWR CIRCUIT		
• Key On, Engine Off. • DVOM on 20V scale. • Measure voltage between VPWR circuit of boost solenoid harness connector and ground.	Voltage reading is less than 10.5V ▶ Voltage reading is 10.5V or greater ▶	SERVICE harness open circuit. REPEAT Quick Test. GO to KN6.
KN6 CHECK CONTINUITY OF BOOST CIRCUIT		
• Key Off, wait 10 seconds. • Disconnect processor 60 Pin connector and inspect for damaged pins, corrosion, loose wires. Service as necessary. • Connect Breakout box to harness. Leave processor disconnected. • DVOM on 200 ohm scale. • Measure resistance between test Pin 32 and boost circuit at harness connector.	Resistance reading is 5 ohms or greater ▶ Resistance reading is less than 5 ohms ▶	SERVICE open circuit. REPEAT Quick Test. GO to KN7.

Turbo Boost — Pinpoint Test — KN

TEST STEP	RESULT ▶	ACTION TO TAKE
KN7 CHECK FOR SHORT TO GROUND		
• Key Off, wait 10 seconds. • Leave Breakout box installed and processor disconnected. • Disconnect boost solenoid. • DVOM on 200,000 ohm scale. • Measure resistance between test Pin 32 and test Pins 40, 46 and 60.	All readings 10,000 ohms or greater ▶ Any reading less than 10,000 ohms ▶	Go to KN8. SERVICE short to ground, REPEAT Quick Test.
KN8 CHECK FOR SHORT TO POWER		
• Key Off, wait 10 seconds. • DVOM on 200,000 ohm scale. • Leave Breakout box installed and processor disconnected. • Boost solenoid disconnected. • Measure resistance between Pin 32 and test Pins 37 and 57.	Resistance reading is less than 10,000 ohms ▶ Resistance reading is 10,000 ohms or greater ▶	SERVICE short to power. REPEAT Quick Test. If symptom is still present, REPLACE processor. REPLACE processor. REPEAT Quick Test.

Exhaust Heat Control (EHC) | Pinpoint Test | KP

KP

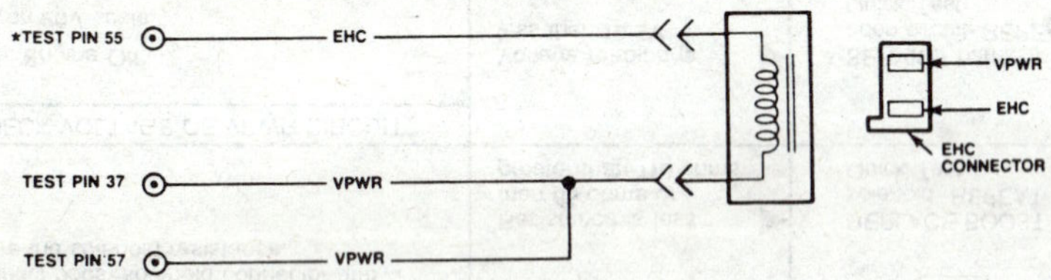

*TEST PINS LOCATED ON BREAKOUT BOX.
ALL HARNESS CONNECTORS VIEWED INTO MATING SURFACE.

STOP-WARNING

You should enter this Pinpoint Test only when a service code 89 is received in Quick Test Step 3.0 or when directed here from Diagnostics by Symptom in the Engine Supplement Section.

This Pinpoint Test is intended to diagnose only the following:

- Exhaust Heat Solenoid
- Vacuum Routing

Exhaust Heat Control (EHC) | Pinpoint Test | KP

TEST STEP	RESULT ▶	ACTION TO TAKE
KP1 ENTER OUTPUT STATE CHECK (REFER TO APPENDIX)		
NOTE: Do not use STAR tester for this Step, use VOM/DVOM. • Key Off, wait 10 seconds. • DVOM on 20V scale. • Connect DVOM negative test lead to STO and positive test lead to battery positive. • Jumper STI to signal return. • Perform Key On, Engine Off Self-Test until the completion of the Continuous Test Codes. • DVOM will indicate zero volts. • Depress and release the throttle. • Did DVOM reading change to a high voltage reading?	Yes ▶	REMAIN in Output State Check. GO to KP2.
	No ▶	DEPRESS throttle to WOT and release. If STO voltage does not go high, GO to Pinpoint Test Step Q40. Leave equipment hooked up.
KP2 CHECK EHC SOLENOID ELECTRICAL OPERATION		
• DVOM on 20V scale. • Measure voltage between VPWR and EHC circuit at EHC solenoid. • While observing DVOM, depress throttle several times to cycle output On and Off. • Does EHC output cycle On and Off?	Yes ▶	GO to KP3.
	No ▶	REMOVE jumper to STI. GO to KP5.
KP3 CHECK EXHAUST HEAT SOLENOID FOR VACUUM CYCLING		
• Install vacuum pump to the exhaust heat solenoid vacuum supply port and install a vacuum gauge to the output port. Apply 6 in Hg minimum. • While cycling outputs On and Off (by depressing and releasing throttle) observe the vacuum gauge at the output. **NOTE: Maintain vacuum at source.**	Vacuum output cycles On and Off ▶	GO to KP4.
	Vacuum output does not cycle On and Off ▶	REPLACE solenoid. REPEAT Quick Test.

Exhaust Heat Control (EHC) — Pinpoint Test — KP

TEST STEP	RESULT ▶	ACTION TO TAKE
KP4 CHECK MANIFOLD VACUUM LINES FOR BLOCKAGE OR LEAKS		
• With vacuum lines disconnected at exhaust heat solenoid, check for vacuum. • Start engine. • Check for vacuum.	Vacuum present ▶	EEC-IV system OK. REFER to Section for exhaust heat diagnostics.
	No vacuum present ▶	SERVICE vacuum source blockage or leak. REPEAT Quick Test.
KP5 MEASURE EHC SOLENOID RESISTANCE		
• Key Off, wait 10 seconds. • DVOM on 200 ohm scale. • Disconnect exhaust heat solenoid connector and measure solenoid resistance.	Resistance is between 50 and 100 ohms ▶	CONNECT exhaust heat solenoid. Go to KP6.
	Resistance is less than 50 ohms or greater than 100 ohms ▶	REPLACE exhaust heat solenoid. REPEAT Quick Test.
KP6 CHECK VOLTAGE OF V PWR CIRCUIT		
• Key On, Engine Off. • DVOM on 20V scale. • Measure voltage at the EHC solenoid harness connector between VPWR circuit and ground.	Voltage reading is less than 10.5V ▶	SERVICE harness open circuit. REPEAT Quick Test.
	Voltage reading is 10.5V or greater ▶	GO to KP7.
KP7 CHECK CONTINUITY OF EHC CIRCUIT		
• Key Off, wait 10 seconds. • Disconnect processor 60 Pin connector and inspect for damaged pins, corrosion, loose wires. Service as necessary. • Connect Breakout box to harness. Leave processor disconnected. • DVOM on 200 ohm scale. • Measure resistance between test Pin 55 and EHC circuit at harness connector.	Resistance readings are 5 ohms or greater ▶	SERVICE harness open circuit. REPEAT Quick Test.
	Resistance reading is less than 5 ohms ▶	GO to KP8.

Exhaust Heat Control (EHC) — Pinpoint Test — KP

TEST STEP	RESULT ▶	ACTION TO TAKE
KP8 CHECK FOR SHORT TO GROUND		
• Key off, wait 10 seconds. • Leave Breakout box installed and processor disconnected. • Disconnect EHC solenoid. • DVOM on 200,000 ohm scale. • Measure resistance between test Pin 55 and test Pins 40, 46 and 60.	Resistance readings are less than 10,000 ohms ▶	GO to KP9.
	Resistance readings are 10,000 ohms or greater ▶	REPLACE processor. REPEAT Quick Test.
KP9 CHECK FOR SHORT TO POWER		
• Key Off, wait 10 seconds. • DVOM on 200,000 ohm scale. • Leave Breakout box installed and processor disconnected. • EHC solenoid disconnected. • Measure resistance between test Pin 55 and test Pins 37 and 57.	Resistance reading is 10,000 ohms or greater ▶	REPLACE processor. REPEAT Quick Test.
	Resistance reading is less than 10,000 ohms ▶	SERVICE short to power. REPEAT Quick Test. If code is repeated, REPLACE processor.

Converter Clutch Override (CCO) — Pinpoint Test — KR

KR

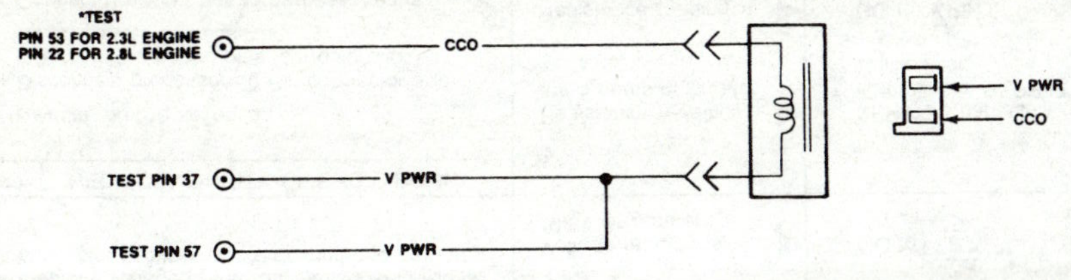

*TEST PINS LOCATED ON BREAKOUT BOX.
ALL HARNESS CONNECTORS VIEWED INTO MATING SURFACE.

STOP-WARNING

You should enter this Pinpoint Test only when a service code 89 is received in Quick Test Step 3.0.

This Pinpoint Test is intended to diagnose only the following:

- Harness Circuits: CCO and V PWR.
- CCO Solenoid.
- Processor Assembly.

TEST STEP	RESULT ▶	ACTION TO TAKE
CODE 89 PRESENT		
KR1 MEASURE CCO SOLENOID RESISTANCE		
• Key Off, wait 10 seconds. • DVOM on 200 ohm scale. • Disconnect CCO solenoid connector and measure solenoid resistance.	Resistance is between 26 and 40 ohms ▶	CONNECT CCO solenoid. GO to KR2
	Resistance is less than 26 ohms or greater than 40 ohms ▶	REPLACE CCO solenoid. REPEAT Quick Test.

Converter Clutch Override (CCO) — Pinpoint Test — KR

TEST STEP	RESULT ▶	ACTION TO TAKE
KR2 CHECK VOLTAGE OF V PWR CIRCUIT		
• Key On, Engine Off. • DVOM on 20V scale. • Measure voltage at the CCO solenoid connector between VPWR and ground.	Voltage reading is less than 10.5V ▶	SERVICE harness open circuit. REPEAT Quick Test.
	Voltage reading is 10.5V or greater ▶	GO to KR3.
KR3 CHECK CONTINUITY OF CCO CIRCUIT		
• Key Off, wait 10 seconds. • Disconnect processor 60 Pin connector and inspect for damaged pins, corrosion, loose wires. Service as necessary. • Connect Breakout box to harness. Leave processor disconnected. • DVOM on 200 ohm scale. • Measure resistance between test Pin 53/22 and CCO circuit at the solenoid harness connector.	Resistance readings are 5 ohms or greater ▶	SERVICE harness open circuit. REPEAT Quick Test.
	Resistance reading is less than 5 ohms ▶	GO to KR4.
KR4 CHECK FOR SHORT TO GROUND		
• Key Off, wait 10 seconds. • Leave Breakout box installed and processor disconnected. • Disconnect CCO solenoid. • DVOM on 200,000 ohm scale. • Measure resistance between test Pin 55 and test Pins 40, 46 and 60.	Resistance readings are less than 10,000 ohms ▶	SERVICE short to ground. REPEAT Quick Test.
	Resistance readings are 10,000 ohms or greater ▶	GO to KR5.
KR5 CHECK FOR SHORT TO POWER		
• Key Off, wait 10 seconds. • DVOM on 200,000 ohm scale. • Leave Breakout box installed and processor disconnected. • CCO solenoid disconnected. • Measure resistance between test Pin 55 and test Pins 37 and 57.	Resistance reading is 10,000 ohms or greater ▶	REPLACE processor. RERUN Quick Test.
	Resistance reading is less than 10,000 ohms ▶	SERVICE short to power. REPEAT Quick Test. If code is still present, REPLACE processor.

Dynamic Response Test — Pinpoint Test — M

M

OPERATOR PERFORMS BRIEF WOT

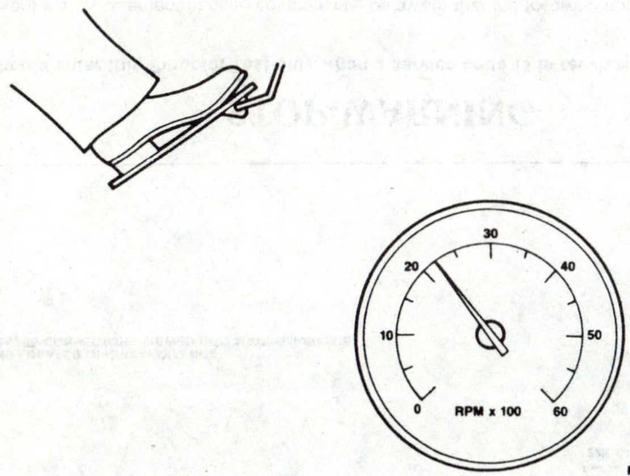

RPM INCREASE
GREATER THAN 2000 RPM

STOP-WARNING

You should enter this Pinpoint Test only when a service code 77 is received in Quick Test Step 5.0.

To prevent the replacement of good components, be aware that the following non-EEC areas may be at fault:

- Operator did not perform a brief WOT after dynamic response code.
- Mechanical engine problems; engine did not achieve greater than 2000 rpm.

This Pinpoint Test is intended to diagnose only the following:

- Throttle movement (greater than 3/4 throttle).
- Vane Air Flow (greater than 50% open).
- RPM increase (greater than 2000 rpm).

Dynamic Response Test — Pinpoint Test — M

TEST STEP	RESULT ▶	ACTION TO TAKE
M1 CODE 77: SYSTEM FAILED TO RECOGNIZE BRIEF WOT		
NOTE: A brief snap of the throttle may not be sufficient to pass this test. Be sure to go to WOT and return.		
• Repeat Engine Running Test of Quick Test. Be sure operator is familiar with the engine running format which proceeds as follows: — With Self-Test activated restart the engine. — ID Code 2 (0) start of test. — Dynamic response Code 1 (0) perform brief WOT. — Testing over. — Service code output begins.	Code 77 still present ▶	REPLACE processor. REPEAT Quick Test.
	No Code 77 ▶	Dynamic Response Test passed. SERVICE any other service code(s) received as necessary.

Ignition Diagnostic Monitor (IDM) — Pinpoint Test — N

N

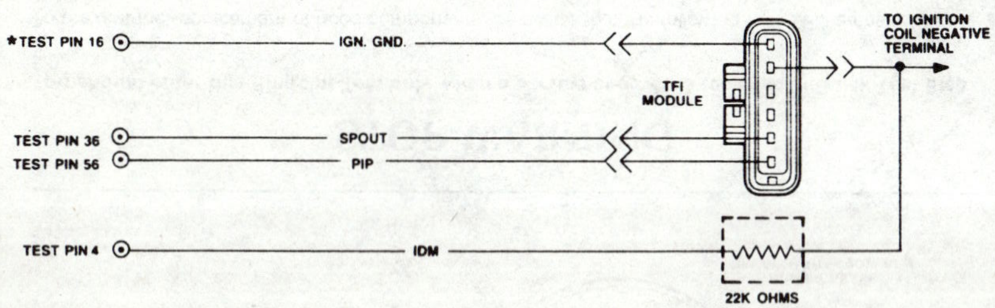

*TEST PINS LOCATED ON BREAKOUT BOX.
ALL HARNESS CONNECTORS VIEWED INTO MATING SURFACE.

STOP-WARNING

You should enter this Pinpoint Test only when a service code 18 is received in Quick Test Step 6.0.

To prevent the replacement of good components, be aware that the following non-EEC areas may be at fault:

- Ignition module.
- Ignition coil.
- Spark plugs and high tension cables.
- Distributor and PIP sensor.

This Pinpoint Test is intended to diagnose only the following:

- Harness circuits: ignition ground, spout, PIP, IDM.

Ignition Diagnostic Monitor (IDM) — Pinpoint Test — N

TEST STEP	RESULT ▶	ACTION TO TAKE
N1 CHECK CONTINUITY OF IDM CIRCUIT		
• Key Off, wait 10 seconds. • Disconnect E-core ignition connector from coil. • Disconnect processor and inspect both 60 Pin connectors for damaged pins, corrosion, loose wires. Service as necessary. • Connect Breakout box to harness. Leave processor disconnected. • DVOM on 200,000 ohm scale. • Measure resistance between test Pin 4 and ignition coil negative terminal.	Resistance reading is less than 20,000 ohms or greater than 24,000 ohms ▶	SERVICE open circuit. REPEAT Quick Test.
	Resistance reading is less than 24,000 ohms but greater than 20,000 ohms ▶	GO to **N2**.
N2 CHECK FOR SHORT TO GROUND		
• Key Off, wait 10 seconds. • Leave Breakout box installed and processor disconnected. • DVOM on 200,000 ohm scale. • Measure resistance between test Pin 4 and test Pins 40, 46 and 60.	Resistance reading is less than 10,000 ohms ▶	SERVICE short to ground. REPEAT Quick Test.
	Resistance reading is 10,000 ohms or greater ▶	GO to **N3**.
N3 CHECK TFI MODULE		
• Key Off, wait 10 seconds. • Deactivate Self-Test. • Connect VOM or STAR per Quick Test Step 2.0. • Enter Engine Running Continuous Monitor Test (as instructed in Quick Test Step 6.0D). • Observe VOM or STAR LED for indication of a fault while performing the following: • Lightly tap on TFI module (simulate road shock). • Wiggle TFI connector. • Is a fault indicated?	Yes ▶	DISCONNECT and INSPECT connectors. If connector and terminals are good, GO to TFI Ignition Diagnostics.
	No ▶	GO to **N4**.

Ignition Diagnostic Monitor (IDM) — Pinpoint Test N

TEST STEP	RESULT ▶	ACTION TO TAKE
N4 CHECK EEC-IV HARNESS		
• While still in continuous monitor test from step N1 observe VOM or STAR LED for a fault indication while performing the following: • While looking for faults listed in the table below, grasp the harness close to the TFI connector. Wiggle, shake or bend a small section of the EEC-IV system harness while working your way to the dash panel. Also wiggle, shake or bend the EEC-IV harness from the dash panel to the processor. Do this test on the circuits listed one at a time if needed to locate a faulty circuit. **FAULT** — **BREAKOUT BOX NO.** PIP shorted to ground or open — Test Pin 56 Spout shorted to ground — Test Pin 36 Ign. ground open — Test Pin 16 IDM open or shorted to ground, power — Test Pin 4 • Is a fault indicated?	Yes ▶	ISOLATE fault and make necessary repairs. REPEAT Quick Test.
	No ▶	GO to [N5].
N5 CHECK PROCESSOR AND HARNESS CONNECTORS		
• Key Off, wait 10 seconds. • Disconnect processor 60 Pin connector and inspect for damaged pins, corrosion, loose wires. • Are connectors and terminals OK? • Reconnect processor when this Step is completed.	No ▶	SERVICE as necessary. REPEAT Quick Test.
	Yes ▶	If unable to DUPLICATE an IDM fault in the EEC-IV System, testing is complete.

Spark Timing Check — Pinpoint Test P

TEST STEP	RESULT ▶	ACTION TO TAKE
P1 CHECK SPARK TIMING		
• Verify that Self-Test is activated during Engine Running Test. • Check timing while performing Engine Running Test. Record value. • Timing base plus 20 degrees ± 3 degrees? **NOTE: Self-Test locks the timing at 20 degrees plus base throughout this test and for two minutes after the last service code is outputted. Timing check must be made during this time period. Self-Test timing is base + 20 degrees (±3 degrees) BTDC. (See VECI decal for base value.)**	Yes ▶	GO to Quick Test Step [5.0].
	No ▶	GO to [P2].
P2 CHECK SPARK OUTPUT (SPOUT) CIRCUIT TO THE TFI MODULE		
• Locate spout connector and open this connection. • Start engine. • Check base timing. • Timing base ±3 degrees BTDC. **NOTE: Base timing must be checked to decal value.**	Yes ▶	GO to [P3].
	No ▶	RECONNECT spout connector. ADJUST timing. After timing is reset, PERFORM Quick Test Step [4].
P3 CHECK FOR POWER TO PROCESSOR		
• Key Off, wait 10 seconds. • Disconnect processor 60 Pin connector and inspect for damaged pins, corrosion, loose wires. Service as necessary. • Install Breakout Box. • Key On, Engine Off. • DVOM on 20V scale. • Measure voltage between test Pin 37 and test Pin 40 and between test Pin 57 and test Pin 60.	Either voltage reading less than 10.5V ▶	GO to Pinpoint Test Step [B1].
	Both voltage readings 10.5V or greater ▶	GO to [P4].

Spark Timing Check — Pinpoint Test P

TEST STEP	RESULT ▶	ACTION TO TAKE
P4 CHECK HARNESS FOR CONTINUITY		
• Key Off, wait 10 seconds. • DVOM on 200 ohm scale. • Measure resistance between test Pin 36 and spout pin (Pin 2) on the TFI harness connector.	DVOM reading 5 ohms or less ▶	GO to P5.
	DVOM reading greater than 5 ohms ▶	REPAIR open circuit. CONNECT spout test connector. CHECK timing per P1.
P5 HARNESS CHECK — IGNITION GROUND		
• Breakout box installed. • Key Off. • DVOM on 200 ohm scale. • Measure resistance between test Pin 16 and TFI connector Pin 6 (ign. grnd.).	Less than 5 ohms ▶	GO to Pinpoint Test Step A7.
	5 ohms or greater ▶	SERVICE harness as necessary. REPEAT Quick Test.

No Codes/Codes Not Listed — Pinpoint Test Q

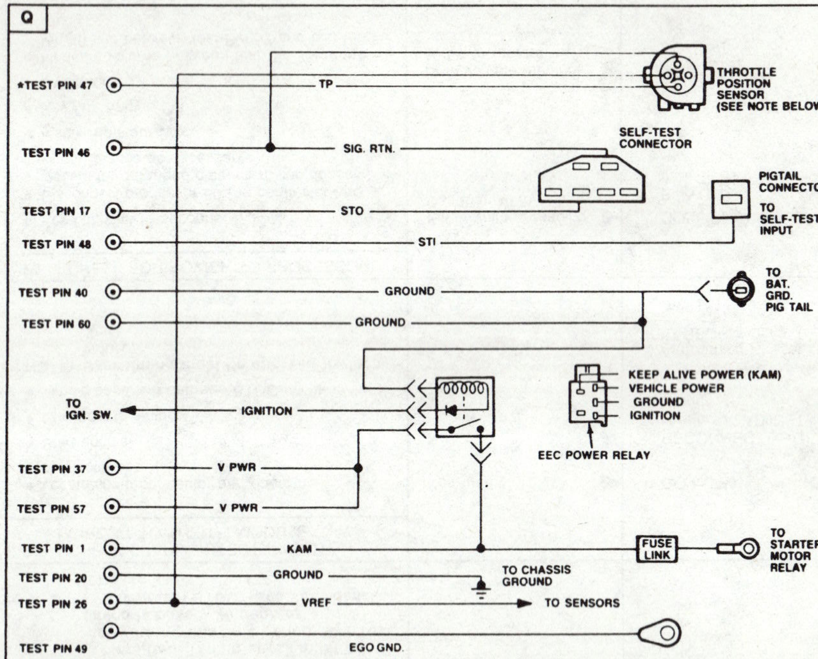

*TEST PIN LOCATED ON BREAKOUT BOX.
ALL HARNESS CONNECTORS VIEWED INTO MATING SURFACE.

STOP-WARNING

You should enter this Pinpoint Test only when directed here from Quick Test Step 3.0, 5.0 or 6.0 or when directed here from Diagnostics by Symptom in the Engine Supplement Section or from a Pinpoint Test Step.

This Pinpoint Test is intended to diagnose only the following:

- Processor.
- Harness Circuits: Signal Return, STO, STI, Ground, V PWR, KAM, VREF, Ignition.

No Codes/Codes Not Listed — Pinpoint Test Q

TEST STEP	RESULT	ACTION TO TAKE
Q1 NO CODES, CODES NOT LISTED		
Refer to illustration Q. • Key Off, wait 10 seconds. • DVOM on 20V scale. • Disconnect TP sensor. • Key On, Engine Off. • Measure voltage at the TP harness connector between VREF and signal return in Self-Test connector.	Less than 4.0V or greater than 6.0V ▶	GO to Pinpoint Test Step C1.
	4.0V to 6.0V ▶	RECONNECT TP sensor. GO to Q2.
Q2 SELF-TEST INPUT CONTINUITY CHECK		
Refer to illustration Q. • Key Off, wait 10 seconds. • Disconnect processor 60 Pin connector and inspect for damaged pins, corrosion, loose wires. Service as necessary. • Install Breakout box, leave processor disconnected. • Set DVOM to 200 ohm scale. • Measure resistance between Self-Test input at the Self-Test pigtail connector and test Pin 48.	Less than 5 ohms ▶	GO to Q3.
	5 ohms or greater ▶	CORRECT open in circuit.
Q3 SELF-TEST OUTPUT CIRCUIT CONTINUITY CHECK		
Refer to illustration Q. • Breakout box installed. • DVOM to 200 ohm scale. • Measure resistance between Self-Test output at Self-Test connector and test Pin 17.	5 ohms or greater ▶	CORRECT open in circuit.
	Less than 5 ohms ▶	GO to Q4.
Q4 EGO SENSOR GROUND CONTINUITY CHECK		
Refer to illustration Q. • Breakout box installed. • Key Off. • Measure resistance between EGO ground point on engine and test Pin 49.	Less than 5 ohms ▶	GO to Q5.
	5 ohms or greater ▶	CHECK and SERVICE EGO sensor ground wire or open circuit bad connection.

No Codes/Codes Not Listed — Pinpoint Test Q

TEST STEP	RESULT	ACTION TO TAKE
Q5 STO SHORT TO GROUND		
• Breakout box installed. • DVOM on 200 ohm scale. • Measure resistance between Self-Test output at Self-Test connector and engine block ground. • Is resistance greater than 5 ohms?	Yes ▶	REPLACE processor. REPEAT Quick Test.
	No ▶	GO to Q6.
Q6 SHIFT LIGHT		
• Is vehicle equipped with shift indicator light?	No ▶	SERVICE STO circuit for short to ground. REPEAT Quick Test.
	Yes ▶	GO to KL7.

Continuous Test Code 15 — Pinpoint Test — Q

TEST STEP	RESULT ▶	ACTION TO TAKE
Q10 CONDITIONS FOR CONTINUOUS CODE 15		
• Power interruption to Keep Alive Memory (KAM) Pin 1 may result in a service code being outputted.* • Clear continuous memory codes (use procedure described in Quick Test Step 6). • Repeat Quick Test Step 3.0 through Continuous memory code output. • Code 15 present on retest? *NOTE: **Anytime power is interrupted to the processor, a code 15 may be outputted the first time Self-Test is run after restoration of power. Repeat Self-Test to ensure correct diagnosis.**	Yes ▶ No ▶	GO to [Q11]. Test complete.
Q11 INSPECT ENGINE COMPARTMENT WIRING FOR PROPER ROUTING		
• Are any EEC components or EEC wiring close to ignition components or wires (High Electrical Energy Sources)? If EEC wiring close, reroute and rerun Self-Test. • Is code 15 still present?	Yes ▶ No ▶	GO to [Q12]. Test complete.
Q12 CHECK POWER CIRCUIT TO KEEP ALIVE MEMORY		
• Check power circuit to Keep Alive Memory for voltage. • Key Off, wait 10 seconds. • Disconnect processor and inspect both 60 Pin connectors. • Connect Breakout box to harness, leave processor disconnected. • DVOM on 20V scale. • Connect positive test lead to test Pin 1 and negative test lead to test Pin 40 or 60. • Key On. • Observe voltage reading.	DVOM reading less than 10 volts ▶ DVOM reading 10V or greater ▶	SERVICE open to KAM circuit. REPEAT Quick Test. REPLACE processor. REPEAT Quick Test.

Engine Stalls During Quick Test — Pinpoint Test — Q

TEST STEP	RESULT ▶	ACTION TO TAKE
Q30 DIAGNOSTIC ROUTINES COMPLETE		
• Have diagnostic routines in Section 2 been completed?	Yes ▶ No ▶	GO to [Q31]. GO to Diagnostic Routines.
Q31 ATTEMPT TO BRING ENGINE TO OPERATING TEMPERATURE		
• Key Off, wait 10 seconds. • Install tachometer. • Try to maintain engine at 2,000 rpm for 2 minutes. • Can rpm be maintained?	Yes ▶ No ▶	GO to Pinpoint Test Step [A13]. GO to [Q32].
Q32 MAINTAIN ENGINE OPERATION DURING ENGINE RUNNING QUICK TEST		
• Engine at operating temperature. • Tachometer installed. • Key Off, wait 10 seconds. • Activate Quick Test. • Perform Engine Running Self-Test while maintaining 2,000 rpm.	Code 11 ▶ Any service code ▶ No codes ▶	GO to Quick Test Step [5.0]. GO to Quick Test Step [5.0B]. GO to Pinpoint Test Step [Q1].

Output State Check Not Functioning — Pinpoint Test Q

TEST STEP	RESULT ▶	ACTION TO TAKE
Q40 CHECK FOR CODES 23, 53, 63 OR 68		
• Key Off, wait 10 seconds. • Perform Key On, Engine Off Self-Test. • Leave Key On to enter Output State Check. • Key On, Engine Off. Are codes 23, 53, 63 or 68 present?	Yes ▶	GO to Quick Test Step 3.0B and SERVICE appropriate code as instructed.
	Code 11 ▶	GO to Q41.
	No Codes ▶	GO to Q1.
Q41 CHECK THROTTLE LINKAGE		
• Check throttle and throttle linkages for sticking and binding. • Throttle OK?	Yes ▶	SERVICE as necessary. REPEAT Quick Test.
	No ▶	REPLACE TP sensor. REPEAT Quick Test.

Erratic Ignition — Pinpoint Test Y

TEST STEP	RESULT ▶	ACTION TO TAKE
Y1 CODE 14: ERRATIC IGNITION		
NOTE: Code 14 indicates two successive erratic profile ignition pick-up (PIP) pulses occurred, resulting in a possible engine miss or stall. • Check EEC-IV and ignition systems harnesses for: — Loose wires connectors. — Arcing secondary ignition components (coil, cap, rotor, wires, plugs, etc.). — On-board transmitter (2-way radio).* • Are any of the above present? *Verify all 2-way radio installations. Carefully follow manufacturer's installation instructions regarding the routing of antenna and power leads.	Yes ▶	SERVICE as necessary. REPEAT Quick Test.
	No ▶	GO to Y2.
Y2 CHECK DISTRIBUTOR		
• Key Off, wait 10 seconds. • Deactivate Self-Test. • Connect VOM or STAR per Quick Test Step 2.0. • Enter Engine Running Continuous Monitor Test (as instructed in Quick Test Step 6.3). • Observe VOM or STAR LED for indication of a fault while performing the following: • Lightly tap on TFI module and distributor. • Wiggle TFI connector. • Is a fault indicated?	Yes ▶	DISCONNECT and INSPECT connectors. If connector and terminals are good, GO to Ignition Diagnostics.
	No ▶	GO to Y3.
Y3 CHECK EEC-IV HARNESS		
• While still in Continuous Monitor Test from Step Y2 observe VOM or STAR LED for a fault indication while performing the following: • While looking for faults grasp the harness close to the TFI connector. Wiggle, shake or bend a small section of the ignition and EEC-IV systems harness while working your way to the other components and dash panel. Also wiggle, shake or bend the EEC-IV harness from the dash panel to the processor. Isolate the PIP circuit if needed for this test. • Is a fault indicated?	Yes ▶	ISOLATE fault and make necessary repairs. REPEAT Quick Test.
	No ▶	GO to Y4.

Erratic Ignition — Pinpoint Test Y

TEST STEP	RESULT	▶	ACTION TO TAKE
Y4 CHECK PROCESSOR AND HARNESS CONNECTORS			
• Key Off, wait 10 seconds. • Disconnect processor 60 Pin connector and inspect for damaged pins, corrosion, loose wires. • Are connectors and terminals OK? • Reconnect processor when this Step is completed.	No	▶	SERVICE as necessary. REPEAT Quick Test.
	Yes	▶	Unable to duplicate an erratic ignition fault in the EEC-IV System, for further diagnosis GO to Ignition System Diagnostics.

MCU System Description

V-EIGHT CYLINDER MCU SYSTEM

The MCU is basically the same on all V8 cylinder engines. The heart of the system is the fuel control system. This is necessary to keep the air-fuel (A/F) ratio at a proper chemical balance (14.7) to obtain maximum catalyst efficiency. The fuel control "loop" consists of an Exhaust Gas Oxygen (EGO) sensor, Microprocessor Control Unit (MCU), and carburetor with Feedback Carburetor Actuator (FBCA).

The EGO sensor senses whether the exhaust gas is rich or lean of a proper chemical balance. This signal is sent to the MCU module, which sends a varying signal to the Feedback Carburetor Actuator (FBCA) to move the air-fuel ratio back to the proper chemical balance. This operation is called "closed loop" operation.

The other mode of operation is called "open loop". In this mode, the MCU module sends out a fixed signal to the Feedback Carburetor Actuator (FBCA). During this time, the input from the EGO sensor is ignored, thus opening the loop.

The determining factor when the system goes into open or closed loop is based upon information from the switch inputs, which senses parameters such as coolant temperature, manifold vacuum, and rpm. Generally, the vehicle will be in closed loop when the vehicle is at operating temperature and at a steady part throttle cruise.

Other functions controlled by the MCU are Canister Purge (CANP), Thermactor Air Diverter (TAD), Thermactor Air Bypass (TAB), Throttle Kicker (TK), and Spark Retard System, or spark retard through the Knock sensor.

Quick Test

1 VISUAL CHECKS

A. CHECK VACUUM LINES FOR CRACKED OR BROKEN HOSES AND PROPER CONNECTIONS.

B. CHECK THE ELECTRICAL HARNESS FOR:

- Broken, detached, or frayed wires/insulation.
- Terminals not completely seated in the connectors.
- Corrosion on terminals.

Quick Test

1 VISUAL CHECKS (Continued)

C. Verify that coolant level is to specification.

2 PRE-TEST CONDITIONS

A. Transmission in Park or Neutral.

B. Set parking brake. Block the wheels.

C. Turn Accessories Off.

Quick Test

2 PRE-TEST CONDITIONS (Continued)

D. Make sure that the vacuum hoses are connected to the air cleaner.

E. When using an analog voltmeter, make sure there is power to the choke (engine must be running).

10 to 12 Volts D.C. Scale

Quick Test

2 PRE-TEST CONDITIONS (Continued)

F. Start and idle the engine until the radiator upper hose is hot and pressurized.

If vehicle will not start, refer to No Start, Pinpoint Test 1.

3 SET-UP PROCEDURE

A. Turn the ignition key Off.

B. Locate the Self-Test connector near MCU module (left or right fender apron as viewed from driver's seat).

Quick Test

3 SET-UP PROCEDURE (Continued)

- If you are using an analog VOM during the Quick Test, connect a jumper wire or paper clip from Self-Test Trigger (Circuit 201) to ground (Circuit 60) on Self-Test connector.
- Then connect an analog voltmeter from battery (+) to Self-Test Output on the Self-Test connector.
- Set voltmeter on DC volt range to read from 0 to 15 VDC.

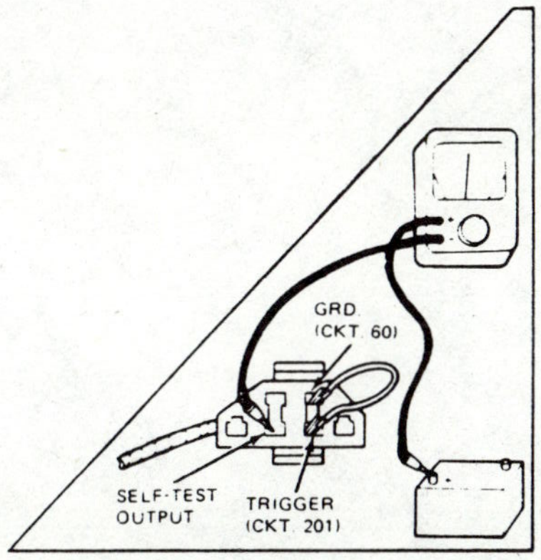

- However, if you are using a STAR tester for the Quick Test:
- Connect black lead to the battery (—) post.
- Connect red lead to the Self-Test Output terminal on the Self-Test connector.
- Connect white lead to the Self-Test Trigger.
- Refer to back of STAR tester for additional instructions.

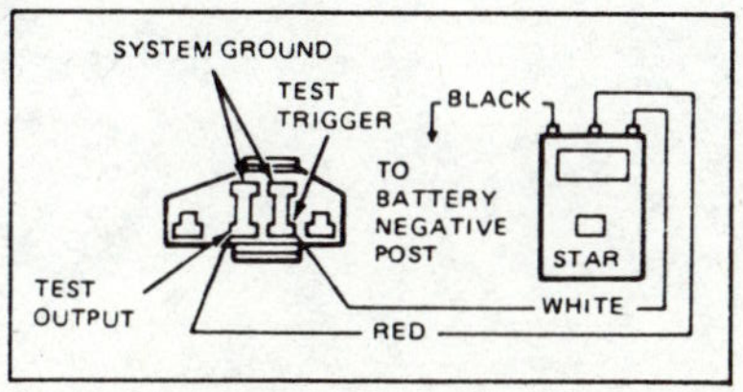

Quick Test

3 SET-UP PROCEDURE (Continued)

C. Remove PCV from breather cap on valve cover.

D. Tee a vacuum gauge tò Canister Purge Solenoid Valve hose on the carbon canister side of the hose.

Quick Test

3	SET-UP PROCEDURE (Continued)

E. On 5.8L engine calibrations with a vacuum delay valve, there is a Tee and restrictor in the Thermactor Bypass Vacuum Control line. The restrictor must be uncapped during the Quick Test. Recap it after the Quick Test is completed.

Quick Test

Test 4

TEST STEP	RESULT ▶	ACTION TO TAKE
4.0 KEY ON, ENGINE OFF QUICK TEST		
• Key Off. • Turn key to On with engine Off. • Observe voltmeter. • Start service with first code received.	Service Code: 1-1 ▶	CONTINUE to Quick Test 5.
	Service Codes: ▶	See note below.
	5-1 ▶	Pinpoint Test 4, open switch.
	6-5 ▶	Pinpoint Test 5, closed switch.
	Steady high or low voltage ▶	Pinpoint Test 14, No Self-Test Output.
	Service Code(s) not listed ▶	Pinpoint Test 14, No Self-Test Output.

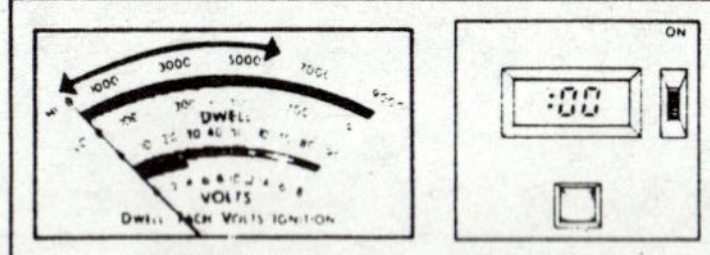

VOLTMETER

STAR TESTER

Display :00

NOTE: ONE QUICK INITIALIZATION PULSE MAY OCCUR ON THE VOLTMETER IMMEDIATELY AFTER TURNING THE IGNITION KEY ON. THE OUTPUT SERVICE CODE WILL OCCUR IN APPROXIMATELY FIVE SECONDS.

*Before proceeding with appropriate Pinpoint Test, read note on Pinpoint Test usage on page 23-21.

Quick Test — Test 5

TEST STEP	RESULT	ACTION TO TAKE
5.0 ENGINE RUNNING QUICK TEST • Verify that engine is at normal operating temperature. • Run engine at 2000 rpm for 2 minutes, then turn key Off. Immediately restart engine and idle. • Observe voltmeter and vacuum gauge for initialization pulses after restarting the engine. The throttle kicker will also extend at this time (rpm increases) and remain on throughout test. • **Vehicles with Knock Sensor:** When 4 initial pulses occur, immediately simulate spark knock by placing a 3/8 inch extension on manifold near base of knock sensor. Tap extension end lightly with light (4 oz.) hammer for approximately 15 seconds. OR • After the 4 initial pulses appear, observe and record voltmeter pulses (service codes). Within 90 seconds, the service codes are complete and the throttle kicker will retract (rpm decreases). • Always start service with first code received.	Service Code: 1-1* for vehicles below 4000 ft. elevation. OR 6-2* for vehicles ABOVE 4000 ft. elevation.	▶ For vehicles with drive complaint of detonation (WOT only) and/or poor performance and equipped with Universal Ignition Modules. CONTINUE to Test 6. MCU system OK.
	For the following results, go to	▶ Pinpoint Test Number:
	No start	▶ Test 1, No Start
	No initialization pulses on voltmeter or service codes not listed	▶ Test 14, No Self-Test Output Test 14, No
	No pulses on vacuum gauge	▶ Test 14, No Self-Test Output
	No Throttle Kicker extension or rpm increase	▶ Test 8, RPM Out of Specification
	Service Codes: 1-2	▶ Test 8, RPM Out of Specification
	2-5	▶ Test 13, Knock Sensor
	4-1	▶ Test 6, Fuel Lean
	4-2	▶ Test 7, Fuel Rich
	4-4	▶ Test 10, Thermactor
	4-5	▶ Test 11, Always Upstream
	4-6	▶ Test 12, Alway Not Bypassing
	5-3 or 5-4	▶ Test 2, Open Switch
	5-1 or 5-5	▶ Test 4, Open Switch
	6-2* below 4000 ft. 1-1* ABOVE 4000 ft.	▶ Test 15.
	6-1 or 6-5	▶ Test 5, Closed Switch

*Before proceeding with appropriate Pinpoint Test, read important note on Pinpoint Test usage

Quick Test — Test 6

TEST STEP	RESULT	ACTION TO TAKE
6.0 SPARK KNOCK, ONLY AT WIDE OPEN THROTTLE, AND/OR POOR PERFORMANCE/MILEAGE. PERFORM STEP 6.3 ON VEHICLES WITH KNOCK SENSOR ONLY		
• Disconnect knock sensor. • Verify initial timing per Emission decal, correct timing if necessary. • With vehicle at normal operating temperature, disconnect the two wire ignition module connector (contains yellow circuit 324 and black/white circuit 60 wires) from vehicle harness, and jumper the wires together at the ignition module. • Read and note ignition timing. • Remove jumper wire, and reread ignition timing.	Retards 16-20 degrees	▶ RECONNECT ignition module, and go to Step 6.1
	Does not retard 16-20 degrees	▶ REPLACE ignition module, and repeat all of Step 6.0

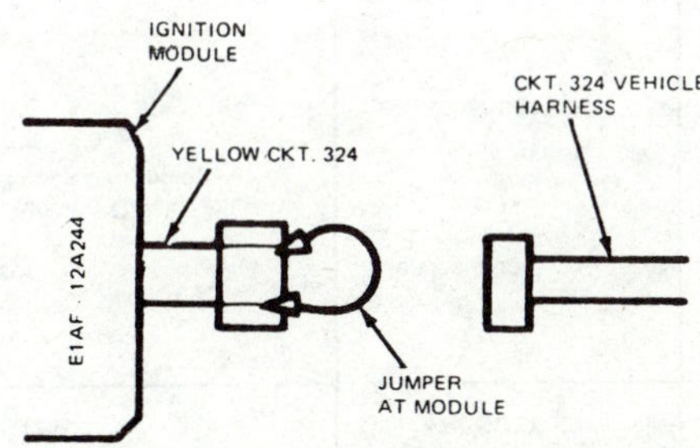

Quick Test — Test 6

TEST STEP	RESULT	▶	ACTION TO TAKE
6.1 • With vehicle at normal operating temperature, set high cam on lowest step that will maintain an engine speed greater than 2400 rpm. • Read and note ignition timing. • Remove and plug the vacuum hose from zone vacuum switch. Reread timing.	Retards more than 5 degrees	▶	If a system has a knock sensor, CONTINUE to Step [6.3], otherwise MCU System, OK; testing completed.
	Retards less than 5 degrees	▶	CONTINUE to Step [6.2]

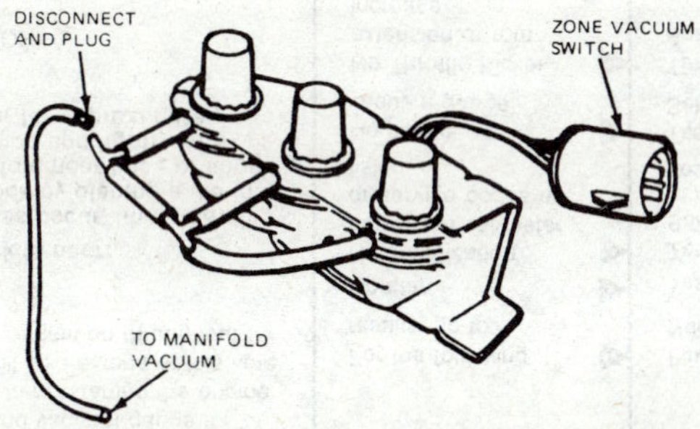

TEST STEP	RESULT	▶	ACTION TO TAKE
6.2 While still at 2400 rpm, disconnect ignition module (connector with yellow and black white wires).	Timing retards	▶	CHECK circuit 324 for short to ground. If OK, REPLACE MCU and retest.
	Timing does not retard	▶	CHECK circuit 324 for open. If OK, REPLACE MCU and retest

Quick Test — Test 6

TEST STEP	RESULT	▶	ACTION TO TAKE
6.3 • Reconnect knock sensor. • Rerun Step 5, Engine Running Quick Test, but DO NOT simulate knock by tapping sensor.	Service Code: 2-5	▶	MCU, OK; testing completed
	Any other service code	▶	REPLACE knock sensor and retest

Pinpoint Test

INSTRUCTIONS FOR USING THE PINPOINT TESTS

Important note on Pinpoint Test usage:

- Do not measure voltage or resistance at the MCU module or connect any test lamps to it, unless otherwise specified.
- The symbol "R" indicates resistance; that is, R = 100 ohms means the resistance is 100 ohms.
- Isolate both ends of a circuit and turn key Off whenever checking for shorts (R less than 1000 ohms) or continuity (R less than 5 ohms), unless specified.
- Connectors shown in schematics are as viewed from the front of the connector. See diagram below.
- Disconnect solenoids and switches from the harness before measuring for continuity or resistance.
- In using the Pinpoint Tests, follow each step in order, starting from the first step in the appropriate routine. Follow each step until the fault is found; correct the problem, then proceed to Quick Test 4 to verify the service. Reconnect all connections before conducting the Quick Test.
- When more than one service code is received, always start service with the first code received.

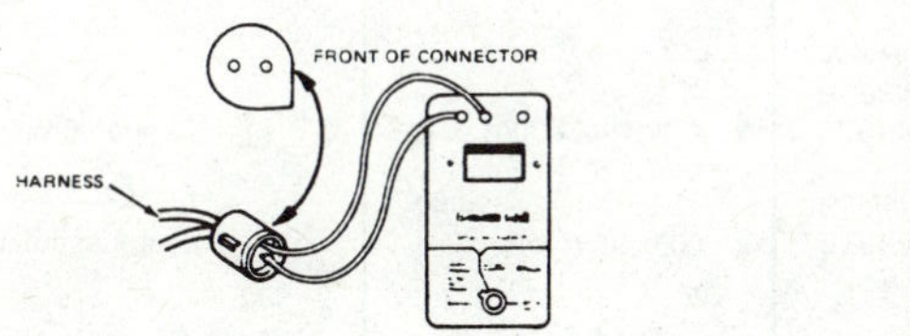

No Start | Pinpoint | Test 1

TEST STEP		RESULT ▶	ACTION TO TAKE
1.0	NO START — DETECTS MCU RELATED FAULTS ONLY		
1.1	• Key Off. • Disconnect the vehicle harness from the ignition module and the horseshoe connector from the coil. • Connect an ohmmeter between circuit 11 and engine block ground. • Measure resistance.	R is more than 1000 ohms. ▶	MCU System OK, refer to appropriate Routine.
		R is 1000 ohms or less. ▶	CONTINUE to Step 1.2

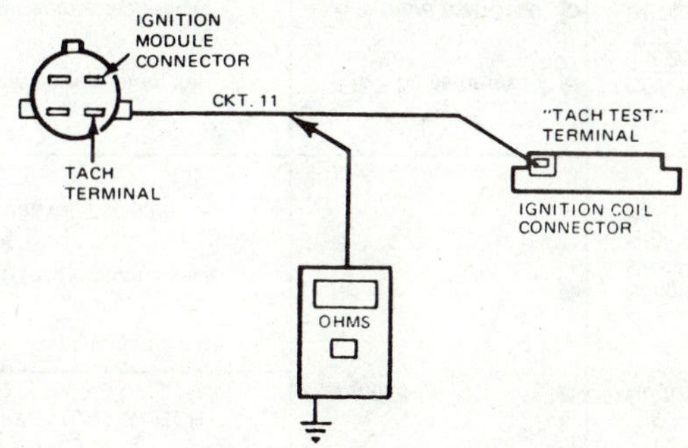

No Start — Pinpoint Test 1

TEST STEP	RESULT	ACTION TO TAKE
1.2 • Disconnect the vehicle harness from the MCU. • Check circuit 11 for short to ground.	R is more than 1000 ohms. ▶	REPLACE MCU, and Retest.
	R is 1000 ohms or less. ▶	SERVICE or REPLACE circuit 11, and Retest.

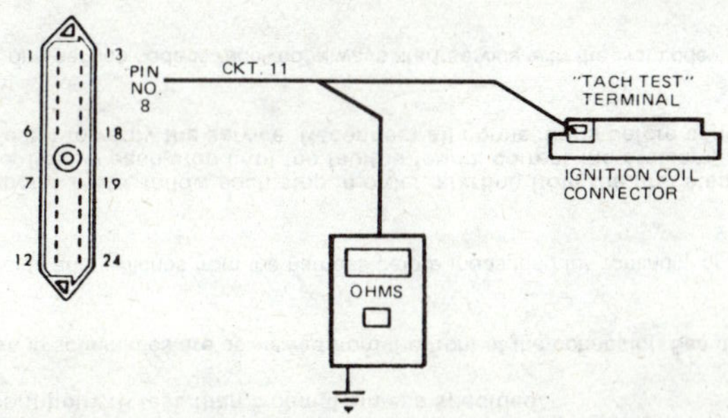

Temperature Switch — Pinpoint Test 2

TEST STEP	RESULT	ACTION TO TAKE
2.0 SERVICE CODE(S) 5-3, 5-4 — ELECTRICAL TEMPERATURE SWITCH, OPEN • CODE 5-3 is for Dual Temperature Switch, circuit 308. • CODE 5-4 is for Mid-Temperature Switch, circuit 309. • Verify coolant temperature is 60-93°C (140-200°F).	Verified ▶	Go to next Step 2.1
	No ▶	Bring engine coolant to specified temperature.
2.1 • Disconnect the vehicle harness from the switch. • Connect an ohmmeter across the switch terminals.	R is 5 ohms or less. ▶	CONTINUE to Step 2.2
	R is more than 1000 ohms. ▶	REPLACE switch
2.2 • Connect one lead of DVOM to Circuit 308 (Dual) or Circuit 309 (MID) and the other lead to engine ground.	R is 1,000 ohms or less. ▶	Service or replace circuit.
	R is more than 1000 ohms. ▶	Continue to next step 2.3
2.3 • Disconnect vehicle harness from MCU. • Check circuits 60 and 308 (Dual), 309 (MID) for continuity.	R is 5 ohms or less. ▶	Replace MCU.
	R is more than 1000 ohms. ▶	Service or repair circuit. Retest

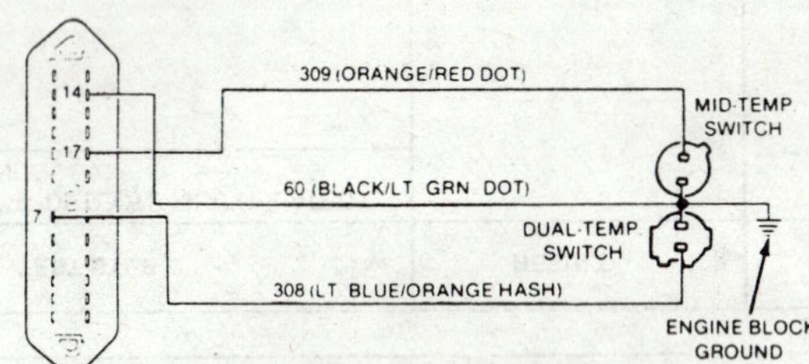

TEMPERATURE SWITCH OPEN
(Service Codes, 5-3 DUAL-TEMPERATURE, and 5-4, MID-TEMPERATURE)

Zoned Vacuum Switch | Pinpoint | Test 4

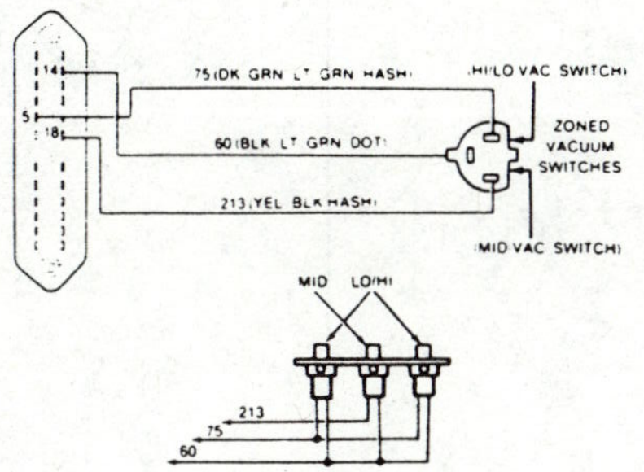

VACUUM SWITCH OPEN (Service Codes 5-1, Hi-Lo, or 5-5, MID VACUUM)

TEST STEP	RESULT ▶	ACTION TO TAKE
4.0 VACUUM CHECK AT THE VACUUM SWITCHES.		
CODE: / SWITCH: / CIRCUIT NUMBER: 5-1 / Hi/Low / 75 5-5 / Mid VAC / 213 • Verify manifold vacuum is present at switches during idle condition.	Vacuum is present. (10 in hg or more) ▶ No vacuum or less than 10. ▶	Continue to next Step 4.1 Check vacuum line for pinched, plugged, cracked conditions, or basic mechanical problem.
4.1 VACUUM SWITCH BLEED-DOWN TEST		
• Key Off. • Apply vacuum to the switch assembly.	Holds vacuum. ▶ Does not hold vacuum. ▶	For a service code 5-1 go to 4.2. For a service code 5-5 go to 4.4 Check vacuum hoses for leaks. If no leaks are found replace the switch assembly and retest.

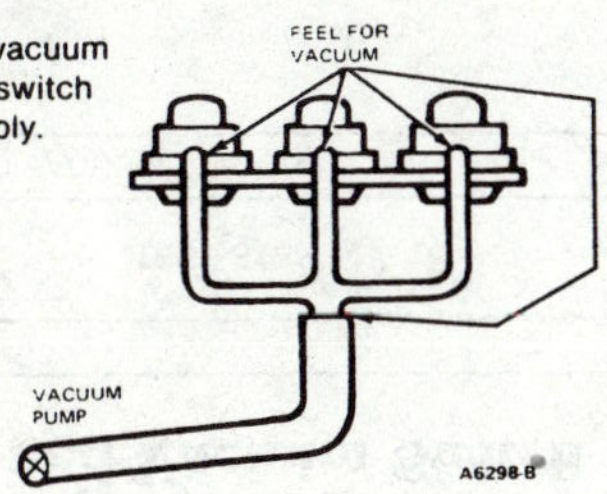

Zoned Vacuum Switch | Pinpoint | Test 4

TEST STEP	RESULT ▶	ACTION TO TAKE
4.2 HI/LOW VACUUM SWITCH		
• Key Off. • Disconnect the vehicle harness from the vacuum switch connector. For Hi/Low, do not apply vacuum to switch. • Check resistance of Hi/Low vacuum switch circuit 75 to circuit 60.	R is 5 ohms or less. ▶ R is more than 5 ohms. ▶	CONTINUE to Step 4.3 REPLACE switch assembly. Retest.
4.3		
• Disconnect the 24 pin connector from MCU module. • Check circuits 60 and 75 of the vehicle harness for continuity.	R is 5 ohms or less. ▶ R is more than 5 ohms. ▶	REPLACE MCU and retest. SERVICE or REPLACE circuit and retest.

Zoned Vacuum Switch — Pinpoint Test 4

TEST STEP	RESULT	▶	ACTION TO TAKE
4.4 MID VACUUM SWITCH			
• Key Off. • Disconnect the vehicle harness from the vacuum switch connector. • For MIDVAC, apply minimum of (15 inches Hg.) of vacuum to switch.	R is 5 ohms or less.	▶	RECONNECT vacuum hose. CONTINUE to Step 4.5
• Check resistance of MIDVAC vacuum switch, circuit 213 to circuit 60.	R is more than 5 ohms.	▶	REPLACE switch assembly. Retest.
4.5			
• Disconnect the 24 pin connector from MCU module.	R is 5 ohms or less.	▶	REPLACE MCU. Retest.
• Check circuits 60 and 213 of the vehicle harness for continuity.	R is more than 5 ohms.	▶	SERVICE or REPLACE circuit. Retest.

Zoned Vacuum Switch — Pinpoint Test 5

TEST STEP	RESULT	▶	ACTION TO TAKE
5.0 SERVICE CODE 6-5, VACUUM SWITCH. CLOSED			
• Verify engine is at normal operating temperature. If not, refer to Section 2, Routines 216 and 217. Verify that the vacuum hoses are connected to the switch; if not, connect and continue.	Verified	▶	GO to Step 5.1
	No	▶	Bring engine to normal operating condition.
5.1			
• Key Off. • Disconnect vehicle harness from vacuum switch connector.	R is more than 1000 ohms.	▶	RECONNECT vacuum hose. CONTINUE to Step 5.2
• Connect ohmmeter across the MIDVAC switch terminals at the switch connector. Must measure from circuit 213 to circuit 60.	R is 1000 ohms or less.	▶	REPLACE switch assembly. Retest.
5.2			
• Disconnect the vehicle harness from the MCU.	R is more than 1000 ohms.	▶	REPLACE MCU. Retest.
• Connect one lead of ohmmeter to the vehicle harness circuit 213 and the other lead to engine block ground.	R is 1000 ohms or less.	▶	SERVICE or REPLACE circuit. Retest.

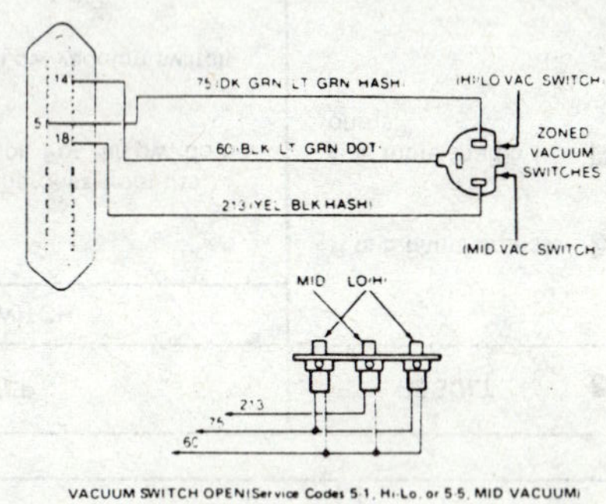

Fuel Always Lean | Pinpoint | Test 6

6.0 SERVICE CODE 4-1

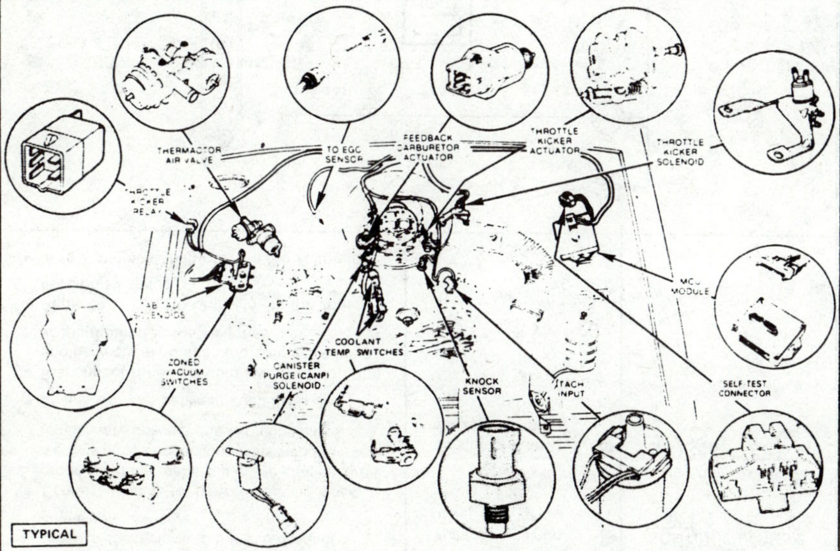

Components controlled by the MCU System in this Pinpoint Test are the Feedback Carburetor Actuator, the Ego Sensor, Thermactor Air Valve, the TAB/TAD Solenoids, and the MCU Module. The components not directly controlled by the MCU System during this Pinpoint Test but could possibly cause a Service Code fault are as follows:

- Malfunctioning Carburetor
- Low Fuel Pressure
  - —Fuel Line Kink
  - —Fuel Pump Malfunction
- Vacuum Leaks
  - —EGR Gasket
  - —Head Gasket
  - —Cracked Hoses, etc.
- Canister Purge Solenoid
- Bowl Vent
- PCV

NOTE: Virtually anything that adds air or restricts fuel should be investigated.

Fuel Always Lean | Pinpoint | Test 6

TEST STEP	RESULT	ACTION TO TAKE
6.0 SERVICE CODE 4-1—FUEL ALWAYS LEAN (CONTINUED)		
• Run engine at 2000 rpm for 60 seconds, to ensure that the EGO sensor is at an operating temperature.	Service code 4-1 still present. ▶	Continue to Pinpoint Test [6.1]
• Rerun Engine Running Self-Test.	No service code 4-1 present. ▶	SERVICE the next service recorded from this self-test.

Fuel Always Lean — Pinpoint Test 6

TEST STEP	RESULT	ACTION TO TAKE
6.1 EGO SENSOR CHECK		
CAUTION: To prevent damage to the EGO sensor, never connect an ohmmeter to the EGO sensor; use a voltmeter with an input impedence of 10 megohms or greater.		
• Key Off. • Deactivate this vehicle self test. • Disconnect the vehicle harness from the EGO sensor. • Connect a jumper wire between the vehicle harness connector and the EGO sensor connector. • Connect a voltmeter between engine block ground and the metal clip of the jumper wire. Verify a good connection exists with the jumper wire between the two connectors. • Engine On, rpm between 2000-2300. • Disconnect zoned vacuum switches from the engine vacuum source, and connect a hand vacuum pump to the switches. • Apply 34-41 kPa (10-12 In.-Hg.) to the zoned vacuum switches. • Wait 2 minutes, then observe the voltmeter.	Voltage is steady, then alternating above and below 0.50 volts ▶	GO to Step 6.14
	Voltage remained under 0.50 volts ▶	CONTINUE to Step 6.2
	Voltage remained above 0.50 volts ▶	GO to Step 6.6

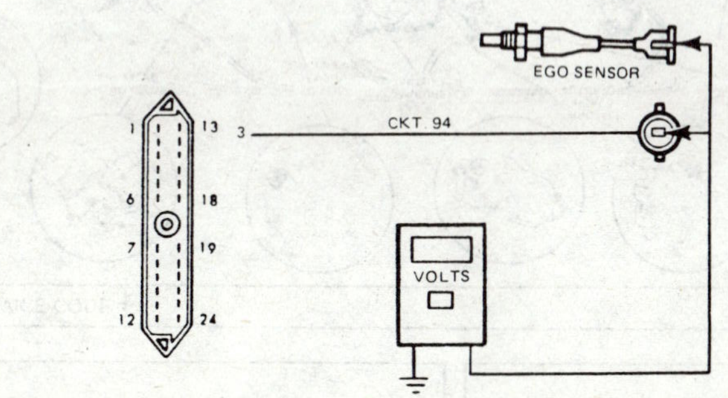

Fuel Always Lean — Pinpoint Test 6

TEST STEP	RESULT	ACTION TO TAKE
6.2		
• Key Off. • Reconnect EGO sensor to to the vehicle harness. • Disconnect the thermactor air hose at the air pump. • Disconnect the vehicle harness connector from the MCU. • Connect a voltmeter between circuit 60 and circuit 94 of the harness connector. • Start engine. Run engine at idle. • Place carburetor on high idle cam. • Force system rich by depressing the CVR rod, and observe voltmeter.	Voltage is 0.5 volts or more ▶	GO to Step 6.6
	Voltage is less than 0.5 volts. ▶	CONTINUE to Step 6.3

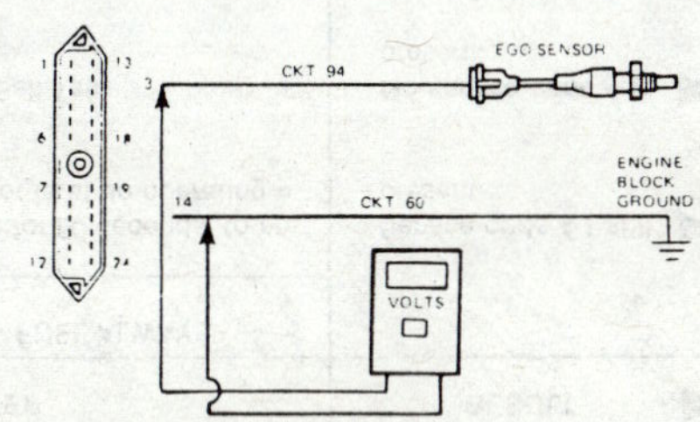

Fuel Always Lean — Pinpoint — Test 6

TEST STEP	RESULT ▶	ACTION TO TAKE
6.3		
• Key Off. • Disconnect the vehicle harness from the EGO Sensor. • Measure resistance between the engine block ground and pin 14 on the MCU harness connector.	R is 5 ohms or less. ▶	Continue to test Step 6.4
	R is more than 5 ohms. ▶	Service or replace circuit. Retest.
6.4		
• Check continuity of circuit 94.	R is 5 ohms or less. ▶	Continue to test Step 6.5
	R is more than 5 ohms. ▶	Service or replace circuit 94. Retest.
6.5		
• Connect an ohmmeter between the engine block ground and circuit 94. • Measure the resistance.	R is 1,000 ohms or less. ▶	SERVICE or replace circuit 94. Retest.
	R is more than 1,000 ohms. ▶	REPLACE EGO sensor. Retest.

Fuel Always Lean — Pinpoint — Test 6

TEST STEP	RESULT ▶	ACTION TO TAKE
6.6 FEEDBACK CARBURETOR ACTUATOR MOTOR (FBCA)		
• Key Off. • Apply 34-41 kPa (10-12 In.-Hg.) to the zoned vacuum switches. • Reconnect the vehicle harness connector to the MCU and EGO sensor. • Disconnect the vehicle harness from the FBCA motor. • Connect a known good FBCA motor, shaft fully depressed, to the vehicle harness. Do not remove existing FBCA motor from carburetor. • Start engine and set on low speed step of fast idle cam. • While observing the FBCA motor shaft, remove the PCV hose from the carburetor.* • Within two minutes, the known good FBCA Motor shaft should fully extend.	FBCA Motor shaft fully extends. ▶	RECONNECT the PCV hose to the carburetor and the thermactor hose to the pump. GO to Step 6.10
	FBCA Motor shaft does not extend ▶	RECONNECT the PCV hose to the carburetor. CONTINUE to Step 6.7

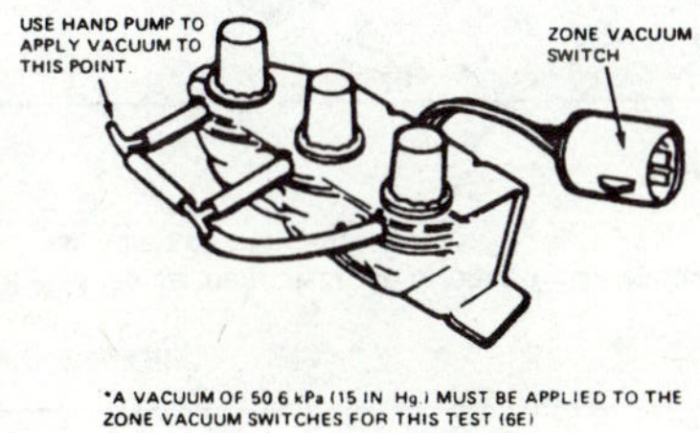

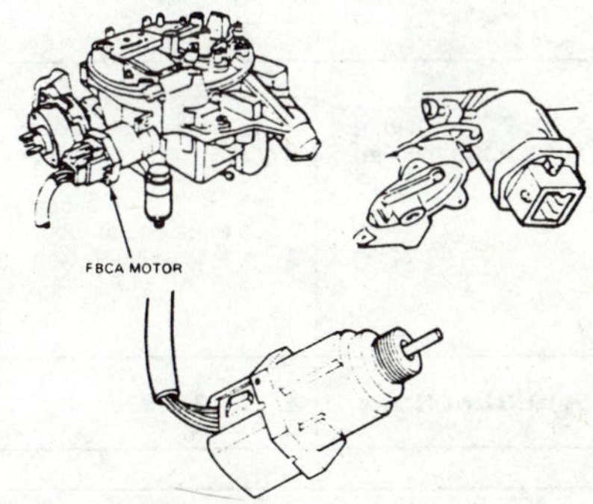

*Because of the vacuum leak caused by this Test Step, the vehicle may run rough.

Fuel Always Lean — Pinpoint Test 6

TEST STEP	RESULT ▶	ACTION TO TAKE
6.7 FBCA MOTOR HARNESS CHECK		
• Key Off. • Disconnect the vehicle harness from the MCU and the FBCA motor. • With an ohmmeter, check the continuity of circuit 687, 95, 96, 97 and 98 of the vehicle harness.	R is 5 ohms or less for each circuit. ▶ R is more than 5 ohms in any one of the circuits. ▶	CONTINUE to Step 6.8 SERVICE or REPLACE the circuit(s). Retest.

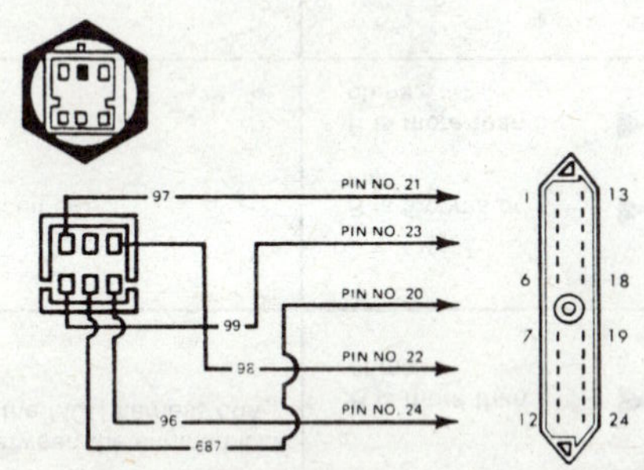

Fuel Always Lean — Pinpoint Test 6

TEST STEP	RESULT ▶	ACTION TO TAKE
6.8 FBCA MOTOR WINDING CHECK		
• Key Off. • Measure the resistance of each FBCA Motor winding on the carburetor.	50-175 ohms at room temperature or above. ▶ Not 50-175 ohms at room temperature or above. ▶	Go to 6.9 Replace FBCA Motor. Retest.

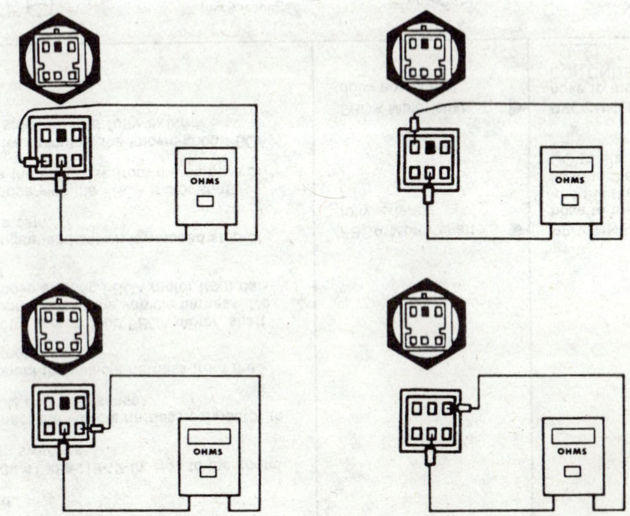

Fuel Always Lean — Pinpoint Test 6

TEST STEP	RESULT	ACTION TO TAKE
6.9		
• Connect one lead of ohmmeter to engine block ground. • With the other lead of the ohmmeter, probe each circuit (95, 96, 97, 98) of the vehicle harness connector, checking each circuit for a short to ground.	R is more than 1000 ohms. ▶	Replace MCU. Retest.
	R is 1000 ohms or less. ▶	SERVICE or REPLACE wire(s). Retest.

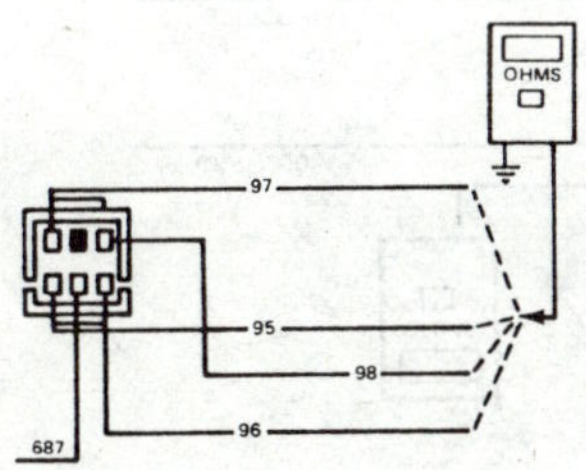

TEST STEP	RESULT	ACTION TO TAKE
6.10 FBCA MOTOR		
• Key Off. • Connect an ohmmeter lead to pin 687 of the original FBCA motor in the carburetor and the other lead to the FBCA motor case.	R is more than 190,000 ohms. ▶	CONTINUE to Step 6.11
	R is 190,000 ohms or less. ▶	REPLACE FBCA motor. Retest.

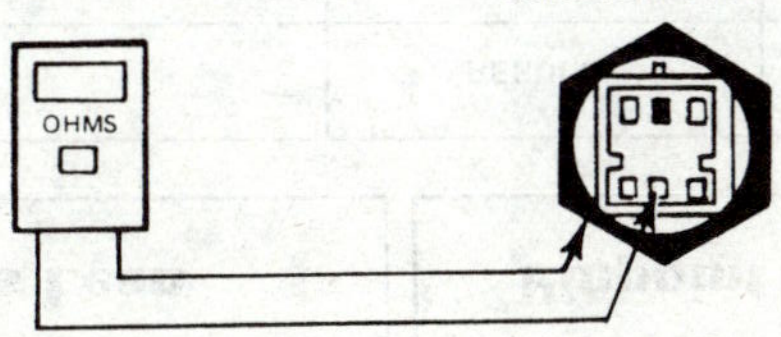

Fuel Always Lean — Pinpoint Test 6

TEST STEP	RESULT	ACTION TO TAKE
6.11		
• Disconnect FBCA motor from the carburetor. • Check the pintle for binding or damage. • Check the FBCA motor for physical damage, and check for movement by turning the key On.	FBCA motor, and pintle, OK. ▶	RECONNECT the FBCA motor to the carburetor and vehicle harness. GO to 6.12
	FBCA pintle, binding. ▶	REPLACE FBCA pintle and retest.
	FBCA motor, physically damaged or no movement. ▶	Replace FBCA motor and retest.
6.12 THERMACTOR AIR DIVERTER SOLENOID (TAD)		
• Key Off. • Disconnect the vehicle harness from the TAD solenoid. • Connect an ohmmeter across the TAD solenoid terminals. • Observe ohmmeter.	R is 50-100 ohms ▶	CONTINUE to Step 6.13
	R less than 50 ohms or greater than 100 ohms. ▶	REPLACE TAD solenoid. Retest.

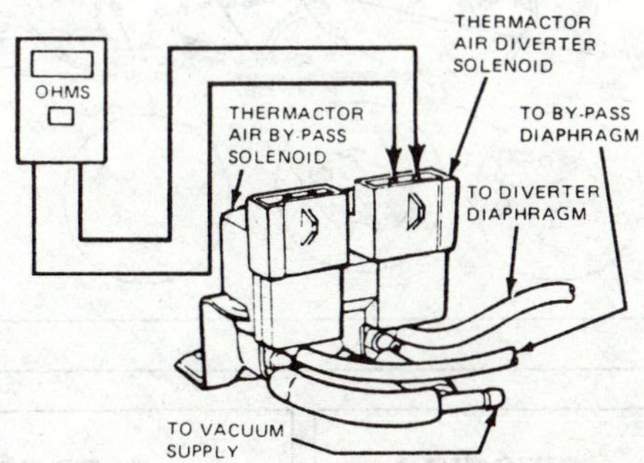

Fuel Always Lean | Pinpoint | Test 6

TEST STEP	RESULT ▶	ACTION TO TAKE
6.13		
• Disconnect the vehicle harness connector from the MCU. • Connect one lead of ohmmeter to circuit 200 of the vehicle harness and the other lead to engine block ground.	R is more than 1000 ohms. ▶	RECONNECT the vehicle harness to the MCU. CONTINUE to Step 6.14
	R is 1000 ohms or less. ▶	SERVICE or REPLACE circuit 200. Retest.

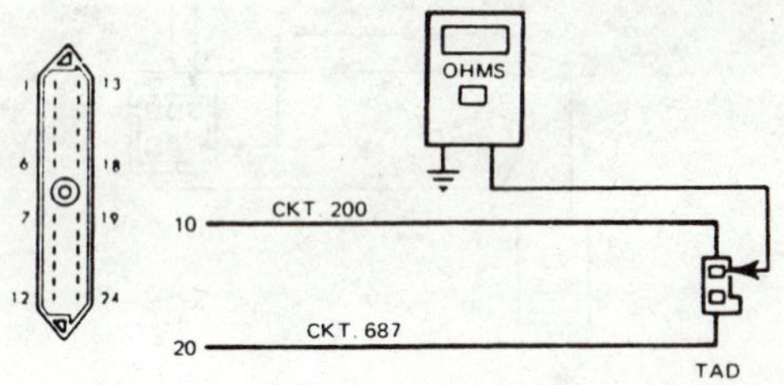

TEST STEP	RESULT ▶	ACTION TO TAKE
6.14		
• Check Thermactor Air Diverter valve, TAD solenoids, and connections. Refer to Section 10. • If Service Code 4-1 and 4-5 were both displayed during the Quick Test, perform Pinpoint Test 11.	No problem exists ▶	CHECK for carburetor leaks, carburetor feedback, bowl vent hose leaks, and the in-line bowl vent solenoid for proper operation.
	Problem exists ▶	CORRECT problems. Retest.

Fuel Always Rich | Pinpoint | Test 7

7.0 SERVICE CODE 4-2

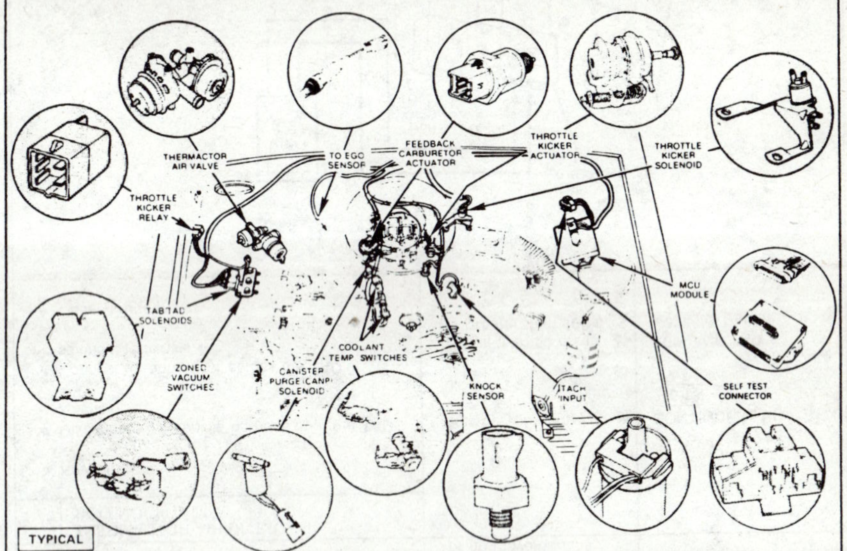

Components controlled by the MCU System in this Pinpoint Test are the EGO Sensor, the Feedback Carburetor Actuator, and the MCU Module. The components not directly controlled by the MCU System during this Pinpoint Test but could possibly cause a Service Code fault are as follows:

- Malfunctioning Carburetor
  - —High Float Level
  - —No Power To Choke Cap
- Misfiring Spark Plugs
  - —Coil
  - —Distributor
- Restricted Air Cleaner
- Restricted PVS
- Air Pump Not Working
  - —Malfunctioning Pulley
  - —Broken Belt(s)
- Throttle Kicker Always On
- PCV

NOTE: Virtually anything that adds fuel or restricts air should be investigated.

Fuel Always Rich — Pinpoint Test 7

TEST STEP	RESULT ▶	ACTION TO TAKE
7.0 SERVICE CODE 4-2—FUEL ALWAYS RICH (CONTINUED)		
• Disconnect the jumper wire from Self-Test Trigger input circuits 201 and 60. • Vehicle must be at normal operating temperature. • Start engine, run at idle. • Verify that there is power to the choke cap, cold enrichment circuit is operating properly, and choke is Off.	System, OK. ▶ System, not OK. ▶	Continue to Step 7.1 Service as required. Retest.
7.1 **CAUTION: To prevent damage to the EGO sensor, never connect an ohmmeter to the EGO sensor; use a voltmeter with an input impedence of 10 megohms or greater.** • Key Off. • Disconnect the thermactor air hose at the pump. • Deactivate self test. • Disconnect the vehicle harness from the EGO sensor. • Connect a voltmeter between the EGO sensor and engine block ground. • Disconnect the PCV hose from the carburetor, creating a vacuum leak. • Run engine at 2400 rpm for 60 seconds, then observe voltmeter.	0.50 volts or less. ▶ Voltage more than 0.50 volts. ▶	RECONNECT PCV hose. RECONNECT the EGO sensor to the vehicle harness. CONTINUE to Step 7.2 REPLACE EGO sensor. Retest.

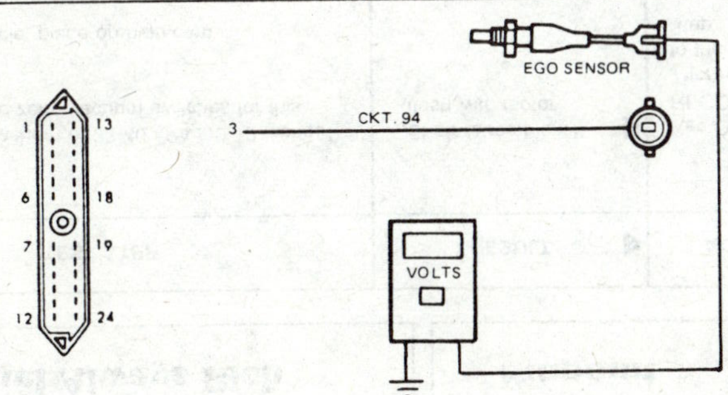

Fuel Always Rich — Pinpoint Test 7

TEST STEP	RESULT ▶	ACTION TO TAKE
7.2 • Key Off. • Disconnect the vehicle harness from the FBCA Motor, connect a known good FBCA motor (shaft fully depressed) to harness, but do not remove the original FBCA motor from the carburetor. • Key to Run.	FBCA motor shaft extends fully, then may retract slightly. ▶ Shaft does not extend fully. ▶	CONTINUE to Step 7.3 CONTINUE to Step 7.4

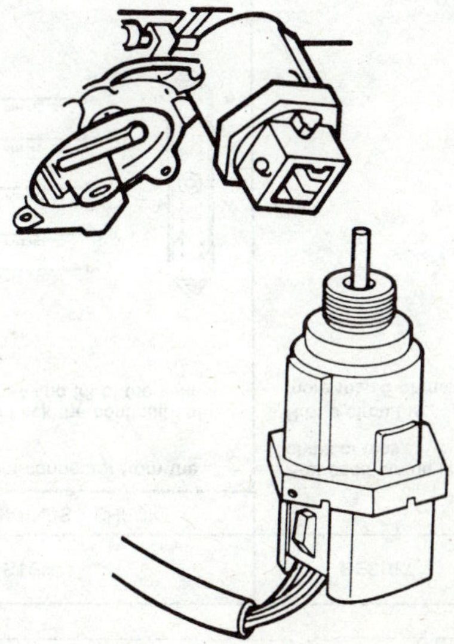

Fuel Always Rich — Pinpoint — Test 7

TEST STEP	RESULT ▶	ACTION TO TAKE
7.3		
• Apply a vacuum of 33-40 kPa (10-12 inches Hg.) to the zone vacuum switches for this test. • Start vehicle, place on high cam. • Allow 120 seconds, then force system rich by depressing CVR rod. • Observe remote FBCA motor shaft.	Shaft retracts until flush with motor ▶	Key Off. RECONNECT Thermactor air hose to the pump. GO to Step 7.6
	Shaft does not retract ▶	Key Off. DISCONNECT the known good FBCA motor from the harness. CONTINUE to Step 7.4

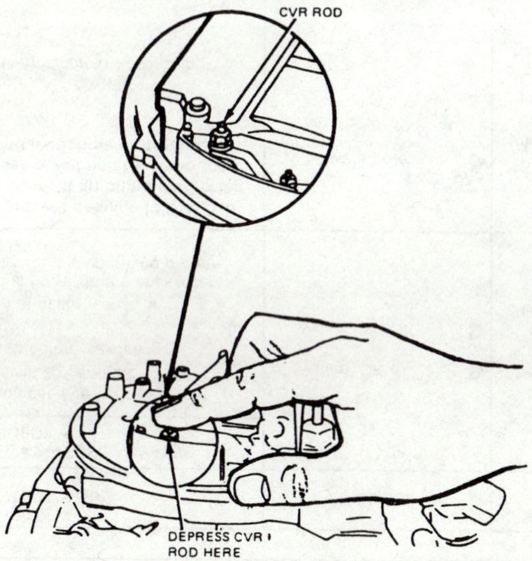

Fuel Always Rich — Pinpoint — Test 7

TEST STEP	RESULT ▶	ACTION TO TAKE
7.4 FBCA MOTOR HARNESS CHECK		
• Disconnect the 24 pin connector from the MCU module. • With an ohmmeter, check the continuity of circuits 687, 95, 96, 97 and 98 of the vehicle harness.	R of each circuit is 5 ohms or less. ▶	CONTINUE to Step 7.5
	R of a circuit is more than 5 ohms. ▶	SERVICE or REPLACE circuit. Retest.

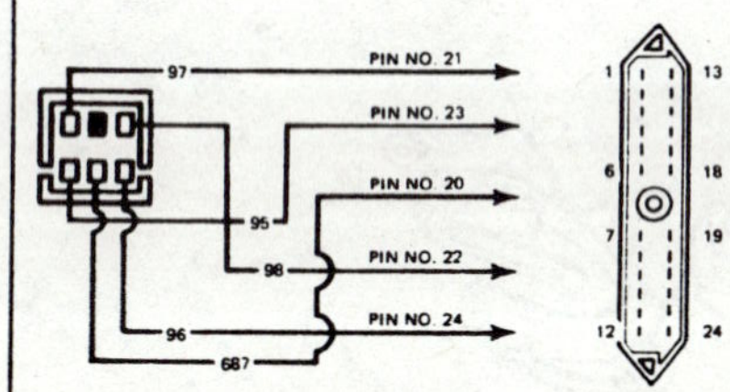

TEST STEP	RESULT ▶	ACTION TO TAKE
7.5 CHECK FBCA MOTOR HARNESS FOR A SHORT TO GROUND		
• Connect one lead of ohmmeter to engine block ground, and with the other lead, probe circuits 95, 96, 97 and 98 of the vehicle harness connector, checking each circuit for a short to ground.	R of each circuit is more than 1000 ohms. ▶	GO to 7.6
	R of each circuit is 1,000 ohms or less. ▶	SERVICE or REPLACE circuit. Retest.

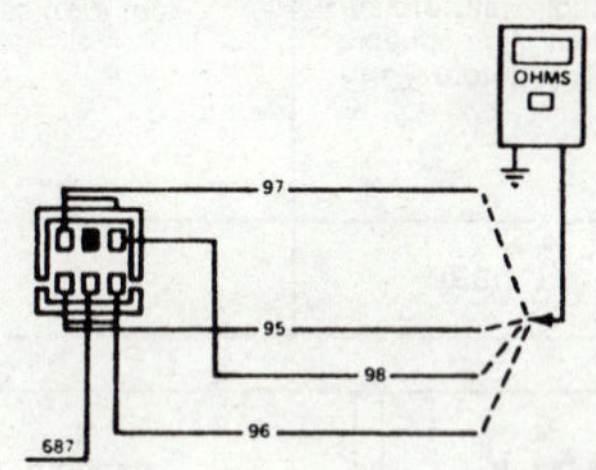

Fuel Always Rich — Pinpoint — Test 7

TEST STEP	RESULT ▶	ACTION TO TAKE
7.6		
• Measure the resistance of each FBCA motor winding to the FBCA in the carburetor.	Not 50-175 ohms at room temperature or above. ▶	GO to 7.7
	50-175 ohms at room temperature or above. ▶	REPLACE FBCA motor and retest.

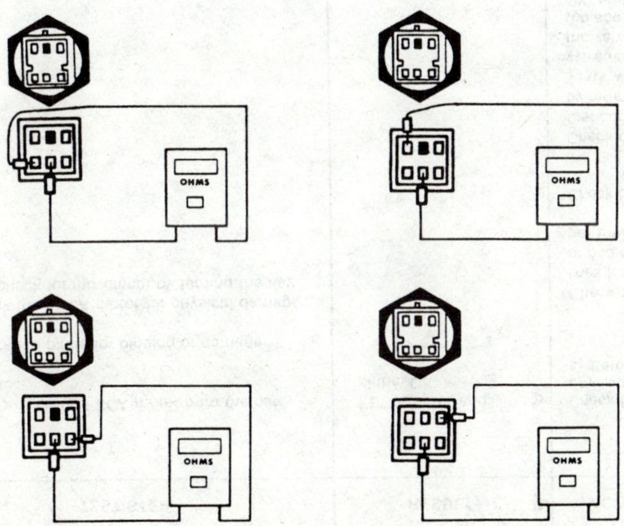

Fuel Always Rich — Pinpoint — Test 7

TEST STEP	RESULT ▶	ACTION TO TAKE
7.7 FBCA MOTOR		
• Key Off. • Connect an ohmmeter lead to pin 687 at the original FBCA motor in the carburetor and the other lead to the FBCA motor outside case.	R is more than 190,000 ohms. ▶	CONTINUE to Step 7.8
	R is 190,000 ohms or less ▶	REPLACE FBCA. Retest.

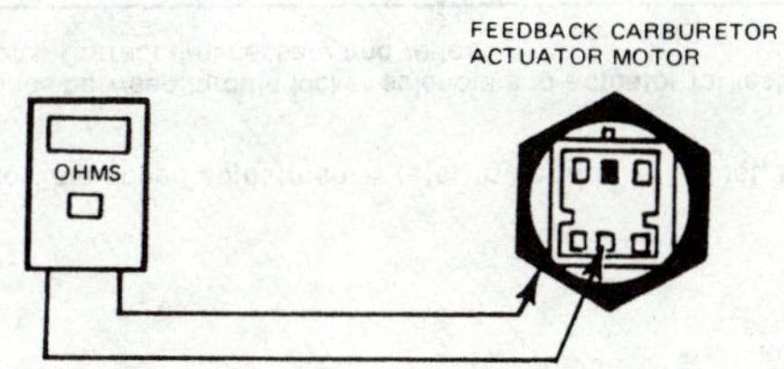

Fuel Always Rich — Pinpoint Test 7

TEST STEP	RESULT ▶	ACTION TO TAKE
7.8 • Disconnect the FBCA motor from the carburetor. • Check the pintle for binding or damage. • Check the FBCA motor for physical damage and check for movement by turning the key On.	FBCA motor and pintle, OK. ▶	Check Evaporative Emission Control system Check all vacuum lines for leaks, wrong or bad connections per Emissions Decal. Check bowl vent per Check crankcase for fuel contamination by rerunning Quick Tests with PCV valve removed from the valve cover. If, the above items are OK, replace MCU and retest.
	FBCA pintle, binding. ▶	REPLACE the FBCA pintle. Retest.
	FBCA motor physically damaged or no movement. ▶	REPLACE FBCA motor and retest.

RPM Check — Pinpoint Test 8

8.0 SERVICE CODE 1-2

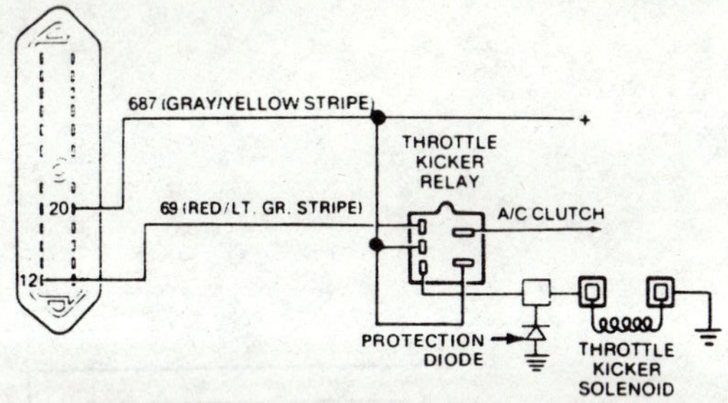

- Check carburetor idle speed adjustments, refer to Section 4. Adjust, if necessary, and re-test.
- Check vacuum lines between throttle kicker solenoid and actuator for leaks, blockage, or improper connections. Correct if necessary and re-test.

TEST STEP	RESULT ▶	ACTION TO TAKE
8.1 • Disconnect self test input. • "T"-in a vacuum gauge to the throttle kicker. • Start engine, run at idle. • Turn A/C On. Vacuum should be present. • Turn A/C Off, no vacuum present.	Vacuum switches. ▶	Key Off. GO to Step 8.2
	No vacuum, or no vacuum switching. ▶	CHECK vacuum line for leaks, blockage, or wrong connections. If OK continue to 8.3

RPM Check — Pinpoint Test 8

TEST STEP	RESULT	ACTION TO TAKE
8.2 • Key Off. • Check throttle kicker actuator for proper function. Apply at least 33 kPa (10 inches Hg.) vacuum to actuator vacuum line. Observe actuator arm.	Arm extends and holds ▶	GO to Step 8.3
	Arm does not extend and hold ▶	CHECK actuator. SERVICE or REPLACE and re-test
8.3 • Check throttle kicker solenoid resistance (harness disconnected).	R is between 45 and 90 ohms ▶	GO to Step 8.4
	R less than 45 ohms or greater than 90 ohms. ▶	REPLACE solenoid. Re-test
8.4 • Check circuits between throttle kicker relay connector and throttle kicker solenoid, and between the solenoid and ground, for continuity.	R is less than 5 ohms ▶	GO to Step 8.5
	R is more than 5 ohms. ▶	Service or replace open circuit in harness. Retest.

RPM Check — Pinpoint Test 8

TEST STEP	RESULT	ACTION TO TAKE
8.5 • MCU and relay disconnected. • Check circuits 69 and 687 for continuity between MCU and throttle kicker relay connector.	R is 5 ohms or less. ▶	GO to Step 8.6
	R is more than 5 ohms. ▶	SERVICE open circuit. Re-test
8.6 • Check circuit 69 for isolation from ground.	R is more than 1,000 ohms. ▶	GO to Step 8.7
	R is 1,000 ohms or less. ▶	SERVICE grounded circuit 69. Re-test
8.7 • Reconnect throttle kicker relay and throttle kicker solenoid. • Key On, Engine Off, A/C Off. • Check throttle kicker relay function when energized. Connect voltmeter across solenoid terminals. Ground circuit 69 at MCU connector, Pin 12, and observe voltage.	Voltage is more than 10 volts. ▶	Verify that circuit 69 is not shorted to 687. REPAIR harness if necessary then replace MCU.
	Voltage is less than 10 volts. ▶	Replace the relay. Retest.

Thermactor Check | **Pinpoint** | **Test 10**

10.0 SERVICE CODE 4-4

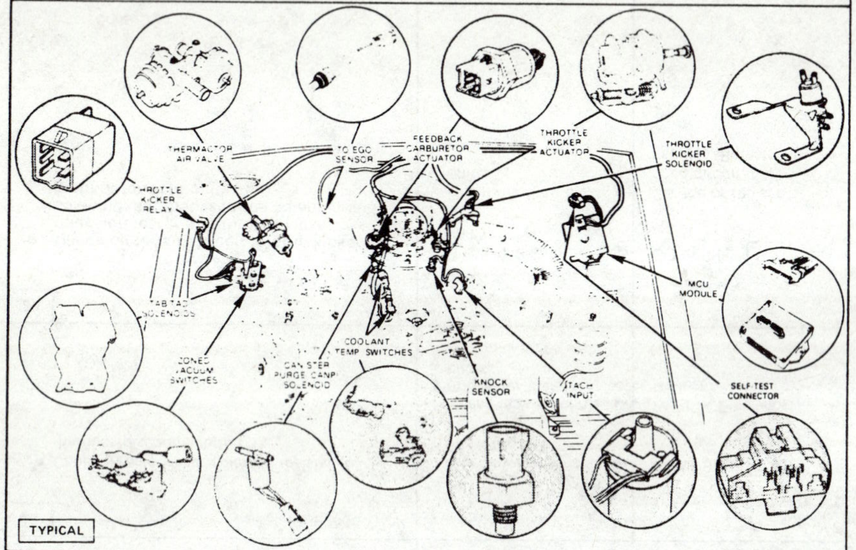

TYPICAL

Components controlled by the MCU System in this Pinpoint Test are the TAB/TAD Solenoids, the Thermactor Air Valve, the EGO Sensor, and the MCU Module. The components not directly controlled by the MCU System during this Pinpoint Test but could possibly cause a Service Code fault are as follows:

- Retard Delay Valve (RDV)
- Temperature Vacuum Switch (TVS)
- Malfunctioning Carburetor
  - —High Float Level
  - —No Power To Choke Plate
- Misfiring Spark Plugs
  - —Coil
  - —Distributor
- Restricted Air Cleaner
- Restricted PVS
- Air Pump Not Working
  - —Malfunctioning Pulley
  - —Broken Belt(s)
- Throttle Kicker Always On
- PCV

NOTE: Virtually anything that adds fuel, restricts air, or inhibits Thermactor System from switching air upstream to the exhaust manifold should be investigated.

Thermactor Check | **Pinpoint** | **Test 10**

10.0 SERVICE CODE 4-4 (CONTINUED)

TYPICAL THERMACTOR SYSTEM

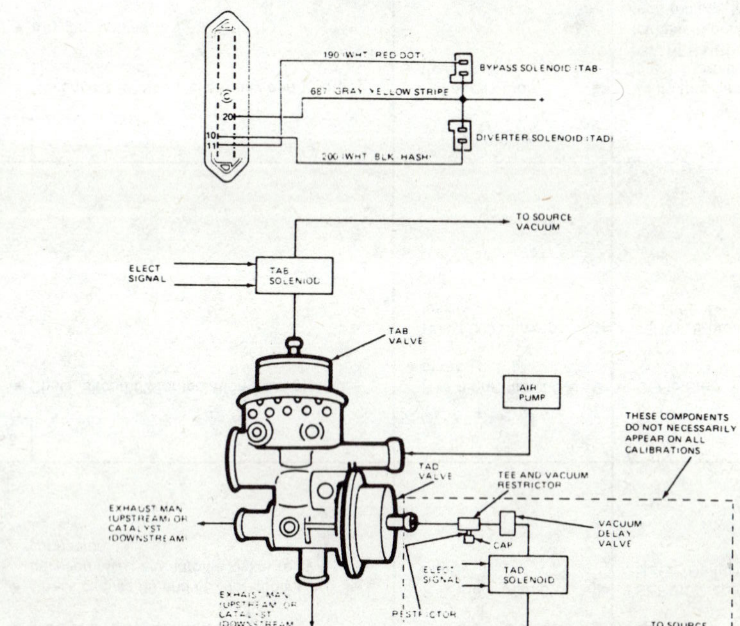

TYPICAL THERMACTOR SYSTEM

Thermactor Check — Pinpoint Test 10

TEST STEP	RESULT ▶	ACTION TO TAKE
10.01 THERMACTOR CHECK This Step is for police vehicles only; all other vehicles, go to Pinpoint Test 10, Step 10.10 • Bring engine to operating temperature. • Verify upper radiator hose is hot and pressurized. • Turn engine off. • Restart engine at curb idle. • Immediately initiate Self-Test within 10 seconds. • Observe DVOM for Service Codes.	Any other Service Codes ▶	Go to the appropriate Pinpoint Test.
	Service Code 4-4 ▶	GO to 10.10
	Service Code 1-1 ▶	Continue with Quick Test.
10.10 • Check for manifold vacuum at input to TAB-TAD solenoids.	Manifold vacuum present (more than 10 inches Hg). ▶	GO to 10.11
	Vacuum not present (10 inches Hg or less). ▶	CORRECT vacuum routing per Vehicle Emission Decal and retest.

Thermactor Check — Pinpoint Test 10

TEST STEP	RESULT ▶	ACTION TO TAKE
10.11 • Remove the vacuum hose from the TAB valve. and connect a vacuum gauge to hose. Start engine. increase rpm to above 2,000 rpm for 2 minutes. Shut off, then restart and initiate Self-Test. Observe gauge during Self-Test.	Vacuum pulses above and below 17 kPa (5 inches Hg.) ▶	GO to 10.12
	Vacuum always above 17 kPa (5 inches Hg.) ▶	GO to 10.20
	Vacuum always below 17 kPa (5 inches Hg.) ▶	GO to 10.30
10.12 • Reconnect hose to TAB valve. Remove vacuum hose from the TAD valve and connect a vacuum gauge to hose. Start re-start engine. increase the rpm to above 2,000 rpm for 2 minutes. Shut off, then restart and initiate Self-Test. Observe gauge during test.	Vacuum pulses above and below 17 kPa (5 inches Hg.) ▶	GO to Thermactor Air System Routine,
	Vacuum always above 17 kPa (5 inches Hg.) ▶	GO to 10.40
	Vacuum always below 17 kPa (5 inches Hg.) ▶	GO to 10.50

Thermactor Check — Pinpoint Test 10

TEST STEP	RESULT	ACTION TO TAKE
10.20 • Key Off. • Check circuit 190 for a short to ground. Measure resistance.	R is more than 1,000 ohms. ▶ R is 1,000 ohms or less. ▶	GO to 10.22 SERVICE short. Retest.
10.21 • Measure the TAB solenoid resistance. Should be between 50-110 ohms.	R within range ▶ R not within range ▶	GO to 10.32 REPLACE TAB solenoid. Retest. If Service Code 4-4 is still present, replace MCU.
10.22 • Check at the TAB solenoid output to make sure vacuum is not present when the solenoid is de-energized (vehicle harness disconnected, engine at idle). • Disconnect the TAB solenoid electrically. • Start engine, run at idle.	Vacuum is present ▶ Vacuum is not present ▶	REPLACE TAB solenoid. Retest. REPLACE MCU module. Retest.

Thermactor Check — Pinpoint Test 10

TEST STEP	RESULT	ACTION TO TAKE
10.30 • Key Off. • Check the vacuum hose between the TAB valve and TAB solenoid for leakage or blockage.	Leakage or blockage found ▶ No problem found ▶	SERVICE as required. Retest. GO to 10.21
10.32 • Check the TAB solenoid output to make sure vacuum is not present when the solenoid is energized. • Jumper the TAB solenoid circuit 190 to ground. • Start engine, run at idle.	Vacuum present. ▶ Vacuum is not present. ▶	GO to 10.33 REPLACE TAB solenoid. Retest if Service Code 4-4 present, continue to 10.33.
10.33 • Key Off. • Check circuits 687 and 190 for continuity between the TAB solenoid and MCU module. Measure resistance.	R is more than 5 ohms. ▶ R is 5 ohms or less. ▶	SERVICE circuit. Retest. REPLACE MCU module. Retest.

Thermactor Check — Pinpoint Test 10

TEST STEP	RESULT	ACTION TO TAKE
10.40		
• Key Off. • Check circuit 200 for a short to ground. Measure resistance.	R is more than 1,000 ohms. ▶	GO to 10.41
	R is 1,000 ohms or less. ▶	SERVICE short
10.41		
• Check at the TAD solenoid output to make sure vacuum is not present when the solenoid is de-energized (vehicle harness disconnected, engine at idle). • Disconnect the TAD solenoid electrically. • Start engine, run at idle.	Vacuum is present ▶	REPLACE TAD solenoid. Retest.
	Vacuum is not present ▶	REPLACE MCU. Retest.
10.50		
• Check the vacuum hoses between the TAD valve and TAD solenoid for leakage or blockage.	Leakage or blockage found ▶	SERVICE as required. Retest.
	OK ▶	GO to 10.52

Thermactor Check — Pinpoint Test 10

TEST STEP	RESULT	ACTION TO TAKE
10.52		
• Key Off. • Measure the TAD solenoid resistance. Should be between 50-110 ohms.	R in range ▶	GO to 10.53
	R not in range ▶	REPLACE solenoid. Retest if Service Code 4-4 is still present go to 10.54.
10.53		
• Check at the TAD solenoid output to make sure vacuum is present when the solenoid is energized by 12V. • Jumper the TAD solenoid circuit 200 to ground. • Start engine, run at idle.	Vacuum present ▶	GO to 10.54
	Vacuum not present ▶	REPLACE solenoid. Retest.
10.54		
• Key Off. • Check circuits 687 and 200 for continuity between TAD solenoid and MCU module. Measure resistance.	R is more than 5 ohms. ▶	SERVICE circuit. Retest
	R is 5 ohms or less. ▶	REPLACE MCU module. Retest.

Thermactor Air Switching	Pinpoint	Test 11

11.0 SERVICE CODE 4-5 THERMACTOR AIR ALWAYS UPSTREAM

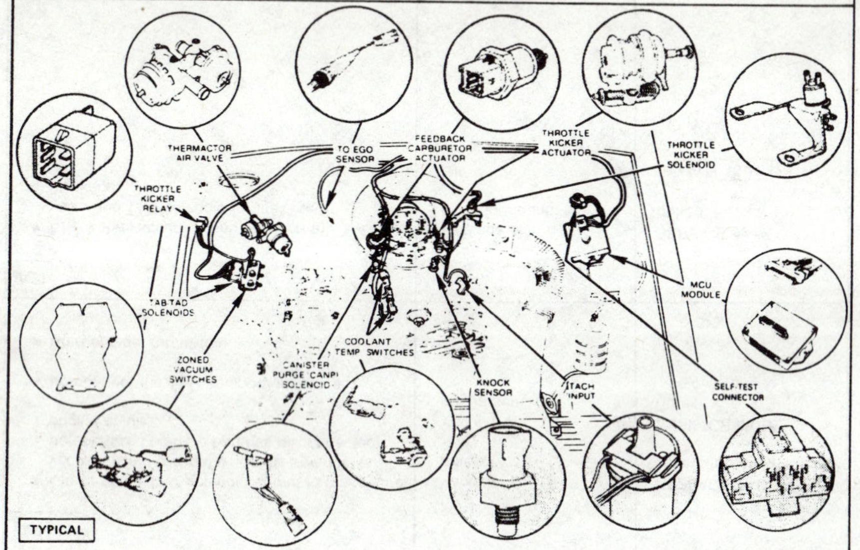

Components controlled by the MCU System in this Pinpoint Test are the TAB/TAD Solenoids, the Thermactor Air Valve, and the MCU Module. The component not directly controlled by the MCU System during this Pinpoint Test but could possibly cause a Service Code fault are as follows:

- Broken, Loose Belts
- Crimped, Pinched or Missing Hoses, Plugged Restrictor
- Malfunctioning Vacuum Delay Valve

NOTE: Virtually anything that inhibits the Thermactor System from switching air downstream to the catalyst should be investigated.

Thermactor Air Switching	Pinpoint	Test 11

11.0 SERVICE CODE 4-5 (Continued)

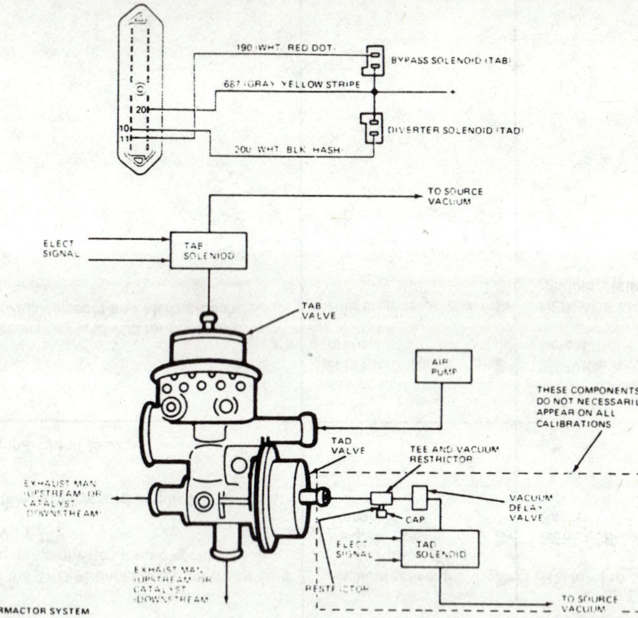

TYPICAL THERMACTOR SYSTEM

TEST STEP	RESULT ▶	ACTION TO TAKE
11.081		
• Verify the restrictor Teed into the TAD vacuum line was uncapped and restrictor was free from contamination.	Restrictor OK ▶	GO to 11.091
	Problem with restrictor ▶	SERVICE as required and retest
11.091		
• Cap vacuum restrictor, and remove vacuum delay valve from TAD vacuum control line. Reconnect system without delay valve and retest.	No Code 4-5 ▶	System OK. SERVICE vacuum delay valve
	Code 4-5 ▶	GO to 11.10

Thermactor Air Switching — Pinpoint Test 11

TEST STEP	RESULT	ACTION TO TAKE
11.10 • Remove vacuum hose from the TAD valve, and connect a vacuum gauge to hose. Start/re-start engine, increase rpm to above 2,000 for 2 minutes. Shut off then restart and initiate self test. Observe vacuum gauge during test.	Vacuum pulses above and below 17 kPa (5 inches Hg.) ▶	MCU system, OK; so CHECK thermactor air.
	Vacuum always above 17 kPa (5 inches Hg.) ▶	GO to 11.30
	Vacuum always below 17 kPa (5 inches Hg.) ▶	GO to 11.20
11.20 • Check vacuum hoses between TAD valve to TAD solenoid and between TAD solenoid and vacuum source for leaks or blockage. Also check the vacuum source per Emission Decal.	Leaks or blockage found ▶	SERVICE as required. Retest.
	No leaks or blockage ▶	GO to 11.21

Thermactor Air Switching — Pinpoint Test 11

TEST STEP	RESULT	ACTION TO TAKE
11.21 • Key Off. • Measure the TAD solenoid resistance. • Should be between 50-110 ohms.	R within range ▶	GO to 11.22
	R not within range ▶	REPLACE solenoid. Retest.
11.22 • Check at the TAD solenoid output that vacuum is present when the solenoid is energized by 12V. • Jumper the TAD solenoid circuit 200 to ground. • Start engine, run at idle.	Vacuum present ▶	GO to 11.23
	Vacuum not present ▶	REPLACE solenoid. Retest.
11.23 • Key Off. • Check continuity at circuits 200 and 687 between the MCU module and the TAD solenoid for continuity. Measure resistance.	R is 5 ohms or less. ▶	REPLACE MCU module. Retest.
	R is more than 5 ohms. ▶	SERVICE circuit. Retest.
11.30 • Check at the TAD solenoid output that vacuum is not present when the solenoid is de-energized. • Disconnect the TAD solenoid electrically. • Start engine, run at idle.	Vacuum present ▶	REPLACE solenoid. Retest.
	Vacuum not present ▶	GO to 11.31
11.31 • Key Off. • Check circuit 200 for a short to ground. Should not be shorted. Measure resistance.	R is 1,000 ohms or less. ▶	SERVICE short. Retest.
	R is more than 1,000 ohms. ▶	REPLACE MCU module. Retest.

Thermactor Air Bypass — Pinpoint Test 12

12.0 SERVICE CODE 4-6 THERMACTOR AIR ALWAYS NOT BYPASSING

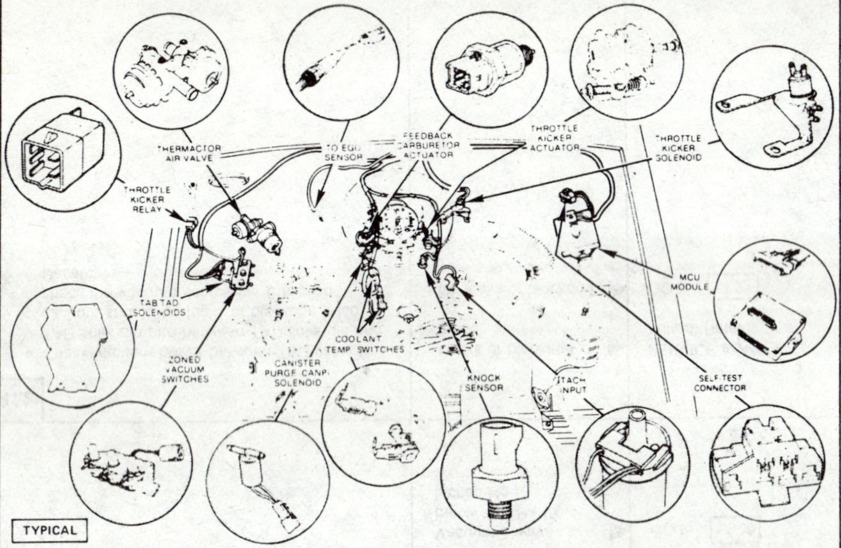

TYPICAL

Components controlled by the MCU System in this Pinpoint Test are the TAB/TAD Solenoids, the Thermactor Air Valve, and the MCU Module. The component not directly controlled by the MCU System during this Pinpoint Test but could possibly cause a Service Code fault are as follows:

- Broken, Loose Belts
- Crimped, Pinched or Missing Hoses, Plugged Restrictor
- Malfunctiong Vacuum Delay Valve

NOTE: Virtually anything that inhibits the Thermactor System from switching air downstream to the catalyst should be investigated.

Thermactor Air Bypass — Pinpoint Test 12

12.0 SERVICE CODE 4-6 (CONTINUED)

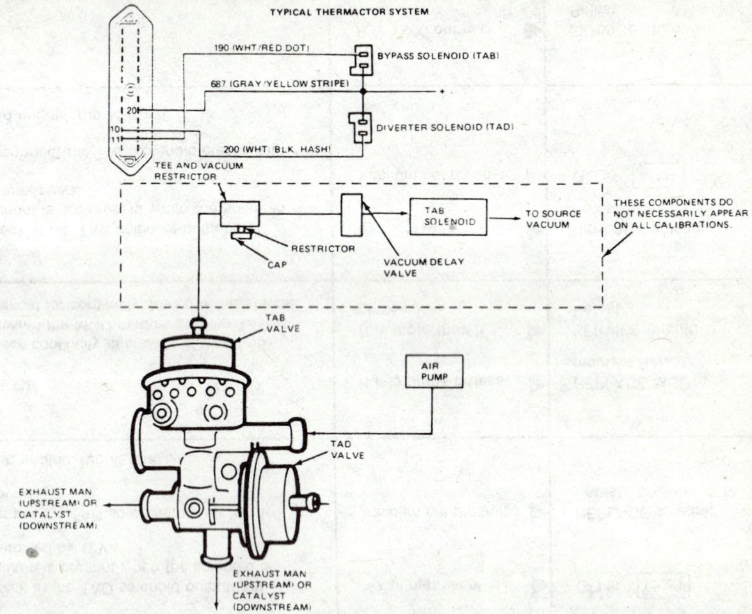

TEST STEP	RESULT	ACTION TO TAKE
12.01		
• Verify the restrictor Teed into the TAB vacuum line was uncapped and restrictor was free from contamination.	Restrictor, OK ▶	GO to 12.02
	Problem with restrictor ▶	SERVICE as required and retest.
12.02		
• Cap vacuum restrictor, and remove vacuum delay valve from TAB vacuum control line. Reconnect system without delay valve and retest.	No Code 4-6 ▶	System, OK. SERVICE vacuum delav valve
	Code 4-6 ▶	GO to 12.10

Thermactor Air Bypass — Pinpoint — Test 12

TEST STEP	RESULT	▶	ACTION TO TAKE
12.03 THERMACTOR CHECK			
This Step is for police vehicles only; all other vehicles, go to Pinpoint Test 12, Step 12.10 • Bring engine to operating temperature. • Verify upper radiator hose is hot and pressurized. • Turn engine Off. • Restart engine at curb idle. • Immediately initiate Self-Test within 10 seconds. • Observe DVOM for Service Codes.	Any other Service Codes.	▶	Go to the appropriate Pinpoint Test.
	Service Code 4-6.	▶	GO to 12.10
	Service Code 1-1.	▶	Continue with Quick Test.
12.10			
• Remove vacuum hose from the TAB valve, and connect a vacuum gauge to hose. Start/re-start engine, increase rpm to above 2,000 for 2 minutes. Shut down, then restart and initiate self test. Observe vacuum gauge during test.	Vacuum pulses above and below 17 kPa (5 inches Hg.)	▶	MCU system, OK; so CHECK thermactor air,
	Vacuum always below 17 kPa (5 inches Hg.)	▶	GO to 12.20
	Vacuum always above 17 kPa (5 inches Hg.)	▶	GO to 12.30

Thermactor Air Bypass — Pinpoint — Test 12

TEST STEP	RESULT	▶	ACTION TO TAKE
12.20			
• Check vacuum hoses between TAB valve to TAB solenoid and between TAB solenoid and vacuum source for leaks or blockage.	Leaks or blockage found	▶	SERVICE as required. Retest.
	OK	▶	GO to 12.21
12.21			
• Check at the TAB solenoid output to make sure vacuum is present when the solenoid is energized. • Jumper the TAB solenoid circuit 190 to ground. • Start engine, run at idle.	Vacuum	▶	GO to 12.22
	No vacuum	▶	REPLACE solenoid. Retest.
12.22			
• Check circuit 190 for continuity. Measure resistance.	R is more than 10 ohms.	▶	SERVICE harness. Retest.
	R is 10 ohms or less.	▶	REPLACE MCU module. Retest.
12.30			
• Check at the TAB solenoid output to make sure vacuum is not present when the solenoid is de-energized. • Disconnect the TAB solenoid electrically. • Start engine, run at idle.	Vacuum not present	▶	GO to 12.31
	Vacuum present	▶	REPLACE solenoid. Retest.
12.31			
• Check circuit 190 for a short to ground. Measure resistance.	R is 1,000 ohms or less.	▶	SERVICE short. Retest.
	R is more than 1,000 ohms.	▶	REPLACE MCU module. Retest.

Knock Detection — Pinpoint Test 13

TEST STEP	RESULT	▶ ACTION TO TAKE
13.0 SERVICE CODE 2-5—KNOCK DETECTION SYSTEM INOPERATIVE		
• Verify that the intake manifold was tapped with a steel object within two inches of the knock sensor.	No	▶ RETEST per Step 5.0 Engine Running Quick Test.
• Disconnect the jumper wire from the Self-Test Trigger input circuit 201.	Yes	▶ CONTINUE to next Step 13.1
13.1		
• Key Off. • Disconnect the vehicle harness from the knock sensor. • Using a torque wrench with a 1⅛ inch deep socket, verify sensor is tightened at 16-24 N•m (12-18 lb-ft). Do not tighten more than 24 N•m (18 lb-ft).	Torque at 16-24 N•m (12-18 lb-ft)	▶ CONTINUE to 13.2
	Torque not at 16-24 N•m (12-18 lb-ft)	▶ TIGHTEN to 16-24 N•m (12-18 lb-ft)

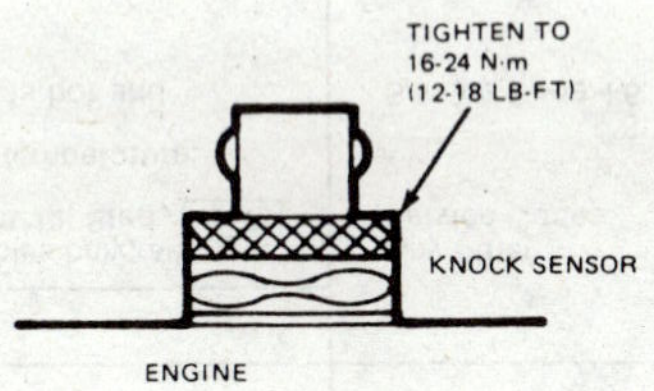

Knock Detection — Pinpoint Test 13

TEST STEP	RESULT	▶ ACTION TO TAKE
13.2		
• Disconnect the vehicle harness from the MCU and knock sensor.	R is more than 1000 ohms.	▶ CONTINUE to Step 13.3
• Connect one lead of ohmmeter to circuit 310 of the vehicle harness and the other lead to engine block.	R is 1,000 ohms or less.	▶ SERVICE or REPLACE circuit 310 for short to ground. Retest.

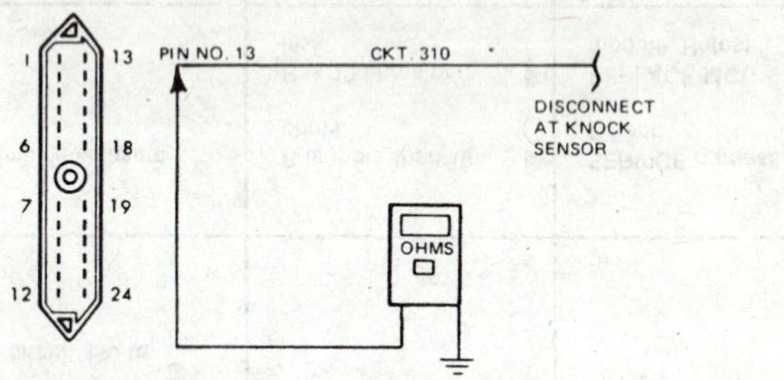

Knock Detection | Pinpoint | Test 13

TEST STEP	RESULT	▶	ACTION TO TAKE
13.3 • Key Off. • Leave knock sensor disconnected. • Check circuit 310 of the vehicle harness for continuity.	R is 5 ohms or less.	▶	CONTINUE to Step 13.4
	R is more than 5 ohms.	▶	SERVICE or REPLACE circuit 310. Retest.

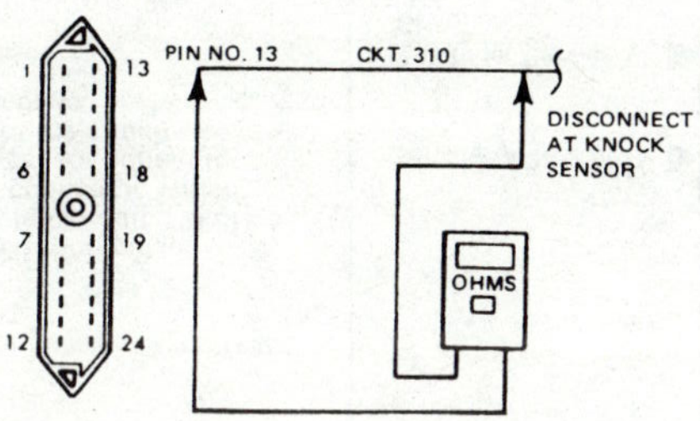

Knock Detection | Pinpoint | Test 13

TEST STEP	RESULT	▶	ACTION TO TAKE
13.4 • Check Circuit 60 of the vehicle harness for continuity between the knock sensor and MCU.	R is 5 ohms or less.	▶	RECONNECT the vehicle harness connector to the MCU. GO to Step 13.5
	R is more than 5 ohms.	▶	SERVICE or REPLACE circuit 60. Retest.

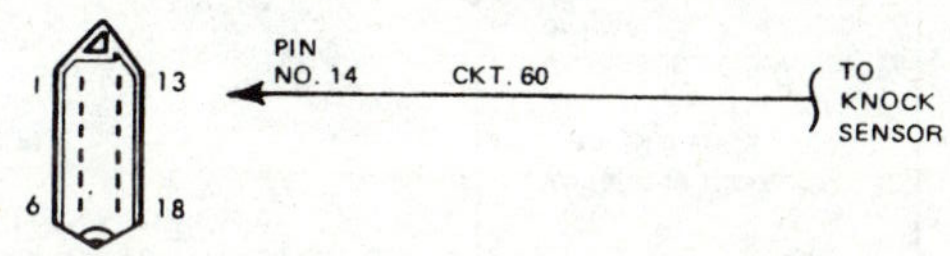

Knock Detection — Pinpoint Test 13

TEST STEP	RESULT ▶	ACTION TO TAKE
13.5		
• Reconnect the jumper to circuits 201 and 60 of the VIP connector. • Connect a circuit test lamp to the positive terminal of the battery. • Disconnect the knock sensor from the vehicle harness. • Go to Step 5.0, Engine Running Quick Test, and follow instructions. • When the four initialization pulses occur, continuously tap circuit 310 of the vehicle harness connector, which mates to the knock sensor connector, with the probe end of the circuit tester lamp for 5 seconds. • Observe voltmeter pulses.	Service Code 2-5 ▶	REPLACE MCU. Retest.
	Any other Service Code ▶	REPLACE Knock Sensor. Retest.

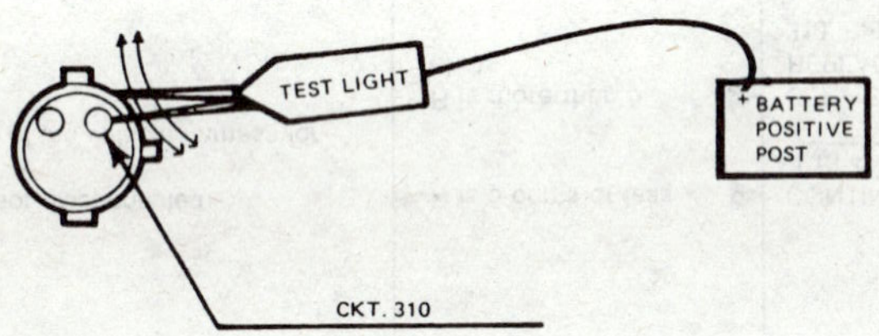

No Self-Test Output — Pinpoint Test 14

TEST STEP	RESULT ▶	ACTION TO TAKE
14.0 NO SELF-TEST OUTPUT		
• Disconnect the jumper wire from the Self-Test Trigger input circuits 201 and 60. • Disconnect the harness connector from the MCU. • Connect a voltmeter between circuit 687 and 60 of the vehicle harness connector. • Key On, engine Off.	Voltage is more than 10.5 volts.	GO to Step 14.2
	Voltage is 10.5 volts or less. ▶	CONTINUE to Step 14.1

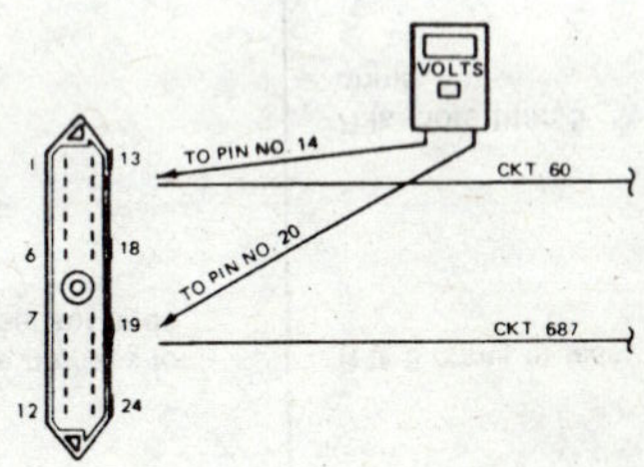

No Self-Test Output | Pinpoint | Test 14

TEST STEP	RESULT ▶	ACTION TO TAKE
14.1 • Leave positive lead of voltmeter connected to circuit 687 of the harness connector, and connect the other lead to engine block ground.	Voltage is more than 10.5 volts. ▶	SERVICE or REPLACE circuit 60. Retest.
	Voltage is less than 10.5 volts. ▶	SERVICE or REPLACE circuit 687. Retest.

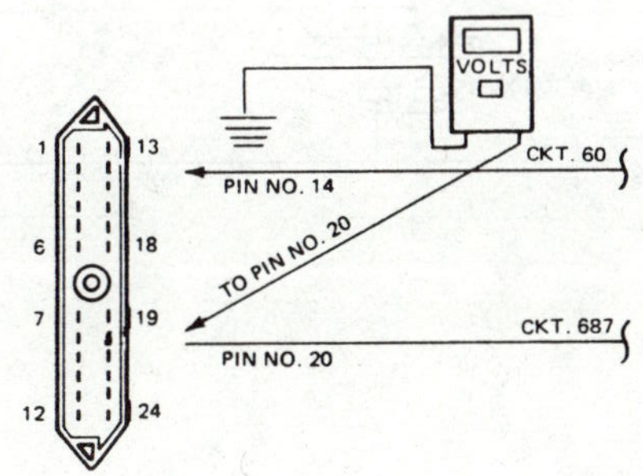

TEST STEP	RESULT ▶	ACTION TO TAKE
14.2 • Key Off. • Check circuit 60 Pin No. 14 for continuity between the Self-Test connector end and engine block ground.	R is 1,000 ohms or less. ▶	CONTINUE to Step 14.3
	R is more than 1,000 ohms. ▶	SERVICE or REPLACE circuit 60. Retest.

No Self-Test Output | Pinpoint | Test 14

TEST STEP	RESULT ▶	ACTION TO TAKE
14.3 • Check circuit 201 for continuity between the Self Test connector and the vehicle harness connector that mates with the MCU, Pin No. 19.	R is 5 ohms or less. ▶	CONTINUE to Step 14.4
	R is more than 5 ohms. ▶	SERVICE or REPLACE circuit 201. Retest.
14.4 TACH LEAD • Connect ohmmeter between Tach terminal of the coil horseshoe connector and circuit 11 of the vehicle harness connector that mates with the MCU connector, Pin No. 8. • Observe ohmmeter.	R is 5 ohms or less. ▶	CONTINUE to Step 14.5
	R is more than 5 ohms. ▶	SERVICE or REPLACE circuit 11. Retest.

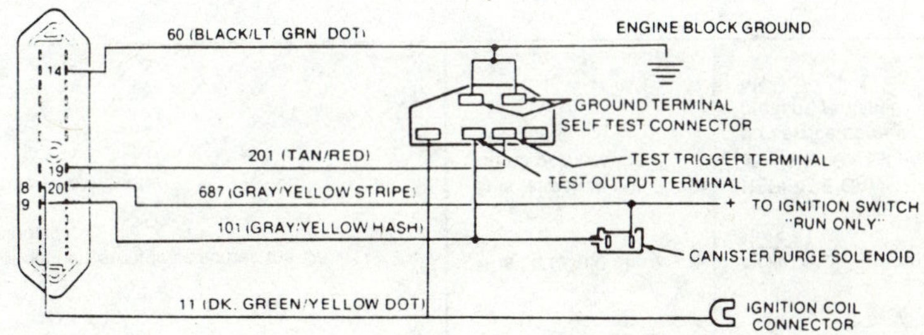

No Self-Test Output — Pinpoint — Test 14

TEST STEP	RESULT ▶	ACTION TO TAKE
14.5 • Disconnect the vehicle harness from the CANP solenoid and the MCU. • Check circuits 687 and 101 for continuity. • Check circuit 101 from Self Test connector to MCU connector, and from CANP to MCU connector.	R of each circuit is 5 ohms or less. ▶	CONTINUE to Step 14.6
	R of a circuit is more than 5 ohms. ▶	SERVICE or REPLACE circuit. Retest.

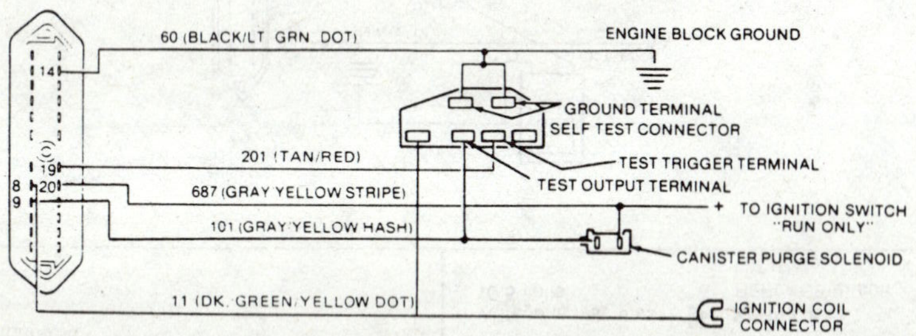

No Self-Test Output — Pinpoint — Test 14

TEST STEP	RESULT ▶	ACTION TO TAKE
14.6 • Connect an ohmmeter across the CANP solenoid. • Observe ohmmeter.	R is 50 to 100 ohms. ▶	CONTINUE to Step 14.7
	R is less than 50 ohms or more than 100 ohms. ▶	REPLACE CANP solenoid. Retest, if no service code is present, replace MCU.

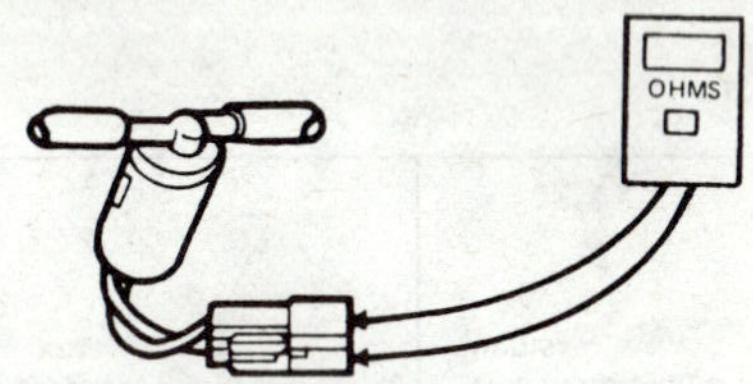

No Self-Test Output — Pinpoint Test 14

TEST STEP	RESULT ▶	ACTION TO TAKE
14.7		
• Connect one lead of ohmmeter to engine block ground and the other lead to circuit 101 of vehicle harness.	R is less than 1000 ohms. ▶	CONTINUE to Step 14.8
	R is 1000 ohms or more. ▶	SERVICE circuit 101 for short to ground. Retest.

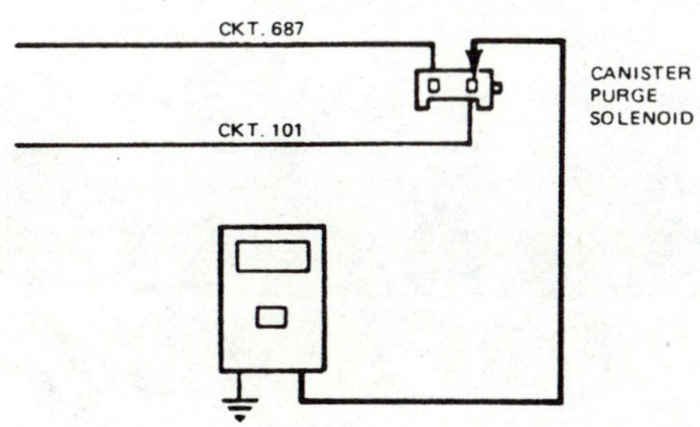

TEST STEP	RESULT ▶	ACTION TO TAKE
14.8		
• Check to make sure the Canister Purge solenoid passes vacuum when it is energized (12V) and blocks vacuum when it is de-energized. Check housing and hose for leaks. • Make sure the CANP hose is not cracked, pinched, or plugged.	Solenoid blocks and passes vacuum as indicated ▶	REPLACE MCU. Retest.
	Solenoid does not pass or block vacuum as indicated. Solenoid housing leaks ▶	REPLACE solenoid. Retest.

Barometric Pressure Switch — Pinpoint Test 15

TEST STEP	RESULT ▶	ACTION TO TAKE
15.0 SERVICE CODE 6-2		
• Make sure the Barometric Pressure switch connector is properly mated with the wiring harness connector. If it is not, connect and retest.	R is 10 ohms or less above 4,000 feet. ▶	Continue to 15.1
• Determine local altitude. If below 3,500 feet, the Barometric Pressure switch is normally open; if above 4,000 feet, the switch is normally closed; and if between 3,500 and 4,000 feet, do not test altitude switch the switch may be in either condition.	R is more than 10 ohms, above 4,000 feet. ▶	Replace switch and retest.
• Disconnect the Barometric Pressure Switch from the vehicle harness. • Connect an ohmmeter across the switch terminals.	R is more than 1,000 ohms, below 3,500 feet. ▶	Continue to 15.1
	R is 1000 ohms or less, below 3,500 feet. ▶	Replace switch and retest.
15.1		
• Disconnect the vehicle harness from the MCU Module.	R is 10 ohms or less. ▶	Replace MCU module and retest.
• Check circuit 60, Pin No. 14 and Pin No. 6 of the MCU module connector, for continuity.	R is more than 10 ohms. ▶	Service or replace circuit. Retest.

Schematics—Electrical

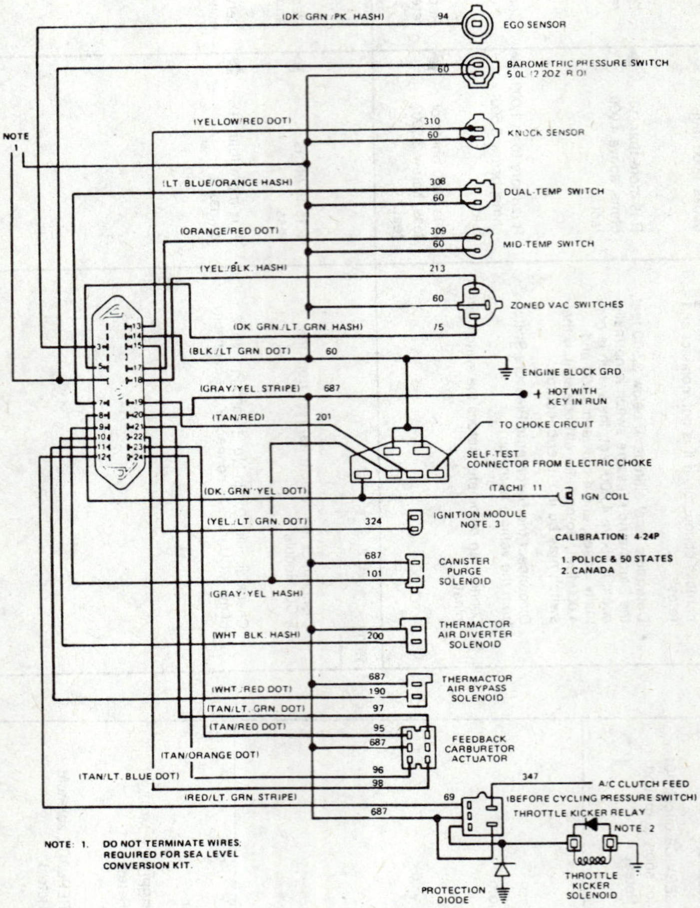

Electronic Diagnosis

1986 FORD CARS & TRUCKS
EEC-IV SYSTEMS

4

When To Use This Diagnostic Procedure

This diagnostic procedure is only used on Electronic Engine Control (EEC-IV) equipped vehicles when the Diagnostic Routines, Section 2 instruct you to perform EEC-IV Diagnostics. It is divided into two test formats. **Quick Test**, a functional EEC-IV system test, and **Pinpoint Tests**, a number of specific component tests.

To test and service the EEC-IV subsystem, perform the Quick Test first. **If the vehicle passes all three phases of the Quick Test: Key On, Engine Off, Engine Running, and Continuous Testing without running any Pinpoint Tests, you will know the EEC-IV subsystem is OK and the vehicle's problem exists somewhere else other than in the EEC-IV subsystem.** However, if a Step of the Quick Test fails, perform only the Pinpoint Tests specified by the failed Step. Do not begin any Pinpoint Test without following the instructions at the beginning of the Section.

After all tests and services have been completed, repeat the entire Quick Test to make sure the EEC-IV subsystem works properly.

Quick Test Description

The Quick Test is a functional test of the EEC-IV system consisting of basic Test Steps (described below). These Steps must be carefully followed in sequence, otherwise mis-diagnosis, or the replacement of non-faulty components may result.

Quick Test Steps

1. Visual Check and Vehicle Preparation
   - Checks for obvious faults.
   - Properly prepares the vehicle for testing.
2. Equipment Hook-up
   - Ensures that the proper equipment for gathering test data is ready, prior to testing.
3. Key On, Engine Off Self-Test
   - Is a static check of processor inputs and outputs.
4. Computed Timing Check
   - Verifies the system ability to compute and maintain a fixed spark timing during Self-Test.
5. Engine Running Self-Test
   - Is a dynamic check with the engine in operation.
6. Continuous Self-Test
   - Checks the sensor inputs for opens and shorts while the vehicle is in operation.

The Key On, Engine Off and Engine Running Self-Tests are intended to detect faults present at the time of testing only, not intermittent faults. Intermittent faults are detected by Continuous Self-Test.

A continuous memory code is information collected and stored from an intermittant fault detected in a previous drive-cycle, the fault may not be present at this time.

WARNING: Anyone who departs from the instruction provided in this publication must first establish that he compromises neither his personal safety nor the vehicle integrity by his choice of methods, tools or parts.

QUICK TEST: Visual Check And Vehicle Preparation — 1.0

Correct test results for the Quick Test are dependent on the proper operation of related non-EEC components systems. It may be necessary to correct faults in these areas before EEC will pass Quick Test. Refer to Diagnostic Routines, Section 2 for service.

Before hooking up any equipment to diagnose the EEC system, make the following checks:

1. Verify the condition of air cleaner and ducting. These components may be removed and reinstalled as necessary for service and/or inspection.
2. Check all engine vacuum hoses for:
   - Leaks or pinched hoses.
   - Proper emission routing per Vehicle Emission Control Information (VECI) Decal.
3. Check the EEC system wiring harness electrical connections for:
   - Proper connections.
   - Loose or detached connectors, wires and terminals.
   - Corrosion.
   - Proper routing of harness.

It may be necessary to disconnect or disassemble the connector assembly to perform some of the inspections. (Note the location of each pin before disassembly.)

4. Check processor, sensors and actuators for physical damage.
5. Perform all safety steps required to start and run operational vehicle tests.
6. Apply the emergency brake. Place shift lever in Park; Neutral for manual transmission.
7. **Turn Off all electrical loads** such as the radio, lamps, air conditioner, etc. Be sure doors are closed whenever readings are made.
8. Verify engine coolant is at the specified level.
9. Start engine and idle until the upper radiator hose is hot and pressurized and the throttle is off fast idle. While the engine is operating, check for leaks around the exhaust manifold, EGO sensor, and vacuum hose connections.

   NOTE: **If engine will not start, starts but stalls, idles rough or runs rough continue through Quick Test Step 3.0. If any of the above conditions are still present after a pass code 11 in the Key On, Engine Off Self-Test, go to Pinpoint Test Step A2 for no start or diagnostics by symptom in the proper Engine Supplement Section for stalls or roughness.**

10. Turn ignition Key Off.
11. Service items as required, and proceed to equipment hook-up, Step 2.0

QUICK TEST: Equipment Hook-Up — 2.0

Using the STAR tester:

- Turn the ignition key Off.
- Connect the color-coded adapter cable leads to the STAR tester.
- Connect the adapter cable's two service connectors to the vehicle's appropriate Self-Test connectors.
- Connect the timing light.

After equipment hook-up, GO to Quick Test Step 3.0

Using analog voltmeter:

- Turn ignition key Off.
- Connect a jumper wire from Self-Test Input (STI) to Pin 2, Signal Return on the Self-Test connector (refer to the diagram below).
- Set analog VOM on a DC voltage range to read from 0 to 15 volts DC. Connect VOM from battery (+) to Pin 4 Self-Test Output (STO) in the Self-Test connector.
- Connect the timing light.

After equipment hook-up, GO to Quick Test Step 3.0

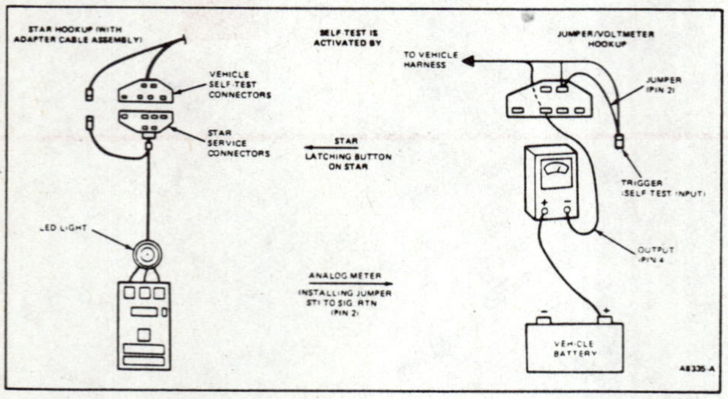

QUICK TEST: Key On, Engine Off Self-Test — 3.0

A CODE OUTPUT

Correct test results for the QUICK TEST are dependent on the proper operation of related non-EEC-IV components/systems. It may be necessary to correct any faults in these areas before EEC-IV will pass the QUICK TEST. Refer to the diagnostic routines, section 2 for proper servicing.

- Verify that the vehicle has been properly prepared per QUICK TEST STEPS 1.0 and 2.0.
- Place transmission in neutral or park. Set parking brake. Place A/C or heater control to off. Place octane switch in premium position on 2.3L SVO TURBO.
- Activate SELF-TEST.
- Place key in run position to start SELF-TEST.

NOTE: Do not depress throttle during SELF-TEST.

- Observe and record all service codes indicated. One of the following outputs will occur:

NOTE: Continuous memory codes recorded in this step will be used for diagnosis later in QUICK TEST Step 6.0!

RESULTS			▶ ACTION TO TAKE
Code Format			
Key On Engine Off	Separator	Continuous Memory	
11	1(0)	11 ▶	KEY ON, ENGINE OFF and CONTINUOUS MEMORY indicate a pass. Go To QUICK TEST STEP 4.0. If vehicle is a no start, go to Pinpoint Test Step A2
ANY CODE(S)	1(0)	11 ▶	KEY ON, ENGINE OFF TEST indicates a fault. Record codes and Go To QUICK TEST STEP 3.0B.
ANY CODE(S)	1(0)	ANY CODE(S) ▶	KEY ON, ENGINE OFF and CONTINUOUS MEMORY indicate a fault. Record ALL codes. DO NOT REPAIR CONTINUOUS MEMORY CODES AT THIS TIME! KEY ON, ENGINE OFF and ENGINE RUNNING codes must be repaired first. Go To QUICK TEST STEP 3.0B.
11	1(0)	ANY CODE(S) EXCEPT 15 ▶	CONTINUOUS MEMORY indicates a fault. DO NOT REPAIR THESE CODES AT THIS TIME! KEY ON, ENGINE OFF and ENGINE RUNNING codes must be repaired first. Go To QUICK TEST STEP 4.0.
11	1(0)	15 ▶	Go To Pinpoint Test Step Q10
NO CODES OUTPUTTED			Repeat SELF-TEST and verify that no service codes are present. Go To Pinpoint Test Step Q1

*Star tester displays a 10 in place of a 1 for the single digit separator code, therefore the separator code is expressed in the results column as 1(0)

QUICK TEST: Key On, Engine Off Self-Test — 3.0

B RESULTS AND ACTION TO TAKE

- Using the KEY ON, ENGINE OFF service codes from QUICK TEST step 3.0A, follow the instructions in the ACTION TO TAKE column in this step.
- When more than one service code is received always start with the first code received.
- Whenever a repair is made REPEAT QUICK TEST!

NOTE: Before proceeding to the specified Pinpoint Test, read the instructions on how to use the Pinpoint Tests at the beginning of the Pinpoint Test section.

PASSENGER CAR

RESULTS ▶	ACTION TO TAKE								
				2.3L CFI ••					
Key On, Engine Off Service Codes	1.9L EFI	2.3L EFI TURBO	2.3L OHC FBC	BASE	PLUS	2.5L CFI	3.0L EFI	3.8L CFI	5.0L SEFI
15 GO to Pinpoint Step ▶	Q12	Q12	Q12	Q12	Q12	Q12	Q12	Q12	Q12
19 GO to Pinpoint Step ▶	•	•	•	•	•	•	Q50	•	Q50
21 GO to Pinpoint Step ▶	DE1	DE1	DE1	DE1	DE1	DE1	DE1	DE1	DE1
22 GO to Pinpoint Step ▶	DF1	DF1	DF1	DF1	DF1	DF1	DF1	DF1	DF1
23 GO to Pinpoint Step ▶	DH1	DH1	DH1	DH1	DH1	DH1	DH1	DH1	DH1
24 GO to Pinpoint Step ▶	DA1	DA1	•	DB1	DB1	DB1	DB1	DB1	DB1
26 GO to Pinpoint Step ▶	DK1	DK1	•	•	•	•	•	•	•
31 GO to Pinpoint Step ▶	•	•	•	DN1	DM1	DN1	DL1	DD2	DN1
32 GO to Pinpoint Step ▶	•	•	•	DN30	•	DN30	•	•	DN30
34 GO to Pinpoint Step ▶	•	•	•	DN35	•	DN35	DL8	•	DN35
35 GO to Pinpoint Step ▶	•	•	•	DN5	•	DN5	DL5	•	DN5
51 GO to Pinpoint Step ▶	DE10	DE10	DE10	DE10	DE10	DE10	DE10	DE10	DE10
52 GO to Pinpoint Step ▶	•	•	•	FF1	FF1	FF1	FF1	•	•
53 GO to Pinpoint Step ▶	DH3	DH3	DH3	DH3	DH3	DH3	DH3	DH3	DH3
54 GO to Pinpoint Step ▶	DA10	DA10		DB10	DB10	DB10	DB10	DB10	DB10
56 GO to Pinpoint Step ▶	DK10	DK10	•	•	•	•	•	•	•
61 GO to Pinpoint Step ▶	DE20	DE20	DE20	DE20	DE20	DE20	DE20	DE20	DE20
62 GO to Pinpoint Step ▶	•	•	•	•	•	•	T61	•	•
63 GO to Pinpoint Step ▶	DH10	DH10	DH10	DH10	DH10	DH10	DH10	DH10	DH10
64 GO to Pinpoint Step ▶	DA20	DA20	•	DB20	DB20	DB20	DB20	DB20	DB20
66 GO to Pinpoint Step ▶	DK20	DK20	•	•	•	•	•	•	•
67 GO to Pinpoint Step ▶	FA1	FA1	FA1	FA1	FA1	FA1	T81	FA1	FA1
68 GO to Pinpoint Step ▶	•	•	KF1	KF1	KF1	KF1	•	KF1	•
81 GO to Pinpoint Step ▶	•	•	•	•	•	•	•	KC8	KC8
82 GO to Pinpoint Step ▶	•	•	•	•	•	•	•	KC8	KC8
83 GO to Pinpoint Step ▶	•	•	•	•	•	X30	X30	DD17	•
84 GO to Pinpoint Step ▶	•	•	•	DN10	DM17	DN10	DL11	DD17	DN10
85 GO to Pinpoint Step ▶	•	•	•	KD4	KD4	KD4	KD4	•	KD4
87 GO to Pinpoint Step ▶	•	•	•	J7	J7	X15	X15	J7	J7
89 GO to Pinpoint Step ▶	•	•	•	•	•	•	T51	KP5	•
No Codes Code Not Listed	GO to Pinpoint Test Step Q1								

••There are two unique Self-Test strategies for 2.3L CFI Tempo/Topaz vehicles, therefore special attention must be given in identifying "base" engine vs "plus" engine on these vehicles to ensure usage of the correct pinpoint test.

QUICK TEST: Key On, Engine Off Self-Test — 3.0

LIGHT TRUCK

RESULTS ▶	ACTION TO TAKE					
Key On, Engine Off Service Codes	2.3L EFI	2.8L FBC	2.9L EFI	3.0L EFI	4.9L FBC	5.0L EFI
15 GO to Pinpoint Step ▶	Q12	Q12	Q12	Q12	Q12	Q12
19 GO to Pinpoint Step ▶	•	•	Q50	Q50	•	•
21 GO to Pinpoint Step ▶	DE1	DE1	DE1	DE1	DE1	DE1
22 GO to Pinpoint Step ▶	DF1	DF1	DF1	DF1	DF1	DF1
23 GO to Pinpoint Step ▶	DH1	DH1	DH1	DH1	DH1	DH1
24 GO to Pinpoint Step ▶	DB1	DB1	DB1	DB1	•	DB1
31 GO to Pinpoint Step ▶	DD2	DD2	DL1	•	•	DD2
34 GO to Pinpoint Step ▶	•	•	DL8	•	•	•
35 GO to Pinpoint Step ▶	•	•	DL5	•	•	•
51 GO to Pinpoint Step ▶	DE10	DE10	DE10	DE10	DE10	DE10
52 GO to Pinpoint Step ▶	FF1	•	•	FF1	•	•
53 GO to Pinpoint Step ▶	DH3	DH3	DH3	DH3	DH3	DH3
54 GO to Pinpoint Step ▶	DB10	DB10	DB10	DB10	•	DB10
61 GO to Pinpoint Step ▶	DE20	DE20	DE20	DE20	DE20	DE20
63 GO to Pinpoint Step ▶	DH10	DH10	DH10	DH10	DH10	DH10
64 GO to Pinpoint Step ▶	DB20	DB20	DB20	DB20	•	DB20
67 GO to Pinpoint Step ▶	FA1	FA1	FA1	FA1	FA1	FA1
68 GO to Pinpoint Step ▶	•	KF1	•	•	KF1	•
81 GO to Pinpoint Step ▶	•	KC8	•	•	•	KC8
82 GO to Pinpoint Step ▶	•	KC8	****	•	•	KC8
83 GO to Pinpoint Step ▶	DD17	DD17	•	•	•	DD17
84 GO to Pinpoint Step ▶	DD17	DD17	DL11	•	•	DD17
85 GO to Pinpoint Step ▶	•	KD4	•	KD4	•	•
87 GO to Pinpoint Step ▶	J7	KK5	J7	J7	•	J7
88 GO to Pinpoint Step ▶	•	KJ6	•	•	•	•
89 GO to Pinpoint Step ▶	KR1	KR1	KR1	KR1	•	•
No Codes Code Not Listed	GO to Pinpoint Test Step Q1					

****Ignore Code 82.

QUICK TEST: TIMING CHECK — 4.0

TEST STEP	RESULT ▶	ACTION TO TAKE
• If engine is a no start, go directly to Pinpoint Test Step A1 NO START. • If engine starts but stalls go directly to DIAGNOSTIC BY SYMPTOM in the proper ENGINE SUPPLEMENT SECTION. • Key off, wait 10 seconds. • Verify SELF-TEST trigger has been activated. • Restart engine and check timing while in SELF-TEST. You have 2 minutes from the time the last code is outputted to check SELF-TEST timing. • Correct SELF-TEST timing equals base plus 20° BTDC ± 3° (see VECI decal for correct base timing). EXAMPLE: If base timing is 10° BTDC then SELF-TEST timing is: 10° + 20° = 30° ± 3° therefore SELF-TEST timing is 27°-33° BTDC.	Engine stalls during timing check ▶	Go To DIAGNOSTIC BY SYMPTOM in the proper ENGINE SUPPLEMENT SECTION
	Timing is not base + 20° BTDC ▶	Go To Pinpoint Test Step P1
	Timing is base plus 20° BTDC ▶	Go To QUICK TEST Step 5.0

QUICK TEST: Engine Running Self-Test — 5.0

A CODE OUTPUT

NOTE: If engine is a no start go directly to Pinpoint Test Step A1. If engine stalls go directly to Diagnostic By Symptom in the proper engine supplement section.

- Deactivate SELF-TEST.
- Start and run engine at greater than 2000 RPM for 2 minutes. This warms up the EGO sensor.
- Turn engine off, wait 10 seconds.
- Activate SELF-TEST.
- **Do not depress throttle during Self-Test unless a dynamic response code occurs.**
- Start engine. The ENGINE RUNNING SELF-TEST will progress as follows:
  - — Engine ID code. 1
    Depress and release the brake pedal, only once, after the ID code has been received. This tests the Brake on off switch on 2.8L and 2.3L EFI trucks.
  - — Run test.
  - — If a dynamic response code occurs, perform a brief WOT 2
  - — ENGINE RUNNING service codes.
- Observe and record all service codes. One of the following outputs will occur.

NOTE: If engine stalls during Self-Test, go directly to Diagnostic By Symptom in the proper engine supplement section.

RESULTS — Code Format: Engine ID	Dynamic Response	Engine Running	▶ ACTION TO TAKE
2(0), 3(0) or 4(0)	1(0) or no dynamic response pulse	11	ENGINE RUNNING SELF-TEST indicates a pass. If the symptom was of an intermittent nature, Go To QUICK TEST STEP 6.0. If the drive symptom is currently present, Go To DIAGNOSTIC BY SYMPTOM in the proper ENGINE SUPPLEMENT SECTION. Otherwise testing is complete. EEC-IV system is OK.
2(0), 3(0) or 4(0)	1(0) or no dynamic response pulse	ANY CODES	ENGINE RUNNING SELF-TEST indicates a fault. Go To QUICK TEST STEP 5.0B
98	No dynamic response pulse	ANY CODES	Rerun QUICK TEST STEP 3.0A and ensure a pass code 11 for KEY ON, ENGINE OFF portion.
NO CODES OUTPUTTED			Repeat SELF-TEST and verify that no service codes are present then Go To Pinpoint Test Step Q1.

1 Engine ID code is equal to half the number of cylinders (a code of 2 equals a 4 cylinder engine). STAR tester adds a zero to all single digit readings, so a 20 code equals a 4 cylinder engine). For 5.0L SEFI a service code 98 may appear instead of the engine ID code indicating the need to ensure a pass code 11 for the KEY ON, ENGINE OFF portion of Quick Test 3.0A.

2 Star displays a 10 in place of a 1 for the single digit separator code, therefore, the dynamic response code is expressed in the results column as 1(0). Some vehicles do not require a brief wot and therefore will not display a 1(0).

QUICK TEST: Engine Running Self-Test — 5.0

B RESULTS AND ACTION TO TAKE

- Using the ENGINE RUNNING service codes from QUICK TEST STEP 5.0A, follow the instructions in the ACTION TO TAKE column in this step.
- When more than one service code is received, always start service with the first code received.
- Whenever a repair is made, REPEAT QUICK TEST!

PASSENGER CAR

RESULTS ▶ Engine Running Service Codes	1.9L EFI	2.3L EFI TURBO	2.3L OHC FBC	2.3L CFI ** BASE	2.3L CFI ** PLUS	2.5L CFI	3.0L EFI	3.8L CFI	5.0L SEFI
12 GO to Pinpoint Step ▶	KE1	KE1	KF16	KF16	KF16	KF16	KE1	KF16	KE1
13 GO to Pinpoint Step ▶	KE3	KE10	KF19	KF19	KF19	KF19	KE10	KF19	KE10
16 GO to Pinpoint Step ▶	·	·	HB21	KF19	·	KF19	KE1	KF19	KE1
17 GO to Pinpoint Step ▶	KE11	·	·	·	·	·	·	·	·
21 GO to Pinpoint Step ▶	DE1	DE1	DE1	DE1	DE1	DE1	DE1	DE1	DE1
22 GO to Pinpoint Step ▶	DF1	DF1	DF7	DF7	DF7	DF7	DF7	DF7	DF7
23 GO to Pinpoint Step ▶	DH1	DH1	DH1	DH1	DH1	DH1	DH1	DH1	DH1
24 GO to Pinpoint Step ▶	DA1	DA1	·	DB1	DB1	DB1	DB1	DB1	DB1
25 GO to Pinpoint Step ▶	·	DG1	·	·	·	·	DG1	·	·
26 GO to Pinpoint Step ▶	DK1	DK1	·	·	·	·	·	·	·
31 GO to Pinpoint Step ▶	·	·	·	DN1	DM1	DN1	DL21	DD1	DN1
32 GO to Pinpoint Step ▶	·	·	·	DN30	DM11	DN30	DL20	DD11	DN30
33 GO to Pinpoint Step ▶	·	·	·	DN40	DM11	DN40	DL30	DD11	DN40
34 GO to Pinpoint Step ▶	KA1	KA1	·	DN35	DM11	DN35	DL25	DD11	DN35
35 GO to Pinpoint Step ▶	·	·	·	DN5	DM30	DN5	DL25	DD30	DN5
41 GO to Pinpoint Step ▶	HA11	HA11	HB1	HF5	HF5	HF5	HG10	HD7	HC1
42 GO to Pinpoint Step ▶	HA8	HA8	HB20	HF9	HF9	HF9	HG7	HD14	HC15
43 GO to Pinpoint Step ▶	·	·	HB40	·	·	·	·	·	·
44 GO to Pinpoint Step ▶	·	·	KC1	·	·	·	·	KC1	KC1
45 GO to Pinpoint Step ▶	·	·	KC1	·	·	·	·	KC1	KC1
46 GO to Pinpoint Step ▶	·	·	KC1	·	·	·	·	KC1	KC1
55 GO to Pinpoint Step ▶	·	·	·	FC3	·	FC3	·	HD1	·
58 GO to Pinpoint Step ▶	·	·	KF12	KF12	KF12	KF12	·	KF12	·
62 GO to Pinpoint Step ▶	·	·	·	·	·	·	T71	·	·
72 GO to Pinpoint Step ▶	·	·	DF10	DF10	DF10	DF10	DF10	DF10	·
73 GO to Pinpoint Step ▶	DH20	DH20	DH20	·	DH20	·	DH20	·	·
74 GO to Pinpoint Step ▶	·	·	·	·	·	FD1	FD1	·	·
75 GO to Pinpoint Step ▶	·	·	·	·	·	FD4	FD4	·	·
76 GO to Pinpoint Step ▶	DK30	DK30	·	·	·	·	·	·	·
77 GO to Pinpoint Step ▶	M1	M1	M1	·	M1	·	M1	·	·
91 GO to Pinpoint Step ▶	·	·	·	·	·	·	·	HD7	HC1
92 GO to Pinpoint Step ▶	·	·	·	·	·	·	·	HD14	HC15
94 GO to Pinpoint Step ▶	·	·	·	·	·	·	·	KC1	KC1
98 GO to Pinpoint Step ▶	·	·	·	·	·	·	·	·	****
99 GO to Pinpoint Step ▶	ISC HAS NOT LEARNED YET. RERUN QUICK TEST STEPS 3.0 THRU 5.0								
No Codes Code Not Listed	GO to Pinpoint Test Step Q1								

****Service Code 98 may appear indicating the need to ensure a pass code 11 for the Key On Engine Off portion of Quick Test Step 3.0A

**There are two unique self test strategies for 2.3L CFI Tempo Topaz vehicles. Therefore special attention must be given in identifying "base" engine vs. "plus" engine on these vehicles to ensure usage of the correct Pinpoint Test.

QUICK TEST: Engine Running Self-Test — 5.0

LIGHT TRUCK

RESULTS ▶ Engine Running Service Codes	2.3L EFI	2.8L FBC	2.9L EFI	3.0L EFI	4.9L FBC	5.0L EFI
12 GO to Pinpoint Step ▶	KE1	KF16	KE1	KE1	KF16	KE1
13 GO to Pinpoint Step ▶	KE10	KF19	KE10	KE10	KF19	KE10
16 GO to Pinpoint Step ▶	KE1	HB21	·	KE1	HB21	KE1
21 GO to Pinpoint Step ▶	DE1	DE1	DE1	DE1	DE1	DE1
22 GO to Pinpoint Step ▶	DF7	DF7	DF7	DF7	DF7	DF7
23 GO to Pinpoint Step ▶	DH1	DH1	DH1	DH1	DH1	DH1
24 GO to Pinpoint Step ▶	DB1	DB1	DB1	DB1	·	DB1
25 GO to Pinpoint Step ▶	DG1	DG1	DG1	DG1	·	DG1
31 GO to Pinpoint Step ▶	DD1	DD1	DL21	·	·	DD1
32 GO to Pinpoint Step ▶	DD11	DD11	DL20	·	·	DD11
33 GO to Pinpoint Step ▶	DD11	DD11	DL30	·	·	DD11
34 GO to Pinpoint Step ▶	DD11	DD11	DL25	·	KA1	DD11
35 GO to Pinpoint Step ▶	DD30	DD30	DL25	·	·	DD30
41 GO to Pinpoint Step ▶	HE11	HB1	HG10	HG10	HB1	HE11
42 GO to Pinpoint Step ▶	HE8	HB20	HG7	HG7	HB20	HE8
43 GO to Pinpoint Step ▶	·	HB40	·	·	HB40	·
44 GO to Pinpoint Step ▶	·	KC1	·	·	KC1	KC1
45 GO to Pinpoint Step ▶	·	KC1	·	·	KC1	KC1
46 GO to Pinpoint Step ▶	·	KC1	·	·	KC1	KC1
55 GO to Pinpoint Step ▶	·	·	FC3	·	·	·
58 GO to Pinpoint Step ▶	·	KF12	·	·	KF12	·
65 GO to Pinpoint Step ▶	·	FC1	FC1	·	·	·
72 GO to Pinpoint Step ▶	DF10	DF10	DF10	DF10	DF10	DF10
73 GO to Pinpoint Step ▶	DH20	DH20	DH20	DH20	DH20	DH20
74 GO to Pinpoint Step ▶	FD1	FD1	FD1	FD1	·	·
75 GO to Pinpoint Step ▶	FD4	FD4	FD4	FD4	·	·
77 GO to Pinpoint Step ▶	M1	M1	M1	M1	M1	M1
No Codes Code Not Listed	GO to Pinpoint Test Step Q1					

QUICK TEST: Continuous Self-Test — 6.0

TEST STEP	RESULT ▶	ACTION TO TAKE
A — CONTINUOUS MEMORY CODES		
• To ensure proper diagnosis of CONTINUOUS MEMORY codes, QUICK TEST STEPS 1.0 through 5.0 must be successfully completed. • Do the results of both KEY ON, ENGINE OFF and ENGINE RUNNING SELF-TEST result in a pass code 11?	Yes ▶	Go To 6.0B
	No ▶	Return to Step 1.0 of QUICK TEST and make the necessary repairs indicated in QUICK TEST STEPS 3.0 and 5.0 before continuing.
B — CLEAR CONTINUOUS MEMORY CODES		
• Prepare the vehicle for KEY ON, ENGINE OFF QUICK TEST per steps 1.0, 2.0, and 3.0. • Perform KEY ON, ENGINE OFF SELF-TEST. When the service codes begin, exit the self-test program (via unlatching STAR or removing the jumper wire from STI to Signal Return). Exiting QUICK TEST during code output will clear all codes stored in the CONTINUOUS MEMORY. • Repeat KEY ON, ENGINE OFF SELF-TEST. • Does KEY ON, ENGINE OFF SELF-TEST output 11-10-11?	Yes ▶	Go To 6.0C
	No ▶	Check STI circuit for short to ground. Repair as necessary. Repeat Step 6.0B.
C — DETERMINE CODES TO BE TESTED		
• Refer to CONTINUOUS MEMORY CODES recorded in QUICK TEST STEP 3.0A. • Have any of the codes been repaired either in KEY ON, ENGINE OFF or ENGINE RUNNING SELF-TESTS?	Yes ▶	Disregard codes already serviced. For remaining codes, Go To 6.0D.
	No ▶	Go To 6.0D

QUICK TEST: Continuous Self-Test — 6.0

LIGHT TRUCK

RESULTS ▶ Continuous Memory Service Codes	ACTION TO TAKE 2.3L EFI	2.8L FBC	2.9L EFI	3.0L EFI	4.9L FBC	5.0L EFI
14 GO to Pinpoint Step ▶	Y1	Y1	Y1	Y1	Y1	Y1
18 GO to Pinpoint Step ▶	N1	N1	N1	N1	N1	N1
21 GO to Pinpoint Step ▶	DE90	DE90	DE90	•	DE90	DE90
22 GO to Pinpoint Step ▶	DF90	DF90	DF90	DF90	DF90	DF90
31 GO to Pinpoint Step ▶	DD90	DD90	DL90	•	•	DD90
32 GO to Pinpoint Step ▶	•	•	DL94	•	•	•
33 GO to Pinpoint Step ▶	•	•	DL97	•	•	•
34 GO to Pinpoint Step ▶	•	•	DL93	•	•	•
35 GO to Pinpoint Step ▶	•	•	DL90	•	•	•
41 GO to Pinpoint Step ▶	•	HB1	•	•	HB1	•
42 GO to Pinpoint Step ▶	•	HB20	•	•	HB20	•
51 GO to Pinpoint Step ▶	DE91	DE91	•	DE91	DE91	DE91
53 GO to Pinpoint Step ▶	•	DH90	DH90	DH90	DH90	DH90
54 GO to Pinpoint Step ▶	DB90	DB90	DB90	DB90	•	DB90
61 GO to Pinpoint Step ▶	DE94	DE94	DE94	DE94	DE94	DE94
63 GO to Pinpoint Step ▶	DH94	DH94	DH94	DH94	DH94	DH94
64 GO to Pinpoint Step ▶	DB93	DB93	DH93	DH93	•	DB93
No Codes Code Not Listed	GO to Pinpoint Test Step Q1					

QUICK TEST: Continuous Self-Test — 6.0

D — TEST RESULTS AND ACTION TO TAKE

- YOU SHOULD NOT BE HERE UNLESS ALL PREVIOUS QUICK TEST STEPS HAVE BEEN SUCCESSFULLY COMPLETED!
- Verify proper test equipment hook up per QUICK TEST STEP 2.0.
- SELF-TEST deactivated. Do not activate SELF-TEST unless instructed.
- Using the service codes determined in STEP 6.0C, follow the action to take instructions. (While in the appropriate PINPOINT TEST, one or both of the continuous monitor modes described below may have to be used to find the intermittent fault.)

CONTINUOUS MONITOR (WIGGLE) MODES

- KEY ON ENGINE OFF
  With SELF-TEST deactivated, turn the ignition key to the ON position. You are now in the continuous monitor mode (wiggle test). STO will be activated whenever a fault is detected. A fault is indicated when the STAR LED turns off or the VOM deflection is 10.5V or greater.
- ENGINE RUNNING
  Activate SELF-TEST and perform ENGINE RUNNING QUICK TEST. After service code output has been completed do not turn engine off or deactivate SELF-TEST. Approximately 2 minutes after code 11 has been outputted the system will enter and remain in continuous monitor mode until SELF-TEST is deactivated or the engine is turned off. STO will be activated whenever a fault is detected. A fault is indicated the same way as KEY ON ENGINE OFF.
- Refer to the appendix for a detailed description of the continuous monitor (wiggle test).

QUICK TEST: Continuous Self-Test — 6.0

PASSENGER CAR

RESULTS ▶ Continuous Memory Service Codes	ACTION TO TAKE 1.9L EFI	2.3L EFI TURBO	2.3L OHC FBC	2.3L CFI** BASE	2.3L CFI** PLUS	2.5L CFI	3.0L EFI	3.8L CFI	5.0L SEFI
13 GO to Pinpoint Step ▶	•	•	•	KF19	KF19	KF19	•	•	•
14 GO to Pinpoint Step ▶	Y1	Y1	Y1	Y1	Y1	Y1	Y1	Y1	Y1
18 GO to Pinpoint Step ▶	N1	N1	N1	N1	N1	N1	N1	N1	N1
21 GO to Pinpoint Step ▶	DE90	DE90	DE90	DE90	DE90	DE90	•	•	•
22 GO to Pinpoint Step ▶	DF90	DF90	DF90	DF90	DF90	DF90	DF90	DF90	DF90
29 GO to Pinpoint Step ▶	•	•	•	•	•	•	T2	•	•
31 GO to Pinpoint Step ▶	•	•	•	DN92	DM90	DN92	DL90	DD90	DN92
32 GO to Pinpoint Step ▶	•	•	•	DN90	•	DN90	DL94	•	DN90
33 GO to Pinpoint Step ▶	•	•	•	DN95	•	DN95	DL97	•	DN95
34 GO to Pinpoint Step ▶	•	•	•	DN98	•	DN98	DL93	•	DN98
35 GO to Pinpoint Step ▶	•	•	•	DN92	•	DN92	DL90	•	DN92
39 GO to Pinpoint Step ▶	•	•	•	•	•	•	T31	•	•
41 GO to Pinpoint Step ▶	HA20	•	HB1	•	•	•	•	•	•
42 GO to Pinpoint Step ▶	HA20	•	HB20	•	•	•	•	•	•
51 GO to Pinpoint Step ▶	DE91	DE91	DE91	DE91	DE91	DE91	DE91	DE91	DE91
53 GO to Pinpoint Step ▶	DH90	DH90	DH90	DH90	DH90	DH90	DH90	DH90	DH90
54 GO to Pinpoint Step ▶	DA90	DA90	•	DB90	DB90	DB90	DB90	DB90	DB90
56 GO to Pinpoint Step ▶	DK90	DK90	•	•	•	•	•	•	•
57 GO to Pinpoint Step ▶	•	•	•	•	•	•	T41	•	•
59 GO to Pinpoint Step ▶	•	•	•	•	•	•	T21	•	•
61 GO to Pinpoint Step ▶	DE94	DE94	DE94	DE94	DE94	DE94	DE94	DE94	DE94
63 GO to Pinpoint Step ▶	DH94	DH94	DH94	DH94	DH94	DH94	DH94	DH94	DH94
64 GO to Pinpoint Step ▶	DA93	DA93	•	DB93	DB93	DB93	DB93	DB93	DB93
66 GO to Pinpoint Step ▶	DK93	DK93	•	•	•	•	•	•	•
69 GO to Pinpoint Step ▶	•	•	•	•	•	•	T11	•	•
78 GO to Pinpoint Step ▶	•	•	•	B10	•	•	•	•	•
No Codes Code Not Listed	GO to Pinpoint Test Step Q1								

**There are two unique Self Test strategies for the 2.3L CFI Tempo Topaz vehicles, therefore special attention must be given in identifying "base" engine vs "plus" engine on these vehicles to ensure usage of the correct Pinpoint Test.

APPENDIX: Self-Test Description

SELF-TEST DESCRIPTION

The Self-Test is divided into three specialized tests: Key On, Engine Off, Engine Running, and Continuous memory. The Self-Test is not a conclusive test by itself, but is used as a part of the functional Quick-Test diagnostic procedure. The processor stores the Self-Test program in its permanent memory. When activated, it checks the EEC-IV system by testing its memory integrity and processing capability, and verifies that various sensors and actuators are connected and operating properly.

The Key On, Engine Off and Engine Running tests are functional tests which only detect faults present at the time of the Self-Test. Continuous testing is an ongoing test that stores fault information for retrieval at a later time (during the Self-Test).

Key On, Engine Off

At this time, a test of the EEC-IV system is conducted with power applied and engine at rest.

For Self-Test to detect errors in this test the fault must be present at the time of testing. For intermittents, refer to Continuous Memory Codes.

Separator Pulse

A single 1/2 second separator pulse is issued 6-9 seconds after the last Functional Test code (Key On, Engine Off only). Then 6-9 seconds after the single 1/2 second separator pulse, the continuous codes will be issued. Refer to the example below:

NOTE: The separator code and continuous memory codes follow Key On, Engine Off testing codes ONLY.

Continuous Memory Codes

Continuous memory codes are issued as a result of information stored during continuous monitor testing, while the vehicle was in normal operation. These codes are displayed only during the Key On, Engine Off testing and after the separator code. These codes should be used for diagnosis only when Key On, Engine Off and Engine Running Self-Tests result in code 11 and all Quick Test steps 1.0 through 5.0 have been successfully completed.

Engine Running Test

At this time, a test of the EEC-IV system is conducted with the engine running. The sensors are checked under actual operating conditions and at normal operating temperatures. The actuators are exercised and checked for corresponding results.

Engine Identification Codes (ID Codes)

Engine ID codes are issued at the beginning of the engine running test and are one-digit numbers represented by the number of pulses sent out. The engine ID code is equal to 1/2 the number of engine cylinder (i.e. 2 pulses = 4 cylinders). These codes are used to verify the proper processor is installed and that the Self-Test has been entered.

Dynamic Response Check

The dynamic response check verifies the movement of the TP, VAF, and MAP sensors during the brief Wide Open Throttle (WOT) performed during the Engine Running Test. The signal for the operator to perform the brief WOT is a single pulse or 10 code on the STAR. Also, during the Engine Running Test, the brake pedal must be depressed and released after the ID code to assure operation of the brake on/off switch for 2.8L and 2.3L EFI light trucks.

APPENDIX: Diagnostic Aids

OUTPUT STATE CHECK

The output state check aids in servicing output actuators associated with the EEC-IV system. It enables the technician to energize and de-energize most of the system output actuators (turn EGR solenoid and Self-Test Output (STO) On and Off) on command. This mode is entered after all codes have been received from Key On, Engine Off and Continuous Testing. At this time, leave Self-Test activated and depress the throttle. Each time the throttle is depressed the output actuators will change state (On to Off or Off to On).

- Enter Self-Test.
- Code Output Ends.
- Do Brief WOT.
- Output Actuators Actuated.
- Next Brief WOT.
- Output Actuators Deactivated.

CYLINDER BALANCE TEST

The cylinder balance test is an aid in servicing the 5.0L SEFI EEC-IV fuel system by energizing and de-energizing each injector separately. This test mode is entered after all service codes have been displayed—following the Engine Running Quick Test. At this time leave the Self-Test activated, and perform a brief Wide Open Throttle (WOT), approximately one second. Total test time is approximately 90 seconds.

APPENDIX: Continuous Self-Test

The Continuous Memory service codes are separated from the Quick Test Key On, Engine Off codes by a single separator pulse, Figure 5.

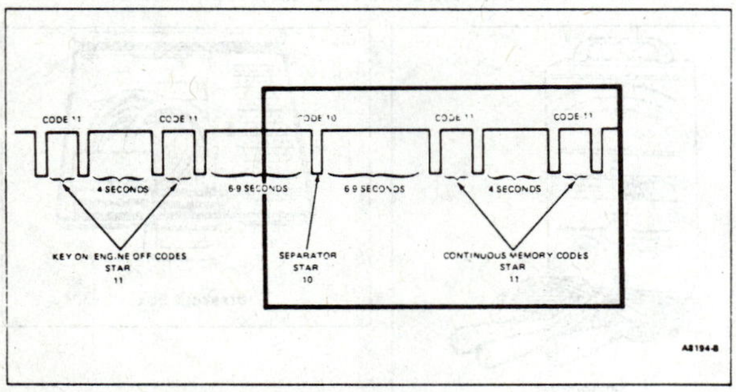

Figure 5 Continuous Memory Self-Test Output Code Format

The Continuous Memory codes should never be used for Diagnosis until the Key On, Engine Off and Engine Running Tests result in a pass code 11.

The service codes that were displayed (repeated) in both the Continuous Memory and Engine Running Test are considered repaired. The unique service code(s) displayed only in Continuous Memory are considered the intermittent service codes.

During this mode of testing we continuously monitor inputs to processor for opens and shorts. The continuous memory codes must be retrieved within twenty engine temperature warm up cycles. On the twenty-first Engine Temperature cycle the service code will be automatically erased. The continuous memory code can also be erased by deactivating Quick Test, while the service codes are being outputted.

Continuous Monitor Test

This test is intended as an aid in diagnosing intermittent failures in the sensor input circuits. The Self-Test output is energized whenever the continuous memory mode senses a fault and de-energized when the system is OK.

- Connect a VOM/STAR to the Self-Test output.

APPENDIX: Continuous Self-Test (Continued)

- **Key On, Engine Off** Continuous Monitor Test. (Wiggle).

  To enter this mode, "verify" Self-Test is not activated and then turn the key to run. Self-Test output will be activated whenever a continuous fault is detected and if the duration is long enough, a fault code will be stored.

- **Engine Running Continuous** Monitor Test. (Wiggle).

  This mode of testing is entered approximately 2 minutes after the service codes from the Engine Running portion of Quick Test (Step 5.0) because the engine running test can be entered only once per ignition cycle. An alternate quick method (eliminates waiting for Self-Test to complete its cycle) is to enter the running test, exit, and re-enter (do not shut the engine off). This will put you in the Engine Running Continuous Monitor (Wiggle) Test.

  The Continuous Monitor (Wiggle) Test will allow the technician to enter this mode of Self-Test and to attempt to re-create the intermittent failure (tap, move, wiggle the harness and/or the suspected sensor).

  If the VOM/STAR indicates a fault (short or open) the corresponding service code will be stored. Now with the knowledge of the affected circuits, a close check of the harness and associated connectors can be made.

NOTE: Remember to keep your eyes on the VOM/STAR for any change which will indicate where the intermittent is located.

EXAMPLE: How to Use the Continuous Monitor Test (Systematically)

If a service code 51 was displayed: Engine Coolant Temperature Sensor:

- Visually inspect the sensor **very closely.**
- Enter Continuous Monitor test.
- Lightly TAP on the sensor.
- Push/pull on the sensor harness connector (do not disconnect, yet).
- Test and Wiggle (shake) harness vigorously working from the sensor connector toward the dash panel and from the dash panel to the processor in short sections.
- If VOM has not given a positive indication of an intermittent, disconnect the sensor from the harness as carefully as possible. Remove terminals from the connector. Visually inspect terminals at both ends for corrosion, bad crimps, improperly seated terminals, etc.
- Reconnect after inspection.
- Disconnect processor from harness as carefully as possible.
- Inspect terminals.
- Only remove terminals associated with the sensor being inspected.
- If the VOM does not give a positive intermittent indication, reconnect the connector and erase the Continuous Test service codes.
- To erase the continuous service codes:
  - —Initiate Key On, Engine Off Self-Test.
  - —Remove jumper from the Self-Test input terminal as soon as the **first** service code is received (even if an "11" is the first code).
  - —Rerun Self-Test with jumper to verify services have been erased.

APPENDIX: Code Output Format

SERVICE CODES

The EEC-IV system communicates service information to the outside world by way of the Self-Test service codes. These service codes are two-digit numbers representing the results of Self-Test.

The service codes are transmitted on the Self-Test output (found in the Self-Test connector) in the form of timed pulses, and read by the technician on a voltmeter or on the STAR tester.

SELF-TEST OUTPUT CODE FORMAT* KEY ON, ENGINE OFF

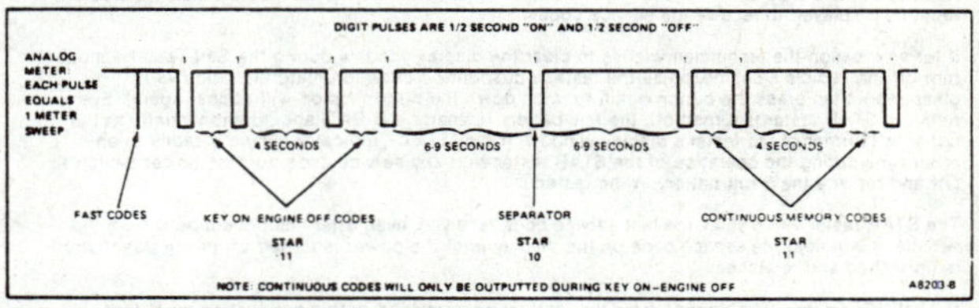

Figure 1 Key On, Engine Off Self-Test Output Code Format

SELF-TEST OUTPUT CODE FORMAT* ENGINE RUNNING

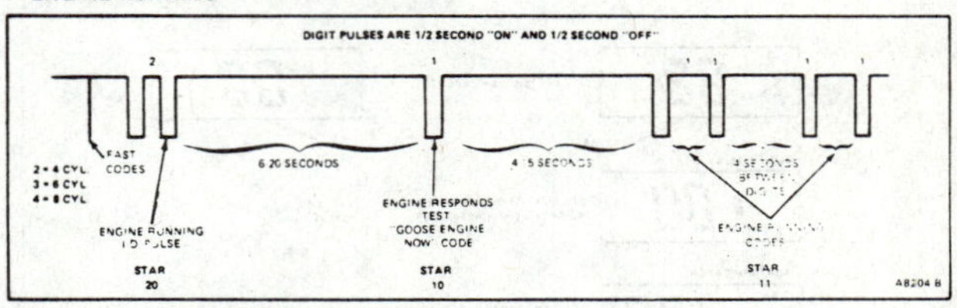

Figure 2 Engine Running Self-Test Output Code Format

Fast Codes

Fast codes are issued prior to regular service codes. These codes contain the identical information as the regular service codes but are transmitted at 100 times the normal rate. These codes are interpreted by special equipment at the Body and Assembly Division. They serve no purpose for the field.

For this reason the STAR has been designed to disregard these codes. Some meters in service detect these codes as a short burst of information (slight meter deflection).

APPENDIX: Self-Test With Analog Voltmeter

Reading Codes—Analog Voltmeter

When a service code is reported on the analog voltmeter for a function test, it will represent itself as a pulsing or sweeping movement of the voltmeter's needle across the dial face of the voltmeter (Figure 3). Therefore, a single-digit number of three will be reported by three needle pulses (sweeps). However, as previously stated, a service code is represented by a two-digit number, such as 2-3. As a result, the Self-Tests service code of 2-3 will appear on the voltmeter as two needle pulses (sweeps), then, after a two-second pause, the needle will pulse (sweep) three times.

The continuous memory codes are separated from the functional codes by a six-second delay, a single half-second sweep, and another six-second delay. They are produced on the voltmeter in the same manner as the functional codes.

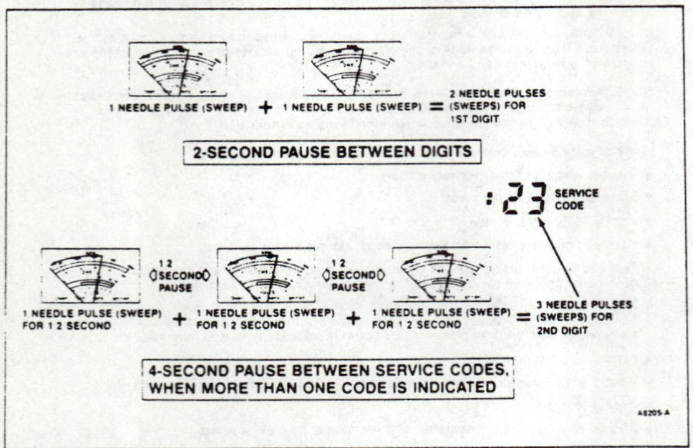

Figure 3 Analog Voltmeter Functional Service Code

APPENDIX: Self-Test With STAR Tester

Reading Codes—Self-Test Automatic Readout (STAR) Rotunda Model 007-00004 or Equivalent

After hooking up the STAR tester and turning on its power switch, the tester will run a display check and the numerals 88 will begin to flash in the display window (Figure 4). A steady 00 will then appear to signify that the STAR tester is ready to start the Self-Test and receive the test's service codes.

To receive the service codes, press the pushbutton at the front of the STAR tester. The button will latch down, and a colon will appear in the display window in front of the 00 numerals. The colon **must** be displayed to receive the service codes.

If for any reason the technician wishes to clear the display window during the Self-Test, he must turn off the vehicle's engine, press the tester's pushbutton once to unlatch it (colon will disappear), then press the button again to latch down the button (colon will appear again). Every time the STAR tester is turned off, the low battery indicator (LO BAT) should show briefly at the upper left corner of the tester's display window. If the LO BAT indicator shows steadily at any other time during the operation of the STAR tester with any service code, turn its power switch to Off and replace the 9-volt battery in the tester.

The STAR tester will display the last service code received, even after disconnecting it from the vehicle. It will hold the service code on the display until the power is turned off or the pushbutton is unlatched and relatched.

WARNING: Anyone who departs from the instructions provided in this publication must first establish that he compromises neither his personal safety nor the vehicle integrity by his choice of methods, tools, or parts.

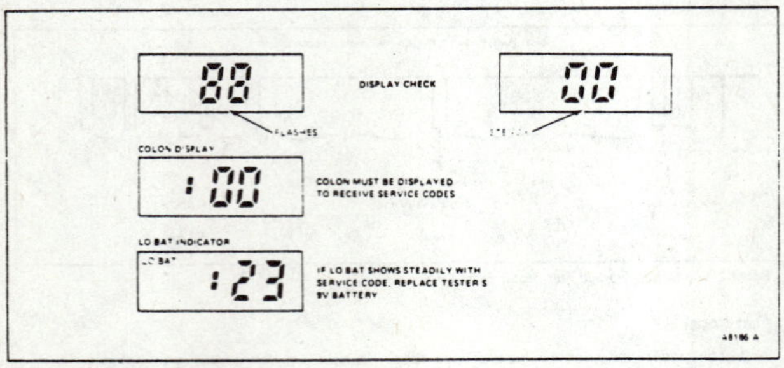

Figure 4 Self-Test Output Code Format

APPENDIX: Test Equipment

EQUIPMENT REQUIRED:

- Rotunda Self-Test Automatic Readout (STAR). No. 007-00004 with cable assembly No. 007-00010. Refer to STAR operation.
- Analog volt-ohmmeter, 0 to 20v DC. (alternate to STAR).

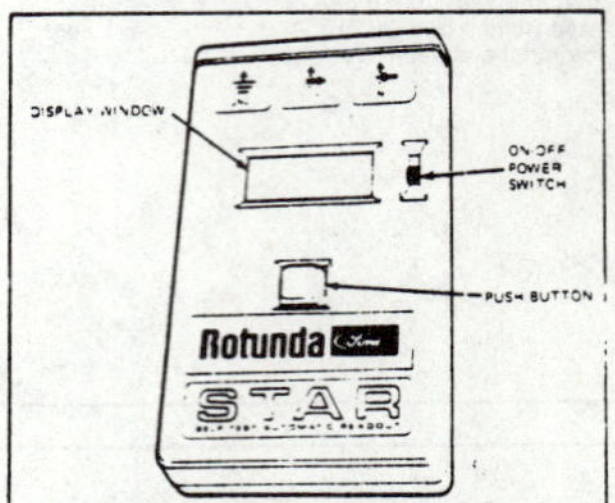

Automatic Readout (STAR) Tester

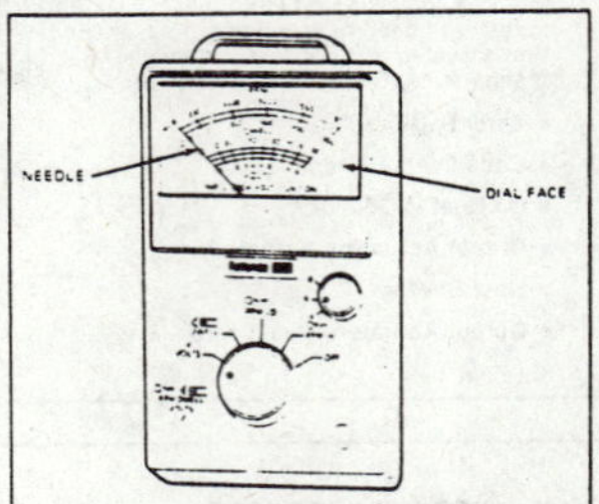

Analog Voltmeter

- Jumper wire.
- Vacuum gauge, Rotunda 059-00008 or equivalent. Range 0-30 in. Hg. Resolution 1 in. Hg.

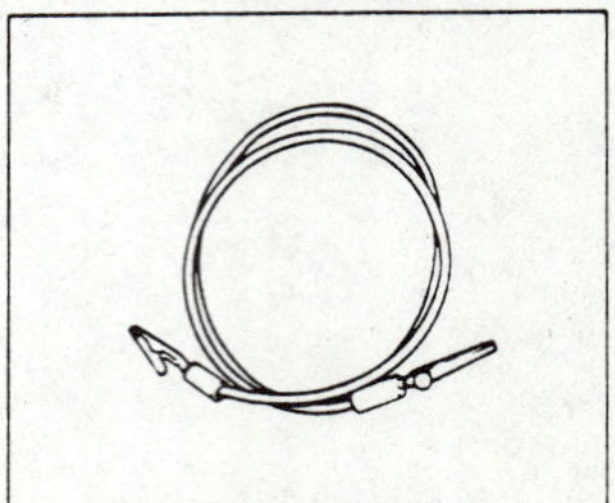

Jumper Wire

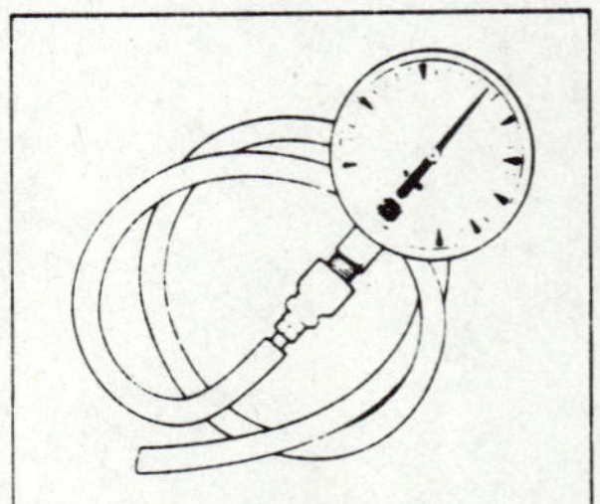

Vacuum Gauge

APPENDIX: Test Equipment (Continued)

- Tachometer. Rotunda No. 059-00010 or equivalent. Range 0-6,000 rpm. Accuracy ± 40 rpm. Resolution 20 rpm.
- Breakout Box. Rotunda 014-00322. Special Service Tool T83L-50-EEC-IV or equivalent.

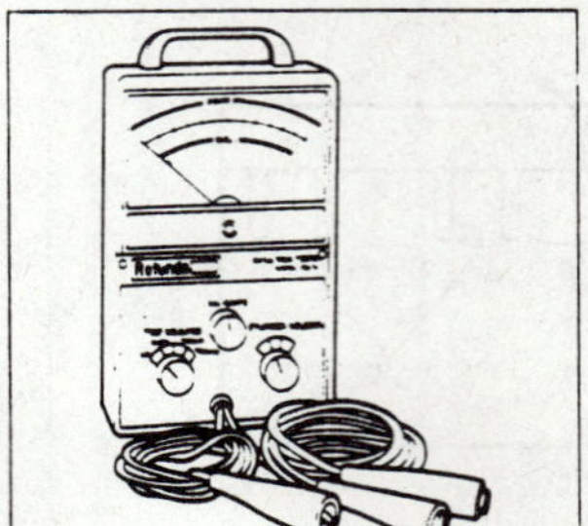

Tachometer

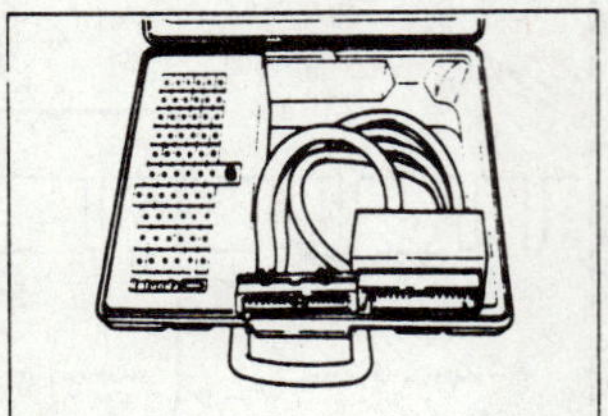

Breakout Box

- Vacuum pump, Rotunda No. 021-00014 or equivalent. Range 0-30 in. Hg.
- Digital volt-ohmmeter, Rotunda No. 014-00407 or equivalent. Input impedance 10 Megaohm minimum.

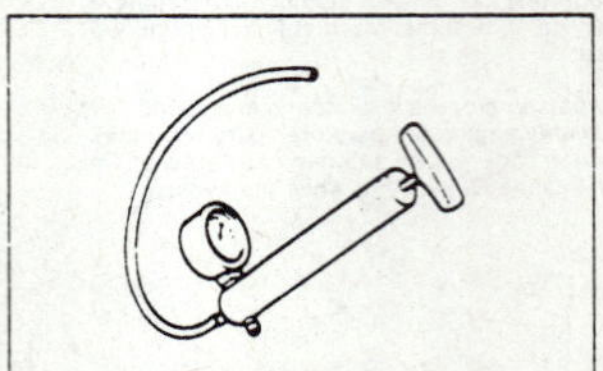

Vacuum Pump

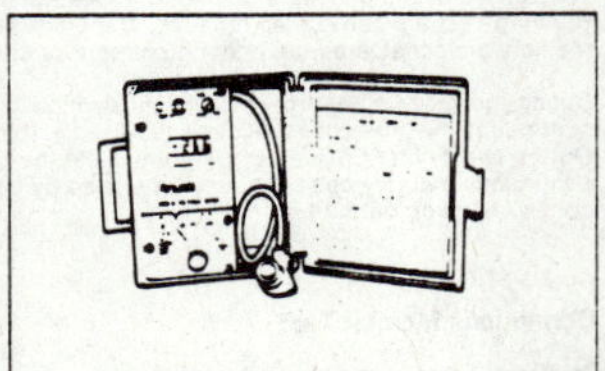

Digital Volt-Ohmmeter

APPENDIX: Test Equipment (Continued)

- Electronic Fuel Injection Pressure Gauge EFI/CFI only. Tool T80L-9974-A or equivalent. (Use instructions. For specific applications, refer to Shop Manual Group 24.)
- Spark tester (optional modified spark plug with side electrode removed). Tool D81P-6666-A or equivalent.

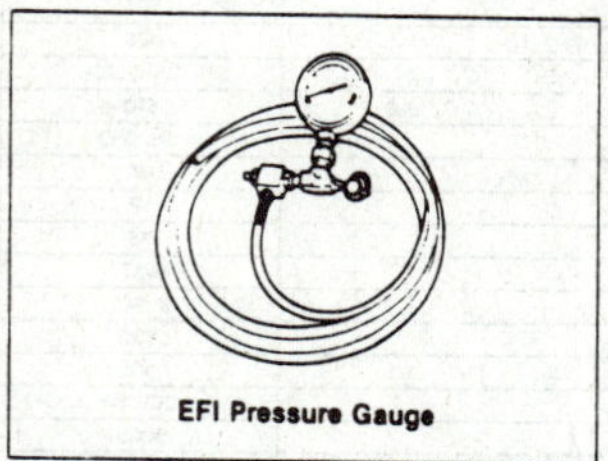

EFI Pressure Gauge

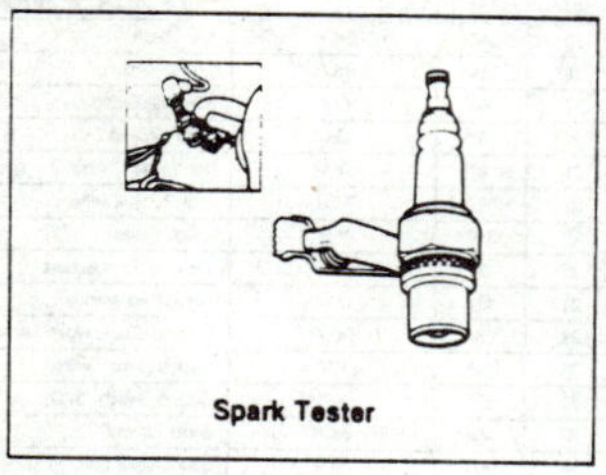

Spark Tester

- Timing light. Rotunda model 059-00006 or equivalent.
- Non-powered Test lamp.

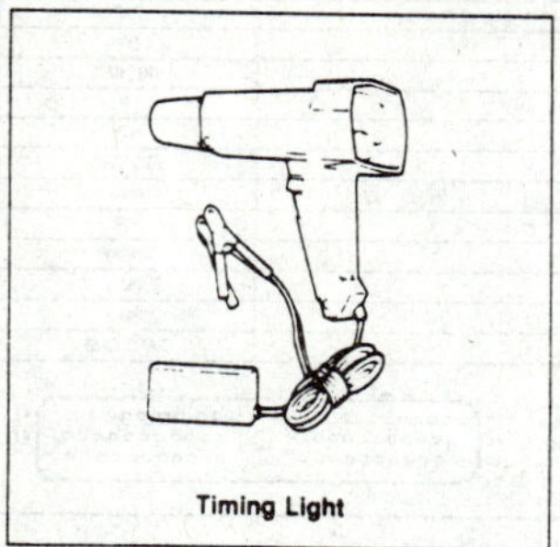

Timing Light

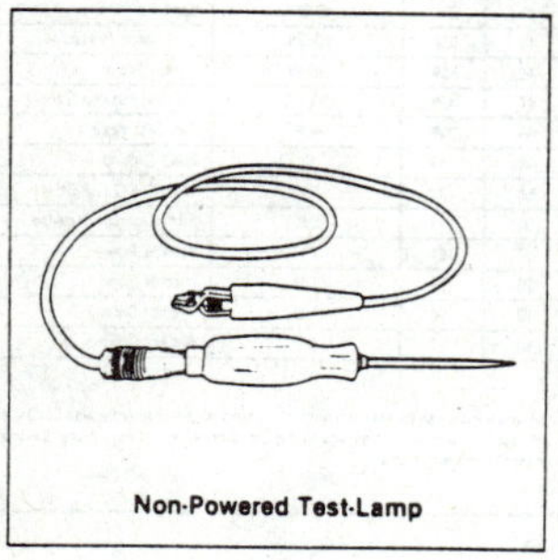

Non-Powered Test-Lamp

Electrical Schematic — 1.9L EFI

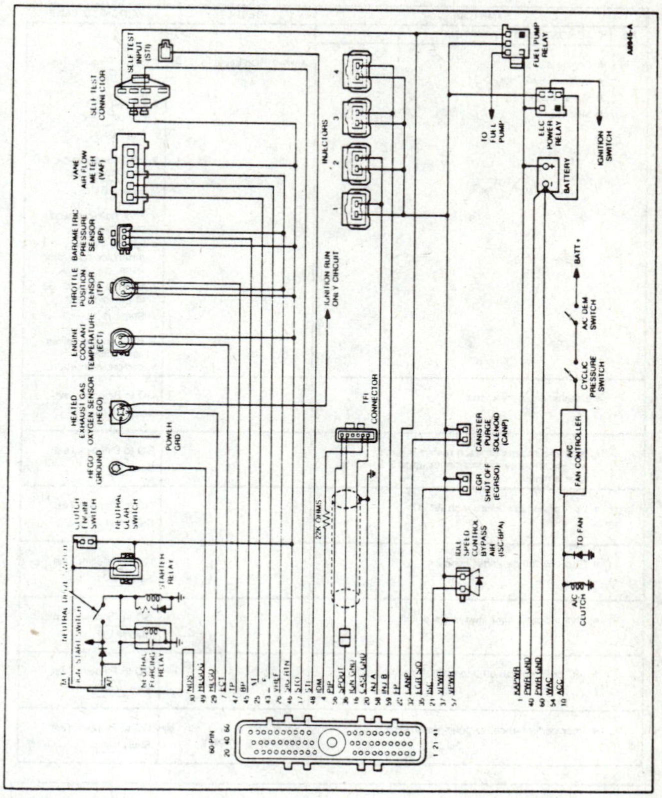

Diagnostic By Symptom — 1.9L EFI

SYMPTOM	RESULT	ACTION TO TAKE
7.0 DIAGNOSTICS BY SYMPTOM		
• Engine stalls. • Stalls in Self-Test • Runs rough • Misses • Always Rich/Lean.		• GO to Pinpoint Test Step S1 • BP test. GO to Pinpoint Test Step DF20 • Poor power/ground connections. • Ignition system distributor cap, rotor, wires, coil, plugs • Base engine valves, cam timing, compression etc.
• Detonation/spark knock.		GO to Pinpoint Test Step KA1
• High idle speeds on each restart may be accompanied by detonation for up to 3-5 minutes after a restart.		GO to Pinpoint Test Step KA1
• Lack of fast idle assist with A/C On.		GO to Pinpoint Test Step FA1
• Gasoline fumes under hood.		GO to Pinpoint Test Step KD1
• Shift indicator light always On or Off		GO to Pinpoint Test Step KL1
• A/C does not cut-off under WOT conditions.		GO to Pinpoint Test Step KM1

EEC-IV Module Connector Pin Usage — 1.9L EFI

Pin	Circuit	Wire Color	Application	Abbreviations
1	37A	Y	Keep Alive Power	KAPWR
4	11B	DG/Y	Ignition Diagnostic Monitor	IDM
7	354	LG/Y	Engine Coolant Temperature	ECT
10	347	BK/Y	A/C Clutch	ACC
16	259	BK/O	Ignition Ground	IGN GND
17	201	T/R	Self-Test Output/Shift Indicator Light	STO/SIL
20	57	BK	Case Ground	CSE GND
21	68	O/BK	Idle Speed Control	ISC
22	97	T/LG	Fuel Pump	FP
25	357	LG/P	Vane Air Temperature	VAT
26	351	O/W	Reference Voltage	V REF
29	94	DG/P	Exhaust Gas Oxygen Sensor	EGO
30	614	GY/O	Neutral Drive Switch (Turbo Grounded)	NDS
32	101	GY/Y	Canister Purge Solenoid	CANP
35	362	Y	Exhaust Gas Recirculation Shut Off	EGR S/O
36	324A	Y/LG	Spark Output	SPOUT
37	361	R	Vehicle Power	V PWR
40	60A	BK/LG	Power Ground	PWR GND
43	200	W/BK	Vane Air Flow	VAF
45	358	LG/BK	Barometric Pressure	BP
46	359	BK/W	Signal Return	SIG RET
47	355	DG/LG	Throttle Position Sensor	TPS
48	100	W/R	Self-Test Input	STI
49	89	O	Exhaust Gas Oxygen Ground	EGOG
54	73	O/LB	Wide Open Throttle A/C Cut-Off	WAC
56	349	DB	Profile Ignition Pickup	PIP
57	361	R	Vehicle Power	V PWR
58	95	T/R	Injector 1 (Bank 1)	INJ 1
59	96	T/O	Injector 2 (Bank 2)	INJ 2
60	60	BK/LG	Power Ground	PWR GND

Pin locations given for reference only. Probing 60 pin connector with DVOM probe will result in permanent damage to the pin connectors. Always probe as directed using the Breakout Box.

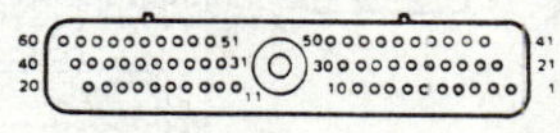

Diagnostic By Symptom

2.3L EFI TURBO

SYMPTOM	RESULT	ACTION TO TAKE
7.0 DIAGNOSTICS BY SYMPTOM		
• Engine stalls. • Stalls in Self-Test. • Runs rough. • Misses.	Always Rich/Lean. ▶	• GO to Pinpoint Test Step S1. • BP test, GO to Pinpoint Test Step DF20. • Idle speed control, GO to Pinpoint Test Step KE1. • Poor power/ground connections. • Ignition system distributor cap, rotor, wires, coil, plugs. • Short to Ground, Go to Pinpoint Test Step J9. • Base engine valves cam timing, compression etc.
• Detonation/spark knock.		GO to Pinpoint Test Step DG1.
• High idle speeds on each restart may be accompanied by detonation for up to 3-5 minutes after a restart.		GO to Pinpoint Test Step DG1.
• Lack of fast idle assist with A/C On.		GO to Pinpoint Test Step FA1.
• Gasoline fumes under hood.		GO to Pinpoint Test Step KD1.
• Shift indicator light always On or Off.		GO to Pinpoint Test Step KL1.
• A/C does not cut-off under WOT conditions.		GO to Pinpoint Test Step KM1.
• Poor performance, sluggish detonation.	Boost control not functioning properly ▶	GO to Pinpoint Test Step KN1.

Electrical Schematic

2.3L EFI TURBO

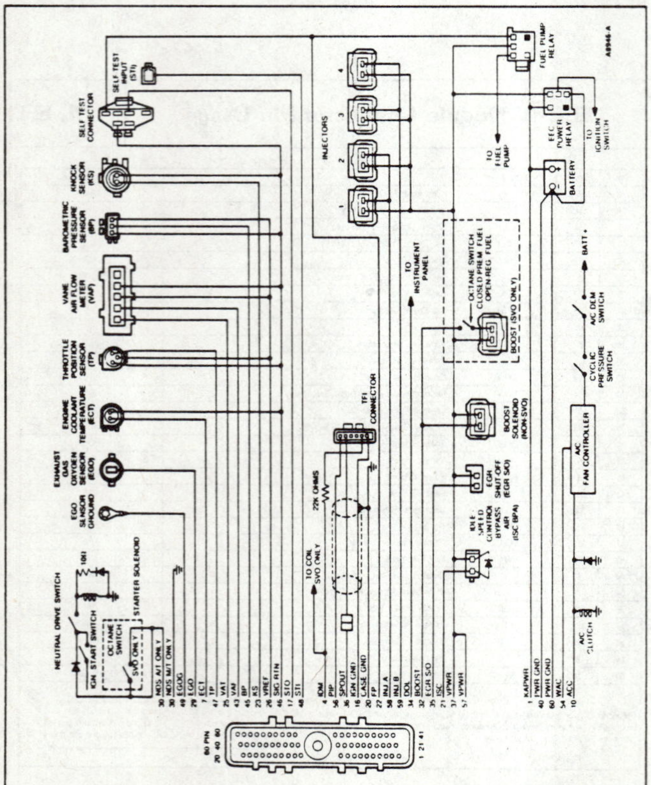

EEC-IV Module Connector Pin Usage

2.3L EFI TURBO

MUSTANG AND SVO MUSTANG

Pin	Circuit	Wire Color	Application	Abbreviations
1	37	Y	Keep Alive Power	KAPWR
4	11	DG/Y	Ignition Diagnostic Monitor	IDM
7	354	LG/Y	Engine Coolant Temperature	ECT
10	347	BK/Y	A/C Clutch	ACC
16	259	BK/O	Ignition Ground	IGN. GND.
17	382	Y/BK	Self-Test Output	STO
20	57	BK	Case Ground	CSE. GND
21	67	GY/W	Idle Speed Control	ISC
22	97	T/LG	Fuel Pump Control	FP
23	310	Y/R	Knock Sensor	KS
25	357	LG/P	Vane Air Temperature	VAT
26	351	O/W	Voltage Reference	V. REF
29	94	DG/P	Exhaust Gas Oxygen Sensor	EGO
30	376	Y/R	Neutral Drive Switch	NDS
30	376	Y/R	Octane Switch (SVO)	OCT. SW.
32	464	BK/P	Boost Control	BOOST
34	305	LB/P	Data Output Link	DOL
35	362	Y	EGR Shut-Off	EGR S/O
36	324	Y/LG	Spark Output	SPOUT
37	361	R	Vehicle Power	V PWR
40	57	BK	Battery Ground	BATT. GND.
43	200	W/BK	Vane Air Flow	VAF
45	358	LG/BK	Barometric Pressure	BP
46	359	BK/W	Signal Return	SIG. RET
47	355	DG/LG	Throttle Position Sensor	TP
48	209	W/R	Self-Test Input	STI
49	89	O	EGO Ground	EGO GND
54	73	O/LB	W.O.T. A/C Cut-Off	WAC
56	349	DB	Profile Ignition Pick-Up	PIP
57	361	R	Vehicle Power	V PWR
58	95	T/R	Injector Bank 1	INJ 1
59	96	T/O	Injector Bank 2	INJ 2
60	57	BK	Battery Ground	BATT. GND.

Pin locations given for reference only. Probing 60 pin connector with DVOM probe will result in permanent damage to the pin connectors. Always probe as directed, using the Breakout Box.

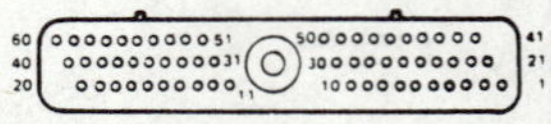

EEC-IV Module Connector Pin Usage

2.3L EFI TURBO

THUNDERBIRD AND XR-7

Pin	Circuit	Wire Color	Application	Abbreviations
1	37	Y	Keep Alive Power	KAPWR
4	11	DG/Y	Ignition Diagnostic Monitor	IDM
7	354	LG/Y	Engine Coolant Temperature	ECT
10	348	LG/P	A/C Clutch	ACC
16	259	BK/O	Ignition Ground	IGN. GND
17	382	Y/BK	Self-Test Output	STO
20	60	BK/LG	Case Ground	CSE. GND
21	68	O/BK	Idle Speed Control	ISC
22	97	T/LG	Fuel Pump Control	FP
23	310	Y/R	Knock Sensor	KS
25	357	LG/P	Vane Air Temperature	VAT
26	351	O/W	Voltage Reference	V. REF.
29	94	DG/P	Exhaust Gas Oxygen Sensor	EGO
30	32	R/LB	Neutral Drive Switch (ATX)	NDS
30	60	BK/LG	Neutral Drive Switch (MTX)	NDS
32	464	BK/P	Boost Control	BOOST
34	305	LB/P	Data Output Link	DOL
35	362	Y	EGR Shut-Off	EGR S/O
36	324	Y/LG	Spark Output	SPOUT
37	175	BK/Y	Vehicle Power	V. PWR
40	60	BK/LG	Battery Ground	BATT. GND.
43	200	W/BK	Vane Air Flow	VAF
45	358	LG/BK	Barometric Pressure	BP
46	359	BK/W	Signal Return	SIG. RET.
47	355	DG/LG	Throttle Position Sensor	TP
48	209	W/R	Self-Test Input	STI
49	89	O	EGO Ground	EGO GND
54	73	O/LB	W.O.T. A/C Cut-Off	WAC
56	349	DB	Profile Ignition Pick-Up	PIP
57	175	BK/Y	Vehicle Power	V. PWR
58	95	T/R	Injector Bank 1	INJ. 1
59	96	T/O	Injector Bank 2	INJ. 2
60	60	BK/LG	Battery Ground	BATT. GND

Pin locations given for reference only. Probing 60 pin connector with DVOM probe will result in permanent damage to the pin connectors. Always probe as directed, using the Breakout Box.

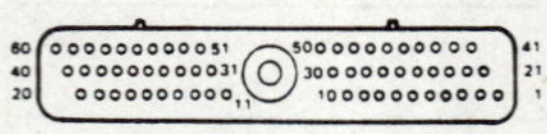

Diagnostic By Symptom

2.3L OHC FBC

SYMPTOM	RESULT	ACTION TO TAKE
• Engine stalls. • Stalls in Self-Test. • Runs rough. • Misses.		• Poor power ground connections. • Ignition system distributor cap, rotor, wires, coil, plugs. Refer to Section 15 • Base engine valves, cam timing, compression etc. Refer to Shop Manual Group 21 • EGR Valve. Refer to Section 6.
• Shift indicator light always On or always Off.		GO to Pinpoint Test Step **KL1**
• A/C does not cut-off during WOT operation. • A/C not functioning.		GO to Pinpoint Test Step **KM1**
• Erratic rpm.	rpm out of specification ▶	• Verify rpm per VECI decal. GO to Idle Adjustment, Section 4
• Gasoline fume odor present under hood.		GO to Pinpoint Test Step **KD1** and/or Section 7
• Lack of fast idle in A/C engine		GO to Pinpoint Test Step **FA1** and/or **KF1**

EEC-IV Module Connector Pin Usage

2.3L OHC FBC

Pin	Circuit	Wire Color	Application	Abbreviations
1	38	BK/O	Keep Alive Memory	KAM
4	11	DG/Y	Ignition Diagnostic Monitor	IDM
5	16	R/LG	Key Power	KEY PWR
7	354	LG/Y	Engine Coolant Temperature	ECT
10	347	BK/Y	A/C Clutch	ACC
11	39	LG/BK	Thermactor Air Diverter	TAD
16	57	BK	Ignition Ground	IGN GND
17	462	P	Self-Test Output Shift Ind. Light	STO SIL
20	60	BK/LG	Case Ground	CASE GND
21	376	Y/R	Idle Speed Control	ISC –
26	351	O/W	Voltage Reference	V REF
28	265	LG/W	Idle Tracking Switch	ITS
29	94	DG/P	Exhaust Gas Oxygen Sensor	EGO
30	199	LB/Y	Neutral Drive Switch	NDS
35	101	GY/Y	Canister Purge Solenoid	CANP
36	324	Y/LG	Spark Output	SPOUT
37	361	R	Vehicle Power	V PWR
40	60	BK/LG	Battery Ground	BATT GND
41	264	W/LB	Idle Speed Control	ISC +
45	358	LG/BK	Manifold Absolute Pressure	MAP
46	359	BK/W	Signal Return	SIG RET
47	355	DG/LG	Throttle Position Sensor	TPS
48	209	W/R	Self-Test Input	STI
49	89	O	EGO Ground	EGO GND
51	100	W/R	Thermactor Air Bypass Solenoid	TAB
53	73	O/LB	Wide Open Throttle A/C Cut-Off	WAC
56	349	DB	Profile Ignition Pick-Up	PIP
57	361	R	Vehicle Power	V PWR
58	95	T/R	Feedback Carburetor Solenoid	FBC
60	60	BK/LG	Battery Ground	BATT GND

Pin locations given for reference only. Probing 60 pin connector with DVOM probe will result in permanent damage to the pin connectors. Always probe as directed using the Breakout Box.

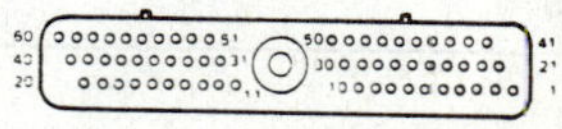

Electrical Schematic

2.3L OHC FBC

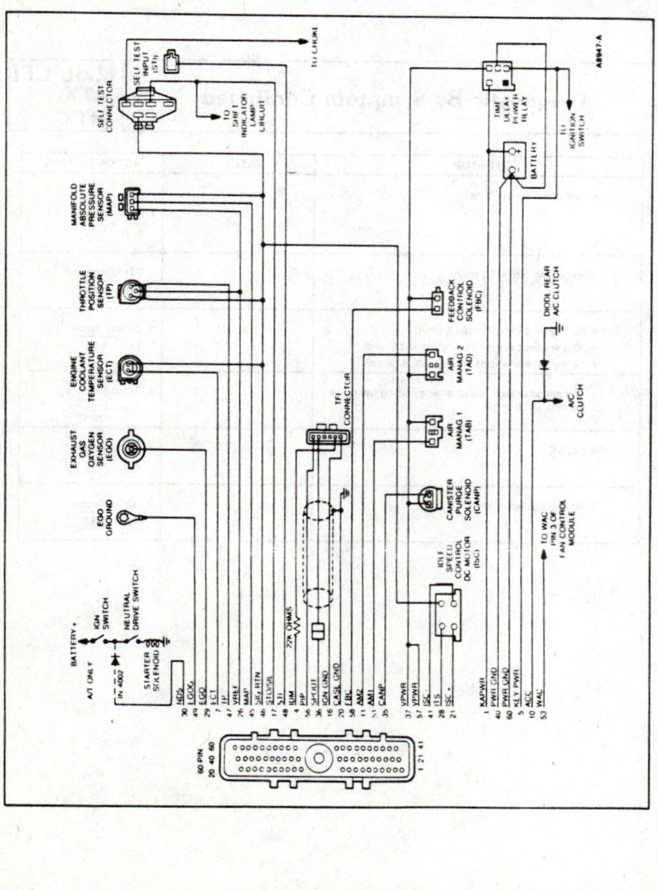

Electrical Schematic

2.3L HSC-CFI

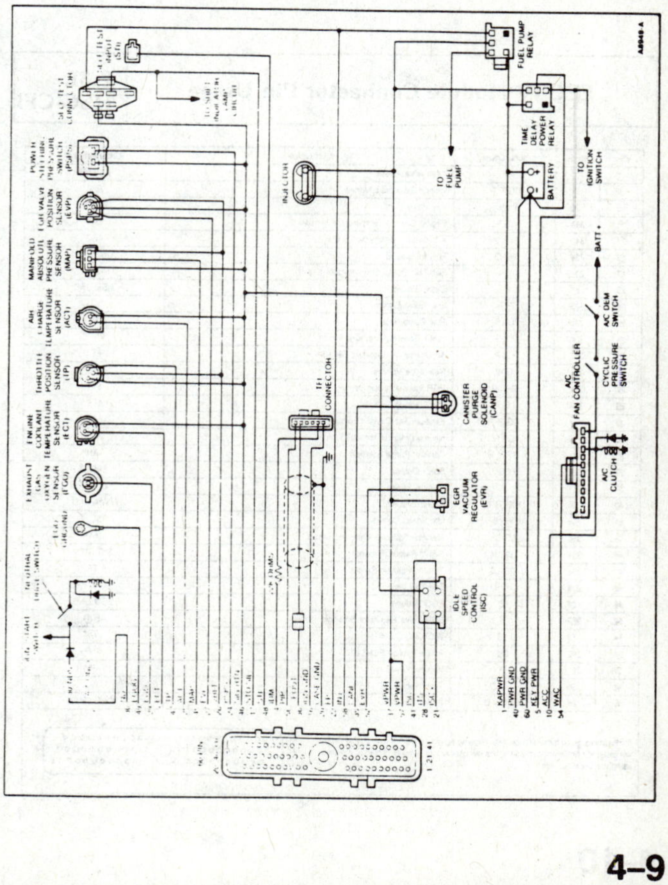

Diagnostic By Symptom — 2.3L HSC-CFI

SYMPTOM	RESULT	ACTION TO TAKE
• Engine runs rough or miss fires. • Engine stalls • Lack of power • Rough idle		• GO to Pinpoint Test Step S2 • GO to Pinpoint Test Step DF20 • EGR Valve Function. • Ignition system distributor cap, rotor, wires, coil, plugs. • Base engine valves, cam timing, compression etc. • Poor power/ground connections.
• Lack of fast idle with A/C On.		• GO to Pinpoint Test Step KF9.
• Gasoline fumes under hood.		• GO to Pinpoint Test Step KD1.
• Stall during parking maneuvers.		• GO to Pinpoint Test Step FF3.
• High idle in Drive, automatic transmission only.		• Place transmission in Drive with engine Off. Test neutral drive circuit (test Pin 30) for short to ground. SERVICE as necessary. **NOTE: Circuit should NOT have continuity to ground in Drive.**

Diagnostic By Symptom — 2.5L CFI ATX, MTX

SYMPTOM	RESULT	ACTION TO TAKE
• Engine runs rough or miss-fires. • Engine stalls. • Lack of power. • Rough idle.		• GO to Pinpoint Test Step S2 • GO to map Pinpoint Test Step DF20 • EGR value function. • Ignition system distributor cap, rotor, wires, coil, plugs. • Base engine valves, cam timing, compression etc. • Poor power/ground connections.
• Lack of fast idle with A/C On.		• GO to Pinpoint Test Step KF9.
• Gasoline fumes under hood.		• GO to Pinpoint Test Step KD1.
• Stalls during parking maneuvers.		• GO to Pinpoint Test Step FF3.
• High idle in Drive, automatic transmission only.		• Place transmission in Drive with engine Off. Test neutral drive circuit (test Pin 30) for short to ground. SERVICE as necessary. **NOTE: Circuit should NOT have continuity to ground in Drive.**

EEC-IV Module Connector Pin Usage — 2.3L HSC-CFI

Pin	Circuit	Wire Color	Application	Abbreviations
1	37	Y	Keep Alive Power	KAPWR
4	11	DG Y D	Ignition Diagnostic Monitor	IDM
5	20	W LB H	Key Power	KPWR
7	354	LG Y	Engine Coolant Temperature Sensor	ECT
10	347	BK Y H	A C Clutch Compressor Signal	ACC
16	259	BK O	Ignition Ground	IGN GND
17	201	T R	Self-Test Output Shift Indicator Lite	STO SIL
20	57	BK	Case Ground	CSE GND
21	376	Y R	Idle Speed Control	ISC -
22	238	O LB	Fuel Pump Relay	FP
24	330	Y LG	Power Steering Pressure Switch	PSPS
25	357	LG P	Air Charge Temperature Sensor	ACT
26	351	O/W	Reference Voltage	V REF
27	352	BR LG	EGR Valve Position Sensor	EVP
28	265	LG W	Idle Tracking Switch	ITS
29	94	DG P H	Exhaust Gas Oxygen Sensor	EGO
30	614	GY O	Neutral Drive Switch (A.T Only)	NDS
35	101	GR Y H	Canister Purge Solenoid	CANP
36	324	Y LG D	Spark Output	SPOUT
37	361	R	Vehicle Power	V PWR
40	60	BK LG D	Power Ground	PWR GND
41	264	W LB	Idle Speed Control	ISC
45	358	LG-BK	Manifold Absolute Pressure Sensor	MAP
46	359	BK W	Signal Return (Ground)	SIG RTN
47	355	DG LG	Throttle Position Sensor	TP
48	209	W R	Self-Test Input	STI
49	89	O	EGO Ground	EGO GND
52	362	Y	EGR Valve Regulator Solenoid	EVR
54	73	O LB H	WOT A C Cutoff	WAC
56	349	DB	Profile Ignition Pick-Up	PIP
57	361	R	Vehicle Power	V PWR
58	95	T R D	Fuel Injector	INJ
60	60	BK LG D	Power Ground	PWR GND

Pin locations given for reference only. Probing 60 pin connector with DVOM probe will result in permanent damage to the pin connectors. Always probe as directed using the Breakout Box.

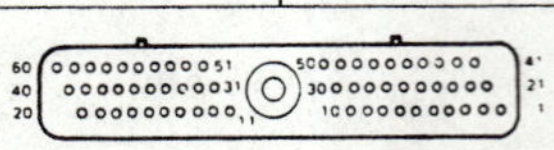

Diagnostic By Symptom-Continued — 2.5L CFI ATX, MTX

SYMPTOM	RESULT	ACTION TO TAKE
• Fuel pump always runs.		• Go to Pinpoint Test X-14
• No cooling fan (MTX only).		• GO to Pinpoint Test X-40
• No engine cooling fan (ATX only). • No low speed engine cooling fan (ATX only). • No high speed engine cooling fan (ATX only).		• GO to Pinpoint Test X-20
• Low speed or high speed engine cooling fan on all the time (ATX only).		• GO to Pinpoint Test X-35
• No A/C.		• GO to Pinpoint Test X-50
• No A/C cut out at WOT.		• GO to Pinpoint Test X-52

Electrical Schematic

2.5L CFI ATX

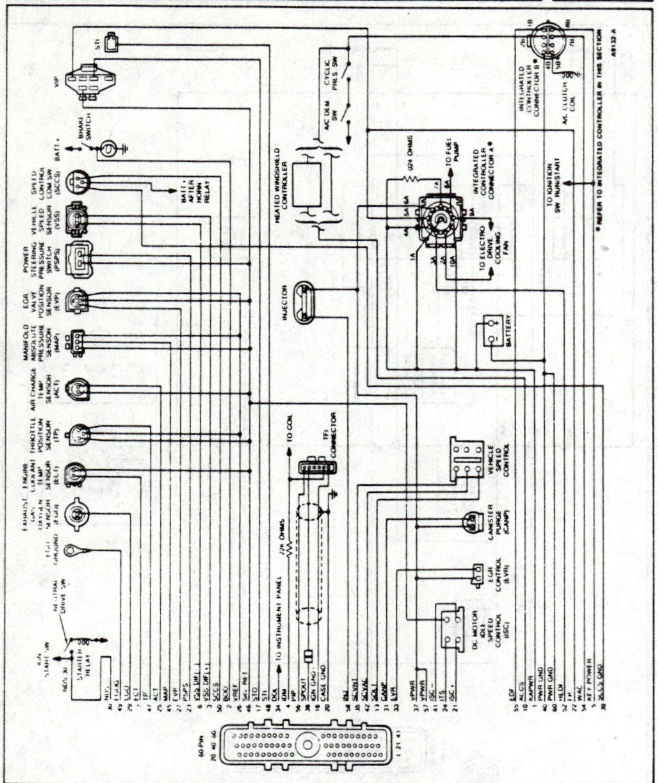

Integrated Controller Schematic

2.5L CFI ATX

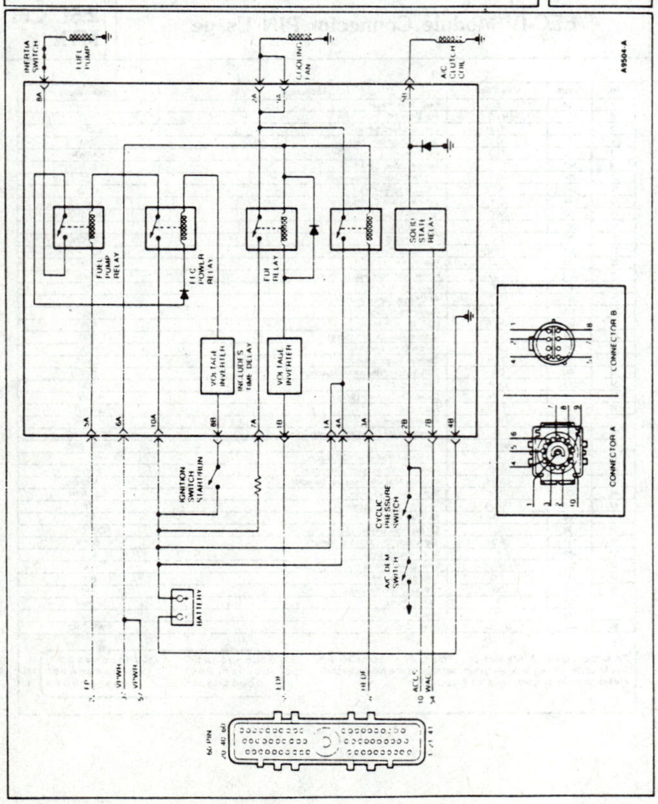

Integrated Controller Usage Pin

2.5L CFI ATX

Pin No.	Circuit	Color	Applications	Abbreviations
Connector A				
1	38	BK-O	HEDF Power into Controller	Batt -
2	228	BY-Y	H EDF Power To Fan	PTF
3	639	PK	HEDF Ground	HEDF
4	38	BK-O	HEDF Power into Controller	Batt -
5	97	T-LG	Fuel Pump Relay Ground	FP
6	361	R	EEC IV Power Out	VPWR
7	181	BR-O	EDF Power Into Controller	Batt +
8	787	PK-BK	Power To The Pump(s)	PTP
9	228	BR-Y	H EDF Power To Fan	PTF
10	37	Y	Battery Voltage (Power Relay)	Batt +
Connector B				
1	197	T-O	EDF Ground	EDF
2	883	PK-LB	A.C Power In	ACCS
3	—	—	Not Used	—
4	60	BK-LG	Vehicle Ground	PWR GRD
5	347	BK-Y	A.C Power To Clutch Coil	PTAC
6	—	—	Not Used	—
7	331	R	WOT A.C Cutoff	WAC
8	16	R-LG	Key Power	KEY PWR

EEC-IV Module Connector PIN Usage

2.5L CFI ATX

Pin	Circuit	Wire Color	Application	Abbreviations
1	37	Y	Keep Alive Power	KA PWR
2	810	R LG	Brake On Off	BOO
3	150	DG W	Vehicle Speed Sensor -	VSSDIF -
4	11	DG Y	Ignition Diagnostic Monitor	IDM
5	16	R LG	Key Power	KAPWR
6	563	O Y	Vehicle Speed Sensor —	VSSDIF -
7	354	LG Y	Engine Coolant Temperature	ECT
10	883	PK LB	A.C Cycling Switch	ACCS
13	144	O Y	Vehicle Speed Control Solenoid	SOL -
16	259	BK O	Ignition Ground	IGN GND
17	201	T R	Self-Test Out	STC
20	57	BK	Case Ground	CSE GND
21	382	Y BK	Idle Speed Control (DC Motor)	ISC +
22	97	T LG	Fuel Pump	FP
23	330	Y LG	Power Steering Pressure Switch	PSPS
24	209	W R	Idle Tracking Switch	ITS
25	357	LG P	Air Charge Temperature	ACT
26	351	O W	Reference Voltage	VREF
27	352	GR LG	EGR Valve Position	EVP
29	94	DG P	Exhaust Gas Oxygen (Sensor)	EGC
30	199	LB Y	Neutral Drive Switch	NDS
31	101	GY Y	Canister Purge Solenoid	CANP
33	360	DG	EGR Vacuum Regulator (Solenoid)	EVR
34	305	LB P	Data Output Link	DOL
35	146	W PK	Speed Control Vent (Solenoid)	SCVNT
36	324	Y LG	Spark Out	SPOUT
37	361	R	Vehicle Power	V PWR
39	461	O	Speed Control Command Switch Ground	SCCS GND
40	60	BK LG	Power Ground	PWR GND
41	377	W	Idle Speed Control — (DC Motor)	ISC -
42	145	GY BK	Speed Control Vacuum (Solenoid)	SCVAC
45	358	LG BK	Manifold Absolute Pressure (Sensor)	MAP
46	359	BK W	Signal Return (Ground)	SIG RTN
47	355	DG LG	Throttle Position (Sensor)	TP
48	200	W BK	Self-Test Input	STI
49	89	O	Exhaust Gas Oxygen (Sensor) Ground	EGO GND
50	151	LB BK	Speed Control Command Switch	SCCS
52	639	PK	High Electro Drive Fan	HEDF
54	331	R	Wide Open Throttle (WOT) A.C Cutoff	WAC
55	197	T O	Electro Drive Fan (Low)	EDF
56	349	DB	Profile Ignition Pick-up	PIP
57	361	R	Vehicle Power	V PWR
58	95	T R	Injector	INJ
60	60	BK LG	Power Ground	PWR GND

Pin locations given for reference only. Probing 60 pin connector with DVOM probe will result in permanent damage to the pin connectors. Always probe as directed, using the Breakout Box.

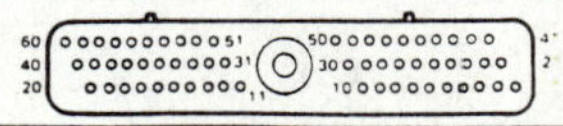

Electrical Schematic — 2.5L CFI MTX

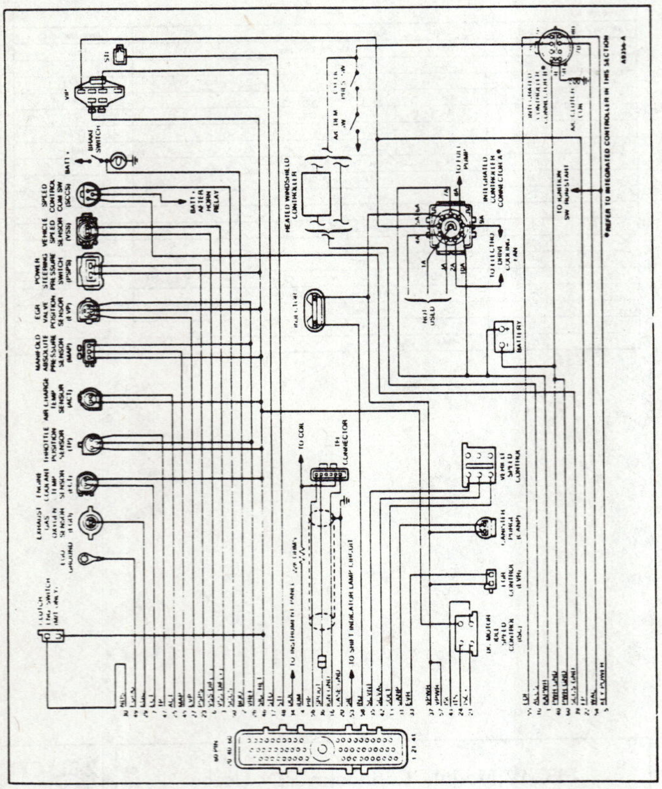

Integrated Controller Schematic — 2.5L CFI MTX

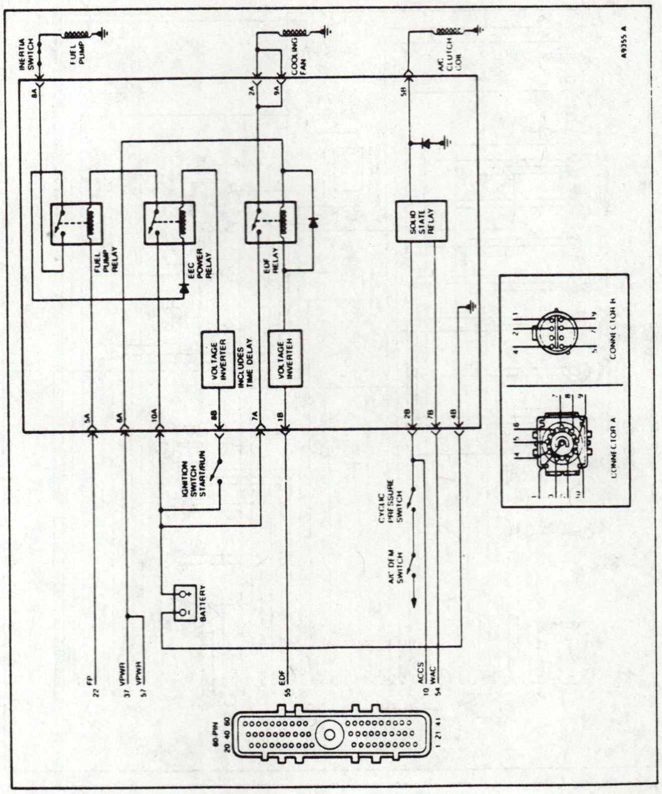

Integrated Controller Usage Pin — 2.5L CFI MTX

Pin No.	Circuit	Color	Applications	Abbreviations
Connector A				
1	—	—	Not Used	—
2	228	BR-Y	H EDF Power To Fan	PTF
3	—	—	Not Used	—
4	—	—	Not Used	—
5	97	T-LG	Fuel Pump Relay Ground	FP
6	361	R	EEC IV Power Out	VPWR
7	181	BR-O	EDF Power Into Controller	Batt +
8	787	PK-BK	Power To The Pump(s)	PTP
9	228	BR-Y	H EDF Power To Fan	PTF
10	37	Y	Battery Voltage (Power Relay)	Batt +
Connector B				
1	197	T-O	EDF Ground	EDF
2	883	PK-LB	A/C Power In	ACCS
3	—	—	Not Used	—
4	60	BK-LG	Vehicle Ground	PWR GRD
5	347	BK-Y	A/C Power To Clutch Coil	PTAC
6	—	—	Not Used	—
7	331	R	WOT A/C Cutoff	WAC
8	16	R-LG	Key Power	KEY PWR

EEC-IV Module Connector PIN Usage — 2.5L CFI MTX

Pin	Circuit	Wire Color	Application	Abbreviations
1	37	Y	Keep Alive Power	KA PWR
2	810	R LG	Brake On Off	BOO
3	150	DG W	Vehicle Speed Sensor +	VSSDIF +
4	11	DG Y	Ignition Diagnostic Monitor	IDM
5	16	R LG	Key Power	KAPWR
6	563	O Y	Vehicle Speed Sensor —	VSSDIF -
7	354	LG Y	Engine Coolant Temperature	ECT
10	883	PR LB	A/C Cycling Switch	ACCS
13	144	O Y	Vehicle Speed Control Solenoid	SOL -
16	259	BK O	Ignition Ground	IGN GND
17	201	T R	Self-Test Out	STO
20	57	BK	Case Ground	CSE GND
21	382	Y BK	Idle Speed Control (DC Motor)	ISC +
22	97	T LG	Fuel Pump	FP
23	330	Y LG	Power Steering Pressure Switch	PSPS
24	209	W R	Idle Tracking Switch	ITS
25	357	LG P	Air Charge Temperature	ACT
26	351	O W	Reference Voltage	VREF
27	352	GR LG	EGR Valve Position	EVP
29	94	DG P	Exhaust Gas Oxygen (Sensor)	EGO
30	199	LB Y	Neutral Drive Switch	NDS
31	101	GY Y	Canister Purge Solenoid	CANP
33	360	DG	EGR Vacuum Regulator (Solenoid)	EVR
34	305	LB P	Data Output Link	DOL
35	146	W PK	Speed Control Vent (Solenoid)	SCVNT
36	324	Y LG	Spark Out	SPOUT
37	361	R	Vehicle Power	V PWR
39	461	O	Speed Control Command Switch Ground	SCCS GND
40	60	BK LG	Power Ground	PWR GND
41	377	W	Idle Speed Control — (DC Motor)	ISC -
42	145	GY BK	Speed Control Vacuum (Solenoid)	SCVAC
45	358	LG BK	Manifold Absolute Pressure (Sensor)	MAP
46	359	BK W	Signal Return (Ground)	SIG RTN
47	355	DG LG	Throttle Position (Sensor)	TP
48	200	W BK	Self-Test Input	STI
49	89	O	Exhaust Gas Oxygen (Sensor) Ground	EGO GND
50	151	LB BK	Speed Control Command Switch	SCCS
53	462	P	Shift Indicator Light	S I L
54	331	R	Wide Open Throttle (WOT) A/C cutoff	WAC
55	197	T O	Electro Drive Fan	EDF
56	349	DB	Profile Ignition Pick-Up	PIP
57	361	R	Vehicle Power	V PWR
58	95	T R	Injector	INJ
60	60	BK LG	Power Ground	PWR GND

Pin locations given for reference only. Probing 60 pin connector with DVOM probe will result in permanent damage to the pin connectors. Always probe as directed, using the Breakout Box.

Diagnostic By Symptom — 3.0L EFI

SYMPTOM	RESULT	ACTION TO TAKE
• Engine runs rough or miss-fires • Engine stalls • Rough idle		• GO to Pinpoint Test Step S1 • GO to map Pinpoint Test Step DF20 • EGR value function • Ignition system distributor cap, rotor, wires, coil, plugs • Base engine valves cam timing, compression etc. • Fuel delivery • Poor power/ground connections
• Detonation spark knock		GO to Pinpoint Test Step DG1
• Stalls during parking maneuvers		GO to Pinpoint Test Step FF3
• Gasoline fumes under hood		GO to Pinpoint Test Step KD1
• Fuel pump always runs		GO to Pinpoint Test Step X-14
• No engine cooling fan • No low speed engine cooling fan • No high speed engine cooling fan		GO to Pinpoint Test Step X-20
• Low speed or high speed engine cooling fan on all the time		GO to Pinpoint Test Step X-35
• Stumble after hot restart (HEGO)		GO to Pinpoint Test Step HG-21
• No A/C		GO to Pinpoint Test X-50
• No A/C cut out at WOT		GO to Pinpoint Test X-52

Electrical Schematic — 3.0L EFI

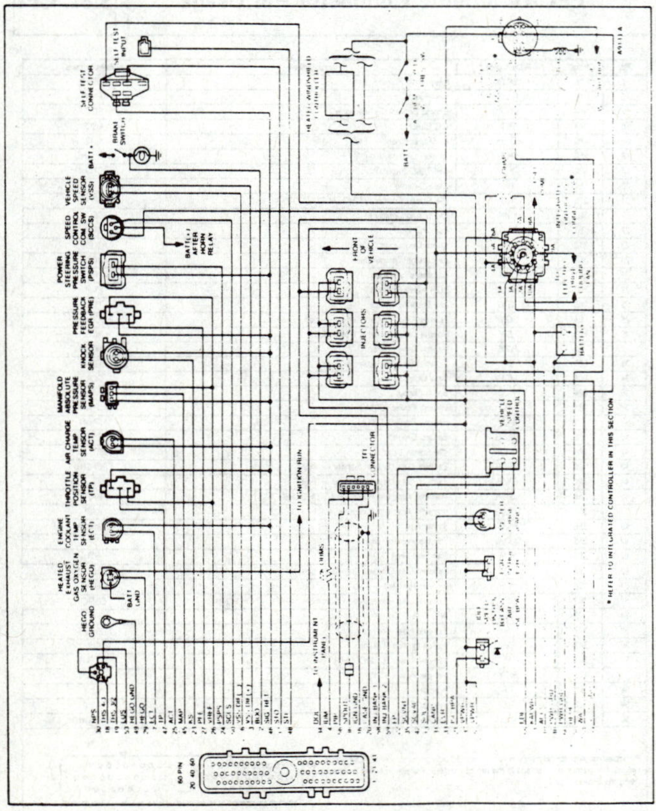

Integrated Controller Schematic — 3.0L EFI

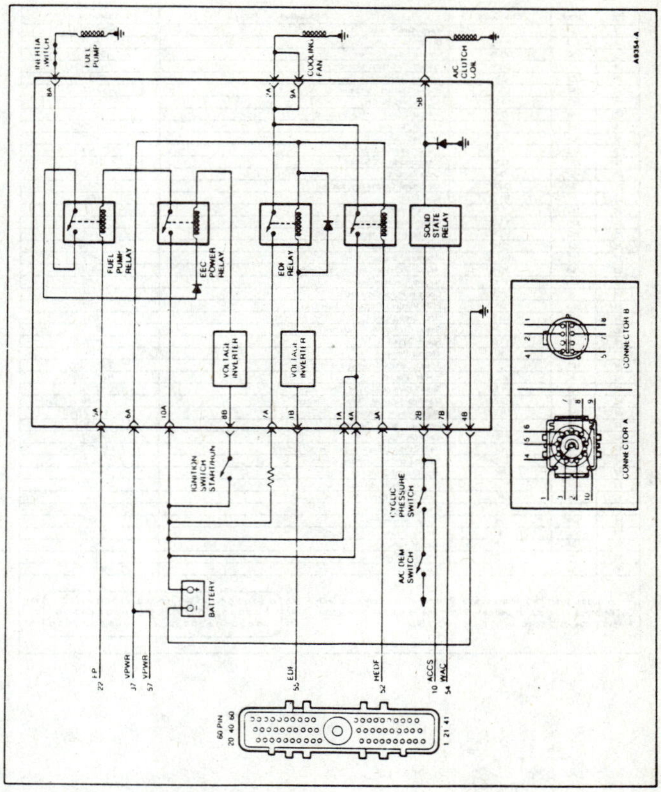

Integrated Controller Usage Pin — 3.0L EFI

Pin No.	Circuit	Color	Applications	Abbreviations
Connector A				
1	38	BK-O	HEDF Power Into Controller	Batt +
2	228	BY-Y	H.EDF Power To Fan	PTF
3	639	PK	HEDF Ground	HEDF
4	38	BK-O	HEDF Power Into Controller	Batt +
5	97	T-LG	Fuel Pump Relay Ground	FP
6	361	R	EEC IV Power Out	VPWR
7	181	BR-O	EDF Power Into Controller	Batt +
8	787	PK-BK	Power To The Pump(s)	PTP
9	228	BR-Y	H/EDF Power To Fan	PTF
10	37	Y	Battery Voltage (Power Relay)	Batt +
Connector B				
1	197	T-O	EDF Ground	EDF
2	883	PK-LB	A/C Power In	ACCS
3	—	—	Not Used	—
4	60	BK-LG	Vehicle Ground	PWR GRD
5	347	BK-Y	A/C Power To Clutch Coil	PTAC
6	—	—	Not Used	—
7	331	R	WOT A/C Cutoff	WAC
8	16	R-LG	Key Power	KEY PWR

EEC-IV Module Connector PIN Usage — 3.0L EFI

Pin	Circuit	Wire Color	Application	Abbreviations
1	37	Y	Keep Alive Power	KA PWR
2	810	R LG	Brake On Off	BOO
3	150	DG W	Vehicle Speed Sensor -	VSSDIF -
4	11	DG Y	Ignition Diagnostic Monitor	DM
6	563	O Y	Vehicle Speed Sensor —	VSSDIF -
7	354	LG Y	Engine Coolant Temperature	ECT
10	883	PR LB	A C Cycling Switch	ACCS
13	144	O Y	Vehicle Speed Control Solenoid	SOL -
16	259	BK O	Ignition Ground	IGN GND
17	201	T R	Self-Test Out	STO
18	315	O W	Transmission 4 3 Switch	THS 4 3
19	237	O Y	Transmission 3 2 Switch	THS 3 2
20	57	BK	Case Ground	CSE GND
21	68	O BK	Idle Speed Control (Bypass Air)	ISC-BPA
22	97	T LG	Fuel Pump	FP
23	310	Y R	Knock Sensor	KS
24	330	Y LG	Power Steering Pressure Switch	PSPS
25	357	LG P	Air Charge Temperature	ACT
26	351	O/W	Reference Voltage	V REF
27	352	BR LG	Pressure Feedback EGR	PFE
29	94	DG P	Heated Exhaust Gas Oxygen (Sensor)	HEGO
30	480	P Y	Neutral Pressure Switch	NPS
31	101	GY Y	Canister Purge Solenoid	CANP
33	380	DG	EGR Vacuum Regulator (Solenoid)	EVR
34	305	LB P	Data Output Link	DOL
35	146	W PK	Speed Control Vent (Solenoid)	SCVNT
36	324	Y LG	Spark Output	SPOUT
37	361	R	Vehicle Power	V PWR
39	461	O	Speed Control Command Switch Ground	SCCS GND
40	60	BK LG	Power Ground	PWR GND
42	145	GY BK	Speed Control Vacuum (Solenoid)	SCVAC
45	358	LG BK	Manifold Absolute Pressure (Sensor)	MAP
46	359	BK W	Signal Return (Ground)	SIG RTN
47	355	DG LG	Throttle Position (Sensor)	TP
48	200	W BK	Self Test Input	STI
49	89	O	Heat Exhaust Gas Oxygen Sensor Ground	HEGO GND
50	151	LB BK	Speed Control Command Switch	SCCS
52	639	PK	High Electro Drive Fan	HEDF
53	224	T LB	Lock-Up Solenoid (Transmission)	LUS
54	331	R	Wide Open Throttle (WOT) A C Cutoff	WAC
55	197	T O	Electro Drive Fan (Low)	EDF
56	349	DB	Profile Ignition Pick-Up	PIP
57	361	R	Vehicle Power	V PWR
58	95	T R	Injector Bank A	INJ A
59	96	T O	Injector Bank B	INJ B
60	60	BK LG	Power Ground	PWR GND

Pin locations given for reference only. Probing 60 pin connector with DVOM probe will result in permanent damage to the pin connectors. Always probe as directed, using the Breakout Box.

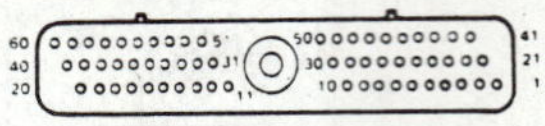

Electrical Schematic — 3.8L CFI

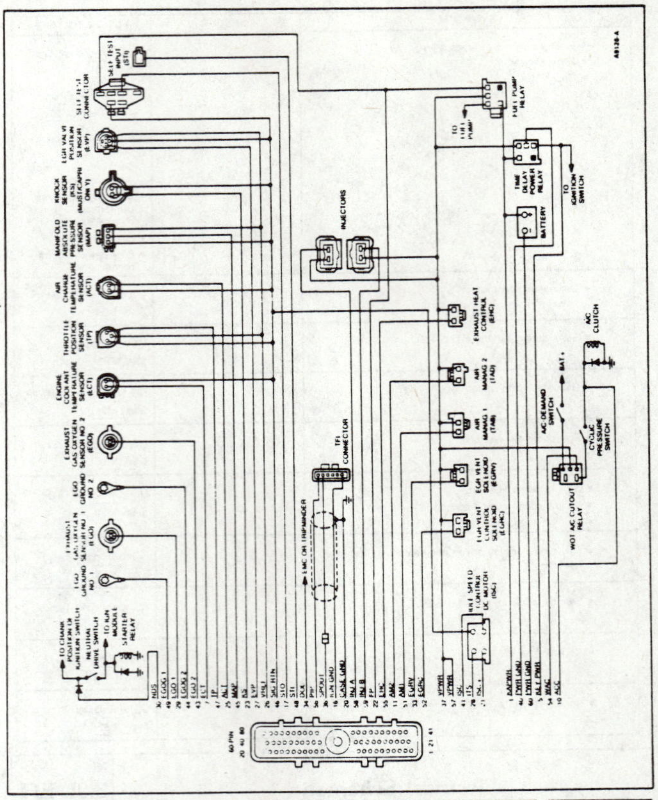

EEC-IV Module Connector Pin Usage — 3.8L CFI

Pin	Circuit	Wire Color	Application	Abbreviations
1	38	BK O	Keep Alive Power	KAPWR
5	16	R LG	Key Power	KPWR
7	354	LG Y	Engine Coolant Temperature	ECT
10	347	BK Y	A C Clutch	ACC
11	99	LG BK	Air Management — TAD	AM2
16	259	BK O	Ignition Ground	IGN GND
17	382	Y B	Self-Test Output	STO
20	57	BK	Case Ground	CSE GND
21	376	Y R	Idle Speed Control (LT MQ, M C)	ISC -
	382	Y BK	Idle Speed Control (T X)	ISC +
22	97	T LG	Fuel Pump	FP
23	310	Y R	Knock Sensor (M C Only)	KS
25	357	LG P	Air Charge Temperature	ACT
26	351	O/W	Reference Voltage	V. REF
27	352	BR LG	EGR Valve Position	EVP
28	265	LG W	Idle Tracking Switch	ITS
29	94	DG/P	Exhaust Gas Oxygen Sensor — Left	EGO 1(L)
30	199	LB Y	Neutral Drive Switch (LT MQ, M/C)	NDS
	32	R LB	Neutral Drive Switch (T/X)	NDS
33	360	DG	Exhaust Gas Recirculation Vent	EGRV
34	305	LB P	Data Output Line	DOL
36	324	Y LG	Spark Output	SPOUT
37	361	R	Vehicle Power (LT MQ, M C)	V PWR
	175	BK/Y	Vehicle Power (T X)	V PWR
40	60	BK LG	Battery Ground	BATT GND
41	264	W LB	Idle Speed Control	ISC -
43	90	DB/LG	Exhaust Gas Oxygen Sensor — Right	EGO 2(R)
44	91	GY/R	Exhaust Gas Oxygen Ground — Right	EGOG 2(R)
45	358	LG BK	Manifold Absolute Pressure (LTD/MQ AND T X)	MAP
	356	DB LG	Manifold Absolute Pressure (M/C)	MAP
46	359	BK W	Signal Return	SIG RTN
47	355	DG LG	Throttle Position Sensor	TP
48	209	W R	Self-Test Input	STI
49	89	O	Exhaust Gas Oxygen Ground — Left	EGOG 1(L)
51	100	W R	Air Management — Tab	AM1
52	362	Y	Exhaust Gas Recirculation Control	EGRC
54	73	O LB	W O T A C Cut Off	WAC
55	377	W	Exhaust Heat Control	EHC
56	349	DB	Profile Ignition Pick-Up	PIP
57	361	R	Vehicle Power (LT MQ, M C)	V PWR
	175	BK Y	Vehicle Power (T X)	V PWR
58	95	T R	Injector 1	INJ 1
59	96	T O	Injector 2	INJ 2
60	60	BL LG	Battery Ground	BATT GND

Pin locations given for reference only. Probing 60 pin connector with DVOM probe will result in permanent damage to the pin connectors. Always probe as directed, using the Breakout Box.

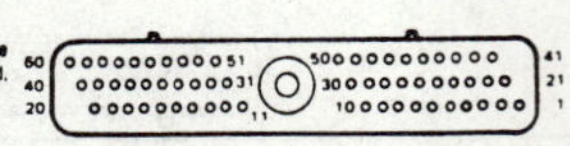

Diagnostic By Symptom — 3.8L CFI

SYMPTOM	RESULT	ACTION TO TAKE
• Engine runs rough or miss-fires. • Engine stalls. • Rough idle.		• GO to Pinpoint Test Step **S2**. • GO to Pinpoint Test Step **DF20**. • EGR valve function. • Ignition system distributor cap, rotor, wires, coil, plugs. • Base engine valves, cam timing, compression etc. • Fuel delivery. • Poor power/ground connections.
• Detonation/spark knock (Mustang/Capri only).		• GO to Pinpoint Test Step **DG1**.
• Lack of fast idle with A/C On.		• GO to Pinpoint Test Step **FA1** and/or **KF1**.
• Gasoline fumes under hood.		• GO to Pinpoint Test Step **KD1**.
• Low idle on cold start-up. • High idle on hot start-up. • Engine diesels.		• GO to Pinpoint Test Step **KF16**.
• High idle in Drive, automatic transmission only.		• Place transmission in Drive with engine Off. Test neutral drive circuit (test Pin 30) for short to ground. SERVICE as necessary. **NOTE: Circuit should NOT have continuity to ground in Drive.**

Diagnostic By Symptom — 5.0L SEFI

SYMPTOM	RESULT COULD BE	ACTION TO TAKE
7.0 DIAGNOSTICS BY SYMPTOM		
• Noise emitting from rear of vehicle with key-off.	Fuel pump always on	▶ GO to Pinpoint Test Step J15.
• Shift indicator light always on or, off.		▶ GO to Pinpoint Test Step KL1.
• Lack of fast idle with A/C ON. • A/C does not cutoff during W.O.T.		▶ GO to Pinpoint Test Step KM1.
• Detonation and, or spark knock.	EGR valve function	▶ GO to EGR Diagnostic Section 6 of Volume "H".
• Engine stalls during quick test.	Idle speed control	▶ GO to Pinpoint Test Step KE3.
• Gasoline fumes under hood.		▶ GO to Canister Purge Pinpoint Test Step KD1.
• Misses. • Engine runs rough. • Engine stalls. • Lack of power. • Erratic RPM.		• Poor power/ground connections. • Ignitions system distributor cap, rotor, wires, coil, plugs. • Base engine valves, cam timing, compression etc. • GO to Pinpoint Test Step S1. (Cylinder Balance Test). • EGR Valve Function.

EEC-IV Module Connector PIN Usage — 5.0L SEFI

Pin	Car Lines: Ford, Mercury, Town Car — Crt.#	Ford, Mercury, Town Car — Wire Color	Thunderbird, Cougar — Crt.#	Thunderbird, Cougar — Wire Color	Application	Abbreviations
1	37	Y	38	BK O	Keep Alive Power	KAPWR
4	11	DG-Y	189	LB-PK	Ignition Diagnostic Monitor	IDM
7	354	LG Y	354	LG Y	Engine Coolant Temperature	ECT
10	883	PK LB	883	PK LB	A C Clutch Signal	ACCS
11	99	LG BK	99	LG BK	Air Management	AM2
12	557	BR Y	557	BR Y	Injector #3	INJ 3
13	558	BR LB	558	BR LB	Injector #4	INJ 4
14	559	T LB	559	T LB	Injector #5	INJ. 5
15	560	LG	560	LG	Injector #6	INJ 6
16	259	BK O	259	BK O	Ignition Ground	ING. GND
17	201	T P	382	Y BK	Self-Test Output	STO
20	57	BK	57	BK	Case Ground	CASE GND
21	264	W LB	264	W LB	Idle Speed Control	ISC
22	97	T LG	97	T LG	Fuel Pump	F P
25	357	LG P	357	LG P	Air Charge Temperature Sensor	ACT
26	351	O W	351	O/W	Reference Voltage	VREF
27	352	BR LG	352	BR/LG	EGR Valve Position Sensor	EVP
29	94	DG P	94	DG P	Heated Exhaust Gas Oxygen Sensor	HEGO
30	33	W PK	376	Y R	Neutral Drive Switch	NDS
31	101	GY Y	101	GY Y	Canister Purge	CANP
33	360	DG	360	DG	EGR Valve Regulator	EVR
34	305	LB P	305	LB P	Data Output Line	DOL
36	324	Y LG	324	Y LG	Spark Output	SPOUT
37	361	R	175	BK Y	Vehicle Power	V PWR
40	60	BK LG	60	BK LG	Power Ground	PWR. GND
42	561	T O	561	T O	Injector #7	INJ. 7
43	90	DB LG	90	DB LG	Heated Exhaust Gas Oxygen Sensor	HEGO
45	358	LG BK	356	DB LG	Manifold Absolute Pressure Sensor	MAP
46	359	BK W	359	BK W	Signal Return	SIG. RTN
47	355	DG LG	355	DG LG	Throttle Position Sensor	T P
48	200	W BK	209	W R	Self-Test Input	S T I
49	89	O	89	O	Heated EGO Ground	HEGO GND
50	—	—	261	O BK		
51	100	W R	100	W R	Air Management	AMI
52	562	LB	562	LB	Injector #8	INJ 8
53	—	—	—	—	Shift Indicator Light	S I L
54	73	O LB	73	O LB	Wide Open Throttle A C Cut Off	WAC
56	349	DB	349	DB	Profile Ignition Pick-Up	PIP
57	361	R	175	BK Y	Vehicle Power	V PWR
58	555	T	555	T	Injector #1	INJ 1
59	556	W	556	W	Injector #2	INJ 2
60	60	BK LG	60	BK LG	Power Ground	PWR GND

Pin locations given for reference only. Probing 60 pin connector with DVOM probe will result in permanent damage to the pin connectors. Always probe as directed, using the Breakout Box.

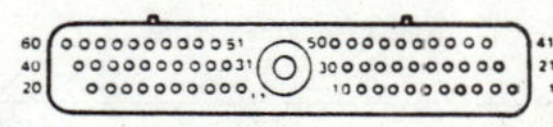

Electrical Schematic — 5.0L SEFI

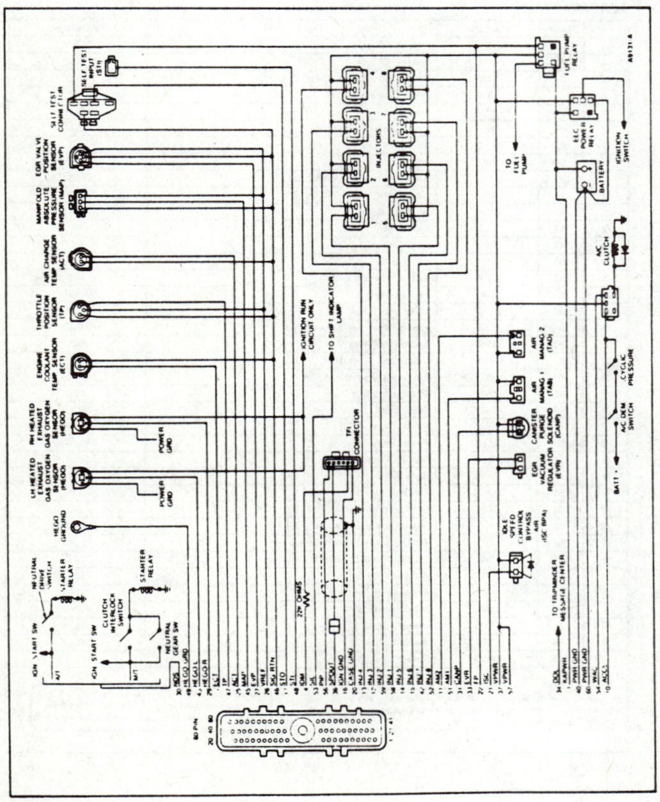

EEC-IV Connector PIN Usage — 5.0L SEFI

Pin	Car Lines: Mark VII, Continental — Crt. #	Mark VII, Continental — Wire Color	Mustang, Capri — Crt. #	Mustang, Capri — Wire Color	Application	Abbreviations
1	38	BK O	38	BK O	Keep Alive Power	KAPWR
4	11	DG Y	11	DG Y	Ignition Diagnostic Monitor	IDM
7	354	LG Y	354	LG Y	Engine Coolant Temperature	ECT
10	348	LG P	883	PK LB	A C Clutch Signal	ACCS
11	99	LG BK	99	LG BK	Air Management	AM2
12	557	T Y	557	BR Y	Injector #3	INJ 3
13	558	T BK	558	BR LB	Injector #4	INJ 4
14	559	T LB	559	T LB	Injector #5	INJ 5
15	560	T LG	560	LG	Injector #6	INJ 6
16	259	BK O	259	BK O	Ignition Ground	ING GND
17	382	Y BK	382	Y BK	Self-Test Output	STO
20	57	BK	57	BK	Case Ground	CASE GND
21	264	W LB	264	W LB	Idle Speed Control	ISC
22	97	T LG	97	T LG	Fuel Pump	F P
25	357	LG P	.357	LG P	Air Charge Temperature	ACT
26	351	O W	351	O W	Reference Voltage	VREF
27	352	BR LG	352	BR LG	EGR Valve Position Sensor	EVP
29	94	DG P	94	DG P	Heated Exhaust Gas Oxygen Sensor	HEGO
30	33	W PK	33	W PK	Neutral Drive Switch	NDS
31	101	GY Y	101	GY Y	Canister Purge	CANP
33	360	DG	360	DG	EGR Valve Regulator	EVR
34	305	LB P	—	—	Data Output Line	DOL
36	324	Y LG	324	Y LG	Spark Output	SPOUT
37	361	R	361	R	Vehicle Power	V PWR
40	60	BK LG	60	BK LG	Power Ground	PWR GND
42	561	T O	561	T O	Injector #7	INJ 7
43	90	DB LG	90	DB LG	Heated EGO	HEGO
45	356	DB LG	356	DB LG	Manifold Absolute Pressure	MAP
46	359	BW W	359	BK W	Signal Return	SIG RTN
47	355	DG LG	355	DG LG	Throttle Position Sensor	T P
48	209	W R	209	W R	Self-Test Input	S T I
49	89	O	89	O	Heated EGO Ground	HEGO GND
50	515	OR	—	—		
51	100	W R	100	W R	Air Management	AMI
52	562	TR	562	LB	Injector #8	INJ 8
53	—	—	462	P	Shift Indicator Light	S I L
54	73	O LB	73	O LB	Wide Open Throttle A C Cut Off	WAC
56	349	DB	349	DB	Profile Ignition Pick-Up	PIP
57	361	R	361	R	Vehicle Power	V PWR
58	555	T	555	T	Injector #1	INJ 1
59	556	T W	556	W	Injector #2	INJ 2
60	60	BK LG	60	BK LG	Power Ground	PWR GND

Pin locations given for reference only. Probing 60 pin connector with DVOM probe will result in permanent damage to the pin connectors. Always probe as directed, using the Breakout Box.

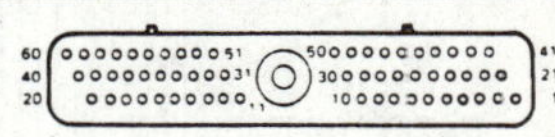

Diagnostic By Symptom — 2.3L EFI

SYMPTOM	RESULT	ACTION TO TAKE
• Engine stalls. • Stalls in Self-Test. • Runs rough. • Misses. • Always rich/lean.		• GO to Pinpoint Test Step S1. • Map test. GO to Pinpoint Test Step DF20. • Idle speed control, GO to Pinpoint Test Step KE1. • Poor power ground connections. • Ignition system distributor cap, rotor, wires, coil, plugs. • Base engine valves, cam timing, compression etc.
• Detonation/spark knock.		GO to Pinpoint Test Step DG1.
• Lack of fast idle assist with A/C On.		GO to Pinpoint Test Step FA1.
• Shift indicator light always On or Off.		GO to Pinpoint Test Step KL1.
• A/C does not cut-off under WOT conditions.		GO to Pinpoint Test Step KM1.
• Surges with A/C on at idle.		GO to Pinpoint Test Step KM20.
• Stumble after hot re-start (HEGO).		GO to Pinpoint Test Step HE21.
• Stalls during parking maneuvers.		GO to Pinpoint Test Step FF3.

EEC-IV Module Connector Pin Usage — 2.3L EFI

RANGER/BRONCO II

Pin	Circuit	Wire Color	Application	Abbreviations
1	37A	Y	Keep Alive Power	KAPWR
2	511	LG	Brake On Off	BOO
3	330	Y-LG	Power Steering Pressure Switch	PSPS
4	11	DG-Y	Ignition Diagnostic Monitor	IDM
5	361C	R	Key Power	KPWR
7	354	LG-Y	Engine Coolant Temperature	ECT
10	347	BK-Y	A/C Clutch	ACC
16	57A	BK	Ignition Ground	IGN GND
17	201	T-R	Self-Test Output/Shift Indicator Light	STO/SIL
20	57	BK	Case Ground	CSE GND
21	67	GY-W	Idle Speed Control	ISC
22	97	T-LG	Fuel Pump	FP
23	99	LG-BK	Knock Sensor	KS
24	347	BK-Y	A/C Demand	ACD
25	310	Y-R	Air Charge Temperature	ACT
26	351	O-W	Reference Voltage	V REF
27	352	BR-LG	EGR Valve Position	EVP
29	94	DG-P	Heated Exhaust Gas Oxygen Sensor	HEGO
30	150	DG-W	Neutral Drive Switch (A/T)	NDS
30	200	W-BK	Neutral Drive Switch (M/T)	NDS
33	360	DG	Exhaust Gas Recirculation Vent	EGRV
34	305	LB-P	Data Output Link	DOL
36	324	Y-LG	Spark Output	SPOUT
37	361	R	Vehicle Power	V. PWR
40	60A	BK-LG	Power Ground	PWR GND
45	356	DB-LG	Manifold Absolute Pressure	MAP
46	359	BK-W	Signal Return	SIG. RTN
47	355	DG-LG	Throttle Position Sensor	T P
48	100	W-R	Self-Test Input	STI
49	89	O	Heated Exhaust Gas Oxygen Sensor Ground	HEGOG
52	362	Y	Exhaust Gas Recirculation Control	EGRC
53	237B	O-Y	Clutch Converter Override (A/T only)	CCO
54	331	R	W.O.T. A/C Cut Off	WAC
56	349	DB	Profile Ignition Pick-Up	PIP
57	361B	R	Vehicle Power	V PWR
58	96	T-O	Injector 1 (Bank 1)	INJ 1
59	95	T-R	Injector 2 (Bank 2)	INJ 2
60	60B	BK-LG	Power Ground	PWR GRD

Pin locations given for reference only. Probing 60 pin connector with DVOM probe will result in permanent damage to the pin connectors. Always probe as directed using the Breakout Box.

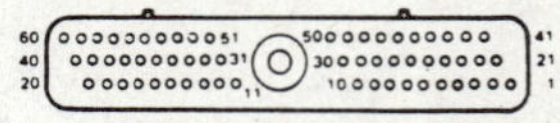

Electrical Schematic — 2.3L EFI

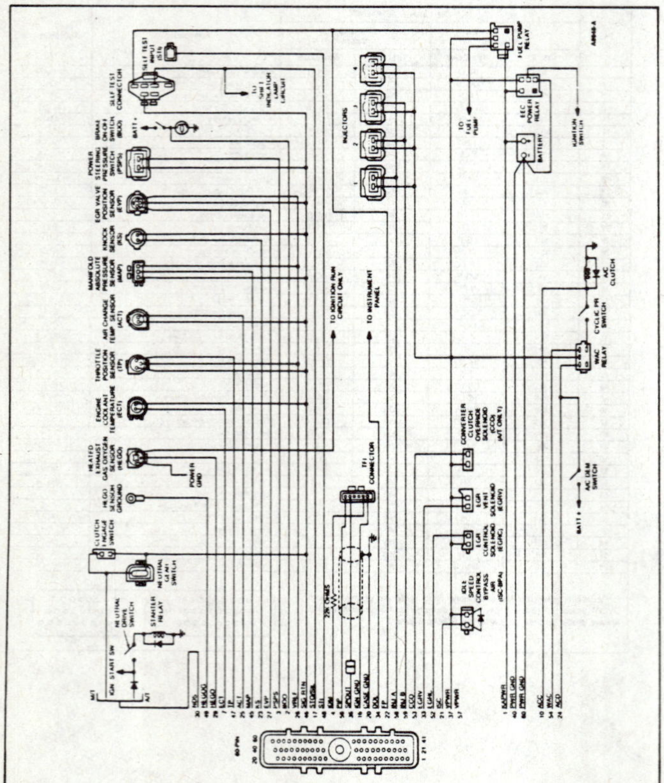

EEC-IV Module Connector PIN Usage — 2.3L EFI

AEROSTAR

Pin	Circuit	Wire Color	Application	Abbreviations
1	37A	Y	Keep Alive Power	KAPWR
2	810	R-LG	Brake On-Off	BOO
3	330	Y-LG	Power Steering Pressure Switch	PSPS
4	11	DG-Y	Ignition Diagnostic Monitor	IDM
5	361C	R	Key Power	KPWR
7	354	LG-Y	Engine Coolant Temperature	ECT
10	347	BK-Y	A/C Clutch	ACC
16	57A	BK	Ignition Ground	IGN. GND
17	201	T-R	Self-Test Output/Shift Indicator Light	STO/SIL
20	57	BK	Case Ground	CSE. GND
21	67	GY-W	Idle Speed Control	ISC
22	97	T-LG	Fuel Pump	FP
23	310	Y-R	Knock Sensor	KS
24	347	BK-Y	A/C Demand	ACD
25	357	LG-P	Air Charge Temperature	ACT
26	351	O-W	Reference Voltage	V REF
27	352	BR-LG	EGR Valve Position	EVP
29	94	DG-P	Heated Exhaust Gas Oxygen Sensor	HEGO
30	150	DG-W	Neutral Drive Switch (A/T)	NDS
30	200	W-BK	Neutral Drive Switch (M/T)	NDS
33	360	DG	Exhaust Gas Recirculation Vent	EGRV
34	305	LB-P	Data Output Link	DOL
36	324	Y-LG	Spark Output	SPOUT
37	361	R	Vehicle Power	V. PWR.
40	60A	BK-LG	Power Ground	PWR. GND
45	356	DB-LG	Manifold Absolute Pressure	MAP
46	359	BK-W	Signal Return	SIG. RTN.
47	355	DG-LG	Throttle Position Sensor	T P
48	100	W-R	Self-Test Input	STI
49	89	O	Heated Exhaust Gas Oxygen Ground	HEGOG
52	362	Y	Exhaust Gas Recirculation Control	EGRC
53	237B	O-Y	Clutch Converter Override (A/T only)	CCO
54	331	R	W.O.T. A/C Cut Off	WAC
56	349	DB	Profile Ignition Pick-Up	PIP
57	361B	R	Vehicle Power	V PWR.
58	96	T-O	Injector 1 (Bank 1)	INJ 1
59	95	T-R	Injector 2 (Bank 2)	INJ 2
60	60B	BK-LG	Power Ground	PWR GRD

Pin locations given for reference only. Probing 60 pin connector with DVOM probe will result in permanent damage to the pin connectors. Always probe as directed, using the Breakout Box.

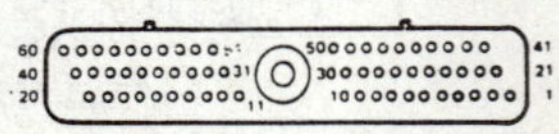

Diagnostic By Symptom — 2.8L FBC

SYMPTOM	RESULT	ACTION TO TAKE
• Engine stalls. • Stalls in Self-Test. • Runs rough. • Misses. • Lack of power.	Carburetor icing ▶	• GO to KH1.
	Minor lean or rich fuel drive ▶	• GO to KK1. • Poor power ground connections. • Ignition system distributor cap, rotor, wires, coil, plugs. • Base engine valves, cam timing, compression etc. • EGR Valve, Section 6. • Fuel Delivery, Section II or Shop Manual, Group 24.
	Choke always On/Off, slow in response ▶	• GO to KJ1.
• Shift indicator light always On or Always Off.		• GO to Pinpoint Test Step KL1.
• Erratic rpm.	rpm out of specification ▶	• Verify rpm per VECI decal. GO to Idle Adjustment, Section 4.
• Gasoline fume odors present under hood.		• GO to Section 7 or Shop Manual, Group 24.
• Lack of fast idle with A/C On.		• GO to Pinpoint Test Step FA1 and or KF1.

EEC-IV Module Connector Pin Usage — 2.8L FBC

Pin	Circuit	Wire Color	Application	Abbreviations
1	37	Y	Keep Alive Memory	KAM
2	810	R/LG	Brake On/Off	BOO
4	11	DG/Y	Ignition Diagnostic Monitor	IDM
5	20	W/LB	Ignition Run Start	—
7	354	LG/Y	Engine Coolant Sensor	ECT
10	347	BK/Y	A/C Clutch	ACC
11	200	W/BK	Thermactor Air Diverter	TAD
16	57	BK	Ignition Ground	IGN. GRD
17	201	T/R	Self-Test Output	STO
20	60	BK/LG	Chassis Ground	GRD.
21	96	T/O	Idle Speed Control	ISC
22	237	O/Y	Converter Clutch Override	CCO
23	310	Y/R	Knock Sensor	KI
24	223	T/LG	Inferred Milage Sensor	IMS
25	357	LG/P	Air Charge Temperature Sensor	ACT
26	351	O/W	Reference Voltage	V. REF.
27	352	BR/LG	EGR Valve Position Sensor	EVP
28	98	T/LB	Idle Tracking Switch	ITS
29	94	DG/P	Exhaust Gas Oxygen Sensor	EGO
30	150	DG/W	Neutral Drive Switch	NDS
33	360	DG	EGR Vent Solenoid	EGRV
35	101	GY/Y	Canister Purge	CANP
36	324	Y/LG	Ignition Module Signal	SPOUT
37	361	R	Vehicle Power	V. PWR.
40	60	BK/LG	Vehicle Ground	GND.
41	95	T/R	Idle Speed Control	ISC
45	356	DB/LG	Manifold Absolute Pressure Sensor	MAP
46	359	BK/W	Signal Return	SIG. RTN.
47	355	DG/LG	Throttle Position Sensor	TPS
48	100	W/R	Self-Test Input	STI
49	89	O	Exhaust Gas Oxygen Sensor Ground	EGO GRD
51	190	W/R	Thermactor Air Bypass	TAB
52	362	Y	EGR Control Solenoid	EGRC
54	256	O/W	Temperature Compensation Pump	TCP
55	462	P	Variable Voltage Choke	VVC
56	349	DB	Profile Ignition Pick-Up	PIP
57	361	R	Vehicle Power	V POWER
58	97	T/LG	Feedback Carburetor	FBC
59	75	DG/LG	Inlet Air Solenoid	IAS
60	60	BK/LG	Vehicle Ground	GRD

Pin locations given for reference only. Probing 60 pin connector with DVOM probe will result in permanent damage to the pin connectors. Always probe as directed, using the Breakout Box.

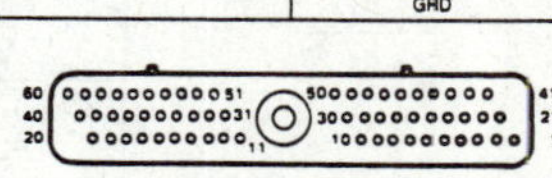

Electrical Schematic — 2.8L FBC

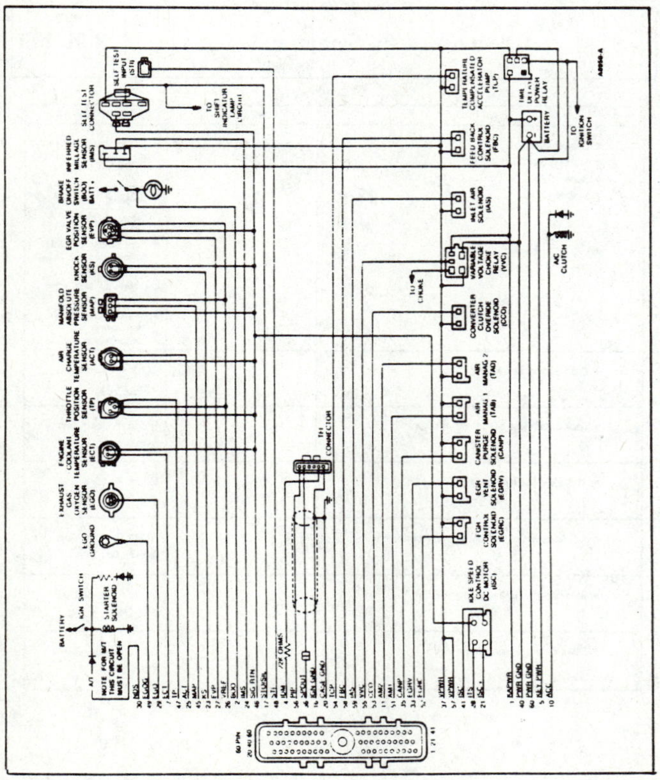

Diagnostic By Symptom — 2.9L EFI

SYMPTOM	RESULT	ACTION TO TAKE
• Engine stalls. • Stalls in Self-Test. • Runs rough. • Misses.	Always rich/lean. ▶	• GO to Pinpoint Test Step S1. • Map test GO to Pinpoint Test Step DF-20. • Poor power/ground connections. • Ignition system distributor cap, rotor, wires, coil, plugs. • Base engine valves cam timing, compression etc.
• Detonation/spark knock.		GO to Pinpoint Test Step DG1.
• Lack of fast idle assist with A/C On.		GO to Pinpoint Test Step FA1.
• A/C does not cut-off under WOT conditions.		GO to Pinpoint Test Step KM1.
• Stumble after hot re-start (HEGO).		GO to Pinpoint Test Step HG2.
• Stalls during parking maneuvers.		GO to Pinpoint Test Step FF3.

Electrical Schematic — 2.9L EFI

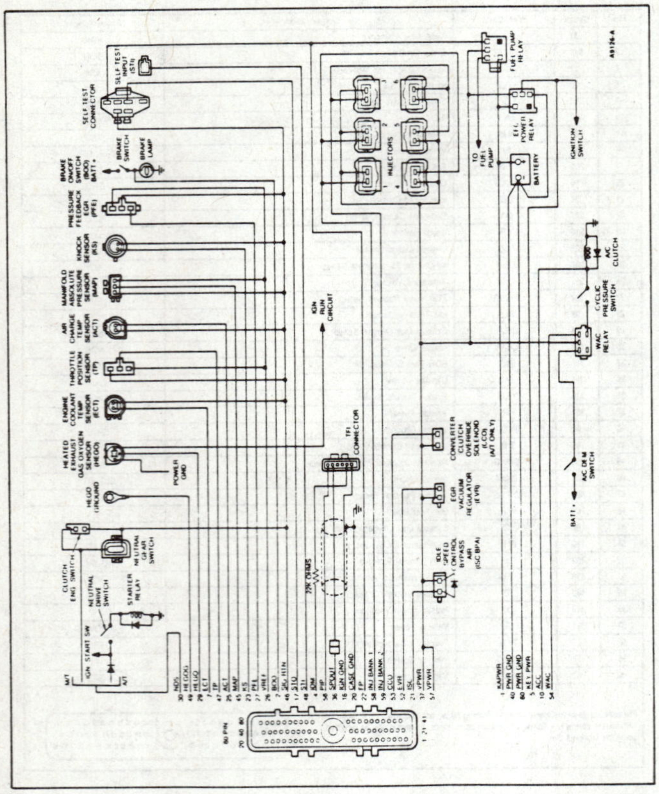

Electrical Schematic — 3.0L EFI

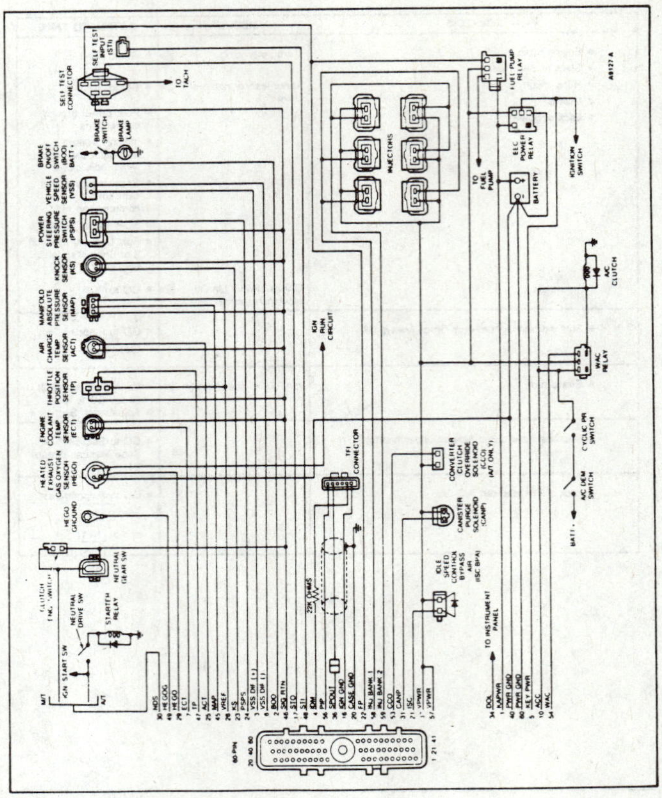

EEC-IV Module Connector PIN Usage — 2.9L EFI

Pin	Circuit	Wire Color	Application	Abbreviations
1	37	Y	Keep Alive Power	KAPWR.
2	511	LG	Brake On-Off	BOO
4	11	DG/Y	Ignition Diagnostic Monitor	IDM
5	16	R/LG	Key Power	KEY PWR.
7	354	LG/Y	Engine Coolant Temp. Sensor	ECT
10	347	BK/Y	A/C Clutch	ACC
16	57	BK	Ignition Ground	IGN. GND
17	201	T/R	Self-Test Output	STO
20	60	B/LG	Case Ground	CASE GND.
21	67	GY/W	Idle Speed Control	ISC
22	97	T/LG	Fuel Pump	FP
23	310	Y/R	Knock Sensor	KS
25	357	LG/P	Air Charge Temp. Sensor	ACT
26	351	O/W	Reference Voltage	V. REF.
27	352	BR/LG	Pressure Feedback Electronic Sensor (EGR)	PFE
29	94	DG/P	Heated Exhaust Gas Oxygen Sensor	HEGO
30	150	DG/W	Neutral Drive Switch	NDS
30	200	W/BK	Neutral Gear & Clutch Switch M/T	NDS
36	324	Y/LG	Spark Output	SPOUT
37	361	R	Vehicle Power	V. PWR.
40	60	BK/LG	Power Ground	PWR. GND.
45	356	DB/LG	Manifold Absolute Pressure Sensor	MAP
46	359	BK/W	Signal Return	SIG. RTN.
47	355	DG/LG	Throttle Position Sensor	TP
48	100	W/R	Self-Test Input	STI
49	89	O	Heated Exhaust Gas Oxygen Sensor Ground	HEGOG
52	362	Y	Electronic Vacuum Regulator	EVR
53	237	O/Y	Clutch Converter Override (Not used on M/T)	CCO
54	331	R	W.O.T. A/C Cut-Off	WAC
56	349	DB	Profile Ignition Pick-up	PIP
57	361	R	Vehicle Power	V. PWR.
58	265	LG/W	Injector Bank 1	INJ 1
59	95	T/R	Injector Bank 2	INJ 2
60	60	BK/LG	Power Ground	PWR. GND.

Pin locations given for reference only. Probing 60 pin connector with DVOM probe will result in permanent damage to the pin connectors. Always probe as directed, using the Breakout Box.

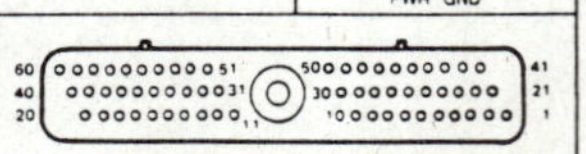

Diagnostics By Symptom — 3.0L EFI

SYMPTOM	RESULT	ACTION TO TAKE
• Engine stalls. • Stalls in Self-Test. • Runs rough. • Misses.	Always rich/lean. ▶	• GO to Pinpoint Test Step S1. • Map test GO to Pinpoint Test Step DF-20. • Poor power ground connections. • Ignition system distributor cap, rotor, wires, coil, plugs. • Base engine valves cam timing, compression etc.
• Detonation/spark knock.		GO to Pinpoint Test Step DG1.
• Lack of fast idle assist with A/C On.		GO to Pinpoint Test Step FA1.
• Shift indicator light always On or Off.		GO to Pinpoint Test Step KL1.
• A/C does not cut-off under WOT conditions.		GO to Pinpoint Test Step KM1.
• Stumble after hot re-start (HEGO).		GO to Pinpoint Test Step HG21.
• Stalls during parking maneuvers.		GO to Pinpoint Test Step FF3.

EEC-IV Module Connector PIN Usage — 3.0L EFI

Pin	Circuit	Wire Color	Application	Abbreviations
1	37	Y	Keep Alive Power	KAPWR
2	810	R LG	Brake ON-OFF	BOO
3	150	DG W	Vehicle Speed Sensor Differential Positive	VSSD F -
4	11	DG Y	Ignition Diagnostic Monitor	IDM
6	397	BK W	Vehicle Speed Sensor Differential Negative	VSSDIF -
7	354	LG Y	Engine Coolant Temperature Sensor	ECT
10	347	BK Y	A C Cycling Pressure Switch	ACCS
16	259	BK O	Ignition Ground	IGN GND
17	201	T R	Self-Test Output	STO
20	57	BK	Case Ground	CASE GND
21	68	O BK	Idle Speed Control	ISC
22	97	T LG	Fuel Pump	FP
23	310	Y R	Knock Sensor	KS
24	330	Y LG	Power Steering Pressure Switch	PSPS
25	357	LG P	Air Charge Temperature Sensor	ACT
26	351	O/W	Reference Voltage	VREF
29	94	DG P	Heated Exhaust Gas Oxygen Sensor	HEGO
30	200	W/BK	Neutral Gear & Clutch Switch	N D S
30	151	LB/BK	A T Neutral Drive Switch	N D S
31	101	GY Y	Canister Purge	CANP
34	305	LB P	Data Output Link	DOL
36	324	Y-LG	Spark Output	SPOUT
37	361	R	Vehicle Power	V PWR
40	60	BK LG	Power Ground	PWR GND
45	356	DB LG	Manifold Absolute Pressure Sensor	MAP
46	359	BK W	Signal Return	SIG RTN
47	355	DG LG	Throttle Position Sensor	T P
48	100	W R	Self-Test Input	STI
49	89	O	Heated Exhaust Gas Oxygen Sensor Ground	HEGOG
53	237	O Y	Converter Clutch Override (Not used on M T)	CCO
54	331	R	W O T A C Cut-Off	WAC
56	349	DB	Profile Ignition Pick-up	PIP
57	361	R	Vehicle Power	V PWR
58	96	T O	Injector Bank 1	INJ 1
59	95	T R	Injector Bank 2	INJ 2
60	60	BL LG	Power Ground	PWR GND

Pin locations given for reference only. Probing 60 pin connector with DVOM probe will result in permanent damage to the pin connectors. Always probe as directed, using the Breakout Box.

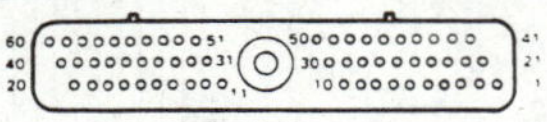

Electrical Schematic — 4.9L FBC

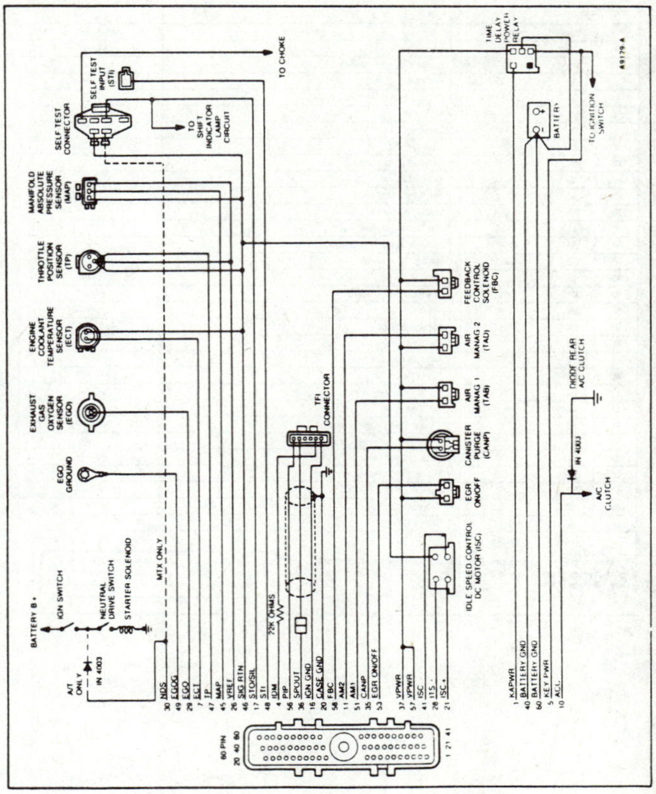

Diagnostic By Symptom — 4.9L FBC

SYMPTOM	RESULT	ACTION TO TAKE
• Engine runs rough or miss-fires. • Engine stalls. • Loss of power. • Misses.		• Poor power ground connections. • Ignition system distributor cap, rotor, wires, coil, plugs, Section 15. • Base engine valves, cam timing, compression etc., refer to Shop Manual Group 21. • Fuel delivery, Section 11 and/or Group 24. • EGR valve function, Section 6.
• Detonation/spark knock. • Rough idle.		• EGR valve function, Section 6. • GO to Pinpoint Test Step KA1.
• Lack of fast idle with A/C On.		• GO to Pinpoint Test Step FA1 and/or KF1.
• Gasoline fumes under hood.		• GO to Pinpoint Test Step KD1 and/or Section 7.
• Shift indicator light always On or always Off.		• GO to Pinpoint Test Step KL1.
• Erratic rpm.	rpm out of specification ▶	• Verify rpm per VECI decal. GO to Idle Adjustment Section 4.

EEC-IV Module Connector Pin Usage — 4.9L FBC

ECONOLINE

Pin	Circuit	Wire Color	Application	Abbreviations
1	37	BK Y	Keep Alive Power	KAPWR
4	11	BK Y	Ignition Diagnostic Module	IDM
5	732	W	Key Power	K PWR
7	354	GR Y	Engine Coolant Temperature	ECT
10	348	GR W	A C Clutch	ACC
11	200	BL	Thermactor Air Diverter	TAD
16	57	BK	Case Ground	CASE GND.
17	201	BL Y	Self-Test Output Shift Ind Light	STO SIL
20	60	BK W	Ignition Ground	IGN GND
21	96	O BK	Idle Speed Control	ISC
26	351	R	Voltage Reference	V REF
28	98	BK	Idle Tracking Switch	ITS
29	94	GR/P	Exhaust Gas Oxygen Sensor	EGO
30	150	R	Neutral Drive Switch	N/D
35	101	GY	Canister Purge	CANP
36	324	Y GR	Spark Output	SPOUT
37	361	R	Vehicle Power	V PWR
40	60	BK W	Battery Ground	BATT GND.
41	95	W	Idle Speed Control	ISC
45	356	BL GR	Manifold Absolute Pressure	MAP
46	359	BK/W	Signal Return	SIG. RTN.
47	355	GR	Throttle Position Sensor	TPS
48	100	BK/R	Self-Test Input	STI
49	89	O	Exhaust Gas Oxygen Ground	EGO GND.
51	190	Y	Thermactor Air Bypass	TAB
53	360	Y BL	Exhaust Gas Recirculate Shut Off	EGR S/O
56	349	BR	Profile Ignition Pick-Up	PIP
57	361	R	Vehicle Power	V PWR
58	97	O	Feedback Control	FBC
60	60	BK W	Battery Ground	BATT GND

Pin locations given for reference only. Probing 60 pin connector with DVOM probe will result in permanent damage to the pin connectors. Always probe as directed, using the Breakout Box.

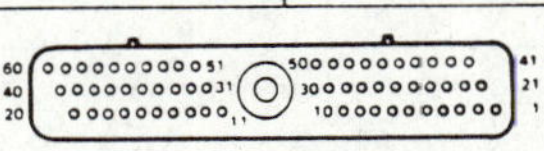

EEC-IV Module Connector Pin Usage — 4.9L FBC

F-SERIES

Pin	Circuit	Wire Color	Application	Abbreviations
1	37	Y	Keep Alive Power	KAPWR
4	11	DG Y	Ignition Diagnostic Module	IDM
5	20	W LB	Key Power	K PWR
7	354	LG Y	Engine Coolant Temperature	ECT
10	348	LG PH	A C Clutch	ACC
11	200	W BK	Thermactor Air Diverter	TAD
16	57	BK	Case Ground	CASE GND
17	201	T R	Self-Test Output Shift Ind Light	STO SIL
20	60	BK LG	Ignition Ground	IGN GND
21	96	T O	Idle Speed Control	ISC
26	351	O W	Voltage Reference	V REF
28	98	T LB	Idle Tracking Switch	ITS
29	94	DG PH	Exhaust Gas Oxygen Sensor	EGO
30	150	DG W	Neutral Drive Switch	N D
35	101	GY Y	Canister Purge	CANP
36	324	Y/LG	Spark Output	SPOUT
37	361	R	Vehicle Power	V PWR
40	60	BK LG	Battery Ground	BATT GND
41	95	T R	Idle Speed Control	ISC
45	356	DB LG	Manifold Absolute Pressure	MAP
46	359	BK W	Signal Return	SIG RTN
47	355	DG LB	Throttle Position Sensor	TPS
48	100	W R	Self-Test Input	STI
49	89	O	Exhaust Gas Oxygen Ground	EGO GND
51	190	W R	Thermactor Air Bypass	TAB
53	360	DG	Exhaust Gas Recirculate Shut Off	EGR S.O
56	349	DB	Profile Ignition Pick-Up	PIP
57	361	R	Vehicle Power	V PWR
58	97	T LG	Feedback Control	FBC
60	60	BK LG	Battery Ground	BATT GND

Pin locations given for reference only. Probing 60 pin connector with DVOM probe will result in permanent damage to the pin connectors. Always probe as directed using the Breakout Box.

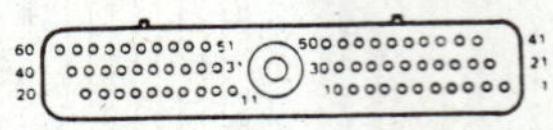

Electrical Schematic — 5.0L EFI

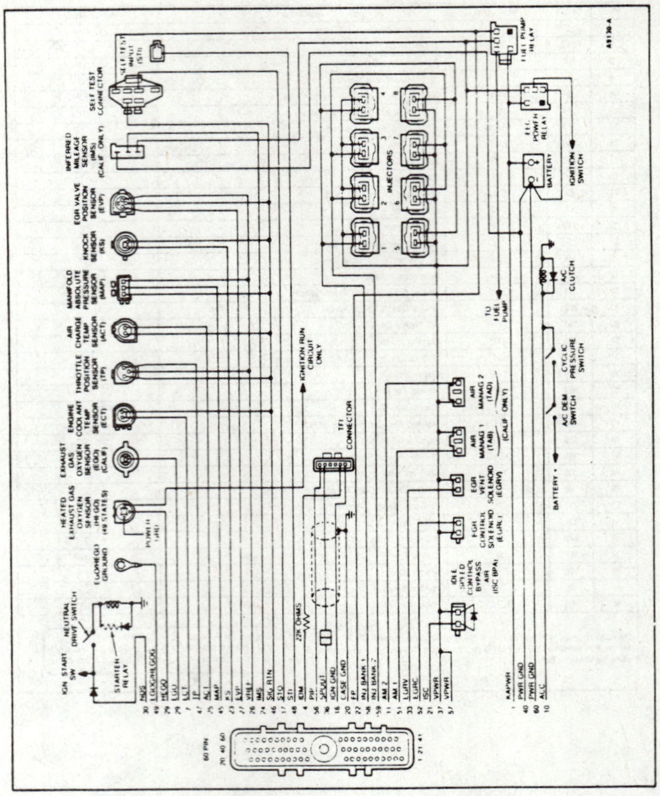

Diagnostic By Symptom — 5.0L EFI

SYMPTOM	RESULT	ACTION TO TAKE
• Engine stalls. • Stalls in Self-Test. • Runs rough. • Misses.	Always rich/lean. ▶	• GO to Pinpoint Test Step S1. • Map test, GO to Pinpoint Test Step DF20. • Idle Speed Control, GO to Pinpoint Test Step KE1. • Poor power ground connections. • Ignition system distributor cap, rotor, wires, coil, plugs. • Base engine valves, cam timing, compression etc.
• Detonation spark knock.		GO to Pinpoint Test Step DG1.
• Lack of fast idle assist with A C On.		GO to Pinpoint Test Step FA1.
• Shift indicator light always On or Off.		GO to Pinpoint Test Step KL1.
• A/C does not cut-off under WOT conditions.		GO to Pinpoint Test Step KM1.
• Surges with A C on at idle.		GO to Pinpoint Test Step KM20.
• Stumble after hot re-start (HEGO).		GO to Pinpoint Test Step HE21.
• Stalls during parking maneuvers.		GO to Pinpoint Test Step FF3.

EEC-IV Module Connector Pin Usage — 5.0L EFI

Pin	Circuit	Wire Color	Application	Abbreviations
1	37D	Y	Keep Alive Power	KAPWR
4	11D	DG/Y	Ignition Diagnostic Monitor	IDM
5	361R	R	Key Power	KPWR
7	354	LG Y	Engine Coolant Temperature	ECT
10	348	LG P	A C Clutch	ACC
11	200	W BK	Air Management 2	AM-2
16	57	BK	Ignition Ground	IGN GND
17	201	T R	Self-Test Output	STO
20	57A	BK	Case Ground	CSE GND
21	67	GY W	Idle Speed Control	ISC
22	97B	T LG	Fuel Pump	FP
23	99	LG BK	Knock Sensor	K S.
24	223	T LG	Inferred Mileage Sensor	IMS
25	310	Y R	Air Charge Temperature	ACT
26	351	O/W	Reference Voltage	V REF
27	352	BR/LG	EGR Valve Position	EVP
29	94	DG/P	Exhaust Gas Oxygen Sensor Heated Exhaust Oxygen Sensor	EGO HEGO
30	150	DG/W	Neutral Drive Switch	NDS
33	360	DG	Exhaust Gas Recirculation Vent	EGRV
36	324A	Y LG	Spark Output	SPOUT
37	361	R	Vehicle Power	V PWR
40	60A	BK LG	Power Ground	PWR GND
45	356	DB/LG	Manifold Absolute Pressure	MAP
46	359	BK W	Signal Return	SIG RTN.
47	355	DG/LG	Throttle Position Sensor	T P
48	100	W/R	Self-Test Input	S T I.
49	89	O	Exhaust Gas Oxygen Ground Heated Exhaust Oxygen Ground	EGOG HEGOG
51	190	W R	Air Management 1	AM-1
52	362	Y	Exhaust Gas Recirculation Control	EGRC
56	349	DB	Profile Ignition Pick-Up	PIP
57	361A	R	Vehicle Power	V PWR
58	96	T O	Injector 1 (Bank 1)	INJ 1
59	95	T R	Injector 2 (Bank 2)	INJ 2
60	60B	BK LG	Power Ground	PWR GND

Pin locations given for reference only. Probing 60 pin connector with DVOM probe will result in permanent damage to the pin connectors. Always probe as directed using the Breakout Box.

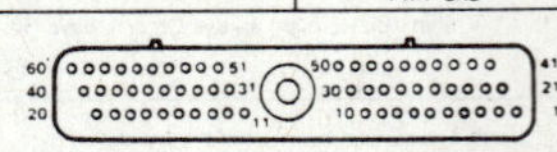

EEC IV No Start — Pinpoint Test — A

A

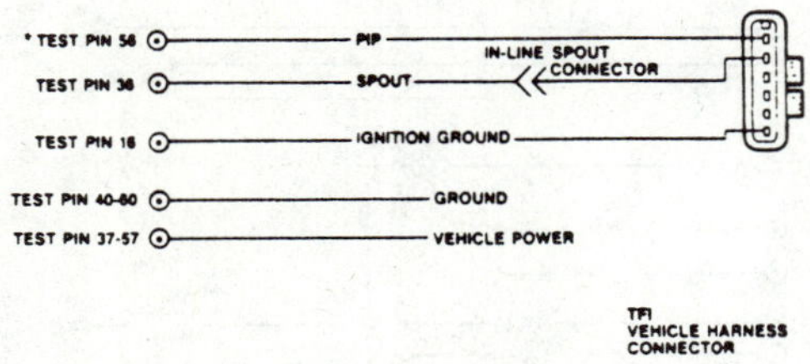

* TEST PIN LOCATED ON BREAKOUT BOX.
ALL CONNECTIONS VIEWED INTO MATING SURFACE.

NOTE: WHEN BREAKOUT BOX IS INSTALLED, ENSURE THAT TIMING SWITCH IS IN "COMPUTED" POSITION UNLESS OTHERWISE NOTED.

STOP-WARNING

You should enter this Pinpoint Test only when Steps 1.0 through 3.0 have been successfully completed and the engine is still a no start. This Pinpoint Test will not diagnose ignition system problems.

To prevent the replacement of good components, be aware that the following non-EEC areas may be at fault:

- Fuel: quantity and quality
- Ignition: general condition, moisture, cracks, damage, etc.
- Engine: internal, valves, timing belt, camshaft.
- Starter and battery circuit

This Pinpoint Test is intended to diagnose only the following:

- Spark (as related to EEC-IV).
- Circuits: pip, spout, ignition ground, vehicle power.

EEC IV No Start — Pinpoint Test — A

WARNING: Stop this test at the first sign of a fuel leak and service as required.
CAUTION: No open flame — No smoking during fuel delivery checks.

TEST STEP	RESULT ▶	ACTION TO TAKE
A1 ATTEMPT TO START ENGINE		
	Engine cranks, but does not start, or stalls out ▶	GO to A2.
	Engine does not crank ▶	REFER to Shop Manual, Group 28.
A2 CHECK FOR VREF AT THROTTLE POSITION SENSOR		
• Key Off, wait 10 seconds. • DVOM on 20V scale • Disconnect TP sensor. • Key On, Engine Off. • Measure voltage at the TP vehicle harness connector between VREF and signal return **NOTE: Refer to electrical schematic in appropriate engine supplement section for connector pin orientation.**	Less than 4.0V or greater than 6.0V ▶ 4.0V to 6.0V ▶	GO to Pinpoint Test Step C1. RECONNECT TP sensor. GO to A3.
A3 CHECK FOR SPARK AT PLUGS		
• Disconnect the spark plug wire to any accessible cylinder • Connect spark tester between spark plug wire and engine ground. • Crank engine and check for spark. • Reconnect the spark plug wire to the spark plug.	Spark ▶ No spark ▶	GO to A13. GO to A4.
A4 CHECK FOR SPARK AT COIL		
• Remove high tension coil wire from distributor and install spark tester. • Check for spark while cranking. • Reconnect high tension coil wire to distributor.	Spark ▶ No spark ▶	REFER to Section 15, Part 2 for TFI Diagnosis for cap, rotor, wires. GO to A5.

EEC IV No Start — Pinpoint Test — A

TEST STEP	RESULT ▶	ACTION TO TAKE
A5 HARNESS CHECK (IGNITION GROUND)		
• Key Off, wait 10 seconds. • Install Breakout box. Leave processor disconnected. • DVOM on 200 ohm scale. • Disconnect TFI. • Measure resistance between test Pin 16 at the Breakout box and TFI harness connector ignition ground.	Less than 5 ohms ▶ 5 ohms or greater ▶	GO to A6. SERVICE harness as necessary. RERUN Quick Test.
A6 ISOLATION OF PROBLEM TO SPOUT CIRCUIT		
• Breakout box installed. • Connect TFI. • Connect processor. • Timing switch to "Dist" position on Breakout box. • Attempt to start vehicle • Does the vehicle start?	Yes ▶ No ▶	GO to A10. GO to A7.
A7 SPOUT SIGNAL CHECK		
• Breakout box installed. • Timing switch to "Computed" position on Breakout box. • DVOM on 20V scale. • Measure voltage between test Pin 36 at the Breakout box and chassis ground, during crank.	Less than 3.0V or greater than 6.0V ▶ Between 3.0V and 6.0V ▶	GO to A8. EEC OK. REFER to Section 15, for TFI diagnosis
A8 CHECK SPOUT FOR SHORTS		
• Key Off, wait 10 seconds. • Breakout box installed. • Disconnect processor. • Disconnect TFI. • DVOM on 200,000 ohm scale. • Measure resistance between test Pin 36 and test Pins 16, 20, 26, 40, 60 (short to ground), 37, 57 (short to power) and 56 (short to pip) at the Breakout box.	All readings 10,000 ohms or greater ▶ Any reading less than 10,000 ohms ▶	GO to A9. SERVICE short in harness. RERUN Quick Test if vehicle does not start. GO to A9.

EEC IV No Start — Pinpoint Test — A

TEST STEP	RESULT ▶	ACTION TO TAKE
A9 ISOLATE SHORT(S) IN PROCESSOR		
• Key Off, wait 10 seconds. • Breakout box installed. • Reconnect processor. • TFI disconnected. • DVOM on 200 ohm scale. • Measure resistance between test Pin 36 and test Pins 37 and 57 (short to power) also, test Pins 40 and 60 (short to ground) at the Breakout box	All readings 5.0 ohms or greater. ▶ Any reading less than 5.0 ohms ▶	Connect TFI. GO to A10. REPLACE processor. RERUN Quick Test.
A10 PIP SIGNAL CHECK		
• Breakout box installed. • DVOM to 20V scale. • Measure voltage between test Pin 56 and test Pin 16 at the Breakout box • Crank engine, record reading.	Between 3.0V and 6.0V ▶ Less than 3.0V or greater than 6.0V ▶	REMOVE Breakout box. REPLACE processor. RERUN Quick Test. GO to A11.
A11 CONTINUITY OF PIP CIRCUIT CHECK		
• Breakout box installed • Key Off, wait 10 seconds. • DVOM on 200 ohm scale. • Disconnect TFI. • Disconnect processor • Measure resistance between test Pin 56 at the Breakout box and TFI harness connector PIP circuit.	Less than 5 ohms ▶ 5 ohms or greater ▶	GO to A12. SERVICE open PIP circuit. RERUN Quick Test.

EEC IV No Start — Pinpoint Test A

TEST STEP	RESULT ▶	ACTION TO TAKE
A12 CHECK PIP CIRCUIT FOR SHORTS		
• Breakout box installed. • Processor disconnected. • Key Off. • Disconnect TFI connector. • DVOM on 200,000 ohm scale. • Measure resistance between test Pin 56 and test Pins 16, 20, 26, 40, 60 (shorts to ground) and test Pins 37 and 57 (shorts to power) and test Pin 36 (short to spout) at the Breakout box.	Any resistance less than 10,000 ohms ▶	SERVICE PIP circuit. RERUN Quick Test.
	All resistance greater than 10,000 ohms ▶	REFER to Section 15 for TFI diagnosis.
A13 SPOUT SIGNAL VERIFICATION		
• Key Off, wait 10 seconds. • Disconnect processors 60 pin connector and inspect for damaged pins, corrosion, loose wires. Service as necessary. • Install Breakout box. • Processor connected. • DVOM on 20V scale. • Measure voltage between test Pin 36 at the Breakout box and chassis ground, during crank. • Ensure timing switch is in "Computed" position on Breakout box.	Between 3.0V and 6.0V ▶	FBC vehicles, REFER to Shop Manual, Group 24. EFI and CFI GO to A21.
	Less than 3.0V or greater than 6.0V ▶	GO to A10.

Vehicle Battery — Pinpoint Test B

B

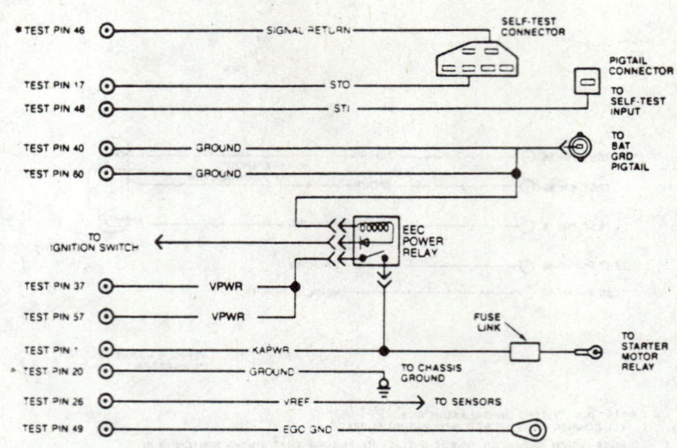

* TEST PINS LOCATED ON BREAKOUT BOX.
ALL HARNESS CONNECTORS VIEWED INTO MATING SURFACE.

STOP-WARNING

You should enter this Pinpoint Test only when directed here from Pinpoint Tests C, J or P or when a continuous memory code 78 is received in Quick Test Step 6.0C.

To prevent the replacement of good components, be aware that the following non-EEC areas may be at fault:

- Ignition switch.
- Battery Cables.
- Alternator.
- Voltage Regulator.
- Ground Straps.

This pinpoint test is intended to diagnose only the following:

- Processor.
- Harness circuits: Signal Return, STO, STI, Ground, VPWR, KAPWR, VREF, Ignition.
- Battery Voltage.
- Power Relay.

EEC IV No Start — Pinpoint Test A

TEST STEP	RESULT ▶	ACTION TO TAKE
A21 FUEL PUMP CHECK		
• **No smoking nearby.** • Disconnect all injector electrical connections at the injectors. • Connect pressure gauge. • Note initial pressure reading. • Observe pressure gauge as you pressurize fuel system. (Turn key to RUN for 1 second, then turn key to OFF. Wait 10 seconds. Repeat 5 times.) • Turn key Off. Wait 10 seconds. • Reconnect all injectors. **WARNING: If fuel starts leaking, turn key OFF immediately. No smoking.**	PRESSURE GAUGE READING: Increased ▶	All EFI Go to Pinpoint Test Step S1. All CFI Go to Pinpoint Test Step S2.
	Did not increase ▶	TURN key Off, and CONTINUE to A22.
A22 INERTIA SWITCH CHECK		
• Key Off. • Fuel pressure gauge installed. • Locate fuel pump inertia switch. Refer to Owner's Manual for location. • Push the button of inertia switch to reset to ON. **NOTE: If switch will not reset to ON, replace Inertia Switch and repeat Step A21. If switch button was on, GO to Step J1, except 2.5L HSC CFI and 3.0L EFI passenger car, GO to Step X-11.** • Observe pressure gauge as you pressurize fuel system. (Turn key to RUN for 1 second, then turn key to OFF. Wait 10 seconds.) Repeat 5 times.	PRESSURE GAUGE READING: Increase ▶	RERUN Quick Test.
	No increase ▶	• 2.5L HSC-CFI and 3.0L EFI passenger car GO to X-11. • All others GO to J1.

Vehicle Battery — Pinpoint Test B

TEST STEP	RESULT ▶	ACTION TO TAKE
B1 BATTERY VOLTAGE CHECK		
• Key On, Engine Off. • DVOM on 20V scale. • Measure voltage across battery terminals.	10.5V or greater ▶	GO to B2.
	Less than 10.5V ▶	SERVICE discharged battery. REFER to Shop Manual, Group 31.
B2 BATTERY POWER GROUND CHECK		
• Key On, Engine Off. • Processor connected. • DVOM on 20V scale. • Measure voltage between battery negative post and Signal Return in the Self-Test connector.	Less than 0.5V ▶	GO to B6.
	0.5V or greater ▶	GO to B3.
B3 GROUND FAULT ISOLATION		
• Breakout box installed. • Key On, Engine Off. • Processor connected. • DVOM on 20V scale. • Measure voltage between battery negative post and test Pins 40 and 60 at the Breakout box.	Both readings less than 0.5V ▶	GO to B4.
	One or both readings 0.5V or greater ▶	Circuit(s) with greater than 0.5V has high resistance or open. CORRECT faulty ground circuit. RERUN Quick Test.
B4 PROCESSOR GROUND FAULT ISOLATION		
• Breakout box installed. • Key Off, wait 10 seconds. • Processor connected. • DVOM on 200 ohm scale. • Measure resistance between test Pin 46 and test Pin 40 and between test Pin 46 and test Pin 60 both at the Breakout box.	Both readings less than 5 ohms ▶	GO to B5.
	One or both readings 5 ohms or greater ▶	DISCONNECT processor connector and INSPECT for corrosion, damaged pins, etc. SERVICE as necessary and RETEST. If fault is still present, REPLACE processor. RERUN Quick Test.

Vehicle Battery — Pinpoint Test B

TEST STEP	RESULT ▶	ACTION TO TAKE
B5 HARNESS CHECK (SIGNAL RETURN)		
• Breakout box installed. • Key Off, wait 10 seconds. • Processor connected. • DVOM on 200 ohm scale. • Measure resistance between test Pin 46 at the Breakout box and Signal Return in the Self-Test connector.	Less than 5 ohms ▶ 5 ohms or greater ▶	System OK. RUN Quick Test. CORRECT cause of resistance in the harness Signal Return circuit. RERUN Quick Test.
B6 12 VOLT BATTERY POWER FAULT ISOLATION		
• Key On, Engine Off. • Processor connected. • DVOM on 20V scale. • Measure voltage between the battery negative post and KAPWR circuit at EEC power relay.	10.5V or greater ▶ Less than 10.5V ▶	GO to B7. CHECK KAPWR and VPWR circuits for shorts to ground and KAPWR circuit from power relay to battery positive post for opens. SERVICE as necessary. RERUN Quick Test.
B7 12 VOLT BATTERY POWER FAULT ISOLATION		
• Key On, Engine Off. • Processor connected. • DVOM on 20V scale. • Measure voltage between the battery negative post and Ignition circuit at EEC power relay	10.5V or greater ▶ Less than 10.5V ▶	GO to B8. CHECK for open in ignition switch circuits. SERVICE as necessary. RERUN Quick Test.
B8 12 VOLT BATTERY POWER FAULT ISOLATION		
• Key On, Engine Off. • Processor connected. • DVOM on 20V scale. • Measure voltage between the battery negative post and ground circuit at EEC power relay.	Less than 0.5V ▶ 0.5V or greater ▶	If you entered this test for a code 78, GO to B10. All others, GO to B9. SERVICE open or ground in Ground Circuit. RERUN Quick Test.

Vehicle Battery — Pinpoint Test B

TEST STEP	RESULT ▶	ACTION TO TAKE
B9 12 VOLT BATTERY POWER FAULT ISOLATION		
• Key On, Engine Off. • Processor connected. • DVOM on 20V scale. • Measure voltage between the battery negative post and VPWR circuit at EEC power relay.	10.5V or greater ▶ Less than 10.5V ▶	SERVICE short to ground or open in VPWR circuit from EEC power relay to processor connector test Pins 37 and 57. RERUN Quick Test. REPLACE power relay. RERUN Quick Test.
B10 WIGGLE TEST VPWR CIRCUITS		
• Key On, Engine Off. • STAR tester or VOM hooked up to Self-Test connector. • Self-Test deactivated. • Observe STAR/VOM for fault indication as explained in Quick Test Step 6.0D • Shake, bend and twist the EEC-IV harness from the EEC time delay power relay to the processor. • Is a fault indicated or does code 78 reappear in continuous memory if Quick Test is rerun?	Yes ▶ No ▶	SERVICE intermittent VPWR circuit. RERUN Quick Test. INSPECT EEC-IV time delay power relay and harness connectors for damaged pins, corrosion, etc. SERVICE as necessary. If OK, REPLACE EEC-IV time delay relay. RERUN Quick Test.

Reference Voltage — Pinpoint Test C

C

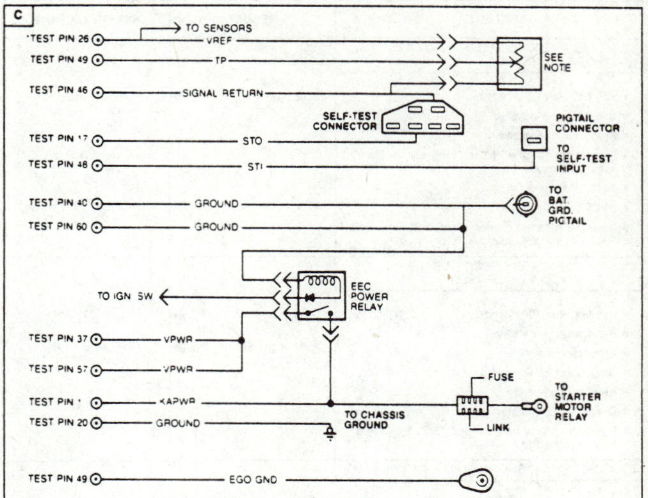

*TEST PINS LOCATED ON BREAKOUT BOX.
ALL HARNESS CONNECTORS VIEWED INTO MATING SURFACE.

NOTE: Refer to the electrical schematic in the appropriate engine supplement section for proper connector and pin orientation.

STOP-WARNING

You should enter this Pinpoint Test only when a check for VREF has failed in the sensor Pinpoint Tests (D-Series) or Pinpoint Tests A or Q.

This pinpoint test is intended to diagnose only the following:

- Processor.
- Sensor harness circuits: Signal Return, STO, STI, Ground, VPWR, KAPWR, VREF, Ignition.

Reference Voltage — Pinpoint Test C

TEST STEP	RESULT ▶	ACTION TO TAKE
C1 VEHICLE BATTERY POWER CIRCUIT CHECK		
• Breakout box installed. • Key On, Engine Off. • Processor connected. • DVOM on 20V scale. • Measure voltage between test Pin 37 at the Breakout box and Signal Return in Self-Test connector.	10.5V or greater ▶ Less than 10.5V ▶	GO to C2. 2.5L HSC CFI and 3.0L EFI passenger car GO to X-1. All others, GO to B1.
C2 VREF VOLTAGE CHECK		
• Breakout box installed. • Key On, Engine Off • Processor connected. • DVOM on 20V scale • Measure voltage between test Pin 26 and test Pin 46 at the Breakout box	6.0V or greater ▶ 4.0V or less ▶ Greater than 4.0V, less than 6.0V ▶	GO to C4. GO to C5. GO to C3.
C3 CHECK VREF AND SIGNAL RETURN FOR CONTINUITY		
• Breakout box installed. • Disconnect 60 Pin connector. Inspect for damaged pins, corrosion, loose wires, etc. Service as necessary. • Processor disconnected. • Key Off • DVOM on 200 ohm scale. • Measure resistance from test Pin 26 at Breakout box to VREF at vehicle harness connector of the sensor that sent you here. • Measure resistance from test Pin 46 at Breakout box to signal return at vehicle harness connector of the sensor that sent you here.	Less than 5 ohms on all readings ▶ 5 ohms or greater on any reading ▶	RECONNECT sensors. Reference voltage OK. RERUN Quick Test. SERVICE open in VREF or Signal Return. RERUN Quick Test.
C4 CHECK FOR SHORT FROM VREF TO VPWR		
• Key Off, wait 10 seconds. • Breakout box installed. • Disconnect processor. • Key On, Engine Off • DVOM on 20V scale • Measure voltage between test Pin 26 at the Breakout box and battery ground.	Less than 0.5V ▶ 0.5V or greater ▶	REPLACE processor. RERUN Quick Test. SERVICE short to battery power in EEC harness. RERUN Quick Test. If condition persists, REPLACE processor.

Reference Voltage — Pinpoint Test C

TEST STEP	RESULT ▶	ACTION TO TAKE
C5 CHECK FOR SHORTED THROTTLE POSITION SENSOR		
• Key Off, wait 10 seconds. • Breakout box installed. • Processor connected. • Disconnect Throttle Position (TP) sensor from vehicle harness. • Key On, Engine Off. • DVOM on 20V scale. • Measure voltage between test Pin 26 and test Pin 46 at the Breakout box.	Less than 4.0V ▶	Vehicles equipped with EVP sensor, GO to C6. All other vehicles, GO to C7.
	4.0V or greater ▶	REPLACE TP sensor. RERUN Quick Test.
C6 CHECK FOR SHORTED EGR VALVE POSITION (EVP) SENSOR		
• Key Off, wait 10 seconds. • Breakout box installed. • Processor connected. • Disconnect EGR valve position (EVP) sensor. • Key On, Engine Off. • DVOM on 20V scale. • Measure voltage between test Pin 26 and test Pin 46 at the Breakout box.	Less than 4.0V ▶	GO to C7.
	4.0V or greater ▶	REPLACE EVP sensor. RERUN Quick Test.
C7 CHECK FOR SHORTED MAP BP SENSOR		
• Key Off, wait 10 seconds. • Breakout box installed. • Processor connected. • Disconnect MAP BP sensor. • Key On, Engine Off. • DVOM on 20V scale. • Measure voltage between test Pin 26 and test Pin 46 at the Breakout box.	Less than 4.0V ▶	Vehicles equipped with VAF sensor, GO to C8. All other vehicles, GO to C9.
	4.0V or greater ▶	REPLACE MAP BP sensor. RERUN Quick Test.

Reference Voltage — Pinpoint Test C

TEST STEP	RESULT ▶	ACTION TO TAKE
C8 CHECK FOR SHORTED VANE AIR METER (VAF) SENSOR		
• Key Off, wait 10 seconds. • Breakout box installed. • Processor connected. • Disconnect vane air meter (VAF) sensor. • Key On, Engine Off. • DVOM on 20V scale. • Measure voltage between test Pin 26 and test Pin 46 at the Breakout box.	Less than 4.0V ▶	GO to C9.
	4.0V or greater ▶	REPLACE VAF sensor and RERUN Quick Test.
C9 SHORT TO GROUND IN VREF		
• Breakout box installed. • Processor disconnected. • Key Off, wait 10 seconds. • Disconnect TP and MAP BP, EVP and VAF, if so equipped. • DVOM on 200 ohm scale. • Measure resistance between test Pin 26 and test Pins 20, 40, 46 and 60 at the Breakout box.	Less than 5 ohms ▶	SERVICE short to ground. CONNECT all sensors. RERUN Quick Test. If original condition still exists, REPLACE processor.
	5 ohms or greater ▶	RECONNECT sensors. REPLACE processor. RERUN Quick Test.

Vane Air Temperature Sensor (VAT) — Pinpoint Test DA

DA

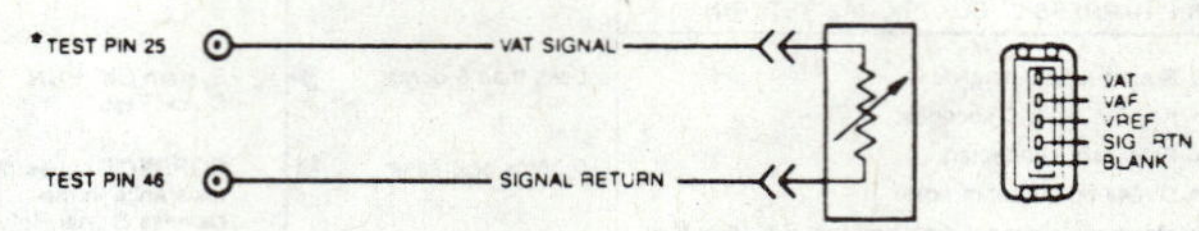

NOTE: AMBIENT TEMP. MUST BE GREATER THAN 50°F TO PASS THIS TEST.

TYPICAL RESISTANCE BETWEEN TEST PINS 25 & 46	5800Ω	2700Ω	300Ω	180Ω	125Ω
AT TEMPERATURE	32°F	65°F	185°F	220°F	240°F

*TEST PIN LOCATED ON BREAKOUT BOX.
ALL CONNECTIONS VIEWED INTO MATING SURFACE.

STOP-WARNING

You should enter this Pinpoint Test only when a service code 24, 54 or 64 is received in Quick Test Step 3.0, 5.0 or 6.0.

To prevent the replacement of good components, be aware that the following non-EEC areas may be at fault:

- Test performed in unusually low (cold) or high (hot) ambient conditions.
- Ambient temperature must be greater than 50°F for this test.

This pinpoint test is intended to diagnose only the following:

- VAT sensor.
- Circuits: VAT, and Signal Return.
- Vehicle harness.
- Processor assembly.

Vane Air Temperature Sensor (VAT) — Pinpoint Test DA

TEST STEP	RESULT ▶	ACTION TO TAKE
DA1 SERVICE CODE 24: CHECK AMBIENT TEMPERATURE		
• Ambient temperature must be greater than 50°F for this test.	Yes ▶	GO to DA2.
	No ▶	RERUN Quick Test.
DA2 CHECK FOR V REF AT THROTTLE POSITION SENSOR		
• Refer to illustration Q. • Key Off, wait 10 seconds. • DVOM on 20V scale. • Disconnect TP sensor. • Key On, Engine Off. • Measure voltage at the TP vehicle harness connector between VREF and signal return.	Less than 4.0V or greater than 6.0V ▶	GO to Pinpoint Test Step C1.
	4.0V to 6.0V ▶	RECONNECT TP sensor. GO to DA3.
DA3 VAT SENSOR CHECK		
NOTE: **Ambient temperature must be greater than 50°F for this test.** • Key Off, wait 10 seconds. • Harness disconnected from the vane meter. • DVOM on 200,000 ohm scale. • Measure resistance at the VAT sensor between VAT signal and Signal Return. • Is reading from 125 ohms (240°F) to 3700 ohms (50°F)?	Yes ▶	REPLACE processor. RECONNECT harness to vane meter. RERUN Quick Test.
	No ▶	REPLACE vane meter. RERUN Quick Test.

Vane Air Temperature Sensor (VAT) — Pinpoint Test — DA

TEST STEP	RESULT ▶	ACTION TO TAKE
DA10 SERVICE CODE 54: INDUCE OPPOSITE CODE		
• Key Off, wait 10 seconds. • Disconnect vehicle harness from vane meter. Inspect for damaged pins, corrosion, loose wires, etc. Service as necessary. • Insert a jumper wire (paper clip) at the vane meter vehicle harness connector between VAT Signal and Signal Return. • Run Key On, Engine Off Quick Test. • Is Code 64 present?	Yes ▶	REPLACE vane meter. REMOVE jumper wire. CONNECT harness to vane meter. RERUN Quick Test.
	No ▶	REMOVE jumper wire. GO to DA11.
DA11 CHECK CONTINUITY OF VAT SIGNAL AND SIGNAL RETURN		
• Key Off, wait 10 seconds. • Harness disconnected from vane meter, jumper wire removed. • Disconnect processor 60 pin connector. Inspect for damaged pins, corrosion, loose wires, etc. Service as necessary. • Install Breakout box leaving processor disconnected. • DVOM on 200 ohm scale. • Measure resistance between VAT signal at the vane meter vehicle harness connector, and test Pin 25 at the Breakout box. • Measure resistance between signal return at the vane meter vehicle harness connector, and test Pin 46 at the Breakout box.	Both readings are less than 5 ohms ▶	REPLACE processor. REMOVE Breakout box. RECONNECT harness to vane meter and processor. RERUN Quick Test.
	Either reading is 5 ohms or greater ▶	CORRECT open circuit. REMOVE Breakout box. RECONNECT harness to vane meter and processor. RERUN Quick Test.

Vane Air Temperature Sensor (VAT) — Pinpoint Test — DA

TEST STEP	RESULT ▶	ACTION TO TAKE
DA20 SERVICE CODE 64: INDUCE OPPOSITE CODE		
• Key off, wait 10 seconds. • Disconnect vehicle harness from vane meter. Inspect for damaged pins, corrosion, loose wires, etc. Service as necessary. • Run Key On, Engine Off Quick Test. • Is code 54 present?	Yes ▶	REPLACE vane meter. RECONNECT harness to vane meter. RERUN Quick Test.
	No ▶	GO to DA21.
DA21 CHECK FOR V REF AT THROTTLE POSITION SENSOR		
• Refer to illustration Q. • Key Off, wait 10 seconds. • DVOM on 20V scale. • Disconnect TP sensor. • Key On, Engine Off. • Measure voltage at the TP vehicle harness connector between VREF and signal return.	Less than 4.0V or greater than 6.0V ▶	GO to Pinpoint Test Step C1.
	4.0V to 6.0V ▶	RECONNECT TP sensor, GO to DA22.
DA22 CHECK VAT SIGNAL FOR SHORTS		
• Key Off, wait 10 seconds. • Harness disconnected from vane meter. • Disconnect processor 60 pin connector. Inspect for damaged pins, corrosion, loose wires, etc. Service as necessary. • Install Breakout box leaving processor disconnected. • DVOM on 200,000 ohm scale. • Measure resistance between Test Pin 25 and Test Pins 40, 46 and 60 at the Breakout box.	Any reading less than 10,000 ohms ▶	CORRECT circuit shorts. REMOVE Breakout box. RECONNECT processor and vane meter. RERUN Quick Test.
	All readings are 10,000 ohms or greater ▶	REPLACE processor. REMOVE Breakout box. RECONNECT processor. RERUN Quick Test.

Vane Air Temperature Sensor (VAT) — Pinpoint Test — DA

TEST STEP	RESULT ▶	ACTION TO TAKE
DA90 SERVICE CODE 54: CONTINUOUS TEST CHECK VAT SENSOR		
• Using continuous monitor mode, observe VOM or STAR LED for indication of a fault while performing the following: • Lightly tap on VAT sensor (simulate road shock). • Wiggle VAT connector. • Is a fault indicated?	Yes ▶	DISCONNECT and INSPECT connectors. If connector and terminals are good, REPLACE VAT sensor. RERUN Quick Test.
	No ▶	GO to DA91.
DA91 CHECK EEC-IV HARNESS		
• Observe VOM or STAR LED for a fault indication while performing the following: • Referring to the illustration in Step DA90, grasp the harness closest to the sensor connector. Wiggle, shake or bend a small section of the EEC-IV system harness while working your way to the dash panel. Also wiggle, shake or bend the EEC-IV harness from the dash panel to the processor. • Is fault indicated?	Yes ▶	ISOLATE fault and make necessary repairs. RERUN Quick Test.
	No ▶	GO to DA92.
DA92 CHECK PROCESSOR AND HARNESS CONNECTORS		
• Key Off, wait 10 seconds. • Disconnect processor 60 pin connector. • Inspect both connectors and connector terminals for obvious damage or faults. • Connectors and terminals are OK.	No ▶	SERVICE as necessary. RERUN Quick Test.
	Yes ▶	Unable to duplicate fault at this time. Continuous code 54 testing complete.

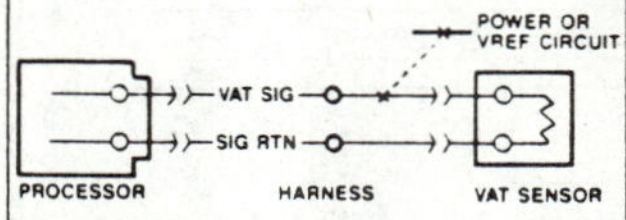

Vane Air Temperature Sensor (VAT) — Pinpoint Test — DA

TEST STEP	RESULT ▶	ACTION TO TAKE
DA93 SERVICE CODE 64: CONTINUOUS TEST: CHECK VAT SENSOR		
• Using continuous monitor mode, observe VOM or STAR LED for indication of a fault while performing the following: • Lightly tap on VAT sensor (simulate road shock). • Wiggle VAT connector. • Is fault indicated?	Yes ▶	DISCONNECT and INSPECT connectors. If connector and terminals are good, REPLACE VAT sensor. RERUN Quick Test.
	No ▶	GO to DA94.
DA94 CHECK EEC-IV HARNESS		
• Observe VOM or STAR LED for a fault indication while performing the following: • Referring to the illustration in Step DA93, grasp the harness closest to the sensor connector. Wiggle, shake or bend a small section of the EEC-IV system harness while working your way to the dash panel. Also wiggle, shake or bend the EEC-IV harness from the dash panel to the processor. • Is fault indicated?	Yes ▶	ISOLATE fault and make necessary service. RERUN Quick Test.
	No ▶	GO to DA95.
DA95 CHECK PROCESSOR AND HARNESS CONNECTORS		
• Key Off, wait 10 seconds. • Disconnect processor 60 pin connector. • Inspect both connectors and connector terminals for obvious damage or faults. • Connectors and terminals are OK.	No ▶	SERVICE as necessary. RERUN Quick Test.
	Yes ▶	Unable to duplicate fault at this time. Continuous code 64 testing complete.

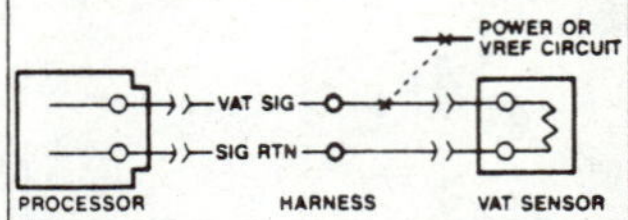

Air Charge Temperature Sensor (ACT) — Pinpoint Test — DB

DB

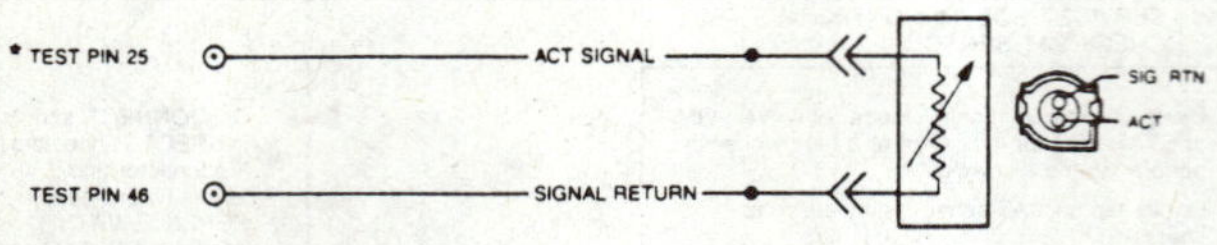

TYPICAL RESISTANCE BETWEEN TEST PINS 25 AND 46	58.750Ω	40 500Ω	3600Ω	1840Ω
AT TEMPERATURE	50°F	65°F	180°F	220°F

* TEST PIN LOCATED ON BREAKOUT BOX.
ALL CONNECTIONS VIEWED INTO MATING SURFACE.

NOTE: Ambient temperature must be greater than 50°F for this test.

STOP-WARNING

You should enter this Pinpont Test only when a service code 24, 54 or 64 is received in Quick Test Step 3.0, 5.0 or 6.0.

To prevent the replacement of good components, be aware that the following non-EEC areas may be at fault:

- Cooling system.
- Air cleaner duct problems.
- Improper engine oil level.

This pinpoint test is intended to diagnose only the following:

- ACT sensor.
- Harness circuits: ACT Signal and Signal Return.
- Processor assembly.

Air Charge Temperature Sensor (ACT) — Pinpoint Test — DB

TEST STEP	RESULT ▶	ACTION TO TAKE
DB1 SERVICE CODE 24: CHECK PROPER INSTALLATION		
• For vehicle with ACT mounted in intake manifold, GO to step DB2. • Is ACT mounted properly in air cleaner?	Yes ▶ No ▶	GO to DB2. INSTALL ACT properly. RERUN Quick Test.
DB2 CHECK FOR V REF AT THROTTLE POSITION SENSOR		
• Key Off, wait 10 seconds. • DVOM on 20V scale. • Disconnect TP sensor. • Key On, Engine Off. • Measure voltage at the TP vehicle harness connector between VREF and signal return.	Less than 4.0V or greater than 6.0V ▶ 4.0V to 6.0V ▶	GO to Pinpoint Test Step C1. RECONNECT TP sensor. GO to DB3.
DB3 CHECK ACT SENSOR — ENGINE OFF		
• Key Off, wait 10 seconds. • Harness disconnected from ACT sensor. • DVOM on 200,000 ohm scale. • Measure resistance of ACT sensor. **NOTE: Make sure engine is warmed up prior to this test.**	Reading is between 1,100 and 58,000 ohms (approximately 50°F) ▶ Reading is less than 1,100 ohms or greater than 58,000 ▶	GO to DB4. CHECK heat stove duct valve operation. If OK, REPLACE ACT sensor. RECONNECT harness to ACT sensor. RERUN Quick Test.
DB4 CHECK ACT SENSOR — ENGINE RUNNING		
• Key Off. Harness disconnected from ACT sensor. • DVOM on 200,000 ohm scale. • Run engine for 2 minutes. • Measure resistance of ACT sensor with engine running.	Reading is between 2,400 and 29,000 ohms ▶ Reading is less than 2,400 ohms or greater than 29,000 ohms ▶	REPLACE processor. RECONNECT harness to ACT sensor. RERUN Quick Test. CHECK heat stove duct valve operation. If OK, REPLACE ACT sensor. RERUN Quick Test.

Air Charge Temperature Sensor (ACT) — Pinpoint Test — DB

TEST STEP	RESULT ▶	ACTION TO TAKE
DB10 SERVICE CODE 54: INDUCE OPPOSITE CODE		
• Key Off, wait 10 seconds. • Disconnect vehicle harness from ACT sensor. Inspect for damaged pins, corrosion, loose wires, etc. Service as necessary. • Insert a jumper wire at the ACT vehicle harness connector between ACT Signal and Signal Return. • Run Key On, Engine Off Quick Test. • Is code 64 present?	Yes ▶ No ▶	REPLACE ACT sensor. REMOVE jumper wire. RECONNECT harness to ACT sensor. RERUN Quick Test. REMOVE jumper wire. GO to DB11.
DB11 CONTINUITY CHECK ACT SIGNAL AND SIGNAL RETURN		
• Key Off, wait 10 seconds. • Harness disconnected from ACT sensor. • Disconnect processor 60 pin connector. Inspect for damaged pins, corrosion, loose wires, etc. Service as necessary. • Install Breakout box. Leave processor disconnected. • DVOM on 200 ohm scale. • Measure resistance between ACT signal, at the ACT vehicle harness connector, and test Pin 25 at the Breakout box. • Measure resistance between Signal Return, at the ACT vehicle harness connector, and test Pin 46 at the Breakout box.	Both readings are less than 5 ohms ▶ Either reading is 5 ohms or greater ▶	REPLACE processor. REMOVE Breakout box. RECONNECT processor and ACT sensor. RERUN Quick Test. CORRECT circuit opens. REMOVE Breakout box. RECONNECT processor and ACT sensor. RERUN Quick Test.

Air Charge Temperature Sensor (ACT) — Pinpoint Test — DB

TEST STEP	RESULT ▶	ACTION TO TAKE
DB20 SERVICE CODE 64. INDUCE OPPOSITE CODE		
• Key Off, wait 10 seconds. • Disconnect vehicle harness from ACT sensor. Inspect for damaged pins, corrosion, loose wires, etc. Service as necessary. • Run Key On, Engine Off Quick Test. • Is code 54 present?	Yes ▶ No ▶	REPLACE ACT sensor. RECONNECT harness to ACT sensor. RERUN Quick Test. GO to DB21.
DB21 CHECK FOR V REF AT THROTTLE POSITION SENSOR		
• Key Off, wait 10 seconds. • DVOM on 20V scale. • Disconnect TP sensor. • Key On. Engine Off. • Measure voltage at the TP vehicle harness connector between VREF and Signal Return.	Less than 4.0V or greater than 6.0V ▶ 4.0V to 6.0V ▶	GO to Pinpoint Test Step C1. RECONNECT TP sensor. GO to DB22.
DB22 CHECK ACT SIGNAL FOR SHORTS TO GROUND		
• Key Off, wait 10 seconds. • Harness disconnected from ACT sensor. • Disconnect processor 60 pin connector. Inspect for damaged pins, corrosion, loose wires, etc. Service as necessary. • DVOM on 200,000 ohm scale. • Measure resistance between test Pin 25 and test Pins 40, 46 and 60 at the Breakout box.	Any reading less than 10,000 ohms ▶ All readings are 10,000 ohms or greater ▶	CORRECT circuit shorts. REMOVE Breakout box. RECONNECT processor and ACT sensor. RERUN Quick Test. REPLACE processor. REMOVE Breakout box. RECONNECT processor and ACT sensor. RERUN Quick Test.

Air Charge Temperature (ACT) — Pinpoint Test DB

TEST STEP	RESULT ▶	ACTION TO TAKE
DB90 SERVICE CODE 54: CONTINUOUS TEST: CHECK ACT SENSOR		
• Using continuous monitor mode, observe VOM or STAR LED for indication of a fault while performing the following: • Lightly tap on ACT sensor (simulate road shock). • Wiggle ACT connector. • Is a fault indicated?	Yes ▶	DISCONNECT and INSPECT connectors. If connector and terminals are good, REPLACE ACT sensor. RERUN Quick Test.
	No ▶	GO to DB91.
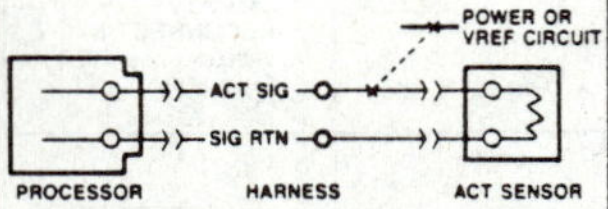		
DB91 CHECK EEC-IV HARNESS		
• Observe VOM or STAR LED for a fault indication while performing the following: • Referring to the illustration in Step DB90, grasp the harness closest to the sensor connector. Wiggle, shake or bend a small section of the EEC-IV system harness while working your way to the dash panel. Also wiggle, shake or bend the EEC-IV harness from the dash panel to the processor. • Is fault indicated?	Yes ▶	ISOLATE fault and make necessary repairs. RERUN Quick Test.
	No ▶	GO to DB92.
DB92 CHECK PROCESSOR AND HARNESS CONNECTORS		
• Key Off, wait 10 seconds. • Disconnect processor 60 pin connector. • Inspect both connectors and connector terminals for obvious damage or faults. • Connectors and terminals are OK.	No ▶	SERVICE as necessary. RERUN Quick Test.
	Yes ▶	Unable to duplicate fault at this time. Continuous code 54 testing complete.

Air Charge Temperature (ACT) — Pinpoint Test DB

TEST STEP	RESULT ▶	ACTION TO TAKE
DB93 SERVICE CODE 64: CONTINUOUS TEST: CHECK ACT SENSOR		
• Using continuous monitor mode, observe VOM or STAR LED for fault while performing the following: • Lightly tap on ACT sensor (simulate road shock). • Wiggle ACT connector. • Is fault indicated?	Yes ▶	DISCONNECT and INSPECT connectors. If connector and terminals are good, REPLACE ACT sensor. RERUN Quick Test.
	No ▶	GO to DB94.
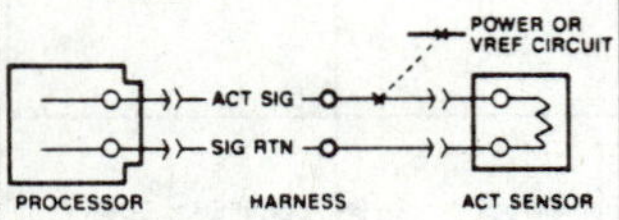		
DB94 CHECK EEC-IV HARNESS		
• Observe VOM or STAR LED for a fault indication while performing the following: • Referring to the illustration in Step DB93, grasp the harness closest to the sensor connector. Wiggle, shake or bend a small section of the EEC-IV system harness while working your way to the dash panel. Also wiggle, shake or bend the EEC-IV harness from the dash panel to the processor. • Is fault indicated?	Yes ▶	ISOLATE fault and make necessary repairs. RERUN Quick Test.
	No ▶	GO to DB95.
DB95 CHECK PROCESSOR AND HARNESS CONNECTORS		
• Key Off, wait 10 seconds. • Disconnect processor 60 pin connector. • Inspect both connectors and connector terminals for obvious damage or faults. • Connectors and terminals are OK.	No ▶	SERVICE as necessary. RERUN Quick Test.
	Yes ▶	Unable to duplicate fault at this time. Continuous code 64 testing complete.

EGR Valve Position Sensor (EVP) Control/Vent (EGRC/EGRV) — Pinpoint Test DD

DD

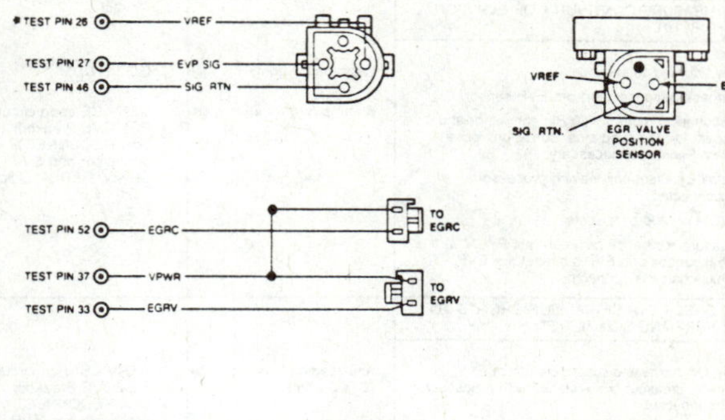

*TEST PINS LOCATED ON BREAKOUT BOX.
ALL HARNESS CONNECTORS VIEWED INTO MATING SURFACE.

STOP-WARNING

You should enter this Pinpoint Test only when a Service Code 31, 32, 33, 34, 35, 83 or 84 is received in Quick Test Step 3.0, 5.0 or 6.0.

To prevent the replacement of good components, be aware that the following non-EEC area may be at fault:

- Damaged EGR valve.

This Pinpoint Test is intended to diagnose only the following:

- EVP sensor.
- Harness circuits: EVP, Signal Return, VREF, EGRV, EGRC, VPWR.
- EGR solenoids.
- EGR Valve assembly.
- Processor assembly.

EGR Valve Position Sensor (EVP) Control/Vent (EGRC/EGRV) — Pinpoint Test DD

TEST STEP	RESULT ▶	ACTION TO TAKE
FAULT CODE 31		
DD1 RUN ENGINE RUNNING QUICK TEST WITH EGR VACUUM SIGNAL LINE DISCONNECTED AT EGR VALVE		
• Key Off, wait 10 seconds. • Disconnect EGR vacuum line at EGR valve and cap EGR vacuum line. • Run Engine Running Quick Test. • Check for code 31.	Code 31 present ▶	GO to DD2.
	No code 31 present, but codes 32, 34 are present ▶	GO to DD11.
DD2 CHECK EVP RESISTANCE WHILE APPLYING VACUUM TO EGR VALVE		
• Key Off, wait 10 seconds. • Vacuum signal line disconnected. • Disconnect vehicle harness at EVP sensor. • DVOM on 200,000 ohm scale. • Connect vacuum pump to EGR valve. • Measure resistance at the EVP sensor between EVP SIG and VREF while gradually increasing vacuum to 33 kPa (10 in. Hg). • Observe resistance as vacuum increases.	Reading gradually decreases from no greater than 5500 ohms to no less than 100 ohms ▶	GO to DD3.
	Reading is less than 100 ohms or greater than 5500 ohms ▶	REPLACE EVP sensor. RECONNECT signal line and harness. RERUN Quick Test.
	Reading does not decrease or unable to hold vacuum ▶	GO to DD16.
DD3 MEASURE VREF TO SIGNAL RETURN VOLTAGE		
• Key On, Engine Off. • Vacuum signal line disconnected and capped, harness disconnected from EVP sensor. • DVOM on 20V scale. • Measure voltage at the EVP vehicle harness connector between VREF and Signal Return.	Reading is between 4 and 6V ▶	GO to DD4.
	Reading is less than 4V or greater than 6V ▶	GO to Pinpoint Test Step C1.

EGR Valve Position Sensor (EVP) Control/Vent (EGRC/EGRV) — Pinpoint Test DD

TEST STEP	RESULT ▶	ACTION TO TAKE
DD4 MEASURE CONTINUITY OF EVP SIGNAL CIRCUIT		
• Key Off, wait 10 seconds. • Harness disconnected from EVP sensor. • Disconnect processor 60 Pin connector and inspect for damaged pins, corrosion, loose wires. Service as necessary. • Install Breakout box leaving processor disconnected. • DVOM on 200 ohm scale. • Measure resistance between test Pin 27 at the Breakout box and EVP signal at the EVP vehicle harness connector.	Less than 5 ohms ▶ 5 ohms or greater ▶	GO to DD5. SERVICE open circuit. REMOVE Breakout box. RECONNECT processor and EVP sensor. RERUN Quick Test.
DD5 CHECK EVP SIGNAL FOR SHORTS TO VREF AND SIGNAL RETURN		
• Key Off, harness disconnected from EVP sensor. Breakout box installed with processor not connected. • DVOM on 200,000 ohm scale. • Measure resistance between test Pin 27 and test Pins 26, 40, 46 and 60 at Breakout box.	Any reading less than 10,000 ohms ▶ All readings 10,000 ohms or greater ▶	SERVICE short circuit. REMOVE Breakout box. RECONNECT processor and EVP sensor. RERUN Quick Test. GO to DD6.
DD6 SUBSTITUTE EVP SENSOR AND EGR VALVE		
• Key Off, wait 10 seconds. • Electrically connect known good EVP sensor and EGR valve assembly. • Remove Breakout box. • Reconnect processor. • Perform Key On, Engine Off Quick Test. • Is code 31 present?	Yes ▶ No ▶	REPLACE processor. Connect original EVP sensor and EGR valve assembly. RERUN Quick Test. GO to DD7.

EGR Valve Position Sensor (EVP) Control/Vent (EGRC/EGRV) — Pinpoint Test DD

TEST STEP	RESULT ▶	ACTION TO TAKE
DD7 EVP SENSOR CHECK		
• Key Off, wait 10 seconds. • Install original EVP sensor on known good EGR valve. • EVP sensor connected. • Rerun Key On, Engine Off Quick Test. • Is code 31 present?	Yes ▶ No ▶	INSTALL new EVP sensor. RERUN Quick Test. REFER to EGR System, Section 6.
FAULT CODES 32, 33 AND 34		
DD11 OUTPUT STATE CHECK (REFER TO APPENDIX)		
NOTE: Do not use STAR tester for this test step. Use VOM/DVOM. • Key Off, wait 10 seconds. • DVOM on 20V scale. • Connect DVOM negative test lead to STO and positive test lead to battery positive. • Jumper STI to signal return. • Perform Key On, Engine Off Self-Test until the completion of the Continuous Test Codes. • DVOM will indicate zero volts. • Depress and release the throttle. • Did DVOM reading change to a high voltage reading?	No ▶ Yes ▶	DEPRESS throttle to WOT and RELEASE. If STO voltage does not go high, GO to Pinpoint Test Step Q40. Remain in output state check and GO to DD12.
DD12 CHECK EGR SOLENOIDS FOR ELECTRICAL CYCLING		
• Key On, Engine Off. • In output state check. • DVOM on 20V scale. • Connect DVOM between EGRV, VPWR and EGRV signal. • While observing DVOM, depress and release the throttle several times to cycle output on and off. • Repeat for EGRC, VPWR and EGRC signal.	Both solenoid outputs cycle on and off ▶ Either output does not cycle on and off ▶	REMAIN in output state check. GO to DD13. Exit output state check. GO to DD17.

EGR Valve Position Sensor (EVP) Control/Vent (EGRC/EGRV) — Pinpoint Test DD

TEST STEP	RESULT ▶	ACTION TO TAKE
DD13 CHECK EGR SOLENOIDS FOR VACUUM CYCLING		
• Key On, Engine Off. • In output state check. • Disconnect and cap vacuum line from bottom port of EGRC solenoid and connect a vacuum pump. • Connect a vacuum gauge in the common output (top) vacuum line to EGR valve. • Disconnect but **do not** cap vacuum vent line from EGRV solenoid. • While cycling outputs on and off (by depressing and releasing throttle), observe vacuum gauge at the output. Maintain vacuum at source.	Vacuum output cycles on and off in less than 2 seconds ▶ Vacuum does not cycle on and off in less than 2 seconds ▶	RECONNECT all vacuum lines. GO to DD14. CHECK filter for obstructions. REPLACE as necessary. If OK, REPLACE solenoid assembly. RECONNECT all vacuum lines. RERUN Quick Test.
DD14 CHECK VACUUM LINES		
• Key Off, wait 10 seconds. • Vacuum lines reconnected. • Check entire EEC vacuum line system per VECI emission schematic decal for kinks, cracks, obstructions or leaks.	Vacuum lines OK ▶ Vacuum lines not OK ▶	GO to DD15. SERVICE as necessary. RERUN Quick Test.

EGR Valve Position Sensor (EVP) Control/Vent (EGRC/EGRV) — Pinpoint Test DD

TEST STEP	RESULT ▶	ACTION TO TAKE
DD15 CHECK EVP RESISTANCE WHILE APPLYING VACUUM TO EGR VALVE		
• Key Off. • Disconnect vehicle harness from EVP sensor. Inspect for damaged pins, corrosion, and pins pushed out. Service as necessary. • DVOM on 200,000 ohm scale. • Disconnect vacuum line at EGR valve. • Connect vacuum pump to EGR valve. • Measure resistance of the EVP sensor between EVP Signal Pin and VREF Pin while increasing vacuum to 33 kPa (10 in. Hg). • Observe resistance as vacuum increases.	Reading does not decrease gradually ▶ Reading gradually decreases from no more than 5,500 ohms to no less than 100 ohms ▶	GO to DD16. REPLACE processor. RECONNECT EVP sensor and EGR vacuum line. RERUN Quick Test.
DD16 MANUALLY EXERCISE EVP SENSOR		
• Key Off, harness disconnected from EVP sensor. • Remove EVP sensor from EGR valve. • Measure resistance of the EVP sensor between EVP Signal Pin and VREF Pin while gradually applying pressure to EVP sensor shaft. • Observe resistance as shaft is slowly pushed in and slowly released. • Look for sudden changes in resistance readings.	Both readings decrease and increase smoothly between 5,500 ohms and 100 ohms ▶ Either reading decreases or increases abruptly between 5,5000 ohms and 100 ohms ▶	REFER to EGR System, Section 6. RECONNECT EVP sensor and EGR supply vacuum line. RERUN Quick Test. REPLACE EVP sensor. RECONNECT harness and EGR supply vacuum line. RERUN Quick Test.
DD17 MEASURE EGRV/EGRC SOLENOID RESISTANCE		
• Key Off, wait 10 seconds. • DVOM on 200 ohm scale. • Disconnect EGRV solenoid connector and measure solenoid resistance. Inspect for damaged pins, corrosion and pins pushed out. Service as necessary. • Disconnect EGRC solenoid connector and measure solenoid resistance. Inspect for damaged pins, corrosion and pins pushed out. Service as necessary.	Both resistances are between 30 and 70 ohms ▶ Either resistance is less than 30 ohms or greater than 70 ohms ▶	CONNECT EGRC/EGRV solenoids. GO to DD18. REPLACE EGRC/EGRV solenoid assembly. RERUN Quick Test.

EGR Valve Position Sensor (EVP) Control/Vent (EGRC/EGRV) — Pinpoint Test — DD

TEST STEP	RESULT ▶	ACTION TO TAKE
DD18 CHECK FOR VOLTAGE ON VEHICLE POWER CIRCUIT		
• Disconnect EGR vent and EGR control solenoids from harness. • Key On, Engine Off. • DVOM on 20V scale. • Measure voltage between battery negative terminal and VPWR circuit on both EGR solenoids.	Either reading is less than 10.5V ▶ Both readings are 10.5V or greater ▶	SERVICE harness circuit open. RERUN Quick Test. GO to DD19.
DD19 CHECK CONTINUITY OF EGRV AND EGRC CIRCUITS		
• Key Off, wait 10 seconds. • EGR vent and EGR control solenoid disconnected from harness. • Disconnect processor 60 Pin connector and inspect for damaged pins, corrosion, loose wires. Service as necessary. • Install Breakout box to processor harness connector. Leave processor disconnected. • DVOM on 200 ohm scale. • Measure resistance between test Pin 33 at the Breakout box and EGRV signal at the EGRV solenoid vehicle harness connector. • Measure resistance between test Pin 52 at the Breakout box and EGRC signal at the EGRC solenoid vehicle harness connector.	Both readings less than 5 ohms ▶ Either reading 5 ohms or greater ▶	GO to DD20. SERVICE open circuit. REMOVE Breakout box. RECONNECT harness to processor. RERUN Quick Test.
DD20 CHECK FOR SHORT TO GROUND		
• Key Off, wait 10 seconds. • DVOM on 200,000 ohm scale. • Leave Breakout box installed and processor disconnected. • EGRV/EGRC solenoids disconnected. • Measure resistance between test Pins 33 and/or 52 and test Pins 40, 46 and 60 at the Breakout box.	Resistance reading is less than 10,000 ohms ▶ Resistance reading is 10,000 ohms or greater ▶	SERVICE short to ground. RERUN Quick Test. GO to DD21.

EGR Valve Position Sensor (EVP) Control/Vent (EGRC/EGRV) — Pinpoint Test — DD

TEST STEP	RESULT ▶	ACTION TO TAKE
DD21 CHECK EGRV AND EGRC SIGNALS FOR SHORTS TO POWER		
• EGR vent and EGR control solenoids disconnected from harness. • Key Off, Breakout box installed. Processor disconnected. • DVOM on 200,000 ohm scale. • Measure resistance between test Pin 33 and test Pins 37 and 57 at the Breakout box. • Measure resistance between test Pin 52 and test Pins 37 and 57 at the Breakout box.	Any reading less than 10,000 ohms ▶ All readings 10,000 ohms or greater ▶	SERVICE circuit short. REMOVE Breakout box. RECONNECT harness to processor. RERUN Quick Test. If code is repeated, REPLACE processor. REPLACE processor. REMOVE Breakout box. RECONNECT harness to processor. RERUN Quick Test.
FAULT CODE 35		
DD30 RPM TOO LOW FOR EGR TEST		
• Check for code 12.	Code 12 present ▶ Code 12 not present ▶	Vehicles equipped with air bypass (EFI), GO to KE1. Vehicles equipped with DC motor control, GO to KF1. GO to DD31.
DD31 RETEST AT 1,500 RPM		
• Key Off, wait 10 seconds. • Install tachometer. • Perform Key On, Engine Running Quick Test while maintaining 1,500 rpm. • Record Engine Running service codes. • Check for code 35.	Code 35 present ▶ Code 35 not present ▶	REPLACE processor. RERUN Quick Test. RERUN Quick Test. SERVICE codes as necessary.

EGR Valve Position Sensor (EVP) Control/Vent (EGRC/EGRV) — Pinpoint Test — DD

TEST STEP	RESULT ▶	ACTION TO TAKE
DD90 SERVICE CODE 31 CONTINUOUS TEST: EXERCISE EVP SENSOR		
• Using continuous monitor mode, observe VOM or STAR LED for indication of a fault while performing the following: • Connect a vacuum pump to the EGR valve. • Very slowly apply 20 kPa (6 in. Hg) vacuum to the EGR valve. • Slowly bleed vacuum off the EGR valve and lightly tap on EVP sensor (simulate road shock). • Wiggle EVP sensor connector. • Is a fault indicated?	Yes ▶ No ▶	GO to DD91. GO to DD92.

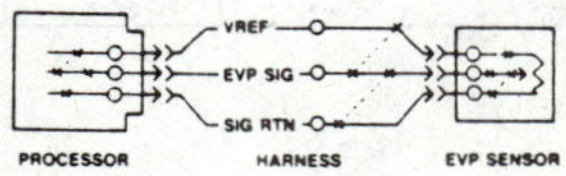

TEST STEP	RESULT ▶	ACTION TO TAKE
DD91 MEASURE EVP SIGNAL VOLTAGE WHILE EXERCISING EVP SENSOR		
• Key Off, wait 10 seconds. • Disconnect processor 60 Pin connector and inspect for damaged pins, corrosion, loose wires. Service as necessary. • Install Breakout box and reconnect processor. • VOM or STAR LED still connected to STO as in previous Step. • Connect a DVOM from test Pin 27 to test Pin 46. • DVOM on 20V scale. • Key On, Engine Off. • While observing DVOM, repeat Step DD90. • Does the fault occur below 4.25V?	Yes ▶ No ▶	DISCONNECT and INSPECT connector. If connector and terminals are good, REPLACE EVP sensor. RERUN Quick Test. EGR valve overshoot may have caused continuous code 31. Sensor service is not required. To verify harness integrity, GO to DD92.

EGR Valve Position Sensor (EVP) Control/Vent (EGRC/EGRV) — Pinpoint Test — DD

TEST STEP	RESULT ▶	ACTION TO TAKE
DD92 CHECK EEC-IV HARNESS		
• Observe VOM or STAR LED for a fault indication while performing the following: • Referring to the illustration in Step DD90, grasp the harness closest to the sensor connector. Wiggle, shake or bend a small section of the EEC-IV system harness while working your way to the dash panel. Also wiggle, shake or bend the EEC-IV harness from the dash panel to the processor. • Is a fault indicated?	Yes ▶ No ▶	ISOLATE fault and SERVICE as necessary. REFER to appropriate figure. RERUN Quick Test. GO to DD93.
DD93 CHECK PROCESSOR AND HARNESS CONNECTORS		
• Key Off, wait 10 seconds. • Disconnect processor 60 Pin connector. • Inspect both connectors and connector terminals for obvious damage or faults. • Are connectors and terminals OK?	No ▶ Yes ▶	SERVICE as necessary. RERUN Quick Test. Unable to duplicate fault at this time. Continuous code 31 testing complete.

Engine Coolant Temperature Sensor (ECT) — Pinpoint Test — DE

DE

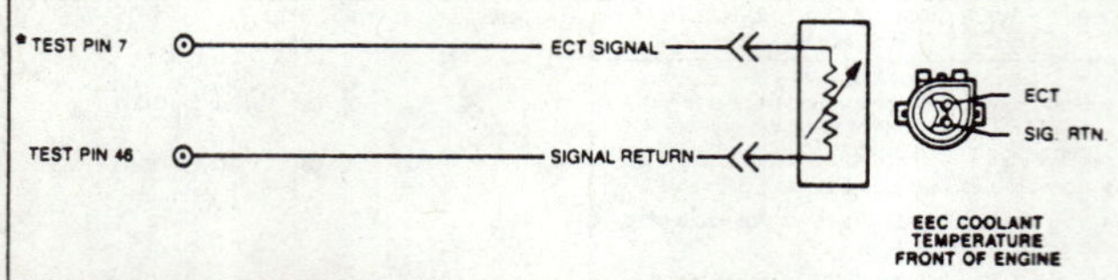

NOTE: TO PASS THIS TEST, ENGINE COOLANT TEMPERATURE MUST BE:
KEY ON, ENGINE OFF (50°F TO 240°F)
ENGINE RUNNING (180°F TO 240°F)

TYPICAL RESISTANCE BETWEEN TEST PINS 7 & 46	58,750Ω	40,500Ω	3600Ω	1840Ω
AT TEMPERATURE	50°F	65°F	180°F	220°F

* TEST PIN LOCATED ON BREAKOUT BOX.
ALL CONNECTIONS VIEWED INTO MATING SURFACE.

STOP-WARNING

You should enter this Pinpont Test only when a service code 21, 51 or 61 is received in Quick Test Step 3.0, 5.0 or 6.0.

To prevent the replacement of good components, be aware that the following non-EEC areas may be at fault:

- Coolant level.
- Oil level.
- Blocked or obstructed air flow.
- Engine not at normal operating temperature.
- Electro drive cooling fan.

This pinpoint test is intended to diagnose only the following:

- ECT sensor.
- Harness sensor circuits: ECT and Signal Return.
- Processor assembly.

Engine Coolant Temperature Sensor (ECT) — Pinpoint Test — DE

TEST STEP	RESULT ▶	ACTION TO TAKE
DE1 SERVICE CODE 21: CHECK ENGINE OPERATING TEMPERATURE		
• Run engine for 2 minutes at 2,000 rpm. • Check that upper radiator hose is hot and pressurized. • Rerun Quick Test.	Vehicle stalls ▶	Do not service code 21 at this time. REFER to diagnosis by symptoms.
	Code 21 present ▶	GO to DE2.
	Code 21 not present ▶	SERVICE other codes as necessary.
DE2 CHECK FOR V REF AT THROTTLE POSITION SENSOR		
• Refer to illustration Q. • Key Off, wait 10 seconds. • DVOM on 20V scale. • Disconnect TP sensor. • Key On, Engine Off. • Measure voltage at the TP vehicle harness connector between VREF and signal return.	Less than 4.0V or greater than 6.0V ▶	GO to Pinpoint Test Step C1.
	4.0V to 6.0V ▶	RECONNECT TP sensor. GO to DE3.
DE3 ECT SENSOR CHECK		
NOTE: Engine may have cooled down. Always warm engine before taking ECT resistance measurement. • Key Off, wait 10 seconds. • Harness disconnected from ECT sensor. • DVOM on 200,000 ohm scale. • Measure resistance of the ECT sensor. • Is the resistance reading — 1300 ohms (240°F) to 7700 ohms (140°F) for engine off? — 1550 ohms (230°F) to 4550 ohms (180°F) for engine running?	Yes ▶	REPLACE processor. RECONNECT harness to ECT sensor. RERUN Quick Test.
	No ▶	REPLACE ECT sensor. RECONNECT harness to ECT sensor. RERUN Quick Test.

Engine Coolant Temperature Sensor (ECT) — Pinpoint Test — DE

TEST STEP	RESULT ▶	ACTION TO TAKE
DE10 SERVICE CODE 51: INDUCE OPPOSITE CODE		
• Key Off, wait 10 seconds. • Disconnect vehicle harness from ECT sensor. Inspect for damaged pins, corrosion, loose wires, etc. Service as necessary. • Insert a jumper wire at the ECT sensor vehicle harness connector between ECT Signal and Signal Return. • Run Key On, Engine Off Quick Test. • Is code 61 present?	Yes ▶	REPLACE ECT sensor. REMOVE jumper wire. RECONNECT ECT sensor. RERUN Quick Test.
	No ▶	GO to DE11.
DE11 CHECK CONTINUITY OF ECT SIGNAL AND SIGNAL RETURN		
• Key Off, wait 10 seconds. • Harness disconnected from ECT sensor, jumper wire removed. • Disconnect processor 60 pin connector. Inspect for damaged pins, corrosion, loose wires, etc. Service as necessary. • Install Breakout box to harness leaving processor disconnected. • DVOM on 200 ohm scale. • Measure resistance between ECT signal at the ECT vehicle harness connector and test Pin 7 at the Breakout box. • Measure resistance between Signal Return at the ECT sensor vehicle harness connector, and test Pin 46 at the Breakout box.	Both readings are less than 5 ohms. ▶	REPLACE processor. REMOVE Breakout box. RECONNECT processor and ECT sensor. RERUN Quick Test.
	Either reading is 5 ohms or greater ▶	SERVICE open circuit(s). REMOVE Breakout box. RECONNECT processor and ECT sensor. RERUN Quick Test.

Engine Coolant Temperature Sensor (ECT) — Pinpoint Test — DE

TEST STEP	RESULT ▶	ACTION TO TAKE
DE20 SERVICE CODE 61: INDUCE OPPOSITE CODE		
• Key Off, wait 10 seconds. • Disconnect vehicle harness from ECT sensor. Inspect for damaged pins, corrosion, loose wires, etc. Service as necessary. • Run Key On, Engine Off Quick Test. • Is code 51 present?	Yes ▶	REPLACE ECT sensor. RECONNECT ECT sensor. RERUN Quick Test.
	No ▶	GO to DE21.
DE21 CHECK FOR V REF AT THROTTLE POSITION SENSOR		
• Refer to illustration Q. • Key Off, wait 10 seconds. • DVOM on 20V scale. • Disconnect TP sensor. • Key On, Engine Off. • Measure voltage at the TP vehicle harness connector between VREF and signal return.	Less than 4.0V or greater than 6.0V ▶	GO to Pinpoint Test Step C1.
	4.0V to 6.0V ▶	RECONNECT TP sensor. GO to DE22.
DE22 CHECK ECT SIGNAL FOR SHORT		
• Key Off, wait 10 seconds. • Harness disconnected from ECT sensor. • Disconnect processor 60 pin connector. Inspect for damaged pins, corrosion, loose wires, etc. Service as necessary. • Install Breakout box, leave processor disconnected. • DVOM on 200,000 ohm scale. • Measure resistance between test Pin 7 and test Pins 40, 46 and 60 at the Breakout box.	Either reading is less than 10,000 ohms ▶	SERVICE circuit shorts. REMOVE Breakout box. RECONNECT processor and ECT sensor. RERUN Quick Test.
	Both readings are 10,000 ohms or greater ▶	REPLACE processor. REMOVE Breakout box. RECONNECT processor and ECT harness. RERUN Quick Test.

Engine Coolant Temperature Sensor (ECT) — Pinpoint Test — DE

TEST STEP	RESULT ▶	ACTION TO TAKE
DE90 SERVICE CODE 21: CONTINUOUS TEST: TEST DRIVE VEHICLE		
• Key Off and wait 10 seconds. • Disconnect all Self-Test equipment and prepare vehicle for test drive. • Drive vehicle. Try to simulate different drive modes or mode in which drive complaint is noticed. Attempt to maintain drive complaint mode for one minute or more, if possible. • Upon completion of drive evaluation, repeat Key On, Engine Off Self-Test. • Is code 21 present in the continuous test results?	Yes ▶	VERIFY thermostat operating properly. If OK, REPLACE ECT sensor. RERUN Quick Test.
	No ▶	Unable to duplicate fault. Code 21 testing complete.

Engine Coolant Temperature (ECT) — Pinpoint Test — DE

TEST STEP	RESULT ▶	ACTION TO TAKE
DE94 SERVICE CODE 61: CONTINUOUS TEST: CHECK ECT SENSOR		
• Using continuous monitor mode, observe VOM or STAR LED for indication of a fault while performing the following: • Lightly tap on ECT sensor (simulate road shock). • Wiggle ECT connector. Voltage greater than 5 volts or STAR LED Off. • Is fault indicated?	Yes ▶	DISCONNECT and INSPECT connectors. If connector and terminals are good, REPLACE ECT sensor. RERUN Quick Test.
	No ▶	GO to DE95.

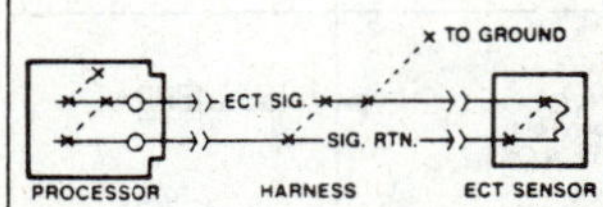

TEST STEP	RESULT ▶	ACTION TO TAKE
DE95 CHECK EEC-IV HARNESS		
• Observe VOM or STAR LED for a fault indication while performing the following: • Referring to the illustration in Step DE94, grasp the harness closest to the sensor connector. Wiggle, shake or bend a small section of the EEC-IV system harness while working your way to the dash panel. Also wiggle, shake or bend the EEC-IV harness from the dash panel to the processor. • Is fault indicated?	Yes ▶	ISOLATE fault and make necessary repairs. RERUN Quick Test.
	No ▶	GO to DE96.
DE96 CHECK PROCESSOR AND HARNESS CONNECTORS		
• Key Off, wait 10 seconds. • Disconnect processor 60 pin connector. • Inspect both connectors and connector terminals for obvious damage or faults. • Connectors and terminals are OK.	No ▶	SERVICE as necessary. RERUN Quick Test.
	Yes ▶	Unable to duplicate fault at this time. Continuous code 61 testing complete.

Engine Coolant Temperature (ECT) — Pinpoint Test — DE

TEST STEP	RESULT ▶	ACTION TO TAKE
DE91 SERVICE CODE 51: CONTINUOUS TEST: CHECK ECT SENSOR		
• Using continuous monitor mode, observe VOM or STAR LED for indication of a fault while performing the following: • Lightly tap on ECT sensor (simulate road shock). • Wiggle ECT connector. • Is a fault indicated?	Yes ▶	DISCONNECT and INSPECT connectors. If connector and terminals are good, REPLACE ECT sensor. RERUN Quick Test.
	No ▶	GO to DE92.

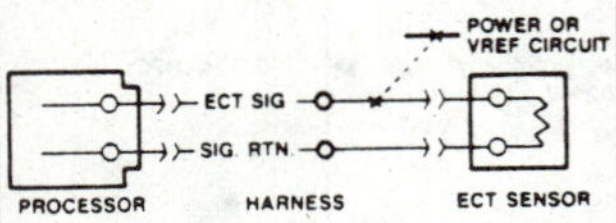

TEST STEP	RESULT ▶	ACTION TO TAKE
DE92 CHECK EEC-IV HARNESS		
• Observe VOM or STAR LED for a fault indication while performing the following: • Referring to the illustration in Step DE91, grasp the harness closest to the sensor connector. Wiggle, shake or bend a small section of the EEC-IV system harness while working your way to the dash panel. Also wiggle, shake or bend the EEC-IV harness from the dash panel to the processor. • Is fault indicated?	Yes ▶	ISOLATE fault and make necessary repairs. RERUN Quick Test.
	No ▶	GO to DE93.
DE93 CHECK PROCESSOR AND HARNESS CONNECTORS		
• Key Off, wait 10 seconds. • Disconnect processor 60 pin connector. • Inspect both connectors and connector terminals for obvious damage or faults. • Connectors and terminals are OK.	No ▶	SERVICE as necessary. RERUN Quick Test.
	Yes ▶	Unable to duplicate fault at this time. Continuous code 51 testing complete.

Manifold Absolute Pressure (MAP)/Barometric Pressure (BP) Sensor — Pinpoint Test — DF

DF

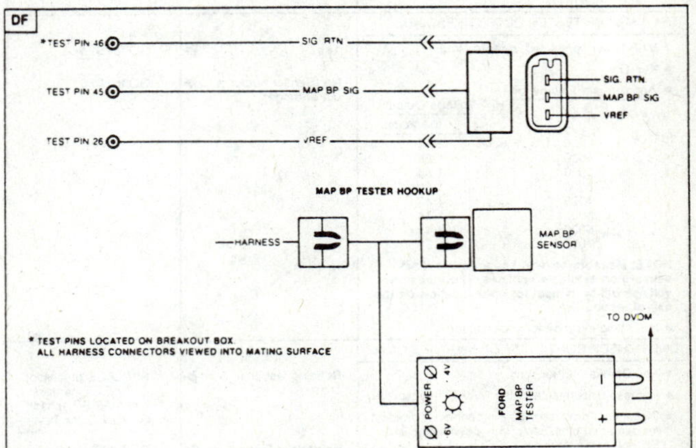

STOP-WARNING

You should enter this Pinpoint Test only when a Service Code 22 or 72 is received in Quick Test Step 3.0, 5.0 or 6.0 or when directed here from Diagnostics by Symptom in the Engine Supplement Section.

To prevent the replacement of good components, be aware that the following non-EEC areas may be at fault:

- Unusually high low atmospheric barometer reading (MAP BP).
- Kinked or obstructed vacuum lines (MAP).
- Basic engine (valves, vacuum leaks, timing, carburetor, EGR valve (MAP), etc.).
- Key On, Engine Off — MAP sensor must not see vacuum.
- Engine Running — MAP sensor must see actual manifold vacuum.

This Pinpoint Test is intended to diagnose only the following:

- MAP BP sensor
- Harness circuits: VREF, MAP BP Signal, and Signal Return
- Processor assembly

Manifold Absolute Pressure (MAP)/Barometric Pressure (BP) Sensor — Pinpoint Test — DF

TEST STEP	RESULT ▶	ACTION TO TAKE
FAULT CODE 22, ENGINE OFF		
DF1 CONNECTING MAP BP TESTER		
• Key Off. • Disconnect the MAP BP sensor from the vehicle harness. • Connect the MAP BP tester between the vehicle harness and the MAP BP sensor. • Insert tester banana plugs into DVOM. • Set DVOM to 20V scale. • Refer to illustration DF	Tester properly hooked up ▶	GO to DF2.
DF2 POWER TO MAP BP SENSOR TEST		
• MAP BP tester connected. • Key On.	(ONLY) Green light, VREF is OK ▶	GO to DF4.
	"Less than 4V" light (RED) or no lights, VREF is too low OR "Greater than 6V" light (Red), VREF is too high ▶	GO to DF3.
DF3 VREF ISOLATION		
• MAP BP tester connected. • Key On. • Disconnect MAP BP • Repeat Test Step DF2.	(ONLY) Green light, VREF is OK ▶	REPLACE MAP BP sensor. RERUN Quick Test.
	"Less than 4V" light (RED) or no lights, VREF is too low OR "Greater than 6V" light (Red), VREF is too high ▶	REMOVE MAP BP tester. GO to Pinpoint Test Step C1.

Manifold Absolute Pressure (MAP)/Barometric Pressure (BP) Sensor — Pinpoint Test — DF

TEST STEP	RESULT ▶	ACTION TO TAKE
DF4 MAP BP TESTER OUTPUT READING		
• MAP tester connected, refer to Note. • Key On. • Approximate Altitude (Ft.) / Voltage Output (+ − .04 Volts) 0 — 1.59 1000 — 1.56 2000 — 1.53 3000 — 1.50 4000 — 1.47 5000 — 1.44 6000 — 1.41 7000 — 1.39 **NOTE: Measure several known good MAP sensors on available vehicles. The measured voltage will be typical for your location on the day of testing.** • Is reading in range for your altitude?	Yes ▶	GO to DF5.
	No (Sensor output is out-of-range) ▶	GO to DF6.
DF5 CHECK CONTINUITY OF MAP BP SIGNAL		
• Key Off, wait 10 seconds • Harness disconnected from MAP BP sensor. • Disconnect processor 60 Pin connector. Inspect for damaged pins, corrosion, loose wires, etc. Service as necessary. • Install Breakout box. Leave processor disconnected. • DVOM on 200 ohm scale • Measure resistance between MAP BP signal at the MAP BP sensor vehicle harness connector and test Pin 45 at the Breakout box.	Reading less than 5 ohms ▶	REPLACE processor. CONNECT harness and MAP BP sensor. RERUN Quick Test.
	Reading is 5 ohms or greater ▶	SERVICE circuit opens. REMOVE Breakout box. RECONNECT processor and MAP BP sensor. RERUN Quick Test.
DF6 CHECK MAP BP SIGNAL FOR SHORTS TO VREF, SIGNAL RETURN AND GROUND		
• Key Off, wait 10 seconds. • Disconnect processor 60 Pin connector and inspect for damaged pins, corrosion, loose wires. Service as necessary. • Install Breakout box, leave processor disconnected. • Harness disconnected from MAP BP sensor. • DVOM on 200,000 ohm scale • Measure resistance between test Pin 45 and test Pins 26, 46, 40 and 60 at the Breakout box.	Any reading less than 10,000 ohms ▶	SERVICE circuit shorts. REMOVE Breakout box. RECONNECT processor and MAP BP Sensor. RERUN Quick Test.
	All readings are 10,000 ohms or greater ▶	REPLACE MAP BP sensor. REMOVE Breakout box. RECONNECT electrical connections. RERUN Quick Test.

Manifold Absolute Pressure (MAP)/Barometric Pressure (BP) Sensor — Pinpoint Test — DF

TEST STEP	RESULT ▶	ACTION TO TAKE
CODE 22 ENGINE RUNNING		
DF7 CHECK MAP SENSOR		
• Key Off, wait 10 seconds. • Disconnect vacuum line from MAP sensor. • Install vacuum pump to MAP sensor. • Apply 18 in. Hg vacuum to MAP sensor. • Does MAP sensor hold vacuum?	Yes ▶	RELEASE vacuum. GO to DF8.
	No ▶	REPLACE MAP sensor. CONNECT vacuum line to MAP sensor. RERUN Quick Test.
DF8 ATTEMPT TO ELIMINATE CODE 22 (ENGINE RUNNING)		
• Key Off, wait 10 seconds. • Plug MAP vacuum supply hose. • Start engine and maintain 1500 ± 100 engine rpm. • Slowly apply 15 in. Hg vacuum to MAP sensor. • While maintaining rpm, perform Engine Running Quick Test. • Is code 22 still present? **NOTE: Disregard any other codes at this time.**	Yes ▶	REPLACE MAP sensor. CONNECT vacuum line to MAP sensor. RERUN Quick Test.
	No ▶	INSPECT vacuum supply hose to MAP sensor. SERVICE as necessary. If OK, SERVICE other engine running codes. If none, GO to Diagnostic Routines, Section 2 for a low vacuum problem.
FAULT CODE 72		
DF10 CHECK THAT VACUUM TO MAP SENSOR DECREASES DURING DYNAMIC RESPONSE		
• Key Off, wait 10 seconds. • Tee a vacuum gauge in the intake manifold vacuum line at the MAP sensor. • Perform Engine Running Quick Test while observing vacuum. • Record engine service codes. • Did vacuum decrease by more than 30 kPa (10 in. Hg) during dynamic response test? • Is code 72 present?	Vacuum decrease is 30 kPa (10 in. Hg) or greater and code 72 is not present ▶	DISCONNECT vacuum equipment and SERVICE other codes as necessary.
	Vacuum decrease is 30 kPa (10 in. Hg) or greater and code 72 is present ▶	REPLACE MAP sensor and RERUN Quick Test.
	Vacuum decrease is less than 30 kPa (10 in. Hg) ▶	GO to DF11.

Manifold Absolute Pressure (MAP)/Barometric Pressure (BP) Sensor — Pinpoint Test — DF

TEST STEP	RESULT ▶	ACTION TO TAKE
DF11 CHECK VACUUM LINES		
• Check vacuum line for proper routing. Refer to VECI decal. Check MAP sensor vacuum line for kinks or blockage. • Vacuum lines are OK	Yes ▶	EEC-IV system OK. REFER to Shop Manual, Group 21 for probable subjects affecting engine vacuum.
	No ▶	SERVICE as necessary and REPEAT DF10.
DF20 CONNECTING MAP BP TESTER		
• Key Off. • Disconnect the MAP BP sensor from the vehicle harness. • Connect the MAP BP Tester between the vehicle harness and the MAP BP sensor. • Plug tester banana plugs into DVOM. • Set DVOM to 20V scale. • Refer to illustration DF	Tester properly hooked up ▶	GO to DF21.
DF21 MAP BP TESTER OUTPUT READING		
• MAP Tester connected, refer to Note. • Key On. • Approximate Altitude (Ft.) / Voltage Output (+ − .04 Volts) 0 — 1.59 1000 — 1.56 2000 — 1.53 3000 — 1.50 4000 — 1.47 5000 — 1.44 6000 — 1.41 7000 — 1.39 **NOTE: Measure several known good MAP sensors on available vehicles. The measured voltage will be typical for your location on the day of testing.** • Is reading in range for your altitude?	Yes ▶	For 1.9L EFI and 2.3L EFI TC engines GO to Diagnostic by Symptom in the Engine Supplement Section. For all others, GO to DF22.
	No (Sensor output is out-of-range) ▶	REPLACE MAP BP sensor.

Manifold Absolute Pressure (MAP)/ Barometric Pressure (BP) Sensor — Pinpoint Test DF

TEST STEP	RESULT	ACTION TO TAKE
DF22 VACUUM LINE CHECK		
• Check MAP sensor vacuum line for holes, disconnections, kinks or blockage. • Are vacuum lines OK?	Yes ▶	GO to Diagnostics by Symptom in the Engine Supplement Section.
	No ▶	SERVICE vacuum lines to MAP sensor. RERUN Quick Test.

Manifold Absolute Pressure (MAP)/ Barometric Pressure (BP) Sensor — Pinpoint Test DF

TEST STEP	RESULT	ACTION TO TAKE
DF90 SERVICE CODE 22: CONTINUOUS TEST: EXERCISE MAP SENSOR		
• Using continuous monitor mode, observe VOM or STAR LED for indication of a fault while performing the following: • Connect a vacuum pump to the MAP sensor. • Slowly apply 84 kPa (25 in. Hg.) vacuum to the sensor. • Slowly bleed vacuum off the MAP sensor. • Lightly tap on MAP sensor (simulate road shock). • Wiggle MAP connector. • Is fault indicated? 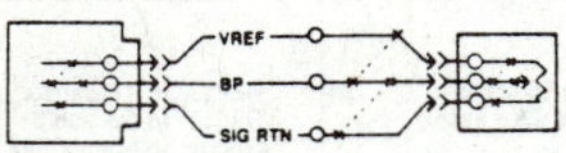 	Yes ▶	DISCONNECT and INSPECT connectors. If connector and terminals are good, REPLACE sensor. RERUN Quick Test.
	No ▶	GO to DF91.
DF91 CHECK EEC-IV HARNESS		
• Observe VOM or STAR LED for a fault indication while performing the following: • Referring to the illustration in Step DF90, grasp the harness closest to the sensor connector. Wiggle, shake or bend a small section of the EEC-IV system harness while working your way to the dash panel. Also wiggle, shake or bend the EEC-IV harness from the dash panel to the processor. • Is a fault indicated?	Yes ▶	ISOLATE fault and SERVICE as necessary. RERUN Quick Test.
	No ▶	GO to DF92.
DF92 CHECK PROCESSOR AND HARNESS CONNECTORS		
• Key Off, wait 10 seconds. • Disconnect processor 60 Pin connector. • Inspect both connectors and connector terminals for obvious damage or faults. • Are connectors and terminals OK?	No ▶	SERVICE as necessary. RERUN Quick Test.
	Yes ▶	Unable to duplicate fault at this time. Continuous code 22 testing complete.

Knock Sensor — Pinpoint Test DG

DG

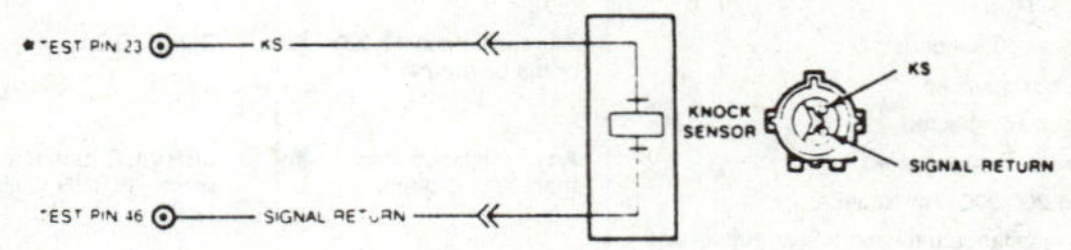

● TEST PINS LOCATED ON BREAKOUT BOX.
ALL HARNESS CONNECTORS VIEWED INTO MATING SURFACE

STOP-WARNING

You should enter this Pinpoint Test only when a service code 25 is received in Quick Test Step 5.0 or you are directed here from Diagnostics by Symptom in the Engine Supplement Section.

To prevent the replacement of good components, be aware that the following non-EEC areas may be at fault:

- Fuel (quality).
- Basic engine.
- Spark timing.

This Pinpoint Test is intended to diagnose only the following:

- Knock sensor.
- Harness circuits: KS and Signal Return.
- Processor assembly.

Knock Sensor — Pinpoint Test DG

TEST STEP	RESULT	ACTION TO TAKE
DG1 SERVICE CODE 25: GENERATE KNOCK MANUALLY		
NOTE: With knock conditions sensitive to fuel, altitude and weather, perform Step DG1 before servicing any components. • Vehicle prepared to run (Engine Running Self-Test). • Equipment needed: 4 oz. hammer. • Prepare to rap-tap on exhaust manifold, directly above the knock sensor, when the Dynamic Response Signal is given. **NOTE: There is no need to actually depress throttle at this point.** • Perform Engine Running Self-Test. • Rap moderately on exhaust manifold when meter indicates Dynamic Response Test is ready. • 15 seconds later a code will be generated. • Check for code 25. **NOTE: Ignore all other codes at this point.**	No ▶	Knock system OK. REPEAT Engine Running Self-Test and SERVICE any other codes from that test.
	Yes ▶ **NOTE: Service code 25 may be received whenever the engine is not tapped**	GO to DG2.
DG2 TEST KNOCK CIRCUIT FOR VOLTAGE		
• Key Off, wait 10 seconds. • Disconnect knock sensor connector and inspect. • Set DVOM on 20V scale. • Key On, Engine Off. • Measure voltage at the vehicle harness connector between KS and signal return.	Voltage is between 1 and 4V ▶	GO to DG6.
	Voltage is less than 1V ▶	GO to DG3.
	Voltage is greater than 4V ▶	GO to DG5.
DG3 CHECK CONTINUITY OF KS AND SIGNAL RETURN CIRCUITS		
• Key Off, wait 10 seconds. • Disconnect processor 60 Pin connector and inspect for damaged pins, corrosion, loose wires. Service as necessary. • Connect Breakout box to harness. Leave processor disconnected. • Knock sensor disconnected. • DVOM on 200 ohm scale. • Measure resistance between Signal Return at the vehicle harness and test Pin 46 at the Breakout box and between KS at the vehicle harness and test Pin 23 at the Breakout box.	Both resistances less than 5 ohms ▶	GO to DG4.
	Either resistance 5 ohms or greater ▶	SERVICE open circuit. RERUN Quick Test.

Knock Sensor — Pinpoint Test — DG

TEST STEP	RESULT ▶	ACTION TO TAKE
DG4 CHECK KS CIRCUIT FOR SHORT TO GROUND		
• Key Off, wait 10 seconds. • Breakout box installed. • Processor disconnected. • Knock sensor disconnected. • DVOM on 200 000 ohm scale. • Measure resistance between KS at the vehicle harness and test Pins 40, 46 and 60 at the Breakout box.	All resistances 10,000 ohms or greater ▶ Any resistance less than 10,000 ohms ▶	GO to DG6. SERVICE harness short. RERUN Quick Test.
DG5 CHECK KS CIRCUIT FOR SHORT TO VOLTAGE		
• Key Off, wait 10 seconds. • Disconnect processor 60 Pin connector and inspect for damaged pins, corrosion, loose wires. Service as necessary. • Connect Breakout box to harness. Leave processor disconnected. • Knock sensor disconnected. • Key On, Engine Off. • DVOM on 20V scale. • Measure voltage between test Pin 23 and test Pin 40 at the Breakout box.	0.5V or greater ▶ Less than 0.5V ▶	SERVICE harness short to power. RERUN Quick Test. GO to DG6.
DG6 TEST PROCESSOR WITH SUBSTITUTE KNOCK SENSOR		
• Key Off, wait 10 seconds. • Remove Breakout box and reconnect processor. • Equipment Required: — Equivalent knock sensor (same part number). — 4 oz. hammer. • Plug substitute sensor in harness (do not install). • Perform Engine Running Self-Test. • Lightly tap knock sensor when Dynamic Test Ready Signal is given. • 15 seconds later, a code will be generated. • Check for code 25. **NOTE: Ignore all other codes at this time.**	No ▶ Yes ▶	INSTALL new knock sensor. RERUN Quick Test. REPLACE processor and REMOVE substitute sensor. RERUN Quick Test with original sensor.

Throttle Position Sensor (TPS) — Pinpoint Test — DH

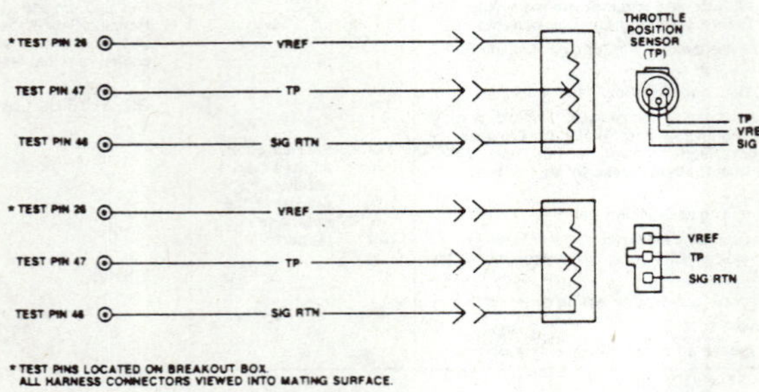

STOP-WARNING

You should enter this Pinpoint Test only when a Service Code 23, 53, 63 or 73 is received in Quick Test Step 3.0, 5.0 or 6.0.

To prevent the replacement of good components, be aware that the following non-EEC areas may be at fault:

- Idle speeds/throttle stop adjustment.
- Binding throttle shaft/linkage or speed control linkage.
- Choke/high cam system, if equipped.

This Pinpoint Test is intended to diagnose only the following:

- TP sensor.
- Sensor harness circuits: VREF, TP Signal, and Signal Return.
- Processor assembly.

Throttle Position Sensor (TPS) — Pinpoint Test — DH

TEST STEP	RESULT ▶	ACTION TO TAKE
DH1 SERVICE CODE 23: THE FOLLOWING CHECK MUST BE MADE BEFORE SERVICING THIS CODE		
• Check for code 68; Key On, Engine Off or codes 58, 31 or 41 Engine Running.	Code(s) present ▶ No codes present ▶	DISREGARD Code 23 at this time and RETURN to Quick Test Step 3.0B for code 68 or Step 5.0B for codes 31, 41 or 58. PROCEED as directed. GO to DH2.
DH2 CHECK FOR STUCK THROTTLE PLATE		
• Visually inspect carburetor/throttle body and throttle linkage for binding or sticking. • Verify the throttle linkage is at mechanical/closed throttle. Check for: binding throttle linkage, speed control linkage, vacuum line/electrical harness interference, etc.	Throttle not stuck ▶ Throttle stuck ▶	GO to DH3. CORRECT faults. RERUN Quick Test.
DH3 GENERATE CODE 63		
• Refer to illustration DH. • Key Off, wait 10 seconds. • Disconnect TP sensor vehicle harness connector at the throttle body. Inspect for damaged pins, corrosion, loose wires, etc. Service as necessary. • Perform Key On, Engine Off self-test and record codes. **NOTE: Ignore all other codes at this time.**	Code 63 present ▶ Code 63 not present ▶	GO to DH4. GO to DH5.
DH4 CHECK VOLTAGE VREF TO SIGNAL RETURN		
• Refer to illustration DH. • Key Off, wait 10 seconds. • Disconnect TP vehicle harness connector at throttle body. Inspect for damaged pins, corrosion, and pins pushed out. Service as necessary. • DVOM on 20V scale. • Key On, Engine Off. • Measure voltage at the TP vehicle harness connector between VREF and Signal Return.	Voltage reading between 4V and 6V ▶ Voltage reading less than 4V or greater than 6V ▶	REPLACE TP sensor. REFER to Section 3 before replacement for adjustment procedures for EFI applications. RERUN Quick Test. GO to Pinpoint Test Step C1.

Throttle Position Sensor (TPS) — Pinpoint Test — DH

TEST STEP	RESULT ▶	ACTION TO TAKE
DH5 CHECK TP SIGNAL FOR SHORT TO POWER		
• Key Off, wait 10 seconds, TP harness disconnected. • DVOM on 200,000 ohm scale. • Disconnect processor 60 Pin connector and inspect for damaged pins, corrosion, loose wires. Service as necessary. • Install Breakout box, leave processor disconnected. • Measure resistance between test Pin 47 and test Pins 26 and 57 at the Breakout box.	Either resistance is less than 10,000 ohms ▶ Both resistances are 10,000 ohms or greater ▶	SERVICE harness short. RERUN Quick Test. REPLACE processor. RERUN Quick Test.
DH10 SERVICE CODE 63: GENERATE CODE 53		
• Key Off, wait 10 seconds, TP harness disconnected. • Jumper VREF to TP signal at TP vehicle harness connector. • Perform Key On, Engine Off self-test. **NOTE: If no codes are generated, immediately remove jumper and go directly to DH13.** • Check for Code 53 (Code 23) **NOTE: Ignore all other codes at this time.**	Code 53 present (Code 23 present) ▶ Code 53 not present (Code 23 not present) ▶	REPLACE TP sensor. REFER to Section 3 before replacement for adjustment procedures for EFI applications and RERUN Quick Test. GO to DH11.
DH11 SERVICE CODE 63: CHECK VOLTAGE VREF TO SIGNAL RETURN		
• Refer to illustration DH. • Key Off, wait 10 seconds. • Disconnect TP vehicle harness connector at throttle body. Inspect for damaged pins, corrosion, and pins pushed out. Service as necessary. • DVOM on 20V scale. • Key On, Engine Off. • Measure voltage at the TP vehicle harness connector between VREF and Signal Return.	Voltage reading between 4V and 6V ▶ Voltage reading less than 4V or greater than 6V ▶	GO to DH12. GO to Pinpoint Test Step C1.

Throttle Position Sensor (TPS) | Pinpoint Test | DH

TEST STEP	RESULT ▶	ACTION TO TAKE
DH12 CHECK CONTINUITY OF TP CIRCUIT		
• Key Off, wait 10 seconds. TP harness disconnected. • DVOM on 200 ohm scale. • Disconnect processor 60 Pin connector and inspect for damaged pins, corrosion, loose wires. Service as necessary. • Connect Breakout box. Processor connected to Breakout box. • Measure resistance between TP Signal at the vehicle harness connector and test Pin 47 at the Breakout box.	Reading 5 ohms or greater ▶	SERVICE faulty circuit. CONNECT throttle position sensor. REMOVE Breakout box and RERUN Quick Test.
	Readings less than 5 ohms ▶	GO to DH13.
DH13 CHECK RESISTANCE OF TP CIRCUIT TO GROUND SIGNAL RETURN		
• Key Off, wait 10 seconds. TP harness disconnected. • Disconnect processor 60 Pin connector and inspect for damaged pins, corrosion, loose wires. Service as necessary. • DVOM on 200,000 ohm scale. • Measure resistance between TP signal at TP vehicle harness connector and test Pin 46 at the Breakout box and between TP signal at TP vehicle harness connector and ground.	Either reading is less than 10,000 ohms ▶	SERVICE circuit short and RERUN Quick Test.
	Both readings are 10,000 ohms or greater ▶	REPLACE processor. CONNECT throttle position sensor. REMOVE Breakout box and RERUN Quick Test.
DH20 SERVICE CODE 73: TP SENSOR MOVES IN ENGINE RESPONSE TEST		
NOTE: Code 73 indicates the TP Sensor did not exceed 25 percent of its rotation in the Engine Response Check. • Key Off. • Install Breakout box. • DVOM on 20V scale. • Connect DVOM to test Pins 47 and 46 at the Breakout box. • Perform Engine Running Quick Test, Step 5.0. • Verify DVOM reading exceeds 3.5V during brief WOT at Engine Response Check.	Reading exceeds 3.5V during Engine Response Check ▶	REPLACE processor. RERUN Quick Test.
	Reading does not exceed 3.5V during Engine Response Check ▶	VERIFY TP Sensor is properly attached to throttle body. If OK, REPLACE TP Sensor. Refer to Section 3 before replacement for adjustment procedures for EFI applications. RERUN Quick Test.

Throttle Position Sensor (TPS) | Pinpoint Test | DH

TEST STEP	RESULT ▶	ACTION TO TAKE
DH90 CONTINUOUS TEST SERVICE CODE 53 EXERCISE TP SENSOR		
• Using continuous monitor mode, observe VOM or STAR LED for indication of a fault while performing the following: • Move throttle slowly to WOT position. • Release throttle slowly to closed position and lightly tap on TP sensor (simulate road shock). • Wiggle TP harness connector. • Is a fault indicated?	Yes ▶	GO to DH91.
	No ▶	GO to DH92.
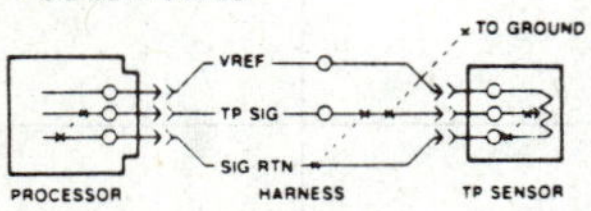		
DH91 MEASURE THROTTLE POSITION SIGNAL VOLTAGE WHILE EXERCISING TP SENSOR		
• Key Off, wait 10 seconds. • Disconnect processor 60 Pin connector and inspect for damaged pins, corrosion, loose wires. Service as necessary. • Install Breakout box and reconnect processor. • VOM or STAR LED still connected to STO as in previous step. • Connect a DVOM from test Pin 47 to test Pin 46. • DVOM on 20V scale. • Key On, engine Off. • While observing DVOM, repeat Step DH90. • Does the fault occur below 4.25V?	Yes ▶	DISCONNECT and INSPECT connectors. If connector and terminals are good, REPLACE TP sensor. REFER to Shop Manual Group 24 and RERUN Quick Test.
	No ▶	Throttle position sensor overtravel may have caused the continuous code 53. Sensor service is not required. To verify harness integrity, GO to DH92.

Throttle Position Sensor (TPS) | Pinpoint Test | DH

TEST STEP	RESULT ▶	ACTION TO TAKE
DH92 CHECK EEC-IV HARNESS		
• Observe VOM or STAR LED for a fault indication while performing the following: • Referring to the illustration in Step DH90, grasp the harness close to the sensor connector. Wiggle, shake or bend a small section of the EEC-IV system harness while working your way to the dash panel. Also wiggle, shake or bend the EEC-IV harness from the dash panel to the processor. • Is a fault indicated?	Yes ▶	ISOLATE fault and make necessary repairs. REFER to appropriate figure. RERUN Quick Test.
	No ▶	GO to DH93.
DH93 CHECK PROCESSOR AND HARNESS CONNECTORS		
• Key Off, wait 10 seconds. • Disconnect processor 60 Pin connector. • Inspect both connectors and connector terminals for obvious damage or faults. • Are connectors and terminals OK?	No ▶	SERVICE as necessary. REPEAT Quick Test.
	Yes ▶	Unable to duplicate fault at this time. Continuous code 53 testing complete.
DH94 CONTINUOUS TEST SERVICE CODE 63: EXERCISE TP SENSOR		
• Using continuous monitor mode, observe VOM or STAR LED for indication of a fault while performing the following: • Move throttle slowly to WOT position. • Release throttle slowly to closed condition. • Lightly tap on TP sensor (simulate road shock). • Wiggle TP harness connector. • Is a fault indicated?	Yes ▶	DISCONNECT and INSPECT connectors. If connector and terminals are good, REPLACE TP sensor. REFER to Shop Manual Group 24 and RERUN Quick Test.
	No ▶	GO to DH95.
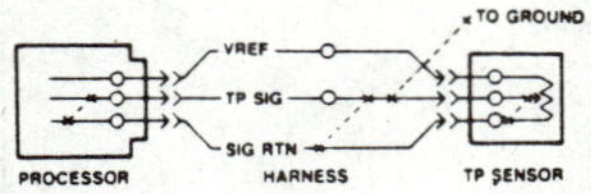		

Throttle Position Sensor (TPS) | Pinpoint Test | DH

TEST STEP	RESULT ▶	ACTION TO TAKE
DH95 CHECK EEC-IV HARNESS		
• Observe VOM or STAR LED for a fault indication while performing the following: • Referring to the illustration in Step DH94 grasp the harness close to the sensor connector. Wiggle, shake or bend a small section of the EEC-IV system harness while working your way to the dash panel. Also wiggle, shake or bend the EEC-IV harness from the dash panel to the processor. • Is a fault indicated?	Yes ▶	ISOLATE fault and make necessary repairs. REFER to appropriate figure. RERUN Quick Test.
	No ▶	Go to DH96.
DH96 CHECK PROCESSOR AND HARNESS CONNECTORS		
• Key Off, wait 10 seconds. • Disconnect processor 60 Pin connector. • Inspect both connectors and connector terminals for obvious damage or faults. • Are connectors and terminals OK?	No ▶	SERVICE as necessary. RERUN Quick Test.
	Yes ▶	Unable to duplicate fault at this time. Continuous code 63 testing complete.

Vane Air Flow Sensor (VAF) — Pinpoint Test — DK

DK

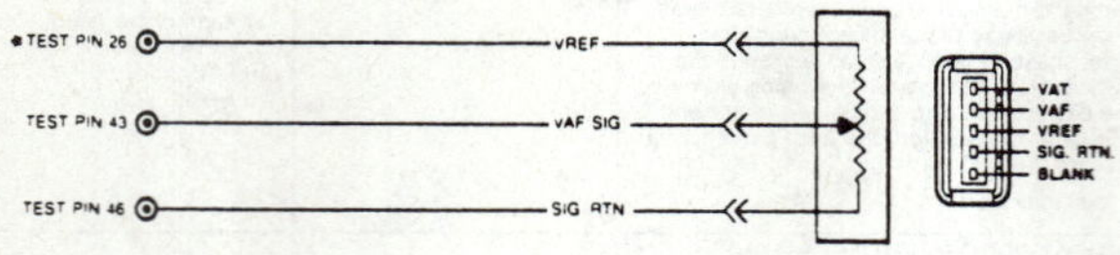

* TEST PIN LOCATED ON BREAKOUT BOX
ALL HARNESS CONNECTORS VIEWED INTO MATING SURFACE.

STOP-WARNING

You should enter this Pinpoint Test only when a service code 26, 56, 66 or 76 is received in Quick Test Step 3.0, 5.0 or 6.0.

To prevent the replacement of good components, be aware that the following non-EEC areas may be at fault:

- Check for unmetered air (air leaks) between VAF meter and throttle body.
- Vacuum leaks.
- Engine sealing (PCV sealing, CANP, valve cover seal dipstick seated).

This Pinpoint Test is intended to diagnose only the following:

- VAF meter.
- Processor.
- Harness circuits: VREF, VAF Sig. and Signal Return.

Vane Air Flow Sensor (VAF) — Pinpoint Test — DK

TEST STEP	RESULT ▶	ACTION TO TAKE
FAULT CODE 26 **DK1** CHECK FOR CONTAMINATION **NOTE: Code 26 (Key On, Engine Off, or Engine Running) indicates the vane air flow input to the processor is out of closed (Engine Not Running) or at idle limits (Engine Off 0.15-0.50V: Engine at idle 1.50-2.70V). There have been no opens or shorts in the VAF circuit or a code 56 (signal always high) or 66 (signal always low) would have been generated.**		
• Key Off, wait 10 seconds. • Remove air cleaner element and check for contamination (oil residue, foreign material, etc.) that may impede VAF sensor vane movement. • Is service code 26 present in Key On, Engine Off portion of Quick Test?	Yes ▶	REINSTALL air cleaner element. REPLACE vane meter. RERUN Quick Test.
	No ▶	GO to DK2.
DK2 VAF SENSOR CHECK • Key Off, air cleaner element reinstalled. • Check for unmetered air leaks between vane meter and throttle body. • Disconnect processor 60 Pin connector. Inspect for damaged pins, corrosion, loose wires, etc. Service as necessary. • Install Breakout box. Connect processor to Breakout box harness. • DVOM on 20V scale. • Key On, engine Off. • Place new unsharpened pencil as shown below. • Measure voltage between test Pins 43 and 46 at the Breakout box. • Does DVOM read between 2.8V and 3.7V?	Yes ▶	Vane meter is capable of outputting an acceptable signal. The VAF code 26 has been caused by incorrect engine speed or an unmetered air leak (vacuum leak). SERVICE as necessary. REMOVE Breakout box. RERUN Quick Test.
	No ▶	REMOVE Breakout box. REPLACE processor. RERUN Quick Test.

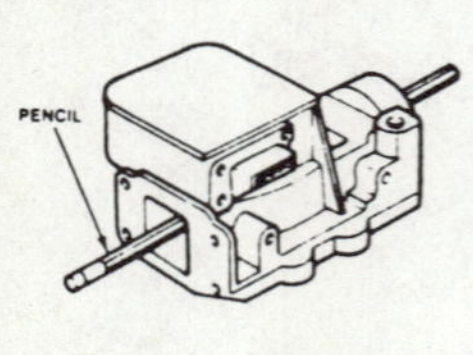

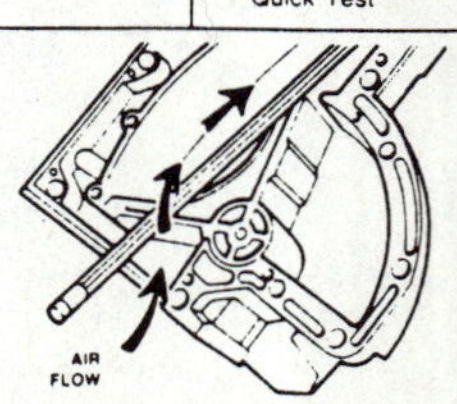

Vane Air Flow Sensor (VAF) — Pinpoint Test — DK

TEST STEP	RESULT ▶	ACTION TO TAKE
FAULT CODE 56 **DK10** INDUCE OPPOSITE CODE • Key Off, wait 10 seconds. • Disconnect vehicle harness from vane meter. Inspect for damaged pins, corrosion, loose wires, etc. Service as necessary. • Run Key On, Engine Off Quick Test. • Is code 66 present? **NOTE: Disconnecting vane meter disconnects both VAF and VAT sensors. A code 54 should also be present. Disregard all codes except VAF codes at this time.**	Yes ▶	GO to DK11.
	No ▶	GO to DK12.
DK11 CHECK VAF TO SIGNAL RETURN VOLTAGE • Key Off, wait 10 seconds. • Harness disconnected from vane meter. • Key On, Engine Off. • DVOM on 20V scale. • Measure voltage at the vane meter vehicle harness connector between VREF and Signal Return.	Reading is between 4.0V and 6.0V ▶	REPLACE vane meter. RECONNECT harness. RERUN Quick Test.
	Reading is less than 4.0V or greater than 6.0V ▶	GO to Pinpoint Test Step C1.
DK12 CHECK VAF SIGNAL FOR SHORT • Key Off, wait 10 seconds. • Harness disconnected from vane meter. • Disconnect processor 60 Pin connector. Inspect for damaged pins, corrosion, loose wires, etc. Service as necessary. • Install Breakout box, leave processor disconnected. • DVOM on 200,000 ohm scale. • Measure resistance between test Pin 43 and test Pins 26 and 57 at the Breakout box.	Either reading is less than 10,000 ohms ▶	SERVICE circuit shorts. REMOVE Breakout box. RECONNECT processor and vane meter. RERUN Quick Test.
	Both readings are 10,000 ohms or greater ▶	REPLACE processor. REMOVE Breakout box. RECONNECT processor and vane meter. RERUN Quick Test.

Vane Air Flow Sensor (VAF) — Pinpoint Test — DK

TEST STEP	RESULT ▶	ACTION TO TAKE
FAULT CODE 66 **DK20** INDUCE OPPOSITE CODE • Key Off, wait 10 seconds. • Disconnect vehicle harness from vane meter. • Install jumper wire in vane meter vehicle harness connector between VREF and VAF signal. • Perform Key On, Engine Off Quick Test. **NOTE: If no codes are generated, immediately remove jumper and go directly to DK23.** • Is code 56 present? **NOTE: Disconnecting vane meter disconnects both VAF and VAT sensors. Code 54 should also be present. Disregard all codes except VAF codes at this time.**	Yes ▶	REPLACE vane meter. REMOVE jumper wire. RECONNECT vane meter. RERUN Quick Test.
	No ▶	REMOVE jumper wire and GO to DK21.
DK21 CHECK VREF TO SIGNAL RETURN VOLTAGE • Key off, wait 10 seconds. • Harness disconnected from vane meter. • Key On, Engine Off. • DVOM on 20V scale. • Measure voltage at the vane meter vehicle harness connector between VREF and Signal Return.	Reading is between 4.0 and 6.0V ▶	GO to DK22.
	Reading is less than 4.0V or greater than 6.0V ▶	GO to Pinpoint Test Step C1.
DK22 CHECK CONTINUITY OF VAF SIGNAL • Key Off, harness disconnected from vane meter. • Disconnect processor 60 Pin connector. Inspect for damaged pins, corrosion, loose wires, etc. Service as necessary. • Install Breakout box, leave processor disconnected. • DVOM on 200 ohm scale. • Measure resistance between VAF signal, at the vane meter vehicle harness connector, and test Pin 43 at the Breakout box.	Reading is less than 5 ohms ▶	GO to DK23.
	Reading is 5 ohms or greater ▶	SERVICE circuit open. REMOVE Breakout box. RECONNECT processor and vane meter. RERUN Quick Test.

Vane Air Flow Sensor (VAF) — Pinpoint Test — DK

TEST STEP	RESULT	ACTION TO TAKE
DK23 CHECK VAF SIGNAL FOR SHORT		
• Key Off, wait 10 seconds. • Processor disconnected. • Harness disconnected from vane meter. • DVOM on 200,000 ohm scale. • Measure resistance at the vane meter vehicle harness between VAF signal and signal return and between VAF signal and negative battery terminal.	Either reading is less than 10,000 ohms ▶	SERVICE circuit shorts. RECONNECT vane meter. RERUN Quick Test.
	Both readings are 10,000 ohms or greater ▶	REPLACE processor. REMOVE Breakout box. RECONNECT processor and vane meter. RERUN Quick Test.
FAULT CODE 76		
DK30 CHECK FOR VOLTAGE INCREASE IN VAF SIGNAL TO SIGNAL RETURN		
NOTE: A sharp snap of the throttle may not be sufficient to pass this test. Be sure to move throttle to WOT and return.		
• Key Off, wait 10 seconds. • Disconnect processor 60 Pin connector. Inspect for damaged pins, corrosion, loose wires, etc. Service as necessary. • Install Breakout box. Connect processor to Breakout box harness. • DVOM on 20V scale. • Connect DVOM to test Pins 43 and 46. • Perform Engine Running Quick Test while monitoring DVOM. • After dynamic response prompt code 1(0) operator does a brief WOT. DVOM should increase more than 2.0V from reading before WOT. • Observe service codes at end of test.	Reading increased more than 2.0V and code 76 is still present ▶	REPLACE processor. REMOVE Breakout box. RERUN Quick Test.
	Reading did not increase more than 2.0V ▶	CHECK air cleaner duct for obstruction. If OK, REPLACE vane meter.
	Reading increased more than 2.0V and code 76 is not present ▶	Vane meter is OK. SERVICE other codes as necessary.

Vane Air Flow Sensor (VAF) — Pinpoint Test — DK

TEST STEP	RESULT	ACTION TO TAKE
DK90 SERVICE CODE 56 CONTINUOUS TEST: CHECK VAF SENSOR		
• Using continuous monitor mode, observe VOM or STAR LED for indication of a fault while performing the following: • Lightly tap on VAF sensor (simulate road shock). • Wiggle VAF connector. • Is a fault indicated?	Yes ▶	DISCONNECT and INSPECT connectors. If connector and terminals are good, REPLACE VAF sensor. RERUN Quick Test.
	No ▶	GO to DK91.
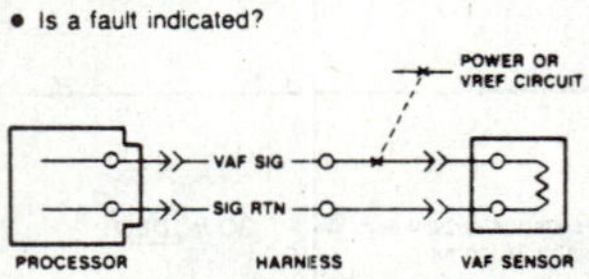		
DK91 CHECK EEC-IV HARNESS		
• Observe VOM or STAR LED for a fault indication while performing the following: • Referring to the illustration in Step DK90, grasp the harness close to the sensor connector. Wiggle, shake or bend a small section of the EEC-IV system harness while working your way to the dash panel. Also wiggle, shake or bend the EEC-IV harness from the dash panel to the processor. • Is a fault indicated?	Yes ▶	ISOLATE fault and make necessary repairs. REFER to appropriate figure. RERUN Quick Test.
	No ▶	GO to DK92.
DK92 CHECK PROCESSOR AND HARNESS CONNECTORS		
• Key Off, wait 10 seconds. • Disconnect processor 60 Pin connector. • Inspect both connectors and connector terminals for obvious damage or faults. • Are connectors and terminals OK?	No ▶	SERVICE as necessary. RERUN Quick Test.
	Yes ▶	Unable to duplicate fault at this time. Continuous code 56 testing complete.

Vane Air Flow Sensor (VAF) — Pinpoint Test — DK

TEST STEP	RESULT	ACTION TO TAKE
DK93 SERVICE CODE 66 CONTINUOUS TEST. CHECK VAF SENSOR		
• Using continuous monitor mode, observe VOM or STAR LED for indication of a fault while performing the following: • Lightly tap on VAF sensor (simulate road shock). • Wiggle VAF connector. • Is a fault indicated?	Yes ▶	DISCONNECT and INSPECT connectors. If connector and terminals are good, REPLACE sensor. RERUN Quick Test.
	No ▶	GO to DK94.
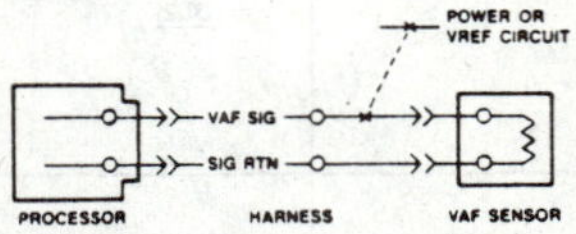		
DK94 CHECK EEC-IV HARNESS		
• Observe VOM or STAR LED for a fault indication while performing the following: • Referring to the illustration in Step DK93 grasp the harness close to the sensor connector. Wiggle, shake or bend a small section of the EEC-IV system harness while working your way to the dash panel. Also wiggle, shake or bend the EEC-IV harness from the dash panel to the processor. • Is a fault indicated?	Yes ▶	ISOLATE fault and make necessary repairs. REFER to appropriate figure. RERUN Quick Test.
	No ▶	Go to DK95.
DK95 CHECK PROCESSOR AND HARNESS CONNECTORS		
• Key Off, wait 10 seconds. • Disconnect processor 60 Pin connector. • Inspect both connectors and connector terminals for obvious damage or faults. • Are connectors and terminals OK?	No ▶	SERVICE as necessary. RERUN Quick Test.
	Yes ▶	Unable to duplicate fault at this time. Continuous code 66 testing complete.

Pressure Feedback EGR (PFE) EGR Valve Regulator (EVR) — Pinpoint Test — DL

DL

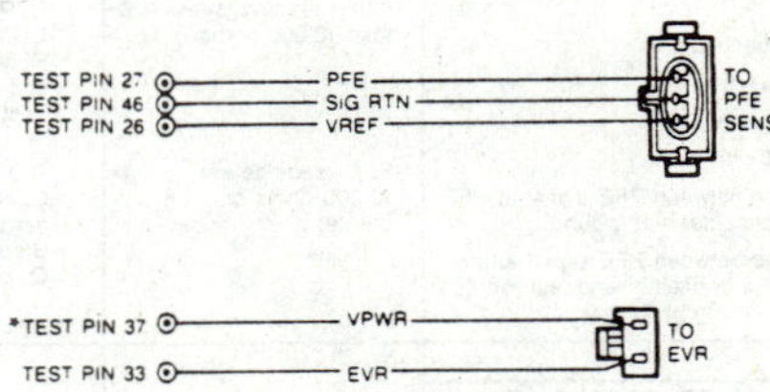

* TEST PINS LOCATED ON BREAKOUT BOX
ALL HARNESS CONNECTORS VIEWED INTO MATING SURFACE

STOP-WARNING

You should enter this Pinpoint Test only when a Service Code 31, 32, 33, 34, 35 or 84 is received in Quick Test Step 3.0, 5.0 or 6.0.

To prevent the replacement of good components, be aware that the following non-EEC area may be at fault:

- Damaged EGR valve.

This Pinpoint Test is intended to diagnose only the following:

- PFE sensor.
- Harness circuits: VREF, PFE, Signal Return, EVR, VPWR
- EVR (EGR valve regulator).
- EGR valve assembly.
- Processor assembly.

Pressure Feedback EGR (PFE) EGR Valve Regulator (EVR) — Pinpoint Test DL

TEST STEP	RESULT ▶	ACTION TO TAKE
SERVICE CODE 31		
DL1 ATTEMPT TO GENERATE OPPOSITE CODE (35)		
• Key Off. • Disconnect PFE vehicle harness at sensor. • Jumper VREF to PFE signal at vehicle harness sensor connector. • Perform Key On, Engine Off Self-Test. **NOTE: If no codes are generated, immediately remove jumper and GO directly to Step DL4.** • Check for code 35. **NOTE: Ignore all other codes at this time.**	Code 35 present ▶	REMOVE Jumper. REPLACE PFE sensor. RERUN Quick Test.
	Code 35 not present ▶	REMOVE jumper. GO to DL2
DL2 MEASURE VREF TO SIGNAL RETURN VOLTAGE		
• Key Off. • PFE harness disconnected. • DVOM on 20V scale. • Key On, Engine Off. • Measure voltage at PFE vehicle harness connector between VREF and Signal Return. • Refer to illustration DL.	Voltage reading between 4V and 6V ▶	GO to DL3.
	Voltage reading less than 4V or greater than 6V ▶	GO to C1.
DL3 CHECK CONTINUITY OF PFE SIGNAL		
• Key Off. • PFE harness disconnected. • DVOM on 200 ohm scale. • Disconnect processor 60 Pin connector and inspect for damaged pins, corrosion, loose wires, etc. Service as necessary. • Connect Breakout box to harness. Processor connected to Breakout box. • Measure resistance between PFE signal at vehicle harness sensor connector and test Pin 27 at the Breakout box.	Reading 5 ohms or greater ▶	SERVICE faulty circuit. CONNECT PFE sensor. REMOVE Breakout box. RERUN Quick Test.
	Reading less than 5 ohms ▶	Go To DL4.

Pressure Feedback EGR (PFE) EGR Valve Regulator (EVR) — Pinpoint Test DL

TEST STEP	RESULT ▶	ACTION TO TAKE
DL4 CHECK RESISTANCE OF PFE SIGNAL TO GROUND AND SIGNAL RETURN		
• Key Off. • PFE harness disconnected. • Breakout box installed. • Processor disconnected. • DVOM on 200,000 ohm scale. • Measure resistance between PFE signal at PFE vehicle harness connector and ground. • Measure resistance between PFE signal at the PFE vehicle harness connector and test Pin 46 (Signal Return) at the Breakout box.	Either reading is less than 10,000 ohms ▶	SERVICE short circuit. CONNECT PFE. REMOVE Breakout box. RERUN Quick Test.
	Both readings are 10,000 ohms or greater ▶	REPLACE processor. CONNECT PFE sensor. REMOVE Breakout box. RERUN Quick Test.
SERVICE CODE 35		
DL5 ATTEMPT TO GENERATE OPPOSITE CODE (31)		
• Key Off. • Disconnect PFE vehicle harness at sensor. Inspect for damaged pins, corrosion, loose wires, etc. Service as necessary. • Perform Key On, Engine Off Self-Test, and record codes. • Check for code 31. **NOTE: Ignore all other codes at this time.**	Code 31 present ▶	GO to DL6.
	Code 31 not present ▶	GO to DL7.
DL6 MEASURE VREF TO SIGNAL RETURN VOLTAGE		
• Key Off. • PFE harness disconnected. • DVOM on 20V scale. • Key On, Engine Off. • Measure voltage at PFE vehicle harness connector between VREF and Signal Return. • Refer to illustration DL.	Voltage reading between 4V and 6V ▶	REPLACE PFE sensor. RERUN Quick Test.
	Voltage reading less than 4V or greater than 6V ▶	GO to C1.

Pressure Feedback EGR (PFE) EGR Valve Regulator (EVR) — Pinpoint Test DL

TEST STEP	RESULT ▶	ACTION TO TAKE
DL7 CHECK PFE CIRCUIT FOR SHORT TO POWER		
• Key Off. • PFE harness disconnected. • Disconnect processor 60 Pin connector and inspect for damaged pins, corrosion, loose wires, etc. Service as necessary. • Install Breakout box, leave processor disconnected. • DVOM on 200,000 ohm scale. • Measure the resistance between test Pin 27 and test Pins 26 and 57 at the Breakout box.	Either resistance is less than 10,000 ohms ▶	SERVICE harness short. REMOVE Breakout box. CONNECT PFE sensor. RERUN Quick Test.
	Both resistances are 10,000 ohms or greater ▶	REPLACE processor. REMOVE Breakout box. CONNECT PFE sensor. RERUN Quick Test.
SERVICE CODE: 34		
DL8 PFE SENSOR OUT OF RANGE		
• PFE system can sense a lack of pressure in the vehicle exhaust system. An efficient garage exhaust ventilation system, installed during Key On, Engine Off Quick Test, may deflect the PFE sensor and generate a code 34. Remove the ventilation system and retest. • Is code 34 present?	No ▶	ADDRESS any other codes in Key On, Engine Off, if none CONTINUE with remaining Quick Test.
	Yes ▶	GO to DL9.
DL9 CHECK PRESSURE FEED TUBE TO PFE SENSOR		
• Remove the pressure feed tube from PFE sensor. • Inspect complete tube, including PFE inlet for blockage. • Is blockage present?	No ▶	GO to DL10.
	Yes ▶	SERVICE as necessary. RERUN Quick Test.

Pressure Feedback EGR (PFE) EGR Valve Regulator (EVR) — Pinpoint Test DL

TEST STEP	RESULT ▶	ACTION TO TAKE
DL10 MEASURE VREF TO SIGNAL RETURN VOLTAGE		
• Key Off. • Disconnect PFE sensor and inspect for damaged pins, corrosion, loose wires, etc. Service as necessary. • DVOM on 20V scale. • Key On, Engine Off. • Measure voltage at PFE vehicle harness connector between VREF and Signal Return. • Refer to Illustration DL.	Voltage reading between 4V and 6V ▶	REPLACE PFE sensor. RERUN Quick Test.
	Voltage reading less than 4V or greater than 6V ▶	GO to Pinpoint Test C1.
SERVICE CODE 84		
DL11 MEASURE EVR SOLENOID RESISTANCE		
• Key Off. • DVOM on 200 ohm scale. • Disconnect EVR solenoid connector and measure solenoid resistance.	Resistance is between 30 and 70 ohms ▶	GO to DL12.
	Resistance is less than 30 ohms or greater than 70 ohms ▶	REPLACE EVR solenoid assembly. RERUN Quick Test.
DL12 CHECK FOR VPWR AT EVR SOLENOID		
• EVR solenoid disconnected from harness. • DVOM on 20V scale. • Key On, Engine Off. • Measure voltage between battery negative terminal and VPWR circuit at EVR solenoid vehicle harness connector.	Reading is less than 10.5V ▶	SERVICE VPWR open circuit. RERUN Quick Test.
	Reading is 10.5V or greater ▶	GO to DL13.
DL13 CHECK CONTINUITY OF EVR CIRCUIT		
• Key Off. • EVR solenoid disconnected from harness. • Disconnect processor 60 Pin connector and inspect for damaged pins, corrosion, loose wires, etc. Service as necessary. • Install Breakout box to processor harness connector. Leave processor disconnected. • DVOM on 200 ohm scale. • Measure resistance between test Pin 33 at the Breakout box and EVR signal at the EVR solenoid vehicle harness connector.	Reading less than 5 ohms ▶	GO to DL14.
	Reading 5 ohms or greater ▶	SERVICE open circuit. REMOVE Breakout box. CONNECT process EVR solenoid. RERUN Quick Test.

Pressure Feedback EGR (PFE) EGR Valve Regulator (EVR) — Pinpoint Test — DL

TEST STEP	RESULT ▶	ACTION TO TAKE
DL14 CHECK EVR CIRCUIT FOR SHORT TO POWER AND GROUND		
• Key Off. • Breakout box installed, processor disconnected. • EVR solenoid disconnected. • DVOM on 200,000 ohm scale. • Measure resistance between test Pin 33 (EVR) and test Pins 37 and 57 (VPWR) and 40, 46 and 60 (GRD) at the Breakout box.	Any reading less than 10,000 ohms ▶ All reading 10,000 ohms or greater ▶	SERVICE short circuit. REMOVE Breakout box. RECONNECT harness to processor and EVR solenoid. RERUN Quick Test. If code is repeated, REPLACE processor. REPLACE processor. REMOVE Breakout box. RECONNECT harness to processor and EVR solenoid. RERUN Quick Test.
SERVICE CODE 32		
DL20 VERIFY ENGINE RUNNING CODES		
The PFE system can sense a lack of pressure in the vehicle exhaust system. An efficient garage exhaust ventilation system installed during Key On, Engine Run Quick Test may, on some calibrations, deflect the PFE sensor and generate a code 32. Temporarily, remove garage forced ventilation system and properly vent to atmosphere. • RERUN Engine Running Quick Test. • Is code 32 present?	Yes ▶ No ▶	GO to DL21. ADDRESS any other codes in Engine Running. If none, CONTINUE with remaining Quick Test.

Pressure Feedback EGR (PFE) EGR Valve Regulator (EVR) — Pinpoint Test — DL

TEST STEP	RESULT ▶	ACTION TO TAKE
DL21 ATTEMPT TO SEPARATE EVR FROM PFE		
• Key Off. • Disconnect EGR valve vacuum line at valve and plug line. • Perform Engine Running Quick Test. • Is code 31 or 32 present?	Yes ▶ No ▶	GO to DL22. GO to DL23.
DL22 CHECK PFE SENSOR SUPPLY TUBE		
• Key Off • Check PFE sensor supply tube for obstructions and or leaks. • Are there any obstructions or leaks?	Yes ▶ No ▶	SERVICE as necessary RECONNECT all lines and RERUN Quick Test. GO to EGR Diagnostic, Section 6
DL23 CHECK EVR FILTER		
• Key Off • Remove and inspect EVR filter for contamination **NOTE: Blockage of filter will cause vacuum to be applied to EGR valve prematurely.** • Is filter contaminated?	Yes ▶ No ▶	REPLACE filter RECONNECT all lines RERUN Quick Test. REPLACE EVR solenoid RERUN Quick Test.
ENGINE RUNNING SERVICE CODE 34 AND 35		
DL25 CHECK FOR EXCESSIVE EXHAUST BACK PRESSURE		
• Service codes 34 and 35 in Engine Running Self-Test indicate excessive exhaust back pressure. There are two possible causes (A) The exhaust system is restricted, and (B) PFE sensor has shifted high. • Key Off • Substitute known good PFE sensor in place of original • Rerun Key On, Engine Running Quick Test • Is code 34 or 35 present?	No ▶ Yes ▶	Original PFE was the cause of the original 34 or 35 REPLACE PFE sensor RERUN Quick Test GO to Section 5, Catalyst and Exhaust Systems Restricted Exhaust System Diagnosis

Pressure Feedback EGR (PFE) EGR Valve Regulator (EVR) — Pinpoint Test — DL

TEST STEP	RESULT ▶	ACTION TO TAKE
SERVICE CODE 33		
DL30 VERIFY VACUUM IS PRESENT AT VALVE		
• Key Off. • Standard vacuum gauge in. Hg (Mercury). • Tee in vacuum gauge at EGR valve. • Perform Engine Running Quick Test while observing vacuum gauge. **NOTE: Disregard code output.**	Vacuum reading 1 in. Hg or less ▶ Vacuum reading greater than 1 in. Hg ▶	GO to DL31. GO to EGR Valve Diagnostic Section 6.
DL31 VACUUM SUPPLY VERIFICATION		
• Key Off. • Check vacuum line from EVR solenoid to EGR valve and source to EVR solenoid for loose or disengaged connections, cracks, etc. • Is vacuum present at EVR solenoid during run?	Yes ▶ No ▶	REPLACE EVR solenoid. RERUN Quick Test. SERVICE as necessary. RERUN Quick Test.
DL90 SERVICE CODE 31 OR 35 CONTINUOUS TEST: EXERCISE PFE SENSOR		
• Using continuous monitor mode, observe VOM or STAR LED for indication of a fault while performing the following: • Connect a vacuum pump to the PFE sensor. • Slowly apply 5 in. Hg vacuum to the sensor. • Slowly bleed vacuum off the PFE sensor. • Lightly tap on PFE sensor (simulate road shock). • Wiggle PFE connector. • Is fault indicated? 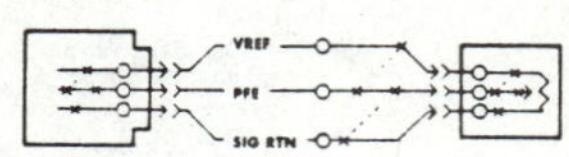	Yes ▶ No ▶	DISCONNECT and INSPECT connectors. If connector and terminals are good, REPLACE sensor RERUN Quick Test. GO to DL91.

Pressure Feedback EGR (PFE) EGR Valve Regulator (EVR) — Pinpoint Test — DL

TEST STEP	RESULT ▶	ACTION TO TAKE
DL91 CHECK EEC-IV HARNESS		
• Observe VOM or STAR LED for a fault indication while performing the following: • Referring to the illustration in Step DL90 grasp the harness closest to the sensor connector. Wiggle, shake or bend a small section of the EEC-IV system harness while working your way to the dash panel. Also wiggle, shake or bend the EEC-IV harness from the dash panel to the processor. • Is a fault indicated?	Yes ▶ No ▶	ISOLATE fault and SERVICE as necessary. RERUN Quick Test. GO to DL92.
DL92 CHECK PROCESSOR AND HARNESS CONNECTORS		
• Key Off, wait 10 seconds • Disconnect processor 60 Pin connector. • Inspect both connectors and connector terminals for obvious damage or faults. • Are connectors and terminals OK?	No ▶ Yes ▶	SERVICE as necessary. RERUN Quick Test. Unable to duplicate and/or identify fault at this time. Continuous code 31 or 35 testing complete.
DL93 SERVICE CODE 34 CONTINUOUS TEST — INSPECT PFE SUPPLY TUBE FOR BLOCKAGE		
• Key Off • Remove PFE sensor and inspect sensor supply inlet for liquids and or any type of blockage • Inspect PFE supply tube to EGR valve base for liquids and or blockage • Is supply tube free of any blockage?	No ▶ Yes ▶	CLEAN and or SERVICE as necessary. RERUN Quick Test. Unable to duplicate and or identify fault at this time. Continuous code 34 testing complete.

Pressure Feedback EGR (PFE) EGR Valve Regulator (EVR) — Pinpoint Test DL

TEST STEP	RESULT ▶	ACTION TO TAKE
DL94 SERVICE CODE 32: CONTINUOUS TEST — INSPECT EGR VALVE FOR SMOOTH OPERATION.		
• Key Off. • Connect a vacuum pump to the EGR valve. • Apply 10 in. Hg of vacuum to EGR valve. • While observing EGR valve, release vacuum. • Does EGR valve function in a smooth manner? **NOTE: Repeat test if necessary to ensure accurate result.**	No ▶	GO to EGR Valve Diagnostic Section 6.
	Yes ▶	GO to DL95.
DL95 INSPECT VACUUM LINES BETWEEN EVR SOLENOID AND EGR VALVE		
• Inspect EGR valve vacuum supply line from EVR solenoid for kinks and/or obstructions. • Is vacuum supply line to EGR valve free of any obstructions?	No ▶	SERVICE as necessary. RERUN Quick Test.
	Yes ▶	GO to DL96.
DL96 EVR REGULATOR FILTER INSPECTION		
• Carefully check EVR filter for contamination and/or obstructions. • Is EVR filter condition acceptable?	No ▶	REPLACE EVR filter. RERUN Quick Test.
	Yes ▶	Unable to duplicate and/or identify fault at this time. Continuous code 34 testing complete.
DL97 SERVICE CODE 33: CONTINUOUS TEST — INSPECT EGR VALVE FOR FREE OPERATION		
• Key Off. • Connect a vacuum pump to the EGR valve. • While observing the EGR valve, slowly apply 10 in. Hg vacuum. **NOTE: EGR valve should begin to open with a very small amount of vacuum, approximately 1 to 1.5 in. Hg and be fully open with about 4 in. Hg vacuum.** • Does EGR valve move freely and smoothly?	No ▶	GO to EGR Valve Diagnostic Section 6.
	Yes ▶	GO to DL98.

Pressure Feedback EGR (PFE) EGR Valve Regulator (EVR) — Pinpoint Test DL

TEST STEP	RESULT ▶	ACTION TO TAKE
DL98 EVR HARNESS CHECK		
• Key Off. • Disconnect processor 60 Pin connector and inspect for damaged pins, corrosion, loose wires, etc. Service as necessary. • Connect Breakout box to harness. Processor connected to Breakout box. • Enter output state check. • DVOM on 20 V scale. • Connect DVOM negative test lead to test Pin 40 at the Brakeout box and DVOM positive test lead to test Pin 33. • Cycle throttle if necessary to indicate 10.5 V or greater. • Remain at this position. • While observing DVOM, grasp the harness closest to the EVR connector. Wiggle, shake or bend a small section of the EEC-IV system harness while working your way to the dash panel. Lightly tap EVR solenoid to simulate road vibration. • Does DVOM indicate less than 10.5 V?	Yes ▶	SERVICE as necessary. RERUN Quick Test.
	No ▶	Unable to duplicate and/or identify fault at this time. Continuous code 33 testing complete.

EGR Valve Position Sensor (EVP) EGR Valve Regulator (EVR) — Pinpoint Test DM

DM

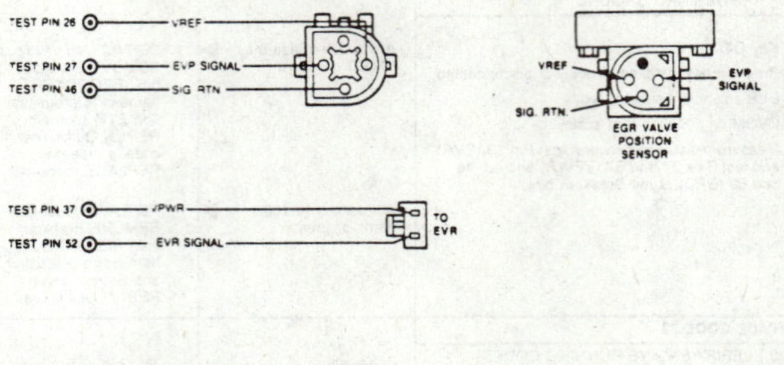

*TEST PINS LOCATED ON BREAKOUT BOX.
ALL HARNESS CONNECTORS VIEWED INTO MATING SURFACE.

STOP-WARNING

You should enter this Pinpoint Test only when a Service Code 31, 32, 33, 34, 35, 83 or 84 is received in Quick Test Step 3.0, 5.0 or 6.0.

To prevent the replacement of good components, be aware that the following non-EEC area may be at fault:

- Damaged EGR valve.

This Pinpoint Test is intended to diagnose only the following:

- EVP sensor.
- Harness circuits: VREF, EVP, Signal Return, EVR, VPWR.
- EVR (EGR valve regulator).
- EGR valve assembly.
- Processor assembly.

EGR Valve Position Sensor (EVP) EGR Valve Regulator (EVR) — Pinpoint Test DM

TEST STEP	RESULT ▶	ACTION TO TAKE
FAULT CODE 31		
DM1 RUN ENGINE RUNNING QUICK TEST WITH EGR VACUUM SIGNAL LINE DISCONNECTED AT EGR VALVE		
• Key Off, wait 10 seconds. • Disconnect EGR vacuum line at EGR valve and cap EGR vacuum line. • Perform Engine Running Quick Test. • Check for code 31.	Code 31 present ▶	GO to DM2.
	No code 31 present, but codes 32 and 34 are present ▶	GO to DM11.
DM2 CHECK EVP RESISTANCE WHILE APPLYING VACUUM TO EGR VALVE		
• Key Off, wait 10 seconds. • Vacuum signal line disconnected and capped. • Disconnect vehicle harness from EVP sensor. • DVOM on 200,000 ohm scale. • Connect vacuum pump to EGR valve. • Measure resistance between EVP signal and VREF at EVP sensor while gradually increasing vacuum to 33 kPa (10 in. Hg). • Observe resistance as vacuum increases.	Reading gradually decreases from no greater than 5,500 ohms to no less than 100 ohms ▶	GO to DM3.
	Reading is less than 100 ohms or greater than 5,500 ohms ▶	REPLACE EVP sensor. RECONNECT signal line and harness. RERUN Quick Test.
	Reading does not decrease or unable to hold vacuum ▶	GO to DM16.
DM3 MEASURE VREF TO SIGNAL RETURN VOLTAGE		
• Key On, Engine Off. Vacuum signal line disconnected and capped, harness disconnected from EVP sensor. • DVOM on 20V scale. • Measure voltage at the EVP vehicle harness connector between VREF and Signal Return.	Reading is between 4V and 6V ▶	GO to DM4.
	Reading is less than 4V or greater than 6V ▶	GO to Pinpoint Test Step C1.

EGR Valve Position Sensor (EVP) EGR Valve Regulator (EVR) — Pinpoint Test DM

TEST STEP	RESULT	ACTION TO TAKE
DM4 MEASURE CONTINUITY OF EVP SIGNAL		
• Key Off, wait 10 seconds. • Harness disconnected from EVP sensor. • Disconnect harness from processor 60 Pin connector and inspect for damaged pins, corrosion, loose wires. Service as necessary. • Install Breakout box leaving processor disconnected. • DVOM on 200 ohm scale. • Measure resistance between test Pin 27 at the Breakout box and EVP signal at the EVP vehicle harness connector.	Less than 5 ohms ▶	GO to DM5.
	5 ohms or greater ▶	SERVICE open circuit. REMOVE Breakout box. RECONNECT processor and EVP sensor. RERUN Quick Test.
DM5 CHECK EVP SIGNAL FOR SHORTS TO VREF AND SIGNAL RETURN		
• Key Off, harness disconnected from EVP sensor. Breakout box installed with processor not connected. • DVOM on 200,000 ohm scale. • Measure resistance between test Pin 27 and test Pins 26 and 46 at the Breakout box.	Any reading less than 10,000 ohms ▶	SERVICE short circuit. REMOVE Breakout box. RECONNECT processor and EVP sensor. RERUN Quick Test.
	All readings 10,000 ohms or greater ▶	GO to DM6.
DM6 SUBSTITUTE EVP SENSOR AND EGR VALVE		
• Key Off, wait 10 seconds. • Electrically connect a known good EVP sensor and EGR valve assembly. • Remove Breakout box. • Reconnect processor. • Perform Key On, Engine Off Quick Test. • Is code 31 present?	Yes ▶	REPLACE processor. CONNECT original EVP sensor and EGR valve assembly. RERUN Quick Test.
	No ▶	GO to DM7.

EGR Valve Position Sensor (EVP) EGR Valve Regulator (EVR) — Pinpoint Test DM

TEST STEP	RESULT	ACTION TO TAKE
DM7 EVP SENSOR CHECK		
• Key Off, wait 10 seconds. • Install original EVP sensor on known good EGR valve. • EVP sensor connected. • Rerun Key On, Engine Off Quick Test. • Is code 31 present?	Yes ▶	REPLACE processor. CONNECT original EVP sensor and EGR valve assembly. RERUN Quick Test.
	No ▶	REFER to EGR Systems, Section 6.
FAULT CODES 32, 33 AND 34		
DM11 OUTPUT STATE CHECK (REFER TO APPENDIX)		
NOTE: Do not use STAR tester for this test Step, use VOM DVOM. • Key Off, wait 10 seconds. • DVOM on 20V scale. • Connect DVOM negative test lead to STO and positive test lead to battery positive. • Jumper STI to signal return. • Perform Key On, Engine Off Self-Test until the completion of the Continuous Test Codes. • DVOM will indicate zero volts. • Depress and release the throttle. • Did DVOM reading change to a high voltage reading?	No ▶	DEPRESS throttle to WOT and RELEASE. If STO voltage does not go high, GO to Pinpoint Test Step Q40.
	Yes ▶	REMAIN in output state check. GO to DM12.
DM12 CHECK EVR SOLENOID FOR ELECTRICAL CYCLING		
• Key On, Engine Off. • In output state check. • DVOM on 20V scale. • Connect DVOM to EVR circuits VPWR and EVR signal. • While observing DVOM, depress and release the throttle several times to cycle output on and off.	Solenoid output cycles on and off ▶	REMAIN in output state check. GO to DM13.
	Ouptut does not cycle on and off ▶	REMOVE jumper. GO to DM17.

EGR Valve Position Sensor (EVP) EGR Valve Regulator (EVR) — Pinpoint Test DM

TEST STEP	RESULT	ACTION TO TAKE
DM13 CHECK EVR SOLENOID FOR VACUUM CYCLING		
• Key On, Engine Off. • In output state check. • Disconnect and cap vacuum line from bottom port of EVR solenoid and connect vacuum pump. • Connect a vacuum gauge in the common output (top) vacuum line to EGR valve. • While cycling outputs on and off (by depressing and releasing throttle) observe vacuum gauge at the output. Maintain vacuum at source.	Vacuum output cycles on and off in less than 2 seconds ▶	RECONNECT all vacuum lines. GO to DM14.
	Vacuum does not cycle on and off in less than 2 seconds ▶	CHECK EVR filter for obstructions. REPLACE as necessary. If OK, REPLACE solenoid assembly. RECONNECT all vacuum lines. RERUN Quick Test.
DM14 CHECK VACUUM LINES		
• Key Off, wait 10 seconds. • Vacuum lines reconnected. • Check entire EEC vacuum line system per VECI schematic decal for leaks, kinks, cracks or obstructions.	Vacuum lines OK ▶	GO to DM15.
	Vacuum lines are not OK ▶	SERVICE as necessary. RERUN Quick Test.
DM15 CHECK EVP RESISTANCE WHILE APPLYING VACUUM TO EGR VALVE		
• Key Off. • Disconnect vehicle harness from EVP sensor. Inspect for damaged pins, corrosion, and pins pushed out. Service as necessary. • DVOM on 200,000 ohm scale. • Disconnect vacuum line at EGR valve. • Connect vacuum pump to EGR valve. • Measure resistance at the EVP sensor between EVP Signal and VREF while increasing vacuum to 33 kPa (10 in. Hg). • Observe resistance as vacuum increases.	Reading does not decrease gradually ▶	GO to DM16.
	Reading gradually decreases from no more than 5,500 ohms to no less than 100 ohms ▶	REPLACE processor. RECONNECT EVP sensor and EGR vacuum line. RERUN Quick Test.

EGR Valve Position Sensor (EVP) EGR Valve Regulator (EVR) — Pinpoint Test DM

TEST STEP	RESULT	ACTION TO TAKE
DM16 MANUALLY EXERCISE EVP SENSOR		
• Key Off, harness disconnected. • Remove EVP sensor from EGR valve. • Measure resistance at the EVP sensor between EVP Signal and VREF while gradually applying pressure to EVP sensor shaft. • Observe resistance as shaft is slowly pushed in and slowly released. • Look for sudden changes in resistance readings.	Both readings decrease and increase smoothly between 5,500 ohms and 100 ohms ▶	REFER to EGR Systems, Section 6. RECONNECT EVP sensor and EGR supply vacuum line. RERUN Quick Test.
	Either reading decreases or increases abruptly between 5,500 and 100 ohms ▶	REPLACE EVP sensor. RECONNECT harness and EGR supply vacuum line. RERUN Quick Test.
DM17 MEASURE EVR SOLENOID RESISTANCE		
• Key Off, wait 10 seconds. • DVOM on 200 ohm scale. • Disconnect EVR solenoid connector and measure solenoid resistance.	Resistance is between 30 and 70 ohms ▶	CONNECT EVR solenoids. GO to DM18.
	Resistance is less than 30 ohms or greater than 70 ohms ▶	REPLACE EVR solenoid assembly. RERUN Quick Test.
DM18 CHECK FOR VOLTAGE ON VEHICLE POWER CIRCUIT		
• Disconnect EVR solenoid from harness. • Key On, Engine Off. • DVOM on 20V scale. • Measure voltage between battery negative terminal and VPWR at the EVR solenoid vehicle harness connector.	Reading is less than 10.5V ▶	SERVICE harness circuit open. RERUN Quick Test.
	Readings are 10.5V or greater ▶	GO to DM19.
DM19 CHECK CONTINUITY OF EVR CIRCUITS		
• Key Off, wait 10 seconds. • EVR solenoid disconnected from harness. • Disconnect processor 60 Pin connector and inspect for damaged pins, corrosion, loose wires. Service as necessary. • Install Breakout box to processor harness connector. Leave processor disconnected. • DVOM on 200 ohm scale. • Measure resistance between test Pin 52 at the Breakout box and EVR signal at the EVR solenoid vehicle harness connector.	Reading less than 5 ohms ▶	GO to DM20.
	Reading 5 ohms or greater ▶	SERVICE open circuit. REMOVE Breakout box. RECONNECT harness to processor. RERUN Quick Test.

EGR Valve Position Sensor (EVP) EGR Valve Regulator (EVR) — Pinpoint Test DM

TEST STEP	RESULT ▶	ACTION TO TAKE
DM20 CHECK FOR SHORT TO GROUND		
• Key Off, wait 10 seconds. • DVOM on 200,000 ohm scale. • Leave Breakout box installed and processor disconnected. • EVR solenoid disconnected. • Measure resistance between test Pin 52 and test Pins 40, 46 and 60 at the Breakout box.	Any resistance reading is less than 10,000 ohms ▶	SERVICE short to ground. RERUN Quick Test.
	All resistance readings are 10,000 ohms or greater ▶	GO to DM21.
DM21 CHECK EVR SIGNAL FOR SHORTS TO POWER		
• EVR solenoid disconnected from harness. • Key Off. Breakout box installed. Processor disconnected. • DVOM on 200,000 ohm scale. • Measure resistance between test Pin 52 and test Pins 37 and 57 at the Breakout box.	Any reading less than 10,000 ohms ▶	SERVICE circuit short. REMOVE Breakout box. RECONNECT harness to processor. RERUN Quick Test. If code is repeated, REPLACE processor.
	All readings 10,000 ohms or greater ▶	REPLACE processor. REMOVE Breakout box. RECONNECT harness to processor. RERUN Quick Test.
FAULT CODE 35		
DM30 RPM TOO LOW FOR EGR TEST		
• Check for code 12.	Code 12 present ▶	Vehicles equipped with air bypass (EFI) GO to KE1. Vehicles equipped with DC motor control GO to KF1.
	Code 12 not present ▶	GO to DM31.

EGR Valve Position Sensor (EVP) EGR Valve Regulator (EVR) — Pinpoint Test DM

TEST STEP	RESULT ▶	ACTION TO TAKE
DM31 RETEST AT 1500 RPM		
• Key Off, wait 10 seconds. • Install tachometer. • Perform Key On, Engine Running Quick Test while maintaining 1500 rpm. • Record engine running service codes. • Check for code 35	Code 35 present ▶	REPLACE processor. RERUN Quick Test.
	Code 35 not present ▶	RERUN Quick Test. SERVICE codes as necessary.
DM90 SERVICE CODE 31 CONTINUOUS TEST EXERCISE EVP SENSOR		
• Using continuous monitor mode, observe VOM or STAR LED for indication of a fault while performing the following: • Connect a vacuum pump to the EGR valve • Very slowly apply 20 kPa (6 in. Hg) vacuum to the EGR valve • Slowly bleed vacuum off the EGR valve and lightly tap on EVP sensor (simulate road shock) • Wiggle EVP sensor connector • Is a fault indicated?	Yes ▶	GO to DM91.
	No ▶	GO to DM92.

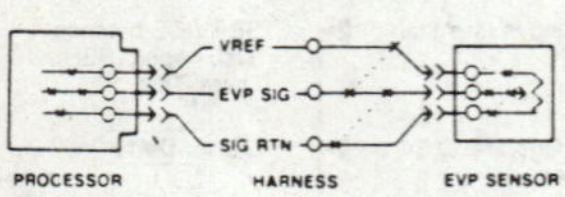

EGR Valve Position Sensor (EVP) EGR Valve Regulator (EVR) — Pinpoint Test DM

TEST STEP	RESULT ▶	ACTION TO TAKE
DM91 MEASURE EVP SIGNAL VOLTAGE WHILE EXERCISING EVP SENSOR		
• Key Off, wait 10 seconds. • Disconnect processor 60 Pin connector and inspect for damaged pins, corrosion, loose wires. Service as necessary. • Install Breakout box and reconnect processor. • VOM or STAR LED still connected to STO as in previous step. • Connect a DVOM from test Pin 27 to test Pin 46. • DVOM on 20V scale. • Key On, Engine Off. • While observing DVOM, repeat Step DM90. • Does the fault occur below 4.25V?	Yes ▶	DISCONNECT and INSPECT connector. If connector and terminals are good, REPLACE EVP sensor. RERUN Quick Test.
	No ▶	EGR valve overshoot may have caused continuous code 31. Sensor service is not required. To VERIFY harness integrity, GO to DM92.
DM92 CHECK EEC-IV HARNESS		
• Observe VOM or STAR LED for a fault indication while performing the following: • Referring to the illustration in Step DM90, grasp the harness close to the sensor connector. Wiggle, shake or bend a small section of the EEC-IV system harness while working your way to the dash panel. Also wiggle, shake or bend the EEC-IV harness from the dash panel to the processor. • Is a fault indicated?	Yes ▶	ISOLATE fault and SERVICE as necessary. REFER to appropriate figure. RERUN Quick Test.
	No ▶	GO to DM93.
DM93 CHECK PROCESSOR AND HARNESS CONNECTORS		
• Key Off, wait 10 seconds. • Disconnect processor 60 Pin connector. • Inspect both connectors and connector terminals for obvious damage or faults. • Are connectors and terminals OK?	No ▶	SERVICE as necessary. RERUN Quick Test.
	Yes ▶	Unable to duplicate fault at this time. Continuous Code 31 testing complete.

EGR Valve Position Sensor (EVP) EGR Valve Regulator (EVR) — Pinpoint Test DN

DN

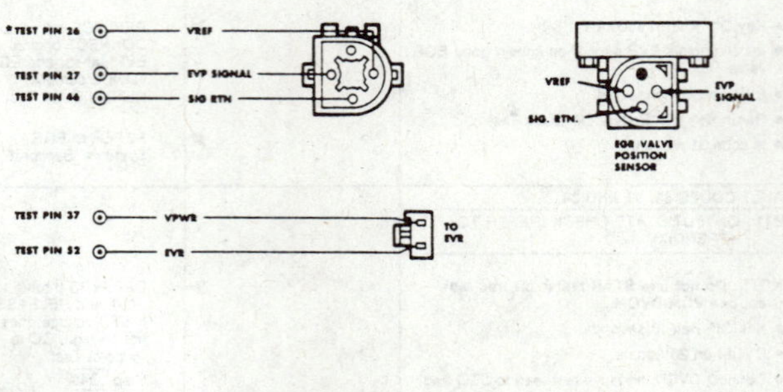

*TEST PINS LOCATED ON BREAKOUT BOX
ALL HARNESS CONNECTORS VIEWED INTO MATING SURFACE

STOP-WARNING

You should enter this Pinpoint Test only when a Service Code 31, 32, 33, 34, 35 or 84 is received in Quick Test Step 3.0, 5.0 or 6.0.

To prevent the replacement of good components, be aware that the following non-EEC area may be at fault:

- Damaged EGR valve

This Pinpoint Test is intended to diagnose only the following:

- EVP sensor
- Harness circuits: VREF, EVP, Signal Return, EVR, VPWR
- EVR (EGR valve regulator)
- EGR valve assembly
- Processor assembly

EGR Valve Position (EVP) EGR Valve Regulator (EVR) — Pinpoint Test — DN

TEST STEP	RESULT ▶	ACTION TO TAKE
SERVICE CODE: 31		
DN1 ATTEMPT TO GENERATE CODE (35)		
• Key Off. • Disconnect EVP vehicle harness at sensor. • Jumper VREF to EVP signal at vehicle harness sensor connector. • Perform Key On, Engine Off Self-Test. • Check for code 35. NOTE: **Ignore all other codes at this time.**	Code 35 present ▶	REPLACE EVP sensor. RERUN Quick Test.
	Code 35 not present ▶	REMOVE jumper. GO to DN2.
DN2 MEASURE VOLTAGE BETWEEN VREF AND SIGNAL RETURN		
• Key Off. • EVP disconnected from harness. • DVOM on 20 V scale. • Key On, Engine Off. • Measure voltage at EVP vehicle harness connector between VREF and Signal Return. • Refer to illustration DL.	Voltage reading between 4V and 6V ▶	GO to DN3.
	Voltage reading less than 4V or greater than 6V ▶	GO to C1.
DN3 CHECK CONTINUITY OF EVP SIGNAL		
• Key Off, EVP harness disconnected. • DVOM on 200 ohm scale. • Disconnect processor 60 Pin connector and inspect for damaged pins, corrosion, loose wires, etc. Service as necessary. • Connect Breakout box to harness. Processor connected to Breakout box. • Measure resistance between EVP signal at vehicle harness connector and test Pin 27 at the Breakout box.	Reading 5 ohms or greater ▶	SERVICE open circuit. CONNECT EVP sensor. REMOVE Breakout box. RERUN Quick Test.
	Reading less than 5 ohms ▶	GO to DN4.

EGR Valve Position (EVP) EGR Valve Regulator (EVR) — Pinpoint Test — DN

TEST STEP	RESULT ▶	ACTION TO TAKE
DN7 CHECK EVP CIRCUIT FOR SHORT TO POWER		
• Key Off. • EVP disconnected from harness. • Disconnect processor 60 Pin connector and inspect for damaged pins, corrosion, loose wires, etc. Service as necessary. • Install Breakout box, leave processor disconnected. • DVOM on 200,000 ohm scale. • Measure the resistance between test Pin 27 and test Pins 26 and 57 at the Breakout box.	Either resistance is less than 10,000 ohms ▶	SERVICE harness short. REMOVE Breakout box. CONNECT EVP sensor. RERUN Quick Test.
	Both resistances are 10,000 ohms or greater ▶	REPLACE processor. REMOVE Breakout box. CONNECT EVP sensor. RERUN Quick Test.
SERVICE CODE: 84		
DN10 MEASURE EVR SOLENOID RESISTANCE		
• Key Off. • DVOM on 200 ohm scale. • Disconnect EVR solenoid connector and measure solenoid resistance.	Resistance is between 30 and 70 ohms ▶	GO to DN11.
	Resistance is less than 30 ohms or greater than 70 ohms ▶	REPLACE EVR solenoid assembly. RERUN Quick Test.
DN11 CHECK FOR VPWR AT EVR SOLENOID		
• EVR solenoid disconnected from harness. • DVOM on 20V scale. • Key On, Engine Off. • Measure voltage between battery negative terminal and VPWR circuit at EVR solenoid vehicle harness connector.	Reading is less than 10.5V ▶	SERVICE VPWR open circuit. RERUN Quick Test.
	Reading is 10.5V or greater ▶	GO to DN12.

EGR Valve Position (EVP) EGR Valve Regulator (EVR) — Pinpoint Test — DN

TEST STEP	RESULT ▶	ACTION TO TAKE
DN4 CHECK RESISTANCE OF EVP SIGNAL TO GROUND AND SIGNAL RETURN		
• Key Off. • EVP harness disconnected. • Breakout box installed. • Processor disconnected. • DVOM on 200,000 ohm scale. • Measure resistance between EVP signal at EVP vehicle harness connector and ground. • Measure resistance between EVP signal at EVP vehicle harness connector and test Pin 46 (Signal Return) at the Breakout box.	Either reading is less than 10,000 ohms ▶	SERVICE short circuit. CONNECT EVP. REMOVE Breakout box. RERUN Quick Test.
	Both readings are 10,000 ohms or greater ▶	REPLACE processor. CONNECT EVP sensor. REMOVE Breakout box. RERUN Quick Test.
SERVICE CODE: 35		
DN5 ATTEMPT TO GENERATE CODE 31		
• Key Off. • Disconnect EVP vehicle harness at sensor. Inspect for damaged pins, corrosion, loose wires, etc. Service as necessary. • Perform Key On, Engine Off Self-Test, and record codes. • Check for code 31. NOTE: **Ignore all other codes at this time.**	Code 31 present ▶	GO to DN6.
	Code 31 not present ▶	GO to DN7.
DN6 MEASURE VREF TO SIGNAL RETURN VOLTAGE		
• Key Off. • EVP harness disconnected. • DVOM on 20V scale. • Key On, Engine Off. • Measure voltage at EVP vehicle harness connector between VREF and Signal Return. • Refer to illustration DN.	Voltage reading between 4V and 6V ▶	REPLACE EVP sensor. RERUN Quick Test.
	Voltage reading less than 4V or greater than 6V ▶	GO to C1.

EGR Valve Position (EVP) EGR Valve Regulator (EVR) — Pinpoint Test — DN

TEST STEP	RESULT ▶	ACTION TO TAKE
DN12 CHECK CONTINUITY OF EVR CIRCUIT		
• Key Off. • EVR solenoid disconnected from harness. • Disconnect processor 60 Pin connector and inspect for damaged pins, corrosion, loose wires, etc. Service as necessary. • Install Breakout box to processor harness connector. Leave processor disconnected. • DVOM on 200 ohm scale. • Measure resistance between test Pin 33 at the Breakout box and EVR signal at the EVR solenoid vehicle harness connector.	Reading less than 5 ohms ▶	GO to DN13.
	Reading 5 ohms or greater ▶	SERVICE open circuit. REMOVE Breakout box. CONNECT process EVR solenoid. RERUN Quick Test.
DN13 CHECK EVR CIRCUIT FOR SHORT TO POWER AND GROUND		
• Key Off. • Breakout box installed, processor disconnected. • EVR solenoid disconnected. • DVOM on 200,000 ohm scale. • Measure resistance between test Pin 33 (EVR) and test Pins 37 and 57 (VPWR) and 40, 46 and 60 (GRD) at the Breakout box.	Any reading less than 10,000 ohms ▶	SERVICE short circuit. REMOVE Breakout box. RECONNECT processor and EVR solenoid. RERUN Quick Test. If code is repeated, REPLACE processor.
	All readings 10,000 ohms or greater ▶	REPLACE processor. REMOVE Breakout box. RECONNECT processor and EVR solenoid. RERUN Quick Test.
DN30 SERVICE CODE 32		
• Replace EVP sensor, rerun Quick Test.		

EGR Valve Position Sensor (EVP) EGR Valve Regulator (EVR) — Pinpoint Test — DN

TEST STEP	RESULT ▶	ACTION TO TAKE
DN35 SERVICE CODE 34: EGR VALVE OPERATION		
• Key Off. • Disconnect vacuum hose from EGR valve and plug hose. • Perform Engine Running Quick Test. • Is code 34 present?	Yes ▶ No ▶	GO to DN36. REPLACE EVR solenoid and RERUN Quick Test.
DN36 EVP OPERATION		
• Key Off, harness disconnected. • Remove EVP sensor from EGR valve. • Measure resistance at the EVP sensor between Signal Return and VREF while gradually applying pressure to EVP sensor shaft. • Observe resistance as shaft is slowly pushed in and slowly released. • Look for sudden changes in resistance readings.	Both readings decrease and increase smoothly between 5,500 ohms and 100 ohms ▶ Either reading decreases or increases abruptly between 5,500 and 100 ohms ▶	REFER to EGR System, Section 6. RECONNECT EVP sensor and EGR supply vacuum line. REPLACE EVP sensor. RECONNECT harness and EGR supply vacuum line. RERUN Quick Test.
DN40 SERVICE CODE 33: CHECK EVP RESISTANCE WHILE APPLYING VACUUM TO EGR VALVE		
• Key Off. • Disconnect harness from EVP sensor. Inspect for damaged pins, corrosion, and pins pushed out. Service as necessary. • DVOM on 200,000 ohm scale. • Disconnect vacuum line at EGR valve. • Connect vacuum pump to EGR valve. • Measure resistance at the EVP sensor between EVP signal and VREF while increasing vacuum to 33 kPa (10 in. Hg). • Observe resistance as vacuum increases.	If resistance reading does not decrease gradually ▶ Reading gradually decreases from no more than 5,500 ohms to no less than 100 ohms ▶	REFER to EGR System, Section 6. GO to DN41.

EGR Valve Position Sensor (EVP) EGR Valve Regulator (EVR) — Pinpoint Test — DN

TEST STEP	RESULT ▶	ACTION TO TAKE
DN41 CHECK VACUUM LINES		
• Key Off, wait 10 seconds. • Vacuum lines reconnected. • Check EGR vacuum line system from EGR valve to EVR, and from EVR to carburetor for obstructions, kicks, leaks, etc.	Vacuum lines OK ▶ Vacuum lines are not OK ▶	CHECK EVR filter for obstructions. REPLACE as necessary. If OK, REPLACE EVR solenoid. RECONNECT all vacuum lines. RERUN Quick Test. SERVICE as necessary. RERUN Quick Test.
DN90 SERVICE CODE 32 CONTINUOUS TEST EXERCISE EVP SENSOR		
• Using continuous monitor mode, observe VOM or STAR LED for indication of a fault while performing the following: • Connect a vacuum pump to the EGR valve. • Very slowly apply 20 kPa (6 in. Hg) vacuum to the EGR valve. • Slowly bleed vacuum off the EGR valve and lightly tap on EVP sensor (simulate road shock). • Wiggle EVP sensor connector. • Is a fault indicated?	Yes ▶ No ▶	GO to DN91. GO to DN92.

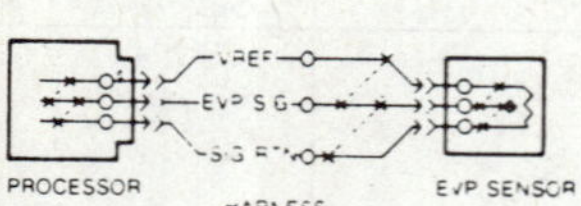

EGR Valve Position Sensor (EVP) EGR Valve Regulator (EVR) — Pinpoint Test — DN

TEST STEP	RESULT ▶	ACTION TO TAKE
DN91 MEASURE EVP SIGNAL VOLTAGE WHILE EXERCISING EVP SENSOR		
• Key Off, wait 10 seconds. • Disconnect processor 60 Pin connector and inspect for damaged pins, corrosion, loose wires, etc. Service as necessary. • Install Breakout box and reconnect processor. • VOM or STAR LED still connected to STO as in previous Step. • Connect a DVOM from test Pin 27 to test Pin 46. • DVOM on 20V scale. • Key On, Engine Off. • While observing DVOM, repeat Step DN90. • Does the fault occur below 4.25V?	Yes ▶ No ▶	DISCONNECT and INSPECT connector. If connector and terminals are good, REPLACE EVP sensor. RERUN Quick Test. EGR valve overshoot may have caused continuous code 31. Sensor service is not required. To VERIFY harness integrity, GO to DN92.
DN92 CHECK EEC-IV HARNESS		
• Observe VOM or STAR LED for a fault indication while performing the following: • Referring to the illustration in Step DN90, grasp the harness close to the sensor connector. Wiggle, shake or bend a small section of the EEC-IV system harness while working your way to the dash panel. Also wiggle, shake or bend the EEC-IV harness from the dash panel to the processor. • Is a fault indicated?	Yes ▶ No ▶	ISOLATE fault and SERVICE as necessary. REFER to appropriate figure. RERUN Quick Test. GO to DN93.
DN93 CHECK PROCESSOR AND HARNESS CONNECTORS		
• Key Off, wait 10 seconds. • Disconnect processor 60 Pin connector. • Inspect both connectors and connector terminals for obvious damage or faults. • Are connectors and terminals OK?	No ▶ Yes ▶	SERVICE as necessary. RERUN Quick Test. Unable to duplicate fault at this time. Continuous Code 31 testing complete.

EGR Valve Position Sensor (EVP) EGR Valve Regulator (EVR) — Pinpoint Test — DN

TEST STEP	RESULT ▶	ACTION TO TAKE
DN95 LEAK TEST		
• Key Off. • Connect a vacuum pump to EGR valve. • Apply 20 in. Hg to EGR valve. • Does valve maintain vacuum and does EGR valve open?	Yes ▶ No ▶	GO to DN96. GO to EGR diagnostic, Section 6.
DN96 EVR CHECK		
• Using continuous monitor mode, observe VOM or STAR LED for indication of a fault while performing the following: • Grasp the harness close to the EVR solenoid connector, wiggle, shake or bend a small section of the harness while working your way to the processor. • Inspect connectors, terminals for obvious damage or faults. • Are any faults detected or indicated?	Yes ▶ No ▶	ISOLATE fault and SERVICE as necessary. RERUN Quick Test. Unable to duplicate fault at this time, testing complete.

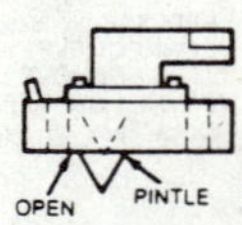

NOTE: TO OPEN, PINTLE SHOULD MOVE WITHIN THE BASE OF VALVE.

EGR Valve Position Sensor (EVP) EGR Valve Regulator (EVR) — Pinpoint Test DN

TEST STEP	RESULT ▶	ACTION TO TAKE
DN98 CHECK EVP RESISTANCE WHILE APPLYING VACUUM TO EGR VALVE		
• Key Off. • Disconnect harness from EVP sensor. • Inspect for damaged pins, corrosion, and pins pushed out. Service as necessary. • DVOM on 200,000 ohm scale. • Disconnect vacuum line at EGR valve. • Connect vacuum pump to EGR valve. • Measure resistance at the EVP sensor between EVP signal Pin and VREF Pin while increasing vacuum to 33 kPa (10 in. Hg). • Observe resistance as vacuum increases.	Reading does not decrease gradually ▶	GO to EGR Diagnostic Section 6.
	Reading gradually decreases from no more than 5,500 ohms to no less than 100 ohms ▶	GO to DN99.
DN99		
• Key Off. • Disconnect vacuum hose from EGR valve and plug hose. • Perform Engine Running Quick Test. • Is code 34 present?	Yes ▶	CHECK EVR filter for obstructions. REPLACE as necessary. If OK, REPLACE EVR solenoid. RECONNECT all vacuum lines. RERUN Quick Test.
	No ▶	Unable to duplicate fault at this time.

Neutral Drive Switch A/C Input — Pinpoint Test FA

TEST STEP	RESULT ▶	ACTION TO TAKE
FA1 CODE 67 SYSTEM IDENTIFICATION		
2.9L TK, 3.0L TK, 5.0L M/T SEFI		GO to FA10.
1.9L M/T, 2.3L EFI M/T Truck, 2.3L CFI M/T		GO to FA10.
2.3L Turbo, M/T		GO to FA15.
2.3L SVO Turbo Octane Switch		GO to FA30.
4.9L M/T, Truck, 2.3L OHC FBC		GO to FA25.
All other systems		GO to FA2.
FA2 NEUTRAL DRIVE INPUT CHECK		

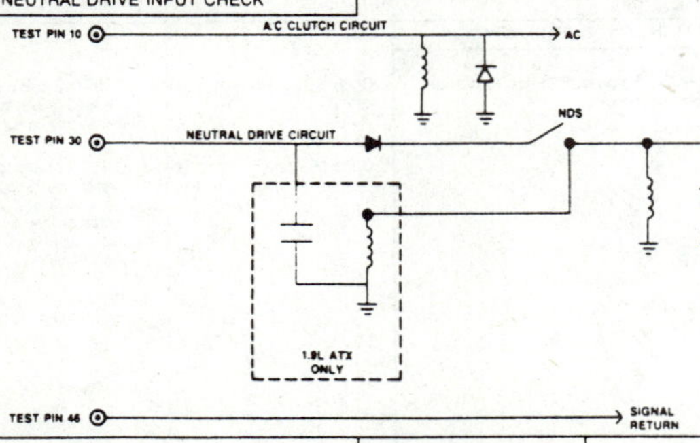

TEST STEP	RESULT ▶	ACTION TO TAKE
• Key Off, wait 10 seconds. • Verify heater control is in OFF position, if so equipped • Verify transmission is in NEUTRAL or PARK. • Disconnect processor 60 Pin connector and inspect for damaged pins, corrosion, loose wires. Service as necessary. • Install Breakout box • Processor connected • Key On, Engine Off • DVOM on 20V scale • Measure voltage between test Pin 30 (Neutral Drive circuit) at the Breakout box and chassis ground	Less than 1.0V ▶	GO to FA4.
	1.0V or greater ▶	GO to FA3.

Neutral Drive Switch A/C Input — Pinpoint Test FA

TEST STEP	RESULT ▶	ACTION TO TAKE
FA3 NEUTRAL DRIVE SWITCH CHECK		
• Key Off, wait 10 seconds. • Breakout box installed. • DVOM on 200 ohm scale. • Locate the Neutral Drive switch. • Disconnect vehicle harness from the Neutral Drive switch and measure resistance across the switch.	Less than 5 ohms ▶	SERVICE open in vehicle harness Neutral Drive circuit. RERUN Quick Test.
	5 ohms or greater ▶	REPLACE Neutral Drive switch. RERUN Quick Test.
FA4 A/C INPUT CHECK		
• Breakout box installed. • Key On, Engine Off. • DVOM on 20V scale. • A/C control Off. Measure voltage between test Pin 10 (A/C clutch circuit) at the Breakout box and chassis ground. **NOTE: On 2.3L Turbo, check for code 18 in continuous memory.**	1.0V or greater ▶	SERVICE short to power in A/C clutch circuit. RERUN Quick Test.
	Less than 1.0V ▶	REPLACE processor. RERUN Quick Test.
FA10 NEUTRAL CLUTCH INPUT CHECK 1.9L, 2.3L EFI TRUCK, 2.3L CFI M/T 2.9L, 3.0L EFI TRUCK, 5.0L M/T SEFI		

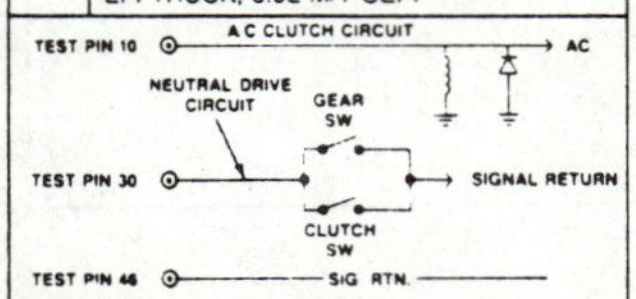

TEST STEP	RESULT ▶	ACTION TO TAKE
• Key Off, wait 10 seconds. • Verify A/C is off, if so equipped. • Verify transmission is in NEUTRAL and clutch is released. • Disconnect processor 60 Pin connector and inspect for damaged pins, corrosion, loose wires. Service as necessary. • Install Breakout box. • Connect processor. • DVOM on 200 ohm scale. • Measure resistance between test Pin 30 (Neutral Drive circuit) and test Pin 46 (Signal Return circuit) at the Breakout box.	Less than 5 ohms ▶	GO to FA4.
	5 ohms or greater ▶	GO to FA11.

Neutral Drive Switch A/C Input — Pinpoint Test FA

TEST STEP	RESULT ▶	ACTION TO TAKE
FA11 NEUTRAL CLUTCH SWITCH CHECK		
• Key Off. • DVOM on 200 ohm scale. • Breakout box installed. • Locate Neutral switch (on transmission) and Clutch switch (under dash). • Disconnect vehicle harness at both switches. • Measure resistance across each switch.	Less than 5 ohms at both switches ▶	GO to FA4.
	5 ohms or greater at either or both switches ▶	REPLACE open switch(es). RECONNECT harness and RERUN Quick Test.
FA15 NEUTRAL INPUT CHECK — 2.3L TC M/T		

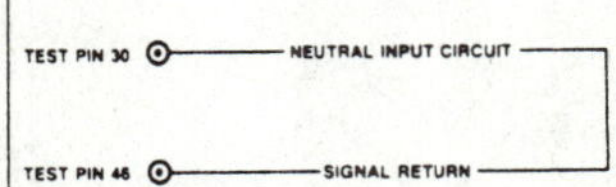

TEST STEP	RESULT ▶	ACTION TO TAKE
• Key Off, wait 10 seconds. • Verify A/C is off, if so equipped. • Disconnect processor 60 Pin connector and inspect for damaged pins, corrosion, loose wires. Service as necessary. • Install Breakout box. • Leave processor disconnected. • DVOM on 200 ohm scale. • Measure resistance between test Pin 30 (Neutral Input circuit) and test Pin 46 (Signal Return circuit) at the Breakout box.	Less than 5 ohms ▶	GO to FA4.
	5 ohms or greater ▶	SERVICE open in Neutral Input or Signal Return circuit. RERUN Quick Test.

Neutral Drive Switch A/C Input — Pinpoint Test FA

TEST STEP	RESULT	▶ ACTION TO TAKE
FA25 NEUTRAL INPUT CHECK — 4.9L M.T. TRUCK, 2.3L OHC FBC		
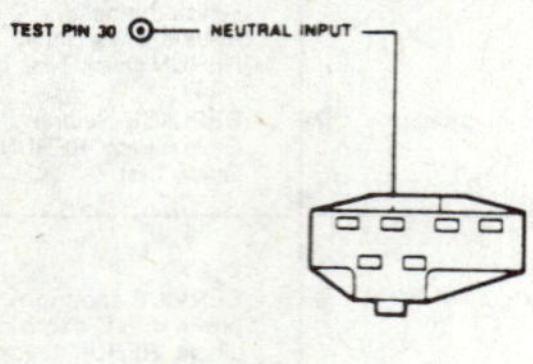 • Key Off, wait 10 seconds. • Verify A/C is off, if so equipped. • Disconnect processor 60 Pin connector and inspect for damaged pins, corrosion, loose wires. Service as necessary. • Install Breakout box. • Connect processor. • DVOM on 200 ohm scale. • Measure resistance between test Pin 30 at the Breakout box (Neutral input circuit) and VIP connector.	Less than 5 ohms ▶ 5 ohms or greater ▶	GO to FA4. SERVICE open in Neutral Input circuit to VIP tester. RERUN Quick Test.
FA30 CHECK OCTANE SWITCH INPUT FOR INPUT CHANGE		
• Key Off, wait 10 seconds. • Disconnect processor 60 Pin connector and inspect for damaged pins, corrosion, loose wires. Service as necessary. • Install Breakout box. Reconnect processor. • DVOM to 20V scale. • Connect positive test lead to test Pin 30 and negative test lead to test Pin 46 at the Breakout box. • Key On, Engine Off. • Cycle octane switch several times while observing DVOM. • Does voltage change from zero volts to 5V?	Yes ▶ No ▶	REPLACE processor. RERUN Quick Test. EEC-IV system OK. REFER to Shop Manual for boost diagnostics.

Key Power Check — Pinpoint Test FC

TEST STEP	RESULT	▶ ACTION TO TAKE
FC1 BATTERY VOLTAGE GREATER THAN 17.5V		
• Key Off, wait 10 seconds. • Install Breakout box with processor connected. • DVOM at 20V range and connected to test Pins 5 and 60 at the Breakout box. • Perform Engine Running Quick Test. Observe DVOM during test and record service codes.	Reading exceeds 17.5V during Quick Test ▶ Reading remains below 17.5V and code 65 is present ▶ Reading remains below 17.5V and code 65 is not present ▶	CORRECT charging system for over-voltage condition. REPLACE processor and RERUN Quick Test. GO to FC2.
FC2 CODE 65 IN CONTINUOUS MEMORY		
• Perform Key On, Engine Off Quick Test and record continuous codes.	Code 65 not present ▶ Code 65 present ▶	TESTING complete. CHARGING system OK at this time. CHECK charging system. REFER to Shop Manual, Group 31 for cause of intermittent overcharging (greater than 17.5V).
FC3 BATTERY VOLTAGE LESS THAN 7.5V		
• Key Off, wait 10 seconds. • Disconnect processor 60 Pin connector and inspect for damaged pins, corrosion, loose wires. Service as necessary. • Install Breakout box to processor harness connector. • Processor connected. • DVOM on 20V scale. • Measure voltage between test Pin 5 and test Pin 60 at the Breakout box. • Perform Engine Running Quick Test.	Voltage below 7.5V during Quick Test ▶ Voltage remains above 7.5V and code 55 is present ▶ Voltage remains above 7.5V and code 55 is not present ▶	GO to FC4. REPLACE processor. RERUN Quick Test. GO to FC5.

Key Power Check — Pinpoint Test FC

FC

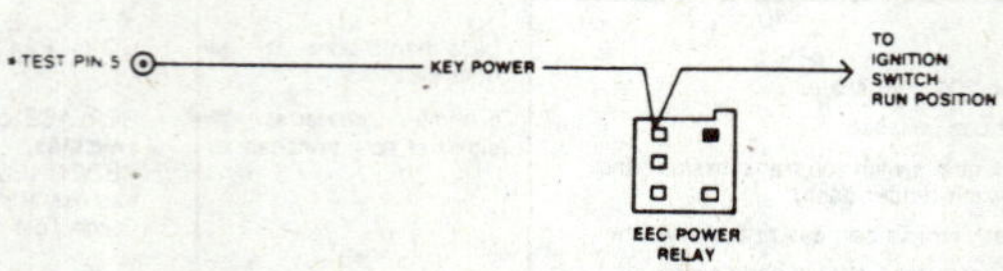

*TEST PINS LOCATED ON BREAKOUT BOX.
ALL HARNESS CONNECTORS VIEWED INTO MATING SURFACE.

STOP-WARNING

You should enter this Pinpoint Test only when a Service Code 55 or 65 is received in Quick Test Step 3.0, 5.0 or 6.0.

To prevent the replacement of good components, be aware that the following non-EEC areas may be at fault:

- Charging system overvoltage.
- Battery charger connected with engine running.
- Jump starting.

This Pinpoint Test is intended to diagnose only the following:

- Harness circuit: key power.
- Processor.

Key Power Check — Pinpoint Test FC

TEST STEP	RESULT	▶ ACTION TO TAKE
FC4 KEY POWER CIRCUIT CHECK		
• Key Off. • Breakout box installed, processor connected. • DVOM at 200 ohm scale. • Measure resistance between test Pin 5 at the Breakout box and key power terminal of EEC power relay.	5 ohms or less ▶ Greater than 5 ohms ▶	CORRECT charging system for under-voltage condition. SERVICE open in key power circuit. RERUN Quick Test.
FC5 CODE 55 IN CONTINUOUS MEMORY		
• Perform Key On, Engine Off Quick Test and record continuous codes.	Code 55 not present ▶ Code 55 present ▶	TESTING complete. CHARGING system OK at this time. CHECK charging system per Shop Manual, Group 31 for cause of intermittent overcharging (greater than 17.5V).

Brake On/Off (BOO) — Pinpoint Test — FD

FD

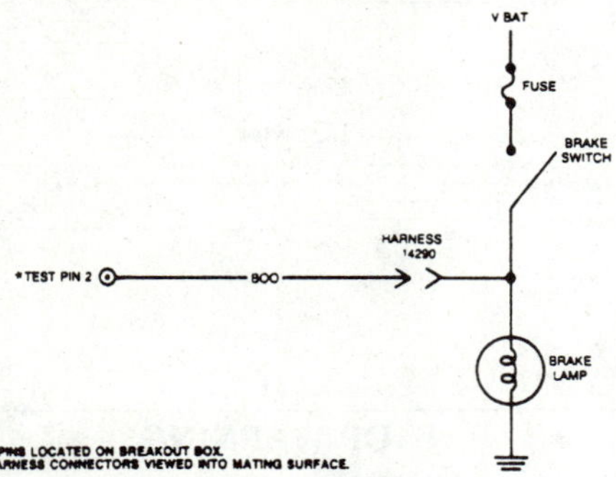

STOP-WARNING

You should enter this Pinpoint Test only when a Service Code 74 or 75 is received in Quick Test Step 5.0.

To prevent the replacement of good components, be aware that the following non-EEC areas may be at fault:

- Brake lamp, brake switch, and fuse.

This pinpoint test is intended to diagnose only the following:

- BOO circuit.
- Processor assembly.

Brake On/Off (BOO) — Pinpoint Test — FD

TEST STEP	RESULT ▶	ACTION TO TAKE
FD1 SERVICE CODE 74		
• Did you press brake during the Engine Running Quick Test?	Yes ▶	GO to FD2.
	No ▶	RERUN Engine Running Quick Test, PRESS brake once during test.
FD2 BOO CIRCUIT CYCLING		
• Key Off, wait 10 seconds. • Disconnect processor 60 Pin connector. Inspect for damaged pins, corrosion, loose wires, etc. Service as necessary. • Install Breakout box, leave processor disconnected. • DVOM on 20V scale. • Measure voltage between test Pin 2 and test Pin 40 at the Breakout box while depressing and releasing brake. • Does the voltage cycle?	Yes ▶	REPLACE processor. RERUN Quick Test.
	No ▶	GO to FD3.
FD3 BOO CIRCUIT SHORT TO GROUND		
• Key Off. • Breakout box installed. • Processor disconnected. • DVOM on 200 Ohm scale. • Disconnect BOO circuit from 14290 harness (12 pin connector). • Measure resistance between test Pin 2 at the Breakout box and ground. • Is resistance reading greater than 5 ohms?	No ▶	SERVICE BOO circuit short to ground.
	Yes ▶	GO to Shop manual, Group 32.

Brake On/Off (BOO) — Pinpoint Test — FD

TEST STEP	RESULT ▶	ACTION TO TAKE
FD4 BOO CIRCUIT CYCLING CODE 75		
• Key Off, wait 10 seconds. • Disconnect processor 60 Pin connector. Inspect for damaged pins, corrosion, loose wires, etc. Service as necessary. • Install Breakout box, leave processor disconnected. • DVOM on 20V scale. • Measure voltage between test Pin 2 and test Pin 40 at the Breakout box while depressing and releasing brake. • Does the voltage cycle?	Yes ▶	REPLACE processor. RERUN Quick Test.
	No ▶	GO to FD5.
FD5 BOO CIRCUIT SHORT TO POWER		
• Key Off. • Breakout box installed. • Processor disconnected. • DVOM on 20V scale. • Disconnect BOO circuit from 14290 harness (12 Pin connector). • Measure voltage between test Pin 2 at the Breakout box and engine block ground. • Is voltage reading greater than 10.5 volts?	Yes ▶	SERVICE BOO circuit short to power.
	No ▶	BOO circuit OK. GO to Shop Manual, Group 32 to SERVICE stoplamp circuit.

Power Steering Pressure Switch (PSPS) — Pinpoint Test — FF

FF

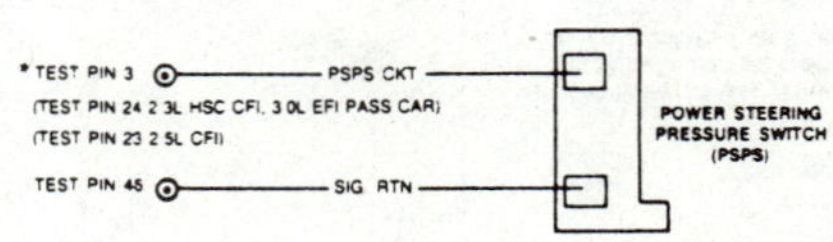

*TEST PINS LOCATED ON BREAKOUT BOX.
ALL HARNESS CONNECTORS VIEWED INTO MATING SURFACE.

STOP-WARNING

You should enter this Pinpoint Test only when a Service Code 52 is received in Quick Test Step 3.0 or if you are directed here from Diagnostics by Symptom in the Engine Supplement Section.

To prevent the replacement of good components, be aware that the following non-EEC areas may be at fault:

- Idle speeds throttle stop adjustment.
- Binding throttle shaft linkage or speed control linkage.

This Pinpoint Test is intended to diagnose only the following:

- Power steering pressure switch.
- Switch harness circuits: PSPS Signal, and Signal Return.
- Processor assembly.

Power Steering Pressure Switch (PSPS) — Pinpoint Test — FF

TEST STEP	RESULT ▶	ACTION TO TAKE
FF1 ATTEMPT TO ELIMINATE CODE 52		
• Key Off, wait 10 seconds. • Disconnect PSPS. • Jumper PSPS circuit to Signal Return at vehicle harness connector. • Rerun Key On, Engine Off Quick Test. • Is code 52 still present?	Yes ▶ No ▶	GO to FF2. REPLACE PSPS. RERUN Quick Test.
FF2 PSPS HARNESS CHECK		
• Key Off, wait 10 seconds. • Disconnect harness from processor 60 Pin connector and inspect for damaged pins, corrosion, loose wires. Service as necessary. • PSPS disconnected. • Processor disconnected. • Breakout box installed. • DVOM on 200 ohm scale. • Measure resistance between test Pin 46 at the Breakout box and Signal Return at the PSPS connector and between test Pin 3 (2.3L HSC), test Pin 23 (2.5L CFI), test Pin 24 (3.0L EFI) at the Breakout box and PSPS circuit at the PSPS harness connector. • Are both readings less than 5 ohms?	Yes ▶ No ▶	REPLACE processor. RERUN Quick Test. SERVICE open in circuit. RERUN Quick Test.
FF3 SWITCH INTEGRITY		
• Install tachometer. • Start engine, allow to idle in Neutral/Park. • Disconnect PSPS at switch. • Does rpm increase?	Yes ▶ No ▶	REPLACE PSPS. GO to FF4.

Power Steering Pressure Switch (PSPS) — Pinpoint Test — FF

TEST STEP	RESULT ▶	ACTION TO TAKE
FF4 PSPS HARNESS CHECK		
• Key Off, wait 10 seconds. • Disconnect harness from processor 60 Pin connector and inspect for damaged pins, corrosion, loose wires. Service as necessary. • PSPS disconnected. • Processor disconnected. • Breakout box installed. • DVOM on 200 ohm scale. • Measure resistance between test Pin 3 (2.3L HSC), test Pin 23 (2.5L CFI), test Pin 24 (3.0L EFI) and test Pin 46 at the breakout box. • Is reading less than 5 ohms?	Yes ▶ No ▶	SERVICE short in harness. REPLACE processor.

Fuel Control — EFI — Pinpoint Test — HA

HA

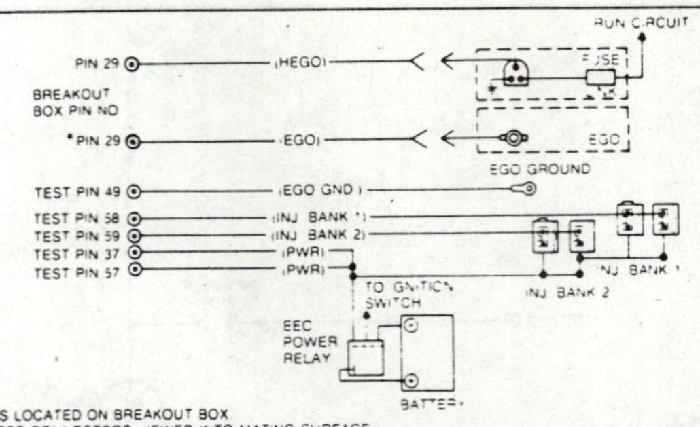

*TEST PINS LOCATED ON BREAKOUT BOX
ALL HARNESS CONNECTORS VIEWED INTO MATING SURFACE

STOP-WARNING

You should enter this Pinpoint Test Only when a Service Code 41 or 42 is received in Quick Test Step 5.0 or 6.0 or when directed here from Pinpoint Test A or Diagnostics by Symptom in the Engine Supplement Section.

To prevent the replacement of good components, be aware that the following non-EEC areas may be at fault:

- Ignition Coil
- Distributor Cap
- Distributor Rotor
- Fouled Spark Plugs
- Spark Plug Wires
- CANP Problems
- PCV Valves (see note below)
- EGR Valve and Gasket
- Air Filter
- Fuel Contamination Engine Oil
- Poor Power Ground
- Fuel Pressure
- Manifold Leaks Intake Exhaust
- Engine Not at Normal Operating Temperatures

This Pinpoint Test is intended to diagnose only the following:

- EGO Sensor
- Harness Circuits EGO Grd., EGO, INJ BANK 1, INJ BANK 2, V PWR
- EGO Sensor Connection
  - — Code 42 start at HA8
  - — Code 41 start at HA11
- Vacuum Systems
- Fuel Injectors
- Processor Assembly

NOTE: Fuel contaminated engine oil may affect 41, 42 Service Codes. If this is suspected, remove the PCV from the valve cover and repeat the Quick Test. If the problem is corrected, then change the engine oil and filter.

Fuel Control — EFI — Pinpoint Test — HA

TEST STEP	RESULT ▶	ACTION TO TAKE
HA1 FUEL PRESSURE CHECK 1		
• Key Off, wait 10 seconds. • Install fuel pressure gauge. • Start and run engine. Fuel pressure must be 241-310 kPa (35-45 psi). **For No Starts:** • If engine will not run, cycle the key from Off to On several times to increase fuel pressure.	Yes ▶ No ▶	GO to HA2. REFER to the Shop Manual Group 24 for electric fuel pump and Section 11 for fuel pressure regulator check.
HA2 FUEL PRESSURE CHECK 2		
• Fuel pressure must remain at 276 ± 34 kPa (40 ± 5 psi) for 60 seconds or longer after final Key Off.	Yes ▶ No ▶	GO to HA3. GO to HA7.
HA3 FUEL DELIVERY TEST		
• Key Off. • Fuel pressure gauge installed. • Pressurize fuel system as per Step HA1. • Locate and disconnect the fuel pump relay. • Crank engine for 5 seconds. • Take pressure reading at the end of 5 second crank. **NOTE: Verify fuel quality: air and/or water will also pressurize and look like acceptable fuel delivery.**	PRESSURE GAUGE READING. Pressure is approximately 10-20 psi at the end of 5 second crank cycle. Refer to note below. ▶ Pressure is greater or less than specified ▶	The EEC system is not the fault of the No Start. Fuel and spark are present. REFER to Section 2 for other No Start Routines. If complaint was runs rough, misses or fuel code GO to HA4. GO to HA4.

NOTE: The colder the engine, the greater the pressure drop (i.e., an engine coolant temperature of 200°F equals approximately a 10 psi drop in 5 seconds. 60°F equals approximately a 20 psi drop in 5 seconds).

Fuel Control — EFI | Pinpoint Test | HA

TEST STEP	RESULT ▶	ACTION TO TAKE
HA4 HARNESS INJECTOR RESISTANCE CHECK		
• Key Off, wait 10 seconds. • Disconnect processor 60 Pin connector and inspect for damaged pins, corrosion, loose wires. Service as necessary. • Connect Breakout box to harness. Leave processor disconnected. • DVOM on 200 ohm scale. • Measure the resistance of injector Bank 1 between test Pin 37 and test Pin 58 at the breakout box. Record reading. • Measure the resistance of injector Bank 2 between test Pin 37 and test Pin 59 at the Breakout box. Record reading. • Are both readings between 1.2 ohms and 1.8 ohms?	Yes ▶ No ▶	GO to HA6. GO to HA5.
HA5 ISOLATE FAULTY INJECTOR CIRCUIT		
• Key Off. • Breakout box installed. • Leave processor disconnected. • DVOM on 200 ohm scale. • Disconnect all injectors on suspect Bank. Measure resistance of each injector by connecting one injector at a time, and reading the resistance between test Pins 37 and 58 at the Breakout box for Bank 1 or test Pins 37 and 59 at the breakout box for Bank 2. • Are all readings between 2.0 and 2.7 ohms?	Yes ▶ No ▶	GO to HA6. SERVICE the harness/connectors on the suspect injector for opens or shorts. If OK, REPLACE injector. RERUN Quick Test.

Fuel Control — EFI | Pinpoint Test | HA

TEST STEP	RESULT ▶	ACTION TO TAKE
HA6 INJECTOR DRIVE SIGNAL CHECK		
Requires standard non-powered 12 volt test lamp. • Key Off. • Breakout box installed. • Connect processor to Breakout box. • Connect test lamp between test Pin 37 and test Pin 58 at the Breakout box. • Crank or start engine. • Repeat above test between test Pin 37 and test Pin 59 at the Breakout box.	Dim glow at light on both tests ▶ No light on one or both tests ▶ Bright light on one or both tests ▶	GO to HA7. VERIFY 12-volt battery power at Pins 37 and 57. If OK, REPLACE processor. RERUN Quick Test. CHECK circuits Bank 1 and Bank 2 for shorts to ground. If OK, REPLACE processor. RERUN Quick Test.
HA7 INJECTOR BALANCE TEST		
• Connect tachometer to engine. Disconnect ISC. Run engine at 2000 rpm. Use throttle body stop screw to set engine speed. • Disconnect and reconnect the injectors one at a time; note rpm drop for each injector. • Does each injector produce at least a 150 rpm drop? • Reconnect ISC. Reset curb idle. Refer to VECI decal.	Yes ▶ No ▶	Fuel delivery OK. Problem is with area common to all cylinders, i.e.: Air/vacuum leak, fuel contamination, EGR. REPLACE faulty injector. RERUN Quick Test.
HA8 FUEL CONTROL — ALWAYS RICH: CODE 42		
• Key Off, wait 10 seconds. • **NOTE: Non-EEC areas could cause a code 42. Check for:** **— Fuel contaminated engine oil** **— Ignition caused misfire** **— CANP problems** • Disconnect vehicle harness at the EGO sensor. Using a jumper wire, ground vehicle harness EGO circuit at the EGO sensor to the engine block. • Repeat Engine Running Self-Test. • Service code 41 results.	Yes ▶ No ▶	GO to HA10. GO to HA9.

Fuel Control — EFI | Pinpoint Test | HA

TEST STEP	RESULT ▶	ACTION TO TAKE
HA9 HARNESS CHECK		
• Key Off, wait 10 seconds. • Install Breakout box. • Measure resistance of EGO ground circuit between test Pin 49 at the Breakout box and EGO ground at engine block. • Measure resistance of EGO circuit between test Pin 29 at the Breakout box and EGO harness connector. • Both circuits less than 5 ohms.	Yes ▶ No ▶	DISCONNECT processor connector. INSPECT for damage or corrosion. If OK, REPLACE processor. RERUN Quick Test. CORRECT harness circuit with resistance greater than 5 ohms. RERUN Quick Test.
HA10 EGO CHECK		
• DVOM on 20V scale. • With EGO sensor disconnected from the vehicle harness, connect a DVOM from EGO sensor to engine ground. • Run the engine at 2000 rpm for 1 minute. While observing DVOM, disconnect the manifold vacuum hose indicated below. • Meter indicates less than 0.4V.	Yes ▶ No ▶	EGO sensor OK. GO to HA1. REPLACE EGO sensor. RERUN Quick Test.

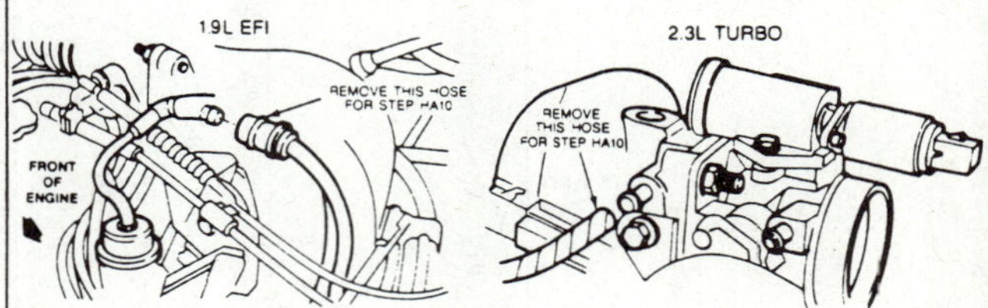

Fuel Control — EFI | Pinpoint Test | HA

TEST STEP	RESULT ▶	ACTION TO TAKE
HA11 FUEL CONTROL — ALWAYS LEAN: CODE 41		
NOTE: Vacuum air leaks in non-EEC areas could cause a code 41. Check for: **— Leaking vacuum actuator (eg: A/C control motor)** **— Engine sealing** **— EGR system** **— PCV system** **— Unmetered air leak between air meter and throttle body** **— Lead contaminated EGO sensor** • Key Off. DVOM on 20V scale. Disconnect EGO sensor from vehicle harness. Connect DVOM to EGO sensor and engine ground. Remove air cleaner to gain access to air meter inlet. Using a standard wood lead pencil, prop the air meter door part-way open. • Start the engine and run at approximately 2000 rpm for 2 minutes. • Does the DVOM read greater than 0.5V at the end of 1 minute?	Yes ▶ No ▶	GO to HA12. REPLACE EGO sensor. RERUN Quick Test.
HA12 HARNESS CHECK EGO CIRCUITS		
• Key Off. • Install Breakout box. Processor disconnected. • Measure resistance of EGO circuit between test Pin 49 at the Breakout box and engine block ground. Also EGO circuit between EGO connector and test Pin 29 at the Breakout box. • Are both circuits less than 5 ohms?	Yes ▶ No ▶	GO to HA13. SERVICE/CORRECT as necessary any circuit greater than 5 ohms. RERUN Quick Test.
HA13 CHECK EGO CIRCUIT FOR SHORT TO GROUND		
• Key Off. • Breakout box installed. Processor disconnected. • DVOM to 200,000 ohms scale. • Measure the resistance between test Pin 29 and test Pin 40 at the Breakout box.	Reading is 10,000 ohms or greater ▶ Reading is less than 10,000 ohms ▶	GO to HA14. CORRECT cause of resistance to ground. RERUN Quick Test.

Fuel Control — EFI | Pinpoint Test | HA

TEST STEP	RESULT ▶	ACTION TO TAKE
HA14 ATTEMPT TO ELIMINATE CODE 41		
• Key Off. • Reconnect EGO sensor. Remove air cleaner to gain access to air meter inlet. Using a standard wood lead pencil, prop the air meter door part-way open. • Start the engine and run at approximately 2000 rpm for 2 minutes. • Perform Engine Running Quick Test. • Is code 41 present?	Yes ▶	INSPECT for corrosion or damaged pins. If OK, REPLACE processor. RERUN Quick Test.
	No ▶	EGO input circuit OK. GO to HA1

Fuel Control (FBC) | Pinpoint Test | HB

HB

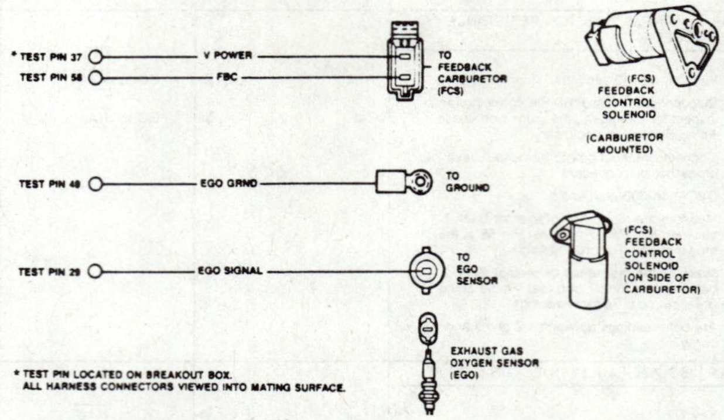

* TEST PIN LOCATED ON BREAKOUT BOX.
ALL HARNESS CONNECTORS VIEWED INTO MATING SURFACE.

STOP-WARNING

You should enter this Pinpoint Test only when a Service Code 16, 41, 42, 43 or 47 in Quick Test Step 5.0 or 6.0 is received or when directed here from Diagnostics by Symptom in the Engine Supplement Section.

To prevent the replacement of good components, be aware that the following non-EEC areas may be at fault:

- Fouled Spark Plugs
- Distributor Cap and Rotor
- Ignition Coil
- Spark Plug Wires
- CANP Problems
- PCV Valve
- Oil Contamination
- EGR Valve and Gasket
- Air Filter
- Fuel Filter
- Fuel Pump and Pressure
- Manifold Leak, Intake Exhaust
- Engine Not at Normal Operating Temperature
- Carburetor
- Choke, Sticking Electrical

This Pinpoint Test is intended to diagnose only the following:

- EGO Sensor Connection
- EGO Sensor
- Harness Circuits: VPWR, FBC, EGO Control, EGO Signal
- Impact of Thermactor Air
- FBC Solenoid
- Vacuum System (EEC Only)
- Choke Circuit
- Processor Assembly

Fuel Control — EFI | Pinpoint Test | HA

HA20 CONTINUOUS TESTING: CODE 41 42

41 — EGO indicated the fuel system was lean for more than 15 seconds when the system should have been in closed loop fuel control.

42 — EGO indicated the fuel system was rich for more than 15 seconds when the system should have been in closed loop fuel control.

*CLOSED LOOP — Fuel control under the influence of the EGO sensor.

*OPEN LOOP — Fuel control NOT under the influence of the EGO sensor.

Before attempting to correct a fuel control code, 41 42, diagnose all other drive complaints first, eg., rough idle, misses, etc.

NOTE: The fuel control code may help in this diagnosis.

Using the fuel control service code, isolate the cause of the fuel control problem.

Some areas to check are:

• Unmetered Air:	Vacuum leaks intake air leaks. — Canister Purge System — PCV System — Engine sealing — Air leaks between VAF meter and throttle body
• EGO Fuel Fouled:	Whenever an over-rich fuel condition has been experienced (fuel fouled spark plugs), make a thorough check of the ignition system. In the event the EGO sensor is suspected of being fuel fouled (low output, slow response), run the vehicle at sustained high speeds (within legal limits) followed by a few hard accelerations. This will burn off EGO contamination and restore proper EGO operation.
• Fuel Pressure:	Perform Pinpoint Test Step HA1
• Ignition System:	Always in default spark (10 degrees). Refer to Quick Test Step 4.0
• Improper Fueling:	Lead fouled EGO sensor
• TP Sensor:	Not moving (mechanical damage). Connect DVOM to test Pin 47 and to test Pin 46. Key to Run. Observe DVOM while moving the throttle. Reading must increase with increase in throttle opening. If not correct, REPLACE as necessary.

• If at this point, the drive concern is still present, perform Steps HA3 through HA6 only.

Fuel Control (FBC) | Pinpoint Test | HB

TEST STEP	RESULT ▶	ACTION TO TAKE
FAULT CODE 41		
HB1 VERIFY VEHICLE PREPARATION		
• Did vehicle run for more than 2 minutes before performing Quick Test?	Yes ▶	GO to HB2.
	No ▶	PREPARE vehicle. RERUN Quick Test.
HB2 TRY TO ELIMINATE CODE 41 BY REMOVING THERMACTOR AIR SUPPLY		
• Key Off. • Reconnect EGO sensor and processor to harness. • Disconnect Thermactor air supply hose at the air pump and cap the hose. • Run engine running Quick Test.	Code 41 is present ▶	GO to HB3. Leave Thermactor disconnected.
	Code 44 is present ▶	RETURN to Quick Test Step 5.0B.
	Code 11 or 42 is present ▶	HANDLE as Code 44. RETURN to Quick Test Step 5.0B.
HB3 CHECK FCS SOLENOID		
• Key Off, wait 10 seconds. • Thermactor disconnected. • Disconnect vehicle harness from FCS solenoid. • Perform Engine Running Quick Test. • Check for Code 41. **NOTE: Ignore all other codes at this time.**	Code 41 is not present ▶	GO to HB5.
	Code 41 is present ▶	GO to HB4.
HB4 FORCE SYSTEM RICH		
• Key Off, wait 10 seconds. • FCS and Thermactor disconnected. • Run for 2 minutes at part throttle. • Key Off, wait 10 seconds. • Perform Engine Running Quick Test while holding choke plate three-quarters closed **(Do not stall engine).** • Check for Code 41	Code 41 is not present ▶	REPLACE FCS. RERUN Quick Test if Code 41 is still present. REFER to Shop Manual, Group 24 for carburetor diagnostics.
	Code 41 is present ▶	GO to HB6. HOOK UP all disconnected components.

Fuel Control (FBC) — Pinpoint Test — HB

TEST STEP	RESULT ▶	ACTION TO TAKE
HB5 CHECK FBC CIRCUIT FOR SHORT TO GROUND		
• Key Off, wait 10 seconds. • Connect Thermactor. • Leave FCS solenoid disconnected. • Disconnect processor 60 Pin connector, inspect for damaged pins, corrosion, loose wires, etc. • Install Breakout box, leave processor disconnected. • DVOM on 200,000 ohm scale. • Measure resistance between test Pin 58 and test Pins 20, 40 and 46 at the Breakout box.	Any reading is less than 10,000 ohms ▶	SERVICE circuit shorts. RERUN Quick Test.
	All readings are 10,000 ohms or greater ▶	REPLACE processor. RERUN Quick Test. HOOK UP all disconnected components.
HB6 FORCE EGO RICH		
• Key Off, EGO sensor disconnected. • Run engine for 2 minutes. • DVOM on 20V scale. • With engine running hold choke plate three-quarters closed **(Do not stall engine).** • Measure voltage between EGO sensor pig tail terminal and battery negative terminal.	Reading is less than .45V ▶	REPLACE EGO sensor. RERUN Quick Test.
	Reading is .45V or greater ▶	GO to HB7.
HB7 CHECK CONTINUITY OF EGO CIRCUIT AND EGO GROUND CIRCUIT		
• Key Off, wait 10 seconds. • EGO sensor disconnected. • Disconnect processor 60 Pin connector, inspect for damaged pins, corrosion, loose wires, etc. Service as necessary. • Install Breakout box, leave processor disconnected. • DVOM on 200 ohm scale. • Measure resistance between test Pin 29 at the Breakout box and EGO circuit at the EGO vehicle harness connector. • Measure resistance between test Pin 49 and test Pins 40 and 60 at the Breakout box.	All readings are less than 5 ohms ▶	GO to HB8.
	Any reading is 5 ohms or greater ▶	SERVICE open in circuit. REMOVE Breakout box. RECONNECT EGO and processor. RERUN Quick Test.

Fuel Control (FBC) — Pinpoint Test — HB

TEST STEP	RESULT ▶	ACTION TO TAKE
HB8 CHECK EGO CIRCUIT FOR SHORT		
• Key Off, wait 10 seconds. • Thermactor and FCS reconnected. • Disconnect vehicle harness from EGO sensor. • DVOM on 200,000 ohm scale. • Measure resistance between EGO circuit at EGO vehicle harness connector and battery negative terminal.	Reading is less than 10,000 ohms ▶	SERVICE circuit short. REMOVE Breakout box. RECONNECT processor and EGO sensor. RERUN Quick Test.
	Reading is 10,000 ohms or greater ▶	REPLACE processor. RECONNECT EGO sensor and processor. RERUN Quick Test.
FAULT CODES 42 47		
HB20 CHECK CHOKE		
• Does choke function properly without sticking or binding?	No ▶	SERVICE choke as required. REFER to Section 4. RERUN Quick Test.
	Yes ▶	GO to HB21.
HB21 GENERATE KNOWN FAULT		
• Run engine for 2 minutes at part throttle. • Key Off, wait 10 seconds. • Disconnect harness from EGO sensor. Jumper vehicle harness EGO circuit to ground. • Rerun Engine Running Quick Test. • Check for Code 41. Ignore all other codes at this time.	Code 41 is present ▶	GO to HB22.
	Code 41 is not present ▶	GO to HB30.
HB22 EGO CHECK		
• Key Off, wait 10 seconds. • Disconnect a 7/32 inch vacuum line (creates a manifold vacuum leak). • DVOM on 20V scale. • With EGO sensor disconnected from the vehicle harness, connect a DVOM from EGO sensor to engine ground. • Run the engine at 2000 rpm for 2 minutes while observing DVOM. • Does meter indicate less than 0.4V?	Yes ▶	EGO sensor OK. GO to HB23.
	No ▶	REPLACE EGO sensor. RECONNECT vacuum hose. RERUN Quick Test.

Fuel Control (FBC) — Pinpoint Test — HB

TEST STEP	RESULT ▶	ACTION TO TAKE
HB23 CHECK FCS SOLENOID		
• Key Off, wait 10 seconds. • Vacuum hose reconnected. • Disconnect FCS solenoid. • DVOM on 200 ohm scale. • Measure resistance between terminals on the FCS solenoid. • **Is resistance within the following ranges:** 15 to 30 ohms: 2150 carburetor 30 to 60 ohms: YFA carburetor	Yes ▶	RECONNECT FCS. GO to HB24.
	No ▶	REPLACE FCS. RERUN Quick Test.
HB24 CHECK V PWR CIRCUIT FOR POWER		
• Key Off, wait 10 seconds. • Deactivate Quick Test. • Disconnect harness from FCS. • DVOM on 20V scale. • Key On, Engine Off. • Measure voltage between VPWR circuit at FCS harness connector and battery negative terminal.	Reading is less than 10V ▶	RECONNECT FCS solenoid. SERVICE open circuit. RERUN Quick Test.
	Reading is 10V or greater ▶	GO to HB25.
HB25 CHECK FOR SHORT TO POWER		
• Key Off, wait 10 seconds. • Disconnect processor 60 Pin connector, inspect connectors for damaged pins, corrosion, loose wires, etc. Service as necessary. • Install Breakout box; leave processor disconnected. • DVOM on 200,000 ohm scale. • Leave FCS solenoid disconnected, measure resistance between test Pin 58 and test Pin 37 at the Breakout box.	Resistance reading is 10,000 ohms or greater ▶	GO to HB26.
	Resistance reading is less than 10,000 ohms ▶	SERVICE short to power. RERUN Quick Test.
HB26 CHECK CONTINUITY OF FBC CIRCUIT		
• Key Off, wait 10 seconds. • Breakout box installed. • FCS disconnected. • DVOM on 200 ohm scale. • Measure resistance between FBC circuit at the FCS harness connector and test Pin 58 at the Breakout box.	Reading is less than 5 ohms ▶	REMOVE Breakout box. RECONNECT FCS solenoid. REPLACE processor and RERUN Quick Test.
	Reading is 5 ohms or greater ▶	SERVICE open circuit. REMOVE Breakout box. RECONNECT processor and FCS solenoid. RERUN Quick Test.

Fuel Control (FBC) — Pinpoint Test — HB

TEST STEP	RESULT ▶	ACTION TO TAKE
HB30 CHECK CANISTER PURGE		
• Key Off, wait 10 seconds. • Disconnect vacuum hose between canister purge solenoid and canister, plug both ends. • Rerun Engine Running Self-Test. • Is service code 42 still present?	Yes ▶	GO to HB31.
	No ▶	GO to KD1.
HB31 HARNESS CHECK		
• Key Off, wait 10 seconds. • Install Breakout box. • Measure resistance between test Pin 49 at the Breakout box and EGO ground at engine block. • Measure resistance between test Pin 29 at the Breakout box and EGO vehicle harness connector. • Both circuits less than 5 ohms.	Yes ▶	DISCONNECT processor connector. INSPECT for damage or corrosion. If OK, REPLACE processor. RERUN Quick Test.
	No ▶	CORRECT harness circuit with resistance greater than 5 ohms. RERUN Quick Test.
FAULT CODE 43		
HB40 WARM ENGINE		
• Run vehicle at 2000 rpm for 2 minutes. • Key Off, wait 10 seconds. • Perform Engine Running Quick Test.	Code 43 is not present ▶	SERVICE other codes as necessary.
	Code 43 is present ▶	GO to HB41.
HB41 CHECK EXHAUST LEAKS		
• Are there any exhaust leaks at manifold or inlet pipe?	Yes ▶	SERVICE as necessary. RERUN Quick Test.
	No ▶	GO to HB42.
HB42 CHECK IDLE QUALITY		
• Start vehicle, let engine idle. • Does engine idle quality deteriorate and remain poor during and after the Quick Test?	Yes ▶	REFER to Section 4, Carburetors.
	No ▶	REPLACE EGO sensor. RERUN Quick Test.

Fuel Control — 5.0L SEFI | Pinpoint Test | HC

HC

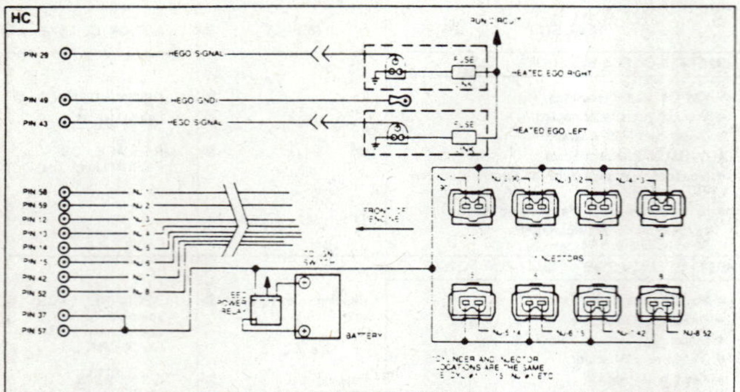

STOP-WARNING

You should enter this Pinpoint Test only when a Service Code 41, 91 or 42, 92 is received in Quick Test Step 5.0 or 6.0 or when directed here from Pinpoint Test A or Diagnostics by Symptom in the Engine Supplement Section.

To prevent the replacement of good components, be aware that the following non-EEC areas may be at fault:

- Ignition Coil
- Distributor Cap
- Distributor Rotor
- Fouled Spark Plugs
- Spark Plug Wires
- CANP Problems
- PCV Valves (see note below)
- EGR Valve and Gasket
- Air Filter
- Fuel Contamination, Engine Oil
- Poor Power Ground
- Fuel Pressure
- Manifold Leaks, Intake Exhaust
- Engine Not at Normal Operating Temperatures

This Pinpoint Test is intended to diagnose only the following:

- HEGO Sensor
- HEGO Sensor Connection
- Vacuum Systems
- Fuel Injectors
- Processor Assembly
- Harness Circuits: HEGO GRD, HEGO, INJ 1-8 and VPWR

NOTE: Fuel-contaminated engine oil may affect 41, 91 and 42, 92 Service Codes, so if it is suspected, remove the PCV from the valve cover, and rerun the Quick Test. If the problem is corrected, then change the engine oil and filter.

Fuel Control — 5.0L SEFI | Pinpoint Test | HC

HC

DESCRIPTION

The cylinder balance test switches each injector "OFF" and "ON," one at a time. Service codes correspond to cylinder number, e.g., service code 30 indicates a problem with cylinder number 3. Causes could include processor, harness and injector, as well as **NON-EEC** items such as fuel, plugs, ignition plug wires, rotor, cap, etc.

SERVICE CODE	90	10	20	30	40	50	60	70	80	77*
Cylinder Number	Pass	1	2	3	4	5	6	7	8	Rerun Test
Breakout Box Pin Number		58	59	12	13	14	15	42	52	
Injector Number		1	2	3	4	5	6	7	8	

***If throttle is touched (moved) during cylinder balance test, service code 77 will appear, indicating test was not completed.**

TEST STEP	RESULT ▶	ACTION TO TAKE
HC1 CYLINDER BALANCE TEST		
• Perform Engine Running Quick Test — to enter cylinder balance test, goose throttle (brief wide-open throttle) immediately after last service code is sent from Quick Test. • Is service code 90 present from cylinder balance test?	Yes ▶	GO to HC10.
	No ▶	GO to HC4.
HC2 FUEL PRESSURE CHECK		
• Key Off, wait 10 seconds. • Install fuel pressure gauge. • Start and run engine. Fuel pressure must be 241-310 kPa (35-45 psi). • Fuel pressure must remain at 276 ± 34 kPa (40 ± 5 psi) for 60 seconds after final key off. **For No Starts:** • If engine will not run, cycle the key from Off to On several times.	Fuel pressure is within specification ▶	GO to HC3.
	Fuel pressure is not 241-310 kPa (35-45 psi) ▶	REFER to Shop Manual Group 24 for electric fuel pump and Section 11 for fuel pressure regulator check.
	Fuel pressure remains at 276 ± 34 kPa (40 ± 5 psi) for 60 sec. or longer ▶	GO to HC1.

Fuel Control — 5.0L SEFI | Pinpoint Test | HC

TEST STEP	RESULT ▶	ACTION TO TAKE
HC3 FUEL DELIVERY TEST		
• Key Off. • Fuel pressure gauge installed. • Pressurize fuel system as per Step HC2. • Locate and disconnect the fuel pump relay. • Crank engine for 5 seconds. • Take pressure reading at the end of 5 second crank. **NOTE: Verify fuel quality; air and/or water will also pressurize and look like acceptable fuel delivery.**	PRESSURE GAUGE READING: Pressure is approximately 10-20 psi at the end of 5 second crank cycle. Refer to note below ▶	The EEC system is not the fault of the No Start; fuel and spark are present; REFER to Section 2 for other No Start Routines. If complaint was runs rough, misses or fuel code, GO to HC4.
	Pressure is greater or less than specified ▶	GO to HC5.
NOTE: The colder the engine, the greater the pressure drop (i.e., an engine coolant temperature of 200°F equals approximately a 10 psi drop in 5 seconds. 60°F equals approximately a 20 psi drop in 5 seconds).		
HC4 RUN CYLINDER BALANCE TEST		
• Key Off. • Perform Engine Running Quick Test. • To enter cylinder balance test; goose throttle (brief wide open throttle) immediately after last code is sent. • Is code 90 present?	Yes ▶	GO to Diagnostic by symptom in appropriate engine supplement section.
	No ▶	GO to HC5.

Fuel Control — 5.0L SEFI | Pinpoint Test | HC

TEST STEP	RESULT ▶	ACTION TO TAKE
HC5 HARNESS INJECTOR RESISTANCE CHECK		
• Key Off, wait 10 seconds. • Disconnect processor 60 Pin connector and inspect for damaged Pins, corrosion, loose wires. Service as necessary. • Connect Breakout box to harness. Leave processor disconnected. • DVOM on 200 ohm scale. • Measure the resistance of the suspect injector circuit between the injector circuit test Pin at the Breakout box and test Pins 37 and 57 at the Breakout box. • Is resistance reading between 15.0 ohms and 19.0 ohms? **NOTE: Refer to chart in test step HC3 for appropriate suspect injector circuit Pin number.**	Yes ▶	GO to HC6.
	No ▶	SERVICE the harness connectors on the suspect injector for OPENS, if OK, replace injector. RERUN Quick Test.
HC6 INJECTOR HARNESS SHORT CHECK		
• Key Off. • Breakout box installed. Processor disconnected. • DVOM to 200,000 ohm scale. • Measure the resistance of the suspect injector circuit between the injector circuit test Pin at the Breakout box and test Pins 40 and 60 at the Breakout box. • Is resistance greater than 10,000 ohms?	Yes ▶	GO to HC7.
	No ▶	Service short to ground. RERUN Quick Test.
HC7 INJECTOR DRIVE SIGNAL CHECK		
Requires standard non-powered 12-volt test lamp. • Key Off. • Breakout box installed. • Connect processor to Breakout box. • Connect test lamp between test Pin 37 and the suspect injector(s) circuit test Pin at the Breakout box. • Crank or start engine. • Is glow on light dim?	Yes ▶	REPLACE injector. RERUN Quick Test and cylinder balance test.
	No (Bright light or no light) ▶	Replace processor. Rerun Quick Test and cylinder balance test.

Fuel Control — 5.0L SEFI | Pinpoint Test | HC

TEST STEP	RESULT ▶	ACTION TO TAKE
HC10 FUEL CONTROL — ALWAYS LEAN: CODE 41 LEFT EGO, OR 91 RIGHT EGO		
NOTE: Vacuum air leaks in non-EEC areas could also cause code 41, or 91. Check for: — Leaking vacuum actuator (e.g.: A/C control motor) — Engine sealing — EGR system — PCV system — Unmetered air leak between air meter and throttle body — Lead contaminated EGO sensor • Key Off. DVOM on 20V scale. Disconnect appropriate EGO sensor from vehicle harness. Connect DVOM to EGO sensor and engine ground. • Apply 14 in. Hg to MAP. • Start the engine. Does the DVOM read greater than 0.5V?	Yes ▶ No ▶	GO to HC11. REPLACE EGO sensor. RERUN Quick Test.
HC11 LEFT EGO, OR RIGHT EGO AND EGO GROUND CONTINUITY CHECK		
• Key Off. • Install Breakout box. Processor disconnected. • DVOM on 200 ohm scale. • Measure resistance between test Pin 49 at Breakout box and engine block ground. • Measure resistance between test Pin 29 at Breakout box and HEGO signal at vehicle harness connector and/or measure resistance between test Pin 43 at Breakout box and HEGO signal at vehicle harness connector. • Are all readings less than 5 ohms?	Yes ▶ No ▶	GO to HC12. SERVICE open in appropriate circuit. RERUN Quick Test.
HC12 CHECK LEFT AND/OR RIGHT HEGO CIRCUIT FOR SHORT TO GROUND		
• Key Off. • Breakout box installed. Processor disconnected. • DVOM to 200,000 ohms scale. • Measure resistance between test Pin 29 (R-HEGO) and test Pin 40 or between test Pin 43 (L-HEGO) and test Pin 40 at the Breakout box.	Reading is 10,000 ohms or greater ▶ Reading is less than 10,000 ohms ▶	GO to HC13. CORRECT cause of resistance to ground. RERUN Quick Test.

Fuel Control — 5.0L SEFI | Pinpoint Test | HC

TEST STEP	RESULT ▶	ACTION TO TAKE
HC13 LEFT OR RIGHT HEGO CIRCUIT SHORT TO HEGO CIRCUIT GROUND		
• Key Off. • Breakout box installed. • DVOM to 200,000 ohm scale. • Disconnect appropriate HEGO. • Measure resistance between HEGO signal and HEGO ground circuit at the HEGO sensor connector. 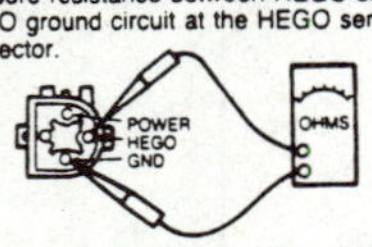 • Is resistance greater than 10,000 ohms?	Yes ▶ No ▶	Start vehicle and run at 2000 rpm for 2 minutes. RERUN Quick Test if code 41 or 91 is present REPLACE processor. If code 41 or 91 is not present GO to HC20. REPLACE appropriate HEGO. RERUN Quick Test.
HC15 FUEL CONTROL — ALWAYS RICH: CODE 42 OR 92		
• Key Off, wait 10 seconds. • NOTE: Non-EEC areas could cause a code 42 or 92. Check for: — Fuel contaminated engine oil — Ignition caused misfire — CANP problems • Disconnect vehicle harness at the HEGO sensor. Using a jumper wire, ground vehicle harness HEGO circuit to the engine block. • Repeat Engine Running Self-Test. • Service code 41 or 91 results.	Yes ▶ No ▶	REPLACE HEGO. RERUN Quick Test if code 42 or 92 is present. GO to HC2. GO to HC16.

Fuel Control — 5.0L SEFI | Pinpoint Test | HC

TEST STEP	RESULT ▶	ACTION TO TAKE
HC16 HEGO HARNESS CIRCUIT CHECK		
• Key Off, wait 10 seconds. • Install Breakout box. • Measure resistance between test Pin 49 (HEGO ground) and engine block. • Measure resistance between test Pin 29 at the Breakout box and HEGO vehicle harness connector for Code 92. • Measure resistance between test Pin 43 at the Breakout box and HEGO vehicle harness connector for Code 42. • Are all circuits less than 5 ohms?	Yes ▶ No ▶	DISCONNECT processor connector. INSPECT for damage or corrosion. If OK, REPLACE processor. RERUN Quick Test. CORRECT harness circuit with resistance greater than 5 ohms. RERUN Quick Test.
HC20 CHECK HEATER ELEMENT RESISTANCE ON BOTH HEATED EGOS		
• Key Off, Engine Off. • DVOM on 200 ohm scale. • Measure resistance between Run circuit and ground at both HEGO sensor connectors at room temperature.	Both readings 2.5 and 5.0 ohms ▶ Either reading less than 2.5 or greater than 5.0 ohms ▶	GO to HC21. REPLACE appropriate HEGO sensor. RERUN Quick Test.
HC21 CHECK FOR POWER AT BOTH HEGO HARNESS CONNECTORS		
• Key On, Engine Off. • DVOM on 20V scale. • Connect positive lead to run circuit and negative lead to ground at both HEGO vehicle harness connectors.	Both readings 10.5V or greater ▶ Either reading less than 10.5V ▶	HEGO's sensor system OK. RETURN to Section 2. GO to HC22.
HC22 CHECK CONTINUITY OF GROUND TO BOTH HEGO CONNECTORS		
• Key Off, wait 10 seconds. • DVOM on 200 ohm scale. • Measure resistance of both ground circuits from HEGO sensor connector to battery ground.	Any reading less than 5.0 ohms ▶ Any reading 5.0 ohms or greater ▶	SERVICE open in run circuit. RERUN Quick Test. SERVICE open in ground circuit. RERUN Quick Test.

Fuel Control — 3.8L CFI | Pinpoint Test | HD

HD REFERENCE DRAWING

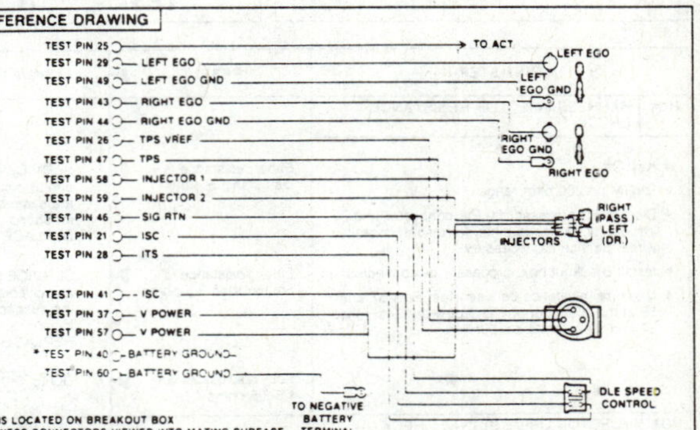

* TEST PINS LOCATED ON BREAKOUT BOX
ALL HARNESS CONNECTORS VIEWED INTO MATING SURFACE

STOP-WARNING

You should enter this Pinpoint Test only when a Service Code 41, 91, 42 or 92 is received in Quick Test Step 5.0 or 6.0 or when directed here from Pinpoint Test A or by Diagnostics by Symptom in the Engine Supplement Section.

To prevent the replacement of good components, be aware that the following non-EEC areas may be at fault:

- Ignition Coil
- Distributor Cap
- Distributor Rotor
- Fouled Spark Plugs
- Spark Plug Wires
- CANP Problems
- PCV Valves (see note below)
- EGR Valve and Gasket
- Air Filter
- Fuel Contamination, Engine Oil
- Poor Power Ground
- Fuel Pressure
- Manifold Leaks, Intake Exhaust
- Engine Not at Normal Operating Temperatures

This Pinpoint Test is intended to diagnose only the following:

- EGO Sensors
- EGO Signals and Ground Circuits
- EGO Sensor Connections
- Fuel Injectors
- Processor Assembly

NOTE: Fuel-contaminated engine oil may affect 41, 91, 42 and 92 Service Codes, so if it is suspected, remove the PCV from the valve cover, and rerun the Quick Test. If the problem is corrected, then change the engine oil and filter.

Fuel Control — 3.8L CFI | Pinpoint Test | HD

TEST STEP	RESULT ▶	ACTION TO TAKE
HD1 KEY POWER VOLTAGE CHECK		
• Key On. • Processor connected. • DVOM on 20V scale. • Measure voltage between test Pins 5 and 60 at the Breakout box.	Reading is less than 10.5V ▶	SERVICE open in key power circuit. RERUN Quick Test.
	Reading is 10.5V or greater ▶	GO to HD2.
HD2 FUEL PRESSURE CHECK		
• Install fuel pressure gauge. • Start and run engine. Fuel pressure must be 241-310 kPa (35-45 psi). • Fuel pressure must remain at 276 ± 34 kPa (40 psi ± 5 psi) for 60 seconds after final Key Off. **For No Starts:** • If engine will not run cycle the key from OFF to ON several times to increase fuel pressure.	Fuel pressure is within specifications ▶	GO to HD3.
	Fuel pressure is not 241-310 kPa (35-45 psi) ▶	REFER to Shop Manual, Group 24 for electric fuel pump.
	Fuel pressure does not remain at 276 ± 34 kPa (40 psi ± 5 psi) for 60 seconds or longer ▶	GO to HD6.
HD3 FUEL DELIVERY TEST		
• Key Off. • Fuel pressure gauge installed. • Pressurize fuel system as per Step HD2. • Locate and disconnect the fuel pump relay. • Crank engine for 5 seconds. • Take pressure reading at the end of 5-second crank. **NOTE: Verify fuel quality: air and or water will also pressurize and look like acceptable fuel delivery.**	Pressure Gauge Reading:	
	Pressure is approximately 10-20 psi at the end of 5-second crank cycle. Refer to Note below ▶	The EEC system is not the fault of the No Start. Fuel and spark are present. REFER to Section 2 for other No Start Routines. If complaint was runs rough, misses or fuel code, GO to HD4.
	Pressure is greater or less than specified ▶	GO to HD4.

NOTE: The colder the engine, the greater the pressure drop (i.e., an engine coolant temperature of 200°F equals approximately a 10 psi drop in 5 seconds. 60°F equals approximately a 20 psi drop in 5 seconds).

Fuel Control — 3.8L CFI | Pinpoint Test | HD

TEST STEP	RESULT ▶	ACTION TO TAKE
HD4 HARNESS INJECTOR RESISTANCE CHECK		
• Key Off. • DVOM on 200 ohm range. • Disconnect processor 60 Pin connector and inspect for damaged pins, corrosion, loose wires. Service as necessary. • Install Breakout box, processor disconnected. • Measure resistance between test Pins 57 and 58 at the Breakout box and between test Pins 57 and 59 at the Breakout box.	Either resistance is less than 1.5 ohms ▶	SERVICE the harness and/or connectors on the suspect injector for shorts. IF OK, REPLACE injector.
	Either resistance is greater than 3.5 ohms ▶	SERVICE the harness and/or connectors on the suspect injector for opens. If OK, REPLACE injector.
	Both resistances are 1.5-3.5 ohms ▶	GO to HD5.
HD5 INJECTOR DRIVE SIGNAL CHECK		
Requires standard non-powered 12V test lamp. • Key Off. • Breakout box installed. • Connect processor to Breakout box. • Refer to illustration HD. • Connect test lamp between test Pin 37 and Pin 58 at the Breakout box. • Crank or start engine. • Repeat above test between test Pin 37 and Pin 59 at the Breakout box.	Dim glow at light on both tests ▶	GO to HD6.
	No light on one or both tests ▶	VERIFY 12V battery power at Pins 37 and 57. If OK, REPLACE processor. RERUN Quick Test.
	Bright light on one or both tests ▶	CHECK injector circuits 1 and 2 for shorts to ground. If OK, REPLACE processor. RERUN Quick Test.
HD6 INJECTOR BALANCE TEST		
• Connect tachometer to engine. Run engine at idle. • Disconnect and reconnect the injectors one at a time, note rpm drop for each injector. • Does each injector produce at least a 100 rpm momentary drop? • Reset curb idle. Refer to VECI decal. Verify ISC connection at end of testing.	Yes ▶	Fuel delivery OK. Problem in area common to all cylinders, i.e., Air vacuum leak, fuel contamination, EGR.
	No ▶	REPLACE faulty injector. RERUN Quick Test.

Fuel Control — 3.8L CFI | Pinpoint Test | HD

TEST STEP	RESULT ▶	ACTION TO TAKE
HD7 CODE 41 AND OR 91 PRESENT		
• Verify codes received during Quick Test. • Are both code 41 and 91 present?	Yes ▶	GO to HD9.
	No ▶	GO to HD8.
HD8 SWAP INJECTOR CONNECTORS		
• Key Off. • Swap injector connections. • Start engine, run engine, run Quick Test. • Does the original code(s) repeat?	Yes ▶	RECONNECT injector harness to original production state. GO to HD9.
	No ▶	RECONNECT injector harness to original production state. GO to HD4.
HD9 FUEL CONTROL — ALWAYS LEAN CODES: 41 LEFT EGO, 91 RIGHT EGO		
NOTE: Vacuum/air leaks in non-EEC areas could cause a code. Check for: **— Leaking vacuum actuator (eg: A/C control motor)** **— Engine sealing** **— EGR system** **— PCV system** **— Lead contaminated EGO sensor** • Refer to illustration HD. • Key Off. DVOM on 20V scale. Disconnect appropriate EGO sensor from vehicle harness. Connect DVOM to EGO sensor and engine ground. Disconnect ACT sensor. • Start the engine and run at approximately 2000 rpm. Does the DVOM read greater than 0.5V within 1 minute?	Yes ▶	GO to HD10.
	No ▶	REPLACE EGO sensor. RERUN Quick Test.

Fuel Control — 3.8L CFI | Pinpoint Test | HD

TEST STEP	RESULT ▶	ACTION TO TAKE
HD10 HARNESS CHECK EGO CIRCUITS		
• Key Off, wait 10 seconds. • Disconnect processor 60 Pin connector and inspect for damaged pins, corrosion, loose wires. Service as necessary. • Install Breakout box. Processor disconnected. • For left EGO code 41, measure resistance between test Pin 49 at the Breakout box and engine block ground and between left EGO connector and test Pin 29 at the Breakout box. • For right EGO code 91, measure between test Pin 44 at the Breakout box and engine block ground and between right test Pin connector 43 at the Breakout box. • Are both circuits less than 5 ohms?	Yes ▶	GO to HD11.
	No ▶	SERVICE/CORRECT open circuit as necessary. RERUN Quick Test.
HD11 CHECK EGO CIRCUIT FOR SHORT TO GROUND		
• Key Off. • Breakout box installed. Processor disconnected. • DVOM to 200,000 ohms range. • Disconnect right EGO sensor at harness. • Measure resistance between test Pin 29 (left EGO) or test Pin 43 (right EGO) to test Pin 40 at the Breakout box.	Reading is 10,000 ohms or greater ▶	GO to HD12.
	Reading is less than 10,000 ohms ▶	CORRECT cause of resistance to ground. RERUN Quick Test.
HD12 ATTEMPT TO ELIMINATE EGO LEAN CODE		
• Key Off. • Reconnect appropriate EGO sensor. • Make certain that the ACT sensor is still disconnected. • Start the engine and run at approximately 2000 rpm for 1 minute. Allow engine to return to idle. • Enter Engine Running Quick Test. • Is code 41 or 91 still present? **NOTE: Disregard any other code output at this time.**	Yes ▶	INSPECT for corrosion or damaged pins. If OK, REPLACE processor. RERUN Quick Test.
	No ▶	EGO input circuit OK. GO to HD13.

Fuel Control — 3.8L CFI | Pinpoint Test | HD

TEST STEP	RESULT ▶	ACTION TO TAKE
HD13 CHECK FOR CODES WITH ATC RECONNECTED		
• Reconnect ACT sensor. • Repeat Quick Test. • Is code 41 or 91 present?	Yes ▶ No ▶	REPLACE processor. RERUN Quick Test. GO to HD2.
HD14 FUEL CONTROL — ALWAYS RICH CODES: 42 LEFT EGO, 92 RIGHT EGO		
• Key Off, wait 10 seconds. • **NOTE: Non-EEC areas could cause a rich code. Check for: — Fuel contaminated engine oil — Ignition caused misfire — CANP problems** • Refer to illustration HD. • Disconnect vehicle harness at the appropriate EGO sensor. Using a jumper wire, ground vehicle harness EGO circuit at the EGO sensor to the engine block. • Repeat Engine Running Self-Test. • Is service code 41 or 91 present?	Yes ▶ No ▶	GO to HD16. GO to HD15.
HD15 HARNESS CHECK		
• Key Off, wait 10 seconds. • Refer to illustration HD. • Disconnect processor 60 Pin connector and inspect for damaged pins, corrosion, loose wires. Service as necessary. • Install Breakout box. • Measure resistance between test Pin 49 at the Breakout box and left EGO ground at Breakout box or Pin 44 and right EGO ground at engine block. • Measure resistance between test Pin 29 and left EGO harness connector or test Pin 43 and right EGO harness connector. • Are both circuits less than 5 ohms?	Yes ▶ No ▶	DISCONNECT processor connector. INSPECT for damage or corrosion. If OK, REPLACE processor. RERUN Quick Test. CORRECT harness circuit with resistance greater than 5 ohms. RERUN Quick Test.

Fuel Control — 3.8L CFI | Pinpoint Test | HD

TEST STEP	RESULT ▶	ACTION TO TAKE
HD16 EGO CHECK		
• Refer to illustration HD. • DVOM on 20V scale. • With appropriate EGO sensor disconnected from the vehicle harness, connect a DVOM from EGO sensor to engine ground. • Disconnect PCV hose • Start engine and run at approximately 2000 rpm. Does the DVOM read less than 0.4V within 1 minute?	Yes ▶ No ▶	EGO sensor OK. GO to HD2. REPLACE EGO sensor. RERUN Quick Test.

Fuel Control — 3.8L CFI | Pinpoint Test | HD

HD17 CONTINUOUS TESTING: CODE 41 42

41 — EGO indicated the fuel system was lean for more than 15 seconds when the system should have been in closed loop fuel control.

42 — EGO indicated the fuel system was rich for more than 15 seconds when the system should have been in closed loop fuel control.

*CLOSED LOOP — Fuel control under the influence of the EGO sensor.

*OPEN LOOP — Fuel control NOT under the influence of the EGO sensor.

Before attempting to correct a fuel control code, 41/42, diagnose all other drive complaints first, eg., rough idle, misses, etc.

NOTE: The fuel control code may help in this diagnosis.

Using the fuel control service code, isolate the cause of the fuel control problem.

Some areas to check are:

- Vacuum Circuits: Vacuum leaks/intake air leaks.
  — Canister Purge System.
  — PCV System.
  — Engine sealing.
- EGO Fuel Fouled: Whenever an over-rich fuel condition has been experienced (fuel fouled spark plugs), make a thorough check of the ignition system. In the event the EGO sensor is suspected of being fuel fouled (low output, slow response), run the vehicle at sustained high speeds (within legal limits) followed by a few hard accelerations. This will burn off EGO contamination and restore proper EGO operation.
- Fuel Pressure: Perform Pinpoint Test Step HD2.
- Ignition System: Always in default spark (10 degrees). Refer to Quick Test Step 4.0.
- Improper Fueling: Lead fouled EGO sensor.
- TP Sensor: Not moving (mechanical damage). Connect DVOM to test Pin 47 and to test Pin 46. Key to Run. Observe DVOM while moving the throttle. Reading must increase with increase in throttle opening. If not correct, SERVICE as necessary.
- If at this point, the drive concern is still present, perform Steps HD3 through HD6 only.

Fuel Control — EFI Truck | Pinpoint Test | HE

HE

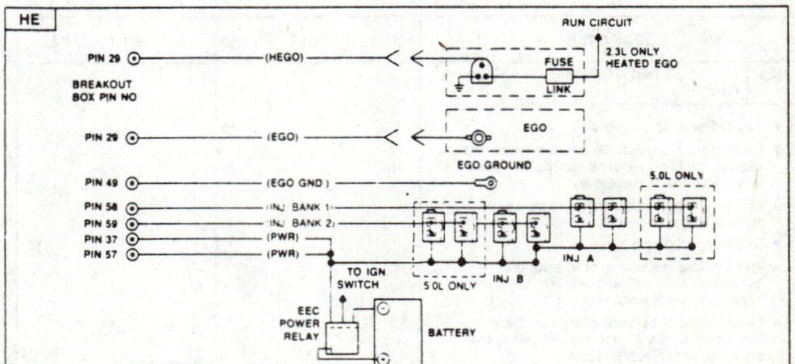

STOP-WARNING

You should enter this Pinpoint Test only when a Service Code 41 or 42 is received in Quick Test Step 5.0 or 6.0 or when directed here from Pinpoint Test A or Diagnostics by Symptom in the Engine Supplement Section.

To prevent the replacement of good components, be aware that the following non-EEC areas may be at fault

- Ignition Coil
- Distributor Cap
- Distributor Rotor
- Fouled Spark Plugs
- Spark Plug Wires
- CANP Problems
- PCV Valves (see note below)
- EGR Valve and Gasket
- Air Filter
- Fuel Contamination, Engine Oil
- Poor Power Ground
- Fuel Pressure
- Manifold Leaks, Intake Exhaust
- Engine Not at Normal Operating Temperatures

This Pinpoint Test is intended to diagnose only the following

- EGO Sensor
- Harness Circuits EGO Grd, EGO, Inj. Bank 1, Inj. Bank 2, V PWR
- EGO Sensor Connection
  — Code 42 start at HE8
  — Code 41 start at HE11
- Vacuum Systems
- Fuel Injectors
- Processor Assembly

NOTE: Fuel contaminated engine oil may affect 41 and 42 Service Codes. If this is suspected, remove the PCV from the valve cover and repeat the Quick Test. If the problem is corrected, then change the engine oil and filter.

Fuel Control — EFI Truck | Pinpoint Test | HE

TEST STEP	RESULT ▶	ACTION TO TAKE
HE1 FUEL PRESSURE CHECK 1		
• Key Off, wait 10 seconds. • Install fuel pressure gauge. • Start and run engine. Fuel pressure must be 241-310 kPa (35-45 psi). **For No Starts:** • If engine will not run, cycle the key from Off to On several times to increase fuel pressure.	Yes ▶ No ▶	GO to HE2. REFER to the Shop Manual, Group 24 for electric fuel pump and Section 11 for fuel pressure regulator check.
HE2 FUEL PRESSURE CHECK 2		
• Fuel pressure must remain at 276 ± 34 kPa (40 ± 5 psi) for 60 seconds or longer after final key Off.	Yes ▶ No ▶	GO to HE3. GO to HE7.
HE3 FUEL DELIVERY TEST		
• Key Off. • Fuel pressure gauge installed. • Pressurize fuel system as per Step HA1. • Locate and disconnect the fuel pump relay. • Crank engine for 5 seconds. • Take pressure reading at the end of 5 second crank. **NOTE: Verify fuel quality; air and or water will also pressurize and look like acceptable fuel delivery.**	PRESSURE GAUGE READING: Pressure is approximately 10-20 psi at the end of 5 second crank cycle. Refer to note below ▶ Pressure is greater or less than specified ▶	The EEC system is not the fault of the No Start. Fuel and spark are present. REFER to Section 2 for other No Start Routines. If complaint was runs rough, misses or fuel code, GO to HE4. GO to HE4.

NOTE: The colder the engine, the greater the pressure drop (i.e., an engine coolant temperature of 200°F equals approximately a 10 psi drop in 5 seconds. 60°F equals approximately a 20 psi drop in 5 seconds).

Fuel Control — EFI Truck | Pinpoint Test | HE

TEST STEP	RESULT ▶	ACTION TO TAKE
HE4 HARNESS INJECTOR RESISTANCE CHECK		
• Key Off, wait 10 seconds. • Disconnect processor 60 pin connector and inspect for damaged pins, corrosion, loose wires. Service as necessary. • Connect Breakout box to harness. Leave processor disconnected. • DVOM on 200 ohm scale. • Measure the resistance of injector Bank 1 between test Pin 37 and test Pin 58 at the Breakout box. Record reading. • Measure the resistance of injector Bank 2 between test Pin 37 and test Pin 59 at the Breakout box. Record reading. • Are both readings between 1.2 ohms and 1.8 ohms?	Yes ▶ No ▶	GO to HE6. GO to HE5.
HE5 ISOLATE FAULTY INJECTOR CIRCUIT		
• Key Off. • Breakout box installed. • Leave processor disconnected. • DVOM on 200 ohm scale. • Disconnect all injectors on suspect Bank. Measure the resistance of each injector by connecting one injector at a time, and reading resistance between test Pin 37 and 58 at the Breakout box for Bank 1 or test Pin 37 and 59 at the Breakout box for Bank 2. • Are all readings between 15.0 and 19.0 ohms?	Yes ▶ No ▶	GO to HE6. SERVICE the harness/connectors on the suspect injector for opens or shorts. If OK, REPLACE injector. RERUN Quick Test.
HE6 INJECTOR DRIVE SIGNAL CHECK		
Requires standard non-powered 12-volt test lamp. • Key Off. • Breakout box installed. • Connect processor to Breakout box. • Connect test lamp between test Pin 37 and test Pin 58 at the Breakout box. • Crank or start engine. • Repeat above test between test Pin 37 and test Pin 59 at the Breakout box.	Dim glow at light on both tests ▶ No light on one or both tests ▶ Bright light on one or both tests ▶	GO to HE7. VERIFY 12-volt battery power at Pins 37 and 57. If OK, REPLACE processor. RERUN Quick Test. CHECK circuits Bank 1 and Bank 2 for shorts to ground. If OK, REPLACE processor. RERUN Quick Test.

Fuel Control — EFI Truck | Pinpoint Test | HE

TEST STEP	RESULT ▶	ACTION TO TAKE
HE7 INJECTOR BALANCE TEST		
• Connect tachometer to engine. Run engine at idle. • Disconnect and reconnect the injectors one at a time; note rpm drop for each injector. • Does each injector produce at least a 100 rpm momentary drop, as ISC will attempt re-establish rpm.	Yes ▶ No ▶	Fuel delivery OK. Problem is with area common to all cylinders, i.e.: Air vacuum leak, fuel contamination, EGR. REPLACE faulty injector. RERUN Quick Test.
HE8 FUEL CONTROL — ALWAYS RICH: CODE 42		
• Key Off, wait 10 seconds. • **NOTE: Non-EEC areas could cause a code 42.** **Check for:** **— Fuel contaminated engine oil** **— Ignition caused misfire** **— CANP problems** • Disconnect vehicle harness at the EGO sensor. Using a jumper wire, ground vehicle harness EGO circuit at the EGO sensor to the engine block. • Repeat Engine Running Self-Test. • Service code 41 results.	Yes ▶ No ▶	GO to HE10. GO to HE9.
HE9 EGO HARNESS CHECK		
• Key Off, wait 10 seconds. • Install Breakout box. • Measure resistance between test Pin 49 at the Breakout box and EGO ground at engine block. • Measure resistance between test Pin 29 at the Breakout box and EGO harness connector. • Are both circuits less than 5 ohms?	Yes ▶ No ▶	DISCONNECT processor connector. INSPECT for damage or corrosion. If OK, REPLACE processor. RERUN Quick Test. CORRECT harness circuit with resistance greater than 5 ohms. RERUN Quick Test.

Fuel Control — EFI Truck | Pinpoint Test | HE

TEST STEP	RESULT ▶	ACTION TO TAKE
HE10 EGO CHECK		
• DVOM on 20V scale. • With EGO sensor disconnected from the vehicle harness, connect a DVOM from EGO sensor to engine ground. • Disconnect PCV hose. • Start engine and run at approximately 2000 rpm. Does the DVOM read less than 0.4V within 1 minute?	Yes ▶ No ▶	EGO sensor OK. GO to HE1. REPLACE EGO sensor. RERUN Quick Test.
HE11 FUEL CONTROL — ALWAYS LEAN: CODE 41		
NOTE: Vacuum/air leaks in non-EEC areas could cause a code 41. Check for: **— Leaking vacuum actuator (eg: A/C control motor)** **— Engine sealing** **— EGR system** **— PCV system** **— Unmetered air leak between air meter and throttle body** **— Lead contaminated EGO sensor** • Key Off. DVOM on 20V scale. Disconnect EGO sensor from vehicle harness. Connect DVOM to EGO sensor and engine ground. • Disconnect ACT sensor. • Start the engine and run at approximately 2000 rpm. Does the DVOM read greater than 0.5V within 1 minute?	Yes ▶ No ▶	GO to HE12. REPLACE EGO sensor. RERUN Quick Test.
HE12 HARNESS CHECK EGO CIRCUITS		
• Key Off. • Install Breakout box. Processor disconnected. • Measure resistance between test Pin 49 at the Breakout box and engine block ground and between EGO connector and test Pin 29 at the breakout box. • Are both circuits less than 5 ohms?	Yes ▶ No ▶	GO to HE13. SERVICE/CORRECT poor connection or open circuit as appropriate. RERUN Quick Test.

Fuel Control — EFI Truck | Pinpoint Test | HE

TEST STEP	RESULT	ACTION TO TAKE
HE13 CHECK EGO CIRCUIT FOR SHORT TO GROUND		
• Key Off. • Breakout box installed. Processor disconnected. • DVOM to 200,000 ohms range. • Measure resistance between test Pin 29 and test Pin 40 at the Breakout box.	Reading is 10,000 ohms or greater ▶	GO to HE14.
	Reading is less than 10,000 ohms ▶	CORRECT cause of resistance to ground. RERUN Quick Test.
HE14 ATTEMPT TO ELIMINATE CODE 41		
• Key Off. • Reconnect EGO sensor. • Make certain that ACT sensor is still disconnected. • Start the engine and run at approximately 2000 rpm for 1 minute. Allow engine to return to idle. • Perform Engine Running Self-Test. • Is code 41 still present? **NOTE: Disregard any other code output at this time.**	Yes ▶	INSPECT for corrosion or damaged pins. If OK, REPLACE processor. RERUN Quick Test.
	No ▶	RECONNECT ACT sensor. EGO input circuit OK. GO to HE1.

Fuel Control — EFI Truck | Pinpoint Test | HE

HE20 CONTINUOUS TESTING: CODE 41/42

41 — EGO indicated the fuel system was lean for more than 15 seconds when the system should have been in closed loop fuel control.

42 — EGO indicated the fuel system was rich for more than 15 seconds when the system should have been in closed loop fuel control.

*CLOSED LOOP — Fuel control under the influence of the EGO sensor.

*OPEN LOOP — Fuel control NOT under the influence of the EGO sensor.

Before attempting to correct a fuel control code, 41/42, diagnose all other drive complaints first, eg., rough idle, misses, etc.

NOTE: The fuel control code may help in this diagnosis.

Using the fuel control service code, isolate the cause of the fuel control problem.

Some areas to check are:

- Vacuum Circuits: Vacuum leaks-intake air leaks.
  - — Canister Purge System.
  - — PCV System.
  - — Engine sealing.
- EGO Fuel Fouled: Whenever an over-rich fuel condition has been experienced (fuel fouled spark plugs), make a thorough check of the ignition system. In the event the EGO sensor is suspected of being fuel fouled (low output, slow response), run the vehicle at sustained high speeds (within legal limits) followed by a few hard accelerations. This will burn off EGO contamination and restore proper EGO operation.
- Fuel Pressure: Perform Pinpoint Test Step HE1.
- Ignition System: Always in default spark (10 degrees). Refer to Quick Test Step 4.0.
- Improper Fueling: Lead fouled EGO sensor.
- TP Sensor: Not moving (mechanical damage). Connect DVOM to test Pin 47 and to test Pin 46. Key to Run. Observe DVOM while moving the throttle. Reading must increase with increase in throttle opening. If not correct, SERVICE as necessary.
- If at this point, the drive concern is still present, perform Steps HE3 through HE6 only.

Fuel Control — EFI Truck | Pinpoint Test | HE

TEST STEP	RESULT	ACTION TO TAKE
HE21 CHECK HEATER ELEMENT RESISTANCE ON HEATED EGO		
• Key Off, Engine Off. • DVOM on 200 ohm scale. • Measure resistance between Run circuit and ground at heated EGO sensor connector at room temperature.	Reading between 2.5 and 5.0 ohms ▶	GO to HE22.
	Reading less than 2.5 or greater than 5.0 ohms ▶	REPLACE HEGO sensor. RERUN Quick Test.
HE22 CHECK FOR POWER AT HEGO HARNESS CONNECTOR		
• Key On, Engine Off. • DVOM on 20V scale. • Connect positive lead to run circuit and negative lead to ground at HEGO vehicle harness connector.	Reading 10.5V or greater ▶	HEGO sensor system OK. RETURN to Section 2.
	Reading less than 10.5V ▶	GO to HE23.
HE23 CHECK CONTINUITY OF GROUND TO HEGO CONNECTOR		
• Key Off, wait 10 seconds. • DVOM on 200 ohm scale. • Measure resistance of ground circuit between HEGO sensor connector and battery ground.	Reading less than 5.0 ohms ▶	SERVICE open in run circuit. RERUN Quick Test.
	Reading 5.0 ohms or greater ▶	SERVICE open in ground circuit. RERUN Quick Test.

Fuel Control — 2.3L & 2.5L (HSC) CFI | Pinpoint Test | HF

HF

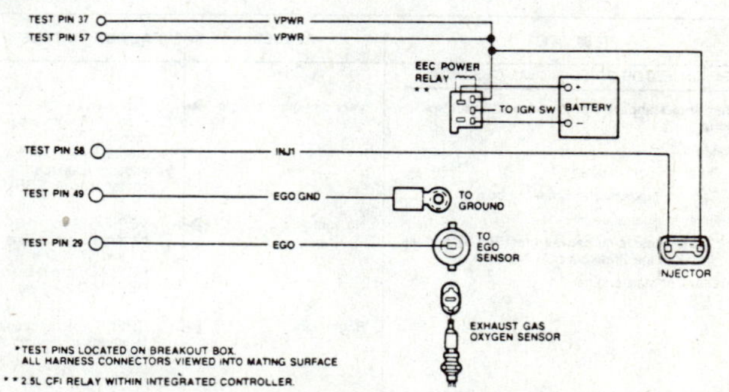

STOP-WARNING

You should enter this Pinpoint Test only when a Service Code 41 or 42 is received in Quick Test Step 5.0 or 6.0 or when directed here from Pinpoint Test A or Diagnostics by Symptom in the Engine Supplement Section.

To prevent the replacement of good components, be aware that the following non-EEC areas may be at fault:

- Ignition Coil
- Distributor Cap
- Distributor Rotor
- Fouled Spark Plugs
- Spark Plug Wires
- CANP Problems
- PCV Valves (see note below)
- EGR Valve and Gasket
- Air Filter
- Fuel Contamination, Engine Oil
- Poor Power Ground
- Fuel Pressure
- Manifold Leaks, Intake Exhaust
- Engine Not at Normal Operating Temperatures

This Pinpoint Test is intended to diagnose only the following:

- EGO Sensor
- EGO Signal and Ground Circuit
- EGO Sensor Connection
- Fuel Injector
- Processor Assembly

NOTE: Fuel contaminated engine oil may affect 41 and 42 Service Codes, so if it is suspected, remove the PCV from the valve cover, and rerun the Quick Test. If the problem is corrected, then change the engine oil and filter.

Fuel Control — 2.3L & 2.5L (HSC) CFI — Pinpoint Test — HF

TEST STEP	RESULT ▶	ACTION TO TAKE
HF1 FUEL PRESSURE CHECK		
• Key Off, wait 10 seconds. • Install fuel pressure gauge. • Start and run engine. Fuel pressure must be 14.5 ± 1.5 psi. • Fuel pressure must remain at 14.5 ± 1.5 psi for 60 seconds after final Key Off. **For No Starts:** • If engine will not run cycle the key from off to on several times.	Fuel pressure is within specification ▶ Fuel pressure is not 14.5 ± 1.5 psi ▶ Fuel pressure does not remain at 14.5 ± 1.5 psi for 60 seconds or longer ▶	GO to HF2. REFER to Shop Manual, Group 24 for electric fuel pump. GO to HF20.
HF2 FUEL DELIVERY TEST		
• Key Off. • Fuel pressure gauge installed. • Pressurize fuel system as per Step HF1. • Locate and disconnect the fuel pump relay. • Crank engine for 5 seconds. • Take pressure reading at the end of 5-second crank. **NOTE: Verify fuel quality: air and or water will also pressurize and look like acceptable fuel delivery.**	Pressure Gauge Reading: Pressure drops approximately 5-8 psi at the end of 5-second crank cycle. Refer to Note below ▶ Pressure is greater or less than specified ▶	The EEC system is not the fault of the No Start. Fuel and spark are present. REFER to Section 2 for other No Start Routines. If complaint was runs rough or misses. CONTINUE to HF3. GO to HF3.
NOTE: The colder the engine, the greater the pressure drop (i.e., an engine coolant temperature of 200°F equals approximately a 5 psi drop in 5 seconds. 60°F equals approximately a 8 psi drop in 5 seconds).		
HF3 HARNESS INJECTOR RESISTANCE CHECK		
• Key Off. • DVOM on 200 ohm range. • Disconnect processor 60 Pin connector and inspect for damaged pins, corrosion, loose wires. Service as necessary. • Install Breakout box, processor disconnected. • Measure resistance between test Pins 57 and 58 at the Breakout box.	Resistance is less than 1.0 ohms or greater than 2.0 ohms ▶ Resistance is between 1.0 and 2.0 ohms ▶	SERVICE the harness and or connectors on the injector for opens or shorts. If OK, REPLACE injector. GO to HF4.

Fuel Control — 2.3L & 2.5L (HSC) CFI — Pinpoint Test — HF

TEST STEP	RESULT ▶	ACTION TO TAKE
HF4 INJECTOR DRIVE SIGNAL CHECK		
Requires standard non-powered 12-volt test lamp. • Key Off. • Breakout box installed. • Connect processor to Breakout box. • Refer to illustration HF. • Connect test lamp between test Pin 37 and test Pin 58 at the Breakout box. • Crank or start engine.	Very dim glow at light ▶ No light ▶ Bright light ▶	Fuel delivery OK. Problem is with area common to all cylinders, i.e., air vacuum leak, fuel contamination, EGR. VERIFY 12-volt battery power at Pins 37 and 57. If OK, REPLACE processor. RERUN Quick Test. CHECK injector circuit for shorts to ground. If OK, REPLACE processor. RERUN Quick Test.
HF5 FUEL CONTROL — ALWAYS LEAN: CODE 41		
NOTE: Vacuum air leaks in non-EEC areas could cause a code 41. Check for: **— Leaking vacuum actuator (eg: A.C control motor)** **— Engine sealing** **— EGR system** **— PCV system** **— Lead contaminated EGO sensor** • Refer to illustration HF. • Key Off. DVOM on 20V scale. Disconnect EGO sensor from vehicle harness. Connect DVOM to EGO sensor and engine ground. Disconnect ACT sensor. • Start the engine and run at approximately 1000 rpm. Does the DVOM read greater than 0.5V within 30 seconds?	Yes ▶ No ▶	GO to HF6. REPLACE EGO sensor. RERUN Quick Test.

Fuel Control — 2.3L & 2.5L (HSC) CFI — Pinpoint Test — HF

TEST STEP	RESULT ▶	ACTION TO TAKE
HF6 HARNESS CHECK EGO CIRCUITS		
• Key Off, wait 10 seconds. • Disconnect processor 60 Pin connector and inspect for damaged pins, corrosion, loose wires. Service as necessary. • Install Breakout box. Processor disconnected. • Measure resistance between test Pin 49 at the Breakout box and engine block ground and between EGO connector and test Pin 29 at the Breakout box. • Are both circuits less than 5 ohms?	Yes ▶ No ▶	GO to HF7. SERVICE CORRECT circuit with resistance greater than 5 ohms, as necessary. RERUN Quick Test.
HF7 CHECK EGO CIRCUIT FOR SHORT TO GROUND		
• Key Off. • Breakout box installed. Processor disconnected. • DVOM to 200,000 ohms range. • Disconnect EGO sensor at harness. • Measure resistance between test Pin 29 and test Pin 40 at the Breakout box.	Reading is 10,000 ohms or greater ▶ Reading is less than 10,000 ohms ▶	GO to HF8. CORRECT cause of resistance to ground. RERUN Quick Test.
HF8 ATTEMPT TO ELIMINATE CODE 41		
• Key Off. • Reconnect EGO sensor. • Make certain that the ACT sensor is still disconnected. • Start the engine and run at approximately 1000 rpm for 1 minute, allow engine to return to idle. • Enter Self-Test. • Is code 41 still present? **NOTE: Disregard any other code output at this time.**	Yes ▶ No ▶	INSPECT for corrosion or damaged pins. If OK, REPLACE processor. RERUN Quick Test. EGO input circuit OK. RECONNECT ACT sensor. GO to HF1.

Fuel Control — 2.3L & 2.5L (HSC) CFI — Pinpoint Test — HF

TEST STEP	RESULT ▶	ACTION TO TAKE
HF9 FUEL CONTROL — ALWAYS RICH: CODE 42		
• Key Off, wait 10 seconds. • **NOTE: Non-EEC areas could cause a code 42.** **Check for:** **— Fuel contaminated engine oil** **— Ignition caused misfire (fouled spark plug)** **— CANP problems** • Refer to illustration HF. • Disconnect vehicle harness at the EGO sensor. Jumper (ground) the EGO circuit at the EGO sensor vehicle harness connector to the engine block. • Repeat Engine Running Self-Test. • Service code 41 results.	Yes ▶ No ▶	GO to HF11. GO to HF10.
HF10 HARNESS CHECK		
• Key Off, wait 10 seconds. • Refer to illustration HF. • Disconnect processor 60 Pin connector and inspect for damaged pins, corrosion, loose wires. Service as necessary. • Install Breakout box. • Measure resistance between test Pin 49 at the Breakout box and EGO ground at engine block. • Measure resistance between test Pin 29 at the Breakout box and EGO circuit at the EGO vehicle harness connector. • Are both circuits less than 5 ohms?	Yes ▶ No ▶	DISCONNECT processor connector. INSPECT for damage or corrosion. If OK, REPLACE processor. RERUN Quick Test. CORRECT harness circuit with resistance greater than 5 ohms. RERUN Quick Test.
HF11 EGO CHECK		
• Refer to illustration HF. • DVOM on 20V scale. • With EGO sensor disconnected from the harness, connect a DVOM from EGO sensor to engine ground. • Disconnect PCV hose. • Start engine and run at approximately 2000 rpm. Does the DVOM read less than 0.4V within 30 seconds?	Yes ▶ No ▶	EGO sensor OK. GO to HF1. REPLACE EGO sensor. RERUN Quick Test.

Fuel Control — 2.3L & 2.5L (HSC) CFI | Pinpoint Test | HF

HF12 CONTINUOUS TESTING: CODE 41/42

41 — EGO indicated the fuel system was lean for more than 15 seconds when the system should have been in closed loop fuel control.

42 — EGO indicated the fuel system was rich for more than 15 seconds when the system should have been in closed loop fuel control.

*CLOSED LOOP — Fuel control under the influence of the EGO sensor.

*OPEN LOOP — Fuel control NOT under the influence of the EGO sensor.

Before attempting to correct a fuel control code, 41/42, diagnose all other drive complaints first, eg., rough idle, misses, etc.

NOTE: The fuel control code may help in this diagnosis.

Using the fuel control service code, isolate the cause of the fuel control problem.

Some areas to check are:

- Vacuum Circuits: Vacuum leaks/intake air leaks.
  - — Canister Purge System.
  - — PCV System.
  - — Engine sealing.
- EGO Fuel Fouled: Whenever an over-rich fuel condition has been experienced (fuel fouled spark plugs), make a thorough check of the ignition system. In the event the EGO sensor is suspected of being fuel fouled (low output, slow response), run the vehicle at sustained high speeds (within legal limits) followed by a few hard accelerations. This will burn off EGO contamination and restore proper EGO operation.
- Fuel Pressure: Perform Pinpoint Test Step HF1.
- Ignition System: Always in default spark (10 degrees). Refer to Quick Test Step 4.0.
- Improper Fueling: Lead fouled EGO sensor.
- TP Sensor: Not moving (mechanical damage). Connect DVOM to test Pin 47 and to test Pin 46. Key to Run. Observe DVOM while moving the throttle. Reading must increase with increase in throttle opening. If not correct, SERVICE as necessary.
- If at this point, the drive concern is still present, perform Steps HF2 through HF3 only.

Fuel Control — 2.3L & 2.5L (HSC) CFI | Pinpoint Test | HF

TEST STEP	RESULT	▶ ACTION TO TAKE
HF20 LEAKING INJECTOR		
• Remove air inlet tube at fuel charging assembly. • Pressurize fuel system as per Test Step HF1. • Key Off. • Observe throughout the air horn inlet for fuel leaking and/or weeping at the fuel injector discharge area.	No sign of fuel leak ▶	GO to HF21
	Fuel leak ▶	SERVICE injector and/or O-rings as necessary. REFER to Shop Manual, Section 24-03 for service procedure. **NOTE: After having completed service RERUN Quick Test.**
HF21 EXTERNAL FUEL LEAKS		
• Pressurize fuel system as per Test Step HF1. • Key Off. • Look for fuel leak external to fuel charging assembly fuel line, fitting, etc.	No sign of fuel leaks ▶	SERVICE fuel pressure regulator as necessary. REFER to Shop Manual, Section 24-03 for service procedure. **NOTE: After having completed service RERUN Quick Test.**
	Fuel leaks ▶	SERVICE as necessary. RERUN Quick Test.

Fuel Control — 2.3L & 2.5L (HSC) CFI | Pinpoint Test | HF

TEST STEP	RESULT	▶ ACTION TO TAKE
HF13 WARM ENGINE, CODE 43		
• Run vehicle at 2000 rpm for 2 minutes. • Key Off, wait 10 seconds. • Perform Engine Running Quick Test.	Code 43 is not present ▶	SERVICE other codes as necessary.
	Code 43 is present ▶	GO to HF14
HF14 CHECK EXHAUST LEAKS, CODE 43		
• Are there any exhaust leaks at manifold or inlet pipe?	Yes ▶	SERVICE as necessary. RERUN Quick Test.
	No ▶	GO to HF15
HF15 CHECK IDLE QUALITY, CODE 43		
• Start vehicle, let engine idle. • Does engine idle quality deteriorate and remain poor during and after the Quick Test?	Yes ▶	REFER to EFI/CFI, Section 4.
	No ▶	REPLACE EGO sensor. RERUN Quick Test.

Fuel Control — 2.9L & 3.0L EFI V-6 | Pinpoint Test | HG

HG

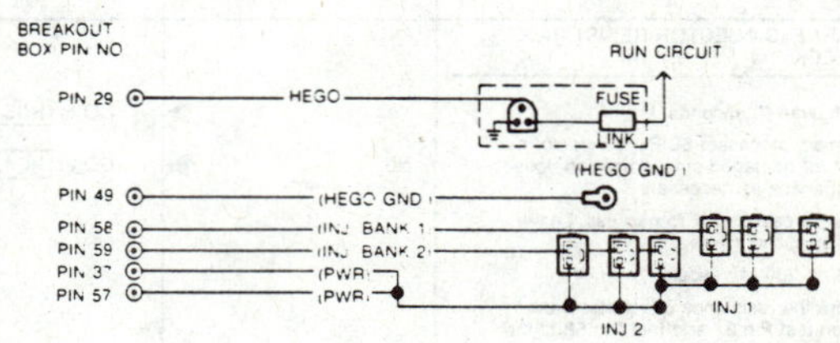

STOP-WARNING

You should enter this Pinpoint Test only when a Service Code 41 or 42 is received in Quick Test Step 5.0 or 6.0 or when directed here from Pinpoint Test S or Diagnostics by Symptom in the Engine Supplement Section.

To prevent the replacement of good components, be aware that the following non-EEC areas may be at fault:

- Ignition Coil
- Distributor Cap
- Distributor Rotor
- Fouled Spark Plugs
- Spark Plug Wires
- CANP Problems
- PCV Valves (see note below)
- EGR Valve and Gasket
- Air Filter
- Fuel Contamination, Engine Oil
- Poor Power Ground
- Fuel Pressure
- Manifold Leaks, Intake Exhaust
- Engine Not at Normal Operating Temperatures

This Pinpoint Test is intended to diagnose only the following:

- HEGO Sensor
- Harness Circuits EGO Grd., HEGO, Inj. Bank 1, Inj. Bank 2, VPWR
- HEGO Sensor Connection
- Vacuum Systems
- Fuel Injectors
- Processor Assembly

NOTE: Fuel contaminated engine oil may affect 41 and 42 Service Codes. If this is suspected, remove the PCV from the valve cover and repeat the Quick Test. If the problem is corrected, then change the engine oil and filter.

Fuel Control — 2.9L & 3.0L EFI V-6 | Pinpoint Test | HG

TEST STEP	RESULT ▶	ACTION TO TAKE
HG1 FUEL PRESSURE CHECK		
• Key Off, wait 10 seconds. • Install fuel pressure gauge. • Start and run engine. Fuel pressure must be 241-310 kPa (35-45 psi). • Fuel pressure must remain at 276 ± 34 kPa (40 ± 5 psi) for 60 seconds after final Key Off. **For No Starts:** • If engine will not run, cycle the key from off to on several times.	Fuel pressure is within all specifications ▶ Fuel pressure is not 241-310 kPa (35-45 psi) ▶ Fuel pressure does not remain at 276 ± 34 kPa (40 ± 5 psi) for 60 seconds or longer ▶	GO to HG2. REFER to the Shop Manual, Group 24 for electric fuel pump and Section 11 for fuel pressure regulator check. GO to HG6.
HG2 FUEL DELIVERY TEST		
• Key Off. • Fuel pressure gauge installed. • Pressurize fuel system as per Step HG1. • Locate and disconnect the fuel pump relay. • Crank engine for 5 seconds. • Take pressure reading at the end of 5 second crank. **NOTE: Verify fuel quality; air and/or water will also pressurize and look like acceptable fuel delivery.**	Pressure Gauge Reading: Pressure is approximately 69-138 kPa (10-20 psi) at the end of 5 second crank cycle. Refer to Note below ▶ Pressure is greater or less than specified ▶	The EEC system is not the fault of the No Start. Fuel and spark are present. REFER to Section 2 for other No Start Routines. If complaint was runs rough, misses or fuel code, GO to HG3. GO to HG3.

NOTE: The colder the engine, the greater the pressure drop (for example, an engine coolant temperature of 200°F equals approximately a 69 kPa [10 psi] drop in 5 seconds. 60°F equals approximately a 138 kPa [20 psi] drop in 5 seconds).

Fuel Control — 2.9L & 3.0L EFI V-6 | Pinpoint Test | HG

TEST STEP	RESULT ▶	ACTION TO TAKE
HG6 INJECTOR BALANCE TEST		
• Connect tachometer to engine. Run engine at idle. • Disconnect and reconnect the injectors one at a time; note rpm drop for each injector. • Does each injector produce at least a 100 rpm momentary drop, as ISC will attempt re-establish rpm?	Yes ▶ No ▶	Fuel delivery OK. Problem is with area common to all cylinders, i.e. Air vacuum leak, fuel contamination, EGR. REPLACE faulty injector. RERUN Quick Test.
HG7 FUEL CONTROL — ALWAYS RICH: CODE 42		
• Key Off, wait 10 seconds. **NOTE: Non-EEC areas could cause a code 42.** **Check for:** **— Fuel contaminated engine oil** **— Ignition caused misfire** **— CANP problems** • Disconnect vehicle harness at the HEGO sensor. • Jumper (ground) the HEGO circuit at the HEGO sensor vehicle harness connector to the engine block. • Repeat Engine Running Self-Test. • Service code 41 results.	Yes ▶ No ▶	GO to HG9. GO to HG8.
HG8 EGO HARNESS CHECK		
• Key Off, wait 10 seconds. • Install Breakout box. • Measure resistance between test Pin 49 at the Breakout box and HEGO ground at engine block. • Measure resistance between test Pin 29 at the Breakout box and HEGO circuit HEGO vehicle harness connector. • Are both circuits less than 5 ohms?	Yes ▶ No ▶	DISCONNECT processor connector. INSPECT for damage or corrosion. If OK, REPLACE processor. RERUN Quick Test. CORRECT harness circuit with resistance greater than 5 ohms. RERUN Quick Test.

Fuel Control — 2.9L & 3.0L EFI V-6 | Pinpoint Test | HG

TEST STEP	RESULT ▶	ACTION TO TAKE
HG3 HARNESS INJECTOR RESISTANCE CHECK		
• Key Off, wait 10 seconds. • Disconnect processor 60 Pin connector and inspect for damaged pins, corrosion, loose wires. Service as necessary. • Connect Breakout box to harness. Leave processor disconnected. • DVOM on 200 ohm scale. • Measure the resistance of injector Bank 1 between test Pin 37 and test Pin 58 at the Breakout box. • Measure the resistance of injector Bank 2 between test Pin 37 and test Pin 59 at the Breakout box. • Are both readings between 5.0 and 6.5 ohms?	Yes ▶ No ▶	GO to HG5. GO to HG4.
HG4 ISOLATE FAULTY INJECTOR CIRCUIT		
• Key Off. • Breakout box installed. • Leave processor disconnected. • DVOM on 200 ohm scale. • Disconnect all injectors on suspect Bank. Measure the resistance of each injector by connecting one injector at a time, and reading resistance between test Pin 37 and 58, for Bank 1, or test Pin 37 and 59, for Bank 2, at the Breakout box. • Are all readings between 16.0 and 18.0 ohms?	Yes ▶ No ▶	GO to HG5. SERVICE the harness and/or connectors on the suspect injector for opens or shorts. If OK, REPLACE injector. RERUN Quick Test.
HG5 INJECTOR DRIVE SIGNAL CHECK		
Requires standard non-powered 12-volt test lamp. • Key Off. • Breakout box installed. • Connect processor to Breakout box. • Connect test lamp between test Pin 37 and test Pin 58 at the Breakout box. • Crank or start engine. • Repeat above test between test Pin 37 and test Pin 59 at the Breakout box.	Dim glow at light on both tests ▶ No light on one or both tests ▶ Bright light on one or both tests ▶	GO to HG6. VERIFY 12V battery power at Pins 37 and 57. If OK, REPLACE processor. RERUN Quick Test. CHECK circuits Bank 1 and Bank 2 for shorts to ground. If OK, REPLACE processor. RERUN Quick Test.

Fuel Control — 2.9L & 3.0L EFI V-6 | Pinpoint Test | HG

TEST STEP	RESULT ▶	ACTION TO TAKE
HG9 HEGO CHECK		
• DVOM on 20V scale. • With HEGO sensor disconnected from the harness, connect a DVOM from HEGO sensor to engine ground. • Disconnect PCV hose. • Start engine and run at approximately 2000 rpm. Does the DVOM read less than 0.4V within 1 minute?	Yes ▶ No ▶	HEGO sensor OK. GO to HG1. REPLACE HEGO sensor. RERUN Quick Test.
HG10 FUEL CONTROL — ALWAYS LEAN: CODE 41 VERIFICATION		
• Run vehicle at 2000 rpm for 2 minutes. • Key Off, wait 10 seconds. • Perform Engine Running Quick Test.	Code 41 is not present ▶ Code 41 is present ▶	GO to HG21. GO to HG11.
HG11 FUEL CONTROL — ALWAYS LEAN		
NOTE: Vacuum/air leaks in non-EEC areas could cause a code 41. **Check for:** **— Leaking vacuum actuator (e.g.: A/C control motor)** **— Engine sealing** **— EGR system** **— PCV system** **— Lead contaminated HEGO sensor** • Key Off. DVOM on 20V scale. Disconnect HEGO sensor from vehicle harness. Connect DVOM to HEGO sensor and engine ground. • Disconnect ACT sensor. • Start the engine and run at approximately 2000 rpm. Does the DVOM read greater than 0.5V within 1 minute?	Yes ▶ No ▶	GO to HG12. REPLACE HEGO sensor. RERUN Quick Test.
HG12 HARNESS CHECK HEGO CIRCUITS		
• Key Off. • Install Breakout box. Processor disconnected. • Measure resistance between test Pin 49 at the Breakout box and engine block ground and between HEGO circuit at the HEGO vehicle harness connector and test Pin 29 at the Breakout box. • Are both circuits less than 5 ohms?	Yes ▶ No ▶	GO to HG13. SERVICE/CORRECT as necessary circuit with greater than 5 ohms resistance. RERUN Quick Test.

Fuel Control — 2.9L & 3.0L EFI V-6 | Pinpoint Test | HG

TEST STEP	RESULT ▶	ACTION TO TAKE
HG13 CHECK HEGO CIRCUIT FOR SHORT TO GROUND		
• Key Off. • Breakout box installed. Processor disconnected. • DVOM to 200,000 ohms range. • Measure resistance between test Pin 29 and test Pin 40 at the Breakout box.	Reading is 10,000 ohms or greater ▶ Reading is less than 10,000 ohms ▶	GO to HG14. CORRECT cause of resistance to ground. RERUN Quick Test.
HG14 ATTEMPT TO ELIMINATE CODE 41		
• Key Off. • Reconnect HEGO sensor. • Make certain that ACT sensor is still disconnected. • Start the engine and run at approximately 2000 rpm for 1 minute. Allow engine to return to idle. • Perform Engine Running Self-Test. • Is code 41 still present? **NOTE: Disregard any other code input at this time.**	Yes ▶ No ▶	INSPECT for corrosion or damaged pins. If OK, REPLACE processor. RERUN Quick Test. RECONNECT ACT sensor. HEGO input circuit OK. GO to HG1.

Fuel Control — 2.9L & 3.0L EFI V-6 | Pinpoint Test | HG

TEST STEP	RESULT ▶	ACTION TO TAKE
HG21 CHECK HEATER ELEMENT RESISTANCE ON HEATED EGO		
• Key Off, Engine Off. • DVOM on 200 ohm scale. • Measure resistance between Run circuit and ground at heated EGO connector at room temperature.	Reading between 2.5 and 5.0 ohms ▶ Reading less than 2.5 or greater than 5.0 ohms ▶	GO to HG22. REPLACE HEGO sensor.
HG22 CHECK FOR POWER AT HEGO HARNESS CONNECTOR		
• Key On, Engine Off. • DVOM on 20V scale. • Connect positive lead to run circuit and negative lead to ground at HEGO vehicle harness connector.	Reading 10.5V or greater ▶ Reading less than 10.5V ▶	HEGO sensor system OK, RETURN to Section 2. GO to HG23.
HG23 CHECK CONTINUITY OF GROUND TO HEGO CONNECTOR		
• Key Off, wait 10 seconds. • DVOM on 200 ohm scale. • Measure resistance of ground circuit from HEGO vehicle harness connector to battery ground.	Reading less than 5.0 ohms ▶ Reading 5.0 ohms or greater ▶	SERVICE open in run circuit. SERVICE open in ground circuit.

Fuel Pump Circuit — (Inertia Switch) | Pinpoint Test | J

J

TYPE I — VEHICLE APPLICATION: LINCOLN TOWN CAR, FORD CROWN VICTORIA/MERCURY GRAND MARQUIS, TEMPO/TOPAZ, ESCORT/LYNX AND EXP

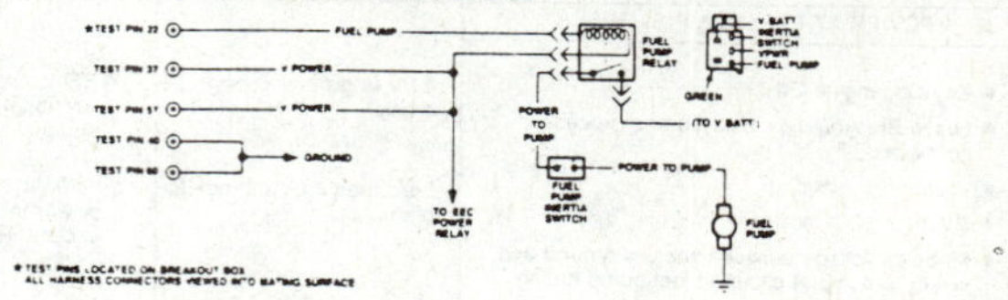

TYPE II — VEHICLE APPLICATION: MARK VII/CONTINENTAL, THUNDERBIRD/COUGAR, LTD/MARQUIS, MUSTANG/CAPRI, RANGER, E-SERIES AND F-SERIES

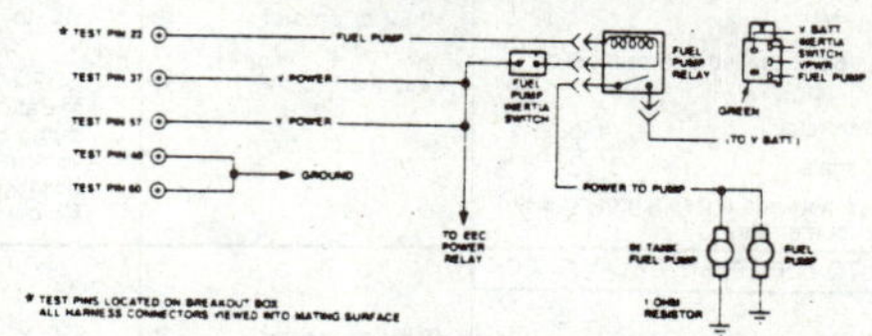

STOP-WARNING

You should enter this Pinpoint Test only when a service code 87 is received in Quick Test Step 3.0 or you are directed here from Pinpoint Test Step A22 or Diagnostics by Symptom in the Engine Supplement Section.

To prevent the replacement of good components, be aware that the following non-EEC areas may be at fault:

- Fuel Lines
- Fuel Filters
- Throttle Body
- Contaminated Fuel
- Fuel Pump

This Pinpoint Test is intended to diagnose only the following:

- Fuel Pump Relay
- Harness Circuits, V Batt., V Power, F.P., Ground and Power to Pump(s)
- Processor Assembly

Fuel Pump Circuit — (Inertia Switch) | Pinpoint Test | J

TEST STEP	RESULT ▶	ACTION TO TAKE
J1 NO FUEL PUMP PRESSURE		
• Fuel pressure gauge installed. • Check if fuel pump runs. • Cycle key from Off to Run, repeat several times (Do not enter start mode.) Fuel pump should run briefly each time the key enters Run.	Yes ▶ No ▶	GO to Shop Manual, Group 24 electric fuel pump. GO to J2.
J2 CHECK FOR V POWER TO PROCESSOR		
• Key Off, wait 10 seconds. • Disconnect processor 60 Pin connector and inspect for damaged pins, corrosion, loose wires. Service as necessary. • Install Breakout box and reconnect processor. • Key On, engine Off. • DVOM on 20V scale. • Measure voltage between test Pin 37 and test Pin 40 at the Breakout box and between test Pin 57 and test Pin 60 at the breakout box.	Either voltage reading is less than 10.5V ▶ Both voltage readings are 10.5V or greater ▶	GO to B1. GO to J3.
J3 RESISTANCE CHECK OF FUEL PUMP INERTIA SWITCH		
• Key Off, wait 10 seconds. • Leave Breakout box installed and processor connected. • Locate and disconnect fuel pump inertia switch. • DVOM on 200 ohm scale. • Measure the resistance of the fuel pump inertia switch.	Less than 5.0 ohms ▶ 5.0 ohms or greater ▶	GO to J4. REPLACE or reset fuel pump inertia switch. RERUN Quick Test.
J4 POWER TO PUMP(S) CHECK		
• Key On, engine Off. • Leave Breakout box installed and processor connected. • Locate fuel pump relay. • DVOM on 20V scale. • Measure voltage between chassis ground and power-to-pump(s) circuit at fuel pump relay during crank mode.	8.0V or greater during crank ▶ Less than 8.0V during crank ▶	GO to J5. GO to J6.

Fuel Pump Circuit — (Inertia Switch) | Pinpoint Test | J

TEST STEP	RESULT ▶	ACTION TO TAKE
J5 POWER AT FUEL PUMP(S) CHECK		
• Key On, engine Off. • Leave Breakout box installed and processor connected. • Locate fuel pump(s). • DVOM on 20V scale. • Measure voltage between chassis ground and power-to-pump(s) circuit at fuel pump during crank mode.	8.0V or greater during crank ▶	GO to Shop Manual, Group 24 electric fuel pump.
	Less than 8.0V during crank ▶	SERVICE open in power to the pump(s) circuit. RERUN Quick Test.
J6 FUEL PUMP CIRCUIT CHECK TO V BATT.		
• Key On, engine Off. • Leave Breakout box installed and processor connected. • Locate fuel pump relay. • DVOM on 20V scale. • Measure voltage between chassis ground and V Batt. at the fuel pump relay.	10.5V or greater ▶	GO to J7.
	Less than 10.5V ▶	SERVICE open in V Batt. between fuel pump relay and vehicle battery positive post. RERUN Quick Test.
J7 V POWER TO FUEL PUMP RELAY CHECK		
• Key On, engine Off. • Leave Breakout box installed and processor connected. • Locate fuel pump relay. • DVOM on 20V scale. • Measure voltage between chassis ground and V Power circuit at the fuel pump relay.	10.5V or greater ▶	GO to J8.
	Less than 10.5V ▶	SERVICE open in V Power circuit between the processor and the fuel pump relay. RERUN Quick Test.
J8 FUEL PUMP CIRCUIT CHECK CONTINUITY		
• Leave Breakout box installed and processor connected. • Key Off, wait 10 seconds. • DVOM on 200 ohm scale. • Measure resistance between fuel pump circuit at the pump relay and test Pin 22 at the Breakout box.	Less than 5 ohms ▶	GO to J9.
	5 ohms or greater ▶	SERVICE open in fuel pump circuit and RERUN Quick Test.

Fuel Pump Circuit — (Inertia Switch) | Pinpoint Test | J

TEST STEP	RESULT ▶	ACTION TO TAKE
J9 CHECK FOR SHORT TO GROUND		
• Key Off. • Leave Breakout box installed and processor disconnected. • Fuel pump relay disconnected. • DVOM on 200,000 ohm scale. • Measure resistance between test Pin 22 and test Pins 40 and 60 at the Breakout box.	10,000 ohms or greater ▶	GO to J10.
	Less than 10,000 ohms ▶	SERVICE short in the fuel pump circuit. RERUN Quick Test.
J10 CHECK FOR SHORT TO POWER		
• Key Off, wait 10 seconds. • Leave Breakout box installed and disconnect processor. • Fuel pump relay disconnected. • DVOM on 200,000 ohm scale. • Measure resistance between test Pin 22 and test Pins 37 and 57 at the Breakout box.	10,000 ohms or greater ▶	RECONNECT fuel pump relay. GO to J11.
	Less than 10,000 ohms ▶	SERVICE short to power in the fuel pump circuit. RECONNECT processor. ATTEMPT to start vehicle. If vehicle fails to start, REPLACE processor. RERUN Quick Test.
J11 FINAL SYSTEM CHECK		
• Leave Breakout box installed and processor disconnected. • Connect jumper wire from test Pin 22 to test Pin 40 or 60 at the Breakout box. • DVOM on 20V scale. • Key On, engine Off. • Measure voltage between chassis ground and power-to-pump(s) circuit at fuel pump relay.	10.5V or greater ▶	REPLACE processor. RERUN Quick Test.
	Less than 10.5V ▶	REPLACE fuel pump relay. RECONNECT processor and RERUN Quick Test.

Fuel Pump Circuit — (Inertia Switch) | Pinpoint Test | J

TEST STEP	RESULT ▶	ACTION TO TAKE
J15 FUEL PUMP RELAY CHECK		
• Key-Off. • Remove fuel pump relay. • Does fuel pump turn off?	Yes ▶	REPLACE fuel pump relay. RERUN Quick Test.
	No ▶	SERVICE short to power to pumps circuit.

EGR On/Off Control | Pinpoint Test | KA

KA

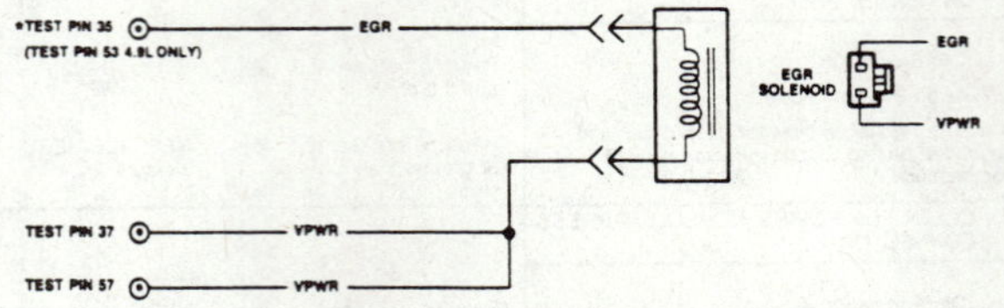

*TEST PINS LOCATED ON BREAKOUT BOX.
ALL HARNESS CONNECTORS VIEWED INTO MATING SURFACE.

STOP-WARNING

You should enter this Pinpoint Test only when a service code 34 is received in Quick Test Step 5.0 or when directed here from Diagnostic by Symptom in the Engine Supplement Section.

To prevent the replacement of good components, be aware that the following non-EEC areas may be at fault:

- Air or Vacuum Leaks
- EGR Flow Restrictions
- EGR Value

NOTE: Code 34 may be the result of high volume exhaust vent system (reduces backpressure). If this is suspected, perform the test in a well-ventilated area without exhaust vent connected.

This Pinpoint Test is intended to diagnose only the following:

- Circuits: EGR and VPWR
- EGR Solenoid
- Presence of Manifold Vacuum
- Processor Assembly

EGR On/Off Control — Pinpoint Test — KA

TEST STEP	RESULT ▶	ACTION TO TAKE
KA1 ENTER OUTPUT STATE CHECK (REFER TO APPENDIX)		
NOTE: Do not use STAR tester for this step, use a VOM DVOM. • Key Off, wait 10 seconds. • DVOM on 20V scale. • Connect DVOM negative test lead to STO at the Self-Test connector and positive test lead to battery positive. • Jumper STI to signal return at the Self-Test connector. • Perform Key On, Engine Off Self-Test until the completion of the Continuous Test Codes. • DVOM will indicate zero volts. • Depress and release the throttle. • Did DVOM reading change to a high voltage reading?	Yes ▶ No ▶	REMAIN in Output State Check. GO to KA2. DEPRESS throttle to WOT and release. If STO voltage does not go high, GO to Pinpoint Test Step Q40. Leave equipment hooked up.
KA2 CHECK EGR ON/OFF CONTROL SOLENOID ELECTRICAL OPERATION		
• DVOM on 20V scale. • Connect DVOM positive test lead to VPWR circuit on EGR solenoid and negative test lead to EGR output circuit. • While observing DVOM, depress and release the throttle several times to cycle output On and Off. • EGR output cycles On and Off.	Yes ▶ No ▶	GO to KA3. REMOVE jumper. GO to KA5.
KA3 CHECK SOLENOID FOR VACUUM CYCLING		
• Install vacuum pump to the solenoid vacuum supply port and install a vacuum gauge to the output port. Apply 6 in. Hg minimum. • While cycling outputs On and Off (by depressing and releasing throttle) observe the vacuum gauge at the output. **NOTE: Maintain vacuum at source.**	Vacuum output cycles On and Off ▶ Vacuum output does not cycle On and Off ▶	GO to KA4. REPLACE solenoid. RERUN Quick Test.

EGR On/Off Control — Pinpoint Test — KA

TEST STEP	RESULT ▶	ACTION TO TAKE
KA4 CHECK MANIFOLD VACUUM LINES FOR BLOCKAGE OR LEAKS		
• Vacuum lines disconnected at solenoid. • Start engine • Check for vacuum	Vacuum present ▶ No vacuum present ▶	EEC-IV system OK. REFER to Section 6. SERVICE vacuum source blockage or leak. RERUN Quick Test.
KA5 MEASURE EGR SOLENOID RESISTANCE		
• Key Off, wait 10 seconds. • DVOM on 200 ohm scale. • Disconnect EGR solenoid connector and measure solenoid resistance.	Resistance is between 65 and 110 ohms ▶ Resistance is less than 65 ohms or greater than 110 ohms ▶	CONNECT EGR solenoid. GO to KA6. REPLACE EGR solenoid. RERUN Quick Test.
KA6 CHECK VOLTAGE OF VPWR CIRCUIT		
• Key On, Engine Off. • DVOM on 20V scale. • Measure voltage between VPWR circuit at the EGR solenoid vehicle harness connector and battery ground.	Voltage reading is less than 10.5V ▶ Voltage reading is 10.5V or greater ▶	SERVICE harness open circuit. RERUN Quick Test. GO to KA7.
KA7 CHECK CONTINUITY OF EGR CIRCUIT		
• Key Off, wait 10 seconds. • Disconnect processor 60 Pin connector and inspect for damaged pins, corrosion, loose wires. Service as necessary. • Connect Breakout box to harness. Leave processor disconnected. • DVOM on 200 ohm scale. • Measure resistance between test Pin 35* at the Breakout box and EGR circuit at vehicle harness connector. *Test Pin 53 for 4.9L engine, truck only.	Resistance reading is 5 ohms or greater ▶ Resistance reading is less than 5 ohms ▶	SERVICE open circuit. RERUN Quick Test. GO to KA8.

EGR On/Off Control — Pinpoint Test — KA

TEST STEP	RESULT ▶	ACTION TO TAKE
KA8 CHECK FOR SHORT TO GROUND		
• Key Off, wait 10 seconds. • Leave Breakout box installed and processor disconnected. • Disconnect EGR solenoid. • DVOM on 200,000 ohm scale. • Measure resistance between test Pin 35* and test Pins 40, 46 and 60 at the Breakout box. *Test Pin 53 for 4.9L engine, truck only.	Resistance reading is less than 10,000 ohms ▶ Resistance reading is 10,000 ohms or greater ▶	SERVICE short to ground. RERUN Quick Test. GO to KA9.
KA9 CHECK FOR SHORT TO POWER		
• Key Off, wait 10 seconds. • DVOM on 200,000 ohm scale. • Leave Breakout box installed and processor disconnected. • EGR solenoid disconnected. • Measure resistance between test Pin 35* and test Pins 37 and 57 at the Breakout box. *Test Pin 53 for 4.9L engine, truck only.	Resistance reading is 10,000 ohms or greater ▶ Resistance reading is less than 10,000 ohms ▶	REPLACE Processor. RERUN Quick Test. SERVICE short to power. RERUN Quick Test. If code is repeated, REPLACE processor.

Air Management System — Pinpoint Test — KC

KC

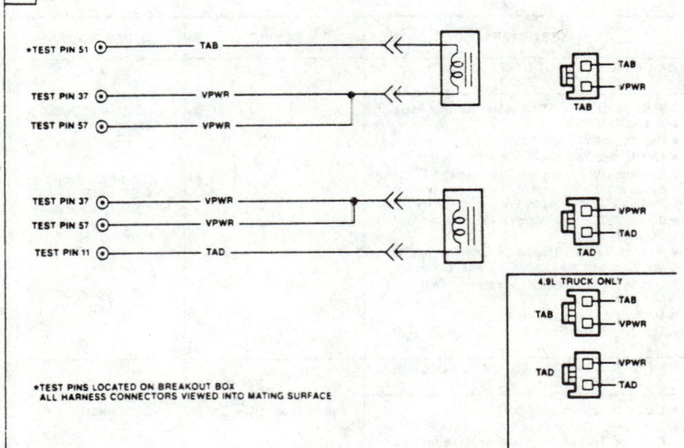

STOP-WARNING

You should enter this Pinpoint Test only when a service code 44, 45, 46, 94, 95, 96, 81 or 82 is received in Quick Test Step 3.0 or 5.0.

To prevent the replacement of good components, be aware that the following non-EEC areas may be at fault:

- Thermactor System
  - — Belt
  - — Pump
  - — Valve

This Pinpoint Test is intended to diagnose only the following:

- TAB and TAD Solenoid Valve Assemblies
- Harness Circuits: TAB, TAD and VPWR
- Vacuum Supply
- Processor Assembly

Air Management System — Pinpoint Test — KC

TEST STEP	RESULT ▶	ACTION TO TAKE
KC1 SERVICE CODES 44 (94), 45 (95) AND 46 (96) VERIFY VACUUM LINE ROUTING		
• Verify proper vacuum line routing to the TAB/TAD solenoids and to the bypass diverter valve. Refer to VECI decal. • Check for kinked or blocked vacuum lines. • Check for kinked or blocked air hoses. • Check for disconnected vacuum lines. • Are visual checks satisfactory?	No ▶	SERVICE routing or faults. RERUN Quick Test
	Yes ▶	Service code 44 (94): GO to KC4 Service code 45 (95): GO to KC2 Service code 46 (96): GO to KC3
KC2 ATTEMPT TO ELIMINATE SERVICE CODE 45 (95) (TAD ONLY)		
• Disconnect vacuum line on diverter valve and cap vacuum line. • Key Off, wait 10 seconds. • Repeat Engine Running Self-Test and record service codes. • Is code 45 (95) present?	Yes ▶	EEC-IV system OK. REFER to Section 3 for diverter valve or check valve diagnostics.
	No ▶	GO to KC4
KC3 ATTEMPT TO ELIMINATE SERVICE CODE 46 (96) (TAB ONLY)		
• Disconnect vacuum line on bypass valve and cap vacuum line. • Key Off, wait 10 seconds. • Repeat Engine Running Self-Test and record codes. • Is code 46 (96) present?	Yes ▶	EEC-IV system OK. REFER to Section 3 for bypass valve diagnostics.
	No ▶	GO to KC4

Air Management System — Pinpoint Test — KC

TEST STEP	RESULT ▶	ACTION TO TAKE
KC4 ENTER OUTPUT STATE CHECK (REFER TO APPENDIX)		
NOTE: Do not use STAR tester for this Step, use a VOM DVOM. • Key Off, wait 10 seconds. • DVOM on 20V scale. • Connect DVOM negative test lead to STO circuit at the Self-Test connector and positive test lead to battery positive. • Jumper STI circuit to signal return at the Self-Test connector. • Perform Key On, Engine Off Self-Test until the completion of the Continuous Test Codes. • DVOM will indicate zero volts when Test is complete. • Depress and release the throttle. • Did DVOM reading change to a high voltage reading?	Yes ▶	REMAIN in Output State Check. GO to KC5
	No ▶	DEPRESS throttle to WOT and RELEASE. If STO voltage does not go high, GO to Pinpoint Test Step Q40. Leave equipment hooked up.
KC5 CHECK TAB/TAD SOLENOID ELECTRICAL OPERATION		
• DVOM on 20V scale. • Connect DVOM positive test lead to VPWR circuit on TAB solenoid and negative test lead to TAB circuit on TAB solenoid. • While observing DVOM depress and release the throttle several times (to cycle output On and Off). • Repeat for TAD solenoid. Connect positive test lead to VPWR circuit on TAD solenoid and negative test lead to TAD circuit on TAD solenoid. • Solenoids cycle On and Off.	Both outputs cycle On and Off ▶	GO to KC6
	Either output does not cycle On and Off ▶	REMOVE jumper. GO to KC8
KC6 CHECK TAB/TAD SOLENOID FOR VACUUM CYCLING		
• Install vacuum pump to the TAB solenoid vacuum supply port and install a vacuum gauge to the output port. • While cycling outputs On and Off (by depressing and releasing throttle), observe the vacuum gauge at the output. **NOTE: Maintain vacuum at source.** • Repeat for TAD solenoid. Connect vacuum pump to the TAD solenoid vacuum supply port and connect a vacuum gauge to the output port. • Cycle output On and Off.	Both vacuum outputs cycle On and Off ▶	GO to KC7
	Either vacuum output does not cycle On and Off ▶	REPLACE solenoid assembly. RERUN Quick Test.

Air Management System — Pinpoint Test — KC

TEST STEP	RESULT ▶	ACTION TO TAKE
KC7 CHECK MANIFOLD VACUUM LINES FOR BLOCKAGE OR LEAKS		
• Vacuum lines disconnected at TAD/TAB solenoids. • Start engine. • Check for vacuum.	Vacuum present ▶	EEC-IV system OK. REFER to Section 3 for Thermactor valve and air pump diagnostics.
	No vacuum present ▶	SERVICE vacuum source blockage or leak. RERUN Quick Test
KC8 MEASURE TAB/TAD SOLENOID RESISTANCE		
• Key Off, wait 10 seconds. • DVOM on 200 ohm scale. • Disconnect TAB solenoid connector and measure solenoid resistance. • Disconnect TAD solenoid connector and measure solenoid resistance.	Both resistances are between 50 and 100 ohms ▶	CONNECT TAB/TAD solenoids. GO to KC9
	Either resistance is less than 50 ohms or greater than 100 ohms ▶	REPLACE TAB/TAD solenoid assembly. RERUN Quick Test.
KC9 CHECK VOLTAGE OF VPWR CIRCUIT		
• Key On, engine Off. • DVOM on 20V scale. • Measure voltage between TAB solenoid VPWR circuit and battery ground. • Repeat for TAD solenoid.	Both voltage readings are greater than 10.5V ▶	GO to KC10
	Either voltage reading is less than 10.5V ▶	SERVICE harness circuit open. RERUN Quick Test.
KC10 CHECK CONTINUITY OF TAB AND TAD CIRCUITS		
• Key Off, wait 10 seconds. • Disconnect processor 60 Pin connector and inspect for damaged pins, corrosion, loose wires. Service as necessary. • Connect Breakout box to harness. Leave processor disconnected. • DVOM on 200 ohm scale. • Measure resistance between test Pin 51 at Breakout box and TAB circuit at vehicle harness connector. • Measure resistance between test Pin 11 at the Breakout box and TAD circuit at vehicle harness connector.	Both resistance readings are less than 5 ohms ▶	GO to KC11
	Either resistance reading is 5 ohms or greater ▶	SERVICE harness open circuit. RERUN Quick Test.

Air Management System — Pinpoint Test — KC

TEST STEP	RESULT ▶	ACTION TO TAKE
KC11 CHECK FOR SHORT TO GROUND		
• Key Off, wait 10 seconds. • DVOM on 200,000 ohm. • Leave Breakout box installed and processor disconnected. • Disconnect TAB/TAD solenoids. • Measure resistance between test Pin 51 and test Pins 37 and 57, and between test Pin 11 and test Pins 37 and 57 at the Breakout box.	All resistance readings 10,000 ohms or greater ▶	GO to KC12
	Any resistance reading less than 10,000 ohms ▶	SERVICE short to power. RERUN Quick Test.
KC12 CHECK FOR SHORT TO GROUND		
• Key Off, wait 10 seconds. • DVOM on 200,000 ohm scale. • Leave Breakout box installed and processor disconnected. • TAB/TAD solenoids disconnected. • Measure resistance between test Pins 51 and test Pins 40, 46, and 60 and between test Pin 11 and test Pins 40, 46, and 60 at the Breakout box.	All resistance readings less than 10,000 ohms ▶	SERVICE short to ground. REMOVE Breakout box. RECONNECT harness to processor. RERUN Quick Test. If code is repeated, REPLACE processor.
	Any resistance readings 10,000 ohms or greater ▶	REPLACE processor. REMOVE Breakout box. RECONNECT harness to processor. RERUN Quick Test.

Canister Purge (CANP) | Pinpoint Test | KD

KD

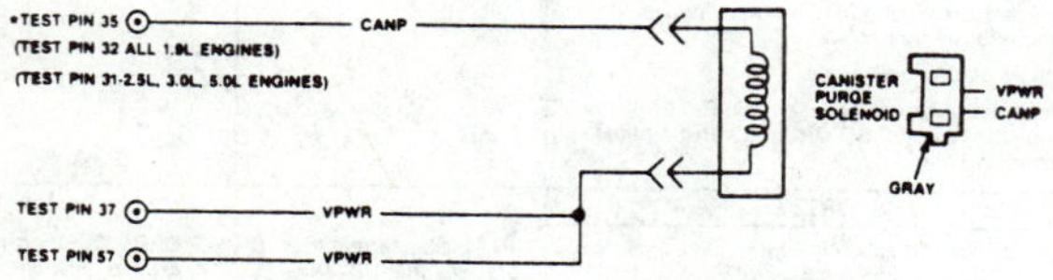

*TEST PINS LOCATED ON BREAKOUT BOX.
ALL HARNESS CONNECTORS VIEWED INTO MATING SURFACE.

STOP-WARNING

You should enter this Pinpoint Test only when a service code 85 is received in Quick Test Step 3.0 or when you are directed here from a fuel control Pinpoint Test and Diagnostics by Symptom in the Engine Supplement Section.

This Pinpoint Test is intended to diagnose only the following:

- Harness Circuits: CANP and VPWR
- Processor Assembly

Canister Purge (CANP) | Pinpoint Test | KD

TEST STEP	RESULT ▶	ACTION TO TAKE
KD1 ENTER OUTPUT STATE CHECK (REFER TO APPENDIX)		
NOTE: Do not use STAR tester for this step, use VOM/DVOM. • Key Off, wait 10 seconds. • DVOM on 20V scale. • Connect DVOM negative test lead to STO circuit at Self-Test connector and positive test lead to battery positive. • Jumper STI circuit to signal return at the Self-Test connector. • Perform Key On, Engine Off Self-Test until the completion of the Continuous Test Codes. • DVOM will indicate zero volts when test is completed. • Depress and release the throttle. • Did DVOM reading change to a high voltage reading?	Yes ▶ No ▶	REMAIN in Output State Check. GO to KD2. DEPRESS throttle to WOT and release. If STO voltage does not go high, GO to Pinpoint Test Step Q40. Leave equipment hooked up.
KD2 CHECK CANISTER PURGE (CANP) SOLENOID ELECTRICAL OPERATION		
• DVOM on 20V scale. • Connect DVOM positive test lead to VPWR on CANP solenoid and negative test lead to CANP output circuit on the solenoid. • While observing DVOM depress and release the throttle several times (to cycle output On and Off). • CANP solenoid cycles On and Off?	Yes ▶ No ▶	GO to KD3. REMOVE jumper. GO to KD4.
KD3 CHECK CANISTER PURGE SOLENOID FOR HOLDING VACUUM		
• Key Off, wait 10 seconds. • Remove jumper from STI to Signal Return. • Disconnect vacuum hose at canister purge solenoid at PCV side. • Apply 16 in. Hg to CANP solenoid. • Does CANP solenoid hold vacuum for 20 seconds?	Yes ▶ No ▶	EEC-IV system OK. Refer to Shop Manual Group 24. REPLACE CANP solenoid. RERUN Quick Test. If code 42 is still present, GO to Section 3.

Canister Purge (CANP) | Pinpoint Test | KD

TEST STEP	RESULT ▶	ACTION TO TAKE
KD4 MEASURE CANP SOLENOID RESISTANCE		
• Key Off, wait 10 seconds. • DVOM on 200 ohm scale. • Disconnect CANP solenoid connector and measure solenoid resistance.	Resistance is between 40 and 90 ohms ▶ Resistance is less than 40 ohms or greater than 90 ohms ▶	CONNECT CANP solenoid. GO to KD5. REPLACE CANP solenoid. RERUN Quick Test.
KD5 CHECK VOLTAGE OF VPWR CIRCUIT		
• Key On, engine Off. • DVOM on 20V scale. • Measure voltage between VPWR at the CANP solenoid vehicle harness connector and battery ground.	Voltage reading is less than 10.5V ▶ Voltage reading is 10.5V or greater ▶	SERVICE harness open circuit. RERUN Quick Test. GO to KD6.
KD6 CHECK CONTINUITY OF CANP CIRCUIT		
• Key Off, wait 10 seconds. • Disconnect processor and inspect both 60 Pin connectors and inspect for damaged pins, corrosion, loose wires. Service as necessary. • Connect Breakout box to harness. Leave processor disconnected. • DVOM on 200 ohm scale. • Measure resistance between test Pin 35* at the Breakout box and CANP vehicle harness connector. *Test Pin 32 for 1.9L engine. *Test Pin 31 for 2.5L, 3.0L, 5.0L engines.	Resistance reading is 5 ohms or greater ▶ Resistance reading is less than 5 ohms ▶	SERVICE open circuit. RERUN Quick Test. GO to KD7.

Canister Purge (CANP) | Pinpoint Test | KD

TEST STEP	RESULT ▶	ACTION TO TAKE
KD7 CHECK FOR SHORT TO GROUND		
• Key Off, wait 10 seconds. • Leave Breakout box to harness. Leave processor disconnected. • Disconnect CANP solenoid. • DVOM on 200,000 ohm scale. • Measure resistance between test Pin 35* and test Pins 40, 46 and 60 at the Breakout box. *Test Pin 32 for 1.9L engine. *Test Pin 31 for 2.5L, 3.0L, 5.0L engines.	All resistance readings are 10,000 ohms or greater ▶ Any resistance reading less than 10,000 ohms ▶	GO to KD8. SERVICE short to power. RERUN Quick Test.
KD8 CHECK FOR SHORT TO POWER		
• Key Off, wait 10 seconds. • DVOM on 200,000 ohm scale. • Leave Breakout box installed and processor disconnected. • CANP solenoid disconnected. • Measure resistance between test Pin 35* and test Pins 37 and 57 at the Breakout box. *Test Pin 32 for 1.9L engine. *Test Pin 31 for 2.5L, 3.0L, 5.0L engines.	All resistance readings are less than 10,000 ohms ▶ Any resistance reading is 10,000 ohms or greater ▶	SERVICE short to ground. REPEAT Quick Test. If code is repeated, REPLACE processor. RERUN Quick Test. REMOVE Breakout box. REPLACE processor. RERUN Quick Test.

Idle Speed Control (By-Pass Air) — Pinpoint Test — KE

KE

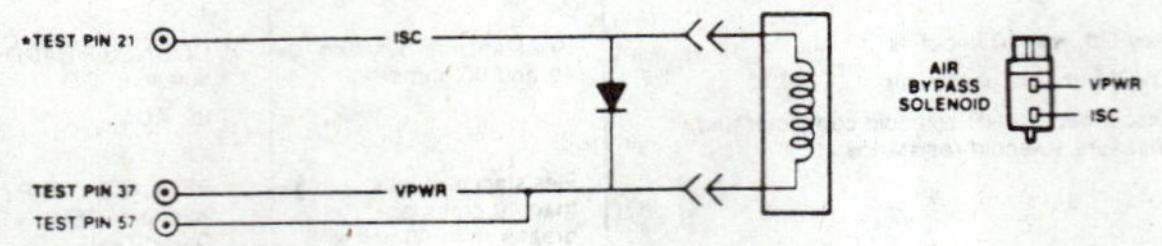

•TEST PINS LOCATED ON BREAKOUT BOX.
ALL HARNESS CONNECTORS VIEWED INTO MATING SURFACE

STOP-WARNING

You should enter this Pinpoint Test only when a service code 12 or 13 is received in Quick Test Step 5.0 or when directed here from Diagnostic by Symptom in the Engine Supplement Section.

CAUTION: If the engine exhibits rough running and/or idle, correct these conditions before diagnosing Idle Speed Control (ISC). Rough running misses may be caused by:

- **Ignition System**
  Refer to Section 15
- **Fuel System**
  Refer to Pinpoint Test Steps HA1 through HA7.
  Refer to Pinpoint Test Steps HE1 through HE7.
  Refer to Pinpoint Test Steps HG1 through HG6.
  Refer to Pinpoint Test Steps HL2 through HL7.
- **EGR System, Section 6**

To prevent the replacement of good components, be aware that the following non-EEC areas may be at fault:

- Engine not up to operating temperature
- Engine over operating temperature
- Improper Idle Speed Throttle Stop Adjustment
- A/C input (electrical problem)
- Throttle Speed Control Linkage

This Pinpoint Test is intended to diagnose only the following:

- RPM in Self-Test only
- ISC Actuator
- Harness Circuits ISC and VPWR
- Processor Assembly

Idle Speed Control (By-Pass Air) — Pinpoint Test — KE

TEST STEP	RESULT ▶	ACTION TO TAKE
KE1 RPM DROP		
• Key Off. • Connect engine tachometer. • Start engine. • Disconnect ISC harness connector. • Does rpm drop or stall?	Yes ▶	GO to KE2
	No ▶	GO to KE3
KE2 CHECK FOR EGR CODES		
• Are service codes 31, 32, 33 or 34 present?	Yes ▶	GO to Quick Test Step 5.0 for appropriate Pinpoint Test
	No ▶	SERVICE next code or if code 12 or 13 only, GO to KE3.
KE3 MEASURE ISC SOLENOID RESISTANCE		
• Key Off. • DVOM on 200 ohm scale. • Disconnect ISC solenoid connector and measure solenoid resistance.	Resistance is between 7 and 13 ohms ▶	GO to KE4
	Resistance is less than 7 ohms or greater than 13 ohms ▶	REPLACE ISC solenoid. RERUN Quick Test.
KE4 ISC SHORT TO CASE (GROUND) CHECK		
• Key Off. • DVOM on 200,000 ohm scale. • ISC solenoid disconnected. • Measure resistance from either ISC pin to ISC housing.	10,000 ohms or greater ▶	GO to KE5
	Less than 10,000 ohms ▶	REPLACE ISC solenoid. RERUN Quick Test.
KE5 CHECK VOLTAGE OF VPWR CIRCUIT		
• Key On, Engine Off. • DVOM on 20V scale. • ISC solenoid disconnected. • Measure voltage between VPWR at the ISC solenoid harness connector and battery ground.	Voltage reading is less than 10.5V ▶	SERVICE harness open circuit. RERUN Quick Test.
	Voltage reading is 10.5V or greater ▶	GO to KE6

Idle Speed Control (By-Pass Air) — Pinpoint Test — KE

TEST STEP	RESULT ▶	ACTION TO TAKE
KE6 CHECK CONTINUITY OF ISC CIRCUIT		
• Key Off, wait 10 seconds. • ISC solenoid disconnected. • Disconnect processor and inspect both 60 Pin connectors for damaged pins, corrosion, loose wires. Service as necessary. • Connect Breakout box to harness. Leave processor disconnected. • DVOM on 200 ohm scale. • Measure resistance between test Pin 21 at the Breakout box and ISC circuit at vehicle harness connector.	Resistance reading is 5 ohms or greater ▶	SERVICE open circuit. RERUN Quick Test.
	Resistance reading is less than 5 ohms ▶	GO to KE7
KE7 CHECK FOR SHORT TO GROUND		
• Key Off, wait 10 seconds. • Leave Breakout box installed and processor disconnected. • ISC solenoid disconnected. • DVOM on 200,000 ohm scale. • Measure resistance between test Pin 21 and test Pins 40, 46 and 60 at the Breakout box.	All resistance readings are less than 10,000 ohms ▶	SERVICE short to ground. RERUN Quick Test.
	Any resistance reading is 10,000 ohms or greater ▶	GO to KE8
KE8 CHECK FOR SHORT TO POWER		
• Key Off, wait 10 seconds. • DVOM on 200,000 ohm scale. • ISC solenoid disconnected. • Leave Breakout box to connected harness. Leave processor disconnected. • Measure resistance between test Pin 21 and test Pin 37 at the Breakout box.	Resistance reading is 10,000 ohms or greater ▶	GO to KE9
	Resistance reading is less than 10,000 ohms ▶	SERVICE short to power. RERUN Quick Test.
KE9 CHECK FOR ISC SIGNAL AT THE PROCESSOR		
• Connect processor. • Connect ISC actuator. • Vehicle prepared for Quick Test. • DVOM on 20V scale. • Connect DVOM between test Pin 21 and test Pin 40. • Start engine and observe the DVOM during Engine Running Quick Test.	Meter reading varies during Quick Test ▶	REPLACE ISC actuator. RERUN Quick Test.
	Meter reading does not vary during Quick Test ▶	GO to KE10

Idle Speed Control (By-Pass Air) — Pinpoint Test — KE

TEST STEP	RESULT ▶	ACTION TO TAKE
KE10 CODE 13: RPM DOES NOT DROP TO LESS THAN 1500 AT END OF TEST		
• Disconnect ISC harness connector. • Connect engine tachometer. • Repeat Engine Running Quick Test. • At end of test, record service codes for future use. • Does rpm remain below 1500 rpm during test?	Yes ▶	REPLACE processor. RERUN Quick Test.
	No ▶	CHECK engine vacuum hoses. REFER to VECI decal. VERIFY curb idle. CHECK that throttle plates are fully closed. CHECK throttle linkage and/or speed control linkage for binding. If OK, REPLACE ISC actuator. RERUN Quick Test.
KE11 CURB IDLE CHECK		
• Key Off, wait 10 seconds. • Deactivate Self-Test. • Run engine at 2000 rpm for 2 minutes or until inlet radiator hose is hot and pressurized. • Key Off, wait 10 seconds. • Activate Self-Test. • Run Engine Running Self-Test. • Is code 17 still present?	No ▶	SERVICE other codes as necessary.
	Yes ▶	INSPECT throttle body and air inlet for contamination. SERVICE as necessary. If OK, ADJUST curb idle (REFER to Section 4 for procedure). RERUN Quick Test.

Idle Speed Control (DC Motor/Idle Tracking Switch Assembly) — Pinpoint Test KF

KF

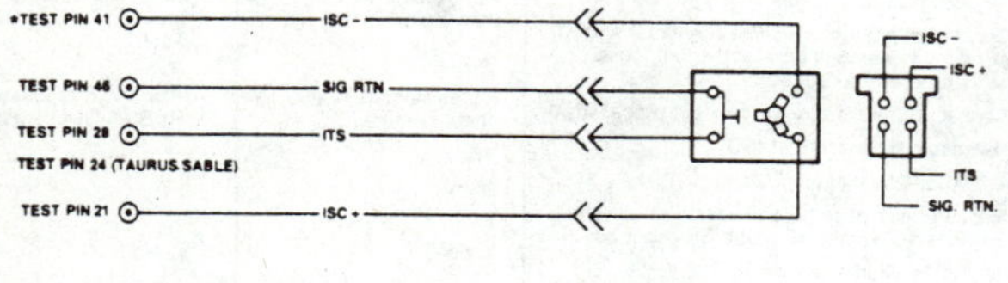

*TEST PINS LOCATED ON BREAKOUT BOX.
ALL HARNESS CONNECTORS VIEWED INTO MATING SURFACE.

STOP-WARNING

You should enter this Pinpoint Test only when a Service Code 12, 13, 58 or 68 is received in Quick Test Step 3.0, 5.0 or 6.0 or when directed here from Diagnostics by Symptom in the Engine Supplement Section.

To prevent the replacement of good components, be aware that the following non-EEC areas may be at fault:

- Anti-Diesel Speed too High
- Basic Engine
- Vacuum Leaks
- Throttle Sticking or on High Cam

This Pinpoint Test is intended to diagnose only the following:

- DC Motor
- Idle Tracking Switch
- Harness circuits: ISC +, ISC –, Signal Return and ITS
- Processor Assembly

Idle Speed Control (DC Motor/Idle Tracking Switch Assembly) — Pinpoint Test KF

TEST STEP	RESULT ▶	ACTION TO TAKE
KF1 SERVICE CODE 68 SIMULATE NO THROTTLE CONTACT TO FORCE A CODE 11 IN KEY ON — ENGINE OFF TEST		
• Key Off, wait 10 seconds • Move throttle away from ISC DC motor shaft • Perform Key On Engine Off Self-Test • Is code 68 present? **NOTE: Ignore all other codes at this point.**	Yes ▶	RELEASE throttle and GO to KF3.
	No ▶	RELEASE throttle and GO to Pinpoint Test Step KF2.

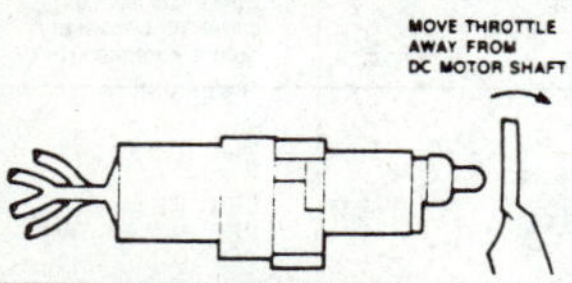

TEST STEP	RESULT ▶	ACTION TO TAKE
KF2 RETRACT ISC DC MOTOR		
• Key Off, wait 10 seconds • Leave Breakout box installed. Leave processor disconnected • Jumper test Pin 41 to test Pin 1 at the Breakout box and test Pin 21* at the Breakout box to battery ground (to retract DC motor) • Observe DC motor **NOTE: Do not leave jumper wire in test pins if the motor will not move; it will damage Breakout box wiring.**	Motor does not retract ▶	REMOVE jumper and GO to KF5.
	Motor retracts and still contacts throttle lever ▶	GO to ISC Adjustments, Section 4.
	Motor retracts and does not contact throttle lever ▶	REMOVE jumper and GO to KF26.

Idle Speed Control (DC Motor/Idle Tracking Switch Assembly) — Pinpoint Test KF

TEST STEP	RESULT ▶	ACTION TO TAKE
KF3 SIMULATE A CLOSED CONTACT CONDITION IN KEY ON — ENGINE OFF TEST		
• Key Off, wait 10 seconds. • Disconnect harness from ISC. • Connect jumper between ISC vehicle harness connector Pins 3 and 4. • Install jumper as shown. • Perform Key On, Engine Off Self-Test. • Is code 68 present?	Yes ▶	GO to KF4.
	No ▶	REPLACE ISC DC motor. RERUN Quick Test.

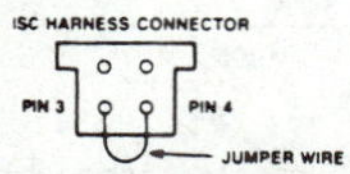

CAUTION: Do not short any other pins.

TEST STEP	RESULT ▶	ACTION TO TAKE
KF4 CHECK ITS AND SIGNAL RETURN CIRCUITS FOR CONTINUITY		
• Key Off, wait 10 seconds. • Disconnect processor 60 Pin connector and inspect for damaged pins, corrosion, loose wires. Service as necessary. • Connect Breakout box to harness. Leave processor disconnected. • DVOM on 200 ohm scale. • Measure resistance between test Pin 46 at the Breakout box and Signal Return circuit at ISC vehicle harness connector and between test Pin 28* at the Breakout box and ITS circuit at ISC vehicle harness connector. *Test Pin 24 on Taurus Sable vehicles.	Either resistance 5 ohms or greater ▶	SERVICE faulty circuit. RERUN Quick Test.
	Both resistances less than 5 ohms ▶	REPLACE processor. RERUN Quick Test.

Idle Speed Control (DC Motor/Idle Tracking Switch Assembly) — Pinpoint Test KF

TEST STEP	RESULT ▶	ACTION TO TAKE
KF5 CHECK TP AND ISC CIRCUITS FOR CONTINUITY		
• Key Off, wait 10 seconds. • Disconnect processor 60 Pin connector and inspect for damaged pins, corrosion, loose wires. Service as necessary. • Connect Breakout box. Leave processor disconnected. • Disconnect harness connector for ISC motor. • DVOM on 200 ohm scale. • Measure resistance between test Pin 41 at the Breakout box and ISC – circuit at vehicle harness connector and between test Pin 21 at the Breakout box and ISC + circuit at vehicle harness connector.	Both resistance readings are less than 5 ohms ▶	GO to KF6.
	Either resistance reading is 5 ohms or greater ▶	SERVICE faulty circuits. RERUN Quick Test.
KF6 CHECK ISC + AND ISC – CIRCUITS FOR SHORTS TO POWER		
• Key On, Engine Off • Leave Breakout box installed. Leave processor disconnected. • Harness disconnected from ISC motor. • DVOM on 20V scale • Measure voltage between test Pin 41 and test Pins 40 and 60 at the Breakout box. Measure voltage between test Pin 21 and test Pins 40 and 60 at the Breakout box.	All voltage readings are less than 1V ▶	GO to KF7.
	Any voltage reading is 1V or more ▶	SERVICE circuit short to power. RERUN Quick Test. If code 12 is still present, REPLACE processor.
KF7 CHECK ISC + AND ISC – CIRCUITS FOR SHORTS TO GROUND		
• Key Off, wait 10 seconds. • Leave Breakout box installed. Leave processor disconnected. • Harness disconnected from ISC motor. • DVOM on 200,000 ohm scale. • Measure resistance between test Pin 41 and test Pins 40, 46 and 60 at the Breakout box. Measure resistance between test Pin 21 and test Pins 40, 46 and 60 at the Breakout box.	All resistance readings greater than 10,000 ohms ▶	GO to KF8.
	Any resistance reading is 10,000 ohms or less ▶	SERVICE faulty circuit. RERUN Quick Test. If code 12 is still present, REPLACE processor.

Idle Speed Control (DC Motor/Idle Tracking Switch Assembly) — Pinpoint Test KF

TEST STEP	RESULT ▶	ACTION TO TAKE
KF8 PREPARE IDLE SPEED CONTROL DC MOTOR FOR OPERATION		
• Key Off, wait 10 seconds. • Leave Breakout box installed. Leave processor disconnected. • Harness disconnected from ISC motor. • DVOM on 20V scale. • Connect DVOM positive test lead to test Pin 1 at the Breakout box and negative test lead to ground.	Voltage reading is 10.5V or greater ▶ Voltage reading is less than 10.5V ▶	GO to KF9. SERVICE open in KAPWR circuit. RERUN Quick Test. If code 12 is still present, GO to KF9.
KF9 ISC DC MOTOR OPERATION CHECK		
• Key Off, wait 10 seconds. • Leave Breakout box installed. Leave processor disconnected. • Connect ISC DC motor to harness connector. • Jumper test Pin 21 to test Pin 1 at the Breakout box and test Pin 41 at the Breakout box to ground (to extend DC motor). • Does DC motor shaft extend 5 cm (2 inches) or more? 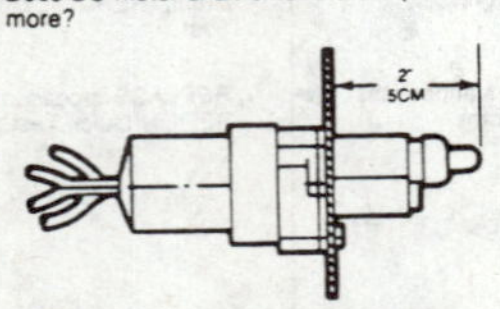	Yes ▶ No ▶	GO to KF10. REPLACE DC motor RERUN Quick Test.
KF10 RETRACT ISC DC MOTOR		
• Key Off, wait 10 seconds. • Leave Breakout box installed. Leave processor disconnected. • Jumper test Pin 41 to test Pin 1 at the Breakout box and test Pin 21 at the Breakout box to ground (to retract DC motor). • Does ISC DC motor retract?	Yes ▶ No ▶	GO to KF11. REPLACE DC motor RERUN Quick Test.

Idle Speed Control (DC Motor/Idle Tracking Switch Assembly) — Pinpoint Test KF

TEST STEP	RESULT ▶	ACTION TO TAKE
KF11 EXTEND ISC DC MOTOR		
• Key Off, wait 10 seconds. • Leave Breakout box installed. Leave processor disconnected. • Jumper test Pin 21 to test Pin 1 at the Breakout box and test Pin 41 at the Breakout box to ground (to extend DC motor). • Does ISC DC motor extend?	No ▶ Yes ▶	REPLACE DC motor. RERUN Quick Test. REPLACE processor. RERUN Quick Test.
KF12 SERVICE CODE 58: SIMULATE THROTTLE CONTACT TO FORCE A SERVICE CODE 68 IN KEY ON — ENGINE OFF TEST		
• Key Off, wait 10 seconds. • Press on ISC DC motor shaft to simulate throttle contact. • With a force pushing on DC motor shaft, perform Key On, Engine Off Self-Test. **NOTE: Ignore all other codes at this point.** • Is code 68 present?	Yes ▶ No ▶	GO to Pinpoint Test Step KF5. GO to KF13.
KF13 RETRACT ISC DC MOTOR		
• Key Off, wait 10 seconds. • Leave Breakout box installed. Leave processor disconnected. • Jumper test Pin 41 to test Pin 1 at the Breakout box and test Pin 21 at the Breakout box to ground (to retract DC motor). • Does ISC DC motor retract?	Yes ▶ No ▶	GO to ISC Adjustments, Section 4. GO to KF14.
KF14 IDLE TRACKING SWITCH ALWAYS "NOT TRACKING" (CLOSED)		
• Key Off, wait 10 seconds. • Reconnect processor. • Disconnect vehicle harness from ISC. • Perform Key On, Engine Off Self-Test. **NOTE: Ignore all other codes at this point.** • Is code 68 present?	Yes ▶ No ▶	REPLACE DC motor RERUN Quick Test. GO to KF15.

Idle Speed Control (DC Motor/Idle Tracking Switch Assembly) — Pinpoint Test KF

TEST STEP	RESULT ▶	ACTION TO TAKE
KF15 CHECK ITS CIRCUIT FOR SHORT TO GROUND		
• Key Off, wait 10 seconds. • Disconnect processor 60 Pin connector and inspect for damaged pins, corrosion, loose wires. Service as necessary. • Connect Breakout box. Disconnect processor. • Harness disconnected at ISC. • DVOM on 200,000 ohm scale. • Measure resistance between test Pin 28* and test Pins 40, 46 and 60 at the Breakout box. *Test Pin 24 Taurus Sable.	Any resistance less than 10,000 ohms ▶ All resistances 10,000 ohms or more ▶	SERVICE short to ground. RERUN Quick Test. REPLACE processor. RERUN Quick Test.
KF16 DC MOTOR CONTROL FAULT ISOLATION: SERVICE CODE 12		
• Are service codes 58, 68, 31 or 41 present?	Yes ▶ No ▶	GO to Quick Test Step 5.0 for appropriate Pinpoint Test. GO to KF17.
KF17 CHECK FOR THROTTLE STICKING		
• Check throttle plates and linkages for binding. • Check speed control for binding (if so equipped).	Yes ▶ No ▶	SERVICE as necessary. RERUN Quick Test. GO to KF18.
KF18 EXTEND ISC DC MOTOR		
• Key Off, wait 10 seconds. • Leave Breakout box installed. Leave processor disconnected. • Jumper test Pin 21 to test Pin 1 at the Breakout box and test Pin 41 at the Breakout box to ground (to extend DC motor). • Does ISC DC motor extend?	No ▶ Yes ▶	REPLACE DC motor. RERUN Quick Test. REPLACE processor. RERUN Quick Test.

Idle Speed Control (DC Motor/Idle Tracking Switch Assembly) — Pinpoint Test KF

TEST STEP	RESULT ▶	ACTION TO TAKE
KF19 ENGINE RUNNING CODES 13 AND/OR 16		
• Prepare system for normal engine operation. • Deactivate Self-Test. • Air conditioner Off. • Run engine. Alternate between 30 seconds at idle and 5 seconds at part throttle modes for 3 minutes. • Is idle speed erratic?	No ▶ Yes ▶	GO to KF20. CHECK for: • Vacuum leaks • Code 22 • Code 31 • Code 41 • Code 58 SERVICE above codes as necessary before continuing.
KF20 CHECK FOR THROTTLE STICKING OR CHOKE POSITION		
• Inspect throttle mechanisms (plate linkage) for sticking or binding. • Inspect choke for sticking or binding. • Inspect for throttle stuck on high cam. • Are above inspections satisfactory?	No ▶ Yes ▶	SERVICE fault. RERUN Quick Test. GO to KF21.
KF21 PERFORM KEY ON — ENGINE OFF TEST. OBSERVE ON DEMAND CODES		
• Perform Key On, Engine Off Test. Record On Demand codes. • Key On, Engine Off. • Is code 68 present?	Yes ▶ No ▶	GO to Pinpoint Test Step KF1. REFER to Section 4 for anti-diesel speed adjustment.
KF25 CHECK ISC – AND ISC – CIRCUITS FOR SHORTS TO GROUND		
• Key Off, wait 10 seconds. • Leave Breakout box installed. Leave processor disconnected. • Harness disconnected from ISC motor. • DVOM on 200,000 ohm scale. • Measure resistance between test Pin 41 and test Pins 40, 46 and 60 at the Breakout box. Measure resistance between test Pin 21 and test Pins 40, 46 and 60 at the Breakout box.	All resistance readings greater than 10,000 ohms ▶ Any resistance reading is 10,000 ohms or less ▶	GO to KF26. SERVICE faulty circuit. RERUN Quick Test. If code 12 is still present, REPLACE processor.

Idle Speed Control (DC Motor/Idle Tracking Switch Assembly) — Pinpoint Test — KF

TEST STEP	RESULT ▶	ACTION TO TAKE
KF26 CHECK ISC + AND ISC − CIRCUITS FOR SHORTS TO POWER		
• Key On, Engine Off. • Leave Breakout box installed. Leave processor disconnected. • Harness disconnected from ISC motor. • DVOM on 20V scale. • Measure voltage between test Pin 41 and test Pins 40 and 60 at the Breakout box. Measure voltage between test Pin 21 and test Pins 40 and 60 at the Breakout box.	All voltage readings are less than 1V ▶ Any voltage reading is 1V or more ▶	GO to KF27. SERVICE circuit short to power. RERUN Quick Test. If code 12 is still present, REPLACE processor.
KF27 EXTEND ISC DC MOTOR		
• Key Off, wait 10 seconds. • Leave Breakout box installed. Leave processor disconnected. • Jumper test Pin 21 to test Pin 1 at the Breakout box and test Pin 41 at the Breakout box to ground (to extend DC motor). • Does ISC DC motor extend?	No ▶ Yes ▶	REPLACE DC motor. RERUN Quick Test. REPLACE processor. RERUN Quick Test.

Inlet Air Solenoid (IAS) — Pinpoint Test — KH

KH

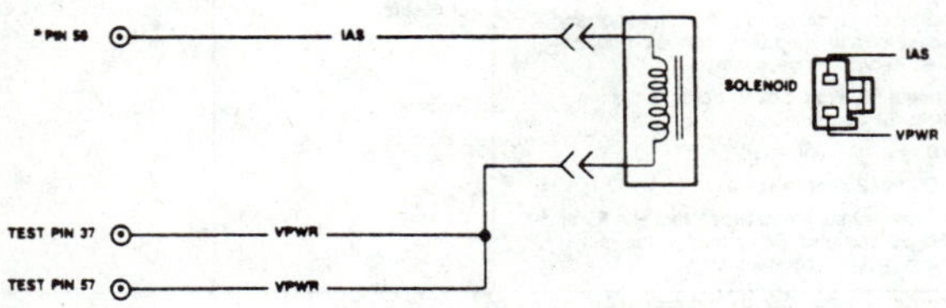

*TEST PINS LOCATED ON BREAKOUT BOX
ALL HARNESS CONNECTORS VIEWED INTO MATING SURFACE

STOP-WARNING

You should enter this Pinpoint Test only when directed here from Diagnostics by Symptom in the Engine Supplement Section.

To prevent the replacement of good components, be aware that the following non-EEC areas may be at fault:

- Air or Vacuum Leaks
- Heat Riser and Stovepipe

This Pinpoint Test is intended to diagnose only the following:

- Circuits IAS and VPWR
- IAS Solenoid
- Presence of Manifold Vacuum
- Processor Assembly

Inlet Air Solenoid (IAS) — Pinpoint Test — KH

TEST STEP	RESULT ▶	ACTION TO TAKE
KH1 SIMULATE SUB-ZERO AMBIENT TEMPERATURE		
• Locate, then disconnect ACT sensor. • Start vehicle. • Does inlet air door close on air cleaner inlet air duct?	Yes ▶ No ▶	IAS is functioning. GO to Diagnostic Routines, Section 2. GO to KH2.
KH2 VACUUM CHECK		
• Key Off. • Disconnect both vacuum hoses from IAS. • Connect both vacuum hoses. • Key On, start vehicle. • Does inlet air door close?	Yes ▶ No ▶	RECONNECT source vacuum hose. GO to KH4. GO to KH3.
KH3 DUCT AND VALVE CHECK		
• Key Off. • Connect vacuum pump to inlet air valve vacuum hose. • Apply vacuum. • Does inlet air door close?	Yes ▶ No ▶	VERIFY proper source vacuum. GO to inlet air duct and valve assembly, Section 8.
KH4 MEASURE IAS SOLENOID RESISTANCE		
• Key Off, wait 10 seconds. • DVOM on 200 ohm scale. • Disconnect IAS solenoid connector. • Measure solenoid resistance.	Resistance is between 65 and 110 ohms ▶ Resistance is less than 65 ohms or greater than 110 ohms ▶	CONNECT IAS solenoid. GO to KH5. REPLACE IAS solenoid. RERUN Quick Test.
KH5 CHECK VOLTAGE OF VPWR CIRCUIT		
• Key On, Engine Off. • DVOM on 20V scale. • Measure voltage between VPWR circuit at IAS solenoid and battery ground.	Voltage reading is less than 10.5V ▶ Voltage reading is 10.5V or greater ▶	SERVICE harness open circuit. RERUN Quick Test. GO to KH6.

Inlet Air Solenoid (IAS) — Pinpoint Test — KH

TEST STEP	RESULT ▶	ACTION TO TAKE
KH6 CHECK CONTINUITY OF IAS CIRCUIT		
• Key Off, wait 10 seconds. • Disconnect processor 60 Pin connector and inspect for damaged pins, corrosion, loose wires. Service as necessary. • Connect Breakout box to harness. Leave processor disconnected. • DVOM on 200 ohm scale. • Measure resistance between test Pin 59 at the Breakout box and IAS circuit at vehicle harness connector.	Resistance reading is 5 ohms or greater ▶ Resistance reading is less than 5 ohms ▶	SERVICE open circuit. RERUN Quick Test. GO to KH7.
KH7 CHECK FOR SHORT TO GROUND		
• Key Off, wait 10 seconds. • Leave Breakout box installed and processor disconnected. • Disconnect IAS solenoid. • DVOM on 200,000 ohm scale. • Measure resistance between test Pin 59 and test Pins 40, 46 and 60 at the Breakout box.	Any resistance reading is less than 10,000 ohms ▶ All resistance readings are 10,000 ohms or greater ▶	SERVICE short to ground. RERUN Quick Test. GO to KH8.
KH8 CHECK FOR SHORT TO POWER		
• Key Off, wait 10 seconds. • DVOM on 200,000 ohm scale. • IAS solenoid disconnected. • Leave Breakout box installed and processor disconnected. • Measure resistance between test Pin 59 and test Pins 37 and 57 at the Breakout box.	All resistance readings are 10,000 ohms or greater ▶ Any resistance reading is less than 10,000 ohms ▶	GO to KH9. SERVICE short to power. RERUN Quick Test.
KH9 PROCESSOR CHECK		
• Key Off. • Leave Breakout box installed and processor disconnected. • Connect a jumper between test Pin 59 and test Pin 40 at the Breakout box. • Key On, start vehicle. • Does inlet air door close?	Yes ▶ No ▶	REPLACE processor. REPLACE IAS.

Variable Voltage Choke (VVC) | Pinpoint Test | KJ

KJ

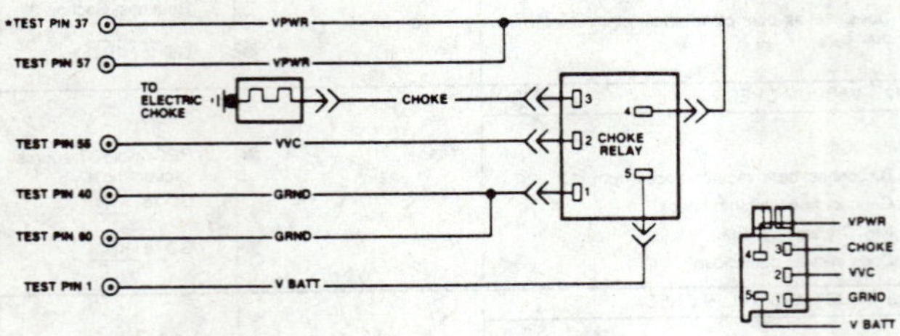

*TEST PINS LOCATED ON BREAKOUT BOX.
ALL HARNESS CONNECTORS VIEWED INTO MATING SURFACE.

STOP-WARNING

You should enter this Pinpoint Test only when a service code 88 is received in Quick Test Step 3.0 or when directed here from Diagnostics by Symptom in the Engine Supplement Section.

This Pinpoint Test is intended to diagnose only the following:

- Harness circuits: VPWR, Choke, VVC, Grnd., and V Batt.
- Choke Relay

Variable Voltage Choke (VVC) | Pinpoint Test | KJ

TEST STEP	RESULT ▶	ACTION TO TAKE
KJ1 ENTER OUTPUT STATE CHECK (REFER TO APPENDIX)		
NOTE: Do not use STAR tester for this step, use VOM/DVOM. • Key Off, wait 10 seconds. • DVOM on 20V scale. • Connect DVOM negative test lead to STO at the Self-Test connector and positive test lead to battery positive. • Jumper STI to signal return at the Self-Test connector. • Perform Key On, Engine Off Self-Test until the completion of the Continuous Test Codes. • DVOM will indicate zero volts when test is complete. • Depress and release the throttle. • Did DVOM reading change to a high voltage reading?	Yes ▶	REMAIN in Output State Check. GO to KJ2.
	No ▶	DEPRESS throttle to WOT and RELEASE. If STO voltage does not go high, GO to Pinpoint Test Step Q40.
KJ2 CHECK VVC OUTPUT ELECTRICAL OPERATION		
• Key On, Engine Off. • DVOM on 20V scale. • Connect DVOM positive test lead to choke circuit on choke cap and negative test lead to battery ground. • While observing DVOM depress and release the throttle several times to cycle output On and Off. • Choke output cycles On and Off?	Yes ▶	GO to Section 4.
	No ▶	REMOVE jumper. GO to KJ3.
KJ3 CHECK CONTINUITY OF GROUND CIRCUIT		
• Key Off, wait 10 seconds. • DVOM on 200 ohm scale. • Measure resistance between battery negative terminal and Pin 1 at the VVC relay vehicle harness connector.	Resistance reading is 5 ohms or greater ▶	SERVICE harness circuit. RERUN Quick Test.
	Resistance reading is less than 5 ohms ▶	GO to KJ4.

Variable Voltage Choke (VVC) | Pinpoint Test | KJ

TEST STEP	RESULT ▶	ACTION TO TAKE
KJ4 CONTINUITY OF CHOKE RELAY		
• Key Off, wait 10 seconds. • Disconnect VVC relay. • DVOM on 200 ohm scale. • Measure resistance between Pin 3 of the VVC relay connector and the choke connector at the choke cap. • Is the reading greater than 5 ohms?	Yes ▶	SERVICE open in choke circuit.
	No ▶	GO to KJ5.
KJ5 CHECK FOR VOLTAGE ON V BATT AND VPWR CIRCUITS		
• Key On, Engine Off. • DVOM on 20V scale. • VVC relay disconnected. • Measure voltage between Pin 1 and Pin 5 and between Pin 1 and between Pin 4 of VVC relay vehicle harness connector.	Voltage is less than 10.5V ▶	SERVICE open in choke power circuit. RERUN Quick Test.
	Voltage is 10.5V or greater ▶	REPLACE choke relay.
KJ6 CHOKE RELAY RESISTANCE		
• Key Off. • DVOM on 200,000 ohm scale. • Disconnect VVC relay. • Measure resistance at the VVC relay between Pin 2 and Pin 4. • Is resistance reading 1,800 ohms ± 200 ohms?	Yes ▶	GO to KJ7.
	No ▶	REPLACE choke relay.
KJ7 VPWR TO CHOKE RELAY		
• Key Off. • DVOM on 20V scale. • VVC relay disconnected. • Key On, engine Off. • Measure voltage between Pin 4 at the VVC relay connector and engine block ground. • Is voltage greater than 10V?	Yes ▶	GO to KJ8.
	No ▶	SERVICE VPWR circuit and RERUN Quick Test.

Variable Voltage Choke (VVC) | Pinpoint Test | KJ

TEST STEP	RESULT ▶	ACTION TO TAKE
KJ8 CHECK CONTINUITY OF VVC CIRCUIT		
• Key Off, wait 10 seconds. • Disconnect processor 60 Pin connector and inspect for damaged pins, corrosion, loose wires. Service as necessary. • Connect Breakout box to harness. Leave processor disconnected. • VVC relay disconnected. • DVOM on 200 ohm scale. • Measure resistance between test Pin 55 at the Breakout box and VVC circuit at the relay vehicle harness connector.	Resistance is 5 ohms or greater ▶	SERVICE harness circuit. RERUN Quick Test.
	Resistance is less than 5 ohms ▶	GO to KJ9.
KJ9 CHECK FOR SHORT TO GROUND ON VVC CIRCUIT		
• Key Off, wait 10 seconds. • DVOM on 200,000 ohm scale. • VVC relay disconnected. • Leave Breakout box installed and processor disconnected. • Measure resistance between Pin 55 to test Pins 40, 46 and 60 at the Breakout box.	All resistance readings are 10,000 ohms or greater ▶	GO to KJ10.
	Any resistance readings less than 10,000 ohms ▶	SERVICE harness short(s). RERUN Quick Test.
KJ10 CHECK FOR SHORT TO POWER		
• Key Off, wait 10 seconds. • VVC relay disconnected. • DVOM on 200,000 ohm scale. • Leave Breakout box installed and processor disconnected. • Measure resistance between test Pin 55 and test Pins 1, 37 and 57 at the Breakout box.	All resistance readings are 10,000 ohms or greater ▶	REPLACE processor. RERUN Quick Test.
	Any resistance reading is less than 10,000 ohms ▶	SERVICE short to power. RERUN Quick Test. If code is repeated, REPLACE processor.

Temperature Compensated Pump (TCP) | Pinpoint Test | KK

KK

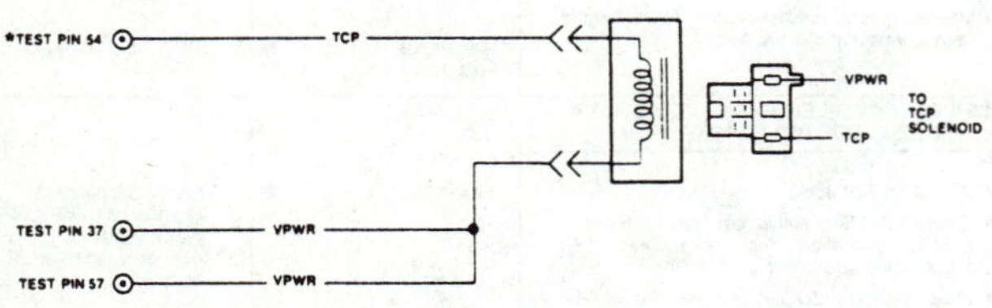

*TEST PINS LOCATED ON BREAKOUT BOX
ALL HARNESS CONNECTORS VIEWED INTO MATING SURFACE.

STOP-WARNING

You should enter this Pinpoint Test only when a service code 87 is received in Quick Test Step 3.0 or when directed here from Diagnostics by Symptom in the Engine Supplement Section.

This Pinpoint Test is intended to diagnose only the following:

- Harness circuits: TCP and VPWR
- TCP Solenoid
- Processor Assembly

Temperature Compensated Pump (TCP) | Pinpoint Test | KK

TEST STEP	RESULT ▶	ACTION TO TAKE
KK1 ENTER OUTPUT STATE CHECK (REFER TO APPENDIX)		
NOTE: Do not use STAR tester for this Step, use VOM/DVOM. • Key Off, wait 10 seconds. • DVOM on 20V scale. • Connect DVOM negative test lead to STO at the Self-Test connector and positive test lead to battery positive. • Jumper STI to signal return at the Self-Test connector. • Perform Key On, Engine Off Self-Test until the completion of the Continuous Test Codes. • DVOM will indicate zero volts when test is complete. • Depress and release the throttle. • Did DVOM reading change to a high voltage reading?	Yes ▶ No ▶	REMAIN in Output State Check. GO to KK2. DEPRESS throttle to WOT and release. If STO voltage does not go high, GO to Pinpoint Test Step Q40. Leave equipment hooked up.
KK2 CHECK TCP SOLENOID ELECTRICAL OPERATION		
• DVOM on 20V scale. • Connect DVOM positive test lead to VPWR circuit on TCP solenoid and negative test lead to TCP circuit at the solenoid. • While observing DVOM depress and release the throttle several times to cycle output On and Off. • Does TCP output cycle On and Off?	Yes ▶ No ▶	GO to KK3. REMOVE jumper to STI. GO to KK5.
KK3 CHECK TCP SOLENOID FOR VACUUM CYCLING		
• Install vacuum pump to the TCP solenoid vacuum supply port and install a vacuum gauge to the output port. Apply 6 in Hg minimum. • While cycling outputs On and Off (by depressing and releasing throttle) observe the vacuum gauge at the output. **NOTE: Maintain vacuum at source.**	Vacuum output cycles On and Off ▶ Vacuum output does not cycle On and Off ▶	GO to KK4. REPLACE solenoid. RERUN Quick Test.

Temperature Compensated Pump (TCP) | Pinpoint Test | KK

TEST STEP	RESULT ▶	ACTION TO TAKE
KK4 CHECK MANIFOLD VACUUM LINES FOR BLOCKAGE OR LEAKS		
• With vacuum lines disconnected at TCP solenoid, check for vacuum. • Start engine. • Check for vacuum.	Vacuum present ▶ No vacuum present ▶	EEC-IV system OK. REFER to Section 11 for TCP diagnostics. SERVICE vacuum source blockage or leak. RERUN Quick Test.
KK5 MEASURE TCP SOLENOID RESISTANCE		
• Key Off, wait 10 seconds. • DVOM on 200 ohm scale. • Disconnect TCP solenoid connector and measure solenoid resistance.	Resistance is between 50-100 ohms ▶ Resistance is less than 50 ohms or greater than 100 ohms ▶	CONNECT TCP solenoid. GO to KK6. REPLACE TCP solenoid. RERUN Quick Test.
KK6 CHECK VOLTAGE OF VPWR CIRCUIT		
• Key On, Engine Off. • DVOM on 20V scale. • Measure voltage between VPWR circuit at the TCP solenoid vehicle harness connector and battery ground.	Voltage reading is 10.5V or less ▶ Voltage reading is 10.5V or greater ▶	SERVICE harness open circuit. RERUN Quick Test. GO to KK7.
KK7 CHECK CONTINUITY OF TCP CIRCUIT		
• Key Off, wait 10 seconds. • Disconnect processor and inspect both 60 Pin connectors. • Connect Breakout box to harness, leave processor disconnected. • DVOM on 200 ohm scale. • Measure resistance between test Pin 54 at the Breakout box and TCP circuit at TCP solenoid vehicle harness connector.	Resistance reading is 5 ohms or greater ▶ Resistance reading is less than 5 ohms ▶	SERVICE harness open circuit. RERUN Quick Test. GO to KK8.

Temperature Compensated Pump (TCP) | Pinpoint Test | KK

TEST STEP	RESULT ▶	ACTION TO TAKE
KK8 CHECK FOR SHORT TO GROUND		
• Key Off, wait 10 seconds. • Leave Breakout box installed and processor disconnected. • Disconnect TCP solenoid. • DVOM on 200,000 ohm scale. • Measure resistance between test Pin 54 and test Pin 40, 46 and 60 at the Breakout box.	Any resistance reading is less than 10,000 ohms ▶ All resistance readings are 10,000 ohms or greater ▶	SERVICE short to ground. RERUN Quick Test. GO to KK9.
KK9 CHECK FOR SHORT TO POWER		
• Key Off, wait 10 seconds. • DVOM on 200,000 ohm scale. • Leave Breakout box installed and processor disconnected. • TCP solenoid disconnected. • Measure resistance between test Pin 54 and test Pins 37 and 57 at the Breakout box.	All resistance readings are 10,000 ohms or greater ▶ Any resistance reading is less than 10,000 ohms ▶	REPLACE processor. RERUN Quick Test. SERVICE short to power. RERUN Quick Test. If code is repeated, REPLACE processor.

Shift Indicator Light (SIL) — Pinpoint Test — KL

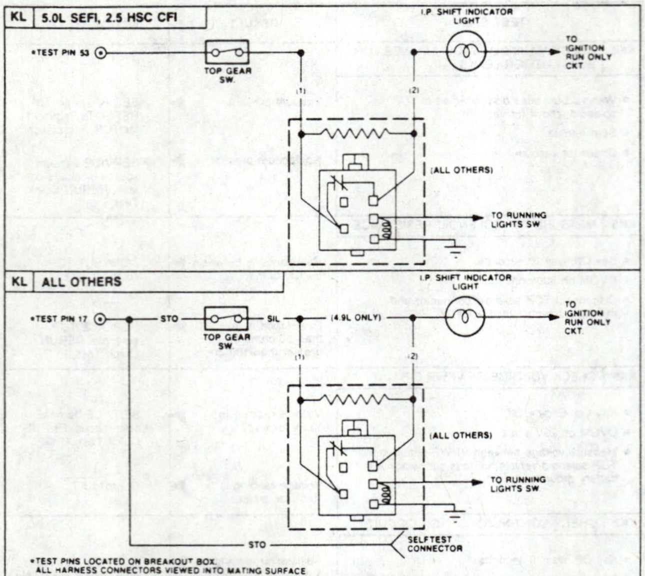

STOP-WARNING

You should enter this Pinpoint Test only when directed here from Diagnostics by Symptom in the Engine Supplement Section or from Pinpoint Test Step Q6.

This Pinpoint Test is intended to diagnose only the following:

- Top Gear Switch
- Shift Indicator Light Bulb and Fuse
- Harness Circuits: SIL and STO

Shift Indicator Light (SIL) — Pinpoint Test — KL

TEST STEP	RESULT ▶	ACTION TO TAKE
KL1 ENTER OUTPUT STATE CHECK IN NEUTRAL (REFER TO APPENDIX)		
NOTE: Do not use STAR tester for this Step, use VOM DVOM. • Key Off, wait 10 seconds. • Transmission in neutral. • DVOM on 20V scale. • Connect DVOM negative test lead to STO at the Self-Test connector and positive test lead to battery positive. • Jumper STI to signal return at the Self-Test connector. • Perform Key On, Engine Off Self-Test until the completion of the Continuous Test Codes. • DVOM will indicate zero volts when test is complete. • Depress and release the throttle. • Did DVOM reading change to a high voltage reading?	Yes ▶ No ▶	REMAIN in output state check. Go to KL2. DEPRESS throttle to WOT and release. If STO voltage does not go high, GO to Pinpoint Test Step KL7. Leave equipment hooked up.
KL2 CHECK FOR ELECTRICAL CYCLING AT DIMMER RELAY INPUT		
• If 4.9L truck go to step KL5. • DVOM on 20V scale. • Connect DVOM positive test lead to battery positive terminal and negative test lead to Pin 1 on dimmer relay. • While observing DVOM, depress and release the throttle several times to cycle Pin 1 On and Off. • Does Pin 1 cycle On and Off?	Yes ▶ No ▶	GO to KL3. GO to KL5.
KL3 CHECK FOR ELECTRICAL CYCLING AT DIMMER RELAY OUTPUT		
• DVOM on 20V scale. • Connect DVOM positive test lead to battery positive terminal and negative test lead to Pin 2 on dimmer relay. • While observing DVOM, depress and release the throttle several times to cycle Pin 2 On and Off. • Does Pin 2 cycle On and Off?	Yes ▶ No ▶	REMOVE jumper. GO to KL4. REMOVE jumper. REPLACE dimmer relay.

Shift Indicator Light (SIL) — Pinpoint Test — KL

TEST STEP	RESULT ▶	ACTION TO TAKE
KL4 CHECK CONTINUITY BETWEEN DIMMER RELAY OUTPUT AND SIL BULB		
• Key Off, wait 10 seconds. • DVOM on 200 ohm scale. • Measure resistance between Pin 2 of dimmer relay connector and SIL bulb.	Resistance reading is 5 ohms or greater ▶ Resistance reading is less than 5 ohms ▶	SERVICE harness open circuit. RERUN Quick Test. REPLACE SIL bulb.
KL5 CHECK FOR ELECTRICAL CYCLING AT SIL SIDE OF TOP GEAR SWITCH		
• DVOM on 20V scale. • Connect DVOM positive test lead to battery positive terminal and negative test lead to SIL circuit of top gear switch. • While observing DVOM, depress and release the throttle several times to cycle SIL circuit On and Off. • Does SIL circuit cycle On and Off?	Yes ▶ No ▶	SERVICE harness open circuit. RERUN Quick Test. 4.9L only, change SIL bulb. GO to KL6.
KL6 CHECK FOR ELECTRICAL CYCLING AT STO SIDE OF TOP GEAR SWITCH		
• DVOM on 20V scale. • Connect DVOM positive test lead to battery positive terminal and negative test lead to STO circuit of top gear switch. • While observing DVOM, depress and release the throttle several times to cycle STO circuit On and Off. • Does STO circuit cycle On and Off?	Yes ▶ No ▶	CHANGE top gear switch. SERVICE harness open circuit. RERUN Quick Test.

Shift Indicator Light (SIL) — Pinpoint Test — KL

TEST STEP	RESULT ▶	ACTION TO TAKE
KL7 ENTER OUTPUT STATE CHECK IN TOP GEAR (REFER TO APPENDIX)		
NOTE: Do not use STAR tester for this Step, use VOM DVOM. • Key Off, wait 10 seconds. • Transmission in top gear. • DVOM on 20V scale. • Connect DVOM negative test lead to STO at the Self-Test connector and positive test lead to battery positive. • Jumper STI to signal return at the Self-Test connector. • Perform Key On, Engine Off Self-Test until the completion of the Continuous Test Codes. • DVOM will indicate zero volts when test is complete. • Depress and release the throttle. • Did DVOM reading change to a high voltage reading?	Yes ▶ No ▶	CHECK SIL bulb and fuse 15. If OK, SERVICE short to ground in SIL circuit. GO to Pinpoint Test Step Q40.

WOT A/C Cutoff (WAC) A/C Demand — Pinpoint Test — KM

KM | 2.3L OHC, 2.3L EFI TC, 2.3L CFI ENGINES

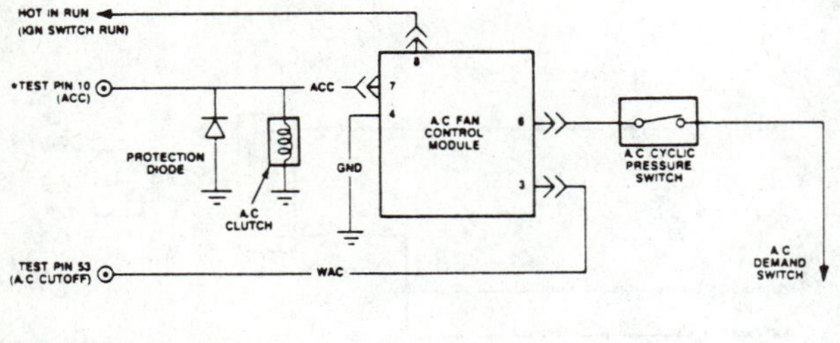

*TEST PINS LOCATED ON BREAKOUT BOX.
ALL HARNESS CONNECTORS VIEWED INTO MATING SURFACE.

KM | 1.9L EFI ENGINE

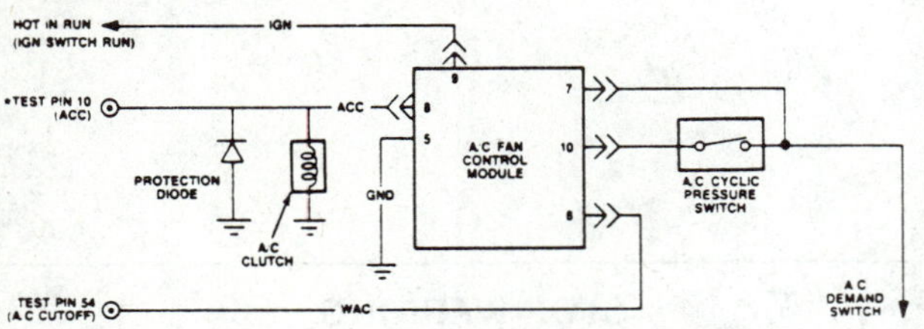

*TEST PINS LOCATED ON BREAKOUT BOX.
ALL HARNESS CONNECTORS VIEWED INTO MATING SURFACE.

WOT A/C Cutoff (WAC) A/C Demand — Pinpoint Test — KM

KM | 3.0L EFI TRUCK AND 5.0L SEFI ENGINES

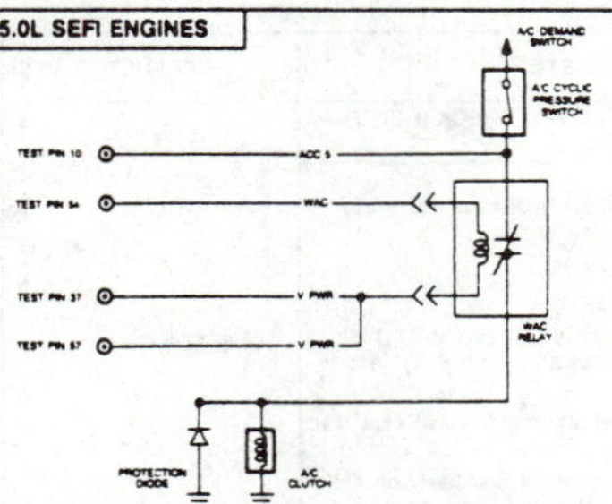

KM | 2.3L, 2.9L EFI TRUCK AND 3.8L CFI ENGINES

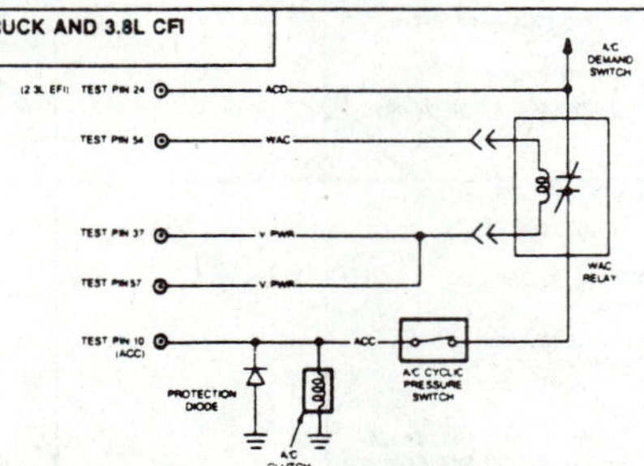

STOP-WARNING

You should enter this Pinpoint Test only when a Service Code 86 is received in Quick Test Step 3.0 or when directed here from Diagnostics by Symptom in the Engine Supplement Section.

This Pinpoint Test is intended to diagnose only the following:

- Harness Circuits: WAC and V PWR
- WOT A/C Cut-Off Relay
- Processor Assembly
- A/C Demand Switch

WOT A/C Cutoff (WAC) A/C Demand — Pinpoint Test — KM

TEST STEP	RESULT ▶	ACTION TO TAKE
KM1 ENTER OUTPUT STATE CHECK (REFER TO APPENDIX)		
NOTE: Do not use STAR tester for this Step, use VOM/DVOM. • Key Off, wait 10 seconds. • DVOM on 20V scale. • Connect DVOM negative test lead to STO at the Self-Test connector and positive test lead to battery positive. • Jumper STI to signal return at the Self-Test connector. • Perform Key On, Engine Off Self-Test until the completion of the Continuous Test Codes. • DVOM will indicate zero volts when test complete. • Depress and release the throttle. • Did DVOM reading change to a high voltage reading?	Yes ▶ No ▶	REMAIN in Output State Check. GO to KM2. DEPRESS throttle to WOT and RELEASE. If STO voltage does not go high, GO to Pinpoint Test Step Q40. Leave equipment hooked up.
KM2 CHECK A/C CLUTCH ELECTRICAL OPERATION		
• Key On, engine Off. • Disconnect processor 60 Pin connector and inspect for damaged pins, corrosion, loose wires. Service as necessary. • Connect Breakout box to harness. Reconnect processor. • A/C switch to A/C. • DVOM on 20V scale. • Connect DVOM positive test lead to test Pin 37 and negative test lead to test Pin 54 at the Breakout box. • While observing DVOM, depress and release the throttle several times (to cycle output On and Off). • Does A/C clutch output cycle On and Off?	Yes ▶ No ▶	EEC-IV system OK. Refer to Shop Manual, Group 36. REMOVE jumper. GO to KM3.
KM3 FAN CONTROL MODULE OR WAC RELAY		
• A/C fan control module applications 2.3L EFI, TC CFI, 1.9L EFI, 2.3L OHC, 2.3L engines. • All other systems with WAC relay.	A/C fan control ▶ WAC relay ▶	GO to KM4. GO to KM10.

WOT A/C Cutoff (WAC) A/C Demand — Pinpoint Test — KM

TEST STEP	RESULT ▶	ACTION TO TAKE
KM4 CHECK CONTINUITY OF GROUND CIRCUIT TO FAN CONTROL MODULE		
• Key Off, wait 10 seconds. • DVOM on 200 ohm scale. • Measure resistance of ground circuit between fan control module connector and battery ground.	Resistance reading is 5 ohms or greater ▶ Resistance reading is less than 5 ohms ▶	SERVICE harness circuit. RERUN Quick Test. GO to KM5.
KM5 CHECK FOR IGN CIRCUIT VOLTAGE		
• Key On, engine Off. • DVOM on 20V scale. • Measure voltage between ignition circuit and ground circuit on the fan control module connector.	Voltage is less than 10.5V ▶ Voltage is 10.5V or greater ▶	SERVICE open in ign. circuit. RERUN Quick Test. GO to KM6.
KM6 CHECK CONTINUITY OF WAC CIRCUIT		
• Key Off, wait 10 seconds. • Disconnect processor 60 Pin connector and inspect for damaged pins, corrosion, loose wires. Service as necessary. • Connect Breakout box to harness. Leave processor disconnected. • DVOM on 200 ohm scale. • Measure resistance between test Pin 54 (test Pin 53 for 2.3L OHC engine) at the Breakout box and WAC circuit at fan control module connector.	Resistance is 5 ohms or greater ▶ Resistance is less than 5 ohms ▶	SERVICE harness WAC circuit. RERUN Quick Test. GO to KM7.
KM7 CHECK FOR SHORT TO GROUND ON WAC CIRCUIT		
• Key Off, wait 10 seconds. • Leave Breakout box installed and processor disconnected. • DVOM on 200,000 ohm scale. • Measure resistance between test Pin 54 (test Pin 53 for 2.3L OHC engine) and test Pins 40, 46 and 60 at the Breakout box.	Resistance reading is less than 10,000 ohms ▶ Resistance reading is 10,000 ohms or greater ▶	GO to KM8. REPLACE processor. RERUN Quick Test.

WOT A/C Cutoff (WAC) A/C Demand — Pinpoint Test — KM

TEST STEP	RESULT	ACTION TO TAKE
KM8 CHECK FOR SHORT TO GROUND WITH FAN CONTROL MODULE DISCONNECTED		
• Key Off, wait 10 seconds. • Leave Breakout box installed and processor disconnected. • DVOM on 200,000 ohm scale. • Disconnect fan control module. • Measure resistance between test Pin 54 (test Pin 53 for 2.3L OHC engine) and test Pins 40, 46 and 60 at the Breakout box.	All resistance readings 10,000 ohms or greater ▶ Any resistance reading is less than 10,000 ohms ▶	REPLACE fan control module. RERUN Quick Test. SERVICE harness short. RERUN Quick Test.
KM10 MEASURE WAC RELAY RESISTANCE		
• Key Off, wait 10 seconds. • DVOM on 200 ohm scale. • Disconnect WAC relay connector and measure relay resistance.	Resistance is between 50 and 70 ohms ▶ Resistance is less than 50 ohms or greater than 70 ohms ▶	CONNECT WAC solenoid. GO to KM11. REPLACE WAC relay. RERUN Quick Test.
KM11 CHECK VOLTAGE OF V PWR CIRCUIT		
• Key On, Engine Off. • DVOM on 20V scale. • Connect DVOM positive test lead to V PWR circuit and negative test lead to ground. • Measure voltage on WAC relay V PWR circuit.	Voltage reading is less than 10.5V ▶ Voltage reading is 10.5V or greater ▶	SERVICE harness open circuit. RERUN Quick Test. GO to KM12.
KM12 CHECK CONTINUITY OF WAC CIRCUIT		
• Key Off, wait 10 seconds. • Disconnect processor 60 Pin connector and inspect for damaged pins, corrosion, loose wires. Service as necessary. • Connect Breakout box to harness. Leave processor disconnected. • DVOM on 200 ohm scale. • Measure resistance between test Pin 54 at the Breakout box and WAC circuit at harness connector.	Resistance readings are 5 ohms or greater ▶ Resistance reading is less than 5 ohms ▶	SERVICE harness open circuit. RERUN Quick Test. GO to KM13.

Turbo Boost — Pinpoint Test — KN

KN

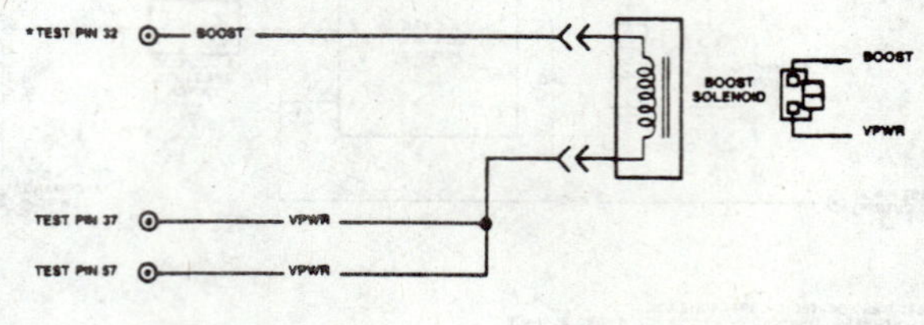

*TEST PINS LOCATED ON BREAKOUT BOX.
ALL HARNESS CONNECTORS VIEWED INTO MATING SURFACE.

STOP-WARNING

You should enter this Pinpoint Test only when directed here from Diagnostics by Symptom in the Engine Supplement Section.

This Pinpoint Test is intended to diagnose only the following:

- Harness circuits: VPWR, Boost

WOT A/C Cutoff (WAC) A/C Demand — Pinpoint Test — KM

TEST STEP	RESULT	ACTION TO TAKE
KM13 CHECK FOR SHORT TO GROUND		
• Key Off, wait 10 seconds. • Leave Breakout box installed and processor disconnected. • DVOM on 200,000 ohm scale. • Measure resistance between test Pin 54 and test Pins 40, 46 and 60 at Breakout box.	Any resistance reading is less than 10,000 ohms ▶ All resistance readings are 10,000 ohms or greater ▶	SERVICE short to ground. RERUN Quick Test. GO to KM14.
KM14 CHECK FOR SHORT TO POWER		
• Key Off, wait 10 seconds. • DVOM on 200,000 ohm scale. • Leave Breakout box installed and processor disconnected. • WAC solenoid disconnected. • Measure resistance between test Pin 54 and test Pins 37 and 57 at Breakout box.	Resistance reading is 10,000 ohms or greater ▶ Resistance reading is less than 10,000 ohms ▶	REPLACE processor. RERUN Quick Test. SERVICE short to power. RERUN Quick Test.
KM20 CYCLE A/C DEMAND SWITCH		
• Key Off, wait 10 seconds. • Disconnect processor 60 Pin connector and inspect for damaged pins, corrosion, loose wires. Service as necessary. • Connect Breakout box to harness. Leave processor disconnected. • DVOM on 20V scale. • Measure voltage between test Pin 10 (and Pin 24 for 2.3L EFI truck) and test Pin 40 at the Breakout box. • Does output cycle 4.0-10.5V when A/C switch is cycled?	Yes ▶ No ▶	REPLACE processor. RERUN Quick Test. GO to KM21.
KM21 CHECK CONTINUITY OF ACC ACD CIRCUIT		
• Key Off, wait 10 seconds. • DVOM on 200 ohm scale. • Measure resistance between ACC test Pin 10 at the Breakout box and A/C clutch. For 2.3L EFI truck, also measure resistance between ACD test Pin 24 at the Breakout box and A/C demand switch.	Resistance reading is 5 ohms or greater ▶ Resistance reading is less than 5 ohms ▶	SERVICE harness circuit. RERUN Quick Test. EEC-IV system OK. REFER to Shop Manual, Group 36.

Turbo Boost — Pinpoint Test — KN

TEST STEP	RESULT	ACTION TO TAKE
KN1 ENTER OUPUT STATE CHECK (REFER TO APPENDIX)		
NOTE: Do not use STAR tester for this Step, use VOM/DVOM. • Key Off, wait 10 seconds. • DVOM on 20V scale. • Connect DVOM negative test lead to STO at the Self-Test connector and positive test lead to battery positive. • Jumper STI to signal return at the Self-Test connector. • Perform Key On, Engine Off Self-Test until the completion of the Continuous Test Codes. • DVOM will indicate zero volts when test is complete. • Depress and release the throttle. • Did DVOM reading change to a high voltage reading?	Yes ▶ No ▶	REMAIN in Output State Check. GO to KN2. DEPRESS throttle to WOT and release. If STO voltage does not go high, GO to Pinpoint Test Step Q40. Leave equipment hooked up.
KN2 CHECK BOOST OUTPUT ELECTRICAL OPERATION		
• Key On, Engine Off. • DVOM on 20V scale. • Connect DVOM positive test lead to circuit VPWR on boost solenoid connector and negative test lead to boost output on boost solenoid connector. • While observing DVOM, depress and release the throttle several times to cycle output On and Off. • Does boost output solenoid cycle On and Off?	Yes ▶ No ▶	GO to KN3. REMOVE jumper. GO to KN4.

Turbo Boost — Pinpoint Test — KN

TEST STEP	RESULT ▶	ACTION TO TAKE
KN3 CHECK BOOST SOLENOID FUNCTION		
• Remain in output state check. • Disconnect black solenoid vacuum hose at the turbocharger end. **(Not the black and yellow stripe hose.)** • Attach a vacuum pump to the hose. • Depress the throttle once to cycle the solenoid closed. • Apply vacuum to the solenoid. • Depress the throttle once again to cycle the solenoid open and release trapped vacuum. • Did the solenoid hold and then release vacuum?	Yes ▶ No ▶	EEC-IV system OK. REFER to Shop Manual for boost diagnostics. REPLACE solenoid. RERUN Quick Test.
KN4 MEASURE BOOST SOLENOID RESISTANCE		
• Key Off, wait 10 seconds. • DVOM on 200 ohm scale. • Disconnect boost solenoid connector and measure the solenoid resistance.	Resistance is between 65 and 110 ohms ▶ Resistance is less than 65 ohms or greater than 110 ohms ▶	CONNECT BOOST solenoid. GO to KN5. REPLACE BOOST solenoid. RERUN Quick Test.
KN5 CHECK VOLTAGE OF VPWR CIRCUIT		
• Key On, Engine Off. • DVOM on 20V scale. • Measure voltage between VPWR circuit of boost solenoid vehicle harness connector and battery ground.	Voltage reading is less than 10.5V ▶ Voltage reading is 10.5V or greater ▶	SERVICE harness open circuit. RERUN Quick Test. GO to KN6.
KN6 CHECK CONTINUITY OF BOOST CIRCUIT		
• Key Off, wait 10 seconds. • Disconnect processor 60 Pin connector and inspect for damaged pins, corrosion, loose wires. Service as necessary. • Connect Breakout box to harness. Leave processor disconnected. • DVOM on 200 ohm scale. • Measure resistance between test Pin 32 at the Breakout box and boost circuit at vehicle harness connector.	Resistance reading is 5 ohms or greater ▶ Resistance reading is less than 5 ohms ▶	SERVICE open circuit. RERUN Quick Test. GO to KN7.

Turbo Boost — Pinpoint Test — KN

TEST STEP	RESULT ▶	ACTION TO TAKE
KN7 CHECK FOR SHORT TO GROUND		
• Key Off, wait 10 seconds. • Leave Breakout box installed and processor disconnected. • Disconnect boost solenoid. • DVOM on 200,000 ohm scale. • Measure resistance between test Pin 32 and test Pins 40, 46 and 60 at the Breakout box.	All readings 10,000 ohms or greater ▶ Any reading less than 10,000 ohms ▶	Go to KN8. SERVICE short to ground. RERUN Quick Test.
KN8 CHECK FOR SHORT TO POWER		
• Key Off, wait 10 seconds. • DVOM on 200,000 ohm scale. • Leave Breakout box installed and processor disconnected. • Boost solenoid disconnected. • Measure resistance between Pin 32 and test Pins 37 and 57 at the Breakout box.	Any resistance reading less than 10,000 ohms ▶ All resistance readings are 10,000 ohms or greater ▶	SERVICE short to power. RERUN Quick Test. If symptom is still present, REPLACE processor. REPLACE processor. RERUN Quick Test.

Exhaust Heat Control (EHC) — Pinpoint Test — KP

KP

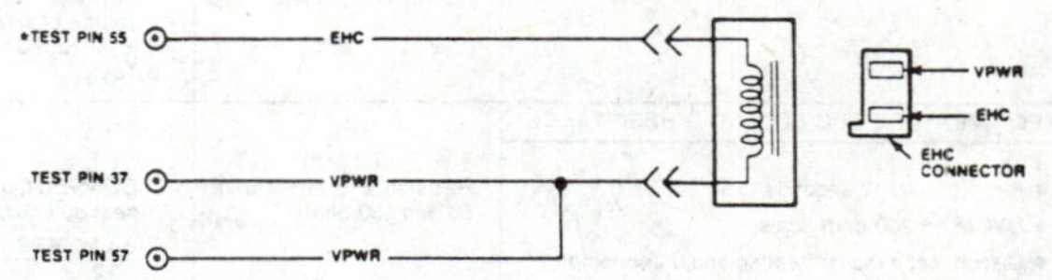

*TEST PINS LOCATED ON BREAKOUT BOX.
ALL HARNESS CONNECTORS VIEWED INTO MATING SURFACE.

STOP-WARNING

You should enter this Pinpoint Test only when a service code 89 is received in Quick Test Step 3.0 or when directed here from Diagnostics by Symptom in the Engine Supplement Section.

This Pinpoint Test is intended to diagnose only the following:

- Exhaust Heat Solenoid
- Vacuum Routing

Exhaust Heat Control (EHC) — Pinpoint Test — KP

TEST STEP	RESULT ▶	ACTION TO TAKE
KP1 ENTER OUTPUT STATE CHECK (REFER TO APPENDIX)		
NOTE: Do not use STAR tester for this Step, use VOM DVOM. • Key Off, wait 10 seconds. • DVOM on 20V scale. • Connect DVOM negative test lead to STO at the Self-Test connector and positive test lead to battery positive. • Jumper STI to signal return at the Self-Test connector. • Perform Key On, Engine Off Self-Test until the completion of the Continuous Test Codes. • DVOM will indicate zero volts when test is complete. • Depress and release the throttle. • Did DVOM reading change to a high voltage reading?	Yes ▶ No ▶	REMAIN in Output State Check. GO to KP2. DEPRESS throttle to WOT and release. If STO voltage does not go high, GO to Pinpoint Test Step Q40. Leave equipment hooked up.
KP2 CHECK EHC SOLENOID ELECTRICAL OPERATION		
• DVOM on 20V scale. • Measure voltage between VPWR and EHC circuit at EHC solenoid. • While observing DVOM, depress throttle several times to cycle output on and off. • Does EHC output cycle on and off?	Yes ▶ No ▶	GO to KP3. GO to KP5.
KP3 CHECK EXHAUST HEAT SOLENOID FOR VACUUM CYCLING		
• Install vacuum pump to the exhaust heat solenoid vacuum supply port and install a vacuum gauge to the output port. Apply 6 in Hg minimum. • While cycling outputs on and off (by depressing and releasing throttle) observe the vacuum gauge at the output. **NOTE: Maintain vacuum at source.**	Vacuum output cycles on and off ▶ Vacuum output does not cycle on and off ▶	GO to KP4. REPLACE solenoid. RERUN Quick Test.

Exhaust Heat Control (EHC) — Pinpoint Test — KP

TEST STEP	RESULT ▶	ACTION TO TAKE
KP4 CHECK MANIFOLD VACUUM LINES FOR BLOCKAGE OR LEAKS		
• With vacuum lines disconnected at exhaust heat solenoid, check for vacuum. • Start engine • Check for vacuum.	Vacuum present ▶	EEC-IV system OK. REFER to Section 3 for exhaust heat diagnostics.
	No vacuum present ▶	SERVICE vacuum source blockage or leak. RERUN Quick Test.
KP5 MEASURE EHC SOLENOID RESISTANCE		
• Key Off, wait 10 seconds. • DVOM on 200 ohm scale. • Disconnect exhaust heat solenoid connector and measure solenoid resistance.	Resistance is between 50 and 100 ohms ▶	CONNECT exhaust heat solenoid. Go to KP6.
	Resistance is less than 50 ohms or greater than 100 ohms ▶	REPLACE exhaust heat solenoid. RERUN Quick Test.
KP6 CHECK VOLTAGE OF VPWR CIRCUIT		
• Key On, Engine Off. • DVOM on 20V scale. • Measure voltage between VPWR at the EHC solenoid vehicle harness connector circuit and ground.	Voltage reading is less than 10.5V ▶	SERVICE harness open circuit. RERUN Quick Test.
	Voltage reading is 10.5V or greater ▶	GO to KP7.
KP7 CHECK CONTINUITY OF EHC CIRCUIT		
• Key Off, wait 10 seconds. • Disconnect processor 60 Pin connector and inspect for damaged pins, corrosion, loose wires. Service as necessary. • Connect Breakout box to harness. Leave processor disconnected. • DVOM on 200 ohm scale. • Measure resistance between test Pin 55 at the Breakout box and EHC circuit at vehicle harness connector.	Resistance readings are 5 ohms or greater ▶	SERVICE harness open circuit. RERUN Quick Test.
	Resistance reading is less than 5 ohms ▶	GO to KP8.

Exhaust Heat Control (EHC) — Pinpoint Test — KP

TEST STEP	RESULT ▶	ACTION TO TAKE
KP8 CHECK FOR SHORT TO GROUND		
• Key off, wait 10 seconds. • Leave Breakout box installed and processor disconnected. • Disconnect EHC solenoid. • DVOM on 200,000 ohm scale. • Measure resistance between test Pin 55 and test Pins 40, 46 and 60 at Breakout box.	Resistance readings are less than 10,000 ohms ▶	SERVICE short to ground. RERUN Quick Test.
	All resistance readings are 10,000 ohms or greater ▶	GO to KP9.
KP9 CHECK FOR SHORT TO POWER		
• Key Off, wait 10 seconds. • DVOM on 200,000 ohm scale. • Leave Breakout box installed and processor disconnected. • EHC solenoid disconnected. • Measure resistance between test Pin 55 and test Pins 37 and 57 at Breakout box.	Resistance reading is 10,000 ohms or greater ▶	REPLACE processor. RERUN Quick Test.
	Resistance reading is less than 10,000 ohms ▶	SERVICE short to power. RERUN Quick Test. If code is repeated, REPLACE processor.

Converter Clutch Override (CCO) — Pinpoint Test — KR

KR

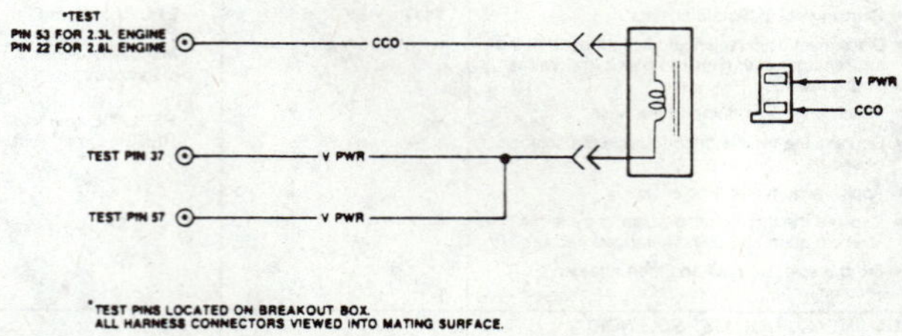

STOP-WARNING

You should enter this Pinpoint Test only when a service code 89 is received in Quick Test Step 3.0.

This Pinpoint Test is intended to diagnose only the following:

- Harness Circuits: CCO and V PWR.
- CCO Solenoid.
- Processor Assembly.

TEST STEP	RESULT ▶	ACTION TO TAKE
CODE 89 PRESENT		
KR1 MEASURE CCO SOLENOID RESISTANCE		
• Key Off, wait 10 seconds. • DVOM on 200 ohm scale. • Disconnect CCO solenoid connector and measure solenoid resistance.	Resistance is between 26 and 40 ohms ▶	CONNECT CCO solenoid. GO to KR2.
	Resistance is less than 26 ohms or greater than 40 ohms ▶	REPLACE CCO solenoid. RERUN Quick Test.

Converter Clutch Override (CCO) — Pinpoint Test — KR

TEST STEP	RESULT ▶	ACTION TO TAKE
KR2 CHECK VOLTAGE OF VPWR CIRCUIT		
• Key On, Engine Off. • DVOM on 20V scale. • Measure voltage at the CCO solenoid connector between VPWR circuit and battery ground.	Voltage reading is less than 10.5V ▶	SERVICE harness open circuit. RERUN Quick Test.
	Voltage reading is 10.5V or greater ▶	GO to KR3.
KR3 CHECK CONTINUITY OF CCO CIRCUIT		
• Key Off, wait 10 seconds. • Disconnect processor 60 Pin connector and inspect for damaged pins, corrosion, loose wires. Service as necessary. • Connect Breakout box to harness. Leave processor disconnected. • DVOM on 200 ohm scale. • Measure resistance between test Pin 53 (test Pin 22 for 2.8L) at the Breakout box and CCO circuit at the solenoid vehicle harness connector.	Resistance readings are 5 ohms or greater ▶	SERVICE harness open circuit. RERUN Quick Test.
	Resistance reading is less than 5 ohms ▶	GO to KR4.
KR4 CHECK FOR SHORT TO GROUND		
• Key Off, wait 10 seconds. • Leave Breakout box installed and processor disconnected. • Disconnect CCO solenoid. • DVOM on 200,000 ohm scale. • Measure resistance between test Pin 55 and test Pins 40, 46 and 60 at Breakout box.	Any resistance readings less than 10,000 ohms ▶	SERVICE short to ground. RERUN Quick Test.
	All resistance readings are 10,000 ohms or greater ▶	GO to KR5.
KR5 CHECK FOR SHORT TO POWER		
• Key Off, wait 10 seconds. • DVOM on 200,000 ohm scale. • Leave Breakout box installed and processor disconnected. • CCO solenoid disconnected. • Measure resistance between test Pin 55 and test Pins 37 and 57 at Breakout box.	All resistance readings are 10,000 ohms or greater ▶	REPLACE processor. RERUN Quick Test.
	Any resistance reading is less than 10,000 ohms ▶	SERVICE short to power. RERUN Quick Test. If code is still present, REPLACE processor.

Dynamic Response Test — Pinpoint Test M

M

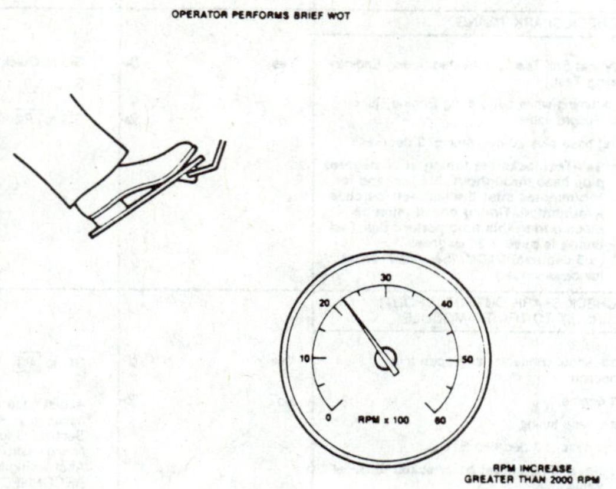

STOP-WARNING

You should enter this Pinpoint Test only when a service code 77 is received in Quick Test Step 5.0.

To prevent the replacement of good components, be aware that the following non-EEC areas may be at fault:

- Operator did not perform a brief WOT after dynamic response code.
- Mechanical engine problems; engine did not achieve greater than 2000 rpm.

This Pinpoint Test is intended to diagnose only the following:

- Throttle movement (greater than 3/4 throttle).
- Vane Air Flow (greater than 50% open).
- RPM increase (greater than 2000 rpm).

Dynamic Response Test — Pinpoint Test M

TEST STEP	RESULT ▶	ACTION TO TAKE
M1 CODE 77: SYSTEM FAILED TO RECOGNIZE BRIEF WOT		
NOTE: A brief snap of the throttle may not be sufficient to pass this test. Be sure to go to WOT and return.		
• Repeat Engine Running Test of Quick Test. Be sure operator is familiar with the engine running format which proceeds as follows: — With Self-Test activated restart the engine. — ID Code 2 (0) start of test. — Dynamic response Code 1 (0) perform brief WOT. — Testing over. — Service code output begins.	Code 77 still present ▶	REPLACE processor. RERUN Quick Test.
	No Code 77 ▶	Dynamic Response Test passed. SERVICE any other service code(s) received as necessary.

Ignition Diagnostic Monitor (IDM) — Pinpoint Test N

N

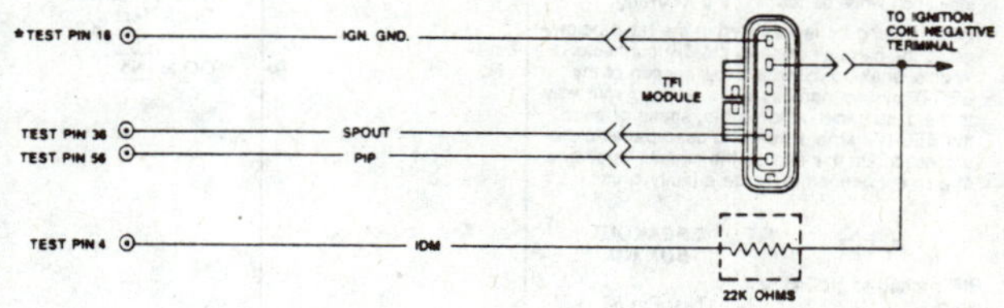

*TEST PINS LOCATED ON BREAKOUT BOX. ALL HARNESS CONNECTORS VIEWED INTO MATING SURFACE.

STOP-WARNING

You should enter this Pinpoint Test only when a service code 18 is received in Quick Test Step 6.0.

To prevent the replacement of good components, be aware that the following non-EEC areas may be at fault:

- Ignition module.
- Ignition coil.
- Spark plugs and high tension cables.
- Distributor and PIP sensor.

This Pinpoint Test is intended to diagnose only the following:

- Harness circuits: ignition ground, spout, PIP, IDM.

Ignition Diagnostic Monitor (IDM) — Pinpoint Test N

TEST STEP	RESULT ▶	ACTION TO TAKE
N1 CHECK CONTINUITY OF IDM CIRCUIT		
• Key Off, wait 10 seconds. • Disconnect E-core ignition connector from coil. • Disconnect processor and inspect both 60 Pin connectors for damaged pins, corrosion, loose wires. Service as necessary. • Connect Breakout box to harness. Leave processor disconnected. • DVOM on 200,000 ohm scale. • Measure resistance between test Pin 4 at the Breakout box and ignition coil negative terminal.	Resistance reading is less than 20,000 ohms or greater than 24,000 ohms ▶	SERVICE open circuit. RERUN Quick Test.
	Resistance reading is less than 24,000 ohms but greater than 20,000 ohms ▶	GO to N2.
N2 CHECK FOR SHORT TO GROUND		
• Key Off, wait 10 seconds. • Leave Breakout box installed and processor disconnected. • DVOM on 200,000 ohm scale. • Measure resistance between test Pin 4 and test Pins 40, 46 and 60 at the Breakout box.	Any resistance reading is less than 10,000 ohms ▶	SERVICE short to ground. RERUN Quick Test.
	All resistance readings are 10,000 ohms or greater ▶	GO to N3.
N3 CHECK TFI MODULE		
• Key Off, wait 10 seconds. • Deactivate Self-Test. • Connect VOM or STAR per Quick Test Step 2.0. • Enter Engine Running Continuous Monitor Test (as instructed in Quick Test Step 6.0D). • Observe VOM or STAR LED for indication of a fault while performing the following: • Lightly tap on TFI module (simulate road shock). • Wiggle TFI connector. • Is a fault indicated?	Yes ▶	DISCONNECT and INSPECT connectors. If connector and terminals are good, GO to Section 15, TFI Ignition Diagnostics.
	No ▶	GO to N4.

Ignition Diagnostic Monitor (IDM) — Pinpoint Test N

TEST STEP	RESULT ▶	ACTION TO TAKE
N4 CHECK EEC-IV HARNESS		
• While still in continuous monitor test from step N1 observe VOM or STAR LED for a fault indication while performing the following: • While looking for faults listed in the table below, grasp the harness close to the TFI connector. Wiggle, shake or bend a small section of the EEC-IV system harness while working your way to the dash panel. Also wiggle, shake or bend the EEC-IV harness from the dash panel to the processor. Do this test on the circuits listed one at a time if needed to locate a faulty circuit. FAULT — BREAKOUT BOX NO. PIP shorted to ground or open — Test Pin 56 Spout shorted to ground — Test Pin 36 Ign. ground open — Test Pin 16 IDM open or shorted to ground, power — Test Pin 4 • Is a fault indicated?	Yes ▶ No ▶	ISOLATE fault and make necessary repairs. RERUN Quick Test. GO to N5.
N5 CHECK PROCESSOR AND HARNESS CONNECTORS		
• Key Off, wait 10 seconds. • Disconnect processor 60 Pin connector and inspect for damaged pins, corrosion, loose wires. • Are connectors and terminals OK? • Reconnect processor when this Step is completed.	No ▶ Yes ▶	SERVICE as necessary. RERUN Quick Test. If unable to DUPLICATE an IDM fault in the EEC-IV System, testing is complete.

Spark Timing Check — Pinpoint Test P

TEST STEP	RESULT ▶	ACTION TO TAKE
P1 CHECK SPARK TIMING		
• Verify that Self-Test is activated during Engine Running Test. • Check timing while performing Engine Running Test. Record value. • Timing base plus 20 degrees ± 3 degrees? **NOTE: Self-Test locks the timing at 20 degrees plus base throughout this test and for two minutes after the last service code is outputted. Timing check must be made during this time period. Self-Test timing is base + 20 degrees (±3 degrees) BTDC. (See VECI decal for base value.)**	Yes ▶ No ▶	GO to Quick Test Step 5.0. GO to P2.
P2 CHECK SPARK OUTPUT (SPOUT) CIRCUIT TO THE TFI MODULE		
• Locate spout connector and open this connection. • Start engine. • Check base timing. • Timing base ±3 degrees BTDC. **NOTE: Base timing must be checked to decal value.**	Yes ▶ No ▶	GO to P3. Adjust base timing if necessary. REFER to Section 15 for engine timing instructions. After timing is reset, RECONNECT spout and PERFORM Quick Test Step 4.
P3 CHECK FOR POWER TO PROCESSOR		
• Key Off, wait 10 seconds. • Disconnect processor 60 Pin connector and inspect for damaged pins, corrosion, loose wires. Service as necessary. • Install Breakout Box. • Key On, Engine Off. • DVOM on 20V scale. • Measure voltage between test Pin 37 and test Pin 40 and between test Pin 57 and test Pin 60 at the Breakout box.	Either voltage reading less than 10.5V ▶ Both voltage readings 10.5V or greater ▶	GO to Pinpoint Test Step B1 for all engines, except 2.5L HSC CFI and 3.0L EFI passenger car. GO to Pinpoint Test Step X1. GO to P4.

Spark Timing Check — Pinpoint Test P

P

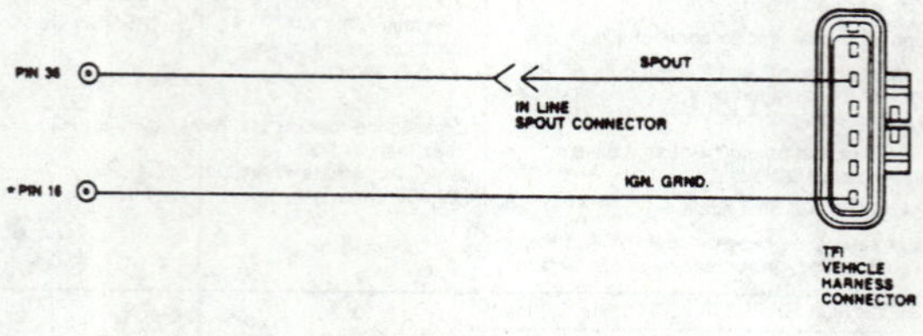

*TEST PINS LOCATED ON BREAKOUT BOX
ALL HARNESS CONNECTORS VIEWED INTO MATING SURFACE.

STOP-WARNING

You should enter this Pinpoint Test only when directed here from Quick Test Step 4.0.

To prevent the replacement of good components, be aware that the following non-EEC areas may be at fault:

- Base Engine.
- PIP Sensor.
- TFI Module.

This Pinpoint Test is intended to diagnose only the following:

- Harness Spout Circuit.
- Base Timing.
- Processor Assembly

Spark Timing Check — Pinpoint Test P

TEST STEP	RESULT ▶	ACTION TO TAKE
P4 CHECK HARNESS FOR CONTINUITY		
• Key Off, wait 10 seconds. • DVOM on 200 ohm scale. • Measure resistance between test Pin 36 at the Breakout box and Pin 2 (spout pin) at the TFI vehicle harness connector.	Reading 5 ohms or less ▶ Reading greater than 5 ohms ▶	GO to P5. SERVICE open circuit. CONNECT spout test connector. CHECK timing per P1.
P5 HARNESS CHECK — IGNITION GROUND		
• Breakout box installed. • Key Off. • DVOM on 200 ohm scale. • Measure resistance between test Pin 16 at the Breakout box and TFI Pin 6 (ign. grnd.) at the TFI vehicle harness connector.	Less than 5 ohms ▶ 5 ohms or greater ▶	GO to Pinpoint Test Step A7. SERVICE harness as necessary. RERUN Quick Test.

No Codes/Codes Not Listed — Pinpoint Test — Q

Q

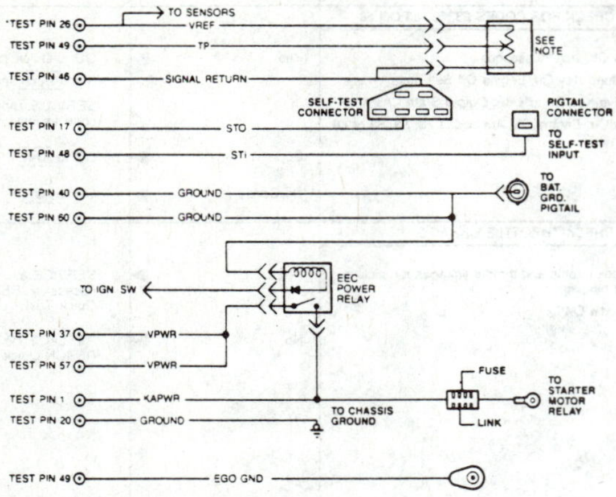

*TEST PINS LOCATED ON BREAKOUT BOX.
ALL HARNESS CONNECTORS VIEWED INTO MATING SURFACE.

NOTE: REFER TO THE ELECTRICAL SCHEMATIC IN THE APPROPRIATE ENGINE SUPPLEMENT SECTION FOR PROPER CONNECTOR AND PIN ORIENTATION.

STOP-WARNING

You should enter this Pinpoint Test only when directed here from Quick Test Step 3.0, 5.0 or 6.0 or when directed here from Diagnostics by Symptom in the Engine Supplement Section or from a Pinpoint Test Step.

This Pinpoint Test is intended to diagnose only the following:

- Processor.
- Harness Circuits: Signal Return, STO, STI, Ground, VPWR, KAM, VREF, Ignition, NDS, EEC Power Relay.

No Codes/Codes Not Listed — Pinpoint Test — Q

TEST STEP	RESULT ▶	ACTION TO TAKE
Q1 NO CODES, CODES NOT LISTED		
Refer to illustration Q. • Key Off, wait 10 seconds. • DVOM on 20V scale. • Disconnect TP sensor. • Key On, Engine Off. • Measure voltage between VREF at the TP harness connector and signal return in Self-Test connector.	Less than 4.0V or greater than 6.0V ▶ 4.0V to 6.0V ▶	GO to Pinpoint Test Step C1. RECONNECT TP sensor. GO to Q2.
Q2 SELF-TEST INPUT CONTINUITY CHECK		
Refer to illustration Q. • Key Off, wait 10 seconds. • Disconnect processor 60 Pin connector and inspect for damaged pins, corrosion, loose wires. Service as necessary. • Install Breakout box, leave processor disconnected. • Set DVOM to 200 ohm scale. • Measure resistance between Self-Test input at the Self-Test single pin connector and test Pin 48 at the Breakout box.	Less than 5 ohms ▶ 5 ohms or greater ▶	GO to Q3. CORRECT open in circuit.
Q3 SELF-TEST OUTPUT CIRCUIT CONTINUITY CHECK		
Refer to illustration Q. • Breakout box installed. • DVOM to 200 ohm scale. • Measure resistance between Self-Test output at the Self-Test connector and test Pin 17 at the Breakout box.	5 ohms or greater ▶ Less than 5 ohms ▶	CORRECT open in circuit. GO to Q4.
Q4 EGO SENSOR GROUND CONTINUITY CHECK		
Refer to illustration Q. • Breakout box installed. • Key Off. • Measure resistance between EGO ground on engine and test Pin 49 at the Breakout box.	Less than 5 ohms ▶ 5 ohms or greater ▶	GO to Q5. CHECK and SERVICE EGO sensor ground wire or open circuit bad connection.

No Codes/Codes Not Listed — Pinpoint Test — Q

TEST STEP	RESULT ▶	ACTION TO TAKE
Q5 STO SHORT TO GROUND		
• Breakout box installed. • DVOM on 200 ohm scale. • Measure resistance between Self-Test output at Self-Test connector and engine block ground. • Is resistance greater than 5 ohms?	Yes ▶ No ▶	REPLACE processor. RERUN Quick Test. For 1.9L EFI, GO to Q6. For 3.0L EFI passenger car GO to Q9. All others GO to Q8.
Q6 HEGO HARNESS SHORT TO POWER		
• Key Off. • Breakout box installed, leave processor disconnected. • Disconnect HEGO. • Measure resistance between test Pin 29 and test Pins 37 and 57 at the Breakout box. • Is reading less than 10,000 ohms?	No ▶ Yes ▶	SERVICE harness short to power. RERUN Quick Test. GO to Q7.
Q7 HEGO SHORT TO POWER		
• Key Off. • HEGO disconnected. • Measure resistance between RUN circuit and EGO circuit on the HEGO connector. • Is reading less than 10,000 ohms? **NOTE: Refer to HA for connector orientation.**	Yes ▶ No ▶	REPLACE HEGO. RERUN Quick Test. GO to Q8.
Q8 INTERMITTENT NDS		
• Key Off. • Breakout box installed. • Connect DVOM between test Pin 30 and test Pin 40 or 60 at the Breakout box. • Key On, engine running Self-Test. • Is voltage greater than 1V? **NOTE: Refer to FA for connector orientation.**	Yes ▶ No ▶	SERVICE intermittent in NDS harness, connector or switch. If OK, GO to Quick Test Step 5.0 for appropriate service codes. SERVICE intermittent in NDS harness, connector or switch. If OK, GO to Q9.

No Codes/Codes Not Listed — Pinpoint Test — Q

TEST STEP	RESULT ▶	ACTION TO TAKE
Q9 POWER RELAY ALWAYS ON		
• Key Off. • Breakout box installed. • Connect DVOM to test Pin 37 or 57 and to test Pin 40 or 60 at the Breakout box. • Turn key on and off. Wait 10 seconds. • Does voltage change from 10.5V or greater to zero volts?	Yes ▶ No ▶	GO to Q9A. REPLACE EEC Power Relay or integrated Relay Controller. RERUN Quick Test.
Q9A SHIFT LIGHT		
• Is vehicle equipped with shift indicator light?	No ▶ Yes ▶	SERVICE STO circuit for short to ground. RERUN Quick Test. GO to KL7.

Continuous Test Code 15 — Pinpoint Test Q

TEST STEP	RESULT ▶	ACTION TO TAKE
Q10 CONDITIONS FOR CONTINUOUS CODE 15		
• Power interruption to Keep Alive Memory (KAM) Pin 1 may result in a service code being outputted.* • Clear continuous memory codes (use procedure described in Quick Test Step 6). • Repeat Quick Test Step 3.0 through Continuous memory code output. • Code 15 present on retest? *NOTE: **Anytime power is interrupted to the processor, a code 15 may be outputted the first time Self-Test is run after restoration of power. Repeat Self-Test to ensure correct diagnosis.**	Yes ▶	GO to Q11.
	No ▶	Test complete.
Q11 INSPECT ENGINE COMPARTMENT WIRING FOR PROPER ROUTING		
• Are any EEC components or EEC wiring close to ignition components or wires (High Electrical Energy Sources)? If EEC wiring close, reroute and rerun Self-Test. • Is code 15 still present?	Yes ▶	GO to Q12.
	No ▶	Test complete.
Q12 CHECK POWER CIRCUIT TO KEEP ALIVE MEMORY		
• Key Off, wait 10 seconds. • Disconnect processor and inspect both 60 Pin connectors. • Connect Breakout box to harness, leave processor disconnected. • DVOM on 20V scale. • Connect positive test lead to test Pin 1 and negative test lead to test Pin 40 or 60 at the Breakout box. • Key On. • Observe voltage reading.	Reading less than 10V ▶	SERVICE open to KAM circuit. RERUN Quick Test.
	Reading 10V or greater ▶	REPLACE processor. RERUN Quick Test.

Engine Stalls During Quick Test — Pinpoint Test Q

TEST STEP	RESULT ▶	ACTION TO TAKE
Q30 DIAGNOSTIC ROUTINES COMPLETE		
• Have diagnostic routines in Section 2 been completed?	Yes ▶	GO to Q31.
	No ▶	GO to Diagnostic Routines, Section 2.
Q31 ATTEMPT TO BRING ENGINE TO OPERATING TEMPERATURE		
• Key Off, wait 10 seconds. • Install tachometer. • Try to maintain engine at 2,000 rpm for 2 minutes. • Can rpm be maintained?	Yes ▶	GO to Pinpoint Test Step A13.
	No ▶	GO to Q32.
Q32 MAINTAIN ENGINE OPERATION DURING ENGINE RUNNING QUICK TEST		
• Engine at operating temperature. • Tachometer installed. • Key Off, wait 10 seconds. • Activate Quick Test. • Perform Engine Running Self-Test while maintaining 2,000 rpm.	Code 11 ▶	GO to Quick Test Step 5.0.
	Any service code ▶	GO to Quick Test Step 5.0B.
	No codes ▶	GO to Pinpoint Test Step Q1.

Output State Check Not Functioning — Pinpoint Test Q

TEST STEP	RESULT ▶	ACTION TO TAKE
Q40 CHECK FOR CODES 23, 53, 63 OR 68		
• Key Off, wait 10 seconds. • Perform Key On, Engine Off Self-Test. • Leave Key On to enter Output State Check. • Key On, Engine Off. Are codes 23, 53, 63 or 68 present?	Yes ▶	GO to Quick Test Step 3.0B and SERVICE appropriate code as instructed.
	Code 11 ▶	GO to Q41.
	No Codes ▶	GO to Q1.
Q41 CHECK THROTTLE LINKAGE		
• Check throttle and throttle linkages for sticking and binding. • Throttle OK?	Yes ▶	SERVICE as necessary. RERUN Quick Test.
	No ▶	REPLACE TP sensor. RERUN Quick Test.

Processor Power Check — Pinpoint Test Q

TEST STEP	RESULT ▶	ACTION TO TAKE
Q50 CHECK FOR POWER TO PROCESSOR		
• Key Off, wait 10 seconds. • Disconnect processor 60 Pin connector and inspect for damaged pins, corrosion, loose wires. Service as necessary. • Install Breakout box. • Key On, Engine Off. • DVOM on 20V scale. • Measure voltage between test Pin 37 and test Pin 40 at the Breakout box and between test Pin 57 and test Pin 60 at the Breakout box.	Either voltage reading less than 10.5V ▶	2.5L CFI and 3.0L EFI passenger car GO to Pinpoint Test X-1. All others GO to Pinpoint Test B-1.
	Both voltage readings 10.5V or greater ▶	REPLACE processor. RERUN Quick Test.

System Check | Pinpoint Test | S

S

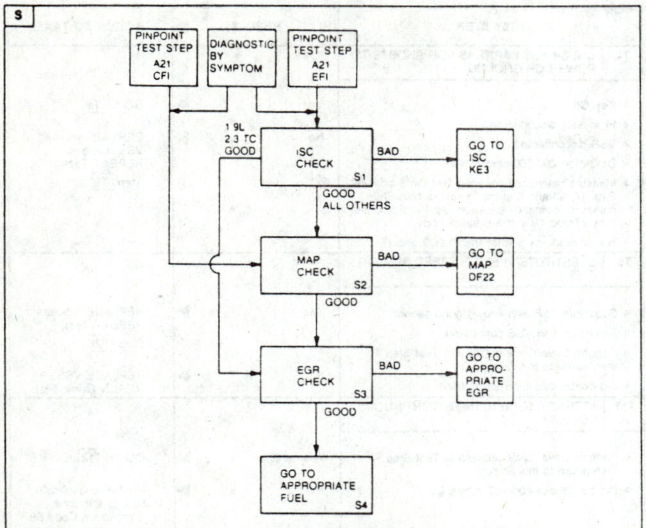

STOP-WARNING

You should enter this Pinpoint Test only after a code 11 is received in Quick Test Step 3.0, and you have been directed here from EEC-IV No-Start Pinpoint Test Step A21 or Diagnostics by symptom.

This Pinpoint Test is intended only as a Quick Check for the basic functioning of the following:

- ISC Bypass Air System
- MAP System
- EGR System

System Check | Pinpoint Test | S

TEST STEP	RESULT ▶	ACTION TO TAKE
S3 EGR CHECK		
• Disconnect and plug vacuum line at EGR valve. • Inspect EGR valve to ensure valve is closed. • Attempt to start vehicle. • Vehicle runs?	Yes ▶	For 1.9L EFI & 2.3L EFI TC, GO to KA1. For 3.8L CFI, 2.3L and 5.0L EFI Truck, GO to DD11. For 2.3L HSC CFI Base, 2.5L HSC CFI, and 5.0L SEFI, GO to DN40. For 2.3L CFI PLUS GO to DM11. For 2.9L EFI Truck and 3.0L EFI, GO to Section 6, EGR Valve Diagnostics.
	No ▶	For 1.9L EFI and 2.3L EFI TC GO to HA1. For 2.3L HSC CFI and 2.5L HSC CFI, GO to HF1. For 2.9L EFI and 3.0L EFI, GO to HG1. For 3.8L CFI, GO to HD2. For 5.0L SEFI, GO to HC1. For 2.3L EFI and 5.0L EFI Truck GO to HE1.

System Check | Pinpoint Test | S

TEST STEP	RESULT ▶	ACTION TO TAKE
S1 ISC-BPA CHECK		
• Attempt to start engine at part throttle. • Can engine run only at part throttle?	Yes ▶	GO to KE3.
	No ▶	1.9L EFI and 2.3L EFI TC GO to S3. All others GO to S2.
S2 MAP CHECK		
• Key Off. • Disconnect the MAP sensor from the vehicle harness. • Connect the MAP tester between the vehicle harness and the MAP sensor. • Plug tester banana plugs into DVOM. • Set DVOM to 20V scale. • Key On. • Observe DVOM. • Approximate Altitude (Ft.) / Voltage Output (+/- .04V): 0 — 1.59; 1000 — 1.56; 2000 — 1.53; 3000 — 1.50; 4000 — 1.47; 5000 — 1.44; 6000 — 1.41; 7000 — 1.39 • Crank engine. • While cranking, does DVOM reading decrease from the appropriate reading for your altitude listed above?	Yes ▶	3.0L EFI Truck, GO to Pinpoint Test Step HG1. all others, GO to S3.
	No ▶	GO to DF22.

Transmission — AXOD | Pinpoint Test | T

T

AXOD — HARNESS CONNECTIONS

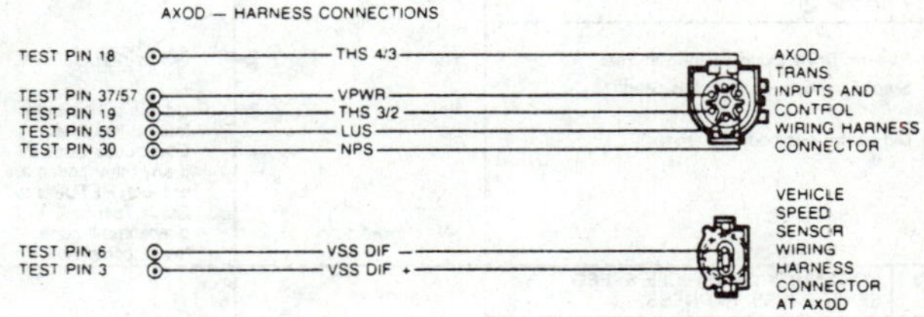

AXOD — FUNCTIONAL

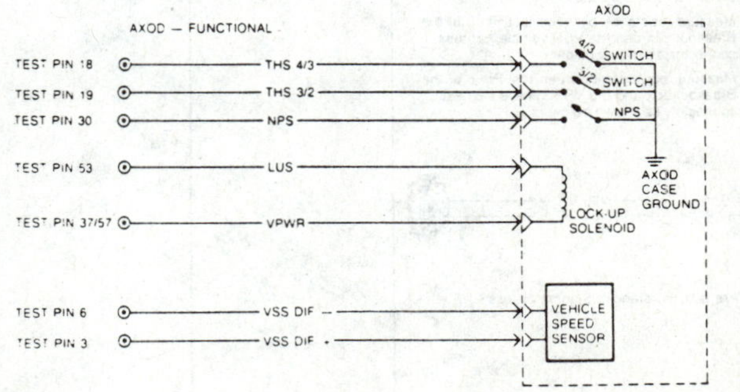

Transmission — AXOD | Pinpoint Test | T

STOP-WARNING

You should enter this Pinpoint Test only when service codes 62, 67 and 89 are received in Quick Test Step 3.0 or a service code 62 is received in Quick Test Step 5.0, and/or service codes 29, 39, 57, 59 and 69 received in Quick Test Step 6.0.

This Pinpoint Test is intended to diagnose only the following:

- Harness Circuits: THS 4-3, THS 3-2, LUS, NPS, VSS +, VSS − and VPWR
- Vehicle Speed Sensor
- Processor Assembly

T1

Drive Cycle For Checking AXOD CONTINUOUS CODES

1. Record and zero EEC-IV Self-Test continuous codes.
2. Warm engine to operating temperature.
3. With transmission in D range, lightly accelerate from a stop to 40 mph to achieve third gear. Hold speed and throttle opening (not closed throttle) steady for 15 seconds minimum (30 seconds above 4000 feet altitude).
4. Shift gear selector to OD range and accelerate lightly from 40 to 50 mph to achieve fourth gear. Hold speed and throttle opening (not closed throttle) steady for 15 seconds minimum in fourth gear.
5. With transmission in fourth gear and steady speed and throttle opening (not closed throttle) lightly apply and release brakes (to light brake lamps) and then hold speed and throttle opening steady for an additional 15 seconds minimum.
6. Brake to a stop and remain stopped for 20 seconds minimum with transmission in OD range.
7. Run EEC-IV Self-Test and record continuous codes.

NOTE: All components must be connected when performing this test.

Transmission — AXOD | Pinpoint Test | T

TEST STEP	RESULT ▶	ACTION TO TAKE
T2 ATTEMPT TO GENERATE CONTINUOUS CODE 29		
• Perform Drive Cycle outlined in Test Step T1, then return to this Step. • Did continuous code 29 repeat?	Yes ▶	GO to T3.
	No ▶	Unable to duplicate fault at this time. Continuous Code 29. If any other codes are present, RETURN to Quick Test for directions. If none, test is completed.
T3 CONTINUITY OF VEHICLE SPEED SENSOR (VSS) HARNESS		
• Key Off, wait 10 seconds. • Disconnect processor 60 Pin connector and inspect for damaged pins, corrosion, loose wires. Service as necessary. • Install Breakout box. • Processor and VSS disconnected. • DVOM on 200 ohm scale. • Measure resistance between test Pin 3 at the Breakout box and the VSS vehicle harness connector as shown below. • Measure resistance between test Pin 6 at the Breakout box and the VSS vehicle harness connector, as shown below. 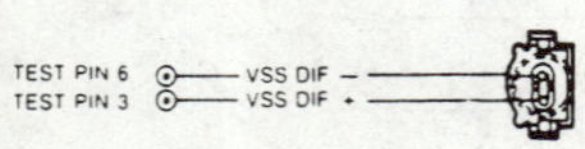 • Are both resistances 5 ohms or less?	Yes ▶	GO to T4.
	No ▶	SERVICE open(s) in VSS harness. REPEAT Test Step T2.

Transmission — AXOD | Pinpoint Test | T

TEST STEP	RESULT ▶	ACTION TO TAKE
T4 CHECK VSS HARNESS FOR SHORTS TO POWER OR GROUND		
• Key Off. • Processor disconnected. • VSS disconnected. • DVOM on 200,000 ohm scale. • Measure resistance between test Pin 3 and test Pins 37, 40 and 6 at the Breakout box. Then measure resistance between test Pin 6 and test Pins 37 and 40 at the Breakout box. • Are all readings greater than 1,000 ohms?	Yes ▶	GO to T5.
	No ▶	SERVICE short(s) in VSS harness. REPEAT Test Step T2.
T5 SUBSTITUTE VEHICLE SPEED SENSOR (VSS)		
• Substitute VSS with known good sensor. • Processor and VSS connected. • Perform Drive Cycle outlined in Test Step T1, then return to this Step. • Did continuous code 29 repeat?	Yes ▶	REPLACE processor. REPEAT Test Step T2.
	No ▶	REPLACE VSS. RERUN Quick Test.
T11 ATTEMPT TO GENERATE CONTINUOUS CODE 69		
• Perform Drive Cycle outlined in Test Step T1, then return to this Step. • Did continuous code 69 repeat?	Yes ▶	GO to T12.
	No ▶	Unable to duplicate fault at this time. Continuous Code 69. If any other codes are present, RETURN to Quick Test for directions. If none, test is completed.

Transmission — AXOD | Pinpoint Test | T

TEST STEP	RESULT ▶	ACTION TO TAKE
T12 CHECK CONTINUITY OF THS 3-2 HARNESS CIRCUIT		
• Key Off, wait 10 seconds. • Disconnect processor 60 Pin connector and inspect for damaged pins, corrosion, loose wires. Service as necessary. • Install Breakout box. • Processor and AXOD harness disconnected. • DVOM on 200 ohm scale. • Measure resistance between test Pin 19 at the Breakout box and the AXOD vehicle harness connector, as shown below. • Is resistance 5 ohms or less? 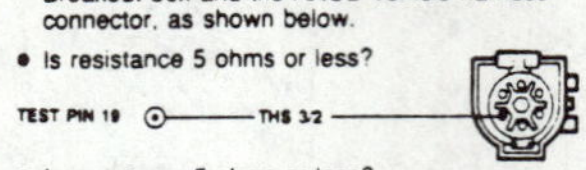 • Is resistance 5 ohms or less?	Yes ▶	GO to T13.
	No ▶	SERVICE open in THS 3/2 harness. REPEAT Test Step T11.
T13 CHECK THS 3-2 HARNESS CIRCUIT FOR SHORTS TO POWER		
• Key Off. • Breakout box installed. • Processor disconnected. • AXOD harness disconnected. • DVOM on 200,00 ohm scale. • Measure the resistance between test Pin 19 and test Pin 37 at the Breakout box. • Is resistance greater than 10,000 ohms?	Yes ▶	GO to T14.
	No ▶	SERVICE short to power in THS 3-2 circuit. REPEAT Test Step T11.
T14 PROCESSOR VERIFICATION		
• Key Off. • Breakout box installed. • Processor connected. • AXOD harness connected. • Jumper test Pin 19 to test Pin 40 at the Breakout box. • Run Key On, Engine Off Quick Test. • Is code 62 present?	No ▶	REMOVE jumper wire. REPLACE processor. REPEAT Test Step T11.
	Yes ▶	REMOVE jumper wire. GO to Taurus/Sable Shop Manual, Section 17-15 for AXOD Transmission Electrical Component Diagnostics.

Transmission — AXOD | Pinpoint Test | T

TEST STEP	RESULT ▶	ACTION TO TAKE
T21 ATTEMPT TO GENERATE CONTINUOUS CODE 59		
• Perform Drive Cycle outlined in Test Step T1, then return to this Step. • Did continuous code 59 repeat?	Yes ▶ No ▶	GO to T22. Unable to duplicate fault at this time. Continuous Code 59. If any other codes are present, RETURN to Quick Test for directions. If none, test is completed.
T22 CHECK CONTINUITY OF THS 4/3 HARNESS CIRCUIT		
• Key Off, wait 10 seconds. • Disconnect processor 60 pin connector and inspect for damaged pins, corrosion, loose wires. Service as necessary. • Install Breakout box. • Processor and AXOD harness disconnected. • DVOM on 200 ohm scale. • Measure resistance between test Pin 18 at the Breakout box and the AXOD vehicle harness connector, as shown below. • Is resistance 5 ohms or less?	Yes ▶ No ▶	GO to T23. SERVICE open in THS 4/3 harness. REPEAT Test Step T21.

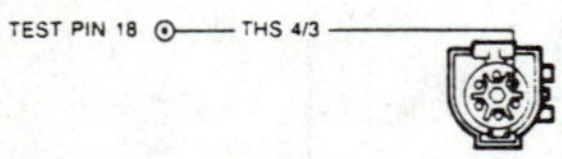

Transmission — AXOD | Pinpoint Test | T

TEST STEP	RESULT ▶	ACTION TO TAKE
T23 CHECK THS 4/3 HARNESS CIRCUIT FOR SHORT TO POWER		
• Key Off. • Breakout box installed. • Processor disconnected. • AXOD harness disconnected. • DVOM on 200,000 ohm scale. • Measure the resistance between Pin 18 and test Pin 37 at the Breakout box. • Is resistance greater than 10,000 ohms?	Yes ▶ No ▶	GO to T24. SERVICE short to power to THS 4/3 circuit. REPEAT Test Step T21.
T24 PROCESSOR VERIFICATION		
• Key Off. • Breakout box installed. • Processor connected. • AXOD harness connected. • Jumper test Pin 18 to test Pin 40 at the Breakout box. • Run Key On, Engine Off Quick Test. • Is code 62 present?	No ▶ Yes ▶	REMOVE jumper wire. REPLACE processor. REPEAT TEST Step T21. REMOVE jumper wire. GO to Taurus/Sable Shop Manual, Section 17-15 for AXOD Transmission Electrical Component Diagnostics.
T31 ATTEMPT TO GENERATE CONTINUOUS CODE 39		
NOTE: If continuous code 59 is also present, go directly to T21. • Perform Drive Cycle outlined in Test Step T1, then return to this Step. • Did continuous code 39 repeat?	Yes ▶ No ▶	GO to Taurus Sable Shop Manual Section 17-15 for AXOD Transmission Electrical Component Diagnostics. Unable to duplicate fault at this time. Continuous Code 39. If any other codes are present, RETURN to Quick Test for directions. If none, test is completed.

Transmission — AXOD | Pinpoint Test | T

TEST STEP	RESULT ▶	ACTION TO TAKE
T41 ATTEMPT TO GENERATE CONTINUOUS CODE 57		
• Perform Drive Cycle outlined in Test Step T1, then return to this Step. • Did continuous code 57 repeat?	Yes ▶ No ▶	GO to T42. Unable to duplicate fault at this time. Continuous Code 57. If any other codes are present, RETURN to Quick Test for directions. If none, test is completed.
T42 CHECK CONTINUITY OF NPS HARNESS CIRCUIT		
• Key Off, wait 10 seconds. • Disconnect processor 60 Pin connector and inspect for damaged pins, corrosion, loose wires, etc. Service as necessary. • Install Breakout box. • Processor and AXOD harness disconnected. • DVOM on 200 ohm scale. • Measure resistance between test Pin 30 at the Breakout box and the AXOD vehicle harness connector, as shown below. • Is resistance 5 ohms or less?	Yes ▶ No ▶	GO to Taurus/Sable Shop Manual, Section 17-15 for AXOD Transmission Electrical Component Diagnostic. SERVICE open in NPS circuit. REPEAT Test Step T41.

Transmission — AXOD | Pinpoint Test | T

TEST STEP	RESULT ▶	ACTION TO TAKE
SERVICE CODE 89		
T51 VPWR HARNESS CONTINUITY CHECK		
• Key Off, wait 10 seconds. • Disconnect processor 60 Pin connector and inspect for damaged pins, corrosion, loose wires. Service as necessary. • Install Breakout box. • Processor and AXOD harness disconnected. • DVOM on 200 ohm scale. • Measure resistance between test Pin 37 at the Breakout box and the AXOD vehicle harness connector, as shown • Is resistance 5 ohms or less?	Yes ▶ No ▶	GO to T52. SERVICE open in VPWR circuit to AXOD. RERUN Quick Test.
T52 LVS HARNESS CONTINUITY CHECK		
• Key Off • Breakout box installed • Processor and AXOD harness disconnected • DVOM on 200 ohm scale • Mesasure resistance between test Pin 53 at the Breakout box and the AXOD vehicle harness connector, as shown • Is resistance 5 ohms or less?	Yes ▶ No ▶	GO to T53. SERVICE open in LUS harness circuit to AXOD. RERUN Quick Test.

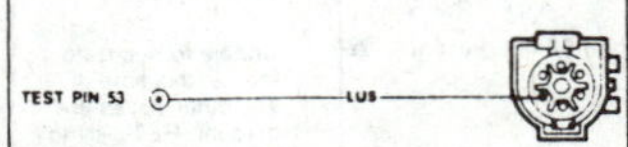

Transmission — AXOD | Pinpoint Test | T

TEST STEP	RESULT ▶	ACTION TO TAKE
T53 CHECK LUS HARNESS FOR SHORTS TO POWER, GROUND		
• Key Off. • Processor disconnected. • AXOD harness disconnected. • DVOM on 200,000 ohm scale. • Measure resistance between test Pin 53 and test Pins 37 and 40 at the Breakout box. • Are both readings greater than 10,000 ohms?	Yes ▶ No ▶	GO to T54. SERVICE shorts in LUS harness. RERUN Quick Test. If code 89 is still present, REPLACE processor. RERUN Quick Test.
T54 TOTAL CIRCUIT RESISTANCE CHECK		
• Key Off. • Breakout box installed. • Processor disconnected. • AXOD transmission harness connected. • DVOM on 200 ohm scale. • Measure the resistance between test Pin 53 and test Pin 57 at the Breakout box. • Is resistance between 20 ohms and 40 ohms?	Yes ▶ No ▶	REPLACE processor. RERUN Quick Test. GO to Taurus Sable Shop Manual, Section 17-15 for AXOD Transmission Electrical Component Diagnostic.
SERVICE CODE 62: KEY ON, ENGINE OFF QUICK TEST		
T61 CHECK FOR CODE 62 IN ENGINE RUNNING SELF-TEST		
• Perform Engine Running Self-Test and record codes. • Is code 62 present?	Yes ▶ No ▶	Fault is in the 4-3 circuit. GO to T62. Fault is in the 3-2 circuit. GO to T62.

Transmission — AXOD | Pinpoint Test | T

TEST STEP	RESULT ▶	ACTION TO TAKE
T81 SERVICE CODE 67 — CHECK VOLTAGE AT NPS INPUT TO PROCESSOR		
• Key On, Engine Off. • Disconnect processor 60 Pin connector and inspect for damaged pins, corrosion, loose wires. Service as necessary. • Install Breakout box and reconnect processor. • DVOM on 20V scale. • Measure voltage between test Pin 30 and test Pin 46 at the Breakout box. • Is voltage less than 4V?	Yes ▶ No ▶	GO to T82. GO to Pinpoint Test FA4.
T82 CHECK NPS HARNESS CIRCUIT FOR SHORT TO GROUND		
• Key Off. • Breakout box installed. • Processor and AXOD harness disconnected. • DVOM on 200,000 ohm scale. • Measure resistance between test Pin 30 and test Pins 40 and 60 at the Breakout box. • Are all readings greater than 10,000 ohms?	Yes ▶ No ▶	GO to T83. SERVICE short to ground in NPS circuit. RERUN Quick Test.

Transmission — AXOD | Pinpoint Test | T

TEST STEP	RESULT ▶	ACTION TO TAKE
SERVICE CODE 62: KEY ON, ENGINE OFF QUICK TEST		
T62 AXOD HARNESS VERIFICATION		
• Key Off. • Disconnect AXOD harness. • Run Key On, Engine Off Quick Test. • Is code 62 still present?	Yes ▶ No ▶	GO to T63. GO to Taurus Sable Shop Manual, Section 17-15 for AXOD Transmission Electrical Component Diagnostics.
T63 CHECK THS 3-2 AND 4-3 HARNESS CIRCUITS FOR SHORT TO GROUND		
• Key Off. • Install Breakout box and disconnect processor. • AXOD harness disconnected. • DVOM on 200,000 ohm scale. • Measure resistance between test Pin 18 and test Pins 40 and 60 at the Breakout box. • Measure resistance between test Pin 19 and test Pins 40 and 60 at the Breakout box. • Are all resistance readings greater than 10,000 ohms?	Yes ▶ No ▶	REPLACE processor. RERUN Quick Test. SERVICE short(s) to ground. RERUN Quick Test.
T71 SERVICE CODE 62 ENGINE RUNNING SELF-TEST		
• Verify code 62 present in Engine Running Self-Test.	Code 62 present ▶ Code 62 not present ▶	GO to Taurus Sable Shop Manual, Section 17-15 for AXOD Transmission Electrical Component Diagnostics. Unable to duplicate fault at this time. If any other codes are present, RETURN to Quick Test for directions. If none, then test is complete.

Transmission — AXOD | Pinpoint Test | T

TEST STEP	RESULT ▶	ACTION TO TAKE
T83 PROCESSOR VERIFICATION		
• Key Off. • Breakout box installed. • Processor connected. • AXOD harness disconnected. • Run Key On, Engine Off Quick Test. • Is code 67 present?	Yes ▶ No ▶	REPLACE processor. RERUN Quick Test. GO to Taurus/Sable Shop Manual, Section 17-15 for AXOD Transmission Electrical Component Diagnostics.

Integrated Relay Controller Module — Pinpoint Test X

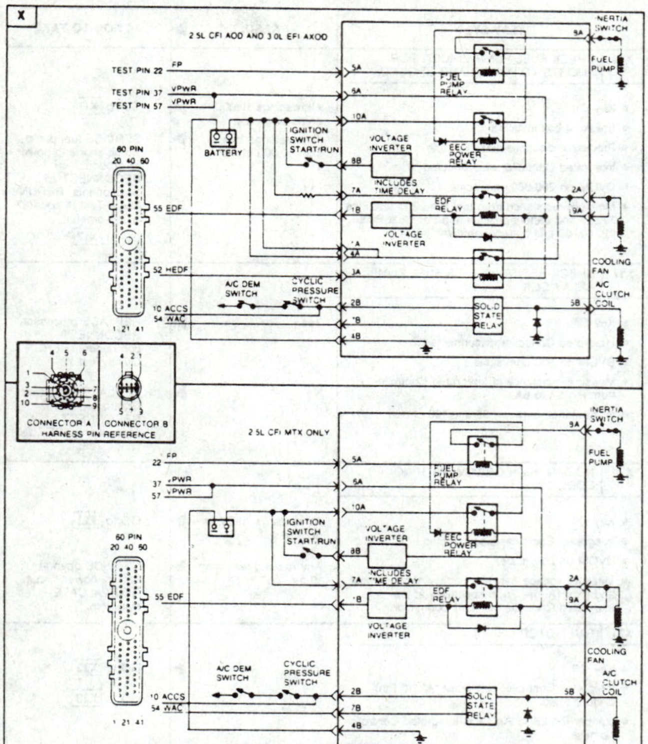

Refer to Engine Supplement for Test Pin Usage Chart.

Integrated Relay Controller Module — Pinpoint Test X

STOP-WARNING

You should enter this Pinpoint Test only when service codes 83 and 87 are received in Quick Test Step 3.0 or you are directed here from Pinpoint Test A, Pinpoint Test C, or Diagnostic By Symptom.

To prevent the replacement of good components, be aware that the following non-EEC area may be at fault:

- Fuel Lines
- Fuel Filters
- Contaminated Fuel
- Fuel Pump
- Ignition Switch
- Battery Cables
- Alternator
- Voltage Regulator
- Ground Straps
- A/C Clutch
- A/C Demand
- Cooling Fan Motor

This pinpoint test is intended to diagnose only the following:

- Integrated Relay Controller Module (within this Pinpoint Named Integrated Controller)
- Battery Voltage
- Power Relay
- EDF Relay
- HEDF Relay (Auto Trans Only)
- WAC Relay
- Fuel Pump Relay
- Harness Circuits: V Batt., VPWR, F.P., Ground and Power to Fuel Pump(s), WAC, ACCS, Cooling Fan Power, A/C Clutch, Key Power, Powers To Integrated Controller
- Processor Assembly
- A/C Demand Switch Input

Integrated Controller — Pinpoint Test X

TEST STEP	RESULT ▶	ACTION TO TAKE
VEHICLE BATTERY		
X1 BATTERY VOLTAGE CHECK		
• Key On, Engine Off. • DVOM on 20V scale. • Measure voltage across battery terminals.	10.5V or greater ▶	GO to X2.
	Less than 10.5V ▶	SERVICE discharged battery. REFER to Shop Manual, Group 31.
X2 BATTERY POWER GROUND CHECK		
• Key On, Engine Off. • Processor connected. • DVOM on 20V range. • Measure voltage between battery negative post and Signal Return circuit in the Self-Test connector.	Less than 0.5V ▶	GO to X6.
	0.5V or greater ▶	GO to X3.
X3 GROUND FAULT ISOLATION		
• Key Off. • Disconnect processor 60 Pin connector and inspect for damaged pins, corrosion, loose wires. Service as necessary. • Install Breakout box. • Key On, Engine Off. • Processor connected. • DVOM on 20V scale. • Measure voltage between battery negative post and test Pins 40 and 60 at the Breakout box.	Both readings less than 0.5V ▶	GO to X4.
	One or both readings 0.5V or greater ▶	Circuit(s) with greater than 0.5V has high resistance or open. CORRECT faulty ground circuit. RERUN Quick Test.

Integrated Controller — Pinpoint Test X

TEST STEP	RESULT ▶	ACTION TO TAKE
X4 PROCESSOR GROUND FAULT ISOLATION		
• Breakout box installed. • Key Off, wait 10 seconds. • Processor connected. • DVOM on 200 ohm scale. • Measure resistance between test Pin 46 and test Pin 40 and between test Pin 46 and test Pin 60.	Both readings less than 5 ohms ▶	GO to X5.
	One or both readings 5 ohms or greater ▶	DISCONNECT processor connector and INSPECT for corrosion, damaged pins, etc. SERVICE as necessary and RETEST. If fault is still present, REPLACE processor. RERUN Quick Test.
X5 HARNESS CHECK (SIGNAL RETURN)		
• Breakout box installed. • Key Off, wait 10 seconds. • Processor connected. • DVOM on 200 ohm scale. • Measure resistance between test Pin 46 and Signal Return circuit at the Self-Test connector.	Less than 5 ohms ▶	System OK. RUN Quick Test.
	5 ohms or greater ▶	CORRECT cause of resistance in the harness Signal Return circuit. RERUN Quick Test.
X6 MEASURE VOLTAGE AND GROUND TO INTEGRATED CONTROLLER		
• Key Off. • Integrated Controller Module disconnected (both connectors.) • DVOM on 20V scale. • Measure voltage at the harness connector to the integrated controller from connector Pin 10A (+) to Pin 4B (−). (Refer to illustration X.)	Less than 10.5V ▶	GO to X9.
	10.5V or greater ▶	GO to X7.
X7 KEY POWER TO INTEGRATED CONTROLLER		
• Integrated Controller disconnected. • DVOM on 20V scale. • Key On. • Measure voltage at the harness connector to the integrated controller from connector Pin 8B (+) to Pin 4B (−). • Refer to illustration X.	Less than 10.5V ▶	SERVICE open between 8B and ignition switch. RECONNECT Integrated Controller. RERUN Quick Test.
	10.5V or greater ▶	GO to X8.

Integrated Controller | Pinpoint Test | X

TEST STEP	RESULT ▶	ACTION TO TAKE
X8 MEASURE CONTINUITY OF VPWR		
• Key Off. • Disconnect processor 60 Pin connector and inspect for damaged pins, corrosion, loose wires, etc. Service as necessary. • Install Breakout box. • Leave processor disconnected. • DVOM on 200 ohm scale. • Measure resistance from test Pin 37 and 57 to Integrated Controller harness Pin 6A.	5 ohms or greater ▶	SERVICE open in VPWR circuit. CONNECT Integrated Controller. RERUN Quick Test.
	Less than 5 ohms ▶	REPLACE Integrated Controller. RERUN Quick Test.
X9 MEASURE CONTINUITY OF POWER GROUND TO INTEGRATED CONTROLLER		
• Key Off. • Integrated Controller disconnected. • DVOM on 200 ohm scale. • Measure resistance from battery negative to Integrated Controller connector Pin 4B.	5 ohms or greater ▶	SERVICE open in battery ground to B4 connector. RERUN Quick Test.
	Less than 5 ohms ▶	SERVICE open in battery positive to 10A (Integrated Controller harness connector). RERUN Quick Test.
X11 POWER AT FUEL PUMP(S) CHECK		
• Key On, engine Off. • Locate fuel pump(s). • DVOM on 20V scale. • Measure voltage between chassis ground and power-to-pump(s) circuit at fuel pump during crank mode.	8.0V or greater during crank ▶	GO to Shop Manual, Group 24, Electric Fuel Pump Diagnosis.
	Less than 8.0V during crank ▶	GO to X12.
X12 RESISTANCE CHECK OF FUEL PUMP INERTIA SWITCH		
• Key Off, wait 10 seconds. • Locate and disconnect fuel pump inertia switch. • DVOM on 200 ohm scale. • Measure the resistance of the fuel pump inertia switch.	Less than 5.0 ohms ▶	GO to X13.
	5.0 ohms or greater ▶	REPLACE fuel pump inertia switch. RERUN Quick Test.

Integrated Controller | Pinpoint Test | X

TEST STEP	RESULT ▶	ACTION TO TAKE
X13 POWER-TO-PUMP CIRCUIT CONTINUITY CHECK		
• Key Off. • DVOM on 200 ohm scale. • Integrated Controller disconnected. • Pump(s) disconnected. • Measure resistance from Integrated Controller harness connector Pin 8A to fuel pump(s) harness connector at fuel pump.	Less than 5 ohms ▶	REPLACE Integrated Controller. RECONNECT all components. RERUN Quick Test.
	5 ohms or greater ▶	SERVICE open in power-to-pump(s) circuit. RECONNECT Integrated Controller. RERUN Quick Test.
X14 CHECK POWER-TO-PUMP(S) FOR SHORTS TO POWER		
• Key Off. • Integrated Controller disconnected. • Fuel pumps disconnected. • DVOM on 200,000 ohm scale. • Measure resistance from Integrated Controller harness connector Pin 8A to Pin 6A (VPWR) and battery – post.	Any readings less than 10,000 ohms ▶	SERVICE short circuit. RECONNECT all components. ATTEMPT to start vehicle. If vehicle runs, RERUN Quick Test. If vehicle will not run, REPLACE Integrated Controller. RERUN Quick Test.
	All readings 10,000 ohms or greater ▶	REPLACE Integrated Controller. RERUN Quick Test.
SERVICE CODE: 87		
X15 FUEL PUMP CIRCUIT CONTINUITY		
• Key Off. • Disconnect processor 60 Pin connector and inspect for damaged pins, corrosion, loose wires, etc. Service as necessary. • Install Breakout box. • Processor disconnected. • Integrated Controller disconnected. • DVOM on 200 ohm scale. • Measure resistance of fuel pump circuit, from test Pin 22 at the Breakout box to Integrated Controller harness connector Pin 5A.	Less than 5 ohms ▶	GO to X16.
	5 ohms or greater ▶	SERVICE open in fuel pump circuit. RECONNECT processor and controller. RERUN Quick Test.

Integrated Controller | Pinpoint Test | X

TEST STEP	RESULT ▶	ACTION TO TAKE
X16 CHECK FUEL PUMP CIRCUIT FOR SHORTS TO POWER AND GROUND		
• Key Off. • Breakout box installed. • Processor disconnected. • Integrated Controller disconnected. • DVOM on 200,000 ohm scale. • Measure resistance from test Pin 22 to test Pins 37, 57 and battery positive and from test Pins 22, 40, 60 and battery negative.	All readings 10,000 ohms or greater ▶	GO to X17.
	Any readings less than 10,000 ohms ▶	SERVICE fuel pump circuit shorts to power or ground. RECONNECT all components. RERUN Quick Test. If code 87 is still present, GO to X17.
X17 CHECK RESISTANCE OF FUEL PUMP RELAY COIL		
• Key Off. • Integrated Controller disconnected. • DVOM on 200 ohm scale. • Measure resistance of Integrated Controller from Pin 5A to 6A. • Is resistance between 65 and 100 ohms?	Yes ▶	REPLACE processor. RECONNECT component. RERUN Quick Test.
	No ▶	REPLACE Integrated Controller. RECONNECT components. RERUN Quick Test.
X20 NO FAN, HIGH OR LOW WITH NO CODE 83		
• Key Off. • Integrated Controller disconnected. • DVOM on 20V scale. • Measure voltage between battery negative post and Pins 1A, 4A, 7A, respectively at the Integrated Controller harness connector.	All readings 10.5V or greater ▶	GO to X21.
	Any reading less than 10.5V ▶	SERVICE open in battery power circuit. RE-EVALUATE symptom.
X21 FAN MOTOR CHECK		
• Key Off. • Integrated Controller connector "A" (10 Pin) disconnected. • Jumper Pin 2A to Pin 1A at Integrated Controller harness. • Does fan run?	Yes ▶	GO to X22.
	No ▶	GO to X23.

Integrated Controller | Pinpoint Test | X

TEST STEP	RESULT ▶	ACTION TO TAKE
X22 CHECK FAN RUNNING MODE (LOW)		
• Key Off. • Disconnect processor. • Integrated Controller connected. • Key On. • Does fan run at low speed?	Yes ▶	GO to X26.
	No ▶	GO to X24.
X23 MEASURE BATTERY VOLTAGE SUPPLY AT FAN — BYPASSING INTEGRATED CONTROLLER		
• Key Off. • Cooling fan disconnected. • Integrated Controller connector "A" (10 Pin) disconnected. • Jumper Pin 2A to Pin 1A at Integrated Controller harness connector. • DVOM on 20V scale. • Measure voltage at cooling fan harness connector.	8.0V or greater ▶	CHANGE fan. RE-EVALUATE symptom.
	Less than 8.0V ▶	GO to X25.
X24 CHECK EDF CIRCUIT FOR SHORT TO GROUND		
• Key Off. • Processor and controller disconnected. • DVOM on 200,000 ohm scale. • Measure resistance from Pin 1B to Pin 4B at Integrated Controller harness connector.	10,000 ohms or greater ▶	CHANGE Integrated Controller. RECONNECT processor and controller. RE-EVALUATE symptom.
	Less than 10,000 ohms ▶	SERVICE short to ground in EDF circuit. RECONNECT processor and Integrated Controller. RE-EVALUATE symptom.

Integrated Controller — Pinpoint Test — X

TEST STEP	RESULT ▶	ACTION TO TAKE
X25 COOLING FAN GROUND VERIFICATION		
• Key Off. • Cooling fan disconnected. • Integrated Controller connector "A" (10 Pin) disconnected. • Jumper Pin 2A to Pin 1A at Integrated Controller harness connector. • DVOM on 20V scale. • Measure voltage positive at cooling fan harness connector, negative to battery negative.	8.0V or greater ▶	SERVICE open in ground circuit to fan. RECONNECT Integrated Controller. RE-EVALUATE symptom.
	Less than 8.0V ▶	SERVICE open in power-to-fan circuit from 2A and 9A of Integrated Controller harness connector to cooling fan connector. RECONNECT controller. RE-EVALUATE symptom.
X26 JUMPER HIGH ELECTRIC-DRIVE SIGNAL (HEDF) TO GROUND		
• Key Off. • Disconnect processor 60 Pin connector and inspect for damaged pins, corrosion, loose wires. Service as necessary. • Install Breakout box. • Integrated Controller connected. • Processor disconnected. • Jumper test Pin 52 to test Pin 40 at Breakout box. • Does fan run at high speed?	Yes ▶	GO to X27.
	No ▶	REPLACE Integrated Controller. RECONNECT processor. RE-EVALUATE symptom.
X27 ECT SENSOR CHECK		
• Reconnect processor. • Check engine coolant level. • Warm engine to operating temperature before taking ECT resistance measurement. • Key Off, wait 10 seconds. • Harness disconnected from ECT sensor. • DVOM on 200,000 ohm scale. • Measure resistance of the ECT sensor. • Is the resistance reading between 1500 ohms and 2000 ohms?	Yes ▶	REPLACE processor. RECONNECT harness to ECT sensor. RECONNECT Integrated Controllers. RE-EVALUATE symptom.
	No ▶	REPLACE ECT sensor. RECONNECT all components. RE-EVALUATE symptom.

Integrated Controller — Pinpoint Test — X

TEST STEP	RESULT ▶	ACTION TO TAKE
X35 LOW SPEED OR HIGH SPEED FAN ALWAYS "ON", NO SERVICE CODE 83 OR 67		
• Key Off. • Disconnect processor 60 Pin connector and inspect for damaged pins, corrosion, loose wires. Service as necessary. • Breakout box installed, leave processor disconnected. • Disconnect the Integrated Controller. • DVOM on 200 ohm scale. • Measure the resistance between test Pin 55 and controller harness Pin 1B. • Is reading 5 ohms or less?	Yes ▶	GO to X36.
	No ▶	SERVICE open in EDF circuit. RECONNECT all components. RE-EVALUATE symptom.
X36 CHECK EDF CIRCUIT FOR SHORTS TO POWER		
• Key Off. • Breakout box installed. • Processor and Integrated Controller disconnected. • DVOM on 200,000 ohm scale. • Measure resistance from test Pin 55 to test Pin 37 and from test Pin 55 to battery positive. • Is resistance 10,000 ohms or less?	No ▶	GO to X37.
	Yes ▶	SERVICE short to power in EDF circuit. GO to X37.
X37 EDF SHORT TO GROUND CHECK		
• Key On. • Breakout box installed. • Processor disconnected. • Integrated Controller connected. • Jumper test Pin 55 to ground test Pin 40 or 60. • Does fan continue to run?	Yes ▶	REPLACE controller. RE-EVALUATE symptom.
	No ▶	REPLACE processor. RE-EVALUATE symptom.

Integrated Controller — Pinpoint Test — X

TEST STEP	RESULT ▶	ACTION TO TAKE
X30 SERVICE CODE 83: RESISTANCE CHECK OF HEDF CONTROLLER CIRCUIT		
• Key Off. • Integrated Controller disconnected. • DVOM on 200 ohm scale. • Measure resistance between Pin 3A and Pin 6A at the Integrated Controller. • Is the resistance reading between 50 ohms and 100 ohms?	Yes ▶	GO to X31.
	No ▶	REPLACE controller. RERUN Quick Test.
X31 CHECK HEDF PROCESSOR SIGNAL TO INTEGRATED CONTROLLER FOR OPEN		
• Key Off • Disconnect processor 60 Pin connector and inspect for damaged pins, corrosion, loose wires. Service as necessary. • Install Breakout box, leave processor disconnected. • Integrated Controller connector "A" (10 Pin) disconnected. • DVOM on 200 ohm scale. • Measure continuity from test Pin 52 at Breakout box to Pin 3A of Integrated Controller harness. • Is resistance 5 ohms or less?	Yes ▶	GO to X32.
	No ▶	SERVICE open in HEDF circuit. RECONNECT all components. RERUN Quick Test.
X32 CHECK FOR SHORTS TO GROUND IN THE HEDF CIRCUIT		
• Key Off • Breakout box installed • Processor and Integrated Controller disconnected. • DVOM on 200,000 ohm scale. • Measure resistance from test Pin 52 to test Pin 40. • Is resistance 10,000 ohms or less?	No ▶	GO to X33.
	Yes ▶	SERVICE short to ground in HEDF circuit. RERUN Quick Test.
X33 CHECK FOR SHORTS TO POWER IN THE HEDF CIRCUIT		
• Key Off. • Breakout box installed. • Processor and Integrated Controller disconnected. • DVOM on 200,000 ohm scale. • Measure resistance from test Pin 52 to test Pin 37. • Is resistance 10,000 ohms or less?	No ▶	REPLACE processor. RECONNECT all components. RERUN Quick Test.
	Yes ▶	SERVICE short to power. RECONNECT all components. RERUN Quick Test. If code 83 is still present, REPLACE processor. RERUN Quick Test.

Integrated Controller — Pinpoint Test — X

TEST STEP	RESULT ▶	ACTION TO TAKE
X40 NO FAN		
• Key Off. • Integrated Controller disconnected. • DVOM on 20V scale. • Measure voltage between battery negative post and Pin 7A at the Integrated Controller harness connector.	10.5V or greater ▶	GO to X41.
	Less than 10.5V ▶	SERVICE open in battery power circuit. RE-EVALUATE symptom.
X41 FAN MOTOR CHECK		
• Key Off. • Integrated Controller connector "A" (10 Pin) disconnected. • Jumper Pin 2A to Pin 7A at Integrated Controller harness. • Does fan run?	Yes ▶	GO to X42.
	No ▶	GO to X43.
X42 CHECK FAN RUNNING MODE		
• Key Off • Disconnect processor • Integrated Controller connected. • Key On • Does fan run?	Yes ▶	GO to X46.
	No ▶	GO to X44.
X43 MEASURE BATTERY VOLTAGE SUPPLY AT FAN — BYPASSING INTEGRATED CONTROLLER		
• Key Off • Cooling fan disconnected. • Integrated Controller connector "A" (10 Pin) disconnected. • Jumper Pin 2A to Pin 7A at Integrated Controller harness connector. • DVOM on 20V scale. • Measure voltage at cooling fan harness connector.	8.0V or greater ▶	CHANGE fan. RE-EVALUATE symptom.
	Less than 8.0V ▶	GO to X45.

Integrated Controller — Pinpoint Test X

TEST STEP	RESULT ▶	ACTION TO TAKE
X44 CHECK EDF CIRCUIT FOR SHORT TO GROUND		
• Key Off. • Processor and controller disconnected. • DVOM on 200,000 ohm scale. • Measure resistance from Pin 1B to Pin 4B at Integrated Controller harness connector.	10,000 ohms or greater ▶	CHANGE Integrated Controller. RECONNECT processor and controller. RE-EVALUATE symptom.
	Less than 10,000 ohms ▶	SERVICE short to ground in EDF circuit. RECONNECT processor and integrated Controller. RE-EVALUATE symptom.
X45 COOLING FAN GROUND VERIFICATION		
• Key Off. • Cooling fan disconnected. • Integrated Controller connector "A" (10 Pin) disconnected. • Jumper Pin 2A to Pin 7A at Integrated Controller harness connector. • DVOM on 20V scale. • Measure voltage positive at cooling fan harness connector, negative to battery negative.	8.0V or greater ▶	SERVICE open in ground circuit to fan. RECONNECT Integrated Controller. RE-EVALUATE symptom.
	Less than 8.0V ▶	SERVICE open in power-to-fan circuit from 2A and 9A of Integrated Controller harness connector to cooling fan connector. RECONNECT controller. RE-EVALUATE symptom.
X46 ECT SENSOR CHECK		
• Reconnect processor. • Check engine coolant level. • Warm engine to operating temperature before taking ECT resistance measurement. • Key Off, wait 10 seconds. • Harness disconnected from ECT sensor. • DVOM on 200,000 ohm scale. • Measure resistance of the ECT sensor. • Is the resistance reading between 1500 ohms and 2000 ohms?	Yes ▶	REPLACE processor. RECONNECT harness to ECT sensor. RECONNECT integrated Controller. RE-EVALUATE symptom.
	No ▶	REPLACE ECT sensor. RECONNECT all components. RE-EVALUATE symptom

Integrated Controller — Pinpoint Test X

TEST STEP	RESULT ▶	ACTION TO TAKE
X50 CHECK FOR VOLTAGE AT A/C CLUTCH		
• Key On, Engine Off. • A/C demand to A/C. • DVOM on 20V scale. • Check voltage at A/C clutch harness connector.	10.5V or greater ▶	GO to Shop Manual, Group 36, A/C Diagnosis.
	Less than 10.5V ▶	GO to X51.
X51 CHECK FOR CONTINUITY FROM INTEGRATED CONTROLLER TO A/C CLUTCH		
• Key Off. • Integrated Controller "B" disconnected, (8 Pin). • DVOM on 200 ohm scale. • Measure resistance between Pin 5B at controller harness to the power circuit at the A/C clutch harness connector. • Is resistance 5 ohms or less?	Yes ▶	GO to X52.
	No ▶	SERVICE open in power to A/C. RE-EVALUATE symptom.
X52 ENTER OUTPUT STATE CHECK (REFER TO APPENDIX)		
NOTE: Do not use STAR tester for this Step, use VOM/DVOM. • Key Off, wait 10 seconds. • DVOM on 20V scale. • Connect DVOM negative test lead to STO and positive test lead to battery positive. • Jumper STI to signal return. • Perform Key On, Engine Off Self-Test until the completion of the Continuous Test Codes. • DVOM will indicate zero volts. • Depress and release the throttle. • Did DVOM reading change to a high voltage reading?	Yes ▶	REMAIN in Output State Check. GO to X53.
	No ▶	DEPRESS throttle to WOT and RELEASE. If STO voltage does not go high, GO to Pinpoint Test Step Q40. LEAVE equipment hooked up.

Integrated Controller — Pinpoint Test X

TEST STEP	RESULT ▶	ACTION TO TAKE
X53 CHECK WAC OUTPUT FOR PROPER ELECTRICAL OPERATION		
• Key On, engine Off. • Disconnect processor 60 Pin connector and inspect for damaged pins, corrosion, loose wires. Service as necessary. • Connect Breakout box to harness. Reconnect processor. • A/C demand switch to A/C on position. • DVOM on 20V scale. • Connect DVOM positive test lead to test Pin 37 and negative test lead to test Pin 54. • While observing DVOM, depress and release the throttle several times (to cycle output on and off). • Does voltage output cycle high and low?	Yes ▶	GO to X54.
	No ▶	GO to X57.
X54 CHECK FOR VOLTAGE AT A/C CLUTCH SWITCH		
• Key On, Engine Off. • A/C demand switch to A/C on position. • DVOM on 20V scale. • Leave Breakout box installed. • Processor and Integrated Controller connected. • Measure voltage between test Pin 10 and test Pin 40 at Breakout box. • Is voltage 10.5V or greater?	No ▶	GO to X56.
	Yes ▶	GO to X55.
X55 CHECK CONTINUITY OF ACCS TO INTEGRATED CONTROLLER		
• Key Off, wait 10 seconds. • Processor disconnected. • Integrated Controller "B" connector disconnected. • DVOM on 200 ohm scale. • Measure resistance between test Pin 10 at Breakout box and test Pin 2B at controller harness connector. • Is resistance 5 ohms or less?	No ▶	SERVICE open in ACCS circuit. RE-EVALUATE symptom.
	Yes ▶	REPLACE Integrated Controller. RE-EVALUATE symptom.

Integrated Controller — Pinpoint Test X

TEST STEP	RESULT ▶	ACTION TO TAKE
X56 CHECK CONTINUITY OF ACCS CIRCUIT		
• Key Off, wait 10 seconds. • A/C demand to A/C. • DVOM on 200 ohm scale. • Measure resistance between test Pin 10 and A/C demand switch. • Is resistance 5 ohms or less?	No ▶	SERVICE open in circuit. RERUN Quick Test.
	Yes ▶	EEC-IV system OK. REFER to Shop Manual, Group 36 A/C Diagnosis.
X57 CHECK CONTINUITY IN WAC TO INTEGRATED CONTROLLER CIRCUIT		
• Key Off, wait 10 seconds. • Disconnect processor 60 Pin connector and inspect for damaged pins, corrosion, loose wires. Service as necessary. • Connect Breakout box to harness. Leave processor and Integrated Controller connect "B" disconnected, (8 Pin). • DVOM on 200 ohm scale. • Measure resistance between test Pin 54 and Pin 7B at Integrated Controller harness. • Is resistance 50 ohms or less?	No ▶	SERVICE open in WAC circuit. RE-EVALUATE symptom.
	Yes ▶	GO to X58.
X58 CHECK WAC CIRCUIT FOR SHORTS TO GROUND		
• Key Off, wait 10 seconds. • Leave Breakout box installed and processor disconnected, (8 Pin). • Integrated Controller connector "B" disconnected. • DVOM on 200,000 ohm scale. • Measure resistance between test Pin 54 and test Pin 40 and between test Pin 54 and test Pin 46 and between test Pin 54 and battery negative post.	All resistance readings 10,000 ohms or greater ▶	GO to X59.
	Any resistance reading less than 10,000 ohms ▶	SERVICE shorts to ground in WAC circuit. RE-EVALUATE symptom.

Integrated Controller — Pinpoint Test X

TEST STEP	RESULT ▶	ACTION TO TAKE
X59 CHECK WAC CIRCUIT FOR SHORTS TO POWER		
• Key Off, wait 10 seconds. • Leave Breakout box installed and processor disconnected. • Integrated Controller connector "B" disconnected. • DVOM on 200,000 ohm scale. • Measure resistance between test Pin 54 and test Pin 37 and between test Pin 54 and battery positive.	All readings 10,000 ohms or greater ▶ Any readings less than 10,000 ohms ▶	GO to X60. SERVICE shorts to power in WAC circuit and then GO to X60.
X60 CHECK FOR VOLTAGE AT A/C CLUTCH		
• Key Off, wait 10 seconds. • Leave Breakout box installed. • Processor disconnected. • Integrated Controller connected. • A/C clutch disconnected. • A/C demand to A/C. • Key On, Engine Off. • DVOM on 20V scale. • Measure voltage at A/C clutch harness connection. • 10.5V or greater?	Yes ▶ No ▶	REPLACE processor. RE-EVALUATE symptom. REPLACE Integrated Controller. RE-EVALUATE symptom.

Erratic Ignition — Pinpoint Test Y

TEST STEP	RESULT ▶	ACTION TO TAKE
Y1 CODE 14 ERRATIC IGNITION		
NOTE: Code 14 indicates two successive erratic profile ignition pickup (PIP) pulses occurred, resulting in a possible engine miss or stall. • Check EEC-IV and ignition systems harnesses for: — Loose wires/connectors. — Arcing secondary ignition components (coil, cap, rotor, wires, plugs, etc.). — On-board transmitter (2-way radio) * • Are any of the above present? *Verify all 2-way radio installations. Carefully follow manufacturer's installation instructions regarding the routing of antenna and power leads.	Yes ▶ No ▶	SERVICE as necessary. RERUN Quick Test. GO to Y2.
Y2 CHECK DISTRIBUTOR		
• Key Off, wait 10 seconds. • Deactivate Self-Test. • Connect VOM or STAR per Quick Test Step 2.0. • Enter Engine Running Continuous Monitor Test (as instructed in Quick Test Step 6.3). • Observe VOM or STAR LED for indication of a fault while performing the following: • Lightly tap on TFI module and distributor. • Wiggle TFI connector. • Is a fault indicated?	Yes ▶ No ▶	DISCONNECT and INSPECT connectors. If connector and terminals are good, GO to Section 15, Ignition System Diagnostics. GO to Y3.
Y3 CHECK EEC-IV HARNESS		
• While still in Continuous Monitor Test from Step Y2 observe VOM or STAR LED for a fault indication while performing the following: • While looking for faults grasp the harness close to the TFI connector. Wiggle, shake or bend a small section of the ignition and EEC-IV systems harness while working your way to the other components and dash panel. Also wiggle, shake or bend the EEC-IV harness from the dash panel to the processor. Isolate the PIP circuit if needed for this test. • Is a fault indicated?	Yes ▶ No ▶	ISOLATE fault and SERVICE as necessary. RERUN Quick Test. GO to Y4.

Erratic Ignition — Pinpoint Test Y

TEST STEP	RESULT ▶	ACTION TO TAKE
Y4 CHECK PROCESSOR AND HARNESS CONNECTORS		
• Key Off, wait 10 seconds. • Disconnect processor 60 Pin connector and inspect for damaged pins, corrosion, loose wires. • Are connectors and terminals OK? • Reconnect processor when this Step is completed.	No ▶ Yes ▶	SERVICE as necessary. RERUN Quick Test. Unable to duplicate an erratic ignition fault in the EEC-IV System, for further diagnosis, GO to Section 15, Ignition System Diagnostics.

Electronic Control Assembly (EFI/CFI/FBC)	12A650	

DESCRIPTION

The center of the EEC IV system is a microprocessor called the Electronic Control Assembly (ECA). The ECA, (Figure 1), receives data from a number of sensors and other electronic components (switches, relays, etc.). The ECA contains a specific calibration for optimizing emissions, fuel economy, and driveability. Based on information received and programmed into its memory, the ECA generates output signals to control various relays, solenoids, and other actuators.

The ECA in the EEC IV system is a microprocessor like the one in the other EEC systems. One significant difference is that this ECA has the calibration module located inside the ECA assembly, unlike the EEC III system.

The ECA is found in different locations, depending on the model. Refer to the chart for locations.

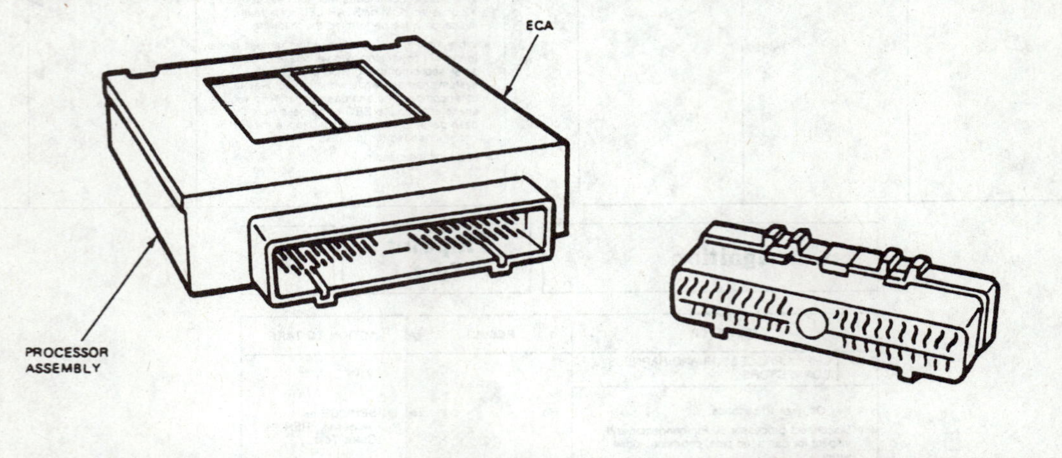

Vehicle	Location
Mark VII/Continental, Thunderbird/Cougar, Mustang/Capri, XR4Ti, Ranger/Bronco II	Passenger compartment under the instrument panel on the RH kick panel
Tempo/Topaz, Escort/Lynx/EXP	Under instrument panel left of steering column
Taurus/Sable	Ahead of glove compartment
Lincoln Town Car, Ford Crown Victoria/Mercury Grand Marquis Aerostar	LH side dash panel in passenger compartment
Econoline with A/C	Rear surface of A/C blower
Econoline without A/C	Under instrument panel on passenger side

Electronic Diagnosis 5

1987 FORD CARS & TRUCKS
EEC-IV SYSTEMS

QUICK TEST: Visual Check Vehicle Preparation	1.0

SPECIAL NOTES:

- Correct results of the QUICK TEST are dependent on the proper operation of related non-EEC-IV components.
- It may be necessary to disconnect or disassemble harness connector assemblies to do some of the inspections. Pin locations should be noted before disassembly.
- If the engine will not start, starts but stalls, idles rough, or runs rough; continue through QUICK TEST STEP 3.0 and follow the instructions in Step 3.0B.

VISUAL CHECK

1. Inspect the air cleaner and inlet ducting.
2. Check all engine vacuum hoses for damage, leaks, cracks, blockage, proper routing, etc.
3. Check EEC-IV system wiring harness for proper connections, bent or broken pins, corrosion, loose wires, proper routing, etc.
4. Check the processor, sensors, and actuators for physical damage.
5. Check the engine coolant for proper level.
6. Make all necessary repairs before continuing with QUICK TEST.

VEHICLE PREPARATION

1. Perform **ALL** safety steps required to start and run vehicle tests - apply parking brake, place shift lever firmly into PARK position (NEUTRAL on manual transmission), block drive wheels, etc.
2. Turn off **ALL** electrical loads — radios, lights, A/C-heater blower fans, etc.
3. Start engine and run until at operating temperature.
4. Turn engine off and proceed to QUICK TEST STEP 2.0.

QUICK TEST: Key On Engine Off Self-Test	3.0

A — PERFORMING THE KEY ON ENGINE OFF SELF-TEST

SPECIAL NOTES:

- It may be necessary to repair Non- EEC-IV faults before running Quick Test. Refer to Section 2.
- Continuous Memory Codes recorded in this step will be used for diagnosis after a PASS code 11 is received in both the Key On Engine Off and the Engine Running Self-Tests.
- On all vehicles equipped with a **4.9L ENGINE** the clutch must be depressed during the Key On Engine Off Self-Test.

HOW TO RUN THE KEY ON ENGINE OFF SELF-TEST

DO

- Verify that the vehicle has been properly prepared per **QUICK TEST STEPS** 1.0 and 2.0.
- Place ignition key in the ON position.
- Activate Self-Test
  —STAR Tester: Latch the center button in the down position.
  —Analog VOM: Jumper STI to SIG RTN at the Self-Test connectors.
- Record all service codes displayed.
- Go to **PART B** of Key On Engine Off Self-Test.

DON'T

- Depress throttle during Key On Engine Off Self-Test.

QUICK TEST: Key On Engine Off Self-Test — 3.0

B CODE OUTPUT

Key On Engine Off	Separator	Continuous Memory	ACTION TO TAKE
11	— 1(0) —	11	• Both tests indicate a PASS. — If engine idles or runs rough, Go to DIAGNOSTIC BY SYMPTOM in the proper Engine Supplement Section. If above condition is not present, Go to Quick TEST STEP 4.0. — If engine is a no start, Go directly to Pinpoint Test Step A1.
ANY CODE(S)	— 1(0) —	11	• Key On Engine Off Self-Test indicates a FAULT. — Go to **PART C** of Key On Engine Off Self-Test. — Always start with the first code displayed.
ANY CODE(S)	— 1(0) —	ANY CODE(S)	• Both tests indicate a FAULT. — DO NOT SERVICE CONTINUOUS MEMORY CODES AT THIS TIME! — Go to **PART C** of Key On Engine Off Self-Test. — Always start with the first code displayed.
11	— 1(0) —	ANY CODE(S) EXCEPT 15	• Continuous Memory indicates a FAULT. — DO NOT SERVICE CONTINUOUS MEMORY CODES AT THIS TIME! — If engine idles or runs rough, Go to DIAGNOSTIC BY SYMPTOM in the proper Engine Supplement Section. If above condition is not present, Go to Quick TEST STEP 4.0.
11	— 1(0) —	15	• Go To Pinpoint Test Step Q10.
NO CODES OUTPUTTED CODES NOT LISTED			• Self-Test did not activate or unlisted codes displayed. — Repeat Key On Engine Off Self-Test to verify the above condition. — If condition still exists Go to Pinpoint Test Step Q1.

QUICK TEST: Key On Engine Off Self-Test — 3.0

C PASSENGER CAR SERVICE CODE CHART

Key On Engine Off Service Code	Pinpoint Test Step Direction								
	1.9L EFI	1.9L CFI	2.3L EFI	2.3L TC	2.3L CFI	2.5L CFI	3.0L EFI	3.8L CFI	5.0L SEFI
13 GO to ▶	—	KB1	—	—	KB1	—	—	—	—
15 GO to ▶	Q12	Q12	Q12	Q12	Q12	Q12	Q12	Q12	Q12
19 GO to ▶	—	—	Q50	—	—	—	Q50	—	Q50
21 GO to ▶	DE1	DE1	DE1	DE1	DE1	DE1	DE1	DE1	DE1
22 GO to ▶	DF1	DF1	DF1	DF1	DF1	DF1	DF1	DF1	DF1
23 GO to ▶	DH1	KB12	DH1	DH1	KB12	DH1	DH1	DH1	DH1
24 GO to ▶	—	DB1	DB1	DB1	DB1	DB1	DB1	DB1	DB1
26 GO to ▶	DK1	—	—	DK1	—	—	—	—	—
28 GO to ▶	DA1	—	—	DA1	—	—	—	—	—
31 GO to ▶	—	DL1	DD2	—	DN1	DN1	DL1	DD2	DN1
32 GO to ▶	—	—	—	—	DN25	DN25	—	—	DN25
34 GO to ▶	—	DL8	—	—	DN20	DN20	DL8	—	DN20
35 GO to ▶	—	DL5	—	—	DN5	DN5	DL5	—	DN5
51 GO to ▶	DE10	DE10	DE10	DE10	DE10	DE10	DE10	DE10	DE10
52 GO to ▶	—	—	FF1	—	FF1	FF1	FF1	—	—
53 GO to ▶	DH3	KB15	DH3	DH3	KB15	DH3	DH3	DH3	DH3
54 GO to ▶	—	DB10	DB10	DB10	DB10	DB10	DB10	DB10	DB10
56 GO to ▶	DK10	—	—	DK10	—	—	—	—	—
58 GO to ▶	DA10	KB5	—	DA10	KB5	—	—	—	—
61 GO to ▶	DE20	DE20	DE20	DE20	DE20	DE20	DE20	DE20	DE20
62 GO to ▶	—	—	—	—	—	—	T61	—	—
63 GO to ▶	DH10	KB18	DH10	DH10	KB18	DH10	DH10	DH10	DH10
64 GO to ▶	—	DB20	DB20	DB20	DB20	DB20	DB20	DB20	DB20
66 GO to ▶	DK20	—	—	DK20	—	—	—	—	—
67 GO to ▶	FA1	FA1	FA1	FA1	FA1	FA1	T81	FA1	FA1
68 GO to ▶	DA20	KB9	—	DA20	KB9	KF1	—	KF1	—
73 GO to ▶	—	KB22	—	—	KB22	—	—	—	—
79 GO to ▶	—	—	—	—	—	—	—	—	FA4
81 GO to ▶	—	—	—	KN1	—	—	—	KC8	KC8
82 GO to ▶	—	—	—	X80	—	—	—	KC8	KC8
83 GO to ▶	—	—	DD17	X30	—	X30	X30	DD17	—
84 GO to ▶	—	DL11	DD17	KA5	DN10	DN10	DL11	DD17	DN10
85 GO to ▶	—	KD6	—	KS10	—	KD6	KD6	—	KD6
87 GO to ▶	—	J7	J7	X15	J7	X15	X15	J7	J7
88 GO to ▶	—	—	—	KS1	—	X80	—	—	—
89 GO to ▶	—	—	KR1	—	—	—	T51	KP5	—
93 GO to ▶	—	KB11	—	—	KB11	—	—	—	—
NO CODES ▶ CODES NOT LISTED ▶	GO to Pinpoint Test Step Q1								

QUICK TEST: Key On Engine Off Self-Test — 3.0

C LIGHT TRUCK SERVICE CODE CHART

Key On Engine Off Service Code	Pinpoint Test Step Direction					
	2.3L EFI	2.9L EFI	3.0L EFI	4.9L EFI	5.0L EFI	7.5L EFI
15 GO to ▶	Q12	Q12	Q12	Q12	Q12	Q12
19 GO to ▶	Q50	Q50	Q50	Q50	Q50	Q50
21 GO to ▶	DE1	DE1	DE1	DE1	DE1	DE1
22 GO to ▶	DF1	DF1	DF1	DF1	DF1	DF1
23 GO to ▶	DH1	DH1	DH1	DH1	DH1	DH1
24 GO to ▶	DB1	DB1	DB1	DB1	DB1	DB1
31 GO to ▶	DD2	DL1	—	DN1	DN1	DN1
32 GO to ▶	—	—	—	DN25	DN25	DN25
34 GO to ▶	—	DL8	—	DN20	DN20	DN20
35 GO to ▶	—	DL5	—	DN5	DN5	DN5
51 GO to ▶	DE10	DE10	DE10	DE10	DE10	DE10
52 GO to ▶	FF1	—	FF1	FF1	FF1	—
53 GO to ▶	DH3	DH3	DH3	DH3	DH3	DH3
54 GO to ▶	DB10	DB10	DB10	DB10	DB10	DB10
61 GO to ▶	DE20	DE20	DE20	DE20	DE20	DE20
63 GO to ▶	DH10	DH10	DH10	DH10	DH10	DH10
64 GO to ▶	DB20	DB20	DB20	DB20	DB20	DB20
67 GO to ▶	FA1	FA1	FA1	FA1	FA1	FA1
81 GO to ▶	—	—	—	KC8	KC8	KC8
82 GO to ▶	—	—	—	KC8	KC8	KC8
83 GO to ▶	DD17	—	—	—	—	—
84 GO to ▶	DD17	DL11	—	DN10	DN10	DN10
85 GO to ▶	—	—	KD6	KD6	KD6	KD6
87 GO to ▶	J7	J7	J7	J7	J7	J7
89 GO to ▶	KR1	KR1	KR1	—	—	—
NO CODES ▶ CODES NOT LISTED ▶	GO to Pinpoint Test Step Q1					

QUICK TEST: Computed Timing Check — 4.0

SPECIAL NOTES:

- If engine is a NO START, go directly to Pinpoint Test Step A1.
- If engine starts but stalls, or stalls during timing check go directly to DIAGNOSTICS BY SYMPTOM in the proper Engine Supplement Section.
- Self-Test timing is equal to Base Timing plus 20 degrees BTDC ± 3 degrees (see VECI decal for correct base timing).

Example

If base timing is 10 degrees BTDC, Self-Test timing is equal to: 10 degrees + 20 degrees = 30 degrees BTDC ± 3 degrees or 27 degrees to 33 degrees BTDC.

HOW TO RUN QUICK TEST TIMING CHECK

DO

- Turn the key off and wait 10 seconds.
- Start engine.
- Activate Self-Test per Quick Test Step 3.0 A.
- Check timing after the last service code has been displayed. The timing will remain fixed for 2 minutes, unless Self-Test is deactivated.

Is Self-Test Timing within specification?

YES	Go To QUICK TEST STEP 5.0.
NO	Go To Pinpoint Test Step P 1.

QUICK TEST: Engine Running Self-Test — 5.0

A — PERFORMING THE ENGINE RUNNING SELF TEST

SPECIAL NOTES:

- If the engine starts but stalls, or stalls during Self-Test, go directly to DIAGNOSTICS BY SYMPTOM in the proper Engine Supplement Section.
- On vehicles equipped with 2.3L TC, 2.3L EFI (Car and Truck), 2.5L CFI, 2.9L EFI, and 3.0L EFI (Car and Truck), the brake pedal MUST be depressed and released AFTER the ID Code.
- On vehicles equipped with 2.3L CFI (Car), 4.9L and 5.0L EFI (Truck), the steering wheel must be turned one-half turn and released after the ID Code.

HOW TO RUN THE ENGINE RUNNING SELF-TEST

DO

- Deactivate Self-Test.
- Start and run engine at 2,000 rpm for 2 minutes. This action warms up the EGO sensor.
- Turn engine off, wait 10 seconds.
- Start engine.
- Activate Self-Test per Quick Test Step 3.0 A.
- After the ID code, depress and release the brake pedal if appropriate. See Special Note above.
- After the ID code, turn the steering wheel one-half turn and release it, if appropriate. See Special Note above.
- If a dynamic response code occurs, perform a brief wide open throttle (WOT).
- Record all service codes displayed.
- Go to **Part B** of Engine Running Self-Test.

DON'T

- Depress the throttle unless a Dynamic Response Code is displayed.

QUICK TEST: Engine Running Self-Test — 5.0

B — CODE OUTPUT

Engine ID	Dynamic Response	Engine Running	ACTION TO TAKE
2(0), 3(0) or 4(0) —	1(0) or no display	— 11	• Engine Running Self-Test indicates a PASS. — If Continuous Memory Codes were present, Go to QUICK TEST STEP 6.0. — If Continuous Memory is a PASS Code 11 and a symptom is present, Go to DIAGNOSTIC BY SYMPTOM in the proper Engine Supplement Section.
2(0), 3(0) or 4(0) —	1(0) or no display —	ANY CODE(S)	• Engine Running Self-Test indicates a FAULT. — Go to **PART C** of Engine Running Self-Test. — Always start with the first code displayed.
98 —	NO DISPLAY —	ANY CODE(S)	• Code 98 in place of the I.D. code indicates that the vehicle DID NOT PASS Key On Engine Off Self-Test. Engine Running Self-Test will not initiate until a PASS Code 11 is obtained in Key On Engine Off Self-Test. — Run Key On Engine Off Self-Test and address all codes displayed.
NO CODE DISPLAYED CODES NOT LISTED			• Self-Test did not activate. — Rerun Engine Running Self-Test to verify the above condition. — If condition is still present, Go to Pinpoint Test Step Q1.

QUICK TEST: Engine Running Self-Test — 5.0

C — PASSENGER CAR SERVICE CODE CHART

Engine Running Service Code	Pinpoint Test Step Direction								
	1.9L EFI	1.9L CFI	2.3L EFI	2.3L TC	2.3L CFI	2.5L CFI	3.0L EFI	3.8L CFI	5.0L SEFI
12 GO to ▶	KE1	KB23	KE1	KE1	KB23	KF16	KE1	KF16	KE1
13 GO to ▶	KE4	KB26	KE15	KE11	KB26	KF19	KE11	KF19	KE11
16 GO to ▶	KE1	KB26	KE1	—	KB26	KF19	KE1	KF19	KE1
17 GO to ▶	KE12	—	—	KE12	—	—	—	—	—
18 GO to ▶	—	—	—	—	—	—	—	—	P1
19 GO to ▶	—	KB26	—	—	—	—	—	—	—
21 GO to ▶	DE1	DE1	DE1	DE1	DE1	DE1	DE1	DE1	DE1
22 GO to ▶	DF1	DF7	DF7	DF1	DF7	DF7	DF7	DF7	DF7
23 GO to ▶	DH1	KB12	DH1	DH1	KB12	DH1	DH1	DH1	DH1
24 GO to ▶	—	DB1	DB1	DB1	DB1	DB1	DB1	DB1	DB1
25 GO to ▶	—	—	DG1	DG1	—	—	DG1	—	—
26 GO to ▶	DK1	—	—	DK1	—	—	—	—	—
28 GO to ▶	DA1	—	—	DA1	—	—	—	—	—
31 GO to ▶	—	DL21	DD1	—	DN1	DN1	DL21	DD1	DN1
32 GO to ▶	—	DL20	DD11	—	DN25	DN25	DL20	DD11	DN25
33 GO to ▶	—	DL30	DD11	—	DN40	DN40	DL30	DD11	DN40
34 GO to ▶	—	DL25	DD11	KA1	DN50	DN50	DL25	DD11	DN50
35 GO to ▶	—	DL25	DD30	—	DN5	DN5	DL25	DD30	DN5
41 GO to ▶	HA14	HF6	HE14	HA14	HF6	HF6	HG15	HD7	HC9
42 GO to ▶	HA8	HF13	HE8	HA8	HF13	HF13	HG8	HD14	HC15
44 GO to ▶	—	—	—	—	—	—	—	KC1	KC1
45 GO to ▶	—	—	—	—	—	—	—	KC1	KC1
46 GO to ▶	—	—	—	—	—	—	—	KC1	KC1
47 GO to ▶	KE11	—	—	—	—	—	—	—	—
48 GO to ▶	KE20	—	—	—	—	—	—	—	—
52 GO to ▶	—	—	—	—	FF5	—	FF5	—	—
55 GO to ▶	—	FC3	—	—	FC3	FC3	—	HD1	—
56 GO to ▶	—	KB5	—	—	KB5	KF12	—	KF12	—
62 GO to ▶	—	—	—	—	—	—	T71	—	—
67 GO to ▶	—	FA1	—	—	—	—	—	—	—
72 GO to ▶	—	—	DF10	—	—	—	DF10	—	—
73 GO to ▶	DH20	—	DH20	DH20	—	—	DH20	—	—
74 GO to ▶	—	—	FD1	FD1	—	FD1	FD1	—	—
75 GO to ▶	—	—	FD4	FD4	—	FD4	FD4	—	—
76 GO to ▶	DK30	—	—	DK30	—	—	—	—	—
77 GO to ▶	M1	—	M1	M1	—	—	M1	—	—
84 GO to ▶	—	DL11	—	—	—	—	—	—	—
85 GO to ▶	—	KD6	—	—	—	—	—	—	—
91 GO to ▶	—	—	—	—	—	—	—	HD7	HC9
92 GO to ▶	—	—	—	—	—	—	—	HD14	HC15
94 GO to ▶	—	—	—	—	—	—	—	KC1	KC1
98 GO to ▶	GO TO QUICK TEST STEP 5.0B								
99 GO to ▶	ISC HAS NOT LEARNED YET. RERUN QUICK TEST STEPS 3 THRU 5								
NO CODES ▶ CODES NOT LISTED ▶	GO to Pinpoint Test Step Q1								

QUICK TEST: Engine Running Self-Test — 5.0

C — LIGHT TRUCK SERVICE CODE CHART

Engine Running Service Code	Pinpoint Test Step Direction					
	2.3L EFI	2.9L EFI	3.0L EFI	4.9L EFI	5.0L EFI	7.5L EFI
12 GO to ▶	KE1	KE1	KE1	KE1	KE1	KE1
13 GO to ▶	KE15	KE11	KE11	KE11	KE11	KE11
16 GO to ▶	KE1	KE1	KE1	KE1	KE1	KE1
21 GO to ▶	DE1	DE1	DE1	DE1	DE1	DE1
22 GO to ▶	DF7	DF7	DF7	DF7	DF7	DF7
23 GO to ▶	DH1	DH1	DH1	DH1	DH1	DH1
24 GO to ▶	DB1	DB1	DB1	DB1	DB1	DB1
25 GO to ▶	DG1	DG1	DG1	DG1	DG1	—
31 GO to ▶	DD1	DL21	—	DN1	DN1	DN1
32 GO to ▶	DD11	DL20	—	DN25	DN25	DN25
33 GO to ▶	DD11	DL30	—	DN40	DN40	DN40
34 GO to ▶	DD11	DL25	—	DN50	DN50	DN50
35 GO to ▶	DD30	DL25	—	DN5	DN5	DN5
38 GO to ▶	—	—	—	DN55	DN55	DN55
41 GO to ▶	HE14	HG15	HG15	HG15	HH15	HH15
42 GO to ▶	HE8	HG8	HG8	HG8	HH8	HH8
44 GO to ▶	—	—	—	KC1	KC1	KC1
45 GO to ▶	—	—	—	KC1	KC1	KC1
46 GO to ▶	—	—	—	KC1	KC1	—
52 GO to ▶	—	—	—	FF5	FF5	—
72 GO to ▶	DF10	DF10	DF10	DF10	DF10	DF10
73 GO to ▶	DH20	DH20	DH20	DH20	DH20	DH20
74 GO to ▶	FD1	FD1	FD1	—	—	—
75 GO to ▶	FD4	FD4	FD4	—	—	—
77 GO to ▶	M1	M1	M1	M1	M1	M1
98 GO to ▶	GO TO QUICK TEST STEP 5.0B					
NO CODES ▶ CODES NOT LISTED ▶	GO to Pinpoint Test Step Q1					

QUICK TEST: Continuous Self-Test — 6.0

A CONTINUOUS MEMORY CODES

SPECIAL NOTES:

- Verify that a **Pass Code 11** was received in both Key On Engine Off and Engine Running Self-Tests before continuing with this test.
- It is necessary to clear the codes in Continuous Memory before continuing with this test.
- Refer to the Appendix for a detailed description of how to use the Continuous Monitor Mode.

HOW TO CLEAR THE CONTINUOUS MEMORY CODES

- Run the Key On Engine Off Self-Test per Quick Test Step 3.0 A.
- When the Service Codes begin to be displayed, deactivate Self-Test:

  —STAR Tester: Unlatching the center button (up position).

  —Analog VOM: Remove the jumper wire from between Self-Test Input (STI) connector and the Signal Return Pin of the Self-Test connector.

DETERMINING THE CONTINOUS MEMORY CODES TO BE TESTED

- Refer to the Continuous Memory Codes recorded in Quick Test Step 3.0 A.
- The cause of some of the Continuous Memory Codes may have been eliminated during either Key On Engine Off or Engine Running Self-Test repairs.
- Address only those Continuous Memory Codes for which a similar code has not been previously serviced.
- Go to Part B of Continuous Self-Test.

QUICK TEST: Continuous Self-Test — 6.0

B HOW TO USE THE CONTINUOUS MONITOR MODE (WIGGLE TEST)

SPECIAL NOTES:

- The Continuous Monitor Modes allow the technician to **ATTEMPT** to recreate an intermittent fault.
- The STAR Tester will flash the LED or sound an alert when a fault is recreated.
- The needle of the VOM will sweep across the face of the meter when a fault is recreated.

KEY ON ENGINE OFF

- Hookup a STAR Tester or VOM as shown in Quick Test Step 2.0.
- Turn the ignition key to the ON position.
- Activate, de-activate, and re-activate Self-Test. **DO NOT** turn the ignition key off.
- You are now in the Continuous Monitor Mode.
- Tap, move, and wiggle the suspect sensor and/or harness. If a fault is detected a Service Code will be stored in memory and will be indicated as explained above depending on the type of equipment being used.
- Go To **PART C** of Continuous Self-Test.

ENGINE RUNNING

- Hookup a STAR Tester or VOM as shown in Quick Test Step 2.0.
- Start the engine.
- Activate Self-Test, wait 10 seconds, deactivate and reactivate Self-Test. **DO NOT** shut the engine off.
- You are now in the Engine Running Continuous Monitor Mode.
- Tap, move and wiggle the suspect sensor and/or harness. If a fault is detected a Service Code will be stored in memory and will be indicated as explained above depending on the type of equipment being used.
- Go to **PART C** of Continuous Self-Test.

QUICK TEST: Continuous Self-Test — 6.0

C PASSENGER CAR SERVICE CODE CHART

Continuous Memory Service Code	Pinpoint Test Step Direction								
	1.9L EFI	1.9L CFI	2.3L EFI	2.3L TC	2.3L CFI	2.5L CFI	3.0L EFI	3.8L CFI	5.0L SEFI
13 GO to ▶	—	KB90	—	—	KB90	—	—	—	—
14 GO to ▶	Y1	Y1	Y1	Y1	Y1	Y1	Y1	Y1	Y1
15 GO to ▶	Q10	Q10	Q10	Q10	Q10	Q10	Q10	Q10	Q10
18 GO to ▶	N1	N1	N1	N1	N1	N1	N1	—	N1
21 GO to ▶	—	—	—	—	DE90	DE90	—	—	—
22 GO to ▶	DF90	DF90	DF90	DF90	DF90	DF90	DF90	DF90	DF90
23 GO to ▶	—	KB97	—	—	KB97	—	—	—	—
27 GO to ▶	—	—	—	DP1	—	—	—	—	—
29 GO to ▶	—	—	—	—	—	—	T2	—	DP1
31 GO to ▶	—	DL90	DD90	—	DN92	DN92	DL90	DD90	DN92
32 GO to ▶	—	DL94	—	—	DN90	DN90	DL94	—	DN90
33 GO to ▶	—	DL97	—	—	DN95	DN95	DL97	—	DN95
34 GO to ▶	—	DL93	—	—	DN98	DN98	DL93	—	DN98
35 GO to ▶	—	DL90	—	—	DN92	DN92	DL90	—	DN92
38 GO to ▶	—	KB91	—	—	KB91	—	—	—	—
39 GO to ▶	—	—	—	—	—	—	T31	—	—
41 GO to ▶	HA30	HF90	—	—	HF90	—	HG90	—	HC90
42 GO to ▶	HA30	—	—	—	—	—	—	—	—
43 GO to ▶	HA30	—	—	HA30	—	—	—	—	—
51 GO to ▶	DE91	DE91	DE91	DE91	DE91	DE91	DE91	DE91	DE91
53 GO to ▶	DH90	KB93	DH90	DH90	KB93	DH90	DH90	DH90	DH90
54 GO to ▶	—	DB90	DB90	DB90	DB90	DB90	DB90	DB90	DB90
56 GO to ▶	DK90	—	—	DK90	—	—	—	—	—
57 GO to ▶	—	—	—	—	—	—	T41	—	—
58 GO to ▶	DA90	—	—	DA90	—	—	—	—	—
59 GO to ▶	—	—	—	—	—	—	T21	—	—
61 GO to ▶	DE94	DE94	DE94	DE94	DE94	DE94	DE94	DE94	DE94
63 GO to ▶	DH94	KB97	DH94	DH94	KB97	DH94	DH94	DH94	DH94
64 GO to ▶	—	DB93	DB93	DB93	DB93	DB93	DB93	DB93	DB93
66 GO to ▶	DK93	—	—	DK93	—	—	—	—	—
67 GO to ▶	FA1	—	—	FA1	—	—	—	—	—
68 GO to ▶	DA93	—	—	DA93	—	—	—	—	—
69 GO to ▶	—	—	—	—	—	—	T11	—	—
71 GO to ▶	Q60	KB92	—	—	KB92	—	—	—	—
72 GO to ▶	B10	—	—	B10	—	—	—	—	—
78 GO to ▶	—	—	—	—	B10	X10	—	—	—
87 GO to ▶	—	J7	—	—	J7	—	J7	—	—
91 GO to ▶	—	—	—	—	—	—	—	—	HC90
NO CODES ▶ / CODES NOT LISTED ▶	GO to Pinpoint Test Step Q1								

QUICK TEST: Continuous Self-Test — 6.0

C LIGHT TRUCK SERVICE CODE CHART

Continuous Memory Service Code	Pinpoint Test Step Direction					
	2.3L EFI	2.9L EFI	3.0L EFI	4.9L EFI	5.0L EFI	7.5I EFI
14 GO to ▶	Y1	Y1	Y1	Y1	Y1	Y1
15 GO to ▶	Q10	Q10	Q10	Q10	Q10	Q10
18 GO to ▶	N1	N1	N1	N1	N1	N1
21 GO to ▶	—	—	—	—	—	—
22 GO to ▶	DF90	DF90	DF90	DF90	DF90	DF90
29 GO to ▶	—	DP1	—	DP1	DP1	—
31 GO to ▶	DD90	DL90	—	DN92	DN92	DN92
32 GO to ▶	—	DL94	—	DN90	DN90	DN90
33 GO to ▶	—	DL97	—	DN95	DN95	DN95
34 GO to ▶	—	DL93	—	DN98	DN98	DN98
35 GO to ▶	—	DL90	—	DN92	DN92	DN92
41 GO to ▶	—	HG90	—	HG90	HH90	HH90
51 GO to ▶	DE91	DE91	DE91	DE91	DE91	DE91
53 GO to ▶	DH90	DH90	HD90	DH90	DH90	DH90
54 GO to ▶	DB90	DB90	DB90	DB90	DB90	DB90
61 GO to ▶	DE94	DE94	DE94	DE94	DE94	DE94
63 GO to ▶	DH94	DH94	DH94	DH94	DH94	DH94
64 GO to ▶	DB93	DB93	DB93	DB93	DB93	DB93
87 GO to ▶	—	J7	—	J7	J7	J7
NO CODES ▶ / CODES NOT LISTED ▶	GO to Pinpoint Test Step Q1					

Self-Test Description

The Self-Test is divided into three specialized tests: Key On Engine Off Self-Test, Engine Running Self-Test, and Continuous Self-Test. The Self-Test is not a conclusive test by itself, but is used as a part of the functional Quick-Test diagnostic procedure. The processor stores the Self-Test program in its permanent memory. When activated, it checks the EEC-IV system by testing its memory integrity and processing capability, and verifies that various sensors and actuators are connected and operating properly.

The Key On Engine Off and Engine Running Self-Tests are functional tests which only detect faults present at the time of the Self-Test. Continuous Self-Test is an ongoing test that stores fault information for retrieval at a later time.

KEY ON ENGINE OFF SELF-TEST

At this time, a test of the EEC-IV system is conducted with power applied and engine at rest.

For Self-Test to detect errors in the Key On Engine Off Self-Test mode, the fault must be present at the time of testing. For intermittents, refer to Continuous Memory Codes.

SEPARATOR PULSE

A single 1/2 second separator pulse is issued 6-9 seconds after the last Key On Engine Off Test code. Then 6-9 seconds after the single 1/2 second separator pulse, the continuous memory codes will be issued.

NOTE: The separator code and continuous memory codes follow Key On Engine Off Testing codes ONLY.

CONTINUOUS MEMORY CODES

Continuous memory codes are issued as a result of information stored during continuous Self-Test, while the vehicle was in normal operation. These codes are displayed only during Key On Engine Off testing and after the separator code. These codes should be used for diagnosis only when Key On Engine Off and Engine Running Self-Tests result in code 11 and all Quick Test Steps 1.0 through 5.0 have been successfully completed.

ENGINE RUNNING SELF-TEST

At this time, a test of the EEC-IV system is conducted with the engine running. The sensors are checked under actual operating conditions and at normal operating temperatures. The actuators are exercised and checked for corresponding results.

ENGINE IDENTIFICATION CODES (ID CODES)

Engine ID codes are issued at the beginning of the Engine Running Self-Test and are one-digit numbers represented by the number of pulses sent out. The engine ID code is equal to 1/2 the number of engine cylinder (i.e. 2 pulses = 4 cylinders). These codes are used to verify the proper processor is installed and that the Self-Test has been entered.

DYNAMIC RESPONSE CHECK

The dynamic response check verifies the movement of the TP, VAF, and MAP sensors during the brief Wide Open Throttle (WOT) performed during the Engine Running Self-Test. The signal for the operator to perform the brief WOT is a single pulse or 10 code on the STAR Tester.

Code Output Format

SERVICE CODES

The EEC-IV system communicates service information to the outside world by way of the Self-Test service codes. These service codes are two-digit numbers representing the results of Self-Test.

The service codes are transmitted on the Self-Test output (STO) line found in the vehicle Self-Test connector. They are in the form of timed pulses, and read by the technician on a voltmeter or on the STAR tester.

SELF-TEST OUTPUT CODE FORMAT
KEY ON ENGINE OFF AND CONTINUOUS MEMORY CODES

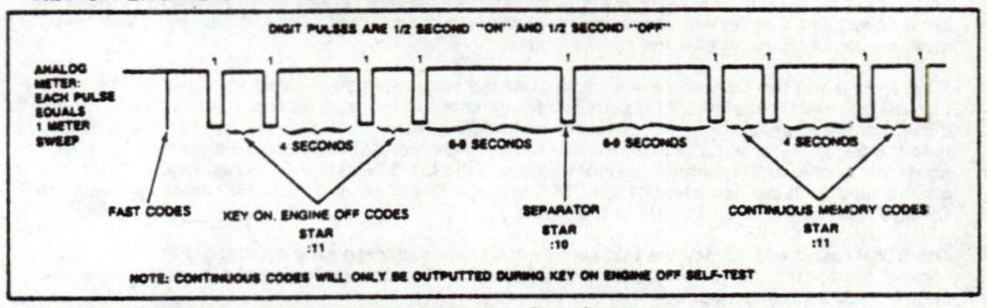

Figure 1 Key On Engine Off and Continuous Memory Code Format

SELF-TEST OUTPUT CODE FORMAT
ENGINE RUNNING CODES

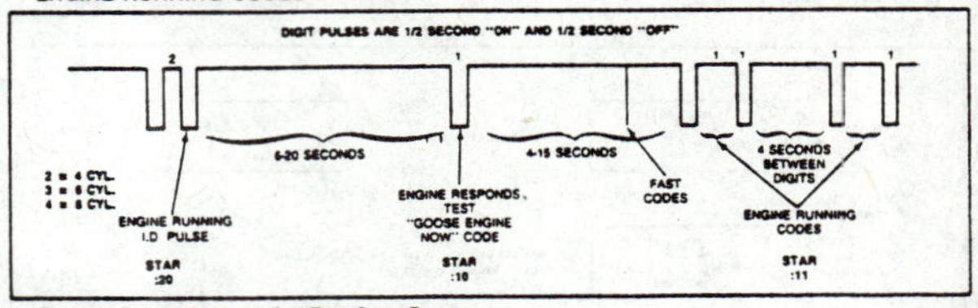

Figure 2 Engine Running Self-Test Code Format

Fast Codes

Fast codes are issued prior to regular service codes. These codes contain the identical information as the regular service codes but are transmitted at 100 times the normal rate. These codes are interpreted by special equipment at the end of the assembly line by the Body and Assembly Division.

Some meters in service detect these codes as a short burst of information (slight meter deflection).

Self-Test Description

POWER STEERING PRESSURE SWITCH TEST

On vehicles equipped with 2.3L CFI (Car), 4.9L and 5.0L EFI (Truck), the steering wheel must be turned one-half turn and released AFTER the ID Code. This tests the ability of the EEC-IV system to detect a change of state in the Power Steering Pressure Switch.

BRAKE ON/OFF SWITCH TEST

On vehicles equipped with 2.3L TC, 2.3L EFI (Car and Truck), 2.5L CFI, 2.9L EFI, and 3.0L EFI (Car and Truck), the brake pedal MUST be depressed and released AFTER the ID Code. This tests the ability of the EEC-IV system to detect a change of state in the Brake ON/OFF Switch.

Continuous Self-Test

The Continuous Memory service codes are separated from the Quick Test Key On Engine Off codes by a single separator pulse, Figure 3.

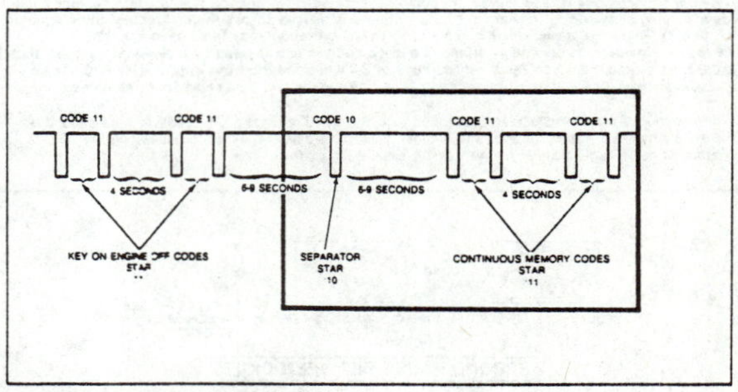

Figure 3 Continuous Memory Code Format

The Continuous Memory codes should never be used for Diagnosis until the Key On Engine Off and Engine Running Self-Tests result in a pass code 11.

During this mode of testing the EEC-IV Processor continuously monitors inputs for opens and shorts. The continuous memory codes must be retrieved within forty engine temperature warm up cycles. On the forty-first Engine Temperature cycle the service code will be automatically erased. The continuous memory codes can also be erased by deactivating Self-Test while the service codes are being outputted.

Self-Test With STAR Tester

READING CODES—SELF-TEST AUTOMATIC READOUT (STAR) TESTER

After hooking up the STAR tester and turning on its power switch, the tester will run a display check and the numerals 88 will begin to flash in the display window (Figure 4). A steady 00 will then appear to signify that the STAR tester is ready to start the Self-Test and receive the test's service codes.

To receive the service codes, press the pushbutton at the front of the STAR tester. The button will latch down, and a colon will appear in the display window in front of the 00 numerals. The colon must be displayed to receive the service codes.

If for any reason the technician wishes to clear the display window during the Self-Test, he must turn off the vehicle's engine, press the tester's pushbutton once to unlatch it (colon will disappear), then press the button again to latch down the button (colon will appear again). Every time the STAR tester is turned off, the low battery indicator (LO BAT) should show briefly at the upper left corner of the tester's display window. If the LO BAT indicator shows steadily at any other time during the operation of the STAR tester with any service code, turn its power switch to Off and replace the 9-volt battery in the tester.

The STAR tester will display the last service code received, even after disconnecting it from the vehicle. It will hold the service code on the display until the power is turned off or the pushbutton is unlatched and relatched.

WARNING: Anyone who departs from the instructions provided in this publication must first establish that he compromises neither his personal safety nor the vehicle integrity by his choice of methods, tools, or parts.

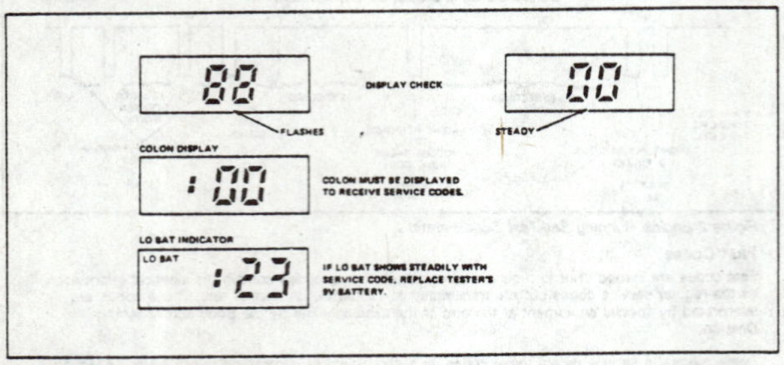

Figure 4 Star Tester Output Code Format

Self-Test With Analog Voltmeter

READING CODES—ANALOG VOLTMETER

When a service code is reported on the analog voltmeter for a function test, it will represent itself as a pulsing or sweeping movement of the voltmeter's needle across the dial face of the voltmeter (Figure 5). Therefore, a single-digit number of three will be reported by three needle pulses (sweeps). However, as previously stated, a service code is represented by a two-digit number, such as 2-3. As a result, the Self-Tests service code of 2-3 will appear on the voltmeter as two needle pulses (sweeps), then, after a two-second pause, the needle will pulse (sweep) three times.

The continuous memory codes are separated from the Key On Engine Off codes by a six-second delay, a single half-second sweep, and another six-second delay. They are produced on the voltmeter in the same manner as the Key On Engine Off codes.

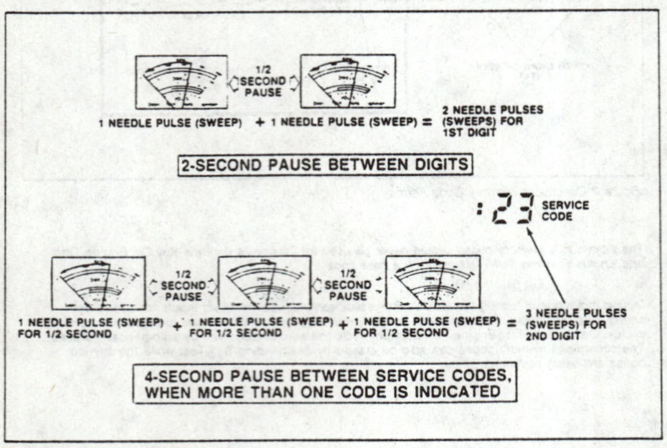

Figure 5 Analog Voltmeter Output Code Format

Diagnostic Aids

CONTINUOUS MONITOR TEST (WIGGLE TEST)

This test is intended as an aid in diagnosing intermittent failures in the sensor input circuits. The Self-Test output is energized whenever the continuous memory mode senses a fault and de-energized when the system is OK.

- Connect a VOM/STAR to the Self-Test output.
- **Key On Engine Off/Engine Running Continuous Monitor Test. (Wiggle Test).**

Continuous Monitor Test is entered by activating, de-activating, and re-activating Self-Test (DO NOT turn key off). This will put you into Continuous Monitor ("Wiggle") Test. The Engine Running Continuous Monitor Test can also be entered by waiting approximately two minutes after the last service codes are received.

The Continuous Monitor (Wiggle) Test will allow the technician to attempt to re-create the intermittent failure (tap, move, wiggle the harness and/or the suspected sensor). The Self-Test output will be activated whenever a continuous fault is detected. If the duration is long enough, a fault code will be stored.

Now with the knowledge of the affected circuits, a close check of the harness and associated connectors can be made.

NOTE: Remember to keep your eyes on the VOM/STAR for any change which will indicate where the intermittent is located.

EXAMPLE: How to Use tr e Continuous Monitor Test (Systematically)

If a service code 51 was displayed: Engine Coolant Temperature Sensor:

- Visually inspect the sensor very closely.
- Enter Continuous Monitor test.
- Lightly TAP on the sensor.
- Push/pull on the sensor harness connector (do not disconnect, yet).
- Test and Wiggle (snake) harness vigorously working from the sensor connector toward the dash panel and from the dash panel to the processor in short sections.
- If VOM has not given a positive indication of an intermittent, disconnect the sensor from the harness as carefully as possible. Remove terminals from the connector. Visually inspect terminals at both ends for corrosion, bad crimps, improperly seated terminals, etc.
- Reconnect after inspection.
- Disconnect processor from harness as carefully as possible.
- Inspect terminals.
- Only remove terminals associated with the sensor being inspected.
- If the VOM does not give a positive intermittent indication, reconnect the connector and erase the Continuous Test service codes.
- To erase the continuous memory service codes:
  - Initiate Key On Engine Off Self-Test.
  - Unlatch the push-button on the Star Tester as soon as the first service code is received (even if an "11" is the first code).
  - Rerun Self-Test to verify services have been erased.

Diagnostic Aids

OUTPUT STATE CHECK

The output state check aids in servicing output actuators associated with the EEC-IV system. It enables the technician to energize and de-energize most of the system output actuators on command. This mode is entered after all codes have been received from Key On Engine Off and Continuous Testing. At this time, leave Self-Test activated and depress the throttle. Each time the throttle is depressed the output actuators will change state from energized to de-energized or from de-energized to energized.

1. Enter Self-Test.
2. Code Output Ends.
3. Do Brief WOT.
4. EEC-IV Output To Actuators Energized.
5. Do Brief WOT.
6. EEC-IV Output To Actuators De-Energized.

CYLINDER BALANCE TEST

All Applications Except Mark VII/Lincoln Town Car Calibration 8-22

The Cylinder Balance Test on the 5.0L SEFI is designed to separately check each injector and injector circuit. By switching each injector off and on, one at a time, Self-Test will verify that an rpm drop of greater than 80 rpm results.

To enter the Cylinder Balance Test:

1. Perform Engine Running Self-Test.
2. After the last REPEATED service code is received wait approximately 10 seconds. (The test must be entered within 2 minutes of the last repeated service code.)
3. Lightly depress throttle. (A two to three degree throttle angle shaft movement is required, NOT wide open throttle.)
4. Test time is approximately 90 seconds.
5. The Cylinder Balance Test can be repeated (entered) as many times as necessary, by lightly depressing the throttle within two minutes after the last repeated service code has been displayed.

The Cylinder Balance Test service codes are listed in Pinpoint Test Step HC4.

The Cylinder Balance Test is designed to aid in the detection of a non-contributing cylinder. The Cylinder Balance Pinpoint Test Steps are designed to isolate only EEC related problems

CYLINDER BALANCE TEST (Continued)
Mark VII/Lincoln Town Car Calibration 8-22

The Cylinder Balance Test on the Mark VII/Lincoln Town Car calibration 8-22 has been modified to test rpm drop at three levels, allowing a more precise check of each cylinder. During first level testing, Cylinder Balance Test will verify that when each injector is separately turned "off and on", an rpm drop of greater than 80 rpm results. When a second test level is requested, an rpm drop of greater than 90 rpm must result. When a third test level is requested, a drop of greater than 100 rpm must result. Any additional tests (ie. fourth test, fifth test, etc.) that are requested will be run at the third level (100 rpm).

The Cylinder Balance Test service codes are listed in Pinpoint Test Step HC4.

1. Perform Engine Running Self-Test.
2. After the last repeated service code is received, wait 5-10 seconds.
3. Lightly depress and release throttle (not wide-open throttle) within two minutes of the last repeated service code.
4. Cylinder Balance Test will be performed at the first test level (80 rpm). Test time is approximately three minutes.
5. After the last repeated Cylinder Balance Test pass code 90, re-enter test within two minutes by lightly depressing and releasing the throttle. The second test level of the Cylinder Balance Test will now be performed (90 rpm).
6. After the last repeated Cylinder Balance Test pass code 90, re-enter test within two minutes by lightly depressing and releasing the throttle. Cylinder Balance Test will now be tested at the third test level (100 rpm).
7. Cylinder Balance Test may be repeated as many times as necessary by repeating Step 6. All additional testing will be performed at the third test level (100 rpm).

The Cylinder Balance Test is designed to aid in the detection of a non-contributing or partially contributing cylinder. The Cylinder Balance Pinpoint Test Steps are designed to isolate only EEC related problems.

Failure Mode Effects Management (FMEM)

DESCRIPTION

FMEM is an alternate system strategy in the ECA designed to allow improved vehicle drive should one or more sensor inputs fail.

When a sensor input is perceived to be out-of-limits by the ECA, an alternative strategy will be initated.

The ECA will substitute a fixed in-limit sensor value and will continue to monitor the faulty sensor input. If the faulty sensor operates within limits, the ECA will return to the normal engine running strategy.

98 — Code 98 will be displayed when FMEM is in effect.

MIL — MIL "Check Engine" Light will remain on when FMEM is in effect.

"Check Engine" Light (Malfunction Indicator Light)

DESCRIPTION

The "Check Engine" light is intended to alert the driver of certain malfunctions in the engine control system.

If such a fault occurs, the EEC-IV processor will substitute a value or values and continue operating. This process is called Failure Mode Effects Management (FMEM). In some cases this action may result in a slight change in driveability.

HOW THE "CHECK ENGINE" LIGHT OPERATES

System OK

The "Check Engine" light will remain on while the key is in the RUN position.

Once the vehicle has started the "Check Engine" light will go out.

System Not OK

If the "Check Engine" light should remain on after the vehicle has started, run Key On Engine Off Self-Test to completion. If the light continues to remain on, go to EEC-IV Diagnostic By Symptom.

If the "Check Engine" light never comes on, go to EEC-IV Diagnostic By Symptom.

If the vehicle is a no start, go to Pinpoint Test Step A1.

NOTE: When in Self-Test the "Check Engine" light will also flash the service codes.

Test Equipment

EQUIPMENT REQUIRED:

- Rotunda Self-Test Automatic Readout (STAR), No. 007-00004 with cable assembly No. 007-00010. Refer to STAR Tester operation.
- Analog volt-ohmmeter, 0 to 20v DC, (alternate to STAR).

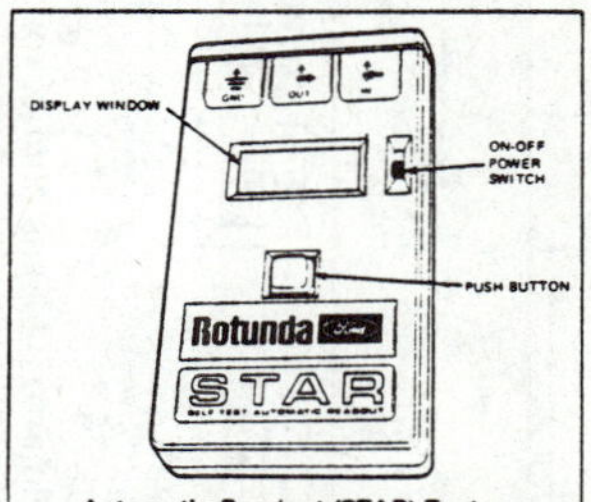

Automatic Readout (STAR) Tester

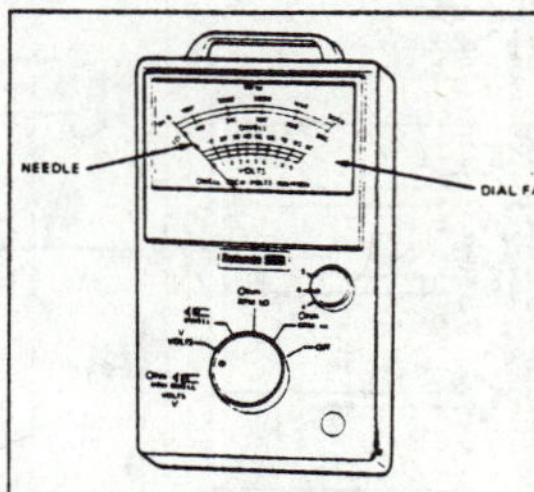

Analog Voltmeter

- Jumper wire.
- Vacuum gauge, Rotunda 059-00008 or equivalent. Range 0-30 in. Hg. Resolution 1 in. Hg.

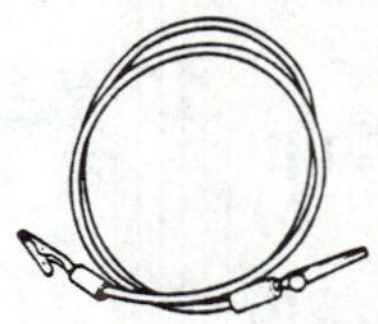
Jumper Wire

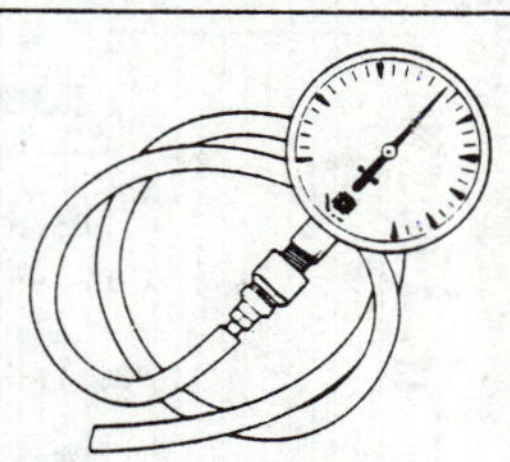
Vacuum Gauge

Test Equipment (Continued)

- Tachometer, Rotunda No. 059-00010 or equivalent. Range 0-6,000 rpm. Accuracy ± 40 rpm. Resolution 20 rpm.
- Breakout Box, Rotunda 014-00322, Special Service Tool T83L-50-EEC-IV or equivalent.

Tachometer

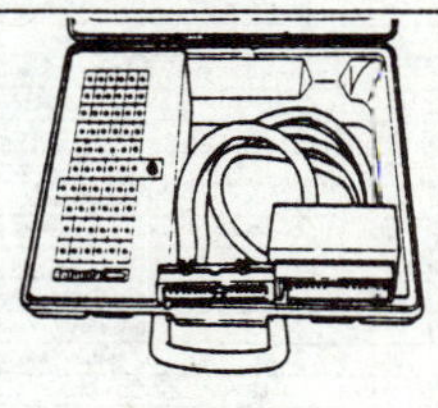
Breakout Box

- Vacuum pump, Rotunda No. 021-00014 or equivalent. Range 0-30 in. Hg.
- Digital volt-ohmmeter, Rotunda No. 014-00407 or equivalent. Input impedance 10 Megaohm minimum.

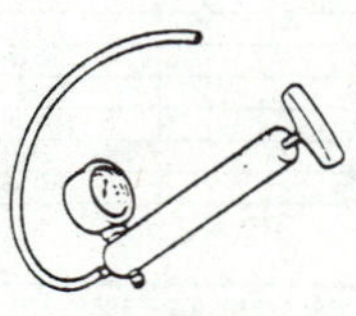
Vacuum Pump

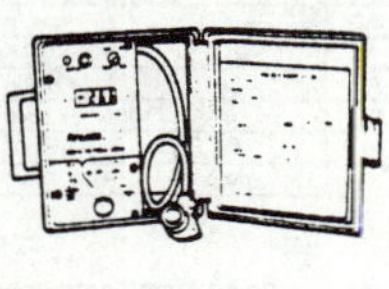
Digital Volt-Ohmmeter

Electrical Schematic — 1.9L CFI

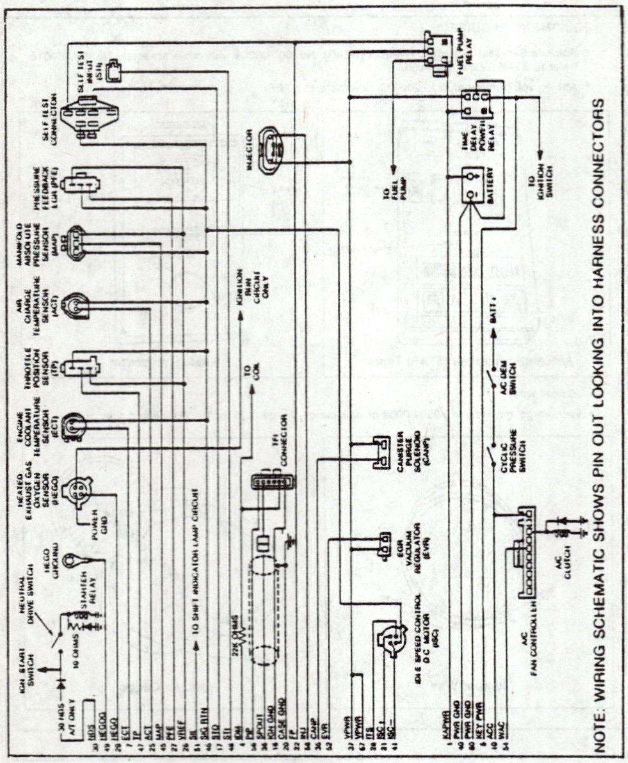

Electrical Schematic — 1.9L EFI

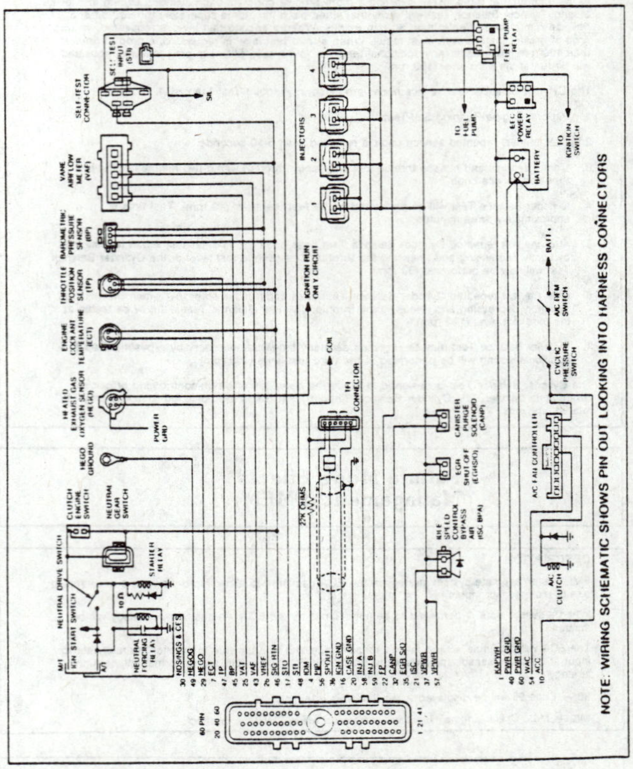

EEC-IV Module Connector Pin Usage — 1.9L CFI

Pin	Circuit	Wire Color	Application	Abbreviations
1	37	Y	Keep Alive Power	KAPWR
4	11	DG Y	Ignition Diagnostic Monitor	IDM
5	16	R LG	Key Power	KPWR
7	354	LG Y	Engine Coolant Temperature Sensor	ECT
10	347	BK Y	A C Clutch Compressor Signal	ACCS
16	259	BK O	Ignition Ground	IGN GND
17	201	T R	Self-Test Output Shift Indicator Light	STO SIL
20	57	BK	Case Ground	CSE GND
21	376	BR W	Idle Speed Control	ISC -
22	97	T LG	Fuel Pump Relay	FP
25	357	LG P	Air Charge Temperature Sensor	ACT
26	351	O W	Reference Voltage	VREF
27	352	BR LG	Pressure Feedback EGR	PFE
28	265	LG W	Idle Tracking Switch	ITS
29	94	DG P	Heated Exhaust Gas Oxygen Sensor	HEGO
30	614	GY O	Neutral Drive Switch (A T Only)	NDS
35	101	GY Y	Canister Purge Solenoid	CANP
36	324	Y LG	Spark Output	SPOUT
37	361	R	Vehicle Power	VPWR
40	60	BK LG	Power Ground	PWR GND
41	264	W LB	Idle Speed Control	ISC -
45	358	LG BK	Manifold Absolute Pressure Sensor	MAP
46	359	BK W	Signal Return (Ground)	SIG RTN
47	355	DG LG	Throttle Position Sensor	TP
48	100	W R	Self-Test Input	STI
49	89	O	Heated Exhaust Gas Oxygen Ground	HEGOG
51	224	T LB	Shift Indicator Light (M T Only)	SIL
52	362	Y	EGR Valve Regulator Solenoid	EVR
54	73	O LB	WOT A C Cutoff	WAC
56	349	DB	Profile Ignition Pick-Up	PIP
57	361	R	Vehicle Power	VPWR
58	95	T R	Fuel Injector	INJ
60	60	BK LG	Power Ground	PWR GND

Pin locations given for reference only. Probing 60 pin connector with DVOM probe will result in permanent damage to the pin connectors. Always probe as directed, using the Breakout Box.

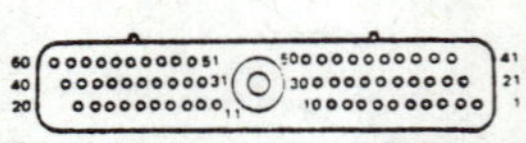

EEC-IV Module Connector Pin Usage — 1.9L EFI

Pin	Circuit	Wire Color	Application	Abbreviations
1	37A	Y	Keep Alive Power	KAPWR
4	11B	DG Y	Ignition Diagnostic Monitor	IDM
7	354	LG Y	Engine Coolant Temperature	ECT
10	347	BK Y	A C Clutch	ACCS
16	259	BK O	Ignition Ground	IGN GND
17	201	T R	Self-Test Output Shift Indicator Light	STO SIL
20	57	BK	Case Ground	CSE GND
21	68	O BK	Idle Speed Control	ISC
22	97	T LG	Fuel Pump	FP
25	357	LG P	Vane Air Temperature	VAT
26	351	O W	Reference Voltage	VREF
29	94	DG P	Heated Exhaust Gas Oxygen Sensor	HEGO
30	614	GY O	Neutral Drive Switch (A T Only)	NDS
32	101	GY Y	Canister Purge Solenoid	CANP
35	362	Y	Exhaust Gas Recirculation Shut Off	EGR S.O
36	324A	Y LG	Spark Output	SPOUT
37	361	R	Vehicle Power	VPWR
40	60A	BK LG	Power Ground	PWR GND
43	200	W BK	Vane Air Flow	VAF
45	358	LG BK	Barometric Pressure	BP
46	359	BK W	Signal Return	SIG RTN
47	355	DG LG	Throttle Position Sensor	TP
48	100	W R	Self-Test Input	STI
49	89	O	Heated Exhaust Gas Oxygen Ground	HEGOG
54	73	O LB	Wide Open Throttle A C Cut-Off	WAC
56	349	DB	Profile Ignition Pickup	PIP
57	361A	R	Vehicle Power	VPWR
58	95	T R	Injector 1 (Bank 1)	INJ 1
59	96	T O	Injector 2 (Bank 2)	INJ 2
60	60B	BK LG	Power Ground	PWR GND

Pin locations given for reference only. Probing 60 pin connector with DVOM probe will result in permanent damage to the pin connectors. Always probe as directed, using the Breakout Box.

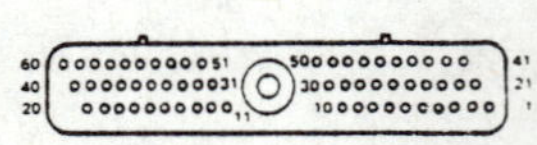

EEC-IV Diagnostic By Symptom — 1.9L CFI

SYMPTOM	RESULT	ACTION TO TAKE
• Engine runs rough or misfires. • Engine stalls. • Lack of power. • Rough idle.		• GO to Pinpoint Test Step S2. • GO to Pinpoint Test Step DF20. • EGR Valve Function. • Ignition system distributor cap, rotor, wires, coil, plugs. • Base engine valves, cam timing, compression etc. • Poor power ground connections.
• Low idle with A/C on.		For A/T: • Place A/C switch to A/C. • Perform KOEO Quick Test. • If code 67 is present GO to FA1. • If code 67 is not present GO to Pinpoint Test Step FA20. For M/T: • GO to Pinpoint Test Step FA20.
• Stumble after hot restart (HEGO).		• GO to Pinpoint Test Step HF20.
• Shift indicator light always on or off.		GO to Pinpoint Test Step KL1.
• Gasoline fumes under hood.		GO to Pinpoint Test Step KD1.
• Stall during parking maneuvers.		GO to Pinpoint Test Step FF3.

EEC-IV Diagnostic By Symptom — 1.9L EFI

SYMPTOM	RESULT	ACTION TO TAKE
• Engine stalls. • Stalls in Self-Test. • Runs rough. • Misses. • Always Rich Lean.		• GO to Pinpoint Test Step S1. • GO to Pinpoint Test Step DF20. • Poor power ground connections. • Ignition system distributor cap, rotor, wires, coil, plugs. • Base engine valves, cam timing, compression etc.
• Stumble after hot restart.		GO to Pinpoint Test Step HA20.
• Fuel pump always runs.		GO to Pinpoint Test Step J15.
• Detonation/spark knock.		GO to Pinpoint Test Step KA1.
• High idle speeds on each restart may be accompanied by detonation for up to 3-5 minutes after a restart.		GO to Pinpoint Test Step KA1.
• Low engine rpm with A/C On.		GO to Pinpoint Test Step FA20.
• Gasoline fumes under hood.		GO to Pinpoint Test Step KD1.
• Shift indicator light always On or Off.		GO to Pinpoint Test Step KL1.
• A/C does not cut-off under WOT conditions.		GO to Pinpoint Test Step KM1.

EEC-IV Diagnostic By Symptom-Continued — 1.9L CFI

SYMPTOM	RESULT	ACTION TO TAKE
• High idle in Drive, automatic transmission only.		• Place transmission in Drive with engine off. Test neutral drive circuit (test Pin 30) for short to ground. SERVICE as necessary. NOTE: Circuit should NOT have continuity to ground in DRIVE.
• Fuel pump always runs.		GO to Pinpoint Test Step J15.
• A/C does not cut off under WOT conditions.		GO to Pinpoint Test Step KM1.

Note: Pin locations given for reference only. Probing 60 pin connector with DVOM probe can result in permanent damage to the pin connectors. Always probe as directed using Breakout Box.

EEC-IV Diagnostic By Symptom — 2.3L EFI

SYMPTOM	RESULT	ACTION TO TAKE
• Engine stalls. • Stalls in Self-Test. • Runs rough. • Misses. • Always rich lean.		• GO to Pinpoint Test Step S1. • MAP test. GO to Pinpoint Test Step DF20. • Idle speed control. GO to Pinpoint Test Step KE1. • Poor power ground connections. • Ignition system distributor cap, rotor, wires, coil, plugs. • Base engine valves, cam timing, compression etc.
• Detonation/spark knock.		GO to Pinpoint Test Step DG1, or EGR Diagnostic Section 6 of Volume H.
• A/C compressor runs continuously.		GO to Pinpoint Test Step FA4.
• Low idle during A/C On.		GO to Pinpoint Test Step FA20.
• "CHECK ENGINE" light always on.		GO to Pinpoint Test ML1.
• "CHECK ENGINE" light never on.		GO to Pinpoint Test ML5.
• A/C does not cut-off under WOT conditions. • A/C not functioning.		GO to Pinpoint Test Step KM1.
• Surges with A/C on at idle.		GO to Pinpoint Test Step KM20.
• Stumble after hot re-start (HEGO).		GO to Pinpoint Test Step HE21.
• Stall during parking maneuvers.		GO to Pinpoint Test Step FF3.
• Fuel pump runs with key off.		GO to Pinpoint Test Step J15.

Electrical Schematic — 2.3L EFI

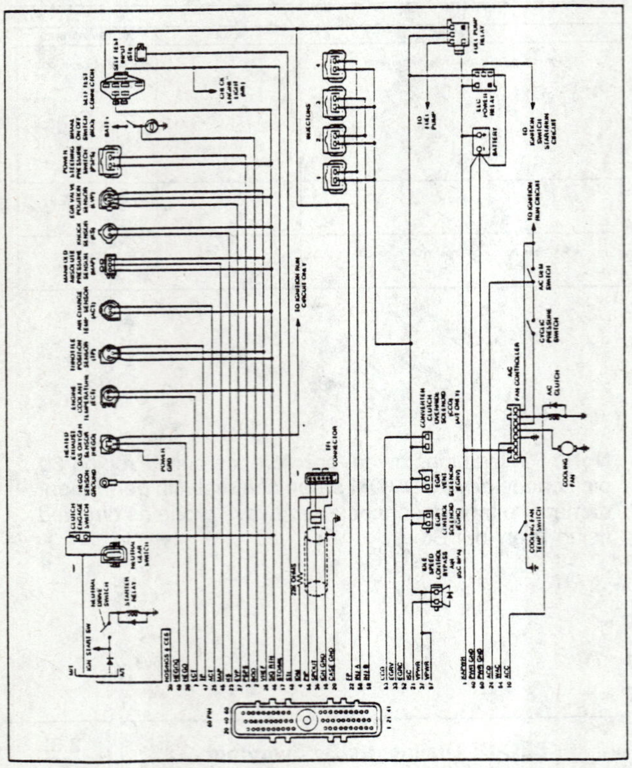

EEC-IV Diagnostic By Symptom — 2.3L EFI TURBO

SYMPTOM	RESULT	ACTION TO TAKE
• Engine stalls. • Stalls in Self-Test. • Runs rough. • Misses.	• Always Rich/Lean ▶	• GO to Pinpoint Test Step S1. • GO to Pinpoint Test Step DF20. • GO to Pinpoint Test Step KE1. • Poor power/ground connections. • Ignition system distributor cap, rotor, wires, coil, plugs. • Short to Ground. Go to Pinpoint Test Step J9. • Base engine valves cam timing, compression etc.
• Stumble after hot restart.		GO to Pinpoint Test Step HA20.
• "CHECK ENGINE" Light always on.		GO to Pinpoint Test Step ML1.
• "CHECK ENGINE" Light never on.		GO to Pinpoint Test Step ML5.
• Detonation/spark knock.		GO to Pinpoint Test Step DG1.
• High idle speeds on each restart may be accompanied by detonation for up to 3-5 minutes after a restart.		GO to Pinpoint Test Step DG1.
• Low engine rpm with A/C On.		GO to Pinpoint Test Step FA20.
• Gasoline fumes under hood.		GO to Pinpoint Test Step KD1.
• Shift indicator light always On or Off.		GO to Pinpoint Test Step KL1.
• Poor performance, sluggish detonation.	Boost control not functioning properly ▶	GO to Pinpoint Test Step KN1.

EEC-IV Module Connector Pin Usage — 2.3L EFI

MUSTANG/CAPRI

Pin	Circuit	Wire Color	Application	Abbreviations
1	38	BK-O	Keep Alive Power	KAPWR
2	810	R-LG	Brake On/Off	BOO
3	330	Y-LG	Power Steering Pressure Switch	PSPS
4	11	DG-Y	Ignition Diagnostic Monitor	IDM
7	354	LG-Y	Engine Coolant Temperature	ECT
10	347	BK-Y	A/C Clutch	ACC
16	259	BK-O	Ignition Ground	IGN GND
17	657	T	Self-Test Output "CHECK ENGINE" Light	STO MIL
20	57	BK	Case Ground	CSE GND
21	264	W-LB	Idle Speed Control	ISC
22	97	T-LG	Fuel Pump	FP
23	310	Y-R	Knock Sensor	KS
24	348	LG-P	A/C Demand	ACD
25	357	LG-P	Air Charge Temperature	ACT
26	351	O-W	Reference Voltage	VREF
27	354	LG-Y	EGR Valve Position	EVP
29	94	DG-P	Heated Exhaust Gas Oxygen Sensor	HEGO
30	771	P-Y	Neutral Drive Switch (A/T Only)	NDS
30	771	P-Y	Neutral Gear Switch and Clutch Engage Switch (M/T)	NGS & CES
33	360	DG	Exhuast Gas Recirculation Vent	EGRV
36	324	Y-LG	Spark Output	SPOUT
37	361	R	Vehicle Power	VPWR
40	60	BK-LG	Power Ground	PWR GND
45	358	LG-BK	Manifold Absolute Pressure	MAP
46	359	BK-W	Signal Return	SIG RTN
47	355	DG-LG	Throttle Position Sensor	TP
48	209	W-R	Self-Test Input	STI
49	89	O	Heated Exhaust Gas Oxygen Sensor Ground	HEGOG
52	362	Y	Exhaust Gas Recirculation Control	EGRC
53	237	O-Y	Clutch Converter Override (A/T only)	CCC
54	73	O-LB	W.O.T. A/C Cut Off	WAC
56	349	DB	Profile Ignition Pick-Up	PIP
57	361	R	Vehicle Power	VPWR
58	96	T-O	Injector 1 (Bank 1)	INJ 1
59	95	T-R	Injector 2 (Bank 2)	INJ 2
60	60	BK-LG	Power Ground	PWR GRD

Pin locations given for reference only. Probing 60 pin connector with DVOM probe will result in permanent damage to the pin connectors. Always probe as directed using the Breakout Box.

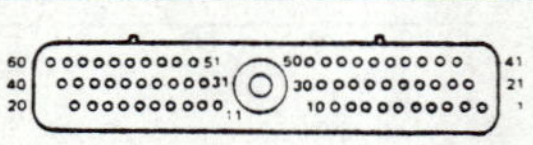

EEC-IV Diagnostic By Symptom-Continued — 2.3L EFI TURBO

SYMPTOM	RESULT	ACTION TO TAKE
• Fuel pump always runs.		GO to Pinpoint Test Step X14.
• Primary fan or secondary fan always on.		GO to Pinpoint Test Step X35.
• No A/C.		GO to Pinpoint Test Step X50.
• No A/C cutout at WOT.		GO to Pinpoint Test Step X52.
• No primary or secondary fan.		GO to Pinpoint Test Step X70.

Note: Pin locations given for reference only. Probing 60 pin connector with DVOM probe can result in permanent damage to the pin connectors. Always probe as directed using Breakout Box.

Electrical Schematic

2.3L EFI TURBO

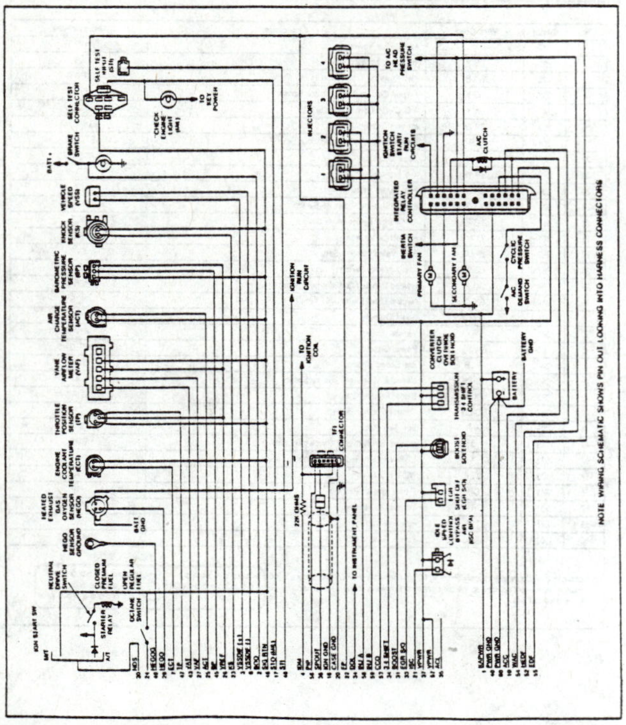

Integrated Controller Usage Pin

2.3L EFI TURBO

Pin	Circuit	Wire Color	Applications	Abbreviations
1	228A	BR Y	Power to Primary Fan	PTPF
2	228B	BR Y	Power to Primary Fan	PTPF
3	37A	Y	Power to Primary and Secondary Relays	PTPSR
4	37B	Y	Power to Primary and Secondary Relays	PTPSR
5	787	PK BK	Power to Fuel Pump	PTP
6	181A	BR O	Power to Secondary Fan	PTSF
7	181B	BR O	Power to Secondary Fan	PTSF
8	37C	Y	Battery to EEC Relay	BATT -
13	16C	R LG	Key Power	KEY PWR
14	197	T O	EDF Circuit (Primary)	EDF
15	60	BK LG	Power Ground	PWR GRD -
16	321	GY W	A C Clutch Ground	PWR GRD
17	539	PK	HEDF Circuit (Secondary)	HEDF
18	97A	T LG	Fuel Pump Circuit	FP
21	883	PK LB	A C Cyclic Switch	ACC
22	331	R	WOT A C Cut Off Circuit	WAC
23	347B	BK Y	Power to A C	PTAC
24	361	R	Vehicle Power	VPWR

Note: Pin locations given for reference only. Probing 60 pin connector with DVOM probe can result in permanent damage to the pin connectors. Always probe as directed using Breakout Box.

Integrated Controller Schematic

2.3L EFI TURBO

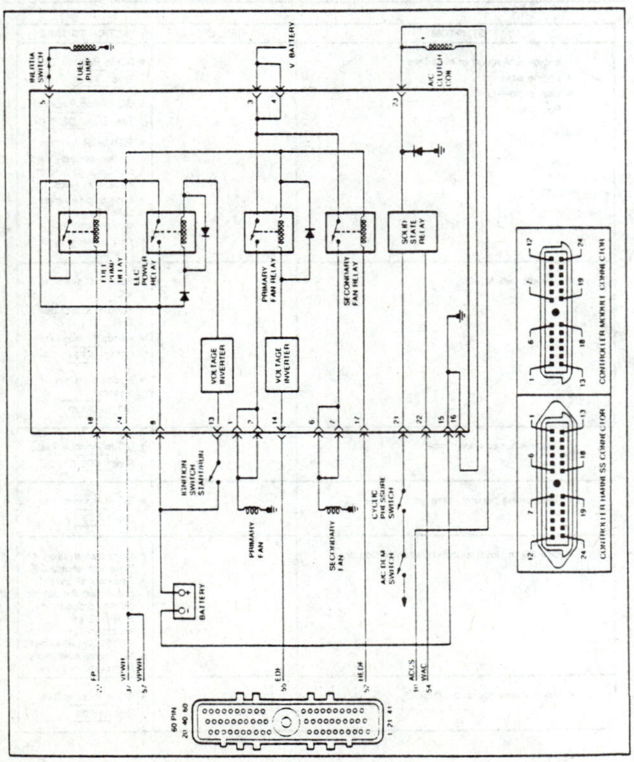

EEC-IV Module Connector Pin Usage

2.3L EFI TURBO

THUNDERBIRD

Pin	Circuit	Wire Color	Application	Abbreviations
1	37	Y	Keep Alive Power	KAPWR
2	511	LG	Brake On Off	BOO
3	150	DG W	Vehicle Speed Sensor -	VSS DIF -
4	11	DG Y	Ignition Diagnostic Monitor	IDM
6	563	O Y	Vehicle Speed Sensor -	VSS DIF
7	354	LG Y	Engine Coolant Temperature	ECT
10	347	BK Y	A C Clutch	ACC
16	259	BK O	Ignition Ground	IGN GND
17	382	Y BK	Self-Test Output "CHECK ENGINE" Light	STO (MIL)
20	60	BK LG	Case Ground	CSE GND
21	68	O BK	Idle Speed Control	ISC
22	97	T LG	Fuel Pump Control	FP
23	310	Y R	Knock Sensor	KS
24	570	R Y	Octane Switch	OCTANE SW
25	470	P BK	Air Charge Temperature Sensor	ACT
26	351	O W	Voltage Reference	VREF
27	200	W BK	Vane Air Flow	VAF
29	34	DG P	Heated Exhaust Gas Oxygen Sensor	HEGO
30	376	BR W	Neutral Drive Switch (A T)	NDS
30	359	BK W	Neutral Gear Switch and Clutch Engage Switch (M T)	NGS and CES
31	464	BK P	Boost Control	BOOST
33	362	Y	Exhaust Gas Recirculation Shut Off	EGR S O
34	305	LB P	Data Output Link	DOL
35	836	O W	Automatic Adjustable Shock Controller	AQL
36	324	Y LG	Spark Output	SPOUT
37	361	R	Vehicle Power	VPWR
40	60	BK LG	Battery Ground	PWR GND
43	357	LG P	Vane Air Temperature	VAT
45	358	LG BK	Barometric Pressure	BP
46	359	BK W	Signal Return	SIG RTN
47	355	DG LG	Throttle Position Sensor	TP
48	209	W R	Self-Test Input	STI
49	89	O	Heated Exhaust Gas Oxygen Ground	HEGOG
52	539	P	High Electric Drive Fan	HEDF
54	331	R	W O T A C Cut-Off	WAC
55	197	T O	Electric Drive Fan	EDF
56	349	DB	Profile Ignition Pick-Up	PIP
57	361	R	Vehicle Power	VPWR
58	95	T R	Injector Bank A	INJ A
59	96	T O	Injector Bank B	INJ B
60	60	BK LG	Battery Ground	PWR GND

Pin locations given for reference only. Probing 60 pin connector with DVOM probe will result in permanent damage to the pin connectors. Always probe as directed using the Breakout Box.

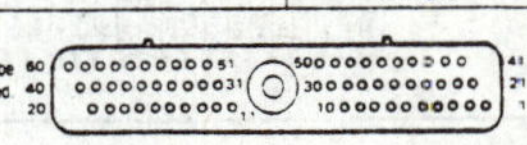

EEC-IV Diagnostic By Symptom — 2.3L HSC-CFI

SYMPTOM	RESULT	ACTION TO TAKE
• Engine runs rough or misfires. • Engine stalls. • Lack of power. • Rough idle.		• GO to Pinpoint Test Step S2. • GO to Pinpoint Test Step DF20. • EGR Valve Function. • Ignition system distributor cap, rotor, wires, coil, plugs. • Base engine valves, cam timing, compression etc. • Poor power/ground connections.
• Low idle with A/C On.		• GO to Pinpoint Test Step FA20.
• Stumble after hot restart (HEGO).		• GO to Pinpoint Test Step HF20.
• Shift indicator light always on or off.		• GO to Pinpoint Test Step KL1.
• Stall during parking maneuvers.		• GO to Pinpoint Test Step FF3.
• High idle in Drive, automatic transmission only.		• Place transmission in Drive with engine Off. Test neutral drive circuit (test Pin 30) for short to ground. SERVICE as necessary. **NOTE: Circuit should NOT have continuity to ground in Drive.**
• Fuel pump always runs.		• GO to Pinpoint Test Step J15.
• A/C does not cutoff under WOT conditions. • No A/C.		• GO to Pinpoint Test Step KM1.

EEC-IV Module Connector Pin Usage — 2.3L HSC-CFI

Pin	Circuit	Wire Color	Application	Abbreviations
1	37	Y	Keep Alive Power	KAPWR
4	[illegible]	DG Y	Ignition Diagnostic Monitor	IDM
5	20	W LB	Key Power	KPWR
7	354	LG Y	Engine Coolant Temperature Sensor	ECT
10	347	BK Y	A C Clutch Compressor Signal	ACC
16	259	BK O	Ignition Ground	IGN GND
17	201	T R	Self-Test Output Shift Indicator Light	STO SIL
20	57	BK	Case Ground	CASE GND
21	376	BR W	Idle Speed Control −	ISC −
22	238	O LB	Fuel Pump Relay	FP
24	330	Y LG	Power Steering Pressure Switch	PSPS
25	357	LG P	Air Charge Temperature Sensor	ACT
26	351	O W	Reference Voltage	VREF
27	352	BR LG	EGR Valve Position Sensor	EVP
28	265	LG W	Idle Tracking Switch	ITS
29	94	DG P	Heated Exhaust Gas Oxygen Sensor	HEGO
30	614	GY O	Neutral Drive Switch (A.T Only)	NDS
36	324	Y LG	Spark Output	SPOUT
37	361	R	Vehicle Power	VPWR
40	60	BK LG	Power Ground	PWR GND
41	264	W LB	Idle Speed Control	ISC −
45	358	LG-BK	Manifold Absolute Pressure Sensor	MAP
46	359	BK W	Signal Return (Ground)	SIG RTN
47	355	DG LG	Throttle Position Sensor	TP
48	209	W R	Self-Test Input	STI
49	89	O	Heated Exhaust Gas Oxygen Ground	HEGOG
52	362	Y	EGR Vacuum Regulator Solenoid	EVR
54	73	O LB	WOT A C Cutoff	WAC
56	349	DB	Profile Ignition Pick-Up	PIP
57	361	R	Vehicle Power	VPWR
58	95	T R	Fuel Injector	INJ
60	60	BK LG	Power Ground	PWR GND

Pin locations given for reference only. Probing 60 pin connector with DVOM probe will result in permanent damage to the pin connectors. Always probe as directed using the Breakout Box.

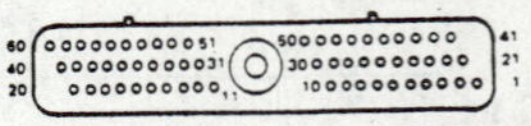

Electrical Schematic — 2.3L HSC-CFI

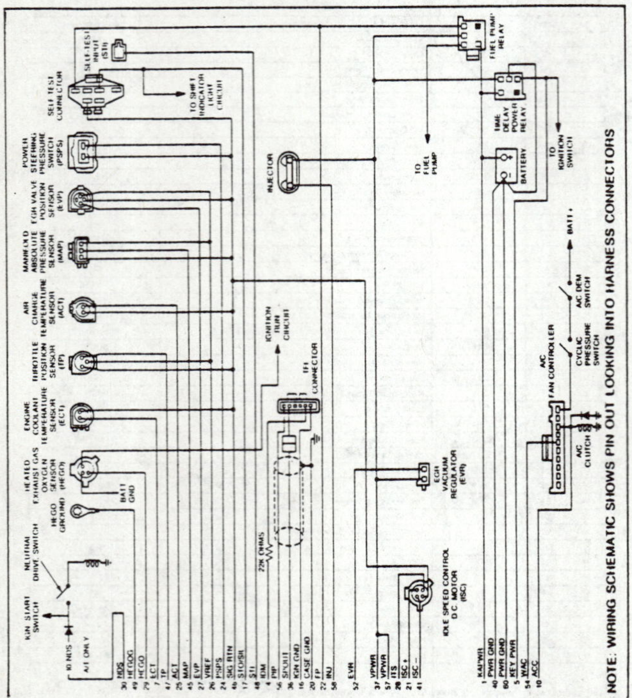

EEC-IV Diagnostic By Symptom — 2.5L CFI CLC, MTX

SYMPTOM	RESULT	ACTION TO TAKE
• Engine runs rough or miss-fires. • Engine stalls. • Lack of power. • Rough idle.		• GO to Pinpoint Test Step S2. • GO to map Pinpoint Test Step DF20. • EGR value function. • Ignition system distributor cap, rotor, wires, coil, plugs. • Base engine valves, cam timing, compression etc. • Poor power/ground connections.
• Low idle with A/C On.		• Place A/C switch to A/C. • Perform KOEO Self Test. • If code 67 is present GO to KF9. • If code 67 is not present GO to FA20.
• Gasoline fumes under hood.		• GO to Pinpoint Test Step KD1.
• Stalls during parking maneuvers.		• GO to Pinpoint Test Step FF3.
• High idle in Drive, automatic transmission only.		• Place transmission in Drive with engine Off. Test neutral drive circuit (test Pin 30) for short to ground. SERVICE as necessary. **NOTE: Circuit should NOT have continuity to ground in Drive.**
• Stumble after hot restart (HEGO).		• GO to Pinpoint Test Step HF20.

EEC-IV Diagnostic By Symptom-Continued

2.5L CFI CLC, MTX

SYMPTOM	RESULT	ACTION TO TAKE
• Shift indicator light always on or off.		• GO to Pinpoint Test Step KL1
• "CHECK ENGINE" light always on.		• GO to Pinpoint Test Step ML1
• "CHECK ENGINE" light never on.		• GO to Pinpoint Test Step ML5
• Fuel pump always runs.		• Go to Pinpoint Test X-14
• No cooling fan (MTX only).		• GO to Pinpoint Test X-40
• No engine cooling fan (CLC only). • No low speed engine cooling fan (CLC only). • No high speed engine cooling fan (CLC only).		• GO to Pinpoint Test X-20
• Low speed or high speed engine cooling fan on all the time (CLC only).		• GO to Pinpoint Test X-35
• No A/C.		• GO to Pinpoint Test X-50
• No A/C cut out at WOT.		• GO to Pinpoint Test X-52

2.5L CFI CLC

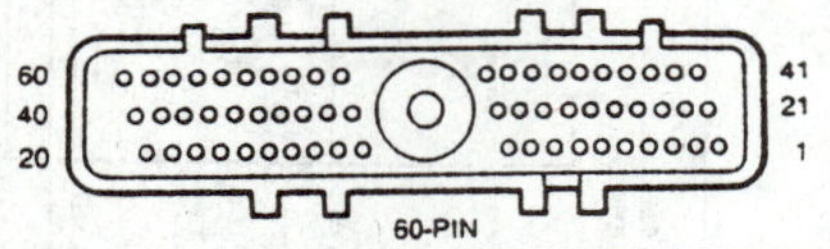

Electrical Schematic

2.5L CFI CLC

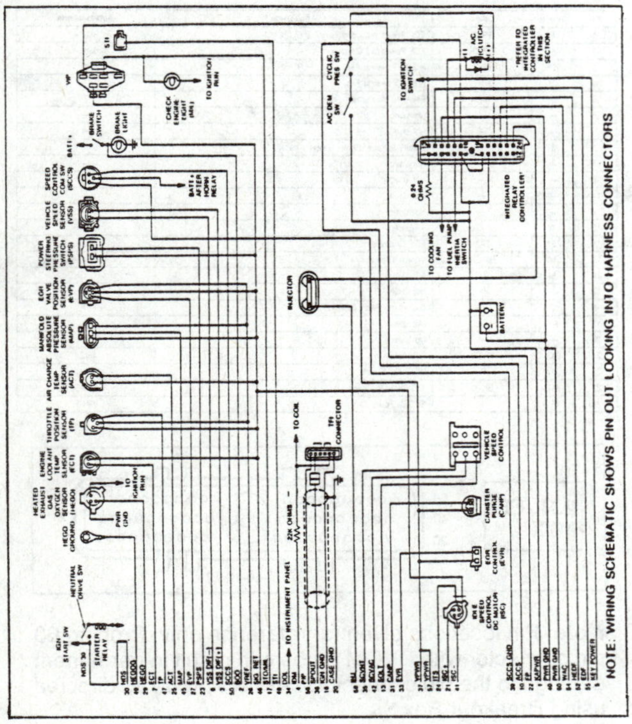

Integrated Controller Schematic

2.5L CFI CLC

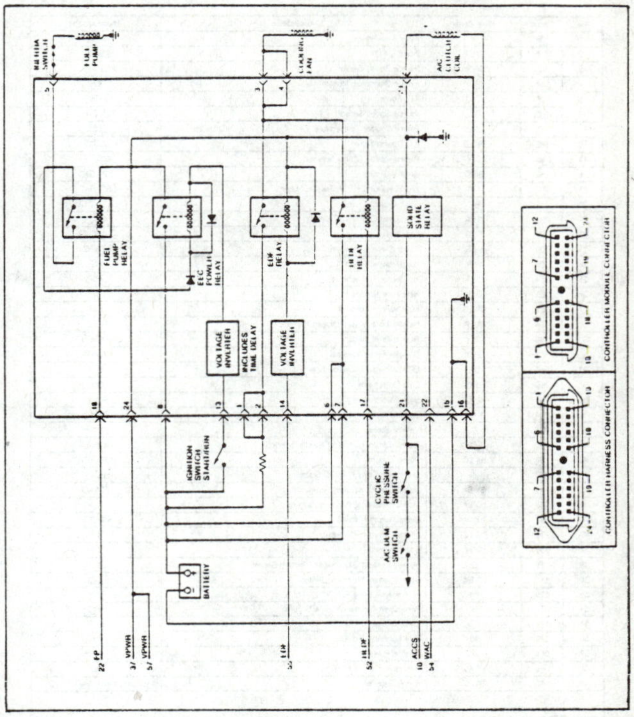

Integrated Controller Usage Pin

2.5L CFI CLC

Pin	Circuit	Color	Application	Abbreviations
1	181	BR-O	EDF Power Into Controller	Batt -
2	181	BR-O	EDF Power Into Controller	Batt -
3	228	BR-Y	H EDF Power to Fan	PTF
4	228	BR-Y	H EDF Power to Fan	PTF
5	787	PK-BK	Power to the Pump	PTP
6	38	BK-O	HEDF Power Into Controller	Batt -
7	38	BK-O	HEDF Power Into Controller	Batt -
8	37	Y	Battery Voltage (Power Relay)	Batt -
9	—	—	Not Used	—
10	—	—	Not Used	—
11	—	—	Not Used	—
12	—	—	Not Used	—
13	16	R-LG	Keypower	KPWR
14	197	T-O	EDF Ground	EDF
15	60	BK-LG	Vehicle Ground	PWR GND
16	57	BK	A C Ground	A C GND
17	639	PK	HEDF Ground	HEDF
18	97	T-LG	Fuel Pump Relay Ground	FP
19	—	—	Not Used	—
20	—	—	Not Used	—
21	883	PK-LB	A C Power In	ACCS
22	331	R	WOT A C Cutoff	WAC
23	347	BK-Y	A C Power to Clutch Coil	PTAC
24	361	R	EEC IV Power Out	VPWR

2.5L CFI CLC

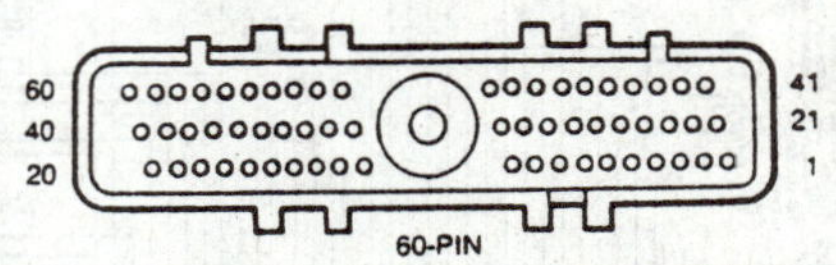

Note: Pin locations given for reference only. Probing 60 pin connector with DVOM probe can result in permanent damage to the pin connectors. Always probe as directed using Breakout Box.

EEC-IV Module Connector Pin Usage — 2.5L CFI CLC

Pin	Circuit	Wire Color	Application	Abbreviations
1	37	Y	Keep Alive Power	KAPWR
2	810	R LG	Brake On Off	BOO
3	150	DG W	Vehicle Speed Sensor -	VSS DIF -
4	11	DG Y	Ignition Diagnostic Monitor	IDM
5	16	R LG	Key Power	KPWR
6	563	O Y	Vehicle Speed Sensor	VSS DIF -
7	354	LG Y	Engine Coolant Temperature	ECT
10	883	PK LB	A C Cycling Switch	ACCS
13	144	O Y	Vehicle Speed Control Solenoid	SOL -
16	259	BK O	Ignition Ground	IGN GND
17	201	T R	Self-Test Out "CHECK ENGINE" Light	STO MIL
20	57	BK	Case Ground	CSE GND
21	382	Y BK	Idle Speed Control (DC Motor)	ISC -
22	97	T LG	Fuel Pump	FP
23	330	Y LG	Power Steering Pressure Switch	PSPS
24	209	W R	Idle Tracking Switch	ITS
25	357	LG P	Air Charge Temperature	ACT
26	351	O W	Reference Voltage	VREF
27	352	BR LG	EGR Valve Position	EVP
29	94	DG P	Heated Exhaust Gas Oxygen (Sensor)	HEGO
30	199	LB Y	Neutral Drive Switch	NDS
31	101	GY Y	Canister Purge Solenoid	CANP
33	360	DG	EGR Vacuum Regulator (Solenoid)	EVR
34	305	LB PK	Data Output Link	DOL
35	146	W PK	Speed Control Vent (Solenoid)	SCVNT
36	324	Y LG	Spark Out	SPOUT
37	361	R	Vehicle Power	VPWR
39	461	O	Speed Control Command Switch Ground	SCCS GND
40	60	BK LG	Power Ground	PWR GND
41	377	W	Idle Speed Control — (DC Motor)	ISC -
42	145	GY BK	Speed Control Vacuum (Solenoid)	SCVAC
45	358	LG BK	Manifold Absolute Pressure (Sensor)	MAP
46	359	BK W	Signal Return (Ground)	SIG RTN
47	355	DG LG	Throttle Position (Sensor)	TP
48	200	W BK	Self-Test Input	STI
49	89	O	Heated Exhaust Gas Oxygen (Sensor) Ground	HEGO GND
50	151	LB BK	Speed Control Command Switch	SCCS
52	639	PK	High Electro Drive Fan	HEDF
54	331	R	Wide Open Throttle (WOT) A C Cutoff	WAC
55	197	T O	Electro Drive Fan (Low)	EDF
56	349	DB	Profile Ignition Pick-up	PIP
57	361	R	Vehicle Power	VPWR
58	95	T R	Injector	INJ
60	60	BK LG	Power Ground	PWR GND

Integrated Controller Schematic — 2.5L CFI MTX

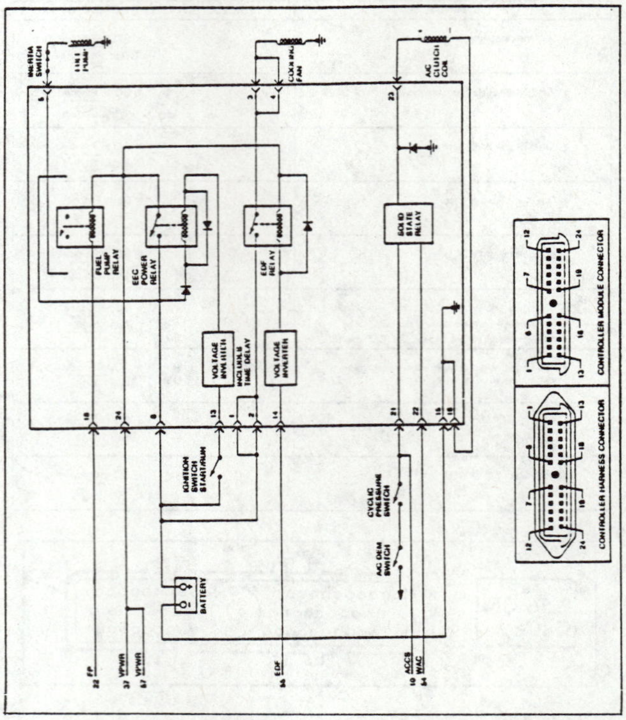

Electrical Schematic — 2.5L CFI MTX

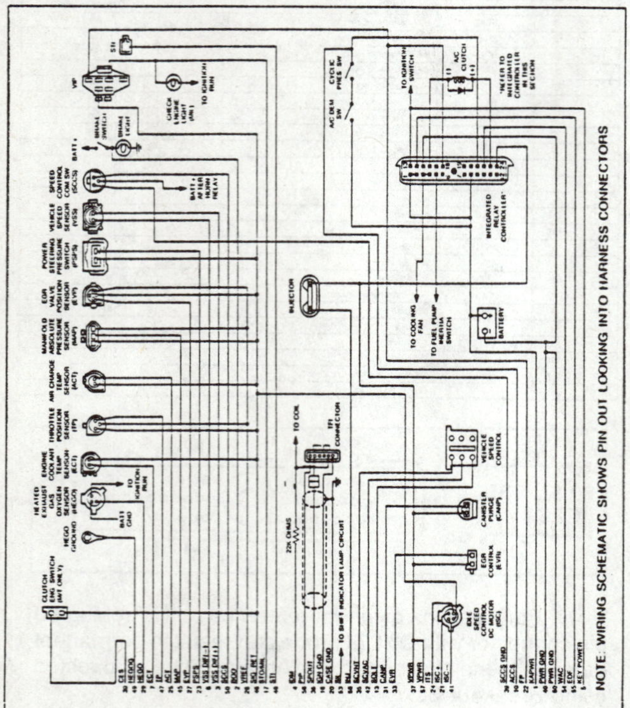

Integrated Controller Usage Pin — 2.5L CFI MTX

Pin	Circuit	Color	Application	Abbreviations
1	181	BR-O	EDF Power Into Controller	Batt -
2	181	BR-O	EDF Power Into Controller	Batt -
3	228	BR-Y	H EDF Power to Fan	PTF
4	228	BR-Y	H EDF Power to Fan	PTF
5	787	PK-BK	Power to the Pump	PTP
6	—	—	Not Used	—
7	—	—	Not Used	—
8	37	Y	Battery Voltage (Power Relay)	Batt -
9	—	—	Not Used	—
10	—	—	Not Used	—
11	—	—	Not Used	—
12	—	—	Not Used	—
13	16	R-LG	Keypower	KPWR
14	197	T-O	EDF Ground	EDF
15	60	BK-LG	Vehicle Ground	PWR GND
16	57	BK	A C Ground	A C GND
17	—	—	Not Used	—
18	97	T-LG	Fuel Pump Relay Ground	FP
19	—	—	Not Used	—
20	—	—	Not Used	—
21	883	PK-LB	A C Power In	ACCS
22	331	R	WOT A C Cutoff	WAC
23	347	BK-Y	A C Power to Clutch Coil	PTAC
24	361	R	EEC IV Power Out	VPWR

2.5L CFI MTX

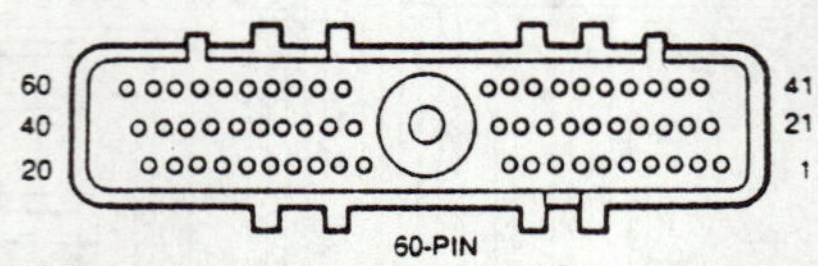

Note: Pin locations given for reference only. Probing 60 pin connector with DVOM probe can result in permanent damage to the pin connectors. Always probe as directed using Breakout Box.

EEC-IV Module Connector Pin Usage — 2.5L CFI MTX

Pin	Circuit	Wire Color	Application	Abbreviations
1	37	Y	Keep Alive Power	KA PWR
2	810	R/LG	Brake On Off	BOO
3	150	DG/W	Vehicle Speed Sensor +	VSS DIF +
4	11	DG/Y	Ignition Diagnostic Monitor	DM
5	16	R/LG	Key Power	KPWR
5	563	O/Y	Vehicle Speed Sensor —	VSS DIF
7	354	LG/Y	Engine Coolant Temperature	ECT
10	383	PK/LB	A/C Cycling Switch	ACCS
13	144	O/Y	Vehicle Speed Control Solenoid	SOL -
16	259	BK/O	Ignition Ground	IGN GND
17	201	T/R	Self-Test Out "CHECK ENGINE" Light	STO MIL
20	57	BK	Case Ground	CSE GND
21	382	Y/BK	Idle Speed Control (DC Motor)	SC -
22	97	T/LG	Fuel Pump	FP
23	330	Y/LG	Power Steering Pressure Switch	PSPS
24	209	W/R	Idle Tracking Switch	ITS
25	357	LG/P	Air Charge Temperature	ACT
26	351	O/W	Reference Voltage	VREF
27	352	BR/LG	EGR Valve Position	EVP
29	94	DG/P	Heated Exhaust Gas Oxygen (Sensor)	HEGO
30	480	P/Y	Clutch Engage Switch	CES
31	101	GY/Y	Canister Purge Solenoid	CANP
33	360	DG	EGR Vacuum Regulator (Solenoid)	EVR
35	148	W/PK	Speed Control Vent (Solenoid)	SCVNT
36	324	Y/LG	Spark Out	SPOUT
37	361	R	Vehicle Power	VPWR
39	461	O	Speed Control Command Switch Ground	SCCS GND
40	60	BK/LG	Power Ground	PWR GND
41	377	W	Idle Speed Control — (DC Motor)	ISC -
42	145	GY/BK	Speed Control Vacuum (Solenoid)	SCVAC
45	358	LG/BK	Manifold Absolute Pressure (Sensor)	MAP
46	359	BK/W	Signal Return (Ground)	SIG RTN
47	355	DG/LG	Throttle Position (Sensor)	TP
48	200	W/BK	Self-Test Input	STI
49	89	O	Heated Exhaust Gas Oxygen (Sensor) Ground	HEGO GND
50	151	LB/BK	Speed Control Command Switch	SCCS
53	462	P	Shift Indicator Light	SIL
54	331	R	Wide Open Throttle (WOT) A/C cutoff	WAC
55	197	T/O	Electro Drive Fan	EDF
56	349	DB	Profile Ignition Pick-Up	PIP
57	361	R	Vehicle Power	VPWR
58	95	T/R	Injector	INJ
60	60	BK/LG	Power Ground	PWR GND

Pin locations given for reference only. Probing 60 pin connector with DVOM probe will result in permanent damage to the pin connectors. Always probe as directed, using the Breakout Box.

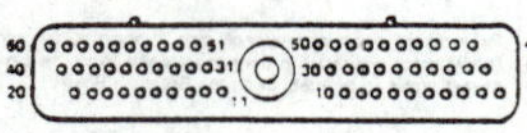

EEC-IV Diagnostic By Symptom — 3.0L EFI

SYMPTOM	RESULT	ACTION TO TAKE
• Engine runs rough or miss-fires. • Engine stalls. • Rough idle.		• GO to Pinpoint Test Step S1. • GO to map Pinpoint Test Step DF20. • EGR value function. • Ignition system distributor cap, rotor, wires, coil, plugs. • Base engine valves cam timing, compression etc. • Fuel delivery. • Poor power ground connections.
• Detonation/spark knock.		GO to Pinpoint Test Step DG1.
• Stalls during parking maneuvers.		GO to Pinpoint Test Step FF3.
• Gasoline fumes under hood.		GO to Pinpoint Test Step KD1.
• Noise emitting from rear of vehicle with Key Off.	Fuel pump always on.	GO to Pinpoint Test Step X-14.
• No engine cooling fan. • No low speed engine cooling fan. • No high speed engine cooling fan.		GO to Pinpoint Test Step X-20.
• Low speed or high speed engine cooling fan on all the time.		GO to Pinpoint Test Step X-35.
• Stumble after hot restart (HEGO).		GO to Pinpoint Test Step HG-21.
• No A/C.		GO to Pinpoint Test X-50.

EEC-IV Diagnostic By Symptom-Continued — 3.0L EFI

SYMPTOM	RESULT	ACTION TO TAKE
• No A/C cut out at WOT.		GO to Pinpoint Test X-52.
• Low idle rpm during A/C On.		GO to Pinpoint Test Step FA20.
• "CHECK ENGINE" light always on.		• GO to Pinpoint Test Step ML1.
• "CHECK ENGINE" light never on.		• GO to Pinpoint Test ML5.

3.0L EFI

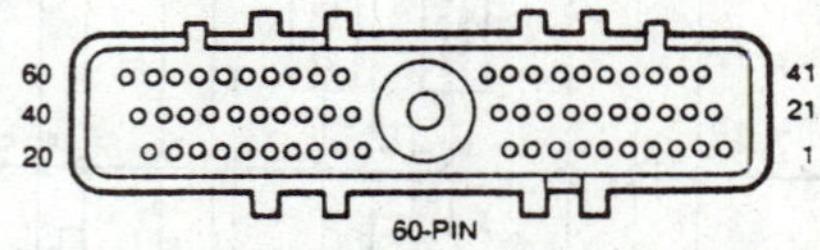

Note: Pin locations given for reference only. Probing 60 pin connector with DVOM probe can result in permanent damage to the pin connectors. Always probe as directed using Breakout Box.

Electrical Schematic — 3.0L EFI

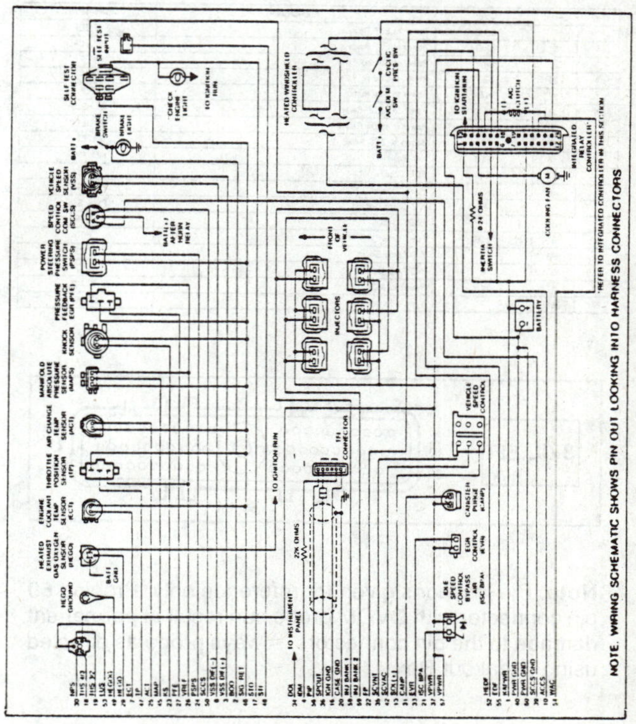

Integrated Controller Schematic — 3.0L EFI

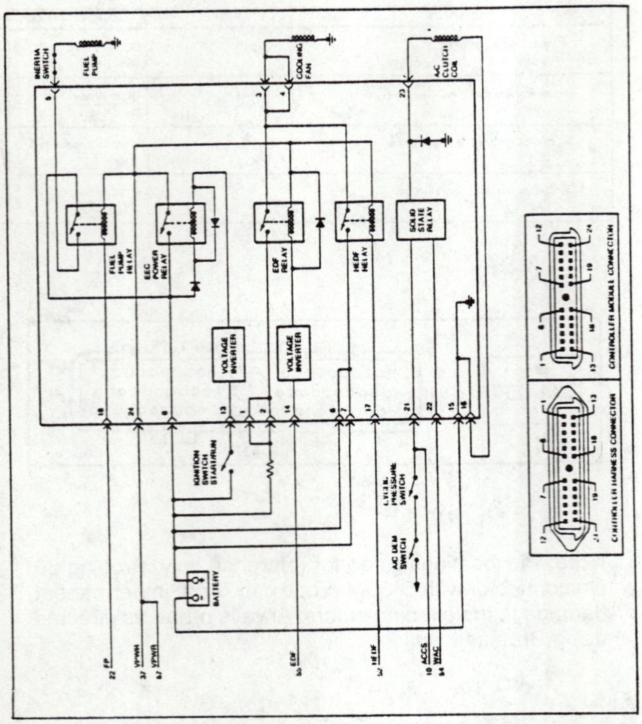

Integrated Controller Usage Pin — 3.0L EFI

Pin	Circuit	Color	Applications	Abbreviations
1	181	BR O	Power to EDF Relay	PT EDF
2	181	BR O	Power to EDF Relay	PT EDF
3	228	BR Y	Power to Cooling Fan	PTF
4	228	BR Y	Power to Cooling Fan	PTF
5	787	PK BK	Power to Fuel Pump	PTP
6	38	BK O	Power to HEDF Relay	PT HEDF
7	38	BK O	Power to HEDF Relay	PT HEDF
8	37	Y	Battery to EEC Relay	BATT +
13	16	R LG	Key Power	KEY PWR
14	197	T O	EDF Circuit	EDF
15	60	BK LG	Power Ground	PWR GRD
16	57	BK	A C Clutch Ground	PWR GRD
17	639	PK	HEDF Circuit	HEDF
18	97	T LG	Fuel Pump	FP
21	883		A C Cyclic Switch	ACCS
22	331	R	WOT A C Cut Off	WAC
23	347	BK Y	Power to A C	PTAC
24	361	R	Vehicle Power	VPWR

3.0L EFI

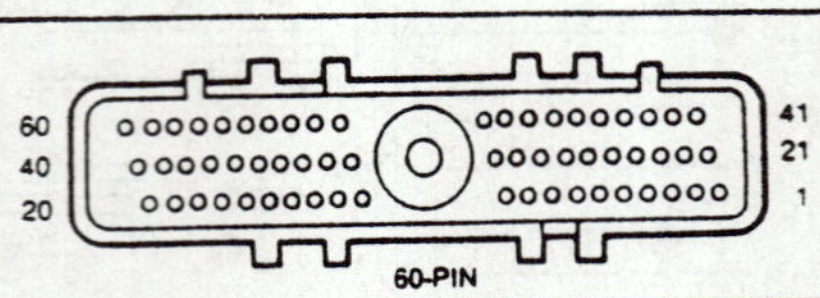

Note: Pin locations given for reference only. Probing 60 pin connector with DVOM probe can result in permanent damage to the pin connectors. Always probe as directed using Breakout Box.

EEC-IV Module Connector Pin Usage — 3.0L EFI

Pin	Circuit	Wire Color	Application	Abbreviations
1	37	Y	Keep Alive Power	KA PWR
2	810	R LG	Brake On Off	BOO
3	150	DG W	Vehicle Speed Sensor +	VSS DIF +
4	11	DG Y	Ignition Diagnostic Monitor	IDM
6	563	O Y	Vehicle Speed Sensor —	VSS DIF −
7	354	LG Y	Engine Coolant Temperature	ECT
10	883	PR LB	A C Cycling Switch	ACCS
13	144	O Y	Vehicle Speed Control Solenoid	SOL −
16	259	BK O	Ignition Ground	IGN GND
17	201	T R	Self-Test Output "CHECK ENGINE" Light	STO MIL
18	315	DG P	Transmission 4/3 Switch	THS 4 3
19	237	O Y	Transmission 3 2 Switch	THS 3 2
20	57	BK	Case Ground	CSE GND
21	68	O BK	Idle Speed Control (Bypass Air)	ISC-BPA
22	97	T LG	Fuel Pump	FP
23	310	Y R	Knock Sensor	KS
24	330	Y LG	Power Steering Pressure Switch	PSPS
25	357	LG P	Air Charge Temperature	ACT
26	351	O W	Reference Voltage	VREF
27	352	BR LG	Pressure Feedback EGR	PFE
29	94	DG P	Heated Exhaust Gas Oxygen (Sensor)	HEGO
30	480	P Y	Neutral Pressure Switch	NPS
31	101	GY Y	Canister Purge Solenoid	CANP
33	360	DG	EGR Vacuum Regulator (Solenoid)	EVR
34	305	LB PK	Data Output Link	DOL
35	146	W PK	Speed Control Vent (Solenoid)	SCVNT
36	324	Y LG	Spark Output	SPOUT
37	361	R	Vehicle Power	VPWR
39	461	O	Speed Control Command Switch Ground	SCCS GND
40	60	BK LG	Power Ground	PWR GND
42	145	GY BK	Speed Control Vacuum (Solenoid)	SCVAC
45	358	LG BK	Manifold Absolute Pressure (Sensor)	MAP
46	359	BK W	Signal Return (Ground)	SIG RTN
47	355	DG LG	Throttle Position (Sensor)	TP
48	200	W BK	Self Test Input	STI
49	89	O	Heat Exhaust Gas Oxygen Sensor Ground	HEGO GND
50	151	LB BK	Speed Control Command Switch	SCCS
52	639	PK	High Electro Drive Fan	HEDF
53	224	T LB	Lock-Up Solenoid (Transmission)	LUS
54	331	R	Wide Open Throttle (WOT) A C Cutoff	WAC
55	197	T O	Electro Drive Fan (Low)	EDF
56	349	DB	Profile Ignition Pick-Up	PIP
57	361	R	Vehicle Power	VPWR
58	95	T R	Injector Bank 1	INJ 1
59	96	T O	Injector Bank 2	INJ 2
60	60	BK LG	Power Ground	PWR GND

Pin locations given for reference only. Probing 60 pin connector with DVOM probe will result in permanent damage to the pin connectors. Always probe as directed using the Breakout Box.

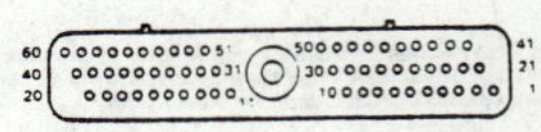

EEC-IV Diagnostic By Symptom — 3.8L CFI

SYMPTOM	RESULT	ACTION TO TAKE
• Engine runs rough or miss-fires. • Engine stalls. • Rough idle.		• GO to Pinpoint Test Step S2. • GO to Pinpoint Test Step DF20. • EGR valve function. • Ignition system distributor cap, rotor, wires, coil, plugs. • Base engine valves, cam timing, compression etc. • Fuel delivery. • Poor power ground connections.
• Detonation spark knock (Mustang Capri only).		• GO to Pinpoint Test Step DG1.
• Low engine rpm with A/C On.		• GO to Pinpoint Test Step FA20. and or KF1.
• Gasoline fumes under hood.		• GO to Pinpoint Test Step KD1.
• Low idle on cold start-up. • High idle on hot start-up. • Engine diesels.		• GO to Pinpoint Test Step KF16.
• High idle in Drive, automatic transmission only		• Place transmission in Drive with engine Off. Test neutral drive circuit (test Pin 30) for short to ground. SERVICE as necessary. **NOTE: Circuit should NOT have continuity to ground in Drive.**
• Fuel pump always runs.		• GO to Pinpoint Test Step J15.
• "CHECK ENGINE" light always on.		• GO to Pinpoint Test Step ML1.
• "CHECK ENGINE" light never on.		• GO to Pinpoint Test Step ML5.

Electrical Schematic — 3.8L CFI

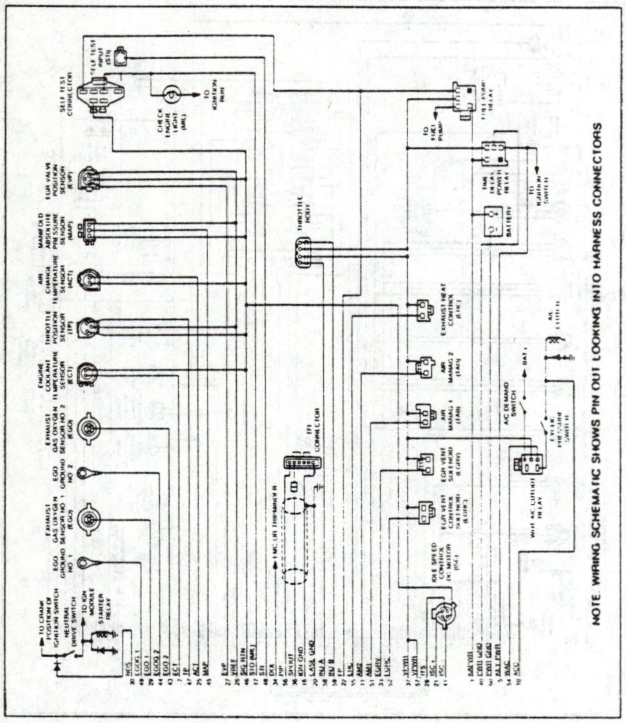

EEC-IV Module Connector Pin Usage — 3.8L CFI

Pin	Circuit	Wire Color	Application	Abbreviations
1	38	BK O	Keep Alive Power	KAPWR
5	16	R LG	Key Power	KPWR
7	354	LG Y	Engine Coolant Temperature	ECT
10	347	BK Y	A C Clutch	ACC
11	99	LG BK	Air Management — TAD	AM2
16	259	BK O	Ignition Ground	IGN GND
17	382	Y B	Self-Test Output "CHECK ENGINE" Light	STO MIL
20	57	BK	Case Ground	CSE GND
21	382	Y BK	Idle Speed Control (T X)	ISC -
22	97	T LG	Fuel Pump	FP
25	357	LG P	Air Charge Temperature	ACT
26	351	O W	Reference Voltage	VREF
27	352	BR LG	EGR Valve Position	EVP
28	265	LG W	Idle Tracking Switch	ITS
29	94	DG P	Exhaust Gas Oxygen Sensor — Left	EGO 1(L)
30	32	R LB	Neutral Drive Switch (T X)	NDS
33	360	DG	Exhaust Gas Recirculation Vent	EGRV
34	305	LB P	Data Output Line	DOL
36	324	Y LG	Spark Output	SPOUT
37	175	BK Y	Vehicle Power (T X)	VPWR
40	60	BK LG	Battery Ground	BATT GND
41	264	W LB	Idle Speed Control	ISC -
43	90	DB LG	Exhaust Gas Oxygen Sensor — Right	EGO 2(R)
44	91	GY R	Exhaust Gas Oxygen Ground — Right	EGOG 2(R)
45	358	LG BK	Manifold Absolute Pressure (T X)	MAP
46	359	BK W	Signal Return	SIG RTN
47	355	DG LG	Throttle Position Sensor	TP
48	209	W R	Self-Test Input	STI
49	89	O	Exhaust Gas Oxygen Ground — Left	EGOG 1(L)
51	100	W R	Air Management — Tab	AM1
52	362	Y	Exhaust Gas Recirculation Control	EGRC
54	73	O LB	W.O.T. A C Cut Off	WAC
55	377	W	Exhaust Heat Control	EHC
56	349	DB	Profile Ignition Pick-Up	PIP
57	175	BK Y	Vehicle Power (T X)	VPWR
58	95	T R	Injector 1	INJ 1
59	96	T O	Injector 2	INJ 2
60	60	BL LG	Battery Ground	BATT GND

Pin locations given for reference only. Probing 60 pin connector with DVOM probe will result in permanent damage to the pin connectors. Always probe as directed, using the Breakout Box.

EEC-IV Diagnostic By Symptom — 5.0L SEFI

SYMPTOM	RESULT COULD BE	ACTION TO TAKE
7.0 DIAGNOSTICS BY SYMPTOM		
• Noise emitting from rear of vehicle with key-off.	Fuel pump always on	▶ GO to Pinpoint Test Step J15
• Shift indicator light always on or, off.		▶ GO to Pinpoint Test Step KL1
• Low idle rpm with A C ON.		▶ GO to Pinpoint Test Step FA20
• A C does not cutoff during W.O.T.		▶ GO to Pinpoint Test Step KM1
• Detonation and, or spark knock.	EGR valve function	▶ GO to EGR Diagnostic Section 6 of Volume "H"
• Engine stalls during quick test.	Idle speed control	▶ GO to Pinpoint Test Step KE4
• Gasoline fumes under hood.		▶ GO to Canister Purge Pinpoint Test Step KD1
• Misses. • Engine runs rough. • Engine stalls. • Lack of power. • Erratic RPM. • Surges.		• Poor power ground connections. • Ignitions system distributor cap, rotor, wires, coil, plugs • Base engine valves, cam timing, compression etc. • GO to Pinpoint Test Step S1 for ISC, MAP, EGR, and Cylinder Balance Test.
• "CHECK ENGINE" light always on.		• GO to Pinpoint Test Step ML1.
• "CHECK ENGINE" light never on.		• GO to Pinpoint Test Step ML5.

Electrical Schematic — 5.0L SEFI

ALL EXCEPT 1987½ MARK VII/LINCOLN TOWN CAR CALIBRATION 8-22

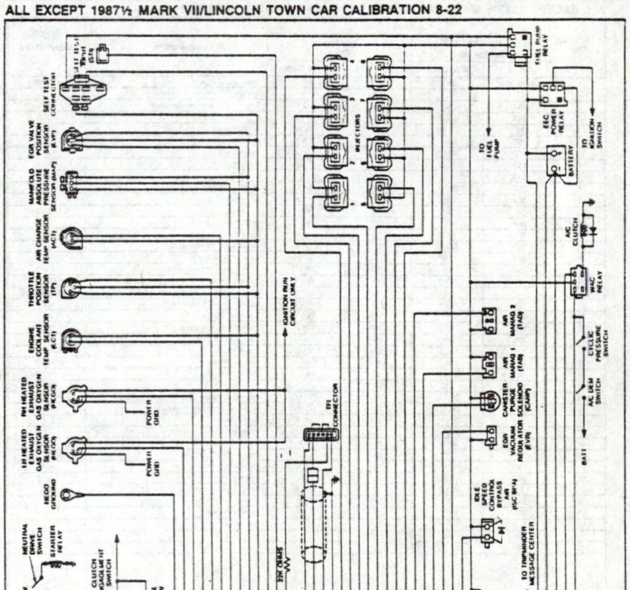

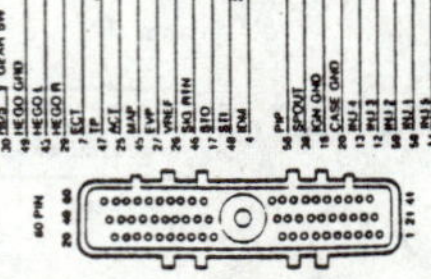

EEC-IV Module Connector Pin Usage — 5.0L SEFI

ALL EXCEPT 1987½ MARK VII/LINCOLN TOWN CAR CALIBRATION 8-22

Pin	Car Lines				Application	Abbreviations
	Ford, Mercury, Town Car		Thunderbird, Cougar			
	Crt.#	Wire Color	Crt.#	Wire Color		
1	37	Y	38	BK O	Keep Alive Power	KAPWR
4	11	DG-Y	189	LB-PK	Ignition Diagnostic Monitor	IDM
7	354	LG Y	354	LG Y	Engine Coolant Temperature	ECT
10	883	PK LB	883	PK LB	A C Clutch Signal	ACCS
11	99	LG BK	99	LG BK	Air Management	AM2
12	557	BR Y	557	BR Y	Injector #3	INJ 3
13	558	BR LB	558	BR LB	Injector #4	INJ 4
14	559	T LB	559	T LB	Injector #5	INJ 5
15	560	LG	560	LG	Injector #6	INJ 6
16	259	BK O	259	BK O	Ignition Ground	IGN GND
17	201	T R	382	Y BK	Self-Test Output	STO
20	57	BK	57	BK	Case Ground	CASE GND
21	264	W LB	264	W LB	Idle Speed Control	ISC
22	97	T LG	97	T LG	Fuel Pump	FP
25	357	LG P	357	LG P	Air Charge Temperature Sensor	ACT
26	351	O W	351	O W	Reference Voltage	VREF
27	352	BR LG	352	BR LG	EGR Valve Position Sensor	EVP
29	94	DG P	94	DG P	Heated Exhaust Gas Oxygen Sensor	HEGO-R
30	33	W/PK	376	BR W	Neutral Drive Switch	NDS
31	101	GY/Y	101	GY/Y	Canister Purge	CANP
33	360	DG	360	DG	EGR Valve Regulator	EVR
34	305	LB/PK	305	LB PK	Data Output Line	DOL
36	324	Y/LG	324	Y/LG	Spark Output	SPOUT
37	361	R	175	BK Y	Vehicle Power	VPWR
40	60	BK/LG	60	BK LG	Power Ground	PWR GND
42	561	T O	561	T O	Injector #7	INJ 7
43	90	DB LG	90	DB LG	Heated Exhaust Gas Oxygen Sensor	HEGO-L
45	358	LG BK	356	DB LG	Manifold Absolute Pressure Sensor	MAP
46	359	BK/W	359	BK W	Signal Return	SIG RTN
47	355	DG LG	355	DG LG	Throttle Position Sensor	TP
48	200	W/BK	209	W R	Self-Test Input	STI
49	89	O	89	O	Heated EGO Ground	HEGO GND
51	100	W/R	100	W R	Air Management	AM1
52	562	LB	562	LB	Injector #8	INJ 8
54	73	O LB	73	O LB	Wide Open Throttle A/C Cut Off	WAC
56	349	DB	349	DB	Profile Ignition Pick-Up	PIP
57	361	R	175	BK Y	Vehicle Power	VPWR
58	555	T	555	T	Injector #1	INJ 1
59	556	W	556	W	Injector #2	INJ 2
60	50	BK LG	60	BK LG	Power Ground	PWR GND

Pin locations given for reference only. Probing 60 pin connector with DVOM probe will result in permanent damage to the pin connectors. Always probe as directed, using the Breakout Box.

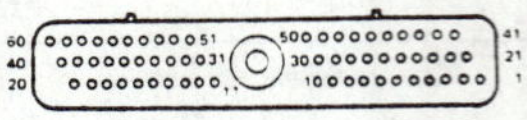

Electrical Schematic — 5.0L SEFI

1987½ MARK VII/LINCOLN TOWN CAR CALIBRATION 8-22

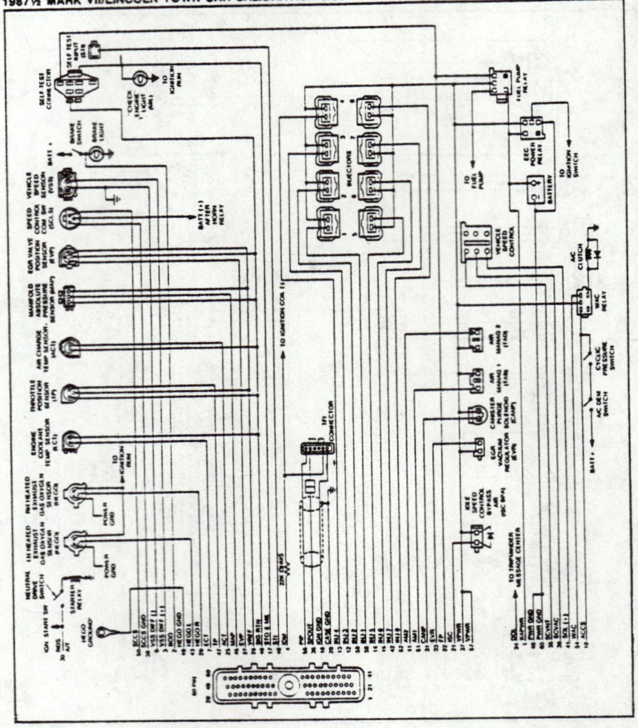

EEC-IV Module Connector Pin Usage — 5.0L SEFI

ALL EXCEPT 1987½ MARK VII/LINCOLN TOWN CAR CALIBRATION 8-22

Pin	Car Lines				Application	Abbreviations
	Mark VII, Continental		Mustang, Capri			
	Crt. #	Wire Color	Crt. #	Wire Color		
1	38	BK O	38	BK O	Keep Alive Power	KAPWR
4	11	DG Y	11	DG Y	Ignition Diagnostic Monitor	IDM
7	354	LG Y	354	LG Y	Engine Coolant Temperature	ECT
10	348	LG P	883	PK LB	A C Clutch Signal	ACCS
11	99	LG BK	99	LG BK	Air Management	AM2
12	557	BR Y	557	BR Y	Injector #3	INJ 3
13	558	BR LB	558	BR LB	Injector #4	INJ 4
14	559	T LB	559	T LB	Injector #5	INJ 5
15	560	LG	560	LG	Injector #6	INJ 6
16	259	BK O	259	BK O	Ignition Ground	IGN GND
17	382	Y BK	382	Y BK	Self-Test Output	STO
20	57	BK	57	BK	Case Ground	CASE GND
21	264	W LB	264	W LB	Idle Speed Control	ISC
22	97	T LG	97	T LG	Fuel Pump	FP
25	357	LG P	357	LG P	Air Charge Temperature	ACT
26	351	O W	351	O W	Reference Voltage	VREF
27	352	BR LG	352	BR LG	EGR Valve Position Sensor	EVP
29	94	DG P	94	DG P	Heated Exhaust Gas Oxygen Sensor	HEGO-R
30	—	—	199	LB Y	Clutch Neutral Gear Switches (M/T)	NGS and CES
30	33	W/PK	33	W PK	Neutral Drive Switch (A/T)	NDS
31	101	GY/Y	101	GY Y	Canister Purge	CANP
33	360	DG	360	DG	EGR Valve Regulator	EVR
34	305	LB/PK	—	—	Data Output Line	DOL
36	324	Y/LG	324	Y LG	Spark Output	SPOUT
37	361	R	361	R	Vehicle Power	VPWR
40	60	BK LG	60	BK LG	Power Ground	PWR GND
42	561	T O	561	T O	Injector #7	INJ 7
43	90	DB LG	90	DB LG	Heated Exhaust Gas Oxygen Sensor	HEGO-L
45	356	DB LG	358	LG BK	Manifold Absolute Pressure	MAP
46	359	BW/W	359	BK W	Signal Return	SIG RTN
47	355	DG LG	355	DG LG	Throttle Position Sensor	TP
48	209	W R	201	T R	Self-Test Input	STI
49	89	O	89	O	Heated EGO Ground	HEGO GND
51	100	W R	100	W R	Air Management	AMI
52	562	LB	562	LB	Injector #8	INJ 8
54	73	O LB	73	O LB	Wide Open Throttle A C Cut Off	WAC
56	349	DB	349	DB	Profile Ignition Pick-Up	PIP
57	361	R	361	R	Vehicle Power	VPWR
58	555	T	555	T	Injector #1	INJ 1
59	556	W	556	W	Injector #2	INJ 2
60	60	BK LG	60	BK LG	Power Ground	PWR GND

Pin locations given for reference only. Probing 60 pin connector with DVOM probe will result in permanent damage to the pin connectors. Always probe as directed, using the Breakout Box.

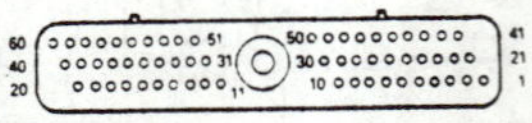

EEC-IV Module Connector Pin Usage — 5.0L SEFI

1987½ MARK VII/LINCOLN TOWN CAR CALIBRATION 8-22

Pin	Car Lines				Application	Abbreviations
	Mark VII		Town Car			
	Crt. #	Wire Color	Crt. #	Wire Color		
1	38	BK O	37	Y	Keep Alive Power	KAPWR
2	511	LG	511	LG	Brake On Off	BOO
3	150	DG W	150	DG W	Vehicle Speed Sensor -	VSS DIF -
4	11	DG Y	11	DG Y	Ignition Diagnostic Monitor	IDM
6	683	P LB	359	BK W	Vehicle Speed Sensor -	VSS DIF
7	354	LG Y	354	LG Y	Engine Coolant Temperature	ECT
10	348	LG P	883	PK LB	A C Clutch Signal	ACCS
11	99	LG BK	99	LG BK	Air Management	AM2
12	557	BR Y	557	BR Y	Injector #3	INJ 3
13	558	BR LB	558	BR LB	Injector #4	INJ 4
14	559	T LB	559	T LB	Injector #5	INJ 5
15	560	LG	560	LG	Injector #6	INJ 6
16	259	BK O	259	BK O	Ignition Ground	IGN GND
17	382	Y BK	201	T R	Self-Test Output "Check Engine" Light	STO & MIL
20	57	BK	57	BK	Case Ground	CASE GND
21	264	W LB	264	W LB	Idle Speed Control	ISC
22	97	T LG	97	T LG	Fuel Pump	FP
25	357	LG P	357	LG P	Air Charge Temperature	ACT
26	351	O W	351	O W	Reference Voltage	VREF
27	352	BR LG	352	BR LG	EGR Valve Position Sensor	EVP
29	94	DG P	94	DG P	Heated Exhaust Gas Oxygen Sensor	HEGO-R
30	33	W PK	33	W PK	Neutral Drive Switch (A/T)	NDS
31	101	GY/Y	101	GY Y	Canister Purge	CANP
33	360	DG	360	DG	EGR Valve Regulator	EVR
34	305	LB PK	305	LB PK	Data Output Line	DOL
35	146	W/PK	146	W PK	Speed Control Vent (Solenoid)	SCVNT
36	324	Y LG	324	Y LG	Spark Output	SPOUT
37	361	R	361	R	Vehicle Power	VPWR
38	145	GY BK	145	GY BK	Speed Control Vacuum (Solenoid)	SCVAC
39	199	LB/Y	679	GY BK	Speed Control Command Switch Ground	SCCS GND
40	60	BK LG	60	BK LG	Power Ground	PWR GND
41	144	O Y	144	O Y	Vehicle Speed Control Solenoid	SOL -
42	561	T O	561	T O	Injector #7	INJ 7
43	90	DB LG	90	DB LG	Heated Exhaust Gas Oxygen Sensor	HEGO-L
45	356	DB LG	358	LG BK	Manifold Absolute Pressure	MAP
46	359	BK W	359	BK/W	Signal Return	SIG RTN
47	355	DG LG	355	DG LG	Throttle Position Sensor	TP
48	209	W R	200	W BK	Self-Test Input	STI
49	89	O	89	O	Heated EGO Ground	HEGO GND
50	151	LB PK	151	LB BK	Speed Control Command Switch	SCCS
51	100	W R	100	W R	Air Management	AMI
52	562	LB	562	LB	Injector #8	INJ 8
54	73	O LB	73	O LB	Wide Open Throttle A C Cut Off	WAC
56	349	DB	349	DB	Profile Ignition Pick-Up	PIP
57	361	R	361	R	Vehicle Power	VPWR
58	555	T	555	T	Injector #1	INJ 1
59	556	W	556	W	Injector #2	INJ 2
60	60	BK LG	60	BK LG	Power Ground	PWR GND

Pin locations given for reference only. Probing 60 pin connector with DVOM probe will result in permanent damage to the pin connectors. Always probe as directed, using the Breakout Box.

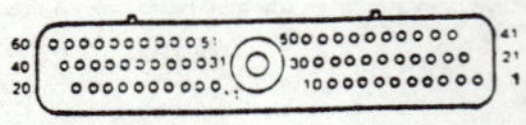

EEC-IV Diagnostic By Symptom — 2.3L EFI

SYMPTOM	RESULT	ACTION TO TAKE
• Engine stalls. • Stalls in Self-Test. • Runs rough. • Misses. • Always rich/lean.		• GO to Pinpoint Test Step S1. • Map test. GO to Pinpoint Test Step DF20. • Idle speed control. GO to Pinpoint Test Step KE1. • Poor power/ground connections. • Ignition system distributor cap, rotor, wires, coil, plugs. • Base engine valves, cam timing, compression etc.
• Detonation/spark knock.		GO to Pinpoint Test Step DG1, or EGR Diagnostic Section 6 of Volume H.
• A/C compressor runs continuously.		GO to Pinpoint Test Step FA4.
• Low idle during A/C On.		GO to Pinpoint Test Step FA20.
• Shift indicator light always On or Off.		GO to Pinpoint Test Step KL1.
• A/C does not cut-off under WOT conditions. • A/C not functioning.		GO to Pinpoint Test Step KM1.
• Surges with A/C on at idle.		GO to Pinpoint Test Step KM20.
• Stumble after hot re-start (HEGO).		GO to Pinpoint Test Step HE21.
• Stalls during parking maneuvers.		GO to Pinpoint Test Step FF3.
• Fuel pump runs with Key Off.		GO to Pinpoint Test Step J15.

Electrical Schematic — 2.3L EFI

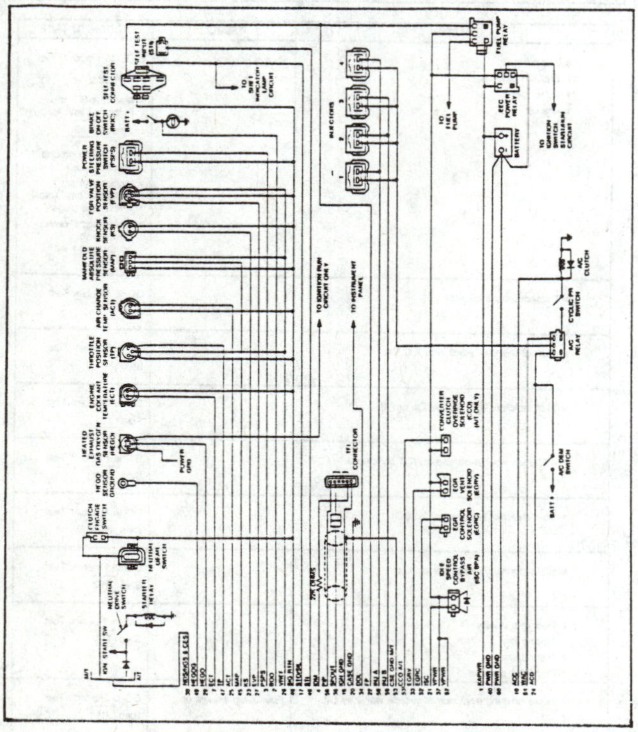

EEC-IV Module Connector Pin Usage — 2.3L EFI

RANGER/BRONCO II

Pin	Circuit	Wire Color	Application	Abbreviations
1	37	Y	Keep Alive Power	KAPWR
2	511	LG	Brake On/Off	BOO
3	330	Y-LG	Power Steering Pressure Switch	PSPS
4	11	DG-Y	Ignition Diagnostic Monitor	IDM
7	354	LG-Y	Engine Coolant Temperature	ECT
10	347	BK-Y	A/C Clutch	ACC
16	57	BK	Ignition Ground	IGN GND
17	201	T-R	Self-Test Output/Shift Indicator Light	STO/SIL
20	57	BK	Case Ground	CSE GND
21	67	GY-W	Idle Speed Control	ISC
22	97	T-LG	Fuel Pump	FP
23	99	LG-BK	Knock Sensor	KS
24	348	LG-P	A/C Demand	ACD
25	310	Y-R	Air Charge Temperature	ACT
26	351	O-W	Reference Voltage	VREF
27	352	BR-LG	EGR Valve Position	EVP
29	94	DG-P	Heated Exhaust Gas Oxygen Sensor	HEGO
30	150	DG-W	Neutral Drive Switch (A/T)	NDS
30	200	W-BK	Neutral Gear Switch and Clutch Engage Switch (M/T)	NGS/CES
33	360	DG	Exhaust Gas Recirculation Vent	EGRV
36	324	Y-LG	Spark Output	SPOUT
37	361	R	Vehicle Power	VPWR
40	60	BK-LG	Power Ground	PWR GND
45	356	DB-LG	Manifold Absolute Pressure	MAP
46	359	BK-W	Signal Return	SIG RTN
47	355	DG-LG	Throttle Position Sensor	TP
48	100	W-R	Self-Test Input	STI
49	89	O	Heated Exhaust Gas Oxygen Sensor Ground	HEGOG
52	362	Y	Exhaust Gas Recirculation Control	EGRC
53	237	O-Y	Clutch Converter Override (A/T only)	CCO
54	331	R	W.O.T. A/C Cut Off	WAC
56	349	DB	Profile Ignition Pick-Up	PIP
57	361	R	Vehicle Power	VPWR
58	96	T-O	Injector 1 (Bank 1)	INJ 1
59	95	T-R	Injector 2 (Bank 2)	INJ 2
60	60	BK-LG	Power Ground	PWR GRD

Pin locations given for reference only. Probing 60 pin connector with DVOM probe will result in permanent damage to the pin connectors. Always probe as directed, using the Breakout Box.

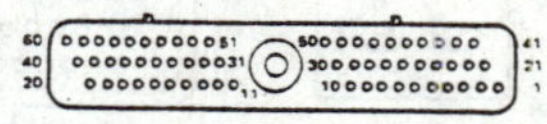

EEC-IV Module Connector Pin Usage — 2.3L EFI

AEROSTAR

Pin	Circuit	Wire Color	Application	Abbreviations
1	37	Y	Keep Alive Power	KAPWR
2	810	R-LG	Brake On-Off	BOO
3	330	Y-LG	Power Steering Pressure Switch	PSPS
4	11	DG-Y	Ignition Diagnostic Monitor	IDM
7	354	LG-Y	Engine Coolant Temperature	ECT
10	347	BK-Y	A/C Clutch	ACC
16	259	BK-O	Ignition Ground	IGN GND
17	201	T-R	Self-Test Output Shift Indicator Light	STO SIL
20	57	BK	Case Ground	CSE GND
21	67	GY-W	Idle Speed Control	SC
22	97	T-LG	Fuel Pump	FP
23	99	LG-BK	Knock Sensor	KS
24	348	LG-P	A/C Demand	ACD
25	357	LG-P	Air Charge Temperature	ACT
26	351	O-W	Reference Voltage	VREF
27	352	BR-LG	EGR Valve Position	EVP
29	94	DG-P	Heated Exhaust Gas Oxygen Sensor	HEGO
30	150	DG-W	Neutral Drive Switch (A/T)	NDS
30	200	W-BK	Neutral Gear Switch and Clutch Engage Switch (M/T)	NGS CES
33	360	DG	Exhaust Gas Recirculation Vent	EGRV
34	305	LB-PK	Data Output Link	DOL
36	324	Y-LG	Spark Output	SPOUT
37	361	R	Vehicle Power	VPWR
40	60	BK-LG	Power Ground	PWR GND
45	356	DB-LG	Manifold Absolute Pressure	MAP
46	359	BK-W	Signal Return	SIG RTN
47	355	DG-LG	Throttle Position Sensor	TP
48	100	W-R	Self-Test Input	STI
49	89	O	Heated Exhaust Gas Oxygen Ground	HEGOG
52	362	Y	Exhaust Gas Recirculation Control	EGRC
53	237	O-Y	Clutch Converter Override (A/T only)	CCO
53	48	Uninsulated	Case Ground (M/T)	CSE GND
54	331	R	W.O.T. A/C Cut Off	WAC
56	349	DB	Profile Ignition Pick-Up	PIP
57	361	R	Vehicle Power	VPWR
58	96	T-O	Injector 1 (Bank 1)	INJ 1
59	95	T-R	Injector 2 (Bank 2)	INJ 2
60	60	BK-LG	Power Ground	PWR GRD

Pin locations given for reference only. Probing 60 pin connector with DVOM probe will result in permanent damage to the pin connectors. Always probe as directed, using the Breakout Box.

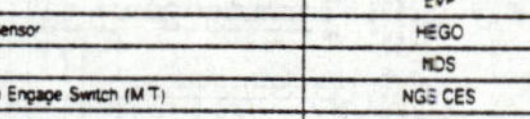

EEC-IV Diagnostic By Symptom — 2.9L EFI

SYMPTOM	RESULT	ACTION TO TAKE
• Engine stalls. • Stalls in Self-Test. • Runs rough. • Misses.	Always rich/lean. ▶	• GO to Pinpoint Test Step S1. • Map test GO to Pinpoint Test Step DF-20. • Poor power/ground connections. • Ignition system distributor cap, rotor, wires, coil, plugs. • Base engine valves cam timing, compression etc.
• Detonation/spark knock.		GO to Pinpoint Test Step DG1.
• Low idle rpm with A/C On.		GO to Pinpoint Test Step FA20.
• A/C does not cut-off under WOT conditions. • A/C not functioning.		GO to Pinpoint Test Step KM1.
• Stumble after hot re-start (HEGO).		GO to Pinpoint Test Step HG21.
• Stalls during parking maneuvers.		GO to Pinpoint Test Step FF3.
• Fuel pump runs with Key Off.		GO to Pinpoint Test Step J15.

3.0L EFI

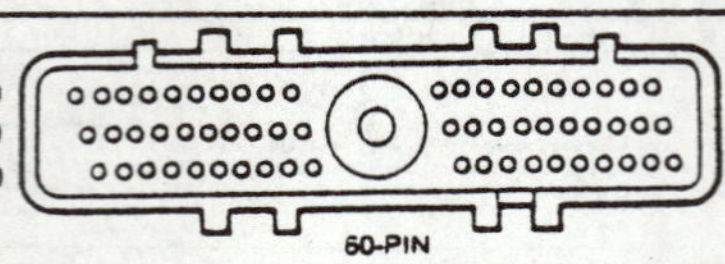

Electrical Schematic — 2.9L EFI

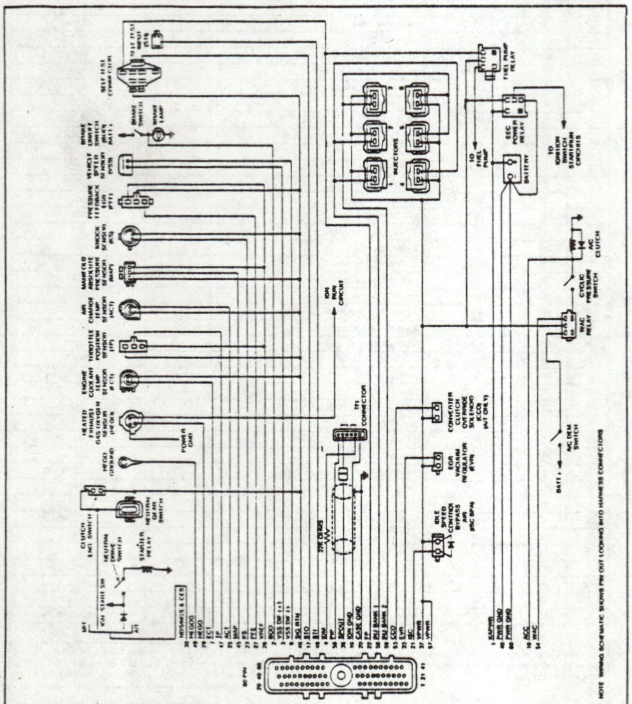

EEC-IV Module Connector PIN Usage — 2.9L EFI

Pin	Circuit	Wire Color	Application	Abbreviations
1	37	Y	Keep Alive Power	KAPWR
2	511	LG	Brake On-Off	BOO
3	150	DG/W	Vehicle Speed Sensor –	VSS DIF –
4	11	DG/Y	Ignition Diagnostic Monitor	IDM
6	397	BK/W	Vehicle Speed Sensor –	VSS DIF –
7	354	LG/Y	Engine Coolant Temp. Sensor	ECT
10	347	BK/Y	A/C Clutch	ACC
16	57	BK	Ignition Ground	IGN GND
17	201	T/R	Self-Test Output	STO
20	60	B/LG	Case Ground	CASE GND
21	67	GY/W	Idle Speed Control	ISC
22	97	T/LG	Fuel Pump	FP
23	310	Y/R	Knock Sensor	KS
25	357	LG/P	Air Charge Temp. Sensor	ACT
26	351	O/W	Reference Voltage	VREF
27	352	BR/LG	Pressure Feedback Electronic Sensor (EGR)	PFE
29	94	DG/P	Heated Exhaust Gas Oxygen Sensor	HEGO
30	33	W/PK	Neutral Drive Switch (A/T Only)	NDS
30	200	W/BK	Neutral Gear Switch and Clutch Engage Switch (M/T)	NGS/CES
33	362	Y	EGR Vacuum Regulator	EVR
36	324	Y/LG	Spark Output	SPOUT
37	361	R	Vehicle Power	VPWR
40	60	BK/LG	Power Ground	PWR GND
45	356	DB/LG	Manifold Absolute Pressure Sensor	MAP
46	359	BK/W	Signal Return	SIG RTN
47	355	DG/LG	Throttle Position Sensor	TP
48	100	W/R	Self-Test Input	STI
49	89	O	Heated Exhaust Gas Oxygen Sensor Ground	HEGOG
53	237	O/Y	Clutch Converter Override (Not used on M/T)	CCO
54	331	R	W.O.T. A/C Cut-Off	WAC
56	349	DB	Profile Ignition Pick-up	PIP
57	361	R	Vehicle Power	VPWR
58	265	LG/W	Injector Bank 1 (Controls Engine Cylinder Numbers 1, 2 and 4)	INJ 1
59	95	T/R	Injector Bank 2 (Controls Engine Cylinder Numbers 3, 5 and 6)	INJ 2
60	60	BK/LG	Power Ground	PWR GND

Pin locations given for reference only. Probing 60 pin connector with DVOM probe will result in permanent damage to the pin connectors. Always probe as directed using the Breakout Box.

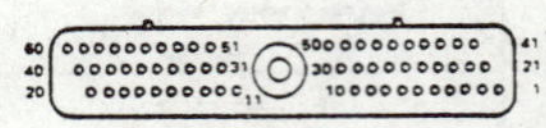

EEC-IV Diagnostic By Symptom — 3.0L EFI

SYMPTOM	RESULT	ACTION TO TAKE
• Engine stalls. • Stalls in Self-Test. • Runs rough. • Misses.	Always rich/lean. ▶	• GO to Pinpoint Test Step S1. • Map test GO to Pinpoint Test Step DF-20. • Poor power/ground connections. • Ignition system distributor cap, rotor, wires, coil, plugs. • Base engine valves cam timing, compression etc.
• Detonation/spark knock.		GO to Pinpoint Test Step DG1.
• Lack of fast idle assist with A/C On.		GO to Pinpoint Test Step FA1.
• Shift indicator light always On or Off.		GO to Pinpoint Test Step KL1.
• A/C does not cut-off under WOT conditions.		GO to Pinpoint Test Step KM1.
• Stumble after hot re-start (HEGO).		GO to Pinpoint Test Step HG21.
• Stalls during parking maneuvers.		GO to Pinpoint Test Step FF3.
• Poor idle quality. • Rolling idle. • Shifts harshly. • Poor fuel economy.		GO to Pinpoint Test Step T90.
• Noise emitting from rear of vehicle with Key Off.	Fuel pump always on.	GO to Pinpoint Test Step J15.

Electrical Schematic — 3.0L EFI

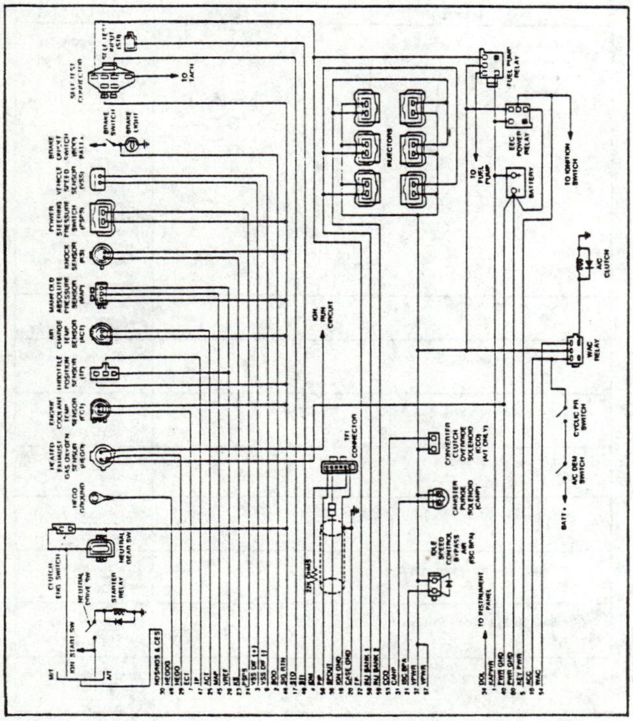

EEC-IV Diagnostic By Symptom — 4.9L EFI

SYMPTOM	RESULT	ACTION TO TAKE
• Engine stalls. • Stalls in Self Test. • Runs rough. • Misses.	Always rich/lean	• GO to Pinpoint Test Step S1. • Map test. GO to Pinpoint Test Step DF20. • Idle Speed Control. GO to Pinpoint Test Step KE1. • Poor power/ground connections. • Ignition system distributor cap, rotor, wires, coil, plugs. • Base engine valves, cam timing, compression etc.
• Detonation/spark knock.		• GO to Pinpoint Test Step DG1.
• Low idle rpm with A/C On.		• GO to Pinpoint Test Step FA20.
• Shift indicator light always On or always Off.		• GO to Pinpoint Test Step KL1.
• Stumble after hot re-start (HEGO).		• GO to Pinpoint Test Step HG21.
• Stalls during parking maneuvers.		• GO to Pinpoint Test Step FF3.
• Fuel pump runs with Key Off.		• GO to Pinpoint Test Step J'5.

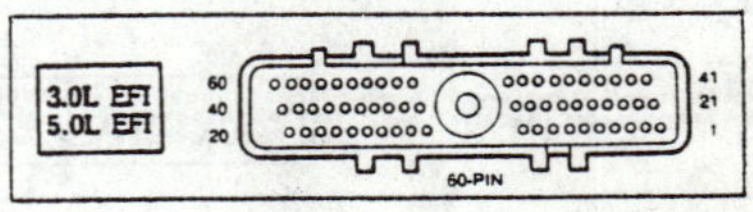

EEC-IV Module Connector Pin Usage — 3.0L EFI

Pin	Circuit	Wire Color	Application	Abbreviations
1	37	Y	Keep Alive Power	KAPWR
2	810	R LG	Brake ON-OFF	BOO
3	150	DG W	Vehicle Speed Sensor Differential Positive	VSSDIF -
4	11	DG Y	Ignition Diagnostic Monitor	IDM
6	397	BK W	Vehicle Speed Sensor Differential Negative	VSSDIF -
7	354	LG Y	Engine Coolant Temperature Sensor	ECT
10	347	BK Y	A C Cycling Pressure Switch	ACCS
16	259	BK O	Ignition Ground	IGN GND
17	201	T R	Self-Test Output	STO
20	57	BK	Case Ground	CASE GND
21	68	O BK	Idle Speed Control (Bypass Air)	ISC-BPA
22	97	T LG	Fuel Pump	FP
23	310	Y R	Knock Sensor	KS
24	330	Y LG	Power Steering Pressure Switch	PSPS
25	357	LG P	Air Charge Temperature Sensor	ACT
26	351	O W	Reference Voltage	VREF
29	94	DG P	Heated Exhaust Gas Oxygen Sensor	HEGO
30	200	W BK	Neutral Gear Switch and Clutch Engage Switch M T	NGS and CES
30	151	LB BK	Neutral Drive Switch A T Only	NDS
31	101	GY Y	Canister Purge	CANP
34	305	LB P	Data Output Link	DOL
36	324	Y-LG	Spark Output	SPOUT
37	361	R	Vehicle Power	VPWR
40	60	BK LG	Power Ground	PWR GND
45	356	DB LG	Manifold Absolute Pressure Sensor	MAP
46	359	BK W	Signal Return	SIG RTN
47	355	DG LG	Throttle Position Sensor	TP
48	100	W R	Self-Test Input	STI
49	89	O	Heated Exhaust Gas Oxygen Sensor Ground	HEGO GND
53	237	O Y	Converter Clutch Override (Not used on M T)	CCO
54	331	P	W O T. A C Cut-Off	WAC
56	349	DB	Profile Ignition Pick-up	PIP
57	361	R	Vehicle Power	VPWR
58	96	T O	Injector Bank 1	INJ 1
59	95	T R	Injector Bank 2	INJ 2
60	60	BL LG	Power Ground	PWR GND

Pin locations given for reference only. Probing 60 pin connector with DVOM probe will result in permanent damage to the pin connectors. Always probe as directed using the Breakout Box.

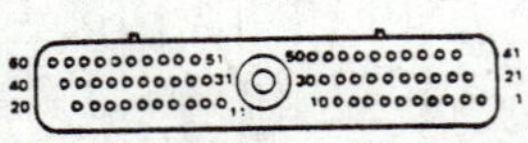

Electrical Schematic — 4.9L EFI

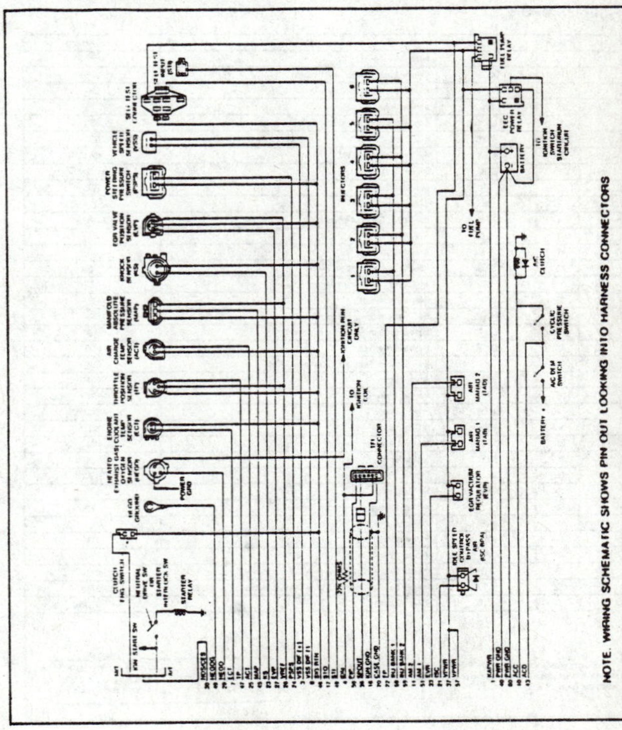

EEC-IV Module Connector Pin Usage — 4.9L EFI

ECONOLINE

Pin	Circuit	Wire Color	Application	Abbreviations
1	37	BK/Y	Keep Alive Power	KAPWR
3	145	BK	Vehicle Speed Sensor −	VSS DIF −
4	11	BK/Y	Ignition Diagnostic Module	IDM
6	146	BK	Vehicle Speed Sensor −	VSS DIF −
7	354	GR/Y	Engine Coolant Temperature Sensor	ECT
10	347	BK	A/C Clutch	ACC
11	200	BL	Air Management 2 (TAD)	AM2
16	259	BK/R	Ignition Ground	IGN GND
17	201	BL/Y	Self-Test Output	STO
20	57	BK	Case Ground	CASE GND
21	264	BL	Idle Speed Control — Bypass Air	ISC — BPA
22	97	O	Fuel Pump	FP
23	99	BK	Knock Sensor	KS
24	330	Y/G	Power Steering Pressure Switch	PSPS
25	310	R/BK	Air Charge Temperature Sensor	ACT
26	351	R	Reference Voltage	VREF
27	352	BR/GR	EGR Valve Position Sensor	EVP
29	94	GR/P	Heated Exhaust Gas Oxygen Sensor	HEGO
30	33	Y	Neutral Drive Switch (A/T)	NDS
30	199	BK/O	Clutch Engage Switch (M/T)	CES
33	360	W/BL	EGR Vacuum Regulator (Solenoid)	EVR
36	324	Y/GR	Spark Output	SPOUT
37	361	R	Vehicle Power	VPWR
40	60	BK/W	Power Ground	PWR GND
43	348	GR/W	A/C Demand	ACD
45	356	BL/GR	Manifold Absolute Pressure	MAP
46	359	BK/W	Signal Return	SIG RTN
47	355	GR	Throttle Position Sensor	TP
48	100	BK/R	Self-Test Input	STI
49	89	O	Heated Exhaust Gas Oxygen Ground	HEGOG
51	190	Y	Air Management 1 (TAB)	AM1
56	349	BR	Profile Ignition Pick-Up	PIP
57	361	R	Vehicle Power	VPWR
58	96	O/BK	Injector Bank 1 (Controls Engine Cylinder Numbers 1, 3 and 5)	INJ BANK 1
59	95	W	Injector Bank 2 (Controls Engine Cylinder Numbers 2, 4 and 6)	INJ BANK 2
60	60	BK/W	Power Ground	PWR GND

Pin locations given for reference only. Probing 60 pin connector with DVOM probe will result in permanent damage to the pin connectors. Always probe as directed, using the Breakout Box.

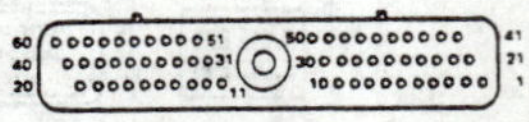

EEC-IV Module Connector Pin Usage — 4.9L EFI

F-SERIES/BRONCO

Pin	Circuit	Wire Color	Application	Abbreviations
1	37	Y	Keep Alive Power	KAPWR
3	150	DG/W	Vehicle Speed Sensor −	VSS DIF −
4	11	DG/Y	Ignition Diagnostic Module	IDM
6	57	BK	Vehicle Speed Sensor −	VSS DIF −
7	354	LG/Y	Engine Coolant Temperature Sensor	ECT
10	347	BK/Y	A/C Clutch	ACC
11	200	W/BK	Air Management 2 (TAD)	AM 2
16	259	BK/O	Ignition Ground	IGN GND
17	201	T/R	Self-Test Output	STO
20	57	BK	Case Ground	CASE GND
21	67	GY/W	Idle Speed Control — Bypass Air	ISC-BPA
22	97	T/LG	Fuel Pump	FS
23	99	LG/BK	Knock Sensor	KS
24	330	Y/LG	Power Steering Pressure Switch	PSPS
25	310	Y/R	Air Charge Temperature Sensor	ACT
26	350	O/W	Reference Voltage	VREF
27	352	BR/LG	EGR Valve Position Sensor	EVP
29	94	DG/P	Heated Exhaust Gas Oxygen Sensor	HEGO
30	481	GY/Y	Neutral Drive Switch	NDS
33	360	DG	EGR Vacuum Regulator (Solenoid)	EVR
36	324	Y/LG	Spark Output	SPOUT
37	361	R	Vehicle Power	VPWR
40	60	BK/LG	Power Ground	PWR GND
43	348	LG/P	A/C Demand	ACD
45	356	DB/LG	Manifold Absolute Pressure	MAP
46	359	BK/W	Signal Return	SIG RTN
47	355	DG/LB	Throttle Position Sensor	TP
48	100	W/R	Self-Test Input	STI
49	89	O	Heated Exhaust Gas Oxygen Ground	HEGOG
51	190	W/R	Air Management 1	AM 1
56	349	DB	Profile Ignition Pick-Up	PIP
57	361	R	Vehicle Power	VPWR
58	96	T/O	Injector Bank 1 (Controls Engine Cylinder Numbers 1, 3 and 5	INJ BANK 1
59	95	T/R	Injector Bank 2 (Controls Engine Cylinder Numbers 2, 4 and 6)	INJ BANK 2
60	60	BK/LG	Power Ground	PWR GND

Pin locations given for reference only. Probing 60 pin connector with DVOM probe will result in permanent damage to the pin connectors. Always probe as directed, using the Breakout Box.

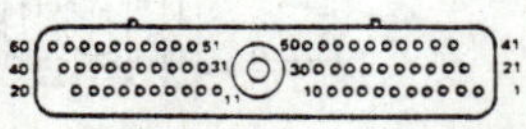

EEC-IV Diagnostic By Symptom — 5.0L EFI

SYMPTOM	RESULT	ACTION TO TAKE
• Engine stalls. • Stalls in Self-Test. • Runs rough. • Misses.	Always rich/lean. ▶	• GO to Pinpoint Test Step S1. • Map test, GO to Pinpoint Test Step DF20. • Idle Speed Control, GO to Pinpoint Test Step KE1. • Poor power/ground connections. • Ignition system distributor cap, rotor, wires, coil, plugs. • Base engine valves, cam timing, compression etc.
• Detonation/spark knock.		GO to Pinpoint Test Step DG1.
• Low idle rpm with A/C On.		GO to Pinpoint Test Step FA20.
• Shift indicator light always On or Off.		GO to Pinpoint Test Step KL1.
• A/C does not cut-off under WOT conditions.		GO to Pinpoint Test Step KM1.
• Surges with A/C on at idle.		GO to Pinpoint Test Step KM20.
• Stumble after hot re-start (HEGO).		GO to Pinpoint Test Step HH21.
• Stalls during parking maneuvers.		GO to Pinpoint Test Step FF3.
• Fuel pump runs with Key Off.		GO to Pinpoint Test Step J15.

Electrical Schematic — 5.0L EFI

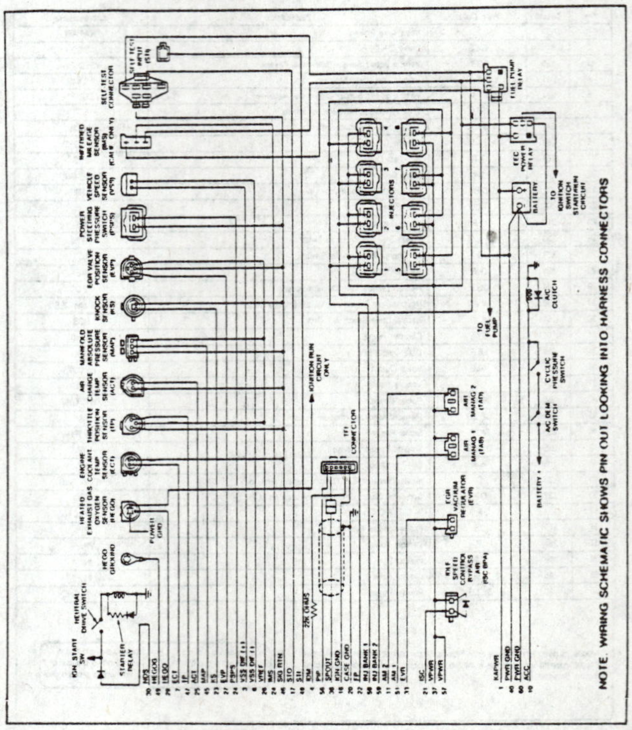

EEC-IV Module Connector Pin Usage — 5.0L EFI

F-SERIES/BRONCO

Pin	Circuit	Wire Color	Application	Abbreviations
1	37	Y	Keep Alive Power	KAPWR
3	150	DG/W	Vehicle Speed Sensor +	VSS DIF +
4	11	DG/Y	Ignition Diagnostic Monitor	IDM
6	57	BK	Vehicle Speed Sensor –	VSS DIF –
7	354	LG/Y	Engine Coolant Temperature Sensor	ECT
10	347	BK/Y	A/C Cycling Switch	ACCS
11	200	W/BK	Air Management 2 (TAD)	AM-2
16	259	BK/O	Ignition Ground	IGN GND
17	201	T/R	Self-Test Output	STO
18	223	T/LG	Inferred Mileage Sensor	IMS
20	57	BK	Case Ground	CASE GND
21	67	GY/W	Idle Speed Control — Bypass Air	ISC — BPA
22	97	T/LG	Fuel Pump	FP
23	99	LG/BK	Knock Sensor	KS
24	330	Y/LG	Power Steering Pressure Switch	PSPS
25	310	Y/R	Air Charge Temperature Sensor	ACT
26	351	O/W	Reference Voltage	VREF
27	352	BR/LG	EGR Valve Position Sensor	EVP
29	94	DG/P	Heated Exhaust Oxygen Sensor	HEGO
30	481	GY/Y	Neutral Drive Switch	NDS
33	360	DG	EGR Vacuum Regulator (Solenoid)	EVR
36	324	Y/LG	Spark Output	SPOUT
37	361	R	Vehicle Power	VPWR
40	60	BK/LG	Power Ground	PWR GND
45	356	DB/LG	Manifold Absolute Pressure	MAP
46	359	BK/W	Signal Return	SIG RTN
47	355	DG/LG	Throttle Position Sensor	TP
48	100	W/R	Self-Test Input	STI
49	89	O	Heated Exhaust Oxygen Ground	HEGOG
51	190	W/R	Air Management 1 (TAB)	AM-1
56	349	DB	Profile Ignition Pick-Up	PIP
57	361	R	Vehicle Power	VPWR
58	96	T/O	Injector Bank 1 (Controls Engine Cylinder Numbers 1, 4, 5 and 8)	INJ BANK 1
59	95	T/R	Injector Bank 2 (Controls Engine Cylinder Numbers 2, 3, 6 and 7)	INJ BANK 2
60	60	BK/LG	Power Ground	PWR GND

Pin locations given for reference only. Probing 60 pin connector with DVOM probe will result in permanent damage to the pin connectors. Always probe as directed using the Breakout Box.

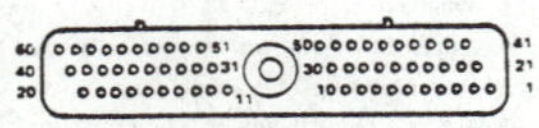

EEC-IV Module Connector Pin Usage — 5.0L EFI

ECONOLINE

Pin	Circuit	Wire Color	Application	Abbreviations
1	38	BK/R	Keep Alive Power	KAPWR
3	145	BK	Vehicle Speed Sensor +	VSS DIF +
4	11	BK-Y	Ignition Diagnostic Monitor	IDM
6	146	BK	Vehicle Speed Sensor –	VSS DIF –
7	354	GR-Y	Engine Coolant Temperature Sensor	ECT
10	347	BK	A/C Cycling Switch	ACCS
11	200	BL	Air Management 2 (TAD)	AM 2
16	57	BK	Ignition Ground	IGN GND
17	201	BL-Y	Self-Test Output	STO
20	60	BK/W	Case Ground	CASE GND
21	67	GR/W	Idle Speed Control — Bypass Air	ISC-BPA
22	97	O	Fuel Pump	FP
23	99	BK	Knock Sensor	KS
24	331	R	Power Steering Pressure Switch	PSPS
25	310	R/BK	Air Charge Temperature Sensor	ACT
26	351	R	Reference Voltage	VREF
27	352	BR/GR	EGR Valve Position Sensor	EVP
29	94	GR/P	Heated Exhaust Gas Oxygen Sensor	HEGO
30	150	R	Neutral Drive Switch	NDS
33	360	Y/BL	EGR Vacuum Regulator (Solenoid)	EVR
36	324	Y/GR	Spark Output	SPOUT
37	361	R	Vehicle Power	VPWR
40	60	BK/W	Power Ground	PWR GND
45	356	BL/GR	Manifold Absolute Pressure	MAP
46	359	BK/W	Signal Return	SIG RTN
47	355	GR	Throttle Position Sensor	TP
48	100	BK/R	Self-Test Input	STI
49	89	O	Heated Exhaust Gas Oxygen Ground	HEGOG
51	190	Y	Air Management I (TAB)	AM I
56	349	BR	Profile Ignition Pick-Up	PIP
57	361	R	Vehicle Power	VPWR
58	96	O/BK	Injector Bank 1 (Controls Engine Cyl. Nos. 1, 4, 5, and 8)	INJ BANK 1
59	95	W	Injector Bank 2 (Controls Engine Cyl. Nos. 2, 3, 6, and 7)	INJ BANK 2
60	60	BK/W	Power Ground	PWR GND

Pin locations given for reference only. Probing 60 pin connector with DVOM probe will result in permanent damage to the pin connectors. Always probe as directed using the Breakout Box.

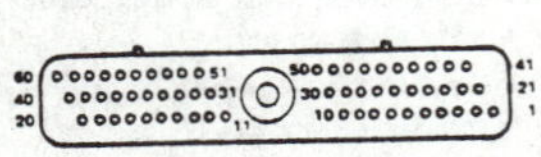

EEC-IV Diagnostic By Symptom — 7.5L EFI

SYMPTOM	RESULT	ACTION TO TAKE
• Engine stalls. • Stalls in Self-Test. • Runs rough. • Misses.	Always rich/lean. ▶	• GO to Pinpoint Test Step [S1]. • Map test, GO to Pinpoint Test Step [DF20]. • Idle Speed Control, GO to Pinpoint Test Step [KE1]. • Poor power/ground connections. • Ignition system distributor cap, rotor, wires, coil, plugs. • Base engine valves, cam timing, compression etc.
• Low idle rpm with A/C On.		GO to Pinpoint Test Step [FA20].
• Shift indicator light always On or Off.		GO to Pinpoint Test Step [KL1].
• A/C does not cut-off under WOT conditions.		GO to Pinpoint Test Step [KM1].
• Surges with A/C on at idle.		GO to Pinpoint Test Step [KM20].
• Stumble after hot re-start (HEGO).		GO to Pinpoint Test Step [HH21].
• Fuel pump runs with Key Off.		GO to Pinpoint Test Step [J15].
• Gasoline fumes under hood.		GO to Pinpoint Test Step [KD1].

Electrical Schematic — 7.5L EFI

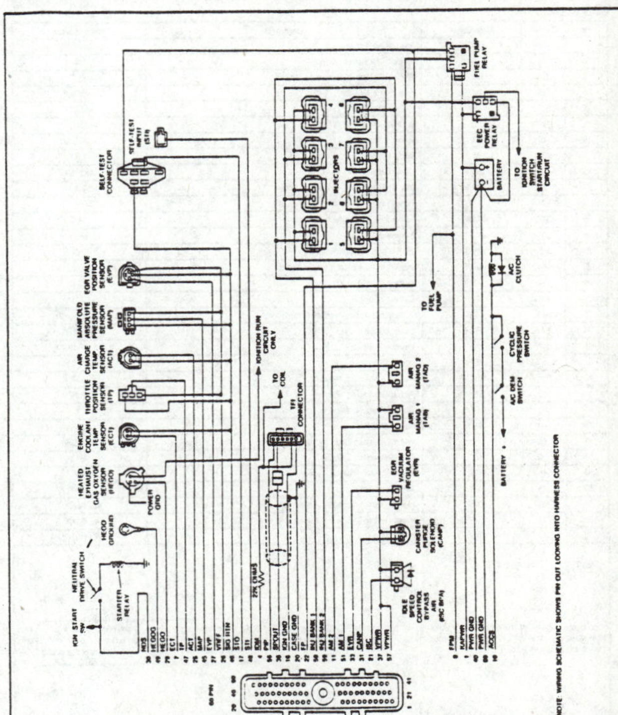

EEC-IV Module Connector Pin Usage — 7.5L EFI

F-SERIES/BRONCO

Pin	Circuit	Wire Color	Application	Abbreviations
1	37	Y	Keep Alive Power	KAPWR
4	11	DG/Y	Ignition Diagnostic Monitor	IDM
7	354	LG/Y	Engine Coolant Temperature Sensor	ECT
8	276	BR	Fuel Pump Monitor	FPM
10	347	BK/Y	A/C Cycling Switch	ACCS
11	200	W/BK	Air Management 2 (TAD)	AM-2
16	259	BK/O	Ignition Ground	IGN GND
17	201	T/R	Self-Test Output	STO
20	57	BK	Case Ground	CASE GND
21	67	GY/W	Idle Speed Control — Bybass Air	ISC — BPA
22	97	T/LG	Fuel Pump	FP
25	310	Y/R	Air Charge Temperature Sensor	ACT
26	351	O/W	Reference Voltage	VREF
27	352	BR/LG	EGR Valve Position Sensor	EVP
29	94	DG/P	Heated Exhaust Oxygen Sensor	HEGO
30	481	GY/Y	Neutral Drive Switch	NDS
31	101	GY/Y	Canister Purge	CANP
33	360	DG	EGR Vacuum Regulator (Solenoid)	EVR
36	324	Y/LG	Spark Output	SPOUT
37	361	R	Vehicle Power	VPWR
40	60	BK/LG	Power Ground	PWR GND
45	356	DB/LG	Manifold Absolute Pressure	MAP
46	359	BK/W	Signal Return	SIG RTN
47	355	DG/LG	Throttle Position Sensor	TP
48	100	W/R	Self-Test Input	STI
49	89	O	Heated Exhaust Oxygen Ground	HEGOG
51	190	W/R	Air Management 1 (TAB)	AM-1
56	349	DB	Profile Ignition Pick-Up	PIP
57	361	R	Vehicle Power	VPWR
58	96	T/O	Injector Bank 1 (Controls Engine Cylinder Numbers 1, 4, 5 and 8)	INJ BANK 1
59	95	T/R	Injector Bank 2 (Controls Engine Cylinder Numbers 2, 3, 6 and 7)	INJ BANK 2
60	60	BK/LG	Power Ground	PWR GND

Pin locations given for reference only. Probing 60 pin connector with DVOM probe will result in permanent damage to the pin connectors. Always probe as directed, using the Breakout Box.

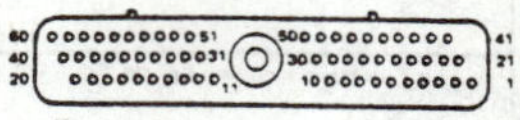

EEC-IV Module Connector Pin Usage — 7.5L EFI

ECONOLINE

Pin	Circuit	Wire Color	Application	Abbreviations
1	38	BK/R	Keep Alive Power	KAPWR
4	11	BK/Y	Ignition Diagnostic Monitor	IDM
7	354	GR/Y	Engine Coolant Temperature Sensor	ECT
8	787	R/BR	Fuel Pump Monitor	FPM
10	347	BK	A/C Cycling Switch	ACCS
11	200	BL	Air Management 2 (TAD)	AM 2
16	57	BK	Ignition Ground	IGN GND
17	201	BL/Y	Self-Test Output	STO
20	60	BK/W	Case Ground	CASE GND
21	67	GR/W	Idle Speed Control — Bypass Air	ISC-BPA
22	97	O	Fuel Pump	FP
25	310	R/BK	Air Charge Temperature Sensor	ACT
26	351	R	Reference Voltage	VREF
27	352	BR/GR	EGR Valve Position Sensor	EVP
29	94	GR/P	Heated Exhaust Gas Oxygen Sensor	HEGO
30	150	R	Neutral Drive Switch	NDS
31	101	GY	Canister Purge	CANP
33	360	Y/BL	EGR Vacuum Regulator (Solenoid)	EVR
36	324	Y/GR	Spark Output	SPOUT
37	361	R	Vehicle Power	VPWR
40	60	BK/W	Power Ground	PWR GND
45	356	BL/GR	Manifold Absolute Pressure	MAP
46	359	BK/W	Signal Return	SIG RTN
47	355	GR	Throttle Position Sensor	TP
48	100	BK/R	Self-Test Input	STI
49	89	O	Heated Exhaust Gas Oxygen Ground	HEGOG
51	190	Y	Air Management I (TAB)	AM I
56	349	BR	Profile Ignition Pick-Up	PIP
57	361	R	Vehicle Power	VPWR
58	96	O/BK	Injector Bank 1 (Controls Engine Cyl. Nos. 1, 4, 5, and 8)	INJ BANK 1
59	95	W	Injector Bank 2 (Controls Engine Cyl. Nos. 2, 3, 6, and 7)	INJ BANK 2
60	60	BK/W	Power Ground	PWR GND

Pin locations given for reference only. Probing 60 pin connector with DVOM probe will result in permanent damage to the pin connectors. Always probe as directed, using the Breakout Box.

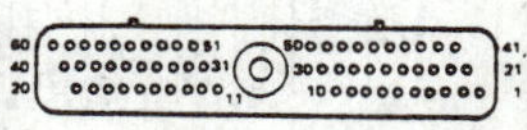

Pinpoint Tests

INSTRUCTIONS FOR USING THE PINPOINT TESTS

- Do not run any of the following Pinpoint Tests unless you are so instructed by the Quick Test. Each Pinpoint Test assumes that a fault has been detected in the system with direction to enter a specific repair routine. Doing any Pinpoint Test without direction from Quick Test may produce incorrect results and replacement of Non-Defective components.
- Correct test results for Quick Test are dependent on the proper operation of related non-EEC components/systems. It may be necessary to correct any defects in these areas before EEC will pass the Quick Test. Refer to the Diagnostic Routines, Section 2 for service.
- Do not replace any parts unless the test result indicates they should be replaced.
- When more than one service code is received, always start service with the first code received.
- Do not measure voltage or resistance at the processor or connect any test lights to it, unless otherwise specified.
- Isolate both ends of a circuit, and turn key Off whenever checking for shorts or continuity, unless specified.
- Disconnect solenoids and switches from the harness before measuring for continuity, resistance, or energizing by way of 12-volt source.
- In using the Pinpoint Tests, follow each Step in order, starting from the first Step in the appropriate test. Follow each Step until the fault is found.
- After completing any repairs to the EEC system, verify all components are properly reconnected and repeat the functional test (Retest).
- An open is defined as any resistance reading greater than 5 ohms unless otherwise specified.
- A short is defined as any resistance reading less than 10,000 ohms to ground, unless otherwise specified.

The standard Ford color abbreviations are:

BK	Black	N	Natural
BL	Blue	O	Orange
BR	Brown	PK	Pink
DB	Dark Blue	P	Purple
DG	Dark Green	R	Red
GY	Gray	T	Tan
GR	Green	W	White
LB	Light Blue	Y	Yellow
LG	Light Green		

Where two colors are shown for a wire, the first color is the basic color of the wire. The second color is the dot, hash, or stripe marking. If D or H is given, the second color is dots or hash marks. If there is no letter after the second color, the wire has a stripe.

For example:

BR/O is a brown wire with an orange stripe.

R/Y D is a red wire with yellow dots.

BK/W H is a black wire with white hash marks.

EEC-IV No Start — Pinpoint Test A

A

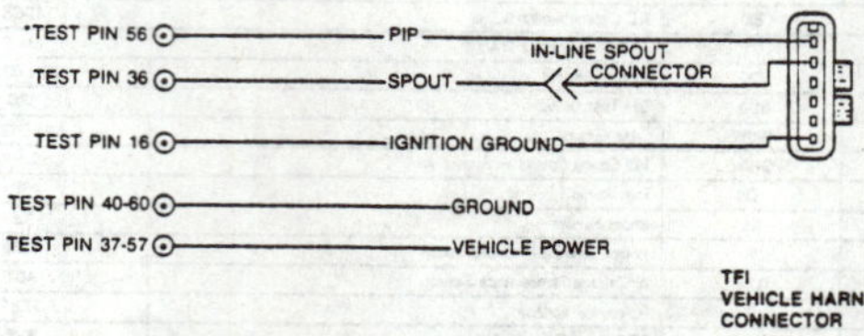

*TEST PINS LOCATED ON BREAKOUT BOX.
ALL HARNESS CONNECTORS VIEWED INTO MATING SURFACE.

NOTE: WHEN BREAKOUT BOX IS INSTALLED, ENSURE THAT TIMING SWITCH IS IN "COMPUTED" POSITION UNLESS OTHERWISE NOTED.

STOP-WARNING

You should enter this Pinpoint Test only when Steps 1.0 through 3.0 have been successfully completed and the engine is still a no start, or when directed here from Pinpoint Test P. This Pinpoint Test will not diagnose ignition system problems.

To prevent the replacement of good components, be aware that the following non-EEC areas may be at fault:

- Fuel: quantity and quality
- Ignition: general condition, moisture, cracks, damage, etc.
- Engine: internal, valves, timing belt, camshaft.
- Starter and battery circuit

This Pinpoint Test is intended to diagnose only the following:

- Spark (as related to EEC-IV).
- Circuits: pip, spout, ignition ground, vehicle power.

EEC-IV No Start — Pinpoint Test A

WARNING: Stop this test at the first sign of a fuel leak and service as required.
CAUTION: No open flame — No smoking during fuel delivery checks.

TEST STEP	RESULT ▶	ACTION TO TAKE
A1 ATTEMPT TO START ENGINE	Engine cranks, but does not start, or stalls out ▶	GO to A2.
	Engine does not crank ▶	REFER to Shop Manual, Group 28.
A2 CHECK FOR VREF AT THROTTLE POSITION SENSOR • Key Off, wait 10 seconds. • DVOM on 20V scale. • Disconnect TP sensor. • Key On, Engine Off. • Measure voltage at the TP vehicle harness connector between VREF and signal return. **NOTE: Refer to electrical schematic in appropriate engine supplement section for connector pin orientation.**	Less than 4.0V or greater than 6.0V ▶	GO to Pinpoint Test Step C1.
	4.0V to 6.0V ▶	RECONNECT TP sensor. GO to A3.
A3 CHECK FOR SPARK AT PLUGS • Disconnect the spark plug wire to any accessible cylinder. • Connect spark tester between spark plug wire and engine ground. • Crank engine and check for spark. • Reconnect the spark plug wire to the spark plug.	Spark ▶	GO to A13.
	No spark ▶	GO to A4.
A4 CHECK FOR SPARK AT COIL • Remove high tension coil wire from distributor and install spark tester. • Check for spark while cranking. • Reconnect high tension coil wire to distributor.	Spark ▶	REFER to Section 15, Part 2 for TFI Diagnosis for cap, rotor, wires.
	No spark ▶	GO to A5.

EEC-IV No Start — Pinpoint Test A

TEST STEP	RESULT ▶	ACTION TO TAKE
A5 HARNESS CHECK (IGNITION GROUND) • Key Off, wait 10 seconds. • Install Breakout box. Leave processor disconnected. • DVOM on 200 ohm scale. • Disconnect TFI. • Measure resistance between test Pin 16 at the Breakout box and TFI harness connector ignition ground.	Less than 5 ohms ▶	GO to A6.
	5 ohms or greater ▶	SERVICE harness as necessary. RERUN Quick Test.
A6 ISOLATION OF PROBLEM TO SPOUT CIRCUIT • Breakout box installed. • Connect TFI. • Connect processor. • Timing switch to "Dist" position on Breakout box. • Attempt to start vehicle. • Does the vehicle start?	Yes ▶	GO to A10.
	No ▶	GO to A7.
A7 SPOUT SIGNAL CHECK • Breakout box installed. • Timing switch to "Computed" position on Breakout box. • DVOM on 20V scale. • Measure voltage between test Pin 36 at the Breakout box and chassis ground, during crank.	Less than 3.0V or greater than 6.0V ▶	GO to A8.
	Between 3.0V and 6.0V ▶	EEC OK. REFER to Section 15, for TFI diagnosis.
A8 CHECK SPOUT FOR SHORTS • Key Off, wait 10 seconds. • Breakout box installed. • Disconnect processor. • Disconnect TFI. • DVOM on 200,000 ohm scale. • Measure resistance between test Pin 36 and test Pins 16, 20, 26, 40, 60 (short to ground), 37, 57 (short to power) and 56 (short to pip) at the Breakout box.	All readings 10,000 ohms or greater ▶	GO to A9.
	Any reading less than 10,000 ohms ▶	SERVICE short in harness. RERUN Quick Test if vehicle does not start. GO to A9.

EEC-IV No Start — Pinpoint Test A

TEST STEP	RESULT ▶	ACTION TO TAKE
A9 ISOLATE SHORT(S) IN PROCESSOR • Key Off, wait 10 seconds. • Breakout box installed. • Reconnect processor. • TFI disconnected. • DVOM on 200 ohm scale. • Measure resistance between test Pin 36 and test Pins 37 and 57 (short to power) also, test Pins 40 and 60 (short to ground) at the Breakout box.	All readings 5.0 ohms or greater ▶	Connect TFI. GO to A10.
	Any reading less than 5.0 ohms ▶	REPLACE processor. RERUN Quick Test.
A10 PIP SIGNAL CHECK • Breakout box installed. • DVOM to 20V scale. • Measure voltage between test Pin 56 and test Pin 16 at the Breakout box. • Crank engine, record reading.	Between 3.0V and 6.0V ▶	REMOVE Breakout box. REPLACE processor. RERUN Quick Test.
	Less than 3.0V or greater than 6.0V ▶	GO to A11.
A11 CONTINUITY OF PIP CIRCUIT CHECK • Breakout box installed. • Key Off, wait 10 seconds. • DVOM on 200 ohm scale. • Disconnect TFI. • Disconnect processor. • Measure resistance between test Pin 56 at the Breakout box and TFI harness connector PIP circuit.	Less than 5 ohms ▶	GO to A12.
	5 ohms or greater ▶	SERVICE open PIP circuit. RERUN Quick Test.

EEC-IV No Start — Pinpoint Test A

TEST STEP	RESULT ▶	ACTION TO TAKE
A12 CHECK PIP CIRCUIT FOR SHORTS • Breakout box installed. • Processor disconnected. • Key Off. • Disconnect TFI connector. • DVOM on 200,000 ohm scale. • Measure resistance between test Pin 56 and test Pins 16, 20, 26, 40, 60 (shorts to ground) and test Pins 37 and 57 (shorts to power) and test Pin 36 (short to spout) at the Breakout box.	Any resistance less than 10,000 ohms ▶	SERVICE PIP circuit. RERUN Quick Test.
	All resistance greater than 10,000 ohms ▶	REFER to Section 15 for TFI diagnosis.
A13 SPOUT SIGNAL VERIFICATION • Key Off, wait 10 seconds. • Disconnect processors 60 pin connector and inspect for damaged pins, corrosion, loose wires. Service as necessary. • Install Breakout box. • Processor connected. • DVOM on 20V scale. • Measure voltage between test Pin 36 at the Breakout box and chassis ground, during crank. • Ensure timing switch is in "Computed" position on Breakout box.	Between 3.0V and 6.0V ▶	GO to A21.
	Less than 3.0V or greater than 6.0V ▶	GO to A10.

EEC-IV No Start — Pinpoint Test A

TEST STEP	RESULT ▶	ACTION TO TAKE
A21 FUEL PUMP CHECK		
• **No smoking nearby.** • Connect pressure gauge. • Note initial pressure reading. • Observe pressure gauge as you pressurize fuel system. (Turn key to RUN for 1 second, then turn key to OFF. Wait 10 seconds. Repeat 5 times.)	PRESSURE GAUGE READING: Increased ▶ Did not increase ▶	All EFI Go to Pinpoint Test Step [S1]. All CFI Go to Pinpoint Test Step [S2]. TURN key Off, and CONTINUE to [A22].
WARNING: If fuel starts leaking, turn key OFF immediately. No smoking.		
A22 INERTIA SWITCH CHECK		
• Key Off. • Fuel pressure gauge installed. • Locate fuel pump inertia switch. Refer to Owner's Manual for location. • Push the button of inertia switch to reset to ON. **NOTE: If switch will not reset to ON, replace Inertia Switch and repeat Step A21. If switch button was on, GO to Step J1, except 2.5L HSC CFI and 3.0L EFI passenger car, GO to Step X-11.** • Observe pressure gauge as you pressurize fuel system. (Turn key to RUN for 1 second, then turn key to OFF. Wait 10 seconds.) Repeat 5 times.	PRESSURE GAUGE READING: Increase ▶ No increase ▶	RERUN Quick Test. • 2.3L EFI TC 2.5L HSC-CFI and 3.0L EFI passenger car GO to [X-11]. • All others. GO to [J1].

Vehicle Battery — Pinpoint Test B

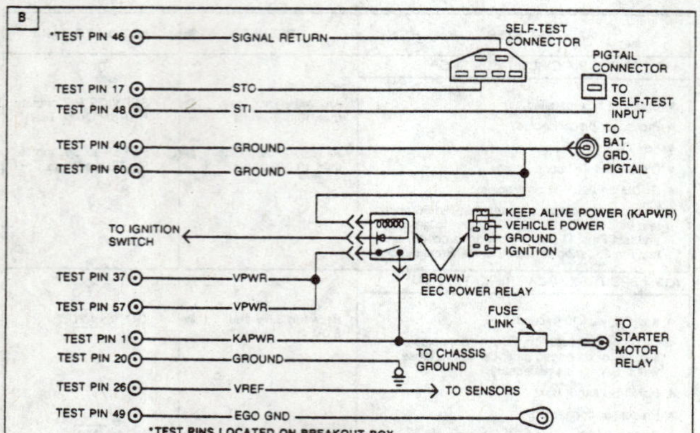

*TEST PINS LOCATED ON BREAKOUT BOX.
ALL HARNESS CONNECTORS VIEWED INTO MATING SURFACE.

STOP-WARNING

You should enter this Pinpoint Test only when directed here from Pinpoint Tests C, J or P or when a continuous memory code 72 or 78 is received in Quick Test Step 6.0C.

To prevent the replacement of good components, be aware that the following non-EEC areas may be at fault:

- Ignition switch.
- Battery Cables.
- Alternator.
- Voltage Regulator.
- Ground Straps.

This pinpoint test is intended to diagnose only the following:

- Processor.
- Harness circuits: Signal Return, STO, STI, Ground, VPWR, KAPWR, VREF, Ignition.
- Battery Voltage.
- Power Relay.

Vehicle Battery — Pinpoint Test B

TEST STEP	RESULT ▶	ACTION TO TAKE
B1 BATTERY VOLTAGE CHECK		
• Key On, Engine Off. • DVOM on 20V scale. • Measure voltage across battery terminals. • Is voltage reading 10.5 volts or greater?	Yes ▶ No ▶	GO to [B2]. SERVICE discharged battery. REFER to Shop Manual, Group 31.
B2 EEC GROUND TO BATTERY GROUND CHECK		
• Key Off, wait 10 seconds. • Install Breakout box. • Processor connected. • DVOM on 200 ohm scale. • Measure resistance between test Pin 40 and negative post of the battery and test Pin 60 and negative post of the battery. • Are both resistance readings less than 5 ohms?	Yes ▶ No ▶	GO to [B3]. CORRECT cause of resistance between negative side and Pin 40, or 60 of the EEC system. RERUN Quick Test.
B3 PROCESSOR GROUND FAULT ISOLATION		
• Breakout box installed. • Key Off, wait 10 seconds. • Processor connected. • DVOM on 200 ohm scale. • Measure resistance between test Pin 46 and test Pin 40 and between test Pin 46 and test Pin 60 both at the Breakout box. • Are both resistance readings 5 ohms or less?	Yes ▶ No ▶	GO to [B4]. DISCONNECT processor connector and INSPECT for corrosion, damaged pins, etc. SERVICE as necessary and RETEST. If fault is still present, REPLACE processor. RERUN Quick Test.
B4 HARNESS CHECK (SIGNAL RETURN)		
• Breakout box installed. • Key Off, wait 10 seconds. • Processor connected. • DVOM on 200 ohm scale. • Measure resistance between test Pin 46 at the Breakout box and Signal Return in the Self-Test connector. • Is resistance reading less than 5.0 ohms?	Yes ▶ No ▶	GO to [B5]. CORRECT cause of resistance in the harness Signal Return circuit. RERUN Quick Test.

Vehicle Battery — Pinpoint Test B

TEST STEP	RESULT ▶	ACTION TO TAKE
B5 12 VOLT BATTERY POWER FAULT ISOLATION		
• Key On, Engine Off. • Processor connected. • DVOM on 20V scale. • Measure voltage between the battery negative post and KAPWR circuit (Test Pin 1) at EEC power relay. • Is voltage reading 10.5V or greater?	Yes ▶ No ▶	GO to [B6]. CHECK KAPWR and VPWR circuits for shorts to ground and KAPWR circuit from power relay to battery positive post for opens. SERVICE as necessary. RERUN Quick Test.
B6 12 VOLT BATTERY POWER FAULT ISOLATION		
• Key On, Engine Off. • Processor connected. • DVOM on 20V scale. • Measure voltage between the battery negative post and Ignition circuit at EEC power relay. • Is the reading 10.5 volts or greater?	Yes ▶ No ▶	GO to [B7]. CHECK for open in ignition switch circuits. SERVICE as necessary. RERUN Quick Test.
B7 12 VOLT BATTERY POWER FAULT ISOLATION		
• Key Off, wait 10 seconds. • Processor connected. • DVOM on 200 ohm scale. • Measure resistance between ground at the EEC power relay and negative post of battery. • Is the resistance reading 5 ohms or less?	Yes ▶ No ▶	GO to [B8]. SERVICE open or ground in Ground Circuit. RERUN Quick Test.
B8 12 VOLT BATTERY POWER FAULT ISOLATION		
• Key On, Engine Off. • Processor connected. • DVOM on 20V scale. • Measure voltage between the battery negative post and VPWR circuit at EEC power relay. • Is the voltage reading 10.5 volts or greater?	Yes ▶ No ▶	SERVICE open in VPWR circuit, if OK, SERVICE short to ground in VPWR circuit. REPLACE power relay. RERUN Quick Test.

Vehicle Battery — Pinpoint Test B

TEST STEP	RESULT ▶	ACTION TO TAKE
B10 WIGGLE TEST VPWR CIRCUITS		
• Key On, Engine Off. • STAR tester or VOM hooked up to Self-Test connector. • Self-Test deactivated. • Observe STAR/VOM for fault indication as explained in Quick Test Step 6.0D. • Shake, bend and twist the EEC-IV harness from the EEC time delay power relay to the processor. • Is a fault indicated or does code 72 or 78 reappear in continuous memory if Quick Test is rerun?	Yes ▶	SERVICE intermittent VPWR circuit. RERUN Quick Test.
	No ▶	INSPECT EEC-IV time delay power relay and harness connectors for damaged pins, corrosion, etc. SERVICE as necessary. If OK, REPLACE EEC-IV time delay relay. RERUN Quick Test.

Reference Voltage — Pinpoint Test C

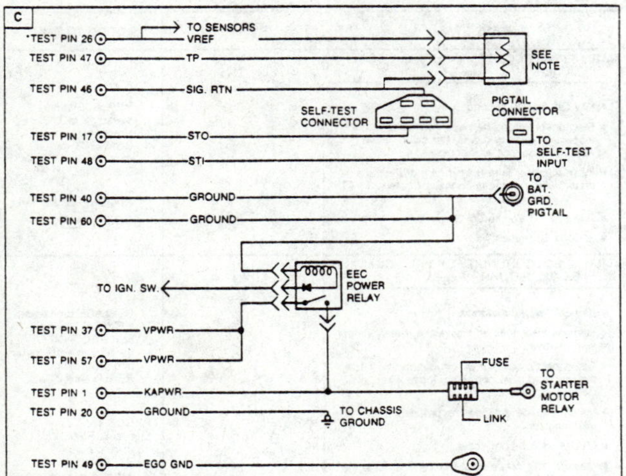

*TEST PINS LOCATED ON BREAKOUT BOX.
ALL HARNESS CONNECTORS VIEWED INTO MATING SURFACE.

NOTE: Refer to the electrical schematic in the appropriate engine supplement section for proper connector and pin orientation.

STOP-WARNING

You should enter this Pinpoint Test only when a check for Vref has failed in the sensor Pinpoint Tests (D-Series) or Pinpoint Tests A or Q.

This pinpoint test is intended to diagnose only the following:

- Processor.
- Sensor harness circuits: Signal Return, STO, STI, Ground, VPWR, KAPWR, VREF, Ignition.

Reference Voltage — Pinpoint Test C

TEST STEP	RESULT ▶	ACTION TO TAKE
C1 VEHICLE BATTERY POWER CIRCUIT CHECK		
• Disconnect 60 Pin connector. Inspect for damaged pins, corrosion, loose wires, etc. Service as necessary. • Breakout box installed. • Processor connected. • Key On, Engine Off. • DVOM on 20V scale. • Measure voltage between test Pin 37 at the Breakout box and Signal Return in Self-Test connector. • Is voltage 10.5 volts or greater?	Yes ▶	GO to C2.
	No ▶	2.3L EFI TC, 2.5L HSC CFI and 3.0L EFI passenger car GO to X-1. All others, GO to B1.
C2 VREF VOLTAGE CHECK		
• Breakout box installed. • Key On, Engine Off. • Processor connected. • DVOM on 20V scale. • Measure voltage between test Pin 26 and test Pin 46 at the Breakout box.	6.0V or greater ▶	GO to C4.
	4.0V or less ▶	GO to C5.
	Greater than 4.0V, less than 6.0V ▶	GO to C3.
C3 CHECK VREF AND SIGNAL RETURN FOR CONTINUITY		
• Breakout box installed. • Processor disconnected. • Sensor that sent you here disconnected. • Key Off. • DVOM on 200 ohm scale. • Measure resistance from test Pin 26 at Breakout box to VREF at vehicle harness connector of the sensor that sent you here. • Measure resistance from test Pin 46 at Breakout box to signal return at vehicle harness connector of the sensor that sent you here. • Are both resistance readings less than 5.0 ohms?	Yes ▶	RECONNECT sensors. Reference voltage OK. RERUN Quick Test.
	No ▶	SERVICE open in Vref or Signal Return. RERUN Quick Test.

Reference Voltage — Pinpoint Test C

TEST STEP	RESULT ▶	ACTION TO TAKE
C4 CHECK FOR SHORT FROM VREF TO VPWR		
• Key Off, wait 10 seconds. • Breakout box installed. • Disconnect processor. • Key On, Engine Off. • DVOM on 20V scale. • Measure voltage between test Pin 26 at the Breakout box and battery ground. • Is voltage less than 0.5 volts?	Yes ▶	REPLACE processor. RERUN Quick Test.
	No ▶	SERVICE short to battery power in EEC harness. RERUN Quick Test. If condition persists, REPLACE processor.
C5 CHECK FOR SHORTED THROTTLE POSITION SENSOR		
• Key Off, wait 10 seconds. • Breakout box installed. • Processor connected. • Disconnect Throttle Position (TP) sensor from vehicle harness. • Key On, Engine Off. • DVOM on 20V scale. • Measure voltage between test Pin 26 and test Pin 46 at the Breakout box. • Is voltage less than 4.0 volts?	Yes ▶	Vehicles equipped with EVP/PFE sensor, GO to C6. All other vehicles, GO to C7.
	No ▶	REPLACE TP sensor. RERUN Quick Test.
C6 CHECK FOR SHORTED EVP/PFE SENSOR		
• Key Off, wait 10 seconds. • Breakout box installed. • Processor connected. • Disconnect EVP/PFE sensor. • Key On, Engine Off. • DVOM on 20V scale. • Measure voltage between test Pin 26 and test Pin 46 at the Breakout box. • Is voltage less than 4.0 volts?	Yes ▶	GO to C7.
	No ▶	REPLACE EVP/PFE sensor. RERUN Quick Test.

Reference Voltage — Pinpoint Test — C

TEST STEP	RESULT ▶	ACTION TO TAKE
C7 CHECK FOR SHORTED MAP/BP SENSOR		
• Key Off, wait 10 seconds. • Breakout box installed. • Processor connected. • Disconnect MAP/BP sensor. • Key On, Engine Off. • DVOM on 20V scale. • Measure voltage between test Pin 26 and test Pin 46 at the Breakout box. • Is reading less than 4.0 volts?	Yes ▶	Vehicles equipped with VAF sensor, GO to C8. All other vehicles, GO to C9.
	No ▶	REPLACE MAP/BP sensor. RERUN Quick Test.
C8 CHECK FOR SHORTED VANE AIR METER (VAF) SENSOR		
• Key Off, wait 10 seconds. • Breakout box installed. • Processor connected. • Disconnect vane air meter (VAF) sensor. • Key On, Engine Off. • DVOM on 20V scale. • Measure voltage between test Pin 26 and test Pin 46 at the Breakout box. • Is voltage less than 4.0 volts?	Yes ▶	GO to C9.
	No ▶	REPLACE VAF sensor and RERUN Quick Test.
C9 SHORT TO GROUND IN VREF		
• Breakout box installed. • Processor disconnected. • Key Off, wait 10 seconds. • Disconnect TP and MAP/BP, EVP/PFE and VAF, if so equipped. • DVOM on 200 ohm scale. • Measure resistance between test Pin 26 and test Pins 20, 40, 46 and 60 at the Breakout box. • Is any resistance less than 5 ohms?	Yes ▶	SERVICE short to ground. CONNECT all sensors. RERUN Quick Test. If original condition still exists, REPLACE processor.
	No ▶	RECONNECT sensors. REPLACE processor. RERUN Quick Test.

Vane Air Temperature Sensor (VAT) — Pinpoint Test — DA

DA

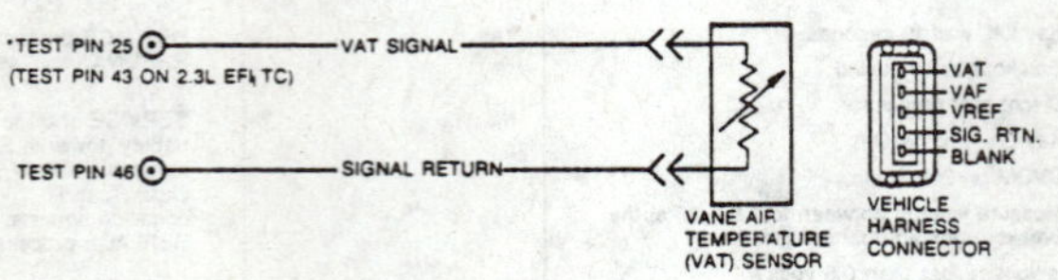

NOTE: AMBIENT TEMPERATURE MUST BE GREATER THAN 50°F TO PASS THIS TEST.

TYPICAL RESISTANCE BETWEEN TEST PINS 25 (OR 43) & 46	5800 ohms	2700 ohms	300 ohms	180 ohms	125 ohms
AT TEMPERATURE	32°F	65°F	185°F	220°F	240°F

*TEST PINS LOCATED ON BREAKOUT BOX.
ALL HARNESS CONNECTORS VIEWED INTO MATING SURFACE.

STOP-WARNING

You should enter this Pinpoint Test only when a service code 28, 58 or 68 is received in Quick Test Step 3.0, 5.0 or 6.0.

To prevent the replacement of good components, be aware that the following non-EEC areas may be at fault:

- Test performed in unusually low (cold) or high (hot) ambient conditions.
- Ambient temperature must be greater than 50°F for this test.

This pinpoint test is intended to diagnose only the following:

- VAT sensor.
- Circuits: VAT, and Signal Return.
- Vehicle harness.
- Processor assembly.

Vane Air Temperature Sensor (VAT) — Pinpoint Test — DA

TEST STEP	RESULT ▶	ACTION TO TAKE
DA1 SERVICE CODE 28: CHECK AMBIENT TEMPERATURE		
• Is ambient temperature greater than 50°F?	Yes ▶	GO to DA2.
	No ▶	RERUN Quick Test.
DA2 CHECK FOR V REF AT THROTTLE POSITION SENSOR		
• Refer to illustration Q. • Key Off, wait 10 seconds. • DVOM on 20V scale. • Disconnect TP sensor. • Key On, Engine Off. • Measure voltage at the TP vehicle harness connector between VREF and signal return. • Is voltage between 4.0 volts and 6.0 volts?	Yes ▶	RECONNECT TP sensor. GO to DA3.
	No ▶	GO to Pinpoint Test Step C1.
DA3 VAT SENSOR CHECK		
NOTE: Ambient temperature must be greater than 50°F for this test. • Key Off, wait 10 seconds. • Harness disconnected from the vane meter. • DVOM on 200,000 ohm scale. • Measure resistance at the VAT sensor between VAT signal and Signal Return. • Is resistance from 125 ohms (240°F) to 3700 ohms (50°F)?	Yes ▶	REPLACE processor. RECONNECT harness to vane meter. RERUN Quick Test.
	No ▶	REPLACE vane meter. RERUN Quick Test.

Vane Air Temperature Sensor (VAT) — Pinpoint Test — DA

TEST STEP	RESULT ▶	ACTION TO TAKE
DA10 SERVICE CODE 58: INDUCE OPPOSITE CODE		
• Key Off, wait 10 seconds. • Disconnect vehicle harness from vane meter. Inspect for damaged pins, corrosion, loose wires, etc. Service as necessary. • Insert a jumper wire (paper clip) at the vane meter vehicle harness connector between VAT Signal and Signal Return. • Run Key On Engine Off Quick Test. • Is Code 68 present?	Yes ▶	REPLACE vane meter. REMOVE jumper wire. CONNECT harness to vane meter. RERUN Quick Test.
	No ▶	REMOVE jumper wire. GO to DA11.
DA11 CHECK CONTINUITY OF VAT SIGNAL AND SIGNAL RETURN		
• Key Off, wait 10 seconds. • Harness disconnected from vane meter, jumper wire removed. • Disconnect processor 60 pin connector. Inspect for damaged pins, corrosion, loose wires, etc. Service as necessary. • Install Breakout box leaving processor disconnected. • DVOM on 200 ohm scale. • Measure resistance between VAT signal at the vane meter vehicle harness connector, and test Pin 25 at the Breakout box. • Measure resistance between VAT signal at the vane meter vehicle harness connector, and test Pin 25 (test Pin 43 on 2.3L EFI TC) at the Breakout box. • Measure resistance between signal return at the vane meter vehicle harness connector, and test Pin 46 at the Breakout box. • Are both resistances less than 5 ohms?	Yes ▶	REPLACE processor. REMOVE Breakout box. RECONNECT harness to vane meter and processor. RERUN Quick Test.
	No ▶	CORRECT open circuit. REMOVE Breakout box. RECONNECT harness to vane meter and processor. RERUN Quick Test.

Vane Air Temperature Sensor (VAT) — Pinpoint Test — DA

TEST STEP	RESULT ▶	ACTION TO TAKE
DA20 SERVICE CODE 68: INDUCE OPPOSITE CODE		
• Key off, wait 10 seconds. • Disconnect vehicle harness from vane meter. Inspect for damaged pins, corrosion, loose wires, etc. Service as necessary. • Run Key On Engine Off Quick Test. • Is code 58 present?	Yes ▶	REPLACE vane meter. RECONNECT harness to vane meter. RERUN Quick Test.
	No ▶	GO to DA21.
DA21 CHECK FOR V REF AT THROTTLE POSITION SENSOR		
• Refer to illustration Q. • Key Off, wait 10 seconds. • DVOM on 20V scale. • Disconnect TP sensor. • Key On, Engine Off. • Measure voltage at the TP vehicle harness connector between VREF and signal return. • Is voltage between 4.0 volts and 6.0 volts?	Yes ▶	RECONNECT TP sensor. GO to DA22.
	No ▶	GO to Pinpoint Test Step C1.
DA22 CHECK VAT SIGNAL FOR SHORTS		
• Key Off, wait 10 seconds. • Harness disconnected from vane meter. • Disconnect processor 60 pin connector. Inspect for damaged pins, corrosion, loose wires, etc. Service as necessary. • Install Breakout box leaving processor disconnected. • DVOM on 200,000 ohm scale. • Measure resistance between test Pin 25 (test Pin 43 on 2.3L EFI TC) and test Pins 40, 46 and 60 at the Breakout box. • Are all resistances greater than 10,000 ohms?	Yes ▶	REPLACE processor. REMOVE Breakout box. RECONNECT processor. RERUN Quick Test.
	No ▶	CORRECT circuit shorts. REMOVE Breakout box. RECONNECT processor and vane meter. RERUN Quick Test.

Vane Air Temperature Sensor (VAT) — Pinpoint Test — DA

TEST STEP	RESULT ▶	ACTION TO TAKE
DA90 SERVICE CODE 58: CONTINUOUS TEST: CHECK VAT SENSOR		
• Using continuous monitor mode, observe VOM or STAR LED for indication of a fault while performing the following: • Lightly tap on VAT sensor (simulate road shock). • Wiggle VAT connector. • Is a fault indicated?	Yes ▶	DISCONNECT and INSPECT connectors. If connector and terminals are good, REPLACE VAT sensor. RERUN Quick Test.
	No ▶	GO to DA91.

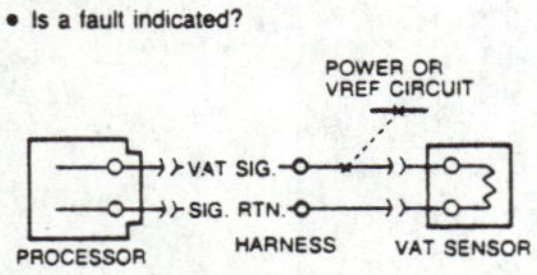

TEST STEP	RESULT ▶	ACTION TO TAKE
DA91 CHECK EEC-IV HARNESS		
• Observe VOM or STAR LED for a fault indication while performing the following: • Referring to the illustration in Step DA90, grasp the harness closest to the sensor connector. Wiggle, shake or bend a small section of the EEC-IV system harness while working your way to the dash panel. Also wiggle, shake or bend the EEC-IV harness from the dash panel to the processor. • Is fault indicated?	Yes ▶	ISOLATE fault and make necessary repairs. RERUN Quick Test.
	No ▶	GO to DA92.
DA92 CHECK PROCESSOR AND HARNESS CONNECTORS		
• Key Off, wait 10 seconds. • Disconnect processor 60 pin connector. • Inspect both connectors and connector terminals for obvious damage or faults. • Are connectors and terminals OK?	No ▶	SERVICE as necessary. RERUN Quick Test.
	Yes ▶	Unable to duplicate fault at this time. Continuous code 58 testing complete.

Vane Air Temperature Sensor (VAT) — Pinpoint Test — DA

TEST STEP	RESULT ▶	ACTION TO TAKE
DA93 SERVICE CODE 68: CONTINUOUS TEST: CHECK VAT SENSOR		
• Using continuous monitor mode, observe VOM or STAR LED for indication of a fault while performing the following: • Lightly tap on VAT sensor (simulate road shock). • Wiggle VAT connector. • Is fault indicated?	Yes ▶	DISCONNECT and INSPECT connectors. If connector and terminals are good, REPLACE VAT sensor. RERUN Quick Test.
	No ▶	GO to DA94.

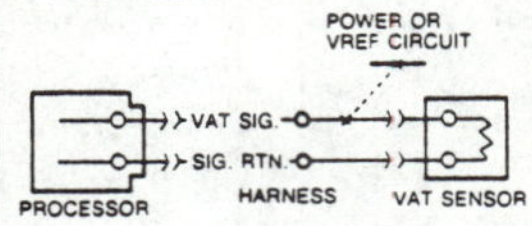

TEST STEP	RESULT ▶	ACTION TO TAKE
DA94 CHECK EEC-IV HARNESS		
• Observe VOM or STAR LED for a fault indication while performing the following: • Referring to the illustration in Step DA93, grasp the harness closest to the sensor connector. Wiggle, shake or bend a small section of the EEC-IV system harness while working your way to the dash panel. Also wiggle, shake or bend the EEC-IV harness from the dash panel to the processor. • Is fault indicated?	Yes ▶	ISOLATE fault and make necessary service. RERUN Quick Test.
	No ▶	GO to DA95.
DA95 CHECK PROCESSOR AND HARNESS CONNECTORS		
• Key Off, wait 10 seconds. • Disconnect processor 60 pin connector. • Inspect both connectors and connector terminals for obvious damage or faults. • Are connectors and terminals OK?	No ▶	SERVICE as necessary. RERUN Quick Test.
	Yes ▶	Unable to duplicate fault at this time. Continuous code 68 testing complete.

Air Charge Temperature Sensor (ACT) — Pinpoint Test — DB

DB

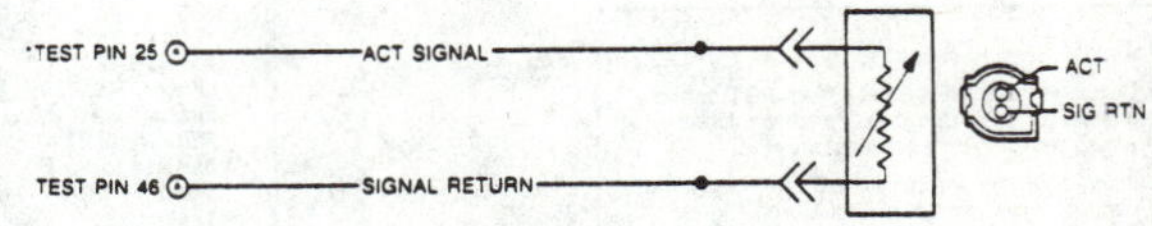

TYPICAL RESISTANCE BETWEEN TEST PINS 25 & 46	58.750 ohms	40.500 ohms	3600 ohms	1840 ohms
AT TEMPERATURE	50°F	65°F	180°F	220°F

*TEST PINS LOCATED ON BREAKOUT BOX.
ALL HARNESS CONNECTORS VIEWED INTO MATING SURFACE.

NOTE: Ambient temperature must be greater than 50°F for this test.

STOP-WARNING

You should enter this Pinpont Test only when a service code 24, 54 or 64 is received in Quick Test Step 3.0, 5.0 or 6.0.

To prevent the replacement of good components, be aware that the following non-EEC areas may be at fault:

- Cooling system.
- Air cleaner duct problems.
- Improper engine oil level.

This pinpoint test is intended to diagnose only the following:

- ACT sensor.
- Harness circuits: ACT Signal and Signal Return.
- Processor assembly.

Air Charge Temperature Sensor (ACT) — Pinpoint Test DB

TEST STEP	RESULT ▶	ACTION TO TAKE
DB1 SERVICE CODE 24: CHECK PROPER INSTALLATION		
• For vehicle with ACT mounted in intake manifold, GO to step DB2. • Is ACT mounted properly in air cleaner?	Yes ▶ No ▶	GO to DB2. INSTALL ACT properly. RERUN Quick Test.
DB2 CHECK FOR VREF AT THROTTLE POSITION SENSOR		
• Key Off, wait 10 seconds. • DVOM on 20V scale. • Disconnect TP sensor. • Key On, Engine Off. • Measure voltage at the TP vehicle harness connector between VREF and signal return.	Less than 4.0V or greater than 6.0V ▶ 4.0V to 6.0V ▶	GO to Pinpoint Test Step C1. RECONNECT TP sensor, GO to DB3.
DB3 CHECK ACT SENSOR — ENGINE OFF		
• Key Off, wait 10 seconds. • Harness disconnected from ACT sensor. • DVOM on 200,000 ohm scale. • Measure resistance of ACT sensor. **NOTE: Make sure engine is warmed up prior to this test.**	Reading is between 1,100 and 58,000 ohms (approximately 50°F) ▶ Reading is less than 1,100 ohms or greater than 58,000 ▶	GO to DB4. CHECK heat stove duct valve operation. If OK, REPLACE ACT sensor. RECONNECT harness to ACT sensor. RERUN Quick Test.
DB4 CHECK ACT SENSOR — ENGINE RUNNING		
• Key Off. Harness disconnected from ACT sensor. • DVOM on 200,000 ohm scale. • Run engine for 2 minutes. • Measure resistance of ACT sensor with engine running.	Reading is between 2,400 and 29,000 ohms ▶ Reading is less than 2,400 ohms or greater than 29,000 ohms ▶	REPLACE processor. RECONNECT harness to ACT sensor. RERUN Quick Test. CHECK heat stove duct valve operation. If OK, REPLACE ACT sensor. RERUN Quick Test.

Air Charge Temperature Sensor (ACT) — Pinpoint Test DB

TEST STEP	RESULT ▶	ACTION TO TAKE
DB10 SERVICE CODE 54: INDUCE OPPOSITE CODE		
• Key Off, wait 10 seconds. • Disconnect vehicle harness from ACT sensor. Inspect for damaged pins, corrosion, loose wires, etc. Service as necessary. • Insert a jumper wire at the ACT vehicle harness connector between ACT Signal and Signal Return. • Run Key On, Engine Off Quick Test. • Is code 64 present?	Yes ▶ No ▶	REPLACE ACT sensor. REMOVE jumper wire. RECONNECT harness to ACT sensor. RERUN Quick Test. REMOVE jumper wire. GO to DB11.
DB11 CONTINUITY CHECK ACT SIGNAL AND SIGNAL RETURN		
• Key Off, wait 10 seconds. • Harness disconnected from ACT sensor. • Disconnect processor 60 pin connector. Inspect for damaged pins, corrosion, loose wires, etc. Service as necessary. • Install Breakout box. Leave processor disconnected. • DVOM on 200 ohm scale. • Measure resistance between ACT signal, at the ACT vehicle harness connector, and test Pin 25 at the Breakout box. • Measure resistance between Signal Return, at the ACT vehicle harness connector, and test Pin 46 at the Breakout box.	Both readings are less than 5 ohms ▶ Either reading is 5 ohms or greater ▶	REPLACE processor. REMOVE Breakout box. RECONNECT processor and ACT sensor. RERUN Quick Test. CORRECT circuit opens. REMOVE Breakout box. RECONNECT processor and ACT sensor. RERUN Quick Test.

Air Charge Temperature Sensor (ACT) — Pinpoint Test DB

TEST STEP	RESULT ▶	ACTION TO TAKE
DB20 SERVICE CODE 64: INDUCE OPPOSITE CODE		
• Key Off, wait 10 seconds. • Disconnect vehicle harness from ACT sensor. Inspect for damaged pins, corrosion, loose wires, etc. Service as necessary. • Run Key On, Engine Off Quick Test. • Is code 54 present?	Yes ▶ No ▶	REPLACE ACT sensor. RECONNECT harness to ACT sensor. RERUN Quick Test. GO to DB21.
DB21 CHECK FOR VREF AT THROTTLE POSITION SENSOR		
• Key Off, wait 10 seconds. • DVOM on 20V scale. • Disconnect TP sensor. • Key On, Engine Off. • Measure voltage at the TP vehicle harness connector between VREF and Signal Return.	Less than 4.0V or greater than 6.0V ▶ 4.0V to 6.0V ▶	GO to Pinpoint Test Step C1. RECONNECT TP sensor, GO to DB22.
DB22 CHECK ACT SIGNAL FOR SHORTS TO GROUND		
• Key Off, wait 10 seconds. • Harness disconnected from ACT sensor. • Disconnect processor 60 pin connector. Inspect for damaged pins, corrosion, loose wires, etc. Service as necessary. • DVOM on 200,000 ohm scale. • Measure resistance between test Pin 25 and test Pins 40, 46 and 60 at the Breakout box.	Any reading less than 10,000 ohms ▶ All readings are 10,000 ohms or greater ▶	CORRECT circuit shorts. REMOVE Breakout box. RECONNECT processor and ACT sensor. RERUN Quick Test. REPLACE processor. REMOVE Breakout box. RECONNECT processor and ACT sensor. RERUN Quick Test.

Air Charge Temperature (ACT) — Pinpoint Test DB

TEST STEP	RESULT ▶	ACTION TO TAKE
DB90 SERVICE CODE 54: CONTINUOUS TEST: CHECK ACT SENSOR		
• Using continuous monitor mode, observe VOM or STAR LED for indication of a fault while performing the following: • Lightly tap on ACT sensor (simulate road shock). • Wiggle ACT connector. • Is a fault indicated?	Yes ▶ No ▶	DISCONNECT and INSPECT connectors. If connector and terminals are good, REPLACE ACT sensor. RERUN Quick Test. GO to DB91.
DB91 CHECK EEC-IV HARNESS		
• Observe VOM or STAR LED for a fault indication while performing the following: • Referring to the illustration in Step DB90, grasp the harness closest to the sensor connector. Wiggle, shake or bend a small section of the EEC-IV system harness while working your way to the dash panel. Also wiggle, shake or bend the EEC-IV harness from the dash panel to the processor. • Is fault indicated?	Yes ▶ No ▶	ISOLATE fault and make necessary repairs. RERUN Quick Test. GO to DB92.
DB92 CHECK PROCESSOR AND HARNESS CONNECTORS		
• Key Off, wait 10 seconds. • Disconnect processor 60 pin connector. • Inspect both connectors and connector terminals for obvious damage or faults. • Connectors and terminals are OK.	No ▶ Yes ▶	SERVICE as necessary. RERUN Quick Test. Unable to duplicate fault at this time. Continuous code 54 testing complete.

DB90 illustration:

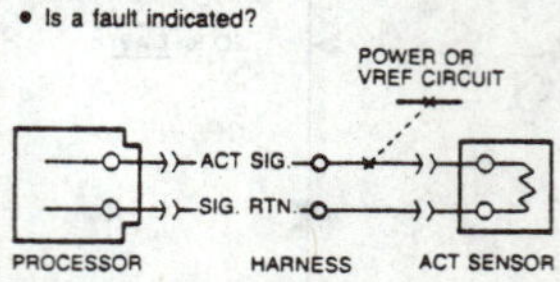

Air Charge Temperature (ACT) — Pinpoint Test — DB

TEST STEP	RESULT ▶	ACTION TO TAKE
DB93 SERVICE CODE 64: CONTINUOUS TEST: CHECK ACT SENSOR		
• Using continuous monitor mode, observe VOM or STAR LED for fault while performing the following: • Lightly tap on ACT sensor (simulate road shock). • Wiggle ACT connector. • Is fault indicated?	Yes ▶	DISCONNECT and INSPECT connectors. If connector and terminals are good, REPLACE ACT sensor. RERUN Quick Test.
	No ▶	GO to DB94.
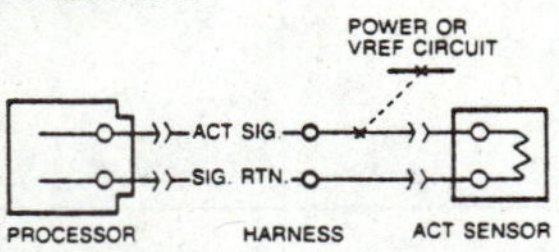		
DB94 CHECK EEC-IV HARNESS		
• Observe VOM or STAR LED for a fault indication while performing the following: • Referring to the illustration in Step DB93, grasp the harness closest to the sensor connector. Wiggle, shake or bend a small section of the EEC-IV system harness while working your way to the dash panel. Also wiggle, shake or bend the EEC-IV harness from the dash panel to the processor. • Is fault indicated?	Yes ▶	ISOLATE fault and make necessary repairs. RERUN Quick Test.
	No ▶	GO to DB95.
DB95 CHECK PROCESSOR AND HARNESS CONNECTORS		
• Key Off, wait 10 seconds. • Disconnect processor 60 pin connector. • Inspect both connectors and connector terminals for obvious damage or faults. • Connectors and terminals are OK.	No ▶	SERVICE as necessary. RERUN Quick Test.
	Yes ▶	Unable to duplicate fault at this time. Continuous code 64 testing complete.

EGR Valve Position Sensor (EVP) Control/Vent (EGRC/EGRV) — Pinpoint Test — DD

DD

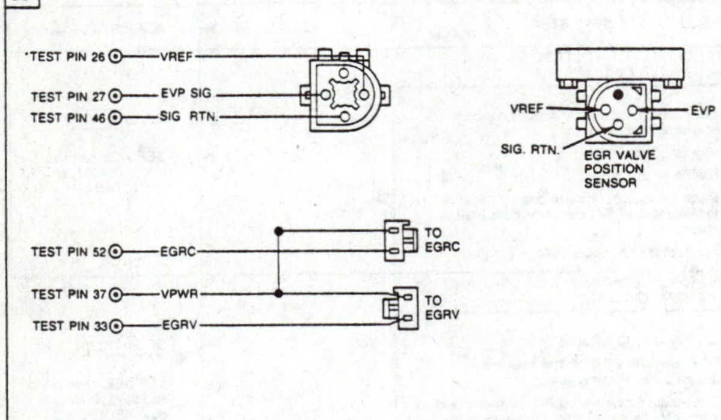

*TEST PINS LOCATED ON BREAKOUT BOX.
ALL HARNESS CONNECTORS VIEWED INTO MATING SURFACE.

STOP-WARNING

You should enter this Pinpoint Test only when a Service Code 31, 32, 33, 34, 35, 83 or 84 is received in Quick Test Step 3.0, 5.0 or 6.0.

To prevent the replacement of good components, be aware that the following non-EEC area may be at fault:

- Damaged EGR valve.

This Pinpoint Test is intended to diagnose only the following:

- EVP sensor.
- Harness circuits: EVP, Signal Return, VREF, EGRV, EGRC, VPWR.
- EGR solenoids.
- EGR Valve assembly.
- Processor assembly.

EGR Valve Position Sensor (EVP) Control/Vent (EGRC/EGRV) — Pinpoint Test — DD

TEST STEP	RESULT ▶	ACTION TO TAKE
FAULT CODE 31		
DD1 RUN ENGINE RUNNING QUICK TEST WITH EGR VACUUM SIGNAL LINE DISCONNECTED AT EGR VALVE		
• Key Off, wait 10 seconds. • Disconnect EGR vacuum line at EGR valve and cap EGR vacuum line. • Run Engine Running Quick Test. • Check for code 31.	Code 31 present ▶	GO to DD2.
	No code 31 present, but codes 32, 34 are present ▶	GO to DD11.
DD2 CHECK EVP RESISTANCE WHILE APPLYING VACUUM TO EGR VALVE		
• Key Off, wait 10 seconds. • Vacuum signal line disconnected. • Disconnect vehicle harness at EVP sensor. • DVOM on 200,000 ohm scale. • Connect vacuum pump to EGR valve. • Measure resistance at the EVP sensor between EVP SIG and Vref while gradually increasing vacuum to 33 kPa (10 in.-Hg.). • Observe resistance as vacuum increases.	Reading gradually decreases from no greater than 5500 ohms to no less than 100 ohms ▶	GO to DD3.
	Reading is less than 100 ohms or greater than 5500 ohms ▶	REPLACE EVP sensor. RECONNECT signal line and harness. RERUN Quick Test.
	Reading does not decrease or unable to hold vacuum ▶	GO to DD16.
DD3 MEASURE VREF TO SIGNAL RETURN VOLTAGE		
• Key On, Engine Off. • Vacuum signal line disconnected and capped, harness disconnected from EVP sensor. • DVOM on 20V scale. • Measure voltage at the EVP vehicle harness connector between VREF and Signal Return. • Is voltage reading between 4.0 and 6.0 volts?	Yes ▶	GO to DD4.
	No ▶	GO to Pinpoint Test Step C1.

EGR Valve Position Sensor (EVP) Control/Vent (EGRC/EGRV) — Pinpoint Test — DD

TEST STEP	RESULT ▶	ACTION TO TAKE
DD4 MEASURE CONTINUITY OF EVP SIGNAL CIRCUIT		
• Key Off, wait 10 seconds. • Harness disconnected from EVP sensor. • Disconnect processor 60 Pin connector and inspect for damaged pins, corrosion, loose wires. Service as necessary. • Install Breakout box leaving processor disconnected. • DVOM on 200 ohm scale. • Measure resistance between test Pin 27 at the Breakout box and EVP signal at the EVP vehicle harness connector. • Is resistance less than 5 ohms?	Yes ▶	GO to DD5.
	No ▶	SERVICE open circuit. REMOVE Breakout box. RECONNECT processor and EVP sensor. RERUN Quick Test.
DD5 CHECK EVP SIGNAL FOR SHORTS TO VREF AND SIGNAL RETURN		
• Key Off, harness disconnected from EVP sensor. Breakout box installed with processor not connected. • DVOM on 200,000 ohm scale. • Measure resistance between test Pin 27 and test Pins 26, 40, 46 and 60 at Breakout box. • Are all readings 10,000 ohms or greater?	Yes ▶	GO to DD6.
	No ▶	SERVICE short circuit. REMOVE Breakout box. RECONNECT processor and EVP sensor. RERUN Quick Test.
DD6 SUBSTITUTE EVP SENSOR AND EGR VALVE		
• Key Off, wait 10 seconds. • Electrically connect known good EVP sensor and EGR valve assembly. • Remove Breakout box. • Reconnect processor. • Perform Key On, Engine Off Quick Test. • Is code 31 present?	Yes ▶	REPLACE processor. CONNECT original EVP sensor and EGR valve assembly. RERUN Quick Test.
	No ▶	GO to DD7.

EGR Valve Position Sensor (EVP) Control/Vent (EGRC/EGRV) — Pinpoint Test DD

TEST STEP	RESULT ▶	ACTION TO TAKE
DD7 EVP SENSOR CHECK		
• Key Off, wait 10 seconds. • Install original EVP sensor on known good EGR valve. • EVP sensor connected. • Rerun Key On, Engine Off Quick Test. • Is code 31 present?	Yes ▶ No ▶	INSTALL new EVP sensor. RERUN Quick Test. REFER to EGR System, Section 6.
FAULT CODES 32, 33 AND 34		
DD11 OUTPUT STATE CHECK (REFER TO APPENDIX)		
NOTE: Do not use STAR tester for this test step. Use VOM/DVOM. • Key Off, wait 10 seconds. • DVOM on 20V scale. • Connect DVOM negative test lead to STO and positive test lead to battery positive. • Jumper STI to signal return. • Perform Key On, Engine Off Self-Test until the completion of the Continuous Test Codes. • DVOM will indicate zero volts. • Depress and release the throttle. • Did DVOM reading change to a high voltage reading?	No ▶ Yes ▶	DEPRESS throttle to WOT and RELEASE. If STO voltage does not go high, GO to Pinpoint Test Step Q40. Remain in output state check and GO to DD12.
DD12 CHECK EGR SOLENOIDS FOR ELECTRICAL CYCLING		
• Key On, Engine Off. • DVOM on 20V scale. • Reconnect DVOM to EGRV solenoid, between VPWR and EGRV signal. • While observing DVOM, depress and release the throttle several times to cycle output on and off. • Repeat for EGRC solenoid, between VPWR and EGRC signal. • Do both solenoid outputs cycle on and off?	Yes ▶ No ▶	REMAIN in output state check. GO to DD13. Exit output state check. GO to DD17.

EGR Valve Position Sensor (EVP) Control/Vent (EGRC/EGRV) — Pinpoint Test DD

TEST STEP	RESULT ▶	ACTION TO TAKE
DD13 CHECK EGR SOLENOIDS FOR VACUUM CYCLING		
• Key On, Engine Off. • Still in output state check. • Disconnect and cap vacuum line from bottom port of EGRC solenoid and connect a vacuum pump. • Connect a vacuum gauge in the common output (top) vacuum line to EGR valve. • Disconnect but do not cap vacuum vent line from EGRV solenoid or remove filter from the top of the EGRV solenoid. • Apply vacuum. • While cycling outputs on and off (by depressing and releasing throttle), observe vacuum gauge at the output. Maintain vacuum at source. • Does the vacuum cycle on and off in less than 2 seconds?	Yes ▶ No ▶	RECONNECT all vacuum lines. GO to DD14. CHECK filter and common output vacuum line for obstructions. REPLACE as necessary. If OK, REPLACE solenoid assembly. RECONNECT all vacuum lines. RERUN Quick Test.
DD14 CHECK VACUUM LINES		
• Key Off, wait 10 seconds. • Vacuum lines reconnected. • Check entire EEC vacuum line system per VECI emission schematic decal for kinks, cracks, obstructions or leaks. • Are vacuum lines OK? **NOTE FOR 3.8L CFI (ONLY): Special attention should be given to the canister purge valve. A canister purge valve that is stuck open will apply vacuum to the EGR valve.**	Yes ▶ No ▶	GO to DD15. SERVICE as necessary. RERUN Quick Test.

EGR Valve Position Sensor (EVP) Control/Vent (EGRC/EGRV) — Pinpoint Test DD

TEST STEP	RESULT ▶	ACTION TO TAKE
DD15 CHECK EVP RESISTANCE WHILE APPLYING VACUUM TO EGR VALVE		
• Key Off. • Disconnect vehicle harness from EVP sensor. Inspect for damaged pins, corrosion, and pins pushed out. Service as necessary. • DVOM on 200,000 ohm scale. • Disconnect vacuum line at EGR valve. • Connect vacuum pump to EGR valve. • Measure resistance of the EVP sensor between EVP Signal Pin and VREF Pin while increasing vacuum to 33 kPa (10 in.-Hg.). • Observe resistance as vacuum increases. • Does the resistance gradually change between 5500 and 100 ohms?	Yes ▶ No ▶	REPLACE processor. RECONNECT EVP sensor and EGR vacuum line. RERUN Quick Test. GO to DD16.
DD16 MANUALLY EXERCISE EVP SENSOR		
• Key Off, harness disconnected from EVP sensor. • Remove EVP sensor from EGR valve. • Measure resistance of the EVP sensor between EVP Signal Pin and VREF Pin while gradually applying pressure to EVP sensor shaft. • Observe resistance as shaft is slowly pushed in and slowly released. • Do either of the readings change suddenly between 5500 and 100 ohms? **NOTE: It is normal for the EVP sensor total resistance to drop below 100 ohms when disconnected from the EGR valve. A defective part will change resistance suddenly between 5500 and 100 ohms.**	Yes ▶ No ▶	REPLACE EVP sensor. RECONNECT harness and EGR supply vacuum line. RERUN Quick Test. REFER to EGR System, Section 6. RECONNECT EVP sensor and EGR supply vacuum line. RERUN Quick Test.
DD17 MEASURE EGRV/EGRC SOLENOID RESISTANCE		
• Key Off, wait 10 seconds. • DVOM on 200 ohm scale. • Disconnect EGRV solenoid connector and measure solenoid resistance. Inspect for damaged pins, corrosion and pins pushed out. Service as necessary. • Disconnect EGRC solenoid connector and measure solenoid resistance. Inspect for damaged pins, corrosion and pins pushed out. Service as necessary. • Are both resistances between 30 and 70 ohms?	Yes ▶ No ▶	CONNECT EGRC EGRV solenoids. GO to DD18. REPLACE EGRC EGRV solenoid assembly. RERUN Quick Test.

EGR Valve Position Sensor (EVP) Control/Vent (EGRC/EGRV) — Pinpoint Test DD

TEST STEP	RESULT ▶	ACTION TO TAKE
DD18 CHECK FOR VOLTAGE ON VEHICLE POWER CIRCUIT		
• Disconnect EGR vent and EGR control solenoids from harness. • Key On, Engine Off. • DVOM on 20V scale. • Measure voltage between battery negative terminal and VPWR circuit on both EGR solenoids. • Are both readings 10.5 volts or greater?	Yes ▶ No ▶	GO to DD19. SERVICE harness circuit open. RERUN Quick Test.
DD19 CHECK CONTINUITY OF EGRV AND EGRC CIRCUITS		
• Key Off, wait 10 seconds. • EGR vent and EGR control solenoid disconnected from harness. • Disconnect processor 60 Pin connector and inspect for damaged pins, corrosion, loose wires. Service as necessary. • Install Breakout box to processor harness connector. Leave processor disconnected. • DVOM on 200 ohm scale. • Measure resistance between test Pin 33 at the Breakout box and EGRV signal at the EGRV solenoid vehicle harness connector. • Measure resistance between test Pin 52 at the Breakout box and EGRC signal at the EGRC solenoid vehicle harness connector. • Are both readings less than 5 ohms?	Yes ▶ No ▶	GO to DD20. SERVICE open circuit. REMOVE Breakout box. RECONNECT harness to processor. RERUN Quick Test.
DD20 CHECK FOR SHORT TO GROUND		
• Key Off, wait 10 seconds. • DVOM on 200,000 ohm scale. • Leave Breakout box installed and processor disconnected. • EGRV/EGRC solenoids disconnected. • Measure resistance between test Pins 33 and/or 52 and test Pins 40, 46 and 60 at the Breakout box. • Is resistance less than 10,000 ohms?	Yes ▶ No ▶	SERVICE short to ground. RERUN Quick Test. GO to DD21.

EGR Valve Position Sensor (EVP) Control/Vent (EGRC/EGRV) — Pinpoint Test DD

TEST STEP	RESULT ▶	ACTION TO TAKE
DD21 CHECK EGRV AND EGRC SIGNALS FOR SHORTS TO POWER		
• EGR vent and EGR control solenoids disconnected from harness. • Key Off, Breakout box installed. Processor disconnected. • DVOM on 200,000 ohm scale. • Measure resistance between test Pin 33 and test Pins 37 and 57 at the Breakout box. • Measure resistance between test Pin 52 and test Pins 37 and 57 at the Breakout box. • Are all readings 10,000 ohms or greater?	Yes ▶	REPLACE processor. REMOVE Breakout box. RECONNECT harness to processor. RERUN Quick Test.
	No ▶	SERVICE circuit short. REMOVE Breakout box. RECONNECT harness to processor. RERUN Quick Test. If code is repeated, REPLACE processor.
FAULT CODE 35		
DD30 RPM TOO LOW FOR EGR TEST		
• Is code 12 also present?	Yes ▶	Vehicles equipped with air bypass (EFI), GO to KE1. Vehicles equipped with DC motor control, GO to KF1.
	No ▶	GO to DD31.
DD31 RETEST AT 1,500 RPM		
• Key Off, wait 10 seconds. • Install tachometer. • Perform Key On, Engine Running Quick Test while maintaining 1,500 rpm. • Record Engine Running service codes. • Is code 35 still present?	Yes ▶	REPLACE processor. RERUN Quick Test.
	No ▶	RERUN Quick Test. SERVICE codes as necessary.

EGR Valve Position Sensor (EVP) Control/Vent (EGRC/EGRV) — Pinpoint Test DD

TEST STEP	RESULT ▶	ACTION TO TAKE
DD90 SERVICE CODE 31 CONTINUOUS TEST: EXERCISE EVP SENSOR		
• Using continuous monitor mode, observe VOM or STAR LED for indication of a fault while performing the following: • Connect a vacuum pump to the EGR valve. • Very slowly apply 20 kPa (6 in.-Hg.) vacuum to the EGR valve. • Slowly bleed vacuum off the EGR valve and lightly tap on EVP sensor (simulate road shock). • Wiggle EVP sensor connector. • Is a fault indicated?	Yes ▶	GO to DD91.
	No ▶	GO to DD92.

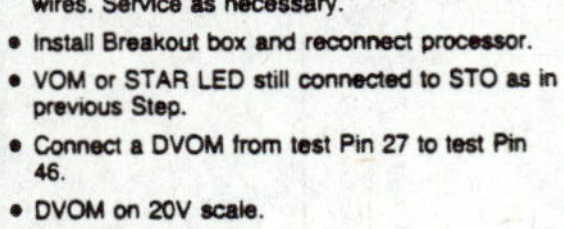

TEST STEP	RESULT ▶	ACTION TO TAKE
DD91 MEASURE EVP SIGNAL VOLTAGE WHILE EXERCISING EVP SENSOR		
• Key Off, wait 10 seconds. • Disconnect processor 60 Pin connector and inspect for damaged pins, corrosion, loose wires. Service as necessary. • Install Breakout box and reconnect processor. • VOM or STAR LED still connected to STO as in previous Step. • Connect a DVOM from test Pin 27 to test Pin 46. • DVOM on 20V scale. • Key On, Engine Off. • While observing DVOM, repeat Step DD90. • Does the fault occur below 4.25V?	Yes ▶	DISCONNECT and INSPECT connector. If connector and terminals are good, REPLACE EVP sensor. RERUN Quick Test.
	No ▶	EGR valve overshoot may have caused continuous code 31. Sensor service is not required. To verify harness integrity, GO to DD92.

EGR Valve Position Sensor (EVP) Control/Vent (EGRC/EGRV) — Pinpoint Test DD

TEST STEP	RESULT ▶	ACTION TO TAKE
DD92 CHECK EEC-IV HARNESS		
• Observe VOM or STAR LED for a fault indication while performing the following: • Referring to the illustration in Step DD90, grasp the harness closest to the sensor connector. Wiggle, shake or bend a small section of the EEC-IV system harness while working your way to the dash panel. Also wiggle, shake or bend the EEC-IV harness from the dash panel to the processor. • Is a fault indicated?	Yes ▶	ISOLATE fault and SERVICE as necessary. REFER to appropriate figure RERUN Quick Test.
	No ▶	GO to DD93.
DD93 CHECK PROCESSOR AND HARNESS CONNECTORS		
• Key Off, wait 10 seconds. • Disconnect processor 60 Pin connector. • Inspect both connectors and connector terminals for obvious damage or faults. • Are connectors and terminals OK?	No ▶	SERVICE as necessary. RERUN Quick Test.
	Yes ▶	Unable to duplicate fault at this time. Continuous code 31 testing complete.

Engine Coolant Temperature Sensor (ECT) — Pinpoint Test DE

DE

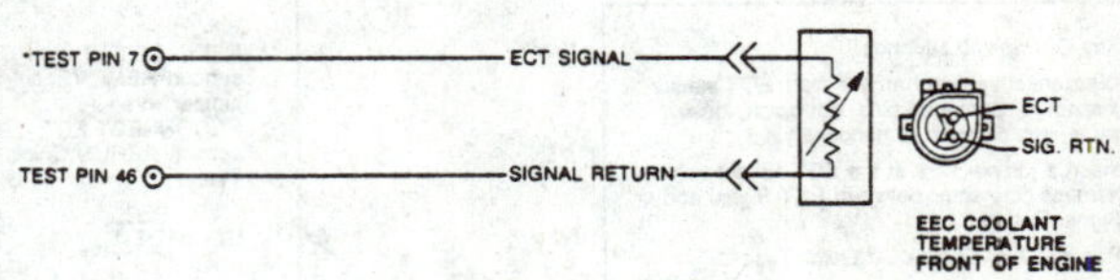

NOTE: TO PASS THIS TEST, ENGINE COOLANT TEMPERATURE MUST BE:
KEY ON, ENGINE OFF (50°F TO 240°F)
ENGINE RUNNING (180°F TO 240°F)

TYPICAL RESISTANCE BETWEEN TEST PINS 7 & 46	58,750 ohms	40,500 ohms	3600 ohms	1840 ohms
AT TEMPERATURE	50°F	65°F	180°F	220°F

*TEST PINS LOCATED ON BREAKOUT BOX.
ALL HARNESS CONNECTORS VIEWED INTO MATING SURFACE.

STOP-WARNING

You should enter this Pinpont Test only when a service code 21, 51 or 61 is received in Quick Test Step 3.0, 5.0 or 6.0.

To prevent the replacement of good components, be aware that the following non-EEC areas may be at fault:

- Coolant level.
- Oil level.
- Blocked or obstructed air flow.
- Engine not at normal operating temperature.
- Electro drive cooling fan.
- Open thermostat.

This pinpoint test is intended to diagnose only the following:

- ECT sensor.
- Harness sensor circuits: ECT and Signal Return.
- Processor assembly.

Engine Coolant Temperature Sensor (ECT) — Pinpoint Test — DE

TEST STEP	RESULT ▶	ACTION TO TAKE
DE1 SERVICE CODE 21: CHECK ENGINE OPERATING TEMPERATURE		
• Run engine for 2 minutes at 2,000 rpm. • Check that upper radiator hose is hot and pressurized. • Rerun Quick Test. • Is code 21 present?	Vehicle stalls ▶	Do not service code 21 at this time. REFER to diagnosis by symptoms.
	Yes ▶	GO to DE2.
	No ▶	SERVICE other codes as necessary.
DE2 CHECK FOR VREF AT THROTTLE POSITION SENSOR		
• Refer to illustration Q. • Key Off, wait 10 seconds. • DVOM on 20V scale. • Disconnect TP sensor. • Key On, Engine Off. • Measure voltage at the TP vehicle harness connector between VREF and signal return. • Is voltage reading between 4.0 and 6.0 volts?	Yes ▶	RECONNECT TP sensor, GO to DE3.
	No ▶	GO to Pinpoint Test Step C1.
DE3 ECT SENSOR CHECK		
NOTE: Engine may have cooled down. Always warm engine before taking ECT resistance measurement. Check for open thermostat. • Key Off, wait 10 seconds. • Harness disconnected from ECT sensor. • DVOM on 200,000 ohm scale. • Measure resistance of the ECT sensor. • Is the resistance reading: — 1300 ohms (240°F) to 7700 ohms (140°F) for engine off? — 1550 ohms (230°F) to 4550 ohms (180°F) for engine running?	Yes ▶	REPLACE processor. RECONNECT harness to ECT sensor. RERUN Quick Test.
	No ▶	REPLACE ECT sensor. RECONNECT harness to ECT sensor. RERUN Quick Test.

Engine Coolant Temperature Sensor (ECT) — Pinpoint Test — DE

TEST STEP	RESULT ▶	ACTION TO TAKE
DE10 SERVICE CODE 51: INDUCE OPPOSITE CODE		
• Key Off, wait 10 seconds. • Disconnect vehicle harness from ECT sensor. Inspect for damaged pins, corrosion, loose wires, etc. Service as necessary. • Insert a jumper wire at the ECT sensor vehicle harness connector between ECT Signal and Signal Return. • Run Key On, Engine Off Quick Test. • Is code 61 present?	Yes ▶	REPLACE ECT sensor. REMOVE jumper wire. RECONNECT ECT sensor. RERUN Quick Test.
	No ▶	GO to DE11.
DE11 CHECK CONTINUITY OF ECT SIGNAL AND SIGNAL RETURN		
• Key Off, wait 10 seconds. • Harness disconnected from ECT sensor, jumper wire removed. • Disconnect processor 60 pin connector. Inspect for damaged pins, corrosion, loose wires, etc. Service as necessary. • Install Breakout box to harness leaving processor disconnected. • DVOM on 200 ohm scale. • Measure resistance between ECT signal at the ECT vehicle harness connector and test Pin 7 at the Breakout box. • Measure resistance between Signal Return at the ECT sensor vehicle harness connector, and test Pin 46 at the Breakout box. • Are both readings less than 5 ohms?	Yes ▶	REPLACE processor. REMOVE Breakout box. RECONNECT processor and ECT sensor. RERUN Quick Test.
	No ▶	SERVICE open circuit(s). REMOVE Breakout box. RECONNECT processor and ECT sensor. RERUN Quick Test.

Engine Coolant Temperature Sensor (ECT) — Pinpoint Test — DE

TEST STEP	RESULT ▶	ACTION TO TAKE
DE20 SERVICE CODE 61: INDUCE OPPOSITE CODE		
• Key Off, wait 10 seconds. • Disconnect vehicle harness from ECT sensor. Inspect for damaged pins, corrosion, loose wires, etc. Service as necessary. • Run Key On, Engine Off Quick Test. • Is code 51 present?	Yes ▶	REPLACE ECT sensor. RECONNECT ECT sensor. RERUN Quick Test.
	No ▶	GO to DE21.
DE21 CHECK FOR VREF AT THROTTLE POSITION SENSOR		
• Refer to illustration Q. • Key Off, wait 10 seconds. • DVOM on 20V scale. • Disconnect TP sensor. • Key On, Engine Off. • Measure voltage at the TP vehicle harness connector between VREF and signal return. • Is voltage reading between 4.0 and 6.0 volts?	Yes ▶	RECONNECT TP sensor, GO to DE22.
	No ▶	GO to Pinpoint Test Step C1.
DE22 CHECK ECT SIGNAL FOR SHORT		
• Key Off, wait 10 seconds. • Harness disconnected from ECT sensor. • Disconnect processor 60 pin connector. Inspect for damaged pins, corrosion, loose wires, etc. Service as necessary. • Install Breakout box, leave processor disconnected. • DVOM on 200,000 ohm scale. • Measure resistance between test Pin 7 and test Pins 40, 46 and 60 at the Breakout box. • Are resistance readings 10,000 ohms or greater?	Yes ▶	REPLACE processor. REMOVE Breakout box. RECONNECT processor and ECT harness. RERUN Quick Test.
	No ▶	SERVICE circuit shorts. REMOVE Breakout box. RECONNECT processor and ECT sensor. RERUN Quick Test.

Engine Coolant Temperature Sensor (ECT) — Pinpoint Test — DE

TEST STEP	RESULT ▶	ACTION TO TAKE
DE90 SERVICE CODE 21: CONTINUOUS TEST: TEST DRIVE VEHICLE		
• Key Off and wait 10 seconds. • Disconnect all Self-Test equipment and prepare vehicle for test drive. • Drive vehicle. Try to simulate different drive modes or mode in which drive complaint is noticed. Attempt to maintain drive complaint mode for one minute or more, if possible. • Upon completion of drive evaluation, repeat Key On, Engine Off Self-Test. • Is code 21 present in the continuous test results?	Yes ▶	VERIFY thermostat operating properly. If OK, REPLACE ECT sensor. RERUN Quick Test.
	No ▶	Unable to duplicate fault. Code 21 testing complete.

Engine Coolant Temperature (ECT) — Pinpoint Test — DE

TEST STEP	RESULT ▶	ACTION TO TAKE
DE91 SERVICE CODE 51: CONTINUOUS TEST: CHECK ECT SENSOR		
• Using continuous monitor mode, observe VOM or STAR LED for indication of a fault while performing the following: • Lightly tap on ECT sensor (simulate road shock). • Wiggle ECT connector. • Is a fault indicated?	Yes ▶	DISCONNECT and INSPECT connectors. If connector and terminals are good, REPLACE ECT sensor. RERUN Quick Test.
	No ▶	GO to DE92.
DE92 CHECK EEC-IV HARNESS		
• Observe VOM or STAR LED for a fault indication while performing the following: • Referring to the illustration in Step DE91, grasp the harness closest to the sensor connector. Wiggle, shake or bend a small section of the EEC-IV system harness while working your way to the dash panel. Also wiggle, shake or bend the EEC-IV harness from the dash panel to the processor. • Is fault indicated?	Yes ▶	ISOLATE fault and make necessary repairs. RERUN Quick Test.
	No ▶	GO to DE93.
DE93 CHECK PROCESSOR AND HARNESS CONNECTORS		
• Key Off, wait 10 seconds. • Disconnect processor 60 pin connector. • Inspect both connectors and connector terminals for obvious damage or faults. • Connectors and terminals are OK.	No ▶	SERVICE as necessary. RERUN Quick Test.
	Yes ▶	Unable to duplicate fault at this time. Continuous code 51 testing complete.

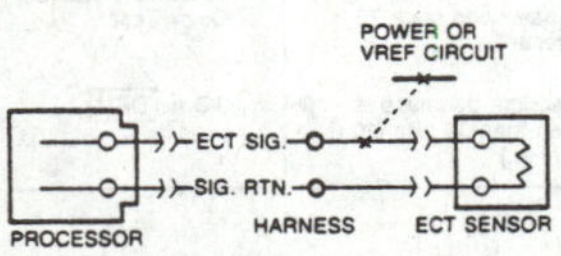

Engine Coolant Temperature (ECT) — Pinpoint Test — DE

TEST STEP	RESULT ▶	ACTION TO TAKE
DE94 SERVICE CODE 61: CONTINUOUS TEST: CHECK ECT SENSOR		
• Using continuous monitor mode, observe VOM or STAR LED for indication of a fault while performing the following: • Lightly tap on ECT sensor (simulate road shock). • Wiggle ECT connector. Voltage greater than 5 volts or STAR LED Off. • Is fault indicated?	Yes ▶	DISCONNECT and INSPECT connectors. If connector and terminals are good, REPLACE ECT sensor. RERUN Quick Test.
	No ▶	GO to DE95.
DE95 CHECK EEC-IV HARNESS		
• Observe VOM or STAR LED for a fault indication while performing the following: • Referring to the illustration in Step DE94, grasp the harness closest to the sensor connector. Wiggle, shake or bend a small section of the EEC-IV system harness while working your way to the dash panel. Also wiggle, shake or bend the EEC-IV harness from the dash panel to the processor. • Is fault indicated?	Yes ▶	ISOLATE fault and make necessary repairs. RERUN Quick Test.
	No ▶	GO to DE96.
DE96 CHECK PROCESSOR AND HARNESS CONNECTORS		
• Key Off, wait 10 seconds. • Disconnect processor 60 pin connector. • Inspect both connectors and connector terminals for obvious damage or faults. • Connectors and terminals are OK.	No ▶	SERVICE as necessary. RERUN Quick Test.
	Yes ▶	Unable to duplicate fault at this time. Continuous code 61 testing complete.

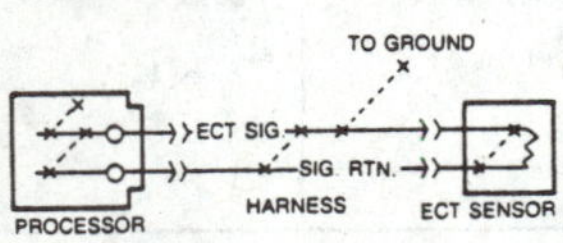

Manifold Absolute Pressure (MAP)/ Barometric Pressure (BP) Sensor — Pinpoint Test — DF

DF

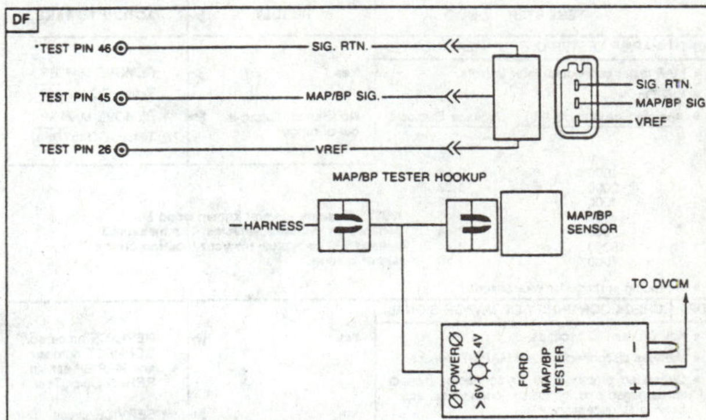

*TEST PINS LOCATED ON BREAKOUT BOX.
ALL HARNESS CONNECTORS VIEWED INTO MATING SURFACE.

STOP-WARNING

You should enter this Pinpoint Test only when a Service Code 22 or 72 is received in Quick Test Step 3.0, 5.0 or 6.0 or when directed here from Pinpoint Test S or Diagnostics by Symptom in the Engine Supplement Section.

To prevent the replacement of good components, be aware that the following non-EEC areas may be at fault:

- Unusually high/low atmospheric barometer reading (MAP/BP).
- Kinked or obstructed vacuum lines (MAP).
- Basic engine (valves, vacuum leaks, timing, EGR valve (MAP), etc.).
- Key On, Engine Off — MAP sensor must not see vacuum.
- Engine Running — MAP sensor must see actual manifold vacuum.

This Pinpoint Test is intended to diagnose only the following:

- MAP/BP sensor.
- Harness circuits: VREF, MAP/BP Signal, and Signal Return.
- Processor assembly.
- MAP vacuum line.

Manifold Absolute Pressure (MAP)/ Barometric Pressure (BP) Sensor — Pinpoint Test — DF

TEST STEP	RESULT ▶	ACTION TO TAKE
FAULT CODE 22, ENGINE OFF		
DF1 CONNECTING MAP/BP TESTER		
• Key Off. • Disconnect the MAP/BP sensor from the vehicle harness. • Connect the MAP/BP tester between the vehicle harness and the MAP/BP sensor. • Insert tester banana plugs into DVOM. • Set DVOM to 20V scale. • Refer to illustration DF.	Tester properly hooked up ▶	GO to DF2.
DF2 POWER TO MAP/BP SENSOR TEST		
• MAP/BP tester connected. • Key On. • Observe red and green lights.	(ONLY) Green light, Vref is OK ▶	GO to DF4.
	"Less than 4V" light (RED) or no lights, Vref is too low OR "Greater than 6V" light (Red), Vref is too high ▶	GO to DF3.
DF3 VREF ISOLATION		
• MAP/BP tester connected. • Key On. • Disconnect MAP/BP • Observe red and green lights.	(ONLY) Green light, Vref is OK ▶	REPLACE MAP/BP sensor. RERUN Quick Test.
	"Less than 4V" light (RED) or no lights, Vref is too low OR "Greater than 6V" light (Red), Vref is too high ▶	REMOVE MAP/BP tester. GO to Pinpoint Test Step C1.

Manifold Absolute Pressure (MAP)/ Barometric Pressure (BP) Sensor — Pinpoint Test — DF

TEST STEP	RESULT ▶	ACTION TO TAKE
DF4 MAP/BP TESTER OUTPUT READING		
• MAP tester connected, refer to Note. • Key On. • Approximate Altitude (Ft.) / Voltage Output (+/− .04 Volts): 0 — 1.59; 1000 — 1.56; 2000 — 1.53; 3000 — 1.50; 4000 — 1.47; 5000 — 1.44; 6000 — 1.41; 7000 — 1.39 **NOTE: Measure several known good MAP sensors on available vehicles. The measured voltage will be typical for your location on the day of testing.** • Is reading in range for your altitude?	Yes ▶ No (Sensor output is out-of-range) ▶	REMOVE MAP/BP Tester. GO to DF5. REMOVE MAP/BP Tester. GO to DF6.
DF5 CHECK CONTINUITY OF MAP/BP SIGNAL		
• Key Off, wait 10 seconds. • Harness disconnected from MAP/BP sensor. • Disconnect processor 60 Pin connector. Inspect for damaged pins, corrosion, loose wires, etc. Service as necessary. • Install Breakout box. Leave processor disconnected. • DVOM on 200 ohm scale. • Measure resistance between MAP/BP signal at the MAP/BP sensor vehicle harness connector and test Pin 45 at the Breakout box. • Is resistance reading 5.0 ohms or less?	Yes ▶ No ▶	REPLACE processor. CONNECT harness and MAP/BP sensor. RERUN Quick Test. SERVICE circuit opens. REMOVE Breakout box. RECONNECT processor and MAP/BP sensor. RERUN Quick Test.
DF6 CHECK MAP/BP SIGNAL FOR SHORTS TO VREF: SIGNAL RETURN AND GROUND		
• Key Off, wait 10 seconds. • Disconnect processor 60 Pin connector and inspect for damaged pins, corrosion, loose wires. Service as necessary. • Install Breakout box, leave processor disconnected. • Harness disconnected from MAP/BP sensor. • DVOM on 200,000 ohm scale. • Measure resistance between test Pin 45 and test Pins 26, 46, 40 and 60 at the Breakout box. • Are all resistance readings 10,000 ohms or greater?	Yes ▶ No ▶	REPLACE MAP/BP sensor. REMOVE Breakout box. RECONNECT electrical connections. RERUN Quick Test. SERVICE circuit shorts. REMOVE Breakout box. RECONNECT processor and MAP/BP Sensor. RERUN Quick Test.

Manifold Absolute Pressure (MAP)/ Barometric Pressure (BP) Sensor — Pinpoint Test — DF

TEST STEP	RESULT ▶	ACTION TO TAKE
CODE 22 ENGINE RUNNING		
DF7 CHECK FOR EGR CODES		
• Are service codes 31, 32, 33, 34 or 35 present?	Yes ▶ No ▶	GO to Quick Test Step 5.0 for appropriate Pinpoint Test. GO to DF8.
DF8 CHECK MAP SENSOR		
• Key Off, wait 10 seconds. • Disconnect vacuum line from MAP sensor. • Install vacuum pump to MAP sensor. • Apply 18 in.-Hg. vacuum to MAP sensor. • Does MAP sensor hold vacuum?	Yes ▶ No ▶	RELEASE vacuum. GO to DF9. REPLACE MAP sensor. CONNECT vacuum line to MAP sensor. RERUN Quick Test.
DF9 ATTEMPT TO ELIMINATE CODE 22 (ENGINE RUNNING)		
• Key Off, wait 10 seconds. • Plug MAP vacuum supply hose. • Start engine and maintain 1500 ± 100 engine rpm. • Slowly apply 15 in.-Hg. vacuum to MAP sensor. • While maintaining rpm, perform Engine Running Quick Test. • Is code 22 still present? **NOTE: Disregard any other codes at this time.**	Yes ▶ No ▶	REPLACE MAP sensor. CONNECT vacuum line to MAP sensor. RERUN Quick Test. INSPECT vacuum supply hose to MAP sensor. SERVICE as necessary. If OK, SERVICE other engine running codes. If none, GO to Diagnostic Routines, Section 2 for a low vacuum problem.

Manifold Absolute Pressure (MAP)/ Barometric Pressure (BP) Sensor — Pinpoint Test — DF

TEST STEP	RESULT ▶	ACTION TO TAKE
FAULT CODE 72		
DF10 CHECK THAT VACUUM TO MAP SENSOR DECREASES DURING DYNAMIC RESPONSE		
• Key Off, wait 10 seconds. • Tee a vacuum gauge in the intake manifold vacuum line at the MAP sensor. • Perform Engine Running Quick Test while observing vacuum. • Record engine service codes. • Did vacuum decrease by more than 30 kPa (10 in.-Hg.) during dynamic response test? • Is code 72 present?	Vacuum decrease is 30 kPa (10 in.-Hg.) or greater and code 72 is not present ▶ Vacuum decrease is 30 kPa (10 in.-Hg.) or greater and code 72 is present ▶ Vacuum decrease is less than 30 kPa (10 in.-Hg.) ▶	DISCONNECT vacuum equipment and SERVICE other codes as necessary. REPLACE MAP sensor and RERUN Quick Test. GO to DF11.
DF11 CHECK VACUUM LINES		
• Check vacuum line for proper routing. Refer to VECI decal. Check MAP sensor vacuum line for kinks or blockage. • Vacuum lines are OK.	Yes ▶ No ▶	EEC-IV system OK. REFER to Shop Manual, Group 21 for probable subjects affecting engine vacuum. SERVICE as necessary and REPEAT DF10.
DF20 CONNECTING MAP/BP TESTER		
• Key Off. • Disconnect the MAP/BP sensor from the vehicle harness. • Connect the MAP/BP Tester between the vehicle harness and the MAP/BP sensor. • Plug tester banana plugs into DVOM. • Set DVOM to 20V scale. • Refer to illustration DF.	Tester properly hooked up ▶	GO to DF21.

Manifold Absolute Pressure (MAP)/ Barometric Pressure (BP) Sensor — Pinpoint Test — DF

TEST STEP	RESULT ▶	ACTION TO TAKE
DF21 MAP/BP TESTER OUTPUT READING		
• MAP Tester connected, refer to Note. • Key On. • Approximate Altitude (Ft.) / Voltage Output (+/− .04 Volts): 0 — 1.59; 1000 — 1.56; 2000 — 1.53; 3000 — 1.50; 4000 — 1.47; 5000 — 1.44; 6000 — 1.41; 7000 — 1.39 **NOTE: Measure several known good MAP sensors on available vehicles. The measured voltage will be typical for your location on the day of testing.** • Is reading in range for your altitude?	Yes ▶ No (Sensor output is out-of-range) ▶	For 1.9L EFI and 2.3L EFI TC engines GO to Diagnostic by Symptom in the Engine Supplement Section. For all others, GO to DF22. REPLACE MAP/BP sensor.
DF22 VACUUM LINE CHECK		
• Check MAP sensor vacuum line for holes, disconnections, kinks or blockage. • Are vacuum lines OK?	Yes ▶ No ▶	GO to Diagnostics by Symptom in the Engine Supplement Section. SERVICE vacuum lines to MAP sensor. RERUN Quick Test.

Manifold Absolute Pressure (MAP)/Barometric Pressure (BP) Sensor — Pinpoint Test DF

TEST STEP	RESULT ▶	ACTION TO TAKE
DF90 SERVICE CODE 22: CONTINUOUS TEST: EXERCISE MAP SENSOR		
• Using continuous monitor mode, observe VOM or STAR LED for indication of a fault while performing the following: • Connect a vacuum pump to the MAP sensor. • Slowly apply 84 kPa (25 in.-Hg.) vacuum to the sensor. • Slowly bleed vacuum off the MAP sensor. • Lightly tap on MAP sensor (simulate road shock). • Wiggle MAP connector. • Is fault indicated? 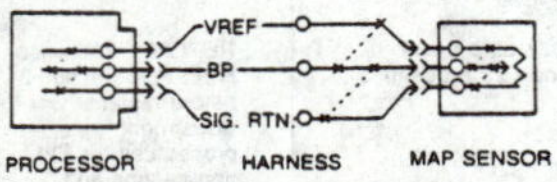	Yes ▶	DISCONNECT and INSPECT connectors. If connector and terminals are good, REPLACE sensor. RERUN Quick Test.
	No ▶	GO to DF91.
DF91 CHECK EEC-IV HARNESS		
• Observe VOM or STAR LED for a fault indication while performing the following: • Referring to the illustration in Step DF90, grasp the harness closest to the sensor connector. Wiggle, shake or bend a small section of the EEC-IV system harness while working your way to the dash panel. Also wiggle, shake or bend the EEC-IV harness from the dash panel to the processor. • Is a fault indicated?	Yes ▶	ISOLATE fault and SERVICE as necessary. RERUN Quick Test.
	No ▶	GO to DF92.
DF92 CHECK PROCESSOR AND HARNESS CONNECTORS		
• Key Off, wait 10 seconds. • Disconnect processor 60 Pin connector. • Inspect both connectors and connector terminals for obvious damage or faults. • Are connectors and terminals OK?	No ▶	SERVICE as necessary. RERUN Quick Test.
	Yes ▶	Unable to duplicate fault at this time. Continuous code 22 testing complete.

Knock Sensor — Pinpoint Test DG

DG

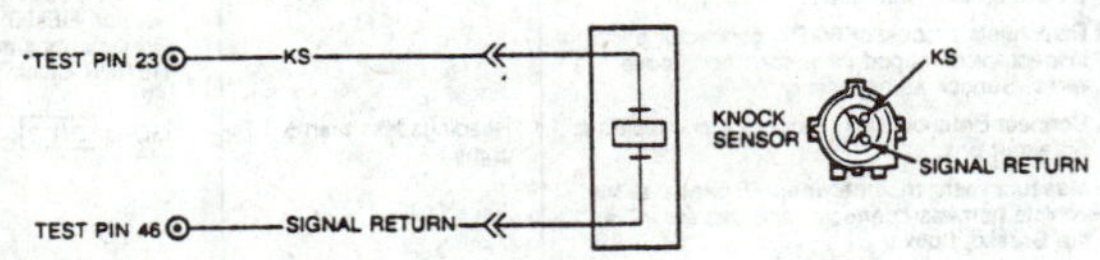

*TEST PINS LOCATED ON BREAKOUT BOX.
ALL HARNESS CONNECTORS VIEWED INTO MATING SURFACE.

STOP-WARNING

You should enter this Pinpoint Test only when a service code 25 is received in Quick Test Step 5.0 or you are directed here from Diagnostics by Symptom in the Engine Supplement Section.

To prevent the replacement of good components, be aware that the following non-EEC areas may be at fault:

- Fuel (quality).
- Basic engine.
- Spark timing.

This Pinpoint Test is intended to diagnose only the following:

- Knock sensor.
- Harness circuits: KS and Signal Return.
- Processor assembly.

Knock Sensor — Pinpoint Test DG

TEST STEP	RESULT ▶	ACTION TO TAKE
DG1 SERVICE CODE 25: GENERATE KNOCK MANUALLY		
NOTE: With knock conditions sensitive to fuel, altitude and weather, perform Step DG1 before servicing any components. • Prepare vehicle to run (Engine Running Self-Test). • Equipment needed: 4 oz. hammer. • Prepare to rap/tap on exhaust manifold, directly above the knock sensor, when the Dynamic Response Signal is given. **NOTE: There is no need to actually depress throttle at this point.** • Perform Engine Running Self-Test. • Rap moderately on exhaust manifold when meter indicates Dynamic Response Test is ready. • 15 seconds later a code will be generated. • Check for code 25. **NOTE: Ignore all other codes at this point.**	No ▶	Knock system OK. REPEAT Engine Running Self-Test and SERVICE any other codes from that test.
	Yes ▶ **NOTE: Service code 25 may be received whenever the engine is not tapped**	GO to DG2.
DG2 TEST KNOCK CIRCUIT FOR VOLTAGE		
• Key Off, wait 10 seconds. • Disconnect knock sensor connector and inspect. • Set DVOM on 20V scale. • Key On, Engine Off. • Measure voltage at the vehicle harness connector between KS and signal return.	Voltage is between 1 and 4V ▶	GO to DG6.
	Voltage is less than 1V ▶	GO to DG3.
	Voltage is greater than 4V ▶	GO to DG5.
DG3 CHECK CONTINUITY OF KS AND SIGNAL RETURN CIRCUITS		
• Key Off, wait 10 seconds. • Disconnect processor 60 Pin connector and inspect for damaged pins, corrosion, loose wires. Service as necessary. • Connect Breakout box to harness. Leave processor disconnected. • Knock sensor disconnected. • DVOM on 200 ohm scale. • Measure resistance between Signal Return at the vehicle harness and test Pin 46 at the Breakout box and between KS at the vehicle harness and test Pin 23 at the Breakout box.	Both resistances less than 5 ohms ▶	GO to DG4.
	Either resistance 5 ohms or greater ▶	SERVICE open circuit. RERUN Quick Test.

Knock Sensor — Pinpoint Test DG

TEST STEP	RESULT ▶	ACTION TO TAKE
DG4 CHECK KS CIRCUIT FOR SHORT TO GROUND		
• Key Off, wait 10 seconds. • Breakout box installed. • Processor disconnected. • Knock sensor disconnected. • DVOM on 200,000 ohm scale. • Measure resistance between KS at the vehicle harness and test Pins 40, 46 and 60 at the Breakout box.	All resistances 10,000 ohms or greater ▶	GO to DG6.
	Any resistance less than 10,000 ohms ▶	SERVICE harness short. RERUN Quick Test.
DG5 CHECK KS CIRCUIT FOR SHORT TO VOLTAGE		
• Key Off, wait 10 seconds. • Disconnect processor 60 Pin connector and inspect for damaged pins, corrosion, loose wires. Service as necessary. • Connect Breakout box to harness. Leave processor disconnected. • Knock sensor disconnected. • Key On, Engine Off. • DVOM on 20V scale. • Measure voltage between test Pin 23 and test Pin 40 at the Breakout box.	0.5V or greater ▶	SERVICE harness short to power. RERUN Quick Test.
	Less than 0.5V ▶	GO to DG6.
DG6 TEST PROCESSOR WITH SUBSTITUTE KNOCK SENSOR		
• Key Off, wait 10 seconds. • Remove Breakout box and reconnect processor. • Equipment Required: — Equivalent knock sensor (same part number). — 4 oz. hammer. • Plug substitute sensor in harness (do not install). • Perform Engine Running Self-Test. • Lightly tap knock sensor when Dynamic Test Ready Signal is given. • 15 seconds later, a code will be generated. • Check for code 25. **NOTE: Ignore all other codes at this time.**	No ▶	INSTALL new knock sensor. RERUN Quick Test.
	Yes ▶	REPLACE processor and REMOVE substitute sensor. RERUN Quick Test with original sensor.

Throttle Position Sensor (TPS) — Pinpoint Test — DH

DH

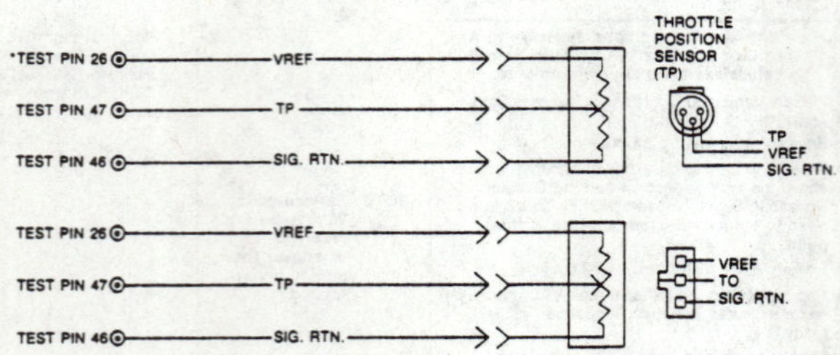

*TEST PINS LOCATED ON BREAKOUT BOX.
ALL HARNESS CONNECTORS VIEWED INTO MATING SURFACE.

STOP-WARNING

You should enter this Pinpoint Test only when a Service Code 23, 53, 63 or 73 is received in Quick Test Step 3.0, 5.0 or 6.0.

To prevent the replacement of good components, be aware that the following non-EEC areas may be at fault:

- Idle speeds/throttle stop adjustment.
- Binding throttle shaft/linkage or speed control linkage.
- Choke/high cam system, if equipped.

This Pinpoint Test is intended to diagnose only the following:

- TP sensor.
- Sensor harness circuits: VREF, TP Signal, and Signal Return.
- Processor assembly.

Throttle Position Sensor (TPS) — Pinpoint Test — DH

TEST STEP	RESULT ▶	ACTION TO TAKE
DH1 SERVICE CODE 23: THE FOLLOWING CHECK MUST BE MADE BEFORE SERVICING THIS CODE		
• Check for code 68: Key On, Engine Off or codes 58, 31 or 41 Engine Running.	Code(s) present ▶	DISREGARD Code 23 at this time and RETURN to Quick Test Step 3.0B for code 68 or Step 5.0B for codes 31, 41 or 58. PROCEED as directed.
	No codes present ▶	GO to DH2.
DH2 CHECK FOR STUCK THROTTLE PLATE		
• Visually inspect carburetor/throttle body and throttle linkage for binding or sticking. • Verify the throttle linkage is at mechanical/closed throttle. Check for: binding throttle linkage, speed control linkage, vacuum line/electrical harness interference, etc.	Throttle not stuck ▶ Throttle stuck ▶	GO to DH3. CORRECT faults. RERUN Quick Test.
DH3 GENERATE CODE 63		
• Refer to illustration DH. • Key Off, wait 10 seconds. • Disconnect TP sensor vehicle harness connector at the throttle body. Inspect for damaged pins, corrosion, loose wires, etc. Service as necessary. • Perform Key On, Engine Off self-test and record codes. **NOTE: Ignore all other codes at this time.**	Code 63 present ▶ Code 63 not present ▶	GO to DH4. GO to DH5.
DH4 CHECK VOLTAGE VREF TO SIGNAL RETURN		
• Refer to illustration DH. • Key Off, wait 10 seconds. • Disconnect TP vehicle harness connector at throttle body. Inspect for damaged pins, corrosion, and pins pushed out. Service as necessary. • DVOM on 20V scale. • Key On, Engine Off. • Measure voltage at the TP vehicle harness connector between VREF and Signal Return.	Voltage reading between 4V and 6V ▶ Voltage reading less than 4V or greater than 6V ▶	REPLACE TP sensor. REFER to Section 3 before replacement for adjustment procedures for EFI applications. RERUN Quick Test. GO to Pinpoint Test Step C1.

Throttle Position Sensor (TPS) — Pinpoint Test — DH

TEST STEP	RESULT ▶	ACTION TO TAKE
DH5 CHECK TP SIGNAL FOR SHORT TO POWER		
• Key Off, wait 10 seconds. TP harness disconnected. • DVOM on 200,000 ohm scale. • Disconnect processor 60 Pin connector and inspect for damaged pins, corrosion, loose wires. Service as necessary. • Install Breakout box, leave processor disconnected. • Measure resistance between test Pin 47 and test Pins 26 and 57 at the Breakout box.	Either resistance is less than 10,000 ohms ▶ Both resistances are 10,000 ohms or greater ▶	SERVICE harness short. RERUN Quick Test. REPLACE processor. RERUN Quick Test.
DH10 SERVICE CODE 63: GENERATE CODE 53		
• Key Off, wait 10 seconds, TP harness disconnected. • Jumper VREF to TP signal at TP vehicle harness connector. • Perform Key On Engine Off self-test. **NOTE: If no codes are generated, immediately remove jumper and go directly to DH13.** • Check for Code 53 (Code 23) **NOTE: Ignore all other codes at this time.**	Code 53 present (Code 23 present) ▶ Code 53 not present (Code 23 not present) ▶	REPLACE TP sensor. REFER to Section 3 before replacement for adjustment procedures for EFI applications and RERUN Quick Test. GO to DH11.
DH11 SERVICE CODE 63: CHECK VOLTAGE VREF TO SIGNAL RETURN		
• Refer to illustration DH. • Key Off, wait 10 seconds. • Disconnect TP vehicle harness connector at throttle body. Inspect for damaged pins, corrosion, and pins pushed out. Service as necessary. • DVOM on 20V scale. • Key On Engine Off. • Measure voltage at the TP vehicle harness connector between VREF and Signal Return.	Voltage reading between 4V and 6V ▶ Voltage reading less than 4V or greater than 6V ▶	GO to DH12. GO to Pinpoint Test Step C1.

Throttle Position Sensor (TPS) — Pinpoint Test — DH

TEST STEP	RESULT ▶	ACTION TO TAKE
DH12 CHECK CONTINUITY OF TP CIRCUIT		
• Key Off, wait 10 seconds, TP harness disconnected. • DVOM on 200 ohm scale. • Disconnect processor 60 Pin connector and inspect for damaged pins, corrosion, loose wires. Service as necessary. • Connect Breakout box. Processor connected to Breakout box. • Measure resistance between TP Signal at the vehicle harness connector and test Pin 47 at the Breakout box.	Reading 5 ohms or greater ▶ Readings less than 5 ohms ▶	SERVICE faulty circuit. CONNECT throttle position sensor. REMOVE Breakout box and RERUN Quick Test. GO to DH13.
DH13 CHECK RESISTANCE OF TP CIRCUIT TO GROUND/SIGNAL RETURN		
• Key Off, wait 10 seconds, TP harness disconnected. • Disconnect processor 60 Pin connector and inspect for damaged pins, corrosion, loose wires. Service as necessary. • DVOM on 200,000 ohm scale. • Measure resistance between TP signal at TP vehicle harness connector and test Pin 46 at the Breakout box and between TP signal at TP vehicle harness connector and ground.	Either reading is less than 10,000 ohms ▶ Both readings are 10,000 ohms or greater ▶	SERVICE circuit short and RERUN Quick Test. REPLACE processor. CONNECT throttle position sensor. REMOVE Breakout box and RERUN Quick Test.
DH20 SERVICE CODE 73: TP SENSOR MOVES IN ENGINE RESPONSE TEST		
NOTE: Code 73 indicates the TP Sensor did not exceed 25 percent of its rotation in the Engine Response Check. • Key Off. • Install Breakout box. • DVOM on 20V scale. • Connect DVOM to test Pins 47 and 46 at the Breakout box. • Perform Engine Running Quick Test, Step 5.0. • Verify DVOM reading exceeds 3.5V during brief WOT at Engine Response Check.	Reading exceeds 3.5V during Engine Response Check ▶ Reading does not exceed 3.5V during Engine Response Check ▶	REPLACE processor. RERUN Quick Test. VERIFY TP Sensor is properly attached to throttle body. If OK, REPLACE TP Sensor. Refer to Section 3 before replacement for adjustment procedures for EFI applications. RERUN Quick Test.

Throttle Position Sensor (TPS) — Pinpoint Test — DH

TEST STEP	RESULT ▶	ACTION TO TAKE
DH90 CONTINUOUS TEST SERVICE CODE 53: EXERCISE TP SENSOR		
• Using continuous monitor mode, observe VOM or STAR LED for indication of a fault while performing the following: • Move throttle slowly to WOT position. • Release throttle slowly to closed position and lightly tap on TP sensor (simulate road shock). • Wiggle TP harness connector. • Is a fault indicated?	Yes ▶ No ▶	GO to DH91. GO to DH92.

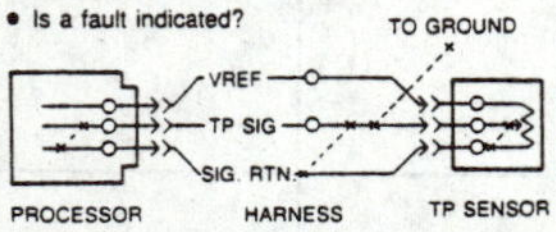

TEST STEP	RESULT ▶	ACTION TO TAKE
DH91 MEASURE THROTTLE POSITION SIGNAL VOLTAGE WHILE EXERCISING TP SENSOR		
• Key Off, wait 10 seconds. • Disconnect processor 60 Pin connector and inspect for damaged pins, corrosion, loose wires. Service as necessary. • Install Breakout box and reconnect processor. • VOM or STAR LED still connected to STO as in previous step. • Connect a DVOM from test Pin 47 to test Pin 46. • DVOM on 20V scale. • Key On engine Off. • While observing DVOM, repeat Step DH90. • Does the fault occur below 4.25V?	Yes ▶ No ▶	DISCONNECT and INSPECT connectors. If connector and terminals are good, REPLACE TP sensor, REFER to Shop Manual Group 24 and RERUN Quick Test. Throttle position sensor overtravel may have caused the continuous code 53. Sensor service is not required. To verify harness integrity, GO to DH92.

Throttle Position Sensor (TPS) — Pinpoint Test — DH

TEST STEP	RESULT ▶	ACTION TO TAKE
DH92 CHECK EEC-IV HARNESS		
• Observe VOM or STAR LED for a fault indication while performing the following: • Referring to the illustration in Step DH90, grasp the harness close to the sensor connector. Wiggle, shake or bend a small section of the EEC-IV system harness while working your way to the dash panel. Also wiggle, shake or bend the EEC-IV harness from the dash panel to the processor. • Is a fault indicated?	Yes ▶ No ▶	ISOLATE fault and make necessary repairs. REFER to appropriate figure. RERUN Quick Test. GO to DH93.
DH93 CHECK PROCESSOR AND HARNESS CONNECTORS		
• Key Off, wait 10 seconds. • Disconnect processor 60 Pin connector. • Inspect both connectors and connector terminals for obvious damage or faults. • Are connectors and terminals OK?	No ▶ Yes ▶	SERVICE as necessary. REPEAT Quick Test. Unable to duplicate fault at this time. Continuous code 53 testing complete.
DH94 CONTINUOUS TEST SERVICE CODE 63: EXERCISE TP SENSOR		
• Using continuous monitor mode, observe VOM or STAR LED for indication of a fault while performing the following: • Move throttle slowly to WOT position. • Release throttle slowly to closed condition. • Lightly tap on TP sensor (simulate road shock). • Wiggle TP harness connector. • Is a fault indicated?	Yes ▶ No ▶	DISCONNECT and INSPECT connectors. If connector and terminals are good, REPLACE TP sensor, REFER to Shop Manual Group 24 and RERUN Quick Test. GO to DH95.

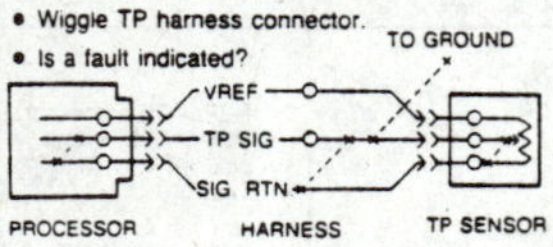

Throttle Position Sensor (TPS) — Pinpoint Test — DH

TEST STEP	RESULT ▶	ACTION TO TAKE
DH95 CHECK EEC-IV HARNESS		
• Observe VOM or STAR LED for a fault indication while performing the following: • Referring to the illustration in Step DH94 grasp the harness close to the sensor connector. Wiggle, shake or bend a small section of the EEC-IV system harness while working your way to the dash panel. Also wiggle, shake or bend the EEC-IV harness from the dash panel to the processor. • Is a fault indicated?	Yes ▶ No ▶	ISOLATE fault and make necessary repairs. REFER to appropriate figure. RERUN Quick Test. Go to DH96.
DH96 CHECK PROCESSOR AND HARNESS CONNECTORS		
• Key Off, wait 10 seconds. • Disconnect processor 60 Pin connector. • Inspect both connectors and connector terminals for obvious damage or faults. • Are connectors and terminals OK?	No ▶ Yes ▶	SERVICE as necessary. RERUN Quick Test. Unable to duplicate fault at this time. Continuous code 63 testing complete.

Vane Airflow Sensor (VAF) — Pinpoint Test — DK

DK

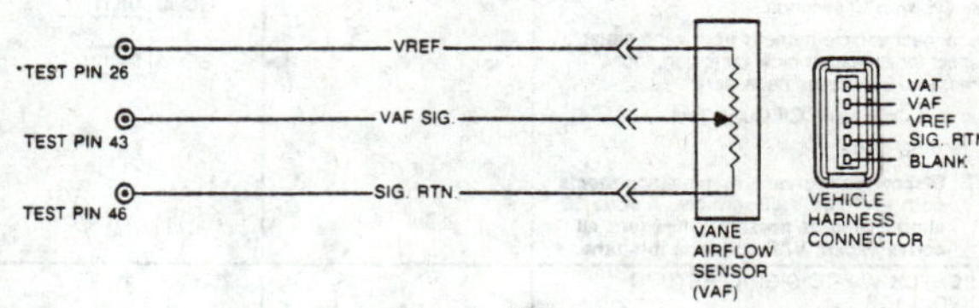

*TEST PINS LOCATED ON BREAKOUT BOX.
ALL HARNESS CONNECTORS VIEWED INTO MATING SURFACE.

STOP-WARNING

You should enter this Pinpoint Test only when a service code 26, 56, 66 or 76 is received in Quick Test Step 3.0, 5.0 or 6.0.

To prevent the replacement of good components, be aware that the following non-EEC areas may be at fault:

- Check for unmetered air (air leaks) between VAF meter and throttle body.
- Vacuum leaks.
- Engine sealing (PCV sealing, CANP, valve cover seal dipstick seated).

This Pinpoint Test is intended to diagnose only the following:

- VAF meter.
- Processor.
- Harness circuits: VREF, VAF Sig. and Signal Return.

Vane Airflow Sensor (VAF) — Pinpoint Test — DK

TEST STEP	RESULT ▶	ACTION TO TAKE
FAULT CODE 26		
DK1 CHECK FOR CONTAMINATION		
NOTE: Code 26 indicates the vane airflow input to the processor is out of engine off or engine idle limits (engine off 0.15-0.50V/engine idle 1.50-2.70V). There have been no opens or shorts in the VAF circuit or a code 56 (signal always high) or 66 (signal always low) would have been generated.		
• Key Off, wait 10 seconds. • Remove air cleaner element and check for contamination (oil residue, foreign material, etc.) that may impede VAF sensor vane movement and service as necessary. • Is service code 26 present in the Key On Engine Off portion of Quick Test?	Yes ▶ No ▶	REPLACE vane meter. RERUN Quick Test. GO to DK2.
DK2 VAF SENSOR CHECK		
• Key Off, air cleaner element reinstalled. • Check for unmetered air leaks between vane meter and throttle body. • Disconnect processor 60 Pin connector. Inspect for damaged pins, corrosion, loose wires, etc. Service as necessary. • Install Breakout box. Connect processor to Breakout box harness. • DVOM on 20V scale. • Key On, engine Off. • Place new unsharpened pencil as shown below. • Measure voltage between test Pins 43 (test Pin 27 on 2.3L EFI TC) and 46 at the Breakout box. • Is voltage between 2.8V and 3.7V?	Yes ▶ No ▶	Vane meter is capable of outputting an acceptable signal. The VAF code 26 has been caused by incorrect engine speed or an unmetered air leak (vacuum leak). SERVICE as necessary. REMOVE Breakout box. RERUN Quick Test. REMOVE Breakout box. REPLACE processor. RERUN Quick Test.

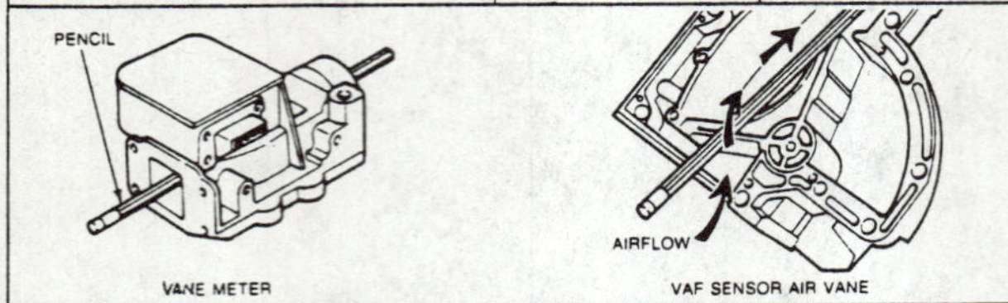

Vane Airflow Sensor (VAF) — Pinpoint Test — DK

TEST STEP	RESULT ▶	ACTION TO TAKE
FAULT CODE 56		
DK10 INDUCE OPPOSITE CODE		
• Key Off, wait 10 seconds. • Disconnect vehicle harness from vane meter. Inspect for damaged pins, corrosion, loose wires, etc. Service as necessary. • Run Key On Engine Off Quick Test. • Is code 66 present? **NOTE: Disconnecting vane meter disconnects both VAF and VAT sensors. A code 58 should also be present. Disregard all codes except VAF codes at this time.**	Yes ▶ No ▶	GO to DK11. GO to DK12.
DK11 CHECK VAF TO SIGNAL RETURN VOLTAGE		
• Key Off, wait 10 seconds. • Harness disconnected from vane meter. • Key On, Engine Off. • DVOM on 20V scale. • Measure voltage at the vane meter vehicle harness connector between VREF and Signal Return. • Is voltage between 4.0 and 6.0 volts?	Yes ▶ No ▶	REPLACE vane meter. RECONNECT harness. RERUN Quick Test. GO to Pinpoint Test Step C1.
DK12 CHECK VAF SIGNAL FOR SHORT		
• Key Off, wait 10 seconds. • Harness disconnected from vane meter. • Disconnect processor 60 Pin connector. Inspect for damaged pins, corrosion, loose wires, etc. Service as necessary. • Install Breakout box, leave processor disconnected. • DVOM on 200,000 ohm scale. • Measure resistance between test Pin 43 (test Pin 27 on 2.3L EFI TC) and test Pins 26 and 57 at the Breakout box. • Are both resistances greater than 10,000 ohms?	Yes ▶ No ▶	REPLACE processor. REMOVE Breakout box. RECONNECT processor and vane meter. RERUN Quick Test. SERVICE circuit shorts. REMOVE Breakout box. RECONNECT processor and vane meter. RERUN Quick Test.

Vane Airflow Sensor (VAF) — Pinpoint Test — DK

TEST STEP	RESULT ▶	ACTION TO TAKE
FAULT CODE 66		
DK20 INDUCE OPPOSITE CODE		
• Key Off, wait 10 seconds. • Disconnect vehicle harness from vane meter. • Install jumper wire in vane meter vehicle harness connector between VREF and VAF signal. • Perform Key On Engine Off Self-Test. **NOTE: If no codes are generated, immediately remove jumper and go directly to DK23.** • Is code 56 present? **NOTE: Disconnecting vane meter disconnects both VAF and VAT sensors. Code 58 should also be present. Disregard all codes except VAF codes at this time.**	Yes ▶ No ▶	REPLACE vane meter. REMOVE jumper wire. RECONNECT vane meter. RERUN Quick Test. REMOVE jumper wire and GO to DK21.
DK21 CHECK VREF TO SIGNAL RETURN VOLTAGE		
• Key off, wait 10 seconds. • Harness disconnected from vane meter. • Key On, Engine Off. • DVOM on 20V scale. • Measure voltage at the vane meter vehicle harness connector between VREF and Signal Return. • Is voltage between 4.0 and 6.0 volts?	Yes ▶ No ▶	GO to DK22. GO to Pinpoint Test Step C1.
DK22 CHECK CONTINUITY OF VAF SIGNAL		
• Key Off, harness disconnected from vane meter. • Disconnect processor 60 Pin connector. Inspect for damaged pins, corrosion, loose wires, etc. Service as necessary. • Install Breakout box; leave processor disconnected. • DVOM on 200 ohm scale. • Measure resistance between VAF signal, at the vane meter vehicle harness connector, and test Pin 43 (test Pin 27 on 2.3L EFI TC) at the Breakout box. • Is resistance greater than 5 ohms?	Yes ▶ No ▶	SERVICE circuit open. REMOVE Breakout box. RECONNECT processor and vane meter. RERUN Quick Test. GO to DK23.

Vane Airflow Sensor (VAF) — Pinpoint Test — DK

TEST STEP	RESULT ▶	ACTION TO TAKE
DK23 CHECK VAF SIGNAL FOR SHORT		
• Key Off, wait 10 seconds. • Processor disconnected. • Harness disconnected from vane meter. • DVOM on 200,000 ohm scale. • Measure resistance at the vane meter vehicle harness between VAF signal and signal return and between VAF signal and negative battery terminal. • Are both resistances greater than 10,000 ohms?	Yes ▶ No ▶	REPLACE processor. REMOVE Breakout box. RECONNECT processor and vane meter. RERUN Quick Test. SERVICE circuit shorts. RECONNECT vane meter. RERUN Quick Test.
FAULT CODE 76		
DK30 CHECK FOR VOLTAGE INCREASE IN VAF SIGNAL TO SIGNAL RETURN		
NOTE: A sharp snap of the throttle may not be sufficient to pass this test. Be sure to move throttle to WOT and return.		
• Key Off, wait 10 seconds. • Disconnect processor 60 Pin connector. Inspect for damaged pins, corrosion, loose wires, etc. Service as necessary. • Install Breakout box. Connect processor to Breakout box harness. • DVOM on 20V scale. • Connect DVOM to test Pins 43 (test Pin 27 on 2.3L EFI TC) and 46. • Perform Engine Running Quick Test while monitoring DVOM. • After dynamic response prompt code 1(0) operator does a brief WOT. DVOM should increase more than 2.0V from reading before WOT.	Yes ▶ No ▶ • Observe service codes at end of test. • Did voltage increase more than 2.0 volts?	GO to DK31. CHECK air cleaner duct for obstruction. If OK, REPLACE vane meter.
DK31 CHECK PROCESSOR		
• Observe Key On Engine Running service codes at end of Pinpoint Test DK30. • Is code 76 present?	Yes ▶ No ▶	REPLACE processor. REMOVE Breakout box. RERUN Quick Test. Vane meter is OK. SERVICE other codes as necessary.

Vane Airflow Sensor (VAF) — Pinpoint Test — DK

TEST STEP	RESULT ▶	ACTION TO TAKE
DK90 SERVICE CODE 56 CONTINUOUS TEST: CHECK VAF SENSOR		
• Using continuous monitor mode, observe VOM or STAR LED for indication of a fault while performing the following: • Lightly tap on VAF sensor (simulate road shock). • Wiggle VAF connector. • Is a fault indicated?	Yes ▶	DISCONNECT and INSPECT connectors. If connector and terminals are good, REPLACE VAF sensor. RERUN Quick Test.
	No ▶	GO to DK91.
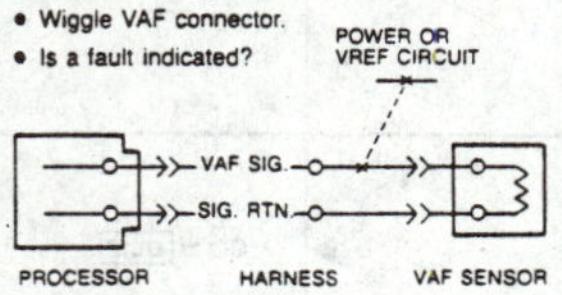		
DK91 CHECK EEC-IV HARNESS		
• Observe VOM or STAR LED for a fault indication while performing the following: • Referring to the illustration in Step DK90, grasp the harness close to the sensor connector. Wiggle, shake or bend a small section of the EEC-IV system harness while working your way to the dash panel. Also wiggle, shake or bend the EEC-IV harness from the dash panel to the processor. • Is a fault indicated?	Yes ▶	ISOLATE fault and make necessary repairs. REFER to appropriate figure. RERUN Quick Test.
	No ▶	GO to DK92.
DK92 CHECK PROCESSOR AND HARNESS CONNECTORS		
• Key Off, wait 10 seconds. • Disconnect processor 60 Pin connector. • Inspect both connectors and connector terminals for obvious damage or faults. • Are connectors and terminals OK?	Yes ▶	Unable to duplicate fault at this time. Continuous code 56 testing complete.
	No ▶	SERVICE as necessary. RERUN Quick Test.

Vane Airflow Sensor (VAF) — Pinpoint Test — DK

TEST STEP	RESULT ▶	ACTION TO TAKE
DK93 SERVICE CODE 66 CONTINUOUS TEST: CHECK VAF SENSOR		
• Using continuous monitor mode, observe VOM or STAR LED for indication of a fault while performing the following: • Lightly tap on VAF sensor (simulate road shock). • Wiggle VAF connector. • Is a fault indicated?	Yes ▶	DISCONNECT and INSPECT connectors. If connector and terminals are good, REPLACE sensor. RERUN Quick Test.
	No ▶	GO to DK94.
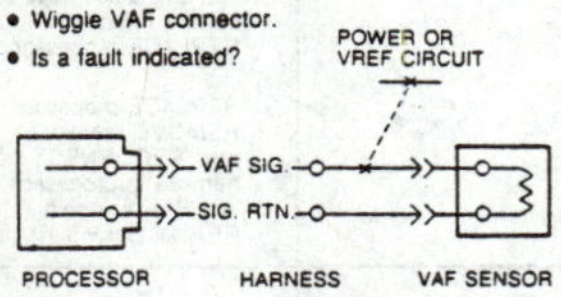		
DK94 CHECK EEC-IV HARNESS		
• Observe VOM or STAR LED for a fault indication while performing the following: • Referring to the illustration in Step DK93 grasp the harness close to the sensor connector. Wiggle, shake or bend a small section of the EEC-IV system harness while working your way to the dash panel. Also wiggle, shake or bend the EEC-IV harness from the dash panel to the processor. • Is a fault indicated?	Yes ▶	ISOLATE fault and make necessary repairs. REFER to appropriate figure. RERUN Quick Test.
	No ▶	Go to DK95.
DK95 CHECK PROCESSOR AND HARNESS CONNECTORS		
• Key Off, wait 10 seconds. • Disconnect processor 60 Pin connector. • Inspect both connectors and connector terminals for obvious damage or faults. • Are connectors and terminals OK?	No ▶	SERVICE as necessary. RERUN Quick Test.
	Yes ▶	Unable to duplicate fault at this time. Continuous code 66 testing complete.

Pressure Feedback EGR (PFE) EGR Valve Regulator (EVR) — Pinpoint Test — DL

DL

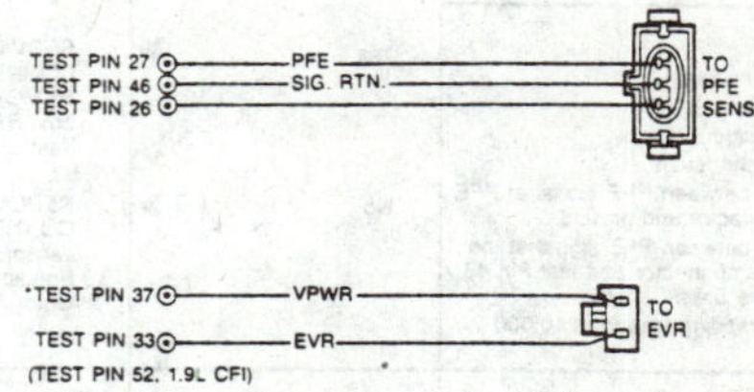

*TEST PINS LOCATED ON BREAKOUT BOX.
ALL HARNESS CONNECTORS VIEWED INTO MATING SURFACE.

STOP-WARNING

You should enter this Pinpoint Test only when a Service Code 31, 32, 33, 34, 35 or 84 is received in Quick Test Step 3.0, 5.0 or 6.0.

To prevent the replacement of good components, be aware that the following non-EEC area may be at fault:

- Damaged EGR valve.

This Pinpoint Test is intended to diagnose only the following:

- PFE sensor.
- Harness circuits: VREF, PFE, Signal Return, EVR, VPWR.
- EVR (EGR valve regulator).
- EGR valve assembly.
- Processor assembly.

Pressure Feedback EGR (PFE) EGR Valve Regulator (EVR) — Pinpoint Test — DL

TEST STEP	RESULT ▶	ACTION TO TAKE
SERVICE CODE 31		
DL1 ATTEMPT TO GENERATE OPPOSITE CODE (35)		
• Key Off. • Disconnect PFE vehicle harness at sensor. • Jumper VREF to PFE signal at vehicle harness sensor connector. • Perform Key On Engine Off Self-Test. **NOTE: If no codes are generated, immediately remove jumper and GO directly to Step DL4.** • Is code 35 present? **NOTE: Ignore all other codes at this time.**	Yes ▶	REMOVE Jumper. REPLACE PFE sensor. RERUN Quick Test.
	No ▶	REMOVE jumper. GO to DL2.
DL2 MEASURE VREF TO SIGNAL RETURN VOLTAGE		
• Key Off. • PFE harness disconnected. • DVOM on 20V scale. • Key On Engine Off. • Measure voltage at PFE vehicle harness connector between VREF and Signal Return. • Is voltage reading between 4 and 6 volts? • Refer to illustration DL.	Yes ▶	GO to DL3.
	No ▶	GO to C1.
DL3 CHECK CONTINUITY OF PFE SIGNAL		
• Key Off. • PFE harness disconnected. • DVOM on 200 ohm scale. • Disconnect processor 60 Pin connector and inspect for damaged pins, corrosion, loose wires, etc. Service as necessary. • Connect Breakout box to harness. Processor connected to Breakout box. • Measure resistance between PFE signal at vehicle harness sensor connector and test Pin 27 at the Breakout box. • Is resistance reading 5 ohms or greater?	Yes ▶	SERVICE faulty circuit. CONNECT PFE sensor. REMOVE Breakout box. RERUN Quick Test.
	No ▶	GO to DL4.

Pressure Feedback EGR (PFE) EGR Valve Regulator (EVR) — Pinpoint Test DL

TEST STEP	RESULT ▶	ACTION TO TAKE
DL4 CHECK RESISTANCE OF PFE SIGNAL TO GROUND AND SIGNAL RETURN		
• Key Off. • PFE harness disconnected. • Breakout box installed. • Processor disconnected. • DVOM on 200,000 ohm scale. • Measure resistance between PFE signal at PFE vehicle harness connector and ground. • Measure resistance between PFE signal at the PFE vehicle harness connector and test Pin 46 (Signal Return) at the Breakout box. • Is either resistance reading less than 10,000 ohms?	Yes ▶ No ▶	SERVICE short circuit. CONNECT PFE. REMOVE Breakout box. RERUN Quick Test. REPLACE processor. CONNECT PFE sensor. REMOVE Breakout box. RERUN Quick Test.
SERVICE CODE 35		
DL5 ATTEMPT TO GENERATE OPPOSITE CODE (31)		
• Key Off. • Disconnect PFE vehicle harness at sensor. Inspect for damaged pins, corrosion, loose wires, etc. Service as necessary. • Perform Key On Engine Off Self-Test, and record codes. • Is code 31 present? **NOTE: Ignore all other codes at this time.**	Yes ▶ No ▶	GO to DL6. GO to DL7.
DL6 MEASURE VREF TO SIGNAL RETURN VOLTAGE		
• Key Off. • PFE harness disconnected. • DVOM on 20V scale. • Key On Engine Off. • Measure voltage at PFE vehicle harness connector between VREF and Signal Return. • Is voltage reading between 4 and 6 volts? • Refer to illustration DL.	Yes ▶ No ▶	REPLACE PFE sensor. RERUN Quick Test. GO to C1.

Pressure Feedback EGR (PFE) EGR Valve Regulator (EVR) — Pinpoint Test DL

TEST STEP	RESULT ▶	ACTION TO TAKE
DL7 CHECK PFE CIRCUIT FOR SHORT TO POWER		
• Key Off. • PFE harness disconnected. • Disconnect processor 60 Pin connector and inspect for damaged pins, corrosion, loose wires, etc. Service as necessary. • Install Breakout box, leave processor disconnected. • DVOM on 200,000 ohm scale. • Measure the resistance between test Pin 27 and test Pins 26 and 57 at the Breakout box. • Is either resistance reading less than 10,000 ohms?	Yes ▶ No ▶	SERVICE harness short. REMOVE Breakout box. CONNECT PFE sensor. RERUN Quick Test. REPLACE processor. REMOVE Breakout box. CONNECT PFE sensor. RERUN Quick Test.
SERVICE CODE: 34		
DL8 PFE SENSOR OUT OF RANGE		
• PFE system can sense a lack of pressure in the vehicle exhaust system. An efficient garage exhaust ventilation system, installed during Key On Engine Off Quick Test, may deflect the PFE sensor and generate a code 34. Remove the ventilation system and retest. • Is code 34 present?	Yes ▶ No ▶	GO to DL9. ADDRESS any other codes in Key On, Engine Off, if none CONTINUE with remaining Quick Test.
DL9 CHECK PRESSURE FEED TUBE TO PFE SENSOR		
• Remove the pressure feed tube from PFE sensor. • Inspect complete tube, including PFE inlet for blockage. • Is blockage present?	Yes ▶ No ▶	SERVICE as necessary. RERUN Quick Test. GO to DL10.

Pressure Feedback EGR (PFE) EGR Valve Regulator (EVR) — Pinpoint Test DL

TEST STEP	RESULT ▶	ACTION TO TAKE
DL10 MEASURE VREF TO SIGNAL RETURN VOLTAGE		
• Key Off. • Disconnect PFE sensor and inspect for damaged pins, corrosion, loose wires, etc. Service as necessary. • DVOM on 20V scale. • Key On Engine Off. • Measure voltage at PFE vehicle harness connector between VREF and Signal Return. • Is voltage reading between 4 and 6 volts? • Refer to illustration DL.	Yes ▶ No ▶	REPLACE PFE sensor. RERUN Quick Test. GO to Pinpoint Test C1.
SERVICE CODE 84		
DL11 MEASURE EVR SOLENOID RESISTANCE		
• Key Off. • DVOM on 200 ohms scale. • Disconnect EVR solenoid connector and measure solenoid resistance. • Is resistance reading between 30 and 70 ohms?	Yes ▶ No ▶	GO to DL12. REPLACE EVR solenoid assembly. RERUN Quick Test.
DL12 CHECK FOR VPWR AT EVR SOLENOID		
• EVR solenoid disconnected from harness. • DVOM on 20V scale. • Key on Engine Off. • Measure voltage between battery negative terminal and VPWR circuit at EVR solenoid vehicle harness connector. • Is voltage reading less than 10.5 volts?	Yes ▶ No ▶	SERVICE VPWR open circuit. RERUN Quick Test. GO to DL13.
DL13 CHECK CONTINUITY OF EVR CIRCUIT		
• Key Off. • EVR solenoid disconnected from harness. • Disconnect processor 60 Pin connector and inspect for damaged pins, corrosion, loose wires, etc. Service as necessary. • Install Breakout box to processor harness connector. Leave processor disconnected. • DVOM on 200 ohm scale. • Measure resistance between test Pin 33 at the Breakout box and EVR signal at the EVR solenoid vehicle harness connector. • Is resistance reading less than 5 ohms?	Yes ▶ No ▶	GO to DL14. SERVICE open circuit. REMOVE Breakout box. CONNECT process EVR solenoid. RERUN Quick Test.

Pressure Feedback EGR (PFE) EGR Valve Regulator (EVR) — Pinpoint Test DL

TEST STEP	RESULT ▶	ACTION TO TAKE
DL14 CHECK EVR CIRCUIT FOR SHORT TO POWER AND GROUND		
• Key Off. • Breakout box installed, processor disconnected. • EVR solenoid disconnected. • DVOM on 200,000 ohm scale. • Measure resistance between test Pin 33 (EVR) and test Pins 37 and 57 (VPWR) and 40, 46 and 60 (GRD) at the Breakout box. • Are any readings less than 10,000 ohms?	Yes ▶ No ▶	SERVICE short circuit. REMOVE Breakout box. RECONNECT harness to processor and EVR solenoid. RERUN Quick Test. If code is repeated, REPLACE processor. REPLACE processor. REMOVE Breakout box. RECONNECT harness to processor and EVR solenoid. RERUN Quick Test.
SERVICE CODE 32		
DL20 VERIFY ENGINE RUNNING CODES		
The PFE system can sense a lack of pressure in the vehicle exhaust system. An efficient garage exhaust ventilation system installed during Key On Engine Run Quick Test may, on some calibrations, deflect the PFE sensor and generate a code 32. Temporarily, remove garage forced ventilation system and properly vent to atmosphere. • RERUN Engine Running Quick Test. • Is code 32 present?	Yes ▶ No ▶	GO to DL21. ADDRESS any other codes in Engine Running. If none, CONTINUE with remaining Quick Test.

Pressure Feedback EGR (PFE) EGR Valve Regulator (EVR) — Pinpoint Test — DL

TEST STEP	RESULT ▶	ACTION TO TAKE
DL21 ATTEMPT TO SEPARATE EVR FROM PFE		
• Key Off. • Disconnect EGR valve vacuum line at valve and plug line. • Perform Engine Running Quick Test. • Is code 31 or 32 present?	Yes ▶ No ▶	GO to DL22. GO to DL23.
DL22 CHECK PFE SENSOR SUPPLY TUBE		
• Key Off. • Check PFE sensor supply tube for obstructions and/or leaks. • Are there any obstructions or leaks?	Yes ▶ No ▶	SERVICE as necessary. RECONNECT all lines and RERUN Quick Test. GO to EGR Diagnostic, Section 6.
DL23 CHECK EVR FILTER		
• Key Off. • Remove and inspect EVR filter for contamination. **NOTE: Blockage of filter will cause vacuum to be applied to EGR valve prematurely.** • Is filter contaminated?	Yes ▶ No ▶	REPLACE filter. RECONNECT all lines. RERUN Quick Test. REPLACE EVR solenoid. RERUN Quick Test.
ENGINE RUNNING SERVICE CODE 34 AND 35		
DL25 CHECK FOR EXCESSIVE EXHAUST BACK PRESSURE		
• Service codes 34 and 35 in Engine Running Self-Test indicate excessive exhaust back pressure; There are two possible causes: (A). The exhaust system is restricted, and (B). PFE sensor has shifted high. • Key Off. • Substitute known good PFE sensor in place of original. • Rerun Key On Engine Running Quick Test. • Is code 34 or 35 present?	Yes ▶ No ▶	GO to Section 5, Catalyst and Exhaust Systems Restricted Exhaust System Diagnosis. Original PFE was the cause of the original 34 or 35. REPLACE PFE sensor. RERUN Quick Test.

Pressure Feedback EGR (PFE) EGR Valve Regulator (EVR) — Pinpoint Test — DL

TEST STEP	RESULT ▶	ACTION TO TAKE
DL90 SERVICE CODE 31 OR 35 CONTINUOUS TEST: EXERCISE PFE SENSOR		
• Using continuous monitor mode, observe VOM or STAR LED for indication of a fault while performing the following: • Connect a vacuum pump to the PFE sensor. • Slowly apply 5 in.-Hg. to the sensor. • Slowly bleed vacuum off the PFE sensor. • Lightly tap on PFE sensor (simulate road shock). • Wiggle PFE connector. • Is fault indicated?	Yes ▶ No ▶	DISCONNECT and INSPECT connectors. If connector and terminals are good, REPLACE sensor. RERUN Quick Test. GO to DL91.

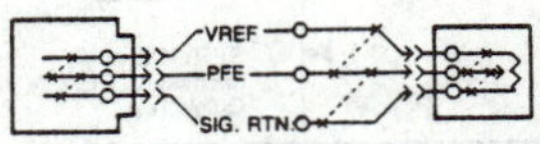

Pressure Feedback EGR (PFE) EGR Valve Regulator (EVR) — Pinpoint Test — DL

TEST STEP	RESULT ▶	ACTION TO TAKE
SERVICE CODE 33		
DL30 VERIFY VACUUM IS PRESENT AT VALVE		
• Key Off. • Standard vacuum gauge in. Hg (Mercury). • Tee in vacuum gauge at EGR valve. • Perform Engine Running Quick Test while observing vacuum gauge. • Is vacuum reading 1 in.-Hg. or less? **NOTE: Disregard code output.**	Yes ▶ No ▶	GO to DL31. GO to DL34.
DL31 VACUUM SUPPLY VERIFICATION		
• Key Off. • Do vacuum lines from EVR solenoid to EGR valve and source to EVR solenoid have loose connections, cracks or obstructions?	Yes ▶ No ▶	SERVICE as necessary. RERUN Quick Test. GO to DL32.
DL32 VERIFY VACUUM TO EVR		
• Key On Engine Running. • Attach vacuum gauge to source line from manifold. • Is vacuum present?	Yes ▶ No ▶	REPLACE EVR solenoid. RERUN Quick Test. REPLACE vacuum line to EVR. RERUN Quick Test.
DL34 CHECK EGR CONTROL PFE SENSOR TUBE		
• Key Off. • Is control pressure input tube to PFE sensor cracked, disconnected or obstructed?	Yes ▶ No ▶	SERVICE as necessary. RERUN Quick Test. REPLACE PFE Sensor. RERUN Quick Test.

Pressure Feedback EGR (PFE) EGR Valve Regulator (EVR) — Pinpoint Test — DL

TEST STEP	RESULT ▶	ACTION TO TAKE
DL91 CHECK EEC-IV HARNESS		
• Observe VOM or STAR LED for a fault indication while performing the following: • Referring to the illustration in Step DL90 grasp the harness closest to the sensor connector. Wiggle, shake or bend a small section of the EEC-IV system harness while working your way to the dash panel. Also wiggle, shake or bend the EEC-IV harness from the dash panel to the processor. • Is a fault indicated?	Yes ▶ No ▶	ISOLATE fault and SERVICE as necessary. RERUN Quick Test. GO to DL92.
DL92 CHECK PROCESSOR AND HARNESS CONNECTORS		
• Key Off, wait 10 seconds. • Disconnect processor 60 Pin connector. • Inspect both connectors and connector terminals for obvious damage or faults. • Are connectors and terminals OK?	Yes ▶ No ▶	Unable to duplicate and/or identify fault at this time. Continuous code 31 or 35 testing complete. SERVICE as necessary. RERUN Quick Test.
DL93 SERVICE CODE 34: CONTINUOUS TEST — INSPECT PFE SUPPLY TUBE FOR BLOCKAGE		
• Key Off. • Remove PFE sensor and inspect sensor supply inlet for liquids and/or any type of blockage. • Inspect PFE supply tube to EGR valve base for liquids and/or blockage. • Is supply tube free of any blockage?	Yes ▶ No ▶	Unable to duplicate and/or identify fault at this time. Continuous code 34 testing complete. CLEAN and/or SERVICE as necessary. RERUN Quick Test.

Pressure Feedback EGR (PFE) EGR Valve Regulator (EVR) — Pinpoint Test DL

TEST STEP	RESULT ▶	ACTION TO TAKE
DL94 SERVICE CODE 32: CONTINUOUS TEST — INSPECT EGR VALVE FOR SMOOTH OPERATION.		
• Key Off. • Connect a vacuum pump to the EGR valve. • Apply 10 in.-Hg. of vacuum to EGR valve. • While observing EGR valve, release vacuum. • Does EGR valve function in a smooth manner? **NOTE: Repeat test if necessary to ensure accurate result.**	Yes ▶ No ▶	GO to DL95. GO to EGR Valve Diagnostic Section 6.
DL95 INSPECT VACUUM LINES BETWEEN EVR SOLENOID AND EGR VALVE		
• Inspect EGR valve vacuum supply line from EVR solenoid for kinks and/or obstructions. • Is vacuum supply line to EGR valve free of any obstructions?	Yes ▶ No ▶	GO to DL96. SERVICE as necessary. RERUN Quick Test.
DL96 EVR REGULATOR FILTER INSPECTION		
• Carefully check EVR filter for contamination and/or obstructions. • Is EVR filter condition acceptable?	Yes ▶ No ▶	Unable to duplicate and/or identify fault at this time. Continuous code 34 testing complete. REPLACE EVR filter. RERUN Quick Test.
DL97 SERVICE CODE 33: CONTINUOUS TEST — INSPECT EGR VALVE FOR FREE OPERATION		
• Key Off. • Connect a vacuum pump to the EGR valve. • While observing the EGR valve, slowly apply 10 in.-Hg. vacuum. **NOTE: EGR valve should begin to open with a very small amount of vacuum, approximately 1 to 1.5 in.-Hg. and be fully open with about 4 in.-Hg. vacuum.** • Does EGR valve move freely and smoothly?	Yes ▶ No ▶	GO to DL98. GO to EGR Valve Diagnostic Section 6.

Pressure Feedback EGR (PFE) EGR Valve Regulator (EVR) — Pinpoint Test DL

TEST STEP	RESULT ▶	ACTION TO TAKE
DL98 EVR HARNESS CHECK		
• Key Off. • Disconnect processor 60 Pin connector and inspect for damaged pins, corrosion, loose wires, etc. Service as necessary. • Connect Breakout box to harness. Processor connected to Breakout box. • Enter output state check. • DVOM on 20 V scale. • Connect DVOM negative test lead to test Pin 40 at the Breakout box and DVOM positive test lead to test Pin 33. • Cycle throttle if necessary to indicate 10.5 V or greater. • Remain at this position. • While observing DVOM, grasp the harness closest to the EVR connector. Wiggle, shake or bend a small section of the EEC-IV system harness while working your way to the dash panel. Lightly tap EVR solenoid to simulate road vibration. • Does DVOM indicate less than 10.5 V?	Yes ▶ No ▶	SERVICE as necessary. RERUN Quick Test. Unable to duplicate and/or identify fault at this time. Continuous code 33 testing complete.

EGR Valve Position Sensor (EVP) EGR Valve Regulator (EVR) — Pinpoint Test DN

DN

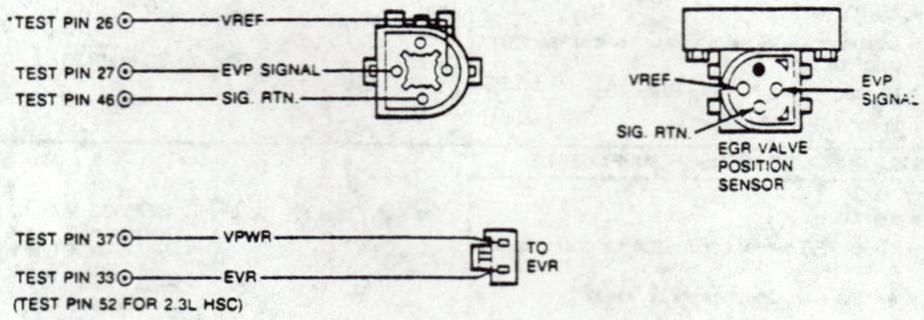

*TEST PINS LOCATED ON BREAKOUT BOX.
ALL HARNESS CONNECTORS VIEWED INTO MATING SURFACE.

STOP-WARNING

You should enter this Pinpoint Test only when a Service Code 31, 32, 33, 34, 35, 38 or 84 is received in Quick Test Step 3.0, 5.0 or 6.0.

To prevent the replacement of good components, be aware that the following non-EEC area may be at fault:

- Damaged EGR valve.

This Pinpoint Test is intended to diagnose only the following:

- EVP sensor.
- Harness circuits: VREF, EVP, Signal Return, EVR, VPWR.
- EVR (EGR valve regulator).
- EGR valve assembly.
- Processor assembly.
- EGR and EVR vacuum lines.

EGR Valve Position Sensor (EVP) EGR Valve Regulator (EVR) — Pinpoint Test DN

TEST STEP	RESULT ▶	ACTION TO TAKE
SERVICE CODE: 31		
DN1 ATTEMPT TO GENERATE CODE (35)		
• Key Off. • Disconnect EVP vehicle harness at sensor. • Jumper VREF to EVP signal at vehicle harness sensor connector. • Perform Key On Engine Off Self-Test. • Is code 35 present? **NOTE: Ignore all other codes at this time.**	Yes ▶ No ▶	REPLACE EVP sensor. RERUN Quick Test. REMOVE jumper. GO to DN2.
DN2 MEASURE VOLTAGE BETWEEN VREF AND SIGNAL RETURN		
• Key Off. • EVP disconnected from harness. • DVOM on 20 V scale. • Key on, engine off. • Measure voltage at EVP vehicle harness connector between VREF and Signal Return. • Is voltage between 4 and 6 volts?	Yes ▶ No ▶	GO to DN3. GO to C1.
DN3 CHECK CONTINUITY OF EVP SIGNAL		
• Key Off, EVP harness disconnected. • DVOM on 200 ohm scale. • Disconnect processor 60 Pin connector and inspect for damaged pins, corrosion, loose wires, etc. Service as necessary. • Connect Breakout box to harness. Processor connected to Breakout box. • Measure resistance between EVP signal at vehicle harness connector and test Pin 27 at the Breakout box. • Is resistance 5 ohms or less?	Yes ▶ No ▶	GO to DN4. SERVICE open circuit. CONNECT EVP sensor. REMOVE Breakout box. RERUN Quick Test.

EGR Valve Position Sensor (EVP) EGR Valve Regulator (EVR) — Pinpoint Test — DN

TEST STEP	RESULT ▶	ACTION TO TAKE
DN4 CHECK RESISTANCE OF EVP SIGNAL TO GROUND AND SIGNAL RETURN		
• Key Off. • EVP harness disconnected. • Breakout box installed. • Processor disconnected. • DVOM on 200,000 ohm scale. • Measure resistance between EVP signal at EVP vehicle harness connector and ground. • Measure resistance between EVP signal at EVP vehicle harness connector and test Pin 46 (Signal Return) at the Breakout box. • Are both resistances greater than 10,000 ohms?	Yes ▶	REPLACE processor. CONNECT EVP sensor. REMOVE Breakout box. RERUN Quick Test.
	No ▶	SERVICE short circuit. CONNECT EVP. REMOVE Breakout box. RERUN Quick Test.
SERVICE CODE: 35		
DN5 ATTEMPT TO GENERATE CODE 31		
• Key Off. • Disconnect EVP vehicle harness at sensor. Inspect for damaged pins, corrosion, loose wires, etc. Service as necessary. Leave EVP sensor disconnected. • Perform Key On Engine Off Self-Test, and record codes. • Is code 31 present? **NOTE: Ignore all other codes at this time.**	Yes ▶	GO to DN6.
	No ▶	GO to DN7.
DN6 MEASURE VREF TO SIGNAL RETURN VOLTAGE		
• Key Off. • EVP harness disconnected. • DVOM on 20V scale. • Key on, engine off. • Measure voltage at EVP vehicle harness connector between VREF and Signal Return. • Is voltage between 4.0 and 6.0 volts?	Yes ▶	REPLACE EVP sensor. RERUN Quick Test.
	No ▶	GO to C1.

EGR Valve Position Sensor (EVP) EGR Valve Regulator (EVR) — Pinpoint Test — DN

TEST STEP	RESULT ▶	ACTION TO TAKE
DN7 CHECK EVP CIRCUIT FOR SHORT TO POWER		
• Key Off. • EVP disconnected from harness. • Disconnect processor 60 Pin connector and inspect for damaged pins, corrosion, loose wires, etc. Service as necessary. • Install Breakout box, leave processor disconnected. • DVOM on 200,000 ohm scale. • Measure the resistance between test Pin 27 and test Pins 26 and 57 at the Breakout box. • Are both resistances greater than 10,000 ohms?	Yes ▶	REPLACE processor. REMOVE Breakout box. CONNECT EVP sensor. RERUN Quick Test.
	No ▶	SERVICE harness short. REMOVE Breakout box. CONNECT EVP sensor. RERUN Quick Test.
SERVICE CODE: 84		
DN10 MEASURE EVR SOLENOID RESISTANCE		
• Key Off. • DVOM on 200 ohm scale. • Disconnect EVR solenoid connector and measure solenoid resistance. • Is resistance between 30 and 70 ohms?	Yes ▶	GO to DN11.
	No ▶	REPLACE EVR solenoid assembly. RERUN Quick Test.
DN11 CHECK FOR VPWR AT EVR SOLENOID		
• EVR solenoid disconnected from harness. • DVOM on 20V scale. • Key on, engine off. • Measure voltage between battery negative terminal and VPWR circuit at EVR solenoid vehicle harness connector. • Is voltage greater than 10.5 V?	Yes ▶	GO to DN12.
	No ▶	SERVICE VPWR open circuit. RERUN Quick Test.

EGR Valve Position Sensor (EVP) EGR Valve Regulator (EVR) — Pinpoint Test — DN

TEST STEP	RESULT ▶	ACTION TO TAKE
DN12 CHECK CONTINUITY OF EVR CIRCUIT		
• Key Off. • EVR solenoid disconnected from harness. • Disconnect processor 60 Pin connector and inspect for damaged pins, corrosion, loose wires, etc. Service as necessary. • Install Breakout box to processor harness connector. Leave processor disconnected. • DVOM on 200 ohm scale. • Measure resistance between test Pin 33 (test Pin 52 for 2.3L HSC) at the Breakout box and EVR signal at the EVR solenoid vehicle harness connector. • Is resistance less than 5 ohms?	Yes ▶	GO to DN13.
	No ▶	SERVICE open circuit. REMOVE Breakout box. CONNECT process EVR solenoid. RERUN Quick Test.
DN13 CHECK EVR CIRCUIT FOR SHORT TO POWER AND GROUND		
• Key Off. • Breakout box installed, processor disconnected. • EVR solenoid disconnected. • DVOM on 200,000 ohm scale. • Measure resistance between test Pin 33 (test Pin 52 for 2.3L HSC) EVR signal and test Pins 37 and 57 (VPWR) and 40, 46 and 60 (GRD) at the Breakout box. • Are all resistances greater than 10,000 ohms?	Yes ▶	REPLACE processor. REMOVE Breakout box. RECONNECT processor and EVR solenoid. RERUN Quick Test.
	No ▶	SERVICE short circuit. REMOVE Breakout box. RECONNECT processor and EVR solenoid. RERUN Quick Test. If code is repeated, REPLACE processor.

EGR Valve Position Sensor (EVP) EGR Valve Regulator (EVR) — Pinpoint Test — DN

TEST STEP	RESULT ▶	ACTION TO TAKE
DN20		
Code 34 Key On Engine Off, indicates that the EGR valve and/or EVP sensor is not fully seated in the closed position. **NOTE: Because of the preload on the installed EVP sensor, it is very difficult to determine whether the problem cause is that the EGR valve is not seated, or that the EVP sensor is not in contact with the EGR valve stem.**		
SUBSTITUTE EVP SENSOR ON ORIGINAL EGR VALVE		
• Key Off, wait 10 seconds. • Install a known good EVP sensor on original EGR valve. • Perform Key On Engine Off Self-Test. • Is code 34 present?	Yes ▶	GO to EGR Valve Diagnostic Section 6.
	No ▶	The original code 34 was the result of the original EVP sensor. SERVICE EVP sensor as necessary. RERUN Quick Test.
DN25		
Code 32 Key On Engine Off or Engine Running: indicates that the EGR valve and/or EVP sensor is lower than normal in the closed position. **NOTE: Because of the preload of the EVP sensor, it is difficult to determine whether the EGR valve has malfunctioned or if the EVP sensor has abnormally high resistance.**		
SUBSTITUTE EVP SENSOR ON ORIGINAL EGR VALVE		
• Key Off, wait 10 seconds. • Install a known good EVP sensor on original EGR valve. • Perform Key On Engine Off Self-Test. • Is code 32 present?	Yes ▶	GO to EGR Valve Diagnostic Section 6.
	No ▶	The original code 32 was the result of the original EVP sensor. SERVICE EVP sensor as necessary. RERUN Quick Test.

EGR Valve Position Sensor (EVP) EGR Valve Regulator (EVR) — Pinpoint Test DN

TEST STEP	RESULT ▶	ACTION TO TAKE
DN40 NOTE: Code 33 Key On Engine Running indicates that the EVP sensor reading did not change after the EVR solenoid was instructed by the processor to open the EGR valve. Items that are known: • EVR solenoid is electrically capable (indicated by a pass code 11 in Key On Engine Off). • EVP sensor was in its expected range when the EGR valve was closed.		
VERIFY VACUUM IS PRESENT AT EGR VALVE • Key Off. • Standard vacuum gauge in.-Hg. • Tee in vacuum gauge at EGR valve. • Perform Engine Running Self-Test while observing vacuum gauge. • Is vacuum greater than 1.5 in.-Hg. (5 kPa)?	Yes ▶	GO to DN43.
	No ▶	GO to DN41.
DN41 VERIFY VACUUM SUPPLY TO EVR SOLENOID • Key Off. • Disconnect the vacuum source to the EVR solenoid. • Install a vacuum gauge at source vacuum line. • Start engine. • Is vacuum greater than 10 in.-Hg. (33 kPa)?	Yes ▶	GO to DN42.
	No ▶	CHECK source vacuum line to EVR solenoid. SERVICE as necessary. RERUN Quick Test.
DN42 CHECK VACUUM LINE FROM EVR SOLENOID TO EGR VALVE • Carefully check vacuum line for cracks, loose connector, blockage, kinks and leaks. • Is vacuum line good?	Yes ▶	CHECK EVR solenoid filter for obstructions. REPLACE as necessary. If OK, REPLACE EVR solenoid assembly. RECONNECT all vacuum lines. RERUN Quick Test.
	No ▶	SERVICE vacuum lines as necessary. RECONNECT vacuum lines and RERUN Quick Test.

EGR Valve Position Sensor (EVP) EGR Valve Regulator (EVR) — Pinpoint Test DN

TEST STEP	RESULT ▶	ACTION TO TAKE
DN43 SUBSTITUTE EVP SENSOR ON ORIGINAL EGR VALVE • Key Off, wait 10 seconds. • Install a known good EVP sensor on original EGR valve. • Perform Engine Running Self-Test. • Is code 33 present?	Yes ▶	GO to EGR Valve Diagnostic Section 6.
	No ▶	The original code 33 was the result of the original EVP sensor. SERVICE EVP sensor as necessary. RERUN Quick Test.
DN50 SERVICE CODE 34: EGR VALVE OPERATION, ENGINE RUNNING SELF-TEST • Key Off. • Disconnect vacuum hose from EGR valve and plug hose. • Perform Engine Running Self-Test. • Is code 34 present?	Yes ▶	GO to DN51.
	No ▶	CHECK EVR filter for obstructions. REPLACE as necessary. If OK, REPLACE solenoid assembly. RECONNECT all vacuum lines. RERUN Quick Test.
DN51 CHECK EVP RESISTANCE WHILE APPLYING VACUUM TO EGR VALVE • Key Off. • Disconnect harness from EVP sensor. Inspect for damaged pins, corrosion, and pins pushed out. Service as necessary. • DVOM on 200,000 ohm scale. • Disconnect vacuum line at EGR valve. • Connect vacuum pump to EGR valve. • Measure resistance at the EVP sensor between EVP signal and VREF while increasing vacuum to 10 in. Hg. (33 kPa). • Observe resistance as vacuum increases. • Does resistance decrease gradually from no more than 5,500 ohms to no less than 100 ohms?	Yes ▶	GO to EGR Valve Diagnostic Section 6.
	No ▶	REPLACE EVP sensor. RECONNECT vacuum lines.

EGR Valve Position Sensor (EVP) EGR Valve Regulator (EVR) — Pinpoint Test DN

TEST STEP	RESULT ▶	ACTION TO TAKE
DN55 Code 38 Engine Running indicates that after opening the EGR valve (during the Engine Running Self Test) the EGR vacuum system was unable to maintain a specific opening (EVP position).		
VERIFY EGR VALVES VACUUM HOLDING CAPABILITY • Key Off. • Disconnect vacuum line at EGR valve. • With a hand vacuum pump, apply 10 in.-Hg. (33 kPa) to EGR valve. • Does vacuum hold for 10 seconds?	Yes ▶	GO to DN56.
	No ▶	GO to EGR Valve Diagnostic Section 6.
DN56 EVR SOLENOID TO EGR VALVE VACUUM LINE VERIFICATION • Key off. • Closely inspect the vacuum line from the EVR solenoid to the EGR valve for any source of leakage, which would include cracks, holes, loose connections etc. • Is the vacuum line good?	Yes ▶	REPLACE the EVR solenoid. RERUN Quick Test.
	No ▶	SERVICE vacuum line as necessary. RERUN Quick Test.
DN60 CHECK VACUUM LINE • Key Off, wait 10 seconds. • Vacuum lines reconnected. • Carefully check EGR vacuum line system from EGR valve to EVR for obstructions, kinks, leaks, etc. • Is vacuum line in good condition?	Yes ▶	CHECK EVR filter for obstructions. REPLACE as necessary. If OK, REPLACE EVR solenoid. RECONNECT all vacuum lines. RERUN Quick Test.
	No ▶	SERVICE as necessary. RERUN Quick Test.

EGR Valve Position Sensor (EVP) EGR Valve Regulator (EVR) — Pinpoint Test DN

TEST STEP	RESULT ▶	ACTION TO TAKE
DN90 CODE 32 CONTINUOUS TEST • **Code 32: The EVP circuit indicated that the EGR valve was closed further than normal with the engine at stabilized operating temperature and at idle.**		
MEASURE EVP SIGNAL VOLTAGE WHILE EXERCISING EVP SENSOR • Key Off, wait 10 seconds. • Disconnect processor 60 Pin connector and inspect for damaged pins, corrosion, loose wires, etc. Service as necessary. • Install Breakout box and reconnect processor. • Connect a vacuum pump to the EGR valve. • Connect a DVOM from test Pin 27 to test Pin 46. • DVOM on 20V scale. • Key on, engine off. • Observe DVOM. • Very slowly apply 6 in.-Hg. (20 kPa) vacuum to the EGR valve. • Slowly bleed vacuum off the EGR valve and lightly tap on EVP sensor (simulate road shock). • Are any voltage readings less than 0.29 volts?	Yes ▶	EGR valve may have caused continuous code 32. Sensor service is not required. GO to EGR Valve Diagnostic Section 6.
	No ▶	Unable to duplicate code 32 fault at this time. Test completed.

EGR Valve Position Sensor (EVP) EGR Valve Regulator (EVR) — Pinpoint Test — DN

TEST STEP	RESULT ▶	ACTION TO TAKE
DN92 CODE 31 AND/OR 35 CONTINUOUS TEST		
• **Code 31: The EVP circuit indicated an open in the EVP signal or VREF, or a short to signal return with the engine at stabilized operating temperature and at idle.** • **Code 35: The EVP circuit indicated a short to VREF and/or V Power, or an open in signal return with the engine at stabilized operating temperature and at idle.**		
CHECK EEC-IV HARNESS		
• Observe VOM or STAR LED for a fault indication while performing the following: • Refer to illustration below by code for possible circuit faults. • Grasp the harness close to the sensor connector. Wiggle, shake or bend a small section of the EEV-IV system harness while working your way to the dash panel. Also wiggle, shake or bend the EEC-IV harness from the dash panel to the processor. • Is a fault indicated?	Yes ▶	ISOLATE fault and SERVICE as necessary. REFER to appropriate figure. RERUN Quick Test.
	No ▶	GO to DN93.

CODE 31

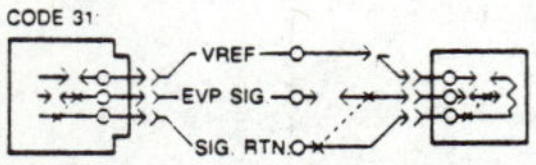

CODE 35

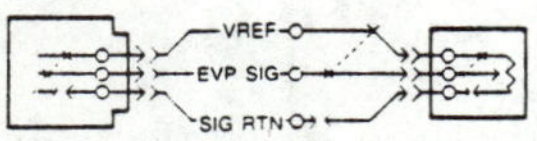

EGR Valve Position Sensor (EVP) EGR Valve Regulator (EVR) — Pinpoint Test — DN

TEST STEP	RESULT ▶	ACTION TO TAKE
DN93 CHECK PROCESSOR AND HARNESS CONNECTORS		
• Key Off, wait 10 seconds. • Disconnect processor 60 Pin connector. • Inspect both connectors and connector terminals for obvious damage or faults. • Are connectors and terminals OK?	Yes ▶	Unable to duplicate fault at this time. Continuous Code 31 or 35 testing complete.
	No ▶	SERVICE as necessary. RERUN Quick Test.
DN95 CODE 33 CONTINUOUS TEST		
• **Code 33: The EVP circuit indicated that the EGR valve did not open with the engine at stabilized temperature and with an EVR solenoid duty cycle present.**		
LEAK TEST		
• Key Off. • Connect a vacuum pump to EGR valve. • Apply 20 in. Hg (66 kPa) to EGR valve. • Does EGR valve open and maintain vacuum?	Yes ▶	GO to DN96.
	No ▶	GO to EGR diagnostic, Section 6.
DN96 EVR CHECK		
• Using continuous monitor mode, observe VOM or STAR LED for indication of a fault while performing the following: • Grasp the harness close to the EVR solenoid connector, wiggle, shake or bend a small section of the harness while working your way to the processor. • Inspect connectors, terminals for obvious damage or faults. • Are any faults detected or indicated?	Yes ▶	ISOLATE fault and SERVICE as necessary. RERUN Quick Test.
	No ▶	Unable to duplicate fault at this time, testing complete.

EGR Valve Position Sensor (EVP) EGR Valve Regulator (EVR) — Pinpoint Test — DN

TEST STEP	RESULT ▶	ACTION TO TAKE
DN98 CODE 34 CONTINUOUS TEST		
• **Code 34: The EVP circuit indicated that the EGR valve was open with the engine at stabilized operating temperature and at idle.**		
CHECK EVP RESISTANCE WHILE APPLYING VACUUM TO EGR VALVE		
• Key Off. • Disconnect harness from EVP sensor. • Inspect for damaged pins, corrosion, and pins pushed out. Service as necessary. • DVOM on 200,000 ohm scale. • Disconnect vacuum line at EGR valve. • Connect vacuum pump to EGR valve. • Measure resistance at the EVP sensor between EVP signal Pin and VREF Pin while increasing vacuum to 10 in. Hg (33 kPa). • Observe resistance as vacuum increases. • Does resistance gradually change from no more than 5,500 ohms to no less than 100 ohms as the vacuum increases?	Yes ▶	GO to DN99.
	No ▶	GO to EGR Diagnostic Section 6.
DN99 EVR CHECK		
• Key Off. • Disconnect vacuum hose from EGR valve and plug hose. • Perform Engine Running Self-Test. • Is code 34 present?	Yes ▶	CHECK EVR filter for obstructions. REPLACE as necessary. If OK, REPLACE EVR solenoid. RECONNECT all vacuum lines. RERUN Quick Test.
	No ▶	Unable to duplicate fault at this time.

Vehicle Speed Sensor — Pinpoint Test — DP

DP

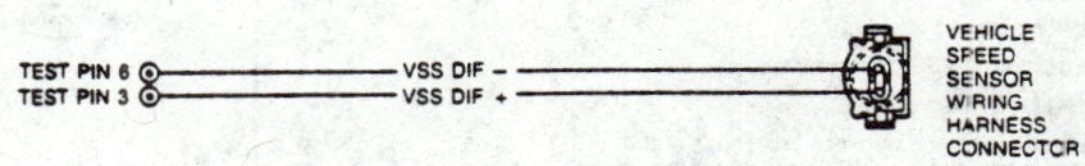

STOP-WARNING

You should enter this Pinpoint Test only when service codes 27 or 29 are received in Quick Test Step 6.0.

This Pinpoint Test is intended to diagnose only the following:

- VSS Harness Circuit
- Vehicle Speed Sensor
- Processor Assembly

Vehicle Speed Sensor — Pinpoint Test — DP

TEST STEP	RESULT ▶	ACTION TO TAKE
DP1 DRIVE CYCLE FOR CHECKING VSS		
AUTOMATIC TRANSMISSION • Record and clear EEC-IV Self-Test continuous codes. • Warm engine to operating temperature. **NOTE: For 2.3L EFI T.C. automatic applications, engine must run without interruption for 5 minutes to enter VSS test.** • With PRNDL selector in 1st gear (low) position, accelerate moderately to 25 mph and coast down to idle and stop. Shut engine off. • Run Key On Engine Off Self-Test. Is code 29 present in continuous? (2.3L EFI T.C. uses code 27.) **MANUAL TRANSMISSION** • Record and clear EEC-IV Self-Test continuous codes. • Warm engine to operating temperature. • Starting in first gear, shift to second gear accelerate moderately to 40 mph, coast down to idle and stop. Shut engine off. • Run Key On Engine Off Self-Test. Is code 29 present in continuous? (2.3L EFI T.C. uses code 27.)	Yes ▶ No ▶	GO to DP2. Unable to duplicate fault at this time. If any other codes are present, return to Quick Test for directions. If codes are not present, test is completed.
DP2 SENSOR CHECK		
• Key Off, wait 10 seconds. • Locate and disconnect vehicle speed sensor. • DVOM on 200,000 ohm scale. • Measure resistance across vehicle speed sensor. • Is resistance between 190 and 240 ohms?	Yes ▶ No ▶	GO to DP3. REPLACE sensor. RERUN Test Step DP1.

Vehicle Speed Sensor — Pinpoint Test — DP

TEST STEP	RESULT ▶	ACTION TO TAKE
DP3 CONTINUITY OF VEHICLE SPEED SENSOR (VSS) HARNESS		
• Key Off, wait 10 seconds. • Disconnect processor 60 Pin connector and inspect for damaged pins, corrosion, loose wires. Service as necessary. • Install Breakout box. • Processor and VSS disconnected. • DVOM on 200 ohm scale. • Measure resistance between test Pin 3 at the Breakout box and the VSS vehicle harness connector as shown below. • Measure resistance between test Pin 6 at the Breakout box and the VSS vehicle harness connector, as shown below. 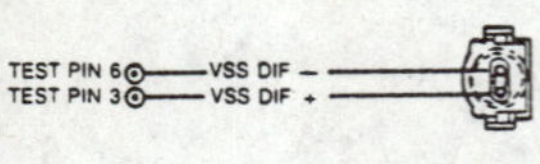 • Are both resistances 5 ohms or less?	Yes ▶ No ▶	GO to DP4. SERVICE open(s) in VSS harness. RERUN Test Step DP1.
DP4 CHECK VSS HARNESS FOR SHORTS TO POWER OR GROUND		
• Key Off. • Processor disconnected. • VSS disconnected. • DVOM on 200,000 ohm scale. • Measure resistance between test Pin 3 and test Pins 37, 40 and 6 at the Breakout box. Then measure resistance between test Pin 6 and test Pins 37 and 40 (except 2.3L EFI T.C.) at the Breakout box. • Are all readings greater than 10,000 ohms?	Yes ▶ No ▶	GO to DP5. SERVICE short(s) in VSS harness. RERUN Test Step DP1.

Vehicle Speed Sensor — Pinpoint Test — DP

TEST STEP	RESULT ▶	ACTION TO TAKE
DP5 SUBSTITUTE VEHICLE SPEED SENSOR (VSS)		
• Substitute VSS with known good sensor. • Processor and VSS connected. • Perform Drive Cycle outlined in Test Step DP1 then return to this Step. • Is code 29 (for 2.3L EFI T.C. vehicles code 27) present in continuous?	Yes ▶ No ▶	REPLACE processor. RERUN Test Step DP1. The original continuous code 29 or 27 was the result of the original VSS. SERVICE VSS as necessary.

Neutral Drive Switch A/C Input — Pinpoint Test — FA

TEST STEP	RESULT ▶	ACTION TO TAKE
FA1 CODE 67 SYSTEM IDENTIFICATION		
2.5L CFI M/T, 5.0L M/T TK		GO to FA4.
2.9L M/T TK, 3.0L M/T TK, 5.0L M/T SEFI		GO to FA10.
1.9L M/T, 2.3L EFI M/T Car and Truck, 2.3L CFI M/T		GO to FA10.
2.3L Turbo		GO to FA15.
2.3L Turbo Octane Switch		GO to FA30.
4.9L M/T, Truck		GO to FA18.
All other systems		GO to FA2.
FA2 NEUTRAL DRIVE INPUT CHECK		
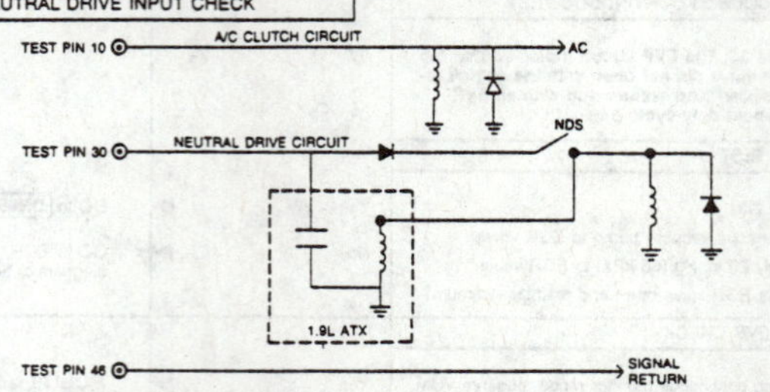 		
• Key Off, wait 10 seconds. • Verify heater control is in OFF position, if so equipped. • Verify transmission is in NEUTRAL or PARK. • Disconnect processor 60 Pin connector and inspect for damaged pins, corrosion, loose wires. Service as necessary. • Install Breakout box. • Processor connected. • Key On, Engine Off. • DVOM on 20V scale. • Measure voltage between test Pin 30 (Neutral Drive circuit) at the Breakout box and chassis ground. • Is reading less than 1.0 volt?	Yes ▶ No ▶	GO to FA4. GO to FA3.

Neutral Drive Switch A/C Input — Pinpoint Test — FA

TEST STEP	RESULT ▶	ACTION TO TAKE
FA3 NEUTRAL DRIVE SWITCH CHECK		
• Key Off, wait 10 seconds. • Breakout box installed. • DVOM on 200 ohm scale. • Locate the Neutral Drive switch. • Disconnect vehicle harness from the Neutral Drive switch and measure resistance across the switch. • Is resistance reading less than 5 ohms?	Yes ▶	SERVICE open in vehicle harness Neutral Drive circuit. RERUN Quick Test.
	No ▶	REPLACE Neutral Drive switch. RERUN Quick Test
FA4 A/C INPUT CHECK		
NOTE: Before entering this test, verify A/C is off. If A/C was on, rerun Quick Test. If code 67 or code 79 is present, continue with this test. • Breakout box installed. • Processor disconnected. • Key on, engine off. • DVOM on 20V scale. • Measure voltage between test Pin 10 at the Breakout box and chassis ground. • Is voltage reading 1.0 volt or greater?	Yes ▶	SERVICE short to power in A/C clutch circuit. RERUN Quick Test.
	No ▶	REPLACE processor. RERUN Quick Test.

Neutral Drive Switch A/C Input — Pinpoint Test — FA

TEST STEP	RESULT ▶	ACTION TO TAKE
FA12 NEUTRAL CLUTCH HARNESS CHECK		
• Key Off. • DVOM on 200 ohm scale. • Breakout box installed. • Vehicle harness disconnected at the Neutral switch and Clutch switch. • Measure resistance between test Pin 30 and the Neutral switch harness connector and between test Pin 30 and the Clutch switch harness connector. • Measure resistance between test Pin 46 and the Neutral switch harness connector and between test Pin 46 and the Clutch switch harness connector. • Are all resistance readings less then 5 ohms?	Yes ▶	GO to FA4.
	No ▶	SERVICE the harness connector. RERUN Quick Test.
FA15 NEUTRAL INPUT CHECK — 2.3L TC M/T		
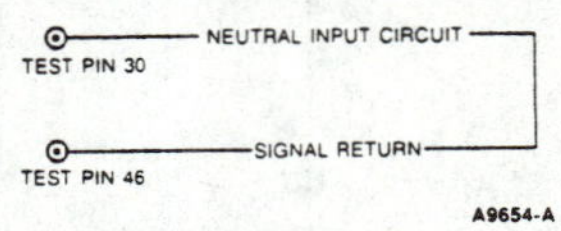 • Key Off, wait 10 seconds. • Verify A/C is off, if so equipped. • Disconnect processor 60 Pin connector and inspect for damaged pins, corrosion, loose wires. Service as necessary. • Install Breakout box. • Leave processor disconnected. • DVOM on 200 ohm scale. • Measure resistance between test Pin 30 (Neutral Input circuit) and test Pin 46 (Signal Return circuit) at the Breakout box. • Is resistance reading less than 5 ohms?	Yes ▶	GO to FA4.
	No ▶	SERVICE open in Neutral Input or Signal Return circuit. RERUN Quick Test.

Neutral Drive Switch A/C Input — Pinpoint Test — FA

TEST STEP	RESULT ▶	ACTION TO TAKE
FA10 NEUTRAL-CLUTCH INPUT CHECK		
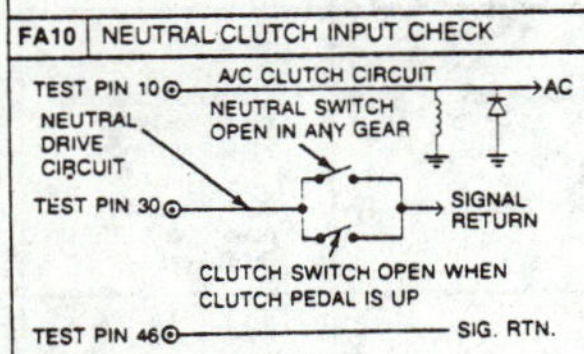 • Key Off, wait 10 seconds. • Verify A/C is off, if so equipped. • Disconnect processor 60 Pin connector and inspect for damaged pins, corrosion, loose wires. Service as necessary. • Install Breakout box. • Connect processor. • DVOM on 200 ohm scale. • Measure resistance between test Pin 30 and test Pin 46. 1. With transmission in NEUTRAL and clutch up. 2. With transmission in GEAR and clutch down. • Are both resistance readings less than 5 ohms?	Yes ▶	GO to FA4.
	No ▶	GO to FA11.
FA11 NEUTRAL/CLUTCH SWITCH CHECK		
• Key Off. • DVOM on 200 ohm scale. • Breakout box installed. • Locate Neutral switch (on transmission) and Clutch switch (under dash). • Disconnect vehicle harness at both switches and inspect connectors for pushed back pins. • Measure resistance across the Neutral switch terminals with transmission in NEUTRAL and across the Clutch switch terminals with the clutch pedal down. • Are both resistance readings less than 5 ohms?	Yes ▶	GO to FA12.
	No ▶	REPLACE open switch(es). RECONNECT harness and RERUN Quick Test.

Neutral Drive Switch A/C Input — Pinpoint Test — FA

TEST STEP	RESULT ▶	ACTION TO TAKE
FA18 CLUTCH PEDAL SWITCH CHECK 4.9L M/T		
NOTE: The clutch pedal must be down during KOEO test; if not, a code 67 will result. • Key Off. • Disconnect processor 60 Pin connector and inspect for damaged pins, corrosion, loose wires. Service as necessary. • Install Breakout box. • DVOM on 200 ohm scale. • Clutch pedal down. • Measure resistance between test Pin 30 and test Pin 46, between test Pin 30 and test Pin 40. • Is resistance reading less than 5 ohms?	Yes ▶	REPAIR open in clutch circuit.
	No ▶	GO to FA4.
FA20 A/C INPUT CIRCUIT CHECK		
NOTE: A low idle with A/C on could be the result of the processor not receiving, or recognizing the A/C input on Pin 10. • Key Off, wait 10 seconds. • Disconnect processor 60 Pin connector and inspect for damage, corrosion or loose wires. Service as necessary. • Install Breakout box. • Leave processor disconnected. • DVOM on 20 volt scale. • Key On. • A/C on. • Check voltage between Pin 10 and Pin 40. • Is voltage reading greater than 10.5 volts?	Yes ▶	REPLACE processor. RERUN Quick Test.
	No ▶	SERVICE open in A/C circuit. Refer to the appropriate engine schematic in Section 17 or 18. RERUN Quick Test.

Neutral Drive Switch A/C Input — Pinpoint Test — FA

TEST STEP	RESULT ▶	ACTION TO TAKE
FA30 CHECK OCTANE SWITCH INPUT FOR INPUT CHANGE		
• Key Off, wait 10 seconds. • Disconnect processor 60 Pin connector and inspect for damaged pins, corrosion, loose wires. Service as necessary. • Install Breakout box. Reconnect processor. • DVOM to 20V scale. • Connect positive test lead to test Pin 30 and negative test lead to test Pin 46 at the Breakout box. • Key On, Engine Off. • Cycle octane switch several times while observing DVOM. • Does voltage change from zero volts to 5V?	Yes ▶	REPLACE processor. RERUN Quick Test.
	No ▶	EEC-IV system OK. REFER to Shop Manual for boost diagnostics.

Key Power Check — Pinpoint Test — FC

FC

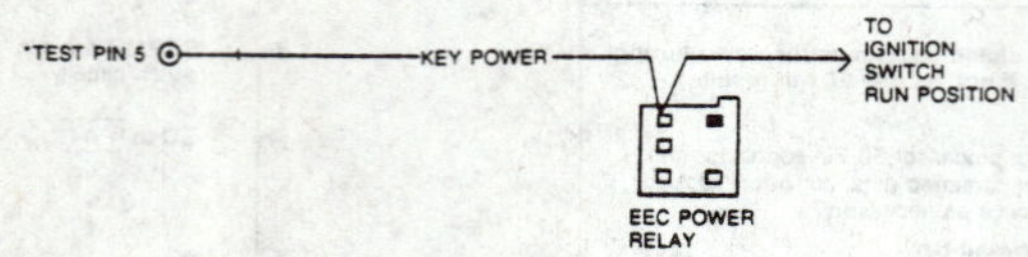

*TEST PINS LOCATED ON BREAKOUT BOX.
ALL HARNESS CONNECTORS VIEWED INTO MATING SURFACE.

STOP-WARNING

You should enter this Pinpoint Test only when a Service Code 55 or 65 is received in Quick Test Step 3.0, 5.0 or 6.0.

To prevent the replacement of good components, be aware that the following non-EEC areas may be at fault:

- Charging system overvoltage.
- Battery charger connected with engine running.
- Jump starting.

This Pinpoint Test is intended to diagnose only the following:

- Harness circuit: key power.
- Processor.

Key Power Check — Pinpoint Test — FC

TEST STEP	RESULT ▶	ACTION TO TAKE
FC1 BATTERY VOLTAGE GREATER THAN 17.5V		
• Key Off, wait 10 seconds. • Install Breakout box with processor connected. • DVOM at 20V range and connected to test Pins 5 and 60 at the Breakout box. • Perform Engine Running Quick Test. Observe DVOM during test and record service codes.	Reading exceeds 17.5V during Quick Test ▶	CORRECT charging system for over-voltage condition.
	Reading remains below 17.5V and code 65 is present ▶	REPLACE processor and RERUN Quick Test.
	Reading remains below 17.5V and code 65 is not present ▶	GO to FC2.
FC2 CODE 65 IN CONTINUOUS MEMORY		
• Perform Key On, Engine Off Quick Test and record continuous codes.	Code 65 not present ▶	TESTING complete. CHARGING system OK at this time.
	Code 65 present ▶	CHECK charging system, REFER to Shop Manual, Group 31 for cause of intermittent overcharging (greater than 17.5V).
FC3 BATTERY VOLTAGE LESS THAN 7.5V		
• Key Off, wait 10 seconds. • Disconnect processor 60 Pin connector and inspect for damaged pins, corrosion, loose wires. Service as necessary. • Install Breakout box to processor harness connector. • Processor connected. • DVOM on 20V scale. • Measure voltage between test Pin 5 and test Pin 60 at the Breakout box. • Perform Engine Running Quick Test.	Voltage below 7.5V during Quick Test ▶	GO to FC4.
	Voltage remains above 7.5V and code 55 is present ▶	REPLACE processor, RERUN Quick Test.
	Voltage remains above 7.5V and code 55 is not present ▶	GO to FC5.

Key Power Check — Pinpoint Test — FC

TEST STEP	RESULT ▶	ACTION TO TAKE
FC4 KEY POWER CIRCUIT CHECK		
• Key Off. • Breakout box installed, processor connected. • DVOM at 200 ohm scale. • Measure resistance between test Pin 5 at the Breakout box and key power terminal of EEC power relay.	5 ohms or less ▶	CORRECT charging system for under-voltage condition.
	Greater than 5 ohms ▶	SERVICE open in key power circuit. RERUN Quick Test.
FC5 CODE 55 IN CONTINUOUS MEMORY		
• Perform Key On, Engine Off Quick Test and record continuous codes.	Code 55 not present ▶	TESTING complete. CHARGING system OK at this time.
	Code 55 present ▶	CHECK charging system per Shop Manual, Group 31 for cause of intermittent overcharging (greater than 17.5V).

Brake On/Off (BOO) — Pinpoint Test — FD

FD

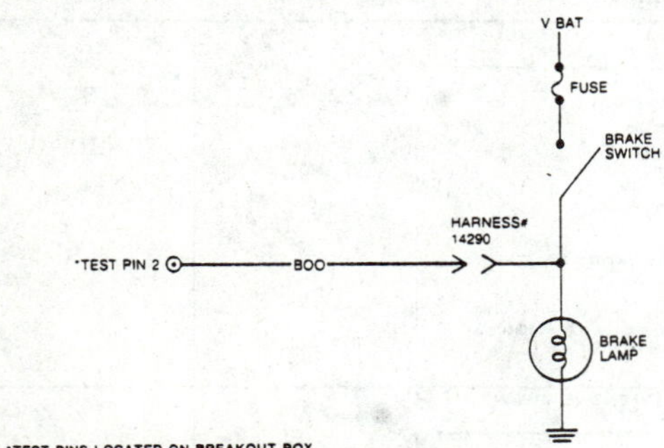

*TEST PINS LOCATED ON BREAKOUT BOX.
ALL HARNESS CONNECTORS VIEWED INTO MATING SURFACE.

STOP-WARNING

You should enter this Pinpoint Test only when a Service Code 74 or 75 is received in Quick Test Step 5.0.

To prevent the replacement of good components, be aware that the following non-EEC areas may be at fault:

- Brake lamp, brake switch, and fuse.

This pinpoint test is intended to diagnose only the following:

- BOO circuit.
- Processor assembly.

Brake On/Off (BOO) — Pinpoint Test — FD

TEST STEP	RESULT ▶	ACTION TO TAKE
FD1 SERVICE CODE 74		
• Did you press brake during the Engine Running Quick Test? **NOTE: On some vehicles it is necessary to depress and release the brake after the dynamic response code 1(0) but before the brief WOT.**	Yes ▶ No ▶	GO to FD2. RERUN Engine Running Quick Test. PRESS brake once during test.
FD2 BOO CIRCUIT CYCLING		
• Key Off, wait 10 seconds. • Disconnect processor 60 Pin connector. Inspect for damaged pins, corrosion, loose wires, etc. Service as necessary. • Install Breakout box, leave processor disconnected. • DVOM on 20V scale. • Measure voltage between test Pin 2 and test Pin 40 at the Breakout box while depressing and releasing brake. • Does the voltage cycle?	Yes ▶ No ▶	REPLACE processor. RERUN Quick Test. GO to FD3.
FD3 BOO CIRCUIT SHORT TO GROUND		
• Key Off. • Breakout box installed. • Processor disconnected. • DVOM on 200 Ohm scale. • Disconnect BOO circuit from 14290 harness (12 pin connector). • Measure resistance between test Pin 2 at the Breakout box and ground. • Is resistance reading greater than 5 ohms?	Yes ▶ No ▶	GO to FD6. SERVICE BOO circuit short to ground. RERUN Engine Running Quick Test.

Brake On/Off (BOO) — Pinpoint Test — FD

TEST STEP	RESULT ▶	ACTION TO TAKE
FD4 BOO CIRCUIT CYCLING CODE 75		
• Key Off, wait 10 seconds. • Disconnect processor 60 Pin connector. Inspect for damaged pins, corrosion, loose wires, etc. Service as necessary. • Install Breakout box, leave processor disconnected. • DVOM on 20V scale. • Measure voltage between test Pin 2 and test Pin 40 at the Breakout box while depressing and releasing brake. • Does the voltage cycle?	Yes ▶ No ▶	REPLACE processor. RERUN Quick Test. GO to FD5.
FD5 BOO CIRCUIT SHORT TO POWER		
• Key Off. • Breakout box installed. • Processor disconnected. • DVOM on 20V scale. • Disconnect BOO circuit from 14290 harness (12 Pin connector). • Measure voltage between test Pin 2 at the Breakout box and engine block ground. • Is voltage reading greater than 10.5 volts?	Yes ▶ No ▶	SERVICE BOO circui short to power. BOO circuit OK. GO to Shop Manual, Group 32 to SERVIC stoplamp circuit.
FD6 BOO HARNESS CONTINUITY CHECK		
• Key Off. • Breakout box installed. • Processor disconnected. • DVOM on 200 Ohm scale. • Disconnect BOO circuit from 14290 harness (12 pin connector). • Measure resistance between test Pin 2 at the Breakout box and BOO circuit at the 14290 harness connector. • Is resistance reading greater than 5 ohms?	Yes ▶ No ▶	SERVICE BOO circuit for open. RERUN Engine Running Self Test. GO to Shop Manual, Group 32.

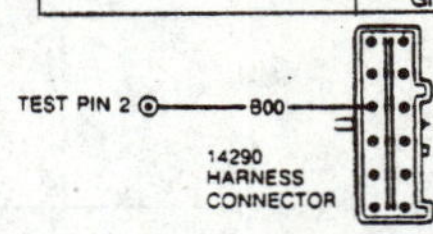

Power Steering Pressure Switch (PSPS) — Pinpoint Test — FF

FF

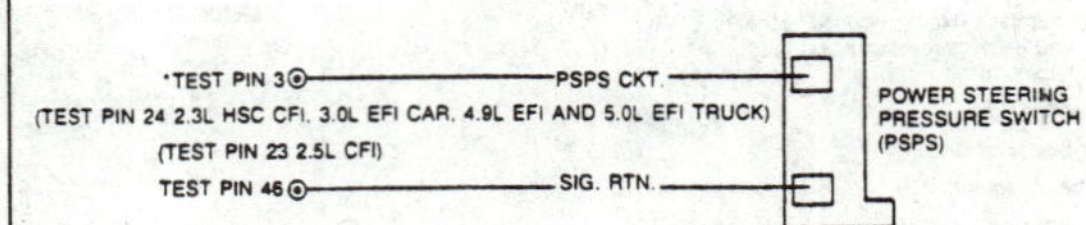

*TEST PINS LOCATED ON BREAKOUT BOX.
ALL HARNESS CONNECTORS VIEWED INTO MATING SURFACE.

A9658-B

STOP-WARNING

You should enter this Pinpoint Test only when a Service Code 52 is received in Quick Test Steps 3.0, 5.0 or if you are directed here from Diagnostics by Symptom in the Engine Supplement Section.

To prevent the replacement of good components, be aware that the following non-EEC areas may be at fault:

- Idle speeds/throttle stop adjustment.
- Binding throttle shaft/linkage or speed control linkage.

This Pinpoint Test is intended to diagnose only the following:

- Power steering pressure switch.
- Switch harness circuits: PSPS Signal, and Signal Return.
- Processor assembly.

Power Steering Pressure Switch (PSPS) — Pinpoint Test — FF

TEST STEP	RESULT ▶	ACTION TO TAKE
FF1 ATTEMPT TO ELIMINATE CODE 52		
• Key Off, wait 10 seconds. • Disconnect PSPS. • Jumper PSPS circuit to Signal Return at vehicle harness connector. • Rerun Key On, Engine Off Quick Test. • Is code 52 still present?	Yes ▶ No ▶	GO to FF2. REPLACE PSPS. RERUN Quick Test.
FF2 PSPS HARNESS CHECK		
• Key Off, wait 10 seconds. • Disconnect harness from processor 60 Pin connector and inspect for damaged pins, corrosion, loose wires. Service as necessary. • PSPS disconnected. • Processor disconnected. • Breakout box installed. • DVOM on 200 ohm scale. • Measure resistance between test Pin 46 at the Breakout box and Signal Return at the PSPS connector and between test Pin 3 (2.3L OHC EFI), test Pin 23 (2.5L CFI), test Pin 24 (2.3L HSC, 3.0L EFI, 4.9L EFI and 5.0L EFI) at the Breakout box and PSPS circuit at the PSPS harness connector. • Are both readings less than 5 ohms?	Yes ▶ No ▶	REPLACE processor. RERUN Quick Test. SERVICE open in circuit. RERUN Quick Test.
FF3 SWITCH INTEGRITY		
• Install tachometer. • Start engine, allow to idle in Neutral/Park. • Disconnect PSPS at switch. • Does rpm increase?	Yes ▶ No ▶	REPLACE PSPS. GO to FF4.

Power Steering Pressure Switch (PSPS) — Pinpoint Test — FF

TEST STEP	RESULT ▶	ACTION TO TAKE
FF4 PSPS HARNESS CHECK		
• Key Off, wait 10 seconds. • Disconnect harness from processor 60 Pin connector and inspect for damaged pins, corrosion, loose wires. Service as necessary. • PSPS disconnected. • Processor disconnected. • Breakout box installed. • DVOM on 200,000 ohm scale. • Measure resistance between test Pin 3 (2.3L OHC EFI), test Pin 23 (2.5L CFI), test Pin 24 (2.3L HSC CFI, 3.0L EFI, 4.9L EFI and 5.0L EFI) and test Pin 46 at the Breakout box. • Is reading less than 10,000 ohms?	Yes ▶ No ▶	SERVICE short in harness. REPLACE processor.
FF5 SERVICE CODE 52 ENGINE RUNNING SELF-TEST		
• Did you turn the steering wheel one-half turn any time after engine ID code, but before the dynamic response code? **NOTE: Make sure the front wheels are centered (no load condition).**	Yes ▶ No ▶	GO to FF6. RERUN Quick Test.
FF6 DETERMINE WHETHER THE PROCESS CAN IDENTIFY AN OPEN CIRCUIT		
• Key Off, wait 10 seconds. • Disconnect PSPS. • Run Key On Engine Off Self-Test. • Is code 52 present?	Yes ▶ No ▶	GO to FF8. GO to FF7.
FF7 PSPS HARNESS CHECK		
• Key Off, wait 10 seconds. • Disconnect harness from processor 60 Pin connector and inspect for damaged pins, corrosion or loose wires. Service as necessary. • PSPS disconnected. • Processor disconnected. • Breakout box installed. • DVOM on 200,000 ohm scale. • Measure resistance between test Pin 46 and test Pin 24 at the Breakout box. • Is resistance reading 10,000 ohms or less?	Yes ▶ No ▶	SERVICE PSPS circuit short to signal return. RERUN Quick Test. REPLACE processor. RERUN Quick Test.

Power Steering Pressure Switch (PSPS) — Pinpoint Test — FF

TEST STEP	RESULT ▶	ACTION TO TAKE
FF8 PSPS POSITION KEY ON ENGINE NOT RUNNING VS. RUNNING		
• Key Off, wait 10 seconds. • Breakout box installed. • Processor connected. • PSPS connected. • DVOM on 200 ohm scale. • Key On. • Measure resistance from test Pin 24 to test Pin 46. • Start engine. • Does resistance remain less than 10 ohms between Key On Engine Off and Engine Running?	Yes ▶ No ▶	GO to FF9. GO to FF11.
FF9 PSPS POSITION ENGINE RUNNING NO LOAD VS. LOAD		
• Breakout box installed. • Processor connected. • PSPS connected. • DVOM on 200 ohm scale. • Engine idling. • Measure the resistance from test Pin 24 to test Pin 46. • Turn the steering wheel at least one-half turn then return. • Does resistance change from less than 10 ohms to infinity (indicating PSPS opening), then returning to 10 ohms or less when steering wheel is returned to center position?	Yes ▶ No ▶	PSPS system OK. REMOVE Breakout box and RETURN to Quick Test Step 5.0 to continue Diagnostics. GO to FF10.

Power Steering Pressure Switch (PSPS) — Pinpoint Test — FF

TEST STEP	RESULT ▶	ACTION TO TAKE
FF10 PSPS VS. POWER STEERING HYDRAULIC PRESSURE (ENGINE RUNNING ALWAYS CLOSED)		
• At this point in the Diagnostics there are only two possible causes for the original code 52 Engine Running: — PSPS (switch) that will not open. — Low available hydraulic pressure. • Key Off, wait 10 seconds. • Substitute original PSPS with a known good PSPS. • Run Engine Running Self Test. (Turn steering wheel one-half turn during test.) • Is code 52 still present?	Yes ▶ No ▶	GO to Power Steering Pressure Diagnostics, Shop Manual, Group 13, looking for low pressure. Original code 52 Engine Running was a result of a bad PSPS (switch). REMOVE all equipment and continue, if necessary, with any other Diagnostics.
FF11 PSPS VS. POWER STEERING HYDRAULIC PRESSURE (ENGINE RUNNING ALWAYS OPEN)		
• At this point in the Diagnostics there are two possible causes for the original code 52 Engine Running: — PSPS (switch) that always remains open during Engine Running. — Excessively high hydraulic pressure. • Key Off, wait 10 seconds. • Substitute original PSPS with a known good PSPS. • Run Engine Running Self Test. (Turn steering wheel one-half turn during test.) • Is code 52 still present?	Yes ▶ No ▶	GO to Power Steering Pressure Diagnostics in Shop Manual, Group 13, looking for high pressure. Original code 52 Engine Running was a result of a bad PSPS (switch). REMOVE all equipment and continue, if necessary, with any other Diagnostics.

Fuel Control — 1.9L EFI and 2.3L EFI TC | Pinpoint Test | HA

HA

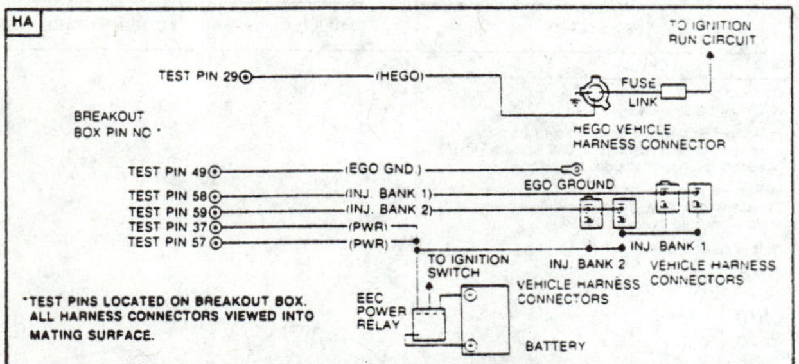

*TEST PINS LOCATED ON BREAKOUT BOX. ALL HARNESS CONNECTORS VIEWED INTO MATING SURFACE.

STOP-WARNING

You should enter this Pinpoint Test Only when a Service Code 41. 42 or 43 is received in Quick Test Step 5.0 or 6.0 or when directed here from Pinpoint Test A or Diagnostics by Symptom in the Engine Supplement Section.

To prevent the replacement of good components, be aware that the following non-EEC areas may be at fault:

- Ignition Coil
- Distributor Cap
- Distributor Rotor
- Fouled Spark Plugs
- Spark Plug Wires
- CANP Problems
- PCV Valves (see note below)
- EGR Valve and Gasket
- Air Filter
- Fuel Contamination, Engine Oil
- Poor Power Ground
- Fuel Pressure
- Manifold Leaks, Intake Exhaust
- Engine Not at Normal Operating Temperatures

This Pinpoint Test is intended to diagnose only the following:

- HEGO Sensor
- Harness Circuits HEGO Grd., HEGO, INJ BANK 1, INJ BANK 2, V PWR
- HEGO Sensor Connection
  - Code 42 start at HA8
  - Code 41 start HA14
  - Code 43 start at HA30
- Vacuum Systems
- Fuel Injectors
- Processor Assembly

NOTE: Fuel contaminated engine oil may affect 41, 42 Service Codes. If this is suspected, remove the PCV from the valve cover and repeat the Quick Test. If the problem is corrected, then change the engine oil and filter.

Fuel Control — 1.9L EFI and 2.3L EFI TC | Pinpoint Test | HA

TEST STEP	RESULT ▶	ACTION TO TAKE
HA1 FUEL PRESSURE CHECK 1		
• Key Off, wait 10 seconds. • Install fuel pressure gauge. • Start and run engine. Is fuel pressure between 35 and 45 psi (241-310 kPa)? **For No Starts:** • If engine will not run, cycle the key from Off to On several times to increase fuel pressure.	Yes ▶ No ▶	GO to HA2. REFER to the Shop Manual Group 24 for electric fuel pump and Section 11 for fuel pressure regulator check.
HA2 FUEL PRESSURE CHECK 2		
• Does fuel pressure remain between 35 and 45 psi (241-310 kPa) for 60 seconds or longer after final Key Off?	Yes ▶ No ▶	GO to HA3. GO to HA7.
HA3 FUEL DELIVERY TEST		
• Key Off. • Fuel pressure gauge installed. • Pressurize fuel system as per Step HA1. • Locate and disconnect the fuel pump relay. • Crank engine for 5 seconds. • Take pressure reading at the end of 5 second crank. • Is pressure between 10 and 20 psi?	Yes ▶ No ▶	The EEC system is not the fault of the No Start. Fuel and spark are present. REFER to Section 2 for other No Start Routines. If complaint was runs rough/ misses or fuel code GO to HA4. GO to HA4.

NOTE: The colder the engine, the greater the pressure drop (i.e., an engine coolant temperature of 200°F equals approximately a 10 psi drop in 5 seconds. 60°F equals approximately a 20 psi drop in 5 seconds).

Verify fuel quality; air and/or water will also pressurize and look like acceptable fuel delivery.

Fuel Control — 1.9L EFI and 2.3L EFI TC | Pinpoint Test | HA

TEST STEP	RESULT ▶	ACTION TO TAKE
HA4 INJECTOR HARNESS RESISTANCE CHECK		
• Key Off, wait 10 seconds. • Disconnect processor 60 Pin connector and inspect for damaged pins, corrosion, loose wires. Service as necessary. • Connect Breakout box to harness. Leave processor disconnected. • DVOM on 200 ohm scale. • Measure the resistance of injector Bank 1 between test Pin 37 and test Pin 58 at the Breakout box. Record resistance. • Measure the resistance of injector Bank 2 between test Pin 37 and test Pin 59 at the Breakout box. Record resistance. • Are both resistances between 1.2 ohms and 1.8 ohms?	Yes ▶ No ▶	GO to HA6. GO to HA5.
HA5 ISOLATE FAULTY INJECTOR CIRCUIT		
• Key Off. • Breakout box installed. • Leave processor disconnected. • DVOM on 200 ohm scale. • Disconnect all injectors on suspect Bank. Measure resistance of each injector by connecting one injector at a time, and reading the resistance between test Pins 37 and 58 at the Breakout box for Bank 1 or test Pins 37 and 59 at the Breakout box for Bank 2. • Are all resistances between 2.0 and 2.7 ohms?	Yes ▶ No ▶	GO to HA6. SERVICE the harness/connectors on the suspect injector for opens or shorts. If OK, REPLACE injector. RERUN Quick Test.

Fuel Control — 1.9L EFI and 2.3L EFI TC | Pinpoint Test | HA

TEST STEP	RESULT ▶	ACTION TO TAKE
HA6 INJECTOR DRIVE SIGNAL CHECK		
Requires standard non-powered 12 volt test lamp. • Key Off. • Breakout box installed. • Connect processor to Breakout box. • Connect test lamp between test Pin 37 and test Pin 58 at the Breakout box. • Crank or start engine. • Repeat above test between test Pin 37 and test Pin 59 at the Breakout box.	Dim glow at light on both tests ▶ No light on one or both tests ▶ Bright light on one or both tests ▶	GO to HA7. VERIFY 12-volt battery power at Pins 37 and 57. If OK, REPLACE processor. RERUN Quick Test. CHECK circuits Bank 1 and Bank 2 for shorts to ground. If OK, REPLACE processor. RERUN Quick Test.
HA7 INJECTOR BALANCE TEST		
• Connect tachometer to engine. Disconnect ISC. Run engine at 2000 rpm. Use throttle body stop screw to set engine speed. • Disconnect and reconnect the injectors one at a time; note rpm drop for each injector. • Does each injector produce at least a 150 rpm drop? • Reconnect ISC. Reset curb idle. Refer to VECI decal and Section 4.	Yes ▶ No ▶	Fuel delivery OK. Problem is with area common to all cylinders, i.e.: Air/vacuum leak, fuel contamination, EGR. REPLACE faulty injector. RERUN Quick Test.

Fuel Control — 1.9L EFI and 2.3L EFI TC | Pinpoint Test | HA

TEST STEP	RESULT ▶	ACTION TO TAKE
HA8 FUEL CONTROL — ALWAYS RICH: CODE 42. CHECK HEGO HARNESS FOR SHORT TO POWER		
• Key Off, wait 10 seconds. • Disconnect the HEGO sensor from the vehicle harness. • DVOM on 20 volt scale. • Key On. • Measure voltage between HEGO signal and ground at the HEGO harness connector. • Is voltage less than .5 volt?	Yes ▶ No ▶	GO to HA9 SERVICE short to power in HEGO signal circuit. RERUN Quick Test.

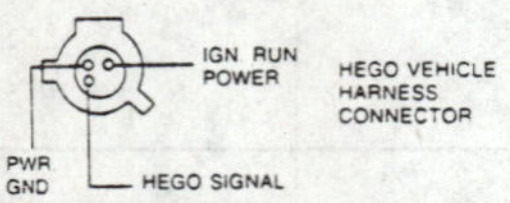

TEST STEP	RESULT ▶	ACTION TO TAKE
HA9 CHECK HEGO FOR SHORT TO KEY POWER		
• Key Off. • HEGO disconnected. • DVOM on 200,000 ohm scale. • Measure resistance between Key Power input and HEGO signal at the HEGO. • Is resistance greater than 10,000 ohms?	Yes ▶ No ▶	GO to HA10 REPLACE HEGO sensor. RERUN Quick Test.

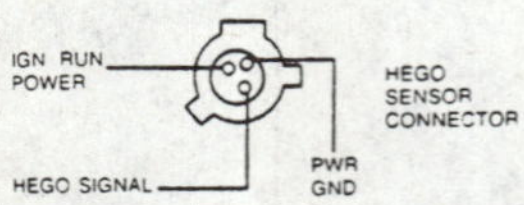

Fuel Control — 1.9L EFI and 2.3L EFI TC | Pinpoint Test | HA

TEST STEP	RESULT ▶	ACTION TO TAKE
HA10 ATTEMPT TO GENERATE CODE 41		
• Key Off, wait 10 seconds. • **NOTE: Non-EEC areas could cause a code 42.** **Check for:** **— Fuel contaminated engine oil** **— Ignition caused misfire** **— CANP problems** • Disconnect vehicle harness at the HEGO sensor. Ground the HEGO circuit at the HEGO vehicle harness connector to the engine block. • Repeat Engine Running Self-Test. • Is service code 41 present?	Yes ▶ No ▶	GO to HA12. GO to HA11.
HA11 HARNESS CHECK		
• Key Off, wait 10 seconds. • Install Breakout box. • Measure resistance of HEGO ground circuit between test Pin 49 at the Breakout box and HEGO ground at engine block. • Measure resistance of HEGO circuit between test Pin 29 at the Breakout box and HEGO harness connector. • Are both circuits less than 5 ohms?	Yes ▶ No ▶	DISCONNECT processor connector. INSPECT for damage or corrosion. If OK, REPLACE processor. RERUN Quick Test. CORRECT harness circuit with resistance greater than 5 ohms. RERUN Quick Test.

Fuel Control — 1.9L EFI and 2.3L EFI TC | Pinpoint Test | HA

TEST STEP	RESULT ▶	ACTION TO TAKE
HA12 EGO CHECK		
• DVOM on 20V scale. • With HEGO sensor disconnected from the vehicle harness, connect a DVOM from HEGO sensor to engine ground. • Run the engine at 2000 rpm for 1 minute. While observing DVOM, disconnect the manifold vacuum hose indicated below. • Is voltage less than 0.4V?	Yes ▶ No ▶	HEGO sensor OK. GO to HA1. REPLACE HEGO sensor. RERUN Quick Test.

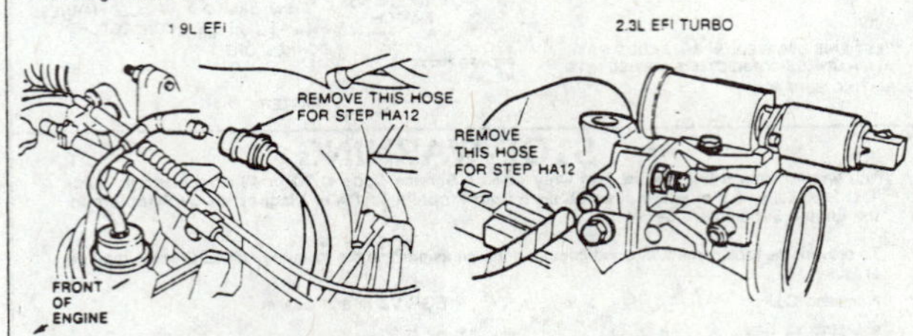

TEST STEP	RESULT ▶	ACTION TO TAKE
HA13 FUEL CONTROL — ALWAYS LEAN: CODE 41 VERIFICATION		
• Run vehicle at 2000 rpm for two minutes. • Key Off, wait 10 seconds. • Perform Engine Running Self-Test. • Is code 41 present?	Yes ▶ No ▶	GO to HA14. GO to HA20.

Fuel Control — 1.9L EFI and 2.3L EFI TC | Pinpoint Test | HA

TEST STEP	RESULT ▶	ACTION TO TAKE
HA14 FUEL CONTROL — ALWAYS LEAN: CODE 41		
NOTE: Vacuum/air leaks in non-EEC areas could cause a code 41. Check for: **— Leaking vacuum actuator (eg: A/C control motor)** **— Engine sealing** **— EGR system** **— PCV system** **— Unmetered air leak between air meter and throttle body** **— Lead contaminated HEGO sensor** • Key Off. DVOM on 20V scale. Disconnect HEGO sensor from vehicle harness. Connect DVOM between HEGO signal at HEGO sensor and engine ground. Remove air cleaner to gain access to air meter inlet. Using a standard wood lead pencil, prop the air meter door part-way open. • Start the engine and run at approximately 2000 rpm for 2 minutes. • Does the DVOM read greater than 0.5V at the end of 1 minute?	Yes ▶ No ▶	GO to HA15. REPLACE HEGO sensor. RERUN Quick Test.

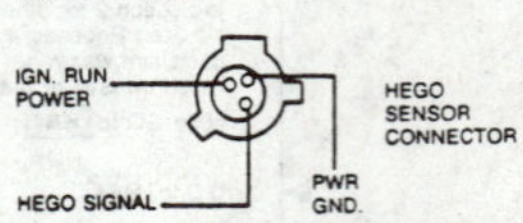

TEST STEP	RESULT ▶	ACTION TO TAKE
HA15 HARNESS CHECK EGO CIRCUITS		
• Key Off. • Install Breakout box. Processor disconnected. • Measure resistance of HEGO circuit between test Pin 49 at the Breakout box and engine block ground. Also HEGO circuit between HEGO connector and test Pin 29 at the Breakout box. • Are both circuits less than 5 ohms?	Yes ▶ No ▶	GO to HA16. SERVICE/CORRECT as necessary any circuit greater than 5 ohms. RERUN Quick Test.

Fuel Control — 1.9L EFI and 2.3L EFI TC | Pinpoint Test | HA

TEST STEP	RESULT ▶	ACTION TO TAKE
HA16 CHECK HEGO CIRCUIT FOR SHORT TO GROUND		
• Key Off. • Breakout box installed. Processor disconnected. • DVOM to 200,000 ohms scale. • Measure the resistance between test Pin 29 and test Pin 40 at the Breakout box. • Is resistance greater than 10,000 ohms?	Yes ▶ No ▶	GO to HA17. CORRECT cause of resistance to ground. RERUN Quick Test.
HA17 CHECK HEGO SENSOR FOR SHORT TO GROUND		
• Key Off. • HEGO disconnected. • DVOM on 200,000 ohm scale. • Measure resistance between ground and HEGO signal at HEGO sensor. • Is resistance greater than 10,000 ohms?	Yes ▶ No ▶	GO to HA18. REPLACE HEGO sensor. RERUN Quick Test.
HA18 ATTEMPT TO ELIMINATE CODE 41		
• Key Off. • Reconnect HEGO sensor. Remove air cleaner to gain access to air meter inlet. Using a standard wood lead pencil, prop the air meter door part-way open. • Start the engine and run at approximately 2000 rpm for 2 minutes. • Perform Engine Running Self-Test. • Is code 41 present?	Yes ▶ No ▶	INSPECT for corrosion or damaged pins. If OK, REPLACE processor. RERUN Quick Test. HEGO input circuit OK. GO to HA1.

Fuel Control — 1.9L EFI and 2.3L EFI TC | Pinpoint Test | HA

TEST STEP	RESULT ▶	ACTION TO TAKE
HA20 CHECK HEATER ELEMENT RESISTANCE ON HEGO		
• Key Off. • DVOM on 200 ohm scale. • HEGO disconnected. • Measure resistance between ignition run circuit and power ground at HEGO sensor. • Hot to warm resistance specification is 5 to 20 ohms. • Is resistance within specification? **NOTE: Room temperature resistance specification is 2 to 5 ohms.**	Yes ▶ No ▶	GO to HA21. REPLACE HEGO sensor.
HA21 CHECK FOR POWER AT HEGO HARNESS CONNECTOR		
• Key On Engine Off. • DVOM on 20V scale. • Measure voltage between ignition run power and power ground at HEGO vehicle harness. • Is voltage greater than 10.5 volts?	Yes ▶ No ▶	HEGO sensor system OK. RECONNECT HEGO and GO to HA14. GO to HA22.
HA22 CHECK CONTINUITY OF GROUND TO HEGO CONNECTOR		
• Key Off, wait 10 seconds. • DVOM on 200 ohm scale. • Measure resistance between battery ground and HEGO ground at HEGO vehicle harness. • Is resistance 5.0 ohms or less?	Yes ▶ No ▶	SERVICE open in ignition run power circuit. SERVICE open in ground circuit.

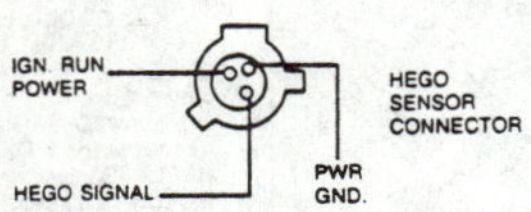

Fuel Control — 1.9L EFI and 2.3L EFI TC | Pinpoint Test | HA

HA30 CONTINUOUS TESTING: CODE 41, 42 OR 43

41 — HEGO indicated the fuel system was lean for more than 15 seconds when the system should have been in closed loop fuel control.

42 — HEGO indicated the fuel system was rich for more than 15 seconds when the system should have been in closed loop fuel control.

43 — HEGO indicated the fuel system was lean at WOT for more than 3 seconds.

CLOSED LOOP — Fuel control under the influence of the HEGO sensor.

OPEN LOOP — Fuel control NOT under the influence of the HEGO sensor.

Before attempting to correct a continuous fuel control code 41, 42 or 43, diagnose all other drive complaints first, eg., rough idle, misses, etc.

NOTE: The fuel control code may help in this diagnosis.

Using the fuel control service code, isolate the cause of the fuel control problem.

Some areas to check are:

- Unmetered Air: Vacuum leaks/intake air leaks.
  — Canister Purge System.
  — PCV System.
  — Engine sealing.
  — Air leaks between VAF meter and throttle body.
  — Crimped fuel line.
  — Plugged fuel filter.
  — Fouled injectors.
- HEGO Fuel Fouled: Whenever an over-rich fuel condition has been experienced (fuel fouled spark plugs), make a thorough check of the ignition system. In the event the HEGO sensor is suspected of being fuel fouled (low output, slow response), run the vehicle at sustained high speeds (within legal limits) followed by a few hard accelerations. This will burn off HEGO contamination and restore proper HEGO operation.
- Fuel Pressure: Perform Pinpoint Test Step HA1.
- Ignition System: Always in default spark (10 degrees). Refer to Quick Test Step 4.0.
- Improper Fueling: Lead fouled HEGO sensor.
- TP Sensor: Not moving (mechanical damage). Connect DVOM to test Pin 47 and to test Pin 46. Key to Run. Observe DVOM while moving the throttle. Reading must increase with increase in throttle opening. If not correct, REPLACE as necessary.
- If at this point, the drive concern is still present, perform Steps HA3 through HA6 only.

Fuel Control — 5.0L SEFI | Pinpoint Test | HC

HC

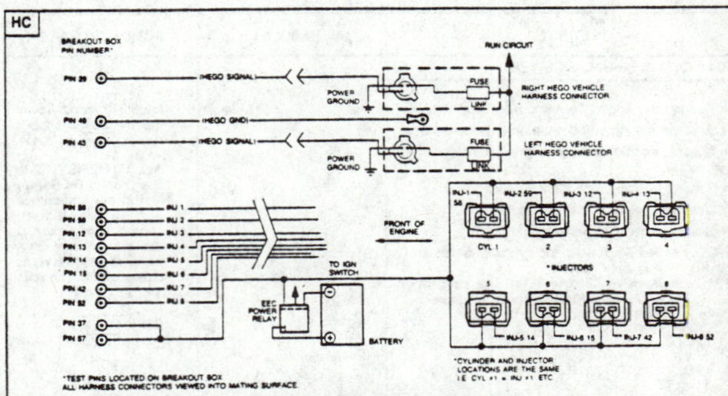

STOP-WARNING

You should enter this Pinpoint Test only when a Service Code 41, 91 or 42, 92 is received in Quick Test Step 5.0 or 6.0 or when directed here from Pinpoint Test S.

To prevent the replacement of good components, be aware that the following non-EEC areas may be at fault:

- Ignition Coil
- Distributor Cap
- Distributor Rotor
- Fouled Spark Plugs
- Spark Plug Wires
- CANP Problems
- PCV Valves (see note below)
- Compression Level
- Valve Problem
- EGR Valve and Gasket
- Air Filter
- Fuel Contamination, Engine Oil
- Poor Power Ground
- Fuel Pressure
- Manifold Leaks, Intake/Exhaust
- Engine Not at Normal Operating Temperatures
- Cracked Piston

This Pinpoint Test is intended to diagnose only the following:

- HEGO Sensor
- HEGO Sensor Connection
- Vacuum Systems
- Fuel Injectors
- Processor Assembly
- Harness Circuits: HEGO GRD, HEGO, INJ 1-8 and VPWR

NOTE: Fuel-contaminated engine oil may affect 41, 91 and 42, 92 Service Codes, so if it is suspected, remove the PCV from the valve cover, and rerun the Quick Test. If the problem is corrected, then change the engine oil and filter.

Fuel Control — 5.0L SEFI | Pinpoint Test | HC

TEST STEP	RESULT ▶	ACTION TO TAKE
HC1 FUEL PRESSURE CHECK 1		
• Key Off, wait 10 seconds. • Install fuel pressure gauge. • Start and run engine. <u>RUNNING FUEL PRESSURE SPECIFICATION</u> 5.0L SEFI = 27-37 psi (186-254 kPa) • Is fuel pressure within specification? **For No Starts:** • If engine will not run, cycle the key from Off to On several times to increase fuel pressure. <u>ENGINE OFF FUEL PRESSURE SPECIFICATION</u> 5.0L SEFI = 35-45 psi (241-310 kPa)	Yes ▶ No ▶	GO to HC2. REFER to the Shop Manual, Group 24 for electric fuel pump and Section 11 for fuel pressure regulator check.
HC2 FUEL PRESSURE CHECK 2		
• Key On. • Does fuel pressure drop more than 5 psi within 60 seconds after Key Off?	Yes ▶ No ▶	GO to HC4. GO to HC3.
HC3 FUEL DELIVERY TEST		
• Key Off. • Fuel pressure gauge installed. • Pressurize fuel system as per Step HC1. • Locate and disconnect inertia switch (which disables fuel pump). • Crank engine for 5 seconds. • Take pressure reading at the end of 5 second crank. • Does pressure drop 5 psi or greater at end of 5 second crank cycle? **NOTE: Verify fuel quality; air and/or water will also pressurize and look like acceptable fuel delivery.**	Yes ▶ No ▶	RECONNECT inertia switch. The EEC system is not the fault of the No Start. Fuel and spark are present. REFER to Section 2 for other No Start Routines. If complaint was runs rough, misses or fuel code, GO to HC4. RECONNECT inertia switch. GO to HC7.

NOTE: The colder the engine, the greater the pressure drop (i.e., an engine coolant temperature of 200°F equals approximately a 10 psi drop in 5 seconds. 60°F equals approximately a 20 psi drop in 5 seconds).

Fuel Control — 5.0L SEFI | Pinpoint Test | HC

TEST STEP	RESULT ▶	ACTION TO TAKE
HC4 CYLINDER BALANCE TEST		
• Perform the Engine Running Self-Test. • After the last repeated code, wait 5-10 seconds. • "Goose" throttle very lightly (not wide-open throttle). • Time of test approximately 90 seconds. • Use table below to interpret codes. • Is service code 90 present? **NOTE: • See Appendix for cylinder balance test definition.** **• Mark VII/Lincoln Town Car calibration 8-22 has added Cylinder Balance Test capability.**	Yes ▶ No ▶	Mark VII/Lincoln Town Car calibration 8-22, GO to HC5. ALL OTHERS, GO to Diagnostic By Symptom in the appropriate engine supplement section. Remember to use conventional diagnostic methods such as compression checks and scoping. GO to HC6.

The cylinder balance test switches each injector "OFF" and "ON," one at a time. Service codes correspond to cylinder number, e.g., service code 30 indicates a problem with cylinder number 3. The cylinder balance test is designed to aid in the detection of a non-contributing cylinder. The Cylinder Balance Pinpoint Test Steps are designed to isolate only EEC related problems.

SERVICE CODE	90	10	20	30	40	50	60	70	80	77*
Cylinder/Injector Number	Pass	1	2	3	4	5	6	7	8	Rerun Test
Breakout Box Pin Number		58	59	12	13	14	15	42	52	

Fuel Control — 5.0L SEFI | Pinpoint Test | HC

TEST STEP	RESULT ▶	ACTION TO TAKE
HC5 PERFORM 2ND/3RD LEVEL CYLINDER BALANCE TEST		
NOTE: Mark VII/Lincoln Town Car calibration 8-22 only, a Cylinder Balance Test service code 90 received in the first test level indicates that the injector harness is not open or shorted and the processor is sending a drive signal to all injectors. The 2nd/3rd level Cylinder Balance Test is intended to aid in the detection of any partially contributing injectors. • Within 2 minutes after the previous Cylinder Balance Test, lightly depress and release throttle to enter 2nd or 3rd level Cylinder Balance Test. • Is service code 90 present for both levels? Refer to appendix for Cylinder Balance Test definition.	Yes ▶ No ▶	GO to Diagnostic By Symptom in the appropriate engine supplement section. Remember to use conventional Diagnostic methods such as engine compression checks and scoping. GO to HC8.
HC6 INJECTOR HARNESS SHORT CHECK		
• Key Off. • Disconnect processor 60 Pin connector and inspect for damaged Pins, corrosion, loose wires. Service as necessary. • Connect Breakout box to harness. Leave processor disconnected. • Suspect injector disconnected. • DVOM to 200,000 ohm scale. • Measure the resistance of the suspect injector circuit between the injector circuit test Pin at the Breakout box and test Pins 40 and 60 at the Breakout box. • Is resistance greater than 10,000 ohms?	Yes ▶ No ▶	RECONNECT injector. GO to HC7. Service short to ground. RECONNECT injector RERUN Quick Test and Cylinder Balance Test.

Fuel Control — 5.0L SEFI | Pinpoint Test | HC

TEST STEP	RESULT ▶	ACTION TO TAKE
HC7 INJECTOR DRIVER SIGNAL CHECK		
Requires standard non-powered 12-volt test lamp. • Key Off. • Breakout box installed. • Connect processor to Breakout box. • Connect test lamp between test Pin 37 and the suspect injector(s) circuit test Pin at the Breakout box. • Crank or start engine. • Is glow on light dim?	Yes ▶ No (Bright light or no light) ▶	GO to HC8. Replace processor. Rerun Quick Test and Cylinder Balance Test.
HC8 HARNESS AND INJECTOR RESISTANCE CHECK		
• Key Off, wait 10 seconds. • Breakout box installed. • Disconnect processor. • DVOM on 200 ohm scale. • Measure the resistance of the suspect injector circuit between the injector circuit test Pin at the Breakout box and test Pins 37 and 57 at the Breakout box. • Is resistance between 13.0 and 19.0 ohms? **For No Starts:** Pick any injector to read resistance.	Yes ▶ No ▶	CHECK for dirty/clogged injector(s), CLEAN as necessary. If OK, REPLACE injector. RERUN Quick-Test and Cylinder Balance Test. SERVICE the harness/connectors on the suspect injector for OPENS or cause of high resistance. If OK, replace the injector. RERUN Quick Test and Cylinder Balance Test. FOR NO STARTS: SERVICE harness for open in VPWR circuit. RERUN Quick Test.

Fuel Control — 5.0L SEFI | Pinpoint Test | HC

TEST STEP	RESULT	ACTION TO TAKE
HC9 FUEL CONTROL — ALWAYS LEAN: CODE 41, 91 VERIFICATION		
• Run vehicle at 2000 rpm for two minutes. • Key Off, wait 10 seconds. • Perform Engine Running Self-Test. • Is code 41 or 91 present?	Yes ▶ No ▶	GO to HC10. GO to HC21.
HC10 FUEL CONTROL — ALWAYS LEAN: CODE 91 LEFT HEGO, OR 41 RIGHT HEGO		
NOTE: Vacuum/air leaks in non-EEC areas could also cause code 41, or 91. Check for: — Leaking vacuum actuator (e.g.: A/C control motor) — Engine sealing — EGR system — PCV system — Unmetered air leak between air meter and throttle body — Lead contaminated HEGO sensor • Key Off. DVOM on 20V scale. Disconnect appropriate HEGO sensor from vehicle harness. Connect DVOM to HEGO sensor and engine ground. • Disconnect and plug vacuum line to MAP sensor. • Apply 14 in.-Hg. to MAP. • Start the engine. Does the DVOM read greater than 0.5V?	Yes ▶ No ▶	GO to HC11. REPLACE HEGO sensor. RECONNECT MAP vacuum line. RERUN Quick Test.
HC11 LEFT HEGO, OR RIGHT HEGO AND HEGO GROUND CONTINUITY CHECK		
• Key Off. • Install Breakout box. Processor disconnected. • DVOM on 200 ohm scale. • Measure resistance between test Pin 49 at Breakout box and engine block ground. • Measure resistance between test Pin 29 at Breakout box and HEGO signal at vehicle harness connector and/or measure resistance between test Pin 43 at Breakout box and HEGO signal at vehicle harness connector. • Are all resistances less than 5 ohms?	Yes ▶ No ▶	GO to HC12. SERVICE open in appropriate circuit. RERUN Quick Test.

Fuel Control — 5.0L SEFI | Pinpoint Test | HC

TEST STEP	RESULT	ACTION TO TAKE
HC15 CHECK HEGO HARNESS FOR SHORT TO POWER		
• Key Off, wait 10 seconds. • Disconnect appropriate HEGO for code 42 or 92. • DVOM on 20 volt scale. • Key On. • Measure voltage between HEGO signal and HEGO power ground at the HEGO harness connector. • Is voltage less than 0.5 volts?	Yes ▶ No ▶	GO to HC16. REPAIR short to power in HEGO circuit. RERUN Quick Test.

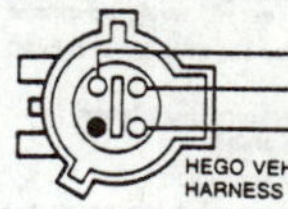

TEST STEP	RESULT	ACTION TO TAKE
HC16 CHECK HEGO SENSOR FOR SHORT TO KEY POWER		
• Key Off. • HEGO disconnected. • DVOM on 200,000 ohm scale. • Measure resistance from Key Power input to HEGO circuit at the HEGO sensor. • Is reading less than 10,000 ohms?	Yes ▶ No ▶	REPLACE HEGO sensor. RERUN Quick Test. GO to HC17.
HC17 ATTEMPT TO GENERATE CODE 41 or 91		
• Key Off, wait 10 seconds. **NOTE: Non-EEC areas could cause a code 42 or 92.** Check for: — Fuel contaminated engine oil. — Ignition caused misfire (fouled spark plug) — CANP problems • Refer to illustration HC. • Disconnect vehicle harness at the appropriate HEGO sensor. Jumper (ground) the HEGO circuit at the HEGO sensor vehicle harness connector to the engine block. • Repeat Engine Running Self Test. • Is service code 41 or 91 present?	Yes ▶ No ▶	REMOVE jumper. GO to HC18. REMOVE jumper. DISCONNECT processor connector. INSPECT for damage or corrosion. If OK, REPLACE processor. RERUN Quick Test.

Fuel Control — 5.0L SEFI | Pinpoint Test | HC

TEST STEP	RESULT	ACTION TO TAKE
HC12 HEGO CIRCUIT SHORT TO GROUND		
• Key Off, processor disconnected. • Disconnect appropriate HEGO. • DVOM on 200,000 ohm scale. • Measure resistance between test Pin 29 or test Pin 43 to test Pin 40 at the Breakout box. • Is resistance 10,000 ohms or less?	Yes ▶ No ▶	CORRECT cause of resistance to ground. RERUN Quick Test. GO to HC13.
HC13 CHECK HEGO SENSOR TO GROUND		
• Key Off. • HEGO disconnected. • DVOM on 200,000 ohm scale. • Measure resistance from HEGO power ground to HEGO signal at HEGO sensor. • Is resistance 10,000 ohms or less?	Yes ▶ No ▶	REPLACE HEGO sensor. RECONNECT MAP vacuum line. RERUN Quick Test. GO to HC14.

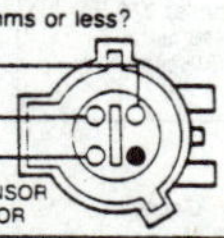

TEST STEP	RESULT	ACTION TO TAKE
HC14 ATTEMPT TO ELIMINATE CODE 41		
• Key Off. • Processor connected. • Reconnect HEGO sensor. • MAP vacuum line still disconnected and plugged. • Apply 10 in.-Hg. (33 kPa) vacuum to MAP sensor. • Start the engine and run at approximately 2000 rpm for one minute. Allow engine to return to idle. • Perform Engine Running Self-Test. • Is code 41 still present? **NOTE: Disregard any other code input at this time.**	Yes ▶ No ▶	INSPECT for corrosion or damaged pins. If OK, REPLACE processor. RECONNECT MAP vacuum line. RERUN Quick Test. RECONNECT MAP vacuum line. HEGO input circuit OK. GO to HC1 (Basic Fuel Delivery).

Fuel Control — 5.0L SEFI | Pinpoint Test | HC

TEST STEP	RESULT	ACTION TO TAKE
HC18 CHECK MAP SENSOR		
NOTE: Due to the MAP sensor's large influence on fuel control, there is a possibility that a code 42 could be a result of a MAP problem, even though a code 22 is not present. Therefore the next two Pinpoint Tests Steps will verify proper vacuum to the MAP sensor and its ability to hold vacuum. • Key Off, wait 10 seconds. • Disconnect vacuum line from MAP sensor. • Install vacuum pump to MAP sensor. • Apply 18 in.-Hg. (59 kPa) vacuum to MAP sensor. • Does MAP sensor hold vacuum?	Yes ▶ No ▶	RELEASE vacuum. GO to HC19. REPLACE MAP sensor. RECONNECT vacuum line to MAP sensor. RERUN Quick Test.
HC19 CHECK FOR LOSS OF MAP VACUUM		
• Tee a vacuum gauge in the intake manifold vacuum line at the MAP sensor. • Start engine and note vacuum reading. • Key Off, wait 10 seconds. • Remove vacuum gauge and tee and reconnect vacuum line to MAP. • Tee in vacuum gauge at a different source of intake manifold vacuum. • Start engine and note vacuum reading. • Do vacuum readings differ greater than 1 in.-Hg. (3.3 kPa)?	Yes ▶ No ▶	INSPECT vacuum lines for leaks, holes, disconnections, kinks, blockage and proper routing. SERVICE as necessary. GO to HC20.
HC20 HEGO CHECK		
• Refer to illustration HC. • DVOM on 20 volt scale. • With HEGO sensor disconnected from the harness, connect a DVOM from HEGO sensor to engine ground. • Disconnect PCV hose. • Start engine and run at approximately 2000 rpm. Does the DVOM read less than 0.4 volts within 30 seconds?	Yes ▶ No ▶	HEGO sensor OK. GO to HC1. REPLACE HEGO sensor. RERUN Quick Test.

Fuel Control — 5.0L SEFI	Pinpoint Test	HC

TEST STEP	RESULT ▶	ACTION TO TAKE
HC21 CHECK HEATER ELEMENT RESISTANCE ON HEGO		
• Key off, engine off. • DVOM on 200 ohm scale. • HEGO disconnected. • Measure resistance between run circuit and ground at HEGO harness connector. — Refer to Pinpoint Test HC wiring schematic. • Hot to warm resistance specification is 5 to 20 ohms. **NOTE: At room temperature, resistance specification is 2 to 5 ohms.** • Is resistance within specification?	Yes ▶ No ▶	GO to HC22. REPLACE HEGO sensor.
HC22 CHECK FOR POWER AT HEGO HARNESS CONNECTOR		
• Key on, engine off. • DVOM on 20V scale. • Connect positive lead to run circuit and negative lead to ground at HEGO vehicle harness connector. • Is voltage greater than 10.5 volts?	Yes ▶ No ▶	HEGO's sensor system OK. RECONNECT HEGO and GO to HC2. GO to HC23.
HC23 CHECK CONTINUITY OF GROUND TO HEGO CONNECTOR		
• Key off, wait 10 seconds. • DVOM on 200 ohm scale. • Measure resistance of ground circuit from HEGO vehicle harness connector to battery ground. • Is resistance less than 5.0 ohms?	Yes ▶ No ▶	SERVICE open in run circuit. SERVICE open in ground circuit.

Fuel Control — 5.0L SEFI	Pinpoint Test	HC

HC90 CONTINUOUS TESTING: CODE 41 RIGHT HEGO/91 LEFT HEGO

CODE 41/91 — Indicates that a HEGO circuit has not switched during closed loop fuel control.

NOTE: In this situation, code 41/91 does not necessarily indicate a lean condition.

CLOSED LOOP —Fuel control under the influence of the HEGO sensors.

OPEN LOOP —Fuel control NOT under the influence of the HEGO sensors.

Before attempting to correct a continuous fuel control code 41/91, diagnose all other drive complaints first., eg., rough idle, misses, etc. in EEC Diagnostic By Symptom in the appropriate engine supplement.

NOTE: The fuel control code may help in this diagnosis.

Using the fuel control service code, isolate the cause of the fuel control problem.

Some areas to check are:

- Vacuum Circuits: Vacuum leaks/intake air leaks.
  - —Canister Purge System.
  - —PCV System.
  - —Engine sealing.
- HEGO Fuel Fouled: Whenever an over-rich fuel condition has been experienced (fuel fouled spark plugs), make a thorough check of the ignition system. In the event a HEGO sensor is suspected of being fuel fouled (low output, slow response), run the vehicle at sustained high speeds (within legal limits) followed by a few hard accelerations. This will burn off HEGO contamination and restore proper HEGO operation.
- Ignition System: Always in default spark (10 degrees). Refer to Quick Test Step 4.0.
- Improper Fueling: Lead fouled HEGO sensor.
- Fuel Pressure: Perform Pinpoint Test Step HC1 and HC2.
- If at this point, the drive concern (if any) is still present, perform Steps HC4 through HC8 only.

Fuel Control — 3.8L CFI	Pinpoint Test	HD

HD REFERENCE DRAWING

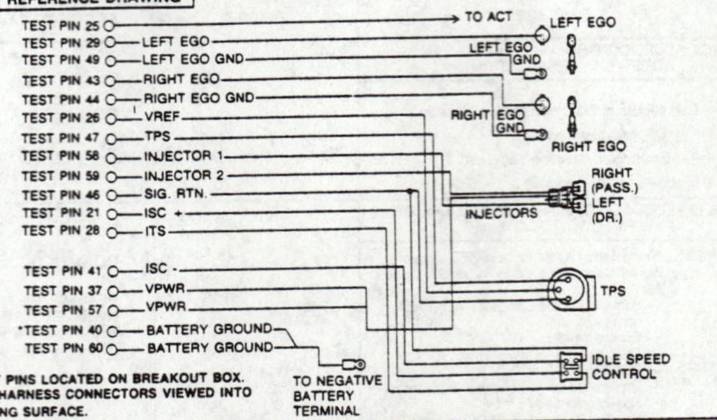

*TEST PINS LOCATED ON BREAKOUT BOX.
ALL HARNESS CONNECTORS VIEWED INTO MATING SURFACE.

STOP-WARNING

You should enter this Pinpoint Test only when a Service Code 41, 91, 42 or 92 is received in Quick Test Step 5.0 or 6.0 or when directed here from Pinpoint Test A or by Diagnostics by Symptom in the Engine Supplement Section.

To prevent the replacement of good components, be aware that the following non-EEC areas may be at fault:

- Ignition Coil
- Distributor Cap
- Distributor Rotor
- Fouled Spark Plugs
- Spark Plug Wires
- CANP Problems
- PCV Valves (see note below)
- EGR Valve and Gasket
- Air Filter
- Fuel Contamination, Engine Oil
- Poor Power Ground
- Fuel Pressure
- Manifold Leaks, Intake/Exhaust
- Engine Not at Normal Operating Temperatures

This Pinpoint Test is intended to diagnose only the following:

- EGO Sensors
- EGO Signals and Ground Circuits
- EGO Sensor Connections
- Fuel Injectors
- Processor Assembly

NOTE: Fuel-contaminated engine oil may affect 41, 91, 42 and 92 Service Codes, so if it is suspected, remove the PCV from the valve cover, and rerun the Quick Test. If the problem is corrected, then change the engine oil and filter.

Fuel Control — 3.8L CFI	Pinpoint Test	HD

TEST STEP	RESULT ▶	ACTION TO TAKE
HD1 KEY POWER VOLTAGE CHECK		
• Key On. • Processor connected. • DVOM on 20V scale. • Measure voltage between test Pins 5 and 60 at the Breakout box.	Reading is less than 10.5V ▶ Reading is 10.5V or greater ▶	SERVICE open in key power circuit. RERUN Quick Test. GO to HD2.
HD2 FUEL PRESSURE CHECK		
• Install fuel pressure gauge. • Start and run engine. Fuel pressure must be 172-310 kPa (25-45 psi). • Fuel pressure must remain at 276 ± 34 kPa (40 psi ± 5 psi) for 60 seconds after final Key Off. **For No Starts:** • If engine will not run cycle the key from OFF to ON several times to increase fuel pressure.	Fuel pressure is within specifications ▶ Fuel pressure is not 172-310 kPa (25-45 psi) ▶ Fuel pressure does not remain at 276 ± 34 kPa (40 psi ± 5 psi) for 60 seconds or longer ▶	GO to HD3. REFER to Shop Manual, Group 24 for electric fuel pump. GO to HD6.
HD3 FUEL DELIVERY TEST		
• Key Off. • Fuel pressure gauge installed. • Pressurize fuel system as per Step HD2. • Locate and disconnect the fuel pump relay. • Crank engine for 5 seconds. • Take pressure reading at the end of 5-second crank. **NOTE: Verify fuel quality; air and/or water will also pressurize and look like acceptable fuel delivery.**	Pressure Gauge Reading: Pressure is approximately 10-20 psi at the end of 5-second crank cycle. Refer to Note below. ▶ Pressure is greater or less than specified ▶	The EEC system is not the fault of the No Start. Fuel and spark are present. REFER to Section 2 for other No Start Routines. If complaint was runs rough, misses or fuel code, GO to HD4. GO to HD4.

NOTE: The colder the engine, the greater the pressure drop (i.e., an engine coolant temperature of 200°F equals approximately a 10 psi drop in 5 seconds. 60°F equals approximately a 20 psi drop in 5 seconds).

Fuel Control — 3.8L CFI | Pinpoint Test | HD

TEST STEP	RESULT ▶	ACTION TO TAKE
HD4 HARNESS INJECTOR RESISTANCE CHECK		
• Key Off. • DVOM on 200 ohm range. • Disconnect processor 60 Pin connector and inspect for damaged pins, corrosion, loose wires. Service as necessary. • Install Breakout box, processor disconnected. • Measure resistance between test Pins 57 and 58 at the Breakout box and between test Pins 57 and 59 at the Breakout box.	Either resistance is less than 1.5 ohms ▶	SERVICE the harness and/or connectors on the suspect injector for shorts. IF OK, REPLACE injector.
	Either resistance is greater than 3.5 ohms ▶	SERVICE the harness and/or connectors on the suspect injector for opens. If OK, REPLACE injector.
	Both resistances are 1.5-3.5 ohms ▶	GO to HD5.
HD5 INJECTOR DRIVE SIGNAL CHECK		
Requires standard non-powered 12V test lamp. • Key Off. • Breakout box installed. • Connect processor to Breakout box. • Refer to illustration HD. • Connect test lamp between test Pin 37 and Pin 58 at the Breakout box. • Crank or start engine. • Repeat above test between test Pin 37 and Pin 59 at the Breakout box.	Dim glow at light on both tests ▶	GO to HD6.
	No light on one or both tests ▶	VERIFY 12V battery power at Pins 37 and 57. If OK, REPLACE processor. RERUN Quick Test.
	Bright light on one or both tests ▶	CHECK injector circuits 1 and 2 for shorts to ground. If OK, REPLACE processor. RERUN Quick Test.
HD6 INJECTOR BALANCE TEST		
• Connect tachometer to engine. Run engine at idle. • Disconnect and reconnect the injectors one at a time; note rpm drop for each injector. • Does each injector produce at least a 100 rpm momentary drop?	Yes ▶	Fuel delivery OK. Problem in area common to all cylinders, i.e.: Air vacuum leak, fuel contamination, EGR.
	No ▶	REPLACE faulty injector. RERUN Quick Test.

Fuel Control — 3.8L CFI | Pinpoint Test | HD

TEST STEP	RESULT ▶	ACTION TO TAKE
HD10 HARNESS CHECK EGO CIRCUITS		
• Key Off, wait 10 seconds. • Disconnect processor 60 Pin connector and inspect for damaged pins, corrosion, loose wires. Service as necessary. • Install Breakout box. Processor disconnected. • Disconnect appropriate EGO. • For left EGO code 41, measure resistance between test Pin 49 at the Breakout box and engine block ground and between left EGO connector and test Pin 29 at the Breakout box. • For right EGO code 91 measure resistance between test Pin 44 and engine block ground and also between right EGO harness connector and test Pin 43 at the Breakout box. • Are both resistances less than 5 ohms?	Yes ▶	GO to HD11.
	No ▶	SERVICE CORRECT open circuit as necessary. RERUN Quick Test.
HD11 CHECK EGO CIRCUIT FOR SHORT TO GROUND		
• Key Off. • Breakout box installed. Processor disconnected. • DVOM to 200,000 ohms range. • Disconnect appropriate EGO sensor at harness. • Measure resistance between test Pin 29 (left EGO) or test Pin 43 (right EGO) to test Pin 40 at the Breakout box.	Reading is 10,000 ohms or greater ▶	GO to HD12.
	Reading is less than 10,000 ohms ▶	CORRECT cause of resistance to ground. RERUN Quick Test.
HD12 ATTEMPT TO ELIMINATE EGO LEAN CODE		
• Key Off. • Reconnect appropriate EGO sensor. • Reconnect processor. • Apply 14 in.-Hg. vacuum to MAP sensor. • Start the engine and run at approximately 2000 rpm for 1 minute. Allow engine to return to idle. • Enter Engine Running Quick Test. • Is code 41 or 91 still present? **NOTE: Disregard any other code output at this time.**	Yes ▶	INSPECT for corrosion or damaged pins. If OK, REPLACE processor. RERUN Quick Test.
	No ▶	EGO input circuit OK. GO to HD13.

Fuel Control — 3.8L CFI | Pinpoint Test | HD

TEST STEP	RESULT ▶	ACTION TO TAKE
HD7 CODE 41 AND/OR 91 PRESENT		
• Verify codes received during Quick Test. • Are both code 41 and 91 present?	Yes ▶	GO to HD9.
	No ▶	GO to HD8.
HD8 SWAP INJECTOR CONNECTORS		
• Key Off. • Swap injector connections. • Start engine, run engine, run Quick Test. • Does the original code(s) repeat?	Yes ▶	RECONNECT injector harness to original production state. GO to HD9.
	No ▶	RECONNECT injector harness to original production state. GO to HD4.
HD9 FUEL CONTROL — ALWAYS LEAN CODES: 41 LEFT EGO, 91 RIGHT EGO		
NOTE: Vacuum/air leaks in non-EEC areas could cause a code. Check for: **— Leaking vacuum actuator (eg: A/C control motor)** **— Engine sealing** **— EGR system** **— PCV system** **— Lead contaminated EGO sensor** • Refer to illustration HD. • Key Off. DVOM on 20V scale. Disconnect appropriate EGO sensor from vehicle harness. Connect DVOM to EGO sensor and engine ground. • Apply 14 in.-Hg. of vacuum to MAP sensor. • Start the engine and run at approximately 2000 rpm. Does the DVOM read greater than 0.5V within 1 minute?	Yes ▶	GO to HD10.
	No ▶	REPLACE EGO sensor. RERUN Quick Test.

Fuel Control — 3.8L CFI | Pinpoint Test | HD

TEST STEP	RESULT ▶	ACTION TO TAKE
HD13 CHECK FOR CODES WITH ACT RECONNECTED		
• Reconnect ACT sensor. • Repeat Quick Test. • Is code 41 or 91 present?	Yes ▶	REPLACE processor. RERUN Quick Test.
	No ▶	GO to HD2.
HD14 FUEL CONTROL — ALWAYS RICH CODES: 42 LEFT EGO, 92 RIGHT EGO		
• Key Off, wait 10 seconds. • **NOTE: Non-EEC areas could cause a code 42.** **Check for:** **— Fuel contaminated engine oil** **— Ignition caused misfire** **— CANP problems** **— Mechanical portion of EGR system** • Refer to illustration HD. • Disconnect vehicle harness at the appropriate EGO sensor. Using a jumper wire, ground vehicle harness EGO circuit at the EGO sensor to the engine block. • Repeat Engine Running Self-Test. • Is service code 41 or 91 present?	Yes ▶	GO to HD16.
	No ▶	GO to HD15.
HD15 HARNESS CHECK		
• Key Off, wait 10 seconds. • Refer to illustration HD. • Disconnect processor 60 Pin connector and inspect for damaged pins, corrosion, loose wires. Service as necessary. • Install Breakout box. • Measure resistance between test Pin 49 at the Breakout box and left EGO ground at engine block or Pin 44 and right EGO ground at engine block. • Measure resistance between test Pin 29 and left EGO harness connector or test Pin 43 and right EGO harness connector. • Are both circuits less than 5 ohms?	Yes ▶	DISCONNECT processor connector. INSPECT for damage or corrosion. If OK, REPLACE processor. RERUN Quick Test.
	No ▶	CORRECT harness circuit with resistance greater than 5 ohms. RERUN Quick Test.

Fuel Control — 3.8L CFI | Pinpoint Test | HD

TEST STEP	RESULT ▶	ACTION TO TAKE
HD16 CHECK MAP SENSOR		
NOTE: Because the MAP sensor influences fuel control, it is possible for a code 42 to occur as a result of a MAP problem even though code 22 is not present. The next 2 steps will verify proper vacuum to the MAP sensor and its ability to hold vacuum.		
• Key Off, wait 10 seconds. • Disconnect vacuum line from MAP sensor. • Install vacuum pump to MAP sensor. • Apply 18 in.-Hg. vacuum to MAP sensor. • Does MAP sensor hold vacuum?	Yes ▶ No ▶	RELEASE vacuum. GO to HD17. REPLACE MAP sensor. CONNECT vacuum line to MAP sensor. RERUN Quick Test.
HD17 CHECK FOR LOSS OF MAP VACUUM		
• Tee a vacuum gauge in the intake manifold vacuum line at the MAP sensor. • Start engine and note vacuum reading. • Key Off, wait 10 seconds. • Remove vacuum gauge and tee and reconnect line to MAP. • Tee in vacuum gauge at a different source of intake manifold vacuum. • Start engine and note vacuum reading. • Do vacuum readings differ greater than 1 in.-Hg.?	Yes ▶ No ▶	INSPECT vacuum lines for leaks, holes, disconnections, kinks, blockage and proper routing. SERVICE as necessary. GO to HD18.
HD18 EGO CHECK		
• Refer to illustration HD. • DVOM on 20V scale. • With appropriate EGO sensor disconnected from the vehicle harness, connect a DVOM from EGO sensor to engine ground. • Disconnect PCV hose. • Start engine and run at approximately 2000 rpm. Does the DVOM read less than 0.4V within 1 minute?	Yes ▶ No ▶	EGO sensor OK. GO to HD2. REPLACE EGO sensor. RERUN Quick Test.

Fuel Control — 3.8L CFI | Pinpoint Test | HD

HD19 CONTINUOUS TESTING, CODE 41/42

41 — EGO indicated the fuel system was lean for more than 15 seconds when the system should have been in closed loop fuel control.

42 — EGO indicated the fuel system was rich for more than 15 seconds when the system should have been in closed loop fuel control.

*CLOSED LOOP — Fuel control under the influence of the EGO sensor.

*OPEN LOOP — Fuel control NOT under the influence of the EGO sensor.

Before attempting to correct a fuel control code 41/42, diagnose all other drive complaints first, eg., rough idle, misses, etc.

NOTE: The fuel control code may help in this diagnosis.

Using the fuel control service code, isolate the cause of the fuel control problem.

Some areas to check are:

- Vacuum Circuits: Vacuum leaks/intake air leaks.
  — Canister Purge System.
  — PCV System.
  — Engine sealing.
- EGO Fuel Fouled: Whenever an over-rich fuel condition has been experienced (fuel fouled spark plugs), make a thorough check of the ignition system. In the event the EGO sensor is suspected of being fuel fouled (low output, slow response), run the vehicle at sustained high speeds (within legal limits) followed by a few hard accelerations. This will burn off EGO contamination and restore proper EGO operation.
- Fuel Pressure: Perform Pinpoint Test Step HD2.
- Ignition System: Always in default spark (10 degrees). Refer to Quick Test Step 4.0.
- Improper Fueling: Lead fouled EGO sensor.
- TP Sensor: Not moving (mechanical damage). Connect DVOM to test Pin 47 and to test Pin 46. Key to Run. Observe DVOM while moving the throttle. Reading must increase with increase in throttle opening. If not correct, SERVICE as necessary.
- If at this point, the drive concern is still present, perform Steps HD3 through HD6 only.

Fuel Control — 2.3L EFI Car/Truck | Pinpoint Test | HE

HE

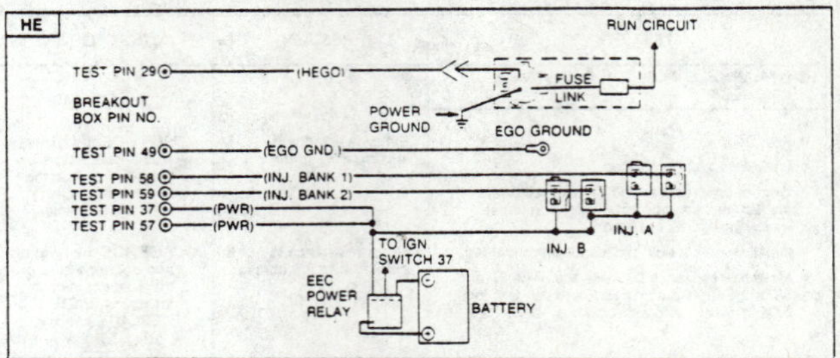

STOP-WARNING

You should enter this Pinpoint Test only when a Service Code 41 or 42 is received in Quick Test Step 5.0 or 6.0 or when directed here from Pinpoint Test A or Diagnostics by Symptom in the Engine Supplement Section.

To prevent the replacement of good components, be aware that the following non-EEC areas may be at fault:

- Ignition Coil
- Distributor Cap
- Distributor Rotor
- Fouled Spark Plugs
- Spark Plug Wires
- CANP Problems
- PCV Valves (see note below)
- EGR Valve and Gasket
- Air Filter
- Fuel Contamination, Engine Oil
- Poor Power Ground
- Fuel Pressure
- Manifold Leaks, Intake Exhaust
- Engine Not at Normal Operating Temperatures

This Pinpoint Test is intended to diagnose only the following:

- HEGO Sensor
- Harness Circuits HEGO Grd., HEGO, Inj. Bank 1, Inj. Bank 2, V PWR
- HEGO Sensor Connection
  — Code 42 start at **HE8**
  — Code 41 start at **HE14**
- Vacuum Systems
- Fuel Injectors
- Processor Assembly

NOTE: Fuel contaminated engine oil may affect 41 and 42 Service Codes. If this is suspected, remove the PCV from the valve cover and repeat the Quick Test. If the problem is corrected, then change the engine oil and filter.

Fuel Control — 2.3L EFI Car/Truck | Pinpoint Test | HE

TEST STEP	RESULT ▶	ACTION TO TAKE
HE1 FUEL PRESSURE CHECK 1		
• Key Off, wait 10 seconds. • Install fuel pressure gauge. • Start and run engine. Fuel pressure must be 206 ± 35 kPa (30 ± 5 psi). **For No Starts:** • If engine will not run, cycle the key from Off to On several times to increase fuel pressure.	Yes ▶ No ▶	GO to HE2. REFER to the Shop Manual, Group 24 for electric fuel pump and Section 11 for fuel pressure regulator check.
HE2 FUEL PRESSURE CHECK 2		
• Fuel pressure must remain at 276 ± 34 kPa (40 ± 5 psi) for 60 seconds or longer after final key Off.	Yes ▶ No ▶	GO to HE3. GO to HE7.
HE3 FUEL DELIVERY TEST		
• Key Off. • Fuel pressure gauge installed. • Pressurize fuel system as per Step HE1. • Locate and disconnect the fuel pump relay. • Crank engine for 5 seconds. • Take pressure reading at the end of 5 second crank.	PRESSURE GAUGE READING: Pressure is approximately 10-20 psi at the end of 5 second crank cycle. Refer to note below ▶ Pressure is greater or less than specified ▶	 The EEC system is not the fault of the No Start. Fuel and spark are present. REFER to Section 2 for other No Start Routines. If complaint was runs rough, misses or fuel code, GO to HE4. GO to HE4.

NOTE: The colder the engine, the greater the pressure drop (i.e., an engine coolant temperature of 200°F equals approximately a 10 psi drop in 5 seconds. 60°F equals approximately a 20 psi drop in 5 seconds).

Fuel Control — 2.3L EFI Car/Truck | Pinpoint Test | HE

TEST STEP	RESULT ▶	ACTION TO TAKE
HE4 HARNESS INJECTOR RESISTANCE CHECK		
• Key Off, wait 10 seconds. • Disconnect processor 60 pin connector and inspect for damaged pins, corrosion, loose wires. Service as necessary. • Connect Breakout box to harness. Leave processor disconnected. • DVOM on 200 ohm scale. • Measure the resistance of injector Bank 1 between test Pin 37 and test Pin 58 at the Breakout box. Record reading. • Measure the resistance of injector Bank 2 between test Pin 37 and test Pin 59 at the Breakout box. Record reading. • Are both resistance readings between 7.0 ohms and 9.5 ohms?	Yes ▶ No ▶	GO to HE6. GO to HE5.
HE5 ISOLATE FAULTY INJECTOR CIRCUIT		
• Key Off. • Breakout box installed. • Leave processor disconnected. • DVOM on 200 ohm scale. • Disconnect all injectors on suspect Bank. Measure the resistance of each injector by connecting one injector at a time, and reading resistance between test Pin 37 and 58 at the Breakout box for Bank 1 or test Pin 37 and 59 at the Breakout box for Bank 2. • Are all resistance readings between 15.0 and 19.0 ohms?	Yes ▶ No ▶	GO to HE6. SERVICE the harness connectors on the suspect injector for opens or shorts. If OK, REPLACE injector. RERUN Quick Test.
HE6 INJECTOR DRIVE SIGNAL CHECK		
Requires standard non-powered 12-volt test lamp. • Key Off. • Breakout box installed. • Connect processor to Breakout box. • Connect test lamp between test Pin 37 and test Pin 58 at the Breakout box. • Crank or start engine. • Repeat above test between test Pin 37 and test Pin 59 at the Breakout box.	Dim glow at light on both tests ▶ No light on one or both tests ▶ Bright light on one or both tests ▶	GO to HE7. VERIFY 12-volt battery power at Pins 37 and 57. If OK, REPLACE processor. RERUN Quick Test. CHECK circuits Bank 1 and Bank 2 for shorts to ground. If OK, REPLACE processor. RERUN Quick Test.

Fuel Control — 2.3L EFI Car/Truck | Pinpoint Test | HE

TEST STEP	RESULT ▶	ACTION TO TAKE
HE7 INJECTOR BALANCE TEST		
• Connect tachometer to engine. Run engine at idle. • Disconnect and reconnect the injectors one at a time; note rpm drop for each injector. • Does each injector produce at least a 100 rpm momentary drop, as ISC will attempt to re-establish rpm?	Yes ▶ No ▶	Fuel delivery OK. Problem is with area common to all cylinders, i.e.: Air/vacuum leak, fuel contamination, EGR. REPLACE faulty injector. RERUN Quick Test.
HE8 FUEL CONTROL — ALWAYS RICH: CODE 42. CHECK HEGO HARNESS FOR SHORT TO POWER		
• Key Off, wait 10 seconds. • Disconnect the HEGO sensor from the vehicle harness. • DVOM on 20 volt scale. • Key On. • Measure voltage between HEGO signal and ground at the HEGO connector. • Is voltage reading less than 0.5 volts? 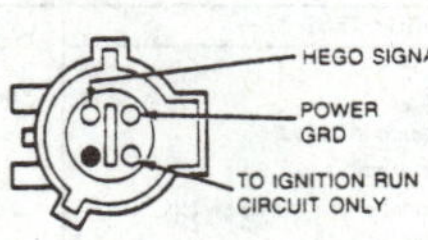	Yes ▶ No ▶	GO to HE9. SERVICE short to power in HEGO circuit. RERUN Quick Test.
HE9 CHECK HEGO FOR SHORT TO KEY POWER		
• Key Off. • HEGO disconnected. • DVOM on 200,000 ohm scale. • Measure resistance from key power input to HEGO circuit at the HEGO. • Is the resistance reading 10,000 ohms or greater?	Yes ▶ No ▶	GO to HE10. REPLACE HEGO sensor. RERUN Quick Test.

Fuel Control — 2.3L EFI Car/Truck | Pinpoint Test | HE

TEST STEP	RESULT ▶	ACTION TO TAKE
HE10 ATTEMPT TO GENERATE CODE 41		
• Key Off, wait 10 seconds. **NOTE: Non-EEC areas could cause a code 42. Check for:** **— Fuel contaminated engine oil** **— Ignition caused misfire** **— CANP problems** • Refer to Illustration HE. • Disconnect vehicle harness at the HEGO sensor. • Jumper (ground) the HEGO circuit at the HEGO sensor vehicle harness connector to the engine block. • Repeat Engine Running Self-Test. • Is service code 41 present?	Yes ▶ No ▶	REMOVE jumper. GO to HE11. REMOVE jumper. DISCONNECT the processor and INSPECT pins for damage or corrosion, if OK, REPLACE processor. RERUN Quick Test.
HE11 CHECK MAP SENSOR		
• Key Off, wait 10 seconds. • Disconnect vacuum line from MAP sensor. • Install vacuum pump to MAP sensor. • Apply 18 in.-Hg. vacuum to MAP sensor. • Does MAP sensor hold vacuum?	Yes ▶ No ▶	RELEASE vacuum. GO to HE12. REPLACE MAP sensor. CONNECT vacuum line to MAP sensor. RERUN Quick Test.
HE12 CHECK FOR LOSS OF MAP VACUUM		
• Install vacuum gauge at the MAP sensor and the manifold source. • Start and run engine. • Compare vacuum readings between the MAP sensor and engine manifold. • Are the vacuum readings the same?	Yes ▶ No ▶	GO to HE13. INSPECT vacuum lines for leaks, holes, disconnects, kinks, blockage and proper routings. SERVICE as necessary.

Fuel Control — 2.3L EFI Car/Truck | Pinpoint Test | HE

TEST STEP	RESULT ▶	ACTION TO TAKE
HE13 HEGO CHECK		
• Refer to Illustration HE. • DVOM on 20V scale. • With HEGO sensor disconnected from the harness, connect a DVOM from HEGO sensor to engine ground. • Disconnect PCV hose. • Start engine and run at approximately 2000 rpm. • Does the DVOM read less than 0.4V within 1 minute?	Yes ▶ No ▶	HEGO sensor OK. GO to HE1. REPLACE HEGO sensor. RERUN Quick Test.
HE14 FUEL CONTROL — ALWAYS LEAN: CODE 41 VERIFICATION		
• Run vehicle at 2000 rpm for 2 minutes. • Key Off, wait 10 seconds. • Perform Engine Running Quick Test. • Is code 41 present?	Yes ▶ No ▶	GO to HE15. GO to HE21.
HE15 FUEL CONTROL — ALWAYS LEAN		
NOTE: Vacuum/air leaks in non-EEC areas could cause a code 41. Check for: **— Leaking vacuum actuator (eg: A/C control motor)** **— Engine sealing** **— EGR system** **— PCV system** **— Unmetered air leak between air meter and throttle body** **— Lead contaminated HEGO sensor** • Key Off. DVOM on 20V scale. Disconnect HEGO sensor from vehicle harness. Connect DVOM to HEGO sensor and engine ground. • Apply 14 in.-Hg. vacuum to MAP. • Start the engine and run at approximately 2000 rpm. Does the DVOM read greater than 0.5V within 1 minute?	Yes ▶ No ▶	GO to HE16. REPLACE HEGO sensor. RERUN Quick Test.

Fuel Control — 2.3L EFI Car/Truck | Pinpoint Test | HE

TEST STEP	RESULT ▶	ACTION TO TAKE
HE16 HARNESS CHECK HEGO CIRCUITS		
• Key Off. • Install Breakout box. Processor disconnected. • Measure resistance between test Pin 49 at the Breakout box and engine block ground and between HEGO circuit at the HEGO vehicle harness connector and test Pin 29 at the Breakout box. • Are both resistance readings less than 5 ohms?	Yes ▶ No ▶	GO to HE17. SERVICE/CORRECT as necessary, circuit with greater than 5 ohms resistance. RERUN Quick Test.
HE17 CHECK HEGO CIRCUIT FOR SHORT TO GROUND		
• Key Off. • HEGO disconnected. • Breakout box installed. Processor disconnected. • DVOM to 200,000 ohms range. • Measure resistance between test Pin 29 and test Pin 40 at the Breakout box. • Is the resistance reading 10,000 ohms or greater?	Yes ▶ No ▶	GO to HE18. CORRECT cause of resistance to ground. RERUN Quick Test.
HE18 CHECK HEGO SENSOR FOR SHORT TO GROUND		
• Key Off. • HEGO disconnected. • DVOM on 200,000 ohm scale. • Measure resistance from ground to HEGO signal at the HEGO sensor. • Is resistance reading 10,000 ohms or greater?	Yes ▶ No ▶	GO to HE19. REPLACE HEGO sensor. RERUN Quick Test.
HE19 ATTEMPT TO ELIMINATE CODE 41		
• Key Off. • Reconnect HEGO sensor. • Apply 14 in.-Hg. vacuum to MAP sensor. • Start the engine and run at approximately 2000 rpm for 1 minute. Allow engine to return to idle. • Perform Engine Running Self-Test. • Is code 41 still present? **NOTE: Disregard any other code output at this time.**	Yes ▶ No ▶	INSPECT for corrosion or damaged pins. If OK, REPLACE processor. RERUN Quick Test. HEGO input circuit OK. GO to HE1.

Fuel Control — 1.9L CFI and 2.3L/2.5L (HSC) CFI | Pinpoint Test | HF

HF

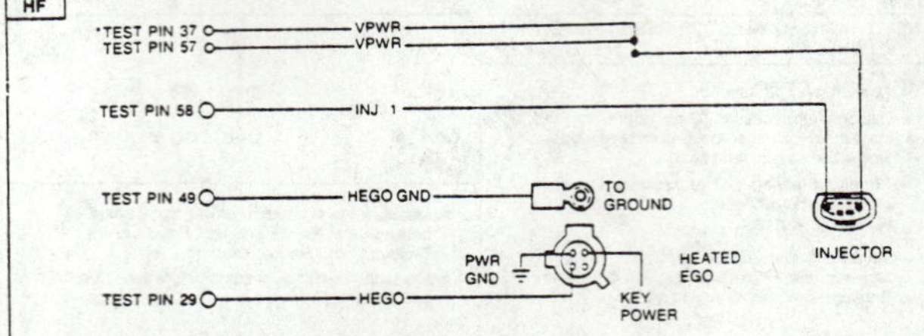

***TEST PINS LOCATED ON BREAKOUT BOX. ALL HARNESS CONNECTORS VIEWED INTO MATING SURFACE.**

****2.5L CFI RELAY WITHIN INTEGRATED CONTROLLER.**

STOP-WARNING

You should enter this Pinpoint Test only when a Service Code 41 or 42 is received in Quick Test Step 5.0 or 6.0 or when directed here from Pinpoint Test S.

To prevent the replacement of good components, be aware that the following non-EEC areas may be at fault:

- Ignition Coil
- Distributor Cap
- Distributor Rotor
- Fouled Spark Plugs
- Spark Plug Wires
- CANP Problems
- PCV Valves (see note below)
- EGR Valve and Gasket
- Air Filter
- Fuel Contamination, Engine Oil
- Poor Power Ground
- Fuel Pressure
- Manifold Leaks, Intake/Exhaust
- Engine Not at Normal Operating Temperatures

This Pinpoint Test is intended to diagnose only the following:

- EGO Sensor
- EGO Signal and Ground Circuit
- EGO Sensor Connection
- Fuel Injector
- Processor Assembly

NOTE: Fuel contaminated engine oil may affect 41 and 42 Service Codes, so if it is suspected, remove the PCV from the valve cover, and rerun the Quick Test. If the problem is corrected, then change the engine oil and filter.

Fuel Control — 2.3L EFI Car/Truck | Pinpoint Test | HE

TEST STEP	RESULT ▶	ACTION TO TAKE
HE21 CHECK HEATER ELEMENT RESISTANCE ON HEATED EGO		
• Key Off, Engine Off. • DVOM on 200 ohm scale. • HEGO disconnected. • Measure resistance between run circuit and ground at HEGO sensor. Refer to Pinpoint Test HE for wiring schematic. • Hot to warm resistance specification is 5 to 20 ohms. **NOTE: At room temperature, resistance specification is 2 to 5 ohms.** • Is resistance within specification?	Yes ▶ No ▶	GO to HE22. REPLACE HEGO sensor.
HE22 CHECK FOR POWER AT HEGO HARNESS CONNECTOR		
• Key On, Engine Off. • DVOM on 20V scale. • Connect positive lead to run circuit and negative lead to ground at HEGO vehicle harness connector. • Is voltage reading 10.5 volts or greater?	Yes ▶ No ▶	HEGO system OK. RECONNECT HEGO and GO to HE1. GO to HE23.
HE23 CHECK CONTINUITY OF GROUND TO HEGO CONNECTOR		
• Key Off, wait 10 seconds. • DVOM on 200 ohm scale. • Measure resistance of ground circuit from HEGO vehicle harness connector to battery ground. • Is resistance reading less than 5.0 ohms?	Yes ▶ No ▶	SERVICE open in run circuit. SERVICE open in ground circuit.

Fuel Control — 1.9L CFI and 2.3L/2.5L (HSC) CFI | Pinpoint Test | HF

TEST STEP	RESULT ▶	ACTION TO TAKE
HF1 FUEL PRESSURE CHECK 1		
• Key Off, wait 10 seconds. • Install fuel pressure gauge. • Start and run engine.* FUEL PRESSURE SPECIFICATION 1.9L and 2.5L CFI = 13-16 psi (14.5 ± 1.5) 2.3L CFI = 14.5-17.5 psi (16 ± 1.5) ***For No Starts:** • If engine will not run, cycle the key from off to on several times. • Is fuel pressure within specification?	Yes ▶ No ▶	GO to HF2. REFER to Shop Manual, Group 24 for electric fuel pump and Section 11 for fuel pressure regulator check.
HF2 FUEL PRESSURE CHECK 2		
• Key On. • Does fuel pressure remain at specification for 60 seconds?	Yes ▶ No ▶	GO to HF3. GO to HF25.
HF3 FUEL DELIVERY TEST		
• Key Off. • Fuel pressure gauge installed. • Pressurize fuel system as per Step HF1. • Locate and disconnect the inertia switch. • Crank engine for 5 seconds. • Take pressure reading at the end of 5-second crank. • Does pressure drop approximately 5-8 psi at the end of 5 second crank cycle. Refer to Note below. **NOTE: Verify fuel quality; air and/or water will also pressurize and look like acceptable fuel delivery.**	Yes ▶ No ▶	The EEC system is not the fault of the No Start. Fuel and spark are present. REFER to Section 2 for other No Start Routines. If complaint was runs rough, misses or fuel code, CONTINUE to HF4. GO to HF4.

NOTE: The colder the engine, the greater the pressure drop (i.e., an engine coolant temperature of 200°F equals approximately a 5 psi drop in 5 seconds. 60°F equals approximately a 8 psi drop in 5 seconds).

Fuel Control — 1.9L CFI and 2.3L/2.5L (HSC) CFI | Pinpoint Test | HF

TEST STEP	RESULT ▶	ACTION TO TAKE
HF4 HARNESS INJECTOR RESISTANCE CHECK		
• Key Off. • DVOM on 200 ohm range. • Disconnect processor 60 Pin connector and inspect for damaged pins, corrosion, loose wires. Service as necessary. • Install Breakout box, processor disconnected. • Measure resistance between test Pins 57 and 58 at the Breakout box. • Is resistance between 1.0 and 2.0 ohms?	Yes ▶ No ▶	GO to HF5. SERVICE the harness and/or connectors on the injector for opens or shorts. If OK, REPLACE injector.
HF5 INJECTOR DRIVE SIGNAL CHECK		
Requires standard non-powered 12-volt test lamp. • Key Off. • Breakout box installed. • Connect processor to Breakout box. • Refer to illustration HF. • Connect test lamp between test Pin 37 and test Pin 58 at the Breakout box. • Crank or start engine.	Very dim glow at light ▶ No light ▶ Bright light ▶	Fuel delivery OK. Problem is with area common to all cylinders, i.e., air/vacuum leak, fuel contamination, EGR. VERIFY 12-volt battery power at Pins 37 and 57. If OK, REPLACE processor. RERUN Quick Test. CHECK injector circuit for shorts to ground. If OK, REPLACE processor. RERUN Quick Test.
HF6 FUEL CONTROL — ALWAYS LEAN: CODE 41 VERIFICATION		
• Run vehicle at 2000 rpm for 2 minutes. • Key Off, wait 10 seconds. • Perform Engine Running Quick Test. • Is code 41 present?	Yes ▶ No ▶	GO to HF7. GO to HF20.

Fuel Control — 1.9L CFI and 2.3L/2.5L (HSC) CFI | Pinpoint Test | HF

TEST STEP	RESULT ▶	ACTION TO TAKE
HF7 FUEL CONTROL — INDICATES LEAN: CODE 41		
NOTE: Vacuum/air leaks in non-EEC areas could cause a code 41. Check for: **— Leaking vacuum actuator (eg: A/C control motor)** **— Engine sealing** **— EGR system** **— PCV system** **— Lead contaminated HEGO sensor** • Refer to illustration HF. • Key Off. DVOM on 20V scale. Disconnect HEGO sensor from vehicle harness. Connect DVOM to EGO signal at HEGO sensor and engine ground. • Apply 14 in.-Hg. vacuum to MAP. • Start the engine and run at approximately 2000 rpm. Does the DVOM read greater than 0.5V within 60 seconds?	Yes ▶ No ▶	GO to HF8. REPLACE HEGO sensor. RERUN Quick Test.
HF8 HARNESS CHECK HEGO CIRCUITS		
• Key Off. • Install Breakout box. Processor disconnected. • HEGO disconnected. • Measure resistance between test Pin 49 at the Breakout box and engine block ground and between EGO circuit at the HEGO vehicle harness connector and test Pin 29 at the Breakout box. • Are both resistance readings less than 5 ohms?	Yes ▶ No ▶	GO to HF9. SERVICE/CORRECT circuit with resistance greater than 5 ohms, as necessary. RERUN Quick Test.
HF9 CHECK HEGO CIRCUIT FOR SHORT TO GROUND		
• Key Off. • HEGO disconnected. • Breakout box installed. Processor disconnected. • DVOM to 200,000 ohms range. • Measure resistance between test Pin 29 and test Pin 40 at the Breakout box. • Is resistance 10,000 ohms or greater?	Yes ▶ No ▶	GO to HF10. CORRECT cause of resistance to ground. RERUN Quick Test.

Fuel Control — 1.9L CFI and 2.3L/2.5L (HSC) CFI | Pinpoint Test | HF

TEST STEP	RESULT ▶	ACTION TO TAKE
HF10 CHECK HEGO SENSOR TO GROUND		
• Key Off. • HEGO disconnected. • DVOM on 200,000 ohm scale. • Measure resistance between power ground and EGO signal at HEGO sensor. • Is resistance 10,000 ohms or greater?	Yes ▶ No ▶	GO to HF11. REPLACE HEGO sensor. RERUN Quick Test.
HF11 ATTEMPT TO ELIMINATE CODE 41		
• Key Off. • Reconnect HEGO sensor. • Apply 14 in.-Hg. vacuum to MAP. • Start the engine and run at approximately 2000 rpm for 1 minute, allow engine to return to idle. • Perform Engine Running Self-Test. • Is code 41 still present? **NOTE: Disregard any other code output at this time.**	Yes ▶ No ▶	INSPECT for corrosion or damaged pins. If OK, REPLACE processor. RERUN Quick Test. HEGO input circuit OK. GO to HF1.
HF13 FUEL CONTROL — ALWAYS RICH: CODE 42 CHECK HEGO HARNESS FOR SHORT TO POWER		
• Key Off, wait 10 seconds. • Disconnect vehicle harness at the HEGO sensor. • DVOM on 20 volt scale. • Key On. • Measure voltage between EGO signal and power ground at the HEGO harness connector. 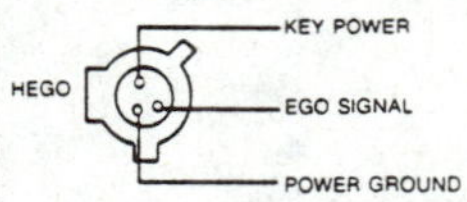 • Is voltage less than .5 volts?	Yes ▶ No ▶	GO to HF14. REPAIR short to power in EGO circuit. RERUN Quick Test. If code 41 is now present, REPLACE HEGO.

Fuel Control — 1.9L CFI and 2.3L/2.5L (HSC) CFI | Pinpoint Test | HF

TEST STEP	RESULT ▶	ACTION TO TAKE
HF14 CHECK HEGO FOR SHORT TO KEY POWER		
• Key Off. • HEGO disconnected. • DVOM on 200,000 ohm scale. • Measure resistance between key power input and EGO circuit at the HEGO sensor. • Is resistance 10,000 ohms or greater.	Yes ▶ No ▶	GO to HF15. REPLACE HEGO sensor. RERUN Quick Test.
HF15 ATTEMPT TO GENERATE CODE 41		
• Key Off, wait 10 seconds. • **NOTE: Non-EEC areas could cause code 42. Check for:** **— Fuel contaminated engine oil** **— Ignition caused misfire (fouled spark plug)** **— CANP problems** **— Mechanical portion EGR system** • Refer to illustration HF. • Disconnect vehicle harness at the HEGO sensor. Jumper (ground) the EGO circuit at the HEGO sensor vehicle harness connector to the engine block. • Repeat Engine Running Self-Test. • Service code 41 results.	Yes ▶ No ▶	DISCONNECT jumper to ground. GO to HF16. DISCONNECT processor connector. INSPECT for damage or corrosion. If OK, REPLACE Processor. RERUN Quick Test.
HF16 CHECK MAP SENSOR		
NOTE: Due to the MAP sensors large influence on fuel control, there is a possibility that a code 42 could be a result of a MAP problem, even though a code 22 is not present. Therefore, in the next two Pinpoint Test Steps we will verify proper vacuum to MAP, and its ability to hold it. • Key Off, wait 10 seconds. • Disconnect vacuum line from MAP sensor. • Install vacuum pump to MAP sensor. • Apply 18 in.-Hg. vacuum to MAP sensor. • Does MAP sensor hold vacuum?	Yes ▶ No ▶	RELEASE vacuum. GO to HF17. REPLACE MAP sensor. CONNECT vacuum line to MAP sensor. RERUN Quick Test.

Fuel Control — 1.9L CFI and 2.3L/2.5L (HSC) CFI | Pinpoint Test | HF

TEST STEP	RESULT ▶	ACTION TO TAKE
HF17 CHECK FOR LOSS OF VACUUM TO MAP		
• Tee a vacuum gauge in the intake manifold vacuum hose at the MAP sensor. • Start engine and let RPM stabilize. NOTE vacuum reading. • Key Off, wait 10 seconds. • Remove vacuum gauge and tee and reconnect vacuum line to MAP. • Tee in vacuum gauge at a different source of intake manifold vacuum. • Start engine and let RPM stabilize. NOTE vacuum reading. • Do the vacuum readings differ greater than 1 in.-Hg.?	Yes ▶	INSPECT vacuum lines for leaks, holes, disconnections, kinks, blockage and proper routing. SERVICE as necessary.
	No ▶	REMOVE vacuum gauge and tee. GO to HF18.
HF18 HEGO CHECK		
• Refer to illustration HF. • DVOM on 20V scale. • HEGO sensor disconnected. • Connect a DVOM from EGO circuit on sensor to engine ground. • Disconnect PCV hose. • Start engine and run at approximately 2000 rpm. Does the DVOM read less than 0.4V within 30 seconds?	Yes ▶	HEGO sensor OK. GO to HF1.
	No ▶	REPLACE HEGO sensor. RERUN Quick Test.
HF20 CHECK HEATER ELEMENT RESISTANCE ON HEATED EGO		
• Key Off Engine Off. • DVOM on 200 ohm scale. • HEGO disconnected. • Measure resistance between run circuit and ground at HEGO sensor. — Refer to pinpoint test HF wiring schematic. • Hot to warm resistance specification is 5 to 20 ohms. NOTE: At room temperature, resistance specification is 2 to 5 ohms. • Is resistance within specification?	Yes ▶	GO to HF21.
	No ▶	REPLACE HEGO sensor.

Fuel Control — 1.9L CFI and 2.3L/2.5L (HSC) CFI | Pinpoint Test | HF

TEST STEP	RESULT ▶	ACTION TO TAKE
HF21 CHECK FOR POWER AT HEGO HARNESS CONNECTOR		
• Key On Engine Off. • DVOM on 20V scale. • Connect positive lead to run circuit and negative lead to ground at HEGO vehicle harness connector. • Is voltage 10.5V or greater?	Yes ▶	HEGO sensor system OK. RECONNECT HEGO. GO to HF1.
	No ▶	GO to HF22.
HF22 CHECK CONTINUITY OF GROUND TO HEGO CONNECTOR		
• Key Off, wait 10 seconds. • DVOM on 200 ohm scale. • Measure resistance of ground circuit from HEGO vehicle harness connector to battery ground. • Is resistance less than 5.0 ohms?	Yes ▶	SERVICE open in run circuit.
	No ▶	SERVICE open in ground circuit.
HF25 LEAKING INJECTOR		
• Remove air inlet tube at fuel charging assembly. • Pressurize fuel system as per Test Step HF1. • Key Off. • Observe throughout the air horn inlet for fuel leaking and/or weeping at the fuel injector discharge area. • Is there a visible leak?	Yes ▶	SERVICE injector and/or O-rings as necessary. REFER to Shop Manual, Section 24-03 for service procedure. **NOTE: After having completed service, RERUN Quick Test.**
	No ▶	GO to HF26.
HF26 EXTERNAL FUEL LEAKS		
• Pressurized fuel system as per Test Step HF1. • Key Off. • Look for fuel leak external to fuel charging assembly fuel line, fitting, etc. • Is there a visible leak?	Yes ▶	SERVICE as necessary. RERUN Quick Test.
	No ▶ **NOTE: After having completed service, RERUN Quick Test.**	SERVICE fuel pressure regulator as necessary. REFER to Shop Manual, Section 24-03 for service procedure.

Fuel Control — 1.9L CFI and 2.3L/2.5L (HSC) CFI | Pinpoint Test | HF

HF90 CONTINUOUS TESTING: CODE 41

CODE 41 — Indicates that the EGO circuit has not switched during closed loop fuel control.

NOTE: In this situation, code 41 does not necessarily indicate a lean condition.

CLOSED LOOP — Fuel control under the influence of the HEGO sensor.

OPEN LOOP — Fuel control NOT under the influence of the HEGO sensor.

Before attempting to correct a continuous fuel control code 41, diagnose all other drive complaints first, eg., rough idle, misses, etc. in EEC Diagnostic By Symptom in the appropriate engine supplement.

NOTE: The fuel control code may help in this diagnosis.

Using the fuel control service code, isolate the cause of the fuel control problem.

Some areas to check are:

- Vacuum Circuits: Vacuum leaks/intake air leaks.
  — Canister Purge System.
  — PCV System.
  — Engine sealing.
- EGO Fuel Fouled: Whenever an over-rich fuel condition has been experienced (fuel fouled spark plugs), make a thorough check of the ignition system. In the event the HEGO sensor is suspected of being fuel fouled (low output, slow response), run the vehicle at sustained high speeds (within legal limits) followed by a few hard accelerations. This will burn off HEGO contamination and restore proper HEGO operation.
- Ignition System: Always in default spark (10 degrees). Refer to Quick Test Step 4.0.
- Improper Fueling: Lead fouled HEGO sensor.
- Fuel Pressure: Perform Pinpoint Test Step HF1 and HF2.
- If at this point, the drive concern (if any) is still present, perform Steps HF3 through HF5 only.

Fuel Control — 2.9L, 3.0L and 4.9L EFI 6-Cylinder Engines | Pinpoint Test | HG

HG

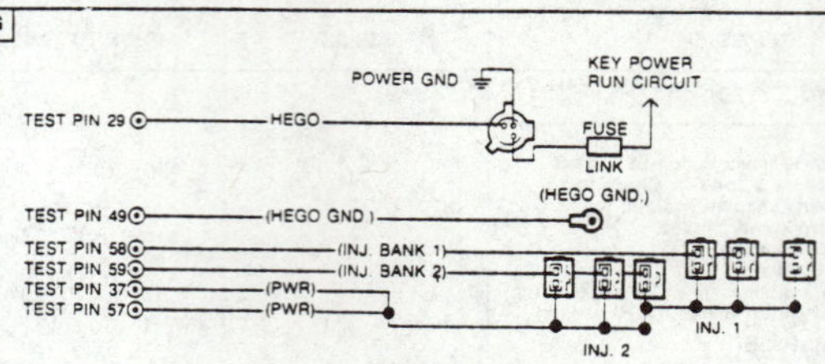

NOTE: REFER TO APPROPRIATE ENGINE SUPPLEMENT FOR PROPER INJECTOR BANK-TO-INJECTOR RELATIONSHIP.

STOP-WARNING

You should enter this Pinpoint Test only when a Service Code 41 or 42 is received in Quick Test Step 5.0 or 6.0 or when directed here from Pinpoint Test S or Diagnostics by Symptom in the Engine Supplement Section.

To prevent the replacement of good components, be aware that the following non-EEC areas may be at fault:

- Ignition Coil
- Distributor Cap
- Distributor Rotor
- Fouled Spark Plugs
- Spark Plug Wires
- CANP Problems
- PCV Valves (see note below)
- EGR Valve and Gasket
- Air Filter
- Fuel Contamination, Engine Oil
- Poor Power Ground
- Fuel Pressure
- Manifold Leaks, Intake/Exhaust
- Engine Not at Normal Operating Temperatures

This Pinpoint Test is intended to diagnose only the following:

- HEGO Sensor
- Harness Circuits EGO Grd., HEGO, Inj. Bank 1, Inj. Bank 2, VPWR
- HEGO Sensor Connection
- Vacuum Systems
- Fuel Injectors
- Processor Assembly

NOTE: Fuel contaminated engine oil may affect 41 and 42 Service Codes. If this is suspected, remove the PCV from the valve cover and repeat the Quick Test. If the problem is corrected, then change the engine oil and filter.

Fuel Control — 2.9L, 3.0L and 4.9L EFI 6-Cylinder Engines | Pinpoint Test | HG

TEST STEP	RESULT ▶	ACTION TO TAKE
HG1 FUEL PRESSURE CHECK 1		
• Key Off, wait 10 seconds. • Install fuel pressure gauge. • Start and run engine. RUNNING FUEL PRESSURE SPECIFICATION 2.9L and 3.0L = 28-38 psi (33 ± 5) 4.9L = 42-52 psi (47 ± 5) ENGINE OFF FUEL PRESSURE SPECIFICATION 2.9L and 3.0L = 35-45 psi (40 ± 5) 4.9L = 50-60 psi (55 ± 5) • Is fuel pressure within specification? For No Starts: • If engine will not run, cycle the key from Off to On several times to increase fuel pressure.	Yes ▶ No ▶	GO to HG2. REFER to the Shop Manual, Group 24 for electric fuel pump and Section 11 for fuel pressure regulator check.
HG2 FUEL PRESSURE CHECK 2		
• Key On. • Does fuel pressure remain at specification for 60 seconds?	Yes ▶ No ▶	GO to HG3. GO to HG7.
HG3 FUEL DELIVERY TEST		
• Key Off. • Fuel pressure gauge installed. • Pressurize fuel system as per Step HG1. • Locate and disconnect the inertia switch which disables fuel pump. • Crank engine for 5 seconds. • Take pressure reading at the end of 5 second crank.	Yes ▶	The EEC system is not the fault of the No Start. Fuel and spark are present. REFER to Section 2 for other No Start Routines. If complaint was runs rough, misses or fuel code, GO to HG4.
• Does pressure drop approximately 5-8 psi at the end of crank cycle? Refer to Notes below. NOTE: Verify fuel quality: air and/or water will also pressurize and look like acceptable fuel delivery.	No ▶	GO to HG4.

NOTE: The colder the engine, the greater the pressure drop (for example, an engine coolant temperature of 200°F equals approximately a 69 kPa [10 psi] drop in 5 seconds. 60°F equals approximately a 138 kPa [20 psi] drop in 5 seconds).

Fuel Control — 2.9L, 3.0L and 4.9L EFI 6-Cylinder Engines | Pinpoint Test | HG

TEST STEP	RESULT ▶	ACTION TO TAKE
HG4 HARNESS INJECTOR RESISTANCE CHECK		
• Key Off, wait 10 seconds. • Disconnect processor 60 Pin connector and inspect for damaged pins, corrosion, loose wires. Service as necessary. • Connect Breakout box to harness. Leave processor disconnected. • DVOM on 200 ohm scale. • Measure the resistance of injector Bank 1 between test Pin 37 and test Pin 58 at the Breakout box. • Measure the resistance of injector Bank 2 between test Pin 37 and test Pin 59 at the Breakout box. • Are both readings between 5.0 and 6.5 ohms?	Yes ▶ No ▶	GO to HG6. GO to HG5.
HG5 ISOLATE FAULTY INJECTOR CIRCUIT		
• Key Off. • Breakout box installed. • Leave processor disconnected. • DVOM on 200 ohm scale. • Disconnect all injectors on suspect Bank. Measure the resistance of each injector by connecting one injector at a time, and reading resistance between test Pin 37 and 58, for Bank 1, or test Pin 37 and 59, for Bank 2, at the Breakout box. • Are all readings between 16.0 and 18.0 ohms?	Yes ▶ No ▶	GO to HG6. SERVICE the harness and/or connectors on the suspect injector for opens or shorts. If OK, REPLACE injector. RERUN Quick Test.
HG6 INJECTOR DRIVE SIGNAL CHECK		
Requires standard non-powered 12-volt test lamp. • Key Off. • Breakout box installed. • Connect processor to Breakout box. • Connect test lamp between test Pin 37 and test Pin 58 at the Breakout box. • Crank or start engine. • Repeat above test between test Pin 37 and test Pin 59 at the Breakout box.	Dim glow at light on both tests ▶ No light on one or both tests ▶ Bright light on one or both tests ▶	GO to HG7. VERIFY 12V battery power at Pins 37 and 57. If OK, REPLACE processor. RERUN Quick Test. CHECK circuits Bank 1 and Bank 2 for shorts to ground. If OK, REPLACE processor. RERUN Quick Test.

Fuel Control — 2.9L, 3.0L and 4.9L EFI 6-Cylinder Engines | Pinpoint Test | HG

TEST STEP	RESULT ▶	ACTION TO TAKE
HG7 INJECTOR BALANCE TEST		
• Connect tachometer to engine. Run engine at idle. • Disconnect and reconnect the injectors one at a time; note rpm drop for each injector. • Does each injector produce at least a 100 rpm momentary drop? ISC will attempt to reestablish rpm.	Yes ▶ No ▶	Fuel delivery OK. Problem is with area common to all cylinders, i.e.: Air vacuum leak, fuel contamination, EGR. REPLACE faulty injector. RERUN Quick Test.
HG8 FUEL CONTROL — ALWAYS RICH: CODE 42. CHECK HEGO HARNESS FOR SHORT TO POWER		
• Key Off, wait 10 seconds. • Disconnect vehicle harness at the HEGO sensor. • DVOM on 20 volt scale. • Key On. • Measure voltage between EGO signal and power ground at the HEGO harness connector. 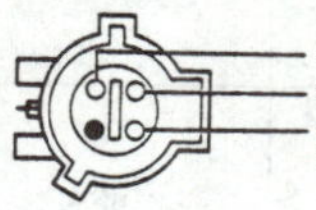 • Is voltage less than .5 volts?	Yes ▶ No ▶	GO to HG9. REPAIR short to power in EGO circuit. RERUN Quick Test.
HG9 CHECK HEGO FOR SHORT TO KEY POWER		
• Key Off. • HEGO sensor disconnected. • DVOM on 200,000 ohm scale. • Measure resistance between key power input to EGO circuit at the HEGO sensor. • Is reading 10,000 ohms or less?	Yes ▶ No ▶	REPLACE HEGO sensor. RERUN Quick Test. GO to HG10.

Fuel Control — 2.9L, 3.0L and 4.9L EFI 6-Cylinder Engines | Pinpoint Test | HG

TEST STEP	RESULT ▶	ACTION TO TAKE
HG10 ATTEMPT TO GENERATE CODE 41		
• Key Off, wait 10 seconds. • NOTE: Non-EEC areas could cause a code 42. Check for: — Fuel contaminated engine oil — Ignition caused misfire (fouled spark plug) — CANP problems • Refer to illustration HF. • Disconnect vehicle harness at the EGO sensor. Jumper (ground) the EGO circuit at the EGO sensor vehicle harness connector to the engine block. • Repeat Engine Running Self-Test. • Service code 41 results.	Yes ▶ No ▶	REMOVE jumper. GO to HG11. REMOVE jumper. DISCONNECT processor connector. INSPECT for damage or corrosion. If OK, REPLACE processor. RERUN Quick Test.
HG11 CHECK MAP SENSOR		
NOTE: Due to the MAP sensor influence on fuel control, there is a possibility that a code 42 could be a result of a MAP problem even though a code 22 is not present. The next two Pinpoint Test steps will verify proper vacuum to the MAP sensor and the ability to hold vacuum. • Key Off, wait 10 seconds. • Disconnect vacuum line from MAP sensor. • Install vacuum pump to MAP sensor. • Apply 18 in.-Hg. vacuum to MAP sensor. • Does MAP sensor hold vacuum?	Yes ▶ No ▶	RELEASE vacuum. GO to HG12. REPLACE MAP sensor. CONNECT vacuum line to MAP sensor. RERUN Quick Test.

Fuel Control — 2.9L, 3.0L and 4.9L EFI 6-Cylinder Engines | Pinpoint Test | HG

TEST STEP	RESULT ▶	ACTION TO TAKE
HG12 CHECK FOR LOSS OF MAP VACUUM		
• Tee a vacuum gauge in the intake manifold vacuum line at the MAP sensor. • Start engine and note vacuum reading. • Key Off, wait 10 seconds. • Remove vacuum gauge and tee and reconnect vacuum line to MAP. • Tee in vacuum gauge at a different source of intake manifold vacuum. • Start engine and note vacuum reading. • Do vacuum readings differ by greater than 1 in.-Hg.?	Yes ▶ No ▶	INSPECT vacuum lines for leaks, holes, disconnections, kinks, blockage and proper routing. SERVICE as necessary. GO to HG13.
HG13 EGO CHECK		
• Refer to Illustration HG. • DVOM on 20V scale. • With EGO sensor disconnected from the harness, connect a DVOM from EGO sensor to engine ground. • Disconnect PCV hose. • Start engine and run at approximately 2000 rpm. Does the DVOM read less than 0.4V within 30 seconds?	Yes ▶ No ▶	EGO sensor OK. GO to HG1. REPLACE EGO sensor. RERUN Quick Test.
HG15 FUEL CONTROL — ALWAYS LEAN: CODE 41 VERIFICATION		
• Run vehicle at 2000 rpm for 2 minutes. • Key Off, wait 10 seconds. • Perform Engine Running Quick Test. • Is code 41 still present?	Yes ▶ No ▶	GO to HG16. GO to HG21.

Fuel Control — 2.9L, 3.0L and 4.9L EFI 6-Cylinder Engines | Pinpoint Test | HG

TEST STEP	RESULT ▶	ACTION TO TAKE
HG16 FUEL CONTROL — INDICATES LEAN 41		
NOTE: Vacuum/air leaks in non-EEC areas could cause a code 41. Check for: **— Leaking vacuum actuator (e.g.: A/C control motor)** **— Engine sealing** **— EGR system** **— PCV system** **— Lead contaminated HEGO sensor** • Key Off. DVOM on 20V scale. Disconnect HEGO sensor from vehicle harness. Connect DVOM to HEGO sensor and engine ground. • Disconnect and plug vacuum line to MAP sensor. • Apply 10 in.-Hg. vacuum to MAP sensor. • Start the engine and run at approximately 2000 rpm. Does the DVOM read greater than 0.5V within 1 minute?	Yes ▶ No ▶	GO to HG17. REPLACE HEGO sensor. RECONNECT MAP vacuum line. RERUN Quick Test.
HG17 HARNESS CHECK HEGO CIRCUITS		
• Key Off. • Install Breakout box. Processor disconnected. • Measure resistance between test Pin 49 at the Breakout box and engine block ground and between HEGO circuit at the HEGO vehicle harness connector and test Pin 29 at the Breakout box. • Are both circuits less than 5 ohms?	Yes ▶ No ▶	GO to HG18. SERVICE/CORRECT as necessary circuit with greater than 5 ohms resistance. RECONNECT MAP vacuum line. RERUN Quick Test.
HG18 CHECK HEGO CIRCUIT FOR SHORT TO GROUND		
• Key Off. • EGO disconnected. • Breakout box installed. Processor disconnected. • DVOM to 200,000 ohms range. • Measure resistance between test Pin 29 and test Pin 40 at the Breakout box. • Is resistance 10,000 ohms or greater?	Yes ▶ No ▶	GO to HG19. CORRECT cause of resistance to ground. RECONNECT MAP vacuum line RERUN Quick Test.

Fuel Control — 2.9L, 3.0L and 4.9L EFI 6-Cylinder Engines | Pinpoint Test | HG

TEST STEP	RESULT ▶	ACTION TO TAKE
HG19 CHECK HEGO SENSOR TO GROUND		
• Key Off. • HEGO disconnected. • DVOM on 200,000 ohm scale. • Measure resistance from power ground to HEGO signal at HEGO sensor. • Is reading 10,000 ohms or less?	Yes ▶ No ▶	REPLACE HEGO sensor. RECONNECT MAP vacuum line. RERUN Quick Test. GO to HG20.
HG20 ATTEMPT TO ELIMINATE CODE 41		
• Key Off. • Processor connected. • Reconnect HEGO sensor. • MAP vacuum line still disconnected and plugged. • Apply 10 in.-Hg. vacuum to MAP sensor. • Start the engine and run at approximately 2000 rpm for 1 minute. Allow engine to return to idle. • Perform Engine Running Self-Test. • Is code 41 still present? **NOTE: Disregard any other code input at this time.**	Yes ▶ No ▶	INSPECT for corrosion or damaged pins. If OK, REPLACE processor. RECONNECT MAP vacuum line. RERUN Quick Test. RECONNECT MAP vacuum line. HEGO input circuit OK. GO to HG1 (Basic Fuel Delivery).
HG21 CHECK HEATER ELEMENT RESISTANCE ON HEATED EGO		
• Key Off. Engine Off. • DVOM on 200 ohm scale. • HEGO sensor disconnected. • Measure resistance between run circuit and power ground at the heated EGO sensor. Refer to Pinpoint Test HG wiring schematic. • Hot to warm resistance is 5-20 ohms. **NOTE: At room temperature, resistance specification is 2-5 ohms.** • Is resistance within specification?	Yes ▶ No ▶	GO to HG22. REPLACE HEGO sensor.

Fuel Control — 2.9L, 3.0L and 4.9L EFI 6-Cylinder Engines | Pinpoint Test | HG

TEST STEP	RESULT ▶	ACTION TO TAKE
HG22 CHECK FOR POWER AT HEGO HARNESS CONNECTOR		
• Key On, Engine Off. • DVOM on 20V scale. • Connect positive lead to run circuit and negative lead to ground at HEGO vehicle harness connector. • Is voltage reading 10.5 volts or greater?	Yes ▶ No ▶	HEGO sensor system OK. RECONNECT HEGO and GO to HG1. GO to HG23.
HG23 CHECK CONTINUITY OF GROUND TO HEGO CONNECTOR		
• Key Off, wait 10 seconds. • DVOM on 200 ohm scale. • Measure resistance of power ground circuit from HEGO vehicle harness connector to battery ground. • Is resistance reading 5.0 ohms or less?	Yes ▶ No ▶	SERVICE open in run circuit. SERVICE open in power ground circuit.

Fuel Control — 2.9L, 3.0L and 4.9L EFI 6-Cylinder Engines | EFI Pinpoint Test | HG

HG90 CONTINUOUS TESTING. CODE 41

Code 41: Indicates that the EGO circuit has not switched during closed loop fuel control.

NOTE: In this situation code 41 does not necessarily indicate a lean condition.

CLOSED LOOP — Fuel control under the influence of the EGO sensor.

OPEN LOOP — Fuel control NOT under the influence of the EGO sensor.

Before attempting to correct continuous code 41, diagnose all other drive complaints first, eg., rough idle, misses, etc. in EEC Diagnostic By Symptom in the appropriate engine supplement.

NOTE: The fuel control code may help in this diagnosis.

Using the fuel control service code, isolate the cause of the fuel control problem.

Some areas to check are:

- Vacuum Circuits: — Vacuum leaks/intake air leaks. — PCV System. — Engine sealing.
- EGO Fuel Fouled: Whenever an over-rich fuel condition has been experienced (fuel fouled spark plugs), make a thorough check of the ignition system. In the event the EGO sensor is suspected of being fuel fouled (low output, slow response), run the vehicle at sustained high speeds (within legal limits) followed by a few hard accelerations. This will burn off EGO contamination and restore proper EGO operation.
- Ignition System: Always in default spark (10 degrees). Refer to Quick Test Step [4.0].
- Improper Fueling: Lead fouled EGO sensor.
- Fuel Pressure: Perform Pinpoint Test Step [HG1] and [HG2].
- If at this point, the drive concern (if any) is still present, perform Steps [HG3] and [HG7] only.

Fuel Control — 5.0L/7.5L EFI Truck | Pinpoint Test | HH

TEST STEP	RESULT ▶	ACTION TO TAKE
HH1 FUEL PRESSURE CHECK 1		
• Key Off, wait 10 seconds. • Install fuel pressure gauge. • Start and run engine. Fuel pressure must be 28-38 psi (193-262 kPa). **For No Starts:** • If engine will not run, cycle the key from Off to On several times to increase fuel pressure. • Is fuel pressure 35-45 psi (241-310 kPa)?	Yes ▶ No ▶	GO to [HH2]. REFER to the Shop Manual, Group 24 for electric fuel pump and Section 11 for fuel pressure regulator check.
HH2 FUEL PRESSURE CHECK 2		
• Key On. • Does fuel pressure remain at 35-45 psi (241-310 kPa) for 60 seconds or longer after final key off?	Yes ▶ No ▶	GO to [HH3]. GO to [HH7].
HH3 FUEL DELIVERY TEST		
• Key Off. • Fuel pressure gauge installed. • Pressurize fuel system as per Step HH1. • Locate and disconnect the fuel pump inertia switch (disables pump). • Crank engine for 5 seconds. • Take pressure reading at the end of 5 second crank. **NOTE: Verify fuel quality; air and/or water will also pressurize and look like acceptable fuel delivery.**	PRESSURE GAUGE READING: Pressure is approximately 10-20 psi at the end of 5 second crank cycle. Refer to note below ▶ Pressure is greater or less than specified ▶	The EEC system is not the fault of the No Start. Fuel and spark are present. REFER to Section 2 for other No Start Routines. If complaint was runs rough, misses or fuel code, GO to [HH4]. GO to [HH4].

NOTE: The colder the engine, the greater the pressure drop (i.e., an engine coolant temperature of 200°F equals approximately a 10 psi drop in 5 seconds. 60°F equals approximately a 20 psi drop in 5 seconds).

Fuel Control — 5.0L/7.5L EFI Truck | Pinpoint Test | HH

HH

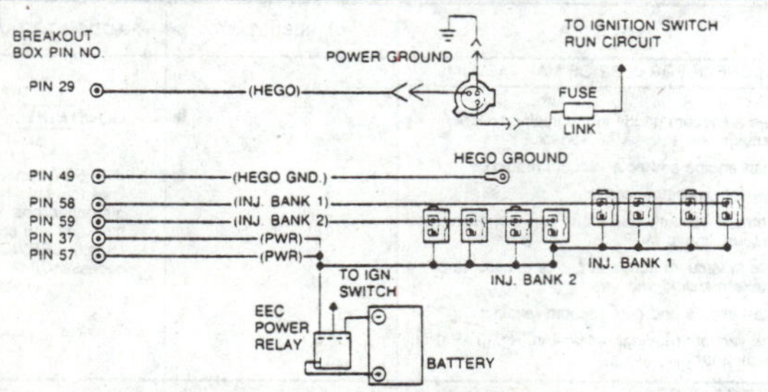

STOP-WARNING

You should enter this Pinpoint Test only when a Service Code 41 or 42 is received in Quick Test Step 5.0 or 6.0 or when directed here from Pinpoint Test S.

To prevent the replacement of good components, be aware that the following non-EEC areas may be at fault:

- Ignition Coil
- Distributor Cap
- Distributor Rotor
- Fouled Spark Plugs
- Spark Plug Wires
- CANP Problems
- PCV Valves (see note below)
- EGR Valve and Gasket
- Air Filter
- Fuel Contamination, Engine Oil
- Poor Power Ground
- Fuel Pressure
- Manifold Leaks, Intake/Exhaust
- Engine Not at Normal Operating Temperatures

This Pinpoint Test is intended to diagnose only the following:

- HEGO Sensor
- Harness Circuits HEGO Grd., HEGO, Inj. Bank 1, Inj. Bank 2, VPWR
- HEGO Sensor Connection
  - Code 42 start at HH8
  - Code 41 start at HH15
- Vacuum Systems
- Fuel Injectors
- Processor Assembly
- MAP Sensor (generating fuel codes)

NOTE: Fuel contaminated engine oil may affect 41 and 42 Service Codes. If this is suspected, remove the PCV from the valve cover and repeat the Quick Test. If the problem is corrected, then change the engine oil and filter.

Fuel Control — 5.0L/7.5L EFI Truck | Pinpoint Test | HH

TEST STEP	RESULT ▶	ACTION TO TAKE
HH4 HARNESS AND INJECTOR RESISTANCE CHECK		
• Key Off, wait 10 seconds. • Disconnect processor 60 pin connector and inspect for damaged pins, corrosion, loose wires. Service as necessary. • Connect Breakout box to harness. Leave processor disconnected. • DVOM on 200 ohm scale. • Measure and record the resistance of injector Bank 1 between test Pin 37 and test Pin 58 at the Breakout box. • Measure and record the resistance of injector Bank 2 between test Pin 37 and test Pin 59 at the Breakout box. • Are both resistances between 3.5 and 5.0 ohms?	Yes ▶ No ▶	GO to [HH6]. GO to [HH5].
HH5 ISOLATE FAULTY INJECTOR CIRCUIT		
• Key Off. • Breakout box installed. • Leave processor disconnected. • DVOM on 200 ohm scale. • Disconnect all injectors on suspect Bank. Measure the resistance of each injector by connecting one injector at a time, and reading resistance between test Pin 37 and 58 at the Breakout box for Bank 1 or test Pin 37 and 59 at the Breakout box for Bank 2. • Are all resistances between 15.0 and 19.0 ohms?	Yes ▶ No ▶	GO to [HH6]. SERVICE the harness/connectors on the suspect injector for opens or shorts. If OK, REPLACE injector. REPEAT Quick Test.

Fuel Control — 5.0L/7.5L EFI Truck | Pinpoint Test | HH

TEST STEP	RESULT ▶	ACTION TO TAKE
HH6 INJECTOR DRIVER SIGNAL CHECK		
Requires standard non-powered 12-volt test lamp. • Key Off. • Breakout box installed. • Connect processor to Breakout box. • Connect test lamp between test Pin 37 and test Pin 58 at the Breakout box. • Crank or start engine. • Repeat above test between test Pin 37 and test Pin 59 at the Breakout box.	Dim glow at light on both tests ▶	GO to HH7.
	No light on one or both tests ▶	VERIFY 12-volt battery power at Pins 37 and 57. If OK, REPLACE processor. REPEAT Quick Test.
	Bright light on one or both tests ▶	CHECK circuits Bank 1 and Bank 2 for shorts to ground. If OK, REPLACE processor. REPEAT Quick Test.
HH7 INJECTOR BALANCE TEST		
• Connect tachometer to engine. Run engine at idle. • Disconnect and reconnect the injectors one at a time: note rpm drop for each injector. • Does each injector produce at least a 100 rpm momentary drop? **NOTE: ISC will attempt to reestablish rpm.**	Yes ▶	Fuel delivery OK. Problem is with area common to all cylinders, i.e.: Air/vacuum leak, fuel contamination, EGR.
	No ▶	REPLACE faulty injector. REPEAT Quick Test.

Fuel Control — 5.0L/7.5L EFI Truck | Pinpoint Test | HH

TEST STEP	RESULT ▶	ACTION TO TAKE
HH8 FUEL CONTROL — ALWAYS RICH: CODE 42 CHECK HEGO HARNESS FOR SHORT TO POWER		
• Key Off, wait 10 seconds. • Disconnect vehicle harness at the HEGO sensor. • DVOM on 20 volt scale. • Key On. • Measure voltage between HEGO signal and ground at the HEGO harness connector. 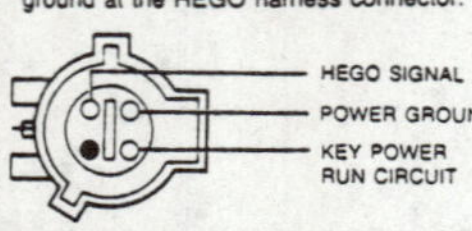 • Is voltage less than 0.5 volts?	Yes ▶	GO to HH9.
	No ▶	REPAIR short to key power in HEGO circuit. RERUN Quick Test.
HH9 CHECK HEGO FOR SHORT TO KEY POWER		
• Key Off. • HEGO disconnected. • DVOM on 200,000 ohm scale. • Measure resistance between key power input and HEGO circuit at the HEGO sensor. • Is resistance less than 10,000 ohms?	Yes ▶	REPLACE HEGO sensor. RERUN Quick Test.
	No ▶	GO to HH10.

Fuel Control — 5.0L/7.5L EFI Truck | Pinpoint Test | HH

TEST STEP	RESULT ▶	ACTION TO TAKE
HH10 ATTEMPT TO GENERATE CODE 41		
• Key Off, wait 10 seconds. • **NOTE: Non-EEC areas could cause a code 42.** **Check for:** **— Fuel contaminated engine oil** **— Ignition caused misfire (fouled spark plug)** **— CANP problems** • Refer to illustration HH. • Disconnect vehicle harness at the HEGO sensor. Jumper (ground) the HEGO circuit at the HEGO sensor vehicle harness connector to the engine block. • Repeat Engine Running Self-Test. • Is service code 41 present?	Yes ▶	REMOVE jumper. GO to HH11.
	No ▶	REMOVE jumper. DISCONNECT processor connector. INSPECT for damage or corrosion. If OK, REPLACE processor. RERUN Quick Test.
HH11 CHECK MAP SENSOR		
NOTE: Due to the MAP sensor's influence on fuel control, there is a possibility that a code 42 could be a result of a MAP problem even though a code 22 is not present. The next 2 Pinpoint Test Steps will verify proper vacuum to the MAP sensor and the ability to hold vacuum. • Key Off, wait 10 seconds. • Disconnect vacuum line from MAP sensor. • Install vacuum pump to MAP sensor. • Apply 18 in.-Hg. (59 kPa) vacuum to MAP sensor. • Does MAP sensor hold vacuum?	Yes ▶	RELEASE vacuum. GO to HH12.
	No ▶	REPLACE MAP sensor. CONNECT vacuum line to MAP sensor. RERUN Quick Test.

Fuel Control — 5.0L/7.5L EFI Truck | Pinpoint Test | HH

TEST STEP	RESULT ▶	ACTION TO TAKE
HH12 CHECK FOR LOSS OF MAP VACUUM		
• Tee a vacuum gauge in the intake manifold vacuum line at the MAP sensor. • Start engine and note vacuum reading. • Key Off, wait 10 seconds. • Remove vacuum gauge and tee and reconnect vacuum line to MAP. • Tee in vacuum gauge at a different source of intake manifold vacuum. • Start engine and note vacuum reading. • Are vacuum readings within 1 in.-Hg. (3 kPa) of each other?	Yes ▶	GO to HH13.
	No ▶	INSPECT vacuum lines for leaks, holes, disconnections, kinks, blockage and proper routing. SERVICE as necessary.
HH13 EGO CHECK		
• Refer to illustration HH. • DVOM on 20V scale. • With HEGO sensor disconnected from the harness, connect a DVOM from HEGO sensor to engine ground. • Disconnect PCV hose. • Start engine and run at approximately 2000 rpm. Does the DVOM read less than 0.4V within 30 seconds?	Yes ▶	HEGO sensor OK. GO to HH1.
	No ▶	REPLACE HEGO sensor. RERUN Quick Test.
HH15 FUEL CONTROL — ALWAYS LEAN: CODE 41 VERIFICATION		
• Run vehicle at 2000 rpm for 2 minutes. • Key Off, wait 10 seconds. • Perform Engine Running Self-Test. • Is code 41 present?	Yes ▶	GO to HH16.
	No ▶	GO to HH21.

Fuel Control — 5.0L/7.5L EFI Truck | Pinpoint Test | HH

TEST STEP	RESULT ▶	ACTION TO TAKE
HH16 FUEL CONTROL — INDICATES LEAN 41		
NOTE: Vaccum/air leaks in non-EEC areas could cause a code 41. **Check for:** — **Leaking vacuum actuator (e.g.: A/C control motor)** — **Engine sealing** — **EGR system** — **PCV system** — **Lead contaminated HEGO sensor** • Key Off. DVOM on 20V scale. Disconnect HEGO sensor from vehicle harness. Connect DVOM to HEGO sensor and engine ground. • Disconnect and plug vacuum line to MAP sensor. • Apply 10 in.-Hg. (33 kPa) vacuum to MAP sensor. • Start the engine and run at approximately 2000 rpm. Does the DVOM read greater than 0.5V within 1 minute?	Yes ▶ No ▶	GO to HH17. REPLACE HEGO sensor. RECONNECT MAP vacuum line. RERUN Quick Test.
HH17 HARNESS CHECK HEGO CIRCUITS		
• Key Off. • Install Breakout box. Processor disconnected. • Measure resistance between test Pin 49 at the Breakout box and engine block ground and between HEGO circuit at the HEGO vehicle harness connector and test Pin 29 at the Breakout box. • Are both circuits less than 5 ohms?	Yes ▶ No ▶	GO to HH18. SERVICE/CORRECT as necessary circuit with greater than 5 ohms resistance. RECONNECT MAP vacuum line. RERUN Quick Test.
HH18 CHECK HEGO CIRCUIT FOR SHORT TO GROUND		
• Key Off. • HEGO disconnected. • Breakout box installed. Processor disconnected. • DVOM to 200,000 ohms range. • Measure resistance between test Pin 29 and test Pin 40 at the Breakout box. • Is resistance greater than 10,000 ohms?	Yes ▶ No ▶	GO to HH19. CORRECT cause of resistance to ground. RECONNECT MAP vacuum line. RERUN Quick Test.

Fuel Control — 5.0L/7.5L EFI Truck | Pinpoint Test | HH

TEST STEP	RESULT ▶	ACTION TO TAKE
HH19 CHECK HEGO SENSOR TO GROUND		
• Key Off. • HEGO disconnected. • DVOM on 200,000 ohm scale. • Measure resistance from power ground to HEGO signal at HEGO sensor. • Is resistance less than 10,000 ohms?	Yes ▶ No ▶	REPLACE HEGO sensor. RECONNECT MAP vacuum line. RERUN Quick Test. GO to HH20.
HH20 ATTEMPT TO ELIMINATE CODE 41		
• Key Off. • Processor connected. • Reconnect HEGO sensor. • MAP vacuum line still disconnected and plugged. • Apply 10 in.-Hg. (33 kPa) vacuum to MAP sensor. • Start the engine and run at approximately 2000 rpm for 1 minute. Allow engine to return to idle. • Perform Engine Running Self-Test. • Is code 41 still present? **NOTE: Disregard any other code input at this time.**	Yes ▶ No ▶	INSPECT for corrosion or damaged pins. If OK, REPLACE processor. RECONNECT MAP vacuum line. RERUN Quick Test. RECONNECT MAP vacuum line. HEGO input circuit OK. GO to HH1 (basic fuel delivery).
HH21 CHECK HEATER ELEMENT RESISTANCE ON HEATED EGO		
• Key off, engine off. • DVOM on 200 ohm scale. • HEGO disconnected. • Measure resistance between run circuit and ground at the HEGO sensor. Refer to illustration HH. • Hot to warm resistance specification is 5 to 20 ohms. **NOTE: At room temperature, resistance specification is 2 to 5 ohms.** • Is resistance within specification?	Yes ▶ No ▶	GO to HH22. REPLACE HEGO sensor.

Fuel Control — 5.0L/7.5L EFI Truck | Pinpoint Test | HH

TEST STEP	RESULT ▶	ACTION TO TAKE
HH22 CHECK FOR POWER AT HEGO HARNESS CONNECTOR		
• Key on, engine off. • DVOM on 20V scale. • Connect positive lead to run circuit and negative lead to ground at HEGO vehicle harness connector. • Is voltage greater than 10.5V?	Yes ▶ No ▶	HEGO sensor system OK. RECONNECT HEGO and GO to HH1. GO to HH23.
HH23 CHECK CONTINUITY OF GROUND TO HEGO CONNECTOR		
• Key Off, wait 10 seconds. • DVOM on 200 ohm scale. • Measure resistance of power ground circuit from HEGO vehicle harness connector to battery ground. • Is resistance less than 5.0 ohms?	Yes ▶ No ▶	SERVICE open in run circuit. SERVICE open in power ground circuit.

Fuel Control — 5.0L/7.5L EFI Truck | Pinpoint Test | HH

HH90 CONTINUOUS TESTING: CODE 41

Code 41: Indicates that the HEGO circuit has not switched during closed loop fuel control.

NOTE: In this situation code 41 does not necessarily indicate a lean condition.

CLOSED LOOP — Fuel control under the influence of the HEGO sensor.

OPEN LOOP — Fuel control NOT under the influence of the HEGO sensor.

Before attempting to correct a continuous fuel control code 41, diagnose all other drive complaints first, eg., rough idle, misses, etc. in EEC Diagnostic By Symptom in the appropriate engine supplement.

NOTE: The fuel control code may help in this diagnosis.

Using the fuel control service code, isolate the cause of the fuel control problem.

Some areas to check are:

- Vacuum Circuits: — Vacuum leaks/intake air leaks. — PCV System. — Engine sealing.
- HEGO Fuel Fouled: Whenever an over-rich fuel condition has been experienced (fuel fouled spark plugs), make a thorough check of the ignition system. In the event the HEGO sensor is suspected of being fuel fouled (low output, slow response), run the vehicle at sustained high speeds (within legal limits) followed by a few hard accelerations. This will burn off HEGO contamination and restore proper HEGO operation.
- Ignition System: Always in default spark (10 degrees). Refer to Quick Test Step 4.0.
- Improper Fueling: Lead fouled HEGO sensor.
- Fuel Pressure: Perform Pinpoint Test Step HH1 and HH2.
- If at this point, the drive concern (if any) is still present, perform Steps HH3 and HH7 only.

Fuel Pump Circuit — (Inertia Switch) | Pinpoint Test | J

J

TYPE I —

VEHICLE APPLICATION: LINCOLN TOWN CAR, FORD CROWN VICTORIA/MERCURY GRAND MARQUIS, MARK VII/CONTINENTAL.THUNDERBIRD/COUGAR, MUSTANG, RANGER, AEROSTAR AND MERKUR

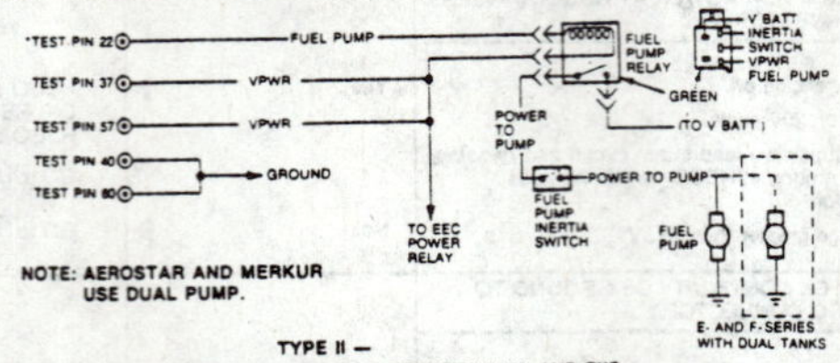

TYPE II —

VEHICLE APPLICATION: TEMPO/TOPAZ, ESCORT/LYNX, AND EXP

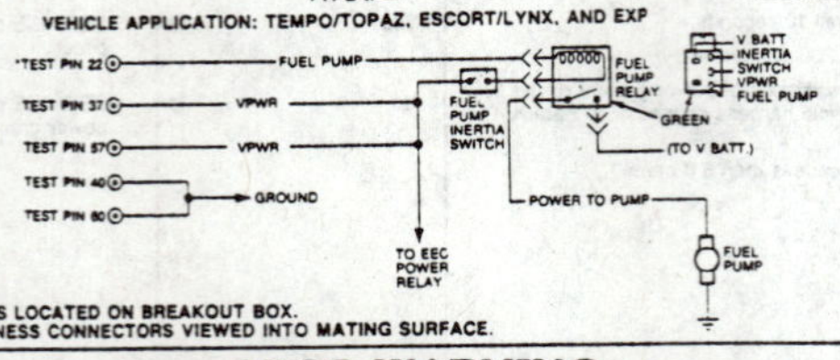

*TEST PINS LOCATED ON BREAKOUT BOX.
ALL HARNESS CONNECTORS VIEWED INTO MATING SURFACE.

STOP-WARNING

You should enter this Pinpoint Test only when a service code 87 is received in Quick Test Step 3.0 or you are directed here from Pinpoint Test Step A22 or Diagnostics by Symptom in the Engine Supplement Section.

To prevent the replacement of good components, be aware that the following non-EEC areas may be at fault:

- Fuel Lines
- Fuel Filters
- Throttle Body
- Contaminated Fuel
- Fuel Pump
- Inertia Switch

This Pinpoint Test is intended to diagnose only the following:

- Fuel Pump Relay
- Harness Circuits: V Batt., V Power, F.P., Ground and Power to Pump(s)
- Processor Assembly

Fuel Pump Circuit — (Inertia Switch) | Pinpoint Test | J

TEST STEP	RESULT ▶	ACTION TO TAKE
J1 NO FUEL PUMP PRESSURE		
• Fuel pressure gauge installed. • To check if fuel pump runs, cycle key from Off to Run, repeat several times. (Do not enter start mode.) Fuel pump should run briefly each time the key enters Run.	Yes ▶	GO to Shop Manual, Group 24 electric fuel pump.
	No ▶	GO to J2.
J2 CHECK FOR V POWER TO PROCESSOR		
• Key Off, wait 10 seconds. • Disconnect processor 60 Pin connector and inspect for damaged pins, corrosion, loose wires. Service as necessary. • Install Breakout box and reconnect processor. • Key On, engine Off. • DVOM on 20V scale. • Measure voltage between test Pin 37 and test Pin 40 at the Breakout box and between test Pin 57 and test Pin 60 at the breakout box.	Either voltage reading is less than 10.5V ▶	GO to B1.
	Both voltage readings are 10.5V or greater ▶	GO to J3.
J3 RESISTANCE CHECK OF FUEL PUMP INERTIA SWITCH		
• Key Off, wait 10 seconds. • Leave Breakout box installed and processor connected. • Locate and disconnect fuel pump inertia switch. • DVOM on 200 ohm scale. • Measure the resistance of the fuel pump inertia switch.	Less than 5.0 ohms ▶	RECONNECT inertia switch and GO to J4.
	5.0 ohms or greater ▶	REPLACE or reset fuel pump inertia switch. RERUN Quick Test.
J4 POWER TO PUMP(S) CHECK		
• Key On, engine Off. • Leave Breakout box installed and processor connected. • Locate fuel pump relay. • DVOM on 20V scale. • Measure voltage between chassis ground and power-to-pump(s) circuit at fuel pump relay during crank mode.	8.0V or greater during crank ▶	GO to J5.
	Less than 8.0V during crank ▶	GO to J6.

Fuel Pump Circuit — (Inertia Switch) | Pinpoint Test | J

TEST STEP	RESULT ▶	ACTION TO TAKE
J5 POWER AT FUEL PUMP(S) CHECK		
• Key On, engine Off. • Leave Breakout box installed and processor connected. • Locate fuel pump(s). • DVOM on 20V scale. • Measure voltage between chassis ground and power-to-pump(s) circuit at fuel pump during crank mode.	8.0V or greater during crank ▶	GO to Shop Manual, Group 24 electric fuel pump.
	Less than 8.0V during crank ▶	SERVICE open in power to the pump(s) circuit. RERUN Quick Test.
J6 FUEL PUMP CIRCUIT CHECK TO V BATT.		
• Key On, engine Off. • Leave Breakout box installed and processor connected. • Locate fuel pump relay. • DVOM on 20V scale. • Measure voltage between chassis ground and V Batt. at the fuel pump relay.	10.5V or greater ▶	GO to J7.
	Less than 10.5V ▶	SERVICE open in V Batt. between fuel pump relay and vehicle battery positive post. RERUN Quick Test.
J7 V POWER TO FUEL PUMP RELAY CHECK		
• Key On, engine Off. • Leave Breakout box installed and processor connected. • Locate fuel pump relay. • DVOM on 20V scale. • Measure voltage between chassis ground and V Power circuit at the fuel pump relay.	10.5V or greater ▶	GO to J8.
	Less than 10.5V ▶	VERIFY inertia switch is reset to On. If switch will not reset, REPLACE switch. If OK, SERVICE open in V Power circuit between the processor and the fuel pump relay. RERUN Quick Test.
J8 FUEL PUMP CIRCUIT CHECK CONTINUITY		
• Leave Breakout box installed and processor connected. • Key Off, wait 10 seconds. • DVOM on 200 ohm scale. • Measure resistance between fuel pump circuit at the pump relay and test Pin 22 at the Breakout box.	Less than 5 ohms ▶	GO to J9.
	5 ohms or greater ▶	SERVICE open in fuel pump circuit and RERUN Quick Test.

Fuel Pump Circuit — (Inertia Switch) | Pinpoint Test | J

TEST STEP	RESULT ▶	ACTION TO TAKE
J9 CHECK FOR SHORT TO GROUND		
• Key Off. • Leave Breakout box installed and processor disconnected. • Fuel pump relay disconnected. • DVOM on 200,000 ohm scale. • Measure resistance between test Pin 22 and test Pins 40 and 60 at the Breakout box.	10,000 ohms or greater ▶	GO to J10.
	Less than 10,000 ohms ▶	SERVICE short in the fuel pump circuit. RERUN Quick Test.
J10 CHECK FOR SHORT TO POWER		
• Key Off, wait 10 seconds. • Leave Breakout box installed and disconnect processor. • Fuel pump relay disconnected. • DVOM on 200,000 ohm scale. • Measure resistance between test Pin 22 and test Pins 37 and 57 at the Breakout box.	10,000 ohms or greater ▶	RECONNECT fuel pump relay. GO to J11.
	Less than 10,000 ohms ▶	SERVICE short to power in the fuel pump circuit. RECONNECT processor, ATTEMPT to start vehicle. If vehicle fails to start, REPLACE processor. RERUN Quick Test.
J11 FINAL SYSTEM CHECK		
• Leave Breakout box installed and processor disconnected. • Connect jumper wire from test Pin 22 to test Pin 40 or 60 at the Breakout box. • DVOM on 20V scale. • Key On, engine Off. • Measure voltage between chassis ground and power-to-pump(s) circuit at fuel pump relay.	10.5V or greater ▶	REPLACE processor. RERUN Quick Test.
	Less than 10.5V ▶	REPLACE fuel pump relay. RECONNECT processor and RERUN Quick Test.
J15 FUEL PUMP RELAY CHECK		
• Key-Off. • Remove fuel pump relay. • Does fuel pump turn off?	Yes ▶	REPLACE fuel pump relay. RERUN Quick Test.
	No ▶	SERVICE short to power to pumps circuit.

EGR On/Off Control — Pinpoint Test — KA

KA

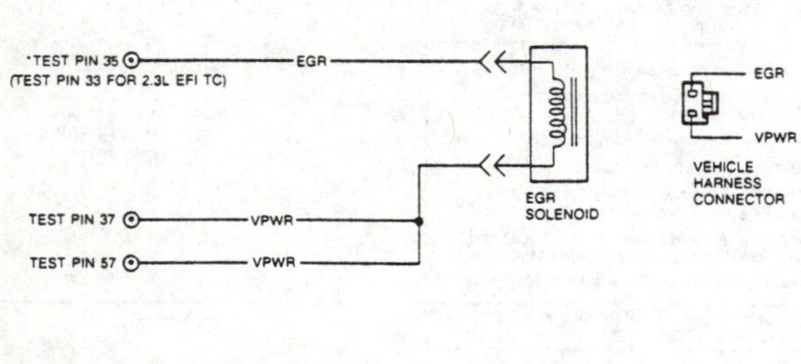

*TEST PINS LOCATED ON BREAKOUT BOX.
ALL HARNESS CONNECTORS VIEWED INTO MATING SURFACE.

STOP-WARNING

You should enter this Pinpoint Test only when a service code 34 or 84 is received in Quick Test Step 3.0, 5.0 or when directed here from Diagnostic by Symptom in the Engine Supplement Section.

To prevent the replacement of good components, be aware that the following non-EEC areas may be at fault:

- Air or Vacuum Leaks
- EGR Flow Restrictions
- EGR Value

NOTE: Code 34 may be the result of high volume exhaust vent system (reduces backpressure). If this is suspected, perform the test in a well-ventilated area without exhaust vent connected.

This Pinpoint Test is intended to diagnose only the following:

- Circuits: EGR and VPWR
- EGR Solenoid
- Presence of Manifold Vacuum
- Processor Assembly

EGR On/Off Control — Pinpoint Test — KA

TEST STEP	RESULT ▶	ACTION TO TAKE
KA1 ENTER OUTPUT STATE CHECK (REFER TO APPENDIX)		
NOTE: Do not use STAR tester for this step, use a VOM/DVOM. • Key Off, wait 10 seconds. • DVOM on 20V scale. • Connect DVOM negative test lead to STO at the Self-Test connector and positive test lead to battery positive. • Jumper STI to signal return at the Self-Test connector. • Perform Key On Engine Off Self-Test until the completion of the Continuous Test Codes. • DVOM will indicate zero volts. • Depress and release the throttle. • Did DVOM reading change to a high voltage reading?	Yes ▶	REMAIN in Output State Check. GO to KA2.
	No ▶	DEPRESS throttle to WOT and release. If STO voltage does not go high, GO to Pinpoint Test Step Q40. Leave equipment hooked up.
KA2 CHECK EGR ON/OFF CONTROL SOLENOID ELECTRICAL OPERATION		
• DVOM on 20V scale. • Connect DVOM positive test lead to VPWR circuit on EGR solenoid and negative test lead to EGR output circuit. • While observing DVOM, depress and release the throttle several times to cycle output On and Off. • Does EGR output cycle On and Off?	Yes ▶	GO to KA3.
	No ▶	REMOVE STI jumper. GO to KA5.
KA3 CHECK SOLENOID FOR VACUUM CYCLING		
• Install vacuum pump to the solenoid vacuum supply port and install a vacuum gauge to the output port. Apply 6 in. Hg minimum. • While cycling outputs On and Off (by depressing and releasing throttle) observe the vacuum gauge at the output. **NOTE: Maintain vacuum at source.** • Does output port vacuum cycle On and Off?	Yes ▶	GO to KA4.
	No ▶	REPLACE solenoid. RERUN Quick Test.

EGR On/Off Control — Pinpoint Test — KA

TEST STEP	RESULT ▶	ACTION TO TAKE
KA4 CHECK MANIFOLD VACUUM LINES FOR BLOCKAGE OR LEAKS		
• Vacuum lines disconnected at solenoid. • Start engine. • Check for vacuum. • Is vacuum present?	Yes ▶	EEC-IV system OK. REFER to Section 6.
	No ▶	SERVICE vacuum source blockage or leak. RERUN Quick Test.
KA5 MEASURE EGR SOLENOID RESISTANCE		
• Key Off, wait 10 seconds. • DVOM on 200 ohm scale. • Disconnect EGR solenoid connector and measure solenoid resistance. • Is resistance between 65 and 110 ohms?	Yes ▶	GO to KA6.
	No ▶	REPLACE EGR solenoid. RERUN Quick Test.
KA6 CHECK VOLTAGE OF VPWR CIRCUIT		
• Key On, Engine Off. • DVOM on 20V scale. • Measure voltage between VPWR circuit at the EGR solenoid vehicle harness connector and battery ground. • Is voltage greater than 10.5 volts?	Yes ▶	GO to KA7.
	No ▶	SERVICE harness open circuit. RERUN Quick Test.
KA7 CHECK CONTINUITY OF EGR CIRCUIT		
• Key Off, wait 10 seconds. • Disconnect processor 60 Pin connector and inspect for damaged pins, corrosion, loose wires. Service as necessary. • Connect Breakout box to harness. Leave processor disconnected. • DVOM on 200 ohm scale. • Measure resistance between test Pin 35 (test Pin 33 for 2.3L EFI TC) at the Breakout box and EGR circuit at vehicle harness connector. • Is resistance less than 5 ohms?	Yes ▶	GO to KA8.
	No ▶	SERVICE open circuit. RERUN Quick Test.

EGR On/Off Control — Pinpoint Test — KA

TEST STEP	RESULT ▶	ACTION TO TAKE
KA8 CHECK FOR SHORT TO GROUND		
• Key Off, wait 10 seconds. • Leave Breakout box installed and processor disconnected. • Disconnect EGR solenoid. • DVOM on 200,000 ohm scale. • Measure resistance between test Pin 35 (test Pin 33 for 2.3L EFI TC) and test Pins 40, 46 and 60 at the Breakout box. • Is resistance greater than 10,000 ohms?	Yes ▶	GO to KA9.
	No ▶	SERVICE short to ground. RERUN Quick Test.
KA9 CHECK FOR SHORT TO POWER		
• Key Off, wait 10 seconds. • DVOM on 200,000 ohm scale. • Leave Breakout box installed and processor disconnected. • EGR solenoid disconnected. • Measure resistance between test Pin 35 (test Pin 33 for 2.3L EFI TC) and test Pins 37 and 57 at the Breakout box. • Is resistance greater than 10,000 ohms?	Yes ▶	REPLACE Processor. RERUN Quick Test.
	No ▶	SERVICE short to power. RERUN Quick Test. If code is repeated, REPLACE processor.

DC Motor Idle Speed Control/Idle Tracking Switch and Throttle Position Sensor — 1.9L CFI and 2.3L (HSC) CFI — Pinpoint Test KB

KB

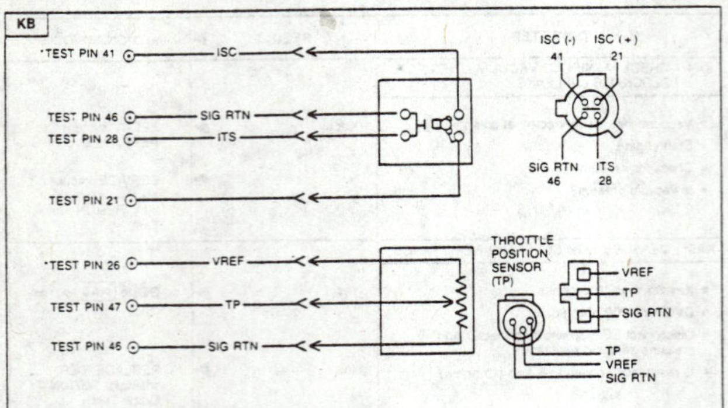

*TEST PINS LOCATED ON THE BREAKOUT BOX.
ALL HARNESS CONNECTORS VIEWED INTO MATING SURFACE.

STOP-WARNING

You should enter this Pinpoint Test only when a Service Code 12, 13, 23, 38, 53, 58, 63, 68, 71, 73 or 93 is received in Quick Test Step 3.0, 5.0, or 6.0 or when directed here from Diagnostics by Symptom in the Engine Supplement Section.

To prevent the replacement of good components, be aware that the following non-EEC areas may be at fault:

- Anti-diesel speed set too high
- Basic engine
- Vacuum leaks
- Throttle sticking

This Pinpoint Test is intended to diagnose only the following:

- DC motor
- Idle Tracking Switch
- Harness circuits ISC+, ISC−, ITS, TP, VREF, and SIG RTN
- Throttle position sensor
- EEC-IV processor assembly

DC Motor Idle Speed Control/Idle Tracking Switch and Throttle Position Sensor — 1.9L CFI and 2.3L (HSC) CFI — Pinpoint Test KB

TEST STEP	RESULT ▶	ACTION TO TAKE
KB1 CHECK DC MOTOR FOR PROPER OPERATION		
• Key Off, wait 10 seconds. • Disconnect harness from DC motor. • Jumper ISC+ circuit of DC motor to battery positive and ISC− circuit of DC motor to battery ground for 4 seconds. • Jumper ISC+ circuit of DC motor to battery ground and ISC− circuit of DC motor to battery positive for 4 seconds. • Does the DC motor shaft extend to greater than 2 inches (5 cm) and retract to less than 1.75 inches (4.4 cm) from mounting bracket (see below)? 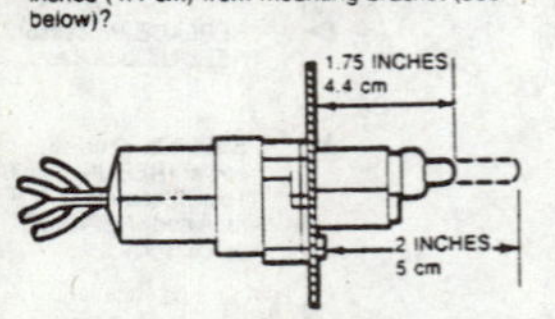	Yes ▶ No ▶	GO to KB2. REPLACE DC MOTOR. RERUN Quick Test.
KB2 CHECK CONTINUITY OF ISC+ AND ISC− CIRCUITS		
• Key Off, wait 10 seconds. • Harness disconnected from DC motor. • Disconnect processor 60 pin connector and inspect for damaged pins, corrosion, loose wires. Service as necessary. • Connect Breakout box. Leave processor disconnected. • DVOM on 200 ohm scale. • Measure resistance between test Pin 41 at the Breakout box and ISC− circuit at the vehicle harness connector and between test Pin 21 at the Breakout box and ISC+ circuit at the vehicle harness connector. • Are both resistances less than 5 ohms?	Yes ▶ No ▶	GO to KB3. SERVICE faulty circuit(s). RERUN Quick Test.

DC Motor Idle Speed Control/Idle Tracking Switch and Throttle Position Sensor — 1.9L CFI and 2.3L (HSC) CFI — Pinpoint Test KB

TEST STEP	RESULT ▶	ACTION TO TAKE
KB3 CHECK FOR SHORTS TO GRND OF ISC+ AND ISC− CIRCUITS		
• Key Off, wait 10 seconds. • Breakout box installed and processor disconnected. • Harness disconnected from DC motor. • DVOM on 200,000 ohm scale. • Measure resistance between test Pin 41 and test Pins 40, 46 and 60 at the Breakout box. • Measure resistance between test Pin 21 and test Pins 40, 46 and 60 at the Breakout box. • Are all resistances greater than 10,000 ohms?	Yes ▶ No ▶	GO to KB4. SERVICE faulty circuit(s). RERUN Quick Test.
KB4 CHECK FOR SHORTS TO PWR. OF ISC+ AND ISC− CIRCUITS		
• Key On engine off. • Breakout box installed and processor disconnected. • Harness disconnected from DC motor. • DVOM on 20V scale. • Measure voltage between test Pin 41 and test Pin 40 and 60 at the Breakout box. • Measure voltage between test Pin 21 and test Pins 40 and 60 at the Breakout-box. • Are all voltages less than 1 volt?	Yes ▶ No ▶	REPLACE processor. RERUN Quick Test. SERVICE faulty circuit(s). RERUN Quick Test.
KB5 CHECK FOR FULL DC MOTOR RETRACTION		
• Key Off, wait 10 seconds. • Disconnect harness from DC motor. • Jumper, at the DC motor connector, ISC− circuit to battery positive and the ISC+ circuit to battery negative for 4 seconds. • Does the DC motor shaft retract away from the throttle lever as shown?	Yes ▶ No ▶ 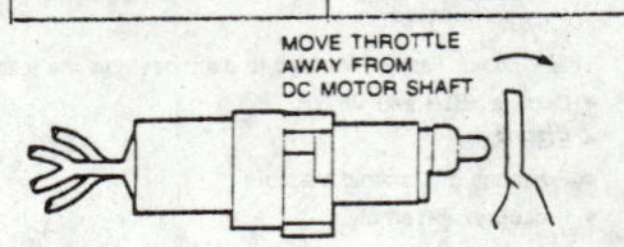	GO to KB7. GO to KB6.

DC Motor Idle Speed Control/Idle Tracking Switch and Throttle Position Sensor — 1.9L CFI and 2.3L (HSC) CFI — Pinpoint Test KB

TEST STEP	RESULT ▶	ACTION TO TAKE
KB6 MEASURE DC MOTOR RETRACTION		
• Key Off, wait 10 seconds. • Harness disconnected from DC motor. • DC motor fully retracted. • Measure the distance from the tip of the DC motor shaft to the mounting bracket. Refer to figure in step KB1. • Is the distance less than 1.75 inches (4.4 cm)?	Yes ▶ No ▶	RECONNECT DC motor. GO to Section 4 for throttle stop adjustment procedure. REPLACE DC motor. RERUN Quick Test.
KB7 CHECK IDLE TRACKING SWITCH STATE		
• Key Off, wait 10 seconds. • Harness disconnected from DC motor. • DC motor fully retracted. • DC motor shaft **NOT** touching the throttle lever. • DVOM on 200 ohm scale. • Measure resistance between ITS circuit and SIG RTN at the DC motor connector. • Is the resistance less than 5 ohms?	Yes ▶ No ▶	GO to KB8. REPLACE DC motor. RERUN Quick Test.
KB8 CHECK CONTINUITY OF ITS AND SIG RTN CIRCUITS		
• Key Off, wait 10 seconds. • Harness disconnected from DC motor. • Disconnect processor 60 pin connector and inspect for damaged pins, corrosion or loose wires. Service as necessary. • Connect Breakout box. Leave processor disconnected. • DVOM on 200 ohm scale. • Measure resistance between test Pin 46 at the Breakout box and SIG RTN circuit at the DC motor vehicle harness connector. • Measure resistance between test Pin 28 at the Breakout box and ITS circuit at the DC motor vehicle harness connector. • Are both resistances less than 5 ohms?	Yes ▶ No ▶	RECONNECT DC motor. REPLACE processor. RERUN Quick Test. RECONNECT DC motor. SERVICE faulty circuit(s) RERUN Quick Test.

DC Motor Idle Speed Control/Idle Tracking Switch and Throttle Position Sensor — 1.9L CFI and 2.3L (HSC) CFI — Pinpoint Test KB

TEST STEP	RESULT	ACTION TO TAKE
KB9 CHECK IDLE TRACKING SWITCH STATE		
• Key Off, wait 10 seconds. • Disconnect harness from DC motor. • DVOM on 200 ohm scale. • Measure resistance between ITS circuit and SIG RTN circuit at the DC motor connector. • Is the resistance greater than 5 ohms?	Yes ▶	GO to KB10.
	No ▶	REPLACE DC motor. RERUN Quick Test.
KB10 CHECK FOR SHORTS TO GRND OF THE ITS CIRCUIT		
• Key Off, wait 10 seconds. • Harness disconnected from DC motor. • Disconnect processor 60 pin connector and inspect for damaged pins, corrosion loose wires. Service as necessary. • Connect Breakout box. Leave processor disconnected. • Measure resistance between test Pin 28 and test Pins 40, 46 and 60 at the Breakout box. • Are all resistances greater than 10,000 ohms?	Yes ▶	RECONNECT DC motor. REPLACE processor. RERUN Quick Test.
	No ▶	RECONNECT DC motor. SERVICE faulty circuit(s). RERUN Quick Test.
KB11 CHECK THROTTLE LEVER AND LINKAGE		
• Key Off, wait 10 seconds. • Inspect throttle for freedom of movement to wide open throttle and for damaged or bent throttle lever. • Is throttle/throttle linkage functioning properly?	Yes ▶	REPLACE DC motor. RERUN Quick Test.
	No ▶	SERVICE as necessary. RERUN Quick Test.
KB12 CHECK THROTTLE PLATE FOR CLOSING		
• Run Key On Engine Off Self Test and disconnect DC motor after it is fully retracted. • Key Off, wait 10 seconds. • Remove air cleaner from throttle body. • Inspect throttle for freedom of movement and proper closure. • Does throttle move freely and close without obstruction?	Yes ▶	RECONNECT DC motor. GO to KB13.
	No ▶	SERVICE as necessary. RERUN Quick Test

DC Motor Idle Speed Control/Idle Tracking Switch and Throttle Position Sensor — 1.9L CFI and 2.3L (HSC) CFI — Pinpoint Test KB

TEST STEP	RESULT	ACTION TO TAKE
KB13 CHECK VOLTAGE OF VREF TO SIGNAL RETURN		
• Key Off, wait 10 seconds. • Disconnect TP vehicle harness connector at the throttle body. Inspect for damaged pins, corrosion, loose wires, etc. Service as necessary. • DVOM on 20V scale. • Key On Engine Off. • Measure voltage between VREF and SIG RTN at the vehicle harness connector. • Is the voltage between 4 and 6 volts?	Yes ▶	RECONNECT TP sensor. GO to KB14.
	No ▶	GO to Pinpoint Test Step C1.
KB14 CHECK THROTTLE STOP RPM		
• Run Key On Engine Off Self Test and disconnect the DC motor after it has fully retracted and exit Self Test. • Start engine and verify that the throttle stop rpm is less than curb idle rpm. • Is the throttle stop set below the curb idle? • Reconnect DC motor.	Yes ▶	REPLACE the TP sensor. RERUN Quick Test.
	No ▶	GO to the adjustment procedure in Section 4. ADJUST throttle stop rpm. RERUN Quick Test.
KB15 GENERATE CODE 63		
• Key Off, wait 10 seconds. • Disconnect harness from the TP sensor at the throttle body. Inspect for damaged pins, corrosion and loose wires. Service as necessary. • Run Key On Engine Off Self Test and record codes. • Is code 63 present? • Ignore all other codes.	Yes ▶	GO to KB16.
	No ▶	GO to KB17.

DC Motor Idle Speed Control/Idle Tracking Switch and Throttle Position Sensor — 1.9L CFI and 2.3L (HSC) CFI — Pinpoint Test KB

TEST STEP	RESULT	ACTION TO TAKE
KB16 CHECK VOLTAGE VREF TO SIG RTN		
• Key Off, wait 10 seconds. • Harness disconnected from TP sensor at throttle body. • DVOM on 20V scale. • Key On Engine Off. • Measure voltage between VREF and SIG RTN at the TP vehicle harness connector. • Is the voltage between 4 and 6 volts?	Yes ▶	REPLACE the TP sensor. RERUN Quick Test.
	No ▶	GO to Pinpoint Test Step C1.
KB17 CHECK TP SIGNAL FOR SHORT TO POWER		
• Key Off, wait 10 seconds. • Harness disconnected from TP sensor. • DVOM on 200,000 ohm scale. • Disconnect the processor 60 pin connector and inspect for damaged pins, corrosion and loose wires. Service as necessary. • Install Breakout box. Leave processor disconnected. • Measure resistance between test Pin 47 and test Pins 26 and 57 at the Breakout box. • Are both resistances 10,000 ohms or greater?	Yes ▶	REPLACE processor. RERUN Quick Test.
	No ▶	SERVICE faulty circuit(s). RERUN Quick Test.
KB18 GENERATE CODE 53		
• Key Off, wait 10 seconds. • Disconnect harness from the TP sensor at the throttle body. Inspect for damaged pins, corrosion and loose wires. Service as necessary. • Jumper VREF to TP signal at the vehicle harness connector. • Run Key On Engine Off Self Test. **NOTE: If no codes are generated, immediately remove jumper and go directly to KB21.** • Is code 53 present? • Ignore all other codes at this time.	Yes ▶	REPLACE TP sensor. RERUN Quick Test.
	No ▶	GO to KB19.

DC Motor Idle Speed Control/Idle Tracking Switch and Throttle Position Sensor — 1.9L CFI and 2.3L (HSC) CFI — Pinpoint Test KB

TEST STEP	RESULT	ACTION TO TAKE
KB19 CHECK VOLTAGE VREF TO SIG RTN		
• Key Off, wait 10 seconds. • Harness disconnected from TP sensor at throttle body. • DVOM on 20V scale. • Key On Engine Off. • Measure voltage between VREF and SIG RTN at the TP vehicle harness connector. • Is the voltage between 4 and 6 volts?	Yes ▶	GO to KB20.
	No ▶	GO to Pinpoint Test Step C1.
KB20 CHECK CONTINUITY OF TP CIRCUIT		
• Key Off, wait 10 seconds. • Harness disconnected from TP sensor at throttle body. • DVOM on 200 ohm scale. • Disconnect the processor 60 pin connector and inspect for damaged pins, corrosion and loose wires. Service as necessary. • Install Breakout box. Leave processor disconnected. • Measure resistance between TP signal at the vehicle harness connector and test Pin 47 at the Breakout box. • Is resistance less than 5 ohms?	Yes ▶	GO to KB21.
	No ▶	SERVICE faulty circuit(s). RECONNECT harness to TP sensor. RERUN Quick Test.
KB21 CHECK TP SIGNAL FOR SHORTS TO GROUND		
• Key Off, wait 10 seconds. • Harness disconnected from TP sensor at throttle body. • Breakout box installed. • DVOM on 200,000 ohm scale. • Measure resistance between TP signal at TP vehicle harness connector and test Pins 40, 46 and 60 at the Breakout box. • Are all resistances greater than 10,000 ohms?	Yes ▶	REPLACE processor. RECONNECT TP sensor. RERUN Quick Test.
	No ▶	SERVICE faulty circuit(s). RERUN Quick Test.

DC Motor Idle Speed Control/Idle Tracking Switch and Throttle Position Sensor — 1.9L CFI and 2.3L (HSC) CFI | Pinpoint Test | KB

TEST STEP	RESULT ▶	ACTION TO TAKE
KB22 SERVICE CODE 73		
• RERUN Key On Engine Off Self Test. • Is code 73 still present?	Yes ▶	REPLACE TP sensor. RERUN Quick Test.
	No ▶	SERVICE other codes.
KB23 CHECK FOR CODES THAT COULD CAUSE CODE 12		
• Are service codes 31, 32, 34, 35, 41, or 58 present in Engine Running Self Test?	Yes ▶	SERVICE these codes first. GO to Quick Test Step 5.0C for direction.
	No ▶	GO to KB24.
KB24 CHECK FOR STICKING THROTTLE LINKAGE		
• Check the throttle plates and/or linkage for sticking or binding. • Check speed control linkage for proper adjustment. • Does throttle open and close properly?	Yes ▶	GO to KB25.
	No ▶	SERVICE as necessary. RERUN Quick Test.
KB25 CHECK DC MOTOR FOR PROPER OPERATION		
• Key Off, wait 10 seconds. • Disconnect harness from DC motor. • Jumper ISC+ circuit of DC motor to battery positive and ISC− circuit of DC motor to battery ground for 4 seconds. • Jumper ISC+ circuit of DC motor to battery ground and ISC− circuit of DC motor to battery positive for 4 seconds. • Does the DC motor shaft extend to greater than 2 inches (5 cm) and retract to less than 1.75 inches (4.4 cm) from mounting bracket (see below)?	Yes ▶	REPLACE processor. RERUN Quick Test.
	No ▶	REPLACE DC MOTOR. RERUN Quick Test.

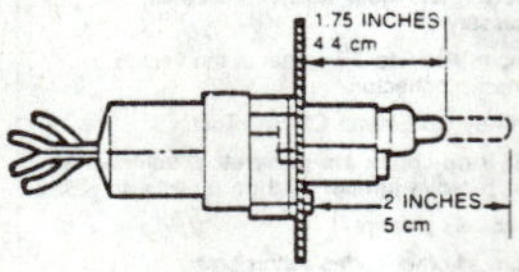

DC Motor Idle Speed Control/Idle Tracking Switch and Throttle Position Sensor — 1.9L CFI and 2.3L (HSC) CFI | Pinpoint Test | KB

TEST STEP	RESULT ▶	ACTION TO TAKE
KB26 CHECK FOR ERRATIC IDLE		
• Prepare system for normal engine operation. • Deactivate Self Test. • A/C off. • Run engine for 3 minutes alternating between 30 second idles and 5 second part throttle modes. • Is the idle erratic at the end of the three minute idle/part throttle test?	Yes ▶	CHECK for • Vacuum leaks • Code 22 • Code 31, 32, 34 or 35 • Code 41 • Code 58 SERVICE above codes before continuing.
	No ▶	GO to KB27.
KB27 CHECK FOR PROPER OPERATION OF THROTTLE		
• Inspect the throttle plates and/or linkage for proper function. • Does the throttle open and close properly?	Yes ▶	REPLACE DC motor. RERUN Quick Test.
	No ▶	SERVICE as necessary. RERUN Quick Test.

DC Motor Idle Speed Control/Idle Tracking Switch and Throttle Position Sensor — 1.9L CFI and 2.3L (HSC) CFI | Pinpoint Test | KB

KB90 CONTINUOUS MEMORY CODE 13

A Continuous Memory Code 13 indicates that sometime in the last 40 warm-up cycles the TP sensor rotation did not follow the reaction of the DC motor when idle speed control was in a dashpot mode. This condition may be caused by:

- The DC motor sticking at part throttle.
- An open in the ITS circuit which, when coupled with other inputs to the processor, causes the EEC-IV system to falsely enter 'dashpot mode'.
- The TP sensor sticking at part throttle.

Each of these areas may generate Key On Engine Off (KOEO) or other continuous memory codes. Therefore, if a repair has been made for KOEO code 13 or 58, the continuous memory code 13 can be considered serviced and erased from memory. If a Continuous Memory Code 38 is present along with the Continuous Memory Code 13, service the 38 first.

If these other codes were not present make the following checks:

- Refer to KB1 and check for FULL travel of the DC motor shaft. Replace the DC motor if full travel is not possible. Leave the motor fully retracted.
- With the DC motor fully retracted and the ITS not touching the throttle lever (ITS closed circuit) check for an intermittent open in the ITS circuit. Turn the ignition key off and install the Breakout box. Make the necessary connector/pin inspections. With the DVOM on the 200 ohm scale, monitor between test Pins 28 and 46 while tapping, wiggling, bending, etc. the DC motor connector and harness.

  DO NOT PUSH IN THE IDLE TRACKING SWITCH.

  The DVOM will change from less than 5 ohms to greater than 5 ohms if an open circuit is created. Service as necessary.
- Check for a sticking TP sensor by monitoring TP voltage while moving the throttle from a wide open position to a closed throttle position. To do this it is necessary to install the Breakout box. Make the necessary connector/pin inspections. It is also necessary to fully retract the shaft of the DC motor by placing a jumper between test Pins 41 and 57. When the motor has fully retracted, disconnect it at the harness and remove the jumper wire from the Breakout box. With the ignition key on and the DVOM on the 20 volt scale, slowly move the throttle wide open to closed throttle. The voltage should move from more than 4 volts to less than 1.5 volts. If the TP sensor hangs up in midrange replace it, otherwise no service should be made.

DC Motor Idle Speed Control/Idle Tracking Switch and Throttle Position Sensor — 1.9L CFI and 2.3L (HSC) CFI | Pinpoint Test | KB

KB91 CONTINUOUS MEMORY CODE 38

A Continuous Memory Code 38 indicates that in the last 40 warm-up cycles the Idle Tracking Switch was open (ITS touching the throttle) when the throttle angle was greater than the MAX extension of the DC motor shaft. This could be caused by:

- An open (either intermittent or hard fault) in the ITS circuit.
- Idle Tracking Switch stuck open (pushed in position).

Either of these conditions may cause a code 58 to appear in Key On Engine Off (KOEO). If a repair has been made for a KOEO code 58, the Continuous Memory Code 38 can be considered serviced and erased from memory.

If KOEO code 58 was not present the following checks can be made:

- With the DC motor fully retracted and the ITS not touching the throttle lever (ITS closed circuit) check for an intermittent open in the ITS circuit. Turn the ignition key off and install the Breakout box. Make the necessary connector/pin inspections. With the DVOM on the 200 ohm scale, monitor between Test Pins 28 and 46 while tapping, wiggling, bending, etc. the DC motor connector and harness.

  DO NOT PUSH IN THE IDLE TRACKING SWITCH.

  The DVOM will change from less than 5 ohms to greater than 5 ohms if an open circuit is created. Service as necessary. If an open circuit cannot be created, no repair should be made.

KB92 CONTINUOUS MEMORY CODE 71

A Continuous Memory Code 71 indicates that sometime in the last 40 warm-up cycles the Idle Tracking Switch was closed (ITS not touching the throttle lever) when the DC motor was in "preposition" — [after the engine has been running and the ignition key is turned off the DC motor fully retracts and then extends to a predetermined position for the next start-up]. This can be caused by:

- The ITS circuit shorted to ground or Signal Return (intermittent or hard fault).
- ITS stuck closed (ITS **NOT** in the pushed in position).

Either of these conditions may cause a Key On Engine Off (KOEO) code 68. If a repair has been made for KOEO code 68, the Continuous Memory Code 71 can be considered serviced and erased from memory.

If KOEO 68 was not present make the following checks:

- Check the ITS circuit for an intermittent short to ground or Signal Return. Turn the ignition key off. Enter the KOEO Continuous Monitor Mode per Quick Test Step 6.0B. Systematically tap, wiggle, or bend the harness while looking for an indication of a fault. If a fault is created, service as necessary, otherwise no repair should be made.

NOTE: Due to the nature of this Test Step, Code 71 will not reappear in memory if a fault is found.

DC Motor Idle Speed Control/Idle Tracking Switch and Throttle Position Sensor — 1.9L CFI and 2.3L (HSC) — Pinpoint Test KB

TEST STEP	RESULT ▶	ACTION TO TAKE
KB93 CONTINUOUS TEST SERVICE CODE 53: EXERCISE TP SENSOR		
• Using continuous monitor mode, observe VOM or STAR LED for indication of a fault while performing the following: • Move throttle slowly to WOT position. • Release throttle slowly to closed position and lightly tap on TP sensor (simulate road shock). • Wiggle TP harness connector. • Is a fault indicated?	Yes ▶ No ▶	GO to KB94. GO to KB95.
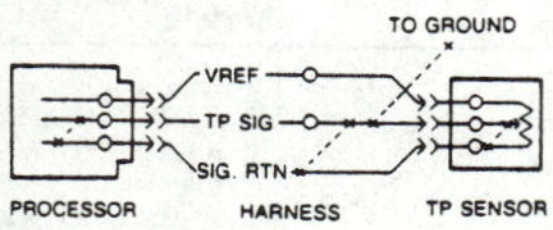		
KB94 MEASURE THROTTLE POSITION SIGNAL VOLTAGE WHILE EXERCISING TP SENSOR		
• Key Off, wait 10 seconds. • Disconnect processor 60 Pin connector and inspect for damaged pins, corrosion, loose wires. Service as necessary. • Install Breakout box and reconnect processor. • VOM or STAR LED still connected to STO as in previous step. • Connect a DVOM from test Pin 47 to test Pin 46. • DVOM on 20V scale. • Key On Engine Off. • While observing DVOM, repeat Step KB93. • Does the fault occur below 4.25V?	Yes ▶ No ▶	DISCONNECT and INSPECT connectors. If connector and terminals are good, REPLACE TP sensor, REFER to Shop Manual Group 24 and RERUN Quick Test. Throttle position sensor overtravel may have caused the continuous code 53. Sensor service is not required. To verify harness integrity, GO to KB95.

DC Motor Idle Speed Control/Idle Tracking Switch and Throttle Position Sensor — 1.9L CFI and 2.3L (HSC) — Pinpoint Test KB

TEST STEP	RESULT ▶	ACTION TO TAKE
KB95 CHECK EEC-IV HARNESS		
• Observe VOM or STAR LED for a fault indication while performing the following: • Referring to the illustration in Step DH90, grasp the harness close to the sensor connector. Wiggle, shake or bend a small section of the EEC-IV system harness while working your way to the dash panel. Also wiggle, shake or bend the EEC-IV harness from the dash panel to the processor. • Is a fault indicated?	Yes ▶ No ▶	ISOLATE fault and make necessary repairs. REFER to appropriate figure. RERUN Quick Test. GO to KB96.
KB96 CHECK PROCESSOR AND HARNESS CONNECTORS		
• Key Off, wait 10 seconds. • Disconnect processor 60 Pin connector. • Inspect both connectors and connector terminals for obvious damage or faults. • Are connectors and terminals OK?	Yes ▶ No ▶	Unable to duplicate fault at this time. Continuous code 53 testing complete. SERVICE as necessary. REPEAT Quick Test.
KB97 CONTINUOUS TEST SERVICE CODES 23 OR 63 EXERCISE TP SENSOR		
• Using continuous monitor mode, observe VOM or STAR LED for indication of a fault while performing the following: • Move throttle slowly to WOT position. • Release throttle slowly to closed condition. • Lightly tap on TP sensor (simulate road shock). • Wiggle TP harness connector. • Is a fault indicated?	Yes ▶ No ▶	DISCONNECT and INSPECT connectors. If connector and terminals are good, REPLACE TP sensor, REFER to Shop Manual Group 24 and RERUN Quick Test. GO to KB98.

DC Motor Idle Speed Control/Idle Tracking Switch and Throttle Position Sensor — 1.9L CFI and 2.3L (HSC) — Pinpoint Test KB

TEST STEP	RESULT ▶	ACTION TO TAKE
KB98 CHECK EEC-IV HARNESS		
• Observe VOM or STAR LED for a fault indication while performing the following: • Referring to the illustration in Step DH94, grasp the harness close to the sensor connector. Wiggle, shake or bend a small section of the EEC-IV system harness while working your way to the dash panel. Also wiggle, shake or bend the EEC-IV harness from the dash panel to the processor. • Is a fault indicated?	Yes ▶ No ▶	ISOLATE fault and make necessary repairs. REFER to appropriate figure. RERUN Quick Test. GO to KB99.
KB99 CHECK PROCESSOR AND HARNESS CONNECTORS		
• Key Off, wait 10 seconds. • Disconnect processor 60 Pin connector. • Inspect both connectors and connector terminals for obvious damage or faults. • Are connectors and terminals OK?	Yes ▶ No ▶	Unable to duplicate fault at this time. Continuous code 63 testing complete. SERVICE as necessary. REPEAT Quick Test.

Air Management System — Pinpoint Test KC

KC

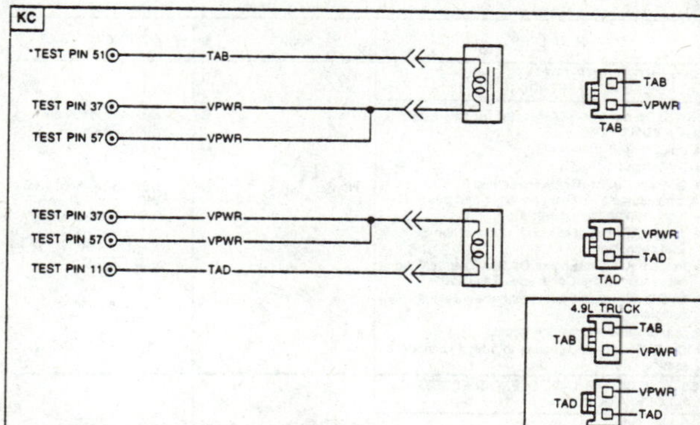

*TEST PINS LOCATED ON BREAKOUT BOX.
ALL HARNESS CONNECTORS VIEWED INTO MATING SURFACE.

STOP-WARNING

You should enter this Pinpoint Test only when a service code 44, 45, 46, 94, 95, 96, 81 or 82 is received in Quick Test Step 3.0 or 5.0.

To prevent the replacement of good components, be aware that the following non-EEC areas may be at fault:

- Thermactor System
  - Belt
  - Pump
  - Valve

This Pinpoint Test is intended to diagnose only the following:

- TAB and TAD Solenoid Valve Assemblies
- Harness Circuits: TAB, TAD and VPWR
- Vacuum Supply
- Processor Assembly

Air Management System — Pinpoint Test — KC

TEST STEP	RESULT ▶	ACTION TO TAKE
KC1 SERVICE CODES 44 (94), 45 (95) AND 46 (96): VERIFY VACUUM LINE ROUTING		
• Verify proper vacuum line routing to the TAB/TAD solenoids and to the bypass diverter valve. Refer to VECI decal. • Check for kinked or blocked vacuum lines. • Check for kinked or blocked air hoses. • Check for disconnected vacuum lines. • Are visual checks satisfactory?	No ▶	SERVICE routing or faults. RERUN Quick Test.
	Yes ▶	Service code 44 (94), GO to KC4. Service code 45 (95), GO to KC2. Service code 46 (96), GO to KC3.
KC2 ATTEMPT TO ELIMINATE SERVICE CODE 45 (95) (TAD ONLY)		
• Disconnect vacuum line on diverter valve and cap vacuum line. • Key Off, wait 10 seconds. • Repeat Engine Running Self-Test and record service codes. • Is code 45 (95) present?	Yes ▶	EEC-IV system OK. REFER to Section 3 for diverter valve or check valve diagnostics.
	No ▶	GO to KC4.
KC3 ATTEMPT TO ELIMINATE SERVICE CODE 46 (96) (TAB ONLY)		
• Disconnect vacuum line on bypass valve and cap vacuum line. • Key Off, wait 10 seconds. • Repeat Engine Running Self-Test and record codes. • Is code 46 (96) present?	Yes ▶	EEC-IV system OK. REFER to Section 3 for bypass valve diagnostics.
	No ▶	GO to KC4.

Air Management System — Pinpoint Test — KC

TEST STEP	RESULT ▶	ACTION TO TAKE
KC4 ENTER OUTPUT STATE CHECK (REFER TO APPENDIX)		
NOTE: Do not use STAR tester for this Step, use a VOM/DVOM. • Key Off, wait 10 seconds. • DVOM on 20V scale. • Connect DVOM negative test lead to STO circuit at the Self-Test connector and positive test lead to battery positive. • Jumper STI circuit to signal return at the Self-Test connector. • Perform Key On, Engine Off Self-Test until the completion of the Continuous Test Codes. • DVOM will indicate zero volts when Test is complete. • Depress and release the throttle. • Did DVOM reading change to a high voltage reading?	Yes ▶	REMAIN in Output State Check. GO to KC5.
	No ▶	DEPRESS throttle to WOT and RELEASE. If STO voltage does not go high, GO to Pinpoint Test Step Q40. Leave equipment hooked up.
KC5 CHECK TAB/TAD SOLENOID ELECTRICAL OPERATION		
• DVOM on 20V scale. • Reconnect DVOM positive test lead to VPWR circuit on TAB solenoid and negative test lead to TAB circuit on TAB solenoid. • While observing DVOM depress and release the throttle several times (to cycle output On and Off). • Repeat for TAD solenoid. Connect positive test lead to VPWR circuit on TAD solenoid and negative test lead to TAD circuit on TAD solenoid. • Solenoids cycle On and Off.	Both outputs cycle On and Off ▶	GO to KC6.
	Either output does not cycle On and Off ▶	REMOVE jumper. GO to KC8.
KC6 CHECK TAB/TAD SOLENOID FOR VACUUM CYCLING		
• Install vacuum pump to the TAB solenoid vacuum supply port and install a vacuum gauge to the output port. • While cycling outputs On and Off (by depressing and releasing throttle), observe the vacuum gauge at the output. **NOTE: Maintain vacuum at source.** • Repeat for TAD solenoid. Connect vacuum pump to the TAD solenoid vacuum supply port and connect a vacuum gauge to the output port. • Cycle output On and Off.	Both vacuum outputs cycle On and Off ▶	GO to KC7.
	Either vacuum output does not cycle On and Off ▶	REPLACE solenoid assembly. RERUN Quick Test.

Air Management System — Pinpoint Test — KC

TEST STEP	RESULT ▶	ACTION TO TAKE
KC7 CHECK MANIFOLD VACUUM LINES FOR BLOCKAGE OR LEAKS		
• Vacuum lines disconnected at TAD/TAB solenoids. • Start engine. • Check for vacuum. • Is vacuum present at the solenoid?	Yes ▶	EEC-IV system OK. REFER to Section 3 for Thermactor valve and air pump diagnostics.
	No ▶	SERVICE vacuum source blockage or leak. RERUN Quick Test.
KC8 MEASURE TAB/TAD SOLENOID RESISTANCE		
• Key Off, wait 10 seconds. • DVOM on 200 ohm scale. • Disconnect TAB solenoid connector and measure solenoid resistance. • Disconnect TAD solenoid connector and measure solenoid resistance. • Are both solenoid resistances between 50 and 100 ohms?	Yes ▶	CONNECT TAB/TAD solenoids. GO to KC9.
	No ▶	REPLACE TAB/TAD solenoid assembly. RERUN Quick Test.
KC9 CHECK VOLTAGE OF VPWR CIRCUIT		
• Key On, engine Off. • DVOM on 20V scale. • Measure voltage between TAB solenoid VPWR circuit and battery ground. • Repeat for TAD solenoid. • Are both voltage readings 10.5 volts or greater?	Yes ▶	GO to KC10.
	No ▶	SERVICE harness circuit open. RERUN Quick Test.

Air Management System — Pinpoint Test — KC

TEST STEP	RESULT ▶	ACTION TO TAKE
KC10 CHECK CONTINUITY OF TAB AND TAD CIRCUITS		
• Key Off, wait 10 seconds. • Disconnect processor 60 Pin connector and inspect for damaged pins, corrosion, loose wires. Service as necessary. • Connect Breakout box to harness. Leave processor disconnected. • DVOM on 200 ohm scale. • Measure resistance between test Pin 51 at Breakout box and TAB circuit at vehicle harness connector. • Measure resistance between test Pin 11 at the Breakout box and TAD circuit at vehicle harness connector. • Are both TAD and TAB circuits resistance 5 ohms or less?	Yes ▶	GO to KC11.
	No ▶	SERVICE harness open circuit. RERUN Quick Test.
KC11 CHECK FOR SHORT TO GROUND		
• Key Off, wait 10 seconds. • DVOM on 200,000 ohm. • Leave Breakout box installed and processor disconnected. • Disconnect TAB/TAD solenoids. • Measure resistance between test Pin 51 and test Pins 40, 46 and 60 and between test Pin 11 and test Pins 40, 46 and 60 at the Breakout box. • Are all resistance readings 10,000 ohms or greater?	Yes ▶	GO to KC12.
	No ▶	SERVICE short to ground. REMOVE Breakout box. RECONNECT harness to processor. RERUN Quick Test.
KC12 CHECK FOR SHORT TO POWER		
• Key Off, wait 10 seconds. • DVOM on 200,000 ohm scale. • Leave Breakout box installed and processor disconnected. • TAB/TAD solenoids disconnected. • Measure resistance between test Pin 51 and test Pins 37 and 57, and between test Pin 11 and test Pins 37 and 57 at the Breakout box. • Are all resistance readings 10,000 ohms or greater?	Yes ▶	REPLACE processor. RERUN Quick Test.
	No ▶	SERVICE short to power. RERUN Quick Test. If code is present, REPLACE processor.

Canister Purge (CANP)	Pinpoint Test	KD

KD

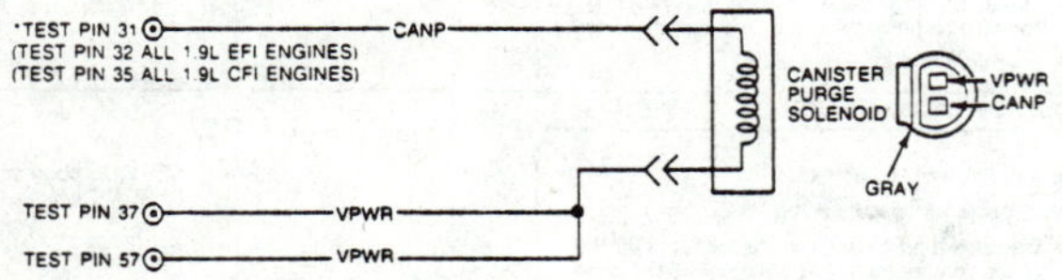

*TEST PINS LOCATED ON BREAKOUT BOX.
ALL HARNESS CONNECTORS VIEWED INTO MATING SURFACE.

STOP-WARNING

You should enter this Pinpoint Test only when a service code 85 is received in Quick Test Step 3.0 or when you are directed here from a fuel control Pinpoint Test and Diagnostics by Symptom in the Engine Supplement Section.

This Pinpoint Test is intended to diagnose only the following:

- Harness Circuits: CANP and VPWR
- Processor Assembly

Canister Purge (CANP)	Pinpoint Test	KD

TEST STEP	RESULT ▶	ACTION TO TAKE
KD1 ENTER OUTPUT STATE CHECK (REFER TO APPENDIX)		
NOTE: Do not use STAR tester for this step, use VOM/DVOM. • Key Off, wait 10 seconds. • DVOM on 20V scale. • Connect DVOM negative test lead to STO circuit at Self-Test connector and positive test lead to battery positive. • Jumper STI circuit to signal return at the Self-Test connector. • Perform Key On, Engine Off Self-Test until the completion of the Continuous Test Codes. • DVOM will indicate zero volts when test is completed. • Depress and release the throttle. • Did DVOM reading change to a high voltage reading?	Yes ▶ No ▶	REMAIN in Output State Check. GO to KD2. DEPRESS throttle to WOT and release. If STO voltage does not go high, GO to Pinpoint Test Step Q40. Leave equipment hooked up.
KD2 CHECK CANISTER PURGE (CANP) SOLENOID ELECTRICAL OPERATION		
• DVOM on 20V scale. • Disconnect CANP solenoid. • Connect DVOM positive test lead to VPWR circuit and negative test lead to CANP output circuit on the vehicle harness connector. • While observing DVOM depress and release the throttle several times (to cycle output On and Off). • CANP circuit cycles On and Off?	Yes ▶ No ▶	GO to KD3. REMOVE jumper. GO to KD6.
KD3 CHECK CANISTER PURGE SOLENOID FOR HOLDING VACUUM		
• Key On. • CANP solenoid disconnected. • Disconnect vacuum hose at canister purge solenoid on PCV side. • Apply 16 in. Hg to CANP solenoid. • Does CANP solenoid hold vacuum for 20 seconds?	Yes ▶ No ▶	Leave vacuum pump setup in place. GO to KD4. REPLACE CANP solenoid. RERUN Quick Test. If symptom is still present, GO to Section 3, Carbon Canister.

Canister Purge (CANP)	Pinpoint Test	KD

TEST STEP	RESULT ▶	ACTION TO TAKE
KD4 CHECK CANISTER PURGE SOLENOID FOR MECHANICAL OPERATION		
• Still in Output State Check. • CANP circuit Off (no voltage). • Reconnect CANP solenoid connector. • Apply 16 in.-Hg. to CANP solenoid as in KD3. • Depress and release throttle. • Is vacuum released?	Yes ▶ No ▶	REMOVE Jumper from STI to signal return. GO to KD5. REPLACE CANP solenoid. REPEAT Quick Test.
KD5 CHECK FOR VACUUM TO CANISTER PURGE SOLENOID		
• Disconnect vacuum hose at canister purge solenoid at PCV side. • Start engine. • Is vacuum present at engine vacuum hose?	Yes ▶ No ▶	EEC-IV system OK. REFER to Shop Manual, Group 24. CHECK vacuum line for proper routing, kinks or blockage. If OK, REFER to Shop Manual, Group 21 for probable subjects affecting engine vacuum.
KD6 MEASURE CANP SOLENOID RESISTANCE		
• Key Off, wait 10 seconds. • DVOM on 200 ohm scale. • Disconnect CANP solenoid connector and measure solenoid resistance. • Is resistance between 40 and 90 ohms?	Yes ▶ No ▶	GO to KD7. REPLACE CANP solenoid. RERUN Quick Test.
KD7 CHECK VOLTAGE OF VPWR CIRCUIT		
• Key On, engine Off. • DVOM on 20V scale. • Measure voltage between VPWR at the CANP solenoid vehicle harness connector and battery ground. • Is voltage reading 10.5 volts or greater?	Yes ▶ No ▶	GO to KD8. SERVICE harness open circuit. RERUN Quick Test.

Canister Purge (CANP)	Pinpoint Test	KD

TEST STEP	RESULT ▶	ACTION TO TAKE
KD8 CHECK CONTINUITY OF CANP CIRCUIT		
• Key Off, wait 10 seconds. • Disconnect processor and inspect both 60 Pin connectors and inspect for damaged pins, corrosion, loose wires. Service as necessary. • Connect Breakout box to harness. Leave processor disconnected. • DVOM on 200 ohm scale. • Measure resistance between test Pin 31 (test Pin 35 for 1.9L CFI engines, test Pin 32 for 1.9L EFI engines) at the Breakout box and CANP on the vehicle harness connector. • Is resistance reading less than 5 ohms?	Yes ▶ No ▶	GO to KD9. SERVICE open circuit. RERUN Quick Test.
KD9 CHECK FOR SHORT TO GROUND		
• Key Off, wait 10 seconds. • Leave Breakout box to harness. Leave processor disconnected. • Disconnect CANP solenoid. • DVOM on 200,000 ohm scale. • Measure resistance between test Pin 31 (test Pin 35 for 1.9L CFI engines, test Pin 32 for 1.9L EFI engines) and test Pins 40, 46 and 60 at the Breakout box. • Are all resistance readings 10,000 ohms or greater?	Yes ▶ No ▶	GO to KD10. SERVICE short to ground. RERUN Quick Test.
KD10 CHECK FOR SHORT TO POWER		
• Key Off, wait 10 seconds. • DVOM on 200,000 ohm scale. • Leave Breakout box installed and processor disconnected. • CANP solenoid disconnected. • Measure resistance between test Pin 31 (test Pin 35 for 1.9L CFI engines, test Pin 32 for 1.9L EFI engines) and test Pins 37 and 57 at the Breakout box. • Are both resistance readings 10,000 ohms or greater?	Yes ▶ No ▶	REMOVE Breakout box. REPLACE processor. RERUN Quick Test. SERVICE short to power. REPEAT Quick Test. If code is repeated, REPLACE processor. RERUN Quick Test.

Idle Speed Control (Bypass Air) — Pinpoint Test — KE

KE

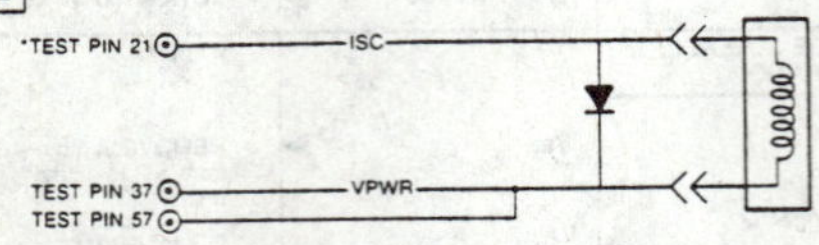

*TEST PINS LOCATED ON BREAKOUT BOX.
ALL HARNESS CONNECTORS VIEWED INTO MATING SURFACE.

STOP-WARNING

You should enter this Pinpoint Test only when a service code 12, 13, 16 and 17 is received in Quick Test Step 5.0 or when directed here from Diagnostic by Symptom in the Engine Supplement Section.

CAUTION: If the engine exhibits rough running and/or idle, correct these conditions before diagnosing Idle Speed Control (ISC). Rough running/misses may be caused by:

- **Ignition System**
  Refer to Section 15
- **Fuel System**
  Refer to Pinpoint Test Steps HA1 through HA7.
  Refer to Pinpoint Test Steps HE1 through HE7.
  Refer to Pinpoint Test Steps HG1 through HG6.
  Refer to Pinpoint Test Steps HC2 through HC9.
- **EGR System, Section 6**

To prevent the replacement of good components, be aware that the following non-EEC areas may be at fault:

- Engine not up to operating temperature
- Engine over operating temperature
- Improper Idle Speed Throttle Stop Adjustment
- A/C input (electrical problem)
- Throttle Speed Control Linkage
- Throttle Sticking or Linkage Binding.

This Pinpoint Test is intended to diagnose only the following:

- RPM in Self-Test only
- ISC Actuator
- Harness Circuits ISC and VPWR
- Processor Assembly

Idle Speed Control (Bypass Air) — Pinpoint Test — KE

TEST STEP	RESULT ▶	ACTION TO TAKE
KE1 RPM DROP		
• Key Off. • Connect engine tachometer. • Start engine. • Disconnect ISC harness connector. • Does rpm drop or stall?	Yes ▶ No ▶	GO to KE2. GO to KE3.
KE2 CHECK FOR EGR CODES		
• Are service codes 31, 32, 33 or 34 present?	Yes ▶	GO to Quick Test Step 5.0 for appropriate Pinpoint Test.
	No ▶	GO to KE3.
KE3 CHECK FOR OTHER EEC CODES		
• Are service codes 22, 41, 42, 91 or 92 present?	Yes ▶	GO to Quick Test Step 5.0 for appropriate Pinpoint Test.
	No ▶	GO to KE4.
KE4 MEASURE ISC SOLENOID RESISTANCE		
• Key Off. • DVOM on 200 ohm scale. • Disconnect ISC solenoid connector and measure solenoid resistance. • Is resistance between 7 and 13 ohms?	Yes ▶ No ▶	GO to KE5. REPLACE ISC solenoid. RERUN Quick Test.
KE5 ISC SHORT TO CASE (GROUND) CHECK		
• Key Off. • DVOM on 200,00 ohm scale. • ISC solenoid disconnected. • Measure resistance from either ISC pin to ISC housing. • Is resistance greater than 10,000 ohms?	Yes ▶ No ▶	GO to KE6. REPLACE ISC solenoid. RERUN Quick Test.

Idle Speed Control (Bypass Air) — Pinpoint Test — KE

TEST STEP	RESULT ▶	ACTION TO TAKE
KE6 CHECK VOLTAGE OF VPWR CIRCUIT		
• Key On, Engine Off. • DVOM on 20V scale. • ISC solenoid disconnected. • Measure voltage between VPWR at the ISC solenoid harness connector and battery ground. • Is voltage less than 10.5 volts?	Yes ▶ No ▶	SERVICE harness open circuit. RERUN Quick Test. GO to KE7.
KE7 CHECK CONTINUITY OF ISC CIRCUIT		
• Key Off, wait 10 seconds. • ISC solenoid disconnected. • Disconnect processor and inspect both 60 Pin connectors for damaged pins, corrosion, loose wires. Service as necessary. • Connect Breakout box to harness. Leave processor disconnected. • DVOM on 200 ohm scale. • Measure resistance between test Pin 21 at the Breakout box and ISC circuit at vehicle harness connector. • Is resistance greater than 5 ohms?	Yes ▶ No ▶	SERVICE open circuit. RERUN Quick Test. GO to KE8.
KE8 CHECK FOR SHORT TO GROUND		
• Key Off, wait 10 seconds. • Leave Breakout box installed and processor disconnected. • ISC solenoid disconnected. • DVOM on 200,000 ohm scale. • Measure resistance between test Pin 21 and test Pins 40, 46 and 60 at the Breakout box. • Is any resistance less than 10,000 ohms?	Yes ▶ No ▶	SERVICE short to ground. RERUN Quick Test. GO to KE9.

Idle Speed Control (Bypass Air) — Pinpoint Test — KE

TEST STEP	RESULT ▶	ACTION TO TAKE
KE9 CHECK FOR SHORT TO POWER		
• Key Off, wait 10 seconds. • DVOM on 200,000 ohm scale. • ISC solenoid disconnected. • Leave Breakout box connected to harness. Leave processor disconnected. • Place positive test lead to test Pin 37 and negative test lead to test Pin 21 to measure resistance. • Is resistance greater than 10,000 ohms?	Yes ▶ No ▶	GO to KE10. SERVICE short to power. RERUN Quick Test. If code or symptom is present, REPLACE processor.
KE10 CHECK FOR ISC SIGNAL FROM THE PROCESSOR		
• Key Off. • Connect processor and ISC actuator. • DVOM on a 20V scale. • Breakout box installed. • Connect DVOM between test Pin 21 and test Pin 40. • Start engine. • Slowly increase and decrease rpm. Does DVOM voltage vary?	Yes ▶ No ▶	GO to KE11. REPLACE processor. RERUN Quick Test.
KE11 VERIFY CURB IDLE		
• Go to Section 4 to verify curb idle. • Is curb idle within specification?	Yes ▶	REPLACE ISC actuator. RERUN Quick Test.
	No ▶	CHECK engine vacuum hoses. REFER to VECI decal. CHECK that throttle plates are fully closed. CHECK throttle linkage and/or speed control linkage for binding. If OK, REPLACE ISC actuator. RERUN Quick Test.

Idle Speed Control (Bypass Air) — Pinpoint Test KE

TEST STEP	RESULT ▶	ACTION TO TAKE
KE12 CURB IDLE CHECK		
• Key Off, wait 10 seconds. • Deactivate Self-Test. • Run engine at 2000 rpm for 2 minutes or until inlet radiator hose is hot and pressurized. • Key Off, wait 10 seconds. • Run Engine Running Self-Test. • Is code 17 still present?	Yes ▶	INSPECT throttle body and air inlet for contamination. SERVICE as necessary. If OK, ADJUST curb idle (REFER to Section 4 for procedure). RERUN Quick Test.
	No ▶	SERVICE other codes as necessary.
KE15 CODE 13 HIGH RPM CHECK		
• Key Off. • Connect tachometer. • Start engine. • Disconnect ISC harness connector. • Does rpm drop or engine stall?	Yes ▶	REPLACE processor. RERUN Quick Test.
	No ▶	CHECK engine vacuum hoses. REFER to VECI decal. VERIFY curb idle. CHECK that throttle plates are fully closed. CHECK throttle linkage and/or speed control linkage for binding. If OK, REPLACE ISC actuator. RERUN Quick Test.
KE20 SERVICE CODE: 48		
• SERVICE holes, cracks, and/or disconnections to air cleaner outlet tube (between vane airflow meter and fuel charging assembly). • Perform Engine Running Self-Test. • Is code 48 present?	Yes ▶	CHECK for other source of excessive unmetered airflow.
	No ▶	EEC system OK for metered air. GO to Quick Test Step 5.0B to service other code if necessary.

Idle Speed Control (DC Motor/Idle Tracking Switch Assembly) — 2.5L CFI, 3.8L CFI — Pinpoint Test KF

KF

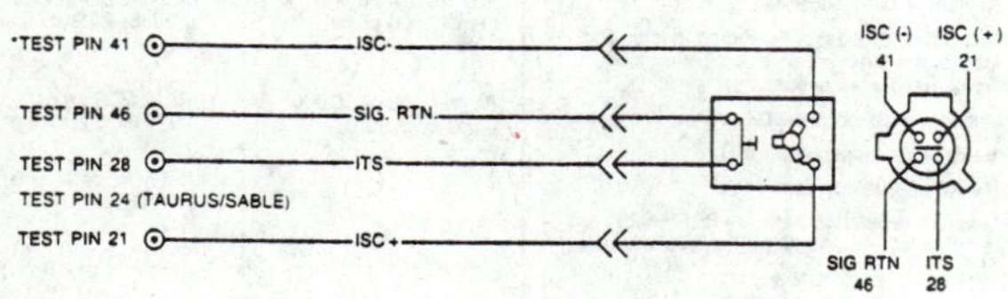

*TEST PINS LOCATED ON BREAKOUT BOX.
ALL HARNESS CONNECTORS VIEWED INTO MATING SURFACE.

STOP-WARNING

You should enter this Pinpoint Test only when a Service Code 12, 13, 58 or 68 is received in Quick Test Step 3.0, 5.0 or 6.0 or when directed here from Diagnostics by Symptom in the Engine Supplement Section.

To prevent the replacement of good components, be aware that the following non-EEC areas may be at fault:

- Anti-Diesel Speed too High
- Basic Engine
- Vacuum Leaks
- Throttle Sticking or on High Cam

This Pinpoint Test is intended to diagnose only the following:

- DC Motor
- Idle Tracking Switch
- Harness circuits: ISC+, ISC−, Signal Return and ITS
- Processor Assembly

Idle Speed Control (DC Motor/Idle Tracking Switch Assembly) — 2.5L CFI, 3.8L CFI — Pinpoint Test KF

TEST STEP	RESULT ▶	ACTION TO TAKE
KF1 SERVICE CODE 68: SIMULATE NO THROTTLE CONTACT TO FORCE A CODE 11 IN KEY ON — ENGINE OFF TEST		
• Key Off, wait 10 seconds. • Move throttle away from ISC DC motor shaft. • Perform Key On Engine Off Self-Test. • Is code 68 present? **NOTE: Ignore all other codes at this point.** 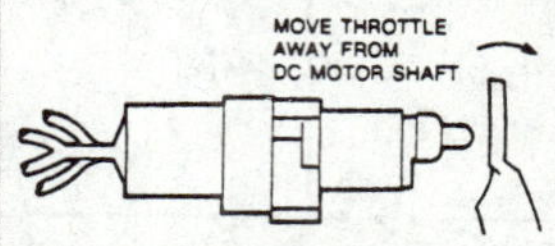	Yes ▶	RELEASE throttle and GO to KF3.
	No ▶	RELEASE throttle and GO to Pinpoint Test Step KF2.
KF2 RETRACT ISC DC MOTOR		
• Key Off, wait 10 seconds. • Connect Breakout box to harness. Leave processor disconnected. • Jumper test Pin 41 to test Pin 1 at the Breakout box and test Pin 21 at the Breakout box to battery ground (to retract DC motor). • Observe DC motor. **NOTE: Do not leave jumper wire in test pins if the motor will not move; it will damage Breakout box wiring.**	Motor does not retract ▶	REMOVE jumper and GO to KF5.
	Motor retracts and still contacts throttle lever ▶	GO to ISC Adjustments, Section 4.
	Motor retracts and does not contact throttle lever ▶	REMOVE jumper and GO to KF25.

Idle Speed Control (DC Motor/Idle Tracking Switch Assembly) — 2.5L CFI, 3.8L CFI — Pinpoint Test KF

TEST STEP	RESULT ▶	ACTION TO TAKE
KF3 SIMULATE A CLOSED CONTACT CONDITION IN KEY ON — ENGINE OFF TEST		
• Key Off, wait 10 seconds. • Disconnect harness from ISC. • Connect jumper between ISC vehicle harness connector Pins 3 and 4. • Install jumper as shown. • Perform Key On Engine Off Self-Test. • Is code 68 present? 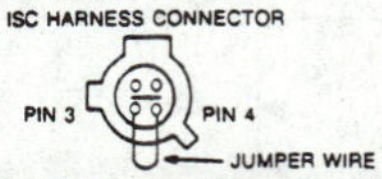 **CAUTION: Do not short any other pins.**	Yes ▶	GO to KF4.
	No ▶	REPLACE ISC DC motor. RERUN Quick Test.
KF4 CHECK ITS AND SIGNAL RETURN CIRCUITS FOR CONTINUITY		
• Key Off, wait 10 seconds. • Disconnect processor 60 Pin connector and inspect for damaged pins, corrosion, loose wires. Service as necessary. • Leave Breakout box installed. Leave processor disconnected. • DVOM on 200 ohm scale. • Measure resistance between test Pin 46 at the Breakout box and Signal Return circuit at ISC vehicle harness connector and between test Pin 28* at the Breakout box and ITS circuit at ISC vehicle harness connector. *Test Pin 24 on Taurus/Sable vehicles.	Either resistance 5 ohms or greater ▶	SERVICE faulty circuit. RERUN Quick Test.
	Both resistances less than 5 ohms ▶	REPLACE processor. RERUN Quick Test.

Idle Speed Control (DC Motor/Idle Tracking Switch Assembly)—2.5L CFI, 3.8L CFI | **Pinpoint Test** | **KF**

TEST STEP	RESULT ▶	ACTION TO TAKE
KF5 CHECK ISC CIRCUITS FOR CONTINUITY		
• Key Off, wait 10 seconds. • Disconnect processor 60 Pin connector and inspect for damaged pins, corrosion, loose wires. Service as necessary. • Connect Breakout box. Leave processor disconnected. • Disconnect harness connector for ISC motor. • DVOM on 200 ohm scale. • Measure resistance between test Pin 41 at the Breakout box and ISC – circuit at vehicle harness connector and between test Pin 21 at the Breakout box and ISC – circuit at vehicle harness connector.	Both resistance readings are less than 5 ohms ▶ Either resistance reading is 5 ohms or greater ▶	GO to KF6. SERVICE faulty circuits. RERUN Quick Test.
KF6 CHECK ISC + AND ISC – CIRCUITS FOR SHORTS TO POWER		
• Key On Engine Off. • Leave Breakout box installed. Leave processor disconnected. • Harness disconnected from ISC motor. • DVOM on 20V scale. • Measure voltage between test Pin 41 and test Pins 40 and 60 at the Breakout box. Measure voltage between test Pin 21 and test Pins 40 and 60 at the Breakout box.	All voltage readings are less than 1V ▶ Any voltage reading is 1V or more ▶	GO to KF7. SERVICE circuit short to power. RERUN Quick Test. If code 12 is still present, REPLACE processor.
KF7 CHECK ISC + AND ISC – CIRCUITS FOR SHORTS TO GROUND		
• Key Off, wait 10 seconds. • Leave Breakout box installed. Leave processor disconnected. • Harness disconnected from ISC motor. • DVOM on 200,000 ohm scale. • Measure resistance between test Pin 41 and test Pins 40, 46 and 60 at the Breakout box. Measure resistance between test Pin 21 and test Pins 40, 46 and 60 at the Breakout box.	All resistance readings greater than 10,000 ohms ▶ Any resistance reading is 10,000 ohms or less ▶	GO to KF8. SERVICE faulty circuit. RERUN Quick Test. If code 12 is still present, REPLACE processor.

Idle Speed Control (DC Motor/Idle Tracking Switch Assembly)—2.5L CFI, 3.8L CFI | **Pinpoint Test** | **KF**

TEST STEP	RESULT ▶	ACTION TO TAKE
KF8 PREPARE IDLE SPEED CONTROL DC MOTOR FOR OPERATION		
• Key Off, wait 10 seconds. • Leave Breakout box installed. Leave processor disconnected. • Harness disconnected from ISC motor. • DVOM on 20V scale. • Connect DVOM positive test lead to test Pin 1 at the Breakout box and negative test lead to ground.	Voltage reading is 10.5V or greater ▶ Voltage reading is less than 10.5V ▶	GO to KF9. SERVICE open in KAPWR circuit. RERUN Quick Test. If code 12 is still present, GO to KF9.
KF9 ISC DC MOTOR OPERATION CHECK		
• Key Off, wait 10 seconds. • Leave Breakout box installed. Leave processor disconnected. • Connect ISC DC motor to harness connector. • Jumper test Pin 21 to test Pin 1 at the Breakout box and test Pin 41 at the Breakout box to ground (to extend DC motor). • Does DC motor shaft extend 5 cm (2 inches) or more? 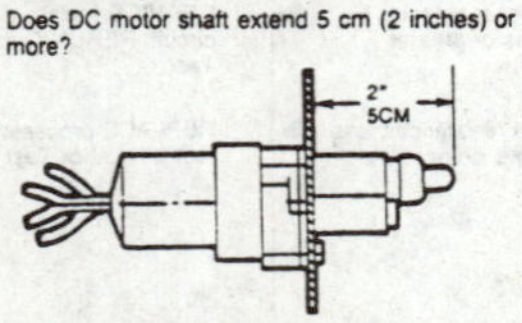	Yes ▶ No ▶	GO to KF10. REPLACE DC motor. RERUN Quick Test.
KF10 RETRACT ISC DC MOTOR		
• Key Off, wait 10 seconds. • Leave Breakout box installed. Leave processor disconnected. • Jumper test Pin 41 to test Pin 1 at the Breakout box and test Pin 21 at the Breakout box to ground (to retract DC motor). • Does ISC DC motor retract?	Yes ▶ No ▶	GO to KF11. REPLACE DC motor. RERUN Quick Test.

Idle Speed Control (DC Motor/Idle Tracking Switch Assembly)—2.5L CFI, 3.8L CFI | **Pinpoint Test** | **KF**

TEST STEP	RESULT ▶	ACTION TO TAKE
KF11 EXTEND ISC DC MOTOR		
• Key Off, wait 10 seconds. • Leave Breakout box installed. Leave processor disconnected. • Jumper test Pin 21 to test Pin 1 at the Breakout box and test Pin 41 at the Breakout box to ground (to extend DC motor). • Does ISC DC motor extend?	No ▶ Yes ▶	REPLACE DC motor. RERUN Quick Test. REPLACE processor. RERUN Quick Test.
KF12 SERVICE CODE 58: SIMULATE THROTTLE CONTACT TO FORCE A SERVICE CODE 68 IN KEY ON — ENGINE OFF TEST		
• Key Off, wait 10 seconds. • Press on ISC DC motor shaft to simulate throttle contact. • With a force pushing on DC motor shaft, perform Key On Engine Off Self-Test. **NOTE: Ignore all other codes at this point.** • Is code 68 present?	Yes ▶ No ▶	GO to Pinpoint Test Step KF5. GO to KF13.
KF13 RETRACT ISC DC MOTOR		
• Key Off, wait 10 seconds. • Leave Breakout box installed. Leave processor disconnected. • Jumper test Pin 41 to test Pin 1 at the Breakout box and test Pin 21 at the Breakout box to ground (to retract DC motor). • Does ISC DC motor retract?	Yes ▶ No ▶	GO to ISC Adjustments, Section 4. GO to KF14.
KF14 IDLE TRACKING SWITCH ALWAYS "NOT TRACKING" (CLOSED)		
• Key Off, wait 10 seconds. • Reconnect processor. • Disconnect vehicle harness from ISC. • Perform Key On Engine Off Self-Test. **NOTE: Ignore all other codes at this point.** • Is code 68 present?	Yes ▶ No ▶	REPLACE DC motor. RERUN Quick Test. GO to KF15.

Idle Speed Control (DC Motor/Idle Tracking Switch Assembly)—2.5L CFI, 3.8L CFI | **Pinpoint Test** | **KF**

TEST STEP	RESULT ▶	ACTION TO TAKE
KF15 CHECK ITS CIRCUIT FOR SHORT TO GROUND		
• Key Off, wait 10 seconds. • Disconnect processor 60 Pin connector and inspect for damaged pins, corrosion, loose wires. Service as necessary. • Connect Breakout box. Disconnect processor. • Harness disconnected at ISC. • DVOM on 200,000 ohm scale. • Measure resistance between test Pin 28* and test Pins 40, 46 and 60 at the Breakout box. *Test Pin 24 Taurus/Sable.	Any resistance less than 10,000 ohms ▶ All resistances 10,000 ohms or more ▶	SERVICE short to ground. RERUN Quick Test. REPLACE processor. RERUN Quick Test.
KF16 DC MOTOR CONTROL FAULT ISOLATION: SERVICE CODE 12		
• Are service codes 58, 68, 31 or 41 present?	Yes ▶ No ▶	GO to Quick Test Step 5.0 B for next appropriate Pinpoint Code. GO to KF17.
KF17 CHECK FOR THROTTLE STICKING		
• Check throttle plates and linkages for binding. • Check speed control for binding (if so equipped).	Yes ▶ No ▶	SERVICE as necessary. RERUN Quick Test. GO to KF18.
KF18 EXTEND ISC DC MOTOR		
• Key Off, wait 10 seconds. • Leave Breakout box installed. Leave processor disconnected. • Jumper test Pin 21 to test Pin 1 at the Breakout box and test Pin 41 at the Breakout box to ground (to extend DC motor). • Does ISC DC motor extend?	No ▶ Yes ▶	REPLACE DC motor. RERUN Quick Test. REPLACE processor. RERUN Quick Test.

Idle Speed Control (DC Motor/Idle Tracking Switch Assembly) — 2.5L CFI, 3.8L CFI | Pinpoint Test | KF

TEST STEP	RESULT ▶	ACTION TO TAKE
KF19 ENGINE RUNNING CODES 13 AND/OR 16		
• Prepare system for normal engine operation. • Deactivate Self-Test. • Air conditioner Off. • Run engine. Alternate between 30 seconds at idle and 5 seconds at part throttle modes for 3 minutes. • Is idle speed erratic?	No ▶ Yes ▶	GO to KF20. CHECK for: • Vacuum leaks • Code 22 • Code 31 • Code 41 • Code 58 SERVICE above codes as necessary before continuing.
KF20 CHECK FOR THROTTLE STICKING OR CHOKE POSITION		
• Inspect throttle mechanisms (plate linkage) for sticking or binding. • Are above inspections satisfactory?	No ▶ Yes ▶	SERVICE fault. RERUN Quick Test. GO to KF21.
KF21 PERFORM KEY ON — ENGINE OFF TEST. OBSERVE ON DEMAND CODES		
• Perform Key On Engine Off Test. Record On Demand codes. • Key On Engine Off. • Is code 68 present?	Yes ▶ No ▶	GO to Pinpoint Test Step KF1. REFER to Section 4 for anti-diesel speed adjustment.
KF25 CHECK ISC – AND ISC – CIRCUITS FOR SHORTS TO GROUND		
• Key Off, wait 10 seconds. • Leave Breakout box installed. Leave processor disconnected. • Harness disconnected from ISC motor. • DVOM on 200,000 ohm scale. • Measure resistance between test Pin 41 and test Pins 40, 46 and 60 at the Breakout box. Measure resistance between test Pin 21 and test Pins 40, 46 and 60 at the Breakout box.	All resistance readings greater than 10,000 ohms ▶ Any resistance reading is 10,000 ohms or less ▶	GO to KF26. SERVICE faulty circuit. RERUN Quick Test. If code 12 is still present, REPLACE processor.

Idle Speed Control (DC Motor/Idle Tracking Switch Assembly) — 2.5L CFI, 3.8L CFI | Pinpoint Test | KF

TEST STEP	RESULT ▶	ACTION TO TAKE
KF26 CHECK ISC – AND ISC – CIRCUITS FOR SHORTS TO POWER		
• Key On Engine Off. • Leave Breakout box installed. Leave processor disconnected. • Harness disconnected from ISC motor. • DVOM on 20V scale. • Measure voltage between test Pin 41 and test Pins 40 and 60 at the Breakout box. Measure voltage between test Pin 21 and test Pins 40 and 60 at the Breakout box.	All voltage readings are less than 1V ▶ Any voltage reading is 1V or more ▶	RECONNECT DC motor. GO to KF27. SERVICE circuit short to power. RERUN Quick Test. If code 12 is still present, REPLACE processor.
KF27 EXTEND ISC DC MOTOR		
• Key Off, wait 10 seconds. • Leave Breakout box installed. Leave processor disconnected. • Jumper test Pin 21 to test Pin 1 at the Breakout box and test Pin 41 at the Breakout box to ground (to extend DC motor). • Does ISC DC motor extend?	No ▶ Yes ▶	REPLACE DC motor. RERUN Quick Test. REPLACE processor. RERUN Quick Test.

Shift Indicator Light (SIL) | Pinpoint Test | KL

KL 1.9L CFI, 2.5 HSC CFI

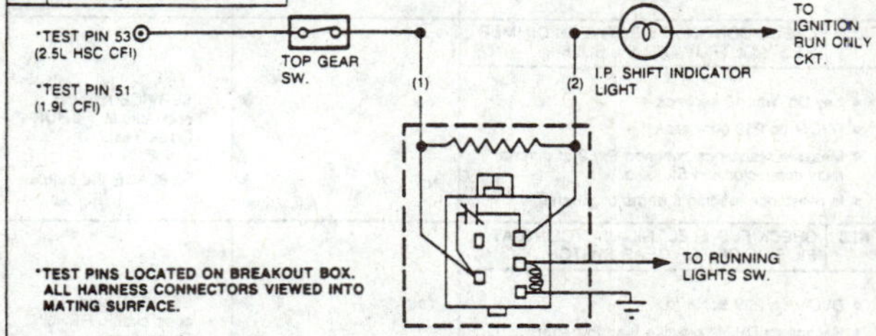

KL ALL OTHERS

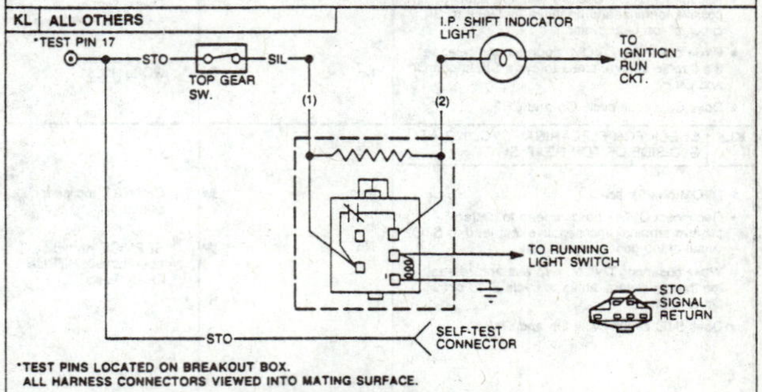

STOP-WARNING

You should enter this Pinpoint Test only when directed here from Diagnostics by Symptom in the Engine Supplement Section or from Pinpoint Test Step Q9A.

This Pinpoint Test is intended to diagnose only the following:

- Top Gear Switch
- Shift Indicator Light Bulb and Fuse
- Harness Circuits: SIL and STO

Shift Indicator Light (SIL) | Pinpoint Test | KL

TEST STEP	RESULT ▶	ACTION TO TAKE
KL1 ENTER OUTPUT STATE CHECK IN NEUTRAL (REFER TO APPENDIX)		
NOTE: Do not use STAR tester for this Step, use VOM/DVOM. • Key Off, wait 10 seconds. • Transmission in neutral. • DVOM on 20V scale. • Connect DVOM negative test lead to STO at the Self-Test connector and positive test lead to battery positive. • Jumper STI to signal return at the Self-Test connector. • Perform Key On, Engine Off Self-Test until the completion of the Continuous Test Codes. • DVOM will indicate zero volts when test is complete. • Depress and release the throttle. • Did DVOM reading change to a high voltage reading?	Yes ▶ No ▶	REMAIN in output state check. Go to KL2. DEPRESS throttle to WOT and release. If STO voltage does not go high, GO to Pinpoint Test Step KL7. Leave equipment hooked up.
KL2 CHECK FOR ELECTRICAL CYCLING AT DIMMER RELAY INPUT		
• DVOM on 20V scale. • Reconnect DVOM positive test lead to battery positive terminal and negative test lead to Pin 1 on dimmer relay. • While observing DVOM, depress and release the throttle several times to cycle Pin 1 On and Off. • Does Pin 1 cycle On and Off?	Yes ▶ No ▶	GO to KL3. GO to KL5.
KL3 CHECK FOR ELECTRICAL CYCLING AT DIMMER RELAY OUTPUT		
• DVOM on 20V scale. • Reconnect DVOM positive test lead to battery positive terminal and negative test lead to Pin 2 on dimmer relay. • While observing DVOM, depress and release the throttle several times to cycle Pin 2 On and Off. • Does Pin 2 cycle On and Off?	Yes ▶ No ▶	REMOVE jumper. GO to KL4. REMOVE jumper. REPLACE dimmer relay.

Shift Indicator Light (SIL) — Pinpoint Test — KL

TEST STEP	RESULT	ACTION TO TAKE
KL4 CHECK CONTINUITY BETWEEN DIMMER RELAY OUTPUT AND SIL BULB		
• Key Off, wait 10 seconds. • DVOM on 200 ohm scale. • Measure resistance between Pin 2 of dimmer relay connector and SIL bulb. • Is resistance reading 5 ohms or greater?	Yes ▶	SERVICE harness open circuit. RERUN Quick Test.
	No ▶	REPLACE SIL bulb.
KL5 CHECK FOR ELECTRICAL CYCLING AT SIL SIDE OF TOP GEAR SWITCH		
• DVOM on 20V scale. • Reconnect DVOM positive lead to battery positive terminal and negative test lead to SIL circuit of top gear switch. • While observing DVOM, depress and release the throttle several times to cycle SIL circuit On and Off. • Does SIL circuit cycle On and Off?	Yes ▶	SERVICE harness open circuit. RERUN Quick Test. 4.9L only, change SIL bulb.
	No ▶	GO to KL6.
KL6 CHECK FOR ELECTRICAL CYCLING AT STO SIDE OF TOP GEAR SWITCH		
• DVOM on 20V scale. • Reconnect DVOM positive lead to battery positive terminal and negative test lead to STO circuit of top gear switch. • While observing DVOM, depress and release the throttle several times to cycle STO circuit On and Off. • Does STO circuit cycle On and Off?	Yes ▶	CHANGE top gear switch.
	No ▶	SERVICE harness open circuit. RERUN Quick Test.

Shift Indicator Light (SIL) — Pinpoint Test — KL

TEST STEP	RESULT	ACTION TO TAKE
KL7 ENTER OUTPUT STATE CHECK IN TOP GEAR (REFER TO APPENDIX)		
NOTE: Do not use STAR tester for this Step, use VOM/DVOM. • Key Off, wait 10 seconds. • Transmission in top gear. • DVOM on 20V scale. • Reconnect DVOM negative test lead to STO at the Self-Test connector and positive test lead to battery positive. • Jumper STI to signal return at the Self-Test connector. • Perform Key On, Engine Off Self-Test until the completion of the Continuous Test Codes. • DVOM will indicate zero volts when test is complete. • Depress and release the throttle. • Did DVOM reading change to a high voltage reading?	Yes ▶	CHECK SIL bulb and fuse 15. If OK, SERVICE short to ground in SIL circuit.
	No ▶	GO to Pinpoint Test Step Q40.

WOT A/C Cutoff (WAC) A/C Demand — Pinpoint Test — KM

KM 2.3L EFI TC, 2.3L EFI CAR, 2.3L CFI ENGINES

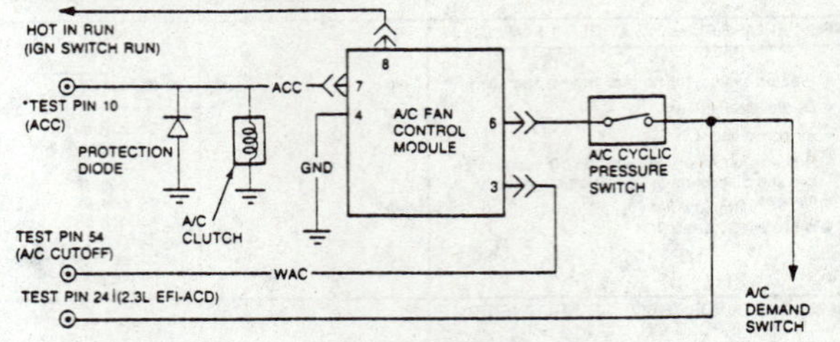

*TEST PINS LOCATED ON BREAKOUT BOX.
ALL HARNESS CONNECTORS VIEWED INTO MATING SURFACE.

KM 1.9L EFI, 1.9L CFI ENGINES

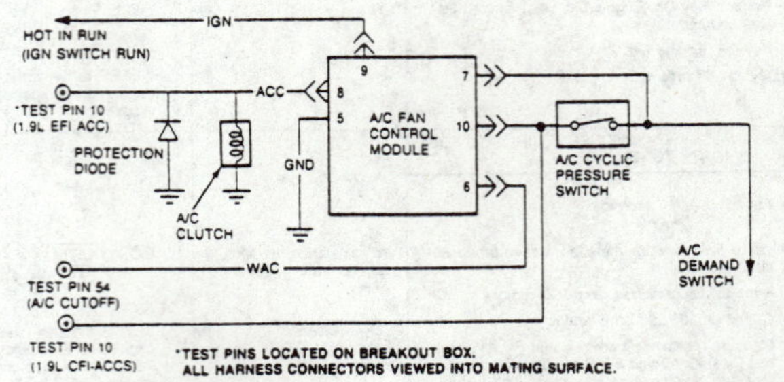

*TEST PINS LOCATED ON BREAKOUT BOX.
ALL HARNESS CONNECTORS VIEWED INTO MATING SURFACE.

WOT A/C Cutoff (WAC) A/C Demand — Pinpoint Test — KM

KM 2.9L EFI, 3.0L EFI TRUCK AND 5.0L SEFI ENGINES

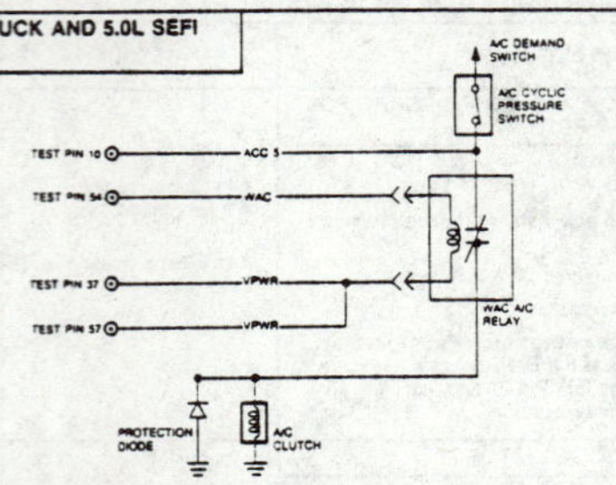

KM 2.3L EFI TRUCK AND 3.8L CFI ENGINES

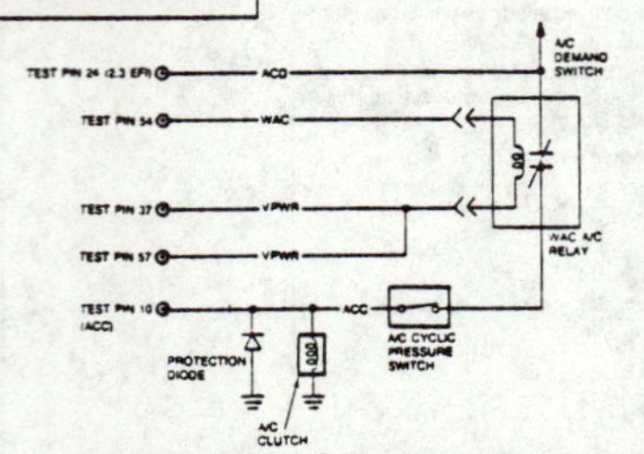

STOP-WARNING

You should enter this Pinpoint Test only when directed here from Diagnostics by Symptom in the Engine Supplement Section.

This Pinpoint Test is intended to diagnose only the following:

- Harness Circuits: WAC and V PWR
- WOT A/C Cut-Off Relay
- Processor Assembly
- A/C Demand Switch

WOT A/C Cutoff (WAC) A/C Demand — Pinpoint Test — KM

TEST STEP	RESULT ▶	ACTION TO TAKE
KM1 ENTER OUTPUT STATE CHECK (REFER TO APPENDIX)		
NOTE: Do not use STAR tester for this Step, use VOM/DVOM. • Key Off, wait 10 seconds. • DVOM on 20V scale. • Connect DVOM negative test lead to STO at the Self-Test connector and positive test lead to battery positive. • Jumper STI to signal return at the Self-Test connector. • Perform Key On, Engine Off Self-Test until the completion of the Continuous Test Codes. • DVOM will indicate zero volts when test complete. • Depress and release the throttle. • Did DVOM reading change to a high voltage reading?	Yes ▶ No ▶	REMAIN in Output State Check. GO to KM2. DEPRESS throttle to WOT and RELEASE. If STO voltage does not go high, GO to Pinpoint Test Step Q40. Leave equipment hooked up.
KM2 CHECK A/C CLUTCH ELECTRICAL OPERATION		
• Key On, engine Off. • Disconnect processor 60 Pin connector and inspect for damaged pins, corrosion, loose wires. Service as necessary. • Connect Breakout box to harness. Reconnect processor. • A/C switch to A/C. • DVOM on 20V scale. • Reconnect DVOM positive test lead to test Pin 37 and negative test lead to test Pin 54 at the Breakout box. • While observing DVOM, depress and release the throttle several times (to cycle output On and Off). • Does A/C clutch output cycle On and Off?	Yes ▶ No ▶	EEC-IV system OK. Refer to Shop Manual, Group 36. REMOVE jumper. GO to KM3.
KM3 FAN CONTROL MODULE OR WAC RELAY		
• A/C fan/control module applications 1.9L EFI, 1.9L CFI, 2.3L EFI, 2.3L CFI engines. • All other systems with WAC relay.	A/C fan control ▶ WAC relay ▶	GO to KM4. GO to KM10.

WOT A/C Cutoff (WAC) A/C Demand — Pinpoint Test — KM

TEST STEP	RESULT ▶	ACTION TO TAKE
KM4 CHECK CONTINUITY OF GROUND CIRCUIT TO FAN CONTROL MODULE		
• Key Off, wait 10 seconds. • DVOM on 200 ohm scale. • Measure resistance of ground circuit between fan control module connector and battery ground. • Is resistance reading 5 ohms or greater?	Yes ▶ No ▶	SERVICE harness circuit. RERUN Quick Test. GO to KM5.
KM5 CHECK FOR IGN. CIRCUIT VOLTAGE		
• Key On, engine Off. • DVOM on 20V scale. • Measure voltage between ignition circuit and ground circuit on the fan control module connector. • Is voltage reading 10.5 volts or greater?	Yes ▶ No ▶	GO to KM6. SERVICE open in ign. circuit. RERUN Quick Test.
KM6 CHECK CONTINUITY OF WAC CIRCUIT		
• Key Off, wait 10 seconds. • Disconnect processor 60 Pin connector and inspect for damaged pins, corrosion, loose wires. Service as necessary. • Connect Breakout box to harness. Leave processor disconnected. • DVOM on 200 ohm scale. • Measure resistance between test Pin 54 at the Breakout box and WAC circuit at fan control module connector. • Is resistance reading 5 ohms or greater?	Yes ▶ No ▶	SERVICE harness WAC circuit. RERUN Quick Test. GO to KM7.
KM7 CHECK FOR SHORT TO GROUND ON WAC CIRCUIT		
• Key Off, wait 10 seconds. • Leave Breakout box installed and processor disconnected. • DVOM on 200,000 ohm scale. • Measure resistance between test Pin 54 and test Pins 40, 46 and 60 at the Breakout box. • Are resistance readings 10,000 ohms or greater?	Yes ▶ No ▶	REPLACE processor. RERUN Quick Test. GO to KM8.

WOT A/C Cutoff (WAC) A/C Demand — Pinpoint Test — KM

TEST STEP	RESULT ▶	ACTION TO TAKE
KM8 CHECK FOR SHORT TO GROUND WITH FAN CONTROL MODULE DISCONNECTED		
• Key Off, wait 10 seconds. • Leave Breakout box installed and processor disconnected. • DVOM on 200,000 ohm scale. • Disconnect fan control module. • Measure resistance between test Pin 54 and test Pins 40, 46 and 60 at the Breakout box. • Are resistance readings 10,000 ohms or greater?	Yes ▶ No ▶	REPLACE fan control module. RERUN Quick Test. SERVICE harness short. RERUN Quick Test.
KM10 MEASURE WAC RELAY RESISTANCE		
• Key Off, wait 10 seconds. • DVOM on 200 ohm scale. • Disconnect WAC relay connector and measure relay resistance. • Is resistance reading less than 50 ohms or greater than 70 ohms?	Yes ▶ No ▶	REPLACE WAC relay. RERUN Quick Test. CONNECT WAC solenoid. GO to KM11.
KM11 CHECK VOLTAGE OF V PWR CIRCUIT		
• Key On, Engine Off. • DVOM on 20V scale. • Connect DVOM positive test lead to V PWR circuit and negative test lead to ground. • Measure voltage on WAC relay V PWR circuit. • Is voltage reading 10.5 volts or greater?	Yes ▶ No ▶	GO to KM12. SERVICE harness open circuit. RERUN Quick Test.
KM12 CHECK CONTINUITY OF WAC CIRCUIT		
• Key Off, wait 10 seconds. • Disconnect processor 60 Pin connector and inspect for damaged pins, corrosion, loose wires. Service as necessary. • Connect Breakout box to harness. Leave processor disconnected. • DVOM on 200 ohm scale. • Measure resistance between test Pin 54 at the Breakout box and WAC circuit at harness connector. • Is resistance reading 5 ohms or greater?	Yes ▶ No ▶	SERVICE harness open circuit. RERUN Quick Test. GO to KM13.

WOT A/C Cutoff (WAC) A/C Demand — Pinpoint Test — KM

TEST STEP	RESULT ▶	ACTION TO TAKE
KM13 CHECK FOR SHORT TO GROUND		
• Key Off, wait 10 seconds. • Leave Breakout box installed and processor disconnected. • DVOM on 200,000 ohm scale. • Measure resistance between test Pin 54 and test Pins 40, 46 and 60 at Breakout box. • Are all resistance readings greater than 10,000 ohms?	Yes ▶ No ▶	GO to KM14. SERVICE short to ground. RERUN Quick Test.
KM14 CHECK FOR SHORT TO POWER		
• Key Off, wait 10 seconds. • DVOM on 200,000 ohm scale. • Leave Breakout box installed and processor disconnected. • WAC solenoid disconnected. • Measure resistance between test Pin 54 and test Pins 37 and 57 at Breakout box. • Are all resistance readings greater than 10,000 ohms?	Yes ▶ No ▶	GO to KM15. SERVICE short to power. RERUN Quick Test. If symptom is still present REPLACE processor.
KM15 CYCLE A/C DEMAND SWITCH		
• Key Off, wait 10 seconds. • Processor disconnected. • DVOM on 20V scale. • Measure voltage between test Pin 10 and test Pin 40 at the Breakout box. • Does output cycle low to high (high being 10.5 volts or greater) when A/C switch is cycled?	Yes ▶ No ▶	REPLACE processor. RERUN Quick Test. REPAIR open in A/C input to test Pin 10. RERUN Quick Test.

WOT A/C Cutoff (WAC) A/C Demand — Pinpoint Test KM

TEST STEP	RESULT ▶	ACTION TO TAKE
KM20 CYCLE A/C DEMAND SWITCH		
• Key Off, wait 10 seconds. • Disconnect processor 60 Pin connector and inspect for damaged pins, corrosion, loose wires. Service as necessary. • Connect Breakout box to harness. Leave processor disconnected. • DVOM on 20V scale. • Measure voltage between test Pin 10 (Pin 24 for 2.3L EFI) and test Pin 40 at the Breakout box. • Does output cycle 4.0-10.5V when A/C switch is cycled?	Yes ▶ No ▶	REPLACE processor. RERUN Quick Test. GO to KM21.
KM21 CHECK CONTINUITY OF ACC/ACD CIRCUIT		
• Key Off, wait 10 seconds. • DVOM on 200 ohm scale. • Measure resistance between ACC test Pin 10 at the Breakout box and A/C clutch. For 2.3L EFI, also measure resistance between ACD test Pin 24 at the Breakout box and A/C demand switch. • Is resistance reading 5 ohms or greater?	Yes ▶ No ▶	SERVICE harness circuit. RERUN Quick Test. EEC-IV system OK. REFER to Shop Manual, Group 36.

Turbo Boost — Pinpoint Test KN

TEST STEP	RESULT ▶	ACTION TO TAKE
KN1 ENTER OUPUT STATE CHECK (REFER TO APPENDIX)		
NOTE: Do not use STAR tester for this Step, use VOM/DVOM. • Key Off, wait 10 seconds. • DVOM on 20V scale. • Connect DVOM negative test lead to STO at the Self-Test connector and positive test lead to battery positive. • Jumper STI to signal return at the Self-Test connector. • Perform Key On Engine Off Self-Test through the completion of the Continuous Test Codes. • DVOM will indicate zero volts when test is complete. • Depress and release the throttle. • Did DVOM reading change to a high voltage reading?	Yes ▶ No ▶	REMAIN in Output State Check. GO to KN2. DEPRESS throttle to WOT and release. If STO voltage does not go high, GO to Pinpoint Test Step Q40. Leave equipment hooked up.
KN2 CHECK BOOST OUTPUT ELECTRICAL OPERATION		
• Key On, Engine Off. • DVOM on 20V scale. • Connect DVOM positive test lead to circuit VPWR on boost solenoid connector and negative test lead to boost output on boost solenoid connector. • While observing DVOM, depress and release the throttle several times to cycle output On and Off. • Does boost output solenoid cycle On and Off?	Yes ▶ No ▶	GO to KN3. REMOVE jumper. GO to KN4.

Turbo Boost — Pinpoint Test KN

KN

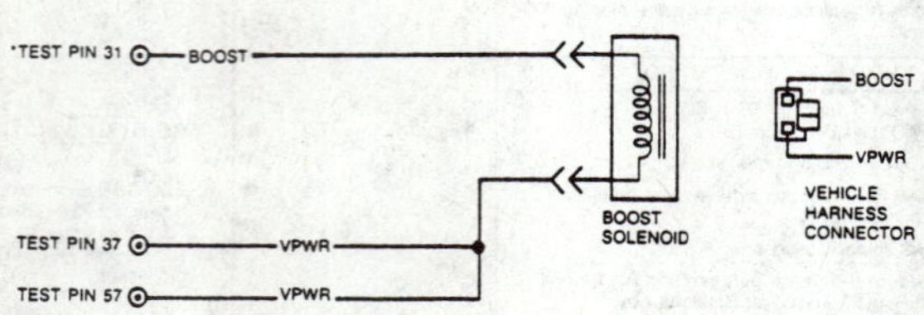

*TEST PINS LOCATED ON BREAKOUT BOX.
ALL HARNESS CONNECTORS VIEWED INTO MATING SURFACE.

STOP-WARNING

You should enter this Pinpoint Test only when a service code 81 is received in Quick Test Step 3.0 or when directed here from Diagnostics by Symptom in the Engine Supplement Section.

This Pinpoint Test is intended to diagnose only the following:

- Harness circuits: VPWR, Boost

Turbo Boost — Pinpoint Test KN

TEST STEP	RESULT ▶	ACTION TO TAKE
KN3 CHECK BOOST SOLENOID FUNCTION		
• Remain in output state check. • Disconnect turbo boost control solenoid vacuum hose at the turbocharger inlet end. • Attach a vacuum pump to the hose. • Depress the throttle once to cycle the solenoid closed. • Apply vacuum to the solenoid. • Depress the throttle once again to cycle the solenoid open and release trapped vacuum. • Did the solenoid hold and then release vacuum? TURBO BOOST CONTROL SOLENOID 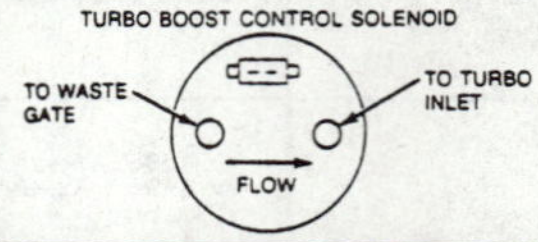	Yes ▶ No ▶	EEC-IV system OK. REFER to Shop Manual for boost diagnostics. REPLACE solenoid. RERUN Quick Test.
KN4 MEASURE BOOST SOLENOID RESISTANCE		
• Key Off, wait 10 seconds. • DVOM on 200 ohm scale. • Disconnect boost solenoid connector and measure the solenoid resistance. • Is resistance between 65 and 110 ohms?	Yes ▶ No ▶	CONNECT BOOST solenoid. GO to KN5. REPLACE BOOST solenoid. RERUN Quick Test.
KN5 CHECK VOLTAGE OF VPWR CIRCUIT		
• Key On, Engine Off. • DVOM on 20V scale. • Measure voltage between VPWR circuit of boost solenoid vehicle harness connector and battery ground. • Is voltage greater than 10.5 volts?	Yes ▶ No ▶	GO to KN6. SERVICE harness open circuit. RERUN Quick Test.

Turbo Boost — Pinpoint Test KN

TEST STEP	RESULT ▶	ACTION TO TAKE
KN6 CHECK CONTINUITY OF BOOST CIRCUIT		
• Key Off, wait 10 seconds. • Disconnect processor 60 Pin connector and inspect for damaged pins, corrosion, loose wires. Service as necessary. • Connect Breakout box to harness. Leave processor disconnected. • DVOM on 200 ohm scale. • Measure resistance between test Pin 31 at the Breakout box and boost circuit at vehicle harness connector. • Is resistance less than 5 ohms?	Yes ▶ No ▶	GO to KN7. SERVICE open circuit. RERUN Quick Test.
KN7 CHECK FOR SHORT TO GROUND		
• Key Off, wait 10 seconds. • Leave Breakout box installed and processor disconnected. • Disconnect boost solenoid. • DVOM on 200,000 ohm scale. • Measure resistance between test Pin 31 and test Pins 40, 46 and 60 at the Breakout box. • Is resistance greater than 10,000 ohms?	Yes ▶ No ▶	GO to KN8. SERVICE short to ground. RERUN Quick Test.
KN8 CHECK FOR SHORT TO POWER		
• Key Off, wait 10 seconds. • DVOM on 200,000 ohm scale. • Leave Breakout box installed and processor disconnected. • Boost solenoid disconnected. • Measure resistance between Pin 31 and test Pins 37 and 57 at the Breakout Box. • Is resistance greater than 10,000 ohms?	Yes ▶ No ▶	REPLACE processor. RERUN Quick Test. SERVICE short to power. RERUN Quick Test. If symptom is still present, REPLACE processor.

Exhaust Heat Control (EHC) — Pinpoint Test KP

KP

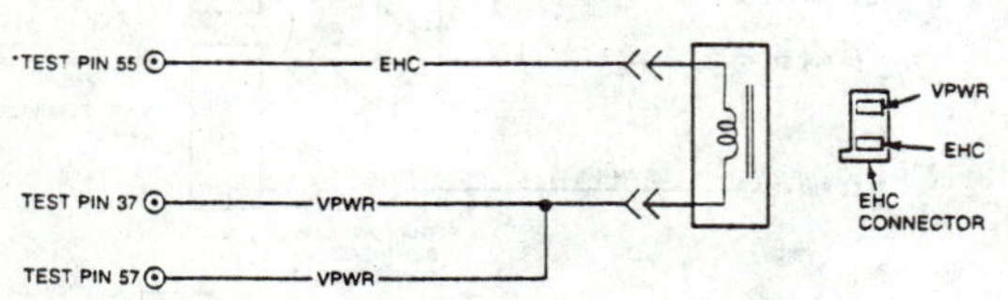

*TEST PINS LOCATED ON BREAKOUT BOX.
ALL HARNESS CONNECTORS VIEWED INTO MATING SURFACE.

STOP-WARNING

You should enter this Pinpoint Test only when a service code 89 is received in Quick Test Step 3.0 or when directed here from Diagnostics by Symptom in the Engine Supplement Section.

This Pinpoint Test is intended to diagnose only the following:

- Exhaust Heat Solenoid
- Vacuum Routing

Exhaust Heat Control (EHC) — Pinpoint Test KP

TEST STEP	RESULT ▶	ACTION TO TAKE
KP1 ENTER OUTPUT STATE CHECK (REFER TO APPENDIX)		
NOTE: Do not use STAR tester for this Step, use VOM/DVOM. • Key Off, wait 10 seconds. • DVOM on 20V scale. • Connect DVOM negative test lead to STO at the Self-Test connector and positive test lead to battery positive. • Jumper STI to signal return at the Self-Test connector. • Perform Key On, Engine Off Self-Test until the completion of the Continuous Test Codes. • DVOM will indicate zero volts when test is complete. • Depress and release the throttle. • Did DVOM reading change to a high voltage reading?	Yes ▶ No ▶	REMAIN in Output State Check. GO to KP2. DEPRESS throttle to WOT and release. If STO voltage does not go high, GO to Pinpoint Test Step Q40. Leave equipment hooked up.
KP2 CHECK EHC SOLENOID ELECTRICAL OPERATION		
• DVOM on 20V scale. • Measure voltage between VPWR and EHC circuit at EHC solenoid. • While observing DVOM, depress throttle several times to cycle output on and off. • Does EHC output cycle on and off?	Yes ▶ No ▶	GO to KP3. GO to KP5.
KP3 CHECK EXHAUST HEAT SOLENOID FOR VACUUM CYCLING		
• Install vacuum pump to the exhaust heat solenoid vacuum supply port and install a vacuum gauge to the output port. Apply 6 in. Hg minimum. • Does the vacuum output cycle On and Off while depressing and releasing the throttle? **NOTE: Maintain vacuum at source.**	Yes ▶ No ▶	GO to KP4. REPLACE solenoid. RERUN Quick Test.

Exhaust Heat Control (EHC) — Pinpoint Test KP

TEST STEP	RESULT ▶	ACTION TO TAKE
KP4 CHECK MANIFOLD VACUUM LINES FOR BLOCKAGE OR LEAKS		
• With vacuum lines disconnected at exhaust heat solenoid, check for vacuum. • Start engine. • Is vacuum present?	Yes ▶ No ▶	EEC-IV system OK. REFER to Section 3 for exhaust heat diagnostics. SERVICE vacuum source blockage or leak. RERUN Quick Test.
KP5 MEASURE EHC SOLENOID RESISTANCE		
• Key Off, wait 10 seconds. • DVOM on 200 ohm scale. • Disconnect exhaust heat solenoid connector and measure solenoid resistance. • Is resistance between 50 and 100 ohms?	Yes ▶ No ▶	CONNECT exhaust heat solenoid. Go to KP6. REPLACE exhaust heat solenoid. RERUN Quick Test.
KP6 CHECK VOLTAGE OF VPWR CIRCUIT		
• Key On, Engine Off. • DVOM on 20V scale. • Measure voltage between VPWR at the EHC solenoid vehicle harness connector circuit and ground. • Is voltage reading 10.5 volts or greater?	Yes ▶ No ▶	GO to KP7. SERVICE harness open circuit. RERUN Quick Test.
KP7 CHECK CONTINUITY OF EHC CIRCUIT		
• Key Off, wait 10 seconds. • Disconnect processor 60 Pin connector and inspect for damaged pins, corrosion, loose wires. Service as necessary. • Connect Breakout box to harness. Leave processor disconnected. • DVOM on 200 ohm scale. • Measure resistance between test Pin 55 at the Breakout box and EHC circuit at vehicle harness connector. • Is resistance reading 5 ohms or greater?	Yes ▶ No ▶	SERVICE harness open circuit. RERUN Quick Test. GO to KP8.

Exhaust Heat Control (EHC) — Pinpoint Test — KP

TEST STEP	RESULT ▶	ACTION TO TAKE
KP8 CHECK FOR SHORT TO GROUND		
• Key Off, wait 10 seconds. • Leave Breakout box installed and processor disconnected. • Disconnect EHC solenoid. • DVOM on 200,000 ohm scale. • Measure resistance between test Pin 55 and test Pins 40, 46 and 60 at Breakout box. • Are resistance readings 10,000 ohms or greater?	Yes ▶ No ▶	GO to KP9. SERVICE short to ground. RERUN Quick Test.
KP9 CHECK FOR SHORT TO POWER		
• Key Off, wait 10 seconds. • DVOM on 200,000 ohm scale. • Leave Breakout box installed and processor disconnected. • EHC solenoid disconnected. • Measure resistance between test Pin 55 and test Pins 37 and 57 at Breakout box. • Is the resistance reading 10,000 ohms or greater?	Yes ▶ No ▶	REPLACE processor. RERUN Quick Test. SERVICE short to power. RERUN Quick Test. If code is repeated, REPLACE processor.

Converter Clutch Override (CCO) — Pinpoint Test — KR

KR

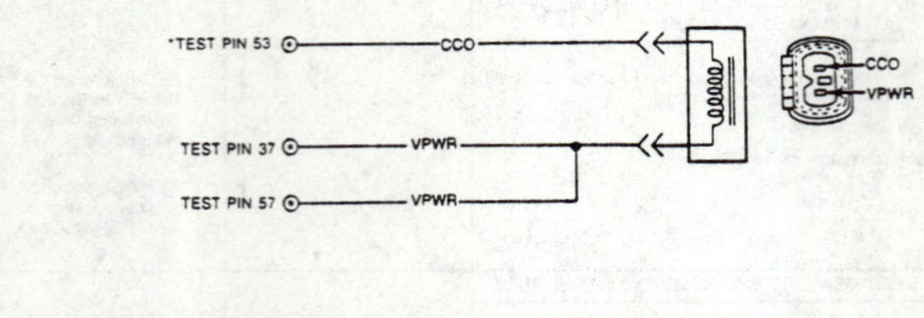

*TEST PINS LOCATED ON BREAKOUT BOX.
ALL HARNESS CONNECTORS VIEWED INTO MATING SURFACE.

STOP-WARNING

You should enter this Pinpoint Test only when a service code 88 or 89 is received in Quick Test Step 3.0.

This Pinpoint Test is intended to diagnose only the following:

- Harness Circuits: CCO and V PWR.
- CCO Solenoid.
- Processor Assembly.

TEST STEP	RESULT ▶	ACTION TO TAKE
CODE 89 PRESENT		
KR1 MEASURE CCO SOLENOID RESISTANCE		
• Key Off, wait 10 seconds. • DVOM on 200 ohm scale. • Disconnect CCO solenoid connector and measure solenoid resistance. • Is resistance reading between 26 and 40 ohms?	Yes ▶ No ▶	CONNECT CCO solenoid. GO to KR2. REPLACE CCO solenoid. RERUN Quick Test.

Converter Clutch Override (CCO) — Pinpoint Test — KR

TEST STEP	RESULT ▶	ACTION TO TAKE
KR2 CHECK VOLTAGE OF VPWR CIRCUIT		
• Key On, Engine Off. • DVOM on 20V scale. • Measure voltage at the CCO solenoid connector between VPWR circuit and battery ground. • Is voltage reading less than 10.5 volts?	Yes ▶ No ▶	SERVICE harness open circuit. RERUN Quick Test. GO to KR3.
KR3 CHECK CONTINUITY OF CCO CIRCUIT		
• Key Off, wait 10 seconds. • Disconnect processor 60 Pin connector and inspect for damaged pins, corrosion, loose wires. Service as necessary. • Connect Breakout box to harness. Leave processor disconnected. • DVOM on 200 ohm scale. • Measure resistance between test Pin 53 at the Breakout box and CCO circuit at the solenoid vehicle harness connector. • Is resistance reading less 5 ohms?	Yes ▶ No ▶	GO to KR4. SERVICE harness open circuit. RERUN Quick Test.
KR4 CHECK FOR SHORT TO GROUND		
• Key Off, wait 10 seconds. • Leave Breakout box installed and processor disconnected. • Disconnect CCO solenoid. • DVOM on 200,000 ohm scale. • Measure resistance between test Pin 53 and test Pins 40, 46 and 60 at Breakout box. • Are all resistance readings 10,000 ohms or greater?	Yes ▶ No ▶	GO to KR5. SERVICE short to ground. RERUN Quick Test.
KR5 CHECK FOR SHORT TO POWER		
• Key Off, wait 10 seconds. • DVOM on 200,000 ohm scale. • Leave Breakout box installed and processor disconnected. • CCO solenoid disconnected. • Measure resistance between test Pin 53 and test Pins 37 and 57 at Breakout box. • Are all resistance readings 10,000 ohms or greater?	Yes ▶ No ▶	REPLACE processor. RERUN Quick Test. SERVICE short to power. RERUN Quick Test. If code is still present, REPLACE processor.

Converter Clutch Override (CCO) and Shift Solenoid 3/4-4/3 (SS 3/4-4/3) — Pinpoint Test — KS

KS

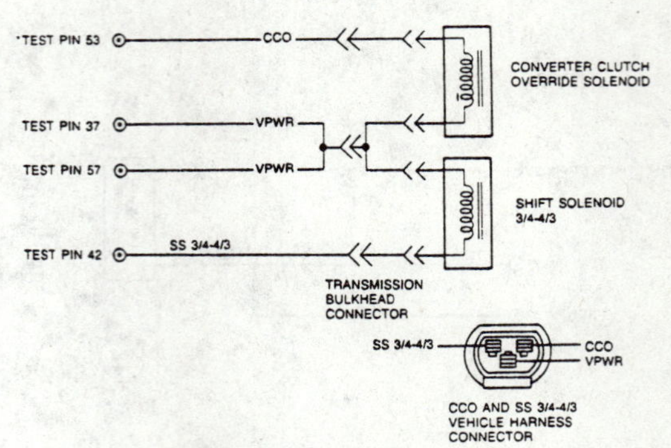

*TEST PINS LOCATED ON BREAKOUT BOX.
ALL HARNESS CONNECTORS VIEWED INTO MATING SURFACE.

STOP-WARNING

You should enter this Pinpoint Test only when a service code 85 or 88 are received in Quick Test Step 3.0.

This Pinpoint Test is intended to diagnose only the following:

- Harness Circuits: CCO SS 3/4-4/3 and V PWR.
- CCO Solenoid.
- Shift Solenoid 3/4-4/3.
- Processor Assembly.

Converter Clutch Override (CCO) and Shift Solenoid 3/4-4/3 (SS 3/4-4/3) — Pinpoint Test KS

TEST STEP	RESULT ▶	ACTION TO TAKE
CODE 88 PRESENT		
KS1 MEASURE CCO SOLENOID RESISTANCE		
• Key Off, wait 10 seconds. • DVOM on 200 ohm scale. • Disconnect CCO solenoid vehicle harness connector and measure solenoid resistance at transmission bulkhead connector. • Is resistance between 26 and 40 ohms.	Yes ▶	CONNECT CCO solenoid. GO to KS2.
	No ▶	GO to Car Shop Manual Volume D, Section 17-08, A4LD transmission diagnosis.

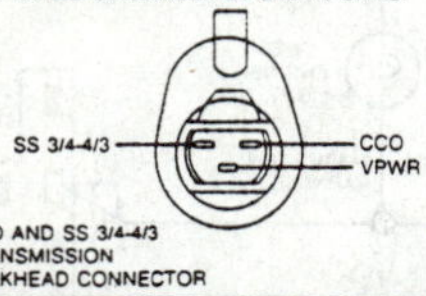

CCO AND SS 3/4-4/3 TRANSMISSION BULKHEAD CONNECTOR

TEST STEP	RESULT ▶	ACTION TO TAKE
KS2 CHECK VOLTAGE OF VPWR CIRCUIT		
• Key On Engine Off. • DVOM on 20V scale. • Measure voltage at the CCO solenoid vehicle harness connector between VPWR circuit and battery ground. • Is voltage greater than 10.5V?	Yes ▶	GO to KS3.
	No ▶	SERVICE harness open circuit. RERUN Quick Test.
KS3 CHECK CONTINUITY OF CCO CIRCUIT		
• Key Off, wait 10 seconds. • Disconnect processor 60 Pin connector and inspect for damaged pins, corrosin, loose wires. Service as necessary. • Connect Breakout box to harness. Leave processor disconnected. • CCO solenoid vehicle harness connector disconnected at transmission bulkhead connector. • DVOM on 200 ohm scale. • Measure resistance between test Pin 53 at the Breakout box and CCO circuit at the solenoid vehicle harness connector. • Is resistance less than 5 ohms?	Yes ▶	GO to KS4.
	No ▶	SERVICE harness open circuit. RERUN Quick Test.

Converter Clutch Override (CCO) and Shift Solenoid 3/4-4/3 (SS 3/4-4/3) — Pinpoint Test KS

TEST STEP	RESULT ▶	ACTION TO TAKE
KS4 CHECK FOR SHORT TO GROUND		
• Key Off, wait 10 seconds. • Leave Breakout box installed and processor disconnected. • CCO solenoid vehicle harness connector disconnected at transmission bulkhead connector. • DVOM on 200,000 ohm scale. • Measure resistance between test Pin 53 and test Pins 40, 46 and 60 at Breakout box. • Is resistance greater than 100,000 ohms?	Yes ▶	GO to KS5.
	No ▶	SERVICE short to ground. RERUN Quick Test.
KS5 CHECK FOR SHORT TO POWER		
• Key Off, wait 10 seconds. • DVOM on 200,000 ohm scale. • Leave Breakout box installed and processor disconnected. • CCO solenoid vehicle harness connector disconnected at transmission bulkhead connector. • Measure resistance between test Pin 53 and test Pins 37 and 57 at Breakout box. • Are all resistances greater than 10,000 ohms?	Yes ▶	REPLACE processor. RERUN Quick Test.
	No ▶	SERVICE short to power. RERUN Quick Test. If code is still present, REPLACE processor.
CODE 85 PRESENT		
KS10 MEASURE SS 3/4-4/3 RESISTANCE		
• Key Off, wait 10 seconds. • DVOM on 200 ohm scale. • Disconnect SS 3/4-4/3 vehicle harness connector and measure solenoid resistance at transmission bulkhead connector. • Is resistance between 26 and 40 ohms?	Yes ▶	CONNECT SS 3/4-4/3. GO to KS11.
	No ▶	GO to Car Shop Manual Volume D, Section 17-08, A4LD transmission diagnosis.

CCO AND SS 3/4-4/3 TRANSMISSION BULKHEAD CONNECTOR

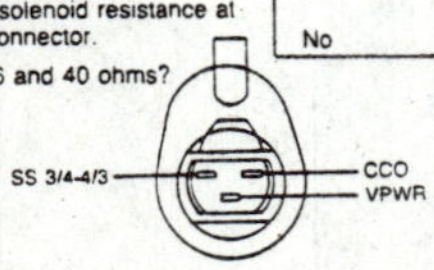

Converter Clutch Override (CCO) and Shift Solenoid 3/4-4/3 (SS 3/4-4/3) — Pinpoint Test KS

TEST STEP	RESULT ▶	ACTION TO TAKE
KS11 CHECK VOLTAGE OF VPWR CIRCUIT		
• Key On Engine Off. • DVOM on 20V scale. • Measure voltage at the SS 3/4-4/3 vehicle harness connector between VPWR circuit and battery ground. • Is voltage greater than 10.5V?	Yes ▶	GO to KS12.
	No ▶	SERVICE harness open circuit. RERUN Quick Test.
KS12 CHECK CONTINUITY OF SS 3/4-4/3 CIRCUIT		
• Key Off, wait 10 seconds. • Disconnect processor 60 Pin connector and inspect for damaged pins, corrosion, loose wires. Service as necessary. • Connect Breakout box to harness. Leave processor disconnected. • SS 3/4-4/3 vehicle harness connector disconnected at transmission bulkhead connector. • DVOM on 200 ohm scale. • Measure resistance between test Pin 53 at the Breakout box and SS 3/4-4/3 circuit at the solenoid vehicle harness connector. • Is resistance less than 5 ohms?	Yes ▶	GO to KS13.
	No ▶	SERVICE harness open circuit. RERUN Quick Test.
KS13 CHECK FOR SHORT TO GROUND		
• Key Off, wait 10 seconds. • Leave Breakout box installed and processor disconnected. • SS 3/4-4/3 vehicle harness connector disconnected at transmission bulkhead connector. • DVOM on 200,000 ohm scale. • Measure resistance between Pin 53 and test Pins 40, 46 and 60 at Breakout box. • Is resistance greater than 100,000 ohms?	Yes ▶	GO to KS14.
	No ▶	SERVICE short to ground. RERUN Quick Test.

Converter Clutch Override (CCO) and Shift Solenoid 3/4-4/3 (SS 3/4-4/3) — Pinpoint Test KS

TEST STEP	RESULT ▶	ACTION TO TAKE
KS14 CHECK FOR SHORT TO POWER		
• Key Off, wait 10 seconds. • DVOM on 200,000 ohm scale. • Leave Breakout box installed and processor disconnected. • SS 3/4-4/3 vehicle harness connector disconnected at transmission bulkhead connector. • Measure resistance between Pin 53 and test Pins 37 and 57 at Breakout box. • Are all resistances greater than 10,000 ohms?	Yes ▶	REPLACE processor. RERUN Quick Test.
	No ▶	SERVICE short to power. RERUN Quick Test. If code is still present, REPLACE processor.

Dynamic Response Test — Pinpoint Test — M

M

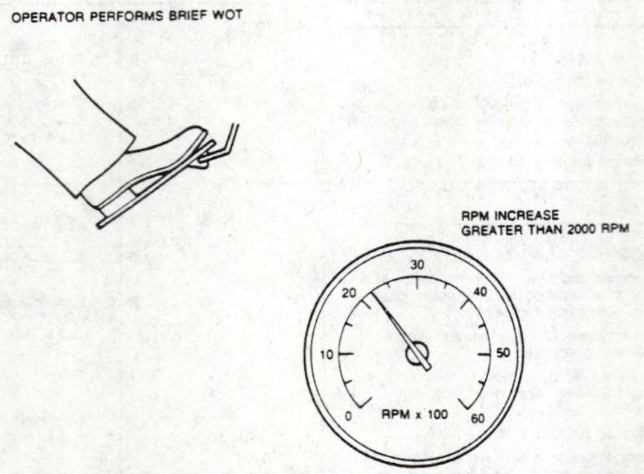

STOP-WARNING

You should enter this Pinpoint Test only when a service code 77 is received in Quick Test Step 5.0.

To prevent the replacement of good components, be aware that the following non-EEC areas may be at fault:

- Operator did not perform a brief WOT after dynamic response code.
- Mechanical engine problems; engine did not achieve greater than 2000 rpm.

This Pinpoint Test is intended to diagnose only the following:

- Throttle movement (greater than 3/4 throttle).
- Vane Air Flow (greater than 50% open).
- RPM increase (greater than 2000 rpm).

"CHECK ENGINE" Light (Malfunction Indicator Light) — Pinpoint Test — ML

ML

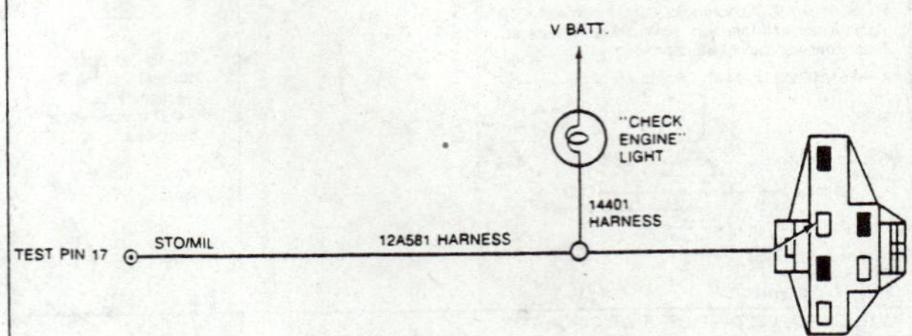

STOP-WARNING

You should enter this Pinpoint Test only from Diagnostic by Symptom.

To prevent the replacement of good components, be aware that the following non-EEC areas may be at fault:

- Fuse, bulb or socket.

This Pinpoint Test is intended to diagnose only the following:

- STO/MIL circuit.
- Processor assembly.

Dynamic Response Test — Pinpoint Test — M

TEST STEP	RESULT ▶	ACTION TO TAKE
M1 CODE 77: SYSTEM FAILED TO RECOGNIZE BRIEF WOT		
NOTE: A brief snap of the throttle may not be sufficient to pass this test. Be sure to go to WOT and return. (If engine is a 5.0L SEFI, Go to Pinpoint Test HC.) • Repeat Engine Running Test of Quick Test. Be sure operator is familiar with the engine running format which proceeds as follows: — With Self-Test activated restart the engine. — ID Code 2 (0) start of test. — Dynamic response Code 1 (0) perform brief WOT. — Testing over. — Service code output begins. • Is code 77 still present?	Yes ▶	REPLACE processor. RERUN Quick Test.
	No ▶	Dynamic Response Test passed. SERVICE any other service code(s) received as necessary.

"CHECK ENGINE" Light (Malfunction Indicator Light) — Pinpoint Test — ML

TEST STEP	RESULT ▶	ACTION TO TAKE
ML1 CHECK FOR SHORT TO GROUND		
NOTE: If vehicle will not start go to Pinpoint Test Step A1. • **If any Key On Engine Off service codes are present, repair before proceeding. If no codes are outputted, continue with this Test Step.** • Key Off. • Disconnect processor. • Inspect 60 Pin connector for damaged pins/corrosion and service as necessary. • Install Breakout box. Leave processor disconnected. • DVOM on 200 ohm scale. • Connect DVOM between test Pin 17 and test Pin 40. • Is resistance less than 5 ohms?	Yes ▶	SERVICE harness for short to ground (from Pin 17 to Self-Test connector or from Pin 17 to the "CHECK ENGINE" light).
	No ▶	REPLACE processor.
ML5 CONTINUITY CHECK		
NOTE: If vehicle will not start go to Pinpoint Test Step A1. • Key Off. • Disconnect processor. • Inspect 60 Pin connector for damaged pins/corrosion and service as necessary. • Install Breakout box. Leave processor disconnected. • DVOM on 200 ohm scale. • Measure resistance between Pin 17 and STO at the Self-Test connector. • Measure resistance between Pin 17 and the "CHECK ENGINE" light. • Are the resistances less than 5 ohms?	Yes ▶	GO to Pinpoint Test Step ML6.
	No ▶	SERVICE OPEN in harness.

"CHECK ENGINE" Light (Malfunction Indicator Light) | Pinpoint Test | ML

TEST STEP	RESULT ▶	ACTION TO TAKE
ML6 POWER TO BULB		
• Check for power to "CHECK ENGINE" light bulb. • Is there power at the light bulb?	Yes ▶	REPLACE bulb or socket. GO to Pinpoint Test Step ML7.
	No ▶	CHECK fuse and input circuit. GO to Pinpoint Test Step ML7.
ML7 CONFIRM CIRCUIT REPAIR		
• Reconnect processor. • Turn key to run. • Is "CHECK ENGINE" light ON? **NOTE: Refer to the appendix for a detailed description of HOW the "CHECK ENGINE" light (malfunction indicator light) operates.**	Yes ▶	System OK.
	No ▶	REPLACE processor.

Ignition Diagnostic Monitor (IDM) | Pinpoint Test | N

N

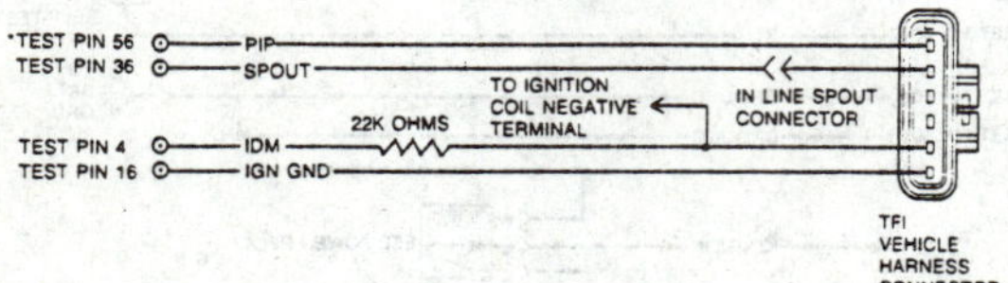

*TEST PINS LOCATED ON BREAKOUT BOX.
ALL HARNESS CONNECTORS VIEWED INTO MATING SURFACE.

STOP-WARNING

You should enter this Pinpoint Test only when a service code 18 is received in Quick Test Step 6.0.

To prevent the replacement of good components, be aware that the following non-EEC areas may be at fault:

- Ignition module.
- Ignition coil.
- Spark plugs and high tension cables.
- Distributor and PIP sensor.

This Pinpoint Test is intended to diagnose only the following:

- Harness circuits: ignition ground, spout, PIP, IDM.

Ignition Diagnostic Monitor (IDM) | Pinpoint Test | N

TEST STEP	RESULT ▶	ACTION TO TAKE
N1 CHECK CONTINUITY OF IDM CIRCUIT		
• Key Off, wait 10 seconds. • Disconnect E-core ignition connector from coil. • Disconnect processor and inspect both 60 Pin connectors for damaged pins, corrosion, loose wires. Service as necessary. • Connect Breakout box to harness. Leave processor disconnected. • DVOM on 200,000 ohm scale. • Measure resistance between test Pin 4 at the Breakout box and ignition coil harness connector negative terminal. • Is resistance between 20,000 and 24,000 ohms?	Yes ▶	GO to N2.
	No ▶	SERVICE open IDM circuit. RERUN Quick Test. RECONNECT E-core ignition connector to coil.
N2 CHECK FOR SHORT TO GROUND		
• Key Off, wait 10 seconds. • Leave Breakout box installed and processor disconnected. • DVOM on 200,000 ohm scale. • Measure resistance between test Pin 4 and test Pins 40, 46 and 60 at the Breakout box. • Are all resistances above 10,000 ohms?	Yes ▶	RECONNECT E-core ignition connector and processor. GO to N3.
	No ▶	SERVICE short to ground in IDM circuit. RECONNECT E-core ignition connector to coil. RERUN Quick Test.
N3 CHECK TFI MODULE		
• Key Off, wait 10 seconds. • Deactivate Self-Test. • Connect VOM or STAR per Quick Test Step 2.0. • Enter Engine Running Continuous Monitor Test (as instructed in Quick Test Step 6.0B). • Observe VOM or STAR LED for indication of a fault while performing the following: • Lightly tap on TFI module (simulate road shock). • Wiggle TFI connector. • Is a fault indicated?	Yes ▶	DISCONNECT and INSPECT connectors. If connector and terminals are good, GO to Section 15, TFI Ignition Diagnostics.
	No ▶	GO to N4.

Ignition Diagnostic Monitor (IDM) | Pinpoint Test | N

TEST STEP	RESULT ▶	ACTION TO TAKE
N4 CHECK EEC-IV HARNESS		
• While still in continuous monitor test from Step N3, observe VOM or STAR LED for a fault indication while performing the following: • While looking for faults listed in the table below, grasp the harness close to the TFI connector. Wiggle, shake or bend a small section of the EEC-IV system harness while working your way to the dash panel. Also wiggle, shake or bend the EEC-IV harness from the dash panel to the processor. Do this test on the circuits listed one at a time if needed to locate a faulty circuit. FAULT — BREAKOUT BOX NO. PIP shorted to ground or open — Test Pin 56 Spout shorted to ground — Test Pin 36 Ign. ground open — Test Pin 16 IDM open or shorted to ground, power — Test Pin 4 • Is a fault indicated?	Yes ▶	ISOLATE fault and make necessary repairs. RERUN Quick Test.
	No ▶	GO to N5.
N5 CHECK PROCESSOR AND HARNESS CONNECTORS		
• Key Off, wait 10 seconds. • Disconnect processor 60 Pin connector and inspect for damaged pins, corrosion, loose wires. • Are connectors and terminals OK?	Yes ▶	REPLACE processor. Start engine and run for about one minute. RERUN Key On Engine Off Self Test observing continuous codes.
	No ▶	SERVICE as necessary. RECONNECT processor. RERUN Quick Test.

Spark Timing Check — Pinpoint Test P

P

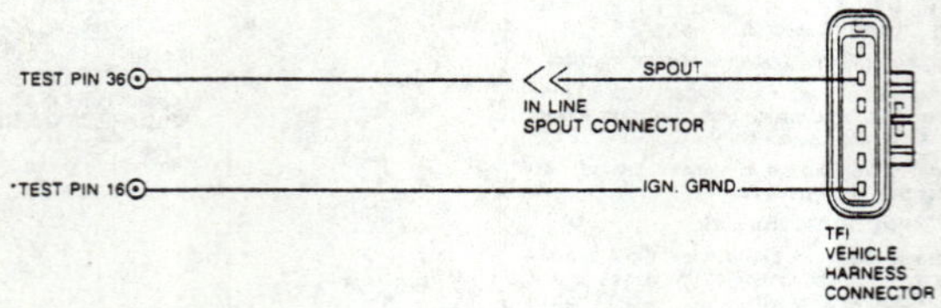

*TEST PINS LOCATED ON BREAKOUT BOX.
ALL HARNESS CONNECTORS VIEWED INTO MATING SURFACE.

STOP-WARNING

You should enter this Pinpoint Test only when directed here from Quick Test Step 4.0 or when a Service Code 18 is received in Quick Test Step 5.0.

To prevent the replacement of good components, be aware that the following non-EEC areas may be at fault:

- Base Engine.
- PIP Sensor.
- TFI Module.

This Pinpoint Test is intended to diagnose only the following:

- Harness Spout Circuit.
- Base Timing.
- Processor Assembly.

Spark Timing Check — Pinpoint Test P

TEST STEP	RESULT ▶	ACTION TO TAKE
P1 CHECK SPARK TIMING		
• If Quick Test cannot be run, Go to Q1, otherwise continue with Test Step. **NOTE: Self-Test locks the timing at 20 degrees plus base during code output and for two minutes after the last service code is outputted. Timing check must be made during this time period. Self-Test timing is base + 20 degrees (±3 degrees) BTDC. (See VECI decal for base value.)** • Check timing. Record value. • Is timing equal to base plus 20 degrees (±3 degrees)?	Yes ▶ No ▶	GO to Quick Test Step 5.0. GO to P2.
P2 CHECK SPARK OUTPUT (SPOUT) CIRCUIT TO THE TFI MODULE		
• Locate spout connector and open this connection. • Start engine. • Check base timing. • Is base timing within ±3 degrees of value on VECI decal?	Yes ▶ No ▶	RECONNECT spout connector. GO to P3. Adjust base timing if necessary. REFER to Section 15 for engine timing instructions. After timing is reset, RECONNECT spout and PERFORM Quick Test Step 4.0.
P3 CHECK FOR POWER TO PROCESSOR		
• Key Off, wait 10 seconds. • Disconnect processor 60 Pin connector and inspect for damaged pins, corrosion, loose wires. Service as necessary. • Install Breakout Box. • Key on, engine off. • DVOM on 20V scale. • Measure voltage between test Pin 37 and test Pin 40 and between test Pin 57 and test Pin 60 at the Breakout box. • Is voltage less than 10.5 volts?	Yes ▶ No ▶	GO to Pinpoint Test Step B1, except 2.3L EFI TC, 2.5L HSC CFI and 3.0L EFI passenger car; GO to Pinpoint Test Step X1. GO to P4.

Spark Timing Check — Pinpoint Test P

TEST STEP	RESULT ▶	ACTION TO TAKE
P4 CHECK HARNESS FOR CONTINUITY		
• Key Off, wait 10 seconds. • DVOM on 200 ohm scale. • Measure resistance between test Pin 36 at the Breakout box and the spout pin at the TFI vehicle harness connector. • Is resistance less than 5 ohms? 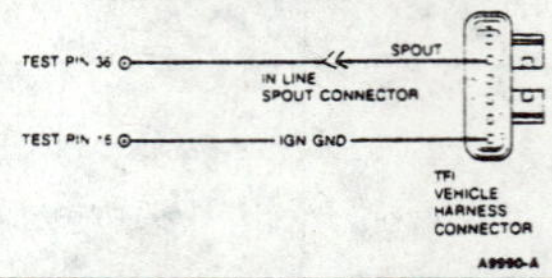	Yes ▶ No ▶	GO to P5. SERVICE open circuit CHECK timing per P1.
P5 HARNESS CHECK — IGNITION GROUND		
• Breakout box installed. • Key Off. • DVOM on 200 ohm scale • Measure resistance between test Pin 16 at the Breakout box and the ignition ground at the TFI vehicle harness connector. • Is resistance less than 5 ohms?	Yes ▶ No ▶	REPLACE processor. GO to Pinpoint Test Step A7. SERVICE harness as necessary. RERUN Quick Test.

No Codes/Codes Not Listed — Pinpoint Test Q

Q

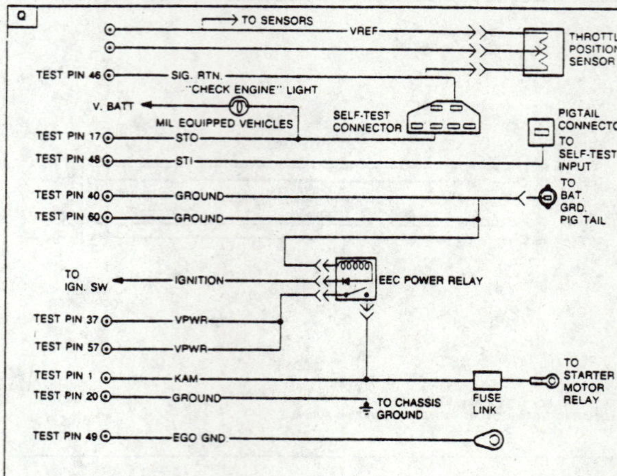

*TEST PINS LOCATED ON BREAKOUT BOX.
ALL HARNESS CONNECTORS VIEWED INTO MATING SURFACE.

STOP-WARNING

You should enter this Pinpoint Test only when directed here from Quick Test Step 3.0, 5.0 or 6.0 or when directed here from Diagnostics by Symptom in the Engine Supplement Section or from a Pinpoint Test Step.

This Pinpoint Test is intended to diagnose only the following:

- Processor.
- Harness Circuits: Signal Return, STO, STI, Ground, VPWR, KAM, VREF, Ignition, NDS, EEC Power Relay.

No Codes/Codes Not Listed — Pinpoint Test Q

TEST STEP	RESULT ▶	ACTION TO TAKE
Q1 CHECK FOR VREF AT TP SENSOR		
Refer to illustration Q. • Key Off, wait 10 seconds. • DVOM on 20V scale. • Disconnect TP sensor. • Key on, engine off. • Measure voltage between VREF at the TP harness connector and signal return in Self-Test connector. • Is voltage between 4.0 and 6.0 volts?	Yes ▶ No ▶	RECONNECT TP Sensor. GO to Q2. GO to Pinpoint Test Step C1.
Q2 SELF-TEST INPUT CONTINUITY CHECK		
Refer to illustration Q. • Key Off, wait 10 seconds. • Disconnect processor 60 Pin connector and inspect for damaged pins, corrosion, loose wires. Service as necessary. • Install Breakout box, leave processor disconnected. • Set DVOM to 200 ohm scale. • Measure resistance between Self-Test input at the Self-Test single pin connector and test Pin 48 at the Breakout box. • Is resistance less than 5 ohms?	Yes ▶ No ▶	GO to Q3. CORRECT open in circuit.
Q3 SELF-TEST OUTPUT CIRCUIT CONTINUITY CHECK		
Refer to illustration Q. • Breakout box installed. • DVOM to 200 ohm scale. • Measure resistance between Self-Test output at the Self-Test connector and test Pin 17 at the Breakout box. • Is resistance less than 5 ohms?	Yes ▶ No ▶	GO to Q4. CORRECT open in circuit.

No Codes/Codes Not Listed — Pinpoint Test Q

TEST STEP	RESULT ▶	ACTION TO TAKE
Q4 EGO SENSOR GROUND CONTINUITY CHECK		
Refer to illustration Q. • Breakout box installed. • Key Off. • DVOM on 200 ohm scale. • Measure resistance between EGO ground on engine and test Pin 49 at the Breakout box. • Is resistance less than 5 ohms?	Yes ▶ No ▶	GO to Q5. SERVICE EGO sensor ground wire or open circuit bad connection.
Q5 STO SHORT TO GROUND		
• Breakout box installed. • DVOM on 200 ohm scale. • Measure resistance between Self-Test output at Self-Test connector and engine block ground. • Is resistance greater than 5 ohms?	No ▶ Yes ▶	REPAIR STO circuit (or "CHECK ENGINE" light circuit on MIL equipped vehicles) for short to ground. RERUN Quick Test. 3.0L EFI passenger car GO to Q7. All others GO to Q6.
Q6 INTERMITTENT NDS		
• Key Off. • Breakout box installed. • Processor connected. • Connect DVOM between test Pin 30 and test Pin 40 or 60 at the Breakout box. • Key On Engine Running Self-Test. • Is voltage greater than 1V? **NOTE: Refer to FA for connector orientation.**	Yes ▶ No ▶	SERVICE intermittent open in NDS harness, connector or switch. If OK, GO to Quick Test Step 5.0 for appropriate service codes. GO to Q7.

No Codes/Codes Not Listed — Pinpoint Test Q

TEST STEP	RESULT ▶	ACTION TO TAKE
Q7 POWER RELAY ALWAYS ON		
• Key Off. • Breakout box installed. • Connect DVOM to test Pin 37 or 57 and to test Pin 40 or 60 at the Breakout box. • Turn key on and off. Wait 10 seconds. • Does voltage change from greater than 10.5V to zero volts?	Yes ▶ No ▶	If vehicle is equipped with MIL (malfunction indicator light displayed as "CHECK ENGINE" light) or SIL (shift indicator light) GO to Q9. If not, REPLACE the processor. RERUN Quick Test. GO to Q8.
Q8 VPWR HARNESS SHORT TO POWER		
• Key Off. • Breakout box installed. • EEC Power Relay or Integrated Relay Controller disconnected. • Connect DVOM to test Pin 37 or 57 and to test Pin 40 or 60 at the Breakout box. • Is voltage greater than 10.5V?	Yes ▶ No ▶	REPAIR VPWR harness short to power. RERUN Quick Test. REPLACE EEC Power Relay or Integrated Relay Controller. RERUN Quick Test.
Q9 MIL AND/OR SIL EQUIPPED VEHICLES		
• Are any of these conditions present?		
• Shift indicator light: — Always ON	▶	GO to KL1.
— Always OFF	▶	GO to KL1.
• Malfunction indicator light: — Always ON	▶	GO to ML1.
— Always OFF	▶	GO to ML5.
• Shift and malfunction indicator lights functioning normally.	▶	REPLACE the processor. RERUN Quick Test.

Continuous Test Code 15 — Pinpoint Test Q

TEST STEP	RESULT ▶	ACTION TO TAKE
Q10 CONDITIONS FOR CONTINUOUS CODE 15		
• Power interruption to Keep Alive Memory (KAM) Pin 1 may result in a service code being outputted.* • Clear continuous memory codes (use procedure described in Quick Test Step 6). • Repeat Quick Test Step 3.0 through Continuous memory code output. • Code 15 present on retest? ***NOTE: Anytime power is interrupted to the processor, for example when installing a Breakout box, a code 15 may be outputted the first time Self-Test is run after restoration of power. Repeat Self-Test to ensure correct diagnosis.**	Yes ▶ No ▶	GO to Q11. Test complete.
Q11 INSPECT ENGINE COMPARTMENT WIRING FOR PROPER ROUTING		
• Are any EEC components or EEC wiring close to ignition components or wires (High Electrical Energy Sources)? If EEC wiring close, reroute and rerun Quick Test. • Is code 15 still present?	Yes ▶ No ▶	GO to Q12. Test complete.
Q12 CHECK POWER CIRCUIT TO KEEP ALIVE MEMORY		
• Key Off, wait 10 seconds. • Disconnect processor and inspect both 60 Pin connectors. • Connect Breakout box to harness, leave processor disconnected. • DVOM on 20V scale. • Connect positive test lead to test Pin 1 and negative test lead to test Pin 40 or 60 at the Breakout box. • Key On. • Is voltage reading less than 10.5 volts?	Yes ▶ No ▶	SERVICE open to KAM circuit. RERUN Quick Test. REPLACE processor. RERUN Quick Test.

Output State Check Not Functioning — Pinpoint Test Q

TEST STEP	RESULT ▶	ACTION TO TAKE
Q40 CHECK FOR CODES 23, 53, 63 OR 68		
• Key Off, wait 10 seconds. • Perform Key On Engine Off Self-Test. • Leave Key On to enter Output State Check. • Key On Engine Off. Are codes 23, 53, 63 or 68 present?	Yes ▶	GO to Quick Test Step 3.0B and SERVICE appropriate code as instructed.
	Code 11 ▶	GO to Q41.
	No Codes ▶	GO to Q1.
Q41 CHECK THROTTLE LINKAGE		
• Check throttle and throttle linkages for sticking and binding. • Throttle OK?	Yes ▶	REPLACE TP sensor. RERUN Quick Test.
	No ▶	SERVICE as necessary. RERUN Quick Test.

Processor Power Check — Pinpoint Test Q

TEST STEP	RESULT ▶	ACTION TO TAKE
Q50 CHECK FOR POWER TO PROCESSOR		
• Key Off, wait 10 seconds. • Disconnect processor 60 Pin connector and inspect for damaged pins, corrosion, loose wires. Service as necessary. • Install Breakout box. • Key On Engine Off. • DVOM on 20V scale. • Measure voltage between test Pin 37 and test Pin 40 at the Breakout box and between test Pin 57 and test Pin 60 at the Breakout box. • Is either voltage reading less than10.5 volts?	Yes ▶	2.3L EFI TC, 2.5L CFI and 3.0L EFI passenger car GO to Pinpoint Test X1. All others GO to Pinpoint Test B1.
	No ▶	REPLACE processor. RERUN Quick Test.

Re-Initialization Check — Pinpoint Test Q

TEST STEP	RESULT ▶	ACTION TO TAKE
Q60 CHECK FOR EEC IV WIRING POSITION		
• Key Off. • Check that the EEC IV wiring and components are greater than 2 inches from secondary ignition wires and ignition coil. • Check that the EEC IV wiring and components are greater than 4 inches from distributor, coil tower, starter motor and its wiring. • Are all above conditions intact?	Yes ▶	GO to Q61.
	No ▶	SERVICE as necessary. RERUN Quick Test.
Q61 HARNESS CHECK — CASE GROUND		
• Key Off. • Install Breakout box. Leave processor disconnected. • Disconnect processor 60 pin connector. Inspect for damaged pins, corrosion, loose wires, etc. Service as necessary. • DVOM on 200 ohm scale. • Measure resistance between test Pin 20 at the Breakout box and chassis ground. • Is the resistance less than 5 ohms?	Yes ▶	GO to Q62.
	No ▶	SERVICE harness as necessary. RERUN Quick Test.
Q62 DISCONNECT HARNESS — CASE GROUND CHECK		
• Key Off. • Reconnect processor to Breakout box, but disconnect harness from Breakout box. • DVOM on 200 ohm scale. • Measure resistance between test Pin 20 at the Breakout box and metal case of processor. • Is the resistance less than 5 ohms?	Yes ▶	GO to Pinpoint Test Step B10.
	No ▶	REPLACE Processor. RERUN Quick Test.

System Check — Pinpoint Test S

S

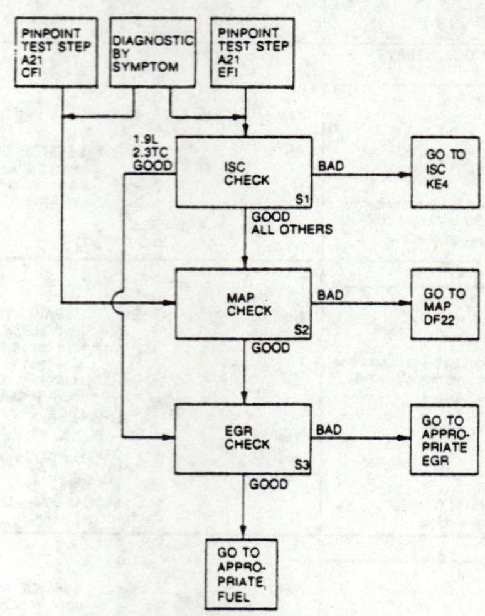

STOP-WARNING

You should enter this Pinpoint Test only after a code 11 is received in Quick Test Step 3.0, and you have been directed here from EEC-IV No-Start Pinpoint Test Step A21 or Diagnostics by symptom.

This Pinpoint Test is intended only as a Quick Check for the basic functioning of the following:

- ISC Bypass Air System
- MAP System
- EGR System

System Check — Pinpoint Test — S

TEST STEP	RESULT	ACTION TO TAKE
S1 ISC-BPA CHECK		
• If you are here for any reason other than stalls or a no start, go to S2. Except 1.9L EFI and 2.3L EFI TC, go to S3. • Attempt to start engine at part throttle. • Will engine run at part throttle?	Yes, but runs rough ▶	GO to S2.
	Yes, and runs smooth ▶	GO to KE4.
	No ▶	1.9 EFI and 2.3L EFI TC GO to S3. All others GO to S2.
S2 MAP CHECK		
• Key Off. • Disconnect the MAP sensor from the vehicle harness. • Connect the MAP tester between the vehicle harness and the MAP sensor. • Plug MAP tester banana plugs into DVOM. • Set DVOM to 20V scale. • Key On. • Observe DVOM.	Yes ▶	3.0L EFI Truck, GO to Pinpoint Test Step HG1. all others, GO to S3.
• Approximate Altitude (Ft.) / Voltage Output (+/−.04V): 0 — 1.59 1000 — 1.56 2000 — 1.53 3000 — 1.50 4000 — 1.47 5000 — 1.44 6000 — 1.41 7000 — 1.39 • If MAP sensor is out of limits (voltage output for altitude) GO to DF21. • Crank engine. • While cranking, does DVOM reading decrease from the appropriate reading for your altitude listed above?	No ▶	GO to DF22.

System Check — Pinpoint Test — S

TEST STEP	RESULT	ACTION TO TAKE
S3 EGR CHECK		
NOTE: This test step will attempt to determine if the EGR system is the cause of the current symptom and/or no start. • Disconnect and plug vacuum line at EGR valve. • Inspect EGR valve to ensure valve is closed. • Start or attempt to start engine. • Is symptom eliminated or is no start resolved?	Yes ▶	For 1.9L EFI & 2.3L EFI TC, GO to KA1. For 3.8L CFI, 2.3L EFI Car and Truck, GO to DD11. For 2.3L HSC CFI, 2.5L HSC CFI, 4.9L EFI, 5.0L EFI, 7.5L EFI Truck and 5.0L SEFI, GO to DN60. For 1.9L CFI, 3.0L EFI Passenger Car, and 2.9L EFI Truck, GO to Section 6, EGR Valve Diagnostics.
	No ▶	For 1.9L EFI and 2.3L EFI TC GO to HA1. For 1.9L CFI, 2.3L HSC CFI and 2.5L HSC CFI, GO to HF1. For 2.9L EFI, 3.0L EFI, and 4.9L EFI, GO to HG1. For 3.8L CFI, GO to HD1. For 5.0L SEFI, GO to HC1. For 2.3L EFI Car and Truck, GO to HE1. For 5.0L EFI, 7.5L EFI Truck, GO to HH1.

Transmission — AXOD — Pinpoint Test — T

T

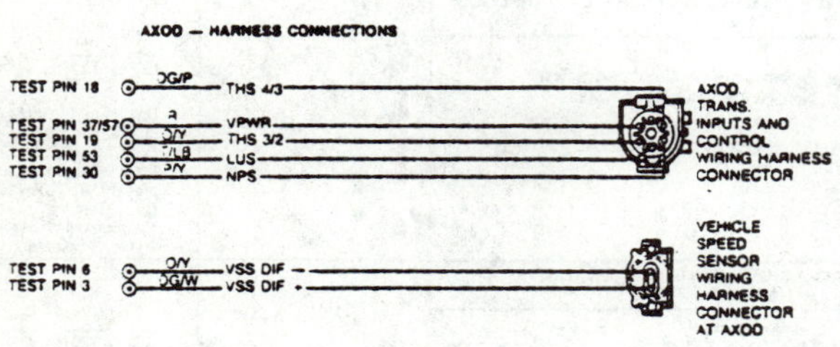

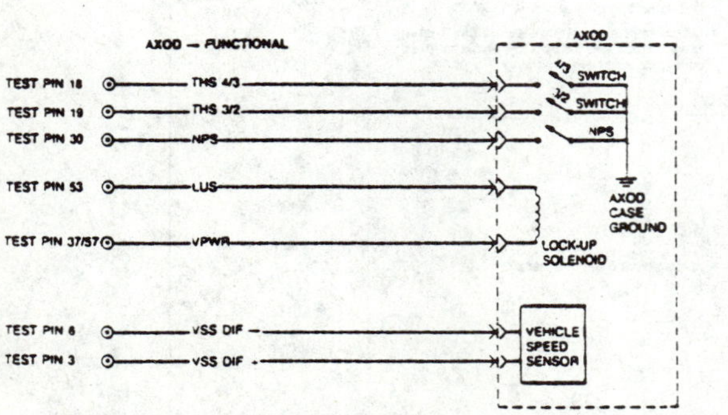

Transmission — AXOD — Pinpoint Test — T

STOP-WARNING

You should enter this Pinpoint Test only when service codes 62, 67 and 89 are received in Quick Test Step 3.0 or a service code 62 is received in Quick Test Step 5.0, and/or service codes 29, 39, 57, 59 and 69 received in Quick Test Step 6.0.

This Pinpoint Test is intended to diagnose only the following:

- Harness Circuits: THS 4/3, THS 3/2, LUS, NPS, VSS+, VSS− and VPWR
- Vehicle Speed Sensor
- Processor Assembly

T1

Drive Cycle For Checking AXOD CONTINUOUS CODES

1. Record and zero EEC-IV Self-Test continuous codes.
2. Warm engine to operating temperature.
3. With transmission in D range, lightly accelerate from a stop to 40 mph to achieve third gear. Hold speed and throttle opening (not closed throttle) steady for 15 seconds minimum (30 seconds above 4000 feet altitude).
4. Shift gear selector to OD range and accelerate lightly from 40 to 50 mph to achieve fourth gear. Hold speed and throttle opening (not closed throttle) steady for 15 seconds minimum in fourth gear.
5. With transmission in fourth gear and steady speed and throttle opening (not closed throttle) lightly apply and release brakes (to light brake lamps) and then hold speed and throttle opening steady for an additional 15 seconds minimum.
6. Brake to a stop and remain stopped for 20 seconds minimum with transmission in OD range.
7. Run EEC-IV Self-Test and record continuous codes.

NOTE: All components must be connected when performing this test.

Transmission — AXOD | Pinpoint Test | T

TEST STEP	RESULT ▶	ACTION TO TAKE
T2 ATTEMPT TO GENERATE CONTINUOUS CODE 29		
• Perform Drive Cycle outlined in Test Step T1, then return to this Step. • Did continuous code 29 repeat?	Yes ▶ No ▶	GO to T3. Unable to duplicate fault at this time. Continuous Code 29. If any other codes are present, RETURN to Quick Test for directions. If none, test is completed.
T3 CONTINUITY OF VEHICLE SPEED SENSOR (VSS) HARNESS		
• Key Off, wait 10 seconds. • Disconnect processor 60 Pin connector and inspect for damaged pins, corrosion, loose wires. Service as necessary. • Install Breakout box. • Processor and VSS disconnected. • DVOM on 200 ohm scale. • Measure resistance between test Pin 3 at the Breakout box and the VSS vehicle harness connector as shown below. • Measure resistance between test Pin 6 at the Breakout box and the VSS vehicle harness connector, as shown below. 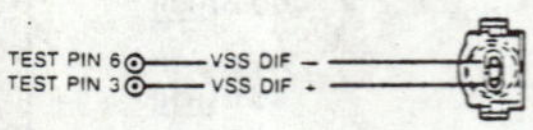 • Are both resistances 5 ohms or less?	Yes ▶ No ▶	GO to T4. SERVICE open(s) in VSS harness. REPEAT Test Step T2.

Transmission — AXOD | Pinpoint Test | T

TEST STEP	RESULT ▶	ACTION TO TAKE
T12 CHECK CONTINUITY OF THS 3/2 HARNESS CIRCUIT		
• Key Off, wait 10 seconds. • Disconnect processor 60 Pin connector and inspect for damaged pins, corrosion, loose wires. Service as necessary. • Install Breakout box. • Processor and AXOD harness disconnected. • DVOM on 200 ohm scale. • Measure resistance between test Pin 19 at the Breakout box and the AXOD vehicle harness connector, as shown below. • Is resistance 5 ohms or less? 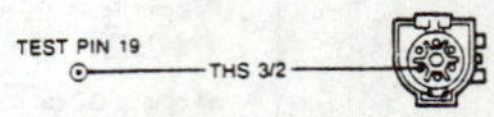 	Yes ▶ No ▶	GO to T13. SERVICE open in THS 3/2 harness. REPEAT Test Step T11.
T13 CHECK THS 3/2 HARNESS CIRCUIT FOR SHORTS TO POWER		
• Key Off. • Breakout box installed. • Processor disconnected. • AXOD harness disconnected. • DVOM on 200,00 ohm scale. • Measure the resistance between test Pin 19 and test Pin 37 at the Breakout box. • Is resistance greater than 10,000 ohms?	Yes ▶ No ▶	GO to T14. SERVICE short to power in THS 3/2 circuit. REPEAT Test Step T11.
T14 PROCESSOR VERIFICATION		
• Key Off. • Breakout box installed. • Processor connected. • AXOD harness connected. • Jumper test Pin 19 to test Pin 40 at the Breakout box. • Run Key On, Engine Off Quick Test. • Is code 62 present?	No ▶ Yes ▶	REMOVE jumper wire. REPLACE processor. REPEAT Test Step T11. REMOVE jumper wire. GO to Taurus/Sable Shop Manual, Section 17-15 for AXOD Transmission Electrical Component Diagnostics.

Transmission — AXOD | Pinpoint Test | T

TEST STEP	RESULT ▶	ACTION TO TAKE
T4 CHECK VSS HARNESS FOR SHORTS TO POWER OR GROUND		
• Key Off. • Processor disconnected. • VSS disconnected. • DVOM on 200,000 ohm scale. • Measure resistance between test Pin 3 and test Pins 37, 40 and 6 at the Breakout box. Then measure resistance between test Pin 6 and test Pin 37 at the Breakout box. • Are all readings greater than 1,000 ohms?	Yes ▶ No ▶	GO to T5. SERVICE short(s) in VSS harness. REPEAT Test Step T2.
T5 SUBSTITUTE VEHICLE SPEED SENSOR (VSS)		
• Substitute VSS with known good sensor. • Processor and VSS connected. • Perform Drive Cycle outlined in Test Step T1, then return to this Step. • Did continuous code 29 repeat?	Yes ▶ No ▶	REPLACE processor. REPEAT Test Step T2. REPLACE VSS. RERUN Quick Test.
T11 ATTEMPT TO GENERATE CONTINUOUS CODE 69		
• Perform Drive Cycle outlined in Test Step T1, then return to this Step. • Did continuous code 69 repeat?	Yes ▶ No ▶	GO to T12. Unable to duplicate fault at this time. Continuous Code 69. If any other codes are present, RETURN to Quick Test for directions. If none, test is completed.

Transmission — AXOD | Pinpoint Test | T

TEST STEP	RESULT ▶	ACTION TO TAKE
T21 ATTEMPT TO GENERATE CONTINUOUS CODE 59		
• Perform Drive Cycle outlined in Test Step T1, then return to this Step. • Did continuous code 59 repeat?	Yes ▶ No ▶	GO to T22. Unable to duplicate fault at this time. Continuous Code 59. If any other codes are present, RETURN to Quick Test for directions. If none, test is completed.
T22 CHECK CONTINUITY OF THS 4/3 HARNESS CIRCUIT		
• Key Off, wait 10 seconds. • Disconnect processor 60 pin connector and inspect for damaged pins, corrosion, loose wires. Service as necessary. • Install Breakout box. • Processor and AXOD harness disconnected. • DVOM on 200 ohm scale. • Measure resistance between test Pin 18 at the Breakout box and the AXOD vehicle harness connector, as shown below. 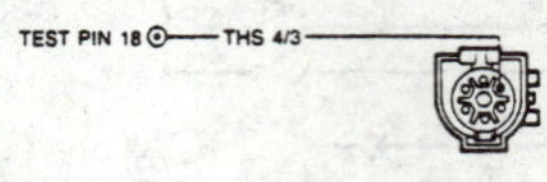 • Is resistance 5 ohms or less?	Yes ▶ No ▶	GO to T23. SERVICE open in THS 4/3 harness. REPEAT Test Step T21.

Transmission — AXOD | Pinpoint Test | T

TEST STEP	RESULT ▶	ACTION TO TAKE
T23 CHECK THS 4/3 HARNESS CIRCUIT FOR SHORT TO POWER		
• Key Off. • Breakout box installed. • Processor disconnected. • AXOD harness disconnected. • DVOM on 200,000 ohm scale. • Measure the resistance between Pin 18 and test Pin 37 at the Breakout box. • Is resistance greater than 10,000 ohms?	Yes ▶	GO to T24.
	No ▶	SERVICE short to power to THS 4/3 circuit. REPEAT Test Step T21.
T24 PROCESSOR VERIFICATION		
• Key Off. • Breakout box installed. • Processor connected. • AXOD harness connected. • Jumper test Pin 18 to test Pin 40 at the Breakout box. • Run Key On Engine Off Quick Test. • Is code 62 present?	No ▶	REMOVE jumper wire. REPLACE processor. REPEAT TEST Step T21.
	Yes ▶	REMOVE jumper wire. GO to Taurus/Sable Shop Manual, Section 17-15 for AXOD Transmission Electrical Component Diagnostics.
T31 ATTEMPT TO GENERATE CONTINUOUS CODE 39		
NOTE: If continuous code 59 is also present, go directly to T21. • Perform Drive Cycle outlined in Test Step T1, then return to this Step. • Did continuous code 39 repeat?	Yes ▶	GO to Taurus/Sable Shop Manual Section 17-15 for AXOD Transmission Electrical Component Diagnostics.
	No ▶	Unable to duplicate fault at this time. Continuous Code 39. If any other codes are present, RETURN to Quick Test for directions. If none, test is completed.

Transmission — AXOD | Pinpoint Test | T

TEST STEP	RESULT ▶	ACTION TO TAKE
T41 ATTEMPT TO GENERATE CONTINUOUS CODE 57		
• Perform Drive Cycle outlined in Test Step T1, then return to this Step. • Did continuous code 57 repeat?	Yes ▶	GO to T42.
	No ▶	Unable to duplicate fault at this time. Continuous Code 57. If any other codes are present, RETURN to Quick Test for directions. If none, test is completed.
T42 CHECK CONTINUITY OF NPS HARNESS CIRCUIT		
• Key Off, wait 10 seconds. • Disconnect processor 60 Pin connector and inspect for damaged pins, corrosion, loose wires, etc. Service as necessary. • Install Breakout box. • Processor and AXOD harness disconnected. • DVOM on 200 ohm scale. • Measure resistance between test Pin 30 at the Breakout box and the AXOD vehicle harness connector, as shown below.	Yes ▶	GO to Taurus/Sable Shop Manual, Section 17-15 for AXOD Transmission Electrical Component Diagnostic.
	No ▶	SERVICE open in NPS circuit. REPEAT Test Step T41.

• Is resistance 5 ohms or less?

Transmission — AXOD | Pinpoint Test | T

TEST STEP	RESULT ▶	ACTION TO TAKE
SERVICE CODE 89		
T51 VPWR HARNESS CONTINUITY CHECK		
• Key Off, wait 10 seconds. • Disconnect processor 60 Pin connector and inspect for damaged pins, corrosion, loose wires. Service as necessary. • Install Breakout box. • Processor and AXOD harness disconnected. • DVOM on 200 ohm scale. • Measure resistance between test Pin 37 at the Breakout box and the AXOD vehicle harness connector, as shown.	Yes ▶	GO to T52.
	No ▶	SERVICE open in VPWR circuit to AXOD. RERUN Quick Test.

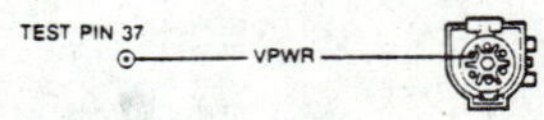

• Is resistance 5 ohms or less?

TEST STEP	RESULT ▶	ACTION TO TAKE
T52 LUS HARNESS CONTINUITY CHECK		
• Key Off. • Breakout box installed. • Processor and AXOD harness disconnected. • DVOM on 200 ohm scale. • Mesasure resistance between test Pin 53 at the Breakout box and the AXOD vehicle harness connector, as shown.	Yes ▶	GO to T53.
	No ▶	SERVICE open in LUS harness circuit to AXOD. RERUN Quick Test.

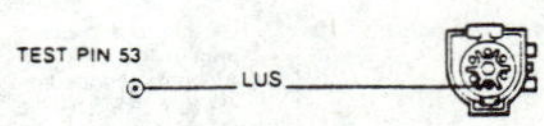

• Is resistance 5 ohms or less?

Transmission — AXOD | Pinpoint Test | T

TEST STEP	RESULT ▶	ACTION TO TAKE
T53 CHECK LUS HARNESS FOR SHORTS TO POWER, GROUND		
• Key Off. • Processor disconnected. • AXOD harness disconnected. • DVOM on 200,000 ohm scale. • Measure resistance between test Pin 53 and test Pins 37 and 40 at the Breakout box. • Are both readings greater than 10,000 ohms?	Yes ▶	GO to T54.
	No ▶	SERVICE shorts in LUS harness. RERUN Quick Test. If code 89 is still present, REPLACE processor. RERUN Quick Test.
T54 TOTAL CIRCUIT RESISTANCE CHECK		
• Key Off. • Breakout box installed. • Processor disconnected. • AXOD transmission harness connected. • DVOM on 200 ohm scale. • Measure the resistance between test Pin 53 and test Pin 57 at the Breakout box. • Is resistance between 20 ohms and 40 ohms?	Yes ▶	REPLACE processor. RERUN Quick Test.
	No ▶	GO to Taurus/Sable Shop Manual, Section 17-15 for AXOD Transmission Electrical Component Diagnostic.
SERVICE CODE 62: KEY ON, ENGINE OFF QUICK TEST		
T61 CHECK FOR CODE 62 IN ENGINE RUNNING SELF-TEST		
• Perform Engine Running Self-Test and record codes. • Is code 62 present?	Yes ▶	Fault is in the 4/3 circuit. GO to T62.
	No ▶	Fault is in the 3/2 circuit. GO to T62.

Transmission — AXOD | Pinpoint Test | T

TEST STEP	RESULT ▶	ACTION TO TAKE
SERVICE CODE 62: KEY ON, ENGINE OFF QUICK TEST		
T62 AXOD HARNESS VERIFICATION		
• Key Off. • Disconnect AXOD harness. • Run Key On Engine Off Quick Test. • Is code 62 still present?	Yes ▶	GO to T63.
	No ▶	GO to Taurus/Sable Shop Manual, Section 17-15 for AXOD Transmission Electrical Component Diagnostics.
T63 CHECK THS 3/2 AND 4/3 HARNESS CIRCUITS FOR SHORT TO GROUND		
• Key Off. • Install Breakout box and disconnect processor. • AXOD harness disconnected. • DVOM on 200,000 ohm scale. • Measure resistance between test Pin 18 and test Pins 40 and 60 at the Breakout box. • Measure resistance between test Pin 19 and test Pins 40 and 60 at the Breakout box. • Are all resistance readings greater than 10,000 ohms?	Yes ▶	REPLACE processor. RERUN Quick Test.
	No ▶	SERVICE short(s) to ground. RERUN Quick Test.
T71 SERVICE CODE 62: ENGINE RUNNING SELF-TEST		
• Verify code 62 present in Engine Running Self-Test.	Code 62 present ▶	GO to Taurus/Sable Shop Manual, Section 17-15 for AXOD Transmission Electrical Component Diagnostics.
	Code 62 not present ▶	Unable to duplicate fault at this time. If any other codes are present, RETURN to Quick Test for directions. If none, then test is complete.

Transmission — AXOD | Pinpoint Test | T

TEST STEP	RESULT ▶	ACTION TO TAKE
T81 SERVICE CODE 67 — CHECK VOLTAGE AT NPS INPUT TO PROCESSOR		
• Key On Engine Off. • Disconnect processor 60 Pin connector and inspect for damaged pins, corrosion, loose wires. Service as necessary. • Install Breakout box and reconnect processor. • DVOM on 20V scale. • Measure voltage between test Pin 30 and test Pin 46 at the Breakout box. • Is voltage less than 4V?	Yes ▶	GO to T82.
	No ▶	GO to Pinpoint Test FA4.
T82 CHECK NPS HARNESS CIRCUIT FOR SHORT TO GROUND		
• Key Off. • Breakout box installed. • Processor and AXOD harness disconnected. • DVOM on 200,000 ohm scale. • Measure resistance between test Pin 30 and test Pins 40 and 60 at the Breakout box. • Are all readings greater than 10,000 ohms?	Yes ▶	GO to T83.
	No ▶	SERVICE short to ground in NPS circuit. RERUN Quick Test.

Transmission — AXOD | Pinpoint Test | T

TEST STEP	RESULT ▶	ACTION TO TAKE
T83 PROCESSOR VERIFICATION		
• Key Off. • Breakout box installed. • Processor connected. • AXOD harness disconnected. • Run Key On Engine Off Quick Test. • Is code 67 present?	Yes ▶	REPLACE processor. RERUN Quick Test.
	No ▶	GO to Taurus/Sable Shop Manual, Section 17-15 for AXOD Transmission Electrical Component Diagnostics.

Transmission — VSS | Pinpoint Test | T

TEST STEP	RESULT ▶	ACTION TO TAKE
T90 SENSOR CHECK		
• Key Off, wait 10 seconds. • Locate, and disconnect vehicle speed sensor. • DVOM on 2,000 ohm scale. • Measure resistance across vehicle speed sensor. • Is resistance between 190 and 230 ohms?	Yes ▶	GO to T91.
	No ▶	REPLACE sensor, verify drive complaint was eliminated.
T91 CONTINUITY OF VEHICLE SPEED SENSOR (VSS) HARNESS		
• Key Off, wait 10 seconds. • Disconnect processor 60 Pin connector and inspect for damaged pins, corrosion, loose wires. Service as necessary. • Install Breakout box. • Processor and VSS disconnected. • DVOM on 200 ohm scale. • Measure resistance between test Pin 3 at the Breakout box and the VSS vehicle harness connector as shown below. • Measure resistance between test Pin 6 at the Breakout box and the VSS vehicle harness connector, as shown below. 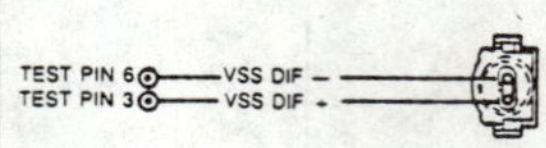 • Are both resistances 5 ohms or less?	Yes ▶	GO to T92.
	No ▶	SERVICE open(s) in VSS harness.

Transmission — VSS | Pinpoint Test | T

TEST STEP	RESULT	▶	ACTION TO TAKE
T92 CHECK VSS HARNESS FOR SHORTS TO POWER OR GROUND			
• Key Off. • Processor disconnected. • VSS disconnected. • DVOM on 200,000 ohm scale. • Measure resistance between test Pin 3 and test Pins 37, 40 and 6 at the Breakout box. Then measure resistance between test Pin 6 and test Pin 37 at the Breakout box. • Are all readings greater than 1,000 ohms?	Yes No	▶ ▶	GO to T93. SERVICE short(s) in VSS harness.
T93 SUBSTITUTE VEHICLE SPEED SENSOR (VSS)			
• Substitute VSS with known good sensor. • Processor and VSS connected. • Drive vehicle to verify drive complaint. • Was drive complaint eliminated?	Yes No	▶ ▶	REPLACE VSS. REPLACE processor.

Integrated Relay Controller Module | Pinpoint Test | X

X

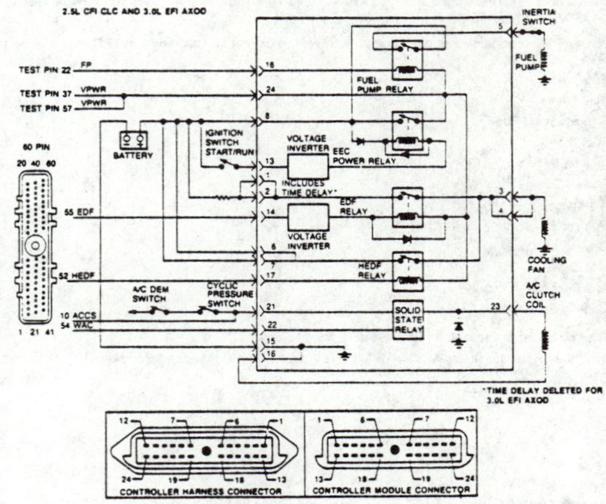

Refer to Engine Supplement for Test Pin Usage Chart.

Integrated Relay Controller Module | Pinpoint Test | X

X

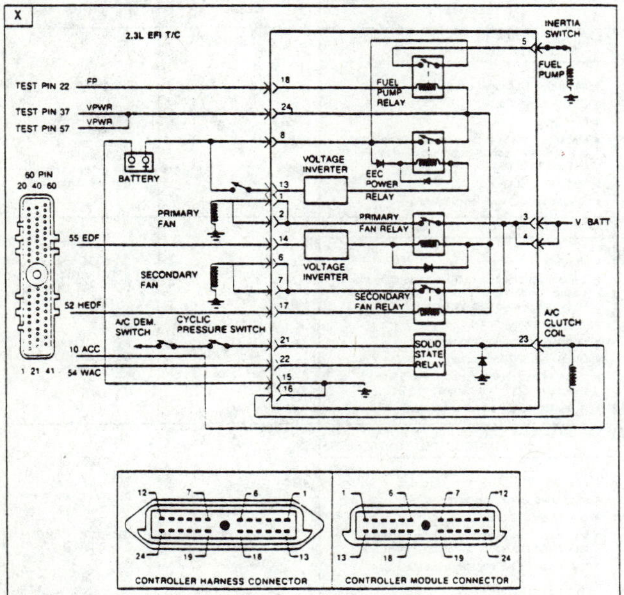

Refer to Engine Supplement for Test Pin Usage Chart.

Integrated Relay Controller Module | Pinpoint Test | X

X

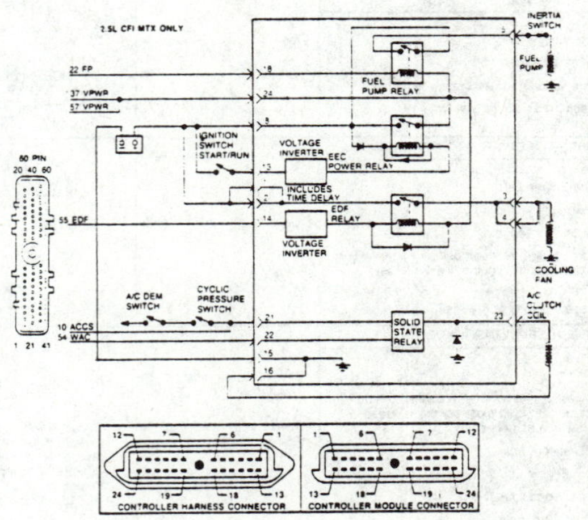

Refer to Engine Supplement for Test Pin Usage Chart.

Integrated Relay Controller Module — Pinpoint Test X

STOP-WARNING

You should enter this Pinpoint Test only when service code 72, 78, 82, 83, 87 and 88 are received in Quick Test Step 3.0 or you are directed here from Pinpoint Test A, Pinpoint Test C, or Diagnostic By Symptom.

To prevent the replacement of good components, be aware that the following non-EEC area may be at fault:

- Fuel Lines
- Fuel Filters
- Contaminated Fuel
- Fuel Pump
- Ignition Switch
- Battery Cables
- Alternator
- Voltage Regulator
- Ground Straps
- A/C Clutch
- A/C Demand
- Cooling Fan Motor

This pinpoint test is intended to diagnose only the following:

- Integrated Relay Controller Module (within this Pinpoint Named Integrated Controller)
- Battery Voltage
- Power Relay
- EDF Relay
- HEDF Relay
- WAC Relay
- Fuel Pump Relay
- Harness Circuits: V Batt., VPWR, F.P., Ground and Power to Fuel Pump(s), WAC, ACC, ACCS, Cooling Fan Power, A/C Clutch, Key Power, Powers To Integrated Controller
- Processor Assembly
- A/C Demand Switch Input

Integrated Controller — Pinpoint Test X

TEST STEP	RESULT	ACTION TO TAKE
VEHICLE BATTERY		
X1 BATTERY VOLTAGE CHECK		
• Key On Engine Off. • DVOM on 20V scale. • Measure voltage across battery terminals. • Is voltage 10.5 volts or greater?	Yes ▶	GO to X2.
	No ▶	SERVICE discharged battery, REFER to Shop Manual, Group 31.
X2 BATTERY POWER GROUND CHECK		
• Key On Engine Off. • Processor connected. • DVOM on 20V range. • Measure voltage between battery negative post and Signal Return circuit in the Self-Test connector. • Is voltage 0.5 volts or greater?	Yes ▶	GO to X3.
	No ▶	GO to X6.
X3 GROUND FAULT ISOLATION		
• Key Off. • Disconnect processor 60 Pin connector and inspect for damaged pins, corrosion, loose wires. Service as necessary. • Install Breakout box. • Key On Engine Off. • Processor connected. • DVOM on 20V scale. • Measure voltage between battery negative post and test Pins 40 and 60 at the Breakout box.	Both readings less than 0.5V ▶	GO to X4.
	One or both readings 0.5V or greater ▶	Circuit(s) with greater than 0.5V has high resistance or open. CORRECT faulty ground circuit. RERUN Quick Test.

Integrated Controller — Pinpoint Test X

TEST STEP	RESULT	ACTION TO TAKE
X4 PROCESSOR GROUND FAULT ISOLATION		
• Breakout box installed. • Key Off, wait 10 seconds. • Processor connected. • DVOM on 200 ohm scale. • Measure resistance between test Pin 46 and test Pin 40 and between test Pin 46 and test Pin 60.	Both reading less than 5 ohms ▶	GO to X5.
	One or both readings 5 ohms or greater ▶	DISCONNECT processor connector and INSPECT for corrosion, damaged pins, etc. SERVICE as necessary and RETEST. If fault is still present, REPLACE processor. RERUN Quick Test.
X5 HARNESS CHECK (SIGNAL RETURN)		
• Breakout box installed. • Key Off, wait 10 seconds. • Processor connected. • DVOM on 200 ohm scale. • Measure resistance between Pin 46 and Signal Return circuit at Self-Test connector. • Is resistance less than 5.0 ohms?	Yes ▶	System OK. RUN Quick Test.
	No ▶	CORRECT cause of resistance in the harness Signal Return circuit. RERUN Quick Test.
X6 MEASURE VOLTAGE AND GROUND TO INTEGRATED CONTROLLER		
• Key Off. • Integrated Controller Module disconnected. • DVOM on 20V scale. • Measure voltage at the harness connector to the integrated cntroller from connector Pin 8 (−) to Pin 15 (−). (Refer to illustration X.) • Is voltage 10.5 volts or greater?	Yes ▶	GO to X7.
	No ▶	GO to X9.

Integrated Controller — Pinpoint Test X

TEST STEP	RESULT	ACTION TO TAKE
X7 KEY POWER TO INTEGRATED CONTROLLER		
• Integrated Controller disconnected. • DVOM on 20V scale. • Key On. • Measure voltage at the harness connector to the integrated controller from connector Pin 13 (−) to Pin 15 (−). • Refer to illustration X. • Is voltage 10.5 volts or greater?	Yes ▶	GO to X8.
	No ▶	SERVICE open between 13 and ignition switch. RECONNECT Integrated Controller. RERUN Quick Test.
X8 MEASURE CONTINUITY OF VPWR		
• Key Off. • Disconnect processor 60 Pin connector and inspect for damaged pins, corrosion, loose wires, etc. Service as necessary. • Install Breakout box • Leave processor disconnected. • DVOM on 200 ohm scale. • Measure resistance from test Pin 37 and 57 to Integrated Controller harness Pin 24. • Is resistance 5.0 ohms or greater?	Yes ▶	SERVICE open in VPWR circuit. CONNECT Integrated Controller. RERUN Quick Test.
	No ▶	REPLACE Integrated Controller. RERUN Quick Test.
X9 MEASURE CONTINUITY OF POWER GROUND TO INTEGRATED CONTROLLER		
• Key Off. • Integrated Controller disconnected. • DVOM on 200 ohm scale. • Measure resistance from battery negative to Integrated Controller connector Pin 15. • Is resistance 5.0 ohms ore greater?	Yes ▶	SERVICE open in battery ground to Pin 15 (Integrated Controller harness connector). RERUN Quick Test.
	No ▶	SERVICE open in battery positive to Pin 8 (Integrated Controller harness connector). RERUN Quick Test.

Integrated Controller — Pinpoint Test X

TEST STEP	RESULT ▶	ACTION TO TAKE
X10 CODE 72 OR 78: INTERMITTENT OPEN IN V POWER CIRCUIT		
NOTE: Code 72 or 78 indicates that while key power was present, V power had an interrupt, resulting in possible stalls, quits on road, or other drive symptoms. Possible Causes: • Intermittent open in V power circuit from integrated controller to processor. • EEC power relay intermittent malfunction. • V battery intermittent open to integrated controller. • Intermittent open in key power circuit to integrated controller. • Using continuous monitor mode (Key On Engine Off) per Quick Step 6D, observe VOM or STAR LED for indication of a fault while performing the following: • Shake, bend and twist harness from integrated controller to the processor, to the ignition switch and to battery positive. • Is a fault indicated or does code 72 or 78 reappear in continuous memory if quick test is rerun?	Yes ▶	SERVICE intermittent V power circuit. RERUN Quick Test.
	No ▶	INSPECT component and harness connectors of integrated controller and processor, for loose or damaged pins, corrosion, etc. SERVICE as necessary. If OK, REPLACE integrated controller. RERUN Quick Test.
X11 POWER AT FUEL PUMP(S) CHECK		
• Key On Engine Off. • Locate fuel pump(s). • DVOM on 20V scale. • Measure voltage between chassis ground and power-to-pump(s) circuit at fuel pump during crank mode. • Is voltage 8.0V or greater during crank?	Yes ▶	GO to Shop Manual, Group 24, Electric Fuel Pump Diagnosis.
	No ▶	GO to X12.
X12 RESISTANCE CHECK OF FUEL PUMP INERTIA SWITCH		
• Key Off, wait 10 seconds. • Locate and disconnect fuel pump inertia switch. • DVOM on 200 ohm scale. • Measure the resistance of the fuel pump inertia switch. • Is resistance less than 5.0 ohms?	Yes ▶	GO to X13.
	No ▶	REPLACE fuel pump inertia switch. RERUN Quick Test.

Integrated Controller — Pinpoint Test X

TEST STEP	RESULT ▶	ACTION TO TAKE
X13 POWER-TO-PUMP CIRCUIT CONTINUITY CHECK		
• Key Off. • DVOM on 200 ohm scale. • Integrated Controller disconnected. • Pump(s) disconnected. • Measure resistance from Integrated Controller harness connector Pin 5 to fuel pump(s) harness connector at fuel pump. • Is resistance less than 5.0 ohms?	Yes ▶	REPLACE Integrated Controller. RECONNECT all components. RERUN Quick Test.
	No ▶	SERVICE open in power-to-pump(s) circuit. RECONNECT Integrated Controller. RERUN Quick Test.
X14 CHECK POWER-TO-PUMP(S) FOR SHORTS TO POWER		
• Key Off. • Integrated Controller disconnected. • Fuel pumps disconnected. • DVOM on 200,000 ohm scale. • Measure resistance from Integrated Controller harness connector Pin 5 to Pin 24 (VPWR) and battery + post.	Any readings less than 10,000 ohms ▶	SERVICE short circuit. RECONNECT all components. ATTEMPT to start vehicle. If vehicle runs, RERUN Quick Test. If vehicle will not run, REPLACE Integrated Controller. RERUN Quick Test.
	All readings 10,000 ohms or greater ▶	REPLACE Integrated Controller. RERUN Quick Test.
SERVICE CODE: 87		
X15 FUEL PUMP CIRCUIT CONTINUITY		
• Key Off. • Disconnect processor 60 Pin connector and inspect for damaged pins, corrosion, loose wires, etc. Service as necessary. • Install Breakout box. • Processor disconnected. • Integrated Controller disconnected. • DVOM on 200 ohm scale. • Measure resistance of fuel pump circuit, from test Pin 22 at the Breakout box to Integrated Controller harness connector Pin 18. • Is resistance less than 5.0 ohms?	Yes ▶	GO to X16.
	No ▶	SERVICE open in fuel pump circuit. RECONNECT processor and controller. RERUN Quick Test.

Integrated Controller — Pinpoint Test X

TEST STEP	RESULT ▶	ACTION TO TAKE
X16 CHECK FUEL PUMP CIRCUIT FOR SHORTS TO POWER AND GROUND		
• Key Off. • Breakout box installed. • Processor disconnected. • Integrated Controller disconnected. • DVOM on 200,000 ohm scale. • Measure resistance from test Pin 22 to test Pins 37, 57 and battery positive post and from test Pin 22 to test Pins 40, 60 and battery negative.	All readings 10,000 ohms or greater ▶	GO to X17.
	Any readings less than 10,000 ohms ▶	SERVICE fuel pump circuit shorts to power or ground. RECONNECT all components. RERUN Quick Test. If code 87 is still present, GO to X17.
X17 CHECK RESISTANCE OF FUEL PUMP RELAY COIL		
• Key Off. • Integrated Controller disconnected. • DVOM on 200 ohm scale. • Measure resistance of Integrated Controller, from Pin 18 to 24. • Is resistance between 65 and 100 ohms?	Yes ▶	REPLACE processor. RECONNECT component. RERUN Quick Test.
	No ▶	REPLACE Integrated Controller. RECONNECT components. RERUN Quick Test.
X20 NO FAN, HIGH OR LOW WITH NO CODE 83		
• Key Off. • Integrated Controller disconnected. • DVOM on 20V scale. • Measure voltage between battery negative post and Pins 1, 2, 6 and 7, respectively at the Integrated Controller harness connector.	All readings 10.5V or greater ▶	GO to X21.
	Any reading less than 10.5V ▶	SERVICE open in battery power circuit. RE-EVALUATE symptom.
X21 FAN MOTOR CHECK		
• Key Off. • Integrated Controller disconnected. • Jumper Pin 3 to Pin 6 at Integrated Controller harness. • Does fan run?	Yes ▶	GO to X22.
	No ▶	GO to X23.

Integrated Controller — Pinpoint Test X

TEST STEP	RESULT ▶	ACTION TO TAKE
X22 CHECK FAN RUNNING MODE (LOW)		
• Key Off. • Disconnect processor. • Integrated Controller connected. • Key On. • Does fan run at low speed?	Yes ▶	GO to X25.
	No ▶	CHANGE Integrated Controller. RECONNECT processor and controller. RE-EVALUATE symptom.
X23 MEASURE BATTERY VOLTAGE SUPPLY AT FAN — BYPASSING INTEGRATED CONTROLLER		
• Key Off. • Cooling fan disconnected. • Integrated Controller disconnected. • Jumper Pin 3 to Pin 6 at Integrated Controller harness connector. • DVOM on 20V scale. • Measure voltage at cooling fan harness connector. • Is voltage 8.0 volts or greater?	Yes ▶	CHANGE fan motor. RE-EVALUATE symptom.
	No ▶	GO to X24.
X24 COOLING FAN GROUND VERIFICATION		
• Key Off. • Cooling fan disconnected. • Integrated Controller disconnected. • Jumper Pin 3 to Pin 6 at Integrated Controller harness connector. • DVOM on 20V scale. • Measure voltage between voltage positive at cooling fan harness connector and negative battery post. • Is voltage 8.0 volts or greater?	Yes ▶	SERVICE Open in ground circuit to fan. RECONNECT Integrated Controller. RE-EVALUATE symptom.
	No ▶	SERVICE open in power-to-fan circuit from 3 and 4 of Integrated Controller harness connector to cooling fan connector. RECONNECT controller. RE-EVALUATE symptom.

Integrated Controller — Pinpoint Test — X

TEST STEP	RESULT	ACTION TO TAKE
X25 JUMPER HIGH ELECTRIC-DRIVE SIGNAL (HEDF) TO GROUND		
• Key Off. • Disconnect processor 60 Pin connector and inspect for damaged pins, corrosion, loose wires. Service as necessary. • Install Breakout box. • Integrated Controller connected. • Processor disconnected. • Key On. • Jumper test Pin 52 to test Pin 40 at Breakout box. • Does fan speed change from low to high?	Yes ▶	GO to X26.
	No ▶	REPLACE Integrated Controller. RECONNECT processor. RE-EVALUATE symptom.
X26 ECT SENSOR CHECK		
• Reconnect processor. • Check engine coolant level. • Warm engine to operating temperature before taking ECT resistance measurement. • Key Off, wait 10 seconds. • Harness disconnected from ECT sensor. • DVOM on 200,000 ohm scale. • Measure resistance of the ECT sensor. • Is the resistance reading between 1500 ohms and 2000 ohms?	Yes ▶	REPLACE processor. RECONNECT harness to ECT sensor. RECONNECT Integrated Controllers. RE-EVALUATE symptom.
	No ▶	REPLACE ECT sensor. RECONNECT all components. RE-EVALUATE symptom.
X30 SERVICE CODE 83: RESISTANCE CHECK OF HEDF CONTROLLER CIRCUIT		
• Key Off. • Integrated Controller disconnected. • DVOM on 200 ohm scale. • Measure resistance between Pin 17 and Pin 24 at the Integrated Controller. • Is the resistance reading between 50 ohms and 100 ohms?	Yes ▶	GO to X31.
	No ▶	REPLACE controller. RERUN Quick Test.

Integrated Controller — Pinpoint Test — X

TEST STEP	RESULT	ACTION TO TAKE
X31 CHECK HEDF PROCESSOR SIGNAL TO INTEGRATED CONTROLLER FOR OPEN		
• Key Off. • Disconnect processor 60 Pin connector and inspect for damaged pins, corrosion, loose wires. Service as necessary. • Install Breakout box, leave processor disconnected. • Integrated Controller disconnected. • DVOM On 200 ohms scale. • Measure continuity from test Pin 52 at Breakout box to Pin 17 of Integrated Controller harness. • Is resistance 5 ohms or less?	Yes ▶	GO to X32.
	No ▶	SERVICE open in HEDF circuit. RECONNECT all components. RERUN Quick Test.
X32 CHECK FOR SHORTS TO GROUND IN THE HEDF CIRCUIT		
• Key Off. • Breakout box installed. • Processor and Integrated Controller disconnected. • DVOM on 200,000 ohm scale. • Measure resistance from test Pin 52 to test Pin 40. • Is resistance 10,000 ohms or less?	No ▶	GO to X33.
	Yes ▶	SERVICE short to ground in HEDF circuit. RERUN Quick Test.
X33 CHECK FOR SHORTS TO POWER IN THE HEDF CIRCUIT		
• Key Off. • Breakout box installed. • Processor and Integrated Controller disconnected. • DVOM on 200,000 ohms scale. • Measure resistance from test Pin 52 to test Pin 37. • Is resistance 10,000 ohms or less?	No ▶	REPLACE Processor. RECONNECT all components. RERUN Quick Test.
	Yes ▶	SERVICE short to power. RECONNECT all components. RERUN Quick Test. If code 83 is still present, REPLACE processor. RERUN Quick Test.

Integrated Controller — Pinpoint Test — X

TEST STEP	RESULT	ACTION TO TAKE
X35 LOW SPEED OR HIGH SPEED FAN ALWAYS "ON", NO SERVICE CODE 83 OR 67		
• Key Off. • Disconnect processor 60 Pin connector and inspect for damaged pins, corrosion, loose wires. Service as necessary. • Breakout box installed, leave processor disconnected. • Disconnect the Integrated Controller. • DVOM on 200 ohm scale. • Measure the resistance between test Pin 55 and controller harness Pin 14. • Is reading 5 ohms or less?	Yes ▶	GO to X36.
	No ▶	SERVICE open in EDF circuit. RECONNECT all components. RE-EVALUATE symptom.
X36 CHECK EDF CIRCUIT FOR SHORTS TO POWER		
• Key Off. • Breakout box installed. • Processor and Integrated Controller disconnected. • DVOM on 200,000 ohm scale. • Measure resistance from test Pin 55 to test Pin 37 and from test Pin 55 to battery positive. • Is resistance 10,000 ohms or less?	Yes ▶	SERVICE short to power in EDF circuit. GO to X37.
	No ▶	GO to X37.
X37 EDF SHORT TO GROUND CHECK		
• Key On. • Breakout box installed. • Processor disconnected. • Integrated Controller connected. • Jumper test Pin 55 to ground test Pin 40 or 60. • Does fan continue to run?	Yes ▶	REPLACE controller RE-EVALUATE symptom.
	No ▶	REPLACE processor. RE-EVALUATE symptom.

Integrated Controller — Pinpoint Test — X

TEST STEP	RESULT	ACTION TO TAKE
X40 NO FAN		
• Key Off. • Integrated Controller disconnected. • DVOM on 20V scale. • Measure voltage between battery negative post and Pin 1 and Pin 2, respectively at the Integrated Controller harness connector. • Is voltage 10.5 volts or greater?	Yes ▶	GO to X41.
	No ▶	SERVICE open in battery power circuit. RE-EVALUATE symptom.
X41 FAN MOTOR CHECK		
• Key Off. • Integrated Controller disconnected. • Jumper Pin 1 to Pin 3 at Integrated Controller harness. • Does fan run?	Yes ▶	GO to X42.
	No ▶	GO to X43.
X42 CHECK FAN RUNNING MODE		
• Key Off. • Disconnect processor. • Integrated Controller connected. • Key On. • Does fan run?	Yes ▶	GO to X46.
	No ▶	GO to X44.
X43 MEASURE BATTERY VOLTAGE SUPPLY AT FAN — BYPASSING INTEGRATED CONTROLLER		
• Key Off • Cooling fan disconnected. • Integrated Controller disconnected. • Jumper Pin 1 to Pin 3 at Integrated Controller harness connector. • DVOM on 20V scale. • Measure voltage at cooling fan harness connector. • Is voltage 8.0 volts or greater?	Yes ▶	CHANGE fan. RE-EVALUATE symptom.
	No ▶	GO to X45.

Integrated Controller — Pinpoint Test X

TEST STEP	RESULT ▶	ACTION TO TAKE
X44 CHECK EDF CIRCUIT FOR SHORT TO GROUND		
• Key Off. • Processor and controller disconnected. • DVOM on 200,000 ohm scale. • Measure resistance from Pin 14 to Pin 15 at Integrated Controller harness connector. • Is resistance 10,000 ohms or greater?	Yes ▶	CHANGE Integrated Controller. RECONNECT processor and controller. RE-EVALUATE symptom.
	No ▶	SERVICE short to ground in EDF circuit. RECONNECT processor and Integrated Controller. RE-EVALUATE symptom.
X45 COOLING FAN GROUND VERIFICATION		
• Key Off. • Cooling fan disconnected. • Integrated Controller disconnected. • Jumper Pin 1 to Pin 3 at Integrated Controller harness connector. • DVOM on 20V scale. • Measure voltage between voltage positive at cooling fan harness connector and negative battery post. • Is voltage 8.0 volts or greater?	Yes ▶	SERVICE open in ground circuit to fan. RECONNECT Integrated Controller. RE-EVALUATE symptom.
	No ▶	SERVICE open in power-to-fan circuit from 3 and 4 of Integrated Controller harness connector to cooling fan connector. RECONNECT controller. RE-EVALUATE symptom.
X46 ECT SENSOR CHECK		
• Reconnect processor. • Check engine coolant level. • Warm engine to operating temperature before taking ECT resistance measurement. • Key Off, wait 10 seconds. • Harness disconnected from ECT sensor. • DVOM on 200,000 ohm scale. • Measure resistance of the ECT sensor. • Is the resistance reading between 1500 ohms and 2000 ohms?	Yes ▶	REPLACE processor. RECONNECT harness to ECT sensor. RECONNECT Integrated Controller. RE-EVALUATE symptom.
	No ▶	REPLACE ECT sensor. RECONNECT all components. RE-EVALUATE symptom.

Integrated Controller — Pinpoint Test X

TEST STEP	RESULT ▶	ACTION TO TAKE
X50 CHECK FOR VOLTAGE AT A/C CLUTCH		
• Key On Engine Off. • A/C demand to A/C. • DVOM on 20V scale. • Check voltage at A/C clutch harness connector. • Is voltage 10.5 volts or greater?	Yes ▶	GO to Shop Manual, Group 36, A/C Diagnosis.
	No ▶	GO to X51.
X51 CHECK FOR CONTINUITY FROM INTEGRATED CONTROLLER TO A/C CLUTCH		
• Key Off. • Integrated Controller disconnected. • DVOM on 200 ohm scale. • Measure resistance between Pin 23 of the controller harness and power side of the A/C clutch harness connector and between Pin 16 of the controller harness and ground side of the A/C clutch harness connector. • Are both resistance readings 5 ohms or less?	Yes ▶	GO to X52.
	No ▶	SERVICE open in power to A/C clutch or ground from A/C clutch. RE-EVALUATE symptom.
X52 ENTER OUTPUT STATE CHECK (REFER TO APPENDIX)		
NOTE: Do not use STAR tester for this Step, use VOM/DVOM. • Key Off, wait 10 seconds. • Disconnect processor 60 pin connector and inspect for damaged pins, corrosion, loose wires. Service as necessary. • Connect Breakout box to harness. Reconnect processor. • DVOM on 20V scale. • Connect DVOM negative test lead to STO and positive test lead to battery positive. • Jumper STI to signal return. • Perform Key On Engine Off Self-Test until the completion of the Continuous Test Codes. • DVOM will indicate zero volts. • Depress and release the throttle. • Did DVOM reading change to a high voltage reading?	Yes ▶	REMAIN in Output State Check. GO to X53.
	No ▶	DEPRESS throttle to WOT and RELEASE. If STO voltage does not go high, GO to Pinpoint Test Step Q40. LEAVE equipment hooked up.

Integrated Controller — Pinpoint Test X

TEST STEP	RESULT ▶	ACTION TO TAKE
X53 CHECK WAC OUTPUT FOR PROPER ELECTRICAL OPERATION		
• Key On engine Off. • A/C demand switch to A/C on position. • DVOM on 20V scale. • Connect DVOM positive test lead to test Pin 37 and negative test lead to test Pin 54. • While observing DVOM, depress and release the throttle several times (to cycle output on and off). • Does voltage output cycle high and low?	Yes ▶	2.3L EFI T/C only, GO to X61. All others, GO to X54.
	No ▶	GO to X57.
X54 CHECK FOR VOLTAGE AT A/C CLUTCH SWITCH		
• Key On Engine Off. • A/C demand switch to A/C on position. • DVOM on 20V scale. • Leave Breakout box installed. • Processor and Integrated Controller connected. • Measure voltage between test Pin 10 and test Pin 40 at Breakout box. • Is voltage 10.5V or greater?	Yes ▶	GO to X55.
	No ▶	GO to X56.
X55 CHECK CONTINUITY OF ACCS TO INTEGRATED CONTROLLER		
• Key Off, wait 10 seconds. • Processor disconnected. • Integrated Controller disconnected. • DVOM on 200 ohm scale. • Measure resistance between test Pin 10 at Breakout box and Pin 21 at controller harness connector. • Is resistance 5 ohms or less?	Yes ▶	REPLACE Integrated Controller. RE-EVALUATE symptom.
	No ▶	SERVICE open in ACCS circuit. RE-EVALUATE symptom.

Integrated Controller — Pinpoint Test X

TEST STEP	RESULT ▶	ACTION TO TAKE
X56 CHECK CONTINUITY OF ACCS CIRCUIT		
• Key Off, wait 10 seconds. • A/C demand to A/C. • DVOM on 200 ohm scale. • Measure resistance between test Pin 10 and A/C demand switch. • Is resistance 5 ohms or less?	No ▶	SERVICE open in circuit. RERUN Quick Test.
	Yes ▶	EEC-IV system OK. REFER to Shop Manual, Group 36 A/C Diagnosis.
X57 CHECK CONTINUITY IN WAC TO INTEGRATED CONTROLLER CIRCUIT		
• Key Off, wait 10 seconds. • Disconnect processor 60 Pin connector and inspect for damaged pins, corrosion, loose wires. Service as necessary. • Connect Breakout box to harness. Leave processor and Integrated Controller disconnected. • DVOM on 200 ohm scale. • Measure resistance between test Pin 54 and Pin 22 at Integrated Controller harness. • Is resistance 50 ohms or less?	No ▶	SERVICE open in WAC circuit. RE-EVALUATE symptom.
	Yes ▶	GO to X58.
X58 CHECK WAC CIRCUIT FOR SHORTS TO GROUND		
• Key Off, wait 10 seconds. • Leave Breakout box installed and processor disconnected. • Integrated Controller disconnected. • DVOM on 200,000 ohm scale. • Measure resistance between test Pin 54 and test Pin 40 and between test Pin 54 and test Pin 46 and between test Pin 54 and battery negative post. • Are all resistance readings 10,000 ohms or greater?	Yes ▶	GO to X59.
	No ▶	SERVICE shorts to ground in WAC circuit. RE-EVALUATE symptom.

Integrated Controller — Pinpoint Test — X

TEST STEP	RESULT ▶	ACTION TO TAKE
X59 CHECK WAC CIRCUIT FOR SHORTS TO POWER		
• Key Off, wait 10 seconds. • Leave Breakout box installed and processor disconnected. • Integrated Controller disconnected. • DVOM on 200,000 ohm scale. • Measure resistance between test Pin 54 and test Pin 37 and between test Pin 54 and battery positive. • Are both readings 10,000 ohms or greater?	Yes ▶ No ▶	GO to X60. SERVICE short to power in WAC circuit and then GO to X60.
X60 CHECK FOR VOLTAGE AT A/C CLUTCH		
• Key Off, wait 10 seconds. • Leave Breakout box installed. • Processor disconnected. • Integrated Controller connected. • A/C clutch disconnected. • A/C demand to A/C. • Key On Engine Off. • DVOM on 20V scale. • Measure voltage at A/C clutch harness connection. • 10.5V or greater?	Yes ▶ No ▶	REPLACE processor. RE-EVALUATE symptom. REPLACE Integrated Controller. RE-EVALUATE symptom.
X61 CHECK FOR VOLTAGE AT A/C INPUT TO CONTROLLER		
• Key Off, wait 10 seconds. • Leave Breakout box installed. • Processor connected. • Integrated controller disconnected. • Key On Engine off. • A/C demand switch to A/C position. • DVOM on 20V scale. • Measure voltage between Pin 21 at controller harness connector and test Pin 40. • Is voltage 10.5V or greater?	Yes ▶ No ▶	REPLACE Integrated Controller. RE-EVALUATE symptom. SERVICE open in A/C input circuit to controller. RE-EVALUATE symptom.

Integrated Controller — Pinpoint Test — X

TEST STEP	RESULT ▶	ACTION TO TAKE
X74 COOLING FAN GROUND VERIFICATION		
• Key Off. • Cooling fan disconnected. • Integrated controller connector disconnected. • Jumper Pin 3 to Pin 1 (for primary fan) and Pin 3 to Pin 6 (for secondary fan) at integrated controller harness connector. • DVOM on 20 volt scale. • Measure voltage between voltage positive at cooling fan harness connector and negative battery post.	One or both 8.0 volts or greater ▶ One or both less than 8.0 volts. ▶	SERVICE open in ground circuit to fan(s). RECONNECT all components. RE-EVALUATE symptom. SERVICE open in power circuit to fan(s). RECONNECT all components. RE-EVALUATE symptom.
X75 JUMPER SECONDARY ELECTRIC DRIVE SIGNAL (HEDF) TO GROUND		
• Key Off. • Disconnect processor 60 Pin connector and inspect for damaged pins, corrosion, loose wires. Service as necessary. • Install Breakout box. • Processor disconnected. • Integrated controller connected. • Jumper test Pin 52 to test Pin 40 at Breakout box. • Key On. • Does secondary fan run?	Yes ▶ No ▶	GO to X76. REPLACE Integrated Controller. RECONNECT processor. RE-EVALUATE symptom.
X76 ECT SENSOR CHECK		
• Processor connected. • Check engine coolant level. • Warm engine to operating temperature before taking ECT resistance measurement. • Key Off, wait 10 seconds. • Harness disconnected from ECT sensor. • DVOM on 200,000 ohm scale. • Measure resistance of ECT sensor. • Is resistance between 1500 and 2000 ohms?	Yes ▶ No ▶	REPLACE processor. RECONNECT all components. RE-EVALUATE symptom. REPLACE ECT sensor. RECONNECT all components. RE-EVALUATE symptom.

Integrated Controller — Pinpoint Test — X

TEST STEP	RESULT ▶	ACTION TO TAKE
X70 NO FAN PRIMARY OR SECONDARY WITH NO CODE 82 OR 88		
• Key Off. • Integrated controller disconnected. • DVOM on 20 volt scale. • Measure voltage between battery negative post and Pins 3 and 4 respectively at controller harness connector. • Are all voltage readings 10.5 volts or greater?	Yes ▶ No ▶	GO to X71. SERVICE open in battery power circuit. RE-EVALUATE symptom.
X71 FAN MOTORS CHECK		
• Key Off. • Integrated controller disconnected. • Jumper Pin 3 to Pin 1 (for primary fan) and Pin 3 to Pin 6 (for secondary fan) at integrated controller harness connector. • Do fans run?	Yes ▶ No ▶	GO to X72. GO to X73.
X72 CHECK FAN RUNNING MODE (LOW)		
• Key Off. • Disconnect processor. • Integrated controller connected. • Key On. • Does primary fan run?	Yes ▶ No ▶	GO to X75. REPLACE Integrated Controller. RECONNECT all components. RE-EVALUATE symptom.
X73 MEASURE BATTERY VOLTAGE SUPPLY AT FANS — BYPASSING INTEGRATED CONTROLLER		
• Key Off. • Cooling fans disconnected. • Integrated controller connector disconnected. • Jumper Pin 3 to Pin 1 (for primary fan) and Pin 3 to Pin 6 (for secondary fan) at integrated controller harness connector. • DVOM on 20 volt scale. • Measure voltage at cooling fan harness connectors.	One or both 8.0 volts or greater. ▶ One or both less than 8.0 volts. ▶	CHANGE fan(s). RE-EVALUATE symptom. GO to X74.

Integrated Controller — Pinpoint Test — X

TEST STEP	RESULT ▶	ACTION TO TAKE
X80 SERVICE CODE 82 OR 88: CHECK EDF PROCESSOR SIGNAL TO INTEGRATED CONTROLLER FOR SHORTS TO GROUND		
NOTE: If fan is always on with Code 82 OR 88, GO to X82 . • Key Off. • Breakout box installed. • Processor and Integrated Controller disconnected. • DVOM on 200,000 ohm scale. • Measure resistance from test Pin 55 to test Pin 40. • Is resistance 10,000 ohms or less?	Yes ▶ No ▶	SERVICE short to ground in EDF circuit. RECONNECT all components. RERUN Quick Test. GO to X81.
X81 CHECK FAN RUNNING MODE		
• Key Off. • Disconnect processor. • Integrated controller connected. • Key On. For 2.5L MTX — Does fan run? For 2.5L and 3.0L ATX — Does fan run at low speed? For 2.3L EFI TC — Does primary fan run?	Yes ▶ No ▶	REPLACE processor. RECONNECT all components. RERUN Quick Test. REPLACE Integrated Controller. RECONNECT all components. RERUN Quick Test.
X82 FAN ALWAYS ON WITH CODE 82 OR 88: CHECK EDF PROCESSOR SIGNAL TO INTEGRATED CONTROLLER FOR OPEN CIRCUIT		
• Key Off. • Disconnect processor 60 Pin connector and inspect for damaged pins, corrosion, and loose wires. Service as necessary. • Breakout box installed. • Processor and integrated controller disconnected. • DVOM on 200 ohm scale. • Measure resistance between test Pin 55 and integrated controller harness Pin 14. • Is resistance 5 ohms or less?	Yes ▶ No ▶	GO to X83. SERVICE open in EDF circuit. RECONNECT all components. RERUN Quick Test.

Integrated Controller — Pinpoint Test X

TEST STEP	RESULT ▶	ACTION TO TAKE
X83 CHECK EDF CIRCUIT FOR SHORTS TO POWER		
• Key Off. • Breakout box installed. • Processor and integrated controller disconnected. • DVOM on 200,000 ohm scale. • Measure resistance from test Pin 55 to test Pin 37, and from test Pin 55 to battery positive. • Is resistance 10,000 ohms or less?	Yes ▶	SERVICE short to power in EDF circuit. GO to X84.
	No ▶	GO to X84.
X84 EDF SHORT TO GROUND CHECK		
• Key Off. • Breakout box installed. • Processor disconnected. • Integrated controller connected. • Key On. • Jumper test Pin 55 to ground test Pin 40 or 60. • Does fan continue to run?	Yes ▶	REPLACE Integrated Controller. RECONNECT all components. RERUN Quick Test.
	No ▶	REPLACE processor. RECONNECT all components. RERUN Quick Test.

Erratic Ignition — Pinpoint Test Y

Y

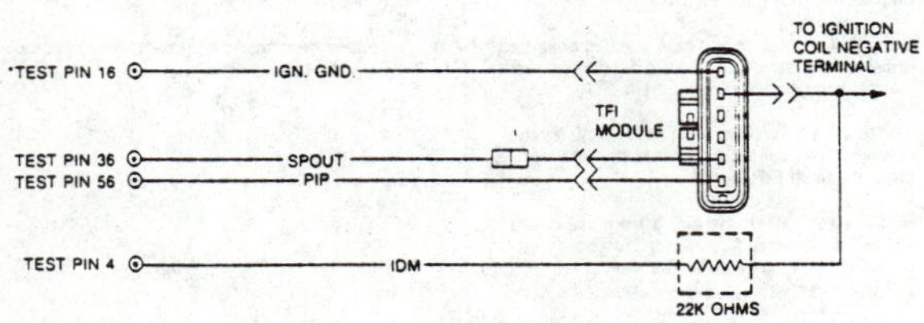

*TEST PINS LOCATED ON BREAKOUT BOX.
ALL HARNESS CONNECTORS VIEWED INTO MATING SURFACE.

STOP-WARNING

You should enter this Pinpoint Test only when a service code 14 is received in Quick Test Step 6.0.

To prevent the replacement of good components, be aware that the following non-EEC areas may be at fault:

- TFI module.
- Arcing secondary ignition components.
  - — Ignition coil.
  - — Spark plugs and high tension cables.
- Distributor and PIP sensor.

This Pinpoint Test is intended to diagnose only the following:

- Harness circuits: ignition ground, spout, PIP, IDM.

Erratic Ignition — Pinpoint Test Y

TEST STEP	RESULT ▶	ACTION TO TAKE
Y1 CODE 14: ERRATIC IGNITION		
NOTE: Code 14 indicates two successive erratic profile ignition pickup (PIP) pulses occurred, resulting in a possible engine miss or stall. • Check EEC-IV and ignition systems harnesses for: — Loose wires/connectors. — Arcing secondary ignition components (coil, cap, rotor, wires, plugs, etc.). — On-board transmitter (2-way radio).* • Are any of the above present? *Verify all 2-way radio installations. Carefully follow manufacturer's installation instructions regarding the routing of antenna and power leads.	Yes ▶	SERVICE as necessary. RERUN Quick Test.
	No ▶	GO to Y2.
Y2 CHECK DISTRIBUTOR		
• Key Off, wait 10 seconds. • Deactivate Self-Test. • Connect VOM or STAR per Quick Test Step 2.0. • Enter Engine Running Continuous Monitor Test (as instructed in Quick Test Step 6.3). • Observe VOM or STAR LED for indication of a fault while performing the following: • Lightly tap on TFI module and distributor. • Wiggle TFI connector. • Is a fault indicated?	Yes ▶	DISCONNECT and INSPECT connectors. If connector and terminals are good, GO to Section 15, Ignition System Diagnostics.
	No ▶	GO to Y3.
Y3 CHECK EEC-IV HARNESS		
• While still in Continuous Monitor Test from Step Y2 observe VOM or STAR LED for a fault indication while performing the following: • While looking for faults grasp the harness close to the TFI connector. Wiggle, shake or bend a small section of the ignition and EEC-IV systems harness while working your way to the other components and dash panel. Also wiggle, shake or bend the EEC-IV harness from the dash panel to the processor. Isolate the PIP circuit if needed for this test. • Is a fault indicated?	Yes ▶	ISOLATE fault and SERVICE as necessary. RERUN Quick Test.
	No ▶	GO to Y4.

Erratic Ignition — Pinpoint Test Y

TEST STEP	RESULT ▶	ACTION TO TAKE
Y4 CHECK PROCESSOR AND HARNESS CONNECTORS		
• Key Off, wait 10 seconds. • Disconnect processor 60 Pin connector and inspect for damaged pins, corrosion, loose wires. • Are connectors and terminals OK? • Reconnect processor when this Step is completed.	Yes ▶	Unable to duplicate an erratic ignition fault in the EEC-IV System, for further diagnosis, GO to Section 15, Ignition System Diagnostics.
	No ▶	SERVICE as necessary. RERUN Quick Test.

Intermittent Fault Diagnosis

WHEN TO USE INTERMITTENT FAULT DIAGNOSIS

It is recommended, that before using this Section for the first time, you view the video program that explains the steps used in this Section. The video program is available in laser disc for viewing in the dealership and a VHS format for home video recorders.

INTERMITTENT FAULT DIAGNOSIS DESCRIPTION

Intermittent fault diagnosis provides a technique that you can use to find an intermittent fault. **A vehicle is diagnosed with an intermittent fault when:**

- **The vehicle has been returned to the dealership with a fault that cannot be duplicated by the technician under normal driving conditions.**

Prior to starting these procedures, the technician **must** have checked out all mechanical systems according to Section 2 of the manual.

Before beginning, the Service Writer should contact the customer and complete the "Customer Contact Sheet." These questions may help you in duplicating the problem, which in turn will allow you to correct the concern.

NOTE: The following procedures for finding intermittent faults DO NOT provide exact answers to problems. They provide a technique that has been proven to correct intermittent faults. At times, it may be necessary to use your own judgement when diagnosing an intermittent fault.

Intermittent No Start And Quits On Road With No Restart

STEPS	POSSIBLE ACTION TO TAKE
• Review the repair order. **If required, contact the driver or drivers of the vehicle to get further information. Refer to the Customer Contact Sheet on page 2 of this Section.**	• If repair order shows any pattern (e.g., no-start occurs whenever the engine is warm or cold), duplicate the same condition and try to start the vehicle.
• Visually inspect the engine compartment. **DO NOT disturb anything.**	• If you determine by a visual inspection that anything is disconnected, **do not reconnect the item. If you move any component or connection, you could make it harder to repeat the intermittent problem. NOTE: Look for areas where electrical wires or vacuum lines could be cut by sharp edges or burned by engine hot spots. Blocked or kinked vacuum lines can also cause trouble.**
• Try to start the vehicle.	• If vehicle does not start, perform routines according to Volume H.
• If vehicle starts, take it for a test drive for "Quits on Road with no Restart" condition. **Take along the spark tester.**	• If vehicle quits on road and can't be restarted, listen for the sound of the fuel pump operating when the ignition key is turned On and check for spark at the coil. Then check TSBs for this concern when you get back to the shop if applicable. 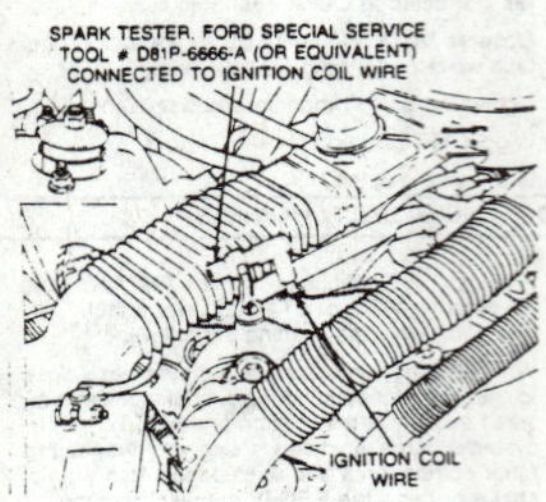

Intermittent No Start And Quits On Road With No Restart

STEPS	POSSIBLE ACTION TO TAKE
• If you cannot verify the concern, contact the driver of the vehicle for additional information.	• If the driver leads you to another vehicle system, work on that system. If not, continue to the next step.
• Check the ignition system secondary wires for damage or loose connections.	• If ignition system secondary wires are not damaged or loose, continue to the next step. • If secondary wires are damaged, replace them and try to start the vehicle. • If secondary wires are loose, repair as necessary and try to start the vehicle. 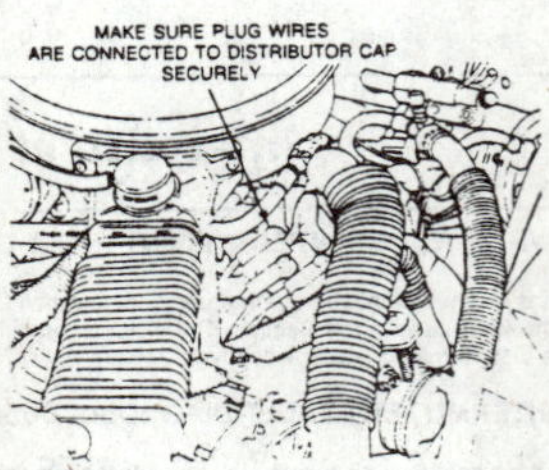

Intermittent No Start And Quits On Road With No Restart

STEPS	POSSIBLE ACTION TO TAKE
• Check distributor cap for moisture, cracks, or salt-like deposits. 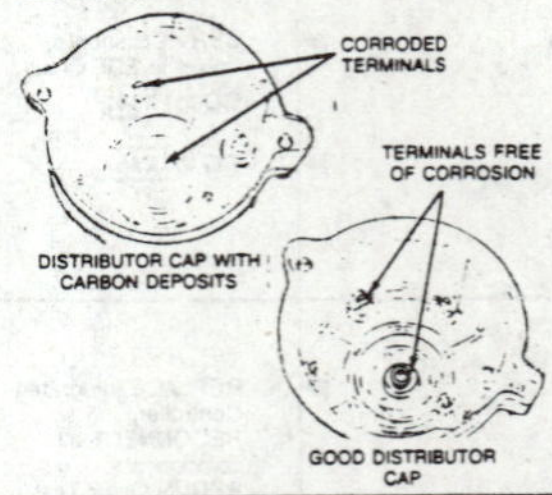	• If distributor cap has moisture, find out why. If necessary, replace the cap. • If distributor cap has cracks or salt-like deposits, replace the cap with a new one. • Check the rotor for damage or burns. • If rotor is damaged or burned, replace with a new one, and try to start the vehicle.
• Check the TFI ignition module for damage or loose connections. **NOTE: Do not use the TFI ignition module as a fulcrum for turning the distributor.** 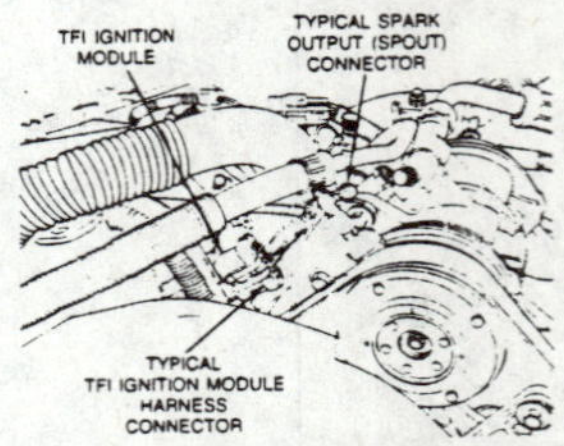	• Replace the TFI ignition module if damaged.
• Check the TFI ignition module six-pin connector to ensure it is installed correctly and that the gasket is not folded over.	• Repair any loose connections. Reinstall connector if not properly installed.

Intermittent No Start And Quits On Road With No Restart

STEPS	POSSIBLE ACTION TO TAKE
• Install a spark tester to the ignition system coil wire. **NOTE: Disconnect the wire at the distributor.** Crank the engine. • Verify that a good, strong blue spark is observed while the engine is cranking. This indicates that the coil and primary circuit are OK. • **NOTE: By connecting the spark tester to the coil wire, you are able to verify that the coil is firing on each PIP signal.** **NOTE: Never allow the spark from the coil wire to go directly to ground. This could cause internal damage to the coil.** 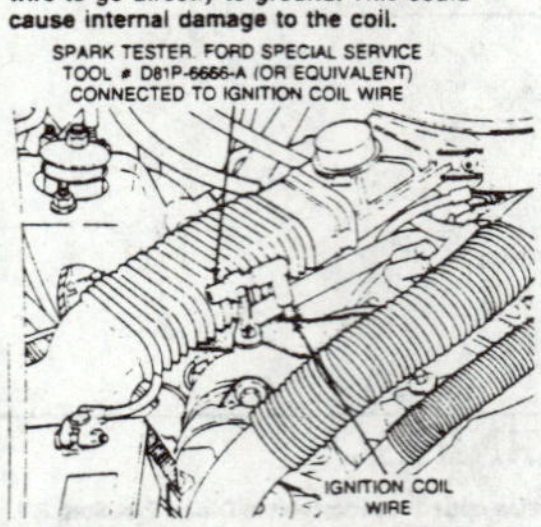	• If a good, strong blue spark is not detected, repair according to ignition procedures located in Section 15 of Volume H. • If an intermittent spark is present, the ignition coil may be damaged internally.
• With the spark tester still connected to the coil wire, wiggle the ignition primary connectors while cranking the engine. • **NOTE: By performing this test while cranking, we check those circuits dedicated to the start mode.**	• If an intermittent loss of spark is detected, there could be an intermittent open in the TFI ignition module connector or harness that could cause an intermittent no-start. • **NOTE: An intermittent open in the TFI ignition module or harness could cause a no-start by preventing the TFI ignition module from firing the coil.** • **NOTE: An intermittent open in the coil primary harness or connecter could cause a no-start because the coil is not receiving battery voltage.**

Intermittent No Start And Quits On Road With No Restart

STEPS	POSSIBLE ACTION TO TAKE
• **NOTE: For the following test the engine MUST be running.** Use a jumper wire with an alligator clip at both ends. Attach one of the clips to a **good** ground location. Attach the other end to the metal part of a screwdriver. While the engine is running, pass the screwdriver around the base of the coil. **WARNING: USE CAUTION NOT TO GET THE SCREWDRIVER OR WIRE ENTANGLED WITH THE DRIVE BELTS OF THE VEHICLE.**	• Arcing indicates a breakdown in the insulation of the coil or coil wire.

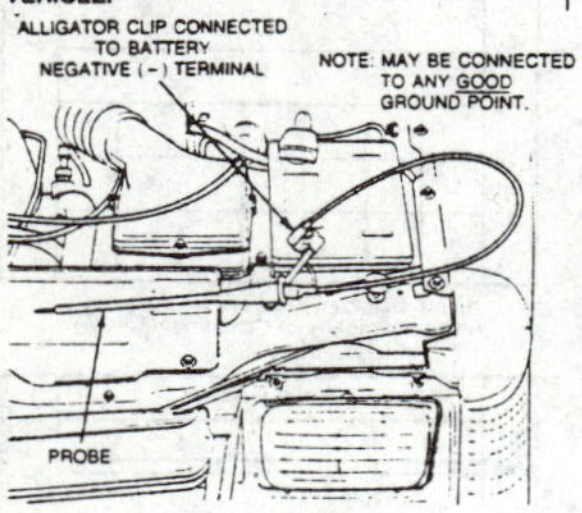

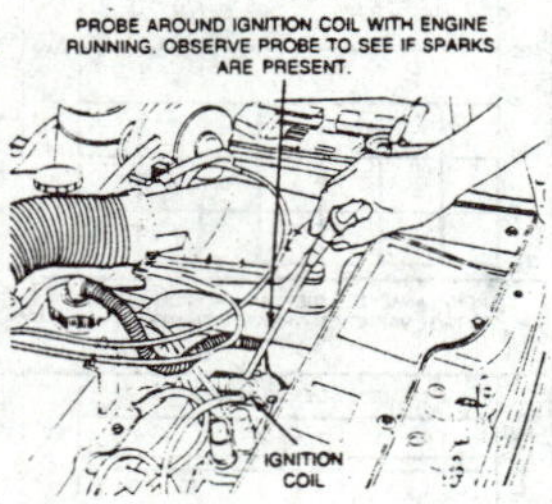

STEPS	POSSIBLE ACTION TO TAKE
• If the vehicle is running, turn it Off. From inside the vehicle, using the ignition switch, **"bump/crank"** the engine approximately ten times.	• If the engine fails to start on any of the tries, the ignition switch may be damaged internally.
• Connect a fuel pressure gauge to the engine with an extension hose long enough to allow the gauge to reach the windshield, as outlined in Group 24 of the Shop Manual. **NOTE: A suitable hose was issued in the 1985 Ford Tool Package.** **NOTE: For 2.3L and 2.5L HSC engines, and other applications without a Schrader valve installed in the fuel system, use Rotunda Tool D85L-9974-B, In-Line Pressure Tester Adapter.**	• If the pressure **does not** increase each time the key is turned to On, there is a problem in the fuel system. • Check the inertia switch, the fuel pump relay wiring connectors, and the fuel pump ground connection for looseness, corrosion, or other factors that could cause an intermittent open circuit. **NOTE: On certain vehicles the fuel pump ground is located beside the driver's seat, under the carpet. Refer to past TSBs for fuel pump ground locations.**

Intermittent No Start And Quits On Road With No Restart

STEPS	POSSIBLE ACTION TO TAKE

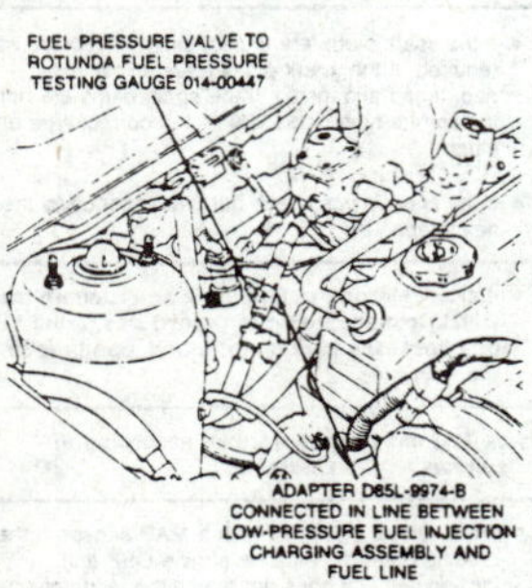

• Tape the pressure gauge to the windshield. This allows you to read the fuel pressure from the inside of the vehicle. **NOTE: Do not tape the pressure gauge where it will interfere with the windshield wiper blades or linkage.**

• Turn the ignition key to the ON position for two seconds, watch the fuel gauge for a pressure increase. Repeat the procedure ten times. There should be a pressure increase each time the key is turned On.

NOTE: On engines equipped with a 10 second delay power relay, pause 10 seconds between turning the key Off and On.

Intermittent No Start And Quits On Road With No Restart

STEPS	POSSIBLE ACTION TO TAKE

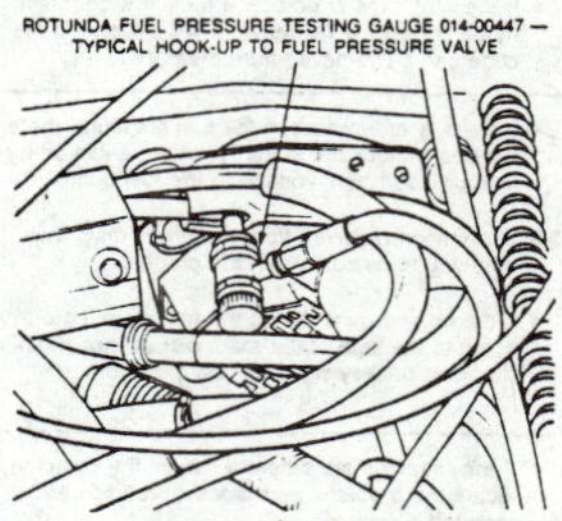

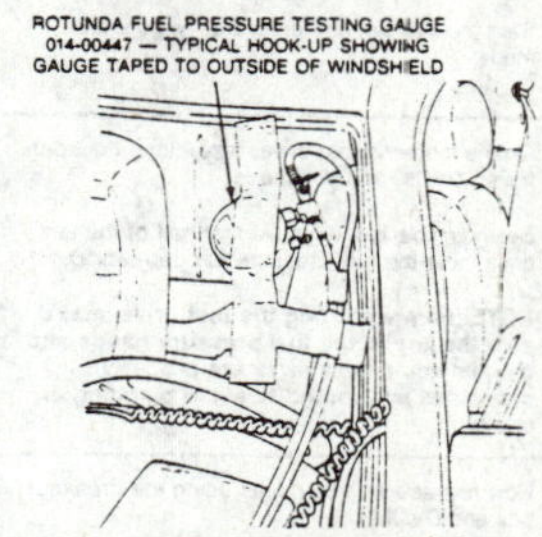

STEPS	POSSIBLE ACTION TO TAKE
• If the fuel and ignition systems are OK, check the spark plugs for soot or fuel fouling.	• If spark plugs are fouled, clean or replace as required.
• **If the intermittent fault is still present, refer to Section 2.**	• **If after referring to Section 2 and running the Quick Test according to procedures found in this book, the fault is not found, continue to the next step.**

NOTE: Do not continue to the next step unless the Quick Test has been performed according to procedures outlined in this book, and the intermittent fault has NOT been found.

STEPS	POSSIBLE ACTION TO TAKE
• If the Quick Test did not find the cause of the intermittent fault, prepare the vehicle for another road test. • Verify that the fuel pressure gauge is still attached securely to the windshield. Hook up the Black STAR tester and connect the breakout box. • Label the following key points on the breakout box: — battery voltage — reference voltage — throttle position sensor — PIP signal — ECT sensor — ACT sensor — SPOUT signal — fuel pump (circuit at ECA)	• Make a copy of the five-column chart located at the end of this section and record the readings taken with the ignition key in the ON position and also while the engine was cranking, and stabilized at idle. • These are "baseline" readings of key voltages while the engine is operating properly. By comparing these baseline readings with the readings when the intermittent symptom occurs, it is possible to isolate the component or circuit that is at fault.

Intermittent No Start And Quits On Road With No Restart

STEPS	POSSIBLE ACTION TO TAKE

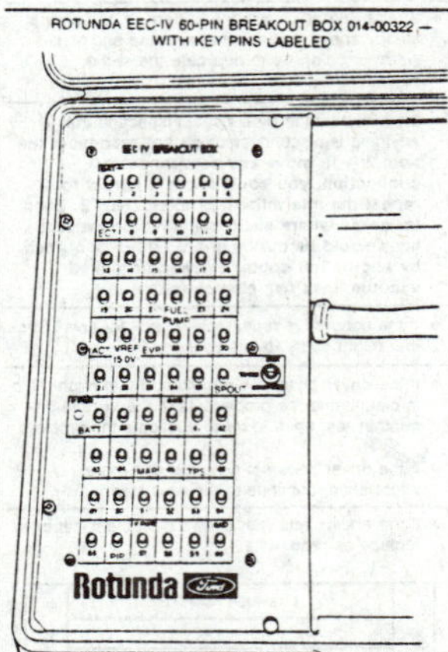

• **NOTE: Refer to the correct vehicle "Engine Supplement" Section for the correct pin numbers for the points listed.**

• Disconnect the coil primary connector so the engine will not start. Turn the ignition key to the ON position and check key points with a DVOM. **DO NOT CRANK THE ENGINE AT THIS TIME.**

• Repeat the procedure while cranking the engine.

• Repeat the procedure at stabilized idle.

STEPS	POSSIBLE ACTION TO TAKE
• After recording the baseline readings, keep the ignition key in the ON position and perform a Key On/Engine Off Quick Test.	• When the Key On/Engine Off Quick Test is completed and everything has checked out properly, unlatch the STAR tester.

Intermittent No Start And Quits On Road With No Restart

STEPS	POSSIBLE ACTION TO TAKE
• Start the vehicle, and enter the "wiggle test" mode.	• If the STAR tester detects a fault in wiggle test mode during the test drive, it will beep, and the code can be retrieved from memory.
• On the following test drive, take along the spark tester, DVOM and Volume H. • Begin vehicle test drive. At first half of the test drive, note the fuel pressure gauge reading. • **NOTE: Before starting the test drive, make sure the line to the fuel pressure gauge and the test equipment wires are properly connected and routed to avoid pinching or crimping.**	• If there is a decrease in the fuel pressure, there may be a blockage in the fuel filter, a kinked fuel line, or insufficient voltage to the fuel pump. • Information received from the driver may now help you to narrow down the problem. • If the engine quits during the test drive, carefully coast to the side of the road, **but do not move the ignition key from the ON position.**
• Now recheck the key points using the breakout box and DVOM.	• If the readings are different, check the circuit to locate the problem, and repair or replace as required. • If the readings are close to the baseline readings, continue to the next step.
• Latch the STAR tester and perform the Key On/Engine Off and Engine Running Quick Tests.	• If the Quick Test steps detect a problem in the circuit, repair or replace as required. • If the Quick Test steps do not detect a problem in the circuits, continue to the next step.
• Check for spark. If no spark is detected, disconnect the SPOUT signal connector, and again check for spark.	• If spark is restored, go to the no-start Pinpoint Test in Volume H. • If spark does not appear, continue to the next step.
• Check for a PIP signal while cranking the engine.	• When the PIP signal is detected, check the TFI and E-core coil circuits for continuity using the DVOM. • If there is no PIP signal, **do not** assume that there is something wrong with the PIP sensor. Check the crank, run, and ground circuits at the TFI connector first.

Intermittent Surging At A Steady Speed

STEPS	POSSIBLE ACTION TO TAKE
• Review the repair order. **If required, contact the driver or drivers of the vehicle to get further information. Refer to the Customer Contact Sheet on page 2 of this Section.**	• If repair order shows any pattern (e.g., surge at steady speed occurs whenever the engine is warm or cold), try to duplicate the same condition.
• Visually inspect the engine compartment. **DO NOT disturb anything.**	• If you determine by a visual inspection that anything is disconnected, **do not reconnect the item. If you move any component or connection, you could make it harder to repeat the intermittent problem. NOTE: Look for areas where electrical wires or vacuum lines could be cut by sharp edges or burned by engine hot spots. Blocked or kinked vacuum lines can also cause trouble.**
• Take the vehicle for a test drive to try to verify the concern.	• If the concern is repeatable, check for any TSBs that might apply to this concern.
• If the concern is not repeatable, get more information; what speed is the intermittent surge at? Are there any certain conditions when the surge always happens?	• If the driver gives you information that might help in diagnosing the problem, take the vehicle on another test drive to try to duplicate the problem. • If the driver does not give you any more information, continue to the next step.
• Connect an engine analyzer to the vehicle. • Check plug firing voltages, cylinder balance, and base timing.	• If the engine analyzer shows a problem, repair or replace as required.

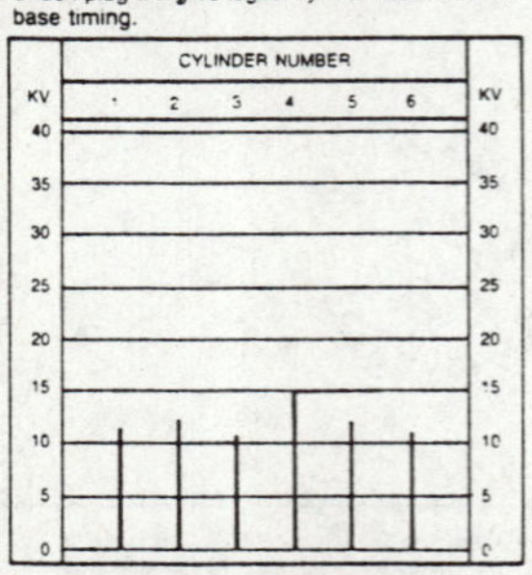

ENGINE ANALYZER DISPLAYS — PROPERLY OPERATING IGNITION SYSTEM

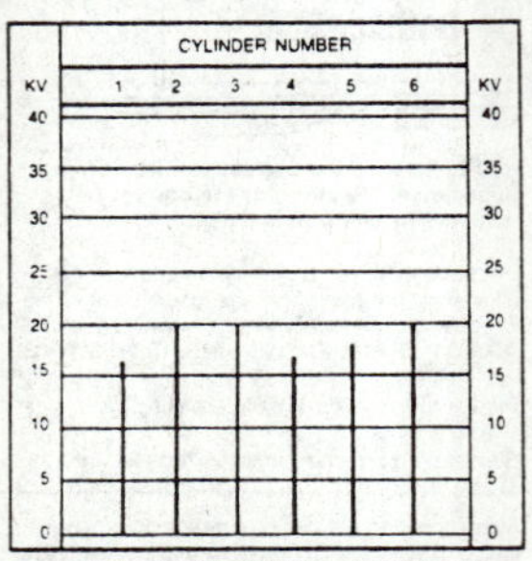

ENGINE ANALYZER DISPLAYS — AVERAGE PLUG FIRING VOLTAGE GREATER THAN 15KV

Intermittent Surging At A Steady Speed

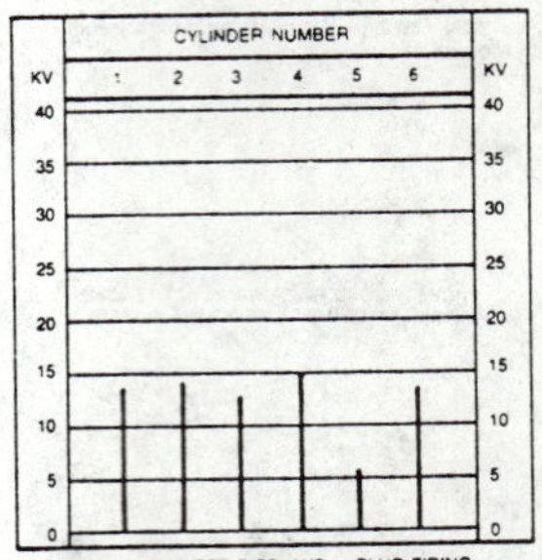

ENGINE ANALYZER DISPLAYS — PLUG FIRING VOLTAGE VARIATION GREATER THAN 5 KV

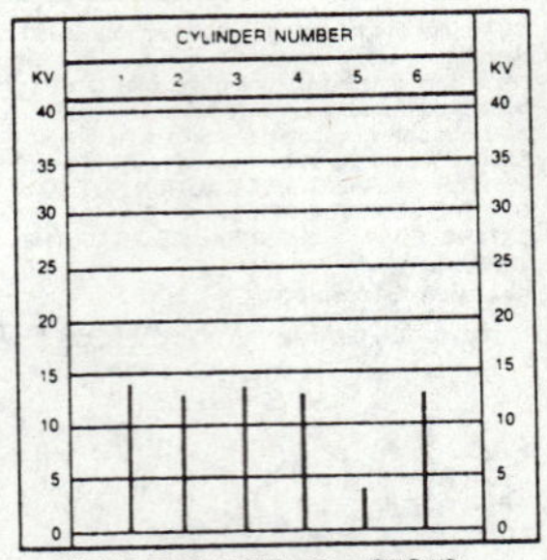

ENGINE ANALYZER DISPLAYS — LOW PLUG FIRING VOLTAGE IN ONE CYLINDER

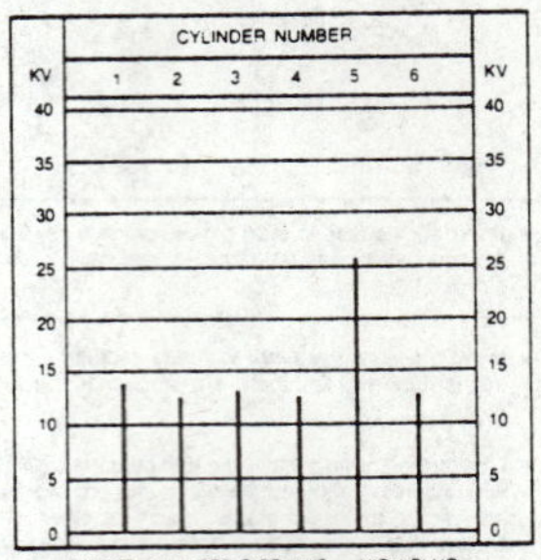

ENGINE ANALYZER DISPLAYS — HIGH PLUG FIRING VOLTAGE IN ONE CYLINDER

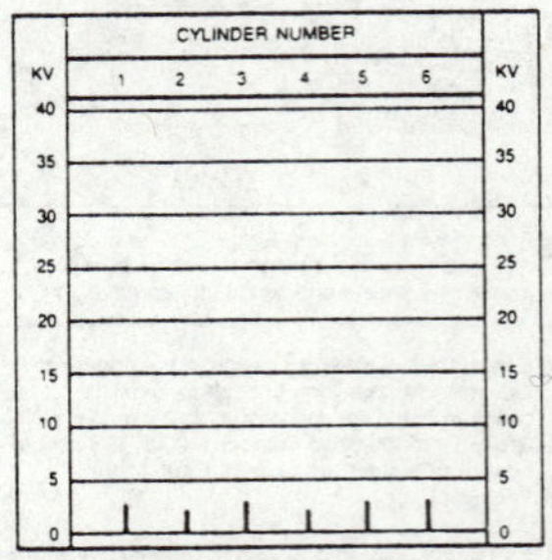

ENGINE ANALYZER DISPLAYS — NEGATIVE FIRING VOLTAGE

Intermittent Surging At A Steady Speed

STEPS	POSSIBLE ACTION TO TAKE
• When all tests have been performed on the engine analyzer, check the spark plugs for ash fouling, proper gap, and to make sure they are the right type of plug for the engine.	• If the spark plugs are fouled, clean or replace as required. If the spark plugs have an improper gap, regap and install. If the spark plugs are not the correct type, replace with the correct type of plugs. • If the spark plugs check out OK, continue to the next step.
• **If the intermittent fault is still present, refer to Section 2.**	• **If after referring to Section 2 and running the Quick Test according to procedures found in this book, the fault is not found, continue to the next step.**
NOTE: Do not continue to the next step unless the Quick Test has been performed according to procedures outlined in this book, and the intermittent fault has NOT been found.	
• Connect a MAP/BP tester to either the manifold absolute pressure (MAP) or barometric pressure (BP) sensor (depending on how vehicle is equipped) and check the readings with the engine Off and the ignition key in the ON position. • Compare the reading to the correct value listed on page 50 of this supplement. • If the vehicle is equipped with a MAP sensor, repeat the MAP test with the engine running. The MAP tester reading should decrease at idle and increase when you bump the throttle. **NOTE: If the vehicle is equipped with a barometric pressure (BP) sensor, the engine running portion of the test does not have to be performed since there is no vacuum hookup to the engine.**	• On a vehicle equipped with a MAP sensor, if the reading obtained with the engine OFF and ignition key ON does not match the value given in the Pinpoint Test, perform the appropriate Pinpoint Test in Volume H. • If the MAP sensor does not change when you bump the throttle with the engine running, substitute a known good hose for the original MAP sensor vacuum hose connected to the sensor and repeat the test. A change in the MAP tester reading indicates that the original hose was damaged, blocked or kinked. • On a vehicle equipped with either a MAP or BP sensor, substitute with a known good component and retest. If readings are now OK, replace the component.

Intermittent Surging At A Steady Speed

STEPS	POSSIBLE ACTION TO TAKE
• If the vehicle is equipped with a vane meter, **carefully** check for unmetered air leaks (i.e., loose air tubes, missing hoses or vacuum caps, etc.)	• Correct air or vacuum leaks as necessary.

SOME TYPICAL PLACES TO CHECK FOR AIR LEAKS ARE SHOWN WITH AN ARROW (2.3L ENGINE SHOWN — 1.9L SIMILAR)

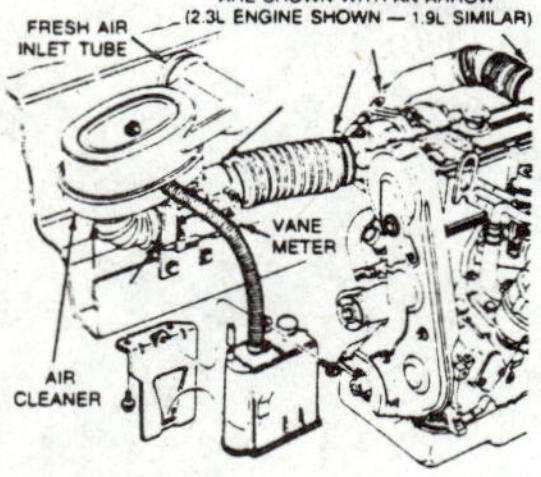

STEPS	POSSIBLE ACTION TO TAKE
• With breakout box installed monitor vane meter signal while you manually move the vane meter door through its entire range.	• Voltage must increase smoothly as door is opened.
• Connect a Rotunda Fuel Injector Tester/Cleaner to the vehicle according to instructions that come with the unit. • Check the flow of the injectors against the specifications given with the Tester/Cleaner and the injector chart.	• If injector flow is not up to specifications, clean the injectors. Follow the instructions that come with the Tester/Cleaner.

Intermittent Surging At A Steady Speed

STEPS	POSSIBLE ACTION TO TAKE

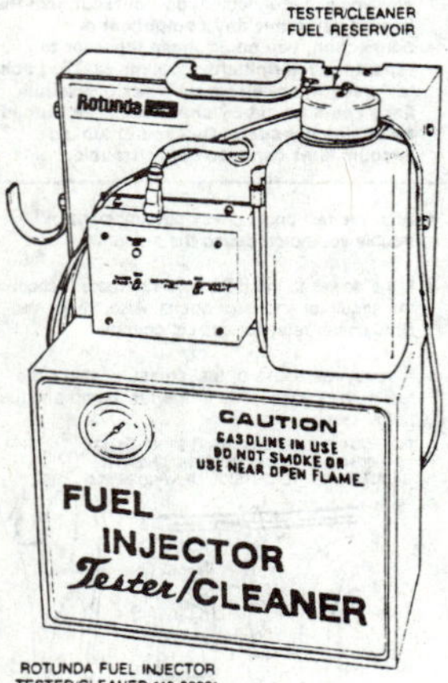

ROTUNDA FUEL INJECTOR TESTER/CLEANER 113-00001

ROTUNDA FUEL INJECTOR TESTER CLEANER 113-00001 — TYPICAL HOOK-UP TO FUEL INJECTORS AND FUEL PRESSURE VALVE

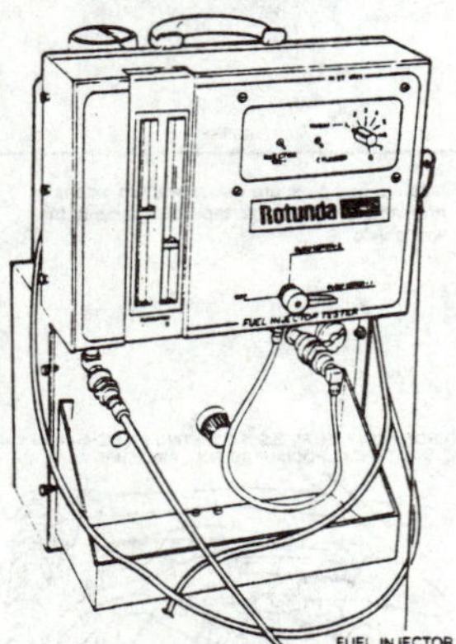

Intermittent Surging At A Steady Speed

The following charts list fuel injector applications, part numbers, source, color code and injector specifications.

HIGH PRESSURE CFI INJECTOR APPLICATION

ENGINE APPLICATION	MODEL YEAR	PART NO. - 9F593 -	SOURCE	COLOR CODE	INJECTOR SPECS. SIZE	RESIST
					gm/sec	Ohms
1.6L	1983	E3EE-BA	Bosch	Blue	1.76 (14# HR)	2.25
	1984	E4EE-AA	N.D.	Blue		2.35
1.6L T C	1984-85	E4EX-AA	Bosch	Black	3.00 (23# HR)	2.40
1.9L	1985-1 2-86	E6EE-AB	Bosch	White	2.45 (19# HR)	2 25
2.3L T/C	1983	E3ZE-BA	Bosch	Green	3.86 (30#/HR)	2.40
	1984	E4ZE-AA	D.K.K.			2.35
	1985	E5ZE-AA	D.K.K.	Brown	4.41 (35#/HR)	2.35
	1986	E5ZE-AB	D.K.K.			
2.3L (Truck)	1985	E59E-AA	N.D.	Gray	1.76 (14#/HR)	16.20
	1986	E59E-AB	N.D.			
2.9L (Car/Truck)	1986	E67C-A1B	Bosch	Gray	1.76 (14# HR)	14.50
3.0L (Car/Truck)	1986	E59E-AB	N.D.	Gray	1.76 (14#/HR)	16.20
		E67E-BB	Bosch			14.50
3.8 CFI	1984-86	E3VE-A1A	Bosch	Green	4.54 (37#/HR)	2.25
		-A2A	N.D.			2.00
5.0L CFI	1980	EOSE-AA	Bosch	Blue	5.81 (46#/HR)	2.40
	1983	EOSE-A1A	Bosch			2.40
		EOSE-A2A	N.D.			2.00
	1984	EOSE-A1A	Bosch			2.25
		EOSE-A2A	N.D.			2.00
5.0L H.O. CFI	1984-85	E4ZE-CA	Bosch	Gray	6.60 (52# HR)	2.25
5.0L SEFI	1986	E59E-AB	N.D.	Gray	1.76 (14# HR)	16.20
		E67E-BB	Bosch			14.50
5.0L H.O. SEFI	1986	E6TE-AB	N.D.	Gold	2.45 (19#/HR)	16.20
5.0L Truck	1985	E5TE-AA	Bosch	Gold	2.45 (19# HR)	15.50
		E5TE-AA	N.D.			16.20
		E5TE-AA	D.K.K.			16.20
	1986	E5TE-AB	D.K.K.			16.20
		E5TE-BB	Bosch			14.50

LOW PRESSURE CFI INJECTOR APPLICATIONS

2.3L HSC	1985-86	E43E-AC	Bosch	Blue	7.00 (56# HR)	1.40
2.3L H.O HSC (Manual Trans. Only)	1985	E53E-AB	Bosch	Green	8.00 (64# HR)	1.40
2.5L HSC	1986	E53E-AB	Bosch	Green	8.00 (64# HR)	1.40

Intermittent Surging At A Steady Speed

STEPS	POSSIBLE ACTION TO TAKE
• After cleaning the injectors, test the injector flow again.	• If the injectors still do not meet specifications, replace them.
• If the problem still exists, and you have not done so yet, check the OASIS system and related TSBs.	• If TSBs or OASIS do not take you in a direction toward fixing the problem, continue to the next step.

NOTE: For 2.3L and 2.5L HSC engines, and other applications without a Schrader Valve installed in the fuel systems, use Rotunda Tool D85L-9974B, In-Line Pressure Tester Adapter.

STEPS	POSSIBLE ACTION TO TAKE
• Install a fuel pressure gauge with an extension hose, so that the fuel pressure can be monitored during the test drive. (Tape the fuel pressure gauge to the windshield for easy viewing.)	• Make a copy of the five-column chart at the end of this section and record the readings on the breakout box, vacuum gauge and fuel pressure gauge taken with the ignition key in the ON position while the engine was cranking. • These are "baseline" readings of key voltages, vacuum condition and fuel pressure while the engine is operating properly. By comparing these baseline readings with the readings when the intermittent symptom occurs, it is possible to isolate the component or circuit that is at fault.

ROTUNDA FUEL PRESSURE TESTING GAUGE 014-00447 — TYPICAL HOOK-UP TO FUEL PRESSURE VALVE

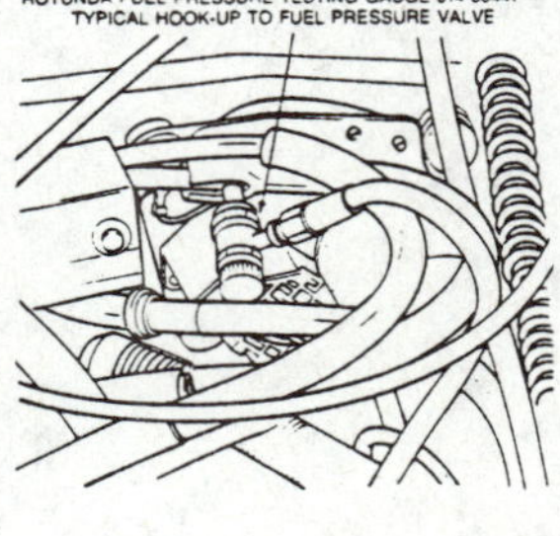

ROTUNDA FUEL PRESSURE TESTING GAUGE 014-00447 — TYPICAL HOOK-UP SHOWING GAUGE TAPED TO OUTSIDE OF WINDSHIELD

Intermittent Surging At A Steady Speed

STEPS	POSSIBLE ACTION TO TAKE
• "Tee" a vacuum gauge and extension hose into the EGR valve vacuum line (use a 0-to-100 inches of water vacuum gauge). Install a black STAR tester. Route the STAR tester and vacuum gauge inside the vehicle so you can read them.	

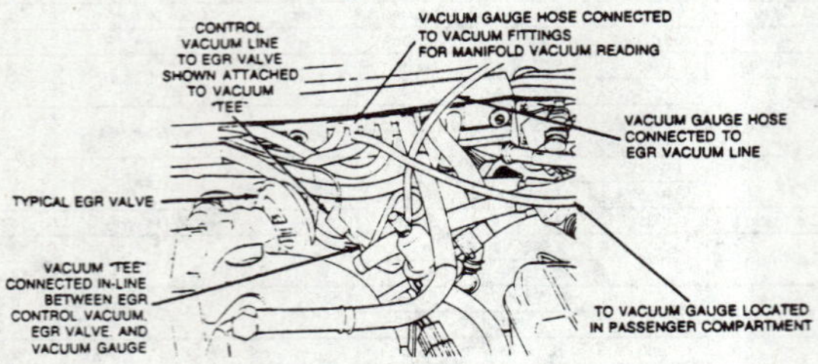

STEPS	POSSIBLE ACTION TO TAKE
• Hook up the breakout box with the following key points labeled on the breakout box: — battery voltage — reference voltage — throttle position sensor — PIP signal — ECT sensor — ACT sensor — SPOUT signal — fuel pump (circuit at ECA)	

Intermittent Surging At A Steady Speed

STEPS	POSSIBLE ACTION TO TAKE
ROTUNDA EEC-IV 60-PIN BREAKOUT BOX 014-00322 — WITH KEY PINS LABELED	

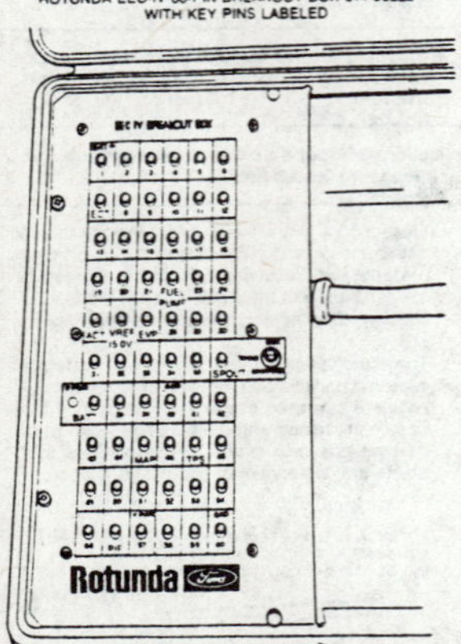

STEPS	POSSIBLE ACTION TO TAKE
• **NOTE: Refer to the correct vehicle "Engine Supplement" Section for the correct pin numbers for the points listed.**	
• **NOTE: Before starting the test drive, make sure the line to the fuel pressure gauge and the test equipment wires are properly connected and routed to avoid pinching or crimping.**	
• Turn the ignition key to the ON position and watch the fuel pressure gauge. Pressure should increase with the key On and the engine cranking.	

Intermittent Surging At A Steady Speed

STEPS	POSSIBLE ACTION TO TAKE
• Run the engine for three minutes. Then, bump the throttle and watch the vacuum gauge. EGR vacuum should increase momentarily when the throttle is bumped. If vacuum doesn't change, the gauge may not be installed correctly. • Before going on the test drive, enter the continuous monitor or "wiggle test" mode with the black STAR tester. • With an assistant take the vehicle on a test drive and operate the vehicle in the same way the customer did to try to repeat the intermittent problem. • Check and record the sensor and gauge readings at the concern speed with the vehicle running properly. When the vehicle surges, have the assistant note the readings on the gauges and breakout box. Then compare them to the good readings.	• If during the test drive the surge is detected and the readings on the breakout box, vacuum gauge, or fuel pressure gauge differ from the baseline readings, this could be where the problem lies. • If the STAR tester "beeps" during the test drive, enter the Quick Test and retrieve the memory code. Follow the procedures located in the Quick Test Section.

Intermittent Quits On Road And Restarts

STEPS	POSSIBLE ACTION TO TAKE
• Review the repair order. **If required, contact the driver or drivers of the vehicle to get further information. Refer to the Customer Contact Sheet on page 2 of this Section.** • Check for any TSBs that might apply, starting with those that relate to what you observed on the test drive.	• If repair order shows any pattern (e.g., quits on road and restarts occurs whenever the engine is warm or cold), try to duplicate the same condition.
• Visually inspect the engine compartment. **DO NOT disturb anything.**	• If you determine by a visual inspection that anything is disconnected, **do not reconnect the item. If you move any component or connection, you could make it harder to repeat the intermittent problem. NOTE: Look for areas where electrical wires or vacuum lines could be cut by sharp edges or burned by engine hot spots. Blocked or kinked vacuum lines can also cause trouble.**
• Install a fuel pressure gauge with an extension hose that allows you to tape the gauge to the windshield.	• After the test drive, check out any signs of trouble you noted during the test drive. • If the power to the ECA was interrupted, check the circuit for shorts or opens. Also, check the EEC power relay for correct operation. • If there was a loss of fuel pressure, check the fuel pump connections at the fuel pump and fuel pump relay.

ROTUNDA FUEL PRESSURE TESTING GAUGE 014-00447 — TYPICAL HOOK-UP TO FUEL PRESSURE VALVE

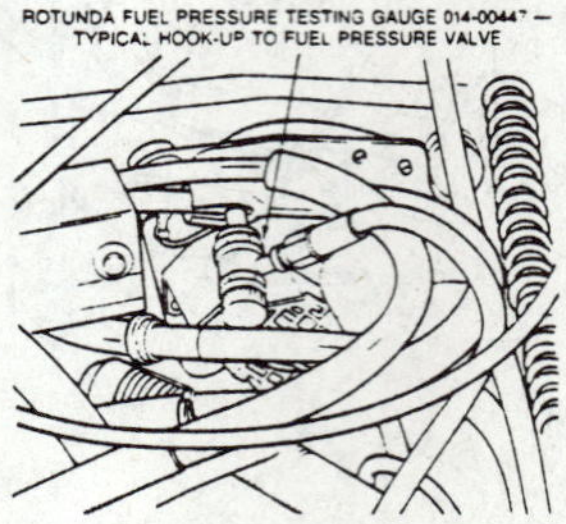

ROTUNDA FUEL PRESSURE TESTING GAUGE 014-00447 — TYPICAL HOOK-UP SHOWING GAUGE TAPED TO OUTSIDE OF WINDSHIELD

Intermittent Quits On Road And Restarts

STEPS	POSSIBLE ACTION TO TAKE
NOTE: For 2.3L and 2.5L HSC engines, and other applications without a Schrader Valve installed in the fuel system, use Rotunda Tool D85L-9974-B, In-Line Pressure Tester Adapter.	
• Install a vacuum gauge and "tee" into the vacuum line to the EGR valve. 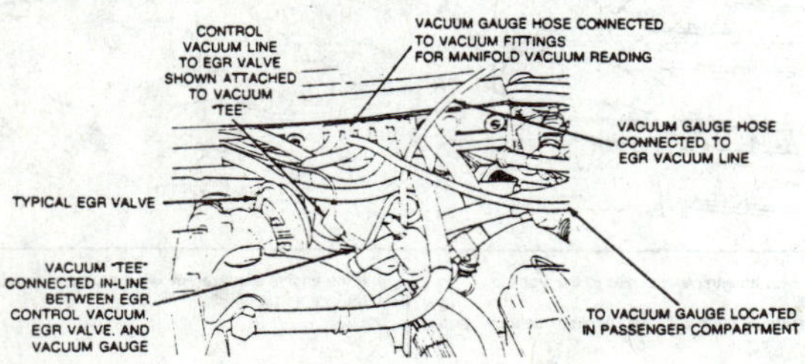 • Connect a DVOM to monitor the battery power to the ECA during the test drive. • Start the vehicle and verify idle speed, following the procedure in Volume H. • Take the vehicle on a short test drive checking the vacuum and fuel pressure gauges and the DVOM occasionally. • If the vehicle quits, check the gauges and DVOM for any signs of trouble with EGR vacuum, fuel pressure, or battery power to the ECA.	
• **If the intermittent fault is still present, refer to Section 2.**	• **If after referring to Section 2 and running the Quick Test according to procedures found in this book, the fault is not found, continue to the next step.**

NOTE: Do not continue to the next step unless the Quick Test has not been performed according to procedures outlined in this book, and the intermittent fault has NOT been found.

Intermittent Quits On Road And Restarts

STEPS	POSSIBLE ACTION TO TAKE
• If the Quick Test did not find the cause of the intermittent fault, prepare the vehicle for another road test. • Verify that the fuel pressure gauge is still attached securely to the windshield. Hook up the Black STAR tester and connect the breakout box. • Label the following key points on the breakout box: — battery voltage — reference voltage — throttle position sensor — PIP signal — ECT sensor — ACT sensor — SPOUT signal — fuel pump (circuit at ECA)	• Make a copy of the five-column chart located at the end of this Section and record the readings taken with the ignition key in the ON position and also while the engine was cranking. • These are "baseline" readings of key voltages while the engine is operating properly. By comparing these baseline readings with the readings when the intermittent symptom occurs, it is possible to isolate the component or circuit that is at fault.

ROTUNDA EEC-IV 60-PIN BREAKOUT BOX 014-00322 — WITH KEY PINS LABELED

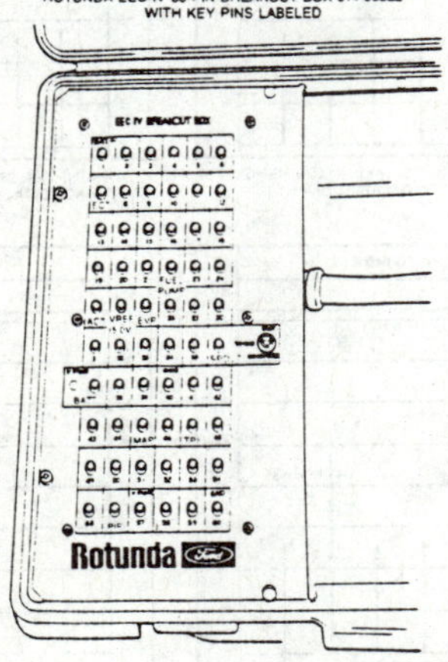

Intermittent Quits On Road And Restarts

STEPS	POSSIBLE ACTION TO TAKE
• **NOTE: Refer to the correct vehicle "Engine Supplement" Section for the correct pin numbers for the points listed.**	
• Test drive the vehicle. If it quits, **do not** touch the ignition key. Coast to the side of the road. • Check the reading obtained on the breakout box with the baseline readings taken before the test drive.	• Perform a Quick Test according to procedures in Volume H.

Intermittent Bucking And Jerking

STEPS	POSSIBLE ACTION TO TAKE
• Review the repair order. **If required, contact the driver or drivers of the vehicle to get further information. Refer to the Customer Contact Sheet on page 2 of this Section.**	• If repair order shows any pattern (e.g., vehicle bucks or jerks at a certain speed when the engine is warm or cold), duplicate the condition and try to start the vehicle.
• Visually inspect the engine compartment. **DO NOT disturb anything.**	• If you determine by a visual inspection that anything is disconnected, **do not reconnect the item. If you move any component or connection, you could make it harder to repeat the intermittent problem. NOTE: Look for areas where electrical wires or vacuum lines could be cut by sharp edges or burned by engine hot spots. Blocked or kinked vacuum lines can also cause trouble.**
• Check the ignition system secondary wires for damage or loose connections.	• If ignition system secondary wires are not damaged or loose, continue to the next step. • If secondary wires are damaged, replace them and try to duplicate the intermittent fault. • If secondary wires are loose, repair as necessary and try to duplicate the intermittent fault.
• Check distributor cap for moisture, cracks, or salt-like deposits. 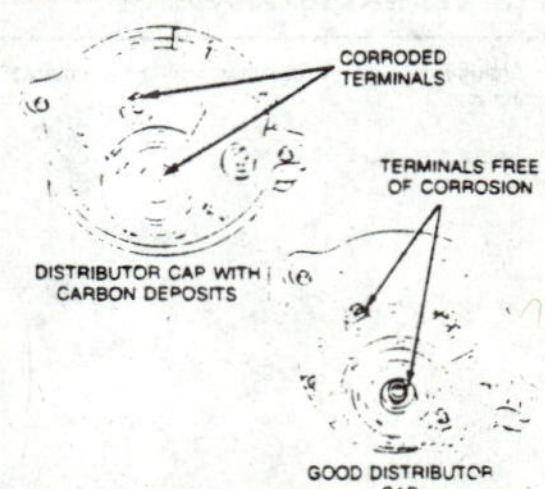 	• If distributor cap has moisture, find the cause. If necessary, replace the cap. • If distributor cap has cracks or salt-like deposits, replace the cap with a new one. • Start the vehicle and try to duplicate the intermittent fault.
• Check the rotor for damage or burns.	• If rotor is damaged or burned, replace with a new one.

Intermittent Bucking And Jerking

STEPS	POSSIBLE ACTION TO TAKE
• Check the TFI ignition module for damage or loose connections. **NOTE: Do not use the TFI ignition module as a fulcrum for turning the distributor.** 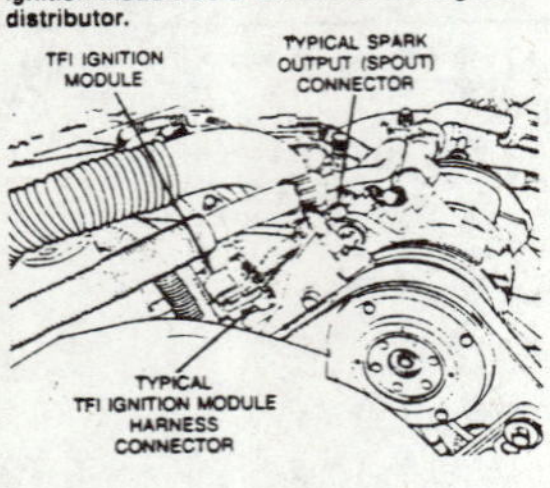	• Replace the TFI ignition module if damaged.
• Check the TFI ignition module six-pin connector to ensure it is installed correctly and that the gasket is properly installed and not folded over.	• Repair any loose connections. Reinstall connector if not properly installed.
• Install a spark tester to the ignition system coil wire. **NOTE: Disconnect the wire at the distributor.** Crank the engine. 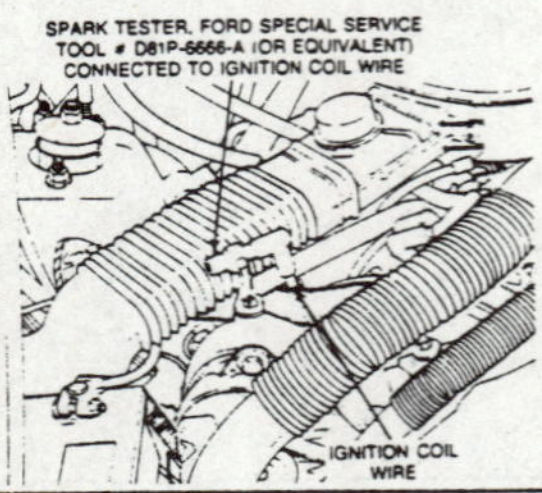	• If a good, strong blue spark is not detected, perform ignition diagnostics according to Ignition Section in Volume H.

Intermittent Bucking And Jerking

STEPS	POSSIBLE ACTION TO TAKE
• Verify that a good, strong blue spark is observed while the engine is cranking. This indicates that the coil and primary circuit are OK. • **NOTE: By connecting the spark tester to the coil wire, you are able to verify that the coil is firing on each PIP signal.** **NOTE: Never allow the spark from the coil wire to go directly to ground. This could cause internal damage to the coil.**	• If an intermittent spark is present, the ignition coil may be damaged internally.
• With the spark tester still connected to the coil wire, wiggle the ignition primary connectors while cranking the engine. **NOTE: By performing this test while cranking, those circuits dedicated to the start mode are checked.**	• If an intermittent loss of spark is detected, there could be an intermittent open in the TFI ignition module connector or harness that could cause an intermittent bucking and jerking. • **NOTE: An intermittent open in the TFI ignition module or harness could cause an intermittent bucking and jerking by preventing the TFI ignition module from firing the coil at certain times.** • **NOTE: An intermittent open in the coil primary harness or connecter could cause an intermittent bucking and jerking because the coil is not receiving battery voltage.**
• **NOTE: For the following test the engine MUST be running.** Use a jumper wire with an alligator clip at both ends. Attach one of the clips to a **good** ground location. Attach the other end to the metal part of a screwdriver. While the engine is running, pass the screwdriver around the base of the coil. **WARNING: USE CAUTION NOT TO GET THE SCREWDRIVER OR WIRE ENTANGLED WITH THE DRIVE BELTS OF THE VEHICLE.**	• Arcing indicates a breakdown in the insulation of the coil. • Replace the coil.

Intermittent Bucking And Jerking

STEPS	POSSIBLE ACTION TO TAKE
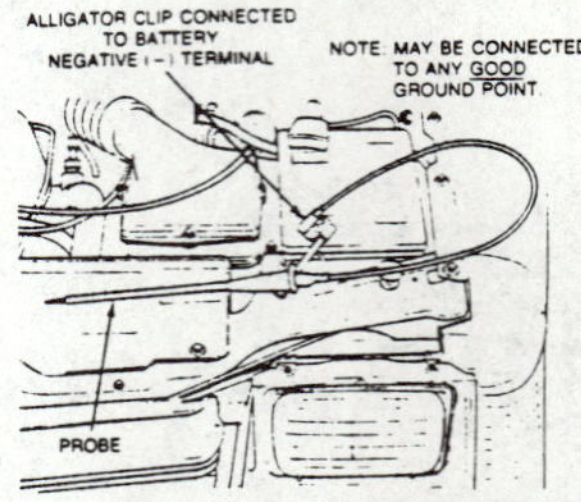	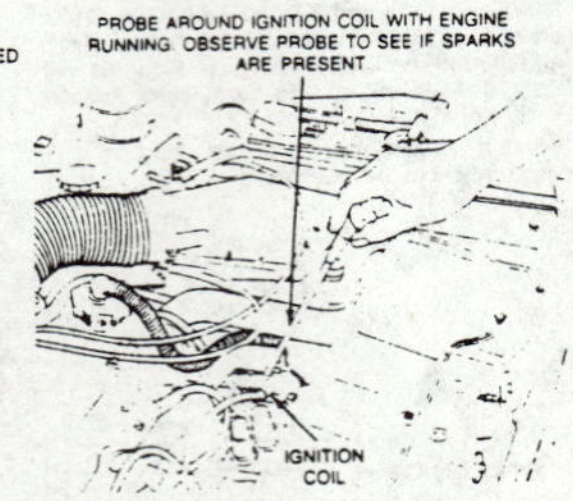
• Connect an engine analyzer to the vehicle. • Check plug firing voltages, cylinder balance, and base timing.	• If the engine analyzer shows a problem, repair or replace as required.

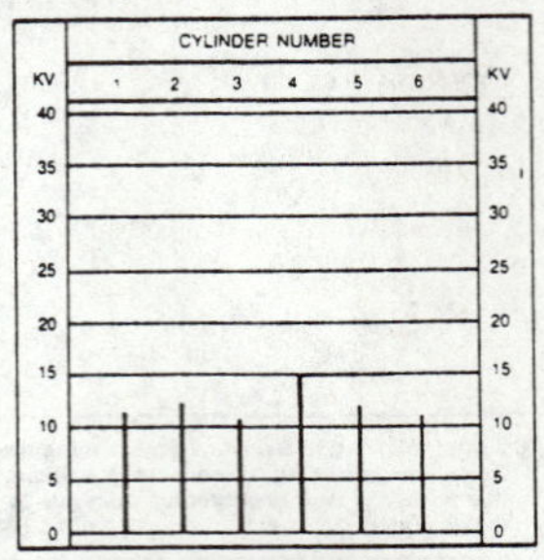

ENGINE ANALYZER DISPLAYS — PROPERLY OPERATING IGNITION SYSTEM

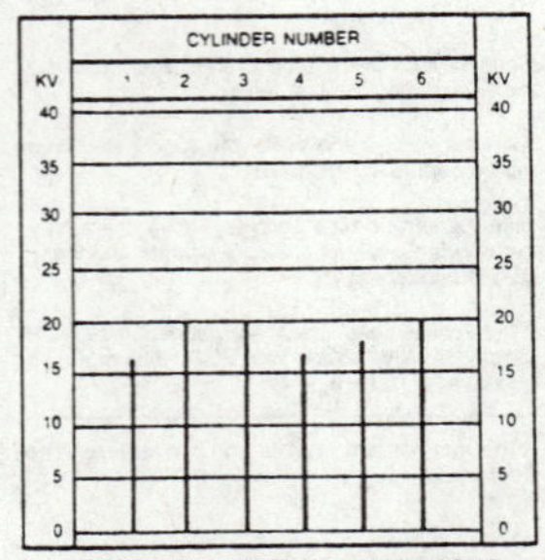

ENGINE ANALYZER DISPLAYS — AVERAGE PLUG FIRING VOLTAGE GREATER THAN 15KV

Intermittent Bucking And Jerking

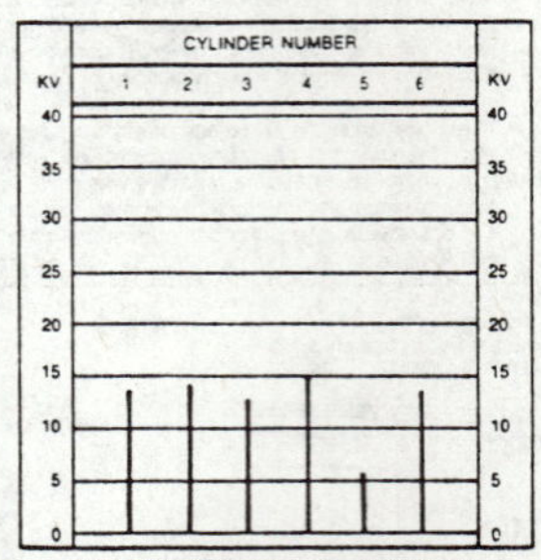

ENGINE ANALYZER DISPLAYS — PLUG FIRING VOLTAGE VARIATION GREATER THAN 5 KV

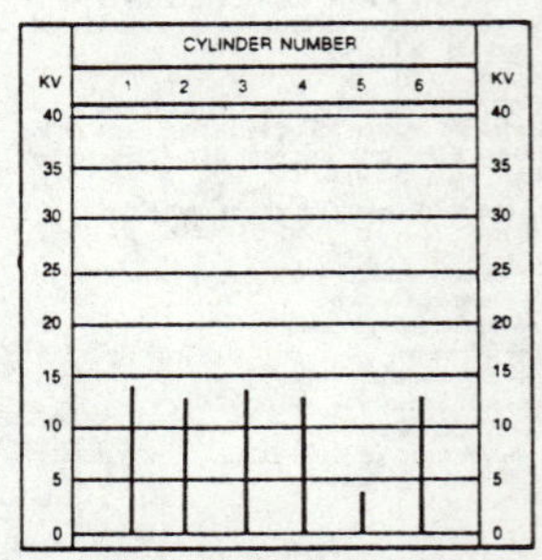

ENGINE ANALYZER DISPLAYS — LOW PLUG FIRING VOLTAGE IN ONE CYLINDER

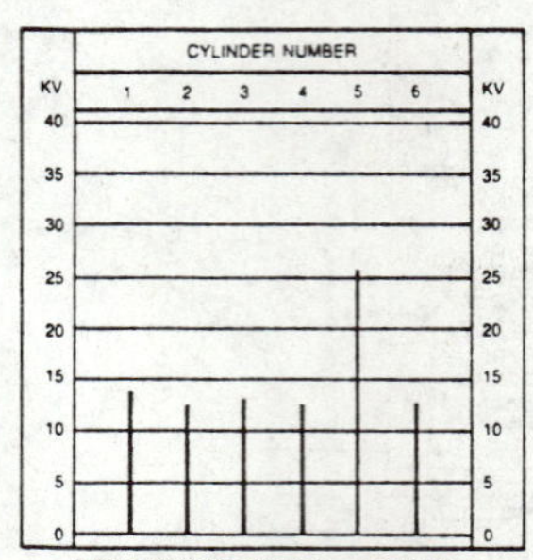

ENGINE ANALYZER DISPLAYS — HIGH PLUG FIRING VOLTAGE IN ONE CYLINDER

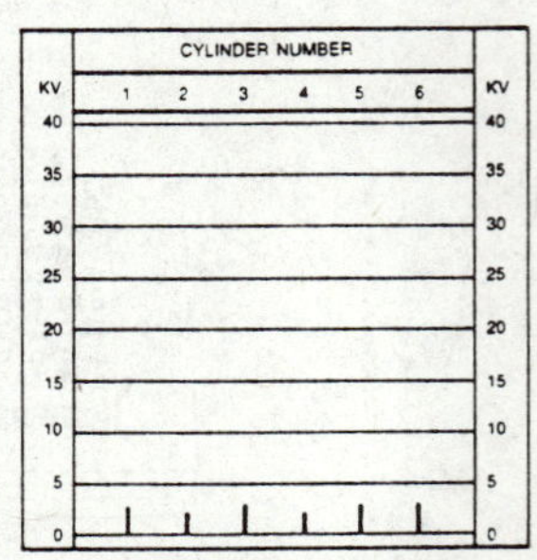

ENGINE ANALYZER DISPLAYS — NEGATIVE FIRING VOLTAGE

Intermittent Bucking And Jerking

STEPS	POSSIBLE ACTION TO TAKE
• When all tests have been performed on the engine analyzer, check the spark plugs for ash fouling, proper gap, and to make sure they are the right type of plug for the vehicle's engine.	• If the spark plugs are fouled, clean or replace as required. If the spark plugs have an improper gap, regap and install. If the spark plugs are not the correct type, replace with the correct type of plugs. • If the spark plugs check out OK, continue on to the next step.
• If the intermittent fault is still present, refer to Section 2.	**• If after referring to Section 2 and running the Quick Test according to procedures found in this book, the fault is not found, continue to the next step.**
NOTE: Do not continue to the next step unless the Quick Test has been performed according to procedures outlined in this book, and the intermittent fault has NOT been found.	
• Connect a MAP/BP tester to either the manifold absolute pressure (MAP) or barometric pressure (BP) sensor (depending on how vehicle is equipped) and check the readings with the engine Off and the ignition key in the ON position. • Compare the reading to the correct value listed on page 50 of this supplement. • If the vehicle is equipped with a MAP sensor, repeat the MAP test with the engine running. The MAP tester reading should decrease at idle and increase when you bump the throttle. **NOTE: If the vehicle is equipped with a barometric pressure (BP) sensor, the engine running portion of the test does not have to be performed since there is no vacuum hookup to the engine.**	• On a vehicle equipped with a MAP sensor, if the reading obtained with the engine OFF and ignition key ON does not match the value given in the Pinpoint Test, perform the appropriate Pinpoint Test in Volume H. • If the MAP sensor does not change when you bump the throttle with the engine running, substitute a known good hose for the original MAP sensor vacuum hose connected to the sensor and repeat the test. A change in the MAP tester reading indicates that the original hose was damaged, blocked or kinked. • On a vehicle equipped with either a MAP or BP sensor, substitute with a known good component and retest. If readings are now OK, replace the component.
NOTE: For 2.3L and 2.5L HSC engines, and other applications without a Schrader Valve installed in the fuel system, use Rotunda Tool D85L-9974-B, In-Line Pressure Tester Adapter. • If the vehicle is equipped with a vane meter, **carefully** check for unmetered air leaks (i.e., loose air tubes, missing hoses or vacuum caps, etc.)	• Correct air or vacuum leaks as necessary.

Intermittent Bucking And Jerking

STEPS	POSSIBLE ACTION TO TAKE
SOME TYPICAL PLACES TO CHECK FOR AIR LEAKS ARE SHOWN WITH AN ARROW (2.3L ENGINE SHOWN — 1.9L SIMILAR) 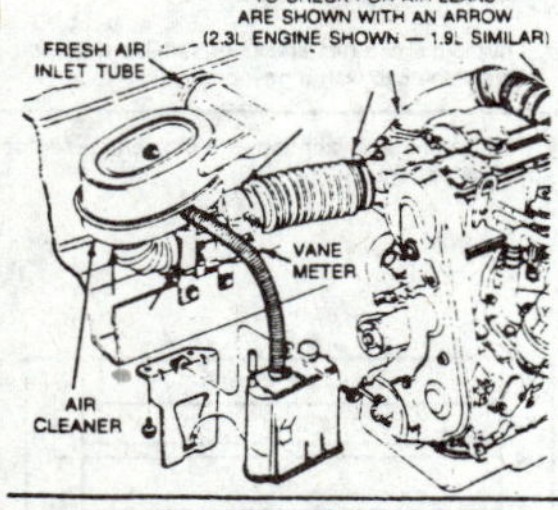	
• With the breakout box installed monitor vane meter signal while you manually move the vane meter door through its entire range.	• Voltage must increase smoothly as door is opened. • If not, repair or replace.
• Install a fuel pressure gauge with an extension hose, so that the fuel pressure can be monitored during the test drive. (Tape the fuel pressure gauge to the windshield for easy viewing.) ROTUNDA FUEL PRESSURE TESTING GAUGE 014-00447 — TYPICAL HOOK-UP TO FUEL PRESSURE VALVE 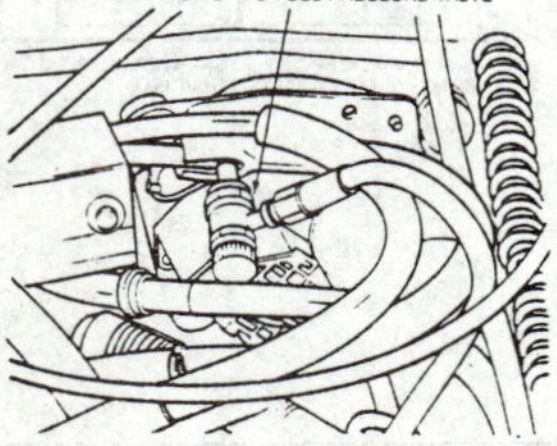	• Make a copy of the five-column chart at the end of this section and record the readings on the breakout box, vacuum gauge and fuel pressure gauge taken with the ignition key in the ON position while the engine was cranking. ROTUNDA FUEL PRESSURE TESTING GAUGE 014-00447 — TYPICAL HOOK-UP SHOWING GAUGE TAPED TO OUTSIDE OF WINDSHIELD 

Intermittent Bucking And Jerking

STEPS	POSSIBLE ACTION TO TAKE
• "Tee" a vacuum gauge and extension hose into the EGR valve vacuum line and manifold vacuum line (use a 0-to-100 inches of water vacuum gauge). Install a black STAR tester. Route the STAR tester and vacuum gauges inside the vehicle so you can read them.	• These are "baseline" readings of key voltages, vacuum condition, and fuel pressure while the engine is operating properly. By comparing these baseline readings with the readings when the intermittent symptom occurs, it is possible to isolate the component or circuit that is at fault.
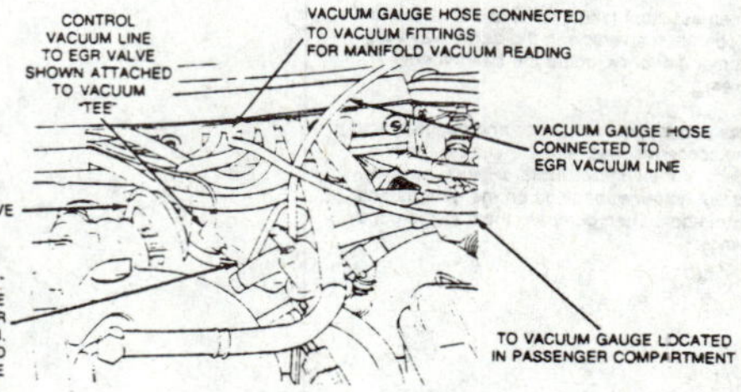	
• Hook up the breakout box with the following key points labeled on the breakout box: — battery voltage — reference voltage — throttle position sensor — PIP signal — ECT sensor — ACT sensor — SPOUT signal — fuel pump (circuit at ECA)	

Intermittent Bucking And Jerking

STEPS	POSSIBLE ACTION TO TAKE
ROTUNDA EEC-IV 60-PIN BREAKOUT BOX 014-00322 — WITH KEY PINS LABELED 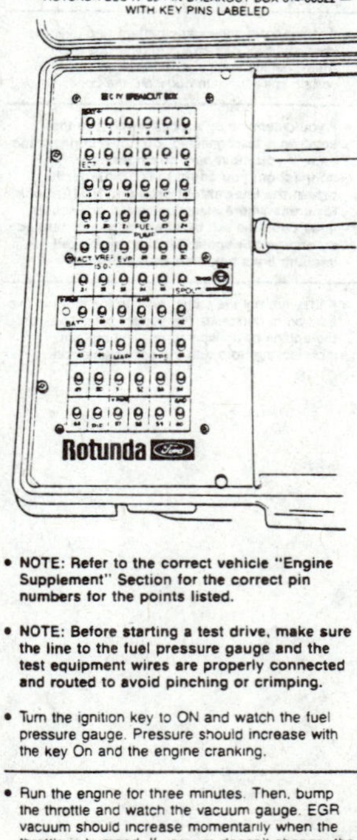 **• NOTE: Refer to the correct vehicle "Engine Supplement" Section for the correct pin numbers for the points listed.** **• NOTE: Before starting a test drive, make sure the line to the fuel pressure gauge and the test equipment wires are properly connected and routed to avoid pinching or crimping.** • Turn the ignition key to ON and watch the fuel pressure gauge. Pressure should increase with the key On and the engine cranking.	
• Run the engine for three minutes. Then, bump the throttle and watch the vacuum gauge. EGR vacuum should increase momentarily when the throttle is bumped. If vacuum doesn't change, the gauge may not be installed correctly.	• If during the test drive the fault is detected and the readings on the breakout box, vacuum gauges, or fuel pressure gauge differ from the baseline readings, this could be where the problem lies.

Intermittent Bucking And Jerking

STEPS	POSSIBLE ACTION TO TAKE
• Before going on the test drive, enter the continuous monitor or "wiggle test" mode with the black STAR tester. • Turn the radio to the low end of the AM band. **This checks for internal arcing in the coil, since the arcing would be picked up as noise on the radio.** • With an assistant take the vehicle on a test drive and operate the vehicle in the same way the customer did to try repeat the intermittent problem. • Check and record the sensor and gauge readings at the concern speed with the vehicle running properly. If the vehicle bucks or jerks, have the assistant note the readings on the gauges and breakout box. Then compare them to the good readings.	• If the STAR tester "beeps" during the test drive, enter the Quick Test and retrieve the memory code. Follow the procedures located in the Quick Test Section.

Intermittent Erratic Running/Rough Idle

STEPS	POSSIBLE ACTION TO TAKE
• Review the repair order. **If required, contact the driver or drivers of the vehicle to get further information. Refer to the Customer Contact Sheet on page 2 of this Section.**	• If repair order shows any pattern (e.g., vehicle idles rough or has erratic engine operation whenever the engine is warm or cold or at a certain speed), try to duplicate the condition.
• Visually inspect the engine compartment. **DO NOT disturb anything.**	• If you determine by a visual inspection that anything is disconnected, **do not reconnect the item. If you move any component or connection, you could make it harder to repeat the intermittent problem. NOTE: Look for areas where electrical wires or vacuum lines could be cut by sharp edges or burned by engine hot spots. Blocked or kinked vacuum lines can also cause trouble.**
• On EFI (multi-point) engines, check the color at the top of the fuel injectors to see that they are the same.	• If they are not the same, refer to the chart in this Section to determine the correct color code for the engine being worked on, and replace incorrect injectors with the correct injectors.

Intermittent Erratic Running/Rough Idle

The following charts list fuel injector applications, part numbers, source, color code and injector specifications.

HIGH PRESSURE CFI INJECTOR APPLICATION

ENGINE APPLICATION	MODEL YEAR	PART NO. - 9F593 -	SOURCE	COLOR CODE	INJECTOR SPECS. SIZE	INJECTOR SPECS. RESIST
					gm/sec	Ohms
1.6L	1983	E3EE-BA	Bosch	Blue	1.76 (14# HR)	2.25
	1984	E4EE-AA	N.D.	Blue		2.35
1.6L T/C	1984-85	E4EX-AA	Bosch	Black	3.00 (23# HR)	2.40
1.9L	1985-1/2-86	E6EE-AB	Bosch	White	2.45 (19# HR)	2.25
2.3L T/C	1983	E3ZE-BA	Bosch	Green	3.86 (30#/HR)	2.40
	1984	E4ZE-AA	D.K.K.			2.35
	1985	E5ZE-AA	D.K.K.	Brown	4.41 (35# HR)	2.35
	1986	E5ZE-AB	D.K.K.			
2.3L (Truck)	1985	E59E-AA	N.D.	Gray	1.76 (14#/HR)	16.20
	1986	E59E-AB	N.D.			
2.9L (Car/Truck)	1986	E67C-A1B	Bosch	Gray	1.76 (14#/HR)	14.50
3.0L (Car/Truck)	1986	E59E-AB	N.D.	Gray	1.76 (14#/HR)	16.20
		E67E-BB	Bosch			14.50
3.8 CFI	1984-86	E3VE-A1A	Bosch	Green	4.54 (37# HR)	2.25
		-A2A	N.D.			2.00
5.0L CFI	1980	EOSE-AA	Bosch	Blue	5.81 (46# HR)	2.40
	1983	EOSE-A1A	Bosch			2.40
		EOSE-A2A	N.D.			2.00
	1984	EOSE-A1A	Bosch			2.25
		EOSE-A2A	N.D.			2.00
5.0L H.O. CFI	1984-85	E4ZE-CA	Bosch	Gray	6.60 (52#/HR)	2.25
5.0L SEFI	1986	E59E-AB	N.D.	Gray	1.76 (14# HR)	16.20
		E67E-BB	Bosch			14.50
5.0L H.O. SEFI	1986	E6TE-AB	N.D.	Gold	2.45 (19# HR)	16.20
5.0L Truck	1985	E5TE-AA	Bosch	Gold	2.45 (19# HR)	15.50
		E5TE-AA	N.D			16.20
		E5TE-AA	D.K.K.			16.20
	1986	E5TE-AB	D.K.K.			16.20
		E5TE-BB	Bosch			14 50

LOW PRESSURE CFI INJECTOR APPLICATIONS

2.3L HSC	1985-86	E43E-AC	Bosch	Blue	7.00 (56# HR)	1.40
2.3L H.O. HSC (Manual Trans. Only)	1985	E53E-AB	Bosch	Green	8.00 (64# HR)	1.40
2.5L HSC	1986	E53E-AB	Bosch	Green	8.00 (64# HR)	1.40

Intermittent Erratic Running/Rough Idle

STEPS	POSSIBLE ACTION TO TAKE
• Check distributor cap for moisture, cracks, or salt-like deposits.	• If distributor cap has moisture, find out the cause. If necessary, replace the cap. • If distributor cap has cracks or salt-like deposits, replace the cap with a new one.
• Connect an engine analyzer to the vehicle. • Check plug firing voltages, cylinder balance, and base timing.	• If the engine analyzer shows a problem, repair or replace as required.

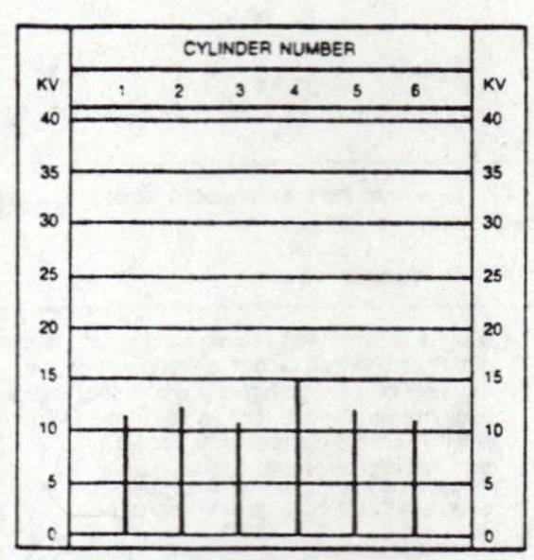

ENGINE ANALYZER DISPLAYS — PROPERLY OPERATING IGNITION SYSTEM

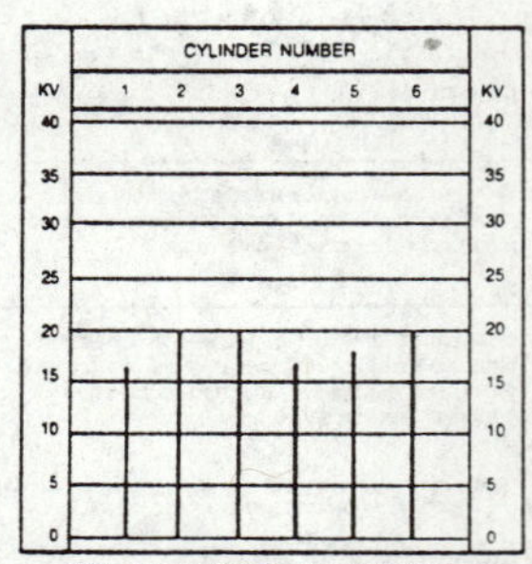

ENGINE ANALYZER DISPLAYS — AVERAGE PLUG FIRING VOLTAGE GREATER THAN 15KV

Intermittent Erratic Running/Rough Idle

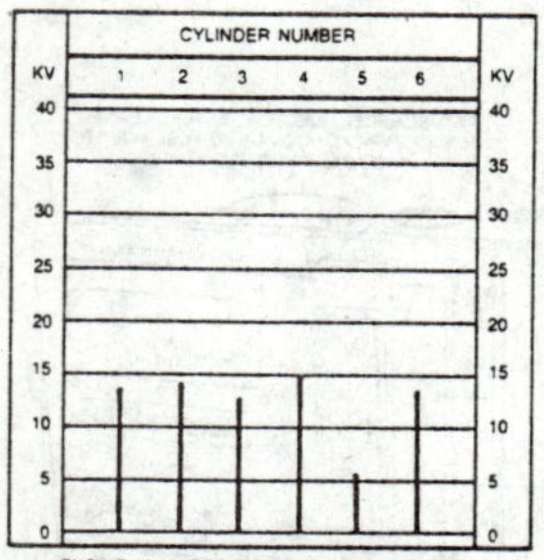

ENGINE ANALYZER DISPLAYS — PLUG FIRING VOLTAGE VARIATION GREATER THAN 5 KV

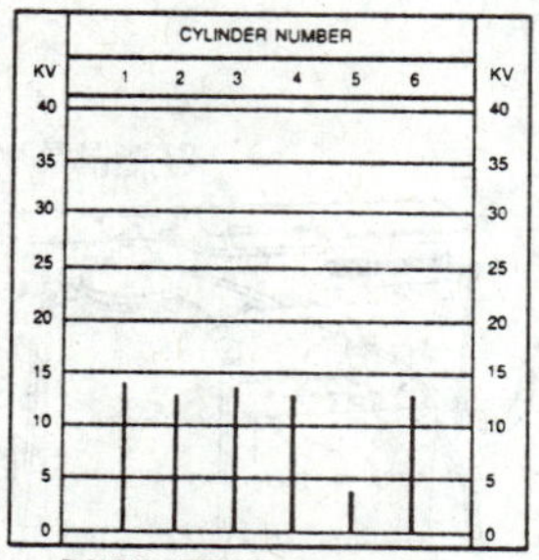

ENGINE ANALYZER DISPLAYS — LOW PLUG FIRING VOLTAGE IN ONE CYLINDER

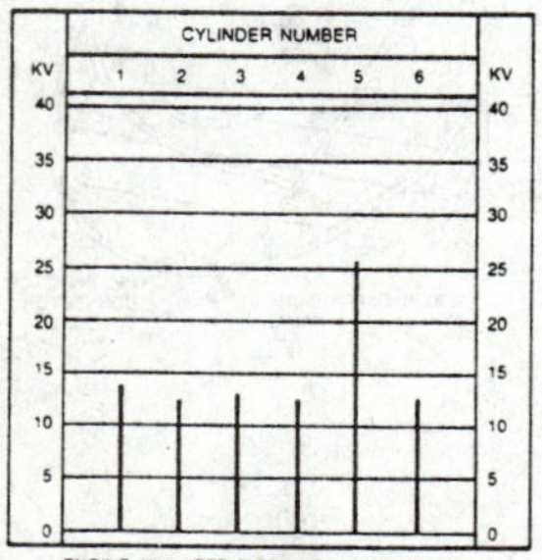

ENGINE ANALYZER DISPLAYS — HIGH PLUG FIRING VOLTAGE IN ONE CYLINDER

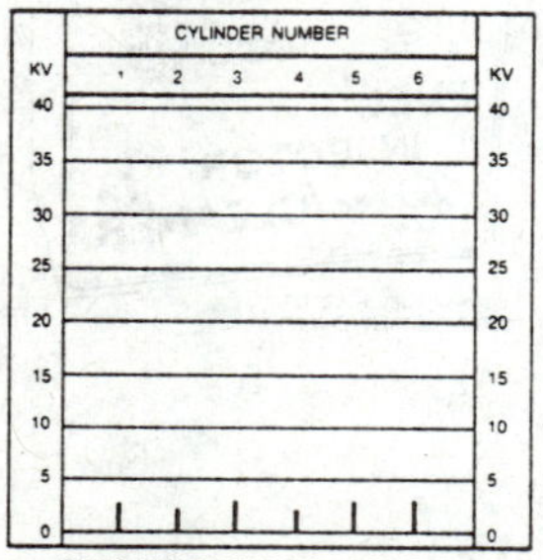

ENGINE ANALYZER DISPLAYS — NEGATIVE FIRING VOLTAGE

Intermittent Erratic Running/Rough Idle

STEPS	POSSIBLE ACTION TO TAKE
• When all tests have been performed on the engine analyzer, check the spark plugs for ash fouling, proper gap, and to make sure they are the right type of plug for the vehicle's engine. • **NOTE: Do not replace spark plugs at this time.**	• If the spark plugs are fouled, clean or replace as required. If the spark plugs have an improper gap, regap and install. If the spark plugs are not the correct type, replace with the correct type of plugs. • If the spark plugs check out OK, continue to the next step.
• Perform a compression test to check for proper engine and valve train operation.	• If compression check shows a problem, repair or replace as required. • If compression check is OK, continue to the next step.
• Check out the TFI ignition module, profile ignition pickup, and coil by following the TFI-IV worksheet provided in this Section.	• If the TFI worksheet leads to the problem, repair or replace as required.

Intermittent Erratic Running/Rough Idle

TFI — WORKSHEET EEC-IV SYSTEMS — TYPICAL

TO PERFORM THESE TESTS YOU WILL NEED: 12V TESTLITE, DVOM AND A SUPPLY OF STRAIGHT PINS.

NOTE: Prior to starting Step 1, disconnect injectors to prevent flooding engine with fuel.

1. Voltage at positive coil — desired 12V, actual ________ (Key On Engine Off).
2. Voltage at negative coil — desired 12V, actual ________ (Key On Engine Off).

   TESTS 1 and 2 PERFORMED WITH IGNITION KEY ON.

3. Using testlite at negative coil, crank engine, watch that testlite flickers OK. If it does not flicker, **problem is indicated. Turn key OFF.**
4. Disconnect coil connector, measure primary and secondary coil resistances, also path to ground.

   Primary coil — .5-1.5 ohms desired, actual ________.

   Secondary coil — 7K-15K ohms desired, actual ________.

   Path to ground — ohms infinity desired, actual ________.

   Reconnect primary and secondary wires.

5. Disconnect TFI module connector at distributor, with DVOM on 20-volt scale.
6. Insert straight pins into connector to measure in wiring harness:

   "Run" voltage — desired 12V, actual ________ (Key On Engine Off).

   "Coil" voltage — desired 12V, actual ________ (Key On Engine Off).

   "Start" voltage — desired 8-10V, actual ________ (engine cranking)

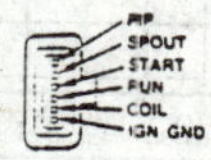

7. Check distributor to engine block ground — desired 0 ohms, actual ________.
8. Check TFI module to distributor ground at Pin 6 — desired 0 ohms, actual ________. Reconnect TFI connector.
9. Check Hall effect device output at SPOUT harness connector while cranking engine — desired 5-7V, actual ________ (distributor side of connector).
10. If #9 is not correct, disconnect ECA and redo test at Pin #56 — desired 5-7V, actual ________. **NOTE: Either a power relay or damaged ECA can cause "no-spark."**
11. If #10 is not correct, check continuity of blue PIP wire to distributor from Pin #56.
12. If #10 is correct, reconnect harness to a substitute processor and redo Step #9.
13. If you do not get 5-7V on SPOUT or PIP wires during crank in any of steps 9-12, Hall effect device in distributor is **probably** shorted.
14. Once you have a good 5-7V PIP signal, if you still do not have pulsating at negative coil the TFI module is bad — provided all the other steps 1-8 are OK. Reconnect injectors.

DO NOT SHORT-CUT THIS TEST — IF YOU DO SO YOU CAN DESTROY A NEW MODULE, A NEW HALL EFFECT DEVICE, A NEW PROCESSOR OR ALL THREE.

Intermittent Erratic Running/Rough Idle

STEPS	POSSIBLE ACTION TO TAKE
• **NOTE: Be sure to use a low-current test light to avoid damaging the electronic control assembly (ECA).**	• If the TFI ignition module, profile ignition pickup, and coil are OK, continue on to the next step.
• Check out the SPOUT connector to ensure it is properly seated.	• If the SPOUT connector is loose, repair or replace as required. • If the SPOUT connector is seated properly, continue to the next step.
• **NOTE: For the following test, the engine MUST be running.** Use a jumper wire with an alligator clip at both ends. Attach one of the clips to a good ground location. Attach the other end to the metal part of a screwdriver. While the engine is running, pass the screwdriver around the base of the coil. **WARNING: USE CAUTION NOT TO GET THE SCREWDRIVER OR WIRE ENTANGLED WITH THE DRIVE BELTS OF THE VEHICLE.**	• Arcing indicates a breakdown in the insulation of the coil. If the coil is malfunctioning, the primary circuit could be open intermittently, causing an intermittent no-start condition.
• **If the intermittent fault is still present, refer to Section 2.**	• **If after referring to Section 2 and running the Quick Test according to procedures found in this book, the fault is not found, continue to the next step.**
NOTE: Do not continue to the next step unless the Quick Test has been performed according to procedures outlined in this book, and the intermittent fault has NOT been found.	
• Connect a MAP/BP tester to either the manifold absolute pressure (MAP) or barometric pressure (BP) sensor (depending on how vehicle is equipped) and check the readings with the engine Off and the ignition key in the ON position. • Compare the reading to the correct value listed on page 50 of this supplement. • If the vehicle is equipped with a MAP sensor, repeat the MAP test with the engine running. The MAP tester reading should decrease at idle and increase when you bump the throttle. **NOTE: If the vehicle is equipped with a barometric pressure (BP) sensor, the engine running portion of the test does not have to be performed since there is no vacuum hookup to the engine.**	• On a vehicle equipped with a MAP sensor, if the reading obtained with the engine OFF and ignition key ON does not match the value given in the Pinpoint Test, perform the appropriate Pinpoint Test in Volume H. • If the MAP sensor does not change when you bump the throttle with the engine running, substitute a known good hose for the original MAP sensor vacuum hose connected to the sensor and repeat the test. A change in the MAP tester reading indicates that the original hose was damaged, blocked or kinked. • On a vehicle equipped with either a MAP or BP sensor, substitute with a known good component and retest. If readings are now OK, replace the component.

Intermittent Erratic Running/Rough Idle

STEPS	POSSIBLE ACTION TO TAKE
• If the vehicle is equipped with a vane meter, **carefully** check for unmetered air leaks (i.e., loose air tubes, missing hoses or vacuum caps, etc.)	• Correct air or vacuum leaks as necessary.

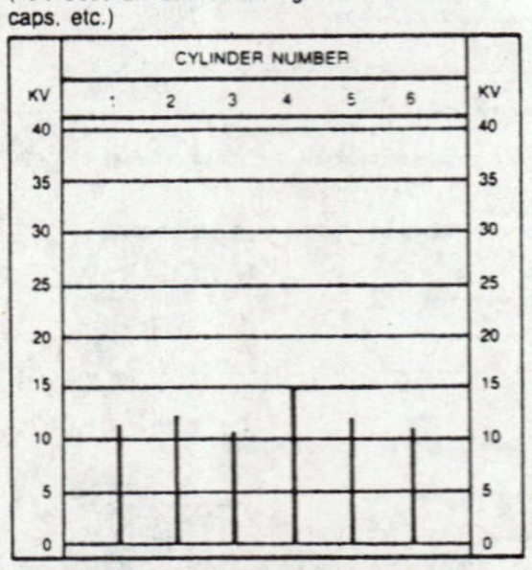

ENGINE ANALYZER DISPLAYS — PROPERLY OPERATING IGNITION SYSTEM

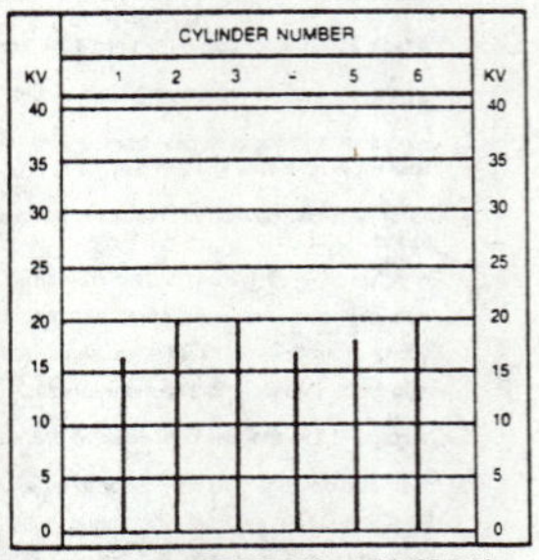

ENGINE ANALYZER DISPLAYS — AVERAGE PLUG FIRING VOLTAGE GREATER THAN 15KV

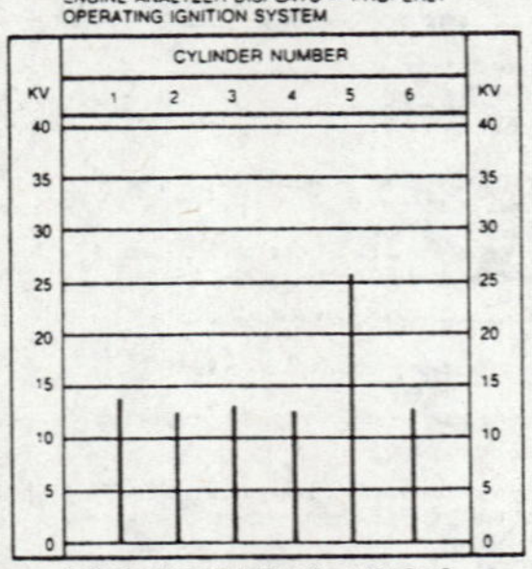

ENGINE ANALYZER DISPLAYS — HIGH PLUG FIRING VOLTAGE IN ONE CYLINDER

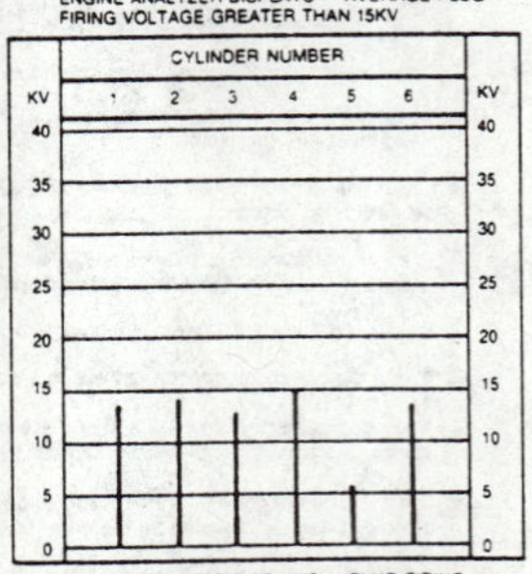

ENGINE ANALYZER DISPLAYS — PLUG FIRING VOLTAGE VARIATION GREATER THAN 5 KV

Intermittent Erratic Running/Rough Idle

STEPS	POSSIBLE ACTION TO TAKE

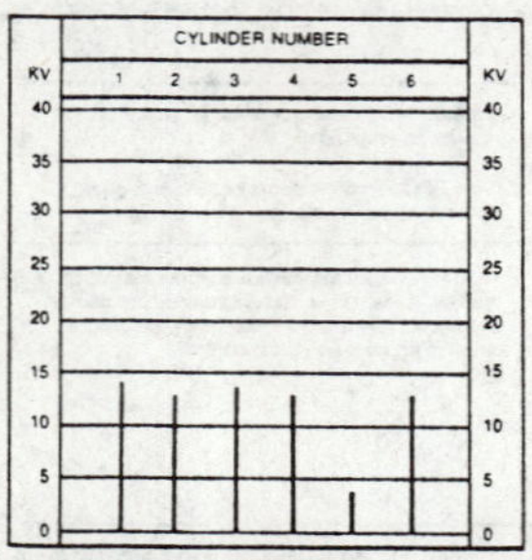

ENGINE ANALYZER DISPLAYS — LOW PLUG FIRING VOLTAGE IN ONE CYLINDER

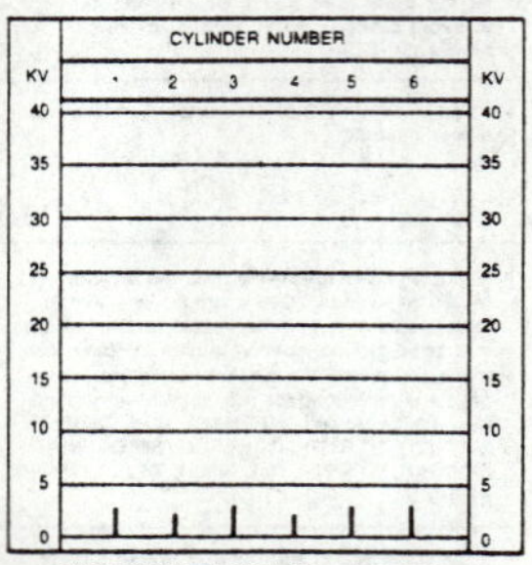

ENGINE ANALYZER DISPLAYS — NEGATIVE FIRING VOLTAGE

STEPS	POSSIBLE ACTION TO TAKE
• With breakout box installed monitor vane meter signal while you manually move the vane meter door through its entire range.	• Voltage must increase smoothly as door is opened.
• Connect a Rotunda Fuel Injector Tester/Cleaner to the vehicle according to instructions that come with the unit. • Check the flow of injectors against the specifications given with the Tester/Cleaner and the injector chart located in this test.	• If injector flow is not up to specifications, clean the injectors. Follow the instructions that come with the Tester/Cleaner.

Intermittent Erratic Running/Rough Idle

STEPS	POSSIBLE ACTION TO TAKE

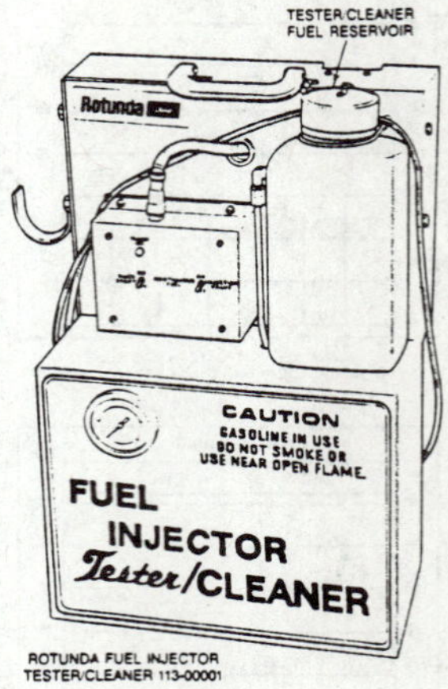

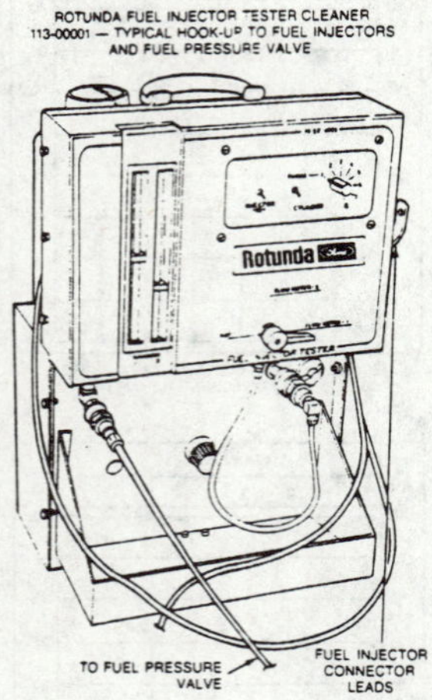

Intermittent Erratic Running/Rough Idle

The following charts list fuel injector applications, part numbers, source, color code and injector specifications.

HIGH PRESSURE CFI INJECTOR APPLICATION

ENGINE APPLICATION	MODEL YEAR	PART NO. - 9F593 -	SOURCE	COLOR CODE	INJECTOR SPECS. SIZE	INJECTOR SPECS. RESIST
					gm/sec	Ohms
1.6L	1983	E3EE-BA	Bosch	Blue	1.76 (14#/HR)	2.25
	1984	E4EE-AA	N.D.	Blue		2.35
1.6L T/C	1984-85	E4EX-AA	Bosch	Black	3.00 (23#/HR)	2.40
1.9L	1985-1/2-86	E6EE-AB	Bosch	White	2.45 (19#/HR)	2.25
2.3L T/C	1983	E3ZE-BA	Bosch	Green	3.86 (30#/HR)	2.40
	1984	E4ZE-AA	D.K.K.			2.35
	1985	E5ZE-AA	D.K.K.	Brown	4.41 (35#/HR)	2.35
	1986	E5ZE-AB	D.K.K.			
2.3L (Truck)	1985	E59E-AA	N.D.	Gray	1.76 (14#/HR)	16.20
	1986	E59E-AB	N.D.			
2.9L (Car/Truck)	1986	E67C-A1B	Bosch	Gray	1.76 (14#/HR)	14.50
3.0L (Car/Truck)	1986	E59E-AB	N.D.	Gray	1.76 (14#/HR)	16.20
		E67E-BB	Bosch			14.50
3.8 CFI	1984-86	E3VE-A1A	Bosch	Green	4.54 (37#/HR)	2.25
		-A2A	N.D.			2.00
5.0L CFI	1980	EOSE-AA	Bosch	Blue	5.81 (46#/HR)	2.40
	1983	EOSE-A1A	Bosch			2.40
		EOSE-A2A	N.D.			2.00
	1984	EOSE-A1A	Bosch			2.25
		EOSE-A2A	N.D.			2.00
5.0L H.O. CFI	1984-85	E4ZE-CA	Bosch	Gray	6.60 (52#/HR)	2.25
5.0L SEFI	1986	E59E-AB	N.D.	Gray	1.76 (14#/HR)	16.20
		E67E-BB	Bosch			14.50
5.0L H.O. SEFI	1986	E6TE-AB	N.D.	Gold	2.45 (19#/HR)	16.20
5.0L Truck	1985	E5TE-AA	Bosch	Gold	2.45 (19#/HR)	15.50
		E5TE-AA	N.D.			16.20
		E5TE-AA	D.K.K.			16.20
	1986	E5TE-AB	D.K.K.			16.20
		E5TE-BB	Bosch			14.50

LOW PRESSURE CFI INJECTOR APPLICATIONS

2.3L HSC	1985-86	E43E-AC	Bosch	Blue	7.00 (56#/HR)	1.40
2.3L H.O. HSC (Manual Trans. Only)	1985	E53E-AB	Bosch	Green	8.00 (64#/HR)	1.40
2.5L HSC	1986	E53E-AB	Bosch	Green	8.00 (64#/HR)	1.40

Intermittent Erratic Running/Rough Idle

STEPS	POSSIBLE ACTION TO TAKE
• Check the injector flow again.	• If the injectors still do not meet specifications, replace them.

NOTE: For 2.3L and 2.5L HSC engines, and other applications without a Schrader Valve installed in the fuel system, use Rotunda Tool D85L-9974-B, In-Line Pressure Tester Adapter.

STEPS	POSSIBLE ACTION TO TAKE
• Install a fuel pressure gauge with an extension hose, so that the fuel pressure can be monitored during the test drive. (Tape the fuel pressure gauge to the windshield for easy viewing.)	• Make a copy of the five-column chart at the end of this section and record the readings on the breakout box, vacuum gauge, and fuel pressure gauge taken with the ignition key in the ON position while the engine was cranking. • These are "baseline" readings of key voltages, vacuum condition and fuel pressure while the engine is operating properly. By comparing these baseline readings with the readings when the intermittent symptom occurs, it is possible to isolate the component or circuit that is at fault.

ROTUNDA FUEL PRESSURE TESTING GAUGE 014-00447 — TYPICAL HOOK-UP TO FUEL PRESSURE VALVE

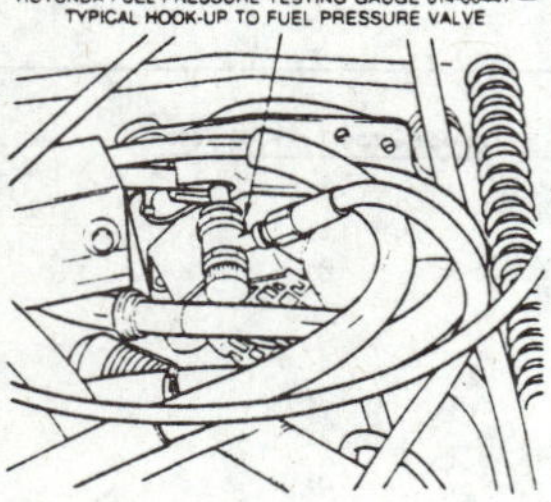

ROTUNDA FUEL PRESSURE TESTING GAUGE 014-00447 — TYPICAL HOOK-UP SHOWING GAUGE TAPED TO OUTSIDE OF WINDSHIELD

STEPS	POSSIBLE ACTION TO TAKE
• "Tee" a vacuum gauge and extension hose into the EGR valve vacuum line and manifold vacuum line (use a 0-to-100 inches of water vacuum gauge). Install a black STAR tester. Route the STAR tester and vacuum gauges inside the vehicle so you can read them.	

Intermittent Erratic Running/Rough Idle

STEPS	POSSIBLE ACTION TO TAKE
ROTUNDA EEC-IV 60-PIN BREAKOUT BOX 014-00322 — WITH KEY PINS LABELED 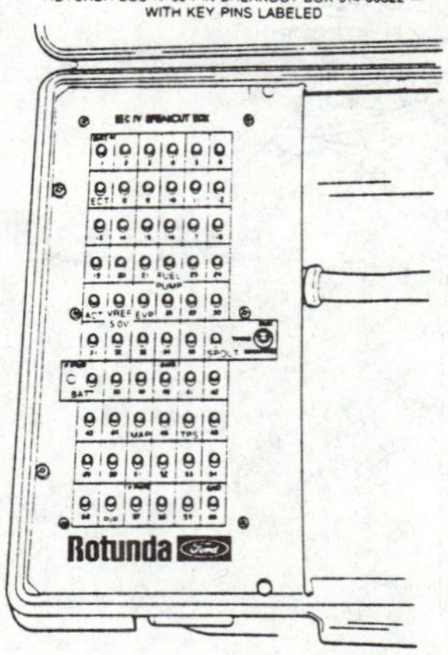 	
• **NOTE: Refer to the correct vehicle "Engine Supplement" Section for the correct pin numbers for the points listed.** • **NOTE: Before starting the test drive, make sure the line to the fuel pressure gauge and the test equipment wires are properly connected and routed to avoid pinching or crimping.** • Turn the ignition key to ON and watch the fuel pressure gauge. Pressure should increase with the key On and the engine cranking.	

Intermittent Erratic Running/Rough Idle

STEPS	POSSIBLE ACTION TO TAKE

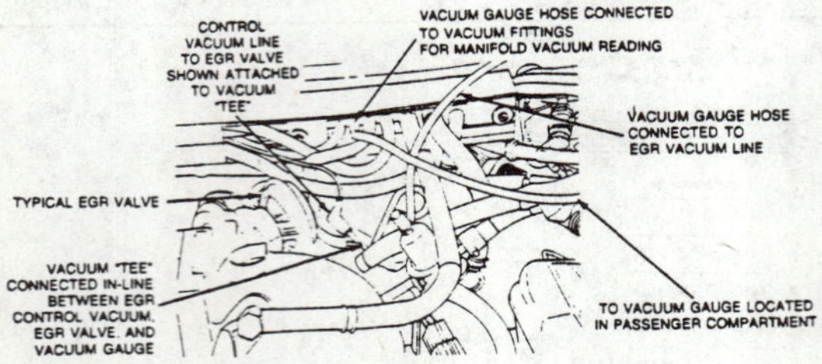

STEPS	POSSIBLE ACTION TO TAKE
• Hook up the breakout box with the following key points labeled on the breakout box: — battery voltage — reference voltage — throttle position sensor — PIP signal — ECT sensor — ACT sensor — SPOUT signal — fuel pump (circuit at ECA)	

Intermittent Erratic Running/Rough Idle

STEPS	POSSIBLE ACTION TO TAKE
• Run the engine for three minutes. Then, bump the throttle and watch the vacuum gauge. EGR vacuum should increase momentarily when the throttle is bumped. If vacuum doesn't change, the gauge may not be installed correctly. • Before going on the test drive, enter the continuous monitor or "wiggle test" mode with the black STAR tester. • Turn the radio to the low end of the AM band. **This checks for internal arcing in the coil, since the arcing would be picked up as noise on the radio.** • With an assistant take the vehicle on a test drive and operate the vehicle in the same way the customer did to try to repeat the intermittent problem. • Check and record the sensor and gauge readings at the concern speed with the vehicle running properly. If the engine starts to idle erratically, have the assistant note the readings on the gauges and breakout box. Then, compare them to the good readings.	• If during the test drive the fault is detected and the readings on the breakout box, vacuum gauges, or fuel pressure gauge differ from the baseline readings, this could be where the problem lies. • If the STAR tester "beeps" during the test drive, enter the Quick Test and retrieve the memory code. Follow the procedures located in the Quick Test Section.

Approximate Altitude (Ft.)	Voltage Output (+/−.04 Volts)
0	1.59
1000	1.56
2000	1.53
3000	1.50
4000	1.47
5000	1.44
6000	1.41
7000	1.39

NOTE: Measure several known good MAP sensors on available vehicles. The measured voltage will be typical for your location on the day of testing.

Electronic Diagnosis

1988 FORD CARS & TRUCKS
EEC-IV SYSTEMS

6

QUICK TEST: Visual Check Vehicle Preparation	1.0

SPECIAL NOTES:

- Correct results of the QUICK TEST are dependent on the proper operation of related non-EEC-IV components.
- It may be necessary to disconnect or disassemble harness connector assemblies to do some of the inspections. Pin locations should be noted before disassembly.
- If the engine will not start, starts but stalls, idles rough, or runs rough; continue through QUICK TEST STEP 3.0 and follow the instructions in Step 3.0B.

VISUAL CHECK

1. Inspect the air cleaner and inlet ducting.
2. Check all engine vacuum hoses for damage, leaks, cracks, blockage, proper routing, etc.
3. Check EEC-IV system wiring harness for proper connections, bent or broken pins, corrosion, loose wires, proper routing, etc.
4. Check the processor, sensors and actuators for physical damage.
5. Check the engine coolant for proper level.
6. Make all necessary repairs before continuing with QUICK TEST.

VEHICLE PREPARATION

1. Perform **ALL** safety steps required to start and run vehicle tests - apply parking brake, place shift lever firmly into PARK position (NEUTRAL on manual transmission), block drive wheels, etc.
2. Turn off **ALL** electrical loads — radios, lights, A/C-heater blower fans, etc.
3. Start engine and run until at operating temperature.
4. Turn engine off and proceed to QUICK TEST STEP 2.0.

QUICK TEST: Equipment Hookup	2.0

SPECIAL NOTES:

- Refer to the illustrations for Self-Test connector pin orientation and VOM and STAR hookup.
- After the equipment is properly hooked up, proceed to QUICK TEST STEP 3.0A.

USING THE STAR TESTER

1. Turn the ignition key off.
2. Connect the color coded adapter cable to the STAR tester.
3. Connect the adapter cable leads to the proper Self-Test connectors.
4. Connect the timing light.

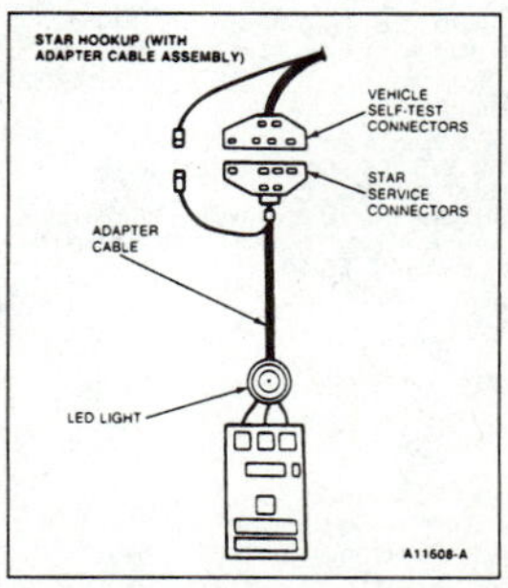

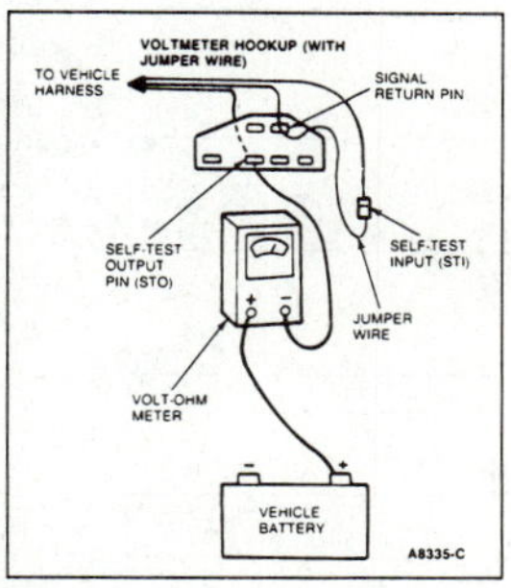

USING AN ANALOG VOLT/OHM METER (VOM)

1. Turn the ignition key off.
2. Set the VOM on a DC voltage range to read from 0 to 15 volts.
3. Connect the VOM from the Battery + terminal to the Self-Test Output pin of the large Self-Test connector.
4. Connect the timing light.

QUICK TEST: Key On Engine Off Self-Test — 3.0

A PERFORMING THE KEY ON ENGINE OFF SELF-TEST

SPECIAL NOTES:

- It may be necessary to service Non-EEC-IV faults before running Quick Test. Refer to Section 2.
- Continuous Memory Codes recorded in this step will be used for diagnosis in Step 6.0 after a PASS code 11 is received in both the Key On Engine Off and the Engine Running Self-Tests.
- On all vehicles equipped with a **4.9L ENGINE**, the clutch must be depressed during the Key On Engine Off Self-Test.

HOW TO RUN THE KEY ON ENGINE OFF SELF-TEST

DO

- Verify that the vehicle has been properly prepared according to **QUICK TEST STEPS** 1.0 and 2.0.
- Place ignition key in the ON position.
- Activate Self-Test.
  - — STAR Tester: Latch the center button in the down position.
  - — Analog VOM: Jumper STI to SIG RTN at the Self-Test connectors.
  - — "Check Engine" Light (MIL): Jumper STI to SIG RTN at the Self-Test connectors. Service Codes will be flashed on the "Check Engine" Light.
  - — Message Center (Continental Applications Only): Refer to Appendix: Self-Test.

NOTE: Refer to Appendix Self-Test for further information on how to read code output.

- Record all service codes displayed.
- Go to **PART B** of Key On Engine Off Self-Test.

DON'T

- Depress throttle during Key On Engine Off Self-Test.

QUICK TEST: Key On Engine Off Self-Test — 3.0

B CODE OUTPUT

Key On Engine Off	Separator	Continuous Memory	ACTION TO TAKE
11	— 1(0) —	11	• Both tests indicate a PASS. — If engine idles rough or runs rough, Go to Pinpoint Test Step S1. If this symptom is not present, Go to QUICK TEST STEP 4.0. — If engine is a no start, Go directly to Pinpoint Test Step A1.
ANY CODE(S)	— 1(0) —	11	• Key On Engine Off Self-Test indicates a FAULT. — Go to **PART C** of Key On Engine Off Self-Test. — Always start with the first code displayed.
ANY CODE(S)	— 1(0) —	ANY CODE(S)	• Both tests indicate a FAULT. — DO NOT SERVICE CONTINUOUS MEMORY CODES AT THIS TIME. — Go to **PART C** of Key On Engine Off Self-Test. — Always start with the first code displayed.
11	— 1(0) —	ANY CODE(S) EXCEPT 15, 56 or 66	• Continuous Memory indicates a FAULT. — DO NOT SERVICE CONTINUOUS MEMORY CODES AT THIS TIME. — If engine idles rough or runs rough, Go to Pinpoint Test Step S1. If this symptom is not present, Go to QUICK TEST STEP 4.0.
11	— 1(0) —	15	• Go To Pinpoint Test Step QB1.
11 11	— 1(0) 1(0) —	56 66	• For 5.0L SEFI MA Mustang only. — Service Code 56 Go to Pinpoint Test Step DC10. — Service Code 66 Go to Pinpoint Test Step DC6.
NO CODES OUTPUTTED CODES NOT LISTED			• Self-Test did not activate or unlisted codes displayed. — Repeat Key On Engine Off Self-Test to verify the above condition. — If condition still exists, Go to Pinpoint Test Step QA1.

QUICK TEST: Key On Engine Off Self-Test — 3.0

C PASSENGER CAR SERVICE CODE CHART

Key On Engine Off Service Code	Pinpoint Test Step Direction										
	1.9L EFI	1.9L CFI	2.3L OHC EFI	2.3L TC EFI	2.3L HSC EFI	2.5L CFI	3.0L EFI	3.8L AXOD EFI	3.8L RWD EFI	5.0L SEFI	5.0L MA SEFI
13 GO to ▶	—	KB1	—	—	—	KB1	—	—	—	—	—
15 GO to ▶	QB3	QB3	QB3	QB3	QB3	QB3	QB3	QB3	QB3	QB3	QB3
19 GO to ▶	—	—	QD1	—	QD1	—	QD1	QD1	QD1	QD1	QD1
21 GO to ▶	DE1	DE1	DE1	DE1	DE1	DE1	DE1	DE1	DE1	DE1	DE1
22 GO to ▶	DF1	DF1	DF1	DF1	DF1	DF1	DF1	DF1	DF1	DF1	DF1
23 GO to ▶	DH1	KB12	DH1	DH1	DH1	KB12	DH1	DH1	DH1	DH1	DH1
24 GO to ▶	—	DB1	DB1	DB1	DB1	DB1	DB1	DB1	DB1	DB1	DB1
26 GO to ▶	DK1	—	—	DK1	—	—	—	—	—	—	DC1
28 GO to ▶	DA1	—	—	DA1	—	—	—	—	—	—	—
31 GO to ▶	—	DL1	DD2	—	DL1	DN1	DL1	DL1	DL1	DN1	DN1
32 GO to ▶	—	—	—	—	—	DN25	—	—	—	DN25	DN25
34 GO to ▶	—	DL8	—	—	DL8	DN20	DL8	DL8	DL8	DN20	DN20
35 GO to ▶	—	DL5	—	—	DL5	DN5	DL5	DL5	DL5	DN5	DN5
51 GO to ▶	DE10	DE10	DE10	DE10	DE10	DE10	DE10	DE10	DE10	DE10	DE10
52 GO to ▶	—	—	FF1	—	FF1	FF1	FF1	FF1	—	—	—
53 GO to ▶	DH3	KB15	DH3	DH3	DH3	KB15	DH3	DH3	DH3	DH3	DH3
54 GO to ▶	—	DB10	DB10	DB10	DB10	DB10	DB10	DB10	DB10	DB10	DB10
56 GO to ▶	DK10	—	—	DK10	—	—	—	—	—	—	DC10
58 GO to ▶	DA10	KB5	—	DA10	—	—	—	—	—	—	—
59 GO to ▶	—	—	—	—	—	—	—	T71	—	—	—
61 GO to ▶	DE20	DE20	DE20	DE20	DE20	DE20	DE20	DE20	DE20	DE20	DE20
62 GO to ▶	—	—	—	—	—	—	T61	—	—	—	—
63 GO to ▶	DH10	KB18	DH10	DH10	DH10	KB18	DH10	DH10	DH10	DH10	DH10
64 GO to ▶	—	DB20	DB20	DB20	DB20	DB20	DB20	DB20	DB20	DB20	DB20
66 GO to ▶	DK20	—	—	DK20	—	—	—	—	—	—	DC6
67 GO to ▶	FA1	FA1	FA1	FA1	FA1	FA1	T81	T81	FA1	FA1	FA1
68 GO to ▶	DA20	KB9	—	DA20	—	KB9	—	—	—	—	—
69 GO to ▶	—	—	—	—	—	—	—	T73	—	—	—
73 GO to ▶	—	KB22	—	—	—	KB22	—	—	—	—	—
79 GO to ▶	—	—	—	—	—	—	—	FA9	FA9	FA9	FA9
81 GO to ▶	—	—	—	KN1	—	—	—	—	—	KC8	KC8
82 GO to ▶	—	—	—	X80	—	—	—	—	—	KC8	KC8
83 GO to ▶	—	—	DD17	X30	—	X30	X30	X30	—	—	—
84 GO to ▶	—	DL11	DD17	KA5	DL11	DN10	DL11	DL11	DL11	DN10	DN10
85 GO to ▶	—	KD6	—	KR10	KD6	KD6	KD6	KD6	KD6	KD6	KD6
87 GO to ▶	—	J7	J7	X15	J7	X15	X15	X15	J7	J7	J7
88 GO to ▶	—	—	—	KR1	—	X80	X80	X80	—	—	—
89 GO to ▶	—	—	KR1	—	—	—	T51	T51	—	—	—
93 GO to ▶	—	KB11	—	—	—	KB11	—	—	—	—	—
95 GO to ▶	J20	J20	—	—	J20	X90	—	X90	J20	J20	J20
96 GO to ▶	J6	J30	—	—	J30	X95	—	X95	J30	J30	J30
NO CODES ▶ CODES NOT LISTED ▶	Go to Pinpoint Test Step QA1										

QUICK TEST: Key On Engine Off Self-Test — 3.0

C LIGHT TRUCK SERVICE CODE CHART

Key On Engine Off Service Code	Pinpoint Test Step Direction						
	2.3L EFI	2.9L EFI	3.0L EFI	4.9L EFI	5.0L EFI	5.8L EFI	7.5L EFI
15 GO to ▶	QB3	QB3	QB3	QB3	QB3	QB3	QB3
19 GO to ▶	QD1	QD1	QD1	QD1	QD1	QD1	QD1
21 GO to ▶	DE1	DE1	DE1	DE1	DE1	DE1	DE1
22 GO to ▶	DF1	DF1	DF1	DF1	DF1	DF1	DF1
23 GO to ▶	DH1	DH1	DH1	DH1	DH1	DH1	DH1
24 GO to ▶	DB1	DB1	DB1	DB1	DB1	DB1	DB1
31 GO to ▶	DD2	—	—	DN1	DN1	DN1	DN1
32 GO to ▶	—	—	—	DN25	DN25	DN25	DN25
34 GO to ▶	—	—	—	DN20	DN20	DN20	DN20
35 GO to ▶	—	—	—	DN5	DN5	DN5	DN5
51 GO to ▶	DE10	DE10	DE10	DE10	DE10	DE10	DE10
52 GO to ▶	FF1	—	FF1	FF1	FF1	—	—
53 GO to ▶	DH3	DH3	DH3	DH3	DH3	DH3	DH3
54 GO to ▶	DB10	DB10	DB10	DB10	DB10	DB10	DB10
61 GO to ▶	DE20	DE20	DE20	DE20	DE20	DE20	DE20
63 GO to ▶	DH10	DH10	DH10	DH10	DH10	DH10	DH10
64 GO to ▶	DB20	DB20	DB20	DB20	DB20	DB20	DB20
67 GO to ▶	FA1	FA1	FA1	FA1	FA1	FA1	FA1
81 GO to ▶	—	—	—	KC8	KC8	KC8	KC8
82 GO to ▶	—	—	—	KC8	KC8	KC8	KC8
83 GO to ▶	DD17	—	—	—	—	—	—
84 GO to ▶	DD17	—	—	DN10	DN10	DN10	DN10
85 GO to ▶	—	—	KD6	KD6	—	KD6	KD6
86 GO to ▶	—	KR10	KR10	—	—	—	—
87 GO to ▶	J7	J7	J7	J7	J7	J7	J7
89 GO to ▶	KR1	KR1	KR1	—	—	—	—
95 GO to ▶	—	J20	J20	J20	J20	J20	J20
96 GO to ▶	—	J30	J30	J30	J30	J30	J30
NO CODES ▶ CODES NOT LISTED ▶	Go To Pinpoint Test Step QA1						

QUICK TEST: Computed Timing Check — 4.0

SPECIAL NOTES:

- If engine is a NO START, go directly to Pinpoint Test Step [A1].
- If engine starts but stalls, or stalls during timing check Go to Pinpoint Test Step [S1].
- If the "Check Engine" Light (MIL) is on, do not run Quick Test timing check. Verify Key On Engine Off Self-Test is a PASS.
- Self-Test timing is equal to Base Timing plus 20 degrees BTDC ± 3 degrees (see VECI decal for correct base timing).

Example

If base timing is 10 degrees BTDC, Self-Test timing is equal to: 10 degrees + 20 degrees = 30 degrees BTDC ± 3 degrees or 27 degrees to 33 degrees BTDC.

HOW TO RUN QUICK TEST TIMING CHECK

1. Turn the key off and wait 10 seconds.
2. Start engine.
3. Activate Engine Running Self-Test.
4. Check timing after the last service code has been displayed. The timing will remain fixed for two minutes, unless Self-Test is deactivated.

Is Self-Test Timing within specification?

[YES] Go To QUICK TEST STEP 5.0.

[NO] Go To Pinpoint Test Step [P1].

QUICK TEST: Engine Running Self-Test — 5.0

A | PERFORMING THE ENGINE RUNNING SELF TEST

SPECIAL NOTES:

- If the engine starts but stalls, or stalls during Self-Test, Go to Pinpoint Test Step [S1].
- On vehicles equipped with the Brake On/Off Switch (BOO), the brake pedal MUST be depressed and released AFTER the ID code.
- On vehicles equipped with the Power Steering Pressure Switch (PSPS), within 1 to 2 seconds after the ID code, the steering wheel must be turned at least one-half turn and released.

HOW TO RUN THE ENGINE RUNNING SELF-TEST

[DO]

- Deactivate Self-Test.
- Start and run engine at 2,000 rpm for two minutes. This action warms up the EGO sensor.
- Turn engine off, wait 10 seconds.
- Start engine.
- Activate Self-Test according to Quick Test Step 3.0 A.
- After the ID code, depress and release the brake pedal if appropriate. See Special Note above.
- After the ID code, within 1 to 2 seconds, turn the steering wheel at least one-half turn and then release it, if appropriate. See Special Note above.
- If a dynamic response code occurs, perform a brief wide-open throttle (WOT).
- Record all service codes displayed.
- Go to **Part B** of Engine Running Self-Test.

[DON'T]

- Depress the throttle unless a Dynamic Response Code is displayed.

QUICK TEST: Engine Running Self-Test — 5.0

B | CODE OUTPUT

Engine ID	Dynamic Response	Engine Running	ACTION TO TAKE
2(0), 3(0) or 4(0)	1(0) or no display	11	• Engine Running Self-Test indicates a PASS. — If Continuous Memory Codes were present, Go to QUICK TEST STEP 6.0. — If Continuous Memory is a PASS Code 11 and a symptom is present, Go to DIAGNOSTIC BY SYMPTOM in the proper Engine Supplement Section.
2(0), 3(0) or 4(0)	1(0) or no display	ANY CODE(S)	• Engine Running Self-Test indicates a FAULT. — Go to **PART C** of Engine Running Self-Test. — Always start with the first code displayed.
.98	NO DISPLAY	ANY CODE(S)	• Code 98 in place of the I.D. code indicates that the vehicle DID NOT PASS Key On Engine Off Self-Test. Engine Running Self-Test will not initiate until a PASS Code 11 is obtained in Key On Engine Off Self-Test. — Run Key On Engine Off Self-Test and address all codes displayed.
NO CODE DISPLAYED CODES NOT LISTED			• Self-Test did not activate. — Rerun Engine Running Self-Test to verify the above condition. — If condition is still present, Go to Pinpoint Test Step [QA1].

QUICK TEST: Engine Running Self-Test — 5.0

C | PASSENGER CAR SERVICE CODE CHART

Engine Running Service Code	Pinpoint Test Step Direction: 1.9L EFI	1.9L CFI	2.3L OHC EFI	2.3L TC EFI	2.3L HSC EFI	2.5L CFI	3.0L EFI	3.8L AXOD EFI	3.8L RWD EFI	5.0L SEFI	5.0L MA SEFI
12 GO to ▶	KE1	KB23	KE1	KE1	KE1	KB23	KE1	KE1	KE1	KE1	KE1
13 GO to ▶	KE4	KB26	KE15	KE11	KE11	KB26	KE11	KE11	KE11	KE11	KE11
16 GO to ▶	KE22	KB31	KE1	—	—	KB31	—	—	—	KE1	KE1
17 GO to ▶	KE13	KB31	—	KE13	KE13	KB31	—	—	—	—	—
18 GO to ▶	P1	—	—	—	P1	—	—	P1	P1	P1	P1
19 GO to ▶	KE12	KB26	—	—	—	—	—	—	—	—	—
21 GO to ▶	DE1	DE1	DE1	DE1	DE1	DE1	DE1	DE1	DE1	DE1	DE1
22 GO to ▶	DF1	DF7	DF7	DF1	DF7	DF7	DF7	DF7	DF7	DF7	DF1
23 GO to ▶	DH1	KB12	DH1	DH1	DH1	KB12	DH1	DH1	DH1	DH1	DH1
24 GO to ▶	—	DB1	DB1	DB1	DB1	DB1	DB1	DB1	DB1	DB1	DB1
25 GO to ▶	—	—	DG1	DG1	—	—	DG1	—	—	—	—
26 GO to ▶	DK1	—	—	DK1	—	—	—	—	—	—	DC4
28 GO to ▶	DA1	—	—	DA1	—	—	—	—	—	—	—
31 GO to ▶	—	DL21	DD1	—	DL21	DN1	DL21	DL21	DL21	DN1	DN1
32 GO to ▶	—	DL20	DD11	—	DL20	DN25	DL20	DL20	DL20	DN25	DN25
33 GO to ▶	—	DL30	DD11	—	DL30	DN40	DL30	DL30	DL30	DN40	DN40
34 GO to ▶	—	DL25	DD11	KA1	DL25	DN50	DL25	DL25	DL25	DN50	DN50
35 GO to ▶	—	DL25	DD30	—	DL25	DN5	DL25	DL25	DL25	DN5	DN5
41 GO to ▶	H11	H11	H11	H11	H11	H11	H11	H11	H11	H11	H1
42 GO to ▶	H23	H23	H23	H23	H23	H23	H23	H23	H23	H23	H1
44 GO to ▶	—	—	—	—	—	—	—	—	—	KC1	KC1
45 GO to ▶	—	—	—	—	—	—	—	—	—	KC1	KC1
46 GO to ▶	—	—	—	—	—	—	—	—	—	KC1	KC1
47 GO to ▶	KE20	—	—	—	—	—	—	—	—	—	—
48 GO to ▶	KE21	—	—	—	—	—	—	—	—	—	—
52 GO to ▶	—	—	—	—	FF5	FF5	FF5	FF5	—	—	—
55 GO to ▶	—	QF1	—	—	—	QF1	—	—	—	—	—
56 GO to ▶	—	—	—	—	—	—	—	—	—	—	DC10
58 GO to ▶	—	KB5	—	—	—	KB5	—	—	—	—	—
66 GO to ▶	—	—	—	—	—	—	—	—	—	—	DC6
67 GO to ▶	—	FA1	—	—	—	—	—	—	—	—	—
68 GO to ▶	—	KB9	—	—	—	KB9	—	—	—	—	—
72 GO to ▶	—	—	DF10	—	DF10	—	DF10	—	—	—	—
73 GO to ▶	DH20	—	DH20	DH20	DH20	—	DH20	—	—	—	—
74 GO to ▶	—	—	FD1	FD1	—	FD1	FD1	FD1	—	—	—
75 GO to ▶	—	—	FD5	FD5	—	FD5	FD5	FD5	—	—	—
76 GO to ▶	DK30	—	—	DK30	—	—	—	—	—	—	—
77 GO to ▶	M1	—	M1	M1	M1	—	M1	—	—	—	—
84 GO to ▶	—	DL11	—	—	—	—	—	—	—	—	—
85 GO to ▶	—	KD6	—	—	—	—	—	—	—	—	—
87 GO to ▶	—	J7	—	—	—	—	—	—	—	—	—
91 GO to ▶	—	—	—	—	—	—	—	—	—	H11	H1
92 GO to ▶	—	—	—	—	—	—	—	—	—	H23	H1
94 GO to ▶	—	—	—	—	—	—	—	—	—	KC1	KC1
98 GO to ▶	GO TO QUICK TEST STEP 5.0B										
99 GO to ▶	—	KB29	—	—	—	KB29	—	—	—	—	—
NO CODES ▶ CODES NOT LISTED ▶	Go to Pinpoint Test Step QA1										

QUICK TEST: Engine Running Self-Test — 5.0

C — LIGHT TRUCK SERVICE CODE CHART

Engine Running Service Code	Pinpoint Test Step Direction 2.3L EFI	2.9L EFI	3.0L EFI	4.9L EFI	5.0L EFI	5.8L EFI	7.5L EFI
12 GO to ▶	KE1	KE1	KE1	KE1	KE1	KE1	KE1
13 GO to ▶	KE15	KE11	KE11	KE11	KE11	KE11	KE11
16 GO to ▶	KE1	—	—	—	—	—	—
17 GO to ▶	KE13	KE13	KE13	KE13	KE13	KE13	KE13
18 GO to ▶	—	—	—	P1	P1	P1	P1
21 GO to ▶	DE1	DE1	DE1	DE1	DE1	DE1	DE1
22 GO to ▶	DF7	DF7	DF7	DF7	DF7	DF7	DF7
23 GO to ▶	DH1	DH1	DH1	DH1	DH1	DH1	DH1
24 GO to ▶	DB1	DB1	DB1	DB1	DB1	DB1	DB1
25 GO to ▶	DG1	DG1	—	DG1	DG1	DG1	—
31 GO to ▶	DD1	—	—	DN1	DN1	DN1	DN1
32 GO to ▶	DD11	—	—	DN25	DN25	DN25	DN25
33 GO to ▶	DD11	—	—	DN40	DN40	DN40	DN40
34 GO to ▶	DD11	—	—	DN50	DN50	DN50	DN50
35 GO to ▶	DD30	—	—	DN5	DN5	DN5	DN5
38 GO to ▶	—	—	—	—	—	—	—
41 GO to ▶	H11	H11	H11	H11	H11	H11	H11
42 GO to ▶	H23	H23	H23	H23	H23	H23	H23
44 GO to ▶	—	—	—	KC1	KC1	KC1	KC1
45 GO to ▶	—	—	—	KC1	KC1	KC1	KC1
46 GO to ▶	—	—	—	KC1	KC1	KC1	—
52 GO to ▶	FF5	—	FF5	FF5	FF5	—	—
72 GO to ▶	DF10	DF10	DF10	DF10	DF10	DF10	DF10
73 GO to ▶	DH20	DH20	DH20	DH20	DH20	DH20	DH20
74 GO to ▶	FD1	FD1	FD1	—	—	—	—
75 GO to ▶	FD5	FD5	FD5	—	—	—	—
77 GO to ▶	M1	M1	M1	M1	M1	M1	M1
98 GO to ▶	GO TO QUICK TEST STEP 5.0B						
NO CODES ▶ CODES NOT LISTED ▶	Go to Pinpoint Test Step QA1						

QUICK TEST: Continuous Self-Test — 6.0

A — CONTINUOUS MEMORY CODES

SPECIAL NOTES:

- Verify that a **Pass Code 11** was received in both Key On Engine Off and Engine Running Self-Tests before continuing with this test.
- Refer to the Appendix for a detailed description of how to use the Continuous Monitor Mode.

DETERMINING THE CONTINUOUS MEMORY CODES TO BE SERVICED

- Refer to the Continuous Memory Codes recorded in Quick Test Step 3.0 A.
- The cause of some of the Continuous Memory Codes may have been eliminated during either Key On Engine Off or Engine Running Self-Test service.
- Address only those Continuous Memory Codes for which a similar code has not been previously serviced.
- Go to **Part B** of Continuous Self-Test.

QUICK TEST: Continuous Self-Test — 6.0

B — PASSENGER CAR SERVICE CODE CHART

Continuous Memory Service Code	Pinpoint Test Step Direction 1.9L EFI	1.9L CFI	2.3L OHC EFI	2.3L TC EFI	2.3L HSC EFI	2.5L CFI	3.0L EFI	3.8L AXOD EFI	3.8L RWD EFI	5.0L SEFI	5.0L MA SEFI
13 GO to ▶	—	KB90	—	—	—	KB90	—	—	—	—	—
14 GO to ▶	Y1	Y1	Y1	Y1	Y1	Y1	Y1	Y1	Y1	Y1	Y1
15 GO to ▶	QB1	QB1	QB1	QB1	QB1	QB1	QB1	QB1	QB1	QB1	QB1
18 GO to ▶	N1	N1	N1	N1	N1	N1	N1	N1	N1	N1	N1
22 GO to ▶	DF90	DF90	DF90	DF90	DF90	DF90	DF90	DF90	DF90	DF90	DF90
23 GO to ▶	—	KB97	—	—	—	KB97	—	—	—	—	—
27 GO to ▶	—	—	—	DP1	—	—	—	—	—	—	—
29 GO to ▶	—	—	—	—	DP1	DP1	T2	T2	DP1	DP1	DP1
31 GO to ▶	—	DL90	DD90	—	DL90	DN92	DL90	DL90	DL90	DN92	DN92
32 GO to ▶	—	DL94	—	—	DL94	DN90	DL94	DL94	DL94	DN90	DN90
33 GO to ▶	—	DL97	—	—	DL97	DN95	DL97	DL97	DL97	DN95	DN95
34 GO to ▶	—	DL93	—	—	DL93	DN98	DL93	DL93	DL93	DN98	DN98
35 GO to ▶	—	DL90	—	—	DL90	DN92	DL90	DL90	DL90	DN92	DN92
38 GO to ▶	—	KB91	—	—	—	KB91	—	—	—	—	—
39 GO to ▶	—	—	—	—	—	—	T31	T31	—	—	—
41 GO to ▶	H30	H29	H29	H30	H29	H29	H29	H29	H29	H29	H1
42 GO to ▶	H30	—	—	H30	—	—	—	—	—	—	—
43 GO to ▶	H30	—	—	H30	—	—	—	—	—	—	—
51 GO to ▶	DE91	DE91	DE91	DE91	DE91	DE91	DE91	DE91	DE91	DE91	DE91
53 GO to ▶	DH90	KB93	DH90	DH90	DH90	KB93	DH90	DH90	DH90	DH90	DH90
54 GO to ▶	—	DB90	DB90	DB90	DB90	DB90	DB90	DB90	DB90	DB90	DB90
56 GO to ▶	DK90	—	—	DK90	—	—	—	—	—	—	DC10
57 GO to ▶	—	—	—	—	—	—	T41	T41	—	—	—
58 GO to ▶	DA90	—	—	DA90	—	—	—	—	—	—	—
59 GO to ▶	—	—	—	—	—	—	T21	T21	—	—	—
61 GO to ▶	DE94	DE94	DE94	DE94	DE94	DE94	DE94	DE94	DE94	DE94	DE94
63 GO to ▶	DH94	KB97	DH94	DH94	DH94	KB97	DH94	DH94	DH94	DH94	DH94
64 GO to ▶	—	DB93	DB93	DB93	DB93	DB93	DB93	DB93	DB93	DB93	DB93
65 GO to ▶	H30	—	—	H30	—	—	—	—	—	—	—
66 GO to ▶	DK93	—	—	DK93	—	—	—	—	—	—	DC6
67 GO to ▶	FA1	—	—	FA1	—	—	—	—	—	—	—
68 GO to ▶	DA93	—	—	DA93	—	—	—	—	—	—	—
69 GO to ▶	—	—	—	—	—	—	T11	T11	—	—	—
71 GO to ▶	QE1	KB92	—	—	—	KB92	—	—	—	—	—
72 GO to ▶	QE4	—	—	QE1	—	—	—	—	—	—	—
78 GO to ▶	—	—	—	—	—	QE1	—	—	—	—	—
85 GO to ▶	H30	—	—	—	—	—	—	—	—	—	—
86 GO to ▶	H30	—	—	—	—	—	—	—	—	—	—
87 GO to ▶	J95	J95	—	—	J95	X104	J95	X104	J95	J95	J95
91 GO to ▶	—	—	—	—	—	—	—	—	—	H29	H1
95 GO to ▶	J90	J90	—	—	J90	X100	—	X100	J90	—	J90
96 GO to ▶	J92	J92	—	—	J92	X102	—	X102	J92	—	J92
NO CODES ▶ CODES NOT LISTED ▶	Go to Pinpoint Test Step QA1										

QUICK TEST: Continuous Self-Test — 6.0

B — LIGHT TRUCK SERVICE CODE CHART

Continuous Memory Service Code	Pinpoint Test Step Direction 2.3L EFI	2.9L EFI	3.0L EFI	4.9L EFI	5.0L EFI	5.8L EFI	7.5L EFI
14 GO to ▶	Y1	Y1	Y1	Y1	Y1	Y1	Y1
15 GO to ▶	QB1	QB1	QB1	QB1	QB1	QB1	QB1
18 GO to ▶	N1	N1	N1	N1	N1	N1	N1
22 GO to ▶	DF90	DF90	DF90	DF90	DF90	DF90	DF90
29 GO to ▶	DP1	DP1	DP1	DP1	DP1	DP1	—
31 GO to ▶	DD90	—	—	DN92	DN92	DN92	DN92
32 GO to ▶	—	—	—	DN90	DN90	DN90	DN90
33 GO to ▶	—	—	—	DN95	DN95	DN95	DN95
34 GO to ▶	—	—	—	DN98	DN98	DN98	DN98
35 GO to ▶	—	—	—	DN92	DN92	DN92	DN92
41 GO to ▶	H29	H29	H29	H29	H29	H29	H29
51 GO to ▶	DE91	DE91	DE91	DE91	DE91	DE91	DE91
53 GO to ▶	DH90	DH90	DH90	DH90	DH90	DH90	DH90
54 GO to ▶	DB90	DB90	DB90	DB90	DB90	DB90	DB90
61 GO to ▶	DE94	DE94	DE94	DE94	DE94	DE94	DE94
63 GO to ▶	DH94	DH94	DH94	DH94	DH94	DH94	DE94
64 GO to ▶	DB93	DB93	DB93	DB93	DB93	DB93	DB93
87 GO to ▶	J95	J95	J95	J95	J95	J95	J95
95 GO to ▶	—	J90	J90	J90	J90	J90	J90
96 GO to ▶	—	J92	J92	J92	J92	J92	J92
NO CODES ▶ CODES NOT LISTED ▶	Go to Pinpoint Test Step QA1						

APPENDIX: Self-Test Description

The Self-Test is divided into three specialized tests: Key On Engine Off Self-Test, Engine Running Self-Test, and Continuous Self-Test. The Self-Test is not a conclusive test by itself, but is used as a part of the functional Quick-Test diagnostic procedure. The processor stores the Self-Test program in its permanent memory. When activated, it checks the EEC-IV system by testing its memory integrity and processing capability, and verifies that various sensors and actuators are connected and operating properly.

The Key On Engine Off and Engine Running Self-Tests are functional tests which only detect faults present at the time of the Self-Test. Continuous Self-Test is an ongoing test that stores fault information for retrieval at a later time.

KEY ON ENGINE OFF SELF-TEST

At this time, a test of the EEC-IV system is conducted with power applied and engine at rest.

For Self-Test to detect errors in the Key On Engine Off Self-Test mode, the fault must be present at the time of testing. For intermittents, refer to Continuous Memory Codes.

SEPARATOR PULSE

A single 1/2 second separator pulse is issued 6-9 seconds after the last Key On Engine Off Test code. Then, 6-9 seconds after the single 1/2 second separator pulse, the Continuous Memory Codes will be issued.

NOTE: The separator code and Continuous Memory Codes follow Key On Engine Testing codes ONLY.

CONTINUOUS MEMORY CODES

Continuous Memory Codes are issued as a result of information stored during continuous Self-Test, while the vehicle was in normal operation. These codes are displayed only during Key On Engine Off testing and after the separator code. These codes should be used for diagnosis only when Key On Engine Off and Engine Running Self-Tests result in code 11 and all Quick Test Steps 1.0 through 5.0 have been successfully completed.

ENGINE RUNNING SELF-TEST

At this time, a test of the EEC-IV system is conducted with the engine running. The sensors are checked under actual operating conditions and at normal operating temperatures. The actuators are exercised and checked for corresponding results.

ENGINE IDENTIFICATION CODES (ID CODES)

Engine ID codes are issued at the beginning of the Engine Running Self-Test and are one-digit numbers represented by the number of pulses sent out. The engine ID code is equal to one-half the number of engine cylinders (i.e. 2 pulses = 4 cylinders). These codes are used to verify the proper processor is installed and that the Self-Test has been entered.

DYNAMIC RESPONSE CHECK

The dynamic response check verifies the movement of the TP, VAF, and MAP sensors during the brief Wide-Open Throttle (WOT) performed during the Engine Running Self-Test. The signal for the operator to perform the brief WOT is a single pulse or 10 code on the STAR Tester.

APPENDIX: Code Output Format

SERVICE CODES

The EEC-IV system communicates service information through the Self-Test service codes. These service codes are two-digit numbers representing the results of Self-Test.

The service codes are transmitted on the Self-Test output (STO) line found in the vehicle Self-Test connector. They are in the form of timed pulses, and read by the technician on a voltmeter, STAR tester, "Check Engine" Light (MIL) or on the Continental message center.

SELF-TEST OUTPUT CODE FORMAT
KEY ON ENGINE OFF AND CONTINUOUS MEMORY CODES

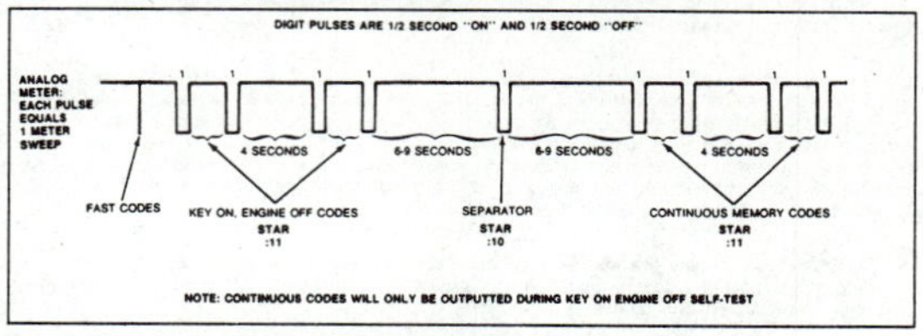

Figure 1 Key On Engine Off and Continuous Memory Code Format

SELF-TEST OUTPUT CODE FORMAT
ENGINE RUNNING CODES

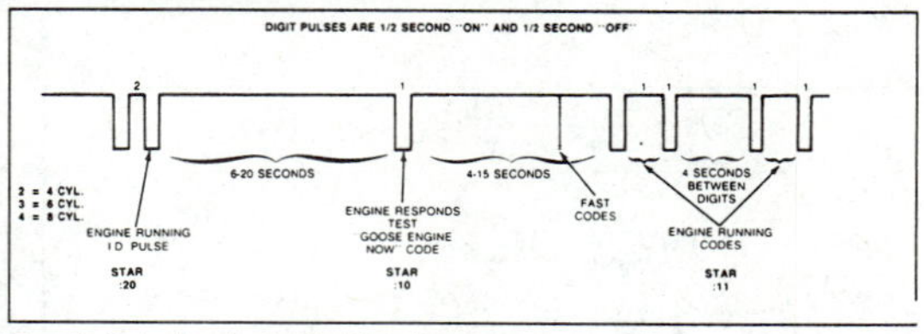

Figure 2 Engine Running Self-Test Code Format

Fast Codes

Fast codes are issued prior to regular service codes. These codes contain the identical information as the regular service codes but are transmitted at 100 times the normal rate. These codes are interpreted by special equipment at the end of the assembly line by the Body and Assembly Division.

Some meters in service detect these codes as a short burst of information (slight meter deflection).

APPENDIX: Self-Test Description

POWER STEERING PRESSURE SWITCH TEST

On vehicles equipped with Power Steering Pressure Switch (PSPS), the steering wheel must be turned one-half turn and released AFTER the ID Code. This tests the ability of the EEC-IV system to detect a change of state in the Power Steering Pressure Switch.

BRAKE ON/OFF SWITCH TEST

On vehicles equipped with Brake ON/OFF Switch (BOO), the brake pedal MUST be depressed and released AFTER the ID Code. This tests the ability of the EEC-IV system to detect a change of state in the Brake ON/OFF Switch.

APPENDIX: Continuous Self-Test

The Continuous Memory service codes are separated from the Quick Test Key On Engine Off codes by a single separator pulse, Figure 3.

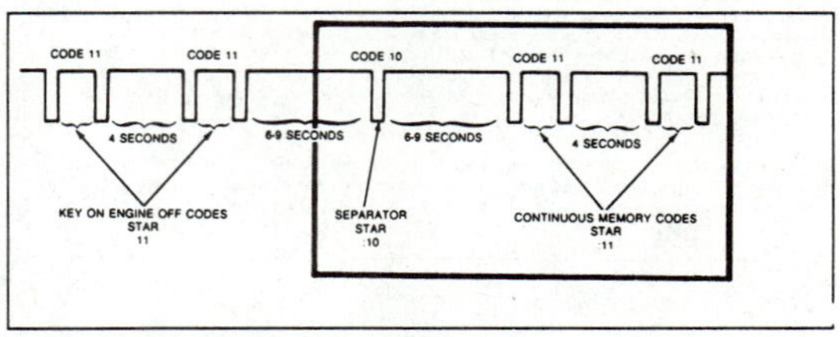

Figure 3 Continuous Memory Code Format

The Continuous Memory codes should never be used for Diagnosis until the Key On Engine Off and Engine Running Self-Tests result in a pass code 11.

During this mode of testing the EEC-IV Processor continuously monitors inputs for opens and shorts. The Continuous Memory Codes must be retrieved within 40 engine temperature warm up cycles. On the 41st Engine Temperature cycle, the service code will be automatically erased. The Continuous Memory Codes can also be erased by deactivating Self-Test while the service codes are being outputted.

APPENDIX: Self-Test With STAR Tester

READING CODES—SELF-TEST AUTOMATIC READOUT (STAR) TESTER

After hooking up the STAR tester and turning on its power switch, the tester will run a display check and the numerals 88 will begin to flash in the display window (Figure [illegible]). A steady 00 will then appear to signify that the STAR tester is ready to start the Self-Test and receive the test's service codes.

To receive the service codes, press the push button at the front of the STAR tester. The button will latch down, and a colon will appear in the display window in front of the 00 numerals. The colon **must** be displayed to receive the service codes.

If, for any reason, the technician wishes to clear the display window during the Self-Test, he must turn off the vehicle's engine, press the tester's push button once to unlatch it (colon will disappear), then press the button again to latch down the button (colon will appear again). Every time the STAR tester is turned off, the low battery indicator (LO BAT) should show briefly at the upper left corner of the tester's display window. If the LO BAT indicator shows steadily at any other time during the operation of the STAR tester with any service code, turn its power switch to Off and replace the 9-volt battery in the tester.

The STAR tester will display the last service code received, even after disconnecting it from the vehicle. It will hold the service code on the display until the power is turned off or the push button is unlatched and relatched.

WARNING: ANYONE WHO DEPARTS FROM THE INSTRUCTIONS PROVIDED IN THIS PUBLICATION MUST FIRST ESTABLISH THAT HE COMPROMISES NEITHER HIS PERSONAL SAFETY NOR THE VEHICLE INTEGRITY BY HIS CHOICE OF METHODS, TOOLS, OR PARTS.

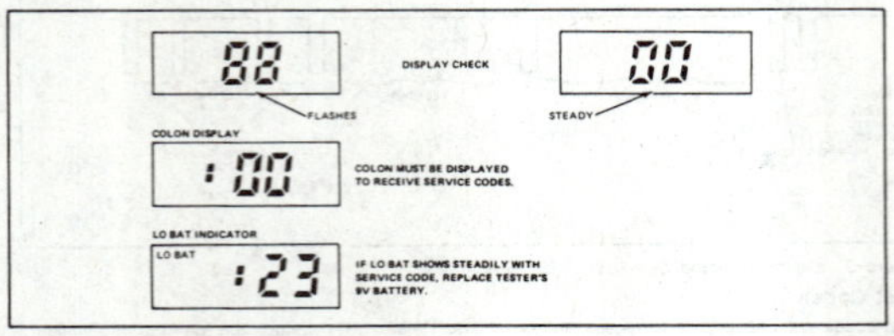

Figure 4 Star Tester Output Code Format

APPENDIX: Self-Test With "Check Engine" Light (MIL)

READING CODES—"CHECK ENGINE" LIGHT (MIL)

The "Check Engine" Light on the front dash panel will remain on when a hard fault (open or short circuit) is present.

During Self-Test a service code is reported by the "Check Engine" Light. It will represent itself as a flash on the "Check Engine" Light display on the dash panel (Figure 6). A single-digit number of three will be reported by three flashes.

However, as previously stated, a service code is represented by a two-digit number, such as 2-3. As a result, the Self-Test service code of 2-3 will appear on the "Check Engine" Light display as two flashes, then, after a two-second pause, the light will flash three times.

The Continuous Memory Codes are separated from the Key On Engine Off codes by a six-second delay, a single half-second flash, and another six-second delay. They are produced on the "Check Engine" Light display in the same manner as the Key On Engine Off codes.

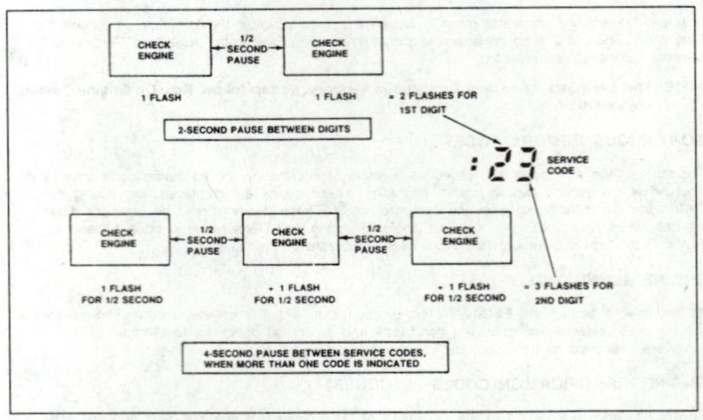

"Check Engine" Light Output Code Format

APPENDIX: Self-Test With Analog Voltmeter

READING CODES—ANALOG VOLTMETER

When a service code is reported on the analog voltmeter for a function test, it will represent itself as a pulsing or sweeping movement of the voltmeter's needle across the dial face of the voltmeter (Figure 5). Therefore, a single-digit number of three will be reported by three needle pulses (sweeps). However, as previously stated, a service code is represented by a two-digit number, such as 2-3. As a result, the Self-Tests service code of 2-3 will appear on the voltmeter as two needle pulses (sweeps), then, after a two-second pause, the needle will pulse (sweep) three times.

The Continuous Memory Codes are separated from the Key On Engine Off codes by a six-second delay, a single half-second sweep, and another six-second delay. They are produced on the voltmeter in the same manner as the Key On Engine Off codes.

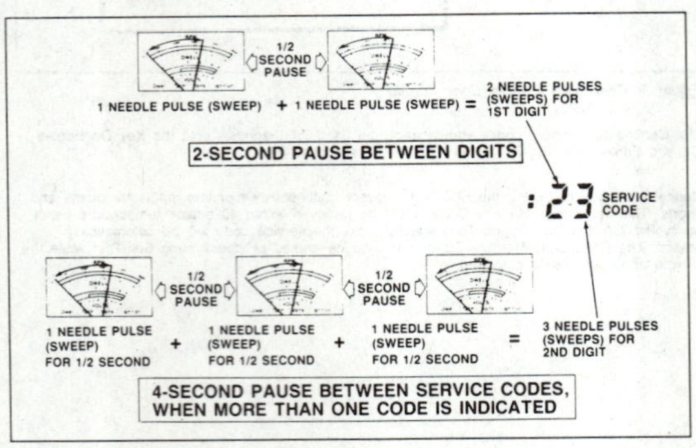

Analog Voltmeter Output Code Format

APPENDIX: Self-Test With Message Center (Continental Only)

HOW TO RUN SELF-TEST USING THE CONTINENTAL MESSAGE CENTER

1. Jumper STI to SIG RTN at the Self-Test connectors.
2. On the Electronic Instrument Cluster, hold in all three buttons (SELECT, CHECKOUT and RESET) at the same time.
3. Key On Engine Off Self-Test
   - While holding in all three buttons, place ignition switch in the ON position. Release buttons.

   — Key On Engine Running Self-Test
   - While holding in all three buttons, place ignition switch in the RUN position. Release buttons.
4. To initiate Self-Test, press the checkout button twice. A base readout of [2] [5] [5] [4] indicates that Self-Test has been entered successfully.
5. Service code output will be displayed in the left side boxes following the base display.
6. To exit Self-Test, turn ignition switch to OFF.

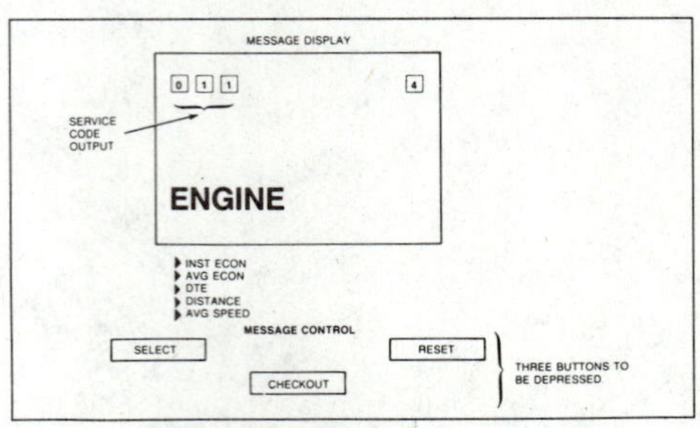

Message Center Output Code Format

APPENDIX: Diagnostic Aids

CONTINUOUS MONITOR MODE (WIGGLE TEST)

SPECIAL NOTES:

- The Continuous Monitor Modes allow the technician to **ATTEMPT** to recreate an intermittent fault.
- The STAR Tester and the "Check Engine" Light (MIL) will flash the LED. The STAR Tester may also sound an alert when a fault is recreated.
- The needle of the VOM will sweep across the face of the meter when a fault is recreated.

KEY ON ENGINE OFF

1. Hook up a STAR Tester or VOM as shown in Quick Test Step 2.0.
2. Turn the STAR Tester On but **DO NOT** latch the center Button. **DO NOT** ground STI if using a VOM or "Check Engine" Light.
3. Turn the ignition key to the ON position.
4. You are now in the Continuous Monitor Mode.
5. Tap, move, and wiggle the suspect sensor and/or harness. If a fault is detected, a Service Code will be stored in memory and will be indicated as explained above depending on the type of equipment being used.

ENGINE RUNNING

1. Hook up a STAR Tester or VOM as shown in Quick Test Step 2.0.
2. Start the engine.
3. Activate Self-Test, wait 10 seconds, deactivate and reactivate Self-Test. **DO NOT** shut the engine off.
4. You are now in the Engine Running Continuous Monitor Mode.
5. Tap, move, and wiggle the suspect sensor and/or harness. If a fault is detected, a Service Code will be stored in memory and will be indicated as explained above depending on the type of equipment being used.

APPENDIX: Diagnostic Aids

CYLINDER BALANCE TEST

The Cylinder Balance test on the 5.0L SEFI and 5.0L SEFI MA vehicles is designed to aid in the detection of a non-contributing cylinder.

The Cylinder Balance test first reads engine rpm, with all injectors activated. Next, each injector is turned "off and on," one at a time. The rpm drop that results, if any, is then read. These two rpm's are compared to verify that the rpm drop was greater than a calibrated level.

The Cylinder Balance test has the capability of comparing rpm drop of each cylinder against three different calibrated levels. The first level checks for gross errors (electrical opens, shorts, etc.). The second and third levels check for partially contributing cylinders. If the appropriate rpm drop is not achieved at any of the levels, then a service code corresponding to that cylinder will be sent.

The Cylinder Balance Test service codes are listed in Pinpoint Test Step H9.

1. Perform Engine Running Self-Test.
2. After the last repeated service code is received, wait 5-10 seconds.
3. Lightly depress and release throttle (not wide-open throttle) within two minutes of the last repeated service code.
4. Cylinder Balance Test will be performed at the first test level. Test time is approximately three minutes.
5. After the last repeated Cylinder Balance Test pass code 90, re-enter test within two minutes by lightly depressing and releasing the throttle. The second test level of the Cylinder Balance Test will now be performed.
6. After the last repeated Cylinder Balance Test pass code 90, re-enter test within two minutes by lightly depressing and releasing the throttle. Cylinder Balance Test will now be tested at the third test level.
7. Cylinder Balance Test may be repeated as many times as necessary by repeating Step 6. All additional testing will be performed at the third test level.

The Cylinder Balance Test is designed to aid in the detection of a non-contributing or partially contributing cylinder. The Cylinder Balance Pinpoint Test Steps are designed to isolate only EEC related problems.

APPENDIX: Diagnostic Aids

HOW TO CLEAR THE CONTINUOUS MEMORY CODES

1. Run the Key On Engine Off Self-Test according to Quick Test Step 3.0A.
2. When the Service Codes begin to be displayed, deactivate Self-Test:
   - STAR Tester: Unlatching the center button (up position).
   - Analog VOM: Remove the jumper wire from between Self-Test Input (STI) connector and the Signal Return Pin of the Self-Test connector.
   - "Check Engine" Light (MIL): Remove the jumper wire from between Self-Test Input (STI) connector and the SIGNAL RETURN pin of the Self-Test connector.
   - Message Center (Continental Only): Remove the jumper wire from between Self-Test input (STI) connector and the SIGNAL RETURN pin of the Self-Test connector.
4. The continuous service codes will be erased from the processor's memory.

OUTPUT STATE CHECK

The output state check aids in servicing output actuators associated with the EEC-IV system. It enables the technician to energize and de-energize most of the system output actuators on command. This mode is entered after all codes have been received from Key On Engine Off and Continuous Testing. At this time, leave Self-Test activated and depress the throttle. Each time the throttle is depressed the output actuators will change state from energized to de-energized or from de-energized to energized.

1. Enter Self-Test.
2. Code Output Ends.
3. Do Brief WOT.
4. EEC-IV Output To Actuators Energized.
5. Do Brief WOT.
6. EEC-IV Output To Actuators De-Energized.

APPENDIX: Diagnostic Aids

FAILURE MODE EFFECTS MANAGEMENT (FMEM)

FMEM is an alternate system strategy in the ECA designed to allow improved vehicle drive should one or more sensor inputs fail.

When a sensor input is perceived to be out-of-limits by the ECA, an alternative strategy will be initiated.

The ECA will substitute a fixed in-limit sensor value and will continue to monitor the faulty sensor input. If the faulty sensor operates within limits, the ECA will return to the normal engine running strategy.

98 — Code 98 will be displayed when FMEM is in effect.

The "Check Engine" Light (MIL)/Message will remain on when FMEM is in effect in all applications except 49 States Continental.

APPENDIX: Diagnostic Aids

"CHECK ENGINE" LIGHT (MALFUNCTION INDICATOR LIGHT) ALL APPLICATIONS EXCEPT CONTINENTAL

The "Check Engine" light is intended to alert the driver of certain malfunctions in the engine control system.

If such a fault occurs, the EEC-IV processor will substitute a value or values and continue operating. This process is called Failure Mode Effects Management (FMEM). In some cases this action may result in a slight change in driveability.

HOW THE "CHECK ENGINE" LIGHT OPERATES

System OK

The "Check Engine" light will remain on while the ignition key is in RUN position.

Once the vehicle has started the "Check Engine" light will go out.

System Not OK

If the "Check Engine" light should remain on after the vehicle has started, run Key On Engine Off Self-Test to completion. If the light continues to remain on, go to EEC-IV Diagnostic By Symptom in the Engine Supplement Section.

If the "Check Engine" light never comes on, go to EEC-IV Diagnostic By Symptom.

If the vehicle is a no start, go to Pinpoint Test Step A1.

NOTE: When in Self-Test the "Check Engine" light will also flash the service codes.

APPENDIX: Diagnostic Aids

"CHECK ENGINE"/"CHECK DCL" MESSAGE (DATA COMMUNICATIONS LINK)

Continental Applications Only

The EEC-IV processor transmits the "Check Engine" message to the message center through the Data Communications Link (DCL). The message center is used to display the "Check Engine" and "Check Engine"/"Check DCL" messages.

HOW THE "CHECK ENGINE" MESSAGE OPERATES

Service Codes: Service codes can be digitally displayed on the message center when running self-test. (See Appendix: Self-Test)

On California applications only, the "Check Engine" message is activated when the EEC-IV processor switches to an alternate strategy of operation. This process is called Failure Mode Effects Management (FMEM). The "Check Engine" message is intended to alert the driver of certain malfunctions in the engine control system.

FMEM Mode (Calif. only): The "Check Engine" message is displayed and accompanied by a one second tone every four seconds. The tone is suppressed after one minute.

On all Continental applications the "Check Engine" message will also be activated during an EEC-IV processor Limited Operating Strategy (LOS) mode. (See Section 12)

LOS/DCL Failure: The "Check Engine" message will be displayed as described in FMEM. In addition, the message "Check DCL" will be displayed alternately with the "Check Engine" message every four seconds.

If the "Check Engine" and/or "Check Engine"/"Check DCL" message should come on after the vehicle has started, Run Key On Engine Off Self-Test to completion. If the message continues to remain on, go to EEC-IV Diagnostic By Symptom in the Engine Supplement Section.

If the vehicle is a no start, go to Pinpoint Test Step A1.

APPENDIX: Test Equipment

EQUIPMENT REQUIRED:

- Rotunda Self-Test Automatic Readout (STAR), No. 007-00004 with cable assembly No. 007-00010. Refer to STAR Tester operation.
- Analog volt-ohmmeter, 0 to 20V DC, (alternate to STAR).

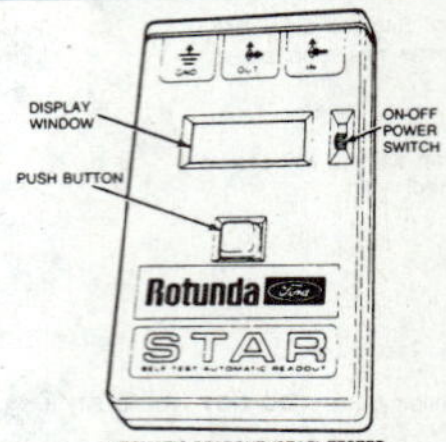

AUTOMATIC READOUT (STAR) TESTER

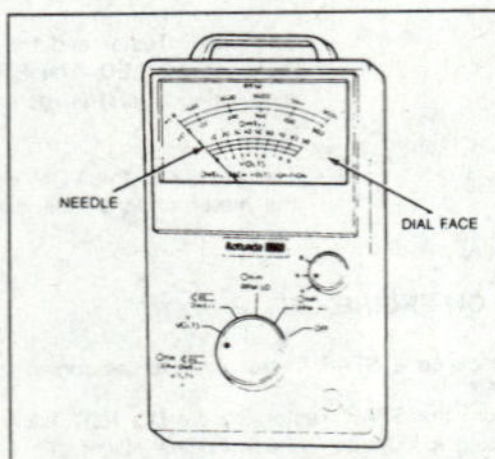

ANALOG VOLTMETER

- Jumper wire.
- Vacuum gauge, Rotunda 059-00008 or equivalent. Range 0-30 in.-Hg. Resolution 1 in.-Hg.

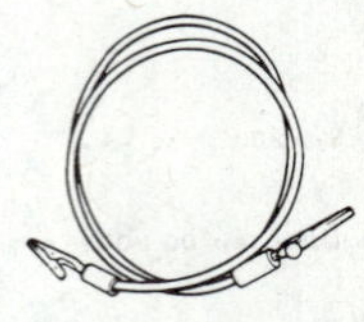
JUMPER WIRE

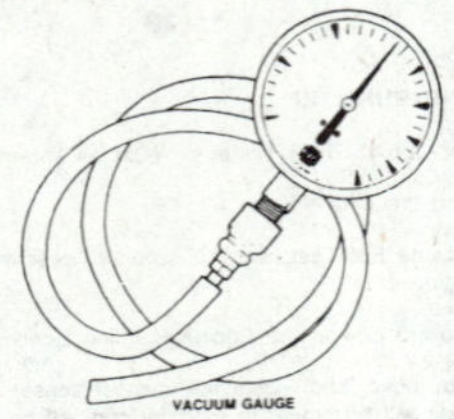
VACUUM GAUGE

APPENDIX: Test Equipment (Continued)

- Tachometer, Rotunda No. 059-00010 or equivalent. Range 0-6,000 rpm. Accuracy ± 40 rpm. Resolution 20 rpm.
- Breakout Box, Rotunda 014-00322, Special Service Tool T83L-50-EEC-IV or equivalent.

TACHOMETER

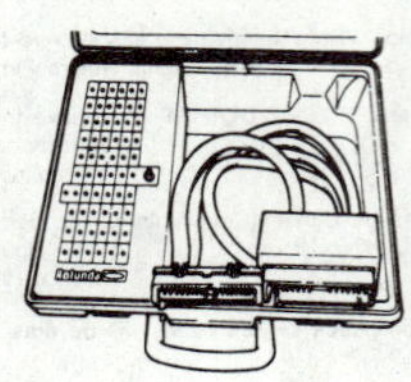
BREAKOUT BOX

- Vacuum pump, Rotunda No. 021-00014 or equivalent. Range 0-30 in. Hg.
- Digital volt-ohmmeter, Rotunda No. 014-00407 or equivalent. Input impedance 10 Megaohm minimum.

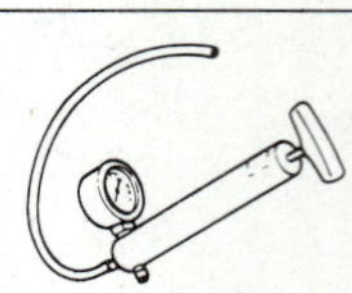
VACUUM PUMP

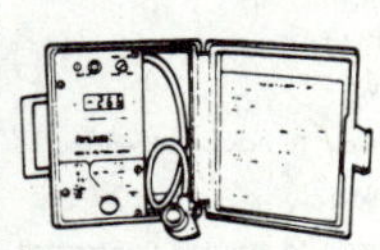
DIGITAL VOLT-OHMMETER

APPENDIX: Test Equipment (Continued)

- Electronic Fuel Injection Pressure Gauge EFI/CFI only, Tool T80L-9974-A or equivalent. (Use instructions. For specific applications, refer to Shop Manual, Group 24.)
- Spark tester (optional modified spark plug with side electrode removed). Tool D81P-6666-A or equivalent.

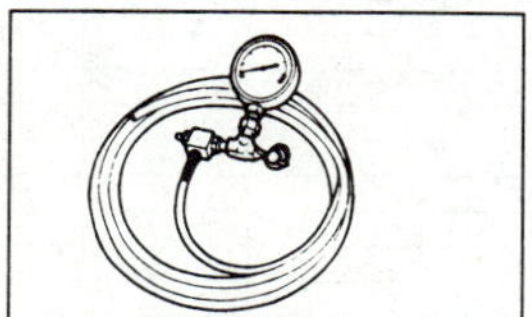
EFI PRESSURE GAUGE

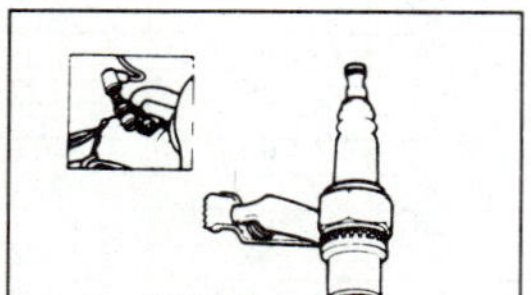
SPARK TESTER

- Timing light, Rotunda model 059-00006 or equivalent.
- Non-powered test lamp.

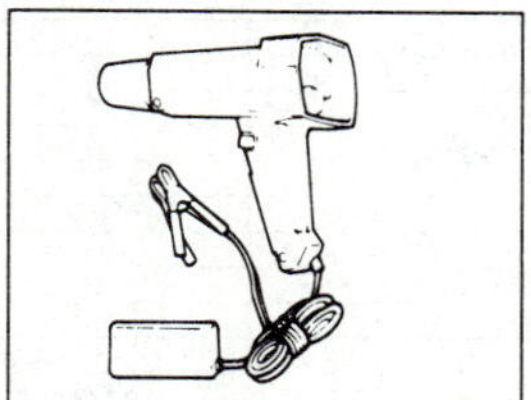
TIMING LIGHT

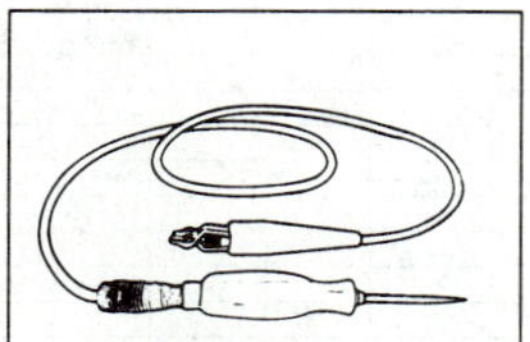
NON-POWERED TEST-LAMP

EEC-IV Diagnostic By Symptom — Car

NOTE: You should only enter Diagnostic By Symptom when directed here from Section 16.

Symptom	Pinpoint Test Step Direction										
	1.9L EFI	1.9L CFI	2.3L OHC EFI	2.3L TC EFI	2.3L HSC EFI	2.5L CFI	3.0L EFI	3.8L AXOD EFI	3.8L RWD EFI	5.0L SEFI	5.0L MA SEFI
Engine runs rough or misses Engine stalls Lack of power Rough idle Erratic rpm Surges Stalls in Self-Test	S1	S2	S1	S1	S1	S2	S1	S1	S1	S1	S1 and verify MAF Sensor is properly connected
Stall during parking maneuvers	—	FF3	FF3	—	—	FF3	FF3	FF3	—	—	—
Surges with A/C on at idle	—	—	KM20	—	KM20	—	—	—	—	—	—
A/C does not cut off under WOT conditions	KM10	KM10	KM10	X52	KM10	X52	X52	X52	KM10	KM10	KM10
A/C not functioning	—	KM1	KM1	X50	KM1	X50	X50	X50	KM1	KM1	KM1
A/C compressor runs continuously	—	—	KM25	—	—	—	—	—	—	—	—
High idle speeds on each restart may be accompanied by spark knock for up to 3-5 minutes after a start	KA1	—	—	DG1	—	—	—	—	—	—	—
Low idle with A/C On	FA1	FA10	FA10	FA10	FA10	FA10	FA10	FA10	FA10	FA10	FA10
High idle in Drive, automatic trans. only	—	—	—	—	—	FA20	—	—	FA20	—	—
Shift indicator light always On or Off	KL1	KL1	—	—	KL1	KL1	—	—	—	—	KL1
"CHECK ENGINE" light always on	ML1	ML1	ML1	ML1	ML1	ML1	ML1	ML1	ML1	ML1	ML1
"CHECK ENGINE" light never on	ML5	ML5	ML5	ML5	ML5	ML5	ML5	ML5	ML5	ML5	ML5
"CHECK ENGINE"/"CHECK DCL" Message On	—	—	—	—	—	—	—	ML15	—	—	—
"CHECK ENGINE" Message On	—	—	—	—	—	—	—	ML10	—	—	—

EEC-IV Diagnostic By Symptom — Car

Symptom	Pinpoint Test Step Direction										
	1.9L EFI	1.9L CFI	2.3L OHC EFI	2.3L TC EFI	2.3L HSC EFI	2.5L CFI	3.0L EFI	3.8L AXOD EFI	3.8L RWD EFI	5.0L SEFI	5.0L MA SEFI
Stumble after hot restart	H20	H20	H20	H20	H20	H20	H20	H20	H20	H20	H20
Engine will not restart	—	QF1	—	—	—	—	—	—	—	—	—
Fuel pump runs with engine off	—	—	J22	X14	—	X14	X14	—	—	J22	—
Gasoline fumes under hood	KD1	KD1	—	—	KD1	KD1	KD1	KD1	KD1	KD1	KD1
Spark knock	KA1 or VOL. H SEC. 6	—	DG1 or VOL. H SEC. 6	DG2 or VOL. H SEC. 6	—	—	DG1 or VOL. H SEC. 6	—	—	—	—
Poor performance, lack of turbo boost	—	—	—	KN1	—	—	—	—	—	—	—
Low boost pressure	—	—	—	KN10	—	—	—	—	—	—	—
Primary fan or secondary fan always on	—	—	—	X35	—	—	—	—	—	—	—
Primary or secondary fan does not operate	—	—	—	X70	—	—	—	—	—	—	—
No engine cooling fan	—	—	—	—	—	X40 MTX ONLY	—	—	—	—	—
Low speed cooling fan does not operate High speed cooling fan does not operate Low or high speed cooling fan does not operate	—	—	—	—	—	X20 CLC ONLY	X20	X20	—	—	—
Low or high speed cooling fan always on	—	—	—	—	—	X35	X35	X35	—	—	—
Engine Lugging Poor Performance Trailer Hitching	—	—	—	—	—	—	—	T75	—	—	—

Electrical Schematic — 1.9L EFI

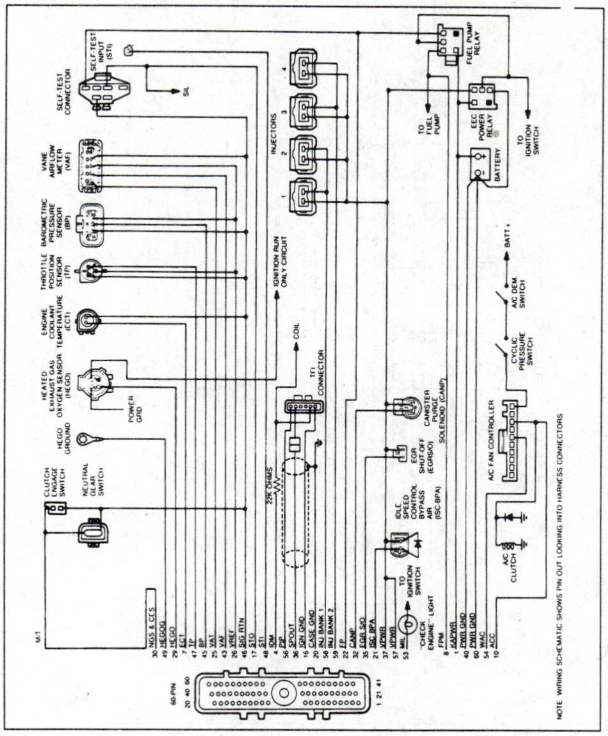

EEC-IV Module Connector Pin Usage — 1.9L EFI

Pin	Circuit	Wire Color	Application	Abbreviations
1	37A	Y	Keep Alive Power	KAPWR
4	11B	DG/Y	Ignition Diagnostic Monitor	IDM
7	354	LG/Y	Engine Coolant Temperature	ECT
8	787A	PK/BK	Fuel Pump Monitor	FPM
10	347	BK/Y	A/C Clutch	ACC
16	259	BK/O	Ignition Ground	IGN GND
17	224A	T/LB	Self-Test Output and Shift Indicator Light	STO and SIL
20	57	BK	Case Ground	CSE GND
21	68	O/BK	Idle Speed Control (Bypass Air)	ISC-BPA
22	97	T/LG	Fuel Pump	FP
25	357	LG/P	Vane Air Temperature	VAT
26	351	O/W	Reference Voltage	VREF
29	94	DG/P	Heated Exhaust Gas Oxygen Sensor	HEGO
30	614	GY/O	Neutral Drive Gear Switch and Clutch Engage Switch	NDS and CES
32	101	GY/Y	Canister Purge Solenoid	CANP
35	362	Y	Exhaust Gas Recirculation Shut Off	EGR S/O
36	324A	Y/LG	Spark Output	SPOUT
37	361	R	Vehicle Power	VPWR
40	60A	BK/LG	Power Ground	PWR GND
43	200	W/BK	Vane Air Flow	VAF
45	358	LG/BK	Barometric Pressure	BP
46	359	BK/W	Signal Return	SIG RTN
47	355	DG/LG	Throttle Position Sensor	TP
48	100	W/R	Self-Test Input	STI
49	89	O	Heated Exhaust Gas Oxygen Ground	HEGOG
53	201	T/R	"Check Engine" Light	MIL
54	73	O/LB	Wide Open Throttle A/C Cut-Off	WAC
56	349	DB	Profile Ignition Pickup	PIP
57	361A	R	Vehicle Power	VPWR
58	95	T/R	Injector (Bank 1)	INJ 1
59	96	T/O	Injector (Bank 2)	INJ 2
60	60B	BK/LG	Power Ground	PWR GND

Pin locations given for reference only. Probing 60 pin connector with DVOM probe will result in permanent damage to the pin connectors. Always probe as directed, using the Breakout Box.

EEC-IV Module Connector Pin Usage — 1.9L CFI

Pin	Circuit	Wire Color	Application	Abbreviations
1	37	Y	Keep Alive Power	KAPWR
4	11	DG/Y	Ignition Diagnostic Monitor	IDM
5	16	R/LG	Key Power	KPWR
7	354	LG/Y	Engine Coolant Temperature Sensor	ECT
8	787	PK/BK	Fuel Pump Monitor	FPM
10	347	BK/Y	A/C Clutch Compressor Signal	ACCS
16	259	BK/O	Ignition Ground	IGN GND
17	201	T/R	Self-Test Ouput/"Check Engine" Light	STO/MIL
20	57	BK	Case Ground	CSE GND
21	376	BR/W	Idle Speed Control	ISC +
22	97	T/LG	Fuel Pump Relay	FP
25	357	LG/P	Air Charge Temperature Sensor	ACT
26	351	O/W	Reference Voltage	VREF
27	352	BR/LG	Pressure Feedback EGR	PFE
28	265	LG/W	Idle Tracking Switch	ITS
29	94	DG/P	Heated Exhaust Gas Oxygen Sensor	HEGO
30	614	GY/O	Neutral Drive Switch (A/T Only)	NDS
35	101	GY/Y	Canister Purge Solenoid	CANP
36	324	Y/LG	Spark Output	SPOUT
37	361	R	Vehicle Power	VPWR
40	60	BK/LG	Power Ground	PWR GND
41	264	W/LB	Idle Speed Control	ISC –
45	358	LG/BK	Manifold Absolute Pressure Sensor	MAP
46	359	BK/W	Signal Return (Ground)	SIG RTN
47	355	DG/LG	Throttle Position Sensor	TP
48	100	W/R	Self-Test Input	STI
49	89	O	Heated Exhaust Gas Oxygen Ground	HEGOG
51	224	T/LB	Shift Indicator Light (M/T Only)	SIL
52	362	Y	EGR Valve Regulator Solenoid	EVR
54	73	O/LB	WOT A/C Cutoff	WAC
56	349	DB	Profile Ignition Pick-Up	PIP
57	361	R	Vehicle Power	VPWR
58	95	T/R	Fuel Injector	INJ
60	60	BK/LG	Power Ground	PWR GND

Pin locations given for reference only. Probing 60 pin connector with DVOM probe will result in permanent damage to the pin connectors. Always probe as directed, using the Breakout Box.

Electrical Schematic — 1.9L CFI

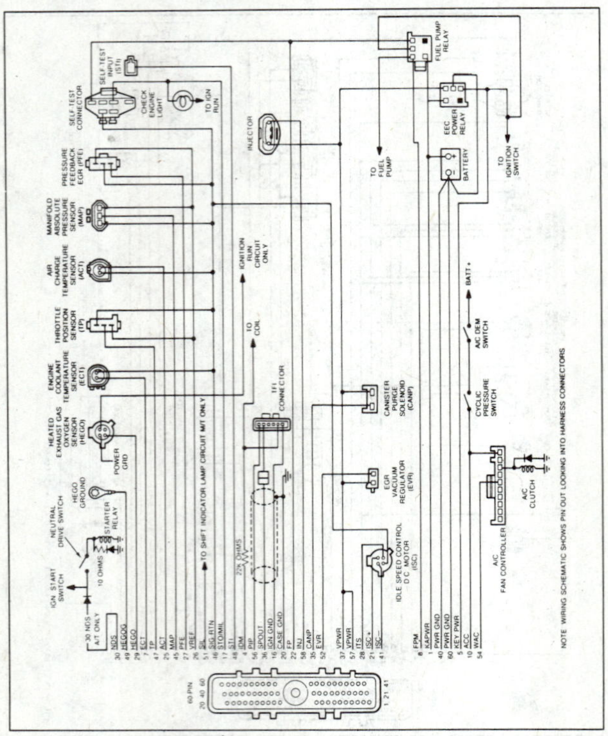

Electrical Schematic — 2.3L OHC EFI

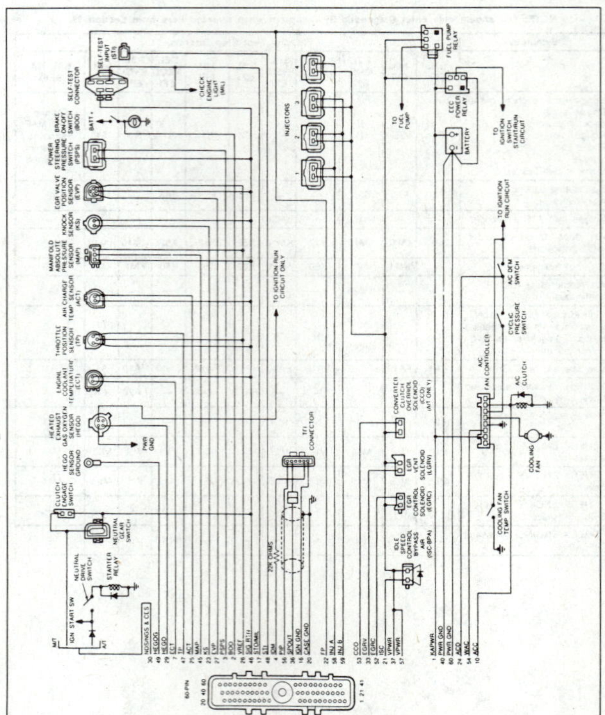

EEC-IV Module Connector Pin Usage — 2.3L OHC EFI

MUSTANG

Pin	Circuit	Wire Color	Application	Abbreviations
1	38	BK/O	Keep Alive Power	KAPWR
2	810	R/LG	Brake On/Off	BOO
3	330	Y/LG	Power Steering Pressure Switch	PSPS
4	11	DG/Y	Ignition Diagnostic Monitor	IDM
7	354	LG/Y	Engine Coolant Temperature	ECT
10	347	BK/Y	A/C Clutch	ACC
16	259	BK/O	Ignition Ground	IGN GND
17	657	T	Self-Test Output/"CHECK ENGINE" Light	STO MIL
20	57	BK	Case Ground	CSE GND
21	264	W/LB	Idle Speed Control	ISC
22	97	T/LG	Fuel Pump	FP
23	310	Y/R	Knock Sensor	KS
24	348	LG/P	A/C Demand	ACD
25	357	LG/P	Air Charge Temperature	ACT
26	351	O/W	Reference Voltage	VREF
27	352	BR/LG	EGR Valve Position	EVP
29	94	DG/P	Heated Exhaust Gas Oxygen Sensor	HEGO
30	771	P/Y	Neutral Drive Switch (A/T Only)	NDS
30	771	P/Y	Neutral Gear Switch and Clutch Engage Switch (M/T)	NGS & CES
33	360	DG	Exhaust Gas Recirculation Vent	EGRV
36	324	Y/LG	Spark Ouput	SPOUT
37	361	R	Vehicle Power	VPWR
40	60	BK/LG	Power Ground	PWR GND
45	358	LG/BK	Manifold Absolute Pressure Sensor	MAP
46	359	BK/W	Signal Return	SIG RTN
47	355	DG/LG	Throttle Position Sensor	TP
48	209	W/R	Self-Test Input	STI
49	89	O	Heated Exhaust Gas Oxygen Sensor Ground	HEGOG
52	362	Y	Exhaust Gas Recirculation Control	EGRC
53	237	O/Y	Clutch Converter Override (A/T Only)	CCO
54	73	O/LB	W.O.T. A/C Cut Off	WAC
56	349	DB	Profile Ignition Pick-Up	PIP
57	361	R	Vehicle Power	VPWR
58	96	T/O	Injector 1 (Bank 1)	INJ 1
59	95	T/R	Injector 2 (Bank 2)	INJ 2
60	60	BK/LG	Power Ground	PWR GRD

Pin locations given for reference only. Probing 60 pin connector with DVOM probe will result in permanent damage to the pin connectors. Always probe as directed, using the Breakout Box.

EEC-IV Module Connector Pin Usage — 2.3L EFI TURBO

THUNDERBIRD

Pin	Circuit	Wire Color	Application	Abbreviations
1	37	Y	Keep Alive Power	KAPWR
2	511	LG	Brake On/Off	BOO
3	150	DG/W	Vehicle Speed Sensor +	VSS DIF +
4	11	DG/Y	Ignition Diagnostic Monitor	IDM
6	563	O/Y	Vehicle Speed Sensor –	VSS DIF -
7	354	LG/Y	Engine Coolant Temperature	ECT
10	347	BK/Y	A/C Clutch	ACC
16	259	BK/O	Ignition Ground	IGN GND
17	382	Y/BK	Self-Test Output/"CHECK ENGINE" Light	STO/(MIL)
20	60	BK/LG	Case Ground	CSE GND
21	68	O/BK	Idle Speed Control - (Bypass Air)	ISC-BPA
22	97	T/LG	Fuel Pump Control	FP
23	310	Y/R	Knock Sensor	KS
24	670	R/Y	Octane Switch	OCTANE SW
25	470	PK/BK	Air Charge Temperature Sensor	ACT
26	351	O/W	Voltage Reference	VREF
27	200	W/BK	Vane Air Flow	VAF
29	94	DG/P	Heated Exhaust Gas Oxygen Sensor	HEGO
30	376	BR/W	Neutral Drive Switch (A/T)	NDS
30	359	BK/W	Neutral Input (M/T)	NI
31	464	BK/PK	Boost Control	BOOST
33	362	Y	Exhaust Gas Recirculation Shut Off	EGR S/O
34	305	LB/PK	Data Output Link	DOL
35	836	O/W	Automatic Adjustable Shock Controller	ACL
36	324	Y/LG	Spark Output	SPOUT
37	361	R	Vehicle Power	VPWR
40	60	BK/LG	Battery Ground	PWR GND
42	435	Y/LB	Shift Solenoid	SS
43	357	LG/P	Vane Air Temperature	VAT
45	358	LG/BK	Barometric Pressure	BP
46	359	BK/W	Signal Return	SIG RTN
47	355	DG/LG	Throttle Position Sensor	TP
48	209	W/R	Self-Test Input	STI
49	89	O	Heated Exhaust Gas Oxygen Ground	HEGOG
52	639	PK	High Electric Drive Fan	HEDF
53	77	DB/Y	Converter Clutch Override	CCO
54	331	R	W.O.T. A/C Cut-Off	WAC
55	197	T/O	Electric Drive Fan	EDF
56	349	DB	Profile Ignition Pick-Up	PIP
57	361	R	Vehicle Power	VPWR
58	555	T	Injector Bank A	INJ A
59	557	BR/Y	Injector Bank B	INJ B
60	60	BK/LG	Battery Ground	PWR GND

Pin locations given for reference only. Probing 60 pin connector with DVOM probe will result in permanent damage to the pin connectors. Always probe as directed, using the Breakout Box.

Electrical Schematic — 2.3L EFI TURBO

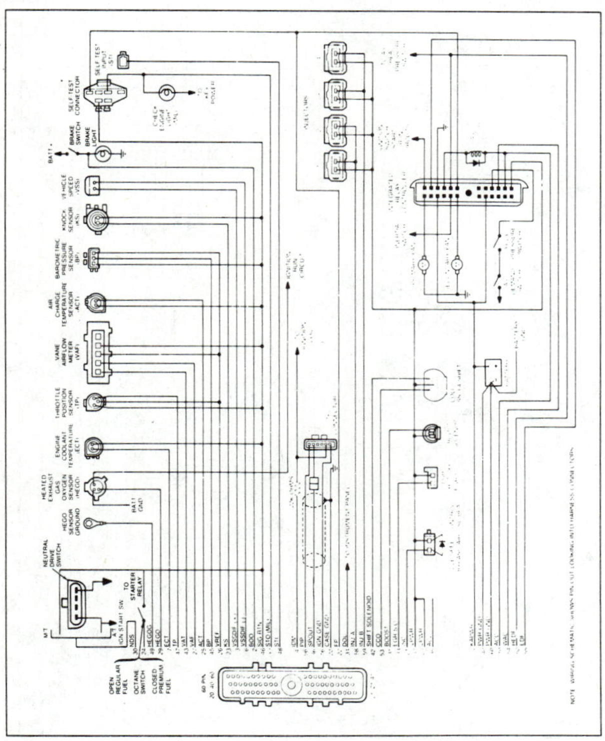

Integrated Controller Schematic — 2.3L EFI TURBO

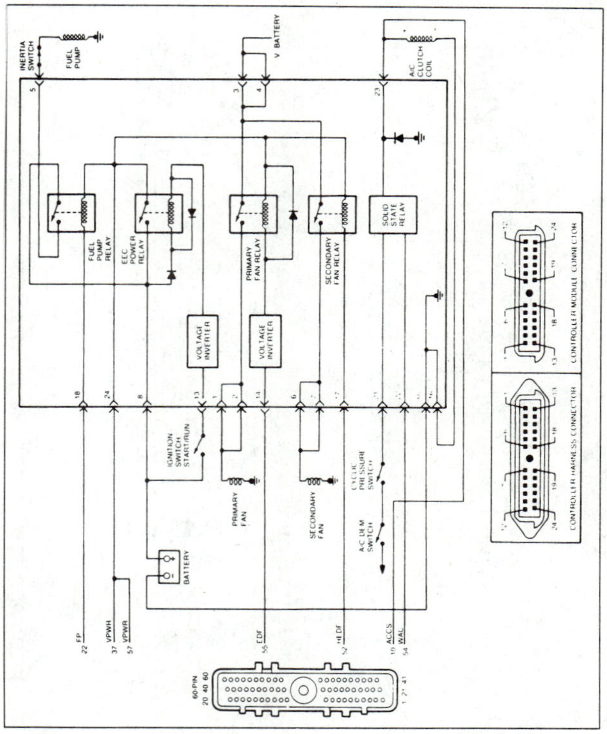

Integrated Controller Pin Usage — 2.3L EFI TURBO

Pin	Circuit	Wire Color	Applications	Abbreviations
1	228A	BR/Y	Power to Primary Fan	PTPF
2	228B	BR/Y	Power to Primary Fan	PTPF
3	37A	Y	Power to Primary and Secondary Relays	PTPSR
4	37B	Y	Power to Primary and Secondary Relays	PTPSR
5	787	PK/BK	Power to Fuel Pump	PTP
6	181A	BR/O	Power to Secondary Fan	PTSF
7	181B	BR/O	Power to Secondary Fan	PTSF
8	37C	Y	Battery to EEC Relay	BATT +
13	16C	R/LG	Key Power	KEY PWR
14	197	T/O	EDF Circuit (Primary)	EDF
15	57G	BK	Power Ground	PWR GRD -
16	321	GY/W	A/C Clutch Ground	PWR GRD
17	639	PK	HEDF Circuit (Secondary)	HEDF
18	97A	T/LG	Fuel Pump Circuit	FP
21	883	PK/LB	A/C Cyclic Switch	ACC
22	331	R	WOT A/C Cut Off Circuit	WAC
23	347B	BK/Y	Power to A/C	PTAC
24	361	R	Vehicle Power	VPWR

EEC-IV Module Connector Pin Usage — 2.3L HSC-EFI

Pin	Circuit	Wire Color	Application	Abbreviations
1	37A	Y	Keep Alive Power	KAPWR
3	150	DG/W	Vehicle Speed Sensor +	VSSDIF +
4	11	DG/Y	Ignition Diagnostic Monitor	IDM
6	563	O/Y	Vehicle Speed Sensor	VSSDIF -
7	354	LG/Y	Engine Coolant Temperature Sensor	ECT
8	787A	PK/BK	Fuel Pump Monitor	FPM
10	198A	T/Y	A/C Clutch Compressor Signal	ACC
16	259	BK/O	Ignition Ground	IGN GND
17	201A	T/R	Self-Test Output and "Check Engine" Light	STO and MIL
20	57	BK	Case Ground	CASE GND
21	376	BR/W	Idle Speed Control (Bypass Air)	ISC(BPA)
22	238	O/LB	Fuel Pump Relay	FP
24	330	Y/LG	Power Steering Pressure Switch	PSPS
25	357	LG/P	Air Charge Temperature Sensor	ACT
26	351	O/W	Reference Voltage	VREF
27	352	BR/LG	Pressure Feedback EGR	PFE
29	94	DG/P	Heated Exhaust Gas Oxygen Sensor	HEGO
30	614	GY/O	Neutral Drive Switch (A/T)	NDS
30	614	GY/O	Clutch Engage Switch	CES
31	101	GY/Y	Canister Purge Solenoid	CANP
33	362	Y	EGR Vacuum Regulator Solenoid	EVR
36	324A	Y/LG	Spark Output	SPOUT
37	361C	R	Vehicle Power	VPWR
40	60A	BK/LG	Power Ground	PWR GND
43	348B	LG/P	A/C Demand	ACD
45	358	LG/BK	Manifold Absolute Pressure Sensor	MAP
46	359	BK/W	Signal Return (Ground)	SIG RTN
47	355	DG/LG	Throttle Position Sensor	TP
48	209	W/R	Self-Test Input	STI
49	89	O	Heated Exhaust Gas Oxygen Ground	HEGOG
53	224	T/LB	Shift Indicator Light	SIL
54	73	O/LB	WOT A/C Cutoff	WAC
56	349	DB	Profile Ignition Pick-Up	PIP
57	361	R	Vehicle Power	VPWR
58	95	T/R	Injector (Bank 1)	INJ 1
59	96	T/O	Injector (Bank 2)	INJ 2
60	60B	BK/LG	Power Ground	PWR GND

Pin locations given for reference only. Probing 60 pin connector with DVOM probe will result in permanent damage to the pin connectors. Always probe as directed, using the Breakout Box.

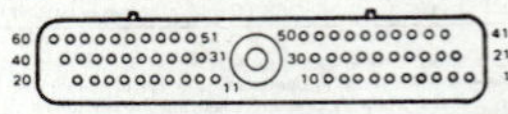

Electrical Schematic — 2.3L HSC-EFI

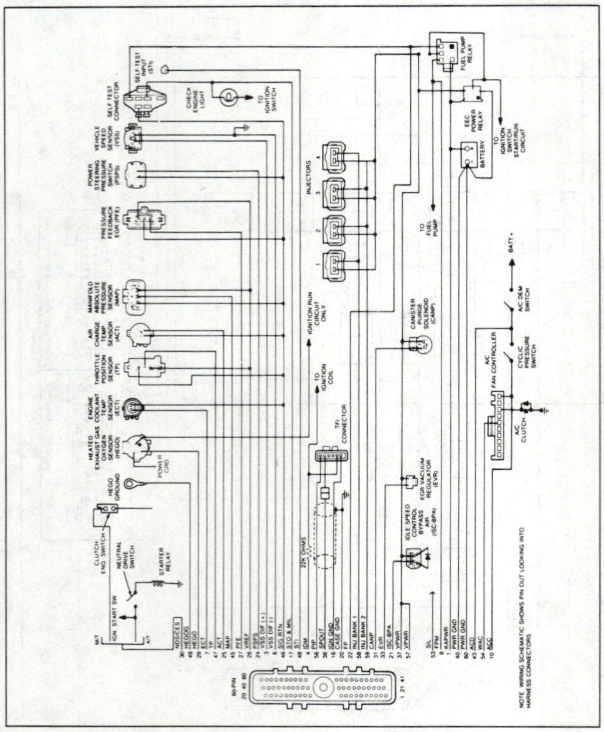

Electrical Schematic — 2.5L CFI CLC

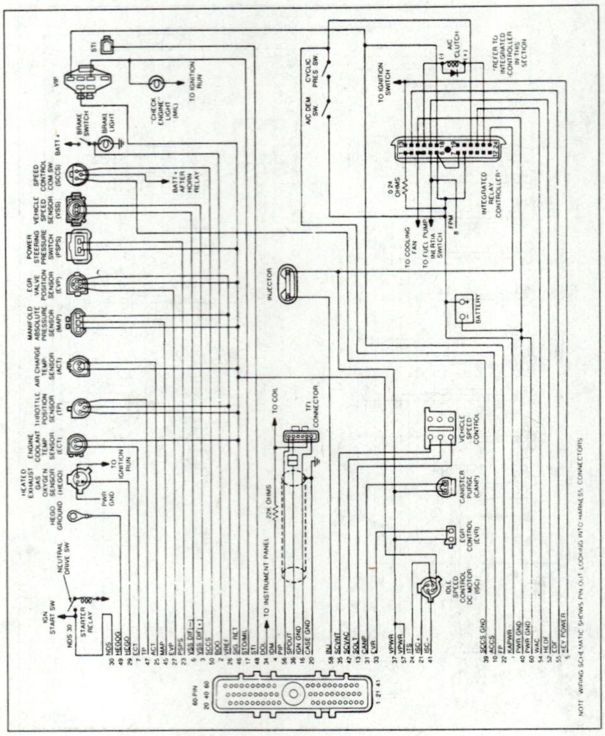

EEC-IV Module Connector Pin Usage

2.5L CFI CLC

Pin	Circuit	Wire Color	Application	Abbreviations
1	37	Y	Keep Alive Power	KAPWR
2	810	R/LG	Brake On/Off	BOO
3	150	DG/W	Vehicle Speed Sensor +	VSS DIF +
4	11	DG/Y	Ignition Diagnostic Monitor	IDM
5	16	R/LG	Key Power	KPWR
6	563	O/Y	Vehicle Speed Sensor –	VSS DIF –
7	354	LG/Y	Engine Coolant Temperature	ECT
8	787	PK/BK	Fuel Pump Monitor	FPM
10	883	PK/LB	A/C Cycling Switch	ACCS
13	144	O/Y	Vehicle Speed Control Solenoid	SOL +
16	259	BK/O	Ignition Ground	IGN GND
17	201	T/R	Self-Test Out/"Check Engine" Light	STO/MIL
20	57	BK	Case Ground	CSE GND
21	382	Y/BK	Idle Speed Control (DC Motor)	ISC +
22	97	T/LG	Fuel Pump	FP
23	330	Y/LG	Power Steering Pressure Switch	PSPS
24	209	W/R	Idle Tracking Switch	ITS
25	357	LG/P	Air Charge Temperature	ACT
26	351	O/W	Reference Voltage	VREF
27	352	BR/LG	EGR Valve Position	EVP
29	94	DG/P	Heated Exhaust Gas Oxygen (Sensor)	HEGO
30	199	LB/Y	Neutral Drive Switch	NDS
31	101	GY/Y	Canister Purge Solenoid	CANP
33	360	DG	EGR Vacuum Regulator (Solenoid)	EVR
34	305	LB/PK	Data Output Link	DOL
35	146	W/PK	Speed Control Vent (Solenoid)	SCVNT
36	324	Y/LG	Spark Out	SPOUT
37	361	R	Vehicle Power	VPWR
39	461	O	Speed Control Command Switch Ground	SCCS GND
40	60	BK/LG	Power Ground	PWR GND
41	377	W	Idle Speed Control — (DC Motor)	ISC –
42	145	GY/BK	Speed Control Vacuum (Solenoid)	SCVAC
45	358	LG/BK	Manifold Absolute Pressure (Sensor)	MAP
46	359	BK/W	Signal Return (Ground)	SIG RTN
47	355	DG/LG	Throttle Position (Sensor)	TP
48	200	W/BK	Self-Test Input	STI
49	89	O	Heated Exhaust Gas Oxygen (Sensor) Ground	HEGO GND
50	151	LB/BK	Speed Control Command Switch	SCCS
52	639	PK	High Electro Drive Fan	HEDF
54	331	R	Wide Open Throttle (WOT) A/C Cutoff	WAC
55	197	T/O	Electro Drive Fan (Low)	EDF
56	349	DB	Profile Ignition Pick-up	PIP
57	361	R	Vehicle Power	VPWR
58	95	T/R	Injector	INJ
60	60	BK/LG	Power Ground	PWR GND

Pin locations given for reference only. Probing 60 pin connector with DVOM probe will result in permanent damage to the pin connectors. Always probe as directed, using the Breakout Box.

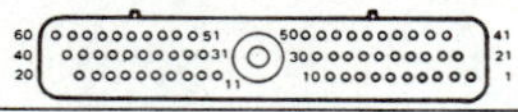

Integrated Controller Pin Usage

2.5L CFI CLC

Pin	Circuit	Color	Application	Abbreviations
1	181	BR/O	EDF Power Into Controller	Batt +
2	181	BR/O	EDF Power Into Controller	Batt +
3	228	BR/Y	H/EDF Power to Fan	PTF
4	228	BR/Y	H/EDF Power to Fan	PTF
5	787	PK/BK	Power to the Pump	PTP
6	38	BK/O	HEDF Power Into Controller	Batt +
7	38	BK/O	HEDF Power Into Controller	Batt +
8	37	Y	Battery Voltage (Power Relay)	Batt +
9	—	—	Not Used	—
10	—	—	Not Used	—
11	—	—	Not Used	—
12	—	—	Not Used	—
13	16	R/LG	Keypower	KPWR
14	197	T/O	EDF Ground	EDF
15	60	BK/LG	Vehicle Ground	PWR GND
16	57	BK	A/C Ground	A/C GND
17	639	PK	HEDF Ground	HEDF
18	97	T/LG	Fuel Pump Relay Ground	FP
19	—	—	Not Used	—
20	—	—	Not Used	—
21	883	PK/LB	A/C Power In	ACCS
22	331	R	WOT A/C Cutoff	WAC
23	347	BK/Y	A/C Power to Clutch Coil	PTAC
24	361	R	EEC IV Power Out	VPWR

Integrated Controller Schematic

2.5L CFI CLC

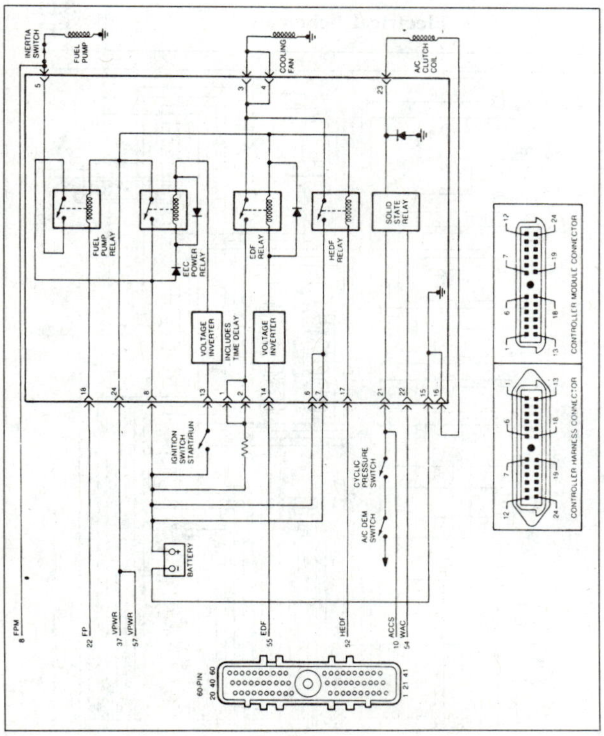

Electrical Schematic

2.5L CFI MTX

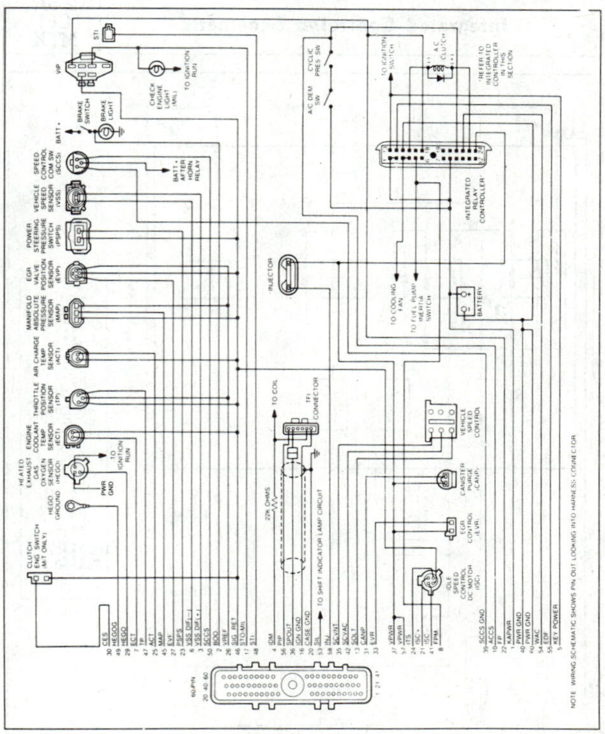

EEC-IV Module Connector Pin Usage — 2.5L CFI MTX

Pin	Circuit	Wire Color	Applications	Abbreviations
1	37	Y	Keep Alive Power	KAPWR
2	810	R/LG	Brake On/Off	BOO
3	150	DG/W	Vehicle Speed Sensor +	VSS DIF +
4	11	DG/Y	Ignition Diagnostic Motor	IDM
5	16	R/LG	Key Power	KPWR
6	563	O/Y	Vehicle Speed Sensor –	VSS DIF –
7	354	LG/Y	Engine Coolant Temperature	ECT
8	787	PK/BK	Fuel Pump Monitor	FPM
10	883	PK/LB	A/C Cycling Switch	ACCS
13	144	O/Y	Vehicle Speed Control Solenoid	SOL +
16	259	BK/O	Ignition Ground	IGN GND
17	201	TR	Self-Test Out/"Check Engine" Light	STO/MIL
20	57	BK	Case Ground	CSE GND
21	382	Y/BK	Idle Speed Control (DC Motor)	ISC +
22	97	T/LG	Fuel Pump	FP
23	330	Y/LG	Power Steering Pressure Switch	PSPS
24	209	W/R	Idle Tracking Switch	ITS
25	357	LG/P	Air Charge Temperature	ACT
26	351	O/W	Reference Voltage	VREF
27	352	BR/LG	EGR Valve Position	EVP
29	94	DG/P	Heated Exhaust Gas Oxygen (Sensor)	HEGO
30	480	P/Y	Clutch Engage Switch	CES
31	101	GY/Y	Canister Purge Solenoid	CANP
33	360	DG	EGR Vacuum Regulator (Solenoid)	EVR
35	146	W/PK	Speed Control Vent (Solenoid)	SCVNT
36	324	Y/LG	Spark Out	SPOUT
37	361	R	Vehicle Power	VPWR
39	461	O	Speed Control Command Switch Ground	SCCS GND
40	60	BK/LG	Power Ground	PWR GND
41	377	W	Idle Speed Control — (DC Motor)	ISC –
42	145	GY/BK	Speed Control Vacuum (Solenoid)	SCVAC
45	358	LG/BK	Manifold Absolute Pressure (Sensor)	MAP
46	359	BK/W	Signal Return (Ground)	SIG RTN
47	355	DG/LG	Throttle Position (Sensor)	TP
48	200	W/BK	Self-Test Input	STI
49	89	O	Heated Exhaust Gas Oxygen (Sensor) Ground	HEGO GND
50	151	LB/BK	Speed Control Command Switch	SCCS
53	462	P	Shift Indicator Light	SIL
54	331	R	Wide Open Throttle (WOT) A/C cutoff	WAC
55	197	T/O	Electro Drive Fan	EDF
56	349	DB	Profile Ignition Pick-Up	PIP
57	361	R	Vehicle Power	VPWR
58	95	T/R	Injector	INJ
60	60	BK/LG	Power Ground	PWR GND

Pin locations given for reference only. Probing 60 pin connector with DVOM probe will result in permanent damage to the pin connectors. Always probe as directed, using the Breakout Box.

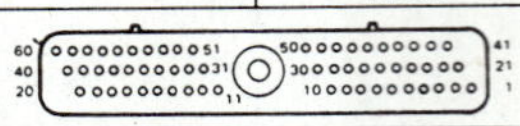

Integrated Controller Pin Usage — 2.5L CFI MTX

Pin	Circuit	Color	Application	Abbreviations
1	181	BR/O	EDF Power Into Controller	Batt +
2	181	BR/O	EDF Power Into Controller	Batt +
3	228	BR/Y	H/EDF Power to Fan	PTF
4	228	BR/Y	H/EDF Power to Fan	PTF
5	787	PK/BK	Power to the Pump	PTP
6	—	—	Not Used	—
7	—	—	Not Used	—
8	37	Y	Battery Voltage (Power Relay)	Batt +
9	—	—	Not Used	—
10	—	—	Not Used	—
11	—	—	Not Used	—
12	—	—	Not Used	—
13	16	R/LG	Keypower	KPWR
14	197	T/O	EDF Ground	EDF
15	60	BK/LG	Vehicle Ground	PWR GND
16	57	BK	A/C Ground	A/C GND
17	—	—	Not Used	—
18	97	T/LG	Fuel Pump Relay Ground	FP
19	—	—	Not Used	—
20	—	—	Not Used	—
21	883	PK/LB	A/C Power In	ACCS
22	331	R	WOT A/C Cutoff	WAC
23	347	BK/Y	A/C Power to Clutch Coil	PTAC
24	361	R	EEC IV Power Out	VPWR

Integrated Controller Schematic — 2.5L CFI MTX

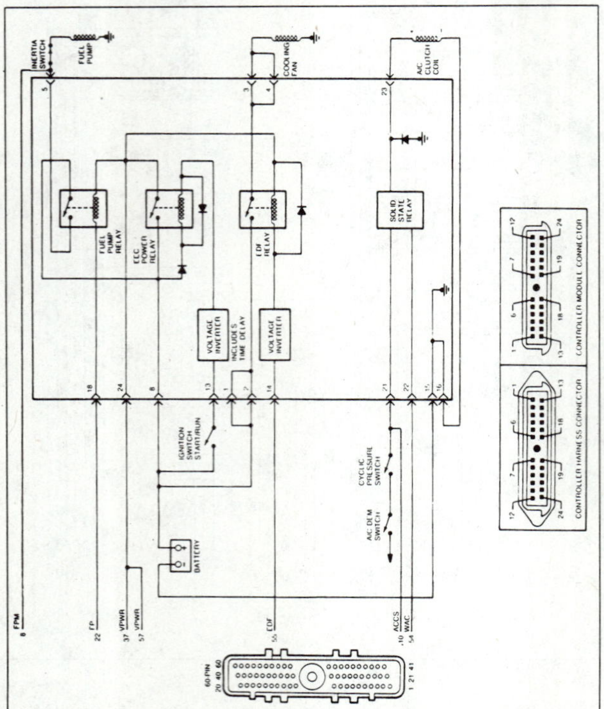

Electrical Schematic — 3.0L EFI

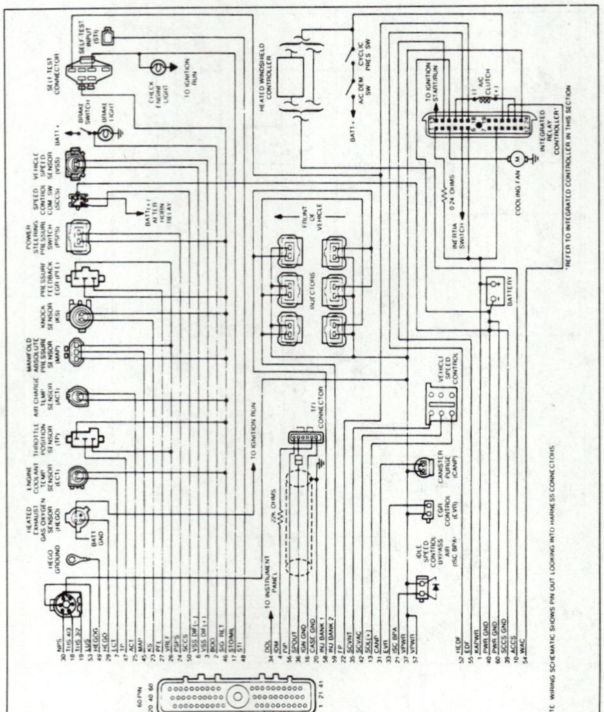

EEC-IV Module Connector Pin Usage — 3.0L EFI

Pin	Circuit	Wire Color	Application	Abbreviations
1	37	Y	Keep Alive Power	KAPWR
2	810	R/LG	Brake On/Off	BOO
3	150	DG/W	Vehicle Speed Sensor +	VSS DIF +
4	11	DG/Y	Ignition Diagnostic Monitor	IDM
6	563	O/Y	Vehicle Speed Sensor –	VSS DIF –
7	354	LG/Y	Engine Coolant Temperature	ECT
10	883	PK/LB	A/C Cycling Switch	ACCS
13	144	O/Y	Vehicle Speed Control Solenoid	SOL +
16	350	GY	Ignition Ground	IGN GND
17	201	T/R	Self-Test Output/ Check Engine Light	STO/MIL
18	315	DG/P	Transmission 4/3 Switch	THS 4/3
19	237	O/Y	Transmission 3/2 Switch	THS 3/2
20	57	BK	Case Ground	CSE GND
21	68	O/BK	Idle Speed Control (Bypass Air)	ISC-BPA
22	97	T/LG	Fuel Pump	FP
23	310	Y/R	Knock Sensor	KS
24	330	Y/LG	Power Steering Pressure Switch	PSPS
25	357	LG/P	Air Charge Temperature	ACT
26	351	O/W	Reference Voltage	VREF
27	352	BR/LG	Pressure Feedback EGR	PFE
29	94	DG/P	Heated Exhaust Gas Oxygen Sensor	HEGO
30	480	P/Y	Neutral Pressure Switch	NPS
31	101	GY/Y	Canister Purge Solenoid	CANP
33	360	DG	EGR Vacuum Regulator Solenoid	EVR
34	305	LB/PK	Data Output Link	DOL
35	146	W/PK	Speed Control Vent Solenoid	SCVNT
36	324	Y/LG	Spark Output	SPOUT
37	361	R	Vehicle Power	VPWR
39	461	O	Speed Control Command Switch Ground	SCCS GND
40	60	BK/LG	Power Ground	PWR GND
42	145	GY/BK	Speed Control Vacuum Solenoid	SCVAC
45	358	LG/BK	Manifold Absolute Pressure Sensor	MAP
46	359	BK/W	Signal Return Ground	SIG RTN
47	355	DG/LG	Throttle Position Sensor	TP
48	200	W/BK	Self Test Input	STI
49	89	O	Heated Exhaust Gas Oxygen Sensor Ground	HEGO GND
50	151	LB/BK	Speed Control Command Switch	SCCS
52	639	PK	High Electro Drive Fan	HEDF
53	224	T/LB	Lock-Up Solenoid (Transmission)	LUS
54	331	R	Wide Open Throttle (WOT) A/C Cutoff	WAC
55	197	T/O	Electro Drive Fan (Low)	EDF
56	349	DB	Profile Ignition Pick-Up	PIP
57	361	R	Vehicle Power	VPWR
58	95	T/R	Injector Bank 1	INJ 1
59	96	T/O	Injector Bank 2	INJ 2
60	60	BK/LG	Power Ground	PWR GND

Pin locations given for reference only. Probing 60 pin connector with DVOM probe will result in permanent damage to the pin connectors. Always probe as directed, using the Breakout Box.

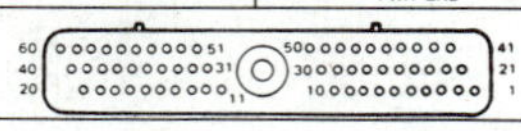

Integrated Controller Pin Usage — 3.0L EFI

Pin	Circuit	Color	Applications	Abbreviations
1	181	BR/O	Power to EDF Relay	PT/EDF
2	181	BR/O	Power to EDF Relay	PT/EDF
3	228	BR/Y	Power to Cooling Fan	PTF
4	228	BR/Y	Power to Cooling Fan	PTF
5	787	PK/BK	Power to Fuel Pump	PTP
6	38	BK/O	Power to HEDF Relay	PT/HEDF
7	38	BK/O	Power to HEDF Relay	PT/HEDF
8	37	Y	Battery to EEC Relay	BATT +
13	16	R/LG	Key Power	KEY PWR
14	197	T/O	EDF Circuit	EDF
15	60	BK/LG	Power Ground	PWR GRD
16	57	BK	A/C Clutch Ground	PWR GRD
17	639	PK	HEDF Circuit	HEDF
18	97	T/LG	Fuel Pump	FP
21	883	PK/LB	A/C Cyclic Switch	ACCS
22	331	R	WOT A/C Cut Off	WAC
23	347	BK/Y	Power to A/C	PTAC
24	361	R	Vehicle Power	VPWR

Integrated Controller Schematic — 3.0L EFI

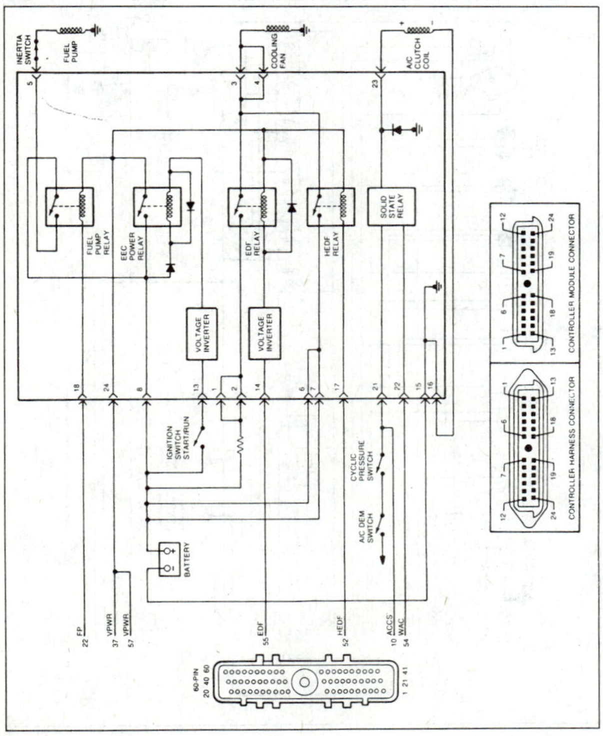

Electrical Schematic — 3.8L EFI AXOD

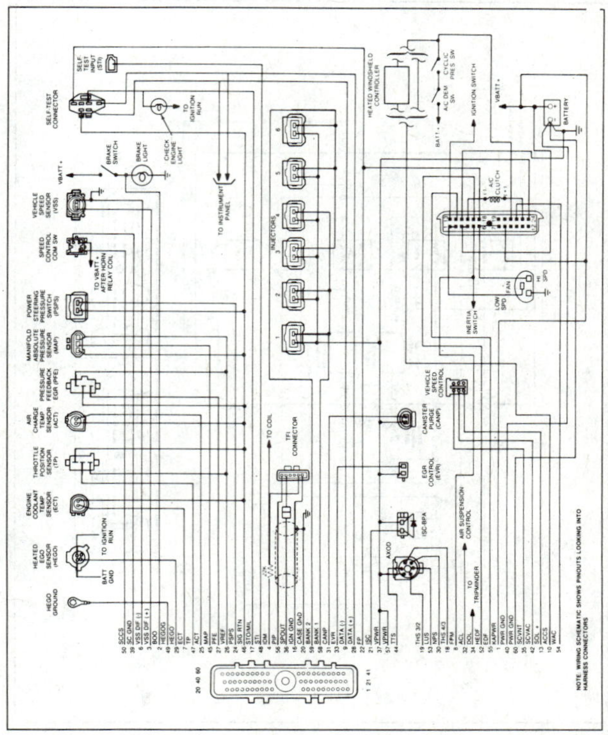

EEC-IV Module Connector Pin Usage

3.8L EFI AXOD

Pin	Circuit	Wire Color	Applications	Abbreviations
1	37	Y	Keep Alive Power	KAPWR
2	810	R/LG	Brake On/Off	BOO
3	150	DG/W	Vehicle Speed Sensor +	VSS DIF +
4	11	DG/Y	Ignition Diagnostic Monitor	IDM
6	563	O/Y	Vehicle Speed Sensor −	VSS DIF −
7	354	LG/Y	Engine Coolant Temperature	ECT
8	787	PK/BK	Fuel Pump Monitor	FPM
9	695	BK/O	Data Communications Link − (Continental Only)	DATA −
10	585	P	A/C Cycling Switch	ACCS
13	144	O/Y	Vehicle Speed Control Solenoid	SOL +
16	350	GY	Ignition Ground	IGN GND
17	201	T/R	Self-Test Output/"Check Engine" Light	STO/MIL
18	315	DG/P	Transmission 4/3 Switch	THS 4/3
19	237	O/Y	Transmission 3/2 Switch	THS 3/2
20	57	BK	Case Ground	CSE·GND
21	68	O/BK	Idle Speed Control (Bypass Air)	ISC-BPA
22	97	T/LG	Fuel Pump	FP
24	330	Y/LG	Power Steering Pressure Switch	PSPS
25	357	LG/P	Air Charge Temperature	ACT
26	351	O/W	Reference Voltage	VREF
27	352	BR/LG	Pressure Feedback EGR	PFE
28	696	O/BK	Data Communications Link + (Continental Only)	DATA +
29	94	DG/P	Heated Exhaust Gas Oxygen Sensor	HEGO
30	480	P/Y	Neutral Pressure Switch	NPS
31	101	GY/Y	Canister Purge Solenoid	CANP
32	637	LG	Air Suspension Control (Continental Only)	ACL
33	360	DG	EGR Vacuum Regulator Solenoid	EVR
34	305	LB/PK	Data Output Link	DOL
35	146	W/PK	Speed Control Vent Solenoid	SCVNT
36	324	Y/LG	Spark Output	SPOUT
37	361	R	Vehicle Power	VPWR
39	461	O	Speed Control Command Switch Ground	SCCS GND
40	60	BK/LG	Power Ground	PWR GND
42	145	GY/BK	Speed Control Vacuum Solenoid	SCVAC
44	854	GY/W	Transmission Temperature Switch (Continental Only)	TTS
45	358	LG/BK	Manifold Absolute Pressure Sensor	MAP
46	359	BK/W	Signal Return (Ground)	SIG RTN
47	355	DG/LG	Throttle Position Sensor	TP
48	200	W/BK	Self Test Input	STI
49	89	O	Heated Exhaust Gas Oxygen Sensor Ground	HEGO GND
50	151	LB/BK	Speed Control Command Switch	SCCS
52	639	PK	High Electro Drive Fan	HEDF
53	224	T/LB	Lock-Up Solenoid (Transmission)	LUS
54	331	R	Wide Open Throttle (WOT) A/C Cutoff	WAC
55	197	T/O	Electro Drive Fan (Low)	EDF
56	349	DB	Profile Ignition Pick-Up	PIP
57	361	R	Vehicle Power	VPWR
58	95	T/R	Injector Bank 1	INJ 1
59	96	T/O	Injector Bank 2	INJ 2
60	60	BK/LG	Power Ground	PWR GND

Pin locations given for reference only. Probing 60 pin connector with DVOM probe will result in permanent damage to the pin connectors. Always probe as directed, using the Breakout Box.

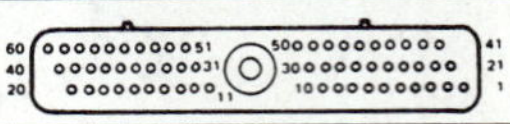

Integrated Controller Pin Usage

3.8L EFI AXOD

Pin	Circuit	Color	Applications	Abbreviations
1	181	BR/O	Power to EDF Relay	PT EDF
2	181	BR/O	Power to EDF Relay	PT EDF
3	38	BK/O	Power to Cooling Fan	PTF
4	38	BK/O	Power to Cooling Fan	PTF
5	787	PK/BK	Power to Fuel Pump	PTP
6	228	BR/Y	Power to HEDF Relay	PT HEDF
7	228	BR/Y	Power to HEDF Relay	PT HEDF
8	37	Y	Battery to EEC Relay	BATT −
13	16	R/LG	Key Power	KEY PWR
14	197	T/O	EDF Circuit	EDF
15	60	BK/LG	Power Ground	PWR GRD
16	321	GY/W	A/C Clutch Ground	PWR GRD
17	639	PK	HEDF Circuit	HEDF
18	97	T/LG	Fuel Pump	FP
21	348	LG/P	A/C Cyclic Switch	ACCS
22	331	R	WOT A/C Cut Off	WAC
23	347	BK/Y	Power to A/C	PTAC
24	361	R	Vehicle Power	VPWR

Integrated Controller Schematic

3.8L EFI AXOD

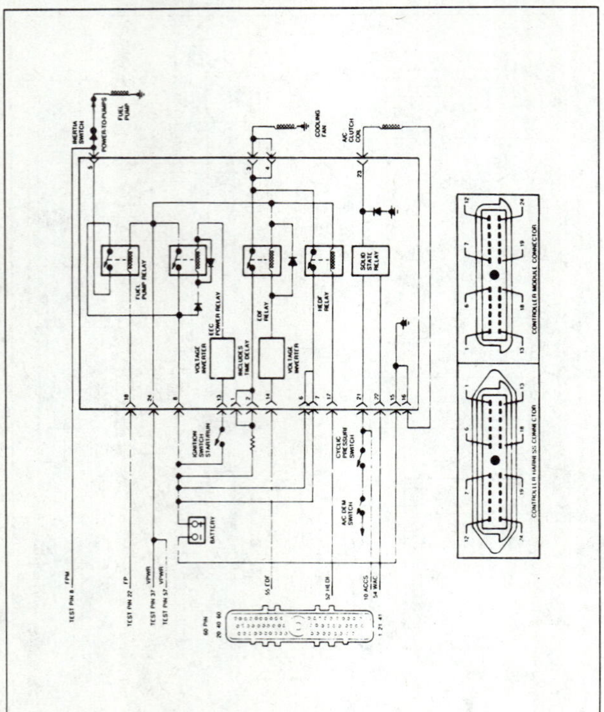

Electrical Schematic

3.8L RWD EFI

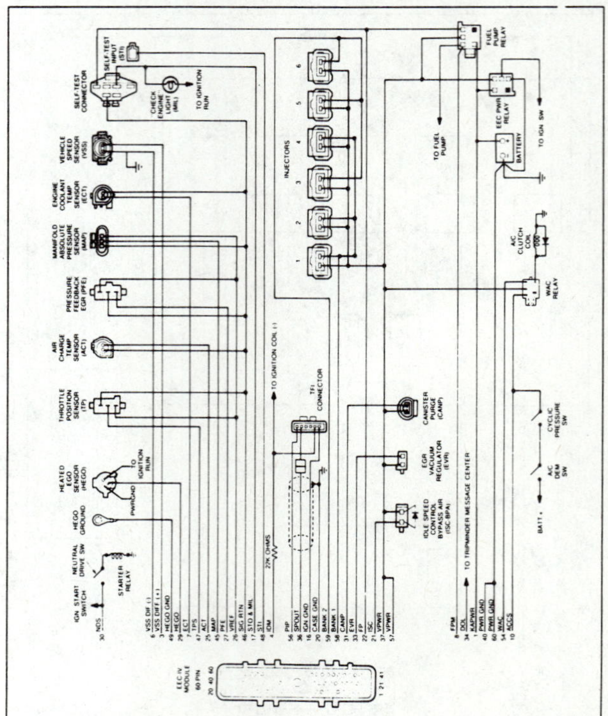

EEC-IV Module Connector Pin Usage — 3.8L RWD EFI

Pin	Circuit	Wire Color	Application	Abbreviations
1	38	BK/O	Keep Alive Power	KAPWR
3	150	DG/W	Vehicle Speed Sensor +	VSS DIF +
4	11	DG/Y	Ignition Diagnostic Monitor	IDM
6	563	O/Y	Vehicle Speed Sensor −	VSS DIF −
7	354	LG/Y	Engine Coolant Temperature	ECT
8	787	PK/BK	Fuel Pump Monitor	FPM
10	883	PK/LB	A/C Cycling Switch	ACCS
16	350	GY	Ignition Ground	IGN GND
17	382	Y/BK	Self-Test Output and "Check Engine" Light	STO/MIL
20	57	BK	Case Ground	CSE GND
21	68	O/BK	Idle Speed Control	ISC
22	97	T/LG	Fuel Pump	FP
25	357	LG/P	Air Charge Temperature	ACT
26	351	O/W	Reference Voltage	VREF
27	352	BR/LG	Pressure Feedback EGR	PFE
29	94	DG/P	Heated Exhaust Gas Oxygen Sensor	HEGO
30	376	BR/W	Neutral Drive Switch	NDS
31	101	GY/Y	Canister Purge	CANP
33	360	DG	EGR Vacuum Regulator	EVR
34	305	LB/PK	Data Output Link	DOL
36	324	Y/LG	Spark Output	SPOUT
37	175	BK/Y	Vehicle Power	VPWR
40	60	BK/LG	Power Ground	PWR GND
45	356	DB/LG	Manifold Absolute Pressure Sensor	MAP
46	359	BK/W	Signal Return	SIG/RTN
47	355	DG/LG	Throttle Position Sensor	TP
48	209	W/R	Self-Test Input	STI
49	89	O	Heated Exhaust Gas Oxygen Sensor Ground	HEGO G
54	73	O/LB	W.O.T. A/C Cutoff	WAC
56	349	DB	Profile Ignition Pick-Up	PIP
57	175	BK/Y	Vehicle Power	VPWR
58	556	W	Injector Bank 1	INJ 1
59	560	LG	Injector Bank 2	INJ 2
60	60	BK/LG	Power Ground	PWR GND

Pin locations given for reference only. Probing 60 pin connector with DVOM probe will result in permanent damage to the pin connectors. Always probe as directed, using the Breakout Box.

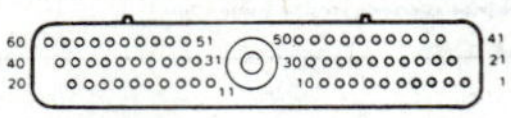

EEC-IV Module Connector Pin Usage — 5.0L SEFI

Pin	Car Lines: Thunderbird, Cougar Crt. #	Thunderbird, Cougar Wire Color	Mustang Crt. #	Mustang Wire Color	Application	Abbreviations
1	38	BK/O	38	BK/O	Keep Alive Power	KAPWR
3	150	DG/W	—	—	Vehicle Speed Sensor +	VSS DIF +
4	189	LB/PK	11	DH/Y	Ignition Diagnostic Monitor	IDM
6	563	O/Y	—	—	Vehicle Speed Sensor −	VSS DIF −
7	354	LG/Y	354	LG/Y	Engine Coolant Temperature	ECT
10	883	PK/LB	883	PK/LB	A/C Clutch Signal	ACCS
11	99	LG/BK	99	LG/BK	Air Management 2	AM2
12	557	BR/Y	557	BR/Y	Injector #3	INJ 3
13	558	BR/LB	558	BR/LB	Injector #4	INJ 4
14	559	T/LB	559	T/LB	Injector #5	INJ 5
15	560	LG	560	LG	Injector #6	INJ 6
16	259	BK/O	259	BK/O	Ignition Ground	IGN GND
17	382	Y/BK	382	Y/BK	Self-Test Output/"Check Engine" Light	STO/MIL
20	57	BK	57	BK	Case Ground	CASE GND
21	264	W/LB	264	W/LB	Idle Speed Control	ISC
22	97	T/LG	97	T/LG	Fuel Pump	FP
25	357	LG/P	357	LG/P	Air Charge Temperature	ACT
26	351	O/W	351	O/W	Reference Voltage	VREF
27	352	BR/LG	352	BR/LG	EGR Valve Position Sensor	EVP
29	94	DG/P	94	DG/P	Heated Exhaust Gas Oxygen Sensor	HEGO-R
30	—	—	199	LB/Y	Clutch/Neutral Gear Switches (M/T)	NGS and CES
30	376	BR/W	33	W/PK	Neutral Drive Switch (A/T)	NDS
31	101	GY/Y	101	GY/Y	Canister Purge	CANP
33	360	DG	360	DG	EGR Valve Regulator	EVR
34	305	LB/PK	—	—	Data Output Link	DCL
36	324	Y/LG	324	Y/LG	Spark Output	SPOUT
37	175	BK/Y	361	R	Vehicle Power	VPWR
40	60	BK/LG	60	BK/LG	Power Ground	PWR GND
42	561	T/O	561	T/O	Injector #7	INJ 7
43	90	DB/LG	90	DB/LG	Heated Exhaust Gas Oxygen Sensor	HEGO-L
45	358	LG/BK	358	LG/BK	Manifold Absolute Pressure Sensor	MAP
46	359	BK/W	359	BK/W	Signal Return	SIG RTN
47	355	DG/LG	355	DG/LG	Throttle Position Sensor	TP
48	209	W/R	201	T/R	Self-Test Input	STI
49	89	O	89	O	Heated EGO Ground	HEGO GND
51	100	W/R	100	W/R	Air Management 1	AM1
52	562	LB	562	LB	Injector #8	INJ 8
54	73	O/LB	73	O/LB	Wide Open Throttle A/C Cut Off	WAC
56	349	DB	349	DB	Profile Ignition Pick-Up	PIP
57	175	BK/Y	361	R	Vehicle Power	VPWR
58	555	T	555	T	Injector #1	INJ 1
59	556	W	556	W	Injector #2	INJ 2
60	60	BK/LG	60	BK/LG	Power Ground	PWR GND

Pin locations given for reference only. Probing 60 pin connector with DVOM probe will result in permanent damage to the pin connectors. Always probe as directed using the Breakout Box.

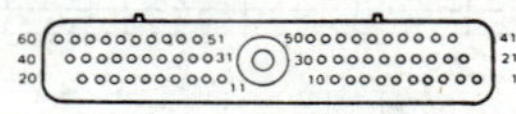

Electrical Schematic — 5.0L SEFI

THUNDERBIRD/COUGAR, MUSTANG

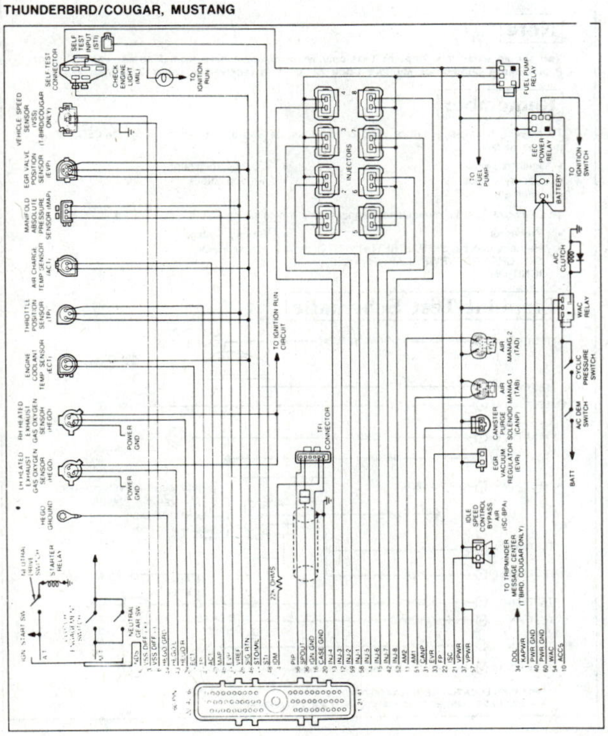

Electrical Schematic — 5.0L SEFI

FORD, MERCURY, MARK VII, TOWN CAR

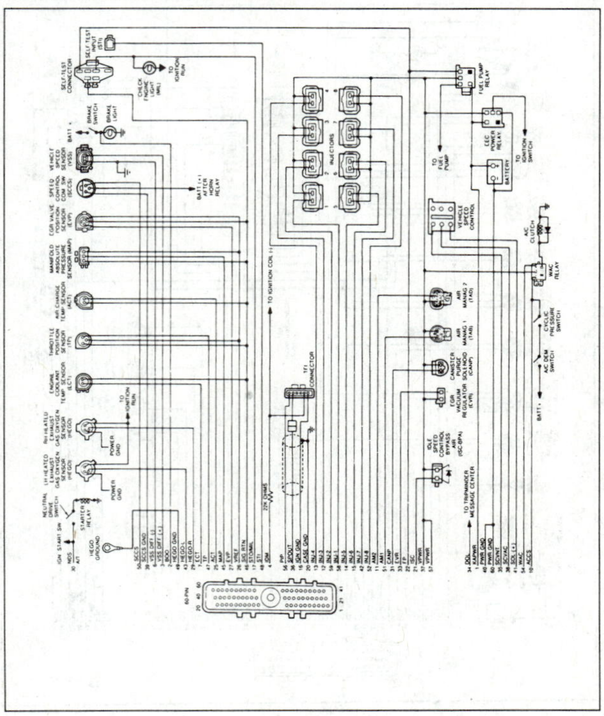

EEC-IV Module Connector Pin Usage — 5.0L SEFI

Pin	Car Lines				Application	Abbreviations
	Mark VII		Ford, Mercury, Town Car			
	Crt.#	Wire Color	Crt.#	Wire Color		
1	38	BK/O	37	Y	Keep Alive Power	KAPWR
2	511	LG	511	LG	Brake On/Off	BOO
3	150	DG/W	150	DG/W	Vehicle Speed Sensor +	VSS DIF +
4	11	DG/Y	11	DG/Y	Ignition Diagnostic Monitor	IDM
6	683	P/LB	359	BK/W	Vehicle Speed Sensor −	VSS DIF −
7	354	LG/Y	354	LG/Y	Engine Coolant Temperature	ECT
10	348	LG/P	883	PK/LB	A/C Clutch Signal	ACCS
11	99	LG/BK	99	LG/BK	Air Management 2	AM2
12	557	BR/Y	557	BR/Y	Injector #3	INJ 3
13	558	BR/LB	558	BR/LB	Injector #4	INJ 4
14	559	T/LB	559	T/LB	Injector #5	INJ 5
15	560	LG	560	LG	Injector #6	INJ 6
16	259	BK/O	259	BK/O	Ignition Ground	IGN GND
17	382	Y/BK	201	T/R	Self-Test Output/"Check Engine" Light	STO/MIL
20	57	BK	57	BK	Case Ground	CASE GND
21	264	W/LB	264	W/LB	Idle Speed Control	ISC
22	97	T/LG	97	T/LG	Fuel Pump	FP
25	357	LG/P	357	LG/P	Air Charge Temperature	ACT
26	351	O/W	351	O/W	Reference Voltage	VREF
27	352	BR/LG	352	BR/LG	EGR Valve Position Sensor	EVP
29	94	DG/P	94	DG/P	Heated Exhaust Gas Oxygen Sensor	HEGO-R
30	33	W/PK	33	W/PK	Neutral Drive Switch	NDS
31	101	GY/Y	101	GY/Y	Canister Purge	CANP
33	360	DG	360	DG	EGR Valve Regulator	EVR
34	305	LB/PK	305	LB/PK	Data Output Line	DOL
35	146	W/PK	146	W/PK	Speed Control Vent (Solenoid)	SCVNT
36	324	Y/LG	324	Y/LG	Spark Output	SPOUT
37	361	R	361	R	Vehicle Power	VPWR
38	145	GY/BK	145	GY/BK	Speed Control Vacuum (Solenoid)	SCVAC
39	199	LB/Y	679	GY/BK	Speed Control Command Switch Ground	SCCS GND
40	60	BK/LG	60	BK/LG	Power Ground	PWR GND
41	144	O/Y	144	O/Y	Vehicle Speed Control Solenoid	SOL +
42	561	T/O	561	T/O	Injector #7	INJ 7
43	90	DB/LG	90	DB/LG	Heated Exhaust Gas Oxygen Sensor	HEGO-L
45	356	DB/LG	358	LG/BK	Manifold Absolute Pressure Sensor	MAP
46	359	BK/W	359	BK/W	Signal Return	SIG RTN
47	355	DG/LG	355	DG/LG	Throttle Position Sensor	TP
48	209	W/R	200	W/BK	Self-Test Input	STI
49	89	O	89	O	Heated EGO Ground	HEGO GND
50	151	LB/PK	151	LB/BK	Speed Control Command Switch	SCCS
51	100	W/R	100	W/R	Air Management 1	AM1
52	562	LB	562	LB	Injector #8	INJ 8
54	73	O/LB	73	O/LB	Wide Open Throttle A/C Cut Off	WAC
56	349	DB	349	DB	Profile Ignition Pick-Up	PIP
57	361	R	361	R	Vehicle Power	VPWR
58	555	T	555	T	Injector #1	INJ 1
59	556	W	556	W	Injector #2	INJ 2
60	60	BK/LG	60	BK/LG	Power Ground	PWR GND

Pin locations given for reference only. Probing 60 pin connector with DVOM probe will result in permanent damage to the pin connectors. Always probe as directed, using the Breakout Box.

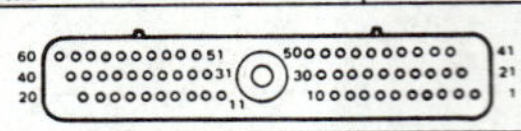

EEC-IV Module Connector Pin Usage — 5.0L SEFI MA

5.0L MASS AIR SEFI MUSTANG

Pin	Circuit	Wire Color	Application	Abbreviations
1	38	BK/O	Keep Alive Power	KAPWR
3	150	DG/W	Vehicle Speed Sensor +	VSS +
4	11	DG/Y	Ignition Diagnostic Monitor	IDM
6	563	O/Y	Vehicle Speed Sensor −	VSS −
7	354	LG/Y	Engine Coolant Temperature	ECT
9	968	T/LB	Mass Air Signal Return	MAF RTN
10	883	PK/LB	A/C Clutch Signal	ACCS
12	557	BR/Y	Fuel Injector #3	INJ 3
13	558	BR/LB	Fuel Injector #4	INJ 4
14	559	T/LB	Fuel Injector #5	INJ 5
15	560	LG	Fuel Injector #6	INJ 6
16	259	BK/O	Ignition Ground	IGN GND
17	657	T	Self-Test Out and Malfunction Light	STO and MIL
19	787	PK/BK	Fuel Pump Monitor	FPM
20	57	BK	Case Ground	CASE GND
21	264	W/LB	Idle Speed Control Bypass Air	ISC-BPA
22	97	T/LG	Fuel Pump	FP
25	357	LG/P	Air Charge Temp	ACT
26	351	O/W	Reference Voltage	VREF
27	352	BR/LG	EGR Valve Position Sensor	EVP
29	94	DG/P	Heated Exhaust Gas Oxygen Sensor	R-HEGO
30	199	LB/Y	Neutral Gear SW. M.T.	NGS
30	33	W/PK	Neutral Drive With Automatics	NDS
31	101	GY/Y	Canister Purge	CANP
32	99	LG/BK	Thermactor Air Diverter	TAD
33	360	DG	EGR Valve Regulator	EVR
36	324	Y/LG	Spark Out	SPOUT
37	361	R	Vehicle Power	VPWR
38	100	W/R	Thermactor Air Bypass	TAB
40	60	BK/LG	Power Ground	PWR GND
42	561	T/O	Fuel Injector #7	INJ 7
43	90	DB/LG	Heated Exhaust Gas Oxygen Sensor	L-HEGO
45	358	LG/BK	Barometric Pressure Sensor	BP
46	359	BG/W	Signal Return	SIG RTN
47	355	DG/LG	Throttle Position Sensor	TPS
48	201	T/R	Self-Test Input	STI
49	89	O	Heated Exhaust Gas Oxygen Sensor Ground	HEGO GND
50	967	DB/O	Mass Air Flow	MAF
52	562	L/B	Fuel Injector #8	INJ 8
54	73	O/LB	W.O.T. A/C Cut Off	WOT A/C
56	349	DB	Profile Ignition Pick Up	PIP
57	361	R	Vehicle Power	VPWR
58	555	T	Fuel Injector #1	INJ 1
59	556	W	Fuel Injector #2	INJ 2
60	60	BK/LG	Power Ground	PWR GND

Pin locations given for reference only. Probing 60 pin connector with DVOM probe will result in permanent damage to the pin connectors. Always probe as directed, using the Breakout Box.

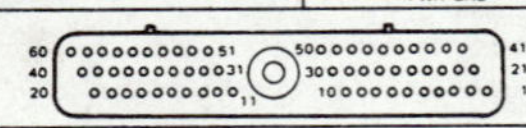

Electrical Schematic — 5.0L SEFI MA

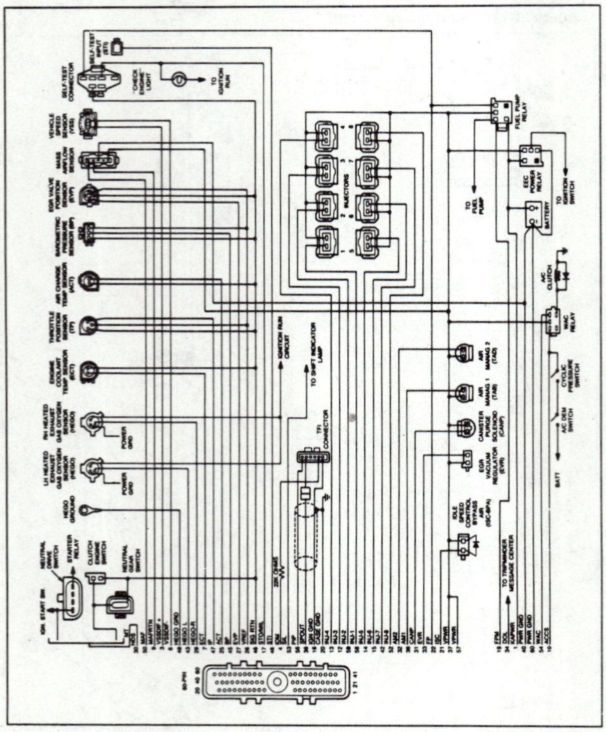

Vehicle Battery — Pinpoint Test B

Note

You should enter this Pinpoint Test only when directed here from Pinpoint Tests C, J or P or when a Continuous Memory Code 72 or 78 is received in Quick Test Step 6.0C.

Remember

To prevent the replacement of good components, be aware that the following non-EEC areas may be at fault:

- Ignition switch.
- Battery Cables.
- Alternator.
- Voltage Regulator.
- Ground Straps.

This Pinpoint Test is intended to diagnose only the following:

- Processor.
- Harness circuits: SIGNAL RETURN, STO, STI, GROUND, VPWR, KAPWR, VREF, IGNITION.
- Battery Voltage.
- Power Relay.

Pinpoint Test Schematic

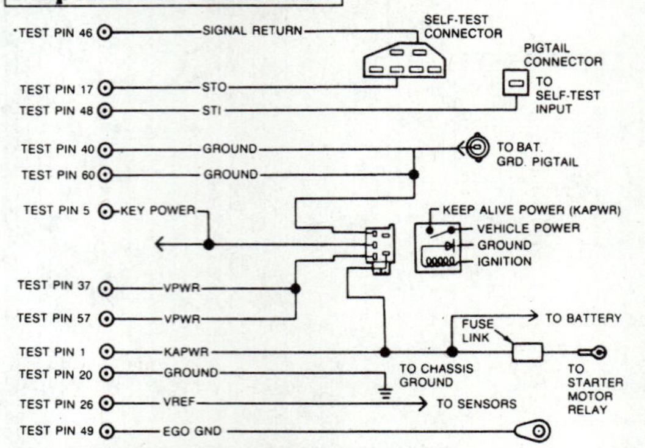

*TEST PINS LOCATED ON BREAKOUT BOX.
ALL HARNESS CONNECTORS VIEWED INTO MATING SURFACE.

Vehicle Battery — Pinpoint Test B

TEST STEP	RESULT ▶	ACTION TO TAKE
B1 BATTERY VOLTAGE CHECK		
• Key on, engine off. • DVOM on 20 volt scale. • Measure voltage across battery terminals. • Is voltage greater than 10.5 volts?	Yes ▶ No ▶	GO to **B2**. SERVICE discharged battery.
B2 CHECK EEC GROUND TO BATTERY GROUND		
• Key off, wait 10 seconds. • Disconnect processor 60 pin connector. Inspect for damaged pins, corrosion, loose wires, etc. Service as necessary. • Install breakout box. • Connect processor to breakout box. • DVOM on 200 ohm scale. • Measure resistance between Test Pin 40 at the breakout box and negative post of the battery and Test Pin 60 at the breakout box and negative post of the battery. • Are both resistances less than 5 ohms?	Yes ▶ No ▶	GO to **B3**. REMOVE breakout box. RECONNECT processor. SERVICE open in EEC ground circuit. RERUN Quick Test.
B3 PROCESSOR GROUND ISOLATION		
• Key off, wait 10 seconds. • Breakout box installed. • Processor connected. • DVOM on 200 ohm scale. • Measure resistance between Test Pin 46 and Test Pin 40 and between Test Pin 46 and Test Pin 60 both at the breakout box. • Are both resistances less than 5 ohms?	Yes ▶ No ▶	GO to **B4**. REMOVE breakout box. REPLACE processor. RERUN Quick Test.

Vehicle Battery — Pinpoint Test B

TEST STEP	RESULT ▶	ACTION TO TAKE
B4 CHECK CONTINUITY OF SIGNAL RETURN CIRCUIT		
• Key off, wait 10 seconds. • Breakout box installed. • Processor connected. • DVOM on 200 ohm scale. • Measure resistance between Test Pin 46 at the breakout box and SIGNAL RETURN in the Self-Test connector. • Is resistance less than 5.0 ohms?	Yes ▶ No ▶	GO to **B5**. REMOVE breakout box. RECONNECT processor. SERVICE open circuit. RERUN Quick Test.
B5 CHECK KEEP ALIVE POWER (KAPWR) CIRCUIT FOR VOLTAGE		
• Key on, engine off. • Breakout box installed. • Processor connected. • DVOM on 20 volt scale. • Measure voltage between Test Pin 1 at the breakout box and the battery negative post. • Is voltage greater than 10.5 volts?	Yes ▶ No ▶	GO to **B6**. CHECK KAPWR and VPWR circuits for shorts to ground and KAPWR circuit from power relay to battery positive post for opens. SERVICE as necessary. REMOVE breakout box. RECONNECT processor. RERUN Quick Test.
B6 CHECK IGNITION CIRCUIT FOR VOLTAGE		
• Key on, engine off. • Breakout box installed. • Processor connected. • DVOM on 20 volt scale. • Measure voltage between the battery negative post and IGNITION circuit at EEC power relay. • Is voltage greater than 10.5 volts?	Yes ▶ No ▶	GO to **B7**. SERVICE open in ignition switch circuits. REMOVE breakout box. RECONNECT processor. RERUN Quick Test.

Vehicle Battery — Pinpoint Test B

TEST STEP	RESULT ▶	ACTION TO TAKE
B7 CHECK CONTINUITY OF EEC POWER RELAY GROUND CIRCUIT		
• Key off, wait 10 seconds. • Breakout box installed. • Processor connected. • DVOM on 200 ohm scale. • Measure resistance between GROUND circuit at the EEC power relay and negative battery post. • Is the resistance less than 5 ohms?	Yes ▶ No ▶	GO to **B8**. SERVICE open circuit. REMOVE breakout box. RECONNECT processor. RERUN Quick Test.
B8 CHECK VOLTAGE OF VPWR CIRCUIT AT EEC POWER RELAY		
• Key on, engine off. • Breakout box installed. • Processor connected. • DVOM on 20 volt scale. • Measure voltage between the battery negative post and VPWR circuit at EEC power relay. • Is the voltage greater than 10.5 volts?	Yes ▶ No ▶	SERVICE open in VPWR circuit, if OK, SERVICE short to ground in VPWR circuit. REMOVE breakout box. RECONNECT processor. RERUN Quick Test. REPLACE power relay. REMOVE breakout box. RECONNECT processor. RERUN Quick Test.

Vehicle Battery — Pinpoint Test B

TEST STEP	RESULT ▶	ACTION TO TAKE
B10 WIGGLE TEST VPWR CIRCUITS		
NOTE: A momentary interrupt ("hiccup") in the vehicle performance may be due to **• EEC-IV wiring harness not properly routed.** **• Electrical or radio noises.** **— High Tension power lines** **— Some CB radio frequencies.** • Key on, engine off. • STAR tester or VOM hooked up to Self-Test connector. • Self-Test deactivated. • Enter key on, engine off continuous monitor mode. • Observe STAR/VOM for indication of a fault while performing the following. — Shake, bend and twist the EEC-IV harness from the EEC time delay power relay to the processor. • Is a fault indicated?	Yes ▶ No ▶	SERVICE VPWR circuit. CLEAR Continuous Memory Code. RERUN Quick Test. INSPECT EEC-IV time delay power relay and harness connectors for damaged pins, corrosion, etc. SERVICE as necessary. If OK, REPLACE EEC-IV time delay relay. CLEAR Continuous Memory Code. RERUN Quick Test.

Reference Voltage — Pinpoint Test C

Note

You should enter this Pinpoint Test only when a check for VREF has failed in the sensor Pinpoint Tests (D-Series) or Pinpoint Tests A or Q.

Remember

This Pinpoint Test is intended to diagnose only the following:

- Processor.
- Sensor harness circuits: SIGNAL RETURN, STO, STI, GROUND, VPWR, KAPWR, VREF, IGNITION.

Pinpoint Test Schematic

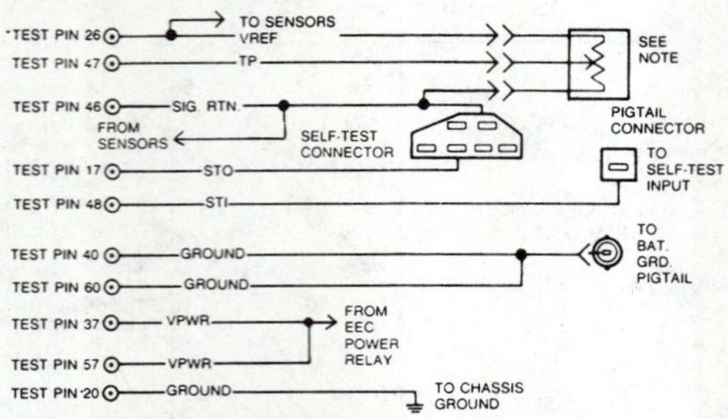

*TEST PINS LOCATED ON BREAKOUT BOX.
ALL HARNESS CONNECTORS VIEWED INTO MATING SURFACE.

Reference Voltage — Pinpoint Test C

TEST STEP	RESULT	▶ ACTION TO TAKE
C1 CHECK VEHICLE BATTERY POWER CIRCUIT		
• Key off, wait 10 seconds. • Disconnect 60 pin connector. Inspect for damaged pins, corrosion, loose wires, etc. Service as necessary. • Breakout box installed. • Processor connected. • Key on, engine off. • DVOM on 20 volt scale. • Measure voltage between Test Pin 37 at the breakout box and SIGNAL RETURN in Self-Test connector. • Is voltage greater than 10.5 volts?	Yes ▶ No ▶	GO to C2. RECONNECT SENSOR. 2.3L EFI TC, 2.5L HSC CFI, 3.8L and 3.0L EFI passenger car GO to X-1. All others, GO to B1.
C2 CHECK VREF VOLTAGE		
• Key on, engine off. • Breakout box installed. • Processor connected. • DVOM on 20 volt scale. • Measure voltage between Test Pin 26 and Test Pin 46 at the breakout box. • What is the voltage?	Greater than 6.0 volts ▶ Less than 4.0 volts ▶ Between 4.0 volts and 6.0 volts ▶	GO to C4. GO to C5. GO to C3.
C3 CHECK VREF AND SIGNAL RETURN FOR CONTINUITY		
• Breakout box installed. • Processor disconnected. • Sensor that sent you here disconnected. • Key off. • DVOM on 200 ohm scale. • Measure resistance from Test Pin 26 at breakout box to VREF at vehicle harness connector of the sensor that sent you here. • Measure resistance from Test Pin 46 at breakout box to signal return at vehicle harness connector of the sensor that sent you here. • Are both resistances less than 5.0 ohms?	Yes ▶ No ▶	RECONNECT sensors. Reference voltage OK. RERUN Quick Test. SERVICE open in VREF or SIGNAL RETURN. REMOVE breakout box. RECONNECT processor and sensor. RERUN Quick Test.

Reference Voltage — Pinpoint Test C

TEST STEP	RESULT	▶ ACTION TO TAKE
C4 CHECK FOR EXCESS VOLTAGE ON VREF CIRCUIT		
• Key off, wait 10 seconds. • Breakout box installed. • Disconnect processor. • Key on, engine off. • DVOM on 20 volt scale. • Measure voltage between Test Pin 26 at the breakout box and battery ground. • Is voltage less than 0.5 volts?	Yes ▶ No ▶	REMOVE breakout box. RECONNECT sensor. REPLACE processor. RERUN Quick Test. SERVICE short to battery power in EEC harness. REMOVE breakout box. RECONNECT processor and sensor. RERUN Quick Test. If condition persists, REPLACE processor.
C5 CHECK FOR SHORTED THROTTLE POSITION SENSOR		
• Key off, wait 10 seconds. • Breakout box installed. • Processor connected. • Disconnect Throttle Position (TP) sensor from vehicle harness. • Key on, engine off. • DVOM on 20 volt scale. • Measure voltage between Test Pin 26 and Test Pin 46 at the breakout box. • Is voltage less than 4.0 volts?	Yes ▶ No ▶	Vehicles equipped with EVP/PFE sensor, GO to C6. All other vehicles, GO to C7. REPLACE TP sensor. REMOVE breakout box. RECONNECT processor. RERUN Quick Test.
C6 CHECK FOR SHORTED EVP/PFE SENSOR		
• Key off, wait 10 seconds. • Breakout box installed. • Processor connected. • Disconnect EVP/PFE sensor. • Key on, engine off. • DVOM on 20 volt scale. • Measure voltage between Test Pin 26 and Test Pin 46 at the breakout box. • Is voltage less than 4.0 volts?	Yes ▶ No ▶	GO to C7. REPLACE EVP/PFE sensor. REMOVE breakout box. RECONNECT processor and sensor(s). RERUN Quick Test.

Reference Voltage — Pinpoint Test C

TEST STEP	RESULT	▶ ACTION TO TAKE
C7 CHECK FOR SHORTED MAP/BP SENSOR		
• Key off, wait 10 seconds. • Breakout box installed. • Processor connected. • Disconnect MAP/BP sensor. • Key on, engine off. • DVOM on 20 volt scale. • Measure voltage between Test Pin 26 and Test Pin 46 at the breakout box. • Is voltage less than 4.0 volts?	Yes ▶ No ▶	Vehicles equipped with VAF sensor, GO to C8. All other vehicles, GO to C9. REPLACE MAP/BP sensor. REMOVE breakout box. RECONNECT processor and sensor(s). RERUN Quick Test.
C8 CHECK FOR SHORTED VANE AIR METER (VAF) SENSOR		
• Key off, wait 10 seconds. • Breakout box installed. • Processor connected. • Disconnect vane air meter (VAF) sensor. • Key on, engine off. • DVOM on 20 volt scale. • Measure voltage between Test Pin 26 and Test Pin 46 at the breakout box. • Is voltage less than 4.0 volts?	Yes ▶ No ▶	GO to C9. REPLACE VAF sensor. REMOVE breakout box. RECONNECT processor and sensor(s). RERUN Quick Test.
C9 SHORT TO GROUND IN VREF		
• Key off, wait 10 seconds. • Breakout box installed. • Processor disconnected. • Disconnect TP and MAP/BP, EVP/PFE and VAF, if so equipped. • DVOM on 200 ohm scale. • Measure resistance between Test Pin 26 and Test Pins 20, 40, 46 and 60 at the breakout box. • Is any resistance less than 5 ohms?	Yes ▶ No ▶	REMOVE breakout box. RECONNECT processor. SERVICE short to ground. CONNECT all sensors. RERUN Quick Test. If original condition still exists, REPLACE processor. REMOVE breakout box. RECONNECT sensors. REPLACE processor. RERUN Quick Test.

Vane Air Temperature Sensor (VAT) — Pinpoint Test DA

Note

You should enter this Pinpoint Test only when a Service Code 28, 58 or 68 is received in Quick Test Step 3.0, 5.0 or 6.0.

Remember

To prevent the replacement of good components, be aware that the following non-EEC areas may be at fault:

- Test performed in unusually low (cold) or high (hot) ambient conditions.
- Ambient temperature must be greater than 50°F for this test.

This Pinpoint Test is intended to diagnose only the following:

- VAT sensor.
- Circuits: VAT, and SIGNAL RETURN.
- Vehicle harness.
- Processor assembly.

Pinpoint Test Schematic

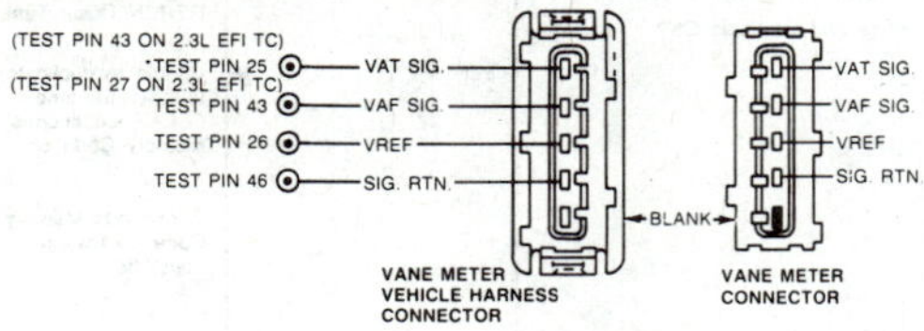

NOTE: AMBIENT TEMPERATURE MUST BE GREATER THAN 50°F TO PASS THIS TEST.

TYPICAL RESISTANCE BETWEEN TEST PINS 25 (OR 43) & 46	5800 ohms	2700 ohms	300 ohms	180 ohms	125 ohms
AT TEMPERATURE	32°F	65°F	185°F	220°F	240°F

*TEST PINS LOCATED ON BREAKOUT BOX.
ALL HARNESS CONNECTORS VIEWED INTO MATING SURFACE.

Vane Air Temperature Sensor (VAT) — Pinpoint Test DA

TEST STEP	RESULT	ACTION TO TAKE
DA1 SERVICE CODE 28: CHECK AMBIENT TEMPERATURE		
NOTE: Ambient temperature must be greater than 50°F for this test. • Is the ambient temperature greater than 50°F?	Yes ▶ No ▶	GO to DA2. RERUN Quick Test.
DA2 CHECK FOR VREF AT THROTTLE POSITION SENSOR		
• Refer to illustration QA. • Key off, wait 10 seconds. • Disconnect TP sensor. • DVOM on 20 volt scale. • Key on, engine off. • Measure voltage between VREF and SIGNAL RETURN at the TP vehicle harness connector. • Is voltage between 4.0 volts and 6.0 volts?	Yes ▶ No ▶	RECONNECT TP sensor, GO to DA3. GO to Pinpoint Test Step C1.
DA3 VAT SENSOR CHECK		
NOTE: Ambient temperature must be greater than 50°F for this test. • Key off, wait 10 seconds. • Disconnect harness from the vane meter. • DVOM on 200,000 ohm scale. • Measure resistance between VAT signal and SIGNAL RETURN at the VAT sensor. • Is resistance between 125 ohms (240°F) and 3700 ohms (50°F)?	Yes ▶ No ▶	REPLACE processor. RECONNECT harness to vane meter. RERUN Quick Test. REPLACE vane meter. RERUN Quick Test.

Vane Air Temperature Sensor (VAT) — Pinpoint Test DA

TEST STEP	RESULT	ACTION TO TAKE
DA10 SERVICE CODE 58: INDUCE OPPOSITE CODE		
• Key off, wait 10 seconds. • Disconnect vehicle harness from vane meter. Inspect for damaged pins, corrosion, loose wires, etc. Service as necessary. • Insert a jumper wire (paper clip) between VAT SIGNAL and SIGNAL RETURN at the vane meter vehicle harness connector. • Rerun Key On Engine Off Self-Test. • Is Code 68 present?	Yes ▶ No ▶	REPLACE vane meter. REMOVE jumper wire. CONNECT harness to vane meter. RERUN Quick Test. REMOVE jumper wire. GO to DA11.
DA11 CHECK CONTINUITY OF VAT SIGNAL AND SIGNAL RETURN		
• Key off, wait 10 seconds. • Harness disconnected from vane meter, jumper wire removed. • Disconnect processor 60 pin connector. Inspect for damaged pins, corrosion, loose wires, etc. Service as necessary. • Install breakout box leaving processor disconnected. • DVOM on 200 ohm scale. • Measure resistance between VAT SIGNAL at the vane meter vehicle harness connector, and Test Pin 25 (Test Pin 43 on 2.3L EFI TC) at the breakout box. • Measure resistance between SIGNAL RETURN at the vane meter vehicle harness connector, and Test Pin 46 at the breakout box. • Are both resistances less than 5 ohms?	Yes ▶ No ▶	REPLACE processor. REMOVE breakout box. RECONNECT harness to vane meter and processor. RERUN Quick Test. CORRECT open circuit. REMOVE breakout box. RECONNECT harness to vane meter and processor. RERUN Quick Test.

Vane Air Temperature Sensor (VAT) — Pinpoint Test DA

TEST STEP	RESULT	ACTION TO TAKE
DA20 SERVICE CODE 68: INDUCE OPPOSITE CODE		
• Key off, wait 10 seconds. • Disconnect vehicle harness from vane meter. Inspect for damaged pins, corrosion, loose wires, etc. Service as necessary. • Run Key On Engine Off Self-Test. • Is Code 58 present?	Yes ▶ No ▶	REPLACE vane meter. RECONNECT harness to vane meter. RERUN Quick Test. GO to DA21.
DA21 CHECK FOR VREF AT THROTTLE POSITION SENSOR		
• Refer to illustration QA. • Key off, wait 10 seconds. • Disconnect TP sensor. • DVOM on 20 volt scale. • Key on, engine off. • Measure voltage at the TP vehicle harness connector between VREF and SIGNAL RETURN. • Is voltage between 4.0 volts and 6.0 volts?	Yes ▶ No ▶	RECONNECT TP sensor, GO to DA22. GO to Pinpoint Test Step C1.
DA22 CHECK VAT SIGNAL FOR SHORTS		
• Key off, wait 10 seconds. • Harness disconnected from vane meter. • Disconnect processor 60 pin connector. Inspect for damaged pins, corrosion, loose wires, etc. Service as necessary. • Install breakout box, leave processor disconnected. • DVOM on 200,000 ohm scale. • Measure resistance between Test Pin 25 (Test Pin 43 on 2.3L EFI TC) and Test Pins 40, 46 and 60 at the breakout box. • Are all resistances greater than 10,000 ohms?	Yes ▶ No ▶	REPLACE processor. REMOVE breakout box. RECONNECT processor. RERUN Quick Test. CORRECT circuit shorts. REMOVE breakout box. RECONNECT processor and vane meter. RERUN Quick Test.

Vane Air Temperature Sensor (VAT) — Pinpoint Test DA

TEST STEP	RESULT ▶	ACTION TO TAKE
DA90 CONTINUOUS MEMORY CODE 58: CHECK VAT SENSOR		
• Enter Key On Engine Off Continuous Monitor mode. Refer to Appendix in Section 16. • Observe VOM or STAR LED for indication of a fault while performing the following: — Lightly tap on vane meter (simulate road shock). — Wiggle connector at vane meter. • Is a fault indicated?	Yes ▶	DISCONNECT and INSPECT connectors. If connector and terminals are good, REPLACE vane meter. CLEAR Continuous Memory Code 58. RERUN Quick Test.
	No ▶	GO to DA91.

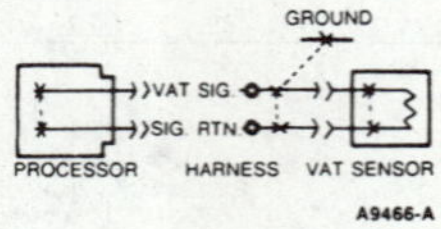

TEST STEP	RESULT ▶	ACTION TO TAKE
DA91 CHECK EEC-IV HARNESS		
• Still in Key On Engine Off Continuous Monitor mode. • Observe VOM or STAR LED for a fault indication while performing the following: — Referring to the illustration in Step DA90, grasp the harness closest to the sensor connector. Wiggle, shake or bend a small section of the EEC-IV system harness while working your way to the dash panel. Also wiggle, shake or bend the EEC-IV harness from the dash panel to the processor. • Is a fault indicated?	Yes ▶	ISOLATE fault and SERVICE as necessary. CLEAR Continuous Memory Code 58. RERUN Quick Test.
	No ▶	GO to DA92.

Vane Air Temperature Sensor (VAT) — Pinpoint Test DA

TEST STEP	RESULT ▶	ACTION TO TAKE
DA92 CHECK PROCESSOR AND HARNESS CONNECTORS		
• Key off, wait 10 seconds. • Disconnect processor 60 pin connector. • Inspect both connectors and connector terminals for obvious damage or faults. • Are connectors and terminals OK?	No ▶	SERVICE as necessary. CLEAR Continuous Memory Code 58. RERUN Quick Test.
	Yes ▶	Unable to duplicate fault at this time. CLEAR Continuous Memory Code 58. Continuous Memory Code 58 testing complete.
DA93 CONTINUOUS MEMORY CODE 68: CHECK VAT SENSOR		
• Enter Key On Engine Off Continuous Monitor mode. Refer to Appendix in Section 16. • Observe VOM or STAR LED for indication of a fault while performing the following: — Lightly tap on vane meter (simulate road shock). — Wiggle connector at vane meter. • Is fault indicated?	Yes ▶	DISCONNECT and INSPECT connectors. If connector and terminals are good, REPLACE vane meter. CLEAR Continuous Memory Code 68. RERUN Quick Test.
	No ▶	GO to DA94.

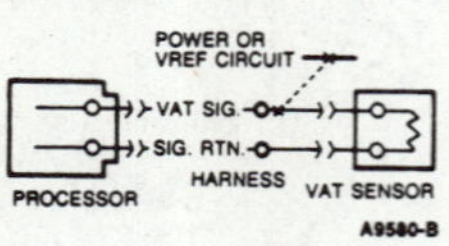

Vane Air Temperature Sensor (VAT) — Pinpoint Test DA

TEST STEP	RESULT ▶	ACTION TO TAKE
DA94 CHECK EEC-IV HARNESS		
• Still in Key On Engine Off Continuous Monitor mode. • Observe VOM or STAR LED for a fault indication while performing the following: — Referring to the illustration in Step DA93, grasp the harness closest to the sensor connector. Wiggle, shake or bend a small section of the EEC-IV system harness while working your way to the dash panel. Also wiggle, shake or bend the EEC-IV harness from the dash panel to the processor. • Is a fault indicated?	Yes ▶	ISOLATE fault and SERVICE as necessary. CLEAR Continuous Memory Code 68. RERUN Quick Test.
	No ▶	GO to DA95.
DA95 CHECK PROCESSOR AND HARNESS CONNECTORS		
• Key off, wait 10 seconds. • Disconnect processor 60 pin connector. • Inspect both connectors and connector terminals for obvious damage or faults. • Are connectors and terminals OK?	No ▶	SERVICE as necessary. CLEAR Continuous Memory Code 68. RERUN Quick Test.
	Yes ▶	Unable to duplicate fault at this time. CLEAR Continuous Memory Code 68. Continuous Memory Code 68 testing complete.

Air Charge Temperature Sensor (ACT) — Pinpoint Test DB

Note

You should enter this Pinpoint Test only when a Service Code 24, 54 or 64 is received in Quick Test Step 3.0, 5.0 or 6.0.

Remember

To prevent the replacement of good components, be aware that the following non-EEC areas may be at fault:

- Cooling system.
- Air cleaner duct problems.
- Improper engine oil level.

This Pinpoint Test is intended to diagnose only the following:

- ACT sensor.
- Harness circuits: ACT SIGNAL and SIGNAL RETURN.
- Processor assembly.

Pinpoint Test Schematic

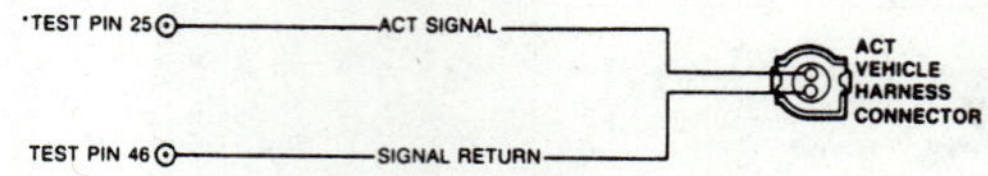

NOTE: AMBIENT TEMPERATURE MUST BE GREATER THAN 50°F.

TYPICAL RESISTANCE BETWEEN TEST PINS 25 & 46	58.750 ohms	40.500 ohms	3600 ohms	1840 ohms
AT TEMPERATURE	50°F	65°F	180°F	220°F

***TEST PINS LOCATED ON BREAKOUT BOX.**
ALL HARNESS CONNECTORS VIEWED INTO MATING SURFACE.

Air Charge Temperature Sensor (ACT) — Pinpoint Test DB

TEST STEP	RESULT	ACTION TO TAKE
DB1 SERVICE CODE 24: CHECK PROPER INSTALLATION OF ACT SENSOR		
• For vehicles with ACT sensor mounted in the intake manifold, GO to step DB2. • Is ACT sensor mounted properly in the air cleaner?	Yes ▶ No ▶	GO to DB2. INSTALL ACT sensor properly. RERUN Quick Test.
DB2 CHECK FOR VREF AT THROTTLE POSITION SENSOR		
• Refer to schematic in Pinpoint Test DH. • Key off, wait 10 seconds. • DVOM on 20 volt scale. • Disconnect TP sensor. • Key on, engine off. • Measure voltage between VREF and SIGNAL RETURN at the TP sensor vehicle harness connector. • Is voltage between 4.0 and 6.0 volts?	Yes ▶ No ▶	RECONNECT TP sensor, GO to DB3. GO to Pinpoint Test Step C1.
DB3 CHECK ACT SENSOR WITH ENGINE OFF		
NOTE: Make sure engine is warmed up prior to this test. • Key off, wait 10 seconds. • Disconnect harness from ACT sensor. • DVOM on 200,000 ohm scale. • Measure resistance of ACT sensor. • Is resistance between 1,100 and 58,000 ohms?	Yes ▶ No ▶	GO to DB4. CHECK heat stove duct valve operation. If OK, REPLACE ACT sensor. RECONNECT harness to ACT sensor. RERUN Quick Test.
DB4 CHECK ACT SENSOR WITH ENGINE RUNNING		
• Key off. • Harness disconnected from ACT sensor. • DVOM on 200,000 ohm scale. • Run engine for 2 minutes. • Measure resistance of ACT sensor with engine running. • Is resistance between 2,400 and 29,000 ohms?	Yes ▶ No ▶	REPLACE processor. RECONNECT harness to ACT sensor. RERUN Quick Test. CHECK heat stove duct valve operation. If OK, REPLACE ACT sensor. RERUN Quick Test.

Air Charge Temperature Sensor (ACT) — Pinpoint Test DB

TEST STEP	RESULT	ACTION TO TAKE
DB10 SERVICE CODE 54: ATTEMPT TO GENERATE CODE 64		
• Key off, wait 10 seconds. • Disconnect vehicle harness from ACT sensor. Inspect for damaged pins, corrosion, loose wires, etc. Service as necessary. • Insert a jumper wire at the ACT vehicle harness connector between ACT SIGNAL and SIGNAL RETURN. • Run Key On Engine Off Self-Test. • Is Code 64 present?	Yes ▶ No ▶	REPLACE ACT sensor. REMOVE jumper wire. RECONNECT ACT sensor. RERUN Quick Test. REMOVE jumper wire. GO to DB11.
DB11 CHECK CONTINUITY OF ACT SIGNAL AND SIGNAL RETURN		
• Key off, wait 10 seconds. • Harness disconnected from ACT sensor. • Disconnect processor 60 pin connector. Inspect for damaged pins, corrosion, loose wires, etc. Service as necessary. • Install breakout box, leave processor disconnected. • DVOM on 200 ohm scale. • Measure resistance between ACT SIGNAL, at the ACT vehicle harness connector, and Test Pin 25 at the breakout box. • Measure resistance between SIGNAL RETURN, at the ACT vehicle harness connector, and Test Pin 46 at the breakout box. • Are both resistances less than 5 ohms?	Yes ▶ No ▶	REPLACE processor. REMOVE breakout box. RECONNECT processor and ACT sensor. RERUN Quick Test. SERVICE open circuit(s). REMOVE breakout box. RECONNECT processor and ACT sensor. RERUN Quick Test.

Air Charge Temperature Sensor (ACT) — Pinpoint Test DB

TEST STEP	RESULT	ACTION TO TAKE
DB20 SERVICE CODE 64: ATTEMPT TO GENERATE CODE 54		
• Key off, wait 10 seconds. • Disconnect vehicle harness from ACT sensor. Inspect for damaged pins, corrosion, loose wires, etc. Service as necessary. • Run Key On Engine Off Self-Test. • Is Code 54 present?	Yes ▶ No ▶	REPLACE ACT sensor. RECONNECT ACT sensor. RERUN Quick Test. GO to DB21.
DB21 CHECK FOR VREF AT THROTTLE POSITION SENSOR		
• Refer to schematic in Pinpoint Test DH. • Key off, wait 10 seconds. • DVOM on 20 volt scale. • Disconnect TP sensor. • Key on, engine off. • Measure voltage at the TP vehicle harness connector between VREF and SIGNAL RETURN. • Is voltage between 4.0 and 6.0 volts?	Yes ▶ No ▶	RECONNECT TP sensor, GO to DB22. GO to Pinpoint Test Step C1.
DB22 CHECK ACT SIGNAL FOR SHORT TO GROUND		
• Key off, wait 10 seconds. • Harness disconnected from ACT sensor. • Disconnect processor 60 pin connector. Inspect for damaged pins, corrosion, loose wires, etc. Service as necessary. • Install breakout box, leave processor disconnected. • DVOM on 200,000 ohm scale. • Measure resistance between Test Pin 25 and Test Pins 40, 46 and 60 at the breakout box. • Are all resistances greater than 10,000 ohms?	Yes ▶ No ▶	REPLACE processor. REMOVE breakout box. RECONNECT processor and ACT sensor. RERUN Quick Test. SERVICE short circuit. REMOVE breakout box. RECONNECT processor and ACT sensor. RERUN Quick Test.

Air Charge Temperature Sensor (ACT) — Pinpoint Test DB

TEST STEP	RESULT	ACTION TO TAKE
DB90 CONTINUOUS MEMORY CODE 54: CHECK ACT SENSOR		
• Enter Key On Engine Off Continuous Monitor mode. Refer to the Appendix in Section 16. • Observe VOM or STAR LED for indication of a fault while performing the following: — Lightly tap on ACT sensor (simulate road shock). — Wiggle ACT connector. • Is a fault indicated? 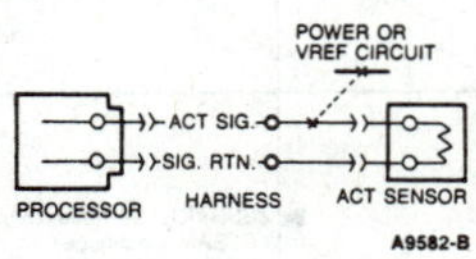	Yes ▶ No ▶	DISCONNECT and INSPECT connectors. If connector and terminals are good, REPLACE ACT sensor. CLEAR Continuous Memory Code 54. RERUN Quick Test. GO to DB91.
DB91 CHECK EEC-IV HARNESS		
• Still in Key On Engine Off Continuous Monitor mode. • Observe VOM or STAR LED for a fault indication while performing the following: — Referring to the illustration in Step DB90, grasp the harness closest to the sensor connector. Wiggle, shake or bend a small section of the EEC-IV system harness while working your way to the dash panel. Also wiggle, shake or bend the EEC-IV harness from the dash panel to the processor. • Is a fault indicated?	Yes ▶ No ▶	ISOLATE fault and SERVICE as necessary. CLEAR Continuous Memory Code 54. RERUN Quick Test. GO to DB92.

Air Charge Temperature Sensor (ACT) — Pinpoint Test DB

TEST STEP	RESULT ▶	ACTION TO TAKE
DB92 CHECK PROCESSOR AND HARNESS CONNECTORS		
• Key off, wait 10 seconds. • Disconnect processor 60 pin connector. • Inspect both connectors and connector terminals for obvious damage or faults. • Are connectors and terminals OK?	No ▶ Yes ▶	SERVICE as necessary. CLEAR Continuous Memory Code 54. RERUN Quick Test. Unable to duplicate fault at this time. CLEAR Continuous Memory Code 54. Continuous Memory Code 54 testing complete.
DB93 CONTINUOUS MEMORY CODE 64: CHECK ACT SENSOR		
• Enter Key On Engine Off Continuous Monitor mode. Refer to the Appendix in Section 16. • Observe VOM or STAR LED for an indication of a fault while performing the following: — Lightly tap on ACT sensor (simulate road shock). — Wiggle ACT connector. • Is a fault indicated? 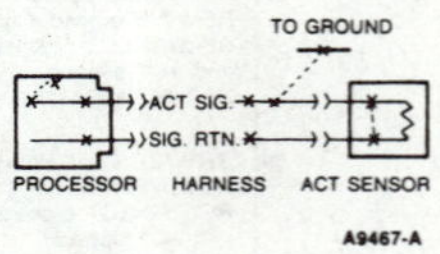	Yes ▶ No ▶	DISCONNECT and INSPECT connectors. If connector and terminals are good, REPLACE ACT sensor. CLEAR Continuous Memory Code 64. RERUN Quick Test. GO to DB94.

Mass Airflow Sensor (MAF) — Pinpoint Test DC

Note

You should enter this Pinpoint Test only when a Service Code 26, 56, or 66 is received in Quick Test Step 3.0, 5.0, or 6.0 or when directed here from Diagnostic By Symptom in the Engine Supplement Section.

Remember

To prevent the replacement of good components, be aware that the following non-EEC areas may be at fault:

- Air cleaner element
- Inlet air duct
- Throttle body

This Pinpoint Test is intended to diagnose only the following:

- Mass Airflow sensor
- Harness circuits: VPWR, POWER GROUND, MAF SIGNAL, and MAF RTN
- Processor assembly

Pinpoint Test Schematic

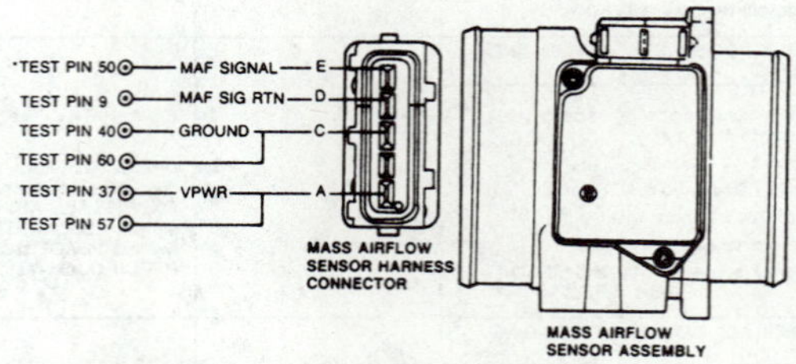

*TEST PINS LOCATED ON THE BREAKOUT BOX.
NOTE: ALL HARNESS CONNECTORS VIEWED INTO MATING SURFACE.

Air Charge Temperature Sensor (ACT) — Pinpoint Test DB

TEST STEP	RESULT ▶	ACTION TO TAKE
DB94 CHECK EEC-IV HARNESS		
• Still in Key On Engine Off Continuous Monitor mode. • Observe VOM or STAR LED for a fault indication while performing the following: — Referring to the illustration in Step DB93, grasp the harness closest to the sensor connector. Wiggle, shake or bend a small section of the EEC-IV system harness while working your way to the dash panel. Also wiggle, shake or bend the EEC-IV harness from the dash panel to the processor. • Is a fault indicated?	Yes ▶ No ▶	ISOLATE fault and SERVICE as necessary. CLEAR Continuous Memory Code 64. RERUN Quick Test. GO to DB95.
DB95 CHECK PROCESSOR AND HARNESS CONNECTORS		
• Key off, wait 10 seconds. • Disconnect processor 60 pin connector. • Inspect both connectors and connector terminals for obvious damage or faults. • Are connectors and terminals OK?	No ▶ Yes ▶	SERVICE as necessary. CLEAR Continuous Memory Code 64. RERUN Quick Test. Unable to duplicate fault at this time. CLEAR Continuous Memory Code 64. Continuous Memory Code 64 testing complete.

Mass Airflow Sensor (MAF) — Pinpoint Test DC

TEST STEP	RESULT ▶	ACTION TO TAKE
DC1 CHECK FOR VOLTAGE AT MAF SENSOR		
• Key off. • Disconnect MAF sensor from vehicle harness. • DVOM on 20 volt scale. • Measure voltage between MAF SIGNAL at the MAF sensor vehicle harness connector and battery negative post. • Is voltage greater than 1.5 volt?	Yes ▶ No ▶	SERVICE short circuit. RECONNECT MAF sensor. RERUN Quick Test. REPLACE MAF sensor. RERUN Quick Test.
DC2 CHECK CONTINUITY OF PWR GND CIRCUIT		
• Key off. • MAF sensor disconnected. • DVOM on 200 ohm scale. • Measure resistance between PWR GND circuit at the MAF sensor vehicle harness connector and battery negative post. • Is resistance less than 5 ohms?	Yes ▶ No ▶	GO to DC3. SERVICE open circuit. RECONNECT MAF sensor. RERUN Quick Test.
DC3 CHECK CONTINUITY OF MAF SIG RTN CIRCUIT		
• Key off. • MAF sensor disconnected. • Disconnect processor 60 pin connector. Inspect for damaged pins, corrosion, loose wires, etc. Service as necessary. • Install breakout box, leave processor disconnected. • DVOM on 200 ohm scale. • Measure resistance between MAF SIG RTN circuit at the MAF sensor vehicle harness connector and Test Pin 9 at the breakout box. • Is resistance less than 5 ohms?	Yes ▶ No ▶	GO to DC8. REMOVE breakout box. RECONNECT all components. SERVICE open circuit. RERUN Quick Test.

Mass Airflow Sensor (MAF) — Pinpoint Test DC

TEST STEP	RESULT ▶	ACTION TO TAKE
DC4 ENGINE RUNNING SERVICE CODE 26: CHECK VOLTAGE OF VPWR CIRCUIT		
• Key off. • Disconnect MAF sensor from vehicle harness. • DVOM on 20 volt scale. • Key on, engine off. • Measure voltage between VPWR circuit at the MAF sensor vehicle harness connector and battery negative post. • Is voltage greater than 10.5 volts?	Yes ▶ No ▶	GO to DC5. GO to Pinpoint Test Step B1.
DC5 CHECK MAF SENSOR GROUND		
• Key on, engine off. • MAF sensor disconnected. • DVOM on 20 volt scale. • Measure voltage between VPWR circuit and PWR GND circuit at the MAF sensor vehicle harness connector. • Is voltage greater than 10.5 volts?	Yes ▶ No ▶	GO to DC6. RECONNECT MAF sensor. SERVICE open PWR GND circuit. RERUN Quick Test.
DC6 CHECK CONTINUITY OF MAF SIGNAL AND VPWR CIRCUITS		
• Key off. • MAF sensor disconnected. • Disconnect processor 60 pin connector. Inspect for damaged pins, corrosion, loose wires, etc. Service as necessary. • Install breakout box, leave processor disconnected. • DVOM on 200 ohm scale. • Measure resistance between MAF SIGNAL at the MAF sensor vehicle harness connector and Test Pin 50 at the breakout box. • Measure resistance between VPWR at the MAF sensor vehicle harness connector and Test Pins 37/57 at the breakout box. • Are both resistances less than 5 ohms?	Yes ▶ No ▶	If vehicle is a no start or starts and stalls, GO to DC2. All others GO to DC7. REMOVE breakout box. RECONNECT all components. SERVICE open circuit. RERUN Quick Test.

EGR Valve Position Sensor (EVP) Control/Vent (EGRC/EGRV) — Pinpoint Test DD

Note

You should enter this Pinpoint Test only when a Service Code 31, 32, 33, 34, 35, 83 or 84 is received in Quick Test Step 3.0, 5.0, or 6.0.

Remember

To prevent the replacement of good components, be aware that the following non-EEC areas may be at fault:

- Damaged EGR valve.

This Pinpoint Test is intended to diagnose only the following:

- EVP sensor.
- Harness circuits: EVP, SIGNAL RETURN, VREF, EGRV, EGRC, VPWR.
- EGRV/EGRC solenoids.
- EGR Valve assembly.
- Processor assembly.
- Vacuum lines (EGRV/EGRC, EGR)

Pinpoint Test Schematic

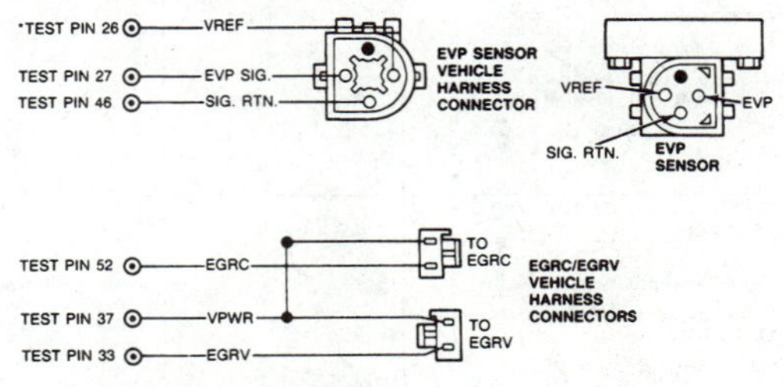

*TEST PINS LOCATED ON BREAKOUT BOX.
ALL HARNESS CONNECTORS VIEWED INTO MATING SURFACE.

Mass Airflow Sensor (MAF) — Pinpoint Test DC

TEST STEP	RESULT ▶	ACTION TO TAKE
DC7 CHECK MAF SIGNAL FOR SHORTS TO GROUND AND MAF SIG RTN		
• Key off. • Breakout box installed. • MAF sensor and processor disconnected. • DVOM on 200,000 ohm scale. • Measure resistance between Test Pin 50 and Test Pins 40, 9, and 60 at the breakout box. • Are all resistances greater than 10,000 ohms?	Yes ▶ No ▶	GO to DC2. REMOVE breakout box. RECONNECT all components. SERVICE short circuit(s). RERUN Quick Test.
DC8 CHECK MAF SENSOR RESISTANCE		
• Key off. • Breakout box installed, processor disconnected. • Reconnect MAF sensor. • DVOM on 20,000 ohm scale. • Measure resistance between Test Pin 50 and Test Pins 9, 40/60, at the breakout box. • Are all resistances less than 10,000 ohms?	Yes ▶ No ▶	REMOVE breakout box. REPLACE processor. RERUN Quick Test. REMOVE breakout box. RECONNECT processor. REPLACE MAF sensor. RERUN Quick Test.
DC10 RUN KEY ON ENGINE OFF SELF-TEST WITH MAF SENSOR DISCONNECTED		
• Key off. • Disconnect MAF sensor from vehicle harness. • Rerun Key On Engine Off Self-Test. • Is Service Code 66 present?	Yes ▶ No ▶	REPLACE MAF sensor. RERUN Quick Test. GO to DC11.
DC11 CHECK MAF SIGNAL FOR SHORT TO VPWR		
• Key off. • MAF sensor disconnected. • Breakout box installed, processor disconnected. • DVOM on 200,000 ohm scale. • Measure resistance between MAF SIGNAL and VPWR at the MAF sensor vehicle harness connector. • Is resistance greater than 10,000 ohms?	Yes ▶ No ▶	REMOVE breakout box. RECONNECT MAF sensor. REPLACE processor. RERUN Quick Test. REMOVE breakout box. RECONNECT all components. SERVICE short circuit. RERUN Quick Test.

EGR Valve Position Sensor (EVP) Control/Vent (EGRC/EGRV) — Pinpoint Test DD

TEST STEP	RESULT ▶	ACTION TO TAKE
DD1 SERVICE CODE 31 RUN ENGINE RUNNING SELF-TEST WITH EGR VACUUM SIGNAL LINE DISCONNECTED AT EGR VALVE		
• Key off, wait 10 seconds. • Disconnect EGR vacuum line at EGR valve and cap EGR vacuum line. • Run Engine Running Self-Test. • Is Code 31 present?	Yes ▶ No ▶	GO to DD2. RECONNECT vacuum line. GO to DD11.
DD2 CHECK EVP RESISTANCE WHILE APPLYING VACUUM TO EGR VALVE		
• Key off, wait 10 seconds. • Disconnect EGR vacuum line at EGR valve and cap the vacuum line. • Disconnect vehicle harness at EVP sensor. • DVOM on 200,000 ohm scale. • Connect vacuum pump to EGR valve. • Measure resistance at the EVP sensor between EVP SIG and VREF while gradually increasing vacuum to 33 kPa (10 in.-Hg.). • Observe resistance as vacuum increases.	Reading gradually decreases from no greater than 5500 ohms to no less than 100 ohms ▶ Reading is less than 100 ohms or greater than 5500 ohms ▶ Reading does not decrease or unable to hold vacuum ▶	RECONNECT vacuum line. GO to DD3. REPLACE EVP sensor. RECONNECT signal line and harness. RERUN Quick Test. GO to DD16.
DD3 CHECK FOR VREF AT THE EVP SENSOR		
• Key on, engine off. • Harness disconnected from EVP sensor. • DVOM on 20 volt scale. • Measure voltage at the EVP vehicle harness connector between VREF and SIGNAL RETURN. • Is voltage between 4.0 and 6.0 volts?	Yes ▶ No ▶	GO to DD4. GO to Pinpoint Test Step C1.

EGR Valve Position Sensor (EVP) Control/Vent (EGRC/EGRV) — Pinpoint Test DD

TEST STEP	RESULT ▶	ACTION TO TAKE
DD4 CHECK CONTINUITY OF EVP SIGNAL CIRCUIT		
• Key off, wait 10 seconds. • Harness disconnected from EVP sensor. • Disconnect processor 60 pin connector. Inspect for damaged pins, corrosion, loose wires, etc. Service as necessary. • Install breakout box, leave processor disconnected. • DVOM on 200 ohm scale. • Measure resistance between Test Pin 27 at the breakout box and EVP SIGNAL at the EVP vehicle harness connector. • Is resistance less than 5 ohms?	Yes ▶ No ▶	GO to DD5. SERVICE open circuit. REMOVE breakout box. RECONNECT processor and EVP sensor. RERUN Quick Test.
DD5 CHECK EVP SIGNAL FOR SHORTS TO VREF AND SIGNAL RETURN		
• Key off. • Harness disconnected from EVP sensor. • Breakout box installed. • Processor disconnected. • DVOM on 200,000 ohm scale. • Measure resistance between Test Pin 27 and Test Pins 26, 40, 46 and 60 at breakout box. • Are all resistances greater than 10,000 ohms?	Yes ▶ No ▶	GO to DD6. SERVICE short circuit. REMOVE breakout box. RECONNECT processor and EVP sensor. RERUN Quick Test.
DD6 SUBSTITUTE EVP SENSOR AND EGR VALVE		
• Key off, wait 10 seconds. • Electrically connect known good EVP sensor and EGR valve assembly. • Connect processor to breakout box. • Perform Key On Engine Off Self-Test. • Is Code 31 present?	Yes ▶ No ▶	REMOVE breakout box. REPLACE processor. CONNECT original EVP sensor and EGR valve assembly. RERUN Quick Test. GO to DD7.

EGR Valve Position Sensor (EVP) Control/Vent (EGRC/EGRV) — Pinpoint Test DD

TEST STEP	RESULT ▶	ACTION TO TAKE
DD7 CHECK EVP SENSOR		
• Key off, wait 10 seconds. • Breakout box installed. • Processor connected. • Install original EVP sensor on known good EGR valve. • Connect harness to EVP sensor. • Rerun Key On Engine Off Self-Test. • Is Code 31 present?	Yes ▶ No ▶	INSTALL new EVP sensor. REMOVE breakout box. RECONNECT processor. RERUN Quick Test. REMOVE breakout box. RECONNECT processor.
DD11 SERVICE CODES 32, 33 and 34: OUTPUT STATE CHECK (REFER TO APPENDIX IN SECTION 16)		
NOTE: Do not use STAR tester for this test step. Use VOM/DVOM. • Key off, wait 10 seconds. • DVOM on 20 volt scale. • Connect DVOM negative test lead to STO and positive test lead to battery positive. • Jumper STI to SIGNAL RETURN. • Perform Key On Engine Off Self-Test until the completion of the Continuous Memory Codes. • DVOM will indicate less than 1.0 volt. • Depress and release the throttle. • Does voltage increase to greater than 10.5 volts?	No ▶ Yes ▶	DEPRESS throttle to WOT and RELEASE. If STO voltage does not go high, GO to Pinpoint Test Step QC1. REMAIN in output state check, GO to DD12.
DD12 CHECK EGRC/EGRV SOLENOIDS FOR ELECTRICAL CYCLING		
• Key on, engine off. • DVOM on 20 volt scale. • Reconnect DVOM to EGRV solenoid, between VPWR and EGRV signal. • While observing DVOM, depress and release the throttle several times to cycle output on and off. • Repeat for EGRC solenoid, between VPWR and EGRC signal. • Do both solenoid outputs cycle on and off?	Yes ▶ No ▶	REMAIN in output state check. GO to DD13. REMOVE jumper. GO to DD17.

EGR Valve Position Sensor (EVP) Control/Vent (EGRC/EGRV) — Pinpoint Test DD

TEST STEP	RESULT ▶	ACTION TO TAKE
DD13 CHECK EGRC/EGRV SOLENOIDS FOR VACUUM CYCLING		
• Key on, engine off. • Still in output state check. • Disconnect and cap vacuum line from bottom port of EGRC solenoid and connect a vacuum pump. • Disconnect vacuum line at EGR valve. Connect vacuum gauge to EGR vacuum line. • Disconnect, but **do not** cap, vacuum vent line from EGRV solenoid or remove filter from the top of the EGRV solenoid. • Apply vacuum. • While cycling outputs on and off (by depressing and releasing throttle), observe vacuum gauge. Maintain vacuum at source. • Does the vacuum cycle on and off in less than 2 seconds?	Yes ▶ No ▶	REMOVE jumper. RECONNECT all vacuum lines. GO to DD14. CHECK filter and common output vacuum line for obstructions. REPLACE as necessary. If OK, REPLACE solenoid assembly. RECONNECT all vacuum lines. RERUN Quick Test.
DD14 CHECK VACUUM LINES		
• Key off, wait 10 seconds. • Vacuum lines reconnected. • Check entire EEC vacuum line system per VECI emission schematic decal for kinks, cracks, obstructions or leaks. • Are vacuum lines OK?	Yes ▶ No ▶	GO to DD15. SERVICE as necessary. RERUN Quick Test.
DD15 CHECK EVP RESISTANCE WHILE APPLYING VACUUM TO EGR VALVE		
• Key off. • Disconnect vehicle harness from EVP sensor. Inspect for damaged pins, corrosion, and pins pushed out. Service as necessary. • DVOM on 200,000 ohm scale. • Disconnect vacuum line at EGR valve. • Connect vacuum pump to EGR valve. • Measure resistance between EVP SIGNAL and VREF at the EVP sensor connector while increasing vacuum to 33 kPa (10 in.-Hg.). • Observe resistance as vacuum increases. • Does the resistance gradually change between 5500 and 100 ohms?	Yes ▶ No ▶	REPLACE processor. RECONNECT EVP sensor and EGR vacuum line. RERUN Quick Test. GO to DD16.

EGR Valve Position Sensor (EVP) Control/Vent (EGRC/EGRV) — Pinpoint Test DD

TEST STEP	RESULT ▶	ACTION TO TAKE
DD16 MANUALLY EXERCISE EVP SENSOR		
• Key off. • Harness disconnected from EVP sensor. • Remove EVP sensor from EGR valve. • Measure resistance between EVP SIGNAL and VREF at the EVP sensor connector while gradually applying pressure to EVP sensor shaft. • Observe resistance as shaft is slowly pushed in and slowly released. • Do either of the readings change suddenly between 5500 and 100 ohms? **NOTE: It is normal for the EVP sensor total resistance to drop below 100 ohms when disconnected from the EGR valve. A defective part will change resistance suddenly between 5500 and 100 ohms.**	Yes ▶ No ▶	REPLACE EVP sensor. RECONNECT harness and EGR supply vacuum line. RERUN Quick Test. RECONNECT EVP sensor and EGR supply vacuum line. RERUN Quick Test.
DD17 CHECK EGRV/EGRC SOLENOID RESISTANCE		
• Key off, wait 10 seconds. • DVOM on 200 ohm scale. • Disconnect EGRV solenoid connector and measure solenoid resistance. Inspect for damaged pins, corrosion and pins pushed out. Service as necessary. • Disconnect EGRC solenoid connector and measure solenoid resistance. Inspect for damaged pins, corrosion and pins pushed out. Service as necessary. • Are both resistances between 30 and 70 ohms?	Yes ▶ No ▶	GO to DD18. REPLACE EGRC/EGRV solenoid assembly. RERUN Quick Test.
DD18 CHECK FOR VPWR at EGRC/EGRV SOLENOIDS		
• Disconnect harness from EGRC/EGRV solenoids. • Key on, engine off. • DVOM on 20 volt scale. • Measure voltage between battery negative post and VPWR circuit on both EGR solenoid vehicle harness connectors. • Are both voltages greater than 10.5 volts?	Yes ▶ No ▶	GO to DD19. SERVICE open circuit. RECONNECT EGRC/EGRV solenoids. RERUN Quick Test.

EGR Valve Position Sensor (EVP) Control/Vent (EGRC/EGRV) — Pinpoint Test DD

TEST STEP	RESULT ▶	ACTION TO TAKE
DD19 CHECK CONTINUITY OF EGRC/EGRV CIRCUITS		
• Key off, wait 10 seconds. • EGRC/EGRV solenoids disconnected from harness. • Disconnect processor 60 pin connector. Inspect for damaged pins, corrosion, loose wires, etc. Service as necessary. • Install breakout box, leave processor disconnected. • DVOM on 200 ohm scale. • Measure resistance between Test Pin 33 at the breakout box and EGRV circuit at the EGRV solenoid vehicle harness connector. • Measure resistance between Test Pin 52 at the breakout box and EGRC circuit at the EGRC solenoid vehicle harness connector. • Are both resistances less than 5 ohms?	Yes ▶	GO to DD20.
	No ▶	SERVICE open circuit. REMOVE breakout box. RECONNECT all components. RERUN Quick Test.
DD20 CHECK EGRC/EGRV CIRCUITS FOR SHORT TO GROUND		
• Key off, wait 10 seconds. • DVOM on 200,000 ohm scale. • Breakout box installed, processor disconnected. • EGRC/EGRV solenoids disconnected. • Measure resistance between Test Pins 33 and Test Pins 40, 46 and 60 at the breakout box. • Measure resistance between Test Pin 52 and Test Pins 40, 46, and 60 at the breakout box. • Are all resistances greater than 10,000 ohms?	Yes ▶	GO to DD21.
	No ▶	SERVICE short circuit. REMOVE breakout box. RECONNECT all components. RERUN Quick Test.

EGR Valve Position Sensor (EVP) Control/Vent (EGRC/EGRV) — Pinpoint Test DD

TEST STEP	RESULT ▶	ACTION TO TAKE
DD90 CONTINUOUS MEMORY CODE 31: EXERCISE EVP SENSOR		
• Enter key on engine off continuous monitor mode. Refer to Appendix in Section 16. • Observe VOM or STAR LED for indication of a fault while performing the following: • Connect a vacuum pump to the EGR valve. • Very slowly apply 20 kPa (6 in.-Hg.) vacuum to the EGR valve. • Slowly bleed vacuum off the EGR valve. Lightly tap on EVP sensor (simulate road shock). • Wiggle EVP sensor connector. • Is a fault indicated?	Yes ▶	GO to DD91.
	No ▶	GO to DD92.

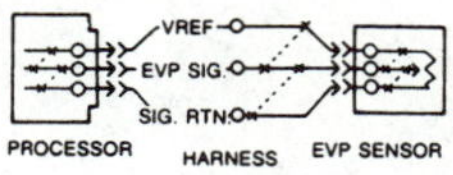

A9584-B

TEST STEP	RESULT ▶	ACTION TO TAKE
DD91 MEASURE EVP SIGNAL VOLTAGE WHILE EXERCISING EVP SENSOR		
• Key off, wait 10 seconds. • Disconnect processor 60 pin connector. Inspect for damaged pins, corrosion, loose wires, etc. Service as necessary. • Install breakout box and reconnect processor. • VOM or STAR LED still connected to STO as in previous Step. • Connect a DVOM between Test Pin 27 and Test Pin 46. • DVOM on 20 volt scale. • Key on, engine off. • While observing DVOM, repeat Test Step DD90. • Does the fault occur below 4.25 volts?	Yes ▶	DISCONNECT and INSPECT connector. If connector and terminals are good, REPLACE EVP sensor. CLEAR Continuous Memory Code 31. RERUN Quick Test.
	No ▶	EGR valve overshoot may have caused Continuous Memory Code 31. Sensor service is not required. To verify harness integrity, GO to DD92.

EGR Valve Position Sensor (EVP) Control/Vent (EGRC/EGRV) — Pinpoint Test DD

TEST STEP	RESULT ▶	ACTION TO TAKE
DD21 CHECK EGRC/EGRV CIRCUITS FOR SHORTS TO POWER		
• Key off. • Breakout box installed, processor disconnected. • EGRC/EGRV solenoids disconnected from harness. • DVOM on 200,000 ohm scale. • Measure resistance between Test Pin 33 and Test Pins 37 and 57 at the breakout box. • Measure resistance between Test Pin 52 and Test Pins 37 and 57 at the breakout box. • Are all resistances greater than 10,000 ohms?	Yes ▶	REPLACE processor. REMOVE breakout box. RECONNECT all components. RERUN Quick Test.
	No ▶	SERVICE short circuit. REMOVE breakout box. RECONNECT all components. RERUN Quick Test. If code is repeated, REPLACE processor.
DD30 SERVICE CODE 35: RPM TOO LOW FOR EGR TEST		
• Is Code 12 also present?	Yes ▶	GO to KE1.
	No ▶	GO to DD31.
DD31 RETEST AT 1,500 RPM		
• Key off, wait 10 seconds. • Install tachometer. • Rerun Engine Running Self-Test while maintaining 1,500 rpm. • Is Code 35 still present? **NOTE: Ignore all other codes at this time.**	Yes ▶	REPLACE processor. RERUN Quick Test.
	No ▶	RERUN Quick Test. SERVICE codes as necessary.

EGR Valve Position Sensor (EVP) Control/Vent (EGRC/EGRV) — Pinpoint Test DD

TEST STEP	RESULT ▶	ACTION TO TAKE
DD92 CHECK EEC-IV HARNESS		
• Observe VOM or STAR LED for a fault indication while performing the following: • Referring to the illustration in Step DD90, grasp the harness closest to the sensor connector. Wiggle, shake or bend a small section of the EEC-IV system harness while working your way to the dash panel. Also wiggle, shake or bend the EEC-IV harness from the dash panel to the processor. • Is a fault indicated?	Yes ▶	ISOLATE fault and SERVICE as necessary. REFER to appropriate figure. CLEAR Continuous Memory Code 31. RERUN Quick Test.
	No ▶	GO to DD93.
DD93 CHECK PROCESSOR AND HARNESS CONNECTORS		
• Key off, wait 10 seconds. • Disconnect processor 60 pin connector. • Inspect both connectors and connector terminals for obvious damage or faults. • Are connectors and terminals OK?	Yes ▶	Unable to duplicate fault at this time. CLEAR Continuous Memory Code 31. Continuous Memory Code 31 testing complete.
	No ▶	SERVICE as necessary. CLEAR Continuous Memory Code 31. RERUN Quick Test.

Engine Coolant Temperature Sensor (ECT) | Pinpoint Test | DE

Note

You should enter this Pinpoint Test only when a Service Code 21, 51 or 61 is received in Quick Test Step 3.0, 5.0 or 6.0.

Remember

To prevent the replacement of good components, be aware that the following non-EEC areas may be at fault:

- Coolant level.
- Oil level.
- Blocked or obstructed airflow.
- Engine not at normal operating temperature.
- Electro drive cooling fan.
- Open thermostat.

This Pinpoint Test is intended to diagnose only the following:

- ECT sensor.
- Harness sensor circuits: ECT and SIGNAL RETURN.
- Processor assembly.

Pinpoint Test Schematic

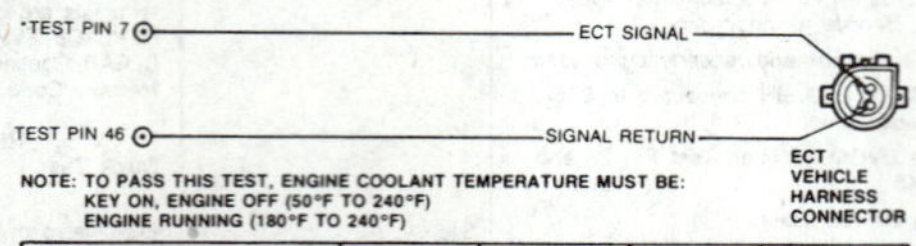

**NOTE: TO PASS THIS TEST, ENGINE COOLANT TEMPERATURE MUST BE:
KEY ON, ENGINE OFF (50°F TO 240°F)
ENGINE RUNNING (180°F TO 240°F)**

TYPICAL RESISTANCE BETWEEN TEST PINS 7 & 46	58,750 ohms	40,500 ohms	3600 ohms	1840 ohms
AT TEMPERATURE	50°F	65°F	180°F	220°F

***TEST PINS LOCATED ON BREAKOUT BOX.
ALL HARNESS CONNECTORS VIEWED INTO MATING SURFACE.**

Engine Coolant Temperature Sensor (ECT) | Pinpoint Test | DE

TEST STEP	RESULT ▶	ACTION TO TAKE
DE10 SERVICE CODE 51: ATTEMPT TO GENERATE CODE 61		
• Key off, wait 10 seconds. • Disconnect vehicle harness from ECT sensor. Inspect for damaged pins, corrosion, loose wires, etc. Service as necessary. • Insert a jumper wire at the ECT sensor vehicle harness connector between ECT SIGNAL and SIGNAL RETURN. • Run Key On Engine Off Self-Test. • Is Code 61 present?	Yes ▶ No ▶	REPLACE ECT sensor. REMOVE jumper wire. RECONNECT ECT sensor. RERUN Quick Test. REMOVE jumper wire. GO to DE11.
DE11 CHECK CONTINUITY OF ECT SIGNAL AND SIGNAL RETURN		
• Key off, wait 10 seconds. • Harness disconnected from ECT sensor. • Disconnect processor 60 pin connector. Inspect for damaged pins, corrosion, loose wires, etc. Service as necessary. • Install breakout box, leave processor disconnected. • DVOM on 200 ohm scale. • Measure resistance between ECT SIGNAL at the ECT vehicle harness connector and Test Pin 7 at the breakout box. • Measure resistance between SIGNAL RETURN at the ECT sensor vehicle harness connector, and Test Pin 46 at the breakout box. • Are both resistances less than 5 ohms?	Yes ▶ No ▶	REPLACE processor. REMOVE breakout box. RECONNECT processor and ECT sensor. RERUN Quick Test. SERVICE open circuit(s). REMOVE breakout box. RECONNECT processor and ECT sensor. RERUN Quick Test.

Engine Coolant Temperature Sensor (ECT) | Pinpoint Test | DE

TEST STEP	RESULT ▶	ACTION TO TAKE
DE1 SERVICE CODE 21: CHECK ENGINE OPERATING TEMPERATURE		
• Run engine for 2 minutes at 2,000 rpm. • Check that upper radiator hose is hot and pressurized. • Rerun Quick Test. • Is Code 21 present?	Vehicle stalls ▶ Yes ▶ No ▶	Do not service Code 21 at this time. REFER to Diagnostic by Symptoms. GO to DE2. SERVICE other codes as necessary.
DE2 CHECK FOR VREF AT THROTTLE POSITION SENSOR		
• Refer to schematic in Pinpoint Test DH. • Key off, wait 10 seconds. • DVOM on 20 volt scale. • Disconnect TP sensor. • Key on, engine off. • Measure voltage between VREF and SIGNAL RETURN at the TP vehicle harness connector. • Is voltage between 4.0 and 6.0 volts?	Yes ▶ No ▶	RECONNECT TP sensor, GO to DE3. GO to Pinpoint Test Step C1.
DE3 CHECK RESISTANCE OF ECT SENSOR		
NOTE: Engine may have cooled down. Always warm engine before taking ECT resistance measurement. Check for open thermostat. • Key off, wait 10 seconds. • Disconnect harness from ECT sensor. • DVOM on 200,000 ohm scale. • Measure resistance of the ECT sensor. • Is resistance: — 1300 ohms (240°F) to 7700 ohms (140°F) for engine off? — 1550 ohms (230°F) to 4550 ohms (170°F) for engine running?	Yes ▶ No ▶	REPLACE processor. RECONNECT harness to ECT sensor. RERUN Quick Test. REPLACE ECT sensor. RECONNECT harness to ECT sensor. RERUN Quick Test.

Engine Coolant Temperature Sensor (ECT) | Pinpoint Test | DE

TEST STEP	RESULT ▶	ACTION TO TAKE
DE20 SERVICE CODE 61: ATTEMPT TO GENERATE CODE 51		
• Key off, wait 10 seconds. • Disconnect vehicle harness from ECT sensor. Inspect for damaged pins, corrosion, loose wires, etc. Service as necessary. • Run Key On Engine Off Self-Test. • Is Code 51 present?	Yes ▶ No ▶	REPLACE ECT sensor. RECONNECT ECT sensor. RERUN Quick Test. GO to DE21.
DE21 CHECK FOR VREF AT THROTTLE POSITION SENSOR		
• Refer to schematic in Pinpoint Test DH. • Key off, wait 10 seconds. • DVOM on 20 volt scale. • Disconnect TP sensor. • Key on, engine off. • Measure voltage between VREF and SIGNAL RETURN at the TP vehicle harness connector. • Is voltage between 4.0 and 6.0 volts?	Yes ▶ No ▶	RECONNECT TP sensor, GO to DE22. GO to Pinpoint Test Step C1.
DE22 CHECK ECT SIGNAL FOR SHORT TO GROUND		
• Key off, wait 10 seconds. • Harness disconnected from ECT sensor. • Disconnect processor 60 pin connector. Inspect for damaged pins, corrosion, loose wires, etc. Service as necessary. • Install breakout box, leave processor disconnected. • DVOM on 200,000 ohm scale. • Measure resistance between Test Pin 7 and Test Pins 40, 46 and 60 at the breakout box. • Are all resistances greater than 10,000 ohms?	Yes ▶ No ▶	REPLACE processor. REMOVE breakout box. RECONNECT processor and ECT harness. RERUN Quick Test. SERVICE short circuit. REMOVE breakout box. RECONNECT processor and ECT sensor. RERUN Quick Test.

Engine Coolant Temperature Sensor (ECT) — Pinpoint Test DE

TEST STEP	RESULT ▶	ACTION TO TAKE
DE90 CONTINUOUS MEMORY CODE 21: TEST DRIVE VEHICLE		
• Key off, wait 10 seconds. • Clear Continuous Memory Code 21. Refer to Appendix in Section 16. • Disconnect all Self-Test equipment and prepare vehicle for test drive. • Drive vehicle. Try to simulate different drive modes or mode in which drive complaint is noticed. Attempt to maintain drive complaint mode for one minute or more, if possible. • Upon completion of drive evaluation, rerun Key On Engine Off Self-Test. • Is Code 21 present in the continuous test results?	Yes ▶	VERIFY thermostat operating properly. If OK, REPLACE ECT sensor. CLEAR Continuous Memory Code 21. RERUN Quick Test.
	No ▶	Unable to duplicate fault. Code 21 testing complete.
DE91 CONTINUOUS MEMORY CODE 51: CHECK ECT SENSOR		
• Enter Key On Engine Off Continuous Monitor mode. Refer to Appendix in Section 16. • Observe VOM or STAR LED for indication of a fault while performing the following: — Lightly tap on ECT sensor (simulate road shock). — Wiggle ECT connector. • Is a fault indicated?	Yes ▶	DISCONNECT and INSPECT connectors. If connector and terminals are good, REPLACE ECT sensor. CLEAR Continuous Memory Code 51. RERUN Quick Test.
	No ▶	GO to DE92.

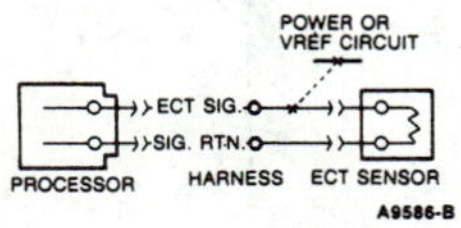

Engine Coolant Temperature Sensor (ECT) — Pinpoint Test DE

TEST STEP	RESULT ▶	ACTION TO TAKE
DE92 CHECK EEC-IV HARNESS		
• Still in Key On Engine Off Continuous Monitor mode. • Observe VOM or STAR LED for a fault indication while performing the following: — Referring to the illustration in Step DE91, grasp the harness closest to the sensor connector. Wiggle, shake or bend a small section of the EEC-IV system harness while working your way to the dash panel. Also wiggle, shake or bend the EEC-IV harness from the dash panel to the processor. • Is a fault indicated?	Yes ▶	ISOLATE fault and SERVICE as necessary. CLEAR Continuous Memory Code 51. RERUN Quick Test.
	No ▶	GO to DE93.
DE93 CHECK PROCESSOR AND HARNESS CONNECTORS		
• Key off, wait 10 seconds. • Disconnect processor 60 pin connector. • Inspect both connectors and connector terminals for obvious damage or faults. • Are connectors and terminals OK?	No ▶	SERVICE as necessary. CLEAR Continuous Memory Code 51. RERUN Quick Test.
	Yes ▶	Unable to duplicate fault at this time. CLEAR Continuous Memory Code 51. Continuous Code 51 testing complete.

Engine Coolant Temperature Sensor (ECT) — Pinpoint Test DE

TEST STEP	RESULT ▶	ACTION TO TAKE
DE94 CONTINUOUS MEMORY CODE 61: CHECK ECT SENSOR		
• Enter Key On Engine Off Continuous Monitor mode. Refer to Appendix in Section 16. • Observe VOM or STAR LED for indication of a fault while performing the following: — Lightly tap on ECT sensor (simulate road shock). — Wiggle ECT connector. • Is a fault indicated?	Yes ▶	DISCONNECT and INSPECT connectors. If connector and terminals are good, REPLACE ECT sensor. CLEAR Continuous Memory Code 61. RERUN Quick Test.
	No ▶	GO to DE95.

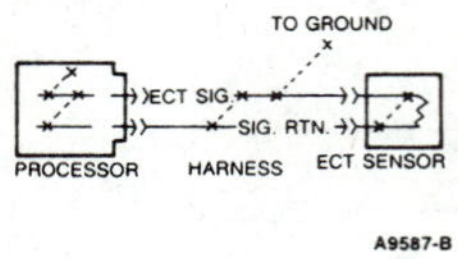

TEST STEP	RESULT ▶	ACTION TO TAKE
DE95 CHECK EEC-IV HARNESS		
• Still in Key On Engine Off Continuous Monitor mode. • Observe VOM or STAR LED for a fault indication while performing the following: — Referring to the illustration in Step DE94, grasp the harness closest to the sensor connector. Wiggle, shake or bend a small section of the EEC-IV system harness while working your way to the dash panel. Also wiggle, shake or bend the EEC-IV harness from the dash panel to the processor. • Is a fault indicated?	Yes ▶	ISOLATE fault and SERVICE as necessary. CLEAR Continuous Memory Code 61. RERUN Quick Test.
	No ▶	GO to DE96.

Engine Coolant Temperature Sensor (ECT) — Pinpoint Test DE

TEST STEP	RESULT ▶	ACTION TO TAKE
DE96 CHECK PROCESSOR AND HARNESS CONNECTORS		
• Key off, wait 10 seconds. • Disconnect processor 60 pin connector. • Inspect both connectors and connector terminals for obvious damage or faults. • Are connectors and terminals OK?	No ▶	SERVICE as necessary. CLEAR Continuous Memory Code 61. RERUN Quick Test.
	Yes ▶	Unable to duplicate fault at this time. CLEAR Continuous Memory Code 61. Continuous Code 61 testing complete.

Manifold Absolute Pressure (MAP)/ Barometric Pressure (BP) Sensor — Pinpoint Test DF

Note

You should enter this Pinpoint Test only when a Service Code 22 or 72 is received in Quick Test Step 3.0, 5.0 or 6.0 or when directed here from Pinpoint Test S or Diagnostic By Symptom in the Engine Supplement Section.

Remember

To prevent the replacement of good components, be aware that the following non-EEC areas may be at fault:

- Unusually high/low barometric pressure.
- Kinked or obstructed vacuum lines (MAP).
- Basic engine (valves, vacuum leaks, timing, EGR valve, etc.).

This Pinpoint Test is intended to diagnose only the following:

- MAP/BP sensor.
- Harness circuits: VREF, MAP/BP SIGNAL, and SIGNAL RETURN.
- Processor assembly.
- MAP vacuum line.

Pinpoint Test Schematic

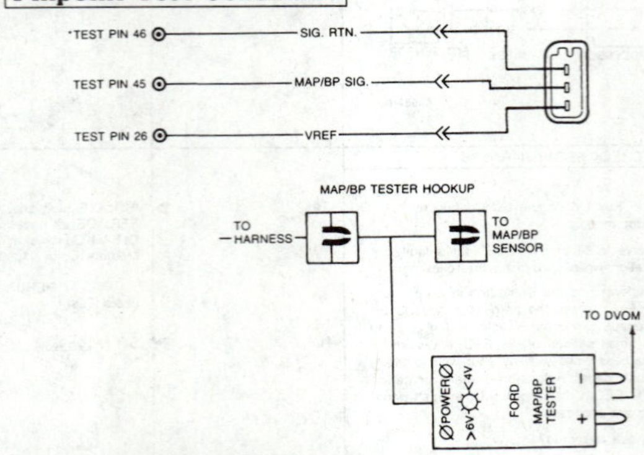

*TEST PINS LOCATED ON BREAKOUT BOX.
ALL HARNESS CONNECTORS VIEWED INTO MATING SURFACE.

Manifold Absolute Pressure (MAP)/ Barometric Pressure (BP) Sensor — Pinpoint Test DF

TEST STEP	RESULT	▶ ACTION TO TAKE
FAULT CODE 22, ENGINE OFF		
DF1 CONNECTING MAP/BP TESTER		
• Key off. • Disconnect the MAP/BP sensor from the vehicle harness. • Connect the MAP/BP tester between the vehicle harness and the MAP/BP sensor. • Insert tester banana plugs into DVOM. • Set DVOM to 20 volt scale. • Refer to schematic in Pinpoint Test DF.	Tester properly hooked up	▶ GO to DF2.
DF2 POWER TO MAP/BP SENSOR TEST		
• MAP/BP tester connected. • Key on. • Observe red and green lights.	(ONLY) Green light, VREF is OK	▶ GO to DF4.
	"Less than 4V" light (RED) or no lights, VREF is too low OR "Greater than 6V" light (Red), VREF is too high	▶ GO to DF3.
DF3 VREF ISOLATION		
• MAP/BP tester connected. • Key on. • Disconnect MAP/BP • Observe red and green lights.	(ONLY) Green light, VREF is OK	▶ REPLACE MAP/BP sensor. RERUN Quick Test.
	"Less than 4V" light (RED) or no lights, VREF is too low OR "Greater than 6V" light (Red), VREF is too high	▶ REMOVE MAP/BP tester. GO to Pinpoint Test Step C1.

Manifold Absolute Pressure (MAP)/ Barometric Pressure (BP) Sensor — Pinpoint Test DF

TEST STEP	RESULT	▶ ACTION TO TAKE
DF4 MAP/BP TESTER OUTPUT READING		
• MAP tester connected, refer to Note. • Key on. • Approximate Altitude (Ft.) / Voltage Output (+/- .04 Volts) 0 — 1.59 1000 — 1.56 2000 — 1.53 3000 — 1.50 4000 — 1.47 5000 — 1.44 6000 — 1.41 7000 — 1.39 **NOTE: Measure several known good MAP sensors on available vehicles. The measured voltage will be typical for your location on the day of testing.** • Is voltage in range for your altitude?	Yes	▶ REMOVE MAP/BP Tester. GO to DF5.
	No (Sensor output is out-of-range)	▶ REMOVE MAP/BP Tester. GO to DF6.
DF5 CHECK CONTINUITY OF MAP/BP SIGNAL		
• Key off, wait 10 seconds. • Harness disconnected from MAP/BP sensor. • Disconnect processor 60 pin connector. Inspect for damaged pins, corrosion, loose wires, etc. Service as necessary. • Install breakout box. Leave processor disconnected. • DVOM on 200 ohm scale. • Measure resistance between MAP/BP signal at the MAP/BP sensor vehicle harness connector and Test Pin 45 at the breakout box. • Is resistance less than 5.0 ohms?	Yes	▶ REPLACE processor. CONNECT harness and MAP/BP sensor. RERUN Quick Test.
	No	▶ SERVICE circuit opens. REMOVE breakout box. RECONNECT processor and MAP/BP sensor. RERUN Quick Test.

Manifold Absolute Pressure (MAP)/ Barometric Pressure (BP) Sensor — Pinpoint Test DF

TEST STEP	RESULT	▶ ACTION TO TAKE
DF6 CHECK MAP/BP SIGNAL FOR SHORTS TO VREF, SIGNAL RETURN AND GROUND		
• Key off, wait 10 seconds. • Disconnect processor 60 pin connector. Inspect for damaged pins, corrosion, loose wires, etc. Service as necessary. • Install breakout box, leave processor disconnected. • Harness disconnected from MAP/BP sensor. • DVOM on 200,000 ohm scale. • Measure resistance between Test Pin 45 and Test Pins 26, 46, 40 and 60 at the breakout box. • Are all resistances greater than 10,000 ohms?	Yes	▶ REPLACE MAP/BP sensor. REMOVE breakout box. RECONNECT electrical connections. RERUN Quick Test.
	No	▶ SERVICE circuit shorts. REMOVE breakout box. RECONNECT processor and MAP/BP Sensor. RERUN Quick Test.
CODE 22: ENGINE RUNNING		
DF7 CHECK FOR EGR CODES		
• Are Service Codes 31, 32, 33, 34 or 35 present?	Yes	▶ GO to Quick Test Step 5.0 for appropriate Pinpoint Test.
	No	▶ GO to DF8.
DF8 CHECK MAP SENSOR		
• Key off, wait 10 seconds. • Disconnect vacuum line from MAP sensor. • Install vacuum pump to MAP sensor. • Apply 18 in.-Hg. vacuum to MAP sensor. • Does MAP sensor hold vacuum?	Yes	▶ RELEASE vacuum. GO to DF9.
	No	▶ REPLACE MAP sensor. CONNECT vacuum line to MAP sensor. RERUN Quick Test.

Manifold Absolute Pressure (MAP)/ Barometric Pressure (BP) Sensor — Pinpoint Test DF

TEST STEP	RESULT ▶	ACTION TO TAKE
DF9 ATTEMPT TO ELIMINATE CODE 22 (ENGINE RUNNING)		
• Key off, wait 10 seconds. • Plug MAP vacuum supply hose. • Start engine and maintain 1500 ± 100 engine rpm. • Slowly apply 15 in.-Hg. vacuum to MAP sensor. • While maintaining rpm, perform Engine Running Self-Test. • Is Code 22 still present? **NOTE: Disregard any other codes at this time.**	Yes ▶	REPLACE MAP sensor. CONNECT vacuum line to MAP sensor. RERUN Quick Test.
	No ▶	INSPECT vacuum supply hose to MAP sensor. SERVICE as necessary. If OK, SERVICE other engine running codes.
DF10 CHECK THAT VACUUM TO MAP SENSOR DECREASES DURING DYNAMIC RESPONSE		
• Key off, wait 10 seconds. • Tee a vacuum gauge in the intake manifold vacuum line at the MAP sensor. • Perform Engine Running Test while observing vacuum. • Did vacuum decrease by more than 30 kPa (10 in.-Hg.) during dynamic response test?	Yes ▶	REMOVE vacuum gauge. RECONNECT all components. REPLACE MAP sensor. RERUN Quick Test
	No ▶	GO to DF11 .
DF11 CHECK VACUUM LINES		
• Check vacuum line for proper routing. Refer to VECI decal. Check MAP sensor vacuum line for kinks or blockage. • Are vacuum lines OK?	Yes ▶	EEC-IV system OK.
	No ▶	SERVICE as necessary and REPEAT DF10 .

Manifold Absolute Pressure (MAP)/ Barometric Pressure (BP) Sensor — Pinpoint Test DF

TEST STEP	RESULT ▶	ACTION TO TAKE
DF20 CONNECTING MAP/BP TESTER		
• Key off. • Disconnect the MAP/BP sensor from the vehicle harness. • Connect the MAP/BP Tester between the vehicle harness and the MAP/BP sensor. • Plug tester banana plugs into DVOM. • Set DVOM to 20 volt scale. • Refer to schematic in Pinpoint Test DF.	Tester properly hooked up ▶	GO to DF21 .
DF21 MAP/BP TESTER OUTPUT READING		
• MAP Tester connected, refer to Note. • Key on. • Approximate Altitude (Ft.) / Voltage Output (+/−.04 Volts): 0 — 1.59 1000 — 1.56 2000 — 1.53 3000 — 1.50 4000 — 1.47 5000 — 1.44 6000 — 1.41 7000 — 1.39 **NOTE: Measure several known good MAP sensors on available vehicles. The measured voltage will be typical for your location on the day of testing.** • Is voltage in range for your altitude?	Yes ▶	For 1.9L EFI, 2.3L EFI TC engines and 5.0L SEFI Mustang GO to Diagnostic by Symptom in the Engine Supplement Section. For all others, GO to DF22 .
	No (Sensor output is out-of-range) ▶	REPLACE MAP/BP sensor.
DF22 VACUUM LINE CHECK		
• Check MAP sensor vacuum line for holes, disconnections, kinks or blockage. • Are vacuum lines OK?	Yes ▶	GO to Diagnostics by Symptom in the Engine Supplement Section.
	No ▶	SERVICE vacuum lines to MAP sensor RERUN Quick Test.

Manifold Absolute Pressure (MAP)/ Barometric Pressure (BP) Sensor — Pinpoint Test DF

TEST STEP	RESULT ▶	ACTION TO TAKE
DF90 SERVICE CODE 22: CONTINUOUS TEST: EXERCISE MAP SENSOR		
• Using Key On Engine Off Continuous Monitor mode, observe VOM or STAR LED for indication of a fault while performing the following: • Connect a vacuum pump to the MAP sensor. • Slowly apply 84 kPa (25 in.-Hg.) vacuum to the sensor. • Slowly bleed vacuum off the MAP sensor. • Lightly tap on MAP sensor (simulate road shock). • Wiggle MAP connector. • Is fault indicated? 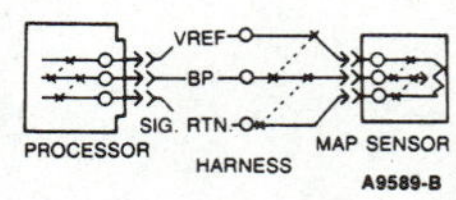	Yes ▶	DISCONNECT and INSPECT connectors. If connector and terminals are good, REPLACE MAP sensor. RERUN Quick Test.
	No ▶	GO to DF91 .
DF91 CHECK EEC-IV HARNESS		
• Remain in Key On Engine Off Continuous Monitor mode. • Observe VOM or STAR LED for a fault indication while performing the following: — Referring to the illustration in Step DF90, grasp the harness closest to the sensor connector. Wiggle, shake or bend a small section of the EEC-IV system harness while working your way to the dash panel. Also wiggle, shake or bend the EEC-IV harness from the dash panel to the processor. • Is a fault indicated?	Yes ▶	ISOLATE fault and SERVICE as necessary. CLEAR Continuous Memory Code. RERUN Quick Test.
	No ▶	GO to DF92 .

Manifold Absolute Pressure (MAP)/ Barometric Pressure (BP) Sensor — Pinpoint Test DF

TEST STEP	RESULT ▶	ACTION TO TAKE
DF92 CHECK PROCESSOR AND HARNESS CONNECTORS		
• Key off, wait 10 seconds. • Disconnect processor 60 pin connector. • Inspect both connectors and connector terminals for obvious damage or faults. • Are connectors and terminals OK?	No ▶	SERVICE as necessary. RERUN Quick Test.
	Yes ▶	Unable to duplicate fault at this time.

Knock Sensor — Pinpoint Test — DG

Note

You should enter this Pinpoint Test only when a Service Code 25 is received in Quick Test Step 5.0 or you are directed here from Diagnostic By Symptom in the Engine Supplement Section.

Remember

To prevent the replacement of good components, be aware that the following non-EEC areas may be at fault:

- Fuel (quality).
- Basic engine.
- Spark timing.

This Pinpoint Test is intended to diagnose only the following:

- Knock sensor.
- Harness circuits: KS and SIGNAL RETURN.
- Processor assembly.

Pinpoint Test Schematic

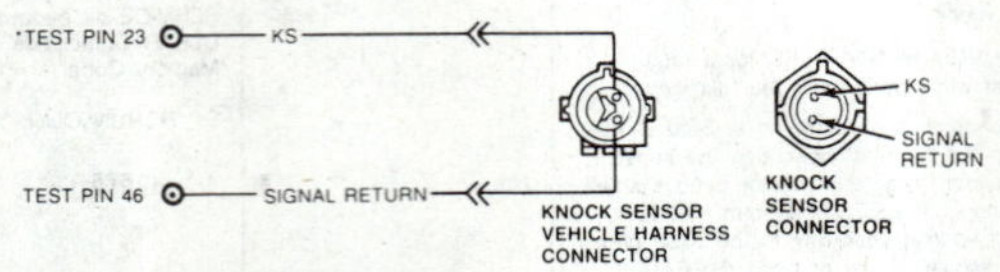

*TEST PINS LOCATED ON BREAKOUT BOX.
ALL HARNESS CONNECTORS VIEWED INTO MATING SURFACE.

Knock Sensor — Pinpoint Test — DG

TEST STEP	RESULT ▶	ACTION TO TAKE
DG1 SERVICE CODE 25: GENERATE KNOCK MANUALLY		
NOTE: With knock conditions sensitive to fuel, altitude and weather, perform Step DG1 before servicing any components. • Locate knock sensor and prepare to rap/tap on exhaust manifold with a 4 oz. hammer. • Run Engine Running Self-Test (engine must be at operating temperature). • Tap exhaust manifold directly above the knock sensor immediately after the dynamic response code is given. **NOTE: It is not necessary to "goose" the throttle. Ignore all other codes except Code 25.** • Is service Code 25 present?	Yes ▶ No ▶	GO to DG2. Knock system OK. RERUN Engine Running Self-Test and SERVICE any other codes from that test.
DG2 TEST KNOCK CIRCUIT FOR VOLTAGE		
• Key off, wait 10 seconds. • Disconnect knock sensor connector. • DVOM on 20 volt scale. • Key on, engine off. • Measure voltage between KS and SIGNAL RETURN at the vehicle harness connector.	Voltage is between 1 and 4 volts ▶ Voltage is less than 1 volt ▶ Voltage is greater than 4 volts ▶	GO to DG6. GO to DG3. GO to DG5.
DG3 CHECK CONTINUITY OF KS AND SIGNAL RETURN CIRCUITS		
• Key off, wait 10 seconds. • Disconnect processor 60 pin connector. Inspect for damaged pins, corrosion, loose wires, etc. Service as necessary. • Connect breakout box to harness, leave processor disconnected. • Knock sensor disconnected. • DVOM on 200 ohm scale. • Measure resistance between SIGNAL RETURN at the knock sensor vehicle harness connector and Test Pin 46 at the breakout box and between KS at the knock sensor vehicle harness connector and Test Pin 23 at the breakout box. • Are both resistances less than 5.0 ohms?	Yes ▶ No ▶	GO to DG4. REMOVE breakout box. RECONNECT processor and knock sensor. SERVICE open circuit. RERUN Quick Test.

Knock Sensor — Pinpoint Test — DG

TEST STEP	RESULT ▶	ACTION TO TAKE
DG4 CHECK KS CIRCUIT FOR SHORT TO GROUND		
• Key off, wait 10 seconds. • Breakout box installed. • Processor disconnected. • Knock sensor disconnected. • DVOM on 200,000 ohm scale. • Measure resistance between KS at the knock sensor vehicle harness connector and Test Pins 40, 46 and 60 at the breakout box. • Are all resistances greater than 10,000 ohms?	Yes ▶ No ▶	REMOVE breakout box. RECONNECT processor. GO to DG6. REMOVE breakout box. RECONNECT processor and knock sensor. SERVICE short circuit. RERUN Quick Test.
DG5 CHECK KS CIRCUIT FOR SHORT TO VOLTAGE		
• Key off, wait 10 seconds. • Disconnect processor 60 pin connector. Inspect for damaged pins, corrosion, loose wires, etc. Service as necessary. • Connect breakout box to harness, leave processor disconnected. • Knock sensor disconnected. • Key on, engine off. • DVOM on 20 volt scale. • Measure voltage between Test Pin 23 and Test Pin 40 at the breakout box. • Is voltage less than 0.5 volts?	Yes ▶ No ▶	REMOVE breakout box. RECONNECT processor. GO to DG6. REMOVE breakout box. RECONNECT processor and knock sensor. SERVICE short circuit. RERUN Quick Test.
DG6 TEST PROCESSOR WITH SUBSTITUTE KNOCK SENSOR		
• Key off, wait 10 seconds. • Connect a known good knock sensor to the vehicle harness. • Do not install this sensor on the engine. • Run Engine Running Self-Test. (Engine must be at operating temperature). • Tap the substitute knock sensor with a 4 oz. hammer immediately after the dynamic response code is given. **NOTE: It is not necessary to "goose" the throttle. Ignore all other codes except Code 25.** • Is service Code 25 present?	Yes ▶ No ▶	REPLACE processor. RECONNECT original knock sensor. RERUN Quick Test. INSTALL new knock sensor. RERUN Quick Test.

Throttle Position Sensor (TPS) — Pinpoint Test — DH

Note

You should enter this Pinpoint Test only when a Service Code 23, 53, 63 or 73 is received in Quick Test Step 3.0, 5.0 or 6.0.

Remember

To prevent the replacement of good components, be aware that the following non-EEC areas may be at fault:

- Idle speeds/throttle stop adjustment.
- Binding throttle shaft/linkage or speed control linkage.
- Choke/high cam system, if equipped.

This Pinpoint Test is intended to diagnose only the following:

- TP sensor.
- Sensor harness circuits: VREF, TP SIGNAL, and SIGNAL RETURN.
- Processor assembly.

Pinpoint Test Schematic

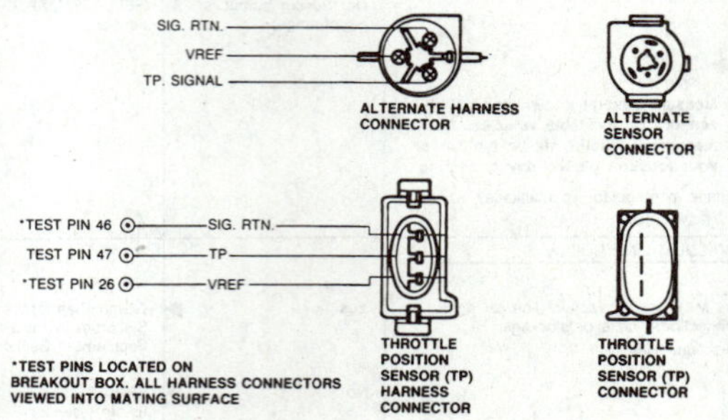

*TEST PINS LOCATED ON BREAKOUT BOX. ALL HARNESS CONNECTORS VIEWED INTO MATING SURFACE

Throttle Position Sensor (TPS) — Pinpoint Test DH

TEST STEP	RESULT ▶	ACTION TO TAKE
DH1 SERVICE CODE 23: THE FOLLOWING CHECK MUST BE MADE BEFORE SERVICING THIS CODE		
• Check for Code 68; Key On Engine Off or Codes 58, 31 or 41 Engine Running. • Are any of the above Codes present?	Yes ▶	RETURN to the Key On Engine Off or Engine Running service code chart as appropriate. PROCEED as directed.
	No ▶	GO to DH2.
DH2 CHECK FOR STUCK THROTTLE PLATE		
• Visually inspect carburetor/throttle body and throttle linkage for binding or sticking. • Verify the throttle linkage is at mechanical/closed throttle. Check for: binding throttle linkage, speed control linkage, vacuum line/electrical harness interference, etc. • Does throttle move freely and return to closed throttle position?	Yes ▶	GO to DH3.
	No ▶	SERVICE as necessary. RERUN Quick Test.
DH3 SERVICE CODE 53: ATTEMPT TO GENERATE CODE 63		
• Refer to schematic in Pinpoint Test DH. • Key off, wait 10 seconds. • Disconnect TP sensor vehicle harness connector at the throttle body. Inspect for damaged pins, corrosion, loose wires, etc. Service as necessary. • RERUN Key On Engine Off Self-Test. • Is Code 63 present? **NOTE: Ignore all other codes at this time.**	Yes ▶	GO to DH4.
	No ▶	GO to DH5.

Throttle Position Sensor (TPS) — Pinpoint Test DH

TEST STEP	RESULT ▶	ACTION TO TAKE
DH11 SERVICE CODE 63: CHECK VOLTAGE VREF TO SIGNAL RETURN		
• Refer to schematic in Pinpoint Test DH. • Key off, wait 10 seconds. • Disconnect TP vehicle harness connector at throttle body. Inspect for damaged pins, corrosion, loose wires, etc. Service as necessary. • DVOM on 20 volt scale. • Key on engine off. • Measure voltage between VREF and SIGNAL RETURN at the TP vehicle harness connector. • Is voltage between 4.0 and 6.0 volts?	Yes ▶	GO to DH12.
	No ▶	GO to Pinpoint Test Step C1.
DH12 CHECK CONTINUITY OF TP CIRCUIT		
• Key off, wait 10 seconds. TP harness disconnected. • DVOM on 200 ohm scale. • Disconnect processor 60 pin connector. Inspect for damaged pins, corrosion, loose wires, etc. Service as necessary. • Connect breakout box. Processor connected to breakout box. • Measure resistance between TP SIGNAL at the vehicle harness connector and Test Pin 47 at the breakout box. • Is the resistance less than 5.0 ohms?	No ▶	SERVICE open circuit. RECONNECT harness to sensor. REMOVE breakout box and RERUN Quick Test.
	Yes ▶	GO to DH13.
DH13 CHECK RESISTANCE OF TP CIRCUIT TO GROUND/SIGNAL RETURN		
• Key off, wait 10 seconds, TP harness disconnected. • Disconnect processor 60 pin connector. Inspect for damaged pins, corrosion, loose wires, etc. Service as necessary. • DVOM on 200,000 ohm scale. • Measure resistance between TP SIGNAL at TP vehicle harness connector and Test Pin 46 at the breakout box and between TP SIGNAL at TP vehicle harness connector and ground. • Are all resistances greater than 10,000 ohms?	No ▶	SERVICE short circuit. REMOVE breakout box. RECONNECT processor and TP sensor. RERUN Quick Test.
	Yes ▶	REMOVE breakout box. REPLACE processor. RECONNECT processor and TP sensor. RERUN Quick Test.

Throttle Position Sensor (TPS) — Pinpoint Test DH

TEST STEP	RESULT ▶	ACTION TO TAKE
DH4 CHECK VOLTAGE VREF TO SIGNAL RETURN		
• Refer to schematic in Pinpoint Test DH. • Key off, wait 10 seconds. • Disconnect TP vehicle harness connector at throttle body. Inspect for damaged pins, corrosion, loose wires, etc. Service as necessary. • DVOM on 20 volt scale. • Key on, engine off. • Measure voltage between VREF and SIGNAL RETURN at the TP vehicle harness connector. • Is voltage between 4.0 and 6.0 volts?	Yes ▶	REPLACE TP sensor. RERUN Quick Test.
	No ▶	GO to Pinpoint Test Step C1.
DH5 CHECK TP SIGNAL FOR SHORT TO POWER		
• Key off, wait 10 seconds, TP harness disconnected. • DVOM on 200,000 ohm scale. • Disconnect processor 60 pin connector. Inspect for damaged pins, corrosion, loose wires, etc. Service as necessary. • Install breakout box, leave processor disconnected. • Measure resistance between Test Pin 47 and Test Pins 26 and 57 at the breakout box. • Are both resistances greater than 10,000 ohms?	No ▶	SERVICE short circuit. REMOVE breakout box. RECONNECT TP sensor and processor. RERUN Quick Test.
	Yes ▶	REMOVE breakout box. REPLACE processor. RECONNECT TP sensor and processor. RERUN Quick Test.
DH10 SERVICE CODE 63: ATTEMPT TO GENERATE CODE 53		
• Key off, wait 10 seconds, TP harness disconnected. • Jumper VREF to TP signal at TP vehicle harness connector. • Perform Key On Engine Off Self-Test. **NOTE: If no codes are generated, immediately remove jumper and go directly to DH13.** • Is Code 53/23 present? **NOTE: Ignore all other codes at this time.**	Yes ▶	REPLACE TP sensor. REMOVE jumper wire. RECONNECT TP sensor. RERUN Quick Test.
	No ▶	GO to DH11.

Throttle Position Sensor (TPS) — Pinpoint Test DH

TEST STEP	RESULT ▶	ACTION TO TAKE
DH20 SERVICE CODE 73: TP SENSOR MOVES IN ENGINE RESPONSE TEST		
NOTE: Code 73 indicates the TP Sensor did not exceed 25 percent of its rotation in the Engine Response Check. • Key off. • Install breakout box. • DVOM on 20 volt scale. • Connect DVOM to Test Pins 47 and 46 at the breakout box. • Perform Engine Running Self-Test, Step 5.0. • Does voltage increase to greater than 3.5 volts during the dynamic response test?	Yes ▶	REMOVE breakout box. REPLACE processor. RERUN Quick Test.
	No ▶	VERIFY TP Sensor is properly installed to throttle body. If OK, REPLACE TP Sensor. RERUN Quick Test.
DH90 CONTINUOUS MEMORY CODE 53: EXERCISE TP SENSOR		
• Enter Key On Engine Off Continuous Monitor mode. Refer to Appendix in Section 16. • Observe VOM or STAR LED for indication of a fault while performing the following: — Move throttle slowly to WOT position. — Release throttle slowly to closed position and lightly tap on TP sensor (simulate road shock). — Wiggle TP harness connector. • Does VOM or STAR LED indicate a fault?	Yes ▶	GO to DH91.
	No ▶	GO to DH92.

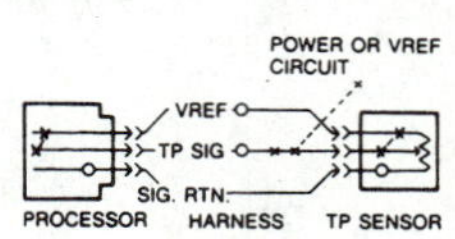

Throttle Position Sensor (TPS) — Pinpoint Test DH

TEST STEP	RESULT ▶	ACTION TO TAKE
DH91 MEASURE THROTTLE POSITION SIGNAL VOLTAGE WHILE EXERCISING TP SENSOR		
• Key off, wait 10 seconds. • Disconnect processor 60 pin connector. Inspect for damaged pins, corrosion, loose wires, etc. Service as necessary. • Install breakout box and reconnect processor. • VOM or STAR LED still connected to STO as in previous step. • Connect a DVOM from Test Pin 47 to Test Pin 46. • DVOM on 20 volt scale. • Key on engine off. • While observing DVOM, repeat Step DH90. • Does the fault occur below 4.25 volts?	Yes ▶	DISCONNECT and INSPECT connectors. If connector and terminals are good, REPLACE TP sensor. CLEAR Continuous Memory Code 53. RERUN Quick Test.
	No ▶	Throttle position sensor overtravel may have caused the Continuous Memory Code 53. VERIFY harness integrity, GO to DH92.
DH92 CHECK EEC-IV HARNESS		
• Still in Key On Engine Off Continuous Monitor mode. • Observe VOM or STAR LED for a fault indication while performing the following: — Referring to the illustration in Step DH90, grasp the harness close to the sensor connector. Wiggle, shake or bend a small section of the EEC-IV system harness while working your way to the dash panel. Also wiggle, shake or bend the EEC-IV harness from the dash panel to the processor. • Does VOM or STAR LED indicate a fault?	Yes ▶	ISOLATE fault. SERVICE as necessary. REFER to appropriate figure. CLEAR Continuous Memory Code 53. RERUN Quick Test.
	No ▶	GO to DH93.

Throttle Position Sensor (TPS) — Pinpoint Test DH

TEST STEP	RESULT ▶	ACTION TO TAKE
DH93 CHECK PROCESSOR AND HARNESS CONNECTORS		
• Key off, wait 10 seconds. • Disconnect processor 60 pin connector. Inspect for damaged pins, corrosion, loose wires, etc. Service as necessary. • Are connectors and terminals OK?	No ▶	SERVICE as necessary. CLEAR Continuous Memory Code 53. RERUN Quick Test.
	Yes ▶	Unable to duplicate fault at this time. CLEAR Continuous Memory Code 53. Continuous Memory Code 53 testing complete.
DH94 CONTINUOUS MEMORY CODE 63: EXERCISE TP SENSOR		
• Enter Key On Engine Off Continuous Monitor mode. Refer to Appendix in Section 16. • Observe VOM or STAR LED for indication of a fault while performing the following: — Move throttle slowly to WOT position. — Release throttle slowly to closed condition. — Lightly tap on TP sensor (simulate road shock). — Wiggle TP harness connector. • Does VOM or STAR LED indicate a fault?	Yes ▶	INSPECT connectors. If connector and terminals are good, REPLACE TP sensor. CLEAR Continuous Memory Code 63. RERUN Quick Test.
	No ▶	GO to DH95.

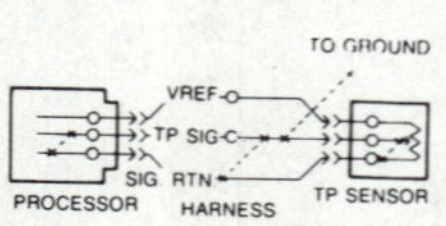

Throttle Position Sensor (TPS) — Pinpoint Test DH

TEST STEP	RESULT ▶	ACTION TO TAKE
DH95 CHECK EEC-IV HARNESS		
• Still in Key On Engine Off Continuous Monitor mode. • Observe VOM or STAR LED for a fault indication while performing the following: — Referring to the illustration in Step DH94 grasp the harness close to the sensor connector. Wiggle, shake or bend a small section of the EEC-IV system harness while working your way to the dash panel. Also wiggle, shake or bend the EEC-IV harness from the dash panel to the processor. • Does VOM or STAR LED indicate a fault?	Yes ▶	ISOLATE fault. SERVICE as necessary. REFER to appropriate figure. CLEAR Continuous Memory Code 63. RERUN Quick Test.
	No ▶	GO to DH96.
DH96 CHECK PROCESSOR AND HARNESS CONNECTORS		
• Key off, wait 10 seconds. • Disconnect processor 60 pin connector. Inspect for damaged pins, corrosion, loose wires, etc. • Are connectors and terminals OK?	No ▶	SERVICE as necessary. CLEAR Continuous Memory Code 63. RERUN Quick Test.
	Yes ▶	Unable to duplicate fault at this time. CLEAR Continuous Memory Code 63. Continuous Memory Code 63 testing complete.

Vane Airflow Sensor (VAF) — Pinpoint Test DK

Note

You should enter this Pinpoint Test only when a Service Code 26, 56, 66 or 76 is received in Quick Test Step 3.0, 5.0, or 6.0.

Remember

To prevent the replacement of good components, be aware that the following non-EEC areas may be at fault:

- Check for unmetered air (air leaks) between VAF meter and throttle body.
- Vacuum leaks.
- Engine sealing (PCV sealing, CANP, valve cover seal dipstick seated).

This Pinpoint Test is intended to diagnose only the following:

- VAF meter.
- Processor.
- Harness circuits: VREF, VAF SIGNAL and SIGNAL RETURN.

Pinpoint Test Schematic

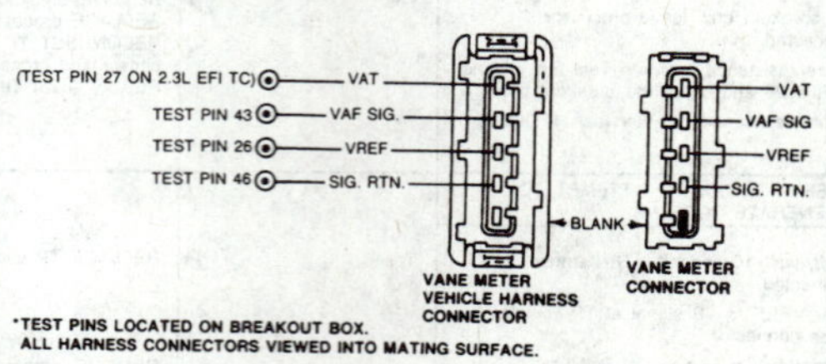

*TEST PINS LOCATED ON BREAKOUT BOX.
ALL HARNESS CONNECTORS VIEWED INTO MATING SURFACE.

Vane Airflow Sensor (VAF) — Pinpoint Test DK

TEST STEP	RESULT ▶	ACTION TO TAKE
DK1 SERVICE CODE 26: CHECK VANE METER FOR CONTAMINATION AND FREEDOM OF MOVEMENT		
NOTE: Code 26 indicates the vane airflow input to the processor is out of engine off or engine idle limits (engine off 0.15-0.50V/engine idle 1.50-2.70V). There have been no opens or shorts in the VAF circuit or a Code 56 (signal always high) or 66 (signal always low) would have been generated. • Key off, wait 10 seconds. • Remove air cleaner element and check for contamination (oil residue, foreign material, etc.) that may impede VAF sensor vane movement and service as necessary. • Was service Code 26 present in Key On Engine Off Self-Test?	Yes ▶ No ▶	REPLACE vane meter. REINSTALL air cleaner. RERUN Quick Test. GO to DK2 .
DK2 CHECK VAF SENSOR		
• Key off. • Check for unmetered air leaks between vane meter and throttle body. • Disconnect processor 60 pin connector. Inspect for damaged pins, corrosion, loose wires, etc. Service as necessary. • Install breakout box. • Connect processor to breakout box. • DVOM on 20 volt scale. • Key on, engine off. • Place new unsharpened pencil as shown. • Measure voltage between Test Pins 43 (Test Pin 27 on 2.3L EFI TC) and 46 at the breakout box. • Is voltage between 2.8 volts and 3.7 volts?	Yes ▶ No ▶	Vane meter is capable of outputting an acceptable signal. The Code 26 has been caused by incorrect engine speed or an unmetered air leak (vacuum leak). SERVICE as necessary. REMOVE breakout box. RERUN Quick Test. REMOVE breakout box. REPLACE processor. RERUN Quick Test.

Vane Airflow Sensor (VAF) — Pinpoint Test DK

TEST STEP	RESULT ▶	ACTION TO TAKE
DK12 CHECK VAF SIGNAL FOR SHORT TO POWER		
• Key off, wait 10 seconds. • Harness disconnected from vane meter. • Disconnect processor 60 pin connector. Inspect for damaged pins, corrosion, loose wires, etc. Service as necessary. • Install breakout box, leave processor disconnected. • DVOM on 200,000 ohm scale. • Measure resistance between Test Pin 43 (Test Pin 27 on 2.3L EFI TC) and Test Pins 26 and 57 at the breakout box. • Are both resistances greater than 10,000 ohms?	Yes ▶ No ▶	REPLACE processor. REMOVE breakout box. RECONNECT processor and vane meter. RERUN Quick Test. SERVICE circuit shorts. REMOVE breakout box. RECONNECT processor and vane meter. RERUN Quick Test.
DK20 SERVICE CODE 66: INDUCE OPPOSITE CODE		
• Key off, wait 10 seconds. • Disconnect vehicle harness from vane meter. • Install jumper wire in vane meter vehicle harness connector between VREF and VAF SIGNAL. • Perform Key On Engine Off Self-Test. **NOTE: If no codes are generated, immediately remove jumper and go directly to DK23 .** • Is Code 56 present? **NOTE: Disconnecting vane meter disconnects both VAF and VAT sensors. Code 58 should also be present. Disregard all codes except VAF codes at this time.**	Yes ▶ No ▶	Replace vane meter. REMOVE jumper wire. RECONNECT vane meter. RERUN Quick Test. REMOVE jumper wire and GO to DK21 .
DK21 CHECK VREF AT THE VANE METER		
• Key off, wait 10 seconds. • Harness disconnected from vane meter. • Key on, engine off. • DVOM on 20 volt scale. • Measure voltage at the vane meter vehicle harness connector between VREF and SIGNAL RETURN. • Is voltage between 4.0 and 6.0 volts?	Yes ▶ No ▶	GO to DK22 . GO to Pinpoint Test Step C1 .

Vane Airflow Sensor (VAF) — Pinpoint Test DK

TEST STEP	RESULT ▶	ACTION TO TAKE

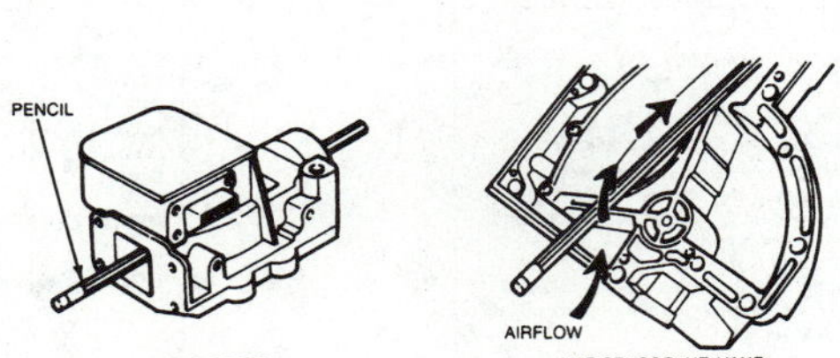

VANE METER

VAF SENSOR AIR VANE

TEST STEP	RESULT ▶	ACTION TO TAKE
DK10 SERVICE CODE 56: INDUCE OPPOSITE CODE		
• Key off, wait 10 seconds. • Disconnect vehicle harness from vane meter. Inspect for damaged pins, corrosion, loose wires, etc. Service as necessary. • Rerun Key On Engine Off Self-Test. • Is Code 66 present? **NOTE: Disconnecting vane meter disconnects both VAF and VAT sensors. A Code 58 should also be present. Disregard all codes except VAF codes at this time.**	Yes ▶ No ▶	GO to DK11 . GO to DK12 .
DK11 CHECK VAF TO SIGNAL RETURN VOLTAGE		
• Key off, wait 10 seconds. • Harness disconnected from vane meter. • Key on, engine off. • DVOM on 20 volt scale. • Measure voltage at the vane meter vehicle harness connector between VREF and SIGNAL RETURN. • Is voltage between 4.0 and 6.0 volts?	Yes ▶ No ▶	REPLACE vane meter. RECONNECT harness. RERUN Quick Test. GO to Pinpoint Test Step C1 .

Vane Airflow Sensor (VAF) — Pinpoint Test DK

TEST STEP	RESULT ▶	ACTION TO TAKE
DK22 CHECK CONTINUITY OF VAF SIGNAL		
• Key off, harness disconnected from vane meter. • Disconnect processor 60 pin connector. Inspect for damaged pins, corrosion, loose wires, etc. Service as necessary. • Install breakout box, leave processor disconnected. • DVOM on 200 ohm scale. • Measure resistance between VAF SIGNAL, at the vane meter vehicle harness connector, and Test Pin 43 (Test Pin 27 on 2.3L EFI TC) at the breakout box. • Is resistance less than 5 ohms?	Yes ▶ No ▶	GO to DK23 . SERVICE open circuit. REMOVE breakout box. RECONNECT processor and vane meter. RERUN Quick Test.
DK23 CHECK VAF SIGNAL FOR SHORT TO GROUND		
• Key off, wait 10 seconds. • Processor disconnected. • Harness disconnected from vane meter. • DVOM on 200,000 ohm scale. • Measure resistance at the vane meter vehicle harness between VAF SIGNAL and SIGNAL RETURN and between VAF SIGNAL and negative battery terminal. • Are both resistances greater than 10,000 ohms?	Yes ▶ No ▶	REPLACE processor. REMOVE breakout box. RECONNECT processor and vane meter. RERUN Quick Test. SERVICE circuit shorts. RECONNECT vane meter. RERUN Quick Test.

Vane Airflow Sensor (VAF) — Pinpoint Test DK

TEST STEP	RESULT ▶	ACTION TO TAKE
DK30 SERVICE CODE 76: CHECK FOR VOLTAGE INCREASE IN VAF SIGNAL DURING DYNAMIC RESPONSE		
NOTE: A sharp snap of the throttle may not be sufficient to pass this test. Be sure to move throttle to WOT and return. • Key off, wait 10 seconds. • Disconnect processor 60 pin connector. Inspect for damaged pins, corrosion, loose wires, etc. Service as necessary. • Install breakout box. • Connect processor to breakout box harness. • DVOM on 20 volt scale. • Connect DVOM to Test Pins 43 (Test Pin 27 on 2.3L EFI TC) and 46. • Perform Engine Running Quick Test while monitoring DVOM. • After dynamic response prompt Code 1(0) operator does a brief WOT. DVOM should increase more than 2.0 volts from reading before WOT. • Observe service codes at end of test. • Did voltage increase more than 2.0 volts?	Yes ▶ No ▶	GO to DK31. CHECK air cleaner duct for obstruction. If OK, REPLACE vane meter.
DK31 CHECK SERVICE CODES FROM STEP DK 30		
• Observe Engine Running service codes outputted in Pinpoint Test Step DK30. • Is Code 76 present?	Yes ▶ No ▶	REPLACE processor. REMOVE breakout box. RERUN Quick Test. Vane meter is OK, SERVICE other codes as necessary.

Vane Airflow Sensor (VAF) — Pinpoint Test DK

TEST STEP	RESULT ▶	ACTION TO TAKE
DK90 CONTINUOUS CODE 56: CHECK VAF SENSOR		
• Enter Key On Engine Off Continuous Monitor mode. Refer to Appendix in Section 16. • Observe VOM or STAR LED for indication of a fault while performing the following: — Lightly tap on VAF sensor (simulate road shock). — Wiggle VAF connector. — Is a fault indicated? 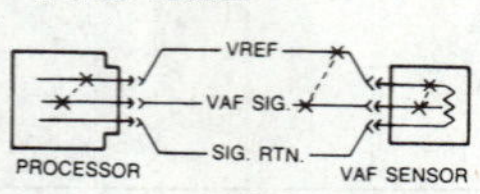	Yes ▶ No ▶	DISCONNECT and INSPECT connectors. If connector and terminals are good, REPLACE VAF sensor. CLEAR Continuous Memory Code. RERUN Quick Test. GO to DK91.
DK91 CHECK EEC-IV HARNESS		
• Still in Key On Engine Off Continuous Monitor mode. • Observe VOM or STAR LED for a fault indication while performing the following: — Referring to the illustration in Step DK90, grasp the harness close to the sensor connector. Wiggle, shake or bend a small section of the EEC-IV system harness while working your way to the dash panel. Also wiggle, shake or bend the EEC-IV harness from the dash panel to the processor. • Is a fault indicated?	Yes ▶ No ▶	ISOLATE fault and SERVICE as necessary. REFER to appropriate figure. CLEAR Continuous Memory Code. RERUN Quick Test. GO to DK92.

Vane Airflow Sensor (VAF) — Pinpoint Test DK

TEST STEP	RESULT ▶	ACTION TO TAKE
DK92 CHECK PROCESSOR AND HARNESS CONNECTORS		
• Key off, wait 10 seconds. • Disconnect processor 60 pin connector. Inspect both connectors and connector terminals for obvious damage or faults. • Are connectors and terminals OK?	Yes ▶ No ▶	Unable to duplicate fault at this time. Continuous Memory Code 56 testing complete. SERVICE as necessary. CLEAR Continuous Memory Code. RERUN Quick Test.
DK93 CONTINUOUS CODE 66: CHECK VAF SENSOR		
• Enter Key On Engine Off Continuous Monitor mode. Refer to Appendix in Section 16. • Observe VOM or STAR LED for indication of a fault while performing the following: — Lightly tap on VAF sensor (simulate road shock). — Wiggle VAF connector. • Is a fault indicated? 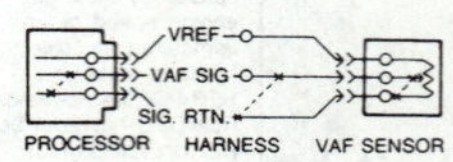	Yes ▶ No ▶	DISCONNECT and INSPECT connectors. If connector and terminals are good, CLEAR Continuous Memory Code. REPLACE sensor. RERUN Quick Test. GO to DK94.

Vane Airflow Sensor (VAF) — Pinpoint Test DK

TEST STEP	RESULT ▶	ACTION TO TAKE
DK94 CHECK EEC-IV HARNESS		
• Still in key on engine off continuous monitor mode. • Observe VOM or STAR LED for a fault indication while performing the following: — Referring to the illustration in Step DK93, grasp the harness close to the sensor connector. Wiggle, shake or bend a small section of the EEC-IV system harness while working your way to the dash panel. Also wiggle, shake or bend the EEC-IV harness from the dash panel to the processor. • Is a fault indicated?	Yes ▶ No ▶	ISOLATE fault and SERVICE as necessary. REFER to appropriate figure. CLEAR Continuous Memory Code. RERUN Quick Test. GO to DK95.
DK95 CHECK PROCESSOR AND HARNESS CONNECTORS		
• Key off, wait 10 seconds. • Disconnect processor 60 pin connector. • Inspect both connectors and connector terminals for obvious damage or faults. • Are connectors and terminals OK?	No ▶ Yes ▶	SERVICE as necessary. CLEAR Continuous Memory Code. RERUN Quick Test. Unable to duplicate fault at this time. CLEAR Continuous Memory Code. Continuous Memory Code 66 testing complete.

Pressure Feedback EGR (PFE) EGR Valve Regulator (EVR) | Pinpoint Test | DL

Note

You should enter this Pinpoint Test only when a Service Code 31, 32, 33, 34, 35 or 84 is received in Quick Test Step 3.0, 5.0 or 6.0.

Remember

To prevent the replacement of good components, be aware that the following non-EEC area may be at fault:

- Damaged EGR valve.

This Pinpoint Test is intended to diagnose only the following:

- Harness circuits: VREF, PFE, SIGNAL RETURN, EVR, VPWR.
- PFE sensor.
- EVR
- EGR valve assembly.
- Processor assembly.
- Vacuum lines/tubes (EVR, PFE)

Pinpoint Test Schematic

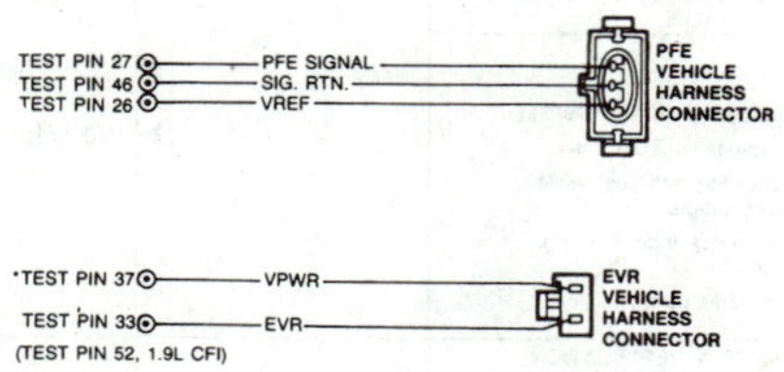

*TEST PINS LOCATED ON BREAKOUT BOX.
ALL HARNESS CONNECTORS VIEWED INTO MATING SURFACE.

Pressure Feedback EGR (PFE) EGR Valve Regulator (EVR) | Pinpoint Test | DL

TEST STEP	RESULT ▶	ACTION TO TAKE
DL1 SERVICE CODE 31: INDUCE CODE 35		
• Key off. • Disconnect PFE vehicle harness at sensor. • Jumper VREF to PFE SIGNAL at vehicle harness sensor connector. • Perform Key On Engine Off Self-Test. **NOTE: If no codes are generated, immediately remove jumper and go directly to Step DL4.** • Is Code 35 present? **NOTE: Ignore all other codes at this time.**	Yes ▶ No ▶	REMOVE Jumper. REPLACE PFE sensor. RERUN Quick Test. REMOVE jumper. GO to DL2.
DL2 MEASURE VREF TO SIGNAL RETURN VOLTAGE		
• Refer to schematic in Pinpoint Test DL. • Key off. • PFE harness disconnected. • DVOM on 20 volt scale. • Key on, engine off. • Measure voltage at PFE vehicle harness connector between VREF and SIGNAL RETURN. • Is voltage between 4 and 6 volts?	Yes ▶ No ▶	GO to DL3. GO to Pinpoint Test Step C1.
DL3 CHECK CONTINUITY OF PFE SIGNAL		
• Key off. • PFE harness disconnected. • DVOM on 200 ohm scale. • Disconnect processor 60 pin connector. Inspect for damaged pins, corrosion, loose wires, etc. Service as necessary. • Install breakout box, leave processor disconnected. • Measure resistance between PFE SIGNAL at vehicle harness sensor connector and Test Pin 27 at the breakout box. • Is resistance greater than 5 ohms?	Yes ▶ No ▶	SERVICE open circuit. RECONNECT PFE sensor. REMOVE breakout box. RERUN Quick Test. GO to DL4.

Pressure Feedback EGR (PFE) EGR Valve Regulator (EVR) | Pinpoint Test | DL

TEST STEP	RESULT ▶	ACTION TO TAKE
DL4 CHECK RESISTANCE OF PFE SIGNAL TO GROUND AND SIGNAL RETURN		
• Key off. • PFE harness disconnected. • Breakout box installed. • Processor disconnected. • DVOM on 200,000 ohm scale. • Measure resistance between PFE SIGNAL at PFE vehicle harness connector and ground. • Measure resistance between PFE SIGNAL at the PFE vehicle harness connector and Test Pin 46 at the breakout box. • Are both resistances greater than 10,000 ohms?	Yes ▶ No ▶	REPLACE processor. RECONNECT PFE sensor. REMOVE breakout box. RERUN Quick Test. SERVICE short circuit. RECONNECT PFE. REMOVE breakout box. RERUN Quick Test.
DL5 SERVICE CODE 35: INDUCE CODE 31		
• Key off. • Disconnect PFE vehicle harness at sensor. Inspect for damaged pins, corrosion, loose wires, etc. Service as necessary. • Rerun Key On Engine Off Self-Test. • Is Code 31 present? **NOTE: Ignore all other codes at this time.**	Yes ▶ No ▶	GO to DL6. GO to DL7.
DL6 MEASURE VREF TO SIGNAL RETURN VOLTAGE		
• Refer to schematic in Pinpoint Test DL. • Key off. • PFE harness disconnected. • DVOM on 20 volt scale. • Key on, engine off. • Measure voltage at PFE vehicle harness connector between VREF and SIGNAL RETURN. • Is voltage between 4 and 6 volts?	Yes ▶ No ▶	REPLACE PFE sensor. RERUN Quick Test. GO to Pinpoint Test Step C1.

Pressure Feedback EGR (PFE) EGR Valve Regulator (EVR) | Pinpoint Test | DL

TEST STEP	RESULT ▶	ACTION TO TAKE
DL7 CHECK PFE CIRCUIT FOR SHORT TO POWER		
• Key off. • PFE harness disconnected. • Disconnect processor 60 pin connector. Inspect for damaged pins, corrosion, loose wires, etc. Service as necessary. • Install breakout box, leave processor disconnected. • DVOM on 200,000 ohm scale. • Measure the resistance between Test Pin 27 and Test Pins 26 and 57 at the breakout box. • Are both resistances greater than 10,000 ohms?	Yes ▶ No ▶	REPLACE processor. REMOVE breakout box. RECONNECT PFE sensor. RERUN Quick Test. SERVICE short circuit. REMOVE breakout box. RECONNECT PFE sensor. RERUN Quick Test.
DL8 SERVICE CODE 34: PFE SENSOR OUT OF RANGE		
• PFE system can sense a lack of pressure in the vehicle exhaust system. An efficient garage exhaust ventilation system installed during Key On Engine Off Self-Test, may deflect the PFE sensor and generate a Code 34. Remove garage forced ventilation system and properly vent to atmosphere. • Rerun Key On Engine Off Self-Test. • Is Code 34 present?	Yes ▶ No ▶	GO to DL9. ADDRESS any other codes in Key On, Engine Off. If none, CONTINUE with remaining Quick Test.
DL9 CHECK PRESSURE FEED TUBE TO PFE SENSOR		
• Remove the pressure feed tube from PFE sensor. • Inspect complete tube, including PFE inlet for blockage. • Is blockage present?	Yes ▶ No ▶	SERVICE as necessary. RERUN Quick Test. GO to DL10.

Pressure Feedback EGR (PFE) EGR Valve Regulator (EVR) — Pinpoint Test — DL

TEST STEP	RESULT ▶	ACTION TO TAKE
DL10 MEASURE VREF TO SIGNAL RETURN VOLTAGE		
• Refer to schematic in Pinpoint Test DL. • Key off. • Disconnect PFE sensor. Inspect for damaged pins, corrosion, loose wires, etc. Service as necessary. • DVOM on 20 volt scale. • Key on, engine off. • Measure voltage between VREF and SIGNAL RETURN at PFE vehicle harness connector. • Is voltage between 4 and 6 volts?	Yes ▶ No ▶	REPLACE PFE sensor. RERUN Quick Test. GO to Pinpoint Test Step [C1].
DL11 SERVICE CODE 84: MEASURE EVR SOLENOID RESISTANCE		
• Key off. • Disconnect EVR solenoid connector. • DVOM on 200 ohm scale. • Measure solenoid resistance. • Is resistance between 30 and 70 ohms?	Yes ▶ No ▶	GO to [DL12]. REPLACE EVR solenoid assembly. RERUN Quick Test.
DL12 CHECK FOR VPWR AT EVR SOLENOID		
• Key on engine off. • EVR solenoid disconnected from harness. • DVOM on 20 volt scale. • Measure voltage between battery negative terminal and VPWR circuit at EVR solenoid vehicle harness connector. • Is voltage less than 10.5 volts?	Yes ▶ No ▶	SERVICE open circuit. RERUN Quick Test. GO to [DL13].
DL13 CHECK CONTINUITY OF EVR CIRCUIT		
• Key off. • EVR solenoid disconnected from harness. • Disconnect processor 60 pin connector. Inspect for damaged pins, corrosion, loose wires, etc. Service as necessary. • Install breakout box, leave processor disconnected. • DVOM on 200 ohm scale. • Measure resistance between Test Pin 33 (Test Pin 52, 1.9L CFI) at the breakout box and EVR SIGNAL at the EVR solenoid vehicle harness connector. • Is resistance less than 5 ohms?	Yes ▶ No ▶	GO to [DL14]. SERVICE open circuit. REMOVE breakout box. RECONNECT processor and EVR solenoid. RERUN Quick Test.

Pressure Feedback EGR (PFE) EGR Valve Regulator (EVR) — Pinpoint Test — DL

TEST STEP	RESULT ▶	ACTION TO TAKE
DL14 CHECK EVR CIRCUIT FOR SHORT TO POWER OR GROUND		
• Key off. • Breakout box installed, processor disconnected. • EVR solenoid disconnected. • DVOM on 200,000 ohm scale. • Measure resistance between Test Pin 33 (Test Pin 52, 1.9L CFI) and Test Pins 37 and 57 at the breakout box. • Measure resistance between Test Pin 33 (Test Pin 52, 1.9L CFI) and Test Pins 40 and 60 at the breakout box. • Are any resistances less than 10,000 ohms?	Yes ▶ No ▶	SERVICE short circuit. REMOVE breakout box. RECONNECT processor and EVR solenoid. RERUN Quick Test. If code is repeated, REPLACE processor. REPLACE processor. REMOVE breakout box. RECONNECT processor and EVR solenoid. RERUN Quick Test.
DL20 SERVICE CODE 32: VERIFY ENGINE RUNNING CODES		
The PFE system can sense a lack of pressure in the vehicle exhaust system. An efficient garage exhaust ventilation system installed during Key On Engine Running Self-Test may, on some calibrations, deflect the PFE sensor and generate a Code 32. Temporarily, remove garage forced ventilation system and properly vent to atmosphere. • Rerun Engine Running Self-Test. • Is Code 32 present?	Yes ▶ No ▶	GO to [DL21]. ADDRESS any other codes in Engine Running. If none, CONTINUE with remaining Self-Test.
DL21 ATTEMPT TO SEPARATE EVR FROM PFE		
• Key off. • Disconnect EGR valve vacuum line at valve and plug line. • Perform Engine Running Self-Test. • Is Code 31 or 32 present?	Yes ▶ No ▶	GO to [DL22]. GO to [DL23].
DL22 CHECK PFE SENSOR SUPPLY TUBE		
• Key off. • Check PFE sensor supply tube for obstructions and/or leaks. • Are there any obstructions or leaks?	Yes ▶ No ▶	SERVICE as necessary. RECONNECT all lines and RERUN Quick Test. GO to EGR Diagnostics

Pressure Feedback EGR (PFE) EGR Valve Regulator (EVR) — Pinpoint Test — DL

TEST STEP	RESULT ▶	ACTION TO TAKE
DL23 CHECK EVR FILTER		
• Key off. • Remove and inspect EVR filter for contamination. **NOTE: Blockage of filter will cause vacuum to be applied to EGR valve prematurely.** • Is filter contaminated?	Yes ▶ No ▶	REPLACE filter. RECONNECT all lines. RERUN Quick Test. REPLACE EVR solenoid. RERUN Quick Test.
DL25 SERVICE CODE 34 AND 35: CHECK FOR EXCESSIVE EXHAUST BACK PRESSURE		
NOTE: Service Codes 34 and 35 in Engine Running Self-Test indicate excessive exhaust back pressure. There are two possible causes: (1). The exhaust system is restricted, and (2). PFE sensor has shifted high. • Key off. • Substitute known good PFE sensor in place of original. • Rerun Engine Running Self-Test. • Is Code 34 or 35 present?	Yes ▶ No ▶	Catalyst and Exhaust Systems Restricted Exhaust System Diagnosis. Original PFE was the cause of the original Service Code 34 or 35. REPLACE PFE sensor. RERUN Quick Test.
DL30 SERVICE CODE 33: VERIFY VACUUM IS PRESENT AT VALVE		
• Key off. • Standard vacuum gauge in.-Hg (Mercury). • Tee in vacuum gauge at EGR valve. • Rerun Engine Running Self-Test while observing vacuum gauge. • Is vacuum reading less than 1 in.-Hg. throughout the test? **NOTE: Disregard code output.**	Yes ▶ No ▶	GO to [DL31]. GO to [DL34].
DL31 VACUUM SUPPLY VERIFICATION		
• Key off. • Do vacuum lines from EVR solenoid to EGR valve and source to EVR solenoid have loose connections, cracks or obstructions?	Yes ▶ No ▶	SERVICE as necessary. RERUN Quick Test. GO to [DL32].

Pressure Feedback EGR (PFE) EGR Valve Regulator (EVR) — Pinpoint Test — DL

TEST STEP	RESULT ▶	ACTION TO TAKE
DL32 VERIFY VACUUM TO EVR		
• Start engine and run at idle. • Attach vacuum gauge to source line from manifold. • Is vacuum present?	Yes ▶ No ▶	REPLACE EVR solenoid. RERUN Quick Test. REPLACE vacuum line to EVR. RERUN Quick Test.
DL34 CHECK EGR CONTROL PFE SENSOR TUBE		
• Key off. • Is control pressure input tube to PFE sensor cracked, disconnected or obstructed?	Yes ▶ No ▶	SERVICE as necessary. RERUN Quick Test. REPLACE PFE Sensor. RERUN Quick Test.
DL90 CONTINUOUS MEMORY CODE 31 OR 35: EXERCISE PFE SENSOR		
• Enter Key On Engine Off Continuous Monitor mode. Refer to the Appendix in Section 16. • Observe VOM or STAR LED for indication of a fault while performing the following: • Connect a vacuum pump to the PFE sensor. • Slowly apply 5 in.-Hg. to the sensor. • Slowly bleed vacuum off the PFE sensor. • Lightly tap on PFE sensor (to simulate road shock). • Wiggle PFE connector. • Is fault indicated? 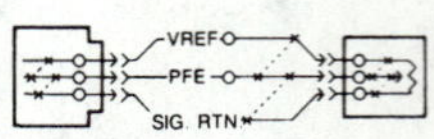	Yes ▶ No ▶	DISCONNECT and INSPECT connectors. If connector and terminals are good, REPLACE PFE sensor. CLEAR Continuous Memory Code 31/35. RERUN Quick Test. GO to [DL91].

Pressure Feedback EGR (PFE) EGR Valve Regulator (EVR) — Pinpoint Test DL

TEST STEP	RESULT	ACTION TO TAKE
DL91 CHECK EEC-IV HARNESS		
• Still in Key On Engine Off Continuous Monitor mode. • Observe VOM or STAR LED for a fault indication while performing the following: — Referring to the illustration in Step DL90 grasp the harness closest to the sensor connector. Wiggle, shake or bend a small section of the EEC-IV system harness while working your way to the dash panel. Also wiggle, shake or bend the EEC-IV harness from the dash panel to the processor. • Is a fault indicated?	Yes ▶	ISOLATE fault and SERVICE as necessary. CLEAR Continuous Memory Code 31/35. RERUN Quick Test.
	No ▶	GO to DL92.
DL92 CHECK PROCESSOR AND HARNESS CONNECTORS		
• Key off, wait 10 seconds. • Disconnect processor 60 pin connector. • Inspect both connectors and connector terminals for obvious damage or faults. • Are connectors and terminals OK?	Yes ▶	Unable to duplicate and/or identify fault at this time. CLEAR Continuous Memory Code 31/35. Continuous Memory Code 31 or 35 testing complete.
	No ▶	SERVICE as necessary. CLEAR Continuous Memory Code 31/35. RERUN Quick Test.

Pressure Feedback EGR (PFE) EGR Valve Regulator (EVR) — Pinpoint Test DL

TEST STEP	RESULT	ACTION TO TAKE
DL93 CONTINUOUS MEMORY CODE 34: INSPECT PFE SUPPLY TUBE FOR BLOCKAGE		
• Key off. • Remove PFE sensor and inspect sensor supply inlet for liquids and/or any type of blockage. • Inspect PFE supply tube to EGR valve base for liquids and/or blockage. • Is supply tube free of any blockage?	Yes ▶	Unable to duplicate and/or identify fault at this time. CLEAR Continuous Memory Code 34. Continuous Code 34 testing complete.
	No ▶	CLEAN and/or SERVICE as necessary. CLEAR Continuous Memory Code 34. RERUN Quick Test.
DL94 CONTINUOUS MEMORY CODE 32: INSPECT EGR VALVE FOR SMOOTH OPERATION.		
• Key off. • Connect a vacuum pump to the EGR valve. • Apply 10 in.-Hg. of vacuum to EGR valve. • While observing EGR valve, release vacuum. • Does EGR valve function in a smooth manner? **NOTE: Repeat test if necessary to ensure accurate result.**	Yes ▶	GO to DL95.
	No ▶	CLEAR Continuous Memory Code 32. GO to EGR Valve Diagnostics, Section 6.
DL95 INSPECT VACUUM LINES BETWEEN EVR SOLENOID AND EGR VALVE		
• Inspect EGR valve vacuum supply line from EVR solenoid for kinks and/or obstructions. • Is vacuum supply line to EGR valve free of any obstructions?	Yes ▶	GO to DL96.
	No ▶	SERVICE as necessary. CLEAR Continuous Memory Code 32. RERUN Quick Test.

Pressure Feedback EGR (PFE) EGR Valve Regulator (EVR) — Pinpoint Test DL

TEST STEP	RESULT	ACTION TO TAKE
DL96 EVR REGULATOR FILTER INSPECTION		
• Carefully check EVR filter for contamination and/or obstructions. • Is EVR filter condition acceptable?	Yes ▶	Unable to duplicate and/or identify fault at this time. CLEAR Continuous Memory Code 32. Continuous Memory Code 34 testing complete.
	No ▶	REPLACE EVR filter. CLEAR Continuous Memory Code 32. RERUN Quick Test.
DL97 CONTINUOUS MEMORY CODE 33: INSPECT EGR VALVE FOR FREE OPERATION		
• Key off. • Connect a vacuum pump to the EGR valve. • While observing the EGR valve, slowly apply 10 in.-Hg. vacuum. **NOTE: EGR valve should begin to open with a very small amount of vacuum, approximately 1 to 1.5 in.-Hg. and be fully open with about 4 in.-Hg. vacuum.** • Does EGR valve move freely and smoothly?	Yes ▶	GO to DL98.
	No ▶	CLEAR Continuous Memory Code 33. GO to EGR Valve Diagnostics,

Pressure Feedback EGR (PFE) EGR Valve Regulator (EVR) — Pinpoint Test DL

TEST STEP	RESULT	ACTION TO TAKE
DL98 EVR HARNESS CHECK		
• Key off. • Disconnect processor 60 pin connector. Inspect for damaged pins, corrosion, loose wires, etc. Service as necessary. • Install breakout box and connect processor to breakout box. • Enter Output State Check. Refer to Appendix in Section 16. • DVOM on 20 volt scale. • Connect DVOM negative test lead to Test Pin 40 at the breakout box and DVOM positive test lead to Test Pin 33. (Test Pin 52 for 1.9L CFI). • Cycle throttle if necessary to indicate greater than 10.5 volts. • Remain at this position. • While observing DVOM, grasp the harness closest to the EVR connector. Wiggle, shake or bend a small section of the EEC-IV system harness while working your way to the dash panel. • Lightly tap EVR solenoid to simulate road shock. • Does DVOM indicate less than 10.5 volts?	Yes ▶	SERVICE as necessary. CLEAR Continuous Memory Code 33. RERUN Quick Test.
	No ▶	Unable to duplicate and/or identify fault at this time. CLEAR Continuous Memory Code 33. Continuous Memory Code 33 testing complete.

EGR Valve Position Sensor (EVP) EGR Valve Regulator (EVR) — Pinpoint Test DN

Note

You should enter this Pinpoint Test only when a Service Code 31, 32, 33, 34, 35, 38 or 84 is received in Quick Test Step 3.0, 5.0 or 6.0 or from Pinpoint Test Step S3.

Remember

To prevent the replacement of good components, be aware that the following non-EEC area may be at fault:

- Damaged EGR valve.

This Pinpoint Test is intended to diagnose only the following:

- EVP sensor.
- Harness circuits: VREF, EVP, SIGNAL RETURN, EVR, VPWR.
- EVR (EGR Valve Regulator).
- EGR valve assembly.
- Processor assembly.
- EGR and EVR vacuum lines.

Pinpoint Test Schematic

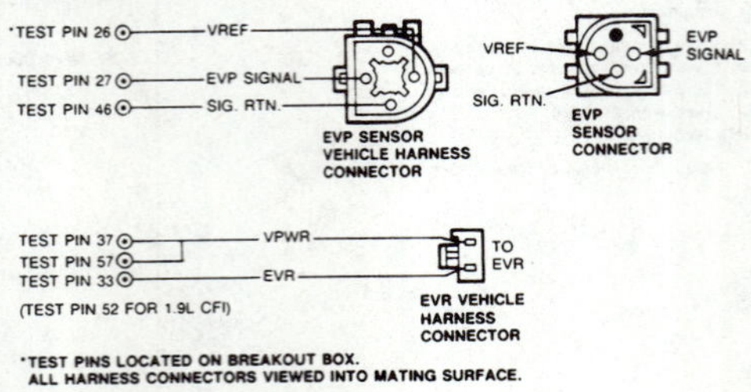

EGR Valve Position Sensor (EVP) EGR Valve Regulator (EVR) — Pinpoint Test DN

TEST STEP	RESULT ▶	ACTION TO TAKE
DN1 SERVICE CODE 31: ATTEMPT TO GENERATE CODE 35		
• Key off, wait 10 seconds. • Disconnect EVP vehicle harness at sensor. • Jumper VREF to EVP signal at vehicle harness connector. • Rerun Key On Engine Off Self-Test. • Is Code 35 present? **NOTE: Ignore all other codes at this time.**	Yes ▶ No ▶	REMOVE Jumper. REPLACE EVP sensor. RERUN Quick Test. REMOVE jumper. GO to DN2.
DN2 CHECK VREF TO SIGNAL RETURN VOLTAGE		
• Key on, engine off. • EVP disconnected from harness. • DVOM on 20 volt scale. • Measure voltage between VREF and SIGNAL RETURN at EVP vehicle harness connector. • Is voltage between 4.0 and 6.0 volts?	Yes ▶ No ▶	GO to DN3. GO to Pinpoint Test Step C1.
DN3 CHECK CONTINUITY OF EVP SIGNAL		
• Key off, wait 10 seconds. • EVP Sensor disconnected. • Disconnect processor 60 pin connector. Inspect for damaged pins, corrosion, loose wires, etc. Service as necessary. • Install breakout box, connect processor to breakout box. • DVOM on 200 ohm scale. • Measure resistance between EVP SIGNAL at vehicle harness connector and Test Pin 27 at the breakout box. • Is resistance less than 5 ohms?	Yes ▶ No ▶	GO to DN4. SERVICE open circuit. REMOVE breakout box. RECONNECT all components. RERUN Quick Test.

EGR Valve Position Sensor (EVP) EGR Valve Regulator (EVR) — Pinpoint Test DN

TEST STEP	RESULT ▶	ACTION TO TAKE
DN4 CHECK EVP SIGNAL FOR SHORT TO GROUND		
• Key off. • EVP harness disconnected. • Breakout box installed. • Processor disconnected. • DVOM on 200,000 ohm scale. • Measure resistance between Test Pin 27 and Test Pins 40, 46 and 60 at the breakout box. • Are all resistances greater than 10,000 ohms?	Yes ▶ No ▶	REPLACE processor. REMOVE breakout box. RECONNECT all components. RERUN Quick Test. SERVICE short circuit. REMOVE breakout box. RECONNECT all components. RERUN Quick Test.
DN5 SERVICE CODE 35: ATTEMPT TO GENERATE CODE 31		
• Key off, wait 10 seconds. • Disconnect EVP sensor. • Rerun Key On Engine Off Self-Test. • Is Code 31 present? **NOTE: Ignore all other codes at this time.**	Yes ▶ No ▶	GO to DN6. GO to DN7.
DN6 CHECK VREF TO SIGNAL RETURN VOLTAGE		
• Key on, engine off. • EVP sensor disconnected. • DVOM on 20 volt scale. • Measure voltage between VREF and SIGNAL RETURN at EVP vehicle harness connector. • Is voltage between 4.0 and 6.0 volts?	Yes ▶ No ▶	REPLACE EVP sensor. RERUN Quick Test. GO to Pinpoint Test Step C1.

EGR Valve Position Sensor (EVP) EGR Valve Regulator (EVR) — Pinpoint Test DN

TEST STEP	RESULT ▶	ACTION TO TAKE
DN7 CHECK EVP SIGNAL FOR SHORT TO POWER		
• Key off. • EVP disconnected from harness. • Disconnect processor 60 pin connector. Inspect for damaged pins, corrosion, loose wires, etc. Service as necessary. • Install breakout box, leave processor disconnected. • DVOM on 200,000 ohm scale. • Measure the resistance between Test Pin 27 and Test Pins 26 and 57 at the breakout box. • Are both resistances greater than 10,000 ohms?	Yes ▶ No ▶	REPLACE processor. REMOVE breakout box. RECONNECT EVP sensor. RERUN Quick Test. SERVICE short circuit. REMOVE breakout box. RECONNECT EVP sensor and processor. RERUN Quick Test.
DN10 SERVICE CODE 84: CHECK RESISTANCE OF EVR SOLENOID		
• Key off, wait 10 seconds. • Disconnect EVR solenoid. • DVOM on 200 ohm scale. • Measure solenoid resistance. • Is resistance between 30 and 70 ohms?	Yes ▶ No ▶	GO to DN11. REPLACE EVR solenoid assembly. RERUN Quick Test.
DN11 CHECK FOR VPWR AT EVR SOLENOID		
• Key on, engine off. • EVR solenoid disconnected. • DVOM on 20 volt scale. • Measure voltage between battery negative post and VPWR circuit at the EVR solenoid vehicle harness connector. • Is voltage greater than 10.5 volts?	Yes ▶ No ▶	GO to DN12. RECONNECT EVR solenoid. SERVICE open circuit. RERUN Quick Test.

EGR Valve Position Sensor (EVP) EGR Valve Regulator (EVR) — Pinpoint Test DN

TEST STEP	RESULT	ACTION TO TAKE
DN12 CHECK CONTINUITY OF EVR CIRCUIT		
• Key off. • EVR solenoid disconnected from harness. • Disconnect processor 60 pin connector. Inspect for damaged pins, corrosion, loose wires, etc. Service as necessary. • Install breakout box to processor, leave processor disconnected. • DVOM on 200 ohm scale. • Measure resistance between Test Pin 33 (Test Pin 52 for 1.9L CFI) at the breakout box and EVR SIGNAL at the EVR solenoid vehicle harness connector. • Is resistance less than 5 ohms?	Yes ▶	GO to DN13.
	No ▶	SERVICE open circuit. REMOVE breakout box. RECONNECT processor and EVR solenoid. RERUN Quick Test.
DN13 CHECK EVR CIRCUIT FOR SHORT TO POWER AND GROUND		
• Key off. • EVR solenoid disconnected. • Breakout box installed, processor disconnected. • DVOM on 200,000 ohm scale. • Measure resistance between Test Pin 33 (Test Pin 52 for 1.9L CFI) and Test Pins 37/57, 40/60 and 46 at the breakout box. • Are all resistances greater than 10,000 ohms?	Yes ▶	REPLACE processor. REMOVE breakout box. RECONNECT processor and EVR solenoid. RERUN Quick Test.
	No ▶	SERVICE short circuit. REMOVE breakout box. RECONNECT processor and EVR solenoid. RERUN Quick Test. If code is repeated, REPLACE processor.

EGR Valve Position Sensor (EVP) EGR Valve Regulator (EVR) — Pinpoint Test DN

TEST STEP	RESULT	ACTION TO TAKE
DN40 SERVICE CODE 33: VERIFY VACUUM IS PRESENT AT EGR VALVE		
NOTE: Engine Running Code 33 indicates that the EVP sensor input did not change after the EVR solenoid was instructed by the processor to open the EGR valve. Because a Code 84 was not received in the Key On Engine Off Self-Test, it is known that the EVR solenoid functions electrically. It is also known that the EVP sensor is in the expected closed valve range because Codes 32 and 34 were not received in either Key On Engine Off or Engine Running Tests.		
• Key off. • Tee in vacuum gauge at EGR valve. • Rerun Engine Running Self-Test while observing vacuum gauge. • Is vacuum greater than 1.5 in.-Hg. (5 kPa)?	Yes ▶	REMOVE vacuum gauge. RECONNECT EGR valve. GO to DN43.
	No ▶	REMOVE vacuum gauge. RECONNECT EGR valve. GO to DN41.
DN41 VERIFY VACUUM SUPPLY TO EVR SOLENOID		
• Key off. • Disconnect the vacuum source to the EVR solenoid. • Install a vacuum gauge at source vacuum. • Start engine and check vacuum. • Is vacuum greater than 10 in.-Hg. (33 kPa)?	Yes ▶	GO to DN42.
	No ▶	CHECK source vacuum hose to EVR solenoid. SERVICE as necessary. RERUN Quick Test.
DN42 CHECK VACUUM HOSE BETWEEN EVR SOLENOID AND EGR VALVE		
• Carefully check EGR vacuum hose from EGR valve to EVR for obstructions cracks, loose connectors, blockage, kinks and leaks, etc. • Is vacuum hose in good condition?	Yes ▶	CHECK EVR solenoid filter for obstructions. REPLACE as necessary. If OK, REPLACE EVR solenoid assembly. RECONNECT vacuum hose. RERUN Quick Test.
	No ▶	SERVICE vacuum hose as necessary. RERUN Quick Test.

EGR Valve Position Sensor (EVP) EGR Valve Regulator (EVR) — Pinpoint Test DN

TEST STEP	RESULT	ACTION TO TAKE
DN20 SERVICE CODE 34: CHECK FOR SERVICE CODE 84.		
• Key off, wait 10 seconds. • Is Code 84 present in Key On Engine Off Self-Test?	Yes ▶	GO to DN10.
	No ▶	GO to DN21.
DN21 SUBSTITUTE EVP SENSOR ON ORIGINAL EGR VALVE		
NOTE: Key On Engine Off Service Code 34 indicates that the EGR valve and/or EVP sensor is not fully seated in the closed position. Because of the preload on the installed EVP sensor, it is very diffficult to determine whether the EGR valve is seated or the EVP sensor is in contact with the EGR valve stem.		
• Key off, wait 10 seconds. • Install a known good EVP sensor on original EGR valve. • Perform Key On Engine Off Quick Test. • Is Code 34 still present?	Yes ▶	GO to EGR Valve Diagnostics, Section 6.
	No ▶	The original Code 34 was the result of the original EVP sensor. SERVICE EVP sensor as necessary. RERUN Quick Test.
DN25 SERVICE CODE 32: SUBSTITUTE EVP SENSOR ON ORIGINAL EGR VALVE		
NOTE: Key On Engine Off and Engine Running Service Code 32 indicates that the EGR valve and/or EVP sensor is lower than normal in the closed position. Because of the preload of the EVP sensor it is very difficult to determine whether EGR valve has malfunctioned or the EVP sensor has abnormally high resistance.		
• Key off, wait 10 seconds. • Install a known good EVP sensor on original EGR valve. • Rerun Key On Engine Off Self-Test. • Is Code 32 present?	Yes ▶	GO to Section 6 for EGR valve diagnostics.
	No ▶	The original Code 32 was the result of the original EVP sensor. SERVICE EVP sensor as necessary. RERUN Quick Test.

EGR Valve Position Sensor (EVP) EGR Valve Regulator (EVR) — Pinpoint Test DN

TEST STEP	RESULT	ACTION TO TAKE
DN43 SUBSTITUTE KNOWN GOOD EVP SENSOR ON ORIGINAL EGR VALVE		
• Key off, wait 10 seconds. • Install a known good EVP sensor on original EGR valve. • Rerun Engine Running Self-Test. • Is Code 33 present?	Yes ▶	GO to EGR valve diagnostics.
	No ▶	The original Code 33 was the result of the original EVP sensor. SERVICE EVP sensor as necessary. RERUN Quick Test.
DN50 SERVICE CODE 34: EGR VALVE OPERATION, ENGINE RUNNING SELF-TEST WITH EGR VACUUM DISCONNECTED		
• Key off. • Disconnect vacuum hose from EGR valve and plug hose. • Rerun Engine Running Self-Test. • Is Code 34 present?	Yes ▶	GO to DN51.
	No ▶	CHECK EVR filter for obstructions. REPLACE as necessary. If OK, REPLACE EVR solenoid assembly. RECONNECT all vacuum hoses. RERUN Quick Test.
DN51 CHECK EVP RESISTANCE WHILE APPLYING VACUUM TO EGR VALVE		
• Key off, wait 10 seconds. • Disconnect harness from EVP sensor. • Disconnect vacuum hose at EGR valve. • Connect vacuum pump to EGR valve. • DVOM on 200,000 ohm scale. • Measure resistance at the EVP sensor between EVP SIGNAL and VREF while increasing vacuum to 10 in. Hg. (33 kPa). • Observe resistance as vacuum increases. • Does resistance decrease gradually from no more than 5,500 ohms to no less than 100 ohms?	Yes ▶	GO EGR valve diagnostics.
	No ▶	REPLACE EVP sensor. RECONNECT vacuum hose. RERUN Quick Test.

EGR Valve Position Sensor (EVP) EGR Valve Regulator (EVR) — Pinpoint Test — DN

TEST STEP	RESULT ▶	ACTION TO TAKE
DN90 CONTINUOUS MEMORY CODE 32: CHECK EVP SIGNAL VOLTAGE WHILE EXERCISING EVP SENSOR		
NOTE: The EVP circuit indicated that the EGR valve was closed further than normal with the engine at stabilized operating temperature and at idle. • Key off, wait 10 seconds. • Disconnect processor 60 pin connector. Inspect for damaged pins, corrosion, loose wires, etc. Service as necessary. • Install breakout box. • Connect processor to breakout box. • Disconnect vacuum hose at EGR valve. • Connect a vacuum pump to the EGR valve. • DVOM on 20 volt scale. • Key on, engine off. • Measure resistance between Test Pin 27 and Test Pin 46 at the breakout box while doing the following. — Slowly increase vacuum at EGR valve to 6 in.-Hg. (20 kPa), then slowly bleed vacuum off the EGR valve and lightly tap on EVP sensor (simulate road shock). • Does voltage drop to less than 0.29 volts?	Yes ▶	EGR valve may have caused Continuous Memory Code 32. CLEAR Continuous Memory Code 32. GO to EGR valve diagnostics.
	No ▶	Unable to duplicate Code 32 fault at this time. CLEAR Continuous Memory Code 32. REFER to Appendix in Section 16. Continuous Memory Code 32 testing complete.

EGR Valve Position Sensor (EVP) EGR Valve Regulator (EVR) — Pinpoint Test — DN

TEST STEP	RESULT ▶	ACTION TO TAKE
DN92 CONTINUOUS MEMORY CODE 31 AND/OR 35: CHECK EEC-IV HARNESS		
NOTE: The EVP circuit indicated an open in the EVP signal or VREF, or a short to SIGNAL RETURN with the engine at stabilized operating temperature and at idle. CODE 31: 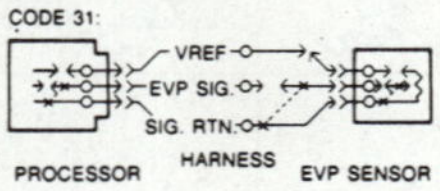		
NOTE: The EVP circuit indicated a short to VREF and/or VPWR, or an open in SIGNAL RETURN with the engine at stabilized operating temperature and at idle. CODE 35: 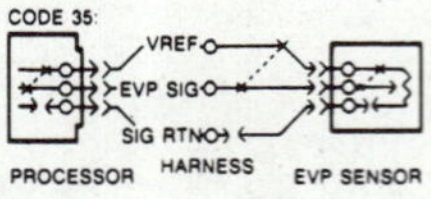 • Still in Key On Engine Off Continuous Monitor mode. • Observe VOM or STAR LED for a fault indication while performing the following: — Refer to illustration above by code for possible circuit faults. — Grasp the harness close to the sensor connector. Wiggle, shake or bend a small section of the EEC-IV system harness while working your way to the dash panel. Also wiggle, shake or bend the EEC-IV harness from the dash panel to the processor. • Is a fault indicated?	Yes ▶	ISOLATE fault and SERVICE as necessary. REFER to appropriate figure. CLEAR Continuous Memory Code 31 and/or 35. RERUN Quick Test.
	No ▶	GO to **DN93**.

EGR Valve Position Sensor (EVP) EGR Valve Regulator (EVR) — Pinpoint Test — DN

TEST STEP	RESULT ▶	ACTION TO TAKE
DN93 CHECK PROCESSOR AND HARNESS CONNECTORS		
• Key off, wait 10 seconds. • Disconnect processor 60 pin connector. Inspect for damaged pins, corrosion, loose wires, etc. on both the processor and harness connectors. • Are connectors and terminals OK?	Yes ▶	Unable to duplicate fault at this time. CLEAR Continuous Memory Code 31 and/or 35. Continuous Memory Code 31 or 35 testing complete.
	No ▶	SERVICE as necessary. CLEAR Continuous Memory Code 31 and/or 35. RERUN Quick Test.
DN95 CONTINUOUS MEMORY CODE 33: LEAK TEST		
NOTE: The EVP circuit indicated that the EGR valve did not open with the engine at stabilized temperature and with an EVR solenoid duty cycle present. • Key off, wait 10 seconds. • Disconnect vacuum hose at EGR valve. • Connect a vacuum pump to EGR valve. • Apply 20 in.-Hg. (66 kPa) to EGR valve. • Does EGR valve open and maintain vacuum?	Yes ▶	REMOVE vacuum pump. RECONNECT EGR valve. GO to **DN96**.
	No ▶	REMOVE vacuum pump. RECONNECT EGR valve. CLEAR Continuous Memory Code 33. GO EGR valve diagnostics.

EGR Valve Position Sensor (EVP) EGR Valve Regulator (EVR) — Pinpoint Test — DN

TEST STEP	RESULT ▶	ACTION TO TAKE
DN96 EVR CHECK		
• Using continuous monitor mode. Refer to Appendix in Section 16. • Observe VOM or STAR LED for indication of a fault while performing the following: — Grasp the harness close to the EVR solenoid connector, wiggle, shake or bend a small section of the harness while working your way to the processor. • Inspect connectors, terminals for obvious damage or faults. • Are any faults detected?	Yes ▶	ISOLATE fault and SERVICE as necessary. CLEAR Continuous Memory Code 33. RERUN Quick Test.
	No ▶	Unable to duplicate fault at this time. CLEAR Continuous Memory Code 33. Testing complete.
DN98 CONTINUOUS MEMORY CODE 34: CHECK EVP RESISTANCE WHILE APPLYING VACUUM TO EGR VALVE		
NOTE: The EVP circuit indicated that the EGR valve was open with the engine at stabilized operating temperature and at idle. • Key off. • Disconnect harness from EVP sensor. • Disconnect vacuum hose at EGR valve. • Connect vacuum pump to EGR valve. • DVOM on 200,000 ohm scale. • Measure resistance between EVP SIGNAL pin and VREF pin at the EVP sensor while increasing vacuum to 10 in.-Hg (33 kPa). • Observe resistance as vacuum increases. • Does resistance gradually change from no more than 5,500 ohms to no less than 100 ohms as the vacuum increases?	Yes ▶	REMOVE vacuum pump. RECONNECT EGR valve. CLEAR Continuous Memory Code 34. GO to **DN99**.
	No ▶	REMOVE vacuum pump. RECONNECT EGR valve. CLEAR Continuous Memory Code 34. GO to EGR valve diagnostics.

EGR Valve Position Sensor (EVP) EGR Valve Regulator (EVR) — Pinpoint Test DN

TEST STEP	RESULT ▶	ACTION TO TAKE
DN99 EVR CHECK		
• Key off. • Disconnect vacuum hose from EGR valve and plug hose. • Rerun Engine Running Self-Test. • Is Code 34 present?	Yes ▶	CHECK EVR filter for obstructions. REPLACE as necessary. If OK, REPLACE EVR solenoid. RECONNECT all vacuum lines. CLEAR Continuous Memory Code 34. RERUN Quick Test.
	No ▶	Unable to duplicate fault at this time. CLEAR Continuous Memory Code 34. Continuous Memory Code 34 testing complete.

Vehicle Speed Sensor — Pinpoint Test DP

TEST STEP	RESULT ▶	ACTION TO TAKE
DP1 DRIVE CYCLE FOR CHECKING VEHICLE SPEED SENSOR (VSS)		
• Record and clear EEC-IV Continuous Memory Codes. • Warm engine to operating temperature. • Perform the drive cycle below as appropriate for the vehicle being tested. **AUTOMATIC TRANSMISSION** **NOTE: On 2.3L EFI TC applications, idle the engine for a full 5 minutes then immediately begin the drive cycle.** Place the gear selector in LOW and moderately accelerate to 25 mph, then coast down to an idle and stop the vehicle. Shut engine off. **MANUAL TRANSMISSION** Starting in first gear, shift to second gear and moderately accelerate to 40 mph, then coast down to an idle and stop vehicle. Shut engine off. • Run Key On Engine Off Self-Test. • Is Code 29 (Code 27 for 2.3L EFI TC vehicles) present in Continuous Memory?	Yes ▶ No ▶	GO to DP2. Unable to duplicate fault at this time. If any other codes are present, return to Quick Test for directions. If codes are not present, test is completed.
DP2 CHECK VEHICLE SPEED SENSOR		
• Key off, wait 10 seconds. • Locate and disconnect Vehicle Speed Sensor. • DVOM on 200,000 ohm scale. • Measure resistance across Vehicle Speed Sensor. • Is resistance between 190 and 240 ohms?	Yes ▶ No ▶	GO to DP3. REPLACE sensor. REPEAT Test Step DP1.

Vehicle Speed Sensor — Pinpoint Test DP

Note

You should enter this Pinpoint Test only when Service Codes 27 or 29 are received in Quick Test Step 6.0.

Remember

This Pinpoint Test is intended to diagnose only the following:

- VSS Harness Circuits
- Vehicle Speed Sensor
- Processor Assembly

Pinpoint Test Schematic

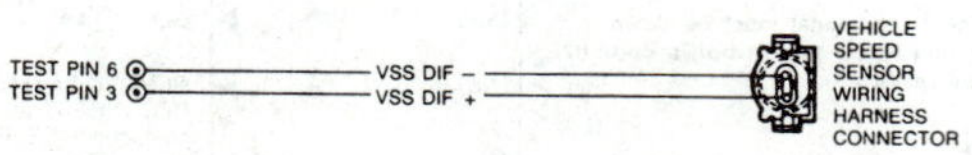

Vehicle Speed Sensor — Pinpoint Test DP

TEST STEP	RESULT ▶	ACTION TO TAKE
DP3 CHECK CONTINUITY OF VEHICLE SPEED SENSOR (VSS) HARNESS		
• Key off, wait 10 seconds. • Disconnect processor 60 pin connector. Inspect for damaged pins, corrosion, loose wires, etc. Service as necessary. • Install breakout box. • Processor and VSS disconnected. • DVOM on 200 ohm scale. • Measure resistance between Test Pin 3 at the breakout box and the VSS vehicle harness connector as shown below. • Measure resistance between Test Pin 6 at the breakout box and the VSS vehicle harness connector, as shown below. 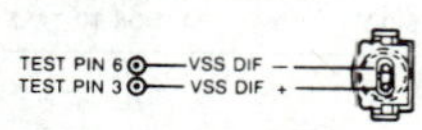 • Are both resistances less than 5 ohms?	Yes ▶ No ▶	GO to DP4. SERVICE open circuit(s). REPEAT Test Step DP1. REMOVE breakout box. RECONNECT processor and VSS.
DP4 CHECK VSS HARNESS FOR SHORTS TO POWER OR GROUND		
• Key off. • Processor disconnected. • VSS disconnected. • DVOM on 200,000 ohm scale. • Measure resistance between Test Pin 3 and Test Pins 37, 40 and 6 at the breakout box. • Measure resistance between Test Pin 6 and Test Pins 37 at the breakout box. • Are all resistances greater than 10,000 ohms?	Yes ▶ No ▶	REMOVE breakout box. RECONNECT processor. GO to DP5. REMOVE breakout box. RECONNECT processor and VSS. SERVICE short circuit(s). REPEAT Test Step DP1.

Vehicle Speed Sensor — Pinpoint Test DP

TEST STEP	RESULT ▶	ACTION TO TAKE
DP5 REPEAT DRIVE CYCLE WITH A KNOWN GOOD VSS INSTALLED		
• Substitute VSS with known good sensor. • Processor and VSS connected. • Perform Drive Cycle outlined in Test Step DP1 then return to this Step. • Is Code 29 (Code 27 for 2.3L EFI TC vehicles) present in continuous memory?	Yes ▶	REMOVE breakout box. REINSTALL original VSS. REPLACE processor. REPEAT Test Step DP1.
	No ▶	The original Continuous Memory Code 29 or 27 was the result of the original VSS. REPLACE VSS. RERUN Quick Test.

Neutral Drive Switch A/C Input — Pinpoint Test FA

Note

You should enter this Pinpoint Test only when a Service Code 67 or 79 is received in Quick Test Step 3.0, 5.0, 6.0, or you have been directed here from Diagnostic By Symptom in the Engine Supplement Section.

Remember

This Pinpoint Test is intended to diagnose only the following:

- A/C input to processor
- Neutral clutch switch
- Neutral drive switch
- Processor
- Harness circuits: NDS, NGS, ACC, ACCS and SIGNAL RETURN

TEST STEP	RESULT ▶	ACTION TO TAKE
FA1 CODE 67 SYSTEM IDENTIFICATION		
1.9L CFI M/T, 5.0L M/T TK		GO to FA9.
2.9L M/T TK, 3.0L M/T TK, 5.8L M/T TK, 5.0L M/T SEFI		GO to FA2.
1.9L EFI M/T, 2.3L EFI M/T Car and Truck		GO to FA2.
2.3L Turbo M/T		GO to FA6.
2.5L M/T, 4.9L M/T TK		GO to FA5.
All other systems		GO to FA7.

Neutral Drive Switch A/C Input — Pinpoint Test FA

FA2 CHECK NEUTRAL GEAR/CLUTCH INPUT

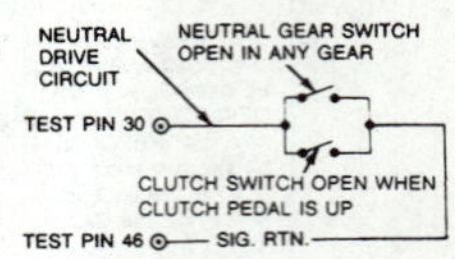

TEST STEP	RESULT ▶	ACTION TO TAKE
FA2 CHECK NEUTRAL GEAR/CLUTCH INPUT		
• Key off, wait 10 seconds. • Verify A/C is off, if so equipped. • Disconnect processor 60 pin connector. Inspect for damaged pins, corrosion, loose wires, etc. Service as necessary. • Install breakout box. • Connect processor. • DVOM on 200 ohm scale. • Measure resistance between Test Pin 30 and Test Pin 46. 1. With transmission in NEUTRAL and clutch up. 2. With transmission in GEAR and clutch down. • Are both resistances less than 5 ohms?	Yes ▶ No ▶	GO to FA9. GO to FA3.
FA3 CHECK NEUTRAL GEAR/CLUTCH SWITCH		
• Key off. • DVOM on 200 ohm scale. • Breakout box installed. • Locate Neutral Gear switch (on transmission) and Clutch switch (at clutch pedal linkage). • Disconnect vehicle harness at both switches and inspect connectors for pushed back pins. • Measure resistance across the Neutral Gear switch terminals with transmission in NEUTRAL and across the Clutch switch terminals with the clutch pedal down. • Are both resistances less than 5 ohms?	Yes ▶ No ▶	GO to FA4. REPLACE open switch(es). REMOVE breakout box. RECONNECT all components. RERUN Quick Test.

Neutral Drive Switch A/C Input — Pinpoint Test FA

TEST STEP	RESULT ▶	ACTION TO TAKE
FA4 CHECK NEUTRAL GEAR/CLUTCH HARNESS		
• Key off. • DVOM on 200 ohm scale. • Breakout box installed. • Vehicle harness disconnected at the Neutral Gear switch and Clutch switch. • Measure resistance between Test Pin 30 and the Neutral Gear switch harness connector and between Test Pin 30 and the Clutch switch harness connector. • Measure resistance between Test Pin 46 and the Neutral Gear switch harness connector and between Test Pin 46 and the Clutch switch harness connector. • Are all resistances less than 5 ohms?	Yes ▶ No ▶	GO to FA9. SERVICE open circuit. REMOVE breakout box. RECONNECT all components. RERUN Quick Test.
FA5 CHECK CLUTCH PEDAL SWITCH		
NOTE: The clutch pedal must be down during KOEO test; if not, a code 67 will result. • Key off. • Disconnect processor 60 pin connector. Inspect for damaged pins, corrosion, loose wires, etc. Service as necessary. • Install breakout box. • DVOM on 200 ohm scale. • Clutch pedal down. • Measure resistance between Test Pin 30 and Test Pin 46, between Test Pin 30 and Test Pin 40. • Are resistances less than 5 ohms?	Yes ▶ No ▶	GO to FA9. REMOVE breakout box. RECONNECT all components. SERVICE open circuit.

Neutral Drive Switch A/C Input — Pinpoint Test FA

TEST STEP	RESULT ▶	ACTION TO TAKE
FA6 CHECK NEUTRAL INPUT — 2.3L TC M/T		
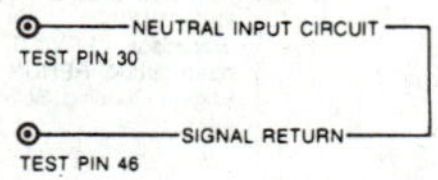 • Key off, wait 10 seconds. • Verify A/C is off, if so equipped. • Disconnect processor 60 pin connector. Inspect for damaged pins, corrosion, loose wires, etc. Service as necessary. • Install breakout box. • Leave processor disconnected. • DVOM on 200 ohm scale. • Measure resistance between Test Pin 30 and Test Pin 46 at the breakout box. • Is resistance less than 5 ohms?	Yes ▶ No ▶	GO to FA9. REMOVE breakout box. RECONNECT all components. SERVICE open circuit. RERUN Quick Test.

Neutral Drive Switch A/C Input — Pinpoint Test FA

TEST STEP	RESULT ▶	ACTION TO TAKE
FA9 CHECK A/C INPUT		
NOTE: Before entering this test, verify A/C is off. If A/C was on, rerun Quick Test. If code 67 or 79 is present, continue with this test. • Breakout box installed. • Disconnect processor. • Key on, engine off. • DVOM on 20 volt scale. • Measure voltage between Test Pin 10 at the breakout box and chassis ground. • Is voltage greater than 1.0 volt? 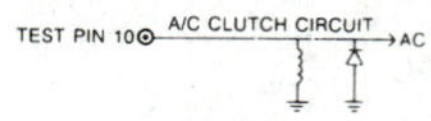	Yes ▶ No ▶	REMOVE breakout box. RECONNECT all components. SERVICE short to power in A/C clutch circuit. RERUN Quick Test. REMOVE breakout box. RECONNECT all components. REPLACE processor. RERUN Quick Test.

Neutral Drive Switch A/C Input — Pinpoint Test FA

TEST STEP	RESULT ▶	ACTION TO TAKE
FA7 CHECK NEUTRAL DRIVE INPUT		
• Key off, wait 10 seconds. • Verify heater control is in OFF position, if so equipped. • Verify transmission is in NEUTRAL or PARK. • Disconnect processor 60 pin connector. Inspect for damaged pins, corrosion, loose wires, etc. Service as necessary. • Install breakout box. • Connect processor to breakout box. • Key on, engine off. • DVOM on 20 volt scale. • Measure voltage between Test Pin 30 at the breakout box and chassis ground. • Is voltage less than 1.0 volt? 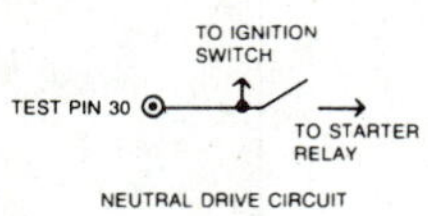	Yes ▶ No ▶	GO to FA9. Go to FA8.
FA8 CHECK NEUTRAL DRIVE SWITCH		
• Key off, wait 10 seconds. • Breakout box installed. • DVOM on 200 ohm scale. • Locate the Neutral Drive switch. • Disconnect vehicle harness from the Neutral Drive switch and measure resistance across the switch. • Is resistance less than 5 ohms?	Yes ▶ No ▶	REMOVE breakout box. RECONNECT all components. SERVICE open in vehicle harness Neutral Drive circuit. RERUN Quick Test. REMOVE breakout box. RECONNECT all components. REPLACE Neutral Drive switch. RERUN Quick Test.

Neutral Drive Switch A/C Input — Pinpoint Test FA

TEST STEP	RESULT ▶	ACTION TO TAKE
FA10 CHECK A/C INPUT CIRCUIT		
NOTE: A low idle with A/C on could be the result of the processor not receiving, or recognizing the A/C input on Pin 10. • Key off, wait 10 seconds. • Disconnect processor 60 pin connector. Inspect for damaged pins, corrosion, loose wires, etc. Service as necessary. • Install breakout box. • Leave processor disconnected. • DVOM on 20 volt scale. • Key on. • A/C on. • Measure voltage between Test Pin 10 and Test Pin 40. • Is voltage greater than 10.5 volts? 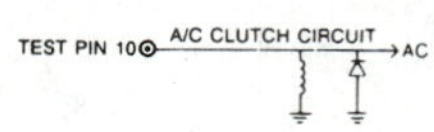	Yes ▶ No ▶	REMOVE breakout box. RECONNECT all components. REPLACE processor. RERUN Quick Test. REMOVE breakout box. RECONNECT all components. SERVICE open in A/C circuit. Refer to the appropriate engine schematic in RERUN Quick Test.
FA20 CHECK NDS CIRCUIT FOR SHORT TO GROUND OR CLOSED NEUTRAL DRIVE SWITCH		
• Key off, wait 10 seconds. • Disconnect processor 60 pin connector. Inspect for damaged pins, corrosion, loose wires, etc. Service as necessary. • Install breakout box. • Connect processor to breakout box. • Place transmission in DRIVE. • DVOM on 200,000 ohm scale. • Measure resistance between Test Pin 30 and Test Pin 40/60 at the breakout box. • Is resistance greater than 10,000 ohms?	Yes ▶ No ▶	GO for high idle. REMOVE breakout box. RECONNECT processor. SERVICE short circuit or closed neutral drive switch. RE-EVALUATE symptom.

Brake On/Off (BOO) — Pinpoint Test FD

Note

You should enter this Pinpoint Test only when a Service Code 74 or 75 is received in Quick Test Step 5.0.

Remember

To prevent the replacement of good components, be aware that the following non-EEC areas may be at fault:

- Brake lamp, Brake switch, and fuse.

This Pinpoint Test is intended to diagnose only the following:

- BOO circuit.
- Processor assembly.

Pinpoint Test Schematic

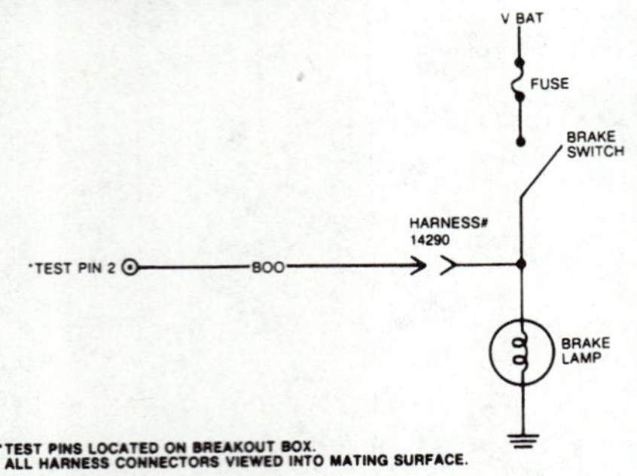

***TEST PINS LOCATED ON BREAKOUT BOX. ALL HARNESS CONNECTORS VIEWED INTO MATING SURFACE.**

Brake On/Off (BOO) — Pinpoint Test FD

TEST STEP	RESULT ▶	ACTION TO TAKE
FD1 SERVICE CODE 74		
• Did you press brake during the Engine Running Self-Test? **NOTE: On some vehicles it is necessary to depress and release the brake after the dynamic response code 1(0) but before the brief WOT.**	Yes ▶	GO to FD2.
	No ▶	RERUN Engine Running Self-Test. PRESS brake once during test.
FD2 CYCLE BOO CIRCUIT		
• Key off, wait 10 seconds. • Disconnect processor 60 pin connector. Inspect for damaged pins, corrosion, loose wires, etc. Service as necessary. • Install breakout box, leave processor disconnected. • DVOM on 20 volt scale. • Measure voltage between Test Pin 2 and Test Pin 40 at the breakout box while depressing and releasing brake. • Does the voltage cycle?	Yes ▶	REMOVE breakout box. REPLACE processor. RERUN Quick Test.
	No ▶	GO to FD3.
FD3 CHECK BOO CIRCUIT FOR SHORT TO GROUND		
• Key off. • Breakout box installed. • Processor disconnected. • DVOM on 200,000 Ohm scale. • Disconnect BOO circuit from 14290 harness (12 pin connector). • Measure resistance between Test Pin 2 and Test Pin 40 at the breakout box. • Is resistance greater than 10,000 ohms?	Yes ▶	GO to FD4.
	No ▶	REMOVE breakout box. SERVICE short circuit. RERUN Engine Running Self-Test.

Brake On/Off (BOO) — Pinpoint Test FD

TEST STEP	RESULT ▶	ACTION TO TAKE
FD4 CHECK CONTINUITY OF BOO CIRCUIT		
• Key off. • Breakout box installed. • Processor disconnected. • DVOM on 200 ohm scale. • Disconnect BOO circuit from 14290 harness (12 pin connector). • Measure resistance between Test Pin 2 at the breakout box and BOO circuit at the 14290 harness connector. • Is resistance greater than 5 ohms?	Yes ▶	REMOVE breakout box. RECONNECT processor. SERVICE open circuit. RERUN Engine Running Self-Test.
	No ▶	
FD5 SERVICE CODE 75: CYCLE BOO CIRCUIT		
• Key off, wait 10 seconds. • Disconnect processor 60 pin connector. Inspect for damaged pins, corrosion, loose wires, etc. Service as necessary. • Install breakout box, leave processor disconnected. • DVOM on 20 volt scale. • Measure voltage between Test Pin 2 and Test Pin 40 at the breakout box while depressing and releasing the brake. • Does the voltage cycle?	Yes ▶	REMOVE breakout box. REPLACE processor. RERUN Quick Test.
	No ▶	GO to FD6.

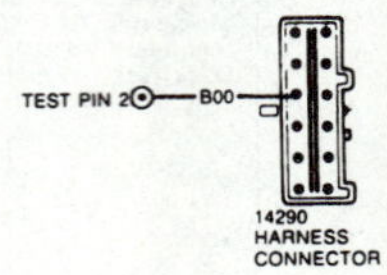

Brake On/Off (BOO) — Pinpoint Test FD

TEST STEP	RESULT ▶	ACTION TO TAKE
FD6 CHECK BOO CIRCUIT FOR SHORT TO POWER		
• Key off. • Breakout box installed. • Processor disconnected. • DVOM on 20 volt scale. • Disconnect BOO circuit from 14290 harness (12 pin connector). • Measure voltage between Test Pin 2 at the breakout box and engine block ground. • Is voltage greater than 10.5 volts?	Yes ▶	REMOVE breakout box. RECONNECT processor. SERVICE short circuit. RERUN Engine Running Self-Test.
	No ▶	BOO circuit OK.

Power Steering Pressure Switch (PSPS) — Pinpoint Test FF

Note

You should enter this Pinpoint Test only when a Service Code 52 is received in Quick Test Steps 3.0, 5.0 or if you are directed here from Diagnostic By Symptom in the Engine Supplement Section.

Remember

To prevent the replacement of good components, be aware that the following non-EEC areas may be at fault:

- Idle speeds/throttle stop adjustment.
- Binding throttle shaft/linkage or speed control linkage.

This Pinpoint Test is intended to diagnose only the following:

- Power steering pressure switch.
- Switch harness circuits: PSPS SIGNAL, and SIGNAL RETURN.
- Processor assembly.

Pinpoint Test Schematic

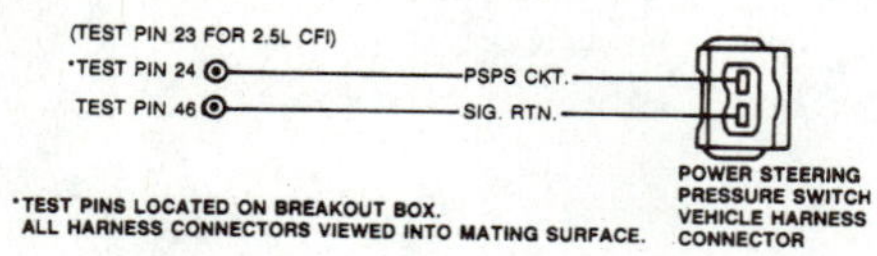

Power Steering Pressure Switch (PSPS) — Pinpoint Test FF

TEST STEP	RESULT	ACTION TO TAKE
FF1 ATTEMPT TO ELIMINATE CODE 52		
• Key off, wait 10 seconds. • Disconnect PSPS. • Jumper PSPS circuit to SIGNAL RETURN at vehicle harness connector. • Rerun Key On Engine Off Self-Test. • Is Code 52 still present?	Yes ▶ No ▶	GO to FF2. REPLACE PSPS. RERUN Quick Test.
FF2 PSPS HARNESS CHECK		
• Key off, wait 10 seconds. • PSPS disconnected. • Disconnect processor 60 pin connector. Inspect for damaged pins, corrosion, loose wires, etc. Service as necessary. • Install breakout box, leave processor disconnected. • DVOM on 200 ohm scale. • Measure resistance between Test Pin 46 at the breakout box and SIGNAL RETURN at the PSPS vehicle harness connector. • Measure resistance between Test Pin 24 (Test Pin 23 for 2.5L CFI) at the breakout box and PSPS circuit at the PSPS vehicle harness connector. • Are both readings less than 5 ohms?	Yes ▶ No ▶	REMOVE breakout box. RECONNECT all components. REPLACE processor. RERUN Quick Test. REMOVE breakout box. RECONNECT all components. SERVICE open circuit. RERUN Quick Test.
FF3 SWITCH INTEGRITY		
• Install tachometer. • Start engine, allow to idle in NEUTRAL/PARK. • Disconnect PSPS at switch. • Does rpm increase?	Yes ▶ No ▶	REPLACE PSPS. GO to FF4.

Power Steering Pressure Switch (PSPS) — Pinpoint Test FF

TEST STEP	RESULT	ACTION TO TAKE
FF4 PSPS HARNESS CHECK		
• Key off, wait 10 seconds. • PSPS disconnected. • Disconnect processor 60 pin connector. Inspect for damaged pins, corrosion, loose wires, etc. Service as necessary. • Install breakout box, leave processor disconnected. • DVOM on 200,000 ohm scale. • Measure resistance between Test Pin 24 (Test Pin 23 for 2.5L CFI) and Test Pin 46 at the breakout box. • Is resistance less than 10,000 ohms?	Yes ▶ No ▶	REMOVE breakout box. RECONNECT all components. SERVICE short in harness. RERUN Quick Test. REMOVE breakout box. RECONNECT all components. REPLACE processor. RERUN Quick Test.
FF5 SERVICE CODE 52 ENGINE RUNNING SELF-TEST		
• Did you turn the steering wheel at least one-half turn within 1 to 2 seconds after engine ID code? **NOTE: Make sure the front wheels are centered (no load condition).**	Yes ▶ No ▶	GO to FF6. RERUN Quick Test.
FF6 DETERMINE WHETHER THE PROCESSOR CAN IDENTIFY AN OPEN CIRCUIT		
• Key off, wait 10 seconds. • Disconnect PSPS. • Run Key On Engine Off Self-Test. • Is Code 52 present?	Yes ▶ No ▶	GO to FF8. GO to FF7.
FF7 PSPS HARNESS CHECK		
• Key off, wait 10 seconds. • PSPS disconnected. • Disconnect processor 60 pin connector. Inspect for damaged pins, corrosion or loose wires, etc. Service as necessary. • Install breakout box, leave processor disconnected. • DVOM on 200,000 ohm scale. • Measure resistance between Test Pin 46 and Test Pin 24 (Test Pin 23 for 2.5L CFI) at the breakout box. • Is resistance 10,000 ohms or less?	Yes ▶ No ▶	REMOVE breakout box. RECONNECT all components. SERVICE short circuit. RERUN Quick Test. REMOVE breakout box. RECONNECT all components. REPLACE processor. RERUN Quick Test.

Power Steering Pressure Switch (PSPS) — Pinpoint Test FF

TEST STEP	RESULT	ACTION TO TAKE
FF8 PSPS POSITION KEY ON ENGINE OFF VS. RUNNING		
• Key off, wait 10 seconds. • Connect PSPS. • Disconnect processor 60 pin connector. Inspect for damaged pins, corrosion, loose wires, etc. Service as necessary. • Processor connected. • Install breakout box, connect processor to breakout box. • DVOM on 200 ohm scale. • Key on. • Measure resistance between Test Pin 24 (Test Pin 23 for 2.5L CFI) and Test Pin 46 at the breakout box. • Start engine. • Does resistance remain less than 10 ohms between key on, engine off and engine running?	Yes ▶ No ▶	GO to FF9. GO to FF11.
FF9 PSPS POSITION ENGINE RUNNING NO LOAD VS. LOAD		
• Engine idling. • Breakout box installed. • Processor connected. • PSPS connected. • DVOM on 200 ohm scale. • Measure the resistance between Test Pin 24 (Test Pin 23 for 2.5L CFI) and Test Pin 46 at the breakout box. • Turn the steering wheel at least one-half turn then return. • Does resistance change from less than 10 ohms to infinity (indicating PSPS opening), then returning to 10 ohms or less when steering wheel is returned to center position?	Yes ▶ No ▶	PSPS system OK, REMOVE breakout box and RETURN to Quick Test Step 5.0 to continue Diagnostics. GO to FF10.

Power Steering Pressure Switch (PSPS) — Pinpoint Test FF

TEST STEP	RESULT ▶	ACTION TO TAKE
FF10 PSPS ALWAYS CLOSED VS. POWER STEERING HYDRAULIC PRESSURE WITH ENGINE RUNNING		
• At this point in the Diagnostics there are only two possible causes for the original Code 52 Engine Running: — PSPS (switch) that will not open. — Low available hydraulic pressure. • Key off, wait 10 seconds. • Substitute original PSPS with a known good PSPS. • Run Engine Running Self-Test. (Turn steering wheel at least one-half turn after engine ID code.) • Is Code 52 still present?	Yes ▶ No ▶	GO to Power Steering Pressure Diagnostics, looking for low pressure. Original Code 52 Engine Running was a result of a bad PSPS (switch). REMOVE all equipment and continue, if necessary, with any other Diagnostics.
FF11 PSPS ALWAYS OPEN VS. POWER STEERING HYDRAULIC PRESSURE WITH ENGINE RUNNING		
• At this point in the Diagnostics there are two possible causes for the original Code 52 Engine Running: — PSPS (switch) that always remains open during Engine Running. — Excessively high hydraulic pressure. • Key Off, wait 10 seconds. • Substitute original PSPS with a known good PSPS. • Run Engine Running Self Test. (Turn steering wheel at least one-half turn after engine ID code.) • Is Code 52 still present?	Yes ▶ No ▶	GO to Power Steering Pressure Diagnostics looking for high pressure. Original Code 52 Engine Running was a result of a bad PSPS (switch). REMOVE all equipment and continue, if necessary, with any other Diagnostics.

Fuel Control — Pinpoint Test H

Note

You should enter this Pinpoint Test only when a Service Code 41, 91 or 42, 92 or 43, 65, 85, 86 is received in Quick Test Step 5.0 or 6.0 or when directed here from Pinpoint Test S or Diagnostic By Symptom in the Engine Supplement Section.

Remember

To prevent the replacement of good components, be aware that the following non-EEC areas may be at fault:

- Ignition Coil
- Distributor Cap
- Distributor Rotor
- Fouled Spark Plugs
- Spark Plug Wires
- CANP Problems
- PCV Valves (see note below)
- EGR Valve and Gasket
- Air Filter
- Fuel Contamination, Engine Oil
- Poor Power Ground
- Fuel Pressure
- Manifold Leaks, Intake/Exhaust
- Engine Not at Normal Operating Temperatures

This Pinpoint Test is intended to diagnose only the following:

- HEGO Sensor
- HEGO Signal and Ground Circuit
- HEGO Sensor Connection
- Vacuum Systems
- Fuel Injector
- Processor Assembly
- Harness Circuits HEGO GRD, HEGO, INJ. 1–8, and VPWR

NOTE: Fuel contaminated engine oil may affect 41, 91 and 42, 92 Service Codes, so if it is suspected, remove the PCV from the valve cover, and rerun Quick Test. If the problem is corrected, then change the engine oil and filter.

Fuel Control — Pinpoint Test H

Pinpoint Test Schematic

All CFI

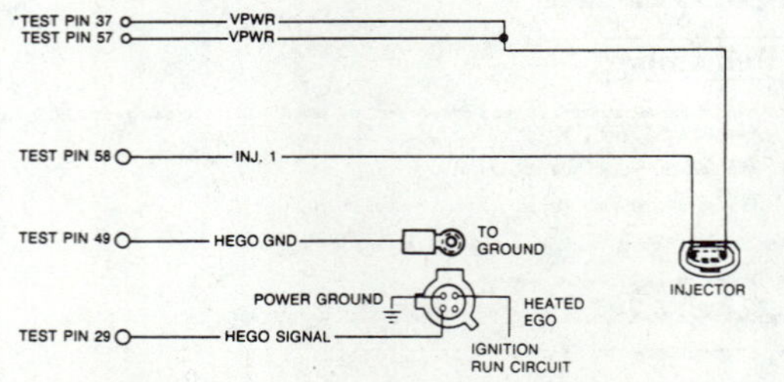

***TEST PINS LOCATED ON BREAKOUT BOX. ALL HARNESS CONNECTORS VIEWED INTO MATING SURFACE.**

Fuel Control — Pinpoint Test H

Pinpoint Test Schematic

All EFI

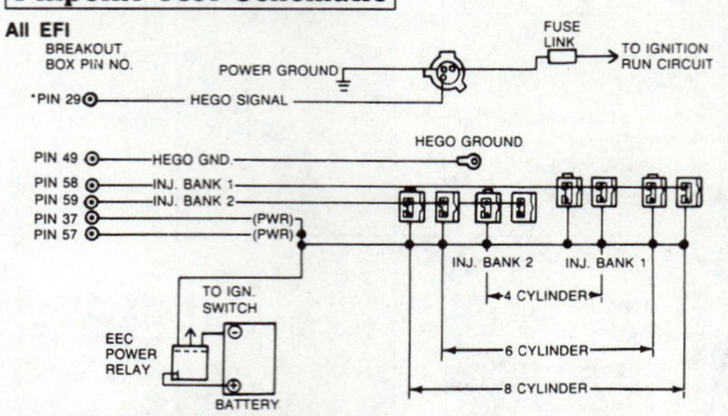

5.0L SEFI and 5.0 SEFI Mass Air

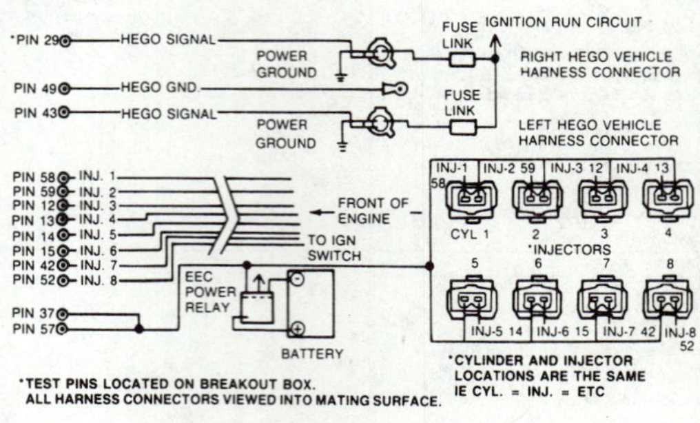

*TEST PINS LOCATED ON BREAKOUT BOX.
ALL HARNESS CONNECTORS VIEWED INTO MATING SURFACE.

Fuel Control — Pinpoint Test H

FUEL PRESSURE SPECIFICATION TABLE

PASSENGER CAR ENGINES

	1.9L EFI	1.9L CFI	2.3L OHC EFI	2.3L TC EFI	2.3L HSC EFI	2.5L CFI	3.0L EFI	3.8L FWD EFI	3.8L RWD EFI	5.0L SEFI	5.0L MA SEFI
ENGINE RUNNING	25 – 35 PSI 172 – 241 kPa	13 – 16 PSI 89 – 110 kPa	25 – 35 PSI 172 – 241 kPa	25 – 35 PSI 172 – 241 kPa	35 – 55 PSI 241 – 379 kPa	13 – 16 PSI 89 – 110 kPa	28 – 38 PSI 193 – 262 kPa	35 – 45 PSI 241 – 310 kPa	25 – 35 PSI 172 – 241 kPa	27 – 37 PSI 186 – 255 kPa	27 – 37 PSI 186 – 255 kPa
KEY ON ENGINE OFF	35 – 45 PSI 241 – 310 kPa	13 – 16 PSI 89 – 110 kPa	35 – 45 PSI 241 – 310 kPa	35 – 45 PSI 241 – 310 kPa	35 – 45 PSI 241 – 310 kPa	13 – 16 PSI 89 – 110 kPa	35 – 45 PSI 241 – 310 kPa	35 – 45 PSI 241 – 310 kPa	35 – 45 PSI 241 – 310 kPa	35 – 45 PSI 241 – 310 kPa	35 – 45 PSI 241 – 310 kPa

LIGHT TRUCK ENGINES

	2.3L EFI	2.9L EFI	3.0L EFI	4.9L EFI	5.0L EFI	5.8L EFI	7.5L EFI
ENGINE RUNNING	25 – 35 PSI 172 – 241 kPa	28 – 38 PSI 193 – 262 kPa	28 – 38 PSI 193 – 262 kPa	42 – 52 PSI 289 – 358 kPa	28 – 38 PSI 193 – 262 kPa	28 – 38 PSI 193 – 262 kPa	28 – 38 PSI 193 – 262 kPa
KEY ON ENGINE OFF	35 – 45 PSI 241 – 310 kPa	35 – 45 PSI 241 – 310 kPa	35 – 45 PSI 241 – 310 kPa	50 – 60 PSI 345 – 413 kPa	35 – 45 PSI 241 – 310 kPa	35 – 45 PSI 241 – 310 kPa	35 – 45 PSI 241 – 310 kPa

Fuel Control — Pinpoint Test H

SINGLE INJECTOR RESISTANCE SPECIFICATION TABLE #1

PASSENGER CAR ENGINES

VALUES ARE IN OHMS

1.9L EFI	1.9L CFI	2.3L OHC EFI	2.3L TC EFI	2.3L HSC EFI	2.5L CFI	3.0L EFI	3.8L FWD EFI	3.8L RWD EFI	5.0L SEFI	5.0L MA SEFI
1.2 TO 1.8	1.0 TO 2.0	7.0 TO 9.5	1.2 TO 1.8	7.0 TO 9.5	1.0 TO 2.0	5.0 TO 6.5	4.0 TO 6.0	4.0 TO 6.0	13.5 TO 19.0	1.5 TO 19.0

LIGHT TRUCK ENGINES

VALUES ARE IN OHMS

2.3L EFI	2.9L EFI	3.0L EFI	4.9L EFI	5.0L EFI	5.8L EFI	7.5L EFI
7.0 TO 9.5	5.0 TO 6.5	5.0 TO 6.5	5.0 TO 6.5	3.5 TO 5.0	2.5 TO 5.0	2.5 TO 5.0

Fuel Control — Pinpoint Test H

TEST STEP	RESULT ▶	ACTION TO TAKE
H1 CHECK FUEL PRESSURE		
• Key off, wait 10 seconds. • Install fuel pressure gauge. • Verify that manifold vacuum is connected to the fuel pressure regulator if applicable. • Start and run engine at idle. • Refer to Fuel Pressure Specification Table. • Is fuel pressure within specification for the engine being tested? **FOR NO STARTS:** • If engine will not run, cycle the key off to on several times. • Refer to Fuel Pressure Specification Table. • Is fuel pressure within specification for the engine being tested?	Yes ▶ No ▶	GO to H2. REMOVE fuel pressure gauge.
H2 CHECK SYSTEM'S ABILITY TO HOLD FUEL PRESSURE		
• Key on, engine off. • Does fuel pressure remain at specification for 60 seconds?	Yes ▶ No ▶	GO to H3. For 5.0L SEFI GO to H9. All others GO to H6.
H3 FUEL DELIVERY TEST		
NOTE: Verify fuel quality; air and/or water will also pressurize and look like acceptable fuel delivery. • Key off. • Fuel pressure gauge installed. • Pressurize fuel system per step H1. • Locate and disconnect the inertia switch. • Crank engine for 5 seconds. • Does pressure drop greater than 5 psi. (34 kPa.) by the end of the 5 second crank cycle?	Yes ▶ No ▶	The EEC-IV system is not the cause of the No Start. REMOVE the fuel pressure gauge. RECONNECT the inertia switch. If the complaint was runs rough, misses or a fuel service code GO to H4. For 5.0L SEFI GO to H9. REMOVE fuel pressure gauge. RECONNECT inertia switch. GO to H4.

Fuel Control — Pinpoint Test H

TEST STEP	RESULT ▶	ACTION TO TAKE
H4 CHECK RESISTANCE OF INJECTOR(S) AND HARNESS		
• Key off, wait 10 seconds. • Disconnect processor 60 pin connector. Inspect for damaged pins, corrosion, loose wires, etc. Service as necessary. • Install breakout box, leave processor disconnected. • DVOM on 200 ohm scale. **For EFI:** — Measure resistance of INJECTOR BANK 1 between Test Pin 37 and Test Pin 58 at the breakout box. Record resistance. — Measure resistance of INJECTOR BANK 2 between Test Pin 37 and Test Pin 59 at the breakout box. Record resistance. **For SEFI:** — From cylinder balance test: Measure resistance between the suspect INJECTOR circuit Test Pin and Test Pin 37 at the breakout box. Record resistance. — For No Starts: Pick any injector and measure resistance between that INJECTOR circuit's Test Pin and Test Pin 37 at the breakout box. Record resistance. **For CFI:** — Measure resistance of INJECTOR circuit between Test Pin 37 and Test Pin 58 at the breakout box. Record resistance. • Refer to Injector Resistance Specification Table #1. • Is/are resistance(s) within specification for the appropriate engine?	Yes ▶ No ▶	GO to H6. **For EFI** GO to H5. **For SEFI:** REMOVE breakout box. RECONNECT processor. SERVICE open or short in harness/connector of suspect injector. If OK, REPLACE injector RERUN Quick Test and Cylinder Balance Test. For No Start: SERVICE open in VPWR circuit. **For CFI:** REMOVE breakout box. RECONNECT processor. SERVICE open or short in harness/connector If OK, REPLACE injector RERUN Quick Test.

Fuel Control — Pinpoint Test H

INJECTOR BANK RESISTANCE SPECIFICATION TABLE #2

PASSENGER CAR ENGINES — VALUES ARE IN OHMS

1.9L EFI	1.9L CFI	2.3L OHC EFI	2.3L TC EFI	2.3L HSC EFI	2.5L CFI	3.0L EFI	3.8L FWD EFI	3.8L RWD EFI
2.0 TO 2.7	1.0 TO 2.0	15.0 TO 19.0	2.0 TO 3.0	13.5 TO 16.0	1.0 TO 2.0	15.0 TO 18.0	13.5 TO 16.0	13.5 TO 16.0

LIGHT TRUCK ENGINES — VALUES ARE IN OHMS

2.3L EFI	2.9L EFI	3.0L EFI	4.9L EFI	5.0L EFI	5.8L EFI	7.5L EFI
13.5 TO 18.0	13.5 TO 18.0	15.0 TO 18.0	13.5 TO 18.0	13.5 TO 18.0	13.5 TO 18.0	13.5 TO 19.0

TEST STEP	RESULT ▶	ACTION TO TAKE
H5 ISOLATE FAULTY INJECTOR CIRCUIT		
• Key off. • Breakout box installed, processor disconnected. • Disconnect all injectors on suspect bank. • DVOM on 200 ohm scale. • Connect one injector and measure resistance between Test Pin 37 and either Test Pin 58 or 59 as appropriate. • Disconnect that injector and repeat process for each of the remaining injectors. • Refer to Injector Resistance Specification Table #2. • Is/are resistance(s) within specification for the appropriate engine?	Yes ▶	GO to H6.
	No ▶	REMOVE breakout box. RECONNECT processor and injectors. SERVICE open/short circuit in injector harness. If OK REPLACE injector. RERUN Quick Test.

Fuel Control — Pinpoint Test H

TEST STEP	RESULT ▶	ACTION TO TAKE
H6 CHECK INJECTOR DRIVER SIGNAL		
Requires standard non-powered 12 volt test lamp. • Key off. • Breakout box installed. • Connect processor to breakout box. **For EFI:** — Connect test lamp between Test Pin 37 and Test Pin 58 at the breakout box. — Connect test lamp between Test Pin 37 and 59 at the breakout box. **For SEFI:** — Connect test lamp between Test Pin 37 and the suspect injectors Test Pin at the breakout box. **For CFI:** — Connect test lamp between Test Pin 37 and Test Pin 58 at the breakout box. • Crank or start engine. • Is glow on lamp dim? **NOTE: Properly operating systems will show a dim glow on the lamp.**	Yes ▶	**For 5.0L SEFI:** REMOVE breakout box. RECONNECT processor. After any servicing, RERUN Quick Test and Cylinder Balance Test. **For all other engines:** GO To H7.
	No ▶	**NO LIGHT:** VERIFY 12 volt battery at Test Pins 37 and 57. **BRIGHT LIGHT:** CHECK injector circuit for shorts to ground. If OK, REMOVE breakout box. REPLACE processor RERUN Quick Test.

Fuel Control — Pinpoint Test H

TEST STEP	RESULT ▶	ACTION TO TAKE
H7 CHECK EXTERNAL SOURCE FOR FUEL PRESSURE PROBLEM		
• Key off. • Pressurize fuel system per Test Step H1. **For EFI:** — Visually look for fuel leaking at fuel injector O-rings, fuel pressure regulator, and fuel rails. **For CFI:** — Remove air inlet tube at the fuel charging assembly. — Visually look for fuel leaking at the air horn inlet, fuel injector O-ring, fuel pressure regulator and fuel line to fuel charging assembly. • Is there a visible leak?	Yes ▶	REMOVE pressure gauge. SERVICE as necessary. After servicing leak, RERUN Quick Test.
	No ▶	**For EFI:** GO to H8. **For CFI:** REMOVE pressure gauge. Fuel delivery system is OK. Problem is in an area common to all cylinders, i.e. air/vacuum leak, fuel contamination, EGR, etc.
H8 INJECTOR BALANCE TEST		
• Connect tachometer to engine. Run engine at idle. • Disconnect and reconnect the injectors one at a time: Note rpm drop for each injector. • Does each injector produce at least a 100 rpm momentary drop? **NOTE: ISC will attempt to re-establish rpm.**	Yes ▶	Fuel delivery OK. Problem is in an area common to all cylinders i.e. air/vacuum leak, fuel contamination, EGR etc.
	No ▶	GO to injector testing and cleaning instructions. After any servicing, RERUN Quick Test.

Fuel Control — Pinpoint Test H

TEST STEP	RESULT ▶	ACTION TO TAKE
H9 CYLINDER BALANCE TEST: SEFI ENGINES ONLY		
• Run the Engine Running Self-Test. • After the last repeated code, wait 5-10 seconds. • "Goose" throttle very lightly (not wide-open-throttle). • Time of test is approximately 90 seconds. • Use table below to interpret codes received from cylinder balance test. • Is Code 90 present? Refer to appendix in the Quick Test Section for detailed information about cylinder balance test.	Yes ▶	GO to H10.
	No ▶	GO to H4.

The Cylinder Balance Test switches each injector OFF and ON one at a time. Service codes correspond to the cylinder number, e.g. Service Code 30 indicates a problem with cylinder No. 3. The Cylinder Balance Test is designed to aid in the detection of a non-contributing cylinder. The cylinder balance Pinpoint Test Steps H9 and H10 are designed to isolate only EEC-IV related problems.

SERVICE CODE	90	10	20	30	40	50	60	70	80	77*
CYLINDER/INJECTOR NUMBER	PASS	1	2	3	4	5	6	7	8	RERUN TEST
BREAKOUT BOX PIN NUMBER		58	59	12	13	14	15	42	52	

* If throttle is touched (moved) during Cylinder Balance Test, Service Code 77 will appear, indicating test was not completed.

Fuel Control — Pinpoint Test H

TEST STEP	RESULT ▶	ACTION TO TAKE
H10 PERFORM 2ND/3RD LEVEL CYLINDER BALANCE TEST		
NOTE: A Cylinder Balance Test Service Code 90 received in the first test level indicates that the injector harness is not open or shorted and the processor is sending a drive signal to all injectors. The 2nd/3rd level Cylinder Balance Test is intended to aid in the detection of any partially contributing injectors.	Yes ▶	For 5.0L SEFI Mass Air vehicles with Service Code 41/91 GO to H11. For Service Code 42/92 GO to H23.
• Within 2 minutes after the previous Cylinder Balance Test, lightly depress and release the throttle to enter 2nd/3rd level Cylinder Balance Test. • Is Code 90 present in all levels?	No ▶	GO to H4.
H11 SERVICE CODE 41/91: FUEL CONTROL ALWAYS LEAN		
NOTE: For 5.0L SEFI and 5.0L SEFI Mass Air Code 41 refers to right HEGO; Code 91 refers to left HEGO. • Run engine at 2000 rpm for 2 minutes. • Key off, wait 10 seconds. • Rerun Engine Running Self-Test. • Is Code 41/91 present?	Yes ▶	**For engines with:** — MAP sensors GO to H12. — Vane Air Meters GO to H13. — Mass Air Meters GO to H14.
	No ▶	GO to H20.

Fuel Control — Pinpoint Test H

TEST STEP	RESULT ▶	ACTION TO TAKE
H13 CHECK HEGO SENSOR ON ENGINES WITH VANE AIR METER		
NOTE: Vacuum/air leaks in non-EEC-IV areas could also cause Code 41. Check for: **— Leaking vacuum actuator (e.g. A/C control motor)** **— Engine sealing** **— EGR system** **— PCV system** **— Lead contaminated HEGO sensor** **— Unmetered air leak between Air Meter and throttle body** • Key off. • Disconnect HEGO sensor from vehicle harness • Remove air cleaner to gain access to air meter inlet. Using a standard wood lead pencil, prop the air meter door partway open. • Connect DVOM to HEGO SIGNAL at the sensor and battery negative post. • DVOM on 20 volt scale. • Start the engine and run at approximately 2000 rpm for 2 minutes. • Does the DVOM indicate greater than 0.5 volts within 1 minute?	Yes ▶ No ▶	GO to H15. REMOVE pencil from Air Meter. REINSTALL air cleaner. REPLACE HEGO sensor. RERUN Quick Test.

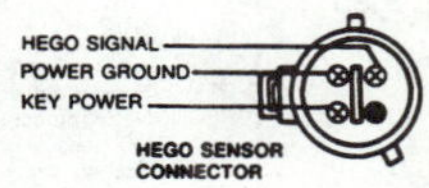

Fuel Control — Pinpoint Test H

TEST STEP	RESULT ▶	ACTION TO TAKE
H12 CHECK HEGO SENSOR ON ENGINES WITH MAP SENSORS		
NOTE: Vacuum/air leaks in non-EEC-IV areas could also cause Code 41/91. Check for: **— Leaking vacuum actuator (e.g. A/C control motor)** **— Engine sealing** **— EGR system** **— PCV system** **— Lead contaminated HEGO sensor** • Key off. • Disconnect appropriate HEGO sensor from vehicle harness. • Connect DVOM to HEGO SIGNAL at the sensor and battery negative post. • Disconnect and plug vacuum line at MAP sensor. • DVOM on 20 volt scale. • Apply 10-14 in. Hg. (33-46 kPa) to MAP sensor. • Start engine. • Does the DVOM indicate greater than 0.5 volts within 1 minute?	Yes ▶ No ▶	GO to H15. RECONNECT MAP sensor vacuum line. REPLACE HEGO sensor. RERUN Quick Test.

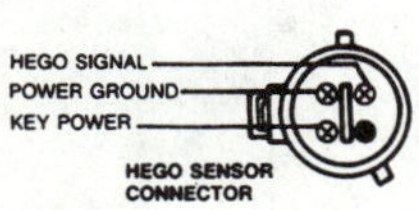

Fuel Control — Pinpoint Test H

TEST STEP	RESULT ▶	ACTION TO TAKE
H14 CHECK HEGO SENSOR ON ENGINES WITH MASS AIR SENSOR		
NOTE: Vacuum/air leaks in non-EEC-IV areas could also cause Code 41/91. Check for: **— Leaking vacuum actuator (e.g. A/C control motor)** **— Engine sealing** **— EGR system** **— PCV system** **— Unmetered air leak between Mass Air Flow sensor and throttle body** **— Lead contaminated HEGO sensor** • Key off. • Disconnect appropriate HEGO sensor from vehicle harness. • Connect DVOM to HEGO SIGNAL at the sensor and battery negative post. • DVOM on 20 volt scale. • Rerun Engine Running Self-Test and monitor HEGO sensor voltage. **NOTE: The purpose of this test is to verify the HEGO sensor can generate greater than 0.5 volts during Engine Running Self-Test.** • Is the voltage greater than 0.5 volts at the end of Self-Test?	Yes ▶ No ▶	GO to H15. REPLACE HEGO sensor. RERUN Quick Test.

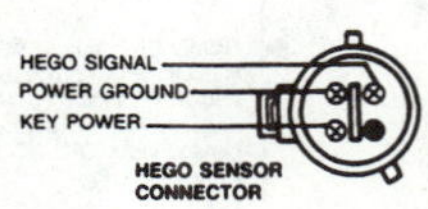

Fuel Control — Pinpoint Test H

TEST STEP	RESULT ▶	ACTION TO TAKE
H15 CHECK CONTINUITY OF HEGO SIGNAL AND HEGO GROUND CIRCUITS		
• Key off. • Disconnect processor 60 pin connector. Inspect for damaged pins, corrosion, loose wires, etc. Service as necessary. • Install breakout box, leave processor disconnected. • HEGO disconnected. • DVOM on 200 ohm scale. • Measure resistance between Test Pin 49 at the breakout box and battery negative post. • Measure resistance between Test Pin 29 at the breakout box and HEGO SIGNAL at the vehicle harness connector. • For vehicles with dual HEGO, also measure resistance between Test Pin 43 at the breakout box and HEGO SIGNAL at the vehicle harness connector. • Are all resistances less than 5.0 ohms?	Yes ▶	GO to H16.
	No ▶	REMOVE breakout box. RECONNECT processor, HEGO sensor, and any other components that are disconnected or removed. SERVICE open circuit. RERUN Quick Test.

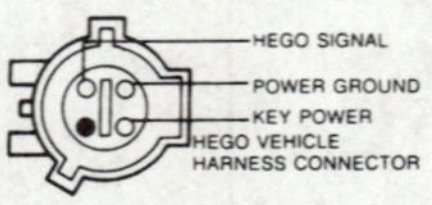

TEST STEP	RESULT ▶	ACTION TO TAKE
H16 CHECK HEGO CIRCUIT FOR SHORT TO GROUND		
• Key off. • Breakout box installed, processor disconnected. • HEGO disconnected. • DVOM on 200,000 ohm scale. • Measure resistance between Test Pin 29 and Test Pin 40 at the breakout box. • For vehicles with dual HEGO also measure resistance between Test Pin 43 and Test Pin 40 at the breakout box. • Is resistance greater than 10,000 ohms?	Yes ▶	GO to H17.
	No ▶	REMOVE breakout box. RECONNECT processor and HEGO sensor. SERVICE short circuit. RERUN Quick Test.

Fuel Control — Pinpoint Test H

TEST STEP	RESULT ▶	ACTION TO TAKE
H17 CHECK HEGO SENSOR FOR SHORT TO GROUND		
• Key off. • Breakout box installed, processor disconnected. • HEGO disconnected. • DVOM on 200,000 ohm scale. • Measure resistance between HEGO PWR GND and HEGO SIGNAL at the HEGO sensor connector. • Is resistance greater than 10,000 ohms?	Yes ▶	**For engines with:** — MAP sensor GO to H18. — Vane Air Meter GO to H19. — Mass Air Meter REMOVE breakout box. RECONNECT HEGO sensor. REPLACE processor. RERUN Quick Test.
	No ▶	REMOVE breakout box. RECONNECT processor. REPLACE HEGO sensor. RERUN Quick Test.

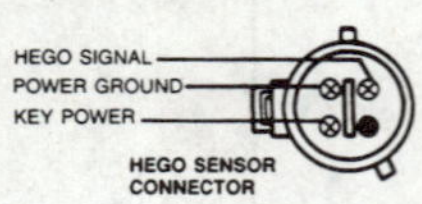

TEST STEP	RESULT ▶	ACTION TO TAKE
H18 ATTEMPT TO ELIMINATE CODE 41 ON ENGINES WITH MAP SENSOR		
• Key off. • Breakout box installed. • MAP vacuum line disconnected and plugged. • Connect processor to breakout box. • Reconnect HEGO sensor. • Apply 10-14 in. Hg. (3-46 kPa) vacuum to MAP sensor. • Start engine and run at approximately 2000 rpm for 2 minutes. Allow engine to return to idle. • Rerun Engine Running Self-Test. • Is Code 41 still present? **NOTE: Disregard other codes received at this time.**	Yes ▶	REMOVE breakout box. RECONNECT MAP sensor vacuum line. REPLACE processor. RERUN Quick Test.
	No ▶	REMOVE breakout box. RECONNECT processor and MAP sensor vacuum line. HEGO sensor input OK. GO to H1.

Fuel Control — Pinpoint Test H

TEST STEP	RESULT ▶	ACTION TO TAKE
H19 ATTEMPT TO ELIMINATE CODE 41 ON ENGINES WITH VANE AIR METER		
• Key off. • Breakout box installed. • Connect processor to breakout box. • Reconnect HEGO sensor. • Air cleaner removed, pencil inserted in vane meter inlet. • Start engine and run at approximately 2000 rpm for 2 minutes. • Rerun Engine Running Self-Test. • Is Code 41 present?	Yes ▶	REMOVE breakout box. REMOVE pencil from vane meter. REINSTALL air cleaner. REPLACE processor. RERUN Quick Test.
	No ▶	REMOVE breakout box. RECONNECT processor. HEGO input circuit OK. GO to H1.
H20 CHECK RESISTANCE OF HEATER ELEMENT ON HEGO		
• Key off. • Disconnect HEGO. • DVOM on 200 ohm scale. • Measure resistance between IGNITION RUN circuit and PWR GND circuit at HEGO sensor connector. • Hot to warm resistance specification is 5.0 to 20.0 ohms. • Is resistance within specification? **NOTE: Room temperature resistance specification is 2.0 to 5.0 ohms.**	Yes ▶	GO to H21.
	No ▶	REPLACE HEGO sensor. RERUN Quick Test.

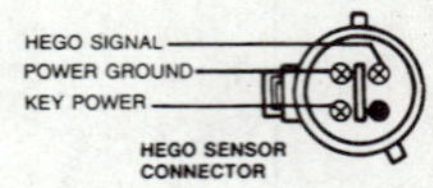

Fuel Control — Pinpoint Test H

TEST STEP	RESULT ▶	ACTION TO TAKE
H21 CHECK FOR POWER AT HEGO HARNESS CONNECTOR		
• Key on, engine off. • HEGO disconnected. • DVOM on 20 volt scale. • Measure voltage between IGNITION RUN circuit and PWR GND circuit at the HEGO vehicle harness connector. • Is voltage greater than 10.5 volts?	Yes ▶	RECONNECT HEGO sensor. HEGO sensor system OK. GO to H1.
	No ▶	GO to H22.

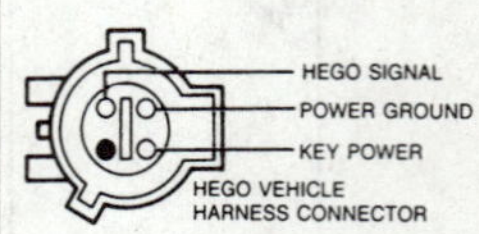

TEST STEP	RESULT ▶	ACTION TO TAKE
H22 CHECK CONTINUITY OF POWER GROUND CIRCUIT		
• Key off, wait 10 seconds. • HEGO disconnected. • DVOM on 200 ohm scale. • Measure resistance between PWR GND circuit at the HEGO vehicle harness connector and battery negative post. • Is resistance less than 5.0 ohms?	Yes ▶	RECONNECT HEGO sensor. SERVICE open in IGNITION RUN circuit. RERUN Quick Test.
	No ▶	RECONNECT HEGO sensor. SERVICE open in PWR GND circuit. RERUN Quick Test.

Fuel Control — Pinpoint Test H

TEST STEP	RESULT ▶	ACTION TO TAKE
H23 SERVICE CODE 42/92: FUEL CONTROL ALWAYS RICH: CHECK HEGO SIGNAL FOR SHORT TO POWER		
NOTE: For 5.0L SEFI and 5.0L SEFI Mass Air Code 42 refers to Right HEGO; Code 92 refers to Left HEGO. • Key off, wait 10 seconds. • Disconnect the appropriate HEGO sensor for Code 42/92. • DVOM on 20 volt scale. • Key on, engine off. • Measure voltage between HEGO SIGNAL and PWR GND at the HEGO vehicle harness connector. • Is voltage less than 0.5 volts? 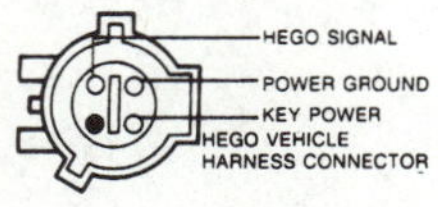	Yes ▶ No ▶	GO to H24. RECONNECT HEGO sensor. SERVICE HEGO circuit short to power. RERUN Quick Test.
H24 CHECK HEGO SENSOR FOR SHORT TO IGNITION RUN CIRCUIT		
• Key off. • HEGO disconnected. • DVOM on 200,000 ohm scale. • Measure resistance between IGNITION RUN circuit and HEGO SIGNAL circuit at the HEGO sensor connector. • Is resistance greater than 10,000 ohms? 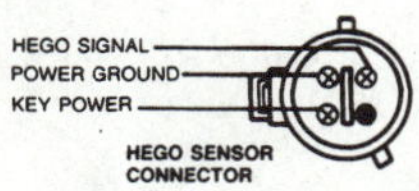	Yes ▶ No ▶	GO to H25. REPLACE HEGO sensor. RERUN Quick Test.

Fuel Control — Pinpoint Test H

TEST STEP	RESULT ▶	ACTION TO TAKE
H27 CHECK FOR LOSS OF VACUUM TO MAP SENSOR		
• Tee a vacuum gauge into the manifold vacuum line at the MAP sensor. • Start the engine and let rpm stabilize. Note vacuum level. • Key off, wait 10 seconds. • REMOVE vacuum gauge and tee and reconnect vacuum line to MAP sensor. • Tee in vacuum gauge at a different source of intake manifold vacuum and restart the engine. Note vacuum level. • Does the vacuum level differ greater than 1 in. Hg.?	Yes ▶ No ▶	REMOVE vacuum gauge and tee. RECONNECT HEGO sensor. INSPECT vacuum lines for leaks, holes, disconnections, kinks, blockages, and proper routing. SERVICE as necessary. RERUN Quick Test. GO to H28.

Fuel Control — Pinpoint Test H

TEST STEP	RESULT ▶	ACTION TO TAKE
H25 ATTEMPT TO GENERATE CODE 41/91		
• Key off, wait 10 seconds. **NOTE: Non-EEC areas could cause a Service Code 42/92. Check for:** **— Fuel contaminated engine oil** **— Ignition caused misfire (fouled spark plug)** **— CANP problems** • HEGO disconnected. • Jumper HEGO SIGNAL circuit at the HEGO vehicle harness connector to battery negative post. • Rerun Engine Running Self-Test. • Is Code 41/91 present?	Yes ▶ No ▶	REMOVE jumper. For engines with MAP sensor GO to H26. All others GO to H28. REMOVE jumper. RECONNECT HEGO sensor. DISCONNECT processor 60 pin connector. Inspect for damaged pins, corrosion, loose wires, etc. SERVICE as necessary. If OK REPLACE processor. RERUN Quick Test.
H26 CHECK MAP SENSOR FOR VACUUM LEAK		
NOTE: Due to the MAP sensor's large influence on fuel control, there is a possibility that a Code 42/92 could be a result of a MAP problem, even though a Code 22 is not present. Therefore the next two Test Steps will verify proper vacuum to the MAP sensor and its ability to hold vacuum. • Key off, wait 10 seconds. • Disconnect vacuum line from MAP sensor. • Connect a vacuum pump to the MAP sensor and apply 18 in. Hg. (60 kPa) vacuum to MAP sensor. • Does MAP sensor hold vacuum?	Yes ▶ No ▶	RELEASE vacuum. GO to H27. REMOVE vacuum pump. RECONNECT HEGO sensor. REPLACE MAP sensor. RERUN Quick Test.

Fuel Control — Pinpoint Test H

TEST STEP	RESULT ▶	ACTION TO TAKE
H28 HEGO SENSOR CHECK		
• Key off, wait 10 seconds. • HEGO sensor disconnected. • Connect DVOM to HEGO SIGNAL at the HEGO sensor connector and to battery negative post. • DVOM on 20 volt scale. • Create a vacuum leak to cause HEGO sensor to go lean. **For 1.9L EFI and 2.3L EFI TC:** — Disconnect the manifold vacuum hose illustrated below. **For 5.0L SEFI MASS AIR:** — Disconnect any vacuum hose from the manifold vacuum tree. **For all other applications:** — Disconnect the PCV valve hose from the PCV valve. • Start engine and run at approximately 2000 rpm. • Does the DVOM indicate less than 0.4 volts within 30 seconds?	Yes ▶ No ▶	RECONNECT HEGO sensor and vacuum lines. HEGO sensor is OK. GO to H1. RECONNECT vacuum hoses. REPLACE HEGO sensor. RERUN Quick Test. 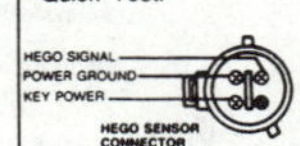

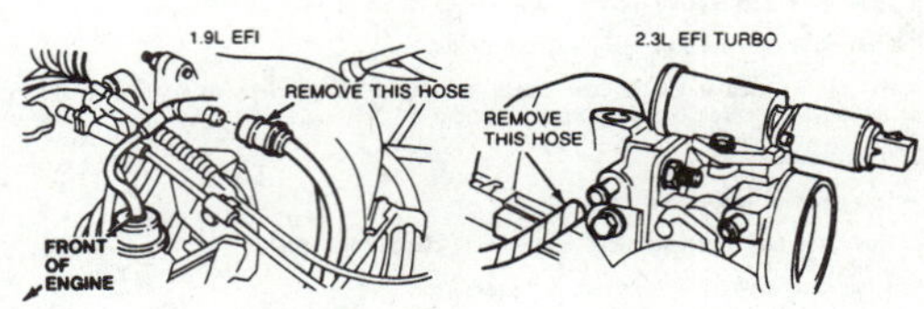

Fuel Control — Pinpoint Test H

H29 CONTINUOUS TESTING: CODE 41, OR 91

CODE 41/91 — Indicates that a HEGO circuit has not switched during closed loop fuel control.

NOTE: In this situation, Code 41/91 does not necessarily indicate a lean condition.

Before attempting to service a Continuous Memory Service Code 41 or 91, DIAGNOSE all other driveability complaints first. E.g., rough idle, misses, etc.

NOTE: The Fuel Service Code may help to isolate the cause of the fuel control problem.

Some areas to check are:

- Unmetered Air (vacuum leaks/intake air leaks):
  - — Canister purge system
  - — PCV system
  - — Engine sealing
  - — Crimped fuel lines
  - — Plugged fuel filter
  - — Fouled fuel injectors
  - — Air leaks between mass air flow sensor and air outlet tube to throttle body
- HEGO Fuel Fouled:

  Whenever an over-rich fuel condition has been experienced (fuel fouled spark plugs), make a thorough check of the ignition system. If a HEGO sensor is suspected of being fuel fouled (low output or slow response), run the vehicle at sustained high speed (within legal limits) followed by a few hard accels. This will burn off the HEGO contamination and restore proper HEGO operation.
- Ignition System:

  If engine is always in DEFAULT spark (base timing) refer to Quick Test Step [4.0].
- Improper Fueling:

  Lead fouled HEGO sensor.
- Fuel Pressure:

  Perform Pinpoint Test Steps [H1] and [H2].
- TP Sensor:

  Turn key to RUN position. While moving throttle slowly toward wide-open position, measure voltage between Test Pins 47 and 46 at the breakout box. If the voltage does not increase with the increase of throttle opening, replace TP sensor or linkage as necessary.
- If at this point the driveability concern is still present, perform Pinpoint Test Steps [H3] through [H6].

Fuel Control — Pinpoint Test H

H30 CONTINUOUS TESTING: CODE 41, 42, 43, 65, 85, OR 86

CODE 41 — HEGO indicated the fuel system was lean for more than 15 seconds when the fuel system should have been in closed loop fuel control.

CODE 42 — HEGO indicated the fuel system was rich for more than 15 seconds when the fuel system should have been in closed loop fuel control.

CODE 43 — HEGO indicated the fuel system was lean at WOT for more than 3 seconds.

CODE 65 — Never went to closed loop fuel control on HEGO switching.

CODE 85 — Adaptive fuel has corrected an excessive rich condition. (Adaptive fuel made the fuel system leaner.)

CODE 86 — Adaptive fuel has corrected an excessive lean condition. (Adaptive fuel made the fuel system richer.)

- Before attempting to service a Continuous Memory Service Code 41, 42, 43, 65, 85, or 86, DIAGNOSE all other drivability complaints first. Examples: rough idle, misses, etc.
- Whenever an over-rich fuel condition has been experienced (fuel fouled spark plugs), make a thorough check of the ignition system. If a HEGO sensor is suspected of being fuel fouled (low output or slow response), after the vehicle service, run the vehicle at sustained high speed (within legal limits) followed by a few hard accels. This will burn off the HEGO contamination and restore proper HEGO operation.
- The fuel Service Code may help to isolate the cause of the fuel control problem. Some areas to check are:

Code 41:

- Intermittant HEGO circuit (SIGNAL or GROUND).
- If Code 65 is also present, service faulty HEGO circuit (SIGNAL or GROUND).
- If Code 43 is also present, service Code 43 first.
- Airflow meter indicates low air flow. Check for vacuum leaks, intake air leaks, or a sticking air meter vane caused by contamination or frost.
- Low fuel pressure at WOT.
  - — Low-pressure fuel pump.
  - — Restricted fuel supply (crimped fuel lines or plugged fuel filter).
- Low fuel flow at WOT with correct fuel pressure.
  - — Clogged fuel injectors.
  - — Low battery (fuel injector voltage less than 11 volts).

Fuel Control — Pinpoint Test H

H30 (CONTINUED)

CODE 42:

- Intermittant HEGO circuit (SIGNAL or GROUND).
- Airflow indicated by the air meter is greater than the actual airflow (causing more fuel to be delivered than necessary). Check for high air meter voltage output due to a sticking air meter vane caused by contamination.
- Excessive fuel pressure. Check for fuel pressure regulator vacuum line disconnected or kinked fuel return line.
- Excessive fuel flow. Check for damaged or stuck open fuel injector(s).

BOTH CODES 41 and 42:

- Intermittant HEGO circuit (SIGNAL or GROUND).
- Sticking air meter vane due to contamination.
- Contaminated HEGO sensor (lead or silicone fouled).

CODE 43:

- Low fuel pressure at WOT.
  - — Low pressure fuel pump.
  - — Restricted fuel supply (crimped fuel lines or plugged filter).
- Low fuel flow at WOT with correct fuel pressure.
  - — Clogged fuel injectors.
  - — Low battery (fuel injector voltage less than 11 volts).

CODE 65:

- Check for faulty HEGO circuit (SIGNAL or GROUND).

CODE 85:

- If Code 42 is also present, service Code 42 first.
- Excessive fuel pressure. Check for fuel pressure regulator vacuum line disconnected or kinked fuel return line.
- Excessive fuel flow. Check for damaged fuel injector pintle or injectors stuck open.

CODE 86:

- If Code 41 is also present, service Code 41 first.
- Low fuel pressure
  - — Low pressure fuel pump.
  - — Restricted fuel supply (crimped fuel lines or plugged filter).
- Low fuel flow with correct fuel pressure
  - — Clogged fuel injectors.
  - — Low battery (fuel injector voltage less than 11 volts).

Fuel Pump Circuit — Pinpoint Test J

Note

You should enter this Pinpoint Test only when a Service Code 87, 95 or 96 is received in Quick Test Step 3.0 or 6.0 or you are directed here from Pinpoint Test Step A or Diagnostic By Symptom in the Engine Supplement Section.

Remember

To prevent the replacement of good components, be aware that the following non-EEC areas may be at fault:

- Fuel Lines
- Fuel Filters
- Throttle Body
- Contaminated Fuel
- Fuel Pump

This Pinpoint Test is intended to diagnose only the following:

- Fuel Pump Relay
- Inertia Switch
- Harness Circuits: V BATT., VPWR, F.P., GROUND and POWER-TO-PUMP(s)
- Processor Assembly

Fuel Pump Circuit — Pinpoint Test J

Pinpoint Test Schematic

1.9L CFI, 1.9L EFI, 2.3L HSC

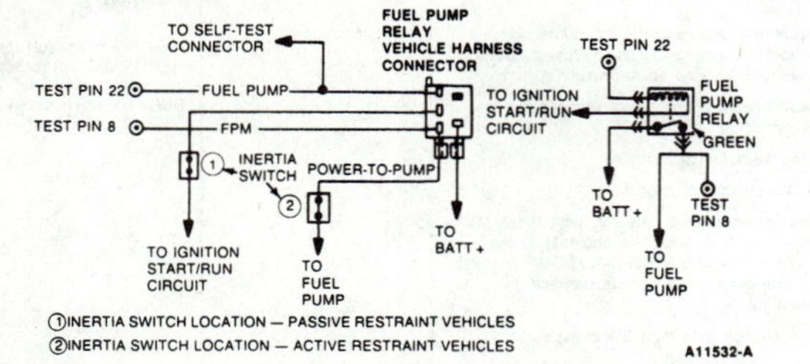

3.8L EFI RWD, 5.0L SEFI MA, 2.3L EFI TRUCK, 2.9L EFI TRUCK, 3.0L EFI TRUCK

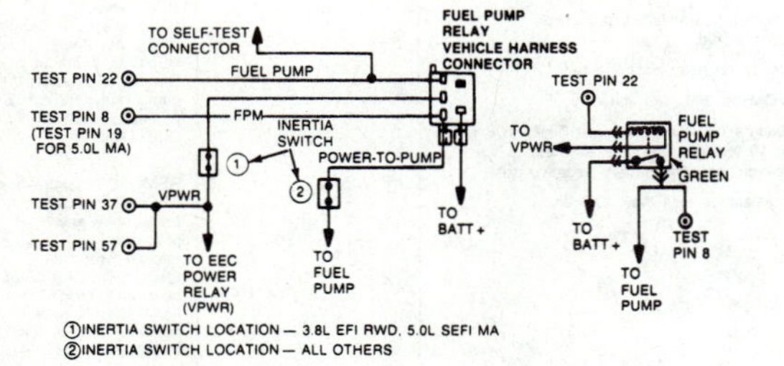

Fuel Pump Circuit — Pinpoint Test J

TEST STEP	RESULT ▶	ACTION TO TAKE
J1 NO FUEL PUMP PRESSURE: CHECK FOR FUEL PUMP ELECTRICAL OPERATION		
• Fuel pressure gauge installed. • To check if fuel pump runs, cycle key from Off to Run, repeat several times, (Do not enter start mode.) Fuel pump should run briefly each time the key enters Run.	Yes ▶	GO to electric fuel pump.
	No ▶	GO to J2.
J2 CHECK FOR VPWR TO PROCESSOR		
• Key off, wait 10 seconds. • Disconnect processor 60 pin connector. Inspect for damaged pins, corrosion, loose wires, etc. Service as necessary. • Install breakout box and reconnect processor. • Key on, engine off. • DVOM on 20 volt scale. • Measure voltage between Test Pin 37 and Test Pin 40 at the breakout box and between Test Pin 57 and Test Pin 60 at the breakout box. • Are both voltages greater than 10.5 volts?	Yes ▶	GO to J3.
	No ▶	GO to B1.
J3 CHECK RESISTANCE OF FUEL PUMP INERTIA SWITCH		
• Key off, wait 10 seconds. • Leave breakout box installed and processor connected. • Locate and disconnect fuel pump inertia switch. • DVOM on 200 ohm scale. • Measure the resistance of the fuel pump inertia switch. • Is resistance less than 5.0 ohms?	Yes ▶	RECONNECT inertia switch and GO to J4.
	No ▶	REPLACE or reset fuel pump inertia switch. RERUN Quick Test.

Fuel Pump Circuit — Pinpoint Test J

Pinpoint Test Schematic

4.9L EFI, 5.0L EFI, 5.8L EFI, 7.5L EFI, TRUCKS

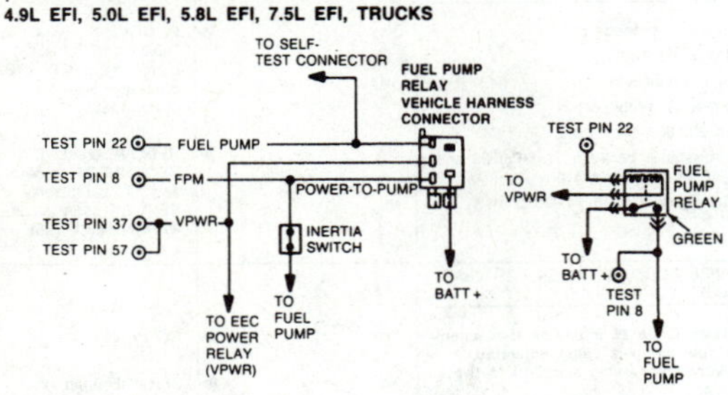

2.3L OHC EFI CAR, 5.0L SEFI

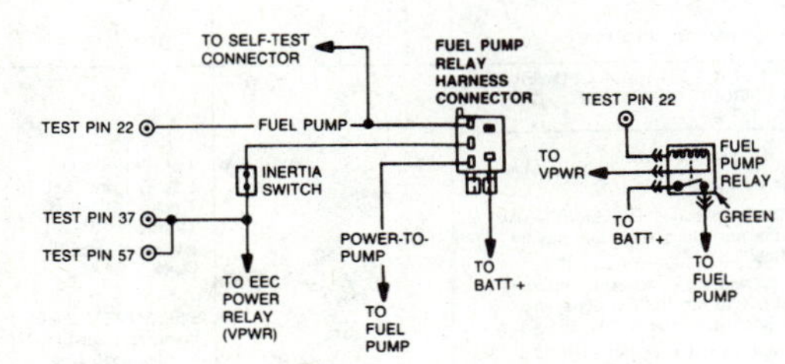

Fuel Pump Circuit — Pinpoint Test J

TEST STEP	RESULT ▶	ACTION TO TAKE
J4 CHECK FOR VOLTAGE TO POWER-TO-PUMP(S) CIRCUIT		
• Key on, engine off. • Leave breakout box installed and processor connected. • Locate fuel pump relay. • DVOM on 20 volt scale. • Measure voltage between chassis ground and POWER-TO-PUMP(s) circuit at fuel pump relay during crank mode. • Is voltage greater than 8.0 volts during crank?	Yes ▶	GO to open in power-to-pump circuit, PWR fuel pump GND, open in pump, etc.
	No ▶	GO to J6.
J6 CHECK FOR V BATT TO FUEL PUMP RELAY		
• Key on, engine off. • Leave breakout box installed and processor connected. • Locate fuel pump relay. • DVOM on 20 volt scale. • Measure voltage between chassis ground and V Batt. at the fuel pump relay. • Is voltage greater than 10.5 volts?	Yes ▶	GO to J7.
	No ▶	SERVICE open in V Batt. between fuel pump relay and vehicle battery positive post. RERUN Quick Test.
J7 CHECK FOR VPWR TO FUEL PUMP RELAY		
• Key on, engine off. • Leave breakout box installed and processor connected. • Locate fuel pump relay. • DVOM on 20 volt scale. • Measure voltage between chassis ground and VPWR circuit (Ignition start/run circuit for 1.9L EFI, 1.9L CFI and 2.3L HSC) at the fuel pump relay. • Is voltage greater than 10.5 volts?	Yes ▶	GO to J8.
	No ▶	VERIFY inertia switch is reset to On. If switch will not reset, REPLACE switch. If OK. — 1.9L EFI, 1.9L CFI and 2.3L HSC, SERVICE open between ignition switch start/run circuit and fuel pump relay. — All others, SERVICE open in VPWR circuit between the EEC power relay and the fuel pump relay. RERUN Quick Test.

Fuel Pump Circuit — Pinpoint Test J

TEST STEP	RESULT ▶	ACTION TO TAKE
J8 CHECK CONTINUITY OF FUEL PUMP CIRCUIT		
• Key off, wait 10 seconds. • Leave breakout box installed and processor connected. • DVOM on 200 ohm scale. • Measure resistance between fuel pump circuit at the pump relay and Test Pin 22 at the breakout box. • Is resistance less than 5.0 ohms?	Yes ▶ No ▶	GO to J9. SERVICE open circuit. RERUN Quick Test.
J9 CHECK FOR SHORT TO POWER		
• Key on. • Breakout box installed. • Disconnect processor. • Disconnect fuel pump relay. • DVOM on 20 volt scale. • Measure voltage between Test Pin 22 and battery negative post. • Is voltage less than 1.0 volt?	Yes ▶ No ▶	GO to J10. SERVICE short circuit. RECONNECT processor, ATTEMPT to start vehicle. If vehicle fails to start, REPLACE processor. RERUN Quick Test.
J10 CHECK FOR SHORT TO GROUND		
• Key off, wait 10 seconds. • Leave breakout box installed and processor disconnected. • Fuel pump relay disconnected. • DVOM on 200,000 ohm scale. • Measure resistance between Test Pin 22 and Test Pins 40 and 60 at the breakout box. • Is resistance greater than 10,000 ohms?	Yes ▶ No ▶	RECONNECT fuel pump relay. GO to J11. SERVICE short circuit. RERUN Quick Test.
J11 CHECK FOR VOLTAGE AT POWER-TO-PUMP(S) CIRCUIT		
• Leave breakout box installed and processor disconnected. • Connect jumper wire from Test Pin 22 to Test Pin 40 or 60 at the breakout box. • DVOM on 20 volt scale. • Key on, engine off. • Measure voltage between chassis ground and POWER-TO-PUMP(s) circuit at fuel pump relay. • Is voltage greater than 10.5 volts?	Yes ▶ No ▶	REPLACE processor. RERUN Quick Test. REPLACE fuel pump relay. RECONNECT processor and RERUN Quick Test.

Fuel Pump Circuit — Pinpoint Test J

TEST STEP	RESULT ▶	ACTION TO TAKE
J20 SERVICE CODE 95: CHECK INERTIA SWITCH		
NOTE: Key On Engine Off Service Code 95 indicates that one of the following has occurred: **— Open circuit in or between fuel pump and FPM circuit at the processor** **— Poor fuel pump ground** **— Fuel pump circuit short to power** **— Fuel pump relay contacts always closed.** • Key off, wait 10 seconds. • Locate and disconnect fuel pump inertia switch (verify that switch is reset). • DVOM on 200 ohm scale. • Measure resistance of the fuel pump inertia switch. • Is resistance less than 5.0 ohms?	Yes ▶ No ▶	RECONNECT inertia switch. GO to J21. REPLACE or RESET inertia switch. RERUN Quick Test.
J21 VERIFY THAT FUEL PUMP IS OFF		
• Key off. • Listen for motor noise from fuel pump. • Is fuel pump off?	Yes ▶ No ▶	GO to J23. GO to J22.
J22 CHECK FOR FUEL PUMP RELAY ALWAYS CLOSED		
• Key off. • Locate and disconnect fuel pump relay. • Does fuel pump shut off when relay is disconnected?	Yes ▶ No ▶	REPLACE fuel pump relay. RERUN Quick Test. SERVICE short to power in POWER-TO-PUMP/FPM circuit. RERUN Quick Test.

Fuel Pump Circuit — Pinpoint Test J

TEST STEP	RESULT ▶	ACTION TO TAKE
J23 CHECK CONTINUITY OF FPM CIRCUIT		
• Key off. • Disconnect processor 60 pin connector. Inspect for damaged pins, corrosion, loose wires, etc. Service as necessary. • Install breakout box, leave processor disconnected. • Disconnect fuel pump relay. • DVOM on 200 ohm scale. • Measure resistance between Test Pin 8 (Test Pin 19 for 5.0L Mass Air engines) at the breakout box and POWER-TO-PUMP circuit at the fuel pump relay vehicle harness connector. • Is resistance less than 5.0 ohms?	Yes ▶ No ▶	GO to J24. REMOVE breakout box. RECONNECT processor and fuel pump relay. SERVICE open circuit. RERUN Quick Test.
J24 CHECK FOR CONTINUITY BETWEEN FPM CIRCUIT AND GROUND		
• Key off. • Breakout box installed, processor disconnected. • Fuel pump relay disconnected. • DVOM on 200 ohm scale. • Measure resistance between Test Pin 8 (Test Pin 19 for 5.0L Mass Air engines) at the breakout box and battery negative post. • Is resistance less than 10.0 ohms?	Yes ▶ No ▶	For 1.9L EFI, RECONNECT fuel pump relay and GO to J25. All others, REMOVE breakout box. RECONNECT fuel pump relay. REPLACE processor. RERUN Quick Test. REMOVE breakout box. RECONNECT fuel pump relay and processor. GO to Shop Manual Group 24, Electric Fuel Pump for open in POWER-TO-PUMP circuit, poor fuel pump GROUND, open in fuel pump, etc.

Fuel Pump Circuit — Pinpoint Test J

TEST STEP	RESULT ▶	ACTION TO TAKE
J25 CHECK FUEL PUMP PRIMARY CIRCUIT FOR SHORT TO GROUND		
• Key off, wait 10 seconds. • Breakout box installed. • Processor disconnected. • Fuel pump relay disconnected. • DVOM on 200,000 ohm scale. • Measure resistance between Test Pin 22 and Test Pin 40 at the breakout box. • Is resistance greater than 10,000 ohms?	Yes ▶ No ▶	REMOVE breakout box. RECONNECT fuel pump relay. REPLACE processor RERUN Quick Test. REMOVE breakout box. RECONNECT processor and fuel pump relay. SERVICE short circuit. RERUN Quick Test.
J30 SERVICE CODE 96: CHECK FOR VBATT TO FUEL PUMP RELAY		
NOTE: Service Code 96 indicates that when the fuel pump is being activated, power is not being supplied to the pump. • Key off, wait 10 seconds. • Locate fuel pump relay. • DVOM on 20 volt scale. • Measure voltage between VBATT circuit at the fuel pump relay and battery negative post. • Is voltage greater than 10.5 volts?	Yes ▶ No ▶	GO to J31. SERVICE open in VBATT circuit. RERUN Quick Test.
J31 CHECK FOR VOLTAGE AT POWER-TO-PUMP CIRCUIT VERIFY FUEL PUMP OPERATION		
• Key off. • DVOM on 20 volt scale. • Connect DVOM between POWER-TO-PUMP circuit at the fuel pump relay and battery negative post. • Observe DVOM as you activate fuel pump relay (turn key to run for 1 second, then to off for 10 seconds. Repeat 5 times. • Does voltage measure greater than 10.5 volts for about 1 second after key is turned to ON position during test?	Yes ▶ No ▶	4.9L EFI, 5.0L EFI, 5.8L EFI and 7.5L EFI Trucks GO to J32. All others REPLACE processor. RERUN Quick Test. DISCONNECT fuel pump relay. INSPECT for damaged pins, corrosion, loose wires, etc. If OK REPLACE fuel pump relay. RERUN Quick Test.

Fuel Pump Circuit — Pinpoint Test J

TEST STEP	RESULT	ACTION TO TAKE
J32 CHECK CONTINUITY OF POWER-TO-PUMP CIRCUIT		
• Key off. • Disconnect processor 60 pin connector. Inspect for damaged pins, corrosion, loose wires, etc. Service as necessary. • Install breakout box, leave processor disconnected. • Disconnect Fuel Pump Relay. • DVOM on 200 ohm scale. • Measure resistance between Test Pin 8 at the breakout box and POWER-TO-PUMP circuit at the fuel pump relay vehicle harness connector. • Is resistance less than 5.0 ohms?	Yes ▶ No ▶	REMOVE breakout box. RECONNECT fuel pump relay. REPLACE processor. RERUN Quick Test. SERVICE open in POWER-TO-PUMP circuit between FPM splice and fuel pump relay. REFER to schematic. RERUN Quick Test.
J90 CONTINUOUS MEMORY CODE 95: CHECK EEC-IV HARNESS		
A Continuous Memory Code 95 indicates that one of the following intermittent conditions has occurred: — Open circuit in or between the fuel pump and pin 8 (pin 19 on 5.0L MA) in the processor (see schematic). — Poor fuel pump ground. • Start engine. • Check for engine stall/stumble while performing the following (also, if possible, listen for fuel pump turning off.) — Shake, wiggle, bend the POWER-TO-PUMP circuit between the POWER-TO-PUMP pin at the fuel pump relay and the fuel pump. — Shake, wiggle, bend the fuel pump ground circuit from the fuel pump to ground. — Lightly tap the fuel pump to simulate road shock. — For vehicles with the inertia switch in the POWER-TO-PUMP circuit (refer to schematic), lightly tap inertia switch to simulate road shock. • Key off. • Inspect the fuel pump harness connector and the fuel pump ground for corrosion, damaged pins, etc. • Is fault indicated/found?	Yes ▶ No ▶	ISOLATE fault and SERVICE as necessary. CLEAR Continuous Memory Code 95. RERUN Quick Test. GO to J91.

Fuel Pump Circuit — Pinpoint Test J

TEST STEP	RESULT	ACTION TO TAKE
J91 CHECK FPM CIRCUIT		
• Key off. • Disconnect processor 60 pin connector. Inspect for damaged pins, corrosion, loose wires, etc. Service as necessary. • Install breakout box, leave processor disconnected. • Key on, engine off. • Connect a TEST LAMP between Test Pin 8 (Test Pin 19 for 5.0L MA.) and Test Pin 37. • Observe test lamp for an indication of a fault while performing the following (The light will go out when a fault is found, indicating an open): — Shake, wiggle, bend the fuel pump monitor circuit between the fuel pump relay (or splice if applicable, see schematic) and the processor. • Is fault indicated?	Yes ▶ No ▶	ISOLATE fault and SERVICE as necessary. REMOVE breakout box. CLEAR Continuous Memory Code 95. RERUN Quick Test. Unable to duplicate fault at this time. REMOVE breakout box. CLEAR Continuous Memory Code 95.
J92 CONTINUOUS MEMORY CODE 96 CHECK FOR CONTINUOUS MEMORY CODE 87		
• Is Continuous Memory Code 87 also present?	Yes ▶ No ▶	GO to J95. GO to J93.

Fuel Pump Circuit — Pinpoint Test J

TEST STEP	RESULT	ACTION TO TAKE
J93 CHECK EEC-IV HARNESS		
A Continuous Memory Code 96, without the presence of a Continuous Memory Code 87, indicates that during vehicle operation, one of the following has occurred: — Open in the VBATT circuit between BATT+ and the fuel pump relay. — Fuel pump relay contacts opened. — Open in the POWER-TO-PUMP circuit from the fuel pump relay to the FPM splice, if applicable (see schematic). • Start engine. • Check for engine stall/stumble while performing the following (also, if possible, listen for fuel pump turning off): — Shake, wiggle, bend the VBATT(+) circuit from BATT(+) to the fuel pump relay. — Lightly tap the fuel pump relay (to simulate road shock). — Shake, wiggle, bend the POWER-TO-PUMP circuit from the fuel pump relay to the FPM splice, if applicable (See schematic). • Key off. • Inspect the fuel pump relay connectors and BATT+ connector terminal for corrosion, damaged pins, etc. • Is fault indicated/found?	Yes ▶ No ▶	ISOLATE fault and SERVICE as necessary. CLEAR Continuous Memory Code 96. RERUN Quick Test. 1.9L EFI: GO to J95. ALL OTHERS: Unable to duplicate fault at this time. CLEAR Continuous Memory Code 96. Continuous Memory Code 96 testing complete.

Fuel Pump Circuit — Pinpoint Test J

TEST STEP	RESULT	ACTION TO TAKE
J95 CONTINUOUS MEMORY SERVICE CODE 87 CHECK EEC-IV HARNESS		
A Continuous Memory Code 87 indicates that one of the following intermittent conditions has occurred: — Open in VPWR circuit between the EEC power relay and the fuel pump relay. — Open coil in fuel pump relay. — Open in fuel pump circuit (pin 22). • Start engine. • Check for engine stall/stumble while performing the following (also, if possible, listen for fuel pump turning off): — Shake, wiggle, bend the VPWR circuit between the EEC power relay and the fuel pump relay. For vehicles with the inertia switch in the VPWR circuit (refer to scheamatic), lightly tap the inertia switch to simulate road shock. — Shake, wiggle, bend the EEC-IV harness fuel pump circuit (Test Pin 22) between the processor and the fuel pump relay. — Lightly tap the fuel pump relay to simulate road shock. • Key off. • Inspect the processor 60 pin connector and them fuel pump relay connectors for corrosion, damaged pins, etc. • Is fault indicated/found?	Yes ▶ No ▶	ISOLATE fault and SERVICE as necessary. CLEAR Continuous Memory Code(s). RERUN Quick Test. Unable to duplicate fault at this time. CLEAR Continuous Memory Code(s).

EGR On/Off Control — Pinpoint Test KA

Note

You should enter this Pinpoint Test only when a Service Code 34 or 84 is received in Quick Test Step 3.0, 5.0 or when directed here from Diagnostic By Symptom in the Engine Supplement Section.

Remember

To prevent the replacement of good components, be aware that the following non-EEC areas may be at fault:

- Air or Vacuum Leaks
- EGR Flow Restrictions
- EGR Value

NOTE: Code 34 may be the result of high volume exhaust vent system (reduces back pressure). If this is suspected, perform the test in a well-ventilated area without exhaust vent connected.

This Pinpoint Test is intended to diagnose only the following:

- Circuits: EGR and VPWR
- EGR Solenoid
- Presence of Manifold Vacuum
- Processor Assembly

Pinpoint Test Schematic

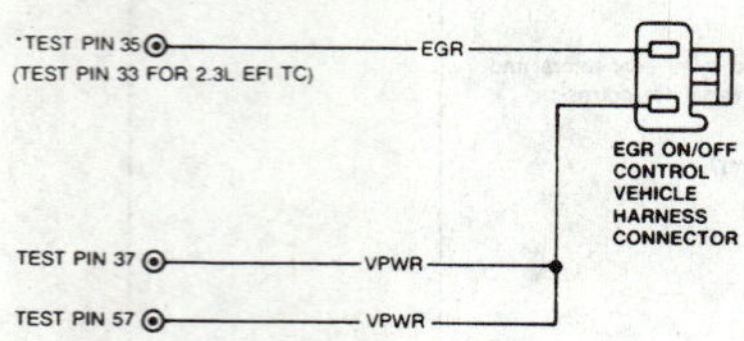

*TEST PINS LOCATED ON BREAKOUT BOX.
ALL HARNESS CONNECTORS VIEWED INTO MATING SURFACE.

EGR On/Off Control — Pinpoint Test KA

TEST STEP	RESULT	▶ ACTION TO TAKE
KA1 ENTER OUTPUT STATE CHECK (REFER TO APPENDIX)		
NOTE: Do not use STAR tester for this step, use a VOM/DVOM. • Key off, wait 10 seconds. • DVOM on 20 volt scale. • Connect DVOM negative test lead to STO at the Self-Test connector and positive test lead to battery positive. • Jumper STI to SIGNAL RETURN at the Self-Test connector. • Rerun Key On Engine Off Self-Test until the completion of the Continuous Test Codes. • DVOM will indicate less than 1.0 volts. • Depress and release the throttle. • Did DVOM reading change to a high voltage reading?	Yes No	▶ REMAIN in Output State Check. GO to KA2. ▶ DEPRESS throttle to WOT and release. If STO voltage does not go high, GO to Pinpoint Test Step QC1. Leave equipment hooked up.
KA2 CHECK EGR ON/OFF CONTROL SOLENOID ELECTRICAL OPERATION		
• DVOM on 20 volt scale. • Connect DVOM positive test lead to VPWR circuit on EGR solenoid and negative test lead to EGR output circuit. • While observing DVOM, depress and release the throttle several times to cycle output On and Off. • Does EGR output cycle On and Off?	Yes No	▶ GO to KA3. ▶ REMOVE STI jumper. GO to KA5.
KA3 CHECK SOLENOID FOR VACUUM CYCLING		
• Install vacuum pump to the solenoid vacuum supply port and install a vacuum gauge to the output port. Apply 6 in. Hg minimum. • While cycling outputs On and Off (by depressing and releasing throttle) observe the vacuum gauge at the output. **NOTE: Maintain vacuum at source.** • Does output port vacuum cycle On and Off?	Yes No	▶ GO to KA4. ▶ REPLACE solenoid. RERUN Quick Test.

EGR On/Off Control — Pinpoint Test KA

TEST STEP	RESULT	▶ ACTION TO TAKE
KA4 CHECK MANIFOLD VACUUM LINES FOR BLOCKAGE OR LEAKS		
• Vacuum lines disconnected at solenoid. • Start engine. • Check for vacuum. • Is vacuum present?	Yes No	▶ EEC-IV system OK. ▶ SERVICE vacuum source blockage or leak. RERUN Quick Test.
KA5 MEASURE EGR SOLENOID RESISTANCE		
• Key off, wait 10 seconds. • DVOM on 200 ohm scale. • Disconnect EGR solenoid. • Measure solenoid resistance. • Is resistance between 65 and 110 ohms?	Yes No	▶ GO to KA6. ▶ REPLACE EGR solenoid. RERUN Quick Test.
KA6 CHECK VOLTAGE OF VPWR CIRCUIT		
• Key on, engine off. • EGR solenoid disconnected. • DVOM on 20 volt scale. • Measure voltage between VPWR circuit at the EGR solenoid vehicle harness connector and battery ground. • Is voltage greater than 10.5 volts?	Yes No	▶ GO to KA7. ▶ RECONNECT EGR solenoid. SERVICE open circuit. RERUN Quick Test.
KA7 CHECK CONTINUITY OF EGR CIRCUIT		
• Key off, wait 10 seconds. • EGR solenoid disconnected. • Disconnect processor 60 pin connector. Inspect for damaged pins, corrosion, loose wires, etc. Service as necessary. • Install Breakout box, leave processor disconnected. • DVOM on 200 ohm scale. • Measure resistance between Test Pin 35 (Test Pin 33 for 2.3L EFI TC) at the breakout box and EGR circuit at vehicle harness connector. • Is resistance less than 5 ohms?	Yes No	▶ GO to KA8. ▶ REMOVE breakout box. RECONNECT all components. SERVICE open circuit. RERUN Quick Test.

EGR On/Off Control — Pinpoint Test KA

TEST STEP	RESULT	▶ ACTION TO TAKE
KA8 CHECK FOR SHORT TO GROUND		
• Key off, wait 10 seconds. • Breakout box installed, processor disconnected. • EGR solenoid disconnected. • DVOM on 200,000 ohm scale. • Measure resistance between Test Pin 35 (Test Pin 33 for 2.3L EFI TC) and Test Pins 40, 46 and 60 at the breakout box. • Is resistance greater than 10,000 ohms?	Yes No	▶ GO to KA9. ▶ REMOVE breakout box. RECONNECT all components. SERVICE short circuit. RERUN Quick Test.
KA9 CHECK FOR SHORT TO POWER		
• Key off, wait 10 seconds. • DVOM on 200,000 ohm scale. • Breakout box installed, processor disconnected. • EGR solenoid disconnected. • Measure resistance between Test Pin 35 (Test Pin 33 for 2.3L EFI TC) and Test Pins 37 and 57 at the breakout box. • Is resistance greater than 10,000 ohms?	Yes No	▶ REMOVE breakout box. RECONNECT all components. REPLACE Processor. RERUN Quick Test. ▶ REMOVE breakout box. RECONNECT all components. SERVICE short to power. RERUN Quick Test. If code is repeated, REPLACE processor.

DC Motor Idle Speed Control/Idle Tracking Switch and Throttle Position Sensor — 1.9L CFI and 2.5L CFI | Pinpoint Test | KB

Note

You should enter this Pinpoint Test only when a Service Code 12, 13, 16, 17, 23, 38, 53, 58, 63, 68, 71, 73 or 93 is received in Quick Test Step 3.0, 5.0, or 6.0 or when directed here from Diagnostic By Symptom in the Engine Supplement Section.

Remember

To prevent the replacement of good components, be aware that the following non-EEC areas may be at fault:

- Throttle stop screw out of adjustment
- Vacuum leaks
- Basic engine
- Throttle sticking

This Pinpoint Test is intended to diagnose only the following:

- DC motor
- Throttle position sensor
- Idle Tracking Switch
- EEC-IV processor assembly
- Harness circuits ISC+, ISC–, ITS, TP, VREF, and, SIG RTN

Pinpoint Test Schematic

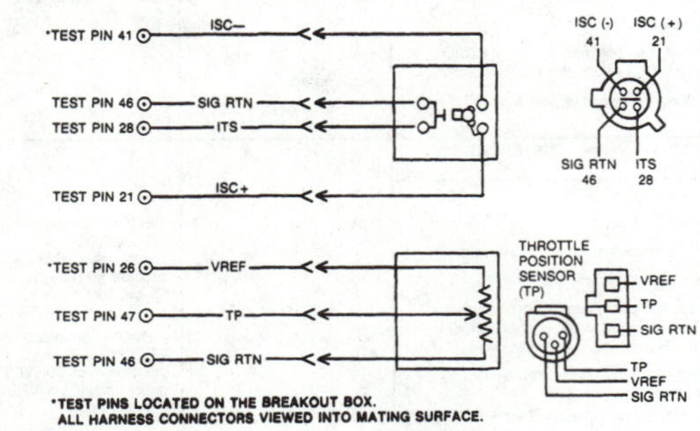

***TEST PINS LOCATED ON THE BREAKOUT BOX.**
ALL HARNESS CONNECTORS VIEWED INTO MATING SURFACE.

DC Motor Idle Speed Control/Idle Tracking Switch and Throttle Position Sensor — 1.9L CFI and 2.5L CFI | Pinpoint Test | KB

TEST STEP	RESULT ▶	ACTION TO TAKE
KB1 CHECK DC MOTOR FOR PROPER OPERATION		
• Key off, wait 10 seconds. • Disconnect harness from DC motor. • Jumper ISC+ circuit of DC motor to battery positive and ISC– circuit of DC motor to battery ground for 4 seconds. • Jumper ISC+ circuit of DC motor to battery ground and ISC– circuit of DC motor to battery positive for 4 seconds. • Does the DC motor shaft extend to greater than 2 inches (5 cm) and retract to less than 1.75 inches (4.4 cm) from mounting bracket (see below)?	Yes ▶ No ▶	GO to KB2. REPLACE DC MOTOR. RERUN Quick Test.
KB2 CHECK CONTINUITY OF ISC+ AND ISC– CIRCUITS		
• Key off, wait 10 seconds. • Harness disconnected from DC motor. • Disconnect processor 60 pin connector. Inspect for damaged pins, corrosion, loose wires, etc. Service as necessary. • Install breakout box, leave processor disconnected. • DVOM on 200 ohm scale. • Measure resistance between Test Pin 41 at the breakout box and ISC– circuit at the vehicle harness connector and between Test Pin 21 at the breakout box and ISC+ circuit at the vehicle harness connector. • Are both resistances less than 5 ohms?	Yes ▶ No ▶	GO to KB3. REMOVE breakout box. RECONNECT all components. SERVICE OPEN circuit(s). RERUN Quick Test.

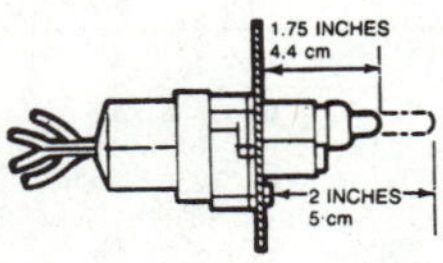

DC Motor Idle Speed Control/Idle Tracking Switch and Throttle Position Sensor — 1.9L CFI and 2.5L CFI | Pinpoint Test | KB

TEST STEP	RESULT ▶	ACTION TO TAKE
KB3 CHECK FOR SHORTS TO GRND OF ISC+ AND ISC– CIRCUITS		
• Key off, wait 10 seconds. • Breakout box installed, processor disconnected. • Harness disconnected from DC motor. • DVOM on 200,000 ohm scale. • Measure resistance between Test Pin 41 and Test Pins 40, 46 and 60 at the breakout box. • Measure resistance between Test Pin 21 and Test Pins 40, 46 and 60 at the breakout box. • Are all resistances greater than 10,000 ohms?	Yes ▶ No ▶	GO to KB4. REMOVE breakout box. RECONNECT all components. SERVICE short circuit(s). RERUN Quick Test.
KB4 CHECK FOR SHORTS TO PWR. OF ISC+ AND ISC– CIRCUITS		
• Key on, engine off. • Breakout box installed, processor disconnected. • Harness disconnected from DC motor. • DVOM on 20 volt scale. • Measure voltage between Test Pin 41 and Test Pin 40 and 60 at the breakout box. • Measure voltage between Test Pin 21 and Test Pins 40 and 60 at the breakout box. • Are all voltages less than 1 volt?	Yes ▶ No ▶	REMOVE breakout box. RECONNECT all components. REPLACE processor. RERUN Quick Test. REMOVE breakout box. RECONNECT all components. SERVICE short circuit(s). RERUN Quick Test.

DC Motor Idle Speed Control/Idle Tracking Switch and Throttle Position Sensor — 1.9L CFI and 2.5L CFI | Pinpoint Test | KB

TEST STEP	RESULT ▶	ACTION TO TAKE
KB5 CHECK FOR FULL DC MOTOR RETRACTION		
• Key off, wait 10 seconds. • Disconnect harness from DC motor. • Jumper, at the DC motor connector, ISC– circuit to battery positive and the ISC+ circuit to battery negative for 4 seconds. • Does the DC motor shaft retract away from the throttle lever as shown?	Yes ▶ No ▶	GO to KB7. GO to KB6.
KB6 MEASURE DC MOTOR RETRACTION		
• Key off, wait 10 seconds. • Harness disconnected from DC motor. • DC motor fully retracted. • Measure the distance from the tip of the DC motor shaft to the mounting bracket. Refer to figure in step KB1. • Is the distance less than 1.75 inches (4.4 cm)?	Yes ▶ No ▶	RECONNECT DC motor. REPLACE DC motor. RERUN Quick Test.
KB7 CHECK IDLE TRACKING SWITCH STATE		
• Key off, wait 10 seconds. • Harness disconnected from DC motor. • DC motor fully retracted. • DC motor shaft **NOT** touching the throttle lever. • DVOM on 200 ohm scale. • Measure resistance between ITS circuit and SIG RTN at the DC motor connector. • Is the resistance less than 5 ohms?	Yes ▶ No ▶	GO to KB8. REPLACE DC motor. RERUN Quick Test.

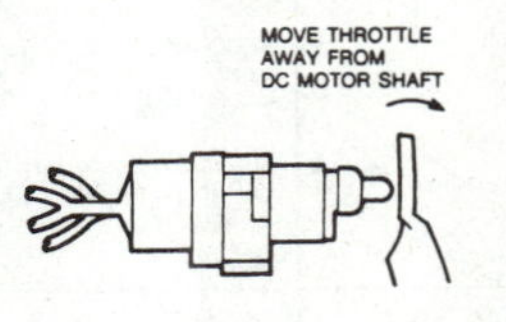

DC Motor Idle Speed Control/Idle Tracking Switch and Throttle Position Sensor — 1.9L CFI and 2.5L CFI — Pinpoint Test KB

TEST STEP	RESULT ▶	ACTION TO TAKE
KB8 CHECK CONTINUITY OF ITS AND SIG RTN CIRCUITS		
• Key off, wait 10 seconds. • Harness disconnected from DC motor. • Disconnect processor 60 pin connector. Inspect for damaged pins, corrosion or loose wires, etc. Service as necessary. • Install breakout box, leave processor disconnected. • DVOM on 200 ohm scale. • Measure resistance between Test Pin 46 at the breakout box and SIG RTN circuit at the DC motor vehicle harness connector. • Measure resistance between Test Pin 28 at the breakout box and ITS circuit at the DC motor vehicle harness connector. • Are both resistances less than 5 ohms?	Yes ▶ No ▶	REMOVE breakout box. RECONNECT all components. RECONNECT DC motor. REPLACE processor. RERUN Quick Test. REMOVE breakout box. RECONNECT all components. RECONNECT DC motor. SERVICE open circuit(s). RERUN Quick Test.
KB9 CHECK IDLE TRACKING SWITCH STATE		
• Key off, wait 10 seconds. • Disconnect harness from DC motor. • DVOM on 200 ohm scale. • Measure resistance between ITS circuit and SIG RTN circuit at the DC motor connector. • Is the resistance greater than 5 ohms?	Yes ▶ No ▶	GO to KB10. REPLACE DC motor. RERUN Quick Test.
KB10 CHECK FOR SHORTS TO GRND OF THE ITS CIRCUIT		
• Key off, wait 10 seconds. • Harness disconnected from DC motor. • Disconnect processor 60 pin connector. Inspect for damaged pins, corrosion, loose wires, etc. Service as necessary. • Install breakout box, leave processor disconnected. • Measure resistance between Test Pin 28 and Test Pins 40, 46 and 60 at the breakout box. • Are all resistances greater than 10,000 ohms?	Yes ▶ No ▶	REMOVE breakout box. RECONNECT all components. RECONNECT DC motor. REPLACE processor. RERUN Quick Test. REMOVE breakout box. RECONNECT all components. RECONNECT DC motor. SERVICE faulty circuit(s). RERUN Quick Test.

DC Motor Idle Speed Control/Idle Tracking Switch and Throttle Position Sensor — 1.9L CFI and 2.5L CFI — Pinpoint Test KB

TEST STEP	RESULT ▶	ACTION TO TAKE
KB15 GENERATE SERVICE CODE 63		
• Key off, wait 10 seconds. • Disconnect harness from the TP sensor at the throttle body. Inspect for damaged pins, corrosion, loose wires, etc. Service as necessary. • Run Key On Engine Off Self-Test and record codes. • Is Code 63 present? • Ignore all other codes.	Yes ▶ No ▶	GO to KB16. GO to KB17.
KB16 CHECK VOLTAGE VREF TO SIG RTN		
• Key off, wait 10 seconds. • Harness disconnected from TP sensor at throttle body. • DVOM on 20 volt scale. • Key on, engine off. • Measure voltage between VREF and SIG RTN at the TP vehicle harness connector. • Is the voltage between 4 and 6 volts?	Yes ▶ No ▶	REPLACE the TP sensor. RERUN Quick Test. GO to Pinpoint Test Step C1.
KB17 CHECK TP SIGNAL FOR SHORT TO POWER		
• Key off, wait 10 seconds. • Harness disconnected from TP sensor. • DVOM on 200,000 ohm scale. • Disconnect the processor 60 pin connector. Inspect for damaged pins, corrosion, loose wires, etc. Service as necessary. • Install breakout box, leave processor disconnected. • Measure resistance between Test Pin 47 and Test Pins 26 and 57 at the breakout box. • Are both resistances greater than 10,000 ohms?	Yes ▶ No ▶	REMOVE breakout box. RECONNECT all components. REPLACE processor. RERUN Quick Test. REMOVE breakout box. RECONNECT all components. SERVICE short circuit(s). RERUN Quick Test.

DC Motor Idle Speed Control/Idle Tracking Switch and Throttle Position Sensor — 1.9L CFI and 2.5L CFI — Pinpoint Test KB

TEST STEP	RESULT ▶	ACTION TO TAKE
KB11 CHECK THROTTLE LEVER AND LINKAGE		
• Key off, wait 10 seconds. • Inspect throttle for freedom of movement to wide-open throttle and for damaged or bent throttle lever. • Is throttle/throttle linkage functioning properly?	Yes ▶ No ▶	REPLACE DC motor. RERUN Quick Test. SERVICE as necessary. RERUN Quick Test.
KB12 CHECK THROTTLE PLATE FOR CLOSING		
• Run Key On Engine Off Self-Test and disconnect DC motor after it is fully retracted. • Key off, wait 10 seconds. • Remove air cleaner from throttle body. • Inspect throttle for freedom of movement and proper closure. • Does throttle move freely and close without obstruction?	Yes ▶ No ▶	RECONNECT DC motor. GO to KB13. SERVICE as necessary. RERUN Quick Test.
KB13 CHECK VOLTAGE OF VREF TO SIGNAL RETURN		
• Key off, wait 10 seconds. • Disconnect TP vehicle harness connector at the throttle body. Inspect for damaged pins, corrosion, loose wires, etc. Service as necessary. • DVOM on 20 volt scale. • Key on engine off. • Measure voltage between VREF and SIG RTN at the TP vehicle harness connector. • Is the voltage between 4 and 6 volts?	Yes ▶ No ▶	RECONNECT TP sensor. GO to KB14. GO to Pinpoint Test Step C1.
KB14 CHECK THROTTLE STOP RPM		
• Run Key On Engine Off Self-Test and disconnect the DC motor after it has fully retracted and exit Self Test. • Start engine and verify that the throttle stop rpm is less than curb idle rpm. • Is the throttle stop set below the curb idle?	Yes ▶ No ▶	RECONNECT DC motor. REPLACE the TP sensor. RERUN Quick Test. RECONNECT DC motor. ADJUST throttle stop rpm. RERUN Quick Test.

DC Motor Idle Speed Control/Idle Tracking Switch and Throttle Position Sensor — 1.9L CFI and 2.5L CFI — Pinpoint Test KB

TEST STEP	RESULT ▶	ACTION TO TAKE
KB18 GENERATE SERVICE CODE 53		
• Key off, wait 10 seconds. • Disconnect harness from the TP sensor at the throttle body. Inspect for damaged pins, corrosion, loose wires, etc. Service as necessary. • Jumper VREF to TP signal at the TP vehicle harness connector. • Run Key On Engine Off Self-Test. **NOTE: If no codes are generated, immediately remove jumper and go directly to KB21.** • Is Code 53 present? • Ignore all other codes at this time.	Yes ▶ No ▶	REPLACE TP sensor. RERUN Quick Test. GO to KB19.
KB19 CHECK VOLTAGE VREF TO SIG RTN		
• Key off, wait 10 seconds. • Harness disconnected from TP sensor at throttle body. • DVOM on 20 volt scale. • Key on, engine off. • Measure voltage between VREF and SIG RTN at the TP vehicle harness connector. • Is the voltage between 4 and 6 volts?	Yes ▶ No ▶	GO to KB20. GO to Pinpoint Test Step C1.
KB20 CHECK CONTINUITY OF TP CIRCUIT		
• Key off, wait 10 seconds. • Harness disconnected from TP sensor at throttle body. • DVOM on 200 ohm scale. • Disconnect the processor 60 pin connector. Inspect for damaged pins, corrosion, loose wires, etc. Service as necessary. • Install breakout box, leave processor disconnected. • Measure resistance between TP signal at the TP vehicle harness connector and Test Pin 47 at the breakout box. • Is resistance less than 5 ohms?	Yes ▶ No ▶	GO to KB21. REMOVE breakout box. SERVICE open circuit. RECONNECT harness to TP sensor. RERUN Quick Test.

DC Motor Idle Speed Control/Idle Tracking Switch and Throttle Position Sensor—1.9L CFI and 2.5L CFI — Pinpoint Test KB

TEST STEP	RESULT	▶ ACTION TO TAKE
KB21 CHECK TP SIGNAL FOR SHORT TO GROUND		
• Key off, wait 10 seconds. • Harness disconnected from TP sensor at throttle body. • Breakout box installed. • Processor disconnected. • DVOM on 200,000 ohm scale. • Measure resistance between TP signal at TP vehicle harness connector and Test Pins 40, 46 and 60 at the breakout box. • Are all resistances greater than 10,000 ohms?	Yes	▶ REMOVE breakout box. REPLACE processor. RECONNECT TP sensor. RERUN Quick Test.
	No	▶ REMOVE breakout box. RECONNECT all components. SERVICE short circuit(s). RERUN Quick Test.
KB22 SERVICE CODE 73		
• Rerun Key On Engine Off Self-Test. • Is Code 73 still present?	Yes	▶ REPLACE TP sensor. RERUN Quick Test.
	No	▶ SERVICE other codes.
KB23 CHECK FOR CODES THAT COULD CAUSE CODE 12		
• Are service codes 31, 32, 34, 35, 41, or 58 present in Engine Running Self-Test?	Yes	▶ SERVICE these codes first. GO to Quick Test Step 5.0C for direction.
	No	▶ GO to KB24.
KB24 CHECK FOR STICKING THROTTLE LINKAGE		
• Check the throttle plates and/or linkage for sticking or binding. • Check speed control linkage for proper adjustment. • Does throttle open and close properly?	Yes	▶ GO to KB25.
	No	▶ SERVICE as necessary. RERUN Quick Test.

DC Motor Idle Speed Control/Idle Tracking Switch and Throttle Position Sensor—1.9L CFI and 2.5L CFI — Pinpoint Test KB

TEST STEP	RESULT	▶ ACTION TO TAKE
KB25 CHECK DC MOTOR FOR PROPER OPERATION		
• Key off, wait 10 seconds. • Disconnect harness from DC motor. **CAUTION: Do not short to other pins when connecting jumper wire.** • Jumper ISC+ circuit of DC motor to battery positive and ISC− circuit of DC motor to battery ground for 4 seconds. • Jumper ISC+ circuit of DC motor to battery ground and ISC− circuit of DC motor to battery positive for 4 seconds. • Does the DC motor shaft extend to greater than 2 inches (5 cm) and retract to less than 1.75 inches (4.4 cm) from mounting bracket (see below)?	Yes	▶ RECONNECT DC motor. REPLACE processor. RERUN Quick Test.
	No	▶ REPLACE DC MOTOR. RERUN Quick Test.
KB26 CHECK FOR ERRATIC IDLE		
• Engine should be at normal operating temperature. • Deactivate Self-Test. • A/C off. • Run engine for 3 minutes alternating between 30 second idles and 5 second part throttle modes. • Is the idle erratic at the end of the three minute idle/part throttle test?	Yes	▶ GO to KB27.
	No	▶ GO to KB28.

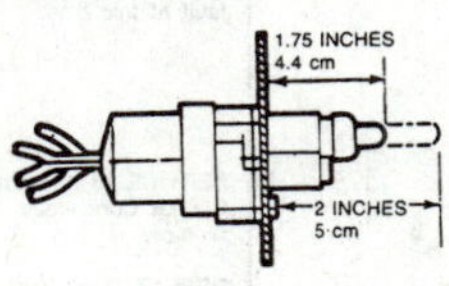

DC Motor Idle Speed Control/Idle Tracking Switch and Throttle Position Sensor—1.9L CFI and 2.5L CFI — Pinpoint Test KB

TEST STEP	RESULT	▶ ACTION TO TAKE
KB27 CHECK FOR ADDITIONAL SERVICE CODES IN ENGINE RUNNING SELF-TEST		
• With engine at idle, check for vacuum leaks. Service as necessary. **NOTE: Extreme caution must be taken when making inspections with Engine Running.** • Rerun Engine Running Self-Test. • Are service codes 22, 31, 32, 34, 35, 41 or 58 present?	Yes	▶ GO to Quick Test Step 5.0C for direction.
	No	▶ GO to KB28.
KB28 CHECK FOR PROPER OPERATION OF THROTTLE		
• Inspect the throttle plates and/or linkage for proper function. • Does the throttle open and close properly?	Yes	▶ REPLACE DC motor. RERUN Quick Test.
	No	▶ SERVICE as necessary. RERUN Quick Test.
KB29 SERVICE CODE 99		
NOTE: Service Code 99 indicates the EEC-IV system has not learned adequately to control the engine idle speed. • With Self-Test deactivated, start the engine. DO NOT touch the throttle. • Let the engine idle for 2 minutes. • Key off, wait 10 seconds. • Run Key On Engine Off Self-Test until the service codes begin to be displayed, then deactivate Self-Test. • Key off, wait 10 seconds. • Rerun Engine Running Self-Test. • Is Service Code 99 present?	Yes	▶ REPLACE processor. RERUN Quick Test.
	No	▶ SERVICE other codes as necessary.

DC Motor Idle Speed Control/Idle Tracking Switch and Throttle Position Sensor—1.9L CFI and 2.5L CFI — Pinpoint Test KB

TEST STEP	RESULT	▶ ACTION TO TAKE
KB30 CHECK FOR OTHER ENGINE RUNNING SELF-TEST SERVICE CODES		
NOTE: Service Code 12 may be the result of a failure in another EEC area. • Check for Service Codes 22, 31, 32, 34, 35, or 41 in the Engine Running Self-Test. • Are any of these codes present?	Yes	▶ GO to Quick Test Step 5.0C for direction.
	No	▶ GO to KB31.
KB31 CHECK FOR RELATED MECHANICAL PROBLEMS OR MISADJUSTMENTS		
NOTE: Service Code 16 could indicate that the accelerator pedal was touched during the Engine Running Self-Test. If Service Code 17 was present, check for electrical loads on the engine, (e.g. A/C or cooling fan on during Self-Test). • Check for vacuum leaks. • Check for throttle plate and/or linkage sticking or binding. • Check for speed control linkage for proper adjustment. • Verify proper base engine timing. • Verify proper throttle stop screw adjustment according to Section 4 of this manual. • Are all of the above areas okay?	Yes	▶ SERVICE other EEC-IV codes as necessary.
	No	▶ SERVICE as necessary. RERUN Quick Test.

DC Motor Idle Speed Control/Idle Tracking Switch and Throttle Position Sensor—1.9L CFI and 2.5L CFI | Pinpoint Test | KB

KB90 CONTINUOUS MEMORY CODE 13

A Continuous Memory Code 13 indicates that sometime in the last 40 warm-up cycles the TP sensor rotation did not follow the reaction of the DC motor when idle speed control was in a dashpot mode. This condition may be caused by:

- The DC motor sticking at part throttle.
- An open in the ITS circuit which, when coupled with other inputs to the processor, causes the EEC-IV system to falsely enter 'dashpot mode'.
- The TP sensor sticking at part throttle.

Each of these areas may generate Key On Engine Off (KOEO) or other Continuous Memory Codes. Therefore, if service has been made for KOEO code 13 or 58, the Continuous Memory Code 13 can be considered serviced and erased from memory. If a Continuous Memory Code 38 is present along with the Continuous Memory Code 13, service the 38 first.

If these other codes were not present make the following checks:

- Refer to KB1 and check for FULL travel of the DC motor shaft. Replace the DC motor if full travel is not possible. Leave the motor fully retracted.
- With the DC motor fully retracted and the ITS not touching the throttle lever (ITS closed circuit) check for an intermittent open in the ITS circuit. Turn the ignition key off and install the breakout box. Make the necessary connector/pin inspections. With the DVOM on the 200 ohm scale, monitor between test Pins 28 and 46 while tapping, wiggling, bending, etc. the DC motor connector and harness.

  DO NOT PUSH IN THE IDLE TRACKING SWITCH.

  The DVOM will change from less than 5 ohms to greater than 5 ohms if an open circuit is created. Service as necessary.
- Check for a sticking TP sensor by monitoring TP voltage while moving the throttle from a wide-open position to a closed throttle position. To do this it is necessary to install the breakout box. Make the necessary connector/pin inspections. It is also necessary to fully retract the shaft of the DC motor by placing a jumper between test Pins 41 and 57. When the motor has fully retracted, disconnect it at the harness and remove the jumper wire from the breakout box. With the ignition key on and the DVOM on the 20 volt scale, slowly move the throttle from wide-open to closed throttle. The voltage should move from more than 4 volts to less than 1.5 volts. If the TP sensor hangs up in midrange replace it; otherwise no service should be made.

DC Motor Idle Speed Control/Idle Tracking Switch and Throttle Position Sensor—1.9L CFI and 2.5L CFI | Pinpoint Test | KB

KB91 CONTINUOUS MEMORY CODE 38

A Continuous Memory Code 38 indicates that in the last 40 warm-up cycles the Idle Tracking Switch was open (ITS touching the throttle) when the throttle angle was greater than the MAX extension of the DC motor shaft. This could be caused by:

- An open (either intermittent or hard fault) in the ITS circuit.
- Idle Tracking Switch stuck open (pushed in position).

Either of these conditions may cause a code 58 to appear in Key On Engine Off (KOEO). If service has been made for a KOEO code 58, the Continuous Memory Code 38 can be considered serviced and erased from memory.

If KOEO code 58 was not present the following checks can be made:

- With the DC motor fully retracted and the ITS not touching the throttle lever (ITS closed circuit) check for an intermittent open in the ITS circuit. Turn the ignition key off and install the breakout box. Make the necessary connector/pin inspections. With the DVOM on the 200 ohm scale, monitor between Test Pins 28 and 46 while tapping, wiggling, bending, etc. the DC motor connector and harness.

  DO NOT PUSH IN THE IDLE TRACKING SWITCH.

  The DVOM will change from less than 5 ohms to greater than 5 ohms if an open circuit is created. Service as necessary. If an open circuit cannot be created, no service should be made.

KB92 CONTINUOUS MEMORY CODE 71

A Continuous Memory Code 71 indicates that sometime in the last 40 warm-up cycles the Idle Tracking Switch was closed (ITS not touching the throttle lever) when the DC motor was in "preposition"—[after the engine has been running and the ignition key is turned off the DC motor fully retracts and then extends to a predetermined position for the next start-up]. This can be caused by:

- The ITS circuit shorted to GROUND or SIGNAL RETURN (intermittent or hard fault).
- ITS stuck closed (ITS **NOT** in the pushed in position).

Either of these conditions may cause a Key On Engine Off (KOEO) code 68. If service has been made for KOEO code 68, the Continuous Memory Code 71 can be considered serviced and erased from memory.

If KOEO 68 was not present make the following checks:

- Check the ITS circuit for an intermittent short to ground or Signal Return. Turn the ignition key off. Enter the KOEO Continuous Monitor Mode per Quick Test Step 6.0B. Systematically tap, wiggle, or bend the harness while looking for an indication of a fault. If a fault is created, service as necessary; otherwise no service should be made.

NOTE: Due to the nature of this Test Step, Code 71 will not reappear in memory if a fault is found.

DC Motor Idle Speed Control/Idle Tracking Switch and Throttle Position Sensor—1.9L CFI and 2.5L CFI | Pinpoint Test | KB

TEST STEP	RESULT ▶	ACTION TO TAKE
KB93 CONTINUOUS TEST SERVICE CODE 53: EXERCISE TP SENSOR		
• Enter Key On Engine Off. • Using continuous monitor mode. Refer to Appendix in Section 16. • Observe VOM or STAR LED for indication of a fault while performing the following: • Move throttle slowly to WOT position. • Release throttle slowly to closed position and lightly tap on TP sensor (simulate road shock). • Wiggle TP harness connector. • Is a fault indicated?	Yes ▶ No ▶	GO to KB94. GO to KB95.

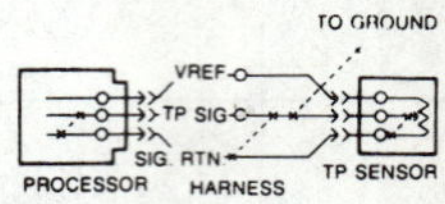

TEST STEP	RESULT ▶	ACTION TO TAKE
KB94 MEASURE THROTTLE POSITION SIGNAL VOLTAGE WHILE EXERCISING TP SENSOR		
• Key off, wait 10 seconds. • Disconnect processor 60 pin connector. Inspect for damaged pins, corrosion, loose wires, etc. Service as necessary. • Install breakout box and reconnect processor. • DVOM on 20 volt scale. • Connect a DVOM from Test Pin 47 to Test Pin 46. • Key on, engine off. • While observing DVOM, perform the following: — Move throttle slowly to closed position and lightly tap on TP sensor (simulating road shock). — Wiggle TP harness and connector. • Does the fault occur below 4.25 volts?	Yes ▶	DISCONNECT and INSPECT connectors. If connector and terminals are good, REPLACE TP sensor. CLEAR Continuous Memory. RERUN Quick Test.
	No ▶	Throttle position sensor overtravel may have caused the Continuous Memory Code 53. Sensor service is not required. To verify harness integrity, GO to KB95.

DC Motor Idle Speed Control/Idle Tracking Switch and Throttle Position Sensor—1.9L CFI and 2.5L CFI | Pinpoint Test | KB

TEST STEP	RESULT ▶	ACTION TO TAKE
KB95 CHECK EEC-IV HARNESS		
• Enter Key On Engine Off Continuous Monitor mode. • Observe VOM or STAR LED for a fault indication while performing the following: Referring to the illustration in Step DH90, grasp the harness close to the sensor connector. Wiggle, shake or bend a small section of the EEC-IV system harness while working your way to the dash panel. Also wiggle, shake or bend the EEC-IV harness from the dash panel to the processor. • Is a fault indicated?	Yes ▶ No ▶	ISOLATE fault and make necessary repairs. CLEAR Continuous Memory. RERUN Quick Test GO to KB96.
KB96 CHECK PROCESSOR AND HARNESS CONNECTORS		
• Key off, wait 10 seconds. • Disconnect processor 60 pin connector. • Inspect both connectors and connector terminals for obvious damage or faults. • Are connectors and terminals OK?	Yes ▶	Unable to duplicate fault at this time.
	No ▶	SERVICE as necessary. CLEAR Continuous Memory. REPEAT Quick Test.
KB97 CONTINUOUS TEST SERVICE CODES 23 OR 63 EXERCISE TP SENSOR		
• Using Key On Engine Off Continuous Monitor mode, observe VOM or STAR LED for indication of a fault while performing the following: • Move throttle slowly to WOT position. • Release throttle slowly to closed condition. • Lightly tap on TP sensor (simulate road shock). • Wiggle TP harness connector. • Is a fault indicated?	Yes ▶	DISCONNECT and INSPECT connectors. If connector and terminals are good, REPLACE TP sensor. CLEAR Continuous Memory. RERUN Quick Test.
	No ▶	GO to KB98.

DC Motor Idle Speed Control/Idle Tracking Switch and Throttle Position Sensor — 1.9L CFI and 2.5L CFI — Pinpoint Test KB

TEST STEP	RESULT ▶	ACTION TO TAKE
KB98 CHECK EEC-IV HARNESS		
• Enter Key On Engine Off Continuous Monitor mode. • Observe VOM or STAR LED for a fault indication while performing the following: Referring to the illustration in Step DH94, grasp the harness close to the sensor connector. Wiggle, shake or bend a small section of the EEC-IV system harness while working your way to the dash panel. Also wiggle, shake or bend the EEC-IV harness from the dash panel to the processor. • Is a fault indicated?	Yes ▶	ISOLATE fault and make necessary repairs. CLEAR Continuous Memory. RERUN Quick Test.
	No ▶	GO to KB99.
KB99 CHECK PROCESSOR AND HARNESS CONNECTORS		
• Key off, wait 10 seconds. • Disconnect processor 60 pin connector. • Inspect both connectors and connector terminals for obvious damage or faults. • Are connectors and terminals OK?	Yes ▶	Unable to duplicate fault at this time.
	No ▶	SERVICE as necessary. CLEAR Continuous Memory. RERUN Quick Test.

Air Management System — Pinpoint Test KC

Note

You should enter this Pinpoint Test only when a Service Code 44, 45, 46, 81, 82 or 94 is received in Quick Test Step 3.0 or 5.0.

Remember

To prevent the replacement of good components, be aware that the following non-EEC areas may be at fault:

- Thermactor System
  - Belt
  - Pump
  - Valve

This Pinpoint Test is intended to diagnose only the following:

- TAB and TAD Solenoid Valve Assemblies
- Harness Circuits: TAB, TAD and VPWR
- Vacuum Supply
- Processor Assembly

Pinpoint Test Schematic

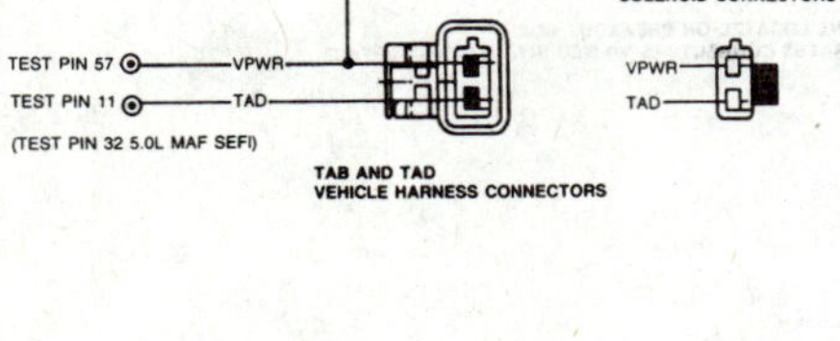

Air Management System — Pinpoint Test KC

TEST STEP	RESULT ▶	ACTION TO TAKE
KC1 SERVICE CODES 44 (94), 45 AND 46: VERIFY VACUUM LINE ROUTING		
• Verify proper vacuum line routing to the TAB/TAD solenoids and to the bypass diverter valve. Refer to VECI decal. • Check for kinked or blocked vacuum lines. • Check for kinked or blocked air hoses. • Check for disconnected or cracked vacuum lines. • Are visual checks satisfactory?	No ▶	SERVICE routing or faults. RERUN Quick Test.
	Yes ▶	Service Code 44 (94), GO to KC4. Service Code 45, GO to KC2. Service Code 46, GO to KC3.
KC2 ATTEMPT TO ELIMINATE SERVICE CODE 45 (TAD ONLY)		
• Disconnect vacuum line on diverter valve and cap vacuum line. • Key off, wait 10 seconds. • Repeat Engine Running Self-Test and record service codes. • Is Code 45 present?	Yes ▶	EEC-IV system OK. REFER to Section 3 for diverter valve or check valve diagnostics.
	No ▶	GO to KC4.
KC3 ATTEMPT TO ELIMINATE SERVICE CODE 46 (TAB ONLY)		
• Disconnect vacuum line on bypass valve and cap vacuum line. • Key off, wait 10 seconds. • Repeat Engine Running Self-Test and record codes. • Is Code 46 present?	Yes ▶	EEC-IV system OK.
	No ▶	GO to KC4.

Air Management System — Pinpoint Test KC

TEST STEP	RESULT ▶	ACTION TO TAKE
KC4 ENTER OUTPUT STATE CHECK (REFER TO APPENDIX IN QUICK TEST)		
NOTE: Do not use STAR tester for this Step, use a VOM/DVOM. • Key off, wait 10 seconds. • DVOM on 20 volt scale. • Connect DVOM negative test lead to STO circuit at the Self-Test connector and positive test lead to battery positive. • Jumper STI circuit to SIGNAL RETURN at the Self-Test connector. • Perform Key On, Engine Off Self-Test until the completion of the Continuous Memory Test Codes. • DVOM will indicate zero volts when Test is complete. • Depress and release the throttle. • Did DVOM change to a high voltage?	Yes ▶	REMAIN in Output State Check. GO to KC5.
	No ▶	DEPRESS throttle to WOT and RELEASE. If STO voltage does not go high, GO to Pinpoint Test Step QC1. Leave equipment hooked up.
KC5 CHECK TAB/TAD SOLENOID ELECTRICAL OPERATION		
• DVOM on 20 volt scale. • Disconnect TAB and TAD solenoids. • Connect DVOM positive test lead to VPWR circuit and negative test lead to TAB circuit on TAB solenoid vehicle harness connector. • While observing DVOM depress and release the throttle several times (to cycle output On and Off). • Repeat for TAD solenoid. Connect positive test lead to VPWR circuit and negative test lead to TAD circuit on TAD solenoid vehicle harness connector. • Do both solenoids cycle On and Off?	Yes ▶	GO to KC6.
	No ▶	REMOVE jumper. GO to KC9.

Air Management System — Pinpoint Test KC

TEST STEP	RESULT ▶	ACTION TO TAKE
KC6 CHECK TAB/TAD SOLENOID FOR VACUUM CYCLING		
• Install vacuum pump to the TAB solenoid vacuum supply port and install a vacuum gauge to the output port. • While cycling outputs On and Off (by depressing and releasing throttle), observe the vacuum gauge at the output. **NOTE: Maintain vacuum at source.** • Repeat for TAD solenoid. Connect vacuum pump to the TAD solenoid vacuum supply port and connect a vacuum gauge to the output port. • Cycle output On and Off. • Do both vacuum outputs cycle On and Off?	Yes ▶ No ▶	GO to KC7. REPLACE solenoid assembly. RERUN Quick Test.
KC7 CHECK MANIFOLD VACUUM LINES FOR BLOCKAGE OR LEAKS		
• Vacuum lines disconnected at TAD/TAB solenoids. • Start engine. • Check for vacuum. • Is vacuum present at the solenoids?	Yes ▶ No ▶	EEC-IV system OK. REFER to Section 3 for Thermactor valve and air pump diagnostics. SERVICE vacuum source blockage or leak. RERUN Quick Test.
KC8 CHECK VOLTAGE OF VPWR CIRCUIT		
• Key on, engine off. • DVOM on 20 volt scale. • Measure voltage between TAB solenoid VPWR circuit and battery ground. • Repeat for TAD solenoid. • Are both voltage greater than 10.5 volts?	Yes ▶ No ▶	GO to KC9. RECONNECT TAB/TAD solenoids. SERVICE harness circuit open. RERUN Quick Test.

Air Management System — Pinpoint Test KC

TEST STEP	RESULT ▶	ACTION TO TAKE
KC11 CHECK FOR SHORT TO GROUND		
• Key Off, wait 10 seconds. • DVOM on 200,000 ohm. • Leave breakout box installed and processor disconnected. • Disconnect TAB/TAD solenoids. • Measure resistance between Test Pin 51 (Test Pin 38 for 5.0L MAF SEFI) and Test Pins 40, 46 and 60 and between Test Pin 11 (Test Pin 32 for 5.0L MAF SEFI) and Test Pins 40, 46 and 60 at the breakout box. • Are all resistances greater than 10,000 ohms?	Yes ▶ No ▶	GO to KC12. SERVICE short to ground. REMOVE breakout box. RECONNECT processor and TAB/TAD solenoids. RERUN Quick Test.
KC12 CHECK FOR SHORT TO POWER		
• Key off, wait 10 seconds. • DVOM on 200,000 ohm scale. • Leave breakout box installed and processor disconnected. • TAB/TAD solenoids disconnected. • Measure resistance between Test Pin 51 (Test Pin 38 for 5.0L MAF SEFI) and Test Pins 37 and 57, and between Test Pin 11 (Test Pin 32 for 5.0L MAF SEFI) and Test Pins 37 and 57 at the breakout box. • Are all resistances greater than 10,000 ohms?	Yes ▶ No ▶	REMOVE breakout box. RECONNECT TAB/TAD solenoid. REPLACE processor. RERUN Quick Test. REMOVE breakout box. RECONNECT processor and TAB/TAD solenoids. SERVICE short to power. RERUN Quick Test. If code is present, REPLACE processor.

Air Management System — Pinpoint Test KC

TEST STEP	RESULT ▶	ACTION TO TAKE
KC9 MEASURE TAB/TAD SOLENOID RESISTANCE		
• Key off, wait 10 seconds. • DVOM on 200 ohm scale. • Disconnect TAB solenoid connector and measure solenoid resistance. • Disconnect TAD solenoid connector and measure solenoid resistance. • Are both solenoid resistances between 50 and 100 ohms?	Yes ▶ No ▶	GO to KC10. REPLACE TAB/TAD solenoid assembly. RERUN Quick Test.
KC10 CHECK CONTINUITY OF TAB AND TAD CIRCUITS		
• Key off, wait 10 seconds. • Disconnect processor 60 pin connector. Inspect for damaged pins, corrosion, loose wires, etc. Service as necessary. • Connect breakout box to harness. Leave processor disconnected. • DVOM on 200 ohm scale. • Measure resistance between Test Pin 51 (Test Pin 38 for 5.0L MAF SEFI) at breakout box and TAB circuit at vehicle harness connector. • Measure resistance between Test Pin 11 (Test Pin 32 for 5.0L MAF SEFI) at the breakout box and TAD circuit at vehicle harness connector. • Are both solenoid resistances less than 5.0 ohms?	Yes ▶ No ▶	GO to KC11. REMOVE breakout box. RECONNECT processor and TAB/TAD solenoids. SERVICE harness open circuit. RERUN Quick Test.

Canister Purge (CANP) — Pinpoint Test KD

Note

You should enter this Pinpoint Test only when a Service Code 85 is received in Quick Test Step 3.0 or when you are directed here from a fuel control Pinpoint Test and Diagnostic By Symptom in the Engine Supplement Section.

Remember

This Pinpoint Test is intended to diagnose only the following:

- Harness Circuits: CANP and VPWR
- Processor Assembly

Pinpoint Test Schematic

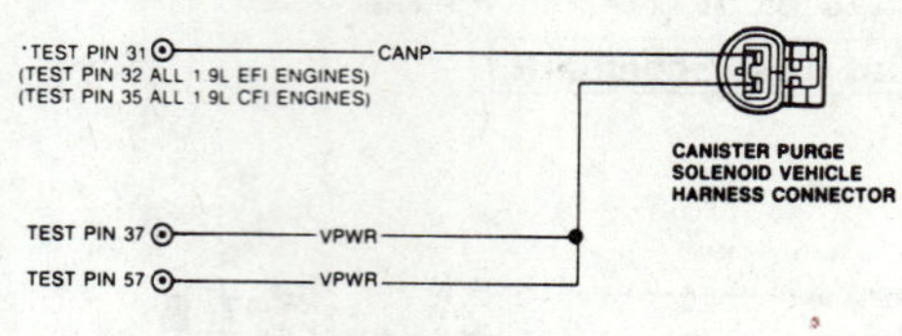

*TEST PINS LOCATED ON BREAKOUT BOX.
ALL HARNESS CONNECTORS VIEWED INTO MATING SURFACE.

Canister Purge (CANP) — Pinpoint Test KD

TEST STEP	RESULT ▶	ACTION TO TAKE
KD1 ENTER OUTPUT STATE CHECK (REFER TO APPENDIX)		
NOTE: Do not use STAR tester for this step, use VOM/DVOM. • Key off, wait 10 seconds. • DVOM on 20 volt scale. • Connect DVOM negative test lead to STO circuit at Self-Test connector and positive test lead to battery positive. • Jumper STI circuit to SIGNAL RETURN at the Self-Test connector. • Perform Key On Engine Off Self-Test until the completion of the Continuous Test Codes. • DVOM will indicate less than 1.0 volt when test is completed. • Depress and release the throttle. • Does voltage increase?	Yes ▶	REMAIN in Output State Check. GO to KD2.
	No ▶	DEPRESS throttle to WOT and release. If STO voltage does not go high, GO to Pinpoint Test Step QC1. Leave equipment hooked up.
KD2 CHECK CANISTER PURGE (CANP) SOLENOID ELECTRICAL OPERATION		
• Key on engine off. • Disconnect CANP solenoid. • Connect DVOM positive test lead to VPWR circuit and negative test lead to CANP output circuit on the vehicle harness connector. • DVOM on 20 volt scale. • While observing DVOM depress and release the throttle several times to cycle output On and Off. • Does CANP circuit cycle On and Off?	Yes ▶	GO to KD3.
	No ▶	REMOVE jumper. GO to KD6.
KD3 CHECK CANISTER PURGE SOLENOID FOR VACUUM LEAKS		
• Key on. • CANP solenoid disconnected. • Disconnect vacuum hose at canister purge solenoid on PCV side. • Apply 16 in. Hg. (53 kPa) of vacuum to PCV side of CANP solenoid. • Does CANP solenoid hold vacuum for 20 seconds?	Yes ▶	REMAIN in output state check. Leave vacuum pump setup in place. GO to KD4.
	No ▶	REPLACE CANP solenoid. RERUN Quick Test. If symptom is still present,

Canister Purge (CANP) — Pinpoint Test KD

TEST STEP	RESULT ▶	ACTION TO TAKE
KD4 CHECK CANISTER PURGE SOLENOID FOR MECHANICAL OPERATION		
• Key on, engine off. • CANP circuit Off (no voltage). • Reconnect CANP solenoid connector. • Apply 16 in.-Hg. (53 kPa) of vacuum to PCV side of CANP solenoid. • Depress and release throttle. • Is vacuum released?	Yes ▶	CHECK hose from solenoid to canister for cracks, leaks, etc. If OK, REMOVE Jumper from STI to SIGNAL RETURN. GO to KD5.
	No ▶	CHECK hose from solenoid to canister for blockage or kinks. If OK, REPLACE CANP solenoid. RERUN Quick Test.
KD5 CHECK FOR VACUUM TO CANISTER PURGE SOLENOID		
• Disconnect vacuum hose at canister purge solenoid at PCV side. • Start engine. • Is vacuum present at engine vacuum hose?	Yes ▶	EEC-IV system OK.
	No ▶	CHECK vacuum line for proper routing, kinks or blockage. If OK,
KD6 MEASURE CANP SOLENOID RESISTANCE		
• Key off, wait 10 seconds. • DVOM on 200 ohm scale. • Disconnect CANP solenoid. • Measure solenoid resistance. • Is resistance between 40 and 90 ohms?	Yes ▶	GO to KD7.
	No ▶	REPLACE CANP solenoid. RERUN Quick Test.
KD7 CHECK VOLTAGE OF VPWR CIRCUIT		
• Key on, engine off. • CANP solenoid disconnected. • DVOM on 20 volt scale. • Measure voltage between VPWR at the CANP solenoid vehicle harness connector and battery ground. • Is voltage greater than 10.5 volts?	Yes ▶	GO to KD8.
	No ▶	RECONNECT CANP solenoid. SERVICE harness open circuit. RERUN Quick Test.

Canister Purge (CANP) — Pinpoint Test KD

TEST STEP	RESULT ▶	ACTION TO TAKE
KD8 CHECK CONTINUITY OF CANP CIRCUIT		
• Key off, wait 10 seconds. • CANP solenoid disconnected. • Disconnect processor 60 pin connectors. Inspect for damaged pins, corrosion, loose wires, etc. Service as necessary. • Install breakout box, leave processor disconnected. • DVOM on 200 ohm scale. • Measure resistance between Test Pin 31 (Test Pin 35 for 1.9L CFI engines, Test Pin 32 for 1.9L EFI engines) at the breakout box and CANP on the vehicle harness connector. • Is resistance less than 5 ohms?	Yes ▶	GO to KD9.
	No ▶	REMOVE breakout box. RECONNECT processor and CANP solenoid. SERVICE open circuit. RERUN Quick Test.
KD9 CHECK FOR SHORT TO GROUND		
• Key off, wait 10 seconds. • Breakout box installed, processor disconnected. • CANP solenoid disconnected. • DVOM on 200,000 ohm scale. • Measure resistance between Test Pin 31 (Test Pin 35 for 1.9L CFI engines, Test Pin 32 for 1.9L EFI engines) and Test Pins 40, 46 and 60 at the breakout box. • Are all resistances greater than 10,000 ohms?	Yes ▶	GO to KD10.
	No ▶	REMOVE breakout box. RECONNECT processor and CANP solenoid. SERVICE short to ground. RERUN Quick Test.
KD10 CHECK FOR SHORT TO POWER		
• Key off, wait 10 seconds. • CANP solenoid disconnected. • Breakout box installed, processor disconnected. • DVOM on 200,000 ohm scale. • Measure resistance between Test Pin 31 (Test Pin 35 for 1.9L CFI engines, Test Pin 32 for 1.9L EFI engines) and Test Pins 37 and 57 at the breakout box. • Are both resistances greater than 10,000 ohms?	Yes ▶	RECONNECT CANP solenoid. REMOVE breakout box. REPLACE processor. RERUN Quick Test.
	No ▶	REMOVE breakout box. RECONNECT processor and CANP solenoid. SERVICE short to power. REPEAT Quick Test. If code is repeated, REPLACE processor. RERUN Quick Test.

Idle Speed Control (Bypass Air) — Pinpoint Test KE

Note

You should enter this Pinpoint Test only when a Service Code 12, 13, 16, 17, 47 or 48 is received in Quick Test Step 5.0 or when directed here from Diagnostic By Symptom in the Engine Supplement Section.

Remember

To prevent the replacement of good components, be aware that the following non-EEC areas may be at fault:

- Engine not up to operating temperature
- Engine over operating temperature
- Improper Idle Speed Throttle Stop Adjustment
- A/C input (electrical problem)
- Throttle Speed Control Linkage
- Throttle Sticking or Linkage Binding.

This Pinpoint Test is intended to diagnose only the following:

- Rpm in Self-Test only
- ISC Solenoid
- Harness Circuits ISC and VPWR
- Processor Assembly

Pinpoint Test Schematic

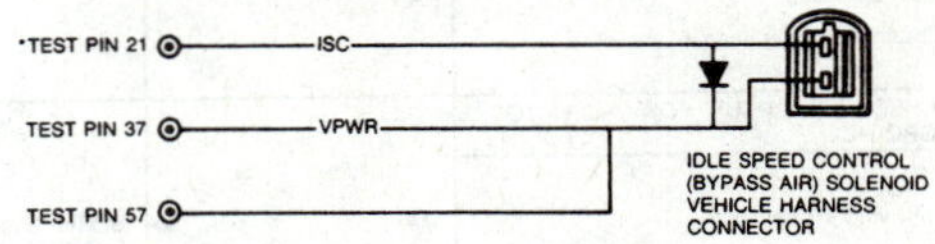

*TEST PINS LOCATED ON BREAKOUT BOX.
ALL HARNESS CONNECTORS VIEWED INTO MATING SURFACE.

Idle Speed Control (Bypass Air) — Pinpoint Test KE

TEST STEP	RESULT	ACTION TO TAKE
KE1 CHECK FOR RPM DROP		
• Key off. • Connect engine tachometer. • Start engine. • Disconnect ISC harness connector. • Does rpm drop or stall?	Yes ▶	GO to KE2.
	No ▶	GO to KE3.
KE2 CHECK FOR EGR CODES		
• Are Service Codes 31, 32, 33 or 34 present?	Yes ▶	RECONNECT ISC solenoid. GO to Quick Test Step 5.0 for appropriate Pinpoint Test.
	No ▶	GO to KE3.
KE3 CHECK FOR OTHER EEC CODES		
• Are Service Codes 22, 41, 42, 91 or 92 present?	Yes ▶	RECONNECT ISC solenoid. For 1.9L EFI with Code 42 present, GO to KE4. All others GO to Quick Test Step 5.0 for appropriate Pinpoint Test.
	No ▶	GO to KE4.
KE4 MEASURE ISC SOLENOID RESISTANCE		
• Key off. • ISC solenoid disconnected. • DVOM on 200 ohm scale. • Measure solenoid resistance. • Is resistance between 7.0 and 13.0 ohms?	Yes ▶	GO to KE5.
	No ▶	REPLACE ISC solenoid. RERUN Quick Test.
KE5 CHECK FOR INTERNAL SHORT TO ISC SOLENOID CASE		
• Key off. • ISC solenoid disconnected. • DVOM on 200,00 ohm scale. • Measure resistance from either ISC solenoid pin to ISC housing. • Is resistance greater than 10,000 ohms?	Yes ▶	GO to KE6.
	No ▶	REPLACE ISC solenoid. RERUN Quick Test.

Idle Speed Control (Bypass Air) — Pinpoint Test KE

TEST STEP	RESULT	ACTION TO TAKE
KE6 CHECK VOLTAGE OF VPWR CIRCUIT		
• Key on, engine off. • ISC solenoid disconnected. • DVOM on 20 volt scale. • Measure voltage between VPWR at the ISC solenoid harness connector and battery ground. • Is voltage less than 10.5 volts?	Yes ▶	SERVICE open circuit. RERUN Quick Test.
	No ▶	GO to KE7.
KE7 CHECK CONTINUITY OF ISC CIRCUIT		
• Key off, wait 10 seconds. • ISC solenoid disconnected. • Disconnect processor and inspect both 60 pin connectors for damaged pins, corrosion, loose wires, etc. Service as necessary. • Install breakout box, leave processor disconnected. • DVOM on 200 ohm scale. • Measure resistance between Test Pin 21 at the breakout box and ISC circuit at vehicle harness connector. • Is resistance greater than 5 ohms?	Yes ▶	SERVICE open circuit. REMOVE breakout box. RECONNECT processor and ISC solenoid. RERUN Quick Test.
	No ▶	GO to KE8.
KE8 CHECK ISC CIRCUIT FOR SHORT TO GROUND		
• Key off, wait 10 seconds. • Breakout box installed and processor disconnected. • ISC solenoid disconnected. • DVOM on 200,000 ohm scale. • Measure resistance between Test Pin 21 and Test Pins 40, 46 and 60 at the breakout box. • Is any resistance less than 10,000 ohms? **NOTE: Placement of probes for DVOM D.**	Yes ▶	SERVICE short circuit. REMOVE breakout box. RECONNECT processor and ISC solenoid. RERUN Quick Test.
	No ▶	GO to KE9.

Idle Speed Control (Bypass Air) — Pinpoint Test KE

TEST STEP	RESULT	ACTION TO TAKE
KE9 CHECK FOR SHORT TO POWER		
• Key off, wait 10 seconds. • ISC solenoid disconnected. • Breakout box installed and processor disconnected. • DVOM on 200,000 ohm scale. • Measure resistance between Test Pin 37 and Test Pin 21 at the breakout box. • Is resistance greater than 10,000 ohms?	Yes ▶	GO to KE10.
	No ▶	SERVICE short circuit. REMOVE breakout box. RECONNECT processor and ISC solenoid. RERUN Quick Test. If code or symptom is present, REPLACE processor.
KE10 CHECK FOR ISC SIGNAL FROM THE PROCESSOR		
• Key off. • Reconnect ISC solenoid. • Breakout box installed. • Connect processor to breakout box. • DVOM on a 20 volt scale. • Connect DVOM between Test Pin 21 and Test Pin 40. • Start engine. • Slowly increase and decrease rpm. • Does DVOM voltage vary?	Yes ▶	GO to KE11.
	No ▶	REMOVE breakout box. REPLACE processor. RERUN Quick Test.
KE11 CHECK CURB IDLE		
• Is curb idle within specification? • Refer to Section 4 for curb idle set procedure.	Yes ▶	REMOVE breakout box. RECONNECT processor. REPLACE ISC solenoid. RERUN Quick Test.
	No ▶	REMOVE breakout box. RECONNECT processor. CHECK engine vacuum hoses. REFER to VECI decal. CHECK that throttle plates are fully closed, CHECK throttle linkage and/or speed control linkage for binding. If OK, REPLACE ISC solenoid. RERUN Quick Test.

Idle Speed Control (Bypass Air) — Pinpoint Test KE

TEST STEP	RESULT	ACTION TO TAKE
KE12 CODE 19: LOW ISC RPM		
• Key off, wait 10 seconds. • Deactivate Self-Test. • Run engine at 2000 rpm for 2 minutes or until inlet radiator hose is hot and pressurized. • Key off, wait 10 seconds. • Rerun Engine Running Self-Test. • Does engine stumble and/or is code 19 still present?	Yes ▶	INSPECT throttle body and air inlet for contamination. SERVICE as necessary. If OK, ADJUST curb idle. RERUN Quick Test.
	No ▶	SERVICE other codes as necessary.
KE13 CODE 17: LOW ISC RPM		
NOTE: Check and correct excessive engine load problems like cooling fan running, lights on, etc. • Run engine at 2000 rpm for 2 minutes or until inlet radiator hose is hot and pressurized. • Key off, wait 10 seconds. • Rerun Engine Running Self-Test. • Is Code 17 still present?	Yes ▶	INSPECT throttle body and air inlet for contamination. SERVICE as necessary. If OK ADJUST curb idle. RERUN Quick Test.
	No ▶	SERVICE other codes as necessary.
KE15 CODE 13: HIGH ISC RPM		
• Key off. • Connect tachometer. • Start engine. • Disconnect ISC harness connector. • Does rpm drop or engine stall?	Yes ▶	INSPECT throttle body and air inlet for contamination. SERVICE as necessary. If OK, REPLACE processor. RERUN Quick Test.
	No ▶	CHECK engine vacuum hoses. REFER to VECI decal. VERIFY curb idle. CHECK that throttle plates are fully closed, CHECK throttle linkage and/or speed control linkage for binding. If OK, REPLACE ISC solenoid. RERUN Quick Test.

Idle Speed Control (Bypass Air) — Pinpoint Test KE

TEST STEP	RESULT ▶	ACTION TO TAKE
KE20 CODE 47: CHECK FOR LOW FLOW UNMETERED AIR		
• Check for holes, cracks, and/or disconnections in fuel charging assembly (manifold gaskets, vacuum lines, vacuum tree, etc). • Check for stuck-open purge solenoid and/or injector O-rings. • Are any faults present?	Yes ▶	SERVICE as necessary. RERUN Quick Test.
	No ▶	EEC system OK for metered air. GO to Quick Test Step 5.0B to service other code if necessary.
KE21 CODE 48: CHECK FOR HIGH FLOW UNMETERED AIR		
• Check for holes, cracks, and/or disconnections in air cleaner outlet tube (between vane airflow meter and fuel charging assembly). • Check for loss of ignition or fuel on one or more cylinder(s).	Yes ▶	SERVICE as necessary. RERUN Quick Test.
	No ▶	EEC system OK for metered air. GO to Quick Test Step 5.0B to SERVICE other code if necessary.
KE22 CODE 16: HIGH ISC RPM		
• Is Code 48 present?	Yes ▶	RESET throttle plate. REFER to Section 4 and VECI decal for curb idle set procedure. RERUN Quick Test. If Code 48 is still present, GO to KE2.
	No ▶	CHECK for vacuum leaks at injector O-rings, purge solenoid, vacuum lines/fittings, excessive PCV, or inlet pin leak between air meter and throttle body. SERVICE as necessary. RERUN Quick Test.

Shift Indicator Light (SIL) — Pinpoint Test KL

Pinpoint Test Schematic

ALL OTHERS

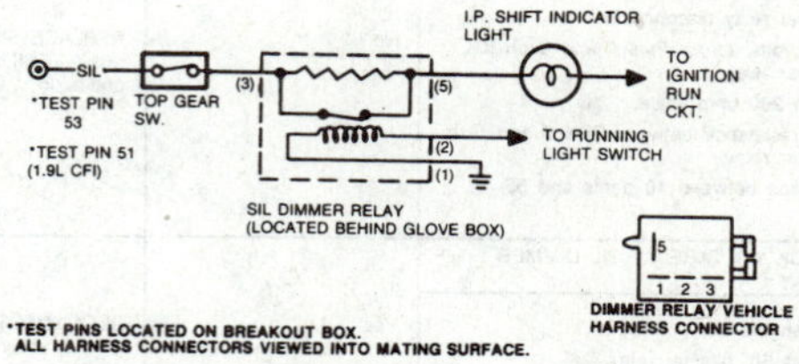

*TEST PINS LOCATED ON BREAKOUT BOX.
ALL HARNESS CONNECTORS VIEWED INTO MATING SURFACE.

Shift Indicator Light (SIL) — Pinpoint Test KL

Note

You should enter this Pinpoint Test only when directed here from Diagnostic By Symptom in the Engine Supplement Section or from Pinpoint Test Step QA9.

Remember

This Pinpoint Test is intended to diagnose only the following:

- Harness Circuits: SIL and STO
- Top Gear Switch
- SIL Dimmer Relay
- Shift Indicator Light Bulb and Fuse

Pinpoint Test Schematic

1.9L EFI

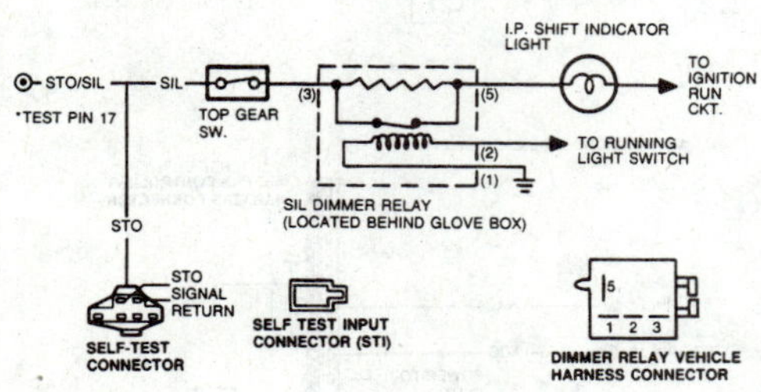

*TEST PINS LOCATED ON BREAKOUT BOX.
ALL HARNESS CONNECTORS VIEWED INTO MATING SURFACE.

Shift Indicator Light (SIL) — Pinpoint Test KL

TEST STEP	RESULT ▶	ACTION TO TAKE
KL1 CHECK SIL OPERATION		
NOTE: To verify SIL operation, inspect the SIL while driving the vehicle. The SIL should turn on when the optimum shift speed is reached in each gear and remain off while in the highest gear. If the SIL is always on, look for a short to ground in the SIL circuit. If the SIL is always off, look for an open in the SIL circuit. • Is SIL on all the time?	Yes ▶	GO to KL6.
	No ▶	GO to KL2.
KL2 CHECK SIL CIRCUIT FUSE		
• Key off, wait 10 seconds. • Remove SIL circuit fuse (#18) and inspect. • Is fuse OK?	Yes ▶	GO to KL3.
	No ▶	SERVICE short to ground between fuse and SIL bulb. REPLACE SIL fuse. VERIFY SIL operation.
KL3 CHECK SIL BULB		
• Key off. • Remove SIL bulb and inspect. • Is bulb OK?	Yes ▶	GO to KL4.
	No ▶	REPLACE SIL bulb. VERIFY SIL operation.
KL4 CHECK SIL DIMMER RELAY CONTINUITY		
• Key off. • Disconnect SIL dimmer relay. • DVOM on 200 ohm scale. • Measure resistance between Pins 3 and 5 on SIL dimmer relay. • Is resistance less than 5 ohms?	Yes ▶	GO to KL5.
	No ▶	REPLACE SIL dimmer relay. VERIFY SIL operation.

Shift Indicator Light (SIL) — Pinpoint Test KL

TEST STEP	RESULT	ACTION TO TAKE
KL5 CHECK SIL DIMMER RELAY FUNCTION		
• Key off. • SIL dimmer relay disconnected. • Apply 12 volts across Pins 1 and 2 on the SIL dimmer relay. • DVOM on 200 ohm scale. • Measure resistance between Pins 3 and 5 on SIL dimmer relay. • Is resistance between 40 ohms and 55 ohms?	Yes ▶ No ▶	GO to KL6. REPLACE SIL dimmer relay. VERIFY SIL operation.
KL6 CHECK VOLTAGE AT SIL DIMMER RELAY		
• Key on, engine off. • Disconnect SIL dimmer relay. • DVOM on 20 volt scale. • Measure voltage between Test Pin 5 on the SIL dimmer relay vehicle harness connector and the battery negative post. • Is voltage greater than 5 volts?	Yes ▶ No ▶	RECONNECT SIL dimmer relay. GO to KL7. SERVICE circuit between SIL dimmer relay and SIL fuse. VERIFY SIL operation.
KL7 CHECK VOLTAGE AT TOP GEAR SWITCH		
• Key on, engine off. • Disconnect top gear switch. • DVOM on 20 volt scale. • Measure voltage between the SIL dimmer relay side of the top gear switch vehicle harness connector and the battery negative post. • Is voltage greater than 5 volts?	Yes ▶ No ▶	GO to KL8. SERVICE circuit between top gear switch and SIL dimmer relay. VERIFY SIL operation.
KL8 CHECK OPERATION OF TOP GEAR SWITCH		
• Key off, wait 10 seconds. • Top gear switch disconnected. • DVOM on 200 ohm scale. • Measure resistance of top gear switch while shifting the transmission from the highest gear to the next lower gear. • Does circuit open and close?	Yes ▶ No ▶	GO to KL9. REPLACE top gear switch. VERIFY SIL operation.

WOT A/C Cutoff (WAC) A/C Demand — Pinpoint Test KM

Note

You should enter this Pinpoint Test only when directed here from Diagnostic By Symptom in the Engine Supplement Section.

Remember

To prevent the replacement of good components, be aware that the following non-EEC areas may be at fault:

- Fuse
- Non-Electrical A/C components
- Refrigerant charge

This Pinpoint Test is intended to diagnose only the following:

- Harness Circuits: WAC, VPWR, GROUND, POWER-TO-CLUTCH, ACD
- WAC Relay or A/C fan controller
- Processor assembly

Pinpoint Test Schematic

APPLICATIONS: 3.8L EFI RWD, 5.0L SEFI, 5.0L SEFI MA, 2.9L EFI TRUCK, 3.0L EFI TRUCK

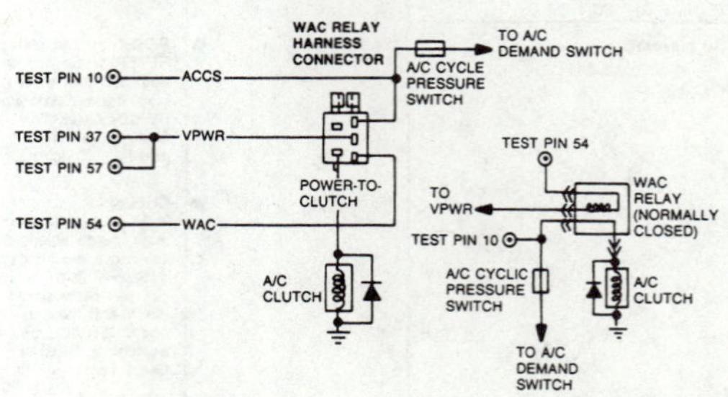

Shift Indicator Light (SIL) — Pinpoint Test KL

TEST STEP	RESULT	ACTION TO TAKE
KL9 CHECK CONTINUITY OF SIL CIRCUIT		
• Key off. • Top gear switch disconnected. • Install breakout box, leave processor disconnected. • DVOM on 200 ohm scale. • Measure resistance between Test Pin 53 (Test Pin 51 on 1.9L CFI or Test Pin 17 on 1.9L EFI) and the processor side of the top gear switch vehicle harness connector. • Is resistance less than 5 ohms?	Yes ▶ No ▶	RECONNECT top gear switch. Shift transmission into highest gear. GO to KL10. SERVICE open circuit between the top gear switch and the processor. VERIFY SIL operation.
KL10 CHECK SIL CIRCUIT FOR SHORT TO GROUND		
• Key off. • Transmission in highest gear. • Breakout box installed and processor disconnected. • DVOM on 200 ohm scale. • Measure resistance between Test Pin 53 (Test Pin 51 on 1.9L CFI or Test Pin 17 on 1.9L EFI) and Test Pin 60. • Is resistance greater than 100,000 ohms?	Yes ▶ No ▶	RERUN Quick Test and CHECK ECT and map circuits for influence on SIL system. If OK, REPLACE processor. SERVICE short to ground between top gear switch and the processor VERIFY SIL operation.

WOT A/C Cutoff (WAC) A/C Demand — Pinpoint Test KM

Pinpoint Test Schematic

APPLICATION: 2.3L EFI TRUCK

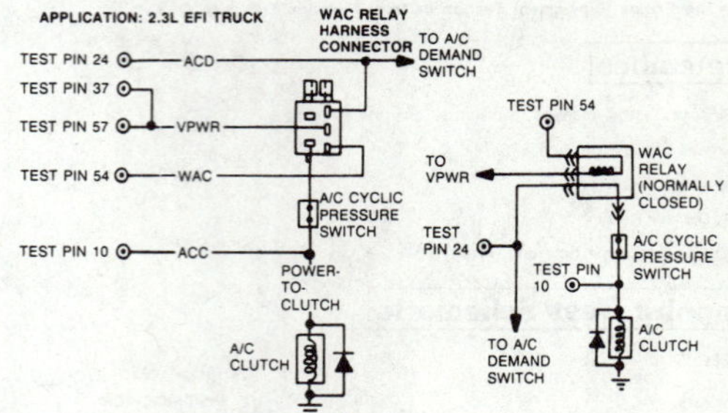

APPLICATIONS: 1.9L CFI, 1.9L EFI

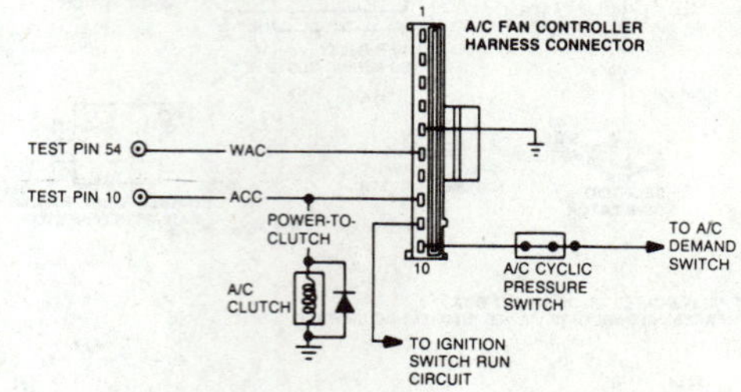

WOT A/C Cutoff (WAC) A/C Demand — Pinpoint Test KM

Pinpoint Test Schematic

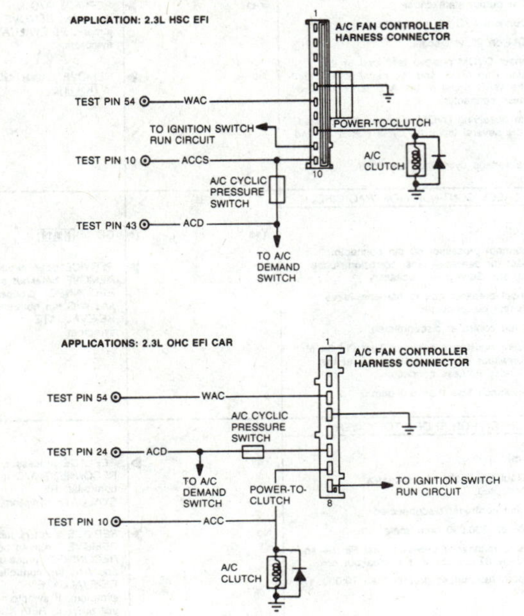

WOT A/C Cutoff (WAC) A/C Demand — Pinpoint Test KM

TEST STEP	RESULT ▶	ACTION TO TAKE
KM1 NO A/C: CHECK FOR VOLTAGE AT A/C CLUTCH		
NOTE: Before proceeding with "NO A/C" diagnostics, verify integrity of related fuses in fuse panel. • Key on, engine off. • Disconnect harness from A/C clutch. • A/C switch to A/C. • DVOM on 20 volt scale. • Measure voltage between the power side of the A/C clutch harness connector and battery negative post. • Is voltage greater than 10.5 volts?	Yes ▶ No ▶	EEC-IV system OK. GO to KM2.
KM2 CHECK CONTINUITY OF POWER-TO-CLUTCH CIRCUIT		
NOTE: Applications with WAC relay: 3.8L EFI RWD, 5.0L SEFI, 5.0L SEFI MA, 2.3L EFI TRK, 2.9L EFI TRK, 3.0L EFI TRK. **Applications with A/C fan controller: 1.9L CFI, 1.9L EFI, 2.3L HSC, 2.3L EFI OHC car.** • Key off. • A/C clutch harness disconnected. • Disconnect harness from WAC relay or A/C FAN controller. • DVOM on 200 ohm scale. • Measure resistance between power side of the A/C clutch harness connector and POWER-TO-CLUTCH pin at the WAC relay or A/C fan controller harness connector. • Is resistance less than 5.0 ohms?	Yes ▶ No ▶	RECONNECT A/C clutch. GO to KM3. 2.3L EFI TRK: VERIFY operation of A/C cyclic pressure switch. If OK, SERVICE open circuit. All others: SERVICE open circuit. RECONNECT A/C clutch and WAC relay or A/C fan controller. RE-EVALUATE symptom.

WOT A/C Cutoff (WAC) A/C Demand — Pinpoint Test KM

TEST STEP	RESULT ▶	ACTION TO TAKE
KM3 CHECK FOR POWER ON A/C DEMAND CIRCUIT		
• Key on, engine off. • WAC relay or A/C fan controller disconnected. • A/C switch to A/C. • DVOM on 20 volt scale. • Measure voltage between A/C demand input pin at WAC relay or A/C fan controller harness connector and chassis ground. • Is voltage greater than 10.5 volts?	Yes ▶ No ▶	GO to KM4. 2.3L EFI TRK: VERIFY operation of A/C demand switch. If OK, SERVICE open circuit. All others: VERIFY operation of A/C cyclic pressure switch and A/C demand switch. If OK, SERVICE open circuit. RECONNECT WAC relay or A/C fan controller. RE-EVALUATE symptom.
KM4 CHECK FOR WAC CIRCUIT SHORT TO GROUND		
• Key off. • WAC relay or A/C fan controller disconnected. • Disconnect processor 60 pin connector. Inspect for damaged pins, corrosion, loose wires, etc. Service as necessary. Leave processor disconnected. • DVOM on 200,000 ohm scale. • Measure resistance between WAC circuit at the WAC relay or A/C fan controller harness connector and chassis ground. • Is resistance greater than 10,000 ohms?	Yes ▶ No ▶	A/C fan controller applications: GO to KM5. WAC relay applications: GO to KM7. SERVICE short circuit. RECONNECT processor and WAC relay or A/C fan controller. RE-EVALUATE symptom.

WOT A/C Cutoff (WAC) A/C Demand — Pinpoint Test KM

TEST STEP	RESULT ▶	ACTION TO TAKE
KM5 CHECK FOR GROUND TO A/C FAN CONTROLLER		
• Key off. • A/C fan controller disconnected. • Processor disconnected. • DVOM on 200 ohm scale. • Measure resistance between ground circuit at A/C fan controller harness connector and chassis ground. • Is resistance less than 5.0 ohms?	Yes ▶ No ▶	GO to KM6. SERVICE open circuit. RECONNECT processor and A/C fan controller. RE-EVALUATE symptom.
KM6 CHECK FOR VOLTAGE TO A/C FAN CONTROLLER		
• Key on. • A/C fan controller disconnected. • Processor disconnected. • DVOM on 20 volt scale. • Measure voltage between Ignition Switch run circuit at the A/C fan controller harness connector and chassis ground. • Is voltage greater than 10.5 volts?	Yes ▶ No ▶	GO to KM7. SERVICE open circuit. RECONNECT processor A/C fan controller. RE-EVALUATE symptom.
KM7 CHECK WAC RELAY, A/C FAN CONTROLLER		
• Key off. • Processor disconnected. • Reconnect WAC relay or A/C fan controller. • Disconnect harness from A/C clutch. • DVOM on 20 volt scale. • Key on, engine off. • A/C switch to A/C. • Measure voltage between the power side of the A/C clutch harness connector and the battery negative post. • Is voltage greater than 10.5 volts?	Yes ▶ No ▶	REPLACE processor. RECONNECT A/C clutch. RE-EVALUATE symptom. REPLACE WAC relay or A/C fan controller. RECONNECT processor and A/C clutch. RE-EVALUATE symptom.

WOT A/C Cutoff (WAC) A/C Demand — Pinpoint Test KM

TEST STEP	RESULT ▶	ACTION TO TAKE
KM10 NO A/C CUTOFF AT WOT: ENTER OUTPUT STATE CHECK (REFER TO APPENDIX IN SECTION 16)		
NOTE: Do not use STAR tester for this Step, use VOM/DVOM. • Key off, wait 10 seconds. • DVOM on 20 volt scale. • Connect DVOM negative test lead to STO at the Self-Test connector and positive test lead to battery positive post. • Jumper STI to SIGNAL RETURN at the Self-Test connector. • Perform Key On Engine Off Self-Test until the completion of the Continuous Memory Codes. • DVOM will indicate less than 1.0 volt when test complete. • Depress and release the throttle. • Does voltage increase to greater than 10.5 volts?	Yes ▶ No ▶	REMAIN in Output State Check. GO to KM11. DEPRESS throttle to WOT and RELEASE. If STO voltage does not go high, GO to Pinpoint Test Step QC2. Leave equipment hooked up.
KM11 WAC RELAY OR A/C FAN CONTROLLER		
• WAC relay applications: 3.8L EFI RWD, 5.0L SEFI, 5.0L SEFI MA, 2.3L EFI TRK, 2.9L EFI TRK, 3.0L EFI TRK. • A/C fan controller applications: 1.9L CFI, 1.9L EFI, 2.3L EFI HSC, 2.3L EFI OHC car.	WAC relay ▶ A/C fan controller ▶	GO to KM12. GO to KM17.
KM12 CHECK FOR VPWR TO RELAY		
• Still in output state check. • Disconnect harness from WAC relay. • DVOM on 20 volt scale. • Measure voltage between VPWR circuit at the WAC relay harness connector and chassis ground. • Is voltage greater than 10.5 volts?	Yes ▶ No ▶	GO to KM13. SERVICE open in VPWR circuit between power relay and WAC relay. RECONNECT WAC relay and REMOVE jumper. RE-EVALUATE symptom.

WOT A/C Cutoff (WAC) A/C Demand — Pinpoint Test KM

TEST STEP	RESULT ▶	ACTION TO TAKE
KM13 CHECK FOR WAC CYCLING		
• Still in output state check. • WAC relay disconnected. • DVOM on 20 volt scale. • Connect DVOM positive test lead to the VPWR circuit and the negative test lead to the WAC circuit at the WAC relay harness connector. • While observing DVOM, depress and release throttle several times (to cycle output on and off). • Does voltage cycle high and low?	Yes ▶ No ▶	REPLACE WAC relay. REMOVE jumper. RE-EVALUATE symptom. REMOVE jumper. GO to KM14.
KM14 CHECK CONTINUITY OF WAC CIRCUIT		
• Key off. • Disconnect processor 60 pin connector. Inspect for damaged pins, corrosion, loose wires, etc. Service as necessary. • Connect breakout box to harness, leave processor disconnected. • WAC relay disconnected. • Measure resistance between Test Pin 54 at the breakout box and WAC circuit at the WAC relay harness connector. • Is resistance less than 5.0 ohms?	Yes ▶ No ▶	GO to KM15. SERVICE open circuit. REMOVE breakout box. RECONNECT processor and WAC relay. RE-EVALUATE symptom.
KM15 CHECK FOR SHORT TO POWER		
• Key off. • Breakout box installed, processor disconnected. • WAC relay disconnected. • DVOM on 200,000 ohm scale. • Measure resistance between Test Pin 54 and Test Pins 37 and 57 at the breakout box. • Are both resistances greater than 10,000 ohms?	Yes ▶ No ▶	REPLACE processor. RECONNECT WAC relay. RE-EVALUATE symptom. SERVICE short circuit. REMOVE breakout box. RECONNECT processor and WAC relay. RE-EVALUATE symptom. IF symptom is still present, REPLACE processor.

WOT A/C Cutoff (WAC) A/C Demand — Pinpoint Test KM

TEST STEP	RESULT ▶	ACTION TO TAKE
KM17 CHECK FOR WAC CYCLING		
• Still in output state check. • Disconnect A/C fan controller. • DVOM on 20 volt scale. • Connect DVOM positive test lead to the Ignition Run circuit and the negative test lead to the WAC circuit at the A/C fan controller harness connector. • While observing DVOM, depress and release throttle several times (to cycle output on and off). • Does voltage cycle high and low?	Yes ▶ No ▶	REPLACE A/C fan controller. REMOVE jumper. RE-EVALUATE symptom. REMOVE jumper. GO to KM18.
KM18 CHECK CONTINUITY OF WAC CIRCUIT		
• Key off. • Disconnect processor 60 pin connector. Inspect for damaged pins, corrosion, loose wires, etc. Service as necessary. • Connect breakout box to harness, leave processor disconnected. • A/C fan controller disconnected. • Measure resistance between Test Pin 54 at the breakout box and WAC circuit at the WAC relay harness connector. • Is resistance less than 5.0 ohms?	Yes ▶ No ▶	GO to KM19. SERVICE open circuit. REMOVE breakout box. RECONNECT processor and A/C fan controller. RE-EVALUATE symptom.
KM19 CHECK FOR SHORT TO POWER		
• Key off. • Breakout box installed, processor disconnected. • A/C fan controller disconnected. • DVOM on 200,000 ohm scale. • Measure resistance between Test Pin 54 and Test Pins 37 and 57 at the breakout box. • Are both resistances greater than 10,000 ohms?	Yes ▶ No ▶	REPLACE processor. RECONNECT A/C fan controller. RE-EVALUATE symptom. SERVICE short circuit. REMOVE breakout box. RECONNECT processor and A/C fan controller. RE-EVALUATE symptom. If symptom is still present, REPLACE processor.

WOT A/C Cutoff (WAC) A/C Demand — Pinpoint Test KM

TEST STEP	RESULT ▶	ACTION TO TAKE
KM20 CYCLE A/C DEMAND SWITCH		
• Key off, wait 10 seconds. • Disconnect processor 60 pin connector. Inspect for damaged pins, corrosion, loose wires, etc. Service as necessary. • Connect breakout box to harness, leave processor disconnected. • DVOM on 20 volt scale. • Key on, engine off. • Connect DVOM positive test lead to Test Pin 24 (Test Pin 43 for 2.3L HSC EFI) and negative test lead to Test Pin 40. • Does voltage cycle high and low when A/C switch is cycled?	Yes ▶ No ▶	REMOVE breakout box. REPLACE processor. RERUN Quick Test. GO to KM21.
KM21 CHECK CONTINUITY OF ACD CIRCUIT		
• Key off, wait 10 seconds. • DVOM on 200 ohm scale. • Measure resistance between Test Pin 24 (Test Pin 43 for 2.3L HSC EFI) at the breakout box and A/C demand switch. • Is resistance greater than 5 ohms?	Yes ▶ No ▶	SERVICE open in ACD circuit. RERUN Quick Test. EEC-IV system OK.
KM25 CHECK ACD CIRCUIT FOR SHORT TO POWER		
• Key off. • Disconnect processor 60 pin connector. Inspect for damaged pins, corrosion, loose wires, etc. Service as necessary. • Connect breakout box to harness, leave processor disconnected. • Disconnect WAC relay (TRK) or A/C fan controller (CAR). • A/C demand switch "OFF". • DVOM on 20 volt scale. • Key on. • Measure voltage between Test Pin 24 at the breakout box and chassis ground. • Is voltage less than 1.0 volt?	Yes ▶ No ▶	EEC-IV system OK. VERIFY operation of A/C demand switch. IF OK, SERVICE short circuit. REMOVE breakout box. RECONNECT processor and WAC relay or A/C fan controller. RE-EVALUATE symptom.

Turbo Boost — Pinpoint Test KN

Note

You should enter this Pinpoint Test only when a Service Code 81 is received in Quick Test Step 3.0 or when directed here from Diagnostic By Symptom in the Engine Supplement Section.

Remember

This Pinpoint Test is intended to diagnose only the following:

- Harness circuits: VPWR, Boost
- Octane switch

Pinpoint Test Schematic

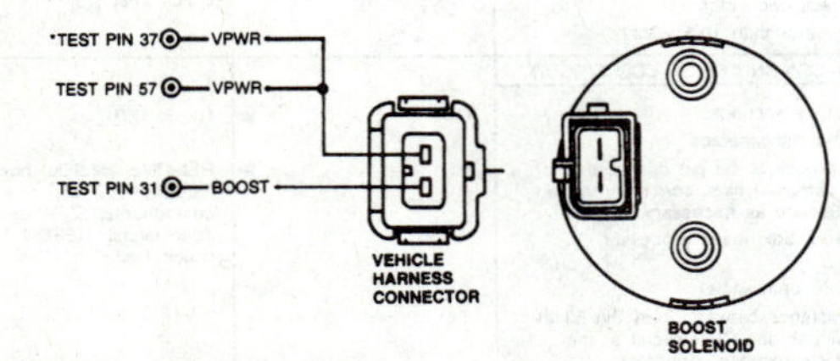

*TEST PINS LOCATED ON BREAKOUT BOX.
ALL HARNESS CONNECTORS VIEWED INTO MATING SURFACE.

Turbo Boost — Pinpoint Test KN

TEST STEP	RESULT	ACTION TO TAKE
KN1 ENTER OUTPUT STATE CHECK (REFER TO APPENDIX)		
NOTE: Do not use STAR tester for this Step, use VOM/DVOM. • Key off, wait 10 seconds. • DVOM on 20 volt scale. • Connect DVOM negative test lead to STO at the Self-Test connector and positive test lead to battery positive. • Jumper STI to SIGNAL RETURN at the Self-Test connector. • Perform Key On Engine Off Self-Test through the completion of the Continuous Test Codes. • DVOM will indicate less than 1.0 volts when test is complete. • Depress and release the throttle. • Did DVOM change to a high voltage?	Yes ▶	REMAIN in Output State Check. GO to KN2.
	No ▶	DEPRESS throttle to WOT and release. If STO voltage does not go high, GO to Pinpoint Test Step QC1. Leave equipment hooked up.
KN2 CHECK BOOST OUTPUT ELECTRICAL OPERATION		
• Key on, engine off. • Disconnect boost solenoid. • DVOM on 20 volt scale. • Connect DVOM positive test lead to VPWR circuit on boost solenoid connector and negative test lead to boost output on boost solenoid connector. • While observing DVOM, depress and release the throttle several times to cycle output On and Off. • Does boost output solenoid cycle On and Off?	Yes ▶	RECONNECT boost solenoid. GO to KN3.
	No ▶	REMOVE jumper. GO to KN4.

Turbo Boost — Pinpoint Test KN

TEST STEP	RESULT	ACTION TO TAKE
KN3 CHECK BOOST SOLENOID FUNCTION		
• Remain in output state check. • Disconnect turbo boost control solenoid vacuum hose at the turbocharger inlet end. • Attach a vacuum pump to the hose. • Depress the throttle once to cycle the solenoid closed. • Apply vacuum to the solenoid. • Depress the throttle once again to cycle the solenoid open and release trapped vacuum. • Did the solenoid hold and then release vacuum? 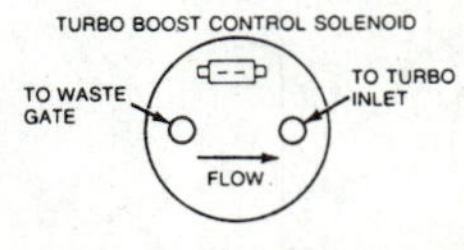	Yes ▶	EEC-IV system OK. for boost diagnostics.
	No ▶	REPLACE solenoid. RERUN Quick Test.
KN4 CHECK VOLTAGE OF VPWR CIRCUIT		
• Key on, engine off. • Boost solenoid disconnected. • DVOM on 20 volt scale. • Measure voltage between VPWR circuit of boost solenoid vehicle harness connector and battery negative post. • Is voltage greater than 10.5 volts?	Yes ▶	GO to KN5.
	No ▶	RECONNECT boost solenoid. SERVICE harness open circuit. RERUN Quick Test.
KN5 MEASURE BOOST SOLENOID RESISTANCE		
• Key off, wait 10 seconds. • Boost solenoid disconnected. • DVOM on 200 ohm scale. • Measure the solenoid resistance. • Is resistance between 65 and 110 ohms?	Yes ▶	GO to KN6.
	No ▶	REPLACE BOOST solenoid. RERUN Quick Test.

Turbo Boost — Pinpoint Test KN

TEST STEP	RESULT	ACTION TO TAKE
KN6 CHECK CONTINUITY OF BOOST CIRCUIT		
• Key off, wait 10 seconds. • Boost solenoid disconnected. • Disconnect processor 60 pin connector. Inspect for damaged pins, corrosion, loose wires etc. Service as necessary. • Install breakout box, leave processor disconnected. • DVOM on 200 ohm scale. • Measure resistance between Test Pin 31 at the breakout box and BOOST circuit at vehicle harness connector. • Is resistance less than 5 ohms?	Yes ▶	GO to KN7.
	No ▶	REMOVE breakout box. RECONNECT all components. SERVICE open circuit. RERUN Quick Test.
KN7 CHECK FOR SHORT TO GROUND		
• Key off, wait 10 seconds. • Breakout box installed, processor disconnected. • Boost solenoid disconnected. • DVOM on 200,000 ohm scale. • Measure resistance between Test Pin 31 and Test Pins 40, 46 and 60 at the breakout box. • Is resistance greater than 10,000 ohms?	Yes ▶	GO to KN8.
	No ▶	REMOVE breakout box. RECONNECT all components. SERVICE short circuit. RERUN Quick Test.
KN8 CHECK FOR SHORT TO POWER		
• Key off, wait 10 seconds. • DVOM on 200,000 ohm scale. • Breakout box installed, processor disconnected. • Boost solenoid disconnected. • Measure resistance between Test Pin 31 and Test Pins 37 and 57 at the breakout box. • Is resistance greater than 10,000 ohms?	Yes ▶	REMOVE breakout box. RECONNECT all components. REPLACE processor. RERUN Quick Test.
	No ▶	REMOVE breakout box. RECONNECT all components. SERVICE short circuit. RERUN Quick Test. If symptom is still present, REPLACE processor.

Turbo Boost — Pinpoint Test KN

TEST STEP	RESULT ▶	ACTION TO TAKE
KN10 CHECK OCTANE SWITCH INPUT FOR INPUT CHANGE		
• Key off, wait 10 seconds. • Disconnect processor 60 pin connector. Inspect for damaged pins, corrosion, loose wires etc. Service as necessary. • Install breakout box. Reconnect processor. • DVOM to 20 volt scale. • Connect positive test lead to Test Pin 24 and negative test lead to Test Pin 46 at the breakout box. • Key on, engine off. • Cycle octane switch several times while observing DVOM. • Does voltage change from less than 1.0 octane switch volts to 5 volts?	Yes ▶ No ▶	REMOVE breakout box. RECONNECT processor. REPLACE processor. RERUN Quick Test. REMOVE breakout box. RECONNECT processor. EEC-IV system OK.

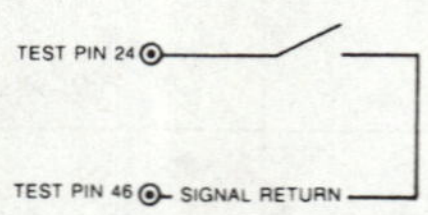

Converter Clutch Override (CCO) and Shift Solenoid 3/4-4/3 (SS 3/4-4/3) — Pinpoint Test KR

TEST STEP	RESULT ▶	ACTION TO TAKE
CODE 88 or 89 PRESENT		
KR1 MEASURE CCO SOLENOID RESISTANCE		
• Key off, wait 10 seconds. • DVOM on 200 ohm scale. • Disconnect CCO solenoid. • Measure resistance of solenoid between CCO and VPWR at the transmission bulkhead connector. • Is resistance reading between 26 and 40 ohms?	Yes ▶ No ▶	GO to **KR2**.
KR2 CHECK VOLTAGE OF VPWR CIRCUIT		
• Key on, engine off. • CCO solenoid disconnected. • DVOM on 20 volt scale. • Measure voltage between the VPWR circuit and battery ground at the CCO solenoid vehicle harness connector. • Is voltage greater than 10.5 volts?	Yes ▶ No ▶	GO to **KR3**. RECONNECT CCO/ shift solenoid. SERVICE open circuit. RERUN Quick Test.
KR3 CHECK CONTINUITY OF CCO CIRCUIT		
• Key off, wait 10 seconds. • CCO solenoid disconnected. • Disconnect processor 60 pin connector. Inspect for damaged pins, corrosion, loose wires, etc. Service as necessary. • Install breakout box, leave processor disconnected. • DVOM on 200 ohm scale. • Measure resistance between Test Pin 53 at the breakout box and CCO circuit at the solenoid vehicle harness connector. • Is resistance reading less than 5 ohms?	Yes ▶ No ▶	GO to **KR4**. REMOVE breakout box. RECONNECT all components. SERVICE open circuit. RERUN Quick Test.
KR4 CHECK CCO CIRCUIT FOR SHORT TO GROUND		
• Key off, wait 10 seconds. • Breakout box installed, processor disconnected. • CCO solenoid disconnected. • DVOM on 200,000 ohm scale. • Measure resistance between Test Pin 53 and Test Pins 40, 46 and 60 at breakout box. • Is resistance greater than 100,000 ohms?	Yes ▶ No ▶	GO to **KR5**. REMOVE breakout box. RECONNECT all components. SERVICE short circuit. RERUN Quick Test.

Converter Clutch Override (CCO) and Shift Solenoid 3/4-4/3 (SS 3/4-4/3) — Pinpoint Test KR

Note

You should enter this Pinpoint Test only when a Service Code 85, 86, 88 or 89 are received in Quick Test Step 3.0.

Remember

To prevent the replacement of good components, be aware that the following non-EEC areas may be at fault:

- Hydraulic brakes
- Emergency brakes
- Internal transmission
- Transmission linkage

This Pinpoint Test is intended to diagnose only the following:

- Harness Circuits: CCO SS 3/4-4/3 and VPWR.
- CCO Solenoid.
- Shift Solenoid 3/4-4/3.
- Processor Assembly.

Pinpoint Test Schematic

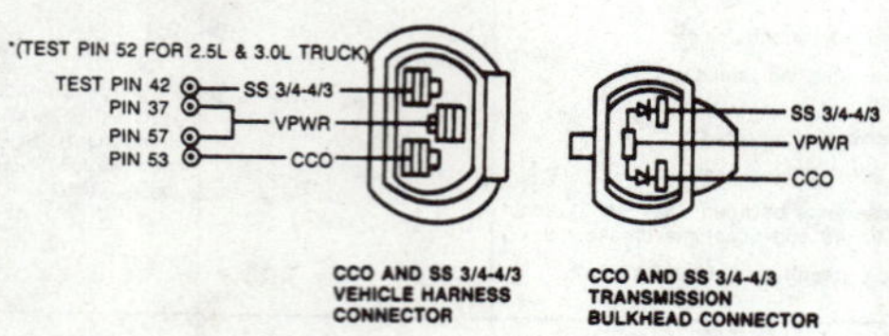

NOTE: TEST PIN 42 (52 ON 2.9L & 3.0L TRUCK) IS NOT USED ON APPLICATIONS EQUIPPED WITH CCO ONLY.

*TEST PINS LOCATED ON BREAKOUT BOX.
ALL HARNESS CONNECTORS VIEWED INTO MATING SURFACE.

Converter Clutch Override (CCO) and Shift Solenoid 3/4-4/3 (SS 3/4-4/3) — Pinpoint Test KR

TEST STEP	RESULT ▶	ACTION TO TAKE
KR5 CHECK CCO CIRCUIT FOR SHORT TO POWER		
• Key off, wait 10 seconds. • DVOM on 200,000 ohm scale. • Breakout box installed, processor disconnected. • CCO solenoid disconnected. • Measure resistance between Test Pin 53 and Test Pins 37 and 57 at breakout box. • Are all resistance readings 10,000 ohms?	Yes ▶ No ▶	REMOVE breakout box. RECONNECT all components. REPLACE processor. RERUN Quick Test. REMOVE breakout box. RECONNECT all components. SERVICE short circuit. RERUN Quick Test. If code is still present, REPLACE processor.
CODE 85 OR 86 PRESENT		
KR10 MEASURE SS 3/4-4/3 RESISTANCE		
• Key off, wait 10 seconds. • Disconnect CCO/shift solenoid. • DVOM on 200 ohm scale. • Measure resistance of solenoid between SS and VPWR at the transmission bulkhead connector. • Is resistance between 26 and 40 ohms?	Yes ▶ No ▶	TO to **KR11**.
KR11 CHECK VOLTAGE OF VPWR CIRCUIT		
• Key on, engine off. • CCO/shift solenoid disconnected. • DVOM on 20 volt scale. • Measure voltage at the SS 3/4-4/3 solenoid vehicle harness connector between VPWR circuit and battery ground. • Is voltage greater than 10.5 volts?	Yes ▶ No ▶	GO to **KR12**. RECONNECT CCO/ Shift solenoid. SERVICE open circuit. RERUN Quick Test.

Converter Clutch Override (CCO) and Shift Solenoid 3/4-4/3 (SS 3/4-4/3) — Pinpoint Test KR

TEST STEP	RESULT ▶	ACTION TO TAKE
KR12 CHECK CONTINUITY OF SS 3/4-4/3 CIRCUIT		
• Key off, wait 10 seconds. • Disconnect processor 60 pin connector. Inspect for damaged pins, corrosion, loose wires, etc. Service as necessary. • Install breakout box, leave processor disconnected. • CCO/shift solenoid disconnected. • DVOM on 200 ohm scale. • Measure resistance between Test Pin 42 (Test Pin 52 for 2.9L and 3.0L truck) at the breakout box and SS 3/4-4/3 circuit at the solenoid vehicle harness connector. • Is resistance less than 5 ohms?	Yes ▶ No ▶	GO to KR13. REMOVE breakout box. RECONNECT all components. SERVICE open circuit. RERUN Quick Test.
KR13 CHECK FOR SHORT TO GROUND		
• Key off, wait 10 seconds. • Breakout box installed, processor disconnected. • CCO/shift solenoid disconnected. • DVOM on 200,000 ohm scale. • Measure resistance between Test Pin 42 (Test Pin 52 for 2.9L and 3.0L truck) and Test Pins 40, 46 and 60 at breakout box. • Is resistance greater than 100,000 ohms?	Yes ▶ No ▶	GO to KR14. REMOVE breakout box. RECONNECT all components. SERVICE short circuit. RERUN Quick Test.
KR14 CHECK FOR SHORT TO POWER		
• Key off, wait 10 seconds. • Breakout box installed, processor disconnected. • CCO/shift solenoid disconnected. • DVOM on 200,000 ohm scale. • Measure resistance between Test Pin 42 (Test Pin 52 for 2.9L and 3.0L truck) and Test Pins 37 and 57 at breakout box. • Are all resistances greater than 10,000 ohms?	Yes ▶ No ▶	REMOVE breakout box. RECONNECT all components. REPLACE processor. RERUN Quick Test. REMOVE breakout box. RECONNECT all components. SERVICE short circuit. RERUN Quick Test. If code is still present, REPLACE processor.

Dynamic Response Test — Pinpoint Test M

Note

You should enter this Pinpoint Test only when a Service Code 77 is received in Quick Test Step 5.0.

Remember

To prevent the replacement of good components, be aware that the following non-EEC areas may be at fault:

- Operator did not perform a brief WOT after dynamic response code.
- Mechanical engine problems; engine did not achieve greater than 2000 rpm.

This Pinpoint Test is intended to diagnose only the following:

- Throttle movement (greater than 3/4 throttle).
- Vane Airflow (greater than 50% open).
- Rpm increase (greater than 2000 rpm).

Pinpoint Test Schematic

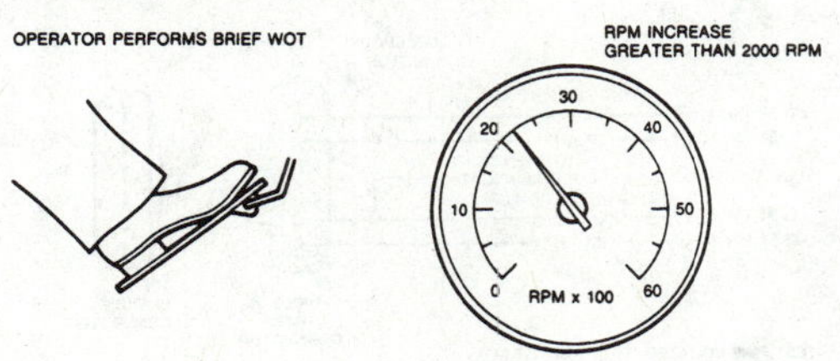

Dynamic Response Test — Pinpoint Test M

TEST STEP	RESULT ▶	ACTION TO TAKE
M1 SERVICE CODE 77: SYSTEM FAILED TO RECOGNIZE BRIEF WOT		
NOTE: A brief snap of the throttle may not be sufficient to pass this test. Be sure to go to WOT and return. • Rerun Engine Running Self-Test. Be sure operator is familiar with the engine running format which proceeds as follows: — With Self-Test activated restart the engine. — ID Code 2 (0) start of test. — Dynamic response Code 1 (0) perform brief WOT. — Testing over. — Service code output begins. • Is Code 77 still present?	Yes ▶ No ▶	REPLACE processor. RERUN Quick Test. Dynamic Response Test passed. SERVICE any other service code(s) received as necessary.

"CHECK ENGINE" Light/Message "CHECK ENGINE"/"CHECK DCL" Message — Pinpoint Test ML

Note

You should enter this Pinpoint Test only when directed here from Pinpoint Test QA or Diagnostic By Symptom in the Engine Supplement Section.

Remember

To prevent the replacement of good components, be aware that the following non-EEC areas may be at fault:

- Fuse, bulb or socket.

This Pinpoint Test is intended to diagnose only the following:

- STO/MIL circuit. (All except 1.9L EFI and 3.8L EFI Continental)
- MIL circuit (1.9L EFI only)
- Processor assembly.
- Data Communications Link (DCL) (3.8L EFI Continental only)

Pinpoint Test Schematic

ALL EXCEPT 1.9L EFI AND 3.8L EFI CONTINENTAL

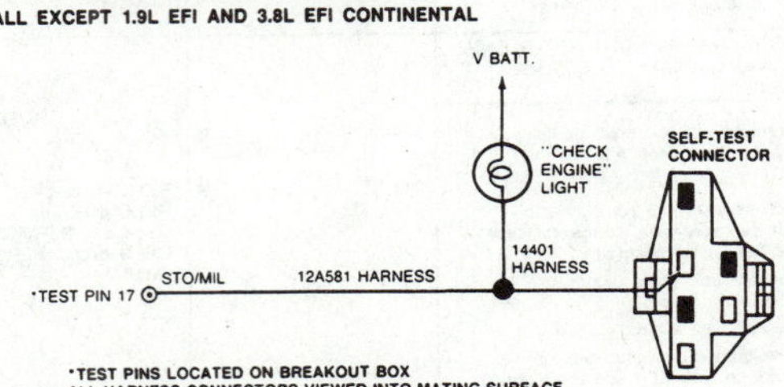

*TEST PINS LOCATED ON BREAKOUT BOX
ALL HARNESS CONNECTORS VIEWED INTO MATING SURFACE

"CHECK ENGINE" Light/Message "CHECK ENGINE"/"CHECK DCL" Message	Pinpoint Test	ML

Pinpoint Test Schematic

1.9L EFI

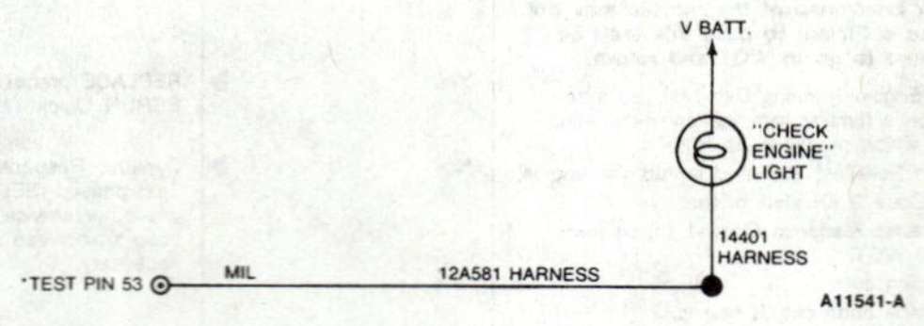

3.8L EFI CONTINENTAL

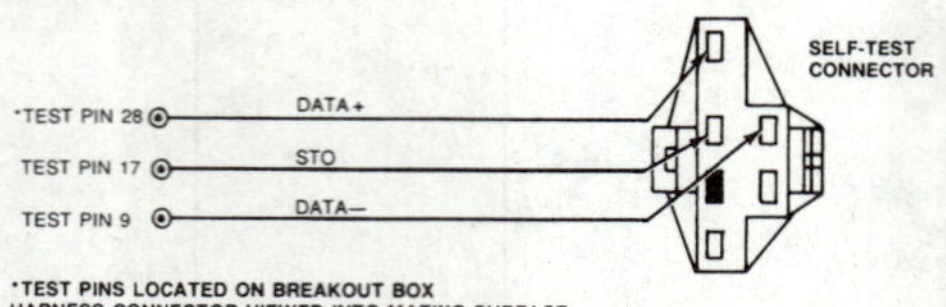

*TEST PINS LOCATED ON BREAKOUT BOX
HARNESS CONNECTOR VIEWED INTO MATING SURFACE.

"CHECK ENGINE" Light/Message "CHECK ENGINE"/"CHECK DCL" Message	Pinpoint Test	ML

TEST STEP	RESULT ▶	ACTION TO TAKE
ML7 CONFIRM CIRCUIT REPAIR		
• Reconnect processor. • Turn key to run. • Is "CHECK ENGINE" Light ON? **NOTE: Refer to Appendix in Section 16 for a detailed description of HOW the "CHECK ENGINE" Light (malfunction indicator light) operates.**	Yes ▶	System OK.
	No ▶	REPLACE processor.
ML10 "CHECK ENGINE" MESSAGE DISPLAYED		
NOTE: If vehicle is a no start, go to Pinpoint Test Step [A1]. • Run Key On Engine Off Self-Test. • Is result 11-10-11 (Pass Codes)? **NOTE: Refer to Appendix in Section 16 for detailed description of how the "CHECK ENGINE" message operates.**	Yes ▶	GO to the Continental Shop Manual.
	No ▶	GO to Quick Test Step 3.0B. PROCEED as directed.
ML15 "CHECK ENGINE"/"CHECK DCL" MESSAGE DISPLAYED		
NOTE: If vehicle is a no start, go to Pinpoint Test Step [A1]. • Run Key On Engine Off Self-Test with a STAR tester or volt/ohmmeter. • Is result 11-10-11 (Pass Codes)? **NOTE: Refer to Appendix in Section 16 for a detailed description of how the "CHECK ENGINE"/"CHECK DCL" message operates.**	Yes ▶	
	No ▶	GO to Quick Test Step 3.0B. PROCEED as directed.

"CHECK ENGINE" Light/Message "CHECK ENGINE"/"CHECK DCL" Message	Pinpoint Test	ML

TEST STEP	RESULT ▶	ACTION TO TAKE
ML1 "CHECK ENGINE" LIGHT ALWAYS ON: CHECK FOR SHORT TO GROUND		
NOTE: If vehicle will not start go to Pinpoint Test Step [A1]. • **If any Key On Engine Off service codes are present, service before proceeding. If no codes are outputted, continue with this Test Step.** • Key off. • Disconnect processor 60 pin connector. Inspect for damaged pins, corrosion, loose wires, etc. Service as necessary. • Install breakout box, leave processor disconnected. • DVOM on 200,000 ohm scale. • Measure resistance between Test Pin 17 (Test Pin 53 on 1.9L EFI) and Test Pin 40 at the breakout box. • Is resistance less than 10,000 ohms?	Yes ▶	REMOVE breakout box. RECONNECT processor. SERVICE short circuit between Test Pin 17/53 and "Check Engine" Light, or between Test Pin 17 and the Self-Test Connector. RERUN Quick Test.
	No ▶	REMOVE breakout box. REPLACE processor. RERUN Quick Test.
ML5 "CHECK ENGINE" LIGHT NEVER ON: CHECK CONTINUITY OF STO/MIL CIRCUIT		
NOTE: If vehicle will not start go to Pinpoint Test Step A1. • Key off. • Disconnect processor 60 pin connector. Inspect for damaged pins, corrosion, loose wires, etc. Service as necessary. • Install breakout box, leave processor disconnected. • DVOM on 200 ohm scale. • Measure resistance between Test Pin 17 (Test Pin 53 on 1.9L EFI) and the "CHECK ENGINE" light. • Is resistance less than 5 ohms?	Yes ▶	GO to [ML6].
	No ▶	REMOVE breakout box. RECONNECT processor. SERVICE OPEN circuit. RERUN Quick Test.
ML6 CHECK FOR POWER TO BULB		
• Check for power to "CHECK ENGINE" light bulb. • Is there power at the light bulb?	Yes ▶	REPLACE bulb or socket. GO to [ML7].
	No ▶	CHECK fuse and VBATT input circuit. GO to [ML7].

Ignition Diagnostic Monitor (IDM)	Pinpoint Test	N

Note

You should enter this Pinpoint Test only when a Service Code 18 is received in Quick Test Step 6.0.

Remember

To prevent the replacement of good components, be aware that the following non-EEC areas may be at fault:

- Ignition module
- Ignition coil
- Spark plugs and high tension cables
- Distributor and PIP sensor

This Pinpoint Test is intended to diagnose only the following:

- Harness circuits: IGNITION GROUND, SPOUT, PIP, IDM.

Pinpoint Test Schematic

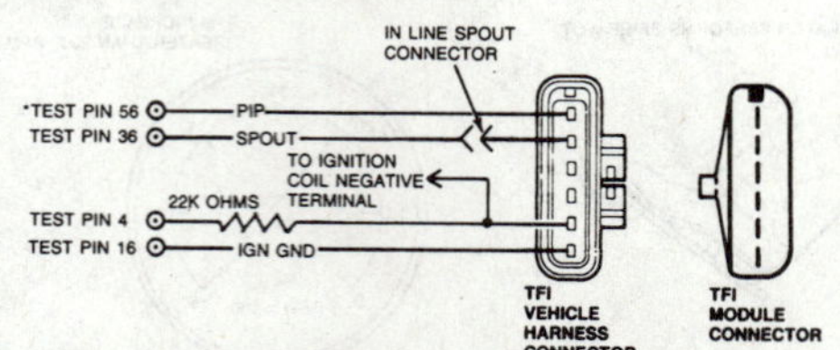

*TEST PINS LOCATED ON BREAKOUT BOX.
ALL HARNESS CONNECTORS VIEWED INTO MATING SURFACE.

Ignition Diagnostic Monitor (IDM) — Pinpoint Test N

TEST STEP	RESULT ▶	ACTION TO TAKE
N1 CHECK CONTINUITY OF IDM CIRCUIT		
NOTE: It is important to know that the IDM circuit has a 20,000 ohm resistor between Test Pin 4 and the Ignition Coil Negative Terminal. • Key off, wait 10 seconds. • Disconnect E-core ignition coil. • Disconnect processor 60 pin connector. Inspect for damaged pins, corrosion, loose wires, etc. Service as necessary. • Install breakout box, leave processor disconnected. • DVOM on 200,000 ohm scale. • Measure resistance between Test Pin 4 at the breakout box and ignition coil harness connector negative terminal. • Is resistance between 20,000 and 24,000 ohms?	Yes ▶ No ▶	GO to N2. REMOVE breakout box. RECONNECT processor. SERVICE open circuit. RERUN Quick Test. RECONNECT E-core ignition coil.
N2 CHECK FOR SHORT TO GROUND		
• Key off, wait 10 seconds. • Breakout box installed, processor disconnected. • DVOM on 200,000 ohm scale. • Measure resistance between Test Pin 4 and Test Pins 40, 46 and 60 at the breakout box. • Are all resistances above 10,000 ohms?	Yes ▶ No ▶	REMOVE breakout box. RECONNECT E-core ignition coil and processor. GO to N3. REMOVE breakout box. SERVICE short to ground in IDM circuit. RECONNECT E-core ignition coil and processor. RERUN Quick Test.

Ignition Diagnostic Monitor (IDM) — Pinpoint Test N

TEST STEP	RESULT ▶	ACTION TO TAKE
N3 CHECK TFI MODULE		
• Key off, wait 10 seconds. • Deactivate Self-Test. • Connect VOM or STAR per Quick Test Step 2.0. • Enter Engine Running Continuous Monitor Test (as instructed in Quick Test Step 6.0B). • Observe VOM or STAR LED for indication of a fault while performing the following: • Lightly tap on TFI module (simulate road shock). • Wiggle TFI connector. • Is a fault indicated?	Yes ▶ No ▶	DISCONNECT and INSPECT connectors. If connector and terminals are good GO to N4.
N4 CHECK EEC-IV HARNESS		
• While still in continuous monitor test from Step N3, observe VOM or STAR LED for a fault indication while performing the following: • While looking for faults listed in the table below, grasp the harness close to the TFI connector. Wiggle, shake or bend a small section of the EEC-IV system harness while working your way to the dash panel. Also wiggle, shake or bend the EEC-IV harness from the dash panel to the processor. Do this test on the circuits listed one at a time if needed to locate a faulty circuit.	Yes ▶ No ▶	ISOLATE fault and make necessary repairs. RERUN Quick Test. GO to N5.

FAULT	BREAKOUT BOX NO.
PIP shorted to ground or open	Test Pin 56
Spout shorted to ground	Test Pin 36
Ign. ground open	Test Pin 16
IDM open or shorted to ground, power	Test Pin 4

• Is a fault indicated?

Ignition Diagnostic Monitor (IDM) — Pinpoint Test N

TEST STEP	RESULT ▶	ACTION TO TAKE
N5 CHECK PROCESSOR AND HARNESS CONNECTORS		
• Key off, wait 10 seconds. • Disconnect processor 60 pin connector. Inspect for damaged pins, corrosion, loose wires, etc. • Are connectors and terminals OK?	Yes ▶ No ▶	REPLACE processor. Start engine and run for about one minute. RERUN Key On Engine Off Self-Test observing continuous codes. SERVICE as necessary. RECONNECT processor. RERUN Quick Test.

Spark Timing Check — Pinpoint Test P

Note

You should enter this Pinpoint Test only when directed here from Quick Test Step 4.0 or when a Service Code 18 is received in Quick Test Step 5.0.

Remember

To prevent the replacement of good components, be aware that the following non-EEC areas may be at fault:

- Base Engine
- PIP Sensor
- TFI Module

This Pinpoint Test is intended to diagnose only the following:

- Harness Spout Circuit
- Base Timing
- Processor Assembly

Pinpoint Test Schematic

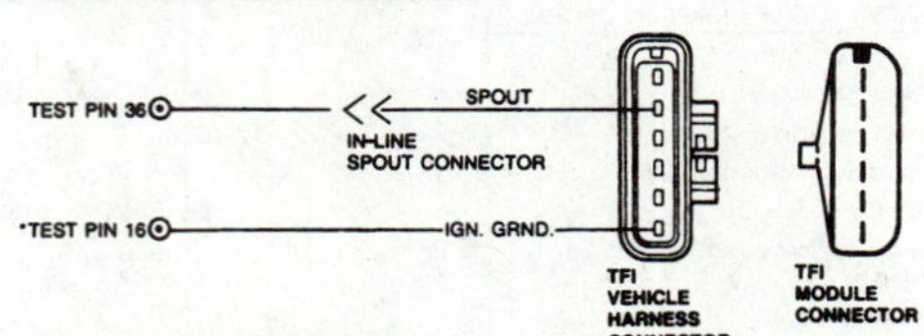

*TEST PINS LOCATED ON BREAKOUT BOX.
ALL HARNESS CONNECTORS VIEWED INTO MATING SURFACE.

Spark Timing Check — Pinpoint Test P

TEST STEP	RESULT ▶	ACTION TO TAKE
P1 CHECK SPARK TIMING		
NOTE: Self-Test locks the timing at 20 degrees plus base during code output and for two minutes after the last service code is outputted. Timing check must be made during this time period. Self-Test timing is base +20 degrees (±3 degrees) BTDC. (See VECI decal for base value.) • Check timing. Record value. • Is timing equal to base plus 20 degrees (±3 degrees)?	Yes No	▶ GO to Quick Test Step [5.0] ▶ GO to [P2]
P2 CHECK SPARK OUTPUT (SPOUT) CIRCUIT TO THE TFI MODULE		
• Locate spout connector and open this connection. • Start engine. • Check base timing. • Is base timing within ±3 degrees of value on VECI decal?	Yes No	▶ RECONNECT spout connector. GO to [P3]. ▶ Adjust base timing if necessary. REFER to Section 15 for engine timing instructions. After timing is reset, RECONNECT spout and PERFORM Quick Test Step [4.0].
P3 CHECK FOR POWER TO PROCESSOR		
• Key off, wait 10 seconds. • Disconnect processor 60 pin connector. Inspect for damaged pins, corrosion, loose wires. Service as necessary. • Install breakout box. • Key on, engine off. • DVOM on 20 volt scale. • Measure voltage between Test Pin 37 and Test Pin 40 and between Test Pin 57 and Test Pin 60 at the breakout box. • Is voltage less than 10.5 volts?	Yes No	▶ GO to Pinpoint Test Step [B1] except 2.3L EFI TC, 2.5L HSC CFI, 3.0L EFI, and 3.8L FWD EFI passenger car; GO to Pinpoint Test Step [X1]. ▶ GO to [P4].

No Codes/Codes Not Listed — Pinpoint Test QA

Note

You should enter this Pinpoint Test only when directed here from Quick Test Step 3.0, 5.0 or 6.0 or when directed here from Diagnostic By Symptom in the Engine Supplement Section or from a Pinpoint Test Step.

Remember

This Pinpoint Test is intended to diagnose only the following:

- Processor.
- EEC Power Relay.
- Harness Circuits: SIGNAL RETURN, STO, STI, GROUND, VPWR, VREF, NDS.

Pinpoint Test Schematic

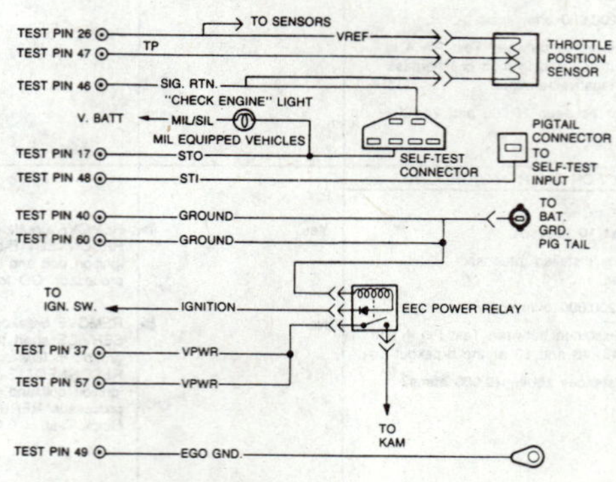

*TEST PINS LOCATED ON BREAKOUT BOX.
ALL HARNESS CONNECTORS VIEWED INTO MATING SURFACE.

Spark Timing Check — Pinpoint Test P

TEST STEP	RESULT ▶	ACTION TO TAKE
P4 CHECK HARNESS FOR CONTINUITY		
• Key off, wait 10 seconds. • Breakout box installed. • Processor disconnected. • Disconnect TFI module. • DVOM on 200 ohm scale. • Measure resistance between Test Pin 36 at the breakout box and the spout pin at the TFI vehicle harness connector. • Is resistance less than 5 ohms? 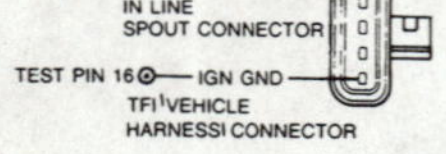 	Yes No	▶ GO to [P5] ▶ SERVICE open circuit. CHECK timing per [P1].
P5 HARNESS CHECK—IGNITION GROUND		
• Key off, wait 10 seconds. • Breakout box installed. • Connect processor to breakout box. • Reconnect TFI module. • Timing switch to "computed" position on breakout box. • DVOM on 20 volt scale. • Measure voltage between Test Pin 36 at the breakout box and battery negative post during Engine Running Self-Test. • Is voltage between 4.0 and 8.0 volts?	Yes No	▶ EEC system OK. ▶ REMOVE breakout box. REPLACE processor. RERUN Quick Test.

No Codes/Codes Not Listed — Pinpoint Test QA

TEST STEP	RESULT ▶	ACTION TO TAKE
QA1 CHECK FOR VREF		
Refer to schematic in Pinpoint Test QA. • Key off, wait 10 seconds. • Disconnect processor 60 pin connector. Inspect for damaged pins, corrosion, loose wires, etc. Service as necessary. • Install breakout box, connect processor to breakout box. • DVOM on 20 volt scale. • Key on, engine off. • Measure voltage between Test Pin 26 and Test Pin 46 at the breakout box. • Is voltage between 4.0 and 6.0 volts?	Yes No	▶ RECONNECT TP Sensor. GO to [QA2]. ▶ GO to Pinpoint Test Step [C1].
QA2 CHECK SELF-TEST INPUT CONTINUITY		
Refer to schematic in Pinpoint Test QA. • Key off, wait 10 seconds. • Breakout box installed, processor disconnected. • Set DVOM to 200 ohm scale. • Measure resistance between SELF-TEST INPUT at the Self-Test single pin connector and Test Pin 48 at the breakout box. • Is resistance less than 5 ohms?	Yes No	▶ GO to [QA3]. ▶ SERVICE open circuit. REMOVE breakout box. RECONNECT processor. RERUN Quick Test.
QA3 CHECK SELF-TEST OUTPUT CIRCUIT CONTINUITY		
Refer to schematic in Pinpoint Test QA. • Breakout box installed, processor disconnected. • DVOM to 200 ohm scale. • Measure resistance between SELF-TEST OUTPUT at the Self-Test connector and Test Pin 17 at the breakout box. • Is resistance less than 5 ohms?	Yes No	▶ GO to [QA4]. ▶ SERVICE open circuit. REMOVE breakout box. RECONNECT processor. RERUN Quick Test.

No Codes/Codes Not Listed — Pinpoint Test QA

TEST STEP	RESULT ▶	ACTION TO TAKE
QA4 CHECK EGO SENSOR GROUND CONTINUITY		
Refer to schematic in Pinpoint Test QA. • Key off. • Breakout box installed, processor disconnected. • DVOM on 200 ohm scale. • Measure resistance between EGO GROUND on engine and Test Pin 49 at the breakout box. • Is resistance less than 5 ohms?	Yes ▶	GO to QA5.
	No ▶	SERVICE open circuit. REMOVE breakout box. RECONNECT processor. RERUN Quick Test.
QA5 STO SHORT TO GROUND		
• Key off. • Breakout box installed, processor disconnected. • DVOM on 200 ohm scale. • Measure resistance between SELF-TEST OUTPUT at Self-Test connector and engine block ground. • Is resistance greater than 5 ohms?	No ▶	REMOVE breakout box. RECONNECT processor. SERVICE STO or MIL/SIL circuit for short to ground. RERUN Quick Test.
	Yes ▶	3.0L EFI passenger car GO to QA7. All others GO to QA6.
QA6 INTERMITTENT NDS		
• Key off. • Breakout box installed. • Connect processor. • Connect DVOM between Test Pin 30 and Test Pin 40 or 60 at the breakout box. • Run Engine Running Self-Test. • Is voltage greater than 1 volt? **NOTE: Refer to proper illustration in Pinpoint-Test FA for connector orientation.**	Yes ▶	SERVICE intermittent open in NDS harness, connector or switch. If OK, REMOVE breakout box. RECONNECT processor. GO to Quick Test Step 5.0 for appropriate service codes.
	No ▶	GO to QA7.

Key On Engine Off and/or Continuous Memory Service Code 15 — Pinpoint Test QB

Note

You should enter this Pinpoint Test only when directed here from Quick Test Step 3.0.

Remember

This Pinpoint Test is intended to diagnose only the following:

- Processor.
- Harness Circuits: GROUND, VPWR, KAM, IGNITION.

Pinpoint Test Schematic

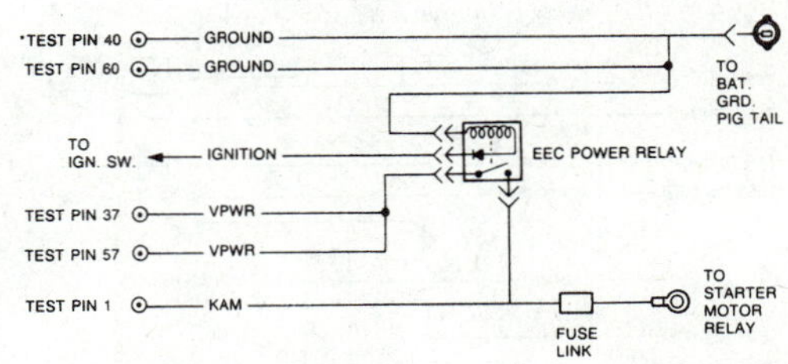

*TEST PINS LOCATED ON BREAKOUT BOX
ALL HARNESS CONNECTORS VIEWED INTO MATING SURFACE

No Codes/Codes Not Listed — Pinpoint Test QA

TEST STEP	RESULT ▶	ACTION TO TAKE
QA7 POWER RELAY ALWAYS ON		
• Key off. • Breakout box installed. • Connect DVOM to Test Pin 37 or 57 and to Test Pin 40 or 60 at the breakout box. • Turn key ON and OFF. Wait 10 seconds. • Does voltage change from greater than 10.5 volts to zero volts?	Yes ▶	If vehicle is equipped with MIL (malfunction indicator light displayed as "CHECK ENGINE" light) or SIL (shift indicator light) GO to QA9. If not, REPLACE the processor. RERUN Quick Test.
	No ▶	GO to QA8.
QA8 VPWR HARNESS SHORT TO POWER		
• Key off. • Breakout box installed. • EEC Power Relay or Integrated Relay Controller disconnected. • Connect DVOM to Test Pin 37 or 57 and to Test Pin 40 or 60 at the breakout box. • Is voltage greater than 10.5 volts?	Yes ▶	SERVICE VPWR harness short to power. RERUN Quick Test.
	No ▶	REPLACE EEC Power Relay or Integrated Relay Controller. RERUN Quick Test.
QA9 MIL AND/OR SIL EQUIPPED VEHICLES		
• Are any of these conditions present?		
• Shift indicator light: — Always ON	▶	GO to KL1.
— Always OFF	▶	GO to KL1.
• Malfunction indicator light: — Always ON	▶	GO to ML1.
— Always OFF	▶	GO to ML5.
• Shift and malfunction indicator lights functioning normally.	▶	REPLACE the processor. RERUN Quick Test.

Key On Engine Off and/or Continuous Memory Service Code 15 — Pinpoint Test QB

TEST STEP	RESULT ▶	ACTION TO TAKE
QB1 CONDITIONS FOR CONTINUOUS CODE 15		
• Power interruption to Keep Alive Memory (KAM) Pin 1 may result in a service code being stored in Continuous Memory.* • Clear continuous memory codes (use procedure described in Quick Test Step 6.0A). • Rerun Quick Test Step 3.0 through Continuous memory code output. • Code 15 present on retest? ***NOTE: Anytime power is interrupted to the processor, for example when installing a breakout box, a Code 15 may be outputted the first time Key On Engine Off Self-Test is run after restoration of power. Rerun Self-Test to ensure correct diagnosis.**	Yes ▶	GO to QB2.
	No ▶	Test complete.
QB2 INSPECT ENGINE COMPARTMENT WIRING FOR PROPER ROUTING		
• Are any EEC components or EEC wiring close to ignition components or wires (High Electrical Energy Sources)? If EEC wiring close, reroute and rerun Step 3.0. • Is Code 15 still present in Continuous Memory?	Yes ▶	GO to QB3.
	No ▶	Test complete.
QB3 CHECK POWER CIRCUIT TO KEEP ALIVE MEMORY		
• Key off, wait 10 seconds. • Disconnect processor 60 pin connector. Inspect for damaged pins, corrosion, loose wires, etc. Service as necessary. • Install breakout box, leave processor disconnected. • DVOM on 20 volt scale. • Connect positive test lead to Test Pin 1 and negative test lead to Test Pin 40 or 60 at the breakout box. • Key on. • Is voltage less than 10.5 volts? **NOTE: If on initial Key On Engine Off Self-Test, no voltage to the processor is observed, a Code 15 will be generated.**	Yes ▶	SERVICE open circuit. REMOVE breakout box. RECONNECT processor. RERUN Quick Test.
	No ▶	REMOVE breakout box. REPLACE processor. RERUN Quick Test.

Output State Check Not Functioning — Pinpoint Test QC

Note

You should enter this Pinpoint Test only when directed here.

Remember

This Pinpoint Test is intended to diagnose only the following:

- Processor
- Harness Circuits: SIGNAL RETURN, STO, STI, GROUND, VPWR, VREF

Pinpoint Test Schematic

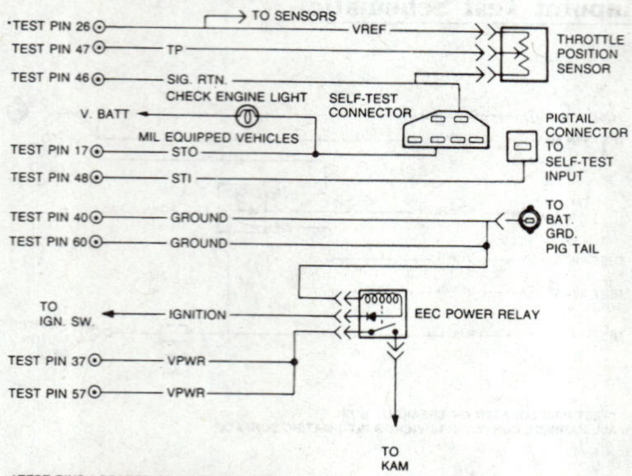

***TEST PINS LOCATED ON BREAKOUT BOX**
ALL HARNESS CONNECTORS VIEWED INTO MATING SURFACE

Processor Power Check — Pinpoint Test QD

Note

You should enter this Pinpoint Test only when directed here from Quick Test Step 3.0 or 6.0 or when directed here from Diagnostic By Symptom in the Engine Supplement Section or from a Pinpoint Test Step.

Remember

This Pinpoint Test is intended to diagnose only the following:

- Processor.
- Harness Circuits: GROUND, VPWR.

Pinpoint Test Schematic

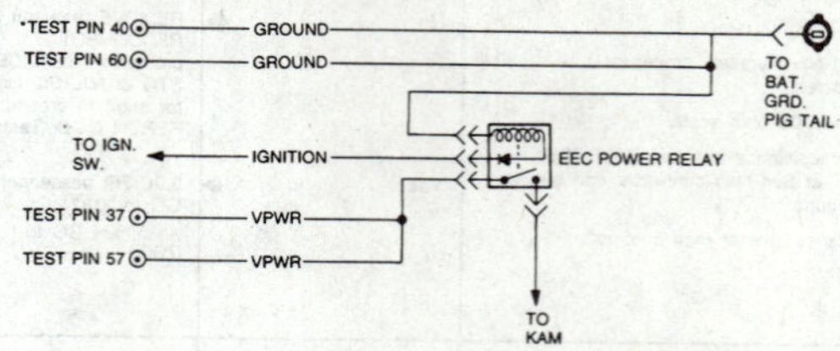

***TEST PINS LOCATED ON BREAKOUT BOX.**
ALL HARNESS CONNECTORS VIEWED INTO MATING SURFACE.

Output State Check Not Functioning — Pinpoint Test QC

TEST STEP	RESULT	▶ ACTION TO TAKE
QC1 CHECK FOR CODES 23, 53, 63 OR 68		
• Key off, wait 10 seconds. • Perform Key On Engine Off Self-Test. • Leave Key On to enter Output State Check. • Key on engine off. • Are any of these codes 23, 53, 63 or 68 present?	Yes ▶	GO to Quick Test Step [3.0B] and SERVICE appropriate code as instructed.
	Code 11 ▶	GO to [QC2].
	No Codes ▶	GO to [QA1].
QC2 CHECK THROTTLE LINKAGE		
• Check throttle and throttle linkages for sticking and binding. • Is throttle OK?	Yes ▶	REPLACE TP sensor. RERUN Quick Test.
	No ▶	SERVICE as necessary. RERUN Quick Test.

Processor Power Check — Pinpoint Test QD

TEST STEP	RESULT	▶ ACTION TO TAKE
QD1 CHECK FOR POWER TO PROCESSOR		
• Key off, wait 10 seconds. • Disconnect processor 60 pin connector. Inspect for damaged pins, corrosion, loose wires, etc. Service as necessary. • Install breakout box. • Key on, engine off. • DVOM on 20 volt scale. • Measure voltage between Test Pin 37 and Test Pin 40 at the breakout box and between Test Pin 57 and Test Pin 60 at the breakout box. • Is either voltage less than 10.5 volts?	Yes ▶	2.3L EFI TC, 2.5L CFI and 3.0L EFI passenger car GO to Pinpoint Test [X1]. All others GO to Pinpoint Test [B1].
	No ▶	REMOVE breakout box. REPLACE processor. RERUN Quick Test.

Re-Initialization Check — Pinpoint Test QE

Note

You should enter this Pinpoint Test only when directed here from Quick Test Step 3.0, 5.0 or 6.0 or when directed here from Diagnostic By Symptom in the Engine Supplement Section.

Remember

This Pinpoint Test is intended to diagnose only the following:

- Processor, EEC Power Relay.
- Harness Circuits: GROUND, VPWR, IGNITION.

Pinpoint Test Schematic

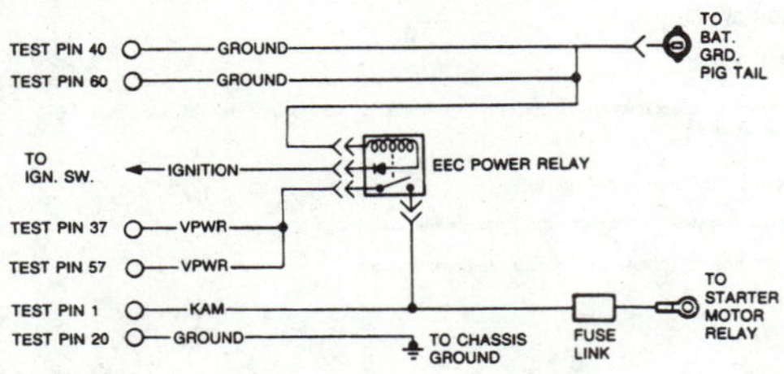

*TEST PINS LOCATED ON BREAKOUT BOX.
ALL HARNESS CONNECTORS VIEWED INTO MATING SURFACE.

Re-Initialization Check — Pinpoint Test QE

TEST STEP	RESULT ▶	ACTION TO TAKE
QE1 CHECK FOR SOURCES OF ELECTRICAL NOISE		
NOTE: Be aware that after-market installed electrical components may influence the driveability of the vehicle. • Key off. • Check that the EEC IV wiring and components are greater than 2 inches from secondary ignition wires and ignition coil. • Check that the EEC IV wiring and components are greater than 4 inches from distributor, coil tower, starter motor and its wiring. • Are all above conditions satisfied?	Yes ▶ No ▶	GO to QE2. SERVICE as necessary. RERUN Quick Test.
QE2 HARNESS CHECK — CASE GROUND		
• Key off. • Disconnect processor 60 pin connector. Inspect for damaged pins, corrosion, loose wires, etc. Service as necessary. • Install breakout box, leave processor disconnected. • DVOM on 200 ohm scale. • Measure resistance between Test Pin 20 at the breakout box and chassis ground. • Is the resistance less than 5 ohms?	Yes ▶ No ▶	GO to QE3. REMOVE breakout box. RECONNECT processor. SERVICE open circuit. RERUN Quick Test.
QE3 DISCONNECT HARNESS — CASE GROUND CHECK		
• Key off. • Reconnect processor to breakout box, but disconnect harness from breakout box. • DVOM on 200 ohm scale. • Measure resistance between Test Pin 20 at the breakout box and metal case of processor. • Is the resistance less than 5 ohms?	Yes ▶ No ▶	For 1.9L EFI GO to QE4. For 2.3L EFI TC and 2.5L CFI GO to X10. REMOVE breakout box. REPLACE processor. RERUN Quick Test.

Re-Initialization Check — Pinpoint Test QE

TEST STEP	RESULT ▶	ACTION TO TAKE
QE4 WIGGLE TEST OF VPWR CIRCUIT		
• Key on, engine off. • Connect STAR or VOM to Self-Test connector. • Self-Test deactivated. • Using Continuous Monitor Mode (Engine Running) per Quick Test Step 6.0B. Observe STAR/VOM for indication of a fault while doing the following: — Shake, bend, and twist the EEC-IV harness from the EEC-IV power relay to the processor. • Is a fault indicated or does Code 72 reappear in continuous memory if the Key On Engine Off Self-Test is rerun?	Yes ▶ No ▶	SERVICE intermittent in the VPWR circuit. RERUN Quick Test. INSPECT EEC-IV power relay and harness connectors for damaged pins, loose wires, corrosion, etc. SERVICE as necessary. If OK, REPLACE EEC-IV power relay. RERUN Quick Test.

Key Power Check — Pinpoint Test QF

Note

You should enter this Pinpoint Test only when a Service Code 55 is received in Quick Test Step 5.0.

Remember

To prevent the replacement of good components, be aware that the following non-EEC areas may be at fault:

- Charging system under voltage
- Battery charger connected with engine running
- Jump starting

This Pinpoint Test is intended to diagnose only the following:

- Harness Circuit: KEY POWER
- Processor

Pinpoint Test Schematic

1.9L CFI

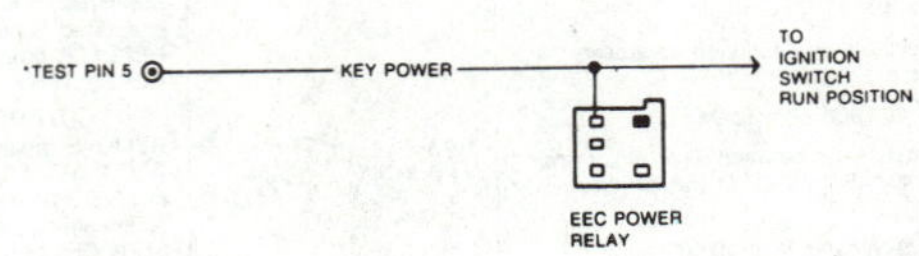

Key Power Check — Pinpoint Test QF

Pinpoint Test Schematic

2.5L CFI WITH INTEGRATED CONTROLLER

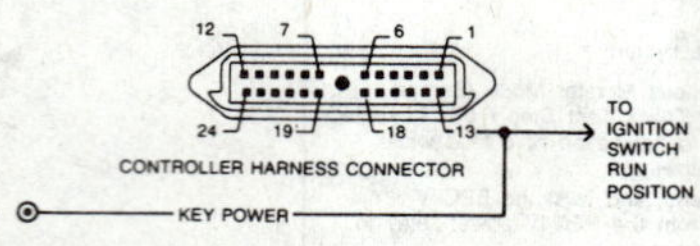

*TEST PINS LOCATED ON BREAKOUT BOX.
ALL HARNESS CONNECTORS VIEWED INTO MATING SURFACE.

System Check — Pinpoint Test S

Note

You should enter this Pinpoint Test only after a Code 11 is received in Quick Test Step 3.0, and you have been directed here from EEC-IV No-Start Pinpoint Test Step [A21] or Diagnostic By Symptom.

Remember

This Pinpoint Test is intended only as a Quick Check for the basic functioning of the following:

- ISC Bypass Air System
- MAP System
- EGR System
- MAF System

To prevent the replacement of good components, be aware that the following Non-EEC areas may be at fault:

- Poor power/ground connections
- Ignition system distributor cap, rotor, wires, coil, plugs
- Base engine valves, cam timing, compression, etc.

Key Power Check — Pinpoint Test QF

TEST STEP	RESULT ▶	ACTION TO TAKE
QF1 CHECK CONTINUITY OF KEY POWER CIRCUIT		
• Key off, wait 10 seconds. • Disconnect processor 60 pin connector. Inspect for damaged pins, corrosion, loose wires, etc. Service as necessary. • Install breakout box, leave processor disconnected. • Disconnect the EEC-IV power relay or integrated controller as appropriate. • DVOM on 200 ohm scale. • Measure resistance between Test Pin 5 at the breakout box and KEY POWER at the EEC power relay or Test Pin 13 at the Integrated Controller vehicle harness connector. • Is resistance less than 5.0 ohms?	Yes ▶ No ▶	GO to [QF2]. REMOVE breakout box. RECONNECT processor and EEC power relay or integrated controller. SERVICE open circuit. RERUN Quick Test.
QF2 CHECK KEY POWER CIRCUIT FOR SHORT TO GROUND		
• Key off. • Breakout box installed, processor disconnected. • EEC power relay or integrated controller disconnected. • DVOM on 200,000 ohm scale. • Measure resistance between Test Pin 5 and Test Pins 40, 46, and 60 at the breakout box. • Is resistance greater than 10,000 ohms?	Yes ▶ No ▶	REMOVE breakout box. RECONNECT EEC power relay or integrated controller. REPLACE processor. RERUN Quick Test. REMOVE breakout box. RECONNECT processor and EEC power relay or integrated controller. SERVICE short circuit. RERUN Quick Test.

System Check — Pinpoint Test S

Pinpoint Test Flow Chart

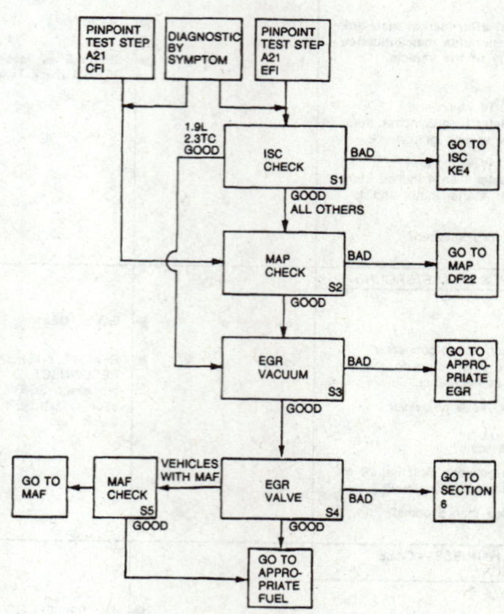

System Check — Pinpoint Test S

TEST STEP	RESULT ▶	ACTION TO TAKE
S1 ISC-BPA CHECK		
NOTE: For 1.9L and 2.5L CFI go directly to S2. • If you are here for any reason other than stalls or a no start, go to S2. Except 1.9L EFI and 2.3L EFI TC, go to S3. • Attempt to start engine at part throttle. • Will engine run at part throttle?	Yes, but runs rough ▶	GO to S2.
	Yes, and runs smooth ▶	GO to KE4.
	No ▶	1.9 EFI and 2.3L EFI TC and 5.0L MA Mustang, GO to S3. All others GO to S2.
S2 MAP CHECK		
• Key off. • Disconnect the MAP sensor from the vehicle harness. • Connect the MAP tester between the vehicle harness and the MAP sensor. • Plug MAP tester banana plugs into DVOM. • Set DVOM to 20 volt scale. • Key on. • Observe DVOM. • Approximate Altitude (Ft.) / Voltage Output (+/ – .04V): 0 – 1.59; 1000 – 1.56; 2000 – 1.53; 3000 – 1.50; 4000 – 1.47; 5000 – 1.44; 6000 – 1.41; 7000 – 1.39 • If MAP sensor is out of limits (voltage output for altitude) GO to DF21. • Crank engine. • While cranking, does DVOM reading decrease from the appropriate reading for your altitude listed above?	Yes ▶	2.9L EFI and 3.0L EFI Truck, GO to Pinpoint Test Step H1, all others, GO to S3.
	No ▶	GO to DF22.

System Check — Pinpoint Test S

TEST STEP	RESULT ▶	ACTION TO TAKE
S3 CHECK EGR VACUUM		
NOTE: The next two Test Steps will attempt to determine if the EGR system is the cause of the current symptom and/or no start. • Disconnect vacuum line at EGR valve. Do not plug the vacuum line. • Start engine. • For Drive Symptom: Is vacuum present at vacuum line? • For No Start: Does engine start?	Yes ▶	For 1.9L EFI and 2.3L EFI TC, GO to KA1. For 2.3L OHC EFI Car and Truck GO to DD11. For 2.5L HSC CFI, 5.0L SEFI Car and 4.9L EFI, 5.0L EFI, 5.8L EFI, 7.5L EFI Truck GO to DN42. For 1.9L CFI, 2.3L HSC, EFI, 3.0L EFI and 3.8L EFI Car GO to DL23.
	No ▶	GO to S4.
S4 CHECK EGR VALVE		
• Inspect EGR valve to ensure proper seating. • Is valve fully seated (closed)?	Yes ▶	Vehicles with MAF sensor GO to S5. All others GO to H1.
	No ▶	
S5 CHECK MAF SENSOR OUTPUT		
• Key off, wait 10 seconds. • Disconnect processor 60 pin connector. Inspect for damaged pins, corrosion, loose wires, etc. Service as necessary. • Install breakout box. • Connect processor to breakout box. • Key on, engine off. • DVOM on 20 volt scale. • Measure voltage between Test Pin 50 at the breakout box and battery negative post. • Is voltage between 0.5 and 1.5 volts?	Yes ▶	If vehicle stalls GO to Pinpoint Test Step DC6. All others GO to H1.
	No ▶	Vehicles with Continuous Memory Code 56 GO to DC10. Vehicles with Continuous Memory Code 66 GO to DC6.

Transmission — AXOD — Pinpoint Test T

Note

You should enter this Pinpoint Test only when Service Codes 59, 62, 67, 69 and 89 are received in Quick Test Step 3.0, and/or Service Codes 29, 39, 57, 59 and 69 received in Quick Test Step 6.0 or when directed here from Diagnostic By Symptom in the Engine Supplement Section.

Remember

This Pinpoint Test is intended to diagnose only the following:

- Harness Circuits: THS 4/3, THS 3/2, TTS, LUS, NPS, VSS+, VSS– and VPWR
- Vehicle Speed Sensor
- Processor Assembly

Transmission — AXOD — Pinpoint Test T

Pinpoint Test Schematic

AXOD — HARNESS CONNECTIONS

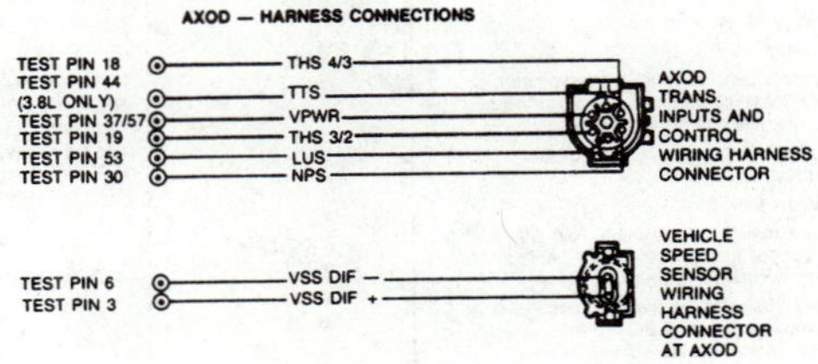

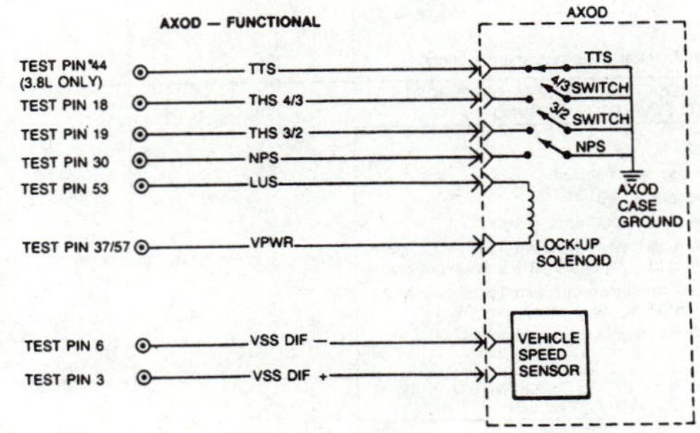

Transmission — AXOD | Pinpoint Test | T

T1 DRIVE CYCLE FOR CHECKING AXOD CONTINUOUS CODES

1. Record and clear Continuous Memory Self-Test codes.
2. Warm engine to operating temperature.
3. With transmission in D range, lightly accelerate from a stop to 40 mph to achieve third gear. Hold speed and throttle opening (not closed throttle) steady for 15 seconds minimum (30 seconds above 4000 feet altitude).
4. Shift gear selector to OD range and accelerate lightly from 40 to 50 mph to achieve fourth gear. Hold speed and throttle opening (not closed throttle) steady for 15 seconds minimum in fourth gear.
5. With transmission in fourth gear and steady speed and throttle opening (not closed throttle) lightly apply and release brakes (to light brake lamps) and then hold speed and throttle opening steady for an additional 15 seconds minimum.
6. Brake to a stop and remain stopped for 20 seconds minimum with transmission in OD range.
7. Turn engine off. Run Key On Engine Off Self-Test and record Continuous Memory Codes.

NOTE: All components must be connected when performing this test.

TEST STEP	RESULT	▶ ACTION TO TAKE
T2 ATTEMPT TO GENERATE CONTINUOUS MEMORY CODE 29		
• Perform Drive Cycle outlined in Test Step [T1], then return to this Step. • Did Continuous Memory Code 29 repeat?	Yes	▶ GO to [T3].
	No	▶ Unable to duplicate Continuous Memory Code 29 at this time. SERVICE other codes as necessary. If none, test is completed.

Transmission — AXOD | Pinpoint Test | T

TEST STEP	RESULT	▶ ACTION TO TAKE
T3 CHECK CONTINUITY OF VEHICLE SPEED SENSOR (VSS) HARNESS		
• Key off, wait 10 seconds. • Disconnect VSS. • Disconnect processor 60 pin connector. Inspect for damaged pins, corrosion, loose wires, etc. Service as necessary. • Install breakout box, leave processor disconnected. • DVOM on 200 ohm scale. • Measure resistance between Test Pin 3 at the breakout box and the VSS vehicle harness connector as shown below. • Measure resistance between Test Pin 6 at the breakout box and the VSS vehicle harness connector, as shown below. 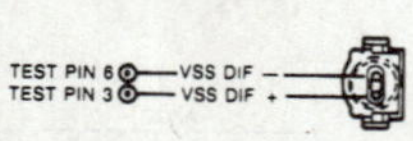 • Are both resistances less than 5 ohms?	Yes	▶ GO to [T4].
	No	▶ REMOVE breakout box. RECONNECT all components. SERVICE open circuit(s). REPEAT Test Step [T2].
T4 CHECK VSS HARNESS FOR SHORTS TO POWER OR GROUND		
• Key off. • Processor disconnected. • VSS disconnected. • DVOM on 200,000 ohm scale. • Measure resistance between Test Pin 3 and Test Pins 37, 40 and 6 at the breakout box. • Measure resistance between Test Pin 6 and Test Pin 37 at the breakout box. • Are all resistances greater than 10,000 ohms?	Yes	▶ REMOVE breakout box. RECONNECT all components. GO to [T5].
	No	▶ REMOVE breakout box. RECONNECT all components. SERVICE short circuit(s). REPEAT Test Step [T2].
T5 REPEAT DRIVE CYCLE WITH A KNOWN GOOD VSS INSTALLED		
• Substitute VSS with known good sensor. • Processor and VSS connected. • Perform Drive Cycle outlined in Test Step [T1], then return to this Step. • Did Continuous Memory Code 29 repeat?	Yes	▶ REPLACE processor. REPEAT Test Step [T2].
	No	▶ REPLACE VSS. RERUN Quick Test.

Transmission — AXOD | Pinpoint Test | T

TEST STEP	RESULT	▶ ACTION TO TAKE
T11 ATTEMPT TO GENERATE CONTINUOUS MEMORY CODE 69		
• Perform Drive Cycle outlined in Test Step [T1], then return to this Step. • Did Continuous Memory Code 69 repeat?	Yes	▶ GO to [T12].
	No	▶ Unable to duplicate Continuous Memory Code 69 at this time. SERVICE other codes as necessary. If none, test is completed.
T12 CHECK CONTINUITY OF THS 3/2 CIRCUIT		
• Key off, wait 10 seconds. • Disconnect AXOD harness. • Disconnect processor 60 pin connector. Inspect for damaged pins, corrosion, loose wires, etc. Service as necessary. • Install breakout box, leave processor disconnected. • DVOM on 200 ohm scale. • Measure resistance between Test Pin 19 at the breakout box and the AXOD vehicle harness connector, as shown below. • Is resistance less than 5 ohms? 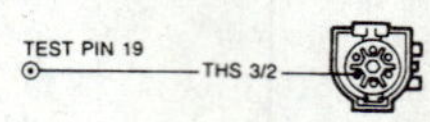	Yes	▶ GO to [T13].
	No	▶ REMOVE breakout box. RECONNECT all components. SERVICE open in THS 3/2 circuit. REPEAT Test Step [T11].
T13 CHECK THS 3/2 CIRCUIT FOR SHORT TO POWER		
• Key off. • Breakout box installed. • Processor disconnected. • AXOD harness disconnected. • DVOM on 200,000 ohm scale. • Measure resistance between Test Pin 19 and Test Pin 37 at the breakout box. • Is resistance greater than 10,000 ohms?	Yes	▶ GO to [T14].
	No	▶ REMOVE breakout box. RECONNECT all components. SERVICE short to power in THS 3/2 circuit. REPEAT Test Step [T11].

Transmission — AXOD | Pinpoint Test | T

TEST STEP	RESULT	▶ ACTION TO TAKE
T14 PROCESSOR VERIFICATION		
• Key off. • Breakout box installed. • Reconnect processor and AXOD harness. • Jumper Test Pin 19 to Test Pin 40 at the breakout box. • Run Key On Engine Off Self-Test. • Is Code 62 or 69 present?	Yes	▶ REMOVE breakout box. REMOVE jumper wire.
	No	▶ REMOVE breakout box. REMOVE jumper wire. REPLACE processor. REPEAT Test Step [T11].
T21 ATTEMPT TO GENERATE CONTINUOUS MEMORY CODE 59		
• Perform Drive Cycle outlined in Test Step [T1], then return to this Step. • Did Continuous Memory Code 59 repeat?	Yes	▶ GO to [T22].
	No	▶ Unable to duplicate Continuous Memory Code 59 at this time. SERVICE other codes as necessary. If none, test is completed.

Transmission — AXOD | Pinpoint Test | T

TEST STEP	RESULT ▶	ACTION TO TAKE
T22 CHECK CONTINUITY OF THS 4/3 CIRCUIT		
• Key off, wait 10 seconds. • Disconnect AXOD harness. • Disconnect processor 60 pin connector. Inspect for damaged pins, corrosion, loose wires, etc. Service as necessary. • Install breakout box, leave processor disconnected. • DVOM on 200 ohm scale. • Measure resistance between Test Pin 18 at the breakout box and the AXOD vehicle harness connector, as shown below. 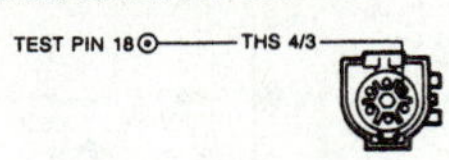 • Is resistance less than 5 ohms?	Yes ▶ No ▶	GO to T23. REMOVE breakout box. RECONNECT all components. SERVICE open in THS 4/3 circuit. REPEAT Test Step T21.

Transmission — AXOD | Pinpoint Test | T

TEST STEP	RESULT ▶	ACTION TO TAKE
T41 ATTEMPT TO GENERATE CONTINUOUS MEMORY CODE 57		
• Perform Drive Cycle outlined in Test Step T1 then return to this Step. • Did Continuous Memory Code 57 repeat?	Yes ▶ No ▶	GO to T42. Unable to duplicate Continuous Memory Code 57 at this time. SERVICE other codes as necessary. If none, test is completed.
T42 CHECK CONTINUITY OF NPS HARNESS CIRCUIT		
• Key off, wait 10 seconds. • Disconnect AXOD harness. • Disconnect processor 60 pin connector. Inspect for damaged pins, corrosion, loose wires, etc. Service as necessary. • Install breakout box, leave processor disconnected. • DVOM on 200 ohm scale. • Measure resistance between Test Pin 30 at the breakout box and the AXOD vehicle harness connector, as shown below. 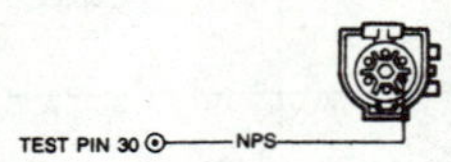 • Is resistance less than 5 ohms?	Yes ▶ No ▶	REMOVE breakout box. RECONNECT all components. REMOVE breakout box. RECONNECT all components. SERVICE open in NPS circuit. REPEAT Test Step T41.

Transmission — AXOD | Pinpoint Test | T

TEST STEP	RESULT ▶	ACTION TO TAKE
T23 CHECK THS 4/3 CIRCUIT FOR SHORT TO POWER		
• Key off. • Breakout box installed. • Processor disconnected. • AXOD harness disconnected. • DVOM on 200,000 ohm scale. • Measure resistance between Test Pin 18 and Test Pin 37 at the breakout box. • Is resistance greater than 10,000 ohms?	Yes ▶ No ▶	GO to T24. REMOVE breakout box. RECONNECT all components. SERVICE short to power in THS 4/3 circuit. REPEAT Test Step T21.
T24 PROCESSOR VERIFICATION		
• Key Off. • Breakout box installed. • Reconnect processor and AXOD harness. • Jumper Test Pin 18 to Test Pin 40 at the breakout box. • Run Key On Engine Off Self-Test. • Is Code 62 or 59 present?	Yes ▶ No ▶	REMOVE breakout box. REMOVE jumper wire. REMOVE breakout box. REMOVE jumper wire. REPLACE processor. REPEAT TEST Step T21.
T31 ATTEMPT TO GENERATE CONTINUOUS MEMORY CODE 39		
NOTE: If Continuous Memory Code 59 is also present, go directly to T21. • Perform Drive Cycle outlined in Test Step T1, then return to this Step. • Did Continuous Memory Code 39 repeat?	Yes ▶ No ▶	 Unable to duplicate Continuous Memory Code 39 at this time. SERVICE other codes as necessary. If none, test is completed.

Transmission — AXOD | Pinpoint Test | T

TEST STEP	RESULT ▶	ACTION TO TAKE
T51 SERVICE CODE 89: CHECK CONTINUITY OF VPWR CIRCUIT		
• Key off, wait 10 seconds. • Disconnect AXOD harness. • Disconnect processor 60 pin connector. Inspect for damaged pins, corrosion, loose wires etc. Service as necessary. • Install breakout box, leave processor disconnected. • DVOM on 200 ohm scale. • Measure resistance between Test Pin 37 at the breakout box and the AXOD vehicle harness connector, as shown. 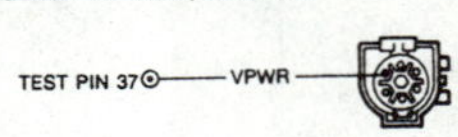 • Is resistance less than 5 ohms?	Yes ▶ No ▶	GO to T52. REMOVE breakout box. RECONNECT all components. SERVICE open in VPWR circuit to AXOD. RERUN Quick Test.
T52 CHECK CONTINUITY OF LUS CIRCUIT		
• Key off. • Breakout box installed. • Processor disconnected. • AXOD harness disconnected. • DVOM on 200 ohm scale. • Measure resistance between Test Pin 53 at the breakout box and the AXOD vehicle harness connector, as shown. 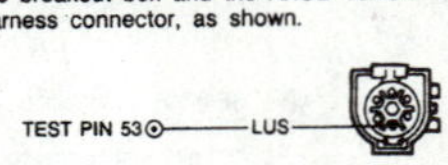 • Is resistance less than 5 ohms?	Yes ▶ No ▶	GO to T53. REMOVE breakout box. RECONNECT all components. SERVICE open in LUS circuit to AXOD. RERUN Quick Test.

Transmission — AXOD — Pinpoint Test T

TEST STEP	RESULT	ACTION TO TAKE
T53 CHECK LUS CIRCUIT FOR SHORT TO POWER OR GROUND		
• Key off. • Breakout box installed. • Processor disconnected. • AXOD harness disconnected. • DVOM on 200,000 ohm scale. • Measure resistance between Test Pin 53 and Test Pins 37 and 40 at the breakout box. • Are both resistances greater than 10,000 ohms?	Yes ▶	GO to T54.
	No ▶	REMOVE breakout box. RECONNECT all components. SERVICE short(s) in LUS circuit. RERUN Quick Test. If code 89 is still present, REPLACE processor. RERUN Quick Test.
T54 CHECK TOTAL CIRCUIT RESISTANCE		
• Key off. • Breakout box installed. • Processor disconnected. • Reconnect AXOD harness. • DVOM on 200 ohm scale. • Measure the resistance between Test Pin 53 and Test Pin 57 at the breakout box. • Is resistance between 20 ohms and 40 ohms?	Yes ▶	REMOVE breakout box. REPLACE processor. RERUN Quick Test.
	No ▶	REMOVE breakout box. RECONNECT processor.
T61 SERVICE CODE 62: AXOD HARNESS VERIFICATION		
• Key off. • Disconnect AXOD harness. • Run Key On Engine Off Self-Test. • Is Code 62 still present?	Yes ▶	GO to T62.
	No ▶	RECONNECT AXOD harness.

Transmission — AXOD — Pinpoint Test T

TEST STEP	RESULT	ACTION TO TAKE
T73 SERVICE CODE 69: AXOD HARNESS VERIFICATION		
• Key off. • Disconnect AXOD harness. • Run Key On Engine Off Self-Test. • Is Code 69 still present?	Yes ▶	GO to T74.
	No ▶	RECONNECT AXOD harness.
T74 CHECK THS 3/2 HARNESS CIRCUIT FOR SHORT TO GROUND		
• Key Off. • Disconnect processor 60 pin connector. Inspect for damaged pins, corrosion, loose wires, etc. Service as necessary. • Install breakout box, leave processor disconnected. • AXOD harness disconnected. • DVOM on 200,000 ohm scale. • Measure resistance between Test Pin 19 and Test Pins 40 and 60 at the breakout box. • Are resistances greater than 10,000 ohms?	Yes ▶	REMOVE breakout box. RECONNECT AXOD harness. REPLACE processor. RERUN Quick Test.
	No ▶	REMOVE breakout box. RECONNECT AXOD harness and processor. SERVICE short to ground. RERUN Quick Test.

Transmission — AXOD — Pinpoint Test T

TEST STEP	RESULT	ACTION TO TAKE
T62 CHECK THS 3/2 AND 4/3 CIRCUITS FOR SHORT TO GROUND		
• Key off. • Disconnect processor 60 pin connector. Inspect for damaged pins, corrosion, loose wires, etc. Service as necessary. • Install breakout box, leave processor disconnected. • AXOD harness disconnected. • DVOM on 200,000 ohm scale. • Measure resistance between Test Pin 18 and Test Pins 40 and 60 at the breakout box. • Measure resistance between Test Pin 19 and Test Pins 40 and 60 at the breakout box. • Are all resistances greater than 10,000 ohms?	Yes ▶	REMOVE breakout box. RECONNECT all components. REPLACE processor. RERUN Quick Test.
	No ▶	REMOVE breakout box. RECONNECT all components. SERVICE short(s) to ground. RERUN Quick Test.
T71 SERVICE CODE 59: AXOD HARNESS VERIFICATION		
• Key off. • Disconnect AXOD harness. • Run Key On Engine Off Self-Test. • Is Code 59 still present?	Yes ▶	GO to T72.
	No ▶	RECONNECT AXOD harness.
T72 CHECK THS 4/3 HARNESS CIRCUIT FOR SHORT TO GROUND		
• Key Off. • Disconnect processor 60 pin connector. Inspect for damaged pins, corrosion, loose wires, etc. Service as necessary. • Install breakout box, leave processor disconnected. • AXOD harness disconnected. • DVOM on 200,000 ohm scale. • Measure resistance between Test Pin 18 and Test Pins 40 and 60 at the breakout box. • Are resistances greater than 10,000 ohms?	Yes ▶	REMOVE breakout box. RECONNECT AXOD harness. REPLACE processor. RERUN Quick Test.
	No ▶	REMOVE breakout box. RECONNECT AXOD harness and processor. SERVICE short to ground. RERUN Quick Test.

Transmission — AXOD — Pinpoint Test T

TEST STEP	RESULT	ACTION TO TAKE
T75 CHECK CONTINUITY OF TTS HARNESS CIRCUIT		
• Key off, wait 10 seconds. • Disconnect AXOD harness. • Disconnect processor 60 pin connector. Inspect for damaged pins, corrosion, loose wires, etc. Service as necessary. • Install breakout box, leave processor disconnected. • DVOM on 200 ohm scale. • Measure resistance between Test Pin 44 at the breakout box and the AXOD vehicle harness connector, as shown below. • Is resistance less than 5 ohms? TEST PIN 44 ⊙—— TTS —— 	Yes ▶	GO to T76.
	No ▶	SERVICE open in TTS circuit. REMOVE breakout box. RECONNECT all components. DRIVE vehicle to verify drive complaint was eliminated.
T76 CHECK TTS HARNESS CIRCUIT FOR SHORT TO POWER OR GROUND		
• Key off. • Breakout box installed. • Processor disconnected. • AXOD harness disconnected. • DVOM on 200,000 ohm scale. • Measure resistance between Test Pin 44 and Test Pin 37 at the breakout box. • Measure resistance between Test Pin 44 and Test Pin 40 at the breakout box. • Are resistances greater than 10,000 ohms?	Yes ▶	GO to T77.
	No ▶	SERVICE short(s) in TTS circuit. REMOVE breakout box. RECONNECT all components. DRIVE vehicle to verify drive complaint was eliminated.

Transmission — AXOD | Pinpoint Test | T

TEST STEP	RESULT ▶	ACTION TO TAKE
T77 PROCESSOR VERIFICATION		
• Key off. • Breakout box installed. • Reconnect processor and AXOD harness. • Jumper Test Pin 44 to Test Pin 40 at the breakout box. • Drive vehicle to verify drive complaint. • Was drive complaint eliminated?	Yes ▶	REMOVE breakout box. REMOVE jumper wire.
	No ▶	**NOTE: To prevent the replacement of good components, be aware that non-EEC areas may cause similar drive symptoms (e.g. driveline, engine running rich).** REMOVE breakout box. REMOVE jumper wire. REPLACE processor.
T81 SERVICE CODE 67: CHECK VOLTAGE AT NPS INPUT TO PROCESSOR		
• Key on, engine off. • Disconnect processor 60 pin connector. Inspect for damaged pins, corrosion, loose wires etc. Service as necessary. • Install breakout box and reconnect processor. • DVOM on 20 volt scale. • Measure voltage between Test Pin 30 and Test Pin 46 at the breakout box. • Is voltage less than 4 volts?	Yes ▶	GO to T82.
	No ▶	GO to Pinpoint Test FA9.

Transmission — VSS | Pinpoint Test | T

TEST STEP	RESULT ▶	ACTION TO TAKE
T90 CHECK VEHICLE SPEED SENSOR		
• Key off, wait 10 seconds. • Locate and disconnect Vehicle Speed Sensor. • DVOM on 200,000 ohm scale. • Measure resistance across Vehicle Speed Sensor. • Is resistance between 190 and 230 ohms?	Yes ▶	GO to T91.
	No ▶	REPLACE sensor. DRIVE vehicle to verify drive complaint was eliminated.
T91 CHECK CONTINUITY OF VEHICLE SPEED SENSOR (VSS) HARNESS		
• Key off, wait 10 seconds. • Disconnect processor 60 pin connector. Inspect for damaged pins, corrosion, loose wires etc. Service as necessary. • Install breakout box, leave processor disconnected. • VSS disconnected. • DVOM on 200 ohm scale. • Measure resistance between Test Pin 3 at the breakout box and the VSS vehicle harness connector as shown below. • Measure resistance between Test Pin 6 at the breakout box and the VSS vehicle harness connector, as shown below. 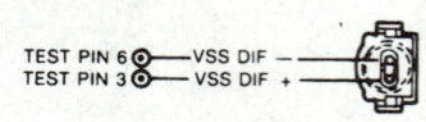 • Are both resistances less than 5 ohms?	Yes ▶	GO to T92.
	No ▶	REMOVE breakout box. RECONNECT all components. SERVICE open circuit(s) in VSS harness. DRIVE vehicle to verify drive complaint was eliminated.

Transmission — AXOD | Pinpoint Test | T

TEST STEP	RESULT ▶	ACTION TO TAKE
T82 CHECK NPS HARNESS CIRCUIT FOR SHORT TO GROUND		
• Key off. • Breakout box installed. • Disconnect processor. • Disconnect AXOD harness. • DVOM on 200,000 ohm scale. • Measure resistance between Test Pin 30 and Test Pins 40 and 60 at the breakout box. • Are all resistances greater than 10,000 ohms?	Yes ▶	GO to T83.
	No ▶	REMOVE breakout box. RECONNECT all components. SERVICE short to ground in NPS circuit. RERUN Quick Test.
T83 PROCESSOR VERIFICATION		
• Key off. • Breakout box installed. • Reconnect processor. • AXOD harness disconnected. • Run Key On Engine Off Self-Test. • Is Code 67 present?	Yes ▶	REMOVE breakout box. RECONNECT all components. REPLACE processor. RERUN Quick Test.
	No ▶	REMOVE breakout box. RECONNECT all components. GO to Taurus/Sable or Continental Shop Manual, Section 17-15 for AXOD Transmission Electrical Component Diagnostics.

Transmission — VSS | Pinpoint Test | T

TEST STEP	RESULT ▶	ACTION TO TAKE
T92 CHECK VSS HARNESS FOR SHORTS TO POWER OR GROUND		
• Key off. • Breakout box installed. • Processor disconnected. • VSS disconnected. • DVOM on 200,000 ohm scale. • Measure resistance between Test Pin 3 and Test Pins 37, 40 and 6 at the breakout box. • Measure resistance between Test Pin 6 and Test Pin 37 at the breakout box. • Are all resistances greater than 10,000 ohms?	Yes ▶	REMOVE breakout box. RECONNECT all components. GO to T93.
	No ▶	REMOVE breakout box. RECONNECT all components. SERVICE short circuit(s) in VSS harness. DRIVE vehicle to verify drive complaint was eliminated.
T93 SUBSTITUTE VEHICLE SPEED SENSOR (VSS)		
• Substitute VSS with known good sensor. • Processor and VSS connected. • Drive vehicle to verify drive complaint. • Was drive complaint eliminated?	Yes ▶	REPLACE VSS.
	No ▶	REPLACE processor.

Integrated Controller	Pinpoint Test	X

Note

You should enter this Pinpoint Test only when service code 72, 78, 82, 83, 87, 88, 95 and 96 are received in Quick Test Steps 3.0 or 6.0, or you are directed here from Pinpoint Test A, Pinpoint Test C, or Diagnostic By Symptom.

Remember

To prevent the replacement of good components, be aware that the following non-EEC area may be at fault:

- Fuel Lines
- Fuel Filters
- Contaminated Fuel
- Fuel Pump
- Ignition Switch
- Battery Cables
- Alternator
- Voltage Regulator
- Ground Straps
- A/C Clutch
- A/C Demand
- Cooling Fan Motor

This Pinpoint Test is intended to diagnose only the following:

- Integrated Relay Controller Module (within this Pinpoint Named Integrated Controller)
- Battery Voltage
- Power Relay
- EDF Relay
- HEDF Relay
- WAC Relay
- Fuel Pump Relay
- Harness Circuits: V Batt., VPWR, F.P., GROUND and POWER to Fuel Pump(s), WAC, ACC, ACCS, COOLING FAN POWER, A/C CLUTCH, KEY POWER, POWERS To Integrated Controller
- Processor Assembly
- A/C Demand Switch Input

Integrated Controller	Pinpoint Test	X

Pinpoint Test Schematic

2.5L CFI CLC; 3.0L EFI AXOD and 3.8L EFI AXOD

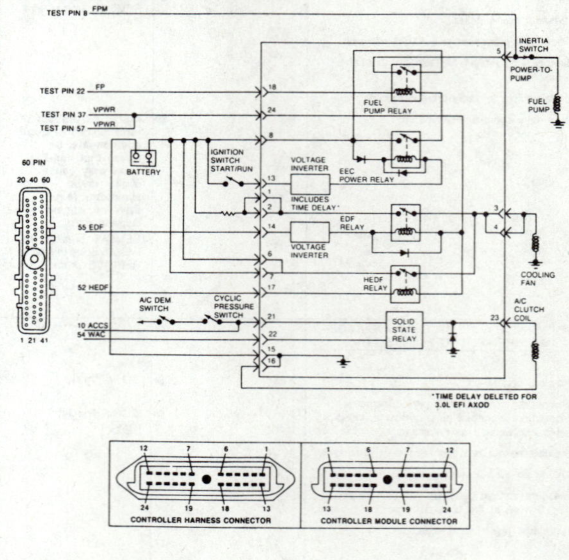

Refer to Engine Supplement for Test Pin Usage Chart.

Integrated Controller	Pinpoint Test	X

Pinpoint Test Schematic

2.3L EFI T/C

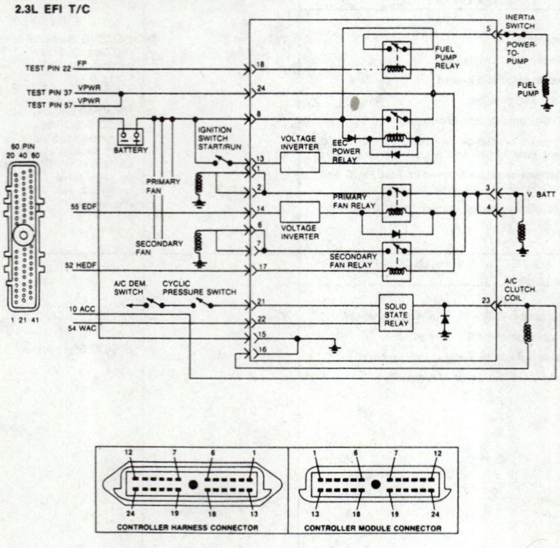

Refer to Engine Supplement for Test Pin Usage Chart.

Integrated Controller	Pinpoint Test	X

Pinpoint Test Schematic

2.5L CFI MTX ONLY

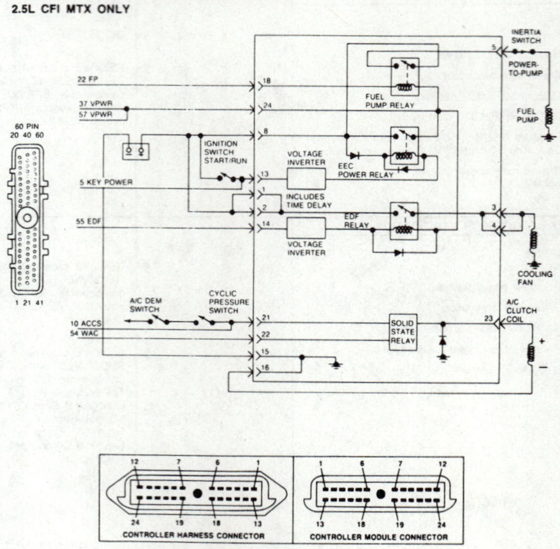

Refer to Engine Supplement for Test Pin Usage Chart.

Integrated Controller — Pinpoint Test X

TEST STEP	RESULT ▶	ACTION TO TAKE
VEHICLE BATTERY		
X1 CHECK BATTERY VOLTAGE		
• Key on, engine off. • DVOM on 20 volt scale. • Measure voltage across battery terminals. • Is voltage greater than 10.5 volts?	Yes ▶	GO to X2.
	No ▶	SERVICE discharged battery
X2 CHECK BATTERY GROUND		
• Key on, engine off. • Processor connected. • DVOM on 20 volt range. • Measure voltage between battery negative post and SIGNAL RETURN circuit in the Self-Test connector. • Is voltage greater than 0.5 volts?	Yes ▶	GO to X3.
	No ▶	GO to X6.
X3 GROUND FAULT ISOLATION		
• Key off. • Disconnect processor 60 pin connector. Inspect for damaged pins, corrosion, loose wires etc. Service as necessary. • Install breakout box. • Key on, engine off. • Processor connected. • DVOM on 20 volt scale. • Measure voltage between battery negative post and Test Pins 40 and 60 at the breakout box. • Are both voltages less than 0.5 volts?	Yes ▶	GO to X4.
	No ▶	Circuit(s) with greater than 0.5 volts has high resistance or open. SERVICE open ground circuit. RERUN Quick Test.
X4 PROCESSOR GROUND FAULT ISOLATION		
• Breakout box installed. • Key off, wait 10 seconds. • Processor connected. • DVOM on 200 ohm scale. • Measure resistance between Test Pin 46 and Test Pin 40 and between Test Pin 46 and Test Pin 60 at the breakout box. • Are both resistances less than 5 ohms?	Yes ▶	GO to X5.
	No ▶	REMOVE breakout box. REPLACE processor. RERUN Quick Test.

Integrated Controller — Pinpoint Test X

TEST STEP	RESULT ▶	ACTION TO TAKE
X5 CHECK CONTINUITY OF SIGNAL RETURN CIRCUIT		
• Breakout box installed. • Key off, wait 10 seconds. • Processor connected. • DVOM on 200 ohm scale. • Measure resistance between Test Pin 46 at the breakout box and SIGNAL RETURN circuit at Self-Test connector. • Is resistance less than 5.0 ohms?	Yes ▶	System OK. RUN Quick Test.
	No ▶	REMOVE breakout box. RECONNECT processor. SERVICE open circuit. RERUN Quick Test.
X6 MEASURE VOLTAGE AND GROUND TO INTEGRATED CONTROLLER		
• Key off. • Disconnect Integrated Controller Module. • DVOM on 20 volt scale. • Measure voltage between Test Pin 8 and Test Pin 15 at the Integrated Controller vehicle harness connector. • Is voltage greater than 10.5 volts?	Yes ▶	GO to X7.
	No ▶	GO to X9.
X7 KEY POWER TO INTEGRATED CONTROLLER		
• Integrated Controller disconnected. • DVOM on 20 volt scale. • Key on. • Measure voltage between Pin 13 and Pin 15 at the Integrated Controller vehicle harness connector. • Refer to schematic in Pinpoint Test X. • Is voltage greater than 10.5 volts?	Yes ▶	GO to X8.
	No ▶	SERVICE open between Pin 13 and ignition switch. RECONNECT Integrated Controller. RERUN Quick Test.

Integrated Controller — Pinpoint Test X

TEST STEP	RESULT ▶	ACTION TO TAKE
X8 MEASURE CONTINUITY OF VPWR		
• Key off. • Integrated Controller disconnected. • Disconnect processor 60 pin connector. Inspect for damaged pins, corrosion, loose wires, etc. Service as necessary. • Install breakout box, leave processor disconnected. • DVOM on 200 ohm scale. • Measure resistance between Test Pin 37 and 57 at the breakout box and Test Pin 24 at the Integrated Controller harness. • Is resistance greater than 5.0 ohms?	Yes ▶	REMOVE breakout box. RECONNECT processor. SERVICE open in VPWR circuit. RECONNECT Integrated Controller. RERUN Quick Test.
	No ▶	REMOVE breakout box. RECONNECT processor. REPLACE Integrated Controller. RERUN Quick Test.
X9 MEASURE CONTINUITY OF POWER GROUND TO INTEGRATED CONTROLLER		
• Key off. • Integrated Controller disconnected. • DVOM on 200 ohm scale. • Measure resistance between battery negative post and at Test Pin 15 at the Integrated Controller connector. • Is resistance greater than 5.0 ohms?	Yes ▶	RECONNECT Integrated Controller. SERVICE open in battery ground to Pin 15 (Integrated Controller harness connector). RERUN Quick Test.
	No ▶	RECONNECT Integrated Controller. SERVICE open in battery positive to Pin 8 (Integrated Controller harness connector). RERUN Quick Test.

Integrated Controller — Pinpoint Test X

TEST STEP	RESULT ▶	ACTION TO TAKE
X10 CODE 72 OR 78: INTERMITTENT OPEN IN VPWR CIRCUIT		
NOTE: Code 72 or 78 indicates that while key power was present, VPWR had an interrupt, or interference from electrical noises caused the processor to reset, resulting in possible stalls, high idle rpm, lack of power on acceleration or other drive symptoms. Possible Causes: • Intermittent open in VPWR circuit from integrated controller to processor. • EEC power relay intermittent malfunction. • Intermittent open in VBATT circuit to integrated controller. • Intermittent open in KEY POWER circuit to integrated controller. • EEC harness to close to the distributor spark plug wires and other vehicle harnesses. • Using Continuous Monitor Mode (Engine Running) per Appendix in Section 16. Observe VOM or STAR LED for indication of a fault while performing the following: • Shake, bend and twist harness from integrated controller to the processor, to the ignition switch and to battery positive. • Is a fault indicated or does Code 72 or 78 reappear in continuous memory if Quick Test is rerun?	Yes ▶	CHECK for proper routing of EEC harness. SERVICE as necessary. If OK SERVICE intermittent VPWR circuit. RERUN Quick Test.
	No ▶	INSPECT component and harness connectors of integrated controller and processor, for loose or damaged pins, corrosion, etc. SERVICE as necessary. If OK, ROAD TEST vehicle through a variety of drive modes to verify if symptom exists. REPLACE integrated controller, otherwise testing complete. RERUN Quick Test.
X11 CHECK POWER-TO-PUMP(S) CIRCUIT		
• Key on, engine off. • Locate and disconnect fuel pump(s). • DVOM on 20 volt scale. • Measure voltage between CHASSIS GROUND and POWER-TO-PUMP(S) circuit at fuel pump during crank mode. • Is voltage greater than 8.0 volts during crank?	Yes ▶	
	No ▶	GO to X12.

Integrated Controller — Pinpoint Test X

TEST STEP	RESULT	ACTION TO TAKE
X12 CHECK RESISTANCE OF FUEL PUMP INERTIA SWITCH		
• Key off, wait 10 seconds. • Fuel pump(s) disconnected. • Locate and disconnect fuel pump inertia switch. • DVOM on 200 ohm scale. • Measure the resistance of the fuel pump inertia switch. • Is resistance less than 5.0 ohms?	Yes ▶ No ▶	GO to X13. REPLACE fuel pump inertia switch. RERUN Quick Test.
X13 POWER-TO-PUMP CIRCUIT CONTINUITY CHECK		
• Key off. • DVOM on 200 ohm scale. • Disconnect Integrated Controller. • Fuel pump(s) disconnected. • Measure resistance between Pin 5 at the integrated controller vehicle harness connector and POWER-TO-PUMP(S) circuit at the fuel pump vehicle harness connector. • Is resistance less than 5.0 ohms?	Yes ▶ No ▶	REPLACE Integrated Controller. RECONNECT all components. RERUN Quick Test. SERVICE open in POWER-TO-PUMP(S) circuit. RECONNECT Integrated Controller. RERUN QuicK Test.
X14 CHECK POWER-TO-PUMP(S) FOR SHORTS TO POWER		
• Key off. • Disconnect Integrated Controller. • Disconnect fuel pumps. • DVOM on 200,000 ohm scale. • Measure resistance between Pin 5 and Pin 24 at the Integrated Controller vehicle harness connector. • Measure resistance between Pin 5 at the Integrated Controller vehicle harness connector and battery positive post. • Is either resistance less than 10,000 ohms?	Yes ▶ No ▶	SERVICE short circuit. RECONNECT all components. ATTEMPT to start vehicle. If vehicle runs, RERUN Quick Test. If vehicle will not run, REPLACE Integrated Controller. RERUN Quick Test. RECONNECT fuel pump. REPLACE Integrated Controller. RERUN Quick Test.

Integrated Controller — Pinpoint Test X

TEST STEP	RESULT	ACTION TO TAKE
SERVICE CODE: 87		
X15 CHECK CONTINUITY OF FUEL PUMP CIRCUIT		
• Key off. • Disconnect processor 60 pin connector. Inspect for damaged pins, corrosion, loose wires, etc. Service as necessary. • Install breakout box, leave processor disconnected. • Disconnect Integrated Controller. • DVOM on 200 ohm scale. • Measure resistance between Test Pin 22 at the breakout box and Pin 18 at the Integrated Controller vehicle harness connector. • Is resistance less than 5.0 ohms?	Yes ▶ No ▶	GO to X16. SERVICE open in fuel pump circuit. REMOVE breakout box. RECONNECT processor and controller. RERUN Quick Test.

Integrated Controller — Pinpoint Test X

TEST STEP	RESULT	ACTION TO TAKE
X16 CHECK FUEL PUMP CIRCUIT FOR SHORTS TO POWER AND GROUND		
• Key off. • Breakout box installed. • Processor disconnected. • Integrated Controller disconnected. • DVOM on 200,000 ohm scale. • Measure resistance between Test Pin 22 and Test Pins 37, 57 and battery positive post and between Test Pin 22 and Test Pins 40, 60 and battery negative. • Are all resistances greater than 10,000 ohms?	Yes ▶ No ▶	GO to X17. REMOVE breakout box. SERVICE fuel pump circuit shorts to power or ground. RECONNECT all components. RERUN Quick Test. If code 87 is still present, GO to X17.
X17 CHECK RESISTANCE OF FUEL PUMP RELAY COIL		
• Key off. • Breakout box installed. • Processor disconnected. • Integrated Controller disconnected. • DVOM on 200 ohm scale. • Measure resistance of Integrated Controller from Pin 18 to 24. • Is resistance between 65 and 100 ohms?	Yes ▶ No ▶	REMOVE breakout box. REPLACE processor. RECONNECT Integrated Controller. RERUN Quick Test. REMOVE breakout box. RECONNECT processor. REPLACE Integrated Controller. RERUN Quick Test.
X20 NO FAN, HIGH OR LOW WITH NO CODE 83		
• Key off. • Disconnect Integrated Controller. • DVOM on 20 volt scale. • Measure voltage between battery negative post and Pins 1, 2, 6 and 7, respectively at the Integrated Controller vehicle harness connector. • Is voltage greater than 10.5 volts?	Yes ▶ No ▶	GO to X21. RECONNECT Integrated Controller. SERVICE open in battery power circuit. RE-EVALUATE symptom.
X21 CHECK FAN MOTOR		
• Key off. • Integrated Controller disconnected. • Jumper Pin 3 to Pin 6 at Integrated Controller harness. • Does fan run?	Yes ▶ No ▶	GO to X22. GO to X23.

Integrated Controller — Pinpoint Test X

TEST STEP	RESULT	ACTION TO TAKE
X22 CHECK FAN RUNNING MODE (LOW)		
• Key off. • Disconnect processor. • Reconnect Integrated Controller. • Key on. • Does fan run at low speed?	Yes ▶ No ▶	GO to X25. CHANGE Integrated Controller. RECONNECT processor and controller. RE-EVALUATE symptom.
X23 MEASURE BATTERY VOLTAGE SUPPLY AT FAN — BYPASSING INTEGRATED CONTROLLER		
• Key Off. • Disconnect cooling fan. • Integrated Controller disconnected. • Jumper Pin 3 to Pin 6 at Integrated Controller vehicle harness connector. • DVOM on 20 volt scale. • Measure voltage at cooling fan vehicle harness connector. • Is voltage greater than 8.0 volts?	Yes ▶ No ▶	RECONNECT Integrated Controller. CHANGE fan motor. RE-EVALUATE symptom. GO to X24.
X24 COOLING FAN GROUND VERIFICATION		
• Key off. • Cooling fan disconnected. • Integrated Controller disconnected. • Jumper Pin 3 to Pin 6 at Integrated Controller vehicle harness connector. • DVOM on 20 volt scale. • Measure voltage between voltage positive at cooling fan harness connector and negative battery post. • Is voltage greater than 8.0 volts?	Yes ▶ No ▶	SERVICE Open in ground circuit to fan. RECONNECT Integrated Controller and cooling fan. RE-EVALUATE symptom. SERVICE open in power-to-fan circuit from 3 and 4 of Integrated Controller harness connector to cooling fan connector. RECONNECT cooling fan and controller, RE-EVALUATE symptom.

Integrated Controller — Pinpoint Test X

TEST STEP	RESULT	ACTION TO TAKE
X25 JUMPER HIGH ELECTRIC-DRIVE SIGNAL (HEDF) TO GROUND		
• Key off. • Inspect processor 60 pin connector for damaged pins, corrosion, loose wires, etc. Service as necessary. • Install breakout box, leave processor disconnected. • Integrated Controller connected. • Key on. • Jumper Test Pin 52 to Test Pin 40 at breakout box. • Does fan speed change from low to high?	Yes ▶ No ▶	GO to X26. REMOVE breakout box. REPLACE Integrated Controller. RECONNECT processor. RE-EVALUATE symptom.
X26 CHECK ECT SENSOR		
• Key off, wait 10 seconds. • Connect processor to breakout box. • Check engine coolant level. • Warm engine to operating temperature before taking ECT resistance measurement. • Key off, wait 10 seconds. • Disconnect harness from ECT sensor. • DVOM on 200,000 ohm scale. • Measure resistance of the ECT sensor. • Is the resistance between 1500 ohms and 2000 ohms?	Yes ▶ No ▶	REMOVE breakout box. REPLACE processor. RECONNECT harness to ECT sensor. RECONNECT Integrated Controllers. RE-EVALUATE symptom. REMOVE breakout box. REPLACE ECT sensor. RECONNECT all components. RE-EVALUATE symptom.
X30 SERVICE CODE 83: CHECK RESISTANCE OF HEDF CONTROLLER CIRCUIT		
• Key off. • Disconnect Integrated Controller. • DVOM on 200 ohm scale. • Measure resistance between Pin 17 and Pin 24 at the Integrated Controller. • Is the resistance reading between 50 ohms and 100 ohms?	Yes ▶ No ▶	GO to X31. REPLACE controller. RERUN Quick Test.

Integrated Controller — Pinpoint Test X

TEST STEP	RESULT	ACTION TO TAKE
X31 CHECK HEDF PROCESSOR SIGNAL TO INTEGRATED CONTROLLER FOR OPEN		
• Key off. • Disconnect processor 60 pin connector. Inspect for damaged pins, corrosion, loose wires, etc. Service as necessary. • Install breakout box, leave processor disconnected. • Integrated Controller disconnected. • DVOM On 200 ohms scale. • Measure resistance between Test Pin 52 at breakout box and Pin 17 of Integrated Controller vehicle harness connector. • Is resistance less than 5 ohms?	Yes ▶ No ▶	GO to X32. REMOVE breakout box. SERVICE open in HEDF circuit. RECONNECT all components. RERUN Quick Test.
X32 CHECK FOR SHORTS TO GROUND IN THE HEDF CIRCUIT		
• Key off. • Breakout box installed. • Processor and Integrated Controller disconnected. • DVOM on 200,000 ohm scale. • Measure resistance between Test Pin 52 and Test Pin 40. • Is resistance less than 10,000 ohms?	No ▶ Yes ▶	GO to X33. REMOVE breakout box. RECONNECT processor and Integrated Controller. SERVICE short to ground in HEDF circuit. RERUN Quick Test.
X33 CHECK FOR SHORTS TO POWER IN THE HEDF CIRCUIT		
• Key off. • Breakout box installed. • Processor and Integrated Controller disconnected. • DVOM on 200,000 ohms scale. • Measure resistance between Test Pin 52 and Test Pin 37. • Is resistance less than 10,000 ohms?	No ▶ Yes ▶	REMOVE breakout box. REPLACE Processor. RECONNECT all components. RERUN Quick Test. REMOVE breakout box. SERVICE short to power. RECONNECT all components. RERUN Quick Test. If code 83 is still present, REPLACE processor. RERUN Quick Test.

Integrated Controller — Pinpoint Test X

TEST STEP	RESULT	ACTION TO TAKE
X35 LOW SPEED OR HIGH SPEED FAN ALWAYS "ON", NO SERVICE CODE 83 OR 67		
• Key off. • Disconnect processor 60 pin connector. Inspect for damaged pins, corrosion, loose wires. Service as necessary. • Install breakout box, leave processor disconnected. • Disconnect the Integrated Controller. • DVOM on 200 ohm scale. • Measure the resistance between Test Pin 55 and controller vehicle harness Pin 14. • Is resistance less than 5 ohms?	Yes ▶ No ▶	GO to X36. REMOVE breakout box. SERVICE open in EDF circuit. RECONNECT all components. RE-EVALUATE symptom.
X36 CHECK EDF CIRCUIT FOR SHORTS TO POWER		
• Key off. • Breakout box installed. • Processor and Integrated Controller disconnected. • DVOM on 200,000 ohm scale. • Measure resistance between Test Pin 55 and Test Pin 37 and between Test Pin 55 and battery positive post. • Is resistance less than 10,000 ohms?	Yes ▶ No ▶	SERVICE short to power in EDF circuit. GO to X37. GO to X37.
X37 CHECK EDF FOR SHORT TO GROUND		
• Key on. • Breakout box installed. • Processor disconnected. • Connect Integrated Controller. • Jumper Test Pin 55 to Test Pin 40 or 60. • Does fan continue to run?	Yes ▶ No ▶	REMOVE breakout box. RECONNECT processor. REPLACE controller. RE-EVALUATE symptom. REMOVE breakout box. RECONNECT controller. REPLACE processor. RE-EVALUATE symptom.

Integrated Controller — Pinpoint Test X

TEST STEP	RESULT	ACTION TO TAKE
X40 NO FAN		
• Key off. • Disconnect Integrated Controller. • DVOM on 20 volt scale. • Measure voltage between battery negative post and Pin 1 and Pin 2, respectively at the Integrated Controller vehicle harness connector. • Is voltage greater than 10.5 volts?	Yes ▶ No ▶	GO to X41. RECONNECT controller. SERVICE open in battery power circuit. RE-EVALUATE symptom.
X41 CHECK FAN MOTOR		
• Key off. • Integrated Controller disconnected. • Jumper Pin 1 to Pin 3 at Integrated Controller harness. • Does fan run?	Yes ▶ No ▶	GO to X42. GO to X43.
X42 CHECK FAN RUNNING MODE		
• Key off. • Disconnect processor. • Connect Integrated Controller. • Key on. • Does fan run?	Yes ▶ No ▶	GO to X46. GO to X44.
X43 MEASURE BATTERY VOLTAGE SUPPLY AT FAN—BYPASSING INTEGRATED CONTROLLER		
• Key off. • Disconnect cooling fan. • Integrated Controller disconnected. • Jumper Pin 1 to Pin 3 at Integrated Controller vehicle harness connector. • DVOM on 20 volt scale. • Measure voltage at cooling fan vehicle harness connector. • Is voltage greater than 8.0 volts?	Yes ▶ No ▶	RECONNECT all components. CHANGE fan. RE-EVALUATE symptom. GO to X45.

Integrated Controller — Pinpoint Test X

TEST STEP	RESULT ▶	ACTION TO TAKE
X44 CHECK EDF CIRCUIT FOR SHORT TO GROUND		
• Key off. • Processor and controller disconnected. • DVOM on 200,000 ohm scale. • Measure resistance from Pin 14 to Pin 15 at Integrated Controller vehicle harness connector. • Is resistance greater than 10,000 ohms?	Yes ▶	REPLACE Integrated Controller. RECONNECT processor and controller. RE-EVALUATE symptom.
	No ▶	SERVICE short to ground in EDF circuit. RECONNECT processor and Integrated Controller. RE-EVALUATE symptom.
X45 COOLING FAN GROUND VERIFICATION		
• Key off. • Cooling fan disconnected. • Integrated Controller disconnected. • Jumper Pin 1 to Pin 3 at Integrated Controller vehicle harness connector. • DVOM on 20 volt scale. • Measure voltage between voltage positive at cooling fan harness connector and negative battery post. • Is voltage greater than 8.0 volts?	Yes ▶	SERVICE open in ground circuit to fan. RECONNECT Integrated Controller, RE-EVALUATE symptom.
	No ▶	SERVICE open in power-to-fan circuit from 3 and 4 of Integrated Controller harness connector to cooling fan connector. RECONNECT controller. RE-EVALUATE symptom.
X46 ECT SENSOR CHECK		
• Reconnect processor. • Check engine coolant level. • Warm engine to operating temperature before taking ECT resistance measurement. • Key off, wait 10 seconds. • Harness disconnected from ECT sensor. • DVOM on 200,000 ohm scale. • Measure resistance of the ECT sensor. • Is the resistance reading between 1500 ohms and 2000 ohms?	Yes ▶	REPLACE processor. RECONNECT harness to ECT sensor. RECONNECT Integrated Controller. RE-EVALUATE symptom.
	No ▶	REPLACE ECT sensor. RECONNECT all components. RE-EVALUATE symptom.

Integrated Controller — Pinpoint Test X

TEST STEP	RESULT ▶	ACTION TO TAKE
X50 CHECK FOR VOLTAGE AT A/C CLUTCH		
• Key on, engine off. • A/C demand switch to A/C ON position. • DVOM on 20 volt scale. • Check voltage at A/C clutch harness connector. • Is voltage greater than 10.5 volts?	Yes ▶	
	No ▶	GO to X51.
X51 CHECK FOR CONTINUITY FROM INTEGRATED CONTROLLER TO A/C CLUTCH		
• Key off. • Disconnect Integrated Controller. • DVOM on 200 ohm scale. • Measure resistance between Pin 23 of the controller harness and power side of the A/C clutch harness connector and between Pin 16 of the controller harness and ground side of the A/C clutch harness connector. • Are both resistances less than 5 ohms?	Yes ▶	GO to X61.
	No ▶	SERVICE open in power to A/C clutch or ground to A/C clutch. RE-EVALUATE symptom.
X52 ENTER OUTPUT STATE CHECK (REFER TO APPENDIX)		
NOTE: Do not use STAR tester for this Step, use VOM/DVOM. • Key off, wait 10 seconds. • Disconnect processor 60 pin connector. Inspect for damaged pins, corrosion, loose wires, etc. Service as necessary. • Install breakout box. Connect processor to breakout box. • DVOM on 20 volt scale. • Connect DVOM negative test lead to STO and positive test lead to battery positive. • Jumper STI to SIGNAL RETURN. • Perform Key On Engine Off Self-Test until the completion of the Continuous Test Codes. • DVOM will indicate zero volts. • Depress and release the throttle. • Did DVOM reading change to a high voltage reading?	Yes ▶	REMAIN in Output State Check. GO to X53.
	No ▶	DEPRESS throttle to WOT and RELEASE. If STO voltage does not go high, GO to Pinpoint Test Step QC1. LEAVE equipment hooked up.

Integrated Controller — Pinpoint Test X

TEST STEP	RESULT ▶	ACTION TO TAKE
X53 CHECK WAC OUTPUT FOR PROPER ELECTRICAL OPERATION		
• Key on, engine off. • A/C demand switch to A/C on position. • Breakout box installed, processor connected. • DVOM on 20 volt scale. • Connect DVOM positive test lead to Test Pin 37 and negative test lead to Test Pin 54. • While observing DVOM, depress and release the throttle several times (to cycle output on and off). • Does voltage output cycle high and low?	Yes ▶	2.3L EFI T/C only, GO to X61. All others, GO to X54.
	No ▶	GO to X57.
X54 CHECK FOR VOLTAGE AT A/C CLUTCH SWITCH		
• Key on, engine off. • A/C demand switch to A/C on position. • DVOM on 20 volt scale. • Breakout box installed. • Processor and Integrated Controller connected. • Measure voltage between Test Pin 10 and Test Pin 40 at breakout box. • Is voltage greater than 10.5 volts?	Yes ▶	GO to X55.
	No ▶	GO to X56.
X55 CHECK CONTINUITY OF ACCS TO INTEGRATED CONTROLLER		
• Key off, wait 10 seconds. • Breakout box installed. • Processor disconnected. • Integrated Controller disconnected. • DVOM on 200 ohm scale. • Measure resistance between Test Pin 10 at breakout box and Pin 21 at controller harness connector. • Is resistance less than 5 ohms?	Yes ▶	REMOVE breakout box. RECONNECT processor. REPLACE Integrated Controller. RE-EVALUATE symptom.
	No ▶	REMOVE breakout box. RECONNECT all components. SERVICE open in ACCS circuit. RE-EVALUATE symptom.

Integrated Controller — Pinpoint Test X

TEST STEP	RESULT ▶	ACTION TO TAKE
X56 CHECK CONTINUITY OF ACCS CIRCUIT		
• Key off, wait 10 seconds. • Breakout box installed. • A/C demand switch to A/C ON position. • Processor and Integrated Controller connected. • DVOM on 200 ohm scale. • Measure resistance between Test Pin 10 and A/C demand switch. • Is resistance less than 5 ohms?	No ▶	SERVICE open in circuit. RERUN Quick Test. REMOVE breakout box. RECONNECT all components.
	Yes ▶	EEC-IV system OK. REMOVE breakout box. RECONNECT all components.
X57 CHECK CONTINUITY IN WAC TO INTEGRATED CONTROLLER CIRCUIT		
• Key off, wait 10 seconds. • Breakout box installed. • Disconnect processor and Integrated Controller. • DVOM on 200 ohm scale. • Measure resistance between Test Pin 54 and Pin 22 at Integrated Controller harness. • Is resistance less than 50 ohms?	No ▶	REMOVE breakout box. RECONNECT all components. SERVICE open in WAC circuit. RE-EVALUATE symptom.
	Yes ▶	GO to X58.
X58 CHECK WAC CIRCUIT FOR SHORTS TO GROUND		
• Key off, wait 10 seconds. • Leave breakout box installed and processor disconnected. • Integrated Controller disconnected. • DVOM on 200,000 ohm scale. • Measure resistance between Test Pin 54 and Test Pin 40 and between Test Pin 54 and Test Pin 46 and between Test Pin 54 and battery negative post. • Are all resistances greater than 10,000 ohms?	Yes ▶	GO to X59.
	No ▶	REMOVE breakout box. RECONNECT all components. SERVICE shorts to ground in WAC circuit. RE-EVALUATE symptom.

Integrated Controller — Pinpoint Test X

TEST STEP	RESULT	ACTION TO TAKE
X59 CHECK WAC CIRCUIT FOR SHORTS TO POWER		
• Key off, wait 10 seconds. • Leave Breakout box installed and processor disconnected. • Integrated Controller disconnected. • DVOM on 200,000 ohm scale. • Measure resistance between Test Pin 54 and Test Pin 37 and between Test Pin 54 and battery positive. • Are both resistances greater than 10,000 ohms?	Yes ▶	GO to X60.
	No ▶	REMOVE breakout box. RECONNECT all components. SERVICE short to power in WAC circuit. GO to X60.
X60 CHECK FOR VOLTAGE AT A/C CLUTCH		
• Key off, wait 10 seconds. • Leave breakout box installed. • Processor disconnected. • Connect Integrated Controller. • A/C clutch disconnected. • A/C demand switch to A/C ON position. • Key on, engine off. • DVOM on 20 volt scale. • Measure voltage at A/C clutch harness connection. • Is voltage greater than 10.5 volts?	Yes ▶	REMOVE breakout box. RECONNECT all components. REPLACE processor. RE-EVALUATE symptom.
	No ▶	REMOVE breakout box. RECONNECT all components. REPLACE Integrated Controller. RE-EVALUATE symptom.
X61 CHECK FOR VOLTAGE AT A/C INPUT TO CONTROLLER		
• Key off, wait 10 seconds. • Leave breakout box installed. • Processor connected. • Integrated controller disconnected. • Key on, engine off. • A/C demand switch to A/C ON position. • DVOM on 20 volt scale. • Measure voltage between Pin 21 at controller harness connector and test Pin 40. • Is voltage greater than 10.5 volts?	Yes ▶	REMOVE breakout box. RECONNECT all components. REPLACE Integrated Controller. RE-EVALUATE symptom.
	No ▶	REMOVE breakout box. RECONNECT all components. SERVICE open in A/C input circuit to controller. RE-EVALUATE symptom.

Integrated Controller — Pinpoint Test X

TEST STEP	RESULT	ACTION TO TAKE
X74 COOLING FAN GROUND VERIFICATION		
• Key off. • Cooling fan disconnected. • Integrated controller disconnected. • Jumper Pin 3 to Pin 1 (for primary fan) and Pin 3 to Pin 6 if equipped (for secondary fan) at integrated controller harness connector. • DVOM on 20 volt scale. • Measure voltage between voltage positive at one or both cooling fan harness connectors and negative battery post as equipped. • Is voltage greater than 8.0 volts?	Yes ▶	SERVICE open in ground circuit to fan(s). RECONNECT all components. RE-EVALUATE symptom.
	No ▶	SERVICE open in power circuit to fan(s). RECONNECT all components. RE-EVALUATE symptom.
X75 JUMPER SECONDARY ELECTRIC DRIVE SIGNAL (HEDF) TO GROUND		
• Key off. • Disconnect processor 60 pin connector. Inspect for damaged pins, corrosion, loose wires, etc. Service as necessary. • Install breakout box, leave processor disconnected. • Integrated controller connected. • Jumper Test Pin 52 to Test Pin 40 at breakout box. • Key on. • Does secondary fan run?	Yes ▶	GO to X76.
	No ▶	REMOVE breakout box. REPLACE Integrated Controller. RECONNECT processor. RE-EVALUATE symptom.
X76 ECT SENSOR CHECK		
• Connect processor. • Check engine coolant level. • Warm engine to operating temperature before taking ECT resistance measurement. • Key off, wait 10 seconds. • Disconnect harness from ECT sensor. • DVOM on 200,000 ohm scale. • Measure resistance of ECT sensor. • Is resistance between 1500 and 2000 ohms?	Yes ▶	REMOVE breakout box. REPLACE processor. RECONNECT all components. RE-EVALUATE symptom.
	No ▶	REMOVE breakout box. REPLACE ECT sensor. RECONNECT all components. RE-EVALUATE symptom.

Integrated Controller — Pinpoint Test X

TEST STEP	RESULT	ACTION TO TAKE
X70 NO FAN PRIMARY OR SECONDARY WITH NO CODE 82 OR 88		
• Key off, wait 10 seconds. • Disconnect Integrated controller. • DVOM on 20 volt scale. • Measure voltage between battery negative post and Pins 3 and 4 respectively at controller harness connector. • Are all voltages greater than 10.5 volts?	Yes ▶	GO to X71.
	No ▶	RECONNECT Integrated Controller. SERVICE open in battery power circuit. RE-EVALUATE symptom.
X71 FAN MOTORS CHECK		
• Key off. • Integrated controller disconnected. • Jumper Pin 3 to Pin 1 (for primary fan) and Pin 3 to Pin 6 (for secondary fan) at integrated controller harness connector. • Do both fans run?	Yes ▶	GO to X72.
	No ▶	GO to X73.
X72 CHECK FAN RUNNING MODE (LOW)		
• Key off. • Disconnect processor. • Connect Integrated controller. • Key on, engine off. • Does primary fan run?	Yes ▶	GO to X75.
	No ▶	REPLACE Integrated Controller. RECONNECT all components. RE-EVALUATE symptom.
X73 MEASURE BATTERY VOLTAGE SUPPLY AT FANS — BYPASSING INTEGRATED CONTROLLER		
• Key off. • Disconnect cooling fans. • Integrated controller disconnected. • Jumper Pin 3 to Pin 1 (for primary fan) and Pin 3 to Pin 6 if equipped (for secondary fan) at integrated controller harness connector. • DVOM on 20 volt scale. • Measure voltage at one or both cooling fan harness connectors as equipped. • Is either voltage greater than 8.0 volts?	Yes ▶	CHANGE fan(s). RE-EVALUATE symptom.
	No ▶	GO to X74.

Integrated Controller — Pinpoint Test X

TEST STEP	RESULT	ACTION TO TAKE
X80 SERVICE CODE 82 OR 88: CHECK EDF PROCESSOR SIGNAL TO INTEGRATED CONTROLLER FOR SHORTS TO GROUND		
NOTE: If fan is always on with Code 82 or 88, GO to X82. • Key off. • Disconnect processor 60 pin connector. Inspect for damaged pins, corrosion, and loose wires, etc. Service as necessary. • Install breakout box, leave processor disconnected. • Disconnect Integrated controller. • DVOM on 200,000 ohm scale. • Measure resistance between Test Pin 55 and Test Pin 40. • Is resistance less than 10,000 ohms?	Yes ▶	SERVICE short to ground in EDF circuit. RECONNECT all components. RERUN Quick Test.
	No ▶	GO to X81.
X81 CHECK FAN RUNNING MODE		
• Key off. • Breakout box installed. • Processor disconnected. • Connect integrated controller. • Key on, engine off. For 2.5L MTX 〉 Does fan run? For 2.5L, 3.0L and 3.8L AXOD 〉 Does fan run at low speed? For 2.3L EFI TC 〉 Does primary fan run?	Yes ▶	REMOVE breakout box. REPLACE processor. RECONNECT all components. RERUN Quick Test.
	No ▶	REMOVE breakout box. REPLACE Integrated Controller. RECONNECT all components. RERUN Quick Test.

Integrated Controller — Pinpoint Test X

TEST STEP	RESULT	ACTION TO TAKE
X82 FAN ALWAYS ON WITH CODE 82 OR 88: CHECK EDF PROCESSOR SIGNAL TO INTEGRATED CONTROLLER FOR OPEN CIRCUIT		
• Key off. • Disconnect processor 60 pin connector. Inspect for damaged pins, corrosion, and loose wires, etc. Service as necessary. • Install breakout box. • Processor and integrated controller disconnected. • DVOM on 200 ohm scale. • Measure resistance between Test Pin 55 and Integrated Controller harness Pin 14. • Is resistance less than 5 ohms?	Yes ▶ No ▶	GO to X83. REMOVE breakout box. SERVICE open in EDF circuit. RECONNECT all components. RERUN Quick Test.
X83 CHECK EDF CIRCUIT FOR SHORTS TO POWER		
• Key off. • Breakout box installed. • Processor and integrated controller disconnected. • DVOM on 200,000 ohm scale. • Measure resistance between Test Pin 55 and Test Pin 37, and between Test Pin 55 and battery positive. • Is resistance less than 10,000 ohms?	Yes ▶ No ▶	REMOVE breakout box. RECONNECT all components. SERVICE short to power in EDF circuit. GO to X84. GO to X84.
X84 CHECK EDF SHORT TO GROUND		
• Key off. • Breakout box installed. • Processor disconnected. • Integrated controller connected. • Key on, engine off. • Jumper test Pin 55 to Test Pin 40 or 60. • Does fan continue to run?	Yes ▶ No ▶	REMOVE breakout box. REPLACE Integrated Controller. RECONNECT all components. RERUN Quick Test. REMOVE breakout box. REPLACE processor. RECONNECT all components. RERUN Quick Test.

Integrated Controller — Pinpoint Test X

TEST STEP	RESULT	ACTION TO TAKE
X90 SERVICE CODE 95: CHECK INERTIA SWITCH		
NOTE: Key On Engine Off Service Code 95 indicates that one of the following has occured: **— Open circuit in/or between the fuel pump and Test Pin 8 (see schematic)** **— Poor fuel pump ground** **— FUEL PUMP circuit short to power** **— Fuel pump relay contacts always closed** • Key off, wait 10 seconds. • Locate and disconnect fuel pump inertia switch. • DVOM on 200 ohm scale. • Measure resistance of the fuel pump inertia switch. • Is resistance less than 5.0 ohms?	Yes ▶ No ▶	RECONNECT inertia switch. GO to X91. REPLACE or RESET inertia switch. RERUN Quick Test.
X91 VERIFY THAT FUEL PUMP IS OFF		
• Key off. • Listen for motor noise from fuel pump. • Is fuel pump off?	Yes ▶ No ▶	GO to X93. GO to X92.
X92 CHECK FOR FUEL PUMP RELAY ALWAYS CLOSED		
• Key off. • Locate and disconnect integrated controller. • Does fuel pump shut off when controller is disconnected?	Yes ▶ No ▶	REPLACE Integrated Controller. RERUN Quick Test SERVICE short to power in POWER-TO-PUMP/FPM circuit. RECONNECT integrated controller. RERUN Quick Test.

Integrated Controller — Pinpoint Test X

TEST STEP	RESULT	ACTION TO TAKE
X93 CHECK CONTINUITY OF FPM CIRCUIT		
• Key off. • Disconnect processor 60 pin connector. Inspect for damaged pins, corrosion, loose wires, etc. Service as necessary. • Install breakout box, leave processor disconnected. • Disconnect integrated controller. • DVOM on 200 ohm scale. • Measure resistance between Test Pin 8 at the breakout box and integrated controller harness connector pin 5. • Is resistance less than 5.0 ohms?	Yes ▶ No ▶	GO to X94. REMOVE breakout box. RECONNECT processor and integrated controller. SERVICE open circuit. RERUN Quick Test.
X94 CHECK FOR CONTINUITY BETWEEN FPM CIRCUIT AND GROUND		
• Key off. • Breakout box installed, processor disconnected. • Integrated controller disconnected. • DVOM on 200 ohm scale. • Measure resistance between Test Pin 8 at the breakout box and battery negative post. • Is resistance less than 5.0 ohms?	Yes ▶ No ▶	REMOVE breakout box. RECONNECT integrated controller. REPLACE processor. RERUN Quick Test. REMOVE breakout box. RECONNECT processor and integrated controller. GO to Shop Manual Group 24, Electric Fuel Pump for open in POWER-TO-PUMP circuit, poor fuel pump ground, open in fuel pump, etc.

Integrated Controller — Pinpoint Test X

TEST STEP	RESULT	ACTION TO TAKE
X95 SERVICE CODE 96: CHECK CONTINUITY OF POWER-TO-PUMP CIRCUIT		
NOTE: Service Code 96 indicates that when the fuel pump is being activated, power is not being supplied to the fuel pump. • Key off, wait 10 seconds. • Disconnect processor 60 pin connector. Inspect for damaged pins, corrosion, loose wires, etc. Service as necessary. • Install breakout box, leave processor disconnected. • Disconnect integrated relay controller. • DVOM on 200 ohm scale. • Measure resistance between Test Pin 8 at the breakout box and integrated controller harness connector pin 5. • Is resistance less than 5.0 ohms?	Yes ▶ No ▶	GO to X96. REMOVE breakout box. RECONNECT processor and integrated controller. SERVICE open in POWER-TO-PUMP circuit between FPM splice and the integrated controller. RERUN Quick Test.
X96 VERIFY FUEL PUMP OPERATION		
• Key off. • Breakout box installed. • Reconnect processor and integrated controller. • DVOM on 20 volt scale. • Connect DVOM between Test Pin 8 and Test Pin 40 at the breakout box. • While observing DVOM, turn key to on. • Does voltage increase to greater than 10.5 volts for about 1 second after key is turned to on?	Yes ▶ No ▶	REMOVE breakout box. REPLACE processor. RERUN Quick Test. REMOVE breakout box. RECONNECT processor. REPLACE integrated controller. RERUN Quick Test.

Integrated Controller — Pinpoint Test X

TEST STEP	RESULT ▶	ACTION TO TAKE
X100 CONTINUOUS MEMORY CODE 95: CHECK EEC-IV HARNESS		
A Continuous Memory Code 95 indicates that one of the following intermittent conditions has occurred: — Open circuit in or between the fuel pump and Pin 8 in the processor (see schematic [X]). — Poor fuel pump ground. • Start engine. • Check for engine stall/stumble while performing the following (also, if possible, listen for fuel pump turning off). — Shake, wiggle, bend the power-to-pump circuit between the Integrated Controller pin 5 and the fuel pump. — Shake, wiggle, bend the fuel pump ground circuit from the fuel pump to ground. — Lightly tap the inertia switch and the fuel pump to simulate road shock. • Key off. • Inspect the fuel pump electrical connector and the fuel pump ground for corrosion, damaged pins, etc. • Is fault indicated/found?	Yes ▶ No ▶	ISOLATE fault and SERVICE as necessary. CLEAR Continuous Memory Code 95. RERUN Quick Test. GO to [X101].
X101 CHECK FPM CIRCUIT		
• Key off. • Disconnect processor 60 pin connector. Inspect for damaged pins, corrosion, loose wires, etc. Service as necessary. • Install breakout box, leave processor disconnected. • Key on, engine off. • Connect a test lamp between Test Pin 8 and Test Pin 37. • Observe test lamp for an indication of a fault while performing the following (The light will go out when a fault is found indicating an open): — Shake, wiggle, bend the fuel pump monitor circuit (Pin 8) between the processor and splice into the POWER-TO-PUMP circuit. • Is fault found/indicated?	Yes ▶ No ▶	ISOLATE fault and SERVICE as necessary. REMOVE breakout box. CLEAR Continuous Memory Code 95. RERUN Quick Test. Unable to duplicate fault at this time. CLEAR Continuous Memory Code 95.

Integrated Controller — Pinpoint Test X

TEST STEP	RESULT ▶	ACTION TO TAKE
X102 CONTINUOUS MEMORY CODE 96 CHECK FOR CONTINUOUS MEMORY CODE 87		
• Is Continuous Memory Code 87 also present?	Yes ▶ No ▶	GO to [X104]. GO to [X103].
X103 CHECK EEC-IV HARNESS		
A Continuous Memory Code 96, without the presence of a Continuous Memory Code 87, indicates that during vehicle operation, one of the following has occurred: — Fuel pump relay contacts opened. — Open in the POWER-TO-PUMP circuit from the integrated relay controller pin 5 to the FPM splice. (See schematic [X]). • Start engine. • Check for engine stall/stumble while performing the following (also, if possible, listen for fuel pump turning off). — Shake, wiggle, bend the POWER-TO-PUMP circuit from the integrated relay controller to the FPM splice. — Lightly tap the integrated relay controller (to simulate road shock). • Key off. • Inspect the integrated relay controller 24 pin connectors for corrosion, damaged pins, etc. • Is fault indicated/found?	Yes ▶ No ▶	ISOLATE fault and SERVICE as necessary. CLEAR Continuous Memory Code. RERUN Quick Test. Unable to duplicate fault at this time. CLEAR Continuous Memory Code 96. Continuous Memory Code 96 testing complete.

Integrated Controller — Pinpoint Test X

TEST STEP	RESULT ▶	ACTION TO TAKE
X104 CONTINUOUS MEMORY SERVICE CODE 87: CHECK EEC-IV HARNESS		
A Continuous Memory Code 87 indicates that one of the following intermittent conditions has occurred: — Open VPWP circuit in the integrated relay controller. — Open coil in fuel pump relay. — Open in fuel pump primary circuit. • Start engine. • Check for engine stall/stumble while performing the following (also, if possible, listen for fuel pump turning off): — Shake, wiggle, bend the EEC-IV Harness fuel pump circuit (pin 22) between the processor and the Integrated Controller (pin 18). — Lightly tap the Integrated Controller (to simulate road shock). • Key off. • Inspect the processor 60 pin connectors and the integrated relay controller 24 pin connectors for corrosion, damaged pins, etc. • Is fault indicated/found?	Yes ▶ No ▶	ISOLATE fault and SERVICE as necessary. CLEAR Continuous Memory Service Code(s). RERUN Quick Test. Unable to duplicate fault at this time. CLEAR Continuous Memory Code(s).

Erratic Ignition — Pinpoint Test Y

Note

You should enter this Pinpoint Test only when a service code 14 is received in Quick Test Step 6.0.

Remember

To prevent the replacement of good components, be aware that the following non-EEC areas may be at fault:

- TFI module.
- Arcing secondary ignition components.
  - — Ignition coil.
  - — Spark plugs and high tension cables.
- Distributor and PIP sensor.

This Pinpoint Test is intended to diagnose only the following:

- Harness circuits: IGNITION GROUND, SPOUT, PIP, IDM.

Pinpoint Test Schematic

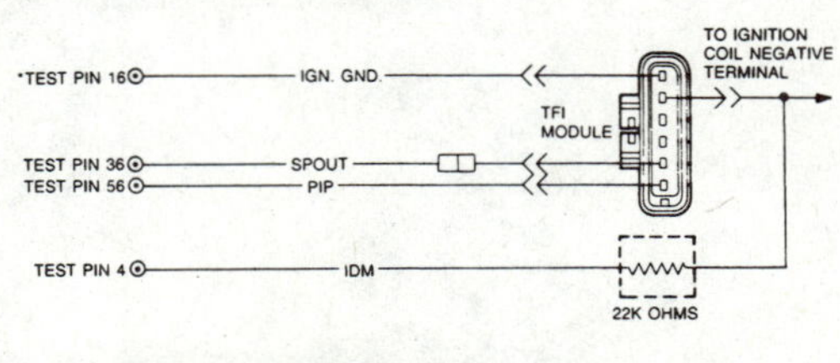

*TEST PINS LOCATED ON BREAKOUT BOX
ALL HARNESS CONNECTORS VIEWED INTO MATING SURFACE

Erratic Ignition — Pinpoint Test Y

TEST STEP	RESULT ▶	ACTION TO TAKE
Y1 SERVICE CODE 14: ERRATIC IGNITION		
NOTE: Code 14 indicates two successive erratic profile ignition pickup (PIP) pulses occurred, resulting in a possible engine miss or stall. • Check EEC-IV and ignition systems harnesses for: — Loose wires/connectors. — Arcing secondary ignition components (coil, cap, rotor, wires, plugs, etc.). — On-board transmitter (2-way radio).* • Are any of the above present? *Verify all 2-way radio installations. Carefully follow manufacturer's installation instructions regarding the routing of antenna and power leads.	Yes ▶ No ▶	SERVICE as necessary. CLEAR Continuous Memory Code 14. RERUN Quick Test. GO to Y2.
Y2 CHECK DISTRIBUTOR		
• Key off, wait 10 seconds. • Enter Engine Running Continuous Monitor mode. Refer to Appendix in Section 16. • Observe VOM or STAR LED for indication of a fault while performing the following: • Lightly tap on TFI module and distributor. • Wiggle TFI connector. • Is a fault indicated?	Yes ▶ No ▶	DISCONNECT and INSPECT connectors. If connector and terminals are good, GO to Y3.

Erratic Ignition — Pinpoint Test Y

TEST STEP	RESULT ▶	ACTION TO TAKE
Y3 CHECK EEC-IV HARNESS		
• While still in Continuous Monitor mode from Step Y2 observe VOM or STAR LED for a fault indication while performing the following: • While looking for faults grasp the harness close to the TFI connector. Wiggle, shake or bend a small section of the ignition and EEC-IV systems harness while working your way to the other components and dash panel. Also wiggle, shake or bend the EEC-IV harness from the dash panel to the processor. Isolate the PIP circuit if needed for this test. • Is a fault indicated?	Yes ▶ No ▶	ISOLATE fault and SERVICE as necessary. CLEAR Continuous Memory Code 14. RERUN Quick Test. GO to Y4
Y4 CHECK PROCESSOR AND HARNESS CONNECTORS		
• Key off, wait 10 seconds. • Disconnect processor 60 pin connector. Inspect for damaged pins, corrosion, loose wires, etc. Service as necessary. • Are connectors and terminals OK? • Reconnect processor when this Step is completed.	Yes ▶ No ▶	Unable to duplicate an erratic ignition fault in the EEC-IV System. For further diagnosis, SERVICE as necessary. CLEAR Continuous Memory Code 14. RERUN Quick Test.